# CELL
# PHYSIOLOGY
## *Source Book*

SECOND EDITION

# CELL PHYSIOLOGY

## Source Book

### SECOND EDITION

Edited by

## NICHOLAS SPERELAKIS

*Department of Physiology and Biophysics*
*University of Cincinnati College of Medicine*
*Cincinnati, Ohio*

### Academic Press

San Diego   London   Boston   New York   Sydney   Tokyo   Toronto

Cover photograph credits: (Left and center) Images © 1997 PhotoDisc, Inc.
(Right) Methods in Enzymology, Volume 207 © 1997 Academic Press

This book is printed on acid-free paper. ∞

Academic Press
*a division of Harcourt Brace & Company*
525 B Street, Suite 1900, San Diego, California 92101-4495, USA
http://www.apnet.com

Academic Press Limited
24-28 Oval Road, London NW1 7DX, UK
http://www.hbuk.co.uk/ap/

Library of Congress Cataloging-in-Publication Data

Cell physiology source book / edited by Nicholas Sperelakis
           p.          cm.
      Includes bibliographical references and index.
      ISBN 0-12-656973-8 (paperback)    ISBN 0-12-656972-X (hardcover)
      1. Cell physiology.    I. Sperelakis, Nick, date.
    QH631.C458    1994
    574.87'6--dc20                              94-31852
                                                CIP

PRINTED IN THE UNITED STATES OF AMERICA
97  98  99  00  01  02  EB  9  8  7  6  5  4  3  2  1

In Memoriam

Professor Hugh Davson
1909–1996

Professor Hugh Davson, to whom the first edition of this book was dedicated, single-handedly authored the classic *A Textbook on General Physiology* through four editions, the last edition being published (in two volumes) in 1970. Davson's textbook was the "bible" for almost all physiology graduate students in the 1950s, 1960s, and early 1970s. As such, Hugh had a tremendous impact and influence on the developing field of cell physiology. He was a pioneer in the rigorous application of the principles of physics and chemistry to biological processes at the cell level. Professor Davson was an outstanding scientist, author, teacher, and human being, and he will be sorely missed by his colleagues and friends worldwide.

The second edition of the **Cell Physiology Source Book** is dedicated to the Planet Earth and its inhabitants, both animals and plants. The destruction of the environment and the polution of the air, soil, and waters are occurring at a tremendous rate. Human population growth is still in an explosive exponential (positive) phase, much like bacteria in a fresh, large culture medium. Our wildlife, forests, parks, and farmland are under increasing pressure. I salute all those national and international organizations, many of which I am proud to be a member, that are dedicated to halting the ruthless destruction of our planet and the inhumane treatment of animals. These organizations are concerned about the environment, wildlife, overpopulation, animal rights, and human rights. I urge all readers of this book to express their serious and urgent concern for the well being of Planet Earth, the environment, plants, animals, and humans, and accordingly to help in educating the public and governments worldwide.

In light of the above, I will donate all royalties due to me as Editor of this book to those courageous organizations on the front lines of the battle to save our planet.

# Contents

## SECTION I
## Biophysical Chemistry, Metabolism, Second Messengers, and Ultrastructure

## SECTION II
## Membrane Potential, Transport Physiology, Pumps, and Exchangers

## SECTION VI
## Muscle and Other Contractile Systems

## SECTION VII
## Protozoa and Bacteria

## SECTION VIII
## Plant Cells, Photosynthesis, and Bioluminescence

## SECTION IX
## Cell Division and Programmed Cell Death

# Contributors

*Numbers in parentheses indicate the pages on which the authors' contributions begin.*

**Thomas K. Akers** (892), 449 Guy Kelly Rd., R.R. #6, Port Angeles, Washington 98362

**Juan Manuel Arias** (1031), Department of Biochemistry, Cinvestav—IPN, 07360 Mexico, DF, Mexico

**David M. Balshaw** (216), Department of Pharmacology and Cell Biophysics, University of Cincinnati College of Medicine, Cincinnati, Ohio 45267-0575

**David W. Barnett** (652), Departments of Medicine and Cell Biology/Physiology, Washington University Medical Center, St. Louis, Missouri 63110

**Clive M. Baumgarten** (253), Department of Physiology, Medical College of Virginia, Virginia Commonwealth University, Richmond, Virginia 23298-0551

**Michael M. Behbehani** (429, 682), Department of Molecular and Cellular Physiology, University of Cincinnati College of Medicine, Cincinnati, Ohio 45267-0576

**Kenneth Blumenthal** (547), Department of Molecular Genetics, Biochemistry, and Microbiology, University of Cincinnati College of Medicine, Cincinnati, Ohio 45267-0524

**John H. B. Bridge** (237), The University of Utah, Salt Lake City, Utah 84112

**Shirley H. Bryant** (574), Department of Pharmacology and Cell Biophysics, University of Cincinnati College of Medicine, Cincinnati, Ohio 45267

**John Cuppoletti** (207), Departments of Molecular and Cellular Physiology, University of Cincinnati College of Medicine, Cincinnati, Ohio 45267-0576

**Alberto Darszon** (456), Instituto de Biotecnología, Departamento de Bioquímica, Universidad Nacional Autónoma de México, Cuernavaca, Morelos 62271, México

**Louis J. DeFelice** (481), Department of Pharmacology, Vanderbilt University, Nashville, Tennessee 37232

**John R. Dedman** (132, 900), Department of Physiology and Biophysics, University of Cincinnati College of Medicine, Cincinnati, Ohio 45267

**Istvan Edes** (225), Department of Heart and Lung Diseases, University Medical School Debrecen, 4004 Debrecen, Hungary

**Joseph J. Feher** (253), Department of Physiology, Medical College of Virginia, Virginia Commonwealth University, Richmond, Virginia 23298-0551

**Donald G. Ferguson** (75), Department of Anatomy, Case Western Reserve University College of Medicine, Cleveland, Ohio 44106

**Darrell Fleischman** (968), Department of Biochemistry and Molecular Biology, Wright State University, Dayton, Ohio 45435-0001

**Jeffrey C. Freedman** (3, 325), Department of Physiology, State University of New York Health Science Center at Syracuse, Syracuse, New York 13210

**Nicole Gallo-Payet** (632), Endocrine Service, Department of Medicine, University of Sherbrooke, Sherbrooke, Québec, Canada J1H 5N4

**Keith D. Garlid** (111), Department of Biochemistry and Molecular Biology, Oregon Graduate Institute of Science and Technology, Portland, Oregon 97291-1000

**W. Gibson-Wood** (61), Department of Physiology and Pharmacology, Texas A&M University, College Station, Texas 77843-4466

**Thomas E. Gorrell** (917), Haskins Laboratory, Pace University, New York, New York 10038

**Anthony L. Gotter** (900), Laboratory of Developmental Chronobiology, Children's Services, Massachusetts General Hospital and Harvard Medical School, Boston, Massachusetts 03114

**Joanna Groden** (1021), Department of Molecular Genetics, Biochemistry and Microbiology, University of Cincinnati College of Medicine, Cincinnati, Ohio 45267-0524

**Dennis W. Grogan** (938), Department of Biological Sciences, University of Cincinnati, Cincinnati, Ohio 45221-0006

**Agustin Guerrero** (1031), Department of Biochemistry, Cinvestav—IPN, 07360 Mexico, DF, Mexico

**Andrés A. Gutiérrez** (1003), Laboratory of Molecular Medicine, Instituto Nacional de Cancerología SS, México DF 14000, Mexico

**Eric J. Hall** (1044), College of Physicians and Surgeons, Columbia University, New York, New York 10032

**J. Woodland Hastings** (984), The Biological Laboratories, Harvard University, Cambridge, Massachusetts 02138-2020

**Christopher D. Heinen** (1021), Department of Molecular Genetics, Biochemistry and Microbiology, University of Cincinnati College of Medicine, Cincinnati, Ohio 45267-0524

**Judith A. Heiny** (805), Department of Molecular and Cellular Physiology, University of Cincinnati College of Medicine, Cincinnati, Ohio 45267-0576

**R. Kent Hermsmeyer** (564, 791), Oregon Regional Primate Research Center, Beaverton, Oregon 97006; and Department of Cell Biology and Anatomy, Oregon Health Sciences University, Portland, Oregon 97201

**Nelson D. Horseman** (153), Department of Molecular and Cellular Physiology, University of Cincinnati, Cincinnati, Ohio 45267

**Ching-hsien Huang** (39), Department of Biochemistry, University of Virginia School of Medicine, Charlottesville, Virginia 22908

**Marcia A. Kaetzel** (132, 900), Department of Physiology and Biophysics, University of Cincinnati College of Medicine, Cincinnati, Ohio 45267

**Edna S. Kaneshiro** (848, 875), Department of Biological Sciences, University of Cincinnati, Cincinnati, Ohio 45221-0006

**James G. Kereiakes** (1044), University of Cincinnati College of Medicine, Cincinnati, Ohio 45267

**Ann B. Kier** (61), Department of Pathobiology, Texas A&M University, College Station, Texas 77843-4467

**Evangelia G. Kranias** (225), Department of Pharmacology and Cell Biophysics, University of Cincinnati College of Medicine, Cincinnati, Ohio 45267

**Yoshihisa Kurachi** (532), Department of Pharmacology II, Osaka University, Faculty of Medicine, Osaka 565, Japan

**William J. Larsen** (467), Department of Anatomy and Cell Biology, University of Cincinnati College of Medicine, Cincinnati, Ohio 45267

**Harold Lecar** (391), Department of Molecular and Cell Biology, University of California at Berkeley, Berkeley, California 94720

**Michael Levandowsky** (917), Haskins Laboratory, Pace University, New York, New York 10038

**Simon Rock Levinson** (406), Department of Physiology, University of Colorado Health Sciences Center, Denver, Colorado 80262

**Michael A. Lieberman** (91, 119), Department of Molecular Genetics, Biochemistry, and Microbiology, University of Cincinnati College of Medicine, Cincinnati, Ohio 45267-0524

**Arturo Liévano** (456), Instituto de Biotecnología, Departamento de Bioqúimica, Universidad Nacional Autónoma de México, Cuernavaca, Morelos 62271, México

**Daniel C. Marcus** (688), Biophysics Laboratory, Boys Town National Research Hospital, Omaha, Nebraska 68131

**Michele Mazzanti** (481), Dipartimento di Fisiologia e Biochimica, Universita Degli Studi di Milano, 20133 Milan, Italy

**Donna McLaren** (1021), Department of Molecular Genetics, Biochemistry and Microbiology, University of Cincinnati College of Medicine, Cincinnati, Ohio 45267-0524

**Gerhard Meissner** (817), Departments of Biochemistry and Biophysics, and Physiology, University of North Carolina, Chapel Hill, North Carolina 27599-7260

**Lauren A. Millette** (216), Department of Biological Sciences, University of Cincinnati, Cincinnati, Ohio 45267

**Stanley Misler** (652), Departments of Medicine and Cell Biology/Physiology, Washington University Medical Center, St. Louis, Missouri 63110

**Catherine E. Morris** (668), Loeb Institute, Ottawa Civic Hospital, Ottawa, Ontario, Canada K1Y 4E9

**Edward F. Nemeth** (142), NPS Pharmaceuticals, Inc., Salt Lake City, Utah 84108

**John C. New** (741), Department of Biology and Parmly Hearing Institute, Loyola University, Chicago, Illinois 60626

**Bernd Nilius** (436), Department of Physiology, KU Leuven, B-3000 Leuven, Belgium

**Richard J. Paul** (830), Department of Molecular and Cellular Physiology, University of Cincinnati College of Medicine, Cincinnati, Ohio 45267

**Marcel Daniel Payet** (632), Department of Physiology and Biophysics, Faculty of Medicine, University of Sherbrooke, Sherbrooke, Québec, Canada J1H 5N4

**J. Wesley Pike** (153), Department of Molecular and Cellular Physiology, University of Cincinnati, Cincinnati, Ohio 45267

**Matthew R. Pincus** (18), Department of Pathology and Laboratory Medicine, Veterans Administration Medical Center, Brooklyn, New York 11235

**David M. Pressel** (652), Departments of Medicine and Cell Biology/Physiology, Washington University Medical Center, St. Louis, Missouri 63110

**Raymund Y. K. Pun** (391), Department of Molecular and Cellular Physiology, University of Cincinnati College of Medicine, Cincinnati, Ohio 45267-0576

**Robert W. Putnam** (293, 312), Department of Physiology and Biophysics, Wright State University, School of Medicine, Dayton, Ohio 45435

**Stephen Roper** (726), Department of Physiology and Biophysics, University of Miami School of Medicine, Miami, Florida 33146

**Friedhelm Schroeder** (61), Department of Physiology and Pharmacology, Texas A&M University, College Station, Texas 77843

**Richard G. Sleight** (91, 119), Yale University Graduate School, New Haven, Connecticut 06520-8236

**Nicholas Sperelakis** (171, 178, 202, 345, 368, 499, 518, 761, 1061), Department of Molecular and Cellular Physiology, University of Cincinnati College of Medicine, Cincinnati, Ohio 45267-0576

**Kira Steigerwald** (1021), Department of Molecular Genetics, Biochemistry and Microbiology, University of Cincinnati College of Medicine, Cincinnati, Ohio 45267-0524

**Richard G. Stout** (953), Department of Biology, Montana State University, Bozeman, Montana 59717-0346

**Janusz B. Suszkiw** (610), Department of Molecular and Cellular Physiology, University of Cincinnati College of Medicine, Cincinnati, Ohio 45267

**Andre Terzic** (532), Division of Cardiovascular Diseases and Medicine, Mayo Clinic, Rochester, Minnesota 55905

**Noritsugu Tohse** (518), Department of Physiology, School of Medicine, Sapporo Medical University, Chuo-Ku, Sapporo 060, Japan

**Timothy C. Tricas** (741), Department of Biological Sciences, Florida Institute of Technology, Melbourne, Florida 32901-6988

**Richard D. Veenstra** (467), Department of Pharmacology, State University of New York Health Science Center at Syracuse, Syracuse, New York 13210

**Gordon M. Wahler** (780), Midwestern University, Downer's Grove, Illinois 60515

**Earl T. Wallick** (216), Department of Pharmacology and Cell Biophysics, University of Cincinnati College of Medicine, Cincinnati, Ohio 45267-0575

**Gary L. Westbrook** (597), Vollum Institute, Oregon Health Sciences University, Portland, Oregon 97201

**Atsuko Yatani** (510), Department of Pharmacology and Cell Biophysics, University of Cincinnati College of Medicine, Cincinnati, Ohio 45267

**Hisashi Yokoshiki** (518), Department of Molecular and Cellular Physiology, University of Cincinnati College of Medicine, Cincinnati, Ohio 45267

**Anita L. Zimmerman** (707, 718), Department of Molecular Pharmacology, Physiology, and Biotechnology, Brown University, Providence, Rhode Island 02912

# Foreword to First Edition

It was kind and generous of my friend Nicholas Sperelakis to relate this excellent book so closely to my own *A Textbook of General Physiology*. In the preface to the first edition of my book, I had expressed the hope that it might be compared with Bayliss' *Principles of General Physiology*. If this comparison is valid, it is very appropriate that the present book be organized and partly written by one who is, along with Sir William Bayliss and myself, associated with University College London (Professor Sperelakis having spent a sabbatical year there). It is a pleasure to recall that it was there that I first met Nicholas, and I remember discussing his pioneering study on the potentials across the crystalline lens of the eye.

For reasons that I think bear no relation to its scope, this book has a different title; I presume it is because the distinction between "ordinary" and "general" physiology has become sufficiently blurred to demand something more appropriate. The definition of general physiology that I had proposed in the preface to the first edition of my textbook was "the study of those aspects of living material that show some immediate prospect of being described in terms of the known laws of physics and chemistry." Later, I had misgivings as to the narrowness of this definition, and I then suggested that it might be replaced by "the study of those features of life that appear to be common to all forms." Whatever definition we choose, however, it is of immense satisfaction to me that this new book, essentially, has been fitted into the same sectional headings that I employed in my own book.

If I may be permitted to reminisce further, I have wondered frequently how a single scientist, actively engaged in research, could write a new book of such wide scope. The answer is that I wrote the book during the 2 years immediately following the end of World War II. Thus, for several years—in Great Britain for as many as 10 years—very little original academic physiology and new research had been published. This made it possible to survey the original literature of a lengthy period without being overwhelmed by a rapid succession of new discoveries that would have rendered my task nearly impossible, a fate similar to that of Sisyphus. Today, this task would be impossible, and it only surprises me that Nicholas has been able to produce this magnificent book with so few collaborators.

*Hugh Davson*
1995

# Foreword to Second Edition

In his Foreword to the first edition of the *Cell Physiology Source Book,* Hugh Davson established it as the lineal descendent of his own well-known and highly respected work *The Textbook of General Physiology.* The second edition of the *Cell Physiology Source Book,* again edited by Nicholas Sperelakis, continues in this same tradition. Although the first edition was enthusiastically received by the cell physiology community because of its depth and breadth of coverage, considerable important progress has been made in this rapidly developing area since its publication. The second edition deals with these new developments by a thorough reworking of topics and by the inclusion of new chapters in all sections covered in the first edition. The new topics introduced into the various sections include lipid structure, mitochondrial physiology, cell responses to hormones, red blood cell transport, neuron physiology, developmental changes in ion channels, sonotransduction, excitation–contraction coupling, and electroplax cells. In addition, the scope of the new edition has been valuably broadened by the inclusion of two entirely new sections. One titled *Protozoa and Bacteria* covers the physiology of these organisms in two chapters. In the other, *Cell Division and Programmed Cell Death,* there are chapters on the regulation of cell division, the cancer cell, apoptosis, and the effects of ionizing radiation. The extensive revisions and the new material in the second edition raise it to a new level.

Cell physiology, an area of central importance in biology, has grown out of a number of more traditional fields, and as a result, the literature continues to be widely dispersed. The great value of the *Cell Physiology Source Book* is that it gathers together under a single cover a broad range of up-to-date chapters that, taken together, define the field. The various chapters exhibit a uniformity of style and level of presentation that are a credit to the editor. Because of this and the scope and clarity of the presentations, this book can serve exceptionally well as an advanced undergraduate- or graduate-level text for cell physiology courses. The broad coverage of this second editon also makes it very attractive for use in cell biophysics, membrane biology, and biomedical engineering courses. It can serve equally well as a textbook for introductory courses in ion channel structure and physiology.

I was pleased, and indeed proud, to be asked by my colleague Nicholas Sperelakis to contribute the Foreword to the second edition of the *Cell Physiology Source Book.* This book clearly sets a new standard of excellence.

*Thomas E. Thompson*
1997

# Preface to First Edition

Hugh Davson's original textbook on cell physiology, *A Textbook of General Physiology,* was a true classic and a huge success. The first edition was published in 1951 and the fourth and final edition in 1970. In the past two decades, cell physiology has advanced by leaps and bounds, as, for example, in the area of ion channels and their regulation. The present book attempts to fill the void left by the discontinuation of Davson's monumental book. This book is dedicated to Professor Davson in recognition of the great impact he has had on the dissemination of the principles of cell physiology.

At present, there is a need for a good text on cell physiology. Several good texts on cell biology are available but these generally treat cell physiology in a superficial and incomplete manner. Therefore, it was our intent to prepare a new work on cell physiology that is high quality, is comprehensive, and covers recent developments (the chapter authors were asked to limit their bibliography to no more than 40 key papers and review-type articles). We hope that the reader will find this book clearly written, thorough, up to date, and worthy of being the successor to Professor Davson's book.

This book focuses on physiology and biophysics at the cellular level. Organ systems and whole organisms are essentially not covered, except for unicellular organisms. The major topics covered include ultrastructure, molecular structure and properties of membranes, transport of ions and nonelectrolytes, ion channels and their regulation, membrane excitability, sensory transduction mechanisms, synaptic transmission, membrane receptors, intracellular messengers, metabolism, energy transduction, secretion, excitation–secretion coupling, excitation–contraction coupling, contraction of muscles, cilia and flagellae, photosynthesis, and bioluminescence. These topics are covered in a comprehensive, but didactic, manner, with each topic beginning in an elementary fashion and ending in a sophisticated and quantitative treatise.

This book is intended primarily for graduate and advanced undergraduate students in the life sciences, including those taking courses in cell physiology, cell biophysics, and cell biology. Selected parts of this book can be used for courses in neurobiology, electrobiology, electrophysiology, secretory biology, biological transport, and muscle contraction. Students majoring in engineering, biomedical engineering, physics, and chemistry should find this book useful to help them understand the living state of matter. Postdoctoral scholars and faculty engaged in biological research would also find this text of immense value for obtaining a better understanding of cell function and as a reference source. Medical, dental, and allied health students could use this book as a valuable comparison text to complement their other textbooks in medical/mammalian physiology. The latter texts often strongly emphasize organ system physiology to the virtual exclusion (in some cases) of cell physiology, membrane biophysics, and mechanisms underlying homeostasis of the entire organism. The chapter authors were asked to include the following aspects wherever appropriate: comparative physiology, developmental changes, pathophysiology, membrane diseases, and molecular biology.

Considering the tremendous amount of new information gained during the past two decades using sophisticated new instruments and techniques, I have recruited a number of outstanding researchers, who are leaders in their respective fields, to contribute to this undertaking. I am delighted that they agreed to participate. It has been my great pleasure and honor to work with them on this important and timely project.

*Nicholas Sperelakis*
1995

Preface to First Edition

# Preface to Second Edition

The first edition of the *Cell Physiology Source Book* was conceived to serve as the replacement for Davson's classic textbook of cell physiology, which was discontinued after the fourth edition was published in 1970. There was a pressing need for a comprehensive and authoritative textbook on cell physiology because the available cell biology textbooks treated cell physiology in a superficial and incomplete manner. To accomplish this Herculean task, I recruited a number of outstanding researchers who were leaders in their respective fields to contribute to the undertaking. The result was a comprehensive source book (738 pages, hardcover) that appeared in February 1995 and rapidly became successful. Within a year, the book had to be reprinted in hardcover, and a softcover version was printed. Book reviews were complimentary to the book, and I received numerous comments from contributors and other scientists about the excellence of the *Source Book*. One of my favorite comments was made by an Israeli physical chemist at a conference in Prague where the book was on display. He said, "Nick, this is the type of book that *speaks* to me." The first edition was selected as one of the outstanding academic books for 1996 by *CHOICE* (current reviews for academic libraries published by the Association of College & Research Libraries, a division of the American Library Association).

I began to formulate plans for the second edition almost immediately after the appearance of the first edition. I believed that it was important for the second edition to appear within 3 years after the first edition. It was especially important to have a relatively short interval because the field is advancing so rapidly. I also wanted to take the opportunity to include some new chapters on topics quite relevant to the discipline of cell physiology. In addition, several of the original chapters were expanded, reorganized, and restructured, and several other chapters were shifted in location, with the goal of making the *Source Book* more useful as a textbook. A total of 18 new chapters were added, and two new sections were organized: Section VII, Protozoa and Bacteria; and Section IX, Cell Division and Programmed Cell Death. The new chapters include lipid structure, physiology of mitochondria, responses to hormones, transport in erythrocytes, developmental changes of ion channels, calcium receptors, cytoskeletal modulation of ion channels, neuronal cells, sonotransduction, excitation–contraction coupling, electroplax cells, electroreceptors, protozoa, bacteria, cell division, cancer cells, apoptosis, and ionizing radiations.

These additional chapters should make the *Source Book* more complete and more useful to graduate students in neuroscience, biology, cell biology, transport physiology, and electrophysiology. Selected chapters in the *Source Book* can be used for courses in cell physiology, cell biophysics, electrophysiology, electrobiology, transport physiology, secretory biology, sensory physiology, neurobiology, and contractile systems. Students majoring in the physical sciences should find this book helpful in understanding the living state of matter. This book is intended primarily for graduate students and advanced undergraduate students majoring in the life sciences.

A new edition also allows the individual contributor the opportunity of adjusting the level at which his or her chapter is pitched (i.e., to aim it at the median level at which the majority of the chapters are pitched). Thus, parts of some chapters were expanded, whereas parts of others were constricted. Certain figures were improved. The subject index was improved by limiting entries to first order and second order.

I sincerely hope that the student will find this book to be clearly written, thorough, up to date, and worthy of being the successor to Davson's monumental book. The book contains several opening chapters and an appendix to help the student review the physical chemistry of solutions, protein structure, lipid structure, and electricity. Hence, within the covers of this one book, the student should have all the information needed to understand the various topics that constitute the discipline of cell physiology. The contributors, publisher, and I want to establish this book as the leading textbook of cell physiology.

*Nicholas Sperelakis*
1997

# Biophysical Chemistry, Metabolism, Second Messengers, and Ultrastructure

*Jeffrey C. Freedman*

# 1

# Biophysical Chemistry of Physiological Solutions

## I. Introduction

All living cells contain proteins, salts, and water enclosed in membrane-bounded compartments. These biochemical and ionic cellular constituents, along with a set of genes, enzymes, substrates, and metabolic intermediates, function to maintain cellular homeostasis and enable cells to perform chemical, osmotic, mechanical, and electrical work. *Homeostasis,* a term introduced by the physiologist Walter Cannon in his book *The Wisdom of the Body* (1932), means that certain parameters, including the cellular volume, the intracellular pH, and the intracellular concentrations of salts are maintained relatively constant in resting cells. Cellular homeostasis depends on a relative constancy of the extracellular fluids that bathe cells. The extracellular fluid compartment was termed the *milieu intérieur,* or internal environment, by Claude Bernard, who recognized around 1865 that "La constance du milieu intérieur est la condition de la vie libre," or the constancy of the internal environment is the condition for an independent life. In *An Introduction the Study of Experimental Medicine* (see 1949 translation, p. 99), Bernard wrote that "only in the physico-chemical conditions of the inner environment can we find the causation of the external phenomena of life," a philosophy that has greatly influenced the development of cell physiology.

Biophysical chemistry concerns the application of the concepts and methods of physical chemistry to the study of biological systems. Physical chemistry includes such physiologically relevant subjects as thermodynamics, chemical equilibria and reaction kinetics, solutions and electrochemistry, properties and kinetic theory of gases, transport processes, surface phenomena, and molecular structure and spectroscopy. It is seen throughout this book that many cellular physiological phenomena are best understood with a rigorous and comprehensive understanding of physical chemistry. Physical chemistry texts that specifically emphasize biological applications include those by Eisenberg and Crothers (1979) and by Tinoco *et al.* (1994). During the past 50 years, outstanding monographs on biophysical chemistry have also been available: Höber (1945), Edsall and Wyman (1958), Tanford (1961), Cantor and Schimmel (1980), Silver (1985), van Holde (1985), and Bergethon and Simons (1990). To begin to understand how cellular homeostasis is achieved, this chapter and Chapter 2 introduce some of the conceptual underpinnings of cell physiology by describing certain physical chemical properties of water, electrolytes, and proteins that are relevant for understanding the structure and function of cells.

## II. Structure and Properties of Water

Biological cells contain a large amount of water, ranging from 0.66 g $H_2O$/g cells in human red blood cells to around 0.8 g $H_2O$/g tissue in skeletal muscle. The amount of water in cells is determined by osmosis (for review, see Dick, 1959, and Chapter 19). Liquid water is a highly polar solvent, with a structure stabilized by extensive intermolecular *hydrogen bonds* (Fig. 1). The hydrogen bonds between adjacent water molecules are linear ($O-H \cdots O$) but are able to bend by about 10°. From X-ray diffraction studies of ice crystals, it is known that the two covalent $O-H$ bonds in water are each 1 Å in length, that the two hydrogen bonds are each about 1.8 Å in length, and that these occur around each oxygen atom at angles of 104°30′ in a tetrahedral array. The work required to break hydrogen bonds in liquid water is 5 to 7 kcal/mol, much less than the 109.7 kcal/mol required to break the covalent $O-H$ bond. A calorie is a unit of work equal to 4.18 J in the SI system of units, where 1 J equals 1 N$-$m. The newton ($-$ kg · m/sec$^2$) is the unit of force in the SI system; the dyne (= g · cm/sec$^2$) is the corresponding unit of force in the older cgs system (1 N = $10^5$ dynes). Recall that, ac-

**3**

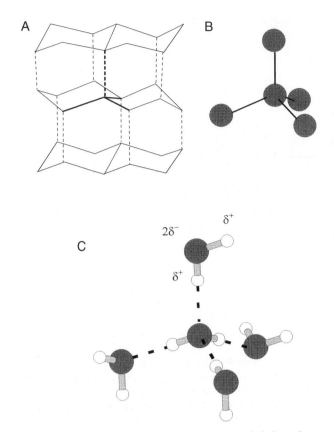

**FIG. 1.** Hydrogen bonds in ice and liquid water. (A) Crystal structure of ice, showing the puckered hexagonal rings of oxygen atoms. (B) Each oxygen atom is connected to four others in a tetrahedron. (C) The linear hydrogen bonds are indicated by dotted lines between adjacent water molecules ($O-H\cdots O$). Two covalent bonds ($O-H$), indicated by solid lines, and two hydrogen bonds occur around each oxygen atom in a tetrahedral array. The partial charge separation of the $O-H$ dipoles is indicated on the upper water molecule.

cording to Newton's first law, the force ($F$) is mass times acceleration, and that work is defined as force times distance. Because of the strength of hydrogen bonds, liquid water has unique physicochemical properties, including a high *boiling point* (100°C), a high *heat capacity* (18 cal/mol · °K), a high *heat of vaporization* (9.7 kcal/mol at 1 atm), and a high *surface tension* (72.75 dynes/cm). As discussed by Bergethon and Simons (1990), all of these thermodynamic parameters are considerably higher for $H_2O$ than for the analogous compound $H_2S$, which boils at −59.6°C. The outer electrons of the S atom shield its nuclear charge and reduce its electronegativity, making the $H-S$ bond weaker, longer, and less polar than the $O-H$ bond. Also, the bond angle of $H_2S$ is only 92°20′, an angle that does not form a tetrahedral array, and thus $H_2S$ does not form a hydrogen-bonded network like water.

The extensive intermolecular association in liquid water is due both to the geometry of the water molecule and to its strong permanent *dipole moment*. Water molecules are electrically neutral and have no net charge, yet possess permanent charge separation along each of the two $O-H$ bonds. Since oxygen is more electronegative than hydro-

gen, the electron density is preferentially greater near oxygen, conferring a partial negative charge ($\delta^-$), thus leaving the hydrogen atoms with a partial positive charge ($\delta^+$) (Fig. 1C). The magnitude of the dipole moment ($\mu$) of a chemical bond is computed as the product of the separated charge ($q$) at either end, and the distance ($d$) between the centers of separated charge ($\mu = q \cdot d$). The dipole moment of a molecule is the vector sum of the dipole moments of each bond. For water in the gaseous phase the dipole moment is 1.85 Debye, where 1 Debye = $10^{-18}$ esu · cm. One esu (electrostatic unit), or statcoulomb, equals $3.336 \times 10^{-10}$ C, the coulomb being a unit of electric charge. In liquid water, the dipole moment becomes even larger (2.5 Debye) because of association with other water molecules. In comparison, the linear molecule carbon dioxide ($O=C=O$) also has separation of charge, with a partial negative charge on each oxygen atom and a partial positive charge on the carbon atom, but the oppositely directed dipole moments of the two $C=O$ bonds sum to a zero dipole moment for the $CO_2$ molecule. The $S-H$ bond has a smaller dipole moment (1.1 Debye) than the $O-H$ bond because sulfur in $H_2S$ is less electronegative than oxygen. Electrostatic attractions between adjacent water dipoles stabilize the hydrogen-bonded structure of liquid water.

Water strongly affects the forces between ions in solution by virtue of its high *dielectric constant*. In a vacuum, the coulombic force, $F_{coul}$ (newtons), between two ions with charges $q^+$ and $q^-$ (coulombs) separated by a distance $d$ (meters) is given by *Coulomb's law*:

$$F_{coul} = \frac{q^+ q^-}{4\pi\varepsilon_o d^2}$$

where $\varepsilon_o$ is the *permittivity constant* ($8.854 \times 10^{-12}$ C²/N · m²). Coulomb's law states that the force between two point charges is directly proportional to the magnitude of the charges, and inversely proportional to the square of the distance between the charges. Note that the term $4\pi r^2$ ($r = d/2$) represents the surface area of a sphere centered halfway between the two charges.

The *dielectric constant $\varepsilon$* of a medium known as a *dielectric* is defined as the ratio of the coulombic force $F_{coul}$ between two charges in a vacuum to the actual force $F$ between the same two charges in the dielectric medium.

$$\varepsilon = \frac{F_{coul}}{F}$$

Substituting for $F_{coul}$ and rearranging yields

$$F = \frac{q^+ q^-}{4\pi\varepsilon_o \varepsilon d^2}$$

The dielectric constant of a vacuum is unity and that of air is close to unity ($\varepsilon_{air} = 1.00054$). Increasing the dielectric constant of the solvent decreases the attractive force between oppositely charged ions. The reason that the force felt by a distant ion is less in a dielectric medium such as water than in a vacuum is that part of the interaction energy is spent in aligning the intervening water dipoles, and in distorting their polarizable electron clouds. The dielectric constant of water at 25°C is 78.5, much greater than that of methanol ($\varepsilon = 32.6$), ethanol (24.0), or methane (1.7). In liquid water, the high dielectric constant weakens the coulombic attractive forces between oppositely charged particles, and thus promotes dissociation and ionization of salts.

Although the thermodynamic properties of liquid water are explicable in terms of extensive hydrogen bonding, the actual structure of water is still unknown (see Eisenberg and Kauzmann, 1969). In the flickering cluster model, groups of 50–70 water molecules, resembling a slightly expanded broken piece of the ice lattice ("icebergs"), are continuously associating and dissociating on a picosecond time scale, in contrast to the extended ordered structure of solid ice. At a given instant, some water molecules are unattached to the clusters and are located in the interstitial regions of the network, but may attach and detach as the clusters continuously form and break down. Other theories treat water as mixtures of distinct states, or as a continuum of states, with considerable short-range order characteristic of the crystalline lattice of ice.

Despite these uncertainties regarding the structure of liquid water, and regarding the influence of macromolecules on its properties in cells (see Cooke and Kuntz, 1975), the ability of water to act as a solvent inside cells closely resembles that of extracellular water. Thus, a variety of nonelectrolytes that are permeant, hydrophilic, and non-metabolized distribute at equilibrium across human red cell membranes with ratios of intracellular-to-extracellular concentrations that deviate from unity by less than 10% (Gary Bobo, 1967). In mouse diaphragm muscle, C. Miller (1974) found that several alcohols, diols, and mono-saccharides exhibit distribution ratios within 2% of unity, whereas certain other sugars appear to be excluded from membrane-bounded intracellular compartments. Some intracellular water is constrained in *gels*, which are cross-linked networks of fibrous macromolecules, but the majority of this water also has normal solubility properties. The diffusion of solutes within gels may be hindered by collisions and interactions with the macromolecules, and by the tortuosity of the diffusion paths.

Another property of water, *viscosity*, contributes to its resistance to flow; a fluid with a greater viscosity exhibits less flow under the influence of a given pressure gradient than a fluid with a lesser viscosity. During *laminar flow*, a frictional force develops between adjacent laminae in the fluid; this force impedes the sliding of one lamina past its neighbor. In a *Newtonian fluid* the frictional force per unit area, or shear stress $\tau$ (dyne/cm$^2$), is proportional to the velocity gradient, or rate of strain $dv/dy$ (sec $^{-1}$) between laminae,

$$\tau = \eta \cdot (dv/dy)$$

where the viscosity $\eta$ is the proportionality constant with units of poise (= 1 dyne · sec/cm$^2$), named after Poiseuille. The viscosity of H$_2$O at 20.3°C is 0.01 poise, or 1 centipoise (cp). In cultured fibroblasts, the fluid-phase cytoplasmic *viscosity,* as determined from rotational motions of fluorescent probes on a picosecond time scale, is only 1.2–1.4 times that of pure water (Fushimi and Verkman, 1991).

Fluorescence studies also show that the viscosity is the same in the cytoplasm and nucleoplasm and is unaffected by large decreases in cell volume or by disruption of the cytoskeleton with cytochalasin B. The fluid-phase viscosity, as determined from fluorophore rotational motions, is affected much less by macromolecules than is bulk viscosity and provides a more accurate view of the physical state of the aqueous domain of the cytoplasm. This and other studies (see Horowitz and Miller, 1984) set limits on the extent to which the physicochemical properties of intracellular water differ from those of extracellular water.

## III. Interactions between Water and Ions

Ions in solution behave as charged hard spheres that interact with and orient water dipoles. When crystals of sodium chloride are dissolved in water, the electrostatic attractive forces between water dipoles and ions in the crystal lattice overcome the interionic attractive forces between oppositely charged ions in the crystal. The dissociated ions then acquire the freedom of translational motion as they diffuse into the solution accompanied by a layer of tightly associated *hydration water* (Fig. 2). *Ion* is the Greek word for "wanderer." The strength of the attraction between ions and water dipoles depends to a large extent on the ionic charge and radius. For the alkali metal cations, the force of attraction and the energy of interaction of ions with water decreases according to the following series:

$$Li^+ > Na^+ > K^+ > Rb^+ > Cs^+$$

Li$^+$, being the smallest of the alkali metal cations, has the strongest interaction with water because its positively charged nucleus can more closely approach the negative side of neighboring water dipoles. As the ionic radius increases with increasing atomic number in the alkali metal series, the filled outer shells of electrons effectively shield the cationic charge and reduce the distance of closest approach to water molecules. The smallest ion thus acquires

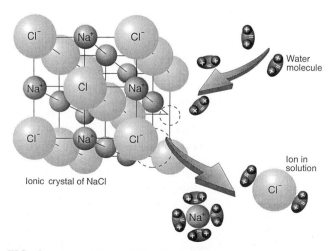

**FIG. 2.** Dissociation of NaCl in water into hydrated Na$^+$ and Cl$^-$ ions.

the greatest degree of hydration and has the largest hydrated ionic radius.

According to Frank and Wen (1957), the orienting influence of ions on water dipoles results in three regions of water structure (Fig. 3). The electric field of an ion is sufficiently strong to remove water dipoles from the bulk water clusters and to attract to itself the oppositely charged ends of one to five water dipoles. A certain number of these water dipoles, called the *hydration number,* then become trapped and oriented in the ion's electric field. The inner hydration shell includes water molecules that are aligned by the force field and that are in direct contact with the ion, $5 \pm 1$ for $Li^+$, $4 \pm 1$ for $Na^+$, $3 \pm 2$ for $K^+$ and $Rb^+$, $4 \pm 1$ for $F^-$, $2 \pm 1$ for $Cl^-$ and $Br^-$, and $1 \pm 1$ for $I^-$. Thus, in physiological saline at 0.15 *M* NaCl in water (55.5 *M*), about 1.6% ($= 100 \times 0.15 \times 6/55.5$) of the water is located in the inner hydration shells of $Na^+$ and of $Cl^-$. Water in the inner hydration shells of $Na^+$, $K^+$, and $Ca^{2+}$ rapidly exhanges with bulk water on a nanosecond time scale. In contrast, the smaller divalent cation $Mg^{2+}$ with its high charge density is some four orders of magnitude slower in exchanging its inner hydration water. The inner hydration sheath of water molecules moves together with the ion as a distinct and single kinetic entity. The mobilities of the alkali cations in water decrease as the nonhydrated ionic radius decreases, but as the hydrated ionic radius increases (Table 1; for discussion, see Hille, 1992). The ionic *mobility* ($u$) is the proportionality constant that relates the velocity ($v$) of ionic migration to the force exerted by an external electric field ($E$) ($v = u \cdot E$); in other words, mobility is the velocity per unit electric field. Farther away from the ion, where the

**TABLE 1**  Radii, Enthalpies of Hydration, and Mobilities of Selected Ions

| Ion | Nonhydrated radius[a] (Å) | $\Delta H°_{\text{hydration}}$[b] (kcal/mol) | Mobility[c] $10^{-4}$ $\dfrac{\text{cm/sec}}{\text{V/cm}}$ |
|---|---|---|---|
| $H^+$ | — | −269 | 36.25 |
| $Li^+$ | 0.60 | −131 | 4.01 |
| $Na^+$ | 0.95 | −105 | 5.19 |
| $K^+$ | 1.33 | −85 | 7.62 |
| $Rb^+$ | 1.48 | −79 | 8.06 |
| $Cs^+$ | 1.69 | −71 | 8.01 |
| $Mg^{2+}$ | 0.65 | −476 | 2.75 |
| $Ca^{2+}$ | 0.99 | −397 | 3.08 |
| $Sr^{2+}$ | 1.13 | −362 | 3.08 |
| $Ba^{2+}$ | 1.35 | −328 | 3.30 |
| $Cl^-$ | 1.81 | −82 | 7.92 |

[a] Radii are from Pauling (1960).
[b] Standard enthalpies of hydration at 25°C are from Edsall and McKenzie (1978).
[c] Mobilities in water at 25°C are from Hille (1992, p. 268).

ion's electric field falls to zero, the water retains the structure of bulk water. In the region between the inner hydration sheath and the bulk water, the orienting influences of the ion and the bulk water network tend to compete and to disrupt the structure of water. In this intermediate structure-breaking region, the water dipoles are partially oriented toward the central ion, yet do not migrate with the ion and join only infrequently with the hydrogen-bonded clusters of the bulk water. If the net effect of an ion is to disorganize more water in the intermediate region than is found in the primary hydration shell, then the ion is termed a "*structure breaker.*" Conversely, water may form a variety of hydrogen-bonded clathrate structures around apolar protein side chains, whose actions resemble that of the class of solutes termed "*structure makers*" (see Klotz, 1970, for review).

The *enthalpy of hydration* of an ion is a measure of the strength of the interaction between ions and water; it is defined as the increase in enthalpy when 1 mole of free ion in a vacuum is dissolved in a large quantity of water. The enthalpy of hydration may be estimated from the heat released upon dissolving salts in water, usually less than 10 kcal/mol, taking into account the energy needed to dissociate the salt crystal and then to hydrate the ions. The enthalpy of hydration may be calculated using either the Born charging method, which estimates the energy needed to transfer a rigid charged sphere into a structureless continuum, or, alternatively, by the Bernal–Fowler structural method, which takes into account the dipolar structure of water (see Bockris and Reddy, 1970). More accurate results are obtained if water is considered to be an electric quadrupole with four centers of charge, two partial positive charges near the hydrogen nuclei and two partial negative charges on the nonbonded electron orbitals near the oxygen nucleus. A further improvement in the calculated enthalpy of hydration is obtained when the *polarizability* ($\alpha$) of the water dipole by the ion is taken

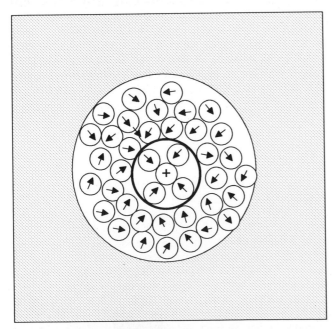

**FIG. 3.**  Three regions of water structure near an ion. A central cation is shown with a primary hydration shell of four oriented water dipoles, surrounded by a partially oriented secondary sheath. The gray region indicates the hydrogen-bonded structure of the bulk water.

into account. The electric field of the ion distorts the electron cloud of the hydration water along its permanent dipole axis, thus inducing an additional increment of charge separation and increasing the dipole moment. For small fields, the *induced dipole moment* ($\mu_{ind}$) is proportional to the electric field strength ($E$), and the constant of proportionality is the polarizability ($\mu_{ind} = \alpha \cdot E$). In Fig. 4, the measured enthalpies of hydration for alkali metal cations and halides are compared with calculated values; the impressive agreement demonstrates the primary importance of electrostatic forces in determining the solvation of ions in water. The enthalpy of hydration for the smallest alkali metal cation $Li^+$ is quite large at −131 kcal/mol. With increasing ionic radius, the enthalpy falls progressively for $Na^+$, $K^+$, and $Rb^+$, reaching −71 kcal/mol for $Cs^+$, the largest ion in the series (see Table 1). For the divalent, alkaline earth metal ions $Mg^{2+}$ and $Ca^{2+}$, the enthalpies of hydration also follow the ionic radius but are considerably larger at −476 and −397 kcal/mol, respectively. As pointed out by Hille (1992), the magnitude of ionic hydration enthalpies approximates the cohesive strength of the ionic bonds in a crystalline salt lattice.

The relationship between the dielectric constant, a macroscopic property of water, and its molecular properties—its dipole moment and polarizability—was discovered by the physical chemist Peter Debye (1929). In a gas consisting of dipoles at concentration $n$, the macroscopic dielectric constant $\varepsilon$ is related to the polarizability $\alpha$ and the molecular dipole moment $\mu$ at temperature $T$ by the Debye equation:

$$\varepsilon - 1 = 4\pi n\alpha + \frac{4\pi n\mu^2}{3kT}$$

where $k$ is Boltzmann's constant ($1.38 \cdot 10^{-23}$ J/°K). In this equation, the first term represents the effect of the polarizability $\alpha$ of the electron clouds, and the second term

represents the effect of reorienting the permanent dipoles, with moments $\mu$, in an electric field, as opposed by the randomizing influence of thermal energy $kT$. Thus, increasing the temperature reduces the dielectric constant by disorienting the dipoles. In a condensed polar dielectric such as water, the Kirkwood equation for the dielectric constant takes into account the formation of molecular groups with $g$ nearest neighbors linked to the central molecule. These groups, which could be a tetrahedral group of water molecules, orient as a unit in the specific local electric fields, as distinct from an externally applied field. The Kirkwood equation is

$$\frac{(\varepsilon - 1)(2\varepsilon + 1)}{9\varepsilon} = \frac{4\pi n}{3}\left[\alpha + \frac{\mu^2(1 + g\overline{\cos\gamma})^2}{3kT}\right]$$

where $\overline{\cos\gamma}$ is the average of the cosines of the angles between the dipole moment of the central water molecule and those of its bonded neighbors. In small spaces that restrict the formation of tetrahedral water clusters such that $g$ is reduced, the Kirkwood equation predicts that the dielectric constant will also be reduced, and electrostatic forces between ions will then be correspondingly increased. In the primary hydration shell of ions, the oriented water is polarizable but cannot be reoriented by applied fields, and so the dielectric constant is reduced from its bulk value of 78 to a value of about 6, with intermediate values in the partially oriented structure-breaking region between the primary hydration shell and the bulk water. For this reason, the dielectric constant of a solution decreases with increasing salt concentration, but in protein-free solutions at physiological salt concentrations the extent of the decrease is small because of the small fraction of hydration water.

Some membrane channels are sufficiently narrow that ions and water permeate by a process of *single-file diffusion* in which water and ions cannot pass each other, at least in the narrowest part of the channel (for review, see Finkelstein and Andersen, 1981, and Chapter 12). In narrow channels, where the diameter of the permeating ion may approximate that of the channel itself, the hydration water is stripped from the ion and is probably replaced by dipolar groups of the proteins lining the channel walls, thus providing electrostatic stabilization of the permeating ion.

## IV. Protons in Solution

The hydrogen-bonded structure of liquid water also contributes to high proton mobility by a mechanism that differs fundamentally from the migration of other hydrated ions. The proton is a highly reactive, positively charged hydrogen nucleus devoid of electrons. Whereas ions with electron shells typically have diameters in angstroms, the diameter of the proton is only about $10^{-5}$ Å. With such a small size, the proton has a strong attraction to electrons, as indicated by the high *ionization energy* of 323 kcal/mol needed to remove an electron from a hydrogen atom to form a proton. By comparison, the ionization energies for the alkali cations are $Li^+$, 124 kcal/mol; $Na^+$, 118 kcal/mol;

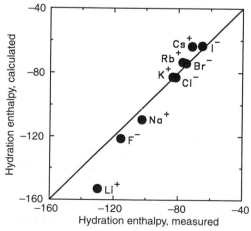

**FIG. 4.** Comparison of measured enthalpies of hydration of alkali metal cations and halides in water (abscissa), with those calculated (ordinate) taking into account the Born charging energy, ion–dipole interactions, ion–quadruple interactions, and ion-induced dipole interactions. (Data from Bockris and Reddy, 1970, p. 107.)

$K^+$, 100 kcal/mol; $Rb^+$, 96 kcal/mol; and $Cs^+$, 90 kcal/mol. The ionization energy decreases with increasing atomic number in the series as more filled electron shells separate and shield the outer shell from the positively charged atomic nucleus, thus making it easier to form a cation by removing an electron. The high affinity of the proton for electrons, and its small size, explains its tendency to form hydrogen bonds with the unshared electrons of oxygen in water. Infrared spectra indicate that protons in solution exist predominantly in the form of *hydronium* ions $H_3O^+$. Nuclear magnetic resonance (NMR) data are consistent with a flattened trigonal pyramidal structure, with $O-H$ bond lengths of 1.02 Å and $H-O-H$ bond angles of 115°, a structure resembling that of $NH_3$ but with a radius similar to that of $K^+$. The free energy of formation of $H_3O^+$ from $H_2O$ and $H^+$ is about −170 kcal/mol, which corresponds to a hypothetical concentration of free protons in solution at room temperature of about $10^{-150}$ $M$—"about as zero as one can get," say Bockris and Reddy (1970). The heat of hydration of $H_3O^+$ is an additional −90 kcal/mol. The change in the density of water with temperature is consistent with the hydration of $H_3O^+$ with an additional three water molecules in the tetrahedral cluster $H_9O_4^+$ (Bockris and Reddy, 1970). The proton mobility in water is 36 · $10^{-4}$ $cm^2$/sec · V, much slower than expected on the basis of hydrodynamic theory if free protons were to carry the current, but curiously about seven times faster than expected if $H_3O^+$ migrates like $K^+$. The abnormally high proton mobility in water (and also in ice) is consistent with a *Grotthus chain mechanism* (Fig. 5) in which the successive breakage and formation of hydrogen bonds, accompanied by proton jumps between neighboring water molecules, effectively results in the passage of $H_3O^+$ along a chain. The transport process is rate limited by the time needed for each successive acceptor water molecule to rotate and reorient its nonbonded orbital into a suitable position to accept the donated proton. The Grotthus mechanism accounts for about 80% of proton mobility, whereas the remaining 20% is due to $H_3O^+$ itself, as a single kinetic entity, undergoing a translational migratory movement through the solvent like other ions.

## V. Interactions between Ions

Interactions between ions may be weak or highly selective. Ions in solution attract ions of the opposite charge in accordance with Coulomb's law. Since the long-range attractive forces are inversely proportional to the square of the distance between charges, the interaction energies are greater in more concentrated solutions. In a uni-univalent salt solution of concentration $c$ (mol/liter), the

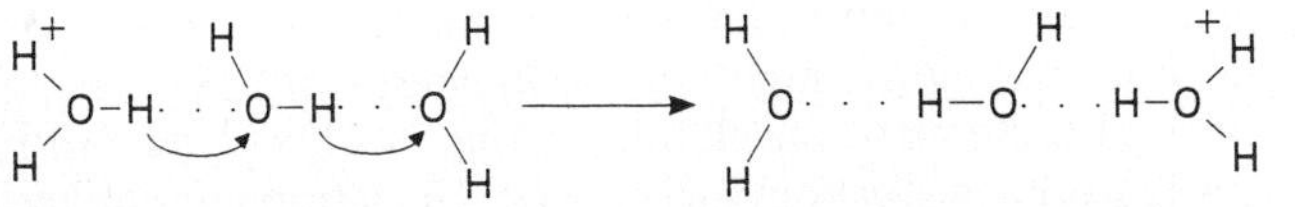

**FIG. 5.**    Grotthus chain mechanism for proton mobility in water.

average distance $d$ in angstroms (Å) between any two ions is given by

$$d = (10^{27}/2N_oc)^{1/3}$$

where $N_o$ is Avogadro's number ($6.023 \cdot 10^{23}$). The factor 2 is taken because both the anions and cations are counted, and the factor $10^{27}$ is Å$^3$/liter. For 0.15 $M$ NaCl, the average distance between $Na^+$ and the nearest $Cl^-$ is 17.7 Å, at which range coulombic forces are highly significant. At pH 7.4, where the concentration of "protons" (really hydronium ions) is only 40 n$M$, the average distance between "protons" is 0.28 $\mu$m.

Attractive ion–ion interactions are stabilizing and lower the chemical potential of an ion from its value in an ideal infinitely dilute solution. The *activity* ($a$) of an ion is defined in terms of its *chemical potential* ($\mu$) as follows:

$$\mu = \mu^o(T, P) + RT \ln a$$

where $\mu^o$ ($T, P$) is the *standard-state chemical potential*, or the chemical potential when the activity is 1. Increases in chemical activity increase the chemical potential. In ideal dilute solutions, the activity equals the chemical concentration ($c$), so that the chemical potential of an ion in an ideal solution is

$$\mu_{ideal} = \mu^o (T, P) + RT \ln c$$

To describe the properties of more concentrated nonideal solutions, G. N. Lewis introduced the *activity coefficient* ($\gamma$), such that the activity is the product of the concentration ($c$) and the activity coefficient ($a = \gamma \cdot c$). The chemical potential of an ion in a nonideal solution is

$$\mu = \mu^o (T, P) + RT \ln c + RT \ln \gamma$$

where $RT \ln \gamma$ includes the effect of ion–ion interactions on the chemical potential.

When a salt ($c_+c_-$) is added to water, the cations ($c_+$) and the anions ($c_-$) both contribute to the free energy of the solution. The chemical potentials of the cations ($c_+$) and anions ($c_-$) are given by

$$\mu_+ = \mu_+^0 + RT \ln c_+ + RT \ln \gamma_+$$

$$\mu_- = \mu_-^0 + RT \ln c_- + RT \ln \gamma_-$$

Adding these two expressions, and taking the average, gives

$$\frac{\mu_+ + \mu_-}{2} = \frac{\mu_+^0 + \mu_-^0}{2} + RT \ln (c_+ + c_-)^{1/2}$$
$$+ RT \ln (\gamma_+ + \gamma_-)^{1/2}$$

The mean ionic activity coefficient ($\gamma_\pm$) is defined as

$$\gamma_\pm = (\gamma_+ + \gamma_-)^{1/2}$$

Mean ionic activity coefficients of salts have been estimated by the *Debye–Hückel theory*, which supposes that central ions attract oppositely charged ions (or *counterions*) in a diffuse and structureless ion cloud (or atmosphere). The forces of coulombic attraction between the central ion and its cloud of counterions are opposed by the randomizing influence of the thermal motion of the ions. In dilute solutions, in which the theory treats central ions as point charges relative to the size of the ion cloud, the mean ionic activity coefficient ($\gamma_\pm$) for salt ions with charges $z_+$ and $z_-$ is given by the limiting law as follows:

$$\log \gamma_\pm = -A|z_+ z_-|I^{1/2}$$

where $I$ is the *ionic strength,* defined as

$$I = \frac{1}{2}\sum_i c_i z_i^2$$

The constant $A$, which equals 0.5108 kg$^{1/2}$mol$^{-1/2}$ is given in terms of $B$, which equals $0.3287 \cdot 10^8$ kg$^{1/2}$mol$^{-1/2}$cm$^{-1}$, both at 25°C, as follows:

$$A = \frac{1}{2.303}\frac{N_A e_0^2}{2\varepsilon RT}B$$

$$B = \sqrt{\frac{8\pi N_A e_0^2}{1000\varepsilon kT}}$$

where $N_A$ is Avogadro's number ($6.023 \cdot 10^{23}$ ions/mol), $e_o$ is the charge on an electron ($4.80 \cdot 10^{-10}$ statcoul = $1.60 \cdot 10^{-19}$ C), $\varepsilon$ is the dielectric constant, $R$ is the gas constant (8.32 J/mol · K), $T$ is the absolute temperature (°K), and $k$ (= $R/N_A$) is Boltzmann's constant ($1.38 \cdot 10^{-23}$ J/°K).

The derivation of the Debye–Hückel limiting law (see Bockris and Reddy, 1970) begins with Gauss' law, one of the four fundamental Maxwell equations of electromagnetic theory, and computes the spherically symmetric electric field around an ion, leading to Poisson's differential equation relating the charge density ($\rho_x$) of the ionic cloud to the electrostatic potential ($\psi_r$) at a distance ($r$) from the central point charge. The counterions, at concentration $n_i$ for ionic species $i$ at a distance $r$ from the central ion, are considered to distribute in the electric field according to a *Boltzmann distribution*

$$n_i = n_i^0 e^{-z_i e_0 \psi_r/kT}$$

where $n_i^0$ is the bulk ion concentration (ions/liter). By linearizing the Boltzmann equation, using a Taylor series expansion and retaining only the first two terms, which assumes that $z_i e_0 \psi_r \ll kT$, the linearized and integrated Poisson–Boltzmann equation shows how the electrostatic potential $\psi$ decreases as a function of distance ($r$) from the central ion (Fig. 6A):

$$\psi_r = \frac{z_i e_0}{\varepsilon}\frac{e^{-\kappa r}}{r}$$

where $\kappa$ is given by

$$\kappa = \sqrt{\frac{4\pi}{\varepsilon kT}\sum_i n_i^0 z_i^2 e_0^2}$$

The charge density ($\rho_r$) of the ionic cloud also decreases with increasing distance from the central ion (Fig. 6B) according to

$$\rho_r = -\frac{z_i e_0}{4\pi}\kappa^2\frac{e^{-\kappa r}}{r}$$

The amount of charge ($dq$) contained in a concentric spherical shell of thickness $dr$ located at a distance $r$ from the central ion (Fig. 6C) is given by

$$dq = z_i e_0 e^{-\kappa r}\kappa^2 r\, dr$$

Furthermore, the maximum amount of charge contained in such a spherical shell occurs at a distance of $\kappa^{-1}$, known as the *Debye length,* which defines the effective radius of the counterion atmosphere. It can also be shown that the Debye length is the distance away from the central ion where an ion of equal and opposite charge would contribute to the electrostatic field by an amount equivalent to that of the dispersed cloud of counterions (Bockris and Reddy, 1970). The Debye length decreases as the ionic concentration increases (Fig. 6D).

The Debye–Hückel limiting law predicts activity coefficients accurately only to concentrations of about 0.01 $M$. In more concentrated solutions, the finite radius ($a$, in cm) of the ion is taken into account in the extended Debye–Hückel equation, given by

$$\log \gamma_\pm = -\frac{A|z_+ z_-|I^{1/2}}{1 + BaI^{1/2}}$$

where the constants $A$ and $B$ are still 0.5108 kg$^{1/2}$mol$^{-1/2}$ and $0.3287 \cdot 10^8$ kg$^{1/2}$mol$^{-1/2}$cm$^{-1}$, respectively, for water at 25°C. The ion-size parameter $a$ is an adjustable parameter, but reasonable values extend the range of concentrations for which the Debye–Hückel theory accurately predicts ionic activity coefficients (Fig. 7).

At still higher salt concentrations, above about 0.7 $M$, the activity coefficient stops decreasing and instead begins to increase with increasing salt concentration. This behavior has been explained by the increasing fraction of hydration water, which both reduces the amount of free water in the solution and raises the effective concentration (i.e., activity of the dissolved ions). Corrections for hydration enable prediction of the activity coefficient as a function of salt concentration over the full range of salt concentrations reaching to several molar.

The activity coefficients calculated by the Debye–Hückel theory indicate that the forces between ions in dilute solutions are weak and nonselective. The predicted activity coefficients depend on the ionic charge and the ionic strength of the solution and are largely independent of the specific ion within, say, the alkali metal series. A very small degree of selectivity is introduced with the ion size parameter. The mean ionic activity coefficients for NaCl and KCl as a function of salt concentration are shown in Fig. 7. At 0.1 $M$ concentrations, the activity coefficient of Na$^+$ is 0.77, whereas those of K$^+$ and Cl$^-$ are each 0.76. These relatively high activity coefficients correspond to ion–ion interaction energies of only about −0.3 kcal/mol (see Hille, 1992).

Agreement between theory and data, such as that seen in Fig. 7 for the Debye–Hückel theory, does not necessarily imply the correctness of the theory, or of the model on which the theory is based. A fundamental criticism of the Debye–Hückel theory concerns the smeared charge model for the cloud of counterions (Frank and Thompson, 1960). Below a concentration of about 0.001 $M$, the Debye length $\kappa^{-1}$ is greater than the average distance between

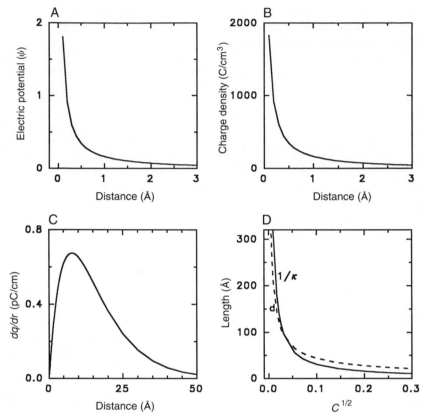

**FIG. 6.** Debye–Hückel theory. (A) Electrostatic potential ($\psi$), (B) charge density ($\rho$) of the counterion cloud, and (C) the charge ($dq$) enclosed in a spherical shell of thickness ($dr$), each calculated and plotted vs. distance ($d$) from the central ion. (D) Debye length $\kappa^{-1}$ (solid line) and average distance ($d$) between ions (dashed line) vs square root of ion concentration.

ions in the solution (Fig. 6D), in which case the counterions could reasonably appear as a cloud of smeared charge from the vantage point of the central ion. Above 0.001 $M$, however, the Debye length incongruously falls below the average distance between ions in the solution. Moreover, at only 0.01 $M$, just one ion is needed to account

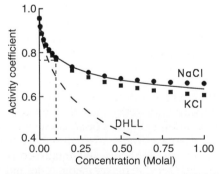

**FIG. 7.** Activity coefficients of NaCl (circles) and KCl (squares). The dashed curve represents the prediction of the Debye–Hückel limiting law. The solid line represents the prediction of the extended Debye–Hückel equation with an ion size parameter of a 4.7 Å. The dashed vertical line indicates that at 0.1 $M$, the activity coefficients of NaCl and KCl are 0.77 and 0.76, respectively. (Data from Robinson and Stokes, 1959.)

for 50% of the effect of the counterion cloud on the central ion, yet this ion must be smeared around the central ion in a cloud of spherically symmetric charge distribution located only 25 Å away from the central ion (Bockris and Reddy, 1970). Thus, well below physiological salt concentrations, the assumption of smeared charge in the Debye–Hückel theory is not in accord with the coarse-grained structure of salt solutions, despite the impressive ability of the theory to predict activity coefficients. A quasilattice approach to salt solutions minimizing the free energy using modern computational power would seem to be a preferable alternative (see Horvath, 1985, for references).

Specific effects of small anions on the solubility, aggregation, or denaturation of proteins often follow the *Hofmeister (lyotropic) series* in the following order of effectiveness:

$$F^- < PO_4^{3-} < SO_4^{2-} < CH_3COO^- < Cl^- < Br^- < NO_3^- < I^- < SCN^-$$

The ions to the right of Cl are referred to as *chaotropic*, since they tend to destabilize proteins (for review, see Collins and Washabaugh, 1985). Ions may also associate with sites on proteins, as exemplified by the binding to albumin of $Ca^{2+}$ (Katz and Klotz, 1953), and of $Cl^-$ (Scatchard *et al.*, 1957; see Chapter 8 in Tanford, 1961). Certain enzymes, ion channels, and membrane transport proteins interact with the alkali cations with a high degree of *ionic selectivity*. Considering the five alkali metal cations, there are 5! (= 120) possible orders of selectivity that might arise. However, only 11 cationic selectivity orders commonly occur in chemical and biological systems, as listed in Table 2. G. Eisenman predicted these selectivity orders by calculating the ion-site interaction energies. If the field strength of a negatively charged site is weak, then an associated ion remains hydrated. In this case, $Cs^+$, having the smallest hydrated ionic radius, is favored (Sequence I) because it can approach most closely to the site, and thus has the strongest coulombic force of attraction. If the field strength of the site is strong, and the interaction energy between the ion and the site is stronger than between the ion and water dipoles, then the ion loses its associated water and becomes dehydrated. In this case, $Li^+$, having the smallest nonhydrated ionic radius, becomes favored (Sequence XI). With intermediate field strengths, ions partially dehydrate, giving rise to the intervening selectivity sequences. Other sequences are possible when the sites are assumed to be polarizable (for review, see Eisenman and Horn, 1983).

## VI. Cell Cations

The predominant biological cations are potassium and sodium (see Ussing, 1960, for review), calcium (see Campbell, 1983, for review), and magnesium. In excitable cells such as nerve and muscle, the inwardly directed concentration gradient of $Na^+$, and the outwardly directed gradient of $K^+$, both created by the Na,K-ATPase (see Chapter 16), are reduced by a very small extent during each action potential, as the cations $Na^+$ and $K^+$ move down their concentration gradients through voltage-gated ion channels (see Chapters 13 and 24). $K^+$ also functions as a specific cofactor for the glycolytic enzyme pyruvate kinase. The maximal catalytic velocity follows Eisenman sequence IV: $K^+ > Rb^+ \gg Cs^+ \geq Na^+ > Li^+$ (Kayne, 1973). During *in vitro* protein synthesis, $K^+$ maintains an active conformation of 50S ribosomal subunits that catalyze peptide bond formation. The order of selectivity for this effect is

Eisenman sequence III: $NH_4^+ \geq Rb^+ > K^+ > Cs^+$; $Na^+$ and $Li^+$ are ineffective (Miskin *et al.*, 1970).

The widespread importance of calcium ion ($Ca^{2+}$) in cellular physiology was first emphasized by L. V. Heilbrunn (see, e.g., *The Dynamics of Living Protoplasm*, 1956). Injections of various salts into frog skeletal muscle fibers revealed that $Ca^{2+}$ is the only intracellular ion that induces muscle contraction (see Chapters 54 and 55). $Ca^{2+}$ is pumped out of the sarcoplasm into the sarcoplasmic reticulum (SR), and also across the plasma membrane into the extracellular solution, by a Ca-ATPase (see Chapter 17). Release of $Ca^{2+}$ from the SR initiates muscle contraction. Using "skinned muscle fibers," in which the sarcolemma has been removed with fine needles, J. Gulati and R. J. Podolski (1978) and others have studied how the tension exerted by myofilaments is directly proportional to the concentration of free $Ca^{2+}$ in the sarcoplasm. Pumping of $Ca^{2+}$ from the cytoplasm of muscle, red cells, and other cells across the plasma membrane normally results in a submicromolar steady-state intracellular concentration of $Ca^{2+}$, whereas plasma $Ca^{2+}$ is around 2.5 m$M$. The ratio of extracellular to intracellular $Ca^{2+}$ concentrations is thus more than 1000, much greater than the ratio of around 25 for $Na^+$. In human mammary glands, $Ca^{2+}$ is secreted by exocytosis via the Golgi system into milk to a total concentration of around 10 m$M$ (Neville *et al.*, 1983).

The classic experiments of S. Ringer (1882) established that plasma $Ca^{2+}$ is essential for the sustained beating of isolated hearts. An inward $Ca^{2+}$ current down its concentration and electrochemical gradient through an ion channel constitutes a major component of the cardiac action potential (see Chapter 51). Transient elevations of intracellular $Ca^{2+}$ also occur in such processes as fertilization, cell division, exocytosis during neurotransmitter release, activation of platelets, neutrophils, and lymphocytes, and the hormonal activation of cells. $Ca^{2+}$ is an essential cofactor in blood clotting and in the activation of complement, and also has a structural role in membranes and in the mineralization of bones, teeth, and other skeletal structures.

In multicellular animals, the concentration of magnesium ion in both the extracellular and intracellular fluids is around 1 m$M$. Consequently, unlike $Na^+$, $K^+$, and $Ca^{2+}$, $Mg^{2+}$ does not act as a carrier of ionic current across cell membranes or as a trigger for the initiation of cellular activities. $Mg^{2+}$ ion, with a divalent positive charge and a nonhydrated diameter of 1.3 Å (see Table 1), has the highest charge density of all of the ions found in cells. Consequently, $Mg^{2+}$ binds readily to anionic sites, particularly to polyphosphates such as ATP. MgATP is the substrate for all kinases and for some phosphatases such as Na,K-ATPase and Ca-ATPase. $Mg^{2+}$ is also a cofactor for the glycolytic enzyme enolase, for glutamine synthetase, and for other enzymes. $Mg^{2+}$ is also tightly bound to porphyrin in chlorophyll, the primary biological molecule for capturing light energy (see Chapter 63).

## VII. Cell Anions

The most abundant permeant anion in cells is chloride, although its intracellular concentration is typically less

**TABLE 2** Eisenman's Selectivity Sequences for Binding of Alkali Cations to Negatively Charged Sites

| Highest field strength of site | | |
|---|---|---|
| Li > Na > K > Rb > Cs | Sequence | XI |
| Na > Li > K > Rb > Cs | Sequence | X |
| Na > K > Li > Rb > Cs | Sequence | IX |
| Na > K > Rb > Li > Cs | Sequence | VIII |
| Na > K > Rb > Cs > Li | Sequence | VII |
| K > Na > Rb > Cs > Li | Sequence | VI |
| K > Rb > Na > Cs > Li | Sequence | V |
| K > Rb > Cs > Na > Li | Sequence | IV |
| Rb > K > Cs > Na > Li | Sequence | III |
| Rb > Cs > K > Na > Li | Sequence | II |
| Cs > Rb > K > Na > Li | Sequence | I |
| Lowest field strength of site | | |

than the extracellular concentration. Macromolecular cellular constituents including nucleic acids and most cytoplasmic proteins carry a net negative charge at physiological pH. Phosphorylated metabolic intermediates and organic acids are also negatively charged. The *condition of macroscopic electroneutrality* requires that the total charge-weighted concentrations of cations and anions be equal in each cellular compartment. However, a very small amount of charge separation occurs across membranes. The intracellular cationic charge, primarily due to $Na^+$ plus $K^+$, is neutralized by intracellular protein, nucleic acids, and organic anions, and by $Cl^-$, bicarbonate ($HCO_3^-$), phosphate, and sulfate. $HCO_3^-$ and phosphate, along with proteins, are the principal buffers that regulate intracellular and extracellular pH (see Chapter 20). Whereas $Cl^-$ functions primarily to maintain bulk electroneutrality, some cells also use $Cl^-$ to carry ionic current. Halobacteria, for example, possess a light-driven $Cl^-$ pump (Oesterhelt and Tittor, 1989).

## VIII.  Trace Elements

In addition to the predominant electrolytes found in the intracellular and extracellular solutions, other ions are tightly associated with certain proteins and enzymes, known as *metalloproteins*. The transport of oxygen ($O_2$) to all of the cells of the organism by red blood cells depends on the reversible association of $O_2$ with ferrous ions ($Fe^{2+}$) in the porphyrin groups of intracellular *hemoglobin*. Heme is an iron porphyrin, whereas chlorophyll is a magnesium porphyrin. Binding of $O_2$ to $Fe^{2+}$ in the muscle heme protein *myoglobin* provides a reservoir of $O_2$ for muscular work. *Catalase* is another Fe–heme protein that protects cells by converting hydrogen peroxide into water and oxygen. Iron is stored in cells in the cavity of the protein *ferritin*, which can accommodate as many as 4500 ferric ($Fe^{3+}$) ions per molecule. Iron ($Fe^{2+}$, $Fe^{3+}$) and copper ($Cu^+$, $Cu^{2+}$) are essential cofactors in *cytochromes* in the respiratory chain of mitochondria, and in the photosystems of chloroplasts. Manganese ions are essential in Photosystem II in plant cells. Selenium is a cofactor in *glutathione peroxidase*, an enzyme that reduces hydrogen peroxide and organic peroxides and helps to prevent oxidative damage in cells. Zinc ($Zn^{2+}$) and copper are cofactors in *superoxide dismutase*, an enzyme that scavenges superoxide anion ($O_2^-$) and helps to prevent oxidative damage from toxic free radicals. $Zn^{2+}$ also functions as a cofactor in some proteolytic digestive enzymes, the zinc proteases, and stabilizes structures known as "zinc fingers" in certain DNA-binding proteins. A notable example of the binding of an ion to a nonprotein compound is that of tervalent cobalt ($Co^{3+}$) in vitamin $B_{12}$ and the $B_{12}$ coenzymes.

## IX.  Free Energy of Ion Transport

When ions cross cell membranes, changes in free energy are involved. Consider a membrane permeant only to cations separating two solutions of KCl of differing concentra-

tions, $c_i$ and $c_o$, where the subscripts i and o represent the intracellular and extracellular compartments, respectively (Fig. 8). If the intracellular concentration $c_i$ is greater than the extracellular concentration $c_o$, then potassium will tend to diffuse out of the cell down its concentration gradient. When only a slight amount of potassium crosses the membrane (which, for this example, is assumed to be impermeant to chloride), the separation of charge creates a membrane potential $E_m$ that is negative inside. The resultant electrical force retards further efflux of potassium. An equilibrium is reached when the diffusion force favoring efflux of potassium exactly balances the electrical force preventing efflux of potassium.

Prior to the attainment of equilibrium, the process of moving $dn$ moles of $K^+$ ions out of the cell from compartment i to compartment o involves a change in free energy ($dG$) of the system. According to the laws of thermodynamics, the process will occur spontaneously only if $dG < 0$, and the system will be at equilibrium if $dG = 0$. Thus, to compute the change in free energy $dG$, we need to review some basic thermodynamics.

The *first law of thermodynamics,* which is a statement of the conservation of energy, states that the increase in energy ($dE$) of a system is the sum of the gain in heat ($dQ$) minus the work ($dW$) done by the system.

$$dE = dQ - dW$$

For reversible processes, the *second law of thermodynamics* defines the change in entropy ($dS$) of a system in terms of the heat gained ($dQ$) and the absolute temperature ($T$) as follows:

$$dS = dQ/T$$

Combining the first and second laws gives

$$dE = TdS - dW$$

Since the difference in pressure between the internal and external solutions for animal cells is negligible, the work ($dW$) done by the system includes *pressure–volume work, pdV,* and *electrical work,* $-zF\varepsilon_m dn$, summed as follows:

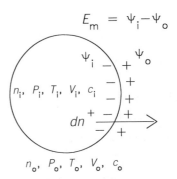

**FIG. 8.**   Free energy of ion transport. A cell membrane separates the intracellular (subscript i) and the extracellular (subscript o) compartments. Each compartment, at pressure $P$ and temperature $T$, has volume $V$, is at electrical potential $\psi$, and contains solutes at concentration $c$. A free energy change is involved when $dn^+$ moles of cation leave the cell. The membrane potential $E_m$ is $\psi_i - \psi_o$.

$$dW = pdV - zF\varepsilon_m dn$$

where $p$ is the pressure, $V$ is the total volume of the system, $z$ is the ionic valence (eq/mol), $F$ is the Faraday constant ($= 96,490$ C/eq), and $dn$ is the number of moles of solute crossing the membrane and leaving the cell. $\varepsilon_m$ is the *transmembrane potential*, or the difference in electrical potential $\psi$, in volts (or J/C), between the two compartments.

$$\varepsilon_m = \psi_i - \psi_o$$

The electrical potential $\psi$ represents the work done in moving a unit positive test charge from infinity to a point in the solution. Note that when the transmembrane potential $\varepsilon_m$ is negative inside, the *electrical* work is positive for pumping $Na^+$ out of the cell, but negative for pumping $K^+$ into the cell. The *electrical* work is negative when $Na^+$ leaks into the cell and positive when $K^+$ leaks out of the cell.

According to the first and second laws, the change in energy $dE$ of the system is

$$dE = TdS - pdV + zF\varepsilon_m dn$$

For systems of variable chemical composition, the *free energy* ($G$) is by definition $G = H - TS + \Sigma\mu_j n_j$, where the *enthalpy* $H = E + PV$. The chemical potential $\mu_j$ of the $j$th solute in the system at constant $T$, $P$, and $n_k$ is defined by

$$\mu_j = (\partial G/\partial n_j)_{n_k}$$

$n_j$ and $n_k$ are the number of moles of the $j$th and $k$th solutes, respectively. The *chemical potential*, also called the *partial molar free energy*, represents the incremental addition of free energy to the system upon incremental addition of a solute. All solutes contribute to the free energy of a solution. While the free energy itself is a parameter of state for the whole system, the chemical potential refers to a particular solute. The free energy is thus given by

$$G = E + PV - TS + \Sigma\mu_j n_j$$

Differentiating,

$$dG = dE + pdV + VdP - TdS - SdT + \Sigma\mu_j dn_j$$

Substituting the expression for $dE$ gives

$$dG = \underline{TdS} - \underline{pdV} + zF\varepsilon_m dn + \underline{pdV} + VdP - \underline{TdS} - SdT + \Sigma\mu_j dn_j$$

Simplifying yields a form of the *Gibbs equation*,

$$dG = -SdT + VdP + zF\varepsilon_m dn + \Sigma\mu_j dn_j$$

which states that the free energy of a system of variable chemical composition is a function of the temperature, the pressure, and the number of moles of each component in the mixture, or $G = G(T,P,n_j)$. For processes that occur at constant temperature and pressure, where $dT = dP = 0$, the Gibbs equation simplifies to

$$dG = zF\varepsilon_m dn + \Sigma\mu_j dn_j$$

which states that the increase in free energy of a system

is equal to the sum of the electrical work done on the system plus the total change in free energy due to changes in chemical composition. Furthermore, when the system is at equilibrium, $dG$ must be zero. The second law of thermodynamics also implies that the change in free energy ($dG$) is negative for all spontaneous processes. Thus, the first and second laws of thermodynamics, when combined with the definitions of free energy and enthalpy, result in the Gibbs equation, which is the fundamental equation permitting estimation of *free energy changes* when water, ions, or other solutes cross cell membranes.

## X. Nernst Equilibrium

A solute is said to be at *equilibrium* when its concentration is constant in time without requiring the continuous input of energy from metabolism or other sources. In human red blood cells, for example, $Cl^-$, $HCO_3^-$, and $H^+$ are at thermodynamic equilibrium (i.e., they are passively distributed). A solute is said to be at *steady state* when its concentration is constant in time, but is dependent on the continuous input of energy from metabolism or other sources. *Active transport* utilizes energy and results in steady-state distributions that represent a deviation from equilibrium. In biological cells, the concentrations of $Na^+$, $K^+$, and $Ca^{2+}$ are at a steady state. Their concentrations are constant in time, but are dependent on the continuous hydrolysis of ATP by the Na,K pump (for review, see Glynn, 1985, and Chapter 16) and the Ca pump (see Chapter 17). With isolated red cell membranes ("ghosts"), active transport of Na and K against their respective concentration gradients has been observed directly by measurements of net fluxes (Freedman, 1976). The active production and maintenance of ion concentration gradients by membrane pumps, such as Na,K-ATPase and Ca-ATPase, represents *chemical work* and *electrical work* done by the cell. The energy required for continuous pumping of $Na^+$ by resting frog sartorious muscles has been estimated to represent some 14–20% of the energy available from the hydrolysis of ATP, and the value is similar in red cells.

*Passive transport* is the movement of solutes toward a state of equilibrium. Passive transport of *hydrophobic* substances across cell membranes usually occurs directly by diffusion across the lipid bilayer. Passive transport of *hydrophilic* substances is usually mediated by specific membrane proteins by a process known as *facilitated diffusion*. Passive ion transport may occur through pores or channels under the influence of concentration gradients and electrical forces by a process known as *electrodiffusion*. The passive flow of ionic currents down their concentration gradients through specific ion channel proteins constitutes *negative work* done by the cell.

The *Nernst equilibrium equation* describes the relationship between voltage across a semipermeable membrane and the equilibrium ion concentrations in the compartments adjacent to the membrane. The Nernst equation provides a simple method of testing whether a particular solute is at equilibrium. Considering the same example described earlier, and using the Gibbs equation, the change in the free energy of the system that occurs when $dn$ moles of $K^+$ ions move from compartment i containing $K^+$ at activity $a_i$ to compartment o containing $K^+$ at activity $a_o$ is given by

$$dG = -zF\varepsilon_m dn + \mu_i dn_i + \mu_o dn_o$$

where the change in free energy has been summed for the two compartments. For this process, the decrease of the number of moles of solute inside the cell ($-dn_i$) equals the increase outside the cell ($dn_o$), so that $-dn_i = dn_o = dn$, and

$$dG = [-zF\varepsilon_m - (\mu_i - \mu_o)]dn$$

In each solution the chemical potential is

$$\mu_i = \mu_i^o + RT \ln a_i$$

$$\mu_o = \mu_o^o + RT \ln a_o$$

For the system under consideration, the standard state chemical potentials are the same in both compartments ($\mu_i^o = \mu_o^o$), and thus,

$$dG = [-zF\varepsilon_m - RT \ln (a_i/a_o)] \, dn$$

At equilibrium, $dG = 0$, and since $dn \neq 0$, $-zF\varepsilon_m - RT \ln (a_i/a_o) = 0$.

For charged solutes, the *electrochemical potential* ($\mu_j$) of the $j$th solute is defined to be the sum of a chemical and an electrical component. The electrical contribution to the electrochemical potential is $zF\psi$, and the chemical contribution is $RT \ln a$.

$$\mu_j = \mu_j^o (T,P) + RT \ln a_j + zF\psi$$

Note that a fundamental condition of equilibrium is that the difference in electrochemical potential between compartments i and o is zero. The electrochemical potential of the solute is the same in each compartment to which that solute has access.

Rearranging yields the Nernst equation for a cationic concentration cell,

$$\varepsilon_m = -\left(\frac{RT}{zF}\right) \ln \left(\frac{a_i}{a_o}\right)$$

Converting the natural logarithm to base 10 yields

$$\varepsilon_m = -\left(2.303 \frac{RT}{zF}\right) \log \left(\frac{a_i}{a_o}\right)$$

The value of $2.303RT/F$ is 58.7 mV at 23°C, and 61.5 mV at 37°C.

The Nernst equation is independent of the mechanism of transport and is often used for ascertaining whether an intracellular ion is at electrochemical equilibrium. If so, then

$$a_i = a_o e^{-zF\varepsilon_m/RT}$$

Another way of understanding the Nernst equation is that at equilibrium, ions distribute across the membrane electric field in accordance with a Boltzmann distribution.

If $z = 0$, as for nonelectrolytes, then $a_i = a_o$, and the activities of the solutes will be the same at equilibrium on both sides of the membrane, as was nearly the case for the distribution of nonelectrolytes across the membranes of human red blood cells (Gary Bobo, 1967). If $z$ does not equal zero, as for electrolytes, then at equilibrium each permeant monovalent ion will reach the same ratio of intracellular-to-extracellular activity. Such is the case for the passive distribution of $Cl^-$, $HCO_3^-$, and $H^+$ in human red blood cells.

$$r = \frac{Cl_i^-}{Cl_o^-} = \frac{HCO_{3i}^-}{HCO_{3o}^-} = \frac{H_o^+}{H_i^+} = e^{F\varepsilon_m/RT}$$

In red cells $Na^+$, $K^+$, and $Ca^{2+}$ deviate from this ratio because of the action of the Na/K pump and the Ca pump. In skeletal muscle, $K^+$ and $Cl^-$ have Nernst equilibrium potentials that are close to the actual measured resting potential, while $Na^+$ is far from equilibrium. In squid axon the Nernst equilibrium potential for $K^+$ is closest to the resting potential, $Cl^-$ is somewhat removed, and, as in muscle and red cells, $Na^+$ is far from equilibrium (see Chapter 13).

The transmembrane potential is usually measured by means of open-tipped microelectrodes, such as those developed and used by G. N. Ling and R. W. Gerard (1949) to obtain accurate and stable measurements of the membrane potential of frog skeletal muscle. With human red blood cells, stable potentials have not been achieved with microelectrodes. As an alternative technique, J. F. Hoffman and P. C. Laris (1974) utilized fluorescent cyanine dyes to monitor and measure red-cell membrane potentials. Fluorescent cyanines, merocyanines, oxonols, styryls, rhodamines, and other dyes have since been used in numerous electrophysiological studies of red cells, neutrophils, platelets, and other nonexcitable cells and organelles that are too small for the use of microelectrodes (for review, see Freedman and Novak, 1989a). The equilibrium distribution of permeant, lipophilic, radioactively labeled ions such as triphenylmethylphosphonium (TPMP$^+$) may also be used to assess the membrane potential using the Nernst equilibrium equation (see Freedman and Novak, 1989b).

## XI. Measurement of Electrolytes

Ion concentrations in biological fluids may be expressed as millimoles per liter of solution (m$M$), or as millimoles per kg of water (mMolal). When a solution containing metallic ions is aspirated into a flame, each type of ion burns with a characteristic color, $Na^+$ giving a yellow flame, $K^+$ giving a violet flame, and $Ca^{2+}$ giving a red flame. In *flame photometry,* the intensity of the emitted light, in comparison with that produced by solutions containing known concentrations of ions, provides a convenient measure of ion concentration in extracellular fluids and in acid extracts of cells. With a uniform rate of aspiration, a flame photometer accurately measures the intensity of the emitted light, which is related linearly to the cation concentrations in suitably diluted standards and unknowns (see, e.g., Funder and Wieth, 1966). *Atomic absorption spectroscopy,* an alternative technique, measures the light absorbed by ions during electronic excitation in a flame. Flame photometry and atomic absorption spectroscopy both measure total ionic concentrations in cell extracts irrespective of any intracellular compartmentation, and are sensitive in the millimolar range of cellular concentrations. $K^+$, $Na^+$, $Ca^{2+}$, and other elements in single cells, or even in single cell organelles, may be measured by *electron probe microanalysis,* a technique that utilizes an electron beam to excite the emission of X-rays with energies characteristic of the various elements in cells.

The development of *ion-specific glass microelectrodes* by G. Eisenman (1967), and then of selective *liquid ion-exchange microelectrodes,* made possible the direct determination of intracellular cation activities. L. G. Palmer and M. M. Civan (1977) found that for $Na^+$, $K^+$, and $Cl^-$ of *Chironomus* salivary gland cells, the ion activities are the same in the nuclear and cytoplasmic compartments. In a related study, Palmer, Civan, and T. J. Century (1978) found that during development of frog oocytes, the ratio of the cytoplasmic concentration of $Na^+$ to $K^+$ increases, while the corresponding ratio of ion activities decreases. This observation could reflect the development of yolk platelets and intracellular vesicles that contain ions at differing concentrations and activities than the bulk cytoplasm. In frog skeletal muscle, the sarcoplasmic reticulum contains a solution resembling that of the cytoplasm, whereas the ionic composition of the solution in the t tubules is extracellular. The extracellular space, however, consists of the interstitial space between the muscle fibers as well as the vascular space; solutes leave these two extracellular compartments with differing rate constants, thus considerably complicating the interpretation of experiments that assess the rate of membrane transport with radioactive isotopes of $Na^+$, $K^+$, and other solutes (Neville, 1979; Neville and White, 1979).

To measure transient changes of intracellular $Ca^{2+}$ in the micromolar and submicromolar range, *fluorescent chelator dyes* such as Quin-2, Fura-2, Indo-1, and Fluo-3 have been developed (Tsien, 1988). Quin-2, Fura-2, and Indo-1 are fluorescent analogues of ethylenediaminetetraacetic acid (EDTA), which contains four carboxylate groups that specifically bind two divalent cations. EGTA is a nonfluorescent analogue with a higher binding affinity for $Ca^{2+}$ as compared with $Mg^{2+}$, and is thus quite useful in experiments where the extracellular concentration of $Ca^{2+}$ is systematically varied. Fluo-3 is a tetracarboxylate fluorescein analogue that exhibits a shift in the emission spectrum upon binding $Ca^{2+}$; in contrast, Fura-2 undergoes a shift in its excitation spectrum. By measuring the ratio of Fura-2 fluorescence upon excitation at two exciting wavelengths, changes in the concentration of $Ca^{2+}$ may be monitored. The cells are incubated with a permeant ester form of the dye to enable the dye to permeate into cells; intracellular esterases then release the $Ca^{2+}$-sensitive chromophore. With video microscopy of cells stained with fluorescent $Ca^{2+}$ indicators, it is also possible to obtain time-resolved and spatially resolved light microscopic images of the changes in intracellular $Ca^{2+}$. Another fluorescent probe (SPQ), developed by Helsey and Verkman (1987), has been used to measure intracellular $Cl^-$ and to study its transport across cell membranes.

## XII. Ionophores

Ionophores are a class of compounds that form complexes with specific ions and facilitate their transport across cell membranes (for review, see Pressman, 1976). An ionophore typically has a *hydrophilic* pocket (or hole) that forms a binding site specific for a particular ion. The exterior surface of an ionophore is *hydrophobic,* allowing the complexed ion in its pocket to cross the hydrophobic membrane. A list of ionophores showing the ion specificity of each is given in Table 3. Ionophores are useful tools in cell physiology. *Nystatin* forms a channel in membranes for monovalent cations and anions and has proved useful for altering the cation composition of cells. *Gramicidin* forms dimeric channels specific for monovalent cations. *Valinomycin* carries $K^+$ across membranes with a high selectivity and has been used extensively to impose a high $K^+$ permeability on cell membranes. *Monensin* is a carrier with specificity for $Na^+$. *Hemisodium* is a new synthetic $Na^+$ ionophore with an even greater degree of selectivity for $Na^+$ (Kaji, 1992). The $Ca^{2+}$ ionophore *A23187* has been used extensively to permit entry of $Ca^{2+}$ into cells that normally have a low native permeability to $Ca^{2+}$, and thereby to activate a variety of cellular processes that are regulated by $Ca^{2+}$. *Nigericin* exchanges $K^+$ for protons and has been used in many studies of mitochondrial bioenergetics to alter electrical and chemical gradients for protons (for review, see Harold, 1986). Ionophores such as *FCCP* and *CCCP* are specific for protons. In addition to their utility in cell-physiology experiments, study of the mechanism of membrane transport mediated by ionophores has provided important conceptual insights (e.g., Stark and Benz, 1971; Finkelstein and Andersen, 1981) relevant to the understanding of ion transport mediated by native transport proteins.

## XIII. Summary

This chapter describes the hydrogen-bonded structure of liquid water and its dipolar and dielectric properties. In salt solutions, water exhibits three regions of structure:

**TABLE 3**　Ionophores and Their Ion Selectivities

| | |
|---|---|
| Conductive carriers | |
| 　Valinomycin | $K^+$ |
| 　Hemisodium | $Na^+$ |
| 　FCCP, CCCP | $H^+$ |
| Electroneutral exchangers | |
| 　A23187 | $Ca^{2+}/Mg^{2+}$; $Ca^{2+}/2H^+$ |
| 　Nigericin | $K^+/H^+$ |
| Channels | |
| 　Gramicidin | $H^+ > Cs^+ \approx Rb^+ > K^+ > Na^+ > Li^+$ |
| 　Nystatin | Monovalent cations and anions |

oriented dipoles near ions, an intermediate structure-breaking region, and flickering clusters with short-range order characteristic of ice. Electrostatic interactions between ions and water based on Coulomb's law account for the enthalpy of hydration, as well as for the sequence of ionic mobilities of the hydrated alkali cations. Intracellular water exhibits thermodynamic properties similar to extracellular water. Salts and nonelectrolytes dissolve in cell water with the same solubility and activity as in extracellular water; even the viscosities are similar. Protons in solution are hydrogen-bonded to water to form $H_3O^+$, which is further hydrated because of electrostatic forces. Protons migrate through water by a Grotthus chain mechanism; in contrast, other hydrated ions migrate through water as hard spheres that interact with water dipoles according to the ionic radius and charge. Nonspecific ion–ion interactions reduce the activity coefficients of ions, as described by the Debye–Hückel theory, which conceptualizes a central ion surrounded by a cloud of counterions with smeared charge. Despite the predictive ability of the Debye–Hückel theory, a quasilattice theory of salt solutions is needed at physiological salt concentrations. Selective ion interactions with sites on enzymes or ion channels occur in certain predictable specific patterns (Eisenman sequences) that depend on the energy of interaction between the ions and water dipoles, relative to that between ions and their binding sites.

Next described are the functions and properties of the predominant cellular electrolytes—$K^+$, $Na^+$, $Ca^{2+}$, $Mg^{2+}$, $Cl^-$, and $HCO_3^-$. Steady-state concentrations of $K^+$, $Na^+$, and $Ca^{2+}$ are established by $Na^+/K^+$-ATPase and $Ca^{2+}$-ATPase, which utilize metabolic energy to pump ions against their electrochemical gradients. The passive flow of $K^+$, $Na^+$, and $Ca^{2+}$ down their electrochemical gradients through voltage-gated ion channels constitutes the ionic currents that form action potentials in excitable cells such as nerve and muscle. Intracellular $K^+$ also activates the glycolytic enzyme pyruvate kinase and is required for peptide bond formation in protein synthesis. Changes in cell $Ca^{2+}$ initiate and modulate a variety of cellular functions, but too much $Ca^{2+}$ inside cells can be harmful. $Mg^{2+}$, with its high charge density, binds tightly to polyphosphates and is a cofactor in all kinases and some phosphatases, including $Na^+/K^+$-ATPase. $HCO_3^-$ and phosphate, along with proteins, are the principal biological buffers that regulate pH. $Cl^-$ mainly serves to maintain bulk electroneutrality. Trace ions, including those of iron, copper, manganese, zinc, selenium, and cobalt, are important cofactors in nutritional and bioenergetic pathways, and in enzymes that protect cells from oxidative and peroxidative damage.

All of the solutes in a cell contribute to the free energy of the intracellular solution. The Gibbs equation, which is derived by combining the first and second laws of thermodynamics, enables estimation of the changes in free energy when solutes cross cell membranes. The change in free energy is zero at equilibrium and negative for spontaneous processes. Solutes will redistribute until the electrochemical potential is the same in every compartment to which that solute has access. The Nernst equilibrium equation, which follows from the Gibbs equation, relates the ratio of intracellular to extracellular ion concentrations to the membrane potential and can be used to test whether a solute is in electrochemical equilibrium across a cell membrane. Cell cations are measured by means of flame photometry, atomic absorption spectroscopy, ion-specific electrodes, electron probe microanalysis, and fluorescent chelator dyes. Membrane-bounded compartments inside cells may contain different ion concentrations than the bulk cytoplasm. Ionophores with a high degree of ion selectivity are useful tools in cell physiology experiments. The study of ion transport mediated by ionophores has provided instructive models for understanding the mechanisms of membrane transport.

## Acknowledgment

The author thanks Dr. Jerry Goodisman of the Chemistry Department of Syracuse University for reviewing parts of this chapter.

## Bibliography

Bergethon, P. R., and Simons, E. R. (1990). "Biophysical Chemistry. Molecules to Membranes." Springer-Verlag, New York.

Bernard, C. (1949). "An Introduction to the Study of Experimental Medicine" (H. C. Green, Trans.). Schuman, New York.

Bockris, J. O'M., and Reddy, A. K. N. (1970). "Modern Electrochemistry," Vol. 1. Plenum Press, New York.

Campbell, A. K. (1983). "Intracellular Calcium. Its Universal Role as Regulator." Wiley, New York.

Cannon, W. B. (1932). "The Wisdom of the Body." Norton, New York.

Cantor, C. R., and Schimmel, P. R. (1980). "Biophysical Chemistry," Parts I, II, III. W. H. Freeman, San Francisco.

Collins, K. D., and Washabaugh, M. W. (1985). The Hofmeister effect and the behavior of water at interfaces. Q. Rev. Biophys. **18**, 323–422.

Cooke, R., and Kuntz, I. D. (1975). The properties of water in biological systems. Annu. Rev. Biophys. Bioeng. **3**, 95–126.

Debye, P. (1929). "Polar Molecules." Dover Publications, New York.

Dick, D. A. T. (1959). Osmotic properties of living cells. Int. Rev. Cytol. **8**, 387–448.

Edsall, J. T., and McKenzie, H. A. (1978). Water and proteins. I. The significance and structure of water: Its interaction with electrolytes and nonelectrolytes. Adv. Biophys. **10**, 137–207.

Edsall, J. T., and Wyman, J. (1958). "Physical Biochemistry." Academic Press, New York.

Eisenberg, D., and Crothers, D. M. (1979). "Physical Chemistry with Applications to the Life Sciences." Benjamin-Cummings, Menlo Park, CA.

Eisenberg, D., and Kauzmann, W. (1969). "The Structure and Properties of Water." Oxford University Press, Oxford.

Eisenman, G. (1967). "Glass Electrodes for Hydrogen and Other Cations." Marcel Dekker, New York.

Eisenman, G., and Horn, R. (1983). Ionic selectivity revisited: The role of kinetic and equilibrium processes in ion permeation through channels. J. Membr. Biol. **76**, 197–225.

Finkelstein, A., and Andersen, O. S. (1981). The gramicidin channel: A review of its permeability characteristics with special reference to the single-file aspect of transport. J. Membr. Biol. **59**, 155–171.

Frank, H. S., and Thompson, P. T. (1960). A point of view on ion clouds. *In* "The Structure of Electrolyte Solutions" (W. J. Hamer, Ed.), pp. 113–134. John Wiley & Sons, New York.

Frank, H. S., and Wen, W.-Y. (1957). Structural aspects of ion–solvent interaction in aqueous solutions: A suggested picture of water structure. *Disc. Faraday Soc.* **24**, 133–140.

Freedman, J. C. (1976). Partial restoration of sodium and potassium gradients by human erythrocyte membranes. *Biochim. Biophys. Acta* **455**, 989–992.

Freedman, J. C., and Novak, T. S. (1989a). Optical measurement of membrane potentials of cells, organelles, and vesicles. *Methods Enzymol.* **172**, 102–122.

Freedman, J. C., and Novak, T. S. (1989b). Use of triphenylmethylphosphonium to measure membrane potentials in red blood cells. *Methods Enzymol.* **173**, 94–100.

Funder, J., and Wieth, J. O. (1966). Determination of sodium, potassium, and water in human red blood cells. Elimination of sources of error in the development of a flame photometric method. *Scand. J. Clin. Lab. Invest.* **18**, 151–166.

Fushimi, K., and A. S. Verkman (1991). Low viscosity in the aqueous domain of cell cytoplasm measured by picosecond polarization microfluorimetry. *J. Cell Biol.* **112**, 719–725.

Gary Bobo, C. M. (1967). Nonsolvent water in human erythrocytes and hemoglobin solutions. *J. Gen. Physiol.* **50**, 2547–2564.

Glynn, I. M. (1985). The $Na^+,K^+$-transporting adenosine triphosphatase. *In* "The Enzymes of Biological Membranes" (A. N. Martonosi, Ed.), 2nd ed., Vol. 3, pp. 35–114. Plenum Press, New York.

Gulati, J., and Podolski, R. J. (1978). Contraction transients of skinned muscle fibers: Effects of calcium and ionic strength. *J. Gen. Physiol.* **72**, 701–716.

Harold, F. M. (1986). "The Vital Force: A Study of Bioenergetics." W. H. Freeman, New York.

Heilbrunn, L. V. (1956). "The Dynamics of Living Protoplasm." Academic Press, New York.

Helsey, N. P., and Verkman, A. S. (1987). Membrane chloride transport measured using a chloride-sensitive fluorescent probe. *Biochemistry* **26**, 1215–1219.

Hille, B. (1992). "Ionic Channels of Excitable Membranes," 2nd ed. Sinauer Associates, Sunderland, MA.

Höber, R. (1945). "Physical Chemistry of Cells and Tissues." The Blakiston Company, Philadelphia.

Hoffman, J. F., and Laris, P. C. (1974). Determination of membrane potentials in human and *Amphiuma* red blood cells by means of a fluorescent probe. *J. Physiol.* (*London*) **239**, 519–552.

Horowitz, S. B., and Miller, D. S. (1984). Solvent properties of ground substance studied by cryomicrodissection and intracellular reference-phase techniques. *J. Cell Biol.* **99**, 172s–179s.

Horvath, A. L. (1985). "Handbook of Aqueous Electrolyte Solutions. Physical Properties, Estimation and Correlation Methods." John Wiley & Sons, New York.

Kaji, D. (1992). Hemisodium, a novel selective Na ionophore. *J. Gen. Physiol.* **99**, 199–216.

Katz, S., and Klotz, I. M. (1953). Interactions of calcium with serum albumin. *Arch. Biochem.* **44**, 351–361.

Kayne, F. J. (1973). Pyruvate kinase. *In* "The Enzymes" (P. D. Boyer, Ed.), pp. 353–382. Academic Press, New York.

Klotz, I. M. (1970). Water: Its fitness as a molecular environment. *In* "Membranes and Ion Transport" (E. E. Bittar, Ed.), Vol. 1, pp. 93–122. Wiley, New York.

Ling, G. N., and Gerard, R. W. (1949). The normal membrane potential of frog sartorius fibers. *J. Cell. Comp. Physiol.* **34**, 383–396.

Miller, C. (1974). Nonelectrolyte distribution in mouse diaphragm muscle. I. The pattern of nonelectrolyte distribution and reversal of the insulin effect. *Biochim. Biophys. Acta* **339**, 71–84.

Miskin, R., Zamir, A., and Elson, D. (1970). Inactivation and reactivation of ribosomal subunits: The peptidyl transferase activity of the 50s subunit of *Escherichia coli. J. Mol. Biol.* **54**, 355–378.

Neville, M. C. (1979). The extracellular compartments of frog skeletal muscle. *J. Physiol.* **288**, 45–70.

Neville, M. C., and White, S. (1979). Extracellular space of frog skeletal muscle *in vivo* and *in vitro*: Relation to proton magnetic resonance relaxation times. *J. Physiol.* **288**, 71–83.

Neville, M. C., Allen, J. C., and Watters, C. (1983). The mechanisms of milk secretion. *In* "Lactation. Physiology, Nutrition, and Breast-Feeding" (M. C. Neville and M. R. Neifert, Eds). Plenum Press, New York.

Oesterhelt, D., and Tittor, J. (1989). Two pumps, one principle: Light driven ion transport in Halobacteria. *Trends Biochem. Sci.* **14**, 57–61.

Palmer, L. G., and Civan, M. M. (1977). Distribution of $Na^+$, $K^+$, and $Cl^-$ between nucleus and cytoplasm in *Chironomus* salivary gland cells. *J. Membr. Biol.* **33**, 41–61.

Palmer, L. G., Century, T. J. and Civan, M. M. (1978). Activity coefficients of intracellular $Na^+$ and $K^+$ during development of frog oocytes. *J. Membr. Biol.* **40**, 25–38.

Pauling, L. (1960). "The nature of the chemical bond and the structure of molecules and crystals: An introduction to modern structural chemistry," 3rd ed. Cornell University Press, Ithaca, NY.

Pressman, B. C. (1976). Biological applications of ionophores. *Ann. Rev. Biochem.* **45**, 501–530.

Ringer, S. (1882). Concerning the influence exerted by each of the constituents of the blood on the contraction of the ventricle. *J. Physiol.* (*London*) **3**, 380–393.

Robinson, R. A., and Stokes, R. H. (1959). "Electrolyte Solutions," 2nd ed. Butterworths, London.

Scatchard, G., Coleman, J. S., and Shen, A. L. (1957). Physical chemistry of protein solutions. VII. The binding of some small anions to serum albumin. *J. Am. Chem. Soc.* **79**, 12–20.

Silver, B. L. (1985). "The Physical Chemistry of Membranes." Allen and Unwin, Boston.

Stark, G., and Benz, R. (1971). The transport of potassium through lipid bilayer membranes by the neutral carriers valinomycin and monactin. *J. Membr. Biol.* **5**, 133–153.

Tanford, C. (1961). "Physical Chemistry of Macromolecules." Wiley, New York.

Tinoco, I., Sauer, K., and Wang, J. C. (1994). "Physical Chemistry: Principles and Applications in the Biological Sciences," 3rd ed. Prentice Hall, Englewood Cliffs, NJ.

Tsien, R. Y. (1988). Fluorescence measurement and photochemical manipulation of cytosolic free calcium. *Trends Neurosci.* **11**, 419–424.

Ussing, H. H. (1960). The alkali metal ions in isolated systems and tissues. *In* "The Alkali Metal Ions in Biology, Handbuch der Experimentellen Pharmakologie, Ergänzungswerk" (O. Eichler and A. Farah, Eds.), Vol. 13, Part 1, pp. 1–195. Springer-Verlag, Berlin.

van Holde, K. E. (1985). "Physical Biochemistry," 2nd ed. Prentice Hall, Englewood Cliffs, NJ.

Matthew R. Pincus

# 2

# Physiological Structure and Function of Proteins

## I. Molecular Structure of Proteins

Proteins are biopolymers that are essential for all plant and animal life. All antibodies, enzymes, and cell receptors are proteins. The basis for all connective tissue are the fibrous proteins such as elastin and collagen. The entire genetic apparatus of every cell, regardless of how simple or complex its functions, is dedicated to the synthesis of proteins. The building blocks for proteins are $\alpha$-amino acids that are linked to one another by peptide (or amide) bonds. There are 20 naturally occurring amino acids (Fig. 1). All of these amino acids have the same backbone structure (i.e., the $N-C^{\alpha}-C=O$ unit) but differ from one another in that they have different side chains attached to the $\alpha$-carbon, as shown in Fig. 1. Also, all of these amino acids are chiral (i.e., are optically active and are in the L-configuration), except for the simplest amino acid, glycine, which has two H atoms attached to the $\alpha$-carbon. Note that cysteine has a sulfhydryl group, giving it the capacity to form disulfide bonds (see Fig. 1). Disulfide bonds are important in stabilizing the three-dimensional structures of proteins as discussed later.

Topographically, proteins are divided into the fibrous proteins, which are long, repeating helical proteins, and the globular proteins, which tend to fold up into spherical or ellipsoidal shapes. As is shown later, how proteins fold is governed uniquely by their linear sequences of amino acids.

### A. Amino Acid Structure

All proteins are composed of linear sequences of amino acids that are unique to each protein and are covalently bound to one another by peptide bonds (Fig. 2). This peptide bond is almost always arranged so that the $C^{\alpha}$ atoms of two successive residues are *trans* to one another and the atoms attached to the C' and N of the peptide all lie in the same plane since the $C'-N$ bond has double

bond character (i.e., the *trans*-planar peptide bond described by Linus Pauling). The linear sequence of amino acids is called the primary structure.

Note that the amino acids in Fig. 1 are classified as polar, nonpolar, and neutral depending on the nature of the side chain. Thus, for example, leucine and valine are nonpolar amino acids, which contain aliphatic side chains that are hydrophobic (i.e., that are nonmiscible with water and other polar solvents). Amino acids such as lysine and glutamic acid, however, contain charged chains that are highly soluble in water and are completely miscible in polar solvents. Amino acids such as glycine and alanine do not contain groups that predispose them to be classified as either polar or nonpolar, and so they are referred to as neutral.

### B. Regular or Secondary Structure in Proteins

The three-dimensional structures of more than 300 proteins have been determined by X-ray crystallography as described later. One striking feature in all of these proteins is the presence of recurrent regular structures, in particular $\alpha$-helices, $\beta$-sheets, and reverse turns (Fig. 3). The existence of the first two of these structural types was predicted on the basis of X-ray diffraction patterns of model poly-$\alpha$-amino acids, by Linus Pauling in the early 1950s.

Note that in the $\alpha$-helix, in Fig. 3A, the residues form a spiral such that at every 3.6 residues, the spiral or helix makes one complete turn. The repeat distance of this helix is 5.4 Å. This structural motif is observed in virtually all proteins, and some proteins such as myoglobin are virtually completely $\alpha$-helical. One very important interaction for stabilizing helices is the $i$-to-$i + 4$ hydrogen-bonding scheme shown in Fig. 3A. The $C=O$ of the $i$th residue accepts a hydrogen bond from the $N-H$ of the $i + 4$th residue. The hydrogen bond can provide up to 3 kcal/mol in stabilization energy. For an amino acid residue in the middle of an $\alpha$-helix, two hydrogen bonds form to both

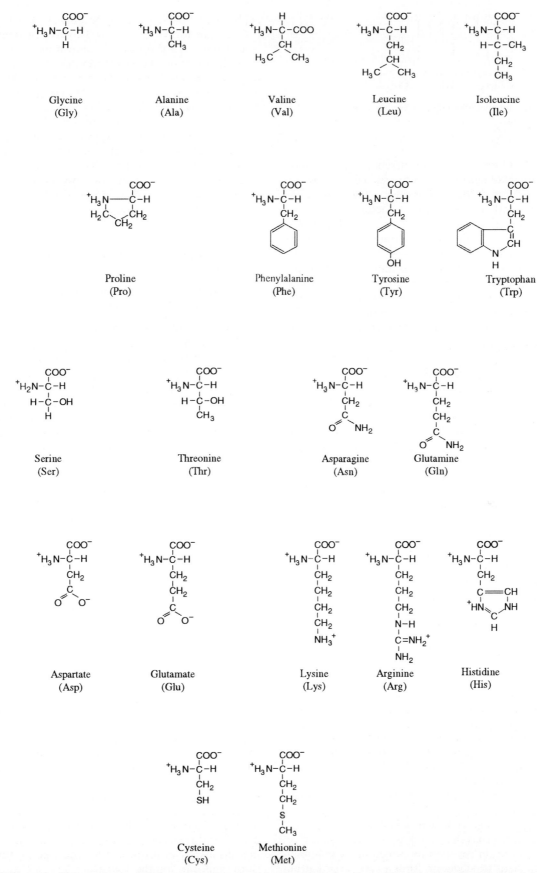

**FIG. 1.** The 20 naturally occurring amino acids grouped by type. Note that Cys residues can form disulfide bonds in the following raction: $R-SH + R-SH \rightarrow R-S-S-R$, where R is the backbone atoms of Cys + the side chain $CH_2$ group.

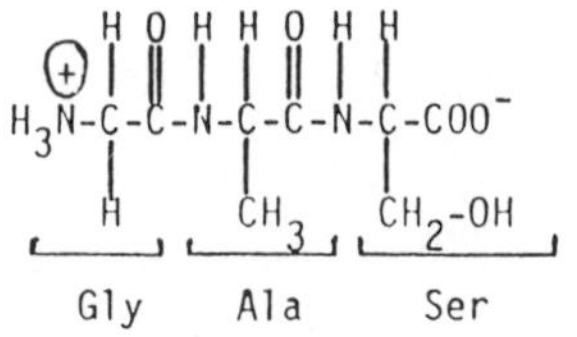

**FIG. 2.** A typical tripeptide, Gly-Ala-Ser, showing the CO—NH peptide bond linkage.

its NH and C=O groups. This occurrence tends to cause helices to propagate because the more such double hydrogen bonds are formed, the greater the stabilization energy provided. Two other longer range factors stabilize $\alpha$-helices. The first is the presence of negatively charged amino acid residues on the amino-terminal end of a helix and positively charged amino acids on the carboxyl-terminal end of the helix. This stabilization results from the dipole moment of an $\alpha$-helix where the center of positive charge is on the amino terminus of the helix and the center of negative charge is on the carboxyl terminus of the helix. The presence of the oppositely charged amino acids stabilizes the helical dipole.

The second stabilizing factor is amphipathicity. This refers to the $i$-to-$i$ + 4 side chain–side chain interactions such that nonpolar or hydrophobic amino acids contact one another on one face of the helix while other polar $i$-to-$i$ + 4 interactions occur on the opposite face of the helix. This segregation of polar and nonpolar faces allows for hydrophobic clustering to occur on one face and hydrogen bonding and charge neutralization to occur on the opposite face. This motif is vital in stabilizing $\alpha$-helices in membrane proteins such as in ion channel proteins.

$\alpha$-Helices are extremely important in membrane proteins, in the transmembrane domains of cellular receptors, and in ion channel proteins, as we discuss in Section IV.

In contrast, the $\beta$-sheet is composed of alternating residues of amino acids that are "flipped" 180° with respect to their nearest neighbors, as shown in Fig. 3B. For single strands in these sheets, we may regard the structure as a flat helix whose repeat is every two residues. Note that within each strand of a $\beta$-sheet, there are minimal interactions between the backbone atoms of different amino acid residues. The hydrogen-bonding stabilization comes from other strands that lie in proximity to one another. As shown in Fig. 3B, the arrangement of the strands is either parallel or antiparallel. Regardless of the arrangement of the two interacting strands, it should be noted from Fig. 3B that two hydrogen bonds between the backbone atoms form for *alternate* residues (i.e., every other residue forms no hydrogen bond to the neighboring strand). Thus, there are twice the number of hydrogen bonds in long $\alpha$-helices as in long $\beta$-sheets. Evidently, other interactions are important in stabilizing $\beta$-sheets besides backbone–backbone hydrogen bonding.

Another type of regular structure is the "reverse turn," the prototype of which is shown in Fig. 3C. In this structure, the polypeptide chain reverses its direction. A minimum of two amino acids is needed to form a reverse or hairpin turn, which is also called a $\beta$-bend. Usually, three amino

acids are involved, however, because the C=O of the first residue can form an $i$-to-$i$ + 3 hydrogen bond with the NH of the $i$ + 3rd residue, as shown in Fig. 3.

Of the three regular structures described, $\alpha$-helices tend to propagate most strongly because of the double hydrogen bonding of the NH and C=O groups of central residues in the helix as discussed earlier and as illustrated in Fig. 3D. This type of "medium-range" interaction is not present for the two other types of regular structures.

The specific sequences of amino acids will adopt specific structures in proteins such as the three basic regular structures just described. Thus, the amino acid sequences of proteins determine their respective three-dimensional structures. This conclusion has been verified as we now describe.

### C. Tertiary or Three-Dimensional Structure of Proteins

#### 1. Amino Acid Sequences of Proteins Determine Their Three-Dimensional Structures

In a classic series of experiments, Anfinsen and his co-workers at the National Institutes of Health showed in the early 1960s that the primary structure (linear sequence of amino acids) of a protein determines the unique three-dimensional structure of that protein (Anfinsen *et al.,* 1961). These investigators took the protein ribonuclease A, which hydrolyzes RNA and whose amino acid sequence had just been determined and was known to contain four disulfide bonds, and denatured it in the denaturing agent 6 $M$ guanidine hydrochloride. This agent and 8 $M$ urea are both known to disrupt the hydrogen bonds in proteins and so to destroy their three-dimensional structures. In addition, using $\beta$-mercaptoethanol, they reduced the four disulfides to 8 sulfhydryl (SH) groups, thereby destroying all of the major determinants of the three-dimensional structure. The denatured protein was completely inactive toward RNA hydrolysis and possessed spectral properties that differed greatly from those of the native protein. They then allowed the sulfhydryl groups to reoxidize to disulfides and dialyzed the 6 $M$ guanidine hydrochloride. Within 2 hours, virtually all of the native enzymatic activity returned, and the spectral properties of the refolded protein were identical to those of the undenatured protein. In addition, all of the disulfides were paired exactly as they were in the native protein.

This experiment illustrated that the primary sequence of amino acids in a protein determines its three-dimensional structure. Further, it demonstrated that the interactions that govern the correct folding of a protein must be highly specific and strong because, of the vast number of possible structures that the polypeptide can adopt, only one such structure actually ultimately forms. This conclusion is reinforced by the following consideration. Once the protein is reduced to the eight-sulfhydryl state, there are many ways in which these eight sulfhydryls can pair to give four disulfides. In particular, there are seven ways in which the first disulfide can form, five for the second, three for the third, and one for the fourth, or 7 × 5 × 3 × 1 = 105 possible ways. Of this total, only one pairing scheme, the

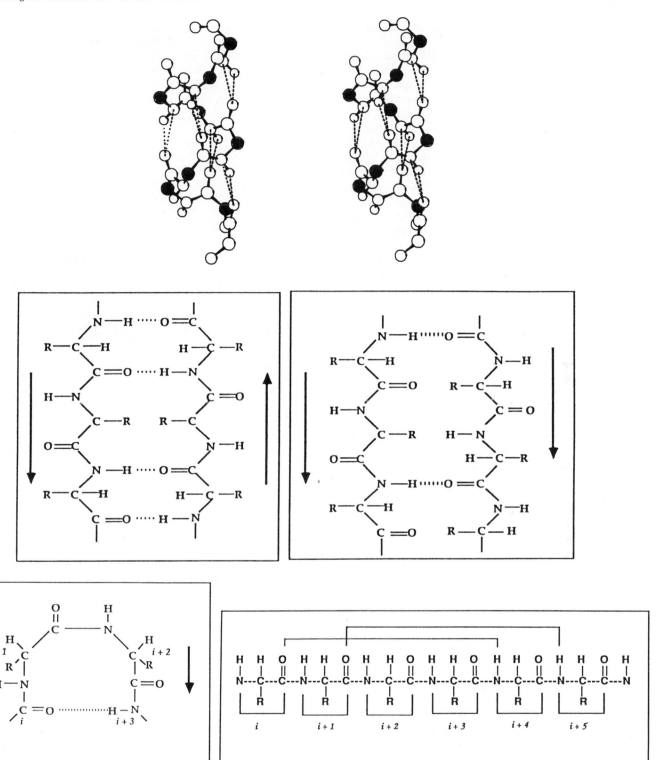

**FIG. 3.** Three types of regular (secondary) structure. (A) Stereo view of an α-helix for a polypeptide backbone showing the $i \rightarrow i + 4$ hydrogen bond. The filled circles are α-carbons; double dashed lines are drawn from C=O to NH (shorter dashed line) and from C=O to N (longer dashed line). (B) Two types of β-sheets. The left side shows the antiparallel arrangement while the right side shows the parallel arrangement, as indicated by the arrows in both figures. (C) The arrangement of backbone atoms in a β-bend showing how an $i$-to-$i + 3$ hydrogen bond stabilizes the reverse turn structure. (D) Illustration of an extended α-helical chain showing hydrogen-bonding propagation.

native pairing, was found ultimately to form in this experiment.

Similar results have been obtained for a number of different proteins. In every case, the native structure is regenerated. Such dramatic results imply that, given the amino acid sequence, it should be possible to infer the three-dimensional structure of the protein.

Methods are being developed that allow us to *predict* the three-dimensional structure of a protein from its amino acid sequence. At present, it is possible to compute the three-dimensional structures of polypeptides from their amino acid sequences for polypeptides of up to about 100 amino acid residues. As might be imagined, the field of structure prediction is a burgeoning one that is developing rapidly. All of the methods used to predict protein structure are based on the physicochemical principle that the observed structure of a molecule is the one of lowest free energy. If we can compute the energies for the possible conformations of a protein, we can pick out the structure of lowest energy, and this is the structure that the protein is most likely to adopt. To perform this task, we must understand the types of energetic interactions that occur in a polypeptide chain that must be optimized for the protein to fold correctly.

### D. Possible Interactions Between Amino Acids in a Protein Chain

As might be expected from the differing nature of the side chains of the naturally occurring amino acids, a vast number of possible interactions between different side chains and backbone are possible for given amino acid sequences. These types of interactions include: (1) electrostatic interactions (given by the coulombic potential $q_1q_2/DR_{12}$, where the $q$s are the charges on two different interacting atoms, $R$ is the distance between them, and $D$ is the dielectric constant of the medium) between positive and negative charges of oppositely charged side chains, such as the $COO^-$ group of glutamic acid and the $-NH_3^+$ group of lysine, and the interactions between the partial charges on all of the atoms of the protein; (2) nonbonded interactions between the individual atoms, explained later; (3) hydrogen-bonding interactions between polar atoms with H-atoms and other polar atoms; and (4) a solvation energy for most proteins.

This solvation energy is such that the nonpolar side chains of nonpolar amino acids tend to avoid water and so tend to "bury" themselves in the core of the protein, whereas the side chains of polar amino acids tend to interact strongly with water and become solvated. This favorable solvation energy is due to water dipole interactions with charged residues such that the charges on the side chain atoms are reduced by favorable interactions with the water dipoles. It is also due to hydrogen bonding between water molecules and the polar atoms of the exposed polar side chains.

### 1. Specific Interactions: Hydrogen-Bonding, Nonbonded, and Hydrophobic Interactions

In the preceding paragraph, we noted that hydrogen-bonding interactions between polar atoms in a protein can

$$
\begin{array}{ccc}
\delta- & \delta+ & \delta- \\
R-N-H & --- & O=C-R' \\
\mid & & \\
R & &
\end{array}
$$

**FIG. 4.**   Illustration of how hydrogen bonds are stabilized by partial sharing of a polar H atom between two polar heavy atoms (N and O in this case).

greatly stabilize the structure of a protein. These interactions are due to the sharing of a hydrogen atom between the donor atom to which the H atom is covalently linked and the receptor polar atom as in Fig. 4. This interaction stabilizes the distribution of charges over the three atoms that are involved in the hydrogen-bonding system (see Fig. 4). For a given protein sequence, there are many possible ways in which hydrogen-bonding patterns can be arranged. However, in every protein whose three-dimensional structure is known, only one or a few hydrogen-bonding schemes are observed, many of these occurring between backbone atoms (NH--O=C) and others occurring between the polar atoms of side chains.

In addition to the hydrogen-bonding interactions between the atoms of a protein, all pairs of atoms of a protein interact with one another so as to attract each other weakly until they become sufficiently close that their electron clouds repel one another. The weak attractive interactions between a pair of atoms is due to the polarization of charges in each atom between the negatively charged electrons and the positively charged nucleus that is induced by the other interacting atom. This induced dipole–induced dipole interaction is attractive because the positive end of the dipole of the first atom interacts attractively with the negative end of the dipole of the other atom of the pair and vice versa. The overall attractive energy for this interaction can be simply expressed as $-B_{ij}/R_{ij}^6$, where $B_{ij}$ represents a constant that characterizes the magnitude of the attractive energy between any two interacting atoms and where $R_{ij}$ is the distance between the two interacting atoms. As shown in Fig. 5, as two interacting atoms approach one another, the attractive energy becomes progressively stronger until the electrons in the outer shells repel one another. This repulsive energy increases rapidly as the internuclear distance between the atoms decreases.

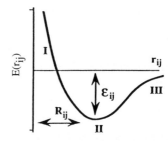

**FIG. 5.**   Interatomic van der Waals energy as a function of distance. Notice that two atoms can approach one another with decreasing energy (regions II and III) until the energy reaches a minimum ($\varepsilon_{ij}$) at a distance of $R_{ij}$. If the atoms become closer, they then repel one another so that the energy rises steeply (region I).

Most proteins fold in an aqueous environment. Changing the solvent, such as by adding nonpolar solvents to water, is known to denature proteins so that they cannot function properly. Evidently, water provides important interactions with proteins, such that they fold correctly to their native conformations.

The three-dimensional structures of more than 300 proteins have been determined, mostly by the technique of X-ray crystallography (see Section II,B). In all of these proteins, the nonpolar amino acid residues tend to "pack" in the interior of the protein, while the polar residues tend to interact with water on the surface of the protein.

Avoidance of interaction of the nonpolar residues with water is due to the hydrophobic (water-avoiding) effect. This behavior is caused predominantly by an entropy effect. Water molecules tend to become highly ordered around the side chains of nonpolar or hydrophobic amino acid residues. In contrast, water molecules can interact in a large number of low-energy complexes with polar side chains and so are less constricted structurally. This ordering of water molecules by nonpolar residues causes the latter to "pack" into the interior of the protein, allowing the water to be more disordered on the outside of the protein.

If one measures the free energy of transfer, $\Delta G$, of nonpolar compounds (i.e., benzene or $n$-hexane) from nonpolar to polar solvents, it is found that the enthalpy of transfer, $\Delta H$, is small while the entropy of transfer, $\Delta S$, is a large, negative value. From the Gibbs free energy expression,

$$\Delta G = \Delta H - T\Delta S \qquad (1)$$

it can be seen that a large negative entropy makes the overall process unfavorable, (i.e., $\Delta G$ becomes more positive). This effect can be attributed to the water structure effect described earlier (see Scheraga, 1984).

All of the preceding interactions have been taken into account in a variety of computer programs that generate the conformations of polypeptides and proteins and calculate their conformational energies. One such program is ECEPP (Empirical Conformational Energies of Peptides Program) developed in the laboratory of Professor Harold A. Scheraga of Cornell University. The energy parameters, such as the constants for the nonbonded interaction energies, have been determined experimentally. This program has been used to compute the structures of many polypeptides and proteins with excellent agreement between predicted and experimentally determined structures (Scheraga, 1984). We give examples of how these methods have been used to predict protein structure and provide insight into the relationship between structure of the protein and its function. First, we describe the properties of the structures of proteins.

## E. Properties of the Structures of Proteins

### 1. Geometry and Dihedral Angles

In the many protein structures that have been determined by crystallography, each of the amino acid residues

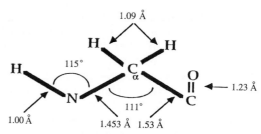

**FIG. 6.** The geometry of a glycine residue showing typical bond lengths and bond angles.

that compose the protein have geometries that are remarkably constant throughout the protein and when compared between any two proteins. Remarkably, this geometry is essentially the same as that found in single crystals of the individual amino acids. Geometry refers to the *bond lengths* and *bond angles* of the individual amino acid. An example of the geometry of an amino acid is illustrated for glycine in Fig. 6. The basic variables that allow for changes in chain conformation are the *dihedral angles*. The dihedral angle is defined as the angle made between two overlapping planes. For example, in Fig. 7, the angle between the planes determined by the $N-C^{\alpha}-C$ atoms and the $C^{\alpha}-C'-N$ atoms of the two amino acid residues shown is a dihedral angle. These angles are generated by rotating around single bonds. As shown in Fig. 7, for the backbone of a single amino acid residue, three dihedral angles exist: $\Phi$, $\Psi$, and $\omega$.

The dihedral angle $\Phi$ is determined by the angle between the planes of $C'-N-C^{\alpha}$ and $N-C^{\alpha}-C'$ whereas the dihedral angle $\Psi$ is determined by the angle between the planes of $N-C^{\alpha}-C'$ and $C^{\alpha}-C'-N$. To determine $\Phi$, one merely has to look down the $N-C^{\alpha}$ axis using the $C'-N$ bond as a reference and to determine the angle made between the $C'-N$ bond and the $C^{\alpha}-C'$ bond, as shown in Fig. 7. To determine $\Psi$, one uses the $N-C^{\alpha}$ bond as a reference and measures its angle with the $C'-N$ bond. The dihedral angle $\omega$ is the interpeptide dihedral angle, which is determined by rotating around the $C'-N$ bond shown in Fig. 7 and can be measured by determining

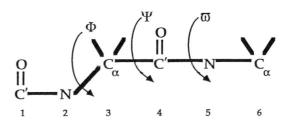

**FIG. 7.** Dihedral angles, $\Phi$, $\Psi$, and $\omega$, for the backbone of an amino acid. $\Phi$ is determined by atoms 1, 2, 3, and 4. Clockwise rotations of the 3—4 bond relative to the 1—2 bond are considered positive, while counterclockwise rotations are considered negative. 0° is the conformation in which the 3—4 bond is eclipsed relative to the 1—2 bond, while 180° is that conformation in which the two bond are *trans* relative to one another. $\Psi$ is determined by the 4—5 bond relative to the 2—3 bond. The same considerations apply to this dihedral angle as to those for $\Phi$. $\omega$ is determined by the angle between the 5—6 bond and the 3—4 bond. This dihedral angle is almost always very close to 180° (i.e., the 5—6 bond is *trans* to the 3—4 bond).

the angle made between the $C^\alpha - C'$ bond and the $N - C^\alpha$ bond when sighting down the $C' - N$ bond. This angle is almost always close to 180°, the *trans*-planar peptide conformation described earlier.

Given the geometry and a complete set of dihedral angles for a given sequence of amino acids in a polypeptide, a unique three-dimensional structure can be generated, called the *conformation of the protein.* All regular structures just described have repeating values for $\Phi$ and $\Psi$ from residue to residue. All $\alpha$-helices have values for $\Phi$ and $\Psi$ that are close to $-60°$, while for the individual extended chains of $\beta$-sheets, the values for both of these dihedral angles lie close to 180°.

In fact, the $\alpha$-helical or A and extended or E conformations are minimum-energy conformations for all of the naturally occurring amino acids. Using conformational energy calculations based on ECEPP, all of the low-energy minimum conformations for the 20 naturally occurring amino acids have been computed as a function of the dihedral angles, $\Phi$ and $\Psi$. In Fig. 8, these energies are plotted as isoenergetic contours in the same way that different elevations are plotted on contour maps. For all of the amino acids, there are at least seven basic low-energy minima, as indicated by the dots in Fig. 7.

Note in this figure that large regions of the map are energetically forbidden and that the allowed (low-energy) regions of the map are relatively restricted (Fig. 7). Analysis of the conformations of the individual amino acids in proteins whose three-dimensional structures are known reveals that virtually all of these amino acids adopt one of these eight basic conformational states, consistent with the results of the energy calculations.

Using programs such as ECEPP, systematic methods, reviewed extensively in Vasquez *et al.* (1994), have been developed that generate representative sets of the low-energy structures for a given polypeptide chain. These methods include the chain buildup procedure (Pincus, 1988), molecular dynamics (Karplus and McCammon, 1986), Monte Carlo procedures, and more specialized methods that allow for "jumping" potential energy barriers between different conformations of the given polypeptide chain (Scheraga, 1989). All of these methods take advantage of the energy map shown in Fig. 8, since the backbone of each amino acid is constricted to adopting one of the low-energy states shown in this figure. Once the low-energy conformations for a given polypeptide are generated and their energies computed, the lowest energy structure is selected and this structure is the one that should be observed experimentally.

Successful computations, using ECEPP, of the three-dimensional structures of a number of polypeptides and proteins have been performed. Examples are shown in Fig. 9 for the cyclic decapeptide gramicidin S, collagen, and melittin (Pincus and Scheraga, 1985). There is close agreement between these computed structures and the experimentally determined ones. Excellent agreement between theory and experiment has also been achieved for other proteins, such as avian pancreatic polypeptide and bovine pancreatic trypsin inhibitor (Scheraga, 1989), and a 100-residue segment of human leukocyte interferon (Gibson *et al.,* 1986). The experimental structures of these polypeptides and proteins have been determined by specific methods [i.e., X-ray crystallography, two-dimensional nuclear magnetic resonance (NMR) spectroscopy, circular dichronism].

## II.  Techniques for Determination of the Structures of Proteins

### A.  Regular (Secondary) Structure: Circular Dichroism

One of the most widely used techniques for determining regular structure is circular dichronism, which is based on the principle that molecules with asymmetric structures absorb light asymmetrically. Light is electromagnetic radiation that can be plane polarized to the right or to the left. Normally, light consists of both types of plane-polarized waves. When a molecule with an asymmetric structure absorbs light, it preferentially absorbs either the left- or right-polarized light wave. The amount of light absorbed, $A$, is equal to $E \times C$, where $E$ is the molar extinction coefficient and $C$ is the concentration of the molecules. For molecules with asymmetric structures, the $E$ values for left and right plane-polarized light differ. The difference $E_L - E_R$ depends on the wavelength of the incident light. A plot of $E_L - E_R$ (called the molar ellipticity) vs wavelength is referred to as the circular dichroism or CD spectrum. Regular structures in proteins are asymmetric both because the amino acids in these structures are them-

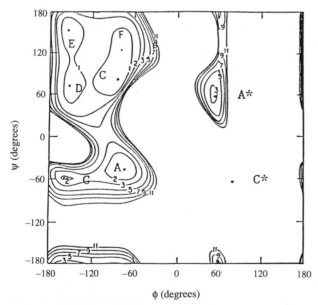

**FIG. 8.**  Contour map for $\Phi$ and $\Psi$ for *N*-acetylalanine-*N'*-methyl amide as computed by ECEPP (Scheraga, 1984). Note that the actual energy minima are denoted by dots on the energy contour map. Seven low-energy minima are shown: A ($\alpha$-helical), C, D (both of which are seen in bends), G, E, and F (the latter two of which are seen in $\beta$-sheets), A* (left-handed $\alpha$-helix), and C*, which is energetically forbidden for all L-amino acids except glycine. Note also that some of these regions are "split" (i.e., more than one minimum can exist for a given conformational region).

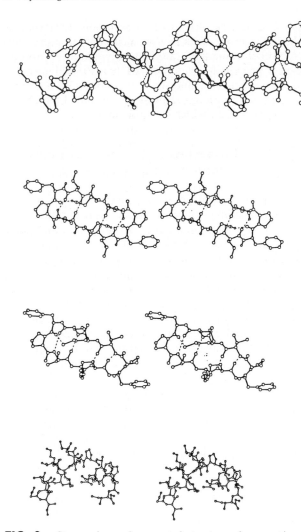

**FIG. 9.** Stereo views of computed structures for a section of a collagen model polypeptide, (Gly-Pro-Pro)$_{20}$, as shown in (A) (Miller and Scheraga, 1976). This structure was directly confirmed by single-crystal X-ray crystallographic studies subsequent to the computations of the structure of this molecule. The computed structure for the cyclic decapeptide gramicidin A (part a of Fig. 9B) and the X-ray structure of the same polypeptide (part b of Fig. 9B) shows the close agreement between the predicted structure and the X-ray crystal structure (Dygert *et al.*, 1975). (C) The computed structure for the membrane-active protein melittin, which consists of two $\alpha$-helical rods, separated by a bend in the middle of the chain (Pincus *et al.*, 1982). These features of have been described in the X-ray crystal structure of this protein. Each polypeptide represents a type of structure seen in many proteins, viz., the collagen helix (A) in fibrous proteins, $\beta$-sheets with reverse turns (B), and $\alpha$-helices (C). (See Pincus and Scheraga, 1985, for a further description of these structures.)

selves asymmetric and because these structures have a "handedness" or a twist-sense as in the case of the $\alpha$-helix, which is right handed. (See van Holde, 1985, for a complete treatment of circular dichroism.) Since the peptide bond has strong absorption in the far UV (i.e., from 230 nm down to about 290 nm), CD spectra for each different type of regular structure should be unique in this range of wavelengths. The patterns in the CD spectra

found for each structure type are shown in Fig. 10. In this figure, the term "random coil" is used and is a misnomer. The term refers collectively to all structures that are not $\alpha$-helices or $\beta$-sheets. These other structures are unique.

For proteins of unknown three-dimensional structures, a knowledge of the contents of regular structure ($\alpha$-helices and $\beta$-pleated sheets) can be obtained by taking their CD spectra. Since patterns for $\alpha$-helix, $\beta$-sheet, and "random coil" are known, these patterns may be combined to reproduce the observed CD spectrum of a particular protein by curve fitting. The combination of regular structure that best fits the observed CD curve is the most likely to exist in the protein. This methodology has been tested on proteins of known three-dimensional structures (so that the percentage of regular structures can be determined) and was found to reproduce the regular structure quite satisfactorily.

## B. Three-Dimensional Structure of Proteins: X-Ray Crystallography

This technique is based on the principle that the electron clouds around atoms diffract incident light in a predictable way. If the atoms are parts of an oriented molecule (i.e., the protein molecules are all oriented in a *crystal*), the same atoms of different molecules diffract the light in the same way. The closer two atoms are in the molecule, the farther apart the diffraction pattern will be. Conversely, the farther away two atoms are in the molecule, the closer the diffraction patterns from each atom will be. Thus, the diffraction pattern is the reciprocal of the atomic pattern. The diffraction pattern can be analyzed with respect to the amplitude of diffraction and the relative positions of the scattering pattern. The light diffracted must be X-rays because the wavelength of light used must be of the order

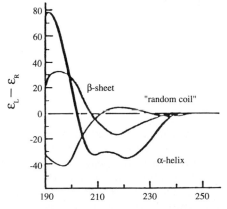

**FIG. 10.** CD spectra for regular structures, $\alpha$-helix, $\beta$-sheet, and so-called "random coil." Note that, for $\alpha$-helices, there is a characteristic "dip" in the region around 210 nm. For a given protein, one can compute the amount of each such structure from its CD spectrum by curve-fitting, using different amounts of each structure type and assuming that the resulting spectum is the sum of the individual spectra for each structure type.

of single bond lengths (i.e., around 1.5–2.0 Å). Only X-rays have this range of wavelengths.

Once a diffraction pattern is obtained, the problem is to relate the positions of the atoms to one another from the diffraction pattern. The relative positions of two atoms can be computed from their (diffraction) amplitudes and a phase factor, which is related to the angular displacement of one atom relative to the other. This phase factor cannot be determined from the diffraction pattern. Since the position of at least one atom is required, in the X-ray diffraction of proteins, a heavy atom bound to the protein is used in the so-called heavy atom replacement method. Once the phase factor is determined, the individual atomic positions can be computed from the diffraction pattern as a Fourier transform that converts from reciprocal (diffraction) space to real (three-dimensional) space.

### C. Two-Dimensional High-Resolution Nuclear Magnetic Resonance Spectroscopy

In one of the most exciting developments in protein chemistry, this technique is being applied to solving the structure of proteins *in solution* without requiring them to be oriented in a crystal. NMR is based on the principle that most nuclei of atoms have spins. If a magnetic field is applied to the atom, the nuclei tend to align their spins with the field. The energy difference between the spins in the absence and the presence of the field is proportional to the magnetic field strength and is given as

$$\Delta E = h\nu = \alpha\beta H \qquad (2)$$

where $\Delta E$ is the energy difference, $h$ is the Planck constant, $\nu$ is the resonance frequency corresponding to the energy absorbed by the spinning nuclei, $g\beta$ is a constant, and $H$ is the magnetic field strength. Nuclei such as hydrogen exist in a protein in different environments, so that some are more shielded by electrons than others, thereby requiring higher $\nu$ values or field strengths to excite them. The frequencies corresponding to the transitions of the spins of individual nuclei to align themselves with the field are in the radio frequency range. Because different H-atoms absorb at different field strengths or values, one can obtain an absorption spectrum for a given compound or protein, where for a given field strength, $H$, different radio frequencies excite the nuclei to change their spins to orient with the field. When two H atoms from different parts of a protein approach one another, their spins either add or subtract from one another and affect the intensity of each other's absorption, in a phenomenon known as the nuclear Overhauser effect or NOE. This change in intensity is inversely proportional to the sixth power of the distance between the two H-atoms. If the individual resonant frequencies for the two interacting H-atoms are known, the distance between them can be calculated.

The specific H-atoms that give rise to different NOE values in a protein can be identified by an elegant technique called two-dimensional NMR (2D-NMR) (Wuthrich, 1986). By irradiating the nuclei at two different frequencies, it is possible to construct a correlation map between the different interacting atoms. A large number

of inter-H distances allows us to fit well-defined structures that have these distances so that a class of structures can be directly determined for the given polypeptide or protein.

Once a protein has adopted its final folded form, it possesses a unique size and shape that confer on it certain bulk properties, among the most important of which are electrical properties, which we now discuss.

## III. Bulk Properties of Proteins: Proteins as Polyelectrolytes

Folded proteins will, in general, have different shapes and sizes depending on their specific sequence of amino acids. Virtually all folded proteins have their nonpolar groups buried in the interior of the protein as a result of hydrophobic interactions and their polar and charged groups on the surface of the protein interacting with the solvent. This distribution of charged groups on the surfaces of proteins strongly affects the manner in which these proteins will be oriented and migrate in electric fields that either are applied in electrophoresis experiments or result from the transmembrane potential in cells. Proteins with net charges also contribute to the fixed charge inside cells. How proteins migrate in electric fields gives much information about their sizes, shapes, and molecular masses. We must therefore consider the properties of proteins as polyelectrolytes dissolved in solution.

### A. Acid–Base Properties of Amino Acids

All amino acids have free $\alpha$-amino ($NH_2$) groups and carboxyl ($COOH$) groups. Amino groups are basic [i.e., *take up* hydrogen ions ($H^+$)], while carboxyl groups are acidic (i.e., *give up* hydrogen ions). Using glycine as an example, at neutral pH (7.0), the only form to exist is the dipolar, or zwitterionic, form, $^+H_3N—CH_2—COO^-$. Figure 11 shows a prototypical titration curve for a dibasic acid such as glycine over the pH range of 1–10. In this figure, note that there are two regions of the titration curve where the change in pH is minimal for added base (i.e., at pHs of around 2.0 and 9.0), the $pK_a$ values for the -COOH and $NH_2$ groups, respectively. At pH 2.0, the buffering capacity of the -COOH group is maximal, while at pH 9.0, the buffering of the $NH_2$ is maximal. The buffering capacity is governed by the Henderson–Hasselbalch equation,

$$pH = pK_a + \log([\text{conjugate base form}]/ \qquad (3)$$
$$[\text{acid form}])$$

where the $pK_a$ is the negative logarithm of the proton dissociation constant for the acid. For the acid segment of the titration curve in Fig. 11, these two forms are $H_3N^+—CH_2—COO^-$ and $H_3N^+—CH_2—COOH$, respectively, while for the high pH portion of the curve, these forms are $H_2N—CH_2—COO^-$ and $H_3N^+—CH_2—COO^-$, respectively. Note also that the following equilibria exist over this range:

$$H_3N^+—CH_2—COOH \leftrightarrows H_3N^+—CH_2—COO^-$$
$$+ H^+ \leftrightarrows H_2N—CH_2—COO^- + H^+ \qquad (4)$$

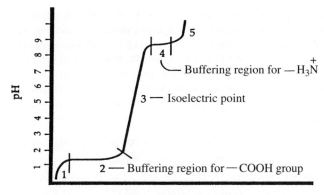

(1) $\overset{+}{H_3N}-CH_2-COOH$

(2) $\overset{+}{H_3N}-CH_2-COOH \longleftrightarrow \overset{+}{H_3N}-CH_2-COO^-$

(3) $\overset{+}{H_3N}-CH_2-COO^-$

(4) $\overset{+}{H_3N}-CH_2-COO^- \longleftrightarrow \overset{+}{H_2N}-CH_2-COO^-$

(5) $\overset{+}{H_2N}-CH_2-COO^-$

**FIG. 11.** Titration curve for a dibasic acid such as glycine to show the two buffering regions for the COOH and $NH_3^+$ groups and the isoelectric point that can be computed as $(pK_{a1} + pK_{a2})/2$. For a protein, there are multiple buffering regions, characteristic of each residue type that is involved in a prototropic dissociation.

We can write two separate sets of equilibrium conditions for the two reactions in Eq. 4 as follows in Eq. 5:

$$K_1 = (H_3N^+ - CH_2 - COO^-)(H^+)/$$
$$(H_3N^+ - CH_2 - COOH)$$

and

$$K_2 = (H_2N - CH_2 - COO^-)(H^+)/$$
$$(H_3N^+ - CH_2 - COO^-) \qquad (5)$$

Solving both of these equations for the common zwitterionic species, $H_3N^+ - CH_2 - COO^-$, we obtain the expression

$$K_1 K_2 = \frac{(H_2N - CH_2 - COO^-)(H^+)^2}{(H_3N^+ - CH_2 - COOH)} \qquad (6)$$

Taking the negative logarithms of both sides of Eq. 6, we obtain the expression

$$pH = (pK_1 + pK_2)/2 + \frac{1}{2}\log[(H_2N - CH_2 - COO^-)/$$
$$(H_3N^+ - CH_2 - COOH)] \qquad (7)$$

where $pK_1$ and $pK_2$ are the negative logarithms of $K_1$ and $K_2$. This equation is the same as the Henderson–Hasselbalch equation for any acid–base equilibrium except, in this case, it is written for two equilibria. The isoelectric point, pI, is defined as the pH at which the total

negative charge of the molecule is equal to its total positive charge (i.e., where only the zwitterionic species exists). In terms of Eq. 7, this condition is met if the numerator and denominator are equal. In this case,

$$pH = (pK_1 + pK_2)/2 \qquad (8)$$

For amino acids with charged side chains, the isoelectric point is easily shown to be that pH equal to $(pK_1 + pK_2 + pK_3)/3$. In general, for multiple prototropic dissociations for a whole protein, the isoelectric point of the protein is simply that point on the titration curve for the whole protein where

$$pH = \Sigma pK_i / N \qquad (9)$$

where the sum is taken over all proton-dissociating groups, i, and $N$ is the number of these groups in the protein.

## B. Protein Charge and Solubility

Since all proteins are charged because of charged side chains and the amino terminal $\alpha$-$NH_3^+$ group and the -$COO^-$ carboxyl terminal groups, there is generally a *net* charge on the protein. These net charges tend to be solvated and interact with counterions in solution. The solvation of these charges makes the protein soluble in $H_2O$. However, interactions of the charged side chains of the protein with surrounding ions and aqueous solvent become minimal at the isoelectric point, where the charges on the protein exactly balance one another. At its isoelectric point, the solubility of the protein in water is minimal, and it will tend to precipitate from solution. Different proteins have different sequences and different charges and isoelectric points so that their tendencies to precipitate will differ from one another and depend on the pH of the solution. Thus, by changing the pH of protein solutions, it is possible to effect differential precipitation of these proteins. A more sophisticated version of isolation of proteins based on their isoelectric points is isoelectric focusing, an electrophoretic method, discussed in Section III,D.

### 1. Isoionic Point

The term *isoionic point* is used when proteins are dissolved in aqueous solutions where the only counterions are $H^+$ and $OH^-$ (i.e., the counterion for the anionic groups is $H^+$, and the counterion for the cationic groups is $OH^-$). Since the total number of positive charges for the system must equal the total number of negative charges for the system,

$(H^+)$ + (total positive charges on protein) = $(OH^-)$
+ (total negative charges on protein) (10)

From Eq. 10, if the isoionic point of a protein is at pH 7 (i.e., $H^+ = OH^-$), *the isoelectric point is the same as the isoionic point*, since from Eq. 10, if $H^+ = OH^-$, then total positive charges on the protein equal total negative charges on the protein.

## C. Titration of Proteins

Because of their large number of exposed acidic and basic groups, proteins behave as buffers. Generally, the pH of the buffering region will depend on the relative numbers of acidic and basic groups on the protein surface. Binding of protons to proteins may be considered to formally be the same as the binding of ligands (small molecules) to proteins. To understand the behavior of proteins as buffers, it is important to understand ligand binding theory.

## I. Ligand Binding Theory

Suppose a protein has $n$ equivalent sites for binding to a ligand and each site is completely independent of any other site. The equilibrium for the binding of a ligand, A, to any one site of the protein can be written as

$$A + P \leftrightarrows AP \tag{11}$$

where A is free ligand concentration, P is the concentration of unbound protein, and AP is the concentration of ligand bound to protein. The equilibrium association constant, $K$, for this process is

$$K = AP/(A)(P) \tag{12}$$

If there are $n$ equivalent binding sites per protein molecule, then the total concentration of sites is $nP_o$, where $P_o$ is the concentration of total protein and is equal to the sum of the concentration of free and bound protein, that is,

$$P_o = P + AP \tag{13}$$

Combining Eqs. 12 and 13, we find that

$$P_o = P + K(A)(P) \tag{14}$$

Solving for P, we obtain,

$$P = P_o/(1 + KA) \tag{15}$$

Since from Eq. 12, $(AP) = K(A)(P)$, the ratio of bound protein to total protein is

$$AP/P_o = KA/(1 + KA) \tag{16}$$

If there are $n$ equivalent binding sites per protein molecule, the fraction of sites bound from Eq. 15 is

$$AP/nP_o = KA/(1 + KA) \tag{17}$$

This equation can be linearized. Defining $R = AP/P_o$,

$$R/A = nK - KR \tag{18}$$

Plots of $R/A$ vs $R$ should give a straight line whose slope is $K$ and whose intercept is $nK$. Thus, both $K$ and $n$ are readily determined. Such plots are called Scatchard plots.

For $n$ equivalent proton binding sites on a protein, we can write Eq. 16 as

$$R = KH^+/(1 + KH^+) \tag{19}$$

In this case, $R = (PH^+)/nP_o$. If we use the dissociation constant $K' = 1/K$, Eq. 19 may be recast as

$$(PH^+)K' = (nP_o - PH^+)(H^+) \tag{20}$$

Rearranging and taking logs of both sides of Eq. 20, we obtain the equation

$$pH = pK' + \log([nP_o - PH^+]/(PH^+)) \tag{21}$$

where $pK'$ is $-\log(K')$. Note that Eq. 21 is the Henderson–Hasselbalch equation for multiple dissociations. The denominator in the logarithmic term in Eq. 21 represents the concentration of $H^+$-bound protein, while the numerator represents non-proton-bound protein. This equation is useful for titration of proteins provided that the $pK_a$ of each group is independent of that any other group for a given set of groups, such as the carboxyl groups of Glu and Asp.

For proteins, it is not generally correct to assume that the foregoing condition holds because, as each carboxyl group dissociates from a protein, the protein becomes progressively negatively charged, so that it becomes progressively more difficult for the next proton to dissociate. In fact, the protein behaves as a polyelectrolyte wherein proton dissociations are strongly dependent on the electric field of the protein.

## 2. The Electrostatic Field Effect

To account for the electrostatic field effect (Bull, 1943) of the charged protein on the dissociation process, we note that the free energy of dissociation per mole of acid group, $\Delta G$, can be written as

$$\Delta G = \Delta G_i + \Delta G_e \tag{22}$$

where $\Delta G_i$ is the intrinsic dissociation free energy for the group, and $\Delta G_e$ is the electrostatic free energy for dissociating a proton from the acid group in the presence of the field of the protein. From electrostatic theory,

$$\Delta G_e = eU \tag{23}$$

where $e = 1$ electrostatic charge unit and $U$ is the electrical potential from the protein. Since the standard free change for dissociation, $\Delta G$, equals $-2.303 \times RT \log K$, and $\Delta G_i$ equals $-2.303 \times RT \log K_i$ (where $R$ is the gas constant and $T$ is the temperature in $°K$), from Eq. 22, we obtain the relation

$$pK = pK_i + eU/2.303RT \tag{24}$$

From Eq. 21, we have

$$pH - \log[(nP_o - PH^+)/(PH^+)] = pK_i + eU/2.303RT \tag{25}$$

It is of obvious importance to obtain an explicit expression for $U$ since it strongly influences the dissociation of protons from proteins.

## 3. Computation of the Electrical Potential, U

To compute this field effect, it is necessary to use a model for the interaction of protein charges with those of small ions interacting with the protein charges. The simplest model is to assume that the protein is a charged sphere and that the counterions, such as protons, surrounding it form a spherical shell around the central charged protein sphere as represented in Fig. 12. The overall arrangement is that of two concentric spheres of radii $R_1$ and $R_2$, the first term being the distance from the center of the protein to its surface, and the second term being

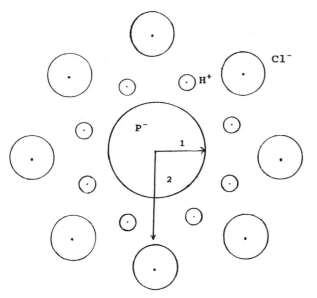

**FIG. 12.** The model for the effect of the charge on a protein ($P^-$) on the dissociation of protons from groups on the protein involved in prototropic dissociations. $R_1$ is the radius of the protein, assuming that it has a spherical shape. $R_2$ is the distance from the center of the protein sphere to the mean center of the ions surrounding the protein. In this figure, $H^+$ ions are shown as the positive ions and $Cl^-$ as the negative ions. Their distributions around the protein will depend on the charge on the protein and the Boltzmann factors for each surrounding ion type. $R_2$ is therefore a mean distance from the center of the protein to the mean center of the surrounding ions.

Solution of this equation for $U$ as a function of the coordinates of the system constitutes the Debye–Hückel theory, and it depends on the assumed configuration of small ions around the central large protein ion, assumed here to be spherical in shape, as in Fig. 12.

Solution of Eq. 30 for this circumstance gives

$$U = Ze(R_2 - R_1)/DR_1R_2 \tag{31}$$

where $Z = \Sigma n_i Z_i$. Letting $1/\chi = R_2 - R_1$,

$$U = Ze/[DR_1(\chi R_1 + 1)] \tag{31A}$$

Combining Eqs. 24 and 31A, we obtain

$$pK + pK_i + (Ze^2/2.303RTD) \times (1/[R_1(\chi R_1 + 1)] \tag{32}$$

Setting

$$W = (e^2/2.303RTD) \times (1/[R_1(\chi R_1 + 1)]) \tag{33}$$

Eq. 25 may now be written as

$$pH - \log[(nP_o - PH^+)/(PH^+)] = pK_i + 0.868WZ \tag{34}$$

This equation now takes into account the field effect of protein and surrounding ion charge provided that the system conforms to the assumptions made for this model, i.e., that $U \ll RT$, that the protein is spherical in shape, and that the ions around the protein form a spherical shell around it. Plots of the left side of Eq. 34 vs $Z$, the overall protein charge, should yield a straight line whose slope is $0.868W$. This term may be regarded as a structure term that depends on the radius of the protein and the position of the center of mass of the surrounding ion cloud.

### 4. More Explicit Models Based on Protein Structures

The preceding theoretical expression is based on several critical assumptions that do not always apply to proteins (i.e., that the charge distribution around the protein is uniform and spherical and that its electrical potential is such that the $ZU$ term $\ll RT$). We know from the many X-ray crystallographically determined protein structures that protein surfaces are far from spherical and that the charge distributions are not uniform. Thus, to understand the effects of protein charge and the effect of the medium on proton dissociations from the protein, more explicit and detailed models are needed. One treatment is based on conformational energy calculations on proteins surrounded explicitly by solvent molecules. These calculations have thus far been aimed at determination of the intrinsic $pK_a$s of specific groups on protein surfaces.

In this approach (Russell and Warshel, 1985), the energy of dissociation of a particular acid group, AH, is computed for the whole protein surrounded by water molecules placed on a grid around the protein. The energy of interaction of the atoms of the protein with one another is computed using energy functions similar to those discussed in Section I,D. Included in this energy is the effect of permanent dipoles of the atoms of the protein that induce dipoles on the other atoms of the protein. The magnitude of this energy depends on the polarizability of the interacting atoms. Further account is taken of the effect of the

the distance from the center of the protein to the center of mass of the ions surrounding the protein. It is clear from this figure that the interactions of protons, and/or any ions, with the central, charged protein molecule is dependent on the distance of the centers of these ions from the center of the protein ion. The distribution of ions around the central charged protein ion is given by the Boltzmann distribution, that is,

$$\Psi = \Sigma e Z_i \exp - (eZ_i U/RT) \tag{26}$$

where $Z_i$ is the valence of the $i$th ion. We can expand the exponential term in a Taylor series as follows:

$$\Psi = \Sigma n_i e Z_i - \Sigma n_i Z_i e(eZ_i U/RT) \tag{27}$$
$$+ (\Sigma n_i Z_i e/2)(Z_i e U/RT)^2 + \dots$$

We make the assumption that the $eZU$ term is much less in value than $RT$, and electrical neutrality must be preserved [i.e., the first sum on the right of Eq. 27 is 0]. Then

$$\Psi \approx - \Sigma n_i Z_i^2 e^2 U/RT \tag{28}$$

This charge distribution must satisfy the Poisson charge distribution equation,

$$\nabla^2 U = -4\Pi\Psi/D \tag{29}$$

where $\nabla^2 U$ is the second derivative of the potential with respect to the coordinates of the system, $\Psi$ is the charge distribution, and $D$ is the dielectric constant of the medium (i.e., water). Combining Eqs. 28 and 29, we obtain

$$\nabla^2 U = (4\Pi e^2)/(DRT) \times \Sigma n_i Z_i^2 U \tag{30}$$

interaction of the dipoles of the atoms of the protein with water molecules, which themselves are represented as dipoles. The calculations are then applied to the dissociation of a proton from a particular group, such as Asp 3 and Glu 7 in bovine pancreatic trypsin inhibitor (BPTI), whose structure has been determined at high resolution by X-ray crystallography (see Section II,B).

In this procedure, a thermodynamic cycle is employed, in which the free energy difference of the system is computed for the acid group in the AH form in the protein and for the AH group dissolved in water (process 1). The group is then allowed to dissociate in water at a given pH (process 2). The free energy difference between AH and A$^-$ in water is known (for Asp residues, this would be the free energy of the dissociation of acetic acid in water at a given pH) and is equal to $2.303RT$ ($pK_a$ − pH). Finally, the free energy difference between solvated A$^-$ in water and A$^-$ in the protein (process 3) is computed. The overall free energy for dissociation is then computed as the difference between the free energy for process 1 − free energy for process 3 + a constant (free energy for dissociation of AH in water).

These calculations have been carried out on a supercomputer using molecular dynamics and iterative procedures (for the polarization energies) and have reproduced ionization constants satisfactorily. The method, which is computer intensive, includes the effects of both solvent and the atoms of the whole protein on the ionization process. Thus, the approach appears to be a fruitful one in the investigation of prototropic dissociations in proteins.

## D. Protein Charge and Electrophoresis

Different proteins have different net charges, as discussed earlier. If these proteins are placed in an electric field, positively charged proteins migrate toward the negatively charged pole, the cathode, while negatively charged proteins migrate to the positively charged poles, the anode. If a protein is present at a pH equal to its isoelectric point, it will generally not move toward either pole. The electrostatic force on a protein present in an electric field is equal to $ZeU$, where $U$ is the electric field, $e$ is the charge of an electron (or a proton), and $Z$ is the number of charges present. There is a drag force on the proteins that is proportional to the velocity at which they move toward either pole, that is, $fv$, where $v$ is the velocity of the protein moving in the electric field and $f$ is a proportionality constant. This force balances the electrostatic force so that

$$ZeU = fv \tag{35}$$

and

$$v = ZeU/f \tag{36}$$

The value of $f$ generally increases with increasing size and molecular weight so that, for proteins of similar charge but different molecular weights, the higher-molecular-weight proteins move more slowly than the lower-molecular-weight proteins.

There are several types of electrophoresis, all of which are extremely effective not only for separating proteins, but also for directly determining their molecular weights.

### 1. Slab Gel Electrophoresis

Proteins are separated on gels made up of such polymers as starch, polyacrylamide, and agarose. The separation is then based mainly on charge and, to an extent, on size.

### 2. Sodium Dodecylsulfate Electrophoresis

Sodium dodecylsulfate (SDS) is a detergent that contains a long aliphatic chain and a sulfate group. This detergent interacts with denatured proteins to form a strongly negatively charged complex (the negative charge arising from the $SO_4^{-2}$ groups of SDS). The proteins are first denatured by heat, and then the SDS is added in large excess. The SDS–protein complexes all contain about the same negative charge because SDS swamps out all of the protein charges. Since the charges are all the same, the proteins all separate from one another strictly on the basis of their sizes. From Eq. (36), the larger polypeptide chains of higher molecular weight migrate the most slowly, while the lower-molecular-weight proteins migrate more rapidly toward the anode. Proteins of known molecular weight can be subjected to this procedure and used as "markers." A plot of the molecular weights vs the log of the distance traveled from the point of application yields a straight line. The log of the distance of migration of a protein of unknown molecular weight can then be plotted on this line and the molecular weight directly determined.

### 3. Isoelectric Focusing

In this elegant technique, a polymer is used in which acidic and basic groups change in density from one end of the polymer to the other. When this polymer is placed in solution, a continuous pH gradient is established along the polymer. A mixture of proteins is then applied to the polymer in a weak buffer. When the proteins migrate in the electric field, they experience local differences in pH on the polymer and eventually reach a local pH equal to their individual pIs. At the pI, the protein no longer migrates in the electric field. This method of separation depends only on the presence of proteins with *different* pI values.

### 4. Two-Dimensional Gel Electrophoresis

Proteins can be separated from one another in one lane as described in the preceding three sections. However, the conditions for separating all of the proteins may not have been met. For example, in agarose gel electrophoresis, two proteins may not separate well at the pH of the buffer used. A buffer of different pH may then be added and the electrophoresis carried out at right angles to the original direction of migration to allow further separation of the proteins.

## 5. Western Blots (Immunoblots)

All proteins are antigens and can provoke the production of antibodies against them when injected into animals. The antibodies are immunoglobulins, which are proteins that bind very specifically to and with high affinity for specific antigens. Antibodies, and, in fact, all proteins, can be conjugated covalently to fluorescent dyes or to enzymes. If we want to identify a particular protein on an electrophoretogram, we can use a "tagged" antibody to the protein and reveal its presence on the electrophoretic gel. The antibody is "tagged" with a covalently labeled enzyme, which catalyzes a reaction that produces a chromophoric (colored) reaction product. For example, alkaline phosphatase catalyzes the hydrolysis of *p*-nitrophenol phosphate, which is colorless, to *p*-nitrophenol, which is bright yellow. Since antibody cannot be added directly to the gel from electrophoresis, the gel itself is blotted onto nitrocellulose or other suitable membrane, which contains all of the separated bands of proteins as they were on the original electrophoretic gel. This nitrocellulose strip is then incubated in a solution containing the antibody. The antibody itself may be conjugated to an enzyme such as alkaline phosphatase, and the band can be identified. Antibodies to the primary antibodies, such as goat anti-rabbit IgG, are then conjugated to the enzyme, and the blot is incubated with these secondary antibodies after being treated with the primary antibody. The secondary antibodies are then markers for the desired protein band.

## IV. Relationship of Protein Structure to Function

In this section, we discuss some of the relationships of the conformations of proteins to their functions. This is a vast field, and so we concentrate on one aspect that is of fundamental importance to cell physiology: membrane polypeptides and proteins.

### A. Membrane Polypeptides and Proteins

Membrane proteins have exceptional physiological importance in a number of different ways. Virtually all extracellular receptors have a transmembrane domain that is essential to the functioning of the receptor protein. Other membrane polypeptides are vital in the transport of secreted proteins across cell membranes. Still other membrane proteins, such as the components of the complement system, are involved in intercalating into the cell membrane, causing cell lysis. Yet other membrane proteins form ion channels that allow for selective entry or exit of specific ions to and from the cell. In this section, we discuss the relationship of the structure of some of these membrane proteins to their functions. It should be emphasized that this entire field of structure–function relationships of membrane proteins is quite new. Most of our knowledge in this field comes from experimental cell biological data on specific proteins and from the results of computations on membrane protein structure.

### 1. Structure and Function of Leader Peptides

Each protein, after it is synthesized on the ribosome, must be transported across the rough endoplasmic reticulum (RER) membranes. If the protein is to be secreted from the cell, or to intercalate into the cell membrane, it must undergo transport across and/or into the cell membrane. For virtually every protein that is synthesized in the cell, there is a polypeptide segment beginning at the N-terminal methionine residue that consists of about 20–30 amino acid residues. Typically, these sequences consist of several hydrophilic amino acids followed by a long stretch of hydrophobic amino acids. When the protein is secreted across the RER membrane, this leader sequence is cleaved off, presumably by intracellular proteases. Absence of these leader sequences results in the inability of the protein to traverse the RER membrane, resulting in ultimate intracellular degradation of the protein. Thus, these leader sequences are of vital importance to cellular protein function in general.

The regulation of transmembrane transport of newly synthesized proteins is delicate. It has been shown in *in vitro* systems that newly synthesized proteins that are secreted across reticulocyte membranes do not undergo leader sequence cleavage. These proteins are nonfunctional because they do not fold correctly. Thus, leader sequences must be attached to the polypeptide chain to enable protein secretion, but must be cleaved off to allow correct protein folding.

To explain how proteins are secreted across membranes, Engelman and Steitz (1981) have proposed the so-called helical hairpin hypothesis, in which the leader sequence adopts an $\alpha$-helical conformation. At the end of this segment there is a hairpin turn, followed by another $\alpha$-helix involving 30–40 residues of the protein itself. Both the amino-terminal end of the leader sequence and the carboxyl-terminal end of the growing polypeptide chain lie on the same side of the membrane, while the hairpin turn lies on the opposite side of the membrane. Because the leader sequence helix and the succeeding helical sequence are both independently stable, they interact minimally with one another, so that the growing polypeptide chain can slide past the hydrophobic leader sequence in the membrane. As the nascent polypeptide chain pushes through the membrane, anchored by the leader sequence, it begins to fold in the aqueous environment of the cytoplasm on the opposite side of the RER membrane. The leader sequence, however, remains in the membrane because of its hydrophobic character. The hairpin connection between the two helices is a signal for intracellular proteases that ultimately results in cleavage of the leader sequence.

Pincus and Klausner (1982), using ECEPP, have computed the low-energy structures for the leader sequence of the $\kappa$-light chain immunoglobulin. This sequence, which contains 16 amino acids, is Asp-Thr-Glu-Thr-Leu-Leu-Leu-Trp-Val-Leu-Leu-Leu-Trp-Val-Pro-Gly. These investigators found that the first four polar residues tend to adopt an extended conformation followed by a long $\alpha$-helix that ends in a hairpin turn at the carboxyl terminal

Val-Pro-Gly sequence. This structure was lower in energy than any other competing structure by at least 10 kcal/mol. Thus, these calculations confirm the essential features of the Engelman–Steitz hypothesis and have been further corroborated experimentally. For example, the CD spectrum of synthetically prepared leader sequences shows high $\alpha$-helical content when the sequence is dissolved in hexafluoroisopropanol, a nonpolar solvent that simulates the low dielectric medium of the membrane. (Rosenblatt *et al.*, 1980).

In site-specific mutagenesis experiments, Silhavy and coworkers at the National Cancer Institute have placed proline residues in the middle of the hydrophobic transmembrane domain of a bacterial leader sequence (Emr and Silhavy, 1982). Proline residues tend to disrupt $\alpha$-helices. These investigators found that if they introduced two proline residues closely together in the sequence, the protein to which the leader sequence was attached was no longer secreted. This result strongly corroborates the conclusion that leader sequences must adopt $\alpha$-helices in the membrane.

In a series of elegant experiments, Blobel and his coworkers (1979) have shown that specific sequences in proteins will insert into membranes but will not lead to cleavage of the inserted sequence from the rest of the protein. These specific sequences contain what are referred to as "stop signals" that do not permit cleavage. Because the transmembrane domain is hydrophobic, it remains inserted in the membrane and there is no impetus for it to move through the membrane, resulting in secretion. This mechanism is extremely important for the function of receptor proteins, as discussed in the later section on transmembrane proteins.

## 2. Melittin

Containing 26 amino acid residues, melittin is perhaps the simplest protein that folds spontaneously. Interestingly, unlike most proteins, it folds in nonpolar environments and actually becomes denatured in water. It is the major component of bee venom and binds to cell membranes at which it causes lysis, resulting in extrusion of the intracellular contents. The sequence of melittin is $\frac{+}{3}$H-Gly-Ile-Gly-Ala-Val-Leu-Lys-Val-Leu-Thr-Thr-Gly-Leu-Pro-Ala-Leu-Ile-Ser-Trp-Ile-Lys-Arg-Lys-Arg-Gln-Gln-NH$_2$. Inspection of this sequence reveals that this protein contains six positive charges and *no* negative charges. At least half of the residues are hydrophobic, as would be expected for a membrane-intercalating protein.

As noted in Section I,E, the structure of melittin has been computed (see Fig. 9) and determined by X-ray crystallography with good agreement between the two structures. The structure of this protein may be thought of as a bent $\alpha$-helical rod with $\alpha$-helices from Gly 1 to Thr 10 and a reverse turn at Thr 11–Gly 12, followed by another $\alpha$-helix from Pro 14 to Gln 26. Leu 13 can adopt an energetically favorable conformation, a D state (see Fig. 8), that results in a compact structure for the monomeric state. In the X-ray structure of tetrameric melittin, this residue adopts an energetically less favorable $\alpha$-helical conforma-

tion, causing it to adopt a more open, "straight" structure. In both structures, melittin forms an amphipathic helix in which the nonpolar groups contact one another on the inside of the structure while the polar groups protrude toward the solvent.

Inspection of the structure for the tetrameric protein reveals that packing between the monomeric units is such that the positive charges of each lie as far away as possible from one another, while the hydrophobic cores pack as tightly as possible with one another. This can be accomplished if the monomeric units become slightly less compact by allowing Leu-13 to adopt the $\alpha$-helical conformation. The sacrifice in conformational energy is about 3 kcal/mol.

Multiple theories about the mechanism by which melittin causes cell lysis have been proposed. One of these is that melittin acts like a detergent. This results from its being amphipathic (i.e., it contains polar positively charged groups and hydrophobic groups that allow it to interact favorably with water and the hydrophobic membrane lipid bilayer, respectively), resulting in solubilization of the membrane and causing its breakdown. This theory does not account for the fact that virtually all biological membranes contain lysolecithins and similar fatty acid derivatives that contain nonpolar and positively charged headgroups that actually stabilize membrane structure. Another theory maintains that melittin forms tetrameric complexes in the membrane, forming pores that allow movement of the intracellular contents into the outside environment.

A third theory (Kempf *et al.*, 1982) proposes that melittin undergoes a conformational change in the presence of a transmembrane electric field such that it adopts a less compact structure and becomes a more rigid rodlike structure that spans the whole membrane and disrupts the lipids in the lipid bilayer. In this theory, monomeric melittin would form pores or channels. Evidence for this theory comes from experimental work in which melittin was added to artificial bilayers. Proteolytic enzymes (pronase) were added to the aqueous phase under the bilayer. If the transmembrane potential was 0 or such that the outside was negative relative to the inside of the bilayer, no proteolysis of melittin occurred. However, if the inside of the membrane was made negative with respect to the outside such that the transmembrane potential was $-60$ millivolts (mV) or less, proteolysis of melittin occurred. A potential of $-60$ mV, found in most resting cells, corresponds to an energy of 2–3 kcal/mol, sufficient to convert compact monomeric melittin to a straightened structure that then spans the membrane and encounters the proteolytic enzymes. This example illustrates the great biological relevance of the relationship of charges on the protein surface to their interactions with electric fields, as in electrophoresis, and to the effects of electric field on the conformations of proteins. This relationship is vital in understanding the functioning of ion channel proteins as described in Section IV,A,4.

## 3. The Function of Transmembrane Proteins

Currently, there appear to be three classes of transmembrane proteins: intercellular adhesion molecules, receptor proteins, and ion channel proteins. Studies on the first category of these proteins are incipient, while the second and third categories of proteins have been better studied. Generally, receptor proteins contain three specific domains: an extracellular, an intramembrane, and an intracytoplasmic domain. This motif has been found to exist for several growth factor receptors and for the T-cell receptor. Although the detailed structures of these proteins have not been elucidated, certain features of their functioning are known.

As an example, the sequence for the growth factor receptor called *neu* or *her-2,* which is a protein of molecular weight 185 kDa, or p185, is known (Padhy *et al.,* 1995). This growth factor receptor is extremely important in the control of mitogenesis in epithelial cells (Padhy *et al.,* 1995). Defects in this receptor protein can result in unregulated cell division and have been found to be highly associated with breast cancer (King *et al.,* 1985; Slamon *et al.,* 1987). It is known that activation of this protein requires that its growth factor bind simultaneously to two receptors (Ben-Levy *et al.,* 1992), thus cross-linking them. As a result of this cross-linking process, tyrosine kinases become activated, resulting in phosphorylation of critical target proteins in a mitogenic signal transduction pathway (Bargmann and Weinberg, 1988).

It is known that substitution of single amino acids *within* the transmembrane domain of this protein activates the protein, presumably by facilitating dimerization in the membrane (Bargmann and Weinberg, 1988). The essential segment of the transmembrane domain contains the sequence (residues 650–683) -Glu-Gln-Arg-Ala-Ser-Pro-Val-Thr-Phe-Ile-Ile-Ala-Thr-Val-XXX-Gly-Val-Leu-Leu-Phe-Leu-Ile-Leu-Val-Val-Val-Val-Gly-Ile-Leu-Ile-Lys-Arg-Arg-. The XXX amino acid residue can be Val, His, Tyr, Lys, or Gly for normal proteins and Gln and Glu for transforming proteins.

Recent calculations (Brandt-Rauf *et al.,* 1989) on the structure of the transmembrane domain of the *neu* protein indicate that the entire transmembrane domain, most of which is hydrophobic, can exist in two states, as shown schematically in Fig. 13. The normal nonmutated transmembrane domain is a bent $\alpha$-helix in which two helices are separated by a $\beta$-turn at residues 664–665. (These conformations are defined in Section I.) In the normal "off" state for the normal protein, not bound to ligand, the $\alpha$-helices are bent and are not favorably disposed to interact with one another. Amino acid substitutions at critical positions around where the bend occurs that result in obliterating the bend between the helices result in the formation of two regular $\alpha$-helices that can easily associate with one another.

A complete analysis (Brandt-Rauf *et al.,* 1990) of the Boltzmann distribution of the low-energy conformations of the p185 protein shows that 90% of the protein is in the bent helical state while 10% is in the all-helical state for *all* normal proteins. For the transforming proteins with Gln or Glu at position 664, this distribution reverses so that 90% of the protein is in the all-helical conformation while 10% is in the bent-helical conformation. These results predict, therefore, that if the normal protein is overex-

pressed by about a factor of 10, then sufficient concentrations of the all-helical form would be present in the cell to cause cell transformation. This computed result has been confirmed in experiments where overexpression of the normal protein by factors of 1–4 causes no effect on the cell (Hudziak *et al.,* 1987) while 10-fold overexpression of the normal protein has been found to cause cell transformation (Di Fiore *et al.,* 1987). They have been further directly confirmed in ligand binding experiments using Scatchard plots (see Section I,C,1); Ben-Levy *et al.,* 1992) in which it was found that, for the normal cellular protein with Val at position 664, 90% of the protein was in a low-affinity state for the cross-linking ligand and only about 10% of the protein was in a high-affinity state. However, with glutamic acid at position 664, a transforming substitution, the reverse was found: More than 90% of the receptors were in the high-affinity state (i.e., were readily cross-linked).

Thus, a change in conformation from a bent to a straight helix causes major changes in the functioning of a transmembrane protein, much like the proposed change in structure of melittin in the presence of an electric field. It is clear from these examples that to understand cell function, it is necessary to understand the conformational properties of proteins, the basis of which has been presented in this chapter.

### 4. Ion Channels

As discussed in Section IV,A,2 cell membranes are polarized resulting in a voltage change across them such that the inside of the cell is negative with respect to the outside. Furthermore, in the resting state of most cells, the extracellular sodium ion concentration is much higher than the intracellular sodium, while the reverse is true for potassium ions. During action potentials in electrically excitable cells, as in axons and dendrites of nerve cells, the membrane becomes depolarized. During this period, there is a large influx of sodium ions, followed by an almost equivalent efflux of potassium ions. These fluxes are then rapidly reversed, during which time the cell becomes refractory to further excitation.

The change in the permeability of the cell to sodium and potassium ions is caused by changes in voltage across the cell membrane. Thus, the motion of these ions across the membrane is said to be voltage-gated. During action potentials, there is a voltage-gated increase in the conductance of the membrane to sodium, followed by a rapid and then a slower inactivation of sodium conductance.

All of these changes in conductance and perm-selectivity of the membrane are mediated by sodium, potassium, and calcium channel proteins, all of which have recently been cloned, purified, and expressed in different cell lines and in artificial lipid bilayers (Catterall, 1995). All of these channel proteins have remarkably similar linear amino acid sequences with differences that cause each to bind selectively to different ions.

All of the channel proteins are composed of an $\alpha$-subunit of molecular mass 260 kDa that is heavily glycosylated. Sequencing of the gene for this protein reveals four do-

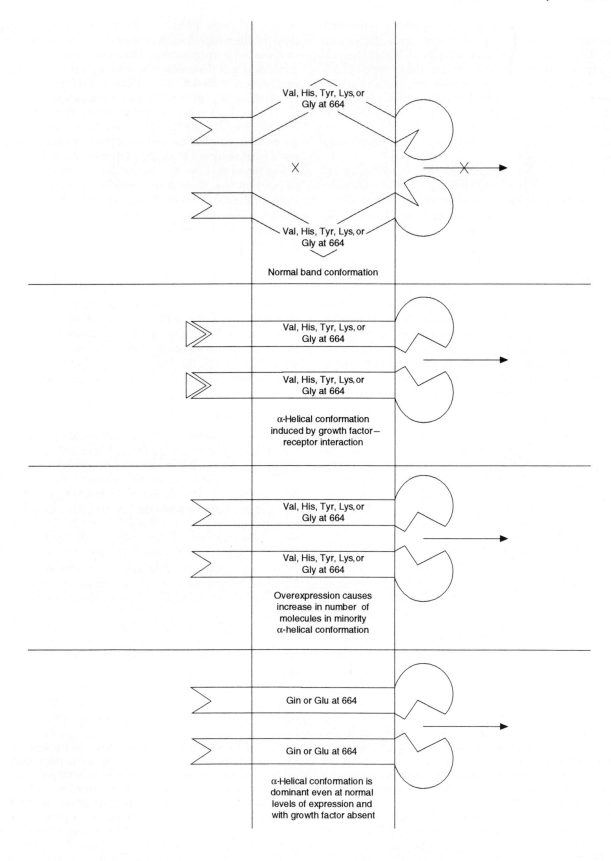

Val, His, Tyr, Lys, or
Gly at 664

Val, His, Tyr, Lys, or
Gly at 664

Normal band conformation

Val, His, Tyr, Lys, or
Gly at 664

Val, His, Tyr, Lys, or
Gly at 664

α-Helical conformation
induced by growth factor—
receptor interaction

Val, His, Tyr, Lys, or
Gly at 664

Val, His, Tyr, Lys, or
Gly at 664

Overexpression causes
increase in number of
molecules in minority
α-helical conformation

Gln or Glu at 664

Gln or Glu at 664

α-Helical conformation is
dominant even at normal
levels of expression and
with growth factor absent

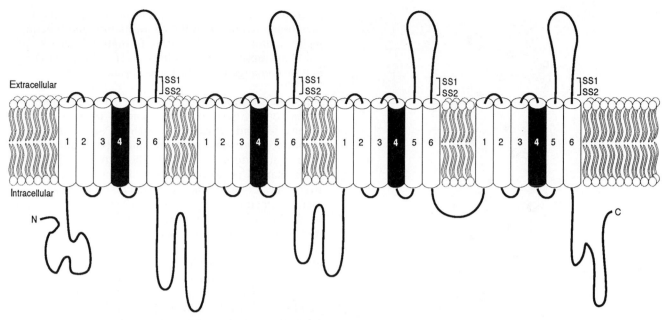

**FIG. 14.** Model for the arrangement of the domains in the $\alpha$-subunit of the sodium ion channel protein. The four domains are labeled I–IV. Each domain is shown to contain six transmembrane helical segments labeled 1–6. Subdomain 4 is highlighted in each domain because it contains the critical residues for voltage-dependent gating. The short segments 1 and 2 that are critical for ion selectivity are labeled SS1 and SS2, respectively. This region contains the critical residues Glu-387 and Phe/Tyr or Cys-385. (This model is based on Hall, 1992, p. 109.)

mains that contain six repeating sequences (called S1–S6). These repeats are characterized by regular spacing in the sequence between hydrophobic amino acid residues and positively charged residues. As with leader sequences, melittin, and the transmembrane domain of the *neu*/HER-2 protein, the six repeat sequences are highly hydrophobic. Therefore, they are believed to be transmembrane domains and have been postulated to be $\alpha$-helical. The basic arrangement of the four domains is shown schematically in Fig. 14. In three dimensions, these four domains would be arranged in a cylindrical fashion and would surround a pore. The $\alpha$-subunit in the sodium channel protein is noncovalently linked to a heavily glycosylated $\beta$-1 subunit of molecular mass 36 kDa and by a disulfide link to a $\beta$-2 subunit.

*a. Voltage Activation.* The actual voltage at which activation of the sodium channel occurs has recently been shown to depend on positively charged and hydrophobic amino acid residues in the S4 amphipathic helical segment of each transmembrane domain. As shown schematically in Fig. 15, there are seven critical positively charged arginine and lysine residues interspersed among many hydrophobic residues. Each of these positive charges must interact with a negative charge from an Asp or Glu residue from another subunit.

In site-specific mutagenesis experiments, each of the positively charged residues has been replaced with neutral amino acid residues. Neutralization of positively charged residues 1, 3, 5, and 7 results in shifting the activation transmembrane potential to more positive values (i.e., the

**FIG. 13.** The effects of transmembrane domain structure of transmembrane proteins on their function. The *neu*-oncogene-encoded p185 protein is a transmembrane protein that has three distinct domains: an extracellular ligand binding domain (left), a transmembrane domain (middle), and an intracellular signaling domain (right). To get signal transduction of a mitogenic signal to the nucleus, two p185 molecules must associate by the binding of their transmembrane domains to one another. This can be accomplished in a variety of ways as shown in this figure. [Reprinted with permission from Brandt-Rauf *et al.* (1990).] (A) The normal protein with Val, His, Tyr, Lys, or Gly at position 664 in the amino acid sequence for the normal protein. The structure of the transmembrane domain is a bent $\alpha$-helix, preventing the two domains from associating with one another. (B) This situation occurs when a growth factor binds to the extracellular domain of the protein. The bend is removed from the middle of the helix, allowing two straight helices to associate. This dimerization allows for activation of mitogenic signaling elements via the intracellular domain in the pathway ending in the nucleus. (C) From the calculations of the distribution of conformations for the normal transmembrane domain (Brandt-Rauf *et al.* 1990) of p185, 90% of the molecules exist in the bent helix state, while 10% are straight helices. Overexpression of the normal protein by a factor of 10 was therefore predicted to produce sufficient straight helices to cause mitogenic cell signaling, leading to cell transformation. (D) Finally, substitution of Glu or Gln at position 664 reverses the distribution of bent and straight helices directly so that 90% of the molecules are straight helices while only 10% are bent helices. This leads to significant levels of dimerization, resulting in permanent mitogenic cell signaling and oncogenic transformation of the cells.

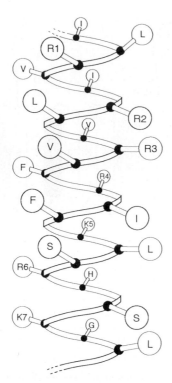

**FIG. 15.** Helical model for the critical segment of the S4 subdomain of the sodium channel. The critical positively charged residues are labeled in numerical order. [Adapted from Catterall (1995).]

influx of sodium ions begins to occur at more depolarized states). Neutralization of charged residues 2 and 4 produces the opposite effect. Thus, residues 1, 3, 5, and 7 are involved with activation of sodium influx, while residues 2 and 4 are involved with pore closing and inactivation of sodium lnflux.

To explain how the S4 segment causes activation of sodium influx, it has been hypothesized that at a certain level of depolarization, the S4 helix rotates such that positive charges on the helix change negatively charged partners in an upward spiral from the inner to the outer membrane, resulting in effective transfer of a positive charge to the outside of the membrane. Since, from the foregoing discussion, not all positive charges perform the same function, this hypothesis may be oversimplified.

Another explanation for charge transfer across the membrane is that, at a critical transmembrane voltage, the S4 helix unfolds into a fully extended conformation, allowing for the shifting of positive charges toward the outer membrane surface. One problem with this explanation is that the energy required to convert an all-helical protein into an all-extended form requires the disruption of a large number of hydrogen bonds, each about 1–2 kcal/mol in energy. Unless the fully extended form can make hydrogen bonds with other segments in a $\beta$-pleated sheet arrangement, the energy barriers for this conformational transition would be very high. Nonetheless, it is clear that the change in the transmembrane potential causes a change in the conformation of the S4 helix, resulting in the opening or closing of the sodium channel. This effect

is strongly analogous to the proposed effect of the transmembrane potential on the conformation and function of melittin, discussed earlier. Pore formation is clearly influenced by transmembrane voltage and results from a conformational change in the transmembrane protein.

*b. Perm-selectivity.* Neural toxins, such as those from snake venom, have long been known to bind with high affinity for and selectivity to specific ion channels. The toxin tetrodotoxin is known to bind to the sodium channel pore at its outermost surface. The tetrodotoxin (and sodium channel pore) binding site has been localized. Between the S5 and S6 helical segments are two short segments, named short segments 1 and 2 (SS1 and SS2), shown in Fig. 14. In SS2, neutralization of Glu-387 by site-specific mutagenesis results in a 10,000-fold decrease in the binding of tetrodotoxin to the mutated sodium channel. In similar experiments, it has further been found that additional residues are vital to perm-selectivity of the sodium ion channel. These include negatively charged residues in domains I and II, a positively charged amino acid residue in domain III, and a neutral amino acid in domain IV.

Position 385 is also a vital amino acid residue in SS2. In brain and skeletal muscle sodium channels, this residue is Phe or Tyr, while in cardiac muscle it contains the nonconservative substitution Cys. The cardiac muscle sodium channel has a several-hundred-fold lower affinity for tetrodotoxin than does brain or skeletal muscle channel protein. If the Cys residue of cardiac muscle channel protein is changed to Tyr or Phe, the high affinity for tetrodotoxin is restored. Thus, Tyr/Phe-385 appears to be vital in tetrodotoxin binding and ion selectivity.

Sodium and calcium channel structure are similar to one another. However, calcium ions bind with only about one-tenth the affinity to the sodium channel protein as that of sodium ions. However, mutation of Lys-1422 and Ala-1714 to Glu residues completely reverses the order of affinity of these two ions to the mutated sodium channel protein. These two residues are therefore implicated as being present at the mouth of the pore of the sodium channel.

*c. Inactivation.* Normally, after the voltage-gated increase in sodium ion permeability, there is a rapid inactivation of the sodium conductance. This inactivation process can be blocked by treatment of the intracellular domain of the sodium channel protein with proteases and by a monoclonal antibody directed against a segment of the protein connecting domains III and IV. Further studies involving site-specific mutagenesis have identified three critical amino acid residues as being vital to the inactivation process: residues 1488–1490, the sequence of which is Ile-Phe-Met, all hydrophobic. Mutation of Phe-1489 to Gln completely blocks inactivation of sodium channel conductance.

To explain the inactivation process, the three-residue segment has been proposed to constitute a "hinged lid," which moves to form a block in the sodium conductance channel after the conformational changes occur that increase sodium ion conductance. Phe-1489 would bind to another as-yet unidentified pocket in the protein, blocking further movement of sodium ions.

Similar experiments on the potassium channel protein have identified a completely different segment of this protein involved in the inactivation of potassium conductance. This constitutes the amino terminal domain consisting of hydrophobic (especially Leu-7) and hydrophilic amino acid residues.

The model for inactivation for the potassium channel protein is the so-called "ball-and-chain" model in which the amino-terminal arm of the protein swings into the pore, blocking further potassium conductance. Convincing evidence for this model has been provided by experiments in which native amino-terminal peptides hybridized with mutant potassium channel proteins that do not inactivate potassium ion conductance completely restore their abilities to inactivate potassium ion conductance. These results compared with those for the sodium channel suggest that inactivation of sodium channels occurs on the intracellular surface, while that for the potassium channels occurs on the extracellular surface.

Our knowledge of the functioning of ion channels is incipient but has been greatly advanced by the cloning of the sodium, potassium, and calcium channel proteins and the site-specific mutagenesis work. These studies all suggest that the functioning of all of these proteins is based on the principles discussed earlier in this chapter. Selective binding of ions is caused by specific ionic (electrostatic) interactions; changes in conductance of the channel proteins are caused by changes in the three-dimensional structure of the proteins induced by changes in the electric field across the cell membrane as proposed for the simpler melittin protein. The basic structure of the channel protein consists of a series of amphipathic $\alpha$-helices containing large numbers of hydrophobic amino acid residues. As found for the transmembrane domain of the *neu*/HER-2 receptor protein, the interaction of the helices with one another is responsible for the functioning of the channel protein.

## V. Summary

We have seen in this chapter that the three-dimensional structures of proteins are dictated by their linear sequences of amino acids. These structures are the one of lowest free energy for the given polypeptide chain, the detailed conformation of which is determined by the dihedral angles of the backbone and side chains. The three-dimensional structures of proteins determine their functions. Proteins are structured in such a way that their polar amino acids tend to be directed toward the aqueous solvent, while the nonpolar or hydrophobic residues tend to point toward the interior of the protein. The three-dimensional structures of proteins can be determined by a variety of techniques that include circular dichroism (for regular structure), X-ray crystallography, two-dimensional NMR, and theoretical techniques.

The three-dimensional structures of proteins determine bulk properties of these proteins such as their behavior

toward titration with acids or bases, their isoionic points, and their migration in electric fields, as in electrophoresis. The ionization of charged groups on the surfaces of proteins can be treated, and thus their titration curves can be predicted, using ligand binding theory combined with electrostatic field theory (using the Debye–Hückel formulation) or using more explicit models based on the actual three-dimensional structure of the protein surrounded by water dipoles.

There is a well-defined relationship between the structure of a protein and its functioning. Membrane proteins, for example, tend to fold into $\alpha$-helices. This structure is of critical importance to the functioning of leader peptides that allow proteins to be secreted both within the cell and outside of the cell. Certain large proteins contain major transmembrane domains, which can adopt several different conformations, some of which allow them to dimerize and thus initiate cellular signal transduction. This process is especially critical for oncogene-encoded proteins that are involved in mitogenic signal transduction to the nucleus causing cell division. Voltage-dependent conformational changes in the transmembrane domains of ion channel proteins are vital to their regulated functioning. The basic functioning of cells is therefore controlled by the proper folding and functioning of cellular proteins, making vital the understanding of protein structure.

## Bibliography

Anfinsen, C. B., Haber, E., Sela, M., and White, F. H., Jr. (1961). The kinetics of formation of native ribonuclease during oxidation of the reduced polypeptide chain. *Proc. Natl. Acad. Sci. USA* **47**, 1309–1314.

Bargmann, C. I., and Weinberg, R. A. (1988). Oncogenic activation of the neu-encoded receptor protein by point mutation and deletion. *EMBO J.* **7**, 2043–2052.

Ben-Levy, R., Peles, E., Goldman-Michael, R., and Yarden, Y. (1992). An oncogenic point mutation confers high affinity ligand binding to the neu receptor. *J. Biol. Chem.* **267**, 17304–17313.

Blobel, G., Walter, P., Chang, C. N., Goldman, B. M., Erickson, A. H., and Lingappa, R. (1979). Translocation of proteins across membranes: The signal hypothesis and beyond. *Symp. Soc. Exp. Biol.* **33**, 9–37.

Brandt-Rauf, P. W., Pincus, M. R., and Chen, J. M. (1989). Conformational changes induced by the transforming amino acid substitution in the transmembrane domain of the *neu*-oncogene-encoded p185 protein. *J. Prot. Chem.* **8**, 749–755.

Brandt-Rauf, P. W., Rackovsky, S., and Pincus, M. R. (1990). Correlation of the transmembrane domain of the *Neu*-oncogene-encoded p185 protein with its function. *Proc. Natl. Acad. Sci. USA* **87**, 8660–8664.

Bull, H. B. (1943). "Physical Biochemistry." John Wiley & Sons, London.

Catterall, W. A. (1995). Structure and function of voltage-gated ion channels. *Annu. Rev. Biochem.* **64**, 493–531.

Di Fiore, P. P., Pierce, J. H., Kraus, M. H., Segatto, O. S., King, R., and Aaronson, S. A. (1987). erbB-2 is a potent oncogene when overexpressed in NIH/3T3 cells. *Science* **237**, 178–182.

Dygert, M., Go, N., and Scheraga, H. A. (1975). Use of a symmetry condition to compute the conformation of gramicidin S. *Macromolecules* **8**, 750–761.

Emr, S. D., and Silhavy, T. J. (1982). Molecular components of the signal sequence that function in the initiation of protein export. *J. Cell Biol.* **95,** 689–696.

Engelman, A. M., and Steitz, T. A. (1981). The spontaneous insertion of proteins into and across membranes: The helical hairpin hypothesis. *Cell* **23,** 411–422.

Gibson, K., Chin, S., Pincus, M. R., Clementi, E., and Scheraga, H. A. (1986). Parallelism in conformational energy calculations on proteins: Partial structure of interferon. *In* "Montreal Symposium on Supercomputer Simulation in Chemistry. Lecture Notes in Chemistry" (M. Dupuis, Ed.), Vol. 44, pp. 198–213.

Hall, Z. W. (1992). Ion channels. *In* Hall, Z. W., ed., Molecular Neurobiology, Sinauer Associates, Inc., Sunderland, Mass., pp. 81–118.

Hudziak, R. M., Schlessinger, J., and Ullrich, A. (1987). Increased expression of the putative growth factor reception p185$^{HER2}$ causes transformation and tumorigenesis of NIH/3T3 cells. *Proc. Natl. Acad. Sci. USA* **84,** 7159–7163.

Karplus, M., and McCammon, A. (1986). The dynamics of proteins. *Sci. Am.* **254,** 42–52.

Kempf, C., Klausner, R. D., Weinstein, J. N., van Renswoude, J., Pincus, M. R., and Blumenthal, R. (1982). Voltage-dependent transbilayer orientation of melittin. *J. Biol. Chem.* **257,** 2469–2475.

King, C. F., Kraus, M. H., and Aaronson, S. A. (1985). Amplification of a novel v-erbB-related gene in a human mammary carcinoma. *Science* **229,** 974–976.

Miller, M. H., and Scheraga, H. A. (1976). Calculation of the structure of collagen models. Role of interchain interactions in determining the triple helical coiled coil conformation. *J. Polymer Sci. Polymer Symp.,* No. 54, 171–200.

Padhy, L. C., Shih, C., Cowing, D., Finkelstein, R., and Weinberg, R. A. (1982). Identification of a phosphoprotein specifically induced by the transforming DNA of rat neuroblastomas. *Cell* **28,** 865–871.

Pincus, M. R. (1988). The chain build-up procedure in computing the structures of biologically active polypeptides and proteins. *Int. J. Quant. Chem.: Quant. Biol. Symp.* **15,** 209–220.

Pincus, M. R., and Klausner, R. D. (1982). Prediction of the three-dimensional structure of the leader sequence of murine pre-kappa light chain, a hexadecapeptide. *Proc. Natl. Acad. Sci. USA* **79,** 3413–3417.

Pincus, M. R., and Scheraga, H. A. (1985). Conformational analysis of biologically active polypeptides, with application to oncogenesis. *Accnt. Chem. Res.* **18,** 372–379.

Pincus, M. R., Klausner, R. D., and Scheraga, H. A. (1982). Calculation of the three dimensional structure of the membrane-bound portion of mellitin from its amino acid sequence. *Proc. Natl. Acad. Sci. USA* **79,** 5107–5110.

Rosenblatt, M., Beaudette, N. V., and Fasman, G. D. (1980). Conformational studies of the synthetic precursor-specific region of preproparathyroid hormone. *Proc. Natl. Acad. Sci. USA* **77,** 3983–3987.

Russell, S. T., and Warshel, A. (1985). Calculations of electrostatic energies in proteins. The energetics of ionized groups in bovine pancreatic trypsin inhibitor. *J. Mol. Biol.* **185,** 389–404.

Scheraga, H. A. (1984). Protein structure and function, from a colloidal to a molecular point of view. *Carlsberg Res. Commun.* **49,** 1–55.

Scheraga, H. A. (1989). Calculations of stable conformations of polypeptides, proteins, and protein complexes. *Chim. Scripta* **29A,** 3–13.

Slamon, D. J., Clark, G. M., Wong, S. G., Levin, W. J., Ullrich, A., and McGuire, W. L. (1987). Human breast cancer: Correlation of relapse and survival with amplification of the HER-2/neu oncogene. *Science* **235,** 177–182.

van Holde, K. E. (1985). "Physical Biochemistry." Prentice Hall, Englewood Cliffs, NJ.

Vasquez, M., Nemethy, G., and Scheraga, H. A. (1994). Conformational energy calculations on polypeptides and proteins. *Chem. Rev.* **94,** 2183–2239.

Wuthrich, K. (1986). "NMR of Proteins and Nucleic Acids." John Wiley & Sons, New York.

Ching-hsien Huang

# 3

# Membrane Lipid
# Structure and Organization

## I. Introduction

Lipids are of fundamental importance as they are a basic structure component of all cell membranes. In addition, many lipids in eukaryotic cell membranes are precursors of lipid-derived second messengers; hence, they are also functionally important. Topologically, membrane lipids are exposed extracellularly as well as intracellularly to an aqueous environment. Consequently, it is relevant and important to know how different membrane lipids are assembled structurally in an aqueous environment and how they interact energetically with each other within the fully hydrated assembly. In this chapter, we first discuss the classification of membrane lipids and the molecular structure of a representative lipid species within each class. We then describe the various organized assemblies of membrane lipids in excess water, with special emphasis on the different forms of lipid bilayers. Subsequently, we turn our attention to the thermally induced polymorphism of the lipid bilayer composed of a single type of phospholipid molecules. In particular, changes in the thermodynamic and conformational properties of the lipid bilayer as it undergoes the gel-to-liquid-crystalline phase transition are illustrated. Finally, in the last part of this chapter, the mixing behavior of two different phospholipids in the bilayer at various temperatures is considered in terms of the temperature–composition phase diagram. Specifically, basic thermodynamic equations applied for simulating the phase diagram for a binary lipid system in excess water are presented. Based on these equations, the simulated and experimentally determined phase diagrams can be compared and matched to determine the nonideality parameters of mixing for the two lipid species, thus leading to a quantitative estimation of the energetics involved in the lipid–lipid interactions in the two-dimensional plane of the lipid bilayer.

## II. Classification and Structures of Membrane Lipids

Membrane lipids are amphipathic molecules composed of polar (or hydrophilic) and nonpolar (or hydrophobic) moieties. From a chemical structure point of view, these amphipathic lipid molecules can be broadly divided into three large groups: (1) *glycerophospholipids*, also loosely called *phospholipids*; (2) *sphingolipids*; and (3) *sterols*. Each membrane lipid group has an enormously wide range of chemically different species; however, there is a common, distinctive structural feature to all lipid species within each group, regardless of their bewildering diversity. All glycerophospholipids, for instance, are well known to contain a common glycerol backbone with L-configuration and a phosphate group ester-linked to the *sn*-3 carbon of the glycerol backbone. The common structural feature to all sphingolipids is the sphingosine backbone. Sterols, however, have four fused rings. The structural features common to all glycerophospholipids and sphingolipids are illustrated in Fig. 1A and B, respectively.

### A. Glycerophospholipids

Glycerophospholipids, also known as phosphoglycerides, are classified on the basis of the polar alcohol esterified with the phosphate group at the *sn*-3 position of the glycerol backbone. The common polar alcohols are choline, ethanolamine, inositol, glycerol, and serine. Each class is further divided into two distinct subclasses based on the type of linkage between the *sn*-1 carbon atom of the glycerol backbone and the nonpolar aliphatic chain that is covalently attached to the *sn*-1 carbon atom. One subclass, called the diacyl phospholipid, has an ester linkage at the *sn*-1 position. The chemical formulas of several commonly occurring diacyl phospholipids are presented in Fig. 2. The

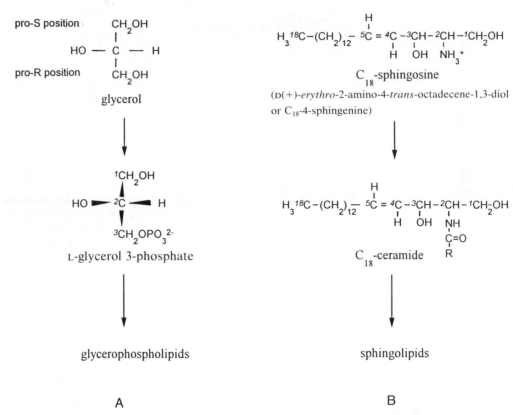

**FIG. 1.** Structural formulas of glycerol and $C_{18}$-sphingosine. All glycerophospholipids have an L-glycerol 3-phosphate backbone, whereas the sphingolipids are built from sphingosine base via ceramide.

other subclass is characterized by an ether linkage at the *sn*-1 carbon of the glycerol backbone (Fig. 3). *Plasmalogens* and *platelet-activating factors* are examples of ether phospholipids belonging to this subclass. In particular, plasmalogens have a vinyl ether linkage, whereas platelet-activating factors have an alkyl ether linkage. In most glycerophospholipids, the aliphatic constituent at the *sn*-2 position is derived from a fatty acid, and this fatty acid is ester-linked to the glycerol backbone (Fig. 3).

In most mammalian cells, diacyl phospholipids constitute the major components of plasma membranes, although ether-linked phospholipids can be present in abundance in some subcellular membranes. A number of other membrane lipids, which are minor constituents of biological membranes, do not belong to either of the subclasses just discussed. These include dialkyl phospholipids, monoacyl phospholipids (e.g., lysophosphatidylcholines), and free fatty acids. The focus of this section is on the structures of various diacyl phospholipids that are found most abundantly in biological membranes.

Diacyl phospholipids isolated from biological membranes of animal cells are primarily mixed-chain lipids with the following general features. With the possible exceptions of mammalian lung and nerve endings, which contain large amounts of dipalmitoylphosphatidylcholine, identical fatty acids esterified at both the *sn*-1 and *sn*-2 positions of the glycerol backbone occur rarely. Instead, the two esterified fatty acids have a different even number

of carbon atoms ranging from 14 to 22. The fatty acids most commonly found in diacyl phospholipids are listed in Table 1. The fatty acids in diacyl phospholipids are nonrandomly distributed. The one that is esterified with the *sn*-1 glycerol hydroxyl group, the *sn*-1 acyl chain, is often derived from a saturated fatty acid, whereas the other, the *sn*-2 acyl chain, is predominantly an unsaturated fatty acid. The unsaturated fatty acids hydrolyzed from the *sn*-2 acyl chains have the structural formula:

$$CH_3-(CH_2)_\alpha-[CH=CH-CH_2]_\beta-(CH_2)_\gamma-COOH,$$

where the subscripts $\alpha = 1, 4, 5,$ and $7, \beta = 1-6,$ and $\gamma = 2-7,$ and the *cis*-double bonds ($\beta = 1-6$) are always separated by one methylene group (Kunau, 1976). Because of the large number of possible ways that permutations of chain length and unsaturation may occur, one can expect that a significant number of molecular species of diacyl phospholipid may be present in any given cell type. Indeed, it has been estimated that there are more than 1000 distinct molecular species of phospholipid in eukaryotic membranes (Raetz, 1986).

Structural features common to all diacyl phospholipids, as shown in Fig. 4, involve the following lipid moieties:

1. The phosphate group ($pK_a \sim 1.3$) in a diacyl phospholipid is completely dissociated at physiological pH and, therefore, bears a negative charge. The four oxygen atoms bonded to the phosphorus atom are tetrahedrally arranged. The two oxygens that do not participate in phosphodiester

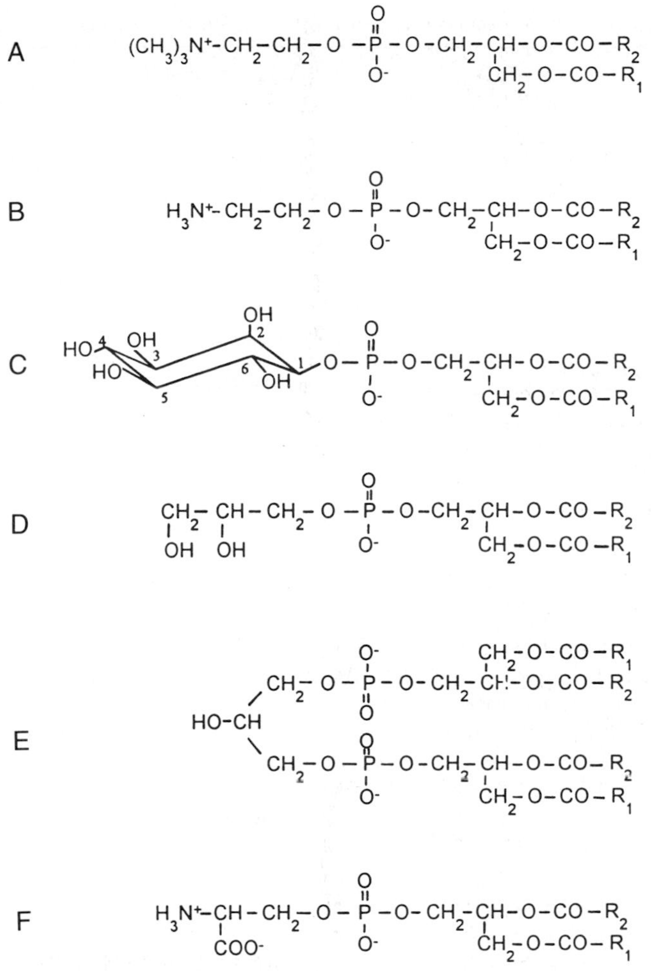

**A**  $(CH_3)_3N^+-CH_2-CH_2-O-\overset{\overset{O}{\|}}{\underset{\underset{O^-}{|}}{P}}-O-CH_2-\underset{\underset{CH_2-O-CO-R_1}{|}}{CH}-O-CO-R_2$

**B**  $H_3N^+-CH_2-CH_2-O-\overset{\overset{O}{\|}}{\underset{\underset{O^-}{|}}{P}}-O-CH_2-\underset{\underset{CH_2-O-CO-R_1}{|}}{CH}-O-CO-R_2$

**C**  (inositol ring) $-O-\overset{\overset{O}{\|}}{\underset{\underset{O^-}{|}}{P}}-O-CH_2-\underset{\underset{CH_2-O-CO-R_1}{|}}{CH}-O-CO-R_2$

**D**  $\underset{\underset{OH}{|}}{CH_2}-\underset{\underset{OH}{|}}{CH}-CH_2-O-\overset{\overset{O}{\|}}{\underset{\underset{O^-}{|}}{P}}-O-CH_2-\underset{\underset{CH_2-O-CO-R_1}{|}}{CH}-O-CO-R_2$

**E**  (cardiolipin structure)

**F**  $H_3N^+-\underset{\underset{COO^-}{|}}{CH}-CH_2-O-\overset{\overset{O}{\|}}{\underset{\underset{O^-}{|}}{P}}-O-CH_2-\underset{\underset{CH_2-O-CO-R_1}{|}}{CH}-O-CO-R_2$

**FIG. 2.** The chemical formulas of diacyl phospholipids that are commonly found in biological membranes. (A) Phosphatidylcholine (PC); (b) phosphatidylethanolamine (PE); (C) phosphatidylinositol (PI); (D) phosphatidylglycerol (PG); (E) diphosphatidylglycerol (cardiolipin); and (F) phosphatidylserine (PS). $R_1$ and $R_2$ refer to hydrocarbon chains of fatty acids at positions 1 and 2 of the glycerol backbone in diacyl phospholipids.

bonds, the pro-R and the pro-S phosphate oxygens, are electrostatically equivalent, each carrying a partially negative charge due to the resonance hybrid effect (Fig. 4A). Because of the electronegativity, the two nonesterified phosphate oxygens are able to serve simultaneously as acceptors in two H-bonds.

2. The two fatty acids esterified with the glycerol backbone of a phospholipid molecule at the *sn*-1 and *sn*-2 positions yield the primary and secondary ester linkages, respectively. The five atoms in the immediate neighborhood of each of the two ester bonds, $[C-O-C_1(=O)-C_2]$, are rigid and planar because of the partial double-bond character of the ester bond $(O-C_1)$ arising from resonance (Fig. 4B), where $C_1$ and $C_2$ are the first and second carbon atoms of the fatty acid, and C denotes the glycerol carbon atom. As a result, the $C-O$ and $C_1-C_2$ bonds adopt a *trans*-configuration at physiological temperature. Moreover, the ester and the carbonyl oxygens bear partial positive and partial negative charges, respectively (Fig. 4B). The electronegative nature of the carbonyl oxygen enables it to serve as a H-bond acceptor.

3. The fatty acid ester-linked at the *sn*-1 position of the glycerol backbone in a diacyl phospholipid usually contains a long polymethylene chain. At low temperature $(<0°C)$ and at thermal equilibrium, the long polymethylene chain adopts the minimum potential energy conformation that corresponds to the fully extended all-*trans*-configuration. Consequently, the consecutive carbon–carbon single bonds along the polymethylene chain are aligned in a zigzag manner (Fig. 4C), with a separation distance of 1.27 Å between two neighboring carbon atoms along the long chain axis. The two-dimensional plane occupied by the polymethylene chain with a zigzag conformation is termed the *zigzag plane*. If the zigzag plane is rotated 90°, the projected successive $C-C$ bonds appear as a straight line with evenly spaced $C-H$ bonds pointing above and below the straight line (Fig. 4D).

Before we turn our attention to the molecular structures of some diacyl phospholipids, it should be mentioned that the basic structure of a diacyl phospholipid molecule has customarily been regarded to consist of three regions: the *polar headgroup,* the *interfacial region,* and the *hydrophobic tail.* The polar headgroup refers to one end of the lipid molecule that is constructed from two polar groups, namely, the alcohol and phosphate groups. In the case of

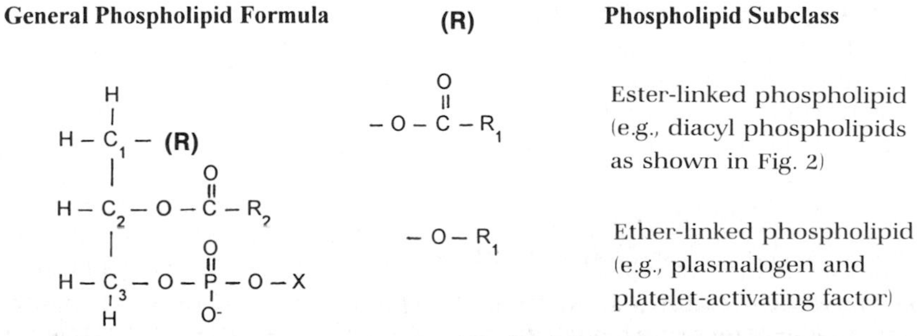

| General Phospholipid Formula | (R) | Phospholipid Subclass |
|---|---|---|
| $\underset{\underset{\underset{\underset{H}{\|}}{H-C_3-O-\overset{\overset{O}{\|}}{\underset{\underset{O^-}{\|}}{P}}-O-X}}{\underset{\underset{O}{\|}}{H-C_2-O-\overset{\overset{O}{\|}}{C}-R_2}}}{H-C_1-(R)}$ | $-O-\overset{\overset{O}{\|}}{C}-R_1$ | Ester-linked phospholipid (e.g., diacyl phospholipids as shown in Fig. 2) |
| | $-O-R_1$ | Ether-linked phospholipid (e.g., plasmalogen and platelet-activating factor) |

**FIG. 3.** The structural formulas of some subclasses of diacyl phospholipids.

**TABLE I** Some of the Common Saturated and Unsaturated Fatty Acids Found in Membrane Phospholipids

| Common name | Systematic name | Abbreviated notation |
|---|---|---|
| Myristic acid | Tetradecanoic acid | C(14) or 14:0 |
| Palmitic acid | Hexadecanoic acid | C(16) or 16:0 |
| Stearic acid | Octadecanoic acid | C(18) or 18:0 |
| Arachidic acid | Eicosanoic acid | C(20) or 20:0 |
| Behanic acid | Docosanoic acid | C(22) or 22:0 |
| Lignoceric acid | Tetracosanoic acid | C(24) or 24:0 |
| Palmitoleic acid | $cis$-9-Hexadecenoic acid | $C(16:1\Delta^9)^a$ or $16:1(n-7)^b$ |
| Oleic acid | $cis$-9-Octadecenoic acid | $C(18:1\Delta^9)$ or $18:1(n-9)$ |
| Vaccenic acid | $cis$-11-Octadecenoic acid | $C(18:1\Delta^{11})$ or $18:1(n-7)$ |
| Nervonic acid | $cis$-15-Tetracosenoic acid | $C(24:1\Delta^{15})$ or $24:1(n-9)$ |
| Linoleic acid | All-$cis$-9, 12-octadecadienoic acid | $C(18:2\Delta^{9,\,12})$ or $18:2(n-6)$ |
| $\alpha$-Linolenic acid | All-$cis$-9, 12, 15-octadecatrienoic acid | $C(18:3\Delta^{9,\,12,\,15})$ or $18:3(n-3)$ |
| Arachidonic acid | All-$cis$-5, 8, 11, 14-eicosatetraenoic acid | $C(20:4\Delta^{5,\,8,\,11,\,14})$ or $20:4(n-6)$ |
| Timnodonic acid | All-$cis$-5, 8, 11, 14, $-17$-eicosapentaenoic acid (EPA) | $C(20:5\Delta^{5,\,8,\,11,\,14,\,17})$ or $20:5(n-3)$ |
| Clupanodonic acid | All-$cis$-7, 10, 13, 16, 19-docosapentaenoic acid (DPA) | $C(22:5\Delta^{7,\,10,\,13,\,16,\,19})$ or $22:5(n-3)$ |
| Cervonic acid | All-$cis$-4, 7, 10, 13, 16, 19-docosahexaenoic acid (DHA) | $C(22:6\Delta^{4,\,7,\,10,\,13,\,16,\,19})$ or $22:6(n-3)$ |

[a] The delta ($\Delta$) refers to the unsaturated double bond and the superscript refers to the position of the double bond from the carboxyl end of the molecule.

[b] The positions of the double bonds are given relative to the methyl end of the molecule. Thus, $18:1(n-9)$ indicates that the double bond occurs at the ninth-from-last carbon atom.

the phosphatidylcholine molecule shown in Fig. 2A, the quaternary base choline (an amino alcohol) is linked to the phosphate group by a phosphoester bond. The resulting phosphorylcholine is thus the polar headgroup of the phosphatidylcholine molecule. Phosphatidylinositol, illustrated in Fig. 2C, differs from phosphatidylcholine in that the polar headgroup has a polyol (inositol) in place of an amino alcohol (choline). The hydrophobic tail refers to the $sn$-1 and $sn$-2 acyl chains excluding the carbonyl groups. Geometrically located in between the polar headgroup and the hydrophobic tail is the third region of the diacyl phospholipid molecule, the interfacial region. Within this region, two rigid and planar elements around the primary and secondary ester bonds exist.

*Phosphatidylcholines* (1,2-diacyl-$sn$-glycero-3-phosphocholines or *lecithins,* often abbreviated as PC) are a structurally diverse class of diacyl phospholipids that are found most abundantly in cell membranes of higher organisms. The repertoire of phosphatidylcholines is extensive, originating from the numerous possible combinations of $sn$-1 and $sn$-2 acyl chains. In plasma membranes, phosphatidylcholine molecules aggregate into the form of the *lipid bilayer* because of their amphipathic nature, thus constituting the basic structural matrix. In addition, some phosphatidylcholine molecules serve as the metabolic precursors of intrinsic signaling elements (Exton, 1994), thus conferring some regulatory properties on eukaryotic cells.

The three-dimensional structure of one molecular species of saturated phosphatidylcholine, dimyristoyl phosphatidylcholine or C(14):C(14)PC, has been determined by X-ray crystallographic approaches (Pascher *et al.,* 1992; Pearson and Pascher, 1979). In this chapter, the saturated PC is abbreviated as C(X):C(Y)PC, where C(X) preceding the colon refers to the $sn$-1 acyl chain with X carbon atoms, and the notation C(Y) succeeding the colon gives the total number of carbon atoms (Y) in the $sn$-2 acyl chain. The single-crystal structure and the energy-minimized structure of C(14):C(14)PC (Huang and Li, 1996) are illustrated in Fig. 5A–D. Several conformational characteristics are revealed by the single-crystal structure. First, the zwitterionic headgroup adopts a bent-down orientation so that the dipole axis or the P–N vector is inclined toward the interfacial region. Second, the primary and secondary ester planes are virtually perpendicular to each other, with the secondary ester plane, $C(2)-O-C_1(=O)-C_2$, running nearly perpendicular to the long molecular axis. Here, C(2) denotes the $sn$-2 carbon atom of the glycerol backbone; $C_1$ and $C_2$ are the first and second carbon atoms of the $sn$-2 fatty acyl chain. Third, the diglyceride moiety exhibits a roughly h-shaped geometry in which the glycerol carbon C(3), the primary ester oxygen, and all carbons in the $sn$-1 acyl chains are arranged in a fully extended conformation, and the $sn$-2 acyl chain is bent 90° at the $C_2$ position. The fully extended $sn$-1 acyl chain and the $sn$-2 acyl chain beyond $C_2$ are aligned in the same direction but nonparallel; however, the two zigzag planes of the two acyl chains are nearly perpendicular to each other (Figs. 5A and B). After force field refinement, the long chain axes of the two acyl chains in the energy-minimized C(14):C(14)PC shown in Fig. 5C and D, are

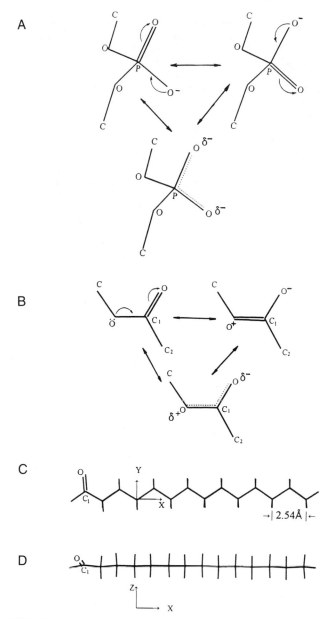

**FIG. 4.** Structural features common to all diacyl phospholipids. (A) The three resonance states of the tetrahedrally arranged phosphate group. (B) The partial double-bond character of the ester bond as represented by the resonance hybrid effect. (C) The zigzag plane of the acyl chain viewed on the x–y plane. (D) The same zigzag plane shown in (C) viewed on the x–z plane.

*Phosphatidylethanolamines* (1,2-diacyl-*sn*-glycero-3-phosphoethanolamines; PE) are also a major zwitterionic lipid component of biological membranes. They usually occur in lesser amounts in higher organisms than do phosphatidylcholines. However, phosphatidylethanolamines are often the principal diacyl phospholipids of microorganisms. Interestingly, phosphatidylethanolamines containing polyunsaturated fatty chains such as docosahexaenoic and arachidonic acids are found at extremely high levels in the retinal rod outer segments.

1,2-Dilauroyl-DL-phosphatidylethanolamine or racemic C(12):C(12)PE is the first molecular species of diacyl phospholipid with which X-ray crystallographic analysis has been performed to determine the structure at the atomic level (Hitchcock *et al.,* 1974). The single-crystal structure, however, is determined in the presence of acetic acid (Fig. 5E and F). The stereochemical features of the polar headgroup may therefore be affected by the contact with acetic acid. Nevertheless, the overall molecular structure of crystalline PE is very similar to that of C(14):C(14)PC shown in Fig. 5A–D. In particular, the diglyceride moiety of C(12):C(12)PE is also characterized by a roughly h-shaped geometry, with a sharp 90° bend occurring at $C_2$ of the *sn*-2 acyl chain. However, there are two notable differences. First, the zwitterionic headgroup of phosphatidylethanolamine has a smaller volume; it bends down slightly more than does the zwitterionic headgroup of phosphatidylcholine. Second, the zigzag planes specified by the long *sn*-1 and *sn*-2 acyl chains of C(12):C(12)PE are nearly parallel to each other, whereas the corresponding zigzag planes are nearly perpendicular in C(14):C(14)PC. It is also worth mentioning that the hydrogen atoms in the terminal -$NH_3^+$ group ($pK_a \sim 9.6$) of a phosphatidylethanolamine are able to form H-bonds with the nonesterified phosphate oxygens of an adjacent phosphatidylethanoamine. Such a H-bonding capability is absent in phosphatidylcholine.

*Phosphatidylinositol* (1,2-diacyl-*sn*-glycero-3-phosphoinositol; PI) is an ester of a cyclic *myo*-inositol with a net negative charge. It is widely distributed among many organisms including plants, but constitutes only about 2 to 8% of all diacyl phospholipids in animal cell membranes. The biological function of phosphatidylinositol has been the subject of intense research in recent years. In eukaryotic cells, phosphatidylinositol on the outer leaflet of the plasma membrane, for instance, can be glycosylated to form glycosylphosphatidylinositol, which is capable of anchoring a large number of proteins to the cell surface (Low, 1989). In addition, two classes of glycosylphosphatidylinositol containing D-*chiro*-inositol and *myo*-inositol, respectively, have been demonstrated to play an important role of mediator in the action of insulin (Huang and Larner, 1993). However, phosphatidylinositol situated on the inner leaflet of the plasma membrane, which faces the cytosol, can be phosphorylated to form phosphatidylinositol bisphosphate. The phosphorylated PI, in turn, can be hydrolyzed by specific enzymes to yield inositol 1,4,5-triphosphate and 1,2-diacylglycerol (Nishizuka, 1984). These hydrolytic products act as second messengers to activate intracellular cascades, thus leading to separate hormonal signaling pathways.

nearly parallel with each other. It should be emphasized that although *sn*-1 and *sn*-2 acyl chains of C(14):C(14)PC have the same total number of methylene units, there is an effective chain length difference between the *sn*-1 and *sn*-2 acyl chains within the lipid molecule. This difference is evident from the clear separation of the two methyl terminal groups along the long molecular axis. In fact, this effective chain-length difference is one of the structural parameters that can be used to characterize the bilayer phase transition behavior, and we discuss it in a later section.

C(14):C(14)PC, crystal structure     C(14):C(14)PC, energy-minimized structure     C(12):C(12)PE, crystal structure

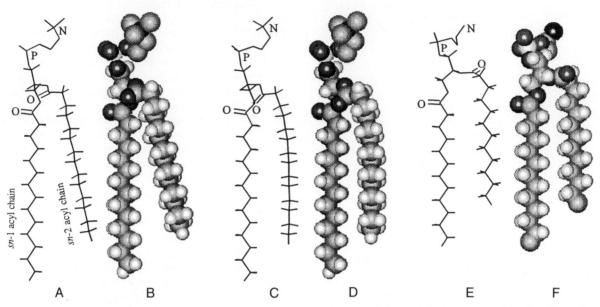

**FIG. 5.** The X-ray single-crystal structure of C(14):C(14)PC is shown in (A) and (B) as represented by the wire and sphere models, respectively. The refined or energy-minimized structure of C(14):C(14)PC is shown in (C) and (D). The refined structure was obtained by the molecular mechanics simulations using the X-ray crystallographic data as the starting data input. Two representations of the X-ray single-crystal structure of C(12):C(12)PE obtained in the presence of acetic acid are shown in (E) and (F).

From a functional point of view, phosphatidylinositol is an extremely important group of diacyl phospholipids.

The orientations of the rigid inositol ring structures of dimyristoyl phosphatidylinositol [C(14):C(14)PI] and dimyristoyl phosphatidylinositol 4-phosphate [C(14):C(14)PI-4P] in the bilayer of C(14):C(14)PC have been determined by neutron diffraction (Bradshaw *et al.,* 1996). According to the neutron diffraction results, the headgroup orientations of these two lipids are strikingly different. Specifically, the inositol ring of C(14):C(14)PI extends almost vertically from the bilayer surface. In addition, the hydroxyl groups at ring positions 2 and 6 are topologically located within the H-bond distance from the two partially charged oxygens of the phosphate group (Fig. 6A and B). In the case of C(14):C(14)PI-4P, the rigid ring is inclined at an angle to the bilayer surface (Fig. 6C and D); moreover, the intramolecular H-bonds are absent. At present, it is not known how the change in the headgroup orientation upon phosphorylation of the ring hydroxyl group is related to the variation of the function of phosphatidylinositol.

*Phosphatidylglycerols* (1,2-diacyl-*sn*-glycerol-3-phosphoryl-1-*sn*-glycerol or 1,2-diacyl-*sn*-3-phosphatidyl-*sn*-1'-glycerol; PG) are negatively charged diacyl phospholipids that are present as minor lipid components in animal cell membranes (largely mitochondrial membranes). PG is also the biosynthetic precursor of *cardiolipin,* shown in Fig. 2E, which is found primarily in mitochondrial membranes. Interestingly, the enzymatic activity of mitochondrial en-

zyme cytochrome c oxidase is stimulated specifically by cardiolipin. Phosphatidylglycerols are major constituents of chloroplast membranes of higher plants and gram-positive bacterial membranes. In a variety of plant cells, saturated identical-chain phosphatidylglycerols are found in chloroplast membranes. The amount of the saturated molecular species of phosphatidylglycerol appears to correlate positively with the susceptibility of the chloroplast membranes to chilling or low-temperature-induced injury (Somerville, 1995).

The polar nonacylated glycerol of naturally occurring phosphatidylglycerol has a stereochemical configuration (*sn*-1 glycerol phosphate) that is opposite to that of the backbone glycerol (*sn*-3-glycerol phosphate). The single-crystal structure of dimyristoyl phosphatidylglycerol, C(14):C(14)PG, has been determined by X-ray diffraction (Pascher *et al.,* 1987, 1992). The energy-minimized structure of C(14):C(14)PG calculated based on the single-crystal coordinates is also known (Jin *et al.,* 1994). The diglyceride moiety of the single-crystal structure, shown in Fig. 7A and B, is essentially identical to that of the energy-minimized conformation (Fig. 7C and D). However, the packing pattern of the two acyl chains in C(14):C(14)PG is distinctly different from that of C(14):C(14)PC or C(12):C(12)PE. Specifically, the *sn*-1 acyl chain of C(14):C(14)PG is initially orientated perpendicular to the long molecular axis of the lipid molecule and then makes a 90° bend at the $C_2$ so that the rest of the *sn*-1 acyl chain axis runs in parallel with the axis of the all-*trans sn*-2 acyl

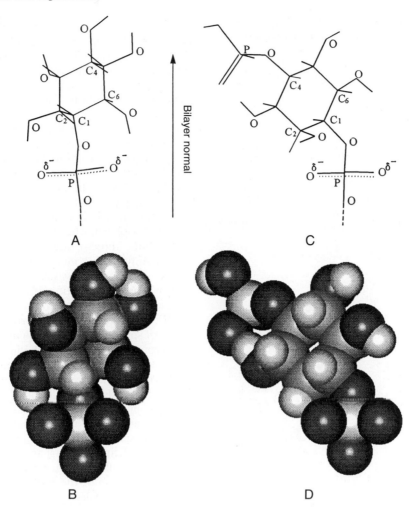

**FIG. 6.** The orientations of the inositol ring structures of dimyristoyl phosphatidylinositol and dimyristoyl phosphatidylinositol 4-phosphate as determined by neutron diffraction are presented in the wire and sphere models as shown in A–B and C–D, respectively.

chain (Fig. 7). The overall acyl chain conformation can thus be considered to be opposite to that of C(14):C(14)PC or C(12):C(12)PE, in which the chain bend occurs at the $C_2$ in the *sn*-2 acyl chain (Fig. 5).

*Phosphatidylserine* (1,2-diacyl-*sn*-glycero-3-phospho-L-serine; PS) is a widely distributed, but minor, class of diacyl phospholipids. It is a negatively charged lipid and is an important constituent of brain cell membranes. In addition to serving as a structural component of membranes, phosphatidylserine is also an activator for all isozymes of protein kinase C. At present, the single-crystal structure of phosphatidylserine is not yet available. It should be mentioned that the serine moiety of phosphatidylserine is the carboxylated ethanolamine; hence, one would expect that the structure of phosphatidylserine may be somewhat similar to that of phosphatidylethanolamine. However, based on the data obtained with $^{31}$P NMR, Browning and Seelig (1980) have suggested that the headgroup of phosphatidylserine is more rigid than that of phosphatidylethanolamine.

### B. Sphingolipids

The second major group of membrane lipids, the sphingolipid, is derived from *sphingosine* (4-sphingenine), a long-chain (18 or 20 carbons in length) amino alcohol with a *trans*-double bond at $C_4$ (Fig. 1). Sphingosine has an amino group on $C_2$; this amino group is jointed in amide linkage with a long chain fatty acid to form *ceramide* (Fig. 1). The fatty acyl moiety of ceramide can be saturated hydrocarbons containing 14–26 carbons and may be monounsaturated and hydroxylated. Ceramides can be further extended to form various kinds of sphingolipids by the addition of different polar headgroups to their terminal hydroxyl group. For instance, if a polar phosphorylcholine molecule is linked to $C_1$ hydroxyl group of ceramide by an ester bond, the resulting molecule, ceramide-1-phosphorylcholine, is *sphingomyelin,* the most commonly occurring sphingolipid. If, however, the reducing end of a carbohydrate residue is linked to the terminal hydroxyl group on $C_1$ of ceramide by the glycosidic bond, then a *glycosphingolipid* (GSL) is formed. The general structure and the

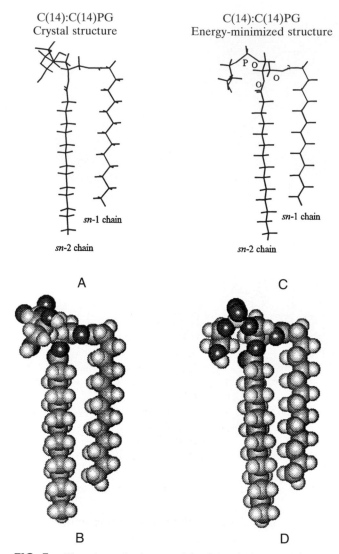

C(14):C(14)PG
Crystal structure

C(14):C(14)PG
Energy-minimized structure

*sn*-1 chain

*sn*-2 chain

A

*sn*-1 chain

*sn*-2 chain

C

B

D

**FIG. 7.** The wire and sphere models of the single-crystal (A and B) and the energy-minimized (C and D) structures of C(14) : C(14)PG.

nomenclature of a few representative GSLs are given in Table 2. These glycosphingolipids are ubiquitously present in animal cell membranes. An excellent review of the structure and the physicochemical properties of individual GSLs is given by Maggio (1994).

Let us now consider the structure of sphingomyelin, the most abundant sphingolipid. In this amphipathic molecule, the hydrophobic chain of the sphingosine backbone has a fixed length of 15 (or 17) carbons that can be considered as one of the two aliphatic chains of the hydrophobic tail. The fatty acid *N*-acylated to the sphingosine backbone constitutes the second chain of the hydrophobic tail. Typical fatty acids *N*-linked to the sphingosine backbone are stearic, behenic, lignoceric, and nervonic acids, with lignoceric acid being the most prevalent. Consequently, the most commonly occurring sphingomyelin is characterized by a marked hydrocarbon chain asymmetry (Fig. 8A).

Because of the identical phosphorylcholine headgroup, sphingomyelin appears structurally to resemble phosphatidylcholine. In fact, these two membrane lipids are quite different in other aspects. As discussed earlier, the most commonly occurring sphingomyelin is a highly asymmetric molecule. In addition, the secondary hydroxyl group on $C_3$ of the sphingosine backbone is an allyl alcohol; hence, it can serve as an excellent H-bond donor. Furthermore, the NH and CO groups of the amide bond linked to $C_2$ of the sphingosine backbone can serve as H-bond donor and acceptor, respectively. These groups can therefore render sphingomyelins in the bilayer to undergo intra- and inter-molecular interactions mediated by H-bonds. In the case of phosphatidylcholine, there is no H-bond donor within the lipid molecule; hence, the intramolecular hydrogen bond and direct intermolecular H-bonds between phosphatidylcholine molecules cannot occur.

The single-crystal structure of sphingomyelin is not known. The conformation of the crystalline galactosylceramide, however, has been reported by Pascher and Sundell (1977). In addition, the conformation of the polar headgroup of sphingomyelin can be approximated by the same headgroup of phosphatidylcholine. A plausible model for the molecular structure of sphingomyelin can thus be simulated based on the crystal structures of galactosylceramide and phosphatidylcholine. This simulated molecular structure of *N*-lignoceryl sphingomyelin is illustrated in Fig. 8A. For comparison, the molecular structure of 1-palmitoyl-2-oleoyl-phosphatidylcholine, obtained by molecular mechanics modeling (Huang and Li, 1996), is presented in Fig. 8B. Parenthetically, 1-palmitoyl-2-oleoyl-phosphatidylcholine is one of the most abundant membrane lipids found in animal cells.

### C. Cholesterol

The third major group of membrane lipids, the sterol, can best be represented by cholesterol. Cholesterols are important components of animal cell membranes and plasma lipoproteins. In addition, cholesterol is the precursor of many steroid hormones (e.g., testosterone, progesterone) and bile acids (e.g., cholate, taurocholate). In plasma membranes, cholesterol concentration is quite high. For instance, in red blood cell membrane, the molar ratio of cholesterol/phospholipid is 0.8–1.0. In contrast, mitochondrial membranes contain little cholesterol.

Cholesterol is characterized structurally by a quasiplanar and frayed conformation. It consists of a fused tetracyclic ring system (the steroid nucleus) and a branched isooctyl side chain. Carbon atoms in the cholesterol molecule are numbered as shown in Fig. 9A. The four fused rings are named A, B, C, and D, with the A–B, B–C, and C–D rings being fused in the *trans*-configuration. For instance, the C-18-angular methyl group is on the opposite side of the hydrogen atom attached at C-14.

The $\alpha$ face or underface of the fused tetracyclic ring system is relatively flat (Fig. 9B), because the seven axial hydrogen atoms at C-1, C-3, C-7, C-9, C-12, C-14, and C-17 are approximately coplanar. The $\beta$ face, however, has greater relief due to the presence of two angular methyl

**TABLE 2**    Names and Structures of Some Common Glycosphingolipids

| Lipid | Chemical structure[a] | Abbreviated nomenclature (IUPC-IUB) | Common name |
|---|---|---|---|
| Galactosylceramide | Gal$\beta$1-1'Cer | GalCer | GalCer |
| Glucosylceramide | Glc$\beta$1-1'Cer | GlcCer | GlcCer |
| Lactosylceramide | Gal$\beta$1-4Glc$\beta$1-1'Cer | LacCer | LacCer |
| Gangliotriaosylceramide | GalNAc$\beta$1-4Gal$\beta$1-4Glc$\beta$1-1'Cer | Gg3Cer | Asialo $G_{M2}$ |
| Gangliotetraosylceramide | Gal$\beta$1-3GalNAc$\beta$1-4Gal$\beta$1-4Glc$\beta$1-1'Cer | Gg4Cer | Asialo $G_{M1}$ |
| Globotetraosylceramide | GalNAc$\beta$1-3Gal$\alpha$1-4Gal$\beta$1-4Glc$\beta$1-1'Cer | Gb4Cer | Globoside |
| $G_{M3}$ ganglioside | NeuAc$\alpha$2-3Gal$\beta$1-4Glc$\beta$1-1'Cer | II$^3$NeuAc-LacCer | Hematoside |
| $G_{M2}$ ganglioside | GalNAc$\beta$1-4Gal(3-2$\alpha$NeuAc)$\beta$1-4Glc$\beta$1-1'Cer | II$^3$NeuAc-Gg3Cer | Tay-Sachs ganglioside |
| $G_{M1}$ ganglioside | Gal$\beta$1-3GalNAc$\beta$1-4Gal$\beta$(3-2$\alpha$NeuAc)$\beta$1-4Glc$\beta$1-1'Cer | II$^3$NeuAc-GgOse$_3$Cer | $G_{M1}$ |

[a] Cer, Ceramide (acylsphingosine); Gal, galactose; Glc, glucose; GalNAc, $N$-acetylgalactosamine; NeuAc, $N$-acetylneuraminic acid (sialic acid).

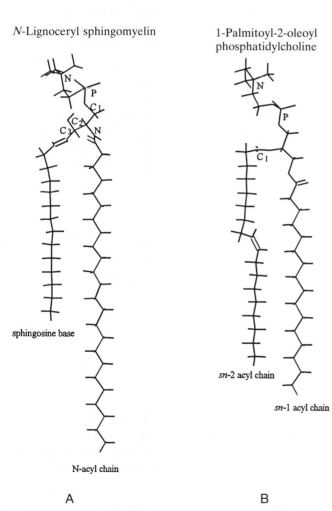

$N$-Lignoceryl sphingomyelin

1-Palmitoyl-2-oleoyl phosphatidylcholine

sphingosine base

N-acyl chain

sn-2 acyl chain

sn-1 acyl chain

A    B

**FIG. 8.**    A structural comparison between $N$-lignoceryl sphingomyelin (A) and 1-palmitoyl-2-oleoyl-phosphatidylcholine (B). In sphingomyelin, $C_1$, $C_2$, and $C_3$ denote the first three carbon atoms in the sphingosine base as indicated in Fig. 1B. In phosphatidylcholine, $C_1$ identifies the carbonyl carbon atom of the $sn$-2 acyl chain. N and P denote the nitrogen and phosphorus atoms, respectively.

groups (C-18 and C-19) attached at C-13 and C-10. In contrast to the alignment of the two angular methyl groups, which are perpendicular to the plane of steroid nucleus, the projected C-21 methyl group lies nearly parallel to the C-16—C-17 bond in the crystal structure of cholesterol monohydrate (Craven, 1976).

As a membrane lipid, cholesterol is unique because the headgroup, the $\beta$-hydroxyl group at C-3, is smaller than a water molecule. Consequently, cholesterols are too weakly amphipathic to self-assemble into the bilayer on their own in water. However, cholesterols can readily dissolve into the lipid bilayer of diacyl phospholipids.

## III. Structural Organizations of Membrane Lipids

Since individual diacyl phospholipid has two long acyl chains, it thus exhibits very limited solubility in water with a critical micellar concentration of about $10^{-10}$ $M$ (Tanford, 1980). Above the critical micellar concentration, diacyl phospholipids self-assemble in water into numerous large organized structures, each characterized by a hydrophobic interior composed of the acyl chains and a hydrophilic surface composed of the headgroups. The driving force for forming spontaneously such a large organized structure is entropic. Specifically, bound water molecules are released into the bulk water from the acyl chains as these hydrophobic chains become sequestered in the interior of the organized structure. Since the motions of bound water are highly restricted, the release of bound water results in a substantial increase in entropy, thus favoring the formation of large organized structures. There are several forms of large organized structures for diacyl phospholipids in water: the micelle, bilayer, inverted hexagonal $H_{II}$ phase, and inverted bicontinuous cubic Pn3m phase. At physiological temperature and ambient pressure, the majority of diacyl phospholipids can self-assemble into *lamellae* or *bilayers* in excess water. The basic feature of a bilayer or lamella is a two-dimensional, bimolecular sheet consisting of two

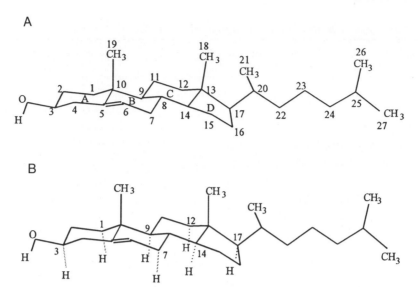

**FIG. 9.**   Conformation of cholesterol molecule. (A) The four fused rings of the steroid nucleus, designated as A, B, C, and D rings, are jointed in a *trans* configuration. The 18 and 19 angular methyl groups at positions 13 and 10 are located above the $\beta$ face of the steroid nucleus. The polar $\beta$-OH group is linked to ring A at position 3. (B) The $\alpha$ face of the steroid nucleus is relatively flat, since the seven axial H-atoms at positions 1, 3, 7, 9, 12, 14, and 17 are approximately coplanar.

opposing leaflets. In this structure, the hydrocarbon chains of the aggregated diacyl phospholipids from the two opposing leaflets are submicroscopically aligned in a nearly parallel manner to form a microscopically two-dimensional hydrocarbon core, and the headgroups are layered on both sides of the hydrocarbon core in a sandwich arrangement.

Because lipid bilayers are structurally the fundamental building matrix of biological membranes (Danielli and Davson, 1935; Blaurock and Wilkins, 1972), studies of lipid bilayers composed of one or two types of pure diacyl phospholipids can provide insights into the properties of these chemically well defined lipid species arranged in two-dimensional arrays of lamellae. Consequently, the roles of these lipids in biological membranes can be inferred. There are two broadly defined model systems for the lipid bilayer that have been widely used in recent years. They are the planar lipid bilayer (or black lipid membrane, BLM) originally developed by Mueller *et al.* (1963) and the liposome admirably pioneered by Bangham *et al.* (1965).

The *planar lipid bilayer* is formed by applying a solution of lipids in an organic solvent such as decane to a small aperture in a hydrophobic sheet of Teflon or polyethylene. This aperture is immersed in an aqueous medium, being located between two chambers. As the lipid bilayer forms spontaneously from the lipid solution under the aqueous medium, it becomes nonreflective to light; hence, it appears as a "black" film. This planar bilayer is an ideal model system for studying simultaneously the electric and transport properties of lipid membranes. An excellent book discussing the water transport across the planar lipid bilayer is that of Finkelstein (1987). Many planar bilayer models using different methods of preparations are also available (Montal and Mueller, 1972; Schindler, 1980). Some can be readily used to incorporate ion channel-forming proteins to allow measurements of electrophysio-

logical events associated with the bilayer (Hartshorne *et al.*, 1985).

*Liposomes* are formed spontaneously by allowing dry powders of phospholipids to undergo hydration in excess aqueous solution at a given temperature. If the temperature is above the main phase transition temperature ($T_m$) of the phospholipid to be used, the dry lipid powders in solution can swell rapidly by vortexing to form concentrically enclosed multilamellar structures like hollowed-out onions, which are heterogeneous in size and shape. Multiple layers of aqueous solutions are partitioned into the multilamellar space between the polar surfaces of enclosed neighboring bilayers. These onion-like liposomes, resembling the myelin sheath of nerve cells, are often called *multilamellar vesicles* (MLVs). If the milky liposome solutions are subjected to ultrasonic irradiation under a $N_2$ atmosphere, the solution becomes translucent as MLVs with diameters up to several micrometers are transformed into small spherical vesicles. Upon gel filtration, small vesicles with an average diameter of about 200 Å can be separated, and each sized vesicle is a single-layered enclosed lamella (Huang, 1969). These are often referred to as *small unilamellar vesicles* (SUVs). Because of the small radius of curvature, diacyl phospholipids packed in the inner and outer leaflets of SUVs are subject to different constraints imposed by the different local surface curvatures; hence, they exhibit distinctly different packing properties. If the large liposomes are subjected to extrusion several times under pressure through a polycarbonate filter of well-defined pore size, single-layered large vesicles with an average diameter of about 0.1–1 $\mu$m, called *large unilamellar vesicles* (LUVs), can be formed (Hope *et al.*, 1985). LUVs can also be formed by first converting the SUVs of saturated diacyl phosphatidylcholines into interdigitated lipid sheets in the presence of ethanol at low temperatures. After ethanol is removed, these fused lipid sheets can be spontaneously vesicularized into LUVs by raising the solution temperature above a critical temperature (Ahl *et al.*, 1994). These LUVs have the potential to serve as drug carriers because of their large internal volumes.

## IV. Thermodynamic and Conformational Properties of Bilayers

### A. Phase Transition Behavior

The lipid bilayer prepared from a single species of diacyl phospholipids can display, in excess water, interesting and unique properties. The thermotropic phase behavior is one such property. This behavior can be determined by using liposomes as the bilayer system and differential scanning calorimetry (DSC) as the detection method (Chapman, 1993). Specifically, the *phase transition temperature* ($T_m$) and the *transition enthalpy* ($\Delta H$) can be determined as the bilayer undergoes the thermally induced phase transitions.

In the case of liposomes prepared from dipalmitoyl phosphatidylcholines of C(16):C(16)PC, which have been preincubated at 0°C for a long time, the first DSC heating scan clearly shows three endothermic phase transitions (Fig. 10). The lower temperature peak is rather broad, centered at 21.5°C ($T_m$) with a $\Delta H$ value of about 6 kcal/mol. Here the $T_m$ value corresponds to the transition-peak position

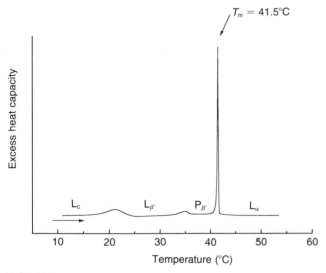

**FIG. 10.** A representative DSC heating curve for the aqueous dispersion of C(16):C(16)PC. Prior to the DSC scan, the sample has been preincubated at 4.0°C under a N$_2$ atmosphere for about 1 year. The DSC experiment was performed with a high-resolution MC-2 differential scanning microcalorimeter (Microcal, Northampton, MA), running at a constant scan rate of 15°C/hour. The buffer solution used for preparing the lipid sample was 50 m*M* NaCl solution containing 5 m*M* phosphate buffer and 1 m*M* EDTA, at pH 7.4, and the final lipid concentration in the aqueous dispersion was 4.5 m*M*. $T_m$ is the main phase-transition temperature. L$_c$, L$_{\beta'}$, P$_{\beta'}$ and L$_\alpha$ are the four lamellar phases described in the text. The arrow indicates the ascending temperature direction of the DSC scan.

at the maximal height, and the $\Delta H$ value corresponds to the integrated area under the peak. This lower temperature transition is called the *subtransition*. The small middle temperature peak, $T_m = 35°C$ and $\Delta H = 1.0$ kcal/mol, is designated as the *pretransition*. The large and sharp upper temperature peak is the *main transition* with $T_m = 41.5°C$ and $\Delta H = 8.9$ kcal/mol. Because of the three discernible transitions, four lamellar phases designated as L$_c$, L$_{\beta'}$, P$_{\beta'}$, and L$_\alpha$ can be defined for the lamellar C(16):C(16)PC within the temperature range of 0 to 50°C (Fig. 10). The sub-, pre-, and main transitions can thus be ascribed to the L$_c \rightarrow$ L$_{\beta'}$, L$_{\beta'} \rightarrow$ P$_{\beta'}$, and P$_{\beta'} \rightarrow$ L$_\alpha$ phase transitions, respectively. Upon immediate reheating, the subtransition is abolished; the pretransition is downshifted by about 1°C; the main transition is the only one that is reproducible upon repeated heatings. Since the pretransition is very small, the main transition is often called the *gel-to-liquid-crystalline phase transition,* referring to the overall L$_{\beta'} \rightarrow$ L$_\alpha$ transition.

Among the multiple phases exhibited by fully hydrated C(16):C(16)PC, the L$_{\beta'}$ and L$_\alpha$ phases have been studied most extensively by X-ray diffraction methods. For instance, in the wide-angle region, a sharp reflection at 4.2 Å with a shoulder at about 4.1 Å is observed for C(16):C(16)PC in the L$_{\beta'}$ or gel phase, whereas the L$_\alpha$ or liquid-crystalline phase is characterized by a diffuse reflection centered at 4.6 Å. These X-ray results thus indicate that in the L$_{\beta'}$ phase the acyl chains of C(16):C(16)PC are highly ordered with a tilted orientation relative to the axis perpendicular to the bilayer surface; however, the chains are melted at $T > T_m$, resembling simple liquid paraffins. Further, refined X-ray analyses of fully hydrated C(16):C(16)PC in the L$_{\beta'}$ (20°C) and L$_\alpha$ (50°C) phases were reported by Nagle *et al.* (1996). Their analyses show that the bilayer thickness decreases from 47.8 Å at 20°C to 39.2 Å at 50°C. In addition, the surface area of C(16):C(16)PC expands from 47.9 to 62.9 Å$^2$ over the same temperature interval. Clearly, the lipid molecule as a whole has increased its volume as it undergoes the thermally induced main phase transition.

### B. *Trans → Gauche* Isomerizations about C—C Bonds in Lipid Acyl Chains

The structural changes of the lipid bilayer associated with the gel-to-liquid-crystalline phase transition can be attributed fundamentally to the following sequence of events: At $T < T_m$, the long acyl chains of C(16):C(16)PC adopt the all-*trans*-conformation. Upon heating to the characteristic temperature of $T_m$, the long acyl chains are "energized"; the tight packing of the long acyl chain will loosen. This thermally induced loosening of the molecular packing is accompanied by a molar volume increase, arising mainly from the *trans → gauche* isomerizations of carbon–carbon single bonds along the acyl chains within each phospholipid molecule.

In Fig. 11A–C, the convention for designating the *trans* and *gauche* conformations of C—C single bonds in an acyl chain is illustrated. The *trans* conformation has a torsion angle of 180°, as shown in Fig. 11A, and is abbreviated as

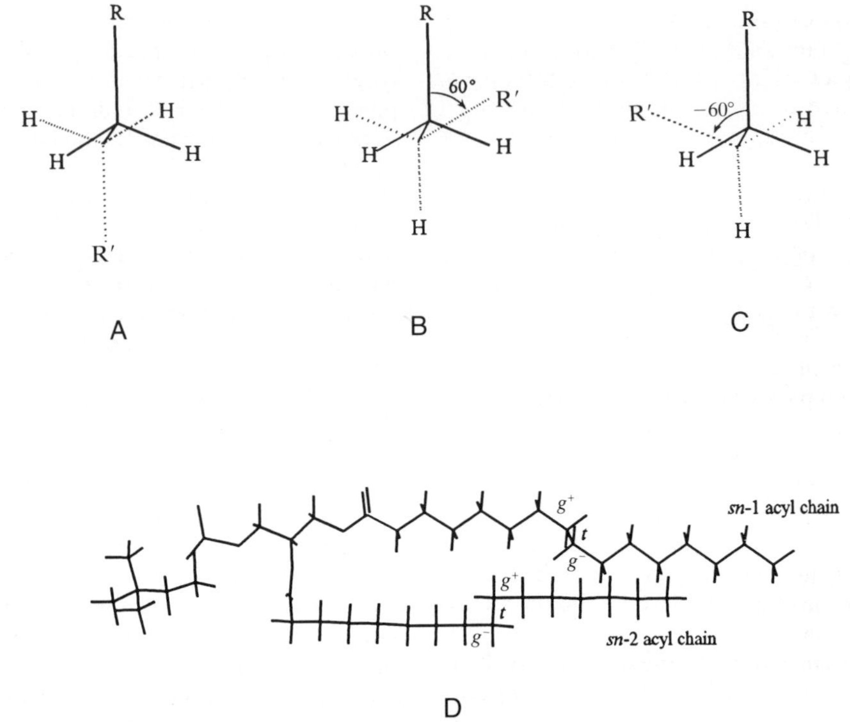

**FIG. 11.** The conformations of *trans, gauche* (+), and *gauche* (−) bonds and $g^+tg^-$ kinks observed in lipid acyl chains. The torsion angle around the C—C bond shown in A–C is defined by the relative positions of R, C, C, and R' atoms. Looking down the C—C bond axis, atom R is placed at 12 o'clock, and atom R' measures the torsional angle; plus if clockwise and minus if counterclockwise. The *trans* or *t* conformation is characterized by a torsion angle of 180° as shown in (A); the *gauche* (±) or $g^\pm$ conformations have torsion angles of ±60° as indicated in (B) and (C), respectively. Energetically, *gauche* (±) conformations are 0.89 kcal/mol higher than the *trans* conformation. In part (D), coupled $g^+tg^-$ kinks are shown in the *sn*-1 and *sn*-2 acyl chains of a PC molecule.

*t*. The two *gauche* conformations shown in Fig. 11B and C are *gauche* (+) with a torsion angle of 60° and *gauche* (−) with a torsion angle of −60°, and they are abbreviated as $g^+$ and $g^-$, respectively. It should be mentioned that *t*, $g^+$, and $g^-$ conformations are positioned in three different minimal troughs in the potential energy diagram, indicating that they all are stable conformations. The *t* conformation is of lower energy than $g^+$ or $g^-$ by about 0.89 kcal/mol (Huang and Li, 1996); hence, *t* is more favored to exist at low temperatures. Upon heating, the conformational changes of the lipid acyl chains in the bilayer occurring abruptly at $T_m$ are thus most likely to involve *trans* → *gauche* isomerizations. Since the *trans* → *gauche* isomerization proceeds with rotations of carbon atoms about the C—C single bond, different chain conformations (*t*, $g^+$, and $g^-$) resulting from the rotated bond are also called rotamers. In this section, some spectroscopic studies are described, suggesting the simultaneous occurrence of coupled and isolated *trans* → *gauche* isomerizations about C—C bonds in lipid acyl chains at the phase transition temperature.

When a single $g^+$ or $g^-$ rotamer is formed, the chain will make a 120° bend. If this pronounced bend is introduced into the center of an acyl chain in the bilayer, it will be inhibited sterically by the neighboring chains. Hence, the single *g* rotamer is most likely to occur, at $T > T_m$, near the methyl ends of the two acyl chains in the C(16):C(16)PC molecule where the steric hindrance due to neighboring chains is small. If, however, a sequence of two *gauche* bonds separated by a *trans* bond ($g^+tg^-$ or $g^-tg^+$) is formed, the resulting acyl chain is then kinked in the shape of a crankshaft. The presence of such a kink leads to a largely parallel packing of the acyl chains in the bilayer. Consequently, the coupled $g^+tg^-$ or $g^-tg^+$ kink is favored in the center of the acyl chain at $T > T_m$. In comparison with an all-*trans* chain, the overall length of a kinked chain is shortened by 1.27 Å along the long-chain axis. In Fig. 11D, this type of kink is graphically illustrated in each of the two acyl chains of a C(16):C(16)PC molecule.

One spectroscopic technique that has been elegantly used to examine the lipid chain order in terms of the *trans* → *gauche* isomerizations about the C—C bonds is $^2$H NMR (Seelig, 1977; Davis, 1983). Specifically, the methylene and terminal methyl groups in the acyl chain are substituted by $CD_2$ and $CD_3$ groups, respectively; the congested powder patterns and the corresponding depacked spectrum with a larger number of resolvable quadrupolar splittings can be obtained from a single $^2$H NMR experiment at $T > T_m$ (Sternin *et al.,* 1983). The experimentally determined quadrupolar splittings can be converted to the *order parameter* ($S_{CD}$) of the C—D bond (Seelig, 1977). Since the order parameter, $S_{CD} = \frac{1}{2} < 3\cos^2 \beta - 1 >$, is a time-average parameter of the angular fluctuation ($\beta$) of the C—D bond relative to the bilayer normal on the $^2$H NMR time scale ($<10^{-5}$ sec), the quantity of $|S_{CD}|$ can thus be used to infer the possible presence of *gauche* rotamers in the acyl chain at $T > T_m$. For instance, if the acyl chain is assumed to adopt an all-*trans*-configuration and is simply reorienting about the long-chain axis, then $\beta = 90°$ and $|S_{CD}| = 0.5$. If, however, the chain is assumed to undergo an isotropic motion over all space, this motional averaging will then lead to $|S_{CD}| = 0$. The *trans* → *gauche* isomerizations of the CD$_2$ groups around the C—C bonds will undoubtedly result in a partially disordered orientation of the acyl chain, which can be expected to give rise to $|S_{CD}|$ values in between the two limiting cases (0.5 and 0). The $|S_{CD}|$ values of the C—D bonds along the acyl chain of C(16):C(16)PC at $T > T_m$ are found experimentally to be approximately constant and are, indeed, equal to about 0.2 over the initial segment near the interface; this part is usually referred to as the *plateau region*. A rapid decrease in $|S_{CD}|$ is observed near the methyl terminus, indicating that the motional averaging of the quadrupolar interactions increases towards the middle of the bilayer. The shape of this orientational order profile can be rationalized by a structural model of the acyl chain at $T > T_m$ as follows: The coupled $g^+tg^{\mp}$ kinks are allowed in the plateau region, whereas the isolated $g^{\pm}$ rotamers exist predominantly near the terminal methyl end.

The relative ratios of *trans* and *gauche* rotamers in the lipid acyl chains at various temperatures can be detected directly by Raman spectroscopy (Levin, 1984). This method has indeed demonstrated that the gel → liquid-crystalline phase transition exhibited by the bilayer of C(16):C(16)PC is accompanied by an abrupt increase in the ratio of *gauche/trans* at the onset of $T_m$ (Gaber and Peticolas, 1977). Moreover, single $g^{\pm}$ rotamers as well as $g^+tg^-$ and $g^-tg^+$ kinks have been detected by FT-infrared spectroscopy in the bilayers of C(16):C(16)PC at $T > T_m$ (Mendelsohn and Scnak, 1993; Casal and McElheney, 1990). Most interestingly, these single $g^{\pm}$ rotamers and coupled $g^{\pm}tg^{\mp}$ kinks are also observed in the computer-simulated C(14):C(14)PC bilayer, in excess water, at $T > T_m$ using the molecular dynamics method (Chiu *et al.*, 1995). Moreover, the $g^{\pm}tg^{\mp}$ kinks shown by molecular dynamics simulations are not fixed at any given preferred positions along the *sn*-1 and *sn*-2 acyl chains of C(14):C(14)PC molecules in the liquid-crystalline bilayer. Based on $^2$H NMR data, it has been suggested earlier that the coupled $g^{\pm}tg^{\mp}$ kinks migrate up and down along the plateau regions of lipid acyl chains in bilayers at $T > T_m$ (Seelig and Seelig, 1980).

Based on the spectroscopic and computational results just discussed, it is obvious that the thickness of the lipid bilayer composed of saturated identical-chain phosphatidylcholines must be reduced at $T > T_m$ because of the presence of $g^{\pm}$ rotamers and $g^{\pm}tg^{\mp}$ kinks along the acyl chain. In addition, the lateral chain–chain van der Waals interactions must also be diminished at $T > T_m$ because of the random distribution and the dynamic nature of the

$g^{\pm}tg^{\mp}$ kinks. Consequently, the average rate of *lateral diffusion* for saturated identical-chain phosphatidylcholine molecules in each of the two monolayers within the two-dimensional plane of the lipid bilayer can be expected to increase appreciably as the bilayer undergoes the thermally induced gel-to-liquid-crystalline phase transition. Indeed, it has been shown by using fluorescent lipid molecules that an averaged lateral diffusion coefficient, typically about $10^{-8}$ cm$^2$/sec is observed in the two-dimensional plane of the liquid-crystalline bilayer, and a much smaller value on the order of $10^{-11}$ cm$^2$/sec is estimated for the same fluorescent lipid molecules in the gel-state bilayer (Clegg and Vaz, 1985).

## C. Factors Affecting the Gel → Liquid-Crystalline Phase Transition Temperature

The phase transition behavior of lipid bilayers composed of a given species of pure diacyl phospholipids can be influenced by many internal and external factors. Internal factors refer to changes in chain length, chain asymmetry, unsaturation, and headgroup structure. External factors include pressure, pH, and chemical composition of the medium. Of the two thermodynamic parameters ($T_m$ and $\Delta H$) that are associated directly with the main phase transition of the lipid bilayer, the value of $T_m$ can be determined with greater accuracy by high-resolution DSC. We shall examine how some of the internal and external factors affect the $T_m$ value associated with the thermally induced main phase transition of diacyl phospholipids in excess water.

Before we discuss the effects of chain length and chain asymmetry on $T_m$, let us first define some *structural parameters* ($\Delta C$, $\Delta C/CL$, and $N$) underlying saturated diacyl phosphatidylcholines or C(X):C(Y)PC packed in the gel-state bilayer. To proceed, we return for a moment to Fig. 5C, in which the effective chain length difference between the *sn*-1 and the *sn*-2 acyl chains of an identical-chain phosphatidylcholine in the crystalline state, $\Delta C_{cry}$, can be clearly identified. This value of $\Delta C_{cry}$ is seen in Fig. 5C to be 3.68 C—C bond lengths along the chain. For the same identical-chain phosphatidylcholine molecule packed in the gel-state bilayer, the effective chain length difference is reduced to about 1.5 C—C bond lengths, and we designate it as the chain length difference of the reference state ($\Delta C_{ref}$). For a saturated mixed-chain C(X):C(Y)PC such as C(18):C(14)PC in the gel-state bilayer, the effective chain length difference $\Delta C$ is related to X and Y as follows (Huang *et al.*, 1994): $\Delta C = |X - Y + \Delta C_{ref}| = |18 - 14 + 1.5| = 5.5$ C—C bond lengths. Schematically, this structural parameter of $\Delta C$ is presented in Fig. 12A. Here, a crystalline structure is conveniently drawn for C(18):C(14)PC; actually, all structural parameters of C(X):C(Y)PC are defined for lipid molecules packed in the gel-state bilayer with shorter chain lengths. A second structural parameter, CL, refers to the effective chain length of the longer of the two acyl chains in C—C bonds, and is defined as CL = X − 1, if the *sn*-1 acyl chain is the longer chain or as CL = Y − 2.5, if the *sn*-2 acyl chain is the longer chain. In the latter case, the effective chain

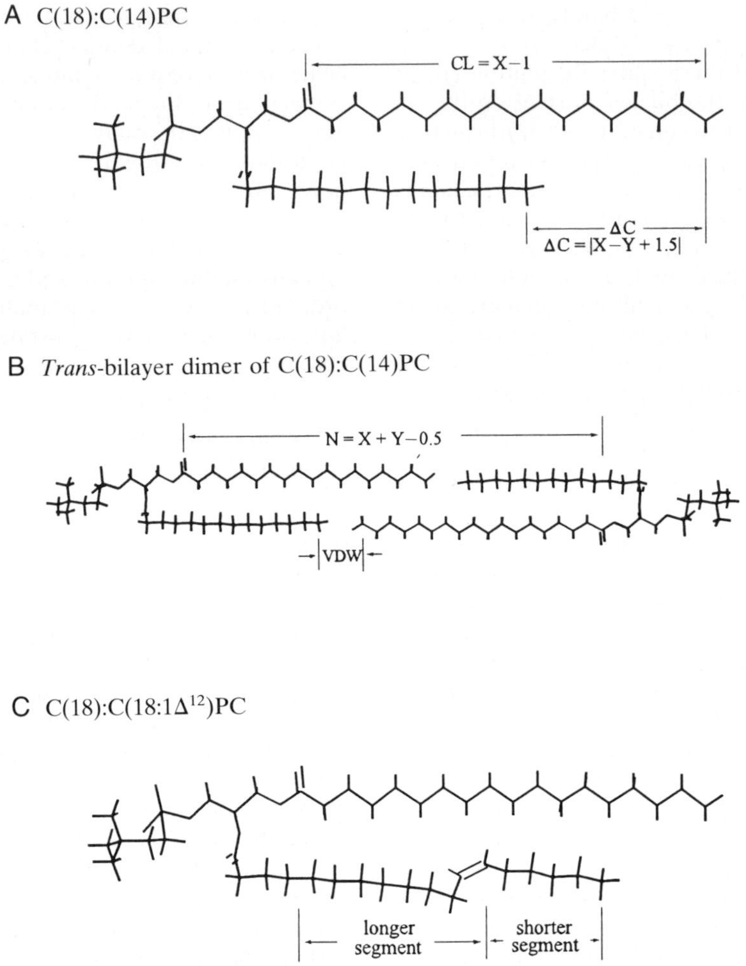

A C(18):C(14)PC

B *Trans*-bilayer dimer of C(18):C(14)PC

C C(18):C(18:1Δ$^{12}$)PC

**FIG. 12.** Molecular graphics representations of phospholipids with various structural parameters. (A) A monomer of saturated C(18):C(14)PC. CL is the effective chain length of the longer of the two acyl chains, and $\Delta C$ is the effective chain-length difference between the two acyl chains along the long-chain axis. For C(X):C(Y)PC in the gel state, the relations between CL, $\Delta C$, and X,Y are indicated. In the case of gel-state C(18):C(14)PC, CL and $\Delta C$ are 17 and 5.5 carbon–carbon bond lengths, respectively. (B) A molecular graphics drawing to illustrate a *trans*-bilayer dimer of C(18):C(14)PC with a partially interdigitated packing motif. $N$ is the hydrophobic thickness of the dimer, corresponding to the distance separating the two carbonyl oxygens of the two opposing *sn*-1 acyl chains. VDW is the van der Waals contact distance between two opposing methyl groups in the bilayer interior. (C) The monomeric C(18):C(18:1Δ$^{12}$)PC. The *sn*-2 acyl chain is shown to adopt a crankshaft-like motif, and hence it consists of two segments. The chain–chain lateral interaction is assumed to be energetically more favorable between the *sn*-1 acyl chain and the longer segment of the *sn*-2 acyl chain.

length extends from the point corresponding to the position of carbonyl carbon of the *sn*-1 acyl chain to the methyl terminus of the *sn*-2 acyl chain. This structural parameter, CL, is also illustrated schematically in Fig. 12A. For C(18):C(14)PC, the CL value is 17 C—C bond lengths. The normalized chain length difference or asymmetry is the ratio of $\Delta C$/CL. For C(18):C(14)PC, the $\Delta C$/CL value is 0.32. Another structural parameter, $N$, is taken to be the distance, in C—C bond lengths, between the two carbonyl oxygens of the *sn*-1 acyl chains in a *trans*-bilayer dimer of C(X):C(Y)PC along the long molecular axis, as shown schematically in Fig. 12B. The $N$ value of a C(X):C(Y)PC dimer is related to X and Y as follows: $N = (X - 1) + VDW + (Y - 2.5) = X + Y - 0.5$, where VDW is

the van der Waals separation distance between the two opposing terminal methyl groups in the *trans*-bilayer dimer (Fig. 12B), which is taken to be 3 C—C bond lengths in the gel-state bilayer.

When the normalized chain asymmetry ratio, $\Delta C$/CL, is less than 0.42, saturated C(X):C(Y)PC molecules, in excess water, can self-assemble into the partially interdigitated bilayer, at $T < T_m$, in which the methyl end of the longer acyl chain of one lipid molecule from one leaflet packs end-to-end with the methyl end of the shorter acyl chain of another lipid molecule from the opposing leaflet and vice versa (Li *et al.*, 1993). This partially interdigitated packing motif is diagrammatically shown in Fig. 12B. If the $\Delta C$/CL value of a phosphatidylcholine molecule is greater than 0.42, this highly asymmetric lipid tends to self-assemble into the mixed interdigitated bilayer at $T < T_m$ (Li et al., 1993). Phosphatidylcholines with $\Delta C/CL$ values greater than 0.42 are rare in biological membranes; hence, they are not discussed in this chapter.

At $\Delta C/CL$ value less than 0.42, DSC measurements on saturated $C(X):C(Y)PC$ have demonstrated that the $T_m$ value of the main phase transition is related to the structural parameters $\Delta C$ and $N$ according to the equations (Huang et al., 1994)

$$T_m = 161.97 - 3716.03\,(1/N) - 292.03\,(\Delta C/N) \\ + 256.46\,\Delta C/(N + \Delta C) \quad (1)$$

for lipids with a longer effective sn-1 acyl chain, and

$$T_m = 155.47 - 3537.16\,(1/N) - 218.50\,(\Delta C/N) \\ + 165.44\,\Delta C/(N + \Delta C) \quad (2)$$

for lipids with a longer effective sn-2 acyl chain. Implicitly, these two equations indicate that if $\Delta C$ is constant, the $T_m$ value increases with increasing $N$; however, the increase tends to level off toward an asymptote. In contrast, the $T_m$ value decreases with increasing $\Delta C$, if $N$ is constant; as a result, $\Delta C$ counteracts the $N$-promoted increase in $T_m$. Thus, it is the delicate balance between the two structural parameters ($N$ and $\Delta C$) underlying the dimeric $C(X):C(Y)PC$ molecules in the gel-state bilayer that determines the unique $T_m$ value of the fully hydrated lipids. Equations (1) and (2) not only show the fundamental antagonistic effect between $N$ and $C$ in determining the $T_m$ value, but also can be employed to predict the unknown value of $T_m$ based on the chemical formula of $C(X):C(Y)PC$ (Huang et al., 1994).

Table 3 lists some representative $T_m$ values obtained calorimetrically with multilamellar vesicles containing diacyl phospholipids. The values of $\Delta C$ and $N$ calculated according to $\Delta C = |X - Y + 1.5|$ and $N = X + Y - 0.5$ are also listed for those saturated lipid species. For the homologous series of saturated identical-chain phosphatidylcholines from $C(14):C(14)PC$ to $C(20):C(20)PC$ with a common $\Delta C$ value of 1.5 C—C bond lengths, the $T_m$ value is seen to increase progressively with a stepwise increase in $N$ that, in fact, is directly related to the chain length. This raises an interesting question as to why the $T_m$ value increases progressively with an increase in the chain length of the saturated lipid molecule. The answer lies in the strength of the lateral chain–chain van der Waals interaction in the gel-state bilayer; this energetic term is directly proportional to the chain length. Specifically, the lateral van der Waals interaction between two all-trans saturated chains ($E_{VDW}$) is related to the number of methylene units in the chain ($N_m$) according to the equation (Salem, 1962)

$$E_{VDW} = A(3\pi/8l)(N_m/D^5) \quad (3)$$

where $l$ is the distance between two methylene units along the long-chain axis (1.27 Å), $D$ is the lateral distance separating the two all-trans chains, and $A$ is a constant ($-1340$ kcal/Å$^{-6}$ mol). If $D$ is taken to be 4.8 Å, then $E_{VDW} = -0.49$ kcal/mol per methylene unit. Clearly, the longer the acyl chain or the larger the number of the methylene units ($N_m$), the greater the attractive van der Waals interaction

between the two saturated acyl chains. As a result, a larger thermal energy and a higher $T_m$ are required to promote the trans → gauche isomerizations within the longer chains as the phospholipid molecules in the bilayer undergo the thermally induced gel-to-liquid-crystalline phase transition.

In Table 3, $C(16):C(16)PC$, $C(14):C(18)PC$, and $C(18):C(14)PC$ are shown to have a common $N$ value of 31.5 C—C bond lengths. The $\Delta C$ and $T_m$ values of these saturated lipids are also shown to be 1.5, 2.5, and 5.5 C—C bond lengths and 41.5, 39.2, and 31.2°C, respectively. This set of saturated $C(X):C(Y)PC$ thus serves to demonstrate that the $T_m$ value decreases with increasing $\Delta C$, when the $N$ value is held constant. Similarly, another pair of the position isomers, $C(16):C(18)PC/C(18):C(16)PC$, exhibits the same inverse relationship between $T_m$ and $\Delta C$ (Table 3). Fundamentally, the structural parameter $\Delta C$ represents the chain asymmetry. It is also a perturbation term that affects the closest lateral chain–chain interaction in the gel-state bilayer. This perturbation stems from the bulky methyl terminal groups.

Most of the naturally occurring phospholipids in eukaryotic cell membranes have a saturated sn-1 acyl chain and an unsaturated sn-2 acyl chain with different numbers and positions of cis double bonds. It is very convenient to abbreviate the names for these enormously complicated sn-1 saturated/sn-2 unsaturated phospholipid species according to the chemical natures of their two acyl chains. These abbreviated names, as specified in Table 3, are used throughout the rest of this chapter. Now, we can proceed to discuss the effects of unsaturation of the sn-2 acyl chain on the gel-to-liquid-crystalline phase transition temperature of phospholipid bilayers as monitored by high-resolution DSC. First, the introduction of a single cis-double bond into the sn-2 acyl chain markedly reduces the $T_m$ value of lamellar phosphatidylcholines. A good example is demonstrated by the bilayer composed of 1-stearoyl-2-oleoyl-phosphatidylcholines or $C(18):C(18:1\Delta^9)PC$, which exhibits calorimetrically a $T_m$ value of 5.6°C. In contrast, a $T_m$ value of 55.3°C is detected for the bilayer composed of the saturated counterparts or $C(18):C(18)PC$. Second, the $T_m$-lowering effect of acyl chain monounsaturation depends critically on the position of the cis-carbon–carbon double bond ($\Delta^n$) in the sn-2 acyl chain. As shown in Table 3, the minimal $T_m$ for a homologous series of $C(18):C(18:1\Delta^n)PC$ occurs at $\Delta^n = \Delta^{11}$ or when the cis-double bond is positioned near the center of the linear segment of the sn-2 acyl chain. Moreover, the $T_m$ value increases progressively as the single cis-double bond moves from $\Delta^{11}$ toward either end of the acyl chain. Third, the progressive increase in the level of acyl chain unsaturation has a nonadditive effect on the $T_m$. Based on the $T_m$ values of $C(18):C(18)PC$, $C(18):C(18:1\Delta^9)PC$, and $C(18):C(18:2\Delta^{9,12})PC$ shown in Table 3, it is evident that the introduction of the first cis-double bond near the chain center gives rise to a drastic decrease of 49.7°C in $T_m$. Although the introduction of a second cis-double bond produces a substantial decrease of 20.2°C in $T_m$, this decrease is not nearly as large as the first one (49.7°C). Most interestingly, the successive incor-

**TABLE 3** Main Phase-Transition Temperatures for Some Diacyl Phospholipids

| Lipid[a] | $\Delta C$ | N | $T_m$ (°C) | Lipid[a] | $T_m$ (°C) |
|---|---|---|---|---|---|
| Saturated phospholipids | | | | | |
| C(14):C(14)PC | 1.5 | 27.5 | 24.1 | C(14):C(14)PE | 49.6 |
| C(16):C(16)PC | 1.5 | 31.5 | 41.5 | C(16):C(16)PE | 63.2 |
| C(18):C(18)PC | 1.5 | 35.5 | 55.3 | C(18):C(18)PE | 74.0 |
| C(20):C(20)PC | 1.5 | 39.5 | 66.4 | C(20):C(20)PE | 82.5 |
| C(14):C(18)PC | 2.5 | 31.5 | 39.2 | C(14):C(18)PE | 61.6 |
| C(18):C(14)PC | 5.5 | 31.5 | 31.2 | C(18):C(14)PE | 54.9 |
| C(16):C(18)PC | 0.5 | 33.5 | 48.8 | C(16):C(18)PE | 70.8 |
| C(18):C(16)PC | 3.5 | 33.5 | 44.4 | C(18):C(16)PE | 66.0 |
| C(20):C(18)PC | 3.5 | 37.5 | 57.5 | C(20):C(18)PE | 76.3 |
| Unsaturated phospholipids | | | | | |
| C(16):C(18:1$\Delta^6$)PC | | | 18.8 | C(16):C(18:1$\Delta^6$)PE | 39.1 |
| C(16):C(18:1$\Delta^9$)PC | | | −2.6 | C(16):C(18:1$\Delta^9$)PE | 26.1 |
| C(18):C(18:1$\Delta^6$)PC | | | 24.8 | C(18):C(18:1$\Delta^6$)PE | 42.6 |
| C(18):C(18:1$\Delta^7$)PC | | | 16.7 | C(18):C(18:1$\Delta^7$)PE | 35.4 |
| C(18):C(18:1$\Delta^9$)PC | | | 5.6 | C(18):C(18:1$\Delta^9$)PE | 31.5 |
| C(18):C(18:1$\Delta^{11}$)PC | | | 3.8 | C(18):C(18:1$\Delta^{11}$)PE | 29.8 |
| C(18):C(18:1$\Delta^{12}$)PC | | | 9.1 | C(18):C(18:1$\Delta^{12}$)PE | 32.3 |
| C(18):C(18:1$\Delta^{13}$)PC | | | 15.9 | C(18):C(18:1$\Delta^{13}$)PE | 35.4 |
| C(18):C(18:2$\Delta^{9,\,12}$)PC | | | −14.6 | C(18):C(18:2$\Delta^{9,\,12}$)PE | 4.4 |

[a] Saturated phosphatidylcholines and phosphatidylethanolamines are abbreviated as C(X):C(Y)PC and C(X):C(Y)PE, respectively. X denotes the total number of carbons in the *sn*-1 acyl chain and Y denotes the total number of carbons in the *sn*-2 acyl chain. For *sn*-1 saturated and *sn*-2 monounsaturated acyl chain with the *cis*-double bond ($\Delta$) at the position $n$ from the carboxyl end it is abbreviated as C(X):C(Y:1$\Delta^n$); similarly, lipid with *sn*-1 saturated and *sn*-2 diunsaturated acyl chain is abbreviated as C(X):C(Y:2$\Delta^{n,m}$), where $m$ is the second *cis*-double bond position at the $m$th carbon atom from the carboxyl end.

poration of a third *cis*-double bond at $\Delta^{15}$ results in a slight increase in $T_m$.

Let us now consider a molecular model that has been invoked to interpret the large $T_m$-lowering effect of acyl chain monounsaturation and the characteristic dependency of $T_m$ on the position of the *cis*-double bond along the acyl chain (Wang *et al.*, 1995). In this molecular model, three basic assumptions have been made. (1) The monoenoic *sn*-2 acyl chain in the *sn*-1 saturated/*sn*-2 monounsaturated phospholipid molecule is assumed to adopt, at $T < T_m$, an energy-minimized crankshaft-like kink motif as shown in Fig. 12C. The *sn*-2 acyl chain thus consists of a longer segment and a shorter segment separated by the $\Delta^n$-containing kink. (2) The longer segment and the neighboring *sn*-1 acyl chain run in a nearly parallel manner with favorable van der Waals attractive distance between them. (3) The shorter segment is considered to be partially disordered at $T < T_m$, thus playing a perturbing role in the lateral chain–chain interactions in the gel-state bilayer. With this molecular model, we are now in a position to interpret first the drastic $T_m$-lowering effect of a single *cis*-double bond. Because the shorter segment is a perturbing element, it behaves similarly as the structural parameter

$\Delta C$ to counteract the effect of $N$, thus lowering $T_m$. Moreover, since this short segment is already partially disordered at $T < T_m$, it cannot contribute significantly to the conformational disordering process of *trans* → *gauche* isomerizations involved in the phase transition. As a result, the total number of *trans* → *gauche* isomerizations during the phase transition is decreased appreciably for the bilayer composed of *cis*-monoenoic phosphatidylcholines relative to that of the saturated counterparts, leading to a significantly lower $T_m$ value.

Similarly, we can use the same molecular model to see how the $T_m$ value varies as the single *cis*-bond migrates along the *sn*-2 acyl chain. Since the longer segment of the kinked chain is assumed to undergo a favorable van der Waals contact interaction with the *sn*-1 acyl chain, this contact interaction energy must then depend on the length of the longer segment. When the single *cis*-double bond is positioned at the center of the *sn*-2 acyl chain, the longer segment of the kinked chain has a minimal length, which is almost equal to that of the shorter segment. Hence, the van der Waals interaction with the *sn*-1 acyl chain is also minimal. As the single *cis*-double bond migrates away successively from the chain center toward either end, the length of the longer segment is progressively increased, leading to a proportionally increased van der Waals interaction and hence a gradual increase in $T_m$. The characteristic change of $T_m$ as a function of the location of the *cis*-double bond, shown in Table 3, can thus be explained by the structural model.

The phospholipid headgroup can also exert a profound effect on the phase transition behavior of the lipid bilayer. For instance, when the headgroup's choline moiety in a saturated identical-chain phosphatidylcholine molecule is replaced by ethanolamine, the resulting phosphatidylethanolamine can self-assemble into the gel-state bilayer, in excess water, after incubating at $T < T_m$ for a brief time. Upon heating, the gel-state PE bilayer transforms directly into the liquid-crystalline ($L_\alpha$) phase without going through the intermediate $P_{\beta'}$ phase that is commonly observed for fully hydrated PC with identical chains. Hence, the main phase transition of saturated identical-chain PE corresponds to the $L_\beta \to L_\alpha$ phase transition, where $L_\beta$ denotes the gel phase with the saturated acyl chains orienting perpendicularly to the bilayer surface. For comparison, the second DSC heating scans obtained with aqueous dispersions of C(16):C(16)PC and C(16):C(16)PE are shown in Fig. 13. Clearly, the $T_m$ value is significantly higher for PE than for PC with the same saturated fatty acyl chain composition, although their $\Delta H$ values are comparable. In Table 3, the $T_m$ values for nine molecular species of saturated diacyl PE as well as PC are given. The $T_m$ values for saturated PE are consistently higher than those for PC with the same fatty acyl chain composition. However, for the homologous series of identical-chain phospholipids from C(14):C(14)PE/PC to C(20):C(20)PE/PC, the difference in $T_m$ between PE and PC ($\Delta T_m$) decreases with increasing acyl chain length (see Table 3). Figure 13 also shows the second DSC heating scans for aqueous dispersions of C(16):C(18:1 $\Delta^6$)PC and C(16):C(18:1$\Delta^6$)PE; here, the $T_m$ value of PE is 20.3°C higher than that of PC. In table 3, the $T_m$ values of a large number of *sn*-1 saturated/*sn*-2 unsaturated PC/PE are presented. Clearly, the $T_m$ values of unsaturated PE are also consistently higher than those of unsaturated PC with the same fatty acyl chain composition. These higher $T_m$ values can thus be taken as evidence to suggest that, in general, the lateral chain–chain interactions in the PE bilayer are stronger than those in the PC bilayer at $T < T_m$. This stronger lateral chain–chain interaction may be attributed to many factors such as the smaller PE headgroup, the interlipid H-bond networks on the two surfaces of the gel-state PE bilayer, and the smaller hydration of the PE bilayer. At the present time, however, the quantitative contribution of each factor to the overall difference in $T_m$ between PE and PC is not well understood and deserves further investigation.

One external factor that has been extensively studied is the pressure (Wong, 1986). Take the C(16):C(16)PC dispersions at 47°C as an example. The lamellar PC in the $L_\alpha$ phase can undergo two isothermal phase transitions upon elevation of pressure up to 1.4 kbar. The first pressure-induced isothermal transition occurs at the critical pressure of 0.17 kbar, and the second occurs at 0.60 kbar.

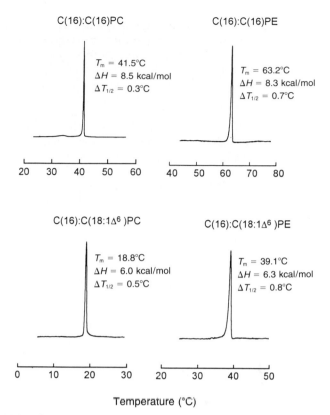

**FIG. 13.** Representative DSC second heating scans for aqueous dispersions of PC and PE with the same saturated and monounsaturated acyl chain compositions. The $T_m$ values of PE bilayers are observed to be consistently higher than those of PC bilayers with the same fatty acyl chain compositions irrespective of acyl chain unsaturation. These DSC experiments were performed with a Microcal Model MC-2 differential scanning microcalorimeter (Microcal). Scan rate: 15°C/hour.

These two phase transitions correspond to the cooling-induced main and pretransitions, respectively. Clearly, an increase in pressure acts in a manner similar to a decrease in temperature by exerting an ordering effect on the acyl chain packing in the bilayer.

## V. Binary Phospholipid Mixtures

Thus far the structure and phase behavior of a limited number of one-component phospholipid systems have been discussed. Attention is now directed to the mixing behavior of two-component (A and B) phospholipid systems at ambient pressure. The mixing behavior of two-component lipids in the bilayer is expected to depend on the difference in the pair interaction energies between the mixed pairs (A–B) and the like pairs (A–A and B–B), which, in turn, depends on the extent of structural similarity between the two-component lipids. Furthermore, the structural characteristics of A and B in the bilayer may change abruptly and significantly as the two-component bilayer undergoes the respective phase transition of components A and B. The relative pair-interaction energy may thus be quite different in the gel and liquid-crystalline states. For instance, the lateral phase separation of a given two-component system may occur in the gel phase because of their structural dissimilarity, whereas a complete miscibility between the same two components may take place in the $L_\alpha$ phase as a result of conformational changes of the component lipids occurring at the $T_m$. The mixing behavior and the interaction energies related to the component lipids in the gel and liquid-crystalline phases are fully contained in the *temperature–composition phase diagram* for the binary lipid systems. In this section, we consider three of the most commonly observed temperature–composition phase diagrams for binary lipid systems. These phase diagrams are shown diagrammatically in Fig. 14A–C. In particular,

each phase diagram is characterized by its position and shape, which are determined by the phase boundaries called the *solidus* and *liquidus* as shown in Fig. 14A. Experimentally, these phase boundaries are commonly constructed from the DSC heating curves for a given binary phospholipid system at various relative compositions. Basically, the onset and completion temperatures for each transition peak, after appropriate correction, are plotted as a function of the mole fraction of the higher melting component lipid ($X_B$). These corrected onset and completion temperature points form the bases for defining the solidus and liquidus, respectively, of the temperature–composition phase diagram (Mabrey and Sturtevant, 1976).

Of all the known phase diagrams constructed on the basis of calorimetric data obtained with binary phospholipids, the mixing behavior of the two lipid components in the bilayer plane is found to be always nonideal. This implies that the phase transition of the component A (or B) is affected by the presence of the second component B (or A) in the binary mixture. The basic thermodynamic equations, derived from the regular solution theory (or Briggs–Williamson approximation), describe phenomenologically the phase diagram for the nonideal system of binary mixtures of phospholipids (A and B) as follows (Lee, 1977):

$$\ln(X_B^G/X_B^L) = \Delta H_B/R\,(1/T - 1/T_B) \\ + 1/RT\,[\rho^L(1 - X_B^L)^2 - \rho^G(1 - X_B^G)^2] \quad (4)$$

$$\ln[(1 - X_B^G)/(1 - X_B^L)] = \Delta H_A/R\,(1/T - 1/T_A) \\ + 1/RT\,[\rho^L(X_B^L)^2 - \rho^G(X_B^G)^2] \quad (5)$$

Here, $X_B^G$ and $X_B^L$, the composition variables, are the mole fraction of the high-melting component B in the gel and liquid-crystalline phases, respectively; $\Delta H_A$ and $\Delta H_B$ are the transition enthalpies (kcal/mol) of pure components A and B, respectively; $T_A$ and $T_B$ are the phase transition temperatures (kelvins) of the bilayers composed of the pure components A and B, respectively; $\rho^G$ and $\rho^L$ are two parameters having units of energy (kcal/mol) that describe the nonideality of mixing between components A and B in the gel and liquid-crystalline phases of the bilayer, respectively; and $R$ is the gas constant. The nonideality parameter, $\rho$, is related to the pair interaction energy between mixed pairs and between like pairs in the two-dimensional plane of the bilayer as follows: $\rho = Z[E_{AB} - \frac{1}{2}(E_{AA} + E_{BB})]$, where $Z$ is the number of nearest neighbors in the plane of the bilayer, and $E_{AB}$, $E_{AA}$, and $E_{BB}$ are the molar interaction energies of A–B, A–A, B–B pairs of nearest neighbor pairs, respectively. If the value of $\rho$ is positive, then there is a tendency for like lipid species to cluster together leading to lateral phase separation; $\rho < 0$ reflects the tendency of compound formation; $\rho = 0$ corresponds to ideal lateral mixing between components A and B in the bilayer.

Equations 4 and 5 allow the position and the shape of the phase diagram for a given binary lipid mixture to be determined, provided that the values of $\rho^L$ and $\rho^G$ and other experimental values ($T_m$ and $\Delta H$) are known. However, if the position and the shape of the phase diagram have already been determined experimentally, Eq. 4 and 5 can

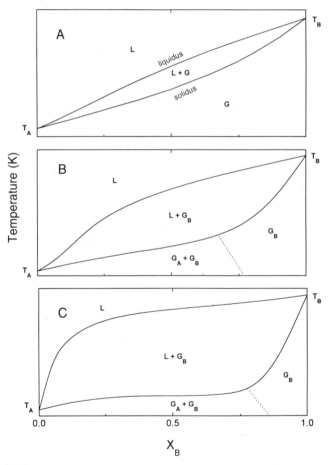

**FIG. 14.** Three commonly observed phase diagrams determined calorimetrically for binary lipid systems. (A) The phase diagram for an isomorphous system as represented by the binary lipid system of C(14):C(14)PC/C(16):C(16)PC. (B) The phase diagram for the binary lipid system of C(14):C(14)PC/C(18):C(18)PC. (C) The peritectic phase diagram of the C(14):C(14)PC/C(20):C(20)PC system. $X_B$ denotes the mole fraction of the higher melting component B. G and L denote the gel and liquid-crystalline phases, respectively. The subscripts A and B represent the lipid components A and B, respectively, in the binary lipid system. The values of $T_A$ and $T_B$, in kelvins, for each of the three binary A/B systems are given in Table 4. The simulated values of the nonideality parameters of mixing ($\rho^L$ and $\rho^G$) for the three phase diagrams are also shown in Table 4.

be used to estimate the $\rho^L$ and $\rho^G$ values by matching the experimental and the simulated phase diagrams with a computer program (Brumbaugh and Huang, 1992). The mutual miscibilities of the two lipid components in the bilayer can thus be compared quantitatively based on the values of $\rho^L$ and $\rho^G$.

## A. Gel-Phase Miscibility and Liquid-Crystalline Phase Miscibility

The simplest phase diagram for binary lipid systems is the type called an *isomorphous system,* in which the two components A and B are mutually miscible over the entire composition range in both the gel and liquid-crystalline states. Each binary mixture with a given $X_B$ value in this isomorphous system exhibits calorimetrically a single endothermic phase transition, corresponding to the gel-to-liquid-crystalline phase transition. The transition curves obtained with various binary mixtures are broadened somewhat in comparison with those of the pure component lipids, particularly in the $X_B$ range of 0.4–0.6. The solidus and liquidus, constructed on the basis of the onset and completion temperatures of these transition curves, are thus smooth curves, giving rise to a *lens-shaped phase diagram* as diagramatically represented in Fig. 14A. Specifically, below the solidus is the single-phase gel region (G); above the liquidus is the single phase liquid-crystalline region (L); between the solidus and the liquidus is a two-phase region in which the G and L phases are in dynamic equilibrium.

Binary mixtures of different saturated phosphatidylcholines, in which one component differs from the other by only two or fewer methylene units in each of their long acyl chains, exhibit complete miscibility in both gel and liquid-crystalline bilayers. The experimentally determined phase diagrams of these systems are characterized by lens-shaped phase boundaries. A good example is provided by the binary lipids of C(14):C(14)PC/C(16):C(16)PC (Mabrey and Sturtevant, 1976). The nonideality parameters $\rho^G$ and $\rho^L$ for the lens-shaped phase diagram of C(14):C(14)PC/C(16):C(16)PC are 0.35 and 0.16 kcal/mol, respectively (Table 4). The value of $\rho^G$ is, in fact, smaller than the average kinetic energy of the thermal environment at room temperature (0.58 kcal/mol); hence, the two lipid components, C(14):C(14)PC and C(16):C(16)PC, are mixed nearly ideally in the gel state

at room temperature. It should be noted that the value of $\rho^L$ is smaller than that of $\rho^G$; consequently, in the two-dimensional plane of the bilayer, these different lipid molecules can mix even better at higher temperatures in the $L_\alpha$ phase. In general, lens-shaped phase diagrams with small values of $\rho^L$ and $\rho^G$ are commonly observed for binary mixtures of different saturated phospholipids with the same headgroup and with a small difference in acyl chain lengths.

## B. Gel-Phase Immiscibility and Liquid-Crystalline Phase Miscibility

We have just seen the lens-shaped phase diagram exhibited by the binary A/B system of C(14):C(14)PC/C(16):C(16)PC. If the higher melting component, C(16):C(16)PC, in the binary system is replaced by C(18):C(18)PC, then the chain-length difference between A and B is increased from two to four methylene units. Upon heating, the DSC curve containing two overlapped or separated phase transitions is noticeably observed over a certain range of $X_B$, indicating that a lateral phase separation of $G_A$ and $G_B$ occurs in the C(14):C(14)PC/C(18):C(18)PC system. Here, $G_A$ and $G_B$ refer to the gel phases of A and B components, respectively. The phase diagram for the binary C(14):C(14)PC/C(18):C(18)PC system constructed from the DSC transition curves is illustrated in Fig. 14B, which deviates considerably from the lens-shaped phase diagram observed for C(14):C(14)PC/C(16):C(16)PC (Fig. 14A). Simulations of the phase boundaries yield the values of $\rho^G$ and $\rho^L$ to be 0.85 and 0.32 kcal/mol, respectively (see Table 4). The $\rho^G$ value is somewhat larger than the thermal energy of the environment; hence, a partial phase separation is expected between C(14):C(14)PC and C(18):C(18)PC in the gel-state bilayer. In contrast, the same two components are mixed nearly ideally in the liquid-crystalline state as indicated by the $\rho^L$ value of 0.32 kcal/mol.

If the chain-length difference between the two saturated identical-chain phosphatidylcholines (A and B) is further increased from four to six methylene units, the increased mismatch will affect the lateral lipid–lipid interactions markedly, leading to $E_{BB} << E_{AB} \approx E_{AA}$ according to the Salem equation (3) discussed earlier. This larger dissimilarity can thus be expected to cause a lower miscibility of C(14):(14)PC and C(20):C(20)PC in the bilayer, particularly the gel-state bilayer. Indeed, the phase diagram for

**TABLE 4**  Values of Nonideality Parameters of Mixing ($\rho$) for Various Binary Lipid Systems[a]

| Binary A/B system | $\rho^L$ | $\rho^G$ | $T_A/T_B$ | $\Delta H_A/\Delta H_B$ |
|---|---|---|---|---|
| C(14):C(14)PC/C(16):C(16)PC | 0.16 | 0.35 | 297.2/314.6 | 6.0/8.5 |
| C(14):C(14)PC/C(18):C(18)PC | 0.32 | 0.85 | 297.2/328.4 | 6.0/10.9 |
| C(14):C(14)PC/C(20):C(20)PC | 0.60 | 1.21 | 297.2/339.5 | 6.0/13.2 |

[a] The units for $\rho$, $T$ (phase transition temperature), and $\Delta H$ (transition enthalpy) are kcal/mol, K, and kcal/mol, respectively; the superscripts L and G denote the liquid-crystalline and gel phases, respectively; the subscripts A and B denote the lipid components A and B in the binary A/B system.

the C(14):C(14)PC/C(20):C(20)PC system is seen in Fig. 14C to deviate drastically from the lens shape. Certain features of the mixing behavior between C(14):C(14)PC and C(20):C(20)PC are evident from this so-called *peritectic phase diagram*. Over a wide $X_B$ range, the solidus is flat, indicating the coexistence of two immiscible gel phases ($G_A$ and $G_B$) below the flat line. The liquidus is not flat, indicating that no obvious liquid-crystalline immiscibility of C(14):C(14)PC and C(20):C(20)PC occurs. However, the shape of the liquidus clearly suggests some degrees of nonideality of mixing in the liquid-crystalline state. Furthermore, the two phase region enclosed by the liquidus and the solidus is considerably larger than the corresponding region in the lens-shaped phase diagram shown in Fig. 14A. The simulation of the phase diagram for C(14):C(14)PC/C(20):C(20)PC yields $\rho^G = 1.21$ and $\rho^L = 0.60$ kcal/mol (see Table 4). The relatively larger $\rho^G$ value is indicative of more extensive gel-phase immiscibility, which is, of course, consistent with the observed extended flat line of the solidus shown in Fig. 14C.

The experimental curves illustrated in Fig. 14A–C and the computational results given in Table 4 may serve as a paradigm to show that the degree of mutual miscibility between two chemically distinct lipid components (A and B) in the bilayer depends on the relative difference in the pair-interaction energies between the mixed pairs (A–B) and the like pairs (A–A and B–B), which, in turn, depends critically on the extent of structural similarity between A and B. In general, the structural similarity should also include the headgroup and the acyl chain unsaturation. If the headgroup of lipid B is different from that of A, this structural dissimilarity is expected to affect the mixing of A and B in the bilayer even if the acyl-chain compositions of A and B are identical. The expectation is indeed borne out by the experimental observation that a peritectic phase diagram is detected for the binary lipid system of C(16):C(16)PC/C(16):C(16)PE. This peritectic phase diagram can be attributed to the stronger headgroup–headgroup interactions among PE molecules, arising mainly from hydrogen bonds between the amine hydrogens of the PE headgroups and the nonesterified phosphate oxygens of the neighboring PE headgroups in the gel-state bilayer.

We have discussed earlier that when a *cis* carbon–carbon double bond is incorporated into the *sn*-2 acyl chain of a phosphatidylcholine molecule, the shorter segment of the monounsaturated chain acts as a perturbing element in the gel-state bilayer. The disordered segments of monounsaturated chains esterified at the *sn*-2 position of lipids (B) therefore can be expected to hinder the close encounters of the highly ordered saturated phospholipids (A) in the gel-state bilayer composed of A/B, thus promoting gel–gel phase separation or immiscibility. Indeed, the experimentally determined phase diagram for the binary lipid system of C(16):C(16)PC/C(16):C(18:1Δ⁹)PC exhibits a shape that is very similar to the one observed for C(14):C(14)PC/C(20):C(20)PC, although the actual chain-length difference between the saturated and the unsaturated PC is less than two C—C bond lengths. Specifically, the shape of this phase diagram is characterized by an extended flat line in

the solidus, indicating an extensive region of immiscibility between C(16):C(16)PC and C(16):C(18:1Δ⁹)PC in the gel-state bilayer.

In summary, the mixing behavior of two different membrane lipids in the bilayer at various temperatures is rather diverse. This diversity arises from the fact that membrane lipids are characterized intrinsically by many different features including the acyl chain length, headgroup structure, and acyl chain unsaturation. Nevertheless, the basic information regarding the miscibility and/or immiscibility of two different lipids in the bilayer can be obtained from the position and shape of the temperature–composition phase diagram. The construction and the simulation of phase diagrams for various binary lipid systems may thus be very useful in establishing the database for characterizing different binary lipid–lipid interactions, from which the more complex multiple lipid–lipid interactions in the two-dimensional plane of cellular membranes may be assessed.

## VI. Summary

In this chapter, we first discuss the classification and structures of membrane lipids with special emphasis on phospholipids. It is important to realize that within each class of phospholipid, a bewildering diversity of molecular species is present because of the complex composition of the fatty acyl chains. Nevertheless, all membrane phospholipids exhibit two common properties: They are amphipathic molecules [i.e., they have a hydrophilic (polar) moiety and a hydrophobic (nonpolar) moiety]; and these molecules, when hydrated, do self-assemble spontaneously into a variety of highly organized structures, including the most commonly observed lipid bilayers. The structural polymorphism of the lipid bilayer composed of a single species of phospholipid as a function of temperature is illustrated. Specifically, the thermally induced phase-transition behavior of the lipid bilayer and the factors that influence the phase-transition behavior are presented. The main phase-transition temperatures of bilayers prepared from various phosphatidylcholines and phosphatidylethanolamines are correlated with structural parameters underlying each lipid species. Finally, the mixing behavior of two different phospholipids in the bilayer at various temperatures is described in terms of phase diagrams. Moreover, simulations of the phase diagrams based on the Briggs–Williamson approximation are also presented. Although lipid bilayers composed of binary lipids are highly simplified compared with the bilayer matrix of biological membranes, the basic information regarding the lipid–lipid interactions obtained with the construction and the simulation of phase diagrams for binary lipid systems may be useful in assessing the more complex lipid–lipid interactions in biological membranes.

## Bibliography

Ahl, P. L., Chen, L., Perkins, W. R., Minchey, S. R., Boni, L. T., Taraschi, T. F., and Janoff, A. S. (1994). Interdigitation-fusion: A new method for producing lipid vesicles of high internal volume. *Biochim. Biophys. Acta* **1195**, 237–244.

Bangham, A. D., Standish, M. M., and Watkins, J. D. (1965). Diffusion of univalent ions across the lamellae of swollen phospholipids. *J. Mol. Biol.* **13**, 238–251.

Blaurock, A. E., and Wilkins, M. H. F. (1972). Structure of frog photoreceptor membranes. *Nature* **236**, 313–314.

Bradshaw, J. P., Bushby, R. J., Giles, C. C. D., Saunders, M. R., and Reid, D. G. (1996). Neutron diffraction reveals the orientation of the headgroup of inositol lipids in model membranes. *Nature Struct. Biol.* **3**, 125–127.

Browning, J., and Seelig, J. (1980). Bilayers of phosphatidylserine: A deuterium and phosphorus nuclear magnetic resonance study. *Biochemistry* **19**, 1262–1270.

Brumbaugh, E. E., and Huang, C. (1992). Parameter estimation in binary mixtures of phospholipids. *Methods Enzymol.* **210**, 521–539.

Casal, L., and McElheney, R. (1990). Quantitative determination of hydrocarbon chain conformational order in bilayer of saturated phosphatidylcholines of various chain lengths by Fourier transform infrared spectroscopy. *Biochemistry* **29**, 5425–5427.

Chapman, D. (1993). Lipid phase transitions. *In* "Biomembranes: Physical Aspects" (M. Shinitzky, Ed.), pp. 29–62. Balaban Publishers, Weinheim.

Chiu, S.-W., Clark, M., Balaji, V. Subramanian, S., Scott, H. L., and Jakobsson, E. (1995). Incorporation of surface tension into molecule dynamics simulation of an interface: A fluid phase lipid bilayer membrane. *Biophys. J.* **69**, 1230–1245.

Clegg, R. M., and Vaz, W. L. C. (1985). Translational diffusion of proteins and lipids in artificial lipid bilayer membranes. A comparison of experiment with theory. *In* "Progress in Protein-Lipid Interactions" (A. Watts, Ed.), pp. 173–229. Elsevier, Amsterdam.

Craven, B. M. (1976). Crystal structure of cholesterol monohydrate. *Nature* **260**, 727–729.

Danielli, J. F., and Davson, H. A. (1935). A contribution to the theory of permeability of thin films. *J. Cell Comp. Physiol.* **5**, 495–508.

Davis, J. H. (1983). The description of membrane lipid conformation, order and dynamics by $^2$H-NMR. *Biochim. Biophys. Acta* **737**, 117–171.

Exton, J. H. (1994). Phosphatidylcholine breakdown and signal transduction. *Biochim. Biophys. Acta* **1212**, 26–42.

Finkelstein, A. (1987). "Water Movement through Lipid Bilayers, Pores, and Plasma Membranes: Theory and Reality." John Wiley & Sons, New York.

Gaber, B. P., and Peticolas, W. L. (1977). On the quantitative interpretation of biomembrane structure by Raman spectroscopy. *Biochim. Biophys. Acta* **465**, 260–274.

Hartshorne, R. P., Keller, B. V., Talvenheimo, J. A., Catterall, W. A., and Montal, M. (1985). Functional reconstitution of the purified brain sodium channel in planar lipid bilayers. *Proc. Natl. Acad. Sci. USA* **82**, 240–244.

Hitchcock, P. B., Mason, R., Thomas, K. M., and Shipley, G. G. (1974). Structural chemistry of 1,2 dilauroyl-DL-phosphatidylethanolamine: Molecular conformation and intermolecular packing of phospholipids. *Proc. Natl. Acad. Sci. USA* **195**, 3036–3049.

Hope, M. J., Bally, M. B., Webb, G., and Cullis, P. R. (1985). Production of large unilamellar vesicles by rapid extrusion procedure. Characterization of size distribution, trapped volume and ability to maintain a membrane potential. *Biochim. Biophys. Acta* **812**, 55–63.

Huang, C. (1969). Studies on phosphatidylcholine vesicles. Formation and physical characteristics. *Biochemistry* **8**, 344–349.

Huang, C., and Li, S. (1996). Computational molecular models of lipid bilayers containing mixed-chain saturated and monounsaturated acyl chains. *In* "Handbook of Nonmedical Applications of Liposomes" (D. D. Lasic and Y. Barenholz, Eds.), pp. 173–194. CRC Press, Boca Raton, FL.

Huang, C., Wang, Z., Lin, H., Brumbaugh, E. E., and Li, S. (1994). Interconversion of bilayer phase transition temperatures between phosphatidylethanolamines and phosphatidylcholines. *Biochim. Biophys. Acta* **1189**, 7–12.

Huang, L. C., and Larner, J. (1993). Inositol phosphoglycan mediators of insulin action: The special role of chiroinositol in insulin resistance. *Adv. Prot. Phosphatases* **7**, 373–392.

Jin, A. Y., Benesch, L. A., and Weaver, D. F. (1994). Computational conformational analysis of neural membrane lipids: Development of force field parameters for phospholipids using semi-empirical molecular orbital calculations. *Can. J. Chem.* **72**, 1596–1604.

Kunau, W.-H. (1976). Chemistry and biochemistry of unsaturated fatty acids. *Angew. Chem. Int. Ed. Engl.* **15**, 61–122.

Lee, A. G. (1977). Lipid phase transitions and phase diagrams. II. Mixtures involving lipids. *Biochim. Biophys. Acta* **472**, 285–344.

Levin, I. W. (1984). Vibrational spectroscopy of membrane assemblies. *In* "Advances in Infrared and Raman Spectroscopy" (R. J. H. Clark and R. E. Hester, Eds.), Vol. 11, pp. 1–48. Heyden, London.

Li, S., Wang, Z.-Q., Lin, H.-N., and Huang, C. (1993). Energy-minimized structures and packing states of a homologous series of mixed-chain phosphatidylcholines: A molecular mechanics study on the diglyceride moieties. *Biophys. J.* **65**, 1415–1428.

Low, M. G. (1989). The glycosyl-phosphatidylinositol anchor of membrane proteins. *Biochim. Biophys. Acta* **988**, 427–454.

Mabrey, S., and Sturtevant, J. M. (1976). Investigation of phase transition of lipids and lipid mixtures by high-sensitivity differential scanning calorimetry. *Proc. Natl. Acad. Sci. USA* **73**, 3862–3866.

Maggio, B. (1994). The surface behavior of glycosphingolipids in biomembranes: A new frontier of molecular ecology. *Progr. Biophys. Mol. Biol.* **62**, 55–117.

Mendelsohn, R., and Senak, L. (1993). Quantitative determination of conformational disorder in biological membranes by FTIR spectroscopy. *In* "Biomolecular Spectroscopy" (R. J. R. Clark and R. E. Heister, Eds.), pp. 339–380. Wiley, New York.

Montal, M., and Mueller, P. (1972). Formation of biomolecular membranes from lipid monolayers and a study of their electrical properties. *Proc. Natl. Acad. Sci. USA* **69**, 3561–3566.

Mueller, P., Rudin, D. O., Tien, H. T., and Wescott, W. C. (1963). Reconstitution of excitable cell membrane structure *in vitro*. *Circulation* **26**, 1167–1171.

Nagle, J. F., Zhang, R., Tristram-Nagle, S., Sun, W., Petrache, H. I., and Suter, R. M. (1996). X-ray structure determination of fully hydrated $L_\alpha$ phase dipalmitoylphosphatidylcholine bilayer. *Biophys. J.* **70**, 1419–1431.

Nishizuka, Y. (1984). The role of protein kinase C in cell surface signal transduction and tumour promotion. *Nature* **308**, 693–698.

Pascher, I., and Sundell, S. (1977). Molecular arrangements on sphingolipids. The crystal structure of cerebroside. *Chem. Phys. Lipids* **20**, 175–191.

Pascher, I., Sundell, S., Harlos, K., and Eibl, H. (1987). Conformation and packing properties of membrane lipid: The crystal structure of sodium dimyristoylphosphatidylglycerol. *Biochim. Biophys. Acta* **896**, 77–88.

Pascher, I., Lundmark, M., Nyholm, P. G., and Sundell, S. (1992). Crystal structures of membrane lipids. *Biochim. Biophys. Acta* **1113**, 339–373.

Pearson, R. H., and Pascher, I. (1979). The molecular structure of lecithin dihydrate. *Nature* **281**, 499–501.

Raetz, C. R. H. (1986). Molecular genetics of membrane phospholipid synthesis. *Annu. Rev. Genet.* **20**, 253–295.

Salem, L. (1962). The role of long-range forces in the cohesion of lipoproteins. *Can. J. Biochem. Physiol.* **40**, 1287–1298.

Schindler, H. (1980). Formation of planar bilayers from artificial or native membrane vesicles. *FEBS Lett.* **122,** 77–79.

Seelig, J. (1977). Deuterium magnetic resonance: Theory and application to lipid membranes. *Q. Rev. Biophys.* **10,** 353–418.

Seelig, J., and Seelig, A. (1980). Lipid conformation in model membranes and biological membranes. *Q. Rev. Biophys.* **13,** 19–61.

Somerville, C. (1995). Direct tests of the role of membrane lipid composition in low-temperature-induced photoinhibition and chilling sensitivity in plants and cyanobacteria. *Proc. Natl. Acad. Sci. USA* **92,** 6215–6218.

Sternin, E., Bloom, M., and MacKay, A. L. (1983). De-Packing of NMR spectra. *J. Magn. Reson.* **55,** 274–282.

Tanford, C. (1980). "The Hydrophobic Effect: Formation of Micelles and Biological Membranes," 2nd ed., John Wiley & Sons, New York.

Wang, G., Lin, H.-N., Li, S., and Huang, C. (1995). Phosphatidylcholines with *sn*-1 saturated and *sn*-2 *cis*-monounsaturated acyl chains: Their melting behavior and structures. *J. Biol. Chem.* **270,** 22738–22746.

Wong, P. T. T. (1986). Phase behavior of phospholipid membranes under high pressure. *Physica* **139–140B,** 847–852.

*Friedhelm Schroeder, W. Gibson Wood, and Ann B. Kier*

# 4

# The Biological Membrane and Lipid Domains

## I. Introduction

Since 1925 it has been recognized that lipids spontaneously form membranes in water (Table 1). A variety of models (reviewed in Yeagle, 1987) eventually evolved into the fluid mosaic lipid bilayer concept popularized over a quarter of a century ago (Singer and Nicolson, 1972). In this structure the fluid lipid bilayer serves as a matrix for embedded proteins functioning as ion channels, receptor–effector coupled systems, transporters, etc. The phenomenal success of this model is attested by its nearly universal acceptance. Unfortunately, its very popularity has overshadowed parallel exciting advances demonstrating that lipids as well as proteins are organized into highly structured domains in biological membranes (reviewed in Schroeder *et al.,* 1996a). Multiple physical techniques [fluorescence photobleaching recovery, fluorescence energy transfer, fluorescence microscopy, differential scanning calorimetry, nuclear magnetic resonance (NMR) spectroscopy, X-ray and neutron diffraction, electron microscopy, elasticity, enzymes, and sterol exchange proteins] show the presence of protein and lipidic domains in biological and/or model membranes. The precise relationship between lipid and protein domains is the subject of intense interest and debate (Parsegian, 1995).

## II. General Structure of Biological Membranes

### A. Surface Membrane

The cell surface plasma membrane is a shared feature of both prokaryotic and eukaryotic cells. Electron microscopy shows the plasma membrane surrounding the cell cytoplasm to have a thickness of about 80 Å. Since lipid bilayers generally have a width less than 50 Å, the remaining thickness is composed of the carbohydrate portion of glycolipids and glycoproteins extending from the exofacial side and the cytoskeleton associated with the cytofacial side of the plasma membrane. The plasma membrane lipids are especially enriched in cholesterol, such that the cholesterol/phospholipid molar ratio ranges from 0.4 to 1.0 (reviewed in Schroeder *et al.,* 1996b). This high cholesterol content orders/rigidifies the plasma membrane phospholipids and condenses their surface area (see Chapter 3), thereby making the plasma membrane lipid bilayer the most rigid in the cell and minimizing permeability of the cell surface lipid bilayer. Thus, the lipid portion of the membrane has primary functions in limiting the diffusion of substances into and out of the cell (see Chapter 11). The relatively poor conductivity of lipids further helps to confer some of the basic electrophysiological properties on the cell membrane (see Chapters 12 and 13). The insertion of specialized proteins (receptors, channels, effectors) into the plasma membrane lipid bilayer confers on the cell the ability to control the intracellular ionic mileu, membrane potential, supply of basic building blocks (sugars, amino acids, nucleotides, lipids), intercellular communication, adhesion, immunogenicity, etc.

### B. Intracellular Membranes

In contrast to prokaryotes, eukaryotic cells contain a highly developed intracellular membrane organization. These intracellular organelles [e.g., endoplasmic reticulum (ER), Golgi, lysosome, peroxisome, mitochondria, nuclear membrane] functionally compartmentalize the eukaryotic cell interior (e.g., macromolecule synthesis, secretion, digestion, fatty acid oxidation, ATP production, genetic replication). This avoids futile cycles and allows subcellular specialization not possible in prokaryotes. For the remainder of this chapter we focus on eukaryotic membranes.

Intracellular membranes differ significantly from the cell surface plasma membrane. Intracellular membranes are less thick than the plasma membranes, have less carbohydrate, have lower levels of cholesterol, and are more fluid.

**TABLE 1**  Progression of Membrane Models

| Year | Author(s) | Technique | Concept |
|---|---|---|---|
| 1925 | Gorter and Grendl | Monolayer area | Plasma membrane is a lipid bilayer |
| 1935 | Davson and Danielli | | Surface protein sandwiches the lipid bilayer |
| 1960 | Robertson | Electron microscopy | Bilayer phospholipid headgroups visualized |
| 1970 | Frye and Edidin | Fluorescence | Membrane components move laterally in bilayer plane |
| 1972 | Singer and Micholson | | Fluid mosaic membrane with integral and surface proteins |

For example, the cholesterol/phospholipid ratio of ER, lysosomal, nuclear, and inner mitochondrial membrane is 0.08–0.2, 0.4–0.6, 0.1–0.2, and 0.01–0.12, respectively (reviewed in Schroeder *et al.*, 1996b). The nuclear envelope, which separates the DNA from the rest of the cell, is actually composed of an inner and outer membrane. The outer nuclear membrane has physical continuity with the ER. Because of the fusion of inner and outer nuclear membranes, the nuclear envelope has pores about 65–75 Å in diameter to allow passage of RNA and other small molecules of molecular weight less than 1000 (see Chapter 30). The ER is abundant in the cell and functions in macromolecule biosynthesis, fatty acid desaturation, detoxification, and some electron transport functions. The principal function of the Golgi is in the posttranslational modification of proteins to glycoproteins and vesicular secretion/trafficking. As such, they may also be important in membrane biogenesis (see later discussion). Because lysosomes are part of the phagocytic and degradative machinery of the cell, the lysosomal membrane surrounds a variety of degradative enzymes that function optimally in the acidic pH of the lysosomal compartment. The very small organelles, peroxisomes, contain degradative enzymes [fatty acid, branched-chain fatty acid (phytol/phytanic acid), amino acid, and certain xenobiotic oxidations] as well as synthetic enzymes [dolichol, cholesterol, isoprenoid (farnesol and geraniol), and plasmalogen biosynthesis] (reviewed in van den Bosch *et al.*, 1992).

Mitochondria, function in lipid oxidation and in ATP production (see Chapters 6 and 7), are surrounded by outer and inner membranes. The lipid composition of the outer membrane resembles that of the ER and contains cholesterol, whereas the inner membrane is essentially devoid of cholesterol. The rate-limiting step in steroidogenesis is the transfer of cholesterol from the outer to the inner mitochondrial membrane. Although there appear to be contact sites between the inner and outer mitochondrial membranes, the lipid complement of the two membranes is distinct.

Perhaps the most important structural feature relating to the lipid components of these biomembranes is that lipids extracted therefrom spontaneously form membrane bilayers when reconstituted in water. This process is energetically highly favorable (see Chapter 3). Moreover, the distribution of lipids in these artificial membranes is generally random. This is in contradistinction to the distribution of lipids in biological membranes as discussed later.

## III. Plasma Membrane Lipid Bilayer: Transbilayer Lipid Distribution

The observation that lipids that make up biomembranes spontaneously self-assemble to form membrane bilayers with a relatively random intramembrane distribution had a dramatic impact on our early understanding of biomembrane structure. It led to the depiction of vectorially organized membrane proteins as being localized in a sea of randomly structured lipids (Fig. 1). This concept underwent a radical transformation in the ensuing two decades (reviewed in Schroeder and Nemecz, 1990; Schroeder *et al.*, 1996b). In the same year that the landmark paper (Singer and Nicolson, 1972) on membrane structure was published, the postulation of a random distribution of lipids in biomembranes was proven to be incorrect (Bretscher, 1972). Exciting new data revealed that the lateral distribution of lipids in hepatocyte plasma membrane and intestinal microvillus plasma membranes was nonrandom. Furthermore, studies with erythrocytes and unilamellar model membrane vesicles (Bretscher, 1972) provided the first convincing and conclusive evidence that the transbilayer distribution of phospholipids in membranes was asymmetric. Other investigators reported an asymmetric transbilayer distribution of cholesterol in erythrocytes and myelin. Thus, a new concept of membrane lipids and proteins grad-

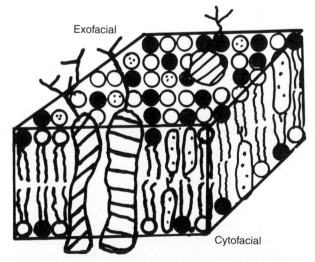

**FIG. 1.**  Random lipid bilayer model of membrane structure.

ually evolved indicating that both were organized into domain structures.

## A. Macroscopic

The transbilayer lipid distribution is not random, but the two leaflets have evolved distinct lipid distributional patterns: the *exofacial (outer) leaflet* is enriched in neutral zwitterionic lipids such as phosphatidylcholine (PC) and sphingomyelin (SM), as well as glycolipids; the *cytofacial (inner) leaflet* is enriched in anionic phospholipids such as phosphatidylethanolamine (PE) and phosphatidylserine (PS), as well as cholesterol (Table 2). Large-scale (macroscopic) transbilayer lipid domains have been demonstrated in plasma membranes of epithelial cells from liver, intestine, and kidney (reviewed in Schroeder *et al.*, 1991). These membranes are the blood sinusoidal, the contiguous (basolateral), and the brush border (microvillar, bile canalicular). Even within the hepatocyte, for example, the different macroscopic plasma membrane domains differ in transbilayer lipid distribution: PC is enriched in the exofacial leaflet of the blood sinusoidal plasma membrane segment; PE is nearly absent in the exofacial leaflet of the contiguous plasma membrane segment; SM is more enriched in the exofacial leaflet of bile canalicular plasma membrane. Individual leaflets of the membrane bilayer can independently regulate permeability (Negrete *et al.*, 1996).

## B. Microscopic

Even when plasma membrane macroscopic domains (segments) cannot be isolated, the plasma membrane has an asymmetric transbilayer distribution of lipids.

### 1. Phospholipid Transbilayer Domains

Phospholipids are asymmetrically distributed across the plane of the plasma membrane lipid bilayer (reviewed in Wood and Schroeder, 1992). With the exception of PS, which appears localized exclusively in the cytofacial leaflet, the transbilayer distribution of the other phospholipids is not absolute. The exofacial leaflet is relatively enriched in the neutral zwitterionic phospholipids having large polar headgroups (e.g., PC and SM) (see Table 2). In contrast, the cytofacial leaflet is enriched in phospholipids that have smaller polar headgroups and are either neutral (PE) or

acidic (PS, PI) (see Table 2). The localization of phospholipids with smaller polar headgroups in the cytofacial leaflet provides for less geometric constraint in this leaflet in highly curved membrane areas (e.g., spicules, microvillar tips, budding viruses). In addition, the distribution of anionic phospholipids primarily toward the cytofacial leaflet and zwitterionic neutral phospholipids toward the exofacial leaflet establishes a transbilayer charge gradient such that the cytofacial leaflet is more negative than the exofacial leaflet. This charge gradient may have an impact on the electrophysiology of the plasma membrane and on intracellular signaling (e.g., the release of cytofacially bound divalent metal counterions).

### 2. Cholesterol Transbilayer Domains

An asymmetric transbilayer distribution of cholesterol has been observed in plasma membranes from both normal (brain synaptosome, sperm, erythrocyte) and tumorigenic cells (L cell). In each case, 3–7 times more cholesterol distributes toward the cytofacial leaflet than the exofacial leaflet (reviewed in Schroeder and Nemecz, 1990; Igbavboa *et al.*, 1996).

### 3. Sphingolipid and Glycolipid Transbilayer Domains

The sphingolipids and glycolipids are localized solely in the exofacial leaflet of the plasma membrane. Sphingolipid–cholesterol complexes are hypothesized to form membrane microdomains differing in thickness from other membrane regions (Smaby *et al.*, 1996). The intracellular trafficking, sorting, and targeting of glycosylphosphatidylinositol-linked proteins is associated with membrane microdomains rich in cholesterol, glycosylceramides, and sphingomyelin (Parton and Simons, 1995).

### 4. Fatty Acid Transbilayer Domains

The fatty acids associated with cytofacial leaflet phospholipids, PE and PS, from L-cell fibroblasts and brain synaptosomes are enriched with unsaturated fatty acids. In contrast, the phospholipids found in the exofacial leaflet are more saturated (reviewed in Fontaine *et al.*, 1980). However, this transbilayer distribution of unsaturated fatty acyl groups is not universal. For example, in the small intestinal brush border membrane, the phospholipid in the outer leaflet contains more unsaturated fatty acids than the inner leaflet (Dudeja *et al.*, 1991).

### 5. Regulation and Pathophysiology

Spontaneous transbilayer migration of phospholipids is a remarkably slow process with a half-time of days to weeks. However, in the plasma membrane the transbilayer distribution of phospholipids is maintained by two plasma membrane integral protein systems: a *phospholipid translocase* and a *scramblase* (reviewed in Williamson *et al.*, 1995; Devaux and Zachowski, 1994). The translocase is specific for anionic phospholipids and rapidly translocates them from the exofacial leaflet to the cytofacial leaflet.

**TABLE 2** Plasma Membrane Leaflet Selectivity of Lipids

| Leaflet | Lipid enriched | Charge |
|---------|----------------|--------|
| Exofacial | Phosphatidylcholine | Zwitterionic |
| | Sphingomyelin | Zwitterionic |
| | Glycolipids | Neutral or anionic |
| Cytofacial | Phosphatidylserine | Anionic |
| | Phosphatidylinositol | Neutral |
| | Phosphatidylethanolamine | Neutral |
| | Cholesterol | |

By exclusion, PC and SM largely remain in the exofacial leaflet. By exclusion, PC and SM largely remain in the exofacial leaflet. The scramblase protein, in contrast, has no specificity for phospholipid polar headgroups, and so randomizes all phospholipid classes across the bilayer when activated. Under normal circumstances, the transbilayer distribution of phospholipids appears to be remarkably stable. However, activation of platelets with thrombin and thapsigargin or with $Ca^{2+}$ and ionophore induces a complete redistribution of all phospholipids across the lipid bilayer (Williamson et al., 1995) (Table 3). In addition, lipid peroxidation, radiation (Schroeder, 1984), and ATP depletion (Devaux and Zachowski, 1994) can lead to loss of transbilayer phospholipid asymmetry. Alterations in phospholipid asymmetry have been observed in human sickle-cell erythrocytes. Chronic ethanol exposure elicits a redistribution of phosphatidylcholine across the bilayer.

The transbilayer distribution of fatty acids occurs by a complex process in which phospholipids appear at the cytofacial leaflet with esterified fatty acids and/or fatty acid remodeling occurring at the cytofacial leaflet of the plasma membrane. The esterified fatty acids are then translocated across the bilayer.

The transbilayer distribution of cholesterol has been examined extensively (reviewed in Schroeder et al., 1996b). Regulation of the transbilayer sterol distribution may occur either by alterations in transbilayer migration rate (flip-flop) or in sterol transbilayer equilibrium. Although little is known regarding factors modulating the cholesterol transbilayer migration rate, chronic ethanol administration to mice significantly increased the sterol transbilayer migration rate (see Table 3). In contrast, at least four factors may contribute to plasma membrane equilibrium transbilayer

sterol distribution: (1) unsaturated fatty acids dramatically alter the plasma membrane transbilayer sterol distribution; (2) chronic exposure of LM cell fibroblasts or mice to ethanol reduced the plasma membrane and synaptosomal plasma membrane transbilayer sterol asymmetry, such that two- to three-fold more sterol was localized in the exofacial leaflet, without any change in the total plasma membrane cholesterol/phospholipid ratio; (3) aging dramatically altered mouse brain synaptic plasma membrane transbilayer sterol distribution such that 2.4-fold more sterol was present in the exofacial leaflet; and (4) although the spontaneous transbilayer migration rate of sterols is fast, on the order of minutes (reviewed in Schroeder and Nemecz, 1990), this cannot account for an asymmetric transbilayer sterol distribution. Several potential mechanisms regulating transbilayer sterol distribution have been proposed:

1. *Sphingomyelin-cholesterol complexes* have been postulated but cannot account for the observed enrichment of cholesterol in the cytofacial leaflet (reviewed in Schroeder and Nemecz, 1990; Schroeder et al., 1991).

2. *Transbilayer cholesterol dimers.* Although transbilayer tail-to-tail interactions of cholesterol exist (reviewed in Harris et al., 1995; Mukherjee and Chattopadhyay, 1996), transbilayer coupled cholesterol dimers would result in a symmetric transbilayer distribution of sterols in plasma membranes. This is not the case, however, as stated earlier.

3. *Plasma membrane integral cholesterol binding proteins.* Such proteins, if asymmetrically distributed across the membrane bilayer, could modulate transbilayer sterol distribution (reviewed in Schroeder and Nemecz, 1990). However, most membrane proteins are localized in cholesterol poor areas (reviewed in Schroeder et al., 1996b).

**TABLE 3**    Plasma Membrane Transbilayer Domain Regulation

| Parameter | Lipid | Plasma membrane | Effector |
|---|---|---|---|
| Transbilayer migration rate | Cholesterol | Synaptosome | Ethanol |
| Transbilayer distribution | Cholesterol | L cell | Ethanol |
| | Cholesterol | L cell | Polyunsaturated fatty acids |
| | Cholesterol | L cell | Sterol carrier proteins |
| | Cholesterol | L cell | Lipid peroxidation |
| | Cholesterol | L cell | Oxidized cholesterol |
| | Cholesterol | Synaptosome | Aging |
| | Cholesterol | Synaptosome | Ethanol |
| Transbilayer distribution | Phospholipid | Platelet | Activation |
| | Phospholipid | Erythrocyte | ATP depletion |
| | Phospholipid | Erythrocyte | Sickle erythrocyte |
| | Phospholipid | Erythrocyte | Ethanol |
| Transbilayer fluidity | | Erythrocyte | Anesthetics |
| | | Synaptosome | Anesthetics |
| | | Intestinal microvillus | Anesthetics |
| | | L cell | Anesthetics |
| | | L cell | Polyunsaturated fatty acids |
| | | L cell | Lipid peroxidation |
| | | Synaptosome | Aging |
| | | Synaptosome | Ethanol |
| | | Synaptosome | Systemic lupus erythematosus |

4. *Cytoplasmic sterol carrier proteins.* At least two cytosolic cholesterol binding proteins with micromolar $K_d$ and 1:1 stoichiometry for cholesterol interact with plasma membranes to function in intermembrane sterol transfer (reviewed in Schroeder *et al.,* 1996b). L-FABP expression in transfected L cells altered plasma membrane transbilayer sterol distribution. Both the expression of these proteins and their interactions with ligand are altered by chronic or acute ethanol treatment of mice or rats.

## IV. Lateral Lipid Microdomains in Membranes

### A. In Model Membranes

Since phospholipids differ in polar headgroups, fatty acyl, and fatty ether composition, it is important to know whether such diverse species cosegregate or mix in membranes. Both phase-miscible and phase-separated mixtures have been observed (see Chapter 3). Coexistence of phospholipid domains is readily demonstrated in phase-separated systems, composed either of pure PCs differing only in fatty acyl chain composition or of phospholipids differing only in fatty acyl chain composition or of phospholipids differing in polar headgroup composition (reviewed in Schroeder and Nemecz, 1990). For example, palmitoyl-oleoyl-phosphatidylcholine (POPC) has a phase transition (determined by differential scanning calorimetry) far below physiological temperature ($-5°C$) and is therefore in the *fluid (liquid-Crystalline) state* at 37°C. In contrast, distearoyl-phosphatidylcholine (DSPC) is in the *solid (gel) phase* up to 52.8°C. When these two phosphatidylcholines are mixed within the same membrane, two separate phase transitions appear near $-5$ and 52.8°C. Thus, between $-5$ and 52.8°C, the fluid-phase POPC and gel-phase DSPC coexist as laterally phase-separated PC domains. Below $-5°C$, both phospholipids are in the gel phase and coexist as separate gel-phase phosphatidylcholine domains. In model membranes composed of two phospholipid species and cholesterol, the cholesterol apparently cosegregates with whichever is the more fluid (lower phase transition temperature) phospholipid species (Schroeder and Nemecz, 1990).

Although of interest, the significance of phase-separated phospholipid lateral domains to biological membranes is unclear. Because of the influence of cholesterol as well as the heterogeneity of phospholipid species, biological membranes are generally in the fluid (liquid-crystalline) state as opposed to solid (gel) state at physiological temperatures. However, a number of investigators using probe approaches (e.g., electron spin resonance, NMR, or fluorescence) have obtained evidence for fluid–fluid immiscibility in model membranes even in the presence of sterols (reviewed in Schroeder and Nemecz, 1990; Sankaran and Thompson, 1990).

Thus, it may be concluded that the appearance of lateral domains may be a property of the phospholipid molecules themselves. In contrast to our understanding of the factors that determine fluid–solid phase separation, little is known regarding determinants of fluid–fluid or solid–solid immiscibility. However, phase separation in mixed bilayers can be induced by ions ($Ca^{2+}$) (Haverstick and Glaser, 1987), drugs (anthracyclines, doxorubicin) (Goormaghtigh *et al.,* 1990; Nicolay *et al.,* 1988), or proteins such as myristoylated alanine-rich protein kinase C substrate (Glaser *et al.,* 1996).

### B. In Biological Membranes

Unlike the simple two- or three-component model membranes discussed in the preceding section, biological membranes (e.g., erythrocyte plasma membrane) are composed of more than 1000 different molecular species. Thus, it is less likely that fluid–solid phase separations would be observed at physiological temperatures in biological membranes. Instead, fluid–fluid phase separations of phospholipids are the norm (reviewed in Bergelson, 1992).

As shown for model membranes, domains of phosphatidylcholines have been demonstrated in biological membranes. The differential action of acidic and basic phospholipases A-2 in human erythrocyte plasma membranes is consistent with the coexistence of at least two PC domains in the surface membrane exofacial leaflet.

Dissimilar phospholipids also form lateral microdomains in cell surface membranes. 9-Anthroylvinyl (AV)-labeled PC and PE partition between different domains. Such lateral microdomains are independent, as evidenced by the differential (opposite) response to the plant toxin ricin of AV-labeled sphingomyelin and PC in Burkitt lymphoma cells. Ricin binds to galactose in glycoproteins and glycolipids. Fluorescence recovery after photobleaching was used to examine lateral diffusion of fluorescent-labeled NBD-C-6-PC and rhodamine-PE of snail (*Aplysia*) neurons. Ethanol significantly increased the diffusion coefficient of rhodamine-PE, but not that of NBD-PC. These observations suggest that the nerve membrane lipids are heterogeneously organized at the submicroscopic scale into microscopic domains.

### C. Plasma Membrane Lateral Microdomains of Sphingolipids and Glycolipids

Glycosphingolipids generally have high phase transition temperatures near 50–85°. In model membranes consisting of stearoyl-oleoyl-phosphatidylcholine, glycosphingolipids phase separate into a solid-phase, glycolipid-rich domain and a stearoyl-oleoyl-phosphatidylcholine-rich fluid-phase domain at greater than 20% glycosphingolipid (Morrow *et al.,* 1992). However, even at low glycolipid concentration the glycosphingolipid and phospholipid exhibit solid-phase immiscibility. Likewise, sphingolipids cluster in the luminal leaflet of the trans Golgi network of MDCK cells where they form the budding site of apical membrane vesicles (reviewed in Parton and Simons, 1995).

### D. Cholesterol Lateral Microdomains

#### 1. Lateral Microdomains of Cholesterol in Model Membranes

Structural studies have demonstrated the coexistence of immiscible cholesterol-rich and cholesterol-poor phases in two-component two-phase lipid bilayers and monolayers. Their existence was inferred from lipid phase diagrams (reviewed in Schroeder *et al.*, 1996b). Fluorescence microscopy and freeze-fracture electron microscopy provided further structure evidence of microscopic cholesterol domains. Fluorescence lifetime analysis of three different fluorescent sterols (dehydroergosterol, cholestatrienol, and naphthylcholesterol) in phosphatidylcholine domains showed the existence of at least two sterol domains.

Studies of cholesterol dynamics in and between model membranes have also revealed multiple cholesterol domains. Findings with NMR indicate that fluid-phase microimmiscibility produces dynamic, short-lived, cholesterol-rich domains. Exchange kinetics of radiolabeled and fluorescent sterols are consistent with the presence of two exchangeable and one very slowly exchangeable sterol domain in phosphatidylcholine/cholesterol membrane vesicles. The size and kinetics of these cholesterol domains are highly dependent on temperature, phospholipid composition, and cholesterol content, and on the presence of proteins, peptides, and drugs (see Table 3). The role of spingomyelin is especially important, since its presence (in excess of 10 mol%) can abolish the rapidly exchangeable domain. Although a considerable literature has postulated preferential interaction of cholesterol with sphingomyelin in 1:1 stoichiometry, other data using fluorescent sterols are not consistent with this possibility (reviewed in Schroeder *et al.*, 1996b).

In model membranes, the cholesterol-induced fluid-phase immiscibility in phospholipid membranes is interpreted in terms of three cholesterol domains. The most dynamic domain, wherein cholesterol is the most exposed to the aqueous phase, is a cholesterol-poor fluid phase, where the cholesterol essentially spans the bilayer in a "time averaged sense" (Mantripragada *et al.*, 1991) (Fig. 2). This cholesterol-poor domain has also been reported as monomer cholesterol randomly distributed in the phospholipids (reviewed in Schroeder *et al.*, 1996b) (Fig. 3). The less dynamic, but still exchangeable, sterol domain is a cholesterol-rich fluid phase in which cholesterol is present in both monolayers (reviewed in Schroeder *et al.*, 1996b) in a rippled network consisting of alternating rows of cholesterol and phospholipid. The very slowly or nonexchangeable sterol domain may represent a highly ordered, nearly pure cholesterol domain within the bilayer (reviewed in Schroeder *et al.*, 1996b). The majority of evidence indicates that the lateral domains of cholesterol in the inner leaflet

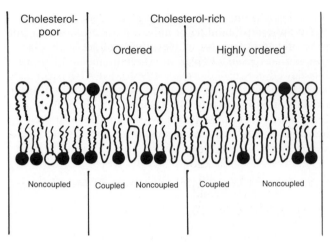

**FIG. 3.** Cholesterol lateral domains in membranes: dynamic exchange model. The cholesterol-poor domain is composed of monomeric cholesterol characterized by random distribution in a highly fluid phospholipid domain; greater exposure to the aqueous environment; rapid transbilayer migration; and rapid desorption into aqueous medium. In contrast, the cholesterol-poor domain is either ordered (rows of cholesterol interspersed with rows of phospholipid) or highly ordered (highly enriched cholesterol patches). In all of these cholesterol domains, the cholesterol in one leaflet may have (coupled) or not have (uncoupled) cholesterol in the opposing leaflet.

and outer leaflet of the model membranes are similar. However, it is not known if the sterol domains in one leaflet of the bilayer are coupled with sterol domains in the opposing leaflet of the bilayer or if they are independently organized (Fig. 3).

### 2. Plasma Membrane Lateral Cholesterol Microdomains

Electron microscopy, cholesterol oxidase, electron spin resonance, and fluorescence techniques, as well as radiolabeled or fluorescent sterol exchange, demonstrated the coexistence of cholesterol-rich and cholesterol-poor domains in cell surface membranes of red blood cells, platelets, L-cell fibroblasts, renal brush border, and human immunodeficiency virus (reviewed in Schroeder *et al.*, 1996b). These surface membranes have up to three different laterally segregated cholesterol domains.

### E. Regulation and Pathophysiology of Plasma Membrane Lateral Domains

The normal regulatory mechanism(s) that determine the lateral lipid domains are not known. The membrane content of sphingomyelin is directly correlated with the amount of nonexchangeable cholesterol domain in model and plasma membranes. Because sphingomyelin is particularly rich in the plasma membrane, the major portion (50–80%) of plasma membrane cholesterol is nonexchangeable (reviewed in Schroeder *et al.*, 1996b). Interaction of specific integral membrane proteins with cholesterol may cause

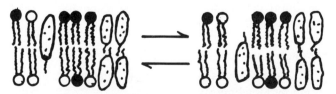

**FIG. 2.** Cholesterol lateral domains in membranes: time-averaged model. In the cholesterol-poor domains the cholesterol migrates rapidly from one leaflet to the other on the NMR time scale.

formation of cholesterol domains. Cytosolic proteins (e.g., sterol carrier protein-2 and the brain fatty acid-binding proteins) alter the size of the sterol domains in plasma membranes *in vitro* (Table 4). Proteins containing basic regions (protein kinase C, myristoylated alanine-rich C kinase substrate, pp 50[src]) bind to membranes and cause formation of negatively charged phospholipid domains. Lateral sterol domains in biomembranes are sensitive to temperature and ionic composition as well as certain drugs and peptides (see Table 4).

Relatively little is known about alterations in plasma membrane lateral lipid domains in pathological states. Pathogens such as Semliki forest virus require cholesterol and/or specific cholesterol domains in target membranes for binding (reviewed in Smaby *et al.,* 1996; Schroeder *et al.,* 1996b). Influenza hemagglutinin and Alzheimer's amyloid peptide fragments induce formation of negatively charged phospholipid domains in membranes. Bacterial cytolysins (hemolysins, $\theta$-toxins) bind to specific cholesterol lateral domains and reveal multiple cholesterol domains in the erythrocyte, lymphoma B, and BALL-1 cell surface plasma membrane, as well as in model membranes. Lateral sterol domains appear to be altered in sickle-cell erythrocyte membranes in response to deoxygenation. Chronic ethanol treatment increased the half-time of exchange for the exchangeable sterol domain in brain synaptosomes. One of the most exciting aspects of plasma membrane lateral cholesterol domains is whether their properties modulate the extent and/or rate of cholesterol efflux from cells. This process, called *reverse cholesterol transport,* may be of importance in atherosclerosis, diabetes, and other lipid disorders.

## V. Transbilayer Asymmetry in Fluidity

Extensive evidence has accumulated that bulk membrane lipid composition is the primary determinant of bulk membrane structure/fluidity, which in turn regulates the activities of membrane lipid-dependent enzymes (reviewed in Schroeder *et al.,* 1996b). However, some integral membrane protein functions are not lipid sensitive. Furthermore, attempts at correlating changes in bulk membrane

lipid composition with functional changes in lipid-sensitive integral membrane proteins are not uniformly successful. Two factors have contributed to resolution of these intriguing observations: (1) evidence for the existence of lipid domains in membranes using new sophisticated methodology (reviewed in Schroeder *et al.,* 1996b), and (2) the recognition that membrane physical properties cannot be simply explained in general terms such as "fluidity" or "microviscosity." It is now recognized that the available physical techniques (fluorescence, NMR, ESR, IR, differential scanning calorimetry, X-ray) generate a multiplicity of terms (fluidity, microviscosity, lateral diffusion constant, rotational correlation time, order parameter, limiting anisotropy, and phase-transition temperature) describing the physical structure of membranes/domains. Moreover, these terms are not synonymous. Instead, they are unique, often-independent parameters that describe different aspects of membrane physical properties. Two classes of lipid structural domains are now recognized in plasma membranes.

### A. Macroscopic Domain Fluidity

The primary determinants of membrane lipid fluidity include phospholipid composition, fatty acid unsaturation, and cholesterol/phospholipid ratio. Phospholipids, fatty acyl chains esterified to the phospholipids, and cholesterol are asymmetrically distributed across the plasma membrane bilayer. Since each of these components has different effects on fluidity, it is not possible to accurately predict the net effect on plasma membrane leaflet fluidity. A rough expectation can be based on the following: PC and SM are generally more fluid phospholipids than PE and PS; unsaturated fatty acids are more fluid than saturated fatty acids; cholesterol orders the membrane structure. Large-scale transbilayer lipid domains have been demonstrated in plasma membranes of epithelial cells from liver, intestine, and kidney (reviewed in Schroeder *et al.,* 1996b). On basis of the transbilayer distribution of phospholipid and fatty acid in intestinal microvillus membrane, one might predict that the cytofacial leaflet might be more fluid than the exofacial leaflet. On the contrary, the microvillus plasma membrane segment from intestine has a much less

**TABLE 4**  Membrane Lateral Domain Regulation

| Lipid | Cell | Membrane | Effector |
|---|---|---|---|
| Cholesterol | L cell | Plasma membrane | Sterol carrier protein |
| | — | Model membrane | Cytolysins |
| | Erythrocyte | Plasma membrane | Cytolysins |
| | Lymphoma | Plasma membrane | Cytolysins |
| | BALL-1 cell | Plasma membrane | Cytolysins |
| | Erythrocyte | Plasma membrane | Deoxygenation |
| | — | Model membrane | Protein kinase C |
| Phospholipid | — | Model membrane | MARCKS |
| | — | Model membrane | pp 50[SRC] |
| | — | Model membrane | VSV |
| | Cell | Plasma membrane | VSV |
| | — | Model membrane | Alzheimer's amyloid |

fluid cytofacial leaflet than exofacial leaflet (Dudeja *et al.*, 1991). This suggests that the cytofacial leaflet is enriched with cholesterol in the microvillus.

### B. Microscopic Domain Fluidity

If the transbilayer distribution of cholesterol is the major determinant of plasma membrane individual leaflet fluidity, one might predict that the cholesterol-rich leaflet is less fluid. This is indeed the case for L-cell, spermatozoa, and erythrocyte plasma membranes (reviewed in Schroeder *et al.*, 1996b). Thus, cholesterol transbilayer distribution is closely associated with transbilayer fluidity gradients in plasma membranes.

### C. Regulation and Pathobiology of Transbilayer Fluidity

Transbilayer fluidity gradients in the plasma membrane are very sensitive to a variety of factors (see Table 3). Acute ethanol administration fluidized the exofacial leaflet, whereas anesthetics were leaflet-selective plasma membrane fluidizers: exofacial leaflet-selective (pentobarbital, phenobarbital, benzyl alcohol, ethanol); cytofacial leaflet-selective (prilocaine, 2-[(2-methoxy-ethoxy)ethyl]-*cis*-8-(2-octylcyclopropyl) octanoate). In contrast, unsaturated fatty acids, aging, chronic ethanol, and systemic lupus erythematosus essentially abolished the plasma membrane transbilayer fluidity. The exofacial fluidizing drugs affected $Na^+,K^+$-ATPase and leucine aminopeptidase activity, whereas cytofacial fluidizing drugs affected $Na^+$-dependent D-glucose uptake and $Na^+,K^+$-dependent L-glutamic acid uptake.

## VI. Lateral Plasma Membrane Lipid Domain Fluidity

### A. Macroscopic Lateral Lipid Domains

The fluidity of large-scale (macroscopic) plasma membrane lipid domains has been demonstrated in plasma membrane segments of epithelial cells from liver, intestine, and kidney (reviewed in Schroeder *et al.*, 1996b; Schroeder *et al.*, 1991). These lateral lipid domains (sinusoidal, basolateral, and brush border or canalicular) are separable by centrifugation techniques. Even from the same cell, each different plasma membrane subfraction is characterized by unique enzymes, transport proteins, receptor/effector coupling systems, lipid composition, and physical structure. It is not understood how lateral intermixing of the lipid components of these macroscopic domains within the same cell is prevented. Lipids are free to diffuse laterally in the plane of the membrane. In the plasma membrane cytofacial leaflet, these lipids freely diffuse laterally across mcrodomain boundaries. Tight junctions prevent the lateral diffusion of lipid molecules in the exofacial leaflet across macrodomain boundaries. How the tight junctions accomplish this sorting of lipids (and proteins) in the exofacial leaflet is not known. Macroscopic lipid domains can also be separated by centrifugation as vesicles shed from cells. Depending on the cellular type, shed vesicle membranes can represent macroscopic cholesterol-rich or cholesterol-poor domains. Shed retinal rod outer segment membrane vesicles are also macroscopic cholesterol-poor domains.

### B. Microscopic Plasma Membrane Lateral Domains

Microscopic lipid domains are not readily separated by techniques such as centrifugation. Their existence is inferred from histochemical staining with filipin to visualize filipin–cholesterol complexes (reviewed in Schroeder *et al.*, 1991, 1996b). However, results obtained with histochemical stains such as filipin must be viewed cautiously. Proteins may interfere with accessibility of the filipin to cholesterol. The major breakthroughs in localizing microscopic lateral lipid domains have come through a variety of probe approaches as well as noninvasive physical techniques. A number of methods, based on cholesterol exchange, cholesterol oxidase, spin-labeled sterol, fluorescent sterol, and freeze-fracture, have provided evidence that such microdomains exist in both model membranes and biomembranes.

### C. Structural Consequences of Lipid Lateral Domains in Model Membranes

Lateral segregation of membrane lipids has important structural consequences. Segregation of phosphatidylcholines differing by >4 methylenes (dimyristoylphosphatidylcholine and distearoylphosphatidylcholine) results in coexistence of fluid and solid phospholipid domains. Segregation of anionic (phosphatidylserine) from neutral zwitterionic (phosphatidylcholine) phospholipids results in lateral charge separation. Cholesterol segregates into cholesterol-poor and cholesterol-rich domains that represent coexisting immiscible fluid phases, with the former relatively less ordered than the latter. In short, lateral lipid domains can differ in phase, in charge, in glycolipid or sphingolipid content, or in cholesterol content. Structural consequences of lipid lateral domains have not been determined, although some evidence indicates that the cholesterol-poor regions of plasma membranes are more fluid than the cholesterol-rich regions.

## VII. Plasma Membrane Lipid Bilayer: Protein Distribution

### A. Coupling of Protein Function to Transbilayer Lipid Distribution

Transbilayer lipid asymmetry is coupled to a variety of plasma membrane functions.

*1. Receptor–effector coupling.* The plasma membrane transbilayer cholesterol distribution is the primary determinant of transbilayer fluidity gradients. This gradient and alterations therein thereby modulate the functions of plasma membrane proteins (review in Schroeder and

Wood, 1995; Schroeder and Sweet, 1988; Sweet and - Schroeder, 1988), such as the glucagon receptor–adenylate cyclase coupled system located in exofacial and cytofacial leaflets, respectively.

2. *Ion transporter coupling.* Ion transporters that are composed of subunits function optimally in an associated/ aggregated state. Plasma membrane leaflet cholesterol content and the associated transbilayer fluidity gradient may define specific ion transporter subunit association. The spontaneous organization of cholesterol into domains may even be responsible for lateral segregation of proteins into specific cholesterol-rich or cholesterol-poor regions in the membrane (Bretscher and Munro, 1993). The $Na^+,K^+$-ATPase in plasma membranes from L-cell fibroblasts transfected with the cDNA encoding liver fatty acid binding protein have an altered plasma membrane transbilayer cholesterol distribution, decreased plasma membrane $Na^+,K^+$-ATPase specific activity, decreased [$^3H$]ouabain binding, and decreased association of specific (but not all) $Na^+,K^+$-ATPase subunits with the plasma membrane. Thus, altered plasma membrane transbilayer sterol distribution may decrease $Na^+,K^+$-ATPase activity by decreasing transbilayer fluidity gradients and/or by causing the recruitment of nonfunctional/inappropriate proportions of $Na^+,K^+$-ATPase subunits to the plasma membrane from intracellular sites.

3. *Cellular cholesterol influx.* Although the major mechanism for cholesterol uptake is via the LDL receptor-mediated endocytic pathway, some LDL free cholesterol may enter by an exchange pathway. This exchange may be facilitated by the low cholesterol content of the plasma membrane exofacial leaflet. The size of the exchangeable cholesterol domain and the transbilayer migration rate of cholesterol may both contribute.

4. *Cellular reverse cholesterol transport.* The reverse process of cholesterol influx is efflux and occurs by molecular cholesterol exchange. This process is very important to the HDL-mediated reverse cholesterol transport for the removal of cholesterol from the cell. The potential effects of cholesterol domain size and transbilayer migration rate are especially significant here.

5. *Translocation of proteins across the plasma membrane.* Not only does protein translocation require a leader sequence in the protein to be secreted and a protein translocase enzyme(s), but the lipid phase is essential. The fatty acid composition, fluidity, phospholipid charge, etc., are all vital, if poorly understood, factors contributing to protein translocation across membranes (Schatz and Dobberstein, 1996).

## B. Transbilayer Coupling of Plasma Membrane Lateral Microdomains: Relation to Protein Function

The subject of the structural properties of lipids in one leaflet of the membrane affecting the structural properties of lipids in the opposing leaflet must be considered. Several points are especially pertinent to lateral domains in opposing leaflets of the membrane bilayer. At least three possibilities can give rise to coupling of laterally segregated lipids

across the bilayer (Fig. 4): (1) Some lipids such as dolichol or dolichol phosphate are extremely long and extend across the bilayer (Fig. 4A). Dolichols are long-chain isoprenoid alcohols with chain lengths generally of 55–115 methylenes. These molecules are extremely potent in fluidizing membranes and in altering $Na^+,K^+$-ATPase activity in synaptosomal membranes. (2) Phospholipid species with fatty acids esterified to the 1 and 2 differing in chain length by four or more methylenes, as well as many sphingomyelin species, can interdigitate (Fig. 4B) in the opposing leaflet acyl chains to varying degrees. (3) A fluorescent-probe approach was recently used to examine the physical state of laterally segregated lipids coupled across opposing leaflets (Almeida *et al.*, 1992) (Fig. 4C). A membrane-spanning NBD-labeled phosphatidylethanolamine in a membrane bilayer containing lignoceryldihydrogalactosylceramide was used to show that the distribution of solid domains in one monolayer of the bilayer was independent of the distribution of solid domains in the opposing bilayer. In contrast, in membrane bilayers lacking lignoceryldihydrogalactosylceramide, the solid domains exactly superimposed on solid domains in the opposing bilayer. These superimposed solid phases were very stable, with half-times of relaxation in excess of 7 hours (Almeida *et al.*, 1992). Such transbilayer-coupled and -uncoupled lateral domains may be highly significant to the functions of transmembrane proteins (i.e., ion channels, pumps, receptor-effector coupling).

## C. Annular Lipid Microdomains

Interfacial or annular lipid exists in the regions between lateral lipid domains (Fig. 5) or around integral proteins (Fig. 6) in cell membranes. At the boundaries of lipid domains such interfacial lipid is believed to be highly disordered and important to the actions of phospholipase enzymes. Integral membrane proteins are primarily localized in fluid, cholesterol-poor lipid domains (Houslay and Stanley, 1982). Moreover, in the fluid lipid microdomains, the proteins are solvated by lipids. The lipids immediately adjacent to the embedded protein are referred to as the lipid annulus, boundary layer lipid, or halo lipid. Although

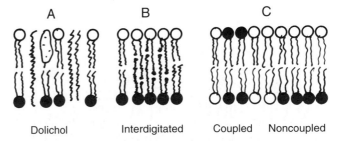

| A | B | C |
|---|---|---|
| Dolichol | Interdigitated | Coupled    Noncoupled |

**FIG. 4.** Transbilayer coupling of lipid domains in membranes. (A) Long-chain dolichol spans the entire bilayer and couples the two leaflets. (B) Phospholipids, glycolipids, and sphingolipids whose acyl chains are dissimilar in length may interdigitate into opposing leaflets. (C) Like lipids may form lateral domains that have identical lipids in the opposing leaflet (coupled) or in different lipids in the opposing leaflet (uncoupled).

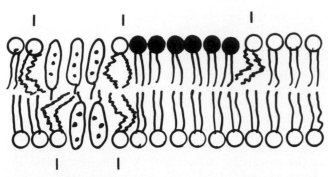

**FIG. 5.** Interfacial lipid in lateral domains of membranes. Interfacial lipid (I) surrounding lipid domains is highly disordered.

the annular lipid may differ from bulk lipid, it exchanges with neighboring lipids much more slowly ($10^{-4}$ to $10^{-6}$ sec) as compared to the extremely rapid hopping frequency, $10^{-7}$ sec, of bulk lipid. The important concept here is that the annular lipid is not static and does turn over, albeit 10 to 1000 times more slowly than bulk lipid. As for interfacial lipid, the annular lipid is highly disordered. One of the best-studied examples is the annular lipid surrounding the $Ca^{2+}$-ATPase from sarcoplasmic reticulum (reviewed in Houslay and Stanley, 1982). This protein spans the membrane and has an annulus of 15 phospholipid molecules in each leaflet of the bilayer. The annulus is essential for activity, since $Ca^{2+}$-ATPase activity is lost if less than 30 mol of lipid remain associated with the protein. When faced with a choice of phospholipids present in a mixed membrane (inhibitory and stimulatory lipids) the $Ca^{2+}$-ATPase segregates lipids necessary for proper function into the lipid annulus. Other examples of integral membrane proteins that segregate annular lipid include cytochrome oxidase, $\beta$-hydroxybutyrate dehydrogenase, glycophorin, $Na^+,K^+$-ATPase, rhodopsin, acetylcholinesterase, and protein kinase C (reviewed in Houslay and Stanley, 1982; Nichols and Roufogalis, 1991). Specific lipid species appear to be selected in the annulus surrounding these proteins: $\beta$-hydroxybutyrate dehydrogenase and phosphatidylcholine; 5'-nucleotidase and sphin-

gomyelin; $Ca^{2+}$-ATPase and nonacidic phospholipids; cytochrome oxidase, glycophorin, $Na^+,K^+$-ATPase, and rhodopsin with acidic phospholipids. Acetylcholinesterase specifically requires associated annular phosphatidylinositol, whose fatty acyl groups differ from those of nonannular phosphatidylinositol.

Cholesterol is generally excluded from the annulus of integral membrane proteins and, in the case of the $Ca^{2+}$-ATPase, may actually be inhibitory (Houslay and Stanley, 1982). One of the rare exceptions to this rule is the acetylcholinesterase receptor, which has both annular and nonannular cholesterol sites around the protein. The nonannular cholesterol sites have a 20 times greater affinity for cholesterol than the annular sites. The nonannular sites appear to be necessary for stabilizing $\alpha$-helical structures in the acetylcholinesterase receptor, which are necessary to support a functional ion channel. However, other neutral lipids such as $\alpha$-tocopherol, squalene, or cholestanol can substitute for cholesterol. Cholesterol enrichment also occurs in the immediate lipid layer surrounding the acetylcholine receptor, as well as vesicular stomatitis virus proteins G and M.

## D. Lateral Coupling of Plasma Membrane Lipid Domains to Protein Function

The function of lateral lipid domains in membranes is not known. Lipids can influence the function of integral membrane proteins through the lipid annulus, lateral diffusion of the proteins, transbilayer domains, and lateral microdomains. Changes in transbilayer or lateral lipid domains can elicit conformational changes as well as vertical or lateral displacement of proteins (Schroeder *et al.,* 1991, 1996b; Bergelson, 1992). It is believed that lateral microdomains can buffer the integral membrane proteins embedded therein from structural changes occurring elsewhere in the membrane. Moreover, insertion of a membrane protein and refolding in the lipid bilayer requires lipid microdomains to be fluid. Lateral lipid domains, especially cholesterol, are coupled to a variety of biological processes.

### 1. Sperm Capacitation

To complete the acrosomal reaction to fertilize an oocyte, sperm must be capacitated. This requires shedding of cholesterol from the periacrosomal region of the sperm plasma membrane head piece. Serum lipid transfer protein-I enhances sperm capacitation by increasing this cholesterol loss.

### 2. Cholesterol Absorption and Efflux

An intrinsic microvillar plasma membrane protein is required to potentiate cholesterol uptake by intestinal microvillus membranes (Schulthess and Hauser, 1995). The extent and direction of this process may be regulated by the size of the exchangeable cholesterol domains and/or the half-time of sterol transfer through the exchangeable domains. Reverse cholesterol transport (reviewed in

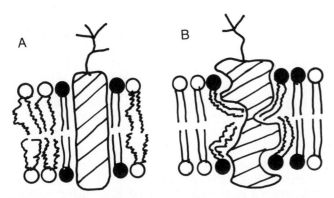

**FIG. 6.** Annular lipid domains. Annular lipid is the first "shell" of lipid surrounding integral membrane proteins. (A) Annular lipid "ordering" by protein. (B) Annular lipid "disordering" by protein.

Schroeder *et al.,* 1996b) is important in atherosclerosis and other serum lipoprotein disorders.

## 3. Membrane Protein Function

Both integral and soluble proteins can modulate the composition and properties of lipid domains. Integral membrane proteins are not randomly distributed in the lateral lipid domains. Lipids form the microenvironment of the proteins, and the immediate annular lipids surrounding the proteins are relatively disordered with respect to bulk lipids. When ligands, agonists, and antagonists interact with proteins, conformational changes occur in the proteins. The binding of hormones, antibodies or lectins (prostaglandins, ricin, muscarinic agents, antibodies) to membrane-associated receptors elicits changes in the fluidity of specific membrane lateral microdomains in which these integral proteins are embedded (Sweet and Schroeder, 1988; Bergelson, 1992). Muscarinic agents decrease the fluidity of the microdomain wherein anthroylvinyl phosphatidylcholine resides in rat brain membranes. One consequence of such lipid microdomain structure changes is that the activities of other proteins that cosegregate into the same lipid microdomain as the receptor protein may be modulated. For example, the acetylcholinesterase receptor is necessary for cholinergic synaptic transmission. The receptor requires both anionic phospholipid and cholesterol for optimal activity (Sunshine and McNamee, 1992). Binding of acetylcholine to the receptor is coupled to the opening of a cation-specific channel. The rate of desensitization of the receptor depends on the type of anionic and neutral lipid surrounding the receptor.

Soluble proteins such as the cytosolic cholesterol carrier proteins (SCP-2 and SCP/L-FABP) can also modulate lipid domains (reviewed in Schroeder *et al.,* 1991, 1996b). *In vitro* studies with model membranes indicate that SCP-2 but not SCP/L-FABP alters both the size and kinetics of the cholesterol domains. In contrast, *in vitro* studies with biological membranes indicate that both SCP-2 and SCP/L-FABP can modulate sterol domains. These results have been interpreted as being consistent with SCP-2 interacting directly with membrane anionic phospholipids, while SCP/L-FABP interacts with a glycoprotein/glycolipid membrane receptor to elicit changes in cholesterol domain structure. These observations with SCP/L-FABP have been confirmed through molecular biological studies performed *in vivo* with intact cultured cells transfected with the cDNA encoding SCP/L-FABP. These alterations in cholesterol lateral microdomain structure dramatically influenced the plasma membrane content and proportion of $Na^+,K^+$-ATPase subunits, as well as the specific activity of the enzyme. The size of the exchangeable domain(s) may relate to the size of cholesterol-"poor" regions of the cell surface membrane. Lipid microdomains appear to regulate in part the activities of transport proteins including the $Ca^{2+}$-ATPase, $Na^+,K^+$-ATPase, Na-H-antiporter, Na dependent glucose transport. Other functions regulated by lipid lateral microdomains include the closure of exocytotic fusion pore (Oberhauser *et al.,* 1992) and protein kinase C activation/inhibition (Epand *et al.,* 1992). Neutral phospholipids such as

phosphatidylcholine as well as phosphatidylserine participate in the activation and differential regulation of the protein kinase C (Chen *et al.,* 1992). It has been shown that autophosphorylation of protein kinase C may require a high order or protein–phospholipid aggregates. The lateral distribution of cholesterol into cholesterol-poor and cholesterol-rich domains may promote the aggregation of proteins into the cholesterol-poor domains. The size of the cholesterol-poor domain in membranes is usually small, and therefore, regulation of the size of this domain by ions, drugs, and cytosolic proteins can have important consequences for integral membrane proteins that are activated in the aggregated state.

In addition to being modulated by proteins, protein functions themselves are modulated by lipid lateral domains. Most proteins reside in relatively cholesterol-poor lateral domains (reviewed in Schroeder *et al.,* 1991, 1996b). Consequently, the size of the exchangeable or sterol-"poor" lateral domain may determine protein concentration in that domain, which may in turn determine the association or aggregation state of protein subunits to form functional receptor–effector coupled systems, ion transporters, etc.

## 4. DNA Replication

In addition to segregating DNA from the rest of the cell, nuclear membrane lipids function in DNA replication. Acidic phospholipids inhibit the replication activity of DnA protein, trigger DNA replication *in vivo,* serve as storage sites for inactive DNA topoisomerase I, bind histones, and influence activity of DNA polymerases (reviewed in Sekimizu, 1994).

## 5. Epithelial Cell Macrodomain Function

Epithelial cells from a variety of tissues (liver, kidney, intestine) have plasma membranes that differ markedly depending on their orientation (basolateral, serosal, microvillar). These different regions of the cell membrane have lipids and other components that are specifically segregated. For example, 5'-nucleotidase, leucine napthylamidase, and alkaline phosphatase are localized predominantly in the bile canalicular region. NaK-ATPase is located primarily in the contiguous region, whereas glucagon activated adenylate cyclase, glutamyl transpeptidase, and CMP-neuraminic acid hydrolase activities are primarily sinusoidal (reviewed in Sweet and Schroeder, 1988).

## VIII. Intracellular Membranes

### A. General Properties

Intracellular membranes generally contain much less cholesterol, sphingomyelin, and glycolipid than the cell surface plasma membrane. The cholesterol/phospholipid ratios in nuclear, microsomal, lysosomal, and mitochondrial membranes are 5–10-fold, 4–5-fold, 1–2-fold, and 4–30-fold, respectively, lower than in plasma membranes.

## B. Organelle Lipid Domains

In general, the cytofacial facing leaflet of those few intracellular membranes that have been examined appears to be enriched with acidic phospholipids (reviewed in Buton et al., 1996). However, the mitochondrial outer membrane has a symmetric transbilayer distribution of phosphatidylcholine (Dolis et al., 1996). Transbilayer motion of glyerolipids across intracellular membranes is protein-mediated and extremely fast, with half-times near 25 sec and 2 min for endoplasmic reticulum and mitochondrial outer membrane, respectively. With the exception of cholesterol domains, almost nothing is known regarding subcellular organelle membrane lateral lipid domains (reviewed in Schroeder et al., 1996b). Similar to plasma membranes from a variety of tissues, fibroblast microsomal membranes had a large nonexchangeable sterol domain composing 59% of total sterol. In contrast, mitochondria had a much larger nonexchangeable lateral sterol domain, 84% of total sterol. At least two and possibly three cholesterol pools exist in adrenal mitochondria.

## C. Protein Functions

To date very few protein functions have been correlated with intracellular membrane lipid domains. One such process is mitochondrial steroidogenesis (reviewed in Schroeder et al., 1996b). In mitochondrial steroidogenesis, the rate-limiting step is the transfer of cholesterol from the outer to the inner mitochondrial membrane. How this transfer occurs is not known, but possible mechanism(s) include (1) a protein that alters the cholesterol transbilayer migration rate and/or equilibrium transbilayer distribution in the outer mitochondrial membrane; (2) a protein that stimulates cholesterol transfer from extramitochondrial sources to mitochondria and/or from outer to inner mitochondrial membranes; and (3) a protein that allows cholesterol to migrate laterally from the outer to the inner mitochondrial membrane via contact sites between the two membranes.

# IX. Membrane Biogenesis

In the past, the topic of membrane biogenesis used to engender thoughts on the origin of life, the primordial membrane enclosing the first protein/DNA, the entrapment of a primordial bacterium to become the progenitor of mitochondria in eukaryotes, etc. Since the mid-1970s, attention has focused away from such global questions to a more pragmatic elucidation of how living cells make membranes. At least two processes are recognized: vesicular-mediated pathways and protein-mediated pathways.

## A. Vesicular Membrane Lipid and Protein Trafficking

Vesicular pathways are important to receptor-mediated endocytosis, phagocytosis, secretions, protein sorting, protein insertion, and membrane partitioning during cell division (reviewed in Pryer et al., 1992; Morris and Frizzell, 1994; Warren, 1993; Schekman and Orci, 1996; Bogdanov et al., 1996; Schatz and Dobberstein, 1996). Phospholipids (e.g., phosphoinositides) may be important regulators of vesicular membrane trafficking in the cell. The role of transbilayer or lateral lipid domains in membranes in vesicular transport is at present unknown.

## B. Protein-Mediated Membrane Lipid Trafficking

In contrast to the paucity of information on the involvement of lipid domains in vesicular trafficking, recent data on intracellular membrane cholesterol domains show that certain cytosolic proteins are capable of interacting with intracellular membrane lipid domains to elicit directional lipid trafficking, in certain instances against the cholesterol concentration gradient (reviewed in Schroeder et al., 1996b)

1. *Spontaneous intermembrane sterol transfer between organelles.* The initial rate of spontaneous sterol transfer among plasma membranes, microsomes, and mitochondria varies over a 3- to 9-fold range, depending on the specific donor/acceptor pair. In addition, the initial rate of spontaneous sterol transfer varied up to 4–13-fold, depending on the direction of sterol transfer examined.

2. *Role of sterol domains in sterol transfer between organelles.* Different acceptor organelles significantly affected the size of the exchangeable sterol domain in plasma membrane, microsomal, and mitochondrial donor membranes. For example, substitution of microsomal acceptor membranes by plasma membrane acceptor membranes decreased the size of the exchangeable sterol domain in donor microsomes by half. Thus, the nature of the acceptor membrane contributes to both the kinetics of spontaneous sterol trafficking and also the domain size in the donor membrane.

3. *Protein-mediated intermembrane sterol transfer between organelles.* SCP-2 (also called nonspecific lipid transfer protein), the liver sterol carrier protein (also called liver fatty-acid binding protein, L-FABP, or sterol and squalene carrier protein), brain fatty-acid binding protein (B-FABP), and heart fatty-acid binding protein (H-FABP) all stimulate sterol transfer, but by different mechanisms (reviewed in Schroeder et al., 1996b). SCP-2 stimulated sterol transfer between all intracellular organelles, whereas L-FABP was more restricted to certain membranes. L-FABP but not SCP-2 decreases the size of the nonexchangeable domain. In contrast, neither brain fatty-acid binding protein (B-FABP) nor heart fatty-acid binding protein (H-FABP) enhance the initial rate of sterol transfer, but instead increase the size of the exchangeable sterol domain. Transfer of cholesterol mediated by SCP-2 and L-FABP requires the interaction of the protein with the membrane and the presence of a sterol binding site. In the case of B-FABP and H-FABP, a sterol binding site was not evident. SCP-2 stimulated the initial rate of sterol transfer from microsomes to plasma membranes 5-fold *in vitro* and in intact cells, suggesting that SCP-2 may participate in reverse cholesterol transport. SCP-2 stimulated the ini-

tial rate of sterol transfer from plasma membrane to microsomes 26-fold, while in the reverse direction it stimulated the initial rate of sterol transfer from microsomes to plasma membranes only 5-fold. Similarly, SCP-2 stimulated the initial rate of sterol transfer 12-fold from plasma membranes to mitochondria, but only 4-fold in the reverse direction from mitochondria, to plasma membranes. SCP-2's effect on interorganelle sterol transport was vectorial and, depending on the donor/acceptor pair, it was also against the cholesterol gradient.

## X. Summary

The landmark fluid–lipid bilayer hypothesis of Singer and Nicholson became the model (before the mid-1970s) for understanding of lipid structure and the way proteins might interact with lipids. However, protein biochemists, biophysicists, and physiologists quickly found that this simplistic model did not consistently fit with their experimental data, especially on protein function. Unfortunately, rather than modify the model, some investigators concluded that lipids might really not be that important to the structure and function of integral membrane proteins. Like a phoenix rising from the ashes, however, the discovery that the lipid bilayer was composed of heterogeneously distributed lipids opened new areas of investigation (Bretscher, 1972). In fact, the growing awareness of lipid domains became a beacon for membranologists willing to embrace a modification of the basic bilayer theory to include the existence of an asymmetric distribution of lipids, that is, lipid domains. Although space limitations do not allow recognition of many of the original contributions in this field, the reader is referred to in-depth citations in the many reviews listed in the references.

In the 1990s, membranologists find themselves with the task of attempting to understand the origin, regulation, and function of lipid domains in cells. Such lipid domains compose macroscopic areas of epithelial cells, microscopic domains in essentially all membranes, and interfacial/boundary lipid in the protein–lipid interface. The challenge for structural biologists will be to find specific probes to explore not only how these domains regualte the tertiary structure and function of proteins, but also how they might regulate lipid trafficking within the cell and lipid efflux/influx from/to the cell. Physiologists will have to modify their concepts of how channels, transport proteins, and receptors function in, and may be regulated by, specific lipid domains in the membrane. The protein biochemists must face the task of isolating and characterizing specific integral membrane proteins that may regulate membrane lipid domains. In addition, since lipids are synthesized in specialized subcellular areas of the cell, a major focus will have to be on how lipids are sorted and targeted toward specific membranes and membrane domains. The molecular biologist must begin identifying specific genetic sequences coding for the proteins that regulate lipid domains in membranes and for proteins that direct lipid trafficking from intracellular sites of synthesis. Contributions from all of these fields are necessary if our understanding of membrane protein function and lipid metabolism in normal physiology and disease states is to advance.

## Bibliography

Almeida, P., Vaz, W., and Thompson, T. (1992). Lateral diffusion and percolation in two-phase, two-component lipid bilayers. Topology of the solid-phase domains in-plane and across the lipid bilayer. *Biochemistry* **31**, 7198–7210.

Bergelson, L. (1992). Lipid domain reorganization and receptor events. Results obtained with new fluorescent lipid probes [letter]. [Review]. *FEBS Letts.* **297**, 212–215.

Bogdanov, M., Sun, J., Kaback, H. R., and Dowhan, W. (1996). A phospholipid acts as a chaperone in assembly of a membrane transport protein. *J. Biol. Chem.* **271**, 11615–11618.

Bretscher, M. S., and Munro, S. (1993). Cholesterol and the Golgi apparatus. *Science* **261**, 1280–1281.

Bretscher, M. (1972). Asymmetrical lipid bilayer structure for biological membranes. *Nature New Biol.* **236**, 11–12.

Buton, X., Morrot, G., Fellman, P., and Seigreuret, M. (1996). Ultrafast glycerophospho lipid selective transbilayer motion mediated by a protein in the endoplasmic reticulum membrane. *J. Biol. Chem.* **271**, 6651–6657.

Chen, S., Kulju, D., Halt, S., and Murakami, K. (1992). Phosphatidylcholine-dependent protein kinase C activation. Effects of *cis*-fatty acid and diacylglycerol on synergism, autophosphorylation and $Ca(2+)$-dependency. *Biochem. J.* **284**, 221–226.

Devaux, P. F., and Zachowski, A. (1994). Maintenance and consequences of membrane phospholipid asymmetry. *Chem. Phys. Lip.* **73**, 107–120.

Dolis, D., De Kroon, A. I. P. M., and De Kruijff, B. (1996). Transmembrane movement of phosphatidylcholine in mitochondrial outer membrane vesicles. *J. Biol. Chem.* **271**, 11879–11883.

Dudeja, P. K., Harig, J. M., Wali, R. K., Knaup, S. M., Ramaswamy, K., and Brasitus, T. A. (1991). Differential modulation of human small intestinal brush-border membrane hemileaflet fluidity affects leucine aminopeptidase activity and transport of D-glucose and L-glutamate. *Arch. Biochem. Biophys.* **284**, 338–345.

Epand, R., Stafford, A., Wang, J., and Epand, R. (1992). Zwitterionic amphiphiles that raise the bilayer to hexagonal phase transition temperature inhibit protein kinase C. The exception that proves the rule. *FEBS Lett.* **304**, 245–248.

Fontaine, R., Harris, R., and Schroeder, F. (1980). Aminophospholipid asymmetry in murine synaptosomal plasma membrane. *J. Neurochem.* **34**, 269–277.

Glaser, M., Wanaski, S., Buser, C. A., Boguslavsky, V., Ashidzada, W., Morris, A., Rebecchi, M., Scarlata, S. F., Runnels, L. W., Prestwich, G. D., Chen, J., Aderem, A., Ahn, J., and McLaughlin, S. (1996). Myristoylated alanine-rich C kinase substrate (MARCKS) produces reversible inhibition of phospholipase C by sequestering phosphatidylinositol 4.5-bisphosphate in lateral domains. *J. Biol. Chem.* **271**, 26187–26193.

Goormaghtigh, E., Huart, P., Praet, M., Brasseur, R., and Ruysschaert, J. (1990). Structure of the adriamycin–cardiolipin complex. Role in mitochondrial toxicity. [Review]. *Biophys. Chem.* **35**, 247–257.

Harris, J. S., Epps, D. E., Davio, S. R., and Kezdy, F. J. (1995). Evidence for transbilayer, tail-to-tail cholesterol dimers in dipalmitoylglycerophosphocholine liposomes. *Biochem.* **34**, 3851–3857.

Haverstick, D., and Glaser, M. (1987). Visualization of $Ca^{2+}$-induced phospholipid domains. *Proc. Nat. Acad. Sci. USA* **84**, 4475–4479.

Houslay, M. D., and Stanley, K. K. (1982). "Dynamics of Biological Membranes." John Wiley & Sons, New York.

Igbavboa, U., Avdulov, N. A., Schroeder, F., and Wood, W. G. (1996). Increasing age alters transbilayer fluidity and cholesterol asymmetry in synaptic plasma membranes of mice. *J. Neurochem.* **66,** 1717–1725.

Mantripragada, B., Sankaram, B., and Thompson, T. E. (1991). Cholesterol-induced third-phase immiscibility in membrane. *Proc. Natl. Acad. Sci. USA* **88,** 8689–8690.

Morris, A., and Frizzell, R. (1994). Vesicle targeting and ion secretion in epithelial cells: Implications for cystic fibrosis. [Review]. *Annu. Rev. Physiol.* **56,** 371–397.

Morrow, M., Singh, D., Lu, D., and Grant, C. (1992). Glycosphingolipid phase behaviour in unsaturated phosphatidylcholine bilayers: A $^2$H-NMR study. *Biochim. Biophys. Acta* **1106,** 85–93.

Mukherjee, S., and Chattopadhyay, A. (1996). Membrane organization at low cholesterol concentrations: A study using NBD-labeled cholesterol. *Biochem.* **35,** 1311–1322.

Negrete, H. O., Rivers, R. L., Gough, A. H., Colombini, M., and Zeidel, M. L. (1996). Individual leaflets of a membrane bilayer can independently regulate permeability. *J. Biol. Chem.* **271,** 11627–11630.

Nichols, C. P., and Roufogalis, B. D. (1991). Influence of associated lipid on the properties of purified bovine erythrocyte acetylcholinesterase. *Biochem. Cell Biol.* **69,** 154–162.

Nicolay, K., Sautereau, A., Tocanne, J., Brasseur, R., Huart, P., Ruysschaert, J., and De Kruijff, B. (1988). A comparative model membrane study on structural effects of membrane-active positively charged anti-tumor drugs. *Biochim. Biophys. Acta* **940,** 197–208.

Oberhauser, A., Monck, J., and Fernandez, J. (1992). Events leading to the opening and closing of the exocytotic fusion pore have markedly different temperature dependencies. Kinetic analysis of single fusion events in patch-clamped mouse mast cells. *Biophys. J.* **61,** 800–809.

Parsegian, V. A. (1995). The cows or the fence? *Mol. Membr. Biol.* **2,** 5–7.

Parton, R. G., and Simons, H. (1995). Digging into caveolae. *Science* **269,** 1398–1399.

Pryer, M., Wuestehube, L., and Schekman, R. (1992). Vesiclemediated protein sorting. [Review]. *Annu. Rev. Biochem.* **61,** 471–516.

Sankaran, M. B., and Thompson, T. E. (1990). Interaction of cholesterol with various glycerophospholipids and sphingomyelin. *Biochem.* **29,** 10670–10675.

Schatz, G., and Dobberstein, B. (1996). Common principles of protein translocation across membranes. *Science* **271,** 1519–1526.

Schekman, R., and Orci, L. (1996). Coat proteins and vesicle budding. *Science* **27,** 1526–1531.

Schroeder, F. (1984). Role of membrane lipid asymmetry in aging. [Review]. *Neurobiol. Aging* **5,** 323–333.

Schroeder, F., and Nemecz, G. (1990). Transmembrane cholesterol distribution. *In* "Advances in Cholesterol Research" (M. Esfahami and J. Swaney, Eds.), p. 47. Telford Press, West Caldwell, NJ.

Schroeder, F., and Sweet, W. D. (1988). The role of membrane lipid and structure asymmetry on transport systems. *In* "Advances in

Biotechnology of Membrane Ion Transport" (P. L. Jorgensen and R. Verna, Eds.), p. 183. Serono Symposia, New York.

Schroeder, F., and Wood, W. G. (1995). Lateral lipid domains and membrane function. *In* "Cell Physiology Source Book" (N. Sperelakis, Ed.), p. 36. Academic Press, New York.

Schroeder, F., Jefferson, J. R., Kier, A. B., Knittell, J., Wood, W. G., Scallen, T. J. A., and Hapala, I. (1991). Membrane cholesterol dynamics: Cholesterol domains and nonexchangeable pools. *Proc. Soc. Exp. Biol. Med.* **196,** 235–252.

Schroeder, F., Frolov, A., Murphy, E., Atshaves, B. P., Jefferson, J. R., Pu, L., Wood, W. G., Foxworth, W. B., and Kier, A. B. (1996a). Recent advances in membrane cholesterol domain dynamics. *Proc. Soc. Exp. Biol. Med.* **213,** 149–176.

Schroeder, F., Frolov, A. A., Murphy, E. J., Atshaves, B. P., Pu, L., Wood, W. G., Foxworth, W. B., and Kier, A. B. (1996b). Recent advances in membrane cholesterol domain dynamics. *Proc. Soc. Exp. Biol. Med.* **213,** 149–176.

Schulthess, G., and Hauser, H. (1995). A unique feature of lipid dynamics in small intestinal brush border membrane. *Mol. Membr. Biol.* **12,** 105–112.

Sekimizu, K. (1994). Interactions between DNA replication-related proteins and phospholipid vesicles *in vitro* [Review]. *Chem. Phys. Lipids* **73,** 223–230.

Singer, S. J., and Nicolson, G. L. (1972). The fluid mosaic model of the structure of cell membranes. *Science* **175,** 720–731.

Smaby, J. M., Momsen, M., Kulkarni, V. S., and Brown, R. E. (1996). Cholesterol-induced interfacial area condensations of galactosylceramides and sphingomyelins with identical acyl chains. *Biochem.* **35,** 5695–5704.

Sunshine, C., and Mcnamee, M. (1992). Lipid modulation of nicotinic acetylcholine receptor, function: The role of neutral and negatively charged lipids. *Biochim. Biophys. Acta* **1108,** 240–246.

Sweet, W. D., and Schroeder, F. (1988). Lipid domains and enzyme activity. *In* "Advances in Membrane Fluidity: Lipid Domains and the Relationship to Membrane Function" (R. C. Aloia, C. C. Cirtain, and L. M. Gordon , Eds.), p. 17. Alan R. Liss, New York.

Van den Bosch, H., Schutgens, R. B. H., Wanders, R. J. A., and Tager, J. M. (1992). Biochemistry of peroxisomes. *Annu. Rev. Biochem.* **61,** 157–197.

Warren, G. (1993). Membrane partitioning during cell division. [Review]. *Annu. Rev. Biochem.* **62,** 323–348.

Williamson, P., Bevers, E., Smeets, E., Comfurius, P., Schlegel, R., and Zwaal, R. (1995). Continuous analysis of the mechanism of activated transbilayer lipid movement in platelets. *Biochem.* **34,** 10448–10455.

Wood, W. G., and Schroeder, F. (1992). A new approach to understanding how ethanol alters brain membranes. *In* "Alcohol and Neurobiology: Receptors, Membranes, and Channels" (W. G. Wood and F. Schroeder, Eds.), p. 161. CRC Press, Boca Raton, FL.

Yeagle, P. (1987). "The Membranes of Cells." Academic Press, New York.

Donald G. Ferguson

# 5

## Ultrastructure of Cells

### I. Introduction

To completely understand the fundamental mechanisms that constitute the normal activities of any organism, it is essential to consider the structural correlates that underlie these activities. Among the earliest observations that form the foundation of our current understanding of morphology was that by Hooke in the 1660s, who examined thin slices of cork using a crude light microscope and realized that the cork was composed of small compartments, or *cells*. These original observations were gradually refined by the work of a large number of light microscopists. One of the key events was the independent postulation by Schleiden and Schwann in the late 1830s that all living organisms are composed of cells. This has come to be recognized as the *cell theory*. Technological developments in sample preparation, staining, and the quality of the microscopes led to a tremendous surge just prior to the turn of the twentieth century. An important question that was under study was the nature of the barrier that separated cells from each other and from the environment. A major breakthrough in resolving this question was made independently by Gorter and Grendel in the 1920s and Davson and Dainelli in the 1930s. Davson and Danielli examined the behavior of phospholipids in aqueous environments and determined that the most favorable configuration energetically was a bilayer with the hydrophilic polar headgroups located at the water interface and the hydrophobic nonpolar tails in contact with each other. Such a cell membrane (*plasmalemma*) formed by a phospholipid bilayer would be impermeable to fluids and charged species such as ions. Thus, it would retain the cell contents separate from the extracellular environment.

It was recognized that proteins were closely associated with the phospholipid bilayer, and in the early models these were proposed to form a layer on either side of the membrane (i.e., a protein–lipid sandwich). This model remained hypothetical until the introduction and refine-

ment of the electron microscope (EM) in the 1930s and 1940s, for which Ruska was awarded the Nobel Prize. This instrument, which uses a beam of electrons rather than electromagnetic radiation in the visible spectrum to produce images, permitted the examination of tissues at magnifications much greater than had been possible with even the best light microscopes. In fact, the resolution was so improved that the morphological details observed using the EM are often referred to as the *ultrastructure* of the cells. In the EM, it was apparent that all cells were bounded by a trilaminar membrane, 70 Å thick, that consisted of two darker lines separated by a relatively clear inner region. This structure was called the *unit membrane* by J. D. Robertson in the early 1950s. It had been recognized even by Davson and Danielli that many of the proteins associated with the cell membrane were tightly attached because they were not extractable by changing the ionic strength of the bathing solutions or even by mild detergent treatment. Thus, the notion of a phospholipid bilayer having a coating of protein evolved in the 1960s to the *fluid mosaic model* proposed by Singer and Nicholson (Fig. 1). In this model, there were *peripheral* (or *extrinsic*) and *integral* (or *intrinsic*) membrane proteins. Peripheral proteins are associated with the inner and outer surfaces of the membrane, but are not anchored in it. Integral membrane proteins are actually embedded in at least one leaflet of the phospholipid bilayer, and some traverse the whole thickness of the membrane.

The fluid mosaic model was more consistent with functional studies that clearly demonstrated that cell membranes were semipermeable, rather than impermeable. The integral membrane proteins included ion channels acting as low-resistance pores, as well as energy-dependent pumps. This could better account for the establishment of ionic gradients and the influx or efflux of ions under various conditions. This fluid mosaic model is now universally accepted.

The extrinsic proteins are membrane-associated rather than embedded in the lipid bilayer. The *extracellular matrix*

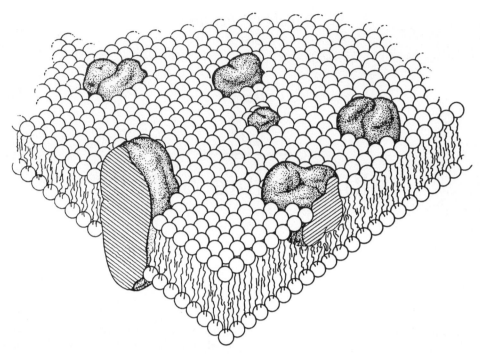

**FIG. 1.** Fluid mosaic model of cell membrane. This diagram illustrates the organization of the phospholipid bilayer with the polar headgroups facing the fluid compartments of the cytoplasm and the extracellular space. The nonpolar tails of the phospholipids form the inner part of the bilayer. Integral membrane proteins are embedded in the bilayer, and some span the whole thickness of the membrane. [Reproduced with permission from Singer, S. J., and Nicholson, G. L. (1972). *Science* **175,** 720–731.]

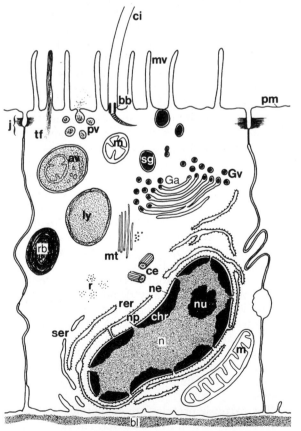

**FIG. 2.** Diagram of the ultrastructure of a typical cell. This diagram provides a two-dimensional representation of the major organelles of an epithelial cell, based on their appearance in electron micrographs. The cell's contents, the cytoplasm and the organelles, are enclosed by the plasma membrane (pm). The central nucleus (n) contains a nucleolus (nu) and clumped chromatin (chr). The nucleus is bounded by a double bilayer membrane that is punctuated by discontinuities called nuclear pores (np). The outer nuclear membrane is studded with small dots representing membrane-attached ribosomes and is continuous with tubules of the rough endoplasmic reticulum (rer). Free ribosomes (r) are also found in the cytoplasm. Elements of smooth endoplasmic reticulum (ser) are observed as membranous profiles with no ribosomes attached to the surface. A Golgi apparatus (Ga) with associated vesicles (Gv) and secretory granules (sg) is located just above the nucleus. Several types of lysosomes are also depicted; a primary lysosome (ly), an autophagic vacuole (av), and a residual body (rb). Several pinocytotic vesicles (pv) are depicted at the apical surface of the cell. Two double-membraned mitochondria (m) are depicted, one above the nucleus and a large one below it. Cytoskeletal elements include actin-containing thin filaments (tf) that form the core of the apical membrane specializations called microvilli (mv). Microtubules (mt) are depicted as cytoskeletal elements and as specialized microtubule organizing structures: the centriole (ce) and the basal body (bb) or a single cilium (ci) at the apical surface of the cell. A complex of junctions (j) that restrict movement of ions and solutes between adjacent cells is depicted at the lateral borders of the cell near the apical surface. The basal lamina (bl) is a layer of fibrous tissue that underlies the cell. It is part of the extracellular matrix that regulates ion flow and provides strong connections to the deeper connective tissue layers. Electron micrographs of each of these organelles are presented in subsequent figures. [Adapted with permission from A. Bubel. (1989). "Microstructure and Function of Cells," Figure 1a. Ellis Horwood Ltd., Chichester, UK.]

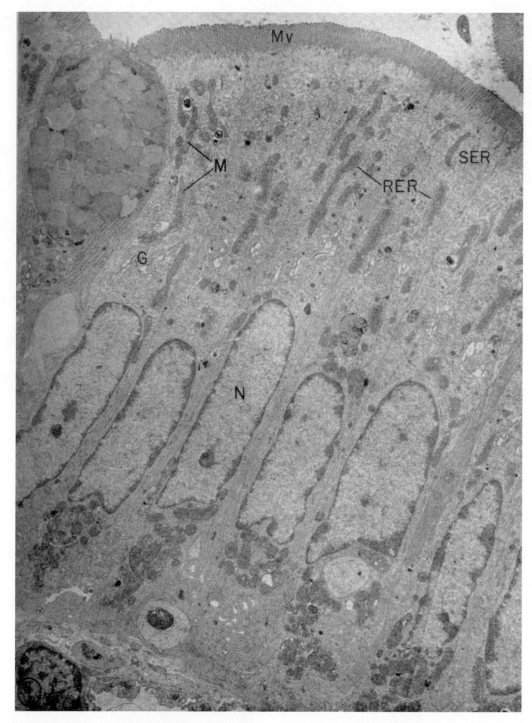

**FIG. 3.** Electron micrograph of epithelial cells lining the lumen of the intestine. This low-magnification electron micrograph illustrates the appearance of an epithelial cell such as the one that is depicted diagrammatically in Fig. 2. Note the nucleus (N), the rough endoplasmic reticulum (RER), the smooth endoplasmic reticulum (SER), the mitochondria (M), and the apical membrane specializations the microvilli (MV) are the most obvious features at this magnification. [Reproduced with permission from R. R. Cardell, J., *et al.* (1957). *J. Cell Biol.* **34,** 123–155.]

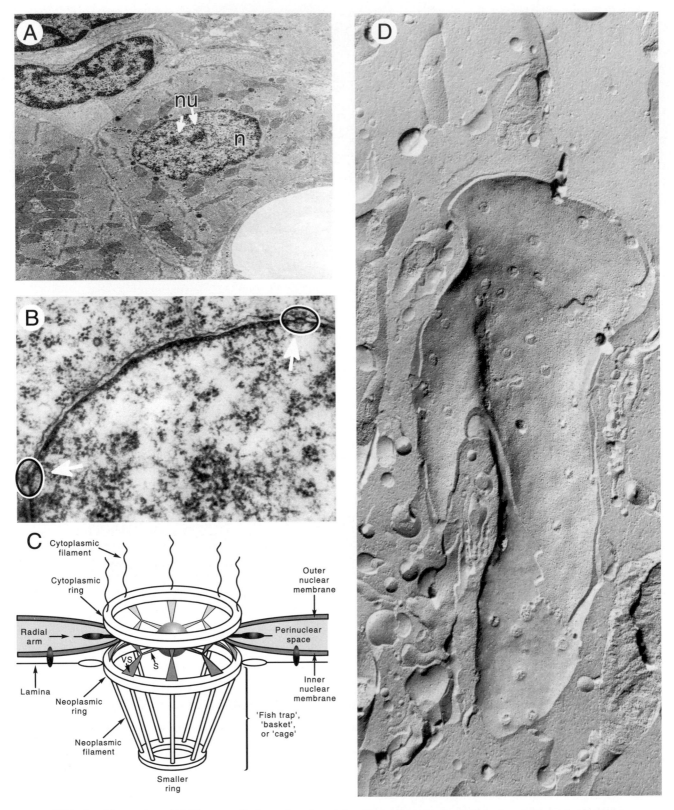

**FIG. 4.** The nucleus. (A) The centrally located nucleus (n), with one or more nucleoli (nu) and chromatin dispersed throughout the nucleoplasm and clumped on the inner surface of the nuclear membrane, is the most obvious feature of most cells. (B) This higher magnification image of the nuclear envelope better illustrates the double membrane enclosing the perinuclear space and shows two circled nuclear pores (arrows). (C) This diagram depicts the details of the nuclear pore complex. There are cytoplasmic and inner (neoplasmic) proteinaceous rings attached to each other by eight vertical

is a layer of varying thickness (depending on the cell type) on the outer surface. Historically, this coating was considered to be an inert structural framework composed of collagen and elastic fibrils. However, recent evidence points to a more active role, with proteins such as fibronectin and laminin active in stimulating a number of cellular activities. This likely is mediated through interactions with intrinsic membrane proteins, which in turn are connected to and influence the extrinsic membrane proteins that form subplasmalemmal networks on the inner surface of the cell membrane. Ultimately, signals from outside the cell may be communicated deep within the cell via the elements of the cytoskeleton (which is discussed in more detail later) and even can influence gene activity in the nucleus.

The contents of the cell which are confined by the plasmalemma are called the *cytoplasm*. The cytoplasm is not an amorphous soup, but rather contains a number of distinct structures known as *organelles* (by analogy to the different organs of the body). Figure 2 is a diagrammatic representation and Fig. 3 is a complementary electron micrograph of a typical animal cell and its organelles. The organelles serve to compartmentalize different cellular activities (as the body organs perform different functions).

## II. Nucleus

The most prominent organelle is readily visible in the light microscope, and since it is more often located centrally, it was named the *nucleus*. The nucleus contains most of the cell's DNA, and it is known to be the compartment that contains the genetic apparatus. In most cells, the nucleus is an ovoid structure, but it can take a variety of shapes and sizes in electron micrographs (Fig. 4A). The earliest EM observations noted that the nucleus was bounded by a double bilayer, the *nuclear envelope,* and often had a dense region within the nucleus that was called the *nucleolus*. In addition, the nucleus contained clumps of darker material called *heterochromatin* that were dispersed throughout the nucleoplasm and also were associated with the inner face of the nuclear envelope. Beyond this, the regional specializations of the nucleoplasm are still relatively poorly understood. Recent studies that have used antibody localization of nuclear-specific proteins demonstrate that there are domains within the nucleoplasm. These domains are not grossly obvious even at the ultrastructural level, but they appear to be organized by attachment to the filamentous proteins of the nuclear matrix. On the basis of the compartmentation of DNA polymerases and RNA splicing factors, it has been suggested that discrete domains appear to be involved in different processes, such as transcription and translation of the genetic material.

The nuclear envelope consists of a proteinaceous lamina inside (composed of cytoskeletal proteins termed *lamins*) and a double bilayer that is interrupted at periodic intervals by numerous discontinuities, the *nuclear pore complexes* (Fig. 4). The nuclear lamina is not simply a scaffolding for the inner membrane. It acts as an attachment site for chromatin and appears to be responsible for regulating the formation of the nuclear domains. Detailed studies have demonstrated that the nuclear pore complexes are rather complicated structures, containing a central aqueous channel 100 mm in diameter that allows passive diffusion of small molecules and ions. The complex is composed mostly of proteins called nucleoporins that are highly conserved and consists of a central spoke assembly with several arms and two peripheral rings: one on the cytoplasmic side and one on the nucleoplasmic side. The central channel complex is attached to the spokes. The nuclear envelope is not a complete barrier, because RNA, proteins, and ions need to pass in and out. It is the nuclear pore complex that is responsible for, and regulates, the trafficking of ions, proteins, and RNA.

As noted earlier, the nuclear envelope is a double bilayer. The outer membrane is often observed to be lined with spherical RNA-containing particles called *ribosomes* and is continuous with a network of cytoplasmic membranes called the *endoplasmic reticulum* (ER).

## III. Endoplasmic Reticulum

The ER was described by light microscopists in the late 1800s using special preparative procedures that selectively infiltrated this organelle with stain. Our understanding of the structure of the ER was greatly advanced by the elegant EM work of Porter and Palade in the late 1940s and early 1950s, who demonstrated that the ER was an extensive network of interconnecting tubules. More recently, the ER has been shown to be organized on a framework of microtu-

---

supports (vs). The central part of the pore is not a simple opening, but contains a central transporter assembly supported by spokes (s) that are attached to each of the vertical supports. The central transporter regulates translocation of proteins and ribonucleic acids across the nuclear envelope. The nucleoplasmic ring is attached by filaments to an inner smaller proteinaceous ring and is also believed to be attached to the inner nuclear lamina. Intermediate filaments are associated with the outer cytoplasmic ring. The edges of the nuclear pore complex have radial arms that often appear to extend into the perinuclear space between the two bilayers of the nuclear envelope. [Reproduced with permission from Figure 2b, L. E. Maquat (1991). *Curr. Opin. Cell Biol.* **3**, 1004–1012. (D) This image of the nuclear envelope was produced by freeze-fracture. This technique involves freezing the tissue, cracking the frozen cells open to expose their contents, and evaporating a heavy metal such as plantinum to form a thin replica of the exposed surfaces. This approach is particularly effective for examining the ultrastructure of membranes and the integral membrane proteins. In this image, the nuclear pores are seen as round structures covering the nucleus. In the lower portion of this nucleus, the fracture plane is in the outer nuclear envelope and the pores appear to drop down into the membrane. In the upper portion of this nucleus, the fracture plane is in the inner nuclear envelope and the pores appear to project up from the surface.

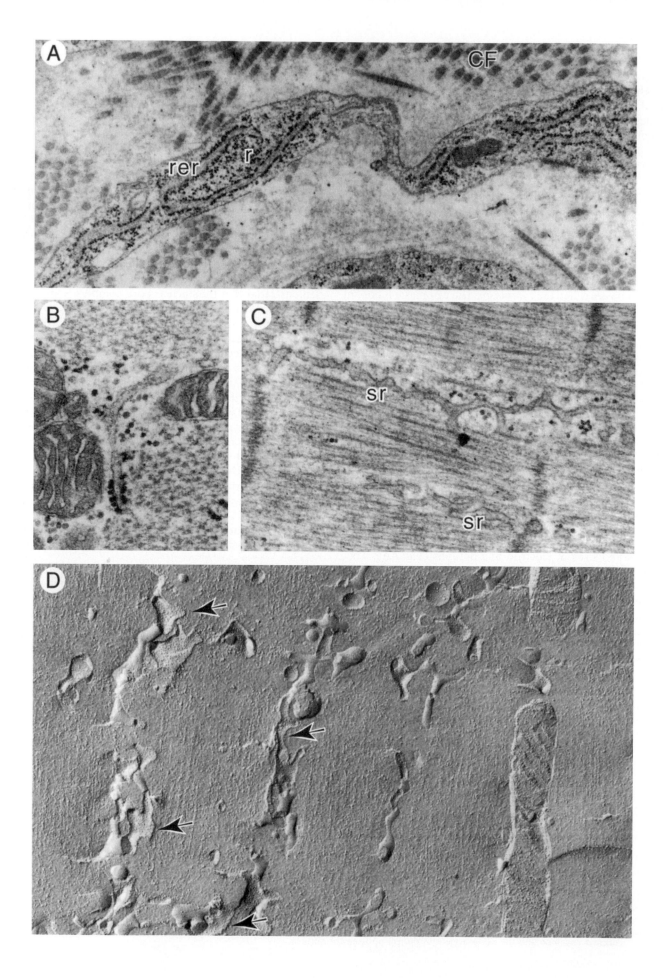

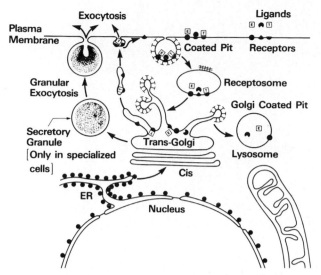

**FIG. 6.** This diagram illustrates involvement of the Golgi apparatus–endoplasmic reticulum–lysosomal system (GERL) in exocytosis and endocytosis. It also depicts the relationship of the Golgi apparatus to the nuclear envelope and the endoplasmic reticulum (ER). The Golgi apparatus modifies secretory products, packages them in vesicles, and exports them by way of exocytosis. The Golgi also generates lysosomes that are involved in the process of receptor-mediated endocytosis. Ligands bind to receptors at the cell surface; the receptor–ligand complexes migrate to specialized coated pits that are internalized and fuse with lysosomes for digestion. [Reproduced with permission from I. Pastan and M. C. Willingham. (1985). "Endocytosis," Chapter 1, Figure 1. Plenum Press, New York.]

bules (elements of the cytoskeleton that are described in more detail later).

Ribosomes (Fig. 5) are compact rounded particles, approximately 300 nm in diameter, that dissociate into two smaller units in the presence of low $Mg^{2+}$. They consist of ribosomal RNA, but in eukaryotes as much of 50% of the ribosomal mass may be composed of associated proteins. Ribosomes are often seen organized into strands or rosettes called *polysomes*. These polysomes are aggregations of ribosomes active in the messenger RNA-directed linkage of amino acids to form peptide chains. Protein synthesis is initiated on ribosomes in the cytoplasm that subsequently bind to the ER to form what is known as *rough endoplasmic reticulum* because of the beaded nature of the decorated membranes (see Fig. 5). The rough ER is presumed to be

involved exclusively in protein synthesis. In contrast, ER that does not have ribosomes on its surface is known as *smooth endoplasmic reticulum* (see Fig. 5) and has been involved in a number of different functions depending on the cell type. In steroid hormone-secreting cells, smooth ER is associated with the production of secretory products. In liver cells, the smooth ER has been involved in glycogen metabolism. In muscle, the smooth ER is known as the sarcoplasmic reticulum (SR), and the SR membrane proteins are responsible for the uptake, sequestration, and release of $Ca^{2+}$ during excitation contraction coupling and relaxation. There is also now evidence that at least some of the ER in the nonmuscle cells is involved in $Ca^{2+}$ handling.

## IV. Golgi Apparatus

Synthesis of membrane proteins and secretory products is completed in a continuation of the ER called the *Golgi apparatus*. This organelle was named for Camillo Golgi, a nineteenth-century microscopist who developed many of the staining techniques and described many of the light-microscopic features of cells. The Golgi apparatus does not stain with dyes that are conventionally used for light microscopy, so it is generally observed as a clear area or "negative" image close to the nucleus. In the electron microscope, this organelle was composed of flat, disc-shaped membranous elements with a hollow lumen that are known as cisternae (singular cisterna). In each Golgi apparatus, several curved cisternae are organized into stacks, with individual cisterna interconnected by networks of tubules. As secretory products pass through the Golgi apparatus, they undergo chemical modifications such as glycosylation, proteolytic cleavage, phosphorylation, and sulfation. Other functions performed by the Golgi apparatus include carbohydrate metabolism, targeting of plasmalemmal proteins (pumps, channels, and receptors), and the condensation of secretory materials.

Although a variety of terminologies have been used to identify different regions of the Golgi apparatus, there are generally four unrecognized morphological subsections (see diagram in Fig. 6 and electrom micrograph in Fig. 7A). Since the cisternae are slightly curved, the Golgi complexes have a concave and a convex face. The convex or *cis face* is

**FIG. 5.** The endoplasmic reticulum. (A) The centrally located cells in this electron micrograph are fibroblasts. One of the major activities of these connective tissue cells is to synthesize and export large quantities of collagen. This extracellular matrix protein is incorporated into the fibrils (CF) that form the robust framework between cells in connective tissue. As might be predicted for cells that are active in protein synthesis, fibroblasts are packed with free ribosomes (r) and rough endoplasmic reticulum (rer). The RER is observed as membrane profiles that have a beaded surface due to the attached ribosomes. (B) Although Fig. 3 and many other diagrams suggest that the rough and smooth endoplasmic reticulum (SER) are distinct compartments, it is clear from this micrograph that elements of the RER and SER may be continuous components of the same membrane system. (C) Smooth endoplasmic reticulum may perform different roles depending on the cell in which it is located (see text). This example of SER is the sarcoplasmic reticulum (sr) in a striated muscle cell. (D) This freeze-fracture image of the SR demonstrates that the smooth ER is not simply a lipid bilayer but may be packed with integral membrane proteins. The particles in the cytoplasmic leaflet of the SR (arrowheads) have been shown to be the $Ca^{2+}$ pump protein that is responsible for removing $Ca^{2+}$ from the cytoplasm during relaxation.

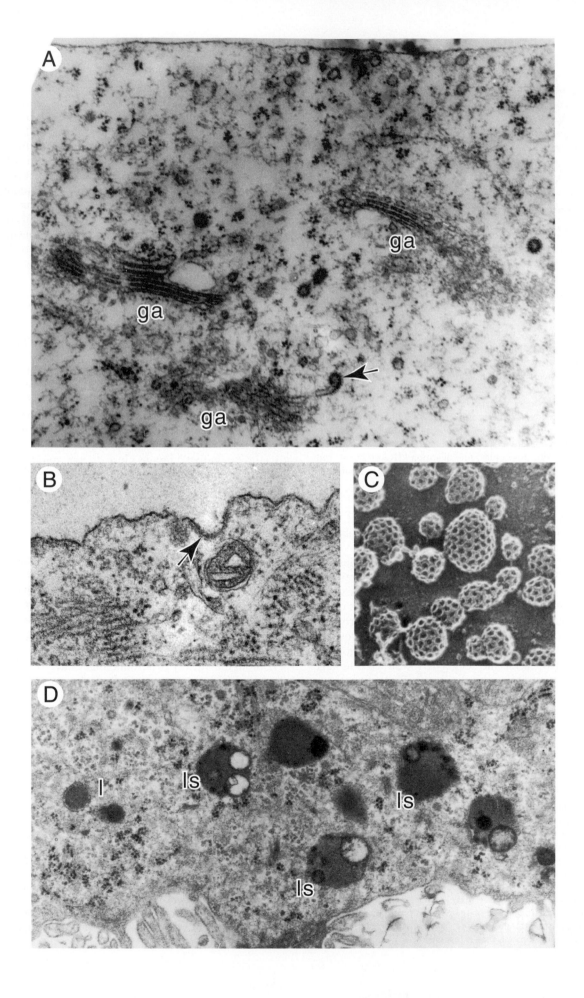

also called the *forming* or *immature face,* because it is the point at which newly synthesized proteins enter the Golgi apparatus. The medial or intermediate region is composed of the variable number of cisternae in the middle of the stack. The concave or *trans face* is also called the *mature* or *secretory face,* because it is the site at which large secretory vesicles bud off after their contents have been modified for export. The *trans Golgi network* is a reticulum of tubules emanating from the *trans* face and is thought to be associated with the lysosomal system (see next section).

This general scheme is common to all Golgi complexes, and transport is generally directional, from the *cis* to the *trans* face. However, more recent evidence suggests that maturation and pinching off is not restricted to the *trans* face. There is likely much more shuttling back and forth of material than was originally believed. The numerous vesicles (of variable shapes and sizes) in the vicinity of the Golgi complex may be transporting proteins along the pathway during maturation and processing. There is also evidence that there is also some retrograde transport from the Golgi apparatus to the ER. The continuity of the Golgi complex with the ER has been clearly established, but whether this occurs through a direct connection or by way of vesicular traffic is not yet certain. The prevailing notion is not that of a rigid morphology, but rather that there is a much more labile, flexible arrangement.

## V. Lysosomes

In addition to packaging and modifying secretory products, the Golgi apparatus plays a role in the formation of another important cellular organelle, the *lysosome* (Fig. 7D). In fact, the apparent continuity of the ER, the Golgi apparatus, and the lysosomes has led to the commonly used terminology *GERL* (Golgi–endoplasmic reticulum–lysosomal network). Lysosomes are a heterogeneous population of membrane-limited vesicles, differing in size and density, that were identified and characterized by the pioneering work of De Duve in the mid-1950s. Lysosomes contain hydrolytic enzymes that break down virtually every form of biological material, and this organelle thus acts as the cell's digestive system. The lysosomal enzymes include proteases (e.g., collagenase, acid phosphatases, cathepsins), lipases (e.g., esterases, phospholipases), glycosidases (e.g., hyaluronidase, lysozyme, galactosidases), and nucleases (e.g., ribo- and deoxyribonucleases). Many of the lysosomal enzymes are more effective at low pH, and the lumen of the lysosome is more acidic than the cytoplasm. Thus, the lysosome acts to recycle intracellular components, such as the cytoskeletal proteins and old organelles (*autophagy*), as well as to digest extracellular material that is trapped in phagocytic vesicles endocytosed as internalized plasmalemma (*heterophagy*). The newly formed primary lysosomes are more uniform in size and have an amorphous electron-dense content when observed in the electron microscope (Fig. 7D). After these organelles fuse with phagocytic vesicles to form secondary lysosomes, their size and density is more variable (Fig. 7D). Lysosomes containing indigestible material often remain in the cytoplasm as residual bodies.

There are several different mechanisms whereby extracellular contents are taken up by cells and directed to the lysosomal compartment. During *pinocytosis* (cellular drinking) the plasmalemma pinches off small ($0.1–0.2$-$\mu$m diameter) vesicles that contain fluid rather than solid material. *Phagocytosis* is the internalization of solid material from the extracellular space and is often mediated by receptor proteins in the plasmalemma. This process is known as *receptor-mediated endocytosis* (see Fig. 6) and can be accomplished for a wide variety of receptor–ligand combinations. The ligands include nutrients too large to diffuse across the plasmalemma (e.g., iron, cholesterol) and extracellular matrix components. Additional ligands include hormones, growth factors, viruses, toxins, and immune complexes. Once receptors bind their specific ligand they migrate to specialized flask-shaped *"coated"* pits in the plasmalemma (Fig. 7B) that are lined with a unique 180-kDa protein called *clathrin.* The coated pits then pinch off to form *endosomes* (Fig. 7D), which are observed to be a complex formed from a small central vesicle with several small tubules attached to it. The lumen of the endosomes is an acid environment that causes the dissociation of receptors and ligands. A certain amount of sorting may be accomplished in the endosomes. Receptors may be recycled to the plasmalemma to be available for binding more ligand. Under certain conditions, receptors may be routed to the lysosomal compartment and this effectively downregulates any activity that may be stimulated by that particular ligand–receptor interaction. Although most cells are capable of endocytosis, this activity is most prevalent in specialized immune cells, such as macrophages, and blood leukocytes, such as neutrophils. Other cells that have higher endocytic activity are those that are exposed to higher levels of foreign material, such as hepatocytes and epithelial cells.

**FIG. 7.** The Golgi apparatus, coated pits, and lysosomes. (A) Three Golgi apparatuses (ga) are present in this electron micrograph. Note the *cis* face (c), the *trans* face (t), tightly stacked cisternae, and the numerous small vesicles at the edges of each golgi complex. A coated vesicle is observed budding off from one of the cisterna (arrowhead). (B) The plasma membrane shown in this electron micrograph contains a coated pit (arrowhead) responsible for receptor-mediated endocytosis. (C) Isolated coated vesicles have been frozen, deep etched, and replicated by shadowing. The clathrin framework of each vesicle is particularly well demonstrated in these images. (Reproduced with permission from B. Alberts *et al.,* "Molecular Biology of the Cell," 2nd edition, Figure 6-73.) (D) The cytoplasm of this liver cell is observed to contain several primary lysosomes (1) that have an amorphous electron dense appearance. Primary lysosomes fuse with cell contents or endocytic vesicles to form secondary lysosomes (ls) that are more variable in size and shape and contain material in various stages of digestion.

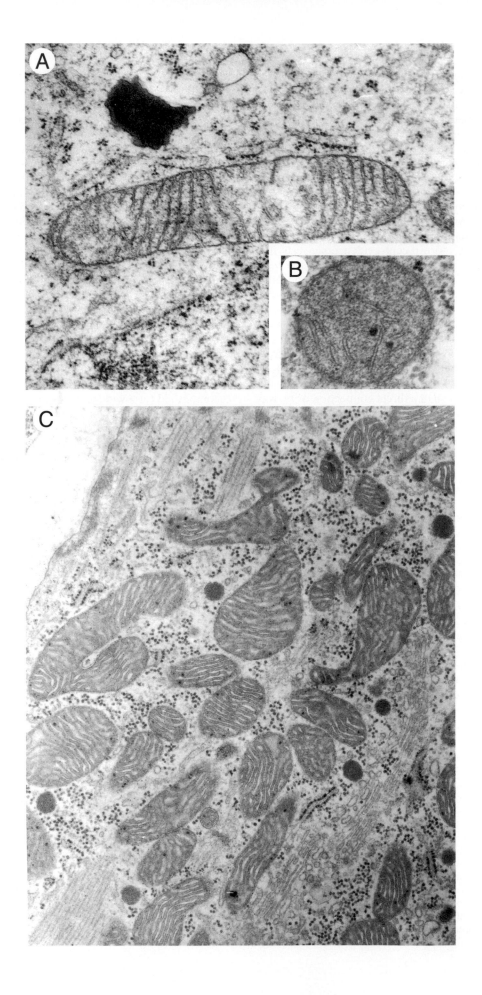

## VI. Mitochondria

Aside from the cell nucleus, the most obvious feature of most cells are the *mitochondria* (Fig. 8). This term was introduced at the turn of the twentieth century and is based on their appearance as threadlike granules in the light microscope. In the EM, each mitochondrion is observed to be limited by a double bilayer membrane. The outer bilayer is a simple membrane that encloses the entire organelle. The inner bilayer has a variable number (several to many) of infoldings called *cristae*, which serve to increase the surface area greatly. At high resolution, the cristae membranes appear to be studded with small raised bumps that have come to be known as *elementary particles.*

It is now clear that the mitochondria are the site of cellular respiration, and thus are the cell's center of energy production. Most of the details concerning the functional activities of the different regions of the mitochondrion has been determined following the development of procedures to isolate intact mitochondria and separate the inner and outer membranes into distinct membrane fractions. It has been shown that the outer membrane contains a large number of molecules of porin, a relatively nonselective channel, that allows the passage of ions and small molecules with a mass less than 10 kDa. This establishes an intermembrane compartment that has essentially the same ionic characteristics as the cytoplasm. In contrast, the folded inner membrane is rather impermeable to ions and encloses an internal compartment often referred to as the *mitochondrial matrix.* The inner mitochondrial membrane contains an abundance of transport proteins responsible for establishing a proton gradient between the intermembrane space and the inner matrix. The proton motive force created by this gradient is essential for driving many of the reactions of *oxidative phosphorylation,* a fundamental metabolic process localized to the inner compartment of the mitochondria. Oxidative phosphorylation generates much of the cell's adenosine triphosphate (ATP), the high-energy compound that is the source of energy for most cellular activities (for more details on the mechanism of this process, see Chapters 6 and 7). The enzymes responsible for the *tricarboxylic acid cycle,* one part of oxidative phosphorylation, are in the mitochondrial matrix, and the enzymes involved in *electron transport,* another major part of oxidative phosphorylation, are associated with the inner mitochondrial membrane. In fact, the elementary particles have been identified as the molecules of the ATPase-synthase complex.

As might be predicted, mitochondria are most abundant at sites within the cell that require high energy utilization.

They are also more abundant in cells that are more highly metabolic. For example, heart muscle tissue has one of the higher metabolic rates, and in myocardial cells, numerous mitochondria are sandwiched in between the contractile myofibrils (Fig. 8C).

Although the mitochondria are the organelles responsible for the bulk of the energy production in animal cells, plant cells contain an additional organelle, the chloroplast. For details on this organelle, please refer to Chapter 63.

## VII. Cytoskeleton

As mentioned earlier, the cytoplasm is not an aqueous solution in which organelles move about freely. Rather, the cell contents are more compartmentalized and there are regions specialized for particular functions. The network of *microfilaments* and *microtubules* that collectively constitute the *cytoskeleton* of each cell is the framework that underlies this organization. Normally 20–35% of the total protein of each cell is cytoskeletal, although this proportion is considerably greater in muscle where the cytoskeletal proteins form the contractile apparatus. In all cells, the cytoskeletal elements are involved in intracellular motility (mitotic movements of chromosomes, organelle translocation within cells, cytoplasmic streaming), cell locomotion, and maintenance of cell shape. More specialized functions are muscle contraction and ciliary and flagellar movement.

### A. Microfilaments

Microfilaments are composed of backbones of single proteins that polymerize to form long, slender filaments. The filaments are helical and polar, and they frequently interact with each other. There are three general categories of microfilaments: *Thin filaments* (Fig. 9A) are composed primarily of *actin* and are 6–8 nm in diameter; *intermediate filaments* (Fig. 9B) are formed from a variety of related proteins, depending on the different cell type, and they are approximately 10 nm in diameter; *thick filaments* (Fig. 9C) are composed primarily of *myosin* and are 12–15 nm in diameter.

### 1. Thin Filaments

As noted earlier, the backbone of these filaments is composed of a cyc oskeletal protein called actin, although regulatory proteins (e.g., *troponins*) and structural proteins (e.g., *tropomyosin*) are associated with the thin filaments in most cell types. Single molecules of *globular (G) actin*

---

**FIG. 8.** Mitochondria. (A and B). These moderately high-magnification electron micrographs of individual mitochondria demonstrate that each mitochondrion is limited by a single outer membrane. Mitochondria also have a second inner membrane that is highly folded into lamellae called cristae. This arrangement serves to enormously increase the surface area of the inner membrane. The darker granules in B are believed to be deposits of $Ca^{2+}$ salts. (C) The cells of highly metabolic tissues, such as the cardiac myocyte in this electron micrograph, are observed to contain numerous mitochondria. These mitochondria have expanded spaces inside the cristae and a relatively electron-dense matrix. Mitochondria having this appearance have been characterized as condensed mitochondria, and it has been suggested that they are in the active state.

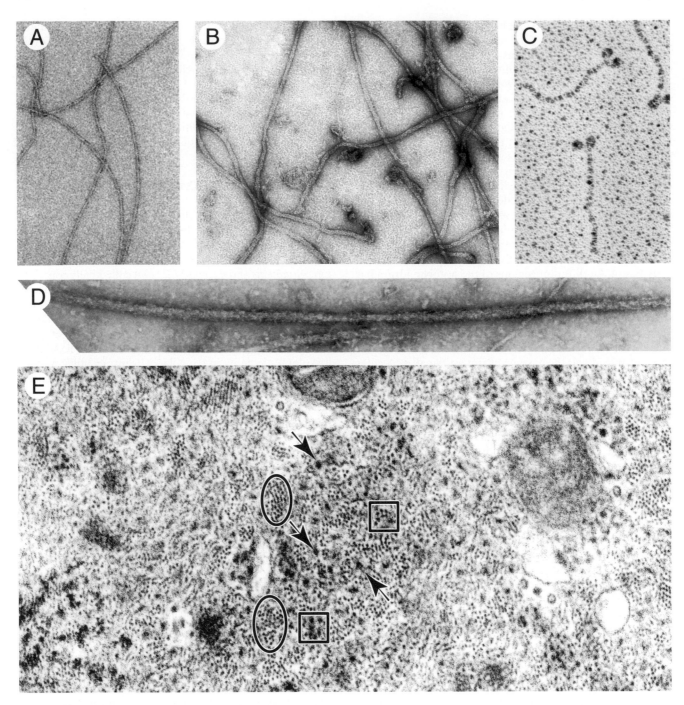

**FIG. 9.** Microfilaments. (A) Thin filaments. (B) Intermediate filaments. (C) Thick filament. The images of the three types of microfilament have been produced by the technique of negative staining. In this procedure, isolated filaments are adsorbed to a support film and covered with a heavy metal stain such as uranyl acetate. The stain dries down, forming an electron dense film that is excluded from space occupied by the proteinaceous filaments. Thus, a negative image is formed that can be examined using an electron microscope. These images clearly demonstrate the filamentous nature and the relative differences in diameter of the thin filaments (A), the intermediate filaments (B), and the thick filaments (C). Although not as obvious, the helical nature of the filaments is also visible. (D) This cross-section of a smooth muscle cell shows the relationship and relative diameters of thin filaments (enclosed in ovals), intermediate filaments (enclosed in squares), and thick filaments (arrowheads) as components of the contractile apparatus *in situ.* [(A) Adapted with permission from A. Bremer and U. Aebi (1992). *Curr. Opin. Cell Biol.* **4,** 20–26. (B) Micrograph provided by Dr. W. Ip. (C) Reproduced with permission from R. W. Kensler and M. Stewart. (1983). *J. Cell Biol.* **96,** 1797–1802. (D) Micrograph provided by Dr. J. D. Strauss.]

have a molecular mass ($M_r$) of 45 kDa and are approximately 55 nm in diameter. Detailed structural studies have demonstrated that the shape of the molecules is not really globular, but consists of a two-lobed dumbbell-shaped entity. G-actin polymerizes to form filaments and in this form is called *filamentous (F) actin*. Using special preparations for the EM, the structure of the thin filament has been determined. It is generally depicted as a double $\alpha$-helix 6–7 nm in diameter, and is of variable length depending on the cell type and the particular cytoskeletal formation with which it is associated. However, there is no evidence that the two strands of this double helix can disassociate or can grow independently. A number of associated proteins may regulate filament initiation and growth as well as the final length. For example, tropomyosin stabilizes thin filaments in muscle cells.

In the cells of different tissues, thin filaments are associated with a number of recognizable intracellular formations. Bundles of filaments all having the same polarity form the cores of the *microvilli* in epithelial cells and are found in smooth muscle and striated muscle cells as part of the contractile apparatus. Bundles of filaments with opposite polarities are also found in muscle cells, on opposite sides of the *Z-disc* (striated muscle) and *dense bodies* (smooth muscle), and in the *stress fibers* of nonmuscle cells. Three-dimensional networks are often found as cortical (subplasmalemmal) networks in many cells.

## 2. Thick Filaments

The most detailed structural information regarding the myosin-containing filaments is available for striated muscle thick filaments. Single myosin molecules consist of dimers of two *heavy chains* with an $M_r$ of approximately 200 kDa. Each heavy chain has a globular head portion and a long slender tail. One or two *light chains* (17–20 kDa) are associated with the head portion of each heavy chain. Thus, the whole myosin complex in striated muscle is observed to be a molecule with two globular heads and a tail approximately 150 nm long (Fig. 9D), having an $M_r$ of approximately 500 kDa. The tails of each myosin molecule are highly organized and form the backbone of the thick filament. The heads of the molecules project out at periodic intervals. These multistranded (three in vertebrate and four in insect striated muscle) $\alpha$-helical filaments are approximately 15 nm in diameter. Thick filaments in smooth muscle and nonmuscle cells have not been as well characterized. Smooth muscle cells clearly contain thick filaments, but the exact organization of the molecules in the filament is still controversial. It is likely that nonmuscle cells also contain thick filaments, but they are shorter and less obvious ultrastructurally.

It is generally accepted that thick filaments are necessary for movement and motility by interactions with actin-containing thin filaments. The details of this interaction are dealt with in considerable detail in Chapter 55.

## 3. Intermediate Filaments

The 10-nm intermediate filaments are formed from a related family of intermediate filament (IF) proteins, and they are found in virtually all cell types (e.g., the lamins are intermediate filament proteins that form the internal lamina of cell nuclei). However, there are tissue-specific IF proteins. *Vimentin* is a 53-kDa protein of mesenchymal tissue (connective tissue, bone, blood, cartilage); *desmin* is a 52-kDa protein of muscle; *neurofilament* (NF) *protein* is found in neurons and is observed as three molecular species: NF light (65 kDa); NF medium (105 kDa); and NF heavy (135 kDa). *Glial fibrillar acidic protein* is a 50-kDa intermediate filament protein found in astroglia cells; *keratins* are IF proteins that are abundant in epithelial cells.

All the IF proteins have common features of molecular structure that give rise to 10-nm filaments when assembled as polymers. Pairs of molecules are the basic subunits that are joined as coiled–coiled dimers. The dimers are assembled into tetrameric complexes about 2 nm in diameter, which then join end-to-end as *protofilaments*. Protofilaments subsequently assemble into 4.5 nm *protofibrils*, which join to form multistranded 10-nm IFs. The actual range of IF diameters is from 8 to 23 nm, but all are classed as 10-nm filaments. IFs are more stable than either thin filaments or thick filaments.

Intermediate filaments are most often observed as loose three-dimensional networks intermixed with other cellular components. For example, IFs form the framework of the inner nuclear membrane, the nuclear lamina. IFs are also attached to the plasma membrane at specialized cell–cell contacts (desmosomes) and are involved in linking actin filaments together into bundles in muscle and nonmuscle cells. Thus, the intermediate filaments are an internal scaffolding that can form networks to link peripheral and central components of the cell as a mechanically integrated complex.

## B. Microtubules

Microtubules are elongated hollow cylinders of larger diameter (25 nm) than other cytoskeletal elements (Fig. 10) and are composed of the protein *tubulin*. Tubulin has $\alpha$- and $\beta$-isoforms that have an $M_r$ of approximately 50 kDa. Each molecule is an $\alpha$–$\beta$ heterodimer of approximately 100 kDa. The walls of the microtubule are formed from approximately 5-nm protofilaments that are composed of linearly arranged pairs of alternating $\alpha$- and $\beta$-subunits. Protofilaments are aligned parallel to the long axis of the microtubule. The number of protofilaments that compose the microtubule cross-section is variable, but 13 is the most frequent. The protofilaments are polar structures and are assembled to assign polarity to the microtubule. The conditions that induce polymerization of the tubulin protofilaments include the presence of certain *microtubule-associated proteins (MAPs)*, GTP, $Mg^{2+}$, and low $Ca^{2+}$.

There are several different MAPs; for example, MAP1, MAP2, and Tau are associated with microtubules in a variety of cell types. The MAPs serve to stabilize the microtubules and they are the targets of regulatory signals. One of the most functionally important MAPs is *dynein*. Dynein is a multimolecular complex with an $M_r$ of approximately 1000–2000 kDa. They form the sidearms of the microtu-

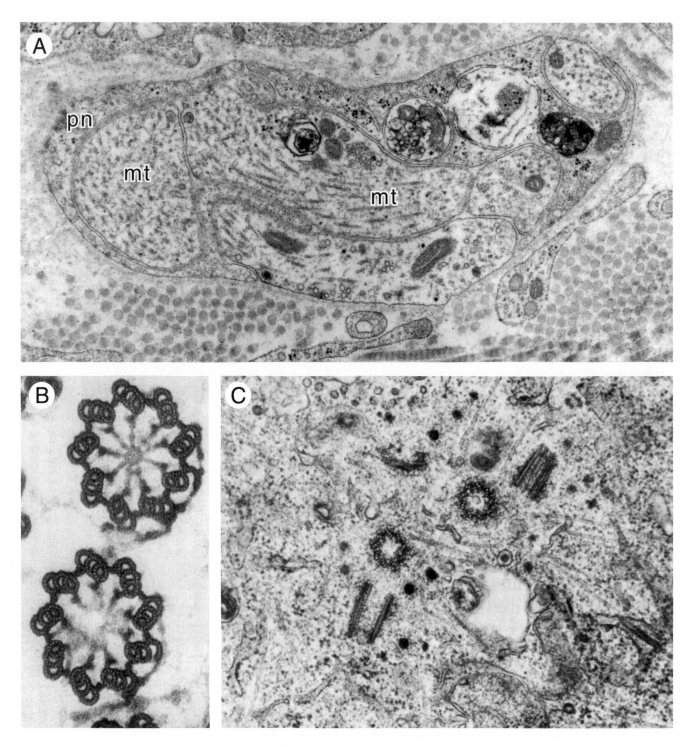

**FIG. 10.** Microtubules. (A) This cross-section of a small peripheral nerve bundle (pn) contains profiles of several nerve axons. The axons contain numerous longitudinally oriented microtubules (mt) seen both in cross-section and at an oblique angle. In neurons, the microtubules are involved in transport of organelles from the nerve cell bodies to the relatively distant synaptic terminals. (B) Two cilia are seen in cross-section. Nine units consisting of three microtubules form the support for this structure. Sliding of the microtubules with respect to each other cause the cilia to bend and are responsible for ciliary motility. [Reproduced with permission from B. Alberts *et al.* "Molecular Biology of the Cell," 2nd edition, Figure 11-59a. Garland Publishing, New York.] (C) This electron micrograph contains two centrosomes. Each centrosome contains a pair of centrioles that are oriented at right angles to each other. The centriole is formed from microtubules organized in the same ninefold triplets observed in the cytoskeleton of cilia and flagellae. Centrioles act as microtubule organizing centers: They give rise to the basal bodies of cilia and flagella, and also serve to organize the microtubules of the mitotic spindle. [Reproduced with permission from B. Alberts *et al.* "Molecular Biology of the Cell," 2nd edition, Figure 11-60. Garland Publishing, New York.]

bules found in the cores of cilia and flagella. These dynein sidearms enable microtubules to slide past each other, causing bending of the whole cilia or flagella, and thus are responsible for ciliary and flagellar motility (for further information, Chapter 56).

In addition to being the central component of cilia and flagella, microtubules are also found in microtubule-organizing centers such as *centrosomes, basal bodies,* and *kinetochores.* Microtubules are involved in movement of chromosomes during cell division and translocation of organelles within the cytoplasm. In some cases, microtubules also can form the framework on which some of the cell compartments are organized, as was noted earlier for the endoplasmic reticulum.

## VIII. Summary

Cell ultrastructure is studied to provide insights into the morphology that underlies the various activities of the cell. These activities are often carried out in distinct regions and are compartmentalized by the cell's organelles. The nucleus is the roughly spherical, centrally located organelle that contains the genetic machinery and initiates RNA transcription. Except during cell division, the nucleus is contained within a double membrane, the nuclear envelope. Pores in the nuclear envelope regulate transport into and out of the nucleus. The outer nuclear envelope is often decorated with small spherical ribosomes, giving it a beaded appearance. The outer nuclear envelope is continuous with the tubules of the endoplasmic reticulum (ER). When the ER is decorated with ribosomes, it also has a beaded appearance in electron micrographs and is thus referred to as rough ER. These membrane systems, largely through the activity of the ribosomes, are involved in linking amino acids together to form polypeptide chains, and inserting the resultant proteins into the membrane or packaging them for export out of the cell. ER that does not have ribosomes attached to it is called smooth ER and may have a variety of functions in the cells of different tissues. In muscle cells the smooth ER is primarily involved in regulating the free $Ca^{2+}$ levels in the cytoplasm, and there is evidence that it is performing this function in non-muscle cells as well. The Golgi complex is a stack of disclike membranes that receives proteins from the ER, processes them further, and then directs them to the appropriate intracellular compartment or exports them as secretory products. Lysosomes are spherical membrane-limited organelles that contain a wide variety of digestive enzymes. Lysosomes function to break down aging or incorrectly synthesized cellular components, and after they have taken up and started to degrade this material, the lysosomal contents have a rather nonhomogeneous appearance. Lysosomes also are involved in digesting extracellular (foreign) substances that enter the cell by exocytosis. Mitochondria are spherical or rodlike organelles composed of two membranes. The simple outer membrane encloses the organelle and serves as the outer boundary of the intermembrane space. The inner membrane is highly folded into lamellae called cristae, which greatly increase the surface area. Mitochondria are the site of the reactions that produce energy in the form of the high-energy phosphate ATP. The cell's cytoskeleton is composed of microfilaments and microtubules. There are three categories of microfilaments: actin-containing thin filaments, myosin-containing thick filaments, and intermediate filaments that are composed of various closely related intermediate filament proteins. Microtubules are long hollow cylinders formed from tubulin. Together the microfilaments and the microtubules of the cytoskeleton provide a scaffolding for the organization and support of other organelles and serve to compartmentalize regions of the cytoplasm. The cytoskeletal elements are also responsible for motility and trafficking of organelles. The cell membranes, both the external plasmalemma and the intracellular membranes, are phospholipid bilayers containing intrinsic proteins that act as ATP-dependent pumps or ion channels. These proteins are responsible for maintaining the ionic gradients that are essential to many of the cell's activities.

## Bibliography

Alberts, B., Bray, D., Lewis, J., Raff, M., Roberts, K., and Watson, J. D. Eds. (1989). "Molecular Biology of the Cell," Garland Publishing, New York.

Andre, J. (1994). Mitochondria. *Biology of the Cell* **80,** 103–106.

Bershadsky, A. D., and Vasiliev, J. M. (1988). "Cytoskeleton." Plenum Press, New York.

Block, S. M. (1996). Fifty ways to love your lever: Myosin motors. *Cell* **87,** 151–157.

Clermont, Y., Rambourg, A., and Hermo, L. (1995). Trans-Golgi network (TGN) of different cell types: Three-dimensional structural characteristics and variability. *Anat. Rec.* **24,** 289–301.

Damke, H. (1996). Dynamin and receptor-mediated endocytosis. *FEBS Lett.* **389,** 48–51.

Davis, L. I. (1995). The nuclear pore complex. *Annu. Rev. Biochem.* **64,** 865–896.

Dessev, G. N. (1992). Nuclear envelope structure. *Curr. Opin. Cell Biol.* **4,** 430–435.

dos Remedios, C. G., and Moens, P. D. (1995). Actin and the actomyosin interface: A review. *Biochim. Biophys. Acta* **1228,** 99–124.

Easterwood, T. R., and Harvey, S. C. (1995). Modeling the structure of the ribosome. *Biochem. Cell Biol.* **73,** 751–756.

Egleman, E. H., and Orlova, A. (1995). Allostery, cooperativity, and different structural states in F-actin. *J. Struct. Biol.* **115,** 159–162.

Gerace, L. (1992). Molecular trafficking across the nuclear pore complex. *Curr. Opin. Cell Biol.* **4,** 637–645.

Gonatas, N. K. (1994). Contributions to the physiology and pathology of the Golgi apparatus. *Am. J. Path.* **145,** 751–761.

Hauri, H. P., and Schweizer, A. (1992). The endoplasmic reticulum–Golgi intermediate compartment. *Curr. Opin. Cell Biol.* **4,** 600–608.

Hughes, T. A., Pombo, A., McManus, J., Hozak, P., Jackson, D. A., and Cook, P. R. (1995). On the structure of replication and transcription factories. *J. Cell Sci. Suppl.* **19,** 59–65.

Kornfeld, S., and Mellman, I. (1989). The biogenesis of lysosomes. *Annu. Rev. Cell Biol.* **5,** 483–525.

Lee, C., Ferguson, M., and Chen, L. B. (1989). Construction of the endoplasmic reticulum. *J. Cell Biol.* **109,** 2045–2055.

Mandelkow, E., and Mandelkow, E. M. (1995). Microtubules and microtubule-associated proteins. *Curr. Opin. Cell Biol.* 7: 72–81.

Mellman, I., and Simons, K. (1992). The Golgi complex: *In vitro veritas? Cell* **68,** 829–840.

Nickerson, J. A., Blencowe, B. J., and Penman, S. (1995). The architectural organization of nuclear metabolism. *Int. Rev. Cytol.* **162A,** 67–123.

Novikoff, A. B. (1976). The endoplasmic reticulum: A cytochemist's view. *Proc. Natl. Acad. Sci. USA* **73,** 2781–2786.

Nunnari, J., and Walter, P. (1992). Protein targeting to and translocation across the membrane of the endoplasmic reticulum. *Curr. Opin. Cell Biol.* **4,** 573–580.

Palade, G. E. (1956). The endoplasmic reticulum. *J. Biophys. Biochem. Cytol.* **2,** 85–98.

Pante, N., and Aebi, U. (1996). Molecular dissection of the nuclear pore complex. *Crit. Rev. Biochem. Mol. Biol.* **31,** 153–199.

Pavelka, M. (1987). Functional morphology of the Golgi apparatus. *Adv. Anat. Embryol. Cell Biol.* **106,** 1–94.

Peters, C., and von Figura, K. (1994). Biogenesis of lysosomal membranes. *FEBS Lett.* **346,** 108–114.

Robinson, M. S., Watts, C., and Zerial, M. (1996). Membrane dynamics in endocytosis. *Cell* **84,** 13–21.

Ruska, E. (1987). Nobel lecture. The development of the electron microscope and electron microscopy. *Biosci. Rep.* **7,** 607–629.

Schafer, D. A., and Cooper, J. A. (1995). Control of actin assembly at filament ends. *Annu. Rev. Cell Dev. Biol.* **11,** 497–518.

Seaman, M. N., Burd, C. G., and Emr, S. D. (1996). Receptor signalling and the regulation of endocytic membrane transport. *Curr. Opin. Cell Biol.* **8,** 549–556.

Shay, J. W. (1986). "Cell and Molecular Biology of the Cytoskeleton." Plenum Press, New York.

Spirin, A. S. (1986). "Ribosome Structure and Protein Biosynthesis." Benjamin/Cummings Publishing Co., Menlo Park, CA.

Tyler, D. (1992). "The Mitochondrion in Health and Disease." VCH Publishers, New York.

Tzagaloff, A. (1982). "Mitochondria." Plenum Press, New York.

Michael A. Lieberman and Richard G. Sleight

# 6

# Energy Production and Metabolism

## I. Introduction

*Energy* must be generated for all living organisms to grow and function. In animals, this energy is derived from the food consumed. The overall function of metabolism is to alter the food into chemical components that aid in cell growth and energy utilization. Two major functions of metabolic pathways are to provide energy to tissues when needed and to store molecules as "potential energy" in times of energy excess. The pathways that *generate* energy are the degradative, or catabolic, pathways. The pathways that *store* energy are the synthetic, or anabolic, pathways. The anabolic pathways also produce many of the compounds required for cells to grow and divide, such as proteins, lipids, and nucleic acids. The synthesis of these compounds requires energy, provided by the catabolic pathways.

Metabolism can be viewed as two opposing sets of pathways, one leading to the biosynthesis of required compounds, the other to the degradation of these compounds. The catabolic pathways are used primarily under three sets of conditions. The first is when food supplies are plentiful. The food is degraded within the digestive tract into basic building blocks, and then converted into the energy storage components (using the anabolic pathways). Degradative pathways are also used in the normal process of turnover of cell components. Many cellular constituents are degraded as they age to protect against the accumulation of damaged material within the cell. These constituents are then replaced by newly synthesized material. The third condition in which degradation is favored is when food supplies are scarce. Examples of this are during dieting and starvation. The degradative pathways are used to provide fuel for energy utilization. Because the biosynthetic pathways oppose the degradative pathways, they will be inactive when the degradative pathways are active. The major catabolic and anabolic pathways are outlined in Fig. 1.

Metabolism is considered in the following manner. First, proteins are discussed in terms of their role as the *catalysts*

of biochemical reactions. The regulation of *enzyme* activity is introduced, along with a discussion of enzyme kinetics. This leads to a more detailed discussion of enzyme regulation, before specific energy-generating or storage pathways are covered. This chapter ends with a discussion of the metabolic changes seen during starvation, including specific changes within the liver, muscles, and fat cells.

## II. Protein Enzymes

Protein molecules are composed of many amino acids linked together through covalent bonds. The sequence of amino acids within the protein determines both its shape and its function. Proteins perform many functions. Many proteins act as *catalysts* for the metabolic reactions that occur within a cell. These protein catalysts are known as *enzymes*. All metabolic reactions are catalyzed by enzymes. Enzymes enhance the rate at which metabolic reactions proceed, without altering the equilibrium point of the reaction. For a reaction to proceed to completion, energy is required. Some of the energy required is used to orient the molecules correctly, such that the reactive groups are adjacent. Achieving the proper *spatial orientation* of the substrates during a reaction is known as the *transition state* of the reaction. Once the transition state has been reached (see Fig. 2), the reaction will proceed. If a large amount of energy is required for the substrates to reach this transition state, the reaction will proceed very slowly, if at all. Enzymes alter the rate at which reactions proceed by *reducing the amount of energy required* to reach the transition state. This means that the rate at which the reaction will proceed is enhanced by the actions of the enzyme.

Because all of the reactions in metabolic pathways are catalyzed by enzymes, *regulation of the pathways* occurs at the enzyme level. The activity of an enzyme can be regulated in both positive and negative ways by specific *modifiers*. Understanding metabolic regulation is the key to understanding metabolism. There are several different levels of regulation. The first, known as *allosteric modifica-*

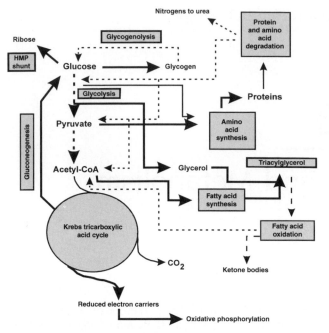

**FIG. 1.** The major pathways of metabolism. The pathway of glycolysis converts glucose to pyruvate and is the entry of all sugars into metabolism. Intermediates along the pathway are used for amino acid biosynthesis and triacylglycerol formation. A key step of metabolism is the conversion of pyruvate to acetyl-CoA. Once acetyl-CoA is formed it can be oxidized by the Krebs tricarboxylic acid cycle (TCA cycle) to produce reduced electron carriers. The reduced electron carriers then generate energy, in the form of ATP, via oxidative phosphorylation through the electron-transport chain. Intermediates of the TCA cycle are also used for amino acid biosynthesis. When glucose is in excess, it is stored in the form of glycogen. When energy is required, glycogen is degraded to glucose via glycogenolysis. Glucose can be synthesized from glycolytic and TCA cycle intermediates via gluconeogenesis. Protein degradation leads to the generation of free amino acids, which are further metabolized to form intermediates of the TCA cycle and glycolysis. The major energy storage form of the cells is triacylglycerol, which is formed from glycerol and fatty acids. The glycerol is obtained from an intermediate of glycolysis, and the fatty acids are synthesized from acetyl-CoA. The pathway of triacylglycerol degradation leads to the formation of glycerol and acetyl-CoA, and derivatives produced from acetyl-CoA known as ketone bodies. Glucose can be metabolized either through glycolysis or through the hexose monophosphate shunt (HMP shunt) pathway to produce the five-carbon sugar ribose. The HMP shunt is necessary to provide sugars for the biosynthesis of nucleotides, which are the building blocks of the nucleic acids DNA (deoxyribonucleic acid) and RNA (ribonucleic acid). In this diagram the solid lines represent catabolic pathways, and the dashed lines anabolic pathways.

tion requires that the modifier bind to a distinct site on the enzyme and alter enzyme activity. Table 1 lists examples of such enzymes and indicates the *positive and negative regulators.* A second level of regulation is by *covalent modification,* in which one of the amino acid residues of the enzyme is chemically modified. This can occur in a number of ways, and examples of this type of modification are also presented in Table 1. In addition to natural modifiers of enzyme activity, many drugs act by inhibiting enzyme ac-

tion. Thus, to understand drug action, it is important to understand how enzymes work and are regulated.

## III. Enzyme Kinetics

The rate at which an enzyme works (Fersht, 1985) can be described by a value known as the *maximal velocity* ($V_{\max}$). The $V_{\max}$ value is obtained by measuring the rate at which the enzyme can form its products at various substrate concentrations. The $V_{\max}$ indicates how fast the enzyme can proceed when all of the available enzyme has bound substrate and is participating in the reaction.

Another kinetic parameter is the *Michaelis constant* (further discussed later), abbreviated as $K_m$. The $K_m$ value is defined as the substrate concentration at which one-half of maximal velocity occurs. The $K_m$ value reflects the *affinity* of the enzyme for its substrate, although it is not a true measure of that affinity. When a $K_m$ value is low, it is an indication that the substrate *binds tightly* to the enzyme, and that half-maximal velocity can be obtained at low substrate concentrations. Conversely, a high $K_m$ value reflects a reduced affinity of the substrate for the enzyme, so that a high substrate concentration is needed to reach half-maximal velocity. A typical enzyme kinetic plot, in which the initial velocity $V$ of the reaction is measured as a function of increasing substrate concentration [S] is shown in Fig. 3.

Before one can understand how to mathematically represent the curve shown in Fig. 3, a number of terms need to be introduced. If one considers the following reaction,

$$E + S \underset{k_2}{\overset{k_1}{\rightleftharpoons}} ES \overset{k_3}{\rightarrow} E \rightarrow P \tag{1}$$

E represents enzyme, S represents the substrate that is being acted upon by the enzyme, and P represents the product of the reaction. ES represents an enzyme–

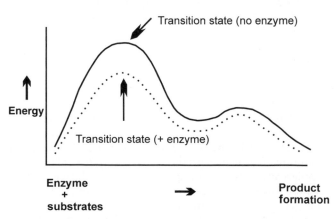

**FIG. 2.** An indication of the energy required to reach the transition state of a reaction. The solid line indicates the energy required for the substrates to reach a transition state. The dotted line indicates how this energy requirement is reduced in the presence of enzyme. By requiring less energy to reach the transition state, the reaction can proceed more rapidly, and the rate of the reaction will be enhanced.

**TABLE I** Examples of Allosteric Modifiers of Key Enzymes in Glycolysis and Glycogen Metabolism[a]

| Enzyme | Positive modifier | Negative modifier | Covalent modification |
|---|---|---|---|
| Phosphofructokinase-I (glycolysis) | AMP, fructose 2,6-bisphosphate | ATP, citrate | None |
| Pyruvate kinase (glycolysis) | Fructose 1,6-bisphosphate | Alanine | Phosphorylation inhibits in the liver |
| Phosphorylase kinase (glycogen metabolism) | Calcium | None | Phosphorylation activates |
| Phosphofructokinase-II (glycolysis) | Fructose 6-phosphate | Fructose 1,6-bisphosphate | Phosphorylation inhibits |
| Hexokinase (glycolysis) | None | Glucose 6-phosphate | None |

[a] This is only a partial list of the enzymes that can be modified either allosterically or by covalent modification, in this case by phosphorylation. Many other enzymes are regulated in this way.

substrate complex, which must form for the reaction to proceed. The velocity of this reaction is the rate at which the product is formed per unit time, such as moles of substrate per minute. The initial velocity of the reaction is the rate of product formation per unit time when the reaction is first initiated. Under initial velocity conditions, there is virtually no back-conversion, that is, conversion of P to S, because of the very high levels of S and the very low levels of P. This is what is represented by Eq. 1.

Two assumptions are made when one derives the mathematical equation that describes the curve shown in Fig. 3 and that represents the reaction shown in Eq. 1. The first is that the reaction rapidly reaches equilibrium, or a steady state. This is true for the majority of biochemical reactions studied. The second assumption is that the product of the reaction does not go back to the starting material. This is not usually true for biochemical systems, but can be approximated by using the initial velocity of the reaction. The curve shown in Fig. 3 can be represented mathematically as

$$V = \frac{V_{max}}{1 + \dfrac{K_m}{[S]}} \qquad (2)$$

where $V$ represents the initial velocity of the reaction, which is measured at different substrate [S] concentrations with a constant enzyme concentration. This equation is known as the *Michaelis–Menten* equation, named after the two scientists who initially derived it.

The curve shown in Fig. 3 is not very useful, as extrapolations need to be made to determine both $V_{max}$ and $K_m$. Equation 2 can be manipulated algebraically into a form in which straight lines are generated. The equation that is derived is known as the *Lineweaver–Burk plot*. It is a *double reciprocal* plot and is represented mathematically as

$$\frac{1}{V} = \frac{1}{[S]}\frac{K_m}{V_{max}} + \frac{1}{V_{max}} \qquad (3)$$

When $1/V$ on the y-axis is plotted against $1/[S]$ on the x-axis, a straight line is obtained, as shown in Fig. 4. The *slope* of the line represents the factor $K_m/V_{max}$, and the *y-intercept* represents the value $1/V_{max}$. It is much easier and more accurate to determine both $K_m$ and $V_{max}$ from the Lineweaver–Burk plot than from the data shown in Fig. 3. It is particularly useful for studying the effects of inhibitors on enzyme function, which is discussed in the next section.

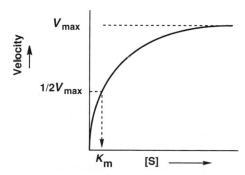

**FIG. 3.** A kinetic plot of reaction velocity $V$ vs substrate concentration [S]. At saturation the addition of more substrate does not increase the rate of the reaction, and the maximal velocity ($V_{max}$) is obtained. Once the $V_{max}$ has been determined, the $K_m$ can be determined, which is the substrate at which one-half maximal velocity occurs.

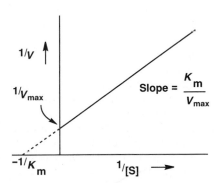

**FIG. 4.** An example of a Lineweaver–Burk plot. This is a more informative plot for analyzing kinetic data than the one shown in Fig. 3.

## IV. Enzyme Inhibitors

Since enzymes carry out the reactions of the metabolic pathways, regulation of the enzymes is very important physiologically. Enzymes can be activated or inhibited by various factors. In addition, many drugs act by inhibiting the action of enzymes. Aspirin, for example, inhibits the enzyme cyclo-oxygenase, which produces mediators of the pain response. Methotrexate, a potent anticancer agent, works by inhibiting the enzyme dihydrofolate reductase, which is needed to produce precursors for DNA synthesis. Inhibitors can work in many ways, but the two most common are termed *competitive inhibition* and *noncompetitive inhibition*. A competitive inhibitor can be overcome by adding excess substrate, as competitive inhibitors compete with the substrate for binding to the enzyme. Since there is competition for the substrate binding site, in the presence of a competitive inhibitor, the amount of substrate required to reach one-half maximal velocity will be increased. This will result in a greater $K_m$ exhibited by the enzyme in the presence of inhibitor than in its absence.

A noncompetitive inhibitor does not affect substrate binding, but binds to the enzyme at a different site, reducing the ability of the enzyme to catalyze the reaction. A noncompetitive inhibitor, therefore, reduces the $V_{max}$ of the enzyme. The $K_m$ will not be affected by a noncompetitive inhibitor. These kinetic differences help to distinguish competitive from noncompetitive inhibitors.

The inhibition patterns exhibited by inhibitors are best represented graphically using the double reciprocal plot established by Lineweaver and Burk. In the presence of an inhibitor, Eq. 3 is altered. As a competitive inhibitor will also bind to the enzyme, in competition with the substrate, the term $K_i$ is used to indicate the *dissociation constant* of the inhibitor from the enzyme. This dissociation constant is a measure of the rate at which bound inhibitor will dissociate from the enzyme. If the effects of the inhibitor (I) are considered in the derivation of the Michaelis–Menten equation, Eq. 2 would become

$$V = \frac{V_{max}}{1 + \dfrac{K_m}{[S]}\left(1 + \dfrac{[I]}{K_i}\right)} \qquad (4)$$

In this equation, the concentration of inhibitor is indicated by [I], and the dissociation constant of the inhibitor from enzyme by $K_i$. Comparing Eq. 4 with Eq. 2 shows that the two expressions are very similar, except that the $K_m/[S]$ term is increased by the factor $(1 + [I]/K_i)$.

In the absence of inhibitor, the expression $(1 + [I]/K_i)$ would be 1, and Eq. 4 would be equivalent to Eq. 2. As the concentration of inhibitor is increased, the $[I]/K_i$ term increases, and the overall velocity of the reaction is reduced. When the reciprocal of Eq. 4 is calculated, the Lineweaver–Burk equivalent for the case of competitive inhibition is obtained as

$$\frac{1}{V} = \frac{1}{V_{max}} + \frac{K_m}{V_{max}}\left(1 + \frac{[I]}{K_i}\right)\frac{1}{[S]} \qquad (5)$$

If $1/V$ is now plotted vs $1/[S]$, the $y$-intercept is $1/V_{max}$ and the slope of the line is given by

$$\text{slope} = \frac{K_m}{V_{max}}\left(1 + \frac{[I]}{K_i}\right) \qquad (6)$$

Compared with the case in which no inhibitor is present, the slope has now been increased by the quantity $(1 + [I]/K_i)$. Thus, when the inhibitor concentration is zero (I is not present), the slope of the line is the same as for a noninhibited enzyme, namely $K_m/V_{max}$. However, as the inhibitor concentration is increased, the slope of the line increases, which results in an increase in the $K_m$. The kinetics of competitive inhibition are illustrated in Fig. 5.

In contrast to competitive inhibition, a noncompetitive inhibitor does not alter substrate binding to the enzyme. Rather, the inhibitor decreases the rate at which the enzyme can catalyze the reaction. Thus, $K_m$ will not be altered, but $V_{max}$ is decreased. The factor by which the velocity is reduced is equal to $(1 + [I]/K_i)$, the same as seen with competitive inhibition. The kinetics of noncompetitive inhibition is shown in Fig. 6. Notice that both lines (one in the presence of inhibitor, the other in the absence of inhibitor) intersect at the $x$-axis. In contrast, for competitive inhibition (see Fig. 5), the two lines intercept on the $y$-axis. This reflects the fact that competitive inhibitors do not alter the $V_{max}$, which is *determined at the y-intercept*, whereas noncompetitive inhibitors do not alter the $K_m$, which is determined at the *extrapolated x-intercept*.

*Regulation* of metabolic pathways occurs primarily at the enzyme level. Enzymes can be regulated by molecules known as *allosteric effectors*, which bind to the enzyme and alter its activity. The binding of the effector leads to a shape change in the protein. The shape change then either makes it *easier* (in the case of an activator) or *more difficult* (in the case of an inhibitor) for the enzyme to catalyze a reaction. The kinetics of such regulated enzymes are shown in Fig. 7. The curves obtained are complex, but the end result is that sigmoidal kinetics are observed in a typical

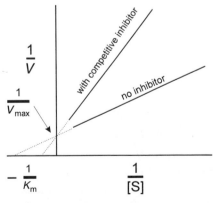

**FIG. 5.** The kinetics of competitive inhibition. Note that the slope of the line will increase as the inhibitor concentration is increased. The lines will intersect on the $y$-axis, indicating that $V_{max}$ is not altered in the presence of the inhibitor. As compared to the line in the absence of inhibitor, the slope of the lines have been increased by the expression $(1 + [I]/K_i)$.

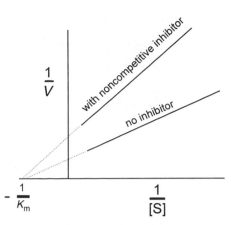

**FIG. 6.** The kinetics of noncompetitive inhibition. Note that as in the case of competitive inhibition (Fig. 5) the slope of the line will increase in the presence of the inhibitor. In the case of noncompetitive inhibition, the lines will intersect at the negative x-axis, indicating that the binding of substrate to the enzyme is not inhibited by the inhibitor. However, the $V_{max}$, as represented by the y-intercept, decreases as the inhibitor concentration increases.

[S] vs V plot. In the presence of an inhibitor, the entire curve gets shifted to the right, indicating that higher levels of substrate are required to reach an equivalent level of activity. In contrast, activators shift the curve to the left, indicating that less substrate is required to obtain maximal activity. Another way of stating this is that activators reduce the $K_m$ of the enzyme, whereas inhibitors increase the $K_m$. *It is important to realize that when a regulated enzyme is inhibited, it means that a higher concentration of substrate is required before the reaction can proceed rapidly;* at normal substrate levels, the reaction will proceed slower than in the absence of the inhibitor. Other means of regulating enzyme activity are discussed later.

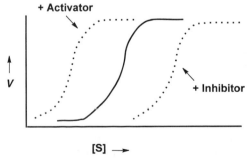

**FIG. 7.** Kinetics of regulated enzymes. Regulated enzymes typically display sigmoidal kinetics, in which the activity of the enzyme increases dramatically over a small substrate concentration range. An inhibitor of the enzyme will shift the entire curve to the right, thereby increasing the level of substrate required to obtain one-half maximal velocity. An activator, however, will shift the curve to the left, in effect decreasing the level of substrate required to reach one-half maximal velocity. The effect of activation, therefore, is to allow the reaction to proceed more rapidly at lower substrate concentrations than in the absence of the activator.

## V. Metabolic Pathways

The major biochemical constituents of a biological cell are *amino acids, carbohydrates, nucleic acids,* and *lipids.* Examples of these four types of compounds are shown in Fig. 8. There are pathways that will either synthesize or degrade these key components. The major function of the degradative pathways is to generate energy. When these pathways are activated, certain energy-storage molecules are degraded so that energy can be obtained. The energy that is obtained from the storage molecules is predominantly in the form of compounds containing *high-energy bonds.* A high-energy bond is one that will release a large amount of energy when the bond is broken. The key energy-containing molecules within the cell are nucleotide triphosphates, such as *adenosine triphosphate (ATP).* The structure of ATP is shown in Fig. 9. Note how this molecule consists of a five-carbon sugar (ribose), three phosphates, and a nitrogenous base, which are the same components as the nucleic acids. The nucleic acids are, in fact, synthesized by the condensation of the nucleotide triphosphates. The high-energy bonds within ATP are between the $\alpha$- and $\beta$-phosphates, and the $\beta$- and $\gamma$-phosphates. Hydrolysis of one of these bonds results in the release of 7.3 kcal/mol of energy, which for a biological system is a large amount. The high-energy bonds of ATP are also found in the other nucleotide triphosphates, such as GTP, CTP, TTP, and UTP.

How is the energy in ATP and other high-energy intermediates used? The primary role of these high-energy compounds is to release the energy at appropriate steps in a pathway to allow reactions to proceed that require energy to reach their transition state. One example of this is muscle contraction (Alberts *et al.,* 1989). For muscles to contract, energy is required. ATP is used to allow molecules (in myofilaments) to slide past one another, which brings about contraction. When ATP levels are low, muscle contraction is inhibited. Another example of energy use is to couple a number of enzyme-catalyzed reactions together, in the form of a pathway, in order to link thermodynamically unfavorable reactions with those that are highly favorable. This is illustrated in Fig. 10. Notice how by coupling an unfavorable reaction with one that releases a large amount of energy, a favorable sequence of reactions occurs despite the presence of the unfavorable reaction.

The key pathways of metabolism were outlined earlier in Fig. 1 (Lehninger *et al.,* 1993). These pathways represent one of the unities of life, as they are common in both bacteria and humans. As stated previously, there are both anabolic and catabolic pathways. One of the major catabolic pathways is *glycolysis,* which is the major pathway for the conversion of carbohydrate (such as glucose and fructose) to pyruvate, a three-carbon intermediate. This pathway does not require oxygen and generates a small amount of energy in the form of ATP during the reaction sequence. Glycolysis is the major pathway of sugar metabolism, as all carbohydrates eventually feed into this pathway.

The metabolism of five-carbon sugars (such as ribose) is connected to glycolysis at a number of points, through a pathway known as the *hexose monophosphate shunt (HMP*

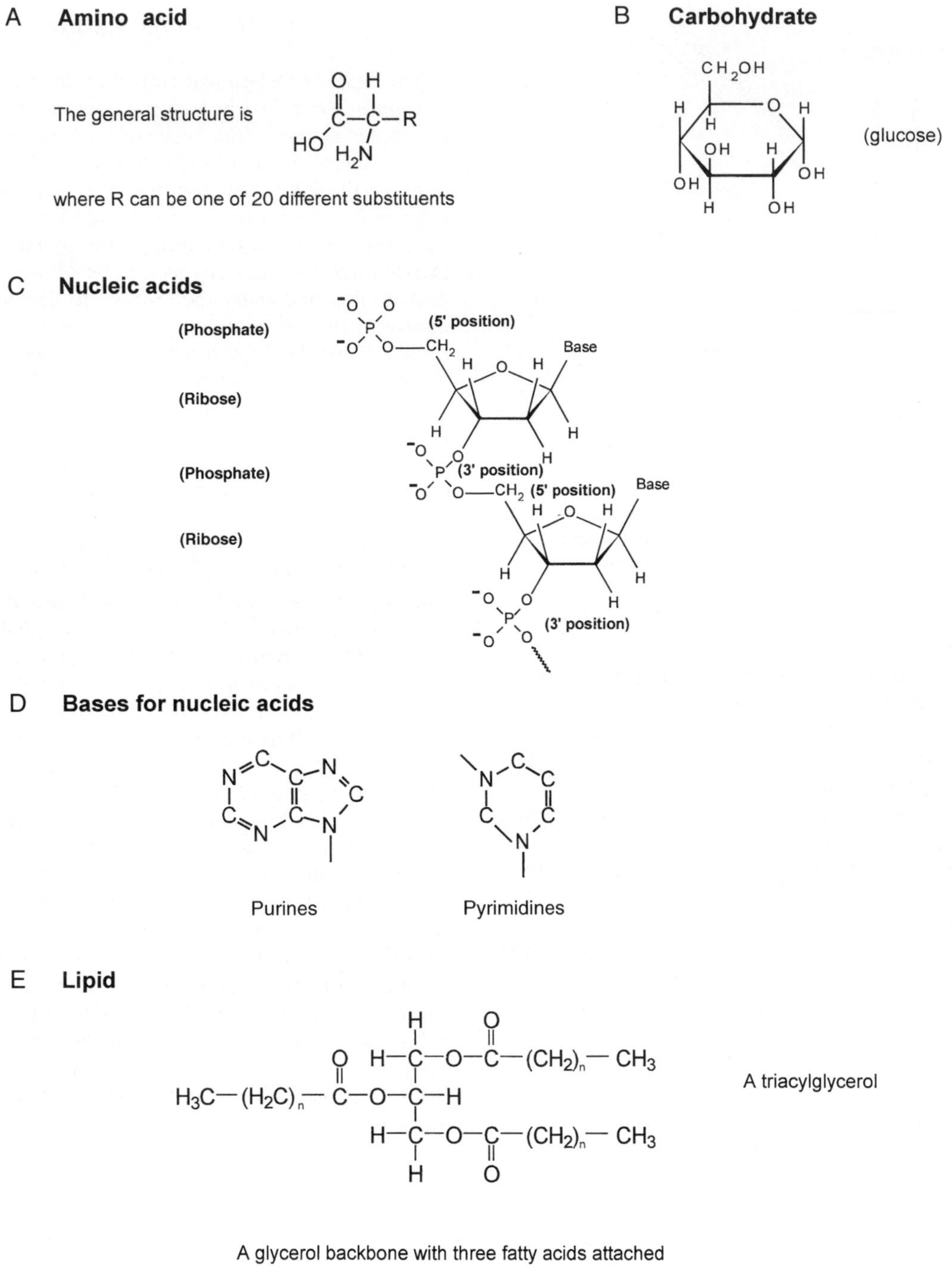

A    **Amino  acid**

The general structure is

where R can be one of 20 different substituents

B    **Carbohydrate**

(glucose)

C    **Nucleic acids**

(Phosphate)

(5' position)

Base

(Ribose)

(Phosphate)

(3' position)

(5' position)    Base

(Ribose)

(3' position)

D    **Bases for nucleic acids**

Purines                          Pyrimidines

E    **Lipid**

A triacylglycerol

A glycerol backbone with three fatty acids attached

**FIG. 8.**    Structures of the major chemical constituents of a cell. (A) The general structure of an amino acid. There are 20 different amino acids found in proteins, each differing in the R group attached to the central carbon atom. The R group can be as simple as a proton (for glycine), or as complex as a conjugated indole ring (for tryptophan). Peptide bonds are formed between the amino group of one amino acid and the carboxyl group of a second amino acid. (B) The structure of glucose as a representative carbohydrate. The molecule consists of a carbon chain with hydroxyl groups on the majority of carbons. (C) A schematic of a nucleic acid. Nucleic acids, such as DNA and RNA, are complex structures consisting of a phosphate, a nitrogenous base, and the five-carbon sugars ribose (for RNA) or deoxyribose (for DNA). The phosphates link different ribose-base units together to form a linear structure. The bases (D) are classified as either purines or pyrimidines, depending on their general structure. (E) The structure of a triacylglycerol. There is a glycerol backbone to which is esterified three fatty acids. Fatty acids consist of long hydrocarbon chains with a terminal carboxylic acid group.

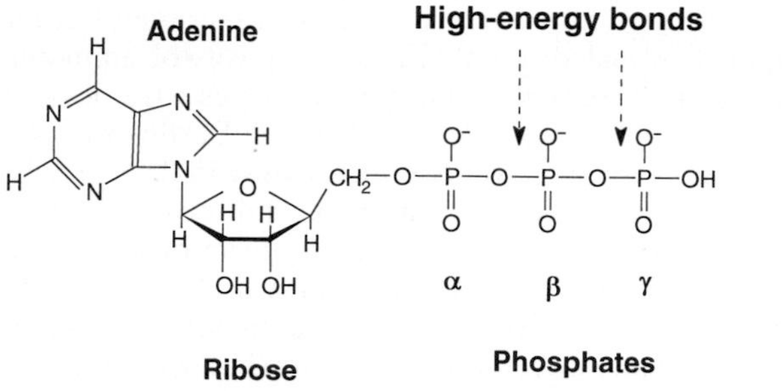

**Adenine**

**High-energy bonds**

**FIG. 9.** The structure of the high-energy compound ATP. ATP consists of the purine base adenine, the sugar ribose, and three phosphates linked in series to the number 5 carbon of the ribose. The high-energy bonds are indicated by the arrows. ATP, and the other nucleotide triphosphates, serve as the precursors of the nucleic acids, in addition to ATP's role as the major energy carrier of the cell.

**A**

**B**

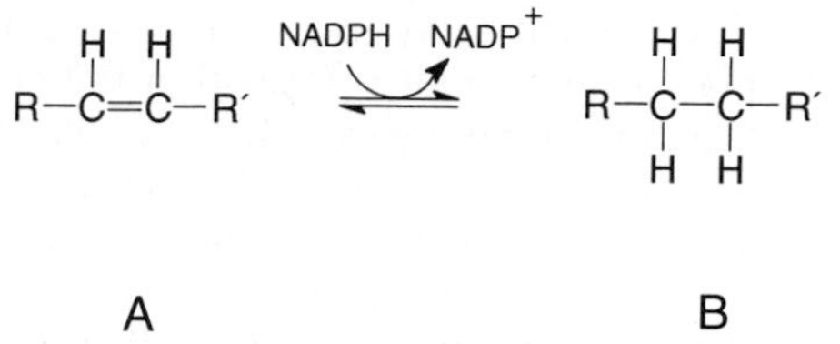

**FIG. 11.** An oxidation reaction. Reactant A is oxidized as compared to reactant B. NADPH can donate its electrons, in the form of a hydride ion, to reactant A to reduce the double bond to a single bond. In the course of this reaction reactant A is reduced to produce reactant B, and the NADPH is oxidized to form $NADP^+$. Compounds that gain electrons are reduced; those that lose electrons are oxidized. Oxidation–reduction reactions are common in biochemical pathways, and almost all require the participation of an electron carrier.

shunt). The HMP shunt pathway performs two major roles. The first is the conversion of five- and six-carbon sugars. Five-carbon sugars are required for the synthesis of nucleoside triphosphates, which are then used for nucleic acid biosynthesis. The second role of the HMP shunt is to generate a molecule known as reduced *nicotinamide adenine dinucleotide phosphate (NADPH)*. NADPH is an *electron carrier* that is required by many biosynthetic pathways. It is used to donate electrons for oxidation–reduction reactions, as illustrated in Fig. 11. Another major electron carrier is reduced *nicotinamide adenine dinucleotide (NADH)*, which has the same structure as NADPH, except it does not contain the phosphate group. As is seen in this chapter, the electrons carried by NADH or NADPH can be donated to oxygen to form both ATP and water.

The three-carbon molecule pyruvate plays an important central role in metabolism. It represents a key point in

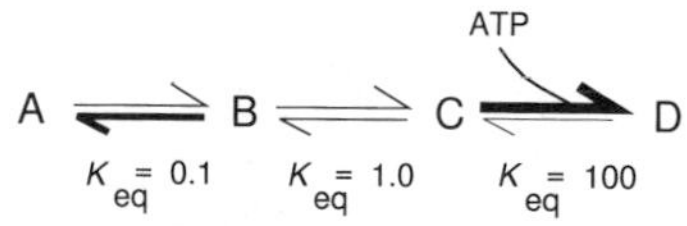

the eventual fate of the carbon atoms. Pyruvate can be converted into glucose, through the process of *gluconeogenesis*. It can also be used as a precursor for the synthesis of various amino acids. In addition, pyruvate can be converted into another central metabolite, *acetyl-CoA*, through an oxidative decarboxylation reaction (Fig. 12). This conversion is a key step in metabolism. Once pyruvate has been converted to acetyl-CoA, the carbons of acetyl-CoA can only be used either to generate energy or to synthesize fatty acids, ketone bodies, or steroids. The carbons of acetyl-CoA cannot be used for amino acid or carbohydrate synthesis. If the acetyl-CoA is used for energy production, it is oxidized to carbon dioxide ($CO_2$) and water through the *Krebs tricarboxylic acid cycle (TCA cycle)*, a pathway that requires oxygen to function. The TCA cycle generates a large amount of NADH, which donates its electrons to the electron transport chain. Through a series of electron transfers the electrons are used to reduce molecular oxygen ($O_2$) to water (HOH), and during this process, energy is generated in the form of ATP. The *electron-transport chain* is the major energy-producing pathway of the cell, and the mechanism of how energy is produced by electron transfer is discussed later in this chapter.

$$A \rightleftharpoons B \rightleftharpoons C \xrightarrow{\text{ATP}} D + ADP$$

$$K_{eq} = 0.1 \qquad K_{eq} = 1.0 \qquad K_{eq} = 100$$

**FIG. 10.** Linking an unfavorable reaction with a favorable reaction to create a pathway. In the example shown the reaction $A \rightarrow B$ is unfavorable, such that at equilibrium 90% of A has not reacted. The reaction $B \rightarrow C$ has an equilibrium constant of 1, indicating that it is a freely reversible reaction. This means that at equilibrium 50% of B will have been converted to C. The reaction $C \rightarrow D$, however, is essentially irreversible because of the release of energy when a high-energy bond of ATP is hydrolyzed. This means that at equilibrium only 1% of the original C will be present, as the other 99% will have been converted to D. However, as C is being converted to D, this affects the equilibrium between B and C. Thus, as C decreases in concentration, more B will be converted to C to maintain the equilibrium constant of 1.0. This has the net effect of decreasing the level of B within the reaction system. As the concentration of B drops, this will affect the equilibrium between A and B. Even though that is an unfavorable reaction, as the concentration of B drops, more A will be converted to B to maintain the A–B equilibrium. The net effect of coupling these three reactions is to push the unfavorable first reaction to completion, because of the hydrolysis of ATP and release of energy at the third step of the pathway.

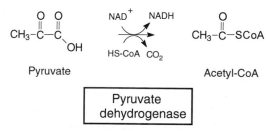

Pyruvate

Acetyl-CoA

Pyruvate
dehydrogenase

**FIG. 12.** The oxidative decarboxylation reaction in which pyruvate is converted into acetyl-CoA. CoA represents coenzyme A, a vitamin-derived cofactor that activates the carboxyl group of acetate to allow it to participate in the subsequent reactions. The enzyme that catalyzes this reaction is pyruvate dehydrogenase. The reaction itself involves both decarboxylation of pyruvate (loss of carbon dioxide) and oxidation of pyruvate to form acetyl-CoA. Since the pyruvate is being oxidized, another molecule needs to accept the electrons that are being lost, and that is done by $NAD^+$ to form NADH (reduced nicotinamide adenine dinucleotide).

*Fatty acids* (FAs) can be synthesized from acetyl-CoA. Fatty acids play a number of roles within the cell. Their main role is as an energy source. In this role, FAs are stored in the form of triacylglycerol, which consists of a glycerol molecule to which three FAs have been esterified (see Fig. 8). A secondary role of FAs is as a *second messenger* (Lands, 1991) in the form of prostaglandins, leukotrienes, and prostacyclins (more information on this topic is provided in Chapter 7). Another secondary role of FAs is as an essential component of cell membranes (in the form of phospholipids), in which two aqueous compartments are separated by a hydrophobic barrier.

The energy from FAs can be obtained by their oxidation and coversion to acetyl-CoA, which is then further oxidized to $CO_2$ and HOH through the TCA cycle and electron-transport chain. In addition, under specialized conditions, the FAs can be degraded to form the *ketone bodies* acetoacetate and $\beta$-hydroxybutyrate, both of which are synthesized when acetyl-CoA levels are high. These compounds are made under conditions in which the body has reduced energy stores. The liver produces the ketone bodies, which are then sent to other tissues for use as alternative energy supplies.

*Diabetes mellitus* is a disease in which glucose metabolism is altered, such that adequate energy cannot be derived from dietary glucose. This can occur when the pancreas stops producing *insulin,* which is required for the proper metabolism of glucose. Because of this problem, untreated diabetics produce excessive levels of ketone bodies to supply energy to all tissues of the body. As the ketone bodies are acids, excessive ketone body production can lead to *diabetic ketoacidosis* (Foster and McGarry, 1983), a condition that leads to a reduced pH in the blood and urine. Immediate treatment of this condition is required to prevent coma and death.

*Glycogen* is the carbohydrate storage form used by the cell. It consists of many glucose molecules organized in a branched structure. Glycogen is synthesized when the carbohydrate content of the cell is high, and is degraded when the cell or body requires more glucose. The pathways of glycogen synthesis and degradation are different, although the manner in which these pathways is regulated is similar. Regulation of these pathways are discussed later in this chapter.

*Amino acids,* in addition to being the building blocks for proteins, can also be used for energy production via degradative pathways. When amino acids are degraded, they can be classified as either glucogenic or ketogenic. *Glucogenic amino acids* can give rise to glucose, through the pathway of gluconeogenesis. *Ketogenic amino acids* can only give rise to acetyl-CoA and ketone bodies, which can generate energy, but not glucose. Upon degradation, the amino acids are converted into acetyl-CoA or intermediates of either glycolysis or of the TCA cycle. Humans can only synthesize 12 of the 20 required amino acids, as key enzymes of the biosynthetic pathways found in other organisms are not expressed in humans. The eight amino acids that cannot be synthesized are referred to as the *essential amino acids,* as they can only be obtained from the diet.

When amino acids are degraded, the amino group requires special disposal. Excessive levels of ammonia are toxic in all species, and different species have devised different methods for nitrogen disposal. Earthworms secrete ammonia directly, which diffuses away in the soil, whereas birds produce uric acid for nitrogen disposal. In mammals the nitrogen is disposed of in the form of urea, which is synthesized in the liver via the *urea cycle.* Through use of this cycle, toxic levels of ammonia do not accumulate, and are instead converted into a nontoxic form, urea.

## VI. Generation of Energy: Mitchell Chemiosmotic Hypothesis

One of the key factors of the catabolic pathways is the generation of energy, usually in the form of ATP. Although small amounts of ATP are produced through the glycolytic pathway, most of the ATP is produced through *oxidative phosphorylation.* These reactions occur within the *mitochondria,* whereas glycolysis occurs in the cytosol. Oxidative phosphorylation refers to the energetically favorable reduction of oxygen, and a coupling of the energy produced by this reduction to the phosphorylation of ADP to form ATP. The electrons used to drive this process are obtained from the TCA cycle through the oxidation of acetyl-CoA to $CO_2$. Electrons are obtained via electron carriers, such as NADPH, NADH, and $FADH_2$. $FADH_2$ is an electron carrier that has not been previously discussed, but is primarily generated during both the oxidation of fatty acids and in one reaction of the TCA cycle. As with NADPH and NADH, $FADH_2$ can donate its electrons to the electron transport chain to generate energy. The process of oxidative phosphorylation is shown schematically in Fig. 13.

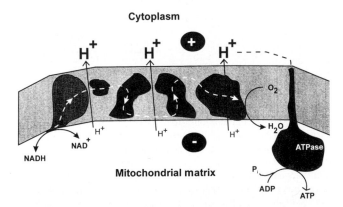

**FIG. 13.** Generation of ATP by the Mitchell chemiosmotic hypothesis. In this model the asymmetric orientation of the membrane-bound electron-transport chain results in the translocation of protons from the inside of the mitochondrion (matrix) to the cytoplasm of the cell as the electrons funnel through the chain. The translocation of protons leads to both a proton and an electrical gradient being established across the mitochondrial inner membrane. The end result of these gradients is that it is now energetically favorable for a proton to go down both its electrical and chemical potentials and enter the mitochondrion. This energy is harnessed by the proton-translocating ATPase to synthesize ATP from ADP and inorganic phosphate.

Mitochondria are intracellular organelles that contain two membranes. The outer membrane is porous to molecules of less than 5000 Da and will allow the small molecules (e.g., electron carriers, ADP, oxygen) required for oxidative phosphorylation to reach the inner mitochondrial membrane. The electron carriers of the electron-transport chain are located within the inner mitochondrial membrane. The inner mitochondrial membrane is impermeable to virtually everything, including protons. The only compounds that can traverse this membrane are those for which specific carriers exist within the membrane, and thereby can transport the molecule into the lumen of the mitochondrion.

Upon acceptance of electrons from either NADH or $FADH_2$, the electrons are transferred in an orderly fashion through the carriers to the terminal *electron acceptor* oxygen, which accepts the electrons and becomes reduced, forming water. As the electrons are transferred from carrier to carrier within the membrane, protons are transferred from the inside to the outside of the inner mitochondrial membrane. As protons are positively charged, the active extrusion of protons during electron transport leads to the loss of positive charge from within the mitochondrion. This leads to a difference in pH across the inner mitochondrial membrane, as well as a difference in charge. These $H^+$ differences across the membrane lead to a *proton-motive force (PMF)* across the membrane. This force consists of two components, $\Delta$pH (change in pH across the membrane) and $\Delta\Psi$, the difference in membrane potential (charge) across the membrane. The PMF (whose units are kcal/mol) can be expressed as

$$PMF = 2.3RT\,\Delta(pH) + zF\,\Delta\Psi \qquad (7)$$

where $r$ is the *gas constant* (1.98 kcal/mol-degree), $T$ is the absolute temperature in kelvins (K), $z$ is the charge on the particle (which in this case is 1, as the particle is a proton), and $F$ is the *Faraday constant* (23 kcal/mol-V). The magnitude of the PMF is directly proportional to both the change in pH across the membrane and the difference in electrical potential across the membrane. Establishing a *proton gradient* in this manner makes it energetically favorable for protons to return to the mitochondrion from the cytoplasm. This is because the protons will be traveling down both a concentration gradient and an electrical gradient, entering the compartment that contains fewer protons and is negatively charged. These two gradients are often lumped together as the *electrochemical gradient.*

Because the entry of protons into the mitochondrion is energetically favorable (i.e., "downhill") when the PMF is established, energy can be generated from proton entry. Protons are normally impermeable to the mitochondrial membrane. If this were not the case, a proton gradient could never be established by the electron-transfer chain in the first place. Thus, for protons to enter the mitochondrion, enzymes are required to facilitate the transport. There exists in the inner mitochondrial membrane an enzyme called the *proton translocating ATP synthase (ATPase).* This enzyme catalyzes the reversible reaction shown as

$$ADP + Pi \leftrightarrows ATP + H_2O \qquad (8)$$

The energy needed to synthesize ATP is derived from the inward transport of the proton down its chemical and electrical gradients. When the enzyme degrades ATP, protons are excluded from the mitochondrion to contribute to the generation and maintenance of the PMF. This process *couples oxidation to phosphorylation.* The process of oxidation (converting oxygen to water) generates a PMF, and this leads to phosphorylation using energy generated by the inward movement of protons down their electrical and chemical gradients, to form a high-energy bond of ATP.

The elucidation of oxidative phosphorylation, and the demonstration of the PMF, was done by Peter Mitchell in the 1960s and 1970s, and the theory is frequently referred to as *Mitchell's chemiosmotic hypothesis* (Mitchell, 1963). Although ridiculed when first proposed, the theory has now gained wide acceptance. All data pertaining to the function of the electron-transfer chain and the ATPase can be explained by this theory. Peter Mitchell was honored for his achievement with the Nobel Prize in Medicine in 1979. Oxidative phosphorylation is the major energy-generating pathway of the cell, and it requires $O_2$. In the absence of $O_2$, ATP is generated only through the glycolytic pathway, but at a much reduced efficiency as compared to oxidative phosphorylation. However, even in the absence of $O_2$, a PMF can still be established by pumping protons out of the cell using the proton-translocating ATPase. This is important, as a number of systems, including the transport of carbohydrates and amino acids into intestinal epithelial cells, require the energy generated by the inward movement of protons down their concentration gradient.

## VII. Food and Energy

As mammals do not eat continuously, mechanisms have been developed to allow energy that is derived from one meal to be stored and used as needed. There are two major energy storage molecules in the body. These are *glycogen,* which is a carbohydrate-based storage medium, and *triacylglycerol,* which is a lipid-based storage medium. Glycogen is stored in the tissues that synthesize it, which are primarily the liver and muscle. Glycogen is stored in small granules, along with the enzymes that both synthesize and degrade it. There are no specialized cells for the storage of glycogen. Lipid, which is stored in the form of triacylglycerol, is stored in specialized cells known as *adipocytes.* As more energy is stored, the size and fat content of the adipocyte increases. The roles of these two storage molecules are different, as is outlined in this chapter.

The energy that is used for cell function and growth is obtained from food in the diet, or from storage molecules that were synthesized when excess food was available. How is food processed such that energy can be obtained from it? In multicellular organisms, this task is carried out by specialized cells found within the digestive tract. In mammals the food enters the stomach, where digestive enzymes from the saliva, stomach, and pancreas act to reduce it to smaller particles. This allows the smaller components to be absorbed by the cells lining the gut and intestines. Certain compounds, such as ethanol, can be absorbed directly by the stomach, but the majority of nutrients are absorbed within the intestine. Once absorbed by the *intestinal epithelial cells,* the small components may or may not be metabolized, before being transported across the cell into either the blood or the lymphatic system, which carries the nutrients to other organs and cells. This process is known as *vectorial transport* (Fig. 14). During vectorial transport, the nutrients, which are at a high concentration in the lumen of the intestine, are transported down their concentration gradient into the epithelial cell. As the concentration of the nutrients increase in the epithelial cell, they are then transported out of the cell into the blood or lymph down their concentration gradient. The transport event can be either *active* or *carrier-mediated facilitated diffusion.* Active transport refers to the process of actively concentrating material within a compartment, such that transport can occur even when the concentration gradient of the nutrients is greater inside the cell than outside. To accomplish active transport, energy must be expended. Carrier-mediated facilitated diffusion is not energy dependent and refers to the use of membrane-bound carriers to transport the nutrients down their concentration gradient. Both mechanisms of transport are used during intestinal absorption of nutrients.

Once food that has been ingested is broken down into constituent sugars, amino acids, and fatty acids, it is transferred to the blood or lymphatic system. Other tissues of the body will use these compounds. They will be used either for storage, such that energy can be obtained at a later time, or for the immediate generation of energy. *These two opposing pathways are never operative at the same time.* To understand why this is so, the basic principles of metabolic regulation need to be addressed.

## VIII. Basic Pathways That Need to Be Regulated

Of key importance is the regulation of pathways that either generate or store energy. The structures of the energy storage molecules glycogen and triacylglycerol are shown in Fig. 15. Glycogen synthesis occurs when blood sugar levels are elevated, indicating adequate energy is available. Glycogen is stored within the tissues that make it, primarily the liver and muscle. As is seen in this chapter, the role of glycogen is different in these two organs. Triacylglycerol is the primary fuel storage compound of mammals. Triacylglycerol, unlike glycogen, is stored in specialized cells known as adipocytes, which have an almost unlimited capacity to store fat in the form of triacylglycerol. Similar to glycogen synthesis, triacylglycerol is produced when blood glucose levels are high, storing energy for later use. The building blocks of triacylglycerol, fatty acids, are synthesized from acetyl-CoA, which is obtained from pyruvate. Fatty acid degradation also leads to the production of acetyl-CoA. As acetyl-CoA cannot form glucose, the oxidation of fatty acids can only lead to energy production, and not to gluconeogenesis. This has important complications for certain physiological conditions, which will be discussed later.

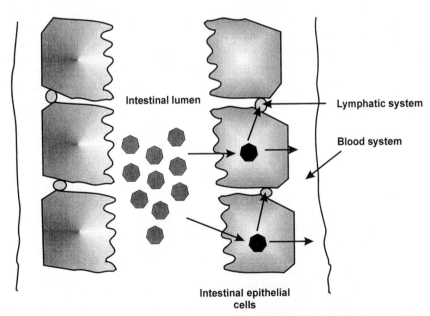

**FIG. 14.**    Vectorial transport by the intestinal epithelial cell. Component X, represented by the hexagons, a required nutrient, is transported into the cell from the intestinal lumen, down its concentration gradient. Once inside the cell X may or may not be metabolized, and then it, or its metabolic product, is transported across the other side of the cell into either the blood or lymph. Removing X from the cell ensures that the concentration of X will be lower within the epithelial cell than in the intestinal lumen, and that as much X as possible can be absorbed from the lumen.

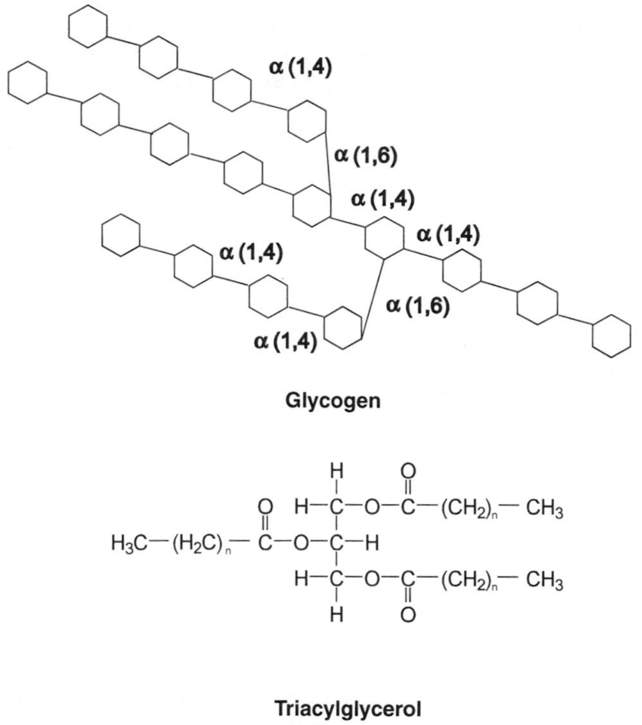

**Glycogen**

**Triacylglycerol**

**FIG. 15.** The structures of the major energy-storage forms of the cell, glycogen and triacylglycerol. Glycogen is a branched polymer of glucose. The basic linkage between glucose residues occurs between carbons 1 and 4 of adjacent glucose molecules. The branches are formed by linkages between the 1 and 6 carbons of adjacent glucose molecules. Triacylglycerol is composed of a glycerol backbone to which three fatty acids are esterified.

## A. Synthesis vs Degradation: Regulatory Principles

The metabolic pathways that synthesize the basic compounds are different from those responsible for their degradation. For example, the pathways of fatty acid synthesis and degradation are different and are compartmentalized. The rationale for this is to prevent the occurrence of *futile cycles,* in which a compound will be synthesized and then immediately degraded by the degradative pathway. Such futile cycles lead to no net gain of material and a useless loss of energy. By having alternate pathways for both the synthesis and degradation of a compound, futile cycles can be avoided by coordinate regulation of the pathways. *Typically, conditions that activate biosynthetic pathways simultaneously inhibit the degradative pathways.*

There are a number of different ways in which pathways can be regulated. Some of these are listed in Table 2. Regulation of pathways occurs by altering the activity of one or more enzymes specific to that pathway, such as controlling the amount of enzyme that is present. Upon activation of the pathway, specific gene transcription is induced that leads to more enzyme being made by the protein synthesis machinery of the cell. This is known as *up-regulating* the enzyme. Producing more enzyme means that increased levels of substrate can be acted on per unit time, and the pathway reaction will proceed at an overall

faster rate. Conversely, if a certain pathway is not used for an extended period of time, the enzymes of that pathway are *down-regulated,* such that less enzyme is made, and the pathway will be less active. Because this type of regulation requires alterations at the genomic level, they do not occur rapidly, but rather require 4 to 7 days before the changes in enzyme levels become apparent. Such regulation of enzyme activity is known as *long-term adaptation.* An example of enzymes regulated by long-term adaptation are five enzymes necessary for synthesizing fatty acids. When food supplies are plentiful over a 4- to 7-day period, the synthesis of these enzymes is increased; under conditions of minimal food intake, the synthesis and levels of these enzymes are decreased.

Long-term adaptation to environmental stimuli, while an important regulatory control, does not provide for a rapid means of altering enzyme activity. This is frequently desired, particularly under conditions of stress and exercise. To rapidly regulate pathways, other control mechanisms are utilized: *allosteric modification* of enzyme activity, and *covalent modification* of the enzyme, which leads to an alteration in enzyme activity. The kinetics of allosteric enzymes have been discussed previously (see Fig. 7). These enzymes have their activities either enhanced (activated) or decreased (inhibited) by small allosteric effectors. Activators tend to *lower* the $K_m$ of the enzyme, and inhibitors tend to *increase* the $K_m$ of the enzyme. An example of an enzyme regulated in this manner is *phosphofructokinase-1 (PFK-1)* (Pilkis *et al.,* 1990), the key regulatory step of glycolysis.

PFK-1 is *activated* by either AMP or fructose 2,6-bisphosphate (F2,6BP), and is *inhibited* by either citrate or ATP. To understand why these compounds are used as regulators, it is necessary to comprehend the role of the pathway being regulated. The major role of glycolysis is to produce energy. This occurs in two ways: anaerobic production of *ATP* by the pathway, or production of *pyruvate,* which will be converted to acetyl-CoA, which, when oxidized by the TCA cycle, will generate energy through oxidative phosphorylation. Thus, under conditions in which energy is required, glycolysis is activated. High levels of ATP indicate that adequate energy levels are available, and glycolysis should not be activated. Thus, ATP inhibits PFK-1 and reduces the glycolytic rate. Alternatively, when ATP levels are low, and AMP levels elevated, it is an indication that energy is required by the cell. AMP will

**TABLE 2** Different Methods for Regulating Enzyme Activity

| Method of regulation | Example |
| --- | --- |
| Change amount of enzyme through alterations in gene transcription | Enzymes required for fatty acid biosynthesis |
| Allosteric modifications | Regulated enzymes of glycolysis |
| Covalent modifications | Regulated enzymes of glycogen metabolism |

therefore activate PFK-1 to increase ATP levels. This activation continues until sufficient ATP is produced, at which point the newly synthesized ATP will inhibit enzyme activity. As AMP is a precursor to ATP, the increase in ATP will also lead to a decrease in the levels of AMP, thereby simultaneously decreasing the concentration of an activator of the enzyme.

*Citrate* is produced in the mitochondrion as an intermediate of the Krebs TCA cycle, whose major function is to produce energy. High energy levels will inhibit the rate at which this cycle functions. When energy levels are high within the mitochondrion, the TCA cycle slows down, and this leads to an increase in the concentration of citrate. Thus, high levels of citrate indicate adequate levels of energy. Citrate in the mitochondrion can diffuse across the mitochondrial membrane and into the cytoplasm. Once in the cytoplasm, the citrate binds to PFK-1 and inactivates it, in a manner similar to ATP. Remember that for citrate levels to increase in the cytoplasm, adequate energy levels must be available. If adequate ATP is available, the rate of glycolysis should decrease.

The role of fructose 2,6-bisphosphate (F2,6BP) is more complex. The levels of F2,6BP increase in response to *hormones* that monitor blood glucose levels and indicate to the liver whether energy needs to be created or stored. It is through the modulation of F2,6BP levels that hormonal signals can regulate glycolysis. The enzyme that produces F2,6BP is activated as the result of specific hormone binding to the cell surface (in this case the hormone is *insulin*), or inhibited when a different hormone (such as *glucagon*) binds to the cell. When the enzyme is activated, F2,6BP levels increase and PFK-1 is activated, thereby enhancing glycolysis. Hormonal control of metabolic pathways is very important and is reemphasized when the regulation of glycogen metabolism is discussed.

A third means of regulating enzyme activity is through covalent modification of the enzyme. An example of this is the enzyme *pyruvate kinase (PK)*, another regulated enzyme in the glycolytic pathway. Pyruvate kinase catalyzes the reaction that produces pyruvate and ATP from phosphoenol pyruvate. Upon appropriate hormonal stimulation, an enzyme that can phosphorylate PK in the liver is activated. Enzymes that phosphorylate other proteins are known as *kinases*. PK is phosphorylated, and its activity in liver is reduced as a result of the covalent modification. A reduction in PK activity results in less pyruvate being produced, and an overall reduction in the glycolytic rate. These changes correlate with the regulation of PFK-2 by F2,6BP. It was stated earlier that the production of F2,6BP is hormonally regulated. The changes in PK activity and F2,6BP production are *coordinately regulated*. The hormones that lead to increases in F2,6BP production (such as insulin) also lead to activation of PK. This ensures that glycolysis is maximally active when insulin levels are high. Thus, hormonal signals can be transmitted into alterations in metabolic pathways through the regulation of a number of key enzymes in the pathway.

A special case of allosteric modification is known as *feedback inhibition* (Fig. 16). Feedback inhibition refers to allosteric modification by an end product(s) of a pathway,

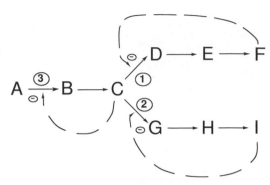

**FIG. 16.**   Examples of feedback inhibition. End products F and I can inhibit the enzymes at the branch points from C (enzymes 1 and 2), whereas product C can inhibit enzyme 3. Product C will only accumulate when both enzymes 1 and 2 are inhibited (i.e., the end products F and I are present at high concentration). This type of feedback inhibition mechanism insures that adequate quantities of all required compounds are available before any step of the pathway is inhibited.

which will feed back and modulate the activity of an enzyme that catalyzes a committed step of that pathway. As seen in Fig. 16, this can occur in a number of ways, particularly when branched pathways are considered. Although feedback inhibition is not universal for all pathways, it does generally occur in biosynthetic pathways to ensure that a particular product is not overproduced.

## B. Hormone Binding to Cells

It is apparent that hormone binding to cells is important in regulating biochemical pathways. A cell will be responsive to a particular hormone if the cell expresses a *receptor* specific for that hormone. For the case of peptide hormones, the receptor is on the cell surface. For steroid hormones, the receptor is intracellular. Hormone binding to the receptor can be analyzed in a manner similar to enzyme kinetics. From these analyses one can determine the affinity of the hormone for the receptor, known as the *dissociation constant, $K_d$*. In addition, the total number of hormone receptors on the cell surface, known as $R_t$, can be determined. This analysis was first worked out in 1949 by Scatchard, who derived the equation

$$\frac{B}{F} = -\frac{1}{K_d}(B) + \frac{R_t}{K_d} \qquad (9)$$

where $B$ represents the concentration of hormone bound to the receptor and $F$ represents the concentration of hormone present at equilibrium that is not bound to receptor. Equation 9 is in the form of a straight line, in which the $y$-axis represents the value of the bound/free ($B/F$) ration, and the $x$-axis represents the bound ($B$) material. The slope of the line defines the dissociation constant ($-1/K_d$), and the $y$-intercept gives $R_t/K_d$. Extrapolating to the $x$-intercept gives $R_t$ and allows the total number of receptors to be obtained. An example of a *Scathard plot* is shown in Fig. 17. The Scatchard equation, although it does have certain limitations, has proven useful for the determination of both binding affinities and types of receptors on a particular cell. Knowing the affinity of a hormone for its receptor means that the concentration of hormone required for a cell to respond to that hormone can be determined.

## C. Regulation of Glycogen Metabolism

The degradative and biosynthetic pathways of glycogen are reciprocally regulated. When one pathway is activated, the other pathway is inhibited (Johnson, 1992). The regulation of these pathways is shown in Fig. 18 and is summarized in the next paragraph.

Glycogen synthesis is catalyzed by the enzyme *glycogen synthase*. The degradation of glycogen is catalyzed by *glycogen phosphorylase*. There are two major ways in which these enzymes are regulated: by covalent modification, and by allosteric modification. We only consider the effects of covalent modification in this discussion. Before discussing the role of covalent modification it is necessary to understand the role of glycogen in metabolism. As stated previously, glycogen is an energy-storage molecule found primarily in the liver and muscle. The job of muscle glycogen is to present the muscle with an immediate source of glucose for energy production. The major function of liver glycogen is to provide glucose to all of the other organs of the body via the circulation. Thus, when blood glucose levels are low, liver glycogen will be degraded into glucose, and the glucose released into the circulation for use by other tissues.

Despite the different functions of glycogen in muscle and liver, the regulation of the metabolic pathways is the same in both tissues. *Both sets of enzymes are regulated by phosphorylation.* Glycogen synthase, when phosphorylated, is inactive. Glycogen phosphorylase, when phosphorylated, is active. This is a simple mechanism to ensure that both pathways are not active at the same time. This leads

to the question of what activates and deactivates the phosphorylation system? The answer is circulating hormones. We have already seen that the major regulatory enzymes of the glycolytic pathway are controlled by hormones. The same hormones that regulate glycolysis also regulate glycogen metabolism. The next section describes how hormones regulate both glycogen metabolism and glycolysis.

## D. cAMP/Phosphorylation Cascade

For a peptide hormone to initiate a cellular response, it must first bind to a specific receptor on the cell surface. Hormones are synthesized by specific tissues and released into the circulation. Once in the blood, the hormone will bind to specific receptors on the surface of appropriate target cells. Of prime importance for our discussion of glycogen metabolism are the hormones *insulin* and *glycogen*. These hormones regulate the level of glucose within the circulation. Specific cells within the pancreas can measure the level of glucose in the blood and release the appropriate hormone. When blood glucose levels are low, the pancreas releases glucagon, which activates glycogen degradation in the liver to raise blood glucose levels. When the blood glucose levels are high, the pancreas releases insulin so that both the muscle and liver will now synthesize glycogen to store excess glucose.

Once a hormone binds to its receptor, the receptor alters its conformation and activates intracellular enzymes to signal the cell. In the case of glucagon secretion, indicating low blood glucose levels, hormone binding will activate the enzyme *adenylate cyclase*, which converts ATP to *cyclic AMP (cAMP,* Fig. 19). The increase in intracellular cAMP levels activates a *cyclic AMP-dependent protein kinase*, which will phosphorylate various enzymes (Sutherland, 1972). Two of the enzymes that are phosphorylated are *glycogen synthase* and *glycogen phosphorylase kinase*. Phosphorylation of glycogen synthase inactivates the enzyme. Phosphorylation of glycogen phosphorylase kinase activates this second kinase. The primary substrate of glycogen phosphorylase kinase is *glycogen phosphorylase*, which upon phosphorylation will be activated. Thus, under conditions of glucagon release the enzyme that makes glycogen is inactivated, and the enzyme that degrades glycogen is activated. This enables the liver to degrade its glycogen to produce glucose and raise blood glucose levels. Muscle does not respond to glucagon, as muscle does not express cell-surface receptors for glucagon.

The release of insulin indicates that blood glucose levels are high. Both muscle and liver express insulin receptors and will therefore respond to this hormone. When this occurs, insulin binds to its specific receptors and reduces intracellular cAMP levels. When this occurs the cAMP-dependent protein kinase is inactivated, and phosphorylation of the key enzymes no longer occurs. This allows enzymes known as *phosphatases* to remove the phosphates from the enzymes, which activates glycogen synthase and inactivates both glycogen phosphorylase kinase and glycogen phosphorylase. Thus, under conditions of insulin release, glycogen synthesis is favored in both muscle

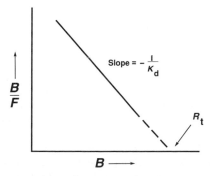

**FIG. 17.** An example of a Scatchard plot. The slope of the line corresponds to $-1/K_d$, and the $x$-intercept will yield the total number of binding sites. If it is known how many binding sites there are per receptor, then the total number of receptors can also be estimated from this analysis.

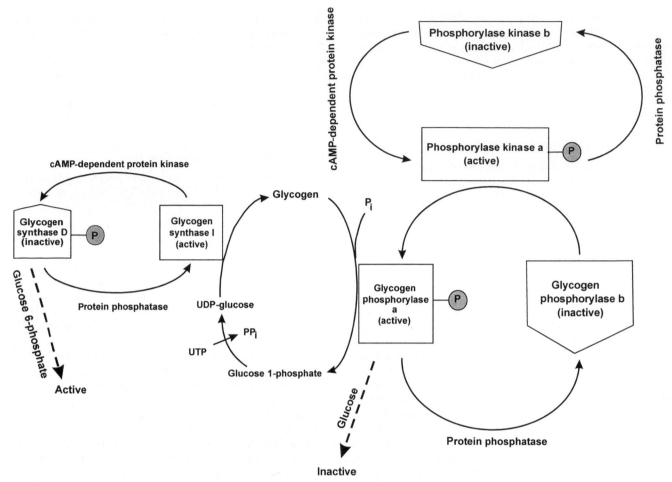

**FIG. 18.**   Regulation of glycogen metabolism. The two key enzymes involved in degrading (glycogen phosphorylase) and synthesizing (glycogen synthase) glycogen are reciprocally regulated by phosphorylation. In addition, the enzymes responsible for phosphorylating these two key enzymes are themselves regulated, either by phosphorylation (phosphorylase kinase) or by allosteric activation (cAMP-dependent protein kinase). This system of regulation ensures that a futile cycle is not established, in that only one pathway is active at any given time. The regulation indicated by the dotted lines is allosteric regulation of the two key enzymes. Allosteric regulation ensures that the enzymes can be activated or inhibited even in the absence of the phosphorylation controls.

and liver, and glycogen degradation is inhibited in the liver.

### E.  Tissue-Specific Regulation

The major role of liver glycogen is to act as a glucose reservoir for the bloodstream. When blood glucose levels drop, liver glycogen is degraded to glucose for use by other tissues. For this to occur, the pathway of liver glycolysis needs to be inhibited when liver glycogen degradation is activated. If it were not, liver glycolysis would use the glucose, and less glucose would be available for export to other tissues. Liver glycolysis is inhibited because two key enzymes of liver glycolysis (one of which is PK) are phosphorylated, and inactivated, by the activated cAMP-dependent protein kinase. This reduces the activity of the glycolytic pathway to a very low level. While the liver is degrading glycogen and releasing glucose for other tissues,

the muscle is also degrading glycogen for its own use. Since the muscle is using the glucose it is producing, *muscle glycolysis needs to be active during periods when liver glycolysis is inhibited.* Thus, *tissue-specific regulation* occurs. In the case of glucagon release, it is relatively clear why muscle glycolysis can continue when liver glycolysis is inhibited. It is because muscle does not contain glucagon receptors. This is actually one way in which differential tissue regulation is accomplished. Thus, when glucagon is released, the cAMP-dependent protein kinase is not activated in muscle.

There is another hormone, however, that will simultaneously activate the cAMP-dependent protein kinase in *both* liver and muscle. This hormone is *epinephrine*. Epinephrine is released upon times of stress or exercise. It indicates that the muscles will be doing a lot of work and will need a lot of energy in the form of glucose. However, even when epinephrine is released, liver glycolysis is inactive while muscle glycolysis remains active. This occurs because muscle contains slightly different forms of the two glycolytic enzymes that are inhibited by phosphorylation in the liver. The muscle enzymes catalyze the same reactions as are catalyzed in the liver; however, the muscle enzymes are not inhibited by phosphorylation (Pilkis *et al.*, 1990). This is an example of *isozymes,* two proteins with different structures that catalyze the same reaction. In most cases the regulation of isozymes is different, even though the reaction catalyzed by each protein is the same. In muscle, phosphorylation of these key glycolytic isozymes does not inhibit their activity. Thus, the muscle isozymes have evolved to be regulated differently than the liver isozymes and lead to tissue-specific regulation.

## F. How Does Fat Storage Fit In?

The major energy storage form of the body is triacylglycerol. Energy can be derived from triacylglycerol by converting it to *glycerol* and *free fatty acids.* The glycerol enters the glycolytic pathway to generate energy, and the fatty acids generate energy by being converted to acetyl-CoA, which is then oxidized by the TCA cycle. Fatty acid oxidation generates considerable energy and is a very efficient energy source. In a manner similar to glycogen metabolism, the pathways of fatty acid synthesis and degradation are reciprocally controlled. Under conditions in which the body requires energy (such as glucagon or epinephrine release) the enzyme that initiates the synthesis of fatty acids (*acetyl-CoA carboxylase*) is phosphorylated and inactive. This results from the activation of the cAMP-dependent protein kinase by the hormone, which then directly phosphorylates acetyl-CoA carboxylase. The enzyme that initiates the degradation of triacylglycerol molecules from the adipocyte is also phosphorylated by the cAMP-dependent protein kinase and is activated. This enzyme, known as *hormone-sensitive lipase,* catalyzes the release of fatty acids from the adipocyte into the circulation, where they travel to other tissues to be used as an energy source. It is important to remember that simultaneous with fatty acid release, the liver has also switched to the degradation of glycogen to provide glucose for the blood. Both of these pathways have been activated by the same mechanism.

*Fatty acids cannot be converted to glucose.* Thus, even though there is more energy available from fatty acids than from glycogen stores in the average individual, and fatty acid breakdown is also activated by glucagon release, the liver must still produce glucose from glycogen if blood glucose levels are low. This is important because the brain has an absolute requirement for glucose and cannot use fatty acids for energy. This metabolic oddity leads to difficulties in times of prolonged fasts or starvation. Under conditions in which food intake is drastically reduced, blood glucose levels are low for extended periods of time. Unfortunately, the liver only contains sufficient glycogen to keep blood glucose levels high for approximately 1 day. After the liver glycogen stores are depleted, the liver will still produce glucose, but will do it via the pathway of gluconeogenesis using amino acids and glycerol as a source of carbon. The liver will also start to produce energy sources known as *ketone bodies,* which are produced from

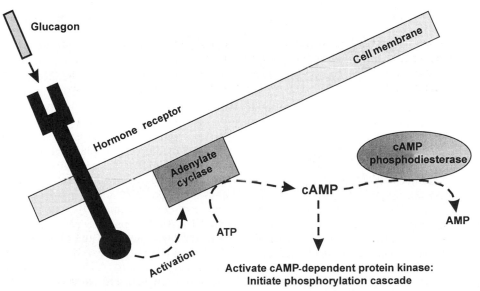

**FIG. 19.** Initiation of the phosphorylation cascade in cells containing the glucagon receptor. Glucagon is released from the α cells of the pancreas in response to low levels of blood glucose and binds to specific cell-surface receptors on the membrane. This binding activates the receptor and a signal is transmitted to the inside of the cell. The nature of this signal is discussed in Chapter 7. This signal activates the enzyme adenylate cyclase such that cAMP is produced. The cAMP produced activates the cAMP-dependent protein kinase, which initiates the regulatory cascade. cAMP is degraded by a phosphodiesterase, which is also activated by opposing hormone action.

high levels of acetyl-CoA. The liver will produce the ketone bodies from fatty acid oxidation and then export them to other tissues to use as an alternative source of energy to glucose. As the other tissues (such as the muscle) use the ketone bodies, there is more glucose available for the brain. The brain can also switch part of its metabolic needs to ketone-body oxidation. Thus, fatty acid oxidation is used to produce ketone bodies, to help reduce the overall body demand for glucose. However, the ketone bodies are acids; too large a buildup of ketone bodies can be detrimental and may lead to acidification of the blood and urine (*keto-acidosis*), which is life threatening.

### G. A Starving Situation

As a unified example of whole-body energy metabolism and regulation, the case of a starving individual is considered. The role of each organ during starvation is considered before integrating all effects. The major function of the "starving" liver is to produce glucose for use by other tissues. To do this, the liver will mobilize glucose from glycogen and produce glucose via gluconeogenesis, and to spare glucose consumption by tissues other than the brain, it will produce ketone bodies from fatty acids. These processes can occur simultaneously, although they usually occur sequentially. *The overriding signal for a switch to these pathways is a reduction in the insulin:glucagon ratio.* The decreased insulin:glucagon ratio brings about the activation of the cAMP-dependent protein kinase. The cAMP-dependent protein kinase will activate glycogen phosphorylase and inactivate glycogen synthase. This results in a rapid mobilization of liver glucose derived from liver glycogen. However, after 12–24 hours of continual glycogen degradation, the stores of liver glycogen are depleted. At this point the liver will switch to gluconeogenesis to maintain blood glucose levels. The pathway of gluconeogenesis has become favored over glycolysis because of the action of the cAMP-dependent protein kinase. This enzyme has been activated by glucagon binding to its receptor and has phosphorylated two key glycolytic enzymes, both of which are inhibited upon phosphorylation. The liver uses as substrates for gluconeogenesis glycerol and amino acids. The glycerol is provided by adipocytes from the degradation of triacylglycerol to glycerol and fatty acids. The amino acids are derived from protein turnover and degradation.

The liver is also beginning to oxidize fatty acids, which have been released from the adipocyte by the activation of hormone-sensitive lipase. The liver is not synthesizing fatty acids because the enzyme neeed to do this, acetyl-CoA carboxylase, is inactivated when the cAMP-dependent enzyme is activated. The oxidation of fatty acids leads to an increase in acetyl-CoA levels in the liver, and this can lead to ketone body formation. The ketone bodies produced are then exported to other tissues to use as an alternative energy source to glucose. The events that occur in the liver under starvation conditions are summarized in Fig. 20.

The role of the *adipocyte* during starvation is to degrade triglycerides to fatty acids and glycerol. This occurs through activation of the hormone-sensitive lipase by the cAMP-dependent protein kinase. The cAMP-dependent protein kinase has been activated by the release of glucagon, brought about by the decrease in blood glucose levels due to the reduced food intake. Glucose metabolism in the adipocyte is reduced because of the limited availability of glucose. As excess accumulation of ketone bodies leads to acidification of the blood, there is a feedback mechanism to try to keep ketone-body formation under control. High levels of ketone bodies will inhibit the degradation of tri-acylglycerol, thus partially regulating the level to which ketone bodies can accumulate. The events that occur in the adipocyte under starvation conditions are summarized in Fig. 21.

The muscle, under starvation conditions, still needs to do work, but it has a problem. During the first day of starvation the muscle glycogen stores are utilized through allosteric activation of glycogen phosphorylase. This is necessary because the muscle does not have glucagon receptors, and the cAMP-dependent protein kinase has not been activated in the muscle. However, since blood glucose levels are low and insulin is not present, the glucose that is in the blood is not entering the muscle at a rapid pace. After 12 to 24 hours, the muscle glycogen stores have been depleted, and the muscle will switch to fatty acid and ketone-body oxidation for energy. Under conditions of extreme starvation the muscle will begin to degrade its own proteins to generate amino acids, which will be sent to the liver for conversion to glucose. This leads to the wasting observed in starved individuals. It should be noted that glycolysis is active in muscle due to allosteric activation. The interactions between all of the organ systems under starvation conditions are shown in Fig. 22.

## IX. Energy Forms Revisited

How is energy stored in the tissues? Energy is in the form of high-energy phosphate bonds in nucleoside triphosphates, such as ATP. The last two phosphoanhydride bonds in ATP, when hydrolyzed, give rise to 7.3 kcal/mol of energy, which for biological systems is a large amount. As previously discussed, ATP is synthesized either through substrate-level phosphorylation (as in glycolysis), or through oxidative phosphorylation. In the muscle, high-energy phosphates are also stored as *creatine phosphate* (the structure and synthesis of creatine phosphate are shown in Fig. 23). Creatine phosphate is generated by the action of *creatine kinase*, in which creatine is phosphorylated by ATP to produce ADP and creatine phosphate, with the high energy of the phosphate bond preserved in the creatine phosphate. The ADP that is produced can be regenerated into ATP by the action of the enzyme *myokinase*, which will convert two ADP molecules into one ATP molecule and one AMP molecule. The ATP can then be used to produce more creatine phosphate, which is used to store energy in muscle. Both ATP and creatine phosphate are energy-rich by virtue of their phosphate–oxygen or phosphate–nitrogen bonds, which when broken release a large amount of energy (7.3 kilocalories per mole

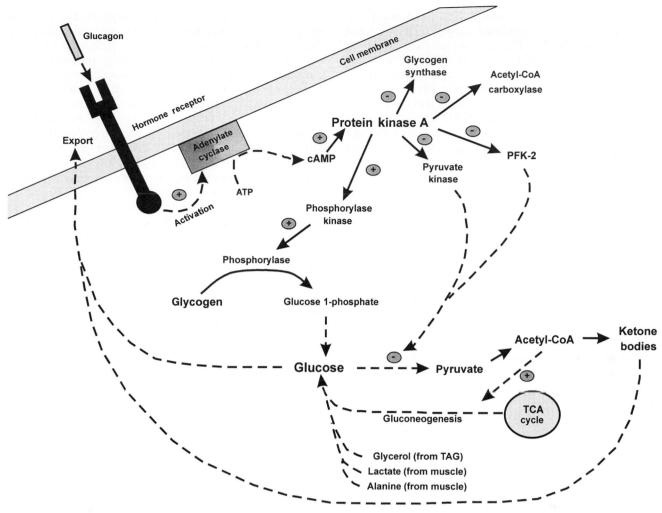

**FIG. 20.**    Regulation of metabolic pathways in the liver under starving conditions. All of the metabolic changes are the result of the activation of the cAMP dependent protein kinase by glucagon binding to its receptor. Once activated, the cAMP-dependent protein kinase phosphorylates the following enzymes: two enzymes of glycolysis (PFK-2 and PK) that are inactivated, one enzyme required for fatty acid synthesis (acetyl-CoA carboxylase) that is inactivated, and two enzymes involved in glycogen metabolism, activating one (phosphorylase kinase) and inactivating the other (glycogen synthase). These enzyme activity changes alter the overall flow of metabolism in the liver to one of energy export, as glycogen is degraded to glucose, and fatty acids are converted to ketone bodies, both of which are exported. Gluconeogenesis is enhanced by the inhibition of glycolysis, allowing glucose to be produced from amino acids and other glucogenic degradation products.

of energy released for ATP, 10.3 kilocalories per mole of energy released for creatine phosphate). Subsequent metabolic reactions will utilize this energy to direct a series of linked reactions, such as a pathway, in the forward direction.

Why does muscle have this alternative energy source? The use of creatine phosphate as an energy source permits the muscle to store only small amounts of ATP. This is important because ATP is often used as a regulator of enzymes involved in the production of ATP. By keeping the levels of ATP low, and storing energy in the form of creatine phosphate, the enzymes that produce ATP will continue to make ATP, and the muscle can store more high-energy compounds than if only ATP itself were the storage form. What happens when the muscle requires energy rapidly? The muscle creatine kinase catalyzes its

reverse reaction and produces ATP from creatine phosphate and ADP. The muscles store creatine phosphate such that when bursts of energy are required there is sufficient energy available to carry out the required task. In fact, during a sprint, which generates energy through anaerobic glycolysis, the initial burst of activity is fueled through the stored creatine-phosphate levels, as the amount of ATP within the muscles is only sufficient for a 2-second burst of activity.

## X. Cori Cycle

The end product of glycolysis in muscle is pyruvate. When muscle needs to work fast and generate ATP at a rapid pace, this is best accomplished using glycolysis in the absence of oxygen. The problem that results from this is that glycolysis needs a continual supply of an oxidized electron carrier, NAD⁺. During glycolysis the NAD⁺ is converted to NADH, and usually the NADH is reconverted to NAD⁺ via the electron-transport chain. In the absence of oxygen, the electron-transport chain is inoperative. This will slow down glycolysis unless there is a means of converting the NADH back into NAD⁺. This does occur through a reduction of pyruvate to lactate, using as the electron donor NADH. The enzyme that catalyzes this reaction is *lactate dehydrogenase.* The lactate that is produced is then sent from the muscle to the liver for its conversion to glucose. If the lactate were to remain in the muscle, the muscle pH would drop, as lactate is an acid. A drop in muscle pH would lead to impaired performance, which is not desirable. Once the liver converts the lactate to glucose, it is sent back to the muscle as an energy source. This cycle is known as the *Cori cycle* and is shown diagrammatically in Fig. 24.

## XI. Summary

All living organisms require energy, which is obtained from the sun and the food that is ingested. The food is

then degraded into smaller components by the action of specific protein molecules, known as enzymes, in the digestive tract. Enzymes are proteins that catalyze specific reactions, primarily by reducing the energy required for the reactants to reach their transition state. Enzyme activity can be studied through enzyme kinetics, which measures the rate at which enzymes catalyze reactions. The rate at which an enzyme works is a function of its binding to the substrate, and this parameter can be approximated by a kinetic parameter, the Michaelis constant, $K_m$. The fastest rate at which the enzyme can produce its product is known as the maximal velocity. Inhibitors of enzyme reactions act either by blocking the substrate from binding to the enzyme, in a process known as competitive inhibition, or by altering the activity of the enzyme without affecting substrate binding, via noncompetitive inhibition. Certain enzymes can be regulated by small effectors. Regulated enzymes display a complex kinetic picture, which is best characterized by sigmoidal kinetics. Regulators either increase or decrease the level of substrate required to reach maximal velocity.

Metabolism is the study of chemical conversions to generate energy. Major pathways required to accomplish this include glycolysis, gluconeogenesis, the TCA cycle, fatty acid synthesis, fatty acid degradation, glycogen synthesis

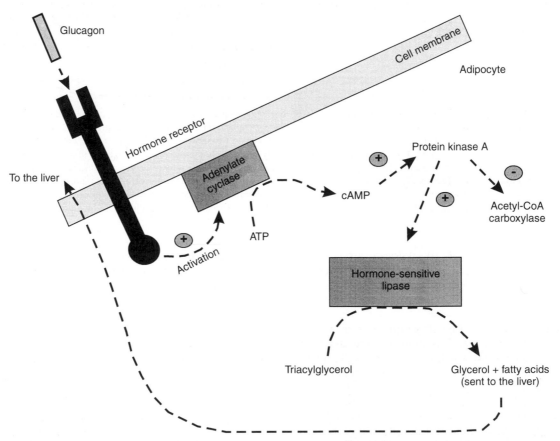

**FIG. 21.** Events that occur in the adipocyte under starvation conditions. As with the liver, all events are initiated by the activation of the cAMP-dependent protein kinase (protein kinase A) and activation or inhibition of the appropriate enzymes within the cell.

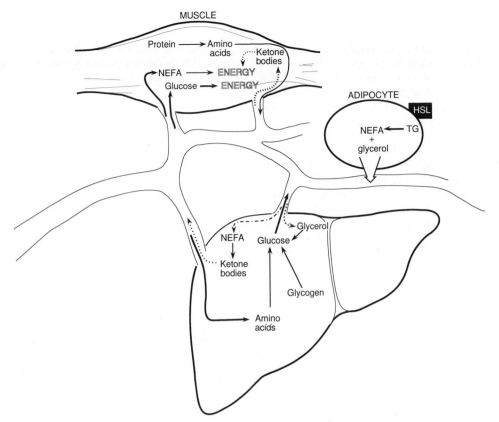

**FIG. 22.** Metabolic interactions between the liver, adipocyte, and muscle under starvation conditions. The liver is producing glucose from glycerol (obtained from the adipocyte) and amino acids (obtained from the muscle) by gluconeogenesis to supply glucose to the brain. The adipocyte is releasing fatty acids into the circulation for use by the muscle and liver as energy sources. The liver is also converting fatty acids to ketone bodies for use by other tissues as an alternative energy source. HSL stands for the enzyme hormone-sensitive lipase, specific for the adipocyte, and NEFA stands for nonesterified fatty acids, that is, fatty acids not attached to glycerol.

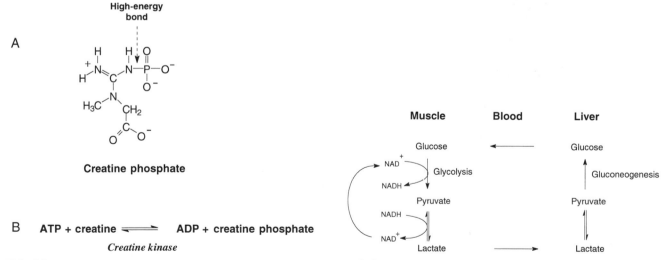

**FIG. 23.** The structure and biosynthesis of creatine phosphate. (A) The structure of creatine phosphate and the location of the high-energy bond. This bond is even more energetic than the one in ATP and will release 10.3 kcal/mol of energy when hydrolyzed. (B) Synthesis of creatine phosphate from ATP, catalyzed by creatine kinase. Creatine kinase catalyzes a reversible reaction; under conditions of high exercise it will generate ATP, and under conditions of rest the direction toward the synthesis of creatine phosphate will be favored.

**FIG. 24.** The Cori cycle. Under conditions of rapid muscle glycolysis, in the absence of oxygen, the supply of the required glycolytic cofactor $NAD^+$ becomes limiting. $NAD^+$ can be regenerated by converting pyruvate to lactate. This, however, leads to an increase in lactate levels within the muscle. This will reduce intracellular pH and inhibit muscle enzyme activity. To overcome the acidification of the muscle, lactate is sent to the liver, where it can be converted to glucose and sent back to the muscle as an energy source.

and degradation, and oxidative phosphorylation. Glycolysis is the process of converting sugars to the three-carbon intermediate pyruvate. Gluconeogenesis refers to the production of glucose from other components, such as amino acids, glycerol, and lactate. The TCA cycle is designed to oxidize acetyl-CoA to $CO_2$ and water, with the generation of a large number of reducing equivalents. The reducing equivalents then donate their electrons to the electron-transfer chain to generate energy through the generation of the proton motive force and oxidative phosphorylation. There are two primary energy storage forms, glycogen (carbohydrate-based) and triacylglycerol (fatty acid-based). Both glycogen and fatty acids are synthesized when energy stores are high, and degraded when energy stores are low. Regulation of these opposing pathways is critical.

It is important to regulate the metabolic pathways such that futile cycles are not maintained, and to accomplish required tissue-specific tasks. Multiple levels of regulation are evident and include regulation of gene transcription, allosteric regulation of enzyme activity, and covalent modification of specific enzymes. Tissue-specific isozymes, which catalyze the same reaction but are subject to different regulatory controls, also are important for providing tissue specificity in the regulatory process. As a general rule, pathways involved in biosynthesis are reciprocally regulated from those that catalyze the degradation of those products. Thus, under conditions of fatty acid and glycogen biosynthesis, the pathways of fatty acid degradation and glycogen degradation are inhibited. By maintaining these rules, the metabolic pathways are directed to accomplish their goals with a minimum of wasted energy.

## Bibliography

Alberts, B., Bray, D., Lewis, J., Raff, M., Roberts, K., and Watson, J. D. (1989). "Molecular Biology of the Cell," 2nd ed. pp. 613–629. Garland Publishing, New York.

Fersht, A. (1985). "Enzyme Structure and Function," pp. 47–120. W. H. Freeman & Co., New York.

Foster, D. W., and McGarry, J. D. (1983). The metabolic derangements and treatment of diabetic ketoacidosis. *New Engl. J. Med.* **309,** 159–169.

Johnson, L. N. (1992). Glycogen phosphorylase: Control by phosphorylation and allosteric effectors. *FASEB J.* **6,** 2274–2282.

Lands, W. E. M. (1991). Biosynthesis of prostaglandins. *Annu. Rev. Nutrition* **11,** 41–60.

Lehninger, A. L., Nelson, D. L., and Cox, M. M. (1993). "Principles of Biochemistry," 2nd ed., pp. 134–159, 198–239, 364–399, 736–788. Worth Publishers, New York.

Mitchell, P. (1963). Molecular, group and electron translocation through natural membranes. *Biochem. Soc. Symp.* **22,** 142–169.

Pilkis, S. J., el Maghrabi, M. R., and Claus, T. H. (1990). Fructose-2,6-bisphosphate in control of hepatic gluconeogenesis. From metabolites to molecular genetics. *Diabetes Care* **13,** 582–599.

Scatchard, G. (1949). The attractions of proteins for small molecules and ions. *Ann. N. Y. Acad. Sci.* **51,** 660–672.

Sutherland, E. W. (1972). Studies on the mechanisms of hormone action. *Science* **177,** 401–408.

Keith D. Garlid

# 7

# Physiology of Mitochondria

## I. Introduction

ATP is the common energy currency of the cell, where it is used to fuel the work of muscle contraction, protein synthesis, and active transport of ions. Most cellular ATP is synthesized by oxidative phosphorylation, in which diverse forms of chemical energy derived from food and body stores are transduced by oxidative phosphorylation into ATP. The complex of enzymes that couple substrate oxidation to ATP synthesis is located in mitochondria. The internal aqueous compartment of mitochondria, called the **matrix,** contains the enzymes of the Krebs tricarboxylic acid cycle. The matrix is enclosed by a highly folded, insulating membrane called the **inner membrane,** and this structure is separated from the cytosol by a more permeable **outer membrane.**

The inner membrane contains the anion carriers responsible for shuttling metabolic substrates between matrix and cytosol. It also contains the cation carriers and channels that regulate mitochondrial volume and matrix calcium. In brown fat cells, the inner mitochondrial membrane contains a unique protein whose function is to dissipate energy both for thermogenesis and to avoid obesity. The vectorial transport enzymes responsible for substrate oxidation, electron transport, and ATP synthesis are all inner membrane proteins. In mitochondria, the action is in the inner membrane.

As one might expect of a membrane-bounded organelle, the mitochondrion has its own transport physiology, which governs its survival and performance within the cytosolic environment. The physiological setting in which these transport cycles operate is provided by the chemiosmotic theory.

## II. Chemiosmotic Theory

The chemiosmotic theory describes, with elegant simplicity, the mechanism by which substrate oxidation is coupled to ATP synthesis. Working far in advance of experimental evidence, Mitchell (1961) postulated that nature uses protonic batteries to perform work, and he proposed that biological energy conservation is a problem in membrane transport. The chemiosmotic theory consists of four postulates, each of which has been validated by experiment.

1. The inner membrane contains electron transport enzymes that are vectorially oriented so that the energy of electron transport drives ejection of protons outward across the membrane. The energy of substrate oxidation is thereby stored as a proton electrochemical potential gradient, called the proton-motive force.
2. The inner membrane contains the F1Fo-ATPase, which is also vectorially oriented so that the energy of ATP **hydrolysis** will drive protons outward across the inner membrane. The ATPase is reversible, so that protons driven inward through the ATPase by the redox-generated proton-motive force will cause ATP synthesis.
3. The inner membrane must have a low diffusive permeability to protons and ions generally. Otherwise, ion leaks would short-circuit the proton-motive batteries, and ATP would not be synthesized.
4. The inner membrane must contain anion-exchange carriers to enable substrates to reach their enzymes in the matrix, and it must contain cation-exchange carriers to remove cations that entered the matrix by diffusion down the very large electrical gradient caused by outward proton pumping.

The consequence of the first three postulates, which are diagrammed in Fig. 1, is the generation of a proton-motive force ($\Delta p$), which is defined as the electrochemical proton gradient divided by the Faraday constant ($\Delta \bar{\mu}_{H^+}/F$):

$$\Delta p = \Delta \Psi - Z \Delta pH \qquad (1)$$

where $Z \equiv (RT \ln 10)/F = 59$ mV at 25°C, and $\Delta \Psi$ is the membrane potential (inside minus outside). The first measurement of pmf was made by Mitchell and Moyle (1969), who obtained a value of $-230$ mV in State 4. Subse-

*111*

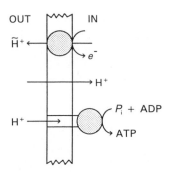

**FIG. 1.** Chemiosmotic energy coupling of electron transport to ATP synthesis. The vectorial enzymes of electron transport and ATP synthesis are coupled indirectly via the proton-motive force. Efficient transfer of proton-motive energy from the redox chain to the ATP synthase is ensured by the low permeability of the inner membrane to protons.

quent measurements have yielded values of about $-200$ mV in isolated rat liver mitochondria respiring in the nonphosphorylating state.

The fourth postulate demonstrates a deep insight into the physiological hazards posed by chemiosmotic coupling. As Mitchell (1966) has written,

> While the introduction of the foregoing sophistication [the chemiosmotic theory of energy coupling] solved one problem, it created another; for, the membrane potential that would now be required to reverse the ATPase reaction would cause the ions of opposite sign of charge to the internal aqueous phase to leak in through the coupling membrane. To prevent swelling and lysis, the ion leakage would have to be balanced by extrusion of ions against the electrical gradient. It was therefore necessary to postulate that the coupling membrane contains exchange-diffusion systems ... that strictly couple the exchange of anions against $OH^-$ ions and of cations against $H^+$ ions (p. 448).

It is noteworthy that the proposal for electroneutral exchange carriers was made at a time when there was no experimental evidence for the existence of ion-exchange carriers in any membrane. (Indeed, the first evidence for the existence of $Na^+/H^+$ exchangers in biological membranes was obtained in Mitchell's laboratory.) Rather, they were born out of physiological necessity, in an elegant application of the scientific method.

Mitchell also introduced a new and useful terminology to distinguish different modes of ion transport. Transport of an ion by itself was called **uniport,** ion transport in exchange with another ion was called **antiport,** and cotransport of two or more ions was called **symport.**

### III. Measurement of Proton-Motive Force

Equilibrium distributions of cations and weak acids are used to measure $\Delta\Psi$ and $\Delta pH$, respectively. Several techniques are available to measure both quantities, and they are well described in Nicholls and Ferguson (1992). All approaches to measuring $\Delta\Psi$ are calibrated to the distribution of $K^+$, which is brought to electrochemical equilibrium

with the ionophore, valinomycin. $\Delta\Psi$ is then given by the Nernst equation:

$$\Delta\Psi = RT/F \ln [K^+]_{in}/[K^+]_{out} \qquad (2)$$

As is discussed later in this chapter, $[K^+]_{in}$ remains relatively constant at 150–180 m$M$, because net $K^+$ transport changes matrix volume rather than $K^+$ concentration. We have measured $\Delta\Psi$ values as high as $-200$ mV in respiring mitochondria, reflecting a 2000-fold concentration gradient of $K^+$, when it is in equilibrium with $\Delta\Psi$. Thus, $[K^+]_{out}$ is maintained at about 70 $\mu M$, and further addition of $K^+$ to the medium is followed by rapid $K^+$ uptake to reestablish equilibrium.

$\Delta pH$ is estimated from the equilibrium distribution of weak acids. If the inner membrane is permeable to the **acid,** but not to the **anion,** the equilibrium distribution is given by

$$(A-)_{in}/(A-)_{out} = 10^{z\,\Delta pH} \qquad (3)$$

where $z$ is the valence of the acid. The standard measurement of $\Delta pH$ in mitochondria uses the distribution of acetate.

Under normal resting conditions, isolated rat liver mitochondria maintain $\Delta\Psi$ at $-170$ to $-180$ mV and $\Delta pH$ at 0.3 to 0.5 units (18 to 30 mV).

### IV. Ion Leaks in Mitochondria

Despite the low permeability of the inner membrane to ions, cation leaks occur at significant rates in respiring mitochondria. This occurs because $\Delta\Psi$ is maintained at very high levels. Ion leaks in mitochondria are physiologically important: Inward potassium leak causes matrix swelling. Inward proton leak dissipates energy and contributes to the basal metabolic rate.

Diffusive transport of ions does not differ fundamentally from transport of polar nonelectrolytes across thin membranes. Thus, the rate of transport is proportional to the concentratiion difference, and the proportionality constant (the permeability coefficient) is a function of the energy barrier that must be overcome to extract the ion from the aqueous phase. The main complication associated with ionic charge derives from long-range effects of the imposed electric field on the local free energy of the diffusing particles.

All of the essential features of current–voltage relationships in biomembrane systems can be derived from a simple model, described by Garlid et al. (1989), in which a single, sharp energy barrier is located at the center of the membrane, as depicted in Fig. 2. Consider the steps taken by a cation crossing the membrane via such a leak pathway. The cation first partitions between the aqueous medium and a surface energy well, created by the phospholipid headgroups. The ion must next overcome an extremely unfavorable Gibbs energy of transfer to move into the hydrophobic interior of the membrane. Only those ions having sufficient energy to reach the barrier peak (center of the membrane) will cross to the energy well on the opposite side, and thence into the aqueous medium. Net

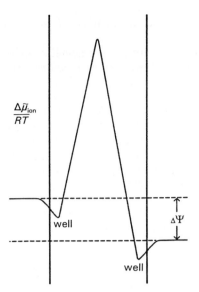

**FIG. 2.** The energy barrier to ion leak across the mitochondrial inner membrane. Experimental data (see Fig. 3) confirm that the energy barrier to cations is best represented by a sharp peak at the center of the membrane. Cations must first partition into the surface energy wells created by the phospholipid headgroups. Those ions having sufficient energy to overcome the unfavorable free energy of transfer from the well to the peak will cross to the other side. Protons and hard cations partition into the same interfacial energy well; however, protons go on to partition into a deeper energy well at the acylglycerol linkages of the phospholipids, where their concentration is near saturation. These surface features cause proton leak to differ *quantitatively* from hard cation leak; however, as shown in Fig. 3, leaks of protons and cations are *qualitatively* identical.

flux will be proportional to the differential probability of getting to the peak from either side.

The major consequence of the high $\Delta\Psi$ in mitochondria is that inward cation flux is no longer ohmic but rather exhibits an exponential dependence on $\Delta\Psi$. To see why this is so, consider the probability of an ion in water having sufficient energy to move to the center of the membrane. This is given by a Boltzmann function, $e^{-\Delta\bar{\mu}_p/RT}$, where $\Delta\bar{\mu}_p \equiv \bar{\mu}_p - \bar{\mu}_{aq}$ is the Gibbs energy of the ion at the peak (p) relative to its value in the aqueous energy well at the surface of the membrane. The exponential dependence of flux on $\Delta\Psi$ derives from the exponential Boltzmann distribution function. Using this approach, it is simple to show (see Garlid *et al.*, 1989) that ion leak across biomembranes is given by

$$J = P(C_1 e^{u/2} - C_2 e^{-u/2}) \tag{4}$$

where $u \equiv -zF\,\Delta\Psi/RT$, and $P$ is the permeability constant, given by

$$P \equiv k e^{-\Delta\mu_p^o/RT} \tag{5}$$

The energy barrier at the center of the membrane splits $\Delta\Psi$ in half when we make the customary assumption that the potential gradient is linear within the hydrophobic interior (the Goldman constant field assumption). This is the origin of the factor $\frac{1}{2}$ in the exponents of Eq. 4. Note in

Eq. 5 that $\Delta\mu_p^o \equiv \mu_p^o - \mu_{aq}^o$ is the standard Gibbs energy of transfer of the ion to the hydrophobic center of the bilayer. $\Delta\mu_p^o$ is the activation energy, and $\Delta\mu_p^o/RT$ represents the height of the barrier that must be overcome for the ion to traverse the membrane at equilibrium.

As is true in all flux equations, $C_1$ and $C_2$ do not refer to the bulk aqueous phase concentrations, but rather to concentrations adjacent to the energy barrier—in this case, to the concentrations in the energy wells. We may define a partition coefficient, $f = C_1/C_{10} = C_2/C_{20}$, where $C_{10}$ and $C_{20}$ are aqueous concentrations, and we assume the membrane to be symmetric with respect to surface potentials and ion extraction. This results in a practical flux equation:

$$J = fP(C_{10} e^{u/2} - C_{20} e^{-u/2}) \tag{6}$$

It is instructive to consider special cases of Eq. 6. When $z = 0$ (transport of a nonelectrolyte) or $\Delta\Psi = 0$ (transport of an ion in the absence of an electric field), Eq. 6 reduces to Fick's law,

$$J = fP(C_{10} - C_{20}) \tag{7}$$

When $\Delta\bar{\mu}$ is sufficiently small, and $C_{10} = C_{20}$, Eq. 6 reduces on expansion of the exponentials to

$$J = -fPC(F\,\Delta\Psi/RT) \tag{8}$$

Equation 8 describes an ohmic (linear) flux–voltage relationship and reveals the important point that Ohm's law is an approximation that is valid only when potential gradients are sufficiently small. Thus, linear flux–voltage dependence should not be viewed as the expected behavior for ion leak. In energy-transducing membranes, electrical gradients greatly exceed the ohmic limit, and flux is exponential with voltage.

The approximation of greatest practical application to mitochondrial bioenergetics is achieved by dropping the second term in Eq. 6. This term represents back-flux of cations from the matrix and becomes negligible at the high values of $\Delta\Psi$ maintained by respiring mitochondria. Thus, Eq. 6 reduces to a simple exponential function of $\Delta\Psi$:

$$J = fPC_{10} e^{u/2} \tag{9}$$

Equation 9 is only valid far from equilibrium, but this is precisely the norm in respiring mitochondria. Measurements of cation leak in mitochondria are in good agreement with the predictions of Eq. 9, as shown by the flux–voltage plots for $H^+$ and $TEA^+$ (tetraethylammonium ion) in Fig. 3.

A great deal of attention has been paid to two "anomalies" of proton leak across lipid bilayers and biomembranes (Deamer and Nichols, 1989). First, the rate constant (flux divided by concentration) for proton transport is about $10^6$ higher than that for $TEA^+$ or alkali cations (see legend to Fig. 3). Second, proton flux exhibits very little dependence on $H^+$ concentration, which appears to contradict Eq. 9. These "anomalies" have led to speculations that protons are transported through the bilayer by a mechanism that differs from that for other cations. This hypothesis is refuted by the data in Fig. 3, which show that there is no difference between $H^+$ and $TEA^+$ in the rate-limiting step of diffusion across the hydrophobic barrier. The

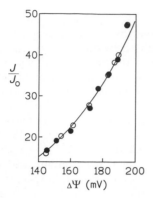

**FIG. 3.** Fluxes of TEA$^+$ and H$^+$ in mitochondria exhibit exponential dependence on $\Delta\Psi$. $J/J_o$ is plotted versus $\Delta\Psi$ over the range to which Eq. 9 applies. $J_o \equiv f \cdot P[\text{C}]_o$, where $[\text{C}]_o$ is the aqueous concentration of TEA$^+$ and H$^+$, respectively, and $f$ and $P$ are defined in the text. $J_o$ was obtained from the extrapolated intercept of a semilogarithmic plot, $\ln J$ vs $F\,\Delta\Psi/RT$. The first point to note from the figure is that the rate-limiting step of crossing the energy barrier is identical for TEA$^+$ and H$^+$, which probably crosses as hydronium ion. That is, the only difference between these ions is in the $J_o$ term, which largely reflects events at the surface of the membrane. The second point is that this quantitative difference is very large indeed. The product, $f \cdot P$, is $2 \times 10^{-10}$ cm/sec for TEA$^+$ ion and $0.85 \times 10^{-3}$ cm/sec for H$^+$ ion. Thus, the apparent permeability constant for hydronium ions is $4.25 \times 10^6$ times greater than that for TEA$^+$. Of this difference, $5.4 \times 10^4$ is accounted for by the fact that the activation enthalpy for proton flux is 27 kJ/mol less than that for TEA$^+$. This is largely a surface effect, reflecting extraction of the ions from their aqueous energy wells into the bilayer. The remainder is attributable to a 79-fold concentration of H$^+$ in its deeper energy well at the acylglycerol linkages. Thus, the apparent "pH" of the proton energy well is about 5.5, which is in reasonable agreement with other data, including the fact that the apparent $pK_a$ values of fatty acids in the bilayer are 2–3 units higher than $pK_a$ values in aqueous solution.

anomalies of proton leak therefore reside in the terms $f$ and $P$ of Eq. 9.

The following observations appear to account for both anomalies of H$^+$ flux. Changing external pH or adding polyvalent cations (e.g., Mg$^{2+}$ or spermine) to the medium profoundly affects TEA$^+$ leak, but has little effect on H$^+$ leak. This shows that the energy well for protons is shielded from the aqueous medium and therefore deeper than the energy well for cations. The observation that $J_H^+$ is relatively unaffected by $[\text{H}^+]_o$ simply means that the proton energy well is nearly saturated with protons. Thus, $C_1 > C_{10}$ for protons, contributing to a higher flux. We also observed that the activation enthalpy for H$^+$ flux is about 27 kJ/mol lower than that for TEA$^+$ ion. Together, these factors can readily account for the apparent anomalies in H$^+$ flux across mitochondrial and bilayer membranes: The hydronium ion equilibrates with an energy well located deeper in the membrane than the energy well for cations. The hydronium energy well is most probably located at the level of the phospholipid acylglycerol linkages. This energy well is not accessible to cations; it is shielded from the aqueous environment; and it is nearly saturated with H$_3^+$O.

## IV. The Mitochondrial Proton Cycle

When inner membrane permeability to H$^+$ ions is increased, electron transport is uncoupled from oxidative phosphorylation. Redox energy is rapidly burned off as heat, because protons leak back into the matrix and no energy is conserved to make ATP. This can readily be demonstrated in the laboratory, using uncouplers such as dinitrophenol. Protonophoretic uncouplers are weak acids that diffuse electroneutrally across the membrane, delivering a proton to the other side. By itself, this has no uncoupling effect, because no charge movement has occurred. The anion, however, contains $\pi$ electrons, enabling charge delocalization in the protonophore structure and consequent membrane permeability to the anion. As diagrammed in Fig. 4, the anion and acid cycle across the membrane until the proton-motive force is dissipated.

Nature has designed a protein expressly for the purpose of uncoupling. Uncoupling protein (UCP) is found only in mammals. Its role is to short-circuit the insulating inner membrane, thereby causing mitochondria to produce heat instead of ATP. UCP plays a major role in providing thermogenesis to hibernating animals and to all mammalian newborns, including humans. UCP is inhibited by cytosolic ATP and released from ATP inhibition by poorly understood mechanisms. As shown in Fig. 5, UCP does not conduct protons, per se; rather, it catalyzes the passive efflux of fatty acid anions from the matrix. Normally, the inner membrane is highly permeable to protonated fatty acid and impermeable to fatty acid anion. When open, UCP therefore permits fatty acids to behave as cycling protonophores (Garlid et al., 1996a). It is of interest that UCP belongs to the family of mitochondrial anion transporters. Its gene has been cloned, and its structure–function is being studied by site-directed mutagenesis.

It has long been suspected that UCP also plays the role of dissipating energy in response to excess fat, thereby dissipating energy and preventing obesity. Recently, a new protein has been discovered that is proposed to provide this function (Fleury et al., 1997). This protein has been designated UCP2 on the basis of 59% amino acid identity

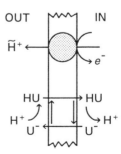

**FIG. 4.** Mechanism of uncoupling by weak acid protonophores. When both the anion and acid can permeate the membrane, as is the case with all weak acid protonophores, their cycling leads to inward transport of H$^+$ ions that is coupled to outward transport of negative charge. This cycling draws current from the proton-motive batteries, diverting energy away from the ATP synthase and dissipating energy as heat. ATP synthesis is said to be *uncoupled* from respiration.

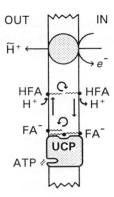

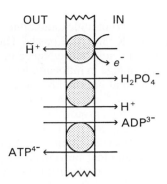

**FIG. 5.** Mechanism of uncoupling by mitochondrial uncoupling protein. Uncoupling protein (UCP) contains a weak binding site for anions near the center of the membrane. This constitutes an energy well for anions and provides a low-resistance pathway for normally impermeant anions to cross the membrane. The physiological substrates of UCP are anions of free fatty acids (FA). The FA headgroups are located at the acylglycerol linkages of the phospholipid bilayer. When the FA is at the surface of UCP, the negative member potential drives the headgroup to the energy well at the center of the membrane. The FA then "flip-flops" and the headgroup is driven to the opposite surface. Here, it diffuses away from the protein and picks up a proton. The protonated FA freely diffuses across the membrane, during which it flip-flops again and delivers a proton to the other side. Note that this uncoupling mechanism is qualitatively identical to that described in Fig. 4. Thus the role of UCP is to enable FA to behave as cycling protonophores. Furthermore, FA cycling is highly regulated: UCP is inhibited from the cytosolic side by nucleotides, including ATP.

**FIG. 6.** Mitochondrial transport of inorganic phosphate, ADP, and ATP. The phosphate carrier catalyzes electroneutral transport, and the ADP/ATP exchanger catalyzes electrophoretic transport. Uptake of $P_i$ and ADP and expulsion of ATP use one electrogenically ejected proton.

with UCP1. Unlike UCP1, which is only expressed in brown adipose tissue, UCP2 is widely expressed in human tissues, including white fat. At this time, the transport function of UCP2 is completely unknown.

and malate are key intermediates. In addition to their normal substrates, the dicarboxylic acid exchanger also catalyzes malate/phosphate exchange, and the tricarboxylic acid exchanger also catalyzes malate/citrate exchange. In this way, both di- and tricarboxylic acids are linked to the phosphate carrier. Since the phosphate carrier transports fully protonated phosphate, the net result is that di- and tricarboxylic acids also enter as if they were fully protonated. Their measured distribution ratios show them to be very close to equilibrium with the $\Delta pH$ across the membrane, as described by Eq. 3.

Many of the anion exchangers have been cloned and sequenced. They are all found to be 30- to 35-kDa proteins and are believed to function as homodimers. Interestingly, all those identified to date contain a tripartite structure and are considered to belong to a common gene superfamily.

## V. The Anion-Exchange Carriers

Mitochondria synthesize ATP in the matrix; consequently, the substrates for the ATP synthase, ADP and phosphate, must be imported across the inner membrane, and ATP must be exported for use by the cell. As shown in Fig. 6, nucleotides are exchanged on the ATP/ADP translocase in a process involving outward movement of one negative charge. Although some of the proton-motive energy is used to export ATP, this is largely compensated by the fact that the F1F0-ATPase can synthesize ATP at a lower phosphorylation free energy than that required by ATP-consuming processes of the cell. The phosphate carrier catalyzes electroneutral $P_i/H^+$ symport or $P_i/OH^-$ antiport, with the net result that it transports phosphoric acid.

The inner membrane also contains a variety of anion-exchange carriers, which are designed to deliver substrates to the tricarboxylic acid cycle and were first described by Chappell (1968) (Fig. 7). The anion-exchange carriers catalyze 1:1 exchange of anions; nevertheless, their distributions follow Eq. 3 exactly, as if they were transported as fully protonated acids. This results from an arrangement of these exchange carriers in a cascade, in which phosphate

## VI. The Mitochondrial Potassium Cycle

The mitochondrial $K^+$ cycle consists of electrophoretic $K^+$ influx and electroneutral $K^+$ efflux across the inner

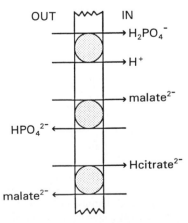

**FIG. 7.** Mitochondrial transport of substrate anions. The dicarboxylic and tricarboxylic exchangers are electroneutral and coupled to the phosphate transporter by malate and citrate. No energy is expended during anion uptake by these mechanisms.

membrane (Fig. 8). When influx and efflux are transiently out of balance, net $K^+$ flux will ensue. The first point to emphasize is that the sole consequence of net $K^+$ flux in mitochondria is a change in mitochondrial volume. To see why this is so, note that both influx and efflux are coupled to proton movements. Electrophoretic $K^+$ influx is driven by the electrogenic redox proton pumps, and electroneutral $K^+$ efflux is obligatorily coupled to $H^+$ influx on the $K^+/H^+$ antiporter. Net $K^+$ flux will therefore perturb the pH gradient, causing corresponding movement of phosphate and other anions on their electroneutral carriers and resulting in net transport of $K^+$ salts. Net salt transport, in turn, will be accompanied by osmotically obligated water. Net $K^+$ flux will not have a significant effect on matrix $K^+$ **concentration,** which is about 180 m$M$ *in vivo*. This is because $K^+$ is transported with anions of valency approximately equal to 2, so the concentration of **transported** $K^+$ is 180–200 m$M$. From these considerations, we may conclude that *any net $K^+$ flux will be accompanied by anions and water and will therefore change mitochondrial volume.*

To maintain the vesicular integrity necessary for oxidative phosphorylation, it is essential that mitochondria regulate net $K^+$ flux to zero in the steady state. Otherwise, $K^+$ leak would cause matrix volume to double within 1–2 min. This regulation is provided by the $K^+/H^+$ antiporter, which ejects exactly the amount of $K^+$ that is taken in. The primary role of the $K^+/H^+$ antiporter is to provide **volume homeostasis** to mitochondria *in vivo*. Note that this comes at some cost (Garlid, 1988). The $K^+$ cycle (see Fig. 8) consumes energy to maintain zero net $K^+$ flux.

Regulation of the $K^+/H^+$ antiporter is designed to sense volume changes, because volume change is the sole consequence of net $K^+$ movement. Regulation is mediated by reversible binding of $Mg^{2+}$ and $H^+$ to the $K^+/H^+$ antiporter on its matrix side. The activity of these ions decreases with uptake of $K^+$ salts, causing a graded, compensatory activation of $K^+$ efflux. The $K^+/H^+$ antiporter has been

identified as an 82-kDa inner membrane protein. It has been purified and reconstituted into proteoliposomes and shown to catalyze electroneutral $K^+/H^+$ antiport (Li *et al.,* 1990).

The primary role of the mitochondrial $K_{ATP}$ channel (mito$K_{ATP}$) is **volume regulation.** Opening mito$K_{ATP}$ will increase electrophoretic $K^+$ influx, causing swelling to a new steady-state volume. We have proposed that volume regulation by mito$K_{ATP}$ plays a central role in cell signaling pathways calling for activation of electron transport and stimulation of fatty-acid oxidation (Garlid, 1996). This hypothesis, which remains to be established, is based on the well-established observation that increasing matrix volume over a fairly narrow range greatly stimulates the rate of electron transport, with a particularly strong effect on fatty acid oxidation. Thus, contracted mitochondria oxidize substrates slowly (and fatty acids hardly at all), and mildly expanded mitochondria oxidize substrates at rapid rates. Volume activation of electron transport has been demonstrated in liver, heart, and brown adipose tissue mitochondria (Halestrap, 1989). It has also been shown that volume regulation of oxidation is mediated strictly by changes in matrix volume, independently of the means used to change it.

Although it possesses certain distinctive characteristics, the mitochondrial $K_{ATP}$ channel (mito$K_{ATP}$) is most striking in its similarities to $K_{ATP}$ channels of the plasma membrane (cell$K_{ATP}$). Specifically, mito$K_{ATP}$ reacts with the same set of biochemical and pharmacological ligands, including adenine and guanine nucleotides, long-chain acyl-CoA esters, sulfonylureas, and $K^+$ channel openers. These similarities suggest that mito$K_{ATP}$, like the pancreatic $\beta$-cell$K_{ATP}$, is a heteromultimeric complex consisting of an inward-rectifying $K^+$ channel, mitoKIR, and a sulfonylurea receptor, mitoSUR. Indeed, we have recently identified and purified these two proteins from the mitochondrial inner membrane. MitoSUR is a 63-kDa sulfonylurea-binding protein, and mitoKIR is a 55-kDa protein. Because of its strong resemblance to cell$K_{ATP}$, we infer that mito-$K_{ATP}$ belongs to the same gene family.

Mito$K_{ATP}$ is regulated by a rich variety of metabolic and pharmacological ligands. It is inhibited with high affinity by ATP, long-chain acyl-CoA esters, and glyburide. The inhibited channel is opened with high affinity by guanine nucleotides and $K^+$ channel openers such as cromakalim and diazoxide. In addition to its metabolic role, there is growing evidence that mito$K_{ATP}$ is the receptor for the cardioprotective action of $K^+$ channel openers in experimental cardiac ischemia. Hearts treated with $K^+$ channel openers maintained higher ATP levels and exhibited reduced infarct size and enhanced postischemic recovery upon reperfusion. All of these effects were blocked by glyburide. These pharmacological effects point to a role of $K_{ATP}$ channels in myocardial protection; however, the receptor for these effects has not been identified. Recent findings indicate that cardiac $K_{ATP}$ channels from mitochondria and plasma membrane can be distinguished pharmacologically: The two channels exhibit similar sensitivities to cromakalim, but mito$K_{ATP}$ is 2000 times more sensitive to diazoxide than cell$K_{ATP}$. In further experiments using

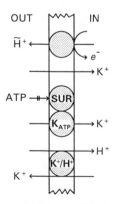

**FIG. 8.**   The mitochondrial $K^+$ cycle. Electrogenic proton ejection drives significant $K^+$ uptake by diffusive leak. In addition, the inner membrane contains a $K_{ATP}$ channel, which is highly regulated by nucleotides, CoA esters, and pharmacological agents. The $K_{ATP}$ channel consists of at least two subunits, a $K^+$-specific channel and a sulfonylurea receptor (SUR). Net $K^+$ flux is regulated to zero in the steady state. Compensatory $K^+$ efflux is provided by the electroneutral $K^+/H^+$ antiporter, which is regulated by matrix $Mg^{2+}$ and $H^+$ and is exquisitely sensitive to changes in matrix volume.

the ischemic model, diazoxide, at doses that had no effect on $cellK_{ATP}$, was cardioprotective over the same dose range as cromakalim. These findings raise the possibility that $mitoK_{ATP}$ is the receptor for the cardioprotective effects of KCOs (Garlid *et al.*, 1996b).

## VII. The Mitochondrial Calcium Cycle

The mitochondrial $Ca^{2+}$ cycle, diagrammed in Fig. 9, consists of three separate processes. $Ca^{2+}$ is taken up by the $Ca^{2+}$ channel at the expense of two ejected protons. $Ca^{2+}$ is ejected via the $Na^+/Ca^{2+}$ antiporter, which catalyzes an electrophoretic $3Na^+/Ca^{2+}$ exchange, using an additional ejected proton (Jung *et al.*, 1995). The three $Na^+$ ions taken up are then ejected via the electroneutral $Na^+/H^+$ antiporter.

The mitochondrial $Ca^{2+}$ channel is characterized by high-capacity, low-affinity, electrophoretic $C^{2+}$ uptake. In experiments with partially purified $Ca^{2+}$ channels reconstituted into liposomes, the $K_m$ for $Ca^{2+}$ ranged between 7 and 20 $\mu M$, and the $V_{max}$ was 130 $\mu$mol/mg protein/min, about 250-fold greater than that observed in intact mitochondria. $La^{3+}$ is a competitive inhibitor of $Ca^{2+}$ uptake, and ruthenium red is a noncompetitive inhibitor. In the reconstituted system $Ca^{2+}$ transport exhibited hyperbolic dependence on $[Ca^{2+}]$, whereas in intact mitochondria, $Ca^{2+}$ appears to be an allosteric activator of transport. This raises the interesting possibility that $Ca^{2+}$ activation of $Ca^{2+}$ uniport in intact mitochondria may be mediated by a separate regulatory protein that is lost upon reconstitution.

Free mitochondrial $[Ca^{2+}]$ is comparable to cytosol $[Ca^{2+}]$ *in vivo*, in spite of the enormous gradient for electrophoretic $Ca^{2+}$ uptake. This disequilibrium is maintained in heart mitochondria by an electrophoretic $Na^+/Ca^{2+}$ antiporter (Jung *et al.*, 1995). The cardiac mitochondrial $Na^+/Ca^{2+}$ antiporter is a 110-kDa protein that has been purified and reconstituted into liposomes. In the presence of $K^+$, the $K_m$ for $Ca^{2+}$ is 300 n$M$, and the $K_m$ for $Na^+$ is 7 $\mu M$. In the absence of $Ca^{2+}$, the reconstituted antiporter catalyzes $Na^+/Li^+$ and $Na^+/K^+$ exchange; however, $K^+/Ca^{2+}$

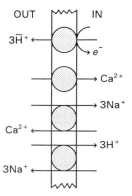

**FIG. 9.** Mitochondrial $Ca^{2+}$ cycle. $Ca^{2+}$ enters the matrix via the electrophoretic $Ca^{2+}$ channel and is ejected by the electrophoretic $Na^+/Ca^{2+}$ antiporter, utilizing three ejected protons per $Ca^{2+}$ taken up. The $Na^+$ is then expelled by the electroneutral $Na^+/H^+$ antiporter.

exchange was not observed (Li *et al.*, 1992). Liver mitochondria appear to have $Na^+$-independent as well as $Na^+$-dependent $Ca^{2+}$ efflux pathways, but these have not been well characterized.

The $Na^+/H^+$ antiporter has been identified as a 59-kDa inner membrane protein, and the purified protein is reconstitutively active (Garlid *et al.*, 1991). Unlike the plasma membrane $Na^+/H^+$ antiporters, the mitochondrial carrier appears to lack regulation, causing $Na^+$ to equilibrate with the $\Delta pH$. Like the plasma membrane carriers, the mitochondrial version is competitively inhibited by $Li^+$ and unaffected by $K^+$.

$Ca^{2+}$ is a second messenger, signaling a need to increase cellular work. Because increased work requires a higher rate of ATP production, this message must be relayed to the mitochondrial matrix. The physiological role of the mitochondrial $Ca^{2+}$ cycle is to regulate matrix $Ca^{2+}$ activity in response to signals from the cytosol. Intramitochondrial $Ca^{2+}$ is required to activate the phosphorylase that converts pyruvate dehydrogenase to its active form, and $\alpha$-ketoglutarate dehydrogenase is allosterically activated by matrix $Ca^{2+}$ in the physiological range (Denton and McCormack, 1990; Gunter *et al.*, 1994).

Studies on isolated hepatocytes strongly suggest that the mitochondrial $Ca^{2+}$ cycle is designed to respond to **oscillations** in cytosolic $Ca^{2+}$. A static elevation in cytosolic $Ca^{2+}$ causes a transient spike in mitochondrial $Ca^{2+}$ and a transient increase in pyridine nucleotide reduction, associated with activation of the $Ca^{2+}$-sensitive mitochondrial dehydrogenases. The energetic effect decayed much more slowly than did matrix $Ca^{2+}$. $IP_3$-mediated oscillations in cytosolic $Ca^{2+}$, however, were matched by oscillations in mitochondrial $Ca^{2+}$ and a sustained activation of the mitochondrial dehydrogenases. Maintaining high activity is believed to be due to the fact that the effects of $Ca^{2+}$ activation decline more slowly than $Ca^{2+}$ itself. From a metabolic point of view, these results indicate that mitochondria can discriminate beween slow increases in cytosolic $Ca^{2+}$, associated with homeostatic mechanisms, and rapid oscillations secondary to $IP_3$ signaling. The kinetics of the $Ca^{2+}$ channel and its proximity to the release sites on the endoplasmic reticulum are well suited to such discrimination. The high-capacity channel will only be activated when local $Ca^{2+}$ is high, because of its allosteric activaiton by $Ca^{2+}$. $Ca^{2+}$ uptake in response to slower oscillations in cytosol $Ca^{2+}$ will rapidly be cleared by the efflux mechanism(s). Similarly, an electrophoretic $Na^+/Ca^{2+}$ exchanger, such as exists in heart mitochondria, would drive rapid $Ca^{2+}$ efflux to permit the observed oscillatory behavior of mitochondrial $Ca^{2+}$ (Hajn'oczky *et al.*, 1995).

## VIII. Summary

Mitochondria have their own unique physiology, which is directed toward maintenance of vesicular integrity and proper functioning in energy transduction. The setting in which this physiology takes place is provided by the elegant postulates of the chemiosmotic theory. Of particular importance is the very high electrical membrane potential required for ATP synthesis. Several transport systems have evolved that either protect against adverse consequences of high membrane potential, or exploit this feature to benefit the organism. The electrical driving force causes significant inward leak of protons, which dissipates energy and reduces the efficiency of oxidative phosphorylation. However, the energy dissipation of proton backflux has also been exploited for thermogenesis in hibernating animals and newborns. Thermogenesis is mediated by uncoupling protein, which permits fatty acids to cycle across the membrane in protonated and deprotonated forms. The high membrane potential also drives significant diffusive uptake of $K^+$ across the membrane, resulting in osmotic swelling of the matrix and threatening the vesicular integrity of the organelle. Excess swelling is prevented, and volume homeostasis is maintained, by an electroneutral $K^+/H^+$ antiporter, which is finely regulated to maintain zero net $K^+$ flux in the face of fluctuations in inward $K^+$ leak. The high driving force for $K^+$ uptake has also been exploited to drive $K^+$ inward through the highly regulated $K_{ATP}$ channel. This additional $K^+$ influx shifts matrix volume to a higher steady-state level at which fatty acid oxidation occurs at much higher rates. Finally, mitochondria have exploited the high membrane potential to drive $Ca^{2+}$ rapidly both inward, via the $Ca^{2+}$ channel, and outward, via the $Na^+/Ca^{2+}$ antiporter, across the inner membrane. The mitochondrial $Ca^{2+}$ cycle thus allows oscillatory cytosolic $Ca^{2+}$ signals to be transmitted to the matrix, where the high frequency of oscillations causes sustained activation of enzymes that synthesize NADH, a major substrate for the electron transport chain. In this way, the $Ca^{2+}$ signal not only directs the cell to increase its ATP-consuming activities, but also directs the mitochondrion to increase its supply of reducing equivalents to meet the demands of increased ATP production.

The reader who wants to explore ion transport and bioenergetics in greater detail is fortunate indeed in the availability of two excellent and highly readable texts by Nicholls and Ferguson (1992) and Gennis (1989). They are highly recommended.

## Acknowledgments

This work relies on major contributions from my colleagues and students who are listed as co-authors of our publications. The research was supported in part by grant GM31086 from the National Institute of General Medical Sciences.

## Bibliography

Chappell, J. B. (1968). Systems used for the transport of substrates into mitochondria. *Brit. Med. Bull.* **24,** 150–157.

Deamer, D. W., and Nichols, J. W. (1989). Proton flux mechanisms in model and biological membranes. *J. Membr. Biol.* **107,** 91–103.

Denton, R. M., and McCormack, J. G. (1990). $Ca^{2+}$ as a second messenger within mitochondria of the heart and other tissues. *Annu. Rev. Physiol.* **52,** 451–456.

Fleury, C., Neverova, M., Collins, S., Raimbault, S., Champigny, O., Levi-Meyrueis, C., Bouillaud, F., Seldin, M. F., Surwit, R. S., Ricquier, D., and Warden, C. H. (1997). Uncoupling protein-2: A novel gene linked to obesity and hyperinsulinemia. *Nature Genetics* **15,** 269–273 .

Garlid, K. D. (1988). Mitochondrial volume control. *In* "Integration of Mitochondrial Function" (J. J. Lemasters, C. R. Hackenbrock, R. G. Thurman, and H. V. Westerhoff, Eds.), pp. 257–276. Plenum Publishing, New York.

Garlid, K. D. (1996). Cation transport in mitochondria—the potassium cycle. *Biochim. Biophys. Acta* **1275,** 123–126.

Garlid, K. D., Beavis, A. D., and Ratkje, S. K. (1989). On the nature of ion leaks in energy-transducing membranes. *Biochim. Biophys. Acta* **976,** 109–120.

Garlid, K. D., Shariat-Madar, Z., Nath, S., and Jezek, P. (1991). Reconstitution and partial purification of the $Na^+$-selective $Na^+/H^+$ antiporter of beef heart mitochondria. *J. Biol. Chem.* **266,** 6518–6523.

Garlid, K. D., Orosz, D.E., Modriansky, M., Vassanelli, S., and Jezek, P. (1996a). On the mechanism of fatty acid-induced proton transport by mitochondrial uncoupling protein. *J. Biol. Chem.* **271,** 2615–2620.

Garlid, K. D., Paucek, P., Yarov-Yarovoy, V., Sun, X., and Schindler, P. A. (1996b). The mitochondrial $K_{ATP}$ channel as a receptor for potassium channel openers. *J. Biol. Chem.* **271,** 8796–8799.

Gennis, R. B. (1989). "Biomembranes: Molecular Structure and Function." Springer-Verlag, New York.

Gunter, T. E., Gunter, K. K., Sheu, S.-S., and Gavin, C. E. (1994). Mitochondrial calcium transport: Physiological and pathological relevance. *Am. J. Physiol.* **267,** C313–C339.

Hajn'oczky, G., Robb-Gaspers, L. D., Seitz, M. B., and Thomas, A. P. (1995). Decoding of cytosolic calcium oscillations in the mitochondria. *Cell* **82,** 415–424.

Halestrap, A. P. (1989). The regulation of the matrix volume mammalian mitochondria *in vivo* and *in vitro* and its role in the control of mitochondrial metabolism. *Biochim. Biophys. Acta* **973,** 355–382.

Jung, D. W., Baysal, K., and Brierley, G. P. (1995). The sodium–calcium antiport of heart mitochondria is not electroneutral. *J. Biol. Chem.* **270,** 672–678.

Li, X., Hegazy, M. G., Mahdi, F., Jezek, P., Lane, R. D., and Garlid, K. D. (1990). Purification of a reconstitutively active $K^+/H^+$ antiporter from rat liver mitochondria. *J. Biol. Chem.* **265,** 15316–15322.

Li, W., Shariat-Madar, Z., Powers, M., Sun, X., Lane, R. D., and Garlid, K. D. (1992). Reconstitution, identification, purification, and immunological characterization of the 110-kDa $Na^+/Ca^{2+}$ antiporter from beef heart mitochondria. *J. Biol. Chem.* **267,** 17983–17989.

Mitchell, P. (1961). Coupling of phosphorylation to electron and hydrogen transfer by a chemiosmotic type of mechanism. *Nature* **191,** 144–148.

Mitchell, P. (1966). Chemiosmotic coupling in oxidative and photosynthetic phosphorylation. *Biol. Rev.* **41,** 445–502.

Mitchell, P., and Moyle, J. (1969). Estimation of membrane potential and pH difference across the cristae membrane of rat liver mitochondria. *Eur. J. Biochem.* **7,** 471–484.

Nicholls, D. G., and Ferguson, S. J. (1992). "Bioenergetics 2." Academic Press, London.

*Richard G. Sleight and Michael A. Lieberman*

# 8

# Signal Transduction

## I. Introduction

Coordination of metabolic activities among cells, tissues, and organs is mediated by the action of extracellular signals. A variety of agents act as extracellular signals, including eicosanoids (e.g., prostaglandins), growth factors, hormones, neurotransmitters, and pheromones (or sex attractants). These agents function by binding to specific receptors located at the cell surface. **Signal transduction** is the process whereby an external chemical signal evokes an intracellular metabolic change. Some agents such as acetylcholine and γ-aminobutyric acid (GABA) bind to receptors that also function as ion channels. Binding to these **ligand-gated channels** has a direct effect on ion flux. However, the majority of extracellular signals work indirectly by complex cascade systems that have the effect of amplification. This allows a small number of extracellular signaling molecules to affect the function of a large number of proteins, which produce the desired physiological response. In this chapter, two mechanisms of signal transduction are discussed: (1) generation of second messengers, and (2) receptor phosphorylation.

## II. Second Messengers

Although some signaling molecules, such as the steroid hormones, diffuse through the plasma membrane and bind to specific receptors present in the cytosol, most extracellular signals interact with target cells by binding to receptors located at the cell surface. Cell surface receptors bind extracellular signaling molecules with high affinity. Molecules that bind specifically to protein receptors are called **ligands.** Structural analogues of natural ligands that bind to receptors are called **agonists** or **antagonists.** Agonists mimic the effect of the natural ligands. Antagonists bind to receptors, but this binding does not lead to a biological response. Thus, antagonists block the effect of natural ligands and agonists.

Binding of ligands (or agonists) to cell surface receptors causes the receptors to undergo a **conformational change.** In some cases, the altered conformation of the receptor initiates a series of events leading to the formation or release of an intracellular signal that alters the metabolism of the target cell. Although the extracellular signaling molecules are rarely called **first messengers,** the intracellular signaling molecules are generally called **second messengers.**

Second messengers may arise by two different mechanisms (Fig. 1). One mechanism involves receptor-mediated activation of an enzyme that catalyzes the production of the second messenger. For example, when glucagon binds to its receptor, adenylate cyclase (see Fig. 3) is activated, catalyzing the synthesis of cyclic AMP (cAMP), a second messenger. Cyclic AMP is synthesized on the cytoplasmic face of the plasma membrane and released into the cytosol where it signals for the activation or inhibition of intracellular enzymes.

A second mechanism for the production of intracellular signals is the opening or closing of ion channels in the plasma membrane. A good example of this is neurotransmitter action at neuromuscular junctions of skeletal muscle. Although in-depth coverage of muscle physiology is not given until a later chapter, this system demonstrates how ions can act as second messengers. Acetylcholine (released from the motor nerve terminal) acts as the extracellular signal at neuromuscular junctions and binds to a specific receptor in the postsynaptic membrane of skeletal muscle fibers. Binding causes a conformational change in the receptor ion-channel complex that leads to an influx of cations (mostly $Na^+$) into the cell.[1] This influx results in depolarization of the plasma membrane (the end-plate potential or excitatory postsynaptic potential), which triggers a propagating action potential. The action potential depolarization activates $Ca^{2+}$ channels that allow $Ca^{2+}$ influx into the cytoplasm. The increased free $Ca^{2+}$ levels trigger muscle contraction.

---

[1] This is an example of a ligand-gated ion channel.

*119*

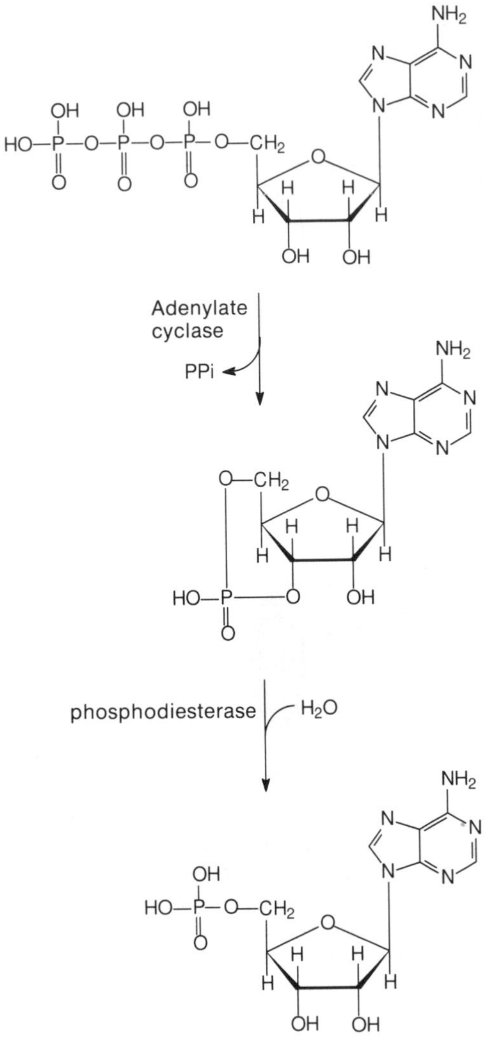

**FIG. 3.**  Production and destruction of cAMP.

distinguished biochemically by the action of bacterial toxins. **Cholera toxin** enzymatically cleaves $NAD^+$ and attaches an ADP–ribose moiety to $G_S$. This **ADP–ribosylation** inhibits the GTPase activity of $G_S$ and thus allows $G_S$ to remain active and permanently activate adenylate cyclase. **Pertussis toxin** permanently inactivates the $G_i$ protein, thereby preventing inhibition of adenylate cyclase. This toxin works by ADP–ribosylation of $G_i$, which prevents GTP binding. Thus, GDP remains bound, prevents GTP activation of the $\alpha_i$-subunit, and $G_i$ cannot inhibit adenylate cyclase.

An increase in intracellular cAMP level causes different responses in different cell types (Table 2). For example, glucagon binds to cells in both liver and fat tissue. In both tissues, the hormone works through  proteins to activate adenylate cyclase and increase cAMP. In hepatocytes, this stimulates glycogen breakdown, whereas in adipocytes the breakdown of triacylglycerols (fat) is stimulated.

Many different extracellular signals may cause an increase in cAMP in a single cell type. The action of four different hormones on adipocytes provides a good exam-

ple. Adrenocorticotropic hormone (ACTH), epinephrine, glucagon, and thyroid-stimulating hormone (TSH) all bind to their respective receptors at the surface of adipocytes. Each hormone causes cAMP to increase, followed by a stimulation of triacylglycerol breakdown. Thus, although the four hormones bind to four different receptors, each receptor reacts with a common pool of G protcins, and the G proteins, in turn, stimulate adenylate cyclase.

The activation (or inhibition) of adenylate cyclase by ligand binding to cell surface receptors only exists while the $\alpha$-subunit of the G protein is transient. Thus, adenylate cyclase activity is quickly shut down when receptor binding ceases. The cAMP produced by the action of adenylate cyclase has a short half-life in the cytoplasm. It is quickly hydrolyzed to 5'-AMP by the action of one or more **cAMP phosphodiesterases** (see Fig. 3). **Caffeine** and **theophylline** are known inhibitors of cAMP phosphodiesterase.

An increase in intracellular cAMP concentration usually elicits a response through the **protein kinase A pathway** (Fig. 4). Cyclic AMP released into the cytoplasm binds to a cAMP-dependent protein kinase (protein kinase A or PKA). This kinase is composed of two catalytic subunits and two regulatory subunits. In its tetrameric form, the enzyme is inactive. When cAMP binds to the regulatory subunits, the tetramer dissociates into two active catalytic subunits and a dimer of regulatory subunits. Active protein kinase A catalyzes the transfer of a phosphoryl group from ATP to target proteins. The proteins are usually phosphorylated at serine or threonine residues. Depending on the cell type, different proteins are phosphorylated. This explains how a common pathway can result in a variety of biological responses.

Phosphorylation of proteins by PKA is the initial unique step in the process of translating receptor binding to biological response. This phosphorylation causes target enzyme activities to change or structural proteins to alter their structure. For example, pyruvate kinase activity is regulated by reversible phosphorylation. When glucagon binds to receptors on hepatocytes, adenylate cyclase is activated, cAMP levels increase, PKA is activated, and pyruvate kinase is phosphorylated. The phosphorylated form of pyruvate kinase is inactive. This is one way that glucagon binding leads to an inhibition of hepatic glycolysis. The process is reversed when the hormone is no longer bound to the receptor. Dephosphorylation of pyruvate kinase and other enzymes phosphorylated by PKA occurs by the action of an enzyme called **protein phosphatase.** The regulation of protein phosphatase is complex and is under intense study.

Hundreds of protein kinases regulate cellular function, and perhaps a thousand or more proteins are targets for these kinases. Dephosphorylation of all of these targets is catalyzed by a set of four protein phosphatases (called PP-1, -2A, -2B, and -2C). Protein phosphatase-1 is responsible for dephosphorylation of many proteins, including muscle glycogen phosphorylase and glycogen synthase. This phosphatase is specifically inhibited by cellular protein inhibitors-1 and -2. The role of these proteins in the regulation of glycogen metabolism is described in Chapter 7. Like PP-1, PP-2A has a broad substrate specificity. The activity

**TABLE 2** Examples of Hormone-Induced Responses Mediated by cAMP

| Hormone | Target tissue | Major response |
| --- | --- | --- |
| Adrenocorticotropic hormone (ACTH) | Adrenal cortex | Cortisol secretion |
| Epinephrine | Muscle, Liver | Glycogen breakdown |
| Epinephrine | Heart | Increase in heart rate |
| Epinephrine, ACTH, glucagon, thyroid-stimulating hormone | Adipose | Triacylglycerol breakdown |
| Follicle-stimulating hormone (FSH) | Ovarian follicle | Secretion of 17$\beta$-estradiol |
| Luteinizing hormone (LH) | Ovarian follicle | Ovulation, progesterone secretion, formation of the corpus luteum |
| Parathormone | Bone | Bone resorption |
| Thyrocalcitonin | Bone | Inhibits bone resorption |
| Thyroid-stimulating hormone | Thyroid | Thyroxin secretion |
| Vasopressin | Kidney | Water resorption |

of PP-2A can be modulated *in vitro* by polyamine; however, it is not known if this occurs *in vivo*. The physiological roles of PP-2B ($Ca^{2+}$-dependent) and PP-2C ($Mg^{2+}$-dependent) are not known.

Signal transduction results in **signal amplification.** A single receptor–ligand complex activates several G proteins, and each $G_\alpha$ protein activates (or inhibits) several adenylate cyclase molecules. Each molecule of adenylate cyclase produces many molecules of cAMP, which in turn activate numerous protein kinase A molecules. Every molecule of PKA then phosphorylates many target proteins and affects

their activities or structure. Theoretically, a single binding event at the cell surface could result in the change in activity of thousands of intracellular molecules, leading to the biological response of the cell.

## C. Cyclic GMP as a Second Messenger

Like cAMP, cGMP can act as a second messenger (Table 3). In many systems, the levels of cAMP and cGMP move in opposite directions in response to ligand–receptor interactions. In these systems, cAMP and cGMP often have

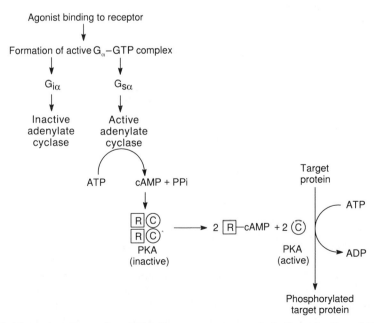

**FIG. 4.** The protein kinase A pathway. Agonist binding to receptors results in the activation of $G_\alpha$ as described in Fig. 2. If the receptor is specific for stimulatory G proteins ($G_s$), then an activated $G_{s\alpha}$–cAMP complex is produced. This complex binds and activates adenylate cyclase, located at the inner leaflet of the plasma membrane. Active adenylate cyclase catalyzes the production of cAMP. cAMP-dependent protein kinase (PKA) normally exists in the cytosol as a tetramer of two regulatory and two inactive catalytic subunits. The catalytic subunits are released from the tetramer and become active when cAMP binds the regulatory subunits. Active PKA catalyzes the transfer of phosphate from ATP to target proteins. Depending on which target protein is phosphorylated, activity of the target protein may increase or decrease.

**TABLE 3**  Examples of Biological Processes Mediated by cGMP

| Target tissue | Effector | Major response |
|---|---|---|
| Eye | Light | Visual processing |
| Kidney | Atrial natriuretic factor (ANF) | Increased Na$^+$ excretion |
| Intestine | *E. coli* endotoxin | Decreased water absorption |
| Heart | Nitric oxide | Decreases forcefulness of contractions |

**GTP**

**cGMP**

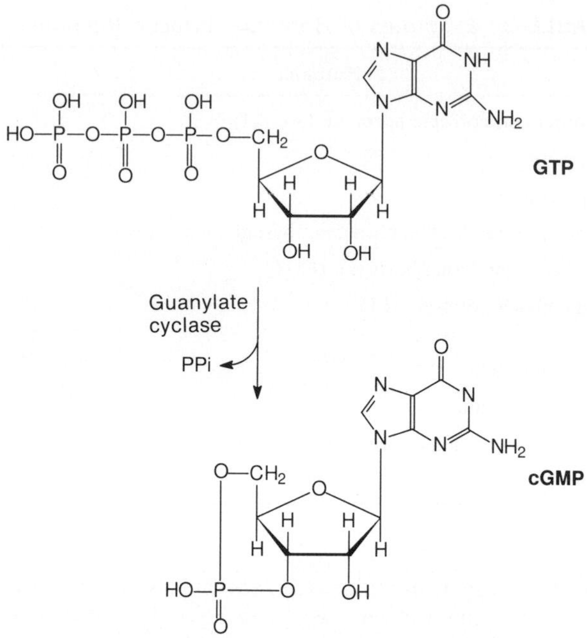

**FIG. 6.**  Production of cGMP by guanylate cyclase.

opposing effects. Although signal transduction in the visual process occurs across membrane discs in rod cells of the retina, the process is remarkably similar to signal transduction across the plasma membrane and provides a good example of cGMP signaling (Fig. 5). The sequence of events in the visual process begins with the absorption of a photon of light by rhodopsin. In this system, the photon may be considered the first messenger and rhodopsin the receptor. Rhodopsin is an integral membrane protein that spans across the membrane thickness of the membrane discs in rod cells. Photoactivation of rhodopsin results in a change in its conformation and allows it to bind a G protein called **transducin** (G$_t$). Like other G proteins, transducin contains $\alpha$-, $\beta$-, and $\gamma$-subunits. While bound to the rhodopsin, the $\alpha$-subunit exchanges GTP for GDP and is released from the complex. As diagrammed in Fig. 5, this subunit then binds to the inhibitory $\gamma$-subunit of a heterotrimeric cyclic nucleotide **phosphodiesterase (PDE)** and removes the inhibitory subunit, thus activating the PDE. The active PDE converts cGMP to 5′-GMP. The reduction

of cytoplasmic cGMP results in the closing of plasma membrane Na$^+$ channels[2] and a hyperpolarization of the membrane. A series of events then occurs, including synaptic transmission between the photoreceptors and retinal bipolar cell, and between bipolar cells and optic neurons. Action potentials in optic nerve fibers transmit a visual signal to the brain.

In rod cells, a sustained high level of cGMP is the cytoplasmic signal that produces steady basal activity of Na$^+$ channels. In this example, the loss (lowering) of the second messenger (cGMP) by a phosphodiesterase leads to a biological response, namely a visual signal. **Guanylate cyclase** (Fig. 6), the enzyme that catalyzes the production of cGMP from GTP, does not appear to be regulated in this process. In other systems, the activity of guanylate cyclase is regulated.

Two isozymes of guanylate cyclase are involved in signal transduction. One of the isozymes is bound to the plasma membrane and responds to receptor binding by increasing cGMP production. The resulting increase in cytoplasmic cGMP activates a **cGMP-dependent protein kinase (PKG).** This kinase phosphorylates intracellular proteins that produce the observed biological effect. For example, when blood volume is increased, the heart releases **atrial natriuretic factor (ANF).**[3] Binding of ANF to the kidney causes an increase in PKG activity and the production of cGMP, which signals the kidney to increase excretion of Na$^+$. Loss of Na$^+$ causes a loss of water (diuresis), which leads to a decrease in blood volume. In addition, PKG phosphorylates L-type Ca$^{2+}$ channels in vascular smooth muscle cells.

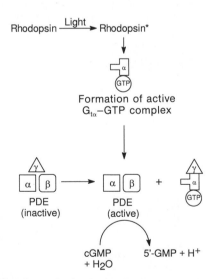

**FIG. 5.**  Signal transduction in the visual process as it occurs in rod cells. In a process similar to that presented in Fig. 2, photoactivated rhodopsin interacts with transducin (G$_t$). When the $\alpha$-subunit of transducin (G$_{t\alpha}$) binds to GTP, it dissociates from the $\beta$- and $\gamma$-subunits and associates with the inhibitory $\gamma$-subunit of cyclic nucleotide phosphodiesterase (PDE). The active $\alpha\beta$ dimer of phosphodiesterase reduces cellular cGMP levels.

[2] Thus, the Na$^+$ channels are open in the dark, giving rise to a steady depolarizing dark current.

[3] ANF is a peptide, and therefore is also known as atrial natriuretic peptide (ANP).

**TABLE 4** Examples of Processes Regulated by Intracellular Free $Ca^{2+}$ Levels

| Tissue | Key protein regulated | Process |
|--------|----------------------|---------|
| Liver | Glycogen phosphorylase kinase | Glycogen degradation |
| Muscle | Troponin C | Muscle contraction |
| Brain | Tyrosine hydroxylase | Production of DOPA |

The phosphorylated channels are inhibited, which causes a decrease in $Ca^{2+}$ influx and leads to vasodilation.

The second isozyme of guanylate cyclase is present in the cytoplasm. This isozyme is activated by nitric oxide (NO). Nitric oxide can be produced by many cell types in response to environmental stimuli. It is also produced by the spontaneous breakdown of the common vasodilators nitroglycerin and nitroprusside. In the heart and systemic circulation, activation of cytoplasmic guanylate cyclase by nitric oxide results in vasodilation and arteriolar relaxation.

### D. Calcium as a Second Messenger

The fact that most intracellular $Ca^{2+}$ is bound to membranes and myofilaments or sequestered into special compartments (organelles) explains the disparity between measurements of total cellular $Ca^{2+}$ ($\sim 10^{-3}$ $M$) and free cytoplasmic $Ca^{2+}$ ($\sim 10^{-7}$ $M$) concentrations. The low level of cytoplasmic $Ca^{2+}$ is maintained by the action of several transporters that move $Ca^{2+}$ either out of the cell or into organelles such as mitochondria, endoplasmic reticulum (ER), or sarcoplasmic reticulum (SR) in muscle. Intracellular $Ca^{2+}$ concentrations can be increased rapidly either by transient opening of plasma membrane $Ca^{2+}$ channels or by release of $Ca^{2+}$ sequestered in the SR (or ER). Both mechanisms are activated by binding of extracellular signals to plasma membrane receptors. Transient changes in intracellular free $Ca^{2+}$ levels regulate diverse processes (Table 4).

Most of the effects of $Ca^{2+}$ as a second messenger are explained by the action of **calmodulin.** Calmodulin is found in all cells, both as a free polypeptide molecule and as a component of some multisubunit enzyme complexes. A typical cell contains more than $10^7$ molecules of calmodulin, constituting as much as 1% of total cell protein. Calmodulin is a single polypeptide chain of 148 amino acids and contains four high-affinity $Ca^{2+}$ binding sites. Each $Ca^{2+}$ binding site consists of a specialized helix–loop–helix structural motif called an EF hand.[4]

The binding of $Ca^{+2}$ to calmodulin causes a large conformational change. In the case of free calmodulin, binding of $Ca^{2+}$ causes it to bind to target proteins and either activate or inhibit them. For example, the $Ca^{2+}$-ATPase of the SR of muscle is normally inactive. When cytoplasmic $Ca^{2+}$ levels are high, $Ca^{2+}$-calmodulin complexes form that bind and activate the $Ca^{2+}$-ATPase. Cytoplasmic $Ca^{2+}$ is then pumped into the lumen of the SR. Some proteins

[4] The term *EF hand* comes from the original studies used to illustrate calcium binding by this motif. These studies were performed using parvalbumin, in which the helices labeled E and F are involved in calcium binding.

have one or more molecules of calmodulin as subunits. Glycogen phosphorylase kinase is a 1.3 million-Da complex composed of 16 subunits ($\alpha_4\beta_4\gamma_4\delta_4$). The four $\delta$-subunits are calmodulin. The binding of $Ca^{2+}$ to the calmodulin subunit of glycogen phosphorylase kinase changes the conformation of the complex and partially activates the kinase.

Calcium- and cAMP-regulated cellular activities often overlap. In some instances, a single protein can be regulated by both $Ca^{2+}$ and cAMP. As described earlier, $Ca^{2+}$ activates glycogen phosphorylase kinase. The same enzyme can also be activated by phosphorylation through the action of a PKA. In some cells, the levels of cAMP and free $Ca^{2+}$ are interdependent. For example, cAMP-dependent kinases can regulate $Ca^{2+}$ channels and pumps by phosphorylation, whereas calmodulin regulates enzymes responsible for cAMP production and degradation.

### E. Second Messengers Generated by Lipid Hydrolysis

#### 1. Phosphatidylinositol Cycle

Phospholipids containing inositol account for 2–8% of total cellular phospholipids. There are three major inositol-containing lipids, **phosphatidylinositol (PI), phosphatidylinositol 4-phosphate (PIP),** and **phosphatidylinositol 4,5-bisphosphate (PIP$_2$).** In response to external signals, cells may hydrolyze PIP$_2$ to form three distinct second messengers, **inositol 1,4,5-trisphosphate** ($I_{1,4,5}P_3$ or **IP$_3$), diacylglycerol (DAG),** and **arachidonic acid** (Figs. 7 and 8). IP$_3$ binds to a specific $Ca^{2+}$ channel in the ER/SR and signals for the release of $Ca^{2+}$ into the cytoplasm. DAG activates a specific protein kinase called PKC.

Arachidonic acid may be converted into a family of biologically active compounds known as the **eicosanoids** (e.g., **leukotrienes, thromboxanes,** prostaglandins). The eicosanoids are local hormones (known as autacoids) because they are extremely short lived. They alter the activities of the cells in which they are synthesized and those of adjoining cells. Eicosanoids have multiple effects, including stimulation of the inflammatory response, regulation of blood flow, control of ion transport, and modulation of synaptic transmission.

The enzyme that hydrolyzes PIP$_2$ to DAG and IP$_3$ is a phosphatidylinositol-specific **phospholipase C (PLC).** At

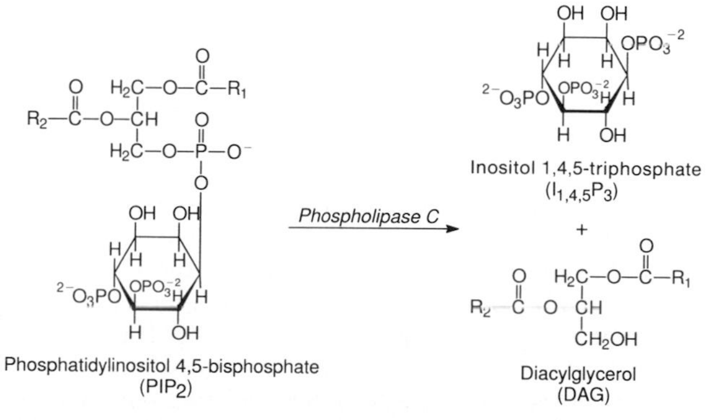

**FIG. 7.** The hydrolysis of PIP$_2$ by phospholipase C produces two second messengers, $I_{1,4,5}P_3$ and diacylglycerol.

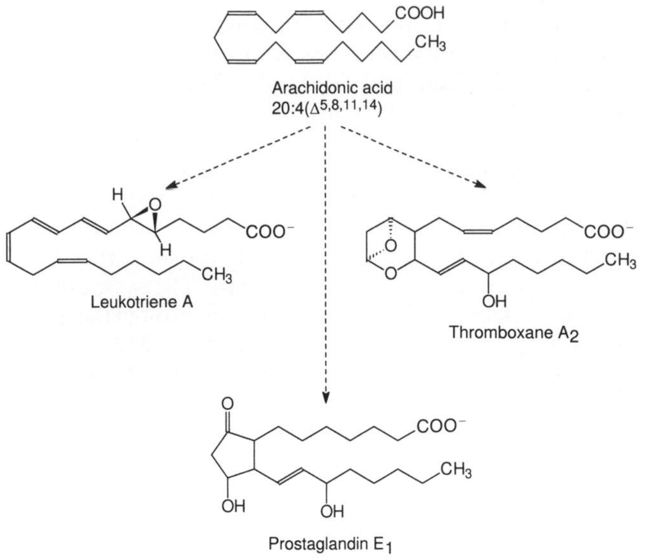

**FIG. 8.** Structures of arachidonic acid and some eicosanoids for which it is a precursor.

$Ca^{2+}$ channels are also present in the nuclear membrane and *cis*-Golgi apparatus. Proteins that bind $IP_4$ and a polyphosphoinositide containing six phosphates have been identified, but their function remains obscure. The more than 100-fold increase in free cytoplasmic $Ca^{2+}$ levels caused by $IP_3$ modulates the activities of several $Ca^{+2}$-sensitive proteins as described earlier.

*b. Diacylglycerol as a Second Messenger.* Diacylglycerol produced by the action of PLC binds to a family of $Ca^{2+}$-phospholipid-dependent protein kinases called protein kinase C (PKC). Binding of DAG to PKC in the presence of $Ca^{2+}$ and phosphatidylserine results in activation of the kinase. PKC contains a **catalytic domain** and a **regulatory domain.** Under some conditions, the regulatory domain is proteolytically cleaved, leaving a constitutively active protein, sometimes called protein kinase M. Active PKC uses ATP to phosphorylate serine or threonine residues of target proteins that are cell specific. Presumably, these phosphorylations affect the proteins' activities and lead to a biological response. A partial list of signal transduction pathways involving PKC is presented in Table 5.

## 2. Phosphatidylcholine Cycle

Two lines of evidence suggest that the hydrolysis of phosphatidylcholine (PC) plays a role in signal transduction. Early studies examining the PI cycle and production of DAG by PLC have often suggested that more DAG was produced than could be accounted for by the breakdown of $PIP_2$. Analysis of the fatty acid composition of the DAG suggested that it may have come from PC. More recently, investigators have prelabeled PC with radiolabeled choline and measured the release of radiolabeled choline and phosphocholine after treating cells with a variety of extracellular signaling agents. Many hormones, neurotransmitters, and growth factors stimulated the breakdown of PC to DAG or phosphatidic acid. This suggests that both phospholipase C and phospholipase D activities were activated. Phosphatidic acid is readily hydrolyzed to DAG and inorganic phosphate by highly active phosphatidate phosphohydrolase(s) present in most cells. Current evidence suggests that depending on the cell type and agonist, either or both phospholipase C and phospholipase D may be activated.

Although the process of signal transduction by PC hydrolysis is not completely understood, current evidence suggests three mechanisms for hormonal stimulation of PC breakdown. Nonhydrolyzable analogues of GTP stimulate PC breakdown in some cell types, whereas similar analogues of GDP inhibit the process. This suggests that G protein may mediate the activation of phospholipases. Since neither cholera toxin nor pertussis toxin affect the hydrolysis of PC, it is unlikely that $G_i$ or $G_s$ are involved in this process.

In some cell types, a rise in intracellular free $Ca^{2+}$ levels stimulates the hydrolysis of PC. For example, when hepatocytes are treated with the $Ca^{2+}$ ionophore A23187, PC hydrolysis is stimulated. It is not known whether $Ca^{2+}$ acts directly on the phospholipase or by a mechanism involving a $Ca^{2+}$-binding protein.

least nine **isoforms** of PLC exist, which are classified into four groups ($\alpha$, $\beta$, $\gamma$, $\delta$) based on structure. There appear to be two mechanisms for activating PLC. In some instances, the binding of an extracellular signaling molecule to a plasma membrane receptor causes the activation of a G protein called $G_q$ or $G_p$. This G protein has a unique $\alpha$-subunit that activates $\beta1$-phospholipase C (PLC$\beta1$) when bound to GTP. Whether other isoforms of PLC are activated by other G proteins is currently unknown.

A second method for PLC activation is **phosphorylation.** PLC$\gamma$ can be activated directly by a **tyrosine kinase** activity present in the receptor–ligand complex. A discussion of receptor kinase activity is presented later in a separate section.

Diacylglycerol and $IP_3$ are recycled to $PIP_2$, as shown in Fig. 9. This constitutes the PI cycle. $IP_3$ is hydrolyzed to inositol by the action of specific **phosphatases.** Alternatively, $IP_3$ is first **phosphorylated** to $IP_4$ ($I_{1,3,4,5}P_4$), or polyphosphoinositides containing five or six phosphates, and then hydrolyzed by phosphatases to inositol. These enzymes ensure that the lifetime of $IP_3$ and $IP_4$ in the cells is short. It has been suggested that the ability of $Li^+$ to inhibit complete removal of phosphate groups from phosphoinositides in brain neurons is correlated with its beneficial therapeutic effect in individuals suffering from manic depression.

*a. $IP_3$ and Other Polyphosphoinositides as Second Messengers.* As stated earlier, $IP_3$ binds and activates a specific $Ca^{2+}$ channel in the SR that is also known as the $Ca^{2+}$ release channel, the $IP_3$-binding protein, or the ryanodine receptor. This channel is a glycoprotein composed of four 260-kDa subunits. At least five different isoforms of the channel are believed to arise by **alternative splicing.** Although it is widely believed that most of the $Ca^{2+}$ released by $IP_3$ binding comes from the ER/SR, it has been demonstrated that

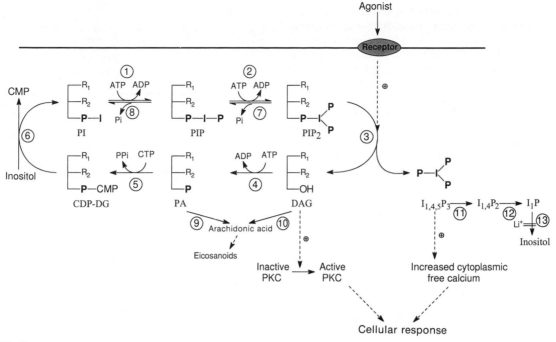

**FIG. 9.** Overview of the PI cycle and activation of protein kinase C. Numbered enzymes: 1, phosphatidylinositol kinase; 2, phosphatidylinositol 4-phosphate kinase; 3, phosphatidylinositol-specific phospholipase C; 4, diacylglycerol kinase; 5, phosphatidate cytidylyltransferase; 6; phosphatidylserine synthase; 7, $PIP_2$ 5-phosphatase; 8, PIP 4-phosphatase; 9, phosphatidate phosphatase; 10, diacylglycerol lipase; 11, $I_{1,4,5}P_3$ 5-phosphatase; 12, $I_{1,4}P_2$ 4-phosphatase, 13, $I_1P$ phosphatase.

A third mechanism of activation of PC hydrolysis may involve PKC. In some cell types, inhibitors of PKC activity also block PC hydrolysis. Conversely, phorbol esters and synthetic DAGs known to activate PKC also stimulate PC breakdown. Vasopressin and $P_2$ purinergic (ATP and analogues) agents appear to cause the activation of a phosphatidylinositol-specific PLC and hydrolysis of $PIP_2$ prior to a more sustained hydrolysis of PC. It has been suggested that the initial transient increase in DAG arising from $PIP_2$ may trigger PC breakdown through activation of PKC. The further production of DAG from PC would then provide a positive feedback to maintain the activation of PKC for prolonged physiological effects.

Hydrolysis of $PIP_2$ is not required for PC turnover. For example, stimulation of growth by interleukins 1 and 3 raise the levels of DAG without the hydrolysis of phosphoinositides.

**TABLE 5** Examples of Physiological Events Mediated by Protein Kinase C

| Agent | Tissue | Effect |
|---|---|---|
| Acetylcholine | Pancreas | Insulin release |
| Angiotensin | Adrenal cortex | Release of aldosterone |
| Thrombin | Platelets | Serotonin release |
| Thyrotropin-releasing hormone | Pituitary | Prolaction secretion |

### 3. Sphingolipids as Second Messengers

Agonist-induced stimulation of sphingomyelinase results in the production of ceramide. Activation of sphingomyelinase has been linked to several cell surface receptors, including the tumor necrosis factor, interleukin-1 (IL-1), and nerve growth factor receptors. During the past few years, an increasing number of cell-specific responses to increased intracellular ceramide levels have been identified, including inhibition of cell division, induction of apoptosis, modulation of prostaglandin secretion, and stimulation of interleukin-2 secretion. Several investigators have identified protein kinases and phosphoprotein phosphatases that are activated by ceramide. It has been hypothesized that through the actions of these enzymes, ceramides may play a role in "cross-talk" between the phosphatidylinositol and phosphatidylcholine cycles.

Sphingosine and sphingosine 1-phosphate are potent regulators of a variety of proteins involved in signal transduction. It has been suggested that sphingolipids play a role in the stimulation of cell proliferation by platelet-derived growth factor (PDGF) and by fetal calf serum. This is a rapidly expanding area of research that holds great promise for our understanding of complex, coordinated signaling systems.

### 4. Phosphatidic Acid as a Second Messenger

Membrane phosphatidic acid (PA) levels rapidly increase in the presence of certain metabolic agonists. Early studies suggested that the increased PA levels affected $Ca^{2+}$ transport across the plasma membrane. It now appears that cellular processes activated by PA induce $Ca^{2+}$ transport. However, PA has been demonstrated to activate PLCγ, Ras, and PIP-4-kinase, and to stimulate respiratory burst in neutrophils independently of diacylglycerol. Thus, it seems likely that, like many other lipids, the regulation of PA levels will play a central role in signal transduction.

Phosphatidic acid phosphohydrolases catalyze the conversion of PA to diacylglycerol. A multifunctional form of the enzyme has been isolated from the plasma membrane. In addition to hydrolyzing PA, this form of the enzyme can dephosphorylate lysophophatidate, sphingosine 1-phosphate, and ceramide 1-phosphate. It has been postulated that phosphatidic acid phosphohydrolases terminate the signaling of bioactive lipids and, at the same time, generate products that are potent signaling molecules. Therefore, PA phosphohydrolases may act as switches between signaling pathways.

## III. Signaling by Receptor Phosphorylation

A family of receptors exists that function by directly phosphorylating target proteins. Binding of high-affinity ligands (e.g., insulin) to these receptors induces a conformational change that activates a **tyrosine kinase** activity located within a cytoplasmic domain. This kinase activity may result in **autophosphorylation** of the receptor and/or phosphorylation of intracellular target proteins.

The response of **growth-factor receptors** is mediated by their tyrosine kinase activity. Mutant receptors missing the tyrosine kianse activity are unable to stimulate their normal biological function. Table 6 lists some of the receptors for growth factors known to be dependent on a tyrosine kinase activity.

Activated receptor tyrosine kinases transmit information by two mechanisms: protein **phosphorylation** and **protein–protein binding.** The action of PDGF receptor requires both of these mechanisms. PDGF binds to a cell surface receptor that possesses cytoplasmic, ligand-activated, tyrosine kinase activity that activates a PI-specific phospholipase C. Mutant receptors that bind the ligands, but lack tyrosine kinase activity, are unable to stimulate the phospholipase activity. Activation of the phospholipase does

**TABLE 6**  Examples of Growth Factors with Receptors having Cytoplasmic Tyrosine Kinase Activities

| |
| --- |
| Colony stimulating factor 1 |
| Epidermal growth factor |
| Insulin-like growth factor 1 |
| Insulin |
| Nerve growth factor |
| Platelet-derived growth factor |
| Transforming growth factor α |

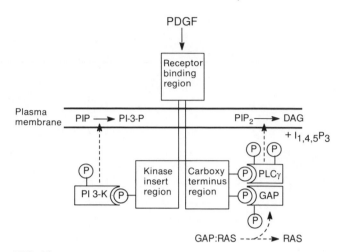

**FIG. 10.**   Interactions between the PDGF receptor and intracellular proteins after binding of PDGF to the receptor. Binding of PDGF to its receptor results in autophosphorylation of the receptor. This phosphorylation allows for the binding and activation of the γ isozyme of PI-specific phospholipase C (PLCγ), GTPase activating protein (GAP), and PI 3-kinase.

not occur through the action of a G protein, but rather by binding of the enzyme directly to the receptor.

A complex set of events is initiated by the binding of PDGF to its receptor and is outlined in Fig. 10. Binding of the ligand to the receptor activates the tyrosine kinase activity. This kinase uses ATP to phosphorylate itself (autophosphorylation) and the γ isoform of phospholipase C. At this time, the receptor and PLCγ bind to each other and the phospholipase becomes activated. Although the exact sequence of these events is not known, it is clear that phosphorylation of both the receptor and PLCγ and association of the receptor and lipase are required for activation to occur.

Activation of PLC by the PDGF receptor is not required for PDGF-induced mitogenesis (cell replication). Although the exact mechanism of PDGF-stimulated mitogenesis is not known, it appears that, in addition to PLC, the receptor binds to two other proteins, namely, phosphatidylinositol 3-kinase and **GTPase activating protein (GAP).** Both of these proteins are phosphorylated by the receptor's kinase. Phosphatidylinositol 4-kinase and phosphatidylinositol 4-phosphate 5-kinase may also bind to the receptor. Thus, it appears that the receptor can assemble the enzymes required to produce $PIP_2$ and the enzyme required to hydrolyze $PIP_2$ to the second messengers $IP_3$ and DAG. GAP and PLCγ bind to the receptor at their carboxyl termini. Phosphatidylinositol 3-kinase binds at the kinase insert region.

Recent work has indicated a pathway through which tyrosine kinase receptors transmit signals to the nucleus, resulting in alterations in transcription, and stimulation of cell proliferation or differentiation. This pathway, known as the **MAP kinase** (mitogen-activated protein kinase) pathway, is outlined in Fig. 11. Binding of a growth factor to a tyrosine kinase receptor results in the activation of the tyrosine kinase activity, one result of which is an auto-

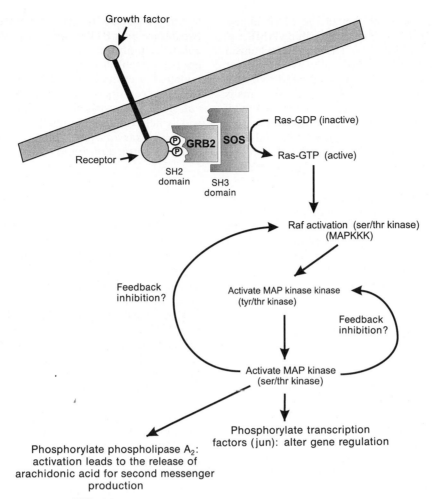

**FIG. 11.** The mitogen-activated protein kinase pathway.

phosphorylation of critical tyrosine residues on the cyto-plasmic portion of the receptor. Once the growth factor receptor has been phosphorylated, an adapter protein, GRB2 (growth-factor receptor binding protein 2), binds to the phosphorylated tyrosine residues on the activated receptor. The domain on GRB2 that allows it to bind to the receptor is known as an SH2 domain. GRB2 also contains a second domain, known as an SH3 domain, that allows it to bind to proteins containing a poly-proline region. One such protein is SOS, which is a guanine nucleotide-releasing factor (GNRF). SOS binds to the SH3 domain of GRB2, which is bound to the activated receptor. This binding event activates SOS, and SOS will accelerate the exchange of GTP for GDP on a small GTP-binding protein known as ras. Ras is similar to the $\alpha$-subunits of the heterotrimeric G-proteins in that it is a GTP binding protein that is active in its GTP-bound form, inactive when bound to GDP, and contains an intrinsic GTPase activity. The GTPase activity of ras is slow acting, but serves to limit the length of time that ras is present in its GTP-bound active state. Thus, so far the effect of receptor activation has been to activate ras, and convert ras–GDP to ras–GTP.

The ras protein, which contains bound fatty acids, is localized to the membrane. When activated, the mem-brane-bound ras recruits a serine/threonine kinase known as raf to the plasma membrane, which activates the raf kinase. The raf kinase is also known as a MAP kinase kinase kinase (MAPKKK). When activated, raf will phos-phorylate other proteins known as the MAP kinase kinases, or MEKS, thereby activating these kinases. The activated MAP kinase kinases then phosphorylate and activate the MAP kinases (also known as ERK). To activate the MAP kinases, both a threonine and a tyrosine residue of these kinases need to be phosphorylated. The MAP kinase kinase catalyzes the phosphorylation of both of these residues. The activated MAP kinases then phosphorylate a group of target proteins, on either serine or threonine residues, which will alter the activity of the target proteins. One such target protein is the transcription factor jun, which when phosphorylated will activate a different set of target genes than when not phosphorylated. Another protein phosphorylated by the MAP kinases is phospholipase D, which when phosphorylated will lead to arachi-donic acid release and allow for the synthesis of additional second messengers.

Recent evidence has indicated that the activation of ade-nyl cyclase via G-protein-linked receptors and the MAP kinase pathway intersect. In certain cells elevation of cAMP levels leads to an inhibition of the MAP kinase pathway, whereas in other cell types elevated cAMP enhances the MAP kinase pathway. The exact mechanism for the cross-talk between the two pathways is not yet known, but for the cells in which elevated cAMP inhibits the MAP kinase pathway, it appears as if activated ras is incapable of activating raf. This effect may be mediated by protein kinase A, although this has yet to be demonstrated. Thus, the entire story of intracellular signaling is still in its infancy, and much more needs to be learned before we can understand how a cell integrates all signals received from the environment.

Recent analyses of signaling pathways, initiated by receptor tyrosine kinases at the plasma membrane, have suggested that signal-transduction pathways play a critical role in the regulation of protein and membrane trafficking. These findings suggest that within the next decade, we will identify many more cellular functions that are controlled by an interlinked group of control processes.

## IV. Other Signaling Mechanisms

### A. Two-Component Systems

To respond to environmental challenges, bacteria must adjust their structure, physiology, and behavior. The simplest signaling systems contain a sensor that monitors an environmental parameter and a response regulator that mediates an adaptive response. Sensors are transmembrane proteins located in the cytoplasmic membrane. Response regulators are cytoplasmic proteins that, when activated, can alter gene expression.

Communication between sensors and response regulators involves phosphorylation and dephosphorylation reactions. Sensors (also called sensory kinases) have an autokinase activity that catalyzes an ATP-dependent phosphorylation of a conserved histidine residue. The high-energy phosphohistidine acts as an intermediate for the subsequent transfer of the phosphate to an aspartate residue present in the response regulator. Phosphorylation of the aspartate residue activates the response regulator. A phosphatase domain within response regulators autocatalyzes the removal of the phosphate from the aspartate residue. This causes inactivation of the response regulator and occurs with half-lives ranging from a few seconds to several minutes.

Two-component systems have been identified in more than 25 distinct types of signal transduction pathways, across a wide variety of prokaryotic genera. Recent studies have indicated that many eukaryotic proteins contain domains with sequences similar to those found in prokaryotic sensors and response regulators, suggesting that similar signaling mechanisms exist in eukaryotes.

### B. Protein Tyrosine Phosphatase Signaling

Transmembrane protein tyrosine phosphatases (PTPases) have been identified in many cell types. These phosphatases contain an extracellular domain of varying architecture connected by a transmembrane peptide to one or two intracellular PTPase domains. Although a direct mechanism for individual mammalian PTPase in signaling has not been established, indirect evidence has begun to accumulate. For example, lymphocyte activation through their antigen receptors is lost in cells deficient in a hematopoietic-specific transmembrane PTPase called CD45. It is widely believed that mammalian signaling systems, analogous to prokaryotic protein tyrosine phosphatase signaling, will be identified during the next few years.

## V. Summary

Signal transduction is the process whereby an external chemical signal elicits an intracellular metabolic change. The process begins with the binding of specific ligands to receptors located at the surface of the plasma membrane. The receptor responds to the binding of agonists in several different ways.

Most signaling appears to involve receptor activation of a GTP-binding protein (G protein). The activated G proteins interact with enzymes that produce second messenger molecules. For example, G protein activation of phosphatidylinositol specific phospholipase C results in the hydrolysis of $PIP_2$ to diacylglycerol and $IP_3$. Both of these act as second messengers transferring information from the plasma membrane to intracellular sites. $IP_3$ binds to receptors at the surface of the endoplasmic reticulum that trigger the release of membrane-bound $Ca^{2+}$ into the cytoplasm. Diacylglycerols activate PKC, which catalyzes the ATP-dependent phosphorylation of intracellular proteins. Both $Ca^{2+}$ release and protein phosphorylation affect the activities of intracellular enzymes. Changes in cellular enzyme activities lead directly to a biological response by the cell. For example, thrombin activation of platelets is mediated by the hydrolysis of phosphoinositides.

Another signaling pathway regulated by the action of G proteins is the PKA pathway. In this pathway, binding of an agonist to a receptor leads to the activation of adenylate cyclase and thus the production of cAMP. Cyclic AMP is a second messenger that activates PKA (cyclic AMP-dependent protein kinase). PKA catalyzes the ATP-dependent phosphorylation of cell-specific target proteins. It is the change in activity of these target proteins that ultimately leads to the biological response of cells bound to an agonist.

A second mechanism of signal transduction involves cell surface receptors that have tyrosine kinase activities associated with their cytoplasmic domains. Binding of agonist to these receptors directly activates their kinase domains. The kinase activity catalyzes the phosphorylation of intracellular proteins and often causes autophosphorylation. This phosphorylation may directly cause changes in enzyme activity or may indirectly affect activity by first promoting binding of proteins to the receptor.

Modulation of ion channels can occur as the result of the binding of extracellular signals to their receptors. In these cases, ions act as second messengers. This process is the least understood mechanism of signal transduction.

# Bibliography

Barik, S. (1996). Protein phosphorylation and signal transduction. *Subcell. Biochem.* **26,** 115–164.

Bell, R., Burns, D., Okazaki, T., and Hannun, Y. (1992). Network of signal transduction pathways involving lipids: Protein kinase C-dependent and -independent pathways. *Adv. Exp. Med. Biol.* **318,** 275–284.

Berridge, M. J. (1993). Inositol trisphosphate and calcium signaling. *Nature* **361,** 315–325.

Brindley, D. N., and Waggoner, E. W. (1996). Phosphatidate phosphohydrolase and signal transduction. *Chem. Phys. Lipids* **80,** 45–57.

Brown, A. M., Yatani, A., VanDongen, M. J., Kirsch, G. E., Codina, J., and Birnbaumer, L. (1990). Networking ionic channels by G proteins. *In* "G Proteins and Signal Transduction" (N. M. Nathanson and T. K. Harden, Eds.), Society of General Physiologists Series, Vol. 45, pp. 1–9. Rockefeller University Press, New York.

Cadena, D. L., and Gill, G. N. (1992). Receptor tyrosine kinases. *FASEB J.* **6,** 2332–2337.

Cobb, M. H., and Goldsmith, E. J. (1995). How MAP kinases are regulated. *J. Biol. Chem.* **270,** 14843–14846.

English, D., Cui, Y., and Siddiqui, R. A. (1996). Messenger functions of phosphatidic acid. *Chem. Phys. Lipids* **80,** 117–132.

Galione, A. (1993). Cyclic ADP-ribose: A new way to control calcium. *Science* **259,** 325–326.

Hidaka, H., and Ishikawa, T. (1992). Molecular pharmacology of calmodulin pathways in the cell functions. *Cell Calcium* **12,** 465–472.

Hoch, J. A., and Thomas, S. J. (Eds.) (1995). "Two-Component Signal Transduction," ASM Press, Washington, DC.

Kleuss, C., Scherübl, Hescheler, J., Schultz, G., and Wittig, B. (1993). Selectivity in signal transduction determined by $\gamma$ heterotrimeric G proteins. *Science* **259,** 832–834.

Kyriakis, J. M., and Avruch, J. (1996). Sounding the alarm: Protein kinase cascades activated by stress and inflammation. *J. Biol. Chem.* **271,** 24313–24316.

Levitzke, A. (1996). Targeting signal transduction for disease therapy. *Curr. Opin. Cell Biol.* **8,** 239–244.

Missiaen, L., Parys, J. B., De Smedt, H., Sienaert, I., Bootman, M. D., and Casteels, R. (1996). Control of the $Ca^{2+}$ release induced by myo-inositol trisphosphate and the implication in signal transduction. *Subcell. Biochem.* **26,** 59–95.

Olivera, A., and Spiegel S. (1993). Sphingosine 1-phosphate as a second messenger in cell proliferation by PDGF and FCS mitogens. *Nature* **365,** 557–560.

Ovadi, J., and Orosz, F. (1992). Calmodulin and dynamics of interactions of cytosolic enzymes. *Curr. Top. Cell. Regul.* **33,** 105–126.

Packer, L., and Wirtz, K. W. A. (Eds.) (1995). "Signalling Mechanisms—from Transcription Factors to Oxidative Stress." Springer-Verlag, Berlin.

Seaman, M. N., Burd, C. G., and Emr, S. D. (1996). Receptor signaling and the regulation of endocytic membrane transport. *Curr. Opin. Cell Biol.* **8,** 549–556.

Spiegel, A. M. (1996). Defects in G protein-coupled signal transduction in human disease. *Annu. Rev. Physiol.* **58,** 143–170.

Spiegel, S., Foster, D., and Kolesnick, R. (1996). Signal transduction through lipid second messengers. *Curr. Opin. Cell Biol.* **8,** 159–167.

Steoli, M. (1996). Protein tyrosine phosphatases in signaling. *Curr. Opin. Cell Biol.* **8,** 182–188.

John R. Dedman and Marcia A. Kaetzel

# 9

# Calcium as an Intracellular Second Messenger: Mediation by Calcium Binding Proteins

## I. Introduction

Sidney Ringer provided the first report relating tissue and cellular function with $Ca^{2+}$ in 1883. He demonstrated that $Ca^{2+}$ was necessary for normal regular contractions of the isolated frog heart. Following this landmark study, $Ca^{2+}$ became an essential component of physiological saline solutions. There have been numerous studies relating $Ca^{2+}$ and cell functions, including fertilization, development, differentiation, adhesion, growth, division, movement, contraction, and secretion. This evidence demonstrates a primary regulatory role for ionized $Ca^{2+}$ in biological systems. $Ca^{2+}$ has also been associated with a number of diseases, particularly those of the muscular and nervous systems, in which this ion plays an important role in contraction and neurotransmitter release.

Understanding the mechanism of $Ca^{2+}$ action has required approaches and expertise from distinct fields. $Ca^{2+}$ is unique compared with other second messengers, which are formed as metabolic intermediates, such as cyclic nucleotides, inositol phosphates, and diacylglycerol. $Ca^{2+}$ is a divalent elemental metal and is not converted to any other form as a part of its cellular regulatory properties. $Ca^{2+}$ is unlike other metal ions such as $K^+$ and $Na^+$, which are involved in membrane potentials and excitability, or $Mg^{2+}$ and $Zn^{2+}$, which act as enzyme cofactors involved in the catalysis of metabolic intermediates. The fact that $Ca^{2+}$ has been associated with a wide variety of cellular functions brings attention to the fact that the blocking of one $Ca^{2+}$-regulated function could very likely affect interdependent secondary and tertiary functions.

## II. Determination of $Ca^{2+}$ Involvement in Physiological Processes

The most direct approach to understanding the involvement of $Ca^{2+}$ in a given physiological activity has been to follow the tradition of Ringer, that is, through the reduction of extracellular $Ca^{2+}$. For example, the reduction of $[Ca]_o$ from 1 to 0.1 m$M$ markedly alters cell growth, adhesion, secretion, and motility. A second approach has been to use various pharmacological agents. As discussed in other chapters, $Ca^{2+}$ channel modulators have also proven useful in probing cellular systems. These distinct chemicals act by binding to the membrane $Ca^{2+}$ channel, thereby blocking ($Ca^{2+}$ channel antagonist) or promoting ($Ca^{2+}$ channel agonist) the influx of extracellular $Ca^{2+}$. The antibiotic A23187 has been shown to cage divalent metal ions such as $Ca^{2+}$ and to thereby act as an ionophore to facilitate their movement across biological membranes (Fig. 1). The A23187 ionophore binds $Ca^{2+}$, $Mn^{2+}$, and $Mg^{2+}$ with respective affinities of $210:2:1$. Under specified conditions, A23187 can provide evidence for a role for $Ca^{2+}$ in a given biological response. The ionophore elevates intracellular $Ca^{2+}$ and, for example, can induce the activation of lymphocytes, platelets, and sea urchin egg development.

## III. $Ca^{2+}$ as an Intracellular Signal

The total intracellular $Ca^{2+}$ concentration has been estimated to be approximately 1 m$M$, a value similar to extracellular $Ca^{2+}$ values. Early studies used model systems such as the skinned muscle fiber, in which the sarcolemma was physically removed. In such a system, the $Ca^{2+}$ ion concentration can be regulated in the bathing solution, while contraction and glycogenolysis are monitored. This approach demonstrated that $Ca^{2+}$ was active at micromolar levels and that it was rapidly sequestered from the cytosol. Conclusions drawn from such studies indicated that contraction and metabolism are regulated at micromolecular $Ca^{2+}$ levels. It was suspected that $Ca^{2+}$ acted by moving from one cellular compartment to another, that is, into and out of the cytosol. $Ca^{2+}$ channels and pumps were identified using

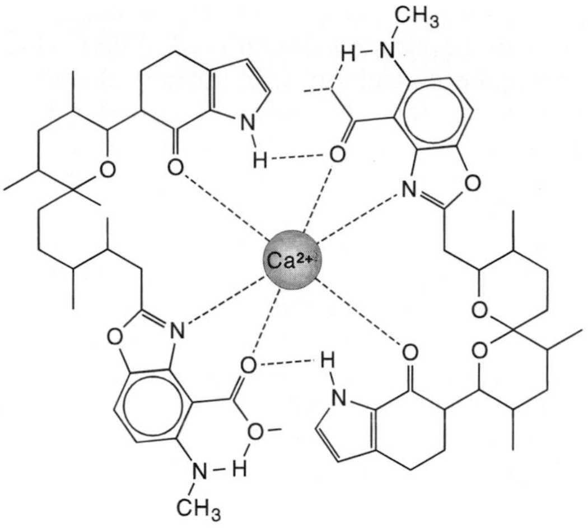

**FIG. 1.** Molecular structure of A23187. The charged $Ca^{2+}$ ions are extremely impermeable to the plasma membrane. Two molecules of ionophore A23187 form a molecular cage around the charged $Ca^{2+}$ ion. The complex is membrane soluble and dissociates in the cytosol. Intracellular $Ca^{2+}$ is then elevated.

isolated intact vesicles through which the uptake of $^{45}Ca^{2+}$ was monitored. Convincing evidence for the movement of $Ca^{2+}$ into and out of cellular compartments using isotopic methods in intact tissues proved difficult, since the ion was in a dynamic state of flux between the various intracellular organelles, the cytoplasm, and the extracellular fluids.

During the past 20 years, however, several generations of $Ca^{2+}$ indicators have been developed. Many coelenterates have the ability to glow by using $Ca^{2+}$-activated luminescent proteins. Aequorin was the first photoprotein to be used to measure intracellular free $Ca^{2+}$. Photoproteins can be used to measure free $Ca^{2+}$ levels within the range of 0.1–10 $\mu M$. The use of aequorin, however, is limited since it is a relatively large molecule, with a molecular weight of approximately 20,000, and hence must be injected into cells. Early studies with aequorin required the use of large cells for microinjections. For example, the premeability of gap junctions to $Ca^{2+}$ ions into large dipteran salivary glands was determined using aequorin luminescence with image intensification techniques. Under these conditions, nanograms of aequorin can be injected into a cell, which gives the cytosol the capacity to produce $10^4$ to $10^8$ photons per second. At these levels, the blue light luminescent output is readily measurable with sensitive photomultipliers.

More recently, a series of fluorescent indicator dyes that bind $Ca^{2+}$ with high affinity and selectivity has been developed. The most commonly used is Fura-2. Since the free acids of these fluorescent indicators are not membrane permeable, the carboxyl acid groups are esterified during their chemical synthesis to make them permeable to the plasma membrane. Intact cells are loaded with the permeable form of the dye and the ester groups are then hydrolyzed in the cytoplasm by cellular esterases, thereby trapping the fluorescent probe within the cells. Fura-2 has been used in many biological systems. The use of Fura-2, in combination with high-resolution fluorescence video

microscopy and computer-assisted image analysis systems, has provided valuable insight into understanding the intracellular $Ca^{2+}$ signal (Berridge, 1990):

1. The resting intracellular ionized $Ca^{2+}$ level is approximately 0.1 $\mu M$. $Ca^{2+}$ ions not complexed to counterions or proteins are frequently referred to as being "free." Cell stimulation by a variety of agents causes a transient rise in intracellular free $Ca^{2+}$; the increase is variable, lasting for a fraction of one second to minutes. The amplitude of the $Ca^{2+}$ spike also varies from tissue to tissue.

2. The elevation in $Ca^{2+}$ may be uniform throughout a cell or group of cells or highly localized to specific regions of individual cells.

3. In many cellular systems the $Ca^{2+}$ signal occurs as a wave, beginning at a discrete initiation site and then moving across the cell. Fertilization is a cellular process that displays this $Ca^{2+}$ wave phenomenon. The $Ca^{2+}$ influx is initiated at the point of sperm–egg contact, from where it spreads as a propagated wave toward the opposite pole. This $Ca^{2+}$ signal initiates cellular reactions to prevent polyspermy (penetration of the ovum by more than one sperm).

4. In many cell systems, the intracellular level of $Ca^{2+}$ oscillates. The frequency depends on several factors, including the cell type and cellular effectors such as hormones, neurotransmitters, growth factors, and cytokines. Most oscillations occur at a periodicity of 20–60 sec when the cell is at "rest." When the cell is stimulated with a hormonal agonist, the frequency can increase to less than 5 sec or result in a sustained elevation of up to 100 sec. The oscillation frequency in $Ca^{2+}$ may represent a periodic code, which can distinguish extracellular effector type and concentration.

## IV. Creation of the $Ca^{2+}$ Signal

Cells are able to maintain a resting level of $Ca^{2+}$ of less than 0.1 $\mu M$ in an environment where the extracellular $Ca^{2+}$ is 1 m$M$ or greater. This gradient is achieved because the plasma membrane is relatively impermeable to $Ca^{2+}$ and also contains ATP-driven $Ca^{2+}$ pumps and $Na^+/Ca^{2+}$ exchangers. In addition, the endoplasmic reticulum sequesters cytosolic $Ca^{2+}$, also by ATP-driven pumps. When $[Ca]_i$ increases to above 1 $\mu M$, mitochondria will internalize $Ca^{2+}$. Mitochondria that are heavily loaded with $Ca^{2+}$ reflect a distressed cellular state: under such conditions, the mitochondria develop dense granules containing complexed $Ca^{2+}$. Collectively, these pump and exchanger systems maintain a resting $[Ca]_i$ of less than 0.1 $\mu M$.

Initiation of the calcium signal is achieved from two primary sources, the extracellular fluid and sequestered internal stores. The plasma membrane can contain multiple species of $Ca^{2+}$-specific channels as determined by electrophysiological or pharmacological methods. In general, there are three channel types: voltage-dependent, ligand-gated, and mechanical (i.e., stretch-activated). The opening of these channels allows a rush of $Ca^{2+}$ to enter the cytoplasm, producing a localized increase in intracellular $Ca^{2+}$. The mechanism of $Ca^{2+}$ release from intracellular stores evaded elucidation for many decades. It was shown that inositol 1,4,5-triphosphate ($IP_3$) causes $Ca^{2+}$ mobilization from internal vesicular stores, primarily the endoplasmic reticulum (Berridge, 1993). $IP_3$ is a phospholipase C (PLC) hydrolysis product of phosphatidylinositol 4,5-bisphosphate ($PIP_2$), which is induced during cell stimulation. Recent evidence indicates that not all of the endoplasmic reticulum-sequestered $Ca^{2+}$ is $IP_3$ sensitive, and release is prompted by cyclic ADP–ribose (Lee, 1993). A second product of PLC hydrolysis of $PIP_2$ is diacylglycerol (DAG).

Stimulation of the membrane phospholipase C is through receptor binding by agonists and is G-protein regulated. GTP$\gamma$S, a nonmetabolizable analogue of GTP, can artificially activate PLC production of $IP_3$ and DAG. G-proteins also appear to regulate several processes, including the exchange of $Ca^{2+}$ from $IP_3$-insensitive to $IP_3$-sensitive pools, the gating properties of plasma membrane $Ca^{2+}$ channels, and the generation of cyclic AMP. The microinjection of GTP$\gamma$S causes a marked increase in $Ca^{2+}$ oscillation frequency and amplitude and stimulation of many cellular processes.

## V. Mediation of $Ca^{2+}$ Signal

Intracellular $Ca^{2+}$, like cyclic AMP, acts as a second messenger. Cyclic AMP is mediated by a limited number of receptor proteins, leading to the activation of specific protein kinases. There are, however, numerous intracellular $Ca^{2+}$ binding proteins; therefore, the intracellular $Ca^{2+}$ signal has many possible bifurcations of action. A simple example is found in skeletal muscle, where elevated $Ca^{2+}$ binds to troponin C to cause myofibrillar contraction at the expense of ATP. An independent, simultaneous $Ca^{2+}$ pathway is mediated by a second $Ca^{2+}$ receptor protein, calmodulin, which activates phosphorylase kinase. This activation initiates glycogenolysis, which leads to the regeneration of expended ATP (Fig. 2).

The most completely described family of intracellular $Ca^{2+}$-binding proteins to date is characterized by a protein structure known as the EF hand. The binding site is achieved through side-chain coordination of $Ca^{2+}$ within a helix–loop–helix composed of precisely spaced amino acids (Fig. 3). This EF hand is found in a number of proteins as determined from the primary amino acid sequence and, in a few cases, has been confirmed by direct $Ca^{2+}$ binding. The sequence data for the $Ca^{2+}$-dependent protease calpain suggest that it may have resulted from the fusion of a gene encoding four EF hand domains with a thiol protease gene (Ohno et al., 1984). Likewise, $\alpha$-actinin may have resulted from the fusion of a gene encoding an actin-binding protein with a gene for a single EF hand domain (Noegel et al., 1987). In many cases the EF hand is nonfunctional. For example, the $Ca^{2+}$ insensitivity of the muscle form of $\alpha$-actinin is due to the imprecise positioning of the amino acids shown in Fig. 3. The identification of a number of putative $Ca^{2+}$-binding proteins is based on sequence similarity to members of the troponin C/calmodulin superfamily. For example, cell cycle yeast mutants, which carry the

temperature-sensitive allele of *CDC31*, are blocked in spindle-pole body duplication (Baum et al., 1986). This gene has 42% sequence similarity with human calmodulin and may contain functional EF hands. A second cell division cycle gene, *CDC24*, is required for bud formation. Sequence analysis of *CDC24* indicates two potential $Ca^{2+}$-binding EF hands (Miyamoto et al., 1987).

Troponin C is present only in skeletal and cardiac muscle; calmodulin is present in all cells. The binding of four $Ca^{2+}$ ions causes conformational changes with the formation of an active state. $Ca^{2+}$-bound calmodulin has very high affinity for its target proteins, which act as additional steps in mediating the original $Ca^{2+}$ signal. The calmodulin target proteins include protein kinases, protein phosphatases, hydrolases, nitric oxide synthetase, and ion channels.

Sequence analysis and site-directed mutagenesis studies have provided a general model for calmodulin regulation of its target proteins (Fig. 4). The target protein contains a pseudosubstrate attached to a flexible region of the polypeptide chain. This sequence acts as an endogenous inhibitor of the enzyme. $Ca^{2+}$-activated calmodulin clamps around an adjacent target site and physically displaces the pseudosubstrate from the active site, causing a derepression or activation of the enzyme. This process is reversed by the reduction of intracellular $Ca^{2+}$ levels, which reduces the affinity of calmodulin for the target site and autoinhibition of the enzyme occurs. A smooth muscle contraction is a paradigm for calmodulin mediation of the $Ca^{2+}$ signal (Fig. 5).

One approach to understanding the cellular role of a protein is to correlate subcellular localization with function. This information can be effectively obtained through the use of antibodies that are monospecific for the targeted protein. As shown in Fig. 6, when affinity-purified antibodies were used to localize calmodulin during mitosis, the spindle poles were stained most brightly. This result suggested that calmodulin is involved in mitotic function. Indeed, microtubules depolymerize in the presence of $Ca^{2+}$-bound calmodulin both *in vitro* and in microinjected cells (Dedman et al., 1982). Many other cellular functions have been assigned to calmodulin through a combination of localization, biochemical, and genetic approaches (Davis, 1992).

## VI. Annexins: Calcium-Dependent Phospholipid-Binding Proteins

A number of laboratories with distinct experimental goals have identified a family of $Ca^{2+}$-dependent phospholipid-binding proteins. These proteins have the common property of binding (annexing) membranes (Crumpton and Dedman, 1990). Sequence data indicate that there are 10 unique mammalian annexins. Annexins are also present in the slime mold *Dictyostelium*, in the sponge, in the coelenterate *Hydra*, in the insect *Drosophila*, and in the mollusk *Aplysia*, as well as in higher plants.

The sequence organization of the family is highly conserved (Fig. 7). All except annexin VI are composed of a core of four repeated domains: annexin VI is composed

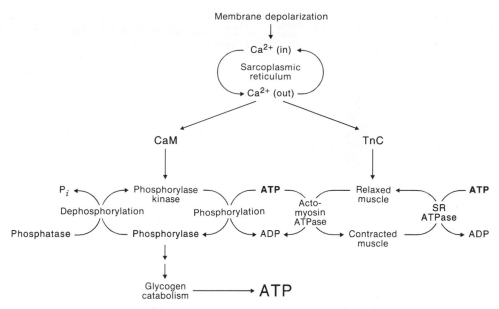

**FIG. 2.** Independent pathways of $Ca^{2+}$ regulation in skeletal muscle. Depolarization of the sarcolemma causes release of $Ca^{2+}$ stored in the sarcoplasmic reticulum. This elevated $Ca^{2+}$ level triggers contraction, an ATP-consuming process. The same signal $Ca^{2+}$ also binds calmodulin (CaM), which initiates glycogenolysis, an ATP-producing process. This parallelism allows for metabolic coordination.

of eight domains. Each domain is approximately 70 amino acids in length. The sequence conservation for each ranges between 40 and 60% when individual annexins are compared. The amino terminus of each protein is unique, suggesting that this region may confer functional differences to the proteins. This property has been confirmed with annexin II, in that the amino terminus binds a subunit, p11, and forms an actin-binding heterotetramer. Calcium-dependent phospholipid binding is a property of the four-domain core of each protein. *In vitro* functions of annexins—which include membrane binding and fusion, ion channel activity, modulation of ion channel activity, inhibition of phospholipase $A_2$, and inhibition of blood coagulation—all require this property. The core does not contain the classic EF hand calcium-binding motif. Coordi-

nation of $Ca^{2+}$ is accomplished through a unique, discontinuous binding loop. Chemical cross-linking data indicate that monomeric annexin in solution self-associates into trimers, which then form higher aggregates when bound to phospholipid vesicles. This observation is consistent with the diffraction pattern obtained from two-dimensional crystals.

Collectively, protein structural data, biophysical reconstitution studies, and subcellular localization results allowed development of the following model to describe the cellular function of the annexins. In the resting cell when free $Ca^{2+}$ concentrations arc low, the annexin exists as a soluble monomer. During cell stimulation, free $Ca^{2+}$ concentrations subjacent to the cell membrane rise to micromolar levels (Llinas *et al.,* 1992). The annexin would bind to target proteins associated with a phospholipid surface, then organize into trimers, hexamers, and higher aggregates, ultimately forming an extended hexagonal array around the target protein (Andree *et al.,* 1992; Concha *et al.,* 1993). The immunofluorescent localization studies of annexins support this model. A sheet of annexin multimers lining the inner membrane leaflet would locally alter membrane properties such as fluidity and sequestration of specific phospholipids. Changes in membrane properties have been shown to modify specific membrane protein function (Sweet and Schroeder, 1988; Bennett, 1985). Such a submembranous scaffolding maintains and stabilizes the membrane. For example, dystrophin is localized to the inner surface of the sarcolemma in normal skeletal muscle. When absent, as in Duchenne muscular dystrophy, the plasma membrane is unstable and the fibers rapidly turn over (Koenig *et al.,* 1988). In addition, the $Ca^{2+}$-dependent lining

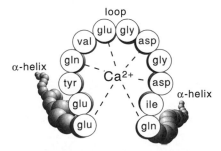

**FIG. 3.** The EF hand $Ca^{2+}$ binding pocket. Many intracellular $Ca^{2+}$-receptor proteins contain precisely positioned amino acids in a loop between highly structured $\alpha$-helical coils. The $Ca^{2+}$ ion is coordinated in this loop, which alters the overall structure of the protein and, in turn, its cellular activity.

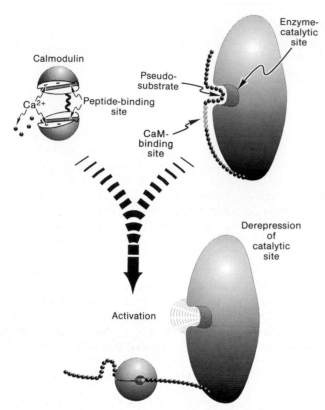

**FIG. 4.** Mechanism of action of calmodulin regulation of target proteins. Calmodulin is molecularly structured as two opposing lobes containing $Ca^{2+}$ binding sites connected by a flexible peptide hinge. $Ca^{2+}$ binding causes the formation of complementary grooves on each which then allows binding to specific regulatory sites on target proteins. $Ca^{2+}$-dependent binding causes derepression of the enzyme active site by displacing an endogenous inhibitory pseudosubstrate.

of the membrane would sterically block the translocation of phospholipid binding proteins such as protein kinase C and cellular phospholipases. The $Ca^{2+}$-dependent self-association on membrane surfaces represents a novel

mechanism of second-messenger coupled cell regulation. This may be the regulatory mechanism by which annexins modify the gating activity of the channels.

Immunolocalization studies have proved valuable in providing insight into evaluating cellular function of the annexins. The individual annexins are associated with secretory granules, the endoplasmic/sarcoplasmic reticulum, actin bundles, and the plasma membrane. Annexin IV, for example, is expressed in many epithelia and is concentrated along the apical membrane, subjacent to the lumen of the organ (Fig. 8) (Kaetzel *et al.*, 1994). This region is the cellular site of fluid secretion into the lumen. Recent studies indicate that this annexin regulates chloride-ion efflux, which produces an electrochemical gradient to draw sodium ions transcellular across the epithelium. This salt causes water to follow because of hyperosmotic pressure. The luminal fluid is required for normal tissue function. Abnormal fluid secretion is involved in the pathologies of cholera and cystic fibrosis.

## VII. Protein Kinase C

The protein kinase C family is a third mediation pathway of intracellular $Ca^{2+}$ action. One class of protein kinase C isozymes is activated by $Ca^{2+}$, which increases the affinity of the enzyme for phosphatidylserine. This ligand binding targets the translocation of protein kinase C to the plasma membrane. Once bound to the membrane surface, the enzyme can be further stimulated by diacylglycerol (DAG) (Fig. 9). This latter metabolite is a product of G-protein-activated phospholipase C hydrolysis of membrane $PIP_2$. Protein kinase C stimulates DNA synthesis and is the cellular mechanism by which many tumor promoters act (Nishizuka, 1992). A major widely distributed cellular substrate for protein kinase C is MARCKS, myristoylated alanine-rich C-kinase substrate (Aderem, 1992). This protein binds to the plasma membrane, calmodulin, and actin and has been associated with secretion, motility, vesicle trafficking, and transformation through the rearrangement of the actin cytoskeleton. Actin filament bundling by MARCKS is reg-

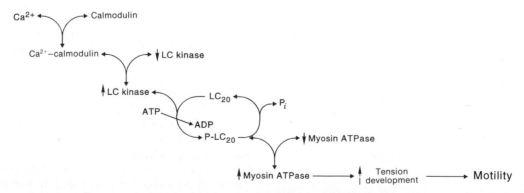

**FIG. 5.** Calmodulin regulation of nonstriated muscle contraction. An increase in intracellular $Ca^{2+}$ binds to calmodulin, causing its increased affinity for the target protein myosin light chain kinase (LC kinase). The activated kinase specifically phosphorylates one of the myosin light chains, which, in turn, increases the activity of myosin and its affinity for actin. Cross-bridge formation develops contraction. This process is reversed after cytosolic reduction in $Ca^{2+}$.

ulated by $Ca^{2+}$-dependent calmodulin binding, which is, in turn, regulated by protein kinase C phosphorylation. The kinase has numerous protein substrates in the cell; however, the precise mechanism of cellular regulation is not fully understood.

There are additional distinct intracellular $Ca^{2+}$ binding proteins that have been well characterized in biochemical terms (see Smith *et al.,* 1990). $Ca^{2+}$ has the responsibility of regulating a large number of unrelated cellular activities such as cell growth, secretion, motility, and transport. The ubiquitous $Ca^{2+}$ signal is discriminated through the individual $Ca^{2+}$-mediator proteins (Fig. 10).

## VIII. Current Perspectives

Scientific progress is not restricted by effort, but is dependent on insight and advances in technology. Technical progress in protein chemistry in the late 1960s and early 1970s led to the identification and characterization of the $Ca^{2+}$ receptors, troponin C and calmodulin. Development of recombinant DNA technology in the mid-1970s provided precise knowledge of the molecular evolution and structural conservation of these proteins. In the early 1980s, $Ca^{2+}$-sensitive dyes and computer-assisted image processing allowed visualization of transient spatial changes in intracellular $Ca^{2+}$. Detailed information is being obtained on the cellular components that are essential for maintaining these levels, including $Ca^{2+}$ channels, $Na^+/Ca^{2+}$ exchangers, and $Ca^{2+}$-ATPases. The original observations of Hokin and Hokin (1953) concerning the turnover of membrane phosphatidylinositols were finally appreciated some 20 years later. New insights in second-messenger action revealed that the hydrolysis of $PIP_2$ to form $IP_3$ and DAG could coordinate, respectively, the release of intracellular $Ca^{2+}$ from the endoplasmic reticulum and the activation of protein kinase C. Although the mechanisms involved in maintaining intracellular $Ca^{2+}$ homeostasis are being defined, understanding the events that couple stimuli to specific cellular responses has resisted biochemical definition. The elucidation of physiological regulation by $Ca^{2+}$ through specific $Ca^{2+}$ mediator proteins is not a trivial undertaking. Initial implications of function can be explored by cellular microinjection of inhibitor peptides, antibodies, and anti-sense RNA or by *in vitro* reconstitution. Information obtained from these inhibition experiments, however, is limited because it does not fully reflect the intact organism. For example, although our knowledge of the molecular aspects and regulatory properties of calmodulin *in vitro* is impressive, little is known of the functional role of calmodulin in the living cell or the whole animal, other than by inference. An attractive approach to obtaining this information is genetic manipulation. Fungi, such as yeast, are powerful models for the introduction and

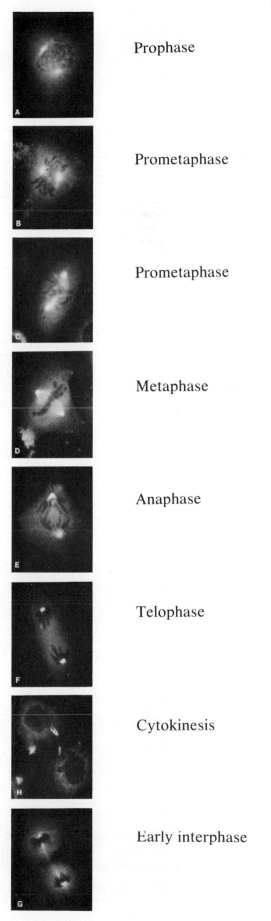

Prophase

Prometaphase

Prometaphase

Metaphase

Anaphase

Telophase

Cytokinesis

Early interphase

**FIG. 6.** Subcellular localization of calmodulin in mitotic cells. Monospecific antibody and immunofluorescence identifies calmodulin to be concentrated at the poles of the spindle apparatus.

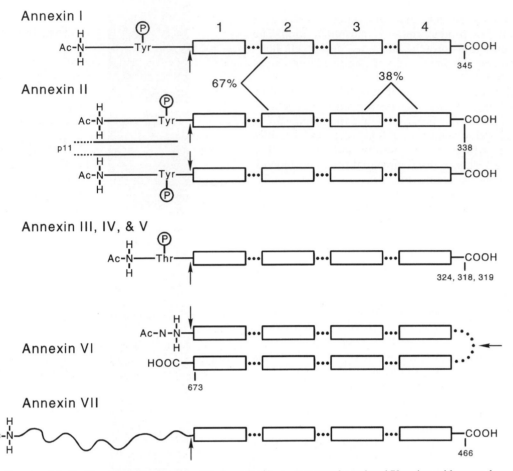

**FIG. 7.**    Structural similarities among the annexins. The boxed regions represent the ordered 70 amino acid repeat domains. The numbers indicate the relative sequence similarities between repeat domains. The arrows indicate highly sensitive sites of proteolysis. The repeat regions form a highly structured, protease resistant, Ca$^{2+}$-phospholipid binding core. The amino terminal "tails" are highly variable in length, sequence, and structure and provide individuality to each annexin family member.

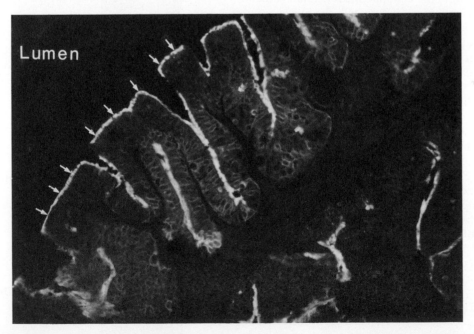

**FIG. 8.**    Localization of annexin IV in the rat fallopian tube. Monospecific antibody identifies annexin IV as concentrated along the membrane bordering the lumen of the oviduct of the ciliated, columnar epithelia (arrows). This region is the cellular site of fluid secretion.

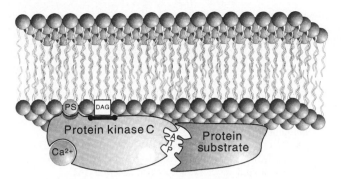

**FIG. 9.** Regulation of protein kinase C. Elevations in intracellular $Ca^{2+}$ cause translocation of PKC to the plasma membrane where phosphatidylserine (PS) and diacylglycerol (DAG) further activate the enzyme.

study of site-directed and temperature-sensitive mutations, and fundamental questions regarding the regulatory role of $Ca^{2+}$ in growth, division, and secretion have been addressed (see Davis, 1992). The usefulness of such haploid systems, however, is limited to genes that are expressed in these organisms and to nonlethal mutations in these genes.

The cellular function of the diverse $Ca^{2+}$-binding proteins that are expressed in highly differentiated cells remains elusive. The respective pathways can be manipulated at many levels. Synthetic genes can be introduced into cultured cells and the resulting phenotypic changes monitored. In addition, specific genes can be deleted or "knocked out" in whole animals using homologous recombination of genes in mouse embryonic stem cells (see Frohmen and Martin, 1989).

Interpretation of targeted gene "knockout" studies is, however, complicated by the fact that the animal lacks the gene in its genome throughout its development. Phenotypic consequences within individual cell types resulting from gene disruptions may be due to defects that occur during embryonic development or inadequate communication with other defective cell types.

An alternative approach to understanding the precise role of proteins associated with the intracellular $Ca^{2+}$ signal is to design and construct dominant-negative genes that, when expressed, neutralize the function of specific proteins. For example, peptides consisting of the calmodulin-binding site on target proteins (see Fig. 4) are extremely potent inhibitors of calmodulin. Wang *et al.* (1995, 1996) constructed a synthetic gene that produced a string of these calmodulin-binding peptides. This peptide inhibitor was

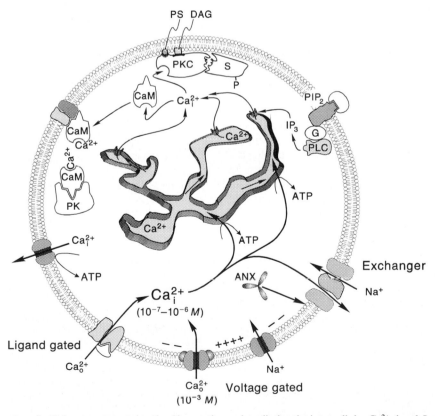

**FIG. 10.** Interplay of cellular components involved in creating and mediating the intracellular $Ca^{2+}$ signal. Low intracellular $Ca^{2+}$ levels are the result of pumps and ion exchangers. $Ca^{2+}$ transients develop from influx through voltage and ligand-gated channels on the plasmalemma and from $IP_3$-gated channels located in the endoplasmic/sarcoplasmic reticulum. $IP_3$ is generated from extracellular stimuli through a G-protein (G) regulated phospholipase C (PLC). The intracellular $Ca^{2+}$ signal is mediated by three primary pathways: calmodulin (CaM), annexins (Anx), and protein kinase C (PKC).

targeted to the nucleus of mouse-lung epithelial cells. By neutralizing the activity of nuclear calmodulin, it was shown that the cells did not synthesize DNA and the embryonic lung did not develop. Dominant-positive genes can also be designed to elucidate $Ca^{2+}$-regulated cellular systems. The amino acid sequence of the pseudosubstrate domain on calmodulin target proteins (see Fig. 4) can be altered to abolish inhibition of the catalytic site. This mutated enzyme becomes autonomous from $Ca^{2+}$–calmodulin regulation and remains active even during periods of low intracellular $Ca^{2+}$. Mayford *et al.* (1996) have shown, for example, that expression of an autonomously active protein kinase in the brain of mice cause deficits in memory.

The understanding of $Ca^{2+}$-coupled stimulus responses has advanced through direct visualization of $Ca^{2+}$ transients, the identification of $Ca^{2+}$-binding proteins, determination of their subcellular localization and their biochemical role through reconstitution of a cellular function, and genetic manipulation of intact cells. Although many candidates for $Ca^{2+}$ mediator proteins are currently under investigation, it is anticipated that more $Ca^{2+}$-binding proteins will be identified as further research continues to solve the stimulus–response coupling puzzle.

## IX. Summary

$Ca^{2+}$ is a ubiquitous intracellular regulator of cellular function. Levels of $Ca^{2+}$ are controlled by a variety of channels, exchangers, and pumps found in the plasmalemma and internal membranes. Cell stimulation causes the intracellular $Ca^{2+}$ to transiently increase. This second-messenger signal is then mediated by $Ca^{2+}$-binding proteins. There are three primary molecular mechanisms of transmitting the signal: calmodulin, annexins, and protein kinase C. Each of these pathways intersects and can be cross-regulatory. $Ca^{2+}$-calmodulin binds to specific sites on target proteins and activates enzymes by derepressing the active site. In the presence of $Ca^{2+}$ the annexins have a strong affinity for phospholipids. Annexins can form an interlocking network along membrane surfaces and alter membrane fluidity. Annexins are important in regulating membrane ion conductances. Protein kinase C is regulated by $Ca^{2+}$, phospholipid and diacylglycerol. Cellular studies using specific activators and inhibitors of protein kinase C have shown this $Ca^{2+}$ pathway to be involved in cell growth, differentiation, and development of tumors. Physiological, cellular, and molecular techniques are being used in combination to define the precise cellular roles of $Ca^{2+}$ binding proteins.

## Bibliography

Aderem, A. (1992). The MARCKS brothers: A family of protein kinase C substrates. *Cell* **71**, 713–716.

Andree, H. A. M., Stuart, M. C. A., Hermens, W. T., Reutelingsperger, C. P. M., Hemker, H. C., Frederik, P. M., and Willems, G. M. (1992). Clustering of lipid-bound annexin V may explain its anticoagulant effect. *J. Biol. Chem.* **267**, 17907–17912.

Baum, P., Furlong, C., and Byers, B. (1986). Yeast gene required for spindle pole body duplication: homology of its product with $Ca^{2+}$-binding proteins. *Proc. Natl. Acad. Sci. USA* **83**, 5512–5516.

Bennett, V. (1985). The membrane skeleton of human erythrocytes and its implications for more complex cells. *Annu. Rev. Biochem.* **54**, 273–304.

Berridge, M. J. (1990). Calcium oscillations. *J. Biol. Chem.* **264**, 9583–9586.

Berridge, M. J. (1993). Inositol trisphosphate and calcium signaling. *Nature* **361**, 315–361.

Concha, N. O., Head, J. F., Kaetzel, M. A., Dedman, J. R., and Seaton, B. A. (1992). Annexin V forms calcium-dependent trimeric units on phospholipid vesicles. *FEBS Letts.* **314**, 159–162.

Crumpton, M. J., and Dedman, J. R. (1990). Protein terminology tangle. *Nature* **345**, 212.

Davis, T. N. (1992). What's new with calcium? *Cell* **71**, 557–564.

Dedman, J. R., Welsh, M. J., Kaetzel, M. A., Pardue, R. L., and Brinkley, B. R. (1982). Localization of calmodulin in tissue culture cells. *In* "Calcium and Cell Function" (W. Y. Cheung, Ed.), Vol. 3, Academic Press, New York.

Frohman, M. A., and Martin, G. R. (1989). Cut, paste, and save: New approaches to altering specific genes in mice. *Cell* **56**, 145–147.

Heizmann, C. W., and Hunziker, W. (1991). Intracellular calcium-binding proteins: More sites than insights. *Trends Biochem. Sci.* **16**, 98–103.

Hodgkin, A. L., and Keynes, R. D. (1957). Movements of labeled calcium in squid giant axons. *J. Physiol.* **138**, 253–281.

Hokin, M. R., and Hokin, L. E. (1953). Enzyme secretion and the incorporation of $^{32}P$ into phospholipids of pancreas slices. *J. Biol. Chem.* **203**, 967–977.

Kaetzel, M. A., Hazarika, P., and Dedman, J. R. (1989). Differential tissue expression of three 35-kDa annexin calcium-dependent phospholipid-binding proteins. *J. Biol. Chem.* **264**, 14463–14470.

Kaetzel, M. A., Chan, H. C., Dubinsky, W. P., Dedman, J. R., and Nelson, D. J. (1994). A role for annexin IV in epithelial cell function: Inhibition of calcium-activated chloride conductance. *J. Biol. Chem.* **269**, 5297–5302.

Koenig, M., Monaco, A. P., and Kunkel, L. M. (1988). The complete sequence of dystrophin predicts a rod-shaped cytoskeletal protein. *Cell* **53**, 219–228.

Llinas, R., Sugimori, M., and Silver, R. B. (1992). Microdomains of high calcium concentration in a presynaptic terminal. *Science* **256**, 677–679.

Lee, H. C. (1993). Potential of calcium- and caffeine-induced calcium release by cyclic ADP–ribose. *J. Biol. Chem.* **268**, 293–299.

Mayford, M., Bach, M. E., Huang, Y. Y., Wang, L., Hawkins, R. D., and Kandel, E. R. (1996). Control of memory formation through regulated expression of a CaMKII transgene. *Science* **274**, 1678–1683.

Meers, P., Daleke, D., Hong, K., and Papahadjopoulos, D. (1991). Interactions of annexins with membrane phospholipids. *Biochem.* **30**, 2903–2908.

Miyamoto, S., Ohya, Y., Ohsumi, Y., and Anraku, Y. (1989). Nucleotide sequence of the *CLS4* (*CDC24*) gene of *Saccharomyces cerevisiae*. *Gene* **54**, 125–132.

Nishizuka, Y. (1992). Intracellular signaling by hydrolysis of phospholipids and activation of protein kinase C. *Science* **258**, 607–614.

Noegel, A., Witke, W., and Schleicher, M. (1987). Calcium-sensitive nonmuscle $\alpha$-actinin contains EF-hand structures and highly conserved regions. *FEBS Lett.* **221**, 391–396.

Ohno, S., Emori, Y., Imajoh, S., Kawaskai, H., Kisaragi, M., and Suzuki, K. (1984). Evolutionary origin of a calcium-dependent protease by fusion of genes for a thiol protease and a calcium-binding. *Nature* **312**, 566–570.

Ringer, S. (1883). A further contribution regarding the influence of the different constituents of the blood of the contraction of the heart. *J. Physiol.* **4**, 29–43.

Smith, V. L., Kaetzel, M. A., and Dedman, J. R. (1990). Stimulus-response coupling: the search for intracellular calcium mediator proteins. *Cell Regul.* **1,** 165–172.

Sweet, W. D., and Schroeder, F. (1988). "Lipid Domains and the Relationship to Membrane Function." p. 1742. R. Liss, New York.

Wang, J., Campos, B., Jamieson, A., Kaetzel, M., and Dedman, J. (1995). Functional elimination of calmodulin with the nucleus by targeted expression of an inhibitor peptide. *J. Biol. Chem.* **270,** 30245–30248.

Wang, J., Moreira, K., Campos, B., Kaetzel, M., and Dedman, J. (1996). Targeted neutralization of calmodulin in the nucleus blocks DNA synthesis and cell cycle progression. *Biochim. Biophys. Acta* **1313,** 223–228.

Edward F. Nemeth

# 10

# Regulation of Cellular Functions by Extracellular Calcium

## I. Introduction

The signaling function of cytoplasmic $Ca^{2+}$ has long been appreciated, and it is now well established that cytoplasmic $Ca^{2+}$ acts to control a variety of cellular responses such as exocytotic secretion, cellular differentiation and proliferation, and muscular contraction (see Chapter 9). These actions of cytoplasmic $Ca^{2+}$ are usually mediated by proteins such as *calmodulin* that bind $Ca^{2+}$ with high affinity, commensurate with the relatively low concentrations of $Ca^{2+}$ within the cytoplasm ($[Ca^{2+}]_i$; 100 n$M$ to 1 $\mu M$). These high-affinity $Ca^{2+}$-binding proteins are essentially *intracellular $Ca^{2+}$ receptors* that transduce changes in $[Ca^{2+}]_i$ into functional cellular responses.

Other important roles for $Ca^{2+}$ are found outside the cell. The levels of extracellular $Ca^{2+}$ in the blood and interstitial fluids modulate a number of vital cellular processes, including thrombosis, muscle and nerve excitability, and proper bone formation. Many of these events are mediated by proteins that bind extracellular $Ca^{2+}$ in concentrations of 0.1 to 1 m$M$, such as proteins involved in cellular adhesion (Clark and Brugge, 1995). In this capacity, extracellular $Ca^{2+}$ functions mostly in a permissive role and cannot by itself elicit a functional cellular response as can cytoplasmic $Ca^{2+}$.

An entirely different role of extracellular $Ca^{2+}$ is found in many of the cells that participate in systemic $Ca^{2+}$ homeostasis, such as those in the parathyroid glands and in the kidney. These cells are capable of detecting very small changes in the concentration of extracellular $Ca^{2+}$. Extracellular $Ca^{2+}$ by itself can profoundly influence cellular activity in these cells. In this capacity, extracellular $Ca^{2+}$ is acting as an intercellular signal much like a neurotransmitter or hormone. It has been suspected for some time that extracellular $Ca^{2+}$ might act in this messenger role, but the molecular mechanism explaining this action of extracellular $Ca^{2+}$ has only recently been discovered. The

novel molecular component of this extracellular $Ca^{2+}$-sensing mechanism is a **G protein-coupled receptor** that uses extracellular $Ca^{2+}$ as its primary physiological ligand. Thus, just as there are intracellular $Ca^{2+}$ receptors responding to changes in $[Ca^{2+}]_i$, so too there are *extracellular $Ca^{2+}$ receptors*, present on the outer surface of some cells, which respond to small changes in the ambient level of $Ca^{2+}$. Not surprisingly, the data revealing this novel mechanism of signal transduction derive from studies of cells that are normally involved in maintaining $Ca^{2+}$ homeostasis in the plasma and extracellular fluids, especially **parathyroid cells.** Because of this, it is appropriate to briefly consider this homeostatic mechanism.

## II. Systemic Calcium Homeostasis

Calcium is the most abundant mineral in the body, and the average human adult contains about 1 kg. Most of the calcium in the body is chemically complexed with phosphate or bound to proteins and deposited in bones and teeth. Only about 0.1% of total body calcium is in the plasma and extracellular fluids. In most mammals, the concentration of total calcium in the plasma is 2.4 m$M$, but only about half of this is free ionized calcium; the remaining calcium is bound to serum proteins and to various inorganic anions such as phosphate and citrate. It is the ionized form of calcium ($Ca^{2+}$) that is biologically active in the extracellular environment, and it is this form that is monitored by the "calciostat" maintaining **systemic $Ca^{2+}$ homeostasis.** The principal factors in this mechanism are **parathyroid hormone** (PTH) and **1,25-dihydroxyvitamin $D_3$.** The former is an 84 amino acid protein secreted by the parathyroid glands and the latter is a steroid-like hormone produced by the kidney. PTH releases $Ca^{2+}$ from bone, increases renal $Ca^{2+}$ absorption, and decreases renal phosphate absorption. An additional action of PTH in the kidney is the stimulation of the synthesis of 1,25-dihydroxyvitamin $D_3$. 1,25-Dihydroxyvitamin $D_3$ then acts in the intestine to increase absorption of dietary $Ca^{2+}$. All these actions tend to increase the level of $Ca^{2+}$ in the circulation. Increased levels of plasma $Ca^{2+}$, in turn, act in a **negative feedback** capacity to depress secretion of PTH. There is a reciprocal relationship between the level of plasma $Ca^{2+}$ and that of PTH, and this simple feedback loop is the primary mechanism regulating systemic $Ca^{2+}$ homeostasis in humans (Mundy, 1989; Brown, 1994; Figs. 1 and 2).

## III. The Calcium Receptor

Because the inhibitory effect of extracellular $Ca^{2+}$ on PTH secretion was readily observed using parathyroid cells *in vitro*, there was never much doubt that extracellular $Ca^{2+}$ acts directly on these cells to regulate secretion. The perplexing issue was how small changes in extracellular $Ca^{2+}$ concentrations (0.05 to 0.1 m$M$) could cause profound changes in hormone secretion. Most cells are not at all

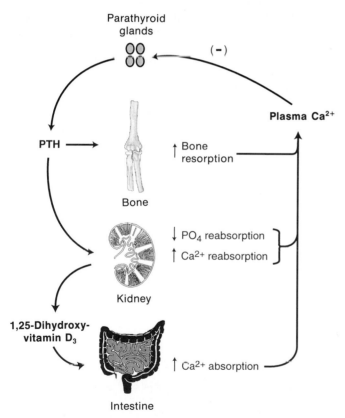

**FIG. 1.** Regulation of systemic $Ca^{2+}$ homeostasis in humans. The key player is PTH, which mobilizes $Ca^{2+}$ from bone and conserves $Ca^{2+}$ by the kidney. It also stimulates the formation of 1,25-dihydroxyvitamin $D_3$ in the kidney. 1,25-Dihydroxyvitamin $D_3$ acts on the intestine to increase $Ca^{2+}$ absorption from ingested food. All these actions tend to increase the plasma level of $Ca^{2+}$. Increases in the circulating levels of $Ca^{2+}$ act on the parathyroid glands (there are four in humans) to depress secretion of PTH. These mechanisms explain the reciprocal relationship between circulating levels of PTH and $Ca^{2+}$.

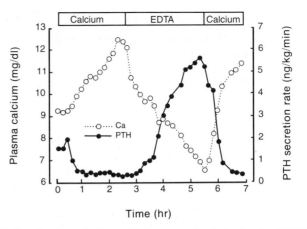

**FIG. 2.** Reciprocal relationship between circulating levels of PTH and $Ca^{2+}$. The data are derived from studies in anesthetized calves where the plasma level of $Ca^{2+}$ was varied either by infusing calcium chloride to raise plasma levels of $Ca^{2+}$ or by infusing the divalent cation chelator EDTA to lower plasma levels of $Ca^{2+}$ (Mayer and Hurst, 1978). Plasma levels of PTH rapidly rise or fall when plasma $Ca^{2+}$ levels decrease or increase.

sensitive to small, physiological changes in the level of extracellular $Ca^{2+}$. It was not even known, for example, if extracellular $Ca^{2+}$ acted at the cell surface or had to enter the parathyroid cell to affect PTH secretion.

A number of studies performed during the mid- to late 1980s provided evidence suggesting that extracellular $Ca^{2+}$ acts at the cell surface and does not have to gain entry into the parathyroid cell to affect secretion of PTH (Nemeth, 1990; Brown, 1991). It was shown that extracellular $Ca^{2+}$ activated **phospholipase C** resulting in the rapid formation of **inositol 1,4,5-trisphosphate** (IP$_3$), which in turn released $Ca^{2+}$ from intracellular stores. Extracellular $Ca^{2+}$ also blocked increases in **cyclic AMP** formation linked to the $\beta$-adrenergic receptor on parathyroid cells. Significantly, this effect on cyclic AMP was blocked by **pertussis toxin,** showing that the action of extracellular $Ca^{2+}$ on cyclic AMP production was mediated by a G protein (in this particular case, a $G_i$ protein). These various findings provided evidence for the existence of an entity in the plasma membrane that is functionally akin to other cell surface receptors linked to the mobilization of intracellular $[Ca^{2+}]_i$. It was hypothesized that parathyroid cells express a cell surface "$Ca^{2+}$ receptor" that enables these cells to detect and respond to small changes in the level of extracellular $Ca^{2+}$.

The ability of extracellular $Ca^{2+}$ to cause the mobilization of intracellular $Ca^{2+}$ in *Xenopus* oocytes injected with parathyroid gland mRNA was used in a functional cloning strategy to isolate cDNA encoding the parathyroid $Ca^{2+}$ receptor (Brown *et al.,* 1993). This $Ca^{2+}$ receptor is structurally and functionally similar to many other cell surface receptors. The major difference is that the primary physiological ligand for the $Ca^{2+}$ receptor is an inorganic ion rather than an organic molecule. As such, it is the first cell surface receptor for an inorganic ion to be structurally identified and functionally characterized.

## A. Structure of the Ca²⁺ Receptor

The nucleotide sequence of the human parathyroid $Ca^{2+}$ receptor encodes a protein of 1078 amino acids; it is 92 and 93% identical, respectively, with the rat kidney (1079 amino acids) and bovine parathyroid (1085 amino acids) $Ca^{2+}$ receptors. Hydropathy analysis of the deduced amino acid sequence reveals a membrane-spanning domain containing seven transmembrane regions (Fig. 3). This serpentine hallmark places the $Ca^{2+}$ receptor within the G protein-coupled receptor superfamily. The $Ca^{2+}$ receptor shares limited homology (25 to 35%) only with **metabotropic glutamate receptors** (mGluRs; Suzdak *et al.*, 1994) and together they define a new receptor subfamily. Despite only limited homology, the topology of the $Ca^{2+}$ receptor and the mGluRs is remarkably similar; both receptors have unusually large extracellular domains and relatively large cytoplasmic domains compared with other G protein-coupled receptors. There is a remarkable conservation in the number and positions of cysteines, again suggesting that the secondary structures of mGluRs and the $Ca^{2+}$ receptor are similar (Brown *et al.*, 1993; Garrett *et al.*, 1995a; Chattopadhyay *et al.*, 1996).

The expressed $Ca^{2+}$ receptor protein has a molecular weight generally around 120 kDa, considerably higher than that predicted solely on the basis of amino acid content. This higher molecular weight results from posttranslational **glycosylation** of the receptor. The human $Ca^{2+}$ receptor contains 11 potential glycosylation residues in the extracellular domain.

It is not surprising that the $Ca^{2+}$ receptor does not possess any **"EF hand" motifs** that bind $Ca^{2+}$ and are characteristic of high-affinity calcium-binding proteins such as *calmodulin* (see Chapter 9). These latter proteins function as *intracellular* $Ca^{2+}$ receptors and sense *micromolar* changes in the cystoplasmic $Ca^{2+}$ concentration, whereas extracellular $Ca^{2+}$ receptors sense *millimolar* changes in the extracellular $Ca^{2+}$ concentration. There are regions of the $Ca^{2+}$ receptor, however, that are enriched with acidic amino acids. Two are in the extracellular domain and one is on the extracellular loop linking transmembrane-spanning segments 4 and 5 (see Fig. 3). Two of these regions are similar to those found in low-affinity calcium-binding proteins such as **calreticulin.** All of these acidic amino acids are conserved among various species homologues of $Ca^{2+}$ receptor proteins. These regions, enriched in negative charges, could conceivably play a role in the binding of extracellular $Ca^{2+}$ to the receptor.

Although the role of these acidic amino acid regions in binding extracellular $Ca^{2+}$ is uncertain, it is certain that the extracellular domain of the $Ca^{2+}$ receptor is crucial for activation by extracellular $Ca^{2+}$ and presumably contains at least one of the sites that bind extracellular $Ca^{2+}$. A similar role for the extracellular domain in binding glutamate has also been shown for mGluRs (Takahashi *et al.*, 1993). Binding of the physiological ligand largely if not exclusively in the extracellular domain is one of the features characterizing this subfamily of G protein-coupled receptors. In most other subfamilies of G protein-coupled receptors, the transmembrane domain plays the dominant or exclusive role in binding the appropriate physiological ligand; the extracellular domain contributes to binding of certain peptide and protein ligands (Beck-Sickinger, 1996).

## B. Regulation of the Ca²⁺ Receptor Gene

The $Ca^{2+}$ receptor gene is located on chromosome 3q and consists of seven exons spanning at least 45 kbp. The human $Ca^{2+}$ receptor promoter has no TATA or CAAT boxes and, in this respect, is similar to other G protein-coupled receptor genes. In human parathyroid cells, an exon 5 to 6 alternative splice isoform has been identified and appears to be functionally equivalent to the major expression product (Garrett *et al.*, 1995a).

Although the effects of altered plasma levels of $Ca^{2+}$ or 1,25-dihydroxyvitamin $D_3$ on expression of the $Ca^{2+}$

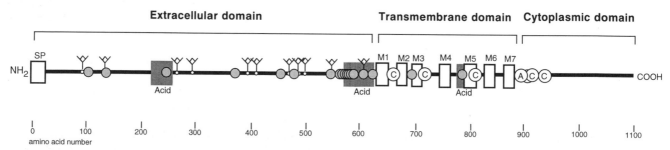

**FIG. 3.** Schematic representation of the human $Ca^{2+}$ receptor protein structure. The primary protein structure is composed of 1078 amino acids and is predicted to be composed of an $NH_2$-terminal extracellular domain ($\approx$620 amino acids), transmembrane domain ($\approx$250 amino acids) and an intracellular carboxy-terminal domain ($\approx$200 amino acids). The hydrophobic transmembrane-spanning regions are indicated by open blocks (M1 to M7). In the cytoplasmic domain and in the cytoplasmic loops linking some transmembrane-spanning domains are the five potential sites for protein kinase C (circled C). On the human receptor, there are two potential sites for protein kinase A phosphorylation (circled A) in the intracellular domain. There are two regions enriched in acidic amino acids in the extracellular domain and one in the extracellular loop linking M4 and M5 (shaded boxes). The locations of 17 cysteine residues, conserved between the $Ca^{2+}$ receptor and metabotropic glutamate receptors, are indicated by shaded circles. Eleven potential sites of *N*-linked glycosylation are present in the extracellular domain (branches), and the signal peptide sequence (SP) is shown at the $NH_2$-terminal.

receptor gene have been studied in rats, a conclusive picture has yet to emerge. These are clearly important factors to assess because chronic states of hypocalcemia and/or lowered 1,25-dihydroxyvitamin $D_3$ (as in **chronic renal failure**) are believed to contribute importantly to parathyroid cell growth and the amount of PTH stored in the cells (Drüeke, 1995; Martin and Slatopolsky, 1994). Increased circulating levels of PTH accompanied by hyperplasia of the parathyroid glands characterize **secondary hyperparathyroidism.** Understanding the actions, if any, of the $Ca^{2+}$ receptor that contributes to the development of secondary hyperparathyroidism clearly has important implications for disease management. Pharmacologically altering the activity of the $Ca^{2+}$ receptor might provide a means of diminishing the severity or preventing the secondary hyperparathyroidism that accompanies renal failure (Nemeth, 1996; Nemeth et al., 1996).

## C. Diseases Caused by Mutations in the $Ca^{2+}$ Receptor Gene

**Familial benign hypocalciuric hypercalcemia (FBHH)** is a rare autosomal dominant disorder characterized by mild hypercalcemia. This disease results from mutations in the $Ca^{2+}$ receptor gene (Nemeth and Heath, 1995; Pearce and Brown, 1996). So far, more than 20 different mutations have been identified in kindreds with FBHH; nearly all are single amino acid substitutions, and most occur within the extracellular domain of the $Ca^{2+}$ receptor. These mutations partially or totally inhibit the ability of extracellular $Ca^{2+}$ to activate the $Ca^{2+}$ receptor (Pollak et al., 1993). This genetic disorder therefore results from the expression of mutated $Ca^{2+}$ receptors that make parathyroid cells less sensitive to regulation by extracellular $Ca^{2+}$. FBHH is the heterozygous manifestation of these $Ca^{2+}$ receptor gene mutations. The homozygous condition is called **neonatal severe hyperparathyroidism (NSHPT)** and is life threatening. In the homozygous disease, the parathyroid gland is completely unresponsive to extracellular $Ca^{2+}$; severe hypercalcemia with greatly elevated levels of PTH ensue, and total parathyroidectomy is usually necessary soon after birth.

The converse of FBHH and NSHPT is **autosomal dominant hypoparathyroidism (ADH).** In this disorder, mutations in the $Ca^{2+}$ receptor gene produce a protein that is more sensitive to activation by extracellular $Ca^{2+}$ (Pollak et al., 1994; Bai et al., 1996). Consequently, normal levels of plasma $Ca^{2+}$ now depress PTH secretion to a much greater extent. Lower circulating levels of PTH will lead to lower levels of $Ca^{2+}$, and this explains the hypocalcemia characteristic of ADH. The deviant $Ca^{2+}$ receptor has essentially lowered the **"set point"** for plasma $Ca^{2+}$ concentration, so that systemic $Ca^{2+}$ concentrations are maintained at a lower value. In this manner, the $Ca^{2+}$ receptor can be considered as the "thermostat" that monitors and maintains the levels of extracellular $Ca^{2+}$. In this particular capacity, the $Ca^{2+}$ receptor serves as the body's "calciostat." Turning the calciostat up (as in ADH) or down (as in FBHH or NSHPT) lowers or raises the circulating levels of $Ca^{2+}$. In the aggregate, these molecular genetic studies

identify the $Ca^{2+}$ receptor as the key entity controlling systemic $Ca^{2+}$ homeostasis.

## D. Other Extracellular $Ca^{2+}$ Receptors?

Curiously, and despite much effort, no subtypes of the $Ca^{2+}$ receptor have yet been identified, and $Ca^{2+}$ receptors expressed in different tissues all seem to be products of a single gene. Usually, related but distinct receptors are identified soon after the first member of a new receptor subfamily is cloned. Some preliminary studies claim to have identified a family of G protein-coupled receptors structurally related to the $Ca^{2+}$ receptor. It is probably just a matter of time before $Ca^{2+}$ receptor subtypes are discovered.

There are reasons for supposing that the parathyroid $Ca^{2+}$ receptor might have related subtypes and/or that structurally distinct proteins might function as $Ca^{2+}$ receptors. One reason is that different cell types sense quite dissimilar ranges of extracellular $Ca^{2+}$. **Keratinocytes** (skin cells), for example, respond to extracellular $Ca^{2+}$ concentrations between 0.05 and 0.5 m$M$, far lower than circulating levels of $Ca^{2+}$ (Bikle et al., 1996). Although the greater sensitivity of keratinocytes to extracellular $Ca^{2+}$ could result from postreceptor mechanisms, it is also possible that a $Ca^{2+}$ receptor subtype or a completely different protein mediates the actions of extracellular $Ca^{2+}$ in this cell type. A second reason is that the pharmacological characteristics of extracellular $Ca^{2+}$ sensing by different cell types is quite variable. Although some of this variability could result from posttranslational modifications of the $Ca^{2+}$ receptor, the pharmacology is often so different as to suggest different receptors entirely.

An issue related to that of $Ca^{2+}$ receptor subtypes is whether the $Ca^{2+}$ receptor is the only kind of protein functioning as a sensor for extracellular $Ca^{2+}$. A 500-kDa glycoprotein, structurally related to **low-density lipoprotein (LDL) receptors,** has been postulated to act as a $Ca^{2+}$ receptor on parathyroid cells (Lundgren et al., 1994). LDL receptors are known to have the ability to bind extracellular $Ca^{2+}$ and some other divalent cations. However, this very large protein has been difficult to express in heterologous systems, and it is not yet known if this protein could function as an extracellular $Ca^{2+}$-sensing protein on parathyroid cells or other types of cells.

A considerable number of studies have investigated the mechanism(s) by which **osteoclasts** respond to changes in the level of extracellular $Ca^{2+}$ (Zaidi et al., 1993). Osteoclasts are cells that resorb bone and, as they do so, $Ca^{2+}$ is released from the bone matrix. The concentration of extracellular $Ca^{2+}$ near the osteoclast is believed to increase as a result of solubilization of the bone matrix. Increases in the concentration of extracellular $Ca^{2+}$ inhibit osteoclastic bone resorption when studied in vitro. The physiological significance of this phenomenon is uncertain. Osteoclasts do not express a parathyroid-like $Ca^{2+}$ receptor and are pharmacologically distinct from parathyroid cells. An alternative mechanism has been suggested to explain the sensing of extracellular $Ca^{2+}$ levels by the osteoclast (Zaidi et al., 1995).

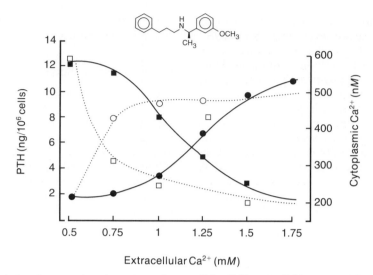

**FIG. 4.** Type II calcimimetics potentiate the actions of extracellular $Ca^{2+}$ at the $Ca^{2+}$ receptor. The data are derived from studies on bovine parathyroid cells. (These cells, when studied *in vitro,* respond to extracellular $Ca^{2+}$ concentrations over a somewhat broader range than human parathyroid cells *in vivo.*) Increasing the concentration of extracellular $Ca^{2+}$ inhibits secretion of PTH (■), and these inhibitory effects are potentiated in the presence of NPS R-467 [(□) 500 n$M$; the *S*-enantiomer is without effect at this concentration]. Likewise, the ability of extracellular $Ca^{2+}$ to increase $[Ca^{2+}]_i$ (●) is potentiated in the presence of NPS R-467 (○). Note that NPS R-467 is without effect when the extracellular $Ca^{2+}$ concentration is very low. Insert: The structure of NPS R-467, the *R*-enantiomer.

## E. Pharmacology of the $Ca^{2+}$ Receptor

The $Ca^{2+}$ receptor is rather promiscuous and senses a variety of positively charged inorganic ions and organic compounds. Many divalent cations and nearly all the trivalent cations activate the $Ca^{2+}$ receptor with a rank order of potency as follows: $La^{3+} > Gd^{3+} > Be^{2+} > Ca^{2+} = Ba^{2+} > Sr^{2+} > Mg^{2+}$. Of these, $Mg^{2+}$ is the only cation likely to have any impact on $Ca^{2+}$ receptor activity under physiological conditions. In addition to inorganic polycations, many organic polycations also activate the $Ca^{2+}$ receptor; these include polyamines (spermine), aminoglycoside antibiotics (neomycin), polyamino acids (polylysine), and proteins (protamine). In general, there is some correlation between net positive charge of these compounds and their potency in activating the $Ca^{2+}$ receptor. Nonetheless, there are exceptions to this rule that indicate that the $Ca^{2+}$ receptor is sensing not just net positive charge. Like $Ca^{2+}$, all these polycations are believed to bind largely, if not exclusively, in the extracellular domain of the $Ca^{2+}$ receptor. It is conceivable that regions of the receptor containing a high density of acidic amino acids, and therefore a high density of negative charge, are involved in the binding of these inorganic and organic polycations.

Although the organic polycations are useful tools for investigating the cellular physiology of parathyroid cells and other extracellular $Ca^{2+}$-sensing cells, they lack the potency and specificity necessary for many *in vitro* studies and all *in vivo* studies. A structurally distinct class of compounds, typified by NPS R-467, comprise superior pharmacological probes that overcome these problems.[1] These phenylalkylamine derivatives carry only one positive charge at physiological pH and, at nanomolar concentrations, act selectively on the $Ca^{2+}$ receptor to inhibit PTH secretion (Nemeth, 1996). These compounds act in a stereoselective manner at the $Ca^{2+}$ receptor, and the *R*-enantiomers are 10- to 100-fold more potent than the corresponding *S*-enantiomers. The mechanism of action of the phenylalkylamine compounds differs from those of the inorganic and organic polycations. Unlike the polycations, these phenylalkylamine compounds require extracellular $Ca^{2+}$ for their activity. Thus, these compounds fail to mobilize intracellular $Ca^{2+}$ in the absence of extracellular $Ca^{2+}$, although a polycation such as neomycin is fully competent in doing so. These phenylalkylamines behave as **positive allosteric modulators** to increase the sensitivity of the $Ca^{2+}$ receptor to activation by extracellular $Ca^{2+}$, thereby shifting the concentration–response curve to the left (Fig. 4). These different mechanisms are also manifest in the concentration–response curves to polycations and phenylalkylamines. That for the former is extremely steep and apparently reflects cooperative binding of the ligand to the receptor, whereas that for the latter is more conventional and suggest a 1 : 1 ligand : receptor complex. The phenylalkylamines bind in the transmembrane domain of the $Ca^{2+}$ receptor. It is not yet known if the phenylalkylamines bind to the $Ca^{2+}$ receptor in the absence of extracellular $Ca^{2+}$ or if binding of extracellular $Ca^{2+}$ to the receptor causes a conformational change in the transmembrane domain that unmasks a cryptic binding site for the phenylalkylamines. Likewise, it is presently unknown whether phenylalkylamines increase the binding affinity of the $Ca^{2+}$ receptor for extracellular $Ca^{2+}$ and/or the efficiency of coupling of the $Ca^{2+}$ receptor to G proteins. Despite these uncertainties regarding the molecular pharmacology of the $Ca^{2+}$

---

[1] NPS R-467 is a small organic compound synthesized at NPS Pharmaceuticals, Inc. Its chemical name is (*R*)-(+)-*N*-(3-phenylpropyl)-$\alpha$-methyl-(3-methoxy)benzylamine hydrochloride.

receptor, it is clear that this receptor can be affected by small organic compounds much like other G protein-coupled receptors.

Compounds that mimic or potentiate the actions of extracellular $Ca^{2+}$ at the $Ca^{2+}$ receptor have been termed **calcimimetics**. As such, they behave as receptor agonists to mobilize intracellular $Ca^{2+}$ and inhibit PTH secretion. Two types of calcimimetics can be distinguished: Type I, exemplified by the inorganic and organic polycations, which bind largely in the extracellular domain and mimic extracellular $Ca^{2+}$, and Type II, typified by the phenylalkylamines, which bind within the transmembrane domain of the $Ca^{2+}$ receptor and potentiate the actions of extracellular $Ca^{2+}$.

Type II calcimimetics have been used to activate the $Ca^{2+}$ receptor *in vivo* and, when administered to animals or humans, cause time- and dose-dependent decreases in plasma levels of PTH and $Ca^{2+}$ (Nemeth *et al.*, 1996). Circulating levels of PTH start to fall sooner than those of $Ca^{2+}$, as expected, since it is this decrease in PTH that leads to hypocalcemia (Fig. 5). These findings, together with those from molecular genetic studies, provide solid evidence that the $Ca^{2+}$ receptor is the essential regulator of systemic $Ca^{2+}$ homeostasis.

## IV. Calcium Receptor-Dependent Regulation of Cellular Functions

There are a number of cells throughout the body whose activity is regulated by changes in the concentration of extracellular $Ca^{2+}$. For many of these cells, it is not certain that the effects of extracellular $Ca^{2+}$ are mediated by a parathyroid-like $Ca^{2+}$ receptor. However, there is clear evidence for the expression of this receptor in certain cells and some understanding of how this receptor regulates cellular responses. Discussed next are the three cell types

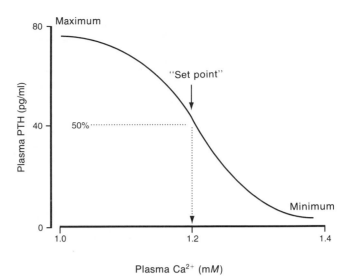

**FIG. 6.** Small changes in the plasma levels of $Ca^{2+}$ cause large changes in the plasma levels of PTH. In the normal adult human, plasma levels vary between 10 and 65 pg/ml and the plasma $Ca^{2+}$ levels are maintained between 1.12 and 1.23 m$M$. The concentration of plasma $Ca^{2+}$ causing a 50% decrease in the maximal levels of plasma PTH (achieved by lowering extracellular $Ca^{2+}$ until no further increase occurs) is called the "set point" for extracellular $Ca^{2+}$. The $Ca^{2+}$ receptor determines the set point and thereby functions as a "calciostat" to maintain systemic $Ca^{2+}$ homeostasis.

that have furnished most of the information about how the $Ca^{2+}$ receptor regulates cellular activity.

### A. Parathyroid Cells

These are the classic cells long known to be responsive to small changes in the concentration of extracellular $Ca^{2+}$. For these cells, extracellular $Ca^{2+}$ is the primary physiological stimulus regulating secretion of PTH. The sensitivity of the parathyroid cell to the ambient $Ca^{2+}$ concentration is impressive: Minimal and maximal rates of PTH secretion are obtained over a concentration difference of only 0.5 m$M$ (range of 1.0 to 1.5 m$M$). The concentration of serum $Ca^{2+}$ that suppresses maximal PTH levels by 50% is called the **"set point"** for extracellular $Ca^{2+}$ (Brown, 1983), and it is usually around the normal serum level of 1.2 m$M$ (Fig. 6). Because of the steep concentration–response relationship between extracellular $Ca^{2+}$ and PTH, very small changes in the concentration of extracellular $Ca^{2+}$ cause large changes in the secretion of PTH (see Fig. 6).

The $Ca^{2+}$ receptor on parathyroid cells is coupled to the activation of phospholipase C through a G protein (Fig. 7). The identity of this G protein is not yet certain, although it is believed to be either $G_{\alpha 11}$ or $G_{\alpha q}$ (Varrault *et al.*, 1995). The G protein linking the $Ca^{2+}$ receptor to phospholipase C is not inhibited by pertussis toxin. Activation of phospholipase C results in the rapid formation of $IP_3$, leading to the mobilization of $Ca^{2+}$ from some nonmitochondrial intracellular store, presumably some component of the endoplasmic reticulum. There is additionally an influx of extracellular $Ca^{2+}$ through **voltage-insensitive channels.** Thus, when the concentration of extracellular $Ca^{2+}$ is increased,

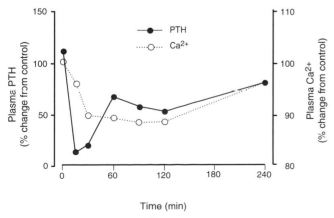

**FIG. 5.** Activation of the $Ca^{2+}$ receptor *in vivo* lowers plasma levels of PTH and $Ca^{2+}$. Normal rats were given 10 mg/kg NPS R-467 orally at time 0. There is a rapid fall in plasma levels of PTH followed by a decrease in plasma $Ca^{2+}$ levels. Both parameters return to normal over several hours.

$Ca^{2+}$ channels, and this confounds the interpretation. Perhaps the most compelling evidence so far derives from the use of selective Type II calcimimetics. Such compounds increase $[Ca^{2+}]_i$ and stimulate calcitonin secretion in various MTC cell lines, but not in TT cells, which lack the $Ca^{2+}$ receptor.

Although the parathyroid cell and the C cell express a $Ca^{2+}$ receptor that is identical at the mRNA level, the pharmacological properties of these cells differ. For example, certain Type I calcimimetics effective on parathyroid cells are without effect on MTC cells. Moreover, Type II calcimimetics are 10- to 40-fold more potent as inhibitors of PTH than as inhibitors of calcitonin secretion. The lower potency in MTC cells is probably a feature of authentic C cells because Type II calcimimetics also show preferential inhibitory effects on PTH secretion *in vivo* (Nemeth *et al.,* 1996). The differential pharmacology of parathyroid cells and C cells might arise from posttranslational modification of the $Ca^{2+}$ receptor, thereby altering its affinity for binding to the phenylalkylamines. Alternatively, the different postreceptor mechanisms in parathyroid cells and in C cells couple to the $Ca^{2+}$ receptor with different efficiencies. The differential pharmacology of parathyroid cells and C cells is significant for medical therapy with drugs that act on $Ca^{2+}$ receptors: Even though no $Ca^{2+}$ receptor subtypes have yet been identified, one can still target selectively two different cell types expressing what appears to be the same $Ca^{2+}$ receptor.

## C. Renal Epithelial Cells

After the parathyroid gland, the $Ca^{2+}$ receptor is most densely expressed in the kidney (Riccardi *et al.,* 1995). mRNA encoding the $Ca^{2+}$ receptor can be detected along most of the nephron, but highest levels are expressed in the **thick ascending limb (TAL)** of the loop of Henle and in the **inner medullary collecting duct (IMCD).** In the TAL, where much of the $Ca^{2+}$ (and $Mg^{2+}$) reabsorption from the nephron takes place, the $Ca^{2+}$ receptor is located largely on the basolateral side of the tubule and is thus positioned to sense changes in the plasma concentration of $Ca^{2+}$. The $Ca^{2+}$ receptor in the TAL couples to mechanisms that control the lumen-positive transepithelial voltage, which, in turn, drives $Ca^{2+}$ (and $Mg^{2+}$) reabsorption by a paracellular pathway (Hebert, 1996). Thus, in hypercalcemic conditions, the TAL $Ca^{2+}$ receptor on the basolateral surface of the epithelial cell is activated and $Ca^{2+}$ reabsorption is diminished. Conversely, hypocalcemic conditions and consequent low $Ca^{2+}$ receptor activity will increase $Ca^{2+}$ reabsorption in the TAL. In this segment of the nephron, the $Ca^{2+}$ receptor acts to control the amount of $Ca^{2+}$ that will be removed in the urine. One of the more convincing pieces of evidence supporting this role of the $Ca^{2+}$ receptor in the TAL derives from molecular genetic studies. Recall that patients suffering from FBHH are hypocalciuric despite their hypercalcemia. These patients continue to reabsorb an inappropriate amount of $Ca^{2+}$. $Ca^{2+}$ receptors in the TAL of patients with FBHH are less sensitive to extracellular $Ca^{2+}$. These mutant receptors do not perceive the hypercalcemia so $Ca^{2+}$ reabsorption is not diminished as it normally would be.

In the IMCD, the $Ca^{2+}$ receptor is localized mostly in vesicles within the epithelial cell and also on the luminal membrane. When luminal $Ca^{2+}$ increases, there is a compensatory decrease in the amount of water reabsorption, and this effect is seemingly mediated by the $Ca^{2+}$ receptor. The model proposed holds that activation of $Ca^{2+}$ receptors on the luminal surface of collecting duct cells inhibits the exocytotic insertion of **aquaporin** (water) channels in the luminal membrane (Brown and Hebert, 1995). This action results in less reabsorption of water, thereby diluting the urine with respect to $Ca^{2+}$. This effect makes sense in the context of systemic $Ca^{2+}$ homeostasis because if the plasma levels rise, then the kidney is called on to remove excess $Ca^{2+}$ from the body. The increased levels of $Ca^{2+}$ appearing in the urine must be diluted to avoid the formation of kidney stones.

## D. Other Extracellular $Ca^{2+}$-Sensing Cells

The $Ca^{2+}$ receptor is expressed in a wide range of cells that have little to do with systemic $Ca^{2+}$ homeostasis. Listed in Table 1 are the cells that express the $Ca^{2+}$ receptor and those that possibly do so. The function of the $Ca^{2+}$ receptor in these cells is often speculative or

**TABLE I**    Extracellular $Ca^{2+}$-Sensing Cells

| Cells known to express the $Ca^{2+}$ receptor | |
| --- | --- |
| Cell type | Function regulated by extracellular $Ca^{2+}$ |
| Parathyroid | PTH secretion and synthesis |
| Parafollicular (C cell) | Calcitonin secretion |
| Kidney<br>  TAL | $Ca^{2+}$ and $Mg^{2+}$ reabsorption |
|   IMCD | Water transport |
|   Macula densa | Renin secretion (?) |
| Pancreas<br>  Endocrine<br>  Exocrine | Unknown |
| Nervous system<br>  Central nervous system<br>  Myenteric plexus neurons | Unknown |

| Cells that respond to extracellular $Ca^{2+}$ | |
| --- | --- |
| Cell type | Function regulated by extracellular $Ca^{2+}$ |
| Osteoclasts | Bone resorption |
| Osteoblasts | Proliferation |
| Keratinocytes | Proliferation/differentiation |
| Mammary | Proliferation |
| Gastrointestinal<br>  G cells<br>  Goblet | Gastrin secretion<br>Proliferation |
| Cytotrophoblasts | Hormone secretion |

completely unknown, although the field does not lack imaginative hypotheses.

The number of different cell types expressing the $Ca^{2+}$ receptor is growing rapidly. This fosters the belief that extracellular $Ca^{2+}$, and perhaps other inorganic ions, act as intercellular signals to affect many different cellular responses.

## V. Summary

A biological phenomenon that is gaining recognition is the peculiar ability of extracellular $Ca^{2+}$ to regulate the activity of certain cells in the body. Many of these cells, such as parathyroid cells, parafollicular cells in the thyroid (C cells), and certain renal tubule cells, are involved in maintaining systemic $Ca^{2+}$ homeostasis. These cell types alter their activity in response to small, physiological changes in the concentration of $Ca^{2+}$ in the plasma or the extracellular fluids, whereas most other cells in the body do not. These "extracellular $Ca^{2+}$-sensing cells" express a $Ca^{2+}$ receptor in the plasma membrane that enables these cells to detect and respond to small changes in the level of extracellular $Ca^{2+}$. The $Ca^{2+}$ receptor is structurally and functionally similar to other plasma membrane receptors that sense changes in the concentration of extracellular ligands and translate these ligand–receptor interactions into intracellular signals that alter cellular activity. The major difference is that the $Ca^{2+}$ receptor binds to an inorganic ion rather than an organic molecule; the $Ca^{2+}$ receptor is the first example of an "inorganic ion receptor."

Extracellular $Ca^{2+}$ can thus act in a messenger capacity much like any number of other extracellular signals, such as neurotransmitters and hormones. This newly discovered function of extracellular $Ca^{2+}$ complements the well-known ability of cytoplasmic $Ca^{2+}$ to act as an intracellular signal. Thus, both extracellular $Ca^{2+}$ and intracellular $Ca^{2+}$ convey information and do so by interacting with $Ca^{2+}$-binding proteins. Intracellular $Ca^{2+}$ receptors, such as calmodulin, bind $Ca^{2+}$ with high affinity (nanomolar to low micromolar levels), whereas the extracellular $Ca^{2+}$ receptor binds $Ca^{2+}$ with much lower affinity (millimolar levels), consistent with the very different concentrations of $Ca^{2+}$ in the cytoplasm and in the extracellular space.

The $Ca^{2+}$ receptor is a G protein-coupled receptor and is structurally homologous to metabotropic glutamate receptors. Both these receptor types possess an unusually large extracellular domain that binds the respective physiological ligand, extracellular $Ca^{2+}$ or glutamate. Although there is no conclusive evidence at present for subtypes of the $Ca^{2+}$ receptor, it seems likely that homologous yet distinct receptors will be discovered.

The $Ca^{2+}$ receptor couples to a number of different transmembrane signaling mechanisms, depending on the particular cell type where it is expressed. In parathyroid cells, the $Ca^{2+}$ receptor couples to phospholipase C and adenylate cyclase to increase $[Ca^{2+}]_i$ and decrease cAMP levels and secretion of PTH. The coupling to adenylate cyclase is mediated by a $G_i$-like protein, whereas that to phospholipase C is mediated either by $G_{11}$ or $G_q$. In C cells, activation of the $Ca^{2+}$ receptor allows the influx of extracellular $Ca^{2+}$

through voltage-sensitive $Ca^{2+}$ channels, leading to an increase in $[Ca^{2+}]_i$ and the secretion of calcitonin. In renal epithelial cells, entirely different intracellular signaling mechanisms are coupled to the $Ca^{2+}$ receptor, and these intracellular signals regulate $Ca^{2+}$ and $Mg^{2+}$ reabsorption and water transport.

Both genetic and pharmacological studies show that the $Ca^{2+}$ receptor is the essential molecular entity maintaining systemic $Ca^{2+}$ homeostasis. Mutations in the $Ca^{2+}$ receptor that decrease its sensitivity to extracellular $Ca^{2+}$ result in hypercalcemic disorders, such as FBHH and NSHPT, and the latter is often fatal if not treated. Mutations in the $Ca^{2+}$ receptor that increase its sensitivity to extracellular $Ca^{2+}$ result in hypocalcemic disorders such as ADH. Moreover, calcimimetic compounds that activate the $Ca^{2+}$ receptor cause decreases in circulating levels of PTH and $Ca^{2+}$. The last studies also suggest that the $Ca^{2+}$ receptor is a novel molecular target for drugs useful in treating a variety of bone and mineral disorders such as hyperparathyroidism and osteoporosis.

Most of what is known about how the $Ca^{2+}$ receptor regulates cellular activity derives from cells involved in systemic $Ca^{2+}$ homeostasis. Yet it is clear that many other types of cells, which have little if any role in systemic $Ca^{2+}$ homeostasis, also express the $Ca^{2+}$ receptor. The function of the $Ca^{2+}$ receptor in most of these cells is far from understood. The widespread distribution of the $Ca^{2+}$ receptor, however, suggests that extracellular $Ca^{2+}$ has an important messenger function throughout the body. These recent discoveries add a new dimension to physiology and show that extracellular $Ca^{2+}$ (and perhaps other inorganic ions) must now be considered a variation on the theme of chemical communication in the body.

## Bibliography

Azria, M. (1989). "The Calcitonins. Physiology and Pharmacology." Karger, New York.

Bai, M., Quinn, S., Trivedi, S., Kifor, O., Pearce, S. H. S., Pollak, M. R., Krapcho, K., Hebert, S. C., and Brown, E. M. (1996). Expression and characterization of inactivating and activating mutations in the human $Ca_o^{2+}$-sensing receptor. *J. Biol. Chem.* **271**, 19537–19545.

Beck-Sickinger, A. G. (1996). Structural characterization and binding sites of G-protein-coupled receptors. *Drug Disc. Today* **1**, 502–513.

Bikle, D. D., Ratnam, A., Mauro, T., Harris, J., and Pillai, S. (1996). Changes in calcium responsiveness and handling during keratinocyte differentiation. Potential role of the calcium receptor. *J. Clin. Invest.* **97**, 1085–1093.

Brown, E. M. (1983). Four parameter model of the sigmoidal relationship between parathyroid hormone release and extracellular calcium concentration in normal and abnormal parathyroid tissue. *J. Clin. Endocrinol. Metab.* **56**, 572–581.

Brown, E. M. (1991). Extracellular $Ca^{2+}$-sensing, regulation of parathyroid cell function, and role of $Ca^{2+}$ and other ions as extracellular (first) messengers. *Physiol. Rev.* **11**, 371–411.

Brown, E. M. (1994). Homeostatic mechanisms regulating extracellular and intracellular calcium metabolism. *In* "The Parathyroids" (J. P. Bilezikian, R. Marcus, and M. A. Levine, Eds.), pp. 15–54. Raven Press, New York.

Brown, E. M., and Hebert, S. C. (1995). A cloned $Ca^{2+}$-sensing receptor: A mediator of direct effects of extracellular $Ca^{2+}$ on renal function? *J. Am. Soc. Nephrol.* **6**, 1530–1540.

Brown, E. M., Gamba, G., Riccardi, D., Lombardi, M., Butters, R., Kifor, O., Sun, A., Hediger, M. A., Lytton, J., and Hebert, S. C. (1993). Cloning and characterization of an extracellular Ca$^{2+}$-sensing receptor from bovine parathyroid. *Nature* **366**, 575–580.

Chattopadhyay, N., Mithal, A., and Brown, E. M. (1996). The calcium-sensing receptor: A window into the physiology and pathophysiology of mineral ion metabolism. *Endocr. Rev.* **17**, 289–307.

Clark, E. A., and Brugge, J. S. (1995). Integrins and signal transduction pathways: The road taken. *Science* **268**, 233–239.

Drüeke, T. B. (1995). The pathogenesis of parathyroid gland hyperplasia in chronic renal failure. *Kidney Intl.* **48**, 259–272.

Garrett, J. E., Capuano, I. V., Hammerland, L. G., Hung, B. C. P., Brown, E. M., Hebert, S. C., Nemeth, E. F., and Fuller, F. (1995a). Molecular cloning and functional expression of human parathyroid calcium receptor cDNAs. *J. Biol. Chem.* **270**, 12,919–12,925.

Garrett, J. E., Tamir, H., Kifor, O., Simin, R. T., Rogers, K. V., Mithal, A., Gagel, R. F., and Brown, E. M. (1995b). Calcitonin-secreting cells of the thyroid express an extracellular calcium receptor gene. *Endocrinology* **136**, 5202–5211.

Hebert, S. C. (1996). Extracellular calcium-sensing receptor: Implications for calcium and magnesium handling in the kidney. *Kidney Intl.* **50**, 2129–2139.

Lundgren, S., Hjälm, G., Hellman, P., Ek, B., Juhlin, C., Rastad, J., Klareskog, L., Åkerström, G., and Rask, L. (1994). A protein involved in calcium sensing of the human parathyroid and placental cytotrophoblast cells belongs to the LDL-receptor protein superfamily. *Exp. Cell Res.* **212**, 344–350.

Martin, K. J., and Slatopolsky, E. (1994). The parathyroids in renal disease. *In* "The Parathyroids" (J. P. Bilezikian, R. Marcus, and M. A. Levine, Eds.), pp. 711–719. Raven Press, New York.

Mayer, G. P., and Hurst, J. G. (1978). Sigmoidal relationship between parathyroid hormone secretion rate and plasma calcium concentration in calves. *Endocrinology* **102**, 1036–1042.

McDermott, M. T., and Kidd, G. S. (1987). The role of calcitonin in the development and treatment of osteoporosis. *Endocr. Rev.* **8**, 377–390.

Muff, R., Nemeth, E. F., Haller-brem, S., and Fischer, J. A. (1988). Regulation of hormone secretion and cytosolic Ca$^{2+}$ by extracellular Ca$^{2+}$ in parathyroid cells and C-cells: Role of voltage-sensitive Ca$^{2+}$ channels. *Arch. Biochem. Biophys.* **265**, 128–135.

Mundy, G. R. (1989). "Calcium Homeostasis: Hypercalcemia and Hypocalcemia." Martin Dunitz, London.

Nemeth, E. F. (1990). Regulation of cytosolic calcium by extracellular divalent cations in C-cells and parathyroid cells. *Cell Calcium* **11**, 323–327.

Nemeth, E. F. (1996). Calcium receptors as novel drug targets. *In* "Principles of Bone Biology" (J. P. Bilezikian, L. G. Raisz, and G. A. Rodan, Eds.), pp. 1019–1035. Academic Press, New York.

Nemeth, E. F., and Heath, H., III. (1995). The calcium receptor and familial benign hypocalciuric hypercalcemia. *Curr. Opin. Endo. Diabetes* **2**, 556–561.

Nemeth, E. F., and Scarpa, A. (1987). Are changes in intracellular free calcium necessary for regulating secretion in parathyroid cells? *Ann. NY Acad. Sci.* **493**, 542–551.

Nemeth, E. F., Steffey, M. E., and Fox, J. (1996). The parathyroid calcium receptor: A novel therapeutic target for treating hyperparathyroidism. *Pediatr. Nephrol.* **10**, 275–279.

Pearce, S. H. S., and Brown, E. M. (1996). The genetic basis of endocrine disease. Disorders of calcium ion sensing. *J. Clin. Endo. Metab.* **81**, 2030–2035.

Petersen, C. C. H. (1996). Store operated calcium entry. *Semin. Neurosci.* **8**, 293–300.

Pollak, M. R., Brown, E. M., Chou, Y.-H. W., Hebert, S. C., Marx, S. J., Steinmann, B., Levi, T., Seidman, C. E., and Seidman, J. G. (1993). Mutations in the human Ca$^{2+}$-sensing receptor gene cause familial hypocalciuric hypercalcemia and neonatal severe hyperparathyroidism. *Cell* **75**, 1297–1303.

Pollak, M. R., Brown, E. M., Estep, H. L., McLaine, P. N., Kifor, O., Park, J., Hebert, S. C., Seidman, C. E., and Seidman, J. G. (1994). Autosomal dominant hypercalcaemia caused by a Ca$^{2+}$-sensing receptor gene mutation. *Nat. Genet.* **8**, 303–307.

Putney, J. W., Jr. (1986). A model for receptor-regulated calcium entry. *Cell Calcium* **7**, 1–12.

Racke, F. K., and Nemeth, E. F. (1993a). Cytosolic calcium homeostasis in bovine parathyroid cells and its modulation by protein kinase C. *J. Physiol.* **468**, 163–176.

Racke, F. K., and Nemeth, E. F. (1993b). Protein kinase C modulates hormone secretion regulated by extracellular polycations in bovine parathyroid cells. *J. Physiol.* **468**, 163–176.

Racke, F. K., and Nemeth, E. F. (1994). Stimulus–secretion coupling in parathyroid cells deficient in protein kinase C activity. *Am. J. Physiol.* **267**, E429–E438.

Raue, F., and Scherübl, H. (1995). Extracellular calcium sensitivity and voltage-dependent calcium channels in C cells. *Endocr. Rev.* **16**, 752–764.

Riccardi, D., Park, J., Lee, W.-S., Gamba, G., Brown, E. M., and Hebert, S. C. (1995). Cloning and functional expression of a rat kidney extracellular calcium-sensing receptor. *Proc. Natl. Acad. Sci. USA* **92**, 131–135.

Suzdak, P. D., Thomsen, C., Mulvihill, E., and Kristensen, P. (1994). Molecular cloning, expression, and characterization of metabotropic glutamate receptor subtypes. *In* "The Metabotropic Glutamate Receptors" (P. J. Conn and J. Patel, Eds.), pp. 1–30. Humana Press, Totowa, NJ.

Takahashi, K., Tsuchida, K., Tanabe, Y., Masu, M., and Nakanishi, S. (1993). Role of the large extracellular domain of metabotropic glutamate receptors in agonist selectivity determination. *J. Biol. Chem.* **268**, 19341–19345.

Varrault, A., Pena, M. S. R., Goldsmith, P. K., Mithal, A., Brown, E. M., and Spiegel, A. M. (1995). Expression of G protein α-subunits in bovine parathyroid. *Endocrinology* **136**, 4390–4396.

Zaidi, M., Alam, A. S. M. T., Huang, C. L.-H., Pazianas, M., Bax, C. M. R., Bax, B. E., Moonga, B. S., Bevis, P. J. R., and Shankar, V. S. (1993). Extracellular Ca$^{2+}$ sensing by the osteoclast. *Cell Calcium* **14**, 271–277.

Zaidi, M., Shankar, V. S., Tunwell, R., Adebanjo, O. A., Mackrill, J., Pazianas, M., O'Connell, D., Simon, B. J., Rifkin, B. R., Venkitaraman, A. R., Huang, C. L.-H., and Lai, F. A. (1995). A ryanodine receptor-like molecule expressed in the osteoclast plasma membrane functions in extracellular Ca$^{2+}$ sensing. *J. Clin. Invest.* **96**, 1582–1590.

Nelson D. Horseman and J. Wesley Pike

# 11

# Cellular Responses to Hormones

## I. Introduction

Hormones are secreted biochemical substances that affect the function of cells within an individual by binding to receptors and eliciting specific reactions from their target cells. Hormones affect cellular functions by altering rates of gene expression and metabolism, but are not themselves substrates for the processes that they regulate. Therefore, hormones are primarily information carriers. The first hormones were identified based on their secretion from specialized endocrine (internal secretion) glands. In recent years, a large number of nonendocrine hormones have been discovered using advanced methods of cell culture, biochemical purification, and recombinant DNA technology. These "paracrine" (locally secreted) hormones include many novel factors, as well as some of the hormones that are also secreted from circumscribed endocrine glands. Homeostasis and development are controlled through the actions of endocrine and paracrine hormones on target cells, which each express receptors for a limited array of hormones.

Because hormones do not directly participate in the processes they control, we conceive of them as initiating "signal transduction" protocols, which lead to the appropriate cellular outcomes. In this chapter, we cover examples of signal transduction mechanisms for two broad classes of hormones: the lipophilic hormones such as steroids, thyroxine, retinoids, and vitamin D, which diffuse into cells and bind to intracellular receptors; and the hydrophilic peptide and amine hormones, which interact with receptors that are exposed on the surface of cells.

## II. Actions of Lipophilic Hormones via Intracellular Receptors

### A. Signal Transduction through Intracellular Receptors: Basic Principles

The sex and adrenal steroids, and the thyroid, retinoic acid, and vitamin D hormones, are members of a chemically

diverse set of endocrine signaling molecules that are produced and regulated in response to both internal and environmental cues. Acting on both distant and local tissue targets, they exert regulatory control over a myriad of specific cellular functions associated with virtually every vertebrate organ system. These actions affect cellular metabolism, reproductive function, nutrient homeostasis, and behavior, to name only several. A common biological feature of each of these hormones is their capacity to regulate cellular differentiation. The mechanism through which lipophilic hormones dictate specific biological responses involves the ability of these hormones to enter the nucleus and to modulate the expression of single genes and gene networks. This action is selective by virtue of the presence in cells of individual receptors for each of these hormones, and these receptors represent primary determinants of tissue response to the cognate hormone. Nuclear receptors belong to a large gene family of transcription factors that function to recognize their respective endocrine signal and to respond accordingly. In this section, we describe the progress that has occurred since the mid-1980s that has enhanced our understanding of how these signaling pathways modify gene expression.

In contrast to peptide hormones, growth factors, and cytokines, the lipophilic hormones are not limited in their ability to gain entry into the cell (Beato, 1989; Katzenellenbogen et al., 1996). Because of their solubility in the membrane and their small size, they are believed to enter the cell by diffusion through the cell membrane; this energy-independent process apparently permits them to ultimately reach the cell nucleus where they associate through a high-affinity interaction with a specific nuclear receptor (Fig. 1). Although the bulk of the members of the nuclear receptor family are believed to reside in the nucleus prior to ligand activation, the receptors for the glucocorticoids (GR) and the mineralocorticoids (MR) appear to be the single exceptions in that they are found in the cytoplasm (Evans, 1988; Beato et al., 1995; O'Malley, 1990). In this specific case, interaction with glucocorticoid or mineralocorticoid ligand induces translocation of the receptors to the nucleus. The

**Steroid, thyroid, retinoid, and 1,25(OH)₂D₃ hormones**

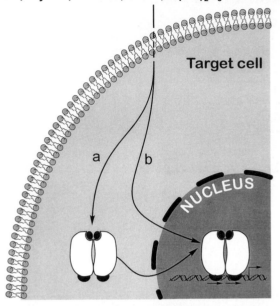

**Target cell**

a    b

NUCLEUS

**FIG. 1.** Model for the molecular mechanism of action of steroid, thyroid, retinoid, and vitamin D hormones. Hormonal ligand enters the cell by diffusion and interacts with its cognate receptor. Activation by the ligand leads to the interaction of receptor with responsive genes and the modulation of gene expression. Active receptors are composed of monomers, homodimers, or heterodimers (see text). (a) Model for glucocorticoid and mineralocorticoid receptors, wherein the receptor is cytoplasmic in the absence of ligand. Upon ligand activation the receptor undergoes nuclear translocation and eventually binds to the regulatory region of a modulated gene. (b) Model for most of the nuclear receptors, wherein the receptor is located in the nucleus and following ligand activation becomes bound to the regulatory region of hormone-responsive genes.

interaction of hormonal ligand with its receptor triggers a series of events that culminate in alterations in gene expression. Allosteric changes in the receptor induced by ligand lead to dissociation of associated proteins that function as inhibitors of receptor DNA-binding ability and transcriptional capacity, formation of functional protein units, binding of the units to specific DNA sequence elements located adjacent to hormone sensitive gene promoters, and activation or repression of transcription (see Fig. 1). The capacity of a hormone to regulate transcription is dependent on the presence of the receptor for that hormone in a cell, but the quantitative and qualitative nature of the response can be both gene-specific and cell-specific depending on the presence and activity of additional cellular protein factors. Cellular mechanisms that serve to inactivate receptor signals within the nucleus are largely unknown, although the induction of enzymes by the hormone-activated receptor that degrade the corresponding hormonal signal is not uncommon.

## B. The Nuclear Receptor Gene Family

### 1. Ligand-Activated Nuclear Receptors

The genes for all of the receptors that mediate the actions of traditional nonpeptide endocrine hormones have been

cloned (Fig. 2). These include the receptor for the glucocorticoids (GR), as well as the mineralocorticoid receptor (MR), estrogen receptors (ERα, β), progesterone receptor (PR), androgen receptor (AR), thyroid hormone receptors (TRα, β), retinoic acid receptors (RARα, β, γ), 9-cis-retinoic acid receptors (RXRα, β, γ) and the vitamin D receptor (VDR) (Gronemeyer, 1993; Truss and Beato, 1993; Mangelsdorf and Evans, 1995; Mangelsdorf et al., 1995; Beato et al., 1995). Single genes encode certain of the receptors, whereas others are derived from multiple genes. In the latter case, these receptors may exhibit differences in tissue distribution as well as differences in gene regulatory activity as a result of slight but significant differences in their primary structure. During transcription, receptor genes can also undergo alternative splicing events that often lead to multiple mRNA transcripts that encode receptor gene products that exhibit different functional activities. The existence of multiple genes for receptors, alternative splicing events within single genes, and differential use of translation start sites within the mRNA transcripts result in a heterogeneous mix of receptor gene products, as well as the possibility of highly diverse responses to ligand within cells, depending on the particular receptor gene product being expressed.

### 2. Orphan Receptors

The cloning of the receptors for known endocrine hormones and the observation that all belonged to a structurally related family of genes precipitated a search for additional gene members in vertebrate tissues as well as in nonvertebrate species. This search has resulted in the identification of numerous additional genes that are clearly members of the nuclear receptor gene family (see Fig. 2) (O'Malley, 1990; Mangelsdorf and Evans, 1995). Although none of these orphan receptors appear currently to mediate the actions of known endocrine hormones, several appear to facilitate the actions of local cellular factors such as 9-cis-retinoic acid and prostaglandin J2 (Forman et al., 1995), as well as cellular metabolic intermediates such as farnesol, arachidonic acid, and perhaps certain cholesterol derivatives. Current efforts are focused on determining whether novel ligands indeed exist for these receptors, although activation pathways independent of ligand are clearly possible. Moreover, in addition to functioning as gene activators, there is evidence that certain of the orphans may play largely negative roles in transcription, acting not only to repress the transcriptional activity of specific genes, but also to function on other genes as repressors of the inducing actions of ligand-activated nuclear receptors. Novel biological roles for several of these orphans receptors have been defined.

### 3. Invertebrate Receptor Genes

A large number of genes from invertebrates have been identified as belonging to the steroid receptor gene family (see Fig. 2) (Mangelsdorf et al., 1995; Thummel, 1995). The genes include those from drosophila, *Chaenorabditis elegans*, and other metazoan species. Perhaps the most interesting member of the nuclear receptor family found

| Genes | Species* | Ligand |
|---|---|---|
| GR | 1 | GLUCOCORTICOIDS |
| MR | 1 | MINERALOCORTICOIDS |
| PR | 1 | PROGESTERONE |
| AR | 1 | TESTOSTERONE |
| EcR | 2 | ECDYSONE |
| FXR | 1 | FARNESOIDS |
| LXR $\alpha,\beta$ | 1 | |
| VDR | 1 | $1,25(OH)_2D_3$ |
| xONR | 1 | |
| MB67 $\alpha,\beta$ | 1 | |
| CeF11A1.3 | 2 | |
| TR $\alpha,\beta$ | 1 | THYROID HORMONE |
| CeB0280.8 | 2 | |
| CeC2B4.2 | 2 | |
| COUP $\alpha,\beta,\gamma$ | 1,2 | |
| RXR $\alpha,\beta,\gamma,(\delta,\varepsilon)$ | 1,2 | 9-cis-RA |
| TR2-11 $\alpha,\beta$ | 1,2 | |
| HNF-4 | 1,2 | |
| TLL | 1,2 | |
| CeF21D12.4 | 2 | |
| GCNF | 1 | |
| PPAR $\alpha,\beta,\gamma,\delta$? | 1 | EICOSANOIDS, PROSTAGLANDINS |
| RAR $\alpha,\beta,\gamma$ | 1 | RETINOIC ACID |
| CNR14 | 2 | |
| E78 | 2 | |
| Rev-Erb $\alpha,\beta$ | 1 | |
| E75A | 2 | |
| ROR $\alpha,\beta,\gamma$ | 1 | |
| DHR3 | 2 | |
| NGFI-B $\alpha,\beta,\gamma$ | 1,2 | |
| FTZ-F1 $\alpha,\beta$ | 1,2 | |
| CeF11C1.6 | 2 | |
| ERR $\alpha,\beta$ | 1 | |
| ER $\alpha,\beta$ | 1 | ESTROGEN |
| KNIRPS $\alpha,\beta,\gamma$ | 2 | |
| CeE2H1.6 | 2 | |
| CEF43C1.4 | 2 | |
| CeKO6A1.4 | 2 | |
| CeODR7 | 2 | |
| CeZK418.1 | 2 | |
| CeF16H9.1 | 2 | |

**FIG. 2.** Members of the nuclear receptor superfamily of genes. The figure documents cloned nuclear receptor family members. Known ligands are indicated on the right. * 1, Vertebrate; 2, invertebrate.

in *Drosophila melanogaster* is the receptor for ecdysone, the metamorphic or molting hormone of insects. This hormone has long been associated with its capacity to induce chromosomal puffing at specific sites on chromatin and to induce complex networks of genes. Interestingly, as described later, the ecdysone receptor functions as a heterodimer with a nuclear protein partner called *ultraspiracle*. Ultraspiracle represents the insect homolog of vertebrate retinoid X receptor (RXR) genes, which in turn function as permissive heterodimer partners with the thyroid receptor, the retinoic acid receptor, the vitamin D receptor, and several others. The presence of nuclear receptors in insects suggests that this evolutionarily highly successful receptor family evolved prior to the divergence of invertebrates and vertebrates.

## C. The Highly Conserved Structural Domains of Nuclear Receptors

The nuclear receptors display a highly modular domain structure composed of a DNA binding domain of 60 to 70 amino acids and an extended carboxy-terminal ligand-binding domain of 300 amino acids separated by a highly flexible hinge of 100 to 150 amino acids, as seen in Fig. 3 (Evans, 1988). Many of the individual functions of each domain are retained following their separation through enzymatic cleavage. The DNA-binding domain is highly conserved among members of the nuclear receptor superfamily and represents the hallmark of this transcription factor family. In addition to containing determinants of specific DNA binding, this domain also contains subregions

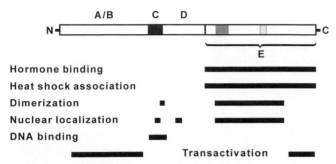

**FIG. 3.** Functional domain structure of the nuclear receptor superfamily. The nuclear receptors (NRs) are separated into four regions designated A/B, C, D, and E. Regions of homology within the nuclear receptor family are shaded. Regions involved in known specific functions of the receptor are indicated below the diagram.

that are involved in directing the protein to the nucleus following synthesis and that stabilize the protein's interaction with DNA through dimerization with a protein partner. The carboxy-terminal domain of the nuclear receptors is more than 300 amino acids in length, exhibits subregions of homology across members of the family, and contains hydrophobic residues that create a high-affinity lipophilic pocket exquisitely selective for the receptor's cognate ligand. Like that of the DNA-binding domain, this region also exhibits additional functional activities, including a dimer interface for interaction with the receptor's protein partner and additional protein/protein interfaces that are integrally involved in the receptor's ability to regulate transcription. Finally, receptors for estrogen, progesterone, androgens, and the glucocorticoids contain an extensive amino-terminal domain that exhibits additional functions associated with transcriptional activation. This complexity in structural organization of the nuclear receptor family implies diverse, highly flexible, and specific activities during the regulation of gene expression.

## 1. The DNA-Binding Domain of Nuclear Receptors

The DNA-binding domain of the nuclear receptor family is composed of two highly conserved zinc finger structures. These structures contain several invariant amino acids that include two sets of four cysteine residues that have been shown to coordinate two zinc ions (Berg, 1989). The two zinc ions hold the DNA-binding domain in an active conformation structurally; in their absence the finger motifs are nonfunctional and sensitive to proteolytic degradation. The three-dimensional structures of the DNA-binding domains of the receptors for the glucocorticoids, estrogen, and 9-*cis*-retinoic acid have been established (Hard *et al.*, 1990; Schwabe *et al.*, 1990; Luisi *et al.*, 1991; Lee *et al.*, 1993; Schwabe *et al.*, 1993). In addition, the three-dimensional structure of the TR and RXR DNA-binding domains bound to DNA as heterodimers has also been determined. The observed structures of these regions of the nuclear receptors confirm and extend earlier predictions that were made based on biochemical evaluations. The two zinc fin-

ger modules are known to serve very different functional roles during association with DNA (Mader *et al.*, 1989). The first functions to dictate the specificity of DNA interaction in the major groove on a core DNA sequence element, whereas the second functions to establish contacts with the receptor's protein partner. These contacts stabilize and strengthen the binding of the two proteins on the DNA sequence. Different regions located within the second zinc module, however, participate in this dimerization function depending on the organization of the DNA-binding site and the orientation of the two receptor molecules that become bound to that site.

## 2. The Ligand-Binding Domain of Nuclear Receptors

In contrast to the DNA-binding domain, the region of the receptor that is required for ligand binding extends to several hundred amino acids. Since the ligand itself is rather small relative to this region, it is clear that the protein pocket that makes direct contact with the ligand is determined through proper folding of different components of the protein domain itself. Indeed, several regions of homology have been defined within this domain, which exist among the nuclear receptor family. These regions of homology however, not only participate in creating a ligand-binding pocket, but also serve several additional functional roles that involve protein/protein interactions; two such interactions are those that lead to homodimer or heterodimer formation with the receptor's protein partner and those that mediate contact with the general transcriptional apparatus. Additional interactions include receptor associations with either inhibitor or repressor proteins that function to prevent receptor DNA-binding and transcriptional activity in the absence of ligand or that facilitate transcriptional repression while bound to DNA in the presence of the activating ligand (Horlein *et al.*, 1995). Studies have led to the elucidation of the crystal structure of ligand-binding domains of TR, RAR, and RXR (Bourguet *et al.*, 1995; Renaud *et al.*, 1995; Wagner *et al.*, 1995; Rastinejad *et al.*, 1995). Crystallization of the first two was achieved in the presence of the corresponding ligands, whereas the structure of the RXR ligand-binding domain was determined in the absence of its cognate ligand. These studies have revealed the three-dimensional structure of the nuclear receptors to be that of an antiparallel α-helical sandwich comprising as many as 12 helices. Importantly, the binding of hormone leads to a significant repositioning of several of these α-helices, including the most carboxy-terminal one that is known to be involved in mediating contacts with the general transcription apparatus. Helices 9 and 10 participate in the formation of receptor homo- or heterodimers. The ligand-binding pocket is composed of extended clusters of residues that include helices 1, 3, 5, a β turn between 5 and 6, a loop between 6 and 7, 11 and 12, and the loop between 11 and 12. Although the crystal structure of a complete receptor has yet to be determined, these studies with both the DNA-binding domain and the ligand-binding domain have provided significant insight into structure–function relationships that exist for this class of transcription factors.

### 3. Additional Nuclear Receptor Domains

Two additional domains are found within the nuclear receptor gene family. The first is a region common to all the receptors that serves to link the DNA-binding domain with the ligand-binding domain. It is not conserved within the receptor gene family. One characteristic of this region is flexibility; depending on the orientation of the DNA sequence elements to which the individual receptors bind, the hinge region allows 180° swiveling of the two domains that surround it. It is likely that this region also subserves other as yet unknown functions. The second domain is a large amino-terminal region found predominantly in receptors for estrogen, progesterone, androgens, and the glucocorticoids. This domain contains additional regions that function to make direct or indirect protein/protein contacts with the general transcriptional apparatus. The three-dimensional structure of this region has not been determined, either independently or in association with the DNA-binding domain or other portions of the receptor molecule.

### D. DNA-Binding Motifs within Nuclear Receptors

Nuclear receptors interact with unique sequences of DNA located within the promoters for hormone-regulated genes. These associations represent the first in a series of steps that result in transcriptional modulation. Research since the mid-1980s on both naturally regulated genes and synthetically derived DNA sequences has revealed the nature of these elements, which serve as selective binding sites for members of the nuclear receptor family. As seen in Fig. 4, three classes of DNA-binding sites have emerged that reflect the three modes of DNA-binding characteristic of members of this gene family—homodimeric DNA binding, heterodimeric binding, and monomeric binding (Umesono and Evans, 1989; Umesono et al., 1991; Perlmann et al., 1993). The core DNA-binding site within promoters is composed of a hexanucleotide sequence. In the case of nuclear receptors that bind to DNA as monomers, this core sequence, together with a relatively well-defined short stretch of nucleotides located immediately 5', make up the DNA-binding site. The sequence of the core as well as the adjacent nucleotides represent determinants of specificity. Numerous orphan receptors interact with DNA as monomers. The nuclear receptors that bind to DNA as dimers interact with similar core elements to those just described. In contrast, however, these core elements are

repeated to provide the two similar and adjacent binding sites necessary to accommodate the association of a dimeric nuclear receptor. Two dissimilar motifs arise as a result of this repeating structure, however: one wherein the core sequence is repeated in a direct fashion, and the second in which the core sequence is inversely repeated in palindromic fashion. Those receptors that function as homodimers, such as the GR, ER, PR, and AR, bind to palindromically repeated core sequences. In contrast, receptors that function as heterodimers in combination with a dissimilar nuclear receptor protein partner, such as TR, RAR, and VDR, bind to directly repeated motifs. The importance of the highly flexible hinge region is highlighted here. Clearly, dimeric protein binding to palindromic DNA sequences requires a symmetric interface between the two protein partners in a head-to-head configuration, whereas dimeric binding to directly repeated DNA sequences requires an asymmetric interface wherein the proteins are arranged head-to-tail. Research has revealed that the DNA-binding domain of the nuclear receptors configure in a head-to-head or head-to-tail arrangement depending on the structure of the DNA-binding site. The carboxy-terminal ligand-binding domains, in contrast, retain their symmetric head-to-head configurations irrespective of whether the DNA-binding site is palindromic or direct in orientation. These two possibilities are accomplished through the hinge region of the nuclear receptors, which allows for an independent swiveling of the DNA-binding domain relative to the ligand-binding domain. Finally, the individual sequences of the core elements determine the ability and therefore the specificity of the interaction by homodimeric receptors such as ER, PR, and GR. In contrast, directly repeated core sequences are generally similar; specificity is determined not through a difference in the sequence, but rather through the number of nucleotides located between the two core halfsites. Thus, binding sites for the VDR, TR, and RAR are determined largely by nucleotide spacings of three, four, and five, respectively; additional arrangements have also been determined.

### E. Modes of Nuclear Receptor DNA Binding

As described earlier, nuclear receptor family members interact with DNA in three different ways: as monomers, homodimers, or heterodimers (Fig. 5) (Glass, 1994). The class of receptors that function as heterdimers, however,

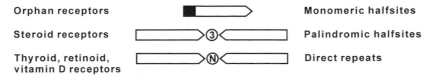

**FIG. 4.** Organization of DNA sequence elements located within gene promoters that mediate the actions of the nuclear receptor family. A core element or monomeric binding site is indicated with an arrow. Receptors that bind to this core element as monomers contact additional DNA sequence located 5' to the core element. Receptors that bind as homodimers interact with two core elements arranged in a palindromic array. These core elements are separated by three base pairs, and receptor specificity is determined by core element sequence. Receptors that bind as heterodimers interact with two core elements arranged in a directly repeated fashion. Receptor binding specificity is determined to a large extent in this motif through the number of nucleotides separating the two core elements.

| Receptors | Mode of DNA binding |
|-----------|---------------------|

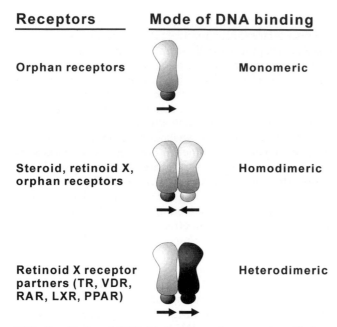

**Orphan receptors**          **Monomeric**

**Steroid, retinoid X, orphan receptors**          **Homodimeric**

**Retinoid X receptor partners (TR, VDR, RAR, LXR, PPAR)**          **Heterodimeric**

**FIG. 5.** Modes of DNA binding by nuclear receptors. Nuclear receptors interact with DNA sequence motifs as monomers, homodimers, and heterodimers.

includes the TRs, RARs, VDR, and several others that mediate the actions of intracellularly produced ligands. Each of these receptors bind to specific hormone response elements in combination with the common nuclear receptor partner RXR. RXR appears to be a generally permissive partner; activation of the heterodimer is achieved through the ligand that binds the signaling partner (Glass, 1994; Kleiwer *et al.*, 1992; Yu *et al.*, 1991). The evolution of the heterodimer mechanism wherein a common protein plays a functional role in the activities of several apparently unrelated biochemical pathways is a repeated theme in cellular biology. The implications of such a mechanism is that competition for the permissive RXR protein pool must arise within cells that express more that one of the RXR-utilizing receptors. This may have profound biological consequences in that apparently unrelated endocrine signaling systems appear to converge and compete during the course of their actions to regulate gene expression.

## F. The Role of the Ligand in Nuclear Receptor Activation

Lipophilic hormonal ligands initiate a series of events that culminate in the modulation of gene expression. These events are largely the result of conformational changes induced by the ligand upon association with its specific nuclear receptor. Considerable evidence exists for this conformation change upon ligand binding. Early studies demonstrated that the sensitivity to endogenous proteases was reduced following association with ligand. The use of limited proteolytic digestion *in vitro* to define the effect of ligand on receptor continues to support the idea that ligand induces selective and specific conformational changes. Although a study of the effect of ligand on the three-dimen-

sional structure of a single receptor has not yet been made, a comparison of the three-dimensional structures of unliganded RXR with liganded RAR and TR reveal a major repositioning of helical structures within the receptors in the liganded state.

### 1. Dissociation of Inhibitory Proteins

Receptors for the sex and adrenal steroids associate in the absence of ligand with a complement of inhibitory proteins that include hsp90 (Pratt *et al.*, 1988). These proteins are believed to dissociate as a result of ligand-induced changes in conformation that occur within the steroid receptors. This dissociation releases free receptor monomers, which in turn associate with other monomers to form active homodimers that bind as high-affinity complexes on DNA (see Fig. 5). Although receptors for T3, retinoic acid, 1,25-dihydroxyvitamin D, and other nonsteroidal ligands do not appear to interact with these inhibitory complexes, it is likely that inhibitor proteins perhaps unrelated to the heat-shock macromolecules exist. If so, a similar ligand-mediated mechanism might release these receptors as well. Since TR, RAR, and VDR form heterodimers with unliganded RXRs, however, an additional ligand-independent mechanism must be proposed that makes available RXR monomers for heterodimer formation. The dissociation of specific inhibitor proteins from TR and the RARs identified as nuclear receptor corepressor (NcoR) (Horlein *et al.*, 1995) have been demonstrated in response to cognate ligands. These inhibitor proteins, however, differ from the typical inhibitor proteins associated with the sex steroid receptors in that they serve a transcriptional repressive function. Considerable evidence exists for additional proteins that function to repress transcription. Disruption of these protein/protein interactions provides additional evidence for ligand-induced conformational changes.

### 2. Formation of Nuclear Receptor Dimers

Ligand-induced conformational changes also promote the formation of nuclear receptor dimers. Thus, binding of ligand to TR, VDR, and perhaps others leads to an increase in the affinity of the ligand-activated receptor for RXR (see Fig. 5). These observations suggest an additional role for the hormone in facilitating the creation of functional receptor units.

### 3. Receptor Contact with the General Transcription Apparatus

Modulation of gene expression requires the interaction of a nuclear receptor with the general transcriptional apparatus. A final and perhaps anticipated role for the ligand appears to be its capacity to regulate this process. The mechanism whereby the nuclear receptors modulate the activity of the general transcriptional apparatus again involves additional modulatory proteins termed coactivators or corepressors (Danielian *et al.*, 1992). Thus, the conformational changes induced by the ligand lead not only to dissociation of inhibitors or repressors and to the formation

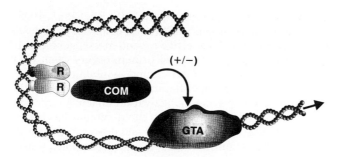

**FIG. 6.** Nuclear receptor contact with the general transcriptional apparatus (GTA). Nuclear receptor dimers (or monomers) contact the GTA via comodulators (COM). These transcription factors enable the nuclear receptor to stimulate (co-activator) or repress (co-repressor) the basal or existing transcription of the gene.

of active receptor dimers, but also facilitate their association with comodulators that are directly able to contact the initiation complex of certain genes (Fig. 6). A variety of comodulators exist that mediate this contact. Most important, since these proteins are not expressed ubiquitously, but rather in tissue-specific patterns, they provide a means whereby cellular response can be graded or altered following induction of the general signaling pathway by ligand.

## III. Cellular Actions of Protein and Amine Hormones via Plasma Membrane Receptors

### A. Signal Transduction from Plasma Membrane: Basic Principles

Because of the selective permeability of the plasma membrane, most hormones cannot traverse the membrane independently; therefore, they bind to integral membrane proteins that serve as sensors of the extracellular hormone concentration. Hormones that bind to membrane receptors include molecules as diverse as large multisubunit proteins, small peptides, and amino acid derivatives. As a general principle, binding of a hormone to its receptor causes the synthesis of one or more diffusible intracellular mediators, which have been termed *second messengers* or *intracellular signal transducers*. The synthesis of these signal transducers is initiated by hormone-induced changes in the conformation of the receptor, or its oligomerization with other membrane-associated proteins. Microbes and single-celled eukaryotes use mechanisms similar to hormonal signal transduction to integrate their responses to nutrients and other environmental cues; therefore, we can infer that hormone response systems evolved from nutrient transport and sensor mechanisms that originated early in evolution. Many of the features of vertebrate signal transduction are conserved in single-celled organisms such as yeast (Herskowitz, 1995), making these simple organisms valuable research tools in contemporary molecular endocrinology.

The intracellular effectors that are controlled by hormonal signal transduction include proteins that reside in a variety of cell compartments. Some, such as the heterotrimeric GTP-binding proteins (G proteins) and various ion channels, are plasma membrane or organellar membrane proteins; others, including numerous protein kinases, are strictly cytosolic proteins; while still others, such as CREB (cAMP-response element binding protein) and Stat proteins (signal transducer and activator of transcription) either are resident in the nucleus or are translocated to the nucleus after their activation. The variety of intracellular processes that are controlled by intracellular signal transduction pathways reflects the capacity of hormones to bring about integrated, adaptive cellular responses that promote either homeostasis or directed developmental changes.

Protein phosphorylation plays a central role in hormonal signal transduction. Since the mid-1980s, it has become clear that protein phosphorylation is a ubiquitous regulatory process that affects all levels of cell signaling. The selectivity of signal-transduction mechanisms is partly accounted for by the substrate specificities of the various protein kinases in signaling pathways. Each signal-transducing kinase is represented by multiple isoforms that can catalyze phosphorylation of different substrates.

Signal-transduction molecules are highly compartmentalized within the cell. Transduction of signals from cell surface-bound receptors to diverse intracellular compartments results in the movement of information from the sensor to ultimate effectors. The means by which this information flow occurs are very poorly understood. What is clear is that most signal-transduction pathways are initiated in multiprotein complexes (receptor signaling complexes), which include not only the receptor, but also a variety of coupling proteins, kinases, and substrates. The addressing of proteins to the receptor signaling complex, and then away from it to effector compartments, provides an important mechanism for conferring signaling selectivity. This addressing phenomenon is accounted for by various conserved protein–protein interaction domains and posttranslational modifications. Since the early 1990s, the functions of some of these addressing sequences have been elucidated. These include the SH2 (Src homology 2) and SH3 domains, which bind to phosphorylated tyrosines and proline-rich motifs, respectively, and carboxy-terminal prenylation signals, which direct the addition of lipid moieties that anchor signaling proteins in the plasma membrane (Pawson, 1995a, b; Zhang and Casey, 1996).

Transduction from the membrane to intracellular compartments occurs by two means. First, small, readily diffusible molecules that are referred to as second messengers carry information from one compartment to others. Examples of these second messengers include cAMP, cGMP, inositol 1,4,5-trisphosphate (IP$_3$), and calcium. After diffusing to their effector compartment, these second messengers bind to proteins (kinases, channels, etc.) and cause conformational alterations that increase or decrease their activation state. Second, a conformational change induced by protein phosphorylation can lead directly to "read-dressing" so that a protein is forwarded to the compartment in which it exerts its effect, such as the nucleus.

These general principles of signal transduction are played out in a rich variety of specific mechanisms. To understand some of these signaling pathways, we focus primarily on those mechanisms that lead to altered control of gene expression. Other chapters in this volume reflect their focus on membrane permeability and transport mechanisms, and cellular metabolism. The reader should refer to them for more information on those topics.

## B. Signaling via Heterotrimeric GTP-Binding Proteins

### 1. Activation of Effector Functions through G Proteins

A large family of receptors share a structure that contains seven membrane-spanning $\alpha$-helices (heptahelical receptors), and all of these receptors interact with and activate GTP-binding proteins that consist of three subunits ($\alpha$, $\beta$, $\gamma$). Hormones that act through heptahelical receptors range from single amino acid derivatives such as epinephrine to multisubunit pituitary hormones, such as luteinizing hormone and follicle-stimulating hormone (Fig. 7). The "heterotrimeric G-proteins" couple receptors to processes that result in the generation of second messengers such as cAMP (Fig. 7). Receptor signaling complexes for heptahelical receptors include the receptor–ligand pair, a trimeric G-protein, and at least one effector, such as adenylyl cyclase.

G-proteins are anchored to the plasma membrane by virtue of a C-terminal prenylation signal in the $\gamma$ subunit. The $\alpha$ subunit binds GTP and hydrolyzes it to GDP. The activation–inactivation cycle of the G proteins begins with the inactive trimeric protein in which GDP is bound to the $\alpha$-subunit and the protein is associated with the membrane through the $\gamma$-subunit. The G-protein complex is capable of binding with receptor proteins that are also in the membrane. Binding of ligand to its receptor induces a conformational shift, and the G protein that is associated with the receptor displaces GDP and binds GTP. The binding of GTP causes the $\alpha$-subunit to be released from the complex, allowing it to interact with an effector molecule, such as

adenylyl cyclase or phospholipase C (Spiegel *et al.*, 1995). Binding of the $\alpha$-subunit to these effector enzymes leads to synthesis of their respective second messengers, cAMP and IP3. Inactivation of the G protein occurs when GTP is hydrolyzed to GDP. GDP-bound $\alpha$-subunit undergoes a conformational change that increases its affinity for the $\beta\gamma$ complex, completing the activation–inactivation cycle. The GTPase activity intrinsic to the $\alpha$-subunit has a slow rate constant, so $\alpha$ stays activated and dissociated from $\beta\gamma$ for an extended period of time, leading to amplification of the hormone signal by virtue of repeated rounds of second-messenger synthesis.

Signal diversity is generated by multiple G-protein $\alpha$-subunit isoforms, which each couple to particular effector proteins (Spiegel *et al.*, 1995). Adenylyl cyclase is regulated by stimulatory ($G_{\alpha s}$) and inhibitory ($G_{\alpha i}$) subunits. $G_{\alpha q}$ activates phospholipase C$\beta$, leading to generation of IP3. The diversity of $\alpha$-subunit isoforms has been realized only in recent years (Table 1), and the details of effector modulation by these isoforms are only partly understood. Signaling by G proteins to the nucleus converges on the transcription factor CREB (cAMP response element binding protein), which is a 341 amino acid protein that is activated by phosphorylation of serine 133 (see Fig. 8). cAMP-dependent protein kinase (PKA) and calcium–calmodulin-dependent protein kinase II (CAM kinase II) each phosphorylate this site. CREB is able to integrate numerous G-protein signals because it responds to changes in both cAMP and calcium, each of which is controlled by multiple G-protein isoforms.

Activation of $G_{\alpha q}$ leads to calcium release from intracellular stores in response to the second messenger $IP_3$. $IP_3$ is a product of the breakdown of the membrane phospholipid phosphatidylinositol 4, 5-bisphosphate ($PIP_2$) by phospholipase C, and diacylglycerol is generated simultaneously. $IP_3$ provokes calcium release from the endoplasmic reticulum and calcium activates numerous targets, including calmodulin and calcium–calmodulin-dependent proteins, and protein kinase C (PKC). PKC is localized to membrane compartments because it binds to diacylglycerol, as well as to calcium. Recent identification of multiple PKC isoforms has allowed a renewed focus on the downstream effectors that are regulated by PKC. It is unclear how PKC participates in nuclear gene regulation, but the presence of hormone-activated PKC in the nucleus suggests that it may be an important regulator of transcription and/or RNA processing (Goss *et al.*, 1994).

The $\alpha$-subunits of G proteins are the most heterogeneous in structure and function, but the $\beta\gamma$ complex is also represented by multiple isoforms. The primary functions of the $\beta\gamma$ complexes are to bind the inactive $\alpha$-subunits and localize the G-protein complex to the membrane. However, liberated $\beta\gamma$ complexes can directly activate the MAP kinase pathway (see later discussion) by interacting with the Ras protein (Luttrell *et al.*, 1995).

### 2. Transcriptional Regulation by G Protein-Activated Pathways

Genes that are regulated by cAMP share a regulatory DNA sequence termed the cAMP response element

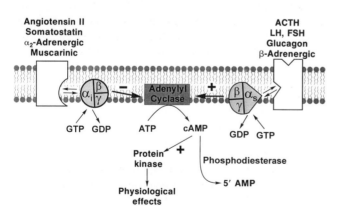

**FIG. 7.** Bivalent regulation of cAMP synthesis by heterotrimeric GTP-binding proteins. Representative ligands are listed, which act through receptors coupled to either stimulatory ($\alpha_s$) or inhibitory ($\alpha_i$) G-protein subunits. Activation of G-protein function mediated by ligand–receptor interaction causes exchange of GTP for GDP, and dissociation of the $\alpha$-subunit from its $\beta\gamma$ partners. The $\alpha$-subunits interact with adenylyl cyclase to either accelerate or inhibit cAMP synthesis. cAMP is degraded by cAMP phosphodiesterase to inactive AMP.

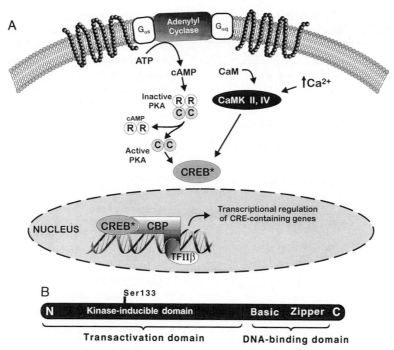

**FIG. 8.** (A) Regulation of nuclear gene transcription by G protein-mediated signaling. Both cAMP, generated in response to $G_{\alpha s}$ coupling, and calcium–calmodulin, generated in response to $G_{\alpha q}$, cause phosphorylation of the transcription factor cAMP response element binding protein (CREB*). Protein kinase A catalytic subunits (C) are released from the regulatory subunits (R) in response to cAMP binding to the R subunits. CREB interacts with DNA and other nuclear proteins, such as CBP and TFIIB, both directly and indirectly. (B). The domain structure of CREB protein. CREB contains a large N-terminal domain that is involved in transcriptional activation and is regulated by phosphorylation on ser133, as well as other sites. The C-terminal DNA-binding domain includes both a region rich in basic amino acids and a leucine zipper motif, which is involved in homo- and heterodimerization.

(CRE) (Pestell and Jameson, 1995). The CRE consists of an octanucleotide sequence that is identical, or closely related, to the palindromic consensus 5'-TGACGTCA. The transcription factor that binds to the CRE is called CRE-binding protein (CREB). CREB is a member of a family of transcription factors termed basic-leucine zipper (bZIP) factors, which bind to related response elements. The bZIP proteins form homodimers and heterodimers by virtue of interactions between their leucine zipper motifs, and they interact with DNA through basic residues located in the C-terminal DNA-binding domain (see Fig. 8) (Meyer and Habener, 1993). CREB can form homodimers that activate transcription. It can also form heterodimers that either activate or inhibit transcription by interacting with isoforms of two closely related bZIP proteins, CREM (CRE modifier) and ATF-1 (activating transcription factor-1) (Habener, 1990). At least seven CREM isoforms are generated by alternative splicing, leading to a rich potential for selective modulation of transcriptional activity.

CREB is phosphorylated on a critical regulatory site (ser 133) by cAMP-dependent protein kinase, calcium–calmodulin (CaM) kinase II, and CaM kinase IV. Other phosphorylation sites in the molecule can down-regulate CREB activity in response to a variety of kinases (Meyer and Habener, 1993).

The analysis of transcriptional activation by CREB led to the identification of a large (molecular weight 300,000)

CREB-binding protein (CBP) that has a high affinity for the phosphorylated CREB dimer. CBP enhances the transcriptional activation driven by CREB (Kwok *et al.*, 1994). CBP was the first identified representative of a wide variety of "coactivator" proteins that form complexes with hormone-inducible transcription factors. These factors enhance, or in some cases inhibit, the transcriptional activity of regulated genes by altering the interactions of upstream transcription factors with the basal transcription machinery. The mechanisms for most of these coactivator proteins are not yet understood. One of the mechanisms by which CBP increases transcriptional activity is by acetylating histones that are bound to the DNA, thereby increasing the ability of proteins to gain access to specific regions of genes (Ogryzko *et al.*, 1996).

CREB is a focal point of regulation by virtue of its ability to be phosphorylated by multiple hormone-regulated kinases, and to interact with other proteins such as CREM and CBP. The commonness of CRE and CRE-like DNA sequences among inducibly regulated genes results in ensembles of coregulated genes in individual cells, as well as diverse cell-specific responses among cell types.

### B. Signaling via Tyrosine Phosphorylation

Tyrosine kinase activity is a property that is shared among certain virus-encoded oncogenes and among pro-

**TABLE 1**  Functional Diversity of α-Subunits among Mammalian Heterotrimeric GTP-Binding Proteins

| α-Subunit | Localization | Effectors |
|-----------|--------------|-----------|
| $G_s$ | Ubiquitous | Adenylyl cylase ↑<br>$Ca^{2+}$ channel ↑ |
| $G_{olf}$ | Olfactory | Adenylyl cyclase ↑ |
| $G_{t1}$ | Rod photoreceptors | cGMP phosphodiesterase ↑ |
| $G_{t2}$ | Cone photoreceptors | cGMP phosphodiesterase ↑ |
| $G_{gust}$ | Taste cells | ? |
| $G_{i1}$ | Neural > others | Adenylyl cyclase ↓<br>$K^+$ channel ↑ |
| $G_{i2}$ | Ubiquitous | Adenylyl cyclase ↓<br>$K^+$ channel ↑ |
| $G_{i3}$ | Others > neural | Adenylyl cyclase ↓<br>$K^+$ channel ↑ |
| $G_o$ | Neural, endocrine | $Ca^{2+}$ channel ↓ |
| $G_z$ | Neural, platelets | ? |
| $G_q$ | Ubiquitous | PLCβ ↑ |
| $G_{11}$ | Ubiquitous | PLCβ ↑ |
| $G_{14}$ | Liver, lung, kidney | PLCβ ↑ |
| $G_{15/16}$ | Blood cells | PLCβ ↑ |
| $G_{12}$ | Ubiquitous | ? |
| $G_{13}$ | Ubiquitous | ? |

teins associated with normal cell signal transduction (Hunter, 1987). This discovery has driven a major advance in our understanding of cellular signal tranduction for hormones and other molecules. Two types of hormone-regulated tyrosine kinase pathways are used by vertebrate cells. In some pathways hormones bind to receptors that have intrinsic tyrosine kinase activity in their intracellular domains, allowing direct activation of tyrosine phosphorylation. Examples of receptors that mediate this type of signaling include those for insulin, insulin-like growth factor I (IGF I), epidermal growth factor (EGF), platelet-derived growth factor (PDGF), fibroblast growth factor (FGF), and others. Other types of receptors, including those for growth hormone (GH), prolactin, erythropoietin, and a variety of hematopoietic cytokines have tyrosine kinases associated with them non-covalently, and the kinase is activated indirectly following ligand binding. Tyrosine phosphorylation leads to association of SH2 domain-containing proteins in the receptor signaling complex (Fig. 9). Tyrosine phosphorylation accounts for only 1% of the protein-bound phosphate in a typical cell, and serine/threonine phosphorylation accounts for the remaining 99%. Phosphorylation on serine and threonine residues is used for many purposes that are not associated with signal transduction, such as creating a highly charged protein that can chelate cations. In contrast, all known tyrosine kinase reactions are directly associated with signal transduction events.

Tyrosine kinases operate in part by stimulating a cascade of serine/threonine kinase reactions that affect numerous

intracellular targets. The prototypic kinase cascade is initiated when tyrosine phosphorylation of a receptor recruits growth factor receptor-binding protein-2 (Grb-2) and a guanine nucleotide exchange factor (GEF) to the receptor signaling complex. GEFs (of which there are several types) convert inactive GDP-bound Ras to active GTP-Ras by catalyzing the exchange of GDP for GTP. Ras-GTP is anchored to the membrane by a C-terminal lipid moiety and binds the serine/threonine kinase Raf-1. Raf-1 is the first of a series of serine/threonine kinases that phosphorylate, in stepwise fashion, other kinases, culminating in phosphorylation of a mitogen-activated protein kinase (MAP kinase) (Campbell et al., 1995). MAP kinases are unusual bifunctional kinases in that they phosphorylate proteins simultaneously both on serine/threonine residues and on tyrosines. Each of the members of this kinase-activating cascade is represented by multiple isoforms that have different substrate specificities and intracellular localizations. A major goal in signal-transduction research is to resolve the mechanisms that allow selectivity amid an array of closely related protein kinases and kinase substrates.

## 1. Hormone Receptors That Have Intrinsic Tyrosine Kinase Activity

The insulin and epidermal growth factor receptors are archetypes for receptors that have intrisic tyrosine kinase activity. These receptors share many features of their signal transduction pathways. The insulin receptor (INS-R) is a heterotetrameric protein that contains two identical α-subunits that are extracellular and bind directly to insu-

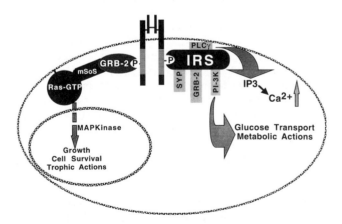

**FIG. 9.**  Insulin-regulated signaling pathways. The insulin receptor is a tetramer consisting of extracellular α-subunits and transmembrane β-subunits that are linked by disulfide bridges. The intracellular domains include a tyrosine kinase region (cross-hatched) and several phosphorylatable tyrosines. Phosphorylated tyrosine residues serve as docking sites for IRS proteins and GRB-2. Each of these is phosphorylated on tyrosine residues also, and recruits SH2-domain-containing effector proteins. The known effector proteins include the guanine nucleotide exchange factor mSoS, the SH2 phosphatase SYP, phosphatidylinositol-3′ kinase (PI-3K), and phospholipase Cγ(PLCγ).

lin, and two identical β-subunits that traverse the plasma membrane. The β-subunits are covalently linked to the α subunit by disulfide bridges (see Fig. 9). The intracellular portion of the INS-R β-subunit has a tyrosine kinase domain and several phosphorylatable tyrosine residues. Binding of insulin to the extracellular domain causes a conformational change that results in autophosphorylation of several tyrosine residues in the β-subunit. The main SH2-domain protein that docks to the INS-R is called insulin receptor substrate, which is represented by two isoforms (IRS-1 and IRS-2).

The function of IRS proteins is to serve as a scaffold on which the INS-R signaling complex is constructed of numerous effector proteins. IRS proteins are large (MW 185,000) proteins that contain several potentially phosphorylatable tyrosines, at least eight of which are phosphorylated by the INS-R (Myers and White, 1996). Association of IRS proteins with the INS-R is mediated by a particular phosphotyrosine residue on the INS-R, but the IRS proteins do not have conventional SH2 domains, so the binding to INS-R is by some alternative, as yet unclear, phosphotyrosine-recognition motif. IRS proteins recruit several SH2 domain proteins that have specific effector activities (Myers and White, 1996). These include the regulatory p85 subunit of phosphotidylinositol-3' kinase (PI-3 kinase), SYP (src-homology phosphotyrosine phosphatase), c src (and related src-family kinases), and Grb2. Grb2 serves as a docking protein for mSoS (mammalian homologue of drosophila *son-of-sevenless*). The mSoS protein represents one of the types of GEF, which, in this case, is employed by the INS-R for activating Ras and a kinase cascade that culminates in MAP kinase activation.

The best-studied effector pathway for receptor tyrosine kinases has been the MAP kinase cascade. There are two main branches of the MAP kinase family: the ERKs (extracellular signal regulated kinases), which activate a ribosomal protein kinase, S6 kinase, and at least two nuclear transcription factors, Elk and Ets; and, the JNKs (Jun N-terminus kinases), which phosphorylate the nuclear transcription factor c-Jun, and related proteins. These substrates are believed to represent the "business end" of the signaling pathway, at which specific changes in the activity of the translation and transcription machinery are imposed. Another important effector pathway, which has been less well studied, is the PI-3 kinase pathway. PI-3 kinase is capable of generating, in collaboration with phospholipases, unique second messengers that mediate some of the rapid metabolic effects of insulin, such as accelerated glucose uptake (Saltiel, 1996). Phospholipid products activated by PI-3 kinase bind to and stimulate the activity of protein kinase B (PKB), which is related to cAMP-dependent protein kinase and protein kinase C. PKB promotes cell survival by preventing the activation of the apoptotic pathway (Frank *et al.*, 1997).

Epidermal growth factor receptor (EGF-R) also contains intrinsic tyrosine kinase activity in its intracellular domain. But unlike INS-R, the EGF-R is a single polypeptide chain that includes both an extracellular ligand-binding domain and an intracellular kinase domain. Binding of EGF to its receptors causes receptor dimerization, bringing the kinase domains of the receptor pair into close proximity where they transphosphorylate tyrosines in the intracellular domain. SH2-domain proteins associate with the receptor through the phosphotyrosine residues (Fig. 10). Some of the proteins that associate with the EGF-R are identical to those that associate with the INS-R, whereas others appear to be unique. For instance, whereas both EGF-R and INS-R recruit Grb-2 and PI-3 kinase, the IRS proteins that are important for insulin signaling do not appear to be involved in EGF signaling.

Receptor tyrosine kinases exert some of their most important actions through activating phospholipid turnover and generating low-molecular-weight second messengers. The SH2 domain protein phospholipase Cγ(PLCγ) is activated as a consequence of recruitment to the signaling complexes and tyrosine phosphorylation in its regulatory domain. PLCγ catalyzes the release of $IP_3$, which generates calcium release from intracellular stores. In this regard the receptor–tyrosine kinases converge with the G-protein coupled receptors, which generate $IP_3$ release by activating the PLCβ isoform. $IP_3$-stimulated calcium release may activate numerous effector proteins, such as PKC, calmodulin, CaM kinase II, CREB, and others. However, it is unclear which, if any, of these known calcium-sensitive proteins participate in physiological effects of the receptor-tyrosine kinases.

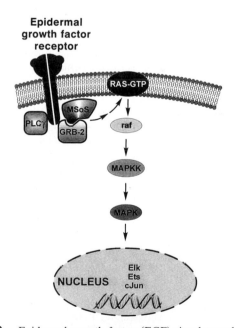

**FIG. 10.** Epidermal growth factor (EGF) signal transduction to the nucleus. EGF binding to its receptor causes dimerization and activates the tyrosine kinase function of the intracellular domain. Binding of GRB-2 to the receptor results in activation of membrane-bound ras through the action of mSoS. Ras recruits raf-family ser/thr kinases, which phosphorylate one of several MAP kinase kinases (MAPKK), leading to phosphorylation and activation of one or more MAP kinases (ERK or JNK). MAP kinases phosphorylate several nuclear transcription factors, as well as cytoplasmic target proteins that are also important for gene expression.

## 2. The Jak–Stat Pathway and Cytokine Receptor Signaling

Growth hormone, prolactin, erythropoietin, interferons, and a variety of interleukins and hematopoietic cytokines bind to receptors that lack intrinsic tyrosine kinase activity and bind to a unique family of nonreceptor kinases called the Janus kinases (Jaks). The Jak proteins were named with the Roman gatekeeper god "Janus" in mind, but the discovery of new tyrosine kinases had become routine enough that many have interpreted Jak to mean "just another kinase." However, although the receptor tyrosine kinases and src-family kinases are closely related in structure and function, the Jaks have distinctive structures and unique functions in cytokine-receptor signal transduction. The primary substrates for Jak kinases are a family of transcription factors named "Stats" (signal transducers and activators of transcription). The currently known members of the Stat family consist of six paralogous proteins in vertebrates, and one insect Stat (Darnell, 1996). Tyrosine phosphorylation of Stats promotes their dimerization and translocation to the nucleus, where they bind to specific DNA sequences on cytokine-responsive genes.

The cytokine receptor superfamily includes two main branches that are distantly related. The receptors for GH and interleukin-6 (IL-6) are representatives of one family, which is large and diverse, and is referred to as the Type I cytokine receptor family; interferon (IFN) receptors make up the other family, which is less diverse. Similar to the INS-R, IFN receptors are made up of covalently coupled subunits that contain extracellular and intracellular portions. Many of the details of Jak–Stat signaling have been initially discovered for interferon signaling, and later found to be widely applicable to other hormones (Darnell, 1996).

Type I cytokine receptors are single transmembrane polypeptides that are characterized by a distinctive signature in the extracellular domain consisting of two pairs of cysteines and a tryptophan–serine–X–tryptophan–serine (WSXWS) motif. The ligands for the Type I cytokine receptors are collectively referred to as "helix-bundle peptide hormones" and they share a similar 3-D structure that consists of four $\alpha$-helices arranged as two antiparallel pairs. The basic features of signaling in the cytokine receptor superfamily are the following: Binding of the hormone causes activation of a Jak-family kinase, which tyrosine-phosphorylates a latent Stat transcription factor. Signaling via Type I cytokine receptors is initiated by the formation of either homodimers (in the case of GH) or heterodimers (in the case of IL-6) (Horseman and Yu-Lee, 1994).

To initiate GH-R signaling a single GH molecule draws together two GH-R molecules. The two binding sites on GH, referred to as sites 1 and 2, have completely different sequences, and the affinity of site 1 is much higher than that of site 2. Although the sequences of the ligand at site 1 and site 2 are different, there is a single binding surface on the GH-R so that the receptors are drawn together in a mirror image conformation (Fig. 11). Homodimerization of the GH-R brings the intracellular domains in close proximity, and receptor-bound Jak2 phosphorylates sites on

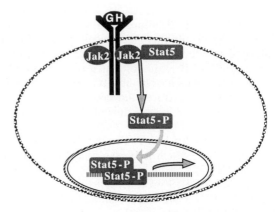

**FIG. 11.** Growth hormone (GH) signal transduction through the Jak–Stat pathway. GH binding to its receptor draws together a receptor pair into a homodimer with which the tyrosine kinase Jak2 is associated. Jak2 phosphorylates tyrosine residues on itself, the GH receptor, and on Stat proteins, represented here by Stat5. Stat5 phosphorylation results in nuclear translocation and dimerization, followed by binding to sites on DNA that accelerate the transcription of GH-inducible genes.

both the receptor and on Jak2 itself. This creates docking sites for SH2 domain proteins. Stat proteins bind to the receptor via the SH2 domains in their N-terminus, and they are phosphorylated by Jak2 on a tyrosine in their C-terminus. Stat5 is the primary GH-activated Stat protein, although Stat1 and Stat3 are also activated. Activation of Jak2 by GH creates docking sites for SH2 domain proteins other than Stats. As with the insulin or EGF receptor, this results in the recruitment of Grb-2, Shc, mSos, and the activation of a MAP kinase cascade. GH-stimulated Stat proteins and MAP kinase substrates participate in transcriptional activation of genes that are involved in GH actions, including insulin-like growth factor 1, c-fos, and the protease inhibitor Spi 2.1 (Carter-Su et al., 1996). Like GH, signal transduction for prolactin and erythropoietin is initiated by homodimerization of their receptors and activation of Jak2.

IL-6 is representative of many cytokines that bind to Type I cytokine receptors that heterodimerize after ligand binding (Hibi et al., 1996) (Fig. 12). The $\alpha$-subunit of the IL-6 receptor is the primary ligand-binding moiety, but it

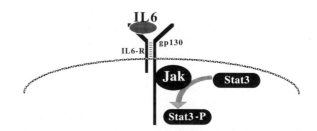

**FIG. 12.** Il6 receptor (IL6-R) heterodimerization with the signaling subunit gp130 activates Jak proteins and Stat phosphorylation.

has only a short intracellular domain and does not have any capability to transduce a signal to the cell's interior. The β-subunit, which is also known as gp130, has no significant affinity for IL-6, but it associates with the liganded α-subunit, and the long intracellular domain of gp130 binds and activates both Jak1 and Jak2 (Ihle *et al.*, 1995; Hibi *et al.*, 1996). The gp130 signaling subunit is shared by several other cytokine receptors, including those for leukemia inhibitory factor (LIF) and oncostatin M (OSM). Stat3 is the primary substrate for the gp130–Jak complex. Activation of the IL-6 receptor in the liver results in the transcription of several acute-phase inflammatory mediators that have Stat3 binding sites in their promoters.

Cytokine receptors induce both terminally differentiated cell functions and cell proliferation during hematopoiesis and organ development. The Stat proteins are connected to the induction of differentiated cell functions, but Jak activation is necessary for both proliferation and differentiation in response to cytokines. Bifurcation of the cytokine signaling pathways downstream of Jak activation leads both to generalized signaling pathways that regulate trophic functions such as growth and cell survival (i.e., MAP kinase, PKC, PKB), and to the differentiation-associated activation of Stat proteins (Horseman and Yu-Lee, 1994; Ihle *et al.*, 1995) (Fig. 13). It is not yet clear how the relative activity in each of these paths is established to allow cells to proliferate or terminally differentiate in response to a cytokine signal.

### C. Signaling via Hormone Receptors That Are Serine/Threonine Kinases

Transforming growth factor β (TGFβ), activin, and inhibin are all members of a family of hormones called "cystine knot proteins," which are vital developmental regula-

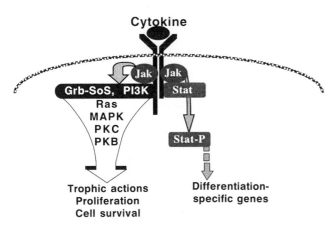

**FIG. 13.** Cytokine receptors activate multiple signaling pathways. Binding to cytokine receptors activates Jak kinases, which in turn phosphorylate both Stat proteins and Grb-2, which recruit multiple SH2-domain proteins. Stat signaling is primarily associated with tissue-specific expression of differentiation-specific genes. Multiple effector molecules (Ras, MAP kinase, PKC, PKB, others) are synthesized in response to cytokines and may contribute to generalized trophic actions, cell proliferation, and cell survival.

tors. The receptors for this family of polypeptide hormones consist of an extracellular ligand binding domain and an intracellular ser/thr kinase domain (Massagué, 1992; ten Dijke *et al.*, 1996). The kinase activity of the receptors is essential for signal transduction by these receptors. TGFβ causes its receptor to aggregate into a tetrameric complex that autophosphorylates sites on several of the receptor-signaling complex proteins. However, unlike tyrosine phosphorylation, which is known to create docking sites for SH2 domain proteins, it is unclear how ser/thr autophosphorylation of the TGFβ complex initiates a signaling process. The cytoplasmic and nuclear targets of this family of receptors are poorly understood at this time.

### D. Signaling via Protein Tyrosine Phosphatases

The pervasive involvement of protein phosphorylation in signal transduction is complemented by a variety of phosphatase mechanisms that are both positive and negative regulators of cell functions. Protein tyrosine phosphatases (PTPases), in particular, appear to include members that are highly specific signal-transducing molecules. PTPases generally have highly specific "addressing domains" that target them to cell compartments, or to associations with other proteins. These include membrane anchors, cytoskeletal association motifs, SH2 domains, transmembrane helices, and fibronectin-like domains (Dixon, 1996; Tonks and Neel, 1996). Two types of PTPases are positive signal transducers in mammalian cells. These are the SH2 domain phosphatases and the receptor-like PTPases.

The SH2 domain phosphatases, for example, Syp in mammals and the *corkscrew* gene product in drosophila, are recruited to phosphotyrosine residues in receptor signaling complexes. These phosphatases are essential for the activation of some, but not all, downstream events following receptor activation. Two mechanisms may account for the involvement of SH2 domain phosphatases in signal transduction. First, the SH2 phosphatase can recruit unique proteins to the complex, which, when dephosphorylated, are active effector proteins. Second, SH2 phosphatases can dephosphorylate inhibitory isoforms of signaling proteins, thereby increasing the effectiveness of positive effector proteins.

The receptor-like PTPases have a single transmembrane domain with the catalytic region oriented to the inside of the cell, and a presumptive ligand-binding domain region outward. The CD45 protein is a T-cell surface antigen that encodes a receptor-like PTPase for which the ligand is unknown. Some of the receptor-like PTPases have extracellular domains that are similar to known cell-adhesion molecules. These contain fibronectin type III repeats or immunoglobulin-like repeats that are known to associate with extracellular matrix factors (Dixon, 1996). Presumably, the activity of these PTPases is regulated by their binding to matrix factors and through direct cell-to-cell interactions. These proteins may, therefore, play important roles in differentiation, and density-dependent regulation of cell growth.

## IV. Integration of Signal Transduction Pathways in Health and Disease

Regulation of cell functions by hormones is necessary for both homeostasis and vectoral differentiation. The actions of endocrine hormones drive changes in cell functions in collaboration with a host of locally secreted hormones and cell matrix factors, providing for the execution of appropriate responses to constantly changing environmental conditions. The widespread use of common signaling reactions, such as phosphorylation–dephosphorylation, and the sharing of signaling proteins, as with the nuclear retinoid X-receptor and various kinases, ensure that cellular responses are highly coordinated and efficient. The responses of cells are determined by the combination of signaling molecules that are activated at a given time, and by the kinetics of their activation. In the case of MAP kinases, transient activation appears to favor cell differentiation, whereas sustained activation favors proliferation (Marshall, 1995).

Human disease states occur when signal pathways are inappropriately activated or disrupted by mutations or infectious agents. A multitude of human diseases exist that are associated with defects in hormone signaling. Diseases arise at every level of signaling, beginning with defects in the control of hormone production, synthesis, secretion, and transport, as well as in the degradation of the hormonal signal in target tissues. Defects occur at various levels within the signaling pathways from receptor proteins themselves through downstream effectors (Latchman, 1996).

Cholera toxin and pertussis (whooping cough) toxin cause disease by catalyzing ADP-ribosylation of $G_{\alpha s}$ and $G_{\alpha i}$, respectively. In the case of cholera toxin, the inappropriate cAMP synthesis in the gut leads to dysregulated water secretion, uncontrollable diarrhea, and death from dehydration. Many virus-encoded oncogenes, such as v-*src* and v-*ras*, are constitutively active forms of normal cellular signaling proteins. Each of the nuclear hormone receptor genes is represented by at least one genetic disease caused by mutation of the receptor. For example, testicular feminization results from mutations that inactivate the androgen receptor, and vitamin D-resistant rickets results from mutations in the DNA-binding region of the VDR (Hughes *et al.*, 1991). Elucidation of the details of signaling pathways will continue to lead to advances in diagnosis, prevention, and therapy for numerous diseases.

It is important to recognize that the phrase "signal transduction" is merely a metaphor for the processes that are initiated by hormones interacting with cells. There are many ways in which this metaphor is inadequate. For example, most everyday signals, such as a red light or a green light, can be interpreted unambiguously. In contrast, the meanings of hormone signals are always very ambiguous except in the context of a myriad of prior events and coregulatory factors. Cells continuously integrate multiple signals by coupling each of the signaling components to multiple receptors. Remarkably, these integrated, combinatorial regulatory systems produce all of the organism's normal physiological responses and disease states, through a highly conserved repertoire of intracellular signaling molecules.

## V. Summary

Hormones are regulatory molecules that are secreted from either circumscribed endocrine glands or cells of nonendocrine organs. Hormones activate signal-transduction protocols that are highly conserved throughout eukaryotic organisms. Lipophilic hormones, such as steroids, thyroxine, retinoids, and vitamin D, enter cells passively and bind to intracellular receptors. These receptors are members of a large family of transcription factors that includes many members that bind known hormones, and others, collectively referred to as orphan receptors, that may bind unknown ligands, or function independent of ligand binding. The nuclear receptors are characterized by a conserved DNA-binding domain flanked by gene activation and ligand-binding domains. Upon interaction with the cognate hormone, or a hormone analogue, the nuclear receptor undergoes significant conformational changes that convert it to an active transcription factor. Hydrophilic protein and amine hormones interact with receptors that are integral membrane proteins. Ligand binding to membrane receptors induces both conformational changes and oligomerization of collaborating signal-transduction proteins. A large family of membrane-bound receptors share a structure that includes seven membrane-spanning $\alpha$-helices (heptahelical receptors). These receptors interact with heterotrimeric GTP-binding proteins, which elicit changes in numerous effector pathways, such as cyclic nucleotide synthesis, phospholipid mobilization, and $Ca^{2+}$ influx. Other membrane receptors activate tyrosine phosphorylation of specialized signaling proteins. The protein–tyrosine kinase enzymatic activity may be either an intrinsic property of the intracellular domain of the hormone receptor, or a property of a receptor-associated protein. Tyrosine phosphorylation creates docking sites for effector proteins that include several families of kinases and latent transcription factors. Physiological integration of signal transduction is essential to normal physiology, and defects in signaling are a common feature of disease states.

## References

Beato, M. (1989). Gene regulation by steroid hormones. *Cell* **56**, 335–344.

Beato, M., Herrliche, P., and Schutz, G. (1995). Steroid hormone receptors: Many actors in search of a plot. *Cell* **83**, 851–857.

Berg, J. M. (1989). DNA binding specificity and steroid receptors. *Cell* **57**, 1065–1068.

Bourguet, W., Ruff, D., Chambon, P., Gronemeyer, H., and Moras, D. (1995). Crystal structure of the ligand binding domain of the human nuclear receptor RXRα. *Nature* **375**, 377–382.

Campbell, J. S., Seger, R., Graves, J. D., Graves, L. M., Jensen, A. M., and Krebs, E. G. (1995). The MAP kinase cascade. *Recent Prog. Horm. Res.* **50**, 131–159.

Carter-Su, C., Schwartz, J., and Smit, L. S. (1996). Molecular mechanisms of growth hormone action. *Annu. Rev. Physiol.* **58**, 187–207.

Danielian, P. S., White, R., Lees, J. A., and Parker, M. G. (1992). Identification of a conserved region required for hormone dependent transcriptional activation by steroid hormone receptors. *EMBO J.* **11**, 1025–1033.

Darnell, J. E., Jr. (1996). The JAK–STAT pathway: Summary of initial studies an recent advances. *Recent Prog. Horm. Res.* **51**, 391–404.

Dixon, J. E. (1996). Protein tyrosine phosphatases: Their roles in signal transduction. *Recent Prog. Horm. Res.* **51**, 405–415.

Evans, R. M. (1988). The steroid and thyroid hormone receptor superfamily. *Science* **240**, 889–895.

Forman, B. M., Tontonoz, P., Chen, J., Brun, R. P., Spiegelman, B. M., and Evans, R. M. (1995). 15-Deoxy-delta$^{12,14}$-prostaglandin J$_2$ is a ligand for the adipocyte determination factor PPARγ. *Cell* **83**, 803–812.

Frank, T. F., Kaplan, D. R., and Cantley, L. C. (1997). PI3K: Downstream AKTion blocks apoptosis. *Cell* **88**, 435–437.

Glass, C. (1994). Differential recognition of target genes by nuclear receptor monomers, dimers, and heterodimers. *Endocr. Rev.* **15**, 391–407.

Goss, V. L., Hocevar, B. A., Thomson, L. J., Stratton, C. A., Burns, D. J., and Fields, A. P. (1994). Identification of nuclear beta II protein kinase C as a mitotic lamin kinase. *J. Biol. Chem.* **269**, 19074–19080.

Gronemeyer, H. (1993). Transcriptional activation by nuclear receptors. *J. Recept. Res.* **13**, 667–691.

Habener, J. F. (1990). Cyclic AMP response element binding proteins: A cornucopia of transcription factors. *Mol. Endocrinol.* **4**, 1087–1094.

Hard, T., Kellenbach, E., Boelens, R., Maler, B. A., Dahlman, K., Freedman, L. P., Carlstedt-Duke, J., Yamamoto K. R., Gustafsson, J. A., and Kaptein, R. (1990). Solution structure of the glucocorticoid receptor DNA binding domain. *Science* **249**, 157–160.

Herskowitz, I. (1995). MAP kinase pathways in yeast: For mating and more. *Cell* **80**, 187–197.

Hibi, M., Nakajima, K., and Hirano, T. (1996). IL-6 cytokine family and signal transduction: A model of the cytokine system. *J. Mol. Med.* **74**, 1–12.

Horlein, A. J., Naar, A. M., Heinsel, T., Torchia, J., Gloss, B., Kurokawa, R., Ryan, A., Kamei, Y., Soderstrom, M., Glass, C. K., and Rosenfeld, M. G. (1995). Ligand-independent repression by the thyroid hormone receptor mediated by a nuclear receptor co-repressor. *Nature* **377**, 397–404.

Horseman, N. D., and Yu-Lee, L.-Y. (1994). Transcriptional regulation by the helix bundle peptides: Growth hormone, prolactin and hematopoietic cytokines. *Endocr. Rev.* **15**, 627–649.

Hughes, M. R., Malloy, P. J., O'Malley, B. W., Pike, J. W., and Feldman, D. (1991). Genetic defects of the 1,25-dihydroxyvitamin D receptor. *J. Recept. Res.* **11**, 699–716.

Hunter, T. (1987). A thousand and one protein kinases. *Cell* **50**, 823–829.

Ihle, J. N., Witthun, B. A., Quelle, F. W., Yamamoto, K., and Silvennoinen, O. (1995). Signaling through the hematopoietic cytokine receptors. *Ann. Rev. Immunol.* **13**, 369–398.

Katzenellenbogen, J. A., O'Malley, B. W., and Katzenellenbogen, B. S. (1996). Tripartite steroid hormone receptor pharmacology: Interaction with multiple effector sites as a basis for the cell- and promoter-specific action of these hormones. *Mol. Endocrinol.* **10**, 119–131.

Kleiwer, S. A., Umesono, K., Mangelsdorf, D. J., and Evans, R. M. (1992). Retinoid X receptor interacts with nuclear receptors in retinoic acid, thyroid, and vitamin D signaling. *Nature* **355**, 446–449.

Kwok, R. P. S., Lundblad, J. R., Chrivia, J. D., Richards, J. P., Bachinger, H. P., Brennan, R. G., Roberts, S. G. E., Green, M. R., and Goodman, R. H. (1994). Nuclear protein CBP is a coactivator for the transcription factor CREB. *Nature* **370**, 223–226.

Latchman, D. S. (1996). Transcription factor mutations and disease. *N. Eng. J. Med.* **334**, 28–33.

Lee, M. S., Kliewer, S. A., Provencal, J., Wright, P. E., and Evans, R. M. (1993). Structure of the retinoid X receptor a DNA binding domain: α helix required for homodimeric DNA binding. *Science* **260**, 1117–1121.

Luisi, B. F., Xu, W., Otwinowski, Z., Freedman, L. P., Yamamoto, K. R., and Sigler, P. B. (1991). Crystallographic analysis of the interaction of the glucocorticoid receptor with DNA. *Nature* **352**, 497–505.

Luttrell, L. M., van Biesen, T., Hawes, B. E., Koch, W. J., Touhara, K., and Lefkowitz, R. J. (1995). G$_{βγ}$ subunits mediate mitogen-activated protein kinase activation by the tyrosine kinase insulin-like growth factor 1 receptor. *J. Biol. Chem.* **270**, 16495–16498.

Mader, S., Kumar, V., deVereneuil, H., and Chambon, P. (1989). Three amino acids of the oestrogen receptor are essential to its ability to distinguish an oestrogen from a glucocorticoid responsive receptor. *Nature* **338**, 271–274.

Mangelsdorf, D. J., and Evans, R. M. (1995). The RXR heterodimer and orphan receptors. *Cell* **83**, 841–850.

Mangelsdorf, D. J., Thummel, C., Beato, M., Herrliche, P., Schutz, G., Umesono, K., Blumberg, B., Kastner, P., Mark, M., Chambon, P., and Evans, R. M. (1995). The nuclear receptor superfamily: The second decade. *Cell* **83**, 835–839.

Marshall, C. J. (1995). Specificity of receptor tyrosine kinase signaling: Transient versus sustained extracellular signal-regulated kinase activation. *Cell* **80**, 179–185.

Massagué, J. (1992). Receptors for the TGFβ family. *Cell* **69**, 1067–1070.

Meyer, T. E., and Habener, J. F. (1993). Cyclic adenosine 3′,5′-monophosphate response element binding protein (CREB) and related transcription-activating deoxyribonucleic acid-binding proteins. *Endocr. Rev.* **14**, 269–290.

Myers, M. G., Jr., and White, M. F. (1996). Insulin signal transduction and the IRS proteins. *Annu. Rev. Pharmacol. Toxicol.* **36**, 615–658.

Ogryzko, V. V., Schiltz, R. L., Russanova, V., Howard, B. H., and Nakatani, Y. (1996). The transcriptional coactivators p300 and CBP are histone acetyltransferases. *Cell* **87**, 953–959.

O'Malley, B. W. (1990). The steroid receptor superfamily: More excitement predicted for the future. *Mol. Endocrinol.* **4**, 363–344.

Pawson, T. (1995a). Protein modules and signalling networks. *Nature* **373**, 573–580.

Pawson, T. (1995b). SH2 and SH3 domains in signal transduction. *Adv. Cancer Res.* **64**, 87–110.

Perlmann, T., Rangarajan, P. N., Umesono, K., and Evans, R. M. (1993). Determinants for selective RAR and TR recognition of direct repeat HREs. *Genes Dev.* **7**, 1411–1422.

Pestell, R. G., and Jameson, J. L. (1995). Transcriptional regulation of endocrine genes by second-messenger signaling pathways. *In* "Molecular Endocrinology" (B. D. Weintraub, Ed.), pp. 59–76. Raven Press, New York.

Pratt, W., Jolly, D. J., Pratt, D. V., Hollenberg, S. M., Giguere, V., Cadepon, F. M., Schweizer-Groyer, G., Cartelli, M. G., Evans, R. M., and Baulieu, E. E. (1988). A region in the steroid-binding domain determines formation of the non-DNA binding glucocorticoid receptor complex. *J. Biol. Chem.* **263**, 267–273.

Rastinejad, F., Perlmann, T., Evans, R. M., and Sigler, P. B. (1995). Structural determinants of nuclear receptor assembly on DNA direct repeats. *Nature* **375**, 203–211.

Renaud, J.-P., Natacha, R., Ruff, M., Vivat, V., Chambon, P., Gronemyer, H., and Moras, D. (1995) Crystal structure of the RARγ ligand binding domain bound to all-*trans* retinoic acid. *Nature* **378**, 681–689.

Saltiel, A. R. (1996). Diverse signaling pathways in the cellular actions of insulin. *Am. J. Physiol.* **270**, E375–E385.

Schwabe, J. W. R., Neuhaus, D., and Rhodes, D. (1990). Solution structure of the DNA binding domain of the oestrogen receptor. *Nature* **348**, 458–461.

Schwabe, J. W. R., Chapman, L., Finch, J. T., and Rhodes, D. (1993). The crystal structure of the oestrogen receptor DNA binding domain bound to DNA: How receptors discriminate between their response elements. *Cell* **75,** 567–578.

Spiegel, A. M., Shenker, A., Simonds, W. F., and Weinstein, L. S. (1995). G protein dysfunction in disease. *In* "Molecular Endocrinology" (B. D. Weintraub, Ed.), pp. 297–318. Raven Press, New York.

ten Dijke, P., Miyazono, K., and Helden, C. H. (1996). Signaling via hetero-oligomeric complexes of type I and type II serine/threonine kinase receptors. *Curr. Opin. Cell Biol.* **8,** 139–145.

Thummel, C. S. (1995). From embryogenesis to metamorphosis: The regulation and function of *Drosophila* nuclear receptor superfamily members. *Cell* **83,** 871–877.

Tonks, N. K., and Neel, B. G. (1996). From form to function: Signaling by protein tyrosine phosphatases. *Cell* **87,** 365–368.

Truss, M., and Beato, M. (1993). Steroid hormone receptors: Interaction with deoxyribonucleic acid and transcription factors. *Endocr. Rev.* **14,** 459–479.

Umesono, K., and Evans, R. M. (1989). Determinants of target gene specificity for steroid/thyroid hormone receptors. *Cell* **57,** 1139–1146.

Umesono, K., Murikami, K. K., Thompson, C. C., and Evans, R. M. (1991). Direct repeats as selective response elements for the thyroid hormone, retinoic acid, and vitamin $D_3$ receptors. *Cell* **65,** 1255–1266.

Wagner, R. L., Apriletti, J. W., McGrath, M. E., West, B. L., Baxter, J. D., and Fletterick, R. J. (1995). A structural role for hormone in the thyroid hormone receptor. *Nature* **378,** 690–697.

Yu, V., Delsert, C., Andersen, B., Holloway, J. M., Devary, O. V., Naar, A. M., Kim, S. Y., Boutin, J.-M., Glass, C. K., and Rosenfeld, M. G. (1991). RXR: A coregulator that enhances binding of retinoic acid, thyroid hormone, and vitamin D receptors to their cognate response elements. *Cell* **67,** 1251–1266.

Zhang, F. L., and Casey, P. J. (1996). Protein prenylation: Molecular mechanisms and functional consequences. *Annu. Rev. Biochem.* **65,** 241–269.

# SECTION

# II

# Membrane Potential, Transport Physiology, Pumps, and Exchangers

Nicholas Sperelakis

# 12

# Diffusion and Permeability

## I. Introduction

It is necessary to learn the basics of diffusion and permeability to understand the mechanisms for development of the resting potential and transport physiology. Therefore, this chapter provides the fundamental principles that are made use of in subsequent chapters in this book.

## II. Diffusion and Diffusion Coefficient

Any substance in solution (or in a gas) tends to move from regions of higher concentration to regions of lower concentration until the substance is uniformly distributed. (The diffusion equations also apply to the diffusion of heat in solids.) Once uniformly distributed (i.e., at equilibrium), the molecules of the substance continue to move about, but the net movement is zero. In the absence of convection currents, the movement of the molecules is by diffusion only. Diffusion occurs because of the random thermal motion of the molecules. If a region of high solute concentration is adjacent to one of low solute concentration, separated by an imaginary plane, it is probable that there will be more molecules per unit of time crossing the plane from the side of higher concentration to the side of lower concentration than there are crossing in the opposite direction. Thus, there is a flux or movement of molecules in both directions (unidirectional fluxes), but the net flux is from the side of higher concentration to the side of lower concentration.

Now if the imaginary plane were replaced with a thin membrane permeable to the molecules, then the same situation would apply; namely, the particles would diffuse from the side of higher concentration to the side of lower concentration across the membrane. Because in a living cell it is often assumed that the inside and outside solutions are relatively well stirred and because diffusion of most substances through the cell membrane is much slower than that through a free solution, our concern is about the diffu-

sion of substances through the cell membrane, that is, the rate-limiting step. We assume for simplicity that the solutions on either side are well stirred; that is, there are no concentration gradients within the bulk solution on either side of the membrane (although there probably are unstirred layers near the membrane). We confine ourselves here to the diffusion of small molecules or ions across membranes.

**Fick's diffusion equation** states that the rate of transfer of uncharged molecules across a membrane is equal to the concentration gradient ($dC/dx$) times the area of the membrane ($A_m$) times a **coefficient of diffusion ($D$)**, expressed as

$$\frac{dQ}{dt} = -DA_m\frac{dC}{dx} \tag{1}$$

where $Q$ is the amount of substance (in moles) transported, $t$ is the time, $C$ is the molecular concentration, and $x$ is the distance. Although concentration is used throughout this chapter for clarity, activity is actually meant (that is, the thermodynamically active concentration), which is the concentration times the activity coefficient. For a thin membrane, when the solution is well stirred on either side, the concentration gradient in the steady state is equal to the difference in the molecular concentration ($\Delta C$) of the solute on both sides of the membrane divided by the thickness of the membrane ($\Delta x$),

$$J' = -DA_m\frac{\Delta C}{\Delta x} \tag{2}$$

where $J' = dQ/dt$ and is the rate of movement (or flux) of the substance across the membrane in mol/sec. Now we can transpose $A_m$ from the right side of the equation to the left side, and letting $J'/A_m = J$, where $J$ is the flux in mol/sec·cm$^2$,

$$J = -D\frac{\Delta C}{\Delta x} \tag{3}$$

$$\frac{mol}{sec \cdot cm^2} = \frac{cm^2}{sec}\frac{mol/cm^3}{cm}$$

**TABLE I** Physicochemical Properties of Selected Ions

| Ion[a] | Limiting equivalent conductivity (cm²/S-eq) 25°C | 35°C | Activity coefficient | Crystallographic radius (Å) | Singly hydrated radius[b] (Å) | Hydrated volume[c] (Å³) | Heat of hydration (kcal g ion⁻¹) | No. of water molecules[c] | | Relative shielding[d] |
|---|---|---|---|---|---|---|---|---|---|---|
| Na⁺ | 50.10 | 61.5 | 0.778 | 0.96 | 3.67 | 150 | −115 | 5 | (4–6) | 0.432 |
| K⁺ | 73.50 | 88.2 | 0.770 | 1.33 | 4.05 | (150) | −90 | (5) | (3–6) | 0.224 |
| Cl⁻ | 76.35 | 92.2 | — | 1.81 | 3.92 | 90 | −59 | (3) | (0–6) | (0.073) |
| Mg²⁺ | 53.05 | | 0.528 | 0.65 | 3.60 | 360 | −501 | 12 | | 2.18 |
| Ca²⁺ | 59.50 | | 0.518 | 0.99 | 3.70 | 310 | −428 | 10 | | 0.812 |
| Sr²⁺ | 59.45 | | 0.515 | 1.13 | 3.85 | 310 | −381 | 10 | | 0.624 |
| Ba²⁺ | 63.63 | | 0.508 | 1.35 | 4.08 | 290 | −347 | 9–10 | | 0.415 |

*Note.* Values taken from R. A. Robinson and R. H. Stokes, *Electrolyte Solutions,* London, Butterworth & Co. (Publishers), Ltd., 1959, and from J. O'M. Bockris and B. E. Conway, *Modern Aspects of Electrochemistry,* London, Butterworth & Co. (Publishers), Ltd., 1954.

[a] Activity coefficient for each cation given for the Cl⁻ salt at 0.1 $M$ and at 25°C.

[b] Singly hydrated radius equals crystal radius + 2.72 Å (diameter of water molecule) for cations with crystal radii between 0.9 and 1.7 Å and crystal radius + 2.23 Å for anions. Values for singly hydrated radius taken from Mullins (1961).

[c] Values in parentheses are estimated values.

[d] Relative shielding was calculated from the number of water molecules per Å² of unhydrated surface area. The order of relative shielding, from least to most shielded, is Cl < K < Na ≅ Ba < Sr < Ca < Mg. This order is similar to that for the heats of hydration: Cl < K < Na ≪ Ba < Sr < Ca < Mg. Thus, Cl⁻ sheds its hydration the easiest and Mg²⁺ the least easiest. The cations that are least shielded and that can shed their hydration the most easily should penetrate the cell membrane most easily if pore diameter is the limiting factor. If so, $P_K$ should be relatively large, and Mg²⁺ should be least active electrophysiologically.

Thus, the net flux is equal to the concentration gradient times the diffusion coefficient. The diffusion coefficient is a constant for a given substance and membrane under a given set of conditions, and the coefficient is given in cm²/sec. The negative sign in Eq. 3 refers to the direction of the net flow, namely, down the concentration gradient.

The diffusion coefficient of substances in free solution is dependent on the molecular size (and shape, for large molecules). For ions, the smaller the unhydrated radius, the greater the charge density, which means that more water molecules are held in the hydration shells (larger hydrated radius); the larger water shell causes diffusion to be slower. The hydrated and unhydrated radii of some relevant ions are given in Table 1.

The time required for diffusion to become 50, 63, or 90% complete varies directly with the square of the distance, and inversely with the diffusion coefficient. Thus, diffusion is extremely fast over short distances (e.g., 10–1000 nm) but exceedingly slow over long distances (e.g., 1 cm). For a small particle like K⁺ or acetylcholine (ACh⁺), with a $D$ value of about $1 \times 10^{-5}$ cm²/sec, the time required for 90% equilibration to be reached for a distance of 1 μm is about 1 msec. Table 2 shows how the time required changes as a function of distance. The chef knows that the time required for a meat roast to cook in the middle is highly dependent on the thickness of the roast, and the physiologist knows that there is a critical thickness (e.g., 0.5–1.0 mm) of a muscle bundle in a bath that allows adequate diffusion of oxygen to the cells within the core of the strip, no matter how vigorously the bath is oxygenated.

The diffusion coefficient across cell membranes for various substances is generally greater when the molecular size of the substance is small and when the lipid solubility is high; that is, small molecules of high lipid solubility (i.e., less polar and nonpolar molecules) penetrate the fastest through the membrane. Most nonpolar molecules pass through the lipid bilayer matrix of the membrane, that is, not through special sites (e.g., water-filled pores or channels). Small charged ions (e.g., Na⁺, K⁺, Cl⁻, Ca²⁺) apparently pass through water-filled channels, some of which have a voltage-dependent gating mechanism; these channels can exhibit a high degree of selectivity for specific ions, the selectivity orders not being solely based on the hydrated or unhydrated sizes of the ions.

**TABLE 2** Calculations of time Required for Diffusion to Become 90% Complete

| Distance | | Time | |
|---|---|---|---|
| 100 | Å | 0.1 | μsec |
| 0.1 | μm | 0.01 | msec |
| 1 | μm | 1 | msec |
| 10 | μm | 100 | msec |
| 100 | μm | 10 | sec |
| 1 | mm | 16.7 | min |
| 1 | cm | 28 | hr |

*Note.* Diffusion time varies with square of diffusion distance. These values are for a substance having a diffusion coefficient ($D$) of about $1 \times 10^{-5}$ cm²/sec in free solution. The diffusion time is inversely proportional to $D$. Einstein's approximation equation is $\lambda = \sqrt{Dt}$, where $\lambda$ is the mean displacement of the diffusing particles along the $x$-axis at any time ($t$).

One type of ionophore (antibiotic) can act as a cage carrier for $K^+$, the $K^+$–ionophore complex being lipid soluble and presumably passing through the lipid bilayer matrix. Carrier molecules are normally present in the cell membrane for transport of substances such as glucose and amino acids. Some of these carriers are for downhill transport only, that is, down the electrochemical gradient. One type of carrier-mediated transport in cells (e.g., for glucose) is known as facilitated diffusion (in contrast with the so-called simple diffusion described in the preceding paragraphs); that is, the downhill movement of the substance across the membrane is enhanced by the presence of the carrier. Another type of carrier-mediated transport is known as exchange diffusion, in which one molecule of substance (or ion) inside the cell is exchanged for one molecule outside the cell; that is, there is no net movement.

One characteristic of carrier-mediated transport systems is that they exhibit saturation kinetics; that is, the rate of transport increases with an increase in the substrate concentration up to some maximum, after which transport levels off, presumably because of a finite number of available carriers. Other characteristics of carrier-mediated transport systems include competitive inhibition (two or more substances may compete for binding to the carrier molecule), noncompetitive inhibition (a nontransported substance may bind to the carrier and prevent the carrier from combining with its usual substrate), specificity (carrier is more or less specific for one physiological substance, different substances having different carriers), and a high temperature coefficient ($Q_{10}$).

## III. Permeability Coefficient

Returning to Eq. 3, we can make one final combination. Because the cell membrane is relatively fixed in thickness, we can combine $\Delta x$ and $D$ into a new coefficient:

$$P = D/\Delta x \tag{3a}$$

Substitution into Eq. 3 gives

$$J = -P \Delta C \tag{4}$$
$$= -P(C_o - C_i) \tag{4a}$$
$$\frac{mol}{sec \cdot cm^2} = \frac{cm}{sec} \frac{mol}{cm^3}$$

where $P$ is the permeability coefficient, $C_i$ is the internal concentration, and $C_o$ is the external concentration. The units of the permeability coefficient are cm/sec, that is, units of velocity.

Hence, the net flux of a nonelectrolyte across a membrane is equal to the permeability coefficient times the difference in concentration across the membrane. The permeability coefficient for $K^+$ ($P_K$) across resting striated muscle membrane is about $1 \times 10^{-6}$ cm/sec. Some $P$ values are given in Table 3 for some selected tissues. Note that although diffusion coefficients apply to diffusion in free solution or in membranes, permeability coefficients only apply to membranes. The higher the diffusion coefficient for movement of a substance across a membrane, the higher

the permeability coefficient. The unidirectional fluxes are given by

$$J_i = -PC_o \tag{5}$$
$$J_o = +PC_i \tag{6}$$

where $J_i$ is the influx of the substance and $J_o$ is the outflux (efflux). Thus,

$$J = J_i - J_o \tag{7}$$
$$= -PC_o - (+PC_i)$$
$$J = -P(C_o - C_i) \tag{7a}$$

Equations 5 and 6 show that the influx of a substance is equal to its $P$ value times the external concentration, whereas the efflux is equal to the $P$ value times the internal concentration. Conversely, the permeability is equal to the ratio of flux to concentration: influx to $C_o$ or efflux to $C_i$.

Equations 4–6 apply only to uncharged molecules. If there is a net charge on the molecule, then the unidirectional and net fluxes are also determined by any electrical field that may exist across the membrane. The equation for the net flux is in the general form

$$J = -Pf(E_m)[C_o - C_i e^{E_m F/RT}] \tag{8}$$

where $f(E_m)$ is some function of the potential difference across the membrane, and $F$, $R$, and $T$ are the Faraday constant, the gas constant, and absolute temperature, respectively. Note the similarity of Eq. 8 to Eq. 4, except for the membrane potential terms. When $f(E_m)$ is equal to $E_m F/RT$, this term is dimensionless ($EF$ is the electrical energy, whereas $RT$ is the thermal energy); $F/RT = (0.026$ V$)^{-1}$ at 37°C. Similarly, the term exp $E_m F/RT$ is dimensionless; therefore, the units are the same as those given in Eq. 4a.

The permeability coefficient (for a charged ion) can be better understood by considering the relationship

$$P = U\frac{\beta}{\delta}\frac{RT}{F} \tag{9}$$

$$\frac{cm}{sec} = \frac{cm/sec}{V/cm}\frac{1}{cm}\frac{V}{1}$$

where $U$ is the mobility of the ion through the membrane and has units of a velocity per unit driving (electrophoresing) voltage gradient (cm/sec per V/cm), $\beta$ is the partition coefficient for the ion between the bulk solution and the edge of the membrane and is dimensionless, $\delta$ is the membrane thickness (in cm), and $RT/F = 0.026$ V at 37°C. Equation 9 indicates that the permeability coefficient is a direct function of the mobility of the ion through the membrane; in other words, it is a velocity in a unit electrical field.

The discussion about Eqs. 5–7 tells us that the unidirectional fluxes of an ion should be proportional to the electrochemical potentials ($\bar{\mu}$) for that ion. That is, the influx should be a direct function of the electrochemical potential of the ion outside ($\bar{\mu}_o$), and the efflux should be a direct function of the electrochemical potential inside $\bar{\mu}_i$):

$$J_i \propto \bar{\mu}_o \tag{10}$$
$$J_o \propto \bar{\mu}_i \tag{11}$$

**TABLE 3**  Summary of Internal $K^+$ and $Na^+$ Concentrations and Activities for Some Selected Heart Tissues

| Preparation | Resting potential (mV) | $P_K$ (cm/sec $\times 10^{-7}$) | $P_{Na}$ (cm/sec $\times 10^{-7}$) | $P_{Na}/P_K$ | $[K]_i$ (mM) | $[Na]_i$ (mM) | Ref. |
|---|---|---|---|---|---|---|---|
| Chick embryonic, 19 days old | | 3.10 | 0.053 | 0.017 | 122 | 15 | Carmeliet et al. (1976) |
| Rabbit papillary | −77.5 | — | — | — | 135 / 82.6[a] | 32.7 / 5.7[a] | Lee and Fozzard (1975) |
| Rabbit papillary | −86 | — | — | — | 119[b] | — | Akiyama and Fozzard (1975) |
| Rabbit ventricular | −76 | — | — | — | 83.1[a] | — | Fozzard and Lee (1976) |
| Cow Purkinje | −75 | 1.66 | — | — | — | — | Carmeliet and Verdonck (1977) |
| Sheep Purkinje | — | — | — | — | 160 | — | Carmeliet and Bosteels (1969) |

*Note.* Unless otherwise specified, all ion concentrations are based on total tissue analyses, and all permeability coefficients are for tissues bathed in normal Ringer's solution.

[a] Measured with ion-selective microelectrodes.

[b] Calculated from the internal $K^+$ concentration and activity coefficient.

The net flux is from the side of greater electrochemical potential to the side of lesser electrochemical potential. The potential difference across the membrane ($E_m$) equals $\Psi_i - \Psi_o$, where $\Psi_i$ is the inside potential and $\Psi_o$ is the outside potential.

The electrochemical potential ($\bar{\mu}$) is a measure of the useful energy, and its units are in joules/mole (just as voltage is in joules/coulomb). Electrochemical potential is composed of a chemical part ($\mu_c$) and an electrical part ($\mu_e$), that is,

$$\bar{\mu} = \mu_c + \mu_e \tag{12}$$

The chemical part is equal to

$$\mu_c = \mu_c^0 + RT \ln a \tag{13}$$

where $\mu_c^0$ is the standard electrochemical potential at standard temperature and pressure, and $a$ is the activity (activity coefficient, $\gamma$, times the concentration). In animal cells, since there is neither a significant temperature difference nor a substantial pressure difference across the cell membrane, the maximal electrochemical potential for $Na^+$ inside is given by

$$\bar{\mu}_i^{Na} = \mu_{Na}^0 + RT \ln(\gamma_i [Na]_i) + zF\Psi_i \tag{14}$$

where $\Psi_i$ is the internal potential, and $zF\Psi_i$ is the electrical energy. The difference in electrochemical potential between inside and outside ($\Delta\bar{\mu} = \bar{\mu}_i - \bar{\mu}_o$) then is

$$\Delta\bar{\mu}^{Na} = RT \ln \frac{[Na]_i}{[Na]_o} + zF\Delta\Psi \tag{15}$$

where $\Delta\Psi = \Psi_i - \Psi_o$ and is the same as $E_m$. The activity coefficients cancel out, assuming $\gamma_i = \gamma_o$.

The electrical current ($i$, in amp) is equal to the flux ($J'$, in mol/sec) times $zF$ or

$$i = J'zF \tag{16}$$

or

$$I = JzF$$
$$\frac{amp}{cm^2} = \frac{mol}{sec \cdot cm^2} \frac{coul}{mol} \tag{17}$$

where $I$ is the current density in amp/cm$^2$, and $J$ is the flux density in mol/sec·cm$^2$.

## IV. Ussing Flux Ratio Equation

As pointed out in the preceding paragraphs, the absolute unidirectional fluxes are determined also by the permeability coefficient. Ussing (1949) developed the so-called flux ratio equation, in which the ratio of influx to efflux is used (permeability cancels out). Specifically, for the ratio of $Na^+$ fluxes (by simple electrodiffusion), the following applies:

$$\frac{J_i^{Na}}{J_o^{Na}} = \frac{[Na^+]_o}{[Na^+]_i} e^{-E_m F/RT} \tag{18}$$

Thus, for $Na^+$, the ratio of influx : efflux (passive) in a resting membrane (assuming a resting potential of −80 mV) is

$$\frac{J_i^{Na}}{J_o^{Na}} = \frac{150 \, mM}{15 \, mM} e^{+80 \, mV/26 \, mV} \tag{18a}$$
$$= 10 \times e^{3.08}$$
$$= 217$$

Thus, the passive influx of $Na^+$ should be 217 times greater than the passive efflux, because of the large electrochemical gradient directed inward.

The flux ratio for $K^+$ would be

$$\frac{J_i^K}{J_o^K} = \frac{4 \, mM}{150 \, mM} e^{+80 \, mV/26 \, mV} \tag{18b}$$
$$= \frac{1}{37.5} e^{3.07}$$
$$= 0.574$$

Thus, the passive influx should be 0.57 times the passive efflux. The K⁺ equilibrium potential ($E_K$) is only slightly greater (more negative by about 14 mV) than the resting potential, and so the passive flux ratio should be close to one.

A modified form of Eq. 18 can be obtained by substituting the ratio of ions with $e^{E_i F/RT}$ (derived from the Nernst equation), giving

$$\frac{J_i^i}{J_o^i} = e^{-(E_m - E_i)F/RT} \tag{19}$$

where $E_i$ is the equilibrium potential for the cation in question, and $J_i^i$ and $J_o^i$ are the inward and outward fluxes, respectively, for ion $i$. The newer sign convention refers inside solution to outside solution. Thus, for Na⁺ we have

$$\frac{J_i^{Na}}{J_o^{Na}} = e^{-[-80\,mV - (+60\,mV)]/26\,mV} \tag{19a}$$

$$= e^{+5.38}$$

$$= 217$$

This value of 217 obtained for the flux ratio is identical to that obtained from Eq. 18. One advantage of Eq. 19 is that it is obvious at a glance that when $E_i = E_m$, the flux ratio is exactly 1.0, because $e^0 = 1$, where $e$ is the base of the natural logarithm. Thus, if Cl⁻ is passively distributed so that $E_{Cl} = E_m$, its flux ratio should be 1.0; that is, influx equals efflux, so there is no net flux. The larger the difference between $E_i$ and $E_m$, that is, the farther the ion is off equilibrium, the greater the flux ratio.

Sometimes Eqs. 18 and 19 do not fit the experimental facts. The data are better fitted if the exponential term contains another factor ($N$), which is an empirical factor:

$$\frac{J_i^i}{J_o^i} = e^{-(E_m - E_i)(F/RT)N} \tag{20}$$

The best fit of the data is when $N$ has a value of 2.5 to 4.0, depending on the membrane under investigation. The interpretation given to $N$ is that if the length of the water-filled pore that the ion must traverse across the membrane is much longer than the ion diameter, as is likely, then for so-called single-file diffusion, $N$ hits on the same side are required for the ion to complete its journey across the membrane. One could consider, for example, that there are three potential energy wells, or chain of reactive sites, along the length of the pore, and the only way for the ion to escape the well is to receive a kinetic bump from an adjacent ion in the file. Complete permeation of an ion through the pore is more likely to happen if the ion is moving in the same direction as the majority, that is, down the electrochemical gradient. Therefore, this factor makes the flux ratio much greater than would otherwise be predicted.

## V. Summary

This chapter described diffusion of uncharged particles and charged ions across membranes and provided the relevant equations that govern such diffusion. The relationship between diffusion coefficient ($D$) and permeability coefficient ($P$) was described. The units for both coefficients were given, and the correspondence of $P$ to mobility of the ion through the membrane under a unit voltage gradient was discussed. Electrochemical potential was defined, and the interconversion between flux and current was developed. Finally, the Ussing flux ratio equation was presented, and examples of its meaning were given. Relating to this, the concept of potential energy wells and barriers was presented, describing the movement of an ion through an ion channel.

# Appendix

The diffusion coefficients ($D$) of different gases are inversely proportional to the square roots of their masses ($M$):

$$DM^{1/2} = \text{constant} \tag{A-1}$$

The molecules have random thermal motion (i.e., Brownian movement). In gases, the molecules are subject to only small intermolecular forces because the molecules are relatively far apart. The kinetic energy of the gas molecules is given by $\frac{1}{2}NM\,\theta^2$, where $N$ is the number of moles, $M$ is their molecular weight, and $\theta$ is their mean velocity. The average velocities of the molecules are inversely proportional to the square root of their masses. Because the velocity of the molecules determines the speed of diffusion, the diffusion coefficient is inversely proportional to the square root of the molecular weight:

$$D \propto \frac{1}{M^{1/2}} \tag{A-1a}$$

In liquids, however, the molecules are close together and within each others' spheres of mutual attraction. Hence, for any molecule to move, it must break away from its surrounding molecules. Diffusion in liquids consequently occurs in a discontinuous manner; a molecule is able to move only when it has acquired sufficient energy by collision to break away from its neighbors and push aside other molecules. Thus, the molecule diffuses in a series of jumps, each jump requiring a certain critical activation energy. When the attractive forces between molecules are weak, diffusion is rapid. Because of the activation energy required to make each jump possible, the diffusion coefficient of a substance in liquid is inversely proportional to the square root of its molecular weight times the temperature coefficient ($Q_{10}$) raised to a power, according to Danielli (1964):

$$DM^{1/2}Q_{10}^{\frac{(T-10)}{10}} = \text{constant} \tag{A-2}$$

$$D \propto \frac{1}{M^{1/2}Q_{10}^{\frac{(T-10)}{10}}} \tag{A-2a}$$

This relationship, derived from kinetic theory, demonstrates that the $Q_{10}$ for diffusion of a substance in water depends on the activation energy necessary for a jump. If a large amount of energy is required (high $Q_{10}$), only a few molecules have the necessary energy to diffuse at any moment; raising the temperature increases the number of molecules with the required energy and thereby has an appreciable effect on the net rate of diffusion. If only a small amount of activation energy is required (low $Q_{10}$), a greater fraction of the molecules possesses this minimal energy at any moment, and so raising the temperature has less of an effect on the net rate of diffusion. Elevation of temperature increases the speed of the molecules and causes an increase in the rate of diffusion. When $Q_{10}$ is close to 1.0, then Eqs. A-2 and A-2a reduce to Eqs. A-1 and A-1a. The $Q_{10}$ for diffusion of $Na^+$ or $K^+$ in water is only about 1.22.

The value of the diffusion coefficient as a function of temperature $[D(T)]$ can be calculated from the following general equation, if the diffusion coefficient value at a given temperature (e.g., at 10°C) and the $Q_{10}$ value are known.

$$D_{T2} = D_{T1}Q_{10}^{\frac{(T_2-T_1)}{10}} \tag{A-3}$$

For example, if the reference temperature ($T_1$) is 10°C, then Eq. A-3 becomes

$$D_{T2} = D_{10°C}Q_{10}^{\frac{(T_2-10)}{10}} \tag{A-3a}$$

Table 4 gives the values of $D$ at temperature $T_2$ calculated from Eq. A-3a for five different temperatures and for $Q_{10}$ values of 1.22 (diffusion through free solution) and 2.0 (diffusion through a membrane), assuming the value of $D_{10°C}$ is $1.0 \times 10^{-6}$ cm$^2$/sec.

**TABLE 4** Calculation of Diffusion Coefficient from Eq. A-3a, Assuming a Diffusion Coefficient of $1.0 \times 10^{-6}$ cm$^2$/sec at 10°C

| Temperature (°C) | $D$(cm$^2$/sec) | |
|---|---|---|
| | $Q_{10} = 1.22$ | $Q_{10} = 2.0$ |
| 0 | $0.82 \times 10^{-6}$ | $0.5 \times 10^{-6}$ |
| 10 | $1.0 \times 10^{-6}$ | $1.0 \times 10^{-6}$ |
| 20 | $1.22 \times 10^{-6}$ | $2 \times 10^{-6}$ |
| 30 | $1.49 \times 10^{-6}$ | $4 \times 10^{-6}$ |
| 40 | $1.82 \times 10^{-6}$ | $8 \times 10^{-6}$ |

Although $Na^+$ has a lower atomic weight than $K^+$, $Na^+$ attracts and holds a larger hydration shell because of its greater charge density (same charge as $K^+$, but smaller unhydrated ion size), and therefore $Na^+$ diffuses in water considerably slower than $K^+$ (ratio of about 1:2).

The cell membrane constitutes a barrier to diffusion. In penetrating through the membrane (through the lipid bilayer matrix), a noncharged small solute molecule must (1) detach itself from its surrounding solvent molecules and jump into the membrane phase; (2) move through the thickness of the membrane, perhaps by a series of small jumps; and (3) detach itself from the membrane environment and jump into the solvent phase again on the opposite side of the membrane. The permeability of the membrane to the solute depends greatly on the activation energy required by the molecule to jump into and out of the membrane. If the activation energy is high, the permeability is low and the $Q_{10}$ is high; if the activation energy is low, the permeability is great and the $Q_{10}$ is low. The rate of permeation of a water-soluble substance is primarily determined by the ease of the substance in passing from water to lipid; its oil–water partition coefficient is a measure of this ease. Equations A-2 and A-2a also apply when the permeability coefficient ($P$) is substituted by the diffusion coefficient ($D$) of the solute in the membrane.

## Bibliography

Akiyama, T., and Fozzard, H. A. (1975). Influence of potassium ions and osmolality on the resting membrane potential of rabbit ventricular papillary muscle with estimation of activity and the activity coefficient of internal potassium. *Cir. Res.* **37,** 621–629.

Bockris, J. O'M., and Conway, B. E. (1954). "Modern Aspects of Electrochemistry." Butterworths, London.

Carmeliet, E., and Bosteels, S. (1969). Coupling between Cl flux and Na or K flux in cardiac Purkinje fibers: influence of pH. *Arch. Int. Physiol. Biochim.* **77,** 57–72.

Carmeliet, E., and Verdonck, F. (1977). Reduction of potassium permeability by chloride substitution in cardiac cells. *J. Physiol. (London)* **265,** 193–206.

Carmeliet, E. E., Horres, C. R., Lieberman, M., and Vereecke, J. S. (1976). Developmental aspects of potassium flux and permeability of the embryonic chick heart. *J. Physiol. (London)* **254,** 673–692.

Crank, J. (1956). "Mathematics of Diffusion." Oxford University Press, New York.

Danielli, J. (1943). The theory of penetration of a thin membrane: Appendix. *In* "The Permeability of Natural Membranes" (H. Davson and D. F. Danielli, Eds.), Cambridge University Press, London.

Davson, H. (1964). "A Textbook of General Physiology," 3rd ed. Little, Brown, Boston.

Fozzard, H. A., and Lee, C. O. (1976). Influence of changes in external potassium and chloride ions on membrane potential and intracellular potassium ion activity in rabbit ventricular muscle. *J. Physiol. (London)* **256,** 663–689.

Hodgkin, A. L., and Huxley, A. F. (1952). Currents carried by sodium and potassium ions through the membrane of the giant axon of Loligo. *J. Physiol. (London)* **116,** 449–472.

Jacobs, M. H. (1967). "Diffusion Processes." Springer-Verlag, New York.

Lee, C. O., and Fozzard, H. A. (1975). Activities of potassium and sodium ions in rabbit heart muscle. *J. Gen. Physiol.* **65,** 695–708.

Mullins, L. J. (1961). The macromolecular properties of excitable membranes. *Ann. N.Y. Acad. Sci.* **94,** 390–404.

Robinson, R. A., and Stokes, R. H. (1959). "Electrolyte Solutions." Butterworths, London.

Sperelakis, N. (1979). Origin of the cardiac resting potential. *In* "Handbook of Physiology, Vol. 1, The Cardiovascular System" (R. M. Berne and N. Sperelakis, Eds.), pp. 187–267. Am. Physiological Society, Bethesda, MD.

Sperelakis, N. (1995). "Electrogenesis of Biopotentials," Kluwer Publishing Co., New York.

Nicholas Sperelakis

# 13

## Origin of Resting Membrane Potentials*

## I. Introduction

The cell membrane exerts tight control over the electrical activity and the contractile machinery during the process of excitation–contraction (electromechanical) coupling. Some drugs and toxins exert primary or secondary effects on the electrical properties of the cell membrane and thereby exert effects, for example, on automaticity, arrhythmias, and force of contraction of the heart. Therefore, for an understanding of the mode of action of therapeutic drugs, toxic agents, neurotransmitters, hormones, and plasma electrolytes on the electrical activity of nerve and muscle, it is necessary to understand the electrical properties and behavior of the cell membrane at rest and during excitation. The first step in gaining such an understanding is to examine the electrical properties of nerve and muscle cells at rest, including the origin of the resting membrane potential ($E_m$). The resting $E_m$ and action potential (AP) result from properties of the cell membrane and the ion distributions across it.

## II. Passive Electrical Properties

### A. Membrane Structure and Composition

As discussed in previous chapters, the cell membrane is composed of a bimolecular leaflet of phospholipid molecules (e.g., phosphatidylcholine, phosphatidylethanolamine) with protein molecules floating in the lipid bilayer. The nonpolar hydrophobic ends of the phospholipid molecules project toward the middle of the membrane, and the polar hydrophilic ends project toward the edges of the membrane bordering on the water phases (Fig. 1). This orientation is thermodynamically favorable. The lipid bilayer membrane is about 50–70 Å thick, and the phospho-

lipid molecules are about the right length (30–40 Å) to stretch across half of the membrane thickness. Cholesterol molecules are in high concentration in the cell membrane (of animal cells) and are inserted between the phospholipid molecules, giving a phospholipid : cholesterol ratio of about 1.0. Some of the large protein molecules, called integral membrane proteins, protrude through the entire membrane thickness; for example, the Na, K-ATPase, Ca-ATPase, and the various ion channel proteins protrude through both leaflets, whereas other proteins are inserted into one leaflet (inner or outer) only. These proteins "float" in the lipid bilayer matrix, and the membrane has fluidity (reciprocal of microviscosity), such that the protein molecules can move around laterally in the plane of the membrane. Some ion channels, for example, fast $Na^+$ channels of the node of Ranvier of myelinated neurons, are tethered in place to the cytoskeleton by anchoring proteins, such as ankyrin.

The outer surface of the cell membrane is lined with strands of mucopolysaccharides (the cell coat or glycocalyx) that endow the cell with immunochemical properties. The cell coat is highly charged negatively and therefore can bind cations, such as $Ca^{2+}$. Treatment with neuraminidase, to remove sialic acid residues, destroys the cell coat.

### B. Membrane Capacitance and Resistivity

Lipid bilayer membranes made artificially have a specific membrane capacitance ($C_m$) of 0.4–1.0 $\mu F/cm^2$, which is close to the value for biologic membranes. The capacitance of natural cell membranes is due to this lipid bilayer matrix. A capacitor consists of two parallel plate conductors separated by a dielectric material of high resistance (e.g., oil). The factors that determine the value of the capacitance of a membrane are given in the equation

$$C_m = \frac{\varepsilon A_m}{\delta} \frac{1}{4\pi k} \qquad (1)$$

where $A_m$ is the membrane area (in $cm^2$) and $k$ is a constant ($9.0 \times 10^{11}$ cm/F). Calculation of membrane thickness ($\delta$)

* Adapted and reprinted by permission from the author's chapter 3 in PHYSIOLOGY (Sperelakis, N. and Banks, R. O., Editors). Copyright © 1993 by Nicholas Sperelakis and Robert O. Banks. Published by Little, Brown and Company.

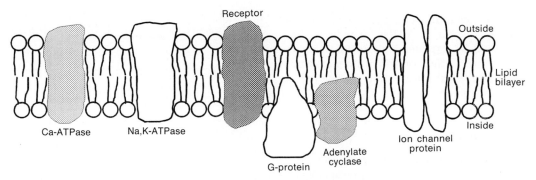

**FIG. 1.** Diagrammatic illustration of cell membrane substructure showing the lipid bilayer. Nonpolar hydrophobic tail ends of the phospholipid molecules project toward the middle of the membrane, and polar hydrophilic heads border on the water phase at each side of the membrane. Lipid bilayer is about 50–70 Å thick. For simplicity, the cholesterol molecules are not shown. Large protein molecules protrude through entire membrane thickness or are inserted into one leaflet only, as depicted. These proteins include various enzymes associated with the cell membrane as well as membrane ionic channels. Membrane has fluidity so that the protein and lipid molecules can move around in the plane of the membrane and fluorescent probe molecules inserted into the hydrophobic region of the membrane have freedom to rotate.

from Eq. 1, assuming a measured membrane capacitance ($C_m$) of 0.7 $\mu$F/cm$^2$ and a dieletric constant ($\varepsilon$) of 5, gives 63 Å. Most oils have dielectric constants of 3–5. The more dipolar the material, the greater the dielectric constant (e.g., water, which is very dipolar, has a value of 81, compared with a value of 1.000 for a vacuum or air). Capacitors are discussed in more detail in the Appendix to this textbook.

The artificial lipid bilayer membrane, on the other hand, has an exceedingly high specific resistance ($R_m$) of $10^6$–$10^9$ $\Omega$-cm$^2$, which is several orders of magnitude higher than that of the biologic cell membrane (about $10^3$ $\Omega$-cm$^2$), with the exception of red blood cells (see Chapter 22). $R_m$ is greatly lowered, however, when the bilayer is doped with certain proteins or substances, such as macrocyclic polypeptide antibiotics (ionophores). The added ionophores may be of the ion-carrier type, such as valinomycin, or of the channel-former type, such as gramicidin. Therefore, the presence of proteins that span across the thickness of the cell membrane must account for the relatively low resistance (high conductance) of the cell membrane. These proteins include those associated with the voltage-dependent gated ion channels of the cell membrane. In summary, the capacitance is due to the lipid bilayer matrix, and the conductance is due to proteins inserted in the lipid bilayer.

The dielectric property of the cell membrane is very good. For a resting $E_m$ of 80 mV and a thickness of 60 Å, the voltage gradient sustained across the membrane is 133,000 V/cm. Thus, the cell membrane tolerates an enormous voltage gradient.

## C. Membrane Fluidity

The electrical properties and the ion transport properties of the cell membrane are determined by the molecular composition of the membrane. The lipid bilayer matrix

even influences the function of the membrane proteins; for example, the Na, K-ATPase activity is affected by the surrounding lipid. A high cholesterol content lowers the fluidity of the membrane. The polar portion of cholesterol lodges in the hydrophilic part of the membrane, and the nonpolar part of the planar cholesterol molecule is wedged between the fatty acid tails, thus restricting their motion and lowering fluidity. A high degree of unsaturation and branching of the tails of the phospholipid molecules raises the fluidity; phospholipids with unsaturated and branched-chain fatty acids cannot be packed tightly because of steric hindrance due to their greater rigidity. Chain length of the lipids also affects fluidity. Low temperature decreases membrane fluidity, as expected. $Ca^{2+}$ and $Mg^{2+}$ may diminish the charge repulsion between the phospholipid head-groups; this allows the bilayer molecules to pack together more closely, thereby constraining the motion of the tails and reducing fluidity. Each phospholipid tail occupies about 20–30 Å$^2$, and each head group about 60 Å$^2$ (Jain, 1972). Membrane fluidity changes occur in muscle development and in certain disease states such as cancer, muscular dystrophy (Duchenne type), and myotonic dystrophy.

The hydrophobic portion of local anesthetic molecules may interpose between the lipid molecules. This separates the acyl chain tails of the phospholipid molecules further, reducing the van der Waals forces of interaction between adjacent tails, and thus increasing the membrane fluidity. Local anesthetics depress the resting conductance of the membrane and the voltage-dependent changes in $g_{Na}$, $g_K$, and $g_{Ca}$. That is, the local anesthetics produce a nonselective depression of all ionic conductances of the membrane. At least part of this depression could be an indirect effect of the anesthetics on the fluidity of the lipid matrix. At the concentration of a local anesthetic required to completely block excitability, its estimated concentration in the lipid bilayer is more than 100,000/$\mu$m$^2$. This should be compared with a density of fast Na$^+$ channels of about 20–100/$\mu$m$^2$, and even less for $K^+$ channels and $Ca^{2+}$ channels. Part of the depression of the Na, K-ATPase activity by local anesthetics also could be explained by an effect on the fluidity, although a direct effect on the protein enzyme is also possible. For additional information on fluidity, the reader is referred to Chapter 4.

### D. Potential Profile across Membrane

The cell membrane has fixed negative charges on its outer and inner surfaces. The charges are presumably due to acidic phospholipids in the bilayer and to protein molecules either embedded in the membrane (islands floating in the lipid bilayer matrix) or tightly adsorbed to the surface of the membrane. Most proteins have an acid isoelectric point, so that a pH near 7.0, they possess a net negative charge. The charge at the outer surface of the cell membrane, with respect to the solution bathing the cell, is known as the **zeta potential.** This charge is responsible for the electrophoresis of cells in an electric field, the cells moving toward the anode (positive electrode), because unlike charges attract. This surface charge affects the true potential difference (PD) across the membrane, as shown in Fig. 2A. At each surface, the fixed charge produces an electric field that extends a short distance into the solution and causes each surface of the membrane to be slightly more negative (by a few millivolts) than the extracellular and intracellular solutions. The potential theoretically recorded by an ideal tiny electrode, as the electrode is driven through the solution perpendicular to the membrane surface, should become negative as the electrode approaches within a few Angstrom units of the surface. The potential difference between the membrane surface and the solution declines exponentially as a function of distance from the surface. The length constant is a function of the ionic strength (or resistivity) of the solution: the lower the ionic strength, the greater the length constant. The magnitude of the PD depends on the density of the charge sites (number of chemical groups per unit of membrane area); the number of ionized charge groups is also affected by the pH and ionic strength.

The membrane potential ($E_m$) measured by an intracellular microelectrode is the potential of the inner solution ($\Psi_i$) minus the potential of the outer solution ($\Psi_o$); that is,

$$E_m = \Psi_i - \Psi_o \qquad (2)$$

The true PD across the membrane ($E_m'$), however, is really that PD directly across the membrane, as shown in Fig. 2A. If the surface charges at each surface of the membrane are equal, then $E_m' = E_m$. If the outer surface charge is decreased to zero by extra binding of protons or cations (such as $Ca^{2+}$), then the membrane becomes slightly hyperpolarized ($E_m' > E_m$), although this is not measurable by the intracellular microelectrode (Fig. 2B). Conversely, if the inner surface charge were neutralized and the outer surface charge restored, then the membrane would become slightly depolarized ($E_m' < E_m$). Again, this change is not measurable by the microelectrode, which can only measure the PD between the two solutions.

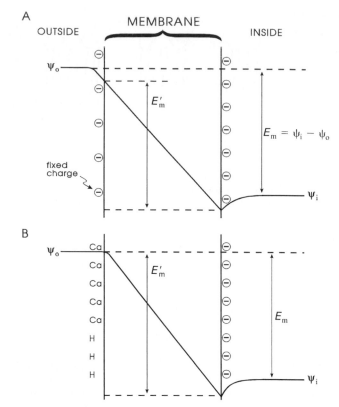

**FIG. 2.** Potential profile across the cell membrane. (A) Because of fixed negative charges (at pH 7.4) at outer and inner surfaces of the membrane, there is a negative potential that extends from the edge of the membrane into the bathing solution on both sides of the membrane. This surface potential falls off exponentially with distance into the solution. Magnitude of the surface potential is a function of the charge density. $\Psi_o$ is the electrical potential of the outside solution, $\Psi_i$ is that of the inside solution, and membrane potential ($E_m$) is the difference ($\Psi_i - \Psi_o$). $E_m$ is determined by the equilibrium potentials and relative conductances. Profile of the potential through the membrane is shown as linear (the constant-field assumption), although this need not be true for the present purpose. If the outer surface potential is exactly equal to that in the inner surface, then the true transmembrane potential ($E_m'$) is exactly equal to the (microelectrode) measured membrane potential ($E_m$). (B) If the outer surface potential is different from the inner potential, for example, by increasing the extracellular $Ca^{2+}$ concentration or decreasing the pH to bind $Ca^{2+}$ or $H^+$ to more of the negative charges, then the $E_m'$ is greater than the measured $E_m$. Diminution of the inner surface charge decreases $E_m'$. The membrane ion channels are controlled by $E_m'$.

Because the membrane ionic conductances are controlled by the PD directly across the membrane (i.e., by $E_m'$ and not by $E_m$), changes in the surface charges (e.g., by drugs, ionic strength, or pH) can lead to apparent shifts in the electrical threshold potential, mechanical threshold potential (the $E_m$ value at which contraction of muscle just begins), activation curve, and inactivation curve (discussed in subsequent chapters). For example, elevation of extracellular $Ca^{2+}$ concentration ($[Ca]_o$) is known to raise the threshold potential (i.e., the critical depolarization required to reach electrical threshold), as expected from the small increase in $E'_m$ that should occur.

## III. Maintenance of Ion Distributions

### A. Resting Potentials and Ion Distributions

The transmembrane potential in resting nerve and muscle cells varies with the type of cell. In myocardial cells, it is about $-80$ mV (Table 1). The resting $E_m$ or maximum diastolic potential in cardiac Purkinje fibers is somewhat greater (about $-90$ mV), whereas that in the nodal cells of the heart is lower (about $-60$ mV). In nerve cells, the resting potential is about $-70$ mV, whereas in skeletal muscle fibers, the value is close to $-80$ mV. In most smooth muscle cells, the resting $E_m$ is lower, about $-55$ mV. The values of the resting $E_m$ for some types of cells are summarized in Table 1.

The ionic composition of the extracellular fluid bathing the muscle cells is similar to that of the blood stream. It is high in $Na^+$ (about 145 m$M$) and $Cl^-$ (about 100 m$M$), but low in $K^+$ (about 4.5 m$M$). The $Ca^{2+}$ concentration is about 2 m$M$, but about half is bound to serum proteins. In contrast, the intracellular fluid has a low concentration of $Na^+$ (about 15 m$M$ or less) and $Cl^-$ (about 6–8 m$M$), but a high concentration of $K^+$ (about 150–170 m$M$). These ion distributions are listed in Table 2. The free intracellular $Ca^{2+}$ concentration ($[Ca]_i$) is about $10^{-7}$ $M$ or less, but during contraction it may rise as high as $10^{-5}$ $M$. The total intracellular $Ca^{2+}$ is much higher (about 2 m$M$/kg), but most of this is bound to molecules such as proteins or is sequestered into compartments such as mitochondria and the sarcoplasmic reticulum (SR). Most of the intracellular $K^+$ is free, and it has a diffusion coefficient only slightly less than that of $K^+$ in free solution, consistent with the tortuosity of the diffusion paths around intracellular structures. Thus, under normal conditions, the cell maintains an internal ion concentration markedly different from that in the medium bathing the cells, and it is these ion concentration differences that underlie the resting potential and excitability. The existence of the resting potential enables

**TABLE 1** Comparison of the Resting Potentials in Different Types of Cells

| Cell type | Resting potential (mV) |
| --- | --- |
| Neuron | $-70$ |
| Skeletal muscle (mammalian) | $-80$ |
| Skeletal muscle (frog) | $-90$ |
| Cardiac muscle (atrial and ventricular) | $-80$ |
| Cardiac Purkinje fiber | $-90$ |
| Atrioventricular nodal cell | $-65$ |
| Sinoatrial nodal cell | $-55$ |
| Smooth muscle cell | $-55$ |
| Red blood cell (human) | $-11$ |

**TABLE 2** Summary of the Ion Distributions in Most Types of Cells and the Equilibrium Potentials Calculated from the Nernst Equation

| Ion | Extracellular distribution (m$M$) | Intracellular distribution (m$M$) | Equilibrium potential (mV) |
| --- | --- | --- | --- |
| $Na^+$ | 145 | 15 | $+60$ |
| $Cl^-$ | 100 | 5[a] | $-80$ |
| $K^+$ | 4.5 | 150 | $-94$ |
| $Ca^{2+}$ | 1.8 | 0.0001 | $+130$ |
| $H^+$ | 0.0001 | 0.0002 | $-18$ |

[a] Assuming $Cl^-$ is passively distributed and resting $E_m$ is $-80$ mV. The extracellular $H^+$ concentration is given for pH 7; it would be 40 n$M$ at pH 7.4.

action potentials (APs) to be produced in those types of cells that have excitability. The ion distributions and related pumps and exchange reactions are depicted in Fig. 3.

Inhibition of the Na–K pump (e.g., by cardiac glycosides such as digitalis) causes the ion concentration gradients to gradually run down or dissipate. The cells lose $K^+$ and gain $Na^+$, $Cl^-$, and water. Therefore, the $K^+$ and $Na^+$ equilibrium potentials ($E_K$ and $E_{Na}$) become smaller and the cells become depolarized (see Section IV). The depolarization causes the cells to gain $Cl^-$ (see Section III,C), and therefore also water because of the resultant gain in osmotic strength, causing the cells to swell.

Although the topic of Na, K-ATPase and the Na–K pump is discussed in detail in Chapter 16, a brief description is given here.

### B. Na$^+$ and K$^+$ Distribution and the Na–K Pump

The intracellular ion concentrations are maintained differently from those in the extracellular fluid by active ion transport mechanisms that expend metabolic energy to transport specific ions against their concentration or electrochemical gradients. These ion pumps are located in the cell membrane at the cell surface and probably also in the transverse tubular membrane of striated muscle cells. The major ion pump is the Na–K-linked pump, which pumps $Na^+$ out of the cell against its electrochemical gradient, while simultaneously pumping $K^+$ in against its electrochemical gradient (Fig. 3). The coupling between $Na^+$ and $K^+$ pumping is obligatory, since in zero $[K]_o$, the $Na^+$ can no longer be pumped out. That is, a coupling ratio of 3 $Na^+$ : 0 $K^+$ is not possible. The coupling ratio of $Na^+$ pumped out to $K^+$ pumped is generally 3:2. The Na–K pump is half-inhibited ($K_i$ value) when $[K]_o$ is lowered to about 2 m$M$.

If the ratio were 3:3, the pump would be electrically neutral or nonelectrogenic. A PD across the membrane would not be produced directly, because the pump would pull in three positive charges ($K^+$) for every three positive charges ($Na^+$) it pushed out. When the ratio is 3:2, the pump is electrogenic and directly produces a PD that causes $E_m$ to be greater (more negative) than it would be other-

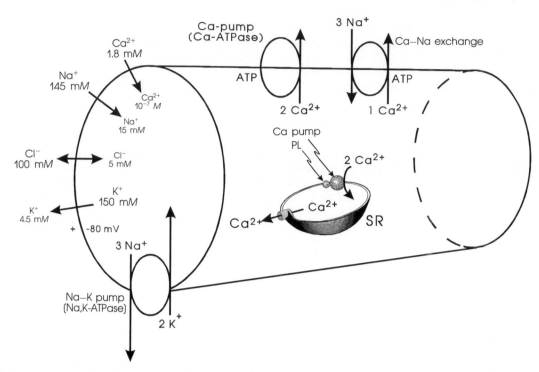

**FIG. 3.** Intracellular and extracellular ion distributions in a myocardial cell or skeletal muscle fiber. Also given are polarity and magnitude of the resting potential. Arrows give direction of the net electrochemical gradient. $Na^+$–$K^+$ pump is located in the cell surface and the T-tubule membranes. A Ca-ATPase/Ca pump, similar to that in the SR, is located in the cell membrane. A Ca–Na exchange carrier is located in the cell membrane.

wise, solely on the basis of the ion concentration gradients and relative permeabilities or net diffusion potential ($E_{diff}$). A coupling ratio of 3 $Na^+$ : 1 $K^+$ would produce a greater electrogenic pump potential ($V_p$). Under normal steady-state conditions, the contribution of the $Na^+$–$K^+$ electrogenic pump potential to $E_m$ ($\Delta V_p$) is only a few millivolts in myocardial cells; the contribution is greater in smooth muscle cells (6–8 mV) and mammalian skeletal muscle fibers (12–16 mV) (see Section VII).

The driving mechanism for the Na–K pump is a membrane ATPase, the Na,K-ATPase, which spans the membrane and requires both $Na^+$ and $K^+$ ions for activation. This enzyme requires $Mg^{2+}$ for activity, and it is actually inhibited by $Ca^{2+}$. ATP, $Mg^{2+}$, and $Na^+$ are thus required at the inner surface of the membrane, and $K^+$ is required at the outer surface. A phosphorylated intermediate (aspartate residue) of the Na,K-ATPase occurs in the transport cycle, its phosphorylation being $Na^+$ dependent and its dephosphorylation being $K^+$ dependent (for references, see Sperelakis, 1979). The pump enzyme usually drives three $Na^+$ ions out and two $K^+$ ions in for each ATP molecule hydrolyzed. The Na,K-ATPase is specifically inhibited by the cardiac glycosides (digitalis drugs) acting on the outer surface. The pump enzyme is also inhibited by vanadate ion and by sulfhydryl (SH) reagents [such as *N*-ethylmaleimide (NEM), mercurial diuretics, and ethacrynic acid], thus indicating that the SH groups are crucial for activity.

Blockade of the Na–K pump produces only a small immediate effect on the resting $E_m$: a small depolarization of about 2–16 mV, depending on cell type, representing the contribution of $V_p$ to $E_m$ ($\Delta V_p$). Because excitability and generation of APs are almost unaffected at short times, excitability is independent of active ion transport. However, over many minutes, depending on the ratio of surface area to volume of the cell, the resting $E_m$ slowly declines because of gradual dissipation of the ionic gradients. The progressive depolarization depresses the rate of rise of the AP, and hence the propagation velocity, and eventually all excitability is lost. Thus, a large resting potential and excitability, although not immediately dependent on the Na–K pump, are ultimately dependent on it.

The rate of Na–K pumping in excitable cells must change with the amount of electrical activity to maintain the intracellular ion concentrations relatively constant. A higher frequency of APs results in a greater overall movement of ions down their electrochemical gradients, and these ions must be repumped to maintain the ion distributions. For example, the cells tend to gain $Na^+$, $Cl^-$, and $Ca^{2+}$ and to lose $K^+$. The factors that control the rate of Na–K pumping include $[Na]_i$ and $[K]_o$. In cells that have a large surface area to volume ratio (such as small-diameter nonmyelinated axons), $[Na]_i$ may increase by a relatively large percentage during a train of APs, and this would stimulate the pumping rate. Likewise, an accumulation of $K^+$ externally occurs and also stimulates the pump. As mentioned above, the $K_m$ value for $K^+$, that is, the concentration for half-maximal rate, is about 2 m$M$. It has been shown that $[K]_o$ is significantly increased during the long AP plateau in cardiac muscle.

## C. Cl⁻ Distribution

In many invertebrate and vertebrate nerve and muscle cells, Cl⁻ ion does not appear to be actively transported; that is, there is no Cl⁻ ion pump (or Cl⁻-ATPase). In such cases, Cl⁻ distributes itself passively (no energy used) in accordance with $E_m$. In such a case, $E_{Cl}$ is equal to $E_m$ in a resting cell. For example, in mammalian myocardial cells, Cl⁻ seems to be close to passive distribution, because $[Cl]_i$ is at, or only slightly above, the value predicted by the Nernst equation from the resting $E_m$ (for references, see Sperelakis, 1979). When passively distributed, $[Cl]_i$ is low because the negative potential inside the cell (the resting potential) pushes out the negatively charged Cl⁻ ion (like charges repel) until the Cl⁻ distribution is at equilibrium with the resting $E_m$. Hence, for a resting $E_m$ of $-80$ mV, and taking $[Cl]_o$ to be 100 m$M$, $[Cl]_i$ calculated from the Nernst equation (see Section IV,B) would be 4.9 m$M$.

$$E_{Cl} = E_m \tag{3}$$

$$E_m = +61 \text{ mV} \log \frac{[Cl]_i}{[Cl]_o}$$

$$= -61 \text{ mV} \log \frac{[Cl]_o}{[Cl]_i}$$

$$\frac{-80 \text{ mV}}{-61 \text{ mV}} = \log \frac{100 \text{ m}M}{[Cl]_i}$$

$$\frac{100 \text{ m}M}{[Cl]_i} = \text{antilog} \frac{-80 \text{ mV}}{-61 \text{ mV}} = \text{antilog } 1.31 = 20.5$$

$$[Cl]_i = \frac{100 \text{ m}M}{20.5} = 4.88 \text{ m}M$$

During the AP, the inside of the cell goes in a positive direction, and a net Cl⁻ influx (outward Cl⁻ current, $I_{Cl}$) will occur and thus increase $[Cl]_i$. The magnitude of the Cl⁻ influx depends on the Cl⁻ conductance ($g_{Cl}$) of the membrane:

$$I_{Cl} = g_{Cl}(E_m - E_{Cl}) \tag{4}$$

This equation is discussed later in Section V. Thus, the average level of $[Cl]_i$ in excitable cells should depend on the frequency and duration of the AP, that is, on the mean $E_m$ averaged over many AP cycles.

In smooth muscle, $[Cl]_i$ is often much higher (ca. 30 m$M$) than the value of 12.5 m$M$ predicted from passive distribution:

$$E_{Cl} = E_m = -55 \text{ mV}$$

$$[Cl]_i = \frac{[Cl]_o}{\text{antilog} \dfrac{E_m}{-61 \text{ mV}}} \tag{3a}$$

$$= \frac{100 \text{ m}M}{\text{antilog} \dfrac{-55 \text{ mV}}{-61 \text{ mV}}}$$

$$= \frac{100}{7.973}$$

$$= 12.5 \text{ m}M$$

The elevated $[Cl]_i$ in smooth muscle could be due to an exchange carrier (e.g., Cl⁻–HCO₃⁻ exchange) or to a co transporter (e.g., Na–K–Cl₂ transport).

## D. Ca²⁺ Distribution

### 1. Need for Calcium Pumps

For the positively charged Ca²⁺ ion, there must be some mechanism for removing Ca²⁺ from the cytoplasm. Otherwise, the cell would continue to gain Ca²⁺ until there was no electrochemical gradient for net influx of Ca²⁺. Ca²⁺ loading would occur until the free $[Ca]_i$ in the cytoplasm was even greater than that outside (ca. 2 m$M$) because of the negative potential inside the cell. Therefore, there must be one or more Ca²⁺ pumps in operation. The SR (or ER) membrane contains a Ca²⁺-activated ATPase (which also requires Mg²⁺) that actively pumps two Ca²⁺ ions from the cytoplasm into the SR lumen at the expense of one ATP. This pump ATPase is capable of pumping down the Ca²⁺ to less than $10^{-7}$ $M$. The Ca-ATPase of the SR is regulated by an associated low-molecular-weight protein, phospholamban. Phospholamban is phosphorylated by cyclic AMP-dependent protein kinase (on serine-16)[1] and, when phosphorylated, stimulates the Ca-ATPase and Ca²⁺ pumping (by a de-repression process). The sequestration of Ca²⁺ by the SR is essential for muscle relaxation. The mitochondria also can actively take up Ca²⁺ almost to the same degree as the SR, but this Ca²⁺ pool probably does not play an important role in normal excitation–contraction coupling processes.

However, the resting Ca²⁺ influx and the extra Ca²⁺ influx that enter with each AP must be returned to the interstitial fluid. Two mechanisms have been proposed for this (for references, see Sperelakis, 1979): (1) a Ca-ATPase, similar to that in the SR, is present in the sarcolemma, and (2) a Ca–Na exchange occurs across the cell membrane. It has been reported that there is a Ca-ATPase in the sarcolemma of myocardial cells (Dhalla et al., 1977; Jones et al., 1980) and smooth muscle (Daniel et al., 1977) that actively transports two Ca²⁺ outward against an electrochemical gradient, using one ATP in the process. Phospholamban is not associated with the sarcolemmal Ca-ATPase.

### 2. Ca/Na Exchange Reaction

The $Ca_i/Na_o$ exchange reaction exchanges one internal Ca²⁺ ion for 3 external Na⁺ ions via a membrane carrier molecule (for references, see Sperelakis, 1979) (see Fig. 3). This reaction is facilitated by ATP, but ATP is not hydrolyzed (consumed) in this reaction. Instead, the energy for the transport of Ca²⁺ against its large electrochemical gradient comes from the Na⁺ electrochemical gradient. That is, the uphill transport of Ca²⁺ is coupled to the downhill movement of Na⁺. Effectively, the energy required for this Ca²⁺ movement is derived from the Na,K-ATPase. Thus, the Na–K pump, which uses ATP to maintain the Na⁺ electrochemical gradient, indirectly helps to maintain the Ca²⁺ electrochemical gradient. Hence, the inward Na⁺ leak is greater than it would be otherwise. A complete discussion of the Ca/Na exchange is given in Chapter 18.

---

[1] Phosphorylation of threonine-17 also occurs in vivo by the Ca–CAM–PK.

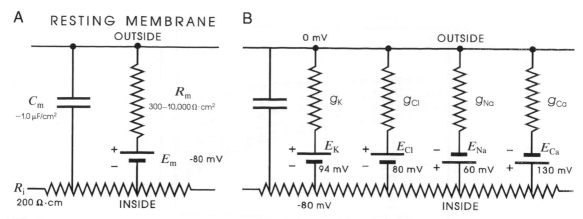

**FIG. 5.** Electrical equivalent circuits for a cell membrane at rest. (A) Membrane as a parallel resistance/capacitance circuit, the membrane resistance ($R_m$) being in parallel with the membrane capacitance ($C_m$). Resting potential ($E_m$) is represented by an 80-mV battery in series with the membrane resistance, the negative pole facing inward. (B) Membrane resistance is divided into its four component parts, one for each of the four major ions of importance: $K^+$, $Cl^-$, $Na^+$, and $Ca^{2+}$. Resistances for these ions ($R_K$, $R_{Cl}$, $R_{Na}$, and $R_{Ca}$) are parallel to one another and represent totally separate and independent pathways for permeation of each ion through the resting membrane. These ion resistances are depicted as their reciprocals, namely, ion conductances ($g_K$, $g_{Cl}$, $g_{Na}$, $g_{Ca}$). Equilibrium potential for each ion (e.g., $E_K$), determined solely by the ion distribution in the steady state and calculated from the Nernst equation, is shown in series with the conductance path for that ion. Resting potential of $-80$ mV is determined by the equilibrium potentials and by the relative conductances.

$$E_m = \frac{Q}{C_m} \tag{9}$$

where $C_m$ is the membrane capacitance. This is further discussed in the next section.

For the ion distributions given previously (see Table 2), the approximate equilibrium potentials are

$$E_{Na} = +60 \text{ mV}$$
$$E_{Ca} = +130 \text{ mV}$$
$$E_K = -93 \text{ mV}$$
$$E_{Cl} = -80 \text{ mV}$$

The sign of the equilibrium potential represents the inside of the cell with reference to the outside (see Fig. 5). Because $Na^+$ is higher outside (ca. 145 m$M$) than inside (ca. 15 m$M$), the positive pole of the $Na^+$ battery ($E_{Na}$) is inside the cell. The concentration gradient for $Ca^{2+}$ is in the same direction as for $Na^+$ (1.8 m$M$ [Ca]$_o$ and about $1 \times 10^{-7}$ $M$ [Ca]$_i$), and so the positive pole of $E_{Ca}$ is inside. $K^+$ is higher inside (ca. 150 m$M$) than outside (ca. 4.5 m$M$), and so the negative pole is inside. Because $Cl^-$ is higher outside (ca. 100 m$M$) than inside (ca. 5 m$M$), the negative pole is inside. Voltages are, by convention, given for the inside with respect to the outside.

### C. Concentration Cell

In a concentration cell (essentially a two-compartment system separated by a membrane), the side of higher concentration becomes negative for cations (positive ions) and positive for anions (negative ions). Any ion whose equilibrium potential is different from the resting potential (e.g., $-80$ mV for a myocardial cell or skeletal muscle fiber) is

off equilibrium and therefore must effectively be pumped at the expense of energy. In many cell types, only $Cl^-$ ion appears to be at or near equilibrium, whereas $Na^+$, $Ca^{2+}$, and $K^+$ are actively transported. Even $H^+$ ion is off equilibrium, $E_H$ being closer to zero potential (see Table 2). If $H^+$ were passively distributed, the negative intracellular potential would pull in more $H^+$ ions, causing [H]$_i$ to increase, making the cell interior more acidic. Extensive discussion of concentration cells and diffusion is given by Sperelakis (1979).

The mechanism for development of the equilibrium potential is depicted in Fig. 6. To show the development of an equilibrium potential, we can use an artificial membrane (e.g., made of celloidin) to separate two solutions, that is, to form a concentration cell. This membrane contains negatively charged pores, which therefore allows cations (like $K^+$) to pass through but prevents anions (like $Cl^-$) from passing. This is because like charges repel one another, and unlike charges attract one another. Therefore, in this particular membrane, $K^+$ is permeable and $Cl^-$ is impermeable. If one side (side 1) contains a salt like KCl at a concentration higher (e.g., 0.10 $M$) than that in the other side (side 2) (e.g., 0.01 $M$), then a steady PD is very quickly built up across the membrane. As can be calculated from the Nernst equation, for a 10-fold difference in concentration of the permeant monovalent cation ($K^+$), the PD would be $-61$ mV at 37°C (or $-59$ mV at a room temperature of 22°C).

This PD is between the two solutions and expressed across the membrane. Side 1 (side of highest $K^+$ concentration) becomes negative with respect to side 2. The PD is developed because of the tendency for diffusion (diffusion force) from high concentration to low concentration. This is

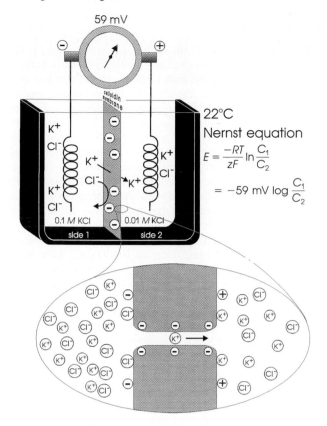

59 mV

celloidin membrane

22°C
Nernst equation

$$E = \frac{-RT}{zF} \ln \frac{C_1}{C_2}$$

$$= -59 \text{ mV} \log \frac{C_1}{C_2}$$

K⁺
Cl⁻
K⁺
Cl⁻
K⁺
Cl⁻

K⁺
Cl⁻
K⁺

0.1 M KCl
0.01 M KCl
side 1
side 2

**FIG. 6.** Upper diagram: Concentration cell diffusion potential developed across artificial membrane containing negatively-charged pores. The membrane is impermeable to Cl⁻ ions, but permeable to cations such as K⁺. Concentration gradient for K⁺ causes a potential to be generated, the side of higher K⁺ concentration becoming negative. Lower diagram: Expanded diagram of a water-filled pore in the membrane, showing the permeability to K⁺ ions, but lack of penetration of Cl⁻ ions. Potential difference is generated by charge separation, a slight excess of K⁺ ions being held close to the right-hand surface of the membrane; a slight excess of Cl⁻ ions is aggregated close to the left surface.

based on the **random thermal motion** of the ions (particles), somewhat related to **Brownian motion** of larger particles. That is, the side of higher concentration has a **greater probability** of K⁺ ions moving from side 1 to side 2 than in the reverse direction, based on the greater number of particles, all moving in random directions. Therefore, there will be a loss of positive charges (K⁺ ions) from side 1 and a gain of positive charges in side 2.

Because negative charges (Cl⁻ ion) cannot accompany the positive charges, as the membrane was made impermeable to anions, a charge separation is built up across the membrane. It can now be readily understood why side 1, the side of higher cation concentration, becomes negatively charged (due to loss of positive charges) and why side 2, the side of lower cation concentration, becomes positively charged (due to gain of positive charges). The charge separation is very tiny, and they stay plastered very close to the membrane. That is, for the example depicted in Fig. 6, side 1 will have the small excess of Cl⁻ ions held very close to the membrane, and side 2 will have the small

excess of K⁺ ions held very close to the membrane. The force holding them there is called the electrostatic or Coulombic force, based on the attraction between unlike charges. This is related to Coulomb's law and the nature of capacitors discussed in the Appendix to the book.

In the two bulk solutions, the **law of electroneutrality** is upheld; that is, for every cation (K⁺) there is a nearby anion (Cl⁻). Thus, the charge separation occurs only directly across the membrane and is very tiny with respect to the total number of charges in the two solutions. In fact, after equilibrium is reached (within a few seconds), the most sensitive chemical analyses would fail to detect the very slight decrease in K⁺ in side 1 or gain of K⁺ on side 2.

Thus, the system comes to equilibrium quickly and with very little charge (K⁺ and Cl⁻) separation. That is, K⁺ does not continue to have a net movement from side 1 to side 2 until the concentrations become equal. Why not? The answer is that the **small charge separation produces a large PD** across the membrane, and this PD is in such a polarity that it antagonizes further net movement of K⁺ from side 1 to side 2. That is, the positive voltage that is developing on side 2 repels the positively charged K⁺ ions because like charges repel. At equilibrium, these two forces become equal and opposite, and there is no further net movement of ions. Unidirectional fluxes of K⁺ ions would still occur (because of their random thermal motion), but these would be equal and opposite and so no further *net flux* would occur.

In the example selected, KCl was used. However, any salt, such as NaCl, CaCl₂, or Na₂SO₄, could have been illustrated. If a divalent cation like Ca²⁺ were used, then from the Nernst equation, the same 10-fold concentration gradient would develop a potential of only half, namely 30.5 mV (37°C) or 29.5 mV (22°C). The reason that this factor is half, rather than double as one might guess from the fact that the charge is double, is that the Nernst equation gives the PD that exactly opposes the diffusion force due to the concentration gradient, as stated in the paragraph above. Therefore, because the charge is double, only half the voltage is necessary to effectively oppose the concentration force. If the cation in question were trivalent, such as La³⁺, then the 2.303 $RT/zF$ factor would be one-third of 61 mV, or about 20.3 mV. It is for this reason that it is convenient to give the Nernst equation in the form shown in Eq. 8c, namely with the factor (at 37°C) being $-61$ mV/$z$. This allows easy calculation of the equilibrium potential for an ion of any charge and sign (polarity).

That is, the sign and charge should be used, for example, $+1$ for K⁺ or Na⁺, $+2$ for Ca²⁺, and $-1$ for Cl⁻. When the ion in question is an anion like Cl⁻, then $-1/-1$ gives a plus ($+$). Because $[Cl]_o > [Cl]_i$, this concentration ratio can be inverted by changing the sign of the 2.30 $RT/zF$ factor back to negative ($-$). That is, changing the sign of the factor in front of a log ratio simply inverts the ratio.

Finally, if the concentration cell depicted in Fig. 6 were made with a membrane that had positively charged pores, then everything would be reversed. The membrane would be permeable to Cl⁻ and impermeable to K⁺, and $E_{Cl}$ would be $+59$ mV (at 22°C) and $+61$ mV (at 37°C). Again, the voltage is given for side 1 with respect to side 2. Thus, in dealing with an anion, the side of higher concentration becomes positive (due to a small loss of $Cl^-$) and the side of lower concentration becomes negative (due to a small gain of $Cl^-$). The separated charges again are plastered very close to the membrane: $K^+$ ions on side 1 and $Cl^-$ ions on side 2.

### D. Activity Coefficient

Thus, an equilibrium potential can be calculated for any speices of ion that is distributed unequally across a membrane. All that one needs to know are the concentrations in the two solutions and the charge (and temperature). Actually, we should use the activity ($a$) of the ion in question in the two solutions instead of concentrations. Thus, the Nernst equation given in Eq. 8c becomes (for $K^+$, for example)

$$E_K = \frac{-61 \, mV}{+1} \log \frac{a_K^i}{a_K^o} \qquad (8d)$$

The activity of an ion (in molar) can be obtained by multiplying the concentration of the ion (in molar) by the activity coefficient ($\gamma$) for the ion:

$$a = c \cdot \gamma \qquad (10)$$

In the biological case, the activity coefficients are relatively close (0.7 to 0.9) to 1.0 for $Na^+$, $K^+$, and $Cl^-$ in *both* the extracellular and the intracellular solutions. Therefore, in these cases, using the concentrations gives a good approximation. However, in the case of $Ca^{2+}$, the activity coefficient in the intracellular solution especially is substantially lower, and so this would affect the calculated value of $E_{Ca}$.

### E. Nernst–Planck Equation

The basic Nernst equation has been modified in several ways for special situations. For example, in the concentration cell depicted in Fig. 6, a cell in which a single salt (both ions of same valence) is distributed across the membrane at two different concentrations, the PD developed across the membrane ($E_m$) can be calculated from the equation

$$E_m = \frac{U_c - U_a}{U_c + U_a} \frac{-61 \, mV}{z} \log \frac{[salt]_1}{[salt]_2} \qquad (11)$$

where $[salt]_1$ and $[salt]_2$ are the concentrations of the salt on side 1 and side 2, and $U_c$ and $U_a$ are the mobilities of the cations and anions, respectively, through the membrane. Thus, when the mobilities (or permeabilities) of the cation and anion are equal ($U_c = U_a$), $E_m$ is zero, regardless of the equilibrium potentials. When the anion is impermeable ($U_a = 0$), the mobility fraction in Eq. 11 becomes 1.0, and the equation reduces to the simple Nernst equation. When the cation is impermeable ($U_c = 0$), the fraction becomes $-1.0$, and the same numerical value of $E_m$ is produced, but of the opposite sign. Equation 11 can be used to calculate $E_m$ for any combination of $U_a$ and $U_c$. For example, if $U_a = 0.5 \, U_c$, then for the problem illustrated in Fig. 6, $E_m$ is about $-20 \, mV$ (side 1 negative). Thus, the membrane potential is related to the relative mobilities.

### F. Energy Wells

Ions do not just "fall" through a water-filled pore in the membrane (protein ion channel) down an electrochemical gradient. Instead, an ion may bind to several charged sites on its journey through the channel pore. The $K^+$ ion depicted in Fig. 6 (bottom), for example, is shown as binding to three negatively charged sites within the pore. These may be considered energy wells, and the ion must gain kinetic energy to become dislodged from this energy well to pass over the next energy barrier and into the next energy well. This energy comes from the ion being hit by another ion just entering the pore, producing a billiard ball effect. Some evidence for this model was presented in Chapter 12, which discussed the Ussing flux ratio equation.

From the measured values of the conductance of single ion channels (e.g., 20 pS, range of 10–300 pS), how many ions that pass through a single channel per second can be estimated. This number is about 6,000,000 ions/sec. Therefore, the average transit time for a single ion to cross the membrane (50–70 Å thick) is about 0.17 $\mu$sec.

### G. Half-Cell Potentials

In the measurement of biological potentials, care must be taken not to introduce artifacts, such as reversible electrode half-cell potentials. See the Appendix (part B) to this chapter for a brief discussion of this topic.

## V. Electrochemical Driving Forces and Membrane Ionic Currents

### A. Electrochemical Driving Forces

The electrochemical driving force for each species of ion is the algebraic difference between its equilibrium potential, $E_i$, and the membrane potential, $E_m$. The total driving force is the sum of two forces: an electrical force (the negative potential in a cell at rest tends to pull in positively charged ions, because unlike charges attract) and a diffusion force (based on the concentration gradient) (Fig. 7); that is,

$$\text{driving force} = E_m - E_i \qquad (12)$$

Thus, in a resting cell, the driving force for $Na^+$ is

$$(E_m - E_{Na}) = -80 \, mV - (+60 \, mV) = -140 \, mV \qquad (12a)$$

The negative sign means that the driving force is directed to bring about net movement of $Na^+$ inward. The driving force for $Ca^{2+}$ is very large and is directed inward:

$$(E_m - E_{Ca}) = -80 \, mV - (+129 \, mV) = -209 \, mV \qquad (12b)$$

The driving force for $K^+$ is

$$(E_m - E_K) = -80 \, mV - (-94 \, mV) = +14 \, mV \qquad (12c)$$

Hence, the driving force for $K^+$ is small and directed outward. The driving force for $Cl^-$ is nearly zero for a cell at rest in which $Cl^-$ is passively distributed (e.g., neuron, myocardial cell, skeletal muscle fiber); that is,

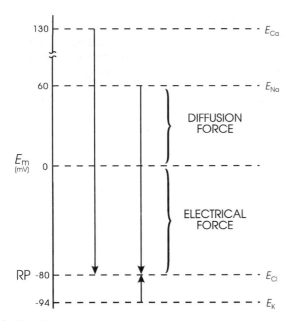

**FIG. 7.** Representation of the electrochemical driving forces for $Na^+$, $Ca^{2+}$, $K^+$, and $Cl^-$. Equilibrium potentials for each ion (e.g., $E_{Na}$) are positioned vertically according to their magnitude and sign: they were calculated from the Nernst equation for a given set of extracellular and intracellular ion concentrations. Measured resting potential (RP) is assumed to be $-80$ mV. Electrochemical driving force for an ion is the difference between its equilibrium potential ($E_i$) and the membrane potential ($E_m$), that is ($E_m - E_i$). Thus, at rest, the driving force for $Na^+$ is the difference between $E_{Na}$ and the resting $E_m$; if $E_{Na}$ is $+60$ mV and resting $E_m$ is $-80$ mV, the driving force is 140 mV; that is, the driving force is the algebraic sum of the diffusion force and the electrical force, and is represented by the length of the arrows in the diagram. Driving force for $Ca^{2+}$ (about 210 mV) is even greater than that for $Na^+$, whereas that for $K^+$ is much less (about 14 mV). Direction of the arrows indicates the direction of the net electrochemical driving force, namely, the direction for $K^+$ is outward, whereas that for $Na^+$ and $Ca^{2+}$ is inward. If $Cl^-$ is passively distributed, then its distribution across the cell membrane can only be determined by the net membrane potential; for a cell sitting a long time at rest, $E_{Cl} = E_m$ and there is no net driving force.

$$(E_m - E_{Cl}) = -80 \, mV - (-80 \, mV) = 0 \qquad (13)$$

However, during the AP, *when $E_m$ is changing, the driving force for $Cl^-$ becomes large,* and there is a net driving force for inward $Cl^-$ movement ($Cl^-$ influx is an outward $Cl^-$ current). Similarly, the driving force for $K^+$ outward movement increases during the AP, whereas those for $Na^+$ and $Ca^{2+}$ decrease.

### B. Membrane Ionic Currents

The net current for each ionic species ($I_i$) is equal to its driving force times its conductance ($g_i$, reciprocal of the resistance) through the membrane. This is essentially Ohm's law.

$$I = \frac{V}{R} = g \cdot V \qquad (14)$$

modified to reflect the fact that, in an electrolytic system, the total force tending to drive net movement of a charged particle must take into account both the electrical force and the concentration (or chemical) force. Thus, for the four ions, the net current can be expressed as

$$I_{Na} = g_{Na}(E_m - E_{Na}) \qquad (15)$$

$$I_{Ca} = g_{Ca}(E_m - E_{Ca}) \qquad (16)$$

$$I_K = g_K(E_m - E_K) \qquad (17)$$

$$I_{Cl} = g_{Cl}(E_m - E_{Cl}) \qquad (18)$$

In a resting cell, $Cl^-$ and $Ca^{2+}$ can be neglected, and the $Na^+$ current (inward) must be equal and opposite to the $K^+$ current (outward) to maintain a steady resting potential:

$$I_K = -I_{Na} \qquad (19)$$

$$g_K(E_m - E_K) = g_{Na}(E_m - E_{Na}) \qquad (19a)$$

Thus, although in the resting membrane the driving force for $Na^+$ is much greater than that for $K^+$, $g_K$ is much larger than $g_{Na}$, so the currents are equal. Hence, there is a continuous leakage of $Na^+$ inward and $K^+$ outward, even in a resting cell, and the system would run down if active pumping were blocked. Because the ratio of the $Na^+/K^+$ driving forces ($-140 \, mV/-14 \, mV$) is 10, the ratio of conductances ($g_{Na}/g_K$) will be about $1:10$. The fact that $g_K$ is much greater than $g_{Na}$ accounts for the resting potential being close to $E_K$ and far from $E_{Na}$.

## VI. Determination of Resting Potential and Net Diffusion Potential ($E_{diff}$)

### A. Determining Factors

For given ion distributions, which normally remain nearly constant under usual steady-state conditions, the resting potential is determined by the relative membrane conductances ($g$) or permeabilities ($P$) for $Na^+$ and $K^+$ ions. That is, the resting potential (of about $-80$ mV in cardiac muscle or skeletal muscle) is close to $E_K$ (about $-94$ mV) because $g_K \gg g_{Na}$ or $P_K \gg P_{Na}$. There is a direct proportionality between $P$ and $g$ at constant $E_m$ and concentrations. From simple circuit analysis (using Ohm's law and Kirchhoff's laws), one can prove that the membrane potential will always be closer to the battery (equilibrium potential) having the lowest resistance (highest conductance) in series with it (see Figs. 5 and 7). In the resting membrane, this battery is $E_K$, whereas in the excited membrane it will be $E_{Na}$ (or $E_{Ca}$), because there is a large increase in $g_{Na}$ and/or $g_{Ca}$ during the AP.

Any ion that is passively distributed cannot determine the resting potential; instead, the resting potential determines the distribution of that ion. Therefore, $Cl^-$ is not considered for myocardial cells, skeletal muscle fibers, and neurons because it seems to be passively distributed. However, transient net movements of $Cl^-$ across the membrane do influence $E_m$; for example, washout of $Cl^-$ (in $Cl^-$-free solution) produces a transient depolarization, and reintroduction of Cl$^-$ produces a small hyperpolarization. Cl$^-$ movement is also involved in the production of inhibitory postsynaptic potentials (IPSPs) (see chapter on synaptic transmission).

Because of its relatively low concentration, coupled with its relatively low resting conductance, the Ca$^{2+}$ distribution has only a relatively small effect on the resting $E_m$, and so it can be ignored.

## B. Constant-Field Equation

A simplified, but most useful, version of the Goldman–Hodgkin–Katz constant-field equation can be given (for 37°C):

$$E_m = -61 \, \text{mV} \log \frac{[K]_i + \frac{P_{Na}}{P_K}[Na]_i}{[K]_o + \frac{P_{Na}}{P_K}[Na]_o} \qquad (20)$$

This equation shows that for a given ion distribution, the resting $E_m$ is determined by the $P_{Na}/P_K$ ratio, the relative permeability of the membrane to Na$^+$ and K$^+$. For myocardial cells and skeletal muscle fibers, the $P_{Na}/P_K$ ratio is about 0.04, whereas for nodal cells of the heart and smooth muscle cells, this ratio is closer to 0.10 or 0.20.

Inspection of the constant-field equation shows that the numerator of the log term will be dominated by the $[K]_i$ term [since the $(P_{Na}/P_K)[Na]_i$ term will be very small], whereas the denominator will be affected by both the $[K]_o$ and $(P_{Na}/P_K)[Na]_o$ terms. This relationship thus accounts for the deviation of the $E_m$ vs log $[K]_o$ curve from a straight line (having a slope of 61 mV/decade) in normal Ringer solution (Fig. 8). When $[K]_o$ is elevated ($[Na]_o$ being reduced by an equimolar amount), the denominator becomes more and more dominated by the $[K]_o$ term, and less and less by the $(P_{Na}/P_K)[Na]_o$ term. Therefore, in bathing solutions containing high K$^+$, the constant-field equation approaches the simple Nernst equation for K$^+$, and $E_m$ approaches $E_K$. As $[K]_o$ is raised stepwise, $E_K$ becomes correspondingly reduced, because $[K]_i$ stays relatively constant; therefore, the membrane becomes more and more depolarized (see Fig. 8).

A more detailed discussion of the constant-field equation and its other variants is given in the Appendix (part C) to this chapter.

When $[K]_o$ is elevated (e.g., to 8 m$M$) in some types of cells, a hyperpolarization of up to about 10 mV may be produced. Such behavior is often observed in cells with a high $P_{Na}/P_K$ ratio (due to low $P_K$) and therefore a low resting $E_m$, such as in young embryonic hearts. This hyperpolarization could be explained by several factors: (a) stimulation of the electrogenic Na$^+$ pump current ($I_p$), (b) an increase in $P_K$ (and therefore $g_K$) due to $[K]_o$ effect on $P_K$, and (c) an increase in $g_K$ (but not $P_K$) due to the concentration effect. A similar explanation may apply to the fall-over in the $E_m$ versus log $[K]_o$ curve when $[K]_o$ is lowered to 1 m$M$ and less, hence depolarizing the cells. This effect is prominent in rat skeletal muscle, for example (see Fig. 7 of Chapter 50).

## C. Chord Conductance Equation

An alternative method of approximating the membrane resting potential ($E_m$) is by use of the chord conductance equation. The word *chord* means a straight line connecting two points on a curve, and here specifically refers to the **average slope** of a nonlinear steady-state voltage/current curve, that is, a straight line from any point on the curve through the origin (zero applied current). (In contrast, slope conductance is the tangent at any point on the curve.) Thus,

$$E_m = \frac{g_K}{\Sigma g}E_K + \frac{g_{Na}}{\Sigma g}E_{Na} + \frac{g_{Cl}}{\Sigma g}E_{Cl} + \frac{g_{Ca}}{\Sigma g}E_{Ca} \qquad (21)$$

where $g_K, g_{Na}, g_{Cl}$, and $g_{Ca}$ are the membrane conductances for K$^+$, Na$^+$, Cl$^-$, and Ca$^{2+}$, respectively, and $\Sigma g$ is the total conductance (sum of all ionic partial conductances). The ratio of $g_K/\Sigma g$, for example, is the relative or fractional conductance for K$^+$.

The chord conductance equation can conveniently take into account all ions, including divalent cations, that are distributed unequally across the membrane. The ions important to membrane potentials (including action potentials, postsynaptic potentials, and receptor potentials) are K$^+$, Na$^+$, Cl$^-$, and Ca$^{2+}$. As discussed previously, Cl$^-$ cannot help in determining the resting potential if it is passively distributed. Thus, Eq. 21 can be rewritten, omitting the Cl$^-$ term, as

$$E_m = \frac{g_K}{\Sigma g}E_K + \frac{g_{Na}}{\Sigma g}E_{Na} + \frac{g_{Ca}}{\Sigma g}E_{Ca} \qquad (21a)$$

For simplicity, we can ignore the Ca$^{2+}$ term also, giving

$$E_m = \frac{g_K}{g_K + g_{Na}}E_K + \frac{g_{Na}}{g_K + g_{Na}}E_{Na} \qquad (21b)$$

where $\Sigma g$ is now equal to $g_K + g_{Na}$.

The chord conductance equation can be derived simply from Ohm's law and from circuit analysis for the condition when net current is zero ($I_{Na} + I_K = 0$) [see Appendix (part D) to this chapter for the derivation]. The equation holds true whenever the net current across the membrane is zero, as for the resting potential.

The chord conductance equation is useful for giving the membrane potential when the ion conductances and distributions are known. For example, at the neuromuscular junction, the neurotransmitter acetylcholine opens the gates of many ionic channels that allow both Na$^+$ and K$^+$ to pass through equally well (that is, $g_{Na} = g_K$). Hence, the potential that the postsynaptic membrane tends to seek when maximally activated (i.e., the equilibrium potential or so-called reversal potential for the end-plate potential, EPP) is

$$E_{EPP} = \frac{1}{2}(-94 \, \text{mV}) + \frac{1}{2}(+60 \, \text{mV})$$
$$= -17 \, \text{mV} \qquad (21c)$$

A disadvantage of the chord conductance equation is that it gives nearly a straight line for the $E_m$ vs log $[K]_o$ plot (actually a slight bend in the opposite direction at low $[K]_o$). In contrast, the constant-field equation gives the

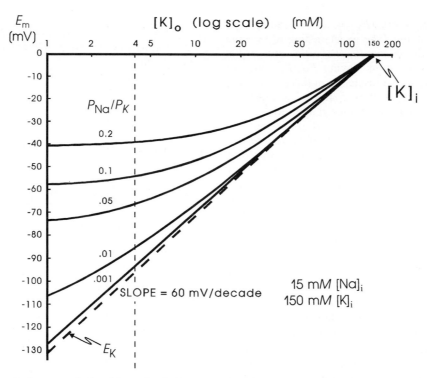

**FIG. 8.** Theoretical curves calculated from the Goldman constant-field equation for resting potential ($E_m$) as a function of $[K]_o$. Family of curves is given for various $P_{Na}/P_K$ ratios (0.001, 0.01, 0.05, 0.1, and 0.2). $K^+$ equilibrium potential ($E_K$) calculated from the Nernst equation (broken straight line). Curves calculated for a $[K]_i$ of 150 m$M$ and a $[Na]_i$ of 15 m$M$. Calculations made holding $[K]_o + [Na]_o$ constant at 154 m$M$; that is, as $[K]_o$ was elevated, $[Na]_o$ was lowered by an equimolar amount. Change in $P_K$ as a function of $[K]_o$ was not taken into account for these calculations. Point at which $E_m$ is zero gives $[K]_i$. The potential reverses in sign when $[K]_o$ exceeds $[K]_i$.

complete bending of the curves (for different $P_{Na}/P_K$ ratios) (see section on Constant-Field Equation).

The chord conductance equation again illustrates the important fact that the $g_K/g_{Na}$ ratio determines the resting potential. When $g_K \gg g_{Na}$, then $E_m$ is close to $E_K$; conversely, when $g_{Na} \gg g_K$ (as during the spike part of the AP), $E_m$ shifts to close to $E_{Na}$ or to $E_{Ca}$ (in the case of many types of smooth muscle cells).

The chord conductance equation can be rewritten, using resistances instead of conductances, and may then be called the **chord resistance equation,**

$$E_m = \frac{R_K}{R_K + R_{Na}} E_{Na} + \frac{R_{Na}}{R_K + R_{Na}} E_K \qquad (21d)$$

where $R_K$ and $R_{Na}$ are the $K^+$ and $Na^+$ resistances, which are the reciprocals of the conductances ($R_K = 1/g_K$ and $R_{Na} = 1/g_{Na}$). Note that in this equation the positions of the two batteries are interchanged. This equation can be derived by simply substituting the two reciprocals given above into the chord conductance equation. It can also be derived by circuit analysis, as discussed in the Appendix (part E) to this chapter. This Appendix section also shows how simple circuit analysis can be used to determine what the resting potential should be, without using either the Goldman constant-field equation or the chord conductance equation.

### D. Net Diffusion Potential, $E_{diff}$

In the presence of ouabain (short-term exposure only) to inhibit the Na–K pump and $V_p$, the resting potential that remains reflects the net diffusion potential, $E_{diff}$. $E_{diff}$ is determined by the ion concentration gradients for $K^+$ and $Na^+$ and by the relative permeability for $K^+$ and $Na^+$. When the Na–K pump is operating, there is normally a small additional contribution of $V_p$ to the resting $E_m$ of about 2–16 mV, depending on cell type (discussed in following section).

Inhibition of the Na–K pump for long periods will gradually run down the ion concentration gradients. The cells lose $K^+$ and gain $Na^+$, and therefore $E_K$ and $F_{Na}$ become smaller. The cells thus become depolarized (even if the relative permeabilities are unaffected), which causes them to gain $Cl^-$ (because $[Cl]_i$ was held low by the large resting potential) and therefore also water (cells swell).

### VII. Electrogenic Sodium Pump Potentials

A brief summary of the previous principles is as follows: The Na–K pump is responsible for maintaining the cation concentration gradients. The equilibrium potentials for $K^+$ ($E_K$) and $Na^+$ ($E_{Na}$) are about −94 mV and +60 mV, respectively. The resting potential value is usually near $E_K$, because the $K^+$ permeability ($P_K$) is much greater than $P_{Na}$ in a resting membrane. The exact resting membrane potential ($E_m$) depends on the $P_{Na}/P_K$ ratio, myocardial cells and skeletal muscle fibers having $P_{Na}/P_K$ ratios of 0.01–0.05, whereas smooth muscle or nodal cells of the heart have a ratio closer to 0.10–0.15. In the various types of cells, the resting $E_m$ has a smaller magnitude (i.e., is less negative) than $E_K$ by 10–40 mV. If there were no electrogenic pump potential contribution to the resting potential (that is, as though the Na–K pump was only indirectly responsible for the resting potential by its role in producing the ionic gradients), $E_m$ would equal $E_{diff}$.

However, a direct contribution of the pump to the resting $E_m$ can be demonstrated. For example, if the Na–K pump is blocked by the addition of ouabain, there usually is an immediate depolarization of 2–16 mV, depending on the type of cell. Thus, the direct contribution of the electrogenic $Na^+$–$K^+$ pump to the measured resting $E_m$ is small under physiologic conditions (but very important).

However, under conditions in which the pump is stimulated to pump at a high rate (e.g., when $[Na]_i$ or $[K]_o$ is abnormally high) the direct electrogenic contribution of the pump to the resting potential can be much greater, and $E_m$ can actually exceed $E_K$ by as much as 20 mV or more. For example, if the ionic concentration gradients are allowed to run down (e.g., by storing the tissues in zero $[K]_o$ and at low temperatures for several hours), then after the tissues are allowed to restart pumping, the measured $E_m$ can exceed the calculated $E_K$ (e.g., by 10–20 mV) for a time (Fig. 9). The $Na^+$ loading of the cells is facilitated by placing them in cold low or zero $[K]_o$ solutions, because external $K^+$ is necessary for the Na–K-linked pump to operate; $K_m$ of the Na,K-ATPase for $K^+$ is about 2 mM. After several hours in such a solution, the internal concentrations of $Na^+$, $K^+$, and $Cl^-$ approach the concentrations in the bathing Ringer solution, and the resting potential is very low ($< -30$ mV) (see Chapter 14). The cells are then transferred to a pumping solution, which is the appropriate Ringer solution containing normal $K^+$ and at normal temperature. Under such conditions, the pump turns over at a maximal rate, because the major control over pump rate is $[Na]_i$ and $[K]_o$. The low initial $E_m$ also stimulates the pump rate, because the energy required to pump out $Na^+$ is less. The measured $E_m$ of such $Na^+$ preloaded cells increases rapidly and more rapidly than $E_K$, as shown in Fig. 9. After this transient phase, however, a crossover of the two curves occurs, so that $E_K$ again exceeds $E_m$, as in the physiologic condition. Cardiac glycosides prevent or reverse the transient hyperpolarization beyond $E_K$. The possibility that ionic conductance changes (e.g., an increase in $g_K$ or a decrease in $g_{Na}$) can account for the observed hyperpolarization can be ruled out whenever $E_m$ exceeds (is more negative than) $E_K$.

Rewarming cells previously cooled leads to the rapid restoration of the normal resting potential (within 10 min), whereas recovery of the intracellular $Na^+$ and $K^+$ concentrations is slower. During prolonged hypoxia, the resting potential of cardiac muscle decreases much less than $E_K$ decreases (a difference of about 25 mV); the electrogenic pump attempts to hold the resting potential constant, de-

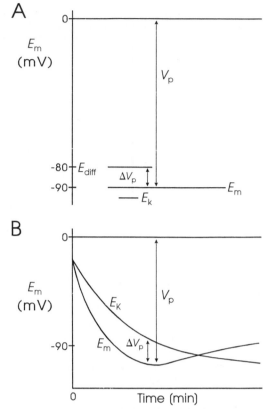

**FIG. 9.** Diagrammatic representation of an electrogenic sodium pump potential. (A) Muscle cell in which the net ionic diffusion potential ($E_{diff}$, function of ion equilibrium potentials and relative conductances) is $-80$ mV, yet exhibits a measured membrane resting potential ($E_m$) that is greater. Difference between $E_m$ and $E_{diff}$ represents the contribution of the electrogenic pump to the resting potential. Usual direct contribution of the pump is only a few millivolts and can be measured by the amount of depolarization produced immediately after complete inhibition of the Na,K-ATPase by cardiac glycosides. Because the pump pathway is separate from and parallel to the ionic conductance pathways, the electrogenic pump potential must be equal to $V_p$. The contribution of the electrogenic pump potential to the resting potential ($E_m - E_{diff}$) is equal to $\Delta V_p$. (B) Cell that was run down (Na loaded, K depleted) over several hours by inhibition of Na–K pumping, resulting in a low resting potential. Returning the muscle cell to a pumping solution allows the resting $E_m$ to rebuild as a function of time. Buildup in $E_m$ occurs faster than buildup in $E_K$, as illustrated. Whenever $E_m$ is greater (more negative) than $E_K$, the difference ($\Delta V_p$) must reflect the contribution of the sodium pump potential.

spite dissipating ionic gradients. It is not known whether the degree of electrogenicity of the pump (e.g., Na/K coupling ratio) might increase to compensate for a slowing pump rate.

Another method used to demonstrate that the pump is electrogenic is to inject $Na^+$ ions into the cell through a micropipette. This procedure rapidly produces a small transient hyperpolarization, which is immediately abolished or prevented by ouabain. The pump current and the rate of $Na^+$ extrusion increase in proportion to the amount of $Na^+$ injected. To prove that the pump is electrogenic, it must be demonstrated that the hyperpolarization produced in an intact muscle is not the result of enhanced pumping of an electroneutral pump. This could cause depletion of external $K^+$ in a restricted diffusion space just outside the cell membrane, leading to a larger $E_K$ and thereby to hyperpolarization. Depletion could occur if the Na–K pump pumped in $K^+$ faster than it could be replenished by diffusion from the bulk interstitial fluid.

The electrogenic $Na^+$ pump is influenced by the membrane potential. From energetic considerations, depolarization should enhance the electrogenic $Na^+$ pumping, whereas hyperpolarization should inhibit it. This is because depolarization reduces the electrochemical gradient (and hence the energy requirements) against which $Na^+$ must be extruded, whereas hyperpolarization increases the gradient. Thus, there should be a distinct potential, more negative than $E_K$, at which $Na^+$ pumping is prevented (e.g., a pump equilibrium potential). A value close to $-140$ mV was reported for cardiac cells and rat skeletal muscle fibers.

Any method used to increase membrane resistance increases the contribution of the pump to the resting potential (Fig. 10); that is, the electrogenic $Na^+$ pump contribution must be augmented under conditions that increase membrane resistance. The contribution of the pump potential to the measured $E_m$ is the difference in $E_m$ when the pump is operating versus that immediately after the pump has been stopped by the addition of ouabain or zero $[K]_o$. Consequently, it appears as though the contribution from the electrogenic pump potential ($\Delta V_p$) were in series with the net cationic diffusion potential ($E_{diff}$),

$$E_m = E_{diff} + R_m I_p = E_{diff} + \Delta V_p \qquad (22)$$

where $I_p$ is the electrogenic component of the pump current, and $E_{diff}$ is the $E_m$ that would exist solely on the basis of the ionic gradients and relative permeabilities in the absence of an electrogenic pump potential (as calculated from the constant-field equation). Equation 22 states that $E_m$ is the sum of $E_{diff}$ and a voltage (IR) drop produced by the electrogenic pump current across $R_m$. The electrogenic pump potential ($V_p$) can be considered to be in parallel with $E_{diff}$ (Fig. 10). Because the density of pump sites is more than 1000-fold greater than that of $Na^+$ and $K^+$ channels in resting membrane, there is no relation between the pump pathway (the active flux path) and $R_m$ (the passive flux paths); that is, the pump path and the passive conductance paths are in parallel. The true pump potential is much greater than the $\Delta E_m$ measured in the absence and presence of ouabain; namely, the pump potential should be considered the full potential between zero and the maximum negative pump potential ($V_p$) while the pump is pumping (see Fig. 9).

One possible equivalent circuit for an electrogenic $Na^+$ pump that takes into account some of the known facts is given in Fig. 10. The pump pathway is in parallel with the resistance pathways. The pump resistance ($R_p$) is estimated to be about 10-fold higher than $R_m$. If so, the pump resistance acts to minimize a short-circuit path to $E_{diff}$ when the pump potential is low or zero (pump inhibited). The pump potential contribution to $E_m$ ($\Delta V_p$) is a function of membrane resistance ($R_m$); the higher the $R_m$ ($R_p$ constant), the more nearly $E_m$ approaches $V_p$. The pump battery is

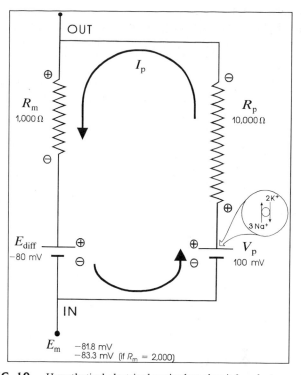

**FIG. 10.** Hypothetical electrical equivalent circuit for electrogenic sodium pump. Model consists of a pump pathway in parallel with the membrane resistance ($R_m$) pathway and the membrane capacitance ($C_m$) pathway. This model fits the evidence that the pump is independent of short-range membrane excitability and that the pump proteins and channel proteins are embedded in the lipid bilayer as parallel elements. Net diffusion potential ($E_{diff}$, determined by the ion equilibrium potentials and relative permeabilities) of $-80$ mV is depicted in series with $R_m$. Pump leg is assumed to consist of a battery in series with a fixed resistor (pump resistance, $R_p$) that does not change with changes in $R_m$ and whose value is 10-fold higher than $R_m$. Pump battery is charged up to some voltage (e.g., $V_p$ of $-100$ mV) by a pump current generator. Net electrogenic pump current is developed by the pumping in of only 2 $K^+$ ions for every 3 $Na^+$ ions pumped out. For the values given in the figure (namely, $R_m$ of 1000 $\Omega$, $E_{diff}$ of $-80$ mV, $R_p$ of 10,000 $\Omega$, and $V_p$ of 100 mV), it may be calculated by circuit analysis that the measured membrane potential ($E_m$) is $-81.8$ mV; that is, the direct electrogenic pump potential contribution to the resting potential is $-1.8$ mV. If $R_m$ were raised to 2000 $\Omega$ (e.g., by placing the membrane in $Cl^-$-free solution, or by adding $Ba^{2+}$ ion to decrease $P_K$, or both), then the calculated $E_m$ would be $-83.3$ mV.

charged to some voltage by a pump current generator. If the pump is stopped by ouabain, $V_p$ goes to zero. Using circuit analysis for the values of the parameters given in Fig. 10, $E_m$ would be $-81.8$ mV, moderately close to $E_{diff}$ ($-80$ mV) (Table 3). If $R_m$ is raised 2-fold (to 2000 $\Omega$), $E_m$ would be $-83.3$ mV. Thus, this circuit clearly gives a pump potential contribution to $E_m$ that is dependent on $R_m$. The higher $R_m$ is relative to $R_p$, the more $E_m$ reflects $V_p$. If $E_{diff}$ is made smaller (e.g., in smooth muscle cells having a higher $P_{Na}/P_K$ ratio), then the relative contribution of the pump potential to $E_m$ becomes greater (see Table 3).

In general, $Cl^-$ ions are known to have a short-circuiting effect on the electrogenic $Na^+$ pump potential. For exam-

**TABLE 3** Summary of Calculations of Resting Potential ($E_m$) for a Model Having an Electrogenic Pump Potential ($V_p$) in Parallel with the Net Diffusion Potential ($E_{diff}$)

| $E_{diff}$ | $R_m$ ($\Omega$-cm$^2$) | $R_p$ ($\Omega$-cm$^2$) | $V_p$ (mV) | Resting $E_m$ (mV) | $\Delta V_p$ ($E_m - E_{diff}$) (mV) |
|---|---|---|---|---|---|
| $-80$ | 1000 | 10,000 | $-100$ | $-81.8$ | $-1.8$ |
| $-80$ | 2000 | 10,000 | $-100$ | $-83.3$ | $-3.3$ |
| $-80$ | 4000 | 10,000 | $-100$ | $-85.7$ | $-5.7$ |
| $-80$ | 8000 | 10,000 | $-100$ | $-88.9$ | $-8.9$ |
| $-80$ | 1000 | 10,000 | 0 | $-72.7$ | $+7.3$ |
| $-80$ | 1000 | 10,000 | 0 | $-79.2$ | $+0.8$ |
| $-40$ | 1000 | 10,000 | $-100$ | $-45.5$ | $-5.5$ |
| $-50$ | 1000 | 10,000 | $-100$ | $-54.5$ | $-4.5$ |
| $-45$ | 2000 | 10,000 | $-100$ | $-54.2$ | $-9.2$ |
| $-45$ | 4000 | 10,000 | $-100$ | $-60.7$ | $-15.7$ |

*Note.* $R_m$, membrane resistivity; $R_p$, pump resistance; $\Delta V_p$, contribution of $V_p$ to the measured $E_m$. $E_m$ was calculated from the equation

$$E_m = \frac{R_m}{R_m + R_p} V_p + \frac{R_p}{R_m + R_p} E_{diff}$$

ple, if the external Cl$^-$ is replaced by less permeant anions, the magnitude of the hyperpolarization produced by the electrogenic Na$^+$ pump is substantially increased. This Cl$^-$ effect could be caused by the lowering of membrane resistance in the presence of Cl$^-$. The greater the $R_m$, the greater the contribution of the electrogenic pump potential to resting $E_m$ (see Fig. 10 and Table 3).

The density of Na–K pump sites, estimated by specific binding of [$^3$H]ouabain, is usually about 700–1000/$\mu$m$^2$. The turnover rate of the pump is generally estimated to be 20–100/sec. The **pump current ($I_p$)** has been estimated as

$$I_p = \frac{\Delta V_p}{R_m} \tag{23}$$

where $\Delta V_p$ is the pump potential contribution. Values of about 20 pmol/cm$^2$-sec were obtained. A density of 1000 sites/$\mu$m$^2$ ($10^{11}$ sites/cm$^2$) times a turnover rate of 40/sec gives $4 \times 10^{12}$ turnovers/cm$^2$-sec. If 3 Na$^+$ are pumped with each turnover, this gives $12 \times 10^{12}$ Na$^+$ ions/cm$^2$-sec; dividing by Avogadro's number ($6.02 \times 10^{23}$ ions/mol) yields $20 \times 10^{-12}$ mol/cm$^2$-sec, which is the same value as the 20 pmol/cm$^2$-sec measured. The net pump current would be less, depending on the amount of K$^+$ pumped in the opposite direction, that is, depending on the coupling ratio (e.g., 3 Na$^+$:2 K$^+$). Whenever the Na–K pump is stimulated to turn over faster, for example, by increasing [Na]$_i$ or [K]$_o$, the electrogenic pump current is increased.

**Ion flux ($J$)** can be converted to current ($I$) by the relationship

$$I = J \cdot zF$$
$$\frac{\text{amp}}{\text{cm}^2} = \frac{\text{mol}}{\text{sec cm}^2} \frac{\text{coul}}{\text{mol}} \tag{24}$$

Thus, a flux of 20 pmol/cm$^2$-sec is equal to approximately 2 $\mu$A/cm$^2$ ($20 \times 10^{-12}$ mol/sec-cm$^2$ $\times$ 0.965 $\times$ $10^5$ coul/mol). Since $\Delta V_p = I_p \times R_m$, if $R_m$ were 1000 ohm-cm$^2$ and

$I_p$ were 2 $\mu$A/cm$^2$, the electrogenic pump contribution to $E_m$ would be 2 mV ($E_m = E_{diff} + I_p R_m$).

Two K$^+$ ions are usually carried in for every 3 Na$^+$ ions moved out. Because the pump is **electrogenic,** that is, produces a net current (and hence potential) across the membrane, then the amount of K$^+$ pumped in must be less than the amount of Na$^+$ pumped out; for example, the Na/K **coupling ratio** must be 3:2 (or 3:1). The coupling ratio cannot be 3:0, because of the well-known fact that external K$^+$ must be present for the pump to operate. The coupling ratio might be increased under some conditions, for example, when [Na]$_i$ is elevated. If the coupling ratio were to increase (e.g., to 3:1), the pump potential contribution would become larger, for a constant pumping rate. The Na–K pump in several cell types can switch to a Na$^+$/Na$^+$ exchanging mode of operation, the mode change being governed by the ATP/$P_i$ ratios.

The pump current may be stimulated by increasing the turnover rate of each pump site and/or by increasing the number of pump sites. In skeletal muscle, insulin has been reported to increase the number of Na–K pump sites in the sarcolemma by increasing the rate of translocation from an internal pool, thereby increasing the pump current. $\beta$-Adrenergic agonists, like isoproterenol, stimulate the pump current by cyclic AMP/protein kinase A phosphorylation of the pump.

The electrogenic pump potential has physiologic importance in cells. Although small, the electrogenic pump potential contribution to the resting potential could have significant effects on the level of inactivation of the fast Na$^+$ channels, and hence on propagation velocity. Further, an electrogenic pump potential could act to delay depolarization under adverse conditions (e.g., ischemia and hypoxia) and would act to speed repolarization of the normal resting potential during recovery from the adverse conditions. It is crucial that the excitable cell maintain its normal resting potential as much as possible, because of the effect of small depolarizations on the AP rate of rise and conduction velocity and the complete loss of excitability with larger depolarizations. For example, the rate of firing of pacemaker nodal cells of the heart is affected significantly by very small potential changes.

In cells in which there are lower resting potentials (e.g., smooth muscle cells and cardiac nodal cells) (see Table 1), the electrogenic pump potential contribution can be considerably larger (see Table 3). Sinusoidal oscillations in the Na–K pumping rate could produce oscillations in $E_m$, which could exert important control over the spontaneous firing of the cell. The period of enhanced pumping hyperpolarizes the cell and suppresses automaticity, whereas slowing of the pump leads to depolarization and consequently to triggering of APs. Oscillation of the pump would be brought about by oscillating changes in [Na]$_i$. For example, the firing of several APs should raise [Na]$_i$ (nodal cells have a small volume to surface area ratio) and stimulate the electrogenic pump. The increased pumping rate, in turn, hyperpolarizes and suppresses firing, thus allowing [Na]$_i$ to decrease again and removing the stimulation of the pump; the latter condition depolarizes and triggers spikes, and the cycle could be repeated. It was concluded that the electrogenic $Na^+$ pump in rabbit sinoatrial nodal cells might be one factor that modulates the heart rate under physiologic conditions. When stimulated at a high rate, cardiac Purkinje fibers and nodal cells undergo a transient period of inhibition of automaticity after cessation of the stimulation, known as **overdrive suppression of automaticity.** Stimulation of the electrogenic pump due to elevation in $[Na]_i$ is the major cause of this phenomenon.

## VIII. Summary

Most of the factors that determine or influence the resting $E_m$ of cells were discussed in this chapter. The structural and chemical composition of the cell membrane was briefly examined and correlated with the resistive and capacitive properties of the membrane. The factors that determine the intracellular ion concentrations in cells were examined. These factors include the Na–K-coupled pump, the Ca/Na exchange reaction, and the sarcolemmal Ca pump. The Na–K pump enzyme, Na,K-ATPase, requires both $Na^+$ and $K^+$ for activity, and transports 3 $Na^+$ ions outward and usually 2 $K^+$ ions inward per ATP hydrolyzed. Cardiac glycosides are specific blockers of this transport ATPase. The Na–K pump is not directly related to excitability, but only indirectly related by its role in maintaining the $Na^+$ and $K^+$ concentration gradients.

The carrier-mediated Ca/Na exchange reaction is driven by the $Na^+$ electrochemical gradient; that is, the energy for transporting out internal $Ca^{2+}$ by this mechanism comes from the Na,K-ATPase. The Ca/Na exchange reaction exchanges one internal $Ca^{2+}$ ion for three external $Na^+$ ions when working in the forward mode in cells at rest. During the AP depolarization, for example, in myocardial cells, the energetics cause the Ca–Na exchanger to operate in reverse mode, allowing $Ca^{2+}$ influx.

The mechanism whereby the ionic distributions give rise to diffusion potentials was discussed, as were the factors that determine the magnitude and polarity of each ionic equilibrium potential. The equilibrium potential for any ion and the transmembrane potential determine the total electrochemical driving force for that ion, and the product of this driving force and membrane conductance for that ion determine the net ionic current or flux. The net ionic movement can be inward or outward across the membrane, depending on the direction of the electrochemical gradient.

The key factor that determines the resting $E_m$—in the absence of any electrogenic pump potential contributions—is the relative permeability of the various ions, particularly of $K^+$ and $Na^+$, that is, the $P_{Na}/P_K$ ratio (or $g_{Na}/g_K$ ratio), as calculated from the Goldman constant-field equation. The major physiologic ions that have some effect on the resting $E_m$ or on the APs are $K^+$, $Na^+$, $Ca^{2+}$, and $Cl^-$. The $Ca^{2+}$ electrochemical gradient has only a small direct effect on the resting $E_m$, although low external $Ca^{2+}$ can affect the permeabilities and conductances for the other ions, such as $Na^+$ and $K^+$. Elevation of internal $Ca^{2+}$ can increase the permeability to $K^+$ by activating $Ca^{2+}$-operated $K^+$-selective $I_{K(Ca)}$ channels.

$Cl^-$ is usually passively distributed according to the membrane potential, that is, not actively transported. However, there is some evidence indicating that $[Cl]_i$ may be about twice as high as that predicted from $E_m$ in some cells like smooth muscle cells; if so, this would give an $E_{Cl}$ value of about 18 mV less negative than the resting $E_m$. Before one can conclude that there is a $Cl^-$ pump directed inward, however, the calculated $E_{Cl}$ (concentrations corrected for activity coefficients) must be proven to be significantly more positive than the mean resting $E_m$ of the cell averaged over time; for example, any spontaneous APs must be taken into account. If $Cl^-$ is passively distributed, it cannot determine the resting $E_m$. However, transient net movements of $Cl^-$ ions, for example, during the AP, can and do affect the $E_m$, particularly when $g_{Cl}$ is high.

Elevation of $[K]_o$ to more than the normal concentration of about 4.5 m$M$ decreases the $K^+$ equilibrium potential ($E_K$), as predicted from the Nernst equation ($[K]_i$ about constant), and depolarization is produced. Sometimes, however, some hyperpolarization is produced at a $[K]_o$ level between 5 and 9 m$M$. In addition, lowering $[K]_o$ to 0.1 m$M$ often produces a prominent depolarization. These effects are usually explained on the basis that (1) $P_K$ is lowered in low $[K]_o$ and elevated in higher $[K]_o$ and (2) an electrogenic Na–K pump potential is inhibited at a low $[K]_o$ ($K_m$ of about 2 m$M$).

Not only is the resting $E_m$ the potential energy storehouse that is drawn upon for production and propagation of the APs, but because the membrane voltage-dependent cationic channels are inactivated with sustained depolarization, the rate of rise of the AP, and hence propagation velocity, is critically dependent on the level of the resting $E_m$. For example, a relatively small elevation of $K^+$ concentration in the blood has dire consequences for functioning of the heart.

The contribution of the Na–K pump to the resting $E_m$ depends on (1) the coupling ratio of $Na^+$ pumped out to $K^+$ pumped in, (2) the turnover rate of the pump, (3) the number of pumps, and (4) the magnitude of the membrane resistance. The electrogenic pump potential is in parallel to the net ionic diffusion potential ($E_{diff}$), determined by the ionic equilibrium potentials and by the relative permeabilities. The contribution of the electrogenic pump potential to the measured resting $E_m$ of cells varies from 2 to 16 mV, depending on the type of cell. Thus, the immediate depolarization produced by complete Na–K pump stoppage with cardiac glycosides is only a few millivolts in cells like myocardial cells. Of course, long-term pump inhibition produces a larger and larger depolarization as the ionic gradients are dissipated. The rate of Na–K pumping, and hence the magnitude of the electrogenic pump contribution to $E_m$, is controlled primarily by $[Na]_i$ and by $[K]_o$. The electrogenic pump potential might be physiologically important to various tissues, particularly the heart, under certain conditions that tend to depolarize the cells, such as transient ischemia or hypoxia. In such cases, the actual depolarization produced may be less because of a relatively constant pump potential in parallel with a diminishing $E_{diff}$. The electrogenic pump potential may also affect automaticity of the nodal cells of the heart as well as other types of cells that exhibit automaticity.

# Appendix

## A. Derivation of Nernst Equation

The Nernst equation may be derived from the general equation for the free energy change ($\Delta G_c$) resulting from both osmotic work and electrical work for transporting 1 mole of cation ($c^+$) across a membrane. Thus,

$$\Delta G_c = RT \ln \frac{[c^+]_i}{[c^+]_o} + zFE_m \qquad \text{(AA-1)}$$

where $R$ is the gas constant; $T$ is absolute temperature; $[c^+]_i$ and $[c^+]_o$ are the internal and external $c^+$ concentrations, respectively; $z$ is the valence; $F$ is the Faraday constant; and $E_m$ is membrane potential. The first term on the right side of this equation, $RT \ln([c^+]_i/[c^+]_o)$, gives the osmotic work for transporting a mole of particles across the membrane against a concentration gradient. The second term, $zFE_m$, gives the electrical work for transporting 1 mole of charged particles across the membrane against an electrical gradient. The sum of these two terms then gives the total work required. At equilibrium, the change in free energy for moving one or only a few particles across the membrane must be zero ($\Delta G = 0$). Therefore,

$$0 = RT \ln \frac{[c^+]_i}{[c^+]_o} + zFE_m \qquad \text{(AA-2)}$$

and

$$zFE_m = -RT \ln \frac{[c^+]_i}{[c^+]_o} \qquad \text{(AA-2a)}$$

or

$$E_m = \left(\frac{-RT}{zF}\right) \ln \frac{[c^+]_i}{[c^+]_o} \qquad \text{(AA-2b)}$$

which is the Nernst equation.

Since the Faraday constant ($F$) is equal to the charge on an electron ($e$, in coulombs) times Avogadro's number ($N_A$, number of ions per mole), then

$$\frac{RT}{F} = \frac{RT}{N_A Q_e} = \frac{kT}{Q_e} \qquad \text{(AA-3)}$$

where $k$ (the Boltzmann constant) is equal to the gas constant divided by Avogadro's number ($N_A$), that is, the energy (in joules) of an ion per degree Kelvin, and $Q_e$ is the charge (in coulombs) on an electron (namely, $1.6 \times 10^{-19}$ coul/e.)

## B. Half-Cell Potentials

In measuring biological potentials, care must be taken not to introduce artifacts, such as half-cell potentials. This section will give a brief description of electrode half-cell potentials. For example, if two beakers containing NaCl at 0.1 and 0.01 $M$ were joined by a salt bridge (agar–NaCl), and if a Ag–AgCl half-cell electrode were placed in each beaker, then a PD of 59 mV would be recorded between the two electrodes, because the potential of each half-cell, reversible to $Cl^-$ ions, would be different (Fig. A-1). In this example, the beaker containing the higher $Cl^-$ concentration would be *negative*, and the one with the lower $Cl^-$ concentration would be *positive*. The AgCl coat of the electrode immersed in the lower $Cl^-$ concentration would have the greater tendency to solubilize and ionize, leaving this electrode positive. Conversely, the AgCl coat of the electrode immersed in the higher $Cl^-$ concentration would have the lower tendency to solubilize and actually would tend to deposit more AgCl, stealing a positive charge from the wire and thus leaving that electrode negatively charged. A positive potential is applied to electroplate the Ag wires with AgCl by electrophoresing $Cl^-$ to the Ag wire, as shown in Fig. A-1.

Note that the resting potential recorded in biological cells by an intracellular microelectrode is not a function of the half-cell potentials (i.e., an artifact), because the solutions bathing the half-cells (e.g., Ag–AgCl wires or calomel half-cells) remain constant; that is, the half-cell potentials stay the same whether the microelectrode is inside or outside the cell. The two half-cell potentials are nearly equal in magnitude and so cancel each other. Any small amount of difference between the two half-cell potentials (e.g., a few millivolts) when the two electrodes are in the same Ringer's solution is arbitrarily called the zero

**A**

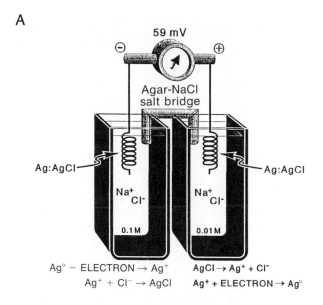

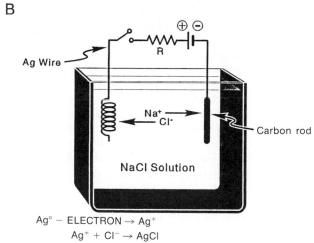

**FIG. A-1.** Half-cell electrode potentials: (A) Two Ag–AgCl half-cell electrodes are bathed in two different solutions containing Cl⁻ ion at different concentrations (0.1 and 0.01 $M$, in figure). The Ag–AgCl half-cells are reversible to Cl⁻, and therefore the half-cell potential depends on the Cl⁻ concentration in which the electrode is bathed. Electrode bathed in the highest Cl⁻ concentration has a more negative value than the electrode in lower Cl⁻ concentration. Thus the two half-cell potentials are not equal to one another, and their potentials do not cancel out, as is normally true. There is a net PD produced by the two unequal half-cell potentials, the electrode in the higher Cl⁻ concentration being negative and the electrode in the lower Cl⁻ concentration being positive. The two solutions are connected by an agar salt bridge to complete the circuit. (B) For electroplating a silver wire electrode with AgCl, the silver wire must be made positive so that Cl⁻ ions can be electrophoresed through the solution to react with silver atoms to plate AgCl.

potential (in practice, with the microelectrode in position, any small microelectrode tip potential, e.g., up to 5 mV, would be included in the zeroing procedure). The resting potential of the cell is added in series with half-cell potential, and thus the recording system gives the true transmembrane resting potential.

## C. Constant-Field Equation

An important modification of the Nernst equation in common use for calculating the membrane potential, or for determining the $P_{Na}/P_K$ ratio, is the Goldman–Hodgkin–Katz constant-field equation (Goldman, 1943; Hodgkin and Katz, 1949),

$$E_m = \frac{-RT}{F} \ln \frac{P_K[K^+]_i + P_{Na}[Na^+]_i + P_{Cl}[Cl^-]_o}{P_K[K^+]_o + P_{Na}[Na^+]_o + P_{Cl}[Cl^-]_i} \quad \text{(AC-1)}$$

where $P_K$, $P_{Na}$, and $P_{Cl}$ are the membrane permeabilities for K⁺, Na⁺, and Cl⁻, respectively. The $P_K$ [K]$_i$ product, for example, is given in the units of a flux (mol/sec per cm²), but the entire right-hand term (fraction) is dimensionless. Plugging in the numerical values for the constants and converting from natural logarithm, or ln, to logarithm to the base 10, or log (ln $N$ = 2.3 log $N$) gives

$$E_m = -61 \text{ mV} \log \frac{P_K[K]_i + P_{Na}[Na]_i + P_{Cl}[Cl]_o}{P_K[K]_o + P_{Na}[Na]_o + P_{Cl}[Cl]_i} \quad \text{(AC-1a)}$$

and $E_m$ is expressed in mV. Dividing the right-hand term by $P_K$ gives

$$E_m = -61 \text{ mV} \log \frac{[K^+]_i + \dfrac{P_{Na}}{P_K}[Na^+]_i + \dfrac{P_{Cl}}{P_K}[Cl^-]_o}{[K^+]_o + \dfrac{P_{Na}}{P_K}[Na^+]_o + \dfrac{P_{Cl}}{P_K}[Cl^-]_i}$$

$$\text{(AC-1b)}$$

Again, the right-hand term (fraction) is dimensionless. Any ion that is passively distributed, that is, not actively pumped, however, cannot determine the resting potential, because the distribution of that ion must follow the resting potential. Therefore, when Cl⁻ is passively distributed, it is not considered because resting potential cannot be determined by Cl⁻. As a result Eq. AC-1b can be reduced to

$$E_m = -61 \text{ mV} \log \frac{[K^+]_i + \dfrac{P_{Na}}{P_K}[Na^+]_i}{[K^+]_o + \dfrac{P_{Na}}{P_K}[Na^+]_o} \quad \text{(AC-1c)}$$

Equation AC-1c is one of the most useful forms of the constant field equation, because if $E_m$ is measured, and if the internal and external ion concentrations are known, the $P_{Na}/P_K$ ratio can be calculated. Thus for given ionic gradients, the resting potential is determined by the $P_{Na}/P_K$ ratio (i.e., the relative permeabilities of the cell membrane to Na⁺ and K⁺) and not by the absolute permeabilities. For simplicity, Ca²⁺ is ignored here as a factor contributing to the resting potential. The relationship between the permeability coefficient for an ion ($P_i$) and the membrane conductance for that ion ($g_i$) is complex and involves several terms, including membrane potential.

Figure 8 gives the expected resting potential as a function of the $P_{Na}/P_K$ ratio for a muscle cell (having a [K]$_i$ of 150 m$M$) bathed in normal Ringer's solution ([K]$_o$ of 4 m$M$, [Na]$_o$ of 150 m$M$), assuming an [Na]$_i$ value of 15 m$M$. As can be seen in the figure, a $P_{Na}/P_K$ ratio of 0.1 gives a resting potential close to −60 mV, whereas a ratio of 0.01

gives a potential close to $-85$ mV. Some muscle cells, for example, smooth muscle and young embryonic myocardial cells, have a low resting potential (at a $[K]_o$ of 4 m$M$) of about $-50$ mV, presumably because of a high $P_{Na}/P_K$ ratio of about 0.15 rather than smaller ionic gradients. The $P_{Na}/P_K$ ratio can be high because of either a high $P_{Na}$ or a low $P_K$, or both; in most cases the main reason appears to be a low $P_K$.

One advantage the constant-field equation has over the chord conductance equation is that it nicely accounts for the bend at low $[K]_o$ in the $E_m$ vs log $[K]_o$ curves (see Fig. 8). As can be seen in Fig. 8, which presents theoretical curves calculated from Eq. AC-1c, the higher the $P_{Na}/P_K$ ratio, the greater the deviation from a straight line as $[K]_o$ is lowered. As mentioned previously, from this equation, one can deduce that as $[K]_o$ is lowered and $[Na]_o$ is concomitantly elevated, the denominator of the right-hand term becomes more and more dominated by the $Na^+$ term, for any given $P_{Na}/P_K$ ratio. Because the numerator is relatively fixed, $E_m$ is more influenced by $E_{Na}$ as $[K]_o$ is lowered more and more. Thus, this relationship accounts for the deviation from the straight line for $E_K$.

Although the $P_{Na}/P_K = 0.05$ curve is almost linear at high $[K]_o$ with a slope of 60 mV/decade, the membrane does not necessarily become "purely $K^+$-selective," as is often stated, because for these theoretical calculations, the $P_{Na}/P_K$ ratio was held constant over the entire $[K]_o$ range. There is some evidence, however, that $P_K$ itself increases as $[K]_o$ increases, but this effect was not taken into consideration in Fig. 8. $g_K$ is a function of $[K]_o$, namely, $g_K \propto P_K [K]_o$. Finally, the increased bending for the higher $P_{Na}/P_K$ ratios can again be seen from Eq. AC-1c; at a given $[K]_o$ (e.g., 4 m$M$) the denominator is more and more dominated by the $Na^+$ term as the $P_{Na}/P_K$ ratio is increased more and more.

The order of selectivity of the resting membrane for the alkali metal ions generally is in the following sequence, from the highest permeability to the lowest: $K^+ > Rb^+ > Cs^+ > Na^+ > Li^+$. For example, the relative permeabilities (assigning $P_K = 1$) in squid giant axon (Baker *et al.*, 1968) are

$$P_K > P_{Rb} > P_{Cs} > P_{Na} > P_{Li}$$
$$1.0 \quad 0.69 \quad 0.19 \quad 0.17 \quad 0.12$$

In frog sartorius (Mullins, 1961) the values are

$$P_K > P_{Rb} > P_{Cs} > P_{Na}$$
$$1.0 \quad 0.54 \quad 0.11 \quad 0.04$$

The $P_{Na}/P_{Ca}$ ratio in frog sartorius is about 3 (Mullins, 1961).

So far in our discussion, $Ca^{2+}$ has been ignored. It can be demonstrated that $Ca^{2+}$ has only a negligible effect on the resting potential, even if it has a permeability equal to, or 10 times greater, than $Na^+$. This is because of the relatively low extracellular and intracellular concentration of free $Ca^{2+}$ ion compared with those of the $K^+$ and $Na^+$ ions. A modified version of the Goldman constant-field equation, which includes a $Ca^{2+}$ term, is[3]

---

[3] This equation was kindly provided by Professor D. E. Goldman.

$$E_m = -60 \text{ mV} \log \frac{(B - A) + \sqrt{y}}{2(A - 4P_{Ca}[Ca]_i)} \quad \text{(AC-2)}$$

where

$$A = P_K[K]_i + P_{Na}[Na]_i$$
$$B = P_K[K]_o + P_{Na}[Na]_o$$
$$y = (B - A)^2 + 4(A + 4P_{Ca}[Ca]_i)(B + 4P_{Ca}[Ca]_o)$$

For simplification, the analogous $Cl^-$ terms ($+P_{Cl}[Ca]_o$ in the definition of $A$ and $+P_{Cl}[Cl]_i$ term in $B$) have been omitted, assuming $Cl^-$ to be passively distributed.

Calculations made from Eq. AC-2 demonstrate some interesting points: (1) For the same permeabilities ($P_{Ca} = P_{Na}$), $Ca^{2+}$ has a much less effect on $E_m$ than does $Na^+$, because of the lower $Ca^{2+}$ concentrations and because of the square root function for the $Ca^{2+}$ concentrations. For example, the depolarization produced by taking into account the $Ca^{2+}$ ion is only $+0.4$ mV for a $P_{Na}/P_K$ ratio of 0.1. (2) Even when $P_{Ca}$ is set equal to 10 times $P_{Na}$, the effect of $Ca^{2+}$ on the resting potential is still relatively small (e.g., $+3.5$ mV for a $P_{Na}/P_K$ ratio of 0.01, and $+8.0$ mV for a $P_{Na}/P_K$ ratio of 0.1). (3) The effect of $Ca^{2+}$ is somewhat greater when the $P_{Na}/P_K$ ratio is higher (as in cardiac nodal cells or smooth muscle). (4) The effect of taking into account $Ca^{2+}$ is considerably less at high $[K]_o$ values.

Thus, these calculations support the view that $Ca^{2+}$ can be virtually ignored in discussion of the ionic basis of the resting potential. This agrees with the well-known fact that a variation in [Ca] throughout a relatively wide range has a negligible effect on the resting potential (see, for example, Sperelakis, 1972). Further these conclusions have implications about the relative importance of $Na^+$ vs $Ca^{2+}$ background currents (inward) during genesis of the pacemaker potential (concomitant with the decrease in $g_K$ and $I_K$). Finally, it should be emphasized that, for example, when $P_{Ca} = P_K$, $g_{Ca}$ does not equal $g_K$, because of the concentration differences. To calculate $g_K$ from a given $P_K$, one must use the appropriate equation that takes into account the concentrations and membrane potential. If $g_{Ca} = g_K$, then the membrane potential is halfway between $E_K$ and $E_{Ca}$, and $Ca^{2+}$ would have a much greater effect on the resting potential.

## D. Derivation of Chord Conductance Equation

Ohm's law states that the current ($I$) is equal to the voltage ($E$) either divided by the resistance ($R$) or multiplied by the conductance ($g = 1/R$):

$$I = \frac{E}{R} = gE \quad \text{(AD-1)}$$

When dealing with solutions, the voltage or driving force must take into account both the concentration force and the electrical force. In this case, the ionic current ($I_i$) is a product of the conductance for a given ion times the total driving force on that ion ($E_m - E_i$) and may be expressed as

$$I_i = g_i(E_m - E_i) \quad \text{(AD-2)}$$

In a resting cell membrane (stable resting potential), the total ionic current must be zero; otherwise, the membrane potential would change. Therefore, the $K^+$ current in the outward direction $(I_K)$ must be equal and opposite to the $Na^+$ current $(I_{Na})$ entering the cell (neglecting $Ca^{2+}$, $Cl^-$, and minor ions) expressed as

$$I_K = -I_{Na} \tag{AD-3}$$

Therefore,

$$I_K + I_{Na} = 0 \tag{AD-3a}$$

Substituting the equations for ionic currents from Eq. AD-2,

$$g_K(E_m - E_K) + g_{Na}(E_m - E_{Na}) = 0 \tag{AD-4}$$

Algebraic manipulations give

$$0 = g_K E_m - g_K E_K + g_{Na} E_m - g_{Na} E_{Na}$$
$$g_K E_m + g_{Na} E_m = g_K E_K + g_{Na} E_{Na} \tag{AD-4a}$$
$$E_m(g_K + g_{Na}) = g_K E_K + g_{Na} E_{Na}$$

Rearrangement gives

$$E_m = \frac{g_K}{g_K + g_{Na}} E_K + \frac{g_{Na}}{g_K + g_{Na}} E_{Na} \tag{AD-5}$$

Equation AD-5 is the chord conductance equation. The ratios $g_K/(g_K + g_{Na})$ and $g_{Na}/(g_K + g_{Na})$ are the fractional conductances (relative) and are dimensionless.

If $Cl^-$ were to be included, the same steps in the derivation would give the chord conductance equation containing a $Cl^-$ term,

$$E_m = \frac{g_K}{\Sigma g} E_K + \frac{g_{Na}}{\Sigma g} E_{Na} + \frac{g_{Cl}}{\Sigma g} E_{Cl} \tag{AD-6}$$

$\Sigma g = g_K + g_{Na} + g_{Cl}$. However, if $Cl^-$ is passively distributed (in equilibrium at the resting $E_m$), then $Cl^-$ cannot be involved in determining the resting potential (although transient movements of $Cl^-$ can affect the membrane potential when $Cl^-$ is shifted off equilibrium during an AP or postsynaptic potential).

On the other hand, the $Ca^{2+}$ ion is actively transported and is off equilibrium, so its conductance influences the resting potential. Thus, the chord conductance equation containing the $Ca^{2+}$ term is

$$E_m = \frac{g_K}{\Sigma g} E_K + \frac{g_{Na}}{\Sigma g} E_{Na} + \frac{g_{Ca}}{\Sigma g} E_{Ca} \tag{AD-7}$$

$\Sigma g = g_K + g_{Na} + g_{Ca}$.

The chord conductance equation, of course, can be written using resistances rather than conductances. For Eq. AD-5 using only $K^+$ and $Na^+$ terms, substitution of $R = 1/g$ in the equation and algebraic manipulation gives the equation

$$E_m = \frac{R_K}{R_K + R_{Na}} E_{Na} + \frac{R_{Na}}{R_K + R_{Na}} E_K \tag{AD-8}$$

Note that the $E_{Na}$ and $E_K$ terms are interchanged from the chord conductance equation. This form of the equation might be termed the **chord resistance equation.**

The chord conductance equation applies only to those situations in which the net ionic current is zero, such as when the membrane is at rest. This equation is derived simply from Ohm's law, and one advantage it has over the constant-field equation is that it can more easily include divalent cations such as $Ca^{2+}$.

### E. Circuit Analysis Applicable to Cell Membrane

Using Ohm's and Kirchhoff's laws and logic, it is possible to see why in the nerve or muscle cell, the $K^+$ battery dominates the resting potential, whereas the $Na^+$ or $Ca^{2+}$ batteries, or both, dominate the peak of the AP. The circuit in Fig. A-2 will be used to show that the battery having the lowest resistance in series with it is the battery that is the most expressed across the network. Before analyzing the circuit rigorously, we can consider three conditions and make some qualitative judgments: (1) If the left resistor $(R_1)$ equals the right resistor $(R_2)$ (regardless of their absolute values), the PD across the network is +150 V (upper terminal positive with respect to the lower terminal), that is, halfway between both batteries because both should be equally expressed. (2) If $R_2$ is made infinite (e.g., open circuit in branch 2) and $R_1$ is finite, then the PD is exactly +100 V, since the right battery $(E_2)$ cannot be expressed at all. (3) If $R_1$ is much less than $R_2$, then the PD approaches +100 V because $E_1$ is dominant.

The circuit in Fig. A-2 can also be analyzed quantitatively. For example, if $R_1 = 10\ \Omega$ and $R_2 = 990\ \Omega$, the exact PD may be reasoned from the following analysis. The current $(I)$ has one magnitude; that is, it is constant throughout this simple closed circuit, but the current flows upward in branch 2 and downward in branch 1. This occurs because the right battery $(E_2)$ is larger than $E_1$, and so the net driving force for the net current is in the direction as indicated in the figure. Therefore, the voltage drops produced across $R_1$ and $R_2$ are in opposite polarities, as shown in the figure. The voltage drop across $R_1$ adds to $E_1$ to make a greater PD across branch 1, like two batteries in series $(+ -, + -)$. In contrast, the voltage drop across $R_2$ subtracts from $E_2$ to make a smaller PD, like two batter-

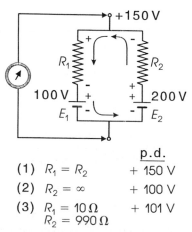

| | | p.d. |
|---|---|---|
| (1) | $R_1 = R_2$ | + 150 V |
| (2) | $R_2 = \infty$ | + 100 V |
| (3) | $R_1 = 10\ \Omega$ $R_2 = 990\ \Omega$ | + 101 V |

**FIG. A-2.** Circuit diagram of the circuit analysis applicable to the cell membrane, showing why the resting potential of a cell is determined by the relative permeabilities (or conductances). Battery having the lowest resistance in series with it is the battery most expressed across such a network.

ies back to back $(- +, + -)$. Therefore, the following two equations can be written for the PD across branch 1 $[(PD)_1]$ and across branch 2 $[(PD)_2]$:

$$(PD)_1 = E_1 - IR_1 \qquad (AE-1)$$

$$(PD)_2 = E_2 - IR_2 \qquad (AE-2)$$

To solve these equations, we must first calculate the current $(I)$. The net driving force for $I$ is equal to $E_2 - E_1$; hence from Ohm's law, the net current is equal to $(E_2 - E_1)$ divided by the total resistance $(R_1 + R_2)$:

$$I = \frac{E_2 - E_1}{R_2 + R_1} \qquad (AE-3)$$

$$I = \frac{200\ V - 100\ V}{990\ \Omega + 10\ \Omega}$$

$$= \frac{100\ V}{1000\ \Omega} = 0.1\ A$$

Now we can enter this value for $I$ into Eq. AE-1:

$$
\begin{aligned}
(PD)_1 &= E_1 - IR_1 \qquad (AE-1)\\
&= 100\ V - (-0.1\ A)(10\ \Omega)\\
&= 100\ V - (-1\ V)\\
&= 100\ V + 1\ V\\
&= 101\ V
\end{aligned}
$$

The negative sign in the current is because the current in branch 1 produces a voltage drop that adds to $E_1$. Because the two branches are connected by zero resistances, that is, they are effectively the same points, $(PD)_1$ must equal $(PD)_2$:

$$(PD)_1 = (PD)_2 \qquad (AE-4)$$

Therefore, we can also calculate the PD by substituting into Eq. AE-2.

$$
\begin{aligned}
(PD)_2 &= E_2 - IR_2 \qquad (AE-2)\\
&= 200\ V - (+0.1\ A)(990\ \Omega)\\
&= 200\ V - (99\ V)\\
&= 101\ V
\end{aligned}
$$

Thus, the two methods check.

To summarize, it has been quantitatively demonstrated that the battery with the lowest series resistance is the battery most expressed across this network. This analysis holds true regardless of the absolute values of the resistances or batteries or the polarity of each battery. Other methods of circuit analysis can be used to calculate the PD across such a network, but this method is one of the simplest.

The following chord resistance equation, analogous to the chord conductance equation, can also be derived and used to calculate the PD across the network:

$$
\begin{aligned}
PD &= \frac{R_1}{R_1 + R_2} E_2 + \frac{R_2}{R_1 + R_2} E_1 \qquad (AE-5)\\
&= \frac{10\ \Omega}{10\ \Omega + 990\ \Omega} 200\ V + \frac{990\ \Omega}{10\ \Omega + 990\ \Omega} 100\ V\\
&= \frac{10}{1000} 200\ V + \frac{990}{1000} 100\ V\\
&= 2\ V + 99\ V\\
&= 101\ V
\end{aligned}
$$

This equation again emphasizes the point that it is the relative resistances that determine which battery is most expressed.

## Bibliography

Baker, P. F. (1968). Nervous conduction: Some properties of the ion-selective channels which appear during the action potential. *Br. Med. Bull.* **24,** 179–182.

Carmeliet, E., and Vereecke, J. (1979). Electrogenesis of the action potential and automaticity. *In* "Handbook of Physiology" (R. M. Berne and N. Sperelakis, Eds.) pp. 269–334. American Physiological Society, Bethesda, MD.

Cole, K. S. (1968). "Membranes, Ions and Impulses: A Chapter of Classical Biophysics." University of California, Berkeley.

Daniel, E. E., Kwan, C. Y., Matlib, M. A., Crankshaw, D., and Kidwai, A. (1977). Characterization and $Ca^{2+}$-accumulation by membrane fractions from myometrium and artery. *In* "Excitation–Contraction Coupling in Smooth Muscle" (R. Casteels, T. Godfraind, and J. C. Ruegg, Eds.), pp. 181–188. Elsevier–North-Holland, Amsterdam.

Dhalla, N. S., Ziegelhoffer, A., and Hazzow, J. A. (1977). Regulatory role of membrane systems in heart function. *Can. J. Physiol. Pharmacol.* **55,** 1211–1234.

Gadsby, D. C., and Nakao, M. (1989). Steady-state current-voltage relationship of the Na/K pump in guinea pig ventricular myocytes. *J. Gen. Physiol.* **94,** 511–537.

Glitsch, H. G. (1972). Activation of the electrogenic sodium pump in guinea-pig auricles by internal sodium ions. *J. Physiol. (London)* **220,** 565–582.

Goldman, D. E. (1943). Potential, impedance, and rectification in membranes. *J. Gen. Physiol.* **27,** 37–60.

Henn, F. A., and Sperelakis, N. (1968). Stimulative and protective action of $Sr^{2+}$ and $Ba^{2+}$ on $(Na^+, K^+)$-ATPase from cultured heart cells. *Biochim. Biophys. Acta* **163,** 415–417.

Hermsmeyer, K., and Sperelakis, N. (1970). Decrease in $K^+$ conductance and depolarization of frog cardiac muscle produced by $Ba^{2+}$. *Am. J. Physiol.* **219,** 1108–1114.

Hodgkin, A. L., and Katz, B. (1949). The effect on sodium ions in electrical activity of the giant axon of the squid. *J. Physiol. (London)* **108,** 37–77.

Irisawa, H. (1978). Comparative physiology of the cardiac pacemaker mechanism. *Physiol. Rev.* **58,** 461–498.

Jain, M. K. (1972). "The Bimolecular Lipid Membrane: A System." Van Nostrand, New York.

Jones, I. R., Maddock, S. W., and Besch, H. R., Jr. (1980). Unmasking effect of alamethicin on the $(Na^+, K^+)$-TPase, beta-adrenergic receptor-coupled adenylate cyclase, and cAMP-dependent protein kinase activities of cardiac sarcolemmal vesicles. *J. Biol. Chem.* **255,** 9971–9980.

McDonald, T. F., and MacLeod, D. P. (1971). Maintenance of resting potential in anoxic guinea pig ventricular muscle: Electrogenic sodium pumping. *Science* **172,** 570–572.

Meech, R. W. (1972). Intracellular calcium injection causes increased potassium conductance in Aplysia nerve cells. *Comp. Biochem. Physiol.* **42A,** 493–499.

New, W., and Trautwein, W. (1972). Inward membrane currents in mammalian myocardium. *Pflugers Arch.* **334,** 1–23.

Noble, D. (1975). "Initiation of the Heartbeat." Oxford University Press (Clarendon), London.

Noma, A., and Irisawa, H. (1974). Electrogenic sodium pump in rabbit sinoatrial node cell. *Pflugers Arch.* **351,** 177–182.

Pelleg, A., Vogel, S., Belardinelli, L., and Sperelakis, N. (1980). Overdrive suppression of automaticity in cultured chick myocardial cells. *Am. J. Physiol.* **238,** H24–H30.

Sperelakis, N. (1972). (Na$^+$, K$^+$)-ATPase activity of embryonic chick heart and skeletal muscles as a function of age. *Biochim. Biophys. Acta* **266,** 230–237.

Sperelakis, N. (1979). Origin of the cardiac resting potential. *In* "Handbook of Physiology, Vol. 1, The Cardiovascular System" (R. M. Berne, and N. Sperelakis, Eds.), pp. 187–267. American Physiological Society, Bethesda, MD.

Sperelakis, N. (1980). Changes in membrane electrical properties during development of the heart. *In* "The Slow Inward Current and Cardiac Arrhythmias." (D. P. Zipes, J. C. Bailey, and V. Elharrar, Eds.), pp. 221–262. Martinus Nijhoff, The Hague.

Sperelakis, N. (1993). Origin of the resting membrane potential. *In* "Physiology" (N. Sperelakis and R. Banks, Eds.), pp. 29–48. Little, Brown, Boston.

Sperelakis, N. (1993). Basis of the resting potential. *In* "Physiology and Pathophysiology of the Heart," 3rd ed. Kluwer Academic Publishers, New York.

Sperelakis, N. (1995). "Electrogenesis of Biopotentials." Kluwer Academic Publishers, Boston.

Sperelakis, N., and Fabiato, A. (1995). Electrophysiology and excitation–contraction coupling in skeletal muscle. *In* "The Thorax: Vital Pump" (C. Roussos, Ed.), 2nd ed. Marcel Dekker, New York.

Sperelakis, N., and Lehmkuhl, D. (1966). Ionic interconversion of pacemaker and nonpacemaker cultured chick heart cells. *J. Gen. Physiol.* **49,** 867–895.

Sperelakis, N., Schneider, M., and Harris, E. J. (1967). Decreased K$^+$ conductance produced by Ba$^{2+}$ in frog sartorius fibers. *J. Gen. Physiol.* **50,** 1565–1583.

Trautwein, W., and Kassebaum, D. G. (1961). On the mechanism of spontaneous impulse generation in the pacemaker of the heart. *J. Gen. Physiol.* **45,** 317–330.

Vassalle, M. (1970). Electrogenic suppression of automaticity in sheep and dog Purkinje fibers. *Circ. Res.* **27,** 361–377.

Nicholas Sperelakis

# 14

# Gibbs–Donnan Equilibrium Potentials

## I. Introduction

Because intracellular cytoplasm contains many colloids, including large nondiffusible polyvalent electrolytes, a **Donnan equilibrium** can be established across the cell membrane with an accompanying transmembrane **Gibbs–Donnan (G–D) potential.** The resting potential of most cells in the body, including nerve and muscle cells, however, is not due to a Donnan equilibrium, and the normal resting potential is not a Gibbs–Donnan potential, as is erroneously stated in some textbooks. *In the true Donnan equilibrium, all diffusible ions are in equilibrium across the membrane.* But many ions—like $Na^+$, $K^+$, $Ca^{2+}$, and $H^+$—in nerve and muscle cells are not in equilibrium; that is,

$$E_{Na} \neq E_m$$

$$E_K \neq E_m$$

$$E_{Ca} \neq E_m$$

and

$$E_H \neq E_m$$

On the other hand, $Cl^-$ is at equilibrium (i.e., passively distributed) in many vertebrate cells; namely,

$$E_{Cl} = E_m$$

In addition, a large internal pressure and concomitant swelling of animal cells would occur if a Donnan equilibrium were allowed to become established. The action of two types of cation pumps keeps the Donnan osmotic pressure from developing and keeps certain cations out of equilibrium. Thus, a second important function of the Na–K pump is the *regulation of cell volume.* The Na–K pump actively pumps 3 $Na^+$ ions out (to 2 $K^+$ ions pumped in) with each cycle. The pump action decreases the osmotic pressure of the cytoplasm and prevents cell swelling. Inhibition of active ion transport by any means leads to osmotic swelling because of the establishment of the Donnan equi-

librium. Under such conditions, the cells gain $Na^+$, $Cl^-$, $Ca^{2+}$, and $H_2O$, and they lose $K^+$.

Thus, the Gibbs–Donnan potential is **passive;** that is, energy is not necessary to its establishment. In contrast, the resting potential is actively generated (indirectly or directly) by the action of the $Na^+$–$K^+$ pump. The Gibbs–Donnan potential is usually less than $-20$ mV, whereas the resting potential is $-40$ to $-100$ mV, depending on the cell type (and its ratio of $P_{Na}$ to $P_K$).

## II. Mechanism for Development of the Gibbs–Donnan Potential

In the **Gibbs–Donnan equilibrium,** a small membrane potential is established even though the biological membrane involved, or the artificial membrane used in a laboratory experiment, may be equally permeable to the small diffusible ions used. For the example illustrated in Fig. 1, where aqueous solutions of 0.1 $M$ $(Na^+)_n$-proteinate$^{n-}$ (side 1) and 0.1 $M$ NaCl (side 2) are initially placed on the two sides of a two-compartment chamber separated by a membrane, and if $g_{Na} = g_{Cl}$ in this membrane, then at equilibrium, $E_{Na} = E_{Cl} = -18$ mV. The side containing the protein anion becomes negative with respect to the other side. Thus, because both diffusion potentials have the same polarity (as well as magnitude), a potential difference (PD) occurs across the membrane, even though conductances for $Na^+$ and $Cl^-$ across the membrane may be equal. The osmotic pressure of the solution on side 1 containing the nonpermeant protein is greater than that on side 2.

In the G–D equilibrium, all permeant ions are in electrochemical equilibrium across the membrane, that is, they are passively distributed, and there is no net electrochemical driving force:

$$(E_m - E_{Na}) = 0$$

$$(E_m - E_{Cl}) = 0$$

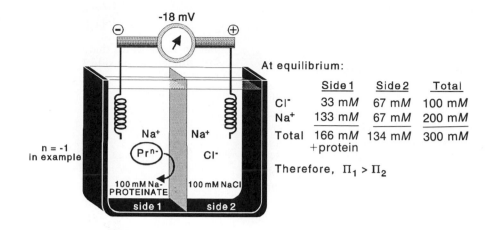

**FIG. 1.** Gibbs–Donnan potential. (A) Gibbs–Donnan experiment. Diagram depicts the experimental arrangement for obtaining Gibbs–Donnan potential. A membrane freely permeable to all small ions, but impermeable to the large protein molecules, is used to separate two solutions, only one of which (side 1) contains protein. Side containing the protein becomes negative, with respect to the other side, by a small voltage ($-18$ mV in example). This membrane potential ($E_m$) does not depend on active ion transport or on selective permeability properties of the membrane, as normal cell resting potential does. The diffusible ions ($Na^+$ and $Cl^-$ in example), however, become unequally distributed across the membrane, and it is their diffusion potentials ($E_{Na} = E_{Cl}$) that produce the Gibbs–Donnan potential. (B) Equivalent circuit for Gibbs–Donnan experiment (depicted in A) demonstrating that $E_m = E_{Na} = E_{Cl}$, the $Na^+$ and $Cl^-$ batteries being of equal magnitude and of the same sign. Therefore, the relative conductances of the membrane to $Na^+$ and $Cl^-$, whether equal or not, are irrelevant to the potential. $g_{Cl}$ and $g_{Na}$ conductance for $Cl^-$ and $Na^+$, respectively.

A more complete explanation for the development of the Gibbs–Donnan potential follows. The Gibbs–Donnan potential (which is an equilibrium PD) does not depend on metabolic energy. Therefore, this discussion applies to a cell that either has no ATP for pumping ions against electrochemical gradients or has had its Na–K pump completely blocked by either ouabain or another agent. The Gibbs–Donnan potential is passively produced by the concentration gradients for diffusible electrolytes (e.g., $Na^+$ and $Cl^-$) across a membrane. These ion gradients are caused by the presence of one or more large nondiffusible (with respect to the membrane) polyvalent electrolytes (e.g., **negatively charged proteins**) on one side of the membrane, as is present in all biological cells. In essence, the negatively charged protein molecules (at pH 7) inside the cell attract cations (e.g., $Na^+$ or $K^+$) and repel anions (e.g., $Cl^-$). Therefore, in the Gibbs–Donnan situation, the inside of the cell has a higher concentration of $Na^+$ (or $K^+$) and a lower concentration of $Cl^-$ than has the solution bathing the cell. The equilibrium potentials for $Na^+$ ($E_{Na}$) and for

Cl$^-$ ($E_{Cl}$) are equal in magnitude and are of the same sign, thereby producing a PD across the membrane. The PD is negative on the inside (side containing the protein) and usually is about $-20$ mV or less.

## III. Gibbs–Donnan Equilibrium

To quantitate the ion distributions produced at equilibrium and the PD developed, let us examine the artificial system shown in Fig. 1A. In this system, a chamber is separated into two compartments by a collodion membrane, which has small uncharged pores that allow Na$^+$ and Cl$^-$ ions to diffuse through, but not large protein molecules. A 100 m$M$ solution of Na proteinate is added to one side (compartment 1), and a 100 m$M$ solution of NaCl to the other side (compartment 2). An electrode is positioned on each side so that the PD across the membrane can be recorded (37°C). Let us assume that the Na proteinate is completely ionized and, for simplicity, that the protein has a net negative charge of only one.

Thus, there is, at the first instant, no diffusion force for Na$^+$, but there is for Cl$^-$, because Cl$^-$ is 100 m$M$ in compartment 2 and 0 m$M$ in compartment 1. Na$^+$ must accompany the diffusion of Cl$^-$ from side 2 to side 1, because the **principle of electroneutrality** in the bulk solution cannot be violated (i.e., there must be an equal number of cations and anions). So one relation that must be true when the system comes to equilibrium is that

$$[Na^+]_2 = [Cl^-]_2 \tag{1}$$

In actuality, there is a small charge separation directly across the membrane to account for the PD; that is, side 2 of the membrane has a small excess of Na$^+$ ions, and side 1 has a small excess of Cl$^-$ ions. Such a charge separation is very small, but is necessary to develop a PD across the membrane ($V = Q/C$), and is discussed in the preceding chapter on the resting potential.

The principle of electroneutrality also requires that the increase in Na$^+$ on side 1 must be exactly equal to the increase in Cl$^-$ on side 1. Thus, the concentration difference of Na$^+$ that is built up at equilibrium must be exactly equal to the final concentration difference for Cl$^-$. This is because the large initial gradient for Cl$^-$ is what drives the Na$^+$ to make its gradient. Therefore, it must also be true that

$$\frac{[Na^+]_1}{[Na^+]_2} = \frac{[Cl^-]_2}{[Cl^-]_1} \tag{2}$$

Cross-multiplying gives

$$[Na^+]_1[Cl^-]_1 = [Na^+]_2[Cl^-]_2 \tag{2a}$$

Another way of considering this is that $E_{Na}$ must equal $E_{Cl}$, and therefore using the respective Nernst equations (see Chapter 13) we can write

$$E_{Na} = E_{Cl}$$

$$\frac{-61\text{ mV}}{+1}\log\frac{[Na^+]_1}{[Na^+]_2} = \frac{-61\text{ mV}}{-1}\log\frac{[Cl^-]_1}{[Cl^-]_2} \tag{3}$$

$$= \frac{-61\text{ mV}}{+1}\log\frac{[Cl^-]_2}{[Cl^-]_1} \tag{3a}$$

Dividing both sides by $-61$ mV and removing the log gives

$$\frac{[Na^+]_1}{[Na^+]_2} = \frac{[Cl^-]_2}{[Cl^-]_1} \tag{2}$$

Equation 2a indicates that, at equilibrium, the product of the diffusible ions on side 1 must be equal to the product of the diffusible ions on side 2. From the Nernst equation, the relationships

$$E_{Na} = \frac{-RT}{zF}\ln\frac{[Na]_1}{[Na]_2} \tag{4}$$

$$= \frac{-61\text{ mV}}{+1}\log\frac{[Na^+]_1}{[Na^+]_2} \tag{4a}$$

and

$$E_{Cl} = \frac{-61\text{ mV}}{-1}\log\frac{[Cl^-]_1}{[Cl^-]_2} \tag{5}$$

can be given because Cl$^-$ is negative ($z = -1$), whereas Na$^+$ is positive ($z = +1$). Equation 5 is the same as (note that a negative sign in front of a log inverts the ratio)

$$E_{Cl} = -61\text{ mV}\log\frac{[Cl^-]_2}{[Cl^-]_1} \tag{5a}$$

Because $[Na]_1/[Na]_2 = [Cl]_2/[Cl]_1$, as Eq. 2 indicates, and from Eqs. 4a and 5a, it is clear that Eq. 3 holds true; that is,

$$E_{Na} = E_{Cl} \tag{3}$$

## IV. Quantitation of the Gibbs–Donnan Potential

For quantitation, let us use $x$ to indicate the amount (in m$M$) of Cl$^-$ or Na$^+$ that shifted from side 2 to side 1 at equilibrium. Then the amount of Na$^+$ on side 2 is 100 m$M - x$ (the original amount minus the amount lost); Cl$^-$ on side 2 is also 100 m$M - x$, because $[Na]_2 = [Cl]_2$. The Na$^+$ on side 1 at equilibrium is 100 m$M + x$ (the original amount plus the amount gained), and the Cl$^-$ on side 1 is simply $x$. These parameters may be listed as follows:

$$[Na]_2 = 100\text{ m}M - x$$
$$[Cl]_2 = 100\text{ m}M - x$$
$$[Na]_1 = 100\text{ m}M + x$$
$$[Cl]_1 = x$$

The value for $x$ can be obtained by substituting these values into Eq. 2a:

$$[Na^+]_1[Cl^-]_1 = [Na^+]_2[Cl^-]_2 \tag{2a}$$
$$(100 + x)x = (100 - x)(100 - x)$$
$$100x + x^2 = 10{,}000 - 200x + x^2$$
$$300x = 10{,}000$$
$$x = 33.3$$

Thus, at equilibrium

$$[Cl]_1 = 33\text{ m}M$$
$$[Na]_1 = (100 + 33) = 133\text{ m}M$$

$$[Cl^-]_2 = (100 - 33) = 67 \text{ m}M$$
$$[Na]_2 = (100 - 33) = 67 \text{ m}M$$

These values are also given in Fig. 1A. Note that all the equations and conditions are obeyed. The Gibbs–Donnan potential produced then may be calculated by substituting into Eqs. 4a and 5:

$$E_{Na} = \frac{-61 \text{ mV}}{+1} \log \frac{133 \text{ m}M}{67 \text{ m}M} \qquad (6)$$
$$= -18 \text{ mV}$$

and

$$E_{Cl} = \frac{-61 \text{ mV}}{-1} \log \frac{33 \text{ m}M}{67 \text{ m}M} \qquad (7)$$
$$= -18 \text{ mV}$$

Hence, $E_{Na} = E_{Cl}$ (Eq. 3). That is, the two diffusion potentials are equal in magnitude and are of the same sign. The PD across the membrane is $-18$ mV; side 1 containing the protein is negative. Therefore, relative permeability of the membrane to $Na^+$ and $Cl^-$ is irrelevant. The equivalent circuit for this example at equilibrium is given in Fig. 1B.

## V. Osmotic Considerations

We should note that, at equilibrium, the sum of $Na^+$ and $Cl^-$ on side 1 (166 m$M$) is greater than that on side 2 (134 m$M$). In addition, there is 100 m$M$ protein on side 1. Thus, the total osmotic concentration on side 1 is 266 mOsm (milliosmolar), compared to 134 mOsm on side 2. Therefore, there is a large osmotic gradient between the two sides. Water moves from side 2 to side 1 (i.e., water accompanies the net movement of $Na^+$ and $Cl^-$) until the hydrostatic pressure head buildup is sufficient to oppose further net movement of water. As expected, the biological cell swells when a Gibbs–Donnan equilibrium is allowed to develop following a blockade of active ion transport for long periods.

The total osmotic concentration ([osm]) on each side, at equilibrium, may be summarized as follows:

$$[osm]_1 = [Na]_1 + [Cl]_1 + [Prot]_1 \qquad (8)$$
$$[osm]_2 = [Na]_2 + [Cl]_2 \qquad (8a)$$

Substitution gives

$$[osm]_1 = 133 \text{ m}M + 33 \text{ m}M + 100 \text{ m}M$$
$$= 266 \text{ m}M$$
$$[osm]_2 = 67 \text{ m}M + 67 \text{ m}M$$
$$= 134 \text{ m}M$$

The osmotic pressure ($\Pi$, in atm) of each solution is equal to the osmotic concentration in osmol/liter ($C$) times the osmotic coefficient ($i$) times the gas constant ($R$, 0.082 liter-atm/mol-K) times the absolute temperature ($T$, in K)

$$\Pi = iCRT \qquad (9)$$

where osmol/liter is the number of osmoles per liter of solution. In the example depicted in Fig. 1 a hydrostatic pressure of 3.17 atm would need to be applied to side 1 to prevent this compartment from gaining water from side 2 (at 20°C and assuming $i = 1.0$)

$$\Pi = i\Delta CRT \qquad (9a)$$
$$= (1.0)(266 \text{ m}M - 134 \text{ m}M)\left(0.082 \frac{\text{liter-atm}}{\text{mol-K}}\right)(273 + 20) \text{ K}$$
$$= \left((0.132 \frac{\text{mol}}{\text{liter}})(0.082 \frac{\text{liter-atm}}{\text{mol-K}})\right)(293 \text{ K})$$
$$= 3.17 \text{ atm}$$

The situation illustrated in Fig. 1 is actually more complex because the net water movement into side 1 acts to dilute the ion concentrations building up there, and therefore a true Gibbs–Donnan equilibrium can become established only if the net water movement is stopped, that is, by allowing an osmotic pressure gradient to develop by making side 1 a closed, or rigid, system. Otherwise, theoretically all of the water and NaCl eventually would move out of side 2.

The example of a G–D equilibrium in Fig. 1 could have been illustrated using another salt, such as KCl, instead of NaCl, or two or more salts.

The extra osmotic pressure in side 1 (or inside a cell) produced by the presence of the negatively charged proteins and other impermeant large charged molecules is known as the **colloid osmotic pressure (COP)**. The COP is also important for water movement across the capillary wall, which separates the blood plasma (containing impermeant proteins) and the interstitial fluid (ISF). At the arterial end of the capillary, the intracapillary hydrostatic blood pressure exceeds the COP, so water moves out of the capillary into ISF space; at the venous end, the COP exceeds the capillary hydrostatic pressure, so water moves into the capillary. In the midregion of the capillary, the two pressures are about equal, and there is no net water flow. Thus, there is a circulation of fluid distributed along the length of the capillary, generally known as the **Starling hypothesis.**

## VI. Summary

In summary, a Gibbs–Donnan equilibrium becomes established and a Gibbs–Donnan potential is developed across the cell membrane of cells under conditions in which metabolism and energy production has been inhibited or the Na–K pump has been inhibited by digitalis. The G–D equilibrium occurs because of the large impermeant charged macromolecules, such as proteins, inside the cell. The G–D equilibrium does not require energy for its establishment; that is, it is passive. This contrasts with the normal resting potential of the cell, which requires active ion transport and use of metabolic energy to establish large ionic electrochemical gradients.

The G–D potential is usually less than $-20$ mV, whereas the resting potential is considerably greater. In the G–D equilibrium, all permeable ions are in equilibrium across the membrane, whereas this is not true for the normal resting potential. The equilibrium potentials for all permeant ions (e.g., $E_K$, $E_{Cl}$) are of equal magnitude and polarity. The G–D potential is developed even if the cell membrane had equal permeability or conductance for all small ions, whereas the normal resting potential requires different permeabilities for $Na^+$ and $K^+$, namely a low $P_{Na}/P_K$ ratio. In the G–D equilibrium, the osmolarity of the cell becomes higher than the interstitial fluid bathing the cell, and so the cell tends to gain water and swell (unless prevented from doing so by a rigid cell wall, such as in plant cells).

When equililbrium is established, the product of the concentrations of the permeant ions inside the cell is equal to that outside the cell, and the concentrations of the anions and cations outside the cell must be equal (law of electroneutrality). The gains in cations and anions inside the cell also must be equal to each other. From these required conditions, the equation can be solved algebraically to give the final concentrations at equilibrium, and from this, the calculated potential difference across the membrane.

## Bibliography

Davson, H. (1964). "A Textbook of General Physiology," 3rd ed. Little, Brown, Boston.

Sperelakis, N. (1979). Origin of the cardiac resting potential. *In* "Handbook of Physiology, Vol. 1, The Cardiovascular system" (R. M. Berne and N. Sperelakis, Eds.), pp. 187–267. American Physiological Society, Bethesda, MD.

Sperelakis, N. (1995). "Electrogenesis of Biopotentials." Kluwer Publishing Co., New York.

John Cuppoletti

# 15

# Transport of Ions and Nonelectrolytes

## I. Introduction

Biological membranes provide selective barriers to the passage of molecules. The barrier properties of biological membranes are due to a thin (4.6 nm) bimolecular leaflet of phospholipids. Selective passage of ions and nonelectrolytes across this barrier is due to the presence of proteins that traverse the membrane and that provide a pathway for the selective movement of solutes. This chapter deals with mechanisms and molecules involved in membrane transport.

## II. Ion Transport by Valinomycin

Many of the characteristics of transport by membrane transport proteins are exhibited by synthetic and natural ionophores. Figure 1 shows the structure of valinomycin, a naturally occurring polypeptide antibiotic that is insoluble in water but is soluble in organic solvents and biological membranes. The structure of valinomycin provides a hydrophobic outer surface and a central pocket lined by negatively charged groups. The pocket is just large enough to accept $K^+$ but is small enough to exclude larger ions. When the valinomycin–$K^+$ complex is placed in one arm of a U-shaped tube with water in both sides of the tube and with an organic solvent in the bottom of the tube, the net movement of potassium from one side of the U through the organic solvent into the other side of the tube will occur much faster than it would occur in the absence of valinomycin. Valinomycin thus facilitates the diffusion of $K^+$. The facilitated diffusion process can be modeled as shown in Fig. 2.

In this model of valinomycin-mediated facilitated diffusion of $K^+$, valinomycin (Val) is soluble in the organic phase. In step 1, valinomycin is oriented toward the water in one arm of the U tube, where it can bind $K^+$. In step 2, the valinomycin–$K^+$ complex then diffuses through the organic phase to the other arm of the U tube, where it reaches the interface on the other side at the tube between the organic and the water phase and orients the release of $K^+$ in step 3. Valinomycin then returns through the organic solvent in step 4 to the interface on side 1. A single cycle of this process results in the movement of a single molecule of $K^+$ from side 1 to side 2 and the return of valinomycin to side 1. If a counterion such as the anionic form of picric acid (which is soluble in both water and organic solvents) were also present, net transport of $K^+$ would be accomplished.

If step 4 were blocked and only the fully loaded $K^+$–valinomycin complex could pass through the organic phase, net $K^+$ flux would not occur because all of the valinomycin would be trapped at the interface on side 2. However, if radioactive[1] $K^+$ were present on one side of the membrane, with nonradioactive $K^+$ on both sides of the membrane, steps 1, 2, and 3 would occur in the forward direction. If the process were reversed through steps 3, 2, and 1, valinomycin would bind nonradioactive $K^+$, pass through the membrane, release the nonradioactive $K^+$, and bind radioactive $K^+$. This process, termed self-exchange, would result in equilibration of radioactive $K^+$ without a change in the concentration $K^+$.

Assuming that the rate of transfer of the valinomycin–$K^+$ complex is slower than the rate of dissociation of the valinomycin–$K^+$ complex, the velocity equation for transport of $K^+$ is given by the expression

$$\frac{v}{V_m} = \frac{[K^+]}{K_s + [K^+]}, \tag{1}$$

where $v$ is the initial velocity of transport, $V_m$ is the maximal velocity for transport, and $K_s$ is the dissociation constant for the valinomycin–$K^+$ complex. This expression is the

---

[1] Radioactive rubidium $^{86}Rb^+$ is used as a substitute for potassium in most experiments. Rubidium behaves similarly to potassium in most cases.

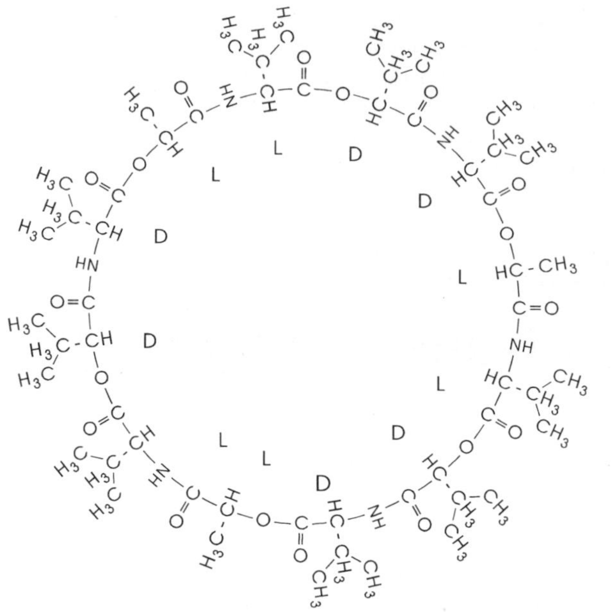

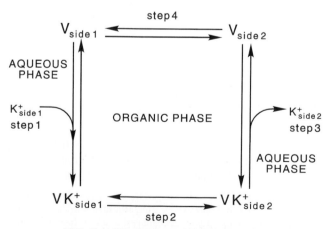

**FIG. 1.**  The sequence of valinomycin. Only the ring and carbonyl atoms are numbered. D and L refer to amino acid isomers.

velocity equation for the valinomycin-induced catalysis of the movement of K$^+$ across the U tube and is identical to the Michaelis–Menten equation for enzyme catalysis. It is similar to that observed for most membrane transport processes.

Consideration of the previous kinetic scheme suggests that the rates of equilibration of K$^+$ will be dependent on the amount of valinomycin present at the interface on side 1 and on side 2 of the apparatus. If the proportion of [Val–K$^+$] on side 1 is increased by increasing the concentration of K$^+$ on side 1, the amount of valinomycin available at the interface on side 2 will be reduced. This could result in inhibition of the initial rate of transport of potassium. This transinhibition phenomenon is characteristic of facilitated diffusion processes and cannot be explained by a simple pore model for membrane transport. An important

characteristic of valinomycin-mediated transport is that the rate of equilibration of other ions is much slower for ions such as sodium and protons than for potassium at any given concentration of cation, because of a higher affinity of valinomycin for K$^+$ than other cations (ion specificity).

## III.  Other Ionophores

Other ionophores are widely employed in studies of membrane transport phenomena and are grouped according to mechanism (whether electrogenic or exchangers) and ion specificity. Valinomycin is an electrogenic ionophore that transports potassium faster than it transports most anions. Other electrogenic ionophores are CCCP and FCCP, both of which act as electrogenic proton ionophores, and the crown ethers, a class of electrogenic ionophores that have been engineered to carry particular cations including sodium. Among the common exchange ionophores are nigericin, which accomplishes the electroneutral exchange of potassium and protons; monensin, which accomplishes the electroneutral exchange of sodium and protons; and ionomycin, a nonfluorescent calcium–proton exchange ionophore and its related compound, A23187. In addition to these cation ionophores, tributyl tin compounds have occasionally been employed as Cl$^-$ hydroxyl exchangers.

## IV.  Membrane Transport Proteins

The processes of net transport and exchange occur through random thermal motions of the ionophore–cation complex across the bilayer membrane. Membrane transport proteins are much larger than simple natural and synthetic ionophores and therefore do not diffuse across the bilayer, but rather accomplish reorientation of binding sites through conformational changes. Despite these differences, the kinetics of membrane transport or protein-mediated transport are similar to the kinetics of ionophore-mediated transport and much more complex than can be explained on the basis of the formation of simple pores through the membrane.

## V.  D-Glucose Carriers

Both electrolytes and nonelectrolytes selectively cross biological membranes on membrane transport proteins. An example of a nonelectrolyte transporter is the D-glucose carrier. Glucose transporters are found in virtually all cells of the body, and these proteins catalyze the entry of glucose, which fuels most of the tissues. The adult human red cell contains a D-glucose carrier that exhibits specificity for D-hexoses. Other substrates of the carrier include 2-deoxyglucose, an analogue that can be taken up by cells and phosphorylated but cannot be metabolized further. In the absence of other energy sources, 2-deoxyglucose will be taken up and phosphorylated to such an extent that cellular ATP will be depleted. 3-*O*-methylglucose can be

**FIG. 2.**  The facilitated diffusion process.

transported, but not phosphorlylated. Since it is not metabolized, 3-*O*-methyl-D-glucose is an ideal substrate for transport studies. L-Glucose, which exhibits a structure that is a mirror image of D-glucose, is not transported by the D-glucose carrier and is therefore ideal for evaluation of the passive permeability of the membrane.

## VI. Identification, Purification, and Molecular Cloning of D-Glucose Carrier Isoforms

Protease treatment of the red cell results in loss of D-glucose transport function without loss of barrier properties, and this finding initially suggested that the D-glucose carrier was a protein, rather than a "pore" in the lipid bilayer. D-Glucose transport can be inhibited by mercuric ions. Other potent inhibitors include forskolin, isolated from decorative plants, and cytochalasin B, a product of fungal metabolism that inhibits D-glucose transport by binding to critical residues within the protein. No apparent relationship exists between the structures of these inhibitors with each other or with D-glucose, yet these toxic substances bind very tightly to the transport protein. One is left to wonder whether these compounds are binding to an important site on the D-glucose carrier that may have physiological significance, such as serving as a binding site for an endogenous compound in the mammalian cell.

Purification of the D-glucose carrier requires detergent solubilization of protein with destruction of the lipid membrane that is essential for transport function. The binding of D-glucose cannot be used to follow the protein through purification, since the affinity of the transport protein for glucose is very low (about a few millimoles/liter) near the concentrations at which glucose is found in the blood and interstitial fluids. The high-affinity inhibitor cytochalasin B was therefore used to identify and follow the D-glucose carrier through the purification process. In these studies, the radioactively labeled inhibitor was bound to the protein near the $K_I$ and was displaced with the unlabeled substrate at concentrations higher than the $K_S$. The purified protein was then reconstituted into liposomes, and glucose transport function was followed. With the purified protein in hand, protein sequencing using traditional cleavage and identification of fragments by protein sequencing and mass spectrometry was performed in a heroic effort to define the complete primary sequence of the red cell D-glucose carrier.

With knowledge of the sequence of the red cell D-glucose carrier, it was possible to search in other tissues for variations in the primary structures of D-glucose carriers using the technique of molecular cloning. In this approach, the amino acid sequence of the red cell D-glucose carrier was used as a guide to generate oligonucleotide probes that bound to small regions of the mRNA or cDNA for the D-glucose carrier from other tissues. Such probes are complementary to the mRNA or cDNA present in the tissues and can be made in a radioactive form. When bacteria are transformed with the mRNA from human tissues, individual bacteria may contain a portion of the nucleic acid regions encoding the D-glucose carrier, and those bacterial colonies that contain mammalian DNA for the D-glucose carriers can then be identified by virtue of binding of the radioactive oligonucleotide "probe." The bacterial colonies that react with the probe can then be grown in quantity. The cDNA that is complementary to the probe can be more easily sequenced than the protein that it encodes. In this way, it was demonstrated that there are at least five different D-glucose carrier isoforms. The distribution of D-glucose carrier isoforms is shown in Table 1. The primary amino acid sequence of the different isoforms varies in some regions, but is conserved in others. The conserved regions apparently serve an important structural or functional role, and the regions of difference exist to serve special functional needs in the tissues in which they are found. Importantly, one of these isoforms is specific for the insulin-sensitive tissues, the heart, skeletal muscle, and the adipocyte.

Despite differences in small regions of the primary structure of the various D-glucose carriers, the majority of the amino acids are conserved, and the small changes do not significantly alter the secondary structure of the carriers. The structure of the protein can be surmised initially by determination of whether a small region of the protein is hydrophobic, and therefore probably in the membrane, or whether the region is hydrophilic and thus probably in the aqueous phase. The consensus structure for the D-glucose carrier is shown in Fig. 3. Other information, such as whether a hydrophobic region reacts with chemicals that dissolve in the membrane or whether a region reacts with antibodies to a region exposed on the outside or inside of the cell, is required to confirm the proposed structure and to determine whether a region is on the outside (extracytosolic) or inside (cytosolic) of the cell. The folding pattern shown in Fig. 3 is the proposed structure of the D-glucose carrier from the insulin-sensitive tissues. The major differences among the various forms of the carrier are found in the extracytosolic loop, which is larger in the carrier from the insulin-sensitive tissues, and in the last 10 amino acids of the C-terminus of the protein. These differences in primary structure are useful for generation of probes that are isoform-specific. For example, monoclonal or polyclonal antibodies prepared against these regions are useful for determination of the amounts of each isoform of the protein in the various tissues or in regions within the tissues.

The D-glucose carrier has a molecular mass of 40–50 kDa, and the extracytosolic face contains a site that is likely glycosylated. The large size of the protein and the

**TABLE I**  D-Glucose Carrier Isoforms

| Designation | Function |
| --- | --- |
| GLUT1 | Basal uptake in placenta, brain, kidney, colon |
| GLUT2 | Primarily found in transport in liver cells |
| GLUT3 | Basal uptake in the brain |
| GLUT4 | Insulin-stimulated uptake in the skeletal muscle, cardiac muscle, and adipose tissue |
| GLUT5 | Absorption in small intestines |

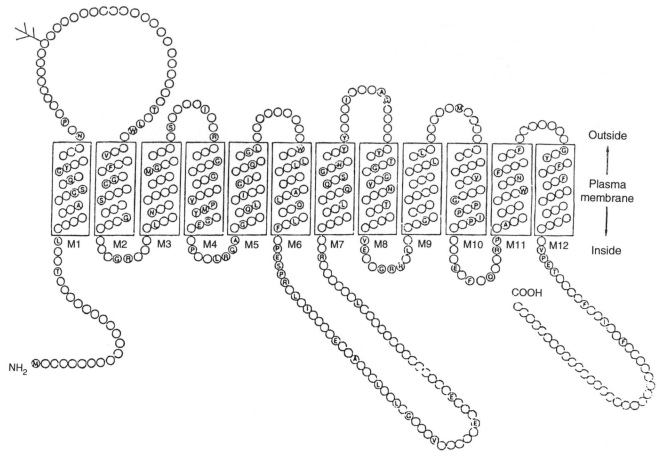

**FIG. 3.**  Consensus structure of the mammalian D-glucose carriers. [Reprinted with permission from Bell (1991). Copyright © 1991 by American Diabetes Association.]

presence of the hydrophilic sugar residues on the cytoplasmic face preclude rotation of the protein from one face of the membrane to the other. Conformational changes that alternatively expose a part of the protein to alternate sides of the membrane must therefore be responsible for binding of the substrate on one face and release on the other. In support of this idea, the D-glucose carrier exhibits transstimulation and transinhibition by substrate, as expected for this alternating conformation hypothesis.

## VII.  Regulation of D-Glucose Transport

The D-glucose carrier is under regulation in the insulin-sensitive tissues. Insulin binding to the insulin receptors in the plasma membrane initiates a series of signal events that lead to increased D-glucose transport (Fig. 4). Concomitant with increases in D-glucose transport is the appearance of one form of the D-glucose carrier in the plasma membrane and a decrease in the complement of D-glucose carriers in intracellular membranes. Increases in D-glucose transport through recruitment occur over a few minutes and do not require synthesis of new D-glucose carriers. Longer term regulation of D-glucose transport, wherein the level of D-glucose carrier protein and mRNA levels change, has also

been observed. In the fasting rat, the intracellular level of insulin-sensitive D-glucose carriers decreases, as demonstrated by testing with isoform-specific antibodies. Refeeding causes in increase in the carrier to prefasting levels. The diabetic state may also play a role in long-term regulation of D-glucose carriers, since experimental diabetes in rats elicits a reduction in the levels of D-glucose carrier, and long-term treatment with insulin leads to a return to normal of carrier levels.

## VIII.  Intestinal D-Glucose Carrier

Another glucose carrier, unrelated to the insulin-sensitive D-glucose carrier, is found in the intestine. Unlike the red cell protein described previously, the intestinal protein is sodium dependent and serves to drive the concentrative uptake of D-glucose from the intestine, where the glucose concentration is low, into the intestinal brush border against a concentration gradient. This 70-kDa protein has binding sites for glucose and for sodium and uses a sodium gradient to drive the uptake of glucose.

Concentrative uptake of molecules is a characteristic of living cells; it requires an input of energy, and some membrane transport proteins act to transduce energy. One

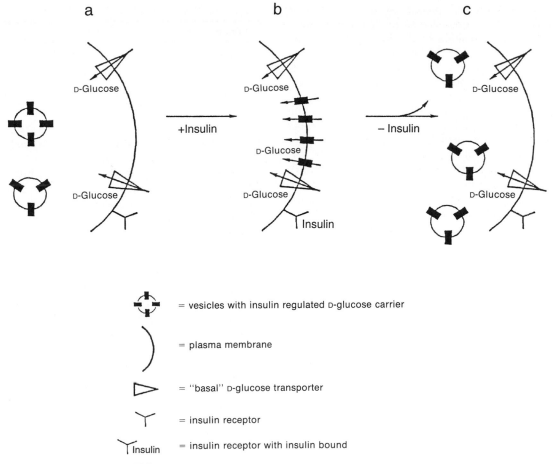

= vesicles with insulin regulated D-glucose carrier

= plasma membrane

= "basal" D-glucose transporter

= insulin receptor

= insulin receptor with insulin bound

**FIG. 4.**    Regulation of the D-glucose carrier by recruitment.

way to harness energy is through coupled processes. For example, harnessing the energy of a 10-fold gradient of sodium moving from one side of the membrane to the other can drive accumulation of a 10-fold gradient of another substance in the opposite direction. Such a process can be modeled for a protein (P) with a binding site for $Na^+$ and a binding site for glucose (S), as shown in Fig. 5.

If the concentration of $Na^+$ on side 1 is higher than that on side 2 (e.g., 10 m$M$ on side 1 vs 1 m$M$ on side 2), the driving force is $2.3RT/F \log[Na^+]_1/[Na^+]_2 = 59$ mV. Binding of sodium in step 1 provides a species that can bind glucose in step 2. Steps 3, 4, 5, and 6 will follow, and the process will be repeated. In the absence of a mechanism for extrusion of sodium, the concentration of $Na^+$ and glucose on side 2 will increase until the sodium gradient is dissipated. In the living cell, the sodium gradient across the brush border membrane is constantly regenerated by the action of a basolateral membrane sodium potassium ATPase ($Na^+$, $K^+$-ATPase). ATP hydrolysis by the $Na^+$, $K^+$-ATPase provides the energy for this primary active transport process, which drives the secondary active transport of D-glucose by the intestinal D-glucose carrier.

It is important to note that the bimolecular lipid leaflet of the biological membrane forms an important component of the process of transduction of energy by membrane

transport proteins. If there were holes in the membrane through which $Na^+$ could flow, the gradient of $Na^+$ would be dissipated, and no useful work such as concentrative uptake of glucose would occur.

## IX. $Cl^-$–Bicarbonate Exchanger

This protein is a 95-kDa protein that spans the membrane approximately 12 times. It is glycosylated at extracellular sites, as are most membrane transport proteins. It functions to exchange $Cl^-$ and bicarbonate. The $Cl^-$–bicarbonate exchanger is found in a variety of tissues, including the human red cell, where it is important in the removal of equivalents of $CO_2$. It has a turnover number of 36,000 per second and is present in the red cell at approximately $10^6$ copies per cell. Structure–function studies have been carried out on this protein. It can be cleaved with proteolytic enzymes, removing 40% of the cytosolic portion of the protein without affecting transport function. In recent studies, it has been shown that this "excess" cytosolic region serves a role in binding other cytosolic proteins, particularly the cytoskeletal proteins. Cleavage of the intact protein into 35- and 60-kDa fragments (albeit without removal of the fragments from the membrane) or cleavage into as many as 12 different pieces also has no affect on function. It is therefore somewhat surprising that the derivatization of a single residue in the intact protein with DIDS, a stilbene derivative, leads to complete loss of transport function. This residue may be essential to the binding of substrate or the conformational changes that are involved in transport of anions.

Determination of the tertiary structure of this protein has been an important goal. Thus, if transport occurs only when the anion carrier is present as a dimer, then a dimer model for transport would be required (for example, Cl⁻ on one subunit, and bicarbonate on the other subunit). Although the Cl–bicarbonate exchanger has been shown to exist as a dimer under many conditions (e.g., in cross-linking experiments), it has not yet been possible to determine whether a multimeric state of the protein is required for its function.

The dynamic aspects of the anion carrier have also been examined. Rotational diffusion of the anion carrier about its own axis has been examined using the technique of fluorescence polarization. The anion carrier rotates about its own axis at a rate approximately equal to that expected for a cylinder of the mass of a dimer of the anion carrier. Indeed, with cross-linking of the cytosolic domain to form covalently linked dimers, no change in rotational diffusion has been observed. Lateral diffusion of the anion carrier through the membrane has also been studied. The anion carrier appears to confined to move within domains of the membrane in normal cells, but restrictions on freedom of movement within the plane of the membrane are removed with treatments that tend to remove the cytoskeleton from the membrane.

## X. Ion Pumps

ATP hydrolysis by the ion pumps, including the Na⁺,K⁺-ATPase, forms the primary source of ion gradients in mammalian cells. Sodium gradients are employed for a variety of cell processes, including coupling net transport of other solutes and nerve conduction. The Na⁺,K⁺-ATPase, or the sodium pump, is found in almost all of the tissues. It is a

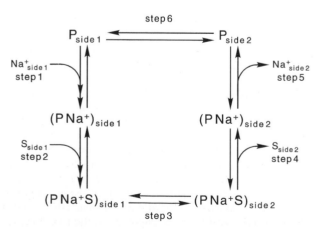

**FIG. 5.** Concentrative uptake of molecules.

**TABLE 2**    Distribution of Na⁺,K⁺-ATPase

| Isoforms | Tissues |
| --- | --- |
| $\alpha$ I and $\beta$ I | Most tissues, highest in kidney and brain |
| $\alpha$ II and $\beta$ II | Brain, heart, skeletal muscle |
| $\alpha$ III and $\beta$ III | Brain, heart |

heterodimer with a 100-kDa alpha ($\alpha$)-subunit and a smaller 55-kDa beta ($\beta$)-subunit. Most of the active sites appear to be located on the $\alpha$-subunit, but the protein will not function without an accompanying $\beta$-subunit. The precise function of the $\beta$-subunit is not clear, but it may play a role in stabilizing the $\alpha$-subunit and/or moving the protein to the plasma membrane. Three isoforms of the Na⁺,K⁺-ATPase $\alpha$ (I, II, and III) with corresponding $\beta$-subunits are known. These are distributed in the various tissues at various levels (Table 2). The $\alpha$ I and $\beta$ I isoform is found in most tissues, highest in the kidney and brain. The $\alpha$ II and $\beta$ II forms are highest in the skeletal muscles, and the $\alpha$ III form is limited to the brain and heart. All of these isoforms are inhibited by the cardiac glycoside ouabain. In the nonepithelial and most excitable cells, the Na⁺,K⁺-ATPase is uniformly distributed in the plasma membrane. In the epithelia, it is exclusively located at the basolateral membrane. The isoform distribution apparently arises from differential expression in the various tissues, rather than differential degradation, as determined by experimental findings of the differential expression of controlling elements plus markers in transgenic animals and cell lines. Thyroid hormone increases the amount of the protein over the course of days in epithelial and nonepithelial cells, presumably acting on genetic controlling elements.

However, the activity of the Na⁺,K⁺-ATPase also increases with a large variety of hormones under conditions that do not seem to require synthesis of the protein. Catecholamines rapidly increase the activity of the Na⁺,K⁺-ATPase in the skeletal muscle, thereby preserving K⁺ during strenuous exercise. Insulin stimulates the Na⁺,K⁺-ATPase rapidly (within minutes) in the liver, muscle, and fat tissues, whereas aldosterone and corticosterones increase the activity over hours in the intestine. The mechanisms of hormone-dependent increases of the Na⁺,K⁺-ATPase in these tissues may arise from either changes in substrate (sodium), recruitment of the protein to the plasma membrane (similar to that seen with recruitment of the D-glucose carrier), and/or covalent modification of the protein.

## XI. Structure–Function Studies

Structure can be divided into a hierarchy, with the primary amino acid sequence containing all of the information necessary to determine the structure of the protein. Unfortunately, it is currently impossible to calculate exactly the folding pattern of the protein on the basis of the primary structure alone. Secondary structures are the structures formed by hydrogen bonding and rotations and restrictions of rotations of the peptide bond. The secondary structure can be $\alpha$-helical (3.6 residues per turn), $\beta$-strand (an extended peptide strand that H-bonds to other strands to form $\beta$-sheets), $\beta$ turns (a characteristic four-amino-acid turn connecting $\alpha$-helices and $\beta$-strands), or random structures. Some success at predicting secondary structure using the Chou–Fasman method can be achieved, but the prediction must be tested by physical and chemical techniques. Tertiary structures are domains formed by folding of secondary structures. Structural predictions based on hydrophobicity as discussed earlier, similarity to other protein domains, or calculation of minimum energy conformations can be carried out, but the predictions must be tested. Quaternary structures are formed from multisubunit interactions. Cross-linking studies, radiation inactivation, rotational diffusion studies, and image reconstruction of electron microscopic data of large proteins provide initial guesses as to quaternary structure. Eventually, determination of the structure of such proteins by high-resolution X-ray crystallography will be required to determine the structure of proteins.

A great deal of effort using molecular biological approaches, covalent modification of critical residues, and physical approaches is currently directed toward definition of the relation between the structure and the function of the ion pumps and other membrane transport proteins. Immediate questions regarding the structure of membrane transport proteins concern the residues involved in formation of "active" sites and delineation of the changes in the structure of the protein during the reaction cycle.

The primary structure of these pumps and most of their isoforms is now known. One helpful key to obtaining the complete structures of these proteins arose from knowledge from protein-sequencing studies that $Ca^{2+}$-ATPase, the $Na^+,K^+$-ATPase, and a related $H^+,K^+$-ATPase from the stomach all reacted with ATP to form a phosphorylated intermediate, which could be isolated and sequenced. The phosphate group was bound to a particular aspartate residue within the sequence of these various ion pumps which was bounded on either side by a number of other conserved residues. Oligonucleotide probes to the conserved regions and other regions of these ion pumps were then used to isolate cDNAs encoding the various pumps and their isoforms, in efforts similar to those described for the isolation of the D-glucose carrier isoforms. The complete sequences provided a testable model for the folding patterns of the pumps in the membrane, which were subsequently validated by antibody localization and other techniques. Among the most powerful techniques have been the mapping of active sites by chemical modification, using proteolytic mapping techniques. For example, a particular lysine residue that is thought to be near the ATP binding region of the $Na^+,K^+$-ATPase can be labeled by fluorescein isothiocyanate and localized on the primary sequence of the protein by labeling the protein, cleaving the protein with proteolytic enzymes, and sequencing the N-terminal residues of the fragments thus obtained. Similarly, binding sites for ATP analogues, inhibitors, and modifiers of the ion pumps have been localized on the ion pumps.

Other approaches that show promise in generating structural information on regions of the ions pumps are NMR (which yields information regarding the orientation and presence of amino acids, sheets, and helical regions with respect to each other) and circular dichroism measurements, which yield information regarding $\alpha$-helical content of regions of the protein. These techniques are most useful for definition of small regions of the proteins and can be carried out on fragments such as those generated by molecular biological techniques.

Low-resolution structural information on the ion pumps has been gained by forming two-dimensional crystals of the protein and placing them in the beam of an electron microscope. Successive images are then obtained by tilting the sample in the beam and reconstructing the images thus obtained (Fig. 6). Sufficient scattering of the regular arrays of ion pump in the samples has provided an idea of the ion pump structure. However, structural information that is detailed enough to allow identification of components as small as amino acid side chains has not yet been acquired. Such information will probably require the preparation of crystals of sufficient quality to obtain X-ray crystal structure.

## XII. Functional Aspects

The $Na^+,K^+$-ATPase moves three sodium ions from the cell and two potassium ions into the cell for each ATP hydrolyzed (under most conditions). Since more sodium is exported than potassium is imported, the operation of the $Na^+,K^+$-ATPase leads to the generation of an inside negative membrane potential. In the process of a single catalytic cycle, the $Na^+,K^+$-ATPase becomes alternatively phosphorylated in the presence of sodium at the internal face of the enzyme; the sodium site is then oriented toward the outside face of the cell, where it is released and potassium then binds. Dephosphorylation and noncovalent binding of (intracellular) ATP promote release of potassium inside of the cell (Fig. 7). Vanadate prevents phosphorylation of the enzyme, presumably by binding to the same aspartyl residue that becomes phosphorylated by ATP. Alternative exposure of the binding sites for sodium and potassium to opposite sides of the membrane during the reaction cycle must be accompanied by changes in affinity of the enzyme for the ions, since sodium, for example, is present at low concentrations in the cell, but is deposited in a region of high concentration outside of the cell. The $K_m$ must change to maintain adequate pump rates consistent with cell growth and maintenance. Conformational changes in the enzyme may be responsible for these changes in orientation and binding affinity. Techniques that presumably monitor changes in the environment of essential residues of the ion pump during the catalytic cycle have developed. In one approach, the pump was modified with the environment-sensitive probe fluorescein isothiocyanate. As the pump changed conformation in small regions, the spectral properties of fluorescein changed. The results of many studies using this probe suggest that the binding of the substrates, $K^+$, $Na^+$, and ATP, as well as binding of

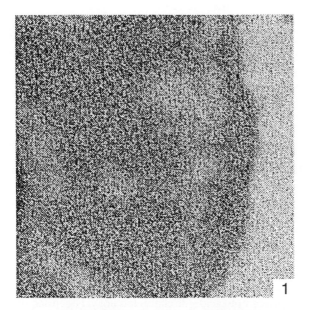

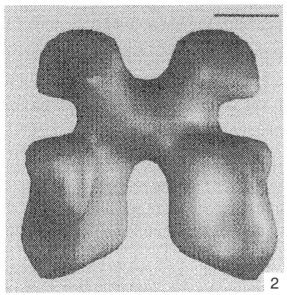

**FIG. 6.** Reconstruction of the structure of the Na⁺,K⁺-ATPase at low resolution from electron microscopic images of two-dimensional crystals. Panel 1 shows the purified Na⁺,K⁺-ATPase in a membrane as viewed perpendicular to the membrane. Panel 2 shows a low-resolution image of a dimer of the Na⁺,K⁺-ATPase from electron micrographs as it would appear when viewed parallel to the plane of the membrane. Bar, 20 Å. (Reproduced from *The Sodium Pump: Recent Developments, 44th Annual Symposium of the Society of General Physiologists*, 1991, by copyright permission of The Rockefeller University Press.)

## XIII. Channels in Secretory Processes

Ion channels play a major role in transport of solutes across biological membranes. In the secretory tissues, the ion channels play a major role in maintaining ion homeostasis and in the movement of salts (and therefore water) across apical (secretory membranes). Several families of ion channels have been identified, including subfamilies of proteins that carry Cl⁻. In the lung, pancreas, and intestinal epithelial cells, basolateral to apical membrane Cl⁻ (and ultimately, water) secretion depends upon a functional apical membrane Cl⁻ channel. The cystic fibrosis transmembrane regulator (CFTR) serves as an apical membrane Cl⁻ channel in these tissues. Cl⁻ transport by the CFTR is ATP dependent, and Cl⁻ transport is activated by cAMP-dependent protein kinase *in vivo* and *in vitro*. Mutations in this protein affect one of 2500 live Caucasian births. It results in a secretory defect that affects salt and water movement, and which ultimately leads to an early death in homozygotes. Many mutations of this protein are known, the most common being at phenylalanine in position 508. This single amino acid in CFTR leads to cystic fibrosis. Repair of the Cl⁻ transport defect in isolated CFTR-negative cells by transfection with the normal CFTR has suggested gene replacement therapies for curing cystic fibrosis in humans. Surprisingly, the major mutant form of the CFTR is equally capable of Cl⁻ transport as the normal protein, as shown in studies of the isolated (synthetic) mutant CFTR. The lack of activity of the mutant protein in the cells of the affected individuals may result from the inability of the mutant protein to be recruited to the plasma membrane or from defects in other intracellular processes that have yet to be defined. Thus, studies of regulation by recruitment, by protein kinase action, and by alteration of the overall ionic homeostasis of the affected cells have been undertaken in an effort to define the mechanisms whereby a functional (mutant) Cl⁻ channel fails to exhibit Cl⁻ channel function *in vivo*. Understanding the mechanisms of control and the various functions of the CFTR will be essential to development of therapies or cures for this serious disease.

An understanding of the physiological significance of molecular details of the function of ion transport proteins requires the combined expertise of biochemistry, molecular biology, biophysics, cellular biology, and whole animal physiology.

## XIV. Summary

The biological membrane is a selective barrier to the movement of molecules. Natural and synthetic ionophores facilitate the movement of small molecules across biological membranes. These ionophores mimic some of the properties of biological transport proteins in that they promote the movement of specific ions by increasing the rates of movements of specific ions across membranes. The larger biological transport proteins are involved in the movement of virtually all substances which cross the membrane (with the exception of some gasses and very hydrophobic sub-

vanadate and ouabain are sufficient to alter the conformation of the protein. It would appear that determination of the regions of the Na⁺,K⁺-ATPase that bind ions and determination of the mechanisms whereby ions become alternatively exposed to the inside and outside face of the protein (with a concomitant change in affinity of the site(s) for ion binding) would be important to understanding the function of the pumps.

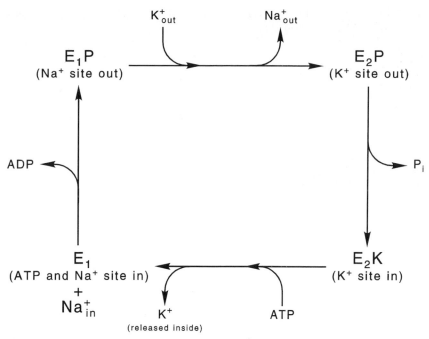

**FIG. 7.** Kinetic scheme for the Na$^+$,K$^+$-ATPase.

stances). Such proteins carry out facilitated diffusion wherein the solute is equilibrated across the membrane; secondary active transport, where solute concentration gradients are produced by coupling to ion or charge gradients; and primary active transport, where the solute is transported against a concentration gradient by using the energy of ATP. Biological transport proteins have been identified at the molecular level using tightly bound inhibitors, protein purification, molecular cloning, and expression of the cloned proteins. The primary structure of biological transport proteins determines the secondary and tertiary structure of the protein. The solute and ATP-binding sites and pores formed by the folding of the amino acids of the transport protein are responsible for the binding and transport of solute. Delineation of these structures and the dynamics of changes in these structures during the reaction cycle will be important next stages in the understanding of the transport process. The transport process is usually the rate-limiting step in the utilization of substrate, and in the case of ion pumps, is responsible for utilization of the bulk of the cellular ATP. Transport proteins are therefore often under regulatory control. Defects in the transport regulation or transport proteins lead to diseases such as diabetes and cystic fibrosis. Correction of defective trans-

port regulation by insulin has long been used as an effective therapy, and new drugs or perhaps gene replacement therapies provide hope for treatment of other transport-protein-related diseases.

## References

Bell, G. I. (1991). Molecular defects in diabetes mellitus. *Diabetes* **40,** 413–422.

Fuller, C. M., and Benos, D. (1992). CFTR! *Am. J. Physiol.* **263** (*Cell Physiol.* **32**), C267–C286.

Pedemonte, C. H., and Kaplan, J. H. (1990). Chemical modification as an approach to elucidation of sodium pump structure–function relationships. *Am J Physiol.* **258** (*Cell Physiol.* **27**), C1–C23.

Pressman, B. C. (1968). Ionophorous antibiotics as models for biological transport. *Fed. Proc. Fed. Am. Soc. Exp. Biol.* **27,** 1283–1288.

Skrivner, E., Hebert, H., Kaveus, U., and Maunsback, A. B. (1991). Three-dimensional structure of CO (NH$_3$)$_4$ ATP-induced membrane crystals of Na,K-ATPase. *In* "The Sodium Pump: Recent Developments, 44th Annual Symposium of The Society of General Physiologists" (J. H. Kaplan and P. De Weer, eds.), Rockefeller University Press, New York.

Stein, W. D. (1990). "Channels, Carriers and Pumps." Academic Press, New York.

David M. Balshaw, Lauren A. Millette, and Earl T. Wallick

# 16

# The Sodium Pump

## I. Introduction

As primordial life forms were evolving, it became necessary to separate the internal milieu of developing cells from their varied, sometimes harsh, external surroundings. For this purpose, biological membranes developed (see Chapter 4). This compartmentalization enabled the evolution of other processes that allowed for functional specialization of cells. One such process that evolved was the active pumping of ions through the phospholipid membrane barrier. Early forms of biological pumps that evolved in prokaryotes were proton pumps that regulated intracellular pH (Gogarten *et al.*, 1989). As biological complexity increased, more elaborate **transport activities** evolved to improve the control of the organism over its environment. As early life forms became exposed to various ionic environments, a need to develop some means of regulating cellular volume and of maintaining cytosolic homeostasis arose. Since water follows the movement of ions as they are transported between compartments, it became imperative to develop mechanisms to control the flow of ions, such as $Na^+$ and $K^+$, through the plasma membrane. In eukaryotic cells, these processes are controlled by the **sodium pump**. The enzyme that fuels the sodium pump is the $Na^+,K^+$-ATPase.

$Na^+,K^+$-ATPase is a member of the P-type ATPase superfamily, which differs structurally and functionally from both the F-type ATPases (ATP-synthases present in prokaryotes, chloroplasts, and mitochondria) and V-type ATPases (e.g., the $H^+$ pump located in vacuolar membranes of eukaryotic cells). The mechanism of action of the P-type ATPases involves formation of a transient, covalently phosphorylated intermediate during their reaction cycle. Structurally, they include at least one catalytic subunit with a mass ranging from 70 to 130 kDa. There are P-type ATPases in both prokaryotes (e.g., copper and cadmium ion pumps) and eukaryotes. The eukaryotic P-type ATPases can be further subdivided into two groups. One group of eukaryotic P-type ATPases consist only of a single subunit, designated $\alpha$, and includes the sarco-endoplasmic

reticulum $Ca^{2+}$-ATPase (SERCA), the plasma membrane $Ca^{2+}$-ATPase, and the $H^+$-ATPase found in yeast and plants. Another grouping of the eukaryotic P-type ATPase family contains an additional subunit ($\beta$) and includes the gastric $H^+,K^+$-ATPase and the $Na^+,K^+$-ATPase. The $\alpha$-subunit of the $H^+,K^+$-ATPase and the $Na^+,K^+$-ATPase show approximately 60% homology in amino acid sequence, and the structure of the genes encoding the two enzymes is very similar. The $\alpha$-subunit of SERCA is approximately 30% homologous in amino acid sequence to the $\alpha$-subunit of the $Na^+,K^+$-ATPase. Not surprisingly, the function and mechanism of action of this family of enzymes is closely related.

## II. Na$^+$/K$^+$ Transport

Figure 1 shows the distribution of $Na^+$, $K^+$, and $Cl^-$ across the plasma membrane of a typical mammalian cell with a resting membrane potential of $-70$ mV. There is little driving force for chloride ions, since their Nernst potential (see Chapter 1) is $-69$ mV, near the resting membrane potential. Potassium ions will tend to flow out of the cell, since their equilibrium potential ($-91$ mV) is more negative than the transmembrane potential. Conversely, $Na^+$ will have a very strong force driving them into the cell, since both the chemical and electrical gradients (equilibrium potential of $+64$ mV) favor $Na^+$ uptake. The pumping of $Na^+$ out of the cell and $K^+$ into the cell, and hence regulation of cytosolic ionic concentrations and cell volume, therefore requires sufficient energy (derived from the hydrolysis of ATP) to overcome these driving forces. This, by definition, is the **sodium pump activity**.

The nonequivalent transport of 3 $Na^+$ for 2 $K^+$ across the membrane maintains transmembrane gradients for these ions, and this became a convenient driving force for the secondary transport of metabolic substrates. In addition, the nonequivalent transport is electrogenic and leads to the generation of a transmembrane electrical potential allowing cells to become excitable. The presence of the

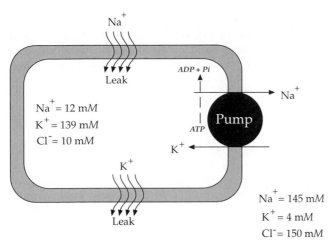

**FIG. 1.** Distribution of sodium, potassium, and chloride ions in a typical mammalian cell.

ionic gradients provided a mechanism for the regulation of many of the processes essential for the viability and stability of the cell. These include maintenance of intracellular pH via the activity of the $Na^+/H^+$ antiporter, contractility by means of $K^+$ activation of the actomyosin ATPase as well as $Na^+/Ca^{2+}$ exchange, and metabolism through symport of $Na^+$ with metabolites, such as sugars, amino acids, and neurotransmitters.

The internal environment has a high concentration of $K^+$ and a low concentration of $Na^+$. Internal biological functions (such as the contractility of muscle cells) have, therefore, evolved to be activated by $K^+$ and inhibited by $Na^+$. If a cell is damaged or becomes starved for energy, the concentration of $K^+$ will decrease and the concentration of $Na^+$ will increase because of either an increased leak of the ions with their gradients or a decreased pumping against the gradient. This decrease in $K^+$ and increase in $Na^+$ leads to inhibition of energy-utilizing processes and allows other compensatory mechanisms to attempt to repair any damage.

## III. Transport Mechanism

Early studies of the pump used intact systems such as giant squid axons and erythrocyte ghosts, enabling both the direct measurement of ion fluxes using radioactive tracers and indirect measurements using electrophysiological techniques. Initial studies found that the appearance of $^{22}Na$ in the bath solution following microinjection into the squid axon required that $K^+$ be present in the extracellular bath solution. This suggested that $Na^+$ efflux (transport out of the cell) was coupled to $K^+$ influx (transport into the cell). Experiments performed in parallel showed that, under physiological conditions, the efflux of $^{22}Na$ and influx of $^{42}K$ occurred at a coupling ratio of 3 $Na^+$:2 $K^+$:1 ATP. More recent studies have demonstrated that this ratio is not fixed, but can vary depending on the concentrations of the ions both within and outside the cell (reviewed in Lauger, 1991; Tepperman *et al.*, 1997).

Transport studies using erythrocytes allowed researchers to determine the ion selectivity of the sodium pump. The intracellular activation of the pump explicitly requires $Na^+$. These sites have an intrinsic affinity[1] of 0.2 m$M$ for $Na^+$. Potassium competes (but does not activate transport) with the $Na^+$ sites with an intrinsic affinity of 9 m$M$ (Garay and Garrahan, 1973). In the presence of normal concentrations of internal $K^+$ (see Fig. 1), the apparent affinity for $Na^+$ at the intracellular site is approximately 3 m$M$. The extracellular transport sites are much less specific for potassium since $Rb^+$, $Cs^+$, $Li^+$, $Tl^+$, $NH_4^+$, and even $Na^+$ can activate ATPase (Bell *et al.*, 1977). The intrinsic affinities of the extracellular transport sites for $K^+$ and $Na^+$ are approximately 0.2 and 9–16 m$M$, respectively (Sachs, 1977). In the presence of normal concentrations of external $Na^+$ (see Fig. 1), the affinity for the extracellular $K^+$ is approximately 2 m$M$. Affinities for cations derived from studies of $Na^+$ and $K^+$ activation of hydrolysis of ATP in fragmented enzyme preparations (Lindenmayer *et al.*, 1974) are in agreement with the studies of intact erythrocytes.

Studies with axons and erythrocytes also demonstrated that the activation of the pump is dependent on the presence of intracellular ATP and that, in particular, it is specific for MgATP. The products of the hydrolysis, ADP and Pi (inorganic phosphate), remain within the cell. The dependence of the pump activity on metabolically derived ATP was confirmed by the finding that metabolic inhibitors such as azide, cyanide, and dinitrophenol greatly reduce the pump activity.

Schatzmann (1953) reported that the sodium pump activity was specifically inhibited by cardiac glycosides, plant-derived steroids that have been used for 200 years to treat heart failure (Withering, 1785). Skou isolated an enzyme found in the plasma membrane whose hydrolysis of ATP was *activated by both $Na^+$ and $K^+$* (Skou, 1957). This is in contrast to the majority of cytosolic enzymes, which are activated by $K^+$ and inhibited by $Na^+$. This $Na^+,K^+$-ATPase is specifically inhibited by cardiac glycosides in the same concentration range that inhibits transport in intact cells, demonstrating that this enzyme is responsible for the sodium pump activity.

Studies using partially purified, fragmented membrane preparations of $Na^+,K^+$-ATPase found that, in the presence of $Na^+$, $Mg^{2+}$ and ATP-$\gamma^{32}P$, the $\gamma P$ was covalently transferred to the enzyme, forming an acid-stable acyl phosphate bond. This established the $Na^+,K^+$-ATPase as a member of the P-type family of ATPases. When $K^+$ is added to the enzyme that has been phosphorylated in the presence of $Na^+$, $Mg^{2+}$, and ATP, the phosphate group is rapidly cleaved and the enzyme is allowed to cycle, continuing to hydrolyze ATP (Fig. 2). Both $N$-ethylmaleimide and oligomycin decrease the sensitivity of the enzyme to $K^+$-induced dephosphorylation and increase the sensitivity of phosphorylated enzyme to ADP (reviewed in Glynn, 1993). This led to the concept that there were two primary enzyme conformations, both of which can exist in either a phos-

---

[1] Intrinsic affinity is the affinity of the ion in the absence of competing ions and is obtained by extrapolation to a concentration of zero competitor.

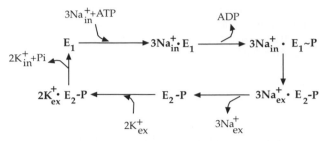

**FIG. 2.** Post–Albers reaction mechanism for the sodium pump.

phorylated or an unphosphorylated form. These are designated $E_1$, which has a high affinity for intracellular $Na^+$, and $E_2$, which has a high affinity for extracellular $K^+$. The cycling of the enzyme between the $E_1$ forms, binding cations at the intracellular face, and $E_2$ forms, binding cations at the extracellular face, results in the transport of the ions through the membrane. This concept, illustrated in Fig. 2, is referred to as the **Post–Albers scheme** in recognition of the contribution of these two scientists.

In the $E_1$ conformation, the enzyme has high affinity for both intracellular $Na^+$ and MgATP. Following the binding of both of these ligands, the $\gamma$ phosphate group of ATP is transferred to an aspartyl residue on the enzyme, forming a "high-energy" phosphoenzyme. In this "high-energy state," the phosphate group can be readily returned to ADP, resynthesizing ATP. In the forward cycle of the enzyme, the "high energy" residing in the phosphate bond is used to drive a change in the conformation of the enzyme, resulting in the exposure of $Na^+$ to the extracellular surface. This "low-energy" phosphoenzyme, with $Na^+$ bound, is now said to be in an $E_2$ conformation. The phosphate group in this state *cannot* be easily transferred back to ADP to form ATP. The $E_2P$ conformation has low affinity for extracellular $Na^+$; therefore, it is released. This conformation has high affinity for extracellular $K^+$. After $K^+$ binds at the extracellular surface, the phosphatase activity of the enzyme is activated, the phosphate group is cleaved, and $K^+$ is translocated and released on the intracellular surface. The enzyme is once again in the $E_1$ conformation and capable of binding $Na^+$ and ATP, completing the cycle. Because of the cycling between these two conformations, the P-type ATPases are also referred to as $E_1E_2$-ATPases.

Additional evidence in support of the two-conformation hypothesis came from studies using controlled proteolysis. Jorgensen found that in the presence of $Na^+$, specific sequences for trypsin proteolysis were exposed, designated T2 and T3 (see later section on $Na^+,K^+$-ATPase structure and Fig. 4). In addition, a site for chymotrypsin, C3, was observed. In the presence of $K^+$, the T2 site is still cleaved, but an additional site, T1, is now exposed, while the T3 and C3 site are now hidden (Jorgensen, 1977). Fluorescence measurements of enzyme labeled at K501 with fluorescein isothiocyanate (FITC) also supported the hypothesis that the enzyme undergoes structural rearrangements dependent on ligand conditions. FITC labeling is competitive with ATP; however, mutation of the labeled residue did not affect the activity of the enzyme. It is likely, therefore,

that the inhibition of ATP binding following FITC modification is due to an indirect, steric effect. Nonetheless, a change in the fluorescence properties of the FITC following the addition of $Na^+$ or $K^+$ indicates that the conformation of the active site is changed (Hegyvary and Jorgensen, 1981). These two types of studies have clearly demonstrated that the enzyme is undergoing structural rearrangement when exposed to different ionic conditions.

It is known that the Post–Albers scheme (see Fig. 2) is overly simplistic. Perhaps a more realistic view of the reaction mechanism is that it consists of a series of steps whereby the ions are **bound, occluded,** and **released** from the enzyme (Forbush, 1987a). The occluded state refers to a form of the enzyme where the ions are trapped within the membrane and are not accessible from either the extracellular or intracellular compartments. This becomes intuitively obvious if one considers the movement of a 2-Å molecule being moved through a 50- to 70-Å thick membrane against its concentration gradient. Figure 3 illustrates a modified version of the Post–Albers scheme that is more consistent with recent findings. First, in the $E_1$ conformation, it has been proposed that the $Na^+$ ions bind in a multistep process, with the rate of binding for the first two sodium ions being so fast as to be indistinguishable, followed by the third $Na^+$ binding at a slower rate (Heyse *et al.*, 1994). In the presence of $Na^+$, MgATP is bound with high affinity and the $\gamma$ phosphate is transferred to the enzyme. This occurs with the simultaneous occlusion of either two or three $Na^+$ ions. The sodium ions are then translocated and released on the extracellular surface; it is believed that the release of sodium is an ordered process. One ion is released first, followed by the remaining two. The release of the sodium ions allows for the binding of the potassium ions, again in an ordered fashion, with positive cooperativity, such that the binding of the first potassium to the extracellular surface increases the affinity for the second (Tepperman *et al.*, 1997). If the concentration of extracellular $K^+$ is low, it is possible that only one $K^+$ will bind and be transported. The binding of potassium results in a rapid dephosphorylation of the enzyme along with occlusion of the ions. In the absence of intracellular ATP, the rate of $K^+$ release to the intracellular surface is very slow. Following the binding of ATP the rate of deocclusion is greatly increased. Occlusion studies by Forbush (1987b) have demonstrated that the release of $K^+$ in the absence of ATP following transport is a multistep, ordered process in which one ion is released at a rapid rate while the second ion is released more slowly. This process is consistent with a flickering gate model, in which the exposure of the cation binding sites limits the release to a single ion at a time. The release of $K^+$ allows $Na^+$ to bind and increase the affinity of the enzyme for ATP, thus completing the enzyme cycle.

Figure 3 does not, however, take into account that internal $K^+$ binds to (competes with) the $Na^+$ transport sites, or that extracellular $Na^+$ is able to bind to (compete with) the external $K^+$ transport sites. Extracellular $Na^+$ has been implicated additionally as an allosteric effector of $K^+$ transport (Sachs, 1977). From the mechanism depicted in Fig. 3, it is clear that the coupling ratio does not have to be

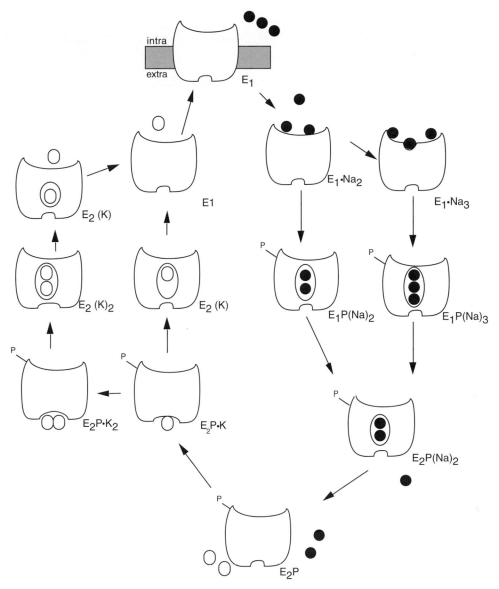

**FIG. 3.** Modified Post–Albers scheme showing binding, occlusion, and release of cations ($Na^+$ and $K^+$) from both the intracellular and extracellular surfaces, and possible noncannonical flux modes. Enzyme forms that contain occluded cations are shown with the cations in parentheses [i.e. $E_1P(Na)_3$]. Enzyme forms with cations bound are shown with a dot (i.e., $E_1 \cdot Na_3$). $Na^+$ is designated by filled circles and $K^+$ by open circles. For simplicity, the binding and hydrolysis of ATP are omitted.

fixed at 3 $Na^+$:2 $K^+$ under all conditions. The study by Tepperman *et al.* demonstrated that the only models that fit their data were models in which translocation could take place when either one or two $K^+$ were bound to the enzyme (Tepperman *et al.*, 1997).

## IV. Na$^+$,K$^+$-ATPase Structure

The Na$^+$,K$^+$-ATPase is composed of two subunits (Fig. 4) in a one-to-one stoichiometry: the **alpha** ($\alpha$) or catalytic subunit and the **beta** ($\beta$) or glycoprotein subunit (Lane *et al.*, 1973). The $\alpha$-subunit contains all of the known binding sites for ligands that regulate pump function. The complete primary structure of the sheep $\alpha$- and $\beta$-subunits has been

determined by the cloning of cDNAs corresponding to the mRNA for both subunits (Shull *et al.*, 1985, 1986a). The sheep kidney $\alpha$-subunit consists of a protein of $M_r$ 112,177 (1016 amino acid residues) and the $\beta$-subunit consists of 302 amino acid residues with a core protein $M_r$ of 34,939 (Shull *et al.*, 1986b). The $\alpha$1-subunit exists as multiple **isoforms**, and the primary sequences of these isoforms from a variety of species, including the $\alpha$1, $\alpha$2, $\alpha$3, and $\alpha$4 from the human and rat, have been determined (reviewed in Lingrel *et al.*, 1990, and see Shamraj and Lingrel, 1994).

The distribution of the four $\alpha$ isoforms is tissue specific (demonstrated in Fig. 5) and is both hormonally and developmentally regulated. The **$\alpha$1 isoform** has been found in all tissues and cell types examined and is considered to be the housekeeping form of the enzyme. The **$\alpha$2 isoform** has

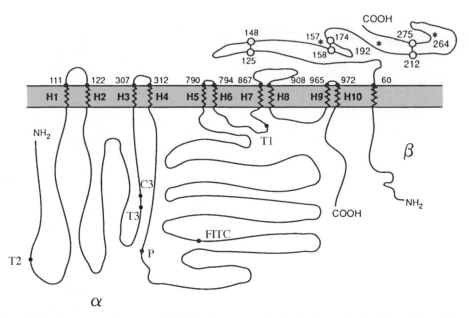

**FIG. 4.** The predicted transmembrane model of the sheep $\alpha_1$- and $\beta_1$-subunits of the Na$^+$,K$^+$-ATPase. The membrane-spanning helices are represented by H1 . . . H10. Asterisks (*) identify the three glycosylation sites, and the open circles denote the position of the disulfide bridges on the $\beta$-subunit. The sites of tryptic digestion have been designated as T1 (R438), T2 (K30), T3 (R262), and chymotryptic digestion, C3 (L266). The FITC binding site is at residue K501 and the phosphorylation site (P) is D369.

been found primarily in skeletal muscle and brain, with smaller amounts in the adult hearts of most species, including human. The **$\alpha$3 isoform** has been found primarily in brain, but also in neonatal rat heart and adult human heart. The **$\alpha$4 isoform** has been found only in rat and human testis, at an mRNA level comparable to the level of $\alpha$1 in the brain. The sequence of the four rat isoforms shows 80–85% homology. The sequence of the $\alpha$1 isoforms is also highly conserved between species. Based on hydropathy plots and analogy with the H$^+$,K$^+$-ATPase and the Ca$^{2+}$-ATPase, the $\alpha$-subunit of Na$^+$,K$^+$-ATPase is predicted to

have 10 membrane-spanning regions labeled H1 to H10, as shown in Fig. 4.

The $\beta$-subunit of sheep kidney contains three $N$-linked oligosaccharide sites and seven cyst(e)ine residues. The one free sulfhydryl, C44, is postulated to be in the membrane-spanning region. The other six residues participate in three disulfide bonds (Kirley, 1989). Based on hydropathy plots, the $\beta$1-subunit possesses only one transmembrane region (see Fig. 4). Three isoforms of the $\beta$-subunit have been described (Shull *et al.*, 1988). The **$\beta$1 isoform** described previously is the major form expressed in virtually all tissues, with $\beta$2 occurring primarily in brain. Although the two mammalian isoforms exhibit only 35% homology, the location of the glycosylation sites, the conserved cysteine residues, and similarity of the hydropathy plots suggest that these two isoforms possess high structural homology. Interestingly, the $\beta$2 isoform appears to be identical to the adhesion molecule on glia (AMOG). A third isoform, $\beta$3, has only been observed during early development in *Xenopus laevis*.

The physiological function of the $\beta$-subunit is unknown. Expression of functional Na$^+$,K$^+$-ATPase activity in yeast, which do not contain a Na$^+$,K$^+$-ATPase, requires that *both* the $\alpha$- and $\beta$-subunits be cotransfected (Horowitz *et al.*, 1990). This indicates that the $\beta$-subunit is required for the proper assembly of the $\alpha$-subunit into the membrane and for the exit of this $\alpha\beta$ complex from the endoplasmic reticulum and proper insertion into the cell membrane (reviewed by Lingrel *et al.*, 1990; McDonough *et al.*, 1990; Sweadner, 1989). The $\beta$-subunit of the enzyme also alters the susceptibility of the $\alpha$-subunit to proteolytic enzymes, further showing its involvement in stabilizing the enzyme's structure.

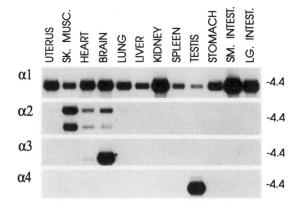

**FIG. 5.** Northern blot analysis of the mRNA expression levels of the four $\alpha$-subunit isoforms in various tissues of rats. [From Shamraj, O. I., and Lingrel, J. B. (1994). A putative fourth Na$^+$,K$^{(+)}$-ATPase alpha-subunit gene is expressed in testis. *Proc. Natl. Acad. Sci. USA* **91**, 12952–12956. Copyright 1994 National Academy of Sciences, U.S.A.]

Furthermore, this subunit has been postulated to affect cation affinity, suggesting that it may undergo conformational changes associated with enzyme function (Chow and Forte, 1995). The $\beta$-subunit of $Na^+,K^+$-ATPase is believed to be in close proximity to the cardiac glycoside binding site on the $\alpha$-subunit because it can be labeled with photoaffinity derivatives of digitoxin. The $\beta$-subunit of $H^+,K^+$-ATPase and $Na^+,K^+$-ATPase are highly homologous, and, in fact the $\beta$-subunit of the $H^+,K^+$-ATPase can replace the native $Na^+,K^+$-ATPase $\beta$-subunit and allow functional expression of $\alpha$ (Horisberger et al., 1991).

The $\alpha$-subunit of the transport ATPases have in common the phosphorylation of an aspartic acid residue (D369,[2] see Fig. 4) as a part of their enzymatic cycle. The large intracellular loop between H4 and H5 has been shown, by chemical modification and ATP analogue labeling, to contain the ATP-binding site (in the region of K501 where FITC is covalently attached; Kirley, et al., 1984) and the phosphorylation site (D369). Immunological data suggest that both the N-terminal and C-terminal regions are on the inside of the cell. Since the ATP site is also located intracellularly, these findings orient the 10 transmembrane regions (see Fig. 4). The N-terminal region contains a large number of both positively and negatively charged residues, including a lysine-rich sequence. It has been argued that this region may act as a cation-selective gate and/or participate in transport via salt bridge formation. Studies investigating the contribution of the N-terminal domain to cation affinity have shown that this region, which varies greatly between the $\alpha$ isoforms, does reduce the apparent affinity for cations, albeit only by a factor of 2 to 3 (Horisberger et al., 1991). It has also been suggested that the sequence between the phosphorylation site (D369) and the H4 transmembrane region may act as an energy-transduction region (Shull et al., 1985).

Another area of research pertains to both the internal and external cation binding sites. It has been difficult to probe the internal $Na^+$ sites because of the difficulty of controlling internal $Na^+$ and the plethora of compensatory $Na^+$ transport mechanisms. Information about the interactions of external $Na^+$ and $K^+$ with the $Na^+,K^+$-ATPase has come from $^{86}Rb$ uptake in whole cells and [$^3H$]ouabain experiments in fragmented membrane preparations. These techniques have yielded apparent affinities for $K^+$ and $Na^+$, and detailed information about their interactions through the analysis of complex mechanistic models. The $\alpha3$ isoform has a higher affinity for extracellular $K^+$ and lower affinity for intracellular $Na^+$ than do $\alpha1$ and $\alpha2$ (Munzer et al., 1994). In addition, site-directed mutagenesis has been used in an effort to determine the amino acids involved in cation binding. Although the initial hypothesis was that negatively charged residues in the predicted transmembrane region would be involved in cation binding and transport, these studies have suggested that the binding site does not consist of a single residue, but rather there is a **binding pocket** composed of a number of residues. The most significant effects have been observed by replacement

of the aspartate residue at position 804 in sheep $\alpha1$, where every replacement that has been made resulted in a nonfunctional enzyme. At residue 808, the only surviving mutation is the conservative replacement of aspartate with glutamate. Other residues (shown in Fig. 6) that affect cation affinity, possibly through indirect mechanisms, are E327, E779, S775, and D926 (summarized in Van Huysse et al., 1996). This has led to the hypothesis that the H5–H6 region, which includes residues S775, E779, D804, and D808, is directly involved in cation binding and transport. It is conceivable that this entire domain may shift in the membrane, facilitating ion transport, similar to what has been postulated for the S4 segment of the voltage-gated ion channels (Sigworth, 1993).

The most convincing evidence regarding the nature of the cardiac glycoside binding site has been made possible from the elegant experiments first performed by Price and Lingrel (1988). These experiments took advantage of the fact that the rodent $\alpha1$ isoform is 1000-fold less sensitive to glycoside inhibition than other isoforms. Expressing an "insensitive" enzyme in HeLa cells, whose endogenous enzyme is glycoside sensitive, allowed growth in media with a high concentration of ouabain, a water-soluble cardiac glycoside. If the cells do not express an insensitive transfected enzyme, they die in 0.2 $\mu M$ ouabain. Expression, however, of mutant enzyme with decreased sensitivity to glycosides will allow the cells to live in 0.2 $\mu M$ ouabain. The amino acid sequence of the H1–H2 region differs substantially between sheep and rat $\alpha1$-subunits. Using chimeric proteins and site-directed mutagenesis, it was determined that the presence of the charged residues, R111 and D122, in the rat $\alpha1$ isoform accounts for the decreased ouabain sensitivity observed for the rat (Price and Lingrel, 1988). Most species have uncharged residues in this position (e.g., sheep $\alpha1$ has Q111 and N122), and these species are sensitive to cardiac glycosides. It has been hypothesized that, following the binding of the cardiac glycosides, these noncharged residues partition into the membrane, resulting in a conformation of the enzyme from which the rate of glycoside release is very slow, and thus the affinity very high.

In addition to these two residues, D121, also in the H1H2 extracellular loop, has been shown to affect the glycoside affinity. Mutagenesis of other regions has revealed that the glycoside binding pocket spans the majority of the extracellular and transmembrane topology (see Fig. 6) (summarized in Palasis et al., 1996). These diverse regions include the H1 transmembrane domain (C104 and Y108), the H5 transmembrane segment (F786), the H5H6 extracellular "hairpin" loop (L793), the H6 transmembrane segment (T797), the H7 transmembrane domain (F863), and the H7H8 extracellular loop (R880) (see Fig. 6). It is interesting to note that the H5H6 region has been implicated in glycoside binding and also appears to be central in the cation-transport processes of the enzyme. Ouabain or $K^+$ is able to protect the enzyme from cleavage under extensive tryptic digestion at the borders of the H5H6 region, suggesting that this region is involved in the conformational transitions of the enzyme (Lutsenko and Kaplan, 1993).

---

[2] The numbering system used in this chapter is for the sheep $\alpha1$ isoform.

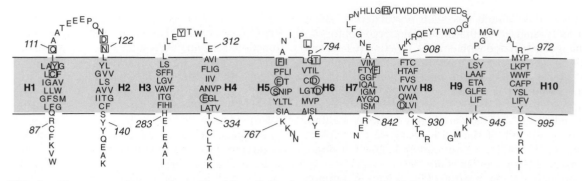

**FIG. 6.** The transmembrane amino acid residues of the sheep $\alpha 1$ sequence. The boxed residues represent amino acids implicated in affecting glycoside sensitivity, and encircled residues are those implicated in altering cation affinity.

## V. Cardiac Glycosides

Plant extracts containing **cardiac glycosides** have been used for many years as medicines (for their therapeutic effect) and as poisons (for their toxic effect) by diverse groups that include ancient Egyptians, Romans, Chinese, and African tribesmen (Hoffman and Bigger, 1985). Withering (1785) noted the use of foxglove (*Digitalis purpurea*) extracts (which contain cardiac glycosides) in treatment of dropsy (pedal edema secondary to heart failure). For the next hundred years, **digitalis** (plant extracts of cardiac glycosides, containing primarily digoxin) was used, frequently at excessive doses, for the treatment of a range of diseases. By the early 20th century, digitalis was known to increase the force of myocardial contractility (**positive inotropic effect**), and thus, to be useful in the treatment of congestive heart failure. The toxic effects of digitalis, however, are frequent and can be severe, even fatal. In spite of this, **digoxin** (the drug used clinically) is still one of the most widely used drugs, being the sixth most commonly prescribed drug in the United States in 1995 (Anon, 1996).

The mechanism of action now generally accepted is termed the **sodium pump lag hypothesis** and postulates that therapeutic concentrations of digoxin inhibit a moderate (30–40%) fraction of the membrane $Na^+,K^+$-ATPase. This, in turn, causes a slight increase in internal $Na^+$ that has two significant effects; (1) it inhibits the forward operation of the $Na^+/Ca^{2+}$ exchanger (Chapter 18), slowing down the rate at which $Ca^{2+}$ exits the cell (note that if $Na^+$ concentration is raised above approximately 20 m$M$, the $Na^+/Ca^{2+}$ exchanger can be forced into reverse mode and drive $Ca^{2+}$ influx); (2) it stimulates the 60–70% fraction of remaining uninhibited pumps, which tend to return the internal $Na^+$ concentration to homeostasis. Thus, therapeutic levels of digoxin produce transient increases in internal $Na^+$ in heart muscle, which lead to transient increases in internal $Ca^{2+}$, which in turn stimulates the release of a larger amount of $Ca^{2+}$ from internal stores in the sarcoplasmic reticulum (Chapter 17). The released calcium interacts with contractile proteins to increase the force of contraction. Toxic concentrations of digoxin lead to extensive (>60%) inhibition of $Na^+/K^+$ transport, so that restoration of normal diastolic levels of $Na^+$ and $Ca^{2+}$ is not possible. Digoxin also indirectly produces sympathetic and parasym-

pathetic effects which can result in cardiac arrhythmias. The first step in both the **therapeutic** and **toxic effects** is clearly mediated via inhibition of $Na^+,K^+$-ATPase (see review by Thomas *et al.*, 1989).

There are three factors that determine the affinity of cardiac glycosides and their analogues for $Na^+,K^+$-ATPase: (1) the nature (and concentration) of the physiological ligands ($Mg^{2+}, Na^+$, ATP, $P_i$, and $K^+$) bound to the enzyme, (2) the structure of the drug, and (3) the structure of the receptor. In the following scheme, LRD represents the ternary complex of drug (D), receptor (R), and physiological ligand(s) (L). The dissociation constant for the rapid binding of the physiological ligand to receptor is represented by $K_L$, and the **association** and **dissociation rate constants** for binding of drug to receptor are represented by $k_1$ and $k_{-1}$, respectively. The **dissociation constant** $K_D$ is equal to $k_{-1}/k_1$.

$$R + L \overset{K_L}{\leftrightarrow} LR + D \underset{k_1}{\overset{k_{-1}}{\rightleftharpoons}} LRD$$

For a given cardiac glycoside, the on rate, and consequently the $K_D$, is very sensitive to the nature of the ligand(s) present in the medium. Inorganic phosphate, $Mg^{2+}$, $Na^+$, and ATP stimulate and $K^+$ inhibits the rate of binding, leading many to hypothesize that phosphorylation is necessary for ouabain binding. By varying the concentration of the ligand, the dissociation constant for the binding of the ligand to the enzyme can be obtained. The dissociation constant ($K_L$) for the $Mg^{2+}$ site is 0.2 m$M$, and when saturated, increases the affinity for ouabain by a factor of 300 over that in presence of buffer alone. The stimulation by $Na^+$ ($K_L = 14$ m$M$) is competitively inhibited by $K^+$ ($K_L = 0.2$ m$M$). The similarity of these $K_L$ values to those derived from transport studies suggest that these monovalent cation regulatory sites for cardiac glycosides are equivalent to the transport sites. In contrast to the on-rate, the off-rate of a specific cardiac glycoside from a specific isoform of the enzyme is not influenced to the same degree by the presence of physiological regulatory ligands. This could be the result of a **ligand-induced conformational change,** such that the enzyme–drug complex initially formed is transformed into a second, more stable complex whose rate of dissociation is independent of the regulatory ligands present.

Mutation of the phosphorylation site, D369N, produces an enzyme that cannot be phosphorylated, cannot hydrolyze ATP, and cannot translocate Na$^+$ and K$^+$. Expression of the sheep $\alpha$1 mutant in NIH 3T2 cells, whose endogenous Na$^+$,K$^+$-ATPase is insensitive, produces a membrane form of the Na$^+$,K$^+$-ATPase that can still bind ouabain and can be used as a probe for the physiological ligands (Kuntzweiler et al., 1995). The mutant enzyme still binds ouabain with high affinity, demonstrating that phosphorylation is not necessary for ouabain binding. The stimulatory effect of phosphate on the binding of ouabain, in the presence of Mg$^{2+}$, is absent in the mutant, showing that the effect of phosphorylation occurs at D369. The fact that ATP and MgATP still bind to the mutant enzyme but inhibit ouabain binding, in contrast to the wild type, in which they stimulate ouabain binding, demonstrates that the Mg-nucleotide site is still present. Three Na$^+$ and two K$^+$ still bind to the mutant enzyme with only slight changes in affinity and still affect ouabain binding (in the presence of Mg$^{2+}$). This indicates that Na$^+$ and K$^+$ can induce conformational changes in the enzyme that affect the affinity for ouabain independent of whether the enzyme is phosphorylated. This is consistent with the hypothesis that ouabain prefers to bind to the E$_2$ conformation. This study also demonstrated that there are distinct sites for Mg$^{2+}$ and MgATP that can be simultaneously occupied.

The structure of the cardiac glycoside also affects their affinity. Classic structure–activity relationship studies of cardiac glycosides, reviewed by Thomas et al. (1989), concluded that there are three primary domains of the cardiac glycosides that contribute to their biological activity. The first of these is a **steroid ring** system with a *cis* configuration of A/B and C/D rings, unique to the cardiac glycosides, as well as 3$\beta$- and 14$\beta$-hydroxyl groups (Fig. 7). The cardiac glycosides also require the presence of an unsaturated $\beta$-lactone on carbon 17. The second domain is a five-membered $\alpha\beta$ unsaturated **lactone** ring in the cardenolide family of the cardiac glycosides, and a six-membered diunsaturated lactone ring in the bufadienolides, a related family of compounds (found primarily in toads) with similar pharmacological action, but significantly higher affinity. Reduction of the double bond(s) in the lactone ring reduces the biological activity. The third domain is a **carbohydrate** moiety. Replacement of the 3$\beta$-hydroxyl group of the **aglycone** with a sugar residue increases affinity by dramatically slowing the rate of dissociation of the cardiac glycoside

from the receptor. The presence of the rhamnose moiety on ouabain, for example, increases the affinity 10- to 30-fold. Cardiac glycosides lacking a carbohydrate group are referred to as genins (i.e., digoxigenin is digoxin without the tridigitoxose moiety on C3, but with a 3$\beta$-hydroxyl group; see Fig 7).

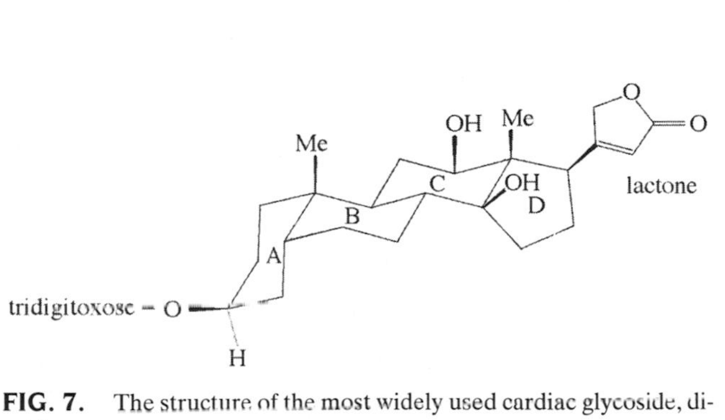

## VI. Summary

The Na$^+$,K$^+$-ATPase is one of the most vital proteins present in animal cells, as demonstrated by the remarkable degree of conservation across evolutionary lines and the fact that it has been found in every animal cell type studied. This enzyme places a high metabolic burden on the cell, using a substantial portion of the total cellular energy production in maintaining the osmotic balance and ionic gradients across the plasma membrane. This enzyme is crucial for the survival of animal cells, as demonstrated by the fact that transgenic knockout (either $\alpha$1 or $\alpha$2 isoform) mice do not survive (Lingrel, personal communication, 1997). In addition to its role in maintaining homeostasis, inhibition of the pump has significant therapeutic relevance by virtue of the secondary increase in Ca$^{2+}$ concentration leading to an increase in cardiac contractility.

The pump activity is due to an integral membrane protein consisting of two subunits, $\alpha$ and $\beta$, and belonging to the P-type family of ATPases. This enzyme hydrolyzes ATP in the presence of Na$^+$ and K$^+$. The $\alpha$-subunit has at least four isoforms and the $\beta$-subunit three, each of which is differentially expressed. The cation affinities for the activation of ion transport, ATP hydrolysis, and effects on glycoside binding agree well with each other, indicating that the same binding sites are involved in all three activities.

Although the enzyme has been studied in great detail over the past 40 years, there remain many unanswered questions that can be placed into three basic categories: (1) What is the three-dimensional structure of the enzyme? (2) What is the mechanism of cation transport? (3) What are the structural domains that make up ligand-binding sites?

## References

Anon. (1996). Top 200 drugs of 1995. *Pharmacy Times,* April, p. 29.

Bell, M. V., Tonduer, F., and Sargent, J. R. (1977). The activation of sodium-plus-potassium ion-dependent adenosine triphosphatase from marine teleost gills by univalent cations. *Biochem. J.* **163,** 185–187.

Chow, D. C., and Forte, J. G. (1995). Functional significance of the beta-subunit for heterodimeric P-type ATPases. *J. Exp. Biol.* **198,** 1–17.

Forbush, B. (1987a). Na$^+$, K$^+$, and Rb$^+$ Movements in a single turnover of the Na/K pump. *Curr. Top. Membr. Transp.* **28,** 19–39.

Forbush, B. (1987b). Rapid release of $^{42}$K or $^{86}$Rb from two distinct transport sites on the Na,K-Pump in the presence of P$_i$ or vanadate. *J. Biol. Chem.* **262,** 11116–11127.

Garay, R. P., and Garrahan, P. J. (1973). The interaction of sodium and potassium with the sodium pump in red cells. *J. Physiol. (London)* **231,** 297–325.

**FIG. 7.** The structure of the most widely used cardiac glycoside, digoxin.

Glynn, I. M. (1993). All hands to the sodium pump. *J. Physiol.* **462**, 1–30.

Gogarten, J. P., Rausch, T., Bernasconi, P., Kibak, H., and Taiz, L. (1989). Molecular evolution of H+-ATPase, I. Methanococcus and Solfolobus are monophyletic with respect to eukaryotes and eubacteria. *Z. Naturforsch.* **44**, 641–650.

Hegyvary, C., and Jorgensen, P. L. (1981). Conformational changes of renal sodium plus potassium ion-transport adenosine triphosphatase labeled with fluorescein. *J. Biol. Chem.* **256**, 6296–6303.

Heyse, S., Wuddel, I., Apell, H.-J., and Sturmer, W. (1994). Partial reactions of the Na,K-ATPase: Determination of rate constants. *J. Gen. Physiol.* **104**, 197–240.

Hoffman, B. F., and Bigger, J. T. (1985). Digitalis and allied cardiac glycosides. *In* "Pharmacological Basis of Therapeutics" (A. G. Gilman, W. S. Goodman, T. W. Rall, and F. Murad, Eds.), pp. 716–747. Macmillan, New York.

Horisberger, J. D., Jaunin, P., Reuben, M. A., Lasater, L. S., Chow, D. C., Forte, J. G., Sachs, G., Rossier, B. C., and Geering, K. (1991). The H,K-ATPase beta-subunit can act as a surrogate for the beta-subunit of Na,K-pumps. *J. Biol. Chem.* **266**, 19131–19134.

Horowitz, B., Eakle, K. A., Scheiner-Bobis, G., Randolph, G. R., Chen, C. Y., Hitzeman, R. A., and Farley, R. A. (1990). Synthesis and assembly of functional mammalian Na,K-ATPase in yeast. *J. Biol. Chem.* **265**, 4189–4192.

Jorgensen, P. L. (1977). Purification and characterization of (Na⁺ + K⁺)-ATPase. VI. Differential tryptic modification of catalytic functions of the purified enzyme in presence of NaCl and KCl. *Biochim. Biophys. Acta* **466**, 97–108.

Kirley, T. L. (1989). Determination of three disulfide bonds and one free sulfhydryl in the β subunit of (Na,K)-ATPase *J. Biol. Chem.* **264**, 7185–7192.

Kirley, T. L., Wallick, E. T., and Lane, L. K. (1984). The amino acid sequence of the fluorescein isothiocyanate reactive site of lamb and rat kidney Na⁺- and K⁺-dependent ATPase. *Biochem. Biophys. Res. Commun.* **125**, 767–773.

Kuntzweiler, T. A., Wallick, E. T., Johnson, C. L., and Lingrel, J. B. (1995). Amino acid replacement of D369 in the sheep α1 isoform eliminates ATP and phosphate stimulation of [H³]ouabain binding to the Na⁺,K⁺-ATPase without altering the cation binding properties of the enzyme. *J.Biol. Chem.* **270**, 16206–16212.

Lane, L. K., Copenhaver, G. H., Lindenmayer, G. E., and Schwartz, A. (1973). Purification and characterization of and [3H]ouabain binding to the transport adenosine triphosphatase from outer medulla of canine kidney. *J. Biol. Chem.* **248**, 7197–7200.

Lauger, P. (1991). "Electrogenic Ion Pumps," Vol. 5. Sinauer Associates, Sunderland, MA.

Lindenmayer, G. E., Schwartz, A., and Thompson, H. K., Jr. (1974). A kinetic description for sodium and potassium effects on (Na⁺ + K⁺)-adenosine triphosphatase: A model for a two-nonequivalent site potassium activation and an analysis of multiequivalent site models for sodium activation. *J. Physiol.* (*London*) **236**, 1–28.

Lingrel, J. B., Orlowski, J., Shull, M. M., and Price, E. M. (1990). Molecular genetics of Na,K-ATPase. *In* "Progress in Nucleic Acids Research and Molecular Biology," Vol. 38, pp. 37–89. Academic Press, London.

Lutsenko, S., and Kaplan, J. H. (1993). An essential role for the extracellular domain of the Na,K-ATPase β-subunit in cation occlusion. *Biochemistry* **32**, 6737–6743.

McDonough, A. A., Geering, K., and Farley, R. A. (1990). The sodium pump needs its β subunit. *FASEB J* **4**, 1598–1605.

Munzer, J. S., Daly, S. E., Jewell-Motz, E. A., Lingrel, J. B., and Blostein, R. (1994). Tissue- and isoform-specific kinetic behavior of the Na,K-ATPase. *J. Biol. Chem.* **269**, 16668–16676.

Palasis, M., Kuntzweiler, T. A., Arguello, J. M., and Lingrel, J. B. (1996). Ouabain interactions with the H5–H6 hairpin of the Na,K-ATPase reveal a possible inhibition mechanism via the cation binding domain. *J. Biol. Chem.* **271**, 14176–14182.

Price, E. M., and Lingrel, J. B. (1988). Structure–function relationships in the Na,K-ATPase α-subunit: Site-directed mutagenesis of glutamine-111 to arginine and asparagine-122 to aspartic acid generates a ouabain-resistant enzyme. *Biochemistry* **27**, 8400–8408.

Sachs, J. R. (1977). Inhibition of the Na–K Pump by External Sodium. *J. Physiol.* (*London*) **264**, 449–470.

Schatzmann, H. J. (1953). Herzglykoside als Hemmstoffe fur den activen Kalium und Natrium-transport durch die Erythrocytenmembran. *Helv. Physiol. Pharmacol. Acta* **11**, 346–354.

Shamraj, O. I., and Lingrel, J. B. (1994). A putative fourth Na⁺,K⁽⁺⁾-ATPase alpha-subunit gene is expressed in testis. *Proc. Natl. Acad. Sci. USA* **91**, 12952–12956.

Shull, G. E., Schwartz, A., and Lingrel, J. B. (1985). Amino-acid sequence of the catalytic subunit of the (Na⁺,K⁺)-ATPase deduced from a complementary DNA. *Nature* **316**, 691–695.

Shull, G. E., Greeb, J., and Lingrel, J. B. (1986a). Molecular cloning of three distinct forms of the Na,K-ATPase alpha subunit from rat brain. *Biochemistry* **25**, 8129–8132.

Shull, G. E., Lane, L. K., and Lingrel, J. B. (1986b). Amino-acid sequence of the β-subunit of the (Na⁺,K⁺)-ATPase deduced from a cDNA. *Nature* **321**, 429–431.

Shull, G. E., Young, R. M., Greeb, J., and Lingrel, J. B. (1988). Amino acid sequences of the α and β subunits of the Na,K-ATPase. *Prog. Clin. Biol. Res.* **268A**, 3–18.

Sigworth, F. J. (1993). Voltage gating of ion channels. *Quart. Rev. Biophys.* **27**, 1–40.

Skou, J. C. (1957). The influence of some cations on an adenosine triphosphatase from peripheral nerves. *Biochim. Biophys. Acta* **23**, 394–401.

Sweadner, K. J. (1989). Isozymes of the Na⁺,K⁺-ATPase. *Biochim. Biophys. Acta* **988**, 185–220.

Tepperman, K., Millette, L. A., Johnson, C. L., Jewell-Motz, E. A., Lingrel, J. B., and Wallick, E. T. (1997). Mutational analysis of glutamate 327 of Na⁺,K⁺-ATPase reveals stimulation of ⁸⁶Rb uptake by external K. *Am. J. Physiol.*, in press.

Thomas, R., Gray, P., and Andrews, J. (1989). Digitalis: Its mode of action, receptor and structure–activity relationship. *In* "Advances in Drug Research," (B. Testa, Ed.) Vol. 19, pp. 311–562. Academic Press, New York.

Van Huysse, J. W., Kuntzweiler, T. A., and Lingrel, J. B. (1996). Critical effects on catalytic function produced by amino acid substitutions at Asp804 and Asp808 of the α1 isoform of Na,K-ATPase. *FEBS Lett.* **389**, 179–185.

Withering, W. (1785). "An Account of the Foxglove and Some of Its Medicinal Uses: With Practical Remarks on Dropsy and Other Diseases," C. G. J. and J. Robinson, London.

*Istvan Edes and Evangelia G. Kranias*

# 17

# $Ca^{2+}$-ATPases/Pumps

## I. Introduction

An important role of $Ca^{2+}$ in muscle contraction was first indicated more than a century ago by Ringer (1883), who demonstrated that the frog's heart would not contract in the absence of extracellular $Ca^{2+}$. Since then, it has been shown that $Ca^{2+}$ is a physiological regulator for the contractile proteins and several other enzymes and processes in muscle. This chapter focuses on the role of the various $Ca^{2+}$-ATPases in maintaining $Ca^{2+}$ homeostasis in the cell, with special emphasis on the sarcoplasmic reticular (SR) $Ca^{2+}$-ATPase(s), which is the primary regulator of the $Ca^{2+}$ levels and thus contractility in muscle.

During the cardiac action potential, $Ca^{2+}$ enters the cell via $Ca^{2+}$ channels, which also act as dihydropyridine receptors (Fig. 1). This $Ca^{2+}$ can either activate the myofilaments directly or produce the release of additional $Ca^{2+}$ from the SR. The SR $Ca^{2+}$-release channel in cardiac and skeletal muscle also acts as a ryanodine receptor and spans the gap between the transverse tubule and the SR ("foot" protein). Furthermore, it was shown that the outer cell membrane $Ca^{2+}$ channel is located close to the SR $Ca^{2+}$ channel. Thus, the excitation–contraction coupling apparently involves the sarcolemmal $Ca^{2+}$ channel and the SR $Ca^{2+}$-release channel with the $Ca^{2+}$ current through the sarcolemmal channel being responsible for the initiation of $Ca^{2+}$ release from the SR (see Fig. 1). In skeletal muscle, the sarcolemmal membrane depolarization itself apparently is responsible for the induction of SR $Ca^{2+}$ release. The relative importance of release from the SR in activation of the cardiac muscle contraction varies from preparation to preparation, but in the heart of mammals it usually accounts for 40–70% of the $Ca^{2+}$ required (Bers, 1991).

The rising cytosolic $Ca^{2+}$ concentration induces contraction through binding to troponin C, which activates a chain of conformational changes, allowing the thin and thick filaments to interact. Subsequently, $Ca^{2+}$ is dissociated from troponin C and is rapidly removed from the cytosol by various systems, resulting in relaxation. At least three processes are responsible for the removal of $Ca^{2+}$ to end con-

traction (see Fig. 1): (1) the SR $Ca^{2+}$ pump, which actively translocates $Ca^{2+}$ at the cost of ATP into the SR system; this is believed to be the most important process in mediating relaxation; (2) the $Na^+/Ca^{2+}$ exchanger, which transports $Ca^{2+}$ out of the cell during diastole; and (3) the sarcolemmal $Ca^{2+}$-ATPase, which also extrudes $Ca^{2+}$ from the cell.

The SR is a tubular network, which seerves as a sink for $Ca^{2+}$ ions during relaxation and as a $Ca^{2+}$ source during contraction. In cardiac muscle, about 60–70% of the intracellular $Ca^{2+}$ released during systole is taken up by the SR (Bers, 1991), and the remaining amount is extruded from the cell by the $Na^+/Ca^{2+}$ exchanger and the sarcolemmal $Ca^{2+}$-ATPase. The SR in both skeletal and cardiac muscles contains an acidic protein, calsequestrin (see Fig. 1), which binds 40–50 mol $Ca^{2+}$/mol protein. The binding and release of $Ca^{2+}$ by calsequestrin is believed to be an integral step of excitation–contraction coupling, but the details of this process are still not fully understood. Mitochondria can also accumulate large amounts of $Ca^{2+}$ under pathological conditions (ischemia, $Ca^{2+}$ overload, etc.) (Bers, 1991).

## II. Sarcoplasmic Reticular $Ca^{2+}$-ATPase

### A. Properties of SR $Ca^{2+}$-ATPase

The major protein in the SR membrane is the $Ca^{2+}$-ATPase ($M_r$ 100,000), representing about 40% of the total protein in cardiac SR. The cardiac SR $Ca^{2+}$-ATPase can create intraluminal $Ca^{2+}$ concentrations of 5–10 m$M$. Recombinant DNA studies revealed that the SR or endoplasmic reticulum (ER) $Ca^{2+}$-ATPase family (SERCA) is the product of at least three alternatively spliced genes, producing a minimum of five different proteins (Burk *et al.*, 1989) (Table 1). SERCA1 is expressed in fast skeletal muscle, and alternative splicing of the 3' end of the primary transcript gives rise to two mRNA forms, which are expressed at different stages of development (Brandl *et al.*, 1986). Alternatively, spliced forms of SERCA2 have been detected in cardiac muscle and slow skeletal muscle (SER-

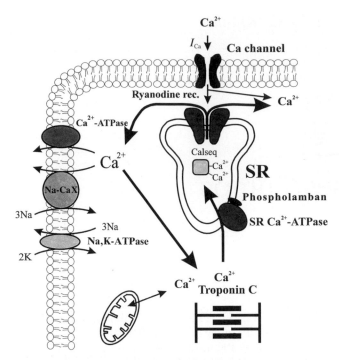

**FIG. 1.** Schematic diagram of $Ca^{2+}$ fluxes in cardiac cell. Na-CaX, $Na^+/Ca^{2+}$ exchanger; Calseq, calsequestrin; Ryanodine rec., ryanodine receptor; $I_{Ca}$, slow inward $Ca^{2+}$ current; SR, sarcoplasmic reticulum.

CA2a) and in adult smooth muscle and nonmuscle tissue (SERCA2b). SERCA3 is expressed in a selective manner, with the highest mRNA levels in intestine, spleen, lung, uterus, and brain. SERCA2 is about 85% identical to SERCA1, whereas SERCA3 is about 75% identical to either SERCA1 or SERCA2. The human SERCA2 gene is localized on chromosome 12 and maps to position 12q23–q24.1.

The proposed general model of the enzyme has three cytoplasmic domains joined to a set of 10 transmembrane helices by a narrow extramembrane pentahelical stalk (Brandl *et al.,* 1986). The cytoplasmic region includes a nucleotide binding site, or a domain to which the MgATP substrate binds, and a phosphorylation domain, which con-

**TABLE I**  Structure and Distribution of the Sarcoplasmic Reticular $Ca^{2+}$-ATPase (SERCA) Isoforms

| Gene | Splice | Tissue |
| --- | --- | --- |
| SERCA1 | a | Adult fast skeletal muscle |
| SERCA1 | b | Neonatal fast skeletal muscle |
| SERCA2 | a | Cardiac/slow skeletal muscle |
| SERCA2 | b | Smooth muscle/nonmuscle |
| SERCA3 | ? | Various tissues |

*Note.* The SERCA numbers identify different gene products, and the letters a and b indicate spliced isoforms.

tains an aspartic acid residue (Asp-351) to which the phosphate is covalently bound (Fig. 2). The third cytoplasmic domain is the $\beta$-strand domain, whose function is still not fully understood. $Ca^{2+}$ has been shown to bind to a region involving several of the membrane-spanning $\alpha$-helices (M4, M5, M6, and M8) on the cytoplasmic side. During the $Ca^{2+}$ transport cycle, the enzyme undergoes a transition from a high-affinity state to a low-affinity state for $Ca^{2+}$, and the ions are translocated from the binding sites into the lumen of the SR (see Fig. 2). This reaction pathway is characterized by the covalent phosphorylated $Ca^{2+}$-ATPase form ($E_1 \sim P$), when the energy of ATP is transferred to an acylphosphoprotein intermediate (Fig. 3). $E_1 \sim P$ rapidly becomes $E_2–P$ when the energy contained originally in the acylphosphoprotein is transduced into the translocation of bound $Ca^{2+}$ into the SR ("marionette" model; see Fig. 2). Subsequently, the acid-labile intermediate ($E_2–P$) decomposes to enzyme ($E_2$) and inorganic phosphate. In this model it is assumed that the phosphorylation of the enzyme at Asp-351 triggers a series of conformational changes in which the high-affinity $Ca^{2+}$-binding sites are disrupted, access to the sites by cytoplasmic $Ca^{2+}$ is closed off, and access to the sites from the luminal surface is gained ($E_2$ conformation) (MacLennan *et al.,* 1992).

The $Ca^{2+}$-free form of the enzyme exists in two different conformational states: one with low affinity for $Ca^{2+}$ ($E_2$) and one with high affinity for $Ca^{2+}$ ($E_1$) (see Fig. 3). The conversion of $E_2$ to $E_1$ is proposed to be the rate-limiting step in the cycle. Thapsigargin (a plant sesquiterpene lactone) has been shown to interact specifically with the $E_2$ form of all members of the SR $Ca^{2+}$-ATPase family and

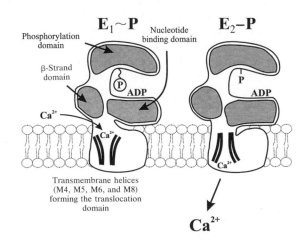

**FIG. 2.** Model illustrating $Ca^{2+}$ translocation by SERCA-type $Ca^{2+}$ pumps. In $E_1 \sim P$ conformation, $Ca^{2+}$ binds to the high-affinity binding sites in the cytosol. The energy of the hydrolyzed ATP triggers a series of conformational changes and transforms the $E_1 \sim P$ intermediate to the $E_2–P$ intermediate. These conformational changes are directly coupled to alterations in the orientation of the transmembrane regions leading to $Ca^{2+}$ release into the lumen of the sarcoplasmic reticulum.

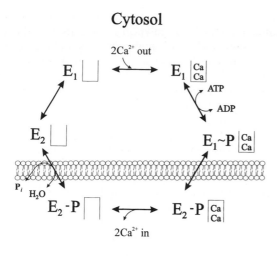

**FIG. 3.**   Reaction scheme of sarcoplasmic reticular $Ca^{2+}$-ATPase.

to inhibit enzyme activity even at subnanomolar concentrations (Lytton *et al.*, 1991). The $E_1$ form of the enzyme has been stabilized and crystallized in the presence of lanthanide ($La^{3+}$) or $Ca^{2+}$ ions (Dux *et al.*, 1985). However, vanadate ions in the absence of $Ca^{2+}$ induced the formation of the $E_2$-type crystals. The $E_1$-type crystals consist of single chains of $Ca^{2+}$-ATPase molecules evenly spaced on the surface of the SR. The $E_2$-type crystals consist of dimer chains of ATPase molecules forming an oblique surface lattice. The transition between $E_1$ and $E_2$ conformation may involve a shift in the monomer–oligomer equilibrium (Dux *et al.*, 1985).

In skeletal muscle, 2 mol $Ca^{2+}$ is transported per mol ATP hydrolyzed. In cardiac muscle, a similar stoichiometry is expected, but this ratio has been generally found to be lower (0.4–1.0 mol $Ca^{2+}$/mol ATP).

The cardiac SR $Ca^{2+}$-ATPase (SERCA2) can be phosphorylated by the $Ca^{2+}$/CAM-dependent protein kinase at $Ser^{38}$ (Toyofuku *et al.*, 1994a). However, the physiological role of this phosphorylation is still not fully understood.

### B. Regulation of SR Ca²⁺-ATPase by Phospholamban

#### 1. Structure of Phospholamban

In cardiac muscle, slow-twitch skeletal muscle, and smooth muscle, the SR contains the low-molecular-weight protein phospholamban, which can be phosphorylated by various protein kinases. The phosphorylation and dephosphorylation of phospholamban, which makes up 3–4% of the SR membrane protein, regulate the $Ca^{2+}$-ATPase activity in the SR membrane. However, the exact stoichiometry of phospholamban to SR $Ca^{2+}$-ATPase is not currently known. In early studies, a stoichiometric relationship of 1 mol phospholamban per 1 mol $Ca^{2+}$-ATPase was proposed for cardiac SR membranes, but subsequent studies, using immunological techniques for identification of the phosphorylated phospholamban species, suggested a relationship of 2 mol of phospholamban per 1 mol of ATPase

(phospholamban monomer vs ATPase monomer) (Colyer and Wang, 1991). In reconstituted systems, a molar ratio of 3:1 of phospholamban/$Ca^{2+}$-ATPase was necessary to obtain the maximal regulatory effects (Reddy *et al.*, 1995).

The complete amino acid sequence of phospholamban has been determined for various tissues and species. There is currently no evidence for the existence of any isoforms for this protein, and the phospholamban gene has been mapped to human chromosome 6 (Fujii *et al.*, 1991). The calculated molecular weight of phospholamban is 6080 (Fujii *et al.*, 1987), and the protein has been proposed to contain two major domains (Fig. 4): a hydrophilic domain (domain I) that has two unique phosphorylatable sites (Ser-16 and Thr-17), and a hydrophobic C-terminal domain (domain II), that is anchored into the SR membrane. The hydrophylic domain (amino acids 1–32) has been suggested to consist of a random coil (amino acids 1–7) and an $\alpha$-helix (amino acids 8–20), followed by a proline residue at position 21 and an antiparallel $\beta$-sheet (residues 21–32), which is located at the membrane–cytosolic interface because of its amphiphilic character (see Fig. 4). The hydrophobic domain (amino acids 33–52) forms an $\alpha$-helix in the SR membrane.

Phospholamban migrates as a 24- to 28-kDa pentamer on sodium dodecyl sulfate (SDS) gels and dissociates into dimers and monomers upon boiling in SDS before electro

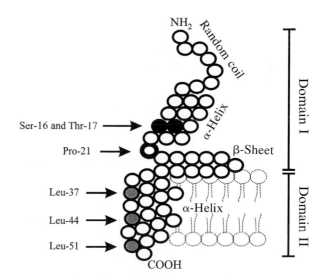

**FIG. 4.**   Molecular model of the structure of phospholamban. The cytoplasmic $\alpha$-helix (residues 8–20) is interrupted by Pro-21 (heavy circle). Residues 22–32 form an antiparallel $\beta$-sheet at the membrane–water interface, and residues 33–52 constitute the transmembrane domain II ($\alpha$-helix). Ser-16 and Thr-17 (black circles) are the adjacent phosphorylation sites. The shaded circles indicate the leucines (Leu-37, Leu-44, and Leu-51) that are important for the phospholamban subunit interactions (pentamer formation).

phoresis. Spontaneous aggregation of phospholamban into pentamers was also observed upon expression of this protein in bacteria or in mammalian cells. Site-specific mutagenesis experiments identified cysteine (Cys-36, Cys-41, and Cys-46), leucine (Leu-37, Leu-44, and Leu-51), and isoleucine (Ile-40 and Ile-47) residues in the hydrophobic transmembrane domain as essential amino acids for phospholamban pentamer formation (Fujii *et al.,* 1989; Simmerman *et al.,* 1996). The leucine and isoleucine amino acids are suggested to form five zippers in the membrane that stabilize the pentameric form of the protein with a central pore, defined by the surface of the hydrophobic amino acids (Simmerman *et al.,* 1996). Based on this pentameric self-association of phospholamban, a channel function for this protein has been proposed (Wegener *et al.,* 1986).

Monoclonal antibodies, raised against phospholamban, stimulate SR $Ca^{2+}$ uptake (Morris *et al.,* 1991). Furthermore, removal of phospholamban from the SR or uncoupling phospholamban from the $Ca^{2+}$-ATPase (using detergents, high-ionic-strength solutions, or polyanions such as heparin sulfate) markedly increased the affinity of the SR $Ca^{2+}$ pump for $Ca^{2+}$. These findings suggest that the dephosphorylated form of phospholamban is an inhibitor of the SR $Ca^{2+}$-ATPase. This "depression hypothesis" has been confirmed by studies using purified $Ca^{2+}$-ATPase and purified phospholamban (or synthetic peptides of phospholamban) in reconstituted systems. Inclusion of phospholamban or the hydrophilic portion of phospholamban resulted in inhibition of the SR $Ca^{2+}$-ATPase activity in reconstituted vesicles (Kim *et al.,* 1990). Cyclic AMP phosphorylation of phospholamban or the synthetic peptide completely reversed its inhibitory effect on the $Ca^{2+}$ pump (Kim *et al.,* 1990). The region of phospholamban interacting with the $Ca^{2+}$-ATPase may involve amino acids 2–18 (Morris *et al.,* 1991; Toyofuku *et al.,* 1994b). Interestingly, phospholamban peptides, corresponding to the hydrophobic membrane-spanning domain, also affect the $Ca^{2+}$-ATPase activity by lowering its affinity for $Ca^{2+}$ (Sasaki *et al.,* 1992; Kimura *et al.,* 1996). These findings indicate that interactions between the $Ca^{2+}$-ATPase and phospholamban within the membrane region may also be important in mediating the regulatory effects of phospholamban (Sasaki *et al.,* 1992; Kimura *et al.,* 1996). The inhibitory role of phospholamban on SR and cardiac function has been directly confirmed using transgenic animal models. Overexpression of the protein (phospholamban-overexpressing mice) was associated with inhibition of SR $Ca^{2+}$ transport, $Ca^{2+}$ transient, and depression of basal left ventricular function (Kadambi *et al.,* 1996). However, partial (phospholamban-heterozygous mice) or complete ablation of the protein (phospholamban-deficient mice) in mouse models was associated with increases in SR $Ca^{2+}$ transport and cardiac function (Luo *et al.,* 1994, 1996). Actually, a close linear correlation between the levels of phospholamban and cardiac contractile parameters was observed, indicating that phospholamban is a prominent regulator of myocardial contractility. These findings suggest that changes in the level of this protein will result in parallel changes in SR function and cardiac contraction.

Based on these reports, the simplest model for the phospholamban-mediated regulation of the SR $Ca^{2+}$-ATPase activity is one in which the highly positively charged region of phospholamban (residues 7–16) interacts directly with a negatively charged region on the surface of the $Ca^{2+}$-ATPase (Lys-Asp-Asp-Lys-Pro-Val$^{402}$) to suppress the pump activity (Fig. 5) (Toyofuku *et al.,* 1994c). This association is disrupted by phosphorylation of Ser-16 or Thr-17 in phospholamban, because the positive charges of the phospholamban cytosolic domain are partially neutralized by the phosphate moiety in this vicinity. Phosphorylation of phospholamban by the cAMP-dependent protein kinase at Ser-16 is associated with local unwinding of the $\alpha$-helix at position 12-16, resulting in conformational changes in the recognition unit of the protein (Mortishire-Smith *et al.,* 1995). Furthermore, phosphorylation also may induce conformational changes in the hydrophobic portion of phospholamban, which could result in changes in the microenvironment of the transmembrane regions of the SR $Ca^{2+}$-ATPase and thus changes in SR $Ca^{2+}$ transport (see Fig. 5).

### 2. In Vitro Studies on Regulation of SR $Ca^{2+}$-ATPase

In the early 1970s, it was suggested that the effects of various catecholamines on cardiac function may be partly attributed to phosphorylation of the SR by cAMP-dependent protein kinase(s). It soon became clear that the substrate for the protein kinase (PK) was not the SR $Ca^{2+}$-ATPase, but phospholamban. Various other high- and low-molecular-weight SR proteins were also identified as minor substrates for cAMP-dependent PK, but only the changes in the phosphorylation of phospholamban were associated with functional alterations of the cardiac SR.

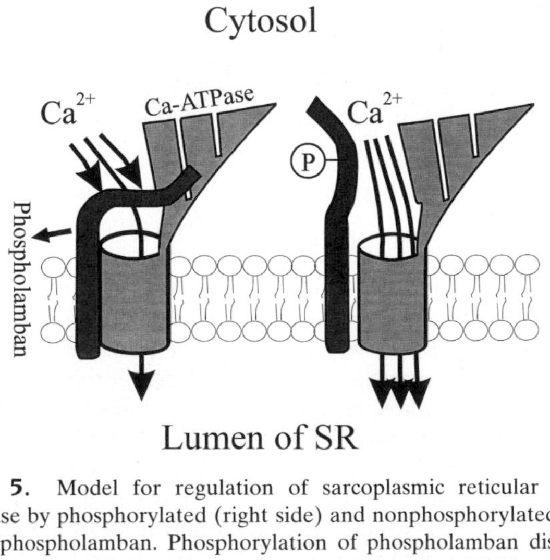

**FIG. 5.** Model for regulation of sarcoplasmic reticular $Ca^{2+}$-ATPase by phosphorylated (right side) and nonphosphorylated (left side) phospholamban. Phosphorylation of phospholamban disrupts the interaction between the two proteins so that the inhibition of the $Ca^{2+}$-ATPase is relieved. Note that both the cytosolic domain and the membrane-spanning region of phospholamban are involved in the phosphorylation-mediated conformational change to relieve the inhibition.

Cardiac SR membranes contain an endogenous cAMP-dependent PK and a $Ca^{2+}$/CaM-dependent PK that have been shown to phosphorylate phospholamban independently of each other (Kranias, 1985a). Phosphorylation by cAMP-dependent PK occurred on Ser-16, whereas $Ca^{2+}$/CaM-dependent PK catalyzed exclusively the phosphorylation of Thr-17 (Simmerman et al., 1986). Phosphorylation by either kinase was shown to result in stimulation of the SR $Ca^{2+}$-ATPase activity and the initial rates of SR $Ca^{2+}$ transport. Stimulation was associated with an increase in the apparent affinity of the SR $Ca^{2+}$-ATPase for $Ca^{2+}$.

In vitro, phospholamban is phosphorylated by two additional PKs: PK-C and a cGMP-dependent PK. Protein kinase C ($Ca^{2+}$-phospholipid-dependent PK) phosphorylated the protein at a site distinct from those phosphorylated by either cAMP-dependent PK or $Ca^{2+}$/CaM-dependent PK (Movsesian et al., 1984). Phosphorylation stimulated the SR $Ca^{2+}$-ATPase activity, and it was suggested that this activity played a role in the action of agents known to stimulate phosphoinositide (PI) hydrolysis, since one product of PI hydrolysis, diacylglycerol, is an activator of PK-C. Cyclic GMP-dependent PK was shown to phosphorylate phospholamban on the same residue (Ser-16) as that phosphorylated by cAMP-dependent PK (Raeymakers et al., 1988). This phosphorylation stimulated cardiac SR $Ca^{2+}$ transport similar to the effects of cAMP-dependent PK. Furthermore, the stimulatory effects on $Ca^{2+}$ transport, mediated by cGMP-dependent phosphorylation of phospholamban, were also observed in smooth muscle, and this may be of particular interest because some vasodilators act by increasing cGMP levels in vascular smooth muscle.

The presence of endogenous PKs in cardiac SR necessitates the presence of phosphoprotein phosphatase(s) for the reversible regulation of the $Ca^{2+}$ pump. Protein phosphatases have been generally classified into type 1 and type 2. Type 1 phosphatase is inhibited by nanomolar concentrations of the protein inhibitor-1 and inhibitor-2, whereas type 2 phosphatases are unaffected. In heart muscle, both types of phosphatase have been reported to be present and both can dephosphorylate phospholamban (Kranias and Di Salvo, 1986; MacDougall et al., 1991). A type I protein phosphatase was shown to be associated with cardiac SR membranes, and this activity could catalyze the dephosphorylation of both the cAMP-dependent PK and the $Ca^{2+}$/CaM-dependent PK phosphorylated sites (Ser-16 and Thr-17) on phospholamban (Steenaart et al., 1992). Dephosphorylation was associated with a reduction in the stimulatory effects of PKs on the SR $Ca^{2+}$ pump (Kranias, 1985b).

The SR phosphatase is similar to the skeletal muscle protein phosphatase $I_G$ ($PPI_G$), which is composed of a catalytic (C)-subunit and a G-subunit (MacDougall et al., 1991). The G-subunit may become phosphorylated by cAMP-dependent PK, and this causes release of the C-subunit from the SR vesicles or glycogen particles into the cytosol, phosphorylating phospholamban, and thus capable of stimulating SR $Ca^{2+}$ transport. In vivo studies have also shown that the phosphatase activity associated with cardiac

SR membranes may be regulated by cAMP-dependent processes. β-Adrenergic stimulation of intact beating hearts was associated with inhibition of the SR phosphatase activity, and this inhibition correlated with increases in the phosphorylation status of inhibitor-1 (Neumann et al., 1991). Thus, regulation of the SR phosphatase activity may be one of the mechanisms by which cells achieve amplification of the cAMP-dependent processes.

### 3. In Vivo Studies on Regulation of SR Ca²⁺-ATPase

The phosphorylation of SR proteins and their regulatory effects on the SR $Ca^{2+}$-ATPase activity have been studied in perfused hearts from various animal species whose ATP pool was labeled with [³²P]orthophosphate. Microsomal fractions enriched in SR were prepared from hearts freeze-clamped during stimulation with different agonists (catecholamines, forskolin, phosphodiesterase inhibitors, phorbol esters) and analyzed by gel electrophoresis and autoradiography for ³²P incorporation. β-Adrenergic agonist (isoproterenol) stimulation of the perfused hearts produced an increase in ³²P incorporation into phospholamban (Kranias and Solaro, 1982; Lindemann et al., 1983). The stimulation of ³²P incorporation into phospholamban was associated with an increased rate of $Ca^{2+}$ uptake into SR membrane vesicles and an increased SR $Ca^{2+}$-ATPase activity (Lindemann et al., 1983; Kranias et al., 1985).

These biochemical changes were associated with increases in left ventricular functional parameters (contractility and relaxation). The in vivo phosphorylation of phospholamban was specific only for inotropic agents that increased the cAMP content of the myocardium (β-adrenergic agonists, forskolin, and phosphodiesterase inhibitors). However, positive inotropic interventions, which increased the intracellular $Ca^{2+}$ level by cAMP-independent mechanisms (α-adrenergic agonists, ouabain, and elevated $[Ca^{2+}]$), failed to stimulate phospholamban phosphorylation and relaxation. Calmodulin inhibitors (fluphenazine) attenuated the isoproterenol-induced phosphorylation of phospholamban (Lindemann and Watanabe, 1985a), and it was shown that at steady-state isoproterenol exposure, phospholamban contains equimolar amounts of phosphoserine (pSer-16) and phosphothreonine (pThr-17).

The muscarinic agonist acetylcholine attenuated the increases in cAMP levels, phosphorylation of phospholamban, and the SR $Ca^{2+}$-ATPase activity produced either by β-adrenergic stimulation or by phosphodiesterase inhibition (using isobutylmethylxanthine) (Lindemann and Watanabe, 1985b). Protein kinase C and cGMP-dependent PK, which have been shown to phosphorylate phospholamban in vitro, failed to demonstrate similar effects in beating guinea pig hearts in response to stimuli that activate PK-C or elevate the cGMP levels (Edes and Kranias, 1990; Huggins et al., 1989). Thus, the physiological relevance of PK-C and PK-G in beating hearts is not clear at present.

The functional alterations in the SR $Ca^{2+}$-ATPase activity may explain, at least partly, the activating and relaxing effects of β-adrenergic agents in cardiac muscle (Figs. 6 and 7). The cAMP-dependent phosphorylation of phos-

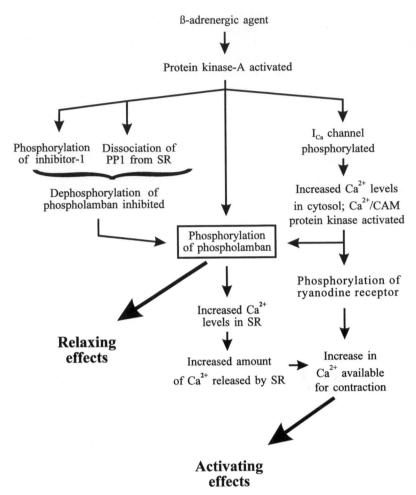

**FIG. 6.**   Schematic diagram of possible relaxing and activating effects of $\beta$-adrenergic agents in the heart. PP1, Protein phosphatase 1; SR, sarcoplasmic reticulum; CAM, calmodulin.

pholamban either *in vitro* or *in vivo* increases the rate of SR $Ca^{2+}$ transport and SR $Ca^{2+}$-ATPase activity. Such an increase in $Ca^{2+}$ transport is expected to contribute primarily to the relaxing effects of catecholamines (see Fig. 6). An additional mechanism, which contributes to the increased phosphorylation of phospholamban upon $\beta$-adrenergic stimulation, is the phosphorylation of the phosphatase inhibitor protein by the stimulated cAMP-dependent kinase. This phosphorylation results in inactivation of protein phosphatase 1 and, thus, inhibition of dephosphorylation of phospholamban during the action of catecholamines (see Fig. 6). The increased phosphorylation of phospholamban and the increased $Ca^{2+}$ levels accumulated by the SR would lead to the availability of higher levels of $Ca^{2+}$ to be subsequently released for binding to the contractile proteins (see Fig. 6). The critical and prominent role of phospholamban in the mediation of $\beta$-adrenergic functional responses was also confirmed in transgenic animal studies. Cardiac myocytes or work-performing heart preparations from phospholamban-deficient mice exhibited largely attenuated responses to $\beta$-adrenergic agonist stimulation (Luo *et al.,* 1996; Wolska *et al.,* 1996), indicating that

phospholamban is a key phosphoprotein in the heart's responses to $\beta$-adrenergic agonists.

Phosphorylation of other myocardial phosphoproteins has also been suggested to be involved in the mediation of positive inotropic and lusitropic effects of $\beta$-adrenergic agonists. Cyclic AMP-dependent protein kinase mediated phosphorylation of the $\alpha_1$-subunit of the $Ca^{2+}$ channel (see Fig. 7) is associated with an increase in the voltage-dependent $Ca^{2+}$ current ($I_{Ca}$), which enhances the $Ca^{2+}$ levels available in the cytosol during $\beta$-adrenergic agonist stimulation. Phosphorylation of troponin I has been shown to decrease the sensitivity of myofilaments for $Ca^{2+}$ both in intact myocardium and in skinned fibers (Kranias *et al.,* 1985). The desensitization of myofibrils is accompanied by an increased off-rate of $Ca^{2+}$ from troponin C, which could contribute to faster relaxation (see Fig. 7). In addition, phosphorylation of the SR $Ca^{2+}$ release channel (ryanodine receptor) by $Ca^{2+}$/CaM-dependent protein kinase may stimulate $Ca^{2+}$ release from the SR vesicles and contribute to the elevation of intracellular $Ca^{2+}$ levels during systole. Thus, the enhanced $Ca^{2+}$ influx across the sarcolemma, together with the increased $Ca^{2+}$ levels to be released from

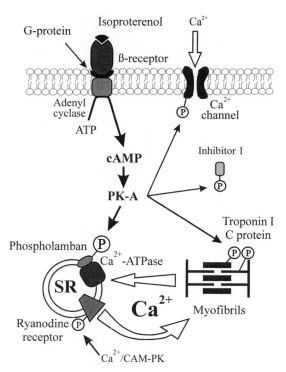

**FIG. 7.** Effects of $\beta$-adrenergic agents on protein phosphorylation in cardiac cells. Increased intracellular cAMP levels activate the cAMP-dependent protein kinase(s), which phosphorylates various proteins (phospholamban, inhibitor-1, Ca²⁺ channel, and myofibrillar proteins) and increases the rates of SR Ca²⁺ uptake and release.

the SR, may result in an elevation of the Ca²⁺ available for the contractile machinery, leading to an increase in the amplitude of contraction (Fig. 7).

## C. SR Ca²⁺-ATPase in Cardiac Diseases

The complex regulation of the SR function clearly indicates that even small disturbances in SR Ca²⁺ handling may result in profound changes and deterioration of normal myocardial function. The fast removal of Ca²⁺ by the SR Ca²⁺-ATPase during diastole and the subsequent rapid release through the SR Ca²⁺ channel (ryanodine receptor) at the beginning of contraction are prerequisites for normal diastolic and systolic function. We briefly outline in the next section the alterations in the SR Ca²⁺-ATPase in the major cardiac diseases.

### 1. SR Ca²⁺-ATPase in Hyperthyroidism and Hypothyroidism

Thyroid hormones are important regulators of myocardial contractility and relaxation. Chronic increases in thyroid hormone levels lead to cardiac hypertrophy, with increases in the heart rate and cardiac output as well as left ventricular contractility and velocity of relaxation. However, opposite effects are associated with a hypothyroid condition. The mechanisms underlying these changes have been the subject of numerous investigations. It is assumed that in hypo- and hyperthyroid hearts the altered gene

expression of the cardiac SR proteins, and hence the changes in the intracellular Ca²⁺ transients, is the most important determinant of the altered myocardial function. It was shown that the velocity of ATP-dependent Ca²⁺ transport and the Ca²⁺-ATPase activity are specifically increased in SR vesicle preparations from hyperthyroid compared with euthyroid hearts (Beekman *et al.*, 1989). Opposite changes were noted for hypothyroid and euthyroid animals (Beekman *et al.*, 1989).

Examination of the steady-state mRNA levels of the cardiac SR Ca²⁺-ATPase and the ryanodine receptor revealed a significant increase (140–190%) in hyperthyroid and a marked decline (40–50%) in hypothyroid animals (Arai *et al.*, 1991). The changes in mRNA levels for the Ca²⁺-ATPase in hypothyroid and hyperthyroid conditions also reflected changes in the protein amounts of the enzyme in these hearts (Kiss *et al.*, 1994). Interestingly, in the case of phospholamban, the regulator of the Ca²⁺-ATPase, and calsequestrin there was no coordinated regulation with respect to the Ca²⁺-ATPase. In fact, both the relative mRNA level and the protein content of phospholamban were reported to decrease in hyperthyroid animals, whereas there was no change noted in the calsequestrin mRNA level upon L-thyroxine treatment. In hypothyroid hearts an opposite trend was noted, since the protein amount of phospholamban was found to be higher than in euthyroid or hyperthyroid animals (Kiss *et al.*, 1994). Consequently, the phospholamban/Ca²⁺-ATPase protein ratio was highest in the hypothyroid (1.82) animals, followed by euthyroid (1.00) and hyperthyroid (0.56) animals (Kiss *et al.*, 1994). These changes in the phospholamban/Ca²⁺-ATPase ratio were associated with coordinate alterations in the SR Ca²⁺ uptake, affinity of the SERCA2 for Ca²⁺, and myocardial function (Kiss *et al.*, 1994; Kimura *et al.*, 1994).

These changes indicate that the SR proteins responsible for Ca²⁺ uptake and release (Ca²⁺-ATPase and ryanodine receptor) are coordinately regulated in hypothyroid and hyperthyroid hearts and provide a simple explanation for the altered Ca²⁺ release and reuptake capacity and hence the myocardial function under these conditions.

### 2. SR Ca²⁺-ATPase in Cardiomyopathies

Dilated cardiomyopathy is a frequent form of cardiac muscle disease and is characterized by an impaired systolic function and dilatation of both ventricles (systolic pump failure). In various animal models of primary and secondary dilated cardiomyopathy, it was shown that both SR Ca²⁺ binding capacity and uptake were depressed because of the decreased activity and protein level of the SR Ca²⁺-ATPase (Edes *et al.*, 1991). In some studies of human idiopathic dilated cardiomyopathy, decreases were noted for both SR Ca²⁺ uptake rates and Ca²⁺-ATPase activity (Limas *et al.*, 1987; Unverferth *et al.*, 1988), as well as myocardial Ca²⁺ handling (Gwathmey *et al.*, 1987; Beuckelmann *et al.*, 1992). Examination of the mRNA levels in left ventricular biopsies from patients with dilated cardiomyopathy revealed a significant decrease in mRNA content for the SR Ca²⁺-ATPase relative to other mRNA forms

(Mercadier *et al.*, 1990; Arai *et al.*, 1993). In contrast, other authors were unable to detect a decrease in SR $Ca^{2+}$ uptake activity (Movsesian *et al.*, 1989) or the immunodetectable levels of the SR $Ca^{2+}$-ATPase protein (Schwinger *et al.*, 1995) in the left ventricular myocardium from patients with idiopathic dilated cardiomyopathy. Furthermore, the gating mechanism of the SR $Ca^{2+}$ release channel was recently reported to be abnormal in dilated cardiomyopathy (D'Agnolo *et al.*, 1992), and it was suggested that defective excitation–contraction coupling is involved in the pathogenesis of this disease.

Another type of cardiomyopathy, hypertrophic cardiomyopathy, has only been recognized in clinical practice for the past three decades. The characteristics of this disease are asymmetric interventricular septal hypertrophy and narrowing of the left ventricular outflow tract, with or without outflow obstruction (outflow tract pressure gradient). It was shown that in the familial form of hypertrophic cardiomyopathy, which accounts for about 60% of all cases, mutations in myofibrillar protein genes ($\beta$-myosin heavy chain, troponin T, $\alpha$-tropomyosin, and C protein) are, associated with the disease (Schwartz *et al.*, 1996). Additionally, prolongation of the $Ca^{2+}$ transient, abnormal $Ca^{2+}$ handling, and a decline in SR $Ca^{2+}$-ATPase mRNA levels are reported to be characteristic for human hypertrophic cardiomyopathy (Gwathmey *et al.*, 1987; Mercadier *et al.*, 1990), which may explain the diastolic function impairment in this disease.

In chronic heart failure due to hemodynamic overload, irrespective of the specific etiology (valvular heart disease, cardiomyopathy, chronic ischemic heart disease, or hypertension), a reduction was observed in both the number and the activity of the SR $Ca^{2+}$ pump (Limas *et al.*, 1987; Unverferth *et al.*, 1988). Furthermore, a close correlation was obtained between the SR $Ca^{2+}$-ATPase mRNA or protein levels and the myocardial function (Mercadier *et al.*, 1990; Hasenfuss *et al.*, 1994). Interestingly, the $Na^+$–$Ca^{2+}$ exchanger gene expression was reported to be increased in failing human hearts, and it was hypothesized that the up-regulation of this protein may compensate for the depressed SR function (Studer *et al.*, 1994).

### 3. SR $Ca^{2+}$-ATPase in Ischemia

A brief period of ischemia (10–20 min) induces reversible tissue damage in cardiac muscle, resulting in the "stunned" myocardium. This condition is characterized by regional contractile abnormalities (declines in both systolic and diastolic function) that persist for several hours, despite the absence of necrosis. These hemodynamic changes are associated with a reduction in SR $Ca^{2+}$ transport (Limbruno *et al.*, 1989; Krause *et al.*, 1989). The maximal activity of the SR $Ca^{2+}$-ATPase was found to be depressed, and the $Ca^{2+}$ sensitivity of this enzyme was decreased (Krause *et al.*, 1989). Furthermore, a decrease in the coupling ratio (mol $Ca^{2+}$/mol ATP) was observed in the SR membranes isolated from the stunned myocardium, which was suggested to be the result of an increase in the $Ca^{2+}$ permeability of the SR membrane. The SR $Ca^{2+}$ release process was also found to be impaired in the stunned myocardium

because of the reduction of the number of ryanodine receptors (Zucchi *et al.*, 1994). These data suggest that complex modifications of the SR function occur in the stunned myocardium, which are at least partly responsible for the contractile impairment found in this condition.

In long-lasting myocardial ischemia, gradual declines in both SR $Ca^{2+}$-ATPase activity and $Ca^{2+}$ uptake were found, which may be due to degradation of the SR $Ca^{2+}$-ATPase (Akiyama *et al.*, 1986; Schoutsen *et al.*, 1989). Ischemia was also shown to result in a gradual decrease in the phosphorylation status of phospholamban under both *in vitro* (Schoutsen *et al.*, 1989; Lamers *et al.*, 1986) and *in vivo* conditions (Bartel *et al.*, 1989), and this correlated with a decrease in the SR $Ca^{2+}$-ATPase activity. Thus, it has been postulated that the long-lasting ischemia-induced progressive inactivation of the SR $Ca^{2+}$ pump not only is a consequence of the specific loss of enzyme activity, but may also be related to the altered characteristics of phospholamban (Schoutsen *et al.*, 1989). A combination of various pathogenic factors has been suggested to be responsible for the reduced SR function and the final tissue necrosis in the ischemic myocardium. These pathological factors include pH reduction (acidosis), activation of intracellular proteolytic enzymes, and increased generation of free radicals.

## III. Other ATPases

The $Ca^{2+}$ regulation in eukaryotic cells involves a complex mechanism that maintains a low background $Ca^{2+}$ concentration (usually 0.1–0.2 $\mu M$) in the cell interior. Eukaryotic cells generally satisfy their $Ca^{2+}$ demands by extracting $Ca^{2+}$ from their own internal stores, but it is also evident that the long-term regulation of the $Ca^{2+}$ gradient across the plasma membrane is a result of the concerted operation of the importing ($Ca^{2+}$ channel) and exporting (SR $Ca^{2+}$ pump; $Na^+$/$Ca^{2+}$ exchanger and $Ca^{2+}$ pump of the surface membrane) $Ca^{2+}$ systems (see Section II.A). The plasma membrane $Ca^{2+}$-ATPase is a low-capacity system possessing a very high $Ca^{2+}$ affinity, which enables the enzyme to interact with $Ca^{2+}$ at low intracellular concentrations. Consequently, its function is continuous and presumably satisfies the fine tuning of $Ca^{2+}$ homeostasis.

### A. General Properties of Plasma Membrane $Ca^{2+}$-ATPase(s)

The plasma membrane $Ca^{2+}$-ATPase has a molecular mass of 140 kDa, and the general kinetic mechanism for this enzyme follows the pattern of the SR $Ca^{2+}$-ATPase. ATP phosphorylates an aspartic acid residue, and the phosphorylated intermediate is acid-stable. The elementary steps of the cycle are probably similar in both SR and plasma membrane $Ca^{2+}$-ATPases (Schatzmann, 1989). The stoichiometry between transported $Ca^{2+}$ and hydrolyzed ATP is only 1.0 for the plasma membrane $Ca^{2+}$ pump. The administration of $La^{3+}$ under various experimental conditions was associated with an increase in the steady-state phosphoenzyme level of the plasma membrane $Ca^{2+}$-ATPase, and this increase possibly results from stabilization of the aspartyl phosphate (inhibition of hydrolysis of the phosphate group). The other classic inhibitor of Ca$^{2+}$ pumps, vanadate, was found to be a potent inhibitor of the plasma membrane Ca$^{2+}$-ATPase even at low concentrations (Bond and Hudgins, 1979).

Calmodulin stimulates the plasma membrane Ca$^{2+}$-ATPase by direct interaction with the enzyme. It was shown that the stimulation results from a combined effect on the affinity for Ca$^{2+}$ ($K_m$) and the maximal transport rate ($V_{max}$) (Carafoli, 1992). The calmodulin-binding domain of the Ca$^{2+}$ pump has been suggested to function as a repressor of the enzymatic activity (autoinhibitory function), and calmodulin may relieve this inhibition (Carafoli, 1992). It has been suggested that a common mechanism exists in the autoinhibition of plasma membrane Ca$^{2+}$-ATPase and phospholamban inhibition of SR Ca$^{2+}$-ATPase. In both proteins, the interacting sites are amphiphilic and located in the cytoplasmic region. The interaction occurs with homologous regions in the SR and plasma membrane Ca$^{2+}$-ATPases close to the phosphorylation sites. In the absence of calmodulin, the plasma membrane Ca$^{2+}$ pump can be activated by several other compounds. Polyunsaturated fatty acids and acidic phospholipids (phosphatidylinositol, phosphatidylinositol 4-phosphate, and phosphatidylinositol 4,5-diphosphate) were reported to be good activators and, since they are present in the plasma membrane, they may be important regulators of the Ca$^{2+}$-ATPase *in vivo* (Carafoli, 1991). Phosphorylation of the enzyme by cAMP-dependent PK or PK-C was also reported to stimulate plasma membrane Ca$^{2+}$-ATPase activity. The cAMP-dependent phosphorylation occurs C-terminally to the calmodulin-binding domain, and the phosphorylation-mediated activation may likewise be significant *in vivo*. The PK-C phosphorylation occurs in the calmodulin binding domain, inhibits the binding of calmodulin to the plasma membrane Ca$^{2+}$-ATPase, and lowers the autoinhibitory potential of this domain (Hofmann *et al.*, 1994). Cyclic GMP-dependent PK has also been reported to stimulate the plasma membrane Ca$^{2+}$ pump in vascular smooth muscle, but the Ca$^{2+}$-ATPase enzyme was not found to be the substrate for this kinase.

### B. Primary Structure and Topography of Plasma Membrane Ca$^{2+}$-ATPase(s)

The complete amino acid sequence of the plasma membrane Ca$^{2+}$ pump has been deduced from rat and human cells (Shull and Greeb, 1988; Verma *et al.*, 1988). It appears that the plasma membrane Ca$^{2+}$-ATPase (PMCA) isoforms are encoded by a multigene family, and additional variability is produced by alternative RNA splicing of each gene transcript (Table 2). The regions that are important for the catalytic function and the transmembrane domains are highly conserved, and there is no diversity observed for these regions. The isoform diversity seems to alter primarily the regulatory characteristics of the enzyme, and it can be regarded as an adaptation to tissue specificity.

The secondary structure of the plasma membrane Ca$^{2+}$-ATPase is similar to that of the SR Ca$^{2+}$-ATPase (Shull and Greeb, 1988). The enzyme contains 10 putative

**TABLE 2**   Distribution of the Plasma Membrane Ca$^{2+}$-ATPase (PMCA) Isoforms

| Gene | Splice | Species |
| --- | --- | --- |
| PMCA1 | a, b, c, d | Rat, human |
| PMCA2 | f, g, h | Rat, human |
| PMCA3 | a | Rat |
| PMCA4 | a, b, g | Human |
| PMCA5 | ? | Bovine |

*Note.* The PMCA numbers identify different gene produts, and the letters a–h indicate spliced isoforms.

transmembrane helices, which are connected on the outside of the plasma membrane by short loops. Three primary domains (about 80% of the pump protein) protrude into the cytoplasm. The first domain corresponds to the transducing unit, which couples ATP hydrolysis to Ca$^{2+}$ translocation. The second protruding domain contains the aspartyl phosphate site (phosphorylation domain). The C-terminal portion of this domain can also be labeled by ATP analogues, and it contains a "hinge" region, which permits the movement of aspartyl phosphate and the ATP-binding site. The third C-terminal protruding domain contains the calmodulin-binding sequence and the phosphorylation sites for protein kinase C and cAMP-dependent protein kinase. The last is not present in all isoforms.

### IV. Summary

The Ca$^{2+}$ levels in muscle are primarily regulated by the sarcoplasmic reticulum network, which serves as a sink for Ca$^{2+}$ ions during relaxation and as a Ca$^{2+}$ source during contraction. In cardiac muscle, most of the intracellular Ca$^{2+}$ released during systole is taken up by the SR through its Ca$^{2+}$-ATPase. This translocation of Ca$^{2+}$ from the cytosol into the SR lumen uses ATP as the energy source, and it is characterized by the formation of a phosphorylated intermediate (E$_1$~P) for the Ca$^{2+}$-ATPase.

In cardiac muscle, slow-twitch skeletal muscle, and smooth muscle, the Ca$^{2+}$-ATPase is regulated by a low-molecular-weight phosphoprotein, called phospholamban. In its dephosphorylated form, phospholamban is an inhibitor of the Ca$^{2+}$-ATPase, and phosphorylation relieves this inhibition. Phosphorylation of phospholamban occurs by cAMP-dependent, cGMP-dependent, Ca$^{2+}$/calmodulin-dependent, and Ca$^{2+}$/phospholipid-dependent protein kinases *in vitro*. However, *in vivo* studies have indicated that phospholamban is phosphorylated only by cAMP-dependent and Ca$^{2+}$/calmodulin-dependent protein kinases in intact beating hearts. A phospholamban phosphatase activity has been reported to be present in SR membranes, which can dephosphorylate this regulatory protein and reverse its stimulatory effects on the Ca$^{2+}$-ATPase.

Alterations in the SR Ca$^{2+}$-ATPase activity and its regulation by phospholamban have been shown to occur in cardiac diseases such as hypothyroidism, hyperthyroidism, hypertrophy, heart failure, and ischemia. In most instances, alterations in Ca$^{2+}$-ATPase activity correlated with alterations in its mRNA levels and ventricular function.

Another Ca$^{2+}$-ATPase, which is also important for maintaining Ca$^{2+}$ homeostasis in muscle, is the plasma membrane Ca$^{2+}$-ATPase. This enzyme transports Ca$^{2+}$ to the extracellular space and uses ATP as its energy source, similar to the SR Ca$^{2+}$-ATPase. The plasmalemmal Ca$^{2+}$-ATPase may be distinguished from the SR Ca$^{2+}$-ATPase primarily by its distinct sensitivity to La$^{3+}$, vanadate, and calmodulin.

The primary structure of the various Ca$^{2+}$ pumps has been published, and there is a growing interest in further use of molecular biological approaches and specifically site-directed mutagenesis for these enzymes to obtain more information about their structural–functional relationships. The ultimate question is, what is the precise mechanism by which Ca$^{2+}$ is transported across the ATPases? Site-directed mutagenesis studies and construction of molecular models will provide the information that will answer this question. Furthermore, in the absence of appropriate crystallographic data, a deeper understanding of the mechanisms that regulate the Ca$^{2+}$-ATPases under normal and pathological conditions may elucidate the structural–functional relationships in these enzymes and their role in maintaining Ca$^{2+}$ homeostasis in the cell.

# References

Akiyama, K., Konno, N., Yanagishita, T., Tanno, F., and Katagiri, T. (1986). Ultrastructural changes in the sarcoplasmic reticulum in acute myocardial ischemia. *Jpn. Circ. J.* **50**, 829–838.

Arai, M., Otsu, K., MacLennan, D. H., Alpert, N. R., and Periasamy, M. (1991). Effect of thyroid hormone on the expression of mRNA encoding sarcoplasmic reticular proteins. *Circ. Res.* **69**, 266–276.

Arai, M., Alpert, N. R., MacLennan, D. H., Barton, P., and Periasamy, M. (1993). Alterations in sarcoplasmic reticulum gene expression in human heart failure. A possible mechanism for alterations in systolic and diastolic properties of the failing myocardium. *Circ. Res.* **72**, 463–469.

Bartel, S., Karczewski, P., and Krause, E.-G. (1989). Phosphorylation of phospholamban and troponin I in the ischemic and reperfused heart: Attenuation and restoration of isoprenaline responsiveness. *Biomed. Biochim. Acta* **48**, 108–113.

Beekman, R. I., Hardeveld, C., and Simonides, W. S. (1989). On the mechanism of the reduction by thyroid hormone of β-adrenergic relaxation rate stimulation in rat heart. *Biochem. J.* **259**, 229–236.

Bers, D. M. (1991). Ca regulation in cardiac muscle. *Med. Sci. Sports Exer.* **23**, 1157–1162.

Beuckelmann, D. J., Näbauer, M., and Erdmann, E. (1992). Intracellular calcium handling in isolated ventricular myocytes from patients with terminal heart failure. *Circulation* **85**, 1046–1055.

Bond, G. H., and Hudgins, P. (1979). Kinetics of inhibition of Na$^+$,K$^+$-ATPase by Mg$^{2+}$, K$^+$ and vanadate. *Biochem.* **18**, 325–331.

Brandl, C. J., Green, N. M., Korczak, B., and MacLennan, D. H. (1986). Two ATPase genes: Homologies and mechanistic implications of deduced amino acid sequences. *Cell* **44**, 597–607.

Burk, S. E., Lytton, J., MacLennan, D. H., and Shull, G. E. (1989).

cDNA cloning, functional expression and mRNA tissue distribution of a third organellar Ca$^{2+}$ pump. *J. Biol. Chem.* **264**, 18561–18568.

Carafoli, E. (1991). The calcium pumping ATPase of the plasma membrane. *Annu. Rev. Physiol.* **53**, 531–547.

Carafoli, E. (1992). The Ca$^{2+}$ pump of the plasma membrane. *J. Biol. Chem.* **267**, 2115–2118.

Colyer, J., and Wang, J. H. (1991). Dependence of cardiac sarcoplasmic reticulum calcium pump activity on the phosphorylation status of phospholamban. *J. Biol. Chem.* **266**, 17486–17493.

D'Agnolo, A., Luciani, G. B., Mazzucco, A., Gallucci, V., and Salviati, G. (1992). Contractile properties and Ca$^{2+}$ release activity of the sarcoplasmic reticulum in dilated cardiomyopathy. *Circulation* **85**, 518–525.

Dux, L., Taylor, K. A., Tin-Beall, H. P., and Martonosi, A. (1985). Crystallization of the Ca$^{2+}$-ATPase of sarcoplasmic reticulum by calcium and lanthanide ions. *J. Biol. Chem.* **260**, 11730–11743.

Edes, I., and Kranias, E. G. (1990). Phospholamban and troponin I are substrates for protein kinase C *in vitro* but not in intact beating guinea pig hearts. *Circ. Res.* **67**, 394–400.

Edes, I., Talosi, L., and Kranias, E. G. (1991). Sarcoplasmic reticulum function in normal heart and in cardiac disease. *Heart Failure* **6**,(6), 221–237.

Fujii, J., Ueno, A., Kitano, K., Tanaka, S., Kadoma, M., and Tada, M. (1987). Complete complementary DNA-derived amino acid sequence of canine cardiac phospholamban. *J. Clin. Invest.* **79**, 301–304.

Fujii, J., Maruyama, K., Tada, M., and MacLennan, D. H. (1989). Expression and site-specific mutagenesis of phospholamban. Studies of residues involved in phosphorylation and pentameric formation. *J. Biol. Chem.* **264**, 12950–12955.

Fujii, J., Zarain-Herzberg, A., Willard, H. F., Tada, M., and MacLennan, D. H. (1991). Structure of the rabbit phospholamban gene, cloning of the human cDNA, and assignment of the gene to human chromosome 6. *J. Biol. Chem.* **266**, 11669–11675.

Gwathmey, J. K., Copelas, L., MacKinnon, R., Schoen, F. J., Feldman, M. D., Grossman, W., and Morgan, J. P. (1987). Abnormal intracellular calcium handling in myocardium from patients with end-stage heart failure. *Circ. Res.* **61**, 70–76.

Hasenfuss, G., Reinecke, H., Studer, R., Meyer, M., Pieske, B., Holtz, J., Holubarsch, C., Posival, H., Just, H., and Drexler, H. (1994). Relation between myocardial function and expression of sarcoplasmic reticulum Ca$^{2+}$-ATPase in failing and nonfailing human myocardium. *Circ. Res.* **75**, 434–442.

Hofmann, F., Anagli, L., and Carafoli, E. (1994). Phosphorylation of the calmodulin binding domain of the plasma membrane Ca$^{2+}$ pump by protein kinase C reduces its interaction with calmodulin and with its pump receptor site. *J. Biol. Chem.* **269**, 24298–24303.

Huggins, J. P., Cook, E. A., Pigott, J. R., Mattinsley, T. J., and England, P. J. (1989). Phospholamban is a good substrate for cGMP-dependent protein kinase *in vitro,* but not in intact cardiac or smooth muscle. *Biochem. J.* **260**, 829–835.

Kadambi, V. J., Ponniah, S., Harrer, J. M., Hoit, B. D., Dorn, G. W., Walsh, R. A., and Kranias, E. G. (1996). Cardiac-specific overexpression of phospholamban alters calcium kinetics and resultant cardiomyocyte mechanics in transgenic mice. *Clin. Invest.* **97**, 533–539.

Kim, H. W., Steenaart, N. A. E., Ferguson, D. G., and Kranias, E. G. (1990). Functional reconstitution of the cardiac sarcoplasmic reticulum Ca$^{2+}$-ATPase with phospholamban in phospholipid vesicles. *J. Biol. Chem.* **265**, 1702–1709.

Kimura, Y., Otsu, K., Nishida, K., Kuzuya, T., and Tada, M. (1994). Thyroid hormone enhances Ca$^{2+}$ pumping activity of the cardiac sarcoplasmic reticulum by increasing Ca$^{2+}$ ATPase and decreasing phospholamban expression. *J. Mol. Cell. Cardiol.* **26**, 1145–1154.

Kimura, Y., Kurzydlowski, K., Tada, M., and MacLennan, D. H. (1996). Phospholamban regulates the Ca$^{2+}$-ATPase through intramembrane interaction. *J. Biol. Chem.* **271**, 21726–21731.

Kiss, E., Jakab, G., Kranias, E. G., and Edes, I. (1994). Thyroid hormone-induced alterations in phospholamban protein expression. Regulatory effects on sarcoplasmic reticulum Ca$^{2+}$ transport and myocardial relaxation. *Circ. Res.* **75,** 245–251.

Kranias, E. G. (1985a). Regulation of Ca$^{2+}$ transport by cyclic 3',5'-AMP-dependent and calcium–calmodulin-dependent phosphorylation of cardiac sarcoplasmic reticulum. *Biochim. Biophys. Acta* **844,** 193–199.

Kranias, E. G. (1985b). Regulation of calcium transport by protein phosphotase activity associated with cardiac sarcoplasmic reticulum. *J. Biol. Chem.* **260,** 11006–11010.

Kranias, E. G., and Di Salvo, J. (1986). A phospholamban protein phosphatase activity associated with cardiac sarcoplasmic reticulum. *J. Biol. Chem.* **261,** 10029–10032.

Kranias, E. G., and Solaro, R. J. (1982). Phosphorylation of troponin I and phospholamban during catecholamine stimulation of rabbit heart. *Nature* **298,** 182–184.

Kranias, E. G., Garvey, J. L., Srivastava, R. D., and Solaro, R. J. (1985). Phosphorylation and functional modifications of sarcoplasmic reticulum and myofibrils in isolated rabbit hearts stimulated with isoprenaline. *Biochem. J.* **226,** 113–121.

Krause, S. M., Jacobus, W. E., and Becker, L. C. (1989). Alterations in cardiac sarcoplasmic reticulum calcium transport in the postischemic "stunned" myocardium. *Circ. Res.* **65,** 526–530.

Lamers, J. M., De Jonge-Stinis, J. T., Hülsman, W. C., and Verdouw, P. D. (1986). Reduced *in vitro* $^{32}$P-incorporation into phospholamban-like protein of sarcolemma due to myocardial ischaemia in anaesthetized pigs. *J. Mol. Cell. Cardiol.* **18,** 115–125.

Limas, C. J., Olivari, M. T., Goldenberg, I. F., Levine, T. B., Benditt, D. G., and Simon, A. (1987). Calcium uptake by cardiac sarcoplasmic reticulum in human dilated cardiomyopathy. *Cardiovasc. Res.* **21,** 601–605.

Limbruno, U., Zucchi, R., Ronca-Testoni, S., Galbani, P., Ronca, G., and Mariani, M. (1989). Sarcoplasmic reticulum function in the "stunned" myocardium. *J. Mol. Cell. Cardiol.* **21,** 1063–1072.

Lindemann, J. P., and Watanabe, A. M. (1985a). Phosphorylation of phospholamban in intact myocardium. Role of Ca$^{2+}$–calmodulin-dependent mechanisms. *J. Biol. Chem.* **260,** 4516–4525.

Lindemann, J. P., and Watanabe, A. M. (1985b). Muscarinic cholinergic inhibition of $\beta$-adrenergic stimulation of phospholamban phosphorylation and Ca$^{2+}$ transport in guinea pig ventricles. *J. Biol. Chem.* **260,** 122–133.

Lindemann, J. P., Jones, L. R., Hathaway, D. R., Henry, B. G., and Watanabe, A. M. (1983). $\beta$-Adrenergic stimulation of phospholamban phosphorylation and Ca$^{2+}$-ATPase activity in guinea pig ventricles. *J. Biol. Chem.* **260,** 4516–4525.

Luo, W., Grupp, I. L., Harrer, J., Ponniah, S., Grupp, G., Duffy, J. J., Doetschman, T., and Kranias, E. G. (1994). Targeted ablation of phospholamban gene is associated with markedly enhanced myocardial contractility and loss of $\beta$-adrenergic stimulation. *Circ. Res.* **75,** 401–409.

Luo, W., Wolska, B. M., Grupp, I. L., Harrer, J. M., Haghighi, K., Ferguson, D. G., Slack, J. P., Grupp, G., Doetschman, T., Solaro, R. J., and Kranias, E. G. (1996). Phospholamban gene dosage effect in the mammalian heart. *Circ. Res.* **78,** 839–847.

Lytton, J., Westlin, M., and Hanley, M. R. (1991). Thapsigargin inhibits the sarcoplasmic or endoplasmic reticulum Ca-ATPase family of calcium pump. *J. Biol. Chem.* **266,** 17067–17071.

MacDougall, L. K., Jones, J. R., and Cohen, P. (1991). Identification of the major protein phosphatases in mammalian cardiac muscle which dephosphorylate phospholamban. *Eur. J. Biochem.* **196,** 725–734.

MacLennan, D. H., Clarke, D. M., Loo, T. W., and Skerjanc, I. S. (1992). Site-directed mutagenesis of the Ca$^{2+}$-ATPase of sarcoplasmic reticulum. *Acta Physiol. Scand.* **146,** 141–150.

Mercadier, J. J., Lompre, A. M., Duc, P., Boheler, K. R., Fraysse, J. B., Wisnewsky, P., Allen, P. D., Komajda, M., and Schwartz, K. (1990). Altered sarcoplasmic reticulum Ca$^{2+}$-ATPase gene expression in the human ventricle during end-stage heart failure. *J. Clin. Invest.* **85,** 305–309.

Morris, G. L., Cheng, H., Colyer, J., and Wang, J. H. (1991). Phospholamban regulation of cardiac sarcoplasmic reticulum (Ca$^{2+}$-Mg$^{2+}$)-ATPase. Mechanism of regulation and site of monoclonal antibody interaction. *J. Biol. Chem.* **266,** 11270–11275.

Mortishire-Smith, R. J., Pitzenberger, S. M., Burke, C. J., Middaugh, C. R., Garsky, V. M., and Johnson, R. G. (1995). Solution structure of the cytoplasmic domain of phospholamban: Phosphorylation leads to a local perturbation in secondary structure. *Biochemistry* **34,** 7603–7613.

Movsesian, M. A., Nishikawa, M., and Adelstein, R. S. (1984). Phosphorylation of phospholamban by calcium-activated, phospholipid-dependent protein kinase. *J. Biol. Chem.* **259,** 8029–8032.

Movsesian, M. A., Bristow, M. R., and Krall, J. (1989). Ca$^{2+}$ uptake by cardiac sarcoplasmic reticulum from patients with idiopathic dilated cardiomyopathy. *Circ. Res.* **65,** 1141–1144.

Neumann, J., Gupta, R. C., Schmitz, W., Scholz, H., Nairn, A. C., and Watanabe, A. M. (1991). Evidence for isoproterenol-induced phosphorylation of phosphatase inhibitor-I in the intact heart. *Circ. Res.* **69,** 1450–1457.

Raeymakers, L., Hofmann, F., and Casteels, R. (1988). Cyclic GMP-dependent protein kinase phosphorylates phospholamban in isolated sarcoplasmic reticulum from cardiac and smooth muscle. *Biochem. J.* **252,** 269–273.

Reddy, L. G., Jones, L. R., Cala, S. E., O'Brian, J. J., Tatulian, S. A., and Stokes, D. L. (1995). Functional reconstitution of recombinant phospholamban with rabbit skeletal Ca$^{2+}$-ATPase. *J. Biol. Chem.* **270,** 9390–9397.

Ringer, S. A. (1883). A further contribution regarding the influence of different constituents of the blood on the contraction of the heart. *J. Physiol.* **4,** 29–42.

Sasaki, T., Inui, M., Kimura, Y., Kuzuya, T., and Tada, M. (1992). Molecular mechanism of regulation of Ca$^{2+}$-ATPase by phospholamban in cardiac sarcoplasmic reticulum. *J. Biol. Chem.* **267,** 1674–1679.

Schatzmann, H. J. (1989). The calcium pump of the surface membrane and of the sarcoplasmic reticulum. *Annu. Rev. Physiol.* **51,** 473–485.

Schoutsen, B., Blom, J. J., Verdouw, P. D., and Lamers, J. M. (1989). Calcium transport and phospholamban in sarcoplasmic reticulum of ischemic myocardium. *J. Mol. Cell. Cardiol.* **21,** 719–727.

Schwartz, K., and Mercadier, J.-J. (1996). Molecular and cellular biology of heart failure. *Curr. Opinion Cardiol.* **11,** 227–236.

Schwinger, R. H. G., Böhm, M., Schmidt, U., Karczewski, P., Bavendiek, U., Flesch, M., Krause, E.-G., and Erdmann, E. (1995). Unchanged protein levels of SERCA II and phospholamban but reduced Ca$^{2+}$ uptake and Ca$^{2+}$-ATPase activity of cardiac sarcoplasmic reticulum from dilated cardiomyopathy patients compared with patients with nonfailing hearts. *Circulation* **92,** 3220–3228.

Shull, G. E., and Greeb, J. (1988). Molecular cloning of two isoforms of the plasma membrane Ca$^{2+}$ transporting ATPase from rat brain. Structural and functional domains exhibit similarity to Na$^+$,K$^+$ and other cation transport ATPases. *J. Biol. Chem.* **263,** 8646–8657.

Simmerman, H. K. B., Collins, J. H., Theibert, J. L., Wegener, A. D., and Jones, L. R. (1986). Sequence analysis of phospholamban. Identification of phosphorylation sites and two major structural domains. *J. Biol. Chem.* **261,** 13333–13341.

Simmerman, H. K. B., Kobayashi, Y. M., Autry, J. M., and Jones, L. R. (1996). A leucine zipper stabilizes the pentameric membrane domain of phospholamban and forms a coiled-coil pore structure. *J. Biol. Chem.* **271,** 5941–5946.

Steenaart, N. A. E., Ganim, J. R., DiSalvo, J., and Kranias, E. G. (1992). The phospholamban phosphatase associated with cardiac sarcoplasmic reticulum is a type 1 enzyme. *Arch. Biochem. Biophys.* **293,** 17–24.

Studer, R., Reinecke, H., Bilger, J., Eschenhagen, T., Böhm, M., Hasenfuss, G., Just, H., Holtz, J., and Drexler, H. (1994). Gene expression of the cardiac $Na^+$–$Ca^{2+}$ exchanger in end-stage human heart failure. *Circ. Res.* **75,** 443–453.

Toyofuku, T., Kurzydlowski, K., Narayanan, N., and MacLennan, D. H. (1994a). Identification of $Ser^{38}$ as the site of cardiac sarcoplasmic reticulum $Ca^{2+}$-ATPase that is phosphorylated by $Ca^{2+}$/calmodulin-dependent protein kinase. *J. Biol. Chem.* **269,** 26492–26496.

Toyofuku, T., Kurzydlowski, K., Tada, M., and MacLennan, D. H. (1994b). Amino acids $Glu^2$ to $Ile^{18}$ in the cytoplasmic domain of phospholamban are essential for functional association with the $Ca^{2+}$-ATPase of sarcoplasmic reticulum. *J. Biol. Chem.* **269,** 3088–3094.

Toyofuku, T., Kurzydlowski, K., Tada, M., and MacLennan, D. H. (1994c). Amino acids Lys-Asp-Asp-Lys-Pro-$Val^{402}$ in the $Ca^{2+}$-ATPase of cardiac reticulum are critical for functional association with phospholamban. *J. Biol. Chem.* **269,** 22929–22932.

Unverferth, D. V., Lee, S. W., and Wallick, E. T. (1988). Human myocardial adenosine triphosphate activities in health and heart failure. *Am. Heart J.* **115,** 139–146.

Verma, A. K., Filoteo, A. G., Stanford, D. R., Wieben, E. D., Penniston, J. T., Strehler, E. E., Fischer, R., Heim, R., Vogel, G., Mathews, S., Strehler-Page, M. A., James, P., Vorherr, T., Krebs, J., and Carafoli, E. (1988). Complete primary structure of a human plasma membrane $Ca^{2+}$ pump. *J. Biol. Chem.* **263,** 14152–14159.

Wegener, A. D., Simmerman, H. K. B., Liepnieks, J., and Jones, L. R. (1986). Proteolytic cleavage of phospholamban purified from canine cardiac sarcoplasmic reticulum. Generation of a low resolution model of phospholamban structure. *J. Biol. Chem.* **261,** 5154–5159.

Wolska, B. M., Stojanovic, M. O., Luo, W., Kranias, E. G., and Solaro, R. J. (1996). Effect of ablation of phospholamban on dynamics of cardiac myocyte contraction and intracellular $Ca^{2+}$. *Am. J. Physiol.* **271,** C391–C397.

Zucchi, R., Ronca-Testoni, S., Yu, G., Galbani, P., Ronca, G., and Mariani, M. (1994). Effect of ischemia and reperfusion on cardiac ryanodine receptors—Sarcoplasmic reticulum $Ca^{2+}$ channels. *Circ. Res.* **74,** 271–280.

*John H. B. Bridge*

# 18

## Na–Ca Exchange Currents

## I. Introduction

The existence of Na–Ca exchange was first postulated by both Repke (1964) and Langer (1964) as a consequence of their studies on the contractility of heart muscle. Three years later Baker *et al.* (1967) provided the first report documenting Na–Ca exchange in giant squid axons. Shortly after this Reuter and Seitz (1968) presented the first complete study describing Na–Ca countertransport in heart. Based on studies of isotopic fluxes, these authors proposed that two $Na^+$ ions were coupled to the extrusion of a single $Ca^{2+}$ ion in a modified exchange diffusion process. Blaustein and Hodgkin then published the results of their studies on squid axons (Blaustein and Hodgkin, 1969). They recognized that the distribution of free $Ca^{2+}$ could not be predicted on simple electrochemical principles. However, cyanide (which was expected to block metabolic processes) failed to prevent the efflux of $^{45}Ca$, so that it seemed unlikely that a metabolic pump was involved in $Ca^{2+}$ extrusion. However, this efflux of $Ca^{2+}$ was (among other things) dependent on external Na. Blaustein and Hodgkin (1969) therefore concluded that, in unpoisoned axons, some or possibly all of the energy for extruding $Ca^{2+}$ ions came from the inward movement of $Na^+$ ions down its electrochemical gradient.

These early studies were seminal and provided impetus for an enormous number of subsequent investigations that have led not only to a study of Na–Ca exchange currents, but also to the molecular cloning and elucidation of the structure of the exchanger molecule itself. This in turn introduced the possibility of studying the relationship between molecular structure and function.

This chapter provides a brief description of current knowledge of Na–Ca exchange currents. The ease with which heart cells can be patch clamped, together with the presence of a vigorous exchange activity, doubtless explains the fact that most of our information on exchange current comes from this tissue. Na–Ca exchange currents have been measured in other cell types, including the squid giant axon (Caputo *et al.,* 1989). In this chapter, we do not deal

with Na–Ca exchange in the vertebrate rod outer segment (ROS). It is now known that this exchange process is different from that found in heart muscle. For example, under physiological conditions the exchange involves not only the Na and Ca gradients, but the K gradient as well. In addition, the structure of the ROS exchanger is quite different from that of the heart exchanger (Achilles *et al.,* 1991). There have been extensive studies of the currents associated with the ROS Na–Ca exchange. The reader interested in the electrogenicity of this exchange can refer to the brief review by Lagnado and McNaughton (1990). Although measurements of exchange current are difficult in many tissues, such as squid axon, barnacle muscle fiber, and smooth muscle, two cell types are likely to prove useful for the study of exchange currents. Cardiac Na–Ca exchangers have been expressed in frog oocytes and SF21 cells from the fall army worm moth *Spodoptera frugipoda.* Reports of exchange currents in these have already been published (see later sections).

## II. Structure and Distribution of Na–Ca Exchanger

The cloned Na–Ca exchanger from canine heart is a protein of molecular weight 110 kDa. Its primary structure has been deduced (Nicoll *et al.,* 1990) and found to consist of a sequence of 970 amino acids. A hydropathy map for the exchanger protein has been obtained and indicates that the protein can be divided into three regions: (1) a hydrophobic $NH_2$-terminal portion containing five potential membrane-spanning segments, (2) a lengthy hydrophilic region, and (3) a hydrophobic COOH-terminal portion containing six potential membrane-spanning segments. Based on the hydropathy data, Philipson has proposed a model of the exchanger (Fig. 1). More than half of the molecule, comprising 520 amino acids (the hydrophilic region), is a large loop on the cytoplasmic surface of the membrane. This loop contains among other things a calmodulin binding site and sites involved in Na-dependent

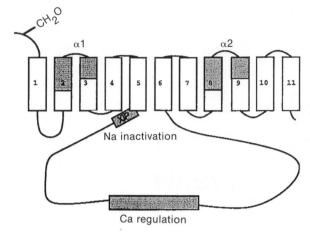

**FIG. 1.** A proposed model of the Na–Ca exchanger. Membrane-spanning segments are labeled 1 to 11. There is a large hydrophilic cytoplasmic domain containing a receptor that binds exchanger inhibitory peptide (XIP), and this site is believed to be associated with the phenomenon of Na-dependent inactivation. In addition, this cytoplasmic domain contains a secondary Ca regulatory site. (This figure was kindly provided by Dr. Kenneth Philipson.)

inactivation and secondary Ca regulation. Potential phosphorylation sites have also been identified, and a region of 23 amino acids with 48% identity to the subunit of the Na,K-ATPase has been noted. It will be of interest to learn whether these regions of identity perform similar functions in the two transporters.

Studies (Levitsky *et al.,* 1994; Matsuoka *et al.,* 1995) of the cardiac sarcolemmal Na–Ca exchanger have revealed some of the relationships between structure and function of the exchange molecule. It is now known that the α1 and α2 repeats (see Fig. 1) are membrane-spanning segments of internal similarity. Mutations in these regions significantly affect exchange activity, that is, they reduce exchange activity or alter current–voltage relationships. Another finding (Nicoll *et al.,* 1996) concerns a region of the exchanger that is known to have sequences that are homologous with sequences in the Na pump. Mutations of two residues in the pumplike region produced nonfunctional exchangers. In addition to these observations on the relationship between structure and function, the distribution of the Na–Ca exchanger has been studied in mammalian cardiac myocytes with immunofluorescence labeling techniques (Frank *et al.,* 1992). It appears that exchangers tend to be concentrated in the vicinity of the t-tubules in this tissue. This places the exchangers in close proximity to the junctional SR, perhaps suggesting an involvement in the events of excitation–contraction coupling.

## III. Energetics of Na–Ca Exchange

Reuter and Seitz in their classic study proposed that the two $Na^+$ ions exchanged for a single $Ca^{2+}$ ion in a modified exchange diffusion process. Based on this proposal, one would not expect steady-state exchange activity to produce a measurable electric current. However, the existence of

exchange currents is a well-established fact, and one that can be appreciated by consideration of the energetic and stoichiometric properties of the exchange as they are currently understood. It is convenient (but not necessarily correct) to represent the transmembranous exchange reaction as a sequential (simultaneous) process.

$$n\mathrm{Na_i} + \mathrm{Ca_o} \rightleftarrows n\mathrm{Na_o} + \mathrm{Ca_i}$$

where *n* is the stoichiometric coefficient of the exchange reaction. If the forward and reverse reaction rates are equal, the exchange reaction is at *equilibrium*. Even if *n* is greater than 2, there can be no net charge movement, and hence no electric current generated at equilibrium. An electric current can only be measured when the exchange is displaced from equilibrium. While the net reaction rate at equilibrium is zero, the unidirectional rates might be substantial (Axelsen and Bridge, 1985). As soon as the forward and reverse exchange rates differ from one another, net ion translocation takes place; provided the stoichiometric coefficient is appropriate (i.e., *n* greater than 2) and no ions of opposite charge are co-transported, one can expect to measure an electric current as a consequence of exchange.

If net movement of ions through the Na–Ca exchanger is solely determined by electrochemical forces, classical thermodynamics may be used to calculate both the direction of exchange and the conditions under which we may expect equilibrium to occur. Having the capacity to do this is of enormous value when designing experiments to measure exchange currents. The electrochemical potential difference or driving force ($n \Delta\mu$) producing exchange is the difference between *n* times the electrochemical potential difference or force producing sodium movement ($n \Delta\mu_{\mathrm{Na}}$) and calcium movement ($\Delta\mu_{\mathrm{Ca}}$). Driving force can be expressed in terms of membrane potential ($E_\mathrm{m}$), $Na^+$ equilibrium potential ($E_{\mathrm{Na}}$), and $Ca^{2+}$ equilibrium potential ($E_{\mathrm{Ca}}$). Thus, we may write

$$\Delta\tilde{\mu} = n \Delta\tilde{\mu}_{\mathrm{Na}} - \Delta\tilde{\mu}_{\mathrm{Ca}} \tag{1}$$

At equilibrium,

$$\Delta\tilde{\mu} = 0 \text{ so that } n \Delta\tilde{\mu}_{\mathrm{Na}} - \Delta\tilde{\mu}_{\mathrm{Ca}} = 0 \tag{2}$$

$$\Delta\tilde{\mu}_{\mathrm{Na}}(\mathrm{mV}) = \frac{RT}{F} \ln \frac{[\mathrm{Na}]_\mathrm{o}}{[\mathrm{Na}]_\mathrm{i}} + E_\mathrm{m} = E_{\mathrm{Na}} + E_\mathrm{m} \tag{3}$$

$$\Delta\tilde{\mu}_{\mathrm{Ca}}(\mathrm{mV}) = \frac{RT}{F} \ln \frac{[\mathrm{Ca}]_\mathrm{o}}{[\mathrm{Ca}]_\mathrm{i}} + 2E_\mathrm{m} = 2E_{\mathrm{Ca}} + 2E_\mathrm{m} \tag{4}$$

Substitution yields

$$nE_{\mathrm{Na}} - 2E_{\mathrm{Ca}} - (n - 2)E_\mathrm{m} = 0 \tag{5}$$

The exponential form of Eq. 5 is

$$\frac{[\mathrm{Ca}]_\mathrm{o}}{[\mathrm{Ca}]_\mathrm{i}} = \left(\frac{[\mathrm{Na}]_\mathrm{o}}{[\mathrm{Na}]_\mathrm{i}}\right)^n \exp - \left((n - 2)\frac{E_\mathrm{m}F}{RT}\right) \tag{6}$$

These are the equations that can be used to predict the equilibrium conditions of the exchanger. Before doing so, one needs to know the stoichiometric coefficient (*n*) of the exchanger. This issue has been the subject of a lengthy debate and numerous investigations. However, there now appears to be broad agreement that three $Na^+$ ions are exchanged for a single $Ca^{2+}$ ion in most mammalian systems studied (Kimura *et al.*, 1987; Bridge *et al.*, 1990; Bridge and Bassingthwaighte, 1983; Crespo *et al.*, 1990). There is also good evidence that exchange stoichiometry is 3:1 in barnacle fibers (Rasgado-Flores and Blaustein, 1987). In squid, the matter is still debated.

Reeves and Hale (1984) provided an elegant (though somewhat indirect) demonstration of exchange stoichiometry. Their study not only produced a value for the exchange stoichiometry, but also provided an excellent example of the way that the foregoing energetic principles may guide experimental design. These authors took advantage of the fact that exchange equilibrium can be achieved simply by appropriate adjustment of the $Na^+$ and $Ca^{2+}$ electrochemical gradients. Bovine sarcolemmal vesicles containing Na–Ca exchanger were equilibrated with solutions of both Na and $^{45}Ca$. Under these circumstances, the equilibrium may be described by Eq. 5. After treating the membrane with valinomycin in the presence of KCl, known membrane potentials were established that caused disequilibrium of the exchanger and either $Ca^{2+}$ entry or exit. By adjusting the $Na^+$ gradient, it was possible to precisely null the tendency of membrane potential to produce $Ca^{2+}$ movement. Thus, if any of the quantities $E_m$, $E_{Na}$, and $E_{Ca}$ are held constant, the relationship between the other two may be found. The point at which Ca movement was nulled by Na gradient is given by:

$$(n - 2)E_m = nE_{Na} \qquad (7)$$

By nulling Ca movement over a range of membrane potentials, the value for $n$ that these authors obtained was 2.97 ± 0.03, which is close to the currently accepted value of 3.0.

Equipped with a value for the stoichiometric coefficient of exchange, some useful parameters can be calculated. Heart muscle is used as an illustrative example to calculate expected reversal potentials for the exchange. It is assumed that a resting ventricular cell maintains intracellular $Na^+$ and $Ca^{2+}$ at concentrations of 10 m$M$ and 100 n$M$, respectively, and extracellular $Na^+$ and $Ca^{2+}$ are, respectively, 140 m$M$ and 2.0 m$M$. For $n = 3$, the reversal potential is given by (see Eq. 5):

$$E_{rev} = 3E_{Na} - 2E_{Ca} = -50 \text{ mV} \qquad (8)$$

For the foregoing conditions $E_{rev}$ may be calculated to be −50 mV.

## IV. Problems Associated with Measurement of Na–Ca Exchange Current

To measure exchange current satisfactorily, it is necessary to be able to control the ionic gradients that are responsible for driving exchange. This implies that one can either control or at least have a good estimate of their magnitudes. The first observations of exchange current were not published unitl nearly 20 year after Reuter and Seitz originally reported Na–Ca exchange activity in heart tissue. It is worth considering some of the problems associated with the measurement of exchange current that might account for this lengthy hiatus.

To stimulate exchange so that it can be measured, one can either change electrochemical gradients (and hence the exchange driving forces) by using a voltage-clamp step, or one can abruptly change external ion composition with a rapid switching device. In the case of the voltage-clamp step, the experimentalist must have a complete understanding of what interfering currents will be activated, as well as some means of inhibiting them. If exchange is to be activated by changing ionic gradients, then these changes must be sufficiently rapid that they are not immediately dissipated by the exchange activity. It is difficult to rapidly change or control extracellular ionic composition in multicellular preparations, because diffusion distances are large and diffusion may be hindered by a surface structure. In multicellular preparations, voltage clamp is usually established with small microelectrodes that are unsuitable for cell dialysis. It is therefore very difficult to control the intracellular ionic composition. This doubtless contributed to the difficulty of isolating exchange currents in multicellular preparations. Despite these difficulties, a measurement of exchange currents have been reported in multicellular preparations (Horackova and Vassort, 1979).

Several techniques have greatly facilitated the isolation of exchange current. The whole-cell ruptured patch voltage-clamp techniques in conjunction with intracellular dialysis (Hamill *et al.*, 1981) have probably contributed most of the initial isolation of exchange currents. It has proved feasible to use the patch-clamp technique in conjunction with wide-tipped microelectrodes (10-$\mu$m diam.) to obtain an isolated patch whose surface area is approximately 75 $\mu m^2$. A conventional isolated patch is on the order of 2–3 $\mu m^2$. These giant patches are sufficiently large to permit the detection of exchange currents (Hilgemann, 1989). Their great virtue is that they permit relatively easy access to either side of the membrane. Thus, solution adjacent to the external surface of the sarcolemma can be changed by changing the pipette solution. The internal surface of the sarcolemma can be changed rapidly by changing the bathing solution. Clearly this method of voltage clamp allows the experimentalist considerable control over the forces that drive exchange.

Several other techniques have become available that improve exchange current isolation. For example, forward exchange current in intact cells can be activated by abruptly elevating intracellular Ca, which is then transported to the cell exterior. This has been accomplished in an elegant fashion by Niggli and Lederer (1991). These workers used the compound DM-dinitrophen, which is commonly referred to as caged Ca. This can be introduced into the cell interior through a patch pipette. Upon appropriate irradiation with UV light, Ca is released from the caged Ca with extremely rapid kinetics. This abruptly elevates cytosolic Ca and transiently stimulates forward Na–Ca exchange as the released Ca is pumped to the cell exterior.

As we have indicated, to activate exchange it is desirable to change external ionic composition (and the electrochemical gradients driving exchange) extremely rapidly in comparison with the time required for the exchange to dissipate these changes. It is now possible to change external solutions surrounding heart cells or isolated patches with a half-time of about 20 msec (this includes exchange of the unstirred layer). One method (Spitzer and Bridge, 1989) consists of placing a cell with attached microelectrodes in one of two adjacent microstreams of solution. The boundary separating these streams is abruptly moved across the cell so that it is placed in the adjacent stream. This rapid switching method has proved valuable in stimulating both inward and outward exchange currents (Bridge *et al.*, 1990; Chin *et al.*, 1993).

Progress in the molecular biology of the Na–Ca exchange has provided additional approaches to the isolation and characterization of Na–Ca exchange. It is possible to prepare baculovirus containing cDNA coding for the canine Na–Ca exchanger (Li *et al.*, 1992). This virus is used to infect SF21 cells (eukaryotic cells from the fall army worm moth *Spodoptera frugipoda*), which results in a high density of expression of Na–Ca exchange molecules in the cell membrane (Li *et al.*, 1992). It is a straightforward matter to record large exchange currents from these cells (Fig. 2). Besides supporting a high level of expression, these cells do not appear to have large numbers of channels that are activated by voltage changes. This greatly simplifies the isolation of current. These cells also introduce the possibility of investigating mutated forms of the exchanger that can in principle be expressed in them.

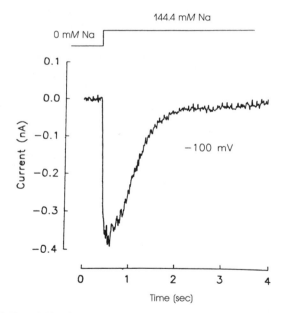

**FIG. 2.** A Na–Ca exchanged current activated in an SF21 cell. The cell was first infected with baculovirus containing the recombinant DNA coding the canine Na–Ca exchanger. The cell was then voltage clamped with a single microelectrode and held at −100 mV while being superfused in a solution containing 2.7 m*M* Ca and no Na (Li replacement). This was designed to reverse exchange and cause Ca to enter the cell. Sudden application of 144 m*M* Na activated forward exchange and presumably caused Ca extrusion in exchange for extracellular Ca. As intracellular Ca declines, so does forward exchange. (Reprinted with permission of the *Annals of the New York Academy of Sciences.*)

It is clear that there are now a variety of methods for measuring Na–Ca exchange current. In the remainder of this chapter, we discover the extent to which the properties of this exchange current have been investigated.

## V. Isolation of Na–Ca Exchange Current

### A. Whole-Cell Patch-Clamp Studies

The first clear evidence that an electric current was generated by Na–Ca exchange was provided simultaneously by Kimura *et al.* (1986) and by Mechmann and Pott (1986). Given the probable stoichiometry of the exchange, electrophysiologists expected to be able to measure a current generated by the exchange. By using the whole-cell patch-clamp technique together with intracellular dialysis, Junko Kimura and her associates were able to isolate the Na–Ca exchange current. These investigators voltage-clamped guinea pig ventricular myocytes with single microelectrodes. The pipette contained a dialyzing solution completely deficient in Na and in which Ca was buffered to a value of 73 n*M* with EGTA. The superfusing external solution contained no Ca. Under these circumstances, no exchange could take place. Very little change in current was observed when external Ca was reapplied and subsequently removed. However, an outward current was generated after the pipette solution was changed for one containing 30 m*M* Na and 1.0 m*M* external Ca was applied. Under these circumstances Ca entered the cell in exchange for internal Na, which was extruded. This current (Fig. 3) is clearly attributable to the operation of electrogenic Na–Ca exchange. This conclusion was strengthened by several additional observations. Upon removal of external Ca the current was turned off. The current was reduced when lower concentrations of dialyzing Na were used and enlarged by increasing external Ca. Thus, the current exhibited the expected dependency on both Na and Ca. The current was blocked by La, which is known to block Na–Ca exchange. The current also exhibited voltage dependence, which we discuss later. One might have expected sustained current under these circumstances. However, this is not observed (see Fig. 3) and the current exhibited a tendency to decay. One plausible explanation for this is that excessive Ca entry displaces protons when it is buffered by the dialyzing EGTA. The resulting acidification in the vicinity of the exchanger should inhibit exchange activity. An alternative possibility is that EGTA fails to buffer incoming Ca, which accumulates in the vicinity of the exchanger. The resulting collapse of the Ca gradient would also slow exchange as observed.

Kimura *et al.* reported the measurement of an outward current corresponding to the entry of Ca and extrusion of Na from the cell. Mechman and Pott (1986) demonstrated the existence of an inward exchange current corresponding to Ca extrusion and Na entry into a cardiac cell. In one experiment, these authors voltage-clamped spherical atrial cells from an adult guinea pig. Transient inward current occurred spontaneously in this preparation. These transient inward currents seemed to depend on intracellular Ca

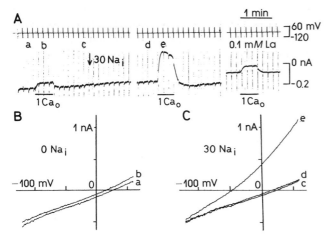

**FIG. 3.** (A) Voltage-clamp record from a single myocyte showing voltage (upper trace) and current (lower trace). A ramp pulse from +60 to −120 mV was given from a holding potential of −30 mV every 10 sec at a ramp speed of about 0.2 V/sec. Current–voltage relationships were then constructed. An application of 1 m$M$ Ca in the absence of intracellular Na (zero pipette Na) produced a small outward current to exchange the pipette solution, a piece of tapered polyethylene tube was inserted into the pipette. The arrow indicates when the solution change was started. The holding current shifted outward slightly on loading Na. The middle current record shows 1 m$M$ Ca superfusion in the presence of 30 m$M$ pipette Na. The Ca-induced outward current appears to decay after reaching a peak. Current undershoot is seen after Ca is washed off. These phenomena may be due to Ca accumulation immediately below the membrane of an acid induced by binding of Ca to EGTA. (B) $I$–$V$ relations measured before (a) and during (b) 1 m$M$ Ca$^{2+}$ superfusion in the absence of Na$_i^+$. (C) $I$–$V$ curves before (c) and after (d) loading Na$_i^+$ and during subsequent 1 m$M$ Ca$_0^{2+}$ application (e). (Reprinted with kind permission of the authors and *Nature*.)

## B. Na–Ca Exchange Current Reversal Potential

It is desirable when studying any ionic current to be able to demonstrate a reversal potential for that current. The existence of reversal potentials is a thermodynamic necessity and provides one of the least ambiguous ways of identifying an ionic current. However, it is not always a straightforward matter to demonstrate reversal potentials. For example, if exchange current was extremely small over a large voltage range in the vicinity of the reversal potential, then the precise potential at which zero current occurred could be extremely difficult to specify. Moreover, subsarcolemmal spaces from which diffusion is restricted can produce changes in ionic concentration that confound measurement of the exchange reversal potential. For the sodium–calcium exchange in heart, the reversal potential is given by Eq. 8. For the conditions of their experiment, Kimura *et al.* (1986; see Fig. 3) calculated that the reversal potential of exchange ought to be −131 mV. Their results indicate that exchange current becomes zero at a potential close to this value. Ehara *et al.* (1989) extended studies of the exchange reversal potential. With the fairly specific exchange inhibitor nickel, these authors were able to show that over a wide range of external Na values, the measured exchange reversal potential conformed to theoretical expectation. Moreover, in view of the wide range over which the reversal potential remains constant, the stoichiometry is unlikely to vary.

## VI. Ionic Dependencies of Na–Ca Exchange Current

It is of considerable physiological significance that the sodium–calcium exchange current is extremely sensitive

because they were abolished if the cell was dialyzed with 1.0 m$M$ EGTA. Moreover, they could not be elicited in the presence of caffeine, which depletes the SR of Ca. Presumably the transient release of SR Ca activated these currents. The currents also showed a dependence on extracellular Na and voltage as expected of a Na–Ca exchange current. The more negative the voltage, the larger the inward current. This is reasonable and simply indicates that at more negative potentials the exchange rate is greater, so that Ca released from the SR is removed rapidly from the cell.

Shortly after these early demonstrations of exchange current, Hume and Uehara (1986a,b) were able to demonstrate Na–Ca exchange currents in isolated frog atrial cells by using the whole-cell patch-clamp technique in combination with intracellular dialysis. It is fair to say that since this time measurement of exchange current has become both reliable and routine. An example of an inward exchange current measured in a ventricular cell is displayed in Fig. 4. Although the majority of exchange currents have been reported in heart cells, measurements of exchange currents have also accomplished in squid axons (Caputo *et al.*, 1989).

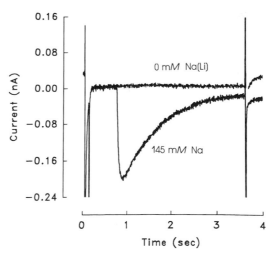

**FIG. 4.** Transient inward Na–Ca exchange. To activate this current, a guinea pig ventricular cell was tetanized with voltage-clamp pulses in the presence of ryanodine and in the absence of external Na. This resulted in an elevation of cytosolic Ca and a sustained contraction (not shown). Abrupt application of 145 m$M$ extracellular Na while the cell was held at −40 mV produced mechanical relaxation and activated a transient inward Na–Ca exchange current. As intracellular Ca declined, the current decayed. (Reprinted with permission of the *Annals of the New York Academy of Science*.)

to internal Na. There are two aspects of this sensitivity to consider. One is purely thermodynamic and the other is kinetic. If we assume that a resting heart cell contains 100 n$M$ free cytosolic Ca and is bathed in 2.0 m$M$ Ca and 140 m$M$ Na, we can use Eq. 8 to calculate the way the exchange reversal potential varies with internal Na. The results are displayed in Fig. 5. It is apparent that with 10 m$M$ intracellular Na, 140 m$M$ extracellular Na, and 100 n$M$ intracellular free Ca, the reversal potential for exchange is −50 mV. Since the resting membrane potential is at least 30 mV negative to this, at rest the exchanger will extrude Ca. However, a modest increase in intracellular Na to 12 m$M$ would change the reversal potential by 14 mV to −64 mV. Were the Na to accumulate to 15 m$M$, then the reversal potential would be −80 mV. Since the resting potential is likely to be close to this value, the exchange would be close to, or at, equilibrium and would be incapable of extruding intracellular Ca. Further Na accumulation would cause the exchange to reverse and transmit Ca into the cell. Na accumulation will tend to bring the exchange closer to its equilibrium potential, and therefore increase the likelihood of Ca entry on the exchanger on membrane depolarization. Cardiac glycosides, which, depending on dose, partially or completely block the sodium pump, tend to produce an accumulation of intracellular Na (Sheu and Fozzard, 1982; Biedert *et al.*, 1979; Lee and Levi, 1991). Part of the basis of their inotropic effect resides in their shifting the reversal potential of the Na–Ca exchange to more negative values. In extreme cases, this will prevent Ca extrusion, which in the face of continued Ca leak will lead to Ca accumulation.

The Na–Ca exchange current also exhibits a physiologically important kinetic dependence on internal Na. This issue has been investigated directly by Miura and Kimura (1989). These authors studied outward exchange current in guinea pig ventricular cells under voltage clamp. Intracellular Ca was buffered to 100 n$M$ and intracellular Na

was varied by varying the pipette Na concentration. External Na was reduced to zero and a unidirectional exchange reaction was activated by applying 0.2 m$M$ Ca as rapidly as possible to the cell exterior. The results demonstrate that exchange current density exhibits a steep and somewhat sigmoid dependence on internal Na. The $K_m$ for this dependency is approximately 20 m$M$, and Hill plots reveal a Hill coefficient of 1.9. It is notable that an application of rate theory revealed that the voltage dependence of the exchange did not depend on intracellular Na, and this topic is discussed more fully later. It is clear from these results that at least Ca influx on the exchange will be extremely sensitive to fluctuations in intracellular Na produced, for example, by increases in stimulation rate or cardiac glycosides. It is unfortunate that we have no information on the intracellular Na dependence of inward exchange current, which corresponds to Ca efflux. However, studies with isotopic fluxes reveal that Na and Ca compete for transport on the exchanger. For example, Na on the same side of the cardiac sarcolemmal membrane as Ca inhibits Ca flux through the exchanger with a $K_i$ of about 15 m$M$ (Reeves and Sutko, 1983). Finally, the dependence of net inward exchange current does depend on external Na. This has been investigated by Kimura *et al.* (1987), who showed that inward current exhibited a sigmoidal dependence on external Na with a $K_{1/2}$ of 87.5 m$M$ and a Hill coefficient of 2.9.

Net inward exchange current (i.e., Ca extrusion in exchange for Na entry) is also dependent on intracellular Ca. The relationship between Ca concentration and current has been measured in two different ways. Miura and Kimura (1989) used whole-cell patch clamp on guinea pig myocytes to determine $K_m$ (Ca). They controlled intracellular Ca by dialyzing the cell with various EGTA-buffered solutions and determined $K_m$ (Ca) to be 0.6 $\mu M$. With a different approach, Barcenas-Ruiz *et al.* (1987) used guinea pig ventricular myocytes to measure exchange current and intracellular Ca simultaneously. Intracellular Ca was measured with the Ca-sensitive indicator Fura-2. This substance was introduced through the dialyzing pipette used to voltage-clamp the cells. Heart cells can be tetanized or caused to go into contracture in the presence of ryanodine. Various clamp pulses up to 120 mV in amplitude (holding potential of −80 mV) produced slow and sustained increases in Ca. This Ca declined upon repolarization, and the decline was accompanied by a Na–Ca exchange current tail. Plots of current against [Ca]$_i$ revealed a linear relationship and no sign of saturation (Fig. 6). It is not possible to extract a $K_m$ value from this data, since the relationship between Ca and current does not saturate. However, it is reasonable to conclude that the $K_m$ value is well above 0.6 $\mu M$ obtained by Kimura *et al.* Discrepancies of this nature are not easy to resolve. It is, however, worth noting that EGTA has the rather curious effect of increasing the affinity of the exchanger for Ca (Trosper and Philipson, 1984).

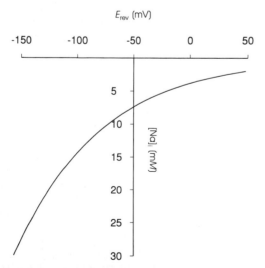

**FIG. 5.** Calculated variation of exchange reversal potential with intracellular Na. Intracellular Ca was set at 100 n$M$ and extracellular Ca was 2.0 m$M$. Extracellular Na was 140 m$M$.

## VII. Regulation of Na–Ca Exchange Current

The regulation of Na–Ca exchange current has largely been studied in the heart. We consider three types of regu-

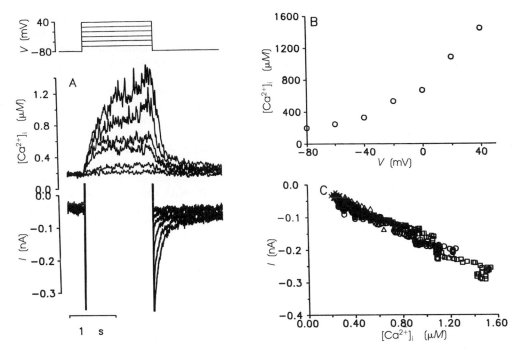

**FIG. 6.** Changes in [Ca]ᵢ and membrane current attributed to Na–Ca exchange in a single guinea pig ventricular myocyte under voltage clamp. Micropipette [Na] was 7.5 m$M$. (A) Simultaneous recordings of [Ca]ᵢ and membrane current. The holding potential was −80 mV, and depolarizing pulses lasting for 1.5 sec were given from −60 to +40 mV. Outward currents during the depolarizing pulse are off scale. (B) Voltage dependence of the change in [Ca]ᵢ. Values plotted are average [Ca]ᵢ over the last 200 msec of the depolarizing pulse. (C) The relationship between [Ca]ᵢ and membrane current after repolarization. The current, $I$, has been plotted as a function of [Ca]ᵢ at 2-msec intervals during the first second after repolarization from +40 mV (circles), 0 mV (triangles), −20 mV (diamonds), and −40 mV (crosses). (Reprinted with kind permission of the authors and *Science*.)

lation. First, the exchange current is regulated by intracellular Ca. The exchange current also exhibits a phenomenon known as Na-dependent inactivation. Finally, the exchange current is known to be regulated by Mg-ATP. The issue of exactly how the exchange current is regulated has recently achieved special significance because regions of the exchange molecule responsible for the Ca-dependent regulation and Na-dependent inactivation have been identified. We first discuss the two forms of ionic regulation and the portion of the exchange molecule likely to contain the regulatory sites. We then briefly describe current understanding of regulation by Mg-ATP.

There are two Ca binding sites on the intracellular surface of the cell membrane. One is the primary transport site, and the other is a high-affinity Ca site that is required for secondary ionic regulation. It is now apparent that Ca influx in exchange for internal Na cannot occur if intracellular Ca is reduced below a certain critical level. The first description of this secondary regulation of Na–Ca exchange by internal Ca was provided by Baker and McNaughton (1976) in their study of squid axons. In their more recent studies of heart cell membrane, both Kimura *et al.* (1986) and Hilgemann (1990) have shown that when internal Ca is sufficiently reduced, exchange current is inactivated. It appears that in dialyzed myocytes the $K_D$ for the regulatory site is about 50 nM. Why the regulatory site exists is unclear, although it may prevent the exchanger from reducing Ca excessively at rest.

Na-dependent inactivation of exchange current is a rather curious regulatory phenomenon first described by Hilgemann *et al.*, 1992). Outward Na–Ca exchange currents were activated in giant sarcolemmal patches by increasing Na on the cytoplasmic side of the patch. Ca was then transported from the pipette to the cytoplasmic side of the patch in exchange for Na. With 60 m$M$ Na on the pipette side of the patch, application of 60 m$M$ Na on the cytoplasmic side of the patch activated outward current in the presence of a large Ca gradient (at 0 mV membrane potential). This outward current declined with a time constant of approximately 1 sec. This decay rate is far too low to represent an initial turnover of exchangers, and it appears to be a true regulatory phenomenon. An example of this sort of behavior is displayed in Fig. 7. Here, application of 100 m$M$ Na to the cytoplasmic surface activated a transient outward current. Some information is now available on the part of the exchange molecule that is involved in Na-dependent inactivation (Matsuoka *et al.*, 1997). By studying mutations in the XIP region of the exchanger molecule, these authors have concluded that the endogenous XIP region is primarily involved in movement of the exchanger into and out of the Na induced inactive state.

Advances in the study of the molecular biology of the Na–Ca exchanger have led to a study of the relationship between ionic regulation and the structure of the exchange molecule. Both Ca secondary regulation and Na-dependent inactivation can be removed if the cytoplasmic surface of

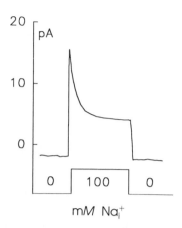

**FIG. 7.** Secondary modulation of outward Na–Ca exchange current in giant excised inside-out patches (37°C, 0 mV). Current transients activated with 100 m$M$ Na$^+$ (substituted for 100 m$M$ Cs$^+$). (This figure was kindly provided by Dr. Donald Hilgemann.)

an excised patch is treated with chymotrypsin (Hilgemann, 1990). From this it may be inferred that these regulatory properties are related in some way to a cytoplasmic domain of the exchange protein. Recent evidence suggests that the sarcolemmal Na–Ca exchanger is regulated by intracellular Ca$^{2+}$ at a high-affinity binding site that is separate from the Ca$^{2+}$ transport site. It was first suggested by Matsuoka *et al.* (1993) that the large hydrophilic loop of the exchanger was the binding site for regulatory Ca. Levitsky *et al.* (1994) have identified a high-affinity Ca binding region comprising amino acids 371–521. This is a large region of the cytoplasmic loop, and Levitsky *et al.* have inferred that substantial secondary structure is necessary for high-affinity Ca binding. Detailed functional studies to measure exchange currents on giant patches from oocytes expressing both wild-type and mutant exchangers have revealed additional information on the Ca regulation of Na–Ca exchange. Mutation of certain aspartic acid residues within two acidic segments markedly decreased Ca binding. These mutations exhibited parallel effects on functional Ca$^{2+}$ regulation. Until recently it has only been possible to study Ca$^{2+}$ regulation when the exchanger operates in the reverse mode. This is because regulatory and transported Ca are on opposite sides of the membrane so that the concentration of regulatory Ca can be manipulated without affecting the concentration of transported Ca. Matsuoka *et al.* (1995) have been able to exploit a class of mutants with low affinity for Ca at the regulatory site. With these mutants it has been possible to demonstrate regulatory effects of Ca on forward-mode exchange. Finally, the affinity of both wild and mutant Na–Ca exchangers for transported Na declines with declining regulatory Ca concentration. This further indicates that Ca regulation modifies transport properties and does not simply control the fraction of exchangers that are available in an active state.

It has been known for some time that in squid giant axons, exchange activity could be regulated by ATP. The main effect of ATP is to increase the affinity of the exchanger for its substrates Na and Ca. Moreover, recent evidence suggests the involvement of Ca-dependent protein kinase. It appears that at least in the squid, Na–Ca exchange that is stimulated by ATP requires intracellular Ca and can be mimicked by hydrolyzable ATP analogues (DiPolo and Beauge, 1987a,b).

It has been difficult to demonstrate regulatory effects of ATP in cardiac tissues. The whole-cell isolated patch technique does not lend itself to studies of this nature because it is extremely difficult if not impossible, to control intracellular ATP concentrations. However, Hilgemann (1990) was able to show that outward exchange current in isolated giant sarcolemmal patches is stimulated by Mg-ATP. With this preparation it is possible to control Mg-ATP on the cytoplasmic side of the cell membrane. This stimulation by Mg-ATP appears to be a different process from that described in squid axon. For example, the process does not appear to involve a protein-dependent kinase (Collins *et al.*, 1992). Hilgemann has also shown that brief treatment of membrane patches with chymotrypsin abolished modulation of exchange current by Mg-ATP. Three mechanisms are likely to explain regulation of Na–Ca exchange by ATP. The first is that an ATP-dependent transport of phosphatidylserine by an amino phospholipid "flippase" leads to an increase in exchange activity. A second possibility is that diacylglycerol is phosphorylated to form phosphatidic acid, which in turn enhances exchange activity. However, recently evidence has been obtained for a third mechanism that seems likely to explain regulation by ATP (Hilgemann and Ball, 1996). It now appears that phosphatidylinositol 4,5-bisphosphate (PIP$_2$) can strongly activate the cardiac Na–Ca exchanger. Apparently, ATP generates PIP$_2$ from phosphatidylinositol (PI). A number of observations suggest that the generation of PIP$_2$ can explain the regulation of Na–Ca exchange by ATP: (1) PI-specific phospholipase C can abolish the action of ATP, which can in turn be restored by the addition of exogenous PI. (2) The effect of ATP can be reversed by a PIP$_2$-specific phospholipase C. However, the stimulatory effect of ATP can be mimicked by exogenous PIP$_2$. (3) Aluminum, which binds with high affinity to PIP$_2$, reversed the effect of ATP on the exchanger. It therefore appears that the generation of PIP$_2$ is an important regulator of Na–Ca exchange activity in the heart.

## VIII. Current–Voltage Relationships and Voltage Dependence of Na–Ca Exchange Current

The origin of the voltage dependence of Na–Ca exchange has yet to be explained. However, current–voltage relationships obtained under a variety of ionic conditions can provide a great deal of information that forms a basis for discussing possible mechanisms of voltage dependence. Here we try to show what can and cannot be concluded about voltage-dependent mechanisms from available current–voltage data. Before interpreting current–voltage relationships, the experimentalist should be confident that he or she is dealing with a pure current. Thus, considerable care in eliminating contaminating currents with appropriate inhibitors and an understanding of which currents (besides exchange current) are activated over the voltage range of interest are essential.

Most Na–Ca exchange current–voltage relationships that have been measured thus far do not show a region of negative slope. This is consistent with the idea that only a single rate-limiting charge translocation step exists in the reaction pathway. Were a second voltage-dependent step to exist in which charge moved in the opposite direction, then one might expect a region of slight negative slope. This is because increasing voltages that stimulated forward exchange would necessarily begin to retard the movement of exchange in the opposite direction, with a resulting decline in net transport and, hence, current. However, conditions under which a region of negative slope in an exchange–current voltage relationship have been detected. Hilgemann et al. (1991b), using the giant patch technique, measured outward exchange current–voltage relationships as a function of extracellular Ca (Fig. 8). When extracellular Ca is 0.1 mM voltage dependence is diminished, and at extreme depolarization the slope of the current–voltage relationship becomes discernibly negative. It is currently believed that most of the exchange voltage dependence is associated with Na translocation. Hilgemann has suggested that the negative slope could be explained if Ca passes through a small fraction of the membrane field before binding to the carrier. This would constitute a second voltage-dependent step associated with Ca translocation in the transport pathway.

A second property one expects of current–voltage relationships for electrogenic exchange reactions is that they

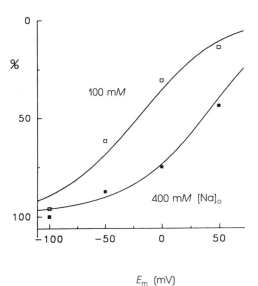

**FIG. 9.** Flattening and saturation of inward $I_{Na-Ca}$–voltage relationships when extracellular $[Na^+]$ is increased from 100 to 400 mM. Na-MES replaced by Ca-MES in pipette, as osmolarity for all extracellular solutions was equal; 20 $\mu M$ free $[Ca^{2+}]$ on cytoplasmic side. $I_{Na-Ca}$ magnitudes in each current–voltage relationship were normalized to values at $-100$ mV.

should saturate at extreme voltages. This is because reaction steps whose rate increases with voltage should become sufficiently rapid that they cease to be rate-limiting for the entire reaction sequence. The reaction may then be rate limited by a voltage-independent step, at which point reaction rate will cease to be responsive to voltage. Many current–voltage relationships do show clear evidence of saturation. For example, the measurements of outward current depicted in Fig. 9 clearly exhibit saturation. Although not so obvious, measurements of inward current by Bridge et al. (1991) and by Miura and Kimura (1989) also show signs of saturation at extremely negative voltages. however, results (from giant patches) under zero trans conditions suggest little saturation of inward exchange current at very negative potentials (Matsuoka and Hilgemann, 1992). The origin of these discrepancies is not clear, but may be related to whether net or unidirectional exchange is producing the inward current. Moreover, current–voltage relationships obtained in whole cells may be complicated by other partial reactions (e.g., Ca dissociation from intracellular buffers) not present in the giant patch experiments.

The way that exchange current–voltage relationships depend on ionic conditions can also yield information about voltage-dependent steps in the reaction sequence that leads to exchange. With voltage-clamped giant patches under zero trans conditions, that is, Ca in the pipette (extracellular) side and Na in the bath solution that superfuses the cytoplasmic surface, outward current–voltage relationships showed a striking dependence on pipette (extracellular) Ca (see Fig. 8). However, when the extracellular (pipette) Ca is reduced to very low values, voltage dependence tends to disappear. An attractive explanation for this result is that as extracellular Ca is reduced, a voltage independent

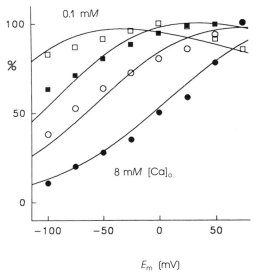

**FIG. 8.** Ion and voltage dependencies of cardiac Na/Ca exchange current conform to consecutive exchange model with voltage dependencies at extracellular $Na^+$ release and binding. Flattening and saturation of outward $I_{Na-Ca}$–voltage relations as extracellular $[Ca^{2+}]$ is lowered from 8 mM (closed circles) to 2 mM (open circles) to 0.4 mM (closed squares) to 0.1 mM (open squares). All results are normalized to the largest current occurring in the current–voltage relation. Note the negative slope with strong depolarization in 0.1 mM.

step or a step with very modest voltage dependence (presumably associated with Ca translocation) starts to become rate-limiting, with the result that exchange loses its voltage dependence.

The dependence of outward current–voltage relationships on sodium is also interesting. Hilgemann *et al.* (1991b) using the giant patch technique have measured outward current–voltage relationships when extracellular Na was 100 and 400 m$M$ (see Fig. 9). At 400 m$M$ there was a pronounced flattening of the current–voltage relationship. There are two possible (and not necessarily unrelated) explanations for this. If we assume that some step in the Na translocation pathway is both rate-limiting and voltage-dependent, it is possible that as extracellular Na increases this step ceases to be rate-limiting because it is accelerated. Were some other voltage-independent step to become rate-limiting, one would expect a flattening of the current–voltage relationship as it became dominated by the voltage-independent step, as is in fact observed. Alternatively, if Na binding to some external site were voltage-dependent, one would expect voltage dependence of this binding to be lost as the site becomes saturated (Lagnado and McNaughton, 1990).

These results seem to suggest that somewhere in the Na translocation pathway a voltage-dependent (charge translocation) step may be involved. In this regard the data obtained with giant patches, where ideal zero trans conditions are most likely to be achieved, are particularly compelling.

## IX. Mechanism of Na–Ca Exchange

A discussion of the evidence for and against various kinetic schemes to account for Na–Ca exchange is beyond the scope of this chapter. The interested reader is referred to the paper by Khananshvili (1991). Here we simply show the way in which measurement of exchange current can be used to infer mechanism. It should, however, be clearly understood that the measurement of isotopic fluxes and rapid mixing techniques provide indispensable tools to investigate mechanism, and that these are not discussed here.

For the Na–Ca exchange, two different basic mechanisms for ion translocation can be considered. These are the Ping-Pong or consecutive mechanism and the sequential or simultaneous mechanism. Although it has been difficult to distinguish between these mechanisms, data using the patch-clamp technique provide strong evidence in favor of the Ping-Pong mechanism. The Ping-Pong mechanism is diagrammed in its simplest form in Fig. 10. There is only one set of binding sites that binds either Ca or Na, and the translocations of Na and Ca are separate events. For example Ca, binds to the carrier E″ at the inner (cytoplasmic) surface and is translocated to the exterior, where it is released. The carrier E′ then can either bind Na or Ca at the external face, which is then transported to the internal surface, where it is in turn released. A basic property of the Ping-Pong mechanism is that Na–Na and Ca–Ca exchange are reversible partial reactions of the exchange. As such, it should be possible to isolate them. These partial

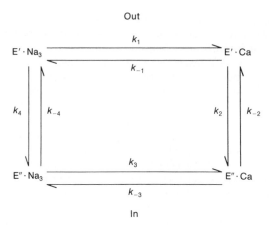

**FIG. 10.** Diagram of the Ping-Pong mechanism of ion translocation.

reactions do not exist as elementary steps in the simultaneous mechanism.

Partial reactions of the exchanger have been observed by Hilgemann *et al.* (1991b). Using giant sarcolemmal patches, rapid application of Na to the intracellular surface caused binding and translocation of intracellular Na to the pipette side. This produced a transient current that could be blocked by exchange blockers, including the recently synthesized exchanger inhibitory peptide (XIP) (Li *et al.*, 1991). Similar currents were measured in giant oocyte patches in which the cloned exchanger had been expressed. Control oocytes did not exhibit this current. It appears that charge translocation is associated with Na translocation, and that the Na translocation step can be isolated, consistent with the idea that a Ping-Pong mechanism is operative.

Further support for a consecutive model of exchange comes from work by Nigli and Lederer (1991). These authors measured very small transient Ca currents induced by the photo release of caged Ca DM-dinitrophen. Insofar as these currents could be inhibited by known blockers of the Na–Ca, exchange they appear to be associated with exchange. Since they are unaffected by Na, they are presumably associated with a partial reaction of the exchange. These authors have speculated that this current represents a charge movement associated with Ca binding and the associated conformational change of the exchange molecule. This suggests that (as we have already seen) a voltage-dependent step might be associated in some way with the partial reactions leading to Ca translocation. It is not yet known exactly how many charge-translocating steps there actually are in the exchange pathway. However, evidence is accruing that the mechanism is consecutive in nature.

Two additional pieces of evidence suggest that a consecutive mechanism could account for Na–Ca exchange. Hilgemann has demonstrated that the apparent affinity of one ion for the exchanger is a function of the concentration of the other (Hilgemann *et al.*, 1991a). This is a requirement of a sequential reaction scheme. Moreover, Kimura has employed classic enzyme kinetics to outward exchange currents under assumed zero trans conditions. Her most recent results suggest that although it is difficult to discriminate between a simultaneous and consecutive reaction, the available evidence favors a consecutive reaction (Li and Kimura, 1991).

## X. Na–Ca Exchange Currents during the Cardiac Action Potential

During every action potential, a Ca current is activated and a modest quantity of Ca enters the cell. As we discuss, it now seems likely that reverse exchange takes place, which also causes some Ca entry. During a steady-state train of action potentials, this continual Ca entry must in some way be compensated. It is therefore likely that forward exchange during the repolarizing phase of the action potential extrudes the Ca entering during the initial phase of the action potential and thus maintains a beat-to-beat homeostasis. Direct evidence that Ca entering the cell through Ca channels can be extruded by the Na–Ca exchange has been obtained by Bridge *et al.* (1990). Guinea pig ventricular cells treated with caffeine can be tetanized in the absence of extracellular Na with a voltage-clamp pulse from −40 to +10 mV. During this clamp an inward Ca current can be measured (Fig. 11). Although the SR is depleted by caffeine, the enlarged Ca current can apparently produce sufficient Ca entry to cause contractures. It should be emphasized that in these experiments the pipette contained no Na, so that Ca entry by reverse exchange was unlikely. At the peak of the contracture, rapid application of extracellular Na produced prompt mechanical relaxation and activated an inward transient current. This current was most likely due to forward Na–Ca exchange, because it could not be activated when intracellular Ca was buffered with EGTA. If the Na–Ca exchange current extruded all the entering Ca, and if we further assume that three Na ions exchange with a single Ca ion, it follows that the integral of the exchange current is half that of the Ca current. The relationship between the integrals of exchange current and Na–Ca exchange current was best explained by assuming that three Na ions exchange with a single Ca ion and that the exchange extruded all the entering Ca. It seems therefore that the Na–Ca exchange does have the capacity to extrude all Ca entering during the duty cycle.

It now seems likely that the activity of Na–Ca exchange is profoundly modified by the cardiac action potential. It is also likely to be the case that the exchange current in part determines the duration of the cardiac action potential. The first comprehensive discussion of this topic was by Mullins (1979). We consider the behavior of the Na–Ca exchange during the ventricular action potential. The guinea pig ventricular action potential is approximately 300 msec in duration. During the initial part of this action potential, Ca is released from the sarcoplasmic reticulum, and this causes a rise of Ca in the cytosol. Peak values for cytosolic free Ca are probably 1–2 $\mu M$. The Ca transient rises to a peak in approximately 50 msec, whereas the peak of the upstroke of the action potential occurs in approximately 2 msec. After the upstroke of the action potential, but before intracellular Ca has risen appreciably, the membrane potential becomes positive to the exchange reversal

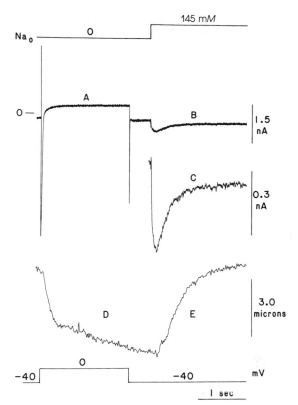

**FIG. 11.** Two-second voltage-clamp pulses in the absence of $Na_o^0$ and presence of 10.0 m$M$ caffeine cause contraction. Rapid application of $Na_+^0$ 500 msec after repolarization causes relaxation. (A) $I_{Ca}$ elicited by membrane depolarization. (B) The application of $Na^+$ produces putative transient inward $I_{Na-Ca}$. This current is displayed on an expanded scale in (C). (D) Contraction (cell shortening) activated by $I_{Ca}$ recorded in (A). (E) After repolarization to −40 mV, relaxation does not occur until 500 msec after the clamp pulse, when $Na^+$ is suddenly applied. (Reprinted with kind permission of the author and *Science.*)

potential, which is about −50 mV. Therefore, an outward exchange current is generated, and this will be accompanied by Ca entry and Na exit. This current has been implicated in the "triggering" of SR Ca release (Leblanc and Hume, 1990). As the Ca released from the SR begins to rise and the membrane begins to repolarize, the membrane potential will become negative to the reversal potential, the exchange current will reverse its direction, and Ca will be extruded in exchange for Na. As the intracellular Ca begins to decline, inward exchange current will also decline to resting values. The same principles presumably govern the behavior of the atrial cell. Earm and Noble (1990) have modeled the time course of Ca current, Na–Ca exchange current, and the Ca transient during the rabbit atrial action potential (Fig. 12). The Na–Ca exchange current is the main depolarizing current during the plateau. Moreover, the exchange activity required to maintain the late plateau is precisely sufficient to balance Ca influx (Ca current + Na–Ca exchange) during the early part of the action potential. It should be appreciated that regardless of the value of intracellular Ca, membrane repolarization will tend to stimulate inward exchange current. On the other hand, the decline of the Ca transient will tend to reduce exchange current. Therefore, the relationship between the Ca transient and membrane repolarization will largely determine the time course of inward exchange current, and therefore the pattern of Ca extrusion.

## XI. Na–Ca Exchange Currents and Excitation–Contraction Coupling

Contractions occur in heart cells when Ca that is released from intracellular stores known as the sarcoplasmic reticulum (SR) activates the contractile elements. For a more detailed account of excitation contraction, the reader is referred to Bers (1991). This release is coupled to electrical excitation at the surface membrane, and the whole process is often referred to as excitation–contraction coupling. The pioneering studies on skinned fibers by Alexander Fabiato led to at least two fundamental findings (for a discussion of these, see Stern and Lakatta, 1992). The first was that Ca can be released from the SR when the concentration of Ca in the vicinity of the SR is abruptly increased. This small increase in the concentration of Ca leads to a much larger release of SR Ca, so the system is one of inherently high gain. This process is usually referred to as Ca-induced Ca release, or CICR. A second and important property of

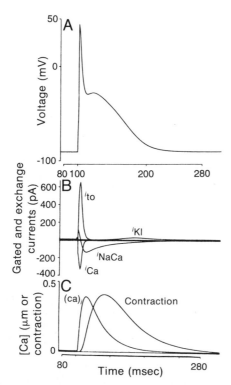

**FIG. 12.** The Earm–Noble model of the single rabbit atrial cell, based on the multicellular model of Hilgemann and Noble (1987). Top: computed action potential. Middle: computed currents. Bottom: [Ca]$_i$ and contraction (Earm and Noble, 1990). (Reprinted with kind permission of the author and the *Annals of the New York Academy of Sciences*.)

CICR is that, under normal circumstances, the release is graded with the size of the Ca increase that induces the release. A priori, such a system might be expected to be regenerative, since the Ca that is released from the SR ought to stimulate further release. A discussion of why SR Ca release in heart is not normally regenerative is beyond the scope of this chapter. However, some of the evidence suggesting that Na–Ca exchange currents are involved in CICR is discussed.

It is now known that large tetrameric Ca release channels (sometimes referred to as ryanodine receptors) are embedded in the sarcoplasmic reticulum (Saito *et al.*, 1988). These apparently respond to elevations of Ca in their vicinity by gating the release of Ca from the SR (Stern and Lakatta, 1992). Therefore, a molecular basis for the early observations by Fabiato has been established. A central question for those studying excitation–contraction in heart is what, in intact cells, produces the rise in intracellular Ca that gates the SR release channel and triggers SR Ca release.

It is now well established that the L-type Ca current is principally involved in triggering SR Ca release in mammalian ventricular cells (London and Krueger, 1986; Beuckelmann and Wier, 1988; Lopez Lopez *et al.*, 1994; Cheng *et al.*, 1993). Because SR Ca release is graded with the size of the increase in Ca concentrations that induces the release, it follows that if L-type Ca channels trigger or induce the release of Ca, the extent of SR Ca release should be graded with the size of the L-type Ca current. Since the size of the Ca current has a bell-shaped dependence on voltage, the finding that the rate or extent of SR Ca release (or the magnitude of triggered contractions) is (under appropriate conditions) also bell-shaped lends strong support to the idea that the Ca current is a trigger for SR Ca release. For example, a detailed study by Beuckelmann and Wier (1988) has clearly established the bell-shaped relationship between triggered Ca transients and voltage.

However, a number of studies in ventricular cells revealed, under certain circumstances, a more complex relationship between the voltage dependence of triggered contractions and Ca current (Nuss and Houser, 1991; Vornanen *et al.*, 1994; Litwin *et al.*, 1996). In particular, tension measurements did not follow a simple bell-shaped relationship with voltage, but showed a rather sigmoid relationship. In particular at positive potentials, shortening did not decline steeply with voltage. Studies by Litwin *et al.* (1996) on guinea pig cells under voltage clamp and dialyzed with various Na solutions indicates that although the shape of the Ca current–voltage relationship was independent of pipette Na, the triggered shortening voltage relationship showed a striking dependence on pipette Na (Fig. 13). At positive potentials the shortening voltage relationship departed from a simple bell shape, and the extent of this departure depended on the concentration of dialyzing Na.

If L-type Ca current is blocked extremely rapidly, triggered contractions are reduced but not abolished (Levi *et al.*, 1996). Thus, if Ca currents are elicited under voltage clamp, it is possible to record a triggered contraction or a triggered Ca transient. If this Ca current is then abruptly blocked with a rapid application of nifedipine a fraction

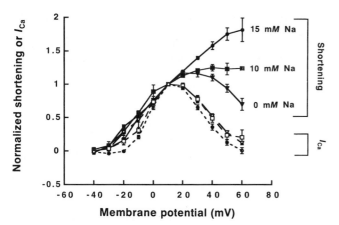

**FIG. 13.** The relationship between voltage and $I_{Ca}$ is bell-shaped regardless of dialyzing Na concentration. However, the relationship between triggered shortening and voltage depends on the concentration of dialyzing Na. When dialyzing Na is nominally 0 m$M$, the relationship between voltage and the extent of shortening approaches a bell shape. (Reprinted with permission of the *Proceedings of the New York Academy of Sciences*.)

of the cell contraction still remains. Similarly, contractions that are elicited during action potentials are not blocked by rapid application of sufficient quantities of nifedipine to abolish Ca current (Levi *et al.,* 1996). It therefore seems likely that another process besides the Ca current is capable of triggering the release of SR Ca. The most likely process is in fact the Na–Ca exchange. Grantham and Cannell (1996) used voltage-clamp pulses shaped like an action potential to infer the magnitude and trajectory of Na–Ca exchange currents during the initial part of an action potential. They found that the magnitude of the exchange current was somewhat less than 30% of the magnitude of the Ca current that occurred during the initial part of the action potential (Fig. 14). As the authors pointed out, this is not a negligible current and is consistent with the idea that at least some of the triggered SR Ca release could be due to the activity of the Na–Ca exchange. If the Na–Ca exchange is capable of contributing part of the trigger for SR Ca release, then this component of the trigger will be extremely sensitive to intracellular Na. In this regard it has been proposed that the initial Na current might provide enough Na accumulation (provided the accumulation takes place in a restricted space) in the vicinity of the Na–Ca exchangers to enhance triggering by reverse Na–Ca exchange current (Leblanc and Hume, 1990; Lipp and Niggli, 1994). It it not yet clear what contribution Na–Ca exchange currents make to triggering SR Ca release under physiological circumstances. However, since the exchanger is regulated by intracellular Na, Ca, and voltage, it seems likely that its contribution to triggering will be both complex and variable.

## XII. Summary

A quarter of a century has elapsed since the Na–Ca exchange was first documented in heart cells. At first, study

of the Na–Ca exchange current was impeded because multicellular preparations did not permit adequate control of the driving forces producing exchange. However, methods for isolating single cells together with the patch-clamp technique made it possible to isolate exchange current and to provide convincing measurements of its properties. Now that the exchange molecule has been cloned, it is possible to express the mammalian exchanger in other cell types, including the frog oocyte and eukaryotic SF21 cells from the fall army worm moth *Spodoptera frugipoda*. Currents may now be measured in these cell types with the giant patch technique or with the whole-cell patch-clamp method. These preparations introduce the possibility of studying mutated forms of the exchanger with a view to

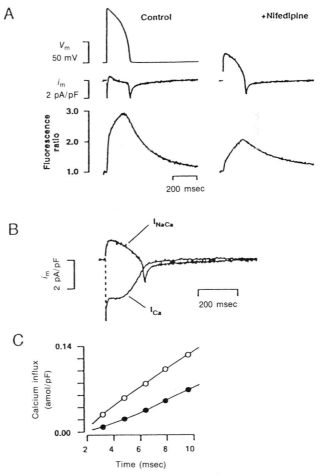

**FIG. 14.** Membrane currents and $Ca^{2+}$ influx in a myocyte with intact $Na^+/Ca^{2+}$ exchange and SR release in response to an AP. The myocyte was at steady state, having been stimulated continuously at 0.2 Hz with a train of at least 20 APs. (A) Top, AP command; middle, membrane current; and bottom, $[Ca^{2+}]$-transient time course in the absence of nifedipine or, on the right, immediately after blockade of $I_{Ca}$ by 10 $\mu$mol/liter nifedipine. (B) Time course of AP-evoked nifedipine-sensitive current ($I_{Ca}$) and calculated $I_{Na/Ca}$. (C) Cumulative $Ca^{2+}$ influx via $I_{Ca}$ (open circles) and the calculated $I_{Na/Ca}$ (closed circles) at the beginning of the AP. Cell capacitance, 150 pF. Similar results were obtained in at least four other cells. (Reprinted with permission of *Circulation Research* from Grantham and Cannell, 1996.)

understanding the relationship between structure and function.

Studies with isolated giant patches have resulted in significant advances in our understanding of both the regulation and mechanism of exchange activity. The phenomenon of Na-dependent inactivation of exchange was first identified in isolated patches, and our understanding of secondary Ca regulation as well as regulation by MgATP has been significantly advanced by this technique. Advances in the molecular biology of the Na–Ca exchanger, which include the elucidation of the primary structure of the exchange molecule, together with the development of recombinant DNAs coding for the Na–Ca exchanger, their expression in systems suitable for current measurement, and, finally, the production of deletion mutants, have already resulted in the first studies of the relationship between structure and function. This has included the elucidation of the sites in the exchange molecule that are responsible for secondary Ca regulation and the regulation of exchange activity by MgATP.

Measurements of both whole-cell currents and currents in giant patches have produced evidence in favor of the idea that a consecutive mechanism can explain exchange activity. Moreover, at least two laboratories have claimed to have isolated currents associated with partial reactions of the exchanger consistent, with the idea that a consecutive (Ping-Pong) reaction scheme is operative. Preliminary results suggest that much of the voltage dependence of the exchange resides on rate-limiting steps associated with Na translocation. With the availability of the techniques of modern molecular biology, it seems likely that we may expect considerable progress in the study of the relationship between the structure and function of the exchange molecule. We may also expect to gain insight into the mechanism of the exchange reaction and, in particular, some understanding of the reaction steps that bestow voltage dependence on the exchange.

Elegant measurements of whole-cell currents have resulted in a plausible model of the behavior of the Na–Ca exchange during the cardiac cycle. The way in which the Na–Ca exchange contributes to excitation–contraction coupling is not understood, but is an active area of research. The advent of peptide inhibitors such as XIP, together with methods for rapidly changing extracellular solutions, suggests that we may soon gain more insight into the contribution of both inward and outward exchange currents to the events of excitation–contraction coupling. Most recently, Na–Ca exchange currents have been proposed as a trigger for SR Ca release under physiological conditions. Thus, in a little more than a quarter of a century the study of exchange currents has expanded enormously to provide insight not only into the way that the Na–Ca exchange controls intracellular Ca at the whole-cell level, but also into the details of the molecular mechanism of the exchange reaction itself.

## References

Achilles, A., Friedel, U., Haase, W., Reilander, H., & Cook, N. (1991). Biochemical and molecular characterization of the sodium–calcium exchange from bovine rod photoreceptors. *Ann. NY Acad. Sci.* **639,** 234–244.

Axelsen, P. H., and Bridge, J. H. B. (1985). Electrochemical ion gradients and the Na/Ca exchange stoichiometry. *J. Gen. Physiol.* **85,** 471–478.

Baker, P. F., and McNaughton, P. A. (1976). Kinetics and energetics of calcium efflux from intact squid giant axons. *J. Physiol.* **259,** 103–144.

Baker, P. F., Blaustein, M. P., Hodgkin, A. L., and Steinhardt, R. A. (1967). Effect of sodium concentration on calcium movements in giant axons of *loligo forbesi. J. Physiol.* (*London*) **192,** 43–44.

Barcenas-Ruiz, L., Beuckelmann, D. J., and Wier, W. G. (1987). Sodium–calcium exchange in heart: Membrane currents and changes in $(Ca^{2+})_i$. *Science* **238,** 1720–1722.

Bers, D. M. (1991). "Excitation–Contraction Coupling and Cardiac Contractile Force." Kluwer Academic Publishers, Dordrecht.

Beuckelmann, D. J., and Wier, W. G. (1988). Mechanism of release of calcium from sarcoplasmic reticulum of guinea-pig cardiac cells. *J. Physiol.* **405,** 233–255.

Biedert, S., Barry, W. H., and Smith, T. W. (1979). Inotropic effects and changes in sodium and calcium contents associated with inhibition of monovalent cation active transport by ouabain in cultured myocardial cells. *J. Gen. Physiol.* **74,** 479–494.

Blaustein, M. P., and Hodgkin, A. L. (1969). The effect of cyanide on the efflux of calcium from squid axons. *J. Physiol.* **200,** 497–527.

Bridge, J. H. B., and Bassingthwaighte, J. B. (1983). Uphill sodium transport driven by an inward calcium gradient in heart muscle. *Science* **219,** 178–180.

Bridge, J. H. B., Smolley, J. R., and Spitzer, K. W. (1990). The relationship between charge movements associated with $I_{Ca}$ and $I_{Na-Ca}$ in cardiac myocytes. *Science* **248,** 376–378.

Bridge, J. H. B., Smolley, J., Spitzer, K. W., and Chin, T. K. (1991). Voltage dependence of sodium–calcium exchange and the control of Ca extrusion in the heart. *Ann. NY Acad. Sci.* **639,** 34–47.

Caputo, C., Bezanilla, F. M., and Dipolo, R. (1989). Currents related to the sodium exchange in squid giant axons. *Biochim. Biophys. Acta* **986,** 250–256.

Cheng, H., Lederer, W. J., and Cannell, M. B. (1993). Calcium sparks: Elementary events underlying excitation–contraction coupling in heart muscle. *Science* **262,** 740–744.

Chin, T. K., Spitzer, K. W., Philipson, K. D., and Bridge, J. H. B. (1993). The effect of exchanger inhibitory peptide (XIP) on sodium–calcium exchange current in guinea pig ventricular cells. *Circ. Res.* **72,** 497–503.

Collins, A., Somlyo, A. V., and Hilgemann, D. W. (1992). The giant cardiac membrane patch method: Stimulation of outward $Na^+$–$Ca^{2+}$ exchange current by Mg-ATP. *J. Physiol.* **454,** 27–57.

Crespo, L. M., Grantham, C. J., and Cannell, M. B. (1990). Kinetics stoichiometry and role of the Na–Ca exchange mechanism in isolated cardiac myocytes. *Nature* **345,** 618–621.

Dipolo, R., and Beauge, L. (1987a). Characterization of the reverse Na–Ca exchange in squid axons and its modulation by $Ca_i$ and ATP. *J. Gen. Physiol.* **90,** 505–525.

DiPolo, R., and Beauge, L. (1987b). In squid axons ATP modulates $Na^+$–$Ca^{2+}$ exchange by a $Ca_i^{2+}$ dependent phosphorylation. *Biochim. Biophys. Acta* **897,** 347–354.

Earm, Y. E., and Noble, D. (1990). A model of the single atrial cell: Relation between calcium current and calcium release. *Proc. Roy. Soc.* (*London*) B **240,** 83–96.

Ehara, T., Matsuoka, S., and Noma, A. (1989). Measurement of reversal potential of $Na^+$–$Ca^{2+}$ exchange current in single guinea-pig ventricular cells. *J. Physiol.* **410,** 227–249.

Frank, J. S., Mottino, G., Reid, D., Molday, R. S., and Philipson, K. D. (1992). Distribution of the $Na^+$–$Ca^{2+}$ exchange protein in mammalian cardiac myocytes: An immunofluorescence and immunocolloidal gold-labeling study. *J. Cell Biol.* **117,** 337–345.

Grantham, C. J., and Cannell, M. B. (1996). $Ca^{2+}$ influx during the cardiac action potential in guinea pig ventricular myocytes. *Circ. Res.* **79**, 194–200.

Hamill, O. P., Marty, E., Neher, E., Sakmann, B., and Sigworth, F. (1981). Improved patch-clamp techniques for high resolution current recording from cells and cell-free membrane patches. *Pflugers Arch.* **391**, 85–100.

Hilgemann, D. W. (1989). Giant excised cardiac sarcolemmal membrane patches: Sodium and sodium–calcium exchange currents. *Pflugers Arch,* 1–3.

Hilgemann, D. W. (1990). Regulation and deregulation of cardiac $Na^+$–$Ca^{2+}$ exchange in giant excised sarcolemmal membrane patches. *Nature* **344**, 242–245.

Hilgemann, D. W., and Ball, R. (1996). Regulation of cardiac $Na^+$, $Ca^{2+}$ exchange and $K_{ATP}$ potassium channels by $PIP_2$. *Science* **273**, 956–959.

Hilgemann, D. W., and Noble, D. (1987). Excitation–contraction coupling and extracellular calcium transients in rabbit atrium: Reconstructions of basic cellular mechanisms. *Proc. Roy. Soc. (London)* B **230**, 163–205.

Hilgemann, D. W., Collins, A., Cash, D. P., and Nagel, G. A. (1991a). Cardiac $Na^+$–$Ca^{2+}$ exchange system in giant membrane patches. *Ann. NY Acad. Sci.* **639**, 126–139.

Hilgemann, D. W., Nicoll, D. A., and Philipson, K. D. (1991b). Change movement during $Na^+$ translocation by native and cloned cardiac $Na^+$/$Ca^{2+}$ exchanger. *Nature* **352**, 715–718.

Hilgemann, D. W., Matsuoka, S., Nagel, G. A., and Collins, A. (1992). Steady-state and dynamic properties of cardiac sodium–calcium exchange. Sodium-dependent inactivation. *J. Gen. Physiol.* **100**, 905–932.

Horackova, M., and Vassort, G. (1979). Sodium–calcium exchange in regulation of cardiac contractility. *J. Gen. Physiol.* **73**, 403–424.

Hume, J. R., and Uehara, A. (1986a). Properties of "creep currents" in single frog atrial cells. *J. Gen. Physiol.* **87**, 833–855.

Hume, J. R., and Uehara, A. (1986b). Creep currents in single frog atrial cells may be generated by electrogenic Na–Ca exchange. *J. Gen. Physiol.* **87**, 857–884.

Khananshvili, D. (1991). Mechanism of partial reactions in the cardiac Na–Ca exchange system. *Ann. NY Acad. Sci.* **639**, 85–95.

Kimura, J., Noma, A., and Irisawa, H. (1986). Na–Ca exchange current in mammalian heart cells. *Nature* **319**, 596–599.

Kimura, J., Miyamae, S., and Noma, A. (1987). Identification of sodium–calcium exchange currents in single ventricular cells of guinea-pig. *J. Physiol.* **384**, 199–222.

Lagnado, L., and McNaughton, P. A. (1990). Electrogenic properties of the Na : Ca exchange. *J. Membr. Biol.* **113**, 177–191.

Langer, G. A. (1964). Kinetic studies of calcium distribution in ventricular muscle of the dog. *Circ. Res.* **15**, 393–405.

Leblanc, N., and Hume, J. R. (1990). Sodium current-induced release of calcium from cardiac sarcoplasmic reticulum. *Science* **248**, 372–376.

Lee, C. O., and Levi, A. J. (1991). The role of intracellular sodium in the control of cardiac contraction. *Ann. NY Acad. Sci.* **639**, 408–427.

Levi, A. J., Li, J., Spitzer, K. W., and Bridge, J. H. B. (1996). Effect on the Indo-1 transient of applying $Ca^{2+}$ channel blocker for a single beat in voltage-clamped guinea-pig cardiac myocytes. *J. Physiol.* **494.3**, 653–673.

Levitsky, D. O., Nicoll, D. A., and Philipson, K. D. (1994). Identification of the high affinity Ca-binding domain of the cardiac Na–Ca exchanger. *J. Biol. Chem.* **269**, 22847–22852.

Li, J., and Kimura, J. (1991). Translocation mechanism of cardiac Na–Ca exchange. *Ann. NY Acad. Sci.* **639**, 48–60.

Li, Z., Nicoll, D. A., Collins, A., Hilgemann, D. W., Filoteo, A. G., Penniston, J. T., Weiss, J. N., Tomich, J. M., and Philipson, K. D. (1991). Identification of a peptide inhibitor of the cardiac sarcolemmal $Na^+$–$Ca^{2+}$ exchanger. *J. Biol. Chem.* **266**, 1014–1020.

Li, Z., Smith, C. D., Smolley, J. R., Bridge, J. H. B., Frank, J. S., and Philipson, K. D. (1992). Expression of the cardiac Na–Ca exchanger in insect cells using a baculovirus vector. *J. Biol. Chem.* **267**, 7828–7833.

Lipp, P., and Niggli, E. (1994). Sodium current-induced calcium signals in isolated guinea-pig ventricular myocytes. *J. Physiol. (London)* **474**, 439–446.

Litwin, S. E., Kohmoto, O., Levi, A. J., Spitzer, K. W., and Bridge, J. H. B. (1996). Evidence that reverse Na-Ca exchange can trigger SR Ca release. *Ann. NY Acad. Sci.* **779**, 451–463.

London, B., and Krueger, J. W. (1986). Contraction in voltage-clamped internally perfused single heart cells. *J. Gen. Physiol.* **88**, 475–505.

Lopez Lopez, J. R., Shacklock, P. S., Balke, C. W., and Wier, W. G. (1994). Local, stochastic release of $Ca^{2+}$ in voltage-clamped rat heart cells: Visualization with confocal microscopy. *J. Physiol. (London)* **480**, 21–29.

Matsuoka, S., and Hilgemann, D. W. (1992). Steady-state and dynamic properties of cardiac sodium–calcium exchange. Ion and voltage dependencies of the transport cycle. *J. Gen. Physiol.* **100**, 963–1001.

Matsuoka, S., Nicoll, D. A., Reilly, R. F., Hilgemann, D. W., and Philipson, K. D. (1993). Initial localization of regulatory regions of the cardiac sarcolemmal $Na^+$–$Ca^{2+}$ exchanger. *Proc. Natl. Acad. Sci. USA* **90**, 3870–3874.

Matsuoka, S., Nicoll, D. A., Hryshko, L. V., Levitsky, D. O., Weiss, J. N., and Philipson, K. D. (1995). Regulation of the cardiac $Na^+$–$Ca^{2+}$ exchanger by $Ca^{2+}$: Mutational analysis of the $Ca^{2+}$-binding domain. *J. Gen. Physiol.* **105**, 403–420.

Matsuoka, S., Nicoll, D. A., He, Z., and Philipson, K. D. (1997). Regulation of the cardiac Na–Ca exchanger by the endogenous XIP region. *J. Gen. Physiol.* **109**, 1–14.

Miura, Y., and Kimura, J. (1989). Sodium–calcium exchange current. *J. Gen. Physiol.* **93**, 1129–1145.

Mullins, L. J. (1979). The generation of electric currents in cardiac fibers by Na/Ca exchange. *Am. J. Physiol.* **236**(3), C103–C110.

Nicoll, D. A., Longoni, S., and Philipson, K. D. (1990). Molecular cloning and functional expression of the cardiac sarcolemmal $Na^+$–$Ca^{2+}$ exchanger. *Science* **250**, 562–565.

Nicoll, D. A., Hryshko, L. V., Matsuoka, S., Frank, J. S., and Philipson, K. D. (1996). Mutation of amino acid residues in the putative transmembrane segments of the cardiac sarcolemmal Na–Ca exchanger. *J. Biol. Chem.* **271**(23), 13385–13391.

Niggli, E., and Lederer, W. J. (1991). Molecular operations of the sodium–calcium exchanger revealed by conformation currents. *Nature* **349**, 621–624.

Nuss, H. B., and Houser, S. R. (1991). Voltage dependence of contraction and calcium current in severely hypertrophied feline ventricular myocytes. *J. Mol. Cell. Cardiol.* **23**, 717–726.

Rasgado-Flores, H., and Blaustein, M. P. (1987). Na/Ca exchange in barnacle muscle cells has a stoichiometry of 3 $Na^+$/1 $Ca^+$. *Am. J. Physiol. Cell Physiol.* **252**(21), C499–C504.

Reeves, J. P., and Hale, C. C. (1984). The stoichiometry of the cardiac sodium–calcium exchange system. *J. Biol. Chem.* **259**, 7733–7739.

Reeves, J. P., and Sutko, J. L. (1983). Competitive interactions of sodium and calcium with the sodium–calcium exchange system of cardiac sarcolemmal vesicles. *J. Biol. Chem.* **285**(5), 3178–3182.

Repke, K. (1964). Ubersichten über den biochemischen Wirkungsmodus von Digitalis. *Klin. Wochenschr.* **41**, 157–165.

Reuter, H., and Seitz, N. (1968). The dependence of calcium efflux from cardiac muscle on temperature and external ion composition. *J. Physiol.* **195**, 451–470.

Saito, A., Inui, M., Radermacher, M., Frank, J., and Fleischer, S. (1988). Ultrastructure of the calcium release channel of sarcoplasmic reticulum. *J. Cell Biol.* **107**, 211–219.

Sheu, S.-S., and Fozzard, H. A. (1982). Transmembrane $Na^+$ and $Ca^{2+}$ electrochemical gradients in cardiac muscle and their relationship to force development. *J. Gen. Physiol.* **319**, 325–351.

Spitzer, K. W., and Bridge, J. H. B. (1989). A simple device for rapidly exchanging solution surrounding a single cardiac cell. *Am. J. Physiol. Cell Physiol.* **256,** C441–C447.

Stern, M. D., and Lakatta, E. G. (1992). Excitation–contraction coupling in the heart: The state of the question. *FASEB J.* **6,** 3092–3100.

Trosper, T. L., and Philipson, K. D. (1984). Stimulatory effect of

calcium chelators on Na–Ca exchange in cardiac sarcolemmal vesicles. *Cell Calcium* **5,** 211–222.

Vornanen, M., Shepherd, N., and Isenberg, G. (1994). Tension–voltage relations of single myocytes reflect Ca release triggered by Na/Ca exchange at 35°C but not 23°C. *Am. J. Physiol. Cell Physiol.* **267,** C623–632.

Clive M. Baumgarten and Joseph J. Feher

# 19

# Osmosis and the Regulation of Cell Volume

## I. Introduction

In whole blood, erythrocytes are biconcave disks about 7 $\mu$m in diameter and 2 $\mu$m thick. When diluted in a solution of 0.9% NaCl (w/v), erythrocytes retain this shape. When diluted with higher concentrations of salt, the erythrocytes shrink, taking on the appearance of a small sphere with spikes all over the surface. Cells in this state are described as **crenated**. If these same cells are diluted with markedly lower concentration of salt, the cells swell. They first become spherical and then, if the solution is sufficiently low in salt, the cells burst and release their contents into the extracellular solution. These elementary observations give rise to the concept of **tonicity**. Tonicity is operationally defined as a measure of the ability of a solution to induce shrinking or swelling when placed in contact with specified cells. Thus, an **isotonic** solution is one that induces no volume change when placed in contact with the cells. The tonicity of the solution is equal to the tonicity of the cell's contents. Exposure of the cell to a **hypertonic** solution results in shrinking. Conversely, a **hypotonic** solution increases cell volume. A solution that is isotonic for one type of cell may or may not be isotonic for others.

Shrinking or swelling on exposure to solutions of differing concentration is ubiquitous in animal cells. Figure 1 shows the volume response of single isolated heart cells on exposure to hypertonic or hypotonic solutions. In hypotonic solution, myocytes swelled to more than 1.5 times their initial volume. Note that the swelling was complete within 2 min, the new volume was stable, and volume returned to normal on return of the isotonic solution. In hypertonic solution, cell volume decreased to about ~0.65 times normal, and the original volume was restored on return of the isotonic solution.

As can be seen from the data of Fig. 1, volume changes upon exposure to solutions of differing tonicity are very rapid. What is moving when the cells swell or shrink? What routes do these substances take? Are there homeostatic mechanisms that limit swelling and shrinking? If so, how

are the compensatory mechanisms engaged? The answers to these questions are not yet complete. This chapter provides the basis for understanding regulation of cell volume through the exchange of water and solutes across the plasma membrane.

## II. Water Movement across Model Membranes

### A. Definition of Osmosis

**Osmosis** refers to the movement of fluid across a membrane in response to differing concentrations of solutes on the two sides of the membrane. Osmosis has been used since antiquity to preserve foods by dehydration with salt or sugar. The removal of water from a tissue by salt was referred to as **imbibition** by the salt. This description comes from the notion that these solutes attracted water from material they touched. In 1748, J. A. Nollet used an animal bladder to separate chambers containing water and wine. He noted that the volume in the wine chamber increased, and, when this chamber was closed, a pressure developed. He named the phenomenon **osmosis** from the Greek $\omega\sigma\mu o\sigma$, meaning thrust or impulse.

Pfeffer (1877) provided early quantitative observations on osmosis. He made an artificial membrane in the walls of an unglazed porcelain vessel by reacting copper salts with potassium ferrocyanide to form a copper ferrocyanide precipitation membrane on the surface of the vessel. He used this membrane to separate a sucrose solution inside the vessel from water outside and found a volume flow from the water side to the sucrose side. Pfeffer observed that the flow was proportional to the sucrose concentration. Further, he observed that a pressure applied inside the vessel produced a filtration flow proportional to the pressure. He found that a closed vessel containing a sucrose solution would develop a pressure proportional to the concentration of sucrose. He recognized this as an equilibrium state in which the pressure balanced the osmosis caused by the sucrose solution. Pfeffer's original data for the osmotic

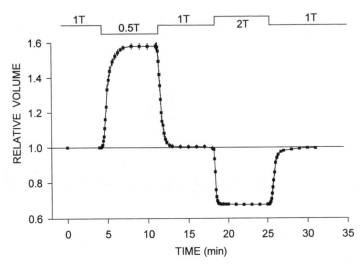

**FIG. 1.** Response of isolated rabbit ventricular myocytes to osmotic stress. Cell volume initially was measured in isotonic solution (1T). Myocytes rapidly swelled 58% in a hypotonic solution with an osmolarity half that of 1T and rapidly shrank 33% in a hypertonic solution with an osmolarity twice that of 1T. Cell volume was stable for the duration of perfusion with either hypotonic or hypertonic media and rapidly returned to its control value when 1T solution was readmitted. Volume was measured by digital video microscopy, and relative volume was calculated as $volume_{test}/volume_{1T}$. Solution osmolarity was adjusted by varying the concentration of mannitol. (From Suleymanian and Baumgarten, 1996; Reproduced from *The Journal of Physiology* by copyright permission of The Rockefeller University Press.)

pressure of sucrose solutions are plotted in Fig. 2. He defined the **osmotic pressure** as the hydrostatic pressure necessary to stop osmotic flow across a barrier (e.g., a membrane) that is impermeable to the solute. This concept is illustrated in Fig. 3. Osmotic pressure is a property intrinsic to the solution and is measured at equilibrium, when the pressure-driven flow exactly balances the osmotic-driven flow. By defining osmotic pressure in this way, we assign a positive value to an apparent reduction in pressure brought about by dissolving the solute. Thus, fluid movement occurs from the solution of low osmotic pressure (water) to the solution of high osmotic pressure, opposite in direction to

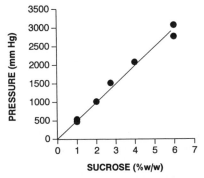

**FIG. 2.** Plot of data from Pfeffer (1877) for the osmotic pressure of sucrose solutions. A copper ferrocyanide precipitation membrane was formed in the walls of an unglazed porcelain cup. The membrane separated a sucrose solution in the inner chamber from water in the outer chamber. The inner chamber was then attached to a manometer and sealed. The linear relation between the pressure measured with this device and the sucrose concentration were the experimental impetus for deriving van't Hoff's law.

the hydraulic flow of water from high to low hydrostatic pressure.

An ideal **semipermeable** membrane is required for determining osmotic pressure. This means that the membrane is permeable to water but absolutely impermeable to the solute. The concept of osmotic pressure differs from tonicity in that tonicity requires comparison of two solutions with reference to a specific nonideal membrane. If the membrane is highly permeable to solute as well as water, no water flow will occur, and therefore the externally applied pressure required to stop osmosis is zero. This observation makes it plain that the **effective** osmotic pressure, the one measured with a real membrane, must be due to some interaction of the membrane with the solute, because pressure depends on both the specific solute and the specific membrane.

### B. van't Hoff's Law

van't Hoff (1887) argued from Pfeffer's data and from thought experiments considering gases in equilibrium with water that the osmotic pressure should be given by

$$\pi = RT \Sigma C_s \qquad (1)$$

where $\pi$ is the usual symbol for osmotic pressure, $R$ is the gas constant, $T$ is the temperature in degrees Kelvin and $C_s$ is the concentration of solute particles in solution. This equation is known as **van't Hoff's law.** Table 1 lists common units for osmotic pressure, along with the values and units of $R$ and $C_s$ needed to make the calculation.

The concentration used in van't Hoff's law, $\Sigma C_s$, refers to the number of osmotically active particles that are formed on dissolution of the solute. For example, organic compounds such as glucose ideally yield one particle,

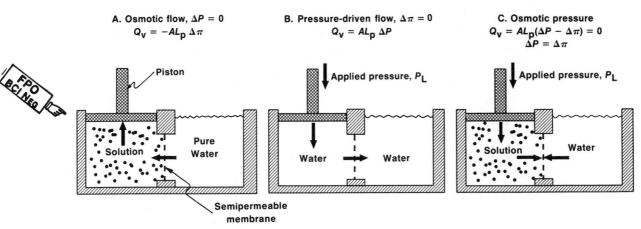

**FIG. 3.**    Equivalence of hydrostatic and osmotic pressures in driving fluid flow across a membrane. (A) An ideal, semipermeable membrane is freely permeable to water, but is impermeable to solute. When the membrane separates pure water on the right from solution on the left, water moves to the solution side. This water flow is *osmosis*. The flow, $Q_v$, in $cm^3$ $sec^{-1}$, is linearly related to the difference in osmotic pressure, $\Delta\pi$, by the area of the membrane, $A$, and the hydraulic conductivity, $L_p$. Positive $Q_v$ is taken as flow to the right. The flow causes expansion of the left compartment and movement of the piston (which is assumed to be weightless). (B) Application of a pressure, $P_L$, to the left compartment forces water out of this compartment, across the semipermeable membrane. The flow is linearly related to the pressure difference between the two compartments. (C) Application of a $P_L$ so that $\Delta P = \Delta\pi$ results in no net flow across the membrane. The osmotic pressure of a solution is defined as the pressure necessary to stop water movement when the ideal, semipermeable membrane separates water from the solution.

whereas strong salts such as NaCl or $CaCl_2$ ideally yield two ($Na^+$ and $Cl^-$) and three ($Ca^{2+}$ and $2\ Cl^-$), respectively. The **osmolarity** of a solution equals $\Sigma C_s$ and is expressed in osmol $liter^{-1}$ to indicate that we are referring to the number of osmotically active particles, termed **osmolytes**, rather than the concentration of the solute. An alternative scale, **osmolality**, defines $\Sigma C_s$ per kilogram of solvent. Although the osmolal scale better describes the osmotic pressure in van't Hoff's equation, the osmolar scale is more generally used in physiological studies. As we shall see, van't Hoff's law is a limiting law that is true only for dilute solutions. In this limit of dilute solutions, both osmolal and osmolar concentration scales converge to the same results.

To illustrate the magnitude of osmotic pressure, ideal solutions of 10 m$M$ glucose or 5 m$M$ NaCl, which dissociates into two particles, both have an osmolarity of 10 mosmol $liter^{-1}$ and an osmotic pressure at 37°C of 0.082 liter atm $mol^{-1}$ $K^{-1}$ × 310 K × 0.01 mol $liter^{-1}$ = 0.254 atm, or 193 mm Hg. Thus, the osmotic pressure of even a dilute solution is large in comparison to normal hydrostatic pressures in physiological systems.

### C. Thermodynamic Derivation of van't Hoff's Law

One of the conclusions of chemistry is that all spontaneous processes are accompanied by a **decrease in free energy**. The total free energy of a solution can be divided among its components. This parceling out of the Gibbs free energy, $G$, is embodied in the concept of **chemical potential**

$$\mu_i = \left(\frac{\partial G}{\partial n_i}\right)_{T,P,n_k} \tag{2}$$

where $\mu_i$ is the chemical potential of component $i$, $n_i$ is the number of moles of component i, and $i \neq k$. The chemical potential of a component of a solution consists of three terms: a standard potential, which refers to the chemical

**TABLE I**    Units for Calculation of Osmotic Pressure[a]

| Pressure units | 1 atm equivalent | Gas constant ($R$) | Solute osmolyte concentration ($\Sigma C_s$) |
|---|---|---|---|
| atm | 1 | 0.082 liter atm $mol^{-1}$ $K^{-1}$ | mol $liter^{-1}$ |
| mm Hg | 760 | 62.36 liter mmHg $mol^{-1}$ $K^{-1}$ | mol $liter^{-1}$ |
| N $m^{-2}$ = Pa | $1.013 \times 10^5$ | 8.314 N m $mol^{-1}$ $K^{-1}$ | mol $m^{-3}$ – mol $(1000\ liter)^{-1}$ |
| dyn $cm^{-2}$ | $1.013 \times 10^6$ | $8.314 \times 10^7$ dyn cm $mol^{-1}$ $K^{-1}$ | mol $cm^{-3}$ |

[a] Osmolarity (osmol $liter^{-1}$) is defined as the concentration of osmotically active particles, osmolytes, in mol $liter^{-1}$. Therefore, the units osmoles and moles cancel in the calculation of osmotic pressure.

energy involved in the formation of the material from standard states; a compositional term, which depends on the presence of other constituents; and a work term, encompassing other work required (per mole) to bring additional material into the solution. The work term in the chemical potential of water is $\int d(\overline{V}_w P)$, where $\overline{V}_w$ is the volume of water per mole and $P$ is the pressure. The **electrochemical potential** of ions in solution requires the inclusion of an electrical work term, $\int z_i F d\psi$, where $z_i$ is the ion's valence, $F$ is Faraday's constant, and $\psi$ is voltage.

In the case of a solution separated from pure water by an ideal semipermeable membrane, water movement will occur when there is a difference in the chemical potential of water on the two sides of the membrane, such that water movement will result in a decrease in free energy. At equilibrium, that is, when the pressure applied to the solution is equal to the osmotic pressure, the chemical potential of water is equal on both sides of the membrane, and no net water movement occurs. This equality of chemical potential is written as

$$\mu^0 + \overline{V}_w P_L + RT \ln a_{w,L} = \mu^0 + \overline{V}_w P_R + RT \ln a_{w,R} \quad (3)$$

where the subscripts L and R refer to the left and right sides of the semipermeable membrane, $\mu^0$ is the chemical potential of liquid water in its standard state (pure water at 1 atm pressure), and $a_w$ is the activity of water. For an ideal solution, the activity of water can be replaced by its **mole fraction,** $X_w$

$$a_w = X_w = \frac{n_w}{n_w + n_s} \quad (4)$$

where $n_w$ and $n_s$ are the moles of water and solute, respectively. The balance of the chemical potential can be written as

$$\mu^0 + \overline{V}_w P_L + RT \ln X_{w,L} = \mu^0 + \overline{V}_w P_R + RT \ln X_{w,R} \quad (5)$$

Consider the situation shown earlier in Fig. 3, where pure water is on the right side of the membrane and a solution is on the left. $X_{w,R} = 1.0$, and thus, $\ln X_{w,R} = 0$. Rearranging, we find

$$\overline{V}_w (P_L - P_R) = -RT \ln X_{w,L} \quad (6)$$

The mole fractions of water and solute in a solution must sum to 1.0. This is expressed as

$$X_{w,L} + X_{s,L} = 1$$
$$\ln X_{w,L} = \ln (1 - X_{s,L}) \quad (7)$$

where $X_{s,L}$ is the mole fraction of solute in the solution on the left. In dilute solutions, $X_{s,L} \ll 1.0$, and thus, $\ln(1 - X_{s,L}) \approx -X_{s,L}$. Substitution of this approximation in Eq. 6 gives

$$P_L - P_R = \frac{RT}{\overline{V}_w} X_{s,L} \quad (8)$$

The left-hand side of Eq. 8 is just the osmotic pressure, $\pi$, which is equal to the extra pressure that must be applied to the solution on the left side in order to establish equality

of the chemical potential of water on the two sides of the membrane. For physiological studies, it is more convenient to express $\pi$ in terms of concentration. From the definition of mole fraction and the assumption of dilute solutions so that $n_s \ll n_w$, we get

$$\pi = \frac{RT}{\overline{V}_w} X_{s,L} = \frac{RT}{\overline{V}_w} \frac{n_s}{n_s + n_w} \approx \frac{RT \, n_s}{n_w \overline{V}_w}$$

$$\pi = RT \frac{n_s}{V} \quad (9)$$

$$\pi = RTC_s$$

where $n_w \overline{V}_w \approx V$, the total volume of solution, and $C_s$ is the concentration of impermeable solute on the solution side of the membrane. This last expression is the van't Hoff equation for the osmotic pressure, Eq. 1. The thermodynamic derivation entails two assumptions: The solution is dilute enough to approach ideality, and the solution is incompressible so that $\int d(\overline{V}_w P) = \overline{V}_w \Delta P$. It is important to recognize that the resulting equation is not exact for physiological solutions. Rather, it is an approximation that is strictly true only for dilute ideal solutions.

The van't Hoff equation is based on thermodynamics, and as such, it tells us nothing about the rates of osmosis or the mechanism by which it occurs. Conceivably, the semipermeable membrane could be like a sieve that allows water to pass freely while blocking solute movement. Alternatively, solvent could dissolve in the membrane, whereas solute is insoluble. Both of these models would exhibit osmotic flow from the region of low osmotic pressure (pure water) to that of high osmotic pressure (impermeant solute solution). The mechanism by which osmosis occurs must be answered by methods of chemical kinetics, and must be answered for every membrane–solvent pair.

### D. Other Colligative Properties of Solutions

As is evident from the thermodynamic derivation given previously, osmotic pressure (and osmotic flow) originates in the lowering of the chemical potential of water by the amount $\sim RTX_s$ when solute is dissolved. Several other properties of solutions also are a consequence of the lowered chemical potential of water because of dissolution of solutes. Together, these are called the **colligative** properties (from the Latin *ligare*, meaning to bind) and include **osmotic pressure, vapor-pressure depression, boiling-point elevation,** and **freezing-point depression.** Consider two open compartments in an enclosed chamber, as shown in Fig. 4. One compartment contains pure water and the other a solution of a nonvolatile solute. The vapor pressure above a solution is defined as the partial pressure of water vapor in equilibrium with solution. Since the vapor pressure of pure water is higher than that of the solution, water vapor above pure water will be at a higher pressure than that above the solution. As a result, water vapor will diffuse from the water side to the solution side. At the surface of the solution, water vapor will condense because the vapor pressure there will be higher than the equilibrium vapor pressure for the solution. Thus, water will move from the

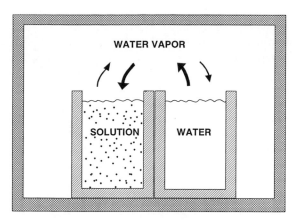

**FIG. 4.** Vapor pressure depression can produce an osmosis-like flow. The vapor pressure is the partial pressure of water in air in equilibrium with a solution. The vapor pressure of a solution of a nonvolatile solute is lower than the vapor pressure of pure water. This is the vapor pressure depression. If open containers of water and a solution are placed in a sealed box, water will evaporate from the water container and condense into the solution container. The surfaces of the fluids and intervening air thus are formally equivalent to an ideal, semipermeable membrane, with a flow of fluid from the water side to the solution side, as in osmosis.

or can be calculated from the parameters tabulated by Pitzer and Mayorga (1973). These osmotic coefficients are corrections to van't Hoff's law due to interactions only for the particular solute. When more than one solute is present, iteractions could occur that are not accounted for by the osmotic coefficients. Therefore, calculations of the osmotic pressure of a mixture of solutes, even when osmotic coefficients are used, are only approximations.

Nonelectrolytes and polyelectrolytes, especially proteins, also show marked departure from van't Hoff's law with increasing concentration. According to Eq. 10, the osmotic coefficient for a single solute can be calculated as

$$\phi_s = \frac{\pi_{observed}}{\pi_{calculated}} = \frac{\pi_{observed}}{RT C_s} \tag{11}$$

The osmotic coefficient for sucrose is plotted against sucrose concentration in Fig. 5. $\phi_s$ is nearly 1.0 in dilute solutions, but approaches 3 in saturated sucrose solutions. Thus, the van't Hoff equation successfully describes the osmotic pressure of dilute solutions, but fails at high solute concentrations. The failure of the van't Hoff equation for highly concentrated solutions is due to the assumptions that solutions are dilute and ideal, which

pure water to the solution side. In short, "osmosis" would occur through the "semipermeable membrane" represented by the surfaces of the two fluids and the intervening air. This serves to illustrate the strong connection among the colligative properties of solutions. Laboratory osmometers typically use either vapor-pressure depression or freezing-point depression to determine the total solute concentration in aqueous solution.

### E. Osmotic Pressure of Nonideal Solutions

As discussed earlier, the van't Hoff equation is an approximation that adequately describes the osmotic pressure for dilute solutions. Its derivation requires the assumptions that the solutions are dilute and that the solutions are ideal. Here "ideal" means that Raoult's law (vapor pressure is proportional to mole fraction of solvent) is valid for the solution (Kiil, 1989; Hildebrand, 1955). Because the behavior of real solutions is not ideal, the van't Hoff equation must be modified to include a correction term, the osmotic coefficient ($\phi_s$)

$$\pi = RT \Sigma \phi_s C_s \tag{10}$$

At physiological concentrations, the osmotic coefficients for NaCl and CaCl₂ are 0.93 and 0.85, respectively. This means the osmolarity of 150 m$M$ NaCl is 0.93 × 2 × 150 = 279 mosmol liter$^{-1}$ and the osmolarity of 150 m$M$ CaCl₂ is 0.85 × 3 × 150 = 382.5 mosmol liter$^{-1}$. The osmotic coefficients for electrolytes vary with temperature, concentration, and the chemical nature of the electrolyte. For most electrolytes, $\phi < 1.0$ for dilute solutions, because of weak attraction of the ions. At higher concentrations, $\phi$ increases to exceed 1.0. Values for the osmotic coefficients for electrolytes can be found in Robinson and Stokes (1959)

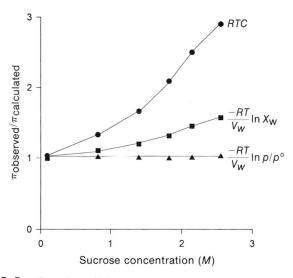

**FIG. 5.** Osmotic coefficients as a function of sucrose concentration. Plotted are the molar osmotic coefficient, defined as $\phi_s$ in Eq. 11 (●), obtained by dividing the observed osmotic pressure by $RTC$; the rational osmotic coefficient, $g$, defined in Eq. 15 (■), obtained by dividing the observed osmotic pressure by $-RT/\overline{V}_w \ln X_w$; and the observed osmotic pressure divided by that predicted by vapor pressure measurements, $-RT/\overline{V}_w \ln p/p^0$ according to Eq. 13 (▲). Deviation of the molar osmotic coefficient from 1.0 means that the van't Hoff law fails to adequately describe the osmotic pressure at high concentrations, but is accurate for dilute solutions. The van't Hoff law requires the assumption of dilute solution and ideal behavior. Deviation of the rational osmotic coefficient from 1.0 means that the solution is not ideal, as the equation requires this assumption. The nearly perfect agreement between the theoretical osmotic pressure predicted from vapor-pressure measurements illustrates the connection between these two colligative properties. [Data from Glasstone (1946).]

were used to derive the equation. $\phi_s$ accounts for these simplifications.

For high solute concentrations, we can calculate the osmotic pressure without assuming a dilute solution from the mole fraction of water by identifying $\pi = P_L - P_R$ in Eq. 6

$$\pi = -\frac{RT}{\overline{V}_w} \ln X_w \qquad (12)$$

This equation still requires the assumption of ideal solution behavior: The activity of water is equal to its mole fraction. The expression for the osmotic pressure without the assumption of either dilute solution or ideality is given by

$$\pi = -\frac{RT}{\overline{V}_w} \ln a_w = -\frac{RT}{\overline{V}_w} \ln \frac{p}{p^0} \qquad (13)$$

where $p$ and $p^0$ are the vapor pressures of the solution and pure water, respectively.

The rational osmotic coefficient, $g$, accounts for nonideal behavior and is defined as

$$\ln a_w = g \ln X_w \qquad (14)$$

Then, from Eqs. 12–14, we find that

$$g = \frac{\pi_{observed}}{\pi_{calculated}} = \frac{\pi_{observed}}{-\dfrac{RT}{\overline{V}_w} \ln X_w} \qquad (15)$$

The rational osmotic coefficient is closer to 1.0, but still deviates significantly at higher sucrose concentrations where solution behavior is further from ideal (see Fig. 5). In contrast, the ratio of the observed osmotic pressure to the theoretical osmotic pressure calculated from vapor pressure measurements, according to Eq. 13, is very close to 1.0 throughout the entire concentration range. This shows the validity of Eq. 13 and the absolute correlation between vapor-pressure depression and osmotic pressure as different measures of the same phenomenon: the lowering of the activity of solvent water by the dissolution of solute. Equation 12 does not adequately describe the variation of $\pi$ with $C_s$ because it requires ideal adherence to Raoult's law (vapor pressure is proportional to $X_w$); van't Hoff's limiting law further deviates from a linear relationship between $\pi$ and $C_s$ because it requires the additional approximation of dilute solutions. Despite these limitations in the high-concentration domain, van't Hoff's law remains a good approximation for electrolyte solutions in the physiological range.

Because of their importance in physiological systems, the nonideality of the osmotic pressure of protein solutions requires special comment. Adair (1928) found that the observed osmotic pressure increased faster than the concentration in hemoglobin solutions, as shown in Fig. 6. Part of the osmotic pressure was due to the unequal distribution of ions across the semipermeable membrane caused by electric charge on the immobile protein molecules. This is the Gibbs–Donnan equilibrium, discussed in more depth later. The contribution of the Gibbs–Donnan distribution to osmotic pressure is small, however, and nearly all of the nonlinearity between $\pi$ and $C_s$ is due to the protein itself.

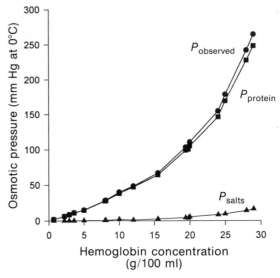

**FIG. 6.** Dependence of the observed osmotic pressure on hemoglobin concentrations. $P_{observed}$ is the observed osmotic pressure (●). $P_{salts}$ is the contribution of the salts to the osmotic pressure as calculated from the Gibbs-Donnan distribution and the van't Hoff equation (▲). $P_{protein}$ is the contribution of the protein itself to the observed osmotic pressure, calculated as $P_{observed} - P_{salts}$ (■). [Data from Adair (1928).]

From the data obtained by Adair (1928), $\phi_{Hb} = 4.03$ at the concentration of hemoglobin within erythrocytes (34.4 g hemoglobin per 100 ml of solution).

The observed osmotic pressure of solutions of plasma proteins also increases more rapidly than concentration, but the degree of deviation from linearity is different for different proteins. Thus, serum albumin shows marked deviation, whereas $\gamma$-globulins are more nearly linear. The empirical fits to the concentration-dependence of osmotic pressure are given by Landis and Pappenheimer (1963) as

$$\pi_{albumin} = 2.8C + 0.18C^2 + 0.012C^3$$

$$\pi_{globulins} = 1.6C + 0.15C^2 + 0.006C^3 \qquad (16)$$

$$\pi_{plasma\ proteins} = 2.1C + 0.16C^2 + 0.009C^3$$

In each of these three equations, the first term represents the limiting law of van't Hoff.

The rather large $\phi_s$ for proteins and polymers is due in part to **excluded volume effects.** That is, proteins and polymers exclude solvent from a larger volume than inorganic ions. The lowering of the free energy of solvent water upon dissolution of solute, which gives rise to osmosis, can be calculated from the increase of entropy on mixing. This entropy of mixing depends on the volume occupied by the solute. From considerations of the excluded volume, it can be shown (Tanford, 1961) that the expected osmotic pressure is given as

$$\pi = RTC \left[ 1 + \frac{\overline{V}_s}{\overline{V}_w} C \right] \qquad (17)$$

## F. Equivalence of Osmotic and Hydrostatic Pressure

As mentioned earlier, Pfeffer originally observed a linear relation between the flow rate and the concentration of solute. This is expressed as

$$J_v = -L_p(\pi_L - \pi_R) = -L_p \Delta\pi \qquad (18)$$

where $J_v$ is the volume flux in $cm^3 \cdot sec^{-1}$ per unit area of membrane, $L_p$ is variously called the **filtration coefficient, hydraulic conductivity,** or **hydraulic permeability**, and $\Delta\pi$ is the osmotic pressure difference. A positive $J_v$ in Eq. 18 represents flux from the left to the right compartment, and this is the order in which the osmotic pressure difference is taken. The minus sign before $L_p$ indicates flux is from the region of low osmotic pressure to the region of high osmotic pressure. In Fig. 3A, $\pi_L > \pi_R$, $\Delta\pi > 0$, and $J_v$ is negative. This means that the flux is from the right to the left compartment. The flow across an extent of membrane is the flux times the area exposed to the driving forces, expressed as

$$Q_v = -A L_p \Delta\pi \qquad (19)$$

where $A$ is the area of the membrane and $Q_v$ is the flow in units of $cm^3 \cdot sec^{-1}$.

In the absence of solute, the volume flow across Pfeffer's artificial membrane was also linearly related to the hydrostatic pressure

$$J_v = L_p(P_L - P_R) = L_p \Delta P$$
$$Q_v = A L_p \Delta P \qquad (20)$$

In a study on collodion membranes, Meschia and Setnikar (1958) found that the proportionality constant for hydrostatic pressure-driven filtration was the same as the constant relating flow and osmotic pressure. That means the $L_p$ in Eq. 19 is the same $L_p$ in Eq. 20. Thus, not only can the osmotic flow be nulled by opposing osmotic pressure with an equal but opposite hydrostatic pressure, but the equivalent proportionality implies that the mechanism of volume flow is also identical for osmotic and hydraulic flow. The equivalence of osmotic and hydrostatic pressures allows us to write

$$Q_v = A L_p [(P_L - P_R) - (\pi_L - \pi_R)]$$
$$= A L_p [\Delta P - \Delta\pi] \qquad (21)$$

This equation describes the net flow that would be observed in the presence of both hydrostatic and osmotic pressure differences across a semipermeable membrane.

## G. Reflection Coefficient

van't Hoff's law, Eq. 1, describes the relation between osmotic pressure and concentration when a solution is separated from water by an ideal semipermeable membrane. Recall that a semipermeable membrane is defined as a membrane that is absolutely impermeable to the solute. Real membranes may not fit this ideal; they may be somewhat permeable to the solute. When membranes are permeable to the solute, the measured osmotic pressure is actually less than that predicted by van't Hoff's law. This phenomenon has led to the definition of a second membrane parameter, $\sigma$, the **reflection coefficient**, which may be defined as

$$\sigma = \frac{\pi_{observed}}{\pi_{theoretical}} = \frac{\pi_{observed}}{\phi_s R T C_s} \qquad (22)$$

The reflection coefficient derives its name from the notion that an absolutely impermeant solute will be totally reflected by the membrane. That is, 100% of collisions with the membrane will result in the solute being reflected back into the solution. The reflection coefficient for an ideal membrane would be 1.0. For a permeable solute, some fraction of the collisions with the membrane will result in permeation of the membrane, so that $\sigma < 1.0$, and the observed osmotic pressure will be less than that predicted by van't Hoff's Law. The value of $\sigma$ is not simply the fraction of collisions that penetrate the membrane. It has more to do with discrimination by the membrane between solvent and solute. Thus, $\sigma$ is a parameter that is different for every membrane–solute pair. A vapor-pressure osmometer or a freezing-point osmometer would still register the proper osmolarity of the solution, however (for nonvolatile solutes). A molecular interpretation of the origin of the reflection coefficient is given later.

Permeation of the solute should reduce osmotic flow along with osmotic pressure. In the presence of both hydrostatic pressure differences and concentration differences across a membrane, the resulting volume flow is given by

$$Q_v = A L_p \left[ (P_L - P_R) - \left( \sum_i \sigma_i \pi_{i,L} - \sum_i \sigma_i \pi_{i,R} \right) \right] \qquad (23)$$

where $\sigma_i$ is the reflection coefficient of solute i, and $\pi_{i,L}$ and $\pi_{i,R}$ are the osmotic pressures of solute i on the left and right sides of the membrane, respectively. The $\pi_i$ in this equation is that given by van't Hoff's law, as the correction to the observed osmotic pressure is accomplished by multiplying by the reflection coefficient, $\sigma$. The effect of a combination of hydrostatic and osmotic pressures on the flow across a membrane is shown in Fig. 7.

## III. Mechanisms of Osmosis

The ultimate cause of osmosis is the reduction of chemical potential of water in a solution. It must be emphasized again that this thermodynamic statement, and equations derived from it tell us nothing about the rates of osmosis or its mechanism. Several possible mechanisms have been investigated. As will be developed, classes of models can be distinguished by comparing the proportionality between applied force, either a pressure or concentration gradient, and water flow.

### A. Microporous Membranes

#### 1. Osmotic and Pressure-Driven Flow through Porous Membranes

The equivalence of $L_p$ for osmotic and pressure-driven flow suggests a common mechanism. At least three differ-

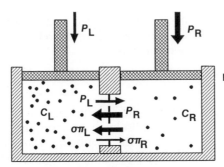

Hydrostatic and osmotic pressure:

$$Q_v = AL_p(\Delta P - \sigma\Delta\pi)$$

**FIG. 7.** Net flow in the presence of osmotic and hydrostatic pressures. For a real membrane, the effective osmotic pressure on the left, $\sigma\pi_L$, causes flow toward the left, while the applied hydrostatic pressure, $P_L$, drives flow to the right. A similar situation occurs on the right. The *net* flow is driven by the balance of the forces, $\Delta P - \sigma\Delta\pi$, and is proportional to the area, $A$, and the hydraulic conductivity, $L_p$.

ent models have been proposed to explain water, volume, and solute flow across membranes: (1) hydrodynamic flow through a porous membrane, (2) diffusion through the membrane, and (3) nonhydrodynamic flow through narrow pores. As we shall see, it is likely that biological membranes are not modeled well by any one of these. Despite this, we shall consider these model membranes because investigators have relied heavily on them to clarify their thinking.

First we consider a porous membrane as a model for understanding osmotic and hydrostatic pressure-driven flow and derive $L_p$, the proportionality constant relating pressure and flow. We assume the membrane is a flat, thin sheet of thickness $\delta$. We imagine that the membrane is pierced by right cylindrical pores of radius $r$, and the number of pores, $N$, per unit area is $n = N/A$. The membrane separates two compartments of water that are at different hydrostatic pressures. If we assume the pores are large enough for laminar flow to occur, then the filtration flow will be given by the Poiseuille equation

$$q_v = \frac{\pi r^4}{8\eta\delta}\Delta P \tag{24}$$

where $q_v$ is the flow per pore in $cm^3 \cdot sec^{-1}$, $\eta$ is the viscosity of the fluid, $\delta$ is the thickness of the membrane (equal to the length of the pore), and $\Delta P$ is the pressure difference across the pore. The $\pi$ in this equation is the geometric ratio, 3.14 . . . , and should not be confused with the symbol for the osmotic pressure. Since the flow through $N$ pores is just $nq_v$, the observed macroscopic flux and flow are

$$J_v = \frac{Nq_v}{A} = nq_v = \frac{n\pi r^4}{8\eta\delta}\Delta P$$
$$Q_v = AJ_v = Nq_v = \frac{N\pi r^4}{8\eta\delta}\Delta P \tag{25}$$

Recall here that $J_v$ is the flux, or flow per unit area of membrane, and $Q_v$ is the flow in units of volume per unit time.

Equation 25 describes the steady-state flow of water across the membrane. Because at steady state there is no buildup or depletion of water, there is no difference in the flow of water at any two points in the pore. Consequently,

pressure changes linearly with distance through the pore, and the gradient of pressure, $\Delta P/\delta$, is constant.

A comparison of Eq. 25 with Eq. 20 indicates that the hydraulic conductivity is

$$L_p = \frac{n\pi r^4}{8\eta\delta} \tag{26}$$

Thus, $L_p$ is a parameter determined by the viscosity of the fluid and by membrane characteristics, including the number of pores per unit area, $n$, the pore radius, and the membrane thickness.

### 2. Diffusional Permeability of Porous Membranes: $P_d$

In the absence of a pressure gradient, solute and water cross a porous membrane by diffusion through the pores. If we assume that the membrane is impermeant at all other points, the permeability is given by Fick's first and second law of diffusion

$$j_s = -D\frac{\partial C}{\partial x}$$
$$\frac{\partial C}{\partial t} = D\frac{\partial^2 C}{\partial x^2} \tag{27}$$

where $j_s$ is the flux of solute through one pore. We will use $J_s$ to signify the macroscopic flux of solute through the entire membrane. The second expression describes the time dependence of the concentration profile over distance, $x$, within the pore. At steady state, the concentration profile no longer changes. This means $\partial C/\partial t = \partial^2 C/\partial x^2 = 0$; the concentration gradient is linear, and $\partial C/\partial x = (C_L - C_R)/(0 - \delta) = -\Delta C/\delta$, where $\delta$ is the thickness of the membrane. The total solute flux across the membrane is given as

$$J_s = \frac{n\pi r^2 D}{\delta}\Delta C \tag{28}$$

where $n$ is the number of pores per unit area of membrane. According to this equation, the observed macroscopic flux of solute across a porous membrane is linearly related to the concentration difference by a coefficient that includes properties of the solute (diffusion coefficient) and the membrane (thickness, pore density, and pore cross-sectional area). The **permeability** of the membrane to solute, $p_s$, includes several parameters in Eq. 28 that are difficult to obtain experimentally; $p_s$ relates solute flux, $J_s$, to the difference in concentration across the membrane

$$J_s = p_s \, \Delta C \qquad (29)$$

From Eqs. 28 and 29, $p_s$ is defined as

$$p_s = \frac{n\pi r^2 D}{\delta} \qquad (30)$$

Isotopic water on one side of a porous membrane is distinguishable from ordinary water and may be viewed as a solute. Thus, water itself will obey these equations. This allows us to define the diffusional permeability of water, $P_d$, for a porous membrane

$$P_d = \frac{n\pi r^2 D_w}{\delta} \qquad (31)$$

where $D_w$ is the diffusion coefficient of water. The units of $P_d$ are cm sec$^{-1}$. Note that multiplication of a permeability by a concentration, as in Eq. 29, gives a flux with units of mol cm$^{-2}$ sec$^{-1}$.

### 3. Evidence for Pores: $P_f/P_d$ Ratio

In the absence of a concentration gradient, pressure-driven water flow gives rise to a second permeability constant termed the **filtration permeability** or **osmotic permeability**, $P_f$, which has units of cm sec$^{-1}$. Mauro (1957) realized that the proportionality constants relating pressure and concentration gradient-driven water flow, $P_f$ and $P_d$, provide evidence for the mechanism of transport. The ratio $P_f/P_d$ should be 1 if water crosses by a dissolution–diffusion process. Mauro (1957) recognized that the flux of water in response to a pressure gradient could be partitioned into two components, diffusional and nondiffusional (e.g., bulk flow), and that the diffusional component of water flux, $J_w$, would obey the Nernst–Planck equation

$$J_w = -\frac{n\pi r^2 D_w}{RT} C_w \frac{d\mu_w}{dx} \qquad (32)$$

where $C_w$ is the concentration of water. In the case where only a hydrostatic pressure is applied, $d\mu_w = \overline{V}_w \, dP$, and $C_w \overline{V}_w = 1$. Assuming steady-state flows and a uniform membrane, $dP/dx = -\Delta P/\delta$, and Eq. 32 becomes

$$J_w = \frac{n\pi r^2 D_w}{RT \delta} \Delta P \qquad (33)$$

This flow of water is in units of moles of water per second per cm$^2$ of membrane. It can be converted to units of volume per second per cm$^2$ (the units of $J_v$) by multiplying by the volume of water per mole, or $\overline{V}_w$

$$J_v = \overline{V}_w J_w = \frac{\overline{V}_w n\pi r^2 D_w}{RT \delta} \Delta P \qquad (34)$$

This equation relates the volume flux to the pressure difference across the membrane.

The total volume flux was earlier given as $J_v = L_p \, \Delta P$ (see Eq. 20). If diffusional flux is the only component of volume flux, Eqs. 20 and 34 may be combined to give

$$L_p = \frac{\overline{V}_w n\pi r^2 D_w}{RT \delta} \qquad (35)$$

Part of this expression for $L_p$ incorporates $P_d$. Insertion of Eq. 31 into Eq. 35 gives

$$P_f \equiv \frac{L_p RT}{\overline{V}_w} = \frac{n\pi r^2 D_w}{\delta} = P_d \qquad (36)$$

This definition of $P_f$ converts $L_p$ into a parameter having the same units as $P_d$, thereby allowing direct comparison of filtration and diffusional permeabilities. The equality of $P_f$ and $P_d$ obtained in Eq. 36 is dependent on the condition that the flow of water in response to a hydrostatic pressure difference is due only to diffusional processes. Thus, for a purely diffusional process, $P_f/P_d = 1$. In contrast, Mauro (1957) found that $P_f/P_d$ was 727 in collodion membranes. That is to say, pressure-driven water movement was much greater than would be expected from a diffusional process. From this he concluded that pressure-driven and osmotic flow across these membranes was predominately nondiffusional.

### 4. Physical Origin of Osmotic Pressure

If a porous membrane separates a solution containing only impermeant solutes from pure water, we observe experimentally that water flows through the membrane from the pure water to the solution side. The flow is proportional to the osmotic pressure of the solution times $L_p$ (see Eq. 20). The question is: What causes this water movement? Because the membrane is impermeable to solute, solute cannot enter the pores, and the fluid in the pore is pure water. Consider a water molecule in the middle of the pore. How does the water "know" to move toward the solution side? It appears there are only two possible answers to this question. Either there is a concentration gradient of water within the pore, or there is a pressure gradient within the pore. These two possibilities are not mutually exclusive, but diffusion-driven water flow and pressure-driven water flow are often thought of as separate mechanisms. The dichotomy reflects the notion that water is an incompressible fluid. Water is not absolutely incompressible, however. The coefficient of compressibility is given as

$$\beta = -\frac{1}{V}\left(\frac{\partial V}{\partial P}\right)_T \qquad (37)$$

and the value of $\beta$ for water is $4.53 \times 10^{-5}$ atm$^{-1}$. The equation for the volume of water is

$$V = V^0 (1 - \beta P) \qquad (38)$$

where $V^0$ is the volume at a standard temperature and pressure (1 atm), and $P$ is the pressure in excess of 1 atm. The coefficient of compressibility is virtually constant in the range $-500$ to $+1000$ atm. Equation 38 indicates that application of a negative pressure of 2.5 atm would expand a pure water solution by 0.01%, which corresponds to a

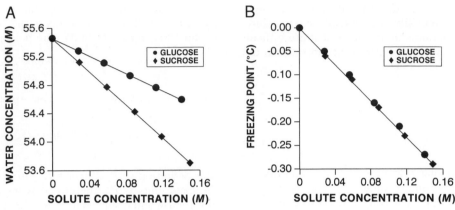

FIG. 8. Effect of solute concentration on water concentration and freezing point depression in glucose and sucrose solutions. (A) Because solute displaces water, the water concentration decreases with increasing solute concentration. Sucrose is nearly twice the size of glucose. Consequently, there is less water in a sucrose solution having the same molarity as a glucose solution. (B) The freezing-point depression, however, is dependent only on the solute concentration. It is the *mole fraction* of water that determines the colligative properties of solutions (the osmotic pressure, vapor-pressure depression, boiling-point elevation, and freezing-point depression). (Data from *The Handbook of Chemistry and Physics*, Chemical Rubber Company, Cleveland, OH, 1965.)

change in the concentration of water of about 5 m$M$. Looked at the other way, an expansion of water of only 0.01% would induce a negative pressure of 2.5 atm, equal to the osmotic pressure of a 0.1 molal solution at 37°C.

Dainty (1965) proposed a model in which he considered the density of water immediately within the pore opening on the solution side. As solute molecules cannot enter the pore, Dainty reasoned that the concentration of water within the pore must be higher than in the solution. Because of this difference in concentration, he argued that water would diffuse into the solution side faster than water could diffuse into the pore. The resulting net movement of water toward the solution side would lower the density of water in the pore, thereby creating a reduced pressure. Bulk movement of water down its pressure gradient would follow.

This explanation of the origin of osmotic pressure supposes that the driving force is actually water diffusing down its concentration gradient. The data in Fig. 8 show, however, that the concentration of water cannot be the major determinant of the colligative properties of solutions. In Fig. 8A, the water concentration in solutions of sucrose and glucose are plotted against the concentration of the solute. The water concentration is indeed decreased by dissolving solute, but sucrose, being almost twice as large as glucose, displaces almost twice as much solvent. As shown in Fig. 8B, however, the colligative properties, represented here by the freezing-point depression, depend *only* on the concentration of solute. Solutions with equal solute but different water concentrations have the same freezing point.

An alternative view of the physical origin of the osmotic pressure begins with the notion of pressure as a force divided by an area. The macroscopic concept of pressure relies on the averaging over time of the myriad of collisions that produce the pressure. By Newton's law, force is the time derivative of the momentum. An elastic collision of a solvent or solute molecule with the walls of the vessel results in a momentum change of $2mv$, where $m$ is the mass of the molecule and $v$ is its velocity, which contributes to the pressure against the vessel wall. At the entrance to the pore, however, solute molecules cannot transfer their momentum to the interior of the pore because they collide with the rim of the pore and are reflected back into the solution. Thus, the water molecules immediately inside the pore experience a momentum deficit that is equal to the component of pressure contributed by the solute molecules in the bulk phase.

Figure 9 shows the one-dimensional concentration profile of solute molecules near a pore opening. Because there

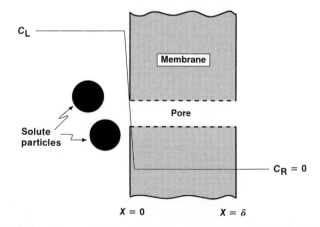

FIG. 9. Concentration profile in the vicinity of a pore in a microporous membrane. The membrane is impermeant at all places except the pores, where water may penetrate but solute particles (solid circles) are too large to enter. The concentration of solute in the bulk solution ($C_L$) must fall to zero upon entering the pore. The steep concentration gradient is accompanied by diffusion toward the pore that is balanced by reflection of solute by collision with the membrane.

is a steep solute gradient, there should be a diffusion of solute toward the pore opening. However, the actual steady-state flux of solute in this direction is zero because of the force exerted on the solute molecules by the membrane. The equation that describes the solute flux, $J_s$, in units of mol cm$^{-2}$ sec$^{-1}$, is

$$J_s = -D\frac{\partial C(x)}{\partial x} + \frac{D}{RT}fC(x) \tag{39}$$

where $C(x)$ is the concentration of solute at position $x$, $D$ is the solute diffusion coefficient in units of cm$^2$ sec$^{-1}$, $f$ is the force per solute molecule, and $R$ and $T$ have their usual meaning.

Villars and Benedek (1974) derived an equation for the drop in pressure immediately inside the pore on the solution side by setting the flux in Eq. 39 to zero and analyzing the net force on a plug of volume near the pore. Under steady-state conditions with zero $J_s$ through the pore, Eq. 39 gives

$$fC(x) = RT\frac{\partial C(x)}{\partial x} \tag{40}$$

where $f C(x)$ is the force per molecule times the number of molecules per unit volume, or the force per unit volume. Figure 10 shows a volume element near the opening of the pore on the solution side. We consider the forces acting on the element of fluid with an area $A$ from a point $x$ well within the bulk solution to a point $x + \Delta x$ just inside the pore. We assume that this element is in mechanical equilibrium: Although it may be moving, it is not accelerated or decelerated. The forces acting on the volume are contact forces on the edges of the volume and additional forces acting on the solute molecules alone to counteract the diffusive flux. At mechanical equilibrium the sum of the forces must be zero. This is written as

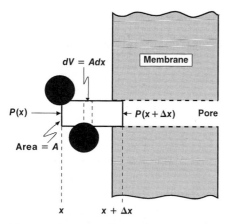

**FIG. 10.** Forces acting on a volume element immediately adjacent to a pore opening. The ideal, porous, semipermeable membrane separates pure water on the right from solution of impermeant solute on the left. The volume element has an area, $A$, equal to the cross-sectional area of the pore. The pressure in the bulk phase at position $x$ is $P(x)$. The pressure at position $x + \Delta x$, just within the pore, is $P(x + \Delta x)$. The net force on the element is the sum of the forces at both ends plus the forces acting only on the solute particles (solid circles) within the element.

$$F_s + F_c = 0 \tag{41}$$

where $F_s$ is the total force acting on the solutes in the volume, and $F_c$ is the net contact force due to the pressure from the adjacent volume elements. The net contact forces are the result of pressure acting over an area

$$F_c = A P(x) - A P(x + \Delta x) \tag{42}$$

The forces acting on the volume due to the solute particles is given by integrating Eq. 40:

$$F_s = \int_x^{x+\Delta x} fC(x)\, dV \tag{43}$$

Inserting the volume element $dV = A\,dx$ and $f\,C(x) = RT\,\partial C(x)/\partial x$ from Eq. 40, we obtain

$$\begin{aligned} F_s &= ART \int_x^{x+\Delta x}\frac{\partial C(x)}{\partial x}\,dx \\ &= ART\,[C(x+\Delta x) - C(x)] \end{aligned} \tag{44}$$

Since $C(x + \Delta x) = 0$ because solute particles are not in the pore, this becomes

$$F_s = -ART\,C(x) \tag{45}$$

where the negative sign indicates that $F_s$ is directed to the left. Inserting Eqs. 42 and 45 into Eq. 41, we have

$$A P(x) - A P(x + \Delta x) = ART\,C(x) \tag{46}$$

or

$$P(x + \Delta x) = P(x) - RT\,C_L \tag{47}$$

This last equation indicates that the pressure experienced by the volume of fluid immediately inside the pore is less than the bulk pressure $[P(x)]$ by the amount $RTC_L$, where $C_L$ is the concentration of impermeant solute in the solution on the left of the membrane. This analysis is consistent with the intuitive idea that water movement from the water side to the solution side of a semipermeable membrane must be due to a real force, which appears in this analysis to be due to the momentum deficit, and thus pressure deficit, within the pore on the solution side.

The theoretical concentration and pressure gradients for the model semipermeable porous membrane separating solution on the left from water on the right are shown in Fig. 11. When $P_L = P_R$, the pressure immediately inside the pore on the left compartment is $P'_L = P_L - RTC_L$, so there is a real pressure gradient along the pore driving flow to the left towards the solution side. When $P_L$ is increased by the addition of an external pressure, the pressure gradient will be less and flow to the left will be less. When the externally applied pressure is equal to $RTC_L$, there is no pressure gradient within the pore, and thus no flow (see Fig. 11C). This is the equilibrium condition that defines the osmotic pressure. If $P_L$ is increased more than $RTC_L$ (i.e., $P_L > P_R + RTC_L$), there will be pressure-driven flow to the right.

## 5. Physical Interpretation of the Reflection Coefficient, $\sigma$

The microporous semipermeable membrane presented previously distinguishes between solvent water and solute

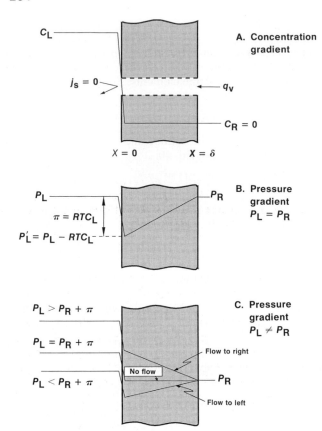

**FIG. 11.** Theoretical concentration and pressure profiles within a porous, semipermeable membrane in response to osmotic and hydrostatic pressure differences across the membrane. (A) The concentration profile shows the bulk solute concentration on the left ($C_L$) drops sharply upon entering the pore. The solute flux through the pore, $j_s$, is zero, and osmotic flow per pore, $q_v$, is from pure water on the right toward the solution. (B) The pressure profile corresponding to the situation in (A), where the pressure in the bulk phase is equal on the left and right compartments ($P_L = P_R$). Immediately upon entering the pore from the left, pressure drops by $RTC_L$. The pressure gradient in the pore is linear, with higher pressure on the right. Flow is down the pressure gradient, from right to left. (C) The pressure profile when $P_L \neq P_R$. When $P_L$ is increased by an amount less than $RTC_L$, the pressure gradient is reduced, but flow remains to the left. When $P_L$ is increased by exactly $RTC_L$, the pressure gradient within the pore disappears, and no flow occurs. This is the definition of osmotic pressure: the pressure that must be added to the solution side to stop water movement to the solution side. When $P_L$ is increased more than $RTC_L$, the pressure within the pore is higher on the left, and there is a net filtration flow of water to the right.

on the basis of pore and solute size. That is, the solute is too large to enter the pore and so cannot cross the membrane. If the pores were somewhat smaller or the solute molecules smaller, the solute could enter the pore, but with a lower probability than water because of the tight fit. In this case, rather than the solute being absolutely impermeant, the membrane would allow its slow passage. How does this affect the situation? Let us suppose that solute molecules that hit the rim of the pore before entry are reflected back into the bulk solution. This is shown diagrammatically in Fig. 12, looking down the axis of the pore

perpendicular to the surface of the membrane. The area of the pore that is accessible to solute is

$$A_s = \pi(r - a)^2 = \pi r^2 \left(1 - \frac{a}{r}\right)^2 \tag{48}$$

where $a$ is the radius of the solute molecule. Assuming that the radius of water molecules (0.75 Å) is negligible compared to the pore's radius, the ratio of areas available to solute and solvent water is

$$\frac{A_s}{A} = \frac{\pi r^2 \left(1 - \dfrac{a}{r}\right)^2}{\pi r^2} = \left(1 - \frac{a}{r}\right)^2 \tag{49}$$

The fraction of collisions of solute molecules with the pore opening that are reflected back, compared with those of water, is approximated by the ratio of the area of the gray annulus in Fig. 12 to the cross-sectional area of the pore. This is identified with the reflection coefficient

$$\sigma = \frac{A - A_s}{A} = 1 - \frac{A_s}{A}$$

$$\sigma = 1 - \left(1 - \frac{a}{r}\right)^2 \tag{50}$$

According to this view, the concentration of solute immediately within the pore would not be zero, as in the case when the solute was impermeant, but instead would be $(1 - \sigma)C$, and the excluded concentration would be $\sigma C$. Thus, the momentum deficit inside the pore would be due only to the excluded solute and would be equal to $\sigma RTC$, which is equal to $\sigma \pi$.

In addition to entrance effects, the concentration profile within the pore will be influenced by the combination of diffusion through the pore, solvent drag due to movement of fluid in response to pressure gradients, and interaction of the solutes with the nonlinear velocity profile within the pore. Laminar flow through long pores is characterized by a parabolic velocity profile, with a motionless layer of fluid adjacent to the pore walls and most rapid flow in the center. Large solute molecules will span several layers of velocity, thereby distorting the velocity profile and changing the solute molecule's velocity. Various equations have been derived to relax the assumption of negligible water radius and to relate the effective filtration area of the solute to the geometric radius of the pore (see Hobbie, 1978; Villars and Benedek, 1974; Renkin, 1954). Although models assuming hydrodynamic flow through pores have been useful, their applicability to osmotic flow across biological membranes remains an open question.

## B. Lipid Bilayer Membranes: The Dissolution–Diffusion Model

There are two types of lipid bilayer membranes that are useful models of membranes: the lipid vesicle and the planar bilayer membrane. The lipid vesicle is a small spherical shell of lipid usually produced by sonicating a dispersion of lipid in water. Planar bilayer membranes consist of a thin film of phospholipids formed over a small hole in a

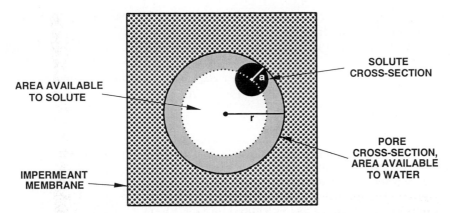

**FIG. 12.** A physical interpretation of the reflection coefficient, $\sigma$, for a solute and a porous membrane. The view is down the pore in a direction perpendicular to the membrane. The pore is modeled as a right circular cylinder of radius, $r$. The model assumes that any contact of the membrane with a solute particle of radius, $a$, (solid circle) will result in reflection of the particle back into the solution. The area available to solute is indicated in white. The total area of the pore, white plus gray annulus, is available for water movement.

partition between two aqueous compartments. The film is formed by "painting" the hole with a nonpolar solvent containing the lipids, which then spontaneously form a bilayer to fully separate the two solutions. This arrangement allows measurement of electrical and permeability properties of the bilayer. In the case of the liposome, the inner compartment is exceedingly small and experimentally inaccessible. In both cases, there are no components of the membranes other than the lipid.

### 1. Osmotic and Pressure-Driven Flow for the Dissolution–Diffusion Model: $P_f$

One model of water permeation through the lipid bilayer supposes that the water dissolves in the lipid phase and crosses the membrane by simple diffusion. Here the solution in contact with the membrane is one phase, while the hydrophobic core of the membrane is a second phase. Equilibrium of water in the solution phase with water in the membrane phase is described by equating the chemical potential of water in the two phases

$$\mu_w^0(\text{solution}) + RT \ln X_w(\text{solution}) + P\overline{V}_w(\text{solution})$$
$$= \mu_w^0(\text{membrane}) + RT \ln X_w(\text{membrane}) \tag{51}$$
$$+ P\overline{V}_w(\text{membrane})$$

where $X_w(\text{solution})$ and $X_w(\text{membrane})$ are the mole fractions of water in the solution in equilibrium with the membrane and in the membrane phase, respectively. The partition coefficient is defined as

$$K_w = \frac{X_w(\text{membrane})}{X_w(\text{solution})}$$
$$= \exp\left[\frac{(\mu_w^0(\text{solution}) - \mu_w^0(\text{membrane}) + \Delta P\overline{V}_w)}{RT}\right] \tag{52}$$

Consider the case where only osmotic pressure drives water flow and hydrostatic pressure across the membrane, $\Delta P$, is 0. Since water generally partitions poorly into hydro-

carbon solvents, we may assume that the mole fraction of water in the membrane phase is low. That is, the water concentration is dilute, and we may replace the mole fraction of water with its concentration

$$C_w(\text{membrane}) \simeq \frac{X_w(\text{membrane})}{\overline{V}_{\text{lipid}}} \tag{53}$$

where $\overline{V}_{\text{lipid}}$ is the partial molar volume of lipid in the membrane. If the concentration of water immediately "inside" the membrane is in equilibrium with the solution in contact with the membrane, we combine Eqs. 52 and 53 to get

$$C_w(\text{membrane}) = K_w \frac{X_w(\text{solution})}{\overline{V}_{\text{lipid}}} \tag{54}$$

For dilute solutions, this is approximated by

$$C_w(\text{membrane}) \simeq K_w \frac{(1 - \overline{V}_w C_s)}{\overline{V}_{\text{lipid}}} \tag{55}$$

where $C_s$ is the solute concentration. The concentration of water immediately inside the left side of the membrane, $C_{w,L}$, is given by Eq. 55, where $C_s$ is the concentration of solute in the solution on the left side of the membrane. A similar expression pertains for the concentration of water immediately inside the membrane on the right side ($C_{w,R}$). Given the concentrations of water at both faces of the membrane, its diffusion across the membrane is, from Fick's law,

$$J_w = -D_w^m \frac{(C_{w,L} - C_{w,R})}{(0 - \delta)} \tag{56}$$

where $D_w^m$ is the diffusion coefficient of water in the membrane phase. Substitution from Eq. 55 into Eq. 56 gives

$$J_w = -\frac{D_w^m K_w \overline{V}_w (C_{s,L} - C_{s,R})}{\overline{V}_{\text{lipid}} \delta}$$
$$= -\frac{D_w^m K_w \overline{V}_w \Delta C_s}{\overline{V}_{\text{lipid}} \delta} \tag{57}$$

The flux of water, $J_w$, is in units of moles of water per second per cm$^2$ of membrane. It is converted to units of volume flow by multiplying by $\overline{V}_w$, as in Eq. 34, to obtain

$$J_v = \overline{V}_w J_w = -\frac{D_w^m K_w \overline{V}_w^2 \Delta C_s}{\overline{V}_{lipid}\,\delta}$$

$$= -\frac{D_w^m K_w \overline{V}_w^2}{\overline{V}_{lipid} RT \delta} RT \Delta C_s \tag{58}$$

The last term on the right is the osmotic pressure difference, $\Delta\pi$. This equation relates the volume flux to the osmotic pressure difference when the mechanism of water flow is dissolution and diffusion. Comparison to the earlier description of osmotic flow by Eq. 18 allows us to identify $L_p$ as

$$L_p = \frac{D_w^m K_w \overline{V}_w^2}{\overline{V}_{lipid}\,\delta RT} \tag{59}$$

From the definition, $P_f = L_p RT/\overline{V}_w$, we get the expression for $P_f$ as

$$P_f = \frac{D_w^m K_w \overline{V}_w}{\overline{V}_{lipid}\,\delta} \tag{60}$$

Equation 58 was derived for an osmotic gradient ($\Delta C_s > 0$) in the absence of a hydrostatic gradient ($\Delta P = 0$). The expression relating volume flux and pressure when $\Delta C_s = 0$ can be derived by returning to Eq. 51 and setting the mole fraction of water equal on the two sides of the membrane, while the pressures differ. The result is that exactly the same $L_p$ is derived for pressure-driven flow and osmotic flow when the mechanism is by rapid dissolution of water followed by slow diffusion through the lipid membrane phase (Finkelstein, 1987).

## 2. Diffusional Water Permeability through Lipid Membranes: $P_d$

The permeability of lipid membranes to a diffusional water flux is expressed as

$$J_w = P_d \Delta C_w \tag{61}$$

where $P_d$ is the diffusional permeability and $\Delta C_w$ is the difference in water concentration across the membrane. The overall permeation of the membrane by water is a consequence of three steps: dissolution into the membrane phase at the left interface; diffusion across the membrane phase; and reversal of dissolution at the right interface. If we assume, as we did in the derivation of $P_f$, that the rate limiting step is diffusion through the membrane phase, then Eq. 61 may be written as

$$J_w = D_w^m \frac{\Delta C_w(\text{membrane})}{\delta} \tag{62}$$

From Eq. 54, this is

$$J_w = \frac{D_w^m K_w}{\overline{V}_{lipid}} \frac{\Delta X_w(\text{solution})}{\delta} \tag{63}$$

Because $\Delta X_w = \overline{V}_w \Delta C_w$, this becomes

$$J_w = \frac{D_w^m K_w \overline{V}_w}{\overline{V}_{lipid}\,\delta} \Delta C_w \tag{64}$$

$P_d$ can be identified by comparing Eqs. 64 and 61:

$$P_d = \frac{D_w^m K_w \overline{V}_w}{\overline{V}_{lipid}\,\delta} \tag{65}$$

## 3. $P_f/P_d$ Ratio for the Dissolution–Diffusion Model

The expressions for $P_f$ in Eq. 60 and $P_d$ in Eq. 65 derived for a lipid membrane under identical assumptions (equilibrium at the interfaces with relatively slow diffusion across the membrane), indicate that the $P_f/P_d$ ratio for diffusive flow of water across lipid membranes should be 1.0. Cass and Finkelstein (1967) measured the osmotic and diffusive permeability of planar lipid bilayers and found that, within experimentally uncertainty, the ratio was 1.0. The uncertainty arose mainly in the determination of $P_d$ because of the presence of unstirred layers adjacent to the planar lipid bilayer. These unstirred layers form an additional diffusional barrier that affects the experimental determination of $P_d$ much more than that of $P_f$. In the flux equations, permeability appears as a conductance relating a flux ($J_w$ or $J_s$) to a driving force ($\Delta C_w$ or $\Delta C_s$). Thus, the inverse of permeability is like a resistance. For a membrane in series with unstirred layers, the total resistance is the inverse of the observed permeability, which is the sum of the resistances offered by the individual barriers, the membrane, and the unstirred layers. We write this as

$$\frac{1}{P_{d_{obs}}} = \frac{1}{P_d} + \frac{1}{P_u} \tag{66}$$

From this equation, it is plain that if $P_u$, the combined permeability of the unstirred layers on both sides of the membrane, is less than infinity, the observed $P_d$ will be less than the actual $P_d$ of the membrane alone. Since diffusion of water through the unstirred layer is given by Fick's law, Eq. 66 may be rewritten as

$$\frac{1}{P_{d_{obs}}} = \frac{1}{P_d} + \frac{\delta_L + \delta_R}{D_w} \tag{67}$$

where $\delta_L$ and $\delta_R$ are the equivalent unstirred layer thickness on the left and right side of the membrane.

Unstirred layers also can affect measurement of the coefficients for pressure gradient-driven flow, $P_f$ and $L_p$, but the error introduced is much less than for concentration gradient-driven flow, $P_d$ (Barry and Diamond, 1984; Finkelstein, 1987) and was safely ignored in evaluating the $P_f/P_d$ ratio (e.g., Cass and Finkelstein, 1967). The osmotic flow sweeps solute towards the membrane on the side with lower osmolarity and away from the membrane on the other side. As long as convection is faster than diffusion, this diminishes the transmembrane osmotic gradient, reducing $J_v$ for the apparent $\Delta\pi$ and causing underestimation of $P_f$ and $L_p$. The observed $P_f$, $P_{f(observed)}$, is given as

$$P_{f(observed)} = P_{f(membrane)} \exp(-J_v\delta/D_s) \tag{68}$$

where $P_{f(\text{membrane})}$ is the true membrane parameter, $\delta$ is the unstirred layer thickness, $J_v$ is the volume flux, and $D_s$ is the diffusion coefficient of the osmolyte. The error can be minimized by determining $P_f$ with a small $\Delta\pi$ so that $J_v$ is small. Because $P_f$ and $L_p$ are proportional (Eq. 36), the errors are also proportional.

The value of $P_d$ and $P_f$ varies with the lipid composition of the membrane and ranges from $1 \times 10^{-5}$ to $5 \times 10^{-3}$ cm sec$^{-1}$ (Deamer and Bramhall, 1986). Cholesterol, which generally reduces the fluidity of lipid bilayers, reduces $P_d$ and $P_f$ progressively with increasing cholesterol content (Finkelstein and Cass, 1968). The *unidirectional* flux across a lipid membrane equals $P_d \times C_w$. Using a typical $P_d$ of $1 \times 10^{-3}$ cm sec$^{-1}$ and a $C_w$ of 55 mol/liter, the unidirectional flux across a membrane is $5.5 \times 10^{-5}$ mol cm$^{-2}$ sec$^{-1}$. By comparison, the unidirectional flux of water across a distance $\delta = 5$ nm (approximately the thickness of the lipid bilayer) can be calculated as $(D_w \times C_w)/\delta$. Using $3 \times 10^{-5}$ cm$^2$ sec$^{-1}$ for $D_w$, the unidirectional flux of water in water is 3.3 mol cm$^{-2}$ sec$^{-1}$. Thus, water flux through the membrane is about 60,000 times slower than that through water. Nevertheless, the water flux is still enormous. Taking 0.7 nm$^2$ as the average area of a typical phospholipid in the bilayer, the unidirectional water flux corresponds to about $2.2 \times 10^5$ water molecules passing each phospholipid molecule each second.

## C. Flow through Narrow Pores: $P_f/P_d$ Ratio

The equations derived earlier for $P_f$ for a porous membrane required the assumption that the pores were large enough to allow laminar flow as described by the Poiseuille equation. Suppose that the pores are so narrow that water passes through the pores in single file. It is clear that laminar flow cannot occur here, and the Poiseuille equation does not apply. As shown in Fig. 13, the narrow pore restricts free diffusion in the pore because diffusion of one water molecule from one position in the pore to the next requires its neighbor to move away to provide a vacancy. In this way, diffusion within the restricted geometry of the pore becomes a collective property of all of the molecules in the pore. The likelihood that a tracer molecule will diffuse all the way through the pore will depend on the number of water molecules in the pore, as movement of tracer water (solid circle) through the pore requires the movement of a vacancy ($\times$) all of the way through the pore.

Suppose that there are $N$ water molecules in a pore. We may assume that the length of the pore, $\delta$, is proportional to $N$, and that the water molecules reside in more or less specific positions separated by $\delta/N$. In the free diffusion of liquid water in the bulk solution, according to Fick's first law of diffusion, the flux of water is proportional to $1/\delta$ or $1/N$. In the case of a single-file diffusion through a narrow pore, the diffusion of a vacancy through the water within the pore looks exactly like the diffusion of water through a series of vacancies in free diffusion. Thus, the flux of the vacancy also is proportional to $1/N$. The movement of tracer from one position to the next in the pore requires the diffusion of a vacancy all the way through the

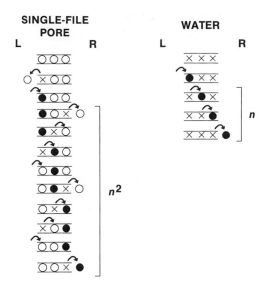

**FIG. 13.** Jumping model of diffusion of water through a narrow pore. In the model of free water diffusion (right), tracer water (solid circles) makes successive jumps from one vacancy ($\times$) to another. To move three places, it must make three jumps. In a narrow pore (left), tracer water cannot move to the next position in the pore unless a vacancy is present because unlabeled water (open circles) cannot move out of the way. Movement of tracer water from one site to the next in the pore requires a vacancy to diffuse all the way through the pore. Since the vacancy requires three steps to diffuse through the pore, tracer water diffusion through three steps in a narrow pore requires nine jumps. The result is that single-file diffusion becomes a collective property of the single file and is slower than free water diffusion.

pore, so that the flux of tracer one step is proportional to $1/N$. To diffuse all the way through the pore, the tracer must make $N$ such jumps. The undirectional flux of tracer over $N$ steps of equal flux is given by (Stein, 1976)

$$J_{0 \rightarrow N} = \frac{J_{ij}^N}{N J_{ij}^{N-1}} = \frac{1}{N} J_{ij} \qquad (69)$$

where $J_{0 \rightarrow N}$ is the unidirectional flux of tracer across $N$ steps (all the way through the membrane) and $J_{ij}$ is the unidirectional flux of tracer over one step. Since $J_{ij}$ requires diffusion of a vacancy all the way through the pore, it is proportional to $1/N$. The unidirectional diffusional flux of tracer through a single-file pore is thus proportional to $1/N^2$, rather than to $1/N$. Pressure-driven flow, on the other hand, remains inversely proportional to pore length. The theoretical analysis of both pressure-driven flow and diffusive flow through a single-file pore has led to the conclusion that the proportionality constants relating $P_f$ to $1/N$ and $P_d$ to $1/N^2$ are identical. The result, for a single-file pore, is

$$\frac{P_f}{P_d} = N \qquad (70)$$

where $N$ is the number of water "binding sites" within the pores (Finkelstein, 1987). A more recent theoretical analysis suggests that this result is true near equilibrium, but the ratio of osmotic to diffusive permeability may exceed $N$ for a membrane far from equilibrium (Hernandez and Fischbarg, 1992).

### D.  Mechanism of Water Transport across Lipid Bilayer Membranes

The previous discussion suggests that water transport across lipid bilayer membranes occurs by rapid dissolution at the membrane interface followed by diffusion across a hydrocarbon-like interior. Three observations strongly support this mechanism: (1) The $P_f/P_d$ ratio, after correction for unstirred layers, appears to be close to 1.0 (Cass and Finkelstein, 1967; Andreoli and Troutman, 1971); (2) insertion of pore-forming antibiotics such as nystatin, amphotericin B, or gramicidin A increases the $P_f/P_d$ ratio to small values of $N$ (between 3 and 5) (Holz and Finkelstein, 1970; Rosenberg and Finkelstein, 1978); and (3) the activation energy for water transport, typically 10–20 kcal mol$^{-1}$, is larger than the activation energy for free water diffusion, ~4 kcal mol$^{-1}$, which should apply if water moves through a pore (Solomon, 1972; Fettiplace and Haydon, 1980). However, experimental studies of water permeation across liposomes appear to give a different answer. The $P_f$ of liposomes, measured by turbidimetric methods, is not affected by the chain length of the lipids or the degree of saturation (Carruthers and Melchior, 1983; Jansen and Blume, 1995), whereas water solubility in hydrocarbons is affected by chain length (Schatzberg, 1963). Further, the activation energy of $P_f$ in liposomes is only 3.2 kcal mol$^{-1}$ (Carruthers and Melchior, 1983), similar to the activation energy for free water diffusion. The $P_f$ of liposomes is discontinuous near phase transitions of the lipid and $P_f/P_d$ ratios from 7 to 23 have been reported (Jansen and Blume, 1995). From these observations, it appears that liposomes do not behave as planar lipid bilayers, perhaps because of membrane defects induced by the marked curvature of the membrane in these small structures. The activation energy has been taken as diagnostic of whether water traverses the membrane through pores or by diffusion through the lipid itself (Finkelstein, 1987; Verkman, 1993). Low activation energies of water transport are associated with a high $P_f$ or $L_p$ in membranes containing water channels or pores, whereas high activation energies are associated with a low $P_f$ or $L_p$, indicating diffusional water transport in membranes without pores or when pores are blocked by mercurials (Table 2).

In the dissolution–diffusion mechanism, water encounters a minimum of three sequential barriers: dissolution on one side of the membrane, transport across the hydrocarbon-like interior, and then removal from the membrane on the opposite side. Assuming rapid equilibration at the interfacial regions is equivalent to assuming that the barriers there are insignificant compared to the barrier of diffusion, but there is no *a priori* reason to make this assumption. The alternative view proposes that the rate-limiting step is lateral movement of the phospholipid headgroups that creates a transient defect required for penetration of water into the interfacial region of the membrane (Trauble, 1971; Haines, 1994). Water transport through the hydrophobic core requires vacancies within the bilayer that form when hydrocarbon chains make *gauche–trans–gauche* kinks caused by the rotation of carbon–carbon bonds in the hydrocarbon tails. Kinks in the hydrocarbon tails propagate rapidly down acyl chains and provide sufficient space for water. Experimental support for this model comes from studies showing that addition of cardiolipin to phosphatidylcholine liposomes decreases $P_f$ without changing bilayer fluidity by stabilizing headgroup interactions (Shibata *et al.*, 1994). These two models make very different assumptions, and the detailed mechanism of water permeation through lipid bilayers remains uncertain.

## IV.  Water Movement across Cell Membranes

In the prior sections, we considered osmotic- or pressure-driven flow and diffusive flow of water across membranes that were characterized as (1) porous membranes with pores large enough to allow laminar flow; (2) lipid membranes with no pores, but allowing water permeation by a dissolution–diffusion mechanism; and (3) membranes containing narrow pores. We found expressions for the osmotic permeability, $P_f$, and the diffusive permeability, $P_d$, for each and found that the membranes could be distinguished in principle by the ratio $P_f/P_d$: Large values of $P_f/P_d$ indicate a porous membrane, $P_f/P_d = 1$ signifies a diffusive mechanism, and small values of $P_f/P_d > 1$ are characteristic of narrow pores. What are the permeabilities of real biological membranes, and what do these permeabilities tell us of the routes of water transport through membranes?

### A.  Rate of Water Exchange: Experimental Measure of $P_d$

Paganelli and Solomon (1957) measured the diffusional exchange of water across erythrocyte membranes by rapidly mixing a suspension of the cells with an isotonic buffer with added tracer $^3H_2O$. The mixture was forced down a tube and at various distances, corresponding to various times of exchange, samples of the extracellular water were obtained by filtration. Paganelli and Solomon found that the half-time for exchange of $^3H_2O$ was 4.2 msec at room temperature. This means that 90% of all of the water within an erythrocyte is exchanged with extracellular water every 14 msec. This is an extraordinarily rapid rate of exchange. Erythrocytes are sufficiently small that the presence of an unstirred layer within the cells does not appreciably affect the determination of $P_d$ [i.e., $(\delta_L + \delta_R)/D_w \ll 1/P_d$; see Eq. 67].

Diffusional exchange of water across erythrocyte membranes also can be measured by nuclear magnetic resonance (NMR) spectroscopy. Relaxation of the nuclear spin states of hydrogen is much slower inside than outside a cell when a relatively impermeant paramagnetic ion such as $Mn^{2+}$ is added to the extracellular solution. This allows calculation of $P_d$ because the relaxation of the spin states is then effectively limited by permeation through the mem-

**TABLE 2** Hydraulic Conductivity and its Apparent Activation Energy[a]

| Tissue | $L_p$ $10^{-12}cm^3dyn^{-1}sec^{-1}$ | $E_a$ kcal mol$^{-1}$ |
|---|---|---|
| **Water channels** | | |
| RBC, human | 18.0 | 3.9 |
| RBC, beef | 18.2 | 4.0 |
| RBC, dog | 23.0 | 4.3 |
| Prox. tubule BLM-V, rabbit | 21.9 | 2.5 |
| Liposomes + AQP1 | 30.8 | 3.1 |
| **Water channels + mercurials** | | |
| RBC, human + PCMBS | 1.3 | 11.6 |
| Prox. tubule BLM-V, rabbit + Hg | 4.4 | 8.2 |
| **Diffusional** | | |
| RBC, chicken | 0.6 | 11.4 |
| Intestinal brush border, rat | 0.9 | 9.8 |
| Ventricle, rabbit | 1.2 | 10.5 |
| Liposomes | 1.9 | 16.0 |
| PC bilayer | 1.6 | 13.0 |
| PC/Chol bilayer | 0.4 | 12.7 |
| Water self-diffusion | — | 4.2 |

[a] Hydraulic conductivity, $L_p$, and activation energy, $E_a$, are characteristic of the mechanism of water transport. High $L_p$ and low $E_a$ are typical of membranes containing functioning water channels, whereas low $L_p$ and high $E_a$ indicate diffusional water transport. RBC, Red blood cell; BLM-V, basolateral membrane vesicle; AQP1, aquaporin-1; PCMBS, *p*-chloromercuribenzene sulfonate; PC, phosphatidylcholine; Chol, cholesterol. For references, see Suleymanian and Baumgarten (1996).

brane. The values of $P_d$ for the erythrocyte determined by isotopic or NMR methods cluster around $4 \times 10^{-3}$ cm sec$^{-1}$ (Solomon, 1989).

### B. Rate of Osmotic Flow: Experimental Measure of $P_f$ and $L_p$

According to Eq. 23, the hydraulic conductivity, $L_p$, can be determined experimentally as

$$L_p = \left(\frac{-Q_v}{A\,\sigma\,\Delta\pi}\right)_{\Delta P=0} \quad (71)$$

The osmotic permeability, $P_f$, can then be calculated as $P_f = L_p RT/\overline{V}_w$ (Eq. 36). The experimentally determined values reflect the rate of water flow, $Q_v$, in the presence of a known osmotic pressure difference, $\Delta\pi$, produced by a solution with a known reflection coefficient, $\sigma \approx 1.0$, over a known surface area, $A$.

The rate of water movement into or out of erythrocytes in response to mixing with hypertonic or hypotonic media has been measured by using light scattering as an index of erythrocyte volume. The experiments are similar to those used to determine $P_d$: A suspension of erythrocytes is mixed with media of defined osmolality and the mixture then flows down a tube passing through an observation cell. Light scattering is monitored at known distances down the tube, and cell volume changes are calculated from changes in light scattering. This and other methods give values for $L_p$ that cluster around $1.8 \times 10^{-11}$ cm$^3$ dyn$^{-1}$ sec$^{-1}$

(Solomon, 1989). Using $R = 8.314 \times 10^7$ dyn cm mol$^{-1}$ K$^{-1}$, $\overline{V}_w = 18$ cm$^3$ mol$^{-1}$, and $T = 298$ K, this average value of $L_p$ corresponds to a $P_f$ of about $2.5 \times 10^{-2}$ cm sec$^{-1}$.

### C. Water Channels in Biological Membranes

The ratio of $P_f/P_d$ for the erythrocyte membrane described earlier is about 5. Although there is some uncertainty in this ratio, it is clearly in excess of the value of 1.0 predicted for a diffusive mechanism. This suggests that there are pores in the erythrocyte membrane. The actual value of $P_f/P_d$ for the pore alone cannot be obtained from just this information, however, because the erythrocyte membrane is actually a mosaic of lipid bilayer and pores, and water moves through both in parallel. That is, $P_f$(observed) = $P_f$(pore) + $P_f$(bilayer) and $P_d$(observed) = $P_d$(pore) + $P_d$(bilayer). Experiments that block the pore (see later discussion) suggest that the $P_f/P_d$ ratio for the pore alone is about 10 and that 90% of water flux is via the pore, whereas 10% crosses the lipid bilayer (Macey, 1979; Finkelstein, 1987).

Two distinct means of inhibiting water transport provide additional evidence for pores and are consistent with the idea that the pores are made of a protein. Mercurial sulfhydryl reagents, such as $HgCl_2$, *p*-chloromercuribenzoate (PCMB), and *p*-chloromercuribenzene sulfonate (PCMBS), decrease the erythrocyte $P_f$ by a factor of 10 and $P_d$ by less than a factor of 2, and the osmotic and diffusional permeabilities become equal (Macey and Farmer, 1970; Macey et al., 1972). Concurrently, the activation energy for permeation is increased from about 4 kcal mol expected for water-filled pores to >10 kcal mol, a value typical of artificial lipid bilayers (see Table 2). Mercurials inhibit water transport primarily by targeting protein SH groups because their effect is fully and rapidly reversed by cysteine. The second inhibitor is radiation. High doses of radiation inhibit water transport in both erythrocytes (van Hoek *et al.*, 1992) and renal brush-border membrane vesicles (van Hoek *et al.*, 1991). The characteristics of radiation inactivation suggest that the target is the size of a 30-kDa protein. Different characteristics would be expected if the entire membrane or transient defects served as the major water pathway.

Although these studies and others, especially in epithelia (e.g., Verkman, 1993), made it clear that water pores or channels must exist, until recently their identity and characteristics remained mysterious. Are what we call "water channels" simply water moving through open ion channels or through an ion exchanger or ion cotransporter? Are water channels specific, admitting water but excluding ions? The application of molecular biological approaches since 1991 has resulted in enormous progress on these issues with the cloning and expression of several members of a family of water channels called the aquaporins (for reviews, see Agre *et al.*, 1993; King and Agre, 1996; Verkman *et al.*, 1996).

### 1. Aquaporins

Water channels are related to the membrane integral protein (MIP) family (20–40% homology). The first water channel to be identified was cloned from a human bone marrow library by Preston and Agre (1991) based on the sequence of a purified protein of uncertain function. Initially called CHIP28 (channel-forming integral protein, 28 kDa), this protein was later redesignated aquaporin-1 (AQP1).

The cloned protein exhibits all of the characteristics of water channels. AQP1 expressed in *Xenopus* oocytes induces up to 30-fold increases in $P_f$ that can be blocked by $HgCl_2$ (Preston *et al.*, 1992). Reconstitution of purified AQP1 protein in liposomes verified that AQP1 itself, rather than modulation of an endogenous oocyte membrane protein, was responsible for mercurial-sensitive water permeation (Zeidel *et al.*, 1992). Furthermore, incorporation of AQP1 into liposomes reduced the activation energy of $P_f$ from 16.1 kcal $mol^{-1}$, characteristic of permeation through the bilayer, to 3.1 kcal $mol^{-1}$, characteristic of water passing through water-filled pores. A closely related channel exhibiting 94% sequence homology with the human erythrocyte channel was cloned from rat renal cortex and expressed (Zhang *et al.*, 1993). Renal cortex contains the proximal tubule, which is highly water permeant.

There are about $2 \times 10^5$ AQP1 molecules per erythrocyte (Zeidel *et al.*, 1992). Taking the erythrocyte's membrane area as $1.35 \times 10^{-6}$ $cm^2$ (Solomon, 1989), this corresponds to a density of about $1.5 \times 10^{11}$ $AQP1/cm^2$ or 1500 $AQP1/\mu m^2$. Despite this remarkable density of pores, the water permeability of a human erythrocyte is only ~10 to 50 times greater than that of a phosphatidylcholine/

cholesterol bilayer (Fettiplace and Haydon, 1980). Thus, 30–150 aquaporin channels are needed to equal the permeability of 1 $\mu m^2$ of bilayer.

AQP1 contains 269 amino acid and is thought to have six membrane-spanning domains based on an analysis of hydrophilicity (Preston and Agre, 1991) and selective proteolysis of protein loops that face the intra- or extracellular side (Preston *et al.*, 1994). The postulated topology for the AQP1 and the other members of the AQP family is shown in Fig. 14. Mercurial-inhibition, *N*-glycosylation, and PKA phosphorylation sites have been identified. There is an internal homology between the halves of AQP designated repeat-1 and -2. The greatest homology among AQPs is in the segments surrounding asparagine–proline–alanine (NPA) motifs in loops B and E. These loops are thought to be arranged antiparallel and are postulated to dip back into the membrane to form an hourglass-shaped pore represented in the cartoon in Fig. 14 (Jung *et al.*, 1994). The biochemical behavior of solubilized erythrocyte AQP1 suggested that it is a noncovalently linked tetrameric structure, and electron microscopy of negatively stained AQP1 confirmed this (Walz *et al.*, 1994). However, coexpression of Hg-sensitive and -insensitive recombinant proteins indicates that each AQP monomer forms an independent functioning pore (Preston *et al.*, 1993). X-ray diffraction patterns at 3.5-Å resolution have been obtained from two-dimensional crystalline AQP1 arrays (Jap and Li, 1995). Electron-density contour maps confirmed a tetrameric arrangement of monomers, and each monomer appeared as a central low-density core, presumably the permeation pathway, surrounded by six (or seven) high-density regions believed to represent membrane-spanning $\alpha$-helices.

Five main mammalian aquaporins, AQP1–AQP5, are known, and additional related proteins have been found in amphibians, *Drosophila*, plants, *E. coli*, and yeast, but not all of these conduct water (Agre *et al.*, 1993; King and Agre, 1996; Verkman *et al.*, 1996). Table 3 lists some of the characteristics of the mammalian family of AQPs. Although water permeates most AQPs, there is disagreement regarding AQP3, which is permeable to glycerol and perhaps to water and urea. All AQPs except AQP4 are blocked by mercurials.

Immunohistochemistry, Western blot (protein determination), and Northern and *in situ* hybridization and RNase protection assays (mRNA determinations) have identified broad but only partially overlapping distributions for four of the AQP homologues in regions where water permeability is high (Hasegawa *et al.*, 1993; Nielsen *et al.*, 1993; Zhang *et al.*, 1993; Umenishi *et al.*, 1996). For example, AQP1 is located in erythrocytes, renal proximal tubules, and the descending thin limb of the loop of Henle (but not in the collecting duct, where water permeability is controlled by vasopressin); the choroid plexus, the iris, ciliary and lens epithelia, and corneal endothelium of the eye; lung alveolar capillaries and epithelium; red splenic pulp (with erythrocyte precursors); colonic crypt epithelium; and nonfenestrated capillary and lymphatic endothelium in a number of organs, including cardiac, skeletal, and smooth muscle. The exception with regard to distribution is AQP2, which has been identified in only one location. AQP2

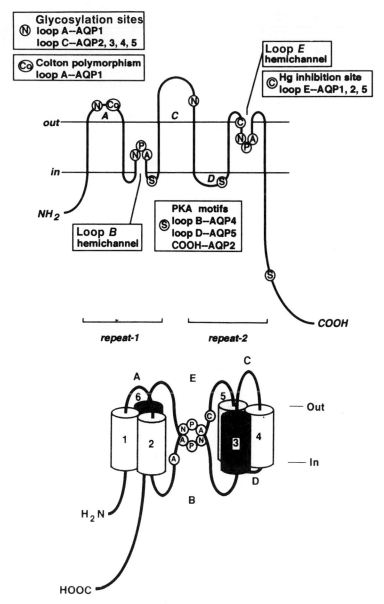

**FIG. 14.** AQP family portrait. Cartoon illustrating the postulated topology of AQP as a structure with six membrane spanning α-helices (top). Cysteine (C) responsible for inhibition by mercurials and consensus PKA phosphorylation (S) and glycosylation (N) sites also are shown. Loops B and E are envisioned to dip into the membrane with the two NPA motifs forming the narrow neck of an hourglass-shaped, water-filled pore (bottom). Single-letter codes are standard amino acid abbreviations. The position of the Colton polymorphic blood group antigen (Co) is noted. [Reproduced with permission from King and Agre (1966) from the *Annual Review of Physiology,* Volume 58, © 1996, by Annual Reviews Inc., and from Jung *et al.* (1994), with permission.]

underlies the vasopressin-regulated water permeation pathway of the renal collecting duct apical membrane (Nielsen *et al.*, 1995).

Several cautions are warranted when drawing physiological interpretations from the localization of AQP. For example, AQP1 mRNA is higher in cardiac homogenate than in homogenates from any other organ (Umenishi *et al.*, 1996), but functionally, the story is different. $L_p$ in isolated myocytes is very low, and its activation energy is high, about 10 kcal mol$^{-1}$ (Suleymanian and Baumgarten, 1996) (see Table 2). Thus, channels do not significantly contribute to water transport across the myocyte membrane.

Organs comprise cells with diverse functions and often different requirements for water transport. Because of its localization in vascular tissue, AQP detected in homogenates should not be attributed to the principal cells of the organ without confirmation. Even careful immunohistochemistry and *in situ* hybridization may lead the physiologist astray. The problem is the high density of AQP necessary to significantly affect $P_f$. If, for example, the density of AQP1 in the erythrocyte membrane was 10 rather than 1500 AQP1/$\mu$m$^2$, AQP1 still would be detected by modern techniques, but it would make a physiologically insignificant contribution to $P_f$.

**TABLE 3**   Characteristics of Mammalian Aquaporins[a]

|  | hAQP1 | hAQP2 | rAQP3 | hAQP4 | hAQP5 |
|---|---|---|---|---|---|
| Synonym | CHIP28 | WCH-CD | GLIP | MIWC |  |
| No. amino acids | 269 | 271 | 285–292 | 341 | 265 |
| Selectivity | water | water | glycerol ±urea, water | water | water |
| Hg-sensitive | + | + | + | − | + |
| N-Glycosylated | + | + | + | − | + |
| PKA site | − | + | − | + | + |
| Vasopressin-stimulated |  | + |  |  |  |
| Location | RBC<br>Kidney<br>Eye<br>Brain<br>Lung<br>Colon<br>Vasc. endothelium, nonfenestrated | Kidney:<br>Coll. duct,<br>apical | Kidney<br>Coll. duct,<br>basolat.<br>Brain<br>Trachea<br>Colon | Brain<br>Lung<br>Kidney<br>Coll. duct,<br>basolat. | Glandular epithelium |

[a] Abbreviations: h, Human; r, rat; AQP, aquaporin; CHIP28, channel-forming integral protein, 28 kDa; WCH-CD, water channel-collecting duct; GLIP, glycerol intrinsic protein; MIWC, mercurial-insensitive water channel. For additional details of tissue distribution and references for AQP characteristics, see King and Agre (1996) and Verkman et al. (1996).

AQP1 appears to be active and selective for water without requiring any modification or cofactors (Preston et al., 1992; Zhang et al., 1993). Voltage-clamp studies showed that AQP1 channels expressed in oocytes do not pass an ionic current. However, evidence for protein kinase A (PKA)-dependent regulation has been reported (Yool et al., 1996). Forskolin, which activates adenylate cyclase and thereby PKA, increases the $P_f$ of AQP1 cDNA-injected oocytes by >60%. Moreover, forskolin, 8-bromo-cyclic AMP, and the catalytic subunit of PKA induce a large, poorly selective cation conductance. Both the increase in $P_f$ and the current are partially suppressed by low concentrations of $HgCl_2$ (20–100 $\mu M$). These results are particularly surprising because AQP1 does not contain a consensus PKA phosphorylation sequence. Nevertheless, an AQP1 COOH-terminal fusion protein was found to be $^{32}P$-labeled by PKA, although indirect phosphorylation by another protein was not excluded. The physiological roles both of the PKA-dependent regulation of AQP1 and of the large conductance induced by PKA remain to be established. [The results of Yool et al. (1996) have been challenged in a series of letters; see Science **275**, 1490–1492, 1997.]

### 2. Other Channels and Transporters

The identification of specific water channels does not exclude the possibility that water flux through ion channels or transporters significantly contributes to the water permeability of the membrane. One interesting example is the cystic fibrosis transmembrane conductance regulator (CFTR). CFTR functions as a cAMP-regulated $Cl^-$ channel. Hasegawa et al. (1992) have found that CFTR expressed in oocytes also acts as a water channel with an estimated single-channel $P_f$ comparable to that of AQP1. Both $P_f$ and $Cl^-$ conductance were increased by cAMP. The effect of the CFTR channel on $P_f$ is large but not unique. Pore-forming antibiotics such as gramicidin, nystatin, and amphotericin are reported to induce a more modest $P_f$. Transporters also may contribute to $P_f$. Fischbarg et al. (1990) found that expression of $Na^+$-independent glucose transporters in oocytes increases water permeability, and Solomon et al. (1983) suggested that some of the water permeability of erythrocytes was contributed by the $Cl^-$–$HCO_3^-$ exchanger. However, water flux mediated by these transporters does not fully account for the macroscopic properties of water transport.

## V. Regulation of Cell Volume under Isosmotic Conditions

### A. Gibbs–Donnan Equilibrium

Because ions exert an effective osmotic pressure, the distribution of ions affects water flow across the cell membrane and thus cell volume. The starting point for understanding these effects is the theoretical ideas of Gibbs that were first demonstrated experimentally by Donnan. The phenomenon is now called **Gibbs–Donnan equilibrium** or simply **Donnan equilibrium**. Macknight and Leaf (1977) elegantly describe the history of how these ideas were applied to cell volume regulation, and Overbeek (1956) provides a detailed derivation and considers nonideal solution behavior.

A Donnan system is illustrated in Fig. 15. The membrane permits the movement of small charged solutes (e.g., $K^+$ and $Cl^-$) between the two compartments but restricts the

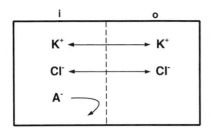

**FIG. 15.** A Donnan equilibrium system. Compartment i represents the intracellular space and contains $K^+$, $Cl^-$ and impermeant anions, $A^-$. Compartment o represents the extracellular space and contains $K^+$ and $Cl^-$. $K^+$ and $Cl^-$ freely cross between the compartments, but $A^-$ is restricted to comparment i. Because of the requirement for macroscopic electroneutrality, $[K^+]_o = [Cl^-]_o$ and $[K^+]_i = [Cl^-]_i + [A^-]_i$. At equilibrium, $\mu_{Ki} = \mu_{Ko}$ and $\mu_{Cli} = \mu_{Clo}$, where $\mu_i$ is the electrochemical potential $K^+$ and $Cl^-$ in compartments i and o, respectively.

movement of large charged species such as proteins, which are usually polyvalent and negatively charged at intracellular pH. Compartment o contains just $K^+$ and $Cl^-$ and for cells represents the extracellular space, and compartment i contains $K^+$, $Cl^-$, and impermeant anionic proteins, $A^-$, as found in the cytoplasm. It is the inability of one (or more) charged species to distribute freely between the compartments that profoundly influences the distribution of the mobile ions and, consequently, water. For a cell, it is the cell membrane that restricts the movement of large ions. A membrane is not required to establish a Donnan equilibrium, however. All that is needed is means of restricting one (or more) of the charged species to a single compartment. Donnan equilibria can arise in gels consisting of charged structural components (e.g., an ion-exchange column) because the charges on the gels are fixed in place. Donnan equilibria also can arise in cells after their membranes have been removed, leaving a cytoplasmic "gel" consisting of interconnected cellular proteins. A Donnan system represents a true equilibrium, although as we will see, an equilibrium is not attained under the usual biological conditions.

There are three important characteristics in a Donnan equilibrium system: (1) an unequal distribution of ions, (2) a potential difference between the compartments, and (3) an osmotic pressure. These characteristic can be understood by application of a basic thermodynamic principle to the system. Species that can cross the membrane, including both mobile ions and water, will distribute themselves so that their electrochemical potential ($\mu_i$) is the same in both compartments.

To understand how a Donnan equilibrium develops, imagine starting an experiment by "filling" the cell and the extracellular space with solutions of concentrations indicated in Fig. 16A. Since ions move down their electrochemical gradients, the first question is: What are the electrochemical gradients? Initially there is no potential difference across the membrane because both internal and external solutions start off with an equal number of positive and negative charges. There is no concentration gradient for $K^+$ because the amount of $K^+$ is the same on both sides of the membrane. Because of the presence of impermeant

anions inside the cell, however, $[Cl^-]_o$ must be greater than $[Cl^-]_i$. Consequently, $Cl^-$ will enter the cell moving down its electrochemical gradient, which initially is just the concentration gradient. Inward movement of $Cl^-$ makes the inside of the cell negative with respect to the outside. The developing inside negative potential has two consequences. First, it slows the rate of further influx of $Cl^-$. Second, the electrical gradient causes the accumulation of $K^+$ inside the cell. As $K^+$ moves down its electrochemical gradient in the direction set by the potential gradient, an opposing concentration gradient is created. Finally, an equilibrium is reached when the electrical and concentration gradients for both $K^+$ and $Cl^-$ are equal in magnitude and opposite in direction. This satisfies the requirement that $\mu_K$ and $\mu_{Cl}$, the electrochemical potential for $K^+$ and $Cl^-$, respectively, are the same in both compartments.

During the evolution of the equilibrium, the influx of $K^+$ and $Cl^-$ are equal on a macroscopic scale and are said to be coupled by the requirement for macroscopic electroneutrality. This is simply a shorthand for the idea that any difference in rates of anion and cation transmembrane flux (a separation of charge) gives rise to a potential that equalizes the fluxes (e.g., if the influx of $K^+$ was greater than that of $Cl^-$, an inside positive potential would develop, slowing $K^+$ influx and accelerating $Cl^-$ influx until the rates

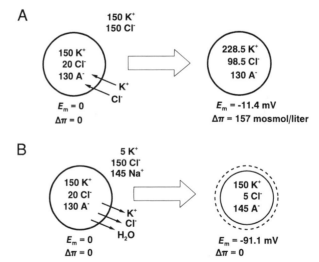

**FIG. 16.** Development of a Donnan equilibrium. (A) Cell bathed in 150 m$M$ KCl is "filled" with 150 m$M$ $K^+$, 20 m$M$ $Cl^-$, and 130 m$M$ $A^-$. Initially, there is no potential or osmotic gradient (left). $K^+$ and $Cl^-$ enters the cell until $[K^+]_o [Cl^-]_o = [K^+]_i [Cl^-]_i$, and the ionic fluxes establish a potential and osmotic gradient (right). An equilibrium is achieved only if the membrane is rigid and cell volume is constant. With a biological membrane, however, water enters the cell, cell volume increases, and the $[K^+]_i [Cl^-]_i$ decreases. This causes further entry of solute and solvent, and the cycle repeats until the cell bursts. (B) Double-Donnan system. Part of the extracellular $K^+$ is replaced by $Na^+$, which is assumed to be impermeant. With impermeant ions on both sides of a membrane (left), equilibrium is attained (right) without an osmotic pressure gradient. In this case, $K^+$, $Cl^-$ and water leave the cell, until at equilibrium $[K^+]_o [Cl^-]_o = [K^+]_i [Cl^-]_i$. The amount of $A^-$ in the cell is fixed, and thus the decrease in cell volume increases $[A^-]_i$.

matched exactly). In the example in Fig. 16A, both $[K^+]_i$ and $[Cl^-]_i$ increased by ~78.5 m$M$. The total entry of $K^+$ and $Cl^-$ cannot be precisely equal, however, because establishing a potential difference implies that a separation of charge, albeit quite small, must have taken place. If this were a spherical cell with a radius of 20 $\mu$m, an inequality of the $K^+$ and $Cl^-$ fluxes of less than 1 ion per 100,000 would cause a potential difference of 100 mV.[1]

The basis for Donnan equilibrium can be expressed in terms of the electrochemical potentials[2] of ions that cross the membrane, here $K^+$ and $Cl^-$. For ideal solutions, $\mu$ is given by

$$\mu_{Ko} = \mu^0 + RT \ln [K^+]_o + z_K F \psi_o$$
$$\mu_{Ki} = \mu^0 + RT \ln [K^+]_i + z_K F \psi_i \qquad (72)$$

and

$$\mu_{Clo} = \mu^0 + RT \ln [Cl^-]_o + z_{Cl} F \psi_o$$
$$\mu_{Cli} = \mu^0 + RT \ln [Cl^-]_i + z_{Cl} F \psi_i$$

where the subscripts o and i represent two compartments such as the outside and inside of a cell, $\mu^0$ is the chemical

---

[1] The amount of charge, $q$, necessary to establish a potential difference, $E_m$, is related to the specific capacitance of the membrane, $C_m$, and its area, $A$:

$$q = C_m A E_m$$

Assuming a specific membrane capacitance of 1 $\mu$F/cm$^2$, as is typical of most biological membranes, the amount of charge necessary to develop a potential of 100 mV in a spherical cell with a radius of 20 $\mu$m is

$$q = (1 \times 10^{-6} \text{ F/cm}^2)[4\pi(20 \times 10^{-4} \text{ cm})^2](0.1 \text{ V})\left(\frac{\text{C/V}}{\text{F}}\right)$$

$$q = 5.0 \times 10^{-12} \text{ C}$$

this can be converted to a change in the concentration, $\Delta C$, of ions in the cell:

$$\Delta C = (q/F)/\vartheta$$

where $F$ is Faraday's constant, and $\vartheta$ is used for the volume of the cell here to distinguish it from voltage. For the same 20-$\mu$-radius cell:

$$\Delta C = (5.0 \times 10^{-12} \text{ C}/96,500 \text{ C/mol})/\left[4/3\,\pi(20 \times 10^{-4} \text{ cm})^3\left(\frac{1\,\text{liter}}{100\,\text{cm}^3}\right)\right]$$

$$\Delta C = 1.5 \times 10^{-6} \text{ mol/liter}$$

Thus, a potential gradient of 100 mV would develop if ion influx were to increased $[K^+]_i$ by only 1.5 $\mu M$ more than it increased $[Cl^-]_i$.

[2] Formally, as presented previously in Eq. (3) for the chemical potential of a neutral species, $\mu_i$ should be defined in terms of the mole fraction rather than the concentration of a component and an additional term, $P\overline{V}$, representing the pressure times the partial molar volume, should be added. For dilute solutions, the mole fraction of a solute closely approximates its concentration (see Eqs. 5 and 9). The $P\overline{V}$ terms are ignored in the derivation of the ion distribution for a Donnan equilibrium because to simplify we initially assume the membrane is rigid (see Fig. 16A), and then we consider a situation without a pressure gradient where the $P\overline{V}$ terms cancel (see Fig. 16B). Derivations of Donnan equilibrium retaining the mole fraction and the pressure–volume terms can be found in Overbeek (1956) and Lakshminarayanaiah (1984).

---

potential of the standard state, $\psi$ is the potential of the compartment, $z_i$ is the valence of species i, and $R$, $T$, and $F$ have their usual meanings. At equilibrium, each permeant species distributes so that $\mu$ is identical inside and outside the cell. Equating the expressions for $\mu$ and simplifying give

$$RT \ln [K^+]_o + z_K F \psi_o = RT \ln [K^+]_i + z_K F \psi_i$$
$$RT \ln [Cl^-]_o + z_{Cl} F \psi_o = RT \ln [Cl^-]_i + z_{Cl} F \psi_i \qquad (73)$$

Membrane potential, $E_m$, is measured as $\psi_i - \psi_o$. Rearranging and substituting in for $E_m$ and $z$ gives

$$E_m = E_K = -RT/F \ln [K^+]_i/[K^+]_o$$
$$E_m = E_{Cl} = -RT/F \ln [Cl^-]_o/[Cl^-]_i \qquad (74)$$

These expressions are the **Nernst equilibrium potentials** for $K^+$ ($E_K$) and $Cl^-$ ($E_{Cl}$). Thus, when the mobile ions attain their equilibrium distribution in a Donnan system, the potential between the compartments, $E_m$, simultaneously equals both $E_K$ and $E_{Cl}$. Equating the Nernst potentials and simplifying gives the ratio of intra- and extracellular ions predicted by Donnan equilibrium:

$$[K^+]_i/[K^+]_o = [Cl^-]_o/[Cl^-]_i \qquad (75)$$

or, expressed another way,

$$[K^+]_i [Cl^-]_i = [K^+]_o [Cl^-]_o \qquad (76)$$

In a Donnan equilibrium, the KCl product inside a cell equals the KCl product outside.

This simple rule implies that increasing extracellular $K^+$ (e.g., by replacing $Na^+$) will cause a cell obeying Donnan equilibrium to take up $K^+$ and $Cl^-$ and swell. Ion movements in a number of tissue appear to follow Donnan equilibrium, at least for some conditions. Boyle and Conway (1941) made careful measurements of $[K^+]_i$, $[Cl^-]_i$, and cell water in frog sartorius muscles while varying extracellular KCl, and some of their results are contained in Table 4. For $[K^+]_o$ greater than 6 m$M$, the experimental ratio of the KCl products, $([K^+]_o [Cl^-]_o)/([K^+]_i [Cl^-]_i)$, was very nearly 1.0, as predicted by Eq. 76. The deviation at low $[K^+]_o$ occurs because $Na^+$ influx causes $E_m$ to deviate from $E_K$. Accompanying changes in cell water in these experiments and in separate experiments in which $K^+$ replaced $Na^+$ also agreed with theory.

The expectations for a Donnan equilibrium are calculated for the system that was shown in Fig. 16A. When the intracellular and extracellular KCl products are equal, $E_m = E_K = E_{Cl} = -11.4$ mV. In addition, because the number of ions inside the cell is much greater than that outside, the Donnan system has established a significant osmotic gradient. Assuming ideal behavior, the osmolarity inside the cell, ~457 mosmol liter$^{-1}$, is about 1.5 times that outside, 300 mosmol liter$^{-1}$. The resulting osmotic pressure can be calculated as follows:

$$\Delta\pi = RT \Delta C = (0.082 \text{ liter atm K}^{-1} \text{ mol}^{-1})(310 \text{ K})$$

$$(0.457 - 0.300 \text{ mol liter}^{-1}) = 3.99 \text{ atm} \quad (77)$$

Because most cells are readily permeable to water, the osmotic pressure generated here would cause water to

**TABLE 4** Intracellular $K^+$ and $Cl^-$ Concentrations in Frog Sartorius Muscle[a,b]

| $[K^+]_o$ (mM) | $[Cl^-]_o$ (mM) | $[K^+]_i$ (mM) | $[Cl^-]_i$ (mM) | $\dfrac{[K^+]_o [Cl^-]_o}{[K^+]_i [Cl^-]_i}$ |
|---|---|---|---|---|
| 3 | 79 | 91 | 7.2 | 0.36 |
| 6 | 82 | 92 | 7.2 | 0.74 |
| 12 | 88 | 101 | 9.9 | 1.05 |
| 18 | 94 | 107 | 16.1 | 0.98 |
| 30 | 106 | 120 | 24.9 | 1.06 |
| 60 | 136 | 142 | 60.6 | 0.94 |
| 90 | 166 | 184 | 86.0 | 0.94 |
| 120 | 196 | 212 | 114.2 | 0.97 |
| 150 | 226 | 240 | 143.1 | 0.99 |
| 210 | 286 | 282 | 186.7 | 1.14 |
| 300 | 376 | 353 | 308 | 1.05 |
| | | | Mean | 1.01 |
| | | | Median | 0.99 |

[a] Adapted from Boyle and Conway (1941).

[b] Sartorius muscles were immersed for 24 hr at 2–3°C in modified Ringer's solution. Solutions were modified by adding KCl, and all other components were unchanged.

enter the cell. This influx of water dilutes the intracellular ion content, and the product $[K^+]_i [Cl^-]_i$ must fall below $[K^+]_o [Cl^-]_o$. As a result, more KCl would enter the cell, followed again by more water, in an endless cycle that would lead to destruction of the cell. That is to say, the simple Donnan system described in Fig. 16A fails to reach equilibrium when typical cell membrane properties are assumed.

### B. Double-Donnan or Pump-Leak Hypothesis

How can we reconcile the failure of the simple Donnan equilibrium model (see Fig. 16A) and the data (see Table 4) demonstrating behavior consistent with Donnan equilibrium in frog skeletal muscle? The Donnan system can be stabilized in two ways. A hydrostatic pressure could be applied to balance the osmotic pressure and arrest transmembrane water movement. In view of the enormous pressures required, this is not a realistic solution for animal cells. Alternatively, equilibrium can be attained by restricting an ionic species to the extracellular compartment just as one is restricted to the intracellular compartment. This arrangement is referred to as a **double-Donnan** (Leaf, 1959) or **pump-leak system** (Tosteson and Hoffman, 1960) and is illustrated diagrammatically in Fig. 16B. Assume that both $Na^+$ and macromolecular anions, $A^-$, are restricted to the extracellular and intracellular compartments, respectively. Filling the cell with the same concentrations of $K^+$, $Cl^-$, and $A^-$ as before, we find now that $[K^+]_i [Cl^-]_i > [K^+]_o [Cl^-]_o$, and thus KCl must leave the cell to establish Donnan equilibrium. In the process, water follows the KCl, adjusting cell volume until the KCl products are equal. A consequence of water flow is that the concentration of impermeant intracellular ion exactly equals the concentration of the impermeant extracellular ions. At equilibrium, $E_m = E_K = E_{Cl} = -91.1$ mV. In contrast to the simple

Donnan system, the osmotic pressure developed by intracellular macromolecules and their counterions (sometimes referred to as **colloid osmotic pressure**) is exactly balanced in the double-Donnan system by the osmotic pressure developed by ions restricted to the extracellular fluid and their counterions. As a consequence, there is no net osmotic pressure across the membrane, and the system is stable.

Another approach for considering the volume change expected in the system in Fig. 16B is to calculate the effective osmotic pressure of each compartment. We can suppose that the reflection coefficients, $\sigma$, are 1.0 for both $A^-$ and $Na^+$ because they are impermeant and are 0 for both $K^+$ and $Cl^-$ because they freely pass the membrane. Extra- and intracellular osmotic pressures are given by

$$\pi = \Sigma \sigma_i C_i$$

$$\pi_o = (1 [Na^+]_o + 0 [K^+]_o + 0 [Cl^-]_o) \tag{78}$$

$$\pi_i = (1 [A^-]_i + 0 [K^+]_i + 0 [Cl^-]_i)$$

Because only the impermeant species contribute to osmotic pressure, the only way for the cell to reach osmotic equilibrium, $\pi_o = \pi_i$, is to alter cell volume until $[Na^+]_o = [A^-]_i$. As a result, the ratio of $\text{volume}_{t=\infty} / \text{volume}_{t=0}$ must equal $[A^-]_i / [Na^+]_o$.

It is important to realize that an ion does not need to be impermeant to be *effectively* restricted to the extracellular space and to counterbalance the osmotic pressure developed by intracellular macromolecules. $Na^+$ is permeant, but it adequately plays this role anyway. As long as the leak of $Na^+$ down its electrochemical gradient into the cell is matched by its transport back out, cell volume will remain stable. Consequently, the existence of a pump that actively extruded $Na^+$ against its concentration gradient was postulated to explain cell volume (Leaf, 1956), and subsequently, the $Na^+$–$K^+$ ATPase, which extrudes three $Na^+$ while taking up two $K^+$ at the cost of ATP hydrolysis, was identified. Because energy is consumed to extrude $Na^+$, the cell is in a *steady state* rather than a true equilibrium.

### C. Modulation of the $Na^+$–$K^+$ Pump

One implication of the double-Donnan or pump-leak model is that the $Na^+$–$K^+$ pump is ultimately responsible for cell volume regulation. Perturbations that alter passive $Na^+$ entry must lead to offsetting changes in the rate of $Na^+$ extrusion by the $Na^+$–$K^+$ pump or cell volume will change. Because the $K_m$ of the $Na^+$–$K^+$ pump for intracellular $Na^+$ is close to the physiological $[Na^+]_i$, alterations in $Na^+$ influx automatically give rise to a compensatory modulation of $Na^+$ efflux. Nevertheless, metabolic or pharmacologic inhibition of the $Na^+$–$K^+$ pump should lead to a net gain of $Na^+$ and anions coupled by macroscopic electroneutrality and result in a swelling of the cell.

The effect of pump inhibition has been examined extensively (Macknight and Leaf, 1977; Macknight, 1988). The predicted cell swelling has been reported in many tissues, including brain slices, kidney slices, renal tubules, hepatocytes, and sheep erythrocytes, when the $Na^+$–$K^+$ pump is inhibited by cardiac glycosides (e.g., ouabain) or by depleting ATP. It is equally clear, however, that swelling after pump inhibition in renal cortex, liver slices, various muscle preparations, lymphocytes, and human erythrocytes is very slow or even absent, perhaps reflecting a low $Na^+$ permeability.

Several processes may affect the response of cell volume to $Na^+$–$K^+$ pump inhibition, and the outcome in some tissues depends on the experimental conditions. Rather than accumulating $Cl^-$ with $Na^+$, cells might instead lose $K^+$ to satisfy macroscopic electroneutrality. An equivalent gain of $Na^+$ and loss of $K^+$ replaces one osmotically active particle with another, and no change in cell volume is anticipated. Closer consideration of this mechanism indicates that it can only be a holding action, however. The loss of intracellular $K^+$ eventually must lead to a reduction of $K^+$ gradient and a less negative $E_m$. This will lead to accumulation of $Cl^-$ with $Na^+$, and cell swelling. Nevertheless, a loss of $K^+$ must slow the swelling that otherwise would have occurred and, in cells with appropriate $Na^+$ and $K^+$ permeabilities, this would explain the absence of swelling over the time course of experiments.

The possibility that a mechanism other than the $Na^+$–$K^+$ pump can extrude $Na^+$ has been considered. Although controversial, a cardiac glycoside-insensitive but metabolically dependent volume regulation mechanism that does not incorporate the $Na^+$–$K^+$ pump has been described in kidney (for review, see Macknight and Leaf, 1977). In addition, circulating erythrocytes from a number of carnivores, including dog, cat, bear, and ferret, lack a functioning $Na^+$–$K^+$ pump and must regulate their volume by a different mechanism (Sarkadi and Parker, 1991). In these cells, $Na^+$ efflux must depend on the gradient of other ions.

### D. Isosmotic Volume Regulation

Although principles of the venerable pump-leak or double-Donnan model are correct and still relevant to the regulation of cell volume, it has become apparent that neither the leak nor the pump is constant. Not only is the control of these fluxes more complex than originally envisioned, but a myriad of other transport processes also contribute to cell volume regulation. Stated simply, constant cell volume under isosmotic conditions implies an equality of intra- and extracellular osmolarity that is perpetuated by a continuous balance of the efflux and influx of osmotic equivalents. The transport processes involved include, but are not limited to (1) ion and organic osmolyte channels; (2) the $Na^+$–$K^+$ pump; (3) the $Na^+$–$K^+$–$2Cl^-$, $K^+$–$Cl^-$, and $Na^+$–$Cl^-$ cotransporters, which transport the specified ions in one direction; (4) $Na^+$-dependent sugar and amino acid cotransport; (5) $Na^+$–$Ca^{2+}$ exchange, which exchanges three $Na^+$ for one $Ca^{2+}$; and (6) osmotically neutral exchangers that indirectly provide a net solute flux. For example, $Na^+$–$H^+$ exchange allows the cell to accumulate $Na^+$, but the $H^+$ removed is replaced by dissociation of $H^+$ from intracellular buffers. $Cl^-$–$HCO_3^-$ and $Na^+$–$H^+$ exchange can operate in parallel to mediate a net influx of $Na^+$ and $Cl^-$ in exchange for $H^+$ and $HCO_3^-$, which are converted to $CO_2$ and $H_2O$ by the action of carbonic anhydrase. (It should be noted that $CO_2$ freely crosses the cell membrane, $\sigma = 0$, and does not directly contribute to solution tonicity. Also, the direct extrusion of water by

these two parallel exchangers is negligible compared to the osmotic water gain caused by the accumulation of $Na^+$ and $Cl^-$.)

Of the number of transporters that participate in volume regulation in one or another type of cell, which are most important in regulating cell volume under isosmotic conditions? No simple answer can be offered. The importance of each process to cell volume regulation depends critically upon the tissue and species under consideration, as well as the conditions. Moreover, transporters operate continually under physiologic and pharmacologic control, and they extensively interact by altering membrane potential or the concentration of the transported species. If the rates of ion transport are not correctly matched, cells will inappropriately shrink or swell. The precise maintenance of cell volume speaks for the need for sensitive and complex regulatory mechanisms. Attempts to integrate the fluxes mathematically and study their interaction have been made based on the cell's requirement for macroscopic electroneutrality and osmotic equilibrium and the equations governing ion fluxes (Jakobsson, 1980). For erythrocytes, the nonideal behavior of hemoglobin has been added (Bookchin et al., 1989). Although these simplified models correctly predict a number of observations, they fail to explain others. In short, we remain a long way from a complete quantitative description of the processes underlying cell volume regulation. In the following sections, we discuss in qualitative detail three examples of isosmotic regulation of cell volume that illustrate some of the underlying principles.

### 1. $Na^+$–$K^+$–$2Cl^-$ Cotransport in the Heart

Recent findings indicate that the $Na^+$–$K^+$–$2Cl^-$ cotransporter plays a critical role in regulating rabbit cardiac myocyte cell volume under isosmotic conditions. As in other tissues, $Na^+$–$K^+$–$2Cl^-$ cotransport conveys osmolytes into cardiac cells under physiological conditions. Because *net* transmembrane fluxes control cell volume, a decreased osmolyte influx is equivalent to increased efflux. Therefore, inhibition of the $Na^+$–$K^+$–$2Cl^-$ cotransport by bumetanide, for example, favors a reduction of cell volume. Consistent with this idea, bumetanide decreases the volume of atrial and ventricular myocytes by about 10% in less than 5 min, and myocyte volume is stable at this new level (Drewnowska and Baumgarten, 1991). The $Na^+$–$K^+$–$2Cl^-$ cotransporter cannot operate without $Na^+$ and $Cl^-$ in the extracellular fluid, and removing either ion renders bumetanide ineffective. These data imply that ion uptake by $Na^+$–$K^+$–$2Cl^-$ cotransport in heart must be responsible for a significant osmolyte flux under isosmotic conditions and that other transport processes are incapable of fully compensating when this flux is removed. In contrast, myocyte volume was unchanged after inhibition of the $Na^+$–$K^+$ pump with 10 $\mu M$ ouabain (Drewnowska and Baumgarten, 1991) or after cooling to 9°C (Drewnowska et al., 1991) for 20 min. In the short term at least, cardiac cell volume in isosmotic solution apparently is influenced more by $Na^+$–$K^+$–$2Cl^-$ cotransport than by the $Na^+$–$K^+$ pump.

Modulation of $Na^+$–$K^+$–$2Cl^-$ cotransport by intracellular messengers such as cGMP may provide a physiological means of modulating cell volume in heart. Figure 17 shows the effects of elevating intracellular cGMP in three ways: (1) by adding 8-Br-cGMP, a membrane-permeant analogue of cGMP; (2) by adding atrial natriuretic factor (ANF), a natriuretic, diuretic, and vasodilatory hormone released by the heart that elevates cGMP by activating guanylate cyclase; and (3) by adding sodium nitroprusside (SNP), a vasodilator, which also activates guanylate cyclase. In each case, cell volume decreased. Furthermore, blocking cGMP-specific phosphodiesterase with zaprinast (M&B22948) augmented the effect of ANF. Based on its sensitivity to bumetanide and the requirement for ions transported by $Na^+-K^+-2Cl^-$ cotransport, cGMP-dependent volume decreases were shown to be due to an inhibition of $Na^+-K^+-2Cl^-$ cotransport by cGMP (Clemo et al., 1992; Clemo and Baumgarten, 1995). Interestingly, *lowering* cGMP levels by inhibiting guanylate cyclase with LY83583 resulted in a small cell swelling. Thus, changing cGMP from its physiological level in either direction altered cell volume. The

mechanism and evidence for isosmotic regulation of cell volume in heart are summarized in Fig. 18.

Does the same mechanism regulate cell volume under isosmotic conditions in other tissues? Perhaps it does in some cells, such as vascular endothelium, in which cGMP inhibits $Na^+-K^+2Cl^-$ cotransport. In other cells $Na^+-K^+-2Cl^-$ cotransport is stimulated by cAMP, cGMP, or a PKC-dependent pathway, and perhaps by a number of other signaling pathways (Palfrey and O'Donnell, 1992; Palfrey, 1994). This diversity in the control of $Na^+/K^+2Cl^-$ cotransport may be related to the variety of transporter isoforms that have been identified and cloned (Haas, 1994; Kaplan et al., 1996). Although the physiological significance of this diversity in control of ion transport is not well understood, it must lead to diversity in the regulation of cell volume.

## 2. Hormones and Substrate Transport in Liver

Another interesting example of isosmotic volume regulation is found in hepatocytes. An impressive number of hormones induce either cell swelling or shrinkage at physiolgoical concentrations, and these actions are related to their control of liver metabolism (Häussinger and Lang, 1991; Häussinger et al., 1994; Agius et al., 1994). $Na^+-H^+$ exchange, $Na^+ K^+-2Cl^-$ cotransport, and the $Na^+-K^+$ pump are stimulated in rat liver cells by insulin. The net effect is that insulin increases $[K^+]_i$, $[Na^+]_i$, and $[Cl^-]_i$ and causes cell swelling by about 12%. This swelling is prevented by bumetanide, a blocker of $Na^+-K^+-2Cl^-$ cotransport, or by amiloride, a blocker of $Na^+-H^+$ exchange. In contrast, glucagon shrinks hepatocytes by about 14%. Instead of directly opposing the action of insulin, glucagon reduces cell volume by increasing $K^+$ and $Cl^-$ efflux through ion channels. Other agents that swell hepatocytes include bradykinin and phenylephrine. Shrinking is initiated by adenosine, 5-HT, vasopressin, and cAMP.

Hepatocytes also swell as a result of $Na^+$-dependent amino acid cotransport in isosmotic media (Häussinger and Lang, 1991; Boyer et al., 1992; Häussinger et al., 1994). Exposure to amino acids that are accumulated with $Na^+$ (e.g., alanine, glutamine, glycine, hydroxyproline, phenylalanine, proline, serine) causes cell swelling. These effects occur at amino acids levels found physiologically in the portal vein. For example, glutamine provokes up to a 10% swelling with a half-maximal effect at approximately 0.7 m$M$. Amino acid cotransport gives rise to an inward current, and $Cl^-$ enters to maintain macroscopic electroneutrality. Instead of loading the cell, $Na^+$ is pumped out mainly by the $Na^+-K^+$ pump, leaving an accumulation of $K^+$ and $Cl^-$. In contrast to these amino acids, substances not accumulated by the liver, such as glucose and leucine, do not affect hepatocyte volume. Cell swelling caused by $Na^+$-dependent amino acid cotransport also has been observed in intestine and renal proximal tubule.

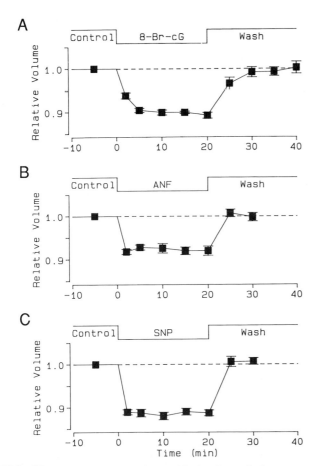

**FIG. 17.** Changes in the volume of isolated ventricular myocytes during a 20-min exposure to (A) 10 $\mu M$ 8-Br-cGMP (8-Br-cG), an permeable cGMP analogue; (B) 1 $\mu M$ atrial natriuretic factor (ANF); and (C) 100 $\mu M$ sodium nitroprusside (SNP). These agents reversibly decreased cell volume 11, 8, and 11%, respectively ($n = 5$ for each time point). The decreases in cell volume were caused by inhibition of $Na^+-K^+-2Cl^-$ cotransport by cGMP. Relative cell volume was measured and calculated as described in the legend of Fig. 1. (From Clemo et al., 1992; Reproduced from *The Journal of General Physiology* by copyright permission of The Rockefeller University Press.)

## 3. $Na^+-Ca^{2+}$ Exchange in Carnivore Erythrocytes

The $Na^+-K^+$ pump in erythrocytes from dogs, cats, ferrets, and bears ceases to function as cells mature. As a

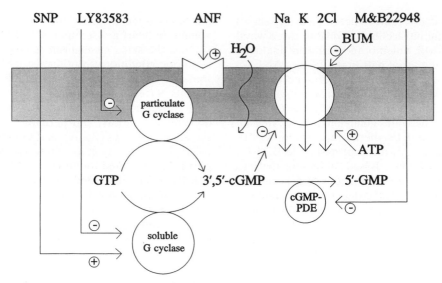

**FIG. 18.** Schematic diagram of the action of ANF and cGMP on cardiac cell volume. Binding of ANF activates guanylate cyclase and increases intracellular cGMP levels. By one or more steps, cGMP inhibits $Na^+$–$K^+$–$2Cl^-$ cotransport. Reducing ion influx by this means is equivalent to increasing *net* ion efflux, and cell shrinkage ensues. LY83583 inhibits guanylate cyclase, thereby blocking the effect of ANF, and zaprinast (M&B22948) potentiates the effect of ANF by inhibiting cGMP-specific phosphodiesterase (PDE). Sodium nitroprusside (SNP) also increases cGMP levels, and bumetanide (BUM) directly inhibits the cotransporter; both cause cell shrinkage. (From Clemo *et al.*, 1992; Reproduced from *The Journal of General Physiology* by copyright permission of The Rockefeller University Press.)

result, $[K^+]_i$ and $[Na^+]_i$ are similar to $[K^+]_o$ and $[Na^+]_o$. This poses a special problem for cell volume regulation. How can these cells offset the osmotic pressure generated by impermeant intracellular molecules and avoid swelling without a $Na^+$–$K^+$ pump to make $Na^+$ effectively impermeant? Carnivore erythrocytes solve their dilemma by extruding three $Na^+$ in exchange for one $Ca^{2+}$ via the $Na^+$–$Ca^{2+}$ exchanger (Parker, 1973; Sarkadi and Parker, 1991). In most cells, the electrochemical gradients for $Na^+$ and $Ca^{2+}$ favor the efflux of $Ca^{2+}$. In these erythrocytes, however, the gradients favor $Ca^{2+}$ entry because the $Na^+$ gradient is reduced, and $Na^+$–$Ca^{2+}$ exchange operates in what is called the "reverse mode." To stabilize their volume, carnivore erythrocytes also must have a means of maintaining the $Ca^{2+}$ gradient (i.e., low $[Ca^{2+}]_i$). This is accomplished by an ATP-dependent $Ca^{2+}$ pump in the plasma membrane. Thus, as in other cells, maintaining cell volume in the face of impermeant intracellular colloids requires the expenditure of energy in a pump-leak mechanism. In this case, ATP is expended by a sarcolemmal $Ca^{2+}$ pump rather than by the $Na^+$–$K^+$ pump.

## VI. Regulation of Cell Volume under Anisosmotic Conditions

### A. Osmometric Behavior of Cells

Because the permeability of most cell membranes to water is much greater than that to solutes, cells swell or shrink when placed in an environment that is hyposmotic or hyperosmotic, respectively. Water rapidly flows to equalize its chemical potential, $\mu_w$, inside and outside the

cell. The initial volume response often is close to that predicted for an ideal osmometer from van't Hoff's law. An example is shown in Fig. 19, which illustrates the response of rabbit ventricular myocytes to solutions with osmolarities ranging from 195 to 825 mosmol/liter (0.60 to

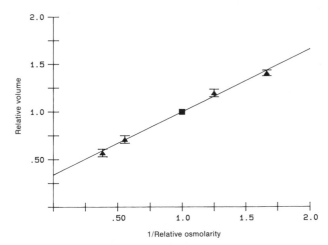

**FIG. 19.** Relationship between relative cell volume and the inverse of relative osmolarity in cardiac ventricular myocytes. Data from 38 measurements of relative cell volume in anisosmotic solutions ($0.6T$, $0.8T$, $1.8T$, and $2.6T$) were fit to a least-squares regression line constrained to pass through 1,1 (square) because relative cell volume is 1 in isotonic solution by definition. The extrapolated intercept on the relative volume axis, 0.34, represents the fraction of cell volume that is osmotically inactive (see Eq. 79). Volume measurements, calculation of relative cell volume, and means of adjusting osmolarity are as described for Fig. 1. [Reproduced from Drewnowska and Baumgarten (1991).]

2.55 times isotonic). Relative cell volume ($V$), calculated as $V_{test}/V_{isosmotic}$, is plotted against the inverse of relative osmolarity, $\pi_{isosmotic}/\pi_{test}$, and the data are fitted to

$$V = (1 - V_b)(\pi_{isosmotic}/\pi_{test}) + V_b \qquad (79)$$

By definition, relative volume is 1.0 at a relative osmolarity of 1.0. Two conclusions can be reached from the data in Fig. 19. First, as expected from van't Hoff's law, the relationship between relative cell volume and the inverse of relative osmolarity is linear. Second, the intercept of the relationship on the volume axis, $V_b$, is 0.34 and is significantly different from 0. This is interpreted as meaning that a fraction of cell volume is **osmotically inactive.** That is, it apparently does not participate in the response to anisosmotic solutions.

Several arguments can be made to justify observing an osmotically inactive volume. The expectation from the simplest model is that *cell water* should vary in proportion to osmolarity. But all of cell volume is not water. The volume of nonaqueous components such as small solutes and proteins, which represent 25–30% of the cell on a weight/weight basis, is unaffected by water movements. Even measurements of cell water show nonideal behavior, however (Macknight and Leaf, 1977; Solomon, 1989), and additional explanations are necessary. One suggestion is that a fraction of cell water is intimately associated with cell proteins or membranes and thereby is *bound* or *structured* and unavailable as solvent (e.g., LeNeveu *et al.*, 1976; Hinke, 1980). Although the state of water molecules adjacent to proteins and membranes must be different from that in the bulk phase of the cytoplasm, in light of NMR, intracellular ion activity, and other data, most investigators believe that virtually all water (~95%) is available as solvent (Shporer and Civan, 1977; Hladky and Rink, 1978). Another possibility is that the behavior of intracellular macromolecules is concentration dependent. For example, the osmotic coefficient and charge on hemoglobin increase upon its concentration as red cells shrink and anions are drawn in to maintain electroneutrality. These phenomena are important in explaining water movement in red cells (Freedman and Hoffman, 1979), but their importance in other tissues remains uncertain. A third possibility is that intracellular compartments such as mitochondria, nucleus, endoplasmic reticulum, and sarcoplasmic reticulum of muscle cells may undergo volume changes that are not proportional to those of the whole cell. Differential responses to an osmotic challenge are expected because the plasmalemma and intracellular membranes possess distinct arrays of transporters and ion channels and each sees a unique environment. Most methods for determining cell water or cell volume fail to distinguish between cytoplasmic and total cell water or volume (for a method that does distinguish, see Reuss, 1985).

A crucial assumption made in determining osmotically inactive volume also may affect the value obtained for $V_b$ in Eq. (79). The analysis assumes that only water has moved at the time volume is measured. That is to say, transmembrane ionic fluxes can be ignored. Although water permeability is many times greater than ionic permeabilities, the net fluxes of both water and ions start at the instant extra-

cellular osmolarity is changed. If ion fluxes significantly affect intracellular osmolarity at the time of measurement, the extrapolated osmotically inactive volume will be imprecise. If in addition the ion fluxes depend on the direction or magnitude of the osmotic gradient, the plot of relative volume vs $\pi_{isosmotic}/\pi_{test}$ can become nonlinear (e.g., Grinstein *et al.*, 1984).

## B. Compensatory Regulation of Cell Volume

Although an osmotic gradient initiates cell swelling or shrinkage, the initial volume response is not maintained in a wide variety of cells. Cell swelling activates compensatory processes leading to an efflux of osmolytes and a reduction of cell volume. This is called a **regulatory volume decrease (RVD).** Similarly, cell shrinking activates an influx of osmolytes in some cells leading to a compensatory swelling referred to as a **regulatory volume increase (RVI).** RVD and RVI nearly restore the original cell volume in some cells, are far less complete in others, and are absent in a few types of cells.

Regulatory volume effects are thought to be adaptive and were first identified in nucleated duck erythrocytes by Kregenow in 1971 (for review, see Kregenow, 1981). Examples of an RVD and RVI taken from work by Grinstein *et al.* (1983) are shown in Fig. 20. Exposure of human peripheral blood lymphocytes to a solution made hypotonic by 50% dilution with water leads to a rapid, 1.6-fold increase in cell volume (panel B). Then, over about 10 min, an RVD returned cell volume almost completely to its initial value. On switching back to isotonic solution, cell volume shrank to less than the control value and then was restored by an RVI. RVDs and RVIs may differ in magnitude, however. A much less complete RVI was observed when lymphocytes in isotonic solution were shrunk in media made hyperosmotic by adding 300 mosmol liter$^{-1}$ of NaCl, and the RVI was absent when the lymphocytes were challenged with 300 mosmol liter$^{-1}$ of sucrose instead of NaCl (panel A). The RVD also could be eliminated, for example, by cooling lymphocytes to 4°C (Grinstein *et al.*, 1984). Thus, regulation of cell volume following an osmotic challenge depends on the particulars of the perturbation as well as the cell under study (compare Figs. 1 and 20).

## C. Transport Processes Responsible for RVD and RVI

How do cells gain or lose osmotic equivalents in anisosmotic medium? The mechanisms underlying RVDs and RVIs have been extensively characterized in a variety of cell types and exhaustively reviewed (Hoffmann and Simonsen, 1989; Chamberlin and Strange, 1989; Grinstein and Foskett, 1990; Sarkadi and Parker, 1991; Häussinger and Lang, 1991; McCarty and O'Neil, 1992; Strange, 1994; Hoffmann and Dunham, 1995). In general, cells undergo RVD or RVI by translocating Na$^+$, K$^+$, and Cl$^-$, and a variety of channels, exchangers, cotransporters, and pumps can participate. In some cases, organic osmolytes (e.g., taurine, betaine, sorbitol, urea) are transported instead of, or in addition to, inorganic ions. Table 5 lists processes

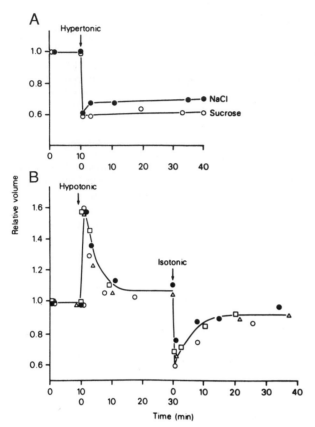

**FIG. 20.** (A) Effect of hypertonic solution ($2T$) on relative cell volume of human peripheral blood mononuclear lymphocytes. Solution osmolarity was increased by adding 300 mosmol liter$^{-1}$ of NaCl or sucrose. Cells shrank 40% in hypertonic solution, and a small regulatory volume increase (RVI) was observed in the NaCl solution. Points are representative of four experiments. (B) Response to hypotonic solution ($0.5T$). After swelling by nearly 60%, a regulatory volume decrease (RVD) virtually restored cell volume to its initial value in 10 min. On returning to isotonic solution, cell volume decreased below its control value, and an RVI led to full recovery. Different symbols represent separate experiments selected from more than 40. Cell volume was measured with a Coulter counter that determines volume from the change in electrical resistance of a column of solution as cells pass through an aperture. (From Grinstein *et al.*, 1983; Reproduced from *The Journal of General Physiology* by copyright permission of The Rockefeller University Press.)

that are activated by altering the volume of various cells, and Fig. 21 illustrates some of the mechanisms diagrammatically. The compensatory mechanisms invoked vary with the tissue, species, and conditions under which osmotic stress is applied.

The primary mechanism of RVD in a number of cell types is activation of conductive pathways, and this process will be discussed in more detail. Cell swelling, acting directly or via a messenger, opens ion channels that allow increased efflux of K$^+$ and Cl$^-$, and H$_2$O follows. Although the openings of cation and anion channels are independent events, K$^+$ and Cl$^-$ efflux are tightly coupled by the need to maintain macroscopic electroneutrality. If only the anion

or cation channel were to open, the resulting change in $E_m$ would rapidly make ion efflux self-limiting and arrest volume regulation. The effect of this coupling of K$^+$ and Cl$^-$ fluxes is illustrated in Fig. 22A, which shows RVDs in Ehrlich ascites cells (Hoffmann *et al.*, 1986; Hoffmann and Dunham, 1995). At time 0, cells were switched from 300 to 150 mosmol liter$^{-1}$ medium. Cells rapidly swelled to about 1.9 times their initial volume, and then underwent an RVD that returned relative cell volume to about 1.3 within 5 min. When K$^+$ conductance was increased by pretreating cells with 0.5 $\mu M$ gramicidin, a K$^+$ ionophore, the compensation by the RVD was more rapid and larger in magnitude, nearly returning cell volume to its control level in 2 min. When K$^+$ channels were blocked with 1 m$M$ quinine, however, only a feeble RVD took place. These data argue that conductive K$^+$ efflux cannot keep up with Cl$^-$ efflux in control cells and limits the rate of RVD (Hoffman *et al.*, 1986). Thus, cell swelling must increase Cl$^-$ conductance more than K$^+$ conductance, leading to greater Cl$^-$ efflux than K$^+$ efflux. Consistent with this idea, a depolarization is observed during the RVD. The activation of Cl$^-$ conductance by swelling is only transient, however. The ability of gramicidin to induce an RVD decays with time, as shown in Fig. 22B. In contrast to the high selectivity of Cl$^-$-dependent cotransporters, RVD also is supported by Br$^-$, NO$_3^-$, and SCN$^-$, suggesting the Cl$^-$ channel responsible for RVD poorly discriminates among these anions. Furthermore, RVD can be suppressed by Cl$^-$ channel blockers such as inacrinone (MK-196) and diphenylamine-2-carboxylate (DPC), but not by the cotransport inhibitors. Taken together, the data in Ehrlich ascites cells provide strong evidence that two independent channels are responsible for RVD instead of, for example, activation of a single K$^+$–Cl$^-$ cotransporter. Organic osmolytes (see next section) also permeate a class of swelling-activated Cl$^-$ channels referred to as volume-sensitive organic osmolyte–anion channels (VSOACs) and are released by a number of cells to regulate volume (Jackson *et al.*, 1994). VSOACs have a high permeability for taurine ($P_{taurine}/P_{Cl} = 0.75$) and glutamate ($P_{glutamate}/P_{Cl} = 0.20$) and other amino acids, and small organic molecules are permeant (Bandarali and Roy, 1992).

Whereas many types of cells exhibit extensive RVDs, fewer cells give robust RVIs. RVIs usually are due to an accumulation of Na$^+$, Cl$^-$, and, in some cases, K$^+$, which occurs over minutes. The primary mechanisms underlying RVIs are acceleration of Na$^+$–K$^+$–2Cl$^-$ or Na$^+$–Cl$^-$ cotransport and coupled Na$^+$–H$^+$ and Cl$^-$–HCO$_3^-$ exchange. Volume recovery can be blocked by application of appropriate inhibitors such as bumetanide for Na$^+$–K$^+$–2Cl$^-$ cotransport, amiloride for Na$^+$–H$^+$ exchange, and SITS (4-acetamido-4′-isothiocyano-stilbene-2,2′-disulfonic acid) for Cl$^-$–HCO$_3^-$ exchange. RVI in most cells is readily observed when simple salt solutions bathe the cells, but in renal cortical collecting duct and proximal tubules, butyrate, acetate, or other metabolizable fatty acids must be in the perfusate to support an RVI. Substrate metabolism may provide H$^+$ and HCO$_3^-$ to support Na$^+$–H$^+$ and Cl$^-$–HCO$_3^-$ exchange in these cells (for review, see McCarty and O'Neil, 1992).

**TABLE 5**  Ionic Mechanisms of Regulatory Responses to Anisosmotic Solutions[a]

| Transport mechanism activated | Cell types |
| --- | --- |
| A. Cell swelling-induced regulatory volume decrease | |
| $K^+$ and $Cl^-$ conductances | Frog urinary bladder |
| | Chinese hamster ovary cells |
| | Ehrlich ascites tumor cells |
| | Frog skin |
| | HeLa carcinoma cells |
| | Human platelets |
| | Human granulocytes |
| | Human lymphocytes |
| | Intestinal 407 cells |
| | Madin–Darby canine kidney (MDCK) cells |
| | *Necturus* enterocytes |
| | *Necturus* gallbladder |
| | Rabbit renal proximal convoluted tubule |
| | Rat hepatocytes |
| $K^+$–$Cl^-$ cotransport | Avian, dog, fish, human, rabbit, and low-$K^+$ sheep erythrocytes |
| | Ehrlich ascites tumor cells ($Ca^{2+}$ depleted) |
| | *Necturus* gallbladder |
| Coupled $K^+$–$H^+$ and $Cl^-$–$HCO_3^-$ exchange | *Amphiuma* erythrocytes |
| $Na^+$–$Ca^{2+}$ exchange | Dog and ferret erythrocytes |
| Organic osmolyte efflux | Crustacean muscle and myocardium |
| | Ehrlich ascites tumor cells |
| | Elasmobranch and molluscan erythrocytes |
| B. Cell shrinkage-induced regulatory volume increase | |
| $Na^+$–$K^+$–$2Cl^-$ cotransport | Astrocytes |
| | C6 glioma cells |
| | Duck, fish, rat, and human erythrocytes |
| | Ehrlich ascites tumor cells |
| | Frog skin |
| | HeLa cells |
| | Rat kidney medullary thick ascending limb |
| | 3T3 cells |
| Coupled $Na^+$–$H^+$ and $Cl^-$–$HCO_3^-$ exchange | Amphibian gallbladder |
| | Dog and amphibian erythrocytes |
| | Human lymphocytes |
| | Mouse medullary thick ascending limb |
| | Rabbit renal proximal straight tubule |
| | Ehrlich ascites tumor cells |
| $Na^+$–$Cl^-$ cotransport | Ehrlich ascites tumor cells |
| | *Necturus* gallbladder |
| $K^+$ and $Cl^-$ conductances, inhibited | Madin–Darby canine kidney (MDCK) cells |
| Organic osmolyte influx | Many animal and plant cells, bacteria, and fungi |

[a] For references, see Yancey *et al.* (1982), Hoffmann and Simonsen (1989), Chamberlin and Strange (1989), Grinstein and Foskett (1990), Wolff and Balaban (1990), Sarkadi and Parker (1991), Häussinger and Lang (1991), McCarty and O'Neil (1992), Boyer *et al.* (1992), Strange (1994), and Hoffmann and Dunham (1995).

## D. Organic Osmolytes

Adaptation to hyperosmolarity over a longer term also occurs in some cells and is mediated by an accumulation of organic osmolytes, including amino acids, polyols, and urea (Yancey *et al.*, 1982; Chamberlin and Strange, 1989; Wolff and Balaban, 1990; Garcia and Berg, 1991; Yancey, 1994; Berg, 1995). Three of these organic osmolytes, taurine, betaine, and inositol, are taken up by $Na^+$- or $Na^+$/$Cl^-$-dependent cotransporters. The taurine and betaine transporters have been cloned and belong to the same family as the $Na^+$- and $Cl^-$-dependent norepinephrine, GABA, dopamine, serotonin, and proline transporters (Uchida *et al.*, 1992). The taurine transporter is found in a variety of cell types. The distribution of mRNA levels is kidney > ileum > brain > liver > heart (Berg, 1995). This

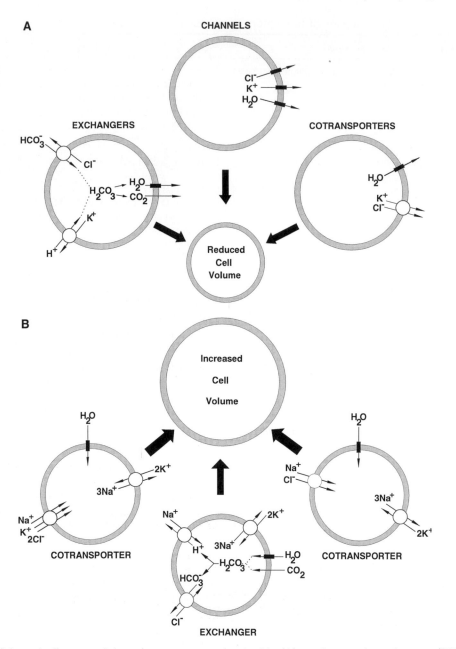

**FIG. 21.** Schematic diagrams of the transport processes involved in (A) regulatory volume decrease (RVD) and (B) regulatory volume increase (RVI). Following cell swelling, a compensatory reduction in cell volume (RVD) may result from increased $K^+$ and $Cl^-$ conductance (channels), activation of functionally coupled $Cl^- - HCO_3^-$ and $K^+ - H^+$ exchange (exchangers), or activation of $K^+ - Cl^-$ cotransport (cotransporters). Following cell shrinking, a compensatory increase in cell volume (RVI) may result from activation of the $Na^+ - K^+ - 2Cl^-$ or $Na^+ - Cl^-$ cotransporters (cotransporters) or activation of functionally coupled $Cl^- - HCO_3^-$ and $Na^+ - H^+$ exchange (exchangers). The $Na^+ - K^+$ pump extrudes the $Na^+$ that enters during RVI, so that cells gain $K^+$ and $Cl^-$.

sequence for taurine transporter message is surprising because heart has the highest intracellular taurine concentration among these tissues. The mRNA levels for the betaine transporter also are highest in the kidney (Yamauchi *et al.*, 1992). A different family of transporters is responsible for $Na^+$-dependent inositol uptake. The inositol transporter is related to those for glucose and nucleo-

sides and is highly expressed in renal medulla (Kwon *et al.*, 1992). A common feature of all three osmolyte transporters is that gene transcription is up-regulated by hypertonic stress. Cells cultured under hypertonic conditions slowly increase their maximum rates of transport, and intracellular concentrations rise. For example, betaine transporter mRNA levels in MDCK cells peak

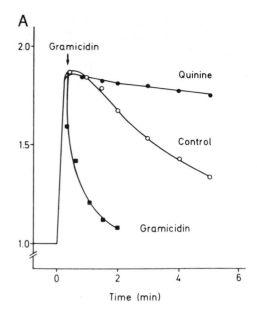

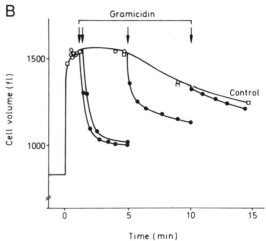

**FIG. 22.** (A) Effects of quinidine (1 m$M$) and gramicidin (5 $\mu M$) on regulatory volume decrease (RVD) in Ehrlich ascites cells suspended in a Na$^+$-free choline medium with 1 m$M$ Ca$^{2+}$. Swelling was induced by diluting the medium to 0.5$T$. Quinidine blocks Ca$^{2+}$-activated K$^+$ channels, and gramicidin increases cation conductance (K$^+$ conductance in the absence of Na$^+$). The data indicate that K$^+$ permeability is rate limiting during the RVD. (B) The swelling-induced increase in Cl$^-$ conductance is transient. Same protocol as in as (A), except that quinidine (1 m$M$) was added to the hypotonic medium to reduce K$^+$ conductance (control). At selected times, 0.5 $\mu M$ gramicidin was added to increase cation permeability. The ability of gramicidin to induce an RVD follows the time course of the swelling-induced Cl$^-$ conductance. Relative cell volumes measured with a Coulter counter, as described for Fig. 20. [Reproduced with permission from Hoffmann *et al.* (1986).]

at about 16 hours after the osmotic challenge, and the transport rate peaks at about 24 hours (Uchida *et al.*, 1993).

Another mechanism for accumulating organic osmolytes while adapting to hypertonicity and cell shrinkage is to modulate their synthesis. For example, renal medullary cells, which experience osmolarities of more than 1200

mosmol liter$^{-1}$ during antidiuresis, accumulate sorbitol and inositol to regulate their volume (Wolff and Balaban, 1990; Garcia and Berg, 1991; Sands, 1994). Sorbitol levels rise 12–24 hours after osmotic challenge of cultured renal medullary cells, and accumulation depends on the induction of aldose reductase by enhanced transcription. Aldose reductase converts exogenous glucose to sorbitol, which in turn is synthesized from stored glycogen in other tissues. This requires coordinated up-regulation of glycogen phosphorylase and hexosekinase to produce glucose 6-phosphate, an intermediate for sorbitol production, and the down-regulation of phosphofructokinase to prevent consumption of substrates for sorbitol production.

Inhibition of degradation is another mechanism for accumulating organic osmolytes when cells are challenged by hypertonicity. This is the main mechanism by which MDCK and medullary collecting duct accumulate glycerophosphocholine (GPC) (Berg, 1995). In response to hypertonicity or elevated urea, GPC–choline phosphodiesterase is inhibited. Under some conditions, synthesis of GPC by phospholipase also may be enhanced.

The accumulation of organic osmolytes is metabolically expensive, especially considering that many cells dump organic osmolytes to respond acutely to swelling (Chamberlin and Strange, 1989; Rasmusson *et al.*, 1993; Jackson *et al.*, 1994; Hoffmann and Dunham, 1995). Why does the cell expend extra energy to use organic compounds rather than inorganic ions? Apparently the reason is that accumulation of inorganic ions may perturb protein structure within cells. The Hofmeister series, first described more than 100 years ago, lists inorganic ions according to their ability to alter the solubility and conformation of proteins. Nonspecific effects of inorganic salts include changes in $V_{max}$, $K_m$, tertiary structure, and subunit assembly of enzymes (Yancey *et al.*, 1982; Somero, 1986; Yancey, 1994). Why are some organic compounds accumulated rather than others? Yancey *et al.* (1982) categorized osmolytes as nonperturbing (stabilizing) or perturbing (destabilizing). Elevated concentrations of nonperturbing osmolytes are compatible with normal enzyme function, whereas high concentrations of perturbing osmolytes are not. This distinction could be due either to direct interaction of osmolytes with enzymes or substrates or to effects on hydration, solubility, or charge interactions of proteins. Organic solutes that are nonperturbing generally are uncharged [e.g., trimethylamine *N*-oxide (TMAO), glycerol] or zwitterionic (e.g., betaine, taurine), although negatively charged octopine is used by some cells. In contrast, most perturbing organic osmolytes are positively charged (e.g., arginine, guanidinium). Neutral urea, however, perturbs proteins. Finally, nonperturbing osmolytes can counteract the destabilizing effects of perturbing osmolytes. Several organisms accumulate perturbing osmolytes, such as urea, in fixed ratios with nonperturbing osmolytes.

## E. Signaling Pathways Underlying RVD and RVI

How do cells detect alterations in their volume and activate the transport processes underlying compensatory volume regulation? Do the transporters themselves sense cell volume and respond by changing their activity? Alternatively, does water movement simply change the concentration of a critical regulatory substance? Answers to these questions are just beginning to emerge. It appears that the signaling pathways modulated by cell volume are as diverse as the transporters that respond.

## 1. Anisosmotic Media

An obvious candidate for signaling a change in cell volume is the composition of the anisosmotic medium itself; that is, the ionic strength or concentrations of ions in the bathing medium might initiate a regulatory volume response. Changing concentration or ionic strength must affect various ion transport processes and transmembrane ion fluxes to some degree. Nevertheless, these factors seem relatively unimportant in volume regulation because comparable responses are observed when nonelectrolytes such as mannitol or sucrose replace a large fraction of the electrolytes in the bathing medium. RVDs also are initiated in isosmotic solutions after swelling caused by sugar or amino acid uptake. Thus, it appears that regulation is initiated by the volume change itself rather than the composition of the bathing medium.

## 2. Membrane Potential

Another possibility is that regulatory volume responses are initiated by dilution or concentration of intracellular $K^+$ via an effect on $E_m$. A twofold change in volume without compensatory $K^+$ fluxes would alter $[K^+]_i$ and cause an 18-mV change in $E_m$ in a cell that conforms to the Nernst equation. Fluxes through $K^+$ and $Cl^-$ channels are voltage dependent, reflecting both the electrochemical driving force and the voltage-dependent conductance. For many $K^+$ channels, the current–voltage relationship is highly nonlinear. Furthermore, the $Na^+$–$K^+$ pump and $Na^+$–$Ca^{2+}$ exchange are voltage dependent because they mediate a net movement of charge across the membrane. Despite this, it is unlikely that changes in $E_m$ are the primary cause of RVDs or RVIs in most cells. Regulatory responses have been observed with change in cell volume of <5% (Hoffmann and Simonsen, 1989), which would directly alter $E_m$ by only ~1 mV. This is too small to have a significant effect. Nevertheless, it is clear that $E_m$ can modulate RVDs through activation of $K^+$ and $Cl^-$ channels. In most instances, the increase in $Cl^-$ conductance is greater than that for $K^+$ (Hoffmann and Simonsen, 1989; Grinstein and Foskett, 1990). This leads to a significant depolarization during the RVD because of a greater passive efflux of $Cl^-$ than of $K^+$, and the depolarization equalizes the anion and cation fluxes (macroscopic electroneutrality). Depolarization may have additional effects that support the RVD. The sensitivity of certain $K^+$ channels to intracellular $Ca^{2+}$ (i.e., $Ca^{2+}$-activated $K^+$ channels) is increased by depolarization, and other $K^+$ channels are directly opened by depolarization.

## 3. Cytoskeleton

The cytoskeleton consists of three main elements: actin filaments (F-actin), microfilaments 5–7 nm in diameter that

are double-stranded $\alpha$-helical polymers of globular actin (G-actin); microtubules, hollow tubes 25 nm in diameter that are made from tubulin monomers arranged in 13 threads; and intermediate filaments (IFs), 10-nm-diameter strands composed of tissue-specific proteins, such as keratin in epithelial cells and desmin in muscle cells (Bershadsky and Vasiliev, 1988; Luna and Hitt, 1992; Mills et al., 1994). In nonmuscle cells, the majority of actin filaments are associated with the cell membrane. F-actin is tied together and to integral membrane proteins by ankyrin, spectrin, MARKS (myristolated acid-rich C-kinase substrate), and other binding proteins forming a structural unit. This scaffolding undergoes constant reorganization as G-actin polymerizes and depolymerizes in response to various stimuli. The main role of microtubules in mature cells is thought to be the transport of vesicles within the cell using specific microtubule-associated proteins (MAPs) to attach to kinesin and dynein, which act as molecular motors. The functions of IF are not fully understood. Because IF binds to ankyrin and to desmosomal plaques, it is thought these filaments have a structural role and link organelles to the membrane cytoskeleton. Intermediate filaments are phosphorylated and dephosphorylated by protein kinases and phosphatases, and it is likely that their phosphorylation stage regulates function.

Alterations in cell volume leads to deformation or reorganization of the cytoskeleton, and ideas as to how these effects may be linked to cell volume regulation have been proposed (Chamberlin and Strange, 1989; Sachs, 1989; Sarkadi and Parker, 1991; Mills et al., 1994; Hoffmann and Dunham, 1995). One possibility is that the cytoskeleton might mechanically resist cell swelling, but the evidence for this is inconclusive.[3] Ion channels, exchangers, and cotransporters are structurally anchored in the membrane by

---

[3] For eukaryotic cells, it is generally assumed that the osmotic gradient across a membrane is negligible. This view arises between the thin bilayer membrane is too fragile to resist the substantial forces developed by even small differences in osmotic pressure. A tension of ~10 dyn cm$^{-1}$ is sufficient to rupture erythrocyte (Evans et al., 1976), protoplast (Wolfe et al., 1986), or lipid bilayer (Needham and Nunn, 1990) membranes. The relationship between tension on the membrane and transmembrane pressure is given by the law of Laplace. For a thin-walled spherical cell,

$$P = 2T/r$$

where $P$ is pressure, $T$ is tension, and $r$ is the cell's radius. If we assume a radius of 10 $\mu$m, a lytic tension is developed by a pressure of $2 \times 10^4$ dyn cm$^{-2}$, which equals ~15 mm Hg. This is the osmotic pressure developed by only ~0.0008 osmol liter$^{-1}$. Hence, the membrane cannot support a sufficient hydrostatic pressure to offset an osmotic gradient.

Before accepting this conclusion, it is necessary to consider the effective radius of the cell in view of the geometry of the cytoskeleton (Jacobson, 1983). For example, there are ~10$^5$ copies of ankyrin per erythrocyte. If all these attach integral membrane proteins to the cytoskeleton and are evenly distributed, the membrane is strengthened by load-bearing cytoskeletal elements at ~40-nm intervals. Even if we assume an effective radius of 100 nm, lytic tension now requires a pressure of $2 \times 10^6$ dyn cm$^{-2}$, which equals ~1500 mm Hg or nearly 2 atm. This is equivalent to the osmotic pressure generated by 0.078 osmol liter$^{-1}$.

specific components of the cytoskeleton (Luna and Hitt, 1992), and their interaction may help regulate cell volume. Agents that disrupt the microfilaments, such as the cytochalasins, modify the regulatory responses to both hypotonic and hypertonic stress by inhibiting $K^+$ and $Cl^-$ channels that promote RVD and stimulating $Na^+-K^+-2Cl^-$ cotransport responsible for RVI (Chamberlin and Strange, 1989; Mills et al., 1994; Hoffmann and Dunham, 1995). In a few cases, cell volume under isosmotic conditions is affected (McCarty and O'Neil, 1992). Microtubules also may be involved in some cell lines. Colchicine, which prevents microtubule polymerization, decreases macrophage volume by 20% (Mills, et al., 1994). The shrinkage is inhibited by SITS, suggesting that either $Cl^--HCO_3^-$ exchange or $Cl^-$ channels are involved. In addition, water and ion channels are inserted reversibly into the membrane in response to volume perturbations in a process that can be blocked by cytochalasin B (Lewis and de Moura, 1982). Furthermore, mechanical deformation of the membrane and supporting structures can modulate biochemical signaling systems such as the cAMP and protein kinase C cascades (Watson, 1991; Richter et al., 1987). Emerging evidence suggests that protein kinases and phosphatases may be specifically localized by cytoskeletal binding proteins to sites adjacent to their target transport proteins. These developments are potentially important because how the cytoskeleton exerts its regulatory effect on ion transporters remains obscure.

Another role for the cytoskeleton involves widely distributed mechanosensitive ion channels (Sachs, 1989; Morris, 1990). The probability of channel opening is increased by osmotic stretch or mechanical deformation, and it has been proposed that mechanical forces are detected by the cytoskeletal protein spectrin. Both stretch-activated cation and anion-selective channels are found. Most stretch-activated channels poorly discriminate between permeant species (e.g., the cation stretch-activated channel admits $Na^+$, $K^+$, and $Ca^{2+}$), but some are highly selective for $K^+$. There also are reports of mechanosensitive channels that are inactivated by stretch (Sachs, 1989; Morris, 1990). Mechanosensitive channels are potentially important in cell volume regulation as both sensors and effectors. The evidence that they are opened by osmotic stretch is convincing, and they carry substantial currents, which might directly or indirectly lead to volume regulation. For example, it has been argued that $K^+$-selective stretch-activated channels in molluscan heart cells pass sufficient current to alter $[K^+]_i$ by 1% in 1 sec when the cell is voltage-clamped away from $E_K$ (Brezden et al., 1986), and that poorly selective cation stretch-activated channels in Necturus choroid epithelium raise $Ca^{2+}$ sufficiently in 100 sec to open $Ca^{2+}$-activated $K^+$ channels and initiate an RVD (Christensen, 1987). Such calculations must be regarded as estimates, however, because they do not reflect $E_m$ under the relevant conditions. Nevertheless, $Ca^{2+}$ entry via stretch-activated channels has been implicated in volume regulation in several types of cells (Foskett, 1994), and arguments favoring the idea that stretch-activated channels participate in cell volume regulation are accumulating (Sackin, 1994). We have found that $Gd^{3+}$, a blocker of

nonselective cation stretch-activated channels, reduced swelling of unclamped rabbit ventricular myocytes in 195 mosmol liter$^{-1}$ solution by 32%, and 9-anthracene carboxylic acid (9-AC), a blocker of anion stretch-activated channels, increased swelling in the same solution by 44% (Suleymanian et al., 1995). Opposite effects of the two blockers are expected because anions and cations travel in opposite directions under these conditions. In contrast, $Gd^{3+}$ and 9-AC have negligible effects on myocyte volume in isosmotic solution when stretch-activated channels are expected to be closed.

### 4. Calcium

A role for $Ca^{2+}$ in cell volume regulation has been recognized for many years (Pierce and Politis, 1990; McCarty and O'Neil, 1992; Foskett, 1994; Hoffmann and Dunham, 1995). RVD is blocked by removing extracellular $Ca^{2+}$ in Amphiuma erythrocytes, Necturus gallbladder, proximal convoluted and straight tubule, intestine 407, and osteosarcoma cells. Extracellular $Ca^{2+}$ is not a requirement for RVD in lymphocytes and Ehrlich ascites cells, but RVD is more rapid in Ehrlich ascites cells when $Ca^{2+}$ is present. In some tissues it appears that the $Ca^{2+}$ involved in RVD enters, at least in part, through dihydropyridine-sensitive, L-type $Ca^{2+}$ channels. Significant $Ca^{2+}$ entry also can occur via nonselective stretch-activated cation channels. Block of RVD by lanthanides and disruption of the cytoskeleton, both of which affect stretch-activated cation channels, is consistent with this possibility. Instead of the entry of extracellular $Ca^{2+}$, release of $Ca^{2+}$ from intracellular stores is critical for RVD in a variety of cells, including lymphocytes, Ehrlich ascites, intestine 407, and opossum kidney cells. Depletion of internal $Ca^{2+}$ stores eliminates RVD in these cells; it is restored by extracellular $Ca^{2+}$ and the $Ca^{2+}$ ionophore, A23187. $[Ca^{2+}]_i$ has been shown to increase during swelling of Amphiuma erythrocytes using arsenazo III as a $Ca^{2+}$ indicator. More recently, fluorescent $Ca^{2+}$ indicators, quin-2 and fura-2, have been used to demonstrate increases in $[Ca^{2+}]_i$ that accompany cell swelling in urinary bladder, osteosarcoma, lymphoma, and proximal and straight convoluted tubule cells. On the other hand, $[Ca^{2+}]_i$ remains unchanged during volume changes in some cells, including human lymphocytes. More evidence for elevated $Ca^{2+}$ comes from patch-clamp studies. In a number of tissues, recordings of single channel activity established that $Ca^{2+}$-activated $K^+$ channels open more frequently during RVD than under isotonic conditions. This suggests an increase in $[Ca^{2+}]_i$ because the probability that these channels open increases as $[Ca^{2+}]_i$ increases. Although studied in less detail, a decrease in $[Ca^{2+}]_i$ has been implicated in RVI in Amphiuma erythrocytes, Ehrlich ascites cells, and lymphocytes.

The mechanism by which $Ca^{2+}$ modulates cell volume appears to vary (Table 6). $Ca^{2+}$ has direct effects on ion channels, but additional signaling mechanisms may be involved. For example, calmodulin inhibitors can block the increased $K^+$ conductance in lymphocytes, Ehrlich ascites cells, and Necturus gallbladder; the increased $Cl^-$ conductance in Ehrlich ascites cells; and increased $K^+-H^+$ exchange in Amphiuma erythrocytes. In other cells, modula-

**TABLE 6**    Intracellular Signaling Pathways for RVD and RVI[a]

| Signal | Effector | Cell type |
|---|---|---|
| $Ca^{2+}$ | $K^+$ conductance | *Necturus* gallbladder |
|  |  | Frog urinary bladder |
|  |  | Ehrlich ascites cells |
|  |  | Human lymphocytes |
|  | $Cl^-$ conductance | Ehrlich ascites cells |
|  | Taurine efflux | Elasmobranch and molluscan erythrocytes |
|  | $Na^+$–$H^+$ exchange | Human lymphocytes |
|  | $K^+$–$H^+$ exchange | *Amphiuma* erythrocytes |
| Phosphorylation | $Na^+$–$H^+$ exchange | Human lymphocytes |
|  | Taurine efflux | Elasmobranch erythrocytes |
|  | $Na^+$–$K^+$–$2Cl^-$ cotransport | Duck erythrocytes |
|  | $K^+$–$Cl^-$ cotransport | Duck, rabbit, and dog erythrocytes |
| Leukotrienes | $K^+$ conductance | Ehrlich ascites cells |
|  | $Cl^-$ conductance | Ehrlich ascites cells |
| cAMP | $Na^+$–$H^+$ exchange | Mouse thick ascending limb of Henle (mTALH) cells |
|  | $Cl^-$–$HCO_3^-$ exchange | Mouse thick ascending limb of Henle (mTALH) cells |
|  | $Na^+$–$Cl^-$ cotransport | *Necturus* gallbladder |
| G proteins | $Na^+$–$H^+$ exchange | Barnacle skeletal muscle |
| Voltage | $K^+$ channels | Human lymphocytes |

[a] For references, see Chamberlin and Strange (1989), Sarkadi and Parker (1991), McCarty and O'Neil (1992), Strange (1994), and Hoffmann and Dunham (1995).

tion by $Ca^{2+}$ of protein kinase C and leukotriene synthesis have been proposed as signals in volume regulation.

## 5. Phosphorylation

The activity of many of the transporters discussed here are modified by phosphorylation (Parker, 1992; McCarty and O'Neil, 1992; Palfrey, 1994; Hoffmann and Dunham, 1995). This raises the possibility that cell volume alterations initiate an RVD or RVI by either increasing or decreasing the fraction of transporters in the phosphorylated state. Until recently, supporting data have been lacking. Over the last few years, however, strong evidence for this idea has come from studies in several tissues. We discuss some of these data from erythrocytes in detail.

Pewitt *et al.* (1990) studied RVI in duck erythrocytes and determined that activation of $Na^+$–$K^+$–$2Cl^-$ cotransport on shrinking is caused by phosphorylation. Both cAMP-dependent and cAMP-independent protein kinase phosphorylate the $Na^+$–$K^+$–$2Cl^-$ cotransporter (or possibly a regulatory protein), but cAMP levels are not affected in duck erythrocytes by osmotic stress. Pewitt *et al.* (1990) found that the protein kinase inhibitors K252a and H-9 prevent transporter activation on shrinking. Conversely, an inhibitor of serine and threonine protein-phosphatases, okadaic acid, which slows protein dephosphorylation, stimulates $Na^+$–$K^+$–$2Cl^-$ cotransport under isotonic conditions. These changes in the activity of the transporter with phosphorylation and with shrinking apparently result largely from a modulation of the number of functioning transporters, as detected by bumetanide binding, rather than from a modulation of their turnover rate. At about the same time, Jennings and al-Rohil (1990) and Jennings and Schulz

(1991) developed evidence that $K^+$–$Cl^-$ cotransport in rabbit erythrocytes, which is responsible for RVD, is activated by dephosphorylation. They discovered that swelling inhibits a protein kinase distinct from protein kinase A and C. An RVD occurred only after a slow dephosphorylation, now identified as due to a type 1 protein phosphatase (PP1) that is blocked by calyculin A (Starke and Jennings, 1993). Parker *et al.* (1991) obtained similar results in dog erythrocytes.

Parker *et al.* (1991) also importantly recognized the reciprocal coordination of $K^+$–$Cl^-$ cotransport and $Na^+$–$H^+$ exchange by phosphorylation and dephosphorylation during both RVDs and RVIs in mammalian erythrocytes. This strategy is illustrated in Fig. 23 and can be summarized as follows. (1) Shrinking activates and swelling inhibits a protein kinase. (2) On shrinking, activated protein kinase rapidly phosphorylates regulatory sites associated with the $K^+$–$Cl^-$ cotransporter and the $Na^+$–$H^+$ exchanger (or the $Na^+$–$K^+$–$2Cl^-$ cotransporter in duck erythrocytes). (3) Phosphorylation inhibits the $K^+$–$Cl^-$ cotransporter, reducing osmolyte efflux, but stimulates the $Na^+$–$H^+$ exchanger or $Na^+$–$K^+$–$2Cl^-$ cotransporter, stimulating osmolyte uptake and leading to an RVI. (4) Conversely, slow dephosphorylation on swelling stimulates $K^+$–$Cl^-$ cotransport and inhibits $Na^+$–$H^+$ exchange or $Na^+$–$K^+$–$2Cl^-$ cotransport, leading to an RVD. Thus, the transporters underlying ion influx and efflux are regulated reciprocally by the activity of a protein kinase that reflects cell volume.

How does erythrocyte volume govern the activity of a protein kinase? Several possible detectors have been considered: cell shape or cytoskeletal deformation; the concentration of an impermeant intracellular cofactor such as $Mg^{2+}$; and a concept referred to as macromolecular

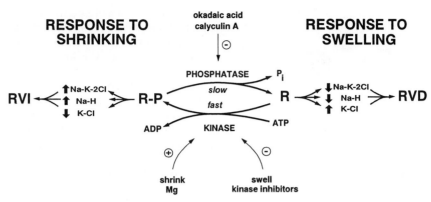

**FIG. 23.** Schematic diagram of the mechanism of coordination of $Na^+-K^+-2Cl^-$ cotransport and $Na^+-H^+$ and $Na^+-K^+$ exchange during cell swelling and shrinking. R–P and R represent phosphorylated and dephosphorylated regulatory sites that modulate the activity of transporters leading to a regulatory volume decrease (RVD) or regulatory volume increase (RVI). Changes in cell volume affect protein kinase activity. Several interventions that inhibit or stimulate the protein kinase and phosphatase are indicated.

crowding. The first possibility has been discussed, but experimental evidence suggests cell shape does not regulate phosphorylation of the relevant transport proteins, at least in erythrocytes. The last two possibilities are considered next.

### 6. Mass Action Model

Increasing intracellular $Mg^{2+}$ activates $Na^+-H^+$ exchange and inhibits $K^+-Cl^-$ cotransport, and it has been suggested that $Mg^{2+}$ might act by inhibiting a phosphatase. Before accepting the idea that $Mg^{2+}$ or another intracellular ion is the volume sensor, it is necessary to explain the steep dependence of ion transport on cell volume. Jennings and Schulz (1990) illustrated one possible answer for $K^+-Cl^-$ cotransport with a theoretical mass action model. They assumed (1) the volume sensor (e.g., $Mg^{2+}$) is an impermeant intracellular species, (2) the kinase and phosphatase are soluble enzymes, and (3) the sensor inhibits dephosphorylation. They then described the model in terms of three first-order Michaelis–Menten expressions. Taking into account that the concentration of sensor, kinase, and phosphatase vary inversely with volume, they were able to reproduce the steep volume dependence of experimental $K^+$ influx data. As the authors emphasized, however, a good fit of the experimental data does not prove that the model is correct. Rather, it illustrates only that a simple dilution mechanism can give rise to a steep volume dependence of transport if dilution has different effects on the activity of enzymes that regulate the transporter (e.g., see Fig. 23).

### 7. Macromolecular Crowding

The concept of **macromolecular crowding** comes from the idea that proteins do not behave ideally in solution at concentrations in the physiological range. We have already mentioned that the osmotic coefficient for hemoglobin and other proteins increases steeply with concentration, and that Freedman and Hoffman (1979) used this fact to explain

water movement in red cells. Nonideal behavior is thought to be a more general phenomenon, however. Minton (1983, 1990, 1994) has argued that the kinetics and equilibria of enzymes (macromolecules) is markedly altered by the presence of inert macromolecules that occupy more than a few percent of the total solution volume. Just as one hemoglobin affects another, macromolecules that are neither substrate nor product affect the behavior of their macromolecular neighbors in solution. This results because "crowding" reduces the solution volume accessible to a macromolecule by excluded-volume effects, as illustrated in Fig. 24. An excluded volume means that solution behavior is nonideal, and the chemical potential, $\mu_i$, and activity, $a_i$, of a macromolecule is increased by crowding. Consequently, reaction rates are affected. On the other hand, small solutes (e.g., ions) are unaffected by the same concentration of macromolecules. Several examples are worth noting. Minton (1983) showed that the specific activity of glyceraldehyde 3-phosphate dehydrogenase decreased dramatically as the concentration of bovine serum albumin, $\beta$-lactoglobulin, poly(ethylene glycol) (PEG), or ribonuclease in the reaction medium was increased. This was explained by suggesting that crowding favored the formation of tetramers of the enzyme that possess a lower catalytic specific activity than monomers. Similarly, the cohesion of complementary ends of $\lambda$ DNA can be increased up to 2000-fold by albumin, Ficoll 70, or PEG (Zimmerman and Harrison, 1985), and the activity of T4 polynucleotide kinase is augmented by PEG (Harrison and Zimmerman, 1986). Protein concentrations within cells are sufficient to give significant excluded-volume effects (Zimmerman and Trach, 1991).

How does macromolecular crowding relate to cell volume regulation? Perhaps the activity of the kinase governing the phosphorylation state of transporters decreases as macromolecular crowding is lessened during cell swelling. This would, for example, activate $K^+-Cl^-$ cotransport and inactive $Na^+-H^+$ exchange (see Fig. 23), and an RVD would ensue. Colclasure and Parker (1991, 1992) provide experimental support for this hypothesis (also see Sarkadi

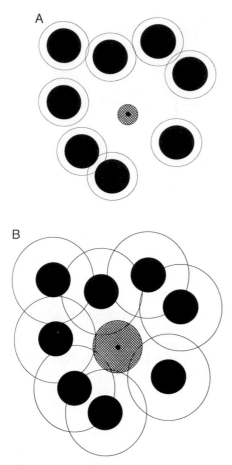

**FIG. 24.** Diagram of macromolecular crowding. Background macromolecules (solid) in solution differentially affect the behavior of small molecules, such as ions (A), and other macromolecules (B). A test molecule (cross-hatched) cannot enter the excluded volume represented by a circle with a radius equal to the sum of the radii of the test molecule and the background macromolecule. The excluded volume is much greater for a test macromolecule (B) than for a test small molecule (A). Restricting the volume that a test molecule can enter increases its activity, $a$, and chemical potential, $\mu$. Consequently, both reaction rates and equilibria are altered. [Reprinted with permission from Minton (1994). Copyright CRC Press, Boca Raton, Florida. © 1994.]

and Parker, 1991; Parker, 1992). They osmotically ruptured dog erythrocytes and then allowed the "ghosts" to reseal in a hypotonic medium. This gave resealed ghosts with about one-fourth the normal volume but with a normal protein (hemoglobin or hemoglobin plus albumin) concentration. During osmotic challenge, $K^+$–$Cl^-$ cotransport and $Na^+$–$H^+$ exchange were regulated at about the same protein concentration as in normal erythrocytes, even though the volumes of the resealed ghosts were vastly different. This hypothesis also is consistent with studies on the kinetics of volume regulation by Jennings and al-Rohil (1990), who concluded that the swelling-sensitive step was a decrease in the rate of phosphorylation rather than an increase in the rate of dephosphorylation.

The new insight that macromolecular crowding may be

a mechanism for volume transduction is an exciting possibility. Minton *et al.* (1992) have presented a model quantitatively accounting for volume-dependent stimulation of ion fluxes on this basis. It is important to recognize that operation of the scheme depends on the components having appropriate sensitivity. In erythrocytes, for example, the activities of both the kinase and phosphatase should be reduced by swelling. Consequently, as the rate of protein phosphorylation falls, so does the rate of protein dephosphorylation. Whether this leads to an increase or decrease in the fraction of transporters in the phosphorylated state must depend both on the relative effects of macromolecular crowding and on the amount of substrate available for each enzyme. With injudicious choices for the parameters (e.g., pathological interventions), this mechanism might lead to an inappropriate RVI rather than an RVD.

## VII. Conclusion

The study of mechanisms underlying osmosis and the regulation of cell volume under both isosmotic and anisosmotic conditions has been fruitful. We understand in substantial detail how water and ions cross the membrane. Where do we go from here? Many directions are possible, and as many details are missing as are known. For example, the identification and cloning of water channels raise several important question. What is it about the protein structure that makes this a water channel? How does water interact with the channel? From a theoretical perspective, how is osmotic pressure sensed, and how does osmosis occur through a channel structure that is different in important ways from the well-explored hydrodynamic models? From the perspective of regulation of ion transport, much remains to be understood about how cells sense swelling and shrinking and how a cell "decides" on its optimal volume. There are also unanswered questions concerning the regulation of volume regulatory ion transporters by cellular messengers, metabolic demands, and pathological states. In short, we can look forward to many more fruitful years of research on these topics.

## Acknowledgment

We thank J. Maghirang for preparing numerous figures. Supported by National Institutes of Health Grants HL-24847 and HL-46764, and Grants-in-Aid from the American Heart Association and its Virginia affiliate.

## References

Adair, G. S. (1928). A theory of partial osmotic pressures and membrane equilibrium with special reference to the application of Dalton's law to haemoglobin solutions in the presence of salts. *Proc. Roy. Soc. London* **120A,** 573–603.
Agius, L., Peak, M., Beresford, G., al-Habori, M., and Thomas, T. (1994). The role of ion content and cell volume in insulin action. *Biochem. Soc. Trans.* **22,** 518–522.

Agre, P., Preston, G. M., Smith, B. L., Jung, J. S., Raina, S., Moon, C., Guggino, W. B., and Nielsen, S. (1993). Aquaporin CHIP: The archetypal molecular water channel. *Am. J. Physiol.* **265**, F463–F476.

Andreoli, T. E., and Troutman, S. L. (1971). An analysis of unstirred layers in series with "tight" and "porous" lipid bilayer membranes. *J. Gen. Physiol.* **57**, 464–478.

Bandarali, U., and Roy, G. (1992). Anion channels for amino acids in MDCK cells. *Am. J. Physiol.* **263**, C1200–C1207.

Barry, P. H., and Diamond, J. M. (1984). Effects of unstirred layers on membrane phenomena. *Physiol. Rev.* **64**, 763–873.

Berg, M. B. (1995). Molecular basis of osmolyte regulation. *Am. J. Physiol.* **268**, F983–F996.

Bershadsky, A. D., and Vasiliev, J. M. (1988). "Cytoskeleton." Plenum Press, New York.

Bookchin, R. M., Ortiz, O. E., Freeman, C. J., and Lew, V. L. (1989). Predictions and tests of a new integrated reticulocyte model: Implications for dehydration of sickle cells. *In* "The Red Cell: Seventh Ann Arbor Conference," pp. 615–625. Alan R. Liss Publishers, New York.

Boyer, J. L., Graf, J., and Meier, P. J. (1992). Hepatic transport systems regulating pH$_i$, cell volume, and bile secretion. *Annu. Rev. Physiol.* **54**, 415–438.

Boyle, P. J., and Conway, E. J. (1941). Potassium accumulation in muscle and associated changes. *J. Physiol.* **100**, 1–63.

Brezden, B. L., Gardner, D. R., and Morris, C. E. (1986). A potassium-selective channel in isolated *Lymnaea stagnalis* heart muscle cells. *J. Exp. Biol.* **123**, 175–190.

Carruthers, A., and Melchior, D. L. (1983). Studies of the relationship between bilayer water permeability and bilayer physical state. *Biochemistry* **22**, 5797–5807.

Cass, A., and Finkelstein, A. (1967). Water permeability of thin lipid membranes. *J. Gen. Physiol.* **20**, 1765–1784.

Chamberlin, M. E., and Strange, K. (1989). Anisosmotic cell volume regulation: A comparative view. *Am. J. Physiol.* **257**, C159–C173.

Christensen, O. (1987). Mediation of cell volume regulation by Ca$^{2+}$ influx through stretch-activated channels. *Nature* **330**, 66–68.

Clemo, H. F., and Baumgarten, C. M. (1995). cGMP and atrial natriuretic factor regulate cell volume of rabbit atrial myocytes. *Cir. Res.* **77**, 741–749.

Clemo, H. F., Feher, J. J., and Baumgarten, C. M. (1992). Modulation of rabbit ventricular cell volume and Na$^+$/K$^+$/2Cl$^-$ cotransport by cGMP and atrial natriuretic factor. *J. Gen. Physiol.* **100**, 89–114.

Colclasure, C. G., and Parker, J. C. (1991). Cytosolic protein concentration is the primary volume signal in dog red cells. *J. Gen. Physiol.* **98**, 881–892.

Colclasure, C. G., and Parker, J. C. (1992). Cytosolic protein concentration is the primary volume signal for swelling-induced [K–Cl] cotransport in dog red cells. *J. Gen. Physiol.* **100**, 1–10.

Dainty, J. (1965). Osmotic flow. *Symp. Soc. Exp. Biol.* **19**, 75–85.

Deamer, D. D., and Bramhall, J. (1986). Permeability of lipid bilayers to water and ionic solutes. *Chem. Phys. Lipids* **40**, 167–188.

Drewnowska, K., and Baumgarten, C. M. (1991). Regulation of cellular volume in rabbit ventricular myocytes: Bumetanide, chlorothiazide, and ouabain. *Am. J. Physiol.* **260**, C122–C131.

Drewnowska, K., Clemo, H. F., and Baumgarten, C. M. (1991). Prevention of myocardial intracellular edema induced by St. Thomas' Hospital cardioplegic solution. *J. Molec. Cell. Cardiol.* **23**, 1215–1221.

Evans, E., Waugh, R., and Melnik, L. (1976). Elastic area compressibility modulus of red cell membrane. *Biophys. J.* **16**, 585–595.

Fettiplace, R., and Haydon, D. A. (1980). Water permeability of lipid membranes. *Physiol. Rev.* **60**, 510–550.

Finkelstein, A. (1987). "Water Movement through Lipid Bilayers, Pores, and Plasma Membranes. Theory and Reality" Wiley-Interscience, New York.

Finkelstein, A., and Cass, A. (1968). Permeability and electrical properties of thin lipid membranes. *J. Gen. Physiol.* **52**, 145s–172s.

Fischbarg, J., Kuang, K., Vera, J. C., Arant, S., Silverstein, S. C., Loike, J., and Rosen, O. M. (1990). Glucose transporters serve as water channels. *Proc. Natl. Acad. Sci. USA* **87**, 3244–3247.

Foskett, J. K. (1994). The role of calcium in the control of volume-regulatory pathways. *In* "Cellular and Molecular Physiology of Cell Volume Regulation" (K. Strange, Ed.), pp. 259–277. CRC Press, Boca Raton, FL..

Freedman, J. C., and Hoffman, J. F. (1979). Ionic and osmotic equilibria of human red blood cells treated with nystatin. *J. Gen. Physiol.* **74**, 157–185.

Garcia, A., and Berg, M. B. (1991). Role of organic osmolytes in adaption of renal cells to high osmolarity. *J. Memb. Biol.* **119**, 1–13.

Glasstone, S. (1946). "Textbook of Physical Chemistry." Van Nostrand, Princeton, NJ.

Grinstein, S., and Foskett, J. K. (1990). Ionic mechanisms of cell volume regulation in leukocytes. *Annu. Rev. Physiol.* **52**, 399–414.

Grinstein, S., Clarke, C. A., and Rothstein, A. (1983). Activation of Na$^+$/H$^+$ exchange in lymphocytes by osmotically-induced volume changes and by cytoplasmic acidification. *J. Gen. Physiol.* **82**, 619–638.

Grinstein, S., Rothstein, A., Sarkadi, B., and Gelfand, E. W. (1984). Responses of lymphocytes to anisotonic media: Volume-regulating behavior. *Am. J. Physiol.* **246**, C204–C215.

Haas, M. (1994). The Na–K–Cl cotransporters. *Am. J. Physiol.* **267**, C869–C885.

Haines, T. H. (1994). Water transport across biological membranes. *FEBS Lett.* **346**, 115–122.

Harrison, B., and Zimmerman, S. B. (1986). T4 polynucleotide kinase: Macromolecular crowding increases the efficiency of reaction at DNA termini. *Anal. Biochem.* **158**, 307–315.

Hasegawa, H., Skach, W., Baker, W., Calayag, M. C., Lingappa, V., and Verkman, A. S. (1992). A multifunctional aqueous channel formed by CFTR. *Science* **258**, 1477–1479.

Hasegawa, H., Zhang, R., Dohrman, A., and Verkman, A. S. (1993). Tissue-specific expression of mRNA encoding rat kidney water channel CHIP28k by *in situ* hydridization. *Am. J. Physiol.* **264**, C237–C245.

Häussinger, D., and Lang, F. (1991). Cell volume in the regulation of hepatic function: A mechanism for metabolic control. *Biochim. Biophys. Acta* **1071**, 331–350.

Häussinger, D., Lang, F., and Gerok, W. (1994). Regulation of cell function by the cellular hydration state. *Am. J. Physiol.* **267**, E343–E355.

Hernandez, J. A., and Fischbarg, J. (1992). Kinetic analysis of water transport through a single-file pore. *J. Gen. Physiol.* **99**, 645–662.

Hildebrand, J. H. (1955). Osmotic pressure. *Science* **121**, 116–119.

Hinke, J. A. M. (1980). Water and electrolyte content of the myofilament phase in the chemically skinned barnacle fiber. *J. Gen. Physiol.* **75**, 531–551.

Hladky, S. B., and Rink, T. J. (1978). Osmotic behaviour of human red blood cells: An interpretation in terms of negative intracellular fluid pressure. *J. Physiol.* **274**, 437–446.

Hobbie, R. K. (1978). "Intermediate Physics for Medicine and Biology," pp. 126–132. John Wiley, New York.

Hoffman, E. K., and Dunham, P. B. (1995). Membrane mechanisms and intracellular signalling in cell volume regulation. *Int. Rev. Cytol.* **161**, 173–262.

Hoffmann, E. K., and Simonsen, L. O. (1989). Membrane mechanisms in volume and pH regulation in vertebrate cells. *Physiol. Rev.* **69**, 315–382.

Hoffmann, E. K., Lambert, I. H., and Simonsen, L. O. (1986). Separate, Ca$^{2+}$-activated K$^+$ and Cl$^-$ transport pathways in Ehrlich ascites tumor cells. *J. Membr. Biol.* **91**, 227–244.

Holz, R., and Finkelstein, A. (1970). The water and nonelectrolyte permeability induced in thin lipid membranes by the polyene antibiotics nystatin and amphotericin B. *J. Gen. Physiol.* **56,** 125–145.

Jackson, P. S., Morrison, R., and Strange, K. (1994). The volume-sensitive organic osmolyte-anion channel VSOAC is regulated by nonhydrolytic ATP binding. *Am. J. Physiol.* **267,** C1203–C1209.

Jacobson, B. S. (1983). Interaction of the plasma membrane with the cytoskeleton: An overview. *Tissue Cell* **15,** 829–852.

Jakobsson, E. (1980). Interactions of cell volume, membrane potential, and membrane transport parameters. *Am. J. Physiol.* **238,** C196–C206.

Jansen, M., and Blume, A. (1995). A comparative study of diffusive and osmotic water permeation across bilayers composed of phospholipids with different head groups and fatty acyl chains. *Biophys. J.* **68,** 997–1008.

Jap, B. K., and Li, H. (1995). Structure of the osmo-regulated $H_2O$-channel, AQP-CHIP, in projection at 3.5 Å resolution. *J. Mol. Biol.* **251,** 413–420.

Jennings, M. L., and al-Rohil, N. (1990). Kinetics of activation and inactivation of swelling-stimulated $K^+/Cl^-$ transport. The volume sensitive parameter is the rate constant for inactivation. *J. Gen. Physiol.* **95,** 1021–1040.

Jennings, M. L., and Schulz, R. K. (1990). Swelling-activated KCl cotransport in rabbit red cells: Flux is determined mainly by cell volume rather than shape. *Am. J. Physiol.* **259,** C960–C967.

Jennings, M. L., and Schulz, R. K. (1991). Okadaic acid inhibition of KCl cotransport. Evidence that protein dephosphorylation is necessary for activation of transport by either cell swelling or *N*-ethylmaleimide. *J. Gen. Physiol.* **97,** 799–817.

Jung, J. A., Preston, G. M., Smith, B. L., Guggino, W. B., and Agre, P. (1994). Molecular structure of the water channel through aquaporin CHIP: The hourglass model. *J. Biol. Chem.* **269,** 14648–14654.

Kaplan, M. R., Mount, D. B., and Delpire, E. (1996). Molecular mechanisms of NaCl cotransport. *Annu. Rev. Physiol.* **58,** 649–668.

Kiil, F. (1989). Molecular mechanisms of osmosis. *Am. J. Physiol.* **256,** R801–R808.

King, L. S., and Agre, P. (1996). Pathophysiology of the aquaporin water channels. *Annu. Rev. Physiol.* **58,** 619–648.

Kregenow, F. M. (1981). Osmoregulatory salt transporting mechanisms: Control of cell volume in anisotonic media. *Ann. Rev. Physiol.* **43,** 493–505.

Kwon, H. M., Yamauchi, A., Uchida, S., Preston, A. S., Garcia-Perez, A., Berg, M. B., and Handler, J. S. (1992). Cloning of a $Na^+/myo$-inositol cotransporter, a hypetonicity stress protein. *J. Biol. Chem.* **267,** 6229–6301.

Lakshminarayanaiah, N. (1984). "Equations of Membrane Biophysics," pp. 107–118. Academic Press, New York.

Landis, E. M., and Pappenheimer, J. R. (1963). Exchange of substances through the capillary walls. *In* "Handbook of Physiology," Vol. 2, Sec. 2, pp. 961–1034. American Physiological Society, Washington, DC.

Leaf, A. (1956). On the mechanism of fluid exchange of tissues *in vitro. Biochem. J.* **62,** 241–248.

Leaf, A. (1959). Maintenance of concentration gradients and regulation of cell volume. *Ann. NY Acad. Sci.* **72,** 396–404.

LeNeveu, D. M., Rand, R. P., and Parsegian, V. A. (1976). Measurement of forces between lecithin bilayers. *Nature* **259,** 601–603.

Lewis, S. A., and de Moura, J. L. (1982). Incorporation of cytoplasmic vesicles into apical membrane of mammalian urinary bladder epithelium. *Nature* **297,** 685–688.

Luna, E. J., and Hitt, A. L. (1992). Cytoskeleton–plasma membrane interactions. *Science* **258,** 955–964.

Macknight, A. D. C. (1988). Principles of cell volume regulation. *Renal Physiol. Biochem.* 3–5, 114–141.

Macknight, A. D. C., and Leaf, A. (1977). Regulation of cellular volume. *Physiol. Rev.* **57,** 510–573.

Macey, R. I. (1979). Transport of water and nonelectrolytes across red cell membranes. *In* "Membrane Transport in Biology," Vol. II, Transport across Single Biological Membranes. (G. Giebisch, D. C. Tosteson, and H. H. Ussing, Eds.), pp. 1–57. Springer Verlag, Berlin.

Macey, R. I., and Farmer, R. E. L. (1970). Inhibition of water and solute permeability in human red cells. *Biochim. Biophys. Acta* **211,** 104–106.

Macey, R. I., Karan, D. M., and Farmer, R. E. L. (1972). Properties of water channels in human red cells. *In* "Biomembranes, Vol. 3, Passive Permeability of Cell Membranes," (F. Kreuzer and J. F. G. Slegers, Eds.), pp. 331–340. Plenum Press, New York.

Mauro, A. (1957). Nature of solvent transfer in osmosis. *Science* **126,** 252–253.

McCarty, N. A., and O'Neil, R. G. (1992). Calcium signaling in cell volume regulation. *Physiol. Rev.* **72,** 1037–1061.

Meschia, G., and Setnikar, I. (1958). Experimental study of osmosis through a collodion membrane. *J. Gen. Physiol.* **42,** 429–444.

Mills, J. W., Schweibert, E. M., and Stanton, B. A. (1994). The cytoskeleton and cell volume regulation. *In* "Cellular and Molecular Physiology of Cell Volume Regulation" (K. Strange, Ed.), pp. 241–258. CRC Press, Boca Raton, FL.

Minton, A. P. (1983). The effect of volume occupancy upon the thermodynamic activity of proteins: Some biochemical consequences. *Mol. Cell Biochem.* **55,** 119–140.

Minton, A. P. (1990). Holobiochemistry: The effect of local environment upon equilibria and rates of biochemical reactions. *Int. J. Biochem.* **22,** 1063–1067.

Minton, A. P. (1994). Influence of macromolecular crowding on intracellular association reactions: Possible role in volume regulation. *In* "Cellular and Molecular Physiology of Cell Volume Regulation" (K. Strange, Eds.), pp. 181–190. CRC Press, Boca Raton, FL.

Minton, A. P., Colclasure, G. C., and Parker, J. C. (1992). Model for the role of macromolecular crowding in regulation of cellular volume. *Proc. Natl. Acad. Sci. USA* **89,** 10504–10506.

Morris, C. E. (1990). Mechanosensitive ion channels. *J. Membr. Biol.* **113,** 93–107.

Needham, D., and Nunn, R. S. (1990) Elastic deformation and failure of lipid bilayer membranes containing cholesterol. *Biophys. J.* **58,** 997–1009.

Nielsen, S., Smith, B. L., Christensen, E. I., Knepper, M. A., and Agre, P. (1993). CHIP28 water channels are localized in constitutively water-permeable segments of the nephron. *J. Cell. Biol.* **120,** 371–383.

Nielsen, S., and Chou, C. L., Marples, D., Christensen, E. I., Kishore, B. K., and Knepper, M. A. (1995). Vasopressin increases water permeability of kidney collecting duct by inducing translocation of aquaporin-CD water channels in rat kidney inner medulla. *Proc. Natl. Acad. Sci. USA* **92,** 1013–1017.

Overbeek, J. T. G. (1956). The Donnan equilibrium. *Prog. Biophys. Biophys. Chem.* **3,** 57–84.

Paganelli, C. V., and Solomon, A. K. (1957). The rate of exchange of tritiated water across the human red cell membrane. *J. Gen. Physiol.* **41,** 259–277.

Palfrey, H. C. (1994). Protein phosphorylation control in the activity of volume-sensitive transport systems. *In* "Cellular and Molecular Physiology of Cell Volume Regulation" (K. Strange, Ed.), pp. 201–214. CRC Press, Boca Raton, FL.

Palfrey, H. C., and O'Donnell, M. E. (1992). Characteristics and regulation of the Na/K/2Cl cotransporter. *Cell Physiol. Biochem.* **2,** 293–307.

Parker, J. C. (1973). Dog red blood cells. Adjustment of density *in vitro. J. Gen. Physiol.* **62,** 147–156.

Parker, J. C. (1992). Volume-activated cation transport in dog red cells: Detection and transduction of the volume stimulus. *Comp. Biochem. Physiol.* **102A,** 615–618.

Parker, J. C., Colclasure, G. C., and McManus, T. J. (1991). Coordinated regulation of shrinkage-induced Na/H exchange and swelling-induced [K–Cl] cotransport in dog red cells. Further evidence from activation kinetics and phosphatase inhibition. *J. Gen. Physiol.* **98,** 869–880.

Pewitt, E. B., Hegde, R. S., Haas, M., and Palfrey, H. C. (1990). The regulation of Na/K/2Cl cotransport and bumetanide binding in avian erythrocytes by protein phosphorylation and dephosphorylation: Effects of kinase inhibitors and okadaic acid. *J. Biol. Chem.* **265,** 20747–20756.

Pfeffer, W. (1877). "Osmotische Untersuchungen. Studien zur Zellmechanik." Wilhelm Engelmann, Leipzig. [Translated by G. R. Kepner and E. J. Tadelmann (1985). "Osmotic Investigations. Studies on Cell Membranes." Van Nostrand Reinhold, New York.]

Pierce, S. K., and Politis, A. D. (1990). $Ca^{2+}$-activated cell volume recovery mechanisms. *Annu. Rev. Physiol.* **52,** 27–42.

Pitzer, K. S., and Mayorga, G. (1973). Thermodynamics of electrolytes. II. Activity and osmotic coefficients for strong electrolytes with one or both ions univalent. *J. Phys. Chem.* **77,** 2300–2308.

Preston, G. M., and Agre, P. (1991). Isolation of the cDNA for erythrocyte integral membrane protein of 28 kilodaltons member of an ancient channel family. *Proc. Natl. Acad. Sci.* **88,** 11110–11114.

Preston, G. M., Carroll, T. P., Guggino, W. B., and Agre, P. (1992). Appearance of water channels in *Xenopus* oocytes expressing red cell CHIP28 protein. *Science* **256,** 385–387.

Preston, G. M., Jung, J. S., Guggino, W. B., and Agre, P. (1993). The mercury-sensitive residue at cysteine-189 in the CHIP28 water channel. *J. Biol. Chem.* **268,** 17–20.

Preston, G. M., Jung, J. S., Guggino, W. B., and Agre P. (1994) Membrane topology of aquaporin CHIP: Analysis of functional epitope scanning mutants by vectorial proteolysis. *J. Biol. Chem.* **269,** 1668–1673.

Rasmusson, R. L., Davis, D. G., and Lieberman, M. (1993). Amino acid loss during volume regulatory decrease in cultured chick heart cells. *Am. J. Physiol.* **264,** C136–C145.

Renkin, E. M. (1954). Filtration, diffusion, and molecular sieving through porous cellulose membranes. *J. Gen. Physiol.* **38,** 225–243.

Reuss, L. (1985). Changes in cell volume measured with an electrophysiological technique. *Proc. Natl. Acad. Sci. USA* **82,** 6014–6018.

Richter, E. A., Cleland, P. J. F., Rattigan, S., and Clark, M. G. (1987). Contraction-association translocation of protein kinase C. *FEBS Lett.* **217,** 232–236.

Robinson, R. A., and Stokes, R. H. (1959). "Electrolyte Solutions," pp. 480–490. Butterworths, London.

Rosenberg, P. A., and Finkelstein, A. (1978). Water permeability of gramicidin A-treated lipid bilayer membranes. *J. Gen. Physiol.* **72,** 341–350.

Sachs, F. (1989). Ion channels as mechanical transducers. *In* "Cell Shape: Determinants, Regulation, and Regulatory Role" (W. D. Stein and F. Bronner, Eds.), pp. 63–92. Academic Press, San Diego.

Sackin, H. (1994). Stretch-activated ion channels. *In* "Cellular and Molecular Physiology of Cell Volume Regulation" (K. Strange, Ed.), pp. 215–240. CRC Press, Boca Raton, FL.

Sands, J. M. (1994). Regulation of intracellular polyols and sugars in response to osmotic stress. *In* "Cellular and Molecular Physiology of Cell Volume Regulation" (K. Strange, Ed.), pp. 133–144. CRC Press, Boca Raton, FL.

Sarkadi, B., and Parker, J. C. (1991). Activation of ion transport pathways by changes in cell volume. *Biochem. Biophys. Acta* **1071,** 407–427.

Schatzberg, P. (1963). Solubilities of water in several normal alkanes from $C_7$ to $C_{16}$. *J. Phys. Chem.* **67,** 776–779.

Shibata, A., Ikawa, K., Shimooka, T., and Terada, H. (1994). Significant stabilization of the phosphatidylcholine bilayer structure by incorporation of small amounts of cardiolipin. *Biochim. Biophys. Acta* **1192,** 71–78.

Shporer, M., and Civan, M. M. (1977). The state of water and alkali cations within the intracellular fluids: The contribution of NMR spectroscopy. *Curr. Top. Membr. Transport* **9,** 1–69.

Solomon, A. K. (1972). Properties of water in red cells and synthetic membranes. In "Biomembranes, Vol. 3, Passive Permeability of Cell Membranes" (F. Kreuzer and J. F. G. Slegers, Eds.), pp. 299–330. Plenum Press, New York.

Solomon, A. K. (1989). Water channels across the red blood cell and other biological membranes. *Methods Enzymol.* **173,** 192–222.

Solomon, A. K., Chasan, B., Dix, J. A., Lukacovic, M. F., Toon, M. R., and Verkman, A. S. (1983). The aqueous pore in the red cell membrane: band 3 as a channel for anions, cations, nonelectrolytes, and water. *Ann. NY Acad.Sci.* **414,** 97–124.

Somero, G. N. (1986). Protons, osmolytes, and fitness of internal milieu for protein function. *Am. J. Physiol.* **251,** R197–R213.

Starke, L. C., and Jennings, M. L. (1993). K–Cl cotransport in rabbit red cells: Further evidence for regulation by protein phosphatase type 1. *Am. J. Physiol.* **264,** C118–C124.

Stein, W. D. (1976). An algorithm for writing down flux equations for carrier kinetics, and its application to co-transport. *J. Theor. Biol.* **62,** 467–478.

Strange, K. (1994). "Cellular and Molecular Physiology of Cell Volume Regulation." CRC Press, Boca Raton, FL.

Suleymanian, M. A., and Baumgarten, C. M. (1996). Osmotic gradient-induced water permeation across the sarcolemma of rabbit ventricular myocytes. *J. Gen. Physiol.* **107,** 503–514.

Suleymanian, M. A., Clemo, H. F., Cohen, N. M., and Baumgarten, C. M. (1995). Stretch-activated channel blockers modulate cell volume in cardiac ventricular myocytes. *J. Mol. Cell. Cardiol.* **27,** 721–728.

Tanford, C. (1961). "Physical Chemistry of Macromolecules." John Wiley, New York.

Tosteson, D. C., and Hoffman, J. F. (1960). Regulation of cell volume by active cation transport in high and low potassium sheep red cells. *J. Gen. Physiol.* **44,** 169–194.

Trauble, H. (1971). The movement of molecules across lipid membranes: A molecular theory. *J. Membr. Biol.* **4,** 193–208.

Uchida, S., Kwon, H., Yamauchi, A., Preston, A., Marumo, F., and Handler, J. (1992). Molecular cloning of the cDNA for an MDCK cell $Na^+$ and $Cl^-$-dependent taurine transporter that is regulated by hypertonicity. *Proc. Natl. Acad. Sci. USA* **89,** 8230–8234.

Uchida, S., Kwon, H., Yamauchi, A., Preston, A., Kwon, H., Handler, J. (1993). Medium tonicity regulates expression of the $Na^+$- and $Cl^-$-dependent betaine transporter in Madin–Darby canine kidney cells by increasing transcription of the transporter gene. *J. Clin. Invest.* **91,** 1604–1607.

Umenishi, F., Verkman, A. S., Gropper, M. A. (1996). Quantitative analysis of aquaporin mRNA expression in rat tissues by RNase protection assay. *DNA Cell Biol.* **15,** 475–480.

van Hoek, A. N., Hom, M. L., Luthjens, L. H., de Jong, M. D., Dempster, J. A., and van Os, C. H. (1991). Functional unit of 30 kDa for proximal tubule water channel as revealed by radiation inactivation. *J. Biol. Chem.* **266,** 16633–16635.

van Hoek, A. N., Luthjens, L. H., Hom, M. L., van Os, C. H., and Dempster, J. A. (1992). A 30 kDa functional size for the erythrocyte water channel determined by *in situ* radiation inactivation. *Biochem. Biophys. Res. Comm.* **184,** 1331–1338.

van't Hoff, J. H. (1887). Die Rolle des osmotischen Druckes in der Analogie zwischen Lösungen und Gasen. *Z. Physik. Chemie* **1,** 481–493. [Translated by G. L. Blackshear (1979). *In* "Cell Membrane Permeability and Transport" (G. R. Kepner, Ed.). Dowden, Hutchinson & Ross, Stroudsburg, PA.]

Verkman, A. S. (1993). "Water Channels." R. G. Landis, Austin, TX.

Verkman, A. S., van Hoek, A. N., Ma, T., Frigeri, A., Skach, W. R., Mitra, A., Tamarappoo, B. K., Farinas, J. (1996). Water transport across mammalian cell membranes. *Am. J. Physiol.* **270,** C11–C30.

Villars, F. M., and Benedek, G. B. (1974). "Physics with Illustrative Examples from Medicine and Biology. Statistical Physics," Vol. 2. Addison-Wesley, Reading, MA.

Walz, T., Smith, B. L., Agre, P., Engel, A. (1994). The three-dimensional structure of human erythrocyte aquaporin CHIP. *EMBO J.* **13,** 2985–2993.

Watson, P. A. (1991). Function follows form: Generation of intracellular signals by cell deformation. *FASEB J.* **5,** 2013–2019.

Wolfe, J., Dowgert, M. F., and Steponkus, P. L. (1986). Mechanical study of the deformation and rupture of the plasma membranes of protoplasts during osmotic expansions. *J. Membr. Biol.* **93,** 63–74.

Wolff, S. D., and Balaban, R. S. (1990). Regulation of the predominant renal medullary organic solutes *in vivo. Annu. Rev. Physiol.* **52,** 727–746.

Yamauchi, A., Uchida, S., Kwon, H., Preston, A., Robey, R., Garcia-Perez, A., Berg, M., and Handler, J. (1992). Cloning of a Na$^+$ and Cl$^-$ dependent betaine transporter that is regulated by hypertonicity. *J. Biol. Chem.* **267,** 649–652.

Yancey, P. H. (1994). Compatible and counteracting solutes. *In* "Cel-

lular and Molecular Physiology of Cell Volume Regulation" (K. Strange, Ed.), pp. 81–177. CRC Press, Boca Raton, FL.

Yancey, P. H., Clark, M. E., Hand, S. C., Bowlus, R. D., and Somero, G. N. (1982). Living with water stress: Evolution of osmolyte systems. *Science* **217,** 1214–1222.

Yool, A. J., Stamer, D., and Regan, J. W. (1996). Forskolin stimulation of water and cation permeability in aquaporin1 water channels. *Science* **273,** 1216–1218.

Zeidel, M. L., Ambudkar, S. V., Smith, B. L., and Agre, P. (1992). Reconstitution of functional water channels in liposomes containing purified red cell CHIP28 protein. *Biochemistry* **31,** 7436–7440.

Zhang, R., Skach, W., Hasegawa, H., van Hoek, A. N., and Verkman, A. S. (1993). Cloning, functional analysis, and cell localization of a kidney proximal tubule water transporter homologous to CHIP28. *J. Cell. Biol.* **120,** 359–369.

Zimmermann, S. B., and Harrison, B. (1985). Macromolecular crowding accelerates the cohesion of DNA fragments with complementary termini. *Nucleic Acids Res.* **13,** 2241–2249.

Zimmerman, S. B., and Trach, S. (1991). Estimation of macromolecular concentrations and excluded volume effects for the cytoplasm of *Escherichia coli. Mol. Biol.* **222,** 599–620.

Robert W. Putnam

# 20

# Intracellular pH Regulation

## I. Introduction

**Intracellular pH** is an important aspect of the intracellular environment. Virtually all cellular processes can potentially be affected by changes in intracellular pH, including metabolism, membrane potential, cell growth, movement of substances across the surface membrane, state of polymerization of the cytoskeleton, and ability to contract in muscle cells. Changes of intracellular pH are also often one of the responses of cells to externally applied agents including growth factors, hormones, and neurotransmitters. Further, many organelles, such as lysosomes, mitochondria, and endosomal vesicles, maintain an organellar pH that is different from the cytoplasmic pH ($pH_i$), and these pH differences have important functional consequences for those organelles. It is thus not surprising to find that cells have elaborated a variety of mechanisms that enable them to regulate their intracellular pH. In this chapter, we will discuss the level of pH in the cytoplasm and various compartments of a cell, the variety of mechanisms available to a cell to regulate its $pH_i$, and the functional consequences of changes in $pH_i$.

## II. pH and Buffering Power

The concept of pH was first introduced in 1909 by Sörensen and defined as the $-\log[H^+]$. This term was a more convenient way to express the concentration of an ion that is present at very low concentrations. Incorporating the concept of activity, pH is now defined as

$$pH = -\log(a_H) = -\log(\gamma_H[H^+]) \qquad (1)$$

where $a_H$ is the activity of $H^+$ and $\gamma_H$ is the activity coefficient of $H^+$. At normal intracellular ionic strength, $\gamma_H$ is about 0.83.

Protons tend to bind to macromolecules and thus are usually present at very low concentrations in biological solutions. This property is the basis for **buffering power.** A variety of weak acids and bases can bind H ions through reversible equilibrium binding reactions. Thus, a weak acid in solution obeys the equilibrium reaction

$$HA \leftrightarrows H^+ + A^- \qquad (2)$$

where HA is the weak acid (e.g., lactic acid) and $A^-$ is the conjugate weak base (e.g., lactate). This equilibrium is described by an apparent equilibrium constant, $K_a'$, as

$$K_a' = a_H \cdot \frac{[A^-]}{[HA]} \qquad (3)$$

This equation is more familiar in its logarithmically transformed expression as

$$pH = pK_a' + \log \frac{[A^-]}{[HA]} \qquad (4)$$

where $pK_a'$ is $-\log K_a'$. This equation, better known as **the Henderson–Hasselbalch equation,** describes the thermodynamic equilibrium that holds for a weak acid in a solution of constant pH. The Henderson–Hasselbalch equation is most commonly used in its specialized form for the total reaction of the hydration of $CO_2$ and the dissociation of the resulting carbonic acid into $H^+$ and bicarbonate as

$$pH = pK_a' + \log \frac{[HCO_3^-]}{\alpha \cdot P_{CO_2}} \qquad (5)$$

where $\alpha$ is the solubility coefficient of $CO_2$ in a given solution and $P_{CO_2}$ is the partial pressure of $CO_2$ in that solution. Two important facts can be deduced from the Henderson–Hasselbalch equation. First, in any weak acid solution, there will be a finite amount of both $[A^-]$ and $[HA]$. For example, if $HCO_3^-$ is added to a solution, $CO_2$ will be generated and thus be present. Conversely, if a solution is bubbled with $CO_2$, $HCO_3^-$ will be produced. Second, this equation can be used to calculate any of the variable parameters if the other two are known. For instance, if a solution is equilibrated with a gas of known $P_{CO_2}$, the $[HCO_3^-]$ in that solution can be calculated from Eq. 5 once the pH has reached a stable value (values for $pK_a'$ are readily available).

On addition of $H^+$ (or $OH^-$) to a solution, the pH will change. However, if the solution contains weak acids (or bases), many of the added protons (or hydroxyl ions) will be bound up, thus minimizing the change in concentration of free $H^+$ (and thereby minimizing the change in pH). Since these substances minimize the change in pH upon addition of acid or base, weak acids and bases are referred to as buffers. The definition of the buffering power ($\beta$) of a solution is

$$\beta = \frac{dB}{dpH} \qquad (6)$$

where $dB$ is the amount of base added to the solution, and $dpH$ is the change in pH of the solution due to that base addition. The addition of acid to the solution is equivalent to a negative addition of base, $-dB$. The units of $\beta$ are $mM$/pH unit.

An example will indicate the importance of buffering power to maintaining the pH of a solution. If 1 $mM$ NaCl is added to a solution, the [Na] and [Cl] increase by 1 $mM$ (for simplicity, the effects of the nonideal activity coefficients will be ignored). However, the addition of 1 $mM$ HCl to a solution that has a pH of 7.0 and a buffering power of 10 $mM$/pH unit will cause that [Cl] to increase by 1 $mM$, but will cause the pH to decrease by only about 0.28 pH unit, to 6.72. Thus, of the added 1 $mM$ $H^+$, only 0.091 $\mu M$ remains free, that is, only 1 of every 11,000 added H ions remains free. The rest are bound to the weak acid buffers. If the buffers had not been present, the same addition of 1 $mM$ HCl would have changed solution pH by about 4 units to pH 3. This clearly demonstrates that the presence of buffers in a solution markedly blunts the effects of added acid or base.

A buffer can act as either a **closed buffer** or an **open buffer**. A closed buffer is one in which the total buffer concentration remains constant. Most of the commonly used laboratory buffers, such as Hepes or Tris, operate as closed buffers in solution. If a buffer is a weak acid (HA $\leftrightarrows A^- + H^+$), then it operates as a closed buffer when the concentration of the total acid ($[A_T] = [HA] + [A^-]$) remains constant. Such a buffer can become protonated or deprotonated, but the total amount of buffer does not change. The following is a mathematical expression for the buffering power of a weak acid acting as a closed buffer,

$$\beta_{closed} = \frac{2.303[A_T]K_a' a_H}{(K_a' + a_H)^2} \qquad (7)$$

where $K_a'$ is the apparent dissociation constant of the weak acid. Several conclusions can be derived from this equation. When pH is very high ($a_H \to 0$) or very low ($a_H \to \infty$), $\beta_{closed}$ approaches 0. $\beta_{closed}$ reaches a maximum when $a_H = K_a'$ (i.e., when pH = $pK_a'$), and $\beta_{closed}^{max} = 0.58[A_T]$. The relationship between pH and $\beta_{closed}$ for a theoretical closed buffer is shown in Fig. 1.

In contrast to the conditions for a closed buffer, if the protonated (or uncharged) form of a buffer remains constant (i.e., [HA] = constant), then the buffer operates as an open buffer. The most common example of an open buffer in solution is $CO_2/HCO_3^-$. Such a solution contains $HCO_3^-$ and is equilibrated with gaseous $CO_2$ (usually by

bubbling). If acid is added to such a solution, H ions combine with $HCO_3^-$ and form additional $CO_2$. Since the solution is in equilibrium with a fixed $P_{CO_2}$, the additional $CO_2$ diffuses from the solution and is removed. Thus, under these conditions, the amount of the uncharged form of the buffer ($CO_2$) remains constant, whereas the total buffer amount decreases because the $[HCO_3^-]$ has decreased. The buffering power of an open buffer is given by

$$\beta_{open} = 2.303[A^-] \qquad (8)$$

where $[A^-] = [HCO_3^-]$ for the open buffering power of $CO_2/HCO_3^-$. The relationship between pH and $\beta_{open}$ for a theoretical weak acid open buffer is shown in Fig. 1.

Open buffers differ from closed buffers in two important respects. Unlike closed buffers, open buffers are not maximal at their $pK_a'$ values, but in fact become better buffers at the more extreme values of pH (Fig. 1). $\beta_{open}$ becomes larger with alkalinization for weak acids and with acidification for weak bases. In addition, open buffers have a much higher buffering power than closed buffers under similar conditions (Fig. 1). Note, however, that neither Eq. 7 nor 8 includes any indication of the nature of the buffer. Thus, all closed buffers are equally potent when the pH is at their $pK_a'$ and all open buffers are equally potent when they have the same $[A^-]$.

Since most uncharged substances are substantially more permeable through cell membranes than charged species, *almost all weak acids and bases can act as open buffers in cells*. For example, a weak acid that has a $pK_a'$ of, say, 4.0, acts like a closed buffer in solution, and since its $pK_a'$ is 4.0, it would be a poor buffer at pH 7.0. However, inside a cell at pH 7.0, this weak acid acts like an open buffer. Any added H ions will bind to the anionic form of the buffer. The uncharged buffer molecule formed will readily diffuse from the cell and be removed by the blood. Thus, although this weak acid in the medium is a poor buffer, it can contribute substantially to the buffering power inside a cell if it is present at sufficient concentration.

Complex solutions, such as blood, contain multiple buffers. The **total buffering power** ($\beta_{total}$) of such solutions will be the sum of the various buffers; that is, buffers operate independently in solution. Thus,

$$\beta_{total} = \Sigma\beta_{closed} + \Sigma\beta_{open} \qquad (9)$$

Finally, it is inherent in the definition of buffering power (Eq. 6) that buffering power is a coefficient that can be used to convert a change of pH into a change in the amount of proton equivalents moved. This relationship has practical application. For instance, in the study of the regulation of intracellular pH, the activity of a transporter that moves H ions across the surface membrane (such as the Na/H exchanger, see Section VI) is determined by measuring the rate of change of $pH_i$ ($dpH_i/dt$). The movements of H ions on this transporter is accompanied by $Na^+$, whose movement is measured as a radioisotopic flux (amount of Na influx/unit time). To compare the flux of $Na^+$ with the flux of $H^+$, the rate of change of $pH_i$ needs to be converted to the amount of H ions moved per time. This is accomplished by the equation

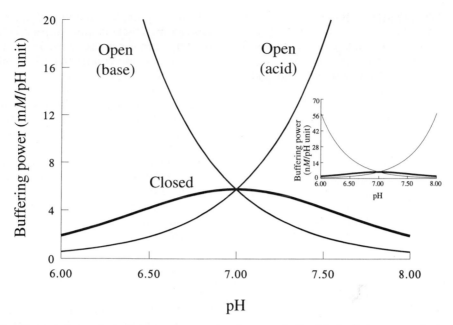

**FIG. 1.** The pH dependence of a buffer operating as a closed or open buffer. The buffering powers for both a weak acid buffer ($H^+ + A^- = HA$) and weak base buffer ($H^+ + B = HB^+$) are plotted. The buffers are assumed to have a $pK_a'$ of 7.0 and to have the same buffering power at pH 7.0 whether operating as an open or closed buffer. Note that while closed buffers have maximal buffering power when $pH = pK_a'$, the buffering power for an open buffer is higher at pH values higher (for weak acid buffers) than $pK_a'$ or at pH values lower (for weak base buffers) than $pK_a'$. Inset: The same plot, only with a different scale for the ordinate. Note how much higher the open buffering power is compared with the closed buffering power at pH values well above (weak acid) or below (weak base) the value of $pK_a'$.

$$J_H = \frac{d\mathrm{pH}}{dt} \cdot \beta_{total} \qquad (10)$$

where $J_H$ is the flux of protons and has units of m$M$ $H^+$ per unit time. Using this equation, the flux of H ions, calculated from the measured rate of pH change, can be directly compared to the flux of another ion determined with radioisotopes.

## III. Intracellular pH

Protons are just like any other cation, except for three distinguishing characteristics: (1) H ions are a dissociation product of water molecules ($H_2O \leftrightarrows H^+ + OH^-$) and thus are always present in aqueous solutions, (2) H ions are present at very low concentrations in most solutions, and (3) H ions have much higher mobility than other cations. However, the equilibrium distribution and movement of $H^+$ across biological membranes are governed by the same principles that govern the movement of all other ions across biological membranes.

Originally, protons were assumed to be at equilibrium across biological membranes because of their very high mobility. Assuming an extracellular pH (pH$_o$) of 7.4, a $V_m$ of $-60$ mV (inside negative) and assuming that H ions are passively distributed across the membrane (i.e., at equilibrium), pH$_i$ would be 6.4 (calculated from the Nernst equation). However, at such an intracellular pH, metabolism and a variety of other cellular functions would be impaired.

With the advent of modern reliable techniques for measuring intracellular pH, including **pH-sensitive glass microelectrodes** and **pH-sensitive fluorescent dyes** (see Appendix), it was shown that in the majority of cells pH$_i$ was between 6.8 and 7.2, well above the calculated value for equilibrium pH. It is now clear that for most cells (with the notable exception of red blood cells), pH$_i$ is considerably more alkaline than it would be if protons were at passive equilibrium across the cell membrane.

The question still remains how pH$_i$ can be well above equilibrium, since H ions should be highly permeable to most cell membranes. In fact, the **permeability of H ions** across biological membranes has been estimated to be between $10^{-4}$ and $10^{-2}$ cm/sec, about four orders of magnitude higher than typical $K^+$ permeabilities. However, it is the conductance and not the permeability of H ions that is crucial. **_Conductance_** is a measure of ion flux and is a function of both the permeability and the free concentration of an ion. H ions have low conductance across biological membranes despite their high permeability, because they are present in such low concentrations ($10^{-7}$ $M$ free concentration for H ions vs $10^{-1}$ $M$ free concentration for K ions). Thus, the acidifying influx of H ions (down their electrochemical gradient) will be small, and these H ions can easily be removed from the cell by membrane transport systems (see Section IV).

In summary, most cells have a cytoplasmic pH that is more alkaline than the value calculated assuming equilibrium of H ions across the cell membrane, and pH$_i$ for most cells is about 6.8–7.2.

## IV. Organellar pH

Several intracellular organelles independently control their internal pH, and it differs from the cytoplasmic pH (Fig. 2). These organelles include **mitochondria** and **acidic intracellular organelles.**

### A. Mitochondria

One of the major roles of mitochondria is the production of ATP. A proton gradient across the inner mitochondrial membrane is required for the production of ATP. The electron transport chain in the mitochondrion translocates protons from the mitochondrion to the cytoplasm across the inner mitochondrial membrane. This proton extrusion creates an electrical and chemical gradient for proton influx into the mitochondrion. This H ion influx occurs through a membrane-bound ATPase, which produces ATP upon passive proton flux back into the mitochondrion. This is known as the **chemiosmotic hypothesis.**

The extrusion of H ions to establish a proton gradient renders mitochondria relatively alkaline to the cytoplasm by about 0.5 pH unit. Thus, intramitochondrial pH in the average cell is about 7.5. In addition, this pH gradient is maintained in the face of considerable acid loads, indicating that mitochondria can probably regulate their internal pH independent of cytoplasmic pH.

### B. Acidic Intracellular Organelles

Organelles with a markedly acidic interior are organelles involved in either the endocytic pathway or the secretory pathway. The acidic organelles involved in endocytosis include **coated pits, endosomes** (i.e., prelysosomal endocytic vesicles), and **lysosomes.** Acidic vesicles in the secretory pathway include the **Golgi apparatus** (or at least part of it),

and **storage granules** for amines (e.g., chromaffin granules involved in catecholamine secretion) and peptides (e.g., secretory granules in the endocrine pancreas). The precise pH in all of these compartments is unknown but can be as low as 4.5–5.0 in lysosomes and 5.0–5.7 in endosomes and secretory granules.

The acidic internal environment of these various organelles is believed to be necessary for their function. Thus, the primary function of lysosomes is the biochemical degradation of macromolecules, and these organelles contain a large number of hydrolytic enzymes whose pH optima are about pH 5.0. This serves as protection for the cell. If the lysosomes should become leaky, the hydrolytic enzymes would be inactivated by the high cytoplasmic pH, thus preventing the indiscriminate degradation of important macromolecules. Receptor-mediated endocytosis involves the internalization of ligand–receptor complexes in endocytic vesicles, or endosomes. Acidification of these vesicles is essential for the dissociation of the ligand from the receptor within the endosome. Once the ligand has dissociated, the internalized receptor is recycled to the surface membrane, while the ligand is delivered to the lysosome. Finally, secretory granules (and perhaps part of the Golgi) serve to accumulate macromolecules to be secreted and often mediate processing or modification of these substances. The maintenance of an acidic environment in these granules can be crucial to both of these functions. There is evidence that the large outward H gradient is used to accumulate biogenic amines in amine secretory granules. This accumulation may be mediated by an $H^+$/amine exchanger in the granule membrane. These accumulated substances can be biochemically modified into their final form, and the enzymes responsible for these modifications often have low pH optima or depend on the availability of organic compounds that will accumulate only in acidic compartments. For all these acidic compartments, then, it is clear that their proper functioning depends on the maintenance of a low internal pH.

### C. Nucleus

The nucleus is separated from the cytoplasm by a double-membrane system that has large nuclear pores. Large macromolecules (up to 5 kDa) readily permeate the nuclear pores and rapidly come to equilibrium between the nucleus and cytoplasm. Given this high degree of permeability of the nuclear membrane, it is unlikely that nuclear pH differs much from cytoplasmic pH.

## V. Maintenance of a Steady-State $pH_i$

If a cell is maintaining a **steady-state** $pH_i$ (i.e., the pH of the cytoplasm), the rate of acid loading must be equal to the rate of acid extrusion from the cell (in these terms it is exactly equivalent if an acid molecule moves in one direction or a base molecule moves in the opposite direction) (Fig. 3). Several processes can contribute to acid loading of a cell, including metabolic production of acid, passive influx of H ions across the cell membrane, leakage

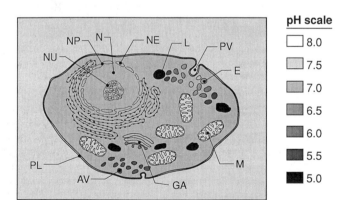

**FIG. 2.** A diagram of an idealized cell with the gray scale representing different values of pH (note calibration scale on the right). The cytoplasm and nucleus have a fairly uniform pH of about 7.0, but note the alkaline mitochondria, the increasing acidification in the Golgi apparatus and in the endosomes, and the very acid lysosomes. AV—acidic vesicles; E—endosomes; GA—Golgi apparatus; L—lysosomes; M—mitochondria; N—nucleus; NE—nuclear envelope; NP—nuclear pore; NU—nucleolus; PL—plasmalemma; PV—pinocytotic vesicle.

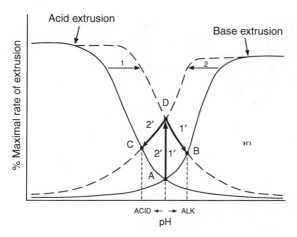

**FIG. 3.** A model for the regulation of steady-state $pH_i$. In the cell, steady-state $pH_i$ is determined as the point where acid extruding processes are balanced by base extruding (or acid influx) pathways (point A). If acid extruding processes are activated (arrow 1—represented as a shift to the right of the pH vs % maximal rate curve), as occurs when growth factors activate Na/H exchange, the rate of acid extrusion exceeds the rate of base extrusion and the cell alkalinizes (arrows 1'). Eventually, the cell alkalinizes to the point where acid and base extrusion are again equal and a new alkaline steady state $pH_i$ is reached (point B). In contrast, if base extrusion is activated (arrow 2—represented as a shift to the left of the pH vs % maximal rate curve), as occurs with an increase in the activity of $Cl/HCO_3$ exchange, the rate of base extrusion exceeds acid extrusion and the cell will acidify (arrows 2'). Eventually, the cell acidifies to the point where acid and base extrusion are again equal and a new acid steady-state $pH_i$ is reached (point C). If acid and base extrusion are activated to the same extent, the steady-state $pH_i$ will remain unchanged (point D).

of H ions from acidic internal compartments, pumping of H ions from alkaline internal compartments, and active influx of H ions (or active extrusion of base). Conversely, acid extrusion includes metabolic consumption of H ions, sequestration of H ions in internal compartments, and active extrusion of acid (or active influx of base).

## A. Metabolic Production and Consumption of Acids

Several metabolic reactions involve the production of H ions (Table 1). These processes include the production of $CO_2$, glycolysis (through the generation of lactic and pyruvic acids), formation of creatine phosphate, ATP hydrolysis, lipolysis, triglyceride hydrolysis, the generation of superoxide, and the operation of the hexose monophosphate shunt. Obviously, when these reactions run in the opposite direction (e.g., creatine phosphate hydrolysis, ATP formation, or consumption of $CO_2$), there is a net consumption of H ions and the cell will alkalinize.

When the cell is at steady state, these reactions must be balanced, and thus the levels of cellular metabolites such as ATP, creatine phosphate, $CO_2$, and lactate will also be at steady state. However, during transient periods, these metabolic reactions can result in a net change in metabolite concentrations and thus contribute markedly to changes in $pH_i$. For instance, during periods of ischemia, cellular

lactate and $CO_2$ may accumulate while ATP is hydrolyzed. All of these reactions can contribute to the observed cellular acidification. The hydrolysis of creatine phosphate during ischemia consumes H ions and will blunt the cellular acidification. Another example of a transient effect of metabolism on cellular pH is the initial acidification seen upon activation of neutrophils with phorbol esters. In this case, stimulation of the production of superoxide and activation of the hexose monophosphate shunt results in an increased production of H ions.

## B. Passive Transmembrane Flux of H Ions

Although the H ion permeability of most biological membranes is quite high ($P_H = 10^{-3}$ cm/sec), the actual H ion flux across the membrane is quite low because of the low free H ion concentration. For example, the putative H ion flux across a frog muscle fiber can be calculated. Assuming a constant electric field across the membrane, a $P_H = 10^{-3}$ cm/sec, $V_m = -90$ mV, $pH_i$ 7.2, $pH_o$ 7.35, and $\beta_{total} = 26$ m$M$, passive H ion influx would result in a cellular acidification of only 0.02 pH unit/hour. Although negligible, this does represent a continued acid load on the cell, and if mechanisms do not exist to remove these H ions, the cell will eventually acidify toward the equilibrium $pH_i$ value.

H ion currents associated with **proton channels** have been reported in several cells, including snail neurons and salamander oocytes. These channels are activated by depolarization and the resulting H ion currents alkalinize the cell. The significance of these channels is not known but they may contribute to the maintenance of pH in a restricted submembrane space within the cell or contribute to $pH_i$ regulation in cells undergoing prolonged depolariza-

**TABLE 1** Examples of Various Metabolic Reactions That Involve the Generation or Consumption of H Ions

| | |
|---|---|
| Glycolysis[a] | Glucose + 2MgADP$^-$ + 2P$_i^{2-}$ → 2(lactate)$^-$ + 2MgATP$^{2-}$ |
| | 2MgATP$^{2-}$ → 2MgADP$^-$ + 2P$_i^{2-}$ + 2H$^+$ |
| Glycogenolysis[b] | Glycogen + 3P$_i^{2-}$ + 3MgADP$^-$ + H$^+$ → 3MgATP$^{2-}$ + 2(lactate)$^-$ + glycogen |
| Creatine phosphate hydrolysis | H$^+$ + creatine phosphate$^{2-}$ + MgADP$^-$ → MgATP$^{2-}$ + creatine |
| Lipolysis | Triglyceride → 3(palmitate)$^-$ + 3H$^+$ |
| | 3(palmitate)$^-$ + 3MgATP$^{2-}$ + 3CoA$^{4-}$ → 3(palmitoyl CoA)$^{4-}$ + 3 AMP$^{2-}$ + 6P$_i^{2-}$ + 3H$^+$ + 3Mg$^{2+}$ |
| Superoxide formation | NADPH + 2O$_2$ → 2O$_2^-$ + NADP$^+$ + H$^+$ |
| Hexose monophosphate shunt | G-6-P$^-$ + 12NADP$^+$ + 6H$_2$O → NADPH + P$_i^-$ + 12H$^+$ + 6CO$_2$ |

[a] At pH 7.2 and high Mg$^{2+}$, 2H$^+$ are always generated but the balance of H$^+$ produced by the two reactions varies with pH and Mg.

[b] Net H$^+$ production when hydrolysis of the 3MgATP is taken into account.

tions, such as the prolonged depolarizing fertilization potential in oocytes.

### C.  Internal Compartments

If the pH values of intracellular compartments, such as mitochondria and lysosomes, are at steady state, then they should have no impact on cytoplasmic pH. However, under pathological conditions where mitochondria or acidic intracellular compartments are rendered leaky or mitochondria take up cytoplasmic calcium in exchange for H ions, these compartments could influence $pH_i$. Under normal conditions, though, these compartments should be contributing little to the maintenance of a steady state $pH_i$, especially owing to their relatively small volume compared with total cell volume.

It is clear that at the very least cells face a continuous acid load from passive H ion influx. In addition, under many conditions of metabolic stress, cells also experience a metabolic acid load. Thus, cells must possess active extrusion mechanisms to maintain a steady-state $pH_i$ well above the equilibrium value for $pH_i$.

## VI.  Active Membrane Transport of Acids and Bases

Several integral proteins within the surface membrane of cells are specialized for the active transport of acids and bases across the membrane. Because of their importance to cellular pH regulation, these transport pathways have been extensively studied and can be divided into five classes: (1) those that move H ions directly in exchange for another cation, (2) those that move $HCO_3^-$ (or an associated species such as $CO_3^{-2}$), (3) H-ATPases (proton pumps) that use energy from ATP hydrolysis to transport H ions, (4) those that cotransport anionic weak bases, with $Na^+$ and (5) those that transport anionic weak bases in exchange for $Cl^-$.

### A.  Cation/H Exchangers

The best characterized of the cation/H exchangers is the **Na/H exchanger (NHE)** (model 1, Fig. 4A). This exchanger responds to cellular acidification by extruding an H ion in exchange for the influx of a Na ion (1:1 stoichiometry means the exchanger is electroneutral, that is, it does not involve net charge movement). There are now known to be at least 5 isoforms of the Na/H exchanger. NHE-1 is found in many different types of cells, is inhibited by the loop diuretic, amiloride, can be activated by growth factors, and has a molecular weight of 91,000. NHE-2, which has only 50% amino acid homology with NHE-1, is found predominantly in the gastrointestinal tract (GI tract), is much less sensitive to amiloride inhibition than NHE-1, and also has a molecular weight of about 91,000. NHE-3 and NHE-4 are recently described isoforms that have about 40% amino acid homology with NHE-1, have molecular weights of 93,000 and 81,000, respectively, and are expressed largely in the GI tract. NHE-4 also shows some expression in

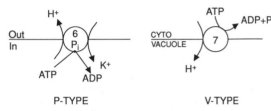

**FIG. 4.** Models of several different types of pH-regulating transporters. (A) Cation/$H^+$ exchangers. Included in this group are the alkalinizing Na/H exchanger (1) and the acidifying K/H exchanger (2). (B) $HCO_3^-$-dependent transporters. This group includes the Cl/$HCO_3$ exchanger (band 3 from red blood cells) (3). Another transporter in this group is the $(Na + HCO_3)/Cl$ exchanger (4). Shown here is merely one possible model for this exchanger. This transporter could involve the influx of 2 $HCO_3^-$ instead of the influx of 1 $HCO_3^-$ and the efflux of an $H^+$. Alternatively, a single $CO_3^{2-}$ or a $NaCO_3$ ion pair could be transported in. These four different variants have the same effect on pH and can be partially distinguished kinetically. The Na–$HCO_3$ cotransporter is shown as model 5. This transporter is electrogenic, mediating the movement of $nHCO_3^-$ for each $Na^+$, and is thus sensitive to membrane potential ($V_m$). The direction of cotransport depends on the value of $V_m$. (C) Proton pumps. Two different types of $H^+$ ATPases. The P-type $H^+$-ATPase (6), typified by the electroneutral gastric H,K-ATPase, involves a phosphorylated intermediate. The V-type $H^+$-ATPase (7), typified by the vacuolar $H^+$-ATPase, does not produce a phosphorylated intermediate.

the uterus, brain, and kidney. NHE-3 and NHE-4 are not readily inhibited by amiloride. The other NHE isoform is $\beta$-NHE, which is activated by cAMP and has so far only been described in trout red blood cells. It has 50–75% homology with NHE-1, but NHE-1 cannot be activated by cAMP. The significance of the properties and distribution of these different isoforms is still largely unclear, but given the variety of functions performed by NHE in the cell, it is not surprising that multiple forms have arisen.

The basic structures of all NHE isoforms are similar (Fig. 5A). NHE is believed to have 12 membrane-spanning domains with two possible glycosylation sites on external loops. This means that NHE is a glycoprotein. Both the N- and C-terminal ends of NHE are cytoplasmic. The C-terminal end represents a large cytoplasmic domain with available serine phosphorylation sites. It is believed that the cytoplasmic C-terminal end is involved in the regulation of NHE while the membrane-spanning regions are respon-

**A**

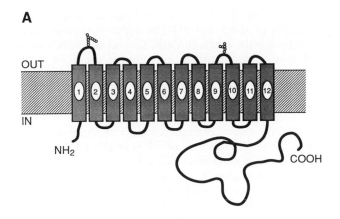

**B**

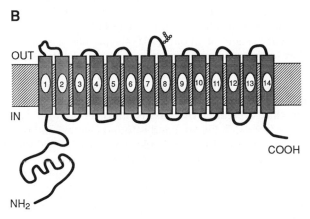

**FIG. 5.** Models of the structure of the Na/H exchanger (NHE1) (A) and the Cl/HCO₃ exchanger (AE1) (B). The Na/H exchanger is believed to have 12 membrane-spanning regions with a large cytoplasmic domain at the C-terminal end. This cytoplasmic domain contains several potential phosphorylation sites that are important to the regulation of the Na/H exchanger. The N-terminal end is also cytoplasmic. Two potential glycosylation sites are shown. The Cl/HCO₃ exchanger is believed to have 14 membrane-spanning regions, and both terminal ends are cytoplasmic. The large cytoplasmic domain is on the N-terminal end in the Cl/HCO₃ exchanger. This cytoplasmic domain (in red blood cells at least) contains many binding sites, including those for the cytoskeletal element ankyrin, for hemoglobin, and for some glycolytic enzymes.

sible for Na and H transport. This is borne out by the expression of NHE mutants that lack the C-terminal end in fibroblasts that do not have NHE. In these cells, Na/H exchange activity is seen but it can no longer be activated by growth factors.

A wide variety of substances have been shown to activate NHE, including hormones, neurotransmitters, growth factors, and the extracellular matrix. Most if not all of these factors are believed to activate the NHE by phosphorylating serine residues on the C-terminal end. These factors may also affect the NHE through the binding of a regulatory protein with the exchanger and/or by debinding of a $Ca^{2+}$/calmodulin complex from a putative autoinhibitory domain on the exchanger. Activation involves an increase in the affinity of the exchanger's internal H binding site

for cytoplasmic H ions (see arrow 1, Fig. 3). Cell acidification also dramatically activates NHE. This activation is due to an internal allosteric H binding site on NHE (distinct from the H transport site) that increases exchange when occupied. Finally, decreased cell volume (cell shrinkage) also activates NHE. This activation does not involve phosphorylation of NHE and is apparently mediated by a unique activation pathway.

Many signaling pathways can activate NHE, depending on the stimulus. For example, growth factor activation of NHE is usually mediated by an elevation of intracellular $Ca^{2+}$ and/or by increased activity of protein kinase C. These pathways commonly involve phosphoinositide breakdown. Activation of NHE by cell shrinkage appears to involve ATP- and GTP-dependent pathways.

The other major cation/H exchanger is the **K/H exchanger** (model 2, Fig. 4A). This transporter exchanges intracellular $K^+$ for extracellular $H^+$ and results in cellular acidification. K/H exchange has been found in nucleated red blood cells and mediates solute efflux during regulatory volume decrease (see Section VII,H). It has also been observed in retinal pigment epithelial cells where it enables the cell to regulate $pH_i$ in the face of an alkaline load.

### B. HCO₃⁻-Dependent Transporters

These transporters are actually a family of related transport proteins that affect $pH_i$ and are characterized by their ability to transport $HCO_3^-$ (or a related species like $CO_3^{2-}$) and by the ability of disulfonic stilbene derivatives to inhibit them. The three major types of $HCO_3^-$-dependent transporters are Cl/HCO₃ exchange (band 3 from red blood cells), (Na + HCO₃)/Cl exchange (often termed Na-dependent Cl/HCO₃ exchange), and electrogenic Na–HCO₃ cotransport.

The operation of **Cl/HCO₃ exchange** (model 3, Fig. 4B) has been studied most extensively in red blood cells. This electroneutral transporter involves a 1:1 exchange of $Cl^-$ for $HCO_3^-$ (although many other ions, such as $SO_4^{2-}$, can also be transported under specific conditions). As with the Na/H exchanger, the Cl/HCO₃ exchanger has several isoforms, denoted AE (for anion exchanger) 1, 2, and 3. These various isoforms have about 80–90% homology. AE 1 is the band 3 transporter from red blood cells and is the smallest isoform ($M_r$ 115,000). This protein has as many as 14 membrane spanning domains (helices) and a large N-terminal cytoplasmic domain that contains various binding sites, including one for ankyrin (Fig. 5B). AE 2 and 3 are larger ($M_r$ 145–165,000). AE 2 is apparently the "housekeeping" anion exchanger and is widely distributed. It is activated by cellular alkalinization and returns $pH_i$ to normal by extruding base ($HCO_3^-$) and thereby reacidifying the cell. This exchanger may also possess an internal allosteric regulatory site that activates the exchanger at *alkaline* values of $pH_i$ (see Base Extrusion curve, Fig. 3). In addition to its role in the regulation of $pH_i$, Cl/HCO₃ exchange may also play a role in regulating intracellular $Cl^-$. The function of AE 3 is not yet clear but it has a far more restricted distribution than AE 2, being found only in the heart and the central nervous system.

The **(Na + HCO₃)/Cl exchanger** (model 4, Fig. 4B) was originally described as the pH-regulating transport system in invertebrate nerve and muscle preparations, but has since been found in a wide variety of cells. Because it transports Na ions, it operates in the opposite way than Na-independent Cl/HCO₃ exchange. Thus, the (Na + HCO₃)/Cl exchanger exchanges 1 external $Na^+$ for 1 internal $Cl^-$ and neutralizes the equivalent of 2 internal protons. As such this exchanger is electroneutral and mediates the alkalinization of the cell in response to an acid load. The ability of this exchanger to neutralize 2 acid equivalents could be achieved by the influx of 2 HCO₃ ions, the influx of a HCO₃ ion in exchange for the efflux of an H ion, the influx of a $CO_3^{2-}$, or the influx of an ion pair ($NaCO_3^-$). In squid axon, this exchanger has been suggested to involve $NaCO_3^-/Cl^-$ exchange, whereas in barnacle muscle fibers it must be another variant, suggesting that this exchanger also has at least two isoforms. The structure of this exchanger is not currently known.

The **Na–HCO₃ cotransporter** (model 5, Fig. 4B) was originally identified in renal epithelial cells but has since been found in a number of other cell types as well. It mediates the movement of 1 Na ion with 1, 2, or 3 HCO₃ ions, depending on the cell type in which the transporter is located. This transporter differs from the other HCO₃-dependent transporters in that it does not require Cl ions and, in the case where the stoichiometry is 1:2 or 1:3, it is electrogenic. In proximal tubule epithelial cells, Na–HCO₃ cotransport is involved in $HCO_3^-$ reabsorption by mediating $HCO_3^-$ efflux across the basolateral membrane. In other cells, such as glial cells, the cotransporter is proposed to mediate $HCO_3^-$ influx and contribute to the regulation of intracellular pH in the face of an acid load. The renal cotransporter (1:3 stoichiometry) has been shown to involve the cotransport of 1 $Na^+$ with 1 $HCO_3^-$ and 1 $CO_3^{2-}$. This cotransporter has recently been cloned. It is a unique glycoprotein that contains over 1000 amino acids. Its structure is similar to other membrane transport proteins, containing 10 putative membrane-spanning regions and large cytoplasmic domains (one each at the N- and C-terminal ends). These cytoplasmic domains probably are the site of cotransport regulation and have a number of potential phosphorylation sites. In addition, this protein has one large extracellular loop. It is not known how the structure of this cotransporter compares to the Na–HCO₃ cotransporters with stoichiometries of 1:1 and 1:2.

### C.  H⁺-ATPases (Proton Pumps)

There are at least three known varieties of **H-ATPases:** (1) F₀–F₁ type ATPase, (2) E₁–E₂ or P-type ATPase, and (3) vacuolar or V-type ATPase. The **F₀–F₁-type ATPase** is found in mitochondria (see Mitochondria). This ATPase has also been called the ATP synthase, because it functions to produce ATP when H ions move down their electrochemical gradient. F₀–F₁-ATPase has a lollipop shape and a membrane-spanning region, F₀, that forms the putative H ion pore through the membrane as well as an extrinsic head region, F₁, that contains the ATPase activity. The F₀ region is quite large, with 5 subunits and a molecular weight

of about 380,000. The F₁ region has 4 subunits and a molecular weight of about 100,000. These two regions are connected by a stalk which contains several subunits, one of which confers sensitivity to oligomycin. The F₀–F₁–type ATPase can be inhibited by azide and N,N'-dicyclohexylcarbodiimide (DCCD) in addition to oligomycin.

The **P-type ATPases** (model 6, Fig. 4C) are characterized by forming a phosphorylated intermediate upon ATP hydrolysis. The classic example of such an ATPase is the Na,K-ATPase. An example of a **P-type H⁺-ATPase** is the H,K-ATPase, best characterized from the apical membrane of gastric glands, where it is responsible for acid secretion into the stomach. This ATPase has also been implicated in the acidification of the urine and reabsorption of $K^+$ by the kidney and in the establishment of an $H^+$ gradient across yeast plasma membrane. In contrast to the F₀–F₁-ATPase, the H,K-ATPase exchanges a K ion for an H ion and is thus electroneutral. The H,K-ATPase has a molecular weight of about 110,000 and has a structure similar to those of other membrane-bound ATPases with several membrane-spanning regions and a large cytoplasmic domain containing the ATP hydrolysing site. Inhibition by vanadate is characteristic for P-type ATPases.

The third type of H⁺-ATPase is the vacuolar, or V-type, ATPase (model 7, Fig. 4C) and is found in yeast and plant vacuoles as well as in several eukaryotic cells (e.g., kidney cells, osteoclasts, macrophages) and organelles (e.g., endosomes, lysosomes, secretory granules, Golgi apparatus). **V-type ATPases** are more like the F₀–F₁-ATPase than the P-type ATPase in that they do not form phosphorylated intermediates, are quite large (>400 kDa), assume a lollipop shape, and are electrogenic. The major function of V-type ATPases is the acidification of intracellular organelles (see Section IV,B), which is so important for proper protein targeting and handling. These ATPases are characterized by their lack of sensitivity to vanadate and oligomycin and by their inhibition by N-ethylmaleimide (NEM), DCCD, and bafilomycin. Whereas P- and F₀–F₁-type ATPases are found in both prokaryotes and eukaryotes, the V-type ATPases are found only in eukaryotes and therefore presumably evolved more recently.

### D.  Transport of Anionic Weak Bases with Na

In a variety of organisms, renal proximal tubule cells have been shown to have **Na⁺-organic anion transporters** (model 1, Fig. 6). These cotransporters mediate the influx of a Na ion with an organic anionic weak base, such as lactate or acetate. Such cotransport would cause cellular alkalinization due to the entry of base. The anionic base would bind a proton upon entry and alkalinize the cell. These transport systems are likely to be part of a mechanism designed for the transepithelial movement of organic molecules and should be considered to *affect* pH$_i$ rather than *regulate* it.

### E.  Chloride/Formate Exchange

Another class of organic anion transporters has been described in renal proximal tubule cells. These transporters

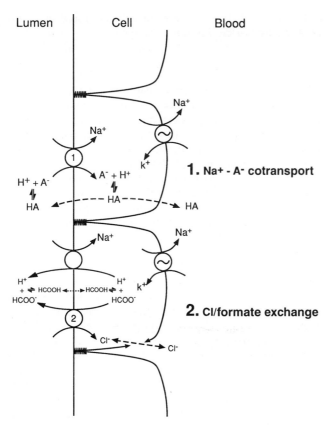

**FIG. 6.** A model of anionic weak base fluxes in the renal proximal tubule mediated by either cotransport with Na$^+$ (1) or in exchange for Cl$^-$ (2). An example of 1 is the Na-acetate cotransporter. The Cl/formate exchanger is depicted by 2. These transporters are apparently involved in the luminal entry of Na$^+$ or Cl$^-$ in NaCl reabsorbing epithelia like the renal proximal tubule. Solid lines represent ion fluxes mediated by membrane transporters. Dashed lines represent passive diffusion of molecules across the membrane. Cl$^-$ movements across the basolateral membrane are mediated by an ion-selective Cl channel.

involve the exchange of organic anions for inorganic anions such as Cl$^-$ or OH$^-$. For example, as part of the mechanism for NaCl reabsorption, an exchanger that mediates Cl$^-$ influx (from the lumen) in exchange for formate efflux resides on the apical membrane of proximal tubule epithelia. This **Cl/formate exchanger** (model 2, Fig. 6) can be inhibited by disulfonic stilbene derivatives and is functionally similar to the Cl/HCO$_3$ exchanger. However, the Cl/formate exchanger is distinct from the Cl/HCO$_3$ exchanger. The operation of this transporter during NaCl reabsorption should result in epithelial cell acidification. Other transporters involve the movement of organic anions including urate and oxalate. When these anions are transported into the cell, they will result in cellular alkalinization.

## VII. Cellular Functions Affected by Intracellular pH

A wide variety of cellular processes and properties are affected by intracellular pH, and perhaps in some way

all cell functions are influenced by the level of pH$_i$. It is impossible to discuss all of the various effects but many of the most important will be highlighted in the following sections (Fig. 7).

### A. Cellular Metabolism

The fact that cellular metabolism can affect pH$_i$ was discussed previously (see Section V,A). It has been appreciated for many years that the converse is also true; that is, changes of pH$_i$ can affect cellular metabolism. Theoretically, because pH will affect the charge on ionizable groups in proteins, it would be anticipated that changes in pH$_i$ could change the configuration of proteins and affect their activity. Such an effect of pH$_i$ has been well documented for two key metabolic enzymes. Phosphofructokinase, a key glycolytic enzyme that converts fructose 6-phosphate (F6P) to fructose 1,6-diphosphate (FDP), has an exquisite pH sensitivity in the physiological range (6.5–7.5), its activity decreasing with a decrease of pH$_i$. The actual pH sensitivity is dependent on the cellular levels of F6P and 5'-AMP. Similarly, the conversion of phosphorylase (which catalyzes the metabolism of glycogen) from its inactive to active form is inhibited by a decrease in pH$_i$.

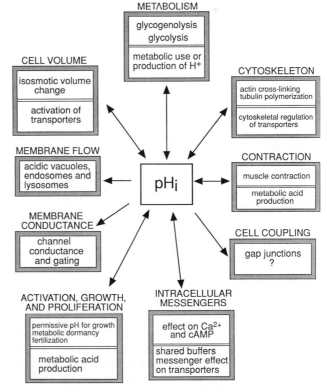

**FIG. 7.** A summary of the cellular processes affected by intracellular pH. Each box represents a different cellular process. A two-headed arrow indicates that pH$_i$ affects the process and that the process can affect pH$_i$. Single-headed arrows suggest that pH affects that cellular function but that the cellular processes probably do not have a major impact on pH$_i$. Within a box, the top half indicates the cellular processes affected by pH$_i$, while the bottom half indicates the mechanisms by which the process can affect pH$_i$.

Two observations can be made from these findings. First, the pH dependence of enzyme activity is often affected by the concentrations of other factors, including substrates and other effectors. Thus, caution must be exercised in relating the *in vitro* pH profile of an enzyme to cellular conditions. Second, the general reaction of metabolic enzymes to changes in pH is a reduction in activity and a decrease in $pH_i$. This suggests that a decrease in $pH_i$ could be used to prevent growth or to put a cell in a dormant state, as has been observed for many cells (see Section VII,G).

## B. Cytoskeleton

Changes in $pH_i$ have been shown to affect the cytoskeleton, which can lead to changes in cell shape or motility. One example of such an effect on the cytoskeleton is the pH dependence of **actin filament** cross-linking to form gels. This cross-linking is mediated by actin-binding proteins and their ability to interact with actin is pH dependent. In some cases, cell alkalinization increases the actin cross-linking to form a gel state in the cytoplasm or to form networks of microfilament bundles. In other cells, alkalinization reduces the cross-linking of actin filaments. The cell-specific responses of cytoskeletal cross-linking are probably due to different pH profiles of actin-binding proteins in different cells.

Changes in cellular pH can also affect the polymerization of cytoskeletal elements, such as **tubulin.** It has been shown in some cells that alkalinization can cause depolymerization of tubulin and disaggregation of microtubules within the cell. There are probably many other pH-dependent components to cytoskeletal assembly and function. It should be pointed out, however, that the conditions that lead to a change in $pH_i$ are often accompanied by changes in intracellular calcium and by phosphorylation of the cytoskeleton, and it is not always clear which is the predominant effector of cytoskeletal changes. Nevertheless, it is clear that changes in $pH_i$ can play a major modulatory role in at least some alterations of the cytoskeleton.

## C. Muscle Contraction

Intracellular acidification is well known to reduce the ability of contractile cells to generate tension. This effect is particularly marked in cardiac muscle, but skeletal muscle and smooth muscle also show a reduced contractility at acidic values of $pH_i$. There are several possible ways by which cellular acidification could influence contractility: Changes in surface channels could reduce cellular excitability, low pH could prevent calcium release from the sarcoplasmic reticulum through the calcium release channel, protons could compete with calcium for binding to the regulatory protein troponin, protons could inhibit the myofibrillar ATPase, or acidification could impair the ability of the cell to generate ATP. The effect of pH on muscle contraction could also be indirect. For example, during intense muscle activity inorganic phosphate ($P_i$) accumulates in the cell. A reduced pH will increase the diprotonated form of $P_i$ ($H_2PO_4^-$), and this has been shown to be particularly effective in inhibiting muscle force development during muscle fatigue.

## D. Cell–Cell Coupling

The coupling of cells through **gap junctions** is apparently affected by intracellular pH. Several studies have suggested that a fall in $pH_i$ uncouples gap junctions, thereby eliminating cell–cell coupling. The direct role of $pH_i$ in the uncoupling process has been questioned, and it appears that changes in intracellular $Ca^{2+}$ levels are responsible for uncoupling in some cells. The control of gap junction conductance may differ from cell to cell, but even in cells where $Ca^{2+}$ is the primary regulator, changes in $pH_i$ probably still have a modulatory effect on gap junctional conductance.

## E. Membrane Conductance

**Ion-selective channels** require the presence of charges within the channel proteins for proper ion conduction and channel gating. If these charges have $pK$ values within the physiological range, these channels could well be affected by changes in pH. Indeed, the conductances of many channels are affected by changes in either $pH_o$ or $pH_i$, including the tetrodotoxin-sensitive Na channel, the delayed rectifying and inward rectifying K channels, and Cl channels. Through the effect on conductance of membrane channels, changes in pH can affect the excitability of nerve and muscle cells and alter the membrane potential in all cells.

## F. Intracellular Messengers

Changes in $pH_i$ can affect the levels of important intracellular signaling molecules, such as **$Ca^{2+}$** and **cAMP.** There are several possible ways by which pH can affect intracellular $Ca^{2+}$. An elevation of cytoplasmic $H^+$ can activate mitochondrial Ca/H exchange, resulting in a sequestering of H ions within the mitochondria and an elevation of cytoplasmic $Ca^{2+}$. A decreased $pH_i$ can reduce $Ca^{2+}$ entry across the plasmalemma. The most direct interaction between cytoplasmic H and Ca ions, however, results from shared buffers. Many molecules that buffer H ions will also bind, and thus buffer, Ca ions. Depending on the relative affinities, an elevation of cytoplasmic $H^+$ can elevate intracellular $Ca^{2+}$ by displacing Ca ions from intracellular buffer sites.

The interaction between changes of $pH_i$ and $Ca_i^{2+}$ can also be indirect. An example of such an interaction is the pH dependence of the binding of $Ca^{2+}$ to calmodulin. Under certain conditions, a decrease in pH can be shown to reduce the binding of $Ca^{2+}$ to calmodulin. Another example of the interaction between pH and $Ca^{2+}$ is the pH dependence of the interaction of the Ca–calmodulin complex with other proteins, the direction of which depends on the protein being considered. Thus, the potential exists for changes of $pH_i$ to alter the effect of $Ca^{2+}$ on cellular function.

Changes of $pH_i$ could affect another important intracellular signaling pathway as well, that involving cAMP. The proposed effects of $pH_i$ on cAMP are based on the pH dependence of adenylyl cyclase (AC—the enzyme that synthesizes cAMP) and the cyclic nucleotide phosphodiesterase (PDE—the enzyme that hydrolyzes cAMP). In most cells, PDE apparently has a rather constant activity over the physiological range of pH (6.5–7.5). However, depending on the cell, an increase in pH can either markedly increase or decrease AC activity. Thus, alkalinization can result in either an increase or a decrease in cellular cAMP levels.

Given the pervasive effect of changes of $pH_i$ on metabolic proteins, it is likely that effects of changes of $pH_i$ on other signaling pathways, such as those mediated by cGMP or phosphoinositide metabolism, also exist. It should be noted, however, that the physiological significance of these effects of $pH_i$ changes on signaling pathways is not often clear, particularly for cells that normally should see only small fluctuations of $pH_i$.

## G. Cell Activation, Growth, and Proliferation

One of the most active areas of research on the role of intracellular pH in cell function has been the study of the role of early changes of $pH_i$ in cell proliferation. These studies grew out of early observations that, shortly after fertilization of sea urchin eggs, egg pH increased markedly (roughly 0.4 pH unit), and this rise in pH was necessary for the initiation of growth by fertilization. These observations were followed by others on mammalian cells showing that a variety of growth-promoting agents, including epidermal growth factor (EGF), platelet-derived growth factor (PDGF), insulin, vasopressin, and serum, similarly induced a cellular alkalinization (of about 0.1–0.2 pH unit) shortly after exposure. All of these alkalinizing effects are mediated by activation of Na/H exchange. These growth promoting agents activate the exchanger by activating cellular signaling pathways, which increase the affinity of the exchanger for internal H ions and increase its activity, thereby alkalinizing the cell (see arrow 1, Fig. 3).

It was initially hypothesized that the growth factor-induced increase in pH was part of a group of early signals that are required for initiation of cell growth and proliferation. Cellular alkalinization was believed to contribute to the initiation of growth by activating key cellular enzymes that then were either direct effectors of growth (e.g., metabolic enzymes) or were activators of other systems.

Recently, the direct signalling role of increases of $pH_i$ in cell activation has been questioned. Changes in $pH_i$ by themselves did not promote cell growth or division. Further, it was shown that a number of cells had a higher $pH_i$ in the presence of 5% $CO_2$ than in its absence. However, most of the initial experiments on the $pH_i$ responses to growth factors had been done in the absence of $CO_2$. Upon repeating a number of these experiments under more physiological conditions (presence of 5% $CO_2$), the initial alkalinization upon exposure to stimulatory factors was not always seen and in some cells the initial response was indeed an *acidification*. Thus, the current view of the role of $pH_i$ in cell activation, growth, and proliferation is that $pH_i$ plays a permissive role. That is, cells will only grow and proliferate if $pH_i$ is above a certain critical value,

regardless of how that value is obtained. If a cell had a $pH_i$ above the critical value before exposure to a stimulatory agent, no change in $pH_i$ would be required for cell growth. In fact, the cell could acidify and still grow as long as its $pH_i$ stayed above the critical value.

In reality, the role of $pH_i$ as a signal to initiate cell growth may depend on the cell and the activating agent. It is likely that a rapid and marked rise in $pH_i$ is one of the critical early steps necessary for the initiation of growth in fertilized sea urchin eggs. Even more dramatic is the over 1 pH unit increase of $pH_i$ in *Artemia* (brine shrimp) embryos upon arousal from anaerobic dormancy by exposure to oxygen. Undoubtedly, this large rise in $pH_i$ is crucial for the translation from metabolic dormancy in these organisms. On the other hand, in many mammalian cells, the rather modest increase in $pH_i$ on exposure to an activating agent is probably of limited physiological significance, especially given the variability in the degree and direction of pH changes observed in different experimental conditions.

Finally, even under conditions where a change of $pH_i$ is not observed in response to a stimulatory agent, pH-regulating transport systems can still be shown to be activated. It has been hypothesized that although a change of $pH_i$ may not be crucial for the stimulation of cell growth, the initiation of this growth may confront the cell with an acid or alkaline load. In this regard, activation of the pH-regulating transport systems by growth-promoting agents could be viewed as preparatory, enabling the cell to better maintain a constant $pH_i$ during a period of high metabolic activity.

## H. Cell Volume Regulation

Most pH-regulating transporters move ions such as $Na^+$ and $Cl^-$ in exchange for a proton equivalent ($H^+$ or $HCO_3^-$). Because the transported proton equivalents are buffered ($H^+$ by $HCO_3^-$ and protein buffers and $HCO_3^-$ by formation of $CO_2$), they are osmotically "invisible." For example, virtually all of the $H^+$ transported by the Na/H exchanger derive from internal buffers and upon efflux from the cell are buffered by external buffers. Thus, the Na/H exchanger mediates the net import of one osmotically active particle (a $Na^+$), and this import of osmolytes will be accompanied by the influx of water and cell swelling. Therefore, the Na/H exchanger, in addition to contributing to $pH_i$ regulation, can mediate the regulation of cell volume.

Other pH-regulating transporters can similarly mediate cell volume changes. For example, the $Cl/HCO_3$ exchanger transports $Cl^-$ into the cell. The $HCO_3^-$ that leaves combines with an $H^+$ and is removed as $CO_2$. Thus, like Na/H exchange, $Cl/HCO_3$ exchange contributes to cell swelling. In fact, these two exchangers often act in concert to result in the net influx of NaCl (and therefore water) into the cell. The Na–$HCO_3$ and Na–anionic weak base cotransporters would be ideally suited to mediate net solute transfer and therefore cell volume change, but such a role for these transport proteins has not been identified yet. Finally, the (Na + $HCO_3$)/Cl exchanger should not contribute to cell volume regulation because it mediates the entry of one osmotically active ion (Na$^+$) for the efflux of another (Cl$^-$).

Cell volume can be rapidly changed by exposure to anisosmotic media, hypertonic media causing cell shrinkage, and hypotonic media causing cell swelling. Many cells respond to shrinkage with a **regulatory volume increase (RVI),** which involves the net uptake of solutes, and therefore water, with cells swelling back toward the initial cell volume (Fig. 8). In several different types of cells, RVI has been shown to involve an activation of Na/H exchange. This exchanger, often in association with the Cl/HCO$_3$ exchanger, results in NaCl influx and a regulatory volume increase. The mechanism by which cell shrinkage activates the Na/H exchanger is not fully understood, but interestingly, unlike most other activation pathways, cell shrinkage apparently does not result in phosphorylation of the exchanger. A possible involvement of the cell cytoskeleton in activating the Na/H exchanger upon cell shrinkage is currently being investigated.

Recently, it has been shown that shrinkage can also activate (Na + HCO$_3$)/Cl exchange in some cells. This observation is interesting since, as stated previously, this exchanger does not mediate any net solute movement and thus should not directly contribute to cell volume regulation. This activation of (Na + HCO$_3$)/Cl exchange by cell shrinkage suggests that pH regulating transporters may be activated by shrinkage to alkalinize the cell regardless of whether they contribute to volume regulation. It is not clear what benefit a shrunken cell derives by becoming alkaline, but it may involve pH-dependent cytoskeletal rearrangements (see Section VII,B).

In response to swelling, most cells exhibit a **regulatory volume decrease (RVD).** RVD involves the efflux of solutes accompanied by water and therefore cell shrinkage back towards the initial cell volume (see Fig. 8). In at least one cell type, the nucleated red blood cell, RVD has been shown to be mediated by K/H exchange (K$^+$ efflux and H$^+$ influx) in association with Cl/HCO$_3$ exchange (Cl$^-$ efflux and HCO$_3^-$ influx).

Cell volume can also be altered under isosmotic conditions by an imbalance of solute influx and efflux. For instance, during periods of active pH recovery from acidification, the Na influx mediated by the Na/H exchanger could result in cell swelling. Thus, changes in pH$_i$ and the response to them could result in an alteration of cell volume.

It is clear that the regulation of intracellular volume and intracellular pH are highly linked in most cells. This linkage is due in part to the use of many of the same membrane transport systems for the regulation of cell pH and volume. In any given cell type, these transporters may respond predominantly either to changes in pH$_i$ or to changes in cell volume.

## I. Intracellular Membrane Flow

The intracellular flow of membranes is affected by changes of pH within acidic vacuolar compartments. In cells, these vacuolar compartments are often involved in the movement of membranes, membrane-bound proteins, and soluble proteins around the cell. In addition, components of the vacuolar system are involved in the synthesis, processing, and degradation of various proteins. This system includes the endoplasmic reticulum, the Golgi apparatus, lysosomes, and endosomes. Movement of materials through this system can be divided into the endocytic and exocytic pathways. The endocytic pathway is involved in the uptake of external macromolecules, the degradation or delivery to the cell of these macromolecules, and the down regulation of surface proteins; it includes coated pits, endosomes, and lysosomes. The exocytic pathway delivers newly synthesized proteins to a variety of sites, including the surface membrane or extracellular space. Many of the compartments within this vacuolar system are acidic (see Section IV,B), and the maintenance of an acidic interior is critical for the functioning of these compartments. This criticality has been shown by the marked disturbance of endocytic and exocytic pathways by a number of agents, such as chloroquine and ammonia, which alkalinize these compartments. In addition, inhibition of vacuolar H$^+$-ATPase results in alkalinization of the acidic compartments and can lead to inhibition of endocytosis/exocytosis. Thus

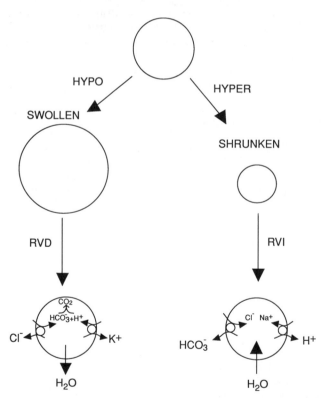

**FIG. 8.** The role of pH-regulating transporters in cellular volume regulation. Upon cell swelling in hypotonic media, KCl has been shown to be removed from nucleated red blood cells by functionally coupled operation of the K/H and the Cl/HCO$_3$ exchangers. The removal of KCl from the cell causes a loss of water from the cell and thus cell shrinkage in a process called regulatory volume decrease (RVD). Upon cell shrinkage in hypertonic media, NaCl enters the cell by parallel operation of Na/H and Cl/HCO$_3$ exchangers. The NaCl entry results in water influx and cell swelling in a process called regulatory volume increase (RVI).

the maintenance of a proper pH in acidic intracellular compartments is crucial for continued and proper intracellular membrane flow.

## VIII. Summary

Virtually all of the H ions within a cell are buffered by reversible binding to weak acids and bases, resulting in a low free H ion concentration. Therefore, the concentration of free H ions within the cytoplasm is usually expressed as cytoplasmic pH ($pH_i$), defined as $pH = -\log(a_H)$, which is a more convenient scale for molecules at low concentrations.

Cytoplasmic pH is an important aspect of the intracellular milieu and can affect nearly all aspects of cell function. In most cells, $pH_i$ is maintained at a value of about 7.0, well alkaline to the equilibrium $pH_i$, calculated on the assumption that H ions are at equilibrium across the membrane. The fact that $pH_i$ is alkaline to its equilibrium value creates a passive acidifying influx of H ions. In fact, most cells face a continuous acid load due not only to this acidifying influx but to metabolic acid production and leakage from internal acidic compartments as well. Such induced changes in $pH_i$ can be blunted by cellular buffers, but the only way to fully regulate $pH_i$ is through the activity of membrane-bound transporters. These transporters fall into five categories: (1) cation/H exchangers, such as the alkalinizing Na/H exchanger and the acidifying K/H exchanger; (2) $HCO_3$-dependent transporters, such as the (Na + $HCO_3$)/Cl and the Cl/$HCO_3$ exchangers and the Na–$HCO_3$ cotransporter; (3) H-ATPases or proton pumps; (4) Na–anionic weak base cotransporters; and (5) Cl/organic anion exchangers.

Changes in $pH_i$ can affect many cellular functions. Cell metabolism can be affected by changes in pH, predominantly because of pH-sensitive metabolic enzymes, such as phosphofructokinase. Changes of $pH_i$ have also been shown to affect the cross-linking and polymerization of cytoskeletal elements such as actin and tubulin. The loss of the ability of muscle cells to generate tension (muscle fatigue) has been correlated with a decrease of $pH_i$. Cell pH is also believed to play a modulatory role in gap junctions and many ion-selective channels. Further, changes of $pH_i$, mediated by activation of the Na/H exchanger, may serve as an intracellular signal for the promotion of cell growth and proliferation. It is significant that changes in $pH_i$ can affect other intracellular signals, such as cellular $Ca^{2+}$ and cAMP levels, suggesting a complex interaction among cellular signalling systems.

Many pH-regulating transporters move an osmotically active ion, such as $Na^+$ or $Cl^-$, in exchange for a buffered (and thus osmotically "invisible") ion like $H^+$ or $HCO_3^-$ and thus mediate the net movement of solute into or out of the cell. This net movement of solute will be accompanied by a net water flow and result in a change in cell volume. Thus, many pH-regulating transporters, in addition to contributing to $pH_i$ regulation, can mediate the regulation of cell volume.

The pH of certain organellar compartments can differ from the value of $pH_i$ and these differences in pH are important for organellar function. For example, mitochondria maintain an internal pH (7.5) about 0.5 unit more alkaline than $pH_i$. This pH gradient across the mitochondrial membrane is essential for the major function of mitochondria, the production of ATP. Further, several intracellular organelles in the vacuolar system (e.g., endosomes, lysosomes, and storage granules) maintain an internal pH (5–6) well below $pH_i$. These organelles contribute to the movement of membranes, membrane-bound proteins, and soluble proteins around the cell and their acidic pH is essential for this function.

Given the importance of pH to so many cellular functions, it is not surprising that cells have elaborated highly regulated mechanisms to control $pH_i$.

# Appendix

The study of intracellular pH and its regulation was enabled in 1974 by the development of the first reliable pH-sensitive microelectrode by Roger Thomas, called the **Thomas recessed-tip microelectrode.** These electrodes are constructed from special glass that is conductive to H ions only. Small capillaries of this pH-sensitive glass are pulled into fine tip electrodes (tip diameter $\approx 1~\mu$m) and these tips are sealed by heating. Such an electrode will respond with a Nernstian slope (about 59 mV/pH unit) to changes in pH. However, to reliably measure $pH_i$, one must assure that the pH-sensitive surfaces of this electrode are exposed to cytoplasm only. This is achieved by carefully lowering the sealed-tip pH electrode into a larger (about 2 $\mu$m) open-tip microelectrode (the shielding microelectrode) constructed from $AlSiO_4$ glass. The tip of the inner pH electrode is brought within a few microns of the open tip of the outer electrode (the distance BT, between tips, in Fig. 9A). The pH electrode is heated, under internal pressure. The pH glass, which melts at a lower temperature than the $AlSiO_4$ glass, is pushed against the inner face of the shielding microelectrode and forms a high resistance glass-glass seal (Fig. 9A). The inner electrode is removed

above the seal and what remains is the tip of the pH electrode sealed near the tip of the shielding electrode (see Fig. 9A). This "recessed-tip" electrode is filled with a buffered conducting solution and connected via a fine chlorided silver wire to an electrometer. The tip of the pH electrode below the seal (exposed length, EXPL in Fig. 9A) will respond to changes in the pH of the fluid trapped within the recess volume (striped area in Fig. 9A). When a cell is impaled with such an electrode, the fluid in the recess volume is replaced (by diffusion) with cytoplasm. A signal is generated across this electrode, which is the sum of the actual membrane potential ($V_m$) and a Nernstian signal based on the difference in pH between the electrode filling solution and the cytoplasm in the recess volume. To derive a signal that is directly proportional to the intracellular pH, a conventional open-tip KCl-filled microelectrode is placed within the same cell to measure $V_m$ and the signals from the two electrodes are subtracted (see Fig. 9B).

The recessed-tip microelectrode gave some of the first continuous, reliable measurements of intracellular pH and initiated 2 decades of intensive study of $pH_i$. However, this electrode has drawbacks. Because of its large size and the need to impale a cell with two electrodes, its use is restricted to fairly large cells. Further, the electrode has a high resistance (often greater than 100 G$\Omega$) and must be used with a high input-impedance amplifier. Finally, because of the need for diffusional exchange between the recess volume and the cytoplasm, these electrodes have a slow response time (time constant no faster than 15 sec).

A different type of pH microelectrode can be constructed using pH-sensitive resins (Fig. 9C). Like pH-sensitive glass, these resins are conductive to H ions only. A bit of this resin is placed in the tip of a conventional microelectrode and the electrodes are back-filled with a buffered conducting medium (see Fig. 9C). When a cell is impaled with these **pH-sensitive resin microelectrodes,** a signal is generated that is the sum of the $V_m$ and a Nernstian signal because of the pH difference between the electrode and the cytoplasm. Thus, a cell must be impaled with a $V_m$

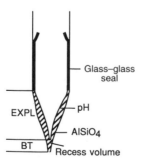

**FIG. 9A.** A diagram of a pH-sensitive recessed-tip glass microelectrode showing the sealed tip of a microelectrode constructed from pH-sensitive glass fused within the tip of a larger shielding aluminosilicate electrode. BT—between tips distance; EXPL—length of exposed pH-sensitive electrode below glass–glass seal.

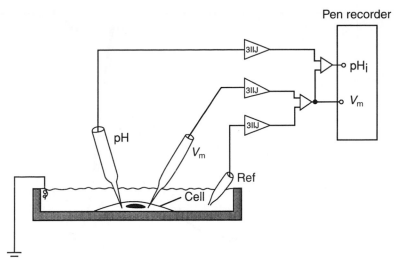

**FIG. 9B.** Measurement of pH$_i$. Diagram of the apparatus used to measure intracellular pH with microelectrodes. The cell is impaled with a pH-sensitive and a $V_m$-sensitive microelectrode. These electrodes are referenced to an open-tip flowing KCl reference electrode. Electrode voltages are amplified by Analog Devices 311J high-input impedance operational amplifiers. The difference in electrode voltage between the $V_m$ and reference electrodes is equal to membrane potential, and the difference in electrode voltage between the pH and $V_m$ electrodes is a signal that is proportional to pH$_i$. These difference signals are plotted on a pen recorder.

microelectrode when using the resin electrodes as well as the recessed-tip electrodes. These pH-sensitive resin microelectrodes can penetrate much smaller cells because of their smaller tip diameters (usually less than 1 $\mu$m) and are faster than the recessed-tip electrodes. Further, pH-sensitive resin microelectrodes can readily be constructed in one barrel of a double-barreled electrode (the other barrel filled with KCl to measure $V_m$) (see Fig. 9C). Thus, the pH-sensitive resin microelectrodes have supplanted the recessed-tip microelectrode for most studies of pH$_i$.

Another reliable method for measuring intracellular pH is based on the fact that a number of dyes bind H ions

reversibly and this binding affects the fluorescence of these dyes. The most common **pH-sensitive dye** in use today is a derivative of fluorescein, biscarboxyethyl carboxyfluorescein (BCECF) (Fig. 9D). An excitation spectrum for this dye can be obtained by shining light of different wavelengths (between 300 and 600 nm) on a solution containing this dye and measuring the fluorescence emitted at 535 nm. When such an excitation spectrum is gathered for the dye at different values of solution pH, the resulting spectra can be superimposed (Fig. 9E). Such curves reveal that when the dye is excited at 440 nm, its fluorescence is the same regardless of pH (called the isoexcitation point), whereas the dye is maximally sensitive to pH when excited at 500 nm, fluorescence increasing as pH increases.

The charged sites on the dye are often rendered neutral by attaching acetoxymethyl ester groups (BCECF/AM). In this form, the dye readily permeates the cell. Once inside, cell esterases cleave the AM groups from the dye, once again creating the charged form, BCECF. This form is impermeable and remains trapped within the cytoplasm. The fluorescence of this dye within cells can be measured using a spectrofluorometer.

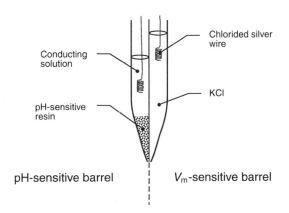

**FIG. 9C.** A diagram of a double-barreled pH-sensitive resin microelectrode. The tip of one barrel of a double-barreled electrode is filled with a pH-sensitive resin and then back-filled with a conducting solution. This is the pH-sensitive barrel. The other barrel, the $V_m$-sensitive barrel, is filled with KCl. Each barrel is connected to an amplifier via a chlorided silver wire immersed in the conducting solution of that barrel.

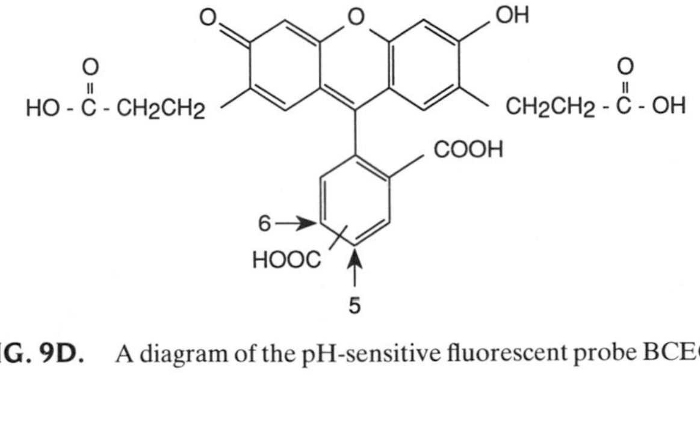

**FIG. 9D.** A diagram of the pH-sensitive fluorescent probe BCECF.

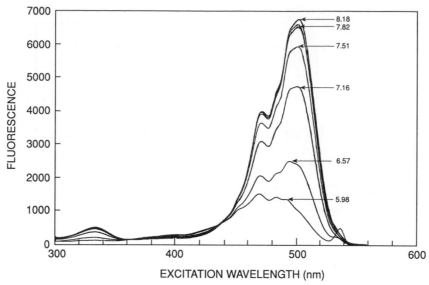

**FIG. 9E.** Excitation spectra of BCECF: the fluorescence emission (at 535 nm) of BCECF when excited at wavelengths from 300 to 600 nm. Each curve represents the excitation spectrum of BCECF at a different pH (indicated to the right of each spectrum). The ordinate represents fluorescence in arbitrary units.

The dye within cells can slowly leak out or photobleach during the course of an experiment. Either of these occurrences would result in a decreased fluorescence signal and appear as a decrease in $pH_i$. To prevent this artifact, the dye fluorescence is often collected at excitation wavelengths of both 440 and 500 nm and a fluorescence ratio ($R_{fl}$) $Fl_{500}/Fl_{440}$, calculated. In this ratio, $Fl_{440}$ is a measure of the amount of dye present and thus serves as a normalizing factor. $R_{fl}$ is proportional to intracellular pH and is not susceptible to leakage or photobleaching artifacts. The fluorescence ratio can be calibrated by exposing cells to high extracellular K (similar to intracellular K) and nigericin, a K/H exchanging ionophore. Under these conditions it is assumed that the nigericin will equilibrate intracellular and extracellular pH. Upon exposing dye-loaded cells to a variety of extracellular pH values and measuring $R_{fl}$, a calibration curve can be constructed (see Fig. 9F). The use of pH-sensitive fluorescent dyes to study $pH_i$ has several advantages over microelectrodes. This technique can be used on cells of any size, including preparations of intracellular organelles. Further, it is a relatively simple and reliable experimental technique. Finally, this technique can be used to visualize pH within single cells or parts of cells, using optical imaging techniques.

To study the ability of cells to actively extrude acid or base from the cell, techniques must be available to alter intracellular pH experimentally. In some large cells, this has been achieved directly by injecting acid into cells, by passing current through microelectrodes, or by internal dialysis through tubing threaded through the cytoplasm of the cell. More recently, changes in $pH_i$ have been accomplished by internal perfusion of cells using whole-cell patch-clamp electrodes. However, the most commonly used experimental method to modify $pH_i$ is by external exposure of cells to weak acids or bases. One of the most popular

of such techniques is the **NH₄Cl prepulse technique.** Cells are exposed to an external solution containing $NH_4Cl$. External $NH_3$, being uncharged, enters the cell (arrow 1, Fig. 9G, part A) far more rapidly than external $NH_4^+$. In the cell, the $NH_3$ combines with an H ion to form $NH_4^+$ (arrow 2, Fig. 9G, part A), thereby alkalinizing the cell. This alkalinization will continue until the internal and external concentrations of $NH_3$ are the same, at which time no more $NH_3$ will enter and the $pH_i$ will reach a new steady state value alkaline to the original $pH_i$. If $NH_4^+$ is unable to enter the cell, no further change in $pH_i$ will occur until external $NH_4Cl$ is removed, at which time cytoplasmic $NH_4^+$ will dissociate to $NH_3$ and $H^+$, and all the $NH_3$ will diffuse from

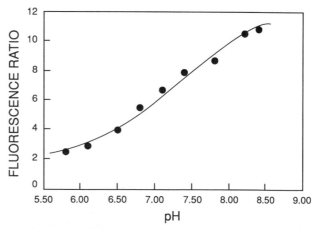

**FIG. 9F.** A calibration curve for BCECF derived from data as shown in Fig. 9E. The ratio of the emitted fluorescence at excitation wavelengths of 500 and 440 nm ($R_{fl} = Fl_{500}/Fl_{440}$) is plotted against the pH of the solution and a sigmoid titration curve is obtained.

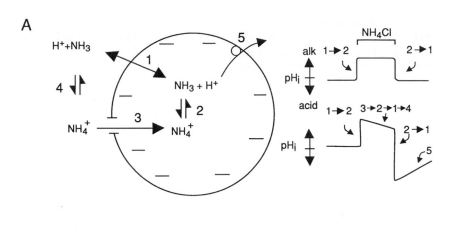

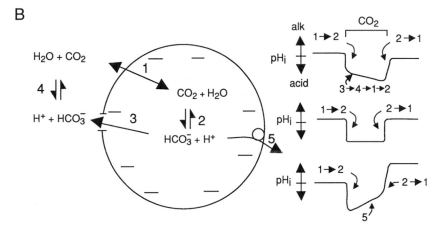

**FIG. 9G.** (A) A diagram of the effects on intracellular pH of a transient exposure of a cell to external $NH_4Cl$. (B) A diagram of the effects on intracellular pH of a transient exposure of a cell to external $CO_2$ and $HCO_3$. The meaning of the various numbers and the pH traces to the right given in the Appendix text.

the cell, returning $pH_i$ back to its original value (see top pH trace from Fig. 9G, part A). However, if $NH_4^+$ has some membrane permeability, it will enter the cell (arrow 3, Fig. 9G, part A), largely driven by the negative internal membrane potential. The $NH_4^+$ that enters will dissociate into $NH_3$ and $H^+$. The newly formed $NH_3$ will diffuse from the cell, leaving an H ion in the cell. Thus, a shuttle is established whereby $NH_4^+$ enters the cell and $NH_3$ leaves the cell (arrows 3,2,1, and 4, Fig. 9G, part A). For each cycle of the shuttle, an H ion is added to the cell and so the cell slowly acidifies in the maintained presence of external $NH_4Cl$ (termed "plateau acidification"). Upon removal of external $NH_4Cl$, all the original $NH_4^+$ formed upon exposure will dissociate and regenerate $NH_3$ (which diffuses from the cell) and $H^+$. However, the cell will contain extra H ions due to the operation of the shuttle and thus $pH_i$ will undershoot its initial value, achieving a more acid $pH_i$ than it had before the $NH_4Cl$ exposure. The degree of this undershoot is dependent on the amount of $NH_4Cl$ initially added, the membrane permeability to $NH_4^+$, and the duration of exposure to $NH_4Cl$. The net effect of an $NH_4Cl$ prepulse is to acidify the cell, and it is entirely analogous to an injection of acid. If a cell possesses mem-

brane transport systems for active $H^+$ extrusion (arrow 5 in Fig. 9G, part A), $pH_i$ will return towards its initial pH value, a process termed **pH recovery** (see lower pH trace in Fig. 9G, part A).

The other common way to alter $pH_i$ is upon exposure of cells to weak acids. The most common weak acid is carbonic acid, in the form of $CO_2$. This process is analogous to the $NH_4Cl$ prepulse. Upon exposure to a solution containing $CO_2$ and $HCO_3^-$, $CO_2$ rapidly enters the cell (arrow 1 of Fig. 9G, part B), hydrates and dissociates to form internal $HCO_3^-$ and $H^+$ (arrow 2, Fig. 9G, part B). The addition of H ions internally acidifies the cell. The cell will continue to acidify until internal and external $CO_2$ are equal. At this point, if $HCO_3^-$ is impermeable, no further change in pH will occur, and pH will return to its initial value upon removal of extracellular $CO_2$ (see middle $pH_i$ trace, Fig. 9G, part B). However, if $HCO_3^-$ can move across the membrane, a shuttle will be established. $HCO_3^-$ will leave the cell (arrow 3, Fig. 9G, part B), mostly driven by the negative membrane potential, internal $CO_2$ will hydrate and dissociate, and more $CO_2$ will enter the cell. Thus, $HCO_3^-$ will leave the cell and $CO_2$ will enter the cell, adding an internal $H^+$ for every cycle of the shuttle (arrows 3,4,1, and 2, Fig. 9G, part B). The cell will slowly acidify (see top $pH_i$ trace in Fig. 9G, part B). If, however, the cell possesses transmembrane H ion extrusion mechanisms (arrow 5, Fig. 9G, part B), pH will recover back toward the initial $pH_i$ value because of active extrusion of internal $H^+$ even in the maintained presence of $CO_2$. If external $CO_2$ is removed after recovery, the H ions initially formed upon $CO_2$ exposure will recombine with $HCO_3^-$ and leave the cell as $CO_2$, thereby alkalinizing the cell. However, the cell pH will overshoot, reaching an alkaline value of pH (see bottom $pH_i$ trace, Fig. 9G, part B), because of the removal of internal H ions during pH recovery. Thus, exposure to or removal of weak acids can induce cellular acidification or alkalinization, respectively.

## Bibliography

Al-Awqati, Q. (1986). Proton-translocating ATPases. *Annu. Rev. Cell Biol.* **2**, 179–199.

Ammann, D., Lanter, F., Steiner, R. A., Schulthess, P., Shijo, Y., and Simon, W. (1981). Neutral carrier based hydrogen ion selective microelectrode for extra- and intracellular studies. *Anal. Chem.* **53**, 2267–2269.

Aronson, P. S. (1989). The renal proximal tubule: A model for diversity of anion exchangers and stilbene-sensitive anion transporters. *Annu. Rev. Physiol.* **51**, 419–441.

Aronson, P. S., and Boron, W. F. (Eds.). (1986). "Na$^+$–H$^+$ Exchange, Intracellular pH, and Cell Function," Vol. 26, "Current Topics in Membranes and Transport." Academic Press, New York.

Aronson, P. S., Nee, J., and Suhm, M. A. (1982). Modifier role of internal H$^+$ in activating the Na$^+$–H$^+$ exchanger in renal microvillus membrane vesicles. *Nature* **299**, 161–163.

Bidani, A., and Brown, S. E. S. (1990). ATP-dependent pH$_i$ recovery in lung macrophages: Evidence for a plasma membrane H$^+$-ATPase. *Am. J. Physiol.* **259**, C586–C598.

Bock, G., and Marsh, J. (Eds.). (1988). "Proton Passage across Cell Membranes," Ciba Foundation Symposium 139. Wiley, New York.

Boron, W. F. (Ed.). (1986). Special topic: Acid/base physiology. *Annu. Rev. Physiol.* **48**, 347–413.

Boron, W. F., and Boulpaep, E. L. (1983). Intracellular pH regulation in the renal proximal tubule of the salamander. Basolateral HCO$_3^-$ transport. *J. Gen. Physiol.* **81**, 53–94.

Boron, W. F., and De Weer, P. (1976). Intracellular pH transients in squid giant axons caused by CO$_2$, NH$_3$, and metabolic inhibitors. *J. Gen. Physiol.* **67**, 91–112.

Busa, W. B., and Nuccitelli, R. (1984). Metabolic regulation via intracellular pH. *Am. J. Physiol.* **246**, R409–R438.

Cala, P. M. (1980). Volume regulation by *Amphiuma* red blood cells. The membrane potential and its implications regarding the nature of the ion-flux pathways. *J. Gen. Physiol.* **76**, 683–708.

Chamberlin, M. E., and Strange, K. (1989). Anisosmotic cell volume regulation: A comparative view. *Am. J. Physiol.* **257**, C159–C173.

Counillon, L., and Pouysségur, J. (1995). Structure–function studies and molecular regulation of the growth factor activatable sodium–hydrogen exchanger (NHE-1). *Cardiovasc. Res.* **29**, 147–154.

DeCoursey, T. E., and Cherney, V. V. (1994). Voltage-activated hydrogen ion currents. *J. Membr. Biol.* **141**, 203–223.

Durham, J. H., and Hardy, M. A. (Eds.). (1989). "Bicarbonate, Chloride, and Proton Transport Systems," Vol. 574, "Annals of the New York Academy of Sciences." New York Academy of Sciences, New York.

Edmonds, B. T., Murray, J., and Condeelis, J. (1995). pH regulation of the F-actin binding properties of *Dictyostelium* elongation factor 1α. *J. Biol. Chem.* **270**, 15222–15230.

Gevers, W. (1977). Generation of protons by metabolic processes in heart cells. *J. Mol. Cell. Cardiol.* **9**, 867–874.

Grinstein, S. (Ed.). (1988). "Na$^+$/H$^+$ Exchange." CRC Press, Boca Raton, FL.

Grinstein, S. (1996). Non-invasive measurement of the luminal pH of compartments of the secretory pathway. *The Physiologist* **39**, 144.

Grinstein, S., and Rothstein, A. (1986). Mechanisms of regulation of the Na$^+$/H$^+$ exchanger. *J. Membr. Biol.* **90**, 1–12.

Häussinger, D. (Ed.). (1988). "pH Homeostasis. Mechanisms and Control." Academic Press, New York.

Hille, B. (1992). "Ionic Channels of Excitable Membranes." Sinauer Associates, Sunderland, MA.

Hochachka, P. W., and Mommsen, T. P. (1983). Protons and anaerobiosis. *Science* **219**, 1391–1397.

Hoffmann, E. K., and Simonsen, L. O. (1989). Membrane mechanisms in volume and pH regulation in vertebrate cells. *Physiol. Rev.* **69**, 315–382.

Karniski, L. P., and Aronson, P. S. (1985). Chloride/formate exchange with formic acid recycling: A mechanism of active chloride transport across epithelial membranes. *Proc. Natl. Acad. Sci. USA* **82**, 6362–6365.

Kopito, R. R., and Lodish, H. F. (1985). Primary structure and transmembrane orientation of the murine anion exchange protein. *Nature* **316**, 234–238.

Kotyk, A., and Slavik, J. (1989). "Intracellular pH and Its Measurement." CRC Press, Boca Raton, FL.

Lowe, A. G., and Lambert, A. (1983). Chloride–bicarbonate exchange and related transport processes. *Biochim. Biophys. Acta* **694**, 353–374.

Madshus, I. H. (1988). Regulation of intracellular pH in eukaryotic cells. *Biochem. J.* **250**, 1–8.

Murer, H., Hopfer, U., and Kinne, R. (1976). Sodium/proton antiport in brush-border membranes isolated from rat small intestine and kidney. *Biochem. J.* **154**, 597–604.

Noël, J., and Pouysségur, J. (1995). Hormonal regulation, pharmacology, and membrane sorting of vertebrate Na$^+$H$^+$ exchanger isoforms. *Am. J. Physiol.* **268**, C283–C296.

Nosek, T. M., Fender, K. Y., and Godt, R. E. (1987). It is diprotonated inorganic phosphate that depresses force in skinned skeletal muscle fibers. *Science* **236**, 191–193.

Nuccitelli, R., and Deamer, D. W. (Eds.). (1981). "Intracellular pH: Its Measurement, Regulation, and Utilization in Cellular Function." A. R. Liss, New York.

Palokangas, H., Metsikkö, K., and Väänänen. (1994). Active vacuolar H$^+$ ATPase is required for both endocytic and exocytic processes during viral infection of BHK-21 cells. *J. Biol. Chem.* **269**, 17577–17585.

Pouysségur, J., Sardet, C., Franchi, A., L'Allemain, G., and Paris, S. (1984). A specific mutation abolishing Na$^+$/H$^+$ antiport activity in hamster fibroblasts precludes growth at neutral and acidic pH. *Proc. Natl. Acad. Sci. USA* **81**, 4833–4837.

Reusch, H. P., Lowe, J., and Ives, H. E. (1995). Osmotic activation of a Na$^+$-dependent Cl$^-$/HCO$_3^-$ exchanger. *Am. J. Physiol.* **268**, C147–C153.

Rink, T. J., Tsien, R. Y., and Pozzan, T. (1982). Cytoplasmic pH and free Mg$^{2+}$ in lymphocytes. *J. Cell Biol.* **95**, 189–196.

Romero, M. F., Hediger, M. A., Boulpaep, E. L., and Boron, W. F. (1996). Expression cloning of the renal electrogenic Na/HCO$_3$ cotransporter (NBC) from *Ambystoma tigrinum*. *The Physiologist* **39**, 144.

Roos, A., and Boron, W. F. (1980). The buffer value of weak acids and bases: Origin of the concept, and first mathematical derivation and application to physico-chemical systems. The work of M. Koppel and K. Spiro (1914). *Respir. Physiol.* **40**, 1–32.

Roos, A., and Boron, W. F. (1981). Intracellular pH. *Physiol. Rev.* **61**, 296–434.

Sardet, C., Franchi, A., and Pouysségur, J. (1989). Molecular cloning, primary structure, and expression of the human growth factor-activatable Na$^+$/H$^+$ antiporter. *Cell* **56,** 271–280.

Sardet, C., Counillon, L., Franchi, A., and Pouysségur, A. (1990). Growth factors induce phosphorylation of the Na$^+$/H$^+$ antiporter, a glycoprotein of 110 kD. *Science* **247,** 723–726.

Seksek, O., Biwersi, J., and Verkman, A. S. (1995). Direct measurement of *trans*-Golgi pH in living cells and regulation by second messengers. *J. Biol. Chem.* **270,** 4967–4970.

Soleimani, M., and Aronson, P. S. (1989). Ionic mechanism of Na$^+$–HCO$_3^-$ cotransport in rabbit renal basolateral membrane vesicles. *J. Biol. Chem.* **264,** 18302–18308.

Thomas, R. C. (1974). Intracellular pH of snail neurones measured with a new pH-sensitive glass microelectrode. *J. Physiol.* (*London*) **238,** 159–180.

Thomas, R. C. (1978). "Ion-Sensitive Intracellular Microelectrodes. How to Make and Use Them." Academic Press, New York.

Thomas, J. A., Buchsbaum, R. N., Zimniak, A., and Racker, E. (1979). Intracellular pH measurements in Ehrlich ascites tumor cells utilizing spectroscopic probes generated *in situ. Biochemistry* **18,** 2210–2218.

Trivedi, B., and Danforth, W. H. (1966). Effect of pH on the kinetics of frog muscle phosphofructokinase. *J. Biol. Chem.* **241,** 4110–4111.

Yun, C. H. C., Tse, C.-M., Nath, S. K., Levine, S. A., Brant, S. R., and Donowitz, M. (1995). Mammalian Na$^+$H$^+$ exchanger gene family: Structure and function studies. *Am. J. Physiol.* **269,** G1–G11.

*Robert W. Putnam*

# 21

# Polarity of Cells and Membrane Regions

## I. Introduction

General discussions of cell physiology often make the simplifying assumption that cells are surrounded by a more or less uniform plasma membrane. In fact, there are numerous examples of highly specialized regions, or **domains,** in the plasma membranes of cells. The most dramatic of which are **epithelial cells,** which contain distinct **apical** and **basolateral membranes.** However, many other cells also exhibit **polarity,** or asymmetry, including neuronal cells, skeletal muscle cells, fibroblasts, eggs, yeast, and bacteria. In addition, this asymmetry can be **spatial,** as with distinct membrane domains, or **temporal,** as with changes in the distribution of cellular components at different stages during development or differentiation.

In addition to the polarization of cell membranes into major domains, surface membranes can exhibit microdomains. **Microdomains** are highly specialized and restricted regions of membrane that have properties that are different from the bulk of the membrane. Examples of microdomains include the end-plate region of skeletal muscle, the nodes of Ranvier of nerves, and "hot spots" on skeletal muscle membrane that are enriched in $Na^+$ channels. Cells can also exhibit intracellular polarity (i.e., nonrandom distribution of intracellular organelles), but in this chapter we deal largely with the phenomenon of specialized major domains in the plasma membrane. Beyond the description of various types of polarity, the pathways that establish polarity, and the mechanisms for the maintenance of polarity are discussed. Much of our understanding of these mechanisms and pathways has been obtained over the last decade, and this continues to be a very active area of research. This chapter is a brief introduction to some of the basic concepts and models of cell polarity, not an exhaustive review of all the recent findings.

## II. Examples of Cell That Are Polarized

The following examples of cell polarity illustrate the wide variety of polarized phenomena in cell biology. In-cluded are examples of polarity in the surface membranes of cells, as well as examples of polarized cell phenomena. These latter examples do not necessarily involve marked polarity of the surface membrane per se, but do involve polarized processes that are often involved in the establishment of distinct surface membrane domains in other cells.

### A. Epithelial Cells

**Epithelia** are composed of closely packed monolayers of cells that separate different tissue compartments. One side of the epithelium often faces a compartment that is "outside" the body, such as the lumen of the gastrointestinal tract, the kidneys, the lungs, or the urinary bladder. The other side of the epithelium faces the "inside" of the body, that is, the blood side.

Epithelia serve two major functions: (1) a **barrier function,** maintaining the integrity of the different fluid compartments that they separate; and (2) a **vectorial transport function,** mediating net transport of solutes between various compartments. The ability to perform both functions depends on the polarized membrane domains. The "outward"-facing membrane, usually referred to as the **apical membrane,** is often covered with microvilli and has a different lipid and integral membrane protein composition than the "inward"-facing membrane, called the **basolateral membrane.** For example, in many epithelial cells, the Na/K-ATPase (the Na pump) is found exclusively on the basolateral membrane, as are cell adhesion molecules and the receptors for blood-borne agents. In contrast, the apical membrane usually contains transporters and channels for the exchange of materials with the external environment. The lipid compositions of these two membrane domains are also distinct, with the apical membrane having a very high content of glycolipids (especially glycosphingolipids). These two distinct membrane domains are separated by **tight junctions** between cells, forming a barrier to diffusion between the apical and lateral membranes through the intercellular space.

The barrier function of epithelial cells is enhanced by the high glycolipid content of the apical membrane, which serves to protect the cells from possible harsh "external" environmental influences. The vectorial movement of solutes across epithelia allow them to maintain concentration gradients between the two different fluid compartments separated by the epithelial cells. Depending on the compartments separated and the distribution of transporters between the apical and basolateral membrane domains, an epithelium may be a **transporting epithelium** (e.g., in kidney), a **secretory epithelium** (e.g., in liver), or an **absorptive epithelium** (e.g., in intestine). An example of each type of epithelium is shown in Fig. 1.

## B. Neuronal Cells

Although polarity is most easily studied in epithelial cells, **neurons** are perhaps the most clearly polarized cells in the body. Neurons are derived embryologically from epithelial cells. Shortly after neurons cease dividing, they form cellular processes called axons and dendrites (Fig. 2). The **axon** extends away from the cell body and is generally quite long. The **dendrites** are usually shorter, tapered, and contain a variety of intracellular organelles such as rough endoplasmic reticula (ER), Golgi bodies, and ribosomes. These two domains, **somatodendritic** (cell body and dendrites) and **axonal,** serve different functions. The dendritic region is adapted to receive input from the axons of other nerve cells and has a large surface area for such contacts, whereas the axon serves as a conducting fiber to carry the action potential rapidly over long distances to the target neuron(s), muscle cells, or endocrine/exocrine cells.

For the neuronal membrane domains to perform their various functions, certain membrane proteins must be specifically localized to either the somatodendritic membrane or the axonal membrane. Unlike epithelial cells, however, neurons have no apparent tight junction-like regions to mark the separation of the two membrane domains. The basis for the maintenance of these domains in neurons is unknown. Interestingly, proteins that have been shown to localize to the apical membrane of epithelial cells [e.g., the viral envelope glycoprotein hemagglutinin (HA) or glycosyl-phosphatidylinositol (GPI)-anchored proteins] are also found in the axonal membrane of neurons. In contrast, proteins localized to the basolateral membrane of epithelial cells (e.g., transferrin receptor or the G protein of vesicular stomatitis virus) are found in the somatodendritic membrane of neurons. This has led to the suggestion that axonal/apical and somatodendritic/basolateral membranes represent analogous membrane domains. This hypothesis is probably too simplistic, however, since other proteins that localize to the basolateral membranes of epithelia are found in both somatodendritic and axonal membranes of neurons (e.g., Na pump) or in axonal membranes (e.g., certain cell adhesion molecules such as integrins).

Within a given neuronal membrane domain, there may be subdomains. For instance, $Na^+$ channels are enriched at the **nodes of Ranvier,** regions of bare axons between

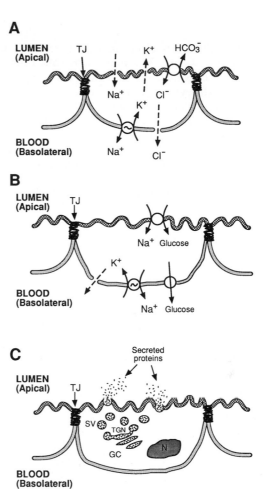

**FIG. 1.** Models of three different types of epithelial cells. (A) Transporting epithelium. Characterized by the asymmetric distribution of ion transporters that confers the ability to conduct vectorial transport of ions and water. This cell is characteristic of a salivary gland ductal cell that reabsorbs Na and Cl and secretes K and $HCO_3$. (B) Absorptive epithelium. Characterized by the ability to reabsorb nutrients, such as this intestinal epithelial cell. The presence on the apical membrane of a Na–glucose cotransporter and on the basolateral membrane of a facilitated glucose transporter enables these cells to move large amounts of glucose from the intestinal lumen into the blood. (C) A secretory epithelium. Proteins that are synthesized within these cells are packaged in secretory vesicles (SVs) in the trans-Golgi network (TGN) and stored for later release from the apical membrane by exocytosis. An example of such a secretory epithelial cell is the exocrine pancreas, which secretes degradative proteins into the intestinal tract. For all three types of epithelia, note that membrane lipids, as well as proteins, are polarized with the apical membrane enriched in glycolipids. Also note the presence of tight junctions (TJs) between the cells that demarcates the apical from the basolateral membranes. GC, Golgi complex.

insulating **Schwann cells** (see Fig. 2). This enrichment is most likely due to interaction between the $Na^+$ channel protein and an anchoring protein (e.g., ankyrin). Fast $Na^+$ channels are also found at higher density in the initial segment of the axon (**axon hillock**) than in the somatodendritic membrane.

## C. Astrocytes

Neuronal cells are not the only polarized cells in the central nervous system. The most numerous cells in the brain are **glial cells,** and the most abundant type of glial cells are the **astrocytes.** The function of astrocytes is not well understood, but they play a role in the development of the nervous system and in controlling the fluid environment surrounding neurons. Differentiated astrocytes have numerous processes, extending from the cell body, that end in terminal enlargements called **end feet,** giving the cell a star-shaped appearance from which the name astrocyte derives. The cellular processes and end feet make connections with neurons and with endothelial cells lining the blood vessels in the brain (Fig. 3). In fact, glial connections with the endothelium release trophic factors that result in brain capillaries being tightly sealed. These tightly sealed capillaries form the **blood–brain barrier.**

The structural polarity of astrocytes results from the membranes of the processes and end feet forming a distinct domain, with certain membrane proteins known to be predominantly localized to these regions. Among these polarized membrane proteins are $K^+$ channels and the electrogenic $Na–HCO_3$ cotransporter (see Chapter 20, Intracellular pH Regulation). The functional significance of this polarity in astrocytes is currently not known.

## D. Endothelial Cells

**Endothelial cells** form a cellular monolayer on the inner face of blood vessels. One side of the endothelial cell faces the blood and the other side faces the basement membrane and smooth muscle cell layer (in arteries and veins) or the basement membrane and interstitial fluid (in capillaries). These cells are clearly polarized and mediate transcellular transport. Generalizations about endothelial cells are hard to make since their properties vary depending on the tissue the vasculature irrigates. For instance, the endothelium from liver capillaries is highly permeable, whereas the endothelium from the brain has a very low permeability (see Fig. 3).

Endothelial cells are similar in structure to epithelial cells, containing a luminal membrane (apical) and an abluminal membrane (basolateral), separated by tight junctions. Receptors for a number of blood-borne agents are localized to the luminal membrane of endothelial cells, including receptors for acetylcholine, bradykinin, histamine, arachidonic acid, thrombin, adenosine, and the adenine nucleotides. Certain transporters have localized distributions in endothelial cells. For example, the Na pump appears to be localized to the basolateral membrane. Finally, endothelial cells secrete a variety of substances, and this secretion is often polarized. Thus, agents such as antithrombin III and plasminogen activator are released into the vascular lumen, whereas collagen IV and V and laminin appear to be largely secreted from the abluminal side. It is likely that many more examples of polarized function will be discovered as we come to understand better the physiology of endothelial cells.

## E. Striated Muscle

Striated muscle fibers (skeletal muscle and cardiac muscle) also exhibit polarity in that they have two distinct plasma membrane domains, the actual **surface membrane**

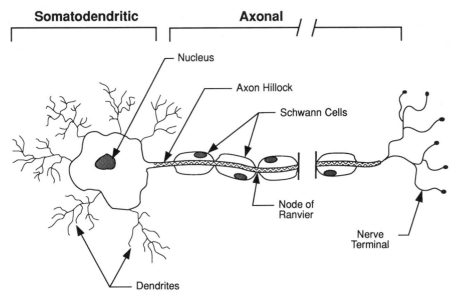

**FIG. 2.** A model of a neuron with the somatodendritic and axonal regions. The somatodendritic region contains the nucleus, the cell body, and the dendrites (membrane extensions of the cell body that receive input). The axonal region is a long conducting region that is covered by insulating Schwann cells. At periodic intervals, the Schwann cell layer is interrupted, leaving bare regions of axon, the nodes of Ranvier. The nerve cell ends at a terminal region that makes synaptic contacts with other cells. The region connecting the axon to the cell body is called the axon hillock and is the region where an action potential is set up.

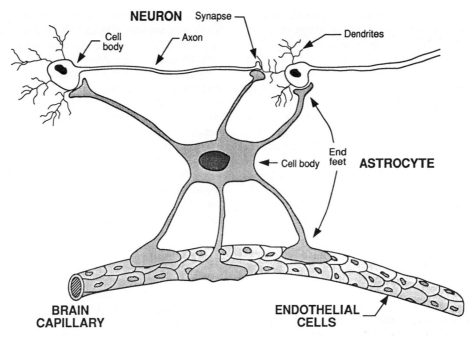

**FIG. 3.**   An astrocyte with a large cell body and processes terminating in end feet. The end feet make contacts with neuronal cell bodies and synapses and with endothelial cells of brain capillaries. The combination of astrocytic end feet and endothelial cells, connected by tight junctions, forms the blood–brain barrier.

(which defines the approximately cylindrical boundaries of the fiber) and an extensive network of radial invaginations of the the surface membrane, the **transverse tubular membrane** (T-tubular membrane). The lipid composition of these two membranes differs, with the T-tubular membrane being enriched in cholesterol compared to the surface membrane. Many differences in integral membrane proteins (proteins that are embedded within the membrane lipids) have also been demonstrated in the two different domains. For example, the T-tubular membrane has been proposed to contain most of the muscle $Cl^-$ and $Ca^{2+}$ conductance, the Na/Ca exchanger, and the glucose carrier, and to contain $Na^+$ channels and Na pumps at a lower density than the surface membrane. That these two membranes represent distinct domains has been further verified by the use of antibodies specific for surface membrane or T-tubular membrane, indicating that each domain contains certain proteins not found in the other domain. As with neurons, there is no apparent junctional marker between the surface and the T-tubular membrane, and thus the mechanism by which these two domains remain isolated is unknown. The significance and functional consequences of having two distinct membrane domains in striated muscle is also unknown.

### F. Embryonic Cardiac Myocytes

In the developing chick heart, the **presumptive myocardium** resides in the outer layer of the forming endocardial tube. The outer layer of this presumptive myocardium appears morphologically as a cuboidal epithelium, with a basal membrane facing inward to a layer of extracellular matrix analogous to a basement membrane (the **cardiac**

**"jelly"**) and an apical membrane facing out from the tube. The presumptive myocardial cells are connected by tight junctions. This myocardial "epithelium" is believed to be involved in the vectorial secretion of the cardiac jelly and to lead to the assembly of myofibrils (rendering the heart contractile) and to the looping of the tubular heart into a C shape.

Upon infection with virus, the presumptive myocardium shows vectorial release of viral envelope proteins. Thus, vesicular stomatitis virus proteins appear in the basolateral membrane of the presumptive myocardium, whereas influenza viral proteins appear predominantly in the apical membrane. These findings clearly show that presumptive myocardial cells have distinct membrane domains and that they have mechanisms to sort newly synthesized proteins and target them to the appropriate surface membrane domain. Indeed, the Na pump has been shown to have a polarized distribution in the surface membrane of these cells. The polarity of the presumptive myocardial cells is probably important for the role of these cells in cardiac morphogenesis (polarized secretion of cardiac jelly, the assembly of myofibrils, and the looping of the endocardial tube). It is not known if these sorting mechanisms, present in embryological cardiac tissue, are maintained into the fully developed cardiac myocyte.

### G. Directed Cell Movement

The ability of a cell to move in a certain direction requires polarity. Examples of such directed movement include **fibroblasts** (migrating or involved in wound healing) and **macrophages** (moving toward a chemoattractant or engaged in phagocytosis). The ability of cells to move in a directed fashion involves the **actin-based cytoskeleton** and **microtubules** as well in some cells. Actin filaments form as a polymer of globular actin subunits, which join together like a string of pearls. Actin filaments are polarized, with a plus end and a minus end, and elongate by a process termed **treadmilling.** New globular actin subunits add to the plus end of the actin filament, while old subunits are preferentially cleaved from the minus end. In this way, the actin filament elongates in the direction of its plus end.

Motile cells express surface projections, called **lamellipodia,** that extend by directed polymerization of actin, creating tension in the surface membrane. If a lamellipodium does not attach firmly to the substrate, the tension causes it to retract in a process termed **ruffling.** However, if the lamellipodium attaches, the same tension will serve to pull the cell in that direction. The initiation of this process probably involves the activation of capping proteins, which reside in the surface membrane and mediate actin polymerization in that region. Thus, the cell becomes polarized into a leading region and a trailing region. This polarity, as stated, is largely due to the polarity of elongation of actin, but may be aided by polarity of the **endocytotic cycle.** Small surface membrane vesicles are continually being endocytosed. In migrating cells, the return of these endocytosed membranes to the surface is polarized, such that membranes are preferentially returned to the leading edge, enabling it to expand. The trafficking of these vesicles occurs along microtubules or actin filaments. Thus, the polarity of migrating cells is largely due to the polarity of the actin-based cytoskeleton and microtubules.

## H. Eggs and Sperm

Eggs are remarkable in that they must contain the information to develop into a complete organism after fertilization. Eggs are generally large, varying from 100 $\mu$m diameter in humans to a few millimeters in fish and amphibians, and several centimeters in birds and reptiles. When the embryo develops outside the body, the egg contains a large store of lipids, the **yolk,** to provide the energy for the initial period of development. These eggs display polarity even before fertilization, with the yolk concentrated at one end of the egg, the **vegetal pole,** and the nucleus at the other end of the egg, the **animal pole.**

In the mature egg, there is polarity beyond the distribution of yolk and nucleus. In some eggs, the animal pole is also pigmented, rendering it visibly darker than the vegetal pole. Of greater significance, there is good evidence that certain cytoplasmic determinants are asymmetrically distributed. Thus, the animal and vegetal pole of eggs contain different complements of mRNA that, upon division, give the daughter cells distinct properties. The egg of the fruit fly, *Drosophila,* is a particularly good example of such asymmetry. The egg is localized at one end of the ovarian follicle and makes direct contact (through structures called **cytoplasmic bridges**) on one end with **nurse cells** in the follicle. These nurse cells produce substances that flow through the cytoplasmic bridges to one pole of the egg. This creates an asymmetric distribution of these substances within the egg, and the pole containing these materials

eventually develops into the head region of the fruit fly. Thus, the asymmetric distribution of materials within an egg can determine the developmental fates of daughter cells from the early divisions of the fertilized egg.

**Sperm** cells also display a marked asymmetry. The sperm cell has three clearly defined regions: (1) a **head** region containing dense nuclear material and the **acrosomal vesicle** (containing hydrolytic enzymes necessary for sperm penetration of the egg coat); (2) a **midpiece** containing mitochondria; and (3) a long **tail** region, which is a **flagellum** and confers the ability to move. This asymmetry arises through differentiation of **spermatids,** the product of the second **meiotic division** of **spermatocytes,** and is ideally suited to the primary function of the sperm, which is to locate, penetrate, and inject nuclear material into an egg.

## I. Yeast

Certain yeast, such as *Saccharomyces cerevisiae,* undergo cell division in a highly polar process called **budding.** Several of the processes in budding are asymmetric. A spherical mother cell develops an outgrowth, or bud, in a specific location (usually along the former axis of division). The yeast cell wall grows preferentially in the bud region, especially at the bud tip, so that the bud grows far more rapidly than the mother cell. Actin bundles and microtubules, oriented from the nucleus into the bud, form and probably serve to support directed transport of vesicles to the bud region. Ultimately, the nucleus migrates from the mother cell to the neck region between mother and bud. The nucleus divides, leaving a nucleus both in the bud and in the mother. Eventually, the bud pinches off as a daughter cell, leaving a **bud scar** on the mother cell.

Proteins assembled during division and remaining as the bud scar (including **septins**) serve as **cortical spatial cues** for attachment of cytoskeletal elements. The rapid growth of the bud undoubtedly involves, at least in part, the organizing movement of cellular components to the bud tip by polarized actin (and perhaps microtubules) in a process similar to directed movement of fibroblasts (see earlier discussion). The microtubules are also clearly involved in the migration and division of the budding yeast nucleus, as they are with the division of the nucleus of other cells.

## J. Bacteria

Polarity is clearly demonstrated by the cell division cycle of a bacterium, *Caulobacter.* In the predivisional cell, one pole contains specialized structures (including a flagellum, pili, receptors for phages, and membrane proteins involved in chemotaxis), while the other pole contains a stalk. Upon division, which divides the cell at its equator, two asymmetric daughter cells are formed (Fig. 4). One cell is a **swarmer cell** that contains the flagellum, the pili, and various specialized receptors and membrane proteins. The other daughter cell is the **sessile stalked cell,** which contains the stalk. The sessile stalked cell is fully capable of DNA replication, but the swarmer cell must differentiate, lose its flagellum and pili, and grow a stalk where the flagellum used to be. Only after this differentiation can the cell undergo DNA

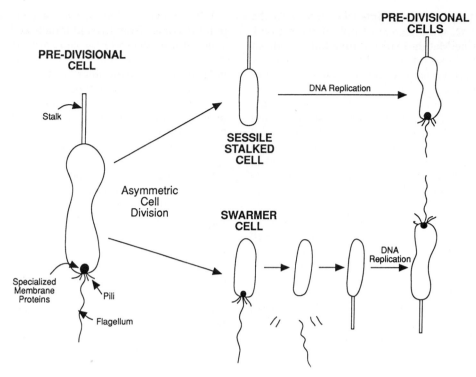

**FIG. 4.** A model of the asymmetric division of *Caulobacter* cells. The predivisional cell is asymmetric, with a stalked pole and a flagellar pole (also containing pili and certain specialized membrane proteins). Division forms a sessile stalked and a swarmer daughter cell. The stalked cell is capable of undergoing DNA replication, synthesizing a new flagellum, and entering into another round of division. The swarmer cell, however, must first lose its flagellum and associated structure and replace it with a stalk before undergoing DNA replication. The former swarmer cell now resynthesizes a new flagellum, pili, and membrane proteins before undergoing another asymmetric division.

replication. Interestingly, in this newly differentiated stalk cell, a new flagellum, pili, phage receptors, and protein machinery for chemotaxis are assembled at the pole *opposite* the stalk. This region becomes the **incipient swarmer cell** portion of the new predivisional cell. Thus, swarmer cells differentiate into stalked cells and then replicate their DNA and regenerate polarity in preparation for the next asymmetric cell division (see Fig. 4).

Since bacteria do not contain cytoskeletal elements per se, the mechanism for establishment of polarity is not known, but several mechanisms for the asymmetric distribution of proteins have been proposed. Newly synthesized proteins themselves may contain the information necessary for specific targeting. For instance, certain bacterial proteins have sequences that target them for the site of the flagellum in the incipient swarmer cell region of the predivisional cell. Such proteins would preferentially localize in the predivisional cell, and thus be found in the swarmer daughter cell. Alternatively, a protein could be present at random in the surface of the predivisional cell, but undergo rapid degradation in only one of the daughter cells. Finally, some evidence exists for polarity deriving from segregation of the mRNA within the predivisional cell. After replication, each pole of the predivisional cell contains one chromosome. Shortly before division, gene products made in the incipient swarmer region do not appear in the incipient stalk region, suggesting preferential gene expression in the

swarmer region and a barrier to protein diffusion between the two regions of the cell. This latter mechanism is a good example of how temporal control of gene expression can contribute to the establishment of polarity.

## III. Establishment and Maintenance of Cell Polarity

The basis for the establishment and maintenance of polarity in various types of cells may well differ. Since these processes are best understood in epithelial cells, we will restrict our discussion to the mechanisms responsible for the establishment and maintenance of cell polarity in these cells. The mechanisms that establish polarity may even vary in different types of epithelial cells. Our discussion will focus on the various mechanisms that have been shown to contribute to the establishment and maintenance of epithelial polarity in some cells.

### A. Cell–Substratum and Cell–Cell Interaction

When fully polarized epithelial cells in culture are separated and put into cell suspension, thus eliminating all **cell–cell** and **cell–substratum interactions,** polarity is lost. Proteins within the plasmalemma become randomly distributed (Fig. 5). If these cells are allowed to reattach to the substratum (largely through **integrin** binding to the extracellular matrix), some degree of polarity is restored. For example, plating Madin–Darby canine kidney cells at low density, which allows for cell–substratum interactions, but not cell–cell interactions, causes a nonrandom distribution of certain proteins. An apical marker protein is excluded from the region of membrane interacting with the substratum while the Na pump, which is usually restricted to the basolateral membrane, is still randomly distributed (see Fig. 5). These studies suggest that cell–substratum interaction is sufficient to establish the polarity of apical proteins, but not of basolateral proteins. Thus, other processes must be involved in the establishment of full cell polarity.

One of these other processes is clearly the development of **cell–cell contacts.** Cell interactions are mediated by **cell adhesion molecules,** which are membrane-spanning glycoproteins. In epithelial cells, adhesion is mediated by molecules such as **L-CAM** and **E-cadherin** (or uvomorulin), both of which require extracellular $Ca^{2+}$ for adhesion. If cultured kidney cells, plated at low density, are allowed to grow to confluence and form cell attachments, they display not only apical polarity, but basolateral polarity as well (see Fig. 5).

Further evidence for a role of cell–cell interaction in polarity comes from thyroid follicle cells grown in suspension culture. When these cells are allowed to associate, they form multicellular hollow spheres that display full polarity (apical membranes facing outward). Interestingly, upon exposure of the outer face of these spherical clusters to collagen, there is a reversal of polarity, with the apical membrane now facing into the center of the sphere and the basolateral membrane facing outward into the collagen-containing solution. Thus, the full development and expression of epithelial polarity is clearly dependent on both cell–cell and cell–substratum interactions.

One of the ways in which cell–cell interactions establish polarity is through the formation of **tight junctions** at the junction between the apical and basolateral membranes. Tight junctions are believed to form a barrier to diffusion, not only externally between the cells, but also internally, keeping the proteins and lipids of the apical membrane distinct from those of the basolateral membrane. This latter function is demonstrated by the loss of epithelial polarity observed upon disruption of the tight junctions in Ca-free solutions.

Formation of tight junctions cannot fully explain the establishment of polarity. The appearance of full basolat-

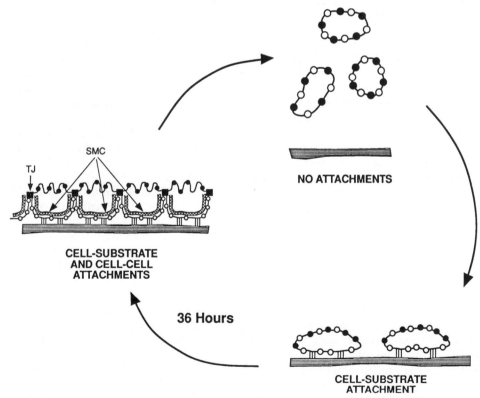

**FIG. 5.** A model for the role of cell–substratum and cell–cell adhesion in the maintenance of polarity. Fully polarized cells lose their polarity if they are removed from their substrate and are not able to make cell–cell contacts. If these cells are allowed to reattach to the substrate (form cell–substrate attachments), but not to form cell–cell contact, apical membranes are polarized to their proper domain but basolateral proteins are still randomly assorted. Within 36 hours of allowing these cells to form cell–cell contacts and tight junctions (TJ), full apical and basolateral polarity is restored. During this time also, a polarized submembranous cytoskeleton (SMC) is formed under the basolateral membrane only, which is believed to help in stabilizing basolateral polarity.

eral polarity is not seen in cultured kidney cells until 36 hours after cell–cell junctions are formed. During this lag period, other processes may mediate the establishment of polarity. Two processes that have been proposed are (1) formation of cytoskeletal elements, and (2) sorting and targeting of proteins to different membrane domains (see later discussion).

Cell adhesion molecules in epithelia also serve as cortical spatial cues, orienting cytoskeletal elements to attach to them and directing the movement of intracellular vesicles to these points of attachment. In this way, cell adhesion molecules also contribute to the establishment and maintenance of cell polarity.

## B. Cytoskeleton

**Cytoskeleton proteins** can form into different structures, including cytoskeletal filaments and the submembranous cytoskeleton. The filaments include microtubules and actin filaments, and these may play a role in directed transport of proteins to specific membrane domains (see later discussion). The submembranous cytoskeleton is a network of structural proteins, including **ankyrin, fodrin** (spectrin), and **adducin,** that forms underneath the plasmalemma and interacts directly with integral membrane proteins. The formation of the submembranous cytoskeleton is correlated with the appearance of cell–cell contacts in cultured kidney cells (see Fig. 5). Thus, when these cells reach confluence and form cell–cell contacts, fodrin molecules assemble to form the submembranous cytoskeleton that is associated with the basolateral membrane, but not the apical membrane. The time of appearance of the submembranous cytoskeleton corresponds to the time of appearance of specific proteins whose distribution is restricted to the basolateral membrane, namely about 36 hours. Further evidence for the relationship between cell–cell contacts and the submembranous cytoskeleton is that the association of fodrin with basal and lateral membranes is lost if cell–cell contacts are disrupted.

The predominant hypothesis for the role of the submembranous cytoskeleton in establishment of cell polarity is based on the direct interaction between these structural elements and integral membrane proteins. For example, the basolaterally distributed Na pump binds to ankyrin with high affinity. It is believed that the submembranous cytoskeleton serves to "corral" or trap integral membrane proteins by binding and immobilizing them. Thus, the polarized distribution of the submembranous cytoskeleton could be responsible for the regional distribution of membrane proteins.

## C. Sorting and Targeting of Proteins

In polarized epithelia, membrane proteins are continually being turned over (half-life between tens of minutes and tens of hours). If polarity is to be maintained, the components of endocytosed membranes must be correctly sorted and returned to the appropriate membrane, and newly synthesized proteins must be correctly targeted to the proper membrane. Several models have been proposed

for how proteins are sorted and targeted to the proper membrane.

Before considering the various models of protein targeting, a short review of protein synthesis is in order. The translation of mRNA into proteins occurs at the ribosomes on the rough endoplasmic reticulum (RER). These proteins undergo a posttranslational modification, **glycosylation,** that involves the addition of sugar groups. From the RER, synthesized proteins are then moved through the Golgi apparatus, from the *cis* face to the *trans* face, where the sugar groups may be further modified. On the *trans* face is the **trans-Golgi network,** in which synthesized proteins are packaged into vesicles for transport to the plasmalemma, lysosomes or secretory granules.

Several models have been proposed for proteins sorting and targeting to either the apical or basolateral membranes. These models include (1) transport of proteins to all membranes with differential removal and retention of proteins in specific domains; (2) transport of all proteins to a default membrane (e.g., basolateral), with later sorting and movement of apical membrane proteins across the cell in vesicles (a process termed **transcytosis**) to the apical membrane; and (3) specific sorting and targeting of proteins at the trans-Golgi network, with vectorial delivery of the appropriate proteins to the proper membrane domain.

### 1. Differential Retention

Polarized epithelial cells are generated from nonpolarized cells. Initially, membrane proteins are distributed randomly, and thus, as domains begin to form, they will contain both apical-specific and basolateral-specific proteins. One mechanism for achieving full polarity from such a random distribution is by **differential** or **selective retention** (Fig. 6). For instance, in subconfluent cultured kidney cells, two randomly distributed adhesion proteins, Dg-1 and E-cadherin, become localized to the basolateral membrane within 8 hours of the formation of cell–cell contacts. Newly synthesized Dg-1 and E-cadherin protein is delivered to both the apical and basolateral membranes during this time, but the residence time of these proteins is only 0.5–2 hours in the apical membrane, in contrast to 8–24 hours in the basolateral membrane. Thus, preferential retention within the basolateral membrane, and rapid removal from the apical membrane, leads to the differential distribution of these proteins. Two days after forming cell–cell contacts, newly synthesized Dg-1 and E-cadherin are almost exclusively *delivered* to the basolateral membrane. Thus, for these two proteins in cultured kidney cells, differential retention is significant for the initial generation of polarity, but polarity is ultimately maintained by the appearance of specific sorting and targeted delivery pathways.

Differential retention is more important for the maintenance of Na pump distribution in cultured kidney cells. For at least 4 days after cell–cell contacts are made, newly synthesized Na pumps are delivered equally to the apical and the basolateral membranes (see Fig. 6). However, the residence time, which is initially 3 hours for both membranes, falls to 1–2 hours for the Na pump in the apical membrane, but increases dramatically to over 36 hours for

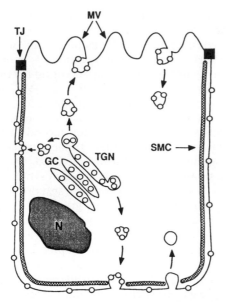

**FIG. 6.** A model of polarity established by differential retention of proteins to a particular membrane domain. In this model, proteins destined for the basolateral membrane are packaged into vesicles in the trans-Golgi network, and these vesicles are inserted into all surface membranes, the apical, lateral, and basal membranes. These proteins have a short residence time in the apical membrane and are rapidly endocytosed. However, these proteins are protected from being endocytosed from the basolateral membranes, probably as a result of interactions with the submembranous cytoskeleton (SMC), and thus have a long retention time in these membranes. Because of this differential retention time, these proteins assume a polarized distribution to the basolateral membrane. GC, Golgi complex; MV, microvilli; N, nucleus; TJ, tight junction.

the pump in the basolateral membrane. This differential retention results in a highly polarized distribution of the Na pump to the basolateral membrane. Unlike Dg-1 and E-cadherin, the asymmetric distribution of the Na pump appears to be both established and maintained by differential retention.

The basis for differential retention of membrane proteins to specific domains may be due to interactions of the protein with the submembranous cytoskeleton (see earlier discussion). The Na pump is known to bind to *ankyrin,* a major component of the submembranous cytoskeleton that associates with the basolateral membrane upon establishment of cell–cell contact. This binding may exclude the Na pump from endocytotic vesicles, thereby markedly increasing its retention time in the basolateral membrane (see Fig. 6).

### 2. Default Pathway/Transcytosis (Indirect Delivery)

Newly synthesized plasma membrane proteins in the trans-Golgi network must ultimately be inserted into either the apical or basolateral membrane domains. One way in which this can be accomplished is for all proteins, both apical and basolateral, to be initially transported together to one membrane domain, say the basolateral membrane

(the **default pathway**). For polarity to be established, the apical proteins would then have to be endocytosed, sorted, and relocated to the apical membrane domain (Fig. 7A). This process of moving an endocytotic vesicle between the apical and basolateral membrane domains in epithelial cells is called **transcytosis.** Transcytosis has been observed in a variety of epithelial cells. An example of the role of transcytosis in cell polarity is in rat hepatocytes, where two newly synthesized apical proteins (aminopeptidase N and dipeptidylpeptidase IV) appear first in the basolateral domain, and only later in the apical domain. In this example, transport from the trans-Golgi network to the basolateral membrane would represent the default pathway, and the polarized distribution of these two proteins arises by transcytosis (see Fig. 7A). A similar pathway from trans-Golgi network to basolateral membrane to apical membrane is followed by the polyimmunoglobulin receptor (pIg-R) in cultured kidney cells. The default pathway domain may vary among different epithelia, but in liver and intestinal epithelia, it is believed to be the basolateral membrane.

The default pathway in many epithelia may involve only some of the newly synthesized proteins. Proteins with certain specialized sorting signals may be directly transported from the trans-Golgi network to the apical membrane (see later discussion), whereas all other proteins, lacking such signals, would be transported to the basolateral membrane. In this case, the apical pathway from the trans-Golgi network would depend on specific sorting signals, whereas the basolateral pathway would be considered a default pathway. Thus, a fully polarized cell with two distinct membrane domains could be maintained with only one set of sorting signals.

### 3. Sorting Signals and Vectorial Transport at the Trans-Golgi Network (Direct Delivery)

The third model for the polarized distribution of newly-synthesized proteins is based on specialized sorting signals for each membrane domain (see Fig. 7B). The best evidence for such a mechanism has come from studies on the distribution of viral envelope glycoproteins in cultured kidney cells (see Section II.B; Neuronal Cells). Hemagglutinin is directly targeted from the trans-Golgi network to the apical membrane, whereas the envelope glycoprotein of vesicular stomatitis virus is targeted to the basolateral membranes. Specific targeting is not just used for these alien proteins, but has been demonstrated for endogenous proteins as well. In cultured kidney cells, an apical sialoglycoprotein, gp114, is delivered directly from the trans-Golgi network to the apical membrane, while the basolateral protein, E-cadherin, is delivered directly from the trans-Golgi network to the basolateral membrane (mistargeting was less than 10%). Clearly, both targeted delivery and default/transcytosis pathways can be used for different proteins within the same cell, since pIg-R reaches the apical membrane of cultured kidney cells only after transport to the basolateral membrane (see previous discussion). These data show that specific targeting signals can exist for transport to the basolateral membrane, as well as to the apical membrane domain.

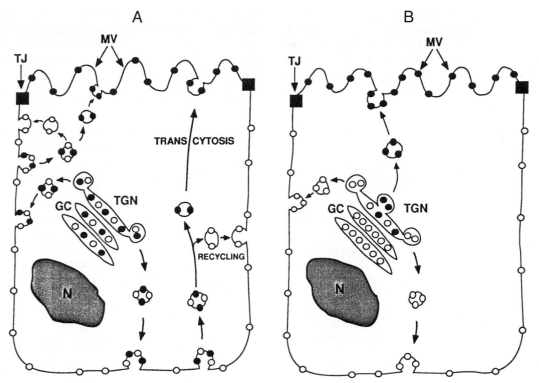

**FIG. 7.** (A) A model for the default/transcytosis mechanism of establishing cell polarity. In this model, all membrane-bound proteins leaving the trans-Golgi network (TGN) are transported to the basolateral membrane by default. These membrane proteins are endocytosed and sorted in the vesicle compartment. The basolateral proteins are recycled to the basolateral membranes while the apical membranes are transcytosed to the apical membrane. In this way, full polarity is established with but a single set of specific signals (those for the apical membrane). (B) A model for the targeted transport of both apical and basolateral membranes. In this model, proteins bound for the apical and basolateral membranes are sorted within the trans-Golgi network and packaged in distinct vesicle pools. One set of vesicles is transported directly to the apical membrane, and the other set is transported to the basolateral membrane. This model requires sorting and targeting signals for both apical and basolateral membrane proteins. GC, Golgi complex; MV, microvilli; N, nucleus; TJ, tight junctions.

## IV. Sorting Signals and Targeted Transport

The presence of distinct basolateral and apical sorting signals within the same cell implies that proteins that are cotransported through the Golgi apparatus become segregated in the trans-Golgi network. These segregated proteins are then incorporated into distinct vesicle pools, one of which is targeted to the apical membrane and the other to the basolateral membrane for either insertion or secretion. Further, the ability of endocytosed membranes to remain segregated in the apical or basolateral membranes, and the presence of transcytosis in cells, suggests that the vesicle pool has both sorting and targeting signals.

Rigorous demonstration of a sorting signal requires that mutation of that signal impairs sorting and that incorporation of that signal into another protein results in sorting. Only a few definitive targeting signals have been described, and they are discussed next. Techniques of molecular biology are ideally suited for characterizing sorting signals, and it is likely that many more such signals will be described in the near future.

## A. Ectodomains and Endodomains of Proteins

Sorted proteins are generally glycoproteins, and thus sorting signals could reside in either the protein itself or in the carbohydrate groups. In one set of proteins, lysosomal hydrolases, phosphorylated mannose side groups bind to a receptor in the Golgi complex and are thereby targeted to the lysosome. For surface-bound glycoproteins (such as hemagglutinin, vesicular stomatitis virus envelope glycoproteins, or endogenous glycoproteins), however, modification of glycosylation does not affect sorting. Thus, sorting information must reside within the structure of the protein.

No apparent signalling structure is evident when domain-specific membrane proteins are compared. However, when chimeras or modified proteins (such as **glycosylated** proteins) are made, it appears that the sorting information for apical proteins, in general, resides on the **ectodomain** (that part of an integral membrane protein that is extracellular), while sorting information for basolateral proteins resides on the **endodomains** (the part that is intracellular). Removal of the transmembrane or the cytoplasmic domain of apically sorted proteins does not affect their sorting. In contrast, removal of the cytoplasmic domain or both the cytoplasmic and transmembrane domains of the poly Ig receptor, which is usually sorted to the basolateral membrane, results in the receptor being directed to the apical domain. Thus, sorting signals seem to exist for both apical and basolateral domains, and both of these signals may reside within a single molecule (Fig. 8).

## B. Lipid Patches

As stated previously, epithelia exhibit polarity in membrane lipids as well as in membrane proteins. The apical membrane is enriched in glycolipids, such as glycosphingolipids, whereas the basolateral membrane is enriched in phosphatidylcholine. Glycolipids, synthesized in the Golgi complex, can form **clusters** or **patches** in the trans-Golgi network. These patches arise by self-association of glycolipids to form enriched regions. It is believed that these glycolipid-enriched patches form into vesicles that are preferentially transported to the apical membrane (see Fig. 8). Other vesicles, not enriched in glycolipids, would be transported to the basolateral membrane by default. In this way, membrane lipid polarity could be established and maintained.

## C. GPI-Anchored Proteins

The presence of glycolipid patches that become apically targeted vesicles offers another pathway for sorting of proteins bound for the apical membrane. Certain glycoproteins are attached to the membrane through a covalent bond to the glycolipid, **glycosyl-phosphatidylinositol (GPI)**. These proteins are exclusively localized at the apical membrane. Foreign proteins that are normally GPI-anchored also localize to the apical membrane when transfected into epithelial cells. Further, proteins normally sorted to the basolateral membrane are redirected to the apical membrane when a GPI anchor is added to their structure (see Fig. 8). These findings clearly show that GPI anchoring is a sorting signal.

The combination of GPI anchoring and glycolipid patches has suggested a new model for apical sorting. GPI-anchored proteins will preferentially attach to patches of trans-Golgi network membrane enriched in glycolipids. When these patches form apically targeted vesicles, the GPI-anchored proteins will be cosorted into these vesicles and thereby targeted to the apical membrane (see Fig. 8). In this way, the sorting of apical lipids and protein is linked in a single pathway.

## D. Endocytotic and Transcytotic Signals

Sorted proteins are endocytosed, and thus must be continually resorted. Many of the initial sorting signals described earlier probably mediate re-sorting as well. For instance, GPI-anchored proteins and glycolipid patches are likely to be present in apically derived endosomes, and

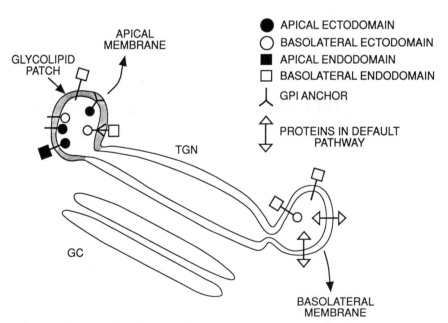

**FIG. 8.** A model of various targeting signals for proteins from the trans-Golgi network to the appropriate membrane domain. Glycolipids form a patch that characterizes a unique vesicle domain that is destined for the apical membrane. GPI-anchored proteins bind preferentially to these patch regions and thus are targeted to the apical membrane. In addition, apically bound proteins have a distinct ectodomain, but the endodomain is not relevant to apical targeting. Basolateral proteins, with their endodomains removed or with GPI anchors, are targeted to the apical membranes as well. The endodomain is essential for basolaterally targeted proteins. It is assumed that proteins and lipids not incorporated into the glycolipid patch form vesicles that target to the basolateral membrane. GC, Golgi complex; GPI, glycosyl-phosphatidylinositol.

these signals would promote return to the apical membrane.

Many proteins contain a **tyrosine (Tyr) residue** in an internalization motif on their cytoplasmic domain that serves as a **signal for internalization.** Deletion of this residue blocks endocytosis. In many such proteins, which are normally localized to the basolateral membrane in epithelial cells, the Tyr deletion changes their sorting to the apical membrane. Thus, a link between endocytosis and basolateral sorting has been proposed. However, in some proteins, deletion of the Tyr residue eliminates endocytosis without affecting basolateral sorting.

Recently, the important role of heterotrimeric and monomeric **G proteins** in controlling vesicle trafficking and the targeting of proteins in these vesicles has become evident. Some of these G proteins are involved in vesicle budding, transport, and fusion and are likely to control vesicle movement to the apical and basolateral membranes, as well as endocytosis, exocytosis, and transcytosis.

## V. Plasticity of Polarity

The establishment and maintenance of cell polarity is a highly involved and integrated process. It is thus of interest that in many cases this polarity can be lost and then the opposite polarity established. As described earlier (see Section III.A), the direction of polarity of thyroid epithelial cells grown in suspension can be switched upon exposure to collagen. The polarity of the chick embryo epiblast epithelium can be switched by reversing the polarity of the transepithelial potential. During the transition, polarity is lost and then re-forms with opposite polarity. The reversal of epithelial polarity may only involve certain transport systems. For example, maintaining rabbits on an acid diet results in the shifting of an $H^+$-ATPase from the basolateral membrane to the apical membrane in intercalated cells from the renal cortical collecting-duct epithelium. This switching of polarity occurs only for the $H^+$ transport systems, and no loss of Na/K ATPase polarity or tight junctions is seen.

Polarity can also be lost. For example, renal proximal tubule epithelia exhibit a loss of polarity upon ischemia. Basolateral and apical membrane lipids and proteins admix during ischemia. The ischemia-induced loss of polarity is attributed to ischemic damage to the actin cytoskeleton. The disruption of the cytoskeleton results in both the loss of the tight-junctional barrier function and the immobilization of membrane proteins, allowing for the loss of polarity due to the random mixing of the two membrane domains.

## VI. Summary

Many cells exhibit asymmetry or polarity. This polarity can be in the form of various membrane domains with distinct properties, asymmetric distribution of intracellular materials, polarized divisions of cells leading to distinct daughter cells, or polarity of cell structure. Such polarity is called **spatial polarity.** Cells can also exhibit **temporal polarity,** with marked changes arising during development or differentiation.

The best examples of polarized cells are epithelial cells, which have distinct **apical membrane and basolateral membrane domains** that differ in their lipid and protein compositions. This polarity of the transport proteins in epithelial cells is essential for the vectorial transport of solute and water across these cells. For example, the presence of $Na^+$ channels on the apical membrane and the Na/K pump on the basolateral membrane is characteristic of $Na^+$-reabsorbing epithelia, such as those in the gallbladder, the renal proximal tubule, and the frog skin.

Neurons are another good example of polarized cells, with distinct dendritic, cell body, and axon regions. The axonal membrane of neurons may be analogous to the apical membrane of epithelia, whereas the somatodendritic membrane may be analogous to the basolateral membrane. Examples of other polarized cells include **astrocytes,** with specialized processes extending from the cell body; **embryonic cardiac myocytes,** which assume an "epithelial-like" morphology early in development; and **egg** and **sperm cells.** Cell division is often asymmetric, as with the meiotic division of vertebrate gametes, the budding of yeast cells, and cell division in the bacterium *Caulobacter*. Finally, directed cell movement depends on the polarized growth of actin microfilaments in a process called **treadmilling.**

The establishment and maintenance of epithelial cell polarity depends on many factors. Cell–substratum attachment can help establish apical polarity by excluding proteins bound for the apical membrane domain from the region of attachment. With cell–cell attachment, in addition to cell–substratum attachment, **tight junctions** are formed and full polarity is observed. The maintenance of this polarity is aided by the appearance of the submembranous cytoskeleton beneath the basolateral membrane.

Several models have been proposed for the sorting and targeting of newly synthesized proteins to the appropriate membrane domain. Proteins could be inserted into both the apical and basolateral domains, but be differentially retained on only one membrane. In contrast, all membrane proteins could be inserted into one membrane (the **default pathway**), for example, the basolateral membrane, and then later the proteins bound for the apical membrane could be endocytosed, sorted, and **transcytosed** to the apical membrane. Finally, proteins could be sorted and targeted to their appropriate membrane domain directly from the trans-Golgi network.

The ability to target proteins to a specific membrane domain implies the existence of **sorting** and **targeting signals.** Membrane proteins contain targeting signals within their own domains. Sorting information to the apical membrane seems to reside on the external face of the membrane protein, whereas sorting information to the basolateral membrane resides on the cytoplasmic face. In addition, **membrane patches,** enriched in glycolipids, appear to cluster in the trans-Golgi network and are preferentially transported to the apical membrane. Certain proteins that are anchored to these patches through **glycosyl-phosphatidyl-inositol (GPI) linkages** are also targeted to the apical membrane. Also, certain amino acid residues within specialized domains on the cytoplasmic face of membrane proteins can serve as signals. Thus, a **tyrosine residue** has been shown to be involved as a signal for internalization. Finally, **G proteins** are believed to play important roles as targeting signals and controllers of endocytosis, exocytosis, and transcytosis.

Finally, cell polarity can be modified. For example, under conditions of ischemia, renal proximal tubule epithelial cells lose their polarity. In contrast, the polarity of chick embryo epiblast epithelial cells can be switched to the opposite polarity by reversing the polarity of the transepithelial potential.

## Bibliography

Alberts, B., Bray, D., Lewis, J., Raff, M., Roberts, K., and Watson, J. D. (1994). "Molecular Biology of the Cell," 3rd ed. Garland, New York.

Ali, S., Hall, J., Hazlewood, G. P., Hirst, B. H., and Gilbert, H. J. (1996). A protein targeting signal that functions in polarized epithelial cells *in vivo*. *Biochem. J.* **315,** 857–862.

Benos, D. J. (1991). "Developmental Biology of Membrane Transport Systems, Vol. 39, Current Topics in Membranes" (D. J. Benos, Ed.). Academic Press, New York.

Breitfeld, P. P., Casanova, J. E., Simister, N. E., Ross, S. A., McKinnon, W. C., and Mostov, K. E. (1989). Sorting signals. *Curr. Opin. Cell Biol.* **1,** 617–623.

Bretscher, M. S. (1996). Moving membrane up to the front of migrating cells. *Cell* **85,** 465–467.

Brown, D., and Stow, J. L. (1996). Protein trafficking and polarity in kidney epithelium: From cell biology to physiology. *Physiol. Rev.* **76,** 245–297.

Casanova, J. E., Apodaca, G., and Mostov, K. (1991). An autonomous signal for basolateral sorting in the cytoplasmic domain of the polymeric immunoglobulin receptor. *Cell* **66,** 65–75.

Drubin, D. G. (1991). Development of cell polarity in budding yeast. *Cell* **65,** 1093–1096.

Drubin, D. G., and Nelson, W. J. (1996). Origins of cell polarity. *Cell* **84,** 335–344.

Eaton, D. C. (1989). Special topic: Polarity of epithelial cells: Intracellular sorting and insertion. *Annu. Rev. Physiol.* **51,** 727–810.

Eaton, S., and Simons, K. (1995). Apical, basal, and lateral cues for epithelial polarization. *Cell* **82,** 5–8.

Gerlach, E., Nees, S., and Becker, B. F. (1985). The vascular endothelium: A survey of some newly evolving biochemical and physiological features. *Basic Res. Cardiol.* **80,** 459–474.

Gober, J. W., Champer, R. Reuter, S., and Shapiro, L. (1991). Expression of positional information during cell differentiation in *Caulobacter. Cell* **64,** 381–391.

Govindan, B., and Novick, P. (1995). Development of cell polarity in budding yeast. *J. Exp. Zool.* **273,** 401–424.

Hammerton, R. W., Krzeminski, K. A., Mays, R. W., Ryan, T. A., Wollner, D. A., and Nelson, W. J. (1991). Mechanism for regulating cell surface distribution of Na$^+$,K$^+$-ATPase in polarized epithelial cells. *Science* **254,** 847–850.

Heilker, R., Manning-Krieg, U., Zuber, J.-F., and Spiess, M. (1996). *In vitro* binding of clathrin adaptors to sorting signals correlates with endocytosis and basolateral sorting. *EMBO J.* **15,** 2893–2899.

Hicks, J. (1985). Yeast cell biology. *In* "UCLA Symposia on Molecular and Cellular Biology" (J. Hicks, Ed.). Alan R. Liss, New York.

Hopkins, C. R. (1991). Polarity signals. *Cell* **66,** 827–829.

Kupfer, A., Kronebusch, P. J., Rose, J. K., and Singer, S. J. (1987). A critical role for the polarization of membrane recycling in cell motility. *Cell Motil. Cytoskeleton* **8,** 182–189.

Lauffenburger, D. A., and Horwitz, A. F. (1996). Cell migration: A physically integrated molecular process. *Cell* **84,** 359–369.

Le Bivic, A., Sambuy, Y., Mostov, K., and Rodriguez-Boulan, E. (1990). Vectorial targeting of an endogenous apical membrane sialoglycoprotein and uvomorulin in MDCK cells. *J. Cell Biol.* **110,** 1533–1539.

Lisanti, M. P., and Rodriguez-Boulan, E. (1990). Glycophospholipid membrane anchoring provides clues to the mechanism of protein sorting in polarized epithelial cells. *Trends Biochem. Sci.* **15,** 113–118.

Matlin, K. S. (1986). The sorting of proteins to the plasma membrane in epithelial cells. *J. Cell Biol.* **103,** 2565–2568.

Matlin, K. S., and Valentich, J. D. (Eds.). (1989). "Functional Epithelial Cells in Culture, Vol. 8, Modern Cell Biology," Alan R. Liss, New York.

Mays, R. W., Siemers, K. A., Fritz, B. A., Lowe, A. W., and Meer, G. (1995). Hierarchy of mechanisms involved in generating Na/K-ATPase polarity in MDCK epithelial cells. *J. Cell Biol.* **130,** 1105–1115.

Meads, T., and Schroer, T. A. (1995). Polarity and nucleation of microtubules in polarized epithelial cells. *Cell Motil. Cytoskeleton* **32,** 273–288.

Molitoris, B. A. (1991). Ischemia-induced loss of epithelial polarity: Potential role of the actin cytoskeleton. *Am. J. Physiol.* **260,** F769–F778.

Mostov, K., Apodaca, G., Aroeti, B., and Okamoto, C. (1992). Plasma membrane protein sorting in polarized epithelial cells. *J. Cell Biol.* **116,** 577–583.

Müsch, A., Xu, H., Shields, D., and Rodriguez-Boulan, E. (1996). Transport of vesicular stomatitis virus G protein to the cell surface is signal mediated in polarized and nonpolarized cells. *J. Cell Biol.* **133,** 543–558.

Pemberton, L. F., Rughetti, A., Taylor-Papadimitrou, J., and Gendler, S. J. (1996). The epithelial mucin MUC1 contains at least two discrete signals specifying membrane localization in cells. *J. Biol. Chem.* **271,** 2332–2340.

Peng, I., Dennis, J. E., Rodriguez-Boulan, E., and Fischman, D. A. (1990). Polarized release of enveloped viruses in the embryonic chick heart. Demonstration of epithelial polarity in the presumptive myocardium. *Dev. Biol.* **141,** 164–172.

Prochiantz, A. (1995). Neuronal polarity: Giving neurons heads and tails. *Neuron* **15,** 743–746.

Putnam, R. W. (1996). Intracellular pH regulation in detubulated frog skeletal muscle fibers. *Am. J. Physiol.* **271,** C1358–C1366.

Rodriguez-Boulan, E., and Nelson, W. J. (1989). Morphogenesis of the polarized epithelial cell phenotype. *Science* **245,** 718–725.

Rodriguez-Boulan, E., and Powell, S. K. (1992). Polarity of epithelial and neuronal cells. *Annu. Rev. Cell Biol.* **8,** 395–427.

Scheiffele, P., Peränen, J., and Simons, K. (1995). N-Glycans as apical sorting signals in epithelial cells. *Nature* **378,** 96–98.

Shapiro, L. (1985). Generation of polarity during *Caulobacter* cell differentiation. *Annu. Rev. Cell Biol.* **1,** 173–207.

Simons, K., and Fuller, S. D. (1985). Cell surface polarity in epithelia. *Annu. Rev. Cell Biol.* **1,** 243–288.

Simons, K., and Wandinger-Ness, A. (1990). Polarized sorting in epithelia. *Cell* **62,** 207–210.

Wollner, D. A., Krzeminski, K. A., and Nelson, W. J. (1992). Remodeling the cell surface distribution of membrane proteins during the development of epithelial cell polarity. *J. Cell Biol.* **116,** 889–899.

Yoshimori, T., Keller, P., Roth, M. G., and Simons, K. (1996). Different biosynthetic transport routes to the plasma membrane in BHK and CHO cells. *J. Cell Biol.* **133,** 247–256.

Zinkl, G. M., Zuk, A., Bijl, P., Meer, G., and Matlin, K. S. (1996). An antiglycolipid antibody inhibits Madin–Darby canine kidney cell adhesion to laminin and interferes with basolateral polarization and tight junction formation. *J. Cell Biol.* **133,** 695–708.

Jeffrey C. Freedman

# 22

## Membrane Transport in Red Blood Cells

### I. Introduction

Red blood cells are prototypical of more complicated cells and have long been a favorite object of study for cellular physiologists. According to Jacobs (1962, p. 1014), "The first serious osmotic study of an animal cell (the mammalian erythrocyte)" was conducted by H. J. Hamburger in 1895, at a time when the importance of the plasma membrane was not widely understood. In their landmark review of membrane permeability, Davson and Danielli (1943) recognized that the ion permeability of red blood cells, together with electrical impedance measurements, lipid extraction studies, and the birefringence of red-cell ghosts, are all consistent with the postulation of a bimolecular lipid membrane, as first proposed for red cells in 1925 by Gorter and Grendel (see Chapter 4). The structural, functional, metabolic, and transport properties of normal and abnormal red blood cells have been well described in detailed monographs (Henderson, 1928; Ponder, 1948; Whittam, 1964; Bishop and Surgenor, 1964; Harris and Kellermeyer, 1970; Surgenor, 1974; Yoshikawa and Rapoport, 1974; Ellory and Lew, 1977; Agre and Parker, 1989; Raess and Tunnicliff, 1990; Ohnishi and Ohnishi, 1994). Methods for studying red cells have also been summarized (Ellory and Young, 1982; Beutler, 1986; Shohet and Mohandas, 1988).

Current research reports indicate that many fundamental unsolved problems remain for further investigation. More than 8000 articles concerning red blood cells were published during the past 5 years alone, including some 325 reviews on the mechanisms of membrane transport, protein associations and genetic defects in the cytoskeleton, oxygen transport and rheology, hematopoiesis and cellular senescence, blood preservation and transfusion medicine, sickle cell anemia and other hemoglobinopathies, metabolic control and enzymopathies, and altered red cells in systemic and infectious diseases, such as hypertension and malaria. Indeed, Ponder's view (1948, p. 1) that "... there is scarcely a fundamental problem in General Physiology which does not have a relation, of one kind or another, to the problems which have arisen in connection with the erythrocyte" remains just as valid today as 50 years ago.

### II. Membrane and Cytoskeleton

Mammalian red blood cells are highly differentiated for their primary function of oxygen and carbon dioxide transport, and from a structural point of view are the simplest of all eukaryotic cells. Devoid of mitochondria, endoplasmic reticulum, ribosomes, Golgi apparatus, and lysosomes, and lacking a nucleus, mammalian red blood cells are free of the complexities associated with intracellular organellar compartments, and thus have served as a classic model system for studying how ions, nutrients, and other solutes cross the plasma membrane. With electron probe microanalysis (Lew et al., 1985) and NMR studies (Murphy et al., 1987), however, the elevated $Ca^{2+}$ in red cells from patients with sickle cell anemia was found sequestered in intracellular vesicles, which were also noted to occur in normal red cells. Observations of $Ca^{2+}$-sequestering vesicles in red cells are reminiscent of the ATP-dependent endocytosis that occurs in intact red cells and in isolated membranes in response to oxidants (Penniston et al., 1979). Another type of compartmentalization proposed in red cells is a membrane pool estimated to contain some 500–600 molecules of ATP (Proverbio et al., 1988). $^{32}P$-labeling experiments in ghosts, and in inside-out vesicles (Mercer and Dunham, 1981), indicated that the membrane-bound glycolytic enzymes glyceraldehyde 3-phosphate dehydrogenase and phosphoglycerokinase form ATP, which comprises a membrane-bound pool that directly provides substrate for the $Na^+,K^+$-ATPase; the structural basis for this membrane pool of ATP is unknown.

Most mammalian red blood cells normally exhibit a biconcave discoidal shape. Under abnormal conditions red cells may be transformed into spiculated forms known as **echinocytes,** or into cup-shaped forms, called **stomatocytes** (for reviews and scanning electron micrographs, see Bessis *et al.*, 1973; Bessis, 1974; and Chapter 14 by Bull and Brailsford in Agre and Parker, 1989). Red cells from camels are an exception, being nonnucleated but with a biconvex ellipsoidal shape, resembling that of nucleated amphibian and avian red cells (for review, see Chapter 3 by Ngai and Lazarides in Agre and Parker, 1989). The red-cell membrane is one of the easiest to isolate (Dodge *et al.*, 1963), has been extensively studied, and is one of the best characterized of all cell membranes. The sidedness and orientation of proteins in the seven major bands found after electrophoresis of human red blood cell membranes in sodium dodecyl sulfate (SDS) polyacrylamide gels was described by Steck (1974). Studies of membrane transport were greatly facilitated by optimizing conditions for "resealing" isolated membranes to make **"resealed ghosts,"** a technique that enabled manipulation of the ionic composition of the intracellular as well as the extracellular solutions (Hoffman *et al.*, 1960). Despite extensive studies of hemolysis, the mechanism of the formation and annealing of a hole or holes in the membrane sufficiently large to permit the passage of $Na^+$, $K^+$, metabolites, and hemoglobin is still not well understood (for review, see Hoffman, 1992). In the red-cell membrane, as in many other eukaryotic cell membranes, the phospholipids are distributed asymmetrically. The neutral phospholipids—phosphatidylcholine and sphingomyelin—are preferentially located in the outer hemileaflet of the lipid bilayer. The neutral phosphatidylethanolamine and the negatively charged phosphatidylinositol and phosphatidylserine are preferentially located in the inner hemileaflet, with the negative charges facing the cytoplasm. An aminophospholipid-specific translocase, or **"flipase,"** uses $Mg^{2+}$-ATP to catalyze the inward transport of phospholipids from the outer to the inner hemileaflet with the following specificity:

phosphatidylserine > phosphatidylethanolamine >
phosphatidylcholine

The flipase exhibits stereospecificity in acting only on L-isomers and is inhibited by vanadate and by oxidation of protein sulfhydryl groups. Elevated intracellular calcium, such as occurs during normal red-cell senescence, as well as in red cells from patients with sickle cell anemia, causes loss of the normal **phospholipid asymmetry,** leading especially to excess phosphatidylserine in the outer hemileaflet. Under these conditions, the external membrane surface acquires procoagulant activity and promotes thrombosis (for review, see Diaz and Schroit, 1996).

The external surface of the red-cell membrane is coated with adsorbed albumin and with some plasma globulins. Because of the sialic acids on **glycophorin,** an integral membrane glycoprotein, the cell surface is negatively charged (for review, see Chapter 27 by Seaman in Surgenor, 1974). The **cytoskeleton** is an organized polygonal fibrous network about 60 nm thick containing spectrin and actin (for reviews, see Chapter 1 by Gardner and Bennett in Agre and Parker, 1989; Bennett and Gilligan, 1993). Sides of the regular five- or six-sided polygons are formed by **spectrin,** a flexible filamentous protein 200 nm in length. Vertices of the polygons are formed by **β-actin** protofilaments 30–40 nm in length consisting of 12–14 actin monomers; both

grooves of the actin filaments contain **tropomyosin,** which binds to **tropomodulin** at junctional complexes. At the mid-region of the spectrin filaments a junctional ternary complex formed by **ankyrin** and band 4.2 protein links spectrin to the membrane-spanning integral band 3 protein. At the ends of the spectrin filaments, an additional linkage site to band 3 and to **glycophorin C** may be provided by protein 4.1. Spectrin–actin junctional complexes also contain the actin-bundling protein **dematin** (band 4.9) and the calmodulin-binding protein **adducin,** which functions to cap and to regulate the length of the actin filaments and which may be involved in $Ca^{2+}$-dependent alterations in cytoskeletal structure (Kuhlman *et al.*, 1996). Measurements of lateral translational diffusion showed that a fraction of band 3 protein is immobile because of its linkage with spectrin, whereas another fraction is capable of slow diffusion in the plane of the membrane (Golan and Veatch, 1980; for review, see Chapter 13 by Golan in Agre and Parker, 1989). A small amount of **myosin** is also found in the red-cell cytoskeleton, but whether active tension is generated or regulated in mature red cells is not known. Presumably, the cytoskeleton is actively involved during **enucleation** in developing red cells, and during **diapedesis,** or egress of red cells from the bone marrow to the peripheral circulation.

## III. Intracellular Environment

The red-cell membrane and associated cytoskeleton enclose a viscous cytoplasmic solution of the oxygen-binding pigment **hemoglobin** at a concentration of 34 g/100 ml cells, corresponding to 5.2 m$M$, 7.3 mMolal, or 44 g Hb/100 g cell water. This concentration is near the threshold for gelation, but hemoglobin is one of the most soluble of all proteins and constitutes more than 98% of red-cell protein by mass. The hemoglobin $\alpha_2\beta_2$ tetramers have a molecular weight of 64,373 Da and are approximately spheroidal with dimensions of 65 Å × 55 Å × 50 Å. In the interior of red blood cells, hemoglobin is nearly close-packed, with the distance between the surfaces of neighboring proteins averaging only about 20 Å (Ponder, 1948, p. 140). With the 86 carboxylates of aspartic and glutamic acids, and with the 98 basic amino and amine groups of arginine, lysine, and histidine per hemoglobin tetramer, the total cellular concentration of the titratable amino acids of hemoglobin is around 0.9 $M$. Thus, hemoglobin contributes significantly, depending on intracellular pH, to a high intracellular ionic strength. Since the charged amino acids are located on the surface of hemoglobin (see Antonini and Brunori, 1971), and taking the radius of hemoglobin to be 28 Å, the average distance between charged sites is only around 7 Å. About 7600 water molecules per hemoglobin tetramer occupy the narrow interstices between the protein molecules. The hydration of hemoglobin in dilute solution is 0.2–0.3 g water/g Hb (see Antonini and Brunori, 1971), representing about 15% of the intracellular water. When the tortuosity of the surface of soluble proteins is taken into account, as much as 30% of red-cell water could reside in the first monolayer around the protein surface.

To determine the **mean ionic activities** of the intracellular KCl and NaCl in this concentrated charged environment, studies were conducted in which the red-cell membrane was rendered permeable to cations by exposure of the cells to the channel-forming antibiotic nystatin, thus allowing $K^+$, $Na^+$, and $Cl^-$ to reach **Gibbs–Donnan equilibrium.** In these experiments sufficient extracellular sucrose was added to prevent cell swelling by balancing the colloid osmotic pressure of hemoglobin and other impermeant cell solutes (see Chapter 14). The **mean ionic activity coefficient** of KCl and NaCl in the concentrated intracellular hemoglobin solution was found to be within 2% of that in the extracellular solution (Freedman and Hoffman, 1979a). Moreover, permeant nonelectrolytes also have equilibrium ratios of intracellular to extracellular concentrations within 10% of unity (Gary-Bobo, 1967). In view of the high intracellular ionic strength and the high volume fraction of cell water in direct contact with hemoglobin, it is both curious and remarkable that intracellular salts and nonelectrolytes appear to behave as if in dilute solution. Either the intracellular solution is indeed like a dilute solution, or alternatively, the expected effect of protein–solvent interactions in altering the activity of intracellular solutes is offset by the effect of interactions between proteins and the solutes themselves.

## IV. Metabolism and Life Span

The red cell also has relatively simple metabolic pathways (Grimes, 1980; Beutler, 1986), at least in comparison with most other cells. Catalogs of red-cell enzymes list about 140 enzymes (see the chapter by Friedemann and Rapaport in Yoshikawa and Rapoport, 1974; and Chapter 3 by Pennell in Surgenor, 1974). **Glycolysis** produces ATP and lactate from glucose, inorganic phosphate, and exogenous purine in the form of adenine, adenosine, or inosine. A mathematical model of red-cell glycolysis was proposed by Rapoport *et al.* (1974). The **pentose shunt** provides reducing equivalents in the form of glutathione, NADH, and NADPH, which, together with catalase, superoxide dismutase, glutathione peroxidase, glutathione reductase, and methemoglobin reductase act to prevent the oxidation of protein sulfhydryl groups and of $Fe^{2+}$ in hemoglobin. No tricarboxylic acid cycle, cytochrome system, or lipid catabolism or utilization are known to occur in red cells, although cholesterol and phospholipids do exchange with plasma lipids, and some fatty acids may be incorporated into membrane phospholipids (for review, see Shohet, 1976). Protein synthesis does not occur in mature red cells, and there is no DNA replication or transcription, and no RNA metabolism or gene action. Red cells in humans make up 40–45% of the blood volume, a fraction known as the **hematocrit;** thus, they are readily accessible, and are easily separable from leukocytes and platelets by centrifugation or filtration. In short-term experiments, red cells may be studied in simple buffered isotonic salt solutions (e.g., 145 m$M$ NaCl, 5 m$M$ KCl, 5 m$M$ Hepes buffer, pH 7.4). For longer experiments, glucose is added to prevent the decline of ATP; for even longer-term experiments, red cells survive *in vitro* at room temperature or at 37°C for many days, albeit with morphological heterogeneity, in a chemically defined culture medium that includes vitamin cofactors, an exogenous purine for synthesis of ATP, and amino acids to support the synthesis of glutathione (Freedman, 1983). *In vitro*, red cells can be studied free from the uncontrollable variables of the intact organism.

Each of the 25 trillion red blood cells in normal adult humans lives for about 120 days. At a turnover rate of about 1%/day, some 250 billion new red cells are released from the bone marrow each day, a rate that corresponds to 3 million cells per second! This may seem like a lot, but 3 million red blood cells occupy less than a microliter of volume, since each biconcave discoidal cell occupies only 87 cubic microns and has a surface area of 133 square microns, a diameter of 8 microns, and a thickness of 2.4 microns at the rim and 1.0 micron at the center. After 120 days, senescent human red cells bind **IgG autoantibody** and are then recognized by macrophages in the initial stage of **erythro-phagocytosis,** a process that leads to the recycling of iron, amino acids, and other essential red-cell constituents. Considerable evidence indicates that the antigenic recognition sites are composed of clusters of an oxidatively denatured form of band 3 protein (for reviews, see Chapter 9 by Low in Agre and Parker, 1989; Kay, 1991). The normal function of band 3 protein, also called **capnophorin** or **AE1 (anion exchange protein 1),** is to mediate the obligate electroneutral exchange of $Cl^-$ for $HCO_3^-$ across the red-cell membrane during gas exchange in the pulmonary and systemic capillaries. The comparative biochemistry and physiology of red blood cells is instructive, with many differences in metabolic and membrane transport properties, as well as oxygen transport properties, known to occur among different animal species (for examples, see Willis, 1992). Whereas human red cells live for 120 days, the **life span** of dog red cells is 60 days, and that of mouse red cells is only 40 days. The life spans $L$ of mammalian red blood cells correlate with body weight $W$ according to the relation $L = 69W^{0.12}$, as illustrated in the log–log plot in Fig. 1, but the physiological basis for this striking correlation is not understood (for discussion, see Vácha and Znojil, 1981). The life span data suggest that red cells contain a biological clock that is obviously not directly determined or controlled in the mature cell by gene transcription or translation.

## V. Membrane Transporters in Red Blood Cells

Among eight mammalian species (Table 1), the intracellular concentration of $K^+$ ranges from 135 m$M$ in human red blood cells to only 8 m$M$ in cat and dog red cells, whereas intracellular $Na^+$ varies from only 17 m$M$ in humans to 142–162 m$M$ in dog and cat; in contrast, the extracellular low $K^+$ and high $Na^+$ are relatively constant among species, as are intracellular and extracelluar $Cl^-$. Cation transport in red blood cells was understood classically in terms of the **pump-leak theory** (Tosteson and Hoffman, 1960), which quantitatively accounted for the differing steady-state $Na^+$ and $K^+$ concentration in red blood cells

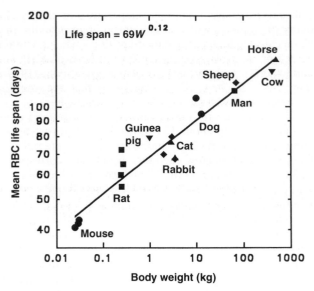

**FIG. 1.**   Mean life span (days) of mammalian red blood cells vs. body weight (kg). [Data from Vácha and Znojil (1981) and various other sources.]

found in two genetic phenotypes of sheep designated as "high-K[+] (HK)" and "low-K[+] (LK)." High-K[+] sheep red cells were found to have a relatively high number of Na[+]–K[+] pumps per cell and high Na[+]–K[+]-pump fluxes with relatively low ouabain-insensitive leakage fluxes. Anions such as Cl[−], HCO$_3^−$, and OH[−] are passively distributed and appear to follow the "**double-Donnan**" equilibrium, as explained later (for reviews on red-cell transport, see Chapter 3 by Passow in Bishop and Surgenor, 1964; Chapter 15 by

Sachs, Knauf, and Dunham in Surgenor, 1974; Ellory and Lew, 1977; Agre and Parker, 1989; Raess and Tunnicliff, 1990). The ionic composition of the intracellular and extracellular solutions may be depicted on bar graphs, as shown for human red blood cells in Fig. 2. In human red cells, the Na[+],K[+]-ATPase (see Chapter 16) specifically selects K[+] from the Na[+]-rich medium, and pumps it against the K[+] concentration gradient (and electrochemical gradient) into the cytoplasmic solution. The same ionic pump specifically selects Na[+] from the K[+]-rich cytoplasmic solution and extrudes it from the cell against the Na[+] concentration gradient. This coupled active transport of three internal Na[+] for two external K[+] uses metabolic energy obtained from the hydrolysis of ATP (for reviews, see Hoffman, 1986, and Chapters 6 and 16 by Mercer *et al.* and by Kaplan in Agre and Parker, 1989). In the steady state, at the same time that K[+] is actively accumulated in the cell, it is also continually leaking out of the cell at the same rate through parallel pathways down its concentration gradient. The same is true for Na[+] leaking into the cell. Steady-state distributions of K[+] and Na[+] are thus achieved by a balance between active pumping and passive leakage. If the Na[+]–K[+]-pump of human red blood cells is completely inhibited, the loss of K[+] and gain of Na[+] is so slow that the leakage fluxes would continue unabated for at least 30 days before Gibbs–Donnan equilibrium is approached (Fig. 3); however, the cells would hemolyze before reaching equilibrium because of the inability of the plasma membrane to withstand any significant osmotic pressure associated with the Gibbs–Donnan equilibrium, as discussed later in this chapter.

**TABLE 1**   Concentrations (m$M$) of Potassium, Sodium, and Chloride in Mammalian Red Blood Cells and Plasma

| Species | Intracellular | | | Extracellular | | |
|---|---|---|---|---|---|---|
| | [K[+]] | [Na[+]] | [Cl[−]] | [K[+]] | [Na[+]] | [Cl[−]] |
| Man[a] | 135 | 17 | 77 | 3.7 | 138 | 116 |
| Baboon[b] | 145 | 24 | 78 | 4.7 | 157 | 115 |
| Rabbit[b] | 142 | 22 | 80 | 5.5 | 150 | 110 |
| Rat[b] | 135 | 28 | 82 | 5.9 | 152 | 118 |
| Horse[b] | 140 | 16 | 85 | 5.2 | 152 | 108 |
| Sheep[c] | | | | | | |
| HK | 124 | 13 | | | | |
| LK | 17 | 119 | | | | |
| Dog[d] | 8 | 162 | 80 | 4.6 | 165 | 123 |
| Cat[b] | 8 | 142 | 84 | 4.6 | 158 | 112 |
| Mean[b] | | | 81 | 4.9 | 153 | 115 |
| SD | | | 3 | 0.7 | 8 | 5 |

[a] Funder, J., and Wieth, J. O. (1966a,b). *Scand. J. Clin. Lab. Invest.* **18**, 151–166, and *Acta Physiol. Scand.* **68**, 234–245.
[b] Bernstein, R. E. (1954). *Science* **120**, 459–460.
[c] Dunham, P. B. (1992). *Comp. Biochem. Physiol.* **102A**, 625–630.
[d] Parker, J. C. (1973). *J. Gen. Physiol.* **61**, 146–157.

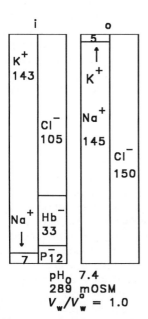

**FIG. 2.**   Ionic composition of the intracellular solution in human red blood cells and of the extracellular solution, designated by i and o, respectively. Organic phosphates are represented by P[−].

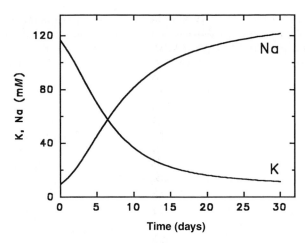

**FIG. 3.** Loss of $K^+$ and gain of $Na^+$ when the $Na^+$–$K^+$ pump of human red blood cells is completely inhibited. Theoretical plot by P. R. Pratap using the integrated model of Lew and Bookchin (1986).

A striking correlation between the properties of the $Na^+$,$K^+$-ATPase and cation fluxes in red blood cells constitutes impressive evidence that the $Na^+$,$K^+$-ATPase is the ion pump that mediates the fluxes of $Na^+$ and $K^+$ across the red-cell membrane: (1) Both the $Na^+$,$K^+$-ATPase and the control of cation transport are located in the membrane; (2) the $Na^+$,$K^+$-ATPase and cation transport are both stimulated specifically by intracellular ATP, by intracellular $Na^+$, and by extracellular $K^+$; (3) the cation concentrations required for half-maximal activation of the $Na^+$,$K^+$-ATPase and cation transport are the same; (4) cardiac glycosides, which are impermeant, specifically inhibit the $Na^+$,$K^+$-ATPase and cation transport without directly affecting other transporters or fluxes; (5) cardiac glycosides inhibit the $Na^+$,$K^+$-ATPase and cation transport half-maximally at the same concentration and with the same molecular specificity in that inhibition of both processes requires an unsaturated lactone group; (6) $K^+$ antagonizes the effect of cardiac glycosides on both the $Na^+$,$K^+$-ATPase and cation transport; and (7) the $Na^+$,$K^+$-ATPase is reversible, and net synthesis of ATP occurs when the $Na^+$ and $K^+$ concentration gradients are reversed. Moreover, the activity of the $Na^+$,$K^+$-ATPase parallels the magnitude of cation fluxes over a 25,000-fold range in a variety of tissues, and active transport of Na and K against their respective concentration gradients has been observed directly by measurement of net fluxes in "resealed ghosts."

P. Läuger (1995) considered that the fluctuating energy barriers in ion pumps and exchangers may be similar to those in ion channels (see also DeFelice and Blakely, 1996). With this idea in mind, it is of considerable interest that **palytoxin,** isolated from a marine soft coral, reversibly increases the cation conductance of red-cell membranes over the native $Cl^-$ conductance, and that the effects of the toxin are totally blocked by external ouabain and prevented by internal vanadate (M. Tosteson *et al.*, 1991). In the presence of the toxin, the conductive cation selectivity becomes $K^+ > Rb^+ > Cs^+ > Na^+ > Li^+$, corresponding to Eisenman

sequence IV (see Chapter 1). It appears that palytoxin "opens a 10 pS channel ... at or near each pump site." Although this channel probably represents a perturbed form of the native cation permeation pathway in the $Na^+$,$K^+$-ATPase, further characterization might provide new insights as to how channels might function within pumps; a related problem in red cells concerns the relationship of DIDS-sensitive $Cl^-$ conductance to the DIDS-sensitive $Cl^-$–$HCO_3^-$ exchanger, as described later in this chapter.

Whereas the $Na^+$–$K^+$-pump is by far the most widely distributed transporter that regulates the monovalent cation composition of cells, the red blood cells of carnivores (e.g., dogs, cats, ferrets, bears) contain high $Na^+$ and low $K^+$ (see Table 1). The immature nucleated red cells of dogs do contain the usual high $K^+$ and low $Na^+$ contents and $Na^+$–$K^+$ pumps, but during maturation primary active transport of $Na^+$ and $K^+$ ceases. In mature dog red cells, intracellular $[Na^+]$ is slightly less than extracellular $[Na^+]$, whereas intracellular $[K^+]$ is slightly greater than extracellular $[K^+]$. These cells, like human red cells, also contain a potent $Ca^{2+}$-ATPase in their membranes that maintains submicromolar concentrations of intracellular $[Ca^{2+}]$ in the face of millimolar extracellular $[Ca^{2+}]$ (for review, see Chapter 17 by Vincenzi in Agre and Parker, 1989, and Chapter 17 of this book). However, dog, ferret, and bear red cells, unlike human red cells, also contain a $Na^+$–$Ca^{2+}$ countertransporter that couples the passive influx of $Ca^{2+}$ to the efflux of $Na^+$, thus resulting in maintenance of a slight reduction in intracellular $[Na^+]$ in the steady state (see Chapter 18). The slight excess of intracellular $K^+$ is consistent with an inside negative membrane potential.

In addition to the many structural and mechanistic studies on active pump fluxes mediated by the $Na^+$,$K^+$-ATPase and the $Ca^{2+}$-ATPase, much progress has also occurred in understanding the mechanisms of passive cation transport (for review, see Chapter 18 by Parker and Dunham in Agre and Parker, 1989). Unlike muscle and nerve, red cells have a low resting permeability to both $Na^+$ and $K^+$, but a high conductive permeability to $Cl^-$ that is about 100 times greater than that to cations. Consequently, $Cl^-$ is in equilibrium with the resting membrane potential $E_m$,

$$E_m = \frac{RT}{\mathscr{F}} \ln \frac{[Cl^-]_i}{[Cl^-]_o}$$

where $R$ is the gas constant, $T$ is the absolute temperature, $\mathscr{F}$ is the Faraday constant, and $RT/\mathscr{F}$ is 25.5 mV at 23°C. For a ratio of $[Cl^-]_i/[Cl^-]_o = 77/116 = 0.66$, the resting potential $E_m$ is −11 mV, inside negative. In the steady state, ionic currents of $K^+$ and $Na^+$ both flow passively across the membrane down their respective concentration gradients. Assuming that the currents of monovalent ions are independent, that the membrane is symmetrical, that the transmembrane electric field is constant, and that ions first partition into the membrane and then diffuse across driven by concentration and electrical gradients, the theory of **electrodiffusion** gives the membrane potential as

$$E_m = -\frac{RT}{\mathscr{F}} \ln \frac{P_K[K^+]_i + P_{Na}[Na^+]_i + P_{Cl}[Cl^-]_o}{P_K[K^+]_o + P_{Na}[Na^+]_o + P_{Cl}[Cl^-]_i}$$

where $P_K$, $P_{Na}$, and $P_{Cl}$ are the constant field permeabilities ($sec^{-1}$). Since $Cl^-$ is at equilibrium, we may substitute $Cl_i = Cl_o e^{\mathscr{F}E_m/RT}$ into the preceding Goldman–Hodgkin–Katz equation. Rearranging and simplifying gives

$$E_m = -\frac{RT}{\mathscr{F}} \ln \frac{P_K[K^+]_i + P_{Na}[Na^+]_i}{P_K[K^+]_o + P_{Na}[Na^+]_o}$$

In other words, all permeant ions—cations and anions—have some relationship to the electrical potential $E_m$ across the membrane. Solving this expression for the ratio $P_K/P_{Na}$ gives

$$\frac{P_K}{P_{Na}} = \frac{[Na^+]_o e^{-\phi} - [Na^+]_i}{[K^+]_i - [K^+]_o e^{-\phi}}$$

where $\phi$ is the reduced potential $\mathscr{F}E_m/RT$. Using standard values for human red cells from Table 1 for $[K^+]_i$, $[K^+]_o$, $[Na^+]_i$, $[Na^+]_o$, and $\phi$ yields $P_K/P_{Na} = 1.5$, much less than the corresponding ratio in nerve and muscle. Thus, the resting cation diffusion potential of human red blood cells is consistent with the membrane having a slightly greater constant-field permeability for $K^+$ than for $Na^+$. However, the measured ground permeabilities to $Na^+$ and $K^+$ appear to be nonselective and more consistent with a single-barrier model than with electrodiffusion (Zade-Oppen *et al.*, 1988). Away from the steady state, such as following cell shrinkage by exogenously added extracellular sucrose, the gradient of the highly permeant $Cl^-$ will dominate and determine the membrane potential of human red blood cells until a new steady state is reached (see Bisognano *et al.*, 1993). The electrogenic current of the $Na^+,K^+$-ATPase is one-third of the active $Na^+$ efflux; this current flows across the membrane electrical resistance, which for normal human red blood cells has been estimated at about $10^6$ ohm $cm^2$ (for review, see Hoffman *et al.*, 1980). The electrogenic potential due to the electrogenic $Na^+$ current is normally less than 1 mV, but is demonstrable (Fig. 4) using fluores-

cent dyes after optimizing the signal by increasing internal $Na^+$, by increasing the membrane resistance through replacing cell $Cl^-$ with $SO_4^{2-}$, and by inhibiting the anion conductance with DIDS (Dissing and Hoffman, 1990).

As illustrated in Fig. 5, red-cell passive fluxes, either conductive or electroneutral, are composed of at least nine phenomenologically separable components, a much more complex situation than assumed in classical theories (Van Slyke *et al.*, 1923; Jacobs and Stewart, 1947; Tosteson and Hoffman, 1960). The pump-leak theory for red cells has been extended to model some of the effects of the additional cotransport pathways (Milanick and Hoffman, 1986; Lew and Bookchin, 1986). Pathways for passive electrolyte transport in human red blood cells include cotransporters for $Na^+$-$K^+$-$2Cl^-$ and $K^+$-$Cl^-$ (for reviews, see Lauf *et al.*, 1992; Hoffmann and Dunham, 1995), countertransporters for $Na^+(Li^+)$–$Na^+$ (e.g., Duhm and Becker, 1977) and $Na^+$–$H^+$ (Escobales and Canessa, 1986), $Ca^{2+}$-activated $K^+$ channels (for review, see Schwarz and Passow, 1983), $Cl^-$–$HCO_3^-$ exchange mediated by capnophorin (AE1 or band 3 protein) (for reviews, see Knauf, 1979; Passow, 1986; Chapter 19 by Gunn *et al.* in Agre and Parker, 1989; Jennings, 1992; Reithmeier, 1993), $Cl^-$ conductance (Freedman *et al.*, 1988, 1994; Freedman and Novak, 1997), and HCl cotransport (Bisognano *et al.*, 1993). Specific red-cell transporters also mediate the transmembrane movement of glucose, nucleosides, lactate and other organic anions, oxidized glutathione, choline, and amino acids (for reviews, see Lefevre, 1961; the chapters by Srivastava and by Martin in Ellory and Lew, 1977; Agre and Parker, 1989; Raess and Tunnicliff, 1990). Water flows through pores made of aquaporin (for review, see Agre, 1996).

Some of the most frequently used inhibitors in studies of red-cell membrane transport are listed in Table 2. In isotope flux studies active transport of $Na^+$ and $K^+$ has often been equated with the "ouabain-sensitive fraction" of $Na^+$ efflux or $K^+$ influx. Similarly, $Na^+$-$K^+$-$2Cl^-$ cotransport is operationally measured as the "ouabain-insensitive, bumetanide-sensitive" fraction of a $Na^+$ flux, whereas $Na^+(Li^+)$–$Na^+$ countertransport is the "ouabain- and bumetanide-insensitive, phloretin-sensitive" fraction. Fluoride and phloretin are relatively nonspecific and act on more than one transporter. Eosin is a relatively new inhibitor for the $Ca^{2+}$-ATPase. Some inhibitors bind reversibly to the transporter (e.g., DNDS to band 3), whereas others inhibit irreversibly by formation of covalent bonds (e.g., DIDS to band 3). Inhibitors may also act indirectly on associated enzymes that modulate transport. For example, the inhibition of a type 1 protein phosphatase by calyculin A or okadaic acid prevents the activation of $K^+$–$Cl^-$ cotransport during cell volume regulation.

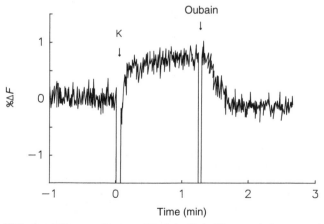

**FIG. 4.** Electrogenic potential of the $Na^+$–$K^+$ pump in human red blood cells, as monitored with the oxonol dye WW781. Intracellular $Cl^-$ has been replaced with $SO_4^{2-}$, followed by treatment with DIDS. In $Na^+$-free medium, the pump is activated by addition of 10 m$M$ $K^+$, and then inhibited with 33 $\mu M$ ouabain, as indicated. An upward deflection indicates hyperpolarization. (Data of P. R. Pratap and J. C. Freedman.)

## VI. Ionic and Osmotic Equilibrium and Cell Volume Regulation

Human red blood cells contain 66% water (100 × g water/g packed cells), as is easily determined by drying a weighed pellet of sedimented cells to constant weight in a vacuum oven, taking into account suitable corrections for

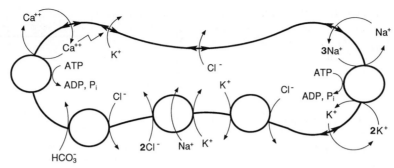

**FIG. 5.** The principal membrane transport systems of human red blood cells. Starting with the $Na^+$–$K^+$ pump and the passive leakage pathways for $Na^+$ and $K^+$ on the right, and proceeding clockwise around the membrane, are $K^+$–$Cl^-$ cotransport, $Na^+$–$K^+$–$2Cl^-$ cotransport, $Cl^-$–$HCO_3^-$ exchange, the $Ca^{2+}$ pump with a $Ca^{2+}$ leakage pathway, the $Ca^{2+}$-activated $K^+$ channel, and $Cl^-$ conductance. The steady-state concentrations of $K^+$, $Na^+$, and $Ca^{2+}$ represent a balance between active pumping and passive leakage. $Cl^-$ and $HCO_3^-$ are passively distributed at equilibrium with a membrane potential of $-11$ mV.

trapped volume. According to the **Boyle–van't Hoff law,** which is derived from the Gibbs equation (see Dick, 1959), **the osmotic pressure ($\pi$)** of a solution containing $N$ moles of solute dissolved in $V_w$ liters of water is given by

$$\pi = \frac{\phi NRT}{V_w} = \phi cRT$$

where $\phi$ is the osmotic coefficient of the solute, $R$ is the gas constant, $T$ is the absolute temperature, and $c$ ($= N/V_w$) is the total concentration of dissolved solutes. The **osmolality** $\phi c$ is the total osmolal concentration of all dissolved solutes (see Chapter 19). The **osmotic coefficient** $\phi$ is unity for ideal dilute solutions; deviations of $\phi$ from unity are due to solute–solvent interactions and other nonidealities. The osmotic pressure $\pi_i$ of the intracellular solution is the sum of the osmotic pressures of each component of the mixture,

$$\pi_i = \sum \frac{\phi_j N_j RT}{V_w} = RT \sum \phi_j C_j$$

where $c_j$ ($= N_j/V_w$) is the concentration of the $j$th solute. Alternatively,

$$\pi_i = \frac{\Phi_i N_T RT}{V_w} = \Phi_i c_T RT$$

where $\Phi_i$ is the osmotic coefficient of the intracellular mixture, a parameter that equals the sum of the osmotic coefficients of the components of the mixture, weighted according to their respective mole fractions.

Water will cross the member until the intracellular osmotic pressure equals the extracellular osmotic pressure, or $\pi_i = \pi_o$, as expressed by

$$\sum \phi_{j,i} C_{j,i} = \sum \phi_{j,o} C_{j,o}$$

An **isosmotic** solution is defined as having an osmolality equal to that of normal plasma, or 289 mOsmolal; an **isotonic** solution will maintain the normal volume of cells incubated during physiological experiments. In isotonic solution, designated by the superscript °,

$$\pi_i^o = \frac{\Phi_i^o N_T^o RT}{V_w^o}$$

Dividing $\pi_i$ by $\pi_i^o$, and setting $\pi_i = \pi_o$, and $\pi_i^o = \pi_o^o$, and rearranging, yields the following expression for the equilib-

**TABLE 2** Some Inhibitors of Membrane Transport in Red Blood Cells

| Transporter | Inhibitor |
| --- | --- |
| Pumps | |
| $Na^+$,$K^+$-ATPase | Ouabain, vanadate |
| $Ca^{2+}$-ATPase | $La^{3+}$, vanadate, eosin |
| GSSG[a] | Fluoride |
| Cotransport | |
| $Na^+$–$K^+$–$2Cl^-$ | Bumetanide, furosemide |
| $K^+$–$Cl^-$ | DIOA[b] |
| | Calcyculin A, okadaic acid |
| $H^+$–$Cl^-$ | DIDS[c] |
| Countertransport | |
| $Cl^-$–$HCO_3^-$ | DIDS[c]; DNDS[d] |
| | Phlorizin; phloretin |
| $Na^+$($Li^+$)–$Na^+$ | Phloretin |
| $Na^+$–$H^+$ | Amiloride |
| Facilitated diffusion | |
| Glucose | Phlorizin; phloretin |
| | Cytochalasin B |
| Nucleosides | Nitrobenzylthioinosine |
| Lactate | PCMBS[e]; DTNB[f] |
| Choline | NEM[g] |
| Channels | |
| $Ca^{2+}$($K^+$) | Charybdotoxin |

[a] Oxidized glutathione.
[b] [(Dihydroindenyl)oxy]alkanoic acid.
[c] 4,4′-Diisothiocyano-2,2′-disulfonic acid stilbene.
[d] 4,4′-Dinitro-2,2′-dinitrosulfonic acid stilbene.
[e] *para*-Chloromercuribenzenesulfonate.
[f] 5,5′-Dithio-bis(2-nitrobenzoate).
[g] $N$-Ethylmaleimide.

rium cell water content relative to that of cells in isotonic solution:

$$\frac{V_{\mathrm{w}}}{V_{\mathrm{w}}^{\mathrm{o}}} = \frac{(\Phi_i N_{\mathrm{T}})}{(\Phi_i^{\mathrm{o}} N_{\mathrm{T}}^{\mathrm{o}})} \frac{\pi_{\mathrm{o}}^{\mathrm{o}}}{\pi_{\mathrm{o}}}$$

This expression quantitates how much cell shrinkage will
occur in **hypertonic** solution ($V_{\mathrm{w}}/V_{\mathrm{w}}^{\mathrm{o}} < 1$ when $\pi_{\mathrm{o}} > \pi_{\mathrm{o}}^{\mathrm{o}}$),
and how much cell swelling will occur in **hypotonic** solution
($V_{\mathrm{w}}/V_{\mathrm{w}}^{\mathrm{o}} > 1$ when $\pi_{\mathrm{o}} < \pi_{\mathrm{o}}^{\mathrm{o}}$).

The osmotic coefficient of hemoglobin $\phi_{\mathrm{Hb}}$ rises approximately with the square of hemoglobin concentration (Fig.
6B; see also Ross and Minton, 1977), resulting in significant
deviations of red cells from ideal osmotic behavior (Dick,
1959; Freedman and Hoffman, 1979a). For an ideal os-
mometer, $\phi_{\mathrm{Hb}}$ and the osmotic coefficients of other solutes
would be unity. At a normal [Hb] of 7.3 mMolal, $\phi_{\mathrm{Hb}}$ is
2.85; in contrast, the osmotic coefficients of KCl and NaCl
decrease with increasing concentration and are 0.93 at
physiological concentrations, whereas that of sucrose rises
slightly with increasing concentration (see Fig. 6A). Mature
human red blood cells maintain stable volumes when
placed in isotonic solution, and rapidly adjust their volumes
to new stable levels that are maintained over time in hypotonic or hypertonic salt solutions (Fig. 7A). In contrast,
human reticulocytes and red cells from many other species,

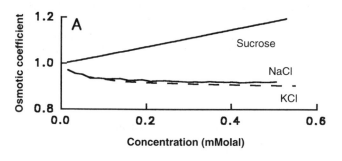

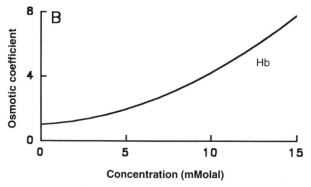

**FIG. 6.** Osmotic coefficients of NaCl, KCl, and sucrose (top panel)
and of hemoglobin (bottom panel). [Data for NaCl, KCl, and sucrose
are from the "CRC Handbook of Chemistry and Physics," 58th Ed.
(R. C. Weast and M. J. Astle, eds.), p. D-261, 1978–79, and from
R. A. Robinson and R. H. Stokes, "Electrolyte Solutions," 2nd Ed.,
Butterworths, London, 1959; data for hemoglobin is from Adair,
adapted from Fig. 7 of Freedman and Hoffman, *The Journal of General
Physiology*, 1979, Vol. 74, p. 177, by copyright permission of The
Rockefeller University Press, and is drawn according to $\phi_{\mathrm{Hb}} = 1 + 0.0645[\mathrm{Hb}] + 0.0258[\mathrm{Hb}]^2$.]

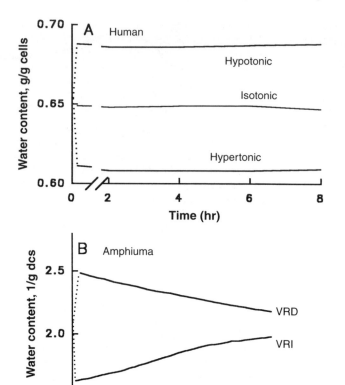

**FIG. 7.** (A) Stable osmotic response of human red blood cells.
Washed human red cells were incubated in hypotonic, isotonic, or
hypertonic media, and the water contents were determined at selected
times by drying packed cells to constant weight in a vacuum oven.
(Data of J. C. Freedman and S. Fazio.) (B) Volume regulatory decrease, VRD (upper trace), or volume regulatory increase, VRI (lower
trace), when *Amphiuma* red cells are incubated in hypotonic or hypertonic media, respectively. For red cells from *Amphiuma,* but not from
humans, the red-cell water content spontaneously returns back toward
its normal value. (Bottom panel adapted from Figs. 1 and 2 of P. M.
Cala, *The Journal of General Physiology*, 1980, Vol. 76, pp. 688–689,
by copyright permission of The Rockefeller University Press.)

including duck, dog, the salamander *Amphiuma*, trout,
skate, rabbit, and a low-$K^+$ genetic variant of sheep, all
possess the capability of altering their membrane permeability in response to an altered extracellular osmolarity
in such a way as to alter their total intracellular solute
concentration $N_{\mathrm{T}}$ and return their cell volumes back to
normal (for review, see Hoffmann and Dunham, 1995,
and Chapter 19 of this volume). Examples of the **volume
regulatory decrease (VRD)** and **volume regulatory increase (VRI)** exhibited by *Amphiuma* red blood cells are
shown in Fig. 7B. The swelling-activated response usually
involves an increase in $K^+$–$Cl^-$ cotransport, as in duck,
sheep, and rabbit red cells, but may also involve formation
of a channel related to band 3 that permits efflux of osmolytes such as taurine, as in red cells from skate and trout.
An exception seems to be net KCl efflux resulting from
activation of $K^+$–$H^+$ exchange coupled with $Cl^-$–$HCO_3^-$

exchange, as proposed for *Amphiuma* red cells. The shrinkage-activated response involves activation of $Na^+-K^+-2Cl^-$-cotransport in duck red cells, $Na^+-H^+$ countertransport in parallel with $Cl^--HCO_3^-$ exchange in *Amphiuma* red cells, and $Na^+-H^+$ exchange in dog red cells. The mechanisms by which cells "sense" their altered volume prior to switching on a transport pathway, and "sense" their normal volume when turning off a transport pathway, are the subjects of current active investigation. Thus, the amount of water in red blood cells is determined by the total concentration (osmolality) of salts (cations and anions), and by proteins and organic metabolites, as well as by the nonideal effects expressed by the osmotic coefficients of hemoglobin and other intracellular solutes.

An understanding of the **Gibbs–Donnan equilibrium** is critically important for properly incubating red cells and for understanding their response to altered pH and osmolarity. Human red blood cells, for example, swell in acid and shrink in alkaline media (Fig. 8). Red cells swell and hemolyze on exposure to compounds such as nystatin that elevate the normal low permeability of the membrane to cations, a phenomenon termed **colloid osmotic hemolysis** by Wilbrandt. As a human red blood cell swells, its membrane surface area remains constant as the biconcave disc gains water and converts to a spherical shape at the **critical hemolytic volume,** which is 1.65 times the original volume (for review, see Hoffman, 1992). The membrane, which has a surface tension of about 1 dyn/cm, but an area expansion modulus of ~450 dyn/cm, is unable to withstand more than about 4% stretching (see Chapter 15 by Berk, Hochmuth,

and Waugh in Agre and Parker, 1989), and consequently ruptures with further swelling. A Gibbs–Donnan equilibrium occurs whenever a membrane separates an aqueous salt solution from a salt solution that also contains impermeant charged electrolytes, such as intracellular charged proteins and organic phosphates (see Chapter 14). For human red cells, the permeability of the membrane to $Na^+$ and $K^+$ is 100 times less than that to $Cl^-$ and $HCO_3^-$, so that certain cellular responses can be understood by considering the cells to approximate a "**double-Donnan**" system. From the principle of bulk electroneutrality in the extracellular and intracellular solutions:

$$Cl_o^- = K_o^+ + Na_o^+$$

and

$$Cl_i^- = Na_i^+ + K_i^+ + zHb$$

where $z$ is the average net charge ($\mu$eq/mol Hb) on impermeant cell solutes including hemoglobin and organic phosphates. The **Donnan ratio** $r_{Cl}$ for $Cl^-$ is

$$r_{Cl} = \frac{Cl_i^-}{Cl_o^-} = \frac{Na_i^+ + K_i^+ + zHb}{K_o^+ + Na_o^+} = e^{FE_m/RT}$$

This expression illustrates most clearly that the double-Donnan equilibrium consists of one Donnan equilibrium being established by the slowly permeant ("functionally impermeant") cations, and the other by hemoglobin (Hb) and organic phosphates. The net charge on hemoglobin depends on intracellular pH as its constituent amino acids are titrated. The titration curve of hemoglobin is approximately given by

$$z = -m(pH_i - pI),$$

where $m$ is the slope of the titration curve, or **buffer capacity** ($m = -12$ eq/mol $Hb_4$/$pH_o$), and $pI$ is the **isoelectric point** of the cell contents, or the $pH_i$ where the net charge is zero ($pI = 6.8$ at 25°C in human red cells). In ectothermic vertebrates, maintaining the constancy of the charge $z$ on intracellular proteins may be more important for homeostasis than the constancy of extracellular pH (for review, see Reeves, 1977). If HCl is added to a suspension of red cells, the intracellular pH will decrease, hemoglobin will become more positively charged, and since $Na^+$ and $K^+$ are so much less permeant than $Cl^-$, the highly permeant anion will enter the cells to balance the increased positive charge on hemoglobin. Since the total intracellular solute concentration has now increased because of entry of $Cl^-$, the red cells swell (see Fig. 8). In contrast, if NaOH is added to a suspension of red cells, the intracellular pH increases, hemoglobin becomes more negatively charged, and $Cl^-$ leaves the cell, resulting in cell shrinkage (see Fig. 8). Model calculations show that at the isoelectric pH, the sum of the cation concentrations is greater in plasma than in the intracellular solution, in accordance with the double-Donnan concept; at the physiological pH of 7.4, however, cell shrinkage raises the intracellular cation concentration so that it nearly equals the plasma cation concentration, thus obscuring the double-Donnan effect. In human red cells, $NO_3^-$, $I^-$, $SCN^-$, and methane sulfonate ($CH_3SO_3^-$),

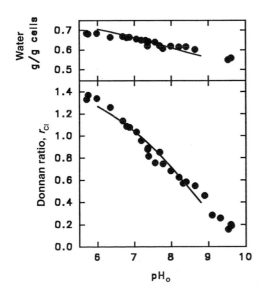

**FIG. 8.** Comparison of measured (points) and predicted (lines) Donnan ratios in human red blood cells. The equilibrium Donnan ratios for $Cl^-$ (bottom panel) and the cell water contents (top panel) were determined at varied extracellular pH. (Adapted from Fig. 3 of R. B. Gunn, M. Dalmark, D. C. Tosteson, and J. O. Wieth, *The Journal of General Physiology*, 1973, Vol. 61, p. 193, by copyright permission of The Rockefeller University Press.) The solid lines are predictions of the program IONIC. (Adapted from Fig. 3 of J. C. Freedman and J. F. Hoffman, *The Journal of General Physiology*, 1979, p. 174, with copyright permission of The Rockefeller University Press.)

but not methyl sulfate ($CH_3SO_4^-$), all bind to hemoglobin and alter its net charge, with a consequent small but significant change in cell volume in accordance with the Gibbs–Donnan equilibrium (Payne *et al.*, 1990). The Donnan equilibrium shifts in the expected manner on experimental alteration of intracellular phosphates (Duhm, 1971) and in abnormal human red cells containing HbC, which has an increased positive charge (Brugnara *et al.*, 1985).

Colloid osmotic hemolysis occurs when the red-cell membrane is exposed to pore-forming ionophores such as nystatin or gramicidin, or when the normal low permeability to cations increases on exposure to detergents or other chemicals. At equilibrium at the isoelectric pH, where the Donnan ratio equals unity and the membrane potential is zero, the intracellular concentrations of $Na^+$, $K^+$, and $Cl^-$ would equal those of the extracellular solution. The osmotic pressure of hemoglobin would be unbalanced by any extracellular solute and would draw water into the cell, leading to swelling and hemolysis. Colloid osmotic hemolysis occurs at any pH when the membrane is permeant both to cations and to anions. The $Na^+$ pump, working against the leakage pathways and any significant cotransport pathways, establishes the steady-state intracellular cation concentrations, while $Cl^-$ and $HCO_3^-$ move to maintain electroneutrality.

In red blood cells, for given steady-state concentrations of $K^+$ and $Na^+$, as set by the balance of pumps and leaks, the equilibrium cellular electrolyte concentrations and water contents are determined by the following three principles:

1. Bulk electroneutrality
2. Osmotic equality of the intracellular and extracellular solutions
3. Equality of the Donnan ratios for all permeant ions of the same charge

The Donnan ratio for monovalent anions is equal to the inverse of the Donnan ratio for protons or any other highly permeant monovalent cations. For human red blood cells, the nonideal equations expressing these constraints, including the activity and osmotic coefficients of the salts and of hemoglobin, have been formulated into a computer program designated IONIC (Freedman and Hoffman, 1979a; Bisognano *et al.*, 1993; see also Raftos *et al.*, 1990). For a given composition of the extracellular medium, and certain cellular parameters, the program predicts the cell volume, the intracellular pH and ion concentrations, the membrane potential, and the charge on hemoglobin and organic phosphates. The predictions of the model generally agree with experimental results within a few percent; for example, the predicted and experimentally determined dependence of the Donnan ratios for $Cl^-$ on pH are shown in Fig. 8. For a model describing blood acid–base status at constant or variable temperature, see Rodeau and Malan (1979). Beginning with a model of the Gibbs–Donnan equilibrium, Lew and Bookchin (1986) developed an integrated model of red-cell transport that includes kinetic parameters describing most of the known transport systems, and begins to describe the time course of the changes in cell volume, intracellular salt concentrations, and membrane potential that occur when the extracellular conditions are changed.

## VII. Anion Exchange and Conductance

The transport of carbon dioxide from the peripheral tissues to the lungs is facilitated by the production of bicarbonate by red blood cells in a cyclic reaction scheme known as the **Jacobs–Stewart cycle.** In systemic tissue capillaries, carbon dioxide diffuses easily across the red-cell membrane into the cells, whereupon **carbonic anhydrase (CA)** catalyzes its hydration to carbonic acid (Fig. 9). The acid then dissociates into a proton and bicarbonate. The proton is buffered by hemoglobin, which when protonated has a reduced affinity for oxygen, a phenomenon known as the **Bohr effect.** The bicarbonate passes across the red-cell membrane into the venous blood by undergoing a rapid electroneutral exchange with $Cl^-$, a reaction known as the **chloride shift.** A small proportion of carbon dioxide forms a covalent **carbamino** compound with amino groups on hemoglobin. Production of $HCO_3^-$ by red blood cells minimizes acidification of venous blood and increases its **carbon dioxide carrying capacity.** Red-cell $Cl^-$–$HCO_3^-$ exchange is mediated by the membrane domain of **capnophorin,** a protein also denoted as anion exchanger AE1. This integral glycoprotein, having a molecular weight of 101,700 Da, makes up 25% of the mass of membrane protein in red blood cells (Steck, 1974) and is the predominant protein located in band 3 of SDS polyacrylamide gels.

DIDS is the most potent of a series of stilbene derivatives originally found by Cabantchik and Rothstein to inhibit red-cell anion transport. DIDS inhibits 99.999% of $Cl^-$–$HCO_3^-$ exchange, but only 67% of net $Cl^-$ efflux. Molecular biological experiments have identified the location of the DIDS binding site. The cDNA from mice and humans that codes for band 3 protein (AE1) was cloned and sequenced (for review, see Alper, 1991). Band 3 protein has also been functionally expressed in *Xenopus laevis* toad oocytes microinjected with mRNA prepared from the human and mouse cDNA clones. Site-directed mutagenesis of the mouse protein indicates that one of the isothiocyanate (NCS) groups on DIDS binds covalently to Lys-558, which is located on the extracellular side of the 65-kDa chymotryptic N-terminal fragment. The stoichiometry of binding is one DIDS per capnophorin monomer (for review, see Passow *et al.*, 1992).

The membrane domain of capnophorin has been reconstituted with lipids and crystallized in two-dimensional arrays (Reithmeier, 1993; Wang *et al.*, 1994). At 20 Å resolution, reconstructed images indicate a dimeric structure with a cavity between the two monomers of band 3. [35]Cl nuclear magnetic resonance studies of capnophorin suggest an hourglass type of structure in which internal and external hemichannels lead to the transport site, which can exist in inward- and outward-facing conformations (Falke and Chan, 1986b). $Cl^-$–$HCO_3^-$ exchange follows Ping-Pong kinetics in which a conformational change occurs only after binding an anion alternately from the inward- and outward-facing conformations (for review, see Fröhlich and Gunn, 1986). DIDS lies in the outer hemichannel between the transport site and the extracellular medium, partially blocking the outward-facing transport site (Falke and Chan, 1986a). Structure–activity studies with a series of

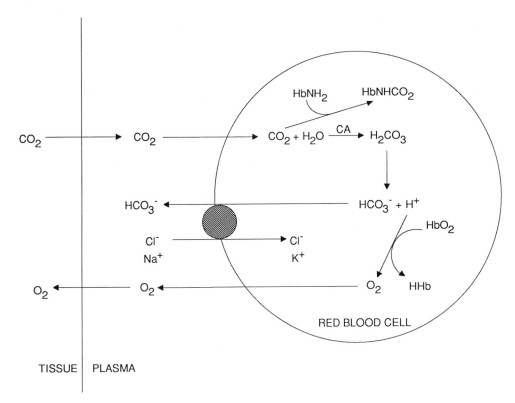

**FIG. 9.** Jacobs–Stewart cycle for red blood cells.

analogues of DIDS suggested a model of the DIDS binding site that included positively charged groups providing electrostatic stabilization of the sulfonates on DIDS, with adjacent hydrophobic and electron-donor centers. Experiments using the technique of fluorescence resonance energy transfer indicate that DIDS is located only 34–42 Å from sulfhydryl reagents bound to cysteine residues on the 40,000-Da amino-terminal cytoplasmic domain of capnophorin, a finding that is also consistent with DIDS residing in a cleft in the outer hemichannel. Moreover, substrate anions such as $Cl^-$ traverse only 10–15% of the full transmembrane electric field when they move from the extracellular medium and bind to the outward-facing transport site (Jennings et al., 1990). This observation is consistent with the hourglass model and indicates a low electrical resistance for the outer hemichannel, because if the resistance of the outer hemichannel were high, the transmembrane voltage would have dropped more than the 10–15% that was observed.

The classical view that the distribution of red-cell $Cl^-$ is simply determined by the "double-Donnan" equilibrium was consistent with the observations that $^{38}Cl$ equilibrates within 1 sec, which is $10^6$ times faster than $K^+$, and that the Donnan ratios for $Cl^-$, $HCO_3^-$, and $OH^-$ are all equal. The concept of rapid net movement of $Cl^-$ across the red-cell membrane changed dramatically when experiments with gramicidin and valinomycin revealed that $Cl^-$ conductance is about $10^4$ times slower than electroneutral $Cl^-$ exchange, yet still about $10^2$ times greater than the cation conductance. Thus net transport of cations normally occurs

over a time course of hours, net movement of $Cl^-$ occurs over minutes, and exchange of $Cl^-$ for $HCO_3^-$ occurs in seconds.

In addition to electroneutral anion exchange, capnophorin is also believed to mediate the net flux of $Cl^-$, which could contribute to loss of salt and water during pathophysiological dehydration of the red cell, such as occurs during the formation of irreversibly sickled cells. Studies have revealed DIDS-sensitive and DIDS-insensitive components of the net $Cl^-$ fluxes, and the DIDS-insensitive component increases with increasing extents of membrane hyperpolarization (Freedman et al., 1988, 1994; Freedman and Novak, 1997).

Because it has not proven possible to use microelectrodes to drive currents or to clamp the voltage across the membrane of human red cells, ionophores have been used instead to increase the permeability to cations, thus permitting ionic currents to flow under the influence of diffusion potentials, which can then be measured indirectly. Consider the equivalent circuit shown in Fig. 10. The concentration gradients of $K^+$, $Na^+$, and $Cl^-$ across the membrane are represented by batteries ($E_K$, $E_{Na}$, and $E_{Cl}$) in parallel with the membrane capacitance $C_m$. Each battery is in series with its respective ionic conductance ($g_K$, $g_{Na}$, and $g_{Cl}$). The electrical potential inside the cell is $E_m$, while that outside is at ground. Treatment of red cells with valinomycin, or with gramicidin in $Na^+$-free medium, increases the normally low value of $g_K$ above that of $g_{Cl}$ and allows the flow of the ionic currents, $i_K$ and $i_{Cl}$, around the loop (see Fig. 10, arrows). At low $[K^+]_o$ the net efflux of $K^+$

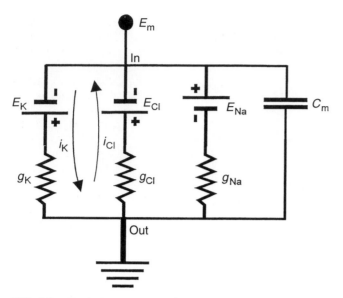

**FIG. 10.** Equivalent electrical circuit for human red blood cells treated with valinomycin, or with gramicidin in Na⁺-free medium. Arrows indicate the direction of positive current flow when the potassium conductance $g_K$ is increased by the addition of valinomycin, or of gramicidin in Na⁺-free medium.

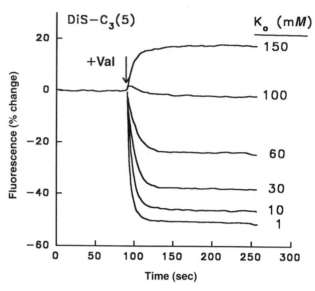

**FIG. 11.** Steady levels of fluorescence of diS-$C_3$(5) following the addition of valinomycin to suspensions of human red blood cells in $x$ m$M$ KCl (as indicated), $150 - x$ m$M$ NaCl, and 5 m$M$ Hepes buffer (pH 7.4 at 23°C). Downward deflections indicate hyperpolarization, whereas upward deflections indicate depolarization. [Adapted from Fig. 2 of Freedman, J. C., and Novak, T. S. (1989). *Methods Enzymol.*, **172**, 108.]

corresponds to the outward current $i_K$, which discharges the potassium battery. The magnitude of $i_K$ depends on $[K^+]_o$, which is easily varied. The efflux of K⁺ hyperpolarizes the membrane potential from its resting value near $-11$ mV to a value of some $-60$ mV, thus driving a net efflux of Cl⁻, corresponding to an inward current $i_{Cl}$ that charges the chloride battery. The K⁺ and Cl⁻ currents are opposite in sign and approximately equal in magnitude; a small disparity of less than 10% is accounted for by a proton current. Whereas the total net current across the membrane must be zero in the absence of an external circuit, the individual ionic currents induced by the addition of ionophores can be either measured directly or inferred from the rate of cell shrinkage. Fluorescent potentiometric indicators that monitor the voltages continuously show that the voltage remains at a steady value after the addition of valinomycin or gramicidin. Steady levels of dye fluorescence are seen with the oxacarbocyanine dye, diO-$C_6$(3) (Hoffman and Laris, 1974), or with the thiadicarbocyanine, diS-$C_3$(5) (Fig. 11), or indodicarbocyanine, diI-$C_3$(5), dyes (Freedman and Hoffman, 1979b; Freedman and Novak, 1989). Human red blood cells treated with valinomycin or with gramicidin thus behave as if voltage-clamped by their intrinsic ion concentration batteries, $E_K$ and $E_{Cl}$, instead of by an external circuit. This alternative voltage-clamp experiment yields current–voltage curves characterizing the conductance to Cl⁻.

The change in voltage upon addition of ionophores has been estimated from the change in the extracellular pH of unbuffered DIDS-treated red-cell suspensions (Macey *et al.*, 1978) using the proton ionophore FCCP, a method that is also useful for calibrating fluorescent potentiometric indicators (Freedman and Novak, 1989; Bifano *et al.*, 1984).

The inward-rectifying current–voltage curve found for the DIDS-insensitive fraction of Cl⁻ net transport (Fig. 12) deviates markedly from the nearly linear outward-rectifying curve predicted by the constant field theory for the inward concentration gradient of Cl⁻. The inward-rectifying current–voltage curve is also inconsistent with certain single-occupancy multiple barrier models, but instead is consistent either with a single barrier located near the center of the transmembrane electric field, or with a voltage-gated mechanism according to which half the channels

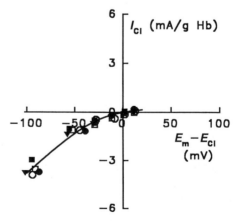

**FIG. 12.** Current–voltage curve for DIDS-insensitive Cl⁻ conductance of human red blood cells. The solid line represents the best fit either of a single barrier model, or of a mechanism involving voltage gating of ionic channels. (Adapted from Fig. 5 of J. C. Freedman and T. S. Novak, 1997, *The Journal of General Physiology*, Vol. 107, p. 207, with copyright permission of The Rockefeller University Press.)

are open at $-27$ mV, and with an equivalent gating charge of $-1.2 \pm 0.3$ (Freedman and Novak, 1997). Further experiments are needed to clarify the relationship of DIDS-insensitive Cl⁻ conductance inferred for red cells voltage-clamped with ionophores to the anion-selective channels recorded with the patch-clamp technique (Schwarz *et al.*, 1989), and to those seen after fusing vesicles from red-cell suspensions into planar lipid bilayers (Freedman and Miller, 1984). It is also important to know if the DIDS-sensitive fraction of Cl⁻ conductance is indeed mediated by the band 3 exchanger, in which case further studies could attempt to discover a unified kinetic scheme for exchange and conductance mediated by the same transporter.

## VIII. Cytotoxic Calcium Cascade

Despite the importance of $Ca^{2+}$ as a trigger and modulator in a variety of cell activities (see Chapter 1), too much intracellular $Ca^{2+}$ is harmful to cells in general and to red blood cells in particular. Elevated intracellular $Ca^{2+}$ stimulates a cytotoxic cascade of pathophysiological and biochemical changes. Some of these events are depicted in Fig. 13. Subjecting human red blood cells to shear stress increases the passive influx of $Ca^{2+}$ as seen in suspension (Larsen *et al.*, 1981), or during flow through synthetic microlattices (Brody *et al.*, 1995). When intracellular $Ca^{2+}$ rises to about 3 $\mu M$, the $K^+$ permeability dramatically increases, a phenomenon first described by Gardos and known as the **Gardos effect.** The increased $K^+$ permeability is due to the opening of a $Ca^{2+}$-activated $K^+$ channel, resulting in loss of intracellular $K^+$ and Cl⁻, accompanied by cell shrinkage (for review, see Schwarz and Passow, 1983). Elevated $[Ca^{2+}]_c$ also stimulates the $Ca^{2+}$-ATPase, causing

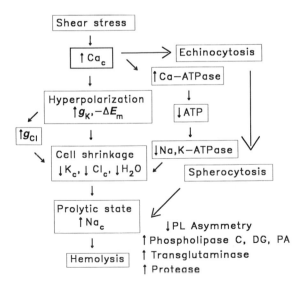

**FIG. 13.** The cytotoxic calcium cascade of human red blood cells. The scheme illustrates some of the effects of elevated intracellular $Ca^{2+}$. (Adapted from Fig. 3 of J. C. Freedman *et al.*, 1988, "Cell Physiology of Blood," p. 222, with copyright permission of The Rockefeller University Press.)

depletion of ATP within minutes (Crespo *et al.*, 1987), which in turn inhibits the $Na^+,K^+$-ATPase. In addition to causing substrate depletion, $Ca^{2+}$ also inhibits the $Na^+,K^+$-ATPase with a potency that depends on a regulatory factor (for review, see Yingst, 1988). At higher levels of intracellular $Ca^{2+}$, the smooth biconcave discoidal form of human red cells converts first to an echinocytic form with spicules protruding from the membrane, and then to a spherocytic form with release of microvesicles. ATP depletion is prevented by inhibiting the ATPases with vanadate, but echinocytosis still develops (Freedman *et al.*, 1988). The echinocytogenic effect of $Ca^{2+}$ is a striking example of the profound influence that $Ca^{2+}$ can exert on cell morphology. At still higher levels, $Ca^{2+}$ activates a transglutaminase that cross-links cytoskeletal proteins and reduces **cell deformability.** $Ca^{2+}$ also activates degradative proteases and a phospholipase C that converts membrane phosphatidylcholine (lecithin) to diacylglycerol.

The various effects of elevated intracellular $Ca^{2+}$ are induced at widely differing concentrations. Addition of A23187 stimulates loss of KCl at 0.1 $\mu M$ free external $Ca^{2+}$, and the membrane hyperpolarizes within 2 sec (Freedman and Novak, 1983). The threshold for $Ca^{2+}$-induced echinocytosis (assessed by the morphologic index) is two orders of magnitude greater, at 10 $\mu M$ $[Ca^{2+}]_f$. It is possible to set external $Ca^{2+}$ at a level sufficient to induce hyperpolarization, but below the threshold for echinocytosis, thus showing that the membrane potential does not directly influence cell shape (Bifano *et al.*, 1984). Under these conditions, the cells are able to crenate because simply increasing extracellular $Ca^{2+}$ produces echinocytes. Under similar conditions, the $Ca^{2+}$-induced decrease in cell deformability, as assessed by a decreased elongation index during flow at set shear stress (measured in an **Ektacytometer**), occurs at 1 m$M$ $Ca^{2+}$. The separate apparent potencies of $Ca^{2+}$, spanning five orders of magnitude, for inducing hyperpolarization, echinocytosis, and reduced deformability are consistent with the idea that $Ca^{2+}$ acts at independent sites to initiate these elements of the cytotoxic calcium cascade (Freedman *et al.*, 1988).

The relationships of the activation by $Ca^{2+}$ of a transglutaminase, a phospholipase C, and a protease to the loss of phospholipid asymmetry (see Lin *et al.*, 1994), and to the changes in permeability, cell volume, and morphology are complex. Lipid alterations, including phosphoinositide hydrolysis and phosphatidic acid production, occur at 1–10 $\mu M$ $Ca^{2+}$. Whereas $\mu M$ $Ca^{2+}$ dramatically affects the viscosity of complexes of spectrin, actin, and band 4.1 (Fowler and Taylor, 1980), the relative contributions of lipid and cytoskeletal alterations to the $Ca^{2+}$-induced shape changes are still unclear.

The final consequence of increased $[Ca^{2+}]_c$ *in vitro*, after prolonged cell shrinkage due to loss of KCl, is the development of prolytic cells leaky to Na that then undergo colloid osmotic hemolysis. Experiments suggest that cell shrinkage, rather than $Ca^{2+}$ itself, results in the gain of $Na^+$ (Crespo *et al.*, 1987). The observation that comparable gains of $Na^+$ occur in cells simply shrunken by sucrose, as upon exposure to valinomycin or to $Ca^{2+}$ plus calcium ionophore A23187 at low external $K^+$, suggests that $Ca^{2+}$

is not essential in causing elevated $[Na^+]_c$. Other flux studies and assessment of relative cell volume distributions indicate that a subpopulation of cells becomes highly permeable to $Na^+$ after shrinkage, induced either osmotically or by $Ca^{2+}$-induced loss of KCl. An apparent steady-state level of elevated $[Na]_c$ in the population of cells is attained because a fraction of the cells become prolytic while another fraction hemolyzes. An interesting question is what determines which cells enter the prolytic state.

A role of the cytotoxic calcium cascade in the formation of dehydrated irreversibly sickle cells (ISCs) is indicated by the high levels of intracellular $Ca^{2+}$ sequestered in sickle cells, and by the ability of charybdotoxin and clotrimazole, two blockers of $Ca^{2+}(K^+)$ channels, to inhibit the formation of ISCs (Ohnishi *et al.*, 1989; Brugnara *et al.*, 1996). It should be noted that $K^+$–$Cl^-$ cotransport has very low activity in normal mature red cells, but has elevated activity in immature reticulocytes and in sickle cells, and is also activated by acidosis (Brugnara *et al.*, 1989). The relative contributions of $K^+$–$Cl^-$ cotransport and net loss of $K^+$ and $Cl^-$ after activation of $Ca^{2+}(K^+)$ channels is an important question pertinent to the development of pharmacological interventions designed to ameliorate the painful vaso-occlusive crises suffered by patients with sickle cell disease.

## IX. SUMMARY

The availability, structural and metabolic simplicity, and ease of manipulating the intracellular as well as the extracellular solutions of red blood cells have made them a favorite subject for the study of membrane transport. The plasma membrane and underlying cytoskeleton have been extensively characterized. A flipase that transports phospholipids from the outer to the inner hemileaflet contributes to phospholipid asymmetry in the membrane. The high concentration of hemoglobin, nearly close-packed and hydrated with a large volume fraction of intracellular water, contributes to a high intracellular ionic strength. Curiously, the intracellular activities of KCl, NaCl, and nonelectrolytes appear to be nearly ideal. Red-cell metabolism supplies ATP as substrate for the $Na^+,K^+$-ATPase and $Ca^{2+}$-ATPase ionic pumps and prevents the oxidation of $Fe^{2+}$ in hemoglobin. The life spans of red blood cells from various mammalian species correlate with body mass, suggesting the existence of a biological clock that is obviously independent of gene transcription or translation in the mature nonnucleated cells.

The red blood cells of all species normally maintain a submicromolar intracellular concentration of $Ca^{2+}$, as set by the $Ca^{2+}$-ATPase pumping against a small passive inward leakage. The red blood cells of most species have high steady-state concentrations of $K^+$ but low $Na^+$, as set by a balance between pumping by the $Na^+,K^+$-ATPase and parallel leaks. An exception is the red cells of carnivores, which, unlike other red cells, have high intracellular $Na^+$ and low intracellular $K^+$. These cells lack a $Na^+,K^+$-ATPase, but do have a $Na^+$–$Ca^{2+}$ exchanger that couples the efflux of $Na^+$ to the influx of $Ca^{2+}$. Whereas human red blood cells are unable to regulate their volume in

anisotonic media, the red cells of many other species use the gradients of $K^+$ and $Na^+$ and activate their $K^+$–$Cl^-$ and $Na^+$–$K^+$–$2Cl^-$ cotransporters to return their volume to normal in hypotonic and hypertonic media, respecitvely. Alternatively, in red cells of trout and skate, an efflux of the osmolyte taurine is used in cell volume regulation.

In human red blood cells, the conductive permeability to $Cl^-$ is 100 times greater than that to $K^+$ or $Na^+$. Consequently, the membrane potential is consistent with the $Cl^-$ equilibrium potential of $-11$ mV, and with a constant field permeability ratio $P_K/P_{Na}$ of 1.5. The electrogenic potential of the $Na^+,K^+$-ATPase is normally less than 1 mV. Away from the steady state, $Cl^-$ will dominate the membrane potential. Treatment of human red blood cells with valinomycin, or with gramicidin in $Na^+$-free media, results in clamping of the voltage by the intrinsic ion concentration batteries. Analysis of individual ionic currents flowing under the influence of diffusion potentials yields current–voltage curves that characterize the conductance to $Cl^-$. An important problem for future study is to understand the "channel-like" properties of exchangers and pumps.

According to the Boyle–van't Hoff law, the total intracellular and extracellular osmolalities are equal at osmotic equilibrium. Significant deviations from osmotic ideality are due to the osmotic coefficient of Hb rising with the square of Hb concentration. For given cation contents of human red cells, the double-Donnan equilibrium describes the passive distribution of $Cl^-$ and water across the red-cell membrane and accounts quantitatively for the effects of pH on cell swelling and shrinkage, as well as for colloid osmotic hemolysis. Computer models are available for (1) red-cell glycolysis, (2) the nonideal double-Donnan equilibrium, (3) acid–base balance, and (4) the kinetics of membrane transport. These models begin to enable prediction of the time course of changes in the cell volume, the intracellular pH, the membrane potential, and the intracellular salt concentrations when the extracellular conditions are altered.

When intracellular $Ca^{2+}$ rises too much for the $Ca^{2+}$-ATPase to maintain the steady state, a cytotoxic calcium cascade leads to depletion of ATP, inhibition of the $Na^+,K^+$-ATPase, activation of $Ca^{2+}(K^+)$ channels, cellular dehydration, loss of phospholipid asymmetry, echinocytosis, membrane vesiculation, activation of phospholipase C, transglutaminase, and protease, and hemolysis. Blockers of $Ca^{2+}(K^+)$ channels reduce the formation of dehydrated, irreversibly sickled cells both *in vitro* and *in vivo* and are currently being tested as a new treatment for sickle cell disease.

### Acknowledgment

The author thanks Dr. Philip B. Dunham of the Biology Department, Syracuse University, for reading this chapter.

## Bibliography

Agre, P. (1996). Pathophysiology of the aquaporin water channels. *Annu. Rev. Physiol.* **58,** 619–648.

Agre, P., and Parker, J. C. (Eds.). (1989). "Red Blood Cell Membranes: Structure, Function, Clinical Implications." Marcel Dekker, New York.

Alper, S. L. (1991). The band 3-related anion exchanger (AE) gene family. *Annu. Rev. Physiol.* **53,** 549–564.

Antonini, E., and Brunori, M. (1971). "Hemoglobin and Myoglobin in Their Reaction with Ligands." North-Holland, Amsterdam.

Bennett, V., and Gilligan, D. M. (1993). The spectrin-based membrane skeleton and micron-scale organization of the plasma membrane. *Annu. Rev. Cell Biol.* **9,** 27–66.

Bessis, M. (1974). "Corpuscles. Atlas of Red Blood Cell Shapes." Springer-Verlag, New York.

Bessis, M., Weed, R. I., and Leblond, P. F. (Eds.). (1973). "Red Cell Shape. Physiology, Pathology, Ultrastructure." Springer-Verlag, New York.

Beutler, E. (1986). "Red Cell Metabolism, Vol. 16. Methods in Hematology." Churchill Livingstone, New York.

Bifano, E. M., Novak, T. S., and Freedman, J. C. (1984). The relationship between the shape and the membrane potential of human red blood cells. *J. Membr. Biol.* **82,** 1–13.

Bishop, C., and Surgenor, D. M. (1964). "The Red Blood Cell. A Comprehensive Treatise." Academic Press, New York.

Bisognano, J. D., Dix, J. A., Pratap, R. R., Novak, T. S., and Freedman, J. C. (1993). Proton (or hydroxide) fluxes and the biphasic osmotic response of human red blood cells. *J. Gen. Physiol.* **102,** 99–123.

Brody, J. P., Han, Y., Austin, R. H., and Bitensky, M. (1995). Deformation and flow of red blood cells in a synthetic lattice: Evidence for an active cytoskeleton. *Biophys. J.* **68,** 2224–2232.

Brugnara, C., Kopin, A. S., Bunn, H. F., and Tosteson, D. C. (1985). Regulation of cation content and cell volume in hemoglobin erythrocytes from patients with homozygous hemoglobin C disease. *J. Clin. Invest.* **75,** 1608–1617.

Brugnara, C., Ha, T. V., and Tosteson, D. C. (1989). Acid pH induces formation of dense cells in sickle erythrocytes. *Blood* **232,** 487–495.

Brugnara, C., Gee, B., Armsby, C. C., Kurth, S., Sakamoto, M., Rifai, N., Alper, S. L., and Platt, O. S. (1996). Therapy with oral clotrimazole induces inhibition of the Gardos channel and reduction of erythrocyte dehydration in patients with sickle cell disease. *J. Clin. Invest.* **97,** 1227–1234.

Crespo, L. M., Novak, T. S., and Freedman, J. C. (1987). Calcium, cell shrinkage, and prolytic state of human red blood cells. *Am. J. Physiol.* **252** (*Cell Physiol.* **21**), C138–C152.

Davson, H., and Danielli, J. F. (1943). "The Permeability of Natural Membranes." Cambridge University Press, Cambridge, UK.

DeFelice, L. J., and Blakely, R. D. (1996). Pore models for transporters? *Biophys. J.* **70,** 579–580.

Diaz, C., and Schroit, A. J. (1996). Role of translocases in the generation of phosphatidylserine asymmetry. *J. Membr. Biol.* **151,** 1–9.

Dick, D. A. T. (1959). Osmotic properties of living cells. *Int. Rev. Cytol.* **8,** 387–448.

Dissing, S., and Hoffman, J. F. (1990). Anion-coupled Na efflux mediated by the human red blood cell Na/K pump. *J. Gen. Physiol.* **96,** 167–193.

Dodge, J. T., Mitchell, C., and Hanahan, D. J. (1963). The preparation and chemical characteristics of hemoglobin-free ghosts of human erythrocytes. *Arch. Biochem. Biophys.* **100,** 119–130.

Duhm, J. (1971). Effects of 2,3-diphosphoglycerate and other organic phosphate compounds on oxygen affinity and intracellular pH of human erythrocytes. *Pflügers Arch.* **326,** 341–356.

Duhm, J., and Becker, B. F. (1977). Studies on the lithium transport across the red cell membrane. IV. Interindividual variations in the Na$^+$-dependent Li$^+$ countertransport system of human erythrocytes. *Pflügers Arch.* **370,** 211–219.

Ellory, J. C., and Lew, V. L. (1977). "Membrane Transport in Red Cells." Academic Press, New York.

Ellory, J. C., and Young, J. D. (1982). "Red Cell Membranes—A Methodological Approach." Biological Techniques Series. Academic Press, New York.

Escobales, N., and Canessa, M. (1986). Amiloride-sensitive Na$^+$ transport in human red cells: Evidence for a Na/H exchange system. *J. Membr. Biol.* **90,** 21–28.

Falke, J. J., and Chan, S. I. (1986a). Molecular mechanism of band 3 inhibitors. I. Transport site inhibitors. *Biochemistry* **25,** 7888–7894.

Falke, J. J., and Chan, S. I. (1986b). Molecular mechanism of band 3 inhibitors. 2. Channel blockers. *Biochemistry* **25,** 7895–7898.

Fowler, V., and Taylor, D. L. (1980). Spectrin plus band 4.1 cross-link actin. Regulation by micromolar calcium. *J. Cell Biol.* **85,** 361–376.

Freedman, J. C. (1983). Partial requirements for *in vitro* survival of human red blood cells. *J. Membr. Biol.* **75,** 225–231.

Freedman, J. C., and Hoffman, J. F. (1979a). Ionic and osmotic equilibria of human red blood cells treated with nystatin. *J. Gen. Physiol.* **74,** 157–185.

Freedman, J. C., and Hoffman, J. F. (1979b). The relation between dicarbocyanine dye fluorescence and the membrane potential of human red blood cells set at varying Donnan equilibria. *J. Gen. Physiol.* **174,** 187–212.

Freedman, J. C., and Miller, C. (1984). Membrane vesicles from human red blood cells in planar lipid bilayers. *Ann. NY Acad. Sci.* **435,** 541–544.

Freedman, J. C., and Novak, T. S. (1983). Membrane potentials associated with Ca-induced K conductance in human red blood cells. Studies with a fluorescent oxonol dye, WW781. *J. Membr. Biol.* **72,** 59–74.

Freedman, J. C., and Novak, T. S. (1989). Optical measurements of membrane potential in cells, organelles, and vesicles. *Methods Enzymol.* **172,** 102–122.

Freedman, J. C., and Novak, T. S. (1997). Electrodiffusion, barrier, and gating analysis of DIDS-insensitive chloride conductance in human red blood cells treated with valinomycin or gramicidin. *J. Gen. Physiol.* **109,** 201–216.

Freedman, J. C., Bifano, E. M., Crespo, L. M., Pratap, P. R., Wallenga, R., Bailey, R. E., Zuk, S., and Novak, T. S. (1988). Membrane potential and the cytotoxic Ca cascade of human red blood cells. *In* "Cell Physiology of Blood" (R. B. Gunn and J. C. Parker, Eds.), pp. 218–231. Society of General Physiologists Series, The Rockefeller University Press, New York.

Freedman, J. C., Novak, T. S., Bisognano, J. D., and Pratap, P. R. (1994). Voltage dependence of DIDS-insensitive chloride conductance in human red blood cells treated with valinomycin or gramicidin. *J. Gen. Physiol.* **104,** 961–983.

Fröhlich, O., and Gunn, R. B. (1986). Erythrocyte anion transport: The kinetics of a single-site obligatory exchange system. *Biochim. Biophys. Acta* **864,** 169–194.

Gary-Bobo, C. M. (1967). Nonsolvent water in human erythrocytes and hemoglobin solutions. *J. Gen. Physiol.* **50,** 2547–2564.

Golan, D. E., and Veatch, W. (1980). Lateral mobility of band 3 in the human erythrocyte membrane studied by fluorescence photobleaching recovery: Evidence for control by cytoskeletal interactions. *Proc. Natl. Acad. Sci. USA* **77,** 2537–2541.

Grimes, A. C. (1980). "Human Red Cell Metabolism." Blackwell, Oxford.

Harris, J. W., and Kellermeyer, R. W. (1970). "The Red Cell. Production, Metabolism, Destruction: Normal and Abnormal," rev. ed. Harvard University Press, Cambridge, MA.

Henderson, L. J. (1928). "Blood. A Study in General Physiology." Yale University Press, New Haven, CT.

Hoffman, J. F. (1986). Active transport of Na$^+$ and K$^+$ by red blood cells. *In* "Physiology of Membrane Disorders," 2nd ed. (T. E. Andreoli, J. F. Hoffman, D. F. Fanestil, and S. G. Schultz, Eds.), pp. 221–231. Plenum Press, New York.

Hoffman, J. F. (1992). On red blood cells, hemolysis and resealed ghosts. *In* "The Use of Resealed Erythrocytes as Carriers and Bioreactors" (M. Magnani and J. R. DeLoach, Eds.). *Adv. Exper. Biol. Med.,* **326,** 1–15. Plenum, New York.

Hoffman, J. F., and Laris, P. C. (1974). Determination of membrane potential in human and *Amphiuma* red blood cells by means of a fluorescent probe. *J. Physiol. (London)* **239,** 519–552.

Hoffman, J. F., Kaplan, J. H., Callahan, T. J., and Freedman, J. C. (1980). Electrical resistance of the red cell membrane and the relation between net anion transport and the anion exchange mechanism. *Ann. NY Acad. Sci.* **341,** 357–360.

Hoffmann, E. K., and Dunham, P. B. (1995). Membrane mechanisms and intracellular signalling in cell volume regulation. *Int. Rev. Cytol.* **161,** 173–262.

Jacobs, M. H. (1962). Early osmotic history of the plasma membrane. *In* "Symposium on the Plasma Membrane, New York Heart Association, Inc." *Circulation* **26,** 1013–1021.

Jacobs, M. H., and Stewart, D. R. (1947). Osmotic properties of the erythrocyte. XII. Ionic and osmotic equilibria with a complex external solution. *J. Cell Comp. Physiol.* **30,** 79–103.

Jennings, M. L. (1992). Inorganic anion transport. *In* "The Structure of Biological Membranes" (P. Yeagle, Ed.), pp. 781–832. CRC Press, Boca Raton, FL.

Jennings, M. L., Schulz, R. K., and Allen, M. (1990). Effects of membrane potential on electrically silent transport. Potential-independent translocation and asymmetric potential-dependent substrate binding to the red blood cell anion exchange protein. *J. Gen. Physiol.* **96,** 991–1012.

Kay, M. M. B. (1991). *Drosophila* to bacteriophage to erythrocyte: The erythrocyte as a model for molecular and membrane aging of terminally differentiated cells. *Gerontology* **37,** 5–32.

Knauf, P. A. (1979). Erythrocyte anion exchange and the band 3 protein: Transport kinetics and molecular structure. *Curr. Top. Membr. Transp.* **12,** 249–363.

Kuhlman, P. A., Hughes, C. A., Bennett, V., and Fowler, V. M. (1996). A new function for adducin. Calcium/calmodulin-regulated capping of the barbed ends of actin filaments. *J. Biol. Chem.* **271,** 7986–7991.

Larsen, R. L., Katz,S., Roufogalis, B. D., and Brooks, D. E. (1981). Physiological shear stresses enhance the $Ca^{2+}$ permeability of human erythrocytes. *Nature* **294,** 667–668.

Lauf, P. K., Bauer, J., Adragna, N. C., Fujise, H., Zade-Oppen, A. M. M., Ryu, K. H., and Delpire, E. (1992). Erythrocyte K–Cl cotransport: Properties and regulation. *Am. J. Physiol.* **263** (*Cell Physiol.* **32**), C917–C932.

Läuger, P. (1995). Conformational transitions of ionic channels. *In* "Single Channel Recording," 2nd ed. (B. Sakmann and E. Neher, Eds.), ch. 22, pp. 651–662. Plenum, New York.

LeFevre, P. G. (1961). Sugar transport in the red blood cell: Structure–activity relationships in substrates and antagonists. *Pharm. Rev.* **13,** 39–70.

Lew, V. L., and Bookchin, R. M. (1986). Volume, pH and ion content regulation in human red cells: Analysis of transient behavior with an integrated model. *J. Membr. Biol.* **92,** 57–74.

Lew, V. L., Hockaday, A., Sipulveda, M.-I., Somlyo, A. P., Somlyo, A. V., Ortiz, O. E., and Bookchin, R. M. (1985). Compartmentalization of sickle-cell calcium in endocytic inside-out vesicles. *Nature* **315,** 586–589.

Lin, S., Yang, E., and Huestis, W. H. (1994). Relationship of phospholipid distribution to shape change in $Ca^{2+}$-crenated and recovered human erythrocytes. *Biochemistry* **33,** 7337–7344.

Macey, R. I., Adorante, J. S., and Orme, F. W. (1978). Erythrocyte membrane potentials determined by hydrogen ion distribution. *Biochim. Biophys. Acta* **512,** 284–295.

Mercer, R. W., and Dunham, P. B. (1981). Membrane-bound ATP fuels the Na/K pump. *J. Gen. Physiol.* **78,** 547–568.

Milanick, M., and Hoffman, J. F. (1986). Ion transport and volume regulation in red blood cells. *Ann. NY Acad. Sci.* **488,** 174–186.

Murphy, E., Berkowitz, L. R., Orringer, E., Levy, L., Gabel, S. A., and London, R. E. (1987). Cytosolic free calcium levels in sickle red blood cells. *Blood* **69,** 1469–1474.

Ohnishi, S. T., and Ohnishi, T. (1994). "Membrane Abnormalities in Sickle Cell Disease and in Other Red Blood Cell Disorders." CRC Press, Boca Raton, FLA.

Ohnishi, S. T., Katagi, H., and Katagi, C. (1989). Inhibition of the *in vitro* formation of dense cells and of irreversibly sickled cells by charybdotoxin, a specific inhibitor of calcium-activated potassium efflux. *Biochim. Biophys. Acta* **1010,** 199–203.

Passow, H. (1986). Molecular aspects of band 3 protein-mediated anion transport across the red blood cell membrane. *Rev. Physiol. Biochem. Pharmacol.* **103,** 61–203.

Passow, H., Wood, P. G., Lepke, S., Müller, H., and Sovak, M. (1992). Exploration of the functional significance of the stilbene disulfonate binding site in mouse band 3 by site-directed mutagenesis. *Biophys. J.* **62,** 98–100.

Payne, J. A., Lytle, C., and McManus, T. J. (1990). Foreign anion substitution for chloride in human red blood cells: Effect on ionic and osmotic equilibria. *Am. J. Physiol.* **259** (*Cell Physiol.* **28**), C819–C827.

Penniston, J. T., Vaughan, L., and Nakamura, M. (1979). Endocytosis in erythrocytes and ghosts: Occurrence at 0°C after ATP preincubation. *Arch. Biochem. Biophys.* **198,** 339–348.

Ponder, E. (1948). "Hemolysis and Related Phenomena." Grune & Stratton, New York.

Proverbio, F., Shoemaker, D. G., and Hoffman, J. F. (1988). Functional consequences of the membrane pool of ATP associated with the human red blood cell Na/K pump. *In* "The $Na^+,K^+$-Pump, Part A: Molecular Aspects" (J. C. Skou, J. G. Norby, A. B. Maunsbach, and M. Esmann, Eds.). *Prog. Clin. Biol. Res.* **268A,** 561–567, Alan R. Liss, New York.

Raess, B. U., and Tunnicliff, G. (Eds.). (1990). "The Red Cell Membrane: A Model for Solute Transport." Humana Press, Clifton, NJ.

Raftos, J. E., Bulliman, B. T., and Kuchel, P. W. (1990). Evaluation of an electrochemical model of erythrocyte pH buffering using $^{31}P$ nuclear magnetic resonance data. *J. Gen. Physiol.* **95,** 1183–1204.

Rapoport, T. A., Heinrich, R., Jacobasch, G., and Rapoport, S. (1974). A linear steady-state treatment of enzymatic chains. A mathematical model of glycolysis of human erythrocytes. *Eur. J. Biochem.* **42,** 107–120.

Reeves, R. B. (1977). The interaction of body temperature and acid–base balance in ectothermic vertebrates. *Annu. Rev. Physiol.* **39,** 559–586.

Reithmeier, R. A. F. (1993). The erythrocyte anion transporter (band 3). *Curr. Opin. Struct. Biol.* **3,** 515–523.

Rodeau, J., and Malan, A. (1979). A two-compartment model of blood acid–base state at constant or variable temperature. *Resp. Physiol.* **36,** 5–30.

Ross, P. D., and Minton, A. P. (1977). Analysis of non-ideal behavior in concentrated hemoglobin solutions. *J. Mol. Biol.* **112,** 437–452.

Salhany, J. (Ed.). (1989). "Erythrocyte Band 3 Protein." CRC Press, Boca Raton, FL.

Schwarz, W., and Passow, H. (1983). $Ca^{2+}$-activated $K^+$ channels in erythrocytes and excitable cells. *Annu. Rev. Physiol.* **45,** 359–374.

Schwarz, W., Grygorczyk, R., and Hof, D. (1989). Recording single-channel currents from human red cells. *Methods Enzymol.* **173,** 112–121.

Shohet, S. B. (1976). Mechanisms of red cell membrane lipid renewal. *In* "Membranes and Disease" (L. Bolis, J. F. Hoffman, and A. Leaf, Eds.), pp. 61–74. Raven Press, New York.

Shohet, S. B., and Mohandas, N. (Eds.). (1988). "Red Cell Membranes, Vol. 19, Methods in Hematology." Churchill Livingstone, New York.

Steck, T. L. (1974). The organization of proteins in the human red blood cell membrane. A review. *J. Cell Biol.* **62,** 1–19.

Surgenor, D. MacN. (Ed.). (1974). "The Red Blood Cell," 2nd ed., Vols. I and II. Academic Press, New York.

Tosteson, D. C., and Hoffman, J. F. (1960). Regulation of cell volume by active cation transport in high and low potassium sheep red cells. *J. Gen. Physiol.* **44,** 169–194.

Tosteson, M. T., Halperin, J. A., Kishi, Y., and Tosteson, D. C. (1991). Palytoxin induces an increase in the cation conductance of red cells. *J. Gen. Physiol.* **98,** 969–985.

Vácha, J., and Znojil, V. (1981). The allometric dependence of the life span of erythrocytes on body weight in mammals. *Comp. Biochem. Physiol.* **69A,** 357–362.

Van Slyke, D. D., Wu, H., and McLean, F. C. (1923). Studies of gas and electrolyte equilibria in the blood. V. Factors controlling the electrolyte and water distribution in the blood. *J. Biol. Chem.* **56,** 765–849.

Wang, D. N., Sarabia, V. E., Reithmeier, R. A. F., and Kühlbrandt, W. (1994). Three-dimensional map of the dimeric membrane domain of the human erythrocyte anion exchanger, Band 3. *EMBO J.* **13,** 3230–3235.

Whittam, R. (1964). "Transport and diffusion in red blood cells." Williams & Wilkins, Baltimore.

Willis, J. S. (1992a). Symposium on diversity of membrane cation transport in vertebrate red blood cells. An overview. *Comp. Biochem. Physiol.* **102A,** 595–596.

Yingst, D. R. (1988). Modulation of the Na,K-ATPase by Ca and intracellular proteins. *Annu. Rev. Physiol.* **50,** 291–303.

Yoshikawa, H., and Rapoport, S. M. (1974). "Cellular and Molecular Biology of Erythrocytes." University Park Press, Baltimore.

Zade-Oppen, A. M., Adragna, N. C., and Tosteson, D. C. (1988). Effects of pH, potential, chloride and furosemide on passive $Na^+$ and $K^+$ effluxes from human red blood cells. *J. Membr. Biol.* **102,** 217–225.

# SECTION
# III

# Membrane Excitability and Ion Channels

*Nicholas Sperelakis*

# 23

## Cable Properties and Propagation of Action Potentials*

### I. Introduction

The resting potential (RP) of cells enables the electrogenesis of action potentials (APs) and excitability. In this chapter, we examine the mechanism for propagation of the APs and excitability from one part of a neuron or muscle to a distal part.

It is imperative that the body be able to transmit a signal from one point to another very rapidly. The only way that this can be accomplished is by an electrical mechanism. Blood flow and diffusion and signaling molecules are much too slow to allow rapid signaling. In contrast, electricity flows very quickly, at the speed of light ($3 \times 10^8$ m/sec) in a copper wire or about one-ninth the speed of light in a water solution (like the composition of the body). Therefore, the body makes use of electricity for rapid signaling in the nervous system, skeletal muscle, heart, and smooth muscles. Propagation velocity is about 120 m/sec in our fastest nerve fibers, about 6 m/sec in skeletal muscle, about 0.5 m/sec in heart, and about 0.05 m/sec in smooth muscle (Table 1).

One example of the need for very fast communication or signaling is the process of walking. Very rapid signals must travel from the motor cortex of the brain, down to the lower spinal cord region, and out the motor axons to the skeletal muscles of the lower extremities (Fig. 1). In this process, the signal crosses one or more synapses, which are regions in which one neuron ends and the next one begins, and in which a special chemical neurotransmitter signal is involved (see Chapter 40) (Fig. 2). At the termination of each branch of a motor nerve axon on the skeletal

muscle fiber, there is another synapse, known as the neuromuscular junction or motor end plate (see Fig. 1). The signal crosses the neuromuscular junction and gives rise to an AP in the muscle fiber that propagates in both directions from the motor end plate. The muscle AP elicits contraction. Receptors in the muscles (e.g., stretch receptors) transmit information (in the form of propagating APs) back into the central nervous system (CNS). Thus, in walking, there is a continuous rapid flow of information and instructions to the muscles in both directions: out of the CNS and into the CNS. Therefore, even a relatively simple skeletal activity such as walking would not be possible without a very rapid signaling system. To illustrate, in various demyelinating diseases (e.g., caused by some viruses, heavy metals, autoimmune reactions), loss of the myelin sheath around the myelinated nerve fibers causes propagation to become slowed and impaired in the affected nerve fibers, with associated uncoordination and partial paralysis.

### II. Frequency-Modulated Signals

Because propagating all-or-none APs are all very similar to each other (in shape, duration, amplitude, rate of rise, and propagation velocity), to make the signal stronger or weaker, the body increases or decreases, respectively, the frequency of the APs. That is, the body uses a frequency-modulated (FM) system, rather than an amplitude-modulated (AM) system (see Fig. 2). It is a digital system composed of on–off identical signals. At each synapse, the signal becomes graded in amplitude rather than all-or-none: The greater the amplitude and duration of the local postsynaptic potential, the higher the frequency of APs triggered. The same is true of the local graded receptor potential generated at some sensory organs/receptors. Stronger signals translate into a higher frequency of im-

---

* Adapted and reprinted by permission from the author's chapter 5 in PHYSIOLOGY (Sperelakis, N. and Banks, R. O., Editors). Copyright © 1993 by Nicholas Sperelakis and Robert O. Banks. Published by Little, Brown and Company.

**TABLE I**  Conduction Velocity as a Function of Fiber
Diameter in Nerve Axons and Muscle Fibers

| Fiber type | Fiber diameter ($\mu$m) | Propagation velocity (m/sec) | Velocity/ diameter (m/sec/$\mu$m) |
|---|---|---|---|
| Myelinated axons | 20 | 120 | 6.0 |
|  | 12 | 70 | 5.8 |
|  | 5 | 30 | 6.0 |
| Nonmyelinated axons | 1.5 | 2.0 | 1.3 |
|  | 1.0 | 1.3 | 1.3 |
| Squid giant axons (20°C) | 500 | 25 | 0.05 |
| Skeletal muscle fibers | 50 | 6 | 0.12 |
| Cardiac muscle fibers | 15 | 0.5 | 0.03 |
| Smooth muscle fibers | 5 | 0.05 | 0.01 |

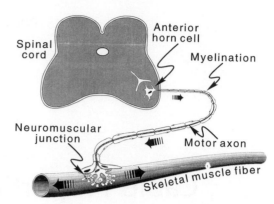

**FIG. 1.**  Schematic diagram of a motor axon, with the cell body
(soma) in the anterior horn of the spinal cord and its terminal branches
ending on skeletal muscle fibers (only one muscle fiber depicted) to
form the neuromuscular junctions (motor end plates). Each motor
axon with its attached skeletal muscle fibers is known as the motor
unit. The motor axons are of large diameter (e.g., 20 $\mu$m) and are
myelinated, and therefore propagate at fast velocities (e.g., 120
m/sec). As a consequence of the chemical synaptic transmission process at the neuromuscular junction, an AP is initiated in the muscle
fiber and propagates in both directions, bringing about contraction.

pulses, and weaker signals correspond to a lower frequency
of impulses.

## III. Cable Properties

### A. Biological Fiber as a Cable

An electrical cable consists of two parallel conductors
separated by insulation material, for example, two copper
wires separated by rubber. Usually one of the conductors
is arranged as a tubular sleeve surrounding a central solid
rod (wire), as depicted in Fig. 3A. The equivalent electrical
circuit for a cable is shown in Fig. 3B: two parallel conductors (wires) separated by a transverse resistance ($R$) shown
distributed along the length of the cable. The resistance
of the conductors is so small compared with the transverse
insulation resistance that it is assumed to be zero. In the
case of the biological cable (a long narrow nerve fiber or
skeletal muscle fiber), one parallel conductor is the inside
fluid (cytoplasm) and the second parallel conductor is the

outside fluid surrounding (bathing) the cell (the interstitial
fluid). Because the conductivity of biological fluid is much
less (i.e., much higher resistance) than that of copper wire,
and because the cross-sectional area of the cell is so small,
the inside longitudinal resistance is high and cannot be
ignored (Fig. 3C). The outside longtudinal resistance is
relatively small, as compared with the inside, because of
the larger volume (cross-sectional area) of fluid available
to carry the outside current, and therefore is assumed to
be negligible.

In addition, there is a stray capacitance distributed along
the length of the cable (Fig. 3D), because a capacitance

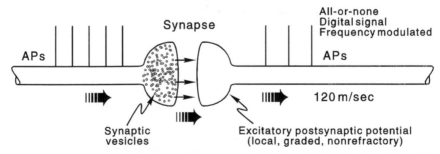

**FIG. 2.**  Diagram of a chemical excitatory synapse between two nerve fibers. An AP in the presynaptic fiber (left) brings
about the release of the neurotransmitter at its nerve terminal. The transmitter molecules rapidly diffuse the short distance
(e.g., 0.1 $\mu$m) across the synaptic cleft, bind to receptor sites on the postsynaptic membrane, and open the associated
nonselective ion channels ($Na^+, K^+$ mixed conductance) complexed to the receptor (i.e., these are ligand-gated ion channels).
The associated synaptic current depolarizes the postsynaptic membrane, producing the excitatory postsynaptic potential
(EPSP); this depolarization spreads passively into the adjacent conductile membrane (excitable) because of cable properties,
thereby triggering one or more APs in the postsynaptic axon. The EPSPs are local, graded in amplitude, and nonrefractory
whereas the APs are all or none (maximal), refractory, and propagated actively. Thus, the AM synaptic process gives rise
to an FM or digital signal. The strength of a biological response (e.g., contraction, secretion, sensation) is a function of the
frequency of the signal. That is, higher AP frequency corresponds to stronger contraction or stronger sensation (in the
sensory nervous system).

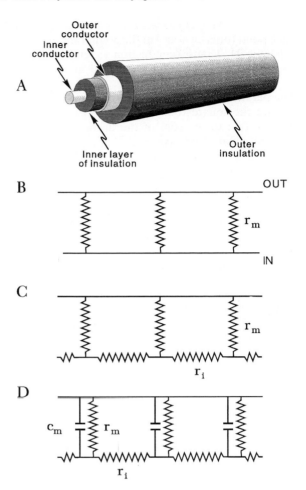

A

B

C

D

FIG. 3. A coaxial cable and associated electrical equivalent circuit (A, B), and the electrical equivalent circuit for a biological cable (C, D). (A) The coaxial cable consists of an inner conductor (e.g., copper wire) and an outer concentric conductor separated by a layer of insulation (e.g., rubber). (B) The equivalent circuit for a coaxial cable. The inner and outer conductors are depicted as having nearly zero resistances, and the transmembrane insulation resistance is shown distributed along the length of the cable. (C) In the biological cable, the inner conductor is the cytoplasm (axoplasm or myoplasm), which is not of negligible resistivity, and therefore is depicted as $r_i$ distributed along the length of the fiber. The transverse insulation resistance is the cell membrane ($r_m$). (D) Addition of the capacitance elements to the biological cable, which arise due to the lipid bilayer matrix of the cell membrane ($c_m$).

occurs when two parallel conductors ("plates") are separated by a high-resistance dielectric material. The dielectric constant of materials is related to vacuum, which is assigned a value of 1.0000; air has a value very close to vacuum, oils have a value of 3–6 and that of pure water is 81. The biological membrane, which has a matrix of phospholipid molecules, has a dielectric constant of about 5, typical of oils. The higher the dielectric constant, the higher the capacitance; the closer the parallel plates, the higher the capacitance. Because the biological membrane is so thin (approx. 70 Å or 7 nm), its capacitance is relatively high: All cell membranes have a membrane capacitance ($C_m$) of about 1.0 $\mu$F/cm$^2$ (where F stands for farads). Capacitors

and dielectric constant are further discussed in the Appendix to this book.

## B. Length Constant

In the electric cable depicted in Figs. 3A and B, a voltage (or signal) applied at one end would be transmitted to a distant end with little or no decrement (diminution or attenuation), and the so-called "length constant" would be very long or nearly infinite. In the biological cable (Figs. 3C and D), however, a signal applied at one end rapidly falls off (decays) in amplitude as a function of distance, with a relatively short length constant ($\lambda$). This decay in voltage is exponential (Fig. 4A). An exponential process gives a straight line on a semilogarithmic plot (log $V$ vs distance) (Fig. 4B). In a cable, the relationship between the voltage at any distance ($x$) from the applied voltage ($V_o$) is

$$V_x = V_o\, e^{-x/\lambda} \tag{1}$$

Thus, when $x = \lambda$

$$V_x = V_o \frac{1}{e}$$

$$= V_o \frac{1}{2.718} \tag{1a}$$

$$= 0.37\, V_o$$

A

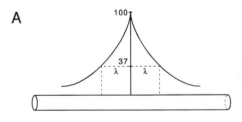

B

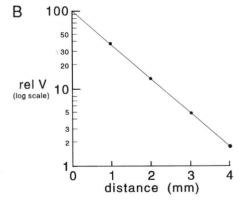

FIG. 4. Length constant ($\lambda$) of the biological cable (fiber). (A) The exponential decay of voltage on both sides of an applied current (voltage) as a function of distance is diagrammed. The distance at which the voltage falls to $1/e$ [or 36.7% of initial voltage (at $x = 0$)] gives the $\lambda$ value. (B) When the voltage is plotted on a log scale against distance, a straight line is obtained.

The mathematical solution to Eq. 1 is

$$V_x = \text{antiln}\left(\ln V_o - \frac{x}{\lambda}\right) \tag{1b}$$

$$V_x = \text{antilog}\,\frac{2.303 \log V_o - \dfrac{x}{\lambda}}{2.303} \tag{1c}$$

Hence, the distance at which the voltage decays to 37% of the initial value gives the length constant, $\lambda$. In nerve fibers and skeletal muscle fibers, $\lambda$ has a value of only about 1–3 mm.

Therefore, the relatively short $\lambda$, compared with the length of the neuron (e.g., over 1.0 m for a lumbar anterior horn cell motor neuron) or skeletal muscle fiber, means that a signal applied at one end (or midpoint) would fall off very quickly with distance along the fiber (Fig. 5). If the length constant were 1.0 mm, then at 4 mm, the signal would become negligible. Hence, the electrical signal cannot be conducted passively in the biological cable, because it would decrement and disappear over relatively short distances. The AP (signal) is amplified to a constant value at each point (or each node) in the membrane, as discussed in Chapter 24. That is, conduction is active, not passive, with energy being put into the signal at each point to prevent any decay of the signal.

The parameters that determine the length constant of a cable are the square root of the ratio of the transverse resistance ($r_m$) to the sum of the inside ($r_i$) and outside ($r_o$) longitudinal resistances,

$$\lambda = \sqrt{\frac{r_m}{r_i + r_o}}$$

$$\text{cm} = \sqrt{\frac{\Omega \cdot \text{cm}}{\dfrac{\Omega}{\text{cm}} + \dfrac{\Omega}{\text{cm}}}} \tag{2}$$

where $r_m$, $r_i$, and $r_o$ are the resistances normalized for a unit length (1 cm) of fiber.

For surface fibers of a nerve or muscle bundle bathed in a large volume conductor, $r_o$ is negligibly small, and Eq. 2 reduces to

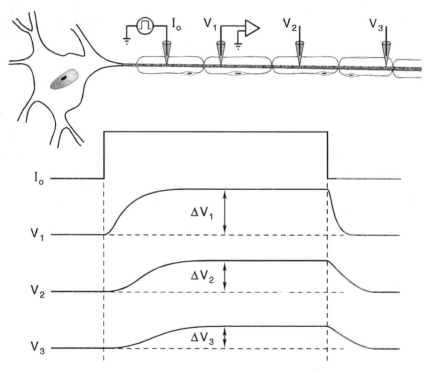

**FIG. 5.** Diagram of a motor axon and how the voltage signal would decay with distance if the neuron were only a passive cable (nonexcitable). The voltage traces illustrate the voltage signals that would be simultaneously recorded at three different points along the axon ($V_1$, $V_2$, and $V_3$) from the site of injection of a rectangular current pulse ($I_o$). As depicted, the amplitude of the steady-state voltage pulse rapidly falls off with distance, because the length constant ($\lambda$) of the biological cable is short (e.g., 1 mm) compared with the length of the axon (e.g., 1000 mm). Therefore, an active response of the cell membrane at each point is required to faithfully propagate the signal. The voltage recorded at point $V_1 (\Delta V_1)$ also illustrates that the membrane potential changes in an exponential manner, both at the beginning of an applied rectangular current pulse and at the end. This exponential change results from the capacitance of the cell membrane, the time constant ($\tau$) being a product of the resistance and capacitance ($R_m C_m$). The time it takes for the voltage to decay to $1/e$ (36.7%) of the initial (maximal) value at the end of the pulse, or to build up to 63.3% ($1 - 1/e$) of the maximal value at the beginning of the pulse, gives the $\tau$ value.

$$\lambda = \sqrt{\frac{r_m}{r_i}}$$

$$= \sqrt{\frac{R_m}{R_i} \frac{a}{2}}$$
(3)

$$cm = \sqrt{\frac{\Omega \cdot cm^2 cm}{\Omega \cdot cm}}$$
(3a)

where $a$ is the fiber radius, and $R_m$ and $R_i$ are the membrane resistance and longitudinal cytoplasmic resistance, respectively, normalized for both length (1 cm) and cell diameter. Thus, the greater the membrane resistance and the smaller the internal longitudinal ressitance (larger cell diameter), the greater the $\lambda$ value. We will see later that the propagation velocity is a function of $\lambda$; that is, larger diameter fibers propagate faster. We will also see that myelination increases the effective membrane resistance ($R_m$) and lowers the effective capacitance, thereby increasing propagation velocity.

## C. Time Constant

Because of the large capacitance of the cell membrane ($C_m$), the membrane potential ($E_m$) cannot change instantaneously upon application of a step current pulse. Instead, $E_m$ changes in an exponenctial (negative) manner (see Fig. 5), on both the charge and the discharge. The membrane time constant ($\tau_m$) is a product of the resistance ($R_m$) and capacitance of the membrane and can be expressed as

$$\tau_m = r_m C_m$$
$$= R_m C_m$$
(4)
$$sec = \Omega \times F$$
(4a)

where $\Omega$ is the resistance in ohms, and $F$ is the capacitance in farads.

The discharge of the membrane (parallel $RC$ network) is given by

$$V_t = V_{max} e^{-t/\tau}$$
(5)

where $V_t$ is the voltage at any time $t$ (at the site of current injection), and $V_{max}$ is the final maximum voltage attained during the pulse. When $t = \tau$,

$$V_t = V_{max} \frac{1}{e}$$

$$= V_{max} \frac{1}{2.718}$$
(5a)
$$= 0.37 \, V_{max}$$

Hence, the time at which the voltage decays to 37% of the initial (maximal) value gives the time constant, $\tau$ (see Fig. 5). In nerve fibers and skeletal muscle fibers, $\tau_m$ has a value of about 1.0 msec.

When the membrane is charging, there is a similar exponential (negative) process, with the identical time constant (see Fig. 5). The corresponding relationship is given by

$$V_t = V_{max} (1 - e^{-t/\tau})$$
(5b)

The time it takes for the voltage to build up to 63% of its final value gives the time constant. When $t = \tau$,

$$V_t = \left(1 - \frac{1}{e}\right) V_{max}$$

$$= (1 - 0.37)V_{max}$$
(5c)
$$= 0.63V_{max}$$

Thus, the time constant can be measured on the build-up of the pulse (time ro reach 63% of the final voltage) or on the decay (time to reach 37% of the initial voltage).

## D. Input Resistance

The **input resistance** ($R_{in}$) of a muscle or nerve fiber is essentially the resistance that an intracellular microelectrode "looks" into when a small current (DC pulse) is injected. Thus, $R_{in}$ is determined by the change in membrane potential ($V_o$ or $V_{x=0}$) at steady state produced at the site of injection ($x = 0$) by the injection of a known (measured) amount of current ($I_o$ or $I_{x=0}$), based on Ohm's law.

$$R_{in} = \frac{V_o}{I_o}$$
(6)

If the microelectrode is near the middle of a very long fiber (or "infinite" cable), then $R_{in}$ is related to the other cable parameters (e.g., to the internal longitudinal resistance $r_i$ and the DC length constant $\lambda_{DC}$):

$$R_{in} = 0.5 \, r_i \, \lambda_{DC}$$

$$\Omega = \frac{\Omega}{cm} \, cm$$
(7)

The factor of 0.5 is present because current flows in both directions from the microelectrode. If the microelectrode is at one end of the long fiber, then the factor of 0.5 is removed. Equation 7 indicates that the $R_{in}$ is greater when $r_i$ and/or $\lambda_{DC}$ are greater.

Combining Eqs. 6 and 7 gives:

$$\frac{V_o}{I_o} = 0.5 \, r_i \, \lambda_{DC}$$
(8)

or

$$V_o = I_o \, 0.5 \, r_i \, \lambda_{DC}$$
(8a)

This equation indicates that the change in membrane potential at $x = 0$ is equal to the applied current times the input resistance.

To obtain the change in membrane potential at any point $x$ along the fiber cable ($V_x$), then an exponential term ($e^{-x/\lambda}$) must be added to account for the exponential decay of voltage over distance:

$$V_x = I_o \, 0.5 \, r_i \, \lambda_{DC} \, e^{-x/\lambda}$$
(9)

This equation is identical to Eq. 1, because the $I_o \, 0.5 \, r_i \, \lambda_{DC}$ term is equal to $V_o$ (see Eq. 8a):

$$V_x = V_o \, e^{-x/\lambda}$$
(1)

Because $\lambda_{DC} = \sqrt{r_m/r_i}$ (Eq. 3), Eq. 7 can be given as:

$$R_{in} = 0.5 \, r_i \sqrt{\frac{r_m}{r_i}} = 0.5 \, \sqrt{r_i r_m} \qquad (7a)$$

and Eq. 9 can also be given as:

$$V_x = I_o \, 0.5 \, \sqrt{r_i} \, \sqrt{r_m} \, e^{-x/\lambda} \qquad (9a)$$

Thus, input resistance is proportional to $\sqrt{r_i r_m}$. This is logical because the higher the longitudinal resistance of the fiber's axoplasm (or myoplasm), which is a function of the resistivity of the axoplasm and fiber diameter (see Appendix to the book, Section XI,A–C), and the higher the membrane resistance, the higher the input resistance should be. The input resistance can be calculated from the electrical equivalent circuit for a cable.

The input impedance can also be given in terms of $R_m$ and $R_i$, the specific resistance of the cell membrane and resistivity of the axoplasm, respectively (see Appendix to book, Section XI,A,B). Since $R_m = 2 \, \pi a \, r_m$ and $R_i = \pi a^2 \, r_i$, then Eq. 7a can be converted to:

$$R_{in} = 0.5 \, \sqrt{r_m r_i} \qquad (7a)$$

$$= 0.5 \, \sqrt{\frac{R_m}{2 \, \pi a} \frac{R_i}{\pi a^2}} \qquad (10)$$

$$= 0.5 \, \sqrt{\frac{R_m R_i}{2 \, \pi^2 a^3}} \qquad (11)$$

Thus, the input resistance is directly proportional to the square root of the membrane-specific resistance and the resistivity of the axoplasm, and inversely proportional to the square root of the fiber radius raised to the third power.

Calculation of the **input impedance** ($Z_{in}$) is more complicated because it also depends on the membrane capacitance ($C_m$) and on the frequency ($f$) of the alternating current (AC) used. According to B. Katz (1966), the input impedance of a long fiber (current injected at one end) is given by:

$$Z_{in} = \sqrt{\frac{R_m R_i}{2 \, \pi^2 a^3 \, \sqrt{1 + 4 \, \pi^2 f^2 R_m^2 C_m^2}}}$$

$$\Omega = \sqrt{\frac{(\Omega\text{-cm}^2) \, (\Omega\text{-cm})}{cm^3 \sqrt{(1/sec^2) \times (sec^2)}}} = \Omega \qquad (12)$$

A discussion of impedance is given in Section VIII of the Appendix to the book and a discussion of the AC length constant ($\lambda_{AC}$) is given in Appendix 2 to this chapter.

### E. Local Potentials

In contrast to the active propagation of APs, synaptic potentials and sensory receptor potentials are not actively propagated. Such potentials decay exponentially (from their source of initiation) along the fiber cable, as described previously. Therefore, postsynaptic potentials and receptor potentials are local potentials. When local potentials are in the depolarizing direction, they can give rise to APs, which are propagated; when hyperpolarizing, they act to inhibit production of APs. These local potentials are similar to the local excitatory response (see Chapter 24), in that both are confined to a local region; however, the electro-

genesis of the two is different. As stated before, the neuromuscular junction is an excitatory type of chemical synapse and produces excitatory postsynaptic potentials (EPSPs), known here as end-plate potentials (EPPs).

Most synaptic potentials are graded, that is, they can add on one another, both in time and in space (temporal summation and spatial summation), to produce larger responses. Larger synaptic potentials exert a greater stimulatory or inhibitory effect on the production of APs.

## IV. Conduction of Action Potentials

### A. Local-Circuit Currents

The generation of APs is described in Chapter 24. This section examines the mechanism for their rapid propagation (conduction). Propagation occurs by means of the local-circuit currents that accompany the propagating APs, as depicted in Fig. 6. Such currents exist because, when two points are at a different potential (voltage) in a conducting medium, current ($I$) will flow between the two points, as governed by Ohm's law ($I = V/R$).

At the peak of the AP in one region of the fiber, the inside of the membrane at that region becomes positive with respect to the outside. The inside is also positive with respect to the inside cytoplasm at a region downstream from the active region. Therefore, current flows through the cytoplasm from the active region (current source) to the adjacent inactive region, then out of the fiber across the cell membrane, then through the interstitial fluid back to the active region (current "sink"), and finally through the membrane of the active region. This completes the closed loop for the current.

The outward current through the membrane of the inactive region produces an $IR$ voltage drop (Ohm's law), positive inside to negative outside. This acts to depolarize this region, because the polarity of the voltage drop is opposite to that of the resting potential (negative inside, positive outside). When the depolarization exceeds the threshold potential, an AP is triggered. Thus, the inactive region becomes converted to an active region. This process is repeated in each segment of fiber, thus resulting in movement (propagation) of the impulse sequentially down the fiber.

If we examine a propagating AP in the middle region of a fiber (Fig. 6), we see that there is also a small backflow of current internally, coupled with a corresponding small forward flow externally, associated with the repolarizing phase of the AP. Thus, as the AP propagates down the fiber, from right to left, there is a simultaneous double flow of local-circuit current: clockwise flow associated with the rising phase of the AP, and counterclockwise flow associated with the repolarizing phase of the AP.

The internal longitudinal current, sweeping past a transverse plane of the fiber, has two phases, first forward (right to left), and then reverse (left to right). The external longitudinal current also has two phases: first left to right and then right to left.

The transverse membrane current, flowing outward in a plane perpendicular to the membrane, has three consecu-

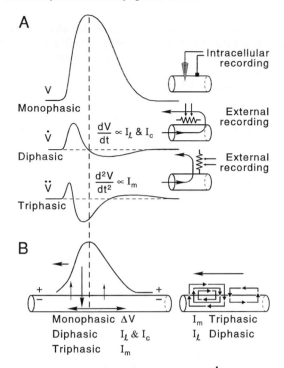

**FIG. 6.** Schematic representation of the first ($\dot{V}$) and second ($\ddot{V}$) time derivatives of the AP spike and the longitudinal and radial currents associated with the propagating spike. The first derivative ($dV/dt$ or $\dot{V}$) is proportional to the capacitive current ($I_c = C_m \cdot dV/dt$) and to the longitudinal (axial) current and is *diphasic*, having an intense forward phase and a less intense backward phase, as depicted in the diagram at the lower right. The second derivative ($d^2V/dt^2$ or $\ddot{V}$) is proportional to the radial transmembrane current ($I_m$), and is *triphasic*, having a moderately intense initial outward phase, then a very intense inward phase (the "current sink"), followed by a least intense second outward phase, as depicted in the lower diagrams. The arrows at the lower left also depict the three phases of the membrane current and the two phases of the axial current. $dV/dt$ can be recorded externally by a pair of closely spaced (relative to the spike wavelength) electrodes arranged parallel to the fiber axis, as illustrated. $d^2V/dt^2$ can be recorded by a pair of electrodes arranged perpendicular to the fiber axis, as depicted. The vertical dashed line indicates that when the slope of the spike goes to zero at the peak of the spike, $dV/dt$ is zero; $dV/dt$ is maximum at about the middle of the rising phase of the spike.

tive phases: first outward (still passive membrane), then inward (active membrane), and finally outward again (still active membrane).

It is the local-circuit current flow that enables the electrocardiogram (ECG), electromyogram (EMG), electroencephalogram (EEG), and electroretinogram (ERG) to be recorded from the body surface over the tissue of interest (heart, skeletal muscle, brain, and eye). The internal longitudinal current is confined to the cytoplasm of the fiber, but the external current can use whatever conducting fluid is available (e.g., the entire torso volume conductor), because of the principle that parallel resistors give a lower total resistance (or current takes the path of least resistance). Thus, this external local-circuit current causes the skin to be at different potentials, and these differences can be recorded (as the ECG, etc.).

## B. Propagation Velocity Determinants

The factors that determine active velocity of propagation ($\theta$) include (1) fiber diameter, (2) length constant ($\lambda$), (3) time constant ($\tau_m$), (4) local-circuit current intensity, (5) threshold potential, and (6) temperature. Some of these factors are interrelated, such as fiber diameter and length constant (since $\lambda$ is proportional to the square root of the radius), and length constant with time constant (since both have a dependence on $R_m$). Propagation velocity is directly proportional to length constant and inversely proportional to time constant as

$$\theta \propto \frac{\lambda}{\tau_m} \tag{13}$$

By substituting Eq. 3 for $\lambda$ and Eq. 4 for the time constant, we have

$$\theta \propto \frac{\sqrt{a}}{c_m \sqrt{R_i} \sqrt{R_m}} \tag{13a}$$

Thus, propagation velocity is directly proportional to the square root of fiber diameter or radius ($a$) and inversely proportional to membrane capacitance ($C_m$). The larger the fiber diameter, the lower the absolute longitudinal resistance of the intracellular cytoplasm (principle of resistors in parallel), and therefore the greater the amount of local-circuit current flowing longitudinally and the greater the length constant. For example, it is well known that the larger the diameter of nerve fibers, the faster they propagate (Table 1). Equation 13a shows that if $C_m$ can be reduced (by myelination), then $\theta$ should increase in proportion. This is discussed in the following section on saltatory conduction.

In addition, $\theta$ depends on the intensity of the local-circuit current, and hence on the rate of rise of the AP. The greater the AP rate of rise (max $dV/dt$), the greater the longitudinal current and the transmembrane capacitive current ($I_c$). Therefore, all other factors being constant, faster rising APs propagate faster. The AP rate of rise depends on the density of the fast Na$^+$ channels that carry inward current, on $C_m$, and on temperature. Max $dV/dt$ decreases with increased $C_m$, with cooling, and with partial depolarization (due to the $h_\infty$ vs $E_m$ relationship discussed in Chapter 24. Cooling slows the rate of all chemical reactions, particularly those with a high $Q_{10}$ (activation energy), such as the ion conductance changes in activated membrane.

Finally, the threshold potential ($V_{th}$) affects propagation velocity. If the threshold were to be shifted to a more positive voltage (more depolarized), then it would take longer for a given point in the membrane to reach threshold (and explode) during propagation of an AP from upstream. A greater critical depolarization (difference between RP and $V_{th}$) would be required to bring the membrane to threshold. Therefore, propagation velocity would be slowed.

As stated earlier, some of these factors are interrelated, and some actually exert opposing effects.

The foregoing discussion applies to nonmyelinated nerve axons and skeletal muscle fibers. An electron micrograph of small bundles of nonmyelinated nerve fibers enveloped by Schwann cells is shown in Fig. 7. In myelinated nerve fibers, propagation velocity is greatly increased by the myelin sheath, as discussed in the following section.

## C. Saltatory Conduction

The nerve cable has been vastly improved by the evolutionary development of myelination in vertebrates. An electron micrograph of a myelinated nerve fiber is shown in Fig. 8. The myelin sheath improves the cable by increasing the effective $R_m$ by about 100-fold and decreasing the effective $C_m$ by about a hundred-fold. This increases the length constant, $\lambda$, and tends to decrease the time constant ($\tau_m$). Because $\tau_m$ tends to increase with the increase in effective $R_m$ (due to the myelin sheath), the decrease in $C_m$ counteracts this effect, thus acting to hold $\tau_m$ almost constant. Thus, Eqs. 13 and 13a predict that propagation velocity should increase with myelination.

One consequence of myelination, therefore, is that propagation velocity is greatly increased (Table 1). Another consequence is that the energy cost of signaling is greatly decreased, because passive ion leaks are limited and active current losses are restricted to the small nodes of Ranvier,

which are spaced relatively far apart. An electron micrograph of a node of Ranvier is shown in Fig. 9. At each node, the length of exposed (naked) cell membrane is only a few micrometers. The internodal distance is about 0.5–2.0 mm (depending on fiber diameter), and the width (length) of each node is only about 0.5–3 $\mu$m. The node forms an annulus around the entire perimeter of the fiber. Therefore, the degree of energy-requiring active ion transport (Na–K and $Ca^{2+}$) required to maintain the steady-state ion distributions and to hold the system in a state of high potential energy is greatly reduced. For example, the amount of $Na^+$ gained and $K^+$ lost per impulse is reduced as a result of myelination. The rate of oxidative metabolism in myelinated fibers reflects this lowered energy requirement.

The myelin sheath is produced by the wrapping of the Schwann cell repeatedly around the nerve fiber in a spiral, forming 20–200 wrappings, depending on axon diameter. That is, the larger axons have a thicker myelin sheath. For purpose of our discussion, we will assume an average of 100 wrappings. The myelin sheath covers the nerve axon as a coat sleeve and is interrupted at each node. The cytoplasm of the Schwann cell in the region of the myelin sheath is nearly completely extruded during its formation, so that the sheath consists essentially of 100 cell membranes

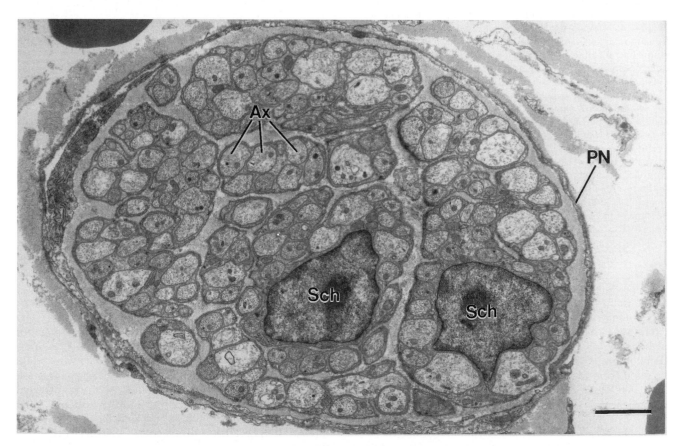

**FIG. 7.** Electron micrograph of an autonomic nerve of mouse heart cut in transverse section. The unmyelinated axons **(Ax)** are arranged in small groups (bundles) that are engulfed by the cytoplasm of a Schwann cell. That is, the nonmyelinated axons are embedded in, and surrounded by, a Schwann cell. Two Schwann cell nuclei **(Sch)** can be seen. The axon bundles are separated and surrounded by a collagenous matrix. The entire nerve is surrounded by a perineurial sheath **(PN)**. The scale bar at lower right equals 2 $\mu$m. (Micrograph courtesy of Dr. Mike Forbes, University of Virginia.)

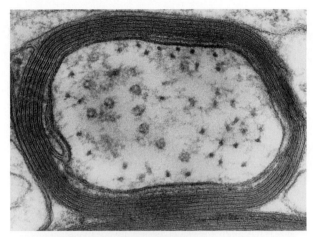

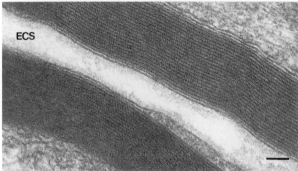

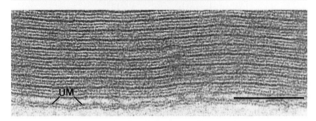

**FIG. 8.** Myelin sheath of myelinated nerve fibers. (Upper panel) Myelinated nerve fiber in spinal cord of rat cut in cross section. There are about seven wrappings of the Schwann cell around the axon, thus giving about 14 membranes in series with the cell membrane of the axon **(axolemma)**. The internal **mesaxon** is visible at the left side (lower) of the axon, and represents the beginning of the spiraling of the Schwann cell membranes (cytoplasm squeezed out) to form the myelin sheath surrounding the axon. On the upper left side, a process of an oligodendrocyte abuts on the myelin sheath. The axon shown is closely plastered up against the myelin sheath of a neighboring axon (bottom of figure). **Microtubules** and **neurofilaments** are visible in the cytoplasm of the axon. Magnification of 168,000X. (From An Atlas of Histology by J. Rhodin. Copyright © 1975 by J. Rhodin. Used by permission of Oxford University Press, Inc.) (Middle panel) Higher magnification of two adjacent nerves fibers (in phrenic nerve of rat) cut in longitudinal section to illustrate axon with a thicker myelin sheath (approx. 23 wrappings = 46 membranes). Note that between the thicker (dense) lines are two thinner lines, which can be seen more clearly in the lower panel. The extracellular space (ECS) between the two axons is labeled. The calibration bar at the lower right equals 0.1 $\mu$m. (Lower panel) Still higher magnification of a portion of one of the myelin sheaths shown in the middle panel to better illustrate the periodicity of the myelin membranes. The dense lines alternate with a pair of thin lines. The space in between the pair of thin lines is extracellular space. The dense lines represent the intracellular (cytoplasmic) region of the Schwann cell process, with

stacked in series. Because resistors in series are added to calculate the total resistance, the effective transmembrane resistance is increased 100-fold, but because the total capacitance of capacitors in series is calculated like resistors in parallel, the effective capacitance is reduced 100-fold. Since $\lambda$ is directly proportional to the $\sqrt{R_m}$, $\lambda$ is increased accordingly. As described in the section on the factors influencing propagation velocity, increasing $\lambda$ and lowering $C_m$ increase $\theta$.

Myelinated nerves usually have an optimal amount of myelin, which is an amount such that the ratio of diameter of axon cylinder (naked axon) to total fiber (including myelin sheath) is about 0.6–0.8. Assuming a maximal total diameter feasible (the body must pack many circuits within a limited space, e.g., sciatic nerve bundle, a greater fraction of myelin, by infringing on the diameter of the axis cylinder, would raise $r_i$ too high, causing $\theta$ to decrease. Thus, there are two opposing factors in determining the degree of optimal thickness of the myelin sheath: the more the myelin, the greater the decrease in $C_m$ and the increase in effective $R_m$, but the smaller the diameter of the axis cylinder and therefore the higher the $r_i$.

In saltatory conduction (Latin *saltere,* to jump), the impulse jumps from one node to the next. The internodal membrane does not fire an AP. This is due to two reasons: (1) The internodal membrane is much less excitable (e.g., much fewer fast Na⁺ channels) and (2) the depolarization of the neuron cell membrane at the internodal region is only about 1/100th of that at the node. The latter occurs because the *IR* voltage drop across the internodal cell membrane is only 1/100th of that across the entire series resistance network (neuron cell membrane plus 100 layers of Schwann cell membrane) (Kirchhoff's laws dealing with voltage drops across resistors in series). Even though the internal potential in the internodal region swings positive (e.g., to +30 mV) when the adjacent nodes fire, the potential at the outer surface of the internodal membrane also swings nearly as positive (e.g., to +29 mV). Therefore, the depolarization of the neuronal membrane at the internode is only about 1 mV, which is well below threshold, and so the internodal membrane does not fire. The potential that controls the membrane conductances (activates the voltage-dependent ion channels) is the pd (potential difference) directly across the membrane and not the absolute potential on either side.

As stated previously, the internodal membrane has only a few fast Na⁺ channels, whereas the nodal membrane has

the cytoplasm extruded, thus allowing the inner (cytoplasmic face) leaflets of the cell membrane of the Schwann cell to come into close contact. The outer (extracellular face) leaflets of the Schwann cell membrane are seen as the thin lines. Thus, each cell membrane has the appearance of a double line, representing the hydrophilic surfaces of the cell membrane; the hydrophobic region of the membrane is the clear region between the dense line and the contiguous thin line on either side. UM = unit cell membrane. The scale bar on the lower right represents 0.1 $\mu$m. [Micrographs (middle and lower panels) courtesy of Dr. Mike Forbes, University of Virginia.]

**FIG. 9.**    Electron micrograph of one node of Ranvier in a single myelinated nerve axon of rat sciatic nerve cut in longitudinal section. As can be seen, the dark myelin bordering the axon on each side is interrupted near the middle of the micrograph, leaving the neuronal cell membrane nude of myelin at the node region. However, cytoplasmic processes of the Schwann cells cover the nodal membrane. Collagenous fibrils of the endoneurium are visible at the region of the node, peripheral to the Schwann cell cytoplasm. The myelin tapers and thins as it approaches the node, and there is a frayed appearance caused by the successive laminae of the myelin sheath terminating as cytoplasmic swellings. The axoplasm contains neurofilaments and microtubules. Magnification of 9000X. (From An Atlas of Histology by J. Rhodin. Copyright © 1975 by J. Rhodin. Used by permission of Oxford University Press, Inc.)

a very high density. Since the cell membrane is fluid and proteins can diffuse (float) laterally in the lipid bilayer matrix, what keeps the fast $Na^+$ channel proteins confined (at high density) in the nodal region? It appears that there are special anchoring proteins (e.g., ankyrin) that anchor the ion channel proteins to the cytoskeletal framework, thus preventing their lateral movement into the internodal membrane.

The effect of myelin and saltatory propagation is to make propagation much faster. For example, a 20-$\mu$m-diameter myelinated nerve fiber conducts even faster than a 1000-$\mu$m (1-mm)-diameter nonmyelinated nerve fiber (e.g., the giant axon in squid, lobster, earthworm): 120 m/sec vs ca. 25–50 m/sec. Thus, invertebrates, to achieve fast conduction in some essential circuits, must resort to giant neurons, resulting in a lower $r_i$ and hence fast conduction. Because of space/size limitations, only a few critical neurons can be made giant in diameter. In vertebrates, on the other hand, a large fraction of the nerve fibers in the peripheral nerves are myelinated for fast propagation.

We saw above that, in nonmyelinated axons and skeletal muscle fibers, $\theta$ should vary with the square root of the cell diameter or radius ($a^{0.5}$). In myelinated axons, $\theta$ varies with the first power of the cell radius ($a^1$) because $\theta$ varies with $\lambda^2$ as indicated by

$$\theta \propto \frac{\lambda^2}{\tau_m} = \frac{a}{2R_iC_m} \qquad (14)$$

The dependence of conduction velocity on diameter of myelinated and nonmyelinated fibers is summarized in Table 1.

### D.  Wavelength of the Impulse

We can calculate the wavelength of the AP, which is the length of the axon simultaneously undergoing some portion of the AP. The wavelength is equal to propagation velocity ($\theta$) times the duration of the AP ($APD_{100}$) thus,

$$Wavelength = \theta \times APD_{100}$$

$$cm = \frac{cm}{sec} \times sec \qquad (15)$$

$$Distance = velocity \times time$$

Note the similarity of this relationship to that for the wavelength of an electromagnetic radiation: wave length = velocity of light/frequency of the radiation. The reciprocal of frequency is the period (duration of one cycle).

The wavelength in a large myelinated nerve axon is about 12 cm: 120 m/sec × 1.0 msec. In a skeletal muscle fiber, it is about 1.8 cm (6 m/sec × 3.0 msec). In a smooth muscle bundle, the wavelength is only about 1.5 mm (5 cm/sec × 30 msec).

## V.  External Recording of Action Potentials

### A.  Monophasic, Diphasic, and Triphasic Recording

As discussed previously, local-circuit currents accompany the propagating AP in each fiber. The intracellular and extracellular longitudinal currents are diphasic; that is, initially they travel in the forward direction intracellularly and then in the reverse direction. The forward direction current is intense (high current density) and the reverse direction current is weak (low current density).

The transmembrane radial currents are triphasic; that is, the first phase is outward (moderate intensity), the second phase is inward (high intensity), and the third phase is outward (low intensity). The first phase (outward) gives rise to the passive exponential foot of the AP, and is due to the passive cable spread of voltage and current. The second phase (inward) corresponds to the large inward fast $Na^+$ current, which occurs during the later portion of the rising phase and peak of the AP. The third phase (outward) corresponds to the net outward current ($K^+$), which occurs during the repolarizing phase.

These longitudinal and radial currents can be recorded by suitably placed external electrodes. The extracellular longitudinal currents can be recorded by two electrodes (bipolar) placed close together along the length of the fiber. If the interelectrode distance is short (relative to the wavelength), an approximate first (time) derivative of the AP is obtained (Figs. 6 and 10C).

The extracellular radial currents can be recorded by two electrodes placed close together in a plane perpendicular to the fiber axis. This gives an approximation of the second (time) derivative of the AP (see Fig. 6).

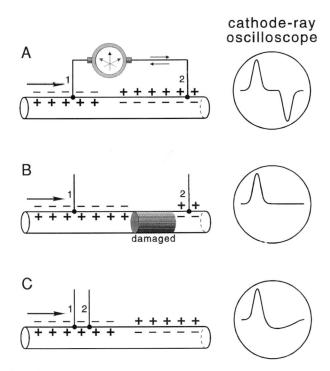

cathode-ray oscilloscope

**FIG. 10.** Diagram of the waveforms that would be recorded externally during propagation of an AP in a single fiber. (A) When the two electrodes are far apart (relative to the AP wavelength), a diphasic recording is obtained, with the two phases being symmetrical and separated by an isopotential segment. The two phases are due to the current flow through the voltmeter recorder being first in one direction and then in the opposite direction. (B) If the fiber between these two electrodes is damaged (e.g., by crushing) or depolarized (by elevated $[K]_o$), so the AP cannot sweep past the second electrode (No. 2), then this second phase is prevented and the recording is monophasic. (C) If the two electrodes depicted in part A are brought progressively closer, then the isopotential segment would shorten and disappear. If electrode No. 2 is brought very close to electrode No. 1, so that the interelectrode distance is short relative to the wavelength, then the second phase is smaller than the first phase, and the record resembles the first derivative of the true AP.

The internal axial currents are confined to the cytoplasm, whereas the external longitudinal currents can use the entire intersitital fluid space of the nerve bundle or muscle or even the entire torso (so-called "volume conductor"), since current takes the path of least resistance (resistors in parallel). As mentioned in the section on local-circuit currents, this allows the recording of the ECG from the body surface and the EMG from the skin overlaying an activated skeletal muscle. The ECG and EMG consist essentially of diphasic potentials, reflecting the external longitudinal currents during propagation of APs.

When the two external electrodes are placed far apart (with respect to the wavelength) along a nerve or muscle fiber, the diphasic recording has the two phases about equal (Fig. 10A). The proximal electrode records the wave of negativity (associated with the propagating AP) first, and then returns to isopotential. When the wave reaches the second electrode, the wave is recorded by it in reversed polarity (because current flow throught the voltmeter is reversed). If the AP were now prevented from reaching the second (distal) electrode by crushing this region of the fiber or elevating $[K]_o$ to depolarize it, then a monophasic recording would be obtained (Fig. 10B). This monophasic recording would most resemble the true AP recorded by a microelectrode impaled into a fiber to record the transmembrane potential, but would be much smaller in amplitude.

### B. Compound Action Potential

When one records the APs externally, the records are graded and not all or none, as in the case of the true APs recorded intracellularly from single fibers (see Chapter 24). That is, the signal recorded becomes larger and larger, up to a maximum amplitude as the intensity of stimulation is increased. This is the so-called "compound action potential." It is graded because, as a greater and greater fraction of the fibers is activated, the external longitudinal currents associated with the all-or-none AP in each fiber cut across the recording electrodes, thereby producing a larger signal. The AP in each indiviual fiber is always all or none. The amplitude of the signal is determined by the resistance between the electrodes multipled by the amount of current flowing through this resistance ($V = IR$). The recording of compound action potentials is diagrammed in Fig. 11.

The compound action potentials can be demonstrated by recording the EMG from a human subject when one electrode is placed on the skin of the ventral forearm and the other (reference) electrode on the wrist of the same arm. Then, as the subject voluntarily produces stronger and stronger contraction to flex the hand, the electrical signals picked up become greater and greater in amplitude and frequency. The amplitude becomes larger because more muscle fibers are simultaneously activated. This is known as **fiber recruitment.** The frequency increases because the motor nerves fire at a higher frequency, causing the muscle fibers to fire at a higher frequency and thus producing a more powerful tetanic contraction.

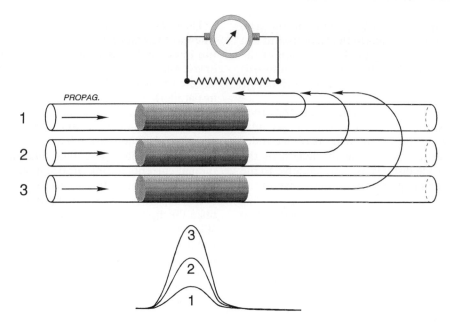

**FIG. 11.** Diagram of a compound action potential in an isolated nerve trunk, such as the frog sciatic nerve, recorded externally by a pair of longitudinal electrodes. The voltmeter (oscilloscope) records the voltage (*IR*) drop across the resistance (fluid) between the two electrodes. If only fiber No. 1 is activated, the current passing between the electrodes is small, and the voltage recorded is small. If fiber No. 2 is simultaneously activated with fiber No. 1, then the amount of current is doubled, and the voltage is doubled. When all three fibers are simultaneously activated, the current is tripled, and the voltage is tripled. Therefore, the externally recorded compound action potential is graded because it reflects the electrical activity of numerous fibers, each of which produces an all-or-none (nongraded) AP.

## VI. Summary

Although the biological cable (i.e., nerve fiber or skeletal muscle fiber) is the best possible, it is a relatively poor cable with a short length constant and relatively long time constant. Therefore, for faithful and rapid signal transmission over long distances, energy must be put into the system at each point along the way. The system evolved is that of AP generation, which are all-or-none signals of constant amplitude and constant propagation velocity, in addition to having refractory periods and sharp thresholds. This is a frequency-modulated system, in which increasing strength of sensation or motor response follows from an increase in frequency of the AP signals.

AP propagation occurs by means of the local-circuit currents. The transmembrane current has three phases: outward, inward, and outward. The internal and external longitudinal currents have two phases: forward and backward (for internal) or backward and forward (for external). The external currents use the path of least resistance, enabling electrograms (e.g., ECG, EMG) to be recorded from the body surface. The compound AP is graded in amplitude, reflecting the summation of the external currents generated from each fiber that is activated; that is, the more fibers simultaneously activated, the greater the amplitude of the electrogram signal.

Propagation velocity is faster the larger the diameter of the fiber, the longer the length constant, and the lower its time constant and capacitance. The myelin sheath evolved by vertebrates enables much faster propagation velocity and at a lower energy cost. Myelination raises the effective membrane resistance and lowers the effective capacitance, and excitability occurs only at the short nodes of Ranvier that periodically interrupt the myelin sheath. Therefore, the AP signal jumps from node to node in a saltatory pattern of conduction.

## Bibliography

Cole, K. S. (1968). "Membranes, Ions and Impulses: A Chapter of Classical Biophysics." University of California Press, Berkeley.

Davis, L., Jr., and Lorente de No, R. (1947). Contribution to the mathematical theory of the electrotonus. *Stud. Rockefeller Inst. Med. Res.* **131,** 442–496.

Hodgkin, A. L., and Rushton, W. A. H. (1946). The electrical constants of a crustacean nerve. *Proc. R. Soc. Lond. (Biol.)* **133,** 444–479.

Jack, J. J. B., Noble, D., and Tsien, R. W. (1975). "Electric Current Flow in Excitable Cells." Clarendon Press, Oxford.

Katz, B. (1966). "Nerve, Muscle, and Synapse." McGraw-Hill, New York.

Lakshminarayanaiah, N. (1984). "Equations of Membrane Biophysics." Academic Press, Orlando, FL.

Rall, W. (1977). *In* "Handbook of Physiology" (J. M. Brookhart and V. B. Mountcastle, Eds.), Vol. 1, pp. 39–97. American Physiological Society, Bethesda, MD.

Sperelakis, N. (1979). Origin of the cardiac resting potential. *In* "Handbook of Physiology," (R. M. Berne and N. Sperelakis, Eds.), Vol. 1, pp. 187–267. American Physiological Society, Bethesda, MD.

Sperelakis, N. (1992). Cable properties and propagation mechanisms. *In* "Physiology" (N. Sperelakis and R. O. Banks, Eds.), pp. 83–97. Little, Brown, Boston.

Sperelakis, N., and A. Fabiato (1985). Electrophysiology and excitation–contraction coupling in skeletal muscle. *In* "The Thorax: Vital Pump" (Ch. Roussous and P. Macklem, Eds.), pp. 45–113. Marcel Dekker, New York.

Taylor, R. E. (1963). Cable theory. *In* "Physical Techniques in Biological Research" (W. L. Nastuk, Ed.), Vol. 6, pp. 219–262. Academic Press, New York.

Nicholas Sperelakis

# Appendix 1: Propagation in Cardiac Muscle and Smooth Muscles

## I. Background

This chapter discusses propagation in cells that are long cables, such as nerve fibers and skeletal muscle fibers. Propagation is more complex in tissues composed of assemblies of short cells, such as cardiac muscle and visceral smooth muscles. These short cells may be considered to be short or truncated cables. In such a truncated cable cell, its true length constant $\lambda$ is much longer than the cell's length. Therefore, there is relatively little voltage fall-off (decay) over the length of each cell. It follows that the entire cell undergoes an action potential (AP) nearly simultaneously. Yet propagation velocity $\theta$ is much slower in cardiac muscle (ca. 0.5 m/sec) and smooth muscles (ca. 0.05 m/sec) than in skeletal muscle (ca. 5 m/sec) or nonmyelinated nerve fibers (ca. 2 m/sec). Part of the reasons for the slower $\theta$ concern fiber diameter (i.e., diameter of cardiac muscle fiber is about 15 $\mu$m compared to about 60 $\mu$m for skeletal muscle fibers). Equation 13a indicates that in a simple cable, $\theta$ is a function of $\sqrt{a}$ (where $a$ is the fiber radius). Thus, the theoretical $\theta$ for cardiac muscle ($\theta_c$) should be:

$$\theta_c = \sqrt{\frac{15\ \mu\text{m}}{60\ \mu\text{m}}} \times \theta_{sk}$$

$$\theta_c = 0.5 \times 5\ \text{m/sec}$$
$$= 2.5\ \text{m/sec}$$

where $\theta_{sk}$ is the propagation velocity for skeletal muscle, and it is assumed that the other parameters given in Eq. 13a are the same in the two types of muscle.

This theoretical velocity of 2.5 m/sec is substantially greater than the actual 0.5 m/sec. Therefore, at least one other factor must determine velocity in cardiac muscle, and that is the high resistance of the junctional membranes (the intercalated disk membranes in the case of cardiac muscle). It is still controversial as to exactly how high the junctional resistance is. When gap junctions are present, the gap junction channels span the intercellular junction, and so serve to lower the cell-to-cell resistance (see Chapter 30). However, the hearts of the lower vertebrates (e.g., amphibians, reptiles) either have an absence of gap junc-

tions or they are very sparse and tiny. In addition, even parts of mammalian hearts and some visceral smooth muscles (e.g., longitudinal muscle layer of intestine) do not appear to contain gap junctions. If so, another mechanism may be involved for cell-to-cell propagation in those muscles in which there is a virtual absence of gap junctions. One of the mechanisms proposed is known as the *electric field model* (Sperelakis *et al.*, 1980). This model is discussed in this appendix.

First, however, let us return to the problem that propagation in cardiac muscle is about five times slower than it should be if the gap junction channels served to interconnect thoroughly (electrically) the cytoplasm of two contiguous cells lying end to end. We also stated previously that the short cardiac muscle cell should show no voltage decay, and so the entire length of the cell should undergo the AP nearly simultaneously (i.e., there is nearly infinite propagation velocity within each cell). Yet the overall propagation velocity in the tissue is relatively slow. Clearly, a large time delay must occur at each cell junction. In fact, most of propagation time is consumed at the cell junctions. This has been demonstrated experimentally. Propagation in cardiac muscle has been shown to be actually discontinuous or "saltatory" in nature (Spach *et al.*, 1981; Rudy and Quan, 1987a,b; Sperelakis *et al.*, 1991). Therefore, the presence of a large number of gap junction channels is insufficient to reduce the junctional resistance enough to allow a chain of cells to behave as a simple cable. This relatively high junctional resistance has relevance to heart block and fibrillation, but this is outside the scope of this chapter.

A similar analysis can be done for visceral smooth muscle. The theoretical $\theta$ for visceral smooth muscle ($\theta_s$) should be

$$\theta_s = \sqrt{\frac{5\ \mu\text{m}}{60\ \mu\text{m}}} \times \theta_{sk}$$

$$= 0.289 \times 5\ \text{m/sec}$$
$$= 1.44\ \text{m/sec}$$

This theoretical velocity of 1.44 m/sec is much greater than the actual velocity of 0.05 m/sec. Thus, propagation velocity in smooth muscle is only about 3.5% of what it should be based on fiber diameter.

The lower maximum rate of rise of the AP (+ max $dV/dt$) in cardiac muscle (ca. 200 V/sec) and smooth muscle (ca. 5 V/sec), as compared with skeletal muscle (ca. 600 V/sec), is another factor that contributes to the lower than expected $\theta$ in cardiac muscle and smooth muscle. In addition, the higher the extracellular resistance, which depends on the tightness of packing of the fibers in the muscle, the slower the velocity (e.g., see Jack *et al.*, 1975).

## II. Experimental Facts

Some of the key experimental facts relevant to the transmission of excitation from one cell to the next in cardiac muscle and visceral smooth muscle can be summarized as follows. These tissues can be enzymatically separated into their individual cells, and the individual single cells are viable and functional. Gap junctions are absent in the hearts of lower vertebrates and in some regions of mammalian hearts, and in some visceral smooth muscles, as stated previously. The length constant $\lambda$ of cardiac muscle and smooth muscle tissues, when measured properly, is relatively short, that is, less than 1 mm (i.e., not much more than about one cell length). The input resistance ($R_{in}$) of cardiac muscle and smooth muscle, when measured properly, is relatively high, about 5 M$\Omega$. The short $\lambda$ and high $R_{in}$ suggest that the cells are not profusely connected by low-resistance pathways (e.g., by gap junction channels). Thus, even when gap junctions are present, the junctional membranes constitute a substantial barrier to current flow from one cell to the next.

As stated in Section I, the true $\lambda$ of individual cells is much greater than the cell length, so that there is almost no voltage decay in a single cell and that the entire cell fires an AP nearly simultaneously. Therefore, most of the propagation time is consumed at the cell junctions, and propagation in these tissues is a discontinuous process, in contrast to a continuous process for skeletal muscle fibers and nonmyelinated nerve fibers. Propagation velocity in cardiac muscle and smooth muscle is slower than what can be accounted for by the smaller fiber diameter and lower +max $dV/dt$. The cell-to-cell transmission process is quite labile, somewhat like that in synaptic transmission.

The reader is referred to several review-type articles by Sperelakis and colleagues for additional details and evidence for some of the statements made here. Considerable data have been published concerning the degree of spread of electrotonic current between neighboring cells in cardiac muscle and visceral and vascular smooth muscles. Only one example is presented here for frog cardiac (ventricular) muscle. In these experiments, electrode 1 was used to record voltage and to inject current (using a bridge circuit), whereas electrode 2 recorded voltage only. As illustrated in Fig. 1, using a pair of microelectrodes whose tips were spaced 11 $\mu$m apart, in some double impalements there was no electrotonic current spread between the two electrodes (e.g., panel A), whereas in other impalements there was substantial spread of current (e.g., panel B). In

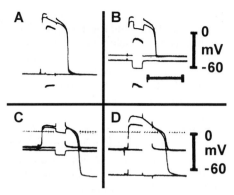

**FIG. 1.** Three typical experiments (A, B, and C–D) measuring the spread of electrotonic current between two closely spaced intracellular microelectrodes in intact frog ventricular trabeculae at rest and during the plateau of the action potential. The interelectrode distance was 11 $\mu$m. Rectangular hyperpolarizing current pulses (approx. 150 ms in duration) were applied. Two successive sweeps of the oscilloscope superimposed in each panel. (A) Interaction at rest and during the plateau was nearly 0%. Capacitive transients are only seen on the trace from electrode 2, whereas a large maintained hyperpolarization occurred at electrode 1. (B) In another impalement in which the cell was injured by the two electrodes, the resting potential was low and the degree of interaction was high. (C–D) In another impalement in which the cell was damaged, the resting potential was low, and the degree of interaction was high (electrode 1 deflection not shown because of bridge imbalance). The contraction accompanying the action potential caused one of the electrodes to become dislodged from that cell and penetrate into a neighboring cell that had a normal resting potential; the degree of interaction then became nearly zero. [Reproduced with permission from Tarr, M., and Sperelakis, N. (1964). *Am. J. Physiol.* **207,** 691–700.

one unusual case (panels C–D), the double impalement first showed good spread (i.e., interaction) between the electrodes (panel C), but then due to muscle contraction, one electrode left that cell and impaled a neighboring cell having a normal resting potential (right portion of panel C); now there was no significant interaction between the two neighboring cells (panel D). In the impalements in which there was substantial interaction between the electrodes (e.g., panels B and C), it was proposed that the two electrodes had impaled the same cell (e.g., both electrodes recorded low resting potentials, perhaps due to damage caused by the two electrodes impaling one cell). In the impalements in which there was little or no interaction (e.g., panels A and D), it was proposed that the electrodes had impaled neighboring cells. This interpretation is most clear in panel D, in which the two electrodes recorded markedly different resting potentials. Thus, there appears to be little or no spread of current between neighboring cells in frog cardiac muscle. If so, then propagation of excitation must occur by some other means.

## III. Electric Field Model

### A. Electric Field Effect

Sperelakis and colleagues developed an electric field hypothesis for propagation of APs in cardiac muscle for situations in which there were no functioning gap junction channels. A computer simulation model for cell-to-cell propagation in cardiac muscle was developed and progressively improved since the mid-1970s (Sperelakis and Mann, 1977; Sperelakis *et al.*, 1985; Sperelakis, 1987; Picone *et al.*, 1991). The model allows electrical transmission to occur between adjacent excitable cells by means of the electrical field effect in the very narrow junctional cleft between the contiguous cells (Sperelakis and Mann, 1977). This electric field model does not require low-resistance channels (gap junctions) between cells. The major requirements of the model are that the pre- and postjunctional membranes (pre-JM and post-JM) be ordinary excitable membranes, and that these membranes be very closely apposed to one another (i.e., the junctional cleft be very narrow, about 10 nm). When the pre-JM fires, the cleft between the cells becomes negative with respect to ground (the interstitial fluid surrounding the cells), and this negative cleft potential (about $-40$ mV) acts to depolarize the post-JM by an equal amount (namely, 40 mV) and brings it to threshold. This, in turn, brings the surface membrane of the postjunctional cell to threshold. The inner surface of the post-JM remains at nearly constant potential with respect to ground.

Figure 2A illustrates propagation of an AP along a chain of 10 cells by the electric field effect. In this computer simulation of cardiac muscle, propagation of overshooting APs occurred at a constant velocity of 32 cm/sec, and the maximum rate of rise of the AP averaged 209 V/sec. As can be seen, the upstroke of each AP exhibited a break or step, reflecting the junctional transmission process.

In the model, a plot of propagation time as a function of distance along a chain of cells has a staircase shape, indicating that almost all propagation time is consumed at the cell junctions and that excitation of each cell is virtually instantaneous (Picone *et al.*, 1991; Sperelakis *et al.*, 1991) (see Fig. 2B). In this figure, it can also be seen that the prejunctional membrane fires a fraction of a millisecond *before* the surface membrane of the same cell, as required by the electric field model. Propagation was found to be strongly dependent on radial cleft resistance ($R_{jc}$) and the junctional membrane properties. There was an optimal $R_{jc}$ for maximum propagation velocity under any given conditions. This model is consistent with many experimental facts about propagation in cardiac muscle, and provides an alternative mechanism for AP propagation that does not require low-resistance pathways to transfer excitation directly between adjacent cells. The electric field model can also account for the fact that propagation in cardiac muscle is actually discontinuous or "saltatory" in nature (Spach *et al.*, 1981; Rudy and Quan, 1987a,b).

## B. High Density of Fast Na⁺ Channels at Intercalated Disks

For this mechanism to work efficiently, there is a requirement for the prejunctional membrane to fire an AP a fraction of a millisecond before the contiguous surface membrane (Sperelakis and Mann, 1977). That is, the prejunctional membrane should be more excitable than the surface membrane, which would cause it to reach threshold

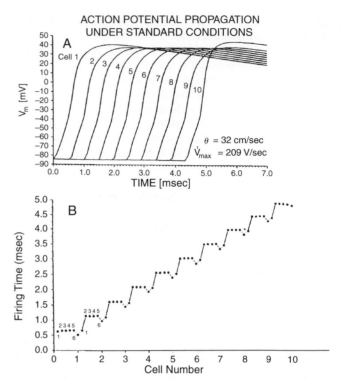

**FIG. 2.** Electric field model for propagation in cardiac muscle using a computer simulation. (A) Successful propagation of an AP at constant velocity along a chain of 10 cells under standard conditions. Propagation velocity was 32 cm/sec; max $dV/dt$ was 209 V/sec. Note the small step, or prepotential, on the AP upstroke. (B) Plot illustrating that junctional delays occupy most of the propagation time. Firing time for an AP to spread along the chain of 10 cells under standard conditions (corresponding to part A) is plotted against distance. Within each cell, the units are numbered 1 through 6 from left to right, 1 corresponding to the post-JM (input), 2–5 the surface membrane, and 6 the pre-JM (output). The AP travels along the surface membrane at a high velocity. At the junctions between cells, there is a significant conduction delay as the AP "jumps" across the cleft. This clearly demonstrates the discontinuous nature of AP propagation. [Adapted with permission from Figs. 2 and 3 of Sperelakis, N., Ortiz-Zuazaga, H., and Picone, J. B. (1991). *Innov. Tech. Biol. Med.* **12**, 404–414.]

first. It was suggested that this situation would be achieved if there were a greater density of fast Na⁺ channels in the junctional membranes (intercalated disks) than in the contiguous surface membranes (Sperelakis and Mann, 1977). This would make the intercalated disks more excitable and give them a lower threshold than the surface membrane. To examine this possibility, a polyclonal antibody, raised against fast Na⁺ channels (from rat brain), was used to immunolocalize the fast Na⁺ channels in rat atrial and ventricular tissues. In immunofluorescence examination, intense labeling was observed associated with the intercalated disks of both atrial and ventricular cells (Ferguson, D., Sperelakis, N., and Angelides, K. J., unpublished observations) (Fig. 3). This fluorescence was more intense than that of the cell surface membrane. These findings are in agreement with results reported by Cohen and Levitt (1993).

Therefore, it is likely that, not only are fast Na⁺ channels present at the intercalated disk membranes, they are present in a higher concentration (density) than in the surface cell membrane. A higher density of fast Na$^+$ channels in the junctional membrane would cause it to have a lower threshold than the surface membrane, and so it would reach threshold and discharge first. Thus, not only are the intercalated disks composed of excitable membranes, but they also may have greater excitability than the contiguous surface membranes. This would be analogous to the initial segment of the axon of the anterior horn neuron having a lower threshold, and therefore discharging before the soma or proximal dendrites (where the excitatory synapses are actually located). Consistent with these findings, it was reported that K$^+$ channels are also localized at the intercalated disks (Mays *et al.*, 1995). Therefore, the electric field model is a plausible mechanism for cell-to-cell transmission of excitability in cardiac muscle and in visceral smooth muscle. Electric field effects may also occur between closely spaced contiguous neurons in the central nervous system.

## IV. Electronic Model for Simulation of Propagation

Sperelakis and colleagues (1990) constructed an electronic model to simulate an excitable membrane, and this

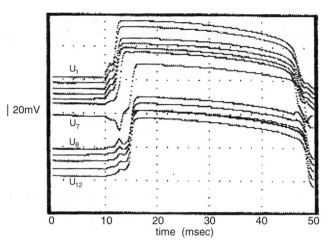

**FIG. 4.** Action potentials produced in an electronic model of two heart cells with a cell junction between them. There were six excitable-circuit units in each cell: four representing the surface cell membrane (U2, 3, 4, 5 and U8, 9, 10, 11) and one for each junctional membrane (U1, 6 and U7, 12). The parameters in the first cell (cell 1) were adjusted so that it fired spontaneously in response to pacemaker potential depolarization. Note the nearly simultaneous firing of all units in cell 1, and after a slight junctional delay time, the nearly simultaneous firing of all units of cell 2. [Reproduced with permission from Fig. 5A of Ge, J., Sperelakis, N., and Ortiz-Zuazaga, H. (1993). *Innov. Tech. Biol. Med.* **14**, 404–420.]

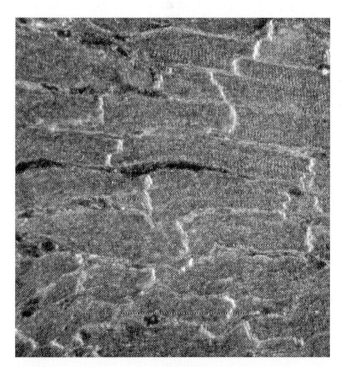

**FIG. 3.** Immunofluorescence localization of fast Na$^+$ channels in adult rat cardiac muscle (ventricle) using polyclonal antibody to the fast Na$^+$ channels of rat brain. The antibody is most densely localized at the cell junctions (intercalated disks), but also stained the surface cell membrane. These results suggest that the intercalated disks are highly excitable membranes, even more excitable than the surface cell membrane, consistent with the requirement of the electric field model for propagation between cells not connected by low-resistance tunnels (gap junction connexons). [From Ferguson, D., Sperelakis, N., and Angelides. K. J., unpublished observations.]

model was used as a new circuit model of propagation. Four such circuits were successfully made to interact with one another through capacitive coupling to simulate propagation over four cells. Adjustment of one parameter either slightly depolarized and caused repetitive spontaneous APs, or hyperpolarized slightly and depressed excitability, and caused partial block (e.g., 2 : 1, 3 : 1, 4 : 1) or complete block of propagation from cell to cell at the cell junctions. When four such cells were connected head to tail in a closed loop, reentry of excitation occurred and could keep going for many seconds before dying out (due to failure at one of the labile junctions). Several configurations of external networks were used in this model to study possible electrical field coupling in cell chains.

In a subsequent study (Ge *et al.*, 1993), 12 such units were arranged to model two adjacent cells: 4 units for each surface membrane and 1 for each junctional membrane. The first unit was stimulated to threshold, and AP propagation spread over the two cells was recorded. The time delay at each junction between two adjacent cells was measured as a function of $R_{jc}$, the radial shunt resistance at the cell junction. The experimental results showed that the time delay was about 1–2 ms when $R_{jc}$ values were changed from infinite down to about 100 k$\Omega$; the time delay increased at lower $R_{jc}$ values, and propagation was blocked when $R_{jc}$ was below 10 k$\Omega$. Excitation of cell 1 caused hyperpolarization of the post-JM (unit 7) and depolarization of the other units in cell 2, due to current flow prior to triggering of the AP in cell 2. Raising the effective coupling resistance ($R_c$) between the two cells increased the conduction delay. As expected, the junctional delay increased when $C_j$ was lowered. Stimulating at the pre-JM (unit 6) resulted in bidirectional propagation. Making the pre-JM inexcitable did not prevent excitation of cell 2, reflecting the presence of longitudinal current flow. However, the post-JM (unit 7) only fired an AP when unit 6 was active, reflecting the electrical field effect across the junction. Thus, in this model, both local-circuit current and the electrical field effect play roles in the transfer of excitation.

Records from an electronic model of two myocardial cells with a cell junction between them are illustrated in Fig. 4. In this example, the first cell was set to be spontaneously active and fire a spontaneous cardiac-like AP. All four surface-membrane units in cell 1 (U2, 3, 4, 5) fired an AP nearly simultaneously. However, there was a short delay (e.g., ca. 1 msec) in the firing of the unit (U6) representing the prejunctional membrane (at the intercalated disk). Firing of U6 led to the firing of all units (U7, 8, 9, 10, 11, 12) of cell 2 after a junctional delay of about 2 msec. As can be seen, the firing of the prejunctional membrane (U6) drove inward hyperpolarizing current through the postjunctional membrane (U7) and outward depolarizing current through the other units (U8, 9, 10, 11, 12) of cell 2.

## Bibliography

Cohen, S. A., and Levitt, L. K. (1993). Partial characterization of the rH1 sodium channel protein from rat heart using subtype-specific antibodies. *Circ. Res.* **73,** 735–742.

Cole, W. C., Picone, J. B., and Sperelakis, N. (1988). Gap junction uncoupling and discontinuous propagation in the heart: A comparison of experimental data with computer simulations. *Biophys. J.* **53,** 809–818.

Ge. J., Sperelakis, N., and Ortiz-Zuazaga, H. (1993). Simulation of action potential propagation with electronic circuits. *Innov. Tech. Biol. Med.* **14,** 404–420.

Jack, J. J. B., Noble, D., and Tsien, R. W. (1975). "Electric Current Flow in Excitable Cells." Oxford University Press, Oxford.

Mann, J. E., Sperelakis, N., and Ruffner, J. A. (1981). Alterations in sodium channel gate kinetics of the Hodgkin–Huxley equations on an electric field model for interaction between excitable cells. *IEEE Trans. Biomed. Eng.* **28,** 655–661.

Mays, D. J., Foose, J. M., Philipson, L. H., and Tamkun, M. M. (1995). Localization of the Kv1.5 K$^+$ channel protein in explanted cardiac tissue. *J. Clin. Invest.* **96,** 282–292.

Picone, J. B., Cole, W. C., and Sperelakis, N. (1989). Discontinuous conduction in cardiac muscle. *In* "Cell Interactions and Gap Junc-

tions" (N. Sperelakis and W. C. Cole, Eds.), Vol. II, pp. 143–154. CRC Press, Boca Raton, FL.

Picone, J. B., Sperelakis, N., and Mann, J. E., Jr. (1991). Expanded model of the electric field hypothesis for propagation in cardiac muscle. *Math. Comp. Mod.* **15,** 17–35.

Rudy, Y., and Quan, W. L. (1987a). A model study of the effects of the discrete cellular structure on electrical propagation in cardiac tissue. *Circ. Res.* **61,** 815–823.

Rudy, Y., and Quan, W.-L. (1987b). Effects of the discrete cellular structure on electrical propagation in cardiac tissue. *In* "Activation, Metabolism, and Perfusion of the Heart—Simulation and Experimental Models" (S. Sideman and R. Beyar, Eds.), pp. 61–76. Martinus Nijhoff, Cordrecht.

Spach, M. S., Miller, W. T., III, Geselowitz, D. B., Barr, R. C. Kootsey, J. M., and Johnson, E. A. (1981). The discontinuous nature of propagation in normal canine cardiac muscle. Evidence for recurrent discontinuity of intracellular resistance that affects the membrane currents. *Circ. Res.* **48,** 39–54.

Sperelakis, N. (1987). Electrical field model for electric interactions between myocardial cells. *In* "Activation, Metabolism, and Perfusion of the Heart—Simulation and Experimental Models" (S. Sideman and R. Beyar, Eds.), pp. 77–113. Martinus Nijhoff, Amsterdam.

Sperelakis, N., and Mann, J. E., Jr. (1977). Evaluation of electric field changes in the cleft between excitable cells. *J. Theor. Biol.* **64,** 71–96.

Sperelakis, N., and Picone, J. (1986). Cable analysis in cardiac muscle and smooth muscle bundles. *Innov. Tech. Biol. Med.* **7,** 433–457.

Sperelakis, N., Marschall, R., and Mann, J. E. (1983). Propagation down a chain of excitable cells by electric field interactions in the junctional clefts: effect of variation in extracellular resistances, including a "sucrose gap" simulation. *IEEE Trans. Biomed. Eng.* **30,** 658–664.

Sperelakis, N., LoBrocco, B., Mann, J. E., and Marschall, R. (1985). Potassium accumulation in intercellular junctions combined with electric field interactions for propagation in cardiac muscle. *Innov. Tech. Biol. Med.* **6**(1), 24–43.

Sperelakis, N., Picone, J. B., and Mann, J. E. (1989). Electric field model for electric interactions between cells: An alternative mechanism for cell-to-cell propagation. *In* "Cell Interactions and Gap Junctions" (N. Sperelakis and W. C. Cole, Eds.), Vol. II, pp. 191–208. CRC Press, Boca Raton, FL.

Sperelakis, N., Rollins, C., and Bryant, S. H. (1990). An electronic analog simulation for cardiac arrhythmias and reentry. *J. Cardiovasc. Electrophysiol.* **1,** 294–302.

Sperelakis, N., Ortiz-Zuazaga, H., and Picone, J. B. (1991). Fast conduction in the electric field model for propagation in cardiac muscle. *Innov. Tech. Biol. Med.* **12**(4), 404–414.

Tarr, M., and Sperelakis, N. (1964). Weak electronic interaction between contiguous cardiac cells. *Am. J. Physiol.* **207,** 691–700.

Richard D. Veenstra

# Appendix 2: Derivation of the Cable Equation and the AC Length Constant

To better understand the conduction process and how the length constant ($\lambda$), time constant ($\tau$), and conduction velocity ($\theta$) relate to the cellular membrane properties, let us consider the simplest case of a uniform biological cable of length $x$ and radius $a$ (Fig. 1A). The cable has a specific membrane resistivity, $R_m$, an axial (internal core, cytoplasmic) resistivity, $R_i$, a specific membrane capacitance, $C_m$, and an external resistivity, $R_o$, of the surrounding medium (extracellular fluid space), all of which are constant along its entire length. Resistivity is defined as the resistance of a 1-cm$^2$ area (or 1-cm length) of membrane and has the units of $\Omega \cdot cm$ ($R_i$ and $R_o$) or $\Omega \cdot cm^2$ ($R_m$), and the specific membrane capacitance is expressed in units of F/cm$^2$. At any given point along the length, $x$, of the cable, the membrane has an equivalent resistive circuit of a membrane resistance, $r_m$, and capacitance, $c_m$, in parallel between the internal, $r_i$, and external, $r_o$, conductive elements. Hence, the simplified one-dimensional biological cable can be represented by a series of discrete resistive–capacitive (RC) elements coupled in series by two longitudinal conductors (Fig. 1B). The resistance and capacitive elements are related to their specific resistivities and capacitance by the following expressions for a 1-cm length of cable:

$$r_m = \frac{R_m}{2\pi a} \qquad (1)$$

$$r_i = \frac{R_i}{\pi a^2} \qquad (2)$$

$$r_o = \frac{R_o}{1\ cm} \qquad (3)$$

$$c_m = 2\pi a C_m \qquad (4)$$

Commonly, the external resistance is assumed to be negligible and $r_o$ is omitted from further consideration. This is not true when restricted extracellular spaces are encountered (e.g., tight junctions or tortuous extracellular pathways in bundles of packed fibers). In this case, significant barriers to ionic diffusion are encountered in the extracellular space. For now, we consider the first case when $r_o = 0$ (Fig. 1C).

From Ohm's law, we know that $V = I \cdot R$, so in order for a voltage to develop across the membrane ($V_m$), current must be flowing across the membrane. For a point source of constant current (DC = direct current) injection at the center ($x = 0$) of an infinitely long cable, $V_m$ will decay uniformly along the length of the cable in both directions. At each point along the cable, the membrane current, $i_m$, will generate a membrane voltage, $V_m$, equivalent to $i_m \cdot r_m$, or $i_m = V_m/r_m$. Since each segment of membrane has a capacitive element, there must also be a capacitive current ($i_c$) component equivalent to $c_m \cdot dV_m/dt$. That is, the total membrane current ($I_m$) is the sum of the ionic current ($I_i$) through the resistance and the capacitive current ($I_c$) through the capacitance: $I_m = I_i + I_c$. Hence,

$$I_m = (V_m/r_m) + c_m \cdot dV_m/dt \qquad (5)$$

However, $V_m$ cannot be constant along the entire length of the cable, because the internal axial current flow, $i_i$, would have to remain constant along the entire cable in order for this to occur, that is, there must be no leakage of internal axial current across the membrane along the length of the cable. Simply, put,

$$I_m = -i_i/dx \qquad (6)$$

where $dx$ is an infinitesimally small distance along the length of the cable. It follows that the change in $V_m$ over the distance $dx$, $dV_m/dx = -i_i \cdot r_i$. Combining this expression with Eqs. 5 and 6 gives

$$I_m = (1/r_i) \cdot d^2V_m/dx^2 \qquad (7)$$

$$= V_m/r_m + c_m \cdot dV_m/dt \qquad (7a)$$

This is referred to as the linear **cable equation.** Multiplying both sides of the equation by $r_m$ yields the following expression:

$$(r_m/r_i) \cdot d^2V_m/dx^2 = V_m + (c_m \cdot r_m) \cdot dV_m/dt \qquad (8)$$

From this version of the cable equation, it is apparent that the value of $V_m$ at the point $x + dx$ ($dV_m/dx$) is directly pro-

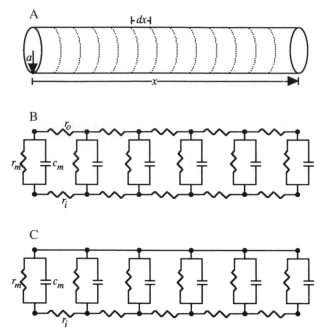

**FIG. 1.** Diagram of a linear cable. (A) Cylindrical model of a linear cable of length $x$ and radius $a$. To derive equations for axial and transverse current flow, the cable is evaluated in small increments of distance ($dx$) along its entire length. (B) Equivalent electrical circuit for the cable illustrated in part A with an axial internal core resistance, $r_i$, an external resistance, $r_0$, and a parallel resistor-capacitor, ($r_m c_m$) circuit, which represents a biological membrane. (C) The equivalent circuit for the cable assuming there is no external voltage gradient (i.e., $r_o = 0$, no resistance to external current flow).

portional to $\sqrt{r_m/r_i}$. It is also apparent that the change in voltage with respect to time at the point $x + dx$ is directly proportional to ($c_m \cdot r_m$). The expression $\sqrt{r_m/r_i}$ is referred to as the **length constant** of the cable, $\lambda$, and ($c_m \cdot r_m$) is the membrane **time constant**, $\tau_m$. These expressions refer to the passive spread of voltage along the length of a linear cable. When $r_o > 0$,

$$\lambda = \sqrt{r_m/(r_i + r_o)} \qquad (9)$$

These conditions apply to the resting muscle fiber or nerve axon, provided that the fiber or axon is several $\lambda$ long and its geometry approximates a linear cable (e.g., no branch points for axial current flow). However, conduction of the action potential (AP) involves the spread of current flow along a passive membrane cable segment from an advancing active membrane segment. The transition from a passive to an active membrane response (i.e., the initiation of an AP) requires that a certain "threshold" voltage level be achieved in order for the segment of membrane undergoing a voltage charge in response to passive current spread (as defined by the cable equation given earlier) to generate an active response. We will see later that the excitation threshold is not actually a voltage value, rather it is equivalent to the amount of charge required to depolarize the membrane and produce an active current response. During the AP, the active portion of the membrane becomes a source of depolarizing current, which contributes to the distal spread of the voltage deflec-

tion. The value $r_m$, or $1/g_m$, is actually the sum of several ionic components resulting from the presence of specific proteins (ion channels and pumps) that permit the selective movement of ions across the membrane. These discrete resistive membrane elements are associated with the activity of ion channels (e.g., $Na^+$, $K^+$, $Ca^{2+}$, and $Cl^-$ channels) and pumps (e.g., $Na^+$,-$K^+$-ATPase, $Na^+$-$Ca^{2+}$ exchanger) associated with the cell membrane. Because each ion possesses a net charge (valence), the translocation of an ion across the membrane produces a current ($I_i$). The activities of the protein channels and transporters are not constant, but are regulated by a variety of mechanisms including membrane voltage ($V_m$). Hence, at different points in time ($dt$), $r_m$ is not constant, but rather is a function of voltage and time, $r_m$ ($V, t$). So a voltage deflection $dV_m$ alters $r_m$, which further affects $V_m$. This constitutes a feedback loop where $r_m$ and $V_m$ are dependent variables (see Chapter 24).

In a passive cable, any voltage deflection caused by the decrement in axial current flow in that particular segment of membrane is initially *opposed* by existing membrane currents, which *counteract* the voltage deflection and maintain the resting (passive) state. This is a **negative feedback** mechanism and is predominant at all times except during the initiation of an action potential. When the voltage deflection exceeds the **threshold** for excitation, certain ion channels are activated, which leads to the generation of more ionic current and a greater voltage deflection in the same direction as the initiating response. Hence, the voltage response of the membrane becomes *additive* to the initial passive $dV_m$ and a **positive feedback** loop is transiently produced. If allowed to persist, a positive feedback loop leads to prolonged excitation, which prevents the return to the resting state and the generation of any further action potentials.[1] Therefore, if the positive feedback loop is to be effective in generating an action potential, it must also be terminated once that action potential is initiated.

The threshold for excitation is the critical value of $V_m$ ($V_{th}$), which must be achieved in order for the active membrane response to commence. However, as observed from the equivalent circuit and cable equation, there are both resistive and capacitive elements to the membrane current. Current is also equivalent to the derivative of charge flowing per unit time (e.g., coulombs/per second). Hence, the resistive and capacitive current components of $i_m$ can be expressed in terms of $dQ/dt$, where $Q$ is the charge in coulombs. From Faraday's law applied to our segment of membrane,

$$Q_m = C_m \cdot V_m \qquad (10)$$

So the threshold for excitation is actually the amount of charge ($Q_{th}$) that must flow across the membrane to achieve the required membrane voltage ($V_{th}$) to initiate the action potential. Once initiated, the action potential propagates down the length of the cable at a rapid rate. The **conduction velocity** ($\theta$) is the distance traveled per unit time or $dx/dt$. Returning to our cable equation (Eq. 8) and solving for $dx/dt$ we obtain:

---

[1] One example of overamplification and positive feedback is the annoying hum of the microphone heard over the audio speaker system of any soundstage production.

$$\lambda^2 \cdot 1/\theta^2 \cdot d^2V_m/dt^2 = V_m. + \tau_m \cdot dV_m/dt \tag{11}$$

It is now evident that the rate at which the action potential propagates along the length of passive cable elements depends on the length and time constants of the membrane or

$$\theta^2 \propto \lambda/\tau \tag{12}$$

To summarize, the cable equation can be expressed in several equivalent forms:

$$I_m = V_m/r_m. + c_m \cdot dV_m/dt \tag{5}$$
$$= (1/r_i) \cdot d^2V_m/dx^2 \tag{7}$$

or

$$(r_m/r_i) \cdot d^2V_m/dx^2 = V_m + (c_m \cdot r_m) \cdot dV/dt \tag{8}$$
$$= \lambda^2 \cdot d^2V_m/dx^2 \tag{13}$$
$$= V_m + (\tau_m) \cdot dV_m/dt \tag{14}$$

or

$$\lambda^2 \cdot 1/\theta^2 \cdot d^2V_m/dt^2 = V_m. + \tau_m \cdot dV_m/dt \tag{11}$$

Several additional derivations are possible that depend on the boundary conditions applied to the cable such as length, extracellular isopotentially ($r_o = 0$), voltage uniformity (i.e., voltage clamp), and uniformity of geometry (e.g., fiber radius $a$ or branch points).

Let us consider further our linear cable of uniform diameter $2a$, length $x$, and external resistance $r_o = 0$. Throughout the passive cable analysis described previously, we have only considered the case for which the voltage displacement along the length of this cable is produced by a constant current (DC) source, $I_0$, located at the exact center ($x = 0$) of the cable. We have shown that $\lambda^2 = (r_m/r_a)$ or

$$\lambda = \frac{(R_m/2\pi a)}{(R_i/\pi a^2)} \tag{15}$$
$$= \left(\frac{aR_m}{2R_i}\right) \tag{15a}$$

or

$$\lambda = \sqrt{(R_m/R_i)(a/2)} \tag{15b}$$

and

$$\tau_m = (c_m \cdot r_m) \tag{16}$$

or

$$\tau_m = \left(\frac{R_m}{2\pi a}\right)(2\pi a C_m) \tag{16a}$$
$$= R_m \cdot C_m \tag{16b}$$

This implies that $\lambda^2$ (and consequently $\theta^2$) $\propto (a/2)$, whereas $\tau_m$ is independent of fiber radius and membrane area (as long as $R_m$, $R_i$, and $C_m$ remain constant). Under steady-state conditions,

$$0 = (r_m/r_a) \cdot d^2V_m/dx^2 - \tau_m \cdot dV_m/dt - V_m \tag{17}$$
$$dV/dt = 0 \tag{18}$$

so the cable equation reduces to

$$0 = \lambda^2 \cdot d^2V_m/dx^2 - V_m. \tag{19}$$

To solve this differential equation, let us define $X = x/\lambda$ (Taylor, 1963; Jack *et al.*, 1983). It follows that $dX/dx = 1/\lambda$, so $\lambda^2 = (dx/dX)^2$ and the equation becomes $0 = d^2V_m/dX^2 - V_m$. The solution to $dX/dx$ is $e^X$ and $1/(dX/dx) = e^{-X}$. The solution to our differential equation therefore becomes

$$V_m = V_0 e^{-X} \tag{20}$$

or

$$V_m = V_0 e^{-(x/\lambda)} \tag{21}$$

where $V_0$ = the value of $V_m$ at $x = 0$. As $x \to 0$, $V_m \to V_0$ and spatial uniformity ($V_m = V_0$) of membrane voltage is approached.

Another consideration is the time required for the membrane to achieve its steady-state value. Again,

$$0 = \lambda^2 \cdot d^2V_m/dx^2 - \tau_m \cdot dV_m/dt - V_m \tag{17a}$$

except in this case $d^2V_m/dx^2 = 0$, since we are observing the same point $x$ at different times during the current injection and $dV_m/dt \neq 0$. Now our equation reduces to

$$0 = \tau_m \cdot dV_m/dt + V_m \tag{22}$$

which is again a differential equation. Let us define $T = t/\tau_m$, and it follows that $dT/dt = 1/\tau_m$ (Taylor, 1963; Jack *et al.*, 1983). Our differential equation takes the form

$$0 = dV_m/dT + V_m \tag{23}$$

which has the solution

$$V_m = V_0 e^{-(t/\tau)} \tag{24}$$

where $V_0$ = the value of $V_m$ when $t = 0$. In conclusion, for a uniform fiber, $V_m$ decays over distance and time by a first-order exponential decay process.

These solutions to the differential equations apply only to the specified conditions described earlier. In the absence of spatial uniformity (but $r_o$ still equals 0), the generalized cable equation $0 = d^2V_m/dX^2 - dV_m/T - V_m$ will have different complex solutions. The solutions to several different boundary conditions are derived in Rall (1977). *In situ*, the condition of spatial uniformity does not apply to a muscle fiber or nerve axon due to a large $r_a$. Hence, for a constant current injection ($I_0$) at $X = 0$ and time $T$, the cable equation has the general solution of (Hodgkin and Rushton, 1946):

$$V_m = \left(\frac{r_a I_0 \lambda}{4}\right) e^{-X} \left[1 - \mathrm{erf}\left(\frac{X}{2\sqrt{T}}\right) - \sqrt{T}\right] \tag{25}$$
$$- e^X \left[1 - \mathrm{erf}\left(\frac{X}{2\sqrt{T}}\right) + \sqrt{T}\right]$$

For $X = x$ and $t = \infty$, this equation reduces to

$$V_m = V_0 e^{-(x/\lambda)} \tag{21}$$

which is the same solution as derived earlier for the decay in $V_m$ as a function of distance. Also, $V_0 = (r_a I_0 \lambda)/2$ since half of the injected current will flow in either direction down the long axis of the cable. However, the special solution for $V_m$ as a function of time at $x = 0$ has the solution

$$V_m = V_0 \text{ erf } \sqrt{t/\tau} \qquad (26)$$

which differs from the previous derivation of the time-dependent voltage response for a space-clamped cable. Because erf(1) = 0.84, when $t = \tau$, $V_m = 0.84V_0$ for the non-space-clamped fiber or axon instead of $0.63V_0$ in the case of the space-clamped cable (Eq. 24; Jack *et al.*, 1983).

Another concept that can be derived from the cable equation is the **input resistance, $R_{in}$**, which is best defined as the resistance of the cable resulting from the internal resistance and membrane resistance over the length of the cable with diameter $2a$. According to Ohm's law,

$$R_{in} = V_0/I_0 \qquad (27)$$

at $t = \infty$ and $x = 0$, or

$$R_{in} = \sqrt{r_m r_a}/2 \qquad (28)$$

$$= \sqrt{(R_m R_a/2)}/(2\pi \sqrt{a^3}) \qquad (28a)$$

Hence, the input resistance of a fiber is proportional to $a^{3/2}$. With these derivations it should be apparent how the length constant, time constant, voltage, conduction velocity, and input resistance of a one-dimensional fiber or axon are related to the parameters of radius, membrane resistance, axial resistance, and current. Other derivations can be derived based on different boundary conditions (e.g., Rall, 1977).

When an **alternating current** (AC) is applied to an RC circuit, the solution is more complex. Capacitive current, $I_c = C_m dV/dt$, and for a sine wave, $dV/dt = 0$ when the current is at its maximum ($I_p$). Simply stated, the voltage across the capacitor ($V_c$) lags behind the current across the resistor by 90°. The current across the resistor is now a sine wave and takes the form of $V_p \sin(\omega t)/r_m$, where $V_p$ = peak voltage of the sine wave. Because $dV_p \sin(\omega t)/dt = V_p \omega \cos(\omega t) = V_p \omega \sin(\omega t - 90°)$, the current equation for our circuit becomes

$$I_p = V_p \sin(\omega t)/r_m + c_m V_p \omega \sin(\omega t - 90°) \qquad (5a)$$

where $\omega = 2\pi f$ and $f$ = frequency of the sine wave (in Hertz). Ohms' law states that $V = IR$, so it follows that $1/\omega C$ has units of resistance (ohms = $\Omega$). Let us define the capacitive **reactance;**

$$X_c = -1/\omega C \qquad (29)$$

Let us further define **impedance;**

$$Z = R + jX \qquad (30)$$

where $R$ is the purely resistive component, $X$ is the reactance of the circuit, and $j = \sqrt{-1}$. Actually, the membrane should always be thought of as having an impedance. In the DC case, $X_c = 0$ because $\omega = 0$ (since there is no frequency component) and the membrane exhibits only a resistance. Since the axial and external resistances in our

linear cable do not contain a capacitive component, $R_a$ and $R_o$ are unaffected. Recall that the steady-state solution for voltage as a function of distance $x$ is

$$V_m = V_0 e^{-(x/\lambda)} \qquad (21)$$

where $V_0 = (r_a I_0 \lambda)/2$, and $\lambda = \sqrt{r_m/r_a}$. We can rewrite Eq. 21 by substituting for $V_0$ and $\lambda$ in the preceding expression, thus obtaining

$$V_m = (I_0/2)\sqrt{r_m r_a}\, e^{-(x/\sqrt{r_m/r_a})} \qquad (31)$$

Substituting $z_m$ for $r_m$ in the preceding expression results in the following expression:

$$V_m = (I_0/2)\sqrt{z_m r_a}\, e^{-(x/\sqrt{z_m/r_a})} \qquad (32)$$

The **membrane impedance, $z_m$**, is a complex number consisting of **real** (resistance) and **imaginary** (reactance) components. So the previous expression should be further subdivided into real and complex numbers. Let us define a **propagation constant, $\gamma$**, which is a complex number composed of an **attenuation factor, $\alpha$**, and a **phase constant, $\beta$**, such that

$$\gamma = \alpha + j\beta \qquad (33)$$

The expression for the peak voltage of our circuit is

$$V_p = (i_p/2)\sqrt{|z_m| r_a}\, e^{-(\gamma x)} \qquad (34)$$

where $|z_m|$ is the magnitude of the impedance $= \sqrt{R^2 + X^2}$ (Eisenberg and Johnson, 1970; see also the Appendix at the back of this book). It follows from Eq. 34 that $\lambda_{AC} = 1/\gamma$. It is easier to solve for the preceding expression by assuming the frequency of the injected current is high and $z_m \approx jX_c = j/\omega c_m$. This occurs when $f > 1/(2\pi r_m c_m)$. Also, $X_c = r_m$ when $f = 1/(2\pi r_m c_m) = f_b$ is the **cutoff** or **corner frequency** for the circuit. Equation 34 now becomes

$$V_p = (i_p/2)\sqrt{jr_a/\omega c_m}\, e^{-(\gamma x)} \qquad (35)$$

and $\gamma = 1/\gamma = 1/\sqrt{z_m/r_a} = 1/\sqrt{j/\omega c_m r_a} = \alpha + j\beta$. Solving for $\alpha$ and $\beta$, we obtain

$$\alpha = 1/\sqrt{2/\omega c_m r_a} = \sqrt{\omega c_m r_a}/2 \qquad (36)$$

and

$$j\beta = j\sqrt{\omega c_m r_a}/2 \qquad (37)$$

Since $\lambda_{AC} = 1/\gamma$, it follows that $\lambda_{AC} = 1/\alpha = \sqrt{2/\omega c_m r_a}$ or

$$\lambda_{AC} = \sqrt{2/2\pi f c_m r_a} = \sqrt{1/\pi f c_m r_a} \qquad (38)$$

When $f < f_b$, $r_m > X_m$ and $z_m \approx r_m$ so $\lambda = \sqrt{r_m/r_a}$, which is equivalent to the DC case. However, even when a DC current is applied, whenever there is a rapid change in the amount of current the DC situation does not apply until the capacitor is charged to its new voltage. So $\lambda_{AC}$ is applicable whenever there is a rapid change in voltage until the steady-state value is achieved. The net effect of the frequency dependence of $\lambda_{AC}$ is that the length constant becomes shorter at higher frequencies (e.g., instantaneous steps in membrane voltage).

# Bibliography

Eisenberg, R. S., and Johnson, E. A. (1970). Three-dimensional electrical field problems in physiology. *Prog. Biophys. Molec. Biol.* **20,** 5–65.

Hodgkin, A. L., and Rushton, W. A. H. (1946). The electrical constants of a crustacean nerve fibre. *Proc. R. Soc.* **B133,** 444–479.

Jack, J. J. B., Noble, D., and Tsien, R. W. (1983). "Electric Current Flow in Excitable Cells." Clarendon Press, Oxford.

Rall, W. (1977). Core conductor theory and cable properties of neurons. *In* "Handbook of Physiology" (J. M. Brookhart and V. B. Mountcastle, Eds.), Vol. 1, Sec. 1: The Nervous System, pp. 39–97. American Physiological Society, Bethesda, MD.

Taylor, R. E. (1963). Cable theory. *In* "Physical Techniques in Biological Research" (W. L. Nastuk, Ed.), Vol. 6B, pp. 219–262. Academic Press, New York.

Nicholas Sperelakis

# 24

# Electrogenesis of Membrane Excitability*

## I. Introduction

Excitability is an intrinsic membrane property that allows a cell to generate an electrical signal or action potential (AP) in response to stimuli of sufficient magnitude. The elongated nerve axon serves to transmit information in the form of APs over long distances. The AP mechanism is required to propagate a uniform signal in a nondecremental manner. In muscle cells, the AP serves to spread excitation over the entire cell surface and is involved in triggering cell contraction.

The energy source for the generation of the AP is stored in the excitable cell itself. An initial depolarization is produced by a stimulus and triggers the intrinsic AP mechanism. The immediate source of energy for the AP comes from the transmembrane ionic gradients for $K^+$ and $Na^+$, which act like a battery. The $K^+$ ion concentration gradient is mainly responsible for the generation of the resting potential, which causes an excess of negative charge to build up on the inner surface of the membrane. Upon depolarization to threshold, the $Na^+$ ion electrochemical driving force, which is directed inward, causes a large and rapid inward $Na^+$ current that generates the AP upstroke. Over a longer time frame, the Na–K pump is responsible for the generation of the $Na^+$ and $K^+$ ionic gradients and for their maintenance and restoration after repetitive AP activity. The Na–K pump derives chemical energy from the hydrolysis of ATP. The bases of the resting potential and active ion transport have been discussed in earlier chapters.

Important technological improvements have led to significant advances in the understanding of the basis of membrane excitability. In the early 1900s, several theories were proposed to explain the mechanism that produces the AP. Julius Bernstein (1902, 1912) proposed that the excitable cell membrane, at rest, was selectively permeable

to $K^+$ ions (producing the resting potential), and that during excitation the membrane became permeable to all ions (producing the AP). About the same time, Overton (1902) had demonstrated that $Na^+$ ions were essential for excitability.

By the 1940s, improvements had been achieved in electronic instrumentation, especially in the high-input impedance amplifiers necessary to record bioelectric phenomena using tiny intracellular microelectrodes. In addition, biophysicists began to study the squid giant axon (500–1000 $\mu$m in diameter), which permitted insertion of relatively large intracellular electrodes, yielding the first measurements of the true transmembrane potential. The transmembrane potential is recorded as the difference between an intracellular and an extracellular electrode (Fig. 1). The findings from the squid giant axon were successfully applied to the smaller diameter (1–20 $\mu$m) neurons found in the vertebrate nervous system (Fig. 2).

## II. Action Potential Characteristics

Action potentials in a given fiber (e.g., myelinated nerve fiber) have the properties of being "all or none," having a sharp threshold, and having a refractory period. All impulses look alike, being very similar in shape, amplitude, and duration. Thus, they constitute a digital system. Strength of the signal is conveyed by changing the frequency of the impulses, as discussed in the preceding chapter on propagation. The all-or-none property means that the single signal is not graded in amplitude (i.e., not an amplitude-modulated system) and that the signal is either zero ("off" or "no") or maximum ("on" or "yes"). These characteristics of the AP are discussed in the next section.

### A. Local-Circuit Currents

During propagation of an AP down a nerve fiber, current flow accompanies the propagating change in membrane voltage. This current is called the **local-circuit current,** and

---

* Adapted and reprinted by permission from the author's chapter 4, with I. R. Josephson, in PHYSIOLOGY (Sperelakis, N. and Banks, R. O., Editors). Copyright © by Nicholas Sperelakis and Robert O. Banks. Published by Little, Brown and Company.

**368**

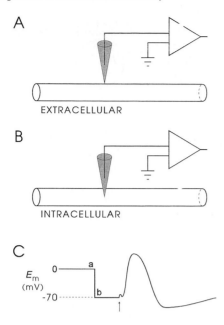

**FIG. 1.** Recording the transmembrane potential of a squid giant nerve axon. The transmembrane potential is measured as the potential difference between the intracellular and extracellular electrodes. (A) The microelectrode is outside the axon, and measures 0 mV [segment a in (C)]. As the electrode is advanced and crosses the membrane (B), the resting potential of $-70$ mV is measured [segment b in (C)]. The cell membrane makes a tight seal around the glass electrode. (C) The membrane potential ($E_m$) is depicted. Stimulation (at arrow) elicits an action potential.

has both longitudinal and radial (transverse) components that make a complete circuit (Figs. 3A and B). Propagation and local-circuit currents have been thoroughly discussed in the preceding chapter. For our purpose here, we focus on the longitudinal component. The intracelluar and the extracellular (external) longitudinal current are exactly equal in amplitude, but flow in opposite directions (Fig. 3A). The intracellular current is, of course, confined to the cross-sectional area of the nerve fiber (neuroplasm), whereas the extracellular current can use the entire extracellular volume (so-called "*volume conductor*"). If a single nerve fiber or nerve bundle is mounted in air, then the extracellular action current is confined to the surface film

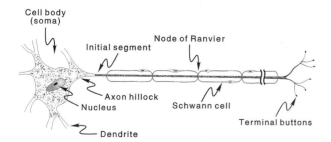

**FIG. 2.** Diagram of a motor neuron with myelinated axon. the major structural elements are diagrammed, including cell body, initial segment, Schwann cell, node of Ranvier.

of fluid adhering to the single fiber or the interstitial fluid space between fibers in the bundle.

To illustrate the principles involved, let us record externally with a pair of electrodes from a single nerve fiber (e.g., squid giant axon) bathed in air (Fig. 3C). As the AP moves from left to right past the pair of recording electrodes, the first electrode becomes negative with respect to the second electrode. Therefore, the voltmeter swings in one direction (an upward deflection is defined as negative in extracellular recording) due to the potential difference (pd). When the wavefront moves to a position between the two electrodes, the voltmeter will record almost zero voltage. When the AP reaches the second electrode, the voltmeter will swing in the opposite direction, because the second electrode is now negative with respect to the first. Thus, the electrodes will record a biphasic change in voltage as the AP sweeps past. Thus, when recording externally, the AP is often described as a wave of negativity sweeping down the fiber; when recording internally, the AP is a wave of positivity.

The reason for the wave of negativity in external recording is that the fluid between the two electrodes constitutes a resistance ($R$), and longitudinal current ($I$) flowing through this fluid resistance produces an $IR$ voltage drop (Ohm's law: $V = IR$). As the wavefront approaches the first electrode, the external local-circuit current is from right to left and produces an $IR$ drop, positive (second electrode) to negative (first electrode). Conversely, when the AP passes beyond the first electrode, the action current is now reversed in direction and produces an $IR$ drop, positive (first electrode) to negative (second electrode). This produces the biphasic voltage recording.

**Propagation velocity** is a function of axon diameter and myelination, such that the larger the diameter, the greater the velocity. **Myelinated axons** conduct much faster than **nonmyelinated axons.** The factors that determine propagation velocity and the mechanism of saltatory conduction are discussed in the preceding chapter.

## B. Threshold and All-or-None Property

Nerve membrane responses near the site of application of brief current pulses vary depending on the magnitude and direction of the pulses (Fig. 4). Anodal (inward) currents produce hyperpolarization, and cathodal (outward) currents produce depolarization. Hyperpolarizing and subthreshold depolarizing responses are graded in magnitude according to the stimulus current. However, as can be seen, a somewhat higher intensity outward current produces a depolarizing response with a different waveform and a longer lasting duration. This condition is referred to as the **local excitatory state** (Figs. 4A and B). It occurs when a small area of the membrane near the stimulus electrode comes close to threshold, but it does not generate an AP. This local membrane activity is not propagated and decays with distance along the axon. A slightly stronger stimulus is, however, sufficient to bring the membrane potential of a large enough membrane area to the threshold potential, thereby initiating an all-or-none AP. At the threshold potential, there is a greater amount of inward (depolarizing)

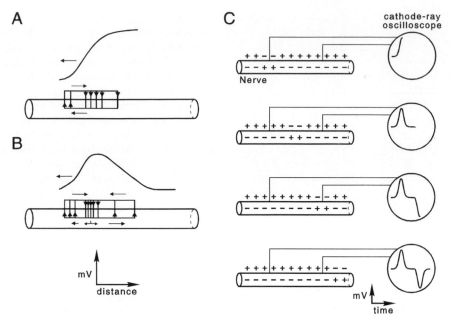

**FIG. 3.** Diagram of the local-circuit current in a nerve fiber during an AP. (A) The current during the rising phase; (B) the current during the entire AP. The AP is depicted as propagating from right to left. (C) The actual registration of the signal is depicted for an AP propagating from left to right past a pair of electrodes spaced relatively far apart.

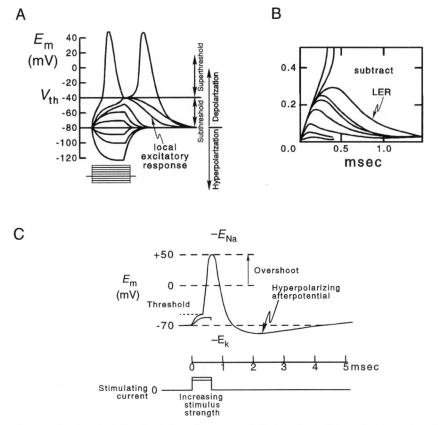

**FIG. 4.** Diagrammatic sketches depicting initiation of the nerve impulse by membrane depolarization. (A) Membrane responses to depolarizing and hyperpolarizing current pulses are shown. The nonlinear local excitatory response (LER) occurs just below the threshold for the all-or-none AP. (B) Illustration of the LER obtained by subtraction of the membrane responses to hyperpolarizing current pulses (passive only) from the responses to depolarizing pulses (passive plus active). (C) A nerve AP is depicted, illustrating the sharp threshold, overshoot, and hyperpolarizing afterpotential.

than outward (repolarizing) current, and the membrane will continue to depolarize.

In regard to the applied stimulus, outward current depolarizes the membrane ($IR$ drop across the resting membrane is positive inside, negative outside), whereas inward current hyperpolarizes ($IR$ drop is negative inside). In contrast, for the active membrane, inward currents are depolarizing (bringing positive charge inside the cell), and outward currents are hyperpolarizing. That is, when the membrane is behaving passively, applied inward current is hyperpolarizing and outward current is depolarizing; this is equivalent to a circuit external to a battery. In contrast, when the membrane is behaving actively, inward current generated is depolarizing, and outward current is hyperpolarizing; this is equivalent to an internal circuit within a battery. This difference is related to the fact that current flows from positive to negative in the external circuit, and from negative to positive within the battery itself.

A subthreshold depolarization is defined as one that does not reach threshold to elicit an AP. It is proportional to the applied stimulus, and it decrements with distance along the nerve axon cable. If a stimulus current is of sufficient magnitude (i.e., enough positive charges are transferred into the cell), then the resulting membrane depolarization reaches a critical value, called the **threshold potential,** at which an AP is initiated (see Fig. 4).

The AP parameters, including overshoot, duration, and rate of rise, are characteristic for each type of excitable cell. For example, the duration of the AP of the squid giant axon is about 1 msec, whereas the cardiac AP lasts for several hundred milliseconds. These differences in the APs subserve the functions performed by the different excitable tissues. The overshoot and a hyperpolarizing afterpotential (following the spike) are illustrated in Fig. 4C.

## C. Refractoriness

Once an AP is initiated, a finite and characteristic time must elapse before a second AP can be generated. This time interval is called the **refractory period,** and its value depends on the type of excitable cell. Cells with long-duration APs (e.g., myocardial cells) have long refractory periods; cells with brief APs (e.g., neurons) have short refractory periods. That is, the refractory periods are proportional to the AP duration.

Two types of refractory periods are usually defined: an **absolute refractory period** and a **relative refractory period** (Fig. 5). The absolute refractory period denotes the interval during which a second AP cannot be elicited, regardless of the intensity of the applied stimulus. During the relative refractory period, a second AP may be elicited, provided that a greater than normal stimulus is applied. The second AP often is subnormal in amplitude and rate of rise. Therefore, the physiologically important refractory period is the **functional (or effective) refractory period,** which is defined by the highest frequency of APs that the excitable cell (e.g., neuron) can propagate. For example, if a myelinated nerve axon can propagate impulses up to 1000/sec, then the functional refractory period is 1.0 msec. The triggering of a second impulse at a given point (or node) is limited by the amount of action current available from an active point (or node) upstream (unlike an electronic stimulator). Therefore, the functional refractory period encompasses all of the absolute refractory period and part of the relative refractory period.

The absolute refractory period extends from when threshold ($V_{th}$) is reached at the initial portion of the rising phase of the AP to when repolarization has reached a level of about −50 mV. During further repolarization beyond −50 mV (e.g., to −70 mV), a larger and larger fraction of the fast $Na^+$ channels recovers from inactivation, and so more fast channels are again available to be reactivated to produce another AP. This is the period that inscribes the relative refractory period. The greater the degree of repolarization (toward the resting potential), the larger the subsequent AP.

In addition to voltage, time is a factor in the recovery of the ion channels. Therefore, the absolute and relative refractory periods persist briefly beyond the theoretical voltages, and the relative refractory period actually slightly exceeds the AP duration.

Membrane excitability is greatly altered during the refractory periods. Excitability is zero during the absolute refractory period, and is depressed during the relative re-

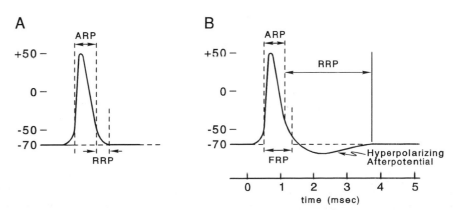

**FIG. 5.** Refractory periods of nerve action potentials, both without (A) and with (B) a hyperpolarizing afterpotential. The absolute (ARP), relative (RRP), and functional (FRP) refractory periods are labeled.

fractory period, becoming less and less depressed as the membrane repolarizes back to the resting potential (Fig. 6).

The afterpotentials that many cells exhibit also affect membrane excitability: hyperpolarizing afterpotentials depress excitability (greater critical depolarization required to reach $V_{th}$) and depolarizing afterpotentials enhance excitability. The latter produces a supernormal period of excitability, and the former blends into and extends the relative refractory period. Thus, for example, since neurons usually exhibit hyperpolarizing afterpotentials, the functional refractory period described previously actually includes part of the afterpotential.

Propagation velocity is slowed during the relative refractory period, achieving the normal value at the end of the relative refractory period. Propagation velocity is slightly faster than normal during the supernormal period of excitability.

### D. Strength–Duration Curve

Whether the threshold potential is reached depends on the amount of charge transferred across the membrane. Figure 7 shows that the total charge transfer across the membrane required to produce excitation is approximately constant (since $Q = I \cdot T$). It is an approximate rectangular hyperbola ($x \cdot y = k$), over the sharply bending region of the curve. The strength–duration (S–D) curve can be derived from the equation for the exponential charge of the membrane capacitance.

The S–D curve deals only with the stimulus parameters (i.e., strength and duration of the applied current pulses) necessary to bring the membrane to threshold. It shows that the greater the duration of the applied pulse, the smaller the current intensity required to just excite the fiber. The asymptote parallel to the $x$ axis is the **rheobase,** which is the lowest intensity of current capable of producing excitation, even when the current is applied for infinite time (practically, >10 msec for myelinated nerve fibers). The asymptote parallel to the $y$ axis is the **minimal stimulation time,** which is the shortest duration of stimulation capable of producing excitation, even when huge currents are applied.

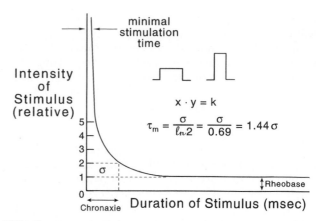

**FIG. 7.** Strength–duration curve for AP initiation in excitable membranes. The intensity of rectangular stimulating pulses is plotted against their duration for stimuli that are just sufficient to elicit an AP. The rheobase current and chronaxie ($\sigma$) are indicated.

The rheobase is useless when comparing the excitability of one nerve with another because only the relative current intensity is meaningful. Furthermore, it is difficult to measure the stimulation time of a current with the intensity of the rheobase because it is an asymptote. Thus, a graphic measurement is made of the *time* during which a stimulus of double the rheobasic strength must act in order to reach threshold. This time is the **chronaxie.** Chronaxie values tend to remain constant regardless of geometry of the stimulating electrodes. The shorter the chronaxie, the more excitable the fiber. The chronaxie value for normal myelinated nerve fibers is about 0.7 msec. Some nerve pathologies in humans can be detected early by changes in their chronaxies.

Measurements of chronaxie in the laboratory is also valuable because it provides an easy method for measuring the value of the membrane time constant ($\tau_m$) (see Chapter 23 on cable properties and the Appendix to the book for a review of electricity). In brief, the relationship between chronaxie ($\sigma$) and time constant ($\tau_m$) is

$$\tau_m = \frac{\sigma}{\ln 2} = \frac{\sigma}{0.69} = 1.44\sigma \tag{1}$$

Thus, $\tau_m$ is 1.44 times the value of $\sigma$. Therefore, $\sigma$ is analogous to a half-time for a first-order reaction, whose rate constant is the reciprocal of the $\tau_m (k = 1/\tau_m)$.

The S–D curve indicates that current pulses of very short duration (e.g., <0.1 msec) are less effective for stimulation. Thus, sinusoidal alternating current (ac) at frequencies above 10,000 Hz is less capable of stimulation. Another way to view this is that, because the membrane impedance decreases greatly at high frequencies (since the cell membrane is a parallel RC network), the potential difference that can be produced across the membrane by current flow across it (*IR* or *IX* drops) is very small. Hence, ac of very high frequency has less tendency to electrocute, but the energy of such currents can be dissipated as heat in body tissues, and thus may be used in diathermy for therapeutic warming of injured tissues.

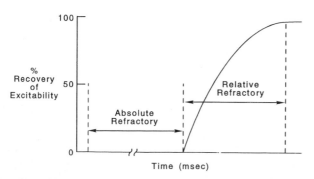

**FIG. 6.** The time course of recovery of nerve excitability during the relative refractory period. As shown, during the absolute refractory period, an AP cannot be elicited, and there is a progressive increase in excitability during the relative refractory period.

## E. Accommodation

**Accommodation,** in electrophysiology, refers to the loss of sensitivity of a cell to an applied stimulus. Sensory organs exhibit the property of accommodation, as do many neurons and other excitable membranes, such as skeletal muscle fibers. When a rectangular (square wave) current pulse is used to depolarize a quiescent motor neuron ("step depolarization") from the resting potential (e.g., $-70$ mV) to the threshold potential ($V_{th}$) or beyond, the neuron quickly responds with an all-or-none AP. However, if the applied pulse is ramp shaped (triangular), the neuron may or may not respond, even if the normal $V_{th}$ is exceeded, depending on the slope of the ramp. If the slope of the ramp is steep, the neuron will respond, but at a higher $V_{th}$ level (more *critical depolarization* is required). If the slope is shallow, the neuron will fail to fire an AP, regardless of the level to which it is depolarized. This is accommodation. That is, when the membrane is depolarized gradually, the stimulus is ineffective in producing an AP response (Fig. 8A).

This is not true of pacemaker cells (Fig. 8B). Automatic (spontaneously discharging) cells do not exhibit accommodation of their membrane to stimuli capable of evoking AP responses (e.g., nodal pacemaker cells of the heart and some sensory neurons). Such cells will discharge an AP no matter how gradually the membrane is brought to the $V_{th}$ point. In fact, the pacemaker potential in cardiac nodal cells is of the ramp type, producing depolarization to $V_{th}$ over a period of about 200–800 msec.

The explanation for this phenomenon of accommodation is as follows. As the membrane is slowly depolarized toward $V_{th}$, the positive feedback cycle between $E_m$, Na$^+$ conductance ($g_{Na}$), and $I_{Na}$ begins to operate (beginning at about 80% of the critical depolarization). Therefore, some of the fast Na$^+$ channels are turned on (activated), only to inactivate spontaneously (I-gates close) within 1–2 msec. If a critical number (*critical mass*) of fast Na$^+$ channels are not activated simultaneously, then the positive feedback cycle does not become explosive, and a regenerative AP is not produced. That is, the slow depolarization does not allow a critical number of fast Na$^+$ channels to be simultaneously in the open conducting state. The channels that opened and spontaneously inactivated cannot be reactivated until they return to the resting state, which requires repolarization to near the resting potential. Hence, they are lost from the pool of available channels.

Added to this is the fact that K$^+$ channels (delayed rectifier type) will open during the slow depolarization, thus increasing K$^+$ conductance ($g_K$). Whenever $g_K$ is increased, excitability becomes depressed, because it tends to repolarize and keep the membrane from depolarizing (to produce an AP); lowering $R_m$ also lowers the effectiveness of the depolarizing current.

Therefore, accommodation to low-slope stimuli occurs for two reasons: (1) spontaneous inactivation of fast Na$^+$ channels that have been activated, and therefore lack of a simultaneously open critical mass and (2) increase in $g_K$, which depresses excitability.

The lack of accommodation in automatic cells (e.g., SA nodal cells of the heart) then may be due to (1) less spontaneous ion channel inactivation and (2) less $g_K$ increase during the applied ramp stimulus or natural ramp pacemaker potential. Both of these conditions apparently apply to cardiac nodal cells. The inward current responsible for the rising phase of the AP is not a fast Na$^+$ current, but rather a slow Ca$^{2+}$ current, which inactivates very slowly, and the kinetics of the turn-on of the delayed rectifier K$^+$ current is also very slow.

As stated previously, accommodation also occurs in sensory organs. For example, some stretch receptors accommodate to a sustained stretch. When the stretch is first applied, there is a burst of APs. But the bursting frequency of discharge gradually slows down and then stops, even though the stretch is maintained.

## F. Anodal-Break Excitation

Excitation occurs on the "make" (the beginning) of a square-wave depolarizing stimulus. If the applied stimulus duration is very long (relative to the AP duration), repetitive firing of APs will occur if the membrane is of the nonaccommodating type. If the membrane is of the accommodating type, then only the initial AP is produced, because accommodation occurs.

If the cathode (negative) and anode (positive) electrodes are placed directly on an isolated single nerve axon (e.g., squid giant axon), then an AP will be triggered at the cathode region on the make of the square-wave stimulus. This happens because, as depicted in Fig. 9A, depolarization occurs under the cathode, whereas hyperpolarization occurs under the anode. However, something unexpected

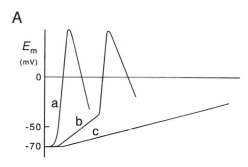

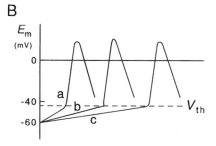

**FIG. 8.** (A) The process of accommodation in response to a ramp stimulus in a motor neuron and (B) contrasted with the lack of accommodation in a nodal pacemaker cell from the heart. In A, as the slope of the ramp stimulus is decreased (from a to c), the action potential is delayed (b), and then fails completely (c). $V_{th}$, threshold potential.

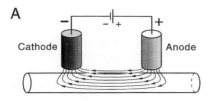

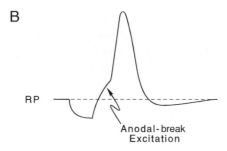

**FIG. 9.** Current flow under the anode and cathode during extracellular stimulation (A) and anodal-break stimulation of an action potential (B). Excitation occurs under the cathode on the "make" of a current pulse and under the anode at the "break" of the rectangular pulse. An intracellularly applied hyperpolarizing current pulse produces an anodal-break response (B). RP, Resting potential.

occurs under the anode on the "break" of the stimulating pulse; namely, an AP is triggered from this hyperpolarized region of the axon (Fig. 9B). The explanation for this is that ion channel changes occur during the hyperpolarization, such that the excitability of that membrane is transiently increased (lower $V_{th}$ point) immediately following cessation of the applied pulse.

The increase in excitability is due to two factors: (1) There is an increase in $h_\infty$ during the hyperpolarization, reflecting that almost 100% of the fast $Na^+$ channels have their I-gates open, and hence are capable of conducting (open state) when the membrane is depolarized on removal of the hyperpolarizing pulse ($g_{Na} = \max g_{Na}\ m^3h$); and (2) there is a decrease in $n_\infty$ during the hyperpolarization, hence decreasing $g_K$ ($g_K = \max g_K\ n^4$). These equations are discussed in the following section. The changes in the $h$ and $n$ parameters persist for a short period after termination of the applied pulse, and hence increase membrane excitability during this brief period and trigger an AP. This is called *anodal-break excitation* and *postanodal enhancement of excitability* (due to the postanodal depolarization).

In contrast, under the cathode, after termination of the applied pulse, an opposite change occurs in $E_m$ and excitability. The membrane is hyperpolarized transiently and excitability is depressed. This is known as *postcathodal hyperpolarization* and *postcathodal depression of excitability*. In Hodgkin–Huxley terms, this phenomenon also is due to two factors: (1) decrease in $h_\infty$ during the depolarization, reflecting a smaller fraction of fast $Na^+$ channels having their I-gates open and hence incapable of conducting and thereby decreasing $g_{Na}$; and (2) increase in $n_\infty$, hence increasing $g_K$. The increased $g_K$ and decreased $g_{Na}$ produce hyperpolarization (refer to chapter on the resting potential), and thereby depress excitability.

## III. Electrogenesis of Action Potential

Early experimentation in electrophysiology focused on determining the mechanism for the generation of the AP. One important finding by Cole and Curtis (1939) was that during the AP, the membrane resistance (but not the capacitance) changed dramatically (Fig. 10). The large reduction in membrane resistance during the AP supported the hypothesis that the AP resulted from a large increase in the ionic permeability of the membrane.

To determine which ionic species might be involved in generating the AP, subsequent experimentation was directed toward varying the concentrations of the different ions bathing the axon. Figure 11 shows a classic experiment in which the concentration of $Na^+$ ions bathing the squid axon was altered. It was found that the overshoot and the rate of rise of the AP were proportional to $[Na]_o$. This result was the first indirect demonstration that the AP resulted from an increase in the membrane permeability to $Na^+$ ions. Several years later, this hypothesis was confirmed directly by the voltage clamp method, as discussed in the next section.

### A. Voltage Clamp Method

The membrane current ($I_m$) that generates the AP is composed of *ionic current* ($I_i$) and *capacitive current* ($I_c$) according to the relationship

$$I_m = I_i + I_c \qquad (2)$$

The flow of ionic currents across their respective resistive membrane pathways (or channels) causes a change in the membrane potential (from Ohm's law: $V = IR$). The

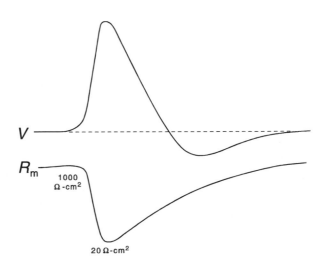

**FIG. 10.** Time course for the decrease in membrane resistance ($R_m$) to the passage of current measured during the action potential ($V$) in a nerve axon. $R_m$ falls from about 1000 $\Omega$-cm$^2$ at rest to about 20 $\Omega$-cm$^2$ near the peak of the AP. [Adapted from *The Journal of General Physiology* by copyright permission of The Rockefeller University Press.]

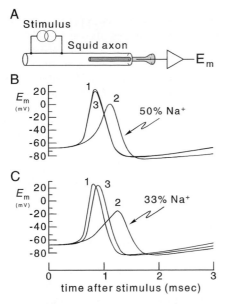

**FIG. 11.** Na$^+$ dependence of the nerve action potential. (A) Diagram depicts the method used to record the transmembrane AP ($E_m$) from a squid axon. (B and C) The voltage traces show the reduction in AP amplitude as the Na$^+$ concentration of the solution bathing the axon was reduced to 50% (B) or to 33% (C) of normal. APs were recorded with 100% of the normal Na$^+$ concentration (traces labeled 1), after reduction of the Na$^+$ concentration (traces 2), and after return to normal Na$^+$ concentration (traces 3). [Adapted from Hodgkin, A. L., and Katz, B. (1949). *J. Physiol.* (*London*) **108**, 37–77.]

change in membrane voltage causes a capacitive current to flow as

$$I_c = C_m \, dV/dt \qquad (3)$$

where $dV/dt$ is the rate of change of the AP. Substituting Eq. 3 into 2 yields

$$I_m = I_i + C_m \, dV/dt \qquad (4)$$

Since the membrane potential during an AP is constantly changing, it would be difficult to separate the contributions of these interacting ionic and capacitive components. In addition, the total ionic current is composed of multiple individual currents carried by specific ions.

To analyze and separate the membrane currents into their capacitive and ionic components, a revolutionary method, called *voltage clamping*, was introduced in the early 1950s by Cole and by Hodgkin and Huxley. A diagram of the voltage clamp method is shown in Fig. 12. During a voltage clamp experiment, the membrane potential is held constant ("clamped") by a negative feedback amplifier, and the amount of current that is necessary to perform this task is recorded. Since the membrane potential ($V_m$) is held constant, the capacitive current is equal to zero. However, at the very beginning of the depolarizing clamp pulse, there is a large transient capacitive current that occurs, and again at the "off" of the pulse; these can be electronically subtracted and thus removed by a special procedure. Since $V_m$ is constant during the clamp pulse,

$$dV/dt = 0 \qquad (4a)$$

and

$$I_c = 0 \qquad (4b)$$

Therefore,

$$I_m = I_i \qquad (4c)$$

The voltage clamp experiment gives the magnitude and time course of the ionic currents at a given clamp potential. By clamping the membrane to many different potentials, information about the flow of ionic currents and the underlying conductance changes during the AP is obtained.

## B. Voltage Clamp Analysis

In the voltage clamp experiments, the various ion currents (such as Na$^+$, Ca$^{2+}$, or K$^+$ currents) can be isolated from the total ionic current and analyzed individually. For example, in the squid axon experiments, the total ionic current consists of an early inward current followed by a delayed outward current (Fig. 13). By varying [Na]$_o$, the early inward current can be shown to be carried by Na$^+$ ions. Similarly, by changing [K]$_o$, the delayed outward current can be shown to be carried by K$^+$ ions. The Na$^+$ and K$^+$ currents can also be separated by blocking their pathways through the membrane. Na$^+$ channels can be blocked with **tetrodotoxin (TTX)**, derived from the ovaries of Japanese puffer fish, and K$^+$ channels can be blocked by several inorganic ions and organic compounds, including tetraethylammonium ions (TEA$^+$) and 4-aminopyridine (4-AP). The current remaining can then be subtracted from the total ionic current to reveal the time course for the current that was blocked.

The *current/voltage relationship* is obtained from measurements of the peak inward Na$^+$ current and peak outward K$^+$ current during a series of voltage clamp steps (Fig. 14). Depolarizing voltage steps to just above the resting potential first produce a small outward current. In this

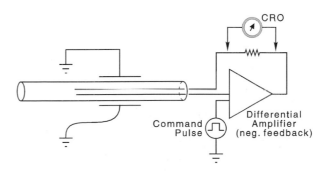

**FIG. 12.** The voltage clamp method used in a squid giant axon. The two wires inserted into the axon are used to measure membrane potential ($V$) and to pass current ($I$). The high-gain negative-feedback amplifier compares the command pulse with the membrane potential, and outputs the amount of current necessary to hold the membrane potential constant (or "clamped"). The magnitude of the feedback current can be measured as the $IR$ voltage drop across a resistor and displayed on a cathode-ray oscilloscope (CRO).

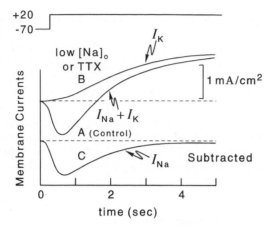

**FIG. 13.** The ionic currents that flow when a squid giant axon in seawater is clamped from its resting potential ($-70$ mV) to a transmembrane potential of $+20$ mV. Trace A shows the net inward Na$^+$ current ($I_{Na}$) and outward K$^+$ current ($I_K$) in normal medium. Trace B shows the net ionic current when the axon is placed in artificial seawater with most of the Na$^+$ replaced by choline$^+$ (an impermeant cation), so that the intracellular and extracellular Na$^+$ concentrations are equal. This current is due to K$^+$ only. TTX also can be used to block $I_{Na}$. Trace C shows the difference between curves A and B, which represents $I_{Na}$. [Redrawn from Hodgkin, A. L., and Huxley, A. F. (1952a). *J. Physiol. (London)* **116**, 449.]

voltage region, the membrane behaves in an ohmic fashion. With greater depolarization, the inward $I_{Na}$ is activated and the I/V relationship displays a negative slope or **negative resistance** region. At potentials above the peak current of the I/V curve, a positive slope is seen, and the current

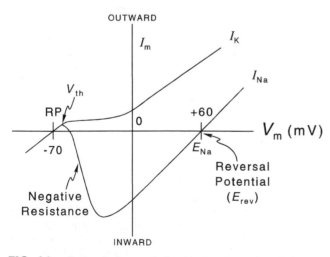

**FIG. 14.** Current/voltage relationship for the peak early inward current and delayed outward current obtained from a squid axon under voltage clamp. The inward current is carried by Na$^+$ ($I_{Na}$), the outward current by K$^+$ ($I_K$). The reversal potential for $I_{Na}$ is the voltage at which the current changes from inward to outward. The region of the $I_{Na}$ curve that has a negative slope is known as negative resistance. RP, Resting potential; $V_{th}$, threshold; $E_{Na}$, Na$^+$ equilibrium potential; $I_m$, total membrane current; $V_m$, membrane potential. (Redrawn from Hodgkin, 1964.)

magnitude decreases as $E_{Na}$ is approached, and actually becomes outward at voltages above $E_{Na}$. The voltage at which the current reverses in direction is the **reversal potential**. The reason that the current diminishes and reverses at potentials approaching $E_{Na}$ and beyond is that the net electrochemical driving force for Na$^+$ ions first becomes smaller and smaller and then outwardly directed, whereas the conductance for Na$^+$ ions remains constant and high over this entire voltage range, as indicated by

$$I_{Na} = g_{Na} (E_m - E_{Na}) \qquad (5)$$

The outward K$^+$ current activates above $-20$ mV and increases with depolarization as

$$I_K = g_K (E_m - E_K) \qquad (6)$$

The voltage clamp experiments have revealed the most fundamental property of the ionic conductances of excitable membrane: namely, that the conductances are both voltage dependent and time dependent (Fig. 15). Both $g_{Na}$ and $g_K$ activate with depolarization, but with different time courses (Figs. 15 and 16). With time, $g_{Na}$ spontaneously turns off, or inactivates. That is, the $I_{Na}$ current shuts off within 1–2 msec.

A number of biological toxins that act on specific ion channels have been discovered. For example, TTX mentioned previously has a very high affinity for the fast Na$^+$ channel of nerve and some types of muscle cells. It binds to the fast Na$^+$ channel and blocks the passage of Na$^+$ ions through the channel. Several different types of toxin, including **batrachotoxin (BTX),** inhibit the inactivation process of the Na$^+$ channel, so that the Na$^+$ currents are greatly prolonged once activated. Such toxins have proven to be valuable tools in analyzing voltage clamp currents and understanding ion channel function. The ion channel toxins are discussed in great detail in a separate chapter.

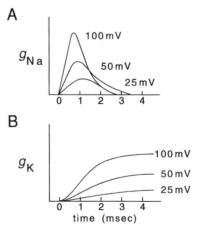

**FIG. 15.** Voltage dependence and time dependence of the changes in Na$^+$ conductance ($g_{Na}$) and K$^+$ conductance ($g_K$) during voltage clamp of the squid giant axon. The numbers refer to the magnitude of depolarization (in millivolts) from the resting potential. The $g_{Na}$ turns on rapidly and then spontaneously declines over a brief time period; $g_K$ turns on more slowly and is sustained in amplitude during the entire clamp pulse. (Redrawn from Hodgkin, 1964.)

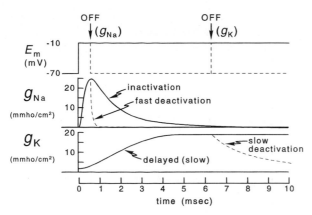

**FIG. 16.** $Na^+$ conductance ($g_{Na}$) and $K^+$ conductance ($g_K$) changes are shown in response to voltage clamp of the squid giant axon from the resting potential to a membrane potential ($E_m$) of $-10$ mV. $g_{Na}$ inactivates after a short time, even when the voltage clamp is maintained; in contrast, $g_K$ remains elevated until the clamp is released. When the clamp pulse is terminated early, the $g_{Na}$ increase is quickly turned off (deactivated); the turn-off of $g_K$ is considerably slower. [Redrawn from Hodgkin, A. L. (1958). *Proc. R. Soc. London (Biol.)* **B148**, 1.]

## C. Whole-Cell Voltage Clamp

A new electrophysiological method, the whole-cell voltage clamp technique, has enabled researchers to examine the basis of excitability at the single cell level, for example, for myocardial cells or smooth muscle cells. The method allows the recording of the macroscopic currents that flow through the assembly of ion channels in the cell membrane. The details of this method are given in the following chapter. In brief, to record the whole-cell current, a small-tipped glass pipette, known as a patch pipette, is pressed against the cell membrane, and negative pressure is applied to the interior of the pipette to draw a small patch of membrane into its tip. A high-resistance seal (e.g., $10^{10}$ $\Omega$) spontaneously forms (Fig. 17). Then, strong suction is applied to the patch pipette to blow out the membrane patch and allow the lumen of the pipette to be continuous with the lumen of the cell. This technique allows recording of the whole-cell current, and a complete voltage clamp analysis can be done for small cells, as described previously for giant axons. This technique also allows some control over the intracellular content of the cell; for example, a substance can be introduced into the cell by diffusion from the patch pipette solution.

Whole-cell voltage clamp records obtained from isolated single uterine smooth muscle cells (18-day pregnant rat) are illustrated in Fig. 18. The condition in Fig. 18 is arranged so that outward $K^+$ currents are blocked (by high $Cs^+$ concentration in the patch pipette), and any inward currents carried by $Ca^{2+}$ or $Na^+$ ions can be recorded. As shown in Fig. 18, there are two inward currents: an initial fast current and a later slow current. The initial fast current is carried by $Na^+$ ion and this $Na^+$ current is blocked by TTX; the later slow current is carried by $Ca^{2+}$ ions. The presence of functional fast $Na^+$ channels is unusual for most smooth muscles, and it was discovered that such channels

developed during pregnancy and reached their maximum close to term.

The information gained from such whole-cell voltage clamp experiments, performed on a variety of excitable cell types, has greatly increased our knowledge of the properties and regulation of the many types of ionic currents. The macroscopic or whole-cell current ($I$) is the product of the number of functional channels in the cell membrane ($N$), times the probability that the "average" channel is open during the clamp step ($p_o$), times the single-channel current ($i$) giving

$$I = iNp_o \qquad (7)$$

The single-channel current (in pA), open probability ($p_o$), and single-channel conductance ($\gamma$) are obtained from patch-clamp experiments, in which the activity of a single channel (or a few channels) can be monitored in a small patch of membrane (e.g., 1–10 $\mu m^2$) that is electrically isolated from the rest of the cell membrane (cell-attached patch) or excised (isolated patch). Opening and closing of the gates of the single channel reflect conformational changes in the channel protein. This sophisticated technique, which earned the Nobel Prize in physiology and medicine in 1991 for E. Neher and B. Sakmann, allows

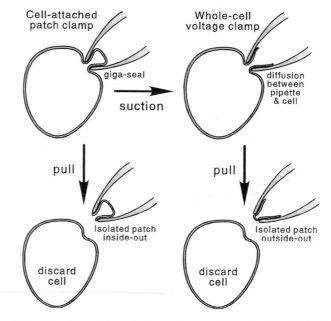

**FIG. 17.** The whole-cell voltage clamp technique in isolated single cells to record the macroscopic current (e.g., $I_{Na}$, $I_{Ca}$, or $I_K$) from the entire cell membrane. (Upper left) Cell-attached patch mode, produced by applying light suction to the patch pipette to produce a "gigaseal" ($10^9$ $\Omega$). This mode is used to record the microscopic currents from only one channel or from a few channels. (Upper right) Whole-cell clamp mode produced by applying strong suction to the patch pipette to blow out the membrane patch and allowing the lumen of the pipette to be continuous with the lumen of the cell. This mode is used to record the macroscopic currents from the entire complement of ion channels in the cell membrane. (Bottom) Preparation of isolated membrane patches. [Reprinted with permission from *Nature*, Hamill, O. P., and Sakmann, B. (1981). **294**, 462–464. Copyright 1981 by Macmillan Magazines Limited.]

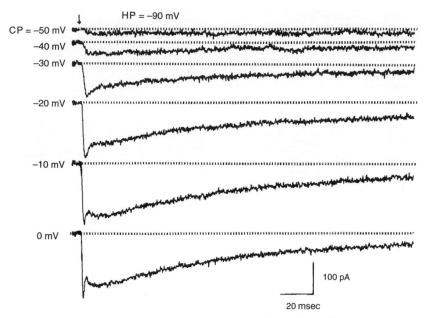

**FIG. 18.**   Two types of inward current recorded from isolated single myometrial cells from pregnant rat using whole-cell voltage clamp with a patch pipette technique. Depolarizing potential steps (−50 to 0 mV) were applied from a holding potential (HP) of −90 mV. Arrow indicates beginning of voltage step that continues to end of trace. Bath contained $K^+$-free solution with 150 m$M$ $Na^+$ and 2 m$M$ $Ca^{2+}$. Pipette solution contained high $Cs^+$. Cell capacitance was 80 pF. CP, command potential. [Reproduced with permission from Ohya, Y., and Sperelakis, N. (1989). *Am. J. Physiol.* **257**, C408–C412.]

the biophysical study of the behavior of a single protein molecule and thus represents electrophysiology at the molecular level. Although, this patch-clamp technique and analysis is discussed in detail in the following chapter, a brief illustration of actual records is given here (Fig. 19).

### D.  Overview of Action Potential Generation

The increase in $g_{Na}$ and resulting increase in inward $I_{Na}$ cause the regenerative depolarization of the AP. The depolarization is limited by the approach of the membrane potential toward $E_{Na}$ and by the Na inactivation process. As the membrane is depolarized, both $g_K$ and the driving force for $I_K$ increase, and the outward $I_K$ repolarizes the membrane. The increase in $g_K$ is *self-limiting;* that is, the increase in $g_K$ produces repolarization, which, in turn, shuts off the increase in $g_K$.

The slow kinetics of the turn-off of $g_K$ results in a transient hyperpolarization, the hyperpolarizing afterpotential. During the hyperpolarizing afterpotential, the membrane potential is brought closer to $E_K$ than at rest. The membrane conductance changes that occur during the AP are shown in Fig. 20.

The time course for the ionic currents during the nerve AP is shown in Fig. 21. The total ionic current ($I_i$) is separated into its two major components, $I_{Na}$ and $I_K$. Since $I_K$ is slower to activate than $I_{Na}$, the inward $I_{Na}$ predominates initially, giving rise to the upstroke of the AP. Later, $I_K$ dominates, causing a net outward current that repolarizes the membrane, that is, is partly responsible for the downstroke of the AP.

As stated in Eqs. 5 and 6, the specific ionic currents are a product of the membrane conductance for the ionic species and the electrochemical driving force exerted on the ion. Thus, the driving forces on $Na^+$ and $K^+$ ions continually change during the time course of the AP, as diagrammed in Fig. 22. At the resting potential, there is a large driving force for $Na^+$ to flow into the cell, since $(E_m - E_{Na})$ is large. Conversely, at the peak of the AP, $(E_m - E_{Na})$ is at its lowest value, and the driving force for $Na^+$ entry is small. In contrast, the driving force for $K^+$ efflux is largest at the peak of the AP, when $(E_m - E_K)$ is maximal. It is important to remember that ionic current flow depends on both conductance and driving force. There is no net current if only one, but not the other, is present.

The biological elements of the excitable membrane may be represented in terms of an electrical equivalent circuit model, as shown in Fig. 23. In the circuit model, currents flow through the individual conductance pathways for each type of ion. The conductances for $Na^+$ and $K^+$ ions are variable and depend on the transmembrane potential and time. Batteries (positive pole directed inwardly for $Na^+$ ions and outwardly for $K^+$ ions) provide the driving forces for current flow. A passive leak conductance for $Cl^-$ ions is also included in the model. If the correct values for the elements are incorporated and varied over time, the model circuit will generate an AP.

### E.  Fast $Na^+$ Channel Activation

During an AP, the increase in $g_{Na}$ is related to the membrane potential ($E_m$) in a *positive-feedback* fashion ("vicious cycle") (Fig. 24A). That is, a small depolarization leads to an increase in $g_{Na}$, which allows a larger inward $I_{Na}$, which causes further depolarization. This greater depolarization produces a greater increase in $g_{Na}$. This positive feedback process is "explosive," with a sharp trigger point (threshold), resulting from the exponential (positive) relationship between $g_{Na}$ and $E_m$ (Fig. 24B). It is this positive-feedback relationship that accounts for the negative resistance (slope) in the current/voltage curve (Fig. 14). As Fig. 24B shows, $g_{Na}$ reaches a maximum (saturates) at positive potentials that gives maximal activation of the population of fast $Na^+$ channels.

The fast $Na^+$ channels (and the slow $Ca^{2+}$ channels) have a double gating mechanism: an inactivation gate (I-gate) and an activation gate (A-gate) (Fig. 25). For a channel to be conducting, both the A-gate and the I-gate must be open; if either one is closed, the channel is nonconducting. The A-gate is located somewhere near the middle of the channel; it is not at the outer surface because even TTX does not prevent the movement of this gate, and it is not at the inner surface because proteases perfused internally do not affect it. The A-gate is closed at the resting $E_m$ and opens rapidly on depolarization; in contrast, the I-gate is open at the resting $E_m$ and closes slowly on depolarization. The time course of the changes in the $m$ and $h$ variables during a depolarizing clamp step is schematized in Fig. 26.

In the Hodgkin–Huxley (1952) analysis, the opening of the A-gate requires simultaneous occupation of three negatively charged sites by three positively charged ($m^+$) particles. The activation variable, $m$, is the probability of one site being occupied, and $m^3$ is the probability that all three sites are occupied: therefore,

$$g_{Na} = \overline{g}_{Na} m^3 h \qquad (8)$$

where $h$ is the inactivation variable and $\overline{g}_{Na}$ is the maximum conductance.

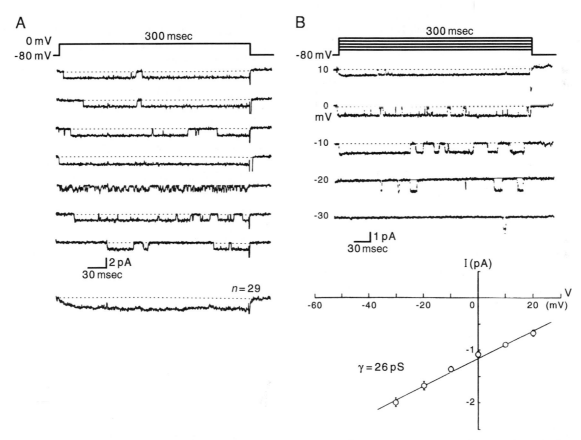

**FIG. 19.** Single-channel currents recorded from a slow (L-type) $Ca^{2+}$ channel, using the cell-attached patch clamp technique, in a single cardiomyocyte isolated from a young (3-day-old) embryonic chick heart (ventricular cell). These $Ca^{2+}$ channels in young embryonic/fetal chick and rat heart cells exhibit a high incidence of long openings. (A) Inward currents were evoked by seven depolarizing voltage step pulses (300-msec duration; 0.5 Hz) to 0 mV from a holding potential (HP) of −80 mV. Pulse protocol is given at the top. The ensemble-averaged current from 29 such pulses is given at the bottom. (B) Similar recordings in another cell illustrating the inward currents recorded at five different command step potentials (−30, −20, −10, 0, and +10 mV) from a HP of −80 mV. Again note the presence of numerous long openings. Also note that the amplitude of the unitary current became smaller at the higher (more positive) command potentials. The current amplitudes are plotted below in the current/voltage curve (each point being the mean ± SE of 5–10 experiments). The data points were fitted by a straight line giving a slope conductance of 26 pS for the single-channel conductance. [Modified from Tohse, N., and Sperelakis, N. (1990). *Am. J. Physiol.* **259,** H639–H642; and Tohse, N., and Sperelakis, N. (1991). *Circ. Res.* **69,** 325–331.]

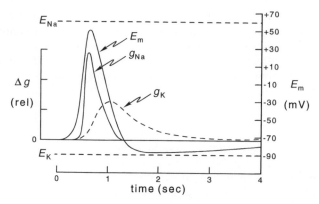

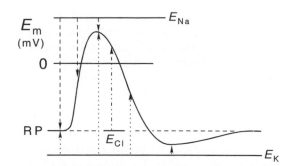

**FIG. 20.**   The relative conductance for Na$^+$ ($g_{Na}$) and K$^+$ ($g_K$) during an action potential in a nerve fiber. The rising phase of the AP is caused by an increase in $g_{Na}$. The falling phase of the AP is due to the rise of $g_K$ (delayed rectification) and to the decrease in $g_{Na}$ (Na$^+$ inactivation). The hyperpolarizing afterpotential is explained by the fact that $g_K$ remains elevated for a short time following repolarization, tending to hold the membrane potential ($E_m$) near the K$^+$ equilibrium potential ($E_K$). $E_{Na}$, Na$^+$ equilibrium potential; $\Delta g$, change in conductance. [Redrawn from Hodgkin, A. L., and Huxley, A. F. (1952). *J. Physiol.* **117**, 500–544.]

**FIG. 22.**   Driving forces for Na$^+$ and K$^+$ currents during the action potential. The total electrochemical driving force is equal to membrane potential ($E_m$) minus the equilibrium potential for the ion ($E_i$): ($E_m - E_i$). As depicted, the driving force for the Na$^+$ current decreases during the AP, whereas that for K$^+$ increases. Even when Cl$^-$ is passively distributed, a net driving force for Cl$^-$ influx (outward Cl$^-$ current) occurs during the AP.

## F.  Gating Current

Since the gating of ionic channels is voltage dependent, a part of the channel protein contains a charged group or dipole that can sense the electric field across the membrane and move in response to a change in transmembrane voltage. When the gating region moves, it causes a shift in the overall conformation of the channel, which allows it to conduct ions. A gating current ($I_g$) that corresponds to the movement of the charged $m^+$ particles (or rotation of an

equivalent dipole) has been measured. The gating current is very small in intensity and is measured by subtracting the linear capacitive current (from a hyperpolarizing clamp step) from the total capacitive current (linear plus nonlinear) that occurs with a depolarizing clamp step beyond threshold. The outward $I_g$ precedes the inward $I_{Na}$. Tetrodotoxin does not block $I_g$, although it does block $I_{Na}$. Thus, the gating current is a nonlinear outward capacitive current (not an ionic current) obtained during depolarizing clamps that reflects movement of the A-gates from the closed to the open configuration. The linear capacitive current results mainly from the lipid bilayer matrix (see Appendix to book), whereas the nonlinear capacitive current arises from the charge movement associated with the A-gates of the protein channels.

Specific charged residues are located along the primary sequence of amino acids that comprise the channel polypeptide, and they can "sense" the transmembrane electric field and move in response to changes in the electric field.

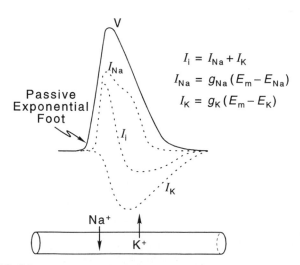

$$I_i = I_{Na} + I_K$$
$$I_{Na} = g_{Na}(E_m - E_{Na})$$
$$I_K = g_K(E_m - E_K)$$

**FIG. 21.**   Ionic currents that flow during the nerve action potential. The total current ($I_i$) is separated into an inward Na$^+$ current ($I_{Na}$) and an outward K$^+$ current ($I_K$). $I_i$ is the algebraic sum of $I_{Na}$ and $I_K$. The appropriate equations for $I_i$, $I_{Na}$, and $I_K$ are given. Also depicted is the fact that a net inward Na$^+$ flux occurs during the rising phase of the AP and a net K$^+$ efflux occurs during the repolarizing phase. (Redrawn from Hodgkin and Huxley, 1952a.)

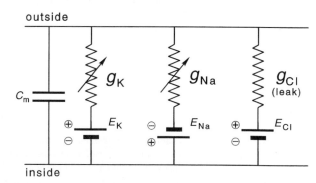

**FIG. 23.**   Hodgkin–Huxley electrical equivalent circuit for the squid giant nerve axon. The K$^+$ conductance ($g_K$) is in series with the K$^+$ equilibrium potential ($E_K$) and $g_{Na}$ is in series with $E_{Na}$. The arrows indicate that $g_{Na}$ and $g_K$ vary with voltage and time. The low conductance for Cl$^-$ ($g_{Cl}$) was termed the *leak conductance*. $C_m$, membrane capacitance. [Redrawn with permission from Hodgkin, A. L. (1964) "The Conduction of the Nervous Impulse." Charles C Thomas, Springfield, IL.]

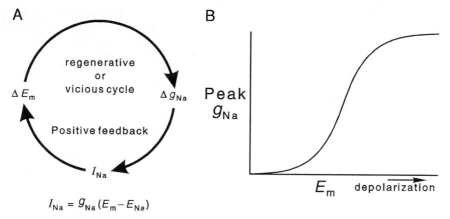

$$I_{Na} = g_{Na}(E_m - E_{Na})$$

**FIG. 24.** The positive feedback relationship between $Na^+$ conductance ($g_{Na}$) and membrane potential ($E_m$), leading to the all-or-none action potential. (A) The increase in $g_{Na}$ allows an increase in the inward $Na^+$ current [$I_{Na} = g_{Na}(E_m - E_{Na})$], which is depolarizing and so triggers a further increase in $g_{Na}$. This explosive feedback cycle is caused by the voltage dependency of the gated fast $Na^+$ channels. (B) Plot of $g_{Na}$ versus depolarization, showing the initial exponential (positive) increase in $g_{Na}$ as a function of voltage, followed by saturation with greater depolarization, thus giving a sigmoidal relationship. [Adapted from Hodgkin, A. L. (1964). "The Conduction of the Nerve Impulse." Charles C Thomas, Springfield, IL.]

The movement of these voltage-sensing or gating charges initiates a conformational change in the channel structure, which then permits ions to flow through the central pore region of the alpha subunit. The movement of the gating charge produces the small, but measurable, gating current. Gating currents have been recorded from several types of ion channels and provide information concerning the steps leading to channel opening.

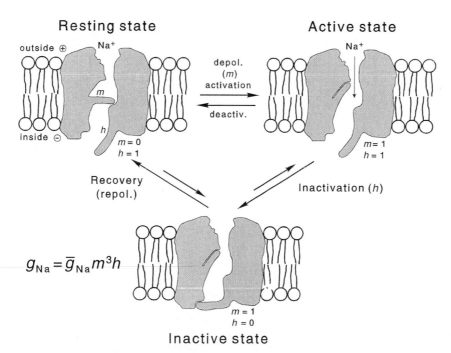

**FIG. 25.** The three hypothetical states of the fast $Na^+$ channel, based on the Hodgkin–Huxley model. In the resting state, the activation gate (A or $m$) is closed and the inactivation gate (I or $h$) is open: $m = 0$, $h = 1$. Depolarization to the threshold or beyond activates the channel to the active state, the A-gate opening rapidly and the I-gate still being open: $m = 1$, $h = 1$. The activated channel spontaneously inactivates to the inactive state because of delayed closure of the I-gate: $m = 1$, $h = 0$. The recovery process on repolarization returns the channel from the inactive state to the resting state, thus making the channel again available for reactivation. $Na^+$ is depicted as being bound to the outer mouth of the channel and poised for entry down its electrochemical gradient when both gates are open (active state of channel). The reaction between resting state and the active states is reversible, whereas the other reactions may be less reversible. The $Ca^{2+}$ slow channels pass through similar states.

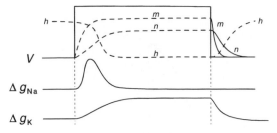

**FIG. 26.**   Sketch of the time course of the changes in the Hodgkin–Huxley $m$, $n$, and $h$ factors during a depolarizing voltage clamp step. The lower traces give the resulting time course for the changes in Na$^+$ and K$^+$ conductances during the step. Because $g_{Na}$ is proportional to the product of the $m$ and $h$ variables, it returns to about the original level within approximately 2 msec after the clamp step is applied.

### G. Na$^+$ Inactivation

The fast $I_{Na}$ lasts only for 1–2 msec because of the spontaneous inactivation of the fast Na$^+$ channels. That is, the fast Na$^+$ channels inactivate quickly, even if the membrane were to remain depolarized (Fig. 26). (In contrast, the slow Ca$^{2+}$ channels inactivate slowly.) Inactivation is produced

in the fast Na$^+$ channels (and in the slow Ca$^{2+}$ channels) by the voltage-sensitive slow closing of the inactivation gate (I-gate) (Fig. 25). The I-gate is located near the inner surface of the membrane, as evidenced by the fact that addition of proteolytic enzyme to the inside of a perfused giant axon chops off the I-gate and eliminates inactivation (protease added outside does not have this effect). The I-gate is presumably charged positively to allow it to move with changes in the membrane potential. During depolarization, the inside of the membrane becomes less negative or more positive, and this causes the I-gate to close. At the normal resting potential, the I-gate is open, but the A-gate is closed.

The voltage dependency of inactivation is given by the $h_\infty$ versus $E_m$ curve (Fig. 27). The inactivation variable $h$ (probability function) varies between 0 and 1.0, perhaps reflecting occupation of a negatively charged site by a positively charged $h$ particle; $h_\infty$ is the value of $h$ at infinite time (>10 msec) or at steady state. When $h = 1.0$, the I-gates of all of the fast Na$^+$ channels are in the open configuration; conversely, when $h = 0$, all the I-gates are closed. Since $g_{Na}$ at any time is equal to the maximal value (max $g_{Na}$) times $m^3h$ (see Eq. 8), when $h = 0$, $g_{Na} = 0$.

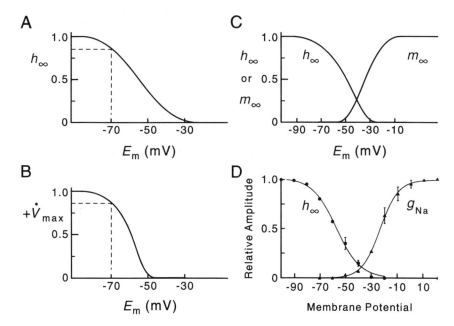

**FIG. 27.**   Voltage inactivation of the fast Na$^+$ channels as a function of the membrane potential ($E_m$). (A) $h_\infty$ plotted against $E_m$, where $h_\infty$ is the inactivation factor of Hodgkin–Huxley. $g_{Na} = \max g_{Na}\, m^3h$, where $g_{Na}$ is the Na$^+$ conductance, $m$ and $h$ are variables, and max $g_{Na}$ is the maximal conductance. The $h_\infty$ represents $h$ at infinite time or steady state (practically, after 10 msec). This graph illustrates that fast Na$^+$ channels begin to inactivate at about $-75$ mV, and nearly complete inactivation occurs at about $-30$ mV. (B) Maximal rate of rise of the AP (max $dV/dt$) as a function of resting $E_m$. Max $dV/dt$ is a measure of the inward current intensity, which is dependent on the number of channels available for activation. Therefore, max $dV/dt$ decreases as $h_\infty$ decreases. (C) Plot of both steady-state inactivation ($h_\infty$) and activation ($m_\infty$) against $E_m$ to illustrate the overlap of the two curves, depicting the "window" current region. (D) Steady-state activation and inactivation curves for the fast Na$^+$ current [$I_{Na(f)}$] recorded using the whole-cell voltage clamp technique from uterine smooth muscle cells isolated from late pregnant (18-day) rats. For the inactivation curve, conditioning pulses of various amplitudes were applied for 2 sec before a test pulse to 0 mV. The activation curve was obtained by measuring peak amplitude of $I_{Na}$ elicited by various command potentials from a HP of $-90$ mV. The two curves were obtained by fitting the data to Boltzmann distributions. Each point represents the mean ± SE of 7 or 5 experiments. Note the overlap between the activation and inactivation curves, resulting in a steady-state window current between about $-50$ and $-30$ mV. (Data from Y. Inoue and N. Sperelakis, unpublished observatons.)

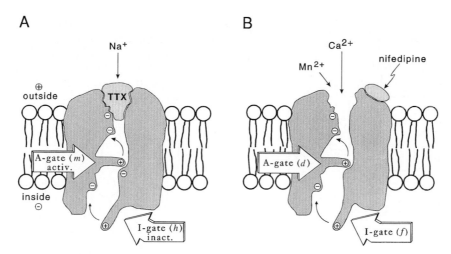

**FIG. 28.** (A) Illustration of a fast Na$^+$ channel and (B) for a Ca$^{2+}$ slow channel. The ionic channels are proteins that float in the lipid bilayer matrix of the cell membrane. These voltage-dependent channels have two gates, as depicted: the activation gate (A or $m$) and the inactivation gate (I or $h$). Conformational changes in the protein could serve as the gates. The I-gate is located near the inner surface of the membrane because proteolytic enzymes added internally destroy this gate. The A-gate is located somewhere in the middle of the channel; TTX does not prevent movement of this gate. TTX binds to the outer mouth of the channel and plugs it physically, as depicted. The gates are presumably charged positively, so that on depolarization [inside going less negative (i.e., in a positive direction)], the gates are electrostatically repelled outward. The A-gate opens relatively quickly (e.g., 0.25 msec), whereas the I-gate closes relatively slowly (e.g., inactivation time constant of 1–2 msec). Therefore, for a short period, both gates are open and the channel is in the conducting mode with Na$^+$ entering the cell down its electrochemical gradient. However, when the I-gate closes (after about 2 msec), the channel again becomes nonconducting. During recovery (reactivation) upon repolarization, the gates return back to their original resting positions. The slow Ca$^{2+}$ channels (B) behave similarly to the fast channels (A), except that their gates apparently move more slowly on a population basis. The slow-channel conductance activates, inactivates, and recovers more slowly. The slow-channel gates operate over a different voltage range than the fast channels (i.e., less negative, more depolarized). TTX does not block the slow channels. Drugs such as nifedipine block the slow channels (but not the fast Na$^+$ channels) by binding to the channel and somehow inactivating it. In addition, the voltage inactivation curve of the Ca$^{2+}$ slow channels is shifted to the right, so that inactivation begins at about −45 mV and is not complete until about −5 mV. The slow channels also have a lower activation (threshold) potential of about −35 mV (compared with about −55 mV for the fast Na$^+$ channel). The activation gate variable is known as $d$, and the inactivation gate variable as $f$.

When $h = 1.0$, $g_{Na} = \max g_{Na}$ (if $m = 1.0$). At the normal resting potential, $h_\infty$ is nearly 1.0 and it diminishes with depolarization, becoming zero at about −30 mV. The maximal rate of rise of the AP (max $dV/dt$) is directly proportional to the net inward $I_{Na}$, which is directly proportional to $g_{Na}$; the decrease in $h_\infty$ is the cause of decrease in $g_{Na}$. At about −30 mV, $h_\infty = 0$, and there is complete inactivation of the fast Na$^+$ channels. Therefore, depolarization by any means (e.g., elevated $[K]_o$ or applied depolarizing current pulses) decreases $g_{Na}$ and excitability disappears at about −50 mV (Fig. 27B). The AP disappears at −50 mV (rather than −30 mV) because of a minimum current density requirement for a regenerative and propagating response.

A plot of the inactivation curve along with the activation curve is given in Fig. 27C to illustrate the overlap in these curves, giving rise to a so-called **"window" current** (or steady-state inward current) over a certain voltage region. The presence of the window current means that, over the voltage range of about −50 to −30 mV, there is a steady inward Na$^+$ current passing through about 10–20% of the fast Na$^+$ channels that are always open (on a rotating population basis).

The slow Ca$^{2+}$ channels (Fig. 28B) behave much the same way as the fast Na$^+$ channels (Fig. 28A) with respect to activation and inactivation, with one main difference being the voltage range over which the slow channels operate. For example, inactivation occurs between −50 and 0 mV for the slow channels, compared with between −100 and −30 mV for the fast Na$^+$ channels. Another major difference is that slow channels inactivate much more slowly than the fast channels; that is, they have a long inactivation time constant ($\tau_{inact}$). In myocardial cells, the $h$ variable for the slow channel is referred to as the $f$ variable, and the $m$ variable as the $d$ variable and the exponent is 2.0:

$$g_{Ca} = \bar{g}_{Ca} d^2 \cdot f \tag{9}$$

The amino acid sequences of Na$^+$ and Ca$^{2+}$ channels are known, as well as their putative tertiary structure (Fig. 29). Na$^+$ and Ca$^{2+}$ channels consist of several subunits, one of which contains the water-filled pore through which the ions pass. Ion channels have one or more sites that can be phosphorylated, and phosphorylation alters their behavior. For example, in cardiac muscle, the slow Ca$^{2+}$ channel activity is increased by adrenaline, a hormone that increases cyclic AMP level and leads to Ca$^{2+}$ channel phosphorylation (see Chapter 51). The molecular structure of ion channels is discussed in more detail in a later chapter.

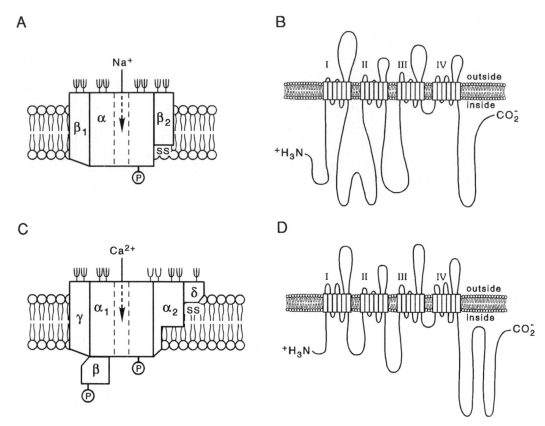

**FIG. 29.** Models of the fast $Na^+$ channel and the slow $Ca^{2+}$ channel proteins based on structural data. (A) The $Na^+$ channel has multiple protein subunits, labeled $\alpha$, $\beta_1$, and $\beta_2$. The $\alpha$ subunit is the one that contains the water-filled pore through which $Na^+$ passes. One site that can be phosphorylated by cyclic AMP-dependent protein kinase (PK) is present on the $\alpha$ subunit. The $\beta_2$ subunit has a disulfide bond (SS). (B) The structure of the central pore-forming $\alpha$ subunit for the $Na^+$ channel. There are four homologous intramembrane polypeptide repeat domains, connected by intracellular polypeptide segments or loops. The four domains are arranged into a circular structure within the plane of the membrane to form a channel. Each domain consists of six units that span the membrane, as depicted. Both the carboxy ($COO^-$) and amino ($NH_3^+$) termini are intracellular. (C) The $Ca^{2+}$ channel also is composed of several subunits: $\alpha_1$, $\alpha_2$, $\beta$, $\gamma$, and $\delta$. The $\alpha_1$ subunit contains the pathway by which $Ca^{2+}$ ions traverse the channel. Two sites can be phosphorylated by the cAMP-PK, one on the $\alpha_1$ subunit and the other on the $\beta$ subunit. As depicted, the $\beta$ and $\delta$ subunits do not span the lipid bilayer, and the $\delta$ subunit possesses a disulfide bond. Ribosylation sites are present on the outer surface of subunits. (D) The polypeptide structure of the $\alpha_1$ subunit of the slow $Ca^{2+}$ channel is similar to that for the fast $Na^+$ channel, with four repeat membrane-spanning domains connected by intracellular and extracellular loops. [Reprinted with permission from Catterall, W. (1988). Structure and function of voltage-sensitive ion channels. *Science* **242,** 50–61. Copyright 1988 American Association for the Advancement of Science.]

## H. Recovery

Any $Na^+$ or $Ca^{2+}$ channel that has been activated and then spontaneously inactivated must go through a recovery process before it can return to the resting state from which it can be reactivated (Fig. 25). The recovery process is dependent on voltage and time. The membrane must be repolarized beyond about $-50$ mV before the recovery process can begin (i.e., traveling up the $h_\infty$ versus $E_m$ curve). At any given $E_m$, time is necessary for the recovery process to occur, namely, the time required for the charged A-gates and I-gates to move back to their resting configuration (A-gate closed, I-gate open) with the electric field. The recovery process is rapid for fast $Na^+$ channels (e.g., 1–10 msec) and less rapid for the slow $Ca^{2+}$ channels. The recovery process of the slow channels is slowed by organic calcium antagonist drugs.

## I. $K^+$ Activation

The $K^+$ channel (outward-going delayed rectifier) is generally believed to have only an A-gate, because it does not inactivate. This gate is thought to be located near the inner surface of the membrane, because $TEA^+$ blocks the $K^+$ channel more readily from the inner surface. The block is use dependent or frequency dependent: As the A-gate opens, the $TEA^+$ molecule can bind in the channel behind the gate. The A-gate is believed to be positively charged, and depolarization (inside going positive) opens the gate. In the Hodgkin–Huxley analysis of squid giant axon, the A-gate opens when four positively charged $n^+$ particles simultaneously occupy four negatively charged sites. If $n$ is the probability that one site is occupied, then $n^4$ is the probability that all four sites are occupied; therefore,

$$g_K = \overline{g}_K n^4 \qquad (10)$$

The power to which $n$ is raised varies in different tissues.

## J. Model for Activation and Inactivation of $Na^+$ and $K^+$ Channels

As stated previously, in the Hodgkin–Huxley analysis, the $Na^+$ conductance $g_{Na}$ is controlled by three charged activating $m$ particles and one blocking $h$ particle, which move with changes in the electric field across the membrane to either occupy or unoccupy certain sites on the $Na^+$ channel protein. If $m$ is the probability of one favorable site being occupied, then $m \times m \times m$ or $m^3$ is the probability that all three sites are occupied simultaneously. The activation A-gate cannot be fully opened unless all three sites are occupied. A simplistic variant is to consider that the A-gate of the $Na^+$ channel actually consists of three separate subgates or subdoors, as illustrated in Fig. 30. For a $Na^+$ ion to pass through the channel, all three subdoors must be open.

The value of $h$ is the probability of one favorable site being occupied which opens the I-gate, and this site is already occupied at the normal resting potential (I-gate open), and becomes unoccupied with depolarization. Therefore

$$g_{Na} = \overline{g}_{Na} m^3 h \qquad (8a)$$

where $\overline{g}_{Na}$ is the maximum $g_{Na}$ possible (primarily function of density of channels), $m$ is the activation variable, and $h$ is the inactivation variable.

The $K^+$ conductance $g_K$ is controlled by four charged activating $n$ particles, which move with depolarization to occupy four sites on the $K^+$ channel protein. If $n$ is the probability of one favorable site being occupied, then $n \times n \times n \times n$ or $n^4$ is the probability that all four sites will be occupied simultaneously. There is no inactivation variable for the $K^+$ channel, because $g_K$ does not exhibit a major decrease during a short depolarizing voltage clamp pulse (e.g., 20 msec). Therefore

$$g_K = \overline{g}_K n^4 \qquad (10a)$$

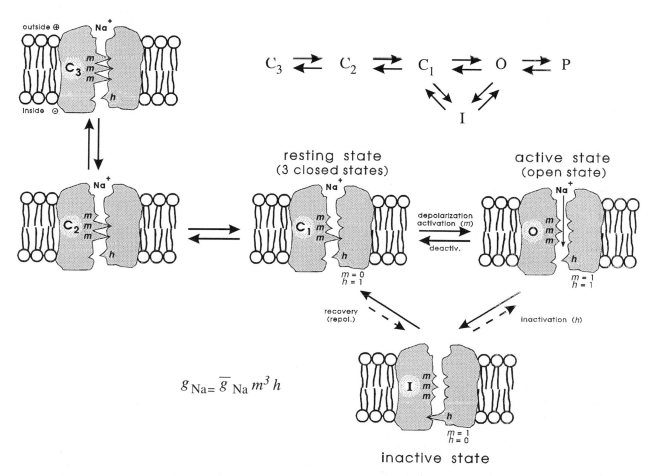

**FIG. 30.** Diagram depicting the modified Hodgkin–Huxley view of the three states of the fast $Na^+$ channel. As shown, there is evidence for the existence of three closed states of the channel ($C_3$, $C_2$, and $C_1$). The activation gate (A-gate) is depicted as consisting of three parts, each one being controlled by one $m^+$ particle. For the A-gate to be open to allow $Na^+$ ions to pass, all three subgates must be open, and therefore all $m$ sites must be occupied. The $C_3$ closed state is depicted as having all three subgates closed, the $C_2$ closed state with two subgates closed (one open), and the $C_1$ closed state with one subgate closed (two open). For the open state, of course, all three subgates must be open. P stands for a plugged channel.

where $\bar{g}_K$ is the maximum $g_K$ and $n$ is the activation variable.

Based on the preceding analysis, Hodgkin and Huxley (1952a,b,c) then gave the differential equations that govern the values of $n$, $m$, and $h$ using a pair of rate constants, $\alpha$ and $\beta$, for the forward reaction and reverse reaction, respectively. The subscripts for the $\alpha$ and $\beta$ rate constants identify which variable they pertain to (i.e., $n$, $m$, or $h$). The $\alpha$ and $\beta$ rate constants depend only on membrane potential (at constant temperature and $[Ca]_o$). A schematic model is given in Fig. 31, and the corresponding equations are

$$\frac{dn}{dt} = \alpha_n (1 - n) - \beta_n n \tag{11}$$

$$\frac{dm}{dt} = \alpha_m (1 - m) - \beta_m m \tag{12}$$

$$\frac{dh}{dt} = \alpha_h (1 - h) - \beta_h h \tag{13}$$

where $n$ is the fraction of sites occupied, and therefore $[1 - n]$ is the fraction of sites unoccupied. Therefore,

$$\text{rate of occupation} = \alpha_n (1 - n) \tag{14}$$

$$\text{rate of unoccupation} = \beta_n n \tag{15}$$

rate of change of $n$ = rate of occupation
$$\quad - \text{rate of unoccupation} \tag{16}$$

or

$$\frac{dn}{dt} = \alpha_n (1 - n) - \beta_n n \tag{11a}$$

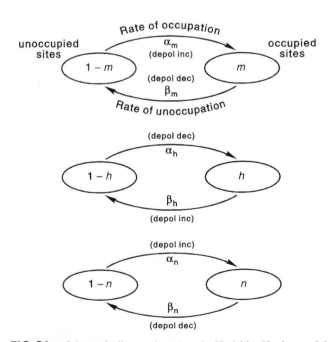

**FIG. 31.** Schematic diagram based on the Hodgkin–Huxley model for activation and inactivation of $Na^+$ channels and $K^+$ channels. $\alpha$ and $\beta$ are the rate constants for occupation and unoccupation, respectively, of sites on the channel protein by the $m^+$, $h^+$, and $n^+$ particles. See text for further explanation.

which is the same as Eq. 11. The same analysis can be applied to the $m$ and $h$ factors. The effect of making the inside of the fiber more positive (i.e., depolarizing) is to increase $\alpha_n$, $\alpha_m$, and $\beta_h$ and to decrease $\beta_n$, $\beta_m$, and $\alpha_h$. One must remember that the $h$ particles occupy the favorable site at the resting potential (I-gate open), and this site becomes unoccupied with depolarization (I-gate closes). This is opposite to the situation for the $m$ and $n$ particles (and A-gate).

In voltage clamp experiments, at a given clamp step, the membrane potential is fixed and the differential equations (Eqs. 11 to 13) lead to exponential expressions for $n$, $m$, and $h$, and the $K^+$ and $Na^+$ conductances can then be calculated. When this was done, there was a good fit of the calculated curves with the experimental data points for the changes in $g_{Na}$ and $g_K$ with time at various clamp steps (see Fig. 15).

### K. Mechanisms of Repolarization

The AP is terminated primarily by the turn-on of $g_K$, the **delayed rectification.** It is called the delayed rectifier because its turn-on is slower and delayed with respect to the turn-on of the fast $Na^+$ channels. The increase in $g_K$ acts to bring $E_m$ toward $E_K$ (about $-90$ mV), since the membrane potential at any time is determined mainly by the ratio of $g_{Na}/g_K$ (see Chapter 13). This type of $g_K$ channel is activated by depolarization and turned off by repolarization. Therefore, this $g_K$ channel is *self-limiting*, in that it turns itself off as the membrane is repolarized by its action.

In addition to the $g_K$ turn-on, there is also some turn-off of $g_{Na}$, which contributes to repolarization. Two reasons for $g_{Na}$ turn-off are (1) spontaneous inactivation of fast $Na^+$ channels that had been activated, that is, closing of their I-gate (inactivation $\tau$ of 1–3 msec); and (2) a reversible shifting of activated channels directly back to the resting state because of the rapid repolarization occurring due to the $g_K$ mechanism. Theoretically, it would be possible to have an AP that would repolarize (but slowly) even if there were no $g_K$ mechanism, because the $g_{Na}$ channels would spontaneously inactivate, and so the $g_{Na}/g_K$ ratio and $E_m$ would slowly be restored to their original resting values. Turn-on of $g_K$ acts to sharpen repolarization of the AP and allows higher frequency of impulses.

### L. Skeletal Muscle Repolarization

The two mechanisms for sharp repolarization of the AP discussed previously for neurons also apply to skeletal muscle. But there is an important third factor involved in repolarization of the skeletal muscle AP, the $Cl^-$ current. The $Cl^-$ permeability ($P_{Cl}$) and conductance ($g_{Cl}$) are very high in skeletal muscle. In fact, $P_{Cl}$ of the surface membrane is much higher than $P_K$, the $P_{Cl}/P_K$ ratio being about 3–7. However, as discussed in Chapter 13, the $Cl^-$ ion is passively distributed, or nearly so, and thus cannot determine the resting potential under steady-state conditions. However, net $Cl^-$ movements inward (hyperpolarizing) or outward (depolarizing) can and do affect $E_m$ transiently until reequilibration occurs. There is no net electrochemical driving force for Cl$^-$ current ($I_{Cl}$) at the resting potential, since

$$E_m = E_{Cl} \tag{17}$$

and

$$(E_m - E_{Cl}) = 0 \tag{17a}$$

However, during AP depolarization, there is a larger and larger driving force for outward $I_{Cl}$ (i.e., Cl$^-$ influx), since

$$I_{Cl} = g_{Cl} (E_m - E_{Cl}) \tag{18}$$

where $E_{Cl}$ is the Cl$^-$ equilibrium potential calculated from the Nernst equation (Fig. 22). In other words, the large electric field that was keeping Cl$^-$ out (i.e., [Cl]$_i$ ≪ [Cl]$_o$) is diminishing during the AP, and so Cl$^-$ ion enters the fiber. This Cl$^-$ entry is hyperpolarizing, and therefore tends to repolarize the membrane more quickly than would otherwise occur. That is, repolarization of the AP is sharpened by the Cl$^-$ mechanism. The higher the $g_{Cl}$, the greater this effect.

To illustrate some of the above points, if skeletal muscle fibers are placed into Cl$^-$-free Ringer solution (e.g., methanesulfonate substitution), depolarization and spontaneous APs and twitches occur for a few minutes until most or all of the [Cl]$_i$ is washed out. After equilibration, the resting $E_m$ returns to the original value (ca. $-90$ mV for frog muscle), clearly indicating that Cl$^-$ does not determine the resting potential and that net Cl$^-$ efflux produces depolarization. Readdition of Cl$^-$ to the bath produces a rapid large hyperpolarization (e.g., to $-120$ mV) due to net Cl$^-$ influx, and $E_m$ then slowly returns to the original value ($-90$ mV) as Cl$^-$ reequilibrates (i.e., redistributes itself passively). These same effects would occur in cardiac muscle, smooth muscle, and nerve, but to a lesser extent, because $P_{Cl}$ is much lower in these tissues.

## M. Types of K$^+$ Channels

The cell membrane of some cells (e.g., myocardial cells) has at least five separate K$^+$ channels. As discussed previously, one type of voltage-dependent K$^+$ channel is the usual K$^+$ channel found in many types of excitable membranes. This channel slowly opens (increasing total $g_K$) on depolarization and is the so-called **delayed rectifier** [$i_K$ or $I_{K(del)}$]. This channel allows K$^+$ to pass readily outward down the usual electrochemical gradient for K$^+$, and so is also known as the *outward-going rectifier*. This delayed rectifier channel in myocardial cells turns on much more slowly than in nerve, skeletal muscle, or smooth muscle, and therefore accounts for the long duration of the AP. The activation of this channel produces the increase in total $g_K$ that terminates the cardiac AP plateau (so-called "phase 3 repolarization") (see Chapter 51).

A second type, known as the $I_{K1}$ channel, allows K$^+$ ion to pass more readily inward than outward, the so-called *inward-going rectifier* or *anomalous rectification*.[1] This ki-

netically fast channel is responsible for the rapid decrease in K$^+$ conductance on depolarization (and increase in conductance with repolarization), helps to set the resting potential, and helps bring about the terminal repolarization of the cardiac AP (phase 3). The $I_{K1}$ channel has been found to appear at a certain stage of development of the heart.

A third type of K$^+$ channel is activated by elevation of [Ca]$_i$ and is therefore known as the *Ca$^{2+}$-activated K$^+$ channel* or $I_{K(Ca)}$. With Ca$^{2+}$ influx and internal release of Ca$^{2+}$ during the AP and contraction, the $I_{K(Ca)}$ channels are activated and help the $I_{K(del)}$ channels to repolarize the AP, that is, to bring the membrane back to the resting potential. The presence of this type of K$^+$ channel has been reported for many types of excitable cells. In actuality, two subtypes of $I_{K(Ca)}$ channel have been found: one that has a high conductance of about 400 pS (big or maxi-K channel) and one that has a lower conductance of about 120 pS (small or mini-K channel).

A fourth type of K$^+$ channel present in some types of cells, such as myocardial cells, is kinetically fast [compared to $I_{K(del)}$] and provides a rapid outward K$^+$ current that produces a small amount of initial repolarization, known as phase 1 repolarization in cardiac cells. This occurs immediately following the rapidly rising spike portion of the AP and is known as the **transient outward current** ($I_{to}$) in myocardial cells and $I_A$ in neurons. There is some evidence that Cl$^-$ current may contribute to $I_{to}$ (Cl$^-$ influx provides an outward $I_{Cl}$, which is repolarizing in direction).

A fifth type of K$^+$ channel is sensitive to ATP, the $K_{ATP}$ *channel,* and provides a current known as $I_{K(ATP)}$. This channel is regulated by ATP, such that in normal myocardial or smooth muscle cells, this K$^+$ channel is inhibited (masked or silent). However, in ischemic or hypoxic conditions, when the ATP level is lowered, the $I_{K(ATP)}$ channels become unmasked and provide a large outward $I_K$ that prematurely shortens the cardiac AP. This channel provides a protection mechanism for the heart; namely, the ischemic region of the heart develops very abbreviated APs, and hence contraction is greatly depressed. This effect acts to conserve ATP in the afflicted cells, enabling full recovery if the blood flow returns to normal after a short time period.

## IV. Effect of Resting Potential on Action Potential

Any agent that affects the resting potential has important repercussions on the AP. Depolarization reduces the rate of rise of the AP, and thereby also slows its velocity of propagation. A slow spread of excitation throughout the nerve or muscle will interfere with its ability to act efficiently. This effect is progressive as a function of the degree of depolarization. If nerve, skeletal muscle fibers, or cardiac muscle cells are depolarized to about $-50$ mV by any means, then the rate of rise goes to zero and all excitability is lost.

Hyperpolarization usually produces only a small increase in the rate of rise. Larger hyperpolarization may actually slow the velocity of propagation, because the critical depo-

---

[1] It is anomalous in the sense that this channel *turns off* with depolarization and *turns on* with hyperpolarization.

larization required to bring the membrane to its threshold potential is increased, and it can cause propagation block.

The explanation for the effect of resting $E_m$ (or takeoff potential) on maximum rate of rise (max $dV/dt$) of the AP is based on the sigmoidal $h_\infty$ versus $E_m$ curve (see Fig. 27B). The I ($h$)-gates are open in a resting membrane and close with depolarization. As discussed previously, $h$, the inactivation variable for the fast $Na^+$ conductance, is a probability factor that deals with the open ($h = 1.0$) versus closed ($h = 0$) positions of the inactivation gate of the channel (see Fig. 25). At the resting potential of $-80$ mV, $h_\infty$ is 0.9–1.0 and diminishes with depolarization, becoming nearly zero at about $-30$ mV.

The resting potential also affects the duration of the AP. With polarizing current, depolarization lengthens the AP, whereas hyperpolarization shortens it. In contrast, when elevated $[K]_o$ levels are used to depolarize the cells, the AP is shortened. One important determinant of the AP duration is $g_K$. Agents or conditions that increase $g_K$, such as elevation of $[K]_o$, tend to shorten the duration. In contrast, agents that decrease $g_K$ or slow its activation, such as $Ba^{2+}$ ion or $TEA^+$, tend to lengthen the AP duration. Because of anomalous rectification (i.e., a decrease in $g_K$ with depolarization and an increase with hyperpolarization), depolarization by current prolongs the AP and hyperpolarization shortens it.

Other factors are also important in determining the AP duration. For example, agents that slow the closing of the I-gates of the fast $Na^+$ channels, such as veratridine, prolong the AP. High rates of activity generally shorten AP duration, for example, by an increase in $[Ca]_i$ (resulting from an increase in $[Na]_i$) producing an increase in $g_K$ [the $g_{K(Ca)}$ channel].

## V. Electrogenesis of Afterpotentials

The APs of nerve and muscle cells usually consist of two components: an initial spike followed by an early afterpotential (Figs. 32A and B). The early afterpotentials may be of two types: depolarizing or hyperpolarizing. In addition, late afterpotentials, both depolarizing or hyperpolarizing, may appear following a brief train of spikes (Figs. 32C and D). The electrogenesis of the early and late afterpotentials is different. The early afterpotentials are due to a conductance change, whereas the late afterpotentials may be due to $K^+$ accumulation or depletion in restricted diffusion spaces and to electrogenic pump stimulation.

### A. Early Depolarizing Afterpotentials

The AP spike in skeletal muscle fibers is immediately followed by a prominent depolarizing afterpotential (also called a "negative" afterpotential, based on the old terminology used in external recording) (Fig. 32A). The early depolarizing afterpotential of frog skeletal fibers is about 25 mV (immediately after the spike component), and gradually decays to the resting potential within 10–20 msec. It results from the fact that the delayed rectifier $K^+$ channel that opens during depolarization to terminate the spike is

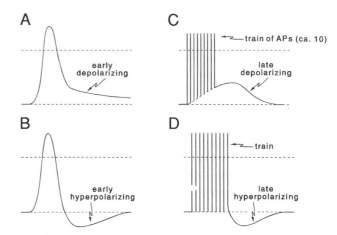

**FIG. 32.** Examples of the different types of afterpotentials. (A) Early depolarizing ("negative") afterpotential recorded after a single AP in a skeletal muscle fiber. (B) Early hyperpolarizing ("positive") afterpotential recorded after a single AP in a nerve fiber. (C) Late depolarizing afterpotential recorded after a train (e.g., 10) of APs in a skeletal muscle fiber. (D) Late hyperpolarizing afterpotential recorded following a train of APs in a nerve terminal.

less selective for $K^+$ (ca. 30:1, $K^+:Na^+$) than is the $K^+$ channel in the resting membrane (ca. 100:1, $K^+:Na^+$). Therefore, from the constant-field equation (see Chapter 13), one can predict that the membrane should be partly depolarized when the membrane is dominated by this delayed rectifier $K^+$ conductance. Thus, the early depolarizing afterpotential is apparently due to the persistence of, and slow decay of, this less-selective $K^+$ conductance.

### B. Early Hyperpolarizing Afterpotentials

Neurons, pacemaker heart cells, and vascular smooth muscle cells often exhibit early hyperpolarizing ("positive") afterpotentials (Fig. 32B). These are due to the delayed rectifier $K^+$ conductance increase (which terminates the spike) persisting after the spike, thereby bringing $E_m$ closer to $E_K$. The maximum amplitude possible for the afterpotential is the difference between $E_K$ and the normal resting potential. The time course of this afterpotential is determined by the decay of the $K^+$ conductance increase.

### C. Late Depolarizing Afterpotentials

The **late depolarizing afterpotential** results from accumulation of $K^+$ ions in the transverse (T) tubules (Fig. 32C). During the AP depolarization and turn-on of the $g_K$ (delayed rectifier), there is a large driving force for $K^+$ efflux from the myoplasm coupled with a large $K^+$ conductance, resulting in a large outward $K^+$ current [$I_K = g_K (E_m - E_K)$] across all surfaces of the fiber, namely, the surface sarcolemma and T-tubule walls. The $K^+$ efflux at the fiber surface membrane can rapidly diffuse away and mix with the relatively large interstitial fluid (ISF) volume, whereas the $K^+$ efflux into the T-tubules is trapped in this restricted diffusion space. The resulting high $[K]_{TT}$ decreases $E_K$ across the T-tubule membrane and thereby depolarizes this membrane. Because of cable properties, part of this depolarization is transmitted to the surface sarcolemma and is recorded by an intracellular microelectrode. The $K^+$ accumulation in the T-tubules can only be dissipated relatively slowly by diffusion out of the mouth of the T-tubules and by active pumping back into the myoplasm across the T-tubule wall. Thus, the decay of the late afterpotential will be a function of these two processes.

The amplitude and duration of the late depolarizing afterpotential of frog skeletal fibers are functions of the number of spikes in the train and their frequency. That is, the greater the spike activity, the greater the amplitude and duration of the late depolarizing afterpotential. If the train consists of 20 spikes at a frequency of 50/sec, a typical value for the amplitude of the afterpotential is about 20 mV. When the diameter of the T-tubules is increased by placing the fibers in hypertonic solutions, the amplitude of the afterpotential decreases, as expected, because of the greater dilution of the $K^+$ ions accumulating in the T-tubule lumen. When the T-tubular system is disrupted and disconnected from the surface membrane by the glycerol osmotic shock method, the late depolarizing afterpotential disappears (whereas the early depolarizing afterpotential persists).

A slow relaxation of a component of the $K^+$ conductance increase may contribute to the late afterpotential.

### D. Late Hyperpolarizing Afterpotentials

Some cells, such as nonmyelinated neurons, exhibit late hyperpolarizing afterpotentials following a train of spikes (Fig. 32D). These hyperpolarizing afterpotentials are due to the Na–K pump, because inhibition of the pump by any means (such as ouabain) abolishes them. Two mechanisms have been proposed for the hyperpolarization: (1) Hyperpolarization occurs because there is an increased electrogenic $Na^+$ pump potential ($V_p$), stimulated both by an increase in the $[Na]_i$ (since these neurons are small in diameter, and hence have a large surface area/volume ratio) and by an increase in $[K]_o$ (since these axons are surrounded by Schwann cells and hence have a narrow intercellular cleft and restricted diffusion space). (2) Hyperpolarization occurs due to an increased $E_K$ caused by $K^+$ depletion in the intercellular cleft because of the stimulated Na–K pump overpumping the $K^+$ back in. It is generally believed that the first mechanism is the most probable.

### E. Importance of Afterpotentials

All afterpotentials have physiological importance because they alter excitability and propagation velocity of the cell. A depolarizing afterpotential should enhance excitability (lower threshold), and a hyperpolarizing afterpotential should depress excitability to a subsequent AP. This is because the critical depolarization required to reach the threshold potential would be decreased or increased, respectively. A large late depolarizing afterpotential, such as that due to $K^+$ accumulation in the T-tubules, can trigger repetitive APs under certain pathological conditions.

The effect of afterpotentials on velocity of propagation is complex because there are two opposing factors: (1) the change in critical depolarization required and (2) the change in maximal rate of rise of the AP, which is a function of the takeoff potential ($h_\infty$ vs $E_m$ curve). For example, during a depolarizing afterpotential in skeletal muscle fibers, the critical depolarization required is decreased, but the maximal rate of rise of the AP is also decreased. Therefore, these two factors exert opposing effects.

## VI. Summary

Membrane excitability is a fundamental property of nerve and muscle cells (skeletal, cardiac, and smooth), as well as certain other cell types, such as some endocrine cells. An excitable cell is one that, in response to certain environmental stimuli (electrical, chemical, or mechanical), generates an all-or-none electrical signal or AP. The AP is sometimes called an "impulse," for example, a nerve impulse. The AP is triggered by a depolarization of the membrane, which is produced by the applied stimulus. The depolarization initiates an increase in the membrane permeability to Na ions, which then flow into the cell, causing a transient reversal in the membrane potential. A slower increase in the permeability of the membrane to K ions contributes to the repolarization of the membrane, in addition to the spontaneous inactivation of the Na channels. Some cells display the property of automaticity; that is, they produce APs spontaneously without any externally applied stimulus.

In some excitable cell types (such as cardiac), a slow inward Ca current contributes to the long plateau of the AP, and in others (such as smooth muscle) the Ca current itself generates the upstroke. The membrane currents that contribute to the AP can be studied by the voltage clamp method, which allows isolation and characterization of each membrane current as a function of membrane potential and time. Ionic currents flow across the membrane by means of numerous ion-specific protein channels. Each channel molecule has a region that senses the transmembrane potential and acts as a gate to open or close the channel to ion passage through its central pore.

Some types of channels (i.e., Na channels) have a second gating system that closes (or inactivates) the channel during a maintained depolarization. The pattern of minute currents that flow through individual voltage-dependent ion channels can be studied using the patch-clamp method. Single-channel current measurement and structural information have given a greater understanding of the molecular basis of membrane excitability.

The APs in vertebrate nerve fibers consist of a spike followed by a hyperpolarizing afterpotential. A large fast inward $Na^+$ current, passing through fast $Na^+$ channels, is responsible for electrogenesis of the spike, which rises rapidly (~1000 V/sec). Subsequently, a small inward $Ca^{2+}$ current, passing through kinetically slower channels, may be involved in excitation–secretion coupling at the nerve terminals. The nerve cell membrane has voltage-dependent $K^+$ channels which allow $K^+$ ions to pass readily outward down the electrochemical gradient for $K^+$. This channel population is responsible for repolarization. This $K^+$ channel, which opens more slowly than $g_{Na}$ upon depolarization, is called the delayed rectifier. The activation of this channel produces the large increase in total $g_K$ that terminates the AP.

The nerve AP amplitude is about 110 mV, from a resting potential of $-70$ mV to a peak overshoot potential of about $+40$ mV. The duration of the AP (at 50% repolarization) ranges between 0.5 and 1 msec, depending on the species and temperature. The threshold potential ($V_{th}$) for triggering of the fast $Na^+$ channels is about $-55$ mV; a critical depolarization of about 15 mV is required to reach $V_{th}$. The turn-on of the fast $g_{Na}$ (fast $I_{Na}$) is very rapid (within 0.2 msec), and $E_m$ is brought rapidly toward $E_{Na}$. There is an explosive (positive exponential initially) increase in $g_{Na}$ caused by a positive-feedback relationship between $g_{Na}$ and $E_m$.

In certain nerve cells, as well as muscle cells, as $E_m$ depolarizes it crosses $V_{th}$ (about $-35$ mV) for the slow Ca channels. Turn-on of the slow Ca conductance ($g_{Ca}$) and $I_{Ca}$ is slow and tends to bring $E_m$ toward $E_{Ca}$. The peak $I_{Ca}$ is considerably smaller than the peak fast $I_{Na}$. From the voltage clamp current versus voltage curves, the maximum inward fast current and slow current occur at an $E_m$ of about $-20$ and $+10$ mV, respectively. The currents decrease at more depolarized $E_m$ levels because of the diminution in electrochemical driving force as the membrane is further depolarized, even though the conductance remains high. At the reversal potential ($E_{rev}$) for the current, the electrochemical driving force goes to zero and reverses direction with greater depolarization.

In some smooth muscle cells, the slow $Ca^{2+}$ current itself is sufficient to depolarize the membrane and generate a regenerative slowly rising AP in the absence of the fast $Na^+$ current.

## Bibliography

Bernstein, J. (1902). Untersuchengen zur Thermodynamik der bio-elektrischen Ströme. *Pflugers Arch.* **92**, 521.

Bernstein, J. (1912). "Elektrobiologie." Braunschweg, Vieweg.

Catterall, W. A. (1988). Structure and function of voltage-sensitive ion channels. *Science* **242**, 50–61.

Cole, K.S. (1949). Dynamic electrical characteristics of the squid axon membrane. *Archs. Sci. Physiol.* **3**, 253–258.

Cole, K. S., and Curtis, H. J. (1939). Electric impedance of the squid giant axon during activity. *J. Gen. Physiol.* **22**, 649–687.

Hamill, O. P., Marty, A., Neher, E., Sakmann, B., and Sigworth, F. J. (1981). Improved patch-clamp technique for high resolution current recording from cells and cell-free membrane patches. *Pflugers Arch.* **391**, 85–100.

Hamill, O. P., and Sakmann, B. (1981). Multiple conductance states of single acetylcholine receptor channels in embryonic muscle cells. *Nature* **294**, 462–464.

Hille, B. (1984). "Ionic Channels of Excitable Membrane." Sinauer Associates, Sunderland, MA.

Hodgkin, A. L. (1958). Ionic movements and electrical activity in giant nerve fibres. *Proc. Roy. Soc. B.* **148**, 1.

Hodgkin, A. L. (1964). "The Conduction of the Nervous Impulse." Liverpool University Press, Liverpool, UK.

Hodgkin, A. L., and Huxley, A. F. (1952a). Currents carried by sodium and potassium ions through the membrane of the giant axon of *Loligo. J. Physiol.* (*London*) **116**, 449–472.

Hodgkin, A. L., and Huxley, A. F. (1952b). The components of membrane conductance in the giant axon of *Loligo. J. Physiol.* (*London*) **116**, 473–496.

Hodgkin, A. L., and Huxley, A. F. (1952c). The dual effect of membrane potential on sodium conductance in the giant axon of *Loligo. J. Physiol.* (*London*) **116**, 497–506.

Hodgkin, A. L., and Huxley, A. F. (1952d). A quantitative description of membrane current and its application to conduction and excitation in nerve. *J. Physiol.* (*London*) **117**, 500–544.

Hodgkin, A. L., and Katz, B. (1949). The effect of sodium ions on the electrical activity of the giant axon of the squid. *J. Physiol.* (*London*) **108**, 37–77.

Katz, B. (1966). "Nerve, Muscle and Synapse." McGraw-Hill, New York.

Ohya, Y., and Sperelakis, N. (1989). Fast $Na^+$ and slow $Ca^{2+}$ channels in single uterine muscle cells from pregnant rat. *Am. J. Physiol. Cell* **257**, C408–C412.

Overton, E. (1902). Beiträge zur allgemeinen Muskel-und Nerven-physiologie. *Pflugers Arch.* **92**, 346.

Sperelakis, N. (1988). Electrical properties of cells at rest and maintenance of the ion distributions. *In* "Physiology and Pathophysiology of the Heart" (N. Sperelakis, Ed.), pp. 59–82. Kluwer, New York.

Sperelakis, N., and Fabiato, A. (1985). Electrophysiology and excitation-contraction coupling in skeletal muscle. *In* "The Thorax: Vital Pump" (C. H. Roussos and P. Macklem, Eds.), pp. 45–113. Marcel Dekker, New York.

Sperelakis, N., and Josephson, I. (1992). Basis of membrane excitability. *In* "Essentials of Physiology" (N. Sperelakis and R. O. Banks, Eds.), pp. 49–97. Little, Brown, Boston.

Tohse, N., and Sperelakis, N. (1990). Long-lasting openings of single slow (L-type) $Ca^{2+}$ channels in chick embryonic heart cells. *Am. J. Physiol.* **259**, H639–H642.

Tohse, N., and Sperelakis, N. (1991). cGMP inhibits the activity of single calcium channels in embryonic chick heart cells. *Circ. Res.* **69**, 325–331.

*Raymund Y. K. Pun and Harold Lecar*

# 25

# Patch-Clamp Techniques and Analysis

## I. Introduction

**Patch-clamp technique** revolutionized electrophysiology by revealing the activity of individual molecular ion channels involved in electrical signaling in excitable cells. Just as the invention of the voltage-clamp technique in the late 1940s led to an understanding of nerve excitation in terms of specific transmembrane ionic currents, the invention of the patch-clamp or **giga-seal technique** provided the impetus for analyzing excitation phenomena at the molecular level in terms of the behavior of specific transport proteins.

This chapter and the several that follow cover the use of the patch-clamp technique in studies of both macroscopic ionic currents and single ion channels. The chapter is divided into three parts: The first focuses on the technique and its application in studies of single channels or unitary conductances; the second discusses the analysis of single-channel data; and the third covers the technique of whole-cell "tight-seal" voltage clamp, an offshoot of the patch-clamp technique that is now the most widely used method of examining membrane excitability.

The ionic basis of the action potential was established in the early 1950s by Hodgkin and Huxley (1952a,b,c) using voltage clamp on the squid giant axon. As described in Chapter 24, they demonstrated that depolarization of the membrane elicits an inward $Na^+$ current, which was followed by an outward $K^+$ current. The flow of $Na^+$ ions commences, following a short delay (within 1.0 msec of membrane depolarization), after which the current follows a sigmoidal time course, reaches a peak within 0.5–1 msec, then declines more slowly (2–5 msec; measurement was made at 6.3°C). The $K^+$ current also follows a sigmoidal time course, but with a longer delay, and is more or less sustained if the membrane remains depolarized. From these and later studies (radioactive isotope measurements), it was inferred that these currents reflect the movement of ions across the membrane via separate ion-specific pathways, which are activated by changes in the transmembrane electric field.

Experiments with neurotoxins showed that ion flux could be blocked by binding to sites of such low density that the rate at which ions flow across a single conducting site was high ($10^6$–$10^7$ ions sec$^{-1}$). Such high transport rates could best be explained by localized channels, which somehow open and close to regulate the rapid movement of ions across the membrane. The random opening and closing of individual ion channels should produce large current fluctuations that are observable as a characteristic form of electrical noise. Pharmacological dissection with site-specific agents showed conclusively that the flow of $Na^+$ ions and the flow of $K^+$ ions are mediated by different species of channels each with its own characteristic ion selectivities and open/close kinetics.

Thus, even before the isolation and cloning of channel proteins, ion channels were thought of as intrinsic membrane proteins, existing as discrete entities, embedded in the phospholipid membrane. When channels switch open, the passage of ions through them causes transient changes in membrane potential that trigger excitability effects, such as muscle contraction, hormone secretion, and all-or-none nerve impulse propagation. In general, an influx of cations (or efflux of anions) will depolarize the membrane potential and increase the generation of action potentials or firing rate, whereas an efflux of cations (or influx of anions) will hyperpolarize and decrease excitability. The $Na^+$ and $K^+$ channels that underlie action potential generation are known as **voltage-gated channels**, because the rates of opening and closing these channels are sharply dependent on changes in the membrane potential. Another class of channels, known as **ligand-gated channels**, are activated when neurotransmitters or hormones bind to them. A prototype of this class of channel is the nicotinic acetylcholine (ACh) channel present at the skeletal neuromuscular junction, which was the first ligand-gated channel to be studied thoroughly (see Chapter 39 for details). Many other types of voltage-gated channels and ligand-gated channels have now been discovered and characterized (discussed in Chapters 26 and 39).

Before the invention of the patch-clamp technique, studies using lipid bilayers doped with various channel-forming peptides and proteins showed how ion channels open in discrete steps or jumps (Ehrenstein and Lecar, 1977). The implication was that channels undergo transitions between states of different conductance and that excitability is the macroscopic concomitant of the random channel gating. In the closed state, channels are nonconducting, whereas in the open state, the channels allow ions to traverse the membrane. Later experiments employing fluctuation analysis or noise analysis on biological membranes supported this idea. Direct confirmation of the existence of microscopic currents passing through single channels in biological membranes came with the advent of the patch-clamp technique.

## II. Patch-Clamp or Giga-Seal Technique

To measure the minute ionic currents carried by a single-channel protein, an experimenter must electrically isolate a small area of membrane containing a small number of channels, thus the term *patch clamp* (clamping the voltage of a membrane patch). The currents recorded from the patch must also have a background noise level sufficiently low for the channel current to be detectable. Neher and Sakmann (1976) first measured single-channel currents in a cell membrane by pressing a blunt-tipped pipette against the surface of a denervated muscle fiber (freed from overlying cells and connective tissue by collagenase treatment), and feeding the signal through a low-noise headstage current-to-voltage converter. A schematic diagram of current measurement through an isolated membrane patch is shown in Fig. 1. The equivalent circuit of this arrangement

is shown in Fig. 2A. Detailed analysis of the circuit and the effects of the various components (e.g., shunt resistance, cell resistance) on the minimization of background noise have been described Hamill *et al.*, 1981; Fenwick *et al.*, 1982; Barry and Lynch, 1991), and the reader is referred to these papers for more in-depth discussion. Several recent books provide a wealth of information on all aspects of the patch-clamp technique (Sakmann and Neher, 1995; Rudy and Iverson, 1992; Ashley, 1995; Boulton *et al.*, 1995).

### A. Noise in Patch Clamp

The main consideration in the design of a patch-clamp experiment is to minimize the background electrical noise that interferes with and obscures the single-channel current fluctuations. The minimum detectable current-jump amplitude in the presence of interfering electrical noise depends not only on the intensity of the noise but also on the duration of the jump. Shorter duration pulses and short-duration flickering channels require broader receiver bandwidth in order to be resolved with some fidelity. The background noise generated in the membrane, the patch pipette, and the recording electronics are spread out in frequency, hence design of the patch clamp requires some compromise between minimizing the bandwidth for noise reduction and increasing it for fidelity of recording. Patch-clamp amplifiers achieve the optimum compromise by employing a first-stage current-to-voltage converter, in which gain and effective noise are minimized at the expense of slow response, and by having a second-stage frequency booster, which reconstructs the sharp channel jumps.

Background noise is generated in all the conductance pathways in parallel with the channels. Thus, particular

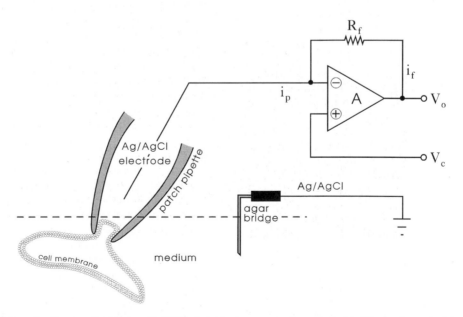

**FIG. 1.** Schematic diagram depicting cell-attached patch-clamp recording. A Ag/AgCl wire immersed in the pipette solution acts as the recording electrode. The ground is another Ag/AgCl electrode that interfaced the bath with an agar bridge (1–3% agar in 0.9% saline). The signal input is into amplifier A. Under voltage-clamp conditions, the voltage to the input (−ve junction of A) is maintained at the same potential as $V_c$, the command potential such that $V_o = -V_c$. Under this condition, $i_f R_f = -i_p R_f$, where $i_f$ is the feedback current, $i_p$ the pipette current, and $R_f$ the feedback resistance.

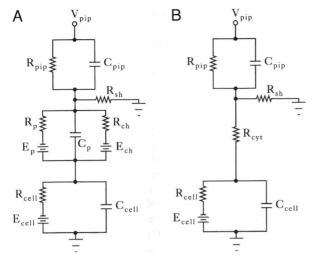

**FIG. 2.** Equivalent circuit diagrams for cell-attached patch-clamp recording (A) and whole-cell tight-seal recording (B). $V_{pip}$, $R_{pip}$, and $C_{pip}$ in the equivalent circuit diagram denote the voltage, resistance, and capacitance of the pipette, respectively; $R_{sh}$ denotes the shunt of seal resistance; $E_p$, $R_p$, and $C_p$ denote the driving potential, resistance, and capacitance of the patch of membrane, respectively; $E_{cell}$, $R_{cell}$, and $C_{cell}$ denote the membrane potential, resistance, and capacitance of the cell, respectively; and $R_{cyt}$ is the resistance of the cytoplasm after establishing whole-cell tight-cell recording. The series or access resistance is the sum of $R_{pip}$ and $R_{cyt}$. The series resistance and the current flowing across this resistor determine the voltage drop across the pipette.

attention must be paid to the thermal noise generated across the pipette-to-membrane seal and across the pipette wall. In early patch-clamp experiments, the current noise generated across the pipette-to-membrane seal, whose rms amplitude is inversely proportional to the square root of seal resistance, was the limiting factor in patch-clamp sensitivity. The multi-giga-ohm seal reduces this seal noise to a minimal level, leaving the noise generated across the pipette walls by several different mechanisms as the main source. Current flowing through the layer of electrolyte clinging to the pipette wall is transmitted across the pipette–wall capacitance, which acts as a high-pass filter. The pipette noise is suppressed by making the electrode of highly resistive glass and/or coating the walls with a hydrophobic sealant. The detailed analysis of the sources of electrical noise operating at different frequencies and the principles underlying design of a patch clamp are discussed by Hamill *et al.* (1981). The introduction of a capacitance feedback headstage coupled with the use of quartz glass electrodes has further lowered the electrical noise level (Levis and Rae, 1993).

When all the sources of background noise are taken into account, a patch with a 100-GΩ seal and 2-kHz recording bandwidth allows the observation of 1-msec channel jumps of amplitude as low as 0.05 pA. This gives some indication of the smallest unit of conductance observable by patch clamp at present (order of 1 pS).

The detection of single-channel jumps is regarded as the *sine qua non* of transport by channels, sites of intense flux,

as opposed to carriers or other transport pathways for which flux is rate limited by the need to translocate binding sites within the membrane. The order of magnitude of the currently minimum detectable channel amplitude implies that detectable single-channel currents are a sufficient condition for identifying channel transport but not a necessary one. There may be membrane-spanning pathways, which transport too slowly to yield observable single-channel jumps. In such cases, the unitary conductances might be inferred from macroscopic membrane noise analysis, as discussed in Section VI,A.

## B. Variations of the Patch-Clamp Technique

Variations in the technique also proved to be successful in determining channel properties (Hamill *et al.*, 1981). The method used in studying the behavior of single channels residing within the electrically isolated patch of membrane described earier is known as the **cell-attached patch,** since the cell is firmly attached to the recording pipette (Fig. 3, left). One drawback of the cell-attached configuration is the inability to determine accurately the cell's membrane potential, which has an effect on the patch potential. To

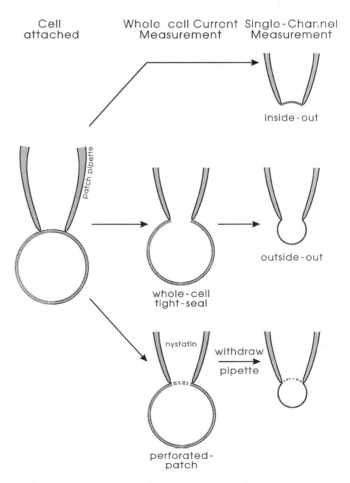

**FIG. 3.** Various configurations of patch-clamp (single-channel current) and whole-cell tight-seal (macroscopic current) recordings. See text for description.

determine the actual potential across the membrane patch, one can measure the cell's potential with another recording pipette and then set the voltage accordingly, or, alternatively, depolarize the cell with high K⁺ solution to "zero" the membrane potential before setting the membrane patch potential. A less accurate means of determining the patch potential is either to use a potential relative to that of the cell (e.g., +20 mV to rest) or to estimate the membrane potential before imposing a voltage across the patch.

When a gigaohm seal is formed, there is a strong bond between the glass surface of the patch pipette and the membrane. The membrane patch is stuck to the pipette and can be excised from the cell by a variety of methods, as illustrated in Fig. 3. The excised patch permits the experimenter to control the voltage across the membrane patch, and thereby to effect the voltage clamp of single-channel currents over a wide voltage range ($\pm 200$ mV). It further allows for the alteration of ionic concentrations and conditions surrounding the patch so that the characteristics and modulation (regulation by second messengers) of the single-channel currents can be studied.

If the pipette is withdrawn and pulled away from the cell, the piece of membrane (estimated area of 1–10 $\mu m^2$) sealed against the pipette is torn away from the cell. Provided this action does not alter seal resistance drastically, channel activities can still be recorded. Occasionally, a membrane vesicle (instead of a simple planar patch) is formed at the tip following excision from the cell, but the vesicle can be broken by passing the pipette rapidly across the fluid/air interface. This variation is known as the isolated **inside-out patch** because the cytoplasmic side of the membrane is now facing the outside or bath solution (Fig. 3, top right). The inside-out configuration can be used to investigate the effects of externally applied second messengers on the behavior of the channel. Another application of the inside-out configuration is the use of excised inside-out macropatches from *Xenopus* oocytes expressing K⁺ channel to study K⁺ gating currents (Stuhmer *et al.*, 1991).

Another variation of the excised patch-clamp technique is the **outside-out patch,** in which the surface of the membrane faces the external bath medium (Fig. 3, middle right). This configuration is comparable to the normal whole-cell tight-seal recording mode (see Section IV,A), with the exception that only a small patch of membrane is being examined. The outside-out patch is obtained by withdrawing the pipette after the establishment of the whole-cell configuration, extruding a tube of membrane, which eventually pinches off to form a membrane patch with the outside surface facing the bath. The outside-out mode of recording is most useful for studying ligand-gated channels, because it is possible to vary the different concentrations of the ligand in the bath and determine their effects on channel activity.

## III.  Single-Channel Analysis

Single-channel recording reveals the discrete steps of ionic current that occur when individual ion channels form nonconducting (closed) and conducting (open) states. Under steady-state conditions, channels fluctuate randomly between different conformations over time periods typically of the order of milliseconds, but ranging from $10^{-5}$ sec to several seconds for different channels. At the single-channel level, the kinetics of the gating process is characterized by statistical distributions of interevent times, which can be related to the rates of state-to-state transition.

The random gating transitions lead to macroscopic excitability phenomena, because the rates of transition between states can be regulated by specific stimuli, such as the change in transmembrane electric field or the binding of ligands. Appropriate stimuli acting on the channel structure are able to alter the free-energy difference between conformational states, and thereby bias the rates of transition between them. The transition rates are highly temperature sensitive, suggesting that the various states are separated by relatively high free-energy barriers. The physical features of the channel structure which allow sensitive coupling to different physical stimuli thus act as sensors for transducing the various types of stimuli into electrical signals.

### A.  Families of Gated Channels

Patch-clamp experiments have revealed families of channels obeying various modes of gating transduction:

1. *Voltage-sensitive gating.* The Na⁺, K⁺, and Ca²⁺ channels of nerve and muscle have very steeply voltage-dependent opening probabilities. These channels have highly charged subunits which must be displaced by the transmembrane electric field in order to effect channel opening. The steepness, an e-fold change in open state probability per 3–6 mV, suggests that the equivalent of four to six elementary (univalent) charges moved across the membrane to bring about channel opening.

2. *Ligand-sensitive gating.* Postsynaptic channels and other chemosensitive channels have extracellular receptors that bind transmitter substances. Transmitter binding alters the free energies of the open and closed states to facilitate opening while the transmitter is bound to its receptor site. Fast synaptic channels, such as the ACh-sensitive channel of the neuromuscular junction, and the GABA$_A$ channel of inhibitory synapses, have two ligand-binding sites leading to some cooperative action of the ligands in activating the channel.

3. *Internal messenger-activated gating.* These channels are sensitive to the binding of intracellular ligands, such as Ca²⁺ and cyclic nucleotides. These channels are generally gated as a steep function of messenger concentration, suggesting that they have multiple binding sites which must be occupied for maximum rate of opening.

4. *G-protein gating.* These are channels whose gating is modified by membrane-bound G-proteins, which in turn are coupled to separate receptor molecules which release them when activated by the appropriate ligand. The G-proteins diffuse laterally within the

membrane and activate the channels they encounter (see Chapter 33).

5. *Stretch-activated gating.* An ubiquitous class of channels can be activated by mechanical tension applied to the membrane. In specific stretch transducer cells, these channels can be coupled to elastic elements of the cytoplasmic structure (see Chapters 35 and 43).

## B. Statistical Nature of Channel Gating

For all of the different gating mechanisms, channel activation is always observed as a random switching between discrete conformational states of the channel protein. The stimulus does not deterministically drive the channel to switch conformations in a fixed time, or to change conductance gradually. Rather, stimuli that induce gating act to bias the rates of an ongoing thermally driven random process. This provides the rationale for modeling channel gating as a stochastic process, in which a system having a fixed number of discrete states passes from one state to another via first-order kinetic transitions.

If the channels reside in any specific metastable state long enough on the average that the transitions are not affected by prior history or the manner of entry into the state, the interstate transitions will be memoryless. Such a process is called a **Markov process** and can be described by a set of equations relating the occupancy probabilities of the states to the transition rates between every pair of states.

The single-channel currents do not represent all the transition among the states but only those that yield a measureable electric current jump. Some of the transitions are electrically silent, although their effect is implicit on the statistics of the channel transitions. A process in which only some of the underlying transitions are directly observable is called an **aggregated Markov process.**

Generally, a gated ion channel is a system with multiple open and closed states; represented by a Markov process with $N$ states, and hence described by a set of $N$ probabilities $P_i$, which obey the set of first-order equations

$$dP_i/dt = \Sigma k_{ij} P_j - (\Sigma k_{ji} P_i) = \Sigma Q_{ij} P_j \qquad (1)$$

Here, the $k_{ij}$ are the transition rates from the $j$th to the $i$th state. The transition rates of a particular model can be gathered in an array that is the $N \times N$ matrix $\mathbf{Q}$. The transitions can be partitioned into groups as observable transitions between an open or closed state or as unobservable transitions between two closed states or two open states. Thus the system can be partitioned into two classes of states and the matrix $\mathbf{Q}$ can be partitioned to show explicitly how the system moves between the two classes of states. These partitioned matrices can be used to predict the distributions of open and closed dwell times, which are multiexponential functions representing sojourns in either of the two collections of states. If all the $Q_{ij}$ were nonzero, so that every state were connected to every other state, the system would be too complex to analyze. However, most of the kinetic schemes that might be suggested on physical or esthetic grounds do not involve transitions between every pair of states.

To describe observable quantities, we divide the states into two classes, open and closed. The matrix of the transition probabilities can be thought of schematically as

$$\mathbf{Q} = \begin{bmatrix} \text{open–open} & \vdots & \text{open–closed} \\ \text{------------} & \vdots & \text{------------} \\ \text{closed–open} & \vdots & \text{closed–closed} \end{bmatrix}$$

Here, the diagonal submatrices represent unobservable transitions that occur during a sojourn in a complex of states having the same conductance, whereas the off-diagonal submatrices represent the observable transitions by which a channel opens or closes.

Figure 4 shows examples of kinetic schemes that have been proposed for different gated channels. Each of the schemes is illustrated by a diagram indicating the states and the allowed transitions between them. The matrix of transition rates shows all the independent rate constants needed to specify a model and the dependence of the rates on external parameters such as membrane potential or ligand concentration. Colquhon and Hawkes (1981, 1982) show how the observable statistical properties of single-channel records are related to the underlying kinetics model.

Three schemes are shown that might describe agonist-activated channels. The first is a hypothetical scheme (Fig. 4A, first proposed by del Castillo and Katz, 1957), in which the binding of an agonist by a closed channel leads to an activated closed state capable of undergoing transitions to the open state. This simple three-state chain fits much of the data on postsynaptic activation.

The major family of postsynaptic channels is known to have two agonist-binding subunits per channel, so that one can hypothesize various ways in which the binding of two agonists leads to channel opening. Two two-agonist schemes are shown in Figs. 4B and C. In B, first one then the other agonist molecule binds before their concerted action leads to channel opening. In C, each bound agonist allows the binding subunit to undergo a conformation change, and the opening results from the sequential effect of both bindings. These more accurate models lead to chains of greater length. Current analysis on chemically activated channels attempts to use single-channel data to distinguish between alternatives of this sort.

A more complex scheme, which has been applied to explain the $Ca^{2+}$-activated $K^+$ channel and glutamate-activated channels, involves allosteric interactions. A ligand-activated channel has ligand-binding subunits, and each binding to a subunit stabilizes open conformations that can exist with any number of bound ligands. This scheme, shown in Fig. 4D, forms a network of connected states for which there are multiple open states as well as closed states.

As examples for schemes representing voltage gating, we can look at the Hodgkin–Huxley kinetic schemes for the $K^+$ and $Na^+$ channels of nerve taken literally. The $K^+$ channel is represented by four identical and independent voltage-activated subunits. The $Na^+$ channel is represented by three independent activation subunits and an indepen-

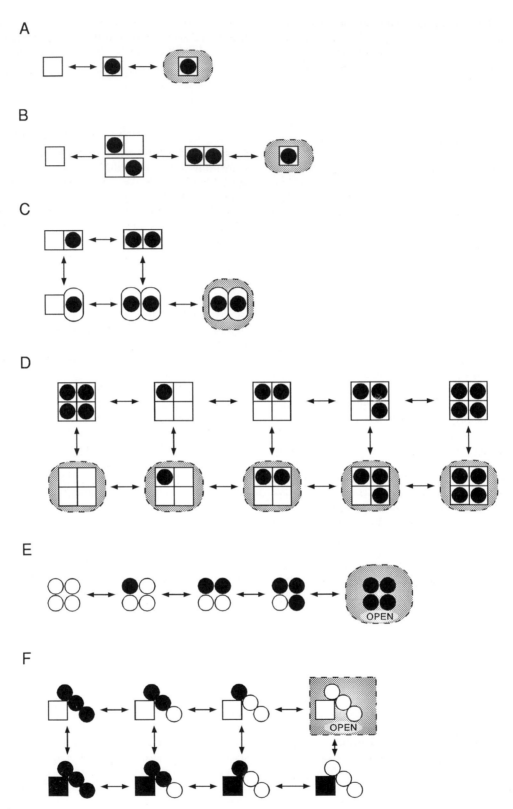

**FIG. 4.** Kinetic diagrams for various models of gating. (A–D) Schemes of increasing complexity for ligand-activated gating. Subunits of the channel protein are shown under transitions to activated states on ligand binding, with further activation leading to channel opening. Open states are indicated by shading. (E and F) Kinetic diagrams for voltage-gated channels, which would follow Hodgkin–Huxley kinetics, literally interpreted. The K$^+$ channel is opened when four activation subunits are simultaneously in the open position. The state of the Na$^+$ channel is determined by the joint activation of three subunits and a separate inactivation subunit.

dent autonomous inactivating subunit. These lead to the kinetic diagrams shown in Figs. 4E and F, a chain of five states for the hypothetical $K^+$ channel and a ladder network of eight states for the $Na^+$ channel, which has the three classes of states closed, open, and inactivated.

More exact modeling that must account for the enormous amount of voltage clamp and gating current data would lead to more complex schemes. Structural data suggest that the $Na^+$ channels have four charged subunits, which are too large and too close to each other to move independently. There is evidence that an inactivation gate exists as a separate unit but that its closing depends on the state of the activation gates. Schemes incorporating all the known information about $Na^+$ channel gating will undoubtedly be complex and somewhat different from the phenomenological Hodgkin–Huxley scheme.

The dynamics of the channel protein in any scheme are summarized by the time-varying vector of the state probabilities

$$\mathbf{p}(t) = [p_1, p_2, \ldots, p_N] \tag{2}$$

In classical kinetic measurements, we observe the entry or exit of the system into the complex of open states in response to some perturbation. The time course of such a relaxation follows a sum of exponentials,

$$\mathbf{p}(t) = \mathbf{p}(\infty) + \mathbf{p}(0) \sum \mathbf{A}_k \exp(\mu_k t) \tag{3}$$

where the $\mu_k$ are the eigenvalues or relaxation rates for a particular kinetic scheme. A system of $N$ states will show relaxations composed of $N - 1$ exponentials. These exponential relaxation times are functions of the rates that connect the substates of the system. In most kinetic schemes, there will be far more than just $N - 1$ rates. There are at least $2(N - 1)$ rates, and there can be as many as $N(N - 1)/2$. Classical kinetic measurements, such as the measurement of current transients by voltage clamp, provide considerable information for specifying the transition scheme of channel gating, but there is not enough information to resolve ambiguities fully in comparing competing schemes. Single-channel measurements by recording more than just the behavior of the ensemble of channels can extract additional information from observations of the detailed fluctuations of the stochastic process itself.

## 1. Dwell-Time Distributions

In most of these schemes, there is a single major open state, but the channel may wander around among its closed configurations before it gets into the open state. The extent of this wandering determines what the distribution of closed dwell times might look like. We can think of this as representative of some free-energy profile that governs the rates of transition. If there is a single dominant barrier, much higher than thermal energy, the open–closed transition is a **Poisson process,** which has a single exponential closed dwell-time distribution. If there are several dominant barriers so that there are a small number of stable closed states, then the transitions among them are governed by a finite-state Markov process, and the distribution of dwell-times is a sum of exponentials.

For a finite-state Markov process, the two observable conditions are represented by two dwell-time distributions, $F_c(t)$, the distribution of closed intervals and $F_o(t)$, the distribution open intervals. These are given by

$$F_c(t) = \sum A_{c,i} \exp(R_{c,i} t) \tag{4}$$

$$F_o(t) = \sum A_{o,i} \exp(R_{o,i} t) \tag{5}$$

where the parameters $A_{c,i}$, $A_{o,i}$, $R_{c,i}$, and $R_{o,i}$ are functions of the rate constants in the Markov model of the gating process (Colquhoun and Hawkes, 1981). Figure 5A shows a channel undergoing a sequence of random transitions among a set of closed states leading to the observable pattern of current jumps depicted at the top of the figure. The underlying sojourn in the set of closed states leads to the multiexponential closed time distribution. In general, the distribution of dwell times can be a complicated function of the rates in the stochastic matrix. Only for relatively simple kinetic models can the distributions $F_c(t)$ and $F_o(t)$ actually be written in terms of the rates, $Q_{ij}$. For a complex of $N_c$ closed states, the dwell time distribution yields $2N_c - 1$ parameters, $N_c$ time constants, and $N_c - 1$ amplitudes. Thus, for a model with $N_o$ open states and $N_c$ closed states, the two dwell-time distributions give $2(N - 1)$ relations among the rates.

A

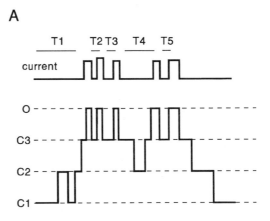

B

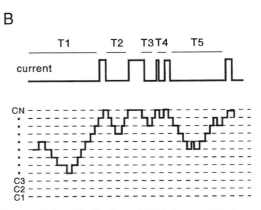

**FIG. 5.** Dwell-time distributions. (A) Finite-state Markov process showing multiexponential dwell-time distribution. (B) Diffusive kinetics showing power-law distribution.

We have assumed that a channel has a small number of closed states through which it passes, so that the dwell-time distribution is characteristic of a Markov process of reasonably low order, and can be fit to a small number of exponentials. Another possibility for the molecular dynamics of excursion through the closed configurations is that the channel takes an enormous number of small steps through an extensive array of closed states that are separated by small free-energy barriers. This would lead to a diffusion-like process for the sojourn in the complex of closed states, which is shown schematically in Fig. 5B. Such a process would lead to a distribution of dwell-times that asymptotically obeys a power law varying between $t^{-3/2}$ and $t^{-1/2}$ (Millhauser $et$ $al.$, 1988).

The long tail distribution expresses the tendency of a diffusion variable to wander away from the channel opening goal and get lost for a time in its vast state space leading to a relatively higher density of long shut intervals. An even higher level of complexity, which has sometimes been suggested for protein conformational transitions, would be a fractal-like free-energy topography, which would lead to chaotic wandering through the energy space and a closed time distribution which falls off as $t^{-1}$ (Leibovitch $et$ $al.$, 1987). There has been some suggestion that the dwell-time distributions of certain channels fit power laws. However, most single-channel data fit well with multiexponential distributions, allowing interpretation in terms of a finite-state Markov process.

## 2. Correlations and Two-Time Distributions

When the distributions of open and closed time are multiexponential, they indicate a multiplicity of closed or open states. In such cases, observations of the correlations between successive open or closed times can yield information about the different routes by which the states of the system interconvert. The Markov assumption limits the extent of correlation since a Markov process starting from a known state has a future evolution completely independent of what happened before entry into the initial state. However, some correlation between successive occupancies can be seen and used to discriminate between kinetic schemes.

Consider an example in which correlation analysis can shed light on the gating mechanism. Say an agonist-activated channel has two distinct open lifetimes. Knowing that there are two agonist receptors, we might attribute the two lifetimes to two different schemes: (1) one in which single-agonist bindings lead to short-lived openings and two-agonist binding leads to more long-lived openings, or (2) a scheme in which two agonists are needed to reach the open state but there are two different open states, one short lived and one long lived. The alternative schemes are shown in Fig. 6.

We can see that although these two schemes might lead to the same two-exponential distribution of open times, they will lead to different bursting patterns or different correlations between the lifetimes of successive openings. For scheme A of Fig. 6, if we consider experiments at low agonist concentration, so that $C_1$–$C_2$ transitions are

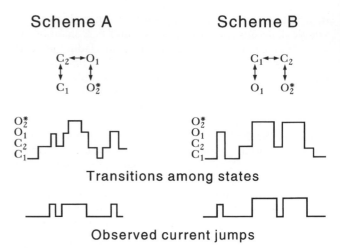

**FIG. 6.** Two kinetic schemes that can be distinguished by measuring correlations between successive open state lifetimes. Both schemes would give a two-exponential open state lifetime distribution, but the successive openings are correlated in scheme A where the two open states are connected and are uncorrelated in scheme B where they are kinetically disjoint.

infrequent compared to $C_1$–$O_1$ or $C_2$–$O_2$, then $C_1$–$O_1$ and $C_2$–$O_2$ transitions will have the opportunity to repeat—these repeated transitions leading to bursting and to correlations between successive openings.

The alternative model (Fig. 6B) does not lead to correlations; every closing of either open state leads back to state $C_2$ and during the sojourn in $C_2$, the channel loses memory of which open state it last came from (Markov definition). A study of correlations can distinguish between these two alternative models.

One measure of correlation is the correlation coefficient for successive lifetimes.

$$K(T_1, T_2) = [\langle T_1 T_2 \rangle - \langle T_1 \rangle \langle T_2 \rangle]/(\langle T_1^2 \rangle - \langle T_1 \rangle^2)(\langle T_2^2 \rangle - \langle T_2 \rangle^2) \tag{6}$$

For the two schemes discussed, scheme A gives $K = 0.3$ to 0.5 depending on the $C_1$–$C_2$ interconversion rate. For scheme B, $K$ must be equal to zero.

All the kinetic information obtainable from stationary patch-clamp data is contained in two-dimensional dwell-time histograms (Fredkin and Rice, 1986). A two-dimensional dwell-time density is defined as $P_2(T_1, T_2)$ equals the probability in a particular condition (e.g., membrane potential) that a channel is in the conducting state $\mathbf{I}$ from $T = 0$ to $T = T_1$ and is in conductance $\mathbf{J}$ from $T = T_1$ to $T = T_1 + T_2$.

Thus, if the two-time distributions can be measured, they can be compared to a model in which the component exponential of a multiexponential distribution can be fit. However, the amount of information available may generally be considerably less than the amount needed to specify the kinetic model fully. For a kinetic model of $N$ states, the system is described by as many as $N(N - 1)$ rates. The number of rates determinable from the patch-clamp data is

$$H = 2 \sum_i^N \sum_j^N N_i N_j \tag{7}$$

Thus, there are $G = N(N - 1) - H$ rates that cannot be determined from stationary patch-clamp data and must be fixed by hypothesis or from other experimental data such as nonstationary voltge clamp. For channels with many closed states and a few open states, it can be seen that $H$ is in fact small compared to $N(N - 1)$.

## C. Nonstationary Analysis

Nonstationary single-channel analysis is analogous to voltage clamp. The experimenter prepares the channel system in a specific set of states by controlling a rate-determining parameter, such as membrane potential. Channels are then suddenly perturbed by a stimulus, such as a step depolarization of potential, which changes some of the transition rates, causing the system of channels to relax to a new equilibrium. The time course of the random transitions during the course of equilibration is what is observed. By repeating pulses, one obtains the statistics of ensemble distributions of outcomes. Voltage-gated channels, which inactivate rapidly, or agonist-gated channels, which desensitize quickly, do not give steady-state responses and are better studied by nonstationary methods.

One example of the kind of information obtainable is the measurement of first-latency distributions—the distribution of times between application of the stimulus and the first opening of the channel (Horn and Vandenberg, 1984). such distributions can be analyzed to obtain the sequence of kinetic steps in a chain of closed states leading to the initial channel opening. So, for example, in the Hodgkin–Huxley $K^+$ channel scheme, a chain of four closed states precedes opening. A hyperpolarizing initial voltage maximizes the population in the leftmost closed states. The probability of arriving at the open state in a particular time interval is equal to the probability of a particular dwell-time for the sojourn in the collection of closed states. Such an experiment is indicated schematically in Fig. 7A. The form of the distribution times to first opening can be predicted from the kinetic model by solving the equations of the modified scheme in which no return is allowed from the open state. The first latency distribution is analogous to the closed state dwell-time distribution but gives information specifically about the sequence of closed states through which the activating channel passes.

Another example of a nonstationary determination is the measurement of the number of times that a channel opens before inactivation (Aldrich *et al.*, 1983; Vandenberg and Horn, 1984). Such a measurement can resolve ambiguities in the macroscopic description of inactivation. In a voltage-clamp experiment on a $Na^+$ channel, the transient conduction increase represents the product of a fast activation process and an independent, slow, voltage-dependent inactivation process. However, the actual shape of the transient is not really sensitive to the choice of activation and inactivation rates and a very similar fit can be obtained with a slow activation and fast inactivation.

At the single-channel level, however, the two situations differ, as illustrated in Fig. 7B (see also Fig. 8). In the first case nearly all channels open on depolarization and then individually slowly fluctuate closed. In the second picture,

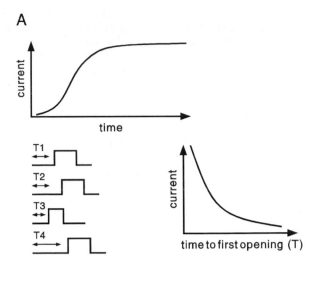

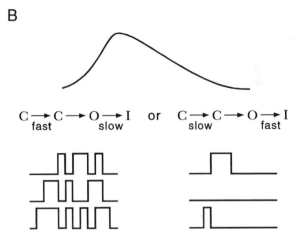

**FIG. 7.** Nonstationary properties of channel lifetime statistics. (A) Distribution of first-latency times. Different kinetic schemes will lead to different excursions through closed states on the way to first opening following a step in membrane potential. (B) Distribution of number of openings per inactivating event. Different kinetic models of the opening and inactivating of a voltage-gated channel which give the same macroscopic transient will lead to different predictions for transient bursts of channel opening following a step change in potential.

only a fraction of the channels opens at any time and the apparent slow abatement comes about because new channels are still entering the open population. Single-channel voltage-clamp experiments show the actual number of openings that an individual channel can undergo before it inactivates and thus make an unambiguous fit to the ratio of activation to inactivation rates.

## D. Model Discrimination

In single-channel analysis, one takes advantage of the ability to observe the actual stochastic process that underlies channel gating. Because even conceptually simple pictures of the molecular conformation changes lead to rather complex kinetic schemes, the single-channel statistics do not yield sufficient information to specify the scheme completely. What they do provide, as illustrated earlier, is accurate criteria for distinguishing between candidate schemes. The detailed picture of channel gating must also be guided by insights derived from the emerging knowledge of channel molecular structure.

## IV. Whole-Cell Current

Single-channel data can provide a detailed description of the kinetic behavior of a single protein. When single-channel data are summed, as shown in Fig. 8, one obtains an ensemble current that reflects the current flowing across the entire cell surface. This current is devoid of any "contamination" from residual currents that are not blocked (e.g., $K^+$ currents when studying $Ca^{2+}$ currents) or any current that might arise from inadequate voltage control (see Section V,B). The tight seal formed in the cell-attached configuration allows the experimenter to breach the patch and gain a low-resistance access to the cell interior. Because the access resistance is much lower than the cell resistance, the cell is in an approximate voltage-clamp configuration without the need for separate voltage and current microelectrodes. This is a boon for voltage clamping small cells grown in tissue culture and has led to an enormous increase in the number and variety of excitable cells that can be studied.

### A. Whole-Cell Tight-Seal Voltage Clamp

The tight-seal patch-clamp technique was invented for investigating the behavior of a single channel in an isolated patch of membrane. A modification of the technique, in which the patch is punctured, provides the basis for a novel voltage-clamp technique for studying currents that flow across the entire membrane of a cell. When more suction is applied to the pipette after sealing onto the membrane, the membrane patch underneath the tip can break, and the pipette solution is now continuous with the inside of the cell. This creates a low electrical resistance pathway between the pipette and cell, making it possible to control the voltage of the cell. This variation is known as the **whole-cell tight-seal voltage clamp** (see Fig. 3; the term "tight seal" is used here to distinguish it from the whole-cell "loose-patch" voltage-clamp technique described by Almers *et al.,* 1987). Figure 2B shows the equivalent circuit for the combination of resting cell membrane and micropipette electrode. To see why the patch technique is so useful, we can compare the recording situation to recording with a conventional microelectrode. For conventional recording, when a microelectrode impales a small cell typical values are $R_{pip} = 100$ M$\Omega$. The voltage drop across the series

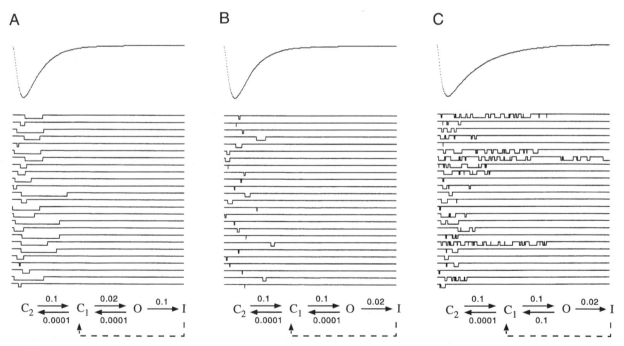

**FIG. 8.** Computer-simulated single-channel records and ensemble current. Simulated single $Na^+$ channel records were generated by a computer program based on a C → C → O → I kinetic scheme. The rates used to calculate channel openings are shown at the bottoms of the figures. Macroscopic currents are obtained by summation of more than 40,000 single-channel records. Scheme A has slow activation and fast inactivation, scheme B is based on the Hodgkin and Huxley model (fast activation and slow inactivation), and scheme C shows a very slow inactivation rate. The macroscopic currents from schemes A and B are identical, except that about a quarter of the number of traces is required to generate the summed current for scheme B. Scheme C has comparable macroscopic current with a slightly slower inactivation. Note that the single-channel behavior for each case is very different. The dashed line indicates a possible connection between the inactivated (I) and closed (C) state. (Figure provided by Dr. Shirley Bryant.)

resistance is comparable to or greater than that across the cell membrane and the shunt current provides a significant leakage current. Thus the membrane potential can only be controlled by a feedback system using either a second microelectrode for monitoring membrane potential or some sort of intermitant switching. These techniques are complicated, are limited in the response time for changing voltage, and generate considerable electrical noise.

In contrast the "patch" junction gives a pipette resistance of 4–10 MΩ and a seal resistance of ≥10 GΩ, so that the voltage drop across the series resistance is small compared to that across the cell membrane and the leakage current is relatively small. In consequence, a patched cell can be voltage clamped with the same type of current-to-voltage converter used in patch clamp.

If a step of voltge is placed across the patch pipette, the voltage rise time can be seen to be (assuming $R_{cell} \gg R_{pip}$)

$$\tau = R_{pip} C_{cell} \tag{8}$$

For a small cell of $R_{pip} = 5$ mΩ and $C_{cell} = 5$ pF, then $\tau = 25$ μsec. The membrane potential, compared to the command potential, is

$$V_m/V_c = R_{cell}/R_{cell} + R_{pip} \tag{9}$$

with $R_{cell} \approx 10$ GΩ, $V_m/V_c = 0.99$. Thus, a small cell can be voltage clamped without any compensation for series resistance or membrane capacitance. In general, $R_{cell}$ scales inversely with cell area and $C_m$ scales proportional to cell area, so that both of these factors will be worse when cell size is increased, limiting the applicability to cells less than ~20 μm in diameter. Patch-clamp circuitry uses capacitance cancellation and series compensation circuits to counteract these limitations. Limitations and design of the whole-cell patch clamp are discussed in Marty and Neher (1983), Lecar and Smith (1985), and Strickholm (1995a,b).

### B. Perforated-Patch Method

The same low-resistance access that permits whole-cell voltage clamp also has the often unwanted effect of allowing the cytoplasmic solution to interdiffuse with the much larger volume of the patch pipette. This often leads to the loss or **rundown** of membrane and cytoplasmic constituents that are vital for maintaining channel activity. The perforated-patch method (Fig. 3; Horn and Marty, 1988) was originally invented to prevent the loss of response due to rundown. This method involves the use of the antibiotic nystatin or amphotericin, which forms ion permeable pores in the membrane without the loss of large molecules from the cytoplasm. The perforated patch creates an electrically low resistance pathway between the pipette and the inside of the cell. The channels created by the antibiotics are more permeable to monovalent cations than monovalent anions, and are thought to be impermeable to divalents. Because of this ion selectivity, it is necessary to include $SO_4^{2-}$ ions in the pipette solution to counteract the Donnan equilibrium potential that arises when only Cl⁻ is present in the pipette solution. More recent use of gramicidin (Kyrozis and Reichling, 1995) which has induced pores that are selectively permeable to monovalent cations and

negligible to chloride ions, appears to ameliorate the problem of changing intracellular free Cl⁻ concentrations. Avoiding the change of chloride equilibrium may be critical when studying ligand-gated or voltage-gated chloride channels.

The tip of the pipette is first filled with antibiotic-free recording solution, then back-filled with the antibiotic-containing recording solution. Following the formation of a high seal resistance, time is allowed for the antibiotic to diffuse down to the tip and form pores in the patch membrane. The time interval (usually between 5–15 min) is dependent on how much antibiotic-free solution is in the tip and the concentration of the antibiotic used, parameters that must be determined experimentally for each cell type. The series resistance under this mode of recording is usually high (>20 MΩ) making adequate control of the voltage in voltage-jump studies difficult. A shift in the current–voltage relation curve to the right following a reduction of current amplitudes (e.g., when studying drug antagonism) and a very sudden or sharp increase in the rising limb of the current–voltage relation curve are indicative of this problem. A slow time-to-peak current is also associated with high series resistance. The perforated-patch method is, therefore, more appropriate for studying regulation of small-amplitude currents or ligand-gated currents.

## V. Problems Associated with Whole-Cell Recording

Three problems are discussed. The first, rundown, is uniquely associated with whole-cell recording of small cells; the second and third are problems associated with any voltage-clamp experiments.

### A. Rundown

One problem that experimenters have encountered when performing whole-cell recordings with small cells (and when using the inside-out mode) is that currents or channel activities often suffer from rundown, that is, a loss of activity with time. This is probably because regulatory component(s) that are not an integral part or tightly bound to the channel protein may diffuse away from the membrane and become lost. It may be possible to sustain channel activities for longer duration by addition of appropriate agents (e.g., nucleotides). In whole-cell recording, the supplementation of the pipette solution with 2–5 m$M$ of ATP has been shown to reduce the rundown rate of $Ca^{2+}$ currents. Moreover, addition of BAY-K 8644 to bath medium in inside-out patch recording prolonged the activities of voltage-dependent $Ca^{2+}$ channels (Ohya and Sperelakis, 1989).

To reduce and minimize the rundown of whole-cell currents, the perforated-patch method can be applied (see Section IV,B). In an attempt to sustain channel activity in patch recording a combination of the perforated-patch and outside-out configuration can be used. After establishing a perforated patch on the whole cell, the membrane is excised such that a resealed vesicle that holds intact cytoplasmic and membrane components can be obtained (see Fig. 3, lower right; Levitan and Kramer, 1990). One can then study the modulatory effects of transmitters on single channels present in the resealed vesicle.

## B. Voltage Control and Space Clamp

Poor or lack of voltage control will give a wrong current–voltage relation and distort the kinetics of the currents. Voltage control will also be inadequate if the cell membrane is not isopotential or "space clamped." In a linear cable, when current is applied to change the potential at a point, the potential measured at a distance from that point will be lower due to the dissipation of charge over the distance. The degree of charge dissipation, and thus the decrement in voltage, is dependent on the electrical properties of the cable. The same principle applies to cells of different geometry. In general, the more electronically compact a cell is (i.e., a longer space constant), the less decrement there is in the voltage over a distance and the more effective the control of voltage. Thus, one will have better voltage control over the membrane potential of a cell that has less surface membrane and a higher input resistance than a cell that has much membrane infolding and low input resistance. Lecar and Smith (1985) have provided estimates of the cell size and contribution of dendritic network when **space clamp** is a problem.

Lack of space clamp alters the apparent kinetics of the currents. The frequency response of a signal is dependent on the location of the conductance change: The further the conductance change is from the recording site (in terms of fraction of space constant), the larger the distortion (Johnston and Brown, 1983). Synaptic currents measured at a site far from the soma (e.g., a distal location on an apical dendrite of a cerebral pyramidal neuron) have a slower decay time compared to the same input measured at the soma.

## C. Junction Potentials

**Junction potentials** between pipette solution and cytoplasm cause problems because they give rise to erroneous potential and erroneous results when determining the permeability ratio for channels from the shift in reversible potential. Junction potentials exist because of (1) the surface potential between the Ag/AgCl surface of the recording electrode and the Ag/AgCl surface of the ground, and (2) the difference in the mobility of the ions in solution. The junction potential that arises from the surface potential of the metal–electrolyte junction can be minimized if the recording silver wire and ground (silver wire or pellet) are chlorided simultaneously, each acting as a half cell, then connected together and stored as a battery in a KCl (1–3 M) solution overnight before use. Any residual potential can be neutralized with the voltage offset of the amplifier after placement of the pipette in the bath.

The junction potential due to different ionic species in the solutions cannot be cancelled in a similar way and requires correction for accuracy. For most recording conditions, the bulk ions present in the bath and pipette solutions consist mainly of $Na^+$, $K^+$, $Cs^+$, and $Cl^-$. The junction potential when the major ions have comparable mobility is usually small (about 2–3 mV maximal) and generally can be ignored. However, when the bulk of the cations is replaced by ions of low mobility (e.g., substituting $Na^+$ with Tris), and particularly when other anions are substituted for $Cl^-$, the junction potential developed can be fairly large. A detailed description for making corrections can be found in Barry and Diamond (1971) and Barry and Lynch (1991). A computer program for making junction potential corrections is available (Barry, 1994).

## VI. Other Applications

In addition to being used for measuring single-channel (unitary conductance) currents, whole-cell voltage, and whole-cell currents, the patch-clamp technique and its variations can also be used to determine electrical parameters of single, isolated cells. In the following two subsections we cover the application of the technique in measuring membrane electrical noise and membrane capacitance.

### A. Membrane Noise Analysis

In whole recording on small cells, the dominant electrical noise observed at frequencies less than 1 kHz is the noise generated by the random opening and closing of ionic channels. Although the individual gating transitions may not be resolvable in such noise records, the unit conductances and some of the kinetic properties of the channels can be obtained from analysis of the noise records.

We can see how the unit conductance can be inferred from a measurement of the fluctuation noise power. We can think of a membrane have $N$ gated channels with unit current $i$ and probability $p$ of being in the open state. The average membrane current will the $I = Npi$ and the variance on the current (or the average fluctuating power in a unit resistor) is

$$\overline{I^2} - \overline{I}^2 = Np(1-p)i^2 \tag{10}$$

for $N$ independent channels. To infer the unit current, $i$, one measures the ratio of variance to mean,

$$\text{Variance/mean} = (1-p)i \tag{11}$$

which approaches the unit current as $p$ is made small.

Kinetic properties of the gating can be obtained by analysis of the frequency spectrum of the noise. For a Markov process, the spectrum by channel fluctuating noise is given by a characteristic spectrum, which is the sum of Lorenzian,

$$P(\omega) = \sum_j (A_j/1 + \omega^2 \tau_j) \tag{12}$$

where the parameters $A_j$ and $\tau_j$ are related to the transition probabilities of the kinetic scheme.

Fluctuation analysis is particularly useful for determining the properties of channels that either cannot be studied by patch clamp or have conductances too small to be resolved by the single-channel method.

## B. Capacitance Measurements

Another application of whole-cell voltage clamp is the monitoring of changes in membrane capacitance during exocytosis or secretion. The regulated secretion of hormones, enzymes, and transmitters is associated with the fusion of cytoplasmic vesicles, which store the substances to be released, to the plasmalemmal membrane. This additon of vesicular membranes following fusion increases the overall surface area of the cell, and since membrane has capacitance, the increase in surface area in turn will lead to an increase in the cell's capacitance. Therefore, by monitoring changes in capacitance one can readily measures secretion from a single cell as it occurs (see Penner and Neher, 1989; Landau and Neher, 1989).

To take advantage of this change in physical parameter, Neher and Marty (1982) first used the increase in the cell's capacitance as an index of monitoring exocytosis in bovine adrenal chromaffin cells. The method involves an injection of a sinusoidal wave of known amplitude into a cell under voltge clamp. The voltage and current response signals are fed into a lock-in amplifier, which measures a phase angle and separates the current response into its resistive and capacitive components. The phase angle is determined by changing the capacitance compensation knob of the amplifier and adjusting the angle until the resistive component output does not alter (Neher and Marty, 1982). The capacitance is then measured from an angle orthogonal to the determined phase angle. An alternative approach is to find the phase angle by changing a resistance in series with the recording ground (Fidler and Fernandez, 1990; note that the phase angle determined by this approach is different from that described by Neher and Marty). A simpler means of determining the phase angle has been proposed (Zierler, 1992). Other approaches to measure membrane capacitance include the use of two sinusoidal frequencies (Donnelly, 1994; Rohlicek and Schmid, 1994) and theoretical calculations (Niu et al., 1995). Capacitance change as a means to measure exocytosis or secretion has been used extensively to study the molecular mechanisms of release in various preparations, including mast cells, pancreatic $\beta$ cells, nerve terminals, neurons, and neuroendocrine cells.

There is a good correlation between capacitance increase and the exocytotic event. Studies with mast cells from the biege mouse showed that capacitance increase occurs following fusion of vesicles (Zimmerberg et al., 1987). Using amperometry, an electrochemical method of detection, to measure release of catecholamines from adrenal chromaffin cells, it was shown that the increase in amperometric signal correlates with capacitance increase (Chow et al., 1992). These results indicate that release of vesicular contents is associated with capacitance increase. The combine use of amperometry and capacitance measurement further revealed that release of vesicular contents does not require full fusion of a vesicle to the plasmalemmal membrane, but that release could occur even when a vesicle transiently interacts with the plasma membrane. These interactions are measured as fluctuations in the capacitance or "foot" of the amperometric signal and reflect the transient formation of a fusion pore (a narrow, aqueous channel; Monck and Fernandez, 1994) between the two membranes, which allows the vesicular contents to leak out of the cell (Chow et al., 1992; Alvarez de Toledo et al., 1993; Zhou et al., 1996). A reduction in cell capacitance following stimulation has also been observed and appears to be related to endocytotic activity (retrieval of membranes after exocytosis; Neher and Marty, 1982; Thomas et al., 1991; Augustine and Neher, 1992; Von Gersdorff and Matthews, 1994a,b). The exact relation between fusion and exocytosis is still not clear (see DeFelice, 1996).

Since the exocytotic process is highly dependent on the entry of $Ca^{2+}$, the use of whole-cell recording offers an advantage by which the experimenter can set the intracellular free $Ca^{2+}$ concentrations and study the relation between $Ca^{2+}$ and secretion (Neher, 1988; Augustine and Neher, 1992). In addition, the use of caged calcium compounds, which release calcium when excited by UV light, has allowed the development of kinetic models for the secretory process (Neher and Zucker, 1993; Heidelberger et al., 1994; Heinemann et al., 1994). Other advantages of using whole-cell recording to measure the exocytotic process include (1) the ability to introduce activators of second messengers, for example, GTP-$\gamma$-S, which activates G-proteins (Fernandez et al., 1984; Penner and Neher, 1989) or antibodies (Schweizer et al., 1989; Tao et al., 1995) to determine the roles of various second messengers and proteins in activating secretion; and (2) the high sensitivity in detecting release down to fusion of a single vesicle; and (3) the ability to examine the variability between individual cells among a population. To circumvent the problem of rundown, that is, the loss of response due to loss or dilution of vital proteins (Neher and Marty, 1982), capacitance measurements can be applied together with the perforated-patch recording technique (Gillis et al., 1991). With the rapid advances in molecular techniques, the ability to generate mutant or variant animals and cells devoid of or with modified vesicular proteins will undoubtedly lead to the unraveling of the molecular mechanisms underlying the once elusive exocytotic process.

## VII. Summary

The technique of patch clamping and its variations in the study of ionic currents at the single channel and whole-cell level have been described. Different variations of the patch-clamp technique—cell-attached, inside-out, and outside-out and its modifications—used in study of single channel currents were described. In addition, the more commonly used application of whole-cell voltage clamp and its variation, the perforated-patch method, to examine macroscopic currents were also described. Although whole-cell recording provides information about the gating properties as well as kinetic characteristics of the channel protein, single-channel recordings allow for a more detailed evaluation of the kinetic properties of the channel protein. As was shown in Fig. 8, similar macroscopic currents can be produced by three entirely different gating and kinetic schemes of channel behavior. Thus, if one were to formulate and obtain a more realistic model for a channel protein it would be more accurate and highly desirable to study the behavior of protein at the single-channel level.

The different variations or modes of patch recordings further permit the experimenter to alter the environment and examine the gating and behavior of a single-channel protein. Under the cell-attached mode it is possible to evaluate the behavior of a channel protein in its more or less native environment. Under the inside-out mode, since the inner membrane is made to face the surrounding medium, one could study whether the behavior of the protein is altered by addition of substances, for example, second messengers, or changing of the ionic conditions, for example, altering the $Ca^{2+}$ concentrations to determine the gating effects of $Ca^{2+}$ on the $Ca^{2+}$-dependent $K^+$ channel. Using the outside-out mode one could examine the detailed gating kinetics of agonist-activated channels with varying concentrations of the agonists.

Another variation of the patch-clamp technique, the whole-cell tight-seal voltage clamp, proves to be particularly useful for the study of macroscopic currents. Because this method is specifically applicable to small cells, it opens up a new vista for experimentations allowing experimenters to examine currents in cells that were previously not accessible to recording with conventional sharp-tipped electrodes. Moreover, the ability to include substances in the pipette solutions allows one to manipulate the cell's interior environment and study the modulation of ionic currents. The advantage of being able to manipulate the cytoplasmic makeup also creates a problem when attempting to record currents for a prolonged duration. Substances that are vital to normal functioning and for sustaining channel functions may be lost following the mixing of the recording solution with the cell's interior. This mixing can result in a rundown or disappearance of the currents. A variation of the whole-cell recording technique, the perforated-patch recording technique, can be used to reduce or overcome this problem.

The fluctuating single-channel current jumps observable with the patch-clamp technique reveal details of the molecular stochastic processes underlying excitability. In equilibrium, channels undergo random transitions through a sequence of conducting and nonconducting states. Distributions of the dwell times in the various states and other statistical distributions can be used to establish and test kinetic schemes of the molecular conformation changes, which lead to channel gating. Excitability is the macroscopic concomitant of gating transition rates that can be altered by an external stimulus.

## Bibliography

Aldrich, R. W., Corey, D. P., and Stevens, C. F. (1983). A reinterpretation of mammalian sodium channel gating based on single channel recording. *Nature (London)* **260**, 436–411.

Almers, W., Stanfield, P. R., and Stuhmer, W. (1987). Lateral distribution of sodium and potassium channels in frog skeletal muscle: Measurements with a patch clamp technique. *J. Physiol. (London)* **336**, 261–284.

Alvarez de Toledo, G., Fernandez-Chacon, R., and Fernandez, J. M. (1993). Release of secretory products during transient vesicle fusion. *Nature (London)* **363**, 554–558.

Ashley, R. H. (Ed.). (1995). "Ion Channels: A Practical Approach." Oxford University Press, New York.

Augustine, G. J., and Neher, E. (1992). Calcium requirements for secretion in bovine chromaffin cells. *J. Physiol. (London)* **450**, 247–271.

Barry, P. H. (1994). JPCalc, a software package for calculating liquid junction potential corrections in patch-clamp, intracellular, epithelial and bilayer measurements and for correcting junction potential measurements. *J. Neurosci. Methods* **51**, 107–116.

Barry, P. H., and Diamond, J. M. (1971). Junction potentials, electrode standard potentials, and other problems in interpreting electrical properties of membranes. *J. Membr. Biol.* **3**, 93–122.

Barry, P. H., and Lynch, J. W. (1991). Liquid junction potentials and small cell effects in patch-clamp analysis. *J. Membr. Biol.* **121**, 101–117.

Boulton, A. A., Baker, G. B., and Walz, W. (1995). Neuromethods. *In* "Patch-Clamp Applications and Protocols," Vol. 26. Humana Press, Totowa, NJ.

Chow, R. H., Von Ruden, L., and Neher, E. (1992). Delay in vesicle fusion revealed by electrochemical monitoring of single secretory events in adrenal chromaffin cells. *Nature (London)* **356**, 60–63.

Colquhoun, D., and Hawkes, A. G. (1981). On the stochastic properties of single ion channels. *Proc. R. Soc. London* **B211**, 205–235.

Colquhoun, D., and Hawkes, A. G. (1982). On the stochastic properties of bursts of single ion channel openings and clusters of bursts. *Proc. R. Soc. London* **B300**, 1–59.

DeFelice, L. J. (1996). Release of secretory vesicles contents: Regulation after fusion? *Biophys. J.* **71**, 1163–1164.

Del Castillo, J., and Katz, B. (1957). Interaction at end-plate receptors between different choline derivatives. *Proc. R. Soc. London* **B146**, 369–381.

Donnelly, D. F. (1994). A novel method for rapid measurement of membrane resistance, capacitance, and access resistance. *Biophys. J.* **66**, 873–877.

Ehrenstein, G., and Lecar, H. (1977). Electrically gated ionic channels in lipid bilayers. *Q. Rev. Biophys.* **10**, 1–34.

Fenwick, E. M., Marty, A., and Neher, E. (1982). A Patch-clamp study of bovine chromaffin cells and of their sensitivity to acetylcholine. *J. Physiol. (London)* **331**, 577–597.

Fernandez, J. M., Neher, E., and Gomperts, B. D. (1984). Capacitance measurements reveal stepwise fusion events in degranulating mast cells. *Nature (London)* **312**, 453–455.

Fidler, N., and Fernandez, J. M. (1989). Phase tracking: An improved phase detection technique for cell membrane capacitance measurements. *Biophys. J.* **56**, 1153–1162.

Fredkin, D., and Rice, J. A. (1986). On aggregated Markov chains. *J. Appl. Prob.* **23**, 208–214.

Gillis, K. D., Pun, R. Y. K., and Misler, S. (1991). Single cell assay of exocytosis from adrenal chromaffin cells using "perforated patch recording." *Pflugers Arch.* **418**, 611–613.

Hamill, O. P., Marty, A., Neher, E., Sakmann, B., and Sigworth, F. J. (1981). Improved patch-clamp techniques for high-resolution current recording from cells and cell-free membrane patches. *Pflugers Arch.* **391**, 85–100.

Heidelberger, R., Heinemann, C., Neher, E., and Matthews, G. (1994). Calcium dependence of the rate of exocytosis in a synaptic terminal. *Nature (London)* **371**, 513–515.

Heinemann, C., Chow, R. H., Neher, E., and Zucker, R. S. (1994). Kinetics of the secretory response in bovine chromaffin cells following flash photolysis of caged $Ca^{2+}$. *Biophys. J.* **67**, 2546–2557.

Hodgkin, A. L., and Huxley, A. F. (1952a). Currents carried by sodium and potassium ions through the membrane of the giant axon of *Loligo. J. Physiol. (London)* **116**, 449–472.

Hodgkin, A. L., and Huxley, A. F. (1952b). The components of membrane conductance in the giant axon of *Loligo. J. Physiol. (London)* **116**, 473–496.

Hodgkin, A. L., and Huxley, A. F. (1952c). A quantitative description of membrane current and its application to conduction and excitation in nerve. *J. Physiol. (London)* **117**, 500–514.

Horn, R., and Marty, A. (1988). Muscarinic activation of ionic currents measured by a new whole cell recording method. *J. Gen. Physiol.* **92**, 145–159.

Horn, R., and Vandenberg, C. A. (1984). Statistical properties of single sodium channels. *J. Gen. Physiol.* **84**, 505–534.

Johnston, D., and Brown, T. H. (1983). Interpretation of voltage-clamp measurements in hippocampal neurons. *J. Neurophysiol.* **50**, 464–486.

Kyrozis, A., and Reichling, D. B. (1995). Perforated-patch recording with gramicidin avoids artifactual changes in intracellular chloride concentration. *J. Neurosci. Methods* **57**, 27–35.

Lecar, H., and Smith, Jr., T. G. (1985). Voltage clamping small cells. *In* "Voltage-Clamping and Patch Clamping with Microelectrodes," (T. G. Smith, Jr., H. Lecar, S. J. Redman, and P. W. Gage, Eds.), pp. 231–256. American Physiological Society.

Leibovitch, L. S., Fishbarg, J., Koniarek, J. P., Todorova, I., and Wang, M. (1987). Fractal model of ion-channel kinetics. *Biochim. Biophys. Acta.* **896**, 173–180.

Levis R. A., and Rae, J. L. (1993). The use of quartz patch pipettes for low noise single channel recording. *Biophys. J.* **65**, 1666–1677.

Levitan, E. S., and Kramer, R. H. (1990). Neuropeptide modulation of single calcium and potassium channels detected with a new patch clamp configuration. *Nature* **348**, 545–547.

Lindau, M., and Neher, E. (1988). Patch-clamp techniques for time-resolved capacitance measurements in single cells. *Pfluegers Arch.* **411**, 137–146.

Marty, A., and Neher, E. (1983). Tight-seal whole-cell recording. *In* "Single-Channel Recording," (B. Sakmann and E. Neher, Eds.), pp. 107–122. Plenum Press, New York.

Millhauser, G. L., Saltpeter, E. E., and Oswald, R. E. (1988). Diffusion models of ion-channel gating and the origin of power-law distributions from single-channel recording. *Proc. Natl. Acad. Sci USA* **85**, 1503–1507.

Monck, J. R., and Fernandez, J. M. (1994). The exocytotic fusion pore and neurotransmitter release. *Neuron* **12**, 707–716.

Neher, E. (1988). The influence of intracellular calcium concentration on degranulation of dialysed mast cells from rat peritoneum. *J. Physiol. (London)* **395**, 193–214.

Neher, E., and Marty, A. (1982). Discrete changes of cell membrane capacitance observed under conditions of enhanced secretion in bovine adrenal chromaffin cells. *Proc. Natl. Acad. Sci. USA* **79**, 6712–6716.

Neher, E., and Sakmann, B. (1976). Single-channel currents recorded from membrane of denervated frog muscle fibres. *Nature* **260**, 779–802.

Neher, E., and Zucker, R. S. (1993). Multiple calcium-dependent processes related to secretion in bovine chromaffin cells. *Neuron* **10**, 21–30.

Niu, A., Zhou, N., Xie, R., and Pun, R. Y. K. (1995). Computational method to determine capacitance changes in secretory cells. *FASEB J. Abstr.* **9**(2), A674 #3910.

Ohya, Y., and Sperelakis, N. (1989). Modulation of single slow (L-type) calcium channels by intracellular ATP in vascular smooth muscle cells. *Pfluegers Arch.* **414**, 257–264.

Penner, R., and Neher, E. (1988). The role of calcium in stimulus–secretion coupling in excitable and non-excitable cells. *J. Exp. Biol.* **139**, 329–345.

Penner, R., and Neher, E. (1989). The patch-clamp technique in the study of secretion. *Trends Neurosci.* **12**, 159–163.

Rohlicek, V., and Schmid, A. (1994). Dual-frequency method for synchronous measurement of cell capacitance, membrane conductance and access-resistance on single cells. *Pfluegers Arch. Eur. J. Physiol.* **428**, 30–38.

Rudy, B., and Iverson, L. E. (Eds.). (1992). *Methods Enzymol.* **207.**

Sakmann, B., and Neher, E. (Eds.). (1995). *Single-Channel Recording,* 2nd ed. Plenum Press, New York.

Schweizer, F. E., Schafer, T., Tapparelli, C., Grob, M., Karli, U. O., Heumann, R., Thonen, H., Bookman, R. J., and Burger, M. M. (1989). Inhibition of exocytosis by intracellularly applied antibodies against a chromaffin granule-binding protein. *Nature* **339**, 709–712.

Strickholm, A. (1995a). A supercharger for single electrode voltage and current clamping. *J. Neurosci. Methods* **61**, 47–52.

Strickholm, A. (1995b). A single electrode voltage, current- and patch-clamp amplifier with complete stable series resistance compensation. *J. Neurosci. Methods* **61**, 53–66.

Stuhmer, W., Conti, F., Stocker, M., Pongs, O., and Heinemann, S. H. (1991). Gating currents of inactivating and non-inactivating potassium channels expressed in Xenopus oocytes. *Pfluegers Arch.* **418**, 423–429.

Tao, J.-X., Howell, M. L., Pun, R. Y. K., Cutler, D., and Dean, G. E. (1995). Different isoforms of synaptotagmin-1 exist in bovine adrenal chromaffin cells. *Soc. Neurosci. Abstr.* **21**, 51, #28.14.

Thomas, P., Suprenant, A., and Almers, W. (1990). Cytosolic $Ca^{2+}$, exocytosis, and endocytosis in single melanotrophs of the rat pituitary. *Neuron* **5**, 723–733.

Vandenberg, C. A., and Horn, R. (1984). Inactivation viewed through single sodium channels. *J. Gen. Physiol.* **84**, 505–534.

Von Gersdorff, H., and Matthews, G. (1994a). Dynamics of synaptic vesicle fusion and membrane retrieval in synaptic terminals. *Nature* **367**, 735–739.

Von Gersdorff, H., and Matthews, G. (1994b). Inhibition of endocytosis by elevated internal calcium in a synaptic terminal. *Nature* **370**, 652–655.

Zhou, Z., Misler, S., and Chow, R. H. (1996). Rapid fluctuations in transmitter release from single vesicles in bovine adrenal chromaffin cells. *Biophys. J.* **70**, 1543–1552.

Zierler, K. (1992). Simplified method for setting the phase angle for use in capacitance measurements in studies of exocytosis. *Biophys. J.* **63**, 854–856.

*Simon Rock Levinson*

# 26

# Structure and Mechanism of Voltage-Gated Ion Channels

## I. Introduction: How Is Ion Channel Structure Studied?

Previous chapters have described how the flow of ions across cell membranes is the basis for electrical excitability and signaling in the nervous system. These flows are controlled by a special class of macromolecules known as **ion channels** that form gated pores in the cell membrane. The structure of ion channels and the current thinking about how these structures form transmembrane pores that gate to open and closed states in response to changes in membrane voltage are discussed in this chapter. Subsequent chapters will describe other important aspects of ion channels, such as interactions with drugs and toxins, modulation by intracellular messengers, and specific ion channel types found in intracellular organelles and involved in synaptic transmission.

What are the questions that should be addressed in this chapter? Most fundamentally, we wish to know how these macromolecular structures give rise to the important functional properties of ion channels, namely, the formation of transmembrane pores, the selective transport of specific ions, and the ability to open and close the pore; that is, how do ion channels function as molecular mechanisms? As we will see, these considerations also lead into a brief consideration of the structural diversity among ion channels and the origin and possible purposes of such diversity.

Cellular ion channels are basically proteins. In the earliest structural studies, biochemical methods were employed to purify channel molecules from excitable tissues. This approach has been essential to the current state of knowledge of ion channel structure. However, ion channels have certain physicochemical properties that have limited the amount of information obtained solely through biochemical characterization. This "bottleneck" has been broken recently through the application of recombinant DNA methodologies, which have uncovered the primary struc-

tures of numerous ion channel types, while providing a powerful means to study higher order structure and function. However, this latter approach has its own limitations, and ultimately we will need to "see" more directly the actual structure of the channels to understand more fully their molecular mechanisms of action.

## II. Biochemistry of Ion Channels: Purification and Characterization of Voltage-Gated Channels

Before the advent of recombinant DNA methods, ion channels were extensively studied by first purifying channels from an excitable tissue and then characterizing the purified molecules by a combination of chemical and physical techniques. The first voltage-gated channel to be purified was the $Na^+$ channel from the electric eel *Electrophorus electricus,* and the account of how this was achieved will serve to illustrate the basic rationale for the biochemical approach (see Miller *et al.,* 1983).

The basic procedure for purification of membrane-associated proteins (Fig. 1) is as follows: First, an enriched fraction of membranes is usually prepared by disrupting the tissue and its cells mechanically and subsequently separating the insoluble fraction containing membranes and connective tissue components from the soluble fraction consisting of cytoplasmic protein. Although relatively nonspecific, this step also often gives substantial enrichment of ion channel preparations. Next, the membranes are solubilized through the use of detergents. This step disperses the membranes into minute droplets of lipid, protein, and detergent known as micelles. This is often the most problematic step of a purification, because the micellar fraction must be dispersed well enough so that no more than one protein molecule is present in each micelle. Too much detergent often results in irreversible denaturation of chan-

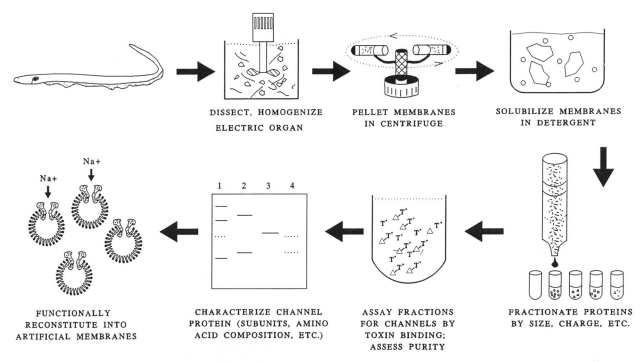

**FIG. 1.** Scheme used for the biochemical isolation of Na⁺ channels from electric tissue of the eel *Electrophorus electricus*. Purification protocols devised for other ion channels utilize similar general principles but differ in details.

nels, hence preventing their identification during the purification process. Thus, the solubilization process must be optimized, and often many different detergents at various concentrations are tested under a plethora of conditions before the right combination is found. Finally, the dispersed micelles are fractionated based on their physical properties or affinities for certain compounds until a pure sample of channel protein is isolated.

### A. Electric Fish Are a Rich Source of Ion Channels

The first requirement for purification of any protein is to have a rich source. Actually, this can be a serious difficulty, since ion channels are present in very small amounts in most excitable tissues. The rarity of channels might at first seem paradoxical in view of the ubiquity of excitability phenomena in nearly all cell membranes. However, the reader should recall that only minute numbers of channels are needed to mediate the flow of ions required to cause the changes in transmembrane voltage during electrical excitation (Table 1).

Fortunately, there exist some highly specialized "freaks" of nature that use large ionic currents for purposes other than signaling and stimulus transduction. These are the strongly electric fish, which are capable of generating powerful electrical discharges to stun prey and defend themselves. The electric organs of these animals contain large numbers of ion channels that are present at high density on the surface of the excitable cells of the electric organ (Fig. 2). As a result, the first two ion channels to be purified were the acetylcholine receptor channel of the electric ray

*Torpedo* and the Na⁺ channel of the electric eel *Electrophorus*.

### B. Toxins and Drugs as Markers for Ion Channels during Purification

Having a rich source of channels does not guarantee a successful isolation of channels. One must also have a very specific and sensitive assay for the presence of solubilized channels to follow the process of fractionation. Unfortunately, unlike enzymes or other chemically active proteins,

**TABLE 1** Sodium Channel Densities in Selected Excitable Tissues

| Tissue | Surface density (channels/$\mu m^2$) | Tissue density ($\mu g$ channel/g tissue) |
|---|---|---|
| Mammalian | | |
| Vague nerve (nonmyelinated) | 110 | 28 |
| Node of Ranvier | 2100 | — |
| Skeletal muscle (various) | 206–557 | 6–14 |
| Other animals | | |
| Squid giant axon | 166–533 | ~1 |
| Frog sartorius muscle | 280 | 6 |
| Electric eel electroplax (excitable surface) | 550 | 38 |
| Garfish olfactory nerve | 35 | 96 |
| Lobster walking leg nerve | 90 | 24 |

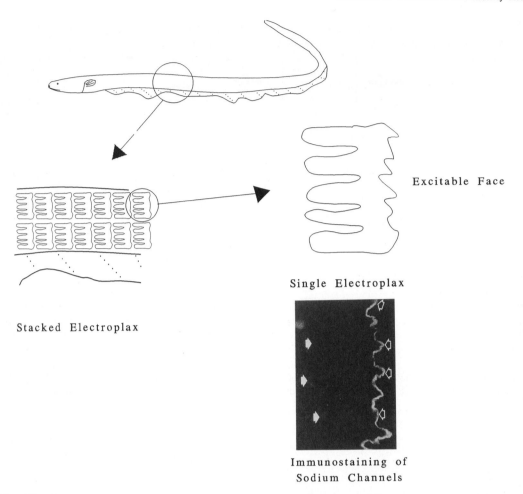

**Excitable Face**

**Single Electroplax**

**Stacked Electroplax**

**Immunostaining of Sodium Channels**

**FIG. 2.** Electric organ as a rich source of Na$^+$ channels. Electric eels are capable of generating powerful electrical discharges using their electric organs, which account for about half the size of these creatures. The organ consists of long columns of excitable cells, or electroplax, which have an excitable face innervated with cholinergic synapses (not shown). These synapses synchronously depolarize the excitable face of each electroplax, opening many Na$^+$ channels in the membrane that further generate a net potential difference across each cell. The geometric stacking of the electroplax thus results in a summation of the transcellular potentials, giving discharges of several hundred volts from a medium-sized animal. Specific antibodies raised to purified eel channels may be used to visualize the distribution of channels in the electroplax. The photograph (lower right, on the same scale as the sketched single cell above) shows that high densities of channels exist on the innervated face (open arrows) but not on the highly convoluted inexcitable surface (white arrows). The combined numbers of Na$^+$ channels from many thousands of such membranes make this electric organ an excellent source of channels for biochemical purification. (Photograph from Ariyasu *et al.*, 1987, with permission.)

the activity of ion channels is to transport ions across membranes. Since this barrier is disrupted during the first stage of purification, the compartments essential to channel function no longer exist, and hence the electrically detectable flow of ions through channels no longer occurs. Instead, one may use the binding of powerful neurotoxins to channels to assess the degree of purity attained during fractionation.

For Na$^+$ channels, there exist highly selective inhibitors of sodium transport known as guanidinium toxins. The most commonly used of these are tetrodotoxin (TTX), obtained from certain species of toxic puffer fish, and saxitoxin (STX), a product of certain dinoflagellates that "bloom" during oceanic "red tides." Either toxin may be radioactively labeled, and these substances very specifically interact with Na$^+$ channels from a wide variety of sources

with high affinity. To assess purity, one may thus compare the number of toxin-binding sites in a fraction with the amount of protein present.

### C. Isolation of Channel Molecules Based on Fractionation Procedures

With a good source of channels, efficient solubilization procedures, and a specific and sensitive assay, one may proceed to attempts at purification. For Na$^+$ channels, fractionation based on charge, followed by size fractionation of the mixed channel/lipid/detergent micelle, has resulted in highly purified preparations of Na$^+$ channels. Alternatively, in some protocols channel proteins have been specifically purified using affinity chromatography, in which channels are selectively retained by a column to which channel-binding substances, such as drugs, toxins, or antibodies, are attached.

Skeletal muscle $Ca^{2+}$ channels have also been extensively purified and characterized using much the same approaches as those described for $Na^+$ channels. In this case, it has been the L-type channels that were isolated, using radiolabeled drugs of the dihydropyridine class to tag channels irreversibly in the intact muscle membrane. This allowed the investigators to isolate the channel protein by following the fractionation of the bound radiolabeled drug (Sharp *et al.*, 1987).

## D. Information about Channel Structure from Channel Purification

There is much to be learned about channel structure from the physicochemical characteristics of the purified molecule. Some of this information forms the basis for separating the molecule from others in membrane extracts (such as size or charge differences), as described previously, whereas other measurements may be done on the purified material itself. Thus, for $Na^+$ channels, it was found that the channel is very negatively charged. Furthermore, in detergent solution the channel appears quite large; in micellar form, its size is on the order of several million molecular weight. Both characteristics are important clues regarding the chemical composition of $Na^+$ channels.

The subunit composition of a channel is an especially important characteristic to be determined from the purified material. For a number of enzymes, separate functions have been attributed to different subunits; hence breaking down the structural entities represents an important basis for further structure/function correlations. In brief, one usually determines the subunit composition of the purified material by analyzing it using denaturing electrophoresis, in which the intersubunit bonds are broken and each subunit separated according to size in an electric field. The most commonly used technique is sodium dodecyl sulfate–polyacrylamide gel electrophoresis (SDS–PAGE), in which each polypeptide species appears as a separate band on staining the polyacrylamide sieve used to separate polypeptides. Unfortunately, this is often more difficult than it appears, since (1) one is often not sure whether different bands on a gel are contaminants or true subunits, and (2) the stoichiometry of subunits is difficult to establish because of variations in the intrinsic staining intensity among various polypeptides. As a result, considerable effort is expended by various researchers to determine (and argue about!) subunit composition using a wide variety of approaches. For the voltage-gated $Na^+$ channel, a single large polypeptide apparently accounts for pore formation, ion selectivity, and gating. This protein appears on SDS–PAGE as a broad band at high molecular weight.

Given the large size of the eel electric organ, enough protein can be purified to allow study of its chemical composition. Not too surprisingly, $Na^+$ channels have an elevated proportion of hydrophobic amino acids compared to soluble globular proteins, likely reflecting domains of the channel that insert into the lipid membrane. However, the channel molecule appears to be much more hydrophobic than can be accounted for from its amino acid composition. Thus, further chemical analysis has found that a large number of lipid molecules are bound to the eel protein. Finally, it has been found that $Na^+$ channels are modified by the presence of extensive domains of carbohydrate. This glycosylation makes up about 30% of the molecule (by weight). The possible significance of such nonprotein lipid and sugar domains is discussed later.

$Na^+$ channels have also been purified from mammalian brain and muscle tissue. Like the eel channel, they consist primarily of heavily glycosylated large polypeptides. However, unlike the electric organ channel, they are found in association with one or several smaller polypeptides that have been thought to be channel subunits (Table 2). The large polypeptide common to $Na^+$ channels from all sources is referred to as alpha ($\alpha$), while smaller subunits have been designated as beta ($\beta$, muscle) or $\beta1$ and $\beta2$ (brain). At present, the role of these accessory polypeptides in channel function is not certain, although recent evidence suggests that they may aid in channel expression and modulate channel gating (Isom *et al.*, 1992).

For the skeletal muscle L-type $Ca^{2+}$ channel, it was found that although a single large polypeptide ($\alpha1$) was affinity labeled by the dihyropyridine (DHP) marker, at least four other polypeptides [$\alpha2$, $\beta$, gamma ($\gamma$), delta ($\delta$)] copurified with the DHP label. As with the $Na^+$ channel, $\alpha1$ seems to have the mechanisms required for ion channel operation. However, there is evidence that the other subunits are important accessory proteins that may be involved in the

**TABLE 2** Subunit Composition of Voltage-Gated Cation Channels

| Channel/tissue | Subunit | Molecular weight[a] |
|---|---|---|
| $Na^+$ | | |
| Eel electric organ | $\alpha$ | 260 (208) |
| Rat brain | $\alpha$ | 260 (220) |
| | $\beta1$ | 36 (23) |
| | $\beta2$ | 33 |
| Rat skeletal muscle | $\alpha$ | 260 (209) |
| | $\beta$ | 36 |
| Chick heart | $\alpha$ | 235 |
| $Ca^{2+}$ | | |
| L-type, rabbit skeletal muscle | $\alpha1$ | 170 (212) |
| | $\alpha2$ | 143 (125)[b] |
| | $\beta$ | 54 (58) |
| | $\gamma$ | 30 (25) |
| | $\delta$ | 24–27 (27)[b] |
| $K^+$ channel | | |
| *Drosophila Shaker* A | — | 65–85 (70) |
| Rat *dkr1* | — | 130 (95) |

[a] First set of values is the apparent molecular weight determined biochemically (SDS–PAGE); weights in parentheses were obtained from cloned DNA sequences.

[b] The biochemically observed calcium channel $\alpha2$ and $\delta$ subunits are derived from proteolysis of a full-length $\alpha2$ translation product during biosynthesis.

key role of this channel as the voltage-sensor in excitation–contraction coupling, which is described in Chapter 41.

Also shown in Table 2 are data for voltage-gated $K^+$ channels. The characterization of $K^+$ channel subunits was achieved primarily through molecular genetic means, as discussed later.

### E. Reconstitution of Purified Proteins Confirms Their Identity as Channels

An important criterion for purification was the reconstitution of the isolated material into artificial membranes and the demonstration that voltage-gated, sodium-selective, toxin-inhibitable channels were formed (see Miller, 1986). Thus, these experiments showed that purified toxin-binding material could be reinserted into small artificial lipid vesicles (liposomes) or planar lipid films (lipid bilayers) where they retained their ability to transport sodium selectively in response to voltage changes. These reconstituted channels had the same ability to respond to different pharmacological compounds, such as toxins and anesthetics, as the channels in natural tissue (see review by Catterall, 1992). Similar experiments have demonstrated that purified DHP receptors can form functional $Ca^{2+}$ channels in bilayers (Catterall, 1988).

### F. Limitations of Biochemical Characterization of Channels

The biochemical approach to channel purification thus gave important information about the size, chemical composition, and polypeptide makeup of $Na^+$ and L-type $Ca^{2+}$ channels. Classically, the next steps would be to sequence the purified material and attempt to obtain a high-resolution structure from crystallographic approaches. However, although these techniques have worked well with a relatively small number of soluble globular proteins, for several reasons they have been unsuccessful with membrane proteins in general and voltage-gated channels in particular. First, sequencing a large polypeptide requires that it be enzymatically or chemically cleaved into overlapping smaller fragments amenable to Edman degradation techniques (which can only sequence from 25 to 50 residues at a time). In such an approach, each small fragment must be separately purified from the fragmented preparation. For the large polypeptides of $Na^+$ and $Ca^{2+}$ channels, the number of fragments generated by such cleavages is just too great for all of them to be purified separately. Furthermore, many of the interesting membrane-spanning segments do not fragment or isolate well because of their high degree of hydrophobicity ("fears water"). Finally, the "amphipathic" (i.e., polarized hydrophilic/hydrophobic) nature of membrane proteins prevents their crystallization; instead, in the absence of lipids or detergents, such purified preparations usually form amorphous aggregates or precipitates with little intrinsic order. Hence, the elaboration of detailed structure of biochemically purified channels has been hampered by the unfavorable physical properties of membrane proteins.

## III. Channel Structure Investigation through Manipulation of DNA Sequences Encoding Channel Polypeptides

Fortunately, the limitations of classic biochemistry may be partly overcome by the application of techniques that allow one to identify and sequence DNA that encodes channel peptides. The next three subsections describe how the powerful methods of molecular biology are being applied to questions of interest to the cell physiologist. These approaches have given important insights into the structure of channels and how they work. This aspect of channel biology is considered in Sections III,A and III,B. However, molecular biology has also shown us that voltage-gated channels exist in a rich diversity of forms even within a single organism. This aspect of channel biology is described in Section III,C. First, a brief description of how recombinant DNA methods are used in such studies is provided for the novice reader (Fig. 3).

### A. Primary Structure of Ion Channels Determined Using Recombinant DNA Technology

One starts with the same tissues shown to be enriched in channel protein because these are also probably enriched in messenger RNAs (mRNAs) encoding these channels. Using the retroviral enzyme reverse transcriptase, one may then make complementary DNA copies (cDNAs) of the mRNAs. Using enzymatic "scissors" (called restriction enzymes), one may then insert each cDNA obtained into a circular DNA plasmid. These plasmids have the ability to be replicated along with other genetic material in bacteria, and also they encode an enzyme that will destroy certain antibiotics that would otherwise kill their bacterial hosts. The recombinant plasmids are next introduced into a bacterial host so that each bacterium will contain no more than a single plasmid with its unique cDNA insert. The bacteria are then diluted and spread over a plate of agar containing an antibiotic. Thus, bacteria without a plasmid are killed, whereas plasmid-containing bacteria can survive and replicate to form visible colony plaques. Since each such colony arises from a single original plasmid-transformed cell, they are known as "clones." Each agar plate may contain many thousands of such clones, each clone having many cells with plasmids containing the identical cDNA insert.

The problem is to identify which of the many colonies has a particular cDNA that encodes for the channel protein of interest, among all of the cloned cDNAs that encode the myriad proteins of the original tissue. The severity of this problem can be appreciated from the fact that even highly enriched tissues such as eel electric organ may have at most only 1 message out of 10,000 or so that encodes a $Na^+$ channel. Hence, there must be a rapid method for screening the colonies to identify the channel cDNA clones. This is where information from biochemical purification proves to be vital.

Despite the near impossibility of totally sequencing large $Na^+$ or $Ca^{2+}$ channel polypeptides, from a proteolytic digest one can readily purify a homogeneous preparation of a particular fragment. This fragment may then be partially

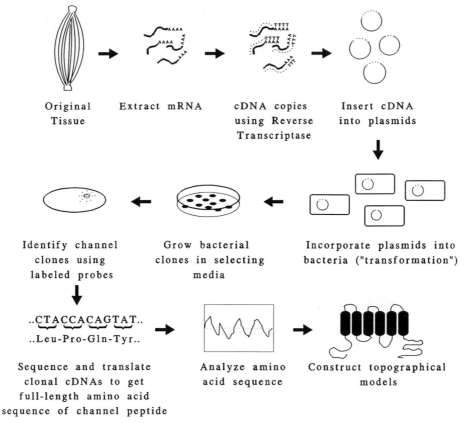

**FIG. 3.**   General strategy used to determine ion channel structure using recombinant DNA approaches.

sequenced using standard Edman degradation techniques. In practice, one only needs a few (perhaps six or so) residues of sequence to be reasonably sure that a unique segment of the protein is encoded. From this short amino acid segment, one can then synthesize (by automated means) a short DNA fragment, called an **oligonucleotide,** that encodes the segment. Further, this synthetic DNA may be radiolabeled, and thus becomes a binding probe to identify cDNA inserts. Identification is done by placing a piece of special paper briefly on top of the agar plate, and lifting off a few bacteria from each colony plaque onto the paper. The paper is then placed in a special solution that lyses the bacteria but binds the released DNA immediately on the surface. The radiolabeled synthetic probe may then be washed over the paper; it will bind by complementary interactions to the antisense strand of clonal plasmids encoding the channel protein containing the original sequence. After unbound probe is washed away, the paper is overlaid in the dark with X-ray film. Channel clones are thus identified as dark spots (usually only a few) on the film, and may be traced back to the plaque on the original agar plate. This plaque is then lifted from the plate with a toothpick and grown in a nutrient broth to produce many bacteria. This technique provides enough cells from which the plasmids are purified by simple chemical means, and the cDNA insert may be readily sequenced. Once the cDNA

sequence is known, the genetic code is applied to derive the primary amino acid sequence of the channel protein.

This description of clonal sequencing omits many details. For example, the length of DNA that one may clone or sequence at one time is limited, and complex strategies are often required to sequence the entire length of long channel polypeptides. In addition, ion channels that are heteromeric complexes will require that each subunit polypeptide be separately sequenced. Despite these apparent complexities, an organized group of researchers using automated equipment can screen hundreds of thousands of clones and sequence thousands of bases of interesting cDNA inserts in a relatively short time.

## B. Analysis of the Primary Structure of Large Polypeptide Voltage-Gated Channels

The electroplax $Na^+$ channel was the first voltage-gated channel to be sequenced (Noda *et al.,* 1984). As suggested by the biochemistry, it was found to be a long polypeptide of 1820 amino acids (Fig. 4). Subsequent cloning of mammalian channels showed them to be similar in size and sequence (see Numa and Noda, 1986). What can one learn about channel structure from such primary sequence data? Typically, one subjects the primary sequence data to several forms of analysis to derive a model of the appearance of the higher order structure. The first striking characteris-

tic revealed in the Na⁺ channel sequence was the presence of four repeats (of about 200 amino acids each) that were highly (but not completely) homologous to one another. Furthermore, these internal repeats were found to have great potential for forming membrane-spanning domains of the channel. From such analyses, the model shown later in Fig. 7 was obtained.

How are such analyses performed? Basically the analyses occur in two or three separate steps. First, one analyzes the sequence of the peptide for regions of high hydrophobicity, on the expectation that membrane-spanning regions should have a high proportion of hydrophobic amino acids. To do this, hydropathy values that reflect its relative hydrophobicity are assigned to each amino acid residue. The assignment of such values is based largely on indirect physical measurements of the hydrophobicity, such as the ability of a given residue to partition from an aqueous solution to an oil phase. Such values may or may not reflect the chemical behavior of a residue in a long polypeptide chain, where chemical properties may be further influenced by neighboring residues. As a result, a number of such hydropathy tables have been developed by various investigators who have significant differences in the way they assign hydropathy weights to a given residue. One of the most popular tables is that of Kyte and Doolittle (1982) shown in Fig. 5. In most cases, the assignments are qualitatively obvious, especially for expected hydrophobic groups (for example, the hydrocarbon chains or leucine or valine) or hydrophilic groups (e.g., the charged residues such as glutamate or lysine).

Figure 6 shows hydropathy plots for the Na⁺ channel sequences given in Fig. 4. The plot is obtained by averaging the hydropathy values of a fixed segment (perhaps 7–15 residues). Starting with the first residues at the N-terminal end, this average is computed and the value is plotted versus the number of the residue in the middle of the segment. This averaging "frame" is then shifted by one residue toward the carboxyl end, the new hydropathy average is recomputed (i.e., the change due to the loss of the N-terminal side residue and the gain of the next C-terminal side residue in the sequence), and the value is plotted by incrementing the residue number by one. This "frameshift" average is thus continued until the end of the protein is reached. The purpose of using the averaging frame is to "smooth out" sharp variations in hydropathy, so that a pattern may be more readily seen.

As can be seen in Fig. 6, there is a repeated pattern of hydropathy for Na⁺ channels that reflects the conserved homology of the four internal repeat domains I–IV. Within each domain, there are five or six peaks of hydrophobicity that are each about 20 amino acids long. This is significant because such a length is considered optimal to span the hydrophobic interior of the lipid bilayer. However, hydro-

phobicity alone is not sufficient to predict a membrane-spanning region; instead, the residues must assume a favorable secondary structure, most commonly assumed to be an α-helix. In the helical conformation the hydrophobic side chains project radially away from the axis of the helix and presumably into the surrounding lipid. Such a configuration prevents the very hydrophilic peptide backbone of the protein from exposure to the hydrophobic lipid environment and thus represents a favorable low free-energy arrangement of the protein/lipid interface.

How are such secondary structures predicted from primary sequence? This too is a highly empirical process (Chou and Fasman, 1978). Basically, investigators have analyzed in detail a number of proteins for which there are high-resolution X-ray structures available at the amino acid level. By cataloging the frequencies with which given amino acids or short combinations of residues are found in helices, β sheets, or turns, a set of empirical rules has been developed to predict the secondary structures that a given sequence of amino acids will most probably form (see Fig. 6). Although the general accuracy of such predictions has been debated, when combined with hydropathy information the method is apparently rather accurate for the prediction of membrane-spanning domains in membrane proteins. This is probably because of the severe thermodynamic constraints confronting polypeptides needing to cross a boundary of high hydrophobicity while interfacing with a highly hydrophilic aqueous phase on either side of the membrane and in the aqueous channel pore.

## C. How Is Channel Structure Formed?

In any case, such primary sequence analyses suggests that each internal domain has six α-helical membrane-spanning segments (see Numa and Noda, 1986). The question to consider next is how such an arrangement might form a transmembrane pore for the passage of ions. The answer is that the four transmembrane regions probably contribute equally to the pore walls in a "staves of a barrel" configuration (Fig. 7; see Guy and Seetharamulu, 1986). In fact, such a model was comforting to the original investigators, because a similar arrangement had previously been shown to exist among the separate subunits of the synaptic acetylcholine-gated channel, the first ion channel to be purified and sequenced via cDNA cloning.

## D. Sequence Homology among the Transmembrane Domains of Ion Channels

Analysis of the DHP-sensitive Ca²⁺ channel α1 peptide revealed that it also has four internal homologous repeats

---

**FIG. 4.** Amino acid sequences of three Na⁺ channels obtained by cDNA cloning techniques. Rats I and II were obtained from rat brain cDNA libraries, while the eel Na⁺ channel is that expressed in the electric organ of *E. electricus*. Boxed residues show homology among the three channels, while dashed lines designate absent segments. Labeled brackets underneath the sequences refer to parts of the highly conserved putative trnsmembrane repeats (see text and Figs. 6 and 7 for details). (From Numa and Noda, 1986.)

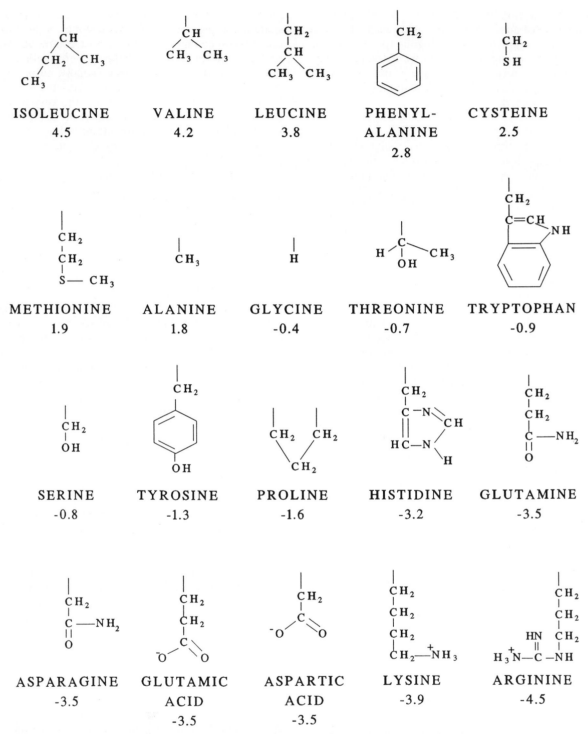

**FIG. 5.** Structure and hydropathy values of amino acid residues. Arranged from most hydrophobic to most hydrophilic according to relative values assigned by Kyte and Doolittle (1982). Only side groups are shown with hydropathy values underneath.

(Fig. 8; see Tanabe *et al.*, 1987). Furthermore, these repeats (but not the intervening loops) were found to be highly homologous to the $Na^+$ channel repeats. The "staves of a barrel" configuration in which either homologous internal repeats (for $Na^+$ and L-type $Ca^{2+}$ channels) or highly homologous subunits (as in the case of transmitter-gated and $K^+$ channels—see later) form a pore seems to be a widespread architecture among ion channels in general.

## E. How Are Topographical Predictions for Channel Structure Tested?

The modeling approach based on primary structure is highly uncertain and needs to be empirically confirmed for each channel. One of the most popular ways this has been done is through the use of "site-directed" antibodies raised to synthetic peptides encoding a short stretch (i.e., 10–20 residues) of the channel of which one wishes to know the topographical orientation. These antibodies may be applied to channels in their native membranes or intact cells, and the relative "sidedness" of the epitope may be determined from the side of the membrane to which the antibody is found. The approach has been used for Na$^+$ channels in several laboratories, and the topographical model shown in Fig. 7 has largely been confirmed (see Catterall, 1992). Similar approaches can be used in which impermeant group-reactive chemical reagents are applied to either intact cells or liposomes with oriented channels, followed by some method to identify the site of interaction on the protein (e.g., sequencing of proteolytic fragments). Alternatively, a synthetic, but highly antigenic, short epi-

tope may be recombinantly inserted into the channel sequence and the mutant expressed in cultured cells. In this case, the same antibody may be used in all experiments to determine on which side of the membrane the epitope "tag" is located as a function of its position in the native channel sequence.

## F. The Genetic Approach Used to Identify Nucleic Acid Clones Coding for K$^+$ Channels

K$^+$ channels are extremely important in excitation and transduction phenomena. Partly because of the way these channels were cloned and partly because their structure is a simplified verison of the long Na$^+$ and Ca$^{2+}$ channels, they are very valuable systems with which to study the general structures and mechanisms of the voltage-gated ion channel class. Such channels were not originally cloned via the usual biochemical purification–oligonucleotide probe approach described previously because there were few good chemical "labels" like tetrodotoxin or DHP.

A highly creative, yet complex, approach was taken independently by several laboratories that used behavioral mu-

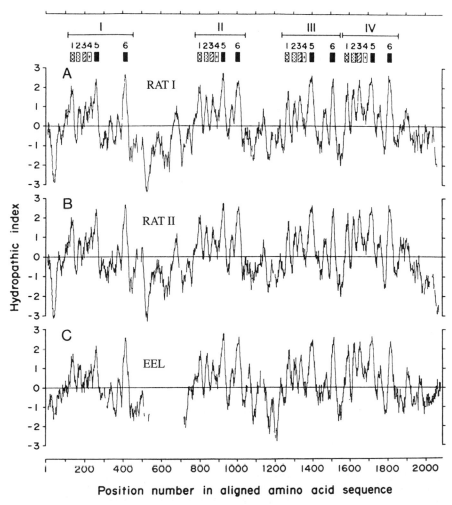

**FIG. 6.** Hydropathy plots for the three Na$^+$ channel sequences of Fig. 4. Brackets and numbered patterns at top refer to location of homologous repeats and transmembrane segments (see text and Fig. 7). (From Numa and Noda, 1986.)

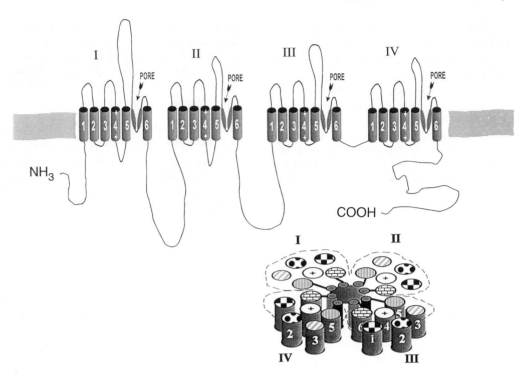

**FIG. 7.** Model of Na⁺ channel structure based on analysis of primary sequence data. Top drawing shows linear placement of four homologous domains, loops, and termini (outer surface of membrane is at the top). Possible pore and selectivity regions are labeled and shown as the angled segments between transmembrane segments 5 and 6. Drawing on bottom right shows hypothetical arrangement of homology repeats in the "staves of a barrel" configuration to form a pore (placement of transmembrane segments 1–6 is somewhat arbitrary).

tants of the fruit fly *Drosophila* (e.g., see Papazian *et al.,* 1987). In brief, mutant strains that had uncontrolled shaking when exposed to ether were identified. Electrophysiology then established that the muscles and certain neurons of such flies had missing or altered K⁺ currents of a specific type known as A-type currents that, like sodium currents, inactivate on depolarization. The locus of the genetic defect causing this behavior could be seen as an actual physical defect (translocation) in the polytene salivary chromosomes of these so-called *Shaker* mutants. Since extensive libraries of cloned identified chromosomal DNA fragments of *Drosophila* have long been available (called "genomic" clones), the investigators were able to obtain the DNA fragment of normal flies on which the mutant defect occurred. This allowed them to sequence such fragments, and by comparison with *Shaker* mutant sequences and by the identification of homologous cDNA frgments (derived as described earlier), the sequence of the A-type K⁺ channel could be inferred. Since the *Shaker* mutation allowed a completely genetic elucidation of the channel primary amino acid sequence, the A-current it manifests has become known as the *Shaker*-type K⁺ channel.

### G. K⁺ Channels Are Homologous to a Single Internal Repeat of Na⁺ Channels

An initially surprising finding when the *Shaker* A-type channel sequence was analyzed was that the length of the

sequence was much shorter than that of the other voltage-gated channels, but contained a single domain that was reminiscent of (but strictly speaking not very homologous to) the internal repeat domains of the Na⁺ or Ca²⁺ channels (Fig. 9; see Pongs, 1992). An obvious hypothesis for channel structure for the *Shaker* channel was that four molecules of this sequence (i.e., subunits) combined to form the "staves of a barrel" structure postulated for the long-channel polypeptides. This tetrameric assembly model has recently received strong support from several elegant experiments (MacKinnon, 1991).

### IV. Molecular Mechanisms of Channel Function: How Does One Investigate Them?

We now discuss the methods that can be used to determine which parts of an elucidated channel sequence are involved in the basic channel properties of pore formation, ion selectivity, and voltage-dependent gating. Here, too, the methods of recombinant DNA technology have proved to be of great use in that they allow the investigator to change the nucleic acid sequence encoding a channel in the manner desired. Using such mutagenesis techniques, the basic idea is to alter the amino acid sequence via deletion, addition, or change in the primary sequence in specific locations and then to test how such changes affect the function of the mutant channels. The hope is that by obtaining many such "structure/function correlations," one might not only infer the part of the sequence that underlies, for example, channel activation, but also infer how this structure works as a mechanism.

## A. Choosing Interesting Sites and Segments as Targets for Mutagenesis

How does one decide which of the many amino acids in a channel polypeptide to change? This is where the modeling approach described previously is so important, for it provides clues or testable hypotheses that focus mutagenesis attempts on reasonable guesses. For example, one might assume that a highly conserved sequence in channels from phylogenetically distant animals might reflect a functionally essential part of the molecule that cannot withstand evolutionary alteration without being deleterious. Another strategy might be to compare the regions of divergence in the sequence of the various voltage-gated channels in the hope of finding candidate regions involved in different functions among channels. For example, the difference in ion selectivity among $Na^+$, $K^+$, and $Ca^{2+}$ channels must lie somewhere in the variations in their amino acid sequence.

Both approaches have had a certain degree of success, although "homology" is a relative term, and in reality large areas of channels are highly variable, even among isotypes of channels from the same organism (see later). This makes it difficult to guess which of these many differences relates to specific functions.

Another approach is to take advantage of certain genetic disorders in which mutations in channel structure occur naturally with functional consequences. A number of human diseases and animal models have been shown to be caused by channel defects. The sequence of such channels can give important clues into structure/function relationships in channels, while directing artificial mutagenesis experiments.

## B. Use of Expression Systems in Mutagenesis Studies

Another essential part of mutagenesis experimentation is the ability to introduce artificially altered genetic material into a cell that is capable of translating it into a protein molecule, processing the molecule appropriately (i.e., folding the polypeptide, assembling subunits, adding sugars or lipids), and inserting the protein into the cell surface. The

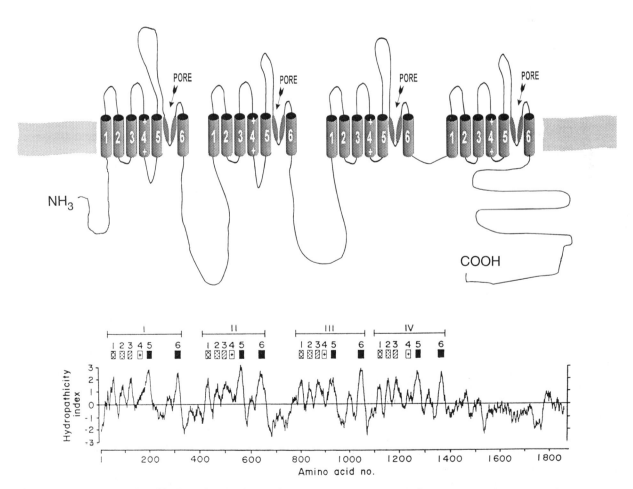

**FIG. 8.** Model (top) and hydropathy plot (bottom) of L-type skeletal muscle $Ca^{2+}$ channel $\alpha 1$ subunit. Compare to Figs. 6 and 7.

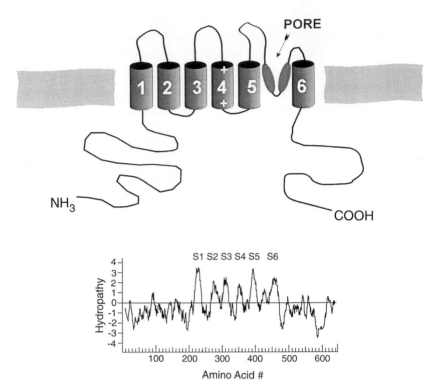

**FIG. 9.** Model (top) and hydropathy plot (bottom) of *Shaker* B-type K⁺ channel. Compare to Figs. 6, 7, and 8.

protein then may be functionally characterized using the biophysical methods described in other chapters. The most popular such "expression" system is the oocyte of the South African frog *Xenopus laevis* (Fig. 10; Leonard and Snutch, 1991). In this system one prepares mRNA from clones of the mutant channel and injects this message into the large (1–2 mm in diameter) oocytes with a micropipette. After waiting a few days for channel synthesis and membrane insertion to occur in sufficient numbers, one observes the expressed ionic currents via voltage-clamp methods, using large microelectrodes to impale the cell. Alternatively, one may also use a very large diameter electrode to patch clamp a large area of membrane; such "macropatch" methods give better voltage control than the impaled electrode technique while allowing macroscopic current characterization. Finally, conventional single-channel analysis may be done in all the usual modes using standard patch pipettes. Thus, the oocyte allows the investigator to compare both the macroscopic and single-channel properties of the mutants with the normal "wild-type" channel.

Although other expression systems in which the altered genetic material may be introduced into cells transiently via viruses or integrated permanently into the host genome (stable transfection) have been developed more recently, the use of such systems is beyond the scope of this chapter. Suffice it to say that the frog oocyte is currently the most popular and productive expression system with which to study structure/function relationships.

Because of its great potential in elucidating channel mechanisms, mutagenesis is being used by a growing num-ber of investigators. As a result, both new and altered concepts of channel function appear frequently in the scientific literature. In the brief account that follows, we can only summarize some of the fairly firm basics of pore formation, selectivity, and gating.

### C. Pore Formation and Ion Selectivity

Even before the advent of structural studies, Hille (see Hille, 1992) hypothesized that ion selectivity and pore formation were performed by the same parts of channel structure. Accordingly, this has influenced investigators to look for short segments that might line a pore with charged groups that would confer ion selectivity properties on the structure. Such structures have been postulated from sophisticated primary sequence analysis (Guy and Seetharamulu, 1986) and from consideration of the sites of binding of pore-blocking neurotoxins (for example, certain scorpion toxins; see Pongs, 1992). Such a candidate segment is a stretch of 20–25 amino acids that lies on a loop connecting the fifth and sixth predicted transmembrane helices in the voltage-gated cation channels (see Figs. 7, 8, and 9).

The role of this structure in ion selectivity has been studied by mutating the residues in the analogous loop in *Shaker*-type K⁺ channels (Yool and Schwarz, 1991). As Fig. 11 shows, discrete changes in this loop can have profound effects on the ability of the resultant mutant to discriminate among cations. Similar findings have been reported for the analogous segments in both Na⁺ and Ca²⁺ channels (Heinemann *et al.*, 1992; Wang *et al.*, 1993). In these latter studies, carboxylate groups have been identified that are

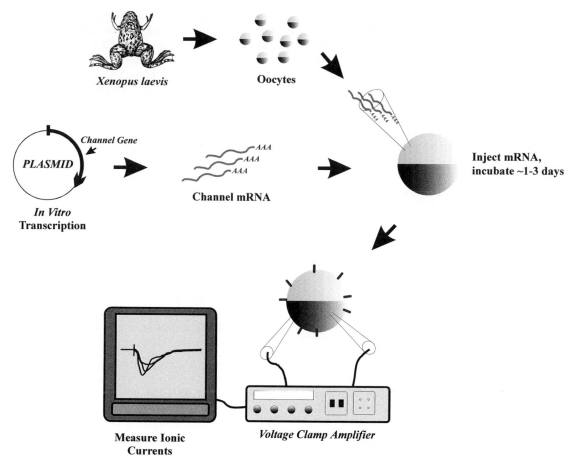

**FIG. 10.** *Xenopus* oocyte expression system. See text for details.

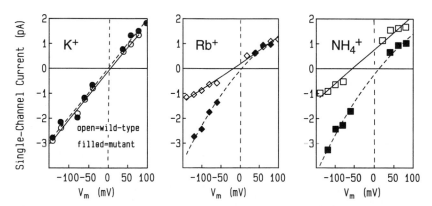

**FIG. 11.** Alteration of *Shaker* B K$^+$ channel ion selectivity caused by a mutation in the putative pore-forming region. In these experiments, the phenylalanine at position 433 was mutated to serine. The oocyte system for expressing both wild-type and mutant channels is described in the text. Single-channel currents were measured using patch-clamp methods in which the ion indicated in each panel was applied to the external surface of the channel with K$^+$ on the inside in all cases. Under these bionic conditions, inward (negative) currents will be carried mainly by the test ion, while outward (positive) currents will be mostly K$^+$ ion fluxes. As can be seen from the increased inward currents, permeability of the channel to rubidium and ammonium ions was greatly enhanced in the mutants, while K$^+$ was unaffected. (Data redrawn from Yool and Schwarz, 1991.)

strong candidates for the negative charges responsible for the selectivity properties of both channels. For $Na^+$ channels, there are two of these groups, one of each in the "pore-forming" segments in domains I and III (see Fig. 7), while for $Ca^{2+}$ channels all four such segments each contributes a negative charge. The effects of altering these groups via mutagenesis have suggested an elegant mechanism of ion selectivity in which paired carboxylates coordinate the favored ions within the pore (Ellinor *et al.,* 1995). In addition to selective ion permeation, this model also accounts for other observations such as "single-file" ion movements, pore saturation, and the effect of $Ca^{2+}$ ions themselves on $Ca^{2+}$ channel selectivity.

### D. Domains Involved in Voltage-Dependent Activation

Here again, investigators looked for clues that might direct their mutagenesis among the large number of residues in cloned voltage-gated channels. In comparing $Na^+$, DHP-sensitive $Ca^{2+}$, and *Shaker*-type $K^+$ channels, the fourth predicted helix in the repeat/subunit motif was seen to display a highly conserved sequence pattern (see Figs. 7, 8, and 9). This segment, known as S4, consists of a repeated triad of a positively charged residue (arginine or lysine), followed by two highly hydrophobic residues (such as valine, leucine, isoleucine; see Fig. 12). While the number of such triads in the various S4 segments ranges from 3 to 8, its structure of both charged and hydrophobic residues suggests an element capable of responding to transmembrane voltage changes (see Catterall, 1992). Accordingly, a number of experiments have now been done in which both the charged and hydrophobic elements have been changed.

In brief, S4 mutagenesis usually causes significant changes in the steady-state activation behavior of the affected channel (Fig. 13; Papazian *et al.,* 1991; Lopez *et al.,* 1991). Thus, shifts of the steady-state activation curve along the voltage axis occur with most changes to either charged or hydrophobic residues. In addition, changes in activation slope are frequently seen.

The question is whether the observed changes can be explained in terms of a mechanism by which S4 participates in gating (instead of the general hypothesis that S4 is some part of a gating structure). Unfortunately, to date the re-sults obtained from mutagenesis do not discriminate very well among the various possible roles for S4 in gating, for example, as a gating sensor, a part of the transduction linkage to a gating element, or simply an unrelated part of channel structure. Thus, the magnitude or direction of changes in activation behavior does not systematically correlate with changes to S4 charge or sequence. In addition, others have questioned whether the charged groups in the S4 segments can completely account for all the *gating charge* (see Chapter 19) that moves when channels activate (see Sigworth, 1995). In general, the mutagenesis approach is inherently limited in its ability to distinguish between alterations that directly affect a mechanistically important part of the molecule and those that indirectly cause functional effects through structural changes that propagate from mechanistically unrelated domains. This problem arises because protein structures are notoriously sensitive to long-range effects of such alterations, generically known as allosteric ("other site") effects.

### E. Limitations to Mutagenesis in the Study of Channel Mechanisms

The naive reader can understand the problem from a crude analogy in which many of us as children tried to understand the workings of a mechanical watch by a similar approach using simple tools. The child understands the function of the mechanism only by the movement of the hands. However, as we all know, a mechanical watch is quite a delicate machine. Thus, even the act of carelessly opening the case may cause the mechanism to cease working. Are we to conclude that the case is therefore part of the mechanism? Given access to the mechanism itself, the situation is highly complex, since the movement of many parts is required for the ultimate movement of the hands. For example, the result of stopping the second hand by interfering with a given gear tells one very little of the role that gear plays in how the watch operates.

For some time it has been known that the interactions of chemical substances or structural alterations with a site on one side of a protein can dramatically affect the function of another site at some distance, for example, on the other side of the protein. This concern and the potentially com-

|  |  | + | + | + | + | + | + | + | + |
|---|---|---|---|---|---|---|---|---|---|
| *Electric eel NaCh IV-S4* | | L F R | V I R | L A R | I A R | V L R | L I R | A A K | G I R |
| *Rat brain NaCh IV-S4* | | L F R | V I R | L A R | I G R | I L R | L I K | G A K | G I R |
| *Rat skeletal muscle NaCh IV-S4* | | L F R | V I R | L A R | I G R | V L R | L I R | G A K | G I R |
| *Drosophila NaCh S4* | | L L R | V V R | V F R | I G R | I L R | L I K | A A K | G I R |
| *Rabbit skel. muscle CaCh IV-S4* | | S S A | F F R | L F R | V M R | L I K | L L S | R A E | G V R |
| *Mouse brain KCh S4* | | I L R | V I R | L V R | V F R | I F K | L S R | H S K | |
| *Drosophila Shaker S4* | | I L R | V I R | L V R | V F R | I F K | L S R | H S K | |

**FIG. 12.** Comparison of S4 segments from different voltage-gated channels. Note the repeating triadic motif of positive charge (R, arginine; K, lysine) followed by two hydrophobic residues. Hatched areas show conservation of charge among channels; all amino acids are designated by single-letter codes.

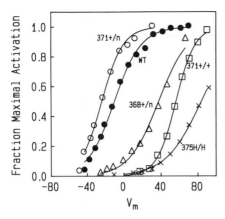

**FIG. 13.** Effect of S4 mutations on activation of *Shaker* K⁺ channels expressed in oocytes. Shown are steady-state activation curves for wild-type (WT) and single amino acid substitutions at various sequence positions indicated by the numbers. +/n, change from a positively charged residue to a neutral one; +/+, substitute arginine or lysine for lysine or arginine; H/H, replace one hydrophobic residue with another. (Data redrawn from Papazian *et al.*, 1991, and Lopez *et al.*, 1991.)

plex nature of channel-gating mechanisms should motivate the responsible practitioner of mutagenesis to display caution in interpreting results in mechanistic terms.

How, then, in the face of such limitations does one make progress in elucidating the mechanisms of channel function? In fact, the approach is philosophically identical to the way one attempts to prove any hypothesis, namely, to establish a circumstantial case by as many independent tests as possible. In the example of S4 and mutagenesis, for example, one might wish to demonstrate that mutations to any other part of the molecule have no effect on voltage-dependent activation as a sufficient (but not necessary) test. Unfortunately, mutations to other parts of channel sequence can have equally profound effects as S4 alterations (for example, to the pore-forming region described previously; see Yool and Schwarz, 1991). Thus, while these other mutations do not disprove a role for S4 as a sensor (or any other role), they do suggest, as in the mechanical watch, that a number of segments of the sequence are either directly involved in a complex mechanism of voltage gating or indirectly affect a gating process through allosteric/structural influences.

## F. The Mechanism of Channel Inactivation

Genetic engineering studies of the inactivation process currently provide a better example of what one may learn mechanistically from the mutagenesis approach. Some of the most important such studies have involved the *Shaker* A-type channel. Here early workers found important clues to the segment involved in the inactivation process by comparing naturally occurring variants of the channel ("isotypes"; see later). Thus, Fig. 14 shows the pattern of four such isotypes, called A, B, C, and D (Timpe *et al.*, 1988). *Shaker* B, C, and D were identified from further testing of cDNA clones using probes constructed from the

original *Shaker* A sequenced as described earlier. When expressed in oocytes, these isotypes displayed very different kinetics of inactivation. Analysis of the regions of homology and differences among the isotypes shown in Fig. 14 indicates that fast inactivation properties correlate most strongly with the nature of the N-terminal segment (thought to be cytoplasmic; see Fig. 9).

Further evidence for the role of this segment in inactivation was obtained by constructing channel "chimeras" in which the *Shaker* A clone was modified by replacing its N-terminal segment with those from other isotypes. In each case the mutant channel assumed the inactivation properties of the isotype donating the N-terminal segment, despite the fact that the rest of the molecule had the sequence of the fast-inactivating Shaker isotype (Aldrich *et al.*, 1990).

Finally, investigators mutagenized specific regions of the fast-inactivating *Shaker* B segment to localize those residues involved in the inactivation process (Hoshi *et al.*, 1990). Both by deleting small segments and by replacing others it was found that the crucial part of the molecule was a stretch of residues from positions 6–83 (i.e., the N-terminal tail of the molecule seen in Fig. 9). In this segment there were distinct effects of the alterations. Thus, mutations to the residues in the 23–83 residue region tended to produce changes in inactivation kinetics, whereas changes to those in the N-terminal side (6–22) of the segment tended to destroy inactivation completely (see Fig. 16).

What might be the mechanism of *Shaker* inactivation? Much earlier Armstrong and coworkers had reported that internal perfusion of proteases in the squid destroyed the ability of Na⁺ channels to inactivate. On the basis of this and other experiments, Armstrong proposed that the inactivation mechanism is a cytoplasmic ball attached to a peptide tether that swings into an open channel to block it from the inside (Fig. 15; Armstrong and Bezanilla, 1977). In this hypothesis the ball blocks the channel by binding to a receptor in the inside of the pore that is accessible only when the channel has opened (activated). The reader may recall the evidence that inactivation is coupled to activation, and hence may be intrinsically voltage independent. Thus, the "ball-and-chain" model also explains these properties since the cytoplasmic ball would experience little of the transmembrane field but could swing into the channel only after it had activated and revealed its receptor (hence the coupling properties).

In the case of the *Shaker* channel, Aldrich and colleagues proposed that the N-terminal segment that they identified was just such a mechanism, since changes to the length of the 23–83 segment affected kinetics (as might be expected from changing the length of the "chain"), while changes to the N-terminal segment (such as alterations of charge) that eliminated inactivation were postulated to prevent the "ball" from binding to its receptor in the activated channel.

How might such a model be further tested? Here the investigators performed a particularly elegant experiment (Fig. 16; Zagottta *et al.*, 1990). First they constructed a mutant *Shaker* channel in which inactivation had been eliminated by a deletion of its own N-terminal segment.

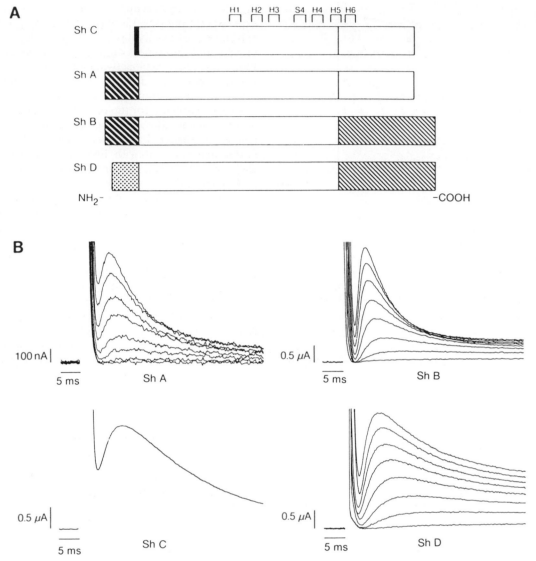

**FIG. 14.** Gating kinetics of alternatively spliced *Shaker* isotypes expressed in oocytes. Top diagram compares sequences, with patterned regions representing alternatively spliced domains that are different among isotypes (other unpatterned regions are identical in sequence). Both current traces obtained from oocyte voltage clamp show currents resulting from each isotype. (From Timpe *et al.,* 1988. Copyright 1988 by Cell Press.)

Next, a synthetic peptide that had the exact sequence of the deleted region was prepared. The mutant channel was then expressed in oocytes, where both excised macropatch and single-channel observations could be made. The investigators found that when the peptide was applied to the solution bathing the inside surface of the membrane, inactivation was restored to these mutant channels. Furthermore, the kinetics of the restored inactivation depended on the concentration of the peptide in the bath, and the inactivation could be reversed by washing away the peptide-containing solution. These observations are thus consistent with a direct blocking interaction of the peptide with a site in the open channel, and constitute strong evidence for the ball-and-chain mechanism of inactivation in A-type K⁺ channels.

### G. Other Gating-Related Domains in Voltage-Sensitive Channels

Analysis of the homology of the various cloned Na⁺ channels revealed that the longest segment of conservation lay in the postulated internal loop between repeats III and IV, even among animals of some phylogenetic distance (see Fig. 4). This high degree of conservation suggested a functionally important role for the segment. This idea was initially tested by constructing mutant Na⁺ channels in which the channel was expressed in oocytes in two pieces, with the genetic "cut" within this segment (Stuhmer *et al.,* 1989). Interestingly, such artificial dimeric channels were assembled and expressed in the oocyte and displayed relatively normal activation and conductance; however, such channels did not inactivate. Further, site-directed antibodies raised to a synthetic peptide encoding this segment were found to slow or eliminate inactivation when they were applied to the cytoplasmic surface of cells expressing normally inactivating channels (Vassilev *et al.,* 1989). Finally, mutagenesis directed to this segment produced channels with altered inactivation (Catterall, 1992).

However, the sequence of this segment has little resemblance to the N-terminal ball-and-chain *Shaker* domain, and it is additionally tethered at both ends. Also, specific correlations of the mutagenesis changes with functional consequences are difficult to interpret in terms of a simple ball-and-chain model. Thus, although it appears that this segment affects channel gating, its mechanistic role in the inactivation process is still unclear.

For DHP-sensitive $Ca^{2+}$ channels, an isotype of the originally cloned skeletal muscle channel has been cloned from the heart. In the parent tissues these channels share some characteristics, but differ greatly in the speed by which their calcium currents activate (heart channel currents being much more rapid in responding to voltage). Although the isotypes share considerable homology, there is still enough difference in the sequences to make it difficult to guess which residues might be responsible for activation kinetics. This problem was elegantly addressed by Beam,

Tanabe, and coworkers, who made chimeras of the two channels (Fig. 17; Tanabe *et al.,* 1991). To focus the mutagenesis it was reasoned that only the predicted transmembrane repeats should be involved in voltage-gated activation. Thus, the chimeras swapped the homologous repeats among constructs. As shown in Fig. 17, the kinetics of activation correlate almost completely with the kinetics of the donor of the first (I) repeat. This finding suggests that despite their approximate homology, the repeats in a given channel contribute differently to gating properties. This is perhaps expected from the observation that the S4 regions among these domains have somewhat different structures and numbers of charges (e.g., see Fig. 4). Having identified the domain responsible, these investigators have been able to narrow their search to shorter segments, and ultimately will identify the residues involved in this activation phenomenon (Nakai *et al.,* 1994).

### H. Nonprotein Domains Possibly Involved in Channel Function

Recall that $Na^+$ channels are extensively modified by the biosynthetic attachment of large numbers of sugars and lipids (Miller *et al.,* 1983). Such "post/translational" domains have also been inferred for some of the $K^+$ channel

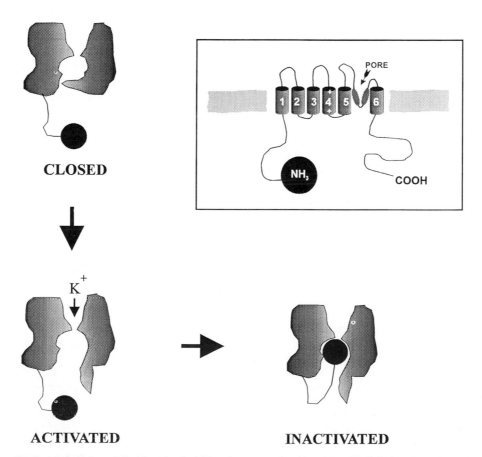

**FIG. 15.** "Ball-and-chain" model of inactivation. Not shown are the three other "balls" that the other subunits of the channel will contribute.

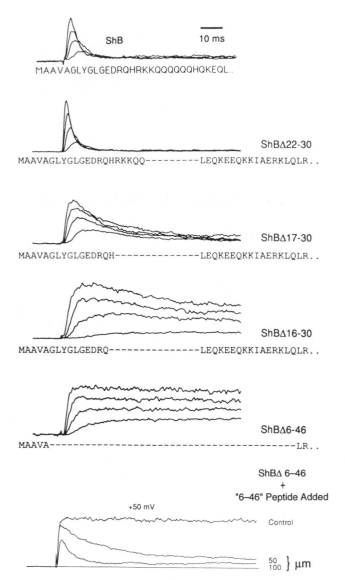

**FIG. 16.** Identification of a putative ball and chain in *Shaker* B channels expressed in oocytes. Panels 2–5 from the top show the effect of deleting increasing lengths of a segment near the N terminus of the *Shaker* protein. Note that the deletion of residues 6–46 results in a noninactivating K+ current. The bottom-most panel shows that inactivation properties of this latter mutant may be restored by application of the synthetic 6–46 peptide to the inner channel surface in a dose-dependent fashion (Redrawn from Hoshi *et al.*, 1990, and Zagotta *et al.*, 1990.)

isotypes (Trimmer, 1991). In general, it has been thought that these modifications might be involved in protein folding or targeting to the plasma membrane during the biosynthetic process. However, it may be that these domains also play a more active role in channel mechanisms (see Levinson *et al.*, 1990).

For example, many of the sugars on Na+ channels are in the form of negatively charged sialic acid residues. It has been considered possible that such charges influence the electrical field around gating elements of the channel. In support of this idea, it has been found that enzymatic removal of these sugars from purified channels causes significant changes in the voltage sensitivity of their activation when they are reconstituted into lipid bilayers (Fig. 18; Recio-Pinto *et al.*, 1990). Furthermore, similar changes in gating result when channel genes are expressed in mutant cell lines unable to add sialyl residues to proteins, or when potential glycosylation sites are removed from channels via mutagenesis (Thornhill *et al.*, 1996; Bennett *et al.*, 1997). Thus, such sugars might represent a means for the cell to control the gating behavior of its channels appropriate to its excitation needs.

The role of covalently attached lipids is less certain. Perhaps they exist to stabilize a microenvironment around gating elements or structurally vital domains of the channel. In any case, their presence might be expected to alter channel structure in a manner not predicted by the modeling approaches described previously.

## V. Isotypes of Voltage-Gated Channels as Part of a Large Superfamily

One of the most striking results of recombinant DNA analysis of channel clones has been the discovery of numerous channel *isotypes* expressed within the same organism and even within the same cell. Naturally, some degree of homology among channels of different organisms was expected, and electrophysiological recordings showed that channels could be functionally diverse at the cellular level. Even a structural similarity among Na+, Ca2+, and K+ channels had been predicted by some investigators. However, the number and diversity of channels within a single organism revealed by molecular biology were rather unexpected. Channels of a functionally similar class (usually based on ion selectivity) are now called *isotypes* or *isoforms* by analogy to the phenomenon of isoenzymes.

We have previously discussed the *Shaker* isotypes. *Shaker*-derived sequences have also been used to construct probes that have discovered other families of K+ channel in flies, mammals, and other species in other animal classes (Salkoff *et al.*, 1992). These families are so extensive that they challenge investigators to develop appropriate nomenclature for their classification (see Table 3; Chandy and Gutman, 1995). In any case, although they are all to some degree homologous with the original *Shaker* family, they have much higher degrees of sequence homology among members of their class. Functionally, they run the gamut from inactivating A-type channels to noninactivating channels of the classic delayed rectifier sort. Interestingly, there is usually a much higher degree of sequence homology among channels of the same class from different animals than there is between channels of separate classes in the same animal. For example, *Shaker* channels from *Drosophila* are much more homologus to *Shaker* channels of mouse brain than they are to the Drosophila *Shab*-type channels.

Considerable diversity is also found among Na+ and Ca2+ channels (Agnew and Trimmer, 1989; Hofmann *et al.*, 1994). For the main channel-forming α-subunits (see Figs. 7 and 8), there are thought to be at least eight Na+

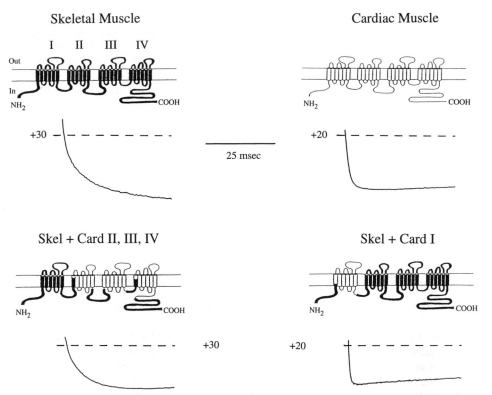

**FIG. 17.** Identification of regions determining activation kinetics in L-type $Ca^{2+}$ channels from skeletal muscle and heart. In these experiments hybrids ("chimeras") of skeletal muscle and cardiac channels were made in various combinations to localize the domains responsible for the differences in $Ca^{2+}$ current activation kinetics in heart and skeletal muscle. cDNA encoding each construct was injected in cultured muscle cells (myotubes) from mutant mice unable to synthesize active L-type channels. Thus these mutant cells would express only currents resulting from the genetically engineered DNA introduced into them. The top two panels show the current kinetics resulting from DNA encoding the wild-type skeletal muscle and cardiac channels. The bottom panels show two chimeras in which current kinetics correlate with the donor of repeat domain I. Most of the other possible combinations of channels were also made, with the same correlation of kinetics with domain I. (Reprinted with permission from *Nature*, Tanabe *et al.*, 1991. Copyright 1991 Macmillan Magazines.)

channel and six $Ca^{2+}$ channel genes in the mammalian genome. Some of these isotypes are expressed very selectively in certain tissues, such as brain, or in only certain cells types or at very specific stages of development.

The term *isotype* is rather nonspecific in that it does not convey a sense of relative relatedness among channels. An evolving system of classification refers to the set of voltage-gated channels as a *superfamily,* while the selectivity types (i.e., $Na^+$, $Ca^{2+}$, or $K^+$ channels) form families. Because of the impressive degree of diversity found for $K^+$ channels (Table 3), investigators have formally proposed a uniform system of nomenclature in which the closely related isoforms such as *Shaker* are referred to as subfamilies. A separate nomenclature for isotypes of $Ca^{2+}$ channel subunits has also been developed. Further consideration of channel classification will undoubtedly be useful in understanding the process of channel evolution.

## A. How Do Isotypes Arise?

In general, two separate mechanisms that create such diversity have been identified. First, there is the expected situation in which distinctly separate genes encode differ-

ent channels in the same beast. Such is largely the situation for mammalian $Na^+$ channel and $Ca^{2+}$ channel isotypes. However, a second mechanism has been found, called *alternative splicing,* in which a given segment of a channel polypeptide may be encoded in several different stretches of genomic DNA known as *exons.* This is the case for *Shaker* in *Drosophilia;* thus, in Fig. 14, the various *Shaker* isotypes are generated by mixing and matching the transcripts of different exons. During transcription some sort of regulatory mechanism exists that decides which one of the alternative coding RNAs for a given segment will be incorporated in the final mRNA. The other undesired homologous segments are then excised from the transcript and the desired RNA segments are spliced together. Currently, there is evidence for the operation of both mechanisms in vertebrates and invertebrates. However, most mammalian systems studied to date appear to rely more heavily on separate genes than alternative splicing, whereas in the fruit fly both mechanisms appear common.

Finally, for all multisubunit channels there is the possibility of diversification through various combinations of each set of subunit isotypes. The best studied examples are the heteromeric formation of channels by different $K^+$ channel

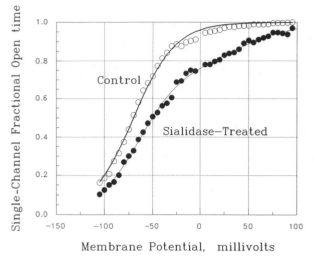

**FIG. 18.** Effect of sialic acid removal on activaton of single Na⁺ channels reconstituted into lipid bilayers. Here purified electric eel sodium channels were treated with a sialidase (neuraminidase) that removes the negatively charged residues of the sugar sialic acid from glycoproteins. Following reincorporation of the treated channel into an artificial lipid bilayer, steady-state activation gating of single channels was measured. Treatment with the enzyme (closed circles) was found to shift the voltage at which channels opened relative to control (open circles). Such a shift is qualitatively consistent with an effect of the sugars on the transmembrane electrical field sensed by the channel-gating mechanism. (Redrawn from Recio-Pinto *et al.,* 1990.)

subunit isotypes (Salkoff *et al.,* 1992). For example, if the same cell synthesized *Shaker* A and *Shaker* B channel subunits, one might find heteromeric complexes of A and B types as well as the homomeric types. In fact, experiments with the oocyte expression system into which both types of mRNA have been injected show that such heteromeric assembly can occur, yielding functional channels with characteristics different from homomultimers of either channel type. Thus, it is theoretically possible that distinct channel properties may be generated by heteromeric assembly of

**TABLE 3** Diversity and Properties of Known Potassium Channels in Fruit Flies and Mammals

| *Drosophila* gene | Homologous mammalian genes (No) | Properties |
|---|---|---|
| *Shaker* | Kv 1.1-1.7 (7) | Extensively spliced in fly; inactivaton fast to slow. |
| *Shab* | Kv 2.1, 2.2 (2) | Limited splicing in fly; intermediate inactivation kinetics. |
| *Shaw* | Kv 3.1-3.4 (4) | Limited splicing in fly and mammal; slow inactivation, delayed rectifier characteristics. |
| *Shal* | Kv 4.1-4.3 (3) | Limited splicing in fly; slow inactivation, delayed rectifier characteristics. |

K⁺ channel subunits. Whether this occurs in nature is currently being investigated.

On the other hand, oocyte experiments show that heteromeric association among subunits from different families (for example, *Shaker* and *Shal*) does not occur. Apparently different families have unique structural domains that allow only homomeric assembly among members of their family. This discovery has led to a recent set of chimeric experiments in which domains among a fly *Shaker* channel and a distantly related mammalian delayed rectifier channel (*drkl*) were exchanged in an attempt to identify the segments specifying assembly specificity (Li *et al.,* 1992). Such a segment was found on the N-terminal part of the peptide. Thus, substitution of the *Shaker* segment into the *drkl* cDNA produced a channel *drkl* subunit that was able to coassemble with native *Shaker*.

## B. Why Are Channels So Diverse in a Given Organism?

One obvious reason for the presence of so many channel isotypes in the same animal is that each isotype has different functional properties that are appropriate to the function of its host cell. For example, high-frequency, repetitive firing requires short-duration action potentials that could be created by rapidly gating Na⁺ channels and fast-inactivating A-type K⁺ channels. In addition, isotypes may be differentially sensitive to intracellular modulators (see Chapter 27) that allow cellular mechanisms to alter the properties of specific channels selectively in response to physiological effectors. On the other hand, many of the discovered isotypes have no significant functional differences among them. In this case it is possible that it is not the differences in the channels themselves that are important, but rather the way the isotypes are expressed. Thus, in a number of cases it has been shown that expression of one isotype over another occurs developmentally or in response to a physiological stimulus (Ribera and Spitzer, 1992). Such specific expression is thought to be controlled by genetic elements outside the coding region for the channel, known as regulatory elements. These stretches of DNA respond specifically to one of many possible soluble factors by increasing (in the case of promoters and enhancers) or inhibiting (silencers) transcription of the channel gene (see Maue *et al.,* 1990). Separate genes, then, potentially allow the cell to express selectively a given channel in response to a given stimulus or at a specific time during development. Finally, there is the possibility that a given isotype sequence encodes information used by the cellular machinery to localize or cluster it in very discrete locations, such as the nodes of Ranvier in myelinated nerve fibers (Dugandzija-Novakovic *et al.,* 1995), neuronal cell bodies, or dendrites (Maletic-Savatic *et al.,* 1994). The diversity of channels and their biological purpose will continue to be a highly active area of cell biology for some time.

## VI. Future Directions

The current molecular biological approaches will no doubt be productive for some time. However, it is likely that better understanding of channel mechanism will require more direct and precise resolution of the higher order structure of channel molecules. The problems inherent in a crystallographic approach have been discussed. However, the problem may be partly solved by recombinantly synthesizing smaller segments of channel proteins that may be amenable to crystal formation. To be useful, these small segments must retain a significant part of the structure they have in the original protein (a serious concern). Alternatively, small segments may be structurally studied using advanced spectroscopic techniques such as magnetic resonance. Finally, some important information has been obtained by direct imaging of purified channels using electron microscopy, and modern instruments that promise greatly improved resolution at the molecular level are being developed. Finally, the study of the genetic regulation of channel expression will be expanded through the further study of regulatory domains in the DNA. This will undoubtedly increase our appreciation of ion channels as important elements in the metabolism of the entire organism.

## VII. Summary

The structure, mechanism, and expression of ion channels are currently intense areas of research interest. The first advances in these areas were made by the application of biochemical techniques to the problems of channel purification. Such studies have told us the size of intact channels and their subunit composition, and have yielded clues regarding functionally important domains (such as pores, gates, modulation sites, and nonprotein modifications). For $Na^+$ and $Ca^{2+}$ channels, attention has been focused on a large polypeptide (so-called alpha peptide) found in each purified preparation. These alpha proteins apparently have all the molecular apparatus required for channel operation, while smaller associated subunits may play other roles in channel modulation, synthesis, or cellular localization. Purified material has also been important in the construction of probes to identify clones of channel-encoding DNA fragments.

Recent work has applied recombinant DNA technology to determine the primary amino acid sequence of channels through cloning. These sequences have revealed widespread sequence homology among the superfamily of voltage-gated ion channels, for example, between $Na^+$ and $Ca^{2+}$ channel alpha polypeptides. Furthermore, the large alpha proteins have four domains of homology within their sequences, suggesting that pore formation occurs via a common "staves of a barrel" architecture. $K^+$ channel sequences, determined via a genetic approach, are relatively shorter, and they partially correspond to a single internal repeat of the larger alpha peptides. Hence, potassium channels are thought to consist of tetramers with each subunit contributing to the pore wall. Other analyses of amino acid sequences have given clues regarding the mechanistically and structurally important domains of ion channels. One analysis was based on thermodynamic (hydropathy) considerations to identify possible membrane-spanning domains. In addition, the sequences were subjected to an empirical analysis that predicts regions of secondary structure, such as helixes and sheets. From such predictions detailed models of channel structure and function have been developed. These models have been tested using immunological and mutagenesis approaches, and candidate domains for ion selectivity, pore wall formation, gating, and modulation have been identified. However, detailed knowledge of the molecular mechanisms of channel function still is limited.

Finally, molecular studies have revealed an astonishing diversity of ion channels at all levels of organization. In particular, the genome of most organisms can express multiple channel isotypes, sometimes coexisting within the same living cell. Much remains to be learned about channel function and the role of channel isotypes in the biology of both simple and complex organisms.

## Bibliography

Agnew, W. S., and Trimmer, J. (1989). Molecular diversity of voltage-sensitive sodium channels. *Annu. Rev. Physiol.* **51,** 401–418.

Aldrich, R. W., Hoshi, T., and Zagotta, W. N. (1990). Differences in gating among amino-terminal variants of *Shaker* potassium channels. *Cold Spring Harbor Symp. Quant. Biol.* **55,** 19–27.

Ariyasu, R. G., Deerinck, T. J., Levinson, S. R., and Ellisman, M. H. (1987). The distribution of ($Na^+ + K^+$) ATPase and Na channels in skeletal muscle and electroplax. *J. Neurocytol.* **16,** 511–522.

Armstrong, C. M., and Bezanilla, F. (1977). Inactivation of the sodium channel. II. Gating current experiments. *J. Gen. Physiol.* **70,** 567–590.

Barchi, R. L. (1983). Protein components of the purified sodium channel from rat skeletal muscle sarcolemma. *J. Neurochem.* **40,** 1377–1385.

Bennett, E., Urcan, M. S., Tinkle, S. S., Koszowski, A. G., and Levinson, S. R. (1997). Contribution of siliac acid to the voltage-dependence of sodium channel gating: A possible electrostatic mechanism. *J. Gen. Physiol.* in press.

Catterall, W. A. (1988). Molecular properties of dihydropyridine sensitive calcium channels in skeletal muscle. *J. Biol. Chem.* **263,** 3535–3538.

Catterall, W. A. (1992). Cellular and molecular biology of voltage-gated sodium channels. *Physiol. Rev.* **72,** Suppl., S15–S48.

Chandy, K. G., and Gutman, G. A. (1995). Voltage-gated $K^+$ channel genes. *In* "CRC Handbook of Receptors and Channels," pp. 1–71. CRC Press, Boca Raton, FL.

Chou, P. Y., and Fasman, G. D. (1978). Empirical predictions of protein conformation. *Annu. Rev. Biochem.* **47,** 251–276.

Dugandzija-Novakovic, S., Koszowski, A. G., Levinson, S. R., and Shrager, P. (1995). Clustering of $Na^+$ channels and node of Ranvier formation in remyelinating axons. *J. Neuroscience* **1,** 492–503.

Ellinor, P. T., Yang, J., Sather, W. A., Zhang, J.-F., and Tsien, R. W. (1995). $Ca^{2+}$ channel selectivity at a single locus for high affinity $Ca^{2+}$ interactions. *Neuron* **15,** 1121–1132.

Guy, H. R., and Seetharamulu, P. (1986). Molecular model of the action potential sodium channel. *Proc. Natl. Acad. Sci. USA* **83,** 508–512.

Heinemann, S. H., Terlau, H., Stuhmer, W., Imoto, K., and Numa, S. (1992). Calcium channel characteristics conferred on the sodium channel by single mutations. *Nature* **356,** 441–443.

Hille, B. (1992). "Ionic channels of Excitable Membranes." Sinauer Associates, Sunderland, MA.

Hofmann, F., Biel, M., and Flockerzi, V. (1994). Molecular basis for $Ca^{2+}$ channel diversity. *Annu. Rev. Neurosci.* **17,** 399–418.

Hoshi, T., Zagotta, W. N., and Aldrich, R. W. (1990). Biophysical and molecular mechanisms of Shaker potassium channel inactivation. *Science* **250,** 533–538.

Isom, L. L., De Jongh, K. S., Patton, D. E., Reber, B. F. X., Offord, J., Charbonneau, H., Walsh, K., Goldin, A. L., and Catterall, W. A. (1992). Primary structure and functional expression of the $\beta_1$ subunit of the rat brain sodium channel. *Science* **256,** 839–842.

Kyte, J., and Doolittle, R. F. (1982). A simple method for displaying the hydropathic character of a protein. *J. Mol. Biol.* **157,** 105–132.

Leonard, J., and Snutch, T. P. (1991). The expression of neurotransmitter receptors and ion channels in *Xenopus* oocytes. *In* "Molecular Neurobiology: A Practical Approach" (D. M. Glover and B. D. Hanes, Eds.), pp. 161–182. Oxford University Press, London/New York.

Levinson, S. R., Thornhill, W. B., Duch, D. S., Recio-Pinto, E., and Urban, B. W. (1990). The role of nonprotein domains in the function and synthesis of voltage-gated sodium channels. *In* "Ion Channels" (T. Narahashi, Ed.), Vol. 2, pp. 33–64. Plenum, New York.

Li, M., Jan, Y. N., and Jan, L. Y. (1992). Specification of subunit assembly by the hydrophilic amino-terminal domain of the shaker potassium channel. *Science* **257,** 1225–1230.

Lopez, G. A., Jan, Y. N., and Jan, L. Y. (1991). Hydrophobic substitution mutations in the S4 sequence alter voltage-dependent gating in Shaker K$^+$ channels. *Neuron* **7,** 327–336.

MacKinnon, R. (1991). Determination of the subunit stoichiometry of a voltage-activated potassium channel. *Nature* **350,** 232–235.

Maletic-Savatic, M., Lenn, N. J., and Trimmer, J. S. (1994). Differential spatiotemporal expression of K$^+$ channel polypeptides in rat hippocampal neurons developing *in situ* and *in vitro*. *J. Neurosci.* **15,** 3840–3851.

Maue, R. A., Kraner, S. D., Goodman, R. H., and Mandel, G. (1990). Neuron-specific expression of the rat brain type II sodium channel gene is directed by upstream regulatory elements. *Neuron* **4,** 223–231.

Miller, C. (Ed.). (1986). "Ion Channel Reconstitution." Plenum, New York.

Miller, J. A., Agnew, W. S., and Levinson, S. R. (1983). Principal glycopeptide of the tetrodotoxin/saxitoxin binding protein from *Electrophorus electricus:* Isolation and partial physical and chemical characterization. *Biochemistry* **22,** 462–470.

Nakai, J., Adams, B. A., Imoto, K., and Beam, K. G. (1994). Critical roles of the S3-segment and S3-S4 linker of repeat I in activation of L-type calcium channels. *Proc. Natl. Acad. Sci. USA* **91,** 1014–1018.

Noda, M., Shimizu, S., Tanabe, T., Takai, T., Kayano, T., Ikeda, T., Takahashi, H., Nakayama, H., Kanaoka, Y., Minamino, N., *et al.* (1984). Primary structure of *Electrophorus electricus* sodium channel deduced from cDNA sequence. *Nature* **312,** 121–127.

Numa, S., and Noda, M., (1986). Molecular structure of sodium channels. *Ann. NY Acad. Sci.* **479,** 338–355.

Papazian, D. M., Timpe, L. C., Jan, Y. N., and Jan, L. Y. (1987). Cloning of genomic and complementary DNA from *Shaker,* a putative potassium channel gene from *Drosophila. Science* **237,** 749–753.

Papazian, D. M., Timpe, L. C., Jan, Y. N., and Jan, L. Y. (1991). Alteration of voltage-dependence of Shaker potassium channel by mutations in the S4 sequence. *Nature* **349,** 305–310.

Pongs, O. (1992). Molecular biology of voltage-dependent potassium channels. *Physiol. Rev.* **72,** Suppl., S69–S88.

Recio-Pinto, E., Duch, D. S., Urban, B. W., Thornhill, W. B., and Levinson, S. R. (1990). Neuraminidase treatment modifies the function of eel sodium channels reconstituted in planar lipid bilayers. *Neuron* **5,** 675–684.

Ribera, A. B., and Spitzer, N. C. (1992). Developmental regulation of potassium channels and the impact on neuronal differentiation. *In* "Ion Channels" (T. Narahashi, Ed.), Vol. 3. Plenum, New York.

Salkoff, L., Baker, K., Butler, A., Covarrubias, M., Pak, M. D., and Wei, A. (1992). An essential set of K$^+$ channels conserved in flies, mice, and humans. *Trends Neurosci.* **15,** 161–166.

Sharp, A. H., Imagawa, T., Leung, A. T., and Campbell, K. P. (1987). Identification and characterization of the dihydropyridine-binding subunit of the skeletal muscle dihydropyridine receptor. *J. Biol. Chem.* **262,** 12,309–12,315.

Sigworth, F. J. (1995). Charge movement in the sodium channel. *J. Gen. Physiol.* **106,** 1047–1051.

Stuhmer, W., Conti, F., Suzuki, H., Wang, X., Noda, M., Yahagi, N., Kubo, H., and Numa, S. (1989). Structural parts involved in the activation and inactivation of sodium channels. *Nature* **339,** 597–603.

Tanabe, T., Adams, B. A., Numa, S., and Beam, K. G. (1991). Repeat I of the dihydropyridine receptor is critical in determining sodium channel activation kinetics. *Nature* **352,** 800–803.

Tanabe, T., Takeshima, H., Mikami, A., Flockerzi, V., Takahashi, H., Kangawa, K., Kojima, M., Matsuo, H., Hirose, T., and Numa, S. (1987). Primary structure of the receptor for calcium channel blockers from skeletal muscle. *Nature* **328,** 313–318.

Thornhill, W. B., Wu, M. B., Wu, X., Morgan, P. T., and Margiotta, J. F. (1996). Expression of Kv1.1 delayed rectifier in *lec* mutant Chinese hamster ovary cells reveals a role for sialidation in channel function. *J. Biol. Chem.* **271,** 19,093–19,098.

Timpe, L. C., Jan, Y. N., and Jan, L. Y. (1988). Four cDNA clones frm the *Shaker* locus of *Drosophila* induce kinetically distinct A-type potassium current in *Xenopus* oocytes. *Neuron* **1,** 659–667.

Trimmer, J. S. (1991). Immunological identification and characterization of a delayed rectifier K$^+$ channel polypeptide in rat brain. *Proc. Natl. Acad. Sci. USA* **88,** 10,764–10,768.

Trimmer, J. S., Cooperman, S. S., Tomiko, S. A., Zhou, J., Crean, S. M., Boyle, M. B., Kallen, R. G., Sheng, Z., Barchi, R. L., Sigworth, F. J., Goodman, R. H., Agnew, W. S., and Mandel, M. (1989). Primary structure and functional expresson of a mammalian skeletal muscle sodium channel. *Neuron* **3,** 33–49.

Vassilev, P., Scheuer, T., and Catterall, W. A. (1989). Inhibition of inactivation of single sodium channels by a site-directed antibody. *Proc. Natl. Acad. Sci. USA* **86,** 8147–8151.

Wang, J., Ellinor, P. T., Sather, W. A., Zhang, J.-F., and Tsien, R. W. (1993). Molecular determinants of Ca$^{2+}$ selectivity and ion permeation in L-type Ca$^{2+}$ channels. *Nature* **366,** 158–161.

Yool, A. J., and Schwarz, T. L. (1991). Alteration of ionic selectivity of a K$^+$ channel by mutation of the H5 region. *Nature* **349,** 700–704.

Zagotta, W. N., Hoshi, T., and Aldrich, R. W. (1990). Restoration of inactivation in mutants of Shaker potassium channels by a peptide derived from ShB [see comments]. *Science* **250,** 568–571.

Michael M. Behbehani

# 27

# Biology of Neurons

## I. Introduction

The general functions of a neuron are to integrate the chemical (and in case of electrical synapses, the electrical) afferent signals and convert the result to action potentials (APs), and to change the AP to chemical release of neurotransmitters at the nerve terminals. To accomplish these tasks, the neuron must be able to (1) synthesize receptors for all the transmitters it responds to and move these receptors to appropriate locations, (2) maintain ion gradient by operating a variety of ion pumps and exchangers, (3) synthesize the neurotransmitters it releases and all the enzymes necessary for their synthesis, (4) transport the transmitters from the cell body to the synaptic terminals, and (5) transport macromolecules and membrane segments to and from the synaptic terminals and the cell body. A close look at the functions previously listed shows that the neuron must manufacture distinct types of molecules. Since the majority of macromolecules used by the neuron are proteins, the cell has developed a complex machinery designed to manufacture specific proteins.

## II. Ultrastructure

A nerve cell consists of a cell body (the *soma*), a *dendritic tree*, an *axon*, and *synaptic terminals*. Although all neurons have the same types of components, there are significant differences in the morphology of nerve cells. Based on the number of processes that originate from the soma, the neurons have been classified as *unipolar*, *bipolar*, or *multipolar* (Fig. 1). The unipolar cells have only a single process that originates from the cell body. Unipolar cells are found in the invertebrate and contain an axon that originates from a specialized dendrite instead of the soma. The sensory neurons in the vertebrates are examples of bipolar cells. These neurons have two processes that originate from the soma: One process, the axon, connects the soma to the central nervous system, and the second process, the sensory nerve, ends in specialized sensory receptors.

The majority of cells within the central nervous system are multipolar. The axons of these neurons initiate from a specialized region of the soma, called the *axon hillock*, and end in multiple branches containing the synaptic terminals.

Nerve cells have a large nucleus and a nucleolus. The nucleus is enclosed by a specific type of membrane called the *nuclear envelope*. This envelope contains pores called *nuclear pores*. These pores are open to the cytosolic compartment of the cell. The nucleus of the cell is in continuity with the *endoplasmic reticulum* (ER), which is a major organelle of the cell. The DNA of the neuron is located in the nucleus and is transcribed in the nucleus to make RNA for the synthesis of a large variety of proteins and macromolecules. The messenger RNA (mRNAs) that encodes macromolecules is released from the nucleus, move through the nuclear pores, are combined with *ribosomes*, and form *polysomes* (Fig. 2). In general, three types of proteins are synthesized by the mRNAs released from the nucleus.

One class of proteins is the *cytosolic* proteins. These proteins are soluble proteins, and their ribosomes remain in the cytosol (the cytoplasm minus organelles). Among the major proteins in this class are those that form the *cytoskeleton* of the cell, including *neurofilaments*, *tubulines*, *actins*, *actin-associated* proteins, and enzymes that catalyze reactions within the cell.

The second class of proteins is the nuclear proteins and mitochondrial proteins. The ribosomes for these proteins are attached to the ER and translocate into the ER lumen as they are synthesized. In this process, a branch made of carbohydrate rich in mannose may be added to each protein to form an N-linked glycosyl chain. The attachment of the ribosomes of these proteins to the ER produces roughness of the ER for this reason, these classes of the ER are called the *rough endoplasmic reticulum* (RER). The ribosomal RNA in the RER stains with several basic dyes including cresyl violet, toluidine blue, and methylene blue. Because of this histological property, the RER is also called the *Nissl substance*.

The third class of proteins encoded by nuclear mRNA is the proteins that are destined to become constituents of

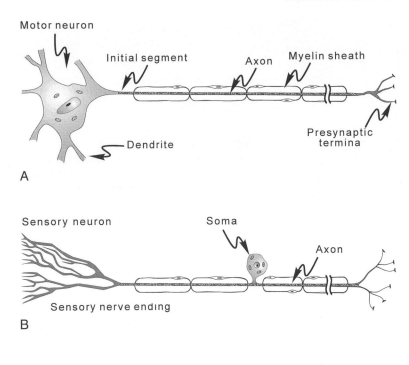

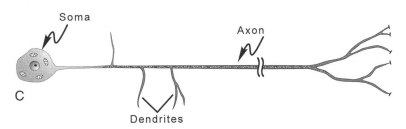

**FIG. 1.**   Morphological characteristics of different types of neurons. (A) A typical motor neuron. This type of cell consists of a large soma, a large myelinated axon, and synaptic terminals. The initial segment is the site where APs are generated. (B) A sensory neuron such as dorsal root ganglion cells. This type of neuron is a bipolar cell with one process located in the periphery and the second branch projecting into the central nervous system. The terminals of the peripheral branch can be sensory receptors, for example, nociceptors, or can innervate a variety of other types of sensory receptors. (C) The type of nerve cell found in invertebrates. The axons of these cells originate from dendrites that are connected to the soma. [Adapted from Hall, Z. W. (1992). *An Introduction to Molecular Neurobiology.* Sinauer Associates, Sunderland, MA.]

the cell membrane and membranes of the organelles within the neuron, and secretory proteins that are released at the nerve terminals. These proteins include neuropeptides that act as neurotransmitters or neuromodulators and growth factors. The polysomes for these proteins attach themselves to the cytoplasmic side of the RER.

## III.  Neuronal Cytoskeleton

The neuronal cytoskeleton is essential for growth of neurons, during development, for arborization of dendritic branches, for maintenance of the shape of the neuron, and for transfer of macromolecules from the cell body to the terminals and from the terminals to the cell body. Consider-

ing the complexity of neurons, each of these processes is highly regulated and involves synthesis and polymerization of a variety of proteins.

The cytoskeleton consists of filaments that cross-link to form a tight meshwork. This network retains its shape even in the absence of the cell membrane. For example, in electrophysiological recording from squid axon, the cytoplasm is sometimes extruded from the axon. The extruded cytoplasm, however, retains its cylindrical shape and many of its function as long as adenosine triphosphate (ATP) can be supplied.

The cytoskeleton contains three major proteins: microfilaments, neurofilaments, and microtubules. These proteins are essential for the axoplasmic flow, which is essential for the survival of the nerve cell.

## A. Microfilaments

*Microfilaments* are formed from *actin*, a 43-kDa globular protein. It is about 7 nm in diameter and is composed of two strands (each 3–5 nm in diameter) of polymerized globular (G) actin monomers that are arranged in a helix. These strands are asymmetric, and each monomer has a pointed tip and a barbed end. Because the globular monomers are asymmetric, these monomers polymerize tip to tail and form polar structures. Some axonal microfilaments are oriented longitudinally. The microfilaments are attached to the plasma membrane through associated proteins linked to actin, including *spectrin, ankyrin, vinculin,* and *talin*. The principal anchoring protein in neurons and other cells of the body is *fodrin* (or neural *spectrin*), some of which is transported down the axon at the same velocity as actin. Microfilaments are also able to interact with proteins in the extracellular matrix (e.g., laminin, fibronectin) through their association with a family of membrane-spanning proteins called *integrins*.

## B. Neurofilaments

*Neurofilaments* are the most abundant fibrillar components in axons and are the "bones" of the cytoskeleton. These filaments are about 10 nm in diameter and are related to a family of proteins that includes *vimentin*, glial fibrillary acidic protein, and *desmin*. Neurofilaments are composed of three polypeptide subunits: NF-H (112 kDa), NF-M (120 kDa), and NG-L (68 kDa). Neurofilaments are built with fibers that twist around each other to produce coils of increasing thickness. The thinnest units are monomers that form coiled-coil heterodimers. These dimers form a tetrameric complex that becomes the protofilament.

Two *protofilaments* become a *protofibril,* and four protofibrils are helically twisted to form the 10-nm neurofilament.

## C. Microtubules

*Microtubules* are long polar polymers about 25 nm in diameter. They are usually constructed of 13 linearly arranged $\alpha$-tubulin and $\beta$-*tubulin* dimers called protofilaments. Each filament is about 5 nm in diameter. The monomeric subunit of microtubules is *tubulin*. Each monomer binds two guanosine triphosphate (GTP) molecules, or one GTP and one guanosine diphosphate (GDP) molecule. In the axon, they are oriented longitudinally with polarity always in the same direction. This arrangement is presumably important for the directional specificities of the two forms of fast axonal transport. Although axonal or dendritic microtubules can be as long as 0.1 mm, they usually do not extend the full length of the axon or dendrite and are not continuous with microtubules in the cell body.

Both monomeric and polymeric components of tubulin and actin are present in the axon. The self-assembly of tubulin and actin requires triphosphates, and depends on the concentration of each component. At a concentration called the *critical concentration*, the rate of subunit added to the polymeric component is the same as the rate at which the polymeric component is disassembled. If the concentration of *monomeric subunits* is larger than its critical concentration, the monomers assemble into polymeric units. However, if the concentration of monomeric subunits is less than the critical concentration, the polymeric subunits disassemble. During *polymerization,* one end of the filament grows faster than the other. This end is called the *plus end*. The other end is called the *minus end.* Under steady state, the net addition of the monomers at the plus

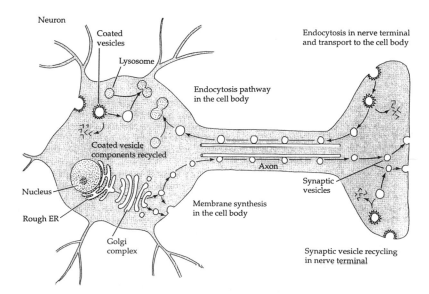

**FIG. 2.** The cell nucleus and structures involved in transportation of macromolecules that are synthesized within the nucleus. The figure shows the nucleus and nucleolus, the rough and smooth endoplasmic reticulum, lysosomes, ribosomes, Golgi complex, and mitochondria.

end is the same as the rate of dissociation from the minus end. This process can occur simultaneously and is called *treadmilling*.

## IV. Axoplasmic Flow

Macromolecules that are needed for the function of the nerve cell are synthesized in the cell body. For the nerve terminals to have access to these macromolecules, they have to be transported from the cell body to the terminals. Typically, cell bodies and nerve terminals are at considerable distances from each other. For example, the cell body of a spinal motor neuron that innervates muscles around the ankle in a 180-cm-tall man is more than 1 m away from its terminals. The separation between cell body and nerve terminals requires complex processes that have specialized components (Fig. 3).

The macromolecules move through the axon by three types of processes: the fast anterograde (forward moving, i.e., moving from cell body to the terminals) axonal transport; fast retrograde (backward moving, i.e., moving from the nerve terminals to the cell body) axonal transport; and slow axoplasmic flow. There are considerable differences in the speed at which these processes transport material through the axon and in the types of material transported by each process.

### A. Fast Axoplasmic Flow

Synaptic vesicles or their precursors are transported to the nerve terminals, in most cases, by the fast axoplasmic flow that moves at a rate of 200–400 mm/day. Particles move in a stop-and-go intermittent or saltatory fashion. Fast anterograde transport can still occur in axons that are severed from their cell bodies, indicating that this type of transport is not dependent on the cell body. In addition, fast axoplasmic flow is independent of protein synthesis. In contrast, this process is critically *dependent on oxidative metabolism*.

The fast anterograde transport is involved in transport of tubulovesicular structures, synaptic vesicles, membrane-associated proteins, neuropeptides, neurotransmitters, and associated enzymes. Fast anterograde transport in the axon is based on *microtubules* that provide a stationary track on which specific organelles move in a saltatory fashion. A major component of these microtubules is a motor molecule called *kinesin*. Kinesin is a rod-shaped tetrameric ATPase consisting of two alpha ($\alpha$) subunits and two beta ($\beta$) subunits. Each of the subunits has a molecular weight of about 115–130 kDa and consists of 340 amino acids and an N-terminal force-generating domain. Each of the light chains ($\beta$ subunit) has a molecular weight of 62–70 kDa. The holoenzyme $\alpha_2\beta_2$ has a molecular weight of about 270,000. High-resolution electron microscopic (EM) analysis indicates that kinesin is an 80-nm rod with two globular

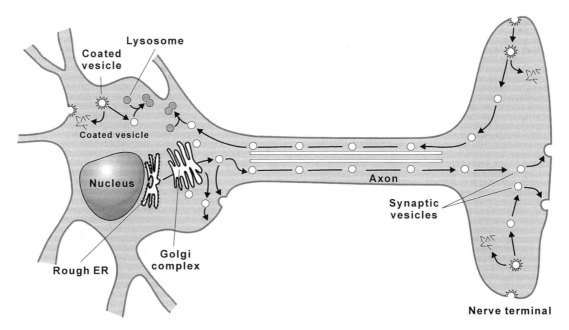

**FIG. 3.** Trafficking of macromolecules within the neuron. Macromolecules that are synthesized by the nucleus are transfered to the endoplasmic reticulum and then to the Golgi complex. Macromolecules that are needed in the axon terminals are transported by anterograde and slow axoplasmic flow. Proteins and macromolecules are transported from the terminals to the cell body by the retrograde axoplasmic flow. [Adapted from Hall, Z. W. (1992). *An Introduction to Molecular Neurobiology*. Sinauer Associates, Sunderland, MA.]

heads approximately 10 nm in diameter. The two heavy chains are arranged in parallel, and there is a kink in the middle of each chain. The light chains form a fan-shaped tail. The ATP-binding site is located in the head region and is also the site of binding to the microtubules. The tail region contains the binding site to the membrane surface. There is evidence that kinesin is anchored to the membrane by a protein called *kinectin*.

Kinesins form the cross-bridges between the moving membranous organelles, which have the appearance of little feet walking along the microtubules (Fig. 4). It has been postulated that the kinesin molecule, which is analogous to myosin in structure, interacts with other proteins such as *actin* or *dynein* to form cross-bridges (see Chapters 55 and 56). It has been proposed that as kinesin hydrolyzes a molecule of ATP, it undergo a series of conformational changes. When the kinesin is attached to a filament, one of the configurational changes produces strain into the proteins. When this strain is relieved, it causes the kinesin to move. Studies of the motility of kinesin indicate that during each of ATP hydrolysis, kinesin moves a distance approximately equal to its length toward the next binding site on the filament. Recent studies have shown that the force generated by a single kinesin molecule is approximately 5 pN.

### B. Fast Retrograde Transport

Materials are transported from nerve terminals to the cell body either for degradation or for restoration and reuse. The rate of fast retrograde transport is 50–100 mm/day. Before being transported these materials are packaged in large membrane-bound organelles that are part of the lysosomal system. The materials that are transported by the fast retrograde transport are prelysosomal vesicles,

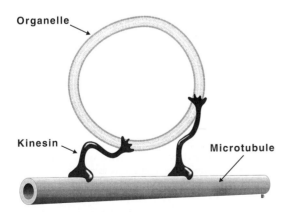

**FIG. 4.** A proposed mechanism by which the motor molecule of the retrograde axoplasmic flow, kinesin, transports materials. Molecules that are transported by anterograde transport system bind to kinesin molecules. Kinesins form cross-bridges with microtubules at the plus end and translocate on the microtubules toward the minus end. [Adapted from Hall, Z. W. (1992). *An Introduction to Molecular Neurobiology.* Sinauer Associates, Sunderland, MA.]

multivesicular bodies, multilamellar bodies, growth factors, and recycled proteins. The retrograde transport also utilizes microtubules. However, the motor molecule for fast retrograde transport is a form of *dynein*, which also is a microtubule-associated ATPase (MAP-1C). Dynein is approximately 40 nm long and is a two-headed structure with two complex polypeptides. It has two heavy chains and several light chains. Dynein is distributed in punctate fashion in neuronal tissue, and it is associated with membrane-bound organelles.

### C. Slow Axonal Transport

Slow axoplasmic flow is used for transport of *cytoskeletal elements* and soluble proteins. This type of transport is more complex than fast anterograde transport. It consists of at least two kinetic components. The slow component travels at a rate of 0.1–2.5 mm/day, and carries the subunits of neurofilaments and the tubulin subunits of the microtubule. The faster component of slow axoplasmic flow travels at a rate of 2–6 mm/day and transports *actin, clathrin* and associated proteins, *spectrin, glycolytic enzymes,* and *calmodulin.* Approximately 2–4% of the protein transported by this process is actin (MW 43,000) that polymerizes to form microfilaments. Other proteins that are transported by this process are calmodulin and neural myosin or a myosin-like protein, clathrin (180,000-MW protein involved in coating of the synaptic vesicles for recycling).

### D. Regulatory Mechanisms for Axonal Transport

The factors that regulate axonal transport and allow the delivery of transported macromolecules to their appropriate location are not well understood. In the case of synaptic vesicles, phosphorylation of *synapsin-I* is an important mechanism that may be involved in the vesicles being transported by the fast axoplasmic flow to localize at the synaptic terminals. Synapsin-I can exist in both phosphorylated and dephosphorylated states. When it is dephosphorylated, *synapsin-I* binds tightly to synaptic vesicles and, upon phosphorylation, it is released from the synaptic vesicles. Phosphorylated synapsin has no effect on axonal transport; however, when synapsin becomes dephosphorylated, it inhibits movement of organelles that are bound to the microfilaments. It has been proposed that dephosphorylated synapsin binds to the passing vesicles and confines them in the presynaptic terminals. It has been suggested that phosphorylation and dephosphorylation are general mechanisms that allow localization of macromolecules to their specialized locations in the neuron.

## V. Summary

The neuron consists of four major components: the soma, the dendritic tree, the axon, and the axon terminals. The DNA of the neuron is located in the nucleus and is transcribed in the nucleus to make RNA for the synthesis of a large variety of proteins and macromolecules. One class of proteins is the cytosolic proteins. Among the major proteins in this class are neurofilaments, tubulines, actins, actin-associated proteins, and enzymes that catalyze reactions within the cell. The second class of proteins is the nuclear proteins and mitochondrial proteins. The third class of proteins encoded by nuclear mRNA is the proteins that are destined to become constituents of the cell membrane and membranes of the organelles within the neuron, and secretory proteins that are released at the nerve terminals. These proteins include neuropeptides that act as neurotransmitters or neuromodulators and growth factors.

Neurons have cytoskeleton that is essential for their growth during development and for arborization of dendritic branches. The cytoskeleton consists of filaments that cross-link to form a tight meshwork. This network retains its shape even in the absence of the cell membrane. The cytoskeleton contains three major proteins: microfilaments, neurofilaments, and microtubules. These proteins are essential for the axoplasmic flow, which is essential for the survival of the nerve cell.

Microfilaments are formed from actin. They are attached to the plasma membrane through associated proteins linked to actin, including spectrin, ankyrin, vinculin, and talin. The principal anchoring protein in neurons and other cells of the body is fodrin (or neural spectrin).

Neurofilaments are the most abundant fibrillar components in axons and are the "bones" of the cytoskeleton. These filaments are about 10 nm in diameter and are related to a family of proteins that includes vimentin, glial fibrillary acidic protein, and desmin. Microtubules are long polar polymers about 25 nm in diameter. In the axon, they are oriented longitudinally with polarity always in the same direction. This arrangement is presumably important for the directional specificities of the two forms of fast axonal transport.

Macromolecules that are needed for the function of the nerve cell are synthesized in the cell body and transported to the other regions of the cell. There are three types of transport mechanisms: the fast anterograde axonal transport, fast retrograde axonal transport, and slow axoplasmic flow.

Fast axoplasmic flow that moves at a rate of 200–400 mm/day is involved in the transport of synaptic vesicles or their precursors. Fast axoplasmic flow is independent of protein synthesis but is critically dependent on oxidative metabolism. The fast anterograde transport is involved in transport of tubulovesicular structures, synaptic vesicles, membrane-associated proteins, neuropeptides, neurotransmitters, and associated enzymes. The fast anterograde transport in the axon is based on microtubules that provide a stationary track on which specific organelles move in a saltatory fashion. A major component of these microtubules is a motor molecule called kinesin.

The fast retrograde transports macromolecules at a rate of 50–100 mm/day. The major macromolecules that are transported by fast retrograde transport are prelysosomal vesicles, multivesicular bodies, multilamellar bodies, growth factors, and recycled proteins. The retrograde transport also uses microtubules. However, the motor molecule for fast retrograde transport is a form of dynein.

Slow axoplasmic flow is used for transport of cytoskeletal elements and soluble proteins. This type of transport is more complex than fast anterograde. It consists of at least two kinetic components. The slow component travels at a rate of 0.1–2.5 mm/day, and carries the subunits of neurofilaments and the tubulin subunits of the microtubule. The faster component of slow axoplasmic flow travels at a rate of 2–6 mm/day and transports actin, clathrin and associated proteins, spectrin, glycolytic enzymes, and calmodulin. Approximately 2–4% of the protein transported by this process is actin that polymerizes to form microfilaments. Other proteins that are transported by this process are calmodulin and neural myosin or a myosin-like protein, clathrin.

The factors that regulate axonal transport and allow the delivery of transported macromolecules to their appropriate location are not well understood. In the case of synaptic vesicles, phosphorylation of synapsin-I is an important mechanism that may be involved in the vesicles being transported by the fast axoplasmic flow to localize at the synaptic terminals.

## References

Bloom, G. S., Wagner, M. C., Pfister, K. K., and Brady, S. T. (1988). Native structure and physical properties of bovine brain kinesin and identification of the ATP-binding subunit polypeptide. *Biochemistry* **27,** 3409–3416.

Bomsel, M., Parton, R., Kuznetsov, S. A., Schroer, T. A., and Gruenberg, J. (1990). Microtubule- and motor-dependent fusion *in vitro* between apical and basolateral endocytic vesicles from MDCK cells. *Cell* **62,** 719–31.

Brady, S. T. (1991). Molecular motors in the nervous system. *Neuron* **7,** 521–533.

Dabora, S. L., and Sheetz, M. P. (1988). Cultured cell extracts support organelle movement on microtubules in vitro. *Cell Motil. Cytoskeleton* **10,** 482–495.

Gelfand, V. I., and Bershadsky, A. D. (1991). Microtubule dynamics: Mechanism, regulation, and function. *Annu. Rev. Cell Biol.* **7,** 93–116.

Gilbert, S. P., Allen, R. D., and Sloboda, R. D. (1985). Translocation of vesicles from squid axoplasm on flagellar microtubules. *Nature* **315,** 245–248.

Hirokawa, N. (1993). Mechanism of axonal transport. Identification of new molecular motors and regulations of transports. *Neurosci. Res.* **18,** 1–9.

Hirokawa, N., Bloom, G. S., and Vallee, R. B. (1985). Cytoskeletal architecture and immunocytochemical localization of microtubule-associated proteins in regions of axons associated with rapid axonal transport: the beta,beta′-iminodipropionitrile-intoxicated axon as a model system. *J. Cell Biol.* **101,** 227–239.

Hirokawa, N., Pfister, K. K., Yorifuji, H., Wagner, M. C., Brady, S. T., and Bloom, G. S. (1989). Submolecular domains of bovine brain kinesin identified by electron microscopy and monoclonal antibody decoration. *Cell* **56,** 867–878.

Hirokawa, N., Funakoshi, T., Sato-Harada, R., and Kanai, Y. (1996). Selective stabilization of tau in axons and microtubule-associated protein 2C in cell bodies and dendrites contributes to polarized localization of cytoskeletal proteins in mature neurons. *J. Cell Biol.* **132,** 667–679.

Jacob, J. M., and O'Donoghue, D. L. (1995). Direct measurement of fast axonal transport rates in corticospinal axons of the adult rat. *Neurosci. Lett.* **197,** 17–20.

Kumar, J., Yu, H., and Sheetz, M. P. (1995). Kinectin, an essential anchor for kinesin-driven vesicle motility. *Science* **267,** 1834–1837.

Li, L., Chin, L. S., Shupliakov, O., Brodin, L., Sihra, T. S., Hvalby, O., Jensen, V., Zheng, D., McNamara, J. O., Greengard, P., *et al.* (1995). Impairment of synaptic vesicle clustering and of synaptic transmission, and increased seizure propensity, in synapsin I-deficient mice. *Proc. Natl. Acad. Sci. USA* **92,** 9235–9239.

McLean, W. G., Kanje, M., and Remgard, P. (1993). An *in vitro* system for the study of slow axonal transport. *Brain Res.* **613,** 295–299.

Rodionov, V. I., Gyoeva, F. K., Tanaka, E., Bershadsky, A. D., Vasiliev, J. M., and Gelfand, V. I. (1993). Microtubule-dependent control of cell shape and pseudopodial activity is inhibited by the antibody to kinesin motor domain. *J. Cell Biol.* **123,** Part 2, 1811–1820.

Terada, S., Nakata, T., Peterson, A. C., Hirokawa, N. (1996). Visualization of slow axonal transport in vivo. *Science* **273,** 784–788.

Terasaki, M., Schmidek, A., Galbraith, J. A., Gallant, P. E., and Reese, T. S. (1995). Transport of cytoskeletal elements in the squid giant axon. *Proc. Natl. Acad. Sci. USA* **92,** 11500–11503.

Vale, R. D. (1987). Intracellular transport using microtubule-based motors. *Annu. Rev. Cell Biol.* **3,** 347–378.

Vale, R. D., Reese, T. S., and Sheetz, M. P. (1985). Identification of a novel force-generating protein, kinesin, involved in microtubule-based motility. *Cell* **42,** 39–50.

Valee, R. B., and Shpetner, H. S. (1991). Mechanisms of fast and slow axonal transport. *Annu. Rev. Neurosci.* **14,** 59–92.

Viancour, T. A., and Kreiter, N. A. (1993). Vesicular fast axonal transport rates in young and old rat axons. *Brain Res.* **628,** 209–217.

Wang, C., Asai, D. J., and Robinson, K. R. (1995). Retrograde but not anterograde bead movement in intact axons requires dynein. *J. Neurobiol.* **27,** 216–226.

Yang, J. T., Laymon, R. A., and Goldstein, L. S. (1989). A three-domain structure of kinesin heavy chain revealed by DNA sequence and microtubule binding analyses. *Cell* **56,** 879–889.

*Bernd Nilius*

# 28

# Ion Channels in Nonexcitable Cells

## I. Introduction

For a long time, electrophysiological methods focused on cells that generate action potentials (APs) by activation of voltage-dependent ion channels such as $Na^+$, $K^+$, and $Ca^{2+}$ channels. These cells—neurons, nerve fibers, cardiac cells, and skeletal muscle fibers—were conventionally named **excitable cells.** Small cells, like most of the blood cells, or extremely flat cells, such as endothelial cells, are not able to evoke regenerative APs. These cells were also not accessible to the hitherto-applied microelectrode techniques. Only with the advent of the patch-clamp technique developed by the 1991 Nobel prize winners Erwin Neher and Bert Sakmann did these cells gain more attention from electrophysiologists. The long-standing dogma, namely, that excitable cells express voltage-gated ion channels and that nonexcitable cells lack those channels, is increasingly challenged by the discovery of nearly all kinds of channels in cells that do not generate APs under physiological conditions. However, nonexcitable cells possess a tremendous variety of functionally important ion channels that make them "exciting" in the best sense of this word.

This chapter focuses only on ion channels in epithelial cells, especially exocrine secretory cells, in endothelium, tumor cells, and endocrine cells, for which a physiological function can be shown.

## II. Different Types of Ion Channels in Nonexcitable Cells

### A. Amiloride-Sensitive $Na^+$ Channels

In epithelial cells that absorb sodium, for example, from renal tubules, distal colon, sweat duct, and tracheal epithelium, the channel that plays the key role is about 20 times more permeable for $Na^+$ than $K^+$. This channel is expressed at the luminal (apical, mucosal) side of these tissues and allows passive $Na^+$ entry (Fig. 1A). The maintenance of $Na^+$ channels at this side of the cell is an essential feature of $Na^+$-reabsorbing epithelial cells and seems to occur via binding of a channel subunit to the apical cytoskeleton, possibly to the cytoskeletal protein **ankyrin.** At the abluminal side (interstitial, serosal, or blood side), $Na^+,K^+$-ATPase pumps the luminally absorbed $Na^+$ out of the cell. Thus, the $Na^+$ channel provides a vectorial transport of $Na^+$ through these cells.

The trademark of the $Na^+$ channel in sodium-absorbing cells, however, is the **amiloride block.** This block results when the open channel is plugged by amiloride at a high-affinity site ($K_i$ between 0.1 and 0.3 $\mu M$, Fig. 1B). However, the situation is more complex: Epithelial $Na^+$ channels can be differentiated into high (H, $K_i < 1$ $\mu M$) and low (L, $K_i > 1$ $\mu M$) affinity. Both types of channels differ also in the block by eicosenoic acids with H channels being insensitive to **EIPA** (pentanoic form).

This $Na^+$ pathway is characterized at the level of single channels, which can be divided into the groups of 5 pS (Na5), 9 pS (Na9), and 28 pS (Na28). The very high-affinity amiloride-sensitive channel, Na5, is the typical epithelial channel that is much more selective for $Na^+$ over $K^+$ ($<10:1$). The conductance is $Na^+$-dependent and saturates with increasing extracellular $Na^+$ concentrations ($K_m$ between 20 and 75 m$M$). However, increased cytosolic $Na^+$ also blocks the channel. $Na^+$ produces a feedback inhibition of the $Na^+$ channel itself. For example, block of the $Na^+$–$K^+$ pump leads to elevation of $[Na^+]_i$ and inhibition of the $Na^+$ channel. Also, a decrease in intracellular pH (pH$_i$) and an increase in the intracellular $Ca^{2+}$ concentration, $[Ca^{2+}]_i$, blocks this channel. This $Ca^{2+}$-induced block, however, is decreased if the intracellular $H^+$ concentration is increased. Thus, intracellular $Ca^{2+}$ blocks the channel by binding to a cytoplasmic site, which is inhibited by protonation (Fig. 1C).

Another functionally important trademark of this channel is its hormonal activation. Hormones such as ADH (**antidiuretic hormone** or **vasopressin**) and **aldosterone** stimulate $Na^+$ entry in $Na^+$-reabsorbing epithelial cells. ADH stimulation is accompanied by an increase in the adenylate cyclase activity and an elevation of intracellular cAMP. The result of hormone action is an increase in

**436**

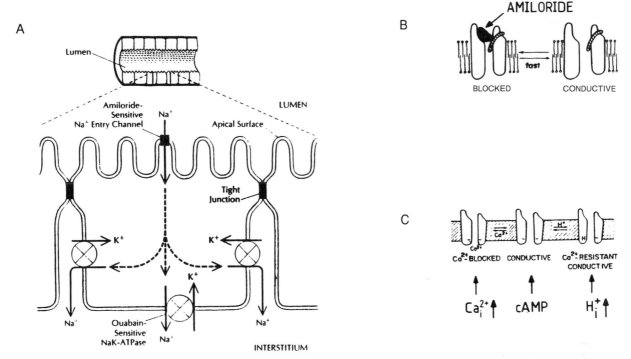

**FIG. 1.** Biology of the amiloride-sensitive Na⁺ channel. (A) The channel is localized at the luminal membrane of Na⁺-absorbing cells. Na⁺–K⁺ pumps are localized opposite at the interstitial side. (B) The trademark of the channel is the fast high-affinity block of the open channel by amiloride. (C) The channel is blocked by intracellular $Ca^{2+}$ in the micromolar range. This $Ca^{2+}$ block is removed by a decrease of the intracellular pH. cAMP activates the channel or transfers it into the conductive state. (Modified from Benos, 1989.)

the number of functional channels in the membrane of epithelial cells by the plasma lemmal insertion of Na⁺ channels from a pool of channel-bearing vesicles from the endoplasmic reticulum or by direct activation of "sleeping channels" in the plasmalemma that do not open. A possible step in this functionally important mechanism is a cAMP-dependent phosphorylation (cyclic-3',5'-adenosine monophosphate: cAMP). Intracellular cAMP activates the channel due to stimulation of a cAMP-dependent protein kinase (protein kinase A: PKA) that catalyzes phosphorylation of the pore-forming channel or of one or more subunits. PKA may also act on regulatory proteins that facilitate insertion of the Na⁺ channels into the plasma membrane. Conversely, protein kinase C (PKC) inhibits up-regulation of the channel.

Modulation of the amiloride-sensitive Na⁺ channel can also occur by a variety of other signals such as hormones **[atrial natriuretic factor (ANF)],** by mechanical forces (stretch), by changes in osmolality of the basolateral fluid, and by PKA, PKC, tyrosine kinases, G proteins, and leukotriens. As discussed for ADH, modulation mainly occurs by up- or down-regulation of the number of available channels but also by changes in the open probability. The effect of G proteins may be mediated by phospholipase $A_2$ and thus via arachidonic acid and its metabolites (leukotrienes).

The molecular biology of this important channel turned out to be very difficult. The breakthrough occurred recently

(1993/1994) and was made by the group of B. Rossier in Switzerland. This group identified three subunits of the channel, αrENaC (α-subunit of the renal epithelial Na⁺ channel), βrENaC, and γrENaC. The α-subunit forms the channel pore. The nucleotide sequence predicted a molecular mass of 79 kDa; the translation product migrates at 92 kDa depending on glycosylation. This channel shows all the functional properties of the H channel (Na5). Although the α-subunit already induced functional channels when expressed in oocytes, coexpression of the other β- (72-kDa), and γ- (75-kDa) subunit resulted in the largest expression and currents. The **amiloride-sensitive Na⁺ channel** is probably formed by a heterotrimer. All subunits are characterized by two membrane-spanning regions, a cysteine-rich long extracellular loop between the two helices. The homology is between 34 and 37%. The structure of the channel is given in Fig. 2. Importantly, this channel shares a very high degree of homology with so-called **degenerins** from the nematode *Caenorhabditis elegans,* which are involved in mechanosensation and may possibly form mechanosensitive ion channels. The three **rENAC**-subunits are very much related in sequence to the degenerins encoded by the three genes, mec-4, mec-6, and mec-10. The mechanosensitive channel Mec-4 (from gene **mec-4**) alone, cannot function normally but might assemble with other "subunits" to become functional. Also this feature very much resembles the normal function of the heterotrimer of the rENaCs. Interestingly, the amiloride-sensitive

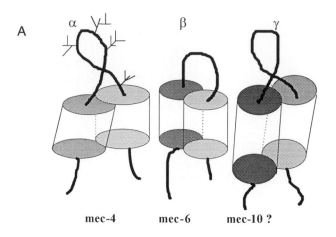

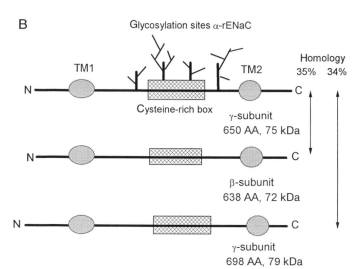

**FIG. 2.** Structure of the amiloride-sensitive Na$^+$ channel. (A) The channel is formed by a heterotrimer. All three subunits have two membrane-spanning helices and a long extracellular loop. They might be related to the mec-4,6,10 encoded mechanosensitive channels from *C. elegans*. (B) Linear models of the three subunits include the conserved cysteine box, glycolysation sites, and the degree of homology.

Na$^+$ channel clearly shows mechanosensitive functional features (see also Section II,K).

It seems to be that the amiloride-sensitive Na$^+$ channel is encoded by human chromosome 7 in a region that flanks the **cystic fibrosis gene** locus (see below). Interestingly, there is strong evidence that in cystic fibrosis in addition to CFRT-Cl$^-$ channels (see Section II,H, and Chapter 36) also amiloride-sensitive Na$^+$ channels are affected.

### B. Ca$^{2+}$-Activated Nonselective Cation Channels

Many nonexcitable cells (epithelial cells, fibroblasts, adipocytes, etc.) express cation channels that are either activated by binding of intracellular Ca$^{2+}$ to the channel or by cyclic nucleotides such as cAMP or cGMP (cyclic-3′,5′-guanosine monophosphate). These channels cannot discriminate between monovalent cations, for example, Na$^+$, Cs$^+$, and K$^+$, and are therefore called **nonselective cation channels (NSC).** Most of these NSCs are also Ca$^{2+}$ permeable and are therefore significantly involved in Ca$^{2+}$ signaling in nonexcitable cells. Ca$^{2+}$-activated NSC (Ca$^{2+}$-NSC) and cyclic nucleotide-gated NSC (cNSC) are impermeable to anions.

A widely distributed Ca$^{2+}$-NSC was first described in epithelial cells. This nonselective channel has a single-channel conductance between 25 and 35 pS and is unable to discriminate between Na$^+$ and K$^+$. In many types of cells (e.g., thyroid follicular cells, pancreatic cells, lacrimal gland cells, insuloma tumor cells), this channel is activated by hormonal stimulation (i.e., by acetylcholine or by **cholecystokinin** in exocrine pancreatic cells). Activation is mediated by an increase in [Ca$^{2+}$]$_i$. This channel is also permeable to Ca$^{2+}$ (e.g., in neutrophils and in endothelial cells) and provides an entry route for Ca$^{2+}$ into the cells.

Figure 3 shows an example of a Ca$^{2+}$-activated nonselective cation channel from pancreatic duct cells. It is activated by 1 $\mu M$ [Ca$^{2+}$]$_i$ (Fig. 3A; control and wash), but 0.1 $\mu M$ Ca$_i^{2+}$ fails to activate the channel. The conductance of this nonrectifying channel is 25 pS (Fig. 2B). It is blocked by quinidine. The channel is activated not only by Ca$^{2+}$ from the cytoplasmic side, but also by depolarization. The Ca$^{2+}$ sensitivity decreases after excision of the patches. The channel is inhibited by intracellular ATP but runs down in the complete absence of intracellular ATP. In this respect, Ca$^{2+}$-NSC shares properties with K$^+$ channels that are opened only by a decrease of intracellular ATP in many cells.

In endothelial cells that line the vessel blood walls a similar class of NSC has been found. These channels are sensitively blocked by nonsteroidal anti-inflammatory drugs (such as **flufenamic acid** or **mefanemic acid**) and are activated by **prostaglandins.**

The physiological role of such a channel (or a family of different Ca$^{2+}$-NSCs) still remains unclear. The channel could provide an entry pathway for Ca$^{2+}$ in stimulated cells. In this case, channel openings would provide positive feedback in a concentration window of [Ca$^{2+}$]$_i$, and thus would provide an amplification mechanism for Ca$^{2+}$ entry. In the case of nonsignificant Ca$^{2+}$ permeation through NSCs, the channel would cause depolarization, and thus could activate or deactivate voltage-dependent ion channels that exist in nonexcitable cells or provide a negative feedback for Ca$^{2+}$ entry via non-voltage-gated channels (see Section II,L).

Interestingly, a variety of calcium-sensitive nonselective ion channels seems to be of rather small conductance and might also be mechanically activated. Other NSCs respond to hypertonicity or are activated by protons. Molecular biology of Ca$^{2+}$-NSCs has not yet given any reliable approach.

### C. Agonist and ATP-Activated Nonselective Cation Channels

In addition to Ca$^{2+}$-activated NSCs (Fig. 4A), nonselective cation channels can be activated by extracellular ago-

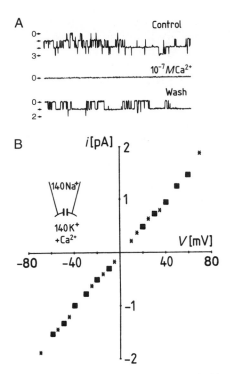

**FIG. 3.** Nonselective cation channels in nonexcitable cells from pancreatic duct of rats. (A) The channel is activated by $Ca^{2+}$ from the cytosolic side (inside-out patch). Conductance levels (0 for the closed channels; 1, 2, 3 indicate one, two, or three open channels, respectively) are indicated at the left-hand side. In "Control" and "Wash," 1 $\mu M$ $[Ca^{2+}]_i$ was used; 0.1 $\mu M$ $[Ca^{2+}]_i$ failed to activate the channel. (B) Current voltage relationship of single $Ca^{2+}$-activated channel. The channel was exposed to 140 m$M$ $Na^+$ in the pipette (external site) and 140 m$M$ $K^+$ from the bath (internal site). In spite of the asymmetric ion composition, the reversal potential is close to 0 mV, indicating a similar permeability of the channel for $Na^+$ and $K^+$. Channel conductance is 25 pS. (Reprinted from *Biochim. Biophys. Acta*, **1029**, Gray, M. A., and Argent, B. E., Nonselective cation channel on pancreatic duct cells, pp. 33–42, Copyright 1990 with kind permission of Elsevier Science–NL, Sara Burgerhartstraat 25, 1055 KV Amsterdam, The Netherlands.)

nist. In many nonexcitable cells the receptor for the agonist is coupled to a G protein. The following possibilities exist: (1) The G protein is activated by the binding of the agonist and directly activates the channel. (2) The activated G protein stimulates an effector system (e.g., adenylate cyclase), which in turn results in the production of an intracellular messenger (e.g., cAMP), which can gate the channel (Fig. 4B). However, in most of the cases, the activating intracellular messenger is not yet known.

One example of such an agonist-gated channel is found in endothelial cells. In these cells, histamine can activate a nonselective cation channel that is also permeable to $Ca^{2+}$. Figures 4E and F show patch-clamp measurements of this nonselective cation channel. The channel is characterized by long openings after activation (Fig. 4E). The size of the unitary currents depends on the membrane potential, that is, the driving force for the permeating cations. Although asymmetric ion compositions were used at both sides of the membrane, the potential at which the current reverses its direction is still close to 0 mV; that is, the channel cannot discriminate between $Na^+$ and $K^+$. From the almost linear current–voltage relationship, a single-channel conductance of 26 pS can be calculated from Ohm's law.

Many excitable cells (i.e., smooth muscle, cardiac cells, neurons) and nonexcitable cells (i.e., parotid and lacrimal acinar cells) express nonselective cation channels that are directly activated by ATP (Fig. 4C, right). These channels are all permeable to $Ca^{2+}$ and provide a receptor-operated influx route for $Ca^{2+}$. For monovalent ions, the single-channel conductance is between 20 and 30 pS, but is only between 4 and 10 pS for $Ca^{2+}$. These results clearly show that $Ca^{2+}$ ions permeate through the channel, but the $Ca^{2+}$ conductance is less than that for monovalents. In some cells (e.g., in pancreatic acinar cells), although there is a smaller conductance of the channel for $Ca^{2+}$, the selectivity for $Ca^{2+}$ is larger than for monovalents, supposedly due to a higher affinity for $Ca^{2+}$ than for monovalents to bind to a site in the conducting pore of the channel.

In some cases, however, nonselective cation channels can be directly activated by extracellular ATP via different types of **purinoceptors** ($P_{2T}$, $P_{2X}$, $P_{2Z}$). It is not yet completely clear to what extent G proteins and elevation of $[Ca^{2+}]_i$ are involved in the gating mechanism. Gating of nonselective cation channels via the receptor-G protein cascade is often coupled to some phosphorylation steps. Figure 4C (left side) shows a scheme for such a mechanism. Binding of the agonist to the receptor activates, for instance, the adenylate cyclase (AC). The resulting generation of cAMP evokes activation of PKA that, in turn, phosphorylates the channel protein (Fig. 4C).

### D. Cyclic Nucleotide-Activated Nonselective Cation Channels

A variety of nonexcitable cells also express **cyclic nucleotide-activated nonselective cation channels (cNSCs).** These channels might be members of a family of ion channels that are well described only in cones, rods, and olfactory cells. Two mechanisms might be involved in gating of the channel: (1) phosphorylation-dependent gating (Fig. 4C) or (2) a direct activation by cAMP or cGMP of the channel (Fig. 4D). An important example for the gating of nonselective cation channels in nonexcitable sensory cells is related to a direct binding of cyclic nucleotides to the channel protein. Phototransduction and signaling in olfactory cells depends on the modulation of such a nonselective cation channel by cGMP and cAMP, respectively. In the absence of $Ca^{2+}$ and $Mg^{2+}$, these channels have a conductance for monovalents of approximately 20 pS. In pineal gland cells, intracellular cGMP (but not cAMP) opens NSC having a single-channel conductance of 15 to 25 pS (for details see Chapter 44). Such cNCSs, which are highly homologous to the photoreceptor and olfactory channels, have also been described in kidney, heart, smooth muscle, and endothelium.

Figure 5 shows the best known members of the cNSC family cloned from rod and cone photoreceptors (cGMP-

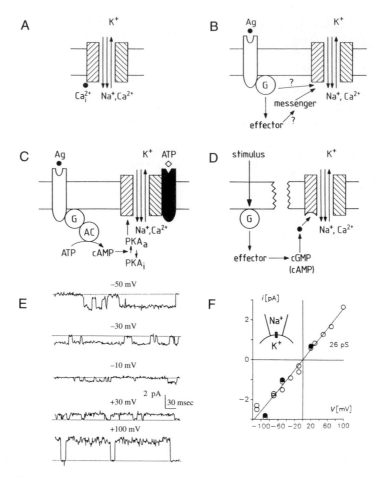

**FIG. 4.** Synopsis of the mechanisms of activation of different nonselective cation channels. (A) A NSC is activated by an increase in the intracellular $Ca^{2+}$ concentration. The channel is permeable for $Na^+$, $K^+$, and $Ca^{2+}$. (B) A NSC is gated by an extracellular agonist (see parts E and F). A G protein is involved. The signaling mechanism from the receptor to the channel is not yet understood (unknown messenger produced by an effector enzyme, gating directly mediated by the G protein?). (C) NSCs can be gated by agonist-mediated phosphorylation of the channel protein (here shown via PKA; $PKA_i$, inactive PKA; $PKA_a$, active PKA). Another well-studied mechanism of NSC gating is the direct activation by extracellular ATP via $P_2$ receptors (right part). (D) An extracellular stimulus such as light (cones and rods) or odor molecules (olfactory cells) can modulate the intracellular cAMP or cGMP concentrations. cGMP and cAMP directly active NSC (see Chapter 44). (E) Example of an agonist-gated NSC in human endothelial cells. Single-channel recordings are shown (cell-attached patch, different potentials: $-50$, $-30$, $-10$, $+30$, $+100$ mV). This channel is activated by application of histamine from the extracellular side. (F) Current–voltage relationship of the same channel. Although the channel is exposed to asymmetric ion compositions, the reversal potential is close to 0 mV, indicating that the channel does not discriminate between $Na^+$ and $K^+$. An increase in $[Ca^{2+}]_i$ is not necessary for channel opening. (From B. Nilius, unpublished, 1991.)

gated channel) and from olfactory cells (cAMP-gated channel). These channels have supposedly six transmembrane helical segments. Both N- and C-terminals are cytoplasmic. The binding site for cyclic nucleotides is located in the cytoplasmic C-terminus of the channel protein. An unexpected feature of this cNSC is the existence of a sequence motif that has until now only been found in voltage-gated ion channels such as $Na^+$, $K^+$, and $Ca^{2+}$ channels. In this motif, four to eight positively charged arginine or lysine residues are located at every third position separated by two hydrophobic amino acids. It is supposed that this motif belongs to the pore-forming region of the channel.

### E. $Ca^{2+}$-Activated $K^+$ Channels

For many functions of nonexcitable cells such as secretory epithelial cells, liver cells, and endothelial cells, release of intracellular $Ca^{2+}$ is associated with hyperpolarization, which increases the inwardly directed driving force for $Ca^{2+}$. Such a physiologically important hyperpolarization is due to $Ca^{2+}$-dependent activation of $K^+$ channels. The existence of high-conductance **maxi $K^+$ channels** with a conductance between 100 and 250 pS was described in a variety of nonexcitable cells including salivary, epithelial, and acrimal acinar cells. Figure 6 shows an example of activation of a high-conductance K$^+$ channel in endothelial cells that were stimulated by the vasoactive agonist ATP.

The high-conductance channel is highly selective for K$^+$. It is selectively blocked by the scorpion venom **charybdotoxin** (CTX) in the nanomaloar range and the still more selective **iberiotoxin** (ITX) but is also sensitive to **tetraethylammonium-chloride (TEA)**. Activation of this channel is both Ca$^{2+}$ dependent and voltage dependent. The open probability of the channel is increased with both increased Ca$^{2+}$ concentration and depolarization. The apparent dissociation constant of Ca$^{2+}$ for activation of the channel is decreased by depolarization. Such a negative feedback of membrane hyperpolarization on the channel provides an ideal tool for fine-tuning of channel activity by both membrane potential and [Ca$^{2+}$]$_i$. This channel will be always activated by a rise of [Ca$^{2+}$]$_i$ that can be induced

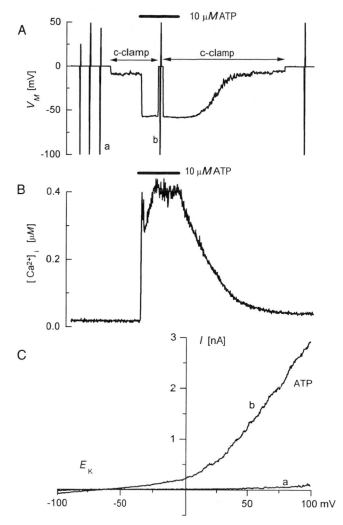

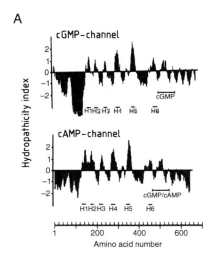

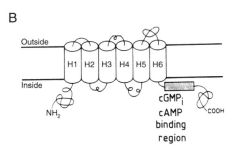

**FIG. 5.** Possible structure of the nucleotide-gated nonselective ion channel in vertebrate rod and cone photoreceptors. (A) Hydropathicity plots. The averaged hydropathicity index of nine amino acids composed of the amino acid residues i$-9$ to i$+9$ is plotted against the number of the amino acid i. These plots indicate the existence of six membrane-spanning regions H1 to H6 of the cGMP- (vertebrate rods and cones) and cAMP-gated channel (olfactory epithelium). (B) Proposed transmembrane structure of the cNSCs. This channel is composed of several copies of a 63 kDa polypeptide. It has four or six helical transmembrane segments. A region of 80 to 100 amino acids near the C-terminus forms the nucleotide (cGMP, cAMP) binding site. This channel is an important representative in the family of cNSC. (From Kaupp, 1991, with permission from *TINS*. Copyright © Elsevier, 1991.)

**FIG. 6.** Activation of high-conductance K$^+$ channels results in hyperpolarization and an increase in the driving force for Ca$^{2+}$-entry. (A) Voltage ramps from $-100$ to $+100$ mV were applied to measure membrane currents (a) before and (b) during stimulation with 10 $\mu M$ ATP. To measure changes in membrane potential, voltage-clamp mode was switched to current clamp as indicated. Application of ATP resulted in a strong hyperpolarization of the cell. Membrane potential does not reach the reversal potential for K$^+$ because of coactivation of a nonselective current. (B) Changes in [Ca$^{2+}$]$_i$. Note the fast peak that results from intracellular Ca$^{2+}$ release. During hyperpolarization the driving force for Ca$^{2+}$ entry is increased. (C) In the voltage-clamp mode, membrane currents were measured. Before agonist application the cells are very tight. Increase in [Ca$^{2+}$]$_i$ activated a current that is outwardly rectifying, is reversed close to the K$^+$ equilibrium potential, and has been identified as current through a maxi K$^+$ channel ($\approx$180 pS in asymmetric K$^+$, whole-cell current-clamp recording from a cultured endothelial cell preincubated in 2 $\mu M$ fura-2 AM for 30 min and the K-aspartate internal solution contained 50 $\mu M$ fura-2). (From F. Viana, G. Droogmans, B. Nilius, unpublished, 1996.)

by a variety of agonists or mechanical responses of stretch-activated membrane channels (SACs; see Section II,K). Ca$^{2+}$-activated K$^+$ channels also play a significant role in modulation of Ca$^{2+}$ influx in nonexcitable cells that are stimulated by agonists which release intracellular Ca$^{2+}$ and open an influx pathway for $Ca^{2+}$ (Section II,L). Because $Ca^{2+}$ influx through open $Ca^{2+}$-permeable channels is determined by the driving force for $Ca^{2+}$, $K^+$ channels will be involved in the fine-tuning of $Ca^{2+}$ influx. Another example is the volume regulation of cells (regulatory volume decrease, RVD; Section III,A). $K^+$ channels initiate a loss of intracellular $K^+$ that results in decrease of cell volume.

**High ("big") conductance $Ca^{2+}$-activated $K^+$ channels (BKCa)** were first cloned from the *Drosophila* slo-gene. The mammalian voltage and $Ca^{2+}$-dependent **maxi $K^+$ channel** (*mslo* or *Kv slo family*) contains six membrane-spanning hydrophobic regions consisting of four segments, S1 to S4, that confer voltage dependence to this channel. The linker between segment S5–S6 forms the $K^+$ channel highly conserved pore region (P region). A large protein domain conferring the $Ca^{2+}$-dependent gating has been appended to the C-terminus. High-conductance $K^+$ channels have a subunit architecture consisting of an $\alpha$-subunit, which is channel forming, and a regulatory $\beta$-subunit.

Other types of $Ca^{2+}$-activated $K^+$ channels are of intermediate conductance (*MKCa*), ranging between 30 and 80 pS in symmetric $K^+$, and are inwardly rectifying. Their conductance at physiological extracellular $K^+$ concentrations is about 15–30 pS. This channel has properties similar to the maxi $K^+$ channel. None of these channels has been cloned yet. It is uncertain whether they belong to the class of Kir 1.0 channels or to the class of $Ca^{2+}$- and voltage-activated $K^+$ channels (slo-like channels). A G-protein-dependent mechanism modulates these channels, probably by increasing their $Ca^{2+}$ sensitivity.

In other nonexcitable cells such as hepatocytes a small-conductance $K^+$ channel has been described that is also activated by an elevation of $[Ca^{2+}]_i$. This **small-conductance channel (SKCa;** between 4 and 12 pS) is highly $K^+$ selective. These channels are selectively blocked by the honeybee venom **apamin** in the nanomolar range, and are resistant to the most common $K^+$ channel blocker, TEA.

Although for all of these channels the intracellular $Ca^{2+}$ concentration and the membrane potential are the main modulators, changes in channel activity by phosphorylation via cAMP-dependent protein kinases have been described. Figure 7 gives some features of $Ca^{2+}$-activated $K^+$ channels coexisting in exocrine epithelial cells.

## F. Inwardly Rectifying $K^+$ Channel

The existence is well documented of **inwardly rectifying $K^+$ channels** in some (but not all) nonexcitable cells, for example, in juxtaglomerular epitheloid cells, in endothelial cells, and in mast cells. In most excitable and nonexcitable cells, these channels are responsible for stabilization of the resting potential near to the $K^+$ equilibrium potential, $E_K$. Inward currents can be measured at potentials negative to $E_K$; at potentials positive to $E_K$, outward currents are almost blocked. Inward rectification depends on intracellular $Mg^{2+}$ being pushed as a plug by membrane depolarization in the open channel. This $Mg^{2+}$ mechanism seems to be less important in some nonexcitable cells (e.g., endothelium). Block by intracellular polyamines such as spermine and spermidine is functionally important. At high extracellular $K^+$ concentrations, the conductance of the channel is between 25 and 30 pS (150 m$M$ $[K^+]_e$). Conductance declines with the square root of $[K^+]_e$. In some cells (e.g., endothelial cells, mast cells, enterocytes, juxtaglomerular cells), the **inward rectifier** is blocked by agonists such as angiotensin II and vasopressin. Block of the inward rectifier depolarizes the cell and decreases the driving force for inwardly transported $Ca^{2+}$, for example, via store-operated channels (SOCs) or $Ca^{2+}$-release-activated channels (CRACs) or via other $Ca^{2+}$-permeable nonselective channels. This block can be mimicked by GTP$\gamma$S and seems to be G-protein mediated.

Inwardly rectifying $K^+$ channels have a much simpler structure than voltage-gated, shaker-type channels or slo-channels. They only have a P domain with two flanking membrane-spanning helices. In nonexcitable cells, two types of inwardly rectifying channels are present. The channel thus lacks the voltage-sensing part of the channel molecule. Gating is thus mainly due to extrinsic block rather than an intrinsic mechanism. The strongly rectifying $Mg^{2+}$ and polyamine-dependent channels belong to the Kir 2.0 subfamily. Other rectifying $K^+$ channels are ATP dependent, are less rectifying, and are members of the Kir 1.0 (ROMK) subfamily of $K^+$ channels. Typical ATP-sensitive $K^+$ channels (KATP; subfamily Kir3.4), which also show inward rectification, have been also described in nonexcitable cells.

## G. $Ca^{2+}$-Activated $Cl^-$ Channels

Chloride channels play an important role in transport functions in nonexcitable cells, particularly in epithelial and endothelial cells. The first evidence for the existence of a $Ca^{2+}$-activated $Cl^-$ channel (ClCa) was published for rat lacrimal glands. Activation of acinar exocrinic cells with secretory agonists (i.e., acetylcholine) induced an increase in $[Ca^{2+}]_i$ up to the micromolar range. This $Ca^{2+}$ concentration opened $Cl^-$ channels. The conductance of these channels in nonexcitable cells is very small, ranging between 1 and 5 pS. Nevertheless, the whole-cell current is large because the estimated number of channels in acinar cells is about 5000 to 20,000 (compared to approximately 150 for maxi $K^+$ channels). This channel is different from other $Cl^-$ channels found in muscle cells. The open probability of the $Cl^-$ channel is greatly increased by $[Ca^{2+}]_i$ between 0.1 and 1 $\mu M$ but is much less sensitive to depolarization than classic voltage-dependent channels.

The functional role of these channels is discussed later. Because $Cl^-$ is passively distributed according to the resting potential, any increase in $Cl^-$ conductance will stabilize this potential.

Big conductance $Cl^-$ channels (<200 pS, BCl) have also been described for a variety of nonexcitable cells. They are modulated by $Ca^{2+}$, cAMP, and PKC. The functional role and incidence of these channels under physiological conditions remain uncertain. A possible molecular candidate for $Ca^{2+}$-activated $Cl^-$ channels has been cloned from the epithelium of bovine trachea. This $Cl^-$ channel consists of 903 amino acids, has four major hydrophobic (trans-

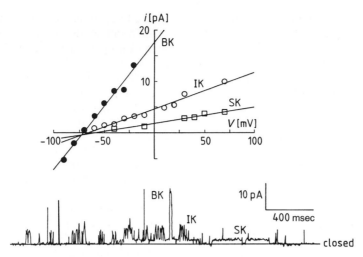

**FIG. 7.**   Examples for different types of $Ca^{2+}$-activated $K^+$ currents coexisting in the same cell from parotid gland. The conductance is 250 pS ("big" conductance maxi $K^+$ channel, BK or BKCa), 75 pS for an intermediate $K^+$ channel (IK or MKCa), 29 pS for a small-conductance $K^+$ channel (SK or SKCa). All currents reverse at $E_K$. SK channels in the patch are almost open during the recording. Two IK channels open in bursts and one BK channel opens occasionally. (From Cook and Young, 1990, with permission. Copyright © Springer-Verlag, 1990.)

membrane) regions, and is $Ca^{2+}$ dependent probably via several $Ca^{2+}$-calmoduline-dependent phosphorylation sites.

## H. cAMP-Activated Cl⁻ Channels

In a variety of nonexcitable cells (airway epithelial cells, rectal gland cells, pancreatic duct epithelial cells, carcinoma cells, but not in endothelium) other $Cl^-$ channels with a low conductance between 3 and 9 pS play an important functional role, especially in the control of secretion. The single-channel conductance scatters substantially within this family. In addition to these low-conductance channels, other populations of channels have conductance as high as 50 pS. All of these $Cl^-$ channels are gated by cAMP rather than $Ca^{2+}$. They are normally closed. In the presence of both cytosolic ATP and the active catalytic subunit of a cAMP-dependent protein kinase, a small conductance channel opens selectively for $Cl^-$ ions. PKA mediates this opening via phosphorylation of the channel protein. Four PKA phosphorylation sites are known in the channel protein. Only one site needs to be phosphorylated to open the channel. Prostaglandins, which stimulate adenylate cyclase, also activate the channel via the same mechanism. These cAMP-activated channels are critically involved in the $Cl^-$ secretion of epithelial cells, stabilization of the resting potential, and probably also in regulation of the cell volume. A direct example of the activation of this current is shown in Figs. 8A and B.

Many symptoms of a widely distributed illness (one in every 20,000 live births is affected), **cystic fibrosis,** are associated with defective **cAMP-regulated Cl⁻ channels** in epithelial cells. A gene that encodes a protein called **cystic fibrosis transmembrane conductance regulator (CFTR,** Fig. 9; for more details see also Chapter 36) has been found. The channel protein consists of 1480 amino acids in two

portions of six helical transmembrane segments (TM1 and TM2; see Fig. 9) and three large cytoplasmic domains of two nucleotide-binding domains (NBF1, NBF2) and another intracellular domain between TM1 and TM2 that contains phosphorylation sites (R-domain). A phenylalanin residue, Phe, is located at site 508 in the NBF1 region. Deletion of Phe-508 (ΔF508 CFTR) and substitution by another amino acid in the CFTR prevent the normal cAMP-activation of the $Cl^-$ channel. This mutation is responsible for a functional defect of the channel that causes failure of exocrine secretion in epithelial cells from respiratory tract, airway epithelium, gastrointestinal tract, exocrine pancreas, and sweat glands. From single-channel data obtained from expression of the wild-type (nonmutated) CFRT and from mutagenesis data, we can conclude that CFTR is a 9-pS low-conductance $Cl^-$ channel rather than only a conductance regulator. Region R is highly charged and has numerous phosphorylation sites for both PKA and PKC. The channel is activated when membrane patches are exposed to either PKA or PKC. PKA activation is potentiated by previous exposure to PKC. Phosphatases inhibit the channel. The first or the sixth helix in TM1 seems to be necessary for anion selectivity of the channel. The channel requires ATP-binding at NBF1 to open (see also Fig. 8B). Phe depletion from site 508 increases strikingly the mean closed time of the channel. Expression of the normal CFRT, but not of the mutant form with the phe508 deletion corrects the channel defect.

However, this situation is still complicated by the finding that ΔF508 CFTR is trapped in the endoplasmic reticulum and cannot travel to the normal cell surface. This defect in the intracellular traffic of the channel causes the functional loss of $Cl^-$ channels in the plasmamembrane.

Other functional defects seem to be involved, for example, changes in the function of the amiloride-sensitive $Na^+$ channels and defects in regulation of cell volume.

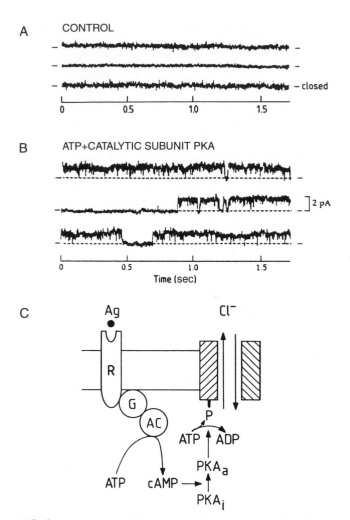

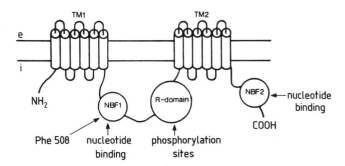

**FIG. 8.** cAMP-gated Cl⁻ channels. (A) ATP and the catalytic subunit of the cAMP-activated protein kinase A are added to the cytoplasmic side of an excised patch from human airway epithelium without channel activity. (B) This experiment shows directly cAMP-dependent gating of Cl⁻ channels. Gating only proceeds in the presence of both active PKA and ATP. [Reprinted with permission from *Nature*, Li *et al.* (1988). Vol. 331, 358–360. Copyright 1988 Macmillan Magazines Limited.] (C) Gating scheme of the cAMP-activated Cl⁻ channel: In intestine, airway, and pancreatic cells, a rise in intracellular cAMP can be induced by receptor stimulation, for instance, by application of noradrenaline, secretin, vasoactive intestinal peptide (VIP), or by a direct G-protein ($G_s$) activation with cholera toxin, which stimulates AC. cAMP activates PKA, which in turn phosphorylates the channel and induces channel openings (PKA$_i$, inactive; PKA, active PKA). The open probability of the channel is increased, mediated by phosphorylation via PKA.

## I. Voltage-Dependent Cl⁻ Channels, the ClC Family

Cl⁻ channels are probably present in every animal cell. Their major functional impact includes regulation of cell volume, regulation of the intracellular pH in cooperation with a variety of transporters such as the band III bicarbonate–chloride exchanger, and setting of the membrane potential and the transport of amino acids. A family of Cl⁻ channels is encoded by a distinct gene family. These chan-

nels, **ClC channels** or **voltage-gated Cl⁻ channels** (because of their voltage-dependent gating behavior) are supposedly ubiquitously expressed in excitable and nonexcitable cells. In nonexcitable cells such as epithelium, a member of this particular family, ClC-2, seems to be the housekeeping Cl⁻ channel. This channel is composed of 907 amino acids and probably 12 membrane-spanning helices. It is activated by hyperpolarization, shows inward rectification, and is more permeable for iodide than chloride. This channel can be also activated by cell swelling and might therefore be a volume sensor in nonexcitable cells. The N-terminus of this channels probably comprise a "ball" structure ("essential" region), which is tethered to the channel by a polypeptide "chain." Normally, the channel is closed by this ball. Mutations in the essential region induce constitutively open channels. One of the intriguing hypotheses predicts that the "ball" closure of the channel can be removed by cell swelling and thus activates a Cl⁻ flux. Many members of the ClC-family might become important Cl⁻ channels in nonexcitable cells.

## J. Volume-Activated Cl⁻ Channels

In many, if not all, nonexcitable cells, an increase in cell volume activates large whole-cell chloride currents (Clvol). These channels are important for the regulation of the cell volume. This function seems to be coupled with other vital events such as regulation of intracellular pH, modulation of resting potential, and also complex cell functions such as cell proliferation. The current is $Ca^{2+}$ independent, but its activation requires both a very low "permissive" intracellular $Ca^{2+}$ concentration and intracellular ATP. In contrast with voltage-gated ClC channels, activation of this current is voltage independent. Only a slow inactivation at very positive potentials is present. The channel shows outward rectification, but much less than the $Ca^{2+}$-activated Cl⁻ channel and is, in contrast to the voltage-gated ClC-2 channel, more permeable for iodide than for chloride. The channel is also permeable for larger molecules such as amino acids (e.g., taurine, glycine, alanine, aspartate, gluta-

**FIG. 9.** Structure of the CFTR. TM1 and TM2 are membrane-spanning helical segments. NBF1 and NBF2 are nucleotide-binding domains, and R contains the phophorylation sites. The cystic fibrosis defect occurs at site 508 in NBF1 (deletion of a phenylalanin). [Reprinted with permission from *Nature*, Ringe, D., and Petsko, G. A. (1990). Vol. 346, 312–313. Copyright 1990 Macmillan Magazines Limited.]

mate), polyols (e.g., sorbitol, myo-inositol), and methylamines (e.g., betaine, choline). The physiological stimulus for its activation is an increase in cell volume, but probably also other mechanical stimuli such as tensile stress (stretch) and changes in cell shape.

The mechanism of gating is unknown. Supposedly, in several cell types (e.g., endothelium) this volume-sensitive channel is already partially activated in resting cells, is inhibited by cell shrinkage, and contributes to the chloride conductance and resting potential of these cells. Four molecular candidates have been proposed for this current: (1) The voltage and volume-dependent ClC-2 channel, which has been already discussed. (2) A protein encoded by the multidrug resistance gene mdr-1, the P-glycoprotein pgp170, which is a member of the ABC-cassette transporter family. This protein has structural similarities to the CFTR-Cl⁻ channel. (3) The putative channel-forming, ATP-sensitive protein pICln (an acid 237 amino acid protein) which consists of possibly four membrane-spanning $\beta$-sheets and extracellular putative bindings sites for nucleotides. (4) Phospholemman, a 72 amino acid intrinsic membrane protein. This small protein was first reported to induce a hyperpolarization-activated chloride channel. Incorporation of phospholemman in planar lipid bilayers resulted in a conductive pathway that was permeable to a whole range of anions ($Cl^- > Br^- > F^-$ selectivity) and to organic molecules such as taurine.

## K. Mechanically Activated Ion Channels

Mechanoelectrical properties of hair cells, sensory mechanoreceptors provide intriguing examples for the function of mechanosensing ion channels (see Chapters 41, 42, and 43). In the nonsensory, nonexcitable cells discussed in this chapter, the ion channels act as mechanosensors. They are activated by **stretch-activated channels (SACs)** or by shear **stress-activated channels (SSCs)** and have been found in kidney cells, mesangial cells, tumor cells (i.e., neuroblastoma cells), astrocytes, endothelial cells, osteoblasts, blood cells, oocytes, etc.

SACs can be divided into channels that are (1) selectively permeable for anions, (2) nonselective for cations, or (3) selectively permeable for K⁺. For SSCs, two entry pathways has been described: (1) a pathway selective for K⁺, and (2) a pathway that is nonselective for cations with mostly a higher permeability for $Ca^{2+}$ than for monovalents.

Reversible block by micromolar gadolinium ($Gd^{3+}$) and lanthanum ($La^{3+}$) is common trademark for all mechano sensitive channels. Most of these channels in nonexcitable cells are also blocked by amiloride.

Cl⁻-selective SACs, which might be different from volume-activated Cl⁻ channels, seem to belong to a class of large-conductance Cl⁻ channels (about 130 pS). K⁺-selective SACs in hepatocytes have low single-channel conductance of approximately 7 pS and conductance in epithelial cells of between 30 and 50 pS. In hepatocytes, nonselective cation SACs have a single-channel conductance of 16 pS. In the basolateral and apical membrane of renal tubule cells and in endothelial cells these channels have a conductance of approximately 16–37 pS for Na⁺ and K⁺ and 15

pS for $Ca^{2+}$. In gallbladder cells, nonselective SACs were described as having conductances of 55 and 116 pS. It seems that all nonselective cation SACs are permeable for $Ca^{2+}$.

Endothelial cells are mechanosensor cells. These cells are subjected to flow-related forces such as biaxial tensile stress (stretch perpendicular to their surface) and shear stress (a tangential force that is generated by friction between blood and endothelial surfaces; see Fig. 10A). Many effects of endothelial cells are related to blood flow, for example, secretion of prostacyclin ($PGI_2$), endothelium-derived relaxing factor [nitric oxide (NO), EDRF], expression of tissue plasminogen activator (tPA), plasminogen activator inhibitor (PAI-1), several adhesion molecules, endothelin, monocytechemoattractant protein (MCP-1), NO synthase, activation of early response genes and small G proteins, cytoskeletal rearrangement, cell cycle entry, long-term responses, such as adaptive changes in cytoskeleton, vessel remodeling, and others. Ion channels may provide a functional link between physical forces and biological responses. Figure 11 gives an overview on mechanically induced cell responses of endothelial cells.

Ion channels are candidates to transfer mechanical forces into cell responses. Therefore, mechanically gated ion channels are of special functional significance in endothelial cells. According to the mode of stimulation, they are SACs or SSCs (Figs. 10 and 11). A typical nonselective cation SAC has been described in endothelium. This channel is permeable for $Ca^{2+}$ having a single-channel conductance of 19 pS in isotonic $Ca^{2+}$ solutions. Single-channel conductance in isotonic K⁺ solutions is 56 pS, for Na⁺ 40 pS. The channel is about six times more permeable for calcium than for sodium (permeation ration $P_{Ca}/P_{Na}$ between 1.2 and 8.4). Modulation of stretch-activated ion channels seems to depend on changes in the cytoskeleton

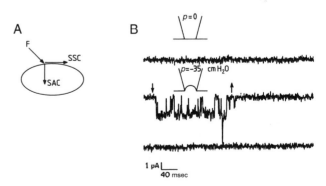

**FIG. 10.** Mechanical-activated ion channels in endothelial cells. (A) A cell is opposed to a mechanical force F that generates at the cell surface a force perpendicular to the surface (stretch that opens SACs) and tangential to the surface (shear stress that opens SSCs). (B) Example of a SAC in endothelial cells. Opening of a nonselective cation channel in an inside-out patch by application of suction pulses (between the arrows) of 1.3–2.7 dyn/cm² through the pipette. The pipette contains 150 m$M$ Na⁺, the bath solution 150 m$M$ K⁺. Under this asymmetric condition, single-channel conductance is approximately 40 pS. [Modified with permission from *Nature*, Lansman, J. B., Hallam, T. J., and Rink, T. J. (1987). Vol. 325, 811–813. Copyright 1987 Macmillan Magazines Limited.]

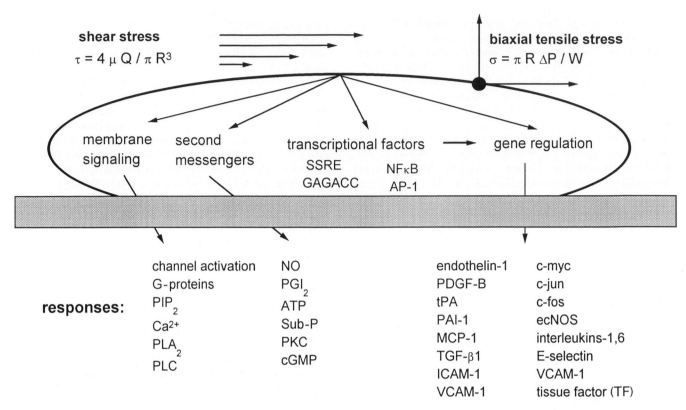

**FIG. 11.** Mechanical activation of endothelial cells and related cell responses. The cell can be activated by shear stress $\tau$, which depends on blood viscosity $\mu$, flow $Q$, and vessel radius $R$. Biaxial tensile stress, $\sigma$, depends on radius $R$, transmural pressure $P$, and wall thickness $W$. Cells respond immediately (in the seconds and minutes range) by channel activation, $Ca^{2+}$ release, activation of phospholipase A and C, $PIP_2$ breakdown, synthesis and release of NO, $PGI_2$, release of different messengers as ATP, and substance P. They react in the range of minutes to a few hours by modulation of the expression of endothelin (ET-1), transcriptional factors (NF$\kappa$B, AP-1, nuclear factors, sensors are probably shear-stress response elements in the gene-promotor region, SSRE with the typical GAGACC-sequence), early response genes (c-myc, c-fos, c-jun), platelet-derived growth factor (PDGF-B), tPA, PAI-1, MCP-1, endothelial cell NO synthase (ecNOS), tissue factor (TF) for initiation of blood clotting, several cell adhesion proteins (ICAM, VCAM), interleukins, and selectins.

that focus membrane distortion from a large area to the site of a single ion channel. An endothelial stretch-activated channel is depicted in Fig. 10B.

Two types of SSCs have been described in endothelial cells. A $K^+$-selective SSC could act as a mechanosensor mediating a hyperpolarizing response due to shear stress in the physiological range between 0.5 and 25 dyn/cm². This channel differs from the classical inwardly rectifying $K^+$ channels (Kir2.0 family) but shares some similarities with ROMK channels (Kir1.0 family). Opening of this channel apparently induces non-EDRF-dependent vasorelaxation via electrical coupling of endothelial cells to smooth muscle cells. Hyperpolarization of endothelial cells would also spread to neighboring smooth muscle cells, causing relaxation. As discussed previously, opening of $K^+$ channels would induce hyperpolarization and therefore an increased driving force for $Ca^{2+}$ into the endothelial cell. By such a mechanism, $[Ca^{2+}]_i$ would rise and could trigger the synthesis and release of vasoactive compounds. Sensi-

tivity of $K^+$ and $Cl^-$ channels to mechanical activation seems to be a more widespread phenomenon including also $Ca^{2+}$-activated channels such as small- (SKCa), intermediate- (MKCa), and high- (big, BKCa) conductance $K^+$ channels.

Other transmembrane currents activated by shear stress appear to be associated with opening of nonselective ion channels that are more permeable for $Ca^{2+}$ than for $Na^+$ or $Cs^+$.

From the molecular point of view, the first revolutionary steps have been done by succeeding in the cloning of mechanically gated channels in the envelope of the bacteria *Escherichia coli*. These channels probably form a multimere structure consisting of four to five 15-kDa subunits of 136 amino acids and probably four transmembrane-spanning regions. The amino acid sequence of these mechanosensitive channels shows no similarities with other putative eukaryotic mechanosensitive channels identified in the nematode *Caenorhabditis elegans* (see Section II,A).

## L. Ca$^{2+}$-Entry and Store-Operated Ca$^{2+}$ Channels

Intracellular Ca$^{2+}$ ions play a fundamental role in linking the information of receptor stimulation at the level of the plasma membrane to various distinct functions in nonexcitable cells, such as enzyme secretion, synthesis of various compounds, control of cell proliferation, and cell differentiation. Different pathways for Ca$^{2+}$ entry have been described. The best studied entry mechanisms are related to voltage-operated Ca$^{2+}$ channels (VOCs; see Chapter 25). Receptor-operated Ca$^{2+}$ channels (ROCs) mediate Ca$^{2+}$ influx as a consequence of occupation of a membrane receptor with an agonist (e.g., P$_2$ receptors with ATP). Other Ca$^{2+}$ channels are indirectly activated by a second messenger via a G-protein or tyrosine-kinase and are often termed **second messenger operated Ca$^{2+}$ channels (SMOCs).** In most nonexcitable cells, the initial rise in [Ca$^{2+}$]$_i$ upon receptor stimulation originates from the release of Ca$^{2+}$ ions from intracellular stores. This release is mediated by Ins(1,4,5)P$_3$. Activator calcium, responsible for the more long-lasting effects on the different cell functions, enters the cell from the extracellular space. Long-lasting Ca$^{2+}$ signals cannot be maintained in Ca$^{2+}$-free extracellular solutions.

Nonexcitable cell types lack voltage-operated Ca$^{2+}$ channels, but have developed another Ca$^{2+}$ entry mechanism that is termed **depletion-operated Ca$^{2+}$ entry** via a putative **depletion-operated channel (DAC), capacitative Ca$^{2+}$ entry** or **Ca$^{2+}$-release-activated Ca$^{2+}$ entry** via a respective channel (CRAC). Another term is **store-operated Ca$^{2+}$ channels (SOCs).** These putative channels appear to be coupled to the state of filling of the internal stores with Ca$^{2+}$. This plethora of terms already points to the uncertainty of the respective mechanisms. However, a basic mechanism seems to be that the mere emptying of these intracellular stores (e.g., by extracellular agonists, intracellular messengers) seems to increase the Ca$^{2+}$ permeability at the plasma membrane (Fig. 12A). The plateau Ca$^{2+}$ entry in nonexcitable cells depends mostly on the extracellular Ca$^{2+}$ concentration. It disappears in Ca$^{2+}$-free solutions (Fig. 12B). This pathway for Ca$^{2+}$ influx could also be involved in the refilling of intracellular stores upon termination of the stimulus.

Ca$^{2+}$ release and CRAC processes are of a more general interest because they are related to the biologically important role of luminal calcium (Ca$^{2+}$ inside the store organelles) in the regulation of many cell functions, such as intracellular protein traffic, secretion of mature processed proteins from the endoplasmic reticulum (ER), retention of the partially processed proteins, and control of cell proliferation.

Evidence for the existence of a Ca$^{2+}$ entry mechanism that depends on store depletion first came from the type of experiment shown in Fig. 12C: Ca$^{2+}$ stores in a nonexcitable cell (here an endothelial cells) are emptied by application of an agonist (e.g., histamine). Refilling of intracellular Ca$^{2+}$ stores can be prevented by Ca$^{2+}$-free external solutions or by use of thapsigargin that selectively blocks the SERCa Ca$^{2+}$ pumps (see Chapter 16) which are responsible for refilling of the stores. Under these conditions, reapplication of Ca$^{2+}$ from outside induces an increase in [Ca$^{2+}$]$_i$ in the absence of an agonist. This increase follows the driving force of Ca$^{2+}$ and is due to Ca$^{2+}$ entry from the extracellular space via a still open channel. Refilling of the stores blocks the entry. An ion current associated with Ca$^{2+}$ release from Ca$^{2+}$ stores in the endoplasmic reticulum was first described in mast cells. Ca$^{2+}$ ions enter the cell via a highly Ca$^{2+}$-selective pathway with a supposedly tiny conductance for Ca$^{2+}$ (presumably in the range of some 20 fS; Fig. 12D).

In other cells, such Ca$^{2+}$ influx following the discharge of Ca$^{2+}$ stores seems to be mediated by a less Ca$^{2+}$-selective pathway. Obviously, as for other channel types, store-operated channels (SOCs, DACs, CRACs?) seem to form a family. The mechanism of signal transduction between the intraluminal Ca$^{2+}$ in a store and the opening of a Ca$^{2+}$ channel in the plasma membrane is not yet well understood. The following mechanisms of cross-talk between Ca$^{2+}$ stores and membrane channels are discussed: (1) Ca$^{2+}$ entry is activated via a direct physical contact between the Ins(1,4,5)P$_3$ receptor in the endoplasmic reticulum and the putative membrane channel. (2) A low molecular weight 500D factor, *Ca$^{2+}$ influx factor* (CIF) is supposed to transfer a signal from the depleted store to the plasmamembrane. CIF possesses a phosphate group. Its action is potentiated by inhibition of phosphatases. (3) A cytosolic 130-kDa signal protein is phosphorylated at a tyrosine residue and is dephosphorylated by a phosphatase. The cytosolic tyrosine kinase is Ca$^{2+}$ dependent. The phosphatase is located in the membrane of the endoplasmic reticulum and is activated by intraluminal Ca$^{2+}$. Thus, phosphorylation and dephosphorylation of the 130-kDa protein would be controlled by Ca$^{2+}$ release and refilling of the store. The phosphorylated protein gates the channel for Ca$^{2+}$ entry. (4) Ca$^{2+}$ entry is controlled via cytochrome P450. The signal for channel gating may be mediated by 5,6-epoxyeicosatrienoic acid (5,6-EET), an arachidonic acid metabolite synthesized by a P450 mono-oxygenase located in the ER membrane that is activated by a decrease of intraluminal Ca$^{2+}$. (5) cGMP is produced by agonist stimulation of a nonexcitable cell and might be the signaling messenger. (6) Ins(1,3,4,5)P$_4$ or Ins(1,4,5)P$_3$ may directly gate the entry channels. All of these mechanisms are still uncertain.

Interestingly, a gene has been cloned in *Drosophila* that encodes a channel-like protein that might be responsible for a store-operated Ca$^{2+}$ influx in *Drosophila* photoreceptors. Similar differences for highly Ca$^{2+}$-selective CRACs and less Ca$^{2+}$-selective Ca$^{2+}$ entry channels have also been reported for these *Drosophila* photoreceptor proteins encoded by the trp and trpl gene (trp from transient receptor potential, trpl is trp-like). The gene products, TRP and TRPL, are Ca$^{2+}$- and Na$^+$-permeable channels. TRP is activated by all known maneuvers that release intracellular Ca$^{2+}$, whereas TRPL is insensitive to store depletion by thapsigargin. TRPL is much less Ca$^{2+}$ selective than TRP.

Both proteins have a putative channel structure with six membrane-spanning (TM) helices, a conserved pore region between TM5 and TM6, and an ankyrin-like (AL) motif that could bind to the ankyrin-binding motif (AB) in the Ins(1,4,5)P$_3$ receptor. The carboxy-terminal domain of TRP contains a lys-pro motif, which is repeated 27 times,

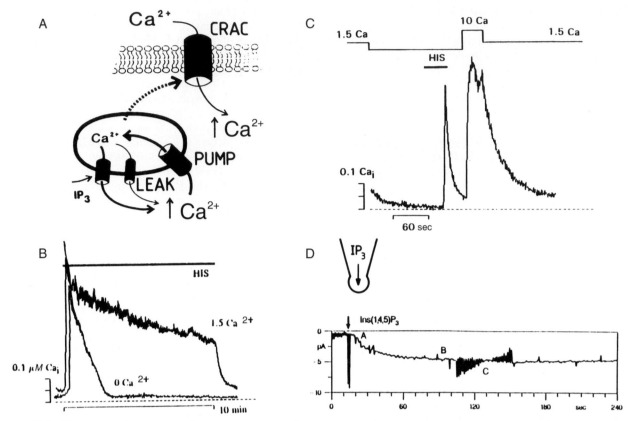

**FIG. 12.** Store depletion, or Ca²⁺-release-operated Ca²⁺ channels. (A) Receptor stimulation of a cell by various agonists induces the production of Ins(1,4,5)P₃ via activation of phospholipase C, PLC. Ins(1,4,5)P₃ opens a Ca²⁺-release channel in Ca²⁺ stores of the endoplasmic reticulum. Discharge of the store signals a Ca²⁺-permeable channel in the plasma membrane to open. Refilling of the store by a Ca²⁺ pump closes CRAC again. (B) Long-lasting Ca²⁺ signals depend on extracellular Ca²⁺. The Ca²⁺ plateau disappears in Ca²⁺-free media. The Ca²⁺ peak is due to Ins(1,4,5)P₃-mediated Ca²⁺ release. Ca²⁺ influx from the extracellular space is necessary. (B. Nilius, unpublished, endothelial cells, 1992.) (C) After depletion of the intracellular Ca²⁺ stores and prevention of refilling (no Ca²⁺ outside), reapplication of Ca²⁺ induces intracellular Ca²⁺ signals that disappear after refilling of the stores. (B. Nilius, unpublished, endothelial cells, 1992.) (D) Example for the first direct measurements of CRAC in mast cells. The cells are loaded via a patch pipette with Ins(1,4,5)P₃. At the arrow, the cell membrane has been perforated under the patch pipette and is then loaded by diffusion from the pipette with Ins(1,4,5)P₃. At A, B, C voltage ramps or steps are applied to measure the voltage dependence of the current activated by Ins(1,4,5)P₃. The holding potential is 0 mV throughout the measurement. Store depletion gates a highly Ca²⁺-selective, low-conductance channel and generates an inward current even at a holding potential of 0 mV. [Reprinted with permission from *Nature*, Hoth, M., and Penner, R. (1992). Vol. 355, 353–356. Copyright 1992 Macmillan Magazines Limited.]

and a highly charged sequence of asp-lys-asp-lys-lys-glu-(arg/gly)-asp (L), which occurs nine times. This C-terminal structure is absent in TRPL and might be responsible for sensing the intraluminal Ca²⁺ concentration in the store. This model would support the hypothesis of gating by physical contact (mechanism 1).

The structural differences in TrP and TRPL might cause TRPL to be much less store operated than TRP. Both proteins also have several calmodulin-binding regions (C), which may mediate the Ca²⁺-dependent modulation of store-related Ca²⁺ entry. The striking differences in ion selectivity and conductance between CRAC channels (highly Ca²⁺ selective, ∼20 fS) and TRP or TRPL channels (Na⁺ and Ca²⁺ permeable, ∼3 pS) point to a family of SOCs rather than to a single class of highly selective CRACs. Recently, two human homologs of TRP, TRPC1 or

HTRP1, and TRPC3 have been described. Figure 13 shows the two proteins that might become very important for Ca²⁺ signaling in nonexcitable cells.

An example of a nonexcitable cell in which a plethora of ion channels has been identified is given in Fig. 14. This figure illustrates the complex organization of ion channels, which might be typical for many nonexcitable cells.

## M. Voltage-Dependent Ion Channels in Nonexcitable Cells

The still useful dogma used to distinguish between excitable cells (which generate APs) and nonexcitable cells (no APs can be generated by these cells under physiological conditions) is not paralleled by the existence of voltage-gated ion channels. As only one example, in lymphocytes,

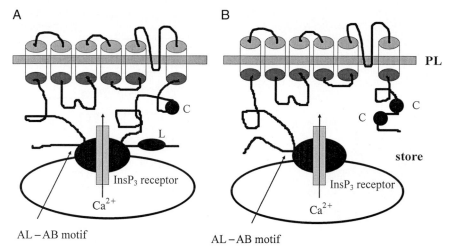

**FIG. 13.** Structure of the two proteins, TRP and TRPL, that may be related to $Ca^{2+}$ entry channels. (A) TRP. Six transmembrane-spanning regions; pore region between S5 and S6. The C-terminus is proline rich. This part of the molecule can bind to the Ins(1,4,5)$P_3$ receptor and may sense $Ca^{2+}$ in the store. Another possible bindings site the Ins(1,4,5)$P_3$ receptor is an ankyrin-like motif in the N-terminus (AL) and an ankyrin-binding motif in the Ins(1,4,5)$P_3$ receptor (AB). L is the lysin-rich region in the C-terminus. C marks both protein calmodulin-binding sites. TRP might be only gated when the store is empty. (B) TRPL. The TRP-like protein is quite different from TRP in the C-terminus. These structural differences might be the reason for the less store-controlled function of TRPL. This protein can function as a less $Ca^{2+}$-selective channel (nonselective) even if the stores are not completely emptied.

TTX-sensitive voltage-operated $Na^+$ channels, voltage-operated $Ca^{2+}$ channels, and different types of voltage-operated $K^+$ channels have been identified. Figure 15 gives another example of a highly voltage-dependent potassium channel in a nonexcitable tumor cell, a human melanoma cell. Expression of this voltage-dependent $K^+$ channel is associated with cell proliferation. In endothelial cells that are dissociated from capillaries, different types of voltage-gated $Ca^{2+}$ channels exist. These channels have some similarities with the classical L- and T-type calcium channels

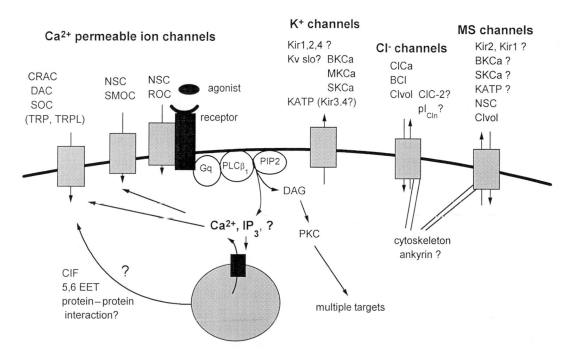

**FIG. 14.** Scheme of an endothelial cell with ion channels identified or discussed in this nonexcitable cell type. MS, mechanosensitive channels; $G_q$, q-type G-protein connected with phospholipase $C\beta_1$ (PLC$\beta_1$), which splits phosphoinositol-diphosphate (PIP2) into Ins(1,4,5)$P_3$ and diacylglcerol (DAG); CIF, $Ca^{2+}$ influx factor. All other terms are explained in the text.

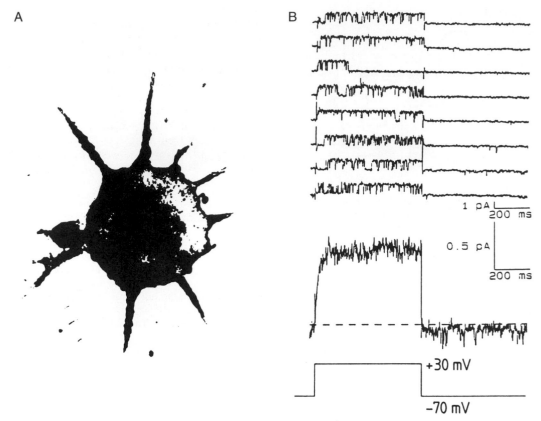

**FIG. 15.** Example of voltage-dependent $K^+$ channels in a nonexcitable tumor cell. (A) Human melanoma cells have dendrite-like processes that contain melanosomes in which the synthesis of melanin occurs. The bar indicates 10 $\mu$m. (B) In these cells using the cell-attached patch-clamp technique, voltage steps induce channel openings of a $K^+$-selective ion channel. This channel is highly voltage dependent, blocked by cAMP and TEA, and seems to be involved in some of the steps of cell proliferation. (Modified from Nilius and Wohlrab, 1992, with permission. Copyright © The Physiological Society, UK, 1992.)

known in excitable tissues such as myocardium or various neurones.

## III. Functional Role of Ion Channels in Nonexcitable Cells

### A. Ion Channels and Cell Volume Regulation

Most if not all nonexcitable cells such as epithelial cells and Ehrlich ascites tumor cells are capable of regulating their volume. Ion channels are critically involved in two mechanisms of volume regulation. First, in response to cell shrinking cells react with a **regulatory volume increase (RVI).** Second, once swollen, cells answer with a **regulatory volume decrease (RVD).** RVI is associated with an increase of $[Na^+]_i$ and $[Cl^-]_i$. This accumulation induces entry of water and cell swelling to compensate for a decrease in volume. RVD is induced by a cellular loss of $K^+$ and $Cl^-$.

Some of the ion channels involved in volume regulation belong to the class of SACs. Cell swelling increases the volume of the cell, which causes opening of these mechanosensitive channels. At least two types of SACs are described: (1) $K^+$-selective ion channels and (2) nonselective cation channels that are permeable for $Ca^{2+}$ and do not

discriminate well between different cations. $K^+$-selective SACs have a single-channel conductance of between 30 and 50 pS in most cells (e.g., in basolateral renal proximal tubule). They are activated by stretch between 0 and $-50$ cm $H_2O$ and are blocked by $Gd^{3+}$. Activation of $K^+$-selective SACs and volume-sensitive $Cl^-$ channels induces an intracellular loss of $K^+$ and $Cl^-$, which is necessary for RVD. Nonselective cation SACs have been demonstrated in a variety of cells. They are $Ca^{2+}$ permeable and therefore provide a pathway for $Ca^{2+}$ influx. This influx leads to an increase in $[Ca^{2+}]_i$ and causes opening of $Ca^{2+}$-activated $K^+$ channels (SKCa, IKCa, BKCa), which can be blocked by quinine and quinidine. These channel blockers also inhibit RVD. Opening of these channels induces a loss of $K^+$. $K^+$ efflux associated with $Cl^-$ efflux induces RVD. Both mechanisms for RVD mediated by SACs are depicted in Fig. 16.

Cooperation of different types of ion channels is necessary for RVD. Currents that can efficiently act in volume regulation must be of different charge and must differ in their respective reversal potentials. An example is shown in Fig. 17. Distal nephron epithelial cells from toad coactivate $K^+$ and $Cl^-$ currents due to swelling in a hypotonic medium. Under isotonic conditions, cells are very tight and

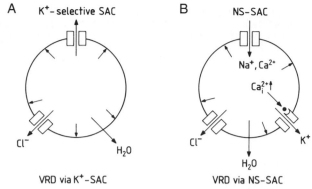

**FIG. 16.** Involvement of ion channels in volume regulation. (A) Swelling of a cell activates $K^+$-selective SACs and possibly also $Cl^-$-permeable SACs. Both events decrease the osmolarity of the cell followed by outflow of water and decrease of the cell volume. (B) Swelling of the cells activates a nonselective SAC that is $Ca^{2+}$ permeable and probably also $Cl^-$ permeable. $Ca^{2+}$ influx activates $K^+$ channels. Both events again decrease the cell volume.

only a small current, $I_{iso}$, is present (Fig. 17A). Swelling in a 40% hypotonic solution activated a large current ($I_{hypo}$), which represents the sum of a $Cl^-$ and a $K^+$ current. The $Cl^-$ component can be inhibited by the $Cl^-$ channel blocker NPPB (a nitro-phenylpropylamino-benzoic acid derivative). The $K^+$ component can be blocked by quinine. The remaining current represents the volume-activated $K^+$ (NPPB) and $Cl^-$ (QUIN) current, which reverse at the respective equilibrium potentials.

During cell swelling, the cell depolarizes and approaches a membrane potential between $E_K$ and $E_{Cl}$. NPPB shifts the potential toward $E_K$, quinine toward $E_{Cl}$ (Fig. 17B). this mix potential reflects the zero-current potential for the swelling-activated current. Indeed, at this potential an inward $Cl^-$ and an outward $K^+$ current reflect the $Cl^-$ and $K^+$ osmolyte efflux, which induces RVD. Blockage of one component shifts the membrane potential toward the equilibrium potential of the other ion and thus prevents net efflux. This must result in blocking of RVD (Figs. 17C and D). Only coactivation of both currents can induce RVD. The described volume-sensitive channel, which is also permeable for other nonionic osmolytes (described in Section

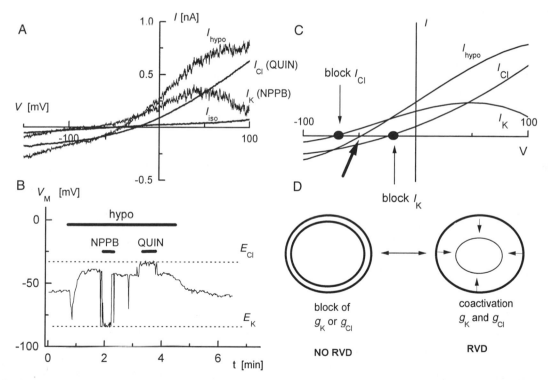

**FIG. 17.** Coactivation of $Cl^-$ and $K^+$ channels is necessary for RVDs. (A) A distal nephron epithelial cell (A6 cell from toad) is challenged by a 40% hypotonic solution. Swelling induces a large current (see change from $I_{iso}$ to $I_{hypo}$). This current can be partially blocked by quinine (500 $\mu M$). The remaining current is a $Cl^-$ current ($I_{Cl}$, QUIN, reversal potential close to $E_{Cl} = -15$ mV). The other component of the current can be blocked by NPPB (100 $\mu M$) and is a $K^+$ current ($I_K$, NPPB, reversal potential close to $E_K = -85$ mV). (B) Changes in membrane potential during cell swelling and after application of NPPB and quinine (for further explanation see text). (C) Schematic presentation of the swelling-activated currents. Block of one component will shift the membrane potential toward the reversal potential of the remaining component. Zero-current represents (thick arrow) inward $Cl^-$ and outward $K^+$ currents, thus efflux of both osmolytes. (D) Block of one conductance will block RVD because of the shift of the membrane potential to the reversal potential of the remaining component and thus inhibition of osmolyte efflux. Only *coactivation* of two currents can induce RVD. (From B. Nilius, unpublished, 1996.)

II,J), is the main player in this complex mechanism of volume regulation.

RVD is not dependent only on activation of ion channels. The functional cooperation between volume-sensitive ion channels, cotransporters, and antiporters (electroneutral $K^+$–$Cl^-$ cotransport, $Na^+$–$H^+$ antiporter, $Cl^-$–$HCO_3^-$ antiporter, $Na^+$–$K^+$–$2\,Cl^-$, $Na^+$–$Cl^-$, and $K^+$–$Cl^-$ cotransporters) in the regulation of cell volume is discussed in Chapter 18.

## B. Role of Ion Channels in Vectorial Transports

Many functions of nonexcitable cells are associated with vectorial secretion and reabsorption. A now well-understood mechanism for vectorial transport is the $Na^+$ reabsorption, for example, in renal cell, colon cells, and frog skin. In these cells, ion channels are immobilized by being anchored to the cytoskeleton at strategically important sites. In the example shown in Fig. 18A for frog skin, the amiloride-blockable $Na^+$ channel is only localised at the outer, mucosal, or apical face of these cells. At the inner, serosal, or basolateral side, however, a high density of $Na^+$–$K^+$ pumps and $K^+$ channels is found. The apical $Na^+$ channels are always open and therefore provide an influx pathway into the cell for $Na^+$ which can be sequestered at the basolateral face by the action of the $Na^+$,$K^+$-ATPase. $K^+$ will be inwardly transported by the pump but will be outwardly directed by the driving force through a variety of $K^+$ channels localized at the basolateral side of the cell. By this concerted action of ion channels and a pump, a sink for $Na^+$ appears that leads the ion through the epithelial cell. Thus, the main task of ion channels in these kinds of nonexcitable cells is to promote transport in the correct direction. This transport can be facilitated by increasing the number of functional amiloride-sensitive $Na^+$ channels in the presence of ADH or aldosterone, but it is inhibited by closing of the channels in the presence of **atrial natriuretic peptide (ANP).**

Another striking example for the functional importance of ion channels is $Cl^-$ channels that control vectorial transport (Fig. 18B). Acinar cells in secretory glands pump $Cl^-$ from the interstitial side (blood side, basolateral side) via a well-understood plasma membrane **$Na^+$–$K^+$–$2\,Cl^-$ cotransporter** into the cell. This cotransporter uses the $Na^+$ gradient generated by the $Na^+$–$K^+$ pump to fuel a $Cl^-$ inward transport against a $Cl^-$ gradient. This transport against the $Cl^-$ gradient will be energetically in equilibrium when the $Na^+$ gradient equals the $Cl^-$ gradient. If the cell is now stimulated, for example, by acetylcholine or cholecystokinine, intracellular $Ca^{2+}$ is released from stores in the endoplasmic reticulum and would open $Cl^-$ channels at the luminal side. Activation of the channel would drain $Cl^-$ into the lumen of the secretory duct, the $Cl^-$ gradient would be increased, and the cotransporter would be activated. The stimulated $K^+$ influx would be compensated by opening of $Ca^{2+}$-activated $K^+$ channels in both the luminal and abluminal membrane. The vectorial transport of $Cl^-$ would be followed by a passive movement of $Na^+$ and $H_2O$ through a paracellular pathway. The inwardly transported $Na^+$ and $K^+$ would be recycled via the $Na^+$–$K^+$ pump and

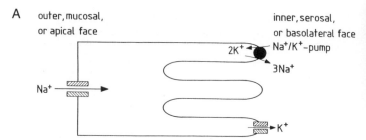

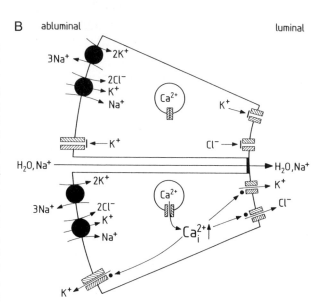

**FIG. 18.** Ion channels are involved in vectorial transport. (A) In $Na^+$-reabsorbing epithelial cells, amiloride $Na^+$ channels are located at the outer, mucosal, and apical faces. $Na^+$–$K^+$ pumps are mainly localized at the inner, serosal, basolateral face. This topology creates a $Na^+$ sink. This scheme describes an absorptive epithelium in frog skin. (B) $Cl^-$ transport occurs due to the polarized distribution of $Ca^{2+}$ or cAMP-gated $Cl^-$ channels. Abluminal $Cl^-$ entry proceeds via a $Na^+$–$K^+$–2 $Cl^-$ cotransporter, luminal secretion via $Cl^-$ channels. The lower cell is stimulated via $Ca^{2+}$ or cAMP that open $Cl^-$ channels, and secretion starts passively followed by water movement through a paracellular pathway for further explanation).

presumably by $Ca^{2+}$-activated $K^+$ channels. $Cl^-$ could escape at the luminal side via $Cl^-$ channels that are not located at the abluminal side.

In other epithelial cells stimulation of $Cl^-$ channels is achieved by cAMP-mediated phosphorylation. Opening of these cAMP-activated $Cl^-$ channels induces the same mechanism as described. If by a defect in the channel protein (CFTR) the channel cannot be activated even though the same amount of cAMP is still available in response to a secretory stimulus, secretion cannot proceed. This channel defect is the main cause of the manifestation of cystic fibrosis, that is, a secretory malfunction including clogging of the airways by mucus and malsecretion in the pancreas and sweat glands—all due mainly to the loss of cAMP regulation of the $Cl^-$ channel.

The opposite effect appears for intact $Cl^-$ channels but an overproduction of cAMP in the presence of the $G_s$-

protein activating cholera toxin. Secretory Cl$^-$ channels are now excessively opened, causing secretion of fluid toward the luminal side.

Another example for the cooperation of ion channels is given in Fig. 19. As observed by intracellular Ca$^{2+}$ imaging, after stimulation of pancreatic acinar cells, [Ca$^{2+}$]$_i$ first rises at the luminal side of the cell. If Cl$^-$ has accumulated in the resting cell and if the membrane potential is more negative than the Cl$^-$ equilibrium potential, $E_{Cl}$, opening of Ca$^{2+}$-activated Cl$^-$ channels at the abluminal side of the membrane would induce a Cl$^-$ efflux out of the cell. The delayed increase in [Ca$^{2+}$]$_i$ at the abluminal side of the cell would open Ca$^{2+}$-activated nonselective cation channels, causing depolarization of the cell, and a Cl$^-$ influx through Ca$^{2+}$-activated Cl$^-$ channels (because the membrane potential is now more positive than the Cl$^-$ equilibrium potential) would occur. These events first "push" Cl$^-$ into the lumen and then "pull" Cl$^-$ into the cell (**push–pull model**). Sequestration of Ca$^{2+}$ would restore the resting state of the cell.

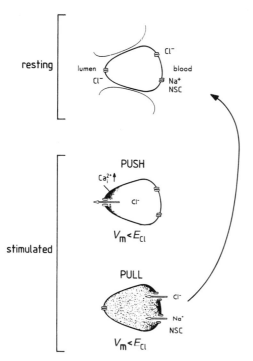

**FIG. 19.** The push–pull hypothesis for vectorial transport. In a resting secretory pancreas acinus cell Cl$^-$ channels and NSC are closed. Stimulation by a secretagogue induces first a luminal increase in [Ca$^{2+}$]$_i$ that activates Cl$^-$ channels and a Cl$^-$ efflux via the channel because the membrane potential is more negative than the Cl$^-$ equilibrium potential. The abluminal channels are still closed. In a later phase of simulation, [Ca$^{2+}$]$_i$ is higher at the abluminal side. Here, Ca$^{2+}$-dependent NSC and Cl$^-$ channels now open, mediating a Cl$^-$ influx because in this phase the membrane potential is more positive than the Cl$^-$ equilibrium potential. The luminal channels are now closed. Sequestration of Ca$^{2+}$ leads again to the resting state. [Modified with permission from *Nature*, Kasai, H., and Augustine, G. J. (1990). Vol. 348, 735–738. copyright 1990 Macmillan Magazines Limited.]

## IV. Function of Ion Channels in Ca$^{2+}$ Signaling

Cell functions are often regulated by changes in [Ca$^{2+}$]$_i$. A transient Ca$^{2+}$ signal that directly follows the activation of nonexcitable cells by an agonist is generated by Ca$^{2+}$ release from Ca$^{2+}$ stores via Ins(1,4,5)P$_3$. However, in many cells a sustained Ca$^{2+}$ signal is required to maintain the secretion or the synthesis of various messengers. An important example is the Ca$^{2+}$-dependent synthesis of **endothelium-derived relaxing factor (EDRF)**. This factor is identical with the chemically simple molecule nitric oxide (NO) and acts as an activator of the soluble guanylate cyclase that increases the intracellular cGMP concentration in neighboring cells. EDRF-NO is released by a long-lasting elevation of [Ca$^{2+}$]$_i$. It is produced by a Ca$^{2+}$-calmodulin-NADPH dependent NO synthase that catalyzes oxidation of the N-guanidine terminal of L-arginine and is released as the easily diffusable NO. An increase in [Ca$^{2+}$]$_i$ in the endothelial cells by release of Ca$^{2+}$ from intracellular stores and by a sustained influx via membrane channels is the trigger event for activation of the NO synthase and the release of EDRF. This sustained Ca$^{2+}$ influx is via transmembrane Ca$^{2+}$-permeable channels (DACs, SOCs, CRACs) or Ca$^{2+}$-permeable nonselective ion channels. For Ca$^{2+}$-activated NSCs positive feedback would be generated: Ca$^{2+}$ influx would induce an increase of [Ca$^{2+}$]$_i$ and would in turn again activate Ca$^{2+}$-NSC. In addition, the driving force for Ca$^{2+}$ could be increased by activation of K$^+$ channels (see also Fig. 6). However, block of K$^+$ channels would decrease the Ca$^{2+}$ influx. Opening of Ca$^{2+}$-activated potassium channels, however, as in endothelial cells, would promote Ca$^{2+}$ influx.

As discussed in detail, most of the exocrine or endocrine cells activate secretion by an increase in [Ca$^{2+}$]$_i$. Another unique area of cooperation between [Ca$^{2+}$]$_i$ and ion channels has been described for renin secretion from juxtaglomerular epitheloid cells. In these cells exocytosis of renin is controlled by the cell volume: Swelling of the cells enhances the renin exocytosis. Release of renin is blocked by an increase of [Ca$^{2+}$]$_i$. Elevation of [Ca$^{2+}$]$_i$ induces activation of Ca$^{2+}$-dependent Cl$^-$ channels and also depolarization of the cells because the membrane potential is shifted toward the Cl$^-$ equilibrium potential. In turn, K$^+$ efflux is promoted through K$^+$ channels. Due to coactivation of both channels, the cells decrease their osmolarity and shrink. This in turn decreases renin secretion. Stimulation of the cells, for example, with angiotensin II (AT II), increases [Ca$^{2+}$]$_i$, which causes negative feedback in the secretion of renin.

## V. Summary

Ion channels in nonexcitable cells have been studied much less than channels in tissues that are involved in fast signaling via voltage-operated ion channels by generation of action potentials. In addition, a considerable lack of information still exists concerning the molecular biology of these channels.

In this chapter, ion channels are described that are involved in the control of biologically important functions.

1. Na$^+$-selective, amiloride-blockable ion channels regulate Na$^+$ absorption and Na$^+$ secretion in epithelial cells.
2. Ca$^{2+}$-activated nonselective cation channels provide in many nonexcitable cells a pathway for the entry of Ca$^{2+}$ and cause depolarization. Other nonselective cation channels can be activated by extracellular agonists via G-protein-dependent mechanisms, and are directly activated by extracellular messengers such as ATP or are directly gated by intracellular cyclic nucleotides (e.g., cGMP, cAMP).
3. K$^+$ channels are widely distributed in nonexcitable cells. In many cells, they are activated by intracellular Ca$^{2+}$ (voltage-dependent small-, intermediate-, and big-conductance channels) and modulate cellular signals that are caused by various agonists. Inwardly rectifying K$^+$ channels may also be targets of modulation.
4. Cl$^-$ channels are involved in secretory functions of nonexcitable cells. They are activated by an elevation of intracellular Ca$^{2+}$, by cAMP via PKA-dependent (CFTR Cl$^-$ channels) changes in cell volume, or are voltage-dependent members of the ClC family.
5. Nonexcitable cells often act as mechanosensors. Various mechanically activated ion channels (K$^+$ channels, nonselective cation channels, Cl$^-$ channels) are involved in volume regulation of many cells and in shear stress-mediated release of EDRF in endothelial cells. Mechanically activated ion channels are widely distributed in nonexcitable cells. The relationship of those channels to the cytoskeleton or to mechanically dependent metabolic pathways is less well understood.
6. Nonexcitable cells lack voltage-operated Ca$^{2+}$ channels, but have developed a Ca$^{2+}$ entry mechanism that depends on intracellular Ca$^{2+}$ stores (SOCs, DACs, and CRACs). Long-lasting Ca$^{2+}$ signals in various nonexcitable cells are due to activation of SOCs. Molecular candidates have recently been described (TRP and TRPL).
7. Some nonexcitable cells also express voltage-operated ion channels, but do not fire action potentials.

The following problems, which are related to functional properties of ion channels in nonexcitable cells, are discussed:

1. Ion channels are involved in the regulation of cell volume. This regulation is coupled to coactivation of different types of ion channels (K$^+$ and Cl$^-$ channels).
2. Ion channels in nonexcitable cells are primarily involved in the vectorial organization of transport rather than in fast signaling. For this function they require localization at strategically important sites of the cells (polarization of the distribution of ion channels). Cytoskeleton may play the major role in establishing polarization. Cooperation of different channel types is responsible for vectorial transport functions.

3. Ca$^{2+}$-signaling is modulated by ion channels that are permeable for Ca$^{2+}$ (SOCs, nonselective cation channels) and channels that control the driving force for Ca$^{2+}$ entry (K$^+$ channels, Cl$^-$ channels).

## Bibliography

Bargman, C. I. (1994). Molecular mechanisms of mechanosensation? *Cell* **78,** 729–731.

Baron, A., Frieden, M., Chaud, F., and Bény, J.-L. (1996). Ca$^{2+}$-activated nonselective cation and potassium channels activated by bradykinin in pig coronary artery endothelial cells. *J. Physiol.* **493,** 691–706.

Bennett, D. L., Petersen, C. C. H., and Cheek, T. R. (1995). Cracking I$_{CRAC}$ in the eye. *Curr. Biol.* **5,** 1225–1228.

Benos, D. J. (1989). The biology of amiloride-sensitive sodium channels. *Hosp. Pract.* **24,** 149–164.

Benos, D. J., Awayda, M. S., Ismailov, I. I., and Johnson, J. P. (1995). Structure and function of amiloride-sensitive Na$^+$ channels. *J. Membrane Biol.* **143,** 1–18.

Berridge, M. J. (1995). Capacitative calcium entry. *Biochem. J.* **312,** 1–11.

Biel, M., Altenhofen, W., Hullin, R., Ludwig, J., Freichel, M., Flockerzi, V., Dascal, N., Kaupp, U. B., and Hofmann, F. (1939). Primary structure and functional expression of a cyclic nucleotide-gated channel in rabbit aorta. *FEBS Lett.* **329,** 134–138.

Butler, A., Tsunoda, S., McCobb, D. P., Wei, A., and Slakoff, L. (1993). mSlo, a complex mouse gene encoding "maxi" calcium activated potassium channels. *Science* **261,** 221–224.

Canessa, C. M., Horisberger, J.-D., and Rossier, B. C. (1993). Epithelial sodium channels related to proteins involved in neurodegeneration. *Nature* **361,** 467–470.

Canessa, C. M., Schild, L., Buell, G., Thorens, B., Gautschi, I., Horisberger, J.-D., and Rossier, B. C. (1994). Amiloride-sensitive epithelial Na$^+$ channel is made of three homologous subunits. *Nature* **367,** 463–467.

Clapham, D. E. (1995). Calcium signaling. *Cell* **80,** 259–268.

Clementi, E., and Meldolesi, J. (1996). Pharmacological and functional properties of voltage-independent Ca$^{2+}$ channels. *Cell Calcium* **19,** 269–279.

Cook, D. I., and Young, J. A. (1990). Cation channels and secretion. *In* "Epithelial Secretion of Water and Electrolytes" (J. A. Young and D. I. Cook, Eds.), pp. 15–38. Springer Verlag, Heidelberg-Berlin.

Cunningham, S. A., Awayda, M. S., Bubien, J. K., Ismailov, I. I., Arrate, M., Berdiev, B. K., Benos, D. J., and Fuller, C. M. (1995). Cloning of an epithelial chloride channel from bovine trachea. *J. Biol. Chem.* **270,** 31016–31026.

Davies, P. F., and Barbee, K. A. (1994). Endothelial cell surface imaging: Insights into hemodynamic force transduction. *News Physiol. Sci.* **9,** 153–157.

Davies, P. F. (1995). Flow-mediated endothelial mechanotransduction. *Physiol. Rev.* **75,** 519–560.

Fasolato, C., Innicenti, B., and Pozzan, T. (1994). Receptor-activated Ca$^{2+}$-influx: How many mechanisms for how many channels? *TIPS* **15,** 77–83.

Gray, M. A., and Argent, B. E. (1990). Nonselective cation channel on pancreatic duct cells. *Biochim. Biophys. Acta* **1029,** 33–42.

Hamill, O. P., and McBridge, D. W., Jr. (1994). The cloning of mechano-gated membrane ion channels. *TINS* **17,** 439–443.

Hille, B. (1992). "Ionic Channels of Excitable Membranes," Chap. 8, pp. 202–235. Sinauer Associates, Sunderland, MA.

Hoffmann, E. K., and Dunham, P. B. (1995). Membrane mechanisms and intracellular signaling in cell volume regulation. *Int. Rev. Cytol.* **161,** 173–262.

Hoth, M., and Penner, R. (1992). Depletion of intracellular calcium stores activates a calcium current in mast cells. *Nature* **355**, 353–356.

Hoyer, J., Popp, R., Meyer, J., Galla, H. J., and Gogelein, H. (1991). Angiotensin II, vasopressin and GTPγS inhibit inward-rectifying K$^+$ channels in porcine cerebral capillary endothelial cells. *J. Membr. Biol.* **123**, 55–62.

Jentsch, T. J. (1993). Chloride channels. *Curr. Opin. Neurobiol.* **3**, 316–321.

Jentsch, T. J. (1994). Trinity of cation channels. *Nature* **367**, 412–413.

Kasai, H., and Augustine, G. J. (1990). Cytosolic Ca$^{2+}$ gradients triggering unidirectional fluid secretion from exocrine pancreas. *Nature* **348**, 735–738.

Kaupp, U. B. (1991). The cyclic nucleotide-gated channel of vertebrate photoreceptors and olfactory epithelium. *TINS* **14**, 150–157.

Kurtz, A. (1990). Do calcium-activated chloride channels control renin secretion? *News Physiol. Sci.* **5**, 43–46.

Kurtz, A., and Penner, R. (1989). Angiotensin II induces oscillations of intracellular calcium and blocks anomalous inward rectifying potassium current in mouse renal juxtaglomerular cells. *Proc. Natl. Acad. Sci. USA* **86**, 3423–3427.

Lansman, J. B., Hallam, T. J., and Rink, T. J. (1987). Single stretch-activated ion channels in vascular endothelial cells as mechanotransducers? *Nature* **235**, 811–813.

Lückhoff, A., and Clapham, D. E. (1992). Inositol 1,3,4,5-tetrakisphosphate activates an endothelial Ca$^{2+}$-permeable channel. *Nature* **355**, 356–358.

Malek, A. M., and Izumo, S. (1994). Molecular aspects of signal transduction of shear stress in the endothelial cell. *J. Hypertens.* **12**, 989–99.

McCarty, N. A., and O'Neil, R. G. (1992). Calcium signaling in cell volume regulation. *Physiol. Rev.* **72**, 1037–1062.

McManus, M. L., Churchwell, K. B., and Strange, K. (1995). Regulation of cell volume in health and disease. *New Engl. J. Med.* **333**, 1260–1266.

Meldolesi, J., Clementi, E., Fasolato, C., Zacchetti, D., and Pozzan, T. (1991). Ca$^{2+}$-influx following receptor activation. *TIPS* **12**, 289–292.

Morris, C. E. (1990). Mechanosensitive ion channels. *J. Membr. Biol.* **113**, 93–107.

Nilius, B. (1991a). Ion channels and regulation of transmembrane Ca$^{2+}$ influx in endothelium. *In* "Electrophysiology and Ion Channels of Vascular Smooth Muscle and Endothelial Cells" (N. Sperelakis and H. Kuriyama, Eds.), pp. 317–325. Elsevier, New York.

Nilius, B. (1991b). Regulation of transmembrane calcium fluxes in endothelium. *News Physiol. Sci.* **6**, 110–114.

Nilius, B., and Castles, R. (1996). Biology of the vascular wall and its interaction with migratory and blood cells. *In* "Comprehensive Human Physiology" (R. Greger and U. Windhorst, Eds.), Vol. II, Chap. 98, pp. 1981–1993. Springer-Verlag, Berlin, Heidelberg.

Nilius, B., Eggermont, J., Voets, T., and Droogmans, G. (1996). Volume-activated Cl$^-$-channels. *Gen. Pharmacol.* **27**, 1137–1140.

Nilius, B., Sehrer, J., De Smet, P., Van Driesche, W., and Droogmans, G. (1995). Volume regulation in a toad epithelial cell line: Role of coactivation of K$^+$ and Cl$^-$ channels. *J. Physiol.* (*London*) **487**, 367–378.

Nilius, B., Viana, F., and Droogmans, G. (1997). Ion channels in vascular endothelium *Annu. Rev. Physiol.* **59**, 145–173.

Nilius, B., and Wohlrab, E. (1992). Potassium channels and regulation of proliferation of human melanoma cells. *J. Physiol.* (*London*) **445**, 537–548.

Oleson, S.-O., Clapham, D. E., and Davies, P. F. (1988). Haemodynamic shear stress activates a K$^+$ current in vascular endothelial cells. *Nature* **331**, 168–170.

Petersen, O. H. (1992). Stimulus-secretion coupling: Cytoplasmic calcium signals and the control of ion channels in exocrine acinar cells. *J. Physiol* (*London*) **448**, 1–54.

Petrov, A. G., and Usherwood, P. N. R. (1994). Mechanosensitivity of cell membranes. *Eur. Biophys. J.* **23**, 1–19.

Pusch, M., and Jentsch, T. J. (1994). Molecular physiology of voltage-gated chloride channels. *Physiol. Rev.* **74**, 813–826.

Putney, J. W., Jr. (1990). Capacitative calcium entry revisited. *Cell Calcium* **11**, 611–624.

Ringe, D., and Petsko, G. A. (1990). A transport problem. *Nature* **346**, 312–313.

Salkoff, L., and Jegla, T. (1995). Surfing the DNA data bases for K$^+$ channels nets yet more diversity. *Neuron* **15**, 489–492.

Sandford, C. A., Sweiry, J. H., and Jenkinson, D. H. (1992). Properties of a cell volume-sensitive potassium conductance in isolated guinea-pig and rat hepatocytes. *J. Physiol.* (*London*) **447**, 133–148.

Schwarz, G., Droogmans, G., Callewaert, G., and Nilius, B. (1992). Shear stress induced calcium transients in human endothelial cells from umbilical cord veins. *J. Physiol.* (*London*) **458**, 527–538.

Schwarz, G., Droogmans, G., and Nilius, B. (1992). Shear stress induced membrane currents and calcium transients in human vascular endothelial cells. *Pflügers Arch.* **421**, 394–396.

Siemen D. (1993). "Nonselective Cation Channels: Pharmacology, Physiology and Biophysics" (D. Siemen and J. Hescheler, Eds.). Birkhäuser Verlag, Germany.

Silver, M. R., and Decoursey, T. E. (1990). Intrinsic gating of inward rectifier in bovine pulmonary artery endothelial cells in the presence or absence of internal Mg$^{2+}$. *J. Gen. Physiol.* **96**, 109–133.

Smith, P. R., Cacomani, G., Joe, E.-H., Angelides, J., and Benos, D. J. (1991). Amiloride-sensitive sodium channel is linked to the cytoskeleton in renal epithelial cells. *Proc. Natl. Acad. Sci. USA* **88**, 6971–6975.

Strange, K., Emma, F., and Jackson, P. S. (1996). Cellular and molecular physiology of volume-sensitive anion channels. *Am. J. Physiol.* **270**, C711–C730.

Sukharev, S. I., Blount, P., Martinac, B., Blattner, F. R., and Kung, C. (1994). A large-conductance mechanosensitive channel in *E. coli* encoded by mscL alone. *Nature* **368**, 265–268.

Swandulla, D., and Partridge, L. D. (1990). Nonspecific cation channels. *In* "Potassium Channels: Structure, Classification, Function and Therapeutic Potential" (N. S. Cook, Ed.). pp. 167–179. Ellis Horwood, Chichester, UK.

Thorn, P., and Peterson, O. H. (1992). Activation of nonselective cation channels by physiological cholecystokinin concentrations in mouse pancreatic acinar cells. *J. Gen. Physiol.* **100**, 11–25.

Tsien, R. W., and Tsien, R. Y. (1990). Calcium channels, stores, and oscillations. *Annu. Rev. Cell Biol.* **6**, 715–760.

Arturo Liévano and Alberto Darszon

# 29

# Sperm Ion Channels

## I. Introduction

Fertilization is one of the most important biological events. It allows not only the generation of a new individual, but also an opportunity for genetic recombination and an increased probability of species preservation over time. It involves a complex communication between gametes, leading to their final encounter and fusion to start a developmental program. Sperm, motile cells that are important protagonists in this process, are very specialized. They lack the machinery for protein or nucleic acid synthesis and have only a nucleus, mitochondria, flagellum, acrosomal vesicle, and a centriole pair. Figure 1A depicts sperm from the sea urchin (upper drawing) and from the mouse (lower drawing).

Sperm, as for other types of cells, have ion channels that participate in fundamental responses to the outer layer of the egg that are required for fertilization. It has been shown that the flow of ions through the plasma membrane of sperm, particularly $Ca^{2+}$, participates crucially in the events leading to fertilization (Schackmann, 1989; Darszon et al., 1996). Indeed, sperm are excitable cells that quickly respond with changes in their plasma membrane ion permeability to components of the outer layer of the egg. During its life span, sperm must respond to different stimuli, thereby changing behavior until fusion is achieved with the egg to create a zygote.

## II. Sperm Responses to Egg Components

### A. Sea Urchin Sperm

Sea urchins are among the most useful and better-known experimental models in fertilization because (1) they undergo external fertilization in a simple medium (seawater) and (2) each male can spawn about $10^{10}$ cells, making it possible to have large amounts of biological material for biochemical and biophysical manipulations. Furthermore, the cells respond to environmental stimuli rapidly, synchronously, and in a compulsory order.

Sea urchin sperm are metabolically arrested in semen. When spawned into seawater, their respiration and motility quickly activate, and chemotaxis to egg components leads them across large distances to the egg (reviewed in Garbers, 1989; Darszon et al. 1996). Contact with the **egg jelly** induces dramatic morphophysiological changes in sperm through a complex process called the **acrosome reaction** (**AR**), which is necessary for fertilization. The most conspicuous morphological change in this reaction is the exocytosis of the **acrosomal vesicle** and the extension of the **acrosomal tubule** (Fig. 1B, upper panel). The AR leads to the release of hydrolytic enzymes present in the acrosome that are required for sperm penetration of the egg's external layers, and to the exposure of new membrane surfaces specialized for fusion with the egg plasma membrane. When sperm undergo the AR, a protein called **bindin** (see Fig. 1B) is exposed and interacts in a species-specific manner with a receptor in the egg that has been recently cloned (Foltz et al., 1993). The AR is triggered by plasma membrane ion permeability changes that lead to modifications in membrane potential ($E_m$), intracellular pH ($pH_i$), and intracellular [Ca] ($[Ca]_i$).

### 1. Responses to Egg Peptides

Small peptides contained in the egg jelly surrounding the sea urchin egg profoundly influence sperm physiology. For instance, picomolar concentrations of **speract**, a decapeptide isolated from *Strougylocentrotus purpuratus* egg jelly, and **resact**, a similar peptide isolated from *Arbacia punctulata*, stimulate sperm phospholipid metabolism and respiration in sperm suspended in acidified seawater. At nanomolar concentrations these peptides stimulate $^{22}Na^+$ and $^{45}Ca^{2+}$ uptake, $H^+$ and $K^+$ efflux, and increases in cGMP, cAMP, $[Ca]_i$, and $pH_i$ (Garbers, 1989; Schackmann, 1989; Ward and Kopf, 1993). They also regulate sperm motility (Ward et al., 1985; Cook et al., 1994) and appear to enhance fertilization (Suzuki and Yoshino, 1992). Chemotaxis has been demonstrated only in *A. punctulata*, where nanomolar concentrations of resact attract sperm in a $Ca^{2+}$-dependent manner (Ward et al., 1985).

456

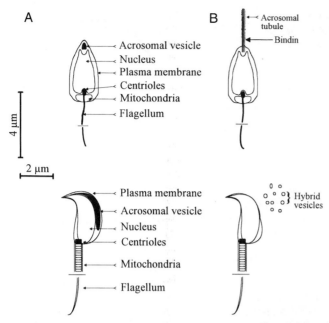

A

B

**FIG. 1.** Schematic diagram of a sea urchin (upper half) and mouse sperm (lower half) (A) before and (B) after acrosome reaction. The sperm head contains all the cell organelles except the flagellum.

Speract triggers a hyperpolarization in *S. purpuratus* sperm flagella (and in flagellar plasma membrane vesicles), probably mediated by $K^+$ channels, which stimulates a voltage-dependent $Na^+/H^+$ exchange (Lee and Garbers, 1986). In these membranes, GTPγS stimulates the speract-induced hyperpolarization, suggesting the possible participation of a G protein (Lee, 1988). Indeed, $G_i$, $G_s$, and several low molecular weight G proteins have been detected in sea urchin sperm (Ward and Kopf, 1993; Cuéllar-Mata *et al.*, 1995).

Speract binds to a 77-kDa sperm plasma membrane protein; this protein was purified, sequenced, and cloned from *S. purpuratus*. The receptor modulates a membrane guanylyl cyclase (Garbers, 1989). In *A. punctulata*, resact binds to a membrane-bound guanylyl cyclase, inducing its dephosphorylation and increasing its activity (Trimmer and Vacquier, 1986). The phosphorylation state of guanylyl cyclase is pH- and $[Na^+]_e$-dependent (Trimmer and Vacquier, 1986; Garbers, 1989). The resact receptor is the first cloned and sequenced member of a family of guanylyl cyclases that are surface receptors participating in a new signal transduction pathway (Drewett and Garbers, 1994).

The relationship between the speract-induced increase in cGMP and cAMP levels and the resulting changes in ionic permeability of the sperm holds the key to understanding how sea urchin gametes manage to meet. Patch-clamping sea urchin sperm is difficult due to their tiny size (Fig. 5C, shown later; Guerrero *et al.*, 1987); thus, to facilitate clamping, they must be swollen in diluted seawater (Fig. 5D, shown later; Babcock *et al.*, 1992). Swollen sperm are spherical (ca. 4 $\mu M$ in diameter) and retain their $E_m$, $pH_i$, and $[Ca]_i$ regulation (Fig. 2, left side). Picomolar concentrations of speract increase a $K^+$-selective perme-

ability in swollen sperm, mediated by $K^+$ channels, as indicated by patch-clamp experiments (Babcock *et al.*, 1992). It is not known how the increase in [cGMP] opens $TEA^+$-insensitive $K^+$-selective channels that hyperpolarize sperm. One of the speract receptors could be coupled to a G protein, which might directly or indirectly activate $K^+$ channels. Higher speract concentrations ($>25$ p$M$) transiently hyperpolarize the cells nearby to the $K^+$ equilibrium potential ($E_K$) and immediately repolarize them toward the resting potential ($E_R$; Fig. 2, upper left panel). As mentioned earlier, the hyperpolarization activates $Na^+/H^+$ exchange (Lee and Garbers, 1986; González-Martinez *et al.*, 1992; Reynaud *et al.*, 1993). The increase, in $pH_i$ inhibits guanylyl cyclase (Trimmer and Vacquier, 1986), and stimulates adenylyl cyclase (Cook and Babcock, 1993a,b), which has also been shown to be sensitive to membrane potential (Beltrán *et al.*, 1996), and $[Ca]_i$ (Garbers, 1989). The decrease in [cGMP] would diminish $K^+$ permeability (Cook and Babcock, 1993a) and repolarize sperm.

In addition, at nanomolar concentrations of speract, $[Ca]_i$ increases and a $Ca^{2+}$-dependent depolarization occurs beyond $E_R$ in swollen sperm (Fig. 2, left records; Babcock *et al.*, 1992; Reynaud *et al.*, 1993; Cook and Babcock, 1993a). These changes are inhibited by $Ca^{2+}$-channel blockers such as $Co^{2+}$, $Ni^{2+}$, and $Zn^{2+}$ (Reynaud *et al.*, 1993; Cook and Babcock, 1993b). These $Ca^{2+}$-permeable channels allow $Mn^{2+}$ to pass through and are regulated by cAMP (Cook and Babcock, 1993b). In normal sperm the depolarizing phase is (1) only partly diminished in the absence of external $Ca^{2+}$, (2) depends on external $Na^{2+}$, (3) is poorly sensitive to $Ca^{2+}$ channel blockers, and (4) is blocked by $TEA^+$ and $Ba^{2+}$ (Labarca *et al.*, 1996). Two (or more) ion channels with distinct selectivity and pharmacology might contribute to the depolarization triggered by speract in normal sea urchin sperm: a cAMP- and/or pH-regulated $Ca^{2+}$ channel (Babcock *et al.*, 1992; Cook and Babcock, 1993b) and a cAMP-regulated $K^+$ channel that allows $Na^+$ flux into sperm (Labarca *et al.*, 1996). A cAMP-modulated $K^+$ channel has been detected in flagellar membranes incorporated into planar lipid bilayers. This channel is blocked by $TEA^+$ (30 m$M$) and $Ba^{2+}$, and has a $P_K/P_{Na}$ of 5. Therefore, its opening would depolarize sperm and could explain part of the $Na^+$ dependence of the speract-induced repolarization (Labarca *et al.*, 1996).

### 2. Acrosome Reaction

Contact of sperm with a glycoprotein–fucose sulfate polymer complex contained in the egg jelly, called **factor (FSG),** triggers the AR (Trimmer and Vacquier, 1986; Keller and Vacquier, 1994). This reaction involves acrosomal vesicle exocytosis, which exposes material required for sperm–egg binding. These events lead to the extension of the acrosomal tubule, the latter being surrounded by the membrane destined to fuse with the egg (Ward and Kopf, 1993; Darszon *et al.*, 1996).

The AR requires external $Ca^{2+}$ and $Na^+$ in seawater at pH 8.0. Exposure of sperm to FSG induces, within seconds, $Na^+$ and $Ca^{2+}$ entry and $H^+$ and $K^+$ efflux (Schackmann, 1989; Ward and Kopf, 1993). These ion fluxes result in

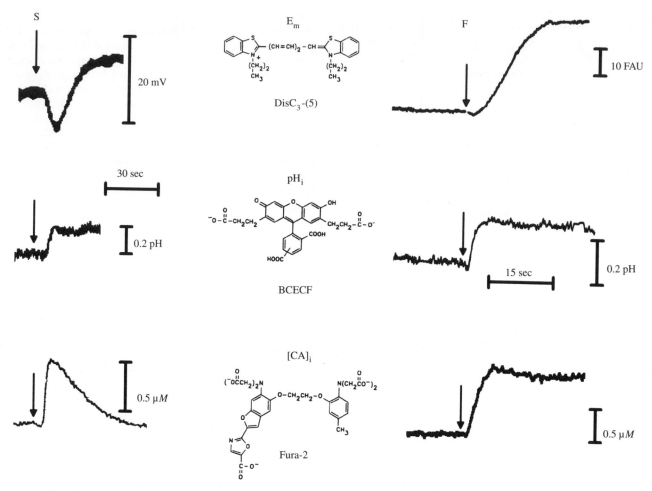

**FIG. 2.** Membrane potential ($E_m$), intracellular pH (pH$_i$), and intracellular Ca$^{2+}$ concentration ([Ca]$_i$) changes in *S. purpuratus* swollen sperm exposed to speract (left-side records) and in normal sperm exposed to the egg factor that triggers AR (FSG) (right-side records). In all records, the arrow indicates 100 nM speract (S) or FSG (F) addition. An upward deflection indicates an increase in the measured parameter. The fluorescent probe used is indicated in the center of the figure: a cyanine dye DisC$_3$-(5) for $E_m$, BCECF for pH$_i$, and fura-2 for [Ca]$_i$. Sperm were swollen in 10-fold diluted artificial seawater containing 20 mM MgCl$_2$ (DASW) (left). For pH$_i$ and [Ca]$_i$ measurements, cells were loaded overnight with the permeant Fura-2-AM or BCECF-AM dyes at 4°C in 0Ca artificial seawater, pH 7.0. For $E_m$ determinations the cells were preequilibrated with 500 nM of the $E_m$-sensitive dye DisC$_3$(5) for 2–3 min (Records kindly provided by Marco González-Martínez, Enrique Reynaud, and Lucia de De La Torre.)

interrelated changes in $E_m$ (González-Martínez and Darszon, 1987; Darszon *et al.*, 1996), [Ca]$_i$ (Schackmann, 1989; Guerrero and Darszon, 1989a), and pH$_i$ (Lee *et al.*, 1983; Guerrero and Darszon, 1989b) (see Fig. 2, records on the right). The FSG also raises cAMP levels, protein kinase A activity, turnover of InsP$_3$, and phospholipase D activity. It is not clear how these changes relate to the FSG-induced permeability changes (Garbers, 1989; Ward and Kopf, 1993).

When exposed to FSG, *Lytechinus pictus* sperm first transiently hyperpolarize and then depolarize (see Fig. 3A). This hyperpolarization is K$^+$ dependent, probably being mediated by K$^+$ channels (González-Martínez and Darszon, 1987), and leads to an increase in pH$_i$, which activates Na$^+$/H$^+$ exchange (Schackmann, 1989). It is not known if speract and FSG modulate the same Na$^+$/H$^+$ exchange. Antagonists to Ca$^{2+}$ channels (verapamil and

dihydropyridines) and K$^+$ channels (TEA$^+$) inhibit Ca$^{2+}$ uptake and the AR in *S. purpuratus* sperm, indicating their mandatory participation in this process (Schackmann, 1989; Darszon *et al.*, 1994). Figure 3B shows the changes in $E_m$ (upper trace), [Ca]$_i$ (middle trace), and pH$_i$ (lower trace) in *L. pictus* sperm suspended in 0K$^+$–seawater associated with a valinomycin-induced hyperpolarization; the latter is followed by a K$^+$-induced depolarization about 60 sec after the hyperpolarization. The first arrow indicates the additon of valinomycin to increase the K$^+$ permeability of the membrane and to bring the $E_m$ close to $E_K$. At this point, only a small percentage of sperm have reacted (at most 16%). However, the hyperpolarization has clearly increased pH$_i$, which at ca. 60 sec reaches the value attained during a normal AR. Addition of KCl to the medium at this time (second arrow) immediately depolarizes and increases [Ca]$_i$ and the percentage of reacted sperm

(>40%). A valinomycin-induced depolarization in high KCl seawater is unable by itself to trigger the AR. Such experiments indicate that it is possible to manipulate $E_m$ and $pH_i$ artificially to induce an increase in $[Ca]_i$, that the induction of the AR requires the coordination between the increase in $[Ca]_i$ and $pH_i$, and that sea urchin sperm have voltage-dependent $Ca^{2+}$ channels (González-Martínez et al., 1992).

Measuring $[Ca]_i$ in sea urchin sperm has provided evidence of the participation of two different $Ca^{2+}$ channels in the AR (Guerrero and Darszon, 1989a,b). Figure 4A depicts the changes in $[Ca]_i$ associated with the FSG-induced AR. After the addition of FSG, $[Ca]_i$ levels increase 10- to 20-fold and remain high. In fact, once the AR has occurred, even though $[Ca]_i$ remains constant, $Ca^{2+}$ influx continues, and $Ca^{2+}$ accumulates in the mitochondria until cell death. Figure 4B shows $[Ca]_i$ changes associated with the addition of FSG to sperm in seawater containing

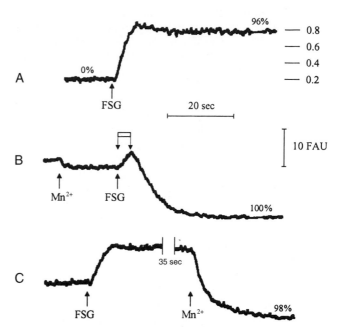

**FIG. 4.** Fura-2 detection of $Mn^{2+}$ influx through $Ca^{2+}$ channels during the *S. purpuratus* sperm FSG-induced AR. (A) Control record showing the profile of $[Ca^{2+}]_i$ change induced by the addition of FSG. $Mn^{2+}$ (3 m*M*) was added (B) before or (C) after FSG. $Mn^{2+}$ influx is indicated by a decrease in fluorescence. Numbers on the right side of (A) indicate the fraction of calcium-bound fura-2. The percentage of FSG-induced AR is shown at the end of each record. (Modified from Guerrero and Darszon, 1989b.)

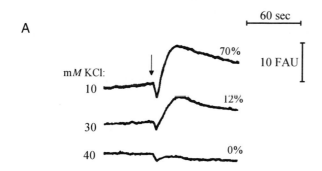

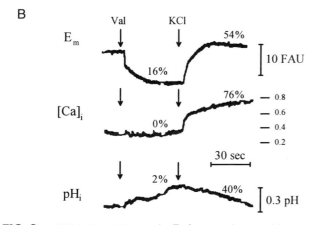

**FIG. 3.** FSG-induced changes in $E_m$ in normal sea urchin sperm. Percent numbers indicate AR; FAU indicates fluorescence arbitrary units. (A) *L. pictus* sperm FSG-induced $E_m$ changes at the indicated KCl concentrations (numbers on the left of each record). (Modified from González-Martínez and Darszon, 1987.) The upper trace shows a control in normal artificial seawater (ASW). (B) Effect of valinomycin (2 $\mu M$) and subsequent KCl addition on $E_m$ (upper trace), $[Ca]_i$ (middle trace), and $pH_i$ (lower trace) in *L. pictus* sperm incubated in 0KASW. Intracellular pH was measured with the fluorescent probe DMCF, and $[Ca]_i$ with QUIN-2. Cells were loaded overnight with 10 $\mu M$ of the permeant form of the dyes at 4°C in 0CaASW, pH 7.0, and $E_m$ was measured as in Fig. 2. (Modified from González-Martínez et al., 1992.)

$Mn^{2+}$.[1] The record illustrates a biphasic behavior: Immediately after FSG addition, a $Ca^+$ channel opens transiently; this channel is very selective for $Ca^{2+}$ and does not allow $Mn^{2+}$ influx. Subsequently, a second type of channel opens that is less selective than the first one, and allows the influx of $Mn^{2+}$ that quenches the fura-2 fluorescence. Thus, the first type of $Ca^{2+}$ channel is a channel that opens by a still unknown mechanism on receptor occupancy. The first type of channel is blocked by verapamil and dihydropyridines (DHPs), and it shows inactivation. The other type is not blocked by these compounds, does not inactivate, and allows $Mn^{2+}$ to permeate. This second type of channel is blocked by conditions that inhibit the increase in $pH_i$ and the AR, but still support a transient increase in intracellular $Ca^{2+}$. Thus, the second channel is modulated by $pH_i$. The opening of the first type of channel is required for the opening of the second, that is, blocking of the first inhibits $Ca^{2+}$ uptake through the second and blocks the AR. How the two types of channels are coupled is still a mystery; however, both are important to fully achieve AR (A. Darszon and M. T. González-Martínez, unpublished, 1997). The first $Ca^{2+}$ channel could be inactivated at the normal sperm resting potential (around −45 mV; González-Martínez and Darszon, 1987), like the low-threshold T-type channels (Hille, 1992). The FSG-induced transient hyperpolariza-

[1] This divalent cation has a 40-fold higher affinity for fura-2 than $Ca^{2+}$ and quenches its fluorescence.

tion could remove this inactivation and subsequently trigger the AR. It is known that rat and mouse spermatogenic cells express T-type $Ca^{2+}$ channels in their membrane (Hagiwara and Kawa, 1984; Liévano et al., 1996; Arnoult et al., 1996; see below), and that AR is blocked by micromolar concentrations of DHPs and $Ni^{2+}$ (Florman et al., 1992), matching the T-type $Ca^{2+}$ channel's sensitivity to those compounds (Hille, 1992; Liévano et al., 1994, 1996).

An approximately 210-kDa plasma membrane protein is the best candidate for the FSG sperm receptor. It is not known how the activated receptor triggers the AR, but it could directly regulate ion channels. This protein has species-specific affinity for egg jelly, and some monoclonal antibodies to it induce the AR (Moy et al., 1996). These antibodies bind to a narrow plasma membrane collar over the acrosome and along the entire flagellum (Trimmer and Vacquier, 1986). Recently this glycoprotein receptor, composed of 1450 residues and approximately 50% carbohydrate, has been cloned (Moy et al., 1996). It has 17 potential sites for N-linked, and 12 for O-linked glycosylation, and only the extreme COOH-terminal appears to have a putative transmembrane region, strongly suggesting that this protein is not an ion channel. The amino-terminal portion has an EGF domain and two contiguous carbohydrate recognition domains with significant relatedness to those of the human macrophage mannose receptor (important in the activation of the classical complement pathway). The receptor contains a novel module (700 residues) that shares extensive homology with the human polycystic kidney disease protein (PKD1). One of the most frequent human genetic diseases, autosomal dominant polycystic kidney disease, is caused by mutations in PKD1 (Moy et al., 1996).

## B. Mammalian Sperm

The **zona pellucida (ZP)**, a thick extracellular glycoprotein coat surrounding the egg, is the main mediator of the sperm AR in mammals. ZP3 is the murine ZP **sulfated glycoprotein**, which displays the sperm-binding and AR-inducing activity of unfertilized eggs. Specific receptors for ZP3 on the plasma membrane of the acrosome-intact sperm overlying the acrosome must mediate its binding and induction of AR. This $Ca^{2+}$-dependent reaction involves the fusion of the plasma and outer acrosomal membranes of the sperm head (Fig. 1B, lower drawing). It exposes the sperm's inner acrosomal membrane, whose surface contains proteases and/or glycosidases believed to allow its penetration through the zona pellucida to reach the egg plasma membrane (Bleil, 1991; Ward and Kopf, 1993).

Identification of the sperm **surface receptor for ZP3** has been attempted by several laboratories. The candidate proteins include these: in mouse sperm, a $\beta$-1,4 galactosyl transferase (Gong et al., 1995), a hexokinase (Leyton and Saling, 1989; Ward and Kopf, 1993), and the lectin sp56 (Bookbinder et al., 1995); and in guinea pig sperm, a hyaluronidase (Gmachl and Kreil, 1993). Multiple concerted and cooperative interactions between ZP3 and the sperm surface, possibly involving receptor aggregation, may be

needed to achieve the signal transduction events that result in AR (Ward and Kopf, 1993; Leyton and Saling, 1989). How these receptors convey information to initiate signal transduction in mammalian sperm is not known.

The Zp-induced AR is inhibited by Pertussis toxin (PTX), a specific inactivator of the $G_i$ class of heterotrimeric G proteins, in mouse, bovine, and human sperm. From the multiple species of G proteins found in mouse sperm, apparently $G_{i1}$ and $G_{i2}$ are preferentially activated by ZP (Ward et al., 1994). Many questions remain to be solved: How does the ZP3-activated receptor turn these $G_i$ proteins on, which activities do they regulate, and are there ion channels among them? It was shown that mouse sperm $\beta$-1,4 galactosyl transferase interacts with a G protein (Gong et al., 1995).

A rise in $[Ca]_i$ is an essential step in the ZP3 signaling path leading to the AR. External $Ca^{2+}$ is required for the physiological AR, and ZP induces $[Ca]_i$ and $pH_i$ increases that precede exocytosis in single sperm loaded with fluorescent dyes as ion indicators. These changes are somehow mediated by $G_i$ proteins since both are inhibited by PTX (Florman et al., 1992).

Evidence has been provided for the presence of voltage-dependent **$Ca^{2+}$ channels** in the plasma membrane of mammalian sperm (Florman et al., 1992). Depolarizing conditions at elevated $pH_i$ that open $Ca^{2+}$ channels sensitive to DHPs bypass the inhibition of the ZP3-induced exocytosis produced by PTX. It is not known how $pH_i$ and $[Ca]_i$ are intimately related and finely tuned in mammalian sperm. The opening of sperm voltage-dependent $Ca^{2+}$ channels appears to be enough to trigger the AR when $pH_i$ is increased. The activation of these channels is a required step in the ZP3 signal transduction pathway (Florman et al., 1992). If $Ca^{2+}$ channels need a depolarization to open, how does ZP depolarize sperm? If not, do second messengers activate the channel possibly by altering its phosphorylation state? Since external $K^+$ depolarizes sperm, $K^+$ channels must be present in the mammalian sperm plasma membrane. Their regulation could be important during the AR and needs to be studied. Because micromolar concentrations of DHPs are needed to block the AR and the increase in $[Ca]_i$, it is difficult to determine the type of $Ca^{2+}$ channels involved: submicromolar concentrations of DHPs block typical L-type $Ca^{2+}$ channels, whereas micromolar concentrations can, as mentioned before, block the low-threshold T-type channels.

It has been suggested that, like in the sea urchin sperm, at least two different $Ca^{2+}$ channels are present in mammalian sperm (Florman, 1994). In addition, in human and mouse sperm, progesterone and other progestins can induce PTX and DHP-insensitive large extracellular $Ca^{2+}$-dependent increases on $[Ca]_i$, which result in AR (Thomas and Meizel, 1989; Darszon et al., 1996). Thus, there are at least two different $Ca^{2+}$-permeable channels in sperm.

Recently, it was shown that the **$IP_3$ receptor**, which is an intracellular $Ca^{2+}$ channel, is present in the acrosomal membrane. A membrane fraction enriched with acrosomal vesicles released $Ca^{2+}$ when exposed to $IP_3$. Thapsigargin ($\sim 3.5 \ \mu M$), an alkaloid that inhibits the endoplasmic reticulum $Ca^{2+}$-ATPase, induced AR in mouse sperm. From these results it was postulated that ZP3 may induce activation of phospholipase C, leading to $IP_3$ generation, which somehow would activate the two plasma membrane $Ca^{2+}$ channels that allow $Ca^{2+}$ entrance during the AR (Walensky and Snyder, 1995). Further experiments are necessary to test this hypothesis.

Progesterone metabolites have been shown to enhance the interaction of $\gamma$-aminobutyric acid (GABA) with the GABA receptor in central nervous system neurons. The **GABA receptor** is a multisubunit protein containing a $Cl^-$ channel, and it has been detected in boar and ram sperm. It has been proposed that the fast progestin-induced human sperm responses may involve steroid interaction with a sperm steroid receptor/$Cl^-$ channel complex, similar to the $GABA_A$/$Cl^-$ channel complex (Wistrom and Meizel, 1993).

## III. Sperm Ion Channels

In spite of the knowledge explosion on ion channels generated by the patch-clamp technique in the last few years, the properties of sperm ion channels are still relatively unknown. The reason for this is their small size and complex geometry (Fig. 1), which has precluded the use of conventional electrophysiological strategies in the characterization of the channels involved in sperm physiology. As mentioned before, one of the main advantages of working with sea urchin sperm is the large quantity of biological material available. This allows the isolation and characterization of different plasma membrane fractions, which can be reassembled to study sperm ion channels in model systmes by different reconstitution strategies (reviewed in Darszon et al., 1994, 1996).

### A. Sea Urchin Sperm Ion Channels

Single $K^+$ **channels** were first recorded in bilayers made at the tip of patch-clamp pipettes from monolayers generated from a mixture of lipid vesicles and isolated sperm flagellar membranes. Three types of $K^+$ channels were identified with conductances of 22, 46, and 88 pS (Fig. 5A). Two of them are blocked by $TEA^+$, which inhibits the AR (Liévano et al., 1985). Although with great difficulty, single channels were recorded directly from sea urchin sperm heads using the patch-clamp technique. Single channel events of 40, 60, and 180 pS were detected, and one of the channels observed was a $K^+$ channel (see Fig. 5C; Guerrero et al., 1987). As mentioned earlier, swelling S. purpuratus sperm improved significantly the success rate of patch formation and allowed the detection of a 2–5 pS $K^+$ channel that is activated by speract. Swollen sperm have opened new possibilities of directly studying the ion channels modulated by egg components and their regulation (Fig. 5D; Babcock et al., 1992).

Two types of $Ca^{2+}$ channels have been detected in S. purpuratus sea urchin sperm by fusing isolated plasma membranes into planar lipid bilayers (Fig. 5B): (1) a voltage-dependent channel of 50-pS conductance observed in 10 mM $Ca^{2+}$ (not shown) and (2) a high-conductance channel having a main conducting state of 172 pS in 50 mM

$CaCl_2$ and several subconductance states (Fig. 5B; Liévano et al., 1990). This channel is strongly voltage dependent, showing a single main-conducting state, with rare closing events at voltages more positive than $-25$ mV, and displaying several subconductance states of lesser conductance at more negative potentials. The main state conductance size sequence is $Ba^{2+} > Sr^2 > Ca^{2+}$, as in many other $Ca^{2+}$ channels (Bean, 1989). The channel discriminates poorly between divalent and monovalent cations: $P_{Ca}/P_{Na} = 5.9$, and is also permeable to $Mg^{2+}$ when added to the cis side ($P_{Ca}/P_{Mg} = 2.8$). In contrast, addition of $Mg^{2+}$ to the trans side blocks the channel in a voltage independent manner (Liévano et al., 1990). Two findings suggest the possible participation of the high-conductance $Ca^{2+}$ channel in the sperm AR. Both $Cd^{2+}$ and $Co^{2+}$ block the channel at concentrations similar to those required to inhibit the AR and the $Ca^{2+}$ uptake induced by egg jelly (Liévano et al., 1990). In addition, $Mg^{2+}$ blocks the high-conductance $Ca^{2+}$ channel only when present in the trans side. This could indicate that $Mg^{2+}$ in seawater may modulate the influx of $Ca^{2+}$ into sperm, either from the outside or after entering the cell. Verapamil and nisoldipine do not significantly modify the kinetics nor the conductance of the high-conductance channel seen in planar bilayers. Therefore, the high-conductance $Ca^{2+}$ channel could be the second type of channel that participates in the AR, allows $Mn^{2+}$ influx, and is $pH_i$-modulated.

Little is known about **anion channels** in these cells. The fusion of sperm plasma membranes into lipid bilayers allowed identification of a 150-pS anion channel (Fig. 5B). This anion channel was enriched from detergent-solubilized sperm plasma memebranes using a wheat germ agglutinin–Sepharose column. Vesicles formed from this preparation were fused into black lipid membranes (BLMs), yielding single-channel anion-selective activity with similar properties as those found in the sperm membranes. The anion selectivity sequence found was: $NO_3^- > CNS^- > Br^- > Cl^-$. This anion channel has a high open probability at the holding potentials tested, it is partially blocked by 4,4′-diisothiocyano-2,2′-stilbendisulfonic acid (DIDS), and often displays substates. DIDS blocks the AR in S. purpuratus sea urchin sperm by a still unknown mechanism. These results suggest that this $Cl^-$ channel could be involved in the events that lead to the AR, or in determining the resting potential of sperm, which modulates this reaction (Morales et al., 1993).

Ion channels are subject to multiple forms of regulation (Hille, 1992), and **cyclic nucleotides** can directly modulate them (Yau, 1994). Spermatozoa respond to the outer layer of eggs with changes in the second messenger levels ($[Ca]_i$, cyclic nucleotides, $IP_3$) (Garbers, 1989; Schackmann, 1989; Ward and Kopf, 1993). Recently a $K^+$-selective channel derived from sea urchin plasma membranes, upwardly regulated by cAMP, has been detected in planar bilayers. Its single-channel conductance is 103 pS in 100 mM KCl. The channel has a low open probability and is weakly voltage dependent. Addition of cAMP in the cis side (the side of membrane addition) up-regulates channel activity in a dose-dependent ($K_D = 200$ $\mu M$) and reversible fashion, increasing the open probability. Millimolar concentrations

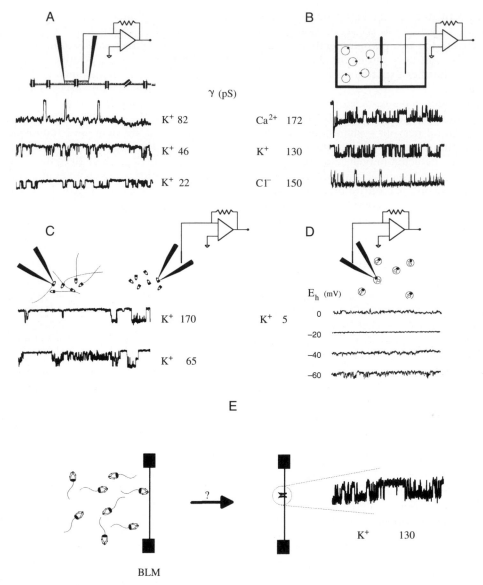

**FIG. 5.** Strategies used for ion channel characterization in sperm cells. The channels detected with each technique are shown, indicating the main ion transported and their single-channel conductance value. (A) Bilayers at the tip of a patch-clamp pipette. (B) Black lipid membranes (BLMs) with fused sperm plasma membrane vesicles. (C) On-cell patch-clamp recordings in sperm (left) and on flagellaless heads (right). (D) On-cell patch-clamp recordings in osmotically swollen sperm. (E) Direct ion channel transfer from cells to the BLM. (Modified from Darszon *et al.*, 1994).

of $Ba^{2+}$ or TEA in the *trans* side blocked the channel in a voltage-dependent fashion. The channel exhibited a low $P_K/P_{Na} \approx 5$, indicating a sizable permeability to $Na^+$; therefore, its opening would depolarize sperm. Hence, this channel could contribute to the depolarizing phase of the response to speract, and perhaps even during the AR as cAMP levels increase (Labarca *et al.*, 1996).

### B. Ion Channels of Mammalian Sperm

There have been fewer reports published on single-channel activity from mammalian sperm than for sea urchin. For example, it was shown that addition of a partially

purified 110-kDa human sperm plasma membrane protein to a preformed lipid bilayer resulted in incorporation of cationic channels having a single-channel conductance of 130 pS (0.1 *M* NaCl). Apparently, the channel was formed from functionally aggregated triplets (Young *et al.*, 1988). Another study described the detection of two types of $Ca^{2+}$ channels in dip-tip bilayers[2] formed from liposomes containing boar sperm plasma membrane. The single-channel conductances detected were between 10 and 20 pS, and 50 and 60 pS, and the smaller channels were partially blocked

[2] Dip-tip bilayers are formed at the tip of patch-clamp electrodes by apposition of two monolayers, see Fig. 5A.

by unknown concentrations of nitrendipine and verapamil, and completely blocked by 0.5 m$M$ La$^{3+}$ (Cox and Peterson, 1989). A nonselective cation channel was also reported, both from cauda epididymal or ejaculated boar sperm plasma membranes incorporated into planar lipid bilayers. Both monovalent and divalent cations permeate through the channel. The channel displays voltage-independent kinetics and is blocked by high concentrations of verapamil or nitrendipine, and by ruthenium red (Cox et al., 1991).

A **cyclic nucleotide-gated** channel from bovine testis is apparently the first sperm channel to be cloned and expressed into *Xenopus* oocytes (Weyand et al., 1994). The channel has 78% identical residues as the cone photoreceptor (for review, see Yau, 1994); in oocytes at 60 mV, it displays a single-channel conductance of 20 pS, allows Ca$^{2+}$ and monovalent cations through, and has a much higher affinity for cGMP (>100 fold) than for cAMP. However, it was not possible to detect cGMP-induced increases in [Ca]$_i$ in fura-2 loaded sperm; but such changes were observed in 10% of vesicles thought to be sperm cytoplasmic droplets. Small cGMP-induced currents associated with single-channel transitions of <10 pS were detected in these vesicles after swelling sperm in a similar fashion as was done for sea urchin sperm (Babcock et al. 1992). Similar cGMP-induced currents were recorded in a small fraction of inside out patches of plasma membrane from human and bovine sperm. What region of the sperm plasma membrane was patch-clamped is unclear, and further work is needed to show that the cloned channel is the one that was recorded.

It is also possible to incorporate ion channels to lipid bilayers directly from sea urchin and mouse spermatozoa (Fig. 5E; Beltrán et al., 1994). The high-conductance Ca$^{2+}$ channel, several K$^+$ channels, and a smaller voltage-dependent 10-pS Ca$^{2+}$ channel (resembling the one from boar sperm described by Tiwari-Woodruff et al., 1995; Fig. 6B) have been recorded with this strategy. This approach opens new avenues to explore cell–cell interactions, such as

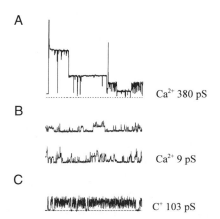

A

Ca$^{2+}$ 380 pS

B

Ca$^{2+}$ 9 pS

C

C$^+$ 103 pS

**FIG. 6.** Mammalian sperm ion channels. (A) A large-conductance Ca$^{2+}$ channel from mouse sperm. (Modified from Beltrán et al., 1994.) (B) A small-conductance Ca$^{2+}$ channel from boar sperm plasma membrane that is blocked by nitrandipine but voltage independent. (Modified from Tiwarí-Woodruff et al., 1995.) (C) A nonselective cationic channel from mouse sperm plasma membranes. (Modified from Labarca et al., 1995.)

sperm–egg fusion, at the single-channel level. More recently several channels were detected by fusing mouse sperm plasma membranes to planar bilayers: (1) an 80-pS anion channel with similar characteristics as the one found in sea urchin sperm plasma membranes, (2) a cation channel ($P_{Na+}/P_{K+}$ = 2.5) with two modes of gating, and (3) a high-conductance Ca$^{2+}$ channel (Fig. 6A; Labarca et al., 1995). This high-conductance Ca$^{2+}$ channel resembles the one from *S. purpuratus* (Liévano et al., 1990). It is attractive to consider that this Ca$^{2+}$ channel may be important in sperm physiology, since it is present in such diverse species (Beltrán et al., 1994). This latter channel was blocked by micromolar concentrations of ruthenium red, which inhibits AR in sea urchin sperm (Labarca et al., 1995).

As mentioned earlier, sperm are differentiated terminal cells lacking the machinery for protein synthesis. Thus, during spermatogenesis, all ion channels required for cell function must be synthesized. Little is known about the biogenesis of ion channels during spermatogenesis and their role in sperm differentiation. Only one report has found that rat spermatogenic cells have only low-threshold, inactivating Ca$^{2+}$ currents that increase their density during rat spermatogenesis. This result could suggest a possible role for Ca$^{2+}$ channels in sperm differentiation (Hagiwara and Kawa, 1984). Recently, the genotypic and phenotypic expression of Ca$^{2+}$ channels was studied in mouse pachytene spermatocytes (PS) and round (RS) and condensing spermatids (CS). These are the cell types in the last phases of spermatogenesis, before the loss of the cytoplasmic vesicle that leads to the formation of mature sperm (Kretzer and Kerr, 1988). It was found that low-threshold T-type Ca$^{2+}$ channels are the only detectable Ca$^{2+}$ channels in the mouse PS (Fig. 7A; Liévano et al., 1996; Santi et al., 1996; Arnoult et al., 1996). RNAs isolated from purified fractions of PS, RS, and CS were explored for the presence of transcripts of the Ca$^{2+}$ channel $\alpha_1$-subunit, the one containing both the pore and the voltage sensor of voltage-dependent Ca$^{2+}$ channels (Stea et al., 1995). Reverse transcription coupled to polymerase chain reaction (RT-PCR) with oligonucleotides specific for $\alpha_{1A}$, $\alpha_{1B}$, $\alpha_{1C}$, $\alpha_{1D}$, and $\alpha_{1E}$ showed that only $\alpha_{1E}$ and to a much lesser extent $\alpha_{1A}$ transcripts are present in PS, RS, and CS (Fig. 7C; Liévano et al., 1996). Consistent with this, the LVA T-type Ca$^{2+}$ currents from primary pachytene spermatocytes are sensitive to nifedipine and Ni$^{2+}$ (Fig. 7B; Liévano et al., 1996; Santi et al., 1996). Since the sperm AR and the uptake of Ca$^{2+}$ that triggers it are also inhibited by these blockers, it is likely that a T-type Ca$^{2+}$ channel is involved in inducing this reaction. Correlating channel gene expression and function in spermatogenic cells will allow for a better understanding as to how these ion channels participate in fertilization.

## IV. Summary

Cell communication involves molecular mechanisms that are at the forefront of research in biology today, since they play a key role in determining the behavior of organisms. Successful gamete interactions requiring cell signaling and

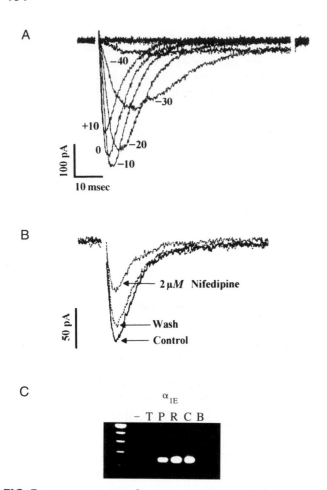

**FIG. 7.**  Low-threshold $Ca^{2+}$ channels are expressed in mouse spermatogenic cells. (A) T-type channel records obtained from pachytene spermatocytes elicited by membrane depolarizations to $-70$ to $+10$ mV in 10-mV steps from a holding potential ($E_h$) of $-80$ mV. Numbers beside the records correspond to the value in millivolts of the applied test pulse. (B) Spermatogenic T-type channel is reversibly blocked by DHPs. Superimposed records show the control, the effect of 2 $\mu M$ nifedipine, and the recovery after washing nifedipine. (C) RT-PCR experiments showing the expression of the $\alpha_{1E}$-subunit gene in testis (T), pachytene spermatocyte (P), round spermatid (R), and condensing spermatid (C). The $\alpha_{1E}$ RNA is detected also in residual bodies, which contain the organelles lost by sperm during spermiogenesis (B). (Modified from Liévano *et al.,* 1996.)

the crucial participation of ion channels determine the propagation of life.

Ionic fluxes play a fundamental role in activation of respiration and motility, chemotaxis, the sperm acrosome reaction, and therefore in fertilization. Indeed, sperm are excitable cells that quickly respond to components from the outer layer of the egg, the jelly, with fast changes in their plasma membrane permeability. Model membranes formed from sperm components and patch-clamp techniques in whole cells have been used to detect, for the first time, the activity of single channels in the plasma membrane of sea urchin sperm. These techniques are now being applied to mammalian sperm, and together with studies of $E_m$, $[Ca]_i$, and $pH_i$ in whole sperm, have allowed

establishing the presence of $K^+$, $Ca^{2+}$, and $Cl^-$ channels in this specialized cell and are helping to unravel their participation in chemotaxis and in the AR.

Sperm are very tiny cells (head diameter ca. 2–3 $\mu$m). This has precluded the characterization of their electrophysiological properties that would shed light on the molecular mechanisms leading to their fascinating, egg-induced, behavioral changes. Sea urchin sperm can be swollen in diluted seawater, maintaining the regulation of their $[Ca]_i$, $pH_i$, and $E_m$. Swollen sperm can be much more easily patch-clamped than normal sperm, thus providing new avenues to study ionic channels and their regulation by egg factors and second messengers. The strategies developed for the sea urchin are now being applied to mammalian sperm. There is presently background information about some of the ion channels present in these cells. Future study will determine the molecular mechanisms that regulate these channels in the cell. An alternative to identify and functionally study ion channels in sperm is to look for their expression in spermatogenic cells, combining molecular biological strategies and electrophysiology. Hopefully, this will allow a deeper understanding of the finely orchestrated events that lead to sperm activation, induction of the acrosome reaction, and in the end to the generation of a new individual.

## Acknowledgments

This work was supported by grants from CONACyT, DGAPA-UNAM, ICGEB, and an International Research Scholar Award to A. D. from the Howard Hughes Medical Institute. The authors thank Carmen Beltrán and Claudia L. Treviño for critically reading the manuscript, and Lucia de De La Torre, Irma Vargas, and Otilia Zapata for their help.

## Bibliography

Arnoult, C., Cardullo, R. A., Leuros, J. R., and Floruian, H. M. (1996). Egg-activation of sperm T-type $Ca^{2+}$ channels regulates acrosome reactions during mammalian fertilization. *Proc. Natl. Acad. Sci. USA.* **93**, 13004–13009.

Babcock, D. F., Bosma, M. M., Battaglia, D. E., and Darszon, A. (1992). Early persistent activation of sperm $K^+$ channels by the egg peptide speract. *Proc. Natl. Acad. Sci. USA* **89**, 6001–6005.

Bean, B. P. (1989). Classes of calcium channels in vertebrate membranes. *Annu. Rev. Physiol.* **51**, 367–389.

Beltrán, C., Darszon, A., Labarca, P., and Liévano, A. (1994). A high-conductance multistate $Ca^{2+}$ channel found in sea urchin and mouse spermatozoa. *FEBS Lett.* **338**, 23–26.

Beltrán, C., Zapata, O., and Darszon, A. (1996). Membrane potential regulates sea urchin sperm adenylylcyclase. *Biochemistry* **35**, 7591–7598.

Bleil, J. D. (1991). Sperm receptors of mammalian eggs. *In* "Elements of Mammalian Fertilization" (P. M. Wassarman, Ed), Vol. 1, pp. 133–152. CRC Press, Boca Raton, FL.

Bookbinder, L. H., Cheng, A., and Bleil, J. D. (1995). Tissue- and species-specific expression of sp56, a mouse sperm fertilization protein. *Science* **269**, 86–89.

Cook, S. P., and Babcock, D. F. (1993a). Selective modulation by cGMP of the $K^+$ channel activated by speract. *J. Biol. Chem.* **268**, 22402–22407.

Cook, S. P., and Babcock, D. F. (1993b). Activation of $Ca^{2+}$ permeability by cAMP is coordinated through the $pH_i$ increase induced by speract. *J. Biol. Chem.* **268**, 22408–22413.

Cook, S. P., Brokaw, C. J., Muller, C. H., and Babcock, D. F. (1994). Sperm chemotaxis: Egg peptides control cytosolic calcium to regulate flegellar response. *Dev. Biol.* **165**, 10–19.

Cox, T., and Peterson, R. N. (1989). Identification of calcium conducting channels in isolated boar sperm plasma membranes. *Biochem. Biophys. Res. Commun.* **161**, 162–168.

Cox, T., Campbell, P., and Peterson, R. N. (1991). Ion channels in boar sperm plasma membranes: Characterization of a cation selective channel. *Mol. Reprod. Dev.* **30**, 135–147.

Cuéllar-Mata, P., Martínez-Cadena, G., Castellano, L. E., Aldana-Velóz, G., Novoa-Martínez, G., Vargas, I., Darszon, A., and García-Soto, J. (1995). Multiple G-binding proteins in sea urchin sperm: Evidence for Gs and small G-proteins. *Dev. Growth Diff.* **37**, 173–181.

Darszon, A., Labarca, P., Beltrán, C., García-Soto, J., and Liévano, A. (1994). Sea urchin sperm: An ion channel reconstitution study case. *Methods: A Companion to Methods in Enzymology* **6**, 37–50.

Darszon, A., Liévano, A., and Beltrán, C. (1996). Ion channels: Key elements in gamete signaling. *Curr. Top. Dev. Biol.* **34**, 117–167.

Drewett, J. G., and Garbers, D. L. (1994). The family of guanylyl cyclase receptors and their ligands. *Endocr. Rev.* **15**, 135–162.

Florman, H. M. (1994). Sequential focal and global elevations of sperm intracellular $Ca^{2+}$ are initiated by the zona pellucida during acrosomal exocytosis. *Dev. Biol.* **165**, 152–164.

Florman, H., Corron, M. E., Kim, T. D.-H., and Babcock, D. F. (1992). Activation of voltage-dependent calcium channels of mammalian sperm is required for zona pellucida-induced acrosomal exocytosis. *Dev. Biol.* **152**, 304–314.

Foltz, K. R., Partin, J. S., and Lennarz, W. J. (1993). Sea urchin egg receptor for sperm: Sequence similarity of binding domain and hsp70. *Science* **259**, 1421–1425.

Garbers, D. L. (1989). Molecular basis of fertilization. *Annu. Rev. Biochem.* **58**, 719–742.

Gmachl, M., and Kreil, G. (1993). Bee venom hyaluronidase to a membrane protein of mammalian sperm. *Proc. Natl. Acad. Sci. USA* **90**, 3569–3573.

Gong, X., Dubois, D. H., Miller, D. J., and Shur, B. D. (1995). Activation of a G protein complex by aggregation of b-1,4-galactosyltransferase on the surface of sperm. *Science* **269**, 1718–1721.

González-Martínez, M. T., and Darszon, A. (1987). A fast transient hyperpolarization occurs during the sea urchin sperm acrosome reaction induced by egg jelly. *FEBS Lett.* **218**, 247–250.

González-Martínez, M. T., Guerrero, A., Morales, E., de De La Torre, L., and Darszon, A. (1992). A depolarization can trigger $Ca^{2+}$ uptake and the acrosome reaction when preceeded by a hyperpolarization in *L. Pictus* sea urchin sperm. *Dev. Biol.* **150**, 193–202.

Guerrero, A., and Darszon, A. (1989a). Egg jelly triggers a calcium influx which inactivates and is inhibited by calmodulin antagonists in the sea urchin sperm. *Biochim. Biophys. Acta* **980**, 109–116.

Guerrero, A., and Darszon, A. (1989b). Evidence for the activation of two different Ca channels during the egg jelly-induced acrosome reaction of sea urchin sperm. *J. Biol. Chem.* **264**, 19593–19599.

Guerrero, A., Sánchez, J. A., and Darszon, A. (1987). Single-channel activity in sea urchin sperm revealed by the patch-clamp technique. *FEBS Lett.* **220**, 295–298.

Hagiwara, N., and Kawa, K. (1984). Calcium and potassium currents in spermatogenic cells dissociated from rat seminiferous tubules. *J. Physiol.* (*London*) **356**, 135–149.

Hille, B. (1992). "Ion Channels of Excitable Membranes," 2nd ed., pp. 83–114. Sinauer Associates, Sunderland, MA.

Keller, S. T., and Vacquier, V. D. (1994). The isolation of acrosome reaction-inducing glycoproteins from sea urchin egg jelly. *Dev. Biol.* **162**, 304–312.

Kretzer, D. M., and Kerr, J. B. (1988). The cytology of the testis. *In* "The Physiology of Reproduction" (E. Knobil and J. D. Neill, Eds), Vol. 1, pp. 837–932. Raven Press, New York.

Labarca, P., Zapata, O., Beltrán, C., and Darszon, A. (1995). Ion channels from the mouse sperm plasma membrane in planar lipid bilayers. *Zygote* **3**, 199–206.

Labarca, P., Santi, C., Zapata, O., Morales, E., Beltrán, C., Liévano, A., and Darszon, A. (1996). A cAMP regulated $K^+$-selective channel from the sea urchin sperm plasma membrane. *Dev. Biol.* **174**, 271–280.

Lee, H. C. (1988). Internal GTP stimulates the speract receptor mediated voltage change in sea urchin spermatozoa membrane vesicles. *Dev. Biol.* **126**, 91–97.

Lee, H. C., and Garbers, D. L. (1986). Modulation of the voltage sensitive $Na^+/H^+$ exchange in sea urchin spermatozoa through membrane potential changes induced by the egg peptide speract. *J. Biol. Chem.* **261**, 16026–16032.

Lee, H. C., Johnson, C., and Epel, D. (1983). Changes in internal pH associated with the initiation of motility and the acrosome reaction of sea urchin sperm. *Dev. Biol.* **95**, 31–45.

Leyton, L., and Saling, P. (1989). 95 kDa sperm proteins bind ZP3 and serve as a tyrosine kinase substrates in response to zona binding. *Cell* **57**, 1123–1130.

Liévano, A., Sánchez, J., and Darszon, A. (1985). Single channel activity of bilayers derived from sea urchin sperm plasma membranes at the tip of a patch-clamp electrode. *Dev. Biol.* **112**, 235–295.

Liévano, A., Vega Saenz de Miera, E. C., and Darszon, A. (1990). $Ca^{2+}$ channels from the sea urchin sperm plasma membrane. *J. Gen. Physiol.* **95**, 273–296.

Liévano, A., Bolden, A., and Horn, R. (1994). Calcium channels in excitable cells: divergent genotypic and genotypic and phenotypic expression of $\alpha_1$-subunits. *Am. J. Physiol.* **267**, C411–C424.

Liévano, A., Santi, C., Serrano, C. J., Treviño, C. L., Bellvé, A. R., Hernádez-Cruz, A., and Darszon, A. (1996). T-type $Ca^{2+}$ channels and $a_{1E}$ expression in spermatogenic cells, and their possible relevance to the sperm acrosome reaction. *FEBS Lett.* **388**, 150–154.

Morales, E., de De la Torre, L., Moy, G., Vacquier, V. D., and Darszon, A. (1993). Anion channels in the sea urchin sperm plasma membrane. *Mol. Reprod. Dev.* **36**, 174–182.

Moy, G., Mendoza, L. M., Schulz, J. R., Swanson, W. J., Glabe, C. G., and Vacquier, V. D. (1996). The sea urchin sperm receptor for egg jelly is a modular protein with extensive homology to the human polycystic kidney disease protein, PKD1. *J. Cell Biol.* **133**, 809–817.

Reynaud, E., de Delatorre, L., Zapata, O., Liévano, A., and Darszon, A. (1993). Ionic bases of the membrane potential changes induced by speract in swollen sea urchin sperm. *FEBS Lett.* **329**, 210–214.

Santi, C. M., Darszon, A., and Hernández, A. (1996). A dihydropyridine-sensitive T-type $Ca^{2+}$ current is the main $Ca^{2+}$ current carrier in mouse primary spermatocytes. *Am. J. Physiol.* **271**, C1583–C1593.

Schackmann, R. W. (1989). Ionic regulation of the sea urchin sperm acrosome reaction and stimulation by egg-derived peptides. *In* "The Cell Biology of Fertilization" (H. Schatten and G. Schatten, Eds.). Academic Press, San Diego.

Stea, A., Wah Soong, T., and Snutch, T. P. (1995). Voltage-gated calcium channels. *In* "Handbook of Receptor and Channels. Ligand and Voltage-Gated Ion Channels" (A. North, Ed.), pp. 112–150. CRC Press, Boca Raton, FL.

Suzuki, N., and Yoshino, K. (1992). The relationship between amino acid sequences of sperm-activating peptides and the taxonomy of equinoids. *Comp. Biochem. Physiol.* **102B**, 679–690.

Thomas, P., and Meizel, S. (1989). Phosphatidyl inositol 4,5-bisphosphate hydrolysis in human sperm stimulated with follicular fluid or progesterone is dependent upon $Ca^{2+}$ influx. *Biochem. J.* **264,** 539–546.

Tiwari-Woodruff, S. K., and Cox, T. (1995). Boar sperm plasma membrane $Ca^{2+}$-selective channels in planar bilayers. *Am. J. Physiol.* **268,** C1284–C1294.

Trimmer, J. S., and Vacquier, V. D. (1986). Activation of sea urchin gametes. *Ann. Rev. Cell Biol.* **2,** 1–26.

Walensky, L. D., and Snyder, S. H. (1995). Inositol 1,4,5-trisphosphate receptors selectively localized to the acrosomes of mammalian sperm. *J. Cell Biol.* **130,** 857–869.

Ward, G. E., Brokaw, C. J., Garbers, D. L., and Vacquier, V. D. (1985). Chemotaxis of *Arbacia punctulata* spermatozoa to resact, a peptide from the egg jelly layer. *J. Cell Biol.* **101,** 2324–2329.

Ward, G. R., and Kopf, G. (1993). Molecular events mediating sperm activation. *Dev. Biol.* **158,** 9–34.

Ward, G. R., Storey, B. T., and Kopf, G. (1994). Selective activation of $G_{i1}$ and $G_{i2}$ in mouse sperm by the zona pellucida, the egg's extracellular matrix. *J. Biol. Chem.* **269,** 13254–13258.

Weyand, I., Godde, M., Frings, S., Welner, J., Muller, F., Altenhofen, W., Hatt, H., and Kaupp, B. (1994). Cloning and functional expression of a cyclic-nucleotide-gated channel from mammalian sperm. *Nature (London)* **368,** 859–863.

Wistrom, C. A., and Meizel, S. (1993). Evidence supporting involvement of a unique human steroid receptor/Cl-channel complex in the progesterone-initiated acrosome reaction. *Dev. Biol.* **159,** 679–690.

Yau, K.-W. (1994). Cyclic nucleotide-gated channels: An expanding new family of ion channels. *Proc. Natl. Acad. Sci. USA* **91,** 3481–3483.

Young, G. P. H., Koide, S. S., Goldstein, M., and Young, J. D. E. (1988). Isolation and partial characterization of an ion channel protein from human sperm membranes. *Arch. Biochem. Biophys.* **262,** 491–500.

*William J. Larsen and Richard D. Veenstra*

# 30

## Gap Junction Channels and Biology

## I. Introduction

In the late 1950s and early 1960s, physiologists who had been poking fine-glass current-injecting and -recording electrodes into neighboring cells within a variety of tissues made an interesting discovery. They found that while the injection of current into a cell caused a predictable shift in its nonjunctional membrane potential, it also caused a similar shift in the nonjunctional membrane potential of immediately adjacent cells (Fig. 1).

One interpretation of this finding was that the ions emanating from the injection electrode were able to flow freely from the injected cell to the adjacent cell and did so in preference to pathways leading to the extracellular medium or to the intercellular space. Moreover, several studies demonstrated that fluorescent dyes were selectively transferred from cell to cell when injected into the cytoplasm through glass injection pipettes. Many observations such as these led to the hypothesis that some cells were "coupled" (electrically or with respect to dye transfer) by **permeable cell junctions.**

In addition, in the neuronal tissues in which this phenomenon had first been witnessed, it was found that action potentials (APs) generated in a presynaptic element could be passed to a postsynaptic element much faster than would occur if the pre- and postsynaptic elements were connected by chemical synapses (Fig. 2). Ironically, this evidence for the presence of permeable cell junctions that could serve as **electrical synapses** in the nervous system came to light soon after the common acceptance of Otto Loewi's findings that supported the idea that neuronal synapses were probably chemical in nature, and not electrical as traditionally believed.

## II. Advantages of Electrical Synapses in Excitable Cells

The utility of electrical synapses in excitable cells is apparent. They may pass APs or subthreshold electrical activity more rapidly from one neuronal element to another than can their chemical counterparts. This advantage is especially obvious in neuronal pathways that serve as **escape mechanisms** such as the tail muscles of crayfish and lobsters and the pectoral fins of fishes. In these cases, the rapidity of the animal's response to imminent danger has selective value. On the other hand, such permeable junctions connecting smooth muscle cells of the uterus or cardiac muscle cells of the heart wall provide a mechanism for the systematic cell-to-cell spread of depolarization that is required for coordinated and effective **contractile activity.**

## III. Ubiquitous Membrane Permeable Junctions

Numerous studies published in the 1960s and 1970s supported the idea that virtually all cells in normal tissues (even those in "inexcitable" tissues) were coupled by permeable cell junctions. Based on studies with native biological molecules and with tracers of different sizes, it was also suggested that the pores within vertebrate cell-to-cell junctions were approximately 1.2 nm in diameter and that molecules larger than about 1 kDa were excluded. Thus, it was suggested that the biological molecules capable of freely moving from cell to cell in normal tissues would include a wide range of ions and small metabolites including sugars, nucleotides and nucleosides, and possible signaling molecules such as cyclic adenosine monophosphate (cAMP).

Because the pores initially appeared to select molecules only with respect to size, it was suggested that one function of permeable cell junctions was to "buffer" the concentrations of small metabolites throughout the tissue. Direct evidence for the cell contact-mediated exchange of metabolites, for example, was provided by "metabolic cooperation" experiments in which it was shown that a normal wild-type cell, able to incorporate thymidine into DNA, could transfer DNA precursor molecules (probably the nucleoside triphosphate form of thymidine) to mutant thy-

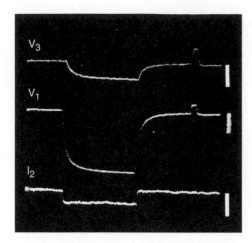

**FIG. 1.** The injection of current into one cultured WI-26 cell (I2) within a monolayer produces a shift in potential within the injected cell (V1) and in a cell two or three cells removed from the injected cell (V3), indicating that ions may move freely between contacting cells. [From Furshpan, E. J., and Potter, D. D. (1968). Low-resistance junctions between cells in embryos and tissue culture. *Curr. Top. Dev. Biol.* **3**, 95, with permission.]

midine kinase-deficient cells only if the wild-type cell and mutant cell were in contact. In these "kiss-of-life" experiments, the mutant cells were then able to incorporate this nucleoside into DNA required for continued proliferation and survival.

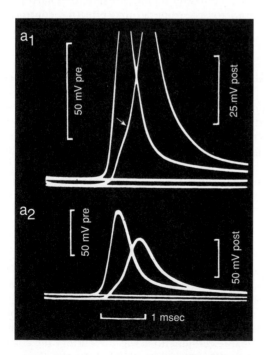

**FIG. 2.** The upper trace is an AP in the presynaptic element of a crayfish septate axon; the lower trace shows the subsequent AP in the postsynaptic element. The short lag of the latter following the former is typical of electrical synapses. The top and bottom traces were recorded at the same synapse at different amplifications. [From Furshpan, E. J., and Potter, D. D. (1959). Transmission at the giant motor synapses of the crayfish. *J. Physiol.* **145**, 289, with permission.]

Functional studies and hypotheses such as these, however, describe relatively general functions of permeable cell-to-cell junctions; functions shared by any number of normal tissues. They do not provide direct evidence or specific models to investigate cell-specific biological functions of cell contact-mediated molecule transfer.

## IV. Structural Candidates for the Permeable Cell Junction

As electron microscopes first came into common use in the late 1950s, several different kinds of cell–cell junctions were discovered. The **tight junction,** or **occludens-type junction,** was characterized by the fusion of membranes of adjacent cells and so held some appeal as a potential permeable cell junction. **Septate junctions** were also good candidates for the permeable cell junction since they were characterized by cell-to-cell bridges or septa covering a large region of cell–cell apposition in many of the tissues shown to be well coupled by electrophysiological or dye-tracing techniques. The basic argument against their function as permeable cell junctions, however, was that their presence in the large catalog of tissues being rapidly characterized by electron microscopy was significantly more limited than was the coupling phenomenon. The septate junction, for example, could not be found in vertebrate cell types that were well coupled.

## V. Ultrastructural Characterization of Gap Junctions

In a pioneering study, Revel and Karnovsky (1967) infiltrated the intercellular spaces of heart and liver with an electron-opaque dye (lanthanum hydroxide) and then examined these preparations in the electron microscope. They found regions where the membranes of adjacent cells were apposed to one another across a uniform lanthanum-infiltrated intercellular space about 2–4 nm in width. Moreover, in cross section, they observed small unstained structures bridging the stain-filled gap between adjacent cells in these regions and in en face views, these bridges were packed hexagonally within the intercellular space. Ultimately, freeze-fracture studies provided evidence that these intercellular bridges were in continuity with structures that spanned the lipid bilayers of both cells and thereby could theoretically provide the structural foundation for the cell-to-cell conduit implied by earlier electrophysiological studies.

It was also very important—for continued progress in this field—that this cell–cell junction, which Revel and Karnovsky called the *gap junction*, was found in many tissues that were ion or dye coupled, including both inexcitable and excitable cells. While well-coupled inexcitable cells such as liver hepatocytes possessed large numbers of gap junctions, it was especially satisfying to find that these structures were prominent features of the cell–cell contact regions of excitable cells such as cardiac myocytes and in neuronal systems that had been shown to possess electrical synapses. In addition, in the metabolic cooperation experiments described earlier, DNA precursor molecules were transferred between test cells only if they were capable of forming gap junctions. Moreover, gap junctions also proved to be a feature of those tissues that possessed large areas of septate junction or tight junction membrane. It is now known that gap junctions are components of virtually all tissues in multicellular organisms of the animal kingdom.

## VI. Molecular Studies of Gap Junction Proteins

Application of biochemical and molecular approaches to the study of gap junctions proved to be more perplexing than originally envisioned, but tenacious investigators in a number of laboratories were able to overcome initial obstacles. It is now known that proteins that make up gap junctions are basic and contain significant stretches of hydrophobic sequence. This is not surprising, given the fact that these are integral membrane proteins. For these reasons, they proved difficult to isolate and purify. Extraction of tissues with boiling sodium dodecyl sulfate (SDS) proved to be the only reliable method for obtaining "morphologically pure" fractions of gap junction membranes for many years, but the proteins extracted from these fractions ran at variable "molecular weights" on polyacrylamide gels. Controversy existed over whether some of these bands were breakdown products or other proteins that were associated with gap junctions in the cell membrane.

It was not until gentler, nondetergent extraction procedures were developed in the early 1980s that a consistent band on polyacrylamide gels with a predicted molecular weight of 26 kDa could be routinely isolated from rodent livers. Once this native protein was obtained and purified, it was possible to produce a polyclonal antibody. This antibody was used to screen an expression cDNA library to isolate and sequence its gene. This first sequence was published in 1986. Hydropathy analysis of the deduced amino acid sequence of this cDNA revealed a 32-kDa protein, which could be interpreted as possessing four potential transmembrane regions, two extracellular loops, and an internal loop and amino and carboxyl termini that extended into the cytoplasm (Fig. 3). The "mapping" of these proteins with antibodies directed against specific sequences and their cutting with specific proteases have largely confirmed this initial interpretation of the relationships of gap junction protein segments to the membrane.

It was also postulated that six gap junction protein molecules (or **connexins**) constitute each gap junction intramembrane particle (**connexon**) thus providing a molecular basis for models of the permeable membrane junction deduced from earlier physiological and structural studies (Figs. 4 and 5).

## VII. A Large Family of Gap Junction Proteins

Early structural studies, particularly those using the freeze-fracture technique, demonstrated structural diversity among the gap junctions distributed throughout the animal kingdom and within a variety of different tissues. The functional significance, however, of such gap junction characteristics as connexon density or connexon packing patterns remains controversial. Variability in these qualities, however, was shown in some cases to depend on the tissue in which the gap junction was located. The packing pattern of connexons in gap junctions of melanocytes or melanoma cells, for example, is recognizable and distinct from the packing pattern of connexons of granulosa, Leydig, and adrenal cortical cells. Although more recent studies have failed to provide a clear-cut molecular basis for these structural differences, it is now known that gap junctions in different tissues and in different species can be formed by a variety of different connexins ranging in molecular weight from 26 to 56 kDa (Fig. 6). Each connexin is named for its deduced molecular weight in kilodaltons so that the 26-kDa connexin in humans, for example, is called human connexin 26 (human Cx26) while the 30.3-kDa connexin in mouse is called mouse connexin 30.3 (mouse Cx30.3). Sequence analysis supports the possibility that two major phylogenetic subfamilies of connexins diverged from each other between 1.3 and 1.9 billion years ago, with one group ranging in size from about 26 to 32 kDa and the other from about 33 to 56 kDa. About a dozen different connexins have been identified in rodents; another dozen or so homologs have been described in other species and the list is growing.

Typically, it appears that only one or a few connexin species may comprise the gap junctions within any given tissue. For example, Cx26 and Cx32 constitute gap junctions of mammalian liver and have been shown to coexist within the same gap junction aggregates. However, the Cx32 and Cx43 documented in thyroid epithelium appear to be segregated into separate gap junctions formed in different regions of the lateral cell membrane. Likewise, the distribution of five different connexins (Cx37, Cx40, Cx43, Cx45, and Cx46) within the human heart appears to be nonrandom, with specific connexins preferentially localized in sinus and AV nodes, bundle branches, and in atrial or ventricular tissues. As many as six different connexin genes (Cx30.1, Cx31, Cx31.1, Cx40, Cx43, and Cx45) are transcribed and translated in mouse embryos as early as the eight-cell stage. These studies imply differential functions for gap junctions composed of different connexins and consequently stimulate the following question: What part of the connexin molecule encodes functional specificity, and how do these specificities differ from connexin to connexin?

It is known that the amino terminus, portions of the third transmembrane helix, cystine-containing regions of the extracellular loops, and serine residues of the carboxyl terminus are well conserved (see Figs. 4 and 6). However, the cytoplasmic loops and cytoplasmic carboxyl termini vary significantly in length and amino acid sequence from one connexin to another. Moreover, heteromeric gap junctions can be formed only with connexons made of particular connexins, suggesting that sequence specificity within the extracellular domains may be relevant to gap junction assembly. Recent studies, therefore, have focused on the

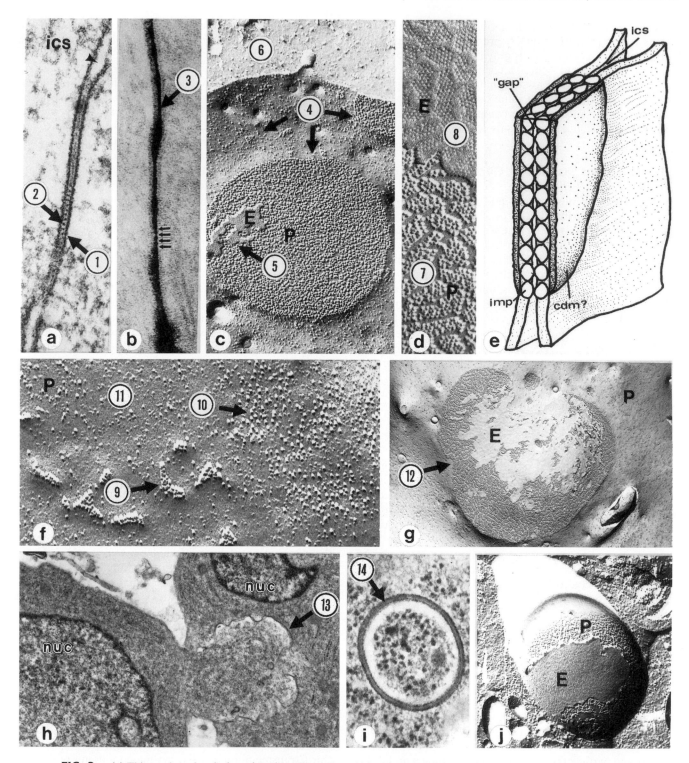

**FIG. 3.** (a) Thin section of typical gap junction. The entire width of both apposed membranes and the intercellular space is about 18 nm. Dense material is often associated with cytoplasmic surface of gap junction (1). Some gap junctions are characterized by stained periodicities evident at level of "gap" (2). Magnification, 157,500×. [From Larsen, W. J., Skowron-Lomneth, C., and Carron, C. (1988). Gap junction modulation: Possible role in tumor cell behavior. *In* "Biochemical Mechanisms and Regulation of Intercellular Communication" (H. A. Milman and E. Elmore, Eds.), Vol. 14, p. 151. Princeton Scientific Publ. Co., Princeton, NJ, with permission.] (b) Lanthanum infiltrated gap junction. The lanthanum infiltrated intercellular "gap" (3) is about 2–4 nm. Magnification, 212,400×. (Reproduced from *The Journal of Cell Biology,* 1975, vol. 67, p. 801, by copyright permission of The Rockefeller University Press.) (c) Freeze-fracture image of a gap junction. Each junctional membrane contains a set of particles and a set of pits. In vertebrate tissues, the particles adhere to the protoplasmic leaflet of the lipid bilayer (protoplasmic fracture face, P), whereas the pits remain associated with the extracellular leaflet

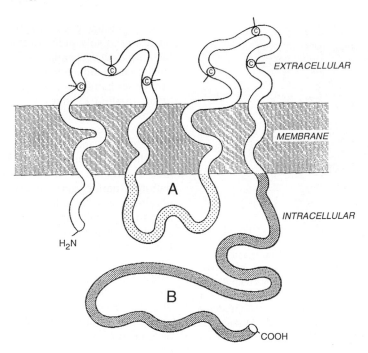

**FIG. 4.** Structure and topology of a connexin. Both amino and carboxy termini (B) are located in the cytoplasm along with an intermediate loop (A). Every connexin contains four membrane-spanning regions and two extracellular loops containing highly conserved cyteine residues (C). [From Beyer, E. C., Paul, D. L., and Goodenough, D. L. (1990). Connexin family of gap junction proteins. *J. Membr. Biol.* **116**, 187, with permission.]

potential functional relevance of specific differences in amino acid sequences within the cytoplasmic loop, hydrophilic carboxyl termini, and extracellular loops of connexin proteins (see Sections IX through XIV).

## VIII. Gap Junctions Contain Channels

Following the identification of the gap junction as the site of cell-to-cell transfer of ions and small hydrophilic

(extracellular fracture face, E). Gap junction particles aggregate within particle-poor zones (4). The fracture may jump between the two membranes, resulting in adhesion of bits of the E-fracture face from the membrane of the adjacent cell (5) on the P-fracture face or vice versa. Occasionally, the fracture plane leaves the membrane and cuts into the cytoplasm (6). Magnification, 64,800×. [From Larsen, W. J. (1977). Structural diversity of gap junctions: A review. *Tissue Cell* **9**, 373, with permission.] (d) Enlargement of P- and E-fracture faces showing details of gap junction particles (7) and corresponding pits (8). Magnification, 139,500×. [From Larsen, W. J. (1977). Structural diversity of gap junctions: A review. *Tissue Cell* **9**, 373, with permission.] (e) Simple model of gap junction showing particle–particle (imp) contact within the "gap" at the level of the intercellular space (ics). Cytoplasmic dense material (cdm) is particularly apparent in gap junctions in excitable tissues. [From Larsen, W. J. (1977). Structural diversity of gap junctions: A review. *Tissue Cell* **9**, 373, with permission.] (f) Gap junctions form by the aggregation (9) of single 11-nm particles (10) in particle-poor regions of the membrane (11). Magnification, 222,300×. [From Larsen, W. J., and Risinger, M. A. (1985). The dynamic life histories of intercellular membrane junctions. *In* "Modern Cell Biology" (B. Satir, Ed.), Vol. 4, p. 151. John Wiley & Sons, New York. Copyright 1985 John Wiley & Sons. Reprinted by permission of Wiley-Liss, Inc., a subsidiary of John Wiley & Sons, Inc.] (g) Very large gap junctions (12) containing thousands of gap junction particles, may represent terminal gap junction caps. Magnification, 21,600×. [From Larsen, W. J. (1977). Structural diversity of gap junctions: A review. *Tissue Cell* **9**, 373, with permission.] (h) The large gap junction caps may invaginate into the cell (13) through an endocytotic mechanism. Magnification, 14,400×. [From Larsen, W. J., and Tung, H. (1978). Origin and fate of gap junction vesicles in rabbit granulosa cells. *Tissue Cell* **10**, 585, with permission.] (i) Bimembranous gap junction vesicles (14) may pinch off from the invaginating junction to fuse with lysosomes before undergoing degradation. Magnification, 52,200×. [From Larsen, W. J., and Tung, H. (1978). Origin and fate of gap junction vesicles in rabbit granulosa cells. *Tissue Cell* **10**, 585, with permission.] (j) Freeze-fracture planes within the cytoplasm may reveal the P-face particles (P) and E-face pits (E) within cytoplasmic gap junction vesicles. Magnification, 38,700×. [From Larsen, W. J., and Tung, H. (1978). Origin and fate of gap junction vesicles in rabbit granulosa cells. *Tissue Cell* **10**, 585, with permission.]

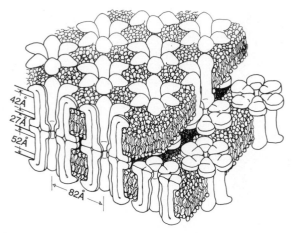

**FIG. 5.** This diagram of a region of a gap junction isolated from mouse liver is based on electron microscope images and X-ray diffraction data. Each gap junction particle or connexon is composed of six gap junction protein molecules called connexins. Connexons in apposed membranes meet within the intercellular space. (Reproduced from *The Journal of Cell Biology*, 1977, vol. 74, p. 449, by copyright permission of The Rockefeller University Press.)

molecules, it was proposed that the pathway for this exchange consisted of an array of parallel aqueous channels. Each imp particle in the freeze-fracture images or hexagonal bridge in the negatively stained micrographs was believed to contain a central aqueous pore. For an aqueous pore with a channel diameter of approximately 14 ü, a

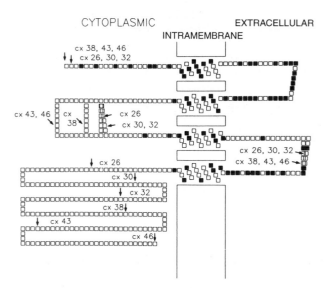

**FIG. 6.** This diagram depicts the predicted structure of rat and *Xenopus* connexins. Each square represents an amino acid. The diagonal arrows mark points within the cytoplasmic and extracellular loops where the connexins differ in length. The vertical arrows indicate the amino and carboxyl termini for each connexin. Closed squares represent amino acids that are the same in all connexins. [From Bennett, M. V. L., Barrio, L. C., Bargiello, T. A., Spray, D. C., Hertzberg, E., and Saez, J. C. (1991). Gap junctions: New tools, new answers, new questions. *Neuron* **6**, 305, with permission. Copyright 1991 Cell Press.]

unitary conductance of $10-10^{-11}$ (= 100 pS) was predicted. Such a relatively high conductance should produce a detectable electrical signal so efforts were undertaken to record quantal changes in electrical coupling as a means to prove the channel hypothesis for gap junction permeability. Small quantal changes in the transfer voltage between paired *Xenopus* embryonic cells were detected as evidence for the existence of discrete coupling elements, but the large cell size and consequential low input resistance limited the resolution of the electrical signals and precluded measurement of the unitary conductance. Precise determination of unitary gap junction channel conductances were obtained only by adapting the two cell voltage clamps to pairs of high input resistance cells. When current is simultaneously recorded from both voltage-clamped cells, the junctional current always appears as "mirror images" since each cell resides on opposite sides of the junction (Fig. 7). This *equal amplitude and opposite polarity* is the criterion used to define junctional current signals in the dual whole-cell recording configuration. Channel conductances of 120 and 160 pS were reasonably close to the original predicted value and were generally regarded as providing supportive evidence for the existence of relatively nonselective aqueous channels. Presently, conductance values range from 30 to 300 pS for different gap junction channels under essentially physiological salt conditions. The same expression used to predict the 100-pS conductance for a 14-ü-diameter gap junction channel requires a pore diameter of 26 ü for a 300-pS channel. It follows that higher conductance gap junction channels should exhibit less ionic selectivity and a higher molecular permeability than channels with smaller conductances.

## IX. Evidence for Charge Selectivity

The apparent permeability to a wide variety of water-soluble (hydrophilic) molecules (ions, cAMP, sugars, ATP, nucleosides, etc.) suggests the presence of a large-diameter aqueous pore with minimal selectivity on the basis of electrical charge (equivalent valence at physiological pH). Several permeability studies on mammalian cell and invertebrate gap junctions were instrumental in assigning the generally accepted molecular permeability limit of 1 kDa or diameter <14 ü. However, two of these same studies used tagged fluorescent tracers with different valences and demonstrated that the molecular permeability limit was lower for molecules with higher negativity than their neutrally charged counterparts. That this selective permeability may occur with molecules of >600 Da and approximately 10 ü in diameter is not surprising given that the size of the permeant molecule is approaching the estimated diameter of the pore, thereby placing any charged surfaces of the pore and the permeant molecule in close proximity to each other. This would enhance any electrostatic attractive or repulsive forces that might be present and suggests that the pore of the gap junction channel contains some electronegative fixed sites within the pore that reduce the permeability of large negatively charged molecules relative to their neutral or less negative counterparts.

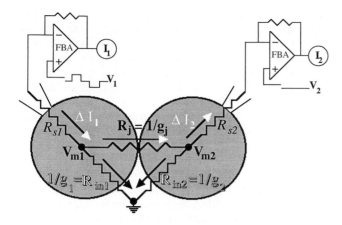

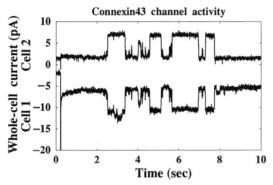

**FIG. 7.** (A) Dual whole-cell recording configuration. Equivalent resistive circuit for the whole-cell patch-clamp recording configuration from two coupled cells. The voltage difference between the command potential ($V_1$ and $V_2$), applied via each negative-feedback patch-clamp amplifier (FBA), and the cell membrane potential ($V_{m1}$ and $V_{m2}$) are minimized by reducing the series resistance ($R_{s1}$ and $R_{s2}$) to <5% of the cell input resistance ($R_{in1}$ and $R_{in2}$) and junctional resistance ($R_j$). Accurate junctional current signals are achieved when a majority of the applied current in response to a transjunctional voltage difference ($V_j = V_{m1} - V_{m2}$) flows across the junction (white arrowheads) instead of the cell membrane (black arrowheads). This condition is readily obtained by using high input resistance cells ($R_{in} > 1$ G$\Omega$). Conductances (g) are the reciprocal of the corresponding resistance element. (B) The whole-cell currents for each amplifier ($I_1$ and $I_2$) appear as simultaneous signals of opposite sign. These are unaltered whole-cell currents digitized at 1 kHz after low-pass filtering at 100 Hz obtained from a Cx43-transfected N2A cell pair. The transjunctional voltage was −30 mV.

Evidence for selectivity at the ionic level was not obtained until the patch-clamp methodology was adapted to the recording of gap junction channel currrents. In two cellular preparations, the junctional membranes of the earthworm septated axon and rat lacrimal gland cells were found to be less permeable to Cl$^-$ relative to K$^+$ by a ratio of 0.52 to 0.69. That this modest selectivity is found among ions with nearly identical aqueous mobilities and diameters of less than 4 ü suggests that there are weak interactions between the permeant ion and wall of the pore and not just between the ions and water as would be true for a simple diffusion-limited pore.

## X. Different Connexins Possess Different Electrical Properties

The functional expression of different connexins by mRNA injection into *Xenopus* oocytes or stable expression in communication-deficient mammalian tumor cell lines has yielded tantalizing results. Single-channel unitary conductances (the ease with which current flows) are connexin specific. Variability in conductance ranges 10-fold from 30 to 300 pS (pS = 1/1012 $\Omega$) with a majority of the channel conductances in the 90–180 pS range. The reporting of channel conductances has become more complicated owing to the presence of multiple conductance states for several of the connexin channels. The physiological relevance of these conductance differences is not yet known.

Connexins also vary in their regulatory responses to intracellular pH, intracellular calcium, protein kinases, and transjunctional (cell-to-cell) voltage. Because the most significant differences in the amino acid sequences of the connexins occur within the cytoplamsic loop and carboxyl terminal domains, it has been proposed that these domains confer the distinct conductance and regulatory properties on the assembled connexin-specific gap junction channel. Because each connexin is encoded for by a specific gene, tissue- and developmental-specific patterns of expression are known to exist. In a few studies, comparisons of the channel conductance and regulatory properties of the connexin channels to the phenotypic properties of the gap junction in native cell types have demonstrated that the presence of at least two distinct connexin channel types can explain the properties of the junctions found in those cells. In one study, the expression of three connexins found in the embryonic chicken heart, Cx42, Cx43, and Cx45, in communication-deficient mouse neuro2A neuroblastoma (N2A) cells demonstrated that each connexin exhibited unique conductance properties that coincided well with the observance of multiple conductance states between cardiac myocytes in culture. Furthermore, by mixing these three connexins in different proportions, the response to transjunctional voltage at different stages of cardiac development could be mimicked closely. Thus, the observed decline in Cx45 expression could significantly explain the decrease in transjunctional voltage conductance sensitivity observed in the developing heart. Developmental changes in connexin expression have been observed in other tissues, suggesting a possible role for this mechanism in the modulation of gap junction communication in a variety of tissues.

## XI. Gating by Ions and Second Messengers

Perhaps of greater physiological relevance than transjunctional voltage is the modulation of junctional communication by highly buffered intracellular cations (protons and calcium) and organic second messengers such as cAMP, diacylglycerol, and inositoltrisphosphate. Whether cations directly bind to a regulatory site (or sites) on the connexin or act through intermediate accessory proteins is not known. The investigation of this issue is complicated by the observation that interventions that lower intracellular pH also increase intracellular free calcium concentrations and vice versa. Conversely, there is direct evidence that several of the connexins are substrates for phosphorylation by protein kinases such as PKA and PKC. Protein kinase-dependent phosphorylation frequently either increases or decreases junctional communication, depending on the tissue or connexin studied. Physiological correlates at the channel level are rare owing to the difficulty of observing channel activity under distinct ionic or phosphorylation conditions. Typically, cAMP is regarded as increasing Cx43 and Cx32 communication and PKC has the opposite effect, at least in Cx43-containing cells. PKA and PKC may both phosphorylate the same serine residue (S233) of Cx32. In contrast, tyrosine phosphorylation by pp60v-src at residue 265 of Cx43 is correlated with a rapid and reversible uncoupling while this tyrosine kinase has no effect on Cx32, which lacks this specific tyrosine residue. It is thought that Y265 is important in the reduction in junctional communication often associated with transformed cells containing Cx43 and related connexins.

Of particular relevance to tissue function is the ability to transfer these chemical signals from cell to cell. The connexin channel permeability to these ions and second messengers is of utmost importance in this regard, and there is evidence that connexins possess different permeabilities to fluorescent tracer molecules. This is indicative of different molecular permeability limits to molecules of similar size and charge to cAMP. The most direct comparison of molecular permeability differences to date comes from the use of fluorescein derivatives with varying valence but essentially constant physical dimensions (+ 10 ü). 2′,7′-Dichlorofluorescein (diCl-F) and 6-carboxyfluorescein (6-CF) were readily permeable through Cx43 channels, yet were only occasionally permeable through Cx40 or Cx37 channels. 6-CF transfer was not observed with Cx45 despite junctional conductances ranging from 0 to 30 nS, while diCl-F was permeable in 60% of the experiments with Cx45. This suggests that the more anionic 6-CF is more restricted in its junctional permeability than diCl-F and that the magnitude of this differential permeability depends on which connexin is also present. This interpretation is consistent with earlier permeability studies, which suggested that gap junction channels possess a fixed anionic charge associated with the pore. In addition, there is no correlation between the maximum conductance of the connexin channel and their molecular permeability since Cx43 has a lower channel conductance than Cx40 and Cx37. Channel conductance measurements during ionic substitution also suggest that a majority of the connexin channels favor cations over anions by a factor of 2:1 or higher. This lack of correlation of charge selectivity with channel conductance suggests that pore size and conductance are not directly related.

## XII. Regulation of Functions of Gap Junctions at Multiple Levels

As suggested earlier, a likely reason for differences in the sequence of the carboxyl tail of connexins may be in the regulation of gap junction-mediated tissue- or cell-

specific biological activities. Indeed, gap junction regulatory mechanisms described so far appear to fall into one of two general categories. On the one hand, rapid changes in cell coupling that occur within seconds or minutes may reflect regulation of gating of preexisting pores. On the other hand, long-term changes that occur within minutes or a few hours to days are likely to involve the modulation of synthesis, assembly, and/or degradation of connexins (Fig. 8). Existence of this latter scheme of modulation is supported by studies that show that the half-lives of several different connexins are relatively short, ranging from 1 to 3 hr.

For example, the possible effect of cAMP on transcription of connexin mRNA and its ultimate translation is supported by studies which show that cAMP-mediated increases in coupling or in junction formation in some cells can be blocked by inhibitors of mRNA or protein synthesis. The potential for control at the level of transcription of connexins 32 and 43 is also implied by the characterization of their 5′ flanking sequences. For example, a specific basal promoter of the rat Cx32 gene has been identified in normal rat hepatocytes called Cx-B2, but the identity of the DNA-binding factor has not been deciphered. In addition, the 5′ flanking sequence of the myometrial Cx43 gene has been shown to possess several consensus activator protein-1 (AP-1) binding sites as well as half-palindromic estrogen response elements. Since the AP-1 proteins, Fos and Jun, are expressed in response to increased estrogen, transcription of Cx43 may be indirectly up-regulated through its

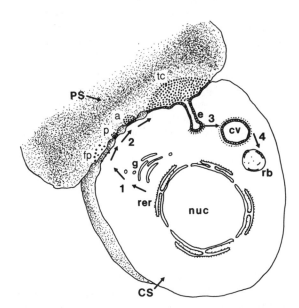

**FIG. 8.** Scheme depicting (1) the synthesis of connexins within rough endoplasmic reticulum (rer), packaging in the Golgi (g), and transport of connexons from the cytoplasm (CS) to the plasma membrane; (2) their assembly in particle-poor regions of the membrane called formation plaques (fp) into small primary plaques (p), functional gap junction aggregates (a), and then into very large terminal caps (tc); (3) their endocytosis, which is facilitated by clathrin and an actin-mediated contractile mechanism; and (4) their degradation, following fusion with lysosomes resulting in the formation of residual bodies (rb).

AP-1, *cis*-acting elements and/or more directly via putative estrogen response elements.

Yet another level of control of gap junction function may be demonstrated by studies that show the up-regulation of coupling or gap junction formation in the absence of transcription or translation, perhaps at a posttranslational level of control. The direct application of dibutyryl cAMP (dbcAMP) to mammary tumor cells in culture, for example, increases the number of gap junctions as well as cell–cell coupling in the absence of an increase in Cx43 mRNA or protein. Likewise, connexin 43 trafficking within several cultured cell lines appears to be regulated by posttranslational mechanisms. Specifically, the trafficking of Cx43 from the cytoplasm to the plasma membrane and its assembly into functional gap junction aggregates may require phosphorylation of two serine phosphorylation consensus sites in the carboxyl tail of this connexin. For example, two cell lines were studied that could not form gap junctions or become coupled, but were able to synthesize significant amounts of Cx43, which was stored within vesicles in the cytoplasm. The Cx43 in these cells was phosphorylated at only one of its serine residues. When one of these lines was transfected with a gene for a cell adhesion molecule, L-CAM, the Cx43 in the transfected line was phosphorylated again, translocated to the plasma membrane, and assembled into gap junctions.

Further evidence for the phosphorylation-dependent trafficking and assembly of Cx43 in cardiomyocytes has been obtained in experiments with monensin, a reagent that inhibits translocation of proteins between Golgi cisternae. These studies suggest that phosphorylation of Cx43 to the mature form occurs in a compartment of the cell between the Golgi apparatus and the plasma membrane. Similar results are obtained in studies of effects of brefeldin A, which blocks both Cx43 phosphorylation and trafficking to the plasma membrane in a mammary tumor cell line.

Finally, it has also been suggested that the process of connexon clustering that gives rise to functional gap junction plaques may be facilitated by the action of microfilaments. Thus the increase in Cx43-positive gap junctions and coupling in the absence of Cx43 mRNA synthesis following treatment of cultured prostatic cells with forskolin may be explained by effects of cAMP on posttranslational processing of connexin43, namely, their assembly into functional plaques and their translocation from the cytoplasm to the membrane.

Numbers of functional gap junctions within the plasma membrane may also be affected by the rate of their removal and degradation. Gap junctions may be removed from the plasma membrane by endocytosis. Following fusion of the endocytotic gap junction vesicles with lysosomes, the gap junctions are degraded (see Fig. 8). For example, the treatment of some cells with phorbol esters may stimulate a decrease in gap junctions. Effects of phorbol esters such as the skin tumor promoter 12-0-tetradecanoylphorbol-13-acetate (TPA), are normally mediated by PKC, which has been demonstrated to induce connexin phosphorylation. Thus, in addition to apparent direct effects on the gating of gap junction channels (see Section XI), some tumor promoters may regulate numbers of gap junction channels at the plasma membrane through posttranslational processing and inhibition of gap junction assembly of by stimulating a process of gap junction endocytosis.

In another example, massive endocytosis of gap junctions in the cumulus cells of the mammalian follicle is precipitated by high titers of follicle-stimulating hormone. An actin-based contractile mechanism and clathrin appear to play a role in this endocytotic process, which leads to the uptake and fusion of the interiorized gap junction vesicles with lysosomes. However, it is not yet known whether the transient decrease in cell coupling in cultured rat liver epithelial cells following treatment with epidermal growth factor is mediated by a decrease in the rate of gap junction synthesis or assembly, modulation of the channel, or an increase in the rate of degradation. Nonetheless, it has been shown that three consensus mitogen-activated protein (MAP) kinase serine phosphorylation sequences (at Ser255, SER279, and Ser282) in the carboxyl tail are concomitantly phosphorylated. On the other hand, it seems likely that decreases in cell coupling associated with the pp60src-mediated phosphorylation of tyrosine residues within the carboxyl tail of Cx43 results from modulation of connexon gating activity (see Section XI).

In summary, these studies support the view that function may be controlled at virtually any point in the gap junction's life history; namely, through regulation of (1) transcription of connexin mRNA, (2) translation of connexin protein, and then through posttranslational modifications, which affect (3) the assembly of functional plaques, (4) gating of junctional channels, and (5) degradation of functional gap junctions.

## XIII. Specific Biological Functions of Gap Junctions

Early speculations regarding the general functional significance of gap junctions, such as the electrical coupling of some neurons and the buffering of metabolites with a tissue mass, have not been seriously challenged since the mid-1960s (see Sections I, II, and IX). Several recent studies, however, have begun to establish even more specific functional relationships between gap junctions and tissue activities and behaviors, most notably with respect to the regulation of smooth muscle contraction and ligand-mediated secretory activity.

Perhaps one of the best studied cases is that of the myometrial connexin, Cx43. It is known, for example, that the myometrium of the nonpregnant uterus in a variety of species is virtually devoid of gap junctions and expresses very little, if any, myometrial Cx43 mRNA or protein. In contrast, however, Cx43, mRNA is rapidly synthesized and translated and then gap junctions appear coincident with rising estrogen and declining progesterone titers just prior to parturition. Experimental manipulations of these hormones also predictably induce or inhibit the appearance of gap junctions in experimental models of preterm labor or tocolysis, respectively (effective labor coincides with their induction, while myometrial quiescence parallels their inhibition). For example, removal of the ovaries or the injection of RU-486 during the latter part of pregnancy effectively increases the estrogen:progesterone ratio, the expression of Cx43 mRNA and protein, the assembly of myometrial gap junctions, and concomitant preterm labor. Conversely, the injection of progesterone several days prior to term inhibits the formation of gap junctions and onset of labor. Thus, it is likely that transcription and translation of Cx43 are required for gap junction formation and coordinated myometrial contractions in this species. Moreover, the evidence for the presence of AP-1 binding sites in 5′ flanking sequences of the myometrial Cx43 gene and the estrogen-mediated expression of Jun and Fos described in Section XII is consistent with this notion.

The function of myometrial gap junctions and contractility may also be regulated at the level of posttranslational processing. For example, high levels of myometrial Cx43 appear a full day prior to delivery in rats (Fig. 9) but most of the Cx43 in these cells is located within perinuclear vesicles that also stain with an antibody against a Golgi-associated protein (Fig. 9b). Six to 12 hr prior to onset of delivery, however, Cx43 immunopositive plaques are observed in the plasma membrane (Fig. 9c). Conversely, cessation of delivery is accompanied by the loss of gap junctions from the cell surface by a process of endocytosis (Fig. 9e).

Gap junctions have also been implicated in the process of secretion and other ligand-mediated functions in many different kinds of cells. For example, gap junctions are well developed in differentiated endocrine and exocrine cells and typically sparse or absent in proliferating stem cells. Moreover, the induction of secretory activity itself appears to be correlated with increased numbers of gap junctions in luteal cells, adrenal cortical cells, pancreatic exocrine cells, mammary alveolar cells, and thyroid cells. Studies of osteoblasts have demonstrated a close correlation between the effects of hormone on the generation of a second messenger, formation of gap junctions, and cell function. In addition, gap junctions appear to be more abundant in cultured confluent osteoblasts than proliferating cells and osteoblast cell coupling in culture and cAMP production are enhanced by the application of parathyroid hormone (PTH). Conversely, an analog of PTH, which binds to PTH receptors but attenuates cAMP accumulation, results in a decrease in cell coupling. Finally, transfection of osteoblasts with antisense Cx43 mRNA results in significant concomitant reductions in coupling and in cAMP synthesis in response to PTH.

It has been suggested that the development of gap junctions may enhance the sensitivity of hormonally responsive cells to their specific ligand as a consequence of the transfer of cAMP through gap junctions. Thus cells not receiving direct stimulation by binding of the secretogogue to its receptor may also respond to the ligand-mediated production of cAMP in neighboring cells that possess the appropriate receptor. Such a hypothesis has been invoked to explain the maintenance of meiotic arrest of mammalian primary oocytes. These germ cells initiate the first meiotic division during embryonic life but are then almost immediately arrested at the first meiotic prophase stage. Typically, a primary oocyte does not resume meiotic maturation until

it responds to an ovulatory surge of gonadotropic hormones following puberty. However, since meiosis spontaneously resumes when the oocyte is removed from the follicle and cultured in medium lacking hormones, it has been postulated that the follicular environment is inhibitory to resumption of meiotic maturation. Meiotic arrest may be maintained by follicle cell-generated cAMP, which enters the oocyte through a well-developed network of gap junctions. Conversely, it has been suggested that meiotic resumption may be signaled by the LH-mediated disruption of the gap junction pathway within the cumulus mass (following an ovulatory surge of gonadotropins), thus preventing cAMP manufactured within the follicle cells from entering the oocyte. Other hypotheses of both negative and positive regulation of meiotic resumption have also been proposed.

## XIV. Gap Junctions in Human Disease

### A. Carcinogenesis

In the early 1970s, the following hypothesis was formulated: The regulated and coordinated growth of cells in normal tissues requires the presence of permeable (communicating) cell junctions, whereas cells incapable of forming "communicating junctions" exhibit uncontrolled growth, such as the uncontrolled growth observed in cancer cells. Studies that compared the presence or absence of coupling and/or gap junctions in normal cells and in several tumors and cancer cell lines initially supported this hypothesis. For example, coupling and/or gap junctions were found to be deficient in certain hepatoma cells and in L-cell derivatives such as the clone-1D cell line in contrast to their normal counterparts. Indeed, more recent studies support these early observations. For example, an extensive study has examined the role of gap junctions in tumor progression in rat liver following initiation by diethylnitrosamine (DEN) and promotion by either phenobarbitol (PB) or 2,3,7,8-dichlorodibenzo-p-dioxin (TCDD). Transcripts and protein levels of Cx26, Cx32, and Cx43 were measured and reductions in Cx26- and Cx32-positive gap junctions were observed in all resulting neoplasms. However, it was also demonstrated that these decreases were not always associated with reductions in specific mRNA transcripts for these connexins, suggesting that the loss of junctions could result from the modulation of transcription and/or translation or from posttranslational modification affecting assembly or degradation (see Section XII). Consistent with these observations, the transfection of HeLa cells with cDNA encoding Cx26, but not Cx40 or Cx43, results in inhibition of tumor formation in nude mice, suggesting that Cx26 gap junctions play a pivotal role in growth control. Similarly, the transfection of communication-deficient hepatoma cells (SKHep1) with cDNA encoding Cx32 results in slowing of the growth rate of tumors in nude mice, compared with tumors arising from communication-deficient parental SKHep1 cells.

In conflict with these findings, however, other highly malignant tumor cells have been shown to be well coupled

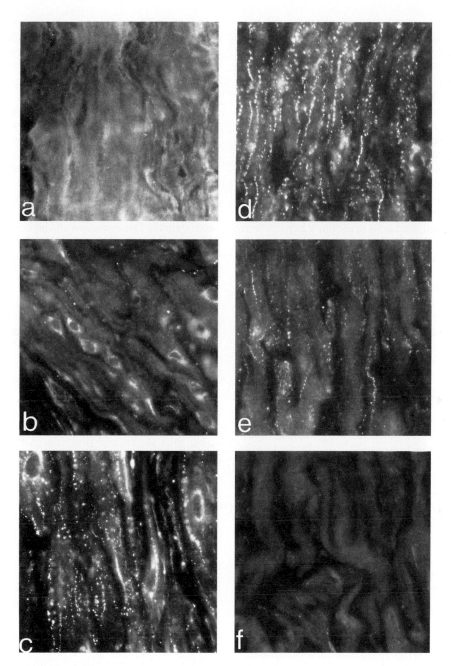

**FIG. 9.** Immunofluorescent staining of myometrium sections incubated with site-specific antibodies to the carboxyl terminus of Cx43 from pregnant rat sacrificed on day 19 (a) or day 21 (b), at midnight on day 21 (c), during delivery on day 22 (d), 6 hr after delivery (e), or 24 hr after delivery (f). Magnification, 500×. [From Hendrix, E. M., Mao, S. J. T., Everson, W., and Larsen, W. J. (1992). Myometrial connexin 43 trafficking and gap junction assembly at term and in preterm labor. *Mol. Reprod. Dev.* **33,** 27, with permission. Copyright 1992 Wiley-Liss, a division of John Wiley & Sons, Inc.]

or to possess large numbers of gap junctions. These include Novikoff hepatoma cells, human SW-13 adrenal cortical carcinoma cells, and murine B16 melanoma cells. Moreover, other studies have analyzed gap junctions during tumor progression in skin tumors and in hepatocarcinoma and have found that while gap junctions or coupling may disappear during the transformation of normal cells by cancer-causing agents (i.e., tumor promoters), their loss

seems to follow, rather than lead the changes that transform these normal cells to tumor cells.

In response to results such as these, a modified "defective communication" hypothesis was proposed to include cancer cells that might not necessarily exhibit an absence of or obvious structural defect in the gap junction itself, but which might be defective with respect to some other part of the gap junction communication mechanism, including the ability of the cell to synthesize the postulated regulatory message or the receptor that translated its activity. Other modifications of the hypothesis have also been advanced, proposing, for example, that some cancer cells (such as cultured, human SW13 cells, and diethylstilbestrol-induced tumors of proximal tubule in Syrian hamster) are not able to maintain a steady-state gap junction level through a regulated balance of synthesis and/or degradation of gap junctions. Under some circumstances, these cells possess normal or even excessive gap junction complements and are mitotically quiescent while under other conditions, the cells lose most or all of their junctions resulting in cancerous growth. A temporal relationship between gap junction formation and degradation and the rate of proliferation is also exhibited by normal liver cells in partially hepatectomized rats. Normal rat liver cells remaining after partial hepatectomy initially lose their gap junctions and proliferate rapidly. By 40 hr after partial hepatectomy the liver cells regain their gap junctions coincident with a reduction in the rate of DNA synthesis and cell division.

Another variation of the hypothesis is based on a series of experiments that demonstrates that normal cells may have the ability to suppress tumor cell behavior in some normal–tumor cell cocultures as long as the tumor cells can also form gap junctions with the cocultured normal cells. In contrast, other cancer cell lines, capable of forming gap junctions between themselves, may lack the capacity to form junctions with normal cells. In these tumor cells, contact with normal cells does not suppress the transformed phenotype of the tumor cell.

In conclusion, it is unlikely that the progression of all cancers is dependent on the activity of gap junctions. It is also unlikely that mechanisms of gap junction-mediated growth control will be identical in all cells or tumors where the activity of gap junctions can be demonstrated to play a pivotal role.

## B. Marie-Charcot-Tooth Disease and Visceral Heterotaxia

Spontaneous mutations of Cx32 and Cx43 result in specific disease states in humans. One of these is the most common peripheral neuropathy, Marie-Charcot-Tooth disease, which affects 1 : 2500 individuals. While this demyelinating disease may arise as a consequence of mutations of a peripheral myelin protein PMP22 (on chromosome 17) or of the myelin constituent Po (on chromosome 1), about 30 families have been identified with an X-linked dominant form of the disease characterized by as many as 25 different mutations of Cx32. Mutations may involve single base substitutions, formation of a premature stop codon, frameshift, or elimination of an amino acid residue. They may occur in amino or carboxyl termini, in cytoplasmic or extracellular loops, or in transmembrane regions of this connexin. The causal relationship between these diverse mutations of Cx32 and demyelination of peripheral nerves remains enigmatic since the mutations are quite diverse and may or may not affect channel properties. Moreover, none of the mutations has discernable effects on other Cx32-containing tissues. Nonetheless, it has been suggested

(1) that Cx32 in the peripheral myelin may either function as ATP-sensitive hemichannels, (2) that reflexive gap junctions comprised of Cx32 may allow exchange of nutrients between the perinuclear region of the Schwann cell and the Schmidt–Lantermann incisures and paranodal processes at the node of Ranvier or (3) that Cx32-positive gap junctions may provide a mechanism for signaling between internodes. Demyelination in this disease, however, could result from some subtle failure of more general properties of the Cx32-positive junctions in peripheral myelin.

A spectrum of human cardiac malformations, which includes anomalies of folding and laterality of the heart, has been related to an apparent autosomal recessive disease characterized by mutations of the gap junction protein, connexin43. The mutation of the connexin43 gene in most of the affected individuals studied so far has resulted in a substitution of proline for serine at amino acid residue 364. This region of the protein is believed to bind phosphate, which may regulate the intracellular trafficking or function of Cx43 (see Section XI). How such mutations influence the mechanism that controls normal situs and the development of handed asymmetry in humans, however, is unknown.

## XV. Summary

A structure serving as an electrical synapse in excitable tissues and as permeable cell-to-cell ion channels in inexcitable cells is the gap junction. Although the function of these structures as electrical synapses in neurons and cardiac muscle is generally understood, new studies relating the activity of specific connexins in particular conduction pathways (for example, in the heart) may begin to specify the functions of gap junctions composed of different connexins. Developmental studies now seem to hold particular promise while studies in the field of cancer cell biology continue to make progress, including analysis of the relationship between connexin expression, processing, assembly, gating, and degradation and transformation to malignancy. In addition, current studies of the properties of gap junction channels and of the cellular regulation of gap junction gating, formation, and turnover in relation to normal cell physiology may yield more specific functional information. Connexin-specific gap junction channels exhibit differences in electrical conductance and molecular permeability that are not directly correlated and dispute the contention that all gap junctions are large aqueous pores. Fortunately, DNAs encoding several connexins have been cloned and significant advances have been made in the correlation of sequence and function. Moreover, the application of this knowledge to the development of transgenic animals exhibiting gain or loss of function of specific connexins should lead to more specific insights into functional differences between connexins.

## Bibliography

Bai, S., Schoenfeld, A., Pietrangelo, A., and Burk, R. D. (1995). Basal promoter of the rat connexin 32 gene: Identification and characterization of an essential element and its DNA-binding protein. *Mol. Cell. Biol.* **15,** 1439–1445.

Bennett, M. V. L., and Verselis, V. K. (1992). Biophysics of gap junctions. *Sem. Cell Biol.* **3,** 29–47.

Bennett, M. V. L., Barrio, L. C., Bargiello, T. A., Spray, D. C., Hertzberg, E., and Saez, J. C. (1991). Gap junctions: New tools, new answers, new questions. *Neuron* **6,** 305–320.

Bennett, M. V. L., Zheng, X., and Sogin, M. L. (1994). The connexins and their family tree. *In* "Molecular Evolution of Physiological Processes," pp. 223–233. The Rockefeller University Press, New York.

Beyer, E. C., Paul, D. L., and Goodenough, D. L. (1990). Connexin family of gap junction proteins. *J. Membr. Biol.* **116,** 187–194.

Brink, P. R., and Dewey, M. M. (1980). Evidence for fixed charge in the nexus. *Nature* **285,** 101–102.

Brink, P. R., and Fan, S.-F. (1989). Patch clamp recordings from membranes which contain gap junction channels. *Biophys. J.* **56,** 579–593.

Budunova, I. V., Carbajal, S., Viaje, A., and Slaga, T. J. (1996). Connexin expression in epidermal cell lines from SENCAR mouse skin tumors. *Mol. Carcinog.* **15,** 190–201.

Chiba, H., Sawada, N., Oyamada, M., Kojima, T., Nomura, S., Ishii, S., and Mori, M. (1993). Relationship between the expression of the gap junction protein and osteoblast phenotype in a human osteoblastic cell line during cell proliferation. *Cell Struct. Func.* **19,** 419–426.

Darrow, B. J., Laing, J. G., Lampe, P. D., Saffitz, J. E., and Beyer, E. C. (1995). Expression of multiple connexins in cultured neonatal rat ventricular myocytes. *Circ. Res.* **76,** 381–387.

Donahue, H. J., McLeod, K. J., Rubin, C. T., Andersen, J., Grinc, E. A., Hertzberg, E. L., and Brink, P. R. (1995). Cell-to-cell communication in osteoblastic networks: Cell line-dependent hormonal regulation of gap junction function. *J. Bone Min. Res.* **10,** 881–889.

Filson, A. J., Azarnia, R., Beyer, E. C., Loewenstein, W. R., and Brugge, J. A. (1990). Tyrosine phosphorylation of a gap junction protein correlated with inhibition of cell-to-cell communication. *Cell Growth Diff.* **1,** 661–668.

Fishman, G. I., Moreno, A. P., Spray, D. C., and Leinwand, L. A. (1991). Functional analysis of human cardiac gap junction channel mutants. *Proc. Natl. Acad. Sci. USA* **88,** 3525–3529.

Flagg-Newton, J., Simpson, I., and Loewenstein, W. R. (1979). Permeability of the cell-to-cell membrane channels in mammalian cell junction. *Science* **205,** 404–407.

Furshpan, E. J., and Potter, D. D. (1959). Transmission at the giant motor synapses of the crayfish. *J. Physiol.* **145,** 289–325.

Furshpan, E. J., and Potter, D. D. (1968). Low-resistance junctions between cells in embryos and tissue culture. *Curr. Top. Dev. Biol.* **3,** 95–127.

Goldberg, G. S., Martyn, K. D., and Lau, A. F. (1994). A connexin 43 antisense vector reduces the ability of normal cells to inhibit the foci formation of transformed cells. *Mol. Cytogen.* **11,** 106–114.

Goodenough, D. A., Goliger, J. A., and Paul, D. (1996). Connexins, connexons, and intercellular communication. *Ann. Rev. Biochem.* **65,** 475–502.

Haeflinger, J.-A., Bruzzone, R., Jenkins, N. A., Gilbert, D. J., Copeland, N. G., and Paul, D. L. (1992). Four novel members of the connexin family of gap junction proteins. *J. Biol. Chem.* **267,** 2057–2064.

Hendrix, E. M., Mao, S. J. T., Everson, W., and Larsen, W. J. (1992). Myometrial connexin 43 trafficking and gap junction assembly at term and in preterm labor. *Mol. Reprod. Develop.* **33,** 27–38.

Hendrix, E. M., Myatt, L., Sellers, S., Russell, P. T., and Larsen, W. J. (1995). Steroid hormone regulation of rat myometrial gap junction formation: Effects on Cx43 levels and trafficking. *Biol. Reprod.* **53,** 547–560.

Hoh, J. H., John, S. A., and Revel, J.-P. (1991). Molecular cloning and characterization of a new member of the gap junction gene family, connexin 31. *J. Biol. Chem.* **266,** 6524–6531.

Kamibayashi, Y., Oyamada, Y., Mori, M., and Oyamada, M. (1995). Aberrant expression of gap junction proteins (connexins) is associated with tumor progression during multistage mouse skin carcinogenesis. *Carcinogenesis.* **16,** 1287–1297.

Kanter, H. L., Saffitz, J. E., and Beyer, E. C. (1992). Cardiac myocytes express multiple gap junction proteins. *Circ. Res.* **70,** 438–444.

Kenne, K., Fransson-Steen, R., Honkasalo, S., and Warngard, L. (1994). Two inhibitors of gap junction intercellular communication, TPA and endosulfan: Different effects on phosphorylation of connexin 43 in the rat liver epithelial cell line, IAR 20. *Carcinogenesis.* **15,** 1161–1165.

Khan-Dawood, F. S., Yang, J., Dawood, M. Y. (1996). Expression of gap junction protein connexin-43 in the human and baboon corpus luteum. *J. Clin. Endocr. Metab.* **81,** 835–842.

Laird, D. W., Puranam, K. L., and Revel, J.-P. (1991). Turnover and phosphorylation dynamics of connexin 43 gap junction protein in cultured cardiac myocytes. *Biochem. J.* **273,** 67–72.

Laird, D. W., Castillo, M., and Kasprzak, L. (1995). Gap junction turnover, intracellular trafficking, and phosphorylation of connexin 43 in brefeldin A-treated rat mammary tumor cells. *J. Cell Biol.* **131,** 1193–1203.

Larsen, W. J., Wert, S. E., and Brunner, G. D. (1987). Differential modulation of follicle cell gap junction populations at ovulation. *Dev. Biol.* **122,** 61–71.

Loewenstein, W. R. (1985). Regulation of cell-to-cell communication by phosphorylation. *Biochem. Soc. Symp.* **50,** 43–58.

Loewenstein, W. R. (1987). The cell-to-cell channel of gap junctions. *Cell* **48,** 725–726.

Loewenstein, W. R., Kanno, Y., and Socolar, S. J. (1978). Quantum leaps of conductance during formation of membrane channels at cell–cell junction. *Nature* **274,** 133–136.

Loo, L. W., Berestecky, J. M., Kanemitsu, M. Y., and Lau, A. F. (1995). pp60src-Mediated phosphorylation of connexin 43, a gap junction protein. *J. Biol. Chem.* **270,** 12751–12761.

Makowski, L., Caspar, D. L. D., Phillips, W. C., and Goodenough, D. A. (1977). Gap junction structures. II. Analysis of the x-ray diffraction data. *J. Cell Biol.* **74,** 629–645.

Mesnil, M., Krutovskikh, V., Piccoli, C., Elfgang, C., Traub, O., Willicke, K., and Yamasaki, H. (1995). Negative growth control of HeLa cells by connexin genes: Connexin species specificity. *Cancer Res.* **55,** 629–639.

Musil, L. S., and Goodenough, D. A. (1990). Gap junctional intercellular communication and the regulation of connexin expression and function. *Curr. Opin. Cell Biol.* **2,** 875–880.

Neveu, M. J., Hully, J. R., Babcock, K. L., Hertzberg, E. L., Nicholson, B. J., Paul, D. L., and Pitot, H. C. (1994). Multiple mechanisms are responsible for altered expression of gap junction genes during oncogenesis in rat liver. *J. Cell Sci.* **107,** 83–95.

Neyton, J., and Trautmann, A. (1985). Single-channel currents of an intercellular junction. *Nature* **317,** 331–335.

Orsino, A., Taylor, C. V., and Lye, S. J. (1996). Connexin-26 and connexin-43 are differentially expressed and regulated in the rat myometrium throughout late pregnancy and the onset of labor. *Endocrinology* **137,** 1545–1553.

Piersanti, M., and Lye, S. J. (1995). Increase in messenger ribonucleic acid encoding the myometrial gap junction protein, connexin-43, requires protein synthesis and is associated with increased expression of the activator protein-1, c-fos. *Endocrinology* **136,** 3571–3578.

Revel, J.-P., and Karnovsky, M. J. (1967). Hexagonal array of subunits in intercellular junctions in the mouse heart and liver. *J. Cell Biol.* **38,** C7–C12.

Revel, J.-P., Hoh, J. H., John, S. A., Laird, D. W., Puranam, K., and Yancey, S. B. (1992). Aspects of gap junction structure and assembly. *Sem. Cell Biol.* **3,** 21–28.

Saez, J., Nairn, A. C., Czernik, A. J., Spray, D. C., Hertzberg, E. L., Greengard, P., and Bennett, M. V. L. (1990). Phosphorylation of connexin32, a hepatocyte gap-junction protein, by cAMP-dependent protein kinase, protein kinase C and Ca2$^+$/calmodulin-dependent protein kinase II. *Eur. J. Biochem.* **192,** 263–273.

Simpson, I., Rose, B., and Loewenstein, W. R. (1977). Size limit of molecules permeating the junctional membrane channels. *Science* **195,** 294–296.

Spray, D. C., and Dermiietzel, R. (1995). X-linked dominant Charcot-Marie-Tooth disease and other potential gap junction diseases of the nervous system. *TINS* **18,** 256–262.

Swenson, K. I., Piwnica-Worms, H., McNamee, H., Paul, D. L. (1990). Tyrosine phosphorylation of the gap junction protein connexin43 is required for the pp60v-src-induced inhibition of communication. *Cell Reg.* **1,** 989–1002.

Veenstra, R. D., and DeHaan, R. L. (1986). Measurement of single channel currents from cardiac gap junctions. *Science* **233,** 972–974.

Veenstra, R. D., Wang, H.-Z., Beblo, D. A., Chilton, M. G., Harris, A. L., Beyer, E. C., and Brink, P. R. (1995). Selectivity of connexin-specific gap junctions does not correlate with channel conductance. *Circ. Res.* **77,** 1156–1165.

Veenstra, R. D., Wang, H.-Z., Westphale, E. M., and Beyer, E. C. (1992). Multiple connexins confer distinct regulatory and conductance properties of gap junctions in developing heart. *Circ. Res.* **71,** 1277–1283.

Warn-Cramer, B. J., Lampe, P. D., Kurata, W. E., Kanemitsu, M. Y., Loo, L. W., Eckert, W., and Lau, A. F. (1996). Characterization of the mitogen-activated protein kinase phosphorylation sites on the connexin-43 gap junction protein. *J. Biol. Chem.* **271,** 3779–3786.

Wohlburg, H., and Rohlmann, A. (1995). Structure-function relationships in gap junctions (review). *Int. Rev. Cytol.* **157,** 315–373.

Yamaguchi, D. T., Huang, J. T., and Ma, D. (1995). Regulation of gap junction intercellular communication by pH in MC3T3-E1 osteoblastic cells. *J. Bone Min. Res.* **10,** 1891–1899.

Yamasaki, H. (1991). Aberrant expression and function of gap junctions during carcinogenesis. *Env. Health Prosp.* **93,** 191–197.

Yu, W., Dahl, G., and Werner, R. (1994). The connexin43 gene is responsive to oestrogen. *Proc. R. Soc. London Series B: Biol. Sci.* **255,** 125–132.

Louis J. DeFelice and Michele Mazzanti

# 31

# Biophysics of the Nuclear Envelope

## I. Introduction

Eukaryotic cells contain a nucleus, that is, a spherical body that encloses characteristic organelles. These organelles include a thin nuclear envelope, one or more nucleoi, chromatin (the chromosomes, DNA attached to proteins, primarily histones), a proteinaceous nuclear lamina (which contacts the inner nuclear membrane, chromosomes, and the nuclear RNA), irregular granules of chromatin material, and linin (fine threads that associate with the chromatin granules). Amorphous nucleoplasm occupies the rest of the nuclear volume.

The nuclear envelope transports macromolecules, and the selectivity of transport changes during the cell cycle and during morphological or developmental stages. Most textbooks describe the envelope as no barrier to small ions, such as Ca, Na, K, and Cl. In this view the nucleus is a sicvclike structure containing large aqueous pores. However, certain data suggest that the nuclear envelope can restrict the movement of intermediate-size molecules, including peptides, amino acids, and sugars. Moreover, although the data are controversial, particular experiments support the idea that inorganic ions such as Ca can accumulate in, or be excluded from, the nucleus. Furthermore, the nucleus has an electrical potential, and it can swell or shrink in response to osmotic forces. One theory is that such phenomena derive from selective absorption by the nucleoplasm. This chapter explores the view that the nuclear envelope can selectively transport small ions and has properties similar to the semipermeable plasma membrane.

The nuclear pore governs the movement of nucleic acids and large molecular weight proteins into and out of the nucleus: Enzymes enter for polymerization of nucleic acids, RNA leaves for translation in the cytosol, and DNA-binding proteins enter for transcription and gene regulation. Selection is precise, because the nucleus eliminates the majority of proteins from its domain. During singular events, such as fertilization or viral infection, or under laboratory manipulations, such as gene transfection, the nuclear pore may establish transitory selection. This chap-

ter explores the possibility that small-ion selection accounts for electric and osmotic properties of the nuclear envelope. The movement of ions across the nuclear envelope may act as regulatory countercurrents for the transport of macromolecules. This assertion derives from the biophysical characteristics of nuclei and the presence of ion-selective channels in the nuclear envelope. Channels in the nuclear envelope have properties similar to channels in the plasma membrane. We further suggest that some channels observed in the nuclear envelope form part of the nuclear pore complex. This proposition introduces a paradox: How could a pore that transports large molecules through a large opening also select small ions? The remainder of this chapter addresses that question.

## II. Permeability of the Nuclear Envelope

Two mechanisms may explain the separation of molecules by the nuclear envelope. One involves solubility in the nucleoplasm or, what amounts to the same thing, exclusion from the cytoplasm. Separation of solutes between adjacent phases falls under the heading of a **Donnan equilibrium.** A Donnan equilibrium relies on the differential absorption of solutes into immiscible phases. If charged molecules partition into adjacent phases, an electric potential could result between the two phases. If the nucleoplasm and the cytoplasm form such domains, and if they connect through open pores, the Donnan equilibrium could explain the nuclear resting potential.

The second mechanism involves a semipermeable membrane that separates the nucleoplasm from the cytoplasm. This mechanism, which characterizes cell plasma membranes, comes under the heading of a **Nernst–Planck regime.** Whereas one membrane surrounds a cell, two membranes surround the nucleus. However, the double-membrane structure is simplified because **nuclear pores** span the envelope and act as pores between nucleoplasm and cytoplasm.

An early model of the double-membrane-spanning nuclear pore depicts it as two parallel rings, one on the cytoplasmic side and the other on the nucleoplasmic side. Each ring consists of eight subunits that surround a central pore. We refer to this structure as the **nuclear pore complex.** This term applies to the entire structure, not the opening in the center. The opening contains an electron-dense **central granule** that connects the two rings. The size and shape of the central pore help determine its selectivity to macromolecules. For example, RNA molecules elongate into cylindrical rods (diameter of 100 Å) as they enter the pore. Proteins larger than 100 Å in diameter cannot pass through the pore. However, the size of the pore cannot explain its selectivity, because 170-Å-diameter gold particles coated with **nucleoplasmin** can enter nuclei. Without nucleoplasmin, or after treatment with trypsin to remove nucleoplasmin, the large gold particles cannot enter. Therefore, proteins may contain a signal that permits nuclear transport. A single amino acid substitution can abolish the nuclear transport of certain tumor antigens, and nucleic acids show a similar specificity. For example, a single base substitution can reduce the rate of tRNA transport by a factor of 20. Some substances apparently block transport by getting in the way, for example, wheat germ agglutinin (WGA). WGA accumulates on the cytoplasmic face of the nucleus and inhibits the uptake of nucleoplasmin. However, blocking cannot explain all the data, because WGA does not influence the rate of dextran entry.

The high rates of tRNA translocation in mammalian cell nuclei ($10^9$ molecules/min) exclude diffusion as the mechanism of transport. Some experiments suggest carrier-mediated transport for macromolecules, augmented by the observation that RNA transport requires adenosine triphosphate (ATP). Although diffusion could account for only small fraction of macromolecular transport, paradoxically, smaller molecules are translocated at rates below diffusion. For example, inulin (5.5 kDa) enters the nucleus at one-fifth its diffusion rate. Another example is sucrose, which passes through the nuclear envelope more readily than it does through the plasma membrane, but at one-third the rate predicted by diffusion. Evidence that nuclei may even restrict small inorganic anions and cations comes from *Drosophila* salivary gland cells. These nuclei maintain an electrical potential of −15 mV with respect to the cytoplasm. Similar potentials exist in other nuclei, although not always of the same sign or amplitude. Furthermore, Na

and K ions may have different concentrations in the cytoplasm than in the nucleoplasm, and the concentration of nuclear Ca changes during cell differentiation and fertilization. These data are controversial, and not everyone agrees that ion gradients exist. Even if they do, we may ask whether the nucleus restricts ions via a Donnan equilibrium or whether the nuclear envelope acts as a semipermeable membrane. If selection occurs via a membrane, how can we reconcile that with the passage of large macromolecules through oversized openings?

## III. Structure of the Nuclear Envelope

Two lipid-bilayer membranes surround the nucleus. In this regard, the nucleus resembles a mitochondrion. In contrast, a single lipid bilayer surrounds other internal organelles, such as the Golgi apparatus, storage vesicles, and the endoplasmic reticulum. The double-membrane structure of isolated nuclei remains intact *in vitro* and, in many ways, the isolated nuclei act similar to cells maintained in culture.

### A. Isolation of Nuclei

*Nuclei from invertebrate cells* are easily dissected from starfish oocytes in the germinal vesicle stage, when the nuclei are large (30 μm in diameter). After the dissection, the nuclei can be kept for hours at cold temperatures (5°C) in an intracellular-like solution (200 m$M$ K$_2$SO$_4$, 20 m$M$ NaCl, 10 m$M$ HEPES, 10 m$M$ EGTA, and surcrose to 1000 mOsm). Nuclei are tolerant to other physiological solutions, such as extracellular media and seawater. **Mammalian cell nuclei** (10 μm in diameter) are readily obtained from mouse oocytes 12 hr after superovulation. On fertilization, pronuclei or early embryo nuclei are collected 12 hr after mating. Prior to removing the nucleus, the oocytes, zygotes, or embryos should be placed in an intracellular-like solution (120 m$M$ KCl, 2 m$M$ MgCl$_2$, 1.1 m$M$ EGTA, 0.1 m$M$ CaCl$_2$, 5 m$M$ glucose, 10 m$M$ HEPES, pH 7.4). Thus when the nucleus comes out of the cell it experiences a normal ionic environment. Mammalian nuclei can also be isolated from differentiated tissue, such as liver cells, by shearing the tissue and centrifuging the homogenate. This results in a pellet of pure nuclei, which can then be resuspended into a physiological solution. These tech-

**FIG. 1.** Structure of the nucleus. Transmission electron micrograph of three mouse oocytes at the germinal vesicle stage. The plasma membrane (pm) surrounds the cytoplasmic compartment (cy), and the nuclear envelope (ne) surrounds the nuclear compartment (nu) within the cytoplasm. The micrograph demonstrates two additional membrane-bound compartments in the cytoplasm, mitochondria (mt) and the endoplasmic reticulum (er). The endoplasmic reticulum has most likely swollen during the fixation procedure. The nuclear envelope consists of a double-membrane structure interrupted by nuclear pores (np). The double membranes, designated the inner membrane and outer membrane (see Fig. 3), define a separate cytoplasmic compartment called the cistern (cs). The cistern may be continuous with the endoplasmic reticulum (see Fig. 3). The nuclei contain darkly stained material adjacent to the inner membrane, called nuclear lamina, and they contain the diffuse darkly stained material called chromatin. Samples were dehydrated with alcohol to propylene oxide, infiltrated and embedded in epon, thin sectioned, and stained in uranyl acetate and lead citrate. The bar length indicates 1 μm, and the bar width indicates 1000 Å. (Reprinted with permission from Mazzanti *et al.*, 1991. Copyright *Journal of Membrane Biology.*)

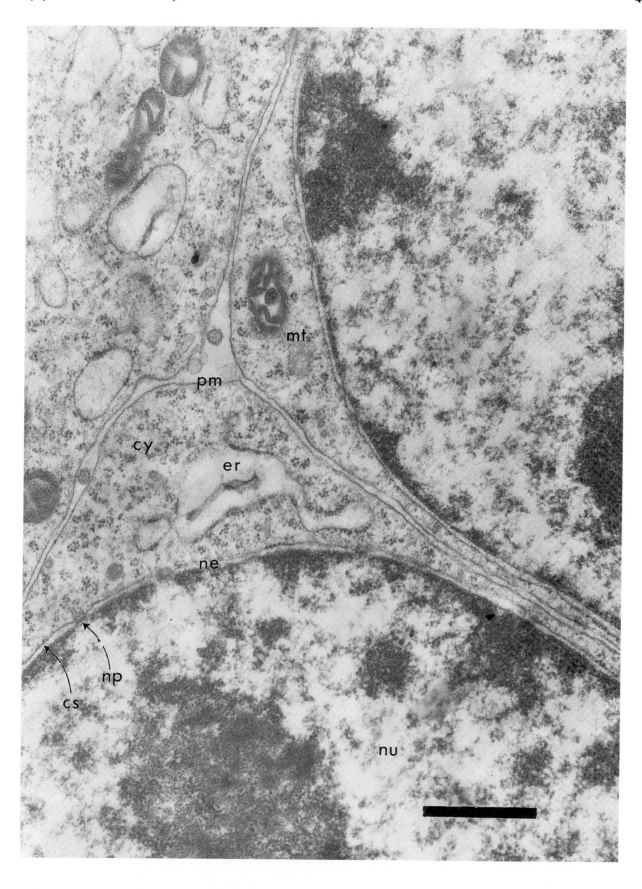

niques result in four categories of nuclei that we have used for comparative studies: germinal vesicles, pronuclei, two-cell embryo nuclei, and adult liver cell nuclei. It is unclear whether nuclei preserve their characteristics in culture conditions; however, zygote enucleation results in pronuclei that have the same resting potential as *in vivo*. Mouse liver nuclei removed by centrifugation have zero resting potential, but it is uncertain whether this condition comes from the isolation procedure.

## B. Electron Microscopy of Nuclei

The procedures for obtaining the photographs and electron micrographs of nuclei become important when evaluating nuclear pore density. The dissected starfish nuclei were fixed in 1% glutaraldehyde, postfixed in 1% osmium tetroxide, dehydrated in alcohol, and dried in $CO_2$. The nuclei were then coated with gold for scanning electron microscopy. The mouse nuclei were fixed in 1% glutaraldehyde and 1% tannic acid, postfixed in $OsO_4$, and dehydrated in alcohol, and 400–500 Å sections were stained with uranyl acetate and lead citrate. These dehydration and fixation procedures may introduce shrinkage and distortion of the nuclear envelope, causing an apparent density of nuclear pores that exceeds the *in vitro* density.

Figure 1 is an electron micrograph of three mouse oocytes packed together. Only a segment of each oocyte appears. The single darkly stained line, designated as **plasma membrane** (pm), represents the bilayer that surrounds the oocyte. The plasma membrane encompasses the **cytoplasm** (cy), which includes internal organelles such as mitochondria (mt) and the endoplasmic reticulum (er). The term **cytosol** (cy) refers to the cytoplasmic volume, excluding the organelles. The designation (cy) will stand either for the cytoplasm or the cytosol, depending on context. In Fig. 2, parallel lines represent the lipid bilayer membrane. Figure 2 depicts the extracellular volume (ex), the cytosol (cy), the single-membrane-enclosed endoplasmic reticulum (er), and the double-membrane-enclosed nucleus (nu). Openings that penetrate the double membrane connect the nucleoplasm with the cytoplasm. The membranes that encircle the nucleus enclose a volume topologi-

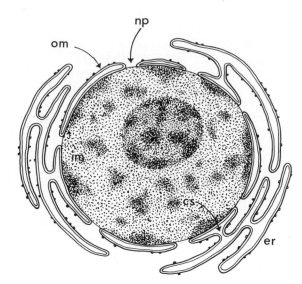

**FIG. 3.** The nuclear envelope and the endoplasmic reticulum. This illustration shows the relationship between the outer membranes (om), the inner membranes (im), the nuclear pore (np), and the endoplasmic reticulum (er). The inner and outer membranes and the nuclear pore complexes define the nuclear envelope. The central egg-shaped structure inside the nucleus represents the nucleolus. The dots along the outer membrane and on the endoplasmic reticulum represent ribosomes. The cistern (cs) of the nuclear envelope and the lumen of the endoplasmic reticulum form a continuous compartment within the cell.

cally equivalent to the lumen of the endoplasmic reticulum. This restricted volume has a particular name, the **cistern** (cs). In effect, cisterns are specialized organelles bound by one membrane. The membrane that defines the cistern is further categorized as the **outer membrane** and the **inner membrane.** Because the inner and outer membranes are close together, they are given the cumulative name **nuclear envelope** (ne). Figure 1 shows the cistern (cs) and the nuclear envelope (ne).

Figure 3 indicates that cistern membranes and endoplasmic reticulum membranes may join one another. However, cisterns may exist without any connection to the endoplasmic reticulum. The anatomy of the nucleus and its associations to cell structures varies with the developmental stage or the specialization of the cell. At some stages, the endoplasmic reticulum is virtually absent. Even in cells that have a well-developed endoplasmic reticulum, large regions of the envelope do not connect with the endoplasmic reticulum. Where they do connect, cisterns adjoin the endoplasmic reticulum via the outer membrane. The endoplasmic reticulum is either smooth or rough, depending on the absence or presence of ribosomes (dots in Fig. 3). The outer membrane of the nuclear envelope is also smooth or rough. The ribosomes on the endoplasmic reticulum are evident in the photograph in Fig. 1; if there are ribosomes on the outer membrane, they are not obvious.

The openings between the cisterns depicted in Figs. 2 and 3 are more than mere breaches in the nuclear envelope. They contain the nuclear pore complex. As already mentioned, the nuclear pore complex is often referred to as the nuclear pore (np). The reader must be careful, because

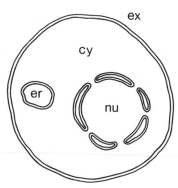

**FIG. 2.** Cell spaces. The diagram defines the volumes within and around a cell: the extracellular space (ex); the cytosol (cy); the endoplasmic reticulum (er); and the nucleus (nu). The double lines represent lipid bilayers.

"nuclear pore" can also mean the *large central opening* in the nuclear pore complex.

## IV. Structure of the Nuclear Pore

The nuclear pore complex has a molecular weight of about 100 MDa and contains more than 100 gene products. It consists of eight symmetrical units that form a large central opening. Before going into details, we note one difference between pore-forming proteins in the nuclear pore and pore-forming proteins in the plasma membrane. Ion channels, gap junction proteins, receptor subunits, and neurotransmitter transporters in the plasma membrane span the bilayer. These we call transmembrane (TM) proteins. Some nuclear pore proteins span the bilayer, but with a different relationship to the flow of material. To illustrate, consider the structure of **gap junctions.** This example has special relevance to our topic because nuclear pore proteins and gap junction proteins both form openings through abutting membranes.

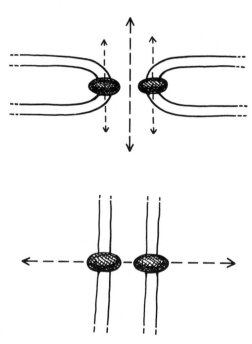

**FIG. 4.** Nuclear pores versus gap junctions. A comparison between the structure and function of the nuclear pore (top) and the gap junction (bottom). The nuclear pore complex contains eight repeating units with TM subunits (two are shown here). Each unit is a multiprotein structure that repeats eight times. Material flows through a central pore (the center arrow in the top figure) defined by the surrounding subunits. In addition, lateral nucleocytoplasmic pathways are located within each unit (see Fig. 5). The shorter arrows in the top drawing illustrate the lateral pathways. The complete nuclear pore complex contains eight lateral pathways surrounding the central pore. Gap junctions have a total of six pairs of abutting membrane-spanning proteins. One such pair is illustrated in the bottom panel. Apart from a different number of repeating units, gap junctions also have a different structural relationship to function when compared to nuclear pore complexes.

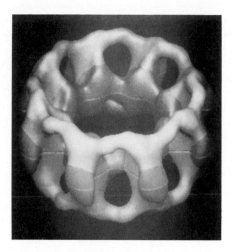

**FIG. 5.** Structure of the nuclear pore. A three-dimensional reconstruction based on an electron density map of the 100-MDa structure that forms the nuclear pore complex. The entire structure has an external diameter of about 1300 Å. The model proposes eight lateral channels that surround a large central pore. The central pore normally contains a proteinaceous plug that occupies most of the opening (see Fig. 6). The central plug is absent in this figures. The reconstruction is based on electron microscopy and image analysis of detergent-released nuclear pore complexes from macronuclei of mature *Xenopus laevis* oocytes. (From Hinshaw *et al., Cell* 69, 1133–1943, 1992. Copyright 1992 Cell Press.)

Figure 4 illustrates that membrane-spanning proteins of the nuclear pore complex insert at the division of the inner and outer membrane. Eight units (only two appear in Fig. 4), each composed of many proteins, form the central opening (center arrow, top drawing). In addition, eight lateral openings exist (side arrows indicate two lateral openings). In gap junctions, six pairs of proteins (only one pair appears in Fig. 4, bottom drawing) define the opening. The distinction may seem artificial, because gap-junction proteins (connexins), as well as the subunits of ion channels and receptors form parallel structures. However, the nuclear envelope exhibits a distinct structure–function relationship between the TM proteins and transport. As far as we know, the functional connections made within the nuclear pore do not include cistern-to-cistern communication, but only cytoplasm-to-nucleoplasm communication.

Figure 5 illustrates the three-dimensional structure in more detail. The dark regions in the reconstruction are the TM domains of the eight repeated units. These units also define eight additional pathways (as indicated in Fig. 4 by the lateral arrows). These pathways surround the central opening. We refer to this structure as an *eight-plus-one configuration,* in analogy with the axoneme nine-plus-two configuration. In a minority of cases, more than eight units can form a nuclear pore complex. The significance of these higher order structures is unknown. Figure 6 defines the proposed relationship between the nuclear pore (np), the outer membrane (om), and the inner membrane (im) of the nuclear envelope. The inner and outer membranes form the connecting loops that define openings between cisterns. The cross-hatched regions in the drawing

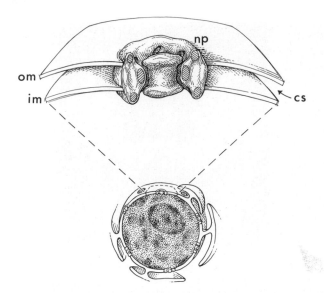

**FIG. 6.** The nuclear pore and the nuclear envelope. Illustration of the position of the nuclear pore (np) within the double-membrane structure that envelops the nucleus. The two planar membranes illustrated in the diagram represent the outer membrane (om) and the inner membrane (im). The interconnecting subunits that make up the supramolecular nuclear pore complex insert into the lipid bilayer at the border between the inner and outer membranes. These subunits form eight channels that surround the main opening of the pore (see Fig. 5). The main opening contains a proteinaceous granule or "plug" (not shown in Fig. 5), which we have pictured as a nearly space-filling, spool-shaped object. Thus, two kinds of pathways connect the cytoplasm and the nucleoplasm: the middle pore that contains the plug and the eight encircling channels that riddle the proteins that form the central pore.

indicate the lateral openings that exist within the complex. Figure 6 shows a feature mentioned earlier: the nuclear granule (or nuclear plug) that resides in the central opening. The nuclear plug is illustrated here as spool-shaped. Recent work indicates that the plug may move within the pore. This movement depends on Ca levels in the cisterns and may be involved in transport. The nuclear plug may also include newly synthesized ribosomes.

The nuclear envelope therefore consists of a large number of proteins, some of which are embedded in bilayers. The nuclear pore complex thus forms an eight-plus-one assembly that spans the double membrane. Some nuclei are virtually covered with nuclear pores. The high density of pores raises a question about the minimum amount of lipid necessary to sustain a dense array of membrane-spanning proteins. In nuclei with a high density of pores, the relative amount of outer lipid membrane diminishes and connections to the endoplasmic reticulum become sparse. The density of pores varies between 1 and 120 per square micrometer depending on the specialization of the cell. The local density of pores may also differ within the same nucleus.

Figure 7 illustrates the patch-clamp technique applied to isolated nuclei. In this technique, a glass electrode isolates a patch of membrane on the outer surface of the nucleus. The applied voltage $V_p$ (the battery in the diagram) appears

at the electrode tip, and the current meter (I) records the flow of ions through the patch. An isolated patch is $1-5 \ \mu m^2$ in area. The voltage across the envelope is the difference between the nuclear resting potential and the pipette potential. This voltage appears across the nuclear pores in the patch. The possibility exists that there are no pores in the patch; in this case, channels in the outer and inner membranes could connect the pipette solution to the nucleoplasm. This cistern channel pathway would be in parallel with the nuclear pore pathway. Can we form a patch that certainly includes nuclear pores? To address this issue, consider an experiment on a nucleus with a very high density of pores.

Figure 8 shows a scanning electron micrograph of a starfish oocyte at the germinal vesicle stage. The nucleus is ~30 $\mu m$ in diameter, larger than the majority of specialized cells. The inset shows an isolated area on the outer membrane that is about 1 $\mu m$ in diameter. The diameter of a nuclear pore complex is ~1000 Å (0.1 $\mu m$), which implies that 100 nuclear pore complexes could fit side by side in 1.0 $\mu m^2$ of membrane. If each doughnut-shaped object in the expanded field in Fig. 8 represents one nuclear pore complex, the pores practically cover the surface. Even if shrinkage has occurred during the fixation procedure, it seems unlikely that a nucleus-attached patch on this nucleus could contain any less than dozens of pores in parallel.

## V. Electrophysiology of the Nucleus

If nuclei like the one shown in Fig. 8 are dispersed in a dish, they can be manipulated and prepared for electrophysiology as if they were cells. As mentioned earlier, some nuclei have resting potentials. For example, in a mouse zygote, the *in situ* pronuclei has a resting potential of −10 mV with respect to the cytoplasm. If we eliminate the resting potential of the zygote plasma membrane (see below), no immediate change occurs in the nuclear potential. We expect this because the nuclear potential appears across the envelope; altering a voltage in series should not affect

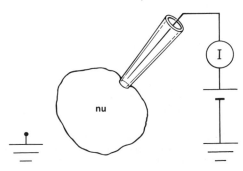

**FIG. 7.** Patch-clamp technique. An illustration of a nucleus-attached patch-clamp experiment. The glass pipette contains a physiological solution connected by a Ag/AgCl electrode to a current meter (I) and a battery. The battery supplies a voltage, $V_p$, to the pipette that ideally appears at the outer membrane of the nucleus. The nucleus bathes in a physiological solution connected to ground. The potential across the nuclear envelope is the resting potential of the nucleus minus $V_p$.

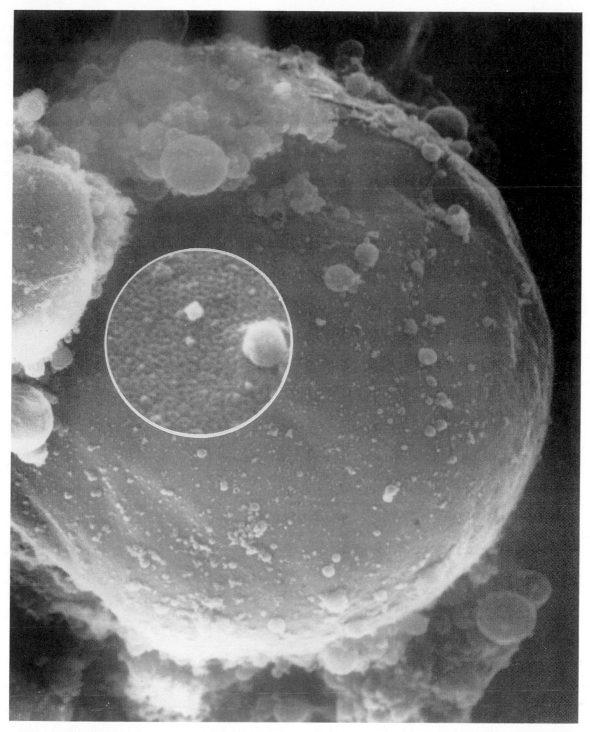

**FIG. 8.** Scanning electron micrograph of a nucleus from a starfish oocyte. The figure shows a scanning electron micrograph of the surface of a germinal vesicle stage nucleus removed from a starfish oocyte (*Mathasterias glacialis*) by manual dissection. The diameter of the nucleus is approximately 30 $\mu$m. The diameter of the inset, which gives a magnified view of the outer membrane, is approximately 1 $\mu$m. Large areas of the external face of the nucleus appear relatively free of extraneous material. The whitish, globular matter is cellular debris stuck to the surface after dissection. The exterior surface of the dissected nucleus appears studded with numerous doughnut-shaped particles that we have interpreted as nuclear pore complexes (average density 60–110 per square micrometer). (Courtesy of Luigia Santella, Stazione Zoologica, Naples, Italy.)

it. For excised pronuclei in an intracellular-like solution (120 m$M$ K), the nuclear potential has the same value it had in the cell. Lowering K concentration *in vitro* to 60 m$M$ K changes the nuclear potential to −20 mV. Table 1 summarizes experiments on *in vivo* and *in vitro* nuclei under different ionic conditions. The data suggest that the nuclear envelope selects K ions. Different conditions may result in other resting potentials. Frog oocyte and adult liver cell nuclei have no resting potential, and starfish nuclei can maintain a positive potential with respect to the bath.

## A. Measurement of Resting Potential

To detect the nuclear resting membrane potentials in intact cells requires a conventional glass microelectrode of high resistance (70–90 M$\Omega$ when filled with 3 $M$ KCl). In such experiments, we penetrate both the outer and inner membranes. Thus the nuclear potential means the voltage across the entire double-membrane nuclear envelope. Ion channels exist in both the outer and inner membranes. For example, Cl channels have been reported on the surface of nuclei as well as ligand-gated channels. However, no measurements exist from isolated cisterns, and we do not know the potential of the cistern with respect to the cytoplasm or nucleoplasm. For *in situ* experiments on the nucleus, cells that are maintained in physiological solutions (133 m$M$ NaCl, 2 m$M$ KCl, 1.5 m$M$ CaCl$_2$, 0.5 m$M$ MgCl$_2$, 10 m$M$ HEPES, 5 m$M$ dextrose, pH 7.4) have normal resting potentials. To eliminate the cell resting potential, we bathe in a solution that mimics the cytoplasm (120 m$M$ KCl, 2 m$M$ MgCl$_2$, 1.1 m$M$ EGTA, 0.1 m$M$ CaCl$_2$, 10 m$M$ HEPES, 5 m$M$ dextrose, pH 7.4).

Among the technical difficulties associated with measuring the nuclear resting potential, electrode tip potential gives the most problems. The structures that compose the nuclear envelope, the inner and outer membranes and the lamina, may stick to the microelectrode and increase resistance and tip potential. To guard against such errors, check

that the tip potential (the electrode voltage with the tip in the bath solution) returns to the baseline on removal. This does not prove that the observed potential comes from the nuclear envelope, but it is an essential first step. Similar technical difficulties related to the value of nuclear potentials arise in measuring plasma membrane potentials.

## B. Measurement of Single Channels from Nuclei

In mouse liver cell nuclei, about 50% of the attempts to patch succeed. Seals up to 50 G$\Omega$ are obtained using pipettes having an external tip diameter of about 1 $\mu$m and a resistance between 5 and 10 M$\Omega$. In the single-channel experiments, 120 m$M$ K (cytoplasmic-like) solutions may be used to fill the electrode and to bathe the nucleus. To investigate the ionic selectivity of the nuclear envelope, NaCl may replace KCl. Single-channel currents can be recorded with a standard current-to-voltage converter (List EPC-7). Positive pressure in the pipette (e.g., 2 cm H$_2$O) keeps the electrode solution from mixing with the bath solution, flattens and cleans the surface, and helps form the seal. For unknown reasons, a negative potential ($V_p = -40$ mV or greater) applied to the electrode tip before touching the nucleus also facilitates seal formation. The negative potential also helps form seals on most cells maintained in cell culture. After touching, notice a slight increase in electrode resistance, then reverse the pressure until the seal occurs, and then release the suction.

A patch-clamp experiment on the nucleus results in single-channel currents similar to those observed in the plasma membrane of most cells. Nuclear envelope channels were first observed in the envelope of mouse oocytes. Figure 9 shows the results of a patch-clamp experiment on a starfish oocyte nucleus (as in Fig. 8). In Fig. 9, $V_p$ is the voltage applied to the pipette (the battery in the external circuit; see Fig. 7). The channel conductance in this experiment was about 100 pS. Not only long-lasting openings but also numerous fast openings occur. These are too brief to resolve in this record. The downward deflections in Fig. 9 indicate electric current that flows into the nucleus. Without additional information concerning ion selectivity, a downward deflection could result from positive ions moving into the nucleus or negative ions moving out of the nucleus.

Mouse oocyte nuclei have a lower pore density than starfish germinal vesicles; nevertheless, they have qualitatively similar ion channel activity. The nuclear pore density depends on the tissue and on the stage of development of the cell. The freeze–fracture images shown in Fig. 10 are from (a) an oocyte germinal vesicle and (b) from an adult liver cell nucleus. The pore density is 3 to 10 per square micrometer in Fig. 10a and 14–18 per square micrometer in Fig. 10b. Recordings from such surfaces show channels similar to those shown in Fig. 9. Figure 11 summarizes results from four different kinds of nuclei at various stages of cellular development: the mouse germinal vesicle, the pronucleus, a two-cell embryo nucleus, and a differentiated adult cell. In Fig. 11, the current is upward because the applied potential is negative. We may conclude from experiments on starfish oocyte nuclei and mammalian cell nuclei

**TABLE I**   Results of Experiments on Nuclei's Subjected to Different Ionic Conditions

| (1) 2 K/133 Na | | (2) 120 K/0 Na | | (3) 120 K/0 Na | (4) 60 K/60 Na |
|---|---|---|---|---|---|
| cy | nu | cy | nu | nu | nu |
| −23 | −11 | 0 | −10 | −10 | −28 |
| −18 | −12 | 2.5 | −9.5 | −7 | −18 |
| −28 | −9 | 3 | −9 | −9 | −23 |
| −25 | −11 | 1 | −11 | −12 | −25 |

*Note:* The ratios of K to Na (in m$M$) refer to extracellular (bath) concentrations. In columns (1) and (2) (zygote pronuclei *in situ*), the voltages (in mV) are as follows: cy = the potential of the cytoplasm with respect to the extracellular solution; nu = the potential of the nucleoplasm with respect to the cytoplasm. In columns (3) and (4) (pronuclei *in vitro*), the voltages are the nuclear potential with respect to the bath.

Vp (mV)

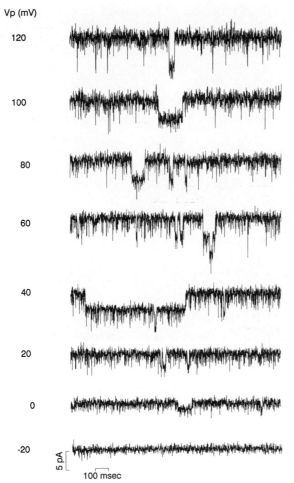

**FIG. 9.** Nucleus-attached patch recordings from starfish germinal vesicle. The figure shows sample currents recorded from the surface of a germinal vesicle stage nucleus isolated from a starfish oocyte similar to that of Fig. 8. Channel openings appear as downward deflections, which signify positive charges flowing into the nucleus. Two categories of openings appear: brief inward currents barely resolved on this timescale and longer openings that last several hundred milliseconds. In some traces (e.g., at 60 mV), simultaneous openings occur giving two current levels of the same amplitude. The voltages ($V_p$) at the side of each trace indicate the pipette potential with respect to the bath. The nucleus-attached electrode contains 200 mM $KSO_4$, 20 NaCl, 200 sucrose, 10 EGTA, 10 HEPES (pH 7.0) and the bath solution comprises natural seawater at room temperature. (From Santella *et al.*, 5th Intl. Conf. on Cell Biol., Madrid, Spain, 1992.)

that channels in the nuclear envelope are an ubiquitous phenomenon. In addition, Fig. 11 shows for embryonic preparations (Figs. 11a–c) only one or two conductance levels, while in the adult preparations there are consistently more levels, as illustrated in the histograms to the right of the figure. Combining morphological information with functional results, we have hypothesized a structural relationship between the nuclear pores and the channels observed by patch clamping the nuclear envelope. Specifically, the channels could represent a pathway through the nuclear pore complex.

## IV. Osmotic Effects in the Nucleus

Figure 12 shows that the nucleus undergoes changes in response to osmotic forces. However, these changes are not the simple osmotic effects expected from solution chemistry. In this experiment, the ionophore nystatin was used to make the plasma membrane permeable to monovalent cations. The extracellular bath and the cytoplasm are in electrical communication through nystatin channels. Adjusting bath osmolarity changes nuclear volume. Such permutations of the volume are difficult to see and even more difficult to quantify. Some responses occur in the way we would expect from osmotic changes: in hypertonic solutions the nucleus shrinks, and in hypotonic solutions the nucleus swells. However, too great an osmotic shock produces different effects, some of which are irreversible: If the nucleus shrinks in external 480 mM KCl, it subsequently swells irreversibly to the configuration shown in Fig. 12d. The swelling that occurs in 480 mM KCl is due in part to internal structural changes in the chromatin. Distended nuclei remain stable for up to 30 min. Returning the nucleus to 120 mM KCl, after swelling with 480 mM KCl, causes the nucleus to disrupt. With $V_p = 0$ and 120 mM KCl in the patch and the bath, channels in the patch carry small inward currents. Perfusing with 480 mM KCl and waiting 5 min for the nucleus to swell, the direction of the current changes from inward to outward, and the amplitude increases. In 120 mM KCl, the conductance of the channels is 230 pS, as in the pronucleus. In 480 mM KCl, the conductance is 510 pS, and their reversal potential shifts negative by 33 mV. The size and direction of the potential shift indicate that the channel is K selective. The existence of K-selective channels would explain the nuclear resting potential discussed earlier. More important, these results imply that ion channels in the patch span the double membrane, because the current pathway must join the pipette with the bath. It is conceivable that channels in the patch reside in the outer membrane and that other channels in the inner membrane complete the circuit. Because the only structure known to connect the cytoplasm to the nucleoplasm is the nuclear pore complex, and because the patch contains nuclear pores, it is likely that the nuclear pores, or some structures within the nuclear pore complex, form channels between the cytoplasm and the nucleoplasm.

## VII. Electrical and Diffusional Forces across the Nuclear Envelope

Solution chemistry is insufficient to explain the complex phases that make up nucleoplasm and cytoplasm. Nevertheless, the data suggest that the nuclear envelope can sustain electric and diffusional gradients similar to those that operate across the plasma membrane. In particular, channels recorded from the nuclear surface respond to trans-envelope voltages, and the nucleus changes shape and volume with osmolarity. Thus, we consider the hypothesis that the nuclear envelope acts as a barrier to the diffusion of small ions. If such forces exist they would play major roles in the transport of macromolecules. We now

**FIG. 10.** Nuclear envelope freeze-fracture from mouse oocyte and adult mouse liver cells. Scanning electron micrographs of a fracture through the outer membrane of the nuclear envelope of (a) mouse oocyte nucleus and (b) adult mouse liver cell nucleus. The doughnut-shaped objects represent the nuclear pore complexes. Note the different pore density in the two preparation. The bar length indicates 1 $\mu$m. (Modified from Innocenti and Mazzanti, 1993.)

apply a well-known approach from cellular physiology to nuclear physiology.

The flux of any substrate through a boundary that sustains diffusional and electrical forces is:

$$-\Phi = ukT(dn/dx) + zenu(dV/dx) \qquad (1)$$

The minus sign comes about because flux occurs down a negative gradient (i.e., downhill). The symbol $\Phi$ represents the flux of the substrate, that is, the number of molecules that move per unit area per unit time. We use the symbol $\multimap$ to stand for "has the units of." Thus $\Phi \multimap$ #/cm$^2$ sec), $n(x)$ is the substrate concentration, and $n \multimap$ #/cm$^3$ at position $x$, $V(x)$ is the voltage (mV) at position $x$, $u$ is the mobility of the substrate through the barrier, and $u \multimap$ (velocity/force), $k$ is the Boltzmann constant, $T$ is the absolute temperature in Kelvins, $z$ is the valence, and $e$ is the electronic charge. These latter quantities will appear as the ratio:

$$a = ze/kT = zF/RT = z/25 \text{ mV}^{-1} \text{ at } 23°C$$

By considering the convention in Fig. 2 and integrating the flux equation between the nucleoplasm (nu) and the cytosol (cy), the expression for the flux becomes:

$$-\Phi = ukT\frac{n_{nu}e^{-aV_{nu}} - n_{cy}e^{-aV_{nu}}}{\int e^{aV(x)} dx} \qquad (2)$$

This assumes that $u$ is constant and flux is in a steady state (no buildup or rundown of substrate; thus $\Phi$ is constant). Note that $D = ukT$, where $D$ is the diffusion constant of the substrate. In the derivation of Eq. 2, no assumption

was made about the spatial distribution of the voltage or concentration gradients, $V(x)$ or $n(x)$, except that they are constant. Only the values of $n$ and $V$ far from the envelope come into play. An unevaluated integral appears in the denominator of Eq. 2. To integrate this term, $V(x)$ must be known, the shape of the voltage gradient across the barrier. $V(x)$ is never known exactly, and it is not constant during transport. Thus the preceding equation is an approximation of the flux driven by diffusional and electrical forces.

The flux equation (in its differential form, Eq. 1, or in its integral form, Eq. 2) is called the **Nernst–Planck equation.** If we assume $V(x)$ to be a linear function of $x$, the Nernst–Planck equation leads to the **Goldman equation.** Rather than assume a particular $V(x)$, we leave the integral undetermined and replace it by the symbol $L$:

$$L = \int e^{aV(x)} dx \qquad (3)$$

$L \multimap$ length (cm) and $ukT/L = D/L \multimap$ velocity (cm/sec); we call this ratio the permeability $\rho$, and $\rho \multimap$ (cm/sec). We can write for the flux of any substrate:

$$-\Phi = \rho(n_{nu}e^{-aV_{nu}} - n_{cy}e^{-aV_{nu}}) \qquad (4)$$

where $n_{nu}$ is the concentration of the substrate in the nucleus and $n_{cy}$ is its concentration in the cytoplasm. For a derivation and discussion, see Chapter 2 in DeFelice (1981).

How would a theory used for cell physiology apply to the transport of macromolecules across the nuclear envelope? To answer this, we overlook the double membrane

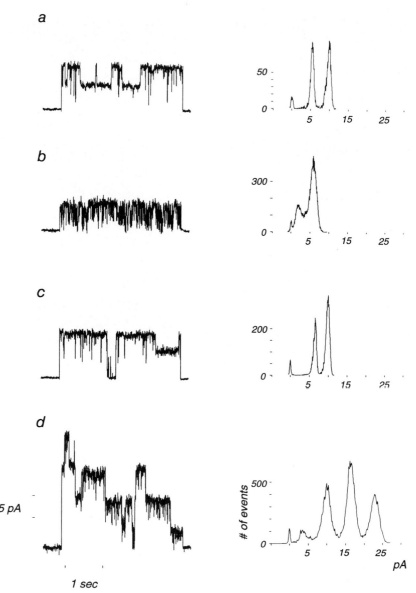

**FIG. 11.** Nucleus-attached patch recordings from mouse nuclei during development. The figure shows large-conductance ion channels that appear on the surface of (a) germinal vesicle stage nucleus from a mouse oocyte, (b) pronucleus from a mouse zygote, (c) nucleus from two-cell mouse embryo, and (d) nucleus from an adult mouse liver cell. Amplitude probability histograms appear to the right of the representative trace. The traces and histograms result from stepping the pipette voltage from 0 to −25 mV. In this experiment, the bath and the nucleus-attached patch pipette contained an intracellular-like solution (120 m$M$ KCl, 2 m$M$ MgCl$_2$, 1.1 m$M$ EGTA, 0.1 m$M$ CaCl$_2$, 5 m$M$ glucose, 10 m$M$ HEPES, pH 7.4) at room temperature. (Reprinted with permission from Mazzanti *et al.*, 1991. Copyright *The Journal of Membrane Biology.*)

and consider the nucleus as a cell within a cell (Fig. 13). This ignores the anatomy of the cisterns and replaces the double membrane with a single boundary penetrated by nuclear pores. First suppose that K ions cannot get across the envelope unless they combine with the nuclear pore complex (np), and that eight ions must interact before transport can occur. We select eight as a multiplier because of the geometry of the complex. Thus we expect eight (or a multiple of eight) interactions between the complex and the ions. Under this assumption, we write:

$$8K + P \Leftrightarrow K_8P$$
$$K_{kp} = [K]^8[P]/[K_8P] \tag{5}$$

where $P$ stands for the nuclear pore complex and $K_{kp}$ is the equilibrium constant for the reaction. Consider a macromolecule, $M$, that undergoes a similar interaction with $P$, but one molecule at a time:

$$M + P = MP$$
$$K_{mp} = [M][P]/[MP] \tag{6}$$

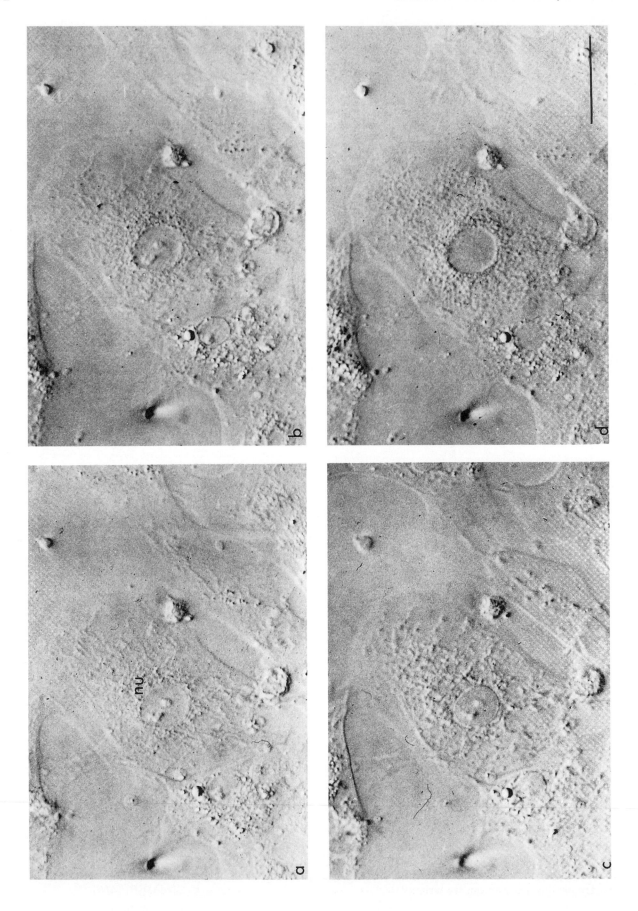

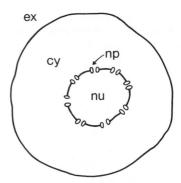

**FIG. 13.** The nucleus as a cell within a cell. This diagram suppresses the double-membrane structure of the nuclear envelope, and it regards the nucleus (nu) as a membrane-bound organelle. In this model, only the nuclear pores (np) connect the nucleoplasm (nu) to the cytoplasm (cy). The nuclear envelope is, therefore, a semipermeable membrane analogous to the cell plasma membrane that separates the cytoplasm from the extracellular space (ex).

where $K_{mp}$ is the equilibrium constant for the reaction. Assume that $K_8P$ and $MP$ can move across the envelope, but that $K$, $M$, and $P$ by themselves cannot. Now consider two fluxes—the flux of $K_8P$ bound to the pore:

$$-\Phi_{kp} = \rho_{kp}([K_8P]_{nu}e^{-aV_{nu}} - [K_8P]_{cy}e^{-aV_{cy}}) \qquad (7)$$

and the flux of $MP$ bound to the pore:

$$-\Phi_{mp} = \rho_{mp}([MP]_{nu}e^{-aV_{nu}} - [MP]_{cy}e^{-aV_{cy}}) \qquad (8)$$

where $\rho_{kp}$ is the permeability of $K_8P$, and $\rho_{mp}$ is the permeability of $MP$ through the envelope. By hypothesis, let the transport of macromolecules and the transport of ions be coupled. We assume this is a property of the nuclear pore. Then the steady-state condition reduces to:

$$\Phi_{kp} + \Phi_{mp} = 0 \qquad (9)$$

This equation treats the nuclear pore complex as a carrier and indicates that the concentration of the carrier does not change: If a certain number of these carriers in the kp state move to the left, than equivalent number in the mp state move to the right. We can solve these equations for $\Phi_{mp}$, which describes the movement of macromolecules, in terms of the concentrations of the substrates, $[K]$ and $[M]$, and the concentration of the pores, $[P]$, using:

$$[K_8P] = [K]^8[P]/K_{kp}$$
$$[MP] = [M][P]/K_{mp} \qquad (10)$$

The values for $K_{kp}$, $K_{mp}$, as well as the concentrations of the substrates, may differ on each side of the pore. The fixed number of pores constrains the equations. Thus, we may write the flux of the macromolecules, $\Phi_{mp}$, in terms of $[K]_{cy}$, $[K]_{nu}$, $[M]_{cy}$, and $[M]_{nu}$.

This outlines the carrier model approach for the transport of ions and macromolecules across the nuclear pore. Now consider a channel model. Assume the nuclear pore complex consists of eight channels for ions and one channel for charged macromolecules, and that no other pathways exist between cytoplasm and nucleoplasm. Then the fluxes of ions and macromolecules are coupled through their electric charge without a separate assumption about the carrier nature of the pore. We may write the flux equations:

$$-\Phi_k = \rho_k([K]_{nu}e^{-aV_{nu}} - [K]_{cy}e^{-aV_{cy}}) \qquad (11)$$
$$-\Phi_m = \rho_m([M]_{nu}e^{-aV_{nu}} - [M]_{cy}e^{-aV_{cy}}) \qquad (12)$$

We have dropped the p in the subscript kp, etc., to distinguish the channel model from the carrier model. In the channel model, the steady-state condition reduces to this:

$$8\Phi_k + z_m\Phi_m = 0 \qquad (13)$$

where $z_m$ is the valence of the macromolecule. Equation 13 leads to an expression for the flux of the macromolecules, $\Phi_m$, in terms of the substrate concentrations on either side of the envelope. The flux of any charged molecule, even a macromolecule, is an electrical current. In general, the relationship between flux, $\Phi \leftrightarrow$ #/cm$^2$ sec, and current, $I \leftrightarrow$ pA/cm$^2$, is:

$$I = zen\Phi \qquad (14)$$

The parameter $z$ in this expression is the valence on the molecule whose concentration is $n$, and $e$ is the electronic charge ($e = 1.6 \times 10^{-19}$ coulombs). For example, if $ze$ were the charge of the K ion, then $ze = +1e$, and $z_me$ would represent the charge on the macromolecule under consideration. Thus we may use Eq. 14 to write an expression for the K current and for the macromolecular current, and these two currents would be constrained by Eq. 13.

The carrier model or the channel model amount to a balance of fluxes and currents. For example, if the charged macromolecule being transported is RNA, then we could write from the above channel model:

$$I_K + I_{RNA} = 0 \qquad (15)$$

A similar equation would hold for charged proteins. It may seem unusual to consider the movement of RNA or proteins as a current. However, we are already familiar with this concept in gel electrophoresis, in which RNA and proteins move under an electric field. Equation 15 merely states that comparable phenomena occur across the nuclear envelope. Assume a rate of $10^9$ RNA molecules/min, one unit charge/base, and $10^4$ bases/molecule. Then $I_{RNA}$ would be the order of 10 nA for the entire nucleus and 1 pA for an individual nuclear pore. The point we want to illustrate is that transport of a charged macromolecule must have a countercurrent. If electrical coupling between macromole-

---

**FIG. 12.** The nucleus as an osmometer. Nuclei (nu) from 3T3 fibroblasts under four osmotic conditions. In all cases, the bath solution contains nystatin, a monovalent cation ionophore, at a concentration of 50–100 $\mu$g/ml. The nystatin pores form holes in the plasma membrane that connect the cytoplasm to the bath. (a) 120 m$M$ KCl bath solution (iso-osmotic), (b) 240 m$M$ KCl (hyper-osmotic), (c) 60 m$M$ KCl (hypo-osmotic), and (d) 480 m$M$ KCl. In (a, b, and c) the effects are reversible, but in (d) the effects are irreversible. (Mazzanti and Innocenti, unpublished data, 1993.)

cules and ions does occur, the ionic current and the voltage across the nuclear envelope could regulate transport of nucleic acids and proteins into and out of the nucleus.

## VIII. Modulation of Ionic Nuclear Permeability by Adenosine Triphosphate

Patch-clamp experiments on isolated nuclei suggest that the channel activity recorded on the nuclear envelope surface represents a direct nucleocytoplasmic communication pathway. As previously stated, several paradoxes arise from this hypothesis. Results obtained from photobleaching techniques indicate that 40-kDa molecules are the upper limit for free diffusion through the pore. In terms of diameter this would mean 9-nm particles. A channel that is 9 nm in diameter, 80 nm in length, and contains a solution of 100 $\Omega$ cm has a conductance of about 1 nS. This value is significantly greater than the channel conductance levels obtained using the patch-clamp recording technique, especially if we consider that the surface isolated by the patch pipette contains many nuclear pores. Pores of 1-nS conductance would shunt the patch and make the observation of smaller conductances impossible. Thus, if the channels are in the outer membrane, the nuclear pores must be closed. On the other hand, if the channels are the nuclear pore, their conductance is too low. Is it possible that, under the experimental conditions of the patch clamp recordings, the pore opening is diminished? As mentioned previously, the nuclear pore contains a plug (see Fig. 6) that may move under altered Ca conditions. Nuclei isolation procedures are traumatic: Cytoplasmic components are lost and nucleocytoplasmic cytoskeletal structures are disrupted. Patch-clamp recordings from nuclei inside the cell help overcome these shortcomings.

Figure 14 shows single-channel events recorded *in situ* from *Xenopus* oocyte nuclei (developmental stage one). Adding 1 m*M* ATP to the pipette solution (a physiological concentration of ATP typical for cytoplasm) transforms sporadic channel openings into a macroscopic current with characteristic kinetics. Figure 15 shows the results of such experiments under three different conditions. The presence of ATP in the recording pipette is essential to record the macroscopic current illustrated in Fig. 15B. In the absence of the nucleotide, either *in situ* or after nuclei isolation, individual channel openings are observed, as illustrated in Figs. 15C and E. However, the ensemble averages of Figs. 15C or E have the same kinetics as B. Furthermore, whole-cell nucleus recordings have resulted in current traces similar to Fig. 15B, although the amplitude of the current is much larger (Dale *et al.*, 1994).

We can estimate the upper limit of the current that should flow if all the pores in the patch were open. Freeze-fracture measurements on first-stage *Xenopus* oocytes indicate 5 to 8 pores per square micrometer. A patch area of 2 $\mu m^2$ would contain 10 to 16 pores. Assuming the maximum expected value of 1 nS per pore, −25 mV would generate a patch current between 250 and 375 pA. In experiments like that of Fig. 15B, the average peak current at −25 mV is 318 ± 34 pA (mean ± SD; $n = 22$). These results show that the channels observed in isolated nuclei are similar to the channels observed *in situ*, that the channel pathways responsible for the currents are the same with or without ATP, and that ATP increases the open probability. Thus, the experimental conditions of *in vitro* patch-clamp recordings apparently do not alter the properties of individual channels, although the open probability may vary. These data, as well as data from whole-nucleus preparations, support the hypothesis that channels recorded from the nuclear envelope represent a direct nucleocytoplasmic communication pathway.

## IX. Summary and Conclusions

The structural intricacy of the nuclear pore complex, the 100 proteins it contains, and the possibility of multiple nucleocytoplasmic pathways indicate that the nuclear pore is not simply a water-filled hole. Rather, it is a coupled transporter for ions and macromolecules. This conclusion seems plausible because the nucleus can partition ions, the nuclear envelope can maintain a resting potential, osmotic forces are manifest, and ion-selective channels exist in patches that contain dozens of nuclear pores.

How can the same structure that transports macromolecules be selective to small ions? One viewpoint would be that the question is misleading, because the observed channels lie not in the pore but in the inner and outer membranes. In this picture, the nuclear pore is the site of macromolecular transport and channels in the cistern membranes are the sites of ion transport. Channels, receptors, and transporters certainly exist in the cistern membranes. However, we hypothesize that separate channels exist within the nuclear pore complex. It may be premature to propose the lateral openings (see Fig. 5) as the site of these channels. Indeed, the only evidence for transport is through the central pore. We nevertheless speculate that the eight lateral structures are pathways for ions, and that the central pore is the pathway for macromolecules. This hypothesis is attractive on several grounds. For example, macromolecular transport requires ATP, and *in situ* patch-clamp experiments indicate that ATP opens channels. If the movement

---

**FIG. 14.**  *In situ* patch-clamp experiment on a *Xenopus* oocyte. Enzymatically isolated first-stage oocytes were transferred into an experimental chamber. The oocytes were held in position by a holding pipette (left) via light suction. After perforating the plasma membrane the patch pipette (right) was pushed gently against the nuclear surface. Before touching the nuclear envelope, positive pressure was applied to the patch-pipette solution. The clear shadow in the oocyte cytoplasm is due to solution outflow, which was used to keep the electrode tip clean and to remove cytoplasmic material from the nuclear surface. (From Mazzanti *et al.*, 1994.)

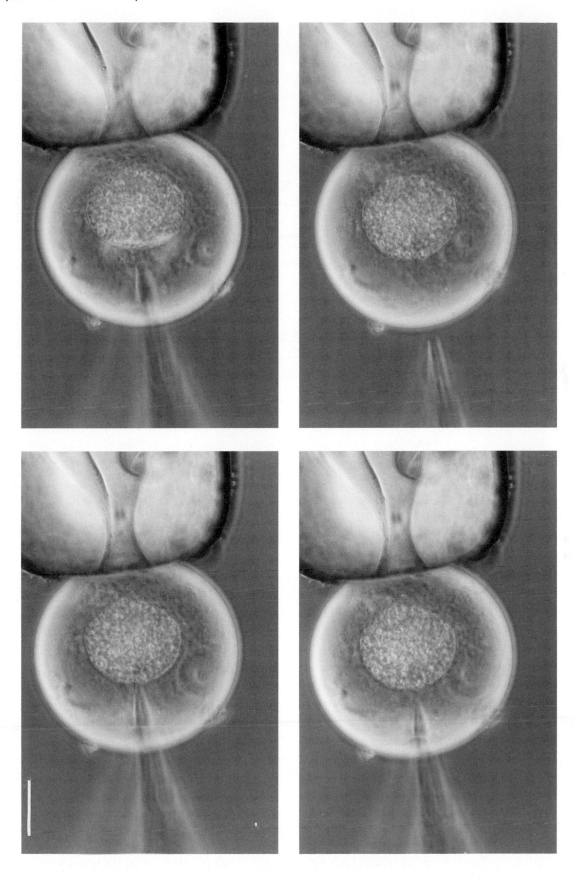

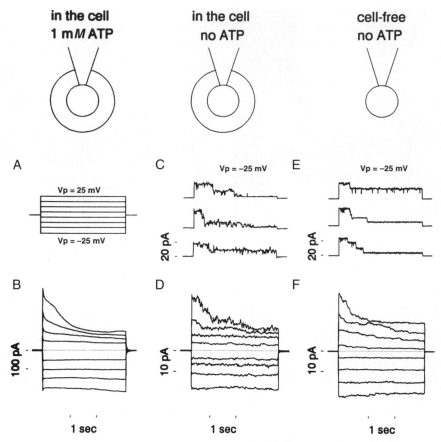

**FIG. 15.** ATP modulation of the ionic permeability of the nuclear envelope. The upper panels in the figure show schematic drawings of the patch-clamp configurations corresponding to the experimental traces plotted below. (A) Voltage protocol used in the experiments. (B) Current traces at eight different test voltages from a nucleus-attached experiment in an intact oocyte. The pipette contained 1 mM ATP. (C) Nucleus-attached single-channel traces with −25 mV in the pipette. The patch electrode did not contain ATP. (D) Average current of 20 single-channel recordings obtained at each voltage step. In (E) and (F) experiments were performed after manual isolation of nuclei. The patch pipette contains no ATP. (E) Single-channel activity at −25 mV. (F) Average currents at different test potentials. (From Mazzanti *et al.*, 1994.)

of proteins and nucleic acids couples to the movement of ions down electrochemical gradients, macromolecular transport would be regulated by ATP through coupled channel openings. It is thus crucial to test whether the nuclear pore complex and the ion channels are linked to the same structure.

The majority of experiments indicate that the ion channels in the envelope select K ions. However, data also exist for Cl channels, ligand-gated ion channels, and pumps. Although channels undoubtedly reside in the outer and inner membranes, this does not preclude the possibility of nuclear-pore-complex-associated ion channels. Furthermore, Donnan equilibria and semipermeable membrane mechanisms may coexist at the nuclear envelope. The electric potentials, dynamic restrictions, and redistribution of small molecules and ions, as well as osmotic gradients across the envelope, probably reflect membrane phenomena. Attributing such properties to *either* a Donnan equilibrium *or* a selective membrane property reflects early controversies concerning the plasma membrane that surrounds the cytoplasmic matrix. Today, a similar debate exists in

another area of cell biology, the release of hormones and neurotransmitters from secretory vesicles. Secretory vesicles apparently contain a gel matrix that restricts the movement of their contents after membrane fusion has occurred. In all such cases, mechanisms that rely on the matrix—or mechanisms that rely on the membrane that surrounds the matrix—are not mutually exclusive. Thus Donnan equilibria and selective nuclear pores may coexist as regulatory transport mechanisms in the nuclear envelope.

It is interesting to speculate whether the structure of the nuclear pore complex resembles other macromolecular assemblies. One similarity exists regarding the central granule (or plug), which represents a protein-within-a-protein motif common to certain macromolecules. Criss Hartzell has noticed that the eight-plus-one structure of the nuclear pore complex is reminiscent of a chaperone molecule, which has a seven-plus-one structure. Mu-ming Poo has asked whether parts of the assembled complex might not exist as independent units in the endoplasmic reticulum, to be reassembled into the nuclear pore complex. K-selective channels and protein-conducting channels reside in the reticulum and bear some similarity to the ionic and macro-molecular pathways that we have discussed. The possibility that channel proteins and transporters in endoplasmic reticulum are components of the nuclear pore complex would have broad implications for pore assembly during development. These possible new directions for inquiry may help resolve the structural basis of nuclear pore function.

## Acknowledgments

We thank B. J. Duke for helping with experiments and preparing the figures. Some unpublished work presented here was done at Stazione Zoologica in Naples, Italy, with Luigia Santella and Brian Dale. NATO-CGR 190025 to B. D. and L. J. D. have supported this work.

## Bibliography

Agutter, P. S. (1991). Role of the cytoskeleton in nucleocytoplasmic RNA and protein distributions. *Biochem. Soc. Trans.* **19**, 1094–1098.

Aidley, D. J. (1989). "The Physiology of the Excitable Cell," 3rd ed. Cambridge University Press, Cambridge, UK.

Akey, C. WE. (1989). Interaction and structure of the nuclear pore complex revealed by cryoelectron microscopy. *J. Cell Biol.* **109**, 955–970.

Alberts, B. *et al.* (1989). "Molecular Biology of the Cell," 2nd ed., Chap. 8. Garland Publishing, New York.

Blobel, G., and Potter, V. R. (1966). Nuclei from rat liver: Isolation method that combines purity with high yield. *Science* **154**, 1662–1665.

Bonner, W. M. (1975a). Protein migration into nuclei: I. Frog oocyte nuclei in vivo accumulate microinjected histones, allow entry to small proteins, and exclude large proteins. *J. Cell Biol.* **64**, 421–430.

Bonner, W. M. (1975b). Protein migration into nuclei: II. Frog oocyte nuclei accumulate a class of microinjected oocyte nuclear proteins and exclude a class of microinjected oocyte cytoplasmic proteins. *J. Cell Biol.* **64**, 431–437.

Bustamente, J. O. (1992a). Nuclear ion channels in cardiac myocytes. *Pflugers Arch.* **421**, 473–485.

Bustamente, J. O. (1992b). Nuclear electrophysiology. *J. Membr. Biol.* **138**, 105–112.

Carmo-Fonesca, M., and Hurt, E. D. (1991). Across the nuclear pores with the help of nucleoporins. *Chromosoma* **101**, 199–205.

Carter, K. C., *et al.* (1993). A 3-D view of precursor messenger RNA metabolism within the mammalian nucleus. *Science* **259**, 1330–1335.

Century, T. J., Fenichel, I. R., and Horowitz, S. B. (1970). The concentration of water, Na and K ions in the nucleus and cytoplasm of amphibian oocytes. *J. Cell Sci.* **7**, 5–13.

Dale, B., DeFelice, L. J., Kyozuka, K., Santella, L., Tosti, E. (1994). Voltage clamp of the nuclear envelope. *Proc. R. Soc. London B* **225**, 119–124.

Dargemont, C., and Kuhn, L. D. (1992). Export of mRNA from microinjected nuclei of *Xenopus laevis* oocytes. *J. Cell Biol.* **118**, 1–9.

DeFelice, L. J. (1981). "Introduction to Membrane Noise." Plenum Press, New York.

DeFelice, L. J. (1997). "Electrical Properties of Cells: Patch Clamp for Biologists." Plenum Press, New York.

Dingwall, C. (1990). Plugging the nuclear pore. *Nature* **346**, 512–514.

Dworetzky, S. I., and Feldherr, C. M. (1988). Translocation of RNA-coated gold particles through the nuclear pores of oocytes. *J. Cell Biol.* **106**, 575–584.

Feldherr, C. M., and Akin, D. (1990). EM visualization of nucleocytoplasmic transport process. *Elec. Micro. Rev.* **3**, 73–86.

Finkelstein, A. (1987). "Water Movement through Lipid Bilayers, Pores, and Plasma Membranes." John Wiley & Sons, New York.

Finlay, D. R., Newmeyer, D. D., Price, T. M., and Forbes, D. J. (1987). Inhibition of *in vitro* nuclear transport by a lectin that binds to nuclear pores. *J. Cell Biol.* **104**, 189–200.

Forbes, D. J. (1992). Structure and function of the nuclear pore complex. *Annu. Rev. Cell Biol.* **8**, 495–527.

Goldberg, M. W., Blow, J. J., and Allen, T. D. (1992). The use of field emission in-lens scanning electron microscopy to study the steps of assembly of the nuclear envelope. *J. Struct. Biol.* **108**, 257–268.

Hernandez-Cruz, A., Sala, F., and Conner, J. A. (1991). Stimulus-induced nuclear Ca signals in fura-2 loaded amphibian neurons. *Ann. NY Acad. Sci.* **635**, 416–420.

Hille, B. (1992). "Ionic Channels of Excitable Membranes," 2nd ed. Sinauer Associated, Sunderland, MA.

Hinshaw, J. E., Carracher, B. O., and Milligan, R. A. (1992). Architecture and design of the nuclear pore complex. *Cell* **69**, 1133–1141.

Horowitz, S. B., and Moore, L. C. (1974). The nuclear permeability intracellular distribution, and diffusion in Inulin in the amphibian oocyte. *J. Cell Biol.* **60**, 405–415.

Innocenti, B., and Mazzanti, M. (1993). Identification of a nucleocytoplasmic ionic pathway by osmotic shock in isolated mouse liver nuclei. *J. Membr. Biol.* **131**, 137–142.

Jarnik, M., and Aebi, U. (1991). Toward a more complete 3-D structure of the nuclear pore complex. *J. Struct. Biol.* **110**, 883–894.

Lauger, P. (1991. "Electrogenic Ion Pumps" 2nd ed. Sinauer Associates, Sunderland, MA.

Loewenstein, W. R., and Kanno, Y. (1963). Some electrical properties of a nuclear membrane examined with a microelectrode. *J. Gen. Physiol.* **46**, 1123–1140.

Matzke, A. J. M., and Matzke, M. A. (1991). The electrical properties of the nuclear envelope, and their possible role in the regulation of eukaryotic gener expression. *Bioelectrochem. Bioenerg.* **25**, 357–370.

Maul, G. G. (1977). The nuclear and cytoplasmic pore complex: Structure, dynamics, distribution, and evolution. *Int. Rev. Cytol.* **6**, Suppl., 75–186.

Mazzanti, M., DeFelice, L. J., Cohen, L., and Malter, H. (1990). Ion channels in the nuclear envelope. *Nature* **343**, 764–767.

Mazzanti, M., DeFelice, L. J., and Smith, E. F. (1991). Ion channels in murine nuclei during early development and in fully differentiated adult cells. *J. Membr. Biol.* **121**, 189–198.

Mazzanti, M., Innocenti, B., and Rigatelli, M. (1994). ATP dependent ionic permeability of nuclear envelope in *in situ Xenopus* oocyte nuclei. *FASEB J.* **8**, 231–236.

Moore, M. S., and Blobel, G. (1992). The two steps of nuclear import, targeting to the nuclear envelope and translocation through the nuclear pore, require different cytosolic factors. *Cell* **69**, 939–950.

Newport, J. W., and Forbes, D. J. (1987). The nucleus: Structure, function and dynamics. *Ann. Rev. Biochem.* **56**, 535–565.

Nicotera, P., McConkey, D. J., Jones, D. P., and Orrenius, S. (1989). ATP stimulates Ca uptake and increases the free Ca concentration in isolated rat liver nuclei. *PNAS* **86**, 453–457.

Nigg, E. A., Baeuerle, P. A., and Luhrmann, R. (1991). Nuclear import-export: In search of signals and mechanisms. *Cell* **66**, 15–22.

Overbeek, J. Th. G. (1956). The donnan equilibrium. *Prog. Biophys. Mol. Biol.* **6**, 57–84.

Paine, P. L., and Horowitz, S. B. (1980). The movement of material between nucleus and cytoplasm. *Cell Biol.* **4**, 299–338.

Paine, P. L., *et al.* (1992). The oocyte nucleus isolated in oil retains in vivo structure and functions. *BioTechniques* **13**, 238–246.

Perez-Terzic, C., Pyle, J., Jaconi, M., Stehno-Bittle, L., and Clapham, D. E. (1996). Conformational states of the nuclear pore complex induced by depletion of nuclear Ca stores. *Science* **273**, 1875–1877.

Richardson, W. D., Mills, A. D., Dilworth, S. M., Laskey, R. A., and Forbes, D. J. (1992). Structure and function of the nuclear pore complex. *Annu. Rev. Cell Biol.* **8,** 495–572.

Santella, L. (1996). The cell nucleus: an Eldorado to future Ca research? *J. Membr. Biol.* **153,** 83–92.

Simon, S. M., and Blobel, G. (1991). A protein-conducting channel in the endoplasmic reticulum. *Cell* **65,** 371–380.

Stehno-Bittel, L., Perez-Terzic, C., and Clapham, D. E. (1995). Diffusion across the nuclear envelope inhibited by depletion of the nuclear Ca store. *Science* **270,** 1835–1838.

Stewart, M., Whytock, S., and Moir, R. D. (1991). Nuclear envelope dynamics and nucleocytoplasmic transport. *J. Cell Sci. Supp.* **14,** 79–82.

Stochaj, U., and Silver, P. (1992). Nucleoplasmic traffic of proteins. *Eur. J. Cell Biol.* **59,** 1–11.

Stricker, S. A., Centonze, V. E., Paddock, S. W., and Schatten, G. (1992). Confocal microscopy of fertilization-induced Ca dynamics in sea urchin eggs. *Dev. Biol.* **149,** 370–380.

Tabares, L. M., Mazzanti, M., and Clapham, D. E. (1991). Cl channels in the nuclear envelope. *J. Membr. Biol.* **123,** 49–54.

Unwin, P. N. T., and Milligan, R. A. (1982). A large particle associated with the perimeter of the nuclear pore complex. *J. Cell Biol.* **93,** 63–75.

Wagner, P., Kunz, J., Koller, A., and Hall, M. N. (1990). Active transport of proteins into the nucleus. *FEBS Lett.* **275,** 1–5.

Waybill, M. M. *et al.* (1991). Nuclear Ca transients in cultured rat hepatocytes. *Am. J. Physiol.* **261,** E94–E57.

Wente, S. R., Rout, M. P., and Blobel, G. (1992). A new family of yeast nuclear pore complex proteins. *J. Cell Biol.* **119,** 705–723.

Williams, D. A., Becker, P., and Fay, F. S. (1987). Regional changes in Ca underlying contraction of single smooth muscle cells. *Science* **235,** 1644–1648.

Yarmola, E. G., Zarudnaya, M. I., and Lazurkin, Yu. S. (1985). Osmotic pressure of DNA solutions and effective diameter of the doulbe helix. *J. Bio. Struct. Dynam.* **2**(5), 981–993.

Zasloff, M. (1983). tRNA transport from the nucleus in a eukaryotic cell: Carrier-mediated translocation process. *Proc. Natl. Acad. Sci. USA* **80,** 6436–6440.

Nicholas Sperelakis

# 32

# Regulation of Ion Channels by Phosphorylation

## I. Introduction

Considerable attention has been given during the past 20 years to phosphorylation of ion channels as a means whereby the activity of the channels can be regulated or modulated. There is evidence for such regulation or modulation of function of $Ca^{2+}$, $K^+$, and $Na^+$ channels by phosphorylation, and biochemical evidence shows that one or a few sites on the channel proteins can be phosphorylated by various protein kinases. Most physiological evidence for such changes in ion channel function is based on slow (L-type) $Ca^{2+}$ channels of nerve, skeletal muscle, cardiac muscle, and vascular smooth muscle (VSM) and for $K^+$ channels (delayed rectifier type) of cardiac muscle and nerve. Therefore, this chapter focuses primarily on the slow $Ca^{2+}$ channel and the delayed rectifier $K^+$ channel of cardiac muscle. These examples should suffice to illustrate the important principles that are involved.

The voltage-dependent slow $Ca^{2+}$ channels in the myocardial cell membrane are the major pathway by which $Ca^{2+}$ ions enter the cell during excitation for initiation and regulation of the force of contraction of cardiac muscle. The slow $Ca^{2+}$ channels have some special properties, including functional dependence on metabolic energy, selective blockade by acidosis, and regulation by the intracellular cyclic nucleotide levels. Because of these special properties of the slow channels, $Ca^{2+}$ influx into the myocardial cell can be controlled by extrinsic factors (such as autonomic nerve stimulation or circulating hormones) and by intrinsic factors [such as cellular pH or adenosine triphosphate (ATP) level].

In myocardial cells, the $Ca^{2+}$ influx that occurs during each cardiac cycle is regulated by cyclic nucleotides. This regulation is presumably mediated by phosphorylation(s) of the slow $Ca^{2+}$ channel protein (L-type). Phosphorylation of the slow $Ca^{2+}$ channels (or of an associated regulatory protein) by cAMP dependent protein kinase (PK-A) (Fig. 1) presumably (1) increases the number of $Ca^{2+}$ slow channels available for voltage activation during the action potential (AP), (2) increases the probability of their opening,

and (3) increases their mean open time. A greater density of available $Ca^{2+}$ channels increases $Ca^{2+}$ influx and inward $Ca^{2+}$ current ($I_{Ca}$) during the AP, and so increases the force of contraction of the heart. Phosphorylation by cGMP-dependent PK (PK-G) depresses the activity of the slow $Ca^{2+}$ channels (Wahler et al., 1990).

## II. Types of $Ca^{2+}$ Channels

Four or five different subtypes of voltage-dependent $Ca^{2+}$ channels have been described for nerve and muscle cells. The first three found in sensory ganglion neurons (rat dorsal root ganglion) of the spinal cord were called L-type (or long-lasting or kinetically slow), T-type (or transient or kinetically fast), and N-type (or neither L-type nor T-type) (Nowycky et al., 1985). More recently, another subtype was initially found in Purkinje neurons of the cerebellum, and hence called P-type (Llinas et al., 1992). The P-type channel has now been identified at the nerve terminals of neuromuscular junctions of both vertebrates and invertebrates. Muscle fibers, in general, apparently possess only the L-type and T-type, with the T-type channels being very sparse or absent in some types of muscles. That is, in muscle cells, the major inward $Ca^{2+}$ current (involved in excitation–contraction coupling) is the current through the L-type slow $Ca^{2+}$ channels. This is true of skeletal muscle, cardiac muscle, and most types of smooth muscles.

The N-type $Ca^{2+}$ channel has a single-channel conductance ($\gamma$) of about 14–18 pS, which is in-between that of the T-type (8–12 pS) and L-type (18–26 pS). The N-type channel has been localized only to neurons so far. The P-type $Ca^{2+}$ channel is high threshold (like the L-type), having an activation voltage of −45 to −35 mV, but is not blocked by the L-type channel blockers. The values reported for single-channel conductance ($\gamma$) in various cells vary from 9 to 20 pS.

Table 1 summarizes the major differences between the slow (L-type) $Ca^{2+}$ channels and the fast (T-type) $Ca^{2+}$ channels. As indicated, the kinetics of activation and inacti-

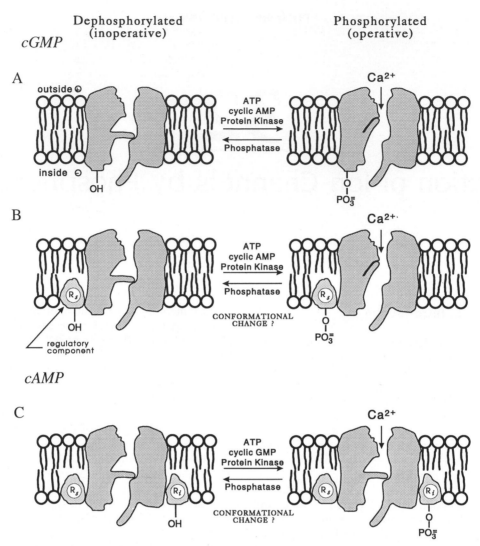

**FIG. 1.** Schematic model for a $Ca^{2+}$ slow channel in myocardial cell membrane in two hypothetical forms: dephosphorylated (or electrically silent) form (left diagrams) and phosphorylated form (right diagrams). The two gates associated with the channel are an activation gate and an inactivation gate. The phosphorylation hypothesis states that a protein constituent of the slow channel itself (A) or a regulatory protein associated with the slow channel (B) must be phosphorylated for the channel to be in a state available for voltage activation. Phosphorylation of a serine or threonine residue occurs by PK-A in the presence of ATP. Phosphorylation may produce a conformational change that effectively allows the channel gates to operate. The slow channel (or an associated regulatory protein) may also be phosphorylated by PK-G (C), thus mediating the inhibitory effects of cGMP on the slow $Ca^{2+}$ channel. (Adapted from Sperelakis and Schneider, 1976.)

vation are slower for the L-type. That is, the slow $I_{Ca(L)}$ turns on (activates) more slowly and turns off (inactivates) more slowly. In addition, the voltage range over which these channels operate is different, the threshold potential and inactivation potential being higher (more positive or less negative) for the slow $Ca^{2+}$ channels. Therefore, the L-type channels are high threshold, and the T-type channels are low threshold. The single-channel conductance is greater for the slow $Ca^{2+}$ channel: 18–26 pS versus 8–12 pS. The slow $Ca^{2+}$ channels are regulated by cyclic nucleotides and phosphorylation, whereas the fast $Ca^{2+}$ channels are not. Finally, the slow $Ca^{2+}$ channels are blocked by

$Ca^{2+}$ antagonist drugs (such as verapamil, diltiazem, and nifedipine) and opened by $Ca^{2+}$ agonist drugs (such as Bay-K-8644, a dihydropyridine that is chemically very close to nifedipine), whereas the fast $Ca^{2+}$ channels are not. In some respects, the fast $Ca^{2+}$ channels behave like fast $Na^{+}$ channels, except that they are $Ca^{2+}$ selective (rather than $Na^{+}$ selective) and are not blocked by TTX.

Table 2 summarizes the blocking or opening action of several drugs and toxins on the various subtypes of $Ca^{2+}$ channels. As indicated, the L-type $Ca^{2+}$ channel is blocked by the prototype $Ca^{2+}$ antagonist drugs (verapamil, diltiazem, and nifedipine) and opened by the dihydropyridine

**TABLE 1**  Summary of Major Differences between the Slow (L-Type) and Fast (T-Type) $Ca^{2+}$ Channels

| Properties | $Ca^{2+}$ channels | |
| --- | --- | --- |
| | Slow (L-type) | Fast (T-type) |
| Duration of current | Long-lasting (sustained) | Transient |
| Inactivation kinetics | Slower | Faster |
| Activation kinetics | Slower | Faster |
| Threshold | High (ca. −35 mV) | Low (ca. −60 mV) |
| Half-inactivation potential | ca. −20 mV | ca. −50 mV |
| Single-channel conductance | High (18–26 pS) | Low (8–12 pS) |
| Regulated by cAMP and cGMP | Yes | No |
| Regulated by phosphorylation | Yes | No |
| Blocked by $Ca^{2+}$ antagonist drugs | Yes | No (slight) |
| Opened by $Ca^{2+}$ agonist drugs | Yes | No |
| Permeation by $Me^{2+}$ | Ba > Ca | Ba ≃ Ca |
| Inactivation by $[Ca]_i$ | Yes | Slight (?) |
| Recordings in isolated patches | Runs down | Relatively stable |

Bay-K-8644. The T-type channels are relatively selectively blocked by tetramethrine and by low concentrations of $Ni^{2+}$ (e.g., 30 $\mu M$). Higher concentrations of $Ni^{2+}$ block the L-type channels as well. Possible blockers of the F-type channel (see below) are not known.

In addition to the $Ca^{2+}$ channel subtypes described previously, a new subtype was discovered in 18-day-old fetal rat ventricular (heart) cells (Tohse *et al.*, 1992). A substan-

**TABLE 2**  Summary of Drugs or Toxins That Block or Open the Various Subtypes of $Ca^{2+}$ Channels

| Channel type | Blockers | Openers |
| --- | --- | --- |
| L-type (slow) | Verapamil, diltiazem, nifedipine | Bay-K-8644 |
| T-type (fast) | Tetramethrin <br> $Ni^{2+}$ (30 $\mu M$) | — |
| N-type | ω-Conotoxin | — |
| P-type | ω-Agatoxin-IVA[a] <br> Polyamine (FTX)[b] | — |
| F-type | — | — |

[a] ω-Agatoxin-IVA, a polypeptide (5202 Da) from funnel-web spider venom (*Agelenopsis aperta*), blocks P-type $Ca^{2+}$ channels of rat Purkinje neurons with a $K_D$ of about 2 n$M$.

[b] A smaller (ca. 254 Da) polyamine (FTX) extracted from venom of this spider also blocks the P-type channels (Mintz *et al.*, 1992). FTX block is antagonized by $Ba^{2+}$ ion. P-type $Ca^{2+}$ channels are involved in presynaptic $Ca^{2+}$ influx and associated neurotransmitter release.

tial fraction (e.g., 30%) of the total $I_{Ca}$ remained in the presence of a high concentration (3 $\mu M$) of nifedipine (nifedipine-resistant $I_{Ca}$), and it was not blocked by diltiazem (another L-type channel blocker) or ω-conotoxin (N-type channel blocker). This novel $Ca^{2+}$ current had a half-inactivation potential about 20 mV more negative than that of the L-type $I_{Ca}$, in this respect being like a T-type $Ca^{2+}$ current. It was called F-type (or fetal-type) $Ca^{2+}$ current ($I_{Ca(F)}$). The single-channel conductance ($\gamma$) is not known.

Another difference discovered during the embryonic or fetal period, first observed in chick and later in rat, is that the slow $Ca^{2+}$ channels (L-type) exhibit an unusually high incidence of very long openings, as observed in single-channel recordings (cell-attached patch) (Tohse and Sperelakis, 1990; Masuda *et al.*, 1995). The incidence of long openings diminished during development and approached the adult channel behavior. The adult behavior primarily consists of bursting patterns (rapid openings and closings of short duration, flickerings).

## III. Cyclic AMP Stimulation of Slow $Ca^{2+}$ Channels

Cyclic AMP (cAMP) modulates the functioning of the slow $Ca^{2+}$ channels (Shigenobu and Sperelakis, 1972; Tsien *et al.*, 1972; Reuter and Scholz, 1977). Histamine and β-adrenergic agonists, after binding to their specific receptors, lead to rapid stimulation of adenylate cyclase with resultant elevation of cAMP levels. Methylxanthines enter the myocardial cells and inhibit the phosphodiesterase, thus causing an elevation of cAMP. These positive inotropic agents also concomitantly stimulate $Ca^{2+}$-dependent slow APs by increasing $I_{Ca}$.

Additional evidence for the regulatory role of cAMP in heart cells includes the following: (1) The GTP analogue GPP(NH)P and forskolin, which directly activate adenylate cyclase, induce $Ca^{2+}$-dependent slow APs (Josephson and Sperelakis, 1978). (2) cAMP microinjection into ventricular muscle cells (by ionotophoresis, pressure, or liposomes) induces slow APs in the injected cells within seconds (Vogel and Sperelakis, 1981; Li and Sperelakis, 1983; Bkaily and Sperelakis, 1985). (3) Injection of cAMP enhances $I_{Ca}$ in isolated single cardiac cells (Irisawa and Kokubun, 1983). (4) A photochemical activation method for suddenly increasing the intracellular cAMP level enhances $I_{Ca}$ in bullfrog atrial cells (Nargeot *et al.*, 1983). (5) Single-channel analysis suggests that cAMP increases the number of functional slow channels available and/or the probability of opening of a given channel (Cachelin *et al.*, 1983; Trautwein and Hofmann, 1983; Bean *et al.*, 1984). Isoproterenol increases the mean open time of single $Ca^{2+}$ channels and decreases the intervals between bursts; the conductance of the single channel is not increased (Reuter *et al.*, 1982). Therefore, the increase in the slow $Ca^{2+}$ current produced by isoproterenol could be produced by the observed increase in mean open time of each channel and the probability of opening, as well as by an increase in the number of available channels.

## IV. Phosphorylation Hypothesis

Because of the relationship between cAMP and the number of available slow $Ca^{2+}$ channels and because of the dependence of the functioning of these channels on metabolic energy, it was postulated that the slow channel protein must be phosphorylated for it to become available for voltage activation (Shigenobu and Sperelakis, 1972; Tsien *et al.*, 1972). Elevation of cAMP by a positive inotropic agent activates PK-A, which phosphorylates a variety of proteins in the presence of ATP. One protein that is phosphorylated might be the slow $Ca^{2+}$ channel protein itself or a contiguous regulatory type of protein (Fig. 1). Agents that elevate cAMP increase the fraction of the channels that are in the phosphorylated form and hence readily available for voltage activation. Phosphorylation could make the slow $Ca^{2+}$ channel available for activation by a conformational change that allowed the activation gate to be opened upon depolarization (or increased the pore diameter).

In this phosphorylation model, the phosphorylated form of the slow $Ca^{2+}$ channel is the active (operational) form, and the dephosphorylated form is the inactive (inoperative) form. The dephosphorylated channels are electrically silent. Thus, phosphorylation increases the probability of channel opening with depolarization. An equilibrium would exist between the phosphorylated and dephosphorylated forms of the channel under a given set of conditions. For example, fluoride ion ($<1$ m$M$) increases the force of contraction of the heart and potentiates the $Ca^{2+}$-dependent slow APs and $Ca^{2+}$ influx ($I_{Ca}$) without increasing the level of cyclic AMP. Fluoride may act by inhibiting the phosphatase, which dephosphorylates the channel protein, thus prolonging the life span of the phosphorylated channel. Some negative inotropic agents or drugs could depress the rate of phosphorylation or stimulate the rate of dephosphorylation. It might be difficult to distinguish between a drug that inhibited phosphorylation of the slow $Ca^{2+}$ channel and one that physically blocked the channel.

Based on the rapid decay of the response to microinjected cAMP (Fig. 2, top), the mean life span of a phosphorylated channel is probably only a few seconds at most, and it is possible that the channels are phosphorylated and dephosphorylated with every cardiac cycle. Agents that affect or regulate the phosphatase that dephosphorylates the channel would affect the life span of the phosphorylated channel. Thus, channel stimulation can be produced either by increasing the rate of phosphorylation (by PK-A activation) or by decreasing the rate of dephosphorylation (inhibition of the phosphatase).

## V. Protein Kinase A Stimulation

To verify that the regulatory effect of cAMP is exerted by means of the PK-A and phosphorylation, intracellular injection of the catalytic subunit of PK-A induces and increases the slow $Ca^{2+}$-dependent APs and potentiates $I_{Ca}$ (Osterrieder *et al.*, 1982; Bkaily and Sperelakis, 1984).

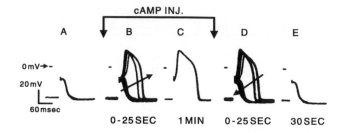

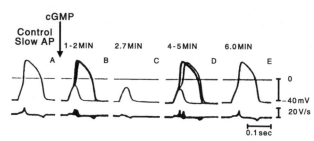

**FIG. 2.** Effects of intracellular injections of cyclic nucleotides. Upper row: Induction of $Ca^{2+}$-dependent slow APs in guinea-pig papillary muscle by intracellular pressure injection of cyclic AMP. The muscle was depolarized in 22 m$M$ $K_o$ to voltage inactivate fast $Na^+$ channels. (A) Small graded response (stimulation rate 30/min). (B) Superimposed records showing the gradual appearance of slow APs upon cAMP injection over a 25-sec period. (C) Presence of stable slow APs after injection for 1 min. (D) Gradual depression of slow APs over a period of 25 sec after the injection is stopped. (E) Complete decay of slow APs 30 sec after cessation of cAMP injection. All records are from one impaled cell. Data taken from Li and Sperelakis, 1983. Lower row: Transient abolition of $Ca^{2+}$-dependent slow APs by pressure injection of cGMP. (A) Control slow AP. (B and C) 1–2 min after the onset of cGMP injection (10-sec duration), the slow APs were depressed and then abolished. (D and E) At 4–6 min, the slow APs recovered spontaneously to control levels. All records from the same cell. (From Wahler and Sperelakis, 1985.)

Injection of an inhibitor (protein) of the PK-A into heart cells inhibits the spontaneous slow $Ca^{2+}$-dependent APs and $I_{Ca}$ (Bkaily and Sperelakis, 1984; Kameyama *et al.*, 1986). Phosphatases decrease the $Ca^{2+}$ current in neurons (Chad and Eckert, 1986) and ventricular myocardial cells (Hescheler *et al.*, 1987). The catalytic subunit of protein phosphatases type 1 and type 2A inhibits $I_{Ca}$ prestimulated by $\beta$-adrenergic agents, whereas okadaic acid, a protein phosphatase inhibitor, stimulates $I_{Ca}$.

Consistent with the phosphorylation hypothesis, the slow $Ca^{2+}$ channel activity disappears within 90 sec in isolated membrane inside-out patches (Reuter, 1983), but can be restored (in neurons) by applying the catalytic subunit of PK-A and Mg-ATP (Armstrong and Eckert, 1987). This is consistent with the washing away of regulatory components of the slow $Ca^{2+}$ channels or of the enzymes necessary to phosphorylate the channel. Even in whole-cell voltage clamp, there is a progressive rundown of the slow $Ca^{2+}$ current, which is slowed or partially reversed by conditions that enhance PK-A phosphorylation.

## VI. Cyclic GMP Inhibition of Slow Ca²⁺ Current

The physiological role played by cyclic GMP on cardiac function is still controversial. It has been proposed that cGMP plays a role antagonistic to that of cAMP, namely, that there was a "Yin–Yang" relationship between cAMP and cGMP (Goldberg *et al.*, 1975). 8Br-cGMP ($10^{-4}$ $M$) shortens the AP duration in rat atria accompanied by a negative inotropic effect, and it was suggested that cyclic GMP might decrease the Ca²⁺ conductance (Nawrath, 1977). ACh and 8Br-cGMP reduce upstroke velocity and duration of the Ca-dependent slow AP in guinea-pig atria (Kohlhardt and Haap, 1978). The abbreviation of AP duration also occurs following injection of cGMP into isolated guinea-pig cardiomyocytes (Trautwein *et al.*, 1982).

Superfusion of isolated ventricular muscle with 8Br-cGMP abolishes the Ca²⁺-dependent slow APs and accompanying contractions (Wahler and Sperelakis, 1985). A similar inhibition by cGMP was shown for the slow APs of atrial muscle and Purkinje fibers (Mehegan *et al.*, 1985). Intracellular pressure injection of cGMP into ventricular cells transiently depresses or abolishes slow APs more quickly (e.g., 1–2 min) (Wahler and Sperelakis, 1985) (Fig. 2, bottom). It was also demonstrated that 8Br-cGMP inhibits the basal $I_{Ca}$ (unstimulated by cAMP) in voltage-clamped ventricular myocytes (Wahler *et al.*, 1990; Haddad *et al.*, 1995) (Figs. 3 and 4).

cGMP inhibition of Ca²⁺ slow channel activity of embryonic chick heart at the single-channel level was demonstrated (Tohse and Sperelakis, 1991) (Fig. 5). Cyclic GMP did not change unit amplitude and slope conductance of the Ca²⁺ channel, but prolonged the closed times and short-

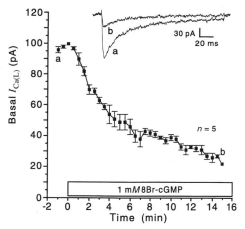

**FIG. 4.** Time course of the inhibition of the basal $I_{Ca(L)}$ by 8Br-cGMP (1 m$M$) in 17-day-old embryonic chick heart cells. Data points plotted are the mean ± standard error. Upper two traces show the original current recordings of $I_{Ca(L)}$ taken at the time points shown by the corresponding letters in the graph. $I_{Ca(L)}$ was elicited by 200 msec depolarizing pulses to +10 mV from a holding potential of −45 mV. Experiments conducted at room temperature. (From Haddad *et al.*, 1995.)

ened the open times. Because 8Br-cGMP is a potent activator of PK-G and does not stimulate cAMP hydrolysis, cGMP-induced inhibition of the basal activity of the Ca²⁺ channels (not prestimulated by cAMP) may be mediated by PK-G.

The Ca²⁺ slow channels of young (3-day-old) embryonic chick heart cells often exhibit long-lasting openings (e.g., for 300 msec) under normal conditions, especially at the more positive command potentials (Tohse and Sperelakis, 1990, 1991). Long-lasting openings were much less frequently observed in 17-day-old embryonic cells. That is, the Ca²⁺ slow channels in early development naturally possess some mode 2 behavior, which is normally produced by Ca²⁺ channel agonists such as the dihydropyridine Bay-K-8644. Addition of 8Br-cGMP to the bath of cells (3-day) exhibiting long openings completely inhibited Ca²⁺ slow channel activity (Fig. 5). Long openings were also observed in fetal (12-day) rat ventricular cardiomyocytes (Masuda *et al.*, 1995).

In whole-cell voltage clamp experiments on single ventricular cardiomyocytes from 17-day embryonic chicks, the stimulation of $I_{Ca(L)}$ produced by 8Br-cAMP added to the bath could be completely reversed by the addition of 8Br-cGMP. Similar results were obtained in experiments on early neonatal (2-day) rat ventricular myocytes; namely, 8Br-cGMP antagonized the stimulation of $I_{Ca(L)}$ by 8Br-cAMP (Fig. 6) (Masuda and Sperelakis, unpublished). Therefore, the ratio of cAMP/cGMP apparently determines the degree of stimulation of $I_{Ca(L)}$, and even the basal $I_{Ca}$ is inhibited by cGMP.

Therefore, cGMP regulates the functioning of the myocardial Ca²⁺ slow channels in a manner that is antagonistic to that of cAMP (Fig. 7). It is possible that the slow Ca²⁺ channel protein has a second site that can be phosphorylated by PK-G and which, when phosphorylated, inhibits

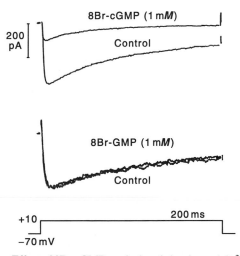

**FIG. 3.** Effect of 8Br-cGMP on the basal slow inward Ca²⁺ current in two cultured embryonic chick ventricular myocytes. Upper traces: Currents elicited by depolarizing pulses from −70 to +10 mV in the control bath solution and after 10 min superfusion with a solution containing 1 m$M$ 8Br-cGMP. Note the large inhibition of $I_{Ca(L)}$. Lower traces: Currents elicited by depolarizing pulses in the control bath solution and after 10 min superfusion with a solution containing 1 m$M$ 8Br-GMP, the noncyclic analog of 8Br-cGMP. (Reproduced from Wahler *et al.*, 1990.)

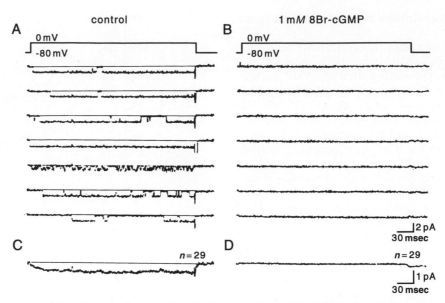

**FIG. 5.** Current recordings from a cell-attached patch showing effect of 8Br-cGMP on the $Ca^{2+}$ slow channel activity in a single myocardial cell isolated from a 3-day-old embryonic chick heart. Single-channel currents were evoked by depolarizing voltage pulses to 0 mV from a holding potential of $-80$ mV, at a duration of 300 msec and repetition rate of 0.5 Hz. (A and B) Examples of original current recordings from the same patch, before (A) and after (B) superfusion with 1.0 mM 8Br-cGMP. (C and D) Ensemble-averaged currents calculated from the current recordings ($n = 29$). (Data from Tohse and Sperelakis, 1991.)

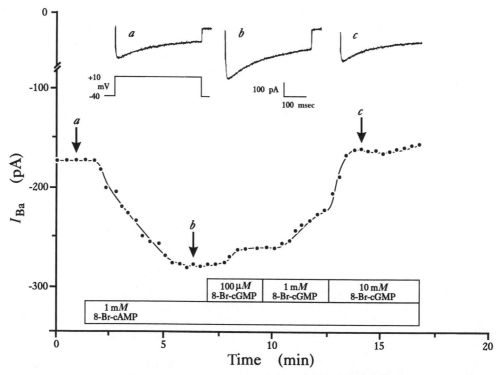

**FIG. 6.** Antagonism of the stimulating effect of 8Br-cAMP on $I_{Ca(s)}$ by 8Br-cGMP in a single young neonatal rat ventricular myocyte. Upper tracings show four original current recordings of $I_{Ca}$ corresponding to the four time points labeled in the lower graph. $I_{Ca(s)}$ was elicited by 300-msec depolarizing pulses to $+10$ mV from a holding potential of $-40$ mV. $Ba^{2+}$ (2.0 mM) was used as the charge carrier. Experiments conducted at room temperature of 25°C. (From Masuda and Sperelakis, unpublished.)

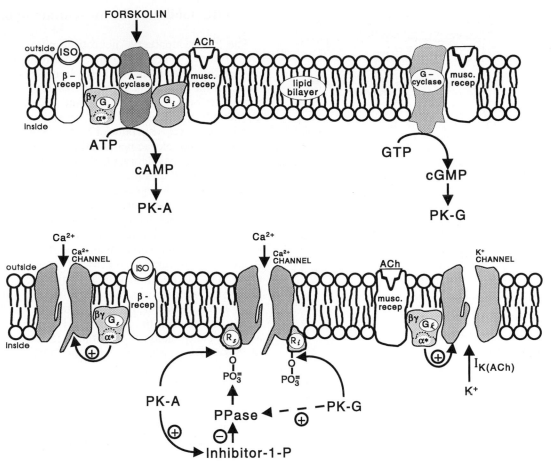

**FIG. 7.** Diagrammatic summary of the regulation of the $Ca^{2+}$ slow channels in the myocardial cell membrane and the mechanisms of action of some inotropic agents. The $\beta$-adrenergic agonists act via their receptor on a GTP-binding protein ($G_s$) to stimulate adenylate cyclase and cAMP production. The voltage-dependent myocardial slow $Ca^{2+}$ channels are stimulated by cAMP, presumably because the channel (or an associated regulatory protein) must be phosphorylated for it to be in a form that is available for voltage activation. cGMP-dependent phosphorylation also regulates the slow channel in a manner antagonistic to cAMP, namely, producing inhibition. Thus, the muscarinic receptor activated by ACh can produce inhibition of $Ca^{2+}$ influx by at least the four mechanisms depicted: (1) reversal of adenylate cyclase stimulation produced by $\beta$-agonists or $H_2$ agonists; (2) stimulation of guanylate cyclase and production of cGMP; (3) activation of a $K^+$ channel ($I_{K(ACh)}$) that produces an outward $K^+$ current, which depresses excitability and terminates the AP earlier, and thereby voltage deactivates the $Ca^{2+}$ slow channels earlier; and (4) stimulation of a phosphatase (PPase) by cGMP and PK-G. Mechanism (3) may be absent in ventricular myocardial cells.

the slow channel. Another possibility is that there is a second type of regulatory protein that is inhibitory when phosphorylated (see Fig. 1).

Another mechanism proposed for cGMP inhibition is based on cGMP depression of the cAMP level. Intracellular application of cGMP inhibited $I_{Ca}$ of frog ventricular myocytes only after the cAMP levels had been increased; that is, there was no effect of cGMP on the basal $I_{Ca}$ (Hartzell and Fischmeister, 1986; Fischmeister and Hartzell, 1987). It was concluded that cGMP inhibited $I_{Ca}$ by stimulating a phosphodiesterase, resulting in increased degradation of cAMP. However, in guinea-pig and rat cardiomyocytes, a direct inhibition of $I_{Ca}$ by cGMP and PK-G occurs (Levi et al., 1989; Mery et al., 1991). In addition, 8Br-cAMP inhibition of slow APs in mammalian

cardiac muscle occurs without a decrease in cAMP levels. However, elevation of intracellular cGMP by photoactivation of a derivative had no effect on the L-type $Ca^{2+}$ current ($I_{Ca}$) in isolated rat ventricular cells (Nargeot et al., 1983).

## VII. Protein Kinase G Inhibition

When PK-G is added to the patch pipette for diffusion into the cell during whole-cell voltage clamp, basal $I_{Ca}$ is inhibited markedly and rapidly by the PK-G, with maximum inhibition reached in about 3–5 min (Figs. 8 and 9). Data from 17-day chick cardiomyocytes are illustrated in Fig. 8. Summary of the effect of PK-G is shown in Fig. 8B.

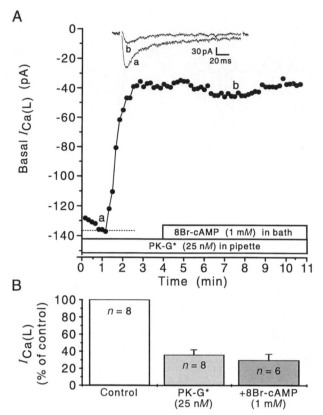

**FIG. 8.** Inhibition of basal $I_{Ca}$ of 17-day embryonic chick cardiomyocytes by PK-G. (A) PK-G (25 nM) was present in the patch pipette for diffusion into the cell during whole-cell voltage clamp. Inhibition of basal $I_{Ca}$ began within 70 sec after breaking into the cell and reached maximum at about 2.5 min. Addition of 1 mM 8Br-cAMP into the bath failed to reverse the inhibition produced by the PK-G. The two current traces illustrated at the top correspond to the time points labeled a and b in the graph. (B) Bar graph summary of the inhibition of basal $I_{Ca(L)}$ by PK-G in 8 cells, and the lack of reversal by 8Br-cAMP in six cells. Experiments were done at room temperature. (From Haddad *et al.*, 1995.)

Note that inhibition of basal $I_{Ca}$ began about 80 sec after breaking into the cell. Similar effects of PK-G infusion were observed in early neonatal rat ventricular myocytes, as illustrated in Fig. 9 (Sumii and Sperelakis, 1995). As can be seen, there is a rapid and prominent inhibition of the basal $I_{Ca}$ by PK-G. Addition of H-8 (a blocker of protein kinases including PK-G) to the bath causes a rapid restoration of $I_{Ca}$ to about the original basal level in some cells. Addition of 1 mM 8Br-cAMP can produce only a small stimulation of $I_{Ca}$ in the continued presence of PK-G (Fig. 8). Therefore, these findings indicate that the inhibitory effects of cGMP on $I_{Ca}$ are mediated by activation of PK-G and resultant phosphorylation, and that the *basal* $I_{Ca}$ is inhibited.

A single protein of approximately 47 kDa has been found to be phosphorylated by PK-G in guinea-pig sarcolemmal preparations (Cuppoletti *et al.*, 1988). Thus, this substrate may be a possible mediator of cGMP-mediated regulation of $Ca^{2+}$ channels of the heart.

## VIII. Inhibition by Muscarinic Agonists

The parasympathetic neurotransmitter acetylcholine (ACh) exerts a negative inotropic effect on ventricular myocardium prestimulated by $\beta$-adrenergic agonists. Activation of the muscarinic receptor by ACh exerts an inhibitory effect on adenylate cyclase and cAMP levels via the $G_i$ (inhibitory) coupling protein, reversing the stimulation of adenylate cyclase produced by the $G_s$ coupling protein due to, for example, activation of the $\beta$-adrenoceptor (Fig. 7). Thus, ACh may depress $Ca^{2+}$ influx and contraction by reversing cAMP elevation produced by various agonists. For example, ACh depressed $I_{Ca}$ of cultured chick ventricular cells that had been prestimulated by isoproterenol (Josephson and Sperelakis, 1978). ACh also reverses the electrophysiological effects of direct adenylate cyclase stimulation by forskolin (Wahler and Sperelakis, 1985). Additionally, in ventricular cells, ACh may inhibit the slow $Ca^{2+}$ channels due partly to elevation of cGMP levels. Muscarinic agonists are known to elevate cGMP (George *et al.*, 1970). However, it was reported that ACh was ineffective in reducing the basal $I_{Ca}$ (Hescheler *et al.*, 1986), and that cGMP actually potentiated the stimulating effect of ISO on $I_{Ca}$, perhaps mediated by PDE inhibition (Ono and Trautwein, 1991).

It has been proposed that muscarinic agonists also act to inhibit the $Ca^{2+}$ slow channel by PK-G stimulation of a phosphatase (type I) that dephosphorylates the channel (Ahmad *et al.*, 1989) (see Fig. 7). This would have the effect of decreasing the fraction of channels in the phosphorylated form, and therefore the $Ca^{2+}$ influx. In this mechanism, the rate of phosphorylation is unaffected, but the rate of dephosphorylation is increased.

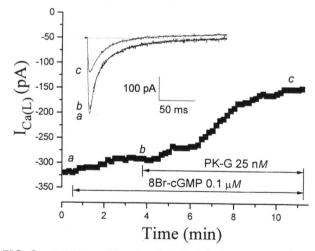

**FIG. 9.** Inhibition of basal $I_{Ca}$ by PK-G (25 nM) in a ventricular myocyte from an early (4-day) neonatal rat heart. Time course of the effect of low doses of 8Br-cGMP (0.1 $\mu$mol/L) and PK-G (25 nmol/L) on basal (not stimulated) $I_{Ca(L)}$. 8Br-cGMP and PK-G were applied by using the perfusion patch-pipette technique. As shown, when 8Br-cGMP was applied in advance of the PK-G, it had only little effect, whereas subsequent addition of PK-G produced rapid inhibition. (Inset) Selected current traces of $I_{Ca(L)}$ (a, b, and c) at points denoted on the time-course curve. (Reproduced with permission from Sumii and Sperelakis, 1995.)

## IX. Protein Kinase C and Calmodulin–Protein Kinase

PK-C is apparently involved in regulation of the myocardial slow $Ca^{2+}$ channels. Angiotensin-II and a high concentration of the $\alpha$-adrenergic agonist phenylephrine cause a positive inotropic effect in cardiac muscle. These agonists stimulate the phosphatidylinositol cycle and generation of inositol tris-phosphate ($IP_3$) and diacyl glycerol (DAG). $IP_3$ acts as a second messenger to release stored $Ca^{2+}$ from the sarcoplasmic reticulum (SR). DAG and $Ca^{2+}$ activate PK-C, which phosphorylates a number of proteins. Phorbol ester (direct activator of PK-C) and angiotensin-II stimulate $I_{Ca}$ in rat and chick hearts, but not in guinea-pig heart.

Inhibitors of calmodulin (e.g., calmidazolium) inhibit the $Ca^{2+}$-dependent slow APs of heart cells, and subsequent injection of calmodulin reverses the inhibition. Apparently, maximal activation of the slow channels requires two separate phosphorylation steps (calmodulin-dependent and cAMP-dependent). These may be on the same protein or on two separate proteins.

## X. Na$^+$, K$^+$, and $I_f$ Channels

The previous discussion dealt primarily with regulation of slow $Ca^{2+}$ channels in cardiac muscle. Table 3 summarizes the effects of the cyclic nucleotides on $I_{Ca(L)}$ in cardiac muscle, VSM, and skeletal muscle (frog). As indicated, in myocardial cells, cAMP and cGMP act in an antagonistic manner, cAMP stimulating and cGMP inhibiting. In contrast, in VSM cells, cAMP and cGMP act in the same direction, both inhibiting $I_{Ca(L)}$. Similarly, in skeletal muscle fibers, both cyclic nucleotides act in the same direction, but here both cAMP and cGMP stimulate $I_{Ca(L)}$. In uterine smooth muscle, neither cyclic nucleotide has any effect on $I_{Ca(L)}$ (Table 4). Therefore, these tissues provide examples of a variety of effects concerning the regulation exerted by cyclic nucleotides.

cAMP also has effects on other types of ion channels (Table 4). For example, the following channels of heart are stimulated by cAMP: (1) delayed rectifier K$^+$ channel (Trautwein *et al.*, 1982; Yazawa and Kameyama, 1990); (2) hyperpolarization-activated Na–K $I_f$ ($I_h$) channel (DiFrancesco and Tromba, 1988); and (3) adrenaline-activated Cl$^-$ channel (Ehara and Ishihara, 1990). The fast Na$^+$ channel was reported to be slightly inhibited by cAMP (Ono *et al.*, 1989), but others have reported no effect or a slight

**TABLE 4** Listing of Some Channels Modulated by cAMP and PK-A Phosphorylation

| Channel type | Tissue | Action |
|---|---|---|
| $I_{Ca(L)}$ | Heart, nerve, skeletal muscle | Stimulated |
| $I_{Ca(L)}$ | Vascular smooth muscle | Inhibited |
| $I_{Ca(L)}$ | Uterine smooth muscle | No effect |
| $I_{K(del)}$ | Heart | Stimulated |
| $I_f(I_h)$ | Heart | Stimulated |
| $I_{Cl(\beta ag)}$ | Heart | Stimulated |
| $I_{Na(f)}$ | Heart | No effect[a] |

[a] Variable subtle effects have been reported for $I_{Na(f)}$, including slight inhibition, slight stimulation, and shifting of the voltage dependency.

stimulation (Ono *et al.*, 1993). cGMP was reported to stimulate, presumably by PDE inhibition, the delayed rectifier K$^+$ current (Ono and Trautwein, 1991) and the Cl$^-$ current activated by $\beta$-adrenergic agonists (Tareen *et al.*, 1991).

The cAMP stimulation of $I_{K(del)}$ in myocardial cells, coupled with its stimulation of $I_{Ca(L)}$, serves to shorten $APD_{50}$ of the cardiac AP, whereas heart rate and force of contraction are increased by agents that elevate cAMP. For example, $\beta$-adrenergic agonists, such as isoproterenol or epinephrine, raise cAMP level, thereby increasing automaticity and force of contraction.

## XI. Summary

The slow $Ca^{2+}$ channels of the heart are stimulated by cAMP. Elevation of cAMP produces a very rapid increase in the number of slow channels available for voltage activation during excitation. The probability of a $Ca^{2+}$ channel opening and the mean open time of the channel are increased. Therefore, any agent that increases the cAMP level of the myocardial cell will tend to potentiate $I_{Ca}$, $Ca^{2+}$ influx, and contraction. The action of cAMP is mediated by PK-A and phosphorylation of the slow $Ca^{2+}$ channel protein or an associated regulatory protein (stimulatory type).

The myocardial slow $Ca^{2+}$ channels are also regulated by cGMP, in a manner that is opposite to that of cAMP. The effect of cGMP is mediated by PK-G and phosphorylation of a protein, as for example, a regulatory protein (inhibitory-type) associated with the $Ca^{2+}$ channel. In addition, cGMP acts to stimulate a phosphatase that dephosphorylates the $Ca^{2+}$ channel.

PK-C and calmodulin-PK may also play roles in regulation of the myocardial slow $Ca^{2+}$ channels, possibly mediated by phosphorylation of some regulatory-type of protein.

Thus, the slow $Ca^{2+}$ channel is apparently a complex structure, including perhaps several associated regulatory proteins, which can be regulated by a number of factors intrinsic and extrinsic to the cell (see Fig. 7).

**TABLE 3** Summary of Effects of Cyclic Nucleotides on Slow $Ca^{2+}$ Channels in Cardiac Muscle, Vascular Smooth Muscle, and Skeletal Muscle

| Cyclic nucleotide | Cardiac muscle | Vascular smooth muscle | Skeletal muscle[a] |
|---|---|---|---|
| cAMP | Stimulation | Inhibition | Stimulation |
| cGMP | Inhibition | Inhibition | Stimulation |

[a] From bullfrog. (From Kokate *et al.*, 1993.)

cAMP and cGMP also have effects on the slow $Ca^{2+}$ channels in cells other than cardiac muscle, including neurons, smooth muscle, and skeletal muscle fibers (see Table 4). In cardiac muscle, the two cyclic nucleotides have opposing effects, cAMP stimulating and cGMP inhibiting. In some smooth muscles (e.g., vascular), both cyclic nucleotides act in the same direction, namely, both inhibit $I_{Ca(L)}$. In skeletal muscle, both cAMP and cGMP act in the same direction on $I_{Ca(L)}$, that is, to stimulate (see Table 3).

The cyclic nucleotides and phosphorylation may also modulate the activity of several other types of ion channels, including $K^+$ channels (delayed rectifier type), $Cl^-$ channels ($\beta$-agonist activated type), fast $Na^+$ channels, and $I_f$ pacemaker channels (see Table 4).

# Bibliography

Ahmad, Z., Green, F. J., Subuhi, H. S., and Watanabe, A. M. (1989). Purification and characterization of an alpha-1,2-mannosidase involved in processing asparagine linked oligosaccharides. *J. Biol. Chem.* **264,** 3859–3863.

Armstrong, D., and Eckert, R. (1987). Voltage-activated calcium channels that must be phosphorylated to respond to membrane depolarization. *Proc. Natl. Acad. Sci. USA* **84,** 2518–2522.

Bean, B. P., Nowycky, M. C., and Tsien, R. W. (1984). Beta-adrenergic modulation of calcium channels in frog ventricular heart cells. *Nature* **307,** 371–375.

Bkaily, G., and Sperelakis, N. (1984). Injection of protein kinase inhibitor into cultured heart cells blocks calcium slow channels. *Am. J. Physiol.* **246,** H630–H634.

Bkaily, G., and Sperelakis, N. (1985). Injection of cyclic GMP into heart cells blocks the $Ca^{2+}$ slow channels. *Am. J. Physiol. (Heart Circ. Physiol.)* **248,** H745–H749.

Bkaily, G., and Sperelakis, N. (1986). Calmodulin is required for a full activation of the calcium slow channels in heart cells. *J. Cyclic Nucleotide Protein Phosphorylation Res.* **11,** 25–34.

Bruckner, R., and Scholz, H. (1984). Effects of alpha-adrenoceptor stimulation with phenylephrine in the presence of propranolol on force of contraction, slow inward current and cyclic AMP content in the bovine heart. *Br. J. Pharmacol.* **82,** 223–232.

Cachelin, A. B., dePeyer, J. E., Kokubun, S., and Reuter, H. (1983). $Ca^{2+}$ channel modulation by 8-bromo-cyclic AMP in culture heart cells. *Nature* **304,** 462–464.

Chad, J. E., and Eckert, R. J. (1986). An enzymatic mechanism for calcium current inactivation in dialysed Helix neurones. *J. Physiol.* **378,** 31–51.

Cuppoletti, J., Thakkar, J., Sperelakis, N., and Wahler, G. (1988). Cardiac sarcolemmal substrate of the cGMP-dependent protein kinase. *Membr. Biochem.* **7,** 135–142.

DiFrancesco, D., and Tromba, C. (1988). Muscarinic control of the hyperpolarization-activated current ($i_f$) in rabbit sino-atrial node myocytes. *J. Physiol. (London)* **405,** 493–510.

Dosemeci, A., Dhalla, R. S., Cohen, N. M., Lederer, W. J., and Rogers, T. B. (1988). Phorbol ester increases calcium current and stimulated the effects of angiotensin II on cultured neonatal rat heart myocytes. *Circ. Res.* **62,** 347.

Ehara, T., and Ishihara, K. (1990). Anion channels activated by adrenaline in cardiac myocytes. *Nature* **347,** 284–286.

Fischmeister, R., and Hartzell, R. C. (1987). Cyclic guanosine 3',5'-monophosphate regulates the calcium current in single cells from frog ventride. *J. Physiol.* **387,** 455–472.

George, W. J., Polson, J. B., O'Toole, A. G., and Goldberg, N. D. (1970). Elevation of guanosine 3',5'-cyclic phosphate in rat heart after perfusion with acetylcholine. *Proc. Natl. Acad. Sci. USA* **66,** 398–403.

Goldberg, N. D., Haddox, M. K., Nicol, S. E., Glass, D. B., Sanford, C. H., Kuehl, F. A., Jr., and Estensen, R. (1975). Biological regulation through opposing influences of cyclic GMP and cyclic AMP: The Yin Yang hypothesis. *Adv. Cyclic Nucleotide* **5,** 307–330.

Haddad, G. E., Sperelakis, N., and Bkaily, G. (1995). Regulation of calcium channel by cyclic GMP-dependent protein kinase in chick heart cells. *Mol. Cell. Biochem.* **148,** 89–94.

Hartzell, H. C., and Fischmeister, R. (1986). Opposite effects of cyclic GMP and cyclic AMP on $Ca^{2+}$ current in single heart cells. *Nature* **323,** 273–275.

Hescheler, J., Kameyama, M., and Trautwein, W. (1986). On the mechanism of muscarinic inhibition of the cardiac Ca current. *Pflugers Arch.* **407,** 182–189.

Hescheler, J., Kameyama, M., Trautwein, W., Mieskes, G., and Soling, H. D. (1987). Regulation of the cardiac calcium channel by protein phosphatases. *Eur. J. Biochem.* **165,** 261–266.

Irisawa, H., and Kokobun, S. (1983). Modulation of intracellular ATP and cyclic AMP of the slow inward current in isolated single ventricular cells of the guinea-pig. *J. Physiol.* **338,** 321–327.

Josephson, I., and Sperelakis, N. (1978). 5'-Guanylimidodiphosphate stimulation of slow $Ca^{2+}$ current in myocardial cells. *J. Mol. Cell Cardiol.* **10,** 1157–1166.

Kameyama, M., Hoffmann, F., and Trautwein, W. (1986). On the mechanism of B-adrenergic regulation of the $Ca^{2+}$ channel in the guinea-pig heart. *Pflugers Arch.* **405,** 285–293.

Kohlhardt, M., and Haap, K. (1978). 8-Bromo-guanosine-3',5'-monophosphate mimics the effect of acetylcholine on slow response action potential and contractile force in mammalian atrial myocardium. *J. Mol. Cell. Cardiol.* **10,** 573–578.

Levi, R. C., Alloatti, G., and Fischmeister, R. (1989). Cyclic GMP regulates the Ca-channel current in guinea pig ventricular myocytes. *Pfluegers Arch.* **413,** 685–687.

Li, T., and Sperelakis, N. (1983). Stimulation of slow action potentials in guinea pig papillary muscle cells by intracellular injection of cAMP, Gpp(NH)p, and cholera toxin. *Circ. Res.* **52,** 111–117.

Llinas, R., Sugimori, D., Hillman, E., and Cherksey, B. (1992). Distribution and functional significance of the P-type, voltage-dependent $Ca^{2+}$ channels in the mammalian central nervous system. *Trends Neurosci.* **15,** 351–355.

Masuda, H., Sumii, K., and Sperelakis, N. (1995). Long openings of calcium channels in fetal rat ventricular cardiomyocytes. *Pflugers Arch. Eur. J. Physiol.* **429,** 595–597.

Mehegan, J. P., Muir, W. W., Unverferth, D. V., Fertel, R. H., and McGuirk, S. M. (1985). Electrophysiological effects of cyclic GMP on canine cardiac Purkinje fibers. *J. Cardiovasc. Pharmacol.* **7,** 30–35.

Mery, P. F., Lohmann, S. M., Walter, U., and Fischmeister, R. (1991). $Ca^{2+}$ current is regulated by cyclic GMP-dependent protein kinase in mammalian cardiac myocytes. *Proc. Natl. Acad. Sci. USA* **88,** 1197–1201.

Mintz, I. M., Venema, V. J., Swiderek, K. M., Lee, T. D., Bean, B. P., and Adams, M. E. (1992). P-type calcium channels blocked by the spider toxin omega-Aga-IVA. *Nature* **355,** 827–829.

Nargeot, J., Nerbonne, J. M., Engels, J., and Lester, H. A. (1983). Time course of the increase in the myocardial slow inward current after a photochemically generated concentration jump of intracellular cAMP. *Proc. Natl. Acad. Sci. USA* **80,** 2395–2399.

Nawrath, H. (1977). Does cyclic GMP mediate the negative inotropic effect of acetylcholine in the heart? *Nature* **267,** 72–74.

Nowycky, M. C., Fox, A. P., and Tsien, R. W. (1985). Three types of neuronal calcium channels with different calcium agonist sensitivity. *Nature* **316,** 440–443.

Ono, K., and Trautwein, W. (1991). Potentiation by cyclic GMP of B-adrenergic effect on $Ca^{2+}$ current in guinea-pig ventricular cell. *J. Physiol.* **443,** 387–404.

Ono, K., Kiyosue, T., and Arita, M. (1989). Isoproterenol, DBcAMP, and forskolin inhibit cardiac sodium current. *Am. J. Physiol.* **256,** C1131–C1137.

Ono, K., Fozzard, H. A., and Nanck, D. A. (1993). Mechanism of cAMP-dependent modulation of cardiac sodium channel current kinetics. *Circ. Res.* **72,** 807–815.

Osterrieder, W., Brum, G., Hescheler, J., Trautwein, W., Flockerzi, V., and Hofmann, F. (1982). Injection of subunits of cyclic AMP-dependent protein kinase into cardiac myocytes modulates $Ca^{2+}$ current. *Nature* **298,** 576–578.

Reuter, H. (1983). Calcium channel modulation by neurotransmitters, enzymes, and drugs. *Nature* **301,** 569–574.

Reuter, H., and Scholz, H. (1977). The regulation of the calcium conductance of cardiac muscle by adrenaline. *J. Physiol. (London)* **264,** 49–62.

Reuter, H., Stevens, C.-F., Tsien, R. W., and Yellen, G. (1982). Properties of single calcium channels in cardiac cell culture. *Nature* **297,** 501–504.

Shigenobu, K., and Sperelakis, N. (1972). $Ca^{2+}$ current channels induced by catecholamines in chick embryonic hearts whose fast $Na^+$ channels are blocked by tetrodotoxin or elevated $K^+$. *Circ. Res.* **31,** 932–952.

Sperelakis, N., and Schneider, J. A. (1976). A metabolic control mechanism for calcium ion influx that may protect the ventricular myocardial cell. *Am. J. Cardiol.* **37,** 1079–1085.

Sumii, K., and Sperelakis, N. (1995). Cyclic GMP-dependent protein kinase regulation of the L-type calcium current in neonatal rat ventricular myocytes. *Circ. Res.* **77,** 803–812.

Tareen, F. M., Ono, K., Noma, A., and Ehara, T. (1991). β-adrenergic and muscarinic regulation of the chloride current in guinea-pig ventricular cells. *J. Physiol.* **440,** 225–241.

Tohse, N., and Sperelakis, N. (1990). Long-lasting openings of single slow (L-type) $Ca^{2+}$ channels in chick embryonic heart cells. *Am. J. Physiol.* **259,** H639–H642.

Tohse, N., and Sperelakis, N. (1991). Cyclic GMP inhibits the activity of single calcium channels in embryonic chick heart cells. *Circ. Res.* **69,** 325–331.

Tohse, N., Kameyama, M., Sakiguchi, K., Shearman, M. S., and Kanno, M. (1990). Protein kinase C activation enhances the delayed rectifier $K^+$ current in guinea-pig heart cells. *J. Mol. Cell Cardiol.* **22,** 725–734.

Tohse, N., Meszaros, J., and Sperelakis, N. (1992). Developmental changes in long-opening behavior of L-type $Ca^{2+}$ (slow) channels in embryonic chick heart cells. *Circ. Res.* **71,** 376–384.

Trautwein, W., and Hoffmann, F. (1983). Activation of calcium current by injection of cAMP and catalytic subunit of cAMP-dependent protein kinase. *Proc. Int. Union Physiol. Sci.* **15,** 75–83.

Trautwein, W., Taniguchi, J., and Noma, A. (1982). The effect of intracellular cyclic nucleotides and calcium on the action potential and acetylcholine response of isolated cardiac cells. *Pfluegers Arch.* **392,** 307–314.

Tsien, R. W., Giles, W., and Greengard, P. (1972). Cyclic AMP mediates the action of adrenaline on the action potential plateau of cardiac Purkinje fibers. *Nature* **240,** 181–183.

Vogel, S., and Sperelakis, N. (1981). Induction of slow action potentials by microiontophoresis of cyclic AMP into heart cells. *J. Mol. Cell Cardiol.* **13,** 51–64.

Vogel, S., Sperelakis, N., Josephson, J., and Brooker, G. (1977). Fluoride stimulation of slow $Ca^{2+}$ current in cardiac muscle. *J. Mol. Cell Cardiol.* **9,** 461–475.

Wahler, G. M., and Sperelakis, N. (1985). Intracellular injection of cyclic GMP depresses cardiac slow action potentials. *J. Cyclic Nucleotide Protein Phosphorylation Res.* **10,** 83–95.

Wahler, G. M., Rusch, N. J., and Sperelakis, N. (1990). 8-Bromo-cyclic GMP inhibits the calcium channel current in embryonic chick ventricular myocytes. *Can. J. Physiol. Pharmacol.* **68,** 531–534.

Yazawa, K., and Kameyama, M. (1990). Mechanism of receptor-mediated modulation of the delayed outward potassium current in guinea-pig ventricular myocytes. *J. Physiol.* **421,** 135–150.

*Atsuko Yatani*

# 33

# Direct Regulation of Ion Channels by G Proteins

## I. Introduction

Heterotrimeric high molecular weight GTP-binding proteins (G proteins) are widely distributed signal-transducing proteins that mediate diverse extracellular signals such as light, odorants, peptide hormones, and neurotransmitters. Signal-transduction systems that use cell surface receptors coupled to G proteins consist of at least three membrane-bound components: (1) a cell membrane receptor, (2) an effector, and (3) a G protein that is coupled to both the receptor and its effector (Stryer, 1986; Gilman, 1987; Birnbaumer *et al.*, 1987, 1990; Dohlman *et al.*, 1991; Strader *et al.*, 1994). Well-characterized G-protein-coupled effector enzymes and their corresponding G proteins include the photoactivated cyclic GMP-phosphodiesterase and transducin ($G_t$), and adenylyl cyclase and its stimulatory regulator, $G_s$, or its inhibitory regulator, $G_i$. Recent combined biochemical, electrophysiological, and molecular cloning studies have shown that G proteins are also involved in the coupling process of membrane receptors to ion channels. Ion channel activity can be modified either by its direct interaction **(direct pathway)** with an activated G protein or with a cytoplasmic second messenger (including cAMP, $Ca^{2+}$, $1P_3$) formed by interaction of activated G protein with membrane-associated enzymes (**indirect pathway;** Fig. 1). This chapter focuses only on the direct pathway.

Using single-channel measurements from (1) cell-free excised patch membranes and (2) plasma membranes vesicles incorporated into planar lipid bilayers, it has been shown that G proteins have effects on some ion channels that are direct (or at least are "membrane-delimited" mechanisms). These are distinct from indirect effects mediated by cytoplasmic second messengers (Yatani *et al.*, 1990). Direct G protein effects take two forms: **obligatory** and **modulatory.** In the obligatory case, the opening probability, $P_o$, of the channel is determined by the G protein. Without agonist occupancy of its corresponding receptor, $P_o$ is low; an example is the cardiac muscarinic $K^+$ ($K_{ACh}$)

channel. In the modulatory case, $P_o$ is primarily determined by membrane voltage and not by receptor agonists. An example is G-protein modulation of voltage-gated dihydropyridine-sensitive L-type $Ca^{2+}$ channels in the heart.

Because direct coupling of G-protein and ion channels is most characterized in the cardiac muscarinic acetylcholine receptor (mAChR) and the $K_{ACh}$ channel to which it is coupled, the development and present understanding of the mAChR–$K_{ACh}$ coupling process is reviewed. In addition, an analogous situation in a $K^+$ channel of clonal pituitary ($GH_3$) cells is reviewed. Other channels that appear to be under direct G-protein regulation are also presented briefly.

## II. General Characteristics of G Proteins

G proteins consist of three subunits, $\alpha$ (39–52 kDa), $\beta$ (35–36 kDa), and $\gamma$ (7–10 kDa). The $\beta\gamma$-subunit is a dimer under native conditions. The $\alpha$-subunits are unique for each G protein, and $\beta$- and $\gamma$-subunits are less unique and are shared by different $\alpha$-subunits. Recent molecular cloning and expression studies have discovered many G-protein subtypes. In mammalian cells, at least 16 $\alpha$-subunits, 4 $\beta$-subunits, and 7 $\gamma$-subunits have been identified to date (Conklin and Bourne, 1993; Strader *et al.*, 1994). The number of G-protein-coupled receptors is also diverse. Molecular cloning studies have revealed the identification and characterization of many new receptor subtypes and sub-subtypes of previously identified receptors. For example, 5 subtypes of muscarinic receptor, 9 adrenergic receptors, and more than 14 subtypes of serotonin receptor are now known to be encoded by distinct genes (Strader *et al.*, 1994). The apparent redundancy of receptor subtypes might reflect differences in their regulation in various tissues, differences over the course of development, or different effector targets.

The mechanism of signal transduction by G proteins has been best studied in adenylyl cyclase and cGMP phosphodiesterase, and the prevailing view of the cycle is summa-

**510**

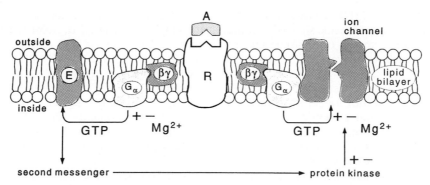

**FIG. 1.** Diagram of the role of G proteins in receptor-mediated regulation of ion channels. When agonist (A) binds receptor (R), in the presence of $Mg^{2+}$, the GDP bound on $G_\alpha$ exchange for GTP is accelerated. The GTP-bound active $G_\alpha$ then regulates the ion channel directly (direct pathway) or indirectly by activating membrane enzyme (E) and cytoplasmic second messenger, which modulate the ion channel (indirect pathway; +, stimulatory; −, inhibitory). In this model, the possibility that a single receptor may interact with more than one type of G protein and/or that a G protein interacts with more than one effector system is not shown.

rized in Fig. 2. The main reactions are as follows: Activation of receptors by ligand promotes the release of GDP, which is tightly associated with the $\alpha$-subunit, to exchange with GTP. This leads to the dissociation of the receptor-associated $\alpha\beta\gamma$ complex of G protein into $\alpha \cdot GTP$ and $\beta\gamma$. $\alpha \cdot GTP$ then activates appropriate effectors (including ion channels). The lifetime of the activated state of the effector depends on the degradation of the activated $\alpha \cdot GTP$. After hydrolysis of GTP to GDP by GTPase, $\alpha \cdot GDP$ is released from the effector. $\alpha \cdot GDP$ reassociates with $\beta\gamma$ to restore the heterotrimeric $G \cdot GDP$ complex, which interacts with the membrane receptors and reinitiates the cycle (see Fig. 2). For most $\alpha$-subunits, the intrinsic rate constant for GTP hydrolysis is low; stimulation of receptor with agonist accelerates this rate by 10- to 50-fold (Birnbaumer *et al.*, 1987; Conklin and Bourne, 1993). Because $\alpha$-subunit binds to and hydrolyzes GTP, and determines receptor specificity, functional characterizations have focused on the $\alpha$-subunit; the function of $\beta\gamma$-subunits is still not completely known. Recent studies have begun to document contribution of $\beta\gamma$-dimer to receptor–G-protein specificity and regulation of effectors (Conklin and Bourne, 1993; Clapham, 1994).

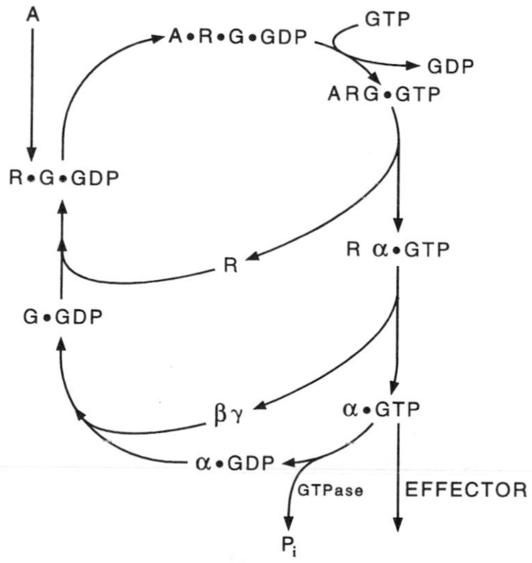

## III. Direct Coupling of Atrial Muscarinic K⁺ Channels to G Proteins

The strongest evidence for direct G-protein coupling to ionic channels was obtained for $K_{ACh}$ channels, which functionally regulate the heart rate by hyperpolarizing the membrane via ACh released from the vagus nerve (Glitsh and Pott, 1978). Muscarinic responses are considerably slower than those mediated by nicotinic receptors, where the nicotinic receptor and ion channel are one and the same protein (e.g., at the skeletal neuromuscular junction). That is, the ligand binding directly produces a conformational change of the channel protein that acts to gate the channel. This suggests that there is a biochemical reaction between the mAChR receptor and the channel. First, single-channel current measurements showed that ACh activates the $K_{ACh}$ channel through a noncytoplasmic pathway (Soejima and Noma, 1984), whereas whole-cell current measurements demonstrated that a GTP-dependent step of mAChR mediates the effects on $K_{ACh}$ current. The mAChR agonist effects were blocked by pretreatment with pertussis toxin (PTX), which prevents receptor interaction with several specific G proteins (e.g., $G_i$ and $G_o$), suggesting the involvement of PTX-sensitive G proteins on the coupling process (Pfaffinger *et al.*, 1985). The ACh response became irreversible after intracellular perfusion of the non-

**FIG. 2.** A simple model of the regulatory cycle of the trimeric G protein in signal transduction. Activation of receptor (R) by agonist (A) activates the release of GDP, tightly associated with the $\alpha$-subunit of G protein to exchange with GTP. This promotes the release of the G protein from receptor and dissociation of the G-protein $\alpha\beta\gamma$ complex into $\alpha \cdot GTP$ and $\beta\gamma$. The GTP-bound active $\alpha$-subunit ($\alpha \cdot GTP$) then modulates the effector. After GTP is hydrolyzed by intrinsic GTPase, $\alpha \cdot GDP$ is released from effector. $\alpha \cdot GDP$ then associates with $\beta\gamma$, which has high affinity for $\alpha \cdot GDP$ to form the $\alpha\beta\gamma$ complex of G protein, which interacts with receptor.

hydrolyzable GTP analog, GMP-P(NH)P (Breitwieser and Szabo, 1985). Single-channel experiments showed that the GTP analog, GTPγS, applied to the cytoplasmic face of inside-out membrane patches, activates $K_{ACh}$ currents (Kurachi et al., 1986a,b).

These results strongly suggested a direct G-protein effect on $K_{ACh}$ channels, but the question remains as to which G protein was involved, and the possibility that the G protein was acting through a membrane-associated enzyme such as protein kinase C (PK-C). To identify which G protein(s) are involved in mAChR-mediated effects on $K_{ACh}$ channels and how G protein interacts with an ion channel, G proteins were applied to the cytoplasmic face of isolated inside-out membrane patches from mammalian atrial cells (Yatani et al., 1987a). Figure 3 summarizes the direct regulation of $K_{ACh}$ channels by G protein. In the presence of $NAD^+$, PTX blocked the carbachol-activated $K_{ACh}$ currents (Fig. 3A). After PTX treatment, GTP (even at 1 mM) could not reactivate the channel, but GTPγS was able to. A PTX-sensitive G protein ($G_k$) purified from human red blood cells (hRBC), preactivated with GTPγS (denoted as $G_k^*$), stimulated the channels in the same way as muscarinic agonists (Fig. 3B). This $G_k$ action was specific. The cholera toxin (CTX)-sensitive $G_s$ from hRBC, which activates adenylyl cyclase, could not activate the $K_{ACh}$ channels. Since all experiments were performed without ATP, and neither AMP-P(NH)P nor phorbol ester altered the $G_k$ effects, the involvement of phosphorylation by kinases (including PK-C) could be excluded. The effective concentrations were picomolar, and the α-subunit of $G_k$, ($α_k$), was equipotent with the holo-G-protein (trimeric) (Fig. 3C).

Amino acid sequence analysis showed that $α_k$ encoded the cDNA designated $α_{i-3}$ and the recombinant $α_{i-3}$ was expressed in bacteria and tested for its effects on $K_{ACh}$ channels (Mattera et al., 1989a). The bacterially expressed recombinant $α_{i-3}$ ($rα_{i-3}$) mimics the effects of $G_k$ and muscarinic activation (Fig. 3D). Neither the antisense recombinant nor recombinant $α_s$ of $G_s$ ($rα_s$) had an effect on $K_{ACh}$ channels. Subsequently, the other α-subunits, $rα_{i-1}$ and $rα_{i-2}$, have been expressed and tested on $K_{ACh}$ channels. All three distinct α-subunits were equally effective in activating $K_{ACh}$ channel current (Yatani et al., 1988b).

Based on these results, it was proposed that the α-subunit mediates the $G_k$ effects. To support this idea, it was also shown that antibodies raised against the α-subunit block muscarinic activation (Yatani et al., 1988a; Okabe et al., 1991). In contrast to the observations that βγ dimers inhibit $K_{ACh}$ currents (Okabe et al., 1990), a dominant role of βγ dimers in activation of the $K_{ACh}$ channels was reported (Logothetis et al., 1987). It is possible that both α- and βγ-subunits cause channel activation in intact cells (Logothetis et al., 1988). Because purified or cloned channel protein was not available until recently, it was not possible to determine if the interaction between G protein and channel is truly direct or mediated by some unknown membrane protein.

Recent molecular cloning and expression studies using Xenopus oocytes identified a G-protein-activated inwardly rectifying $K^+$ channel, termed GIRK (Kubo et al., 1993), although subsequent reports indicate that $K_{ACh}$ channels expressed in heart cells is a hetero-multimeric channel composed of GIRK and a second related protein termed the cardiac inward rectifier (CIR) (Krapivinsky et al., 1995).

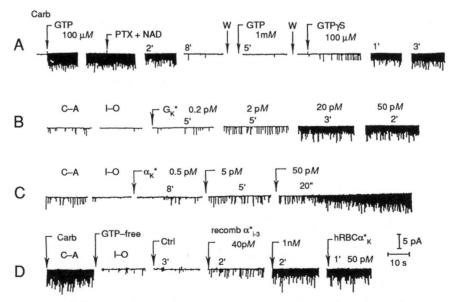

**FIG. 3.** Reactivation of $K_{ACh}$ currents by GTPγS (A) after uncoupling of endogenous G protein by activated PTX plus $NAD^+$ at a concentration of 10 μg/ml and 1 mM, respectively. Activation of $K_{ACh}$ currents by hRBC $G_k^*$ (B), hRBC $α_k^*$ (C) and recombinant $α_{i-3}$ (D). In (B), (C), and (D), G proteins are preactivated with GTPγS, as denoted by the asterisk. Single $K^+$ channel currents were recorded as cell-attached (C-A) and inside-out (I-O) patch configurations as noted. Carbachol (10 μM) was present in patch solution throughout in (A) and (D). The recording potential was −80 mV. Numbers above records denote time elapsed in min (′) or second (″) between solution change and the beginning of the record.

Nevertheless, all components necessary to reconstitute $K_{ACh}$ channel activation are now available, and the role of G-protein subunits in cloned GIRK activation is the subject of intense study (Kofuji *et al.*, 1995). Initial experiments on the structural basis of GIRK regulation by G proteins using selected mutagenesis and protein–protein interaction suggest that $\beta\gamma$ dimers bind to both the N-terminal hydrophilic domain and C-terminal domain of GIRK (Huang *et al.*, 1995). Expression studies have shown a modulatory role for $\alpha$-subunit in the activation of $K_{ACh}$ similar to adenylyl cyclase regulation by G proteins (Schreibmayer *et al.*, 1996).

## IV. Direct Coupling of Somatostatin and Muscarinic Receptor Activated K⁺ Channel by G Proteins

The evidence for direct coupling of G proteins to ionic channels has been extended to muscarinic- and somatostatin-regulated $K^+$ channels in clonal GH₃ pituitary cells. Somatostatin (SST) inhibits hormone secretion in normal and clonal pituitary cells. The mechanism by which SST inhibits hormone secretion is complex, and has been suggested to be linked to its ability to reduce cAMP levels. The latter is due to a GTP-dependent inhibition of adenylyl cyclase, which involves the PTX-sensitive $G_i$ protein. ACh acting through a mAChR has the same effects as SST (Schlegel *et al.*, 1985). All the effects of SST and ACh are inhibited by PTX, suggesting the involvement of a common signal transduction mechanism.

Although reduction of cAMP levels, and consequent inhibition of phosphorylation, could be the mechanism of action of SST and ACh in these cells, there is evidence that decreased cAMP cannot solely account for their inhibitory actions. SST also inhibits hormone secretion induced by agents that bypass the adenylyl cyclase system, such as 8Br-cAMP, or by depolarization with $K^+$ (Reisine, 1984).

Tests were conducted to determine whether these cells also contain a $K^+$ channel that can be activated by $G_k$, SST, or ACh, and whether this receptor-mediated activation is sensitive to PTX, and, if so, whether a re-PTX-inactivated system could be reactivated by exogenously applied $G_k$ (Yatani *et al.*, 1987b). This was done as follows. To characterize receptor- and/or $G_k$-gated $K^+$ channels in membranes of GH₃ cells, $K^+$ currents from membrane patches were analyzed before and after excision. Figure 4 illustrates typical single-channel recordings from a patch (inside-out configuration) in the presence of agonist in the pipette solution and GTP in the bath (intracellular face of the patch). An amplitude histogram of openings obtained at $-80$ mV showed the presence of one major group of unitary openings having a mean current amplitude of 4 pA. Analysis of the current–voltage relations of the ligand-induced openings showed a slope conductance of $55 \pm 2$ pS. No differences between SST-induced and carbachol (CCh)-induced currents could be detected, either in terms of conductance, amplitude histogram, or mean open time.

The involvement of a G-protein coupling occupancy of receptors by SST and ACh (or CCh) to activation

of $K^+$ channels is demonstrated in Fig. 5: $K^+$ channel activity was dependent on the presence of GTP. Addition of GTP$\gamma$S to excised patches (without prior activation by ligands) produced activation of $K^+$ channels. These channels had the same conductance as seen with receptor ligands plus GTP, and addition of purified $G_k$ activated the $K^+$ channels.

The experiments described earlier defined the existence in GH₃ cell membrane of a GTP-dependent step in the receptor-mediated activation of $K^+$ channels (Fig. 5C). They also indicated that these $K^+$ channels can be activated by $G_k$ (Fig. 5B). After inactivation of the channel by PTX, the channel became reactivated by the addition of a nonactivated $G_k$ plus GTP (Fig. 5D). At the single-channel level, the unit conductance of the $K^+$ channel activity restored after PTX treatment by $G_k$ was the same as before PTX inactivation of the endogenous $G_k$. As in the case of atrial cells, $\alpha_k$ was equipotent with trimeric G protein (Codina *et al.*, 1987). These experiments demonstrate that PTX uncouples endogenous G protein, and that exogenously added $G_k$ reconstitutes the activity.

## V. Direct Activation of Ca²⁺ Channels by G Proteins

G protein also directly regulates voltage-gated dihydropyridine-sensitive, L-type $Ca^{2+}$ channels from skeletal muscle T-tubules (Yatani *et al.*, 1988c) and from cardiac sarcolemma (Yatani *et al.*, 1987c; Imoto *et al.*, 1988; Wang *et al.*, 1993). In the case of $Ca^{2+}$ channels, G-protein activation does not, *by itself,* produce $Ca^{2+}$ channel opening, but rather modulates the voltage-dependent activity.

These are pharmacological and biochemical indications that a G protein interacts (is interposed) between receptors and $Ca^{2+}$ channels (Holtz *et al.*, 1986; Hescheler *et al.*, 1987). To test whether G protein(s) have direct effects on cardiac $Ca^{2+}$ channels (Yatani *et al.*, 1987c; Imoto *et al.*, 1988), GTP$\gamma$S and G proteins were applied to patches excised from ventricular myocytes. However, the L-type $Ca^{2+}$ channel quickly runs down following patch excision. Two approaches can be used to deal with this problem: (1) Ventricular $Ca^{2+}$ channels can be stimulated prior to patch excision by the $\beta$-adrenoreceptor agonist isoproterenol (Iso) or by the dihydropyridine $Ca^{2+}$ channel agonist Bay-K-8644. Iso phosphorylates $Ca^{2+}$ channels via protein kinase A (PK-A) (Kameyama *et al.*, 1985, 1986) and Bay-K-8644 activates the channels directly. Both produced some prolongation of survival time after patch excision. (2) Purified $Ca^{2+}$ channels can be incorporated into planar lipid bilayers (from skeletal muscle T-tubule). In the presence of Bay-K-8644, single-channel currents are stable for relatively long periods (Yatani *et al.*, 1988c; Rosenberg *et al.*, 1986).

In inside-out membrane patches (approach 1), GTP or GTP$\gamma$S had slowed rundown. Iso in the patch pipette was an absolute requirement, presumably to increase the off rate (see Fig. 2) of GDP (from endogenous G protein) to allow GTP$\gamma$S activation to occur before rundown became irreversible. The effect was not blocked by PK-A inhibitor.

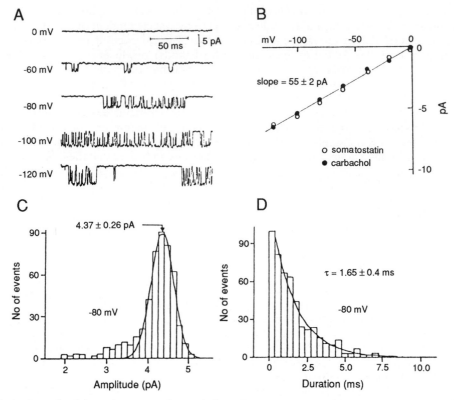

**FIG. 4.** Single-channel activity and current–voltage relations of receptor-activated $K^+$ channels in $GH_3$ cells. (A) Representative records of 100-msec duration of activity in a patch with 10 $\mu M$ CCh in the pipette side and 100 $\mu M$ GTP on the bath side of the membrane, obtained at the indicated holding potentials. (B) Current–voltage relations for $K^+$ channels activated by 10 $\mu M$ CCh ($\cdot$) or 0.5 $\mu M$ SST ($\bigcirc$). (C) Amplitude histogram of channel openings obtained in the presence of CCh in the pipette and GTP in the bath at a holding potential of $-80$ mV. (D) Open-time duration histogram of the 55-pS channel openings at $-80$ mV holding potential. (Reproduced from A. Yatani, J. Codina, R. D. Sekura, L. Birnbaumer, and A. M. Brown, Reconstitution of somatostatin and muscarinic receptor mediated simulation of $K^+$ channels by Isolated $G_k$ protein in clonal rat anterior pituitary cell membranes. *Mol. Endocr.* **4,** 283–289, Copyright 1987 The Endocrine Society.)

Combination of ATP, cAMP, PK-A, and forskolin (direct adenylyl cyclase activator) failed to mimic the effects of GTP or GTPγS.

Several lines of evidence indicate that the $Ca^{2+}$ channel regulatory protein is $G_s$. First, for GTPγS to be effective in excised patches, there was a requirement for the adrenergic receptor to be occupied and coupled to $G_s$. Second, only activated $G_s$ and its α-subunit mimicked the effect of GTPγS. Third, the active G protein is a CTX substrate (Yatani *et al.*, 1988c).

In the reconstitution experiments on skeletal muscle T-tubule $Ca^{2+}$ channels (approach 2), $G_s$ stimulation was more straightforward, since the control $Ca^{2+}$ channel currents were not running down. Neither ATP nor Bay-K-8644 were required, and addition of exogenous $G_s$ (activated either with GTPγS or CTX) was effective. Activation of endogenous $G_s$ by addition of either GTPγS or GTP plus Iso also increased $Ca^{2+}$ channel currents. As expected, Iso was effective only from the extracellular side, and GTP, GTPγS, and $G_s$ were effective only from the intracellular side. Prior addition of GDPβS prevented the stimulation produced by GTPγS. The effect was specific for $G_s$; $G_k$

had no effect. Single $Ca^{2+}$-channel currents in lipid bilayers were also stimulated by phosphorylation catalyzed by PK-A, and such phosphorylated channels were stimulated still further by addition of $G_s$ (Yatani *et al.*, 1988c).

In the case of $\alpha_s$, there are four closely related splice variants (Robishaw *et al.*, 1986), and the $\alpha_s$ from hRBC is of the short type. To determine whether single $\alpha_s$ can act on both adenylyl cyclase and $Ca^{2+}$ channels, three of the four splice variants of $\alpha_s$ were expressed in *Escherichia coli*. All forms of α-subunit, when preactivated, stimulated both adenylyl cyclase and $Ca^{2+}$ channels. The results indicate that a single G protein α-subunit can regulate two distinct effector functions (Mattera *et al.*, 1989b) (see Fig. 1).

The physiological significance of the direct $G_2$ effects on $Ca^{2+}$ channels is still unclear (Hartzell and Fischmeister, 1992). One proposal is the direct pathway, which produces a faster response to receptor agonists, might prime the $Ca^{2+}$ channels to respond to the phosphorylation-mediated indirect pathway (Yatani and Brown, 1989; Cavalié *et al.*, 1991). That is, the fast direct pathway primes the channel for the slower indirect pathway.

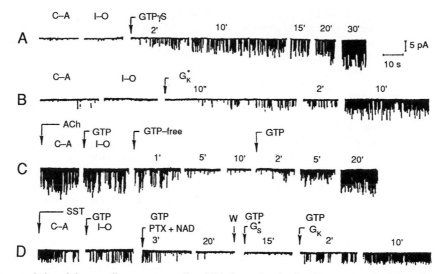

**FIG. 5.** Characteristics of the coupling system mediated $K^+$ channel activation by receptors in $GH_3$ cell membranes. $GH_3$ cell $K^+$ currents were recorded at a holding potential of $-80$ mV, first in the cell-attached (C-A) mode and then, in the inside-out (I-O) configuration. Times on top of the traces refer to minutes (or seconds) elapsed between the last addition of test material and the recording of the trace shown. (A) Basal channel activities in cell-attached and inside-out patches and activation of $K^+$ currents by addition of 100 $\mu M$ GTP$\gamma$S. (B) Effect of hRBC $G_k^*$ (200 p$M$) to activate $GH_3$ $K^+$ channels in the absence of either receptor ligands (in the pipette) or GTP (in the bath). (C) Dependence of receptor-induced $K^+$ currents in isolated patches on GTP. GTP-containing (100 $\mu M$) and GTP-free media were exchanged by perfusion; the patch pipette contained 10 $\mu M$ ACh throughout. (D) Dependence of SST-stimulated $K^+$ currents on a PTX-sensitive step and reconstitution of receptor–$K^+$ channel coupling by readdition of $G_k$, but not $G_s$. Throughout, the pipette contained 0.5 $\mu M$ SST and the bath solutions 100 $\mu M$ GTP. Preactivated PTX and NAD$^+$ were added to give 10 $\mu$g/ml and 1 m$M$, respectively; washing occurred after PTX treatment (W) with the same GTP-containing medium but without PTX and NAD$^+$. Preactivated $G_s^*$ and, $G_k^*$ were added to give final concentrations of 2000 p$M$ each. (Reproduced from A. Yatani, J. Codina, R. D. Sekura, L. Birnbaumer, and A. M. Brown, Reconstitution of somatostatin and muscarinic receptor mediated simulation of $K^+$ channels by Isolated $G_k$ protein in clonal rat anterior pituitary cell membranes. *Mol. Endocr.* **4**, 283–289, Copyright 1987 The Endocrine Society.)

## VI. Neuronal Ca²⁺ Channels That May be Under Direct Regulation by G Proteins

Although firm evidence from single-channel data in a cell-free system have not yet been reported, there is some evidence for G-protein modulation of voltage-gated $Ca^{2+}$ channels of neuronal cells. Several comprehensive reviews are available (Dolphin, 1991; Anwyl, 1991; Hille, 1992).

SST inhibits $Ca^{2+}$ channel currents in the anterior pituitary AtT-20 cell line by a PTX-sensitive mechanism and was mimicked by GTP$\gamma$S (Lewis *et al.*, 1986). GABA and the GABA$_B$ analogue baclofen and $\alpha_2$-adrenergic agonist noradrenaline (NE) inhibited $Ca^{2+}$ channel currents in dorsal root ganglion (DRG) neurons; PTX and GDP$\beta$S prevented these effects (Holtz *et al.*, 1986; Dolphin and Scott, 1987). GTP$\gamma$S inhibited the transient $Ca^{2+}$ current in DRG cells and potentiated the inhibition of transient and sustained components of $Ca^{2+}$ current caused by baclofen (Dolphin and Scott, 1987; Hescheler *et al.*, 1987). Since the block of $Ca^{2+}$ currents by GTP$\gamma$S was relieved by depolarization, a voltage-dependent interaction between G protein and $Ca^{2+}$ channel has been proposed (Dolphin, 1991).

An opioid peptide agonist D-Ala-D-Leu-encephalin (DADLE) inhibits $Ca^{2+}$ currents in neuroblastoma X glioma hybrid cells (Hescheler *et al.*, 1987). The effect was

PTX sensitive, and was restored by injection of $G_o$ or $G_i$ proteins. $G_o$ or its $\alpha$-subunit ($\alpha_o$) restored the inhibitory effects. $G_o$ was at least 10 times more effective than $G_i$ protein.

Neuropeptide Y (NPY) and bradykinin (BK) inhibited $Ca^{2+}$ channel currents in DRG cells, and these effects were blocked by PTX (Ewald *et al.*, 1989). Although complete restoration of NPY effects was obtained with $\alpha_o$, a combination of $\alpha_o$ and $\alpha_{i-2}$ was required for full restoration of BK action. Using a neuroblastoma cell line that expresses a mutant PTX-resistant $\alpha$-subunit of $G_{oA}$ (subtype of $G_o$ protein), it was demonstrated that $G_{oA}$ mediates the blocking effects of Leu-enkephalin, and NE but does not transduce the SST block (Taussing *et al.*, 1992). Studies on the coupling processes between specific receptors and G proteins in $GH_3$ cells (Kleuss *et al.*, 1991) also demonstrated functional specificity in the physiological cellular responses to external signals. CCh and SST inhibit $Ca^{2+}$ currents by different $G_o$ subtypes. The $Ca^{2+}$ channel inhibition by CCh (but not SST) is eliminated by injection of antisense DNA to subtype $G_{o1}$ into $GH_3$, and the response to SST (but not CCh) is eliminated by antisense DNA to $G_{o2}$.

These data suggest that several G proteins are involved in receptor-mediated inhibition of neuronal $Ca^{2+}$ channels. Direct G-protein action might be responsible for these effects, but $Ca^{2+}$ currents are also reduced by activators of PK-C such as OAG and phorbol esters (Rane and Dunlap, 1986; Anwyl, 1991), and the G protein could be acting on phosphatidylinositol-specific PK-C rather than directly.

## VII. Summary

G proteins play an essential role in the signal transduction pathway between a wide range of membrane receptors and ion channels. All G proteins share a common heteromeric $\alpha\beta\gamma$-type structure, and the number of components of G-protein-mediated signaling pathways is increasing rapidly. It is now recognized that ion channels comprise a major class of effector. The mechanism by which G proteins regulate ion channels was believed to be indirect through cytoplasmic agents such as cyclic AMP, $IP_3$, or $Ca^{2+}$. Results obtained from combined approaches of biochemistry, molecular biology, and single-channel recordings provide evidence that ion channels are regulated, not only by ligand and voltage, but also by G proteins in a direct (or at least "membrane-delimited") manner. The direct G-protein gating of ionic channels provides a fast pathway to respond to external signals quickly. The best characterized example of direct G-protein gating is the $K_{ACh}$ channels. Reconstitution studies using excised membrane patches demonstrate that $K_{ACh}$ channels are activated by a purified (or recombinant) G protein independently from cytoplasmic agents. Other ion channels that appear to be regulated by G proteins were also briefly summarized in this chapter.

The molecular cloning studies revealed the existence of many receptor subtypes. Likewise, the number of G proteins is rapidly increasing. It is known that various G-protein subtypes can interact with a single effector. Furthermore, a single transmitter or hormone can activate more than one closely related receptor subtype, and a single G protein can modulate several ion channels. Given the complexity of signal-transduction systems, it is necessary to perform experiments with pure component of receptor, G-protein subunit, and ion channel to differentiate functional specificity in the physiological cellular responses to external signals. The development of recombinant model cell systems in which the interactions of a defined receptor subtype with a specific ion channel or a specific G protein can be studied will be needed to answer detailed questions. Alternatively, the development of transgenic model systems that overexpress (or selective knockout) of genes in the G-protein systems may be useful to provide unique insight into a variety of the G-protein regulation of ion channels under physiological conditions.

## Bibliography

Anwyl, R. (1991). Modulation of vertebrate neuronal calcium channel by transmitters. *Brain Res. Rev.* **16,** 265–281.

Birnbaumer, L., Codina, J., Mattera, R., Yatani, A., Scherer, N., Toro, M.-J., and Brown, A. M. (1987). Signal transduction by G proteins. *Kidney Int.* **32,** Suppl. 23, S14–S37.

Birnbaumer, L., Abramowitz, J., Yatani, A., Okabe, K., Mattera, R., Graf, R., Sanford, J., Codina, J., and Brown, A. M. (1990). Roles

of G proteins in coupling of receptors to ionic channels and other effector systems. *In* "Critical Reviews in Biochemistry and Molecular Biology" (G. D. Fasman, Ed.), Vol. 25, pp. 225–244. CRC Press, Boca Raton, FL.

Breitwieser, G. E., and Szabo, G. (1985). Uncoupling of cardiac muscarinic and $\beta$-adrenergic receptors from ion channels by a guanine nucleotide analogue. *Nature* **317,** 538–540.

Cavalié, A., Allen, T. J. A., and Trautwein, W. (1991). Role of the GTP binding protein $G_s$ in the $\beta$-adrenergic modulation of cardiac Ca channels. *Pflugers Arch.* **419,** 433–443.

Clapham, D. E. (1994). Direct G protein activation of ion channels? *Annu. Rev. Neurosci.* **17,** 441–464.

Codina, J., Grenel, D., Yatani, A., Birnbaumer, L., and Brown, A. M. (1987). Hormonal regulation of pituitary $GH_3$ cell $K^+$ channels by $G_k$ is mediated by its $\alpha$-subunit. *FEBS Lett.* **216,** 104–106.

Conklin, B. R., and Bourne, H. R. (1993). Structural elements of $G\alpha$ subunits that interact with $G\beta\gamma$ receptors, and effectors. *Cell* **73,** 631–641.

Dohlman, H. G., Thorner, J., Caron, M. G., and Lefkowitz, R. J. (1991). Model systems for the study of seven-transmembrane-segment receptors. *Annu. Rev. Biochem.* **60,** 653–688.

Dolphin, A. (1991). Regulation of calcium channel activity by GTP binding proteins and second messengers. *Biochem. Biophys. Acta* **1091,** 68–80.

Dolphin, A. C., and Scott, R. H. (1987). Calcium channel currents and their inhibition by (−)-baclofen in rat sensory neurones: Modulation by guanine nucleotides. *J. Physiol. (London)* **386,** 1–17.

Ewald, D. A., Pang, I.-H., Sternweis, P. C., and Miller, R. J. (1989). Differential G protein-mediated coupling of neurotransmitter receptors to $Ca^{2+}$ channels in rat dorsal root ganglion neurons in vitro. *Neuron* **2,** 1185–1193.

Gilman, A. G. (1987). G proteins: Transducers of receptor-generated signals. *Annu. Rev. Biochem.* **56,** 615–649.

Glitsch, H. G., and Pott, L. (1978). Effects of acetylcholine and parasympathetic nerve stimulation on membrane potential in quiescent guinea pig atria. *J. Physiol. (London)* **279,** 655–668.

Hartzell, H. C., and Fischmeister, R. (1992). Direct regulation of cardiac $Ca^{2+}$ channels by G proteins: Neither provennor necessary? *Trends Pharmacol. Sci.* **13,** 380–385.

Hescheler, J., Rosenthal, W., Trautwein, W., and Schultz, G. (1987). The GTP-binding protein, $G_o$, regulates neuronal calcium channels. *Nature* **325,** 445–447.

Hille, B. (1992). G protein-coupled mechanisms and nervous signalling. *Neuron* **9,** 187–195.

Holtz, G. G. VI, Rane, S. G., and Dunlap, K. (1986). GTP-binding proteins mediate transmitter inhibition of voltage-dependent calcium channels. *Nature* **319,** 670–672.

Huang, C.-L., Slesinger, P. A., Casey, P. J., Jan, Y. N., and Jan, L. Y. (1995). Evidence that direct binding of Gbg to the GIK1 G protein-gated inwardly rectifying $K^+$ channel is important for channel activation. *Neuron* **15,** 1133–1143.

Imoto, Y., Yatani, A., Reeves, J. P., Codina, J., Birnbaumer, L., and Brown, A. M. (1988). The $\alpha$ subunit of $G_s$ directly activates cardiac calcium channels in lipid bilayers. *Am. J. Physiol.* **255,** H722–H728.

Kameyama, M., Hofmann, F., and Trautwein, W. (1985). On the mechanism of beta-adrenergic regulation of the Ca channel in the guinea-pig heart. *Pflugers Arch.* **405,** 285–293.

Kameyama, M., Hescheler, J., Hofmann, F., and Trautwein, W. (1986). Modulation of Ca current during the phosphorylation cycle in the guinea pig heart. *Pflugers Arch.* **407,** 123–128.

Kofuji, P., Davidson, N., and Lester, H. A. (1995). Evidence that neuronal G-protein-gated inwardly rectifying $K^+$ channels are activated by $G\beta\gamma$ subunits and function as heteromultimers. *Proc. Natl. Acad. Sci. USA* **92,** 6542–6546.

Kleuss, C., Hescheler, J., Ewel, C., Rosenthal, W., Schultz, G., and Wittig, B. (1991). Assignment of G protein subtypes to specific receptors inducing inhibition of calcium currents. *Nature* **353**, 43–48.

Krapivinsky, G., Gordon, E. A., Wickman, K., Velimirovic, B., Krapivinsky, L., and Clapham, D. E. (1995). The G-protein-gated atrial $I_{KACh}$ is a heteromultimer of two inwardly rectifying $K^+$ channel proteins. *Nature* **374**, 135–141.

Kubo, Y., Reuveny, E., Slesinger, P. A., Jan, Y. N., and Jan, L. Y. (1993). Primary structure and functional expression of a rat G-protein-coupled muscarinic potassium channel. *Nature* **364**, 802–806.

Kurachi, Y., Nakajima, T., and Sugimoto, T. (1986a). On the mechanism of activation of muscarinic $K^+$ channels by adenosine in isolated atrial cells: Involvement of GTP-binding proteins. *Pflugers Arch.* **407**, 264–274.

Kurachi, V., Nakajima, T., and Surgimoto, T. (1986b). Role of intracellular $Mg^{2+}$ in the activation of muscarinic $K^+$ channel in cardiac atrial cell membrane. *Pflugers Arch.* **407**, 572–574.

Lewis, D. W., Weight, F. F., and Luini, A. (1986). A guanine nucleotide binding protein mediates the inhibition of voltage-dependent calcium current by somatostatin in a pituitary cell line. *Proc. Natl. Acad. Sci. USA* **83**, 9035–9041.

Logothetis, D. E., Kurachi, Y., Galper, J., Neer, E. J., and Clapham, D. E. (1987). The $\beta\gamma$ subunits of GTP-binding proteins activate the muscarinic $K^+$ channel in heart. *Nature* **325**, 321–326.

Logothetis, D. E., Kim, D., Northup, J. K., Neer, E. J., and Clapham, D. E. (1988). Specificity of action of guanine nucleotide-binding regulatory protein subunits on the cardiac muscarinic $K^+$ channels. *Proc. Natl. Acad. Sci. USA* **85**, 5814–5818.

Mattera, R., Yatani, A., Kirsh, G. E., Graf, R., Olate, J., Codina, J., Brown, A. M., and Birnbaumer, L. (1989a). Recombinant $\alpha_{i-3}$ subunit of G protein activates $G_k$-gated $K^+$ channels. *J. Biol. Chem.* **264**, 465–471.

Mattera, R., Graziano, M. P., Yatani, A., Zhou, Z., Graf, R., Codina, J., Birnbaumer, L., Gilman, A. G., and Brown, A. M. (1989b). Splice variants of the $\alpha$ subunit of the G protein, $G_s$ activate both adenylyl cyclase and calcium channels. *Science* **243**, 804–807.

Okabe, K., Yatani, A., Evans, T., Ho, Y.-K., Codina, J., Birnbaumer, L., and Brown, A. M. (1990). $\beta\gamma$ dimers of G proteins inhibit atrial muscarinic $K^+$ channels. *J. Biol. Chem.* **265**, 12854–12858.

Okabe, K., Yatani, A., and Brown, A. M. (1991). The nature and origin of spontaneous noise in G protein-gated ion channels. *J. Gen. Physiol.* **97**, 1279–1293.

Pfaffinger, P. J., Martin, J. M., Hunter, D. D., Nathanson, N. M., and Hille, B. (1985). GTP-binding proteins couple cardiac muscarinic receptors to a K channel. *Nature* **317**, 536–538.

Rane, S. O., and Dunlap, K. (1986). Kinase C activator 1,2-oleylacetylglycerol attenuates voltage-dependent calcium current in sensory neurones. *Proc. Natl. Acad. Sci. USA* **83**, 184–188.

Reisine, T. D. (1984). Cellular mechanisms of regulating adrenocorticotropin release. *J. Recep. Res.* **4**, 291–300.

Robishaw, J. D., Smigel, M. D., and Gilman, A. G. (1986). Molecular basis for two forms of the G protein that stimulates adenylate cyclase. *J. Biol. Chem.* **261**, 9587–9590.

Rosenberg, R. L., Hess, P., Reeves, J. P., Smilowitz, H., and Tsien,

R. W. (1986). Calcium channels in planar lipid bilayers: Insights into mechanisms of ion permeation and gating. *Science* **231**, 1564–1566.

Schlegel, W., Wuarin, F., Zbaren, C., Wolheim, C. B., and Zahnd, G. R. (1985). Pertussis toxin selectively abolishes hormone induced lowering of cytosolic calcium in $GH_3$ cells. *FEBS Lett.* **189**, 27–32.

Schreibmayer, W., Dessauer, C. W., Vorobiov, K., Gilman, A. G., Lester, H. A., Davidson, N., and Dascal, N. (1996). Inhibition of an inwardly rectifying $K^+$ channel by G-protein $\alpha$-subunits. *Nature* **380**, 624–627.

Soejima, M., and Noma, A. (1984). Mode of regulation of the ACh-sensitive K-channel by the muscarinic receptor in rabbit atrial cells. *Pflugers Arch.* **400**, 424–431.

Strader, C. D., Fong, T. M., Tota, M. R., Underwood, D., and Dixon, R. A. F. (1994). Structure and function of G protein-coupled receptors. *Annu. Rev. Biochem.* **63**, 101–132.

Stryer, L. (1986). Cyclic GMP cascade of vision. *Annu. Rev. Neurosci.* **9**, 87–119.

Taussing, R., Sanchez, S., Rifo, M. F., Gilman, A. G., and Belavdetti, F. (1992). Inhibition of the $\omega$-conotoxin-sensitive calcium current by distinct G proteins. *Neuron* **8**, 799–809.

Wang, Y., Townsend, C., and Rosenberg, R. L. (1993). Regulation of cardiac L-type Ca channels in planar lipid bilayers by G-proteins and protein phosphorylation. *Am. J. Physiol.* **264**, C1473–C1479.

Yatani, A., and Brown, A. M. (1989). Rapid $\beta$-adrenergic modulation of cardiac calcium channel currents by a fast G protein pathway. *Science* **245**, 71–74.

Yatani, A., Codina, J., Brown, A. M., and Birnbaumer, L. (1987a). Direct activation of mammalian atrial muscarinic K channels by a human erythrocyte pertussis toxin-sensitive G protein, $G_k$. *Science* **235**, 207–211.

Yatani, A., Codina, J., Sekura, R. D., Birnbaumer, L., and Brown, A. M. (1987b). Reconstitution of somatostatin and muscarinic receptor mediated stimulated of $K^+$ channels by isolated $G_k$ protein in clonal rat anterior pituitary cell membranes. *Mol. Endocrinol.* **1**, 283–289.

Yatani, A., Codina, J., Imoto, Y., Reeves, J. P., Birnbaumer, L., and Brown, A. M. (1987c). Direct regulation of mammalian cardiac calcium channels by a G protein. *Science* **238**, 1288–1292.

Yatani, A., Hamm, H., Codina, J., Mazzoni, M. R., Birnbaumer, L., and Brown, A. M. (1988a). A monoclonal antibody to the alpha subunit of $G_k$ blocks muscarinic activation of atrial $K^+$ channels. *Science* **241**, 828–831.

Yatani, A., Mattera, R., Codina, J., Graf, R., Okabe, K., Padrell, E., Iyengar, R., Brown, A. M., and Birnbaumer, L. (1988b). The G protein-gated atrial $K^+$ channel is stimulated by three distinct $G_{i\alpha}$-subunits. *Nature* **336**, 680–682.

Yatani, A., Imoto, Y., Codina, J., Hamilton, S., Brown, A. M., and Birnbaumer, L. (1988c). The stimulatory G protein of adenylyl cyclase $G_s$, also stimulates dihydropyridine-sensitive $Ca^{2+}$ channels. *J. Biol. Chem.* **263**, 9887–9895.

Yatani, A., Codina, J., and Brown, A. M. (1990). G protein-mediated effects on ionic channels. *In* "G Proteins" (R. Iyenger and L. Birnbaumer, Eds.), pp. 241–266. Academic Press, New York.

Noritsugu Tohse, Hisashi Yokoshiki, and Nicholas Sperelakis

# 34

## Developmental Changes in Ion Channels

## I. Introduction

Cellular functions and tissue structures change dramatically during development. Ion channels are responsible for cellular signaling and maintenance of the intracellular environment. For example, the $Ca^{2+}$ channels allow $Ca^{2+}$ influx into the cell, and the $Ca^{2+}$ acts as a second messenger that affects several structures: activation of enzymes, activation of some ion channels, and activation of the contractile proteins. The ion channels change during both the embryonic/fetal period and the neonatal period. These developmental changes include changes in the types of, number of, and kinetic properties of the ion channels. The developmental changes of ion channels are clearly observed in excitable cells (i.e., cardiomyocytes, skeletal muscle fibers, neurons) because their resting potential and action potential are greatly altered progressively during the developmental stages. For example, resting potential increases in amplitude during development, and large changes occur in the action potential rate of rise, overshoot, and duration. In general, the rate of rise increases markedly, the overshoot increases, and the duration decreases during development. This chapter focuses primarily on the ion channels of cardiomyocytes, skeletal muscle fibers, and neurons where most is known about the developmental changes.

## II. Cardiomyocytes

### A. Resting Potential

In the early embryonic period, the heart is tubular in shape. In the heart tube, there is no distinction yet between atria and ventricles. In the middle embryonic period, the heart tube twists and the cardiac loop is constructed. In this period, the ventricular portion becomes distinguished from the atrial portion.

The electrophysiological properties are also altered during development. The resting potential (RP) of the ventric-

ular cells in the early embryonic/fetal period is low (e.g., $-40$ to $-50$ mV), and there is a gradual hyperpolarization during development. Finally, in the late embryonic period, the RP becomes nearly the level of adult cells (around $-80$ mV). During the hyperpolarization of the RP, a decrease in the permeability ratio for $Na^+$ and $K^+$ ($P_{Na}/P_K$ ratio) was observed (Sperelakis and Shigenobu, 1972; Sperelakis and Haddad, 1995). However, the developmental changes in the RP cannot be accounted for by changes in the intracellular ion concentrations because $[K^+]_i$ is already high in the early embryonic period (Sperelakis and Shigenobu, 1972; Sperelakis and Haddad, 1995). Although the Na,K-ATPase pump specific activity was found to be low in the early embryonic period (Sperelakis and Lee, 1971), the Na–K pump is sufficient to maintain a high $[K^+]_i$ and low $[Na^+]_i$ because of the less leaky membrane (i.e., high in resistance) (Sperelakis and Shigenobu, 1972). Therefore, the developmental change in the RP is due to changes in membrane permeability (conductance) of the ions.

In the early embryonic period, the low RP of the ventricular portion is not stable, but exhibits a spontaneous depolarization, the pacemaker potential (phase 4 diastolic depolarization). The maximum diastolic potential increases (hyperpolarized) and the slope of the pacemaker potential progressively decreases during embryonic development. When the RP has attained the adult level in the late embryonic period, the pacemaker potential disappears. Thus, automaticity of the ventricular cells is lost by the middle embryonic period. Possible factors in the loss of automaticity are the decrease in the $P_{Na}/P_K$ ratio and the resultant hyperpolarization. These factors are closely related to the increase in the inward rectifier $K^+$ current ($I_{K(IR)}$) and the loss of the hyperpolarization-activated inward current ($I_h$ or $I_f$) (see discussion later).

### B. Action Potential

The action potentials (APs) get larger and rise faster during embryonic development. These changes are caused

**518**

by the hyperpolarization of the RP in the middle embryonic period and by an increase in overshoot to about $+30$ mV. The maximal rate of depolarization (max $dV/dt$) progressively increases during development, from about 20 to about 200 V/sec in the late embryonic stage, which is about the adult level. However, the time course of the increase in max $dV/dt$ was not parallel to the increase in RP. The increase in RP preceded the increase in max $dV/dt$ by several days. Therefore, this increase in max $dV/dt$ is not simply due to the hyperpolarization, but is produced by a much greater number (density) of tetrodotoxin (TTX)-sensitive fast $Na^+$ channels (see discussion later).

The duration of the AP (e.g., at 50% repolarization, $APD_{50}$) was hardly changed in the chick during development. The same is true of the guinea-pig heart as well as many other mammalian species. However, in human atrial cells, the $APD_{50}$ is significantly shortened (e.g., about 20%). In addition, in the rat, there is a marked decrease in $APD_{50}$, beginning in the late fetal period and extending through the first 3 weeks of the neonatal period, after which the adult-like brief APs are attained (Kojima *et al.*, 1990). Several factors contribute to this marked abbreviation of the AP in the rat, including increase in the transient outward current ($I_{to}$) and loss of the sustained component of fast $Na^+$ current (discussed later).

## C. Na⁺ Channels

Slow $Ca^{2+}$ channel current makes a major contribution to the upstroke of the AP in the early embryonic period. The TTX-sensitive fast $Na^+$ current in ventricular cells increases markedly during development by a factor of about 10-fold in both chick and rat hearts (Fujii *et al.*, 1988; Sada *et al.*, 1988, 1995; Conforti *et al.*, 1993). Saxitoxin (STX, a specific blocker of the fast $Na^+$ channels) binding reveals a marked increase in density of the fast $Na^+$ channel protein during development of embryonic chick hearts (Renaud *et al.*, 1981). This increase in number (density) of fast $Na^+$ channels accounts for the large increase in max $dV/dt$ of the AP that occurs during development. Therefore, the contribution of fast $Na^+$ channel to the AP is progressively increased during development.

The heart greatly enlarges in size during development. Thus, the excitation wave must travel over longer distances in the larger hearts during the late embryonic period and adult. A fast propagation velocity of excitation is required to allow a synchronized contraction of ventricle, that is, to allow the heart to serve as an effective pump. The increase in max $dV/dt$ during development would contribute to the required increase in propagation velocity. Another factor involved in propagation velocity is that the cell size (i.e., diameter) becomes much greater. It is well known that propagation velocity is a function of the square root of cell diameter (see Chapter 23).

The TTX sensitivity of the fast $Na^+$ channels in avian cardiomyocytes (in the nanomolar range) was about 1000-fold greater than that for adult mammalian hearts (such as guinea pig), which are in the micromolar range (Sada *et al.*, unpublished observation). This finding is in agreement with previous reports of high sensitivity to TTX of embry-

onic chick hearts (Iijima and Pappano, 1979; Marcus and Fozzard, 1981; Fujii *et al.*, 1988). The fast $Na^+$ channels are completely blocked by 10 $\mu M$ TTX in fetal rat cardiomyocytes (Conforti *et al.*, 1993), and by 30 $\mu M$ TTX in adult rat cardiomyocytes (Brown *et al.*, 1981). It is not clear whether the high TTX sensitivity of chick embryonic hearts is due to a different isoform of the channel.

The TTX-sensitive fast $Na^+$ current has a slow inactivating or sustained component. This component is small, but gives a relatively larger contribution in the embryonic period than in adult (Conforti *et al.*, 1993). Sustained $Na^+$ current, which is blocked by TTX, is observed in the early embryonic period of chick. Reopening of some of the fast $Na^+$ channels during a long depolarizing clamp step is one explanation for the small sustained component (Josephson and Sperelakis, 1989). The sustained component may reflect the "window current" produced by a balance between the activating (m) gate and the inactivating (h) gate (Sada *et al.*, 1995). (Window current is discussed in Chapter 24.) In rat heart cells, the fast $Na^+$ current has a slow inactivating component, and the time constant of the slow component decreased in the neonatal cells compared to the fetal cells (Conforti *et al.*, 1993). Although the slow component of the $Na^+$ current is small, inward current produced by the slow component helps to maintain the longer duration of the AP plateau in the fetal period. TTX, which does not affect types of ion channels other than the fast $Na^+$ channels, shortens the AP duration in rat fetal cardiomyocytes (Fig. 1) (Conforti *et al.*, 1993). A key factor that may contribute to the shortening of the AP duration during development of rat heart is the loss of the slow component of $Na^+$ current. However, in adult heart, it appears that the sustained component of $Na^+$ current persists in the Purkinje fiber, because the AP plateau is substantially shortened by TTX. Another factor responsible is an increase in $I_{to}$ carried primarily by $K^+$ (see discussion later).

## D. Ca²⁺ Channels

Changes in the slow (L-type) $Ca^{2+}$ current also occur during development of the heart. However, the direction of the change is opposite in avian versus mammalian hearts (Fig. 2). In rat heart, the L-type $Ca^{2+}$ current increases during development (Masuda *et al.*, 1995), whereas in chick heart it actually decreases (Tohse *et al.*, 1992b). In early embryonic period of chick, the current density of the L-type channels is 8 $\mu A/cm^2$, which is comparable to that in other adult animals (about 10 $\mu A/cm^2$). The current density decreases during development to about 5 $\mu A/cm^2$ in the late embryonic period (Fig. 2B). However, current density of the L-type $Ca^{2+}$ channel of rat cardiomyocytes increases through the middle fetal, late fetal, and neonatal period (Masuda *et al.*, 1995) (Fig. 2A). Another investigation demonstrated that the current density in the neonatal period is actually larger than that in adult rat (Cohen and Lederer, 1988). That is, in development of rat heart, there is an increase of the current density, followed by a decrease. In contrast, in rabbit (Osaka and Joyner, 1991) and guinea-pig (Kato *et al.*, 1996) cardiomyocytes, the current density in the neonatal period is smaller than that in adult. Thus,

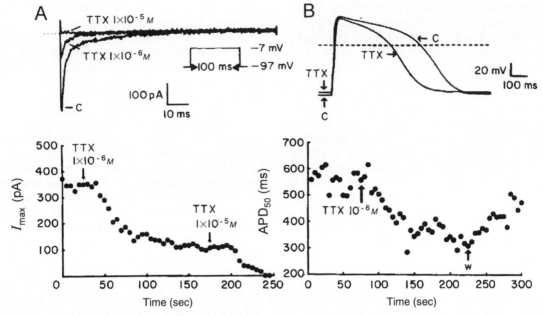

**FIG. 1.** Effect of TTX on the $Na^+$ current and the AP configuration recorded from a fetal rat cardiomyocyte. (A) Top panel: Superimposed current traces showing the $Na^+$ current recorded before (C: Ctrl) and after a 90-sec exposure to 1 and 10 $\mu M$ TTX. Holding potential (HP) was $-97$ mV, and the test potential was $-7$ mV. Bottom panel: Time course of the TTX effect; steady-state responses were attained at about 1–2 min. (B) Effect of TTX on the AP configuration of a fetal cell recorded in current-clamp mode. Top panel: Superimposed traces (averaged from 10 consecutive records) showing APs before (C: Ctrl) and after a 90-sec exposure to 1 $\mu M$ TTX. Bottom panel: Time course of the change in $APD_{50}$ produced by TTX. Arrows indicate points of introduction and washout (w) of TTX. (Reproduced with permission from Conforti *et al.*, 1993.)

the changes in the L-type $Ca^{2+}$ channel density that occur during development are complex and vary from one species to another.

Other types of $Ca^{2+}$ channels are also observed during development. In chick embryonic heart cells, there is a

report showing that the T-type channel is dominant in the early embryonic period, but that the L-type current is dominant in the late embryonic period (Kawano and DeHaan, 1991). However, other reports indicate that the L-type current is also dominant in the early embryonic

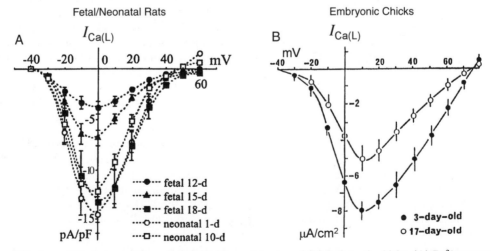

**FIG. 2.** Developmental changes of $I_{Ca(L)}$ in (A) fetal/neonatal rats and (B) embryonic chicks. (A) $Ba^{2+}$ currents through L-type $Ca^{2+}$ channels ($I_{Ba(L)}$) were elicited by depolarizing steps from a HP of $-40$ mV (22°C). Current–voltage curves [normalized as current density (in pA/pF) (mean $\pm$ SE)] shown for the different developmental stages (from day 12 fetal to day 10 neonatal). (B) Changes in the density of $I_{Ca(L)}$ in isolated embryonic chick heart cells; 1.8 m$M$ [$Ca^{2+}$]$_o$, 35°C. [Modified with permission from (panel A) Masuda *et al.*, 1995. Long openings of calcium channels in fetal rat ventricular cardiomyocytes. *Pfluger Arch.* **429,** 595–597, copyright Springer-Verlag, and (panel B) Tohse *et al.*, 1992b. Developmental changes in long-opening behavior of L-type $Ca^{2+}$ channels in embryonic chick heart cells. *Circ. Res.* **71,** 376–384.]

period (Tohse *et al.*, 1992a,b). In rat fetal cardiomyocytes, it was found that a substantial fraction of the total $Ca^{2+}$ current is resistant to nifedipine (a relatively selective blocker of L-type $Ca^{2+}$ channels) and other $Ca^{2+}$ channel blockers (Fig. 3) (Tohse *et al.*, 1992a). This nifedipine-resistant current is not blocked by an N-type $Ca^{2+}$ channel blocker, ω-conotoxin, and is only partially inhibited by 30 μM $Ni^{2+}$, which is a blocker of T-type $Ca^{2+}$ channels. Therefore, this channel is called a *fetal-type* (F-type) $Ca^{2+}$ channel. The F-type $Ca^{2+}$ current is absent in adult heart cells. That is, in the fetal period, the total $Ca^{2+}$ current has two main components: the L-type current, which is blocked by nifedipine, and the F-type current, which is not blocked by nifedipine.

In chick embryonic heart, unit conductance of the L-type $Ca^{2+}$ channel is 26 pS (using 50 mM $Ba^{2+}$ in the pipette), which is comparable to that in adult heart cells (Tohse and Sperelakis, 1990). The single-channel activity of the L-type $Ca^{2+}$ channel in the embryonic cells was completely blocked by nifedipine.

The kinetics of opening of the L-type $Ca^{2+}$ channels in embryonic heart cells is different from that in adult heart cells. Long-lasting openings of the channels occur relatively frequently, in addition to the more usual brief-bursting openings as observed in adult heart cells (Fig. 4). These long-lasting openings are similar to the mode 2 openings produced by $Ca^{2+}$ agonists, such as the dihydropyridine Bay-K-8644. For example, in Fig. 4, one can see long openings that persist over the entire duration of the clamp pulse (i.e., 300 msec); the long openings are sometimes punctuated by brief closures. As can be seen, in many sweeps, the open probability ($P_o$) is close to 1.00. The long-lasting openings gradually disappear during development (Tohse *et al.*, 1992b; Masuda *et al.*, 1995).

### E. Inward-Rectifier $K^+$ Channels

It was demonstrated that the inward-rectifier $K^+$ current ($I_{K(IR)}$) of ventricular cells increase markedly during development in both embryonic chick (Josephson and Sperelakis, 1990) and fetal/neonatal rat hearts (Masuda and Sperelakis, 1993). The increase in $I_{K(IR)}$ is likely to be a major factor responsible for the increase in RP (hyperpolarization) that occurs during development, concomitant with a decrease in membrane resistivity and in membrane time constant ($\tau_m = R_m C_m$). The increase in $I_{K(IR)}$ channels can also account for the decrease in the $P_{Na}/P_K$ ratio that occurs during development (Sperelakis and Shigenobu, 1972); namely, it results in an increase in the $K^+$ permeability ($P_K$). Similar change in the $P_{Na}/P_K$ ratio during development is shown in the skeletal muscle fibers (see discussion later).

The increase in $I_{K(IR)}$ during development may be due to two factors: (1) an increase in the number of channel molecules and (2) an increase in single-channel conductance (Masuda and Sperelakis, 1993). The single-channel conductance in young fetal rat is much less than that in old fetus and neonate (Fig. 5). However, the mean open time of the channel is longer in young fetal cells than in old fetal and neonatal cells. These observations suggest that the structure (i.e., a different isoform) of the inward-rectifier $K^+$ channel changes dramatically during development.

### F. Voltage-Gated $K^+$ Channels

Transient outward current ($I_{to}$) density, mainly carried by $K^+$ ion, has been reported to increase during development. A substantial amount of transient outward current was observed in early embryonic chick heart cells (Satoh

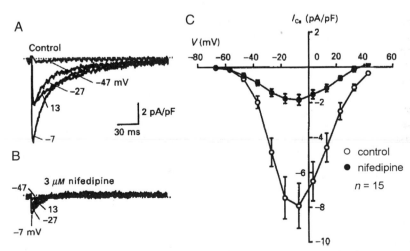

**FIG. 3.** Presence of fetal-type $Ca^{2+}$ channels in fetal (18-day) rat cardiomyocytes. (A) Currents elicited by 300 msec (only 150 msec shown) depolarizing pulses to −47, −27, −7, and 13 mV from a HP of −87 mV. (B) In the presence of 3 μM nifedipine, a significant inward current remained at each potential. (C) Current–voltage relationship; data points given as mean ± SE. Nifedipine did not completely block the $Ca^{2+}$ current, indicating the presence of a nifedipine-resistant $Ca^{2+}$ current. (Reproduced with permission from Tohse *et al.*, 1992a.)

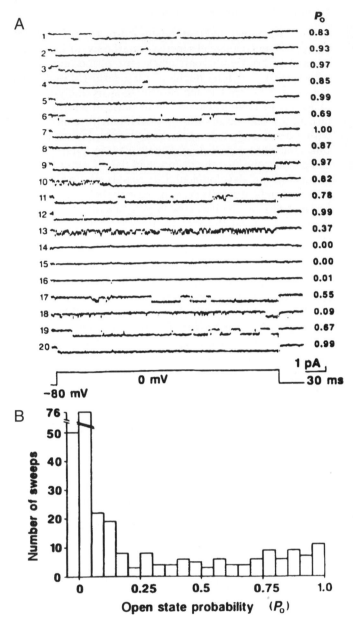

**FIG. 4.** Presence of long openings of the slow (L-type) Ca$^{2+}$ channels in young embryonic (3-day) chick heart cell. (A) Single-channel activity elicited by consecutive command pulses to 0 mV (from a HP of −80 mV) every 2 sec. Sweep-to-sweep variations of the probability of the channel opening ($P_o$) are given in the right-hand column. (B) A histogram of $P_o$ data from nine cells (30 sweeps each). Note that many sweeps showed long openings and high $P_o$. (Reproduced with permission from Tohse and Sperelakis, 1990.)

and Sperelakis, 1994). In neonatal rat ventricular cells, the density of $I_{to}$ was reported to increase about fourfold between day 1 and day 10 (Kilborn and Fedida, 1990). This increase in $I_{to}$ contributes to the abbreviation of the AP that occurs in the neonatal period. Other reports also demonstrate that the current density in adults is markedly larger than that in the neonatal period for both rat (Wahler *et al.*, 1994) and canine ventricular cells (Jeck and Boyden, 1992).

Recently, several genes coding for cardiac K$^+$ channels have been found using the techniques of molecular biology. The subunits Kv1.2 and Kv1.4, cloned from cardiomyocytes, probably contribute to formation of the $I_{to}$ channel. Subunits encoded by these two genes may bind together to make a heteromultimer for the $I_{to}$ channel (Po *et al.*, 1993). The level of Kv1.2 mRNA expression in adult rats was significantly higher than that in neonatal rats (Nakamura and Iijima, 1994). Therefore, the increase in $I_{to}$ that occurs during the postnatal period may be due to the increase in expression of the ion channel genes.[1]

The developmental change of the delayed-rectifier K$^+$ channels has been examined in guinea-pig ventricular myocytes (Kato *et al.*, 1996). The density of delayed-rectifier K$^+$ current ($I_K$) is larger in neonatal cells than in fetal cells. In the postnatal period, the density of $I_K$ exhibited no further change. No substantial changes in the kinetics and voltage dependency of $I_K$ are observed during development.

### G. Hyperpolarization-Activated Inward Current

The hyperpolarization-activated inward current ($I_h$ or $I_f$), which is mainly carried by Na$^+$ and K$^+$ ions, is observed in early chick embryonic cardiomyocytes (Satoh and Sperelakis, 1993). This current progressively decreases during development and essentially disappears in the late embryonic period (Fig. 6).

The $I_h$ is called the *pacemaker current* in adult cardiomyocytes. In Purkinje fibers, $I_h$ plays a key role for pacemaker depolarization during the diastolic phase. In sinoatrial node cells, contribution of $I_h$ to pacemaker potential is still controversial (Irisawa *et al.*, 1993). This is because that the time course of activation of $I_h$ is too slow to account for the high frequency of the pacemaker, and that the threshold potential for activation of $I_h$ (close to −70 mV) is beyond the maximum diastolic potential (−60 to −70 mV) for the nodal cells. In chick embryonic cardiomyocytes, although the time course of decrease in $I_h$ parallels the disappearance of the pacemaker potential, the contribution of $I_h$ to the pacemaking may still be small (Satoh and Sperelakis, 1993).

### H. Excitation–Contraction Coupling

Changes in the excitation–contraction coupling process also occur during development of the heart. Especially, the source of Ca$^{2+}$ for producing contraction is altered during development (Fig. 7A) (Nakanishi *et al.*, 1988). In the fetal heart cells, the role of the SR is minimal, so that most of the Ca$^{2+}$ required for contraction is derived from Ca$^{2+}$ influx through the voltage-dependent Ca$^{2+}$ channels (L-type and T-type) (i.e., originates from the extracellular

---

[1] It was demonstrated by immunofluorescence that the Kv1.5 channel protein is more strongly localized in the intercalated disk membrane than in the cell surface membrane (Mays *et al.*, 1995) (See Appendix 1 of Chapter 23). Similarly, using immunofluorescence and antibody to fast Na$^+$ channels, there was more intense localization at the intercalated disk membrane than at the cell surface membrane. Therefore, junctional membranes of myocardial cells are likely to be highly excitable membranes.

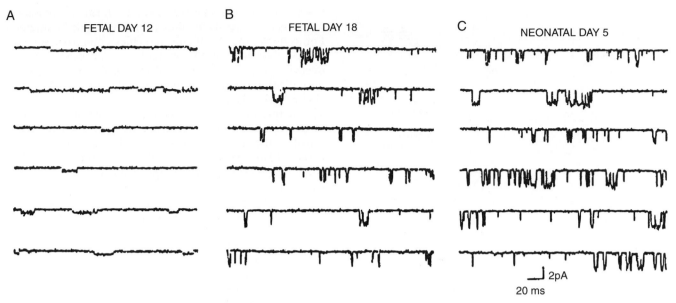

**FIG. 5.** Developmental increase in $I_{K(IR)}$ in maturing rats. Single-channel activities illustrated for (A) 12-day fetal, (B) 18-day fetal, and (C) 5-day neonatal ventricular myocytes. Single-channel activities were recorded at −80 mV. (Reproduced with permission from Masuda and Sperelakis, 1993.)

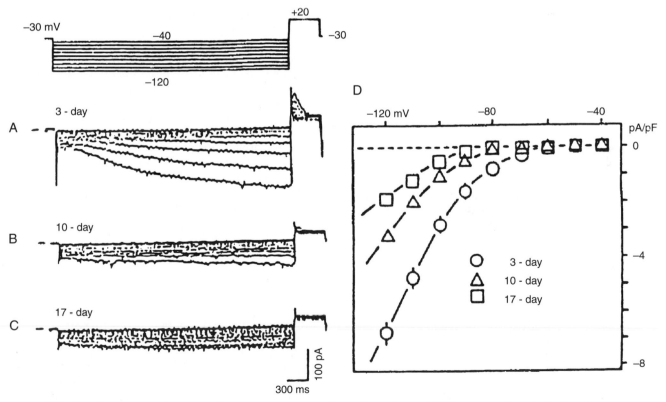

**FIG. 6.** Developmental changes of the hyperpolarization-activated inward current ($I_h$) in young embryonic chick ventricular cells. Test pulses were applied between −40 and −120 mV, in 10-mV increments, from a HP of −30 mV. (A) A large inward current was slowly activated by hyperpolarization in a 3-day-old cell. (B) Smaller $I_h$ in a 10-day-old cell. (C) Further reduced $I_h$ in a 17-day-old cell. (D) Current–voltage relations for $I_h$ current density at the three developmental stages (mean ± SE). (Reproduced from *Eur. J. Pharmacol.,* **240,** H. Satoh and N. Sperelakis, Hyperpolarization-activated inward current embryonic chick cardiac myocytes: Developmental changes and modulation by isoproterenol and carbachol, pp. 283–290, Copyright 1993 with kind permission of Elsevier Science–NL, Sara Burgerhartstraat 25, 1055 KV Amsterdam, The Netherlands.)

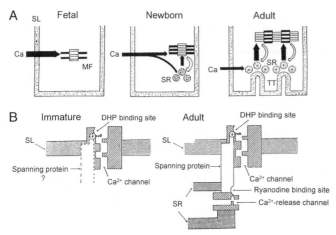

**FIG. 7.** Schematic diagrams showing developmental changes in the excitation–contraction (E-C) coupling. (A) In the early fetus (left), $Ca^{2+}$ entering the cell across the sarcolemma (SL) directly activates the myofilaments (MF). In the newborn (middle), $Ca^{2+}$ influx across the SL induces some $Ca^{2+}$ release from the sarcoplasmic reticulum (SR), which then contributes to the activation of the MFs. In the adult (right), the $Ca^{2+}$ released from SR is greater and more important. TT, T-tubule. (B) Model of E-C coupling mechanisms in immature and adult heart cells. In the immature cells (left), the protein linking the $Ca^{2+}$ channel in the SL with the $Ca^{2+}$-release channel in the SR is absent. In the adult cells (right), the spanning protein is present, allowing $Ca^{2+}$ release from the SR. The $Ca^{2+}$ channel in the SL contains the dihydropyridine (DHP) binding site. The sarcolemmal sensor responsible for intramembrane charge movement is shown as a plus ($+$) sign. [Reproduced in modified form, with permission, from (panel A) Nakanishi *et al.*, 1988, and (panel B) Cohen and Lederer, 1988.]

space). In the neonatal heart cells, the SR matures, and plays a main role as the source of $Ca^{2+}$ for contraction. Therefore, the $Ca^{2+}$-induced $Ca^{2+}$ release from the SR compartment (Fabiato and Fabiato, 1978) becomes the more important system for contraction. That is, in adult heart cells, most of the $Ca^{2+}$ for contraction comes from the internal SR stores. However, $Ca^{2+}$ influx through the sarcolemma is still the determining factor for contractile force, because the $Ca^{2+}$ influx controls the amount of $Ca^{2+}$ released.

The spanning protein (foot protein) couples the L-type $Ca^{2+}$ channel in the T-tubule wall membrane to the $Ca^{2+}$-release channel in the SR membrane. This protein has been reported to be absent in the immature fetal heart (Fig. 7B) (Cohen and Lederer, 1988). If so, then the release of $Ca^{2+}$ is prevented from any SR that may be present. Thus, $Ca^{2+}$ influx would be the main source of $Ca^{2+}$ for contraction in the immature heart.

## III. Skeletal Muscle Fibers

### A. Resting Potential and Action Potential

In skeletal muscle fibers, RP also increases during development and differentiation, from the individual myoblast stage, to the multinucleated myotube stage, to the mature fiber. (The myoblasts fuse end to end to form the myotubes, which get progressively larger in length and diameter.) For example, the RP of cultured chick embryonic skeletal myocytes is low in the immature stage (short mononucleated myoblasts) and is dramatically hyperpolarized during differentiation (to long myotubes or mature fibers) (Fig. 8A) (Fischbach *et al.*, 1971; Spector and Prives, 1977). The mature fibers in older cultures consist of multinucleated myotubes with cross-striations (i.e., aligned myofibrils). In the immature myoblasts, the RP generally was about $-40$ mV. However, the RP of the maturing myotubes is about $-60$ mV, which is approximately equal to that from adult skeletal muscles bathed in the same culture medium (Fig. 8A). A similar change in RP occurs in rat skeletal myoblasts/myotubes during development (Ritchie and Fambrough, 1975). A progressive decrease in the $P_{Na}/P_K$ ratio, which sets the RP closer to the equilibrium potential for $K^+$ ($E_K$), accounts for the developmental changes in RP (Fig. 8B). The family of curves shown in Fig. 8B is similar to that reported in developing chick heart (Sperelakis and Shigenobu, 1972).

In human skeletal muscle (from biopsy), the RP increases (hyperpolarization) during culture (Iannaccone *et al.*, 1987). Fetal myocytes exhibit RPs of about $-35$ mV in early period of culture, and then hyperpolarize to about $-50$ mV in later culture. Therefore, the RP of human myocytes increases during development *in vitro*.

The AP configuration of skeletal muscle fibers also changes during development and differentiation. For example, in one study of chick embryonic myotubes, prolonged APs that persist for more than 500 msec, with prolonged contractions, were exhibited on day 5 in culture (Spector and Prives, 1977). By culture day 7, the myotubes exhibited primarily brief APs (less than 10 msec) and brief twitch contractions. (On day 5, small regions of the myotubes sometimes displayed brief APs with localized twitches in those regions.) Thus, the brief AP becomes more dominant during development that proceeds in culture.

A second study on cultured chick skeletal myoblasts/myotubes is depicted in Fig. 9. As can be seen, at the early stage, AP is small and does not overshoot (the zero membrane potential level) (Fig. 9A, upper record). In the myotubes formed by fusion of myoblasts after a few more days in culture (days 7 and 11), the amplitude and maximum rate of the spikelike APs increased markedly during development while in culture (Fig. 9A, middle and lower records, and Fig. 9B). A plot of max $dV/dt$ versus the number of days in culture is given in Fig. 9B. Since the APs were blocked by TTX in almost all myotubes tested (Fig. 9A, right side, and Fig. 9B), differentiation of the spikelike APs is due to progressively increased intensity of inward current through fast $Na^+$ channels (Kano and Yamamoto, 1977). That is, an increase in the number (density) of fast $Na^+$ channels allows a regenerative AP to be produced, and their maximum rate of rise to become faster and faster.

In myocytes of a marine tunicate (ascidian), AP duration abbreviates during the embryonic period (Greaves *et al.*, 1996). There is a progressive decrease in AP duration from the middle embryonic to the late embryonic period, and

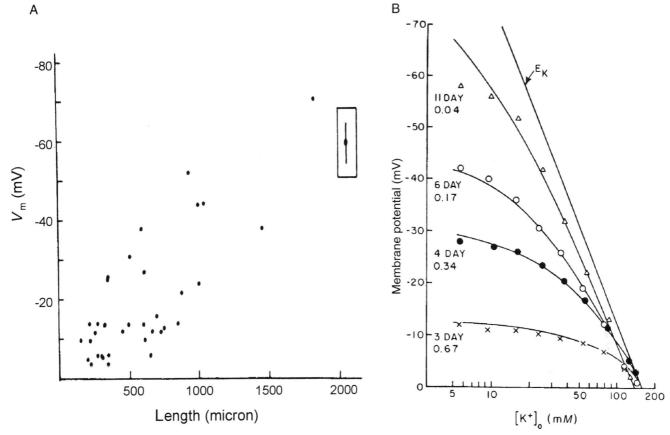

**FIG. 8.** Developmental changes of resting potential in the skeletal myoblasts/myotubes from (A) embryonic chick and (B) fetal rat. (A) The relation between RP and length of the chick skeletal myoblasts/myotubes in culture. The filled circle enclosed within the box is the mean (±2 SE) resting potential of 20 myotubes. (B) The relationship between RP and external K$^+$ ion concentration ([K$^+$]$_o$) for rat skeletal myotubes of different periods in culture. The Nernst equilibrium potential for K$^+$ is indicated by the straight line labeled E$_K$. The solid line for each myotube is the theoretical curve predicted by the Goldman equation for the $P_{Na}/P_K$ ratios given at the left and the extrapolated [K$^+$]$_i$ value. [Reproduced with permission from (panel A) Fischbach *et al.*, 1971. *J. Cell. Physiol.* **78**, 289–300. Copyright 1971 John Wiley & Sons, Inc. Reprinted by permission of Wiley-Liss, Inc., a subsidiary of John Wiley and Sons, Inc., and (panel B) Ritchie and Fambrough, 1975; Reproduced from *The Journal of General Physiology* by copyright permission of The Rockefeller University Press.]

a corresponding increase in the rate of rise and fall of the AP. Spontaneous firing of APs is observed in most ascidian myocytes in the middle embryonic period. The automaticity progressively disappears during development.

### B. Inward-Rectifier K$^+$ Channels

The RP of maturing myotubes gets closer to $E_K$ because the $P_{Na}/P_K$ ratio is gradually decreased during development (see Fig. 8B). These characteristics are also observed in chick embryonic cardiomyocytes during development (Sperelakis and Shigenobu, 1972) and are produced by marked expression of the inward-rectifier K$^+$ channels in the surface membrane of the myocytes. Therefore, it seems likely that the hyperpolarization of the RP of skeletal myocytes in culture is produced by a similar developmental change of inward-rectifier K$^+$ channels.

The $I_{K(IR)}$ is present in skeletal muscle cells from early embryonic amphibian (Linsdell and Moody, 1995). Although there is a brief period (approximately 4 hr) during

which its density decreases, the overall trend is an increase during development.

In ascidian myocytes, $I_{K(IR)}$ exhibited dramatic changes during development (Fig. 10C) (Greaves *et al.*, 1996). In the ascidian, the inward-rectifier K$^+$ current is gained after fertilization of the egg. (The same change also occurs in other species.) When gastrulation ends (at 16 hr after fertilization), the current density suddenly decreases from 4 to 0.5 pA/pF. After the tailbud stage (22 hr after fertilization), the current density progressively increases again and reaches a value of 5 pA/pF before hatching. Because $I_{K(IR)}$ is one of the most important resting conductances (which stabilizes and helps to set the RP), this transient decrease and subsequent increase in the current density parallels the generation of spontaneous APs.

### C. Ca$^{2+}$ Channels and Na$^+$ Channels

High-voltage-activated Ca$^{2+}$ channels have also been observed in ascidian embryo (Fig. 10A) (Greaves *et al.*, 1996).

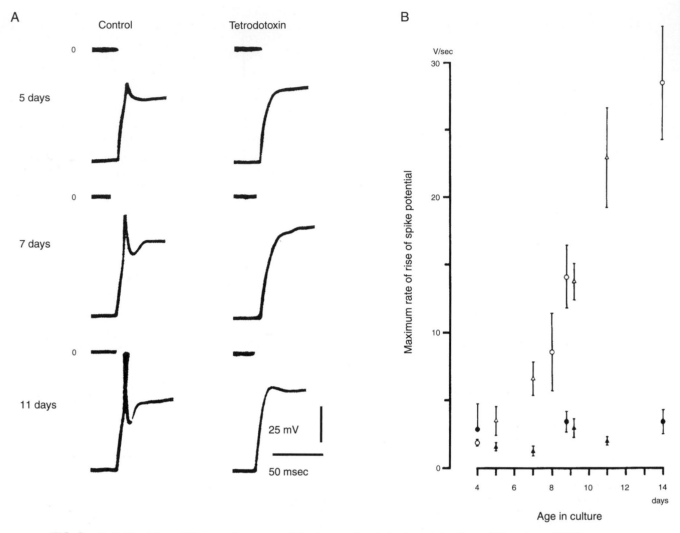

**FIG. 9.** Spikelike APs and their maximum rate of rise in maturing skeletal myotubes from chick embryo. (A) Responses of three myotubes at different ages in culture [5 days (upper), 7 days (middle), and 11 days (lower)]. Each pair of records is taken from the same myotube before (left) and after (right) application of TTX ($10^{-7}$ $M$). Depolarizing current pulses were applied after the membrane potential was hyperpolarized to a standard level of $-80$ mV. The zero potential level is indicated. (B) Maximum rate of rise of spike potentials as a function of the age in culture (mean $\pm$ SE). Circles and triangles represent two different batches of cultures; filled symbols indicate presence of TTX. (From Kano and Yamamoto, 1977. Development of spike potentials in skeletal muscle cells differentiated in vitro from chick embryo. *J. Cell. Physiol.* **90,** 439–444. Copyright © 1977 *Journal of Cellular Physiology*. Reprinted by permission of Wiley-Liss, Inc., a subsidiary of John Wiley & Sons, Inc.)

These $Ca^{2+}$ channels exhibit inactivation, and may be an N-type $Ca^{2+}$ channel because the current was blocked by conotoxin. The channels increased after the neurula stage (16 hr after fertilization). After the tailbud stage, sustained $Ca^{2+}$ channel activity (probably L-type) began to increase, and was dominant at time of hatching. Low-voltage-activating, rapidly inactivating $Ca^{2+}$ channels (T-type) are detected in about 50% of the cells at each stage of development. The contribution of the T-type channels to total $Ca^{2+}$ influx is relatively small in comparison with the L-type and N-type channels.

In early embryonic skeletal myocytes of amphibian, substantial $Ca^{2+}$ currents and $Na^+$ currents appeared almost at the same time (after 10–12 hr in culture) (Linsdell and Moody, 1995). These channel currents continued to increase steadily over an observation period of 10–28 hr in culture.

### D. Delayed-Rectifier K⁺ Channels

The delayed-rectifier $K^+$ current ($I_K$) in skeletal myocytes of early embryonic amphibian progressively increases during development in culture (Linsdell and Moody, 1995). A similar increase in $I_K$ occurs in ascidian myocytes after the neurula stage (16 hr postfertilization) (Fig. 10B) (Greaves *et al.,* 1996). However, $I_{K(Ca)}$ progressively in-

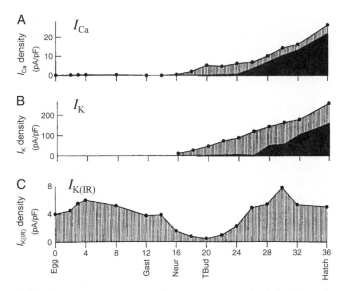

**FIG. 10.** Development of $Ca^{2+}$ and $K^+$ currents in skeletal muscle of a marine tunicate (ascidian). The plots start at fertilization (0 hr). (A) Total $Ca^{2+}$ current density (filled circles), with inactivating (hatched) and sustained (solid) components (B) Total $K^+$ current density (filled circles), with voltage-dependent (hatched) and $Ca^{2+}$-dependent (solid) components. (C) Inward-rectifier $K^+$ current density. Gast, Neur, and TBud indicate the stages of gastrula, neurula, and tailbud, respectively. (Modified with permission from Greaves *et al.*, 1996.)

creased after 26 hr postfertilization. The increase in $I_{K(Ca)}$ closely parallels the increase in sustained L-type $Ca^{2+}$ current (Figs. 10A and B). These developmental changes in $K^+$ channels contribute to the abbreviation of AP duration during the late developmental period.

### E. Acetylcholine Receptor/Channel

The nicotinic acetylcholine receptor/channel (nAChR) is essential to transmission at the neuromuscular junction (see Chapter 40). During development, the nAChR channels in embryonic muscles are converted to adult-type nAChR channels around the time of birth (Mishina *et al.*, 1986). The fetal nAChR channel is composed of $\alpha$-, $\beta$-, $\gamma$-, and $\delta$-subunits, and in the adult channel, the $\gamma$-subunit is substituted by an $\varepsilon$-subunit. In functional characteristics, the fetal channel exhibits a low conductance and long openings, compared with those of the adult channel. This conversion of the nAChR channel may be related to innervation of the muscles that occurs during development, because the fetal type of nAChR channel has also been observed in denervated muscles and the extrajunctional nAChR channel of fetal muscles.

## IV. Neurons

### A. Action Potential

The ionic dependence of the neuronal AP is altered during the early stages of embryonic development (Spitzer

and Baccaglini, 1976; Spitzer *et al.*, 1994). Initially, the AP exhibits a prominent $Ca^{2+}$ dependence (i.e., $Ca^{2+}$-dependent AP) and its duration is prolonged. Later in development, the AP duration becomes brief, and most of the inward current during the depolarizing phase is carried by $Na^+$ (i.e., $Na^+$-dependent AP). The $Na^+$-dependent APs continue until maturation of the neuron. For example, in embryonic amphibian neurons *in vivo*, the AP is prolonged and the rate of rise is slow at the relatively early stage (Fig. 11A). Removal of $Na^+$ ion does not affect the AP configuration at this stage, whereas it is almost abolished by $Co^{2+}$ ion, an inorganic blocker of voltage-dependent $Ca^{2+}$ channels. At the late embryonic stage, the AP becomes greatly abbreviated and loses the shoulder on its falling phase (Fig. 11B). This AP is completely blocked by removal of $Na^+$ ion (or by TTX), but is unaffected by $Co^{2+}$ (Spitzer and Baccaglini, 1976).

### B. $Ca^{2+}$ Transient

Spontaneous transient elevation of intracellular $Ca^{2+}$ is observed in developing neurons (Holliday and Spitzer, 1990). The spontaneous $Ca^{2+}$ transient (recorded by use of fluorescent dyes, e.g., fura-2, fluo-3, indo-1, etc.) is exclusively dependent on $Ca^{2+}$ influx, because it is abolished either by removal of extracellular $Ca^{2+}$ or by agents that block $Ca^{2+}$ channels (Holliday and Spitzer, 1990; Spitzer, 1994). Therefore, the observed intracellular $Ca^{2+}$ transient is due to $Ca^{2+}$ influx from the extracellular space.

Two classes of spontaneous $Ca^{2+}$ transients have been detected: rapid events, termed *$Ca^{2+}$ spikes,* and slow events,

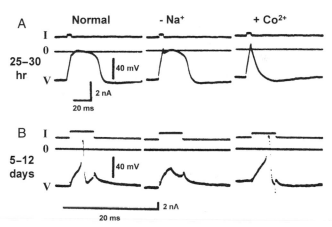

**FIG. 11.** Developmental changes in APs of embryonic amphibian neurons at two different stages: (A) 25–30 hr and (B) 5–12 days after fertilization of the egg. Depolarizing current (I) is applied to evoke the AP (V). The zero potential is shown by a solid line (0). (A) The AP is of long duration, and the rate of rise is slow at this early stage. This AP is little affected by removal of $Na^+$, but is abolished by $Co^{2+}$. (B) The AP at this late stage is brief and its amplitude is large. This AP is blocked by removal of $Na^+$ and is unaffected by $Co^{2+}$. (Modified from *Brain Res.,* **107**, N. C. Spitzer and P. I. Baccaglini, Development of the action potential in embryo amphibian neurons in vivo, pp. 610–616, Copyright 1976 with kind permission of Elsevier Science–NL, Sara Burgerhartstraat 25, 1055 KV Amsterdam, The Netherlands.)

termed *Ca²⁺ waves* (Gu *et al.*, 1994). The incidence of these Ca²⁺ transients changes during development in culture. Ca²⁺ spikes in the cell body (soma) are triggered by spontaneous APs and are rapidly propagated to the growth cone. Ca²⁺ spikes also use the intracellular Ca²⁺ store, because depletion of this store with caffeine substantially reduces their amplitude. Ca²⁺ spikes may be required for the normal appearance of transmitter GABA, since blocking of Ca²⁺ spikes by a Ca²⁺ channel blocker prevents the acquisition of GABA immunoreactivity. The normal developmental increase in the activation kinetics of K⁺ currents is also prevented by the blocking of Ca²⁺ spikes.

Ca²⁺ waves occur often (about 10/hr) in the growth cone, and they are not generally propagated to the soma. The Ca²⁺ waves in the soma occur at a lower frequency (about 2/hr). Therefore, the Ca²⁺ waves are local and occur independently in separate growth cones of the same neuron. Because there seems to be some relation between external Ca²⁺ and the length of the neurite, Ca²⁺ waves in growth cones are likely to regulate neurite extension (Gu *et al.*, 1994; Spitzer, 1994; Spitzer *et al.*, 1994).

## C. Voltage-Gated Ion Channels

In mature excitable cell membranes, the major inward currents consist of two ions, Na⁺ and Ca²⁺, which are carried through voltage-gated Na⁺ channels and Ca²⁺ channels, respectively. Na⁺ and Ca²⁺ channels exhibit two patterns of development (Gottmann *et al.*, 1988; O'Dowd *et al.*, 1988). In the first pattern, Ca²⁺ channels appear earlier and develop faster than the Na⁺ channels. This explains, at least in part, the conversion of the Ca²⁺-dependent AP to the Na⁺-dependent one (see earlier discussion). In the second pattern, Ca²⁺ channels and Na⁺ channels become expressed almost at the same time. For example, in cultured embryonic amphibian neurons, Ca²⁺ currents are large even at the early stage of development, and the peak current density does not change from the early stage to the late stage (Fig. 12A). Na⁺ currents are also present, but the current density is small at the early stage. The peak density of the Na⁺ current approximately doubles between the early and late stages of development in culture (Fig. 12B).

As described earlier, the Ca²⁺ channel allows Ca²⁺ influx, and is responsbile for spontaneous Ca²⁺ transient in developing neurons. However, regulation of Ca²⁺ influx during development is not due to change in the density of the Ca²⁺ channels, because the degree of expression of the Ca²⁺ channels may not change. The ratio of the K⁺ current to the Ca²⁺ current ($I_K/I_{Ca}$) affects the configuration of the APs, because the inward $I_{Ca}$ is deactivated earlier when

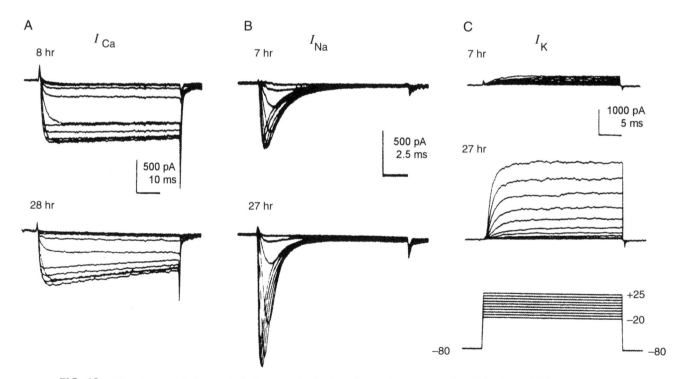

**FIG. 12.**   Developmental changes in ionic currents of cultured maturing neurons of embryonic amphibian. Records from cells in early stage (7–8 hr in culture) (upper traces) and late stage (27–28 hr) (lower traces). Records in (A) and (B) were obtained in presence of Cs⁺ (internal solution) and TEA (external solution) to block outward K⁺ currents. (A) Ca²⁺ currents ($I_{Ca}$) recorded in Na⁺-free solution. Amplitude of $I_{Ca}$ was nearly equal at early and late stages. (B) Na⁺ current ($I_{Na}$) recorded in presence of Co²⁺ to block $I_{Ca}$. $I_{Na}$ was smaller at the early stage than at the late stage. (C) K⁺ currents ($I_K$) recorded in presence of Co²⁺ and TTX to block the inward currents ($I_{Ca}$ and $I_{Na}$). $I_K$ is small at the early stage, and gets larger in amplitude and faster in rate of activation at the late stage. Pulse protocols for parts A, B, and C are given in the bottom of panel C. [Modified with permission from O'Dowd *et al.*, Development of voltage-dependent calcium, sodium ... *The Journal of Neuroscience* 8(3):796–800 (1988).]

the outward $I_K$ is augmented, resulting in a decrease in the net inward current. That is, any outward ($K^+$) current subtracts from the inward ($Ca^{2+}$) current, to give a lower *net* inward current. In addition, the change in the amplitude and the kinetics of $I_K$ that occurs during development greatly affects the AP configuration and apparent ionic dependence (Barish, 1986; Lockery and Spitzer, 1992). Therefore, the $I_K/I_{Ca}$ ratio determines whether $Ca^{2+}$-independent APs are exhibited, and the ratio influences the $Ca^{2+}$ influx.

The developmental increase in $K^+$ currents and change in their activation kinetics has been investigated in amphibian neurons in intact (isolated) spinal cord (Desarmenien *et al.*, 1993) and in dissociated cell culture (Fig. 12C) (Barish, 1986; O'Dowd *et al.*, 1988). The general rule regarding the order of appearance of ionic conductances in developing neurons is that $K^+$ currents appear first, before functional expression of the voltage-dependent inward currents (Spitzer *et al.*, 1994). However, the type of $K^+$ currents regulating the AP duration and the order of their maturation vary among different neuronal cells at early stages of development. In some neurons, delayed-rectifier $K^+$ current precedes the appearance of the transient outward $K^+$ current ($I_A$ or $I_{to}$). In contrast, in other neurons, the inactivating $K^+$ current ($I_A$) precedes expression of the delayed-rectifier $K^+$ current (Bader *et al.*, 1985; Aguayo, 1989; Beck *et al.*, 1992).

## D. Ligand-Gated Channels

During development, activity of the N-methyl-D-aspartate (NMDA) receptor channels decreases, and neuronal differentiation is supported by $Ca^{2+}$ influx through the receptor channels. In some cases, NMDA receptors are expressed transiently during early stages of development, restricting an increase in the intracellular $Ca^{2+}$ at the early stages (Garthwaite *et al.*, 1987).

Glutamate receptors gated by kinate/AMPA ($\alpha$-amino-3-hydroxy-methyl-4-isoxazolcpropionic acid) allow $Ca^{2+}$ influx and elevation of intracellular $Ca^{2+}$. The $Ca^{2+}$ permeability of glutamate receptors is governed by their subunit composition, as well as by single amino acid substitutions generated by RNA editing; both factors may be developmentally regulated (Hume *et al.*, 1991).

## V. Summary

Action potentials and resting potentials in excitable cells, such as cardiomyocytes, skeletal muscle fibers, and neurons, are greatly altered during development. In general, the rate of rise increases, the overshoot increases, the duration decreases, and the RP is hyperpolarized. These electrophysiological alterations are mainly produced by developmental changes in ion channels, that is, by changes in the types of, number of, and kinetic properties of the ion channels.

In cardiomyocytes, the density of inward-rectifier $K^+$ currents *increases* during development. This increase may result from changes in the single-channel conductance, as well as in the number and open probability of the channel, because the conductance of early fetal myocytes is much smaller than that of neonate. The hyperpolarization of the RP during development can be accounted for by the *increase* in the density of $I_{K(IR)}$ and the resultant decrease in the $P_{Na}/P_K$ ratio. In contrast, the hyperpolarization-activated inward current ($I_h$), which may affect automaticity, is dominant in the early embryonic chick cardiomyocytes and disappears during development.

The fast $Na^+$ current also increases markedly during development. That is, there are few or no functional fast $Na^+$ channels present at the earliest stages, and the density of these channels increases progressively during development. The fast $Na^+$ current is responsible for the increase in the AP rate of rise (independent of the hyperpolarization of RP). The max $dV/dt$ increases dramatically during development, for example, in chick heart, from 20 to 200 V/sec.

The fast $Na^+$ current in embryonic/fetal hearts has a significant sustained (i.e., slow inactivating or steady-state) component. The sustained component of the embryonic/fetal $Na^+$ current *decreases* during development, and this decrease contributes, at least in part, to the abbreviation of AP duration.

Development of $Ca^{2+}$ channels seems more complex. The density of total $Ca^{2+}$ current in chick cardiomyocytes decreases during the developmental period from fetal to neonate. In rat, however, it increases from the fetal to the neonatal period, followed by a substantial decrease in adult. In rabbit and guinea-pig cardiomyocytes, the current density in the neonatal period is smaller than that in adult. The total $Ca^{2+}$ current is composed of currents through several different types of channels: L-type, T-type, and F-type. (1) The proportion of the T-type $Ca^{2+}$ channel current in immature cells is generally more than that in mature cells, and it may actually disappear in adult. That is, the L-type $Ca^{2+}$ channel current becomes more dominant in mature cells. (2) A nifedipine-resistant F-type $Ca^{2+}$ channel current is also present in early fetal cardiomyocytes of rats. (3) Long-lasting openings of the L-type $Ca^{2+}$ channels are relatively frequently observed in embryonic chick and fetal rat cardiomyocytes, which are quite unusual in adult cells.

$Ca^{2+}$ influx through the $Ca^{2+}$ channels is especially important for the excitation–contraction coupling process of fetal cardiomyocytes. This is because the SR function and/or the essential component for $Ca^{2+}$ release from the SR is immature, and so $Ca^{2+}$ influx from the extracellular space is the main source of $Ca^{2+}$ for contraction.

The density of transient outward current ($I_{to}$) increases during development. An increase in the density of the delayed-rectifier $K^+$ current ($I_K$) also occurs during early development, and there is no further change during the postnatal period. The change of these voltage-gated outward currents (i.e., $I_{to}$ and $I_K$) helps to abbreviate the AP duration during development.

In skeletal muscle fibers, there is an overall trend toward an increase in the density of $I_{K(IR)}$ during the early developmental period. However, there is a transient period during which the current density decreases. This transient period parallels the increased incidence of spontaneous firings presumably due to a less stable RP. The major voltage-gated inward currents, Na$^+$ and Ca$^{2+}$ currents, are already present in the early embryonic skeletal myocytes in culture, and they increase in intensity during development. These changes contribute to the increases in the AP rate of rise, overshoot, and propagation velocity. Prolonged APs and brief APs can be elicited in chick embryonic skeletal myotubes, depending on the stage of development. The prolonged APs are associated with long-lasting contractions of the myotubes. The brief APs become more dominant during development in culture, and they are associated with myotube twitches. The nicotinic acetylcholine receptor/channel, which is essential to transmission at the neuromuscular junction, is converted from fetal-type to adult-type, and this conversion may be related to innervation of the muscles that occurs during development.

In neuronal cells, the ionic dependence of the AP is altered from the Ca$^{2+}$-dependent (prolonged AP duration) to the Na$^+$-dependent (brief AP duration) during development of amphibian embryo. The patterns of ion channel development vary among different types of neuronal cells, with faster development of Ca$^{2+}$ channels in some cells. Another important fator that determines the ionic dependence of the AP is the developmental increase in $I_K$. Therefore, the $I_K/I_{Ca}$ ratio is the major determinant of the conversion of the AP configuration, and influences the Ca$^{2+}$ influx during development. Two types of Ca$^{2+}$ transients, Ca$^{2+}$ spikes and Ca$^{2+}$ waves, are present in developing neurons. Ca$^{2+}$ spikes in the cell body (soma) are generated by spontaneous APs and propagate to the growth cone. The incidence of these Ca$^{2+}$ transients changes in developing neurons in culture. The activity of ligand-gated Ca$^{2+}$-permeable channels (such as NMDA receptor/channels and kinate/AMPA-gated receptor channels) are also altered during development. Therefore, the Ca$^{2+}$ influx through the voltage-gated and the ligand-gated Ca$^{2+}$ channels, and the subsequent effects on intracellular Ca$^{2+}$, may affect the structural changes of developing neurons and help the establishment of the neuronal network.

As described earlier, ion channels exhibit dramatic changes during development in their type, structure, function, and distribution. These dynamic alterations are controlled by expression of genes coding ion channels, and may be essential to cellular growth and differentiation by affecting several structures such as intracellular Ca$^{2+}$ concentration and excitability. Thus, molecular biology, combined with electrophysiology, has enabled large advances in our understanding of cell function.

## Bibliography

Aguayo, A. G. (1989). Post-Natal development of K$^+$ currents studied in isolated rat pineal cells. *J. Physiol. (London)* **414**, 283–300.

Bader, C. R., Bertrand, D., and Dupin, E. (1985). Voltage-dependent potassium currents in developing neurons from quail mesencephalic neural crest. *J. Physiol. (London)* **366**, 129–151.

Barish, M. E. (1986). Differentiation of voltage-gated potassium current and modulation of excitability in cultured amphibian spinal neurons. *J. Physiol. (London)* **375**, 229–250.

Beck, H., Ficker, E., and Heinemann, U. (1992). Properties of two voltage-activated potassium currents in acutely isolated juvenile rat dentate gyrus granule cells. *J. Neurophysiol.* **68**, 2086–2099.

Brown, A. M., Lee, K. S., and Powell, T. (1981). Sodium current in single rat heart muscle cells. *J. Physiol. (London)* **318**, 479–500.

Cohen, N. M., and Lederer, W. J. (1988). Changes in the calcium current of rat heart ventricular myocytes during development. *J. Physiol. (London)* **406**, 115–146.

Conforti, L., Tohse, N., and Sperelakis, N. (1993). Tetrodotoxin-sensitive sodium current in rat fetal ventricular myocytes—Contribution to the plateau phase of action potential. *J. Mol. Cell. Cardiol.* **25**, 159–173.

Desarmenien, M. G., Clendening, B., and Spitzer, N. C. (1993). *In vivo* development of voltage-dependent ionic currents in embryonic *Xenopus* spinal neurons. *J. Neurosci.* **13**, 2575–2581.

Fabiato, A., and Fabiato, F. (1978). Calcium-induced release of calcium from the sarcoplasmic reticulum of skinned cells from adult human, dog, cat, rabbit, rat, and frog hearts and from fetal and new-born rat ventricles. *Ann. N.Y. Acad. Sci.* **307**, 491–522.

Fischbach, G. D., Nameroff, M., and Nelson, P. G. (1971). Electrical properties of chick skeletal muscle fibers developing in cell culture. *J. Cell. Physiol.* **78**, 289–300.

Fujii, S., Ayer, R. K., Jr., and DeHaan, R. L. (1988). Development of the fast sodium current in early embryonic chick heart cells. *J. Membr. Biol.* **101**, 209–223.

Garthwaite, G., Yamini, B., Jr., and Garthwaite, J. (1987). Selective loss of Purkinje and granule cell responsiveness to N-methyl-D-aspartate in rat cerebellum during development. *Brain Res.* **433**, 288–292.

Gottmann, K., Dietzel, I. D., Lux, H. D., Huck, S., and Rohrer, H. (1988). Development of inward currents in chick sensory and autonomic neuronal precursor cells in culture. *J. Neurosci.* **8**, 3722–3732.

Greaves, A. A., Davis, A. K., Dallman, J. E., and Moody, W. J. (1996). Co-ordinated modulation of Ca$^{2+}$ and K$^+$ currents during ascidian muscle development. *J. Physiol. (London)* **497**, 39–52.

Gu, X., Olson, E. C., and Spitzer, N. C. (1994). Spontaneous neuronal calcium spikes and waves during early differentiation. *J. Neurosci.* **14**, 6325–6335.

Holliday, J., and Spitzer, N. C. (1990). Spontaneous calcium influx and its roles in differentiation of spinal neurons in culture. *Dev. Biol.* **141**, 13–23.

Hume, R. I., Dingledine, R., and Heinemann, S. F. (1991). Identification of a site in glutamate receptor subunits that controls calcium permeability. *Science* **253**, 1028–1031.

Iannaccone, S. T., Li, K. X., and Sperelakis, N. (1987). Transmembrane electrical characteristics of cultured human skeletal muscle cells. *J. Cell. Physiol.* **133**, 409–413.

Iijima, T., and Pappano, A. J. (1979). Ontogenetic increase of the maximal rate of rise of the chick embryonic heart action potential: Relationship to voltage, time and tetrodotoxin. *Circ. Res.* **44**, 358–367.

Irisawa, H., Brown, H. F., and Giles, W. (1993). Cardiac pacemaking in the sinoatrial node. *Physiol. Rev.* **73**, 197–227.

Jeck, C. D., and Boyden, P. A. (1992). Age-related appearance of outward currents may contribute to developmental differences in ventricular depolarization. *Circ. Res.* **71**, 1390–1403.

Josephson, I. R., and Sperelakis, N. (1989). Tetrodotoxin differentially blocks peak and steady-state sodium channel currents in early embryonic chick ventricular myocytes. *Pflugers Arch.* **414**, 354–359.

Josephson, I. R., and Sperelakis, N. (1990). Developmental increases in the inwardly-rectifying K$^+$ current of embryonic chick ventricular myocytes. *Biochim. Biophys. Acta* **1052**, 123–127.

Kano, M., and Yamamoto, M. (1977). Development of spike potentials in skeletal muscle cells differentiated in vitro from chick embryo. *J. Cell. Physiol.* **90**, 439–444.

Kato, Y., Masumiya, H., Agata, N., Tanaka, H., and Shigenobu, K. (1996). Developmental changes in action potential and membrane currents in fetal, neonatal and adult guinea-pig ventricular myocytes. *J. Mol. Cell. Cardiol.* **28**, 1515–1522.

Kawano, S., and DeHaan, R. L. (1991). Developmental changes in the calcium currents in embryonic chick ventricular myocytes. *J. Membr. Biol.* **120**, 17–28.

Kilborn, M. J., and Fedida, D. (1990). A study of the developmental changes in outward currents of rat ventricular myocytes. *J. Physiol. (London)* **430**, 37–60.

Kojima, M., Sada, H., and Sperelakis, N. (1990). Developmental changes in beta-adrenergic and cholinergic interactions on calcium-dependent slow action potentials in rat ventricular muscles. *Br. J. Pharmacol.* **99**, 327–333.

Lindsdell, P., and Moody, W. (1995). Electrical activity and calcium influx regulate ion channel development in embryonic *Xenopus* skeletal muscle. *J. Neurosci.* **15**, 4507–4514.

Lockery, S. R., and Spitzer, N. C. (1992). Reconstruction of action potential development from whole-cell currents of differentiating spinal neurons. *J. Neurosci.* **12**, 2268–2287.

Marcus, N. C., and Fozzard, H. (1981). Tetrodotoxin sensitivity in the developing and adult chick heart. *J. Mol. Cell. Cardiol.* **13**, 335–340.

Masuda, H., and Sperelakis, N. (1993). Inwardly-rectifying potassium current in rat fetal and neonatal ventricular cardiomyocytes. *Am. J. Physiol.* **265**, H1107–H1111.

Masuda, H., Sumii, K., and Sperelakis, N. (1995). Long openings of calcium channels in fetal rat ventricular cardiomyocytes. *Pflugers Arch.* **429**, 595–597.

Mays, D. J., Foose, J. M., Philipson, L. H., and Tamkun, M. M. (1995). Localization of the Kv1.5 $K^+$ channel protein in explanted cardiac tissue. *J. Clin. Invest.* **96**, 282–292.

Mishina, M., Takai, T., Imoto, K., Noda, T., Takahashi, S., Numa, S., Methfessel, C., and Sakmann, B. (1986). Molecular distinction between fetal and adult forms of muscle acetylcholine receptor. *Nature* **321**, 406–411.

Nakamura, K., and Iijima, T. (1994). Postnatal changes in mRNA expression of the $K^+$ channel in rat cardiac ventricles. *Jpn. J. Pharmacol.* **66**, 489–492.

Nakanishi, T., Sebuchi, M., and Takao, A. (1988). Development of the myocardial contractile system. *Experientia* **44**, 936–944.

O'Dowd, D. K., Ribera, A. B., and Spitzer, N. C. (1988). Development of voltage-dependent calcium, sodium and potassium currents in *Xenopus* spinal neurons. *J. Neurosci.* **8**, 792–805.

Osaka, T., and Joyner, R. W. (1991). Developmental changes in calcium currents of rabbit ventricular cells. *Circ. Res.* **68**, 788–796.

Po, S., Snyders, D. J., Tamkun, M. M., and Bennett, P. B. (1993). Heteromultimeric assembly of human potassium channels. *Circ. Res.* **72**, 1326–1336.

Renaud, J. F., Romey, G., Lombet, A., and Lazdunski, M. (1981). Differentiation of the $Na^+$ channel in embryonic heart cells: Interaction of the channel with neurotoxin. *Proc. Natl. Acad. Sci. USA* **78**, 5348–5352.

Ritchie, A. K., and Fambrough, D. M. (1975). Electrophysiological properties of the membrane and acetylcholine receptor in developing rat and chick myotubes. *J. Gen. Physiol.* **66**, 327–355.

Sada, H., Kojima, M., and Sperelakis, N. (1988). Fast inward current properties of voltage-clamped ventricular cells of embryonic chick heart. *Am. J. Physiol.* **255**, H540–H553.

Sada, H., Ban, T., Fujita, T., Ebina, Y., and Sperelakis, N. (1995). Developmental change in fast $Na^+$ channel properties in embryonic chick ventricular heart cells. *Can. J. Physiol. Pharmacol.* **73**, 1475–1484.

Satoh, H., and Sperelakis, N. (1993). Hyperpolarization-activated inward current in embryonic chick cardiac myocytes: Developmental changes and modulation by isoproterenol and carbachol. *Eur. J. Pharmacol.* **240**, 283–290.

Satoh, H., and Sperelakis, N. (1994). Identification of and developmental changes in transient outward current in embryonic chick cardiomyocytes. *J. Dev. Physiol.* **20**, 149–154.

Spector, I., and Prives, J. M. (1977). Development of electrophysiological and biochemical membrane properties during differentiation of embryonic skeletal muscle in culture. *Proc. Natl. Acad. Sci. USA* **74**, 5166–5170.

Sperelakis, N., and Haddad, G. E. (1995). Developmental changes in membrane electrical properties of the heart. In "Physiology and Pathology of the Heart," 3rd ed. (N. Sperelakis, Ed.), Chap. 35, pp. 669–700. Kluwer Academic Publishers, New York.

Sperelakis, N., and Lee, E. C. (1971). Characterization of $(Na^+, K^+)$-ATPase isolated from embryonic chick hearts and cultured chick heart cells. *Biochem. Biophys. Acta* **233**, 562–579.

Sperelakis, N., and Shigenobu, K. (1972). Changes in membrane properties of chick embryonic hearts during development. *J. Gen. Physiol.* **60**, 430–453.

Spitzer, N. C. (1994). Spontaneous $Ca^{2+}$ spikes and waves in embryonic neurons: Signaling systems for differentiation. *Trends Neurosci.* **17**, 115–118.

Spitzer, N. C., and Baccaglini, P. I. (1976). Development of the action potential in embryo amphibian neurons in vivo. *Brain Res.* **107**, 610–616.

Spitzer, N. C., Gu, X., and Olson, E. (1994). Action potentials, calcium transients and the control of differentiation of excitable cells. *Curr. Opin. Neurobiol.* **4**, 70–77.

Tohse, N., and Sperelakis, N. (1990). Long-lasting openings of single slow (L-type) $Ca^{2+}$ channels in chick embryonic heart cells. *Am. J. Physiol.* **259**, H639–H642.

Tohse, N., Masuda, H., and Sperelakis, N. (1992a). Novel isoform of $Ca^{2+}$ channel in rat fetal cardiomyocytes. *J. Physiol. (London)* **451**, 295–306.

Tohse, N., Maszaros, J., and Sperelakis, N. (1992b). Developmental changes in long-opening behavior of L-type $Ca^{2+}$ channels in embryonic chick heart cells. *Circ. Res.* **71**, 376–384.

Wahler, G. M., Dallinger, S. J., Smith, J. M., and Flemal, K. L. (1994). Time course of postnatal changes in rat heart action potential and in transient outward current is different. *Am. J. Physiol.* **267**, H1157–H1166.

*Andre Terzic and Yoshihisa Kurachi*

# 35

## Cytoskeleton Effects on Ion Channels

### I. Introduction

**Ion channels** are integral membrane proteins that mediate ion permeation through cellular membranes. As movement of charged ions creates electrical currents, ion channels are the molecular structures responsible for the electrical properties of a cell. The probability of channel opening can be regulated by several gating mechanisms including voltage changes, ligand binding, or covalent modifications. In addition to conventional regulators of ion channel gating, more recently it became apparent that the cytoskeleton itself, or proteins associated with the cellular cytoskeleton, may also provide a major regulator of ion channel behavior.

The **cytoskeleton** forms fibrillar network structures throughout the cytosol, including the microenvironment surrounding ion channel proteins within the plasma membrane. The cytoskeleton is made of microfilaments, microtubules, intermediate filaments, and associated proteins (Gallo-Payet and Payet, 1995). Microfilaments are composed of G-actin monomers, which assemble into actin polymers (F-actin). Actin polymers link into three-dimensional frameworks, which cross-link myosin, or form cortical networks at the cell periphery. There is a dynamic equilibrium between polymeric forms of actin and G-actin monomers (Fig. 1A). The status of actin and the state of myofilament organization is set by diverse actin-binding proteins, including profilin (which binds G-actin), gelsolin (which caps F-actin), and filamin and $\alpha$-actinin (which cross-link microfilaments). Microtubules are composed of $\alpha$- and $\beta$-tubulin heterodimers that form tubular filaments. Polymerization, stabilization, and modulation of microtubule function is regulated by microtubule-associated proteins. The cytoskeleton is essential not only in the maintenance of cell shape and motility, but also in the distribution, stability, and function of integral membrane proteins (Bennett and Gilligan, 1993; Hitt and Luna, 1994).

Interactions between various ion channel proteins and cytoskeletal structures are numerous, and have been impli-

cated in mediating spatial sorting of surface channel proteins, and in the regulation of channel activity (Cantiello, 1995; Gomperts, 1996; Smith and Benos, 1996). This chapter provides a synopsis on the effects of cytoskeleton of ion channels.

### II. Site-Directed Ion Channel Distribution by Cytoskeleton-Associated Proteins

A critical role for the cytoskeleton, and associated proteins, lies in the governance of ion channel distribution within specialized regions of plasma membranes. Such regulated distribution of ion channels at the cell surface is necessary for proper intra- and intercellular signaling, in particular within and between excitable cells, such as neurons (Sheng, 1996). In the nervous system, the electrical signaling is driven by the synchronized function of ion channels, which are typically localized at specific locations, such as the neuromuscular junction, nodes of Ranvier, or postsynaptic sites. In addition to the nervous system, association of cytoskeletal proteins, such as ankyrin and spectrin, with ion channels and ion transporters, including the $Cl/HCO_3$ exchanger and the $\alpha$-subunit of the Na–K pump has been reported in other tissues such as the erythrocyte or epithelial cells (Cantiello, 1995; Smith and Benos, 1996). This section provides an overview on the interaction between cytoskeletal proteins and the ion channel responsible for site-directed ion channel distribution in the nervous system (Table 1).

#### A. Rapsyn and Clustering of Nicotinic Acetylcholine Receptors at the Neuromuscular Junction

It is well established that site-specific ion channel clustering of **nicotinic acetylcholine receptors** (nAChRs) at the **neuromuscular junction** is essential for synaptic efficacy during neurotransmission. At the neuromuscular junction, nAChRs are localized at the motor end plate opposite the presynaptic nerve terminal. In this specialized membrane

A

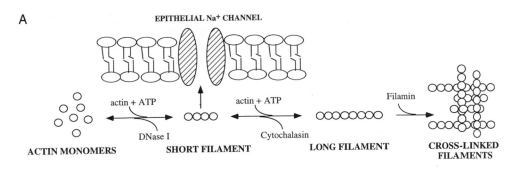

EPITHELIAL Na⁺ CHANNEL

ACTIN MONOMERS  SHORT FILAMENT  LONG FILAMENT  CROSS-LINKED FILAMENTS

B

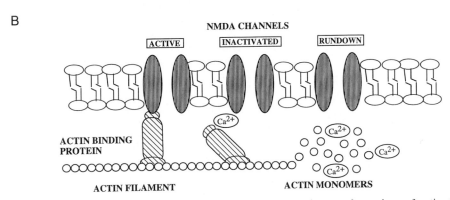

NMDA CHANNELS

**FIG. 1.** (A) A simplified scheme of dynamic interactions among various conformations of actin. In the presence of MgATP, monomeric G-actin nucleates into short actin filaments (e.g., tetramers) that anneal to produce long actin filaments. DNAse I inhibits polymerization by binding with G-actin. Cytochalasins favor the transition from long to short filaments. Cross-linking proteins, such as filamin, induce the gelation of actin filaments. The length of actin filaments appears to be a determinant of channel activation. For example, an increase in the availability of short actin filaments enhances the probability for renal epithelial Na⁺ channels to be open. (Modified from Cantiello, 1995; see also Gallo-Payet and Payet, 1995.) (B) Proposed model of actin-dependent regulation of NMDA channels. This model (Rosenmund and Westbrook, 1993) includes an actin-binding protein that dissociates from the NMDA channel in a Ca²⁺-dependent manner, leading the channel to inactivation. Calcium-dependent actin filament depolymerization results in removal of the actin-binding protein and channel rundown.

domain, nAChRs are tightly clustered at a density of 10,000/$\mu$m², in contrast to the thousand-fold lower density of nAChRs found just outside the motor end plate. The spatially localized aggregation of receptor/channel complexes ensures a rapid and robust response to released acetylcholine molecules, while a change in the cluster density of nAChRs strongly impacts postsynaptic response (Gomperts, 1996). The 43-kDa protein, **rapsyn,** which is associated with the inner face of the postsynaptic membrane, clusters and localizes nAChRs and links them to the subsynaptic cytoskeleton (Phillips *et al.*, 1991; Froehner, 1993). Mutational studies on rapsyn have identified binding domains for nAChRs, as well as for the cytoskeleton, within the primary structure of rapsyn.

Rapsyn can also cluster dystroglycan, a member of the **dystrophin–glycoprotein complex** (DGC) found along the

TABLE 1  Ion Channel Distribution Directed by Cytoskeleton-Associated Proteins

| Channel | Location | Cytoskeleton-associated protein |
|---|---|---|
| Na⁺ channel | Node of Ranvier | Ankryin |
| Glycine receptor | Postsynaptic membrane | Gephyrin |
| NMDA receptor | Postsynaptic density | PSD-95 |
| NMDA receptor | Postsynaptic domains | $\alpha_2$-actinin |
| Shaker type K⁺ channel (Kv1.4) | Various neural microdomains | PSD-95 |
| Inwardly rectifying K⁺ channel (Kir 2.3) | Hippocampus | PSD-95 |

*Note:* See text for further details and abbreviations.

sarcolemma of skeletal muscle and associated with the cytoskeleton, extracellular matrix, and with synapse-specific proteins, utrophin and $\beta_2$-syntrophin (Apel and Merlie, 1995). By binding to DGC, rapsyn may, in turn, localize nAChR clusters to the synapse. Targeted disruption of the rapsyn gene abolishes nAChRs clustering in the neuromuscular junction, directly implicating rapsyn as essential for the immobilization of nAChRs in the postsynaptic membrane (Gautam *et al.*, 1995). The importance of protein association with the DGC is further underscored by the demonstration that several muscular dystrophies, which cause improper function of the neuromuscular junction with progressive muscle weakness and premature death, derive from genetic errors within members of the DGC, and associated loss of the cytoskeleton–extracellular matrix linkage (Campbell, 1995).

## B. Ankyrin and Na⁺ Channels in the Node of Ranvier

Segregation of ion channels to specialized regions of neurons is also crucial for the propagation of action potentials (APs). Studies of lateral mobility and topography indicate that Na⁺ channels are freely mobile on the neuronal cell body, but are immobile at the axon hillock, presynaptic terminal, and at focal points along the axon. In particular, a high density in **voltage-gated Na⁺ channels** is found at the **node of Ranvier,** a specialized membrane domain in myelinated axons essential for fast propagation of nerve impulses. Such domain-dependent expression of Na⁺ channels has been ascribed to an interaction of channel proteins with a specific isoform of **ankyrin,** a cytoskeletal linker protein (Srinivasan *et al.*, 1988). Specifically, during the process of myelinization, ankyrin clusters at the ends of Schwann cell processes and flanks the site of nodal formation. As myelinization proceeds, clustered zones fuse to form mature nodes of Ranvier. During nodal development, Na⁺ channels and adhesion molecules (i.e., neurofascin) colocalize with ankyrin cluster zones and define the molecular composition of the node of Ranvier. Thus, the multivalent properties of ankyrin (i.e., a capacity to bind not only to membranes but also to ion channel proteins and adhesion molecules) plays an important role in the coordinated recruitment and targeted distribution of Na⁺ channels to the node of Ranvier, and thereby in AP conduction (Davis *et al.*, 1996).

## C. Gephyrin and Postsynaptic Ion Channel/ Receptor Mosaic

The molecular mechanisms underlying the postsynaptic ion channel/receptor mosaic within central nervous system neurons has been related, at least in part, to cytoskeleton binding proteins. In particular, clustering of inhibitory **glycine receptors** (GlyR) at postsynaptic membranes, underneath glycine-releasing nerve terminals, has been ascribed to the associations of these pentameric receptor/ion channel complexes with **gephyrin,** a 93-kDa, tubulin-binding protein (Kuhse *et al.*, 1995). Specifically, a gephyrin-binding domain has been identified in the cytoplasmic loop of the $\beta$-subunit of GlyR, between the third and fourth transmem-

brane segment (Meyer *et al.*, 1995). Antisense inhibition of the expression of gephyrin prevents GlyR accumulation at postsynaptic membrane specialization sites (Kirsch *et al.*, 1993). Furthermore, depolymerization of microtubules in spinal neurons reduces the percentage of cells with postsynaptic gephyrin clusters and disperses postsynaptic GlyR clusters. This supports the notion that postsynaptic localization of GlyR is regulated by the gephyrin-mediated anchoring of receptor polypeptides to the subsynaptic cytoskeleton through interactions with microtubules and possibly microfilaments (Kirsch and Betz, 1995).

The essential role for gephyrin in setting postsynaptic receptor/channel topology may not be restricted to GlyR. Gephyrin may also anchor another inhibitory receptor–ion channel complex, the **GABA receptor/channel,** to postsynaptic membranes in hippocampal neurons (Craig *et al.*, 1996).

## D. Postsynaptic Density Protein and NMDA Receptor/Channels

At central excitatory synapses, the N-methyl-D-aspartate (NMDA) receptor is highly concentrated at postsynaptic sites. These ionotropic neurotransmitter-gated receptor/channels are actually imbedded in the postsynaptic density (PSD), a specialized compartment of the submembrane cytoskeleton. Recently, it has been established that one of the subunits of the NMDA receptor complex (NR2) specifically interacts with the **postsynaptic density protein, PSD-95** (Kornau *et al.*, 1995). PSD-95 (also known as the synapse-associated protein 90 kDa, or SAP90) is a 95-kDa, cytoskeleton-associated protein abundant in the postsynaptic synaptosomal fraction (Gomperts, 1996). It has been proposed that members of the PSD-95 family serve to anchor NMDA receptors to the submembrane cytoskeleton and to aid in the assembly of signal transduction complexes at postsynaptic sites (Ehlers *et al.*, 1996; Gomperts, 1996).

## E. PSD-95-Related Proteins and Voltage-Gated K⁺ Channels

This class of channel-associated proteins, collectively named the PSD-95/SAP90 family (which in addition to PSD-95 also includes related proteins such as hdlg/SAP97), also interacts with subunits of several K⁺ channels (Kim *et al.*, 1995; Sheng, 1996). Voltage-gated K⁺ channels are concentrated at various neuronal microdomains, including presynaptic terminals, nodes of Ranvier, and dendrites, where they regulate local membrane excitability. Evidence has been obtained that cell-surface clustering of **Shaker type-K⁺ channels,** such as Kv1.4, is mediated by the PSD-95 family of membrane- and cytoskeleton-associated proteins. This occurs through direct and specific binding of the C-terminal cytoplasmic tails of K⁺ channel subunits to two characteristic, so-called PDZ domains, in the PSD-95 protein (see later discussion).

## F. PDZ Domains and Protein–Protein Interaction

PDZ domains are viewed as modular protein-binding sites that recognize a short consensus peptide sequence, making a small domain of a much larger protein responsible for specific protein–protein association (Sheng, 1996). PDZ domains are 90 amino acid repeats in the N-terminal half of PSD-95-related proteins (PSD-95, hdlg/SAP97, chapsyn-110/PSD-93). In addition to the presence of three PDZ domains, these proteins, which also belong to the membrane-associated guanylate-kinase (MAGUK) superfamily of proteins, are characterized by the existence of Src 3 homology (SH3) and guanylate-like domains within the C-terminal region (Gomperts, 1996; Sheng, 1996). Both the NR2 subunit of the NMDA receptor and the C-terminal region of *shaker* K$^+$ channels possess four highly conserved amino acids (-E-S/T-D-V-motif) that are specifically recognized by the PDZ domains in PSD-95 (Doyle *et al.*, 1996; Gomperts, 1996; Sheng, 1996).

The ability of PDZ domains to function as independent modules for protein–protein interaction suggests that PDZ-domain-containing polypeptides may be widely involved in the organization of proteins at sites of membrane specialization (Kim *et al.*, 1995; Sheng, 1996). While *shaker* and NR2 proteins do not cluster in the absence of PSD-95, coexpression of *shaker*-type K$^+$ channels (or NR2 subunits of NMDA receptors) with PSD-95 (or chapsyn-110) results in the coclustering of both proteins (Kim *et al.*, 1995, 1996; Sheng, 1996). This emphasizes the importance of PSD-95 proteins in directing the distribution of NMDA and voltage-dependent K$^+$ channels.

### G. PSD-95 and Inwardly Rectifying K$^+$ Channels

In addition to promoting clustering of NMDA receptors and voltage-dependent K$^+$ channels, the cytoskeletal protein PSD-95 also binds **inwardly rectifying K$^+$ channels,** including Kir 2.1 and 2.3. Furthermore, Kir 2.3 colocalizes with PSD-95 in neuronal populations in the forebrain, whereas a PSD-95/Kir 2.3 complex occurs in the hippocampus (Cohen *et al.*, 1996). Based on the multiple protein–protein interactions, and the observation that PSD-95 and related proteins forms oligomers, a scaffolding role for PSD-95 in organizing signaling cascades at the PSD has been proposed (Cohen *et al.*, 1996; Gomperts, 1996).

### H. $\alpha_2$-Actinin and Ion Channel Expression in Postsynaptic Domains

Other cytoskeleton-related proteins, such as **$\alpha_2$-actinin,** have also been related to the regulation of NMDA receptor expression within the postsynaptic domains. The integrity of postsynaptic actin filaments is, indeed, important for NMDA receptor function. A specific biochemical association between $\alpha_2$-actinin and the cytoplasmic tail of the NMDA receptor subunit NR1 has been recently demonstrated. It has, therefore, been suggested that the actin-binding protein, $\alpha_2$-actinin, could serve as an anchor protein to mediate NMDA receptor association with the postsynaptic actin cytoskeleton (Wyszynski and Sheng, 1996). The interaction of $\alpha_2$-actinin with NMDA receptor proteins may, in turn, affect the plasticity of excitatory synapses.

In summary, evidence has been obtained to indicate that cytoskeleton-associated proteins (including rapsyn, ankyrin, gephyrin, PSD-95, $\alpha_2$-actinin) direct and maintain the site-specific distribution of several ion channels (such as nAChRs and Na$^+$ and K$^+$ channels, as well as glycine and NMDA receptor/channel complexes) within microdomains of the plasma membrane (see Table 1).

## III. Role for Phosphorylation in Cytoskeletal Protein-Directed Clustering of Ion Channels

Targeting of ion channels to discrete plasma membrane sites is a dynamic process and is believed to be commonly regulated by additional enzymatic processes. Most commonly, evidence for protein phosphorylation has been obtained.

### A. Tyrosine Kinase Activity and Clustering of Nicotinic Acetylcholine Receptors

During development, nAChRs are clustered to postsynaptic muscular junctions in response to release of the nerve-derived factor, agrin (Wallace, 1992). Agrin-mediated signal transduction is mediated by receptor protein tyrosine kinases (Glass *et al.*, 1996). Targeted disruption of the gene encoding **MuSK,** a receptor tyrosine kinase selectively localized to the postsynaptic muscle surface, disrupts neuromuscular synapse formation (DeChiara *et al.*, 1996). Thus, nAChR clustering requires not only the cytoskeleton-associated protein, rapsyn, as discussed earlier, but also protein–tyrosine kinase activity. Moreover, it has also been reported that rapsyn may induce autophosphorylation of MuSK. This could, in turn, lead to a MuSK-specific phosphorylation of the $\beta$-subunit of nAChRs (Gillespie *et al.*, 1996). Thus, rapsyn may mediate the synaptic localization of MuSK in muscle, which may play an important role in the rapsyn-induced clustering of nAChRs within the neuromuscular junction.

### B. Protein Kinase A Activity and Synaptic Channel Density

It has recently been established that phosphorylation by **protein kinase A** (PKA) is important for the interaction between inward rectifying K$^+$ channels and the cytoskeletal protein, PSD-95. Stimulation of PKA in intact cells causes rapid dissociation of the inwardly rectifying K$^+$ channel, Kir 2.3, from PSD-95 (Cohen *et al.*, 1996). A serine residue (Ser-440), located within the C-terminal tail of the inwardly rectifying K$^+$ channel, Kir 2.3, is critical not only for interaction with PSD-95, but also serves as a substrate for phosphorylation by PKA. In turn, phosphorylation and dephosphorylation of such amino acid residue may regulate the dynamic interaction between K$^+$ channels and the cytoskeleton. Rapid cyclic AMP-mediated changes in the structure of the PSD may, in turn, determine the postsynaptic channel density, and mediate synaptic plasticity (Cohen *et al.*, 1996).

Anchoring of PKA also appears important in the regulation of synaptic function. Specifically, it has been shown that anchoring of PKA by A-kinase-anchoring proteins (AKAPs) is required for the modulation of alpha-amino-3-hydroxy-5-methyl-4-isoxazole-propionic acid (AMPA)/kainate channels. Intracellular perfusion of hippocampal neurons with peptides derived from the conserved kinase-binding region of AKAPs prevents PKA-mediated regulation of AMPA/kainate currents as well as fast excitatory synaptic currents. Thus, positioning of kinases by anchoring proteins near their substrates, including ion channel complexes, may be essential in the regulation of the electrical properties of a cellular membrane (Rosenmund et al., 1994).

## IV. Regulation of Ion Channel Function by Cytoskeletal Proteins

As described earlier, ion channels and other ion transport molecules are integral to the plasma membrane and are surrounded by cytoskeletal strands, in particular actin filaments (Ruknudin et al., 1991; Horber et al., 1995). Actin accounts for more than 20% of total cell proteins, and actin and actin-binding proteins couple to several ion channels and ion transport molecules (Cantiello, 1995). This is of importance in epithelia and neurons, where maintenance of ion channels and transporters to specific membrane domains is vital for the normal function of epithelial and neuronal tissues (Smith and Benos, 1996). In addition to structural interactions, there is growing evidence for functional interactions between ion channels and the adjacent actin microfilament network both in epithelia and nervous tissues, as well as in the heart (Cantiello and Prat, 1996).

### A. Epithelial Ion Channel Function

#### 1. Regulation of Epithelial $Na^+$ Channels by the Cytoskeleton

It has been established that **epithelial $Na^+$ channels** are linked to cytoskeleton structures, including ankyrin, spectrin, and actin (Smith and Benos, 1996). Using antibodies generated against the purified renal epithelial $Na^+$ channel, epithelial $Na^+$ channels were found to colocalize to the apical membrane with actin and apically associated isoforms of ankyrin and spectrin (Smith et al., 1991; Cantiello and Prat, 1996). It is a proline-rich region in the epithelial $Na^+$ channel that mediates binding to an SH3 region of $\alpha$-spectrin, which in turn maintains the polarized distribution of the channel to the apical membrane of renal epithelial (Rotin et al., 1994). While such interaction may serve to determine the spatial distribution of $Na^+$ channels, colocalization of actin filaments with apical $Na^+$ channels has also been related to the functional regulation of $Na^+$ channel activity (Cantiello, 1995). Agents that depolymerize actin filaments, such as cytochalasin D (see Fig. 1A), enhance the open probability of a 9-pS $Na^+$ channel activity in renal epithelial cells (Cantiello et al., 1991). In contrast, DNAse I, which stabilizes the pool of monomeric actin (see Fig.

1A), lacked such effect. The **length of the actin filaments** (see Fig. 1A) appears to be a determinant of channel activation. Addition of short actin filaments to excised membrane patches enhanced the probability for $Na^+$ channels to be open, yet whenever actin was added after being polymerized to achieve predominantly long filaments, no $Na^+$ channel activation was observed (Cantiello et al., 1991; Cantiello, 1995). This suggests that short actin filaments, in contrast to G-actin or long actin filaments, are responsible for channel activation (Cantiello and Prat, 1996).

The actin-dependent regulation of channel activity may participate in the stretch-dependent activation of the renal epithelial $Na^+$ channel (Awayda et al., 1995). Under basal condition, stabilized actin filaments may contribute to maintaining $Na^+$ channels in the closed state, whereas actin depolymerization, by stretch or a hormone (e.g., vasopressin), may result in channel activation, either by affecting the membrane environment or by interacting with other membrane–cytoskeleton proteins associated with the channel (Cantiello, 1995; Cantiello and Prat, 1996; Smith and Benos, 1996). The effect of actin on $Na^+$ channels is modulated by phosphorylation through PKA (Prat et al., 1993). Actin filaments have also been implicated in hormone (vasopressin)-mediated insertion of $Na^+$ channels into the apical membrane (Verrey et al., 1995).

#### 2. Regulation of Epithelial $Cl^-$ Channels by the Cytoskeleton

Chloride channels play an important role in fluid movement across epithelia. The actin cytoskeleton appears to regulate the behavior of the **cystic fibrosis transmembrane regulator (CFTR),** a low-conductance $Cl^-$ channel predominantly expressed in the **apical** membrane of epithelia (Cantiello and Prat, 1996; Smith and Benow, 1996). Severing of the endogenous actin-cytoskeleton by cytochalasin D, or direct addition of exogenous actin, induces activation of $Cl^-$ current in adenocarcinoma cells transfected with the human CFTR gene (Prat et al., 1995). Thus, increase in the availability of "short" filaments (see Fig. 1A) by either disruption of preexisting filaments or de novo formation of new ones activates CFTR. In contrast, decrease in the number of "short" actin filaments by preventing actin polymerization using DNAse I, for example, or by bundling filaments with filamin, inhibits CFTR-associated channel activity (Cantiello and Prat, 1996). In this regard the actin-dependent regulation of CFTR activity appears similar to that of epithelial $Na^+$ channels (see earlier discussion; Cantiello, 1995), since long, filamentous actin maintains the channels in the closed state while severing of actin into "short" filaments activates channels (Cantiello, 1995; Cantiello and Prat, 1996; Smith and Benos, 1996). Comparison of amino acid sequences of CFTR with known actin-binding proteins, such as severin and filamin, revealed putative actin-binding domain(s) within the nucleotide-binding folds of CFTR (Prat et al., 1995).

It should be pointed out, however, that the effect of the F-actin network-dependent regulation of $Cl^-$ channels may not be uniform. For example, opening of a different $Cl^-$ conductance, the 33-pS $Cl^-$ channel present in renal proximal tubule epithelia or of the Cl⁻ conductances in bronchial epithelia, is actually inhibited by cytochalasin D (Suzuki *et al.*, 1993; Hug *et al.*, 1995). Signaling pathways that include disruption of F-actin have been suggested to mediate the activation of a 305-pS Cl⁻ channel, during cell swelling, in renal collecting duct cells (Schwiebert *et al.*, 1994). Regardless of the outcome on channel activity resulting from the modification in actin microfilament structure, these experimental data point toward a functional interaction between the actin cytoskeleton and epithelial Cl⁻ channels.

Further evidence for an interaction between actin and a Cl⁻ channel or associated protein has been obtained from coimmunoprecipitation studies. Actin does coimmuno-precipitate with a 27-kDa protein, named $pI_{Cln}$, which gives rise to an outwardly rectifying Cl⁻ channel activity (Paulmichl *et al.*, 1992; Krapivinsky *et al.*, 1994). Thus, in epithelia the actin network interacts with channel-related proteins and regulates Cl⁻ channel behavior.

### 3. Regulation of Epithelial K⁺ Channels by the Cytoskeleton

The membrane-cytoskeleton is involved in the modulation of the low-conductance K⁺ channel present in the apical membrane of the **cortical collecting duct** (Wang *et al.*, 1994). This K⁺ channel is inactivated by application of known disrupters of actin filaments, such as cytochalasins. Phalloidin, however, which stabilizes actin filaments, prevents cytochalasin-induced K⁺ channel inactivation. Based on such findings, it has been proposed that the actin cytoskeleton is critically involved in the interaction between epithelial K⁺ channel proteins and the lipid phase of the cell membrane (Wang *et al.*, 1994).

Taken together, these findings indicate that the activity of various epithelial Na⁺, K⁺ and Cl⁻ channels is modulated by agents acting on cytoskeletal structures (Table 2). This, in turn, suggests a role for the submembrane cytoskeleton in the regulation of ion channel function in epithelial tissues.

### B. Neuronal Ion Channel Function

In addition to compartmenting and anchoring integral membrane proteins, the neuronal cytoskeleton has also been suggested to modulate neuronal excitability and synaptic plasticity trough regulation of ion channel function (Fukuda *et al.*, 1981; Rosenmund and Westbrook, 1993). The initial observation was that cytoskeleton breakdown

decreases AP upstroke in dorsal root ganglion neurons and axons, apparently through regulation of Na⁺ and Ca²⁺ channels (Matsumoto and Sakai, 1979; Fukuda *et al.*, 1981). Thereafter, cytoskeletal breakdown in neuronal tissue was also shown to affect membrane excitability through regulation of Ca²⁺ and NMDA channels (Fukuda *et al.*, 1981; Johnson and Byerly, 1993; Rosenmund and Westbrook, 1993). Here, we summarize the studies that relate to the cytoskeleton-dependent regulation of ion channel activity.

### 1. Regulation of Neuronal Ca²⁺ Channels by the Cytoskeleton

It was first shown that the cytoskeletal disrupter, colchicine, causes a reduction of the upstroke velocity of action potentials. From these data, it was inferred that Ca²⁺ channels in neurons interact with microtubules (Fukuda *et al.*, 1981). The **metabolic dependence** and **inactivation** by intracellular Ca²⁺ of Ca²⁺ channels were found to be mediated by an allosteric interaction between channel proteins and the cytoskeleton (Johnson and Byerly, 1993). Cytoskeletal disruption (by colchicine and cytochalasin B) prevents ATP from preserving Ca²⁺ channel activity, whereas cytoskeletal stabilizers (taxol and phalloidin) reduce both channel dependence on ATP and inactivation by intracellular Ca²⁺. An allosteric interaction between the cytoskeleton and Ca²⁺ might trigger a conformational change in the cytoskeleton rapidly closing the adjacent Ca²⁺ channels (Johnson and Byerly, 1993). Cytoskeletal stabilizers would reduce Ca²⁺-induced channel inactivation by restricting the Ca²⁺-dependent conformational change in the cytoskeleton. Thus, it is proposed that Ca²⁺-dependent inactivation of Ca²⁺ current in neurons may be related to cytoskeleton integrity (Johnson and Byerly, 1993).

### 2. Regulation of Neuronal NMDA Channels by the Cytoskeleton

F-actin is a major component of the cytoskeleton in postsynaptic densities and dendritic spines, and is under dynamic regulation of both Ca²⁺, which rapidly induces depolymerization, and of adenosine triphosphate (ATP), which promotes repolymerization. **Actin depolymerization** influences NMDA channel activity in whole-cell recordings of cultured hippocampal neurons (Rosenmund and Westbrook, 1993). Specifically, the ATP- and Ca²⁺-dependent "rundown" of NMDA channels (a parameter used to probe channel regulation) was prevented when actin depolymeri-

**TABLE 2**  Effects of Actin Cytoskeleton on Selected Epithelial Channels

| Channel | Tissue | Effect |
|---|---|---|
| 9-pS Na⁺ channel | Renal epithelial (A6) cells | Activation by short actin filaments |
| CFTR | Transfected adenocarcinoma cell line | Activation by long actin filaments |
| 33-pS Cl⁻ channel | Proximal tubular cell line | Activation by long actin filaments |
| 30-pS K⁺ channel | Principal cells in cortical collecting duct | Inactivation by actin filament disrupters |

*Note:* See text for further details. See also Smith and Benos (1996).

zation was blocked by phalloidin. This agent binds to F-actin and shifts the equilibrium between F-actin and actin monomers (G-actin) toward the polymerized state. Cyto-chalasins, which enhance actin-ATP hydrolysis, induced NMDA channel rundown, whereas taxol or colchicine, which stabilize or disrupt microtubule assembly, had no effect (Rosenmund and Westbrook, 1993). These results were interpreted to suggest that $Ca^{2+}$ and ATP can influence NMDA channel activity by altering the state of actin polymerization (Rosenmund and Westbrook, 1993; see Fig. 1B). Thus, actin dynamics may contribute to calcium-dependent postsynaptic events, such as long-term depression.

### C. Cardiac Ion Channel Function

In addition to epithelial and neuronal ion channels, more recently, indications were obtained for a functional interaction between the cytoskeleton and $Na^+$, $Ca^{2+}$, as well as ATP-sensitive $K^+$ channels expressed in cardiac cells. These data are based primarily on the ability of agents known to affect the cytoskeleton to modulate ion channel activity.

#### 1. Cytoskeleton Modulates Gating of Voltage-Dependent Cardiac $Na^+$ Channels

Agents that interfere with actin polymerization, such as cytochalasin D, reduce whole-cell peak $Na^+$ current, and slow current decay in ventricular cardiac myocytes (Undrovinas et al., 1995). Application of cytochalasin on the cytoplasmic side of inside-out patches results in reduction of peak open probability, accompanied with long bursts of $Na^+$ channel openings (Undrovinas et al., 1995). These results were interpreted to indicate that cytochalasin D, through effects on the cytoskeleton, induces cardiac $Na^+$ channels to enter a mode characterized by a lower peak open probability but a greater persistent activity as if the inactivation rate were slowed (Undrovinas et al., 1995).

#### 2. Cytoskeleton Disrupters Regulate L-Type Cardiac $Ca^{2+}$ Channels

Initially, it was observed that cardiac excitability can be modulated by agents that target microtubules, such as tubulin a depolymerizing agent (Lampidis et al., 1992). Colchicine, which dissociates microtubules into tubulin, and taxol, which stabilizes microtubules, strongly influence the kinetics of L-type $Ca^{2+}$ channels in intact cardiac cells (Galli and DeFelice, 1994). Colchicine increases the probability that $Ca^{2+}$ channels are in the closed state, whereas taxol increases the probability that $Ca^{2+}$ channels are in the open state. Moreover, taxol lengthens the mean open time of $Ca^{2+}$ channels. Neither taxol nor colchicine affects the number of $Ca^{2+}$ channels (Galli and DeFelice, 1994). Several interpretations were proposed for these findings, including a direct interaction of tubulin with $Ca^{2+}$ channels, or alternatively an action of taxol and/or colchicine through the buffering ability of the cytoskeleton to regulate the effective concentration of inactivating ions near the mouths of channels (Galli and DeFelice, 1994). This relates to the concept that the dynamics of current-induced inactivation are dictated by restricted and heterogeneous spaces near the membrane, as well as by a transient local buffering within cells. In this regard, the structure of the cytoskeleton surrounding the mouths of channels could contribute to both compartmentalization and buffering. Thus, alterations in the structure of the cytoskeleton within the $Ca^{2+}$ channel's microenvironment could participate in the phenomenon of channel inactivation, and thereby in the regulation of cell excitability (Galli and DeFelice, 1994).

#### 3. Actin Filaments Regulate Cardiac ATP-Sensitive $K^+$ Channel Activity

The defining property of ATP-sensitive $K^+$ ($K_{ATP}$) channels is their inhibition by intracellular ATP, whereby these channels are viewed as a link between the metabolic state and electrical excitability of a cardiac cell (Terzic et al., 1995). Opening of **$K_{ATP}$ channels** in the myocardium is sensitive to the mechanical distortion of the membrane (Van Wagoner, 1993) suggesting that the integrity of the microenvironment surrounding $K_{ATP}$ channels may play a role in modulating channel activity. Indeed, cytoskeletal disrupters, DNAse I (Fig. 2) and cytochalasin B (but not antimicrotubule agents), were found to antagonize the ATP-induced inhibition of cardiac $K_{ATP}$ channels, that is, they produced an apparent decrease in the sensitivity of $K_{ATP}$ channels toward ATP blockade (Terzic and Kurachi, 1996). When denatured by boiling (Fig. 2) or co-incubated with purified actin subunits (Fig. 3A), DNAse could not antagonize the ATP-induced inhibition of $K_{ATP}$ channels. Conversely, the DNAse-induced decrease in the sensitivity of $K_{ATP}$ channels toward ATP-induced inhibition was partially restored by addition of purified actin subunits (Fig. 3B; Terzic and Kurachi, 1996). Taken together, these findings may fulfill the established criteria for a disrupter of actin microfilaments to regulate a specific ion channel (Cantiello, 1995), and support the notion that DNAse I acts on actin filaments to modulate $K_{ATP}$ channel activity (Terzic and Kurachi, 1996).

The subsarcolemmal actin microfilament network may be of importance in governing not only the ATP-dependent gating of the channel, but also the sulfonylurea-dependent $K_{ATP}$ channel regulation (Fig. 4). In addition to ATP, a major pharmacological property of $K_{ATP}$ channels is their sensitivity to sulfonylurea drugs that are considered among the more specific $K_{ATP}$ channel ligands to inhibit channel activity. DNAse I, when applied to the internal surface of excised membrane patches, impaired the action of sulfonylurea drugs on myocardial $K_{ATP}$ channel activity (Brady et al., 1996). Specifically, this high-affinity actin-sequestering protein, which depolymerizes actin filaments, decreased the apparent sensitivity of $K_{ATP}$ channels toward inhibition by a prototype sulfonylurea, glyburide (Fig. 4; Brady et al., 1996). The effect of DNAse appeared mediated through binding to actin molecules since cytoskeletal strands are present in excised-membrane patches (Brady et al., 1996), whereas other known targets of DNAse are absent (Ruknudin et al., 1991; Horber et al., 1995). Denatured DNAse I could not antagonize glyburide-induced $K_{ATP}$ channel

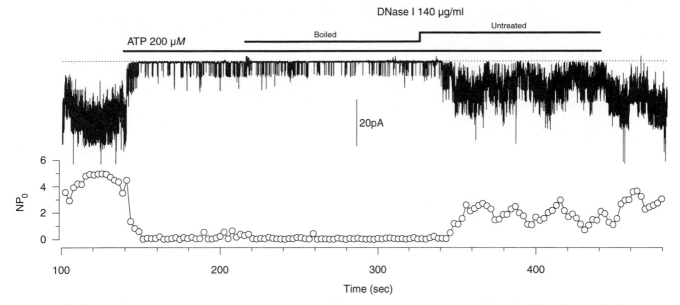

**FIG. 2.** The actin microfilament disrupter, DNAse I, enhances $K_{ATP}$ channel opening. Untreated DNAse I antagonized ATP-induced $K_{ATP}$ channel inhibition. By contrast, DNAse that has been denatured by boiling had no effect. Upper trace: Original trace record from an inside-out patch excised from a guinea-pig ventricular cardiac cell. Lower trace: Channel open probability calculated over 2.5-sec intervals. (From Terzic and Kurachi, 1996.)

inhibition (Brady *et al.*, 1996), which is consistent with the requirement that the native structure of the DNAse protein be intact for it to form complexes with actin molecules and interfere with actin filament formation. Co-incubation of DNAse I with excess purified actin, which forms 1:1 molar complexes with DNAse, prevented DNAse action on $K_{ATP}$ channels, suggesting that unoccupied binding sites for actin binding on the DNAse molecule are important for the modulation of $K_{ATP}$ channel regulation (Terzic and Kurachi, 1996; Brady *et al.*, 1996). The finding that the intraburst kinetic properties and conductance of the channel were not affected by DNAse could suggest that the loss of responsiveness of the channel to sulfonylurea, as well as ATP, was secondary to alterations remote from the pore region of the channel (Brady *et al.*, 1996; Terzic and Kurachi, 1996).

The coupled impairment in ligand sensitivity induced by DNAse supports the notion that sulfonylurea drugs and ATP bind to a common subunit, within the so-called sulfonylurea-binding protein, distinct from the presumed pore-forming components of the cardiovascular $K_{ATP}$ channel (Inagaki *et al.*, 1996; Isomoto *et al.*, 1996). Further evidence for a functional linkage of $K_{ATP}$ channels to the actin cytoskeleton was obtained from the observation that phalloidin, an actin filament stabilizing agent, could maintain channel activity and partially restore rundown channel activity (Furukawa *et al.*, 1996). It was thus proposed that for "fully activated" channels long, polymerized F-actin filaments are required. "Partially rundown" channels were associated with short actin filaments capped by actin-binding proteins. "Completely rundown" channels were related to depolymerized G-actin (Furukawa *et al.*, 1996). Taken together, these results could be interpreted to indicate that

cardiac $K_{ATP}$ channels can be regulated by the assembly and disassembly of the actin cytoskeletal network (Brady *et al.*, 1996; Furukawa *et al.*, 1996; Terzic and Kurachi, 1996).

## V. Mechanosensitive Gating of Ion Channels and Cytoskeleton

The interaction of protein channels with the cytoskeleton has also been suggested in the gating of **mechanosensitive** and **transduction channels.** This is the case, for example, in certain specialized hair cells (Sachs, 1988). Also, in skeletal muscle, an absence of normal dystrophin (a spectrin-like component of the cortical cytoskeleton) is associated with altered mechanosensitive gating, thus implicating cortical cytoskeleton in the mechanism of strech sensitivity (Morris, 1995).

Gating of stretch-activated channels is thought to rely on forces between the cytoskeleton and the attached membrane channels (Morris, 1995). Although the biochemical basis of this interaction is uncertain, disruption of actin by cytochalasin alters the behavior of mechanosensitive channels (Sachs, 1986). In the case of vertebrate hair cells, adaptation of the transduction current involves a $Ca^{2+}$- and actin-dependent mechanism. $Ca^{2+}$ influx is believed to activate a molecular motor that maintains gating spring tension by moving along the actin core of the stereocilia (Hudspeth, 1989).

## VI. Summary

The cytoskeleton regulates ion channel function through integrated interactions with cytoskeleton-associated pro-

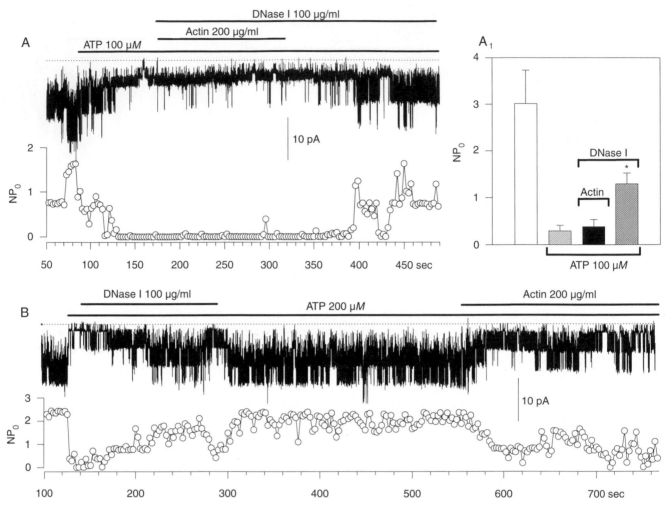

**FIG. 3.** Exogenous actin prevents the effect of DNAse I on $K_{ATP}$ channel activity. (A) Co-incubation and coapplication of purified actin subunits prevented the effect of DNAse I on $K_{ATP}$ channel activity. Original record with calculated open channel probability on the left and summarized data on the right. (B) Purified actin subunits partially restored the left and summarized data on the right. (B) Purified actin subunits partially restored the sensitivity of $K_{ATP}$ channels toward ATP-induced inhibition. (From Terzic and Kurachi, 1996.)

teins, and dynamic regulation of its own state. Two main roles for the cytoskeleton have been recognized in the regulation of ion channel function: (1) targeted distribution of ion channel proteins within specialized domains of the plasma membranes and (2) modulation of ion channel activity.

It is now established that cytoskeleton-associated proteins, such as rapsyn, ankyrin, gephyrin, PSD-95, and $\alpha_2$-actinin, target the distribution of Na$^+$ and K$^+$ channels, nAChRs, glycine, and NMDA receptor/channel complexes to specialized membrane domains including the postsynaptic membranes or the nodes of Ranvier (see Table 1). Cytoskeleton-dependent targeting and maintenance of ion channels at discrete plasma membrane sites, in turn, are regulated by catalytic processes, including protein phosphorylation by tyrosine kinases and PKA. Thus, structural interactions between the cytoskeleton, cytoskeleton-associated proteins, and channel/receptor subunits determine the highly specialized distribution of ion channel proteins

within domains of plasma membranes. Such site-directed distribution and anchoring of ion channels is required for proper intra- and intercellular signaling.

In addition to structural interactions, functional interactions between ion channel and colocalized subplasmalemmal cytoskeletal networks have been described in epithelia, neurons, and cardiac cells. Apparently, the open probability of epithelial Na$^+$, K$^+$, and Cl$^-$ channels is regulated by the length of actin filament networks (see Table 2). Also, the cytoskeleton has been suggested to modulate neuronal excitability and synaptic plasticity through modulation of ion channel activity, in particular through regulation of the open probability and kinetics of Ca$^{2+}$ and NMDA channels. In the heart, more recently, the activities of Na$^+$, Ca$^{2+}$, and $K_{ATP}$ channels have been shown to depend on the integrity of the cytoskeleton. Therefore, based on the current understanding of the relationship between the cytoskeleton and ion channels, it has become apparent that modulation of the cytoskeleton and associated proteins may represent

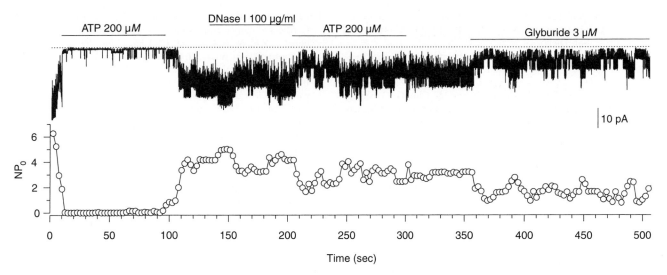

**FIG. 4.** DNAse I-induced loss of ATP sensitivity coupled to a loss of sulfonylurea sensitivity of $K_{ATP}$ channels. Original record with calculated open channel probability from an inside-out patch excised from a guinea-pig ventricular myocyte. Not illustrated is the ability of glyburide, a prototype sulfonlyurea drug, to inhibit fully myocardial $K_{ATP}$ channel activity when applied at micromolar concentration in the absence of DNAse I treatment. (From Brady *et al.*, 1996.)

important means of regulating the physiology of ion channels, and thereby cellular functions, including signaling and excitability.

Moreover, disturbances of the cytoskeleton or associated proteins can occur under disease conditions, including muscular dystrophies (Campbell, 1995), as well as under pathophysiological conditions, such as ischemia and hypoxia (Ganote and Armstrong, 1993). Thus, it is conceivable that under such conditions the distribution and behavior of ion conductances, which depend on the integrity of cytoskeleton networks, could be dramatically altered. The effects of cytoskeleton and ion channel function in disease states await to be elucidated.

## Bibliography

Apel, E. D., and Merlie, J. P. (1995). Assembly of the postsynaptic apparatus. *Curr. Opin. Neurobiol.* **5**, 62–67.

Awayda, M. S., Ismailov, I. I., Berdiev, B. K., and Benos, D. J. (1995). A cloned renal epithelial Na⁺ channel proteins display stretch activation in planar lipid bilayers. *Am. J. Physiol.* **268**, C1450–C1459.

Bennett, V., and Gilligan, D. M. (1993). The spectrin-based membrane skeleton and micron-scale organization of the plasma membrane. *Annu. Rev. Cell Biol.* **9**, 27–66.

Brady, P. A., Alekseev, A. E. A., Aleksandrova, L. A., Gomez, L. A., and Terzic, A. (1996). A disrupter of actin microfilaments impairs sulfonylurea-inhibitory gating of cardiac $K_{ATP}$ channels. *Am. J. Physiol.* **271**, H2710–H2716.

Campbell, K. P. (1995). Three muscular dystrophies: Loss of cytoskeleton–extracellular matrix linkage. *Cell* **80**, 675–679.

Cantiello, H. F. (1995). Role of the actin cytoskeleton on epithelial Na⁺ channel regulation. *Kidney Int.* **48**, 970–984.

Cantiello, H. F., and Prat, A. G. (1996). Role of actin filament organization in ion channel activity and cell volume regulation. *Curr. Top. Membr.* **43**, 373–396.

Cantiello, H. F., Stow, J. L., Prat, A. G., and Ausiello, D. A. (1991). Actin filaments regulate epithelial Na⁺ channel activity. *Am. J. Physiol.* **261**, C882–C888.

Cohen, N. A., Brenman, J. E., Snyder, S. H., and Bredt, D. S. (1996). Binding of the inward rectifier K⁺ channel Kir 2.3 to PSD-95 is regulated by protein kinase A phosphorylation. *Neuron* **17**, 759–767.

Craig, A. M., Banker, G., Chang, W., McGrath, M. E., and Serpinskaya, A. S. (1996). Clustering of gephyrin at GABAergic but not glutamatergic synapses in cultured rat hippocampal neurons. *J. Neurosci.* **16**, 3166–3177.

Davis, J. Q., Lambert, S., and Bennett, V. (1996). Molecular composition of the node of Ranvier—Identification of ankyrin-binding cell adhesion molecules neurofascin and NrCAM at nodal axon segments. *J. Cell. Biol.* **135**, 1355–1367.

DeChiara, T. M., Bowen, D. C., Valenzuela, D. M., Simmons, M. V., Poueymirou, W. T., Thomas, S., Kinetz, E., Compton, D. L., Rojas, E., Park, J. S., Smith, C., DiStefano, P. S., Glass, D. J., Burden, S. J., and Yancopoulos, G. D. (1996). The receptor tyrosine kinase MuSK is required for neuromuscular junction formation *in vivo*. *Cell* **85**, 501–512.

Doyle, D. A., Lee, A., Lewis, J., Kim, E., Sheng, M., and MacKinnon, R. (1996). Crystal structures of a complexed and peptide-free membrane domain—Molecular basis of peptide recognition by PDZ. *Cell* **85**, 1067–1076.

Ehlers, M. D., Mammen, A. L., Lau, L. F., and Huganir, R. L. (1996). Synaptic targeting of glutamate receptors. *Curr. Opin. Cell Biol.* **8**, 484–489.

Froehner, S. C. (1993). Regulation of ion channel distribution at synapses. *Annu. Rev. Neurosci.* **16**, 347–368.

Fukuda, J., Kameyama, M., and Yamaguchi, K. (1981). Breakdown of cytoskeletal filaments selectively reduces Na and Ca spikes in cultured mammal neurones. *Nature* **294**, 82–85.

Furukawa, T., Yamane, T., Terai, T., Katayama, Y., and Hiraoka, M. (1996). Functional linkage of the cardiac ATP-sensitive K⁺ channel to the actin cytoskeleton. *Pflugers Arch.* **431**, 504–512.

Galli, A., and DeFelice, L. J. (1994). Inactivation of L-type Ca channels in embryonic chick ventricle cells: Dependence on the cytoskeletal agents colchicine and taxol. *Biophys. J.* **67**, 2296–2304.

Gallo-Payet, N., and Payet, M. D. (1995). Excitation-secretion coupling. *In* "Cell Physiology" (N. Sperelakis, Ed.), pp. 465–482. Academic Press, San Diego.

Ganote, C., and Armstrong, S. (1993). Ischaemia and the myocyte cytoskeleton: Review and speculation. *Cardiovasc. Res.* **27**, 1387–1403.

Gautam, M., Noakes, P. G., Mudd, J., Nichol, M., Chu, G. C., Sanes, J. R., and Merlie, J. P. (1995). Failure of postsynaptic specialization to develop at neuromuscular junctions of rapsyn-deficient mice. *Nature* **377**, 195–196.

Gillespie, S. K., Balasuramanian, S., Fung, E. T., and Huganir, R. L. (1996). Rapsyn clusters and activates the synapse-specific receptor tyrosine kinase MuSK. *Neuron* **16**, 953–962.

Glass, D. J., Bowen, D. C. Stitt, T. N., Radziejewski, C., Bruno, J., Ryan, T. E., Gies, D. R. Shah, S., Mattsson, K., Burden, S. J., DiStefano, P. S., Valenzuela, D. M., Dechiara, T. M., and Yancopoulos, G. D. (1996). Agrin acts via a MuSK receptor complex. *Cell* **85**, 513–523.

Gomperts, S. N. (1996). Clustering membrane proteins; It's all coming together with the PSD-95/SAP90 protein family. *Cell* **84**, 659–662.

Hitt, A. L., and Luna, E. A. (1994). Membrane interactions with the actin cytoskeleton. *Curr. Opin. Cell Biol.* **6**, 120–130.

Horber, J. K. H., Mosbacher, J., Haberele, W., Ruppersberg, J. P., and Sackmann, B. (1995). A look at membrane patches with a scanning force microscope. *Biophys. J.* **68**, 1687–1693.

Hudspeth, A. J. (1989). How the ear's works work. *Nature* **341**, 397–401.

Hug, T., Koslowsky, T., Ecke, T., Greger, R., and Kunzelmann, K. (1995). Actin-dependent activation of ion conductances in bronchial epithelial cells. *Pflugers Arch.* **429**, 682–690.

Inagaki, N., Gonoi, T., Clement, J. P., Wang, C. Z., Aguilar-Bryan, L., Bryan, J., and Seino, S. (1996). A family of sulfonylurea receptors determines the pharmacological properties of ATP-sensitive K$^+$ channels. *Neuron* **16**, 1011–1017.

Isomoto, S., Kondo, C., Yamada, M., Matsumoto, S., Higashiguchi, O., Horio, Y., Matsuzawa, Y., and Kurachi, Y. (1996). A novel sulfonylurea receptor forms with BIR (Kir 6.2) a smooth muscle type ATP-sensitive K$^+$ channel. *J. Biol. Chem.* **271**, 24321–24324.

Johnson, B. D., and Byerly, L. (1993). A cytoskeletal mechanism for Ca$^{2+}$ channel metabolic dependence and inactivation by intracellular Ca$^{2+}$. *Neuron* **10**, 797–804.

Kim, E., Niethammer, M., Rothschild, A., Jan, Y. N., and Sheng, M. (1995). Clustering of Shaker-type K$^+$ channels by interaction with a family of membrane-associated guanylate kinases. *Nature* **378**, 85–88.

Kim, E., Cho, K. O., Rothschild, A., and Sheng, M. (1996). Heteromultimerization and NMDA receptor-clustering activity of chapsyn-110, a member of the PSD-95 family of proteins. *Neuron* **17**, 103–113.

Kirsch, J., and Betz. H. (1995). The postsynaptic localization of the glycine receptor-associated protein gephyrin is regulated by the cytoskeleton. *J. Neurosci.* **15**, 4148–4156.

Kirsch, J., Wolters, I., Triller A., and Betz, H. (1993). Gephyrin antisense oligonucleotides prevent glycine receptor clustering in spinal neurons. *Nature* **366**, 745–748.

Kornau, H. C., Schenker, L. T., Kennedy, M. B., and Seeburg, P. H. (1995). Domain interaction between NMDA receptor subunits and the postsynaptic density protein PSD-95. *Science* **269**, 1737–1740.

Krapivinsky, G. B., Ackerman, M., Gordon, E., Krapivinsky, L., and Clapham, D. E. (1994). Molecular characterization of a swelling-induced chloride conductance regulatory protein, pI$_{Cln}$. *Cell* **76**, 439–448.

Kuhse, J., Betz, H., and Kirsch, J. (1995). The inhibitory glycine receptor: Architecture, synaptic localization and molecular pathology of a postsynaptic ion-channel complex. *Curr. Opin. Neurobiol.* **5**, 318–323.

Lampidis, T. J., Kolonias, D., Savara, J. N., and Rubin, R. (1992). Cardiostimulatory and antiarrhythmic activity of tubulin-binding agents. *Proc. Natl. Acad. Sci. USA* **86**, 1256–1260.

Matsumoto, G., and Sakai, H. (1979). Microtubules inside the plasma membrane of squid giant axons and their possible physiological function. *J. Membr. Biol.* **50**, 1–14.

Meyer, G., Kirsh, J., Betz, H., and Langosch, D. (1995). Identification of a gephyrin binding motif on the glycine receptor beta subunit. *Neuron* **15**, 563–572.

Morris, C. E. (1995). Strech-sensitive ion channels. *In* "Cell Physiology" (N. Sperelakis, Ed.), pp. 483–489. Academic Press, San Diego.

Phillips, W. D., Kopta, C., Blount, P., Gardner, P. D, Steinbach, J. H., and Merlie, J. P. (1991). ACh receptor-rich membrane domains organized in fibroblasts by recombinant 43-kilodalton protein. *Science* **251**, 568–570.

Paulmichl, M., Li, Y., Wickman, K., Ackerman, M., Peralta, E., and Clapham, D. (1992). New mammalian chloride channel identified by expression cloning. *Nature* **356**, 238–241.

Prat, A. G., Bertorello, A. M., Ausiello, D. A., and Cantiello, H. F. (1993). Activation of epithelial Na$^+$ channels by protein kinase A requires actin filaments. *Am J. Physiol.* **265**, C224–C233.

Prat, A. G., Xiao, Y.-F., Ausiello, D. A., and Cantiello, H. F. (1995). c-AMP-independent regulation of CFTR by the actin cytoskeleton. *Am. J. Physiol.* **268**, C1522–C1561.

Rosenmund, C., and Westbrook, G. L. (1993). Calcium-induced actin depolymerization reduces NMDA channel activity. *Neuron* **10**, 805–814.

Rosenmund, C., Carr, D. W., Bergeson, S. E., Nilaver, G., Scott, J. D., and Westbrook, G. L. (1994). Anchoring of protein kinase A is required for modulation of AMPA/kainate receptors on hippocampal neurons. *Nature* **368**, 853–856.

Rotin, D., Bar-Sagi, D., O'Brodovich, H., Merilainen, J., Lehto, V. P., Canessa, C. M., Rossier, B. C., and Downey, G. P. (1994). An SH3 binding region in the epithelial Na$^+$ channel ($\alpha$rENaC) mediates its localization at the apical membrane. *EMBO J.* **13**, 4440–4450.

Ruknudin, A., Song, M. J., and Sachs, F. (1991). The ultrastructure of patch-clamped membranes: A study using high voltage electron microscopy. *J. Cell Biol.* **112**, 125–134.

Sachs, F. (1986). Biophysics of mechanoreception. *Membrane Bio. Chem.* **6**, 173–195.

Sachs, F. (1988). Mechanical transduction in biological systems. *Crit. Rev. Biomed. Engineer* **16**, 141–169.

Schwiebert, E. M., Mills, J. W., and Stanton, B. A. (1994). Actin-based cytoskeleton regulates a chloride channel and cell volume in a renal cortical collecting duct cell line. *J. Biol. Chem.* **269**, 7081–7089.

Sheng, M. (1996). PDZs and receptor/channel clustering: Rounding up the latest suspects. *Neuron* **17**, 575–578.

Smith, P. R., and Benos, D. J. (1996). Regulation of epithelial ion channel activity by the membrane-cytoskeleton. *Curr. Top. Membr.* **43**, 345–372.

Smith, P. R., Saccomani, G., Joe, E.-H., Angelides, K. J., and Benos, D. J. (1991) Amiloride-sensitive sodium channel is linked to the cytoskeleton in renal epithelial cells. *Proc. Natl. Acad. Sci. USA* **88**, 6971–6975.

Srinivasan, Y., Elmer, L., Davis, J., Bennett, V., and Angelides, K. (1988). Ankyrin and spectrin associate with voltage-dependent sodium channels in brain. *Nature* **333**, 177–180.

Suzuki, M., Miyazaki, K., Ikeda, M., Kawaguchi, Y., and Sakai, O. (1993). F-actin network may regulate a Cl$^-$ channel in proximal tubule cells. *J. Membr. Biol.* **134**, 31–39.

Terzic, A., and Kurachi, Y. (1996). Actin microfilament disrupters enhance K$_{ATP}$ channel opening in patches from guinea-pig cardiomyocytes. *J. Physiol.* **492**, 395–404.

Terzic, A., Jahangir, A., and Kurachi, Y. (1995). Cardiac ATP-sensitive K$^+$ channels: Regulation by intracellular nucleotides and K$^+$ channel-opening drugs. *Am. J. Physiol.* **269,** C525–C545.

Undrovinas, A. I., Shander, G. S., and Makielski, J. C. (1995). Cytoskeleton modulates gating of voltage-dependent sodium channel in heart. *Am. J. Physiol.* **269,** H203–H214.

Van Wagoner, D. R. (1993). Mechanosensitive gating of atrial ATP-sensitive potassium channels. *Circ. Res.* **72,** 973–983.

Verrey, F., Groscurth, P., and Bolliger, U. (1995). Cytoskeletal disruption of A6 kidney cells: Impact on endo/exocytosis and NaCl transport regulation by antidiuretic hormone. *J. Membr. Biol.* **145,** 193–294.

Wallace, B. G. (1992). Mechanism of agrin-induced acetylcholine receptor aggregation. *J. Neurobiol.* **23,** 592–604.

Wang, W.-H., Cassola, A., and Giebisch, G. (1994). Involvement of actin cytoskeleton in modulation of apical K channel activity in rat collecting duct. *Am. J. Physiol.* **267,** F592–F598.

Wyszynski, M., and Sheng, M. (1996). NMDA receptor anchoring to the actin cytoskeleton mediated by $\alpha_2$-actinin in dendritic spines. *J. Gen. Physiol.* **108,** 32a.

# Ion Channels as Targets for Toxins, Drugs, and Genetic Diseases

Kenneth Blumenthal

# 36

# Ion Channels as Targets for Toxins

## I. Introduction

Since the 1970s, a wide variety of neurotoxins have become potent tools in the armementarium of the biochemist, biophysicist, physiologist, or pharmacologist interested in identifying, purifying, and characterizing voltage-sensitive ion channels of excitable membranes and in understanding the molecular details of their structure and function. This chapter provides a brief description of some of these toxins, focusing on their chemical natures, similarities and differences, target macromolecules, and effects on transmembrane ionic fluxes. In this chapter, the ion channels are used as a framework, and in those cases where multiple toxin binding sites have been demonstrated with a single channel, the relationships among these sites are discussed. A number of excellent reviews on this subject have been published (Stephan and Agnew, 1991; Catterall, 1995; Fozzard and Hanck, 1996; Goldstein, 1996).

The toxins are diverse both chemically and functionally. They include an ever-expanding array of polypeptides derived from marine invertebrates and terrestrial arthropods, alkaloids of diverse structures, heterocyclic compounds of the tetrodotoxin family and synthetic insecticides. Functionally, these molecules are equally diverse. Toxins have been characterized that shift the voltage dependence of activation, delay channel inactivation, or block ion fluxes. Some have multiple actions. Thus, in addition to being useful probes for purification of channel constituents, they have contributed significantly to our understanding of how these various processes are coupled to one another.

## II. Voltage-Sensitive Sodium Channels

The neuronal sodium channel consists of three polypeptides having molecular masses of 260 ($\alpha$), 36 ($\beta_1$), and 33 kDa ($\beta_2$) (Tamkun et al., 1984; Hartshorne and Catterall, 1984); whether the smaller subunits are present in muscle or cardiac channels remains unresolved. All of the toxin binding sites characterized to date appear to be associated with the $\alpha$-subunit, reconstitution of which into planar lipid bilayers recapitulates many of the activities of the native channel (Tamkun et al., 1984). Microinjection of mRNA encoding this subunit into Xenopus oocytes results in expression of sodium channels that are functionally similar, though not identical, to those seen in nature. The $\alpha$-subunit is organized into four repeated structurally homologous domains, each containing 300 to 400 amino acid residues. Each domain is predicted to include six transmembrane helices (designated S1–S6). There is also a seventh region (SS1–SS2) having a length that varies among the four domains and that lies at least partially within the membrane. Recent data are consistent with the conclusion that these SS1–SS2 sequences contribute to the inner lining of the conducting pore and also provide its outer vestibule. A variety of mutually incompatible models for the secondary structure of this region exist. Whatever its structure, the binding sites for both tetrodotoxin/saxitoxin and the $\alpha$-scorpion toxins have been associated with this region by both biochemical and mutagenic analyses. Figure 1 depicts the putative transmembrane organization of the $\alpha$-subunit of this channel; as described in subsequent sections, this organization is shared by known $K^+$ channels and by the $\alpha_1$-subunit of the $Ca^{2+}$ channel as well.

A series of elegant analyses, carried out mainly by Catterall, have clearly established the presence of four independent classes of neurotoxin binding sites on the rat brain $Na^+$ channel (Catterall, 1977; also, see summary on Table 1). These include sites for (1) the classic channel inhibitors **tetrodotoxin** and **saxitoxin;** (2) for activating alkaloids, such as **batrachotoxin** and **veratridine;** (3) for **polypeptide ($\alpha$) toxins** from scorpion and sea anemone venoms, which delay channel inactivation; and (4) a distinct set of **scorpion toxins ($\beta$-toxins),** which alter the voltage dependence of channel inactivation. Each of these four classes is discussed next.

### A. Site 1: Tetrodotoxin, Saxitoxin, and $\mu$-Conotoxin

The presence of a paralytic toxin in newts, pufferfish, and other organisms has been known for well over 50

**547**

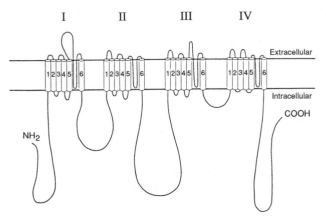

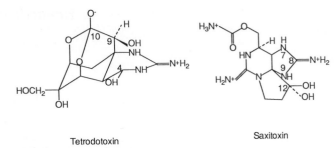

Tetrodotoxin                          Saxitoxin

**FIG. 2.** The structures of tetrodotoxin (left) and saxitoxin (right). For TTX, the guanidinium group and the C4, 9, and 10 hydroxyls are important for activity, whereas in saxitoxin, the C12 hydroxyl and the 7, 8, 9 guanidinium group are essential.

**FIG. 1.** Proposed structure of the voltage-sensitive sodium channel. The channel consists of four homologous domains (I–IV), each consisting of six transmembrane segments (S1–S6). Cytoplasmic loop III–IV has been implicated in both activation and inactivation of the channel, while residues lying between S5 and S6 are involved in both ionic selectivity and binding of specific channel blockers such as tetrodoxin. [Reproduced from McClatchey *et al.* (1992), *Cell* **68,** 769–774. Copyright 1992 Cell Press.]

years, and the structure of this compound (Fig. 2), called **tetrodotoxin,** was first described in 1964. Narahashi, Moore, and collaborators demonstrated that the toxicity of tetrodotoxin (TTX) was attributable to its ability to block the increase in sodium conductance associated with the rising phase of the action potential (AP) in frog muscle. Because TTX and the structurally related, dinoflagellate-derived saxitoxin (STX) bind to the channel with high affinity (1–5 n*M* for the neuronal isoform), a property that is essentially unaltered by detergent solubilization, their discovery provided biochemists with an invaluable probe for purification of the channel protein. Because of the availability of TTX and STX, purification, biochemical

characterization, and molecular cloning of voltage-dependent sodium channels from a variety of sources has been accomplished during the past 10 years.

Tetrodotoxin and saxitoxin are heterocyclic guanidinium compounds that bind reversibly and with high affinity ($K_d = 1$–5 n*M*) to a site accessible from the external face of neuronal and muscle Na$^+$ channels. This $K_d$ agrees well with that obtained from analysis of the effects of these toxins on Na$^+$ currents in these tissues. Binding of TTX and STX to cardiac Na$^+$ channels is of much lesser affinity, with $K_d$ values on the order of micromolars. The cardiac channels are thus said to be TTX resistant. Analyses of channels by site-directed mutagenesis (Noda *et al.*, 1989; Satin *et al.*, 1992) have implicated Cys-374 and Arg-377 of the cardiac channel as being important for their decreased TTX sensitivity (*vide infra*).

Binding of TTX and STX blocks the Na$^+$ current, and was originally believed to involve interaction of toxin with the ionized form of a channel carboxyl group, which Hille (1975) proposed was associated with the **selectivity filter** of the channel. However, the notion that the carboxylate(s)

**TABLE I**  Toxin Effectors of Voltage-Sensitive Sodium Channels

| Toxin | Binding affinity | Functional effect |
|---|---|---|
| Tetrodotoxin | 1–5 n*M* (nerve) | Blockade |
|  | 1–10 μ*M* (heart) | Blockade |
| Saxitoxin | 1 n*M* (nerve) | Blockade |
|  | 100 n*M* (heart) | Blockade |
| Alkaloids |  |  |
|   Batrachotoxin | 0.25 μ*M* | Persistent activation |
|   Grayanotoxin | >1 m*M* | Persistent activation |
|   Veratridine | 50 μ*M* | Persistent activation |
| Polypeptides |  |  |
|   α-Scorpion toxins | 1–2 n*M* | Delayed inactivation |
|   β-Scorpion toxins | 0.5 n*M* | Shifts voltage dependence of activation |
|   Anemone toxins | 10–1000 n*M* | Delayed inactivation |
| Dinoflagellate toxins |  |  |
|   Brevetoxins | μ*M* range | Shifts voltage dependence of activation |
|   Ciguatoxin | 0.6 nG/mL | Induces depolarization |
|   *Goniopora* | 30–50 n*M* | Delayed inactivation |

important in TTX binding was also a part of the selectivity filter was abandoned when it was shown that chemical modification of channel carboxylates, which abolished binding, was without effect on ion selectivity. Subsequent structure–function analysis of TTX and STX, using both natural and synthetic analogues (Mosher, 1986; Shimizu, 1986), clearly delineated the importance of the guanidinium group of TTX (and the corresponding 7, 8, 9 guanidinium of the bifunctional STX) for activity. In addition, the C4, C9, and C10 hydroxyl groups of TTX and the C12 hydroxyl of STX are required. It is assumed that these hydroxyl groups form hydrogen bonds to as yet unknown acceptor sites on the channel. Recent data suggest that the TTX binding site is located near the external face of the channel and may lie outside of the transmembrane field.

In the absence of direct information on the three-dimensional structure of the channel, precise localization of the TTX binding site is impossible. Although current models for the binding of both toxins still emphasize the importance of fixed negative charges at or near the channel mouth, it is clear that both molecules make multiple contacts to channel residues lying within the linker connecting the putative S5 and S6 helices. Site-directed mutagenesis of both $Na^+$ channels and $K^+$ channels indicates that this region is important both for ion selectivity and binding of channel-specific toxins. Thus, replacement of arginine-377 of the TTX-resistant cardiac ($RH_I$) channel by asparagine results in a small increase in TTX sensitivity (Stuhmer et al., 1989a), whereas the mutation cysteine-374 to tyrosine renders the channel approximately 700-fold more sensitive to TTX (Satin et al., 1992). This latter result raises the possibility that the positively charged guanidinium groups of TTX and STX are actually bound not by ionic interactions but rather interact with the $\pi$-clouds contributed by a cluster of electron-rich aromatic residues located near the channel mouth. Analogous models have been proposed to account for acetylcholine binding to the nicotinic acetylcholine receptor, and tetraethylammonium binding to a variety of potassium channels. However, it is clear that interaction with aromatic residues is not the *sole* binding determinant since mutagenesis has also identified a carboxyl group, glutamate-387, essential for high-affinity TTX binding in the brain $R_{II}$ channel (Noda et al., 1989). Like Cys-374 and Arg-377, this residue is also located in the SS2 region. Thus, whatever the nature of the specific interactions involved in TTX binding, important determinants are clearly found in the SS1–SS2 region. A very recent development has utilized the ability of TTX to protect mutagenically created channel cysteine residues from reaction with sulfhydryl-specific reagents to develop topographic maps of the channel pore.

The corrsponding S5–S6 linker regions from domains III and IV are also important determinants of ionic selectivity, as shown by the recent finding that the mutations lysine-1422 to glutamate and alanine-1714 to glutamate give the $R_{II}Na^+$ channel an ionic selectivity similar to that of a $Ca^{2+}$ channel. Interestingly, mutagenesis has also been used to show that the S5–S6 region of *shaker* $K^+$ channels is directly involved in both ion transit and interaction of inhibitors such as tetraethylammonium (TEA) and charybdotoxin.

Tyrosine-379 of the $K^+$ channel RBK-1 has been identified as an important determinant of TEA binding (MacKinnon and Yellen, 1990), while the introduction of a tyrosine at position 449 of the *shaker* H4 channel greatly increases its sensitivity to TEA. Thus, a leitmotif stressing the importance of aromatic residues in the binding of cationic ligands to ion channels is emerging (Dougherty, 1996).

A structurally unrelated toxin, designated **$\mu$-conotoxin** or **geographutoxin,** is a competitive inhibitor of STX binding to muscle, but not nerve, $Na^+$ channels ($K_d < 100$ n$M$ in muscle vs $> 1$ $\mu M$ in brain; Cruz et al., 1985; Ohizumi et al., 1986). $\mu$-Conotoxin is a 22 residue, hydroxyproline-containing peptide, tightly cross-linked by three disulfide bonds. A model for its three-dimensional structure, based on two-dimensional nuclear magnetic resonance (NMR) measurements and simulated annealing protocols, has been proposed. The overall structure (Fig. 3) includes a series of tight turns in the N-terminal region and a short stretch of right-handed helix, while the core of the peptide encompasses what the authors refer to as a **disulfide cage.** A key feature of the structure is that the cationic side chains of arginines 1, 13, and 19 and lysines 8 and 16 all project into solution, away from the core of the molecule as defined by the disulfide bonds. Analysis of synthetic variants of $\mu$-conotoxin has shown that arginine-13 is essential for activity: Even conservative replacement by lysine reduces potency by about 10-fold. In contrast, arginine-19 can be replaced by lysine with essentially no change in activity, while substitution at arginine-1 yields an intermediate result (Cruz et al., 1989). It is interesting that binding of both TTX/STX and the $\mu$-conotoxins is at least partially dependent on guanidinium functions, despite the disparate chemistries of these molecules.

Unlike TTX and STX, the binding of $\mu$-conotoxin is dependent on membrane potential, with $K_d$ decreasing $e$-fold for every 34 mV of depolarization (Cruz et al., 1989). TTX/STX-resistant (i.e., cardiac) channels are likewise resistant to $\mu$-conotoxins. In addition, a mutated form of $\mu$-conotoxin has been used as a monitor of distance between the voltage sensor and channel pore (French et al., 1996).

## B. Site 2: Batrachotoxin, Grayanotoxin, Veratridine, and Related Compounds

The existence of low molecular weight substances having dramatic effects on the gating properties of voltage-dependent $Na^+$ channels has been known for more than 20 years. The most commonly used of these alkaloids are **batrachotoxin, grayanotoxin,** and **veratridine.** Despite their different biological origins (respectively, skin secretions of the frog *P. aurotaenia, Lilaceae,* and *Ericaceae* sp.) and dose-response curves ($IC_{50}$: approximately 0.5, 10, and 25 $\mu M$ for batrachotoxin, grayanotoxin, and veratridine, respectively), these molecules have very similar effects on channel function. They all appear to interact at a common binding site that is allosterically coupled to both the TTX binding site described in the previous section and the $\alpha$-scorpion toxin site to be discussed later (Catterall, 1977). Binding of any of the alkaloids to this site causes persistent activation of the channel and leads to membrane depolarization.

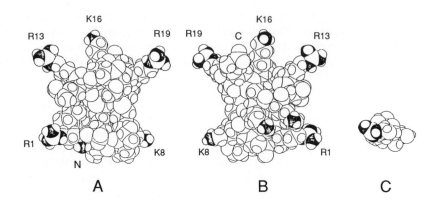

| Variant | Sequence |
|---------|----------|
| GIIIA   | RDCCTPPKKC KDRQCKPQRC CA |
| GIIIB   | RDCCTPPRKC KDRRCKPMKC CA |
| GIIIC   | RDCCTPPKKC KDRRCKPLKC CA |
| | |
| Consensus | RDCCTPP+KC KDROCKPX+C CA |

**FIG. 3.** $\mu$-Conotoxin structures. Top: A space-filling model of the three-dimensional structure of $\mu$-GIIIA, as deduced from two-dimensional NMR data. Black spheres indicate side chain nitrogens of Arg, Lys, and the N-terminal group. A and B depict different faces of the molecule, while the structure of tetrodotoxin is shown in scale in C for comparison. [Reproduced from Lancelin *et al.* (1991), *Biochemistry* **30**, 6908–6916. Copyright 1991 American Chemical Society.] Bottom: Amino acid sequences of three $\mu$-conotoxin variants are shown in the single-letter code with P designating hydroxyproline. Disulfide bonds link Cys3–Cys15, Cys4–Cys20, and Cys10–Cys21. In the consensus sequence +, cationic residue; X, ambiguous character.

In addition, alkaloid-treated channels display a more relaxed ionic selectivity than do their naive counterparts. Thus, while the permeability order of untreated Na$^+$ channels is Na$^+$:guanidinium:K$^+$:Rb$^+$ = 1:0.13:0.09:0.01, veratridine increases the relative permeability for guanidinium, K$^+$, and Rb$^+$ by factors of 2.7-, 4.5-, and 10-fold, respectively. Qualitatively similar results are obtained with the other alkaloids of this class and with combinations of alkaloid and polypeptide toxins from scorpion or sea anemone venoms.

The steroidal alkaloid batrachotoxin (BTX) was purified and characterized by Witkop, Daly, and their coworkers (1971). In collaboration with Albuquerque, the effects of BTX have been measured in a variety of tissues including axons, brain slices and isolated synaptosomes, neuromuscular junctions, heart papillary muscle, and cardiac Purkinje fibers. Batrachotoxin binding has been measured directly in rat brain synaptosomes using a tritiated derivative, and the measured density of binding sites correlates well with that for TTX and STX. Binding of BTX to this site is competitively inhibited by veratridine with a $K_i$ very similar to the $K_{0.5}$ for its activation of the channel. Because BTX acts from either side of the membrane, it had been speculated that this binding site might lie within the lipid bilayer, and a very recent study has demonstrated that a photoactivatable derivative of batrachotoxin specifically labels a site within the S6 transmembrane segment of domain I of the rat brain channel (Trainer *et al.*, 1996). As discussed later, fluorescent derivatives of BTX have also

been important for measuring distances between the various toxin-binding sites by resonance energy transfer experiments. Because the effects of BTX on membrane potential are dependent on the presence of extracellular Na$^+$ and are blocked by treatment with TTX, they are attributed to the toxin's ability to cause voltage-dependent Na$^+$ channels to activate at the resting potential, and to persist in the activated state. BTX also causes increased permeability to larger cations, consistent with allosteric coupling between the BTX binding site and the selectivity filter of the Na$^+$ channel.

Other alkaloids that appear to interact with the BTX site with similar functional effects include veratridine, grayanotoxin (GTX), and aconitine. While structurally distinct from one another, these highly hydrophobic compounds do have certain common features, such as the bridged ring systems and esterfield aromatic moieties found in BTX, veratridine, and aconitine (Fig. 4). In the case of BTX, both the bridge and the esterified group are essential for activity.

For all of these toxins, the available data strongly suggest that interaction of a single toxin molecule per channel is sufficient for activation. The known biological effects of these toxins are ascribed to their ability to interact with the voltage-dependent Na$^+$ channel, although only BTX binding has been measured directly. Catterall (1977) proposed that the alkaloids bind preferentially to the activated conformation of the Na$^+$ channel, shifting a preexisting equilibrium between active and inactive states toward the former. Thus, toxin action would best be approximated by

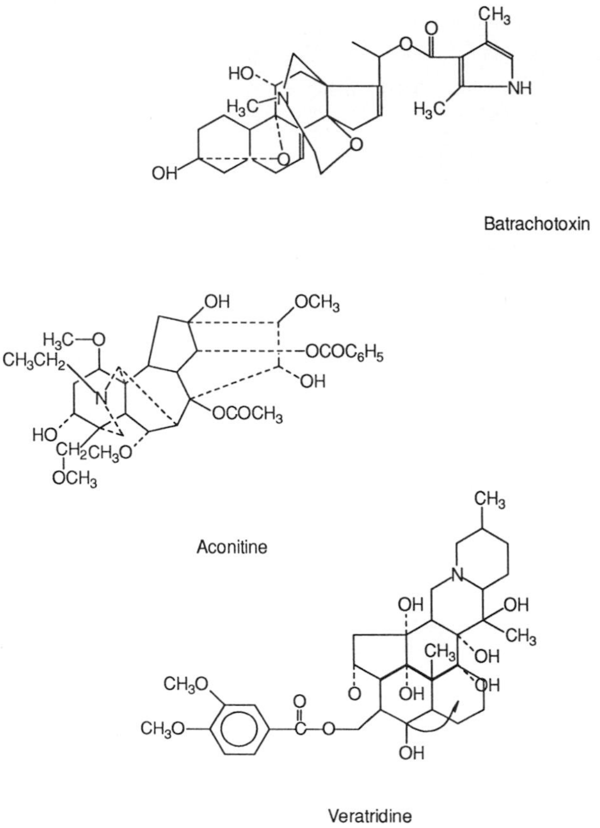

Batrachotoxin

Aconitine

Veratridine

**FIG. 4.** Structures of some major alkaloid agonists of the voltage-sensitive sodium channel. Top, batrachotoxin; center, aconitine; bottom, veratridine.

the allosteric model of Monod, Wyman, and Changeaux (1965) in which the binding energy for a given toxin is expressed as a shift in the voltage dependence for channel activation. As discussed later, conformational changes in the channel induced by binding of toxins have been directly demonstrated, thus supporting this model.

## C. Sites 3 and 4: Scorpion and Sea Anemone Toxins

An effect of scorpion venom on frog $Na^+$ channels was first demonstrated almost 25 years ago, and individual toxins from *Androctonus* venom were purified and sequenced beginning in the mid-1970s. Subsequently, toxins active on the sodium channel have been purified from the venoms of *Androctonus, Leiurus, Centruroides, Tityus,* and *Buthus* sp. While these toxins exhibit certain common structural features, the pharmacology of the *Androctonus* and *Leiurus* toxins is quite distinct from that of the corresponding polypeptides from *Centruroides* and *Tityus*. The former (α-toxin) group delays or abolishes channel inactivation and thus delays repolarization of the membrane through binding to neurotoxin site 3, whereas the latter group (β toxins) shifts the voltage dependence of activation to more negative potentials as a result of binding to site 4 (in the Catterall nomenclature), resulting in spontaneous activity. Binding of the α-toxins is dependent on membrane potential, with $K_D$ values in the nanomolar range at the resting

potential. In contrast, β-toxin binding is potential-independent and generally of slightly lesser affinity. The pharmacologic distinction between the α- and β-toxins should be emphasized, in view of the close relationship among these toxins at the level of covalent and tertiary structure.

The primary structures of representative scorpion toxins are depicted in Fig. 5. All scorpion toxins characterized to date are cationic polypeptides of approximate molecular mass 7000 Da and contain four disulfide bonds whose integrity is essential for activity; in addition, they display a significant degree of sequence homology, with approximately 30% of all positions being conserved among the known α-toxins. The β-toxins, exemplified by *Centruroides* toxin 3, are generally similar to those of the α group, but display less extensive homology. It is, however, possible that much of this sequence conservation within a group is essential simply for allowing proper folding of these toxins, similar to results obtained recently with the conotoxin family; but until residues important for interaction with the channel binding site have been definitively identified, this question will remain unresolved. Nonetheless, three-dimensional structures of representatives of each group show a good deal of similarity in the peptide backbones. With respect to structure–activity correlations, it has been shown that alkylation of lysine-56 of AaH-I is correlated with its inactivation, suggesting the importance of a cationic group at this position in the α-scorpion toxins, and that modification of arginines-2 and -60 likewise results in inactivation. While other residues have been implicated in activity by treatment of one or another of the scorpion toxins with group-selective reagents, the results are somewhat fragmentary, and no overall patterns have as yet emerged. While at first glance this paucity of information might appear surprising, given the small size of the toxins and their ready availability, it must be remembered that prior to the advent of molecular biological approaches, the most accessible techniques (e.g., chemical modification using group-specific reagents) for identifying essential residues in proteins were developed to probe substrate binding to enzymes and their catalytic mechanisms. In general, these methods do not adapt well to proteins lacking superreactive residues, such as the toxins now under consideration. The ability to mutagenize peptide neurotoxins has in the past few years led to more extensive knowledge of the relationship between their structure and function, as exemplified by analysis of anemone toxins described in a later section.

The three-dimensional structures of a number of scorpion toxins are now known, including members of both the α- and β-classes (Fig. 6). As might be anticipated from the highly cross-linked nature of these proteins, both possess a dense core of secondary structure connected by two of the four disulfides: This core contains three strands of antiparallel β-sheet (residues 1–4, 37–41, and 46–50) and a short strand of α-helix (residues 23–32). In addition to several loops that protrude from this core region, a prominent feature of both structures is a relatively nonpolar surface region that contains a number of conserved aromatic residues. Because chemical modification of sites external to this surface region is, in general, not deleterious to toxin function, and because Lysine-56, referred to pre-

### Representative α-Toxins

```
AaH-I     KRDGYIVYPN  NCVYHCVPP-  -CDGLCKKNG  GSSGSSCFLV
AaH-II    VKDGYIVDDV  NCTYFCGRNA  YCNEECTKLK  GESGY-CQWA
Lqq-v     LKDGYIVDDK  NCTFFCGRNA  YCNDECKKKG  GESGY-CQWA
BoT-1     GRDAYIAQPE  NCVYECAQNS  YCNDLCTKNG  ATSGY-CQWL
BoT-XI    VKDGYIVDDV  NCTYFCGRNA  YCHEECTKLK  GESGY-CQWA

consensus N+DGYIUOXX  NCßΦXCXOOO  YCOOXCOKXO  GOSGO-CXUU


AaH-I     -PSGLACWC-  KDLPDNVPIK  DTSRKCT
AaH-II    SPYGNACYCY  K-LPDHVRTK  GPGR-CH
LqQ-V     SPYGNACWCY  K-LPDRVSIK  EKGR-CN
BoT-I     GKYGNACWC-  KDLPDNVPIR  IPGK-CHF
BoT-XI    SPYGNACYCY  K-LPDHVRTK  GSPGRCH

consensus OXOGXACΦCY  KDLPDOVOß+  OXOO+CO
```

### Representative ß-Toxins

```
CsE-I     KEGYLVEKTG  CKKTCYKLGE  NDFCNRECKW  KHIGGSYGYC
TsS-G     KEGYLMDHEG  CKLSCFIRPS  -GYCGRECGI  KK--GSSGYC

consensus KEGYLU-+OG  CKXOCΦXXXO  XOΦCORECXU  K+XXGSOGYC


CsE-I     YGFGCYCEGL  PDSTQTWPLP  -NKCT
TsS-G     AWPACYCYGL  PNWVKVWDRA  TNKC

consensus XXXOCYCXGL  POXßOßWXXX  XNKC
```

**FIG. 5.** Primary structures of representative α- and β-scorpion toxins. Sequences are shown in the single-letter code, and disulfide bonds link half-cystines 12 and 63, 16 and 36, 22 and 46, and 26 and 48 in both groups. Numbering is based on the sequence of AaH-II. Highlighted positions are either invariant or occupied by functionally equivalent residues within the α and β subgroup. In the consensus sequence, the coding is +, cationic residue; O, polar residue; X, character undefined; U, nonpolar residue; Φ, aromatic residue; β, branched side chain. AaH, *Androctonus australis* Hector: Lqq, *Leiurus quinquestriatus quinquestriatus*; BoT, *Buthus occitanus Tunetanus*; CsE, *Centruroides sculpturatus* Ewing; TsS, *Tityus serrulatus*.

viously, lies within it, Fontecilla-Camps and coworkers (1980) first suggested that this region was essential for receptor binding. Interestingly, this surface is fairly well conserved in both the α- and β-toxins, despite their distinct pharmacologies and nonoverlapping binding sites (*vide infra*). Indeed, all of the structure features discussed are common among scorpion toxins regardless of binding site or species specificity.

In addition to the α-scorpion toxins, site 3 also binds a variety of polypeptide toxins derived from marine invertebrates. The best characterized of these are a diverse family of sea anemone toxins derived from *Anemonia, Anthopleura, Stichodactyla,* and *Radianthus* sp. (Catterall and Beress, 1978). Under appropriate conditions, certain of these polypeptides can function as cardiac stimulants. While also basic and rich in disulfides, the anemone toxins differ from those discussed previously in being somewhat smaller (46–50 residues versus 60–65) and having three, rather than four, disulfides. Moreover, despite the wealth of structural information now available, no sequence homology has been found between neurotoxins from scorpi-

ons and anemones. Representative anemone toxin sequences are shown in Fig. 7.

Within the past few years, the solution structures of four anemone toxins have been solved (Fig. 8). These structures, all having a core of twisted, four-stranded antiparallel β-pleated sheet (residues 2–4, 18–23, 31–34, and 42–47 in anthopleurin A numbering), demonstrate unequivocally that anemone and scorpion toxins are structurally unrelated. Nonetheless, anemone and α-scorpion toxins have very similar pharmacologies and display mutually competitive binding to the Na⁺ channel, suggesting that at least a limited degree of relatedness may yet remain to be found. Biophysical analyses have provided evidence for the exposure of a number of hydrophobic residues on the surface of homologous anemone toxins. In anthopleurin B, the functional roles played by a subset of these residues have been analyzed using site-directed mutagenesis of a synthetic gene capable of encoding the toxin polypeptide. These studies have clearly demonstrated that both leucine-18 and tryptophan-33 play important roles in regulating high-affinity binding of this toxin to both neuronal and

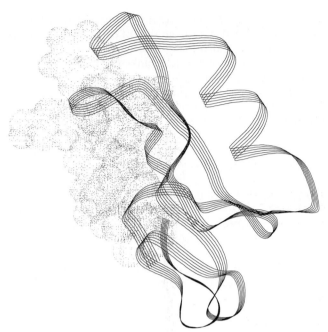

**FIG. 6.** Three-dimensional structure of *Centruroides* toxin variant 3. Structural coordinates were obtained from the Brookhaven protein database, with permission. Although variant 3 is a β-toxin, the overall folding patterns of the α- and β-toxins are highly similar. [Reproduced from Fontecilla-Camps *et al.* (1988), *Proc. Natl. Acad. Sci. USA* **85,** 7443–7447.] The backbone of the molecule is depicted as a ribbon, and the hydrophobic surface proposed to be important in toxin/receptor interactions is shown as a dot surface. All side chains have been omitted for the sake of clarity.

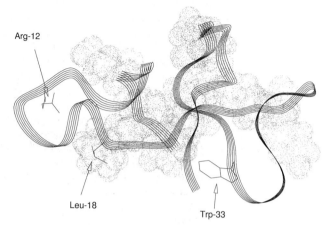

**FIG. 8.** Three-dimensional structure of anthopleurin B, a representative anemone toxin. The structure shown represents a model developed in the author's laboratory (Khera *et al.*, 1995) based on the coordinates of the homologous toxin from *S. helianthus*. [From Fogh *et al.* (1990), *J. Biol. Chem.* **265,** 13016.] The overall folding patterns for anemone toxins of known structure are similar. The model is also consistent with the recently determined solution structure of ApB [Monks *et al.* (1995), *Structure* **3,** 791]. The backbone is shown in ribbon format, and the side chains of residues which have been shown to be important determinants of binding affinity are annotated. The van der Waals surface depicts a region of partially exposed hydrophobic residues, including the required Leu-18 and Trp-33, which may be the functional equivalent of the hydrophobic surface involved in scorpion toxin binding to the sodium channel.

cardiac sodium channels, and further show that two other exposed hydrophobic sites, isoleucine-43 and tryptophan-45, are external to this binding epitope. Thus, the former pair of residues may act as the functional equivalent of the

```
AsI       GAPCKCKSDG  PNTRGNSMSG  TIWV--FGCP  SGWNNCEGRA
AsII      GVPCLCDSDG  PSVRGNTLSG  IIWL--AGCP  SGWHNCKKHG
AsV       GVPCLCDSDG  PSVRGNTLSG  ILWL--AGCP  SGWHNCKKHK
ApA       GVSCLCDSDG  PSVRGNTLSG  TLWLYPSGCP  SGWHNCKAHG
ApB       GVPCLCDSDG  PRPRGNTLSG  ILWFYPSGCP  SGWHNCKAHG
ShI       -AACKCDDEG  PDIRTAPLTG  --TVDLGSCN  AGWEKCASYY
RpII      -ASCKCDDDG  PDVRSATGTG  --TVDFWNCN  EGWEKCTAVY


consensus XUXCXCOODG  POXROOOUOG  ßUXUXXXOCX  OGWOOCOOXX


AsI       --HGYCCKQ
AsII      PTIGWCCKQ
AsV       PTIGWCCK
ApA       PTIGWCCKQ
ApB       PNIGWCCKK
ShI       TIIADCCRKKK
RpII      TPVASCCRKK
```

**FIG. 7.** Primary structures of representative sea anemone toxins. Sequences are shown in the single-letter code, with the numbering based on the sequence of anthopleurin A. Disulfide bonds link Cys4–Cys46, Cys6–Cys36, and Cys29–Cys47. Highlighted positions are either invariant or contain functionally conservative replacements. In the consensus sequence, the coding is +, cationic residue; O, polar residue; X, character undefined; U, nonpolar residue; β, branched side chain. As, *A. sulcata*, Ap, *A. xanthogrammica*; Sh, *S. helianthus*; Rp, *R. paumotensis*.

hydrophobic surface implicated in scorpion toxin binding (Dias-Kadambi *et al.*, 1996a,b). Indeed, charge neutralizing mutations of all the ionizable groups in this toxin reveal that the contribution to affinity made by hydrophobic contacts is far more significant than any involving electrostatic interactions. Surprisingly, even arginine-14, which is conserved in all known anemone toxins, is revealed by mutagenesis to play at most a minor role in defining binding affinity (Khera and Blumenthal, 1994).

While sea anemone and α-scorpion toxins bind to the channel in a membrane-potential-dependent manner and their binding sites overlap at least in part, β-scorpion toxins bind to a demonstrably distinct site on the same protein and their binding is potential-independent. All of the known polypeptide toxin-binding sites are structurally distinct from those for the alkaloid activators and guanidinium inhibitors. Affinity labeling experiments, using derivatives of either TTX or *Leiurus* toxin have demonstrated that the binding sites for both toxins are found on the α-subunit of the channel. Like the TTX/STX binding site discussed previously, a photoactivatable derivative of *Leiurus* toxin labels residues between the S5 and S6 helices (i.e., SS1–SS2) of domains I and IV of the rat brain channel. In some cells, the β₁-subunit of the channel is also photolabeled, leading to the suggestion that the scorpion toxin binding site may be located near a subunit interface. Labeling of both subunits is specific, being blocked in the presence of excess unlabeled ligand, or by membrane depolarization. While these experiments demonstrate convincingly that scorpion toxin is capable of labeling the channel in the SS1–SS2 region of domains I and IV, they do not prove that interaction at this site is responsible for the physiologic consequences of scorpion toxin binding. Interestingly, recent genetic studies have highlighted the role of other extracellular sequences in modulating channel inactivation kinetics. The dominantly heritable disorder paramyotonia congenita, which displays an electrophysiologic phenotype identical to that of an anemone or scorpion toxin-treated, wild-type channel, has been mapped to a set of missense mutations in the S3–S4 linker region of domain IV of the muscle sodium channel (Cannon and Corey, 1993; Chahine et al., 1994; Yang et al., 1994). As such, the recent report that *Leiurus* toxin may also interact with this region is of great interest.

### D.  Other Toxins Specific for the Na⁻ Channel

A variety of nonproteinaceous substances of diverse (and in some cases, unknown) structure have been shown to interact with the $Na^+$ channel at sites distinct from those discussed above. These include lipid-soluble dinoflagellate toxins (brevetoxins, ciguatoxin, and maitotoxin), the *Goniopora* toxins, and the pyrethoid insecticides. The brevetoxins induce the $Na^+$ channel to open at membrane potentials of $-80$ to $-160$ mV, and also abolish raid inactivation; these effects lead to both generation of repetitive action potentials and ultimately membrane depolarization. Both effects are blocked by TTX. Analyses of brevetoxin binding suggest the existence of a fifth ligand site on the channel, since brevetoxin binding fails to displace toxins bound to sites 1–4. The physiologic effects of ciguatoxin are very similar to those of the brevetoxins, while the mechanism of maitotoxin toxicity is unresolved at this time. *Goniopora* toxins, while not completely characterized, appear to be protein in nature, with molecular weights between 12,000 and 19,000 being reported from different preparations. The mode of action of at least one of these proteins appears to be similar to that of sea anemone toxins, but there is evidence that multiple *Goniopora* toxins with distinct molecular targets exist. Finally, the pyrethroids delay inactivation via interaction with a site distinct from that for polypeptide toxins having the same effect. Physiologic effects of pyrethroids are antagonized by TTX.

### E.  Relationships among Sites 1 through 4

The best data available from direct binding studies support the notion that each channel molecule contains single sites for each class of toxins. The next obvious question relates to identification of these sites and characterization of interactions among them. Angelides and coworkers (1986) have estimated the distances between the various toxin sites using resonance energy transfer between fluorescent derivatives of the bound ligands. In addition to confirming that the $\alpha$ and $\beta$ sites are unique and separated by about 22 Å, these studies indicate the TTX site to be

some 33 Å removed from that for $\beta$-toxins and the BTX site to lie about 37 Å distant from that of $\alpha$-toxins. Analogous experiments have also been used to verify that the sites involved are conformationally linked, with distances between a given pair of ligands changing on binding of a third.

Because multiple subtypes of the $Na^+$ channel have been characterized, it is natural to wonder whether different toxins are directed against specific channel isoforms, either in a tissue-specific manner or not. As discussed later, this may be the case for $K^+$ channels. However, to date no convincing evidence has been presented that a given, for example, scorpion toxin distinguishes among $Na^+$ channel subtypes within an individual tissue, although it is clear that toxins do differ substantially in their affinities for channels that are expressed in a tissue-specific manner (Khera et al., 1995). Obvious examples of this phenomenon would include the specificity of $\mu$-conotoxins for muscle $Na^+$ channels, and the 10-fold affinity preference displayed by some sea anemone toxins for cardiac, as compared to neuronal, $Na^+$ channels.

## III.  Voltage-Activated and Ca-Activated Potassium Channels

Our knowledge of the structure, function, and diversity of potassium channels has increased dramatically since the early 1990s, with the application of combined molecular genetic and electrophysiologic approaches (Miller, 1991). As seen later, despite the enormous power of the genetic approach, the ability of polypeptide toxins specific for this family of channels to distinguish among subtypes is proving to be an invaluable tool. In addition, both polypeptides and nonproteinaceous blockers, such as tetraethylammonium ions and 4-aminopyridines, are proving to be valuable tools for development of structural models of the K channel pore. These models may well apply to certains aspects of the homologous Na and Ca channels as well. In the case of $K^+$ channels, toxins from both arthropods and snakes have provided a wealth of pharmacologically significant data (summarized in Table 2), and very recently a new

**TABLE 2**  Polypeptide Blockers of Potassium Channels

| Toxin | Channel type | $K_d$ |
|---|---|---|
| Charybdotoxin | Ca-activated K | 3.5 n$M$ |
| | Voltage-sensitive K | 140 p$M$ |
| Iberiotoxin | Ca-activated K (BK) | 250 p$M$ |
| Kaliotoxin | Ca-activated K (BK) | 20 n$M$ |
| Noxiustoxin | Ca-activated K (BK) | >300 n$M$ |
| Scyllatoxin | Ca-activated K (SK) | 80 p$M$ |
| Apamin | SK | 10–400 p$M$ |
| Dendrotoxins | Voltage-sensitive K | 100 p$M$ |

class of polypeptide $K^+$ channel blockers has been purified from sea anemone cnidocytes.

While voltage-dependent $Na^+$ channels exist in a relatively small number of molecular forms, the diversity seen with $K^+$ channels is vastly greater: Some are ligand-coupled; others are activated by either $Ca^{2+}$ or depolarization. A large number of channels seem to exist even within this last category. In *Drosophila*, $K^+$ channels are encoded by genetic loci designated as *shaker, shal, shab, shaw,* and *eag,* and mammalian counterparts for each of these channel types have been identified. *Shaker* apparently encodes channels of the rapidly inactivating (or A) type, and transcripts from this locus display alternative splicing. It seems likely that this gives rise to channel forms that differ with respect to virtually all important properties: pharmacology, voltage dependence, and unit conductance (Stuhmer *et al.*, 1989b). Multiple transcripts from the *shab* locus have also been observed, and the gene products have many of the properties associated with the delayed rectifier channel. The *shal* and *shaw* genes seem to encode A-type and delayed rectifier channels, respectively.

In addition to this multiplicity of voltage-gated $K^+$ channels, there exists an array of channels gated by $Ca^{2+}$ as well. Because the unit conductances within this group vary quite widely, it is generally divided into high-conductance (BK or maxi-K; 130–300 pS) and small-conductance (SK, 10–50 pS) subtypes. The amount of structural information presently available on these channels is significantly less than that for the A-type channels, and because low-stringency hybridization screening using known $K^+$ channel probes has failed to identify members of this group, it is inferred that significant differences exist at the primary structure level despite (*vide infra*) some toxin cross-reactivity. Toxin binding may thus provide a means for purification of, for example, BK channels, and partial sequence analysis would then allow design of oligonucleotide probes, accessing additional structural information using classic molecular biological approaches.

The channels encoded by *shaker* and related genes have many structural features in common with the $Na^+$ channels discussed earlier, and these proteins display a limited degree of homology at the amino acid level. In both cases, the channel "subunit" contains six putative transmembrane helices with the N-terminus of the first being cytoplasmic. Helix S4, which contains a series of cationic groups separated by two hydrophobic residues, is postulated to serve as the voltage sensor, perhaps with some participation by charged residues in S2 and S3. As with the $Na^+$ channel, sequences joining the helices designated S5 and S6 are thought to provide the inner lining of the pore, and a great deal of evidence that these residues also provide the binding sites for channel-specific toxins has accumulated in the past few years. The most obvious structural difference between these channels is that whereas the $Na^+$ channel $\alpha$-subunit has a molecular mass of about 260 kDa and contains some 1800 amino acid residues grouped into four homologous domains, the $K^+$ channel monomer is approximately one-quarter this size. Whereas the monomeric unit of the $K^+$ channel is thus small relative to its $Na^+$ and $Ca^{2+}$ counterparts, coexpression of mRNAs encoding distinct channel isoforms in *Xenopus* oocytes results in the appearance of functional channels having intermediate properties, strongly suggesting that the physiologically relevant form of this channel is a noncovalent tetramer. The possibility of functional diversity arising from both alternative splicing and formation of hetero-oligomeric channels thus arises.

Most of the known $K^+$-channel-specific toxins have been purified from scorpion, honeybee, and snake venoms. The first such agent to be characterized was purified by Miller from *Leiurus* venom, and given the colorful, if nondescriptive, name **charybdotoxin** (Miller *et al.*, 1985). Homologs of this polypeptide have been found in venoms from *Leiurus* (Scyllatoxin and Agitoxin), *Androctonus* (Kaliotoxin), *Buthus* (Iberiotoxin), and *Centruroides* (Noxiustoxin) species. In addition, the honeybee peptides apamin and mast cell degranulating peptide and homologous family of snake (*Dendroaspis*) dendrotoxins have been shown to have $K^+$ channel blocking activity. Analogous to the case for sodium channels, the potassium channel toxins from scorpion venom comprise a homologous group in which essentially the only *absolutely* conserved residues are involved in disulfide bond formation (Fig. 9). Thus, one would reasonably expect these toxins to fold into a generally similar overall structure, but to differ in detail. It is therefore not surprising to realize that, as discussed later, at least some of the $K^+$ channel toxins are able to distinguish among channel subtypes. The honeybee toxins and dendrotoxins display no homology to any of the scorpion polypeptides.

Charybdotoxin (ChTX) was first identified in 1985 as a polypeptide from *Leiurus* venom capable of inhibiting $Ca^{2+}$-activated $K^+$ channels from skeletal muscle. Amino acid sequence analysis of the purified toxin revealed a primary structure completely distinct from that of other known scorpion toxins. The three-dimensional structure of ChTX was solved by NMR spectroscopy in 1991 (Fig. 10; Bontems *et al.*, 1991, 1992). Its major features include a small, triple-stranded $\beta$-sheet encompassing residues 1–2, 25–29, and 32–36 linked by disulfide bonds ($Cys_{13}$–$Cys_{33}$ and $Cys_{17}$–$Cys_{35}$) to an $\alpha$-helix including residues 10–19. Surprisingly, however, the overall fold of this molecule displays significant similarities to both the *Centruroides* and *Androctonus* $Na^+$-channel-specific toxins for which crystal structures had been obtained during the 1980s. It is important to recall that comparison of the sequences of the three toxins shows that only the positions of the disulfide bonds are conserved. However, in three dimensions, additional common features of these toxins are revealed, including three disulfide bonds occurring in similar positions and orientations, the triple-stranded $\beta$-pleated sheet (residues 1–4, 37–41, and 46–50 of *Centruroides* toxin vs residues 1–2, 25–29, and 32–36 of charybdotoxin), the single strand of $\alpha$-helix (residues 23–32 vs 10–19), and a significant proportion of extended structure (Fig. 11). The differences in comparing the structures of these long and short neurotoxins are most pronounced in loop regions connecting the elements of the defined secondary structure. Thus, it would appear that the scorpion neurotoxins as a group share a common structural motif, apparent only at the level of three-dimensional structure, regardless of the nature of

```
ChTX            QFTNVSCTTS KECWSVCQRL HNTSRGK-CMN KKCRCYS
ScTX            AFCNL- RMCQLSCRSL GLL--GK-CIG DKCECVKH
KaTX            GVEINVKCSGS PQCLKPCKDA GM-RFGK-CMN RKCHCTP
IbTX            QFTDVDCSVS KECWSVCKDL FGVDRGK-CMG KKCRCYQ
NxTX            TIINVKCTSP KQCSKPCKEL YGSSAGAKCMN GKCKCYNN


consensus       XXßOUOCOXX XXCXXXCOOU XXXOXGO CUO OKCOCXX




Apamin          CNCKAPETAL CARRCQQH




αDtx        ZPRRKLCILH RNPGRCYDKI PAFYYNQKKK QCERFDWSGC
            GGNSNRFKTI EECRRTCIG
```

**FIG. 9.**    Primary structures of polypeptide toxins specific for K channels. Sequences are depicted in the single-letter code, with disulfide bonds linking Cys7–Cys28, Cys13–Cys33, and Cys17–Cys35 in charybdotoxin. Highlighted residues are invariant within a class. In the consensus sequence, the coding is +, cationic residue; O, polar residue; X, character undefined; U, nonpolar residue; $\beta$, branched side chain. ChTX, charybdotoxin; ScTX, scyllatoxin; KaTX, kaliotoxin; IbTX, iberiatoxin; NxTX, noxiustoxin; $\alpha$DTx, $\alpha$-dendrotoxin; Dtx$_1$, dendrotoxin I; kalicludine, K channel blocker from *Anemonia sulcata*.

their target channel. The apparent similarities in channel folding thus seem to be mirrored in the structures of their "ligands." Interestingly, this motif appears to be similar to one observed in insect defensins, small peptides produced in response to injury, which are thought to play a role in protecting insects against bacterial infections (Bontems *et al.*, 1992). It seems almost counterintuitive that the scorpion toxins, despite their vastly different pharmacologies, display conservation of tertiary structure, while anemone and scorpion $\alpha$-toxins, which show very similar pharmacologies, do not. Perhaps this is simply indicative of

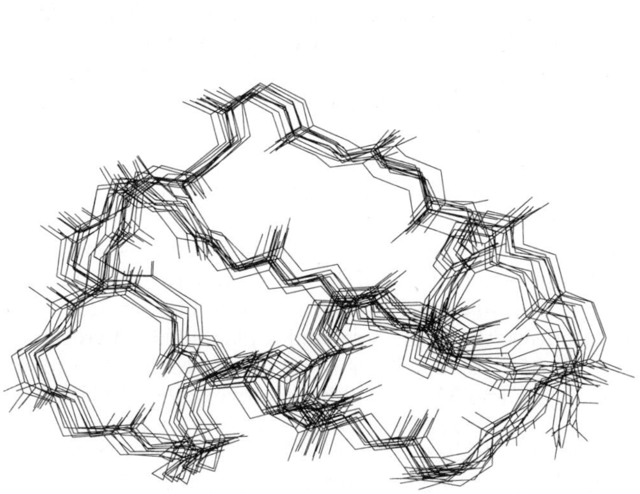

**FIG. 10.**    Three-dimensional structure of charybdotoxin as deduced from NMR experiments. Positions of the backbone atoms of each of the 12 contributing structures in the Brookhaven database file pdb2crn.ent are depicted. [Reproduced from Bontems *et al.* (1991), *Eur. J. Biochem.* **196,** 19–28.]

the fact that our knowledge of the molecular interactions important for productive binding of these toxins to their target sites remains incomplete (however, see later discussion). However, note that the very recently charactrized K$^+$ channel blockers from anemones (**kalicludines** and **kaliseptines,** for examples), which are functional homologs of charybdotoxin and dendrotoxin (*vide infra*), appear to share neither this structural motif, sequence homolgy, or even disulfide connectivities.

Because ChTX possesses a net charge of +5, it was predicted that ionic interactions with the channel would be important in its binding. This prediction is supported by the observations that (1) ChTX binding affinity is decreased 100-fold in medium of increased ionic strength (300 m$M$ salt); this reduction in binding is due to a decrease in toxin association and (2) chemical modification of the channel with the carboxylate-specific reagent trimethyloxonium (TMO) ion greatly reduces ChTX binding. Affinity also depends on membrane potential, with ChTX preferring the open conformation by approximately 7-fold. More recent studies with charybdotoxin mutants are consistent with hydrophobic interactions between channel sites and Met-29 and Tyr-36 of ChTX as also being important for binding (MacKinnon and Miller, 1989).

Although charybdotoxin was discovered because of its ability to block BK channels, evidence has since accumulated that its specificity is less well defined. It now seems clear that voltage-dependent K$^+$ channels in frog ganglia, and SK channels from *Aplysia* among others, are also blocked by ChTX. However, because it affords us the ability to map precisely ChTX : channel interactions, the most interesting new target for ChTX is the *shaker* channel. To test the importance of ionic interactions in ChTX binding to this target, acidic residues located in extracellular sequences of the channel were altered by site-directed muta-

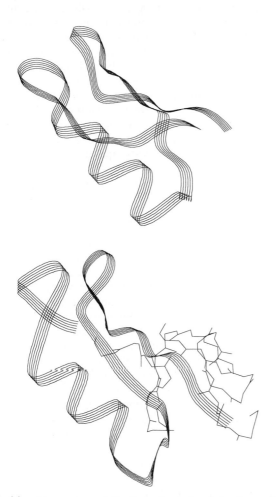

**FIG. 11.** Comparison of the three-dimensional structures of *Centruroides* variant 3 (lower) and charybdotoxin (upper). Backbone structures are shown as ribbons, within the region of high similarity: α-helices 10–20 (charybdotoxin) and 22–32 (variant 3), and β-strands 1–2, 25–29, and 32–36 (charybdotoxin) and 1–4, 37–42, and 44–48 (variant 3). Images were created from the Brookhaven database files pdb2crd.ent and pdb2sn3.ent, respectively, using InsightII. [Original data from Bontems *et al.* (1991), *Eur. J. Biochem.* **196,** 19–28.]

genesis, and the kinetics of ChTX block measured. This strategy identified Glu-422, located in the SS1 region, as important for binding; substitution by aspartic acid being without effect while conversion to glutamine or lysine decreased blocking activity by 3- and 12-fold, respectively. Very recent data have identified another point mutation (Phe-425 to Gly) whose major consequence is to render the *shaker* channel approximately 2000 times more sensitive to ChTX. From both the nature of the substitution made, and the magnitude of the change observed, one might reasonably ascribe this increase to a major conformational change in the toxin binding site.

More recently, application of charybdotoxin and other homologous peptide blockers of potassium channels, notably agitoxin-2, as molecular calipers capable of providing

topographical information on the vestibule of the potassium channel, has provided exciting results. For the sake of clarity, description of these results will be restricted to the charybdotoxin/agitoxin:*shaker* system, although analyses of other toxin:channel pairs have also proven quite useful. Miller's laboratory first systematically mutated all solvent-exposed residues of charybdotoxin, allowing identification of a subset of residues whose alteration had major effects on its affinity for the *shaker* channel (Goldstein *et al.,* 1994). Subsequently, MacKinnon and coworkers (Hidalgo and MacKinnon, 1995; Ranganathan *et al.,* 1996) demonstrated by comutagenesis of agitoxin and the S5–S6 region of the *shaker* channel the existence of interactions between Arg-24 (toxin) and Asp-431 (channel); Ser-10 and Gly-425; Lys-27 and Thr-449; and Arg-25 and Met-448; while Miller's group have found that Met-29 and Thr-449 and Lys-31 and Asp-427 also form interacting pairs (Naranjo and Miller, 1996; Naini and Miller, 1996). Interestingly, this last interaction was first predicted based on previous models of the charybdotoxin:*shaker* complex and knowledge of the toxin's solution, structure, and considerations of symmetry in the tetrameric channel. From such studies as these, a low-resolution structural map of the outer vestibule of the *shaker* channel is beginning to emerge (Fig. 12). Because nonpeptide pore blockers that function at the internal mouth of the channel are available, and channel chimeras and mutagenesis have identified residues at or near the cytoplasmic ends of S5 and S6 as essential to their

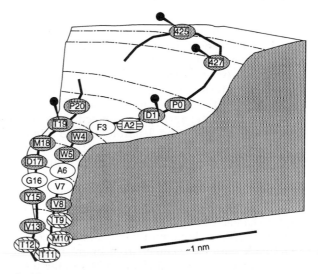

**FIG. 12.** Proposed topography of the external pore and vestibule of the *shaker* K⁺ channel. A hypothetical single *shaker* subunit, a 90° sector of the outer pore region and vestibule, is displayed. Residue numbers given are referenced to the P-region (P0–P20) or to *shaker* numbering. Push pins indicate residues whose positions relative to the pore axis were established by mapping with pore-blocking polypeptides. Shaded residues are predicted to project into the aqueous pore or vestibule, while unshaded residues are proposed to project away from the aqueous phase. No predictions have been made regarding side chain orientations for those residues having diagonal striping. [Reproduced from Lu and Miller (1995), *Science* **268,** 304. Copyright 1995 American Association for the Advancement of Science.]

binding, an analogous approach to mapping its structure may also prove fruitful. That the high-affinity polypeptide K$^+$ channel blockers from sea anemone venom are structurally distinct from the charybdotoxin family makes them yet another potentially fruitful probe of pore three-dimensional structure (Tudor *et al.*, 1996).

The effects of ChTX on *shaker* homologs cloned from rat brain (RCK channels) have also been examined. These experiments indicate a correlation between toxin affinity and the presence of negative charges in the S3–S4 loop. While this correlation remains to be verified experimentally, involvement of the S3–S4 loop in charybdotoxin binding would be of special interest in view of the demonstrated involvement of this region of the Na channel in inactivation (see discussion in Section II,C).

Because ChTX displays high-affinity interactions with so many different K$^+$ channels, the search for more specific probes has continued and a number of groups have purified and characterized new toxins whose targeting may be more stringent. **Iberiotoxin (IbTX),** isolated from *Buthus* venom on the basis of its ability to inhibit ChTX biding, is 70% homologous to ChTX but significantly less basic. IbTX inhibits ChTX binding to smooth muscle noncompetitively, consistent with its having a distinct binding site, and data available to date are consistent with IbTX recognizing only Ca$^{2+}$-activated channels in a variety of tissues (Galvez *et al.*, 1990). Studies with chemically synthesized chimeras of the two toxins suggest that C-terminal sequences of each toxin dictate channel specificity to a large degree: a construct containing the N-terminal 19 residues of ChTX and the C-terminal 18 from IbTX has IbTX-like properties and vice versa.

Two additional scorpion toxins have been shown to target the BK-type channel, at least preferentially. **Kaliotoxin,** recently purified and characterized from *Androctonus* venom, is approximately 50% homologous to ChTX, IbTX, and **noxiustoxin (vide infra)** and has been shown to suppress the whole-cell Ca$^{2+}$-activated K$^+$ current in mollusk and rabbit sympathetic nerve. The affected channel is insensitive to both apamin and dendrotoxins and is blocked by ChTX and TEA (Crest *et al.,* 1992). Noxiustoxin (NxTX), purified from *Centruroides* venom, is 50% identical to ChTX, but due to its relatively low affinity in many systems, and its apparent interaction with a multiplicity of channel types, it seems unlikely that NxTX will be widely useful for channel purification and/or discrimination studies. NxTX appears to bind most tightly to Jurkat cells and to the ChTX-sensitive channel from brain. In contrast, its affinity for Ca$^{2+}$-activated channels lies in the 0.1–1.0 $\mu$m range.

Two distinct, nonhomologous toxins directed at the SK channel are now known: **apamin** and **scyllatoxin.** The latter is 25% identical to ChTX, although most of this relatedness derives from alignment of the cysteine residues. Apamin is an 18-residue polypeptide containing two disulfide bonds; it has an $\alpha$-helical core comprising residues 9–15 flanked by regions rich in $\beta$-turns. Analysis of synthetic analogues of apamin highlights the importance of arginines-13 and -14 for activity. Electrophysiologic analyses and ion flux experiments indicate that apamin is directed against the SK

channel in guinea pig hepatocytes, murine neuroblastoma cells, and cultured rat muscle cells. Iodinated apamin has been used to demonstrate the existence of high affinity ($K_d = 10$–400 p$M$) receptors in rat brain, with affinity increasing with [K$^+$]. The chemical nature of the apamin binding protein has been studied by radiation inactivation and chemical cross-linking, but the results have been inconclusive, with multiple polypeptides being labeled.

Scyllatoxin seems to be directed against the same target as apamin, despite the fact that the two polypeptides display no sequence homology. In all systems analyzed to date, scyllatoxin competes with apamin for binding and/or exhibits the same functional effects as apamin. Although the latter toxin appears to bind more tightly ($K_D$ is 15 p$M$ for apamin and 80 p$M$ for scyllatoxin), both proteins display very high affinities (Auguste *et al.*, 1992).

While the positions of defined helix and sheet structures are conserved in charybdotoxin and scyllatoxin (with the exception of the first $\beta$-strand, which is absent in the latter polypeptide), very recent results indicate that these toxins, which act on different K$^+$ channel subtypes, are structurally distinct. In addition to lacking strand $\beta$-1, scyllatoxin also lacks many of the residues known to form an important hydrophobic cluster in charybdotoxin. Sequence homology suggests that this hydrophobic cluster is also maintained in noxiustoxius and iberiotoxins. In addition, scyllatoxin can be distinguished from the others based on positioning of cationic side chains. Three positively charged residues (arginines-25 and -34 and lysine-31) of charybdotoxin are replaced by leucine, aspartic acid, and glutamic acid, respectively, in scyllatoxin. Two of these residues, arginines-25 and -34, have been shown to be important for interaction of charybdotoxin with the rat muscle Ca-activated K channel: Mutagenesis to glutamine causes a large increase in the rate of toxin dissociation. Thus, scyllatoxin, which targets the SK channel, lacks these cationic sites but contains alternative ones that appear to be important for channel interaction. Modification of arginine-18 and histidine-38 of scyllatoxin appears to abolish its binding to rat brain channels. One possible interpretation of these data is that the charybdotoxin subgroup utilizes mainly positive residues in the $\beta$-sheet in its primary binding interaction, whereas scyllatoxin uses cationic sites in both the $\beta$-sheet and the $\alpha$-helix.

**Dendrotoxins** comprise yet another family of toxins that target the K$^+$ channel and at present are the only snake (*Dendroaspis* sp.) toxins known to do so. Dendrotoxins (Dtx; four homologs known) are grossly similar to the toxins discussed previously, being highly cationic molecules containing 57–59 amino acid residues cross-linked by three disulfide bonds. While there is no apparent sequence homology between the Dtx's and the other toxins described earlier, the Dtx's are clearly related to the small subunit of $\beta$-bungarotoxin and to bovine pancreatic trypsin inhibitor (BPTI). Very recently, the crystal structure of $\alpha$-Dtx has been solved at 2-Å resolution, showing the similarity between Dtx and BPTI. Although the author (Starzynski, 1992) failed to discuss this structure in the context of those of other polypeptide neurotoxins, examination of the relevant structures does not reveal any obvious relationships.

The similarity between Dtx and BPTI is even more curious in view of the fact that the latter protein binds to an internal site of the BK channel from skeletal muscle and that this binding is associated with the generation of subconducting states.

Dendrotoxins were first identified by their ability to stimulate neurotransmitter release at both central and peripheral neurons. Available electrophysiologic data now indicate that $\alpha$-Dtx recognizes a variety of voltage-dependent $K^+$-channels. The Dtx sensitivity of RCK channels expressed in *Xenopus* oocytes has been analyzed, showing that rapidly inactivating channels are less sensitive than slowly inactivating ones, although all four types were responsive. To date, no data demonstrating an effect of the Dtx's on $Ca^{2+}$-activated currents have been published. Based on immunologic cross-reactivity, the $\alpha$-Dtx receptor in mammalian brain is likely related to proteins encoded by the *shaker* locus, and mutations in the S5–S6 linker of RBK1 have been shown to strongly influence Dtx binding to this *shaker* homolog.

## IV. Voltage-Dependent Calcium Channels

Voltage-dependent $Ca^{2+}$ channels are a good deal more complex than those discussed earlier, containing five subunits ranging in size from approximately 30–170 kDa (Catterall, 1991). Nonetheless, certain structural features of the $Ca^{2+}$ channel are strikingly similar to those of the $Na^+$ and $K^+$ channels. Most importantly, the $\alpha_1$-subunit of the calcium channel is 29% identical to the $\alpha$-subunit of the rat brain $R_{II}$ $Na^+$ channel, and the two proteins have functionally related but chemically distinct residues at about the same frequency. Application of the usual predictive structural algorithms to $\alpha_1$-subunit suggests the protein to be organized into four homologous structural domains, each containing the now familiar six transmembrane helices. In addition, the sequence of the S4 helix suggests that, like its counterpart in $Na^+$ and $K^+$ channels, this region functions as the voltage sensor. Heterologous expression experiments have shown that the $\alpha_1$-subunit is essential to the properties of dihydropyridine-sensitive $Ca^{2+}$ channels.

Knowledge of the toxinology of $Ca^{2+}$ channels is not extensive, at least with regard to polypeptide effectors. There are no known scorpion, anemone, or snake toxins directed at this channel molecule. As of this writing, the only known peptide blockers of the $Ca^{2+}$ channel are the omega conotoxins and a family of peptides from funnel web spiders, which is structurally unrelated to the conotoxins; both groups prevent transmitter release at presynaptic terminals (Venema et al., 1992). At least seven **ω-conotoxins** have been described. Their common features include their cationic nature, the presence of three disulfide bonds, amidated C-termini, and a large number of hydroxylated amino acids, including in some cases hydroxyproline (Fig. 13). Two basic classes of these conotoxins are known, and homology across these classes amounts to about 40%. However, we again emphasize that this value may be misleading in view of the fact that approximately half the identical residues are involved in disulfide bonds. At least two classes

of spider toxins, distinguishable by size and amino acid sequences, are now known and have been designated either as **agatoxins** (not to be confused with agitoxin, discussed previously) or curtatoxins. The shorter group ($\mu$-toxins) contains 36 to 38 residues and is cross-linked by four disulfides; in addition, these $\mu$-toxins are distinguished from the polypeptides discussed previously in that they are generally less cationic. A second group of spider toxins ($\omega$-toxins) contains 65 to 75 residues and as many as six disulfides, and is more cationic than the shorter group. Sequences of representative spider toxins are also shown in Fig. 13, emphasizing their lack of relatedness to the conotoxins at the level of primary structure.

$\omega$-Conotoxins irreversibly block stimulus-evoked transmitter release in frog neuromuscular preparations; because the muscle retains the ability to respond to a direct stimulus, a presynaptic target for the toxin is indicated. That this target is the $Ca^{2+}$ channel has since been confirmed in chick dorsal root ganglia (DRG). The effect of this toxin seems to be irreversible: in experiments in which DRG was treated with 100 n$M$ toxin, then washed for 2 hr, no recovery of $Ca^{2+}$ activity was observed.

Toxins isolated from funnel web spider venom are in general less well characterized than the conotoxins with regard to their target channel. The unfractionated venom has been shown to block a dihydropyridine and $\omega$-CgTX-insensitive $Ca^{2+}$ current detectable in *Xenopus* oocytes following microinjection of rat brain mRNA. However, because these spider venoms are known to contain both peptide and nonpeptide toxins, the chemical nature of the active substance in these experiments is unknown. More recent analyses using purified $\omega$-agatoxins support the conclusion that these peptides target the $Ca^{2+}$ channel. $\omega$-Aga-IIA has been shown to inhibit binding of CgTX GVIA to chick brain membranes, while $\omega$-Aga-IA and -IB do not, suggesting that the spider toxins may be able to discriminate among channel subtypes. Additional evidence consistent with this hypothesis arises from analysis of the effects of Aga-IA, -IIA, and -IIIA on $Ca^{2+}$ entry into depolarized chicken synaptosomes. In this system, the latter two polypeptides block entry, with 50% effective doses in the nanomolar range, while $\omega$-Aga-IA is ineffective at 1 $\mu M$. Very recently, a new $\omega$-agatoxin, $\omega$-Aga-IVA, has been characterized and found to target P-type $Ca^{2+}$ channels in synaptosomes: these channels are distinct from the N type, and are resistant to both dihydropyridine and $\omega$-CgTX blockade. Based on its sequence, $\omega$-Aga-IVA may represent yet another class of spider toxins.

Because many distinct $Ca^{2+}$ channel subtypes have been characterized electrophysiologically, it is natural to ask whether all are equally susceptible to the $\omega$-conotoxins. Clearly, from the experiments described earlier, muscle channels are unaffected. However, even among neuronal channels, at least three subtypes are distinguishable by a variety of criteria, including gating kinetics, unit conductance, and antagonist pharmacology: Nowycky and co-workers, who studied them in chick DRG, designated these channels as T, N and L types. The L-type channel is distinguished by its sensitivity to dihydropyridines (DHPs), whereas T types are not affected by these agents. $\omega$-Cono-

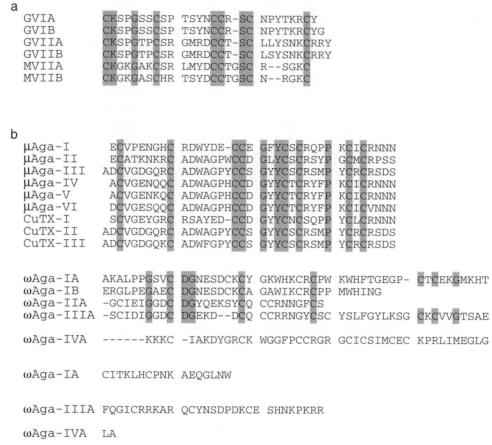

a

| | | | |
|---|---|---|---|
| GVIA | CKSPGSSCSP | TSYNCCR-SC | NPYTKRCY |
| GVIB | CKSPGSSCSP | TSYNCCR-SC | NPYTKRCYG |
| GVIIA | CKSPGTPCSR | GMRDCCT-SC | LLYSNKCRRY |
| GVIIB | CKSPGTPCSR | GMRDCCT-SC | LSYSNKCRRY |
| MVIIA | CKGKGAKCSR | LMYDCCTGSC | R--SGKC |
| MVIIB | CKGKGASCHR | TSYDCCTGSC | N--RGKC |

b

| | | | |
|---|---|---|---|
| μAga-I | ECVPENGHC | RDWYDE-CCE | GFYCSCRQPP | KCICRNNN |
| μAga-II | ECATKNKRC | ADWAGPWCCD | GLYCSCRSYP | GCMCRPSS |
| μAga-III | ADCVGDGQRC | ADWAGPYCCS | GYYCSCRSMP | YCRCRSDS |
| μAga-IV | ACVGENQQC | ADWAGPHCCD | GYYCTCRYFP | KCICRNNN |
| μAga-V | ACVGENKQC | ADWAGPHCCD | GYYCTCRYFP | KCICRNNN |
| μAga-VI | DCVGESQQC | ADWAGPHCCD | GYYCTCRYFP | KCICVNNN |
| CuTX-I | SCVGEYGRC | RSAYED-CCD | GYYCNCSQPP | YCLCRNNN |
| CuTX-II | ADCVGDGQRC | ADWAGPYCCS | GYYCSCRSMP | YCRCRSDS |
| CuTX-III | ADCVGDGQKC | ADWFGPYCCS | GYYCSCRSMP | YCRCRSDS |

| | | | | |
|---|---|---|---|---|
| ωAga-IA | AKALPPGSVC | DGNESDCKCY | GKWHKCRCPW | KWHFTGEGP- | CTCEKGMKHT |
| ωAga-IB | ERGLPEGAEC | DGNESDCKCA | GAWIKCRCPP | MWHING |
| ωAga-IIA | -GCIEIGGDC | DGYQEKSYCQ | CCRNNGFCS |
| ωAga-IIIA | -SCIDIGGDC | DGEKD--DCQ | CCRRNGYCSC | YSLFGYLKSG | CKCVVGTSAE |

| | | | | |
|---|---|---|---|---|
| ωAga-IVA | ------KKKC | -IAKDYGRCK | WGGFPCCRGR | GCICSIMCEC | KPRLIMEGLG |

| | |
|---|---|
| ωAga-IA | CITKLHCPNK | AEQGLNW |

| | |
|---|---|
| ωAga-IIIA | FQGICRRKAR | QCYNSDPDKCE | SHNKPKRR |

| | |
|---|---|
| ωAga-IVA | LA |

**FIG. 13.** (a) Primary structures of ω-conotoxin variants. The sequences are shown in the single-letter code, with invariant residues or functionally conservative residues highlighted. Disulfide bonds link Cys1–Cys16, Cys15–Cys27, and Cys8–Cys20. (b) Primary structures of representative spider toxins specific for Ca channels. Highlighted positions are invariant within a group or occupied by functionally conservative replacements. Where complete sequences have not been determined, N-terminal sequences are shown. Aga, agatoxins; CuTX, curtatoxins.

toxins block T-type channels only weakly, whereas inhibition of neuronal L-type channels is essentially complete and irreversible in almost all cases; the irreversibility precludes an accurate determination of $K_d$.

Binding of ω-conotoxins to a variety of tissues has been analyzed, with $K_d$ values in the subnanomolar range commonly reported. However, the validity of these values may fairly be questioned, given the irreversible nature of the toxin's effects. What is most clear is the lack of competition between the conotoxins and any class of organic channel blockers: DHP, benzothiazepine, and phenylalkylamine binding is unaffected in the presence of high concentrations of any of these drugs. Reported densities of ω-conotoxin binding sites are in the range of picomoles per milligram membrane protein in a variety of vertebrate neuronal preparations. Binding analyses also suggest that channels have either a DHP or a conotoxin site, although since $L_n$ channels can be blocked by both agents, exceptions to this rule must exist. In interpreting the literature on ω-conotoxin binding, remember that distinct forms of this toxin are made by different *Conus* species, and these toxins do not invariably have the same specificity.

ω-Conotoxins have been cross-linked to putative receptors in rat and chick brain. The variation in the molecular weight of this complex on reduction suggests that the ω-conotoxin "receptor" is associated with the $\alpha_2$-$\delta$-subunit of the $Ca^{2+}$ channel. However, given the ability of the ω-conotoxins to block this channel, and the fact that the $\alpha_1$-subunit is integral to ion permeation, it is reasonable to postulate interactions of the toxins with $\alpha_1$ as well.

## V. Other Toxins and Channels

### A. α-Conotoxins and the Nicotinic Acetylcholine Receptor-Associated Channel

Although not targeted to ion channels, the α-conotoxins are briefly discussed here in the interest of covering all of the conotoxin classes. The α-conotoxins are the smallest of the conotoxin peptides known, comprising 13 to 15 amino acid residues of which 4 are half-cystines involved in disulfide bonds. All of the known α-conotoxins are cationic, with formal charges of +1.5 to +3.5 at neutral pH.

Sequences of the α-conotoxins are shown in Fig. 14. Conservative changes are tolerated at most positions without significant loss of activity, so that relatively little information on structure–function relationships can be derived from synthetic analogues. It is clear that the presence of two disulfides in such a short peptide imposes relatively severe constraints on its three-dimensional structure. The proposed structure for α-conotoxin, based on NMR measurements, includes β-turns at residues 5–8 and 9–12.

α-Conotoxins are directed at the nicotinic acetylcholine receptor channel (nAChR), and mechanistically act like to the snake toxins exemplified by α-bungarotoxin. In fact, the most widely accepted model for α-conotoxin structure places several of its side chain functional groups into positions very similar to those seen in the "active site" of bungarotoxin. The postsynaptic site of action of α-conotoxin is consistent with its observed ability to compete with other nAChR antagonists like d-tubocurarine and α-BgTX. On this basis, α-conotoxin is, strictly speaking, not truly directed against the ion channel, and will not be discussed further.

## B. Chlorotoxin and Chloride Channels

The toxinology of anion-selective channels is not nearly as well understood as that of the cation channels described previously and therefore represents a fertile ground for future enquiry. The recent literature is replete with reports of chloride channels in a variety of tissues, which are activated by diverse stimuli including membrane potential, cyclic AMP, volume, pH, etc. (Cuppoletti et al., 1993; Pusch et al., 1995a,b; Malinowska et al., 1995). These channels, which are all members of the family designated ClC, are formed by proteins of $M_r$ 90–100 kDa, although the oligomeric state of the functional channel remains undefined at present. Hydropathy-based models predict that both their N- and C-termini are located in the cytoplasm, and are consistent with the presence of 12 transmembrane domains, which are modeled as α-helices. Structure–function analysis of these channels is still at a rather early stage, and although recent reports have identified N-terminal cytoplasmic sequences as essential for channel inactivation, identification of residues involved in forming the conducting pore has thus far proven elusive. Clearly, high-affinity toxins analogous to TTX or charybdotoxin would be an invaluable tool in such analyses. However, to date only a single peptide blocker of chloride channels has been identified. This polypeptide, isolated from *Leiurus* venom and designated as **chlorotoxin,** contains 37 amino acid residues cross-linked by four disulfide bonds, and displays no

sequence homology to toxins from the same venom that target Na or K channels (DeBin et al., 1993). It is, however, highly homologous to the scorpion insectotoxins derived from *Buthus eupeus* venom. Interestingly, although chlorotoxin blocks enterocyte chloride conductances with an affinity in the low micromolar range, it does so only when applied to the interior aspect of the channel; thus, its biological role is unlikely to involve chloride channel blockade. The homologous insectotoxins have been suggested to target the insect postsynaptic glutamate receptor, although supporting data for this proposal have not been presented, and the effects of chlorotoxin on this receptor are unknown.

The solution structure of chlorotoxin was recently solved by Lippens et al. (1995) using multidimensional NMR. Not surprisingly, the averaged structure, which contains a small three-stranded β-sheet packed against a single α-helix, is very similar to that of the insectotoxin. More unexpected is the observation that the overall fold of the molecule is rather similar to that of charybdotoxin, despite the lack of significant sequence homology between the two proteins. While structure–function relationships have not been addressed experimentally for either the insectotoxins or chlorotoxin, our recent ability to express active toxin in bacteria opens up this area for further exploitation.

## VI. Summary

Neurotoxins have proven to be invaluable tools in the purification of many of the best characterized cation channels, and in addition have contributed significantly to our understanding of how these macromolecules function in excitable tissues. As should be clear from the foregoing outline, this collection of molecules displays a great deal of diversity in both a chemical and a functional sense. Although many of the compounds discussed in the preceeding sections are cationic, additional unifying structural themes are, in essence, conspicuous by their absence. Even within the subgroup of polypeptide toxins, the most well-studied groups (i.e., anemone, scorpion, snail, and snake toxins) fail to display any intergroup relatedness at the level of primary structure. Indeed, this is rather surprising in view of the ability of, for example, anemone and α-scorpion toxins to compete for a common receptor. While all of the polypeptides do share a cationic character and are rich in disulfide bonds, it remains to be proven that the former property is intimately related to their activity; the latter is almost certainly a consequence of the necessity for stabilizing small peptides in an extracellular environment. Perhaps the most striking observation is the relatedness between charybdotoxin and the $Na^+$-channel-directed scorpion toxins at the level of their three-dimensional structures. This clearly raises the possibility of common binding mechanisms having evolved.

Given the vast diversity of the venoms from which each of these toxins is derived, and the growing array of ion channels purified and/or channel activities characterized, it is likely that new and interesting toxins will continue to emerge in the future, and that their characterization will add significantly to our knowledge of channels and recep-

```
G1      ECCNPACGRH  YSC
G1A     ECCNPACGRH  YSCGK
G11     ECCHPACGKH  FSC
MI      GRCCHPACGKN YSC
```

**FIG. 14.** Primary structures of α-conotoxins. The sequences are shown in the single-letter code, with highlighted residues either invariant of allowing only functionally conservative substitution.

tors in excitable tissues. Two areas of research that might prove especially fruitful involve using existing (or newly discovered) toxins to probe for specific channel isoforms in different tissues, and using behavioral assays to identify new venom components likely to have neuronal activities. This latter approach has already been applied to components of *Conus geographus* venom with very intriguing results, among them being the discovery of the so-called "sleeper" and "King Kong" peptides. It is likely that this surface has as yet only been scratched, and that future studies will unveil additional toxins that interact with new, and perhaps presently unknown, target sites.

## Bibliography

Albuquerque, E. X., Daly, J. W., and Witkop, B. (1971). Batrachotoxin: chemistry and pharmacology. *Science* **172**, 995–1002.

Angelides, K., Terakawa, S., and Brown, G. B. (1986). Spatial relations of the neurotoxin binding sites on the sodium channel. *Ann. NY Acad. Sci.* **479**, 221–237.

Auguste, P., Hugues, M., Mourre, C., Moinier, D., Tartar, A., and Lazdunski, M. (1992). Scyllatoxin, a blocker of Ca-activated K-channels: Structure–function relationships and brain localization of the binding sites. *Biochemistry* **31**, 648–654.

Bontems, F., Roumestand, C., Gilquin, B., Menez, A., and Toma, F. (1991). Refined structure of charybdotoxin: Common motifs in scorpion toxins and defensins. *Science* **254**, 1521–1523.

Bontems, F., Gilquin, B., Roumestand, C., Menez, A., and Toma, F. (1992). Analysis of side-chain organization on a refined model of charybdotoxin: Structural and functional implications. *Biochemistry* **31**, 7757–7764.

Cannon, S. C., and Corey, D. P. (1993). Loss of Na-channel inactivation by anemone toxin mimics the myotonic state in HPP. *J. Physiol.* **466**, 501.

Catterall, W. A. (1977). Activation of the action potential Na-ionophore by neurotoxins: An allosteric model. *J. Biol. Chem.* **252**, 8669–8676.

Catterall, W. A. (1991). Functional subunit structure of voltage-gated calcium channels. *Science* **253**, 1499–1500.

Catterall, W. A. (1995). Structure and function of voltage-gated ion channels. *Annu. Rev. Biochem.* **64**, 493–531.

Catterall, W. A., and Beress, L. (1978). Sea anemone toxin and scorpion toxin share a common receptor site associated with the action potential sodium ionophore. *J. Biol. Chem.* **253**, 7393–7396.

Chahine, M., George, A. L. Jr., Zhou, M., Ji, S., Sun, W., Barchi, R. L., and Horn, R. (1994). Na channel mutation in paramyotonia congenita uncouple inactivation from activation. *Neuron* **12**, 281.

Crest, M., Jacquet, G., Gola, M., Zerrouk, H., Benslimane, A., Rochat, H., Mansuelle, P., and Martin-Eauclaire, M.-F. (1992). Kaliotoxin, a novel peptidyl inhibitor of neuronal BK-type Ca-activated K-channels characterized from *Androctonus mauretanicus mauretanicus* venom. *J. Biol. Chem.* **267**, 1640–1647.

Cruz, L. J., Gray, W. R., Olivera, B. M., Zeikus, R. D., Kerr, L., Yoshikami, D., and Moczydlowski, E. (1985). *Conus geographus* toxins that discriminate between neuronal and muscle sodium channels. *J. Biol. Chem.* **260**, 9280–9288.

Cruz, L. J., Kupryszewski, G., LeCheminant, G. W., Gray, W. R., Olivera, B. M., and Rivier, J. (1989). μ-Conotoxin GIIIA, a peptide ligand for muscle sodium channels: Chemical synthesis, radiolabeling and receptor characterization. *Biochemistry* **28**, 3437–3442.

Cuppoletti, J., Baker, A. M., and Malinowska, D. H. (1993). Chloride channels of the gastric parietal cell that are active at low pH. *Am. J. Physiol.* **264**, C1609.

DeBin, J. A., Maggio, J. E., and Strichartz, G. R. (1993). Purification and characterization of chlorotoxin, a chloride channel ligand from the venom of the scorpion, *Am. J. Physiol.* **264**, C361–C369.

Dias-Kadambi, B. L., Combs, K. A., Drum, C. L., Hanck, D. A., and Blumenthal, K. M. (1996a). The role of exposed tryptophan residues in the activity of the cardiotonic polypeptide anthopleurin B. *J. Biol. Chem.* **271**, 23828–23835.

Dias-Kadambi, B. L., Drum, C. L., Hanck, D. A., and Blumenthal, K. M. (1996b). Leucine-18, a hydrophobic residue essential for high affinity binding of anthopleurin-B to the voltage sensitive sodium channel. *J. Biol. Chem.* **271**, 9422–9429.

Doughtery, D. A. (1996). Cation-π interactions in chemistry and biology: A new view of benzene, Phe, Tyr and Trp. *Science* **271**, 163–168.

Fontecilla-Camps, J. C., Almassy, R. J., Suddath, F. L., Watt, D. D., and Bugg, C. E. (1980). Three-dimensional structure of a protein from scorpion venom: a new structural class of neurotoxins. *Proc. Natl. Acad. Sci. USA* **77**, 6496–6500.

Fozzard, H. A., and Hanck, D. A. (1996). Structure and function of voltage-dependent sodium channels: Comparison of brain II and cardiac isoforms. *Physiol. Rev.* **76**, in press.

French, R. J., Sochaczewski, E. P., Zamponi, G. W., Becker, S., Kularatna, A. S., and Horn, R. (1996). Interactions between a pore-blocking peptide and the voltage sensor of the sodium channel: An electrostatic approach to channel geometry. *Neuron* **16**, 407–413.

Galvez, A., Gimenez-Gallego, G., Reuben, J. P., Roy-Contancin, L., Feigenbaum, P., Kaczorowski, G. J., and Garcia, M. L. (1990). Purification and characterization of a unique, potent, peptidyl probe for the high conductance calcium-activated potassium channel from venom of the scorpion *Buthus tamulus*. *J. Biol. Chem.* **265**, 11083–11090.

Goldstein, S. A. N. (1996). A structural vignette common to voltage sensors and conduction pores: Canaliculi. *Neuron* **16**, 717–722.

Goldstein, S. A. N., Pheasant, D. J., and Miller, C. (1994). The charybdotoxin receptor of a *shaker* K+ channel: Peptide and channel residues mediating molecular recognition. *Neuron* **12**, 1377–1388.

Hartshorne, R. P., and Catterall, W. A. (1984). The sodium channel from rat brain: Purification and subunit composition. *J. Biol. Chem.* **259**, 1667–1675.

Hidalgo, P., and MacKinnon, R. (1995). Revealing the architecture of a K+ channel pore through mutant cycles with a peptide inhibitor. *Science* **268**, 307–310.

Hille, B. (1975). An essential ionized acid group in sodium channel. *Federation Proc.* **34**, 1318–1321.

Hoshi, T., Zagotta, W. N., and Aldrich, R. W. (1990). Biophysical and molecular mechanisms of *shaker* channel inactivation. *Science* **250**, 533–538.

Kerr, L. M., and Yoshikami, D. (1984). A venom peptide with a novel presynaptic blocking action. *Nature* **308**, 282–284.

Khera, P. K., and Blumenthal, K. M. (1994). Role of the cationic residues arginine-14 and lysine-48 in the function of the cardiotonic polypeptide anthopleurin B. *J. Biol. Chem.* **269**, 921–926.

Khera, P. K., Benzinger, G. R., Lipkind, G., Drum, C. L., Hanck, D. A., and Blumenthal, K. M. (1995). Multiple cationic residues of anthopleurin B that determine high affinity and channel isoform discrimination. *Biochemistry* **34**, 8533–8541.

Lippens, G., Najib, J., Wodak, S. J., and Tartar, A. (1995). NMR sequential assignments and solution structure of chlorotoxin, a small scorpion toxin that blocks chloride channels. *Biochemistry* **34**, 13–21.

MacKinnon, R., and Miller, C. (1989). Mutant potassium channels with altered binding of charybdotoxin, a pore-blocking inhibitor. *Science* **245**, 1382–1385.

MacKinnon, R., and Yellen, G. (1990). Mutations affecting TEA blockade and ion permeation in voltage-activated K-channels. *Science* **250**, 276–279.

Malinowska, D. H., Kupert, E. Y., Bahinski, A., Sherry, A. M., and Cuppoletti, J. (1995). Cloning, functional expression and characterization of a PKA-activated gastric Cl⁻ channel. *Am. J. Physiol.* **268,** C191–200.

Miller, C. (1991). 1990: Annus mirabilis of potassium channels. *Science* **252,** 1092–1096.

Miller, C., Moczydlowski, E., Latorre, R., and Phillips, M. (1985). Charybdotoxin, a protein inhibitor of single Ca-activated K-channels from mammalian skeletal muscle. *Nature* **313,** 316–318.

Monod, J., Wyman, J., and Changeux, J. P. (1965). On the nature of allosteric transitions: a plausible model. *J. Mol. Biol.* **12,** 88–118.

Mosher, H. S. (1986). The chemistry of tetrodotoxin. *Ann. NY Acad. Sci.* **479,** 32–43.

Naini, A. A., and Miller, C. (1996). "A symmetry-driven search for electrostatic interaction partners in charybdotoxin and a voltage-gated K⁺ channel. *Biochemistry* **35,** 6181–6187.

Naranjo, D., and Miller, C. (1996). A strongly interacting pair of residues on the contact surface of charybdotoxin and a *shaker* K⁺ channel. *Neuron* **16,** 123–130.

Noda, M., Suzuki, H., Numa, S., and Stuhmer, W. (1989). A single point mutation confers tetrodotoxin and saxitoxin insensitivity on the sodium channel II. *FEBS Lett.* **259,** 213–216.

Ohizumi, Y., Nakamura, H., Kobayashi, J., and Catterall, W. A. (1986). Specific inhibition of saxitoxin binding to skeletal muscle sodium channels by geographutoxin II, a polypeptide channel blocker. *J. Biol. Chem.* **261,** 6149–6152.

Pusch, M., Ludewig, U., Rehfeldt, A., and Jentsch, T. J. (1995a). Gating of the voltage-dependent chloride channel ClC-0 by the permeant anion. *Nature* **373,** 527–531.

Pusch, M., Steinmeyer, K., Koch, M. C., and Jentsch, T. J. (1995b). Mutations in dominant human myotonia congenita drastically alter the voltage dependence of the ClC-1 chloride channel. *Neuron* **15,** 1455–1463.

Ranganathan, R., Lewis, J. H., and MacKinnon, R. (1996). Spatial localization of the K⁺ channel selectivity filter by mutant cycle-based structure analysis. *Neuron* **16,** 131–139.

Satin, J., Kyle, J. W., Chen, M., Bell, P., Cribbs, L. L., Fozzard, H. A., and Rogart, R. B. (1992). A mutant of TTX-resistant cardiac sodium channels with TTX-sensitive properties. *Science* **256,** 1202–1205.

Shimizu, Y. (1986). Chemistry and biochemistry of saxitoxin analogues and tetrodotoxin. *Ann. NY Acad. Sci.* **479,** 24–31.

Starzynski, T. (1992). Crystal structure of α-dendrotoxin from the green mamba venom and its comparison with the structure of bovine pancreatic trypsin inhibitor. *J. Mol. Biol.* **224,** 671–683.

Stephan, M., and Agnew, W. S. (1991). Voltage-sensitive Na-channels: Motifs, modes and modulation. *Curr. Opin. Cell Biol.* **3,** 676–684.

Stuhmer, W., Conti, F., Suzuki, H., Wang, X., Noda, M., Yahagi, N., Kubo, H., and Numa, S. (1989a). Structural parts involved in activation and inactivation of the sodium channel. *Nature* **339,** 597–603.

Stuhmer, W., Ruppersberg, J. P., Schroter, K. H., Sakmann, B., Stocker, M., Giese, K. P., Perschke, A., Baumann, A., and Pongs, O. (1989b). Molecular basis of functional diversity of voltage-gated potassium channels in mammalian brain. *EMBO J.* **8,** 3235–3244.

Tamkun, M. M., Talvenheimo, J. A., and Catterall, W. A. (1984). The sodium channel from rat brain: Reconstitution of neurotoxin-activated ion flux and scorpion toxin binding from purified components. *J. Biol. Chem.* **259,** 1676–1687.

Trainer, V. L., Brown, G. B., and Catterall, W. A. (1996). Site of covalent labeling by a photoreactive batrachotoxin derivative near transmembrane segment IS6 of the sodium channel α subunit. *J. Biol. Chem.* **271,** 11261–11267.

Tudor, J. E., Pallaghy, P. K., Pennington, M. W., and Norton, R. S. (1996). Solution structure of ShK toxin, a novel potassium channel inhibitor from a sea anemone. *Nature Struct. Biol.* **3,** 317–320.

Venema, V. J., Swiderek, K. M., Lee, T. D., Hathaway, G. M., and Adams, M. E. (1992). Antagonism of synaptosomal calcium channels by subtypes of omega agatoxins. *J. Biol. Chem.* **267,** 2610–2615.

Yang, N., Ji, S., Zhou, M., Ptáček, L. J., Barchi, R. L., Horn, R., and George, A. L. Jr. (1994). Na channel mutations in paramyotonia congenita exhibit similar biophysical phenotypes *in vitro. Proc. Natl. Acad. Sci. USA* **91,** 12785.

R. Kent Hermsmeyer

# 37

# Ion Channels as Targets for Drugs

## I. Introduction

Creative thinking is a valuable asset that has in many instances allowed great conceptual advances in physiology. The process to be explained can yield a solution that would not have been possible without developing whole new concepts, in many ways analogous to similar breakthroughs in chemistry, engineering, and physics. An example of development of new concepts based largely on correct intuition is the prediction made more than 50 years ago to explain how ions pass through cell membranes that are composed of lipids and thus nearly impermeable to charged substances (Hodgkin, 1964). This insightful intuition was based on a mechanistic picture of water-filled, minute channels through cell membranes that provide a route for charged particles to move through the insulating lipid bilayer of cell membranes. Pioneering electrophysiologists introduced the concept of ion channels to account for the electrical charge movement, which was measured at a rate many orders of magnitude greater than could occur through the fastest of known enzyme reactions. As they correctly hypothesized, water-filled ion pores, or channels, of approximately the dimension of an ion in solution would be sufficient to account for the currents measured in experiments. Increasing sophistication in these experiments, especially the development of voltage-clamp techniques, allowed the isolation of individual types of ionic currents, and pointed strongly to the existence of ion channels with characteristic voltage and kinetic properties.

During recent years, a virtual explosion in information about ion channels has resulted from the insightful development of recording techniques allowing measurement of currents through individual ion channels. Improved technical quality and isolations that permit experiments on single cells have led to important discoveries. The history of electrophysiology has been a consistent wealth of the most advanced scientific thinking from the early days of recording action potentials with 1-mm-diameter wire electrodes through the present high-resolution recording technique that uses glass pipettes that have an internal diameter of about 0.1 $\mu$m. The Nobel Prize for physiology and medicine was awarded to Bert Sakmann and Erwin Neher in 1991 for the development of **patch-clamp techniques,** which have allowed investigation of very small cells that were not otherwise accessible (Neher and Sakmann, 1992). These patch-clamp techniques, which literally allow the submillisecond resolution of the action of an individual protein molecule in the membrane of a living cell, have led to the establishment of an entire field of investigation that is the subject of this chapter.

Ion channels are extraordinarily important because they represent the **molecular basis for excitability** of neurons, muscle cells, and secretory cells (Sakmann, 1992). These and other excitable cells depend on similar ion channel mechanisms for signaling. The essential importance of these elements of **membrane signals** indicates that this field is permanently established as an integral component of both biology and medical research. By studying ion channels, we find answers to fundamental questions that strongly influence our working knowledge of living cells. Because ion channels are targets for a growing list of transmitters and drugs, $Ca^{2+}$, $K^+$, and $Na^+$ channels are being considered individually, with examples of agents acting on them.

## II. Ion Channels as Receptors

A major thesis of this chapter is that the distinction between **ion channels** and **membrane receptors** is diminishing as more details become clear about membrane function. The earliest recognition of membrane proteins triggering events in response to drugs was with membrane receptors. From the classical work of Alquist in the adrenergic and others in the cholinergic parts of the autonomic nervous system, and with substantial development from the beautiful model systems offered by frog neuromuscular junctions, giant synapses in squid, and other convenient animal models, the concept of **receptor occupation** as **ligand binding** provided important insight that virtually revolutionized the

**564**

way in which one perceives how hormones and neurotransmitters act (Sakmann, 1992). These perceptions have given considerable impetus to the development of the discipline of pharmacology.

The concept of the function of ion channels is much more recent. Ion channels as an experimental subject awaited the decade of the 1980s for widespread acceptance. The hypothetical construct, that **water-filled ion channels in the cell membrane** provide the mechanism by which the most rapid electrical signaling underlying instantaneous responsiveness occurs, was visionary. This brilliant hypothesis has only been established in the last few years. The similarity between ion channels and the membrane receptors for adrenergic (and other) hormones and transmitters was recognized as increasing details of their nature were revealed (Venter and Triggle, 1987). Both receptors and ion channels consist of proteins that have several regions that are ideal for passing through the cell membrane. In the case of ion channels, there is coiling to form water-filled channels. The **aqueous channel** is an ideal tunnel by which highly charged ions can pass through the lipid environment of the cell membrane. Receptors consist of a single, membrane-spanning molecule that contains both extracellular sensors and intracellular signals (Triggle *et al.*, 1989). Often a protein will dissociate to activate (or inhibit) a regulatory cascade with contraction or secretion as the endpoint. Because there are so many similarities between the membrane-spanning ion channels and membrane receptors, it becomes more understandable that interactions of drugs with ion channels and receptors are similar.

The resemblances of regulation between ion channels and receptors are the rule rather than the exception. The interaction of $Ca^{2+}$ antagonists with $Ca^{2+}$ channels occurs with high affinity (in the range of nanomolar dissociation constants), which is in virtually the same realm as hormones and neurotransmitters with their membrane receptors. Furthermore, these interactions are reversible, stereospecific, subject to competition, and can be characterized with virtually every concept developed for receptors. For example, competitive and noncompetitive inhibition, allosteric interactions, and partial agonists all are found in ion channels, as well as with receptors (Glossmann *et al.*, 1989).

## III. $Ca^{2+}$ Channels and Their Ligands

One of the outstanding characteristics of $Ca^{2+}$ channels that has been revealed by ion channel studies is that stereoisomers of a given drug (i.e., dihydropyridine) can have opposite effects on an ion channel, with one acting as **agonist** and one as **antagonist.** The most famous of these combinations is the compound produced by Bayer, known by the company number Bay-K-8644. In this Bay-K-8644 dihydropyridine (DHP), the agonist (plus isomer) action dominates the racemic mixture, and enhancement of $Ca^{2+}$ current by several-fold is produced. With the pure agonist isomer (Bay-R-5417), there is even greater enhancement, up to $10\times$, as shown in Fig. 1 (Bean *et al.*, 1986). In contrast, the minus isomer is an antagonist. Similarly for the analogous Sandoz compound known as SDZ 202-791, the plus

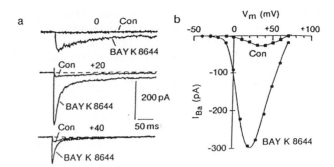

**FIG. 1.** Augmentation of calcium channel current by Bay-K-8644. (a) Currents shown were recorded before or 2 min after superfusing the cell with solution containing 200 n$M$ Bay-K-8644. (b) Peak current vs test potential before and after drug. Cell B49B, 115 Ba//Cs glutamate, cell capacity 35 pF. (Reproduced with permission from Mishra and Hermsmeyer, 1994, *Circ. Res.* **75**, 144–148. Copyright 1994 American Heart Association.)

isomer is the agonist and the minus isomer is the antagonist. This kind of **stereospecificity** is probably the result of a binding site that has several three-dimensional constraints, which result in agonist or antagonist action depending on binding to multiple facets of the channel. Strong binding to only a fraction of these sites might produce block of the channel and occupation of all sites might activate the ion channel analogous to a receptor. On the other hand, ion channel block could be a result of all the binding facets and agonist action might be the result of occupying only part of the sites.

Additional classes of ion channels also need to be described, although documentation of their function as drug targets is less clear. Chloride channels exist in several kinds of cells and can provide an important depolarizing current. The distribution of intracellular chloride is not in electrochemical equilibrium, but is in fact higher than would be predicted by the potential across membranes of many kinds of cells. Thus, there is an electrochemical gradient for chloride to move out of cells when ion channels are opened, and since the chloride has a charge of $-1$, this outward movement of a negative ion is a depolarizing influence. Such currents are known to be activated in vascular muscle by norepinephrine (Loirand *et al.*, 1990; Pacaud *et al.*, 1991, 1992) or thrombin (Baron *et al.*, 1993), and might contribute to activation of vascular muscle cells under certain circumstances.

## IV. $K^+$ Channels and Their Ligands

Another important type of channel as a drug target is the $K^+$ channel. Some $K^+$ channels respond to a new class of drugs called **$K^+$ channel openers** that cause hyperpolarization and relaxation. Such $K^+$ channel openers form a new major class of vasodilator, with a mechanism of action similar in certain respects to that discovered for minoxidil. Like minoxidil, which has been used as a potent vasodilator for decades, the newer compounds, cromakalim and pinacidil, act primarily by increasing the open probability of $K^+$ channels. It appears that ATP-sensitive $K^+$ channels that are inhibited by glyburide may be involved in this vasodilator action (Meisheri *et al.,* 1993). The glyburide inhibition, presumably of ATP-sensitive $K^+$ channels, demonstrates how agents can block relaxation by decreasing $K^+$ efflux in vascular muscle cells. This depolarization by $K^+$ efflux inhibition was first demonstrated in the $\beta$ cells of the endocrine pancreas (Ashcroft *et al.,* 1984). Glyburide (and related sulfonylureas) cause inhibition of this fraction of $K^+$ channels and appear to inhibit the primary action of $K^+$ channel openers. When studied as vascular muscle relaxation, there is a shift to the right by up to two orders of magnitude by glyburide of the relaxation dose–response curve for pinacidil and other $K^+$ channel openers (Meisheri *et al.,* 1993).

Other types of $K^+$ channels, such as $Ca^{2+}$-dependent $K^+$ channels, have even larger conductances, and can therefore dominate membrane potential in certain cells. The largest conductance is found in $Ca^{2+}$-dependent $K^+$ channels, which are inhibited by the scorpion toxin, charybdotoxin. This toxin was relatively ineffective in inhibiting the hyperpolarizations caused by $K^+$ channel openers. Thus, $K^+$ channel openers may act preferentially on the ATP-sensitive $K^+$ channels, and perhaps on other types of $K^+$ channels. Conversely, there are agents that cause block of $K^+$ channels. The most common experimentally used $K^+$ channel blockers are $Ba^{2+}$ and $Cs^+$. Both of these agents have been used often, especially in voltage-clamp experiments, to allow isolation of the much smaller $Ca^{2+}$ currents, which can only be detected under conditions where the $K^+$ currents are blocked. Thus, $K^+$ channels may be either made to spend more time in an open (or closed) state by exposure to agents useful for experimental conditions. These discoveries will be conceptionally useful in the design of new drugs.

## V. $Na^+$ Channels and Their Ligands

Even $Na^+$ channels provide a useful target for block by drugs. The class I antiarrhythmics, such as lidocaine, effectively block $Na^+$ channels, and this blockade is use dependent (Hille, 1992). That is to say, the higher the frequency of stimulation, the more potently lidocaine blocks the $Na^+$ channels. This is believed to be the result of the preferential block of $Na^+$ channels in the inactivated state, which occurs more rapidly and for a greater percentage of the time at high frequencies of stimulation. As an antiarrhythmic, this is highly desirable because it is the high-frequency stimulation in the form of cardiac action potentials that causes ventricular fibrillation, which is life-threatening. On the other hand, $Na^+$ channels operating at normal (lower) heart rates are less inhibited, and thus the normal heart rhythm is much less affected. Lidocaine also blocks $Na^+$ channels in neurons and various kinds of muscle cells, making it a useful experimental tool (Hille, 1992). Even more dramatic block of $Na^+$ channels is found with the specific toxins that have a high affinity for $Na^+$ channels. Tetrodotoxin is the pufferfish poison that is well known (at least in James Bond fiction) as a deadly poison because it blocks $Na^+$ channels in neurons with high affin-

ity. The affinity is much lower in cardiac muscle and even lower in vascular muscle cells. However, tetrodotoxin and the closely related toxin, saxitoxin, are very useful for identification of currents through $Na^+$ channels. Much more detail about toxins, including more details and examples of how and where they act on $Na^+$ channels, can be found in Chapter 36.

## VI. Examples of Drugs Acting on Multiple Ion Channels

Thus, a number of ion channels provide relatively specific targets for drugs, and an increasing number of developing classes (by mechanism of action and molecular structure). Some well-known drugs even act on ion channels in combination. An example is veratradine, which appears to open both $Na^+$ and $K^+$ channels, and in the dual action to cause a depolarization (the relative increase in $Na^+$ channel opening is much larger). Veratradine opening of $Na^+$ channels can be occasionally used to search for the presence of $Na^+$ channels that are not otherwise detectable. This example shows that a combination of ion channel effects can occur, which emphasizes the relative complexity that is involved in the development of accurate concepts of modulation of ion channels by drugs.

## VII. Conceptual Basis for Modulation of Ion Channels by Drugs

A large number of molecules act on ion channels, and this rich, almost unlimited, array reveals considerable insight into membrane function at the molecular level. However, the number of important types of drug or regulatory molecule actions on ion channels presently known is limited, so the reader should not be concerned by having to learn a vast array of pharmacological mechanisms. In fact, the list of well-understood mechanisms is relatively short. The $Ca^{2+}$ channel is treated as a prototype in this chapter because it has been intensively studied and is of fundamental importance to the function of most cells. The nearly ubiquitous distribution of $Ca^{2+}$ channels among different types of cells and organs probably also contributes to universal interest in $Ca^{2+}$ channels. For $Ca^{2+}$, $K^+$, and $Na^+$ channels, there is vigorous study of structure and functions, gating mechanisms, and cell biology, yielding important new insights.

Evidence for modulation of the $Ca^{2+}$ channel by phosphorylation has been reported with respect to growth factors. Platelet-derived growth factor (PDGF) is a potent mitogen for cells of mesenchymal origin that is produced by platelets, endothelial cells, macrophages, and vascular muscle cells, which acts by a protein tyrosine kinase to enhance $Ca^{2+}$ channels, as indicated by reduction of rabbit ear artery muscle cells by the selective inhibitor of tyrosine kinases, tryphostin-23, bistyrophostin, or genistein (Wijetunge and Hughes, 1995). Furthermore, protein kinase C (PKC) activation of L-type $Ca^{2+}$ channels in vascular muscle cells in rat tail artery by pituitary adenylate cyclase activating polypeptide was furthermore modulated by protein kinase (PKA), providing an example of integration of modulatory actions at phosphorylation sites on a single molecule (Chik *et al.,* 1996). Type 2A phosphatase inhibition by okadaic acid increased the opening of $Ca^{2+}$ channels in vascular muscle cells from human umbilical vein, an action opposite to exposure of the cytoplasmic side of the cell membrane to the purified catalytic subunit of the phosphatase (Groschner *et al.,* 1996). Thus, the type 2A phosphatase dephosphorylates a regulatory site that appears to determine fast gating of the L-type $Ca^{2+}$ channels. Several important phenomena related to growth and remodeling of blood vessels that were previously unexplained are likely to be addressed by further research along these conceptual lines.

## VIII. $Ca^{2+}$ Channel Modulation

A number of the properties of regulation by drugs and hormones have been inferred from $Ca^{2+}$ channels. This $Ca^{2+}$ channel research is literally dependent on the picomolar affinity of $Ca^{2+}$ channels for DHPs. In particular nitrendipine affinity is sufficient to allow separation of the proteins on chromatography affinity columns that have nitrendipine permanently attached. The high affinity of DHPs for $Ca^{2+}$ channels has been used in research to un-

derstand $Ca^{2+}$ channel function. Resulting concepts have also been extended to other types of ion channels (Catterall *et al.,* 1989). More characterization, especially of differences in ion channels, will be possible when drugs have been discovered that allow similar, high-affinity separation of specific subtypes of ion channel proteins from the large number of other proteins that exist in the cell membrane. In fact, separation of key signal molecules, exemplified by receptors and ion channels, can be expected to be one of the most important challenges for the limiting step in understanding cell function for the foreseeable future.

Examples given here illustrate important unifying features and how differences occur. Emphasis is placed on $Ca^{2+}$ channels, in which the conditions necessary for study of modulation and fulfilling criteria of isolation of the protein have been met. $Ca^{2+}$ channels from several tissues have been sequenced, cloned, and expressed as subunits in cultured host cells. Figure 2 provides an overview that simplifies important functional features of the $Ca^{2+}$ channel. Both extracellular and intracellular modulation would occur based on target sites on both sides of the cell membrane. The interaction of most drugs with a $Ca^{2+}$ channel might have been expected to be in the extracellular phase, but several examples of drugs seem to act within the cell membrane or even at the cytoplasmic end of the $Ca^{2+}$ channel. Another counterintuitive discovery about the $Ca^{2+}$ channel is the similarity of $Ca^{2+}$ antagonists to an

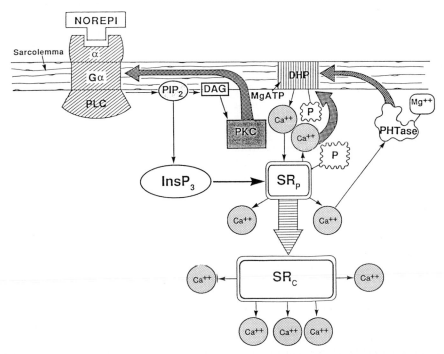

**FIG. 2.** Excitation and membrane signal mechanisms in vascular muscle (VM) show multiple sites for modulation of both the $Ca^{2+}$ channels and $Ca_i^{2+}$ release. The two signal mechanisms diagrammed here are $Ca^{2+}$ and phosphorylation. This diagram indicates (by the different processes activated at different times but with the same $Ca^{2+}$ or phosphorylation signal mechanisms) the necessity for careful study of the spatial localization of intracellular signaling. Abbreviations: DHP, dihydropyridine receptor; P, phosphorylation; G, guanine nucleotide binding protein; PLC, phospholipase C; $PIP_2$, phosphatidyl inositol 4,5-bisphosphate; DAG, 1,2-diacylglycerol; $InsP_3$, inositol-1,4,5-trisphosphate; PKC, protein kinase C, $SR_p$, peripheral sarcoplasmic reticulum; $SR_c$, central sarcoplasmic reticulum; PHTase, phosphatase.

excess of extracellular $Ca^{2+}$ in stabilizing (decreasing excitability of) the cell membrane. One useful way of thinking about $Ca^{2+}$ antagonists is to postulate that they may act like extracellular $Ca^{2+}$ to increase the threshold for excitation. The result of this decreased excitability has been termed *membrane stabilization*. If $Ca^{2+}$ antagonists do stabilize the cell membrane, they actually aid the normal function of $Ca^{2+}$ and in this sense behave contrary to what their name might have implied. Some authors have chosen **$Ca^{2+}$ entry blockers** to more accurately connote what is currently regarded as the major action of the many drugs called $Ca^{2+}$ antagonists. There is growing evidence that a definition of $Ca^{2+}$ antagonists should imply that $Ca^{2+}$ regulation is aided (not antagonized) by this action.

Figure 3 shows structures of some $Ca^{2+}$ antagonists that represent different groups. Like certain kinds of receptors, there are a number of different molecular types. This multiplicity suggests an array of successful avenues of interaction for modulating the function of the $Ca^{2+}$ channel. At least five chemical classes of compounds acting on $Ca^{2+}$ channels should be recognized. Several more classes are likely to be developed in the coming years. The multiple subunits of the $Ca^{2+}$ channel, and consequently the number of potential sites for interaction that might modulate channel function, make it likely that increasingly specialized classes of $Ca^{2+}$ antagonists will be developed. The evidence for at least two isoforms of $Ca^{2+}$ channels, and the likely additional isoforms to be discovered, shows how selective action of drugs on particular kinds of cells can be predicted.

Table 1 provides examples of several kinds of cells that show sensitivity to $Ca^{2+}$ antagonists. While the smooth muscle cells, especially those in blood vessels, are superbly sensitive to $Ca^{2+}$ antagonists, there are also significant actions on the other kinds of muscle. Cardiac muscle, skeletal muscle, neurons, and secretory cells all can be affected by $Ca^{2+}$ antagonists. It is not only the vasodilator action that is beneficial in consideration of $Ca^{2+}$ antagonists. Asthma attacks have been effectively relieved using selected $Ca^{2+}$ antagonists. Cinnarazine, in particular, has been reported to decrease the incidence of asthma attacks. Table 1 shows that $Ca^{2+}$ channels can be blocked in a wide variety of cells, but that there are differences in potency for actions of these compounds among the various cell types.

Another important comparison is the action of $Ca^{2+}$ antagonists of widely different chemical classes. A number of compounds for which the mechanism of action is unknown, or complex enough to be said to be poorly defined or undefined, are given in Table 2. The actions of different chemical classes on various smooth muscles are shown. Explanations for actions of drugs that had previously not been characterized, at least by mechanisms of action, have been found in $Ca^{2+}$ channel block. A number of drugs developed for purposes other than $Ca^{2+}$ channel block are included to illustrate, for example, SKF 525A, reserpine, diazepam, gentamicin, cyproheptadine, pimozide, and lidoflazine.

Examples of drugs that were thought to act by a different mechanism (than $Ca^{2+}$ antagonist) include cinnarizine, which was considered an antihistaminic with antiemetic actions, chlorpromazine, an antipsychotic, antianxiety diaz-

epam, SKF 525A, another central nervous system drug, and cyproheptadine, the serotonergic antagonist. All of these drugs show $Ca^{2+}$ antagonist actions in inhibiting contraction, especially the tonic portion of smooth muscle contraction. Also, reserpine, the psychiatric drug, is a $Ca^{2+}$ antagonist with 1 $\mu M$ potency. Pimozide, a specific dopamine antagonist, is also an effective $Ca^{2+}$ channel blocker.

While block by ions is under discussion, the especially effective inhibition of $Ca^{2+}$ channels by the trivalent ion, $La^{3+}$, should be mentioned. $La^{3+}$ has been used experimentally to inhibit the influx and efflux of $Ca^{2+}$ through virtually all membrane ion channels. While the earliest experiments were carried out with approximately 30–50 $\mu M$ $La^{3+}$ (in electrophysiological studies), the use of $La^{3+}$ for radioisotope flux studies has commonly been at concentrations from 10–50 m$M$, which can be shown to inhibit all movement of $Ca^{2+}$ across cell membranes. $La^{3+}$ continues to be used at a range of concentrations in electrophysiological studies for inhibiting $Ca^{2+}$ currents. $La^{3+}$ might also point the way to development of an additional class of $Ca^{2+}$ antagonists, perhaps acting at new sites.

## IX. $Ca^{2+}$ Channel Subunits and Molecular Biology

A model to show the allosteric interactions between different sites of a subunit deduced from the action of different classes of $Ca^{2+}$ antagonists is shown in Fig. 4. The interaction of DHPs, often used to quantify binding in $Ca^{2+}$ channels, is increased by diltiazem and decreased by verapamil. These drug interactions, either interference or enhancement, among different classes of $Ca^{2+}$ antagonists, suggest actions at different sites on $Ca^{2+}$ channels. While these studies have been carried out primarily in rabbit skeletal muscle and in rabbit heart, they may be representative; although in some details they may well be different from those found in other smooth muscles. For example, interactions between diltiazem and DHPs are difficult to demonstrate in intestinal muscle.

Considerable progress in reconstituting the several $Ca^{2+}$ channels now recognized has allowed more detailed insights, and L, T, N, P, Q, and R classes are now recognized. A useful hierarchical classification has been proposed by Snutch *et al.* (1990) based on gene product and classified by DHP sensitivity, as shown in Table 3. The T-type $Ca^{2+}$ channel (also known as the low-voltage-activated $Ca^{2+}$ channel because only a small depolarization is necessary to cause its activation) has not yet been identified with a gene product, and is likely to be a completely separate entity on a different gene, although homology with other ion channel proteins is anticipated.

Although research on the T-type $Ca^{2+}$ channel has lagged behind that on all other classes, the recent discovery of the new tetraline $Ca^{2+}$ antagonist, mibefradil, which is selective for T-type $Ca^{2+}$ channels (Mishra and Hermsmeyer, 1994b), at least in vascular muscle cells, promises to improve understanding of the functional significance of this ion channel. Mibefradil blocks $Ca^{2+}$ channels, even in the resting state (Mishra and Hermsmeyer, 1994a), and inter-

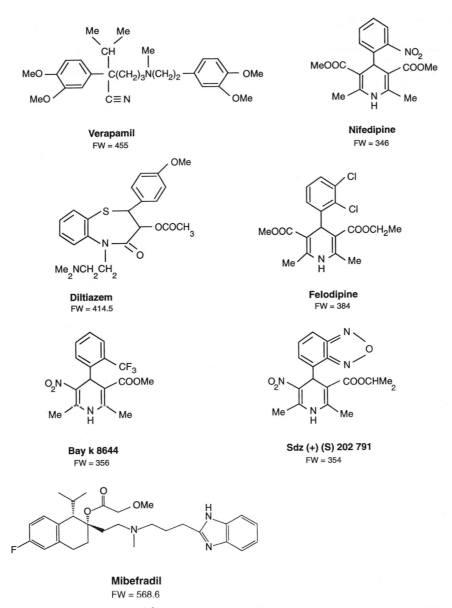

**FIG. 3.** Structures of drugs that act as $Ca^{2+}$ antagonists form a diverse array of chemical structures. The most familiar in order of experimental usefulness are phenylalkylamines (e.g., verpamil and D600), DHPs (e.g., nitrendipine, felodipine, amlodipine), and benzothiazepines (e.g., diltiazem). Stereoisomers of a few of these compounds act as $Ca^{2+}$ agonists (i.e., increase the probability of $Ca^{2+}$ channel opening), providing the chemical tool for increases in $Ca^{2+}$ channel currents (e.g., Bay-K-8644 or Bay-R-5417, Sandoz (+) 202-791). Mibefradil is a tetralin existing as a single enantiomer that blocks T-type $Ca^{2+}$ channels with >30 times selectivity over L-type $Ca^{2+}$ channels in vascular muscle cells.

fcres with $Ca^{2+}$ release and activation of PKC (Hermsmeyer and Miyagawa, 1996). Block of vascular muscle cell proliferation and blood vessel wall thickening suggest a beneficial effect and an important mechanism by which T-type $Ca^{2+}$ channel block, as exemplified by mibefradil, may be useful (Vacher *et al.*, 1996). Selective block of T-type $Ca^{2+}$ channels in the sinoatrial node results in a slowing of the heart rate that is uniquely associated with this novel $Ca^{2+}$ antagonist, as was discussed as part of a review of the evidence for putative T-type $Ca^{2+}$ channel functions (Hermsmeyer, 1997).

The classes are formed according to the $\alpha_1$-subunit, which is where all drug binding occurs, and are thus classi-fied as $\alpha_{1A}$, $\alpha_{1B}$, $\alpha_{1C}$, $\alpha_{1D}$, and $\alpha_{1E}$ as discussed in recombination reports. Both the $\alpha_{1B}$- and $\alpha_{1E}$-subunits respond to PKC stimulation by phorbol esters with a 30–40% increase in currents, whereas the $\alpha_{1A}$- and $\alpha_{1C}$-subunits are insensitive to modulation by phorbol stimulation (Stea *et al.*, 1995). The $\beta$-subunits, of which several have now been cloned, affect current inactivation when coexpressed with the $\alpha$ types. The $\beta_3$-type, isolated from vascular muscle cells, expressed together with the matching $\alpha_{1C}$-subunit, selectively modulated the interaction of $Ca^{2+}$ antagonists with the $Ca^{2+}$ channel, left-shifting the verapamil and gallopamil curves at 80 and −40 mV, but not shifting the isradipine or mibefradil curves significantly (Lacinová *et*

**TABLE 1**   Examples of Muscle Cells Sensitive
to Ca²⁺ Antagonists

| Cell type<br>Ca²⁺ antagonist | $IC_{50}$<br>(n$M$) | Ref. |
|---|---|---|
| Aorta (rabbit) | | |
| Nitrendipine | 3.1 | Towart, 1982 |
| Verapamil | 140 | Sperelakis and Mras, 1983 |
| Gallopamil (D600) | 100 | Sperelakis and Mras, 1983 |
| Diltiazem | 1,200 | Sperelakis and Mras, 1983 |
| Bepridil | 3,000 | Mras and Sperelakis, 1981 |
| Trachea (guinea pig) | | |
| Nifedipine | 6 | Weichman *et al.*, 1983 |
| Verapamil | 6 | Weichman *et al.*, 1983 |
| Diltiazem | 6.3 | Cerrina *et al.*, 1982 |
| Vas deferens (rat) | | |
| Nifedipine | 50 | Triggle *et al.*, 1979 |
| Verapamil | 940 | Triggle *et al.*, 1979 |
| Gallopamil (D600) | 880 | Triggle *et al.*, 1979 |
| Flunarizine | 2,800 | Hay and Wadsworth, 1982 |

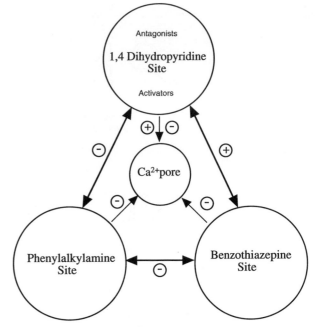

**FIG. 4.**   Schematic view of the three distinct drug receptor sites of the calcium channel. The receptor sites (named after the prototype of drug) communicate with each other by reciprocal allosterism. Calcium-binding sites have a key role in these regulatory phenomena. The receptor sites are diagrammed here in a circular model, connected by bidirectional arrows, but the model neither excludes the existence of more sites nor other possible interactions. The plus and minus signs on these arrows indicate whether a drug receptor can exert a positive or negative allosteric effect on the neighboring receptor site. Each drug receptor site can exist in low- and high-affinity state. According to one hypothesis, the 1,4-dihydropyridine channel antagonists stabilize a ternary high-affinity calcium–calcium channel–drug complex, consequently raising the energy barrier for the divalent cation. The 1,4-dihydropyridine channel agonists would stabilize the low-affinity state, with a lower energy barrier. The distribution of states of the receptor site would then be modulated by temperature, respective ligand, drugs binding to allosteric sites, and a variety of other experimental conditions such as pH and ion concentrations. The model is based on *in vitro* binding experiments. (Modified from Glossmann *et al.*, 1985.)

*al.*, 1995). The $\alpha_2/\delta$-subunit also significantly increased Ca²⁺ channel kinetics because both activation and deactivation were more rapid while amplitude increased (Bangalore *et al.*, 1996). Coexpression of the $\alpha_2/\delta$-subunit along with the $\alpha_{1C}$-subunit may increase expression of the L-type Ca²⁺ channels and enhance transitions between open and closed states.

## X. Drug Binding and Blocking of Ion Channels

There is a strong correlation between binding to ion channels and pharmacological activity of diverse drugs, toxins, and Ca²⁺ antagonists, at least for several kinds of cells (Fleckenstein, 1983). As previously stated, smooth muscle cells show the highest sensitivity to Ca²⁺ antagonists and are most strongly affected by this class of drugs (Table 4). Cardiac muscle is much less sensitive, but still has moderate sensitivity. Inhibition of cardiac function occurs at concentrations only a few-fold higher than those causing vasodilation. Skeletal muscle shows relatively little sensitivity to Ca²⁺ antagonists, probably based primarily on storage of large quantities of Ca²⁺ that allow skeletal muscle to function without rapid entry of Ca²⁺ through the

surface membrane. Nonetheless, in Ca²⁺-limited conditions such as low Ca²⁺ solution, the dependence of skeletal muscle tension on Ca²⁺ channels can be demonstrated. The selectivity for smooth muscle results from differences in

**TABLE 2**   Other Drugs That Block Ca²⁺ Channels

| Drug | Tissue | $IC_{50}$ (n$M$) | Ref. |
|---|---|---|---|
| Proadifen (SKF 525A) | Intestinal | 20,000 | Triggle *et al.*, 1989 |
| Reserpine | *Taenia coli* | 1,000 | Casteels and Login, 1983 |
| Diazepam | Intestinal | 30,000 | Ishii *et al.*, 1982 |
| Pimozide | *Taenia coli* | 220 | Quintana, 1978 |
| Cyproheptadine | Intestinal | 1,000 | Lowe *et al.*, 1981 |
| Lanthanum (La³⁺) | Intestinal | 800 | Goodman and Weiss, 1971 |

**TABLE 3** Classes of $Ca^{2+}$ Channel by Gene Product

| Class of $Ca^{2+}$ channel | Found in | Type |
|---|---|---|
| $S^a$ | Skeletal muscle | L |
| A | Neurons, endocrine cells | P, Q |
| B | Neurons | N |
| $C^a$ | Cardiovascular, neurons | L |
| $D^a$ | Endocrine cells, neurons | L |
| E | Neurons | R |

$^a$ Dihydropyridine sensitive.

the $Ca^{2+}$ channel proteins, and the $Ca^{2+}$ channel proteins have taught us how ion channels modulated by drugs can alter function. The ion channels are, in conclusion, excellent targets for existing drugs and for new drug design.

## XI. Summary

Ion channels have regulatory sites which bind drugs that powerfully influence their function. In this way, ion channels provide a membrane signal function that integrates multiple transmitter and modulator substance inputs, thus behaving very much like drug receptors. Ion channels thus can be thought of as a special type of drug receptor. Because ion channels are by definition water-filled pathways that allow ions to move through the otherwise impermeable lipid bilayer of the cell membrane, they are a perfect mech-

anism for integrating rapid, localized events that are directly translated into electrical signals to act quickly and effectively as triggers for specific functions of the cell. Several important technical advances in the last few years have allowed the details of the function of single ion channels in the cell membrane to be identified. These so-called patch-clamp techniques for measurement of single-channel ion currents now allow the routine investigation of various kinds of channels in virtually all kinds of cells with submillisecond resolution. In the most fortunate instances, live functional studies of the single protein molecules that constitute ion channels are possible, and even can be accomplished in the live cell membrane under physiological conditions.

Single-channel studies have verified the molecular basis of a water-filled channel through the membrane, and the remarkable control by intracellular and extracellular modulators and drugs of the movement of ions through the channels. Rapid opening and closing characterize single ion channels and spontaneously reveal the nature of the underlying channel kinetics, often with resolution with milliseconds for hypothesized charge movement and molecular conformational changes that constitute the gating events. When drugs interact with these channels, the affinity can be as high as the nanomolar range, allowing for extreme specificity and selectivity of action on these fundamentally important proteins. $Ca^{2+}$ channels have been favored for study because of their central importance in cell signaling and because drugs with very high affinity for the $Ca^{2+}$ channels (the DHPs) have been discovered. These compounds, which are called $Ca^{2+}$ antagonists and $Ca^{2+}$ agonists, are capable of blocking or increasing the opening

**TABLE 4** Diverse Array of Ion Channels as Targets for Drug and Toxin High-Affinity Binding

| Action | Ion channel agents active on smooth and striated muscle | | |
|---|---|---|---|
| | $Ca^{2+}$ channel | $K^+$ channel | $Na^+$ channel |
| Blockers | Verapamil (PAA) | Tetraethylammonium | Local anesthetics |
| | Nitrendipine (DHP) | 4-Aminopyridine | Tetrodotoxin |
| | Diltiazem (BTZ) | 9-Aminoacridine | Barbiturates |
| | Flunarizine | Quinidine | Alcohol |
| | $\alpha$-Conotoxin | Apamin | Chlorpromazine |
| | $La^{3+}$ | Charybdotoxin | Allethrin |
| | $Co^{2+}$, $Mn^{2+}$, $Co^{2+}$ | Glyburide | Scorpion toxins |
| | Aminoglycoside antibiotics | $Ba^{2+}$ | |
| | SDZ (−) 202-791 | $Cs^+$, $Rb^+$ | |
| | | Local anesthetics | |
| | | Alcohol | |
| | | Nicotine | |
| | | Barbiturates | |
| | | DDT | |
| Openers | Bay-K-8644 | Acetylcholine | Veratridine |
| | SDZ (+) 202-791 | Chromakalim | Batrachotoxin |
| | Anemone toxin | Pinacidil | Grayanotoxin |
| Inhibit inactivation | | | Condylactis toxin |
| | | | DDT |
| | | | Veratridine |

of $Ca^{2+}$ channels. The remarkable specificity of drug binding is demonstrated by the fact that stereoisomers of the same compound exist as $Ca^{2+}$ antagonist and $Ca^{2+}$ agonist pairs. This stereospecificity suggests precise three-dimensional constraints on binding of these drugs to the $Ca^{2+}$ channel.

In addition to $Ca^{2+}$ channels, sodium ($Na^+$), potassium ($K^+$), and chloride ($Cl^-$) channels are known to be important in cell signaling. $Na^+$ channels have been isolated using a compound that allows their identification with high affinity, namely, tetrodotoxin, the poison of the pufferfish. $Na^+$ channels were the first to be characterized and sequenced, and have important homologies with other ion channels that have allowed them to be used as a prototype, and are now recognized to act as a drug target. Based on fundamental similarities with $Na^+$ channels, multiple $K^+$ channels have also been identified at the molecular level and studied extensively. Important contributions to understanding the function of $K^+$ channels have been made by studying drugs that act on them. Drugs that act on $K^+$ channels have been central in the recognition that multiple types exist. Furthermore, we now realize that there are multiple types of all ion channels so far explored. The largest number of channels studied have been in the $K^+$ channel group. $K^+$ channel blockers and $K^+$ channel openers both exist and are needed for understanding $K^+$-channel-dependent cell functions, especially the regulation of membrane potential, which is dominated in both nerve and muscle cells by $K^+$ channels. $Cl^-$ channels can also make important contributons to determining membrane potential, and there are examples of drugs that act on $Cl^-$ channels. Though $Cl^-$ channels are among the least explored, their conductances can be very high, similar to those of $K^+$ channels, and both $K^+$ channels and chloride channels can provide the long-term, steady-state membrane potentials in many cell types.

One of the earliest types of ion channels to be thoroughly investigated was the cholinergic receptor, which is in fact a ligand-gated channel. All ligand-gated channels studied to date are relatively nonspecific, allowing multiple kinds of ions to pass through, and driving the cell to a membrane potential that is an average of their electrochemical gradients. Cholinergic receptors of the type found in skeletal muscle, and well characterized by electrophysiological and molecular biological studies, have long been known to conduct $K^+$, $Na^+$, and $Cl^-$, and this is typical of ligand-gated ion channels. Other examples are the gamma aminobutyric acid channel in the brain and the adenosine triphosphate ligand-gated channel found in various excitable cells. This class of ligand-gated channels is an exquisite example of the direct modulation of ion movements through cell membranes by synaptic transmitter, released in localized areas to carry out vital functions rapidly and reliably. As research reveals additional details of the action of drugs on ligand-gated and voltage-gated channels, one can expect continued progress in understanding and selectively modifying cell membrane function. The concept of ion channels as drug receptors has emerged as valid and productive in cellular physiology and pharmacology.

## Bibliography

Ashcroft, F. M., Harrison, D. E., and Ashcroft, S. J. H. (1984). Glucose induces closures of single potassium channels in isolated rat pancreatic $\beta$-cells. *Nature* **312**, 446–448.

Bangalore, R., Mehrke, G., Gingrich, K., Hofmann, F., and Kass, R. S. (1996). Influence of L-type $Ca^{2+}$ channel $\alpha_2/\delta$-subunit on ionic and gating current in transiently transfected HEK 293 cells. *Am. J. Physiol.* **39**, H1521–H1528.

Baron, A., Loirand, G., Pacaud, P., Mironneau, C., and Mironneau, J. (1993). Dual effect of thrombin on voltage-dependent $Ca^{2+}$ channels of portal vein smooth muscle cells. *Circ. Res.* **72**, 1317–1325.

Bean, B. P., Sturek, M., Puga, A., and Hermsmeyer, K. (1986). Calcium channels in vascular muscle cells isolated from rat mesenteric arteries: Modulation by dihydropyridine drugs. *Circ. Res.* **59**, 229–235.

Casteels, R., and Login, I. S. (1983). Reserpine has a direct action as a calcium antagonist on mammalian smooth muscle cells. *J. Physiol. (London)* **340**, 403–414.

Catterall, W. A., Seagar, M. J., Takahashi, M., and Nunoki, K. (1989). Molecular properties of dihydropyridine-sensitive calcium channels. *In* "Calcium Channels. Structure and Function" (D. W. Wray, R. I. Norman, and P. Hess, Eds.), pp. 1–14. Annals of the New York Academy of Sciences, New York.

Cerrina, J., Renier, A., Floch, A., Duroux, P., and Advenier, C. (1982). Effects of Ca antagonists on guinea pig tracheal contraction induced by various agonists. *Am. Rev. Respir. Dis.* **125**, 226.

Chik, C. L., Li, B., Okiwara, T., Ho, A. K., and Karpinski, E. (1996). PACAP modulates L-type $Ca^{2+}$ channel currents in vascular smooth muscle cells: Involvement of PKC and PKA. *FASEB J.* **10**, 1310–1317.

Fleckenstein, A. (1983). "Calcium Antagonism in Heart and Smooth Muscle. Experimental Facts and Therapeutic Prospects." John Wiley & Sons, New York.

Glossmann, H., Ferry, D. R., Goll, A., Streissnig, J., and Schober, M. (1985). Calcium channels: Basic properties as revealed by radioligand binding studies. *J. Cardiovasc. Pharmacol.* **7**, S20–S30.

Glossmann, H., Striessnig, J., Knaus, H.-G., Muller, J., Grassegger, A., Holtje, H.-D., Marrer, S., Hymel, L., and Schindler, H. G. (1989). Structure of calcium channels. *In* "Calcium Channels. Structure and Function" (D. W. Wray, R. I. Norman, and P. Hess, Eds.), pp. 198–214. Annals of the New York Academy of Sciences, New York.

Goodman, F. R., and Weiss, G. B. (1971). Dissociation by lanthanum of smooth muscle responses to potassium and acetylcholine. *Am. J. Physiol.* **220**, 759–766.

Groschner, K., Schumann, K., Mieskes, G., Baumgartner, W., and Ramanin, C. (1996). A type 2A phosphatase-sensitive phosphorylation site controls modal gating of L-type $Ca^{2+}$ channels in human vascular smooth muscle cells. *Biochem. J.* **318**, 503–517.

Hay, D. W. P., and Wadsworth, R. M. (1982). Effects of some organic calcium antagonists and other procedures affecting $Ca^{2+}$ translocation on KCl-induced contractions in the rat vas deferens. *Br. J. Pharmacol.* **76**, 103–113.

Hermsmeyer K. (1997). Block of L and T types of $Ca^{2+}$ channels in vascular muscle cells by mibefradil. *In Proceedings of the Low Voltage Activated T-Type $Ca^{2+}$ Channel Meeting*, (J. Nargeot, R. W. Tsien, and J. P. Clozel, Eds.). Adis International, Chester, England.

Hermsmeyer, K., and Miyagawa, K. (1996). Protein kinase C mechanism enhances vascular muscle relaxation by the $Ca^{2+}$ antagonist, Ro 40-5967. *J. Vasc. Res.* **33**, 71–77.

Hille, B. (1992). "Ionic Channels of Excitable Membranes," 2nd ed. Sinauer Associates, Sunderland, MA.

Hodgkin, A. L. (1964). "The Conduction of the Nervous Impulse," the Sherrington Lectures VII. Liverpool University Press, Liverpool, England.

Ishii, K., Kano, T., Akutagawa, M., Makino, M., Tanaka, T., and Ando, J. (1982). Effects of flurazepam and diazepam in isolated guinea-pig *Taenia coli* and longitudinal muscle. *Eur. J. Pharmacol.* **83**, 329–333.

Lacinová, L., Ludwig, A., Bosse, E., Flockerzi, V., and Hofmann, F. (1995). The block of expressed L-type Ca$^{2+}$ channel is modulated by the $\beta_3$ subunit. *FEBS Lett.* **373**, 103–107.

Loirand, G., Pacaud, P., Mironneau, C., and Mironneau, J. (1990). GTP-binding proteins mediate noradrenaline effects on calcium and chloride currents in rat portal vein myocytes. *J. Physiol. (London)* **428**, 517–529.

Lowe, D. A., Matthews, E. K., and Richardson, B. P. (1981). The calcium antagonistic effects of cyproheptadine on contraction, membrane electrical events, and calcium influx in the guinea-pig *Taenia coli*. *Br. J. Pharmacol.* **74**, 641–663.

Meisheri, K. D., Khan, S. A., and Martin, J. L. (1993). Vascular pharmacology of ATP-sensitive K$^+$ channels: Interactions between glyburide and K$^+$ channel openers. *J. Vasc. Res.* **30**, 2–12.

Mishra, S. K., and Hermsmeyer, K. (1994a). Resting state block and use independence of rat vascular muscle Ca$^{2+}$ channels by Ro 40-5967. *J. Pharmacol. Exp. Ther.* **269**, 178–183.

Mishra, S. K., and Hermsmeyer, K. (1994b). Selective inhibition of T-type Ca$^{2+}$ channels by Ro 40-5967. *Circ. Res.* **75**, 144–148.

Mras, S., and Sperelakis, N. (1981). Bepridil (CERM-1978) and verapamil depression of contractions of rabbit aortic rings. *Blood Vessels* **18**, 196–205.

Neher, E., and Sakmann, B. (1992). The patch clamp technique. *Sci. Am.* **266**, 44–51.

Pacaud, P., Loirand, G., Baron, A., Mironneau, C., and Mironneau, J. (1991). Ca$^{2+}$ channel activation and membrane depolarization mediated by Cl$^-$ channels in response to noradrenaline in vascular myocytes. *Br. J. Pharmacol.* **104**, 1000–1006.

Pacaud, P., Loirand, G., Grégoire, G., Mironneau, C., and Mironneau, J. (1992). Calcium-dependence of the calcium-activated chloride current in smooth muscle cells of rat portal vein. *Pflugers Arch.* **421**, 125–130.

Quintana, A. (1978). Effects of pimozide on the response of smooth muscle to nondopamine agonists and calcium. *Eur. J. Pharmacol.* **53**, 113–116.

Sakmann, B. (1992). Elementary steps in synaptic transmission revealed by currents through single ion channels. *Science* **256**, 503–512.

Sather, W. A., Tanabe, T., Zhang, J. F., and Tsien, R. W. (1994). Biophysical and pharmacological characterization of a class A calcium channel. *Ann. NY Acad. Sci.* **747**, 294–301.

Snutch, T. P., Leonard, J. P., Gilbert, M. M., Lester, H. A., and Davidson, N. (1990). Rat brain expresses a heterogeneous family of Ca$^{2+}$ channels. *Proc. Natl. Acad. Sci. USA* **87**, 3391–3395.

Sperelakis, N. (1992). Chemical agent actions on ion channels and electrophysiology of the heart. *In* "Cardiovascular Toxicology" (D. Acosta, Jr., Ed.), 2nd ed., pp. 283–338. Raven Press, New York.

Sperelakis, N., and Mras, S. (1983). Depression of contractions of rabbit aorta and guinea pig vena cava by mesudipine and other slow channel blockers. *Blood Vessels* **20**, 172–183.

Stea, A., Soong, T. W., and Snutch, T. P. (1995). Determinants of PKC-dependent modulation of a family of neuron Ca$^{2+}$ channels. *Neuron* **15**, 929–940.

Towart, R. (1982). Effects of nitrendipine (Bay e 5009), nifedipine, verapamil, phentolamine, papaverine, and minoxidil on contractions of isolated rabbit aortic smooth muscle. *J. Cardiovasc. Pharmacol.* **4**, 895–902.

Triggle, C. R., Swamy, V. C., and Triggle, D. J. (1979). Calcium antagonists and contractile responses in rat vas deferens and guinea pig ileal smooth muscle. *Can. J. Physiol. Pharmacol.* **57**, 804–818.

Triggle, D. J., Zheng, W., Hawthorn, M., Kwon, Y. W., Wei, X.-Y., Joslyn, A., Ferrante, J., and Triggle, A. M. (1989). Calcium channels in smooth muscle. Properties and regulation. *In* "Calcium Channels: Structure and Function" (D. W. Wray, R. I. Norman, and P. Hess, Eds.), pp. 215–229. Annals of the New York Academy of Sciences, New York.

Vacher, E., Richer, C., Fornes, P., Clozel, J.-P., and Giudicelli, J.-F. (1996). Mibefradil, a selective calcium T-channel blocker, in stroke-prone spontaneously hypertensive rats. *J. Cardiovasc. Pharmacol.* **27**, 686–694.

Van Breemen, C., and Siegel, B. (1980). The mechanism of alpha-adrenergic activation of the dog coronary artery. *Circ. Res.* **46**, 426–429.

Venter, J. C., and Triggle, D. (Eds.) (1987). "Structure and Physiology of the Slow Inward Calcium Channel. Receptor Biochemistry and Methodology," Vol. 9 (J. C. Venter and L. C. Harrison, Series Eds.). Alan R. Liss, New York.

Weichman, B. M., Muccitelli, R. M., Tucker, S. S., and Wasserman, M. A. (1983). Effect of calcium antagonists on leukotriene D$_4$-induced contractions of the guinea-pig trachea and lung parenchyma. *J. Pharmacol. Exp. Ther.* **225**, 310–315.

Wijetunge, S., and Hughes, A. D. (1995). Effect of platelet-derived growth factor on voltage-operated Ca$^{2+}$ channels in rabbit isolated ear artery cells. *Br. J. Pharmacol.* **115**, 534–538.

*Shirley H. Bryant*

# 38

# Ion Channels as Targets for Disease

## I. Introduction

Ion channels are involved in many critical cellular processes as varied as the transmission of impulses in excitable tissues to the triggering of contraction. Altered function of membrane ion channels then would be expected to have serious consequences for the survival of the organism. The ion channels can be affected in a variety of ways by disease, both directly and indirectly. Direct action on the channel protein structure occurs as a result of genetic mutations of the channel gene. Many of the genetic mutations that markedly alter the channel function are point mutations affecting only a single amino acid. Most spontaneous genetic mutations of ion channels or of their regulators are probably lethal, but some alterations allow survival. We recognize these as the hereditary diseases whose target is ion channels. Indirect action includes any of the following: (1) abnormalities in regulatory mechanisms such as phosphorylation required for channels to function, (2) presence of any number of kinds of toxic materials that block channels or prevent their synthesis, and (3) development of autoimmune disease directed at the channel protein or its regulators.

Naturally occurring channel diseases have been reported within the general families of $Cl^-$, $Na^+$, $Ca^{2+}$, $K^+$, and transmitter-activated channels. With the recent introduction of rapid cloning methods and improvements in functional analysis of channels, especially the patch-clamp methodology, the rate of identification of dysfunctional channels causing disease has mushroomed. The diseases themselves, now being referred to as *channelopathies*, are becoming too numerous to discuss with any depth. It is our task in this chapter to make the reader aware of pertinent examples of disease for each of these general families of channels. The chapter is organized by the channel type (i.e., $Cl^-$, $Na^+$, $Ca^{2+}$, $K^+$, and neurotransmitter gated) rather than by disease. Some disease entities (e.g., myotonia) may be caused by mutations in different types of channels. Such diseases are discussed under each of the channel type headings.

## II. Ion Channel Diseases

### A. Mutations of Ion Channel Genes

A gene mutation can affect channels in one of the following ways: (1) A defective channel protein is expressed or its expression is blocked or (2) a nonchannel protein on which channel function depends is expressed but defective or its expression is blocked. Defective and malfunctioning proteins may be expressed as a result of improper coding. However, certain mutations can completely block the expression of a product. It is possible further for a defect to occur in a subunit or in a regulator shared by several types of channels. This may offer an explanation of how a single defective gene could lead to malfunction of more than one type of channel.

Identification of genes involved in ion channel diseases involves two principal search strategies: (1) functional or expression cloning and (2) positional cloning. In expression cloning the gene is isolated on the basis of information about the expressed protein product, for example, its amino acid sequence or antibody reactivity, or its function (e.g., receptor/ligand reactivity or ion channel characteristics). cDNA libraries are screened using different types of probes including antibodies and oligonucleotides. The polymerase chain reaction (PCR) technique can be employed in several ways including the amplification of cDNA using oligonucleotides derived from the protein sequence. Functional cloning was used to characterize the genes for the CLC-1 $Cl^-$ channel and the SkM-1 $Na^+$ channel discussed later. In contrast, positional cloning is the process of isolating the gene starting from information about its genetic or physical location in the genome. Often little or nothing may be known about the function of the product. The work involved is enormous since it relies on the method of **chromosome walking** and identification of expressed sequences. In spite of the difficulties, 19 disease genes were identified in this way between 1989 and 1993 (Ballabio, 1993), including myotonic dystrophy and cystic fibrosis, which are discussed in this chapter.

## B. Autoimmune Disease and Pleotropic Effects

Ion channel diseases can be due to the production of antibodies directed at the ion channel protein. Three examples in which autoimmune disease affects channels are discussed. Two of these are **myasthenia gravis** and the **Lambert–Eaton syndrome.** These conditions lead to bouts of skeletal muscle weakness because neuromuscular transmission is compromised. The third example is a voltage-gated $K^+$ channel, which is the target of an autoimmune disease called **neuromyotonia.**

Mutations on one gene may influence the functions of channels coded by different genes; these are referred to as **pleotropic effects.** In **myotonic dystrophy** (discussed later), skeletal muscle $Na^+$ channel dysfunction leads to the abnormality excitability known as **myotonia.** The disease, however, is due to a mutant gene coding for a protein kinase called **myotonin.** The mechanisms for pleotropic effects are not well understood.

## C. Diseases, Linkage, Candidate Genes, and Candidate Channels

Common methods are evolving for the study of familial diseases. Two basic strategies are used: **positional cloning** and the **candidate gene.** The first approach, positional cloning, is used to identify the genetic locus even when there is no knowledge of the biophysical or biochemical abnormalities underlying the disease. This is done by examining affected individuals for genetic linkage to polymorphic markers located throughout the genome. In this way a region of the gene is found that is **linked** to the disease. Linkage is established when there is a low rate of recombination between the marker locus and the disease locus. From the gene one can express the protein it encodes and then determine its function and how the disease mutations disturb this function. The cystic fibrosis transmembrane regulator channel was discovered in this way.

The second approach, or candidate gene strategy, requires knowledge of functional abnormalities brought about by the disease. Genes responsible for maintaining the normal function of the process affected by the disease are then considered to be candidate genes for the disease. Genetic linkage to the disease is then sought, and if one is found the candidate gene is screened for mutations. These mutations can be incorporated into the wild-type DNA and expressed in heterologous systems. Then one can examine the biophysical and biochemical behavior and determine if the abnormality could account for the disease. The new rapid cloning techniques are largely responsible for the increased pool of known channel genes having known functions. This makes it easier to identify candidate channel genes, thus the candidate gene approach has been very successful for the study of the channelopathies. Most of the new channelopathies including the CLC-5 mutations causing kidney stones, the episodic ataxias, and the long Q-T interval diseases were discovered using the candidate gene approach.

From the small number of channelopathies now reported we can begin to make certain generalizations with respect to excitable cells. In excitable cells channels often assume both excitatory and stabilizing roles. Therefore blocking the stabilizing $Cl^-$ channel (e.g., the low $Cl^-$ channel myotonias) is tantamount to blocking the inactivation of the excitatory $Na^+$ channels (e.g., the $Na^+$ channel myotonias). We see further examples. Similarly, in the human heart the long Q-T syndrome is linked both to $K^+$ channels that have a decreased $P_{open}$ or to $Na^+$ channels that have an increased $P_{open}$. In the central nervous system the phenomenon of episodic ataxia is linked both to $K^+$ channels that have a decreased $P_{open}$ (*episodic ataxia type-1*) or to $Ca^{2+}$ channels that have an increased $P_{open}$ (*episodic ataxia type-2*).

Sometimes a linkage to a mutant gene is present but the expressed channels containing the mutation do not behave abnormally. This is the case for hypokalemic periodic paralysis (HoPP) discussed later. The linkage to a mutation in the gene coding for a skeletal muscle $Ca^{2+}$ channel is clear, but heterologously expressed channels do not display behavior that would explain the disease. In another example, some mutant $Cl^-$ channels linked to myotonia in patients when expressed in heterologous systems have normal behavior. One explanation for discrepancies of this type is that important factors necessary for phenotypic expression may be absent in the expressions system, be it a cultured mammalian cell or a *Xenopus* oocyte. The missing factors could include regulatory paths or chemical modifications (e.g., disulfide linkages) necessary for proper folding of the channel protein (George, 1995).

## D. Phenomenon of Myotonia

Myotonia is a clinical sign of certain diseases that target skeletal muscle membrane ion channels. It is a relatively rare phenomenon, usually not life threatening, yet it has held the interest of scientists and physicians for more than a century after it was first described by Dr. J. Thomsen in 1875 (see Rüdel and Lehmann-Horn, 1985, for early references). Myotonia is defined as an abnormal contraction of skeletal muscles following a willed contraction. In a typical example, a patient with myotonia is asked to grip your hand for a few seconds and then asked to release his grip. In spite of the fact that the patient is no longer willing a contraction, i.e., causing nerve action potentials to excite the muscle, the contraction continues for several more seconds. The myotonic muscle fibers are hyperexcitable (i.e., have low threshold potential) and tend to fire repetitive action potentials on their own. The **after discharge** of action potentials leading to the **after contraction** has been shown to result from a slowly decaying depolarization of the transverse tubules by accumulated $K^+$ in their lumen (Adrian and Bryant, 1974). Myotonic fibers fire repetitively for many seconds to steady depolarization, whereas normal fibers fire a short burst for about one-tenth of a second. This characteristic produces a diagnostically distinct electromyographic recording, which sounds like a "dive-bomber" when played through a sound system.

Myotonic disease occurs as a result of ion channel dysfunction in the skeletal muscle excitable membranes (*surface* plus *tubular*). Hereditary myotonias occur indirectly as a result of mutations in enzymes regulating the ion channels, or directly through mutations of genes coding for ion channels. Although normal muscle fiber excitability requires $Na^+$, $K^+$, and $Cl^-$ channels, to date, only errors in $Na^+$ and $Cl^-$ channels have been associated with myotonia. $K^+$ channels play an important role in action potential generation so it is surprising that there are no natural $K^+$ channel myotonias; however, toxins affecting these channels produce myotonia.

The most frequently encountered myotonic disease is human **myotonia dystrophica.** This autosomal dominant disease (actually, the most common inherited neuromuscular disorder) is caused by the expression of an abnormal kinase known as myotonin. The mutant gene contains a large number of CTG repeats, which may increase with each generation, worsening the disease in the process (Harley *et al.,* 1993). The disease is generalized to many organ systems and the skeletal muscles undergo dystrophy. Although ion channels are dysfunctional here, mutations of these genes are not involved. Less frequently occurring, but better understood, are several nondystrophic muscle diseases due to mutations in the skeletal muscle membrane ion channels. These can be classified into (1) **$Cl^-$ channel type** and (2) **$Na^+$ channel type.** The membrane properties characteristic of these two types are illustrated in Fig. 1.

### 1. $Cl^-$ Channel Type

The main $Cl^-$ channel type myotonias include human *dominant myotonia congenita* (DMC, Thomsen's disease), human **recessive myotonia congenita** (RMC; also called recessive generalized myotonia), and among animals there is the dominant myotonia of goats (which we also consider a DMC) and the recessive myotonic **adr mouse.** The basic physiological problem in these myotonias is a low membrane $Cl^-$ conductance ($g_{Cl}$) due to dysfunctional CLC-1 $Cl^-$ channels. Normally, the mammalian CLC-1 $Cl^-$ channels provide the major stabilizing steady conductance (i.e., $g_{Cl}$) necessary for normal excitability. When $g_{Cl}$ falls below 25% of normal, the skeletal muscle membrane becomes hyperexcitable and myotonia results. $Cl^-$ channel myotonia can be induced by agents that specifically block the CLC-1 channel such as anthracene-9-carboxylic acid (9-AC), and simulated by computer models of action potentials in which $g_{Cl}$ is lowered (Bryant and Morales-Aguilera, 1971; Furman and Barchi, 1978; Adrian and Marshall, 1976). See Figs. 1A–C and G.

### 2. $Na^+$ Channel Type

The main $Na^+$ channel myotonias, which are all dominant, include human and equine **hyperkalemic periodic paralysis,** human **paramyotonia congenita,** and human **$Na^+$ channel myotonia.** The common physiological factor in this group is a small fraction of $Na^+$ channels that inactivate more slowly and/or show abnormal reopenings compared to the wild type. The abnormally increased $Na^+$ currents result in hyperexcitability because a smaller stimulus can initiate an action potential and can then cause reexcitation with myotonic repetitive firing. $Na^+$ channel myotonia can be induced by agents that block $Na^+$ channel inactivation

like the sea anemone toxin, ATX-II, and, like the low $g_{Cl}$ myotonia, $Na^+$ channel myotonia has been extensively simulated by the computer (Cannon *et al.,* 1993). See Figs. 1D–F and H.

## III. $Cl^-$ Channels

### A. General Comments

$Cl^-$ channels and nonspecific anion channels have only recently gained popularity, yet the number of distinctly different types of $Cl^-$ channels may in time exceed that of other channels (Vaughan and French, 1989; Pusch and Jentsch, 1994). Unfortunately, many of the newly described channels have not yet been assigned physiological functions. The functions already identified include osmotic pressure and stretch sensitivity, $Cl^-$ concentration regulation, and membrane potential stabilization. With respect to the latter function, the channels can furnish a constant stabilizing membrane conductance in skeletal muscle, a transient agonist-activated stabilizing conductance in neurons (known as inhibition), and a hormonally activated stabilizing conductance in heart fibers. Based on structural similarity and function we can identify many classes of $Cl^-$ channels. Examples of genetic diseases are given for three: (1) the CFTR channel important for normal airway function whose dysfunction produces cystic fibrosis, (2) the CLC-1 channel present mainly in skeletal muscle membranes that controls the excitability of muscle fibers and whose dysfunction leads to forms of myotonia, and (3) the CLC-5 channel of the kidney whose dysfunction leads to formation of kidney stones.

### B. Cystic Fibrosis Transmembrane Regulator

Cystic fibrosis (CF) is an autosomal recessive disease that affects about 1 in 2000 Caucasians. The recessive gene is present in about 5% of that population, and it is the most common inherited lethal disease. It is caused by mutations (more than 100 have been identified) in the gene for the **cystic fibrosis transmembrane regulator,** which is a $Cl^-$ channel known as **CFTR.** The primary structure of CFTR and site of the major (70%) mutation, $\Delta508f$, a deletion of a single phenylalanine at position 508, is shown in Fig. 2A. The primary sequence suggests 12 transmembrane helices, two hydrophilic nucleotide-binding sites, and a large cytosolic regulatory (or "R") domain, which has many potential phosphorylation sites. This channel is regulated by phosphorylation and by nucleoside triphosphates. These agents act at the proposed regulatory areas.

Normal operation is assumed to be that the channel opens when the sites are occupied. Levels of cAMP determined by physiological needs of the cell would then regulate normal function. When mutant CFTR is present in the pulmonary epithelia it does not respond to regulation by cAMP and remains closed, not allowing $Cl^-$ ions to enter the cell, which prevents normal secretion in airway passages. This leads to accumulation of a thick mucus, which blocks the passages and leads to death in 90% of the

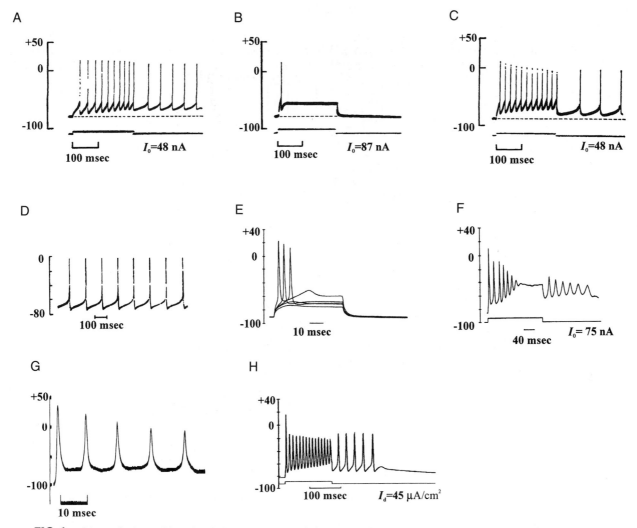

**FIG. 1.** Myotonia due to Cl⁻ and Na⁺ channels. Tracings (A) through (F) are actual microelectrode recordings of membrane potentials from single mammalian skeletal muscle fibers. Except for trace (D), in which the fiber was spontaneously active, the fibers were stimulated by long square current pulses passed through a second microelectrode inserted into the same fiber close to the voltage recording microelectrode. A lower trace representing this depolarizing current is shown in several of the traces and the value $I_0$ is the intensity of this pulse in nanoamperes. *Top row:* Low $g_{Cl}$ myotonia. (A) From a myotonic mutant goat fiber. Note the repetitive action potential and an after discharge when the current is turned off. (B) From a normal goat fiber for control. Note that a larger current than in trace A produced no myotonia. (C) Impermeant sulfate ion replaced Cl⁻ ion bathing a normal fiber, thus blocking $g_{Cl}$. Myotonic responses are induced identical to those of the mutant fiber in trace A. Other $g_{Cl}$ blockers such as 9-AC at 50 $\mu M$ would give the same result. These goat fibers were studied at 38°C (Adrian and Bryant, 1974). *Center row:* Na⁺ channel myotonia. (D) From a human patient with the PC mutation (Lehmann-Horn *et al.,* 1981). PC patients have normal $g_{Cl}$ yet the fibers are myotonic. Cooling this fiber from 37 to 30°C caused the depolarization, the small action potentials, and myotonia. (E and F) From normal rat fibers (Cannon *et al.,* 1993). In trace E are five superimposed control measurements with 50-msec current pulses ranging from 90 to 175 nA (threshold at 115 nA); there is no myotonia. In trace F the fiber was exposed to 10 $\mu M$ ATX-II, the sea anemone toxin that blocks Na⁺ channel inactivation, and stimulated with a long pulse. Note the induction of myotonia, the baseline depolarization, and an after discharge. *Bottom row:* Computer simulations of myotonia. (G) Simulates low $g_{Cl}$ myotonia. Shown are the first five surface membrane potentials calculated for propagating the action potential model, which has a simulated T-tubular system that accumulates K⁺. These action potentials are from a myotonic train during a long depolarizing current pulse. After the pulse was turned off there was a short after discharge. The model was made myotonic by lowering the $g_{Cl}$ term in a Hodgkin–Huxley mathematical model of the action potential adapted to skeletal muscle. The control calculation (not shown) with normal $g_{Cl}$ gave only one action potential for a large stimulating current. (From Adrian and Marshall, 1976.) (H) Computer simulation of Na⁺ channel myotonia (Cannon *et al.,* 1993). The mathematical model was similar to that used for the low $g_{Cl}$ myotonia except in this case the Na⁺ channel currents were modified by adding 2% of the slowly inactivating HYPP mutant-type Na⁺ currents to the normal Na⁺ currents. Trace II shares many details with trace F from the toxin model of Na⁺ channel myotonia; for example, the tendency for the average potential to move toward a depolarizing plateau during the pulse, followed by an after discharge after cessation of the current pulse. Both low $g_{Cl}$ and Na⁺ myotonias require an intact T-tubular system for an after discharge. Detubulated real or simulated myotonic fibers remain myotonic but lack the after discharge.

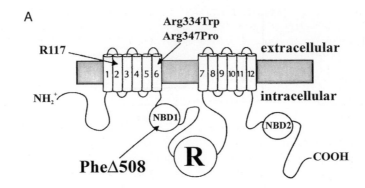

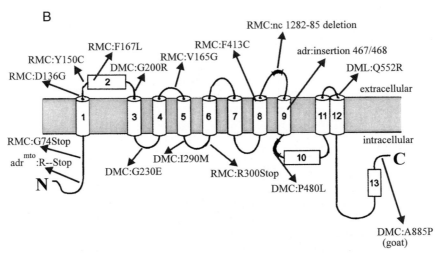

**FIG. 2.** Cl⁻ channels. The primary structures as deduced from hydropathy plots and other information are given for the following Cl⁻ channels: (A) The CFTR channel located in pulmonary and other epithelia. The lower arrow points to NBD1 (nucleotide-binding domain 1), the location of the principal mutation PheΔ508, a deletion of phenylalanine at position 508, which prevents the channel from being regulated by cAMP and leads to 70% of the cases of severe CF. M6 is indicated as the site of two mutations at which arginine (334 and 347) is substituted by tryptophan and proline, respectively. M2 is the site of a similar mutation at position 117. These latter three mutations decrease the Cl⁻ conductance of the channel and lead to mild CF. (B) The CLC-1 channel of mammalian skeletal muscle produces the large stabilizing $g_{Cl}$ necessary for normal excitability. Shown are some of the known mutations with the theoretical structural alterations of the CLC-1 channel leading to the following phenotypes exhibiting low Cl⁻ conductance myotonia: DMC, RMC, DML (dominant myotonia levior), adr, and adr^mto (recessive myotonia in the myotonic mouse). All of the diseases indicated occur in human patients except for those conditions labeled as adr mouse or myotonic goat. Following each disease designation is the mutation given in standard single-letter amino acid codes and the residue position. For example, DMC:G230E is interpreted as dominant myotonia congenita in which at residue 230 a glutamate (E) is substituted for the wild-type glycine (G). The adr-mouse syndrome is due to the insertion of a transposon into the gene, which prevents expression of the channel. Shown here is the position (between 467 and 468) on the protein corresponding to the region on the gene affected by the transposon. This is for illustration only since the protein is never completely expressed. For more complete information on mutations of the CLCN1 gene that encodes for the human CLC-1 channel, see Lehmann-Horn and Rüdel (1996). (Redrawn and modified from Lehmann-Horn and Rüdel, 1996, using data from their Table 2.)

patients before reaching adulthood. There are additional problems with secretion by the epithelia of the pancreas, gut, salivary glands, and other organs. One diagnostic feature of CF is a high NaCl content of sweat. A mutant CFTR can account for most of the pathological signs of the disease but some additional factors may also be involved. (For a review see Warner, 1992.) It has been recently argued that in spite of the fact that CFTR is a channel it may not be the actual Cl⁻ channel that furnishes the ion

transport. CFTR may indeed act as a regulatory protein, which controls this channel as originally thought.

The discovery that an abnormal Cl⁻ channel was ultimately at fault came from resting potential measurements in epithelial cells from airways or sweat glands that always showed the CF cells to be hyperpolarized. Perfusing the luminal side of the cells with a Cl⁻-free (sulfate) medium increased the potential of normal cells as predicted, but had no effect on the diseased cells, i.e., Cl⁻ channels were not conducting properly. The major breakthrough occurred when molecular biologists sequenced the gene responsible for CF. The initial analysis of the proposed product suggested that it was a regulatory enzyme, hence, it received the name cystic fibrosis transmembrane regulatory factor or CFTR. Patch-clamp studies later showed that CFTR codes for a nonrectifying cAMP-activated channel of about 8 pS. The $\Delta$508f mutation prevents the channel from responding to regulation by cAMP, although the defective CFTR protein is expressed. Impressive proof that CFTR was a channel came from constructing a CFTR mutant that exhibited an altered permeability sequence to various anions, suggesting that the CFTR product was indeed a channel. As mentioned earlier it is not completely clear that CFTR and the responsible channel are one and the same. Gene therapy by transfecting genes for the normal channel/regulator into diseased pulmonary epithelia is under investigation. It is too early to evaluate these results.

## C. Skeletal Muscle CLC-1 Channels

Skeletal muscle fiber membranes in their resting or unstimulated state have a relatively high $g_{Cl}$. This amounts to about 80–90% of the resting conductance in mammals, the remainder being $K^+$ and less than 1% $Na^+$ and other ions. The resting $g_{Cl}$ has an equilibrium potential ($E_{Cl}$) about equal to the resting potential (ca. $-82$ mV), and thus it acts to stabilize the membrane toward this potential, which in effect prevents myotonia.

The $g_{Cl}$ of skeletal muscle fibers has been studied for a long time, but identification of the responsible channel, known as CLC-1, has only occurred recently. The CLC family of $Cl^-$ channels cloned by Jentsch and his colleagues (1995) presently includes about a dozen diverse members, and the number appears to be still increasing. The first member cloned was CLC-0 from electrolytes of the marine electric fish, *Torpedo,* where it stabilizes the membrane on one side of the electric cells thus allowing the cells to act as electric batteries (see Chapter 59). Because these electrocytes were derived embryologically from skeletal muscle, polypeptides from short sequences of the CLC-0 channel were used to probe a rat skeletal muscle cDNA library, which resulted in cloning of the CLC-1 channel (Steinmeyer *et al.,* 1991b). The CLC-1 channel, whose primary structure is shown in Fig. 2B, is controlled by development and innervation. The channel is not seen in embryonic muscle or in cultured muscle cells, or following denervation of the adult fiber. The corresponding mRNA was seen only in trace amounts in other tissues.

Monoclonal antibodies to CLC-1 react only with sarcolemmal membrane fragments, suggesting a purely sarcolemmal location of this channel (Gurnett *et al.,* 1995). This is in contrast to earlier functional studies, which placed mammalian $g_{Cl}$ in both the T-tubular and surface membranes (e.g., Dulhunty, 1979). Possibly, there is a T-tubular isoform that does not bind the antibody, or the channels are not accessible in the T-tubular fragment. The location problem awaits further resolution.

The single conductance of CLC-1 is around 1 pS (Pusch *et al.,* 1993), as determined from noise analysis (for an explanation of this method, see Heineman and Conti, 1992), and therefore single-channel currents are too small (i.e., below 1 pA) for practical study. The CLC-1 channel whether expressed in oocytes, mammalian cells, or insect cells (Astill *et al.,* 1996) is sensitive to 9-AC and behaves essentially like the macroscopic $g_{Cl}$ of mammalian skeletal muscle.

The CLC $Cl^-$ channel family has continued to expand with the reports of new members. At present this family, in addition to CLC-0 and CLC-1, consists of CLC-2, CLC-3, and CLC-4, which are ubiquitous mammalian channels of little understood function, possibly involved in cell volume regulation; CLC-K1 and CLC-K2, which are involved in reabsorption in the mammalian kidney; and CLC-5, an additional kidney channel. So far, only two members of the CLC family are associated with genetic disease. Mutations of the CLC-1 channel account for several classical hereditary myotonias in man, mouse, and goat; and mutations in the CLC-5 channel produce human hereditary hypercalcuric nephrolithiasis (kidney stones). These conditions are discussed next.

### 1. Myotonia Congenita

Thomsen's disease or dominant myotonia congenita (DMC) is an autosomal dominant condition in which there is myotonia due to loss of the stabilizing $g_{Cl}$. As shown in Fig. 2B (Koch *et al.,* 1992; George *et al.,* 1993), the defect producing the Thomsen family form of DMC is a single-point mutation where the conserved glycine (G) (residue 230) is replaced by glutamic acid (E), shown on the figure as DMC:G230E. Some of the other known dominant mutations are also shown in Fig. 2B. The fact that a dominant mutation in a channel gene can cause a large reduction in the total number of functional normal channels requires explanation. In the **heterozygous** individual where only half the subunits expressed are abnormal, the $g_{Cl}$ is much less than 50%. One of the mechanisms proposed is that the CLC-1 channel functions as a multimer, probably a tetramer, and that the presence of even a single mutant subunit out of the four makes the channel dysfunctional. From the binomial theorem the conductance works out to be about $\frac{1}{16}$ of normal, that is, this is the probability of getting four wild subunits in a tetramer assembly. The remaining $\frac{15}{16}$ of the combinations of tetramers would have at least one mutant subunit and therefore be inactive. Experimental proof of this possibility is that coexpression of a DMC mutant CLC-1 with the wild-type CLC-1 produces a reduction in $g_{Cl}$ to about $\frac{1}{16}$ of the predicted $g_{Cl}$, although the protein channels were expressed (Steinmeyer *et al.,* 1994). It has now been shown that the CLC-0 channel is a dimer (Ludewig *et al.,* 1996; Middleton *et al.,* 1996), and because of the large identity between CLC-0 and CLC-1 it is reasonable to propose that CLC-1 is also a dimer. Recent coexpression experiments support a dimer model for CLC-1 (Fahlke *et al.,* 1997), and these authors question the methodology and use of binomial model in the earlier work of Steinmeyer *et al.* (1994), which was discussed earlier. It is also of great theoretical interest that the gating process of CLC-0 and CLC-1 is fairly unique in that these channels utilize the permeating chloride ion itself as the gating particle (Rychkov *et al.*, 1996).

## 2. Recessive Myotonia Congenita

A large population of nondystrophic myotonic patients was studied, and two separate diseases corresponding to their dominant or recessive inheritance were distinguished (Becker, 1973). The smaller group (30%) were the autosomal dominant DMC patients who showed less myotonia on the average and whose physiology we discussed previously. The larger group (70%) were the autosomal recessive RMC patients who were generally more severely myotonic. RMC is also called **recessive generalized myotonia.** In RMC there is a reduced $g_{Cl}$ due to a variety of mutations, some of which are shown in Fig. 2B. Additionally, some dysfunctional $Na^+$ channels may contribute to the myotonia in certain patients (Franke *et al.*, 1991). This $Na^+$ channel abnormality may be a pleotropic effect consequent to the CLC-1 mutation.

## 3. Myotonic Mouse

A mouse that had difficulties in righting when placed on its back was discovered at the Jackson laboratories (Mehrke *et al.*, 1988). This condition was due to a recessive mutation, and at first was thought to be a neurological condition affecting the righting mechanism and thus received the name of *arrested development of righting* response or *adr* mouse. Later it became clear that the major problem was the myotonia of the skeletal muscles, and thus the term *myotonic mouse* is also used. The repetitive firing and other defects in excitability were shown to be due to a specific loss of $g_{Cl}$. The heterozygous animals that had nearly normal $g_{Cl}$ behaved normally, which indicates that the normal allele in the heterozygous mouse is capable of coding for and expressing a normal density of $Cl^-$ channels. Jentsch and colleagues (Steinmeyer *et al.*, 1991a,b) have shown that the CLC-1 channel (Fig. 2B) is also the major skeletal muscle $Cl^-$ channel in the mouse. They further showed that the mutation was caused by a **transposon** (i.e., a nonsense code) inserted into this gene at a region that would otherwise code for an essential part of the channel protein. Actually, the defect is so severe that no mRNA for this channel was detected in the myotonic mouse. In addition to the transposon defect of the *adr* mouse, other missense and nonsense mutations have been reported in the mouse CLC-1 that lead to myotonic conditions (Gronemeier *et al.*, 1994). One of these is the *adr*[mto] mouse (Bryant *et al.*, 1987).

## 4. Myotonic Goat

A dominant mutation in goats in which myotonia is the only clinical sign has been known for more than a century (see Bryant, 1979). These animals (so-called "stiff," "falling," or "nervous" goats) are still being raised for their interesting behavior on sudden movement, which often causes bizarre stiffness and sometimes falling. Based on human experience it is believed that the muscle effects are not painful. Electrophysiological studies on biopsied

muscle fibers from these animals advanced our knowledge of the biophysics of myotonia, and in particular led to the discovery of myotonia due to low $g_{Cl}$ discussed earlier. Also shown was the necessity of $K^+$ accumulation in the T-tubular system for maintaining an after discharge (Adrian and Bryant, 1974). The myotonia can be simulated by blocking $g_{Cl}$ with 9-AC or other $Cl^-$ channel blockers (Bryant and Morales-Aguilera, 1971), and by computer modeling of action potentials when the $Cl^-$ conductance term is made low (Adrian and Marshall, 1976).

The mutation responsible for CLC-1 dysfunction in myotonic goat muscle was very recently identified by Beck (Beck *et al.*, 1996) to be A885P, a replacement of a highly conserved alanine by proline in the carboxyterminus region. The functional effect of this mutation, determined in a human CLC-1 construct, is to shift the curve relating open probability ($P_{open}$) to membrane potential to the right by 45 mV. This is illustrated in Fig. 3. These results imply that at the physiological resting potential of $-82$ mV, if only mutant channels were present, the $Cl^-$ conductance would be roughly 25% of a membrane having wild-type channels; this is similar to the reduction of $Cl^-$ conductance recorded in native myotonic goat membrane. An exact comparison with the older studies would be difficult, since these data were usually averaged from mixed genetic populations having both mutant and wild-type alleles. The pattern of shifting the open probability curves toward more depolarized voltages also occurs in four dominant human myotonias (I290M, R317C, P480L and Q552R), and may represent a "biophysical phenotype" for dominant $Cl^-$ channel myotonic disease (Beck *et al.*, 1996). This pattern of shifting the center potential (or $V_{1/2}$) of activation curves (e.g., $P_{open}$ vs potential) occurs also in mutant $K^+$ channels, and it is discussed in that section. This shift may represent

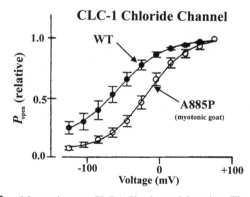

**FIG. 3.** Myotonic goat CLC-1 $Cl^-$ channel function. The relative open probabilities at the different membrane voltages were determined with a two-electrode voltage clamp from from *Xenopus* oocytes expressing wild-type (WT) or mutant (A885P) CLC-1 channel currents. The curves were fitted to single Boltzmann functions. The principal change observed is the right shift of the center potential, $V_{1/2}$, of the mutant channel by 45 mV from the wild-type control. This kind of shift is typical of shifts observed in several dominant human myotonias involving the CLC-1 channel. At the resting potential of $-80$ mV the mutant channels would have a lower probability of being open, and this would account for the low $g_{Cl}$ and myotonia in the affected goat. (Redrawn from Beck *et al.*, 1996.)

a general mechanism accounting for the malfunction of mutant voltage-dependent channels.

## D. Kidney CLC-5 Channel: Kidney Stone Diseases

The second member of the CLC family of Cl⁻ channels to be associated with genetic disease is CLC-5. This channel is strongly **outwardly rectifying,** unlike CLC-1, which is an **inward rectifier.** Kidney stones affect about 12% males and 5% females in the Western world, and in 45% of these patients, the disease is inherited. Dent's disease and two other types of kidney stone disease were linked to chromosome Xp11.22. A microdeletion was deleted in one of the Dent's disease patients that allowed the identification of a candidate gene (*CLCN5*) that codes for the renal Cl⁻ channel, CLC-5. Many types of mutations of this gene were then detected; these errors include nonsense, missense, donor splice mutations, and various deletions (Steinmeyer *et al.*, 1995; Lloyd *et al.*, 1996).

A speculative explanation for the pathological effects of the mutant CLC-5 is that the protein leakage through the glomerulus is taken up by endocytosis in the proximal tubules, and subsequently degraded in a lysosomal compartment. The essential endocytotic uptake is postulated to require acidification of the endosomes, and the CLC-5 channel provides the Cl⁻ transport necessary to maintain the low pH in this compartment. Because the mutant CLC-5 does not transport Cl⁻, the endosomal pH rises, endocytosis is impaired, protein accumulates in the tubule, and kidney stone formation is promoted.

## IV. Na⁺ Channels

### A. General Comments

Voltage-gated Na⁺ channels play an essential role in nerve tissue where they are responsible for the rapidly conducting nerve impulse; and in the heart, they account for the rapid rising phase of the cardiac AP. Voltage-gated Na⁺ channels belong to the general family of voltage-gated cation channels, which includes Ca²⁺ and K⁺ channels. Basically, the main or α-subunit (see Fig. 4A) consists of about 2000 amino acids. There are four repeats, each containing six putative transmembrane helices. Site-directed mutagenesis studies have determined that the fourth transmembrane segment, called S4, acts as the voltage sensor. The cytoplasmic loop connecting repeats III and IV is necessary for inactivation, and the region between S5 and S6 contains the permeable pore. Ca²⁺ and K⁺ channels share a similar molecular design. The K⁺ channel, however, departs from the scheme in that genes encode for only a single repeat; the channel is assembled from four identical or differing polypeptides. Non-voltage-gated Na⁺ transporters are also to be found. One of them, the amiloride-sensitive Na⁺/glucose transporter, has undergone mutations causing human metabolic disease (see Ashcroft and Röper, 1993 for discussion). Well-described disease entities involving Na⁺ channel mutations include those of the human SCN4A gene on chromosome 17q, which codes for skeletal muscle Na⁺ channels. These are discussed as a group next.

## B. Skeletal Muscle SkM-1 and SkM-2 Channels

There are multiple subtypes of Na⁺ channels in skeletal muscle based on sensitivity to toxins and to antibodies, and these subtypes may be located preferentially in the T-tubules or surface membrane. Two Na⁺ channels, specific to skeletal muscle, known as SkM-1 (George *et al.*, 1992) and SkM-2 (Kallen *et al.*, 1990), have been cloned and sequenced. These channels are sometimes designated $\mu_1$ and $\mu_2$, respectively. Skeletal muscle Na⁺ channels are heterodimers having a 260-kDa α-subunit and a 38-kDa β-subunit (Kraner *et al.*, 1985). Type SkM-1 is expressed in both innervated and denervated adult muscle and is blocked by nanomolar concentrations of tetrodotoxin (TTX) and μ-conotoxin. The TTX-insensitive SkM-2 is not found in innervated adult mammalian muscle, but appears within hours following denervation, reaching a maximum at 48 hr and then decreasing. The type SkM-2 is expressed in early development and disappears as type SkM-1 increases, thus SkM-1 is the only adult form. Type SkM-2 is also seen *in vitro* in skeletal muscle culture (absence of innervation) where it accounts for about 30% of the functioning channels; its sequence is essentially the same as the normal TTX-insensitive Na⁺ channel of heart (Rogart *et al.*, 1989). The discussion of hyperkalemic periodic paralysis has been combined with the related myotonia syndromes, because these two conditions appear to be due to different mutations of the SCN4A gene, which codes for the SkM-1-type Na⁺ channel (Fig. 4A).

### 1. Periodic Paralysis and Paramyotonia

In a number of inherited diseases, periodic bouts of paralysis occur. Some of these conditions are associated with either hyperkalemia or hypokalemia, and myotonia may also be present. Recent studies have brought considerable light to understanding many of these conditions. Several genetic forms of myotonia and periodic paralysis may be due to failure of a small fraction of the skeletal muscle voltage-gated Na⁺ channels to inactivate normally (Rüdel and Lehmann-Horn, 1985). Hyperkalemic periodic paralysis (HYPP) and paramyotonia congenita (PC) are related because they are due to autosomal *dominant* mutations of the α-subunit of the SkM-1 Na⁺ channel located in the surface membrane and T-tubular membrane of mammalian skeletal muscle fibers. These channels are normally responsible for conducting the skeletal muscle action potential (AP) throughout the tubular membranes, which is necessary for excitation–contraction coupling. Figure 4A shows the locations of frequent PC and HYPP mutations of the α-subunit of the Na⁺ channel (Fontaine *et al.*, 1990). In Fig. 4B, the lack of effect on activation and the profound effect on inactivation are clearly seen in mutant constructs having point mutations in these regions (residue numbers differ slightly from Fig. 4A due to species). Note in the single-channel records the longer openings and repeated reopenings of the mutant constructs. This accounts for the

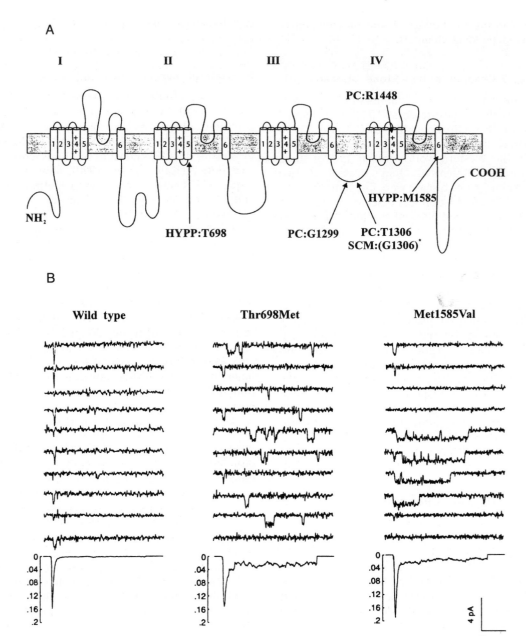

**FIG. 4.** Skeletal muscle Na$^+$ channel. (A) Primary structure of the $\alpha$-subunit of the SkM-1 skeletal muscle Na$^+$ channel coded by the gene SCN4A. Mutations are indicated that lead to PC (paramyotonia congenita), HYPP (hyperkalemic periodic paralysis), and SCM (Na$^+$ channel myotonia). *The amino acid position numbers for PC and HYPP follow those of Cannon and Strittmatter (1993); however, for the mutations at the conserved glycine causing SCM, the numbering system of Lerche *et al.* (1993) was used. Unfortunately, the position numbers overlap here although the real amino acid locations are different. (B) Single Na$^+$ channel currents recorded in HEK293 cells after expressing the wild type, and two genetically engineered constructs, Thr6698Met and Met1585Val, which correspond to natural mutations indicated in part A that cause HYPP. Ten single-channel current traces are shown above, and the averaged ensemble currents are shown below. Note that the wild-type channel shows brief single openings and a transient ensemble current. In contrast, the mutant Na$^+$ channels display long openings and reopenings giving averaged Na$^+$ currents with long noninactivating tails. These prolonged depolarizing inward currents lead to the myotonic behavior and, if sufficiently large, to the block of excitation and a flaccid paralysis. The HYPP mutant channels, unlike the PC channels, show a sensitivity to K$^+$ ions that further decreases their ability to inactivate. The PC channels, on the other hand, are sensitive to lowering the temperature, which has the same effect.

longer persistence of the averaged mutant sodium currents. Another way of stating this is to say that the mutant Na$^+$ channels exhibit a decreased tendency to inactivate.

HYPP is an autosomal dominant mutation in which one sees recurrent episodes of weakness with small elevations in the serum K$^+$ concentration. In this particular form of the disease, there is marked depolarization of skeletal muscle fibers during an attack, which makes the fiber inexcitable and leads to a flaccid paralysis of the muscle. Under voltage-clamp conditions, diseased biopsied fibers produce a noninactivating Na$^+$ current that can be blocked by TTX, and further this has been shown to be the mechanism for the depolarization (Lehmann-Horn et al., 1987). In turn, single-channel recordings have shown that the Na$^+$ channels giving rise to the depolarizing current lack normal inactivation. These channels have prolonged open times and tend to open repetitively (Cannon et al., 1991). This is shown in Figs. 4A and B. The wild SkM-1-type channels very rarely show this noninactivating behavior as do the mutant channels when exposed to low extracellular K$^+$. However, in high K$^+$ concentrations (around 10 m$M$ compared with a normal of 4–5 m$M$) a small fraction (5–10%) of the mutant channels does not inactivate and this produces a constant open probability of between .02 and .05, yielding a steady depolarizing current large enough to account for the depolarization block (Cannon et al., 1993). As mentioned when discussing the mechanisms of myotonia, delayed inactivation of Na$^+$ channels can lead to myotonia, and myotonia has been observed in HYPP patients. A condition closely resembling HYPP of man has also been described in mutant quarter horses (Pickar et al., 1991).

To demonstrate that lack of Na$^+$ channel inactivation is the possible cause of some forms of myotonia and periodic paralysis, two approaches have been taken. In the first approach, an in vitro model was created in rat muscle exposed to a polypeptide toxin (ATX-II) obtained from a sea anemone. The toxin at 10 $\mu$m produced a noninactivating open probability at $-10$ mV of approximately .02, similar to that reported for myotubes from HYPP patients, and myotonia was apparent (Cannon and Corey, 1993). In the second approach, a computer simulation was developed based on modifications of the Hodgkin–Huxley equations adapted to skeletal muscle fibers. This approach is similar to that discussed for computer simulation of low-$g_{Cl}$ myotonia. The simulated fiber, like the low-Cl$^-$ model, required a T-tubular compartment to act as a diffusion-limited space in which activity-induced K$^+$ can accumulate. The computed APs for increasing degrees of incomplete inactivation of Na$^+$ channels effectively simulated normal, myotonic, and paralytic muscle. The simulated APs with altered Na$^+$ inactivation compared favorably with the abnormal APs produced pharmacologically with ATX-II (Cannon et al., 1993).

PC, an autosomal dominant condition, has been also referred to as paradoxical myotonia because the signs of myotonia increase with use of the skeletal muscles, rather than diminish, and there is an increase in myotonia and a dramatic increase in the response of a muscle to percussion with local cooling. At least in vitro, low-Cl$^-$ conductance

myotonias are quite different since the signs are diminished with cooling below 27°C (Furman and Barchi, 1978). If one replaces the sensitivity to cooling with the worsening of symptoms with elevated serum K$^+$ levels, there is a physiological resemblance between PC and HYPP. Both of these conditions have myotonia with episodic weakness and the similarities appear to be explained by the findings that both diseases are due to different mutations but on the same skeletal muscle Na$^+$ channel gene (Ptacek et al., 1993; McClatchey et al., 1992). Figure 4B diagrams the relation of some of the common mutations which have been identified. The most highly conserved region of the various Na$^+$ channels, which as a family show remarkable homology, are the S4 segments of each domain. This region is commonly believed to function in Na$^+$ gating and inactivation. These segments contain positively charged arginines and glycines at every third position, with principally neutral amino acids interdigitated between. Figure 4B shows single-channel recordings from cells expressing constructed mutants similar to those seen in PC patients. Note the lack of inactivation as evidenced by repeated openings and longer open times compared with the wild type (WT). The ensemble averages shown immediately below the single-channel records of the constructs in Fig. 4B display the prolonged Na$^+$ currents capable of causing depolarization and myotonia.

### 2. Na$^+$ Channel Myotonia

Some mutations in the SCN4A gene for the SkM-1 channel can produce in patients a relatively pure form of myotonia known as human **Na$^+$ channel myotonia (SCM)** (Lerche et al., 1993). This syndrome is devoid of dystrophy and weakness, and does not resemble PC or HYPP, although the latter pair are also due to mutations in the same gene (Fig. 4A). Glycine at position 1306 is completely conserved in all Na$^+$ channel $\alpha$-subunits in all species. This is an important region for control of fast Na$^+$ channel inactivation; for example, substituting a glutamine for the phenylalanine at position 1311 completely removes fast inactivation (West et al., 1992). It has even been proposed that the two glycines at 1306 and 1307 may act as a "hinge" to block the channel pore. In the families having SCM the glycine 1306 is substituted by glutamine, valine, or alanine. The glutamine substitution produces the most severe myotonia. None of the mutations in the patients affects inactivation as much as the substitution for phenylalanine at 1311, but the degree is sufficient to account for the myotonia. Patch-clamp studies of the SCM channel show an increase in the time constant of fast inactivation and an increased frequency of late channel openings. These parameters lead to an increase in the ratio of late current to peak current from control of 0.4 to 6%. The simulation experiments using ATX-II anemone toxin (Cannon and Corey, 1993) with computer modeling showed that myotonia occurred when only 2% of Na$^+$ channels fail to inactivate, which corresponds to a late over-peak ratio of 5%. It should be clear that a larger late depolarizing current can result in restimulation and repetitive firing, that is, myotonia. Thus, SCM and its severity can be explained by the length, ramification, and charge of the group substituted for glycine 1306.

### 3. Denervation

Denervation of mammalian skeletal muscle produces some interesting changes in the electrical excitability of the cell membrane. Within 2–3 days after cutting the motor nerve to the muscle fibers in rats, the fast SkM-1 Na$^+$ channels are partly replaced by newly synthesized, fast SkM-2 Na$^+$ channels that are 400 times less sensitive to TTX (Redfern and Thesleff, 1971). Interestingly, the SkM-2 channels retain sensitivity to saxitoxin (STX), a poison that is structurally related to TTX, but has two guanidinium groups instead of one. Thus, TTX at concentrations that usually block SkM-1 channels now only reduces the maximal rate of rise of the AP (max $dV/dt$), but does not block excitability. The APs are said to be "TTX-resistant," but this is, of course, relative because higher concentrations of TTX do cause block. In addition, the presence of SkM-2 channels reduces the max $dV/dt$ to about 65% of the control (innervated) value, in the absence of TTX.

The SkM-2 channel is normally inhibited from expression by both activity and trophic influences from the nerve, and none of them is detectable in innervated muscle. Following denervation the new SkM-2 channels reach a maximal density around 4 days and slowly disappear over the next 10 days. When the SkM-2 channel is studied in planar bilayers it shows pharmacological sensitivity and electrical properties similar to the normal heart channel (Guo *et al.*, 1987). SkM-2 channels are also seen in cultured muscle cells in the absence of nerve influence. Both the heart and denervation channels have been cloned and sequenced and these two channels differ only very slightly as might be expected (Rogart *et al.*, 1989; Kallen *et al.*, 1990).

### C. Cardiac Na$^+$ Channel: Long QT Interval Disease

Na$^+$ currents flow during the initial upward stroke or phase 0 of the ventricular cardiac action potential. The Na$^+$ channels quickly inactivate but the action potential remains depolarized for its duration due to the flow of Ca$^{2+}$ currents. The action potential is repolarized by the activation of K$^+$ currents. The cardiac Na$^+$ channel is coded by the *SCN5A* gene, and mutations in this gene produce aberrant Na$^+$ channels in which fast inactivation is delayed (Roden *et al.*, 1995). The end result is a cardiac action potential that has a prolonged repolarization, which is seen as a distinctive long Q-T interval of the electrocardiogram. Associated with the repolarization defect is a serious arrhythmia leading to syncope (fainting), seizures, and sudden death from a distinctive ventricular tachycardia known as *torsade de pointes*. The long Q-T syndrome is genetically linked to sites other than the Na$^+$ channel gene (Keating and Sanguinetti, 1996). One of these linkages is also brought about by mutations of K$^+$ channels that do not open properly. (See the K$^+$ channel section.)

## V. Ca$^{2+}$ Channels

### A. General Comments

There are several types of Ca$^{2+}$ channels, both agonist activated and voltage controlled. The primary structure of the $\alpha$-subunit of the L-type Ca$^{2+}$ channel of skeletal muscle is shown in Fig. 5A. This channel is blocked by dihydropyridines (DHPs) so it is also known as the skeletal muscle DHP receptor; in addition it is the T-tubule voltage sensor. (For a review of the important voltage-gated Ca$^{2+}$ channels see Tsien *et al.*, 1991.) The Ca$^{2+}$ release channel of the sarcoplasmic reticulum (Fig. 5B) can be reviewed from the references given in MacLennan and Philips (1992).

### B. Dihydropyridine Receptor

#### 1. Muscular Dysgenesis

Mice that are homozygous for the autosomal recessive dysgenic gene mutation (mdg/mdg) lack excitation–contraction (EC) coupling and the slow Ca$^{2+}$ current. These animals are therefore incapable of muscular movement and survival. The fetuses do not move *in utero* and die at birth, presumably from respiratory paralysis. Cultured myotubes that carry the mdg/mdg phenotype can also be produced for experimentation. The physiological problem is related to lack of function of the DHP receptor. In the mutant dysgenic animals there is a fivefold decrease in DHP binding in skeletal muscle, there is a low level of mRNA encoding the DHP receptor, and monoclonal antibodies detect no DPH receptors. On the other hand, cardiac DHP binding is normal, illustrating the fact that the L-type Ca$^{2+}$ channels (or DHP receptors) are distinctly different in these two tissues. Muscular dysgenesis appears to have been the first documented example of a genetic disease in vertebrates that is produced by a defect in a structural gene coding for an ion channel (Tanabe *et al.*, 1988).

Although a human counterpart has not been reported for this mutation, basic studies with the dysgenic mouse model have helped confirm the role of the DHP receptor as both a voltage sensor and channel in the T-tubular membrane. A critical experiment was done by Tanabe and coworkers (Tanabe *et al.*, 1988) in which both EC coupling and slow Ca$^{2+}$ current were restored in the dysgenic muscle by microinjection of the expression plasmid (pCAC6) carrying cDNA for the missing DHP receptor. The restoration of both voltage sensing and ion channel functions with the single cDNA confirmed the dual role of this protein in the T-tubule. See also the discussion of muscular dysgenesis in the chapter by Levinson, and the discussion in Lehmann-Horn and Rüdel (1996).

#### 2. Hypokalemic Periodic Paralysis

Hypokalemic periodic paralysis is a disease of skeletal muscle in which episodes of weakness occur in association with lowered serum K$^+$ levels. It is an autosomal dominant trait, but in one-third of the cases it appears spontaneously.

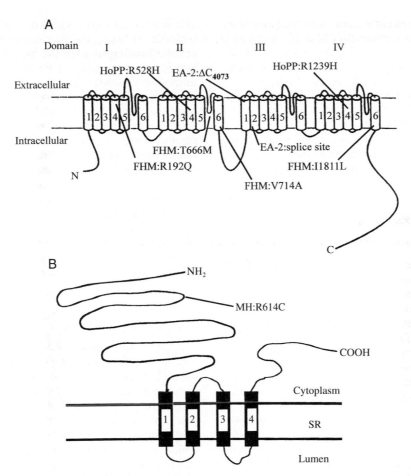

**FIG. 5.** $Ca^{2+}$ channels. (A) The primary structure of the $\alpha_1$-subunit of the human skeletal muscle type-1 voltage-sensor/ channel (i.e., the DHP receptor) of the T-tubular membrane, and the neuronal P/Q-type $Ca^{2+}$ channel are shown here superimposed on the same illustration. In the skeletal muscle two mutations linked to human hypokalemic periodic paralysis (HoPP) are indicated. Both mutations occur in segment 4 of domains II and IV, which is a region believed to be involved in gating of the channel. Using the same schematic, the primary structure is superimposed on the $\alpha_1$-subunit of the neuronal P/Q-type $Ca^{2+}$ channel. The mutations causing familial hemiplegic migraine (FHM) and EA-2 are indicated. Note that the mutations causing FHM are of the missense type, resulting in single amino acid substitutions, whereas the two mutations for EA-2 both cause premature stops. (Redrawn and modified from Ophoff *et al.*, 1996.) (B) The primary structure of the human ryanodine receptor indicating the main mutation responsible for malignant hyperthermia (MH). The mutation at the conserved arginine-614 is a substitution for cystine in the cytoplasmic region of the molecule. In porcine MH arginine-615 is substituted in the same way. One should note that except for the four membrane-spanning segments M1 through M4 (indicated here as simple 1 through 4) 90% of this molecule (5037 amino acids) is cytoplasmic. The functioning $Ca^{2+}$ release channel consists of four of these units (i.e., a homotetramer) in the SR membrane. Activation to the T-tubular DHP receptor causes activation of the $Ca^{2+}$ release channel in the process of excitation–contraction coupling. One of many speculations as to how this is accomplished is that a third molecule, triadin, interacts with the connector between domains II and III of the DHP receptor (shown in Part A) giving a "mechanical" coupling between the two systems.

A typical attack of weakness and low serum $K^+$ can begin a few hours after a high carbohydrate meal, often when asleep, and can last for 4–24 hr. Other precipitating factors include rest after exercise and situations where $K^+$ moves into muscle cells. For years this disease was thought to be involved somehow with abnormal pump ($Na^+, K^+$-ATPase) or cation channel function and the syndrome can be reasonably modeled in rats that have their skeletal muscle $K^+$ contents altered through insulin and/or low $K^+$ diet (Bond and Gordon, 1993). It came as a surprise then when linkages were discovered to the gene that encodes the skeletal muscle L-type $Ca^{2+}$ channel. Two mutations were described and are shown in Fig. 5A. In each mutation there is a replacement of a highly conserved arginine by histidine in the S4 segment, a region believed to be involved in channel gating. One replacement occurs in the D4 domain at arginine 1239 (Ptacek *et al.*, 1994), and the other occurs in the D2 domain at arginine 528 (Jurkat-Rott *et al.*, 1994). Recent studies of these mutations expressed in hererologous systems have thus far been disappointing in that they have not revealed abnormal functions of this channel which can explain HoPP. (For further discussion see Lehmann-Horn and Rüdel, 1996.)

## C. Ca²⁺ Release Channel

### 1. Malignant Hyperthermia

Malignant hyperthermia (MH) is a rare autosomal dominant condition in humans that predisposes these individuals to react to anesthesia with muscle rigidity, hypermetabolism, and high fever in which the muscles become highly stimulated metabolically, with a consequent rise in body temperature (MacLennan and Phillips, 1992). A popular form of anesthesia involving use of halothane in conjunction with a depolarizing neuromuscular blocker, succinylcholine, can trigger MH. If not treated immediately these patients may die within a few minutes from ventricular fibrillation or within hours to days from neurological or renal complications. Although MH is rare, it is serious enough to be fatal in apparently healthy individuals undergoing anesthesia. MH is also seen in animals where it is often triggered by heat stress. MH-susceptible pigs have been recognized for many years (Lucke *et al.,* 1979) and genetic and physiological studies of the MH pigs have hastened our understanding of the comparable condition in human patients.

The mechanism of MH is the presence of a mutated Ca²⁺ release channel located in the sarcoplasmic reticulum (SR) of skeletal muscle (Fig. 5B). The specific mutations that account for the dysfunction of this channel are different in man and pigs. The Ca²⁺ release channel is a key element of the EC coupling process. This is the chain of events beginning with the muscle AP that invades the T-tubular membrane causing movement of gating charge in the DHP-sensitive Ca²⁺ channel/voltage sensor, which in turn activates the Ca²⁺ release channel on the SR to release Ca²⁺ from stores from within the SR to initiate contraction. The exact means by which the voltage sensor of the T-tubular membrane couples with the release channel of SR is not yet fully understood. In addition, pumps and regulators exist that accumulate and store the Ca²⁺ for release. These events and a scheme to explain the pathophysiology of a MH attack are depicted in Figs. 6A and B.

The Ca²⁺ release channel is specifically opened at nanomolar concentrations and blocked at micromolar concentrations by the toxin ryanodine. Thus, this channel is also referred to as the *ryanodine receptor*. Using this high-affinity ligand, the protein was purified and several full-length cDNAs from different species were subsequently cloned. The channel is coded by two different genes. RYR1 on chromosome 19 codes for the skeletal muscle channel, and RYR2 on chromosome 1 codes for heart and brain channel. In pigs the cDNA sequences for RYR1 between normal and MH pigs predict a single amino acid difference, a cysteine for an arginine (Fig. 5B). In human studies two mutations of RYR1 are associated with MH, but some MH patients have no linkage with this gene; further, mutations of this gene may cause unrelated forms of muscle disease. Human MH may therefore involve other gene mutations

and interactions (see Gillard *et al.,* 1992). These major isoforms of the ryanodine receptor are the only known cellular binding sites for the ligand. The protein is a large homotetrameric complex constructed from 565-kDa subunits. The channel region is probably located in the membrane-spanning segments, which are 20% of the subunit from the carboxyterminus. The remainder of each subunit is cytoplasmic and the four subunits together form a cytoplasmic structure holding the four extended channels that empty into lateral vestibules. This cytoplasmic region also must contain the sites for coupling to the DHP receptor.

All of the signs of MH as seen in Figs. 6A and B can be explained by a defect in the regulation of intracellular Ca²⁺. Sustained Ca²⁺ levels in the fiber would cause contracture, increased glycolytic and anaerobic metabolism with depletion of ATP, glucose, and oxygen, and overproduction of CO₂, lactic acid, and heat. A specific antagonist for the MH reaction is dantrolene Na⁺, which is capable of blocking release of intracellular Ca²⁺ from the SR.

### 2. Central Core Disease

Another autosomal dominant disease of the RYR1 gene is central core disease (CCD). It is characterized by hypotonia and proximal weakness first seen in infants, but appears to be nonprogressive. The severity of the disease is variable within families. To date, three mutations causing the disease have been identified among affected families. The substitution of histidine for a conserved arginine (R2434H), which is absent in the general population, was identified in a large Canadian family. Two other missense mutations, I404M and R163C, have also been identified. One might presume that these channel defects should cause EC coupling malfunction and be able to explain for the observed pathology; however, this is still an open question (George, 1995).

## D. Neuronal Ca²⁺ Channels

The neuronal Ca²⁺ channels are like the L-type skeletal muscle channels discussed earlier. They are multimeric entities consisting of one each of $\alpha_1$-, $\alpha_2$-, $\delta$-, and $\beta$-subunits. The central $\alpha$1-subunit has the voltage sensor and the conducting pore. The neuronal channels discussed here differ from the skeletal muscle L-type channel in lacking the DHP receptor on the $\alpha_1$-subunit. Two different classes of neuronal Ca²⁺ channel diseases are considered. The first, Lambert–Eaton syndrome, is due to autoimmune block of Ca²⁺ channels in motor nerve terminals leading to skeletal muscle weakness. The second class is a pair of neurological diseases related by the fact that they are caused by different types of mutations in the same gene that codes for a central nervous system Ca²⁺ channel. The types of mutations in the same channel determine the resultant form of the disease.

### 1. Lambert–Eaton Syndrome

In *Lambert–Eaton syndrome* autoimmune-generated antibodies reduce the number of voltage-activated Ca²⁺ channels in the presynaptic nerve terminal. This decreases

A

B

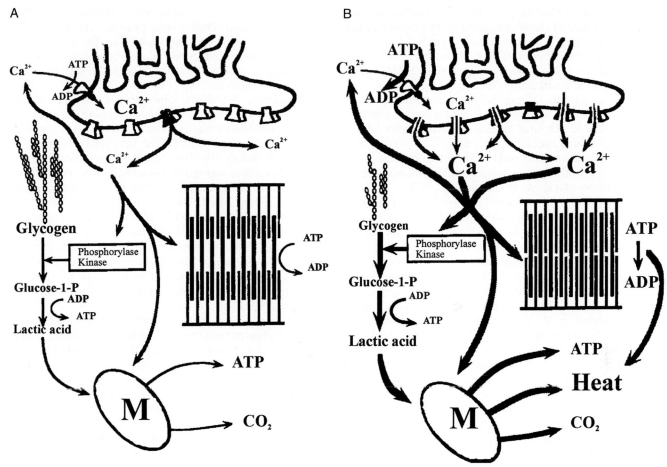

**FIG. 6.** Malignant hyperthermia. (A) Normal physiological mechanism of EC coupling. The $Ca^{2+}$ release channel (shown here as square elements) is a key element of the process known as excitation–contraction coupling. (B) MH muscle contraction cycle. Mutant $Ca^{2+}$ release channels can explain the basic mechanism of MH. The mutations produce a release channel that is more sensitive to opening stimuli and fails to close rapidly, thus leading to abnormally high $Ca^{2+}$ levels. Sustained $Ca^{2+}$ levels in the fiber then would cause contracture, increased glycolytic and anaerobic metabolism with depletion of ATP, glucose and oxygen, and there is overproduction of $CO_2$, lactic acid and heat. M indicates a mitochondrion. (Modified from MacLennan and Phillips, 1992; copyright © 1992 AAAS.)

the amount of $Ca^{2+}$ entering the nerve terminal with each nerve impulse. Because $Ca^{2+}$ entry is necessary for the release of transmitter by presynaptic terminal, we find a decrease in acetylcholine released to the postsynaptic AChR channels with reduced neuromuscular transmission and weakness. There are many subtypes of neuronal $Ca^{2+}$ channels; however, the type involved in the Lambert–Eaton syndrome is specifically labeled by the marine snail toxin, $\omega$-conotoxin (see Penn et al., 1993). The Lambert–Eaton syndrome has a presynaptic origin that is to be contrasted with *myasthenia gravis* (another form of neuromuscular weakness, discussed later), which has a postsynaptic origin. Patients with Lambert–Eaton syndrome often

have an accompanying carcinoma and it has been asserted that the antibodies to the cancer cells also recognize the $Ca^{2+}$ channel.

### 2. Hemiplegic Migraine and Episodic Ataxia Type-2

Migraine is a common neurological disorder characterized by recurrent severe headaches accompanied by phonophobia, photophobia, and nausea and vomiting. It occurs in about 24% of females and 12% of males, and varies greatly with regard to frequency, duration, and severity. Familial hemiplegic migraine (FHM) is a rare autosomal dominant form of migraine, which in addition to the usual migraine symptoms has an accompanying hemiparesis (paralysis of one side of the body) and sometimes progressive cerebellar atrophy. Episodic ataxia type-2 (EP-2) is an autosomal dominant neurological disorder characterized by ataxia of cerebellar origin, and resembles FHM in also having migraine attacks and cerebellar atrophy. The episodic ataxia component responds to treatment with acetazolamide (a carbonic anhydrase inhibitor), otherwise it is similar to the $K^+$ channel defect, episodic ataxia type-1 (see the $K^+$ channel section).

We are discussing FHM and EP-2 together because these diseases have both been mapped to chromosome 19p13 and are due to mutations in the gene (*CACNL1A4*) that codes for a brain $Ca^{2+}$ channel known as the P/Q-type $Ca^{2+}$ channel. Four missense mutations causing FHM and two premature stops causing EP-2 have been identified. The primary structure of the brain P/Q-type $Ca^{2+}$ channel and the sites of the mutations causing the aforementioned diseases are shown in Fig. 5A. The electrophysiological dysfunctions of the mutant channels in FHM and EP-2 have not been yet been identified for these very recent studies (Ophoff *et al.*, 1996), but some suggestions may be offered based on homology to the $Na^+$ channel and its defects.

The four missense mutations associated with FHM could produce a situation analogous to the missense mutations of the skeletal muscle $Na^+$ channel that cause hyperkalemic periodic paralysis, paramyotonia, and $Na^+$ channel myotonia (see $Na^+$ channel section). In these latter conditions, the $Na^+$ channels have impaired inactivation, and by analogy in FHM the $Ca^{2+}$ channels would exhibit increased activity. The cellular mechanism proposed is that increased $Ca^{2+}$ activity would promote the process known as **spreading cortical depression,** which would in turn facilitate initiation of a migraine attack.

On the other hand, the mutations occurring in EP-2 are classed as premature stops (i.e., frame shifts and splice-site mutations), and these would lead to truncated $\alpha$-subunits consisting only of repeats I and II and part of III. Thus, the EP-2 channels would not be expected to function and probably would be present in lower density, the end result being decreased $Ca^{2+}$ channel function in EP-2 with consequent dysfunction of the cerebellum, which would account for the ataxia.

## VI. $K^+$ Channels

### A. General Comments

Voltage-dependent $Na^+$ and $Ca^{2+}$ ion channels are sometimes thought of as **pseudotetramers** since their main functioning subunit consists of four interconnected domains, each having six membrane-spanning segments (see Figs. 7A and B). $K^+$ channels in contrast, are true tetramers with each subunit coded by a single gene. The membrane-spanning segments can vary from six as in the $Na^+$ or $Ca^{2+}$ channels, or only two in the case of the inward rectifying channels associated with the sulfanylurea receptor (SUR). There are three main functional types, the voltage-gated,

the $Ca^{2+}$-activated, and the inward rectifying types. Many have been cloned. They consist of homotetramers or heterotetramers within the same class. Thus, there is the possibility for mixing subunits that would allow for the evolution of "designer" channels to match particular different functional circumstances.

Mutations induced in voltage-gated $K^+$ channels in *Drosophila* have been very useful in determining the molecular basis for gating and inactivation. Many colorful terms have been introduced to describe the families of $K^+$ channels that came out of the *Drosophila* studies. For example, **shaker** and ether **a-go-go** refer to phenotypic behaviors of the flies having these channel gene mutations in response to the challenge of ether vapor. In essence, some mutants would shake their legs, while others would wriggle their tails provocatively. The *Drosophila* studies have become increasingly useful as more structural similarities are being found to *Drosophila*-related channels in vertebrates. There is a systematic classification of mammalian $K^+$ channels, and for greater detail one should consult Gutman and Chandy (1993).

$K^+$ channels serve many important functions in excitable tissues such as nerve or heart. Voltage-dependent $K^+$ channels often function to shorten APs or quickly return the membrane to its resting state after activity. Nerve cells depend more on $K^+$ conductance for resting stability, unlike mammalian skeletal muscle, which depends on $Cl^-$ channels. Dysfunction of $K^+$ channels might therefore be expected in neurological mutants and, in fact, the very first mammalian $K^+$ channel mutant to be described causes a human neurological condition, episodic ataxia, discussed later. We note that the defective Kv1.1 $K^+$ channel in episodic ataxia is a member of the **shaker K** $K^+$ channel family, and one form of human heart long Q-T syndrome (also discussed later) is due to mutations in a $K^+$ channel expressed by HERG (human ether a-go-go related gene). Although the *Drosophila ether a-go-go* channel was not found in the mammalian heart, the distantly related HERG channel was found.

The inward rectifying HERG $K^+$ channel may even prove to be a major repolarizing mechanism in the human heart. It is an interesting channel in several ways. Like the CLC-1 $Cl^-$ channel it is activated by depolarization, yet it conducts very little current in the outward direction. This is accounted for by a very rapid inactivation and slow activation when depolarized (Smith *et al.*, 1996). The HERG channel has six membrane-spanning regions (Fig. 7B), whereas other inward rectifying $K^+$ channels contain only two, the only other exception being the KAT1 channel. Also the inactivation mechanism appears to be of the C type, which is the slow type in other $K^+$ channels. Note that N-type inactivation utilizes the intracellular "ball-and-chain" mechanism, and C-type is due to conformational change at the extracellular mouth of the channel. Before the recent discovery of the inherited $K^+$ channel diseases, the major $K^+$ channel disease known was *acquired neuromyotonia*. This syndrome is not due to mutations, but rather to autoimmune destruction of a particular $K^+$ channel. This condition is included for comparison.

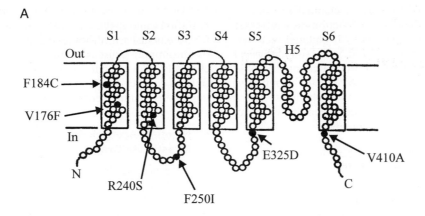

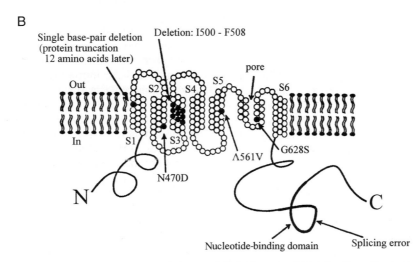

**FIG. 7.** K⁺ channels. (A) Structure of the Kv1.1 K⁺ channel (*KCNA1* gene), which is implicated in the human neurological condition EA-1. Point mutations from six families having this condition are indicated. (Redrawn and used with permission of Browne *et al.*, 1994.) (B) Structure of the *HERG* (human ether a-go-go related gene) inward rectifying K⁺ channel of the human heart, which is implicated in the K⁺ channel type of the long Q-T syndrome. Several types of mutations found in this condition are indicated. There are three point missense mutations, N470D, A561V, and G628S. There is also a deletion between I500 and F508, a protein truncation due to a single base-pair deletion in S1, and a splicing error in the carboxy-terminus. The errors in S1 and S3 are unlikely to form functional channels, whereas the point mutations might do so. (Redrawn and used with permission of Roden *et al.*, 1995.)

## B. Episodic Ataxia Type-1

Hereditary *episodic ataxia type-1* (also called myokymia) is an autosomal dominant disease characterized by periodic attacks of motor imbalance, incoordination, and involuntary tremor. These attacks are precipitated by emotional or physical stress. The condition is caused by point mutations in the Kv1.1 (or *KCNA1*) gene (located on chromosome 12p) of the *shaker K* subfamily, which codes for a delayed outward rectifying K⁺ channel present in neurons. The primary structure of the Kv1.1 channel and the mutations causing EA-1 are shown in Fig. 7A. This Kv1.1 channel displays the C-type inactivation (conformational change at the mouth of the pore) as contrasted with the N-type inactivation (classical ball-and-chain mechanism) seen, for example, in the Kv1.4 channel. A common defect

shared by many of the mutant channels thus far examined is shift of the $V_{1/2}$ for activation in the hyperpolarizing direction, and of the $V_{1/2}$ for inactivation in the depolarizing direction. Because heterozygous individuals show neurological signs and no homozygous individuals have yet been found, the many other types of K⁺ channels that might be present are unable to take over the function of the Kv1.1 channel. This speaks against a current notion that the large number of different types of K⁺ channels that have evolved is to prevent just such a dependence. Coexpression studies suggest that the mutant subunits, in addition to forming homotetramers characteristic of the Kv1.1 channel, probably mix with those from the wild-type allele and assemble to form a variety of functionally defective heterotetramers *in vivo* (Browne *et al.*, 1994; Adelman *et al.*, 1995; Boland *et al.*, 1997). This is the same mechanism invoked to explain dominance in the case of the CLC-1 channel myotonias, and to support the idea that CLC-1 functions as a tetramer or similar order oligomultimer (see Cl⁻ channel section).

The cellular mechanism suggested by these biophysical findings is that nerve cells having mutant channels would not effectively repolarize after each AP due to the impaired delayed rectifier function of the mutant channels. These cells would become temporarily depolarized and dysfunctional as the late AP effects accumulate.

### C. HERG: Long Q-T Interval Disease

The inherited *long Q-T syndrome* is cardiac disorder characterized by fainting (syncope), seizures, and sudden death due to a particular ventricular tachycardia called *torsade de pointes*. Patients having the disease have a characteristically prolonged Q-T interval, which indicates dysfunction of ventricular repolarization. Mutations in both a Na⁺ channel (discussed in that section) and the HERG K⁺ channel have been implicated in the human long Q-T interval syndrome (Keating and Sanguinetti, 1996). The Na⁺ channel stays open too long and the K⁺ channel opens less efficiently. Both conditions produce a similar physiological effect, namely, a prolonged cardiac AP, as reflected in the long Q-T interval of the electrocardiogram. The primary structure of the HERG channel and mutations causing the long Q-T interval syndrome are shown in Fig. 7B.

### D. Acquired Neuromyotonia

In addition to the two human diseases due to mutations of K⁺ channels, there is an autoimmune disease involving a gated K⁺ channel known as *acquired neuromyotonia* (Sinha *et al.*, 1991). In this disease the peripheral motor nerves are hyperexcitable, resulting in repetitive firing and abnormal contractions of skeletal muscle. This must be clearly distinguished from "true" myotonia discussed previously in which the muscle fibers but *not* the nerve fibers are hyperexcitable. The two syndromes can be distinguished by administration of a neuromuscular blocker such as tubocurarine and, with direct muscle stimulation, the "true" myotonic fibers remain hyperexcitable, whereas in the neural type of myotonia the muscle fibers have normal excitability.

## VII. Neurotransmitter-Gated Channels

### A. General Comments

Several types of channels are gated open by neurotransmitters. The major ones can be classed by the neurotransmitter effective at their receptors. These include acetylcholine (ACh), glutamate, 5-HT, adenosine triphosphate (ATP), gamma-aminobutyric acid (GABA), and glycine. The selectivity of the channels can be for nonspecific cations, such as the ACh nicotinic channel, the K⁺ selective ATP channels, or the Cl⁻ selective glycine or GABA channels. This class of channels usually has five subunits, with

one of the units occurring more than once. Only a few mutations among this class of channels have been described for human disease, although the channels have been targets of autoimmune disease. Our disease examples include a clinically relevant one involving the skeletal muscle nicotinic receptor channel, and a rare but instructive one involving the strychnine receptor glycine channel.

### B. Nicotinic Acetylcholine Receptor: Myasthenia Gravis

In the human disease *myasthenia gravis,* the number of nicotinic acetylcholine receptors (AChRs) at the postsynaptic side of the neuromuscular junction is decreased, resulting in smaller postsynaptic currents and the tendency to block neuromuscular transmission. As a result patients experience weakness of skeletal muscles. The AChR is a membrane-bound glycoprotein that exists as a pentamer. Mammals have two main isoforms: the mature or innervated form having two α- and one each of β-, δ-, and ε-subunits, and the immature or denervated form in which a γ-subunit replaces the ε-subunit. The α-subunit has the ACh receptor and all five subunits are necessary for the ligand-gated channel function during neuromuscular transmission. When ACh is released from the nerve terminal it diffuses to the AChR, and when two molecules occupy each α receptor, the AChR channel can open causing both Na⁺ and K⁺ currents to flow, resulting in depolarization of the postsynaptic membrane and excitation of the muscle fiber. Most of the myasthenia gravis syndromes are apparently due to a T-lymphocyte-dependent serum autoantibody against AChR (see Penn *et al.,* 1993). Experimentally produced monoclonal antibodies have been shown to cause the channel to make kinetic transitions, leading to desensitization rather than activation. Myasthenia gravis may also occur very rarely due to heredity, reportedly by a mutation in the ε-subunit.

### C. Glycine Receptor: Hyperekplexia

Glycine-activated Cl⁻ channels are found in many central neurons where they function to produce inhibition. Activation of these channels causes an increase in $g_{Cl}$, which tends to clamp the membrane potential to the Cl⁻ equilibrium potential, which is often in the hyperpolarizing direction. The effects of the excitatory depolarizing channels are thus reduced causing inhibition. The mammalian glycine-activated channel is a hetero-oligomeric Cl⁻ channel that is inhibited by the plant toxin strychnine. The adult receptor is a pentamer consisting of three 48-kDa α-subunits and two 58-kDa β-subunits. The α-subunit of the channel protein has the strychnine binding site and behaves as a channel when expressed in *Xenopus* oocytes. There are several subtypes of this channel and a large degree of sequence homology to the GABA, glutamate, and ACh ligand-gated ion channels.

In mammals small doses of strychnine cause highly exaggerated startle reactions to unexpected acoustic or mechanical stimulation. A response pattern resembling strychnine poisoning is seen in patients afflicted with the autosomal

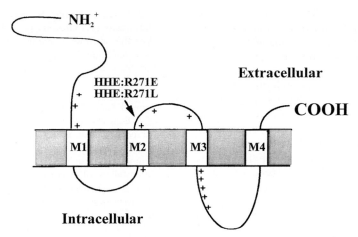

**FIG. 8.** Glycine receptor Cl⁻ channel. The α-subunit of the glycine-activated, strychnine-blocked, Cl⁻ channel (or GLRA1 channel) present in the mammalian nervous system. The functional pentameric channel is assembled from three of these α-subunits and two β-subunits (not shown). The arrow points to position 271 near the presumed mouth of the assembled channel where autosomal dominant mutations cause a charged lysine to be replaced by either an uncharged leucine or glutamine. The result is a blocked inhibitory channel leading to the nervous hyperexcitability called hereditary hyperekplexia (HHE).

dominant disorder known as **hereditary hyperekplexia** (HHE) or familial startle disease (STHE) (Shiang *et al.*, 1993). Mutations in the α-subunit of the glycine-activated Cl⁻ channels have been shown to be the cause. The proposed structure of the α-subunit seen in Fig. 8 shows position 271 between domains M2 and M3 where a positively charged arginine is substituted by a neutral leucine or glutamine in the major mutations. These mutations lie at the presumed mouth of the channel where altering the charge has a significant effect on channel function.

The fact that three α-subunits are required in the functional adult receptor helps explain the dominance of the inheritance. If the channel is rendered dysfunctional if only one mutant subunit is present, then even if there is only one mutant allele (the other producing normal subunits) the heterozygous individual would have only one-eighth or 12.5% of the normal functioning channels. This theory was discussed earlier for dominant MC.

## VIII. Summary

Ion channels are often targets of disease, either directly through a mutation in the gene coding for some critical part of the channel protein structure, or indirectly through autoimmune disease or by defective coding of a critical ion channel regulator. We have seen examples of each case. When ion channel function is abnormal, we may get malfunction of excitable cells as occurs in the myotonias, periodic paralyses and episodic ataxias, or problems with ion transport as occurs in cystic fibrosis and malignant hyperthermia. No region of the channel protein appears to be immune to the effects of mutation. Effects are seen on gating, selectivity, and unit conductance. Often a change in a single amino acid accounts for the altered function of

a channel or of its regulator. Ion channel diseases are often called **channelopathies.**

Modern electrophysiological methods, including the patch-clamp method, have made the monitoring of channel function very direct and convenient. Combining the new electrophysiology with the new molecular biology has been successful in offering us a staggering amount of information on normal and diseased channels in a remarkably short period of time.

Common methods are evolving for the study of familial diseases. Two basic strategies are used, positional cloning and the candidate gene. The first approach, positional cloning, is used to identify the genetic locus even when there is no knowledge of the biophysical or biochemical abnormalities underlying the disease. This is done by examining affected individuals for genetic linkage to polymorphic markers located throughout the genome. In this way a region of the gene is found that is linked to the disease. Linkage is established when there is a low rate of recombination between the marker locus and the disease locus. From the gene one can express the protein it encodes and then determine its function and how the disease mutations disturb this function. The CFTR channel of cystic fibrosis was discovered in this way.

The second approach, or candidate gene strategy, requires knowledge of functional abnormalities brought about by the disease. Genes responsible for maintaining the normal function of the process affected by the disease are then considered to be candidate genes for the disease. Genetic linkage to the disease is then sought, and if one is found the candidate gene is screened for mutations. These mutations can be incorporated into the wild-type DNA and expressed in heterologous systems. Then one can thus examine the biophysical and biochemical behavior and determine if the abnormality could account for the disease. The new rapid cloning techniques are largely responsible for the increased pool of known channel genes having known functions. This makes it easier to identify candidate channel genes, thus the candidate gene approach has been very successful for the study of the channelopathies. Most of the new channelopathies including the CLC-5 mutations causing kidney stones, the episodic ataxias, and the long Q-T interval diseases were discovered using the candidate gene approach.

With the increase in the number of channelopathies now described we can begin to make certain generalizations. We have seen that if more than one channel type is involved in control of cellular excitability, then either decreasing the $P_{open}$ of the stabilizing channel or increasing the $P_{open}$ of the depolarizing channel will lead to hyperexcitability of the excitable cell. Examples to date are $Cl^-$ channel block or decreased $Na^+$ channel inactivation in the myotonias, $K^+$ channel block or decreased $Na^+$ channel inactivation in the long Q-T syndromes, and $K^+$ channel block or increased activation of $Ca^{2+}$ channels in the two types of episodic ataxias.

## Bibliography

Adelman, J. P., Bond, C. T., Pessia, M., and Maylie, J. (1995). Episodic ataxia results from voltage-dependent $K^+$ channels with altered functions. *Neuron* **15**, 1445–1454.

Adrian, R. H., and Bryant, S. H. (1974). On the repetitive discharge in myotonic muscle fibres. *J. Physiol.* **240**, 505–515.

Adrian, R. H., and Marshall, M. W. (1976). Action potentials reconstructed in normal and myotonic muscle fibres. *J. Physiol. (London)* **258**, 125–143.

Ashcroft, F. M., and Röper, J. (1993). Transporters, channels and human disease. *Curr. Opin. Cell Biol.* **5**, 677–683.

Astill, D. S., Rychkov, G., Clarke, J. D., Hughes, B. P., Roberts, M. L., and Bretag, A. H. (1996). Characteristics of skeletal muscle chloride channel ClC-1 and point mutant R304E expressed in Sf-9 insect cells. *Biochim. Biophys. Acta Bio-Membr.* **1280**, 178–186.

Ballabio, A. (1993). The rise and fall of positional cloning? *Nature Genet.* **3**, 277–279.

Beck, C. L., Fahlke, C., and George, A. L. (1996). Molecular basis for decreased muscle chloride conductance in the myotonic goat. *Proc. Natl. Acad. Sci. USA* **93**, 11248–11252.

Becker, P. E. (1973). Generalized non-dystrophic myotonia: The dominant (Thomsen) type and the recently identified recessive type. *New Devel. Electromyography Clin. Neurophysiol.* **1**, 407–412.

Boland, I. M., Price, D. L., and Jackson, K. A. (1997). Functional consequences of potassium channel mutations in families with inherited episodic ataxia. *Biophys. J.* **72**, A140.

Bond, E. F., and Gordon, A. M. (1993). Insulin-induced membrane changes in $K^+$-depleted rat skeletal muscle. *Am. J. Physiol. Cell Physiol.* **265**, C257–C265.

Browne, D. L., Gancher, S. T., Nutt, J. G., Brunt, E. R. P., Smith, E. A., Kramer, P., and Litt, M. (1994). Episodic ataxia/myokymia syndrome is associated with point mutations in the human potassium channel gene, KCNA1. *Nature Genet.* **8**, 136–140.

Bryant, S. H. (1979). Myotonia in the goat. *Ann. NY Acad. Sci.* **317**, 314–325.

Bryant, S. H., and Morales-Aguilera, A. (1971). Chloride conductance of normal and myotonic goat fibres and the action of monocarboxylic aromatic acids. *J. Physiol. (London)* **219**, 367–383.

Bryant, S. H., Mambrini, M., and Entrikin, R. K. (1987). Chloride and potassium conductances are decreased in skeletal muscle fibers from the (mto) myotonic mouse. *Neuroscience* **465**, 17.

Cannon, S. C., and Corey, D. P. (1993). Loss of $Na^+$ channel inactivation by anemone toxin (ATX II) mimics the myotonic state in hyperkalaemic periodic paralysis. *J. Physiol. (London)* **466**, 501–520.

Cannon, S. C., and Strittmatter, S. M. (1993). Functional expression of sodium channel mutations identified in families with periodic paralysis. *Neuron* **10**, 317–326.

Cannon, S. C., Brown, R. H., and Corey, D. P. (1991). A sodium channel defect in hyperkalemic periodic paralysis: Potassium-induced failure of inactivation. *Neuron* **6**, 619–626.

Cannon, S. C., Brown, R. H., Jr., and Corey, D. P. (1993). Theoretical reconstruction of myotonia and paralysis caused by incomplete inactivation of sodium channels. *Biophys. J.* **65**, 270–288.

Dulhunty, A. F. (1979). Distribution of potassium and chloride permeability over the surface and T-tubule membranes of mammalian skeletal muscle. *J. Membr. Biol.* **45**, 293–310.

Fahlke, C., Knittle, T., Gurnett, C. A., Campbell, K. P., and George, A. L., Jr. (1997). Subunit stoichiometry of human muscle chloride channels. *J. Gen. Physiol.* **109**, 93–104.

Fontaine, B., Khurana, T. S., Hoffman, E. P., Bruns, G. A. P., Haines, J. L., Trofatter, J. A., Hanson, M. P., Rich, J., McFarlane, H., Yasek, D. M., Romano, D., Gusella, J. F., and Brown, R. H. (1990). Hyperkalemic periodic paralysis and the adult muscle sodium channel alpha-subunit gene. *Science* **250**, 1000–1002.

Franke, C., Iaizzo, P. A., Hatt, H., Spittlemeister, W., Ricker, K., and Lehmann-Horn, F. (1991). Altered Na channel activity and reduced Cl conductance cause hyperexcitability in recessive generalized myotonia (Becker). *Muscle Nerve* **14**, 762–770.

Furman, R. E., and Barchi, R. L. (1978). The pathophysiology of myotonia produced by aromatic carboxylic acids. *Ann. Neurol.* **4**, 357–365.

George, A. L., Jr. (1995). Molecular genetics of ion channel diseases. *Kidney Int.* **48**, 1180–1190.

George, A. L., Kamisarof, J., Kallen, R. G., and Barchi, R. L. (1992). Primary structure of adult human skeletal muscle voltage-dependent $Na^+$ channel. *Ann. Neurol.* **31**, 131–137.

George, A. L., Crackower, M. A., Abdalla, J. A., Hudson, A. J., and Ebers, G. C. (1993). Molecular basis of Thomsen's disease (autosomal dominant myotonia congenita). *Nature Genet.* **3**, 305–310.

Gillard, E. F., Otsu, K., Fujii, J., Duff, C., DeLeon, S., Khanna, V. K., Britt, B. A., Warton, R. G., and MacLennan, D. H. (1992). Polymorphisms and deduced amino acid substitutions in the coding sequence of the ryanodine receptor (RYR1) gene in individuals with malignant hyperthermia. *Genomics* **13**, 1247–1254.

Gronemeier, M., Condie, A., Prosser, J., Steinmeyer, K., Jentsch, T. J., and Jockusch, H. (1994). Nonsense and missense mutations in the muscular chloride channel gene *Clc-1* of myotonic mice. *J. Biol. Chem.* **269**, 5963–5967.

Guo, X., Uehara, A., Ravindran, A., Bryant, S. H., Hall, S., and Moczydlowski, E. (1987). Kinetic basis for insensitivity to tetrodotoxin and saxitoxin in sodium channels of canine heart and denervated rat skeletal muscle. *Biochemistry* **26**, 7546–7556.

Gurnett, C. A., Kahl, S. D., Anderson, R. D., and Campbell, K. P. (1995). Absence of the skeletal muscle sarcolemma chloride channel CLC-1 in myotonic mice. *J. Biol. Chem.* **270**, 9035–9038.

Gutman, G. A., and Chandy, K. G. (1993). Nomenclature of mammalian voltage-dependent potassium channel genes. *Neurosciences* **5**, 101–106.

Harley, H. G., Rundle, S. A., MacMillan, J. C., Myring, J., Brook, J. D., Crow, S., Reardon, W., Fenton, I., Shaw, D. J., and Harper, P. S. (1993). Size of the unstable CTG repeat sequence in relation to phenotype and parental transmission in myotonic dystrophy. *Human Genet.* **52,** 1164–1174.

Heinemann, S. H., and Conti, F. (1992). Nonstationary noise analysis and application to patch clamp recordings. *Methods Enzymol.* **207,** 131–148.

Jentsch, T. J., Günther, W., Pusch, M., and Schwappach, B. (1995). Properties of voltage-gated chloride channels of the CLC gene family. *J. Physiol. (London)* **482,** Suppl. P, 19S–25S.

Jurkat-Rott, K., Lehmann-Horn, F., Elbaz, A., Heine, R., Gregg, R. G., Hogan, K., Powers, P. A., Lapie, P., Vale-Santos, J. E., Weissenbach, J., and Fontaine, B. (1994). A calcium channel mutation causing hypokalemic periodic paralysis. *Human Mol. Genet.* **3,** 1415–1419.

Kallen, R. G., Sheng, Z.-H., Yang, J., Chen, L., Rogart, R. B., and Barchi, R. L. (1990). Primary structure and expression of a sodium channel characteristic of denervated and immature rat skeletal muscle. *Neuron* **4,** 233–242.

Keating, M. T., and Sanguinetti, M. C. (1996). Molecular insights into cardiovascular disease. *Science* **272,** 681–685.

Koch, M. C., Steinmeyer, K., Lorenz, C., Ricker, K., Wolf, F., Otto, M., Zoll, B., Lehmann-Horn, F., Grzeschik, K.-H., and Jentsch, T. J. (1992). The skeletal muscle chloride channel in dominant and recessive human myotonia. *Science* **257,** 797–800.

Kraner, S. D., Tanaka, J. C., and Barchi, R. L. (1985). Purification and functional reconstitution of the voltage-sensitive sodium channel from rabbit T-tubular membranes. *J. Biol. Chem.* **25,** 6341–6347.

Lehmann-Horn, F., and Rüdel, R. (1996). Molecular pathophysiology of voltage-gated ion channels. *Rev. Physiol. Biochem. Pharmacol.* **128,** 195–268.

Lehmann-Horn, F., Rüdel, R., Dengler, R., Lorkovic, H., Haass, A., and Ricker, K. (1981). Membrane defects in paramyotonia congenita with and without myotonia in a warm environment. *Muscle Nerve* **4,** 496–406.

Lehmann-Horn, F., Kuther, G., Ricker, K., Grafe, P., Ballanyi, K., and Rüdel, R. (1987). Adynamia episodica hereditaria with myotonia: A non-inactivating sodium current and the effect of extracellular pH. *Muscle Nerve* **10,** 363–374.

Lerche, H., Heine, R., Pika, U., George, A. L., Jr., Mitrovic, N., Browatzki, M., Weiss, T., Rivet-Bastide, M., Franek, C., Lomonaco, M., Ricker, K., and Lehmann-Horn, F. (1993). Human sodium channel myotonia: Slowed channel inactivation due to substitutions for a glycine within the III-IV linker. *J. Physiol. (London)* **470,** 13–22.

Lloyd, S. E., Pearce, S. H. S., Fisher, S. E., Steinmeyer, K., Schwappach, B., Scheinman, S. J., Harding, B., Bolino, A., Devoto, M., Goodyer, P., Rigden, S. P. A., Wrong, O., Jentsch, T. J., Craig, I. W., and Thakker, R. V. (1996). A common molecular basis for three inherited kidney stone diseases. *Nature* **379,** 445–449.

Lucke, J. N., Hall, G. M., and Lister, D. (1979). Malignant hyperthermia in the pig and the role of stress. *Ann. NY Acad. Sci.* **317,** 326–337.

Ludewig, U., Pusch, M., and Jentsch, T. J. (1996). Two physically distinct pores in the dimeric ClC-0 chloride channel. *Nature* **383,** 340–343.

MacLennan, D. H., and Phillips, M. S. (1992). Malignant hyperthermia. *Science* **256,** 789–794.

McClatchey, A. I., Van den Bergh, P., Pericak-Vance, M. A., Raskind, W., Verellen, C., McKenna-Yasek, D., Rao, K., Haines, J. L., Bird, T., Brown, R. H., and Gusella, J. F. (1992). Temperature-sensitive mutations in the III-IV cytoplasmic loop region of the skeletal muscle sodium channel gene in paramyotonia congenita. *Cell* **68,** 769–774.

Mehrke, G., Brinkmeier, H., and Jockusch, H. (1988). The myotonic mouse mutant ADR: Electrophysiology of the muscle fiber. *Muscle Nerve* **11,** 440–446.

Middleton, R. E., Pheasant, D. J., and Miller, C. (1996). Homodimeric architecture of a ClC-type chloride ion channel. *Nature* **383,** 337–340.

Ophoff, R. A., Terwindt, G. M., Vergouwe, M. N., Van Eijk, R., Oefner, P. J., Hoffman, S. M. G., Lamerdin, J. E., Mohrenweiser, H. W., Bulman, D. E., Ferrari, M., Haan, J., Lindhout, D., Van Ommen, G. J. B., Hofker, M. H., Ferrari, M. D., and Frants, R. R. (1996). Familial hemiplegic migraine and episodic ataxia type-2 are caused by mutations in the $Ca^{2+}$ channel gene CACNL1A4. *Cell* **87,** 543–552.

Penn, A. S., Richman, D. P., Ruff, R. L., and Lennon, V. A. (Eds.). (1993). Myasthenia gravis and related disorders. *Ann. NY Acad. Sci.* **681.**

Pickar, J. G., Spier, S. J., Snyder, J. R., and Carlsen, R. C. (1991). Altered ionic permeability in skeletal muscle from a horse with hyperkalemic periodic paralysis. *Am. J. Physiol.* **260,** C926–C933.

Ptacek, L. J., Gouw, L., Kwiecinski, H., McManis, P., Mendell, J. R., Barohn, R. J., George, A. L., Barchi, R. L., Robertson, M., and Leppert, M. F. (1993). Sodium channel mutations in paramyotonia congenita and hyperkalemic periodic paralysis. *Ann. Neurol.* **33,** 300–307.

Ptacek, L. J., Tawil, R., Griggs, R. C., Engel, A. G., Layzer, R. B., Kwiecinski, H., McManis, P. G., Santiago, L., Moore, M., Fouad, G., Bradley, P., and Leppert, M. F. (1994). Dihydropyridine receptor mutations cause hypokalemic periodic paralysis. *Cell* **77,** 863–868.

Pusch, M., and Jentsch, T. J. (1994). Molecular physiology of voltage-gated chloride channels. *Physiol. Rev.* **74,** 813–828.

Pusch, M., Steinmeyer, K., and Jentsch, T. J. (1993). Low single channel conductance of the major skeletal muscle chloride channel, CLC-1. *Biophys. J.* **66,** 149–152.

Redfern, P., and Thesleff, S. (1971). Action potential generation in denervated rat skeletal muscle: II. The action of tetrodotoxin. *Acta. Physiol. Scand.* **82,** 70–78.

Roden, D. M., George, A. L., Jr., and Bennett, P. B. (1995). Recent advances in understanding the molecular mechanisms of the long QT syndrome. *J. Cardiovasc. Electrophysiol.* **6,** 1023–1031.

Rogart, R. B., Cribbs, L. L., Muglia, L. K., Kephart, D. D., and Kaiser, M. W. (1989). Molecular cloning of a putative tetrodotoxin-resistant rat heart $Na^+$ channel isoform. *Proc. Natl. Acad. Sci. USA* **86,** 8170–8174.

Rüdel, R., and Lehmann-Horn, F. (1985). Membrane changes in cells from myotonia patients. *Physiol. Rev.* **65,** 310–356.

Rychkov, G. Y., Pusch, M., Astill, D. S., Roberts, M. L., Jentsch, T. J., and Bretag, A. H. (1996). Concentration and pH dependence of skeletal muscle chloride channel ClC-1. *J. Physiol. (London)* **497,** 423–435.

Shiang, R., Ryan, S. G., Zhu, Y. Z., Hahn, A. F., O'Connell, P., and Wasmuth, J. J. (1993). Mutations in the $\alpha_1$ subunit of the inhibitory glycine receptor cause the dominant neurological disorder, hyperekplexia. *Nature Genet.* **5,** 351–358.

Sinha, S., Newson-Davies, J., Mills, K., Byrne, N., Lang, B., and Vincent, A. (1991). Autoimmune aetiology for acquired neuromyotonia (Isaac's Syndrome). *Lancet* **338,** 75–77.

Smith, P. L., Baukrowitz, T., and Yellin, G. (1996). The inward rectification mechanism of the HERG cardiac channel. *Nature* **379,** 833–836.

Steinmeyer, K., Klocke, R., Ortland, C., Gronemeier, M., Jockusch, H., Grunder, S., and Jentsch, T. (1991a). Inactivation of muscle chloride channel by transposon insertion in myotonic mice. *Nature* **354,** 304–308.

Steinmeyer, K., Ortland, C., and Jentsch, T. (1991b). Primary structure and functional expression of a developmentally regulated skeletal muscle chloride channel. *Nature* **354,** 301–304.

Steinmeyer, K., Lorenz, C., Pusch, M., Koch, M. C., and Jentsch, T. J. (1994). Multimeric structure of ClC-1 chloride channel revealed by mutations in dominant myotonia congenita (Thomsen). *EMBO J.* **13,** 737–743.

Steinmeyer, K., Schwappach, B., Bens, M., Vandewalle, A., and Jentsch, T. J. (1995). Cloning and functional expression of rat CLC-5, a chloride channel related to kidney disease. *J. Biol. Chem.* **270,** 31172–31177.

Tanabe, T., Beam, K. G., Powell, J. A., and Numa, S. (1988). Restoration of excitation–contraction coupling and slow calcium current in dysgenic muscle by dihydropyridine receptor complementary DNA. *Nature* **336,** 134–139.

Tsien, R. W., Ellinor, P. T., and Horne, W. A. (1991). Molecular diversity of voltage-dependent calcium channels. *Trends Pharmacol. Sci.* **12,** 349–354.

Vaughan, P. C., and French, A. S. (1989). Non-ligand activated chloride channels of skeletal muscle and epithelia. *Prog. Biophys. Mol. Biol.* **54,** 59–79.

Warner, J. O. (Ed.). (1992). Cystic fibrosis. *Br. Med. Bull.* **48,** 717–978.

West, J. W., Patton, D. E., Scheuer, T., Wang, Y., Goldin, A. L., and Catterall, W. A. (1992). A cluster of hydrophobic amino acid residues required for fast Na+-inactivation. *Proc. Natl. Acad. Sci. USA* **89,** 10910–10914.

# SECTION

# V

# Synaptic Transmission and Sensory Transduction

Gary L. Westbrook

# 39

## Ligand-Gated Ion Channels

## I. Introduction

Ion channels are fundamental signaling molecules in virtually all cells. Although channels are present on intracellular membranes of organelles, our current understanding of ion channel function originated with studies of ion channels in the plasmalemma of excitable cells. In the nervous system, voltage-gated ion channels mediate action potentials (APs) and trigger transmitter release, whereas ligand-gated channels are responsible for chemical signaling mediated by classical fast-acting neurotransmitters. Neurotransmitters also trigger slower synaptic responses that are mediated by G-protein-coupled receptors, as discussed in Chapter 40.

Until the 1980s, knowledge about the molecular properties of ion channel proteins was largely inferred from physiological and biophysical studies (reviewed in Hille, 1992). In retrospect, many of the predictions from these biophysical studies of channels have been confirmed in dramatic fashion with the elucidation of the amino acid sequence of many voltage- and ligand-gated ion channels by molecular cloning. However, the increasing knowledge of the molecular structure has also revealed a number of unexpected findings and has begun to permit sophisticated correlations of protein structure with function (see, e.g., Miller, 1989; Montal, 1990; Unwin, 1993). These advances have led to an explosive increase in our understanding of the molecular operation of this important class of membrane proteins. These interesting molecules have captured the interest of many outstanding scientists whose studies of ion channels have led to several Nobel Prizes (see, e.g., Neher, 1992; Sakmann, 1992).

Studies of ion channels were pioneered by studies of the voltage-gated ion channels in squid axon. These studies established that a conductance change was associated with the AP, and that this conductance change could be attributed to selective increases in the membrane permeability to $Na^+$ and $K^+$ ions, with the energy for the process derived from the transmembrane ion gradients created by the Na,K-ATPase. Both the $Na^+$ and $K^+$ conductances in-

creased with membrane depolarization. These findings suggested that there were discrete pores in the membrane that accounted for the $Na^+$ and $K^+$ conductance. However it was not until the 1970s that the existence of ion channels was confirmed, first by measurements of the statistical properties of currents through populations of ion channels, a technique called *fluctuation* or *noise analysis*. Later, this was refined by the introduction of patch-clamp recording, which allowed measurement of current through single ion channels (for review, see Neher, 1992; Sakmann, 1992). Patch-clamp recording revolutionized the study of ion channels, because it allowed detailed biophysical studies of the electrical activity of single molecules. Finally, protein purification and subsequent molecular cloning have confirmed that ion channels are a large and heterogeneous family of membrane proteins. Although the ligand-gated ion channels vary in their structural features, they share a common basic structure in that they are composed of multiple subunits arranged around a central water-filled pore. Each subunit is a polypeptide encoding by a separate gene. Generally, there are a large number of possible subunit combinations; thus the particular subunits expressed in a certain class of neurons can result in channels with distinct functional characteristics.

This chapter is divided into three topic areas. In the first, the categories of ligand-gated ion channels are briefly reviewed. The basic physiological properties and general molecular structure of the ligand-gated channels are then discussed using examples from the muscle nicotinic acetylcholine receptor (AChR) and the *N*-methyl-D-aspartate (NMDA)-type glutamate receptor. In the third section, the features of neuronal ligand-gated channels that are responsible for fast synaptic transmission are reviewed with an emphasis on the distinctive characteristics of each class of channel. Channels gated by acetylcholine (ACh), $\gamma$-aminobutyric acid (GABA), glycine, and glutamate are included as representative examples. Related topics on ion channels are covered extensively in other chapters in Sections III through V of this book. The reader is also referred to a large number of excellent monographs and reviews listed in the bibliography for further details and original citations in this rapidly advancing field.

## II. Classes of Ligand-Gated Ion Channels

Ligands that activate ion channels in nerve cell membranes can be divided into two major categories, neurotransmitters and intracellular ligands, as listed in Table 1. The best studied are neurotransmitters that are involved in fast chemical synaptic transmission. These include ACh, which is the transmitter at the vertebrate neuromuscular junction and in autonomic ganglia; and the amino acids, L-glutamate and GABA, which mediate the majority of fast excitatory and inhibitory synaptic transmission, respectively, in the vertebrate central nervous system (Betz, 1990). Each of these ligands also activates receptors that are coupled to second messengers via GTP-binding (G) proteins. G-protein-coupled receptors generally mediate slower and neuromodulatory transmembrane signaling (reviewed in Nicoll et al., 1990). Other neurotransmitters that activate ligand-gated ion channels in various cell types include serotonin, glycine, histamine, and adenosine triphosphate (ATP). For example, the 5HT-3 receptor activated by serotonin is a ligand-gated ion channel (reviewed in Julius, 1991). ATP is released from nerve terminals in several pathways and activates an ATP-gated channel in some spinal neurons that process incoming sensory information (reviewed in Bean, 1992).

In addition to neurotransmitter-gated ion channels, increasing attention has focused on ligand-gated ion channels that are activated by intracellular ligands. These ligands include $Ca^{2+}$, cyclic nucleotides, ATP, and inositol trisphosphate (IP3). Intracellular ligands can increase ($Ca^{2+}$, cAMP, inositol trisphosphate) or decrease (cGMP, ATP) activity of the associated channel. These channels play important roles in cell function. For example, $Ca^{2+}$-dependent potassium, chloride, and nonspecific cation channels

### TABLE 1 Ligand-Gated Ion Channels

| Ligand | Receptor | Ion selectivity[a] |
|---|---|---|
| Neurotransmitters | | |
| Acetylcholine | Muscle, neuronal AChRs | NS |
| Glutamate | AMPA, kainate, NMDA | NS |
| GABA | $GABA_A$, $GABA_C$ | Cl |
| Glycine | GlyR | Cl |
| Serotonin | $5HT_3$ receptor | NS |
| ATP | ATP receptor | NS |
| Intracellular ligands | | |
| Calcium | Calcium-dependent channels | K, Cl, NS |
| Cyclic nucleotides | cGMP and cAMP receptors | NS |
| ATP | ATP-dependent channel | K |
| IP3 | Calcium release channel | Ca |

[a] Abbreviations for ion selectivity are NS (nonselective cation permeability, some with calcium permeability); Cl (permeable to chloride); K (permeable to potassium); and Ca (permeable to calcium).

modify the excitability of the nerve cell membrane and thus can alter the AP and release of transmitter from nerve terminals. The cyclic nucleotide (cAMP, cGMP)-gated channels mediate sensory transduction in the visual and olfactory system. Light results in the closing of cGMP channels in vertebrate photoreceptors while odors trigger the opening of cAMP-gated channels in olfactory receptor neurons. See Chapters 46 and 48 for more discussion of channels gated by cyclic nucleotides.

## III. Basic Physiological Features

Modern patch-clamp techniques now allow investigators to define the three fundamental properties of any ion channel—conductance, selective permeability, and gating—in molecular terms. In addition to these fundamental properties, modulation, channel block, and desensitization are processes that affect the activity of ligand-gated ion channels and are important in shaping the impact of channel activity on membrane excitability. The biophysical basis of these properties is discussed first and then the structural features of ligand-gated channels that control these properties are considered. For a more detailed discussion, see Hille (1992).

Conductance reflects the flux of charged ions through the channel, and is measured in pico-Siemens (pS). Most ligand-gated ion channels have a conductance of 5–50 pS, which corresponds to the movement of more than $1 \times 10^6$ ions per second through the channel pore. These high flux rates initially suggested that ligand-gated receptors contained a water-filled pore, because this rate is much higher than that predicted for other transport mechanisms, such as pumps or exchangers. The size of the single-channel conductance ($\gamma$) is characteristic for a given channel, although many ligand-gated channels also have subconductance levels (states). An example of a glutamate-activated channel in a hippocampal neuron with several subconductance states is shown in Fig. 1. Note that the conductance of the open channel is usually near 50 pS, but drops occasionally to 8, 35, 40, and 45 pS. The conductance of a channel increases as the permeant ion concentration is raised, but eventually saturates at concentrations well above physiological levels. This behavior can be described by Michaelis–Menten kinetics, suggesting that ions bind to sites within the pore rather than simply obeying the laws of free diffusion. Most permeant ions have a low $K_m$ of approximately 100 m$M$ ($K_m$ is the concentration at which the conductance is half-maximal as defined by the Michaelis–Menten equation); thus the ions bind for only a microsecond or so before continuing through the pore.

The current carried by the opening of a single channel is measured in picoamperes (pA) with the single channel current:

$$i = \gamma(V - V_{eq}) \qquad (1)$$

where $V$ is the membrane potential and $V_{eq}$ is the equilibrium potential at which no net current is measured. This equation is simply Ohm's law, where $(V - V_{eq})$ is the electrochemical driving force for ions that can pass through

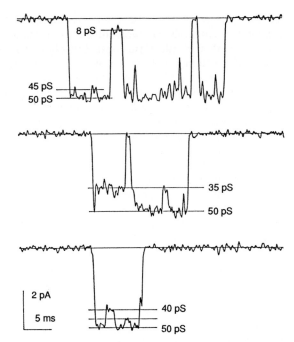

**FIG. 1.** Ligand-gated channels can open to multiple conductance levels. Examples of NMDA-type glutamate channel recording in an outside-out membrane patch obtained from a cultured hippocampal neuron. Downward deflections indicate opening of the channel in the presence of ligand (in this case, 20 $\mu M$ NMDA). The channel usually opens to the 50-pS level, but may switch to lower levels of 8, 35, 40, or 45 pS before closing to the baseline level. [Modified with permission from Jahr and Stevens (1987). *Nature* **325**, 522–525. Copyright 1987 Macmillan Magazines Limited.]

the channel. For example, if the extracellular and cytoplasmic concentrations of permeant ions are equal, then $V_{eq}$ is 0 mV. Thus, as the membrane potential changes, the size of the single-channel current also changes; a plot of this relationship is called a current/voltage or I/V plot. For channels, such as the muscle AChR, which are permeable to $Na^+$ and $K^+$, the reversal potential is 0 mV and the I/V curve is nearly linear. However some channels deviate from linear behavior and show inward or outward rectification (that is, the channel passes current in one direction better than the other), analogous to the behavior of some voltage-gated $K^+$ channels.

Channels are not equally permeable to all ions; that is, they exhibit *selective permeability*. As listed in Table 1, most ligand-gated ion channels are permeable to either monovalent cations or anions. The selective permeability of channels is partly due to the physical size of the pore. For example, the AChR is permeable to cations with diameters up to about 6.5 Å. In general, cationic ligand-gated ion channels are less selective than voltage-gated ion channels, which are highly selective for $Na^+$, $K^+$ or $Ca^{2+}$, but more selective than gap junction channels, which are permeable to small molecular weight molecules such as cyclic nucleotides as well as ions. Because anions and cations are approximately the same size, it is obvious that the pore dimensions cannot totally account for selective permeability. As discussed later, positively and negatively charged

amino acid residues in the entrance to the channel, and within the pore, provide a means for this discrimination. Selective permeability is extremely important in considering the function of an ion channel. For example, channels that are permeable to $Na^+$ and $K^+$ do not significantly alter the ion concentrations on either side of the membrane, but instead provide an electrical signal that depolarizes the neuron and brings the membrane potential closer to the threshold for AP generation. However, channels with $Ca^{2+}$ permeability can transiently increase the cytoplasmic $Ca^{2+}$ concentration, and thus act as a biochemical signal.

Channel *gating* refers to the conformational change in the ion channel that is triggered by ligand binding. The conformational change results in all all-or-none switch between conducting ("open") and nonconducting ("closed") states of the channel. The gating behavior of channels has many parallels with the allosteric behavior of enzymes, with the binding of ligand providing the free energy necessary to maintain the channel in the open conformation. Gating between the open and closed configuration is extremely rapid (ca. 10 $\mu$sec), and thus is beyond the resolution of standard recording methods. Each channel opening usually lasts a few milliseconds, but can, on some occasions, last up to several hundred milliseconds. Binding of more than one agonist molecule is usually necessary to open a channel. The binding of multiple agonist molecules creates a sigmoidal dose–response relationship, characteristic of the cooperative binding of substrates to enzymes. The steep activation created by sigmoid activation kinetics prevents the channel from opening in the presence of a low concentration of agonist. This may be quite important in preventing desensitization of receptors at synapses. In general, the higher the binding affinity of the agonist, the longer the channel will remain open. This is because the channel can open (or reopen) until the agonist dissociates.

The gating of ligand-gated ion channels can be described by multistate kinetic diagrams as shown in Fig. 2A. Such state diagrams have been developed for many ligand-gated ion channels, based on the analysis of open and closed time distributions obtained in single-channel recording. For example, the presence of two exponentials in an open time histogram implies the existence of two open states. The minimal kinetic model for a ligand-gated channel usually has at least four states including closed and unbound, closed and bound, open and bound, and nonconducting (desensitized). For the case shown in Figs. 2A and B, note that the channel opens only when two agonist molecules are bound. For fast-acting neurotransmitters, the concentration of transmitter in the synaptic cleft is typically very high for a brief period. At high concentrations of agonist, the receptors quickly reach the fully liganded but closed state; this transition is determined by the product of the binding rate $r_b$ and the concentration of agonist. Thus the rising and falling phase of the synaptic response largely reflect the unbinding rate $r_u$, the channel opening and closing rates ($\beta$ and $\alpha$), and in some cases the rates in and out of the desensitized state, $R_d$. Classical pharmacological terms can be interpreted in terms of these kinetic schemes. Agonists open the channel, while antagonists bind, but apparently do not cause the conformation change in the

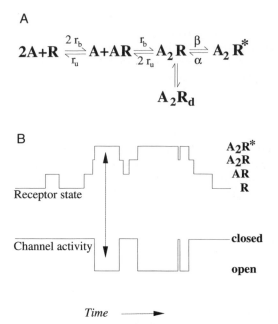

**FIG. 2.** Kinetics of ligand-gated channels. (A) Example of a four-state kinetic scheme used to describe the behavior of ligand-gated ion channels. In this example, two agonist molecules (A) must bind before the channel can open ($A_2R^*$) or enter a nonconducting desensitized state ($A_2R_d$). The rate constants for agonist binding and unbinding are designated $r_b$ and $r_u$, and the rate constants for channel opening and closing are designated as $\beta$ and $\alpha$. (B) Relationship of the state of the receptor to the open and closing of the channel. The channel opens only when the receptor is in the $A_2R^*$ state as indicated by the arrows.

channel protein that opens the channel. The fraction of time the bound channel spends in the open state is called the *open probability*, and may differ between agonists—an agonist that activates the channel with higher probability has a higher "efficacy" in the nomenclature of classical receptor pharmacology.

Studies of the activity of ligand-gated channels have generally been limited to equilibrium measurements due to the limits in the speed of application of agonists to cells or cell-free patches. Such equilibrium measurements are difficult to interpret in terms of state diagrams such as that shown in Fig. 2. In addition, the duration of fast-acting transmitters such as glutamate is generally brief, suggesting that the synapse itself operates under nonequilibrium conditions. However, new rapid solution exchange techniques now allow agonists to be applied within several hundred microseconds to membrane patches, usually in the outside-out configuration to allow agonist in the bath to access the extracellular face of the receptor. Many invesigators have now used these methods to analyze the gating of ligand-gated channels. Such approaches also appear to more closely resemble conditions at synapses and are particularly useful in evaluating the role of channel modulation, channel block, and desensitization on synaptic responses. An example of the response of a membrane patch from a hippocampal neuron to a brief application of glutamate is shown in Fig. 3.

In general, the conductance and selective permeability of ligand-gated channels are not subject to regulation. However, a number of allosteric control mechanisms can profoundly affect the gating. These can be categorized as *desensitization, channel block*, and allosteric *modulation*. *Desensitization* refers to the loss of the response during the continued presence of agonist, and was first noted in studies of the muscle AChR. At the single-channel level, desensitization reflects a nonconducting state of the channel. This phenomenon has now been seen for most ligand-gated channels, and can be described by including an extra bound, but nonconducting, state in the kinetic scheme (see Fig. 2A). Depending on the channel, the nonconducting state may be accessible from either the closed state ($A_2R$) or the open state ($A_2R^*$). Desensitization is thought to represent a distinct conformation of the receptor, but this has yet to be demonstrated at the molecular level. Very rapid desensitization may play a role in terminating the neuronal response to neurotransmitters at some synapses, whereas slower desensitization may actually prolong synaptic responses as channels reenter open states after transiting through desensitized states (see, e.g., Jones and Westbrook, 1996).

Ligand-gated channels can also be plugged by ions or drugs, a phenomenon referred to as *channel block*. If the blocking particle is charged, such as a large ion, the block will be influenced by the voltage across the membrane. Thus the reduction in channel activity will be more pronounced at some membrane potentials. The voltage-dependent block of NMDA-type glutamate channels by extracellular $Mg^{2+}$ is an important example of this phenomenon (see Ascher and Nowak, 1987). At the single-channel level, channel block is usually seen as rapid interruptions ("flickers") of the open state as illustrated in Figs. 4A and B. A characteristic feature of most forms of channel block is that the blocker is only effective when the channel is open. Thus the binding site for the blocker is thought to be within the pore of the channel, or at least at a site that is only available in the open conformation of the channel. Channel block is the mechanism of action for a number of drugs and psychoactive compounds such as phencyclidine (PCP).

Channel gating is also subject to *modulation* by either covalent modifications such as phosphorylation, or via the noncovalent binding of modulators to the channel (e.g., Swope *et al.*, 1992; Changeux *et al.*, 1990). Most ligand-gated channels undergo phosphorylation of intracellular portions of the channel protein; however, the resulting effect on channel function is dependent on the specific channel and the type of kinase involved. Phosphate groups are highly charged, and thus can modify interactions in local regions. For example, phosphorylation of the AChR enhances desensitization, and may also be important in clustering of AChR molecules at sites of innervation. The list of noncovalent modulators is long and includes $Ca^{2+}$ toxins, drugs, and some endogenous substances such as protons and steroid hormones. Most of these reagents affect gating by altering the rate of channel opening, or the time spent in the open state. For example, glycine binds to the NMDA channel, and results in an increased proba-

## Hippocampal Neuron

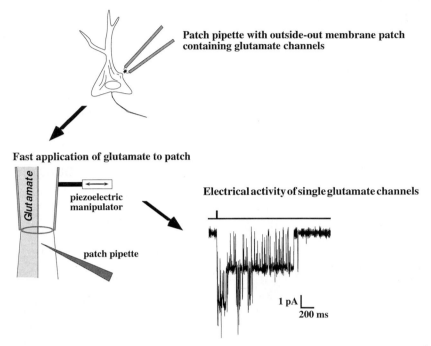

**Patch pipette with outside-out membrane patch containing glutamate channels**

**Fast application of glutamate to patch**

piezoelectric manipulator

patch pipette

**Electrical activity of single glutamate channels**

1 pA
200 ms

**FIG. 3.** Use of rapid application methods to study kinetics of ligand-gated channels. Use of the patch-clamp technique allows small membrane patches to be removed from a cell such as a hippocampal neuron with the extracellular side of the membrane facing the bath (outside-out configuration). The tip of the pipette is then placed in the interface of two rapidly moving streams of solution. A piezoelectric manipulator moves the solutions a few microns to change the solution flow over the patch. Such methods allow solution exchange in less than 1 msec. Thus the response to short applications of ligand, similar as is thought to occur at synapses, can be recorded as openings of single channels. Note, in the response at the right, that the opening of single glutamate channels of the NMDA type greatly outlasts the brief application of agonist. The duration of the glutamate application is indicated by the bar above the channel activity.

bility of opening (Figs. 4C and D). This is a dramatic example of allosteric regulation because the channel will not open unless both glycine and the transmitter (glutamate) are bound (for review see McBain and Mayer, 1994). A less profound, but equally important, example is the upregulation of GABA$_A$ receptor activity by the benzodiazepines (Olsen et al., 1991).

## IV. Molecular Structure

The AChR from skeletal muscle is the prototypic ligand-gated ion channel, and its physiological properties have been extensively studied during the past 35 years (Changeux et al., 1990; Karlin, 1991). The AChR also led the way in determining the molecular structure of ligand-gated channels. This analysis took advantage of the high density of AChRs in the electric organ of the ray (*Torpedo californica*), and the high-affinity snake toxin, α-bungarotoxin, that binds to muscle-type AChRs. Protein purification studies revealed an approximately 250-kDa complex with two α-bungarotoxin binding sites per complex, and separation on denaturing gels revealed polypeptides of 40, 50, 60, and 65 designated as the α-, β-, γ-, and δ-subunits.

Because of the molecular weight of the complex, and the binding of toxin molecules to the α-subunit, a pentameric structure with two α-subunits and single copies of β, γ, and δ was proposed. Using the partial amino sequence of the purified subunits, the cDNAs encoding AChRs in the *Torpedo* electric organ were cloned and sequenced in the early 1980s. Incorporation of the purified protein complex in lipid bilayers or expression of AChR mRNA in *Xenopus* oocytes demonstrated channels activated by acetylcholine, confirming the identity of the molecule. Consistent with the proposed combination of subunits in intact receptors, the expression of AChRs in oocytes was much greater when all four subunit mRNAs were included. The α-subunit was essential in order to obtain responses to acetylcholine. The $\alpha_2\beta\gamma\delta$ pentamer occurs in embryonic muscle, but later in development the γ-subunit is replaced by an ε-subunit. This substitution results in an increase in the single-channel conductance and a decrease in the time the channel is open. A much greater heterogeneity of receptor subunits appears to be involved in the expression of ligand-gated channels at central synapses (Sections V through VII, this chapter).

During the past 15 years, great strides have been made in understanding the structural characteristics of the AChR

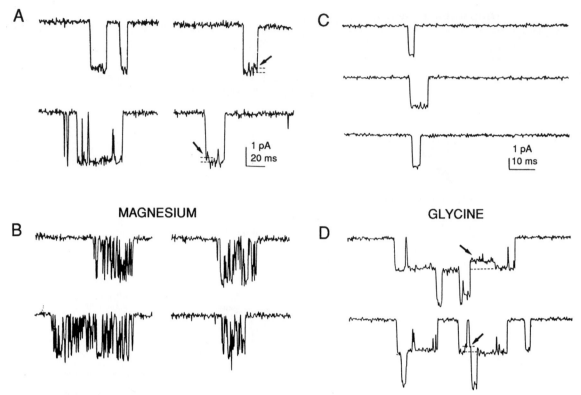

**FIG. 4.** NMDA-type glutamate channels illustrate two common mechanisms of channel regulation: channel block and allosteric modulation. (A and B) Channels activated by NMDA in a membrane patch from a hippocampal neuron in the absence of extracellular $Mg^{2+}$ remain open for several milliseconds before closing (A). However in the presence of $Mg^{2+}$ (B), the openings are interrupted by brief closures due to plugging of the pore by $Mg^{2+}$ ions. Downward deflections are due to channel opening. (C and D) The frequency of NMDA channel opening is increased in the presence of extracellular glycine (D) compared to glycine-free solutions (C). The stepwise increase in current in part D indicates that at least two channels were present in the membrane patch. Because glycine is now known to be required for NMDA channel opening, the occasional openings in part C probably reflect contamination of the solution with low levels of glycine. [Modified with permission from Ascher, R., and Nowak, L. (1987). *Trends Neurosci.* **10,** 284–288.]

channel. Based on the hydrophobicity of amino acid residues in the primary sequence, a transmembrane topology has been proposed that includes four putative transmembrane domains. A large N-terminal domain and a short C-terminal domain extend from the extracellular side of the membrane. This configuration predicts two cytoplasmic loops—a short one between M1 and M2 and a longer loop containing phosphorylation sites between M3 and M4. For the AChR, several residues in the N-terminal domain have been shown to be involved in forming the ACh binding site. The N-terminal domain may also determine subunit interactions during assembly of the pentamer, and contains carbohydrate residues whose function is largely unknown. The structures of the neuronal nicotinic and GABA/glycine receptor subunits have the same general characteristics (Fig. 5A), and have therefore been grouped together as a gene "superfamily", perhaps suggesting a common ancestral gene (reviewed in Betz, 1990). The length of the transmembrane domains (approximately 20 aa) is consistent with an $\alpha$-helical configuration, although structural studies suggest that the actual configuration of the transmembrane domains is not strictly $\alpha$-helical. The M2 domains of the five subunits make a major contribution to

the lining of the channel pore as shown schematically in Fig. 5B.

The glutamate receptor subunits differ substantially in their primary amino acid sequence (Nakanishi, 1992), Sommer and Seeburg, 1992; Hollmann and Heinemann, 1994). Notable differences include the much longer N-terminal domain (approximately 500 aa) of the glutamate subunits (Fig. 5A). The glutamate subunit transmembrane topology was initially modeled by homology with the AChR receptor family with both the N- and C-terminal domains as extracellular. However, subsequent analysis has revealed marked differences between the transmembrane topology of the AChR family and the glutamate receptor family. For example, the second hydrophobic domain of the glutamate subunits (M2 in the AChR subunits) appears to be a reentrant loop that enters and exits the membrane from the cytoplasmic side without fully traversing the membrane. This "P" loop shows striking homology to the P loop of voltage-gated channels that line the channel pore. Secondly, the region between the third and fourth hydrophobic domains is extracellular in glutamate subunits, whereas it is cytoplasmic in the AChR subunits. Finally the C-terminus of the glutamate subunits is cytoplasmic. These unique

## A
### AChR family

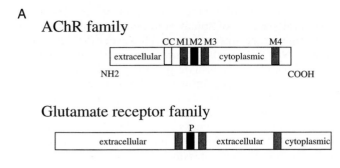

### Glutamate receptor family

**FIG. 5.** Schematic topology of ligand-gated channels. (A) The proposed membrane arrangement of polypeptide subunits for the AChR and glutamate receptors, diagrammed from the amino-terminal (NH2) to carboxyl-terminal (COOH) ends. The hydrophobic segments M1–M4 span the membrane with a long cytoplasmic loop between M3 and M4. The region marked CC in the AChR subunits contains cysteines that form disulfide bridges, and is involved in ligand binding. Note the much larger extracellular domain of the glutamate receptors. (B) Cross-section through the membrane showing the proposed distribution of five subunits around a central pore of the AChR. The M2 domain lines the pore.

features provide important clues to the location of the transmmitter binding site(s), the behavior of the channel pore, and sites of interaction with other proteins that anchor and/or localize glutamate receptor subunits at synapses (see Section VII, this chapter).

The analysis of the primary amino acid sequence gives limited information about the three-dimensional structure of the channel. Recently crystals of AChRs from the *Torpedo* electric organ have been analyzed with the electron microscope (Unwin, 1993). The resolution of these studies is not yet adequate to resolve the location of individual amino acid residues, but a good representation of the overall shape of the channel has been achieved (Fig. 6). Several features are striking and include the large component of the receptor that protrudes out of the membrane into the extracellular space. In addition, the entrance to the channel (vestibule) is rather long and shaped more like a cylinder than a funnel. This vestibule is ≈25 Å in diameter, and ion flux may be influenced by electrostatic interactions with residues in the wall of the pore. It is also interesting that a cytoplasmic protein remains attached to the base of the channel protein. The attached protein is likely to be the

43-kDa protein that copurifies with AChRs, and is involved in clustering of receptors. Similar proteins linking ligand-gated ion channels to the cytoskeleton may be involved in localizing, anchoring, or influencing the gating of other channels, including the glycine receptor and the NMDA receptor. The evidence for M2 as the pore lining comes from several lines of evidence. For example, a 23 amino acid peptide fragment from the M2 region of the *Torpedo* δ AChR subunit forms a cation channel when expressed in lipid bilayers. More direct evidence comes from site-directed mutagenesis of conserved residues in this putative membrane-spanning segment. It has been suggested that the M2 region forms an α-helix with 3.6 residues per turn. However, studies using a technique known as scanning cysteine mutagenesis (see Karlin and Akabas, 1995) suggest that some portions of M2 are in a β-sheet configuration. This method has great application in structure–function studies of ion channels because substitution of other amino acid residues by cysteines usually does not affect channel properties. The cysteine residue can then form disulfide bonds with applied reagents, allowing investigators to systematically "map" residues that are exposed in particular conformations of the molecule, for example, during channel opening.

Rings of negatively charged residues (glutamate or aspartate) at the ends of M2 for the cationic AChR and glutamate receptor subunits are thought to attract permeant cations, because mutation of these residues to neutral or positively charged amino acids results in a decrease in

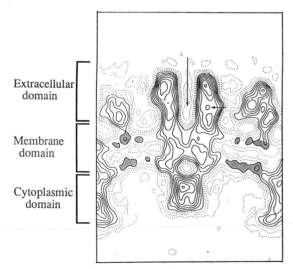

**FIG. 6.** Cross-section of AChR obtained from crystallized postsynaptic membrane of *Torpedo* electric organ at 9-Å resolution. Parts of several receptors are shown, with a complete profile of one receptor along the axis of the pore. The dark lines indicate the channel protein, with arrows identifying the pore and a presumptive area ca. 30 Å above the membrane that may be the site of ACh binding. The pore narrows in the membrane-spanning region. The square contour lines at the cytoplasmic face of the receptor may be the 43-kDa protein that is involved in receptor clustering. (Modified with permission from Unwin, 1993. Copyright 1993 Cell Press.)

the single-channel conductance. Consistent with this hypothesis, glycine and GABA subunits have positively charged residues (lysine or arginine) near the ends of the M2 region that may attract permeant anions (Fig. 7B). The M2 region was assumed to be membrane spanning because the amino acid residues are hydrophobic, that is, they have no charged side chains (except of course for the charged residues at the ends of M2). Several small uncharged residues near the middle of the bilayer appear to line the channel lumen, as they are involved in the binding of some open channel blockers to the AChR. Mutations of the conserved leucine residue (position 12 in Fig. 7B) also cause a profound change in the response to ACh in the neuronal AChR subunit $\alpha_7$, and is involved in the binding of the noncompetitive antagonist chlorpromazine (Section V, this chapter).

## V. Neuronal Acetylcholine Receptor Channels

Similar to muscle AChRs, neuronal AChRs (NAChRs) are ligand-gated ion channels that are permeable to $Na^+$ and $K^+$, and in some cases $Ca^{2+}$ (for review, see Luetje *et al.*, 1990; McGehee and Role, 1995). ACh has long been known to activate channels mediating excitatory synaptic transmission in autonomic ganglia, and at the recurrent collateral synapse of spinal motoneurons onto Renshaw cells (inhibitory interneurons). However, the functional role of NAChRs in the central nervous system has been more difficult to establish despite the well-known effect of nicotine (via cigarette smoke) on human behavior (Dani and Heinemann, 1996). NAChRs differ from prototypic muscle AChRs both pharmacologically and structurally. First, $\alpha$-bungarotoxin blocks muscle AChRs, but not most types of NAChRs. However, binding studies revealed $\alpha$-bungarotosin-binding sites on neurons whose function was unclear for many years. It is now accepted that the neuronal $\alpha$-bungarotoxin receptor is a subtype of the NAChRs that has faster kinetics and a higher calcium permeability than other NAChR subtypes. The $Ca^{2+}$ permeability of some NAChRs and changes in channel properties during synaptogenesis suggest that various NAChR subtypes play a role in synaptic plasticity and development. The activity of NAChRs may also be modulated by phosphorylation. The kinase activation may be due to the effects of neuropeptides such as vasoactive intestinal peptide and substance P that are coreleased with acetylcholine at some synapses.

Perhaps even more than for other ligand-gated channels, the cloning and characterization of subunit genes has greatly accelerated studies of both the structure and func-

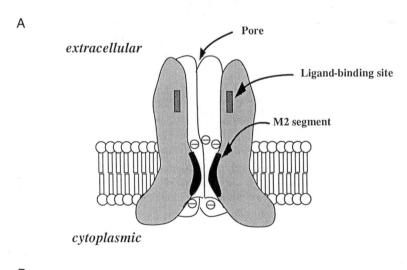

**FIG. 7.** The M2 segment lines the channel pore. (A) Schematic of a multiple subunit AChR channel, as suggested by electron micrographs (see Fig. 6) and physiological data. The membrane-spanning region of the pore is lined by the M2 segment of each subunit, with negatively charged residues at either end of M2. (B) Amino acid sequence of the M2 region of the $\alpha_1$-subunit of the AChR that forms a cation channel is compared with the $\alpha_1$-subunit of the GABA$_A$ receptor that forms an anion channel. The charged residues are in bold, and conserved small uncharged residues are highlighted by gray bars. Note that the negatively charged residues (glutamate, Glu) in the AChR subunit are replaced by positively charged residues (arginine, Arg) in the GABA$_A$ subunit. These differences may account for the cation and anion selectivity, respectively.

tion of NAChRs (Steinbach, 1989; Role, 1992). The structure of NAChRs is comprised of a pentamer of subunits assembled around a central pore. But unlike the muscle AChRs, the NAChR pentamer can consist of only one or two classes of subunits, aptly named $\alpha$ and $\beta$. The subunits with cysteines at residues 192 and 193 (involved in binding of ACh to the $\alpha$-subunit of muscle AChRs) are designated as $\alpha$; subunits lacking these cysteines are called $\beta$. To date, the genes for eight $\alpha$ ($\alpha_{2-9}$) and three $\beta$ ($\beta_{2-4}$) subunits have been cloned from rat and chick brain cDNA libraries. The stoichiometry of each channel molecule appears to be two $\alpha$- and three $\beta$-subunits, at least for channels comprised of $\alpha_4$- and $\beta_2$-subunits. Expression of single subunits and combination of subunits in *Xenopus* oocytes (by injection of subunit mRNA) has begun to narrow the possible combinations of subunits in functional channels. For example, $\alpha_{2-4}$ do not express as homo-oligomeric (e.g., five $\alpha_2$-subunits) channels, whereas $\alpha_7$, $\alpha_8$, and $\alpha_9$ do. On the other hand, $\alpha_{5-6}$ do not express in oocytes even in combination with any of the current $\beta$-subunits, although it is of course possible (even likely) that these subunits contribute to functional receptors *in vivo*. Likewise, combinations of subunits differ in their single-channel properties, ligand affinity, and sensitivity to antagonists such as neuronal bungarotoxin. Homo-oligomeric $\alpha_7$ channels produce a rapidly desensitizing response to ACh. This subunit is being used to probe the structural substrates of channel properties such as desensitization, ion flux, and selectivity. Homo-oligomeric $\alpha_7$, $\alpha_8$, and $\alpha_9$ channels are also sensitive to $\alpha$-bungarotoxin, suggesting that $\alpha_7$ (which is the predominant subunit of these three subunits in brain) is at least a component of the $\alpha$-bungarotoxin receptor. Although it is still controversial the $\alpha$-bungarotoxin receptor may be composed exclusively of $\alpha_7$ subunits. Recent studies suggest that presynaptic $\alpha$-bungarotoxin receptors, particularly of the $\alpha_7$ type with their high calcium permeability, can enhance the release of glutamate and other transmitters. The recently cloned $\alpha_9$-subunit is unique due to its limited expression in the cochlea as well as its mixed muscarinic/nicotinic pharmacology.

Although many central neurons have responses to exogenous application of nicotinic agonists, the distribution of NAChR subunit genes revealed by *in situ* hybridization is wider than might have been predicted based on physiological studies. However, there is a reasonable correlation between pathways with functional NAChRs (e.g., the cholinergic input to the hippocampus and medial hebenula from the medial septum and diagonal band), the presence of nicotinic binding sites, and the expression of NAChR subunits. Recent studies of NAChRs in different neuronal cell types has revealed a spectrum of channel properties and pharmacological characteristics, suggesting that individual neurons may express several combinations of NAChR subunits. However, the simplest scenario—that each cell type expresses only one combination of NAChR subunits—appears to be incorrect. For example, *in situ* hybridization studies of neurons in the medial habenula nucleus have revealed two $\alpha$- and three $\beta$-subunit genes. Thus, the particular selection of subunits comprising a NAChR channel in a particular cell type is still uncertain.

## VI. $\gamma$-Aminobutyric Acid and Glycine Receptor Channels

Synaptic inhibition in the central nervous system is mediated by channels gated by GABA and glycine. Virtually all neurons have GABA channels, while the distribution of glycine channels is restricted primarily to the brainstem and spinal cord. These two channels are selectively permeable to anions, principally $Cl^-$ under physiological conditions. GABA-gated $Cl^-$ channels are designated as $GABA_A$ receptors to distinguish them from the G-protein-coupled $GABA_B$ receptor (reviewed by Bormann, 1988). $GABA_A$ channels are often localized on proximal dendrites of central neurons. Because the $Cl^-$ equilibrium potential in many neurons is more negative than the resting potential, the opening of $GABA_A$ channels hyperpolarizes the cell, and thus reduces excitability. In addition to the hyperpolarization, the opening of large numbers of $GABA_A$ channels lowers the resistance of the membrane, and effectively "shunts" excitation traveling down the dendrite from excitatory synapses on more distal dendritic branches. However, in some neurons, particularly during early development, the $Cl^-$ equilibrium is more positive than the resting potential, resulting in "excitatory" $GABA_A$ responses. Such excitatory responses may act as a development signal in some cases.

The behavior of single $GABA_A$ and glycine channels can be described by a kinetic scheme similar to that of the AChR, with the binding of two agonist molecules required for channel opening (see Macdonald and Twyman, 1992). Three conductance levels are apparent. These are approximately 19, 30, and 45 pS when the $Cl^-$ concentration is equal on both sides of the membrane. The 30-pS level is the most frequently observed level for $GABA_A$ channels, and 45 pS is generally more common for glycine channels. Analysis of the openings and closings of single $GABA_A$ channels suggests that the channel may open briefly following the binding of a single $GABA_A$ molecule, and into two longer lived open states from the doubly liganded configuration. Receptors may close and reenter the longer lived open states before the agonist dissociates, so-called *bursts* where short closings interrupt a series of openings. These bursts last tens of milliseconds. Desensitization of $GABA_A$ channels results in long closed intervals that are grouped with bursts into "clusters" lasting up to several hundred milliseconds. These clusters are important in determining the duration of inhibitory postsynaptic potentials at some synapses (see Jones and Westbrook, 1996). By measuring the permeability of anions of different diameters, the pore size has been estimated to be 5.6 Å. $GABA_A$ channels also become less active during prolonged recording. This process has been called *rundown* and appears, at least in part, to be due to channel dephosphorylation, and may also be sensitive to intracellular $Ca^{2+}$ These are themes that are common to several ligand-gated ion channels.

The drugs that act on $GABA_A$ and glycine channels comprise a fascinating assortment of clinically important compounds (Olsen *et al.*, 1991). Because these channels are so important in synaptic inhibition, either enhancement or reduction in their activity can lead to profound changes in brain function including amnesia (increased $GABA_A$ activity) or seizures (decreased $GABA_A$ activity). Antagonists for these receptor channels include strychnine that blocks glycine receptors, bicuculline that inhibits $GABA_A$ channels, and picrotoxin that inhibits both channel types. The $GABA_A$ channel is also the target of sedative-hypnotic drugs, such as the benzodiazepines and barbiturates. Benzodiazepines (BDZ) increase the probability of channel opening, whereas barbiturates appear to act by prolonging long channel openings ("bursts"). The pharmacology of benzodiazepine modulation of the $GABA_A$ receptor is particularly interesting as compounds can either enhance channel opening (BDZ agonists), reduce channel opening (BDZ inverse agonists), or block the effects of BDZ agonists (BDZ antagonists). $GABA_A$ receptor activity is also modulated by alcohol, volatile anesthetics such as isoflurane, and some steroid anesthetics (so-called "neurosteroids").

Using benzodiazepines and strychnine as selective ligands, $GABA_A$ and glycine receptors were purified as several polypeptides, each with molecular weights of approximately 50–60 kDa (reviewed in Betz, 1992; Delorey and Olsen, 1992). The solubilized receptor complex had a molecular weight of approximately 250 kDa, suggesting that, as for the AChR, multiple (four or five) subunits constitute a receptor. Molecular cloning has now identified a series of receptor subunits for both the $GABA_A$ and glycine receptor. The glycine subunits include the strychnine-binding subunit ($\alpha$) and a $\beta$-subunit, with a proposed stoichiometry of $\alpha_3\beta_2$. A larger number of $GABA_A$ subunits (seventeen) have been identified including six $\alpha$, three $\beta$, three $\gamma$, $\delta$, $\varepsilon$, $\pi$, and two $\rho$ (see Schofield et al., 1990; Wisden and Seeburg, 1992; Macdonald and Olsen, 1994). The benzodiazepine-binding site appears to reside on the $\alpha$-subunit, but the presence of a $\gamma$-subunit (usually $\gamma_2$) is necessary for high-affinity benzodiazepine binding. Interestingly, the $\alpha_6$-subunit has a low affinity for BDZ agonists, but still can bind BDZ inverse agonists or antagonists. This observation may explain the occurrence of benzodiazepine-insensitive $GABA_A$ receptors in some neurons. As for the NAChRs and glutamate receptors, the large number of $GABA_A$ subunits provides a formidable challenge in determining which combinations are the predominant ones in neurons. The combinations of subunits that form functional receptors are just beginning to be explored, but it is clear that different combinations can alter receptor properties. Subunit expression also varies during development and with neuronal cell type.

There is now increasing evidence for a distinct GABA-mediated chloride channel termed the $GABA_C$ receptor in some retinal neurons. This receptor can be activated by several conformationally restricted GABA analogues including cis-4-aminocrotonic acid. The $GABA_C$ receptor is bicuculline insensitive, weakly antagonized by picrotoxin, and not modulated by BDZs, barbiturates, or neurosteroids. These channels show distinct gating properties and conductance compared to $GABA_A$ receptors. Because the $GABA_C$ receptor shares some features with the $\rho_1$-subunit that is highly expressed in retina, $\rho_1$ may be one of the $GABA_C$ receptor subunits.

## VII. Glutamate Receptor Channels

Although it was known from the early 1950s that L-glutamate was a neuroexcitant, it was not until the 1980s that the role of glutamate-gated ion channels in central synaptic transmission was widely accepted (see Collingridge and Lester, 1989; Westbrook and Jahr, 1989). It is now clear that glutamate receptors mediate a substantial portion of fast excitatory transmission in the brain and spinal cord through the simultaneous activation of two types of ion channels colocalized at excitatory synapses. Characterization of this family of ligand-gated ion channels was initially based on selective activation by the exogenous amino acid ligands: NMDA, kainate, and quisqualate. Because quisqualate also activates a G-protein-coupled receptor, AMPA ($\alpha$-amino-3-hydroxy-5-methyl-4-isoxazole proprionic acid) has replaced quisqualate as one of the prototypic ligands. It soon became apparent that NMDA receptors were distinct from kainate and AMPA receptors, but for some time it was debated whether kainate and AMPA (i.e., non-NMDA) responses were due to the same or different receptors.

Initial studies of native AMPA/kainate receptor channels in neurons demonstrated many of the same features as AChRs. In hippocampal pyramidal neurons, AMPA channel activation leads to brief openings (1–5 msec) of monovalent cation channels that show little or no voltage dependence. However, in some neurons, particularly interneurons, AMPA channels have a higher calcium permeability and are inwardly rectifying. AMPA receptors rapidly desensitize when gated by glutamate, and it appears that both agonist dissociation (unbinding) and desensitization (bound, but nonconducting) contribute to the decay of the synaptic current. The behavior of AMPA channels seems well suited to the fast relay of information that characterizes the fast component of excitatory postsynaptic potentials in the brain. However, this simple picture of AMPA channels has been complicated by recent molecular studies (see later discussion), and it appears likely that at least some modifications of this scheme will be necessary as more information becomes available.

The NMDA channel has been intensively investigated by neuroscientists in recent years, and provides perhaps the best example of the linkage between fundamental properties of ion channels and the electrical activity of single cells and complex behavioral phenomena (for review see Bekkers and Stevens, 1990; McBain and Mayer, 1994). This linkage is based on three features of the NMDA channels: (1) voltage-dependent block of the open channel by extracellular $Mg^{2+}$, (2) a relatively high permeability to $Ca^{2+}$, and (3) slow channel kinetics resulting in long-lasting channel activity (see, e.g., Jahr and Lester, 1992). Because $Mg^{2+}$ is positively charged, these ions sense the membrane electric field, and are drawn into the open channel at negative membrane potentials. $Mg^{2+}$ binds in the channel pore and impedes the flow of permeant ions, even though the ligand (glutamate) remains bound. However, as the cell membrane depolarizes during synaptic activity, $Mg^{2+}$ falls out of the channel and permeant cations flow through the channel. Thus ion flux through the channel is voltage dependent.

This mechanism differs from other voltage-dependent ion channels such as the sodium or calcium channel, where voltage dependence results from an intrinsic conformational change in the channel protein. However, the end result is the same; the NMDA channel becomes more effective with depolarization, and thus acts as a positive feedback on synaptic activation.

Also important to the cellular function of NMDA receptors is ion permeability. Unlike many AMPA channels, $Ca^{2+}$ contributes a substantial fraction of the flux through NMDA channels. Based on measurements of the equilibrium (reversal) potential in different ionic solutions, the permeability ratio of $Ca^{2+}$ to $Na^+$ has been estimated to be approximately 10. However, the rate of flux of $Ca^{2+}$ through the NMDA channel is slower than $Na^+$ because $Ca^{2+}$ transiently binds with higher affinity to sites within the channel ("energy wells"). Thus most of the current through open NMDA channels is due to $Na^+$. Nonetheless, the approximately 5–10% of the current that is carried by $Ca^{2+}$ is sufficient to act as a biochemical signal for such processes as the induction of long-term potentiation in the hippocampus, an experimental model of associative learning. Excessive activation of glutamate receptors may also cause neuronal damage, presumably through elevations of intracellular $Ca^{2+}$, activation of proteases, and free radical formation. This mechanism, called *excitotoxicity*, has been implicated in brain damage due to prolonged seizures, following strokes, and may also contribute to loss of neurons in several degenerative neurological diseases.

Because no high-affinity ligands were available for purification of glutamate channel proteins, the structure of this class of ligand-gated channels defied analysis for a number of years. Thus molecular biologists were forced to use the somewhat laborious task of expression cloning to isolate cDNA clones for glutamate receptor subunits. The first AMPA receptor subunit (GluR1) was isolated by this technique in 1989, and the first NMDA receptor clone (NMDAR1) was also isolated by expression cloning in 1991. As would be predicted from analysis of other ligand-gated ion channels, the proposed structure of glutamate channels involves multisubunit complexes surrounding a central pore. Using the initial AMPA and NMDA clones to screen for homologous sequences, approximately 17 related glutamate receptor subunits have now been isolated from rat and mouse cDNA libraries. Surprisingly, the sequence of glutamate channel subunits is not highly homologous to AChR, GABA, or glycine channels. In particular, glutamate channels have a very large extracellular domain constituting approximately 50% of the protein. Likewise, as discussed in Section IV of this chapter, the topology of glutamate subunits is very different than the AChR family of receptor channels. The two major extracellular domains of the glutamate subunits, the N-terminus and the loop between the third and fourth hydrophobic domains, are homologous to a group of bacterial proteins that bind amino acids. This homology is also shared by the N-terminus of the single polypeptide metabotropic glutamate receptors, which led to a proposed model for the glutamate-binding site based on the crystal structure of the periplasmic binding proteins (O'Hara *et al.*, 1993). Using chimeric

receptors composed of domains from two glutamate subunits with distinct pharmacology, the role of these two domains in binding of glutamate has now been confirmed (Stern-Bach *et al.*, 1994).

On the basis of sequence homology and the characteristics of the expressed receptors, the subunits can be grouped into three categories: Four AMPA subunits (GluR1–4 or A–D), five high-affinity kainate subunits (GluR5–7, KA-1, KA-2), and five NMDA subunits (NMDA1, NMDA2A–D). Several other related glutamate subunits (delta 1, delta 2, and chi-1 or NMDAR-L) have been isolated but their function remains unclear. AMPA receptors have a low affinity for glutamate and kainate, compared to the so-called "high-affinity" kainate subunits. Some of the subunits can combine to form functional homo-oligomeric receptors, and this has revealed some interesting and curious phenomena. For example, in the M2 region of GluR1–4 (which forms at least part of the channel pore), the genes all code for a glutamine residue at one location. However, the RNA for GluR2, 5, and 6 is edited, such that the expressed protein has an arginine residue at this site. This switch has a profound effect on the behavior of the channel. The current evoked by homo-oligomeric expression of unedited subunits shows marked inward rectification and an increased permeability to $Ca^{2+}$. The edited versions have a linear current/voltage relationship and are permeable only to monovalent cations ($Na^+$ and $K^+$). Because the synaptic response mediated by AMPA receptors in hippocampal pyramidal cells has a linear current/voltage relationship, it seems likely that most AMPA receptors in these neurons contain one or more edited copies of GluR2. However, in interneurons in cortex and hippocampus as well as in dorsal horn neurons of the spinal cord, AMPA receptors lacking edited GluR2 subunits show inward rectification and increased permeability to $Ca^{2+}$. Physiological studies have demonstrated a more restricted distribution of high-affinity kainate responses, including dorsal root ganglion neurons and some hippocampal neurons. Expression studies and *in situ* immunohistochemistry suggest that combinations of GluR5 or 6 with KA-2 may account for at least some of the high-affinity kainate responses.

Glutamate channels, particularly NMDA channels, are regulated both by phosphorylation and by a variety of allosteric mechanisms. The most dramatic is the action of glycine as a *coagonist* that is required for the opening of NMDA channels, and extracellular $Mg^{2+}$ that blocks NMDA channels (Fig. 4). It appears that two molecules of glutamate and two molecules of glycine are required to activate an NMDA channel. Because there are four or five subunits per channel, it would appear that each subunit can contribute to agonist binding. Although AMPA channels do not require a coagonist, given the homology between AMPA and NMDA channels it may be that four molecules of glutamate are also necessary to activate AMPA channels. Other regulator factors that have been shown to affect NMDA channel activity include extracellular protons, $Zn^{2+}$, polyamines, redox potential, and intracellular $Ca^{2+}$. NMDA receptors can be regulated by protein kinase A and C and tyrosine kinases, whereas AMPA channel activity is up-regulated by cAMP-dependent protein kinase. The impact of these regulatory mechanisms on synaptic function is an important area for future investigation.

As for all synaptic receptor channels, AChR and glutamate channels are clustered at synaptic sites on muscle and neurons. For AchRs on muscle, the clustering/anchoring involves a complex cascade involving factors released from the neuron as well as a number of protein–protein interactions in the postsynaptic membrane including the 43 KDa protein, rapsyn (for review, see Froehner, 1993). Recent studies have revealed a number of candidate proteins in central neurons that are involved in the clustering and localization of glycine and glutamate receptor channels. Glutamate channels are imbedded in the postsynaptic density that contains receptors, regulatory proteins, and cytoskeletal proteins. Biochemical studies suggest that calmodulin and cytoskeletal proteins can interact with C-terminal domains of several NMDA receptor subunits at domains that are involved in the regulation of NMDA receptor desensitization, suggesting an intriguing link between dynamic regulation of channel activity (e.g., by compartmentalized regulatory proteins) and structural features such as channel anchoring and clustering. A family of cytoskeletal proteins with so-called PDZ domains appear to be involved in interactions with AMPA and NMDA subunits (for review see Sheng, 1996; Ehlers *et al.*, 1996). Studies of these interactions may be important in synapse development as well as in the dynamic regulation of activity of ligand-gated channels at synapses.

## VIII. Summary

Ligand-gated ion channels are membrane proteins that are fundamental signaling molecules in neurons. These molecules are localized in the plasmalemma and on intracellular organelles and can be gated by both intracellular and extracellular ligands. The neurotransmitter-gated ion channels discussed in this chapter mediate fast excitation and inhibition in the nervous system, and have now been well characterized by physiological and molecular studies. Studies using the technique of voltage and patch-clamp recording have examined the three basic features of an ion channel: gating, conductance, and selective permeability. In general, ligand-gated ion channels can be described by kinetic models involving binding of two or more ligand molecules that induce a conformational change in the protein. As a result, a central water-filled pore opens and conducts ions at very high rates of up to $10^7$ ions per second. Channel activity is terminated when either the channel closes or enters a nonconducting (desensitized) state. The ACh and glutamate receptors are cation channels, whereas GABA and glycine receptors are anion channels. Many of the genes coding for these channels have been cloned. The primary structure of the ACh, GABA, and glycine channel subunits, deduced from the primary amino acid sequence, is highly homologous. Thus, this group of cDNAs constitutes a gene "superfamily" that probably evolved from a common ancestral gene. Glutamate channels comprise a separate but related family of channel subunits. The ligand-

binding domain of glutamate channels shares homology with bacterial amino acid binding proteins whereas the pore shares homology with the pore of voltage-dependent ion channels.

Each of the neurotransmitter-gated channels appears to be comprised of a hetero-oligomeric complex comprised of five closely related subunits surrounding a central ion pore. For each receptor type there are a number of possible subunit combinations that can form functional channels. AChR subunits have four transmembrane domains with the second transmembrane domain lining the pore. Charged amino acid residues near the mouth of the pore are important in determining the selectivity and conductance of the individual channel types. The glutamate subunits have three transmembrane domains and a reentrant loop domain that forms a portion of the pore. A number of important drugs act by binding to the walls of the channel pore and obstructing the flow of ions through the open channel.

Molecular studies of these subunits have begun to reveal the structural features of the channel proteins that determine their gating behavior, conductance, and selective ion permeability. Ligand-gated ion channels are also highly regulated by allosteric and covalent modifications. Allosteric regulation also constitutes an important mechanism for drug action in the brain. Recent progress in the study of ligand-gated ion channel function and structure has answered many long-standing questions, but has also raised many new and interesting issues concerning the function of this important class of membrane proteins.

## Bibliography

Ascher, P., and Nowak, L. (1987). Electrophysiological studies of NMDA receptors. *Trends Neurosci.* **10**, 284–288.

Bean, B. P. (1992). Pharmacology and electrophysiology of ATP-activated ion channels. *Trends Pharmacol. Sci.* **13**, 87–90.

Bekkers, J. M., and Stevens, C. E. (1990). Computational implications of NMDA receptor channels. *Cold Spring Harbor Symp. Quant. Biol.* **55**, 131–135.

Betz, H. (1990). Ligand-gated ion channels in the brain: The amino acid receptor superfamily. *Neuron* **5**, 383–392.

Betz, H. (1992). Structure and function of inhibitory glycine receptors. *Q. Rev. Biophys.* **25**, 381–394.

Bormann, J. (1988). Electrophysiology of GABA_A and GABA_B receptor subtypes. *Trends Neurosci.* **11**, 112–116.

Changeux, J. P., Benoit, P., Bessis, A., Cartaud, J., Devillers-Thiery, A., Fontaine, B., Galzi, J. L., Klarsfeld, A., Laufer, R., Mulle, C., Nghiem, H. O., Osterlund, M., Piette, J., and Rehav, F. (1990). The acetylcholine receptor: Functional archhitecture and regulation. *Adv. Second Messenger Phosphoprotein Res.* **24**, 15–19.

Collingridge, G. L., and Lester, R. A. (1989). Excitatory amino acid receptors in the vertebrate central nervous system. *Pharmacol. Rev.* **41**, 143–210.

Dani, J. A., and Heinemann, S. (1996). Molecular and cellular aspects of nicotine abuse. *Neuron* **16**, 905–908.

DeLorey, T. M., and Olsen, R. W. (1992). γ-Aminobutyric acid_A receptor structure and function. *J. Biol. Chem.* **267**, 16747–16750.

Ehlers, M. D., Mammen, A. L., Lau, L. F., and Huganir, R. L. (1996). Synaptic targeting of glutamate receptors. *Curr. Opin. Cell Biol.* **8**, 484–489.

Froehner, S. (1993). Regulation of ion channel distribution at synapses. *Annu. Rev. Neurosci.* **16**, 347–368.

Hille, B. (1992). "Ionic Channels of Excitable Membranes," 2nd ed. Sinauer Associates, Sunderland, MA.

Hollmann, M., and Heinemann, S. (1994). Cloned glutamate receptors. *Annu. Rev. Neurosci.* **17**, 31–108.

Jahr, C. E., and Lester, R. A. J. (1992). Synaptic excitation mediated by glutamate-gated ion channels. *Curr. Opin. Neurobiol.* **2**, 270–274.

Jones, M. V., and Westbrook, G. L. (1996). The impact of receptor desensitization on fast synaptic transmission. *Trends Neurosci.* **19**, 96–101.

Julius, D. (1991). Molecular biology of serotonin receptors. *Annu. Rev. Neurosci.* **14**, 335–360.

Luetje, C. W., Patrick, J., and Seguela, P. (1990). Nicotine receptors in the mammalian brain. *FASEB J.* **4**, 2753–2760.

Karlin, A. (1991). Explorations of the nicotinic acetylcholine receptor. *Harvey Lectures* **85**, 71–107.

Karlin, A., and Akabas, M. H. (1995). Toward a structural basis for the function of nicotinic acetylcholine receptors and their cousins. *Neurons* **15**, 1231–1244.

Macdonald, R. L., and Olsen, R. W. (1994). GABA$_A$ receptor channels. *Annu. Rev. Neurosic.* **17**, 569–602.

Macdonald, R. L., and Twyman, R. E. (1992). Kinetic properties and regulation of GABA$_A$ receptor channels. *Ion Channels* **3**, 315–343.

McBain, C., and Mayer, M. (1994). *N*-methyl-D-aspartic acid receptor structure and function. *Physiol. Rev.* **74**, 723–759.

McGehee, D. A., and Role, L. W. (1995). Physiological diversity of nicotinic acetylcholine receptors expressed by vertebrate neurons. *Annu. Rev. Physiol.* **57**, 521–546.

Miller, C. (1989). Genetic manipulation of ion channels: A new approach to structure and mechanism. *Neuron* **2**, 1195–1205.

Montal, M. (1990). Molecular anatomy and molecular design of channel proteins. *FASEB J.* **4**, 2623–2635.

Nakanishi, S. (1992). Molecular diversity of glutamate receptors and implications for brain function. *Science* **258**, 597–603.

Neher, E. (1992). Ion channels for communication between and within cells. *Science* **256**, 498–502.

Nicoll, R. A., Malenka, R. C., and Kauer, J. A. (1990). Functional comparison of neurotransmitter receptor subtypes in mammalian central nervous system. *Physiol. Rev.* **70**, 513–565.

O'Hara, P. J., Sheppard, P. O., Thogersen, H., Venezia, D., Haldeman, B. A., McGrane, V., Houamed, K. M., Thomsen, C., Gilbert, T. L., and Mulvihill, E. R. (1993). The ligand-binding domain in metabotropic glutamate receptors is related to bacterial periplasmic binding proteins. *Neuron* **11**, 41–52.

Olsen, R. W., Sapp, D. M., Bureau, M. H., Turner, D. M., and Kokka, N. (1991). Allosteric actions of central nervous system depressants including anesthetics on subtypes of the inhibitory gamma-aminobutyric acid$_A$ receptor-chloride channel complex. *Ann. NY Acad. Sci.* **625**, 145–154.

Role, L. W. (1992). Diversity in primary structure and function of neuronal nicotinic acetylcholine receptor channels. *Curr. Opin. Neurobiol.* **2**, 254–262.

Sakmann, B. (1992). Nobel Lecture. Elementary steps in synaptic transmission revealed by currents through single ion channels. *Neuron* **8**, 613–629.

Schofield, R. R., Shivers, B. D., and Seeburg, P. H. (1990). The role of receptor subtype diversity in the CNS. *Trends Neurosci.* **13**, 8–11.

Sheng, M. (1996). PDZs and receptor/channel clustering: Rounding up the latest suspects. *Neuron* **17**, 575–578.

Sommer, B., and Seeburg, P. H. (1992). Glutamate receptor channels: Novel properties and new clones. *Trends Pharmacol. Sci.* **13**, 291–296.

Steinbach, J. H. (1989). Structural and functional diversity in vertebrate skeletal muscle nicotinic acetylcholine receptors. *Ann. Rev. Physiol.* **51**, 353–365.

Stern-Bach, Y., Bettler, B., Hartley, M., Sheppard, P. O., O'Hara, P. I., and Heinemann, S. F. (1994). Agonist selectivity of glutamate receptors is specified by two domains structurally related to bacterial amino acid–binding proteins. *Neuron* **13**, 1345–1357.

Swope, S. L., Moss, S. J., Blackstone, C. D., and Huganir, R. L. (1992). Phosphorylation of ligand-gated ion channels: A possible mode of synaptic plasticity. *FASEB J.* **6**, 2514–2523.

Unwin, N. (1993). Neurotransmitter action: Opening of ligand-gated ion channels. *Neuron* **10**, Suppl., 31–41.

Westbrook, G. L., and Jahr, C. E. (1989). Glutamate receptors in excitatory neurotransmission. *Sem. Neurosci.* **1**, 103–114.

Wisden, W., and Seeburg, P. H. (1992). GABA$_A$ receptor channels: From subunits to functional entities. *Curr. Opin. Neurobiol.* **2**, 263–269.

Janusz B. Suszkiw

# 40

# Synaptic Transmission

## I. Introduction

Transfer of signals between neurons or from neurons to effector cells takes place at morphologically and functionally specialized sites called **synapses.** A characteristic feature of most synaptic junctions in vertebrates is a 20- to 100-nm-wide extracellular space called the **synaptic cleft,** which separates the presynaptic and postsynaptic elements of the synapse. This physical discontinuity prevents direct flow of current between the cells, and the transmission of signals across the synapse is accomplished by means of a chemical neurotransmitter substance.

The fundamental aspects of chemically mediated transmission is the transformation of voltage signal into the release of neurotransmitter from the presynaptic neuron and transduction of transmitter binding to a specific receptor into voltage change in the postsynaptic cell. The transmitter–receptor interaction alters the permeability of the postsynaptic membrane to certain ions, resulting in depolarizing or hyperpolarizing currents and generation of the associated excitatory postsynaptic potentials or inhibitory postsynaptic potentials, respectively.

One of the best studied excitatory synapses is the vertebrate skeletal neuromuscular junction (NMJ). Here, the action potential (AP) in the presynaptic motor neuron triggers the release of acetylcholine (ACh), which binds to the postsynaptic nicotinic acetylcholine receptor (nAChR) and produces a large increase in the permeability of the postsynaptic membrane to sodium and potassium ions. This results in a depolarizing synaptic potential that is normally large enough to elicit a muscle AP and muscle twitch.

In contrast to the relatively simple one-to-one relationship between the presynaptic axon and the postsynaptic muscles fiber at the NMJ, neurons in the central nervous system (CNS) receive multiple excitatory and inhibitory synaptic inputs. The ionic mechanisms of EPSPs at central synapses and NMJ are similar. The inhibitory synaptic potentials are associated with the increase in the permeability to chloride or potassium ions. This tends to counteract depolarization produced by excitatory synapses and move membrane potential away from the threshold. The postsynaptic neuron integrates the excitatory and inhibitory inputs and the balance between the excitatory and inhibitory influences determines whether or not an AP is initiated.

The basic principles of chemical neurotransmission have been elaborated in a series of seminal experiments conducted on the NMJ by Katz and his collaborators in the 1950s and 1960s. At about the same time, the pioneering work by Eccles and his coworkers on the spinal motorneurons showed that the general principles of chemical transmission established at the vertebrate neuromuscular junction also apply to central synapses (Eccles, 1964; Katz, 1966). Since then, considerable insights into the mechanism of synaptic transmission have been gained on the molecular level, through the application of contemporary techniques of molecular biology and electrophysiology (Jessell *et al.*, 1993).

Although this chapter focuses on chemical synaptic transmission, the reader should be cognizant of another form of signal transfer, referred to as **electrotonic** or **ephatic transmission.** Electrotonic transmission is accomplished at the gap junctions, which provide electrical continuity between the communicating cells. Unlike the chemical synapses, gap junctions and the electrical coupling they mediate are not unique to the nervous system but are also found in many other tissues. For additional information on the subject of electrotonic transmission, the interested reader is referred to the article by Bennett *et al.* (1991).

## II. Structural Features of Chemical Synapses

A synaptic junction may be defined as a morphofunctional unit consisting of specialized presynaptic and postsynaptic cell membranes and the intervening synaptic cleft. The presynaptic element, frequently the axon terminal, is the transmitter output region of a neuron making the synaptic connection. A *sine qua non* of a chemical synapse is the presence in the presynapse of submicroscopic spherical vesicles that function to store and release transmitter, and are called **synaptic vesicles.** The postsynaptic structure is the receptive membrane, which contains specific receptors for the transmitter released from the presynaptic terminal.

## A. Skeletal Neuromuscular Junction

The vertebrate NMJ (Fig. 1) is an example of an excitatory neuroeffector synapse designed for secure transfer of APs from the spinal and brainstem motor neurons to skeletal muscles. As a rule each muscle fiber is singly innervated but owing to axon branching a single motoneuron can innervate several to several hundred individual muscle fibers. The junction is formed by naked (unmyelinated) terminal arborizations of the axons that expand into bulbous structures or terminal **boutons.** The shape and size of contact between the axon terminals and the muscle fiber vary among species. In mammals the contact between the axon terminal boutons and the muscle fiber has a discoidal platelike shape, giving rise to the term **end plate.** In amphibians, the terminals spread longitudinally forming the so-called **end brushes.** However, the term *end plate* is commonly applied to all vertebrate NMJs.

The presynaptic nerve terminal is separated from the postsynaptic muscle fiber by about a 50-nm-wide synaptic cleft, and the junction is insulated by a Schwann cell. The terminal contains mitochondria and a large number of synaptic vesicles which contain the transmitter ACh. The synaptic vesicles tend to cluster at the presynaptic active zones. These are regions of presynaptic membrane specialized for directional exocytotic release of transmitter directly opposite the postjunctional folds that contain the receptors for ACh. The active zone is characterized by dense parallel bars on the internal face of presynaptic plasmalemma. The bars are defined by intramembranous particles that are distributed along the margins of each bar and thought to represent the calcium channels. Synaptic vesicles are arranged in rows along each presynaptic bar. The ACh receptors are concentrated in the crests of postjunctional folds, formed by the sarcolemma beneath the nerve ending. The dense material within the synaptic cleft is the basal lamina, which presumably provides an adherent for stabilizing the synapse. The material also contains the enzyme acetylcholinesterase (AChE), which hydrolyzes the re-

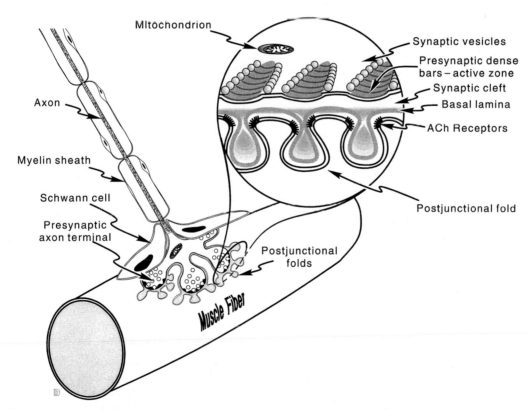

**FIG. 1.** Anatomy of motor end plate. The end plate is a specialized region of the muscle fiber in contact with the nerve terminal boutons. The approximately 50-nm-wide space between the presynaptic and postsynaptic membranes is the synaptic cleft. Terminals are covered with a thin layer of Shwann cells, which insulate the synapse. The nerve endings contain many synaptic vesicles that serve as storage sites for the transmitter ACh. The enlargement of a nerve terminal segment shows that some vesicles are aligned along the so-called "presynaptic" dense bars, which define the active zone of nerve terminal. Receptors for ACh are clustered at the crests of the postjunctional folds, opposite the presynaptic active zones. The basal lamina within the synaptic cleft is a connective tissue material that contains AChE, an enzyme that degrades ACh released into the synaptic cleft.

leased ACh and terminates its synaptic action (Couteaux and Pecot-Dechavassine, 1970; Dreyer *et al.*, 1973; Hall and Sanes, 1993).

## B. Neuro-Neuronal Synapses

Neuro-neuronal synapses are classified morphologically with respect to the parts of the presynaptic and postsynaptic neuron that form the junction (Fig. 2A). The axo-dendritic and axo-somatic synapses refer to junctions between the axon terminal of the transmitting neuron and a dendrite or somatic membrane of the receiving neuron, respectively. Other synaptic configurations include axo-axonic, dendro-dendritic, and soma-somatic connections. Typical ultrastructural features of a central synapse are illustrated in Fig. 2B. The presynaptic membrane frequently contains dense material forming the so-called "presynaptic vesicular grid." This structure is functionally equivalent to the active zone at the neuromuscular synapse and is thought to provide apparatus for anchoring synaptic vesicles at the plasmalemmal release sites. A thickening of postsynaptic membrane, the so-called "postsynaptic density," is also frequently observed. Although its precise function is not

known, it probably plays a role in aggregating specific neurotransmitter receptors at the subsynaptic membrane. The synaptic cleft at neuro-neuronal synapses is about 20 nm wide and usually contains fibrous material, which probably serves as an adherent for stabilizing the synapse. For a detailed discussion of synaptic ultrastructure, the reader may want to consult the articles in Pappas and Purpura (1972).

## III. Overview of Chemical Neurotransmission Process

### A. Neurotransmitter Substances

Several criteria define a chemical substance as a neurotransmitter: (1) The biosynthetic enzymes for the synthesis of the substance must be present in the identified presynaptic neuron to catalyze the synthesis of transmitter in the nerve terminals. (2) The substance must be released by stimulation of the presynaptic neuron in a $Ca^{2+}$-dependent manner. (3) Exogenous application of the substance to the postsynaptic cell must mimic the actions of a neurally released substance. (4) A specific mechanism for inactivation of the transmitter substance, such as a selective uptake system in the presynaptic terminals and presence of degradative enzyme(s), must be present at the synapse investigated.

In the CNS, the major transmitter at the excitatory synapses is the amino acid glutamate. The amino acids γ-aminobutyric acid (GABA) and glycine are the common neurotransmitters associated with inhibitory synapses. In addition to amino acids, ACh, dopamine (DA), norepinephrine (NE), epinephrine, serotonin (5-HT), histamine, and adenosine triphosphate (ATP) are all recognized as neurotransmitters or neurotransmitter candidates. Acetylcholine and NE are the well-established transmitters at peripheral nervous system. In addition there is good evidence that in the vascular smooth muscle, ATP is coreleased and acts as cotransmitter with NE. Neurons may also costore and corelease various peptide hormones. These may act as cotransmitters or modulate the synaptic actions of conventional transmitters. Common low molecular weight transmitter substances are listed in Table 1 and their chemical structures are shown in Fig. 3. Examples of neuroactive peptides are given in Table 2.

### B. Transmitter Receptors

Neurotransmitter receptors are integral membrane proteins containing the transmitter recognition site and the transducer site. Binding of transmitter to the receptor activates the transducer site, which in turn initiates the physiological response. Two major receptor classes have been distinguished (Table 1). Conventional, fast synaptic transmission characterized by fast-onset brief-duration potentials, is mediated by ionophoric or channel-forming receptors. When the receptor forms part of the channel complex, the gating is said to be direct and the ionic channels are referred to as **ligand-gated** or **receptor-gated channels.** The

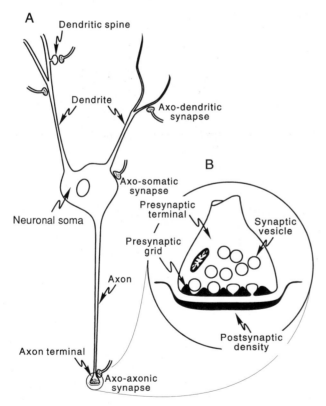

**FIG. 2.** Synapses of the CNS. (A) Examples of various synaptic configurations in the CNS include axo-dendritic, axo-somatic, and axo-axonic synaptic contacts. (B) Submicroscopic organization of nerve terminal bouton includes the characteristic presence of synaptic vesicles within the terminal cytosol and docked at the presynaptic "vesicular grid," which constitutes the active zone of the terminal. Postsynaptic density is illustrated by a thickening of the postsynaptic membrane. The synaptic cleft, which separates the presynaptic and postsynaptic elements, is about 20 nm wide.

**TABLE 1**  Common Neurotransmitters and Major Receptor Types

| Amino acidergic | Glutamic acid | **Kainate, AMPA, NMDA,** mGluR |
| | Gamma-aminobutyric acid (GABA) | **GABA$_A$,** GABA$_B$, **GABA$_c$** |
| | Glycine | **Gly-R** |
| Cholinergic | Acetylcholine | **nAChR,** mAChR, |
| Aminergic | Dopamine | D$_1$, D$_2$ |
| | Norepinephrine | $\alpha_1$, $\alpha_2$, $\beta_1$, $\beta_2$ |
| | Serotonin | 5HT$_1$, 5HT$_2$, **5HT$_3$** |
| | Histamine | H$_1$, H$_2$, H$_3$ |
| Purinergic | ATP, adenosine | P$_1$, **P$_{2x}$**, P$_{2y}$ |

Ionophoric receptors are indicated in boldface letters. Regular lettering indicates G-protein-coupled receptors.

**TABLE 2**  Examples of Neuroactive Peptides

| |
| --- |
| Angiotensin |
| Cholecystokinin |
| Enkephalin |
| Neuropeptide Y (NPY) |
| Neurotensin |
| Oxytocin |
| Somatostatin |
| Substance P (SP) |
| Thyrotropin-releasing hormone (TRH) |
| Vasoactive intestinal peptide (VIP) |
| Vasopressin |

ionophoric, cation channel-forming receptors include the muscle and neural nicotinic acetylcholine receptors; three pharmacologically distinguishable glutamate receptors that are selectively activated by analogues of glutamate, that is, kainate, $\alpha$-amino-3-hydroxy-5-methylisoxazole-4-propionic acid (AMPA), or $N$-methyl-D-aspartic acid (NMDA);

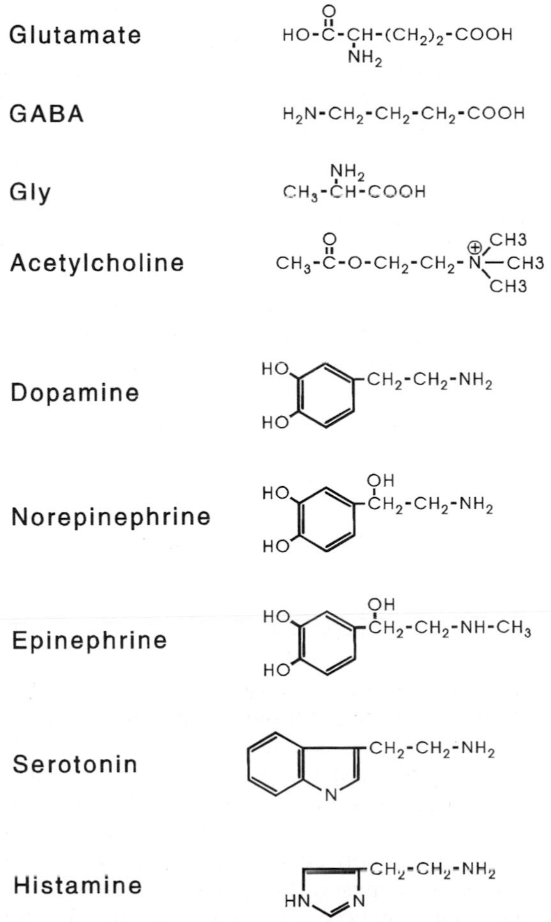

Glutamate

GABA

Gly

Acetylcholine

Dopamine

Norepinephrine

Epinephrine

Serotonin

Histamine

**FIG. 3.**  Structural formulas of common neurotransmitters.

the serotonin 5-HT$_3$ receptor; and the P$_{2x}$ purinoceptor for ATP. The anion (Cl$^-$) channel-forming receptors include the GABA$_A$ and GABA$_C$ receptors and the glycine receptors (Gly-R).

Slow synaptic transmission, characterized by slow-onset, long-lasting potentials, is mediated by distinct family proteins that do not contain intrinsic channel activity but transduce the transmitter binding into physiological response via G-protein-coupled mechanisms and second messenger systems. The synaptic actions of biogenic amines, norepinephrine, dopamine, serotonin, and histamine are exerted via the G-protein/second-messenger-coupled mechanisms systems. The G-protein-coupled receptors also include the muscarinic ACh receptors, metabotropic glutamate receptor (mGluR), GABA$_B$, and several receptors for adenosine and ATP.

### C. Functional Characteristics of Chemical Synapses

The sequence of events in chemical neurotransmission is summarized in Fig. 4. Transmission at chemical synapses is unidirectional from the presynaptic to postsynaptic element. The presynaptic terminals are specialized for transmitter biosynthesis, packaging of transmitter into the synaptic vesicles, and vesicular transmitter exocytosis. The transmitter receptors in the postsynaptic membrane transduce transmitter binding into ionic current, which generates voltage signal, the postsynaptic potential (PSP). The generation of the postsynaptic potentials is not instantaneous but rather registers 0.3–0.5 msec after the arrival of the action potential at the presynaptic terminal. This synaptic delay is a characteristic feature of chemical synapses and reflects, for the most part, the time required for the molecular events associated with the transmitter release.

As indicated earlier, transmitters are synthesized in the nerve ending of the presynaptic neuron and are concentrated and stored prior to release in synaptic vesicles. The intravesicular packet or quantum of transmitter is released by the process of Ca$^{2+}$-dependent exocytosis. This process is normally triggered by arrival of an AP and depolarization

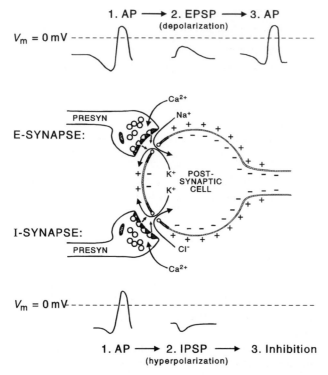

**FIG. 4.** Sequence of events in excitatory and inhibitory neurotransmission. (1) Presynaptic AP depolarizes nerve terminal. Depolarization-activated influx of $Ca^{2+}$ triggers exocytosis of transmitter from synaptic vesicles into the synaptic cleft. (2) Combination of transmitter with receptors in the postsynaptic membrane causes transient alteration of membrane permeability to ions and generation of a postsynaptic potential (PSP). In excitatory (E) synapses, change in the postsynaptic membrane permeability to $Na^+$ and $K^+$ ions results in a local depolarizing potential called *excitatory postsynaptic potential* (EPSP). In inhibitory (I) synapses, change in the postsynaptic membrane permeability to either $Cl^-$ or $K^+$ ions results in membrane hyperpolarization called an *inhibitory postsynaptic potential* (IPSP). Direction of ionic fluxes is indicated by arrows. (3) Depolarization of postsynaptic membrane to threshold will trigger a propagated AP. The likelihood of AP is diminished by activation of I synapses, because IPSPs hold or move membrane potential of the postsynaptic cell away from the threshold. (Modified from Goodman and Gilman's *The Pharmacological Basis of Therapeutics*, 1980.)

of the nerve terminal membrane. Depolarization opens voltage-dependent Ca channels in the nerve terminal membrane, allowing influx of $Ca^{2+}$ and activation of synaptic vesicle exocytosis. Following its release into the synaptic cleft, transmitter combines with specific receptors in the postsynaptic membrane, causing a change in its permeability to specific ions. Change in membrane permeability to ions gives rise to synaptic current, which, depending on the ions involved, depolarizes or hyperpolarizes the postsynaptic membrane. When the synaptic potential is depolarizing it is called an **excitatory postsynaptic potential (EPSP),** because it tends to bring the membrane potential toward the threshold for an AP. EPSPs are associated with a change in the postsynaptic membrane permeability to

$Na^+$ and $K^+$ ions and a net influx of positive charges (inward current carried by $Na^+$). When the synaptic potential is hyperpolarizing it is called an **inhibitory postsynaptic potential (IPSP),** because it tends to move or hold the membrane potential away from the threshold, thus decreasing the likelihood of an AP being fired. IPSPs are associated with an increase in the postsynaptic membrane permeability to $Cl^-$ or, less frequently, $K^+$ ions.

Transmitter release terminates within less than 1 msec, as AP decays and the terminal repolarizes. As the terminal repolarizes back to the resting potential, the depolarization-activated Ca channels reclose, Ca-influx ceases, and $[Ca]_i$ is rapidly lowered to the prestimulus level. Transmitter action on the receptors in the postsynaptic membrane is terminated by diffusion, reuptake, or enzymatic degradation. Reuptake via high-affinity transporters in the plasma membrane of the nerve terminals is the primary route of transmitter inactivation for most transmitters. The exception is acetylcholine, in which case, inactivation is primarily by means of degradation of ACh to inactivate acetate and choline by AChE.

### D. Integration versus Amplification

Central neurons usually receive several dozens to several thousand excitatory (E) and inhibitory (I) synaptic connections, which converge on the target neuron from a variety of other neurons in brain. Whether or not a neuron discharges a propagated AP is determined by the number and type of synapses active at any one time. The postsynaptic neurons integrate the postsynaptic potentials by adding the EPSPs and subtracting the IPSPs from the membrane potential at any instant of time, and APs are discharged when the summated membrane potential exceeds the threshold. Note that even in the absence of IPSPs, activation of a single excitatory synapse would be insufficient to discharge an AP, because individual EPSPs generated at any one synapse are always subthreshold. Therefore, summation of several EPSPs is usually necessary to bring the membrane of the postsynaptic neuron to the threshold potential.

In contrast to the integrative activity at central synapses, the function of neuromuscular junction is to transfer without failure the AP from the presynaptic motor neuron to the postsynaptic muscle fiber. In this case, the excitatory postsynaptic potential, called the end-plate potential (EPP) at the neuromuscular junction, is always suprathreshold and sufficient to trigger the muscle AP. Note that because the nerve terminal is very small in diameter compared with the muscle fiber it innervates, the AP cannot simply be transferred from nerve to muscle membrane. Even if these two membranes were continuous, the current generated during the invasion of presynaptic terminal by an AP would be insufficient to depolarize the postsynaptic membrane to threshold due to an impedance mismatch between the two membranes; that is, the small nerve terminal cannot provide enough action current to depolarize the large-diameter skeletal muscle fiber much more than about 1 mV. Thus, one function of transmitter at the neuromuscular junction is to amplify the presynaptic signal.

## IV. Presynaptic Processes

### A. Neurotransmitter Synthesis

Transmitters are synthesized locally in the nerve terminals and packaged into synaptic vesicles prior to release. The local synthesis of transmitter in the nerve terminals ensures that transmitter is available for refilling the vesicles after they have released their content of transmitter into the synaptic cleft.

The key enzyme(s) involved in transmitter synthesis are selectively expressed in the neurons that utilize the particular transmitter. The neurons that use ACh as the transmitter and the synapses formed by these neurons are called **cholinergic.** Acetylcholine is synthesized from two precursors, acetyl-coenzyme A (AcCoA) and choline (Ch), in a reaction catalyzed by choline acetyltransferase (ChAT):

$$AcCoA + Choline \xrightarrow{ChAT} ACh$$

Acetyl-coenzyme A derives from general cell metabolism, but Ch is not synthesized in neurons and is ultimately derived from the diet and delivered to neurons from the liver via the blood supply. The transport of Ch into nerve terminals is mediated by a specific high-affinity carrier system in the presynaptic plasma membrane. Transport of choline and synthesis of ACh are tightly coupled. The rate of Ch uptake and ACh synthesis increases during activity, ensuring adequate supply of the transmitter. Much of the Ch used in ACh synthesis is recovered from the synaptic cleft after the released ACh is enzymatically hydrolyzed by AChE to acetate and Ch.

Catecholaminergic neurons and their synaptic connections comprise dopaminergic, noradrenergic, and adrenergic systems. Dopamine (DA), norepinephrine (NE) and epinephrine (adrenaline) are synthesized from tyrosine by tyrosine hydroxylase (TH) in a reaction requiring $O_2$ and reduced pteridine as cofactors. The reaction product is L-dihydroxyphenylalanine (L-dopa), which is decarboxylated by a pyridoxal phosphate-dependent aromatic L-amino acid decarboxylase to dopamine. Dopamine is taken up into synaptic vesicles and serves as transmitter at dopaminergic synapses. In noradrenergic nerve endings, dopamine is further converted to norepinephrine by dopamine-$\beta$-hydroxylase (D$\beta$H). Methylation of norepinephrine to epinephrine is catalyzed by phenylethanolamine-$N$-methyltransferase (PNMT) in a reaction requiring S-adenosylmethionine as methyl donor:

$$Tyrosine \xrightarrow{TH} \text{L-dopa}$$
$$\text{L-dopa} \xrightarrow{Decarboxylase} DA$$
$$DA \xrightarrow{D\beta H} NE$$
$$NE \xrightarrow{PNMT} Epinephrine$$

Serotonergic neurons utilize serotonin as the transmitter. Serotonin is formed by hydroxylation of tryptophan to 5-hydroxytryptamine (5-HTP), followed by decarboxylation of 5-HTP to 5-hydroxytryptamine (5-HT, serotonin). The hydroxylation is catalyzed by serotonergic neuron-specific tryptophan hydroxylase:

$$Tryptophan \rightarrow 5\text{-HTP} \rightarrow 5\text{-HT}$$

Histamine serves as the transmitter at histaminergic synapses. It is formed by decarboxylation of histidine. The decarboxylase that catalyzes this process appears to be characteristic of histaminergic neurons.

Amino acidergic transmission includes the excitatory glutamatergic and inhibitory GABAergic and glycinergic systems. Both glutamic acid and glycine are common constituents of a large amino acid pool found in all cells. Therefore, their selective use as neurotransmitters at glutamatergic and glycinergic synapses implies a subcompartmentation in nerve terminals of respective amino acidergic neurons dedicated for the neurotransmitter function. This subcompartment in all likelihood corresponds to synaptic vesicles, which presumably selectively accumulate glutamic acid and glycine from the cytosol and release them during synaptic transmission.

GABA is synthesized by decarboxylation of glutamate in a reaction catalyzed by glutamic acid decarboxylase (GAD), which is selectively localized in the GABAergic neurons and their terminals:

$$Glutamate \xrightarrow{GAD} GABA$$

### B. Storage of Neurotransmitters in Synaptic Vesicles

Neurotransmitters are taken up into synaptic vesicles by specific transporters. The uptake of transmitters is energetically coupled to electrochemical proton gradients, generated by electrogenic proton pump in the vesicular membrane. Selectivity of the vesicular uptake may be particularly important at amino acidergic nerve terminals, where transporter in synaptic vesicles must distinguish between the amino acid that is utilized as the neurotransmitter at the particular synapse from the general metabolic pool of amino acids that are not (Maycox *et al.*, 1990).

Cholinergic vesicles have been purified from the electric organ of torpedinid fish (rays) and from mammalian brain. The vesicles from mammalian tissue are approximately 40–50 nm in diameter and contain an estimated 2000 to 10,000 molecules of ACh. The vesicles isolated from *Torpedo* are larger (ca. 85 nm) and contain a correspondingly greater amount of ACh. In addition to taking up ACh against large concentration gradient, vesicles take up and store ATP, which may serve as a counterion for ACh to prevent its loss from vesicles. ATP may also act as cotransmitter with ACh. The major functional proteins identified in the cholinergic vesicles include ACh carrier, proton-ATPase, Ca/Mg-activated ATPase, and ATP carrier (Whittaker, 1987).

Catecholamine-storing vesicles have been isolated from mammalian sympathetic nerves and adrenal glands. The sympathetic nerve endings contain two types of vesicles: large dense-cored vesicles (LDVs) and small dense-cored vesicles (SDVs). The LDVs from sympathetic nerves closely resemble adrenal chromaffin granules in function and composition. They contain catecholamines, ATP, ascorbic acid, chromogranins, and several neuroactive peptides. The membranes of catecholaminergic vesicles contain proton pump, amine carrier, nucleotide carrier, D$\beta$H, and cytochrome $b_{561}$. Unlike the LDVs, the soluble contents of SDVs consist of catecholamines and nucleotides, but do not contain peptides. The SDVs may be derived from LDV and may participate in multiple cycles of transmitter exocytosis and reuptake within the terminal (Winkler *et al.*, 1988).

The amino acid-containing synaptic vesicles have been less well characterized. Nevertheless, accumulating evidence supports the notion that these vesicles selectively accumulate the amino acid neurotransmitters and release them by exocytosis.

## C. Transmitter Release

### 1. Quantal-Vesicular Hypothesis of Transmitter Release

The evolution of the current understanding of the mechanism of transmitter release began with the formulation of the quantal-vesicular hypothesis of transmitter release by Katz and his collaborators (Fatt and Katz, 1952; delCastillo and Katz, 1954). Katz and colleagues observed that in resting NMJ, that is, in the absence of stimulation, nerve terminals spontaneously release ACh, giving rise to small depolarizations that occur at random intervals and average about 0.5 mV in amplitude (Fig. 5A). These small potentials behaved in all respects like miniature replicas of the

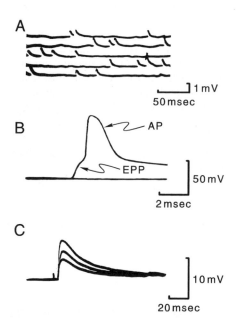

**FIG. 5.** Intracellular recordings from the end plate at the frog neuromuscular junction. (A) Spontaneous MEPPs recorded from a resting (not stimulated) junction. Note that these are small depolarizing potentials less than 1 mV in amplitude and occur randomly. (B) Postsynaptic response to presynaptic action potential. The initial hump on the recorded waveform is the EPP elicited by ACh released by the presynaptic AP. Note that the EPP is a large depolarization (>40 mV), sufficient to bring the end plate to threshold and trigger a muscle AP. (C) Fluctuations in EPPs when output of transmitter has been reduced by adding 10 m*M* Mg$^{2+}$ to bathing medium. (Adapted from Fatt and Katz, 1952; and del Castillo and Katz, 1954.)

EPPs evoked by presynaptic APs (Fig. 5B) and therefore were called **miniature end-plate potentials (MEPPs).** The crucial insight into the relationship between MEPPs and EPPs was provided by the observation that when evoked release was reduced in low Ca$^{2+}$/high Mg$^{2+}$ solutions, the size of EPPs fluctuated in a random manner (Fig. 5C) such that the EPP amplitudes appeared to be made up of integral multiples of the average MEPP amplitude and could be described by a Poisson distribution (Fig. 6). Katz concluded that transmitter release is a stochastic process, consisting of random release of multimolecular packets or quanta of ACh each producing a unit response (MEPPs), and that the EPP is a summation of many quantal units released nearly synchronously by presynaptic AP.

For detailed discussion of the statistics of transmitter release the interested reader is referred to Martin (1977). For the present purpose it suffices to state that transmitter release can be described by a simple statistical expression:

$$m = nP \tag{1}$$

The parameter $m$ is the average number of quanta released per presynaptic impulse when a large number of trials are performed and is called the **quantal content** of the EPP. The parameter $n$ represents the number of quanta immediately available for release and most likely corresponds to either the population of synaptic vesicles associated with the presynaptic active zones or the number of release sites. Parameter $P$ is the probability of any single quantum being released and primarily reflects the probability of a productive Ca-dependent vesicle fusion with the plasmalemma as a function of Ca$^{2+}$ concentration at the release sites. Reducing the availability of Ca$^{2+}$ to enter terminals would reduce $P$, thus accounting for a reduction of transmitter release in low Ca$^{2+}$/high Mg$^{2+}$ media. Conversely, increasing the concentration of Ca$^{2+}$ at or near the release sites would tend to enhance the probability of exocytosis, that is, facilitate transmitter release.

### 2. Essential Role of Ca$^{2+}$ in Depolarization-Release Coupling

The voltage-gated Ca channels in the presynaptic plasma membrane couple membrane depolarization to transmitter exocytosis. The essential role of Ca$^{2+}$ influx in transmitter release was demonstrated by Katz and Miledi (1967) who showed that depolarization of presynaptic terminals fails to evoke transmitter release when Ca$^{2+}$ is absent or its entry into nerve terminals is prevented by high Mg concentration in the extracellular medium (Fig. 7). In subsequent experiments carried out at the squid giant synapse, where it is possible to make intracellular recordings from both pre- and postsynaptic cells simultaneously, Miledi (1973) showed that injection of Ca$^{2+}$ into presynaptic terminals elicited transmitter release, thus providing direct evidence that a rise in intracellular Ca$^{2+}$ alone is sufficient for activation of the release process. Furthermore, using voltage clamping to control the membrane potential of the presynaptic terminals at the squid giant synapse, Llinas (1977) showed that the quantity of synaptic transmitter released is a function of the size of the presynaptic Ca current,

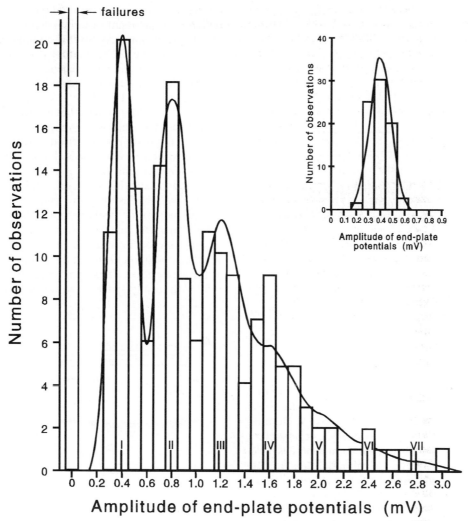

**FIG. 6.** Distribution of EPP amplitudes recorded from mammalian NMJ. Recordings of evoked potentials were obtained under conditions of reduced transmitter release in high (12.5 m*M*) $Mg^{2+}$. Inset shows a histogram of spontaneous potentials (MEPPs) recorded from a resting junction. Note that EPP amplitudes group around multiples of mean MEPP amplitude; the number of failures (0 quanta released) and the distribution of single, double, triple, etc., quantal responses observed experimentally fit the theoretical distribution (solid curve) calculated from the Poisson equation. (From Boyd and Martin, 1956.)

which in turn depends on the extent of nerve terminal depolarization (Fig. 8).

### 3. Exocytosis and Recycling of Synaptic Vesicles

The active zones in the presynaptic nerve terminals provide plasmalemmal specializations for synaptic vesicle docking and exocytosis. In ingenious experiments, Heuser *et al.* (1979) employed specially constructed quick-freeze apparatus that enabled them to capture images of vesicle exocytosis at the active zone of the NMJ. The comparison of the number of exocytotic pores captured in quick-freeze experiments correlated with the estimated number of quanta released under similar conditions of stimulation. Moreover, statistical analysis of synaptic vesicle discharge indicated that individual vesicles fuse independently of

each other, as predicted by the quantal hypothesis of transmitter release. These experiments provided the most direct evidence that exocytosis of synaptic vesicles at the active zone region is the likely mechanism of quantal transmitter release (Fig. 9).

Electrophysiological and ultrastructural studies indicate the existence of two distinct, the so-called "readily available" and "reserve" pools of synaptic vesicles. The readily available pool is thought to correspond to the vesicles docked at the active zone, whereas the reserve pool comprises the vesicles distributed within the terminal at some distance from the active zone. It is evident that to maintain the availability of quanta for release, exocytosis must be accompanied by mobilization of new vesicles from the reserve pool within the terminal to the plasmalemmal release sites. Mobilization of vesicles is thought to involve alter-

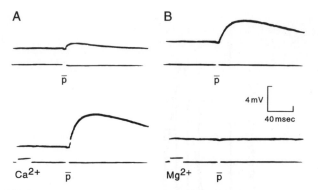

**FIG. 7.** Calcium is required for transmitter release evoked by depolarization of nerve terminals. (A) In the absence of $Ca^{2+}$, application of a depolarizing pulse ($p$) to the presynaptic nerve endings fails to evoke transmitter release and postsynaptic response is nearly absent (top). Application of $Ca^{2+}$ to the neuromuscular junction shortly before the stimulus ($p$) to the presynaptic nerve (lower trace) supports transmitter release as shown by appearance of postsynaptic response. (B) Normal postsynaptic response in $Ca^{2+}$-containing media (top trace) is blocked by applying high $Mg^{2+}$ before nerve stimulation. The failure to elicit transmitter release in the presence of high $Mg^{2+}$ is due to block of $Ca^{2+}$ influx into the terminals. (Adapted from Katz and Miledi, 1967.)

ation in vesicle cytoskeleton interactions. One of the proteins that may play a role in this process is the vesicle-associated family of phosphoprotein synapsins. Dephosphorylated synapsin seems to stabilize vesicle–cytoskeletal interactions and inhibit vesicle mobilization, whereas phosphorylation of synapsin by the Ca-calmodulin-dependent protein kinase II is thought to promote vesicle mobilization (Llinas *et al.*, 1991).

The mechanism of vesicle attachment and initiation of exocytosis by $Ca^{2+}$ has not yet been completely worked out; however, remarkable progress has been made in this direction during the past few years. Studies in model systems such as mast cells or chromaffin cells indicate that secretion is accompanied by a stepwise increase in cell capacitance, consistent with fusion of secretory granules with the cell membrane. Electrical measurements further suggest that the first event in exocytosis may be formation of a pore that connects vesicle lumen with the extracellular space and may provide a channel for the release of soluble contents into the synaptic cleft (Almers, 1990). Several proteins have been identified that are likely to be critically involved in vesicle docking, activation, and fusion (Südhof and Jahn, 1991; Bennett and Scheller, 1993; Söllner *et al.*, 1993; Niemann *et al.*, 1994; Südhof, 1995). The vesicle membrane proteins synaptobrevin and plasma membrane proteins syntaxin and a 25-kDa synaptosomal-associated protein (SNAP-25), have been implicated in formation of a macromolecular core complex, which may function to anchor the vesicles near the calcium channels at the active zone. Selective proteolytic cleavage of either protein by clostridial neurotoxins prevents assembly of the complex and abolishes transmitter release. The core complex is thought to be primed for exocytosis through its interactions with the soluble constitutive fusion factors NSF (*N*-ethyl-

maleimide-sensitive factor) and the associated SNAPs (soluble NSF attachment proteins). Synaptotagmin, a synaptic vesicle membrane protein that contains two C2 motifs homologous to the C2 regulatory domain of protein kinase C, may serve as the $Ca^{2+}$ receptor of exocytosis. Another protein that has been implicated in exocytosis is neurexin, which provides a target for $\alpha$-latrotoxin, a component of black widow spider venom that induces massive transmitter exocytosis. Together with synaptotagmin with which it associates, neurexin may function to transduce the Ca signal into activation of exocytosis. Finally, the synaptic vesicle membrane protein synaptophysins have the capacity to form channels in lipid bilayers and could be involved in making vesicle fusion pore, perhaps in association with plasma membrane proteins physophilins (Fig. 10).

Following the release of transmitter, the process of exocytosis may be terminated in at least two ways. One is simply through reclosure of the pore and fission of vesicles at the active zone. These postexocytotic vesicles may then refill with locally synthesized transmitter and by virtue of

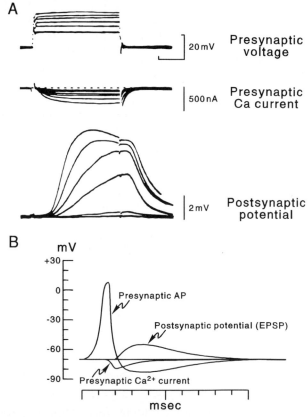

**FIG. 8.** Relationship between $Ca^{2+}$ influx and transmitter release. (A) Experiments in squid giant synapse illustrating the relationship between magnitude of presynaptic $Ca^{2+}$ current and transmitter release monitored by recording postsynaptic potentials. Graded depolarizations of presynaptic terminals (top trace) are associated with graded inward $Ca^{2+}$ currents (middle trace) and correlated with graded postsynaptic potentials (bottom trace) that reflect the amount of transmitter released. (B) Relationship between presynaptic AP, presynaptic Ca current, and the postsynaptic potential, reconstructed from experiments at the squid giant synapse. (Adapted from Llinas, 1977.)

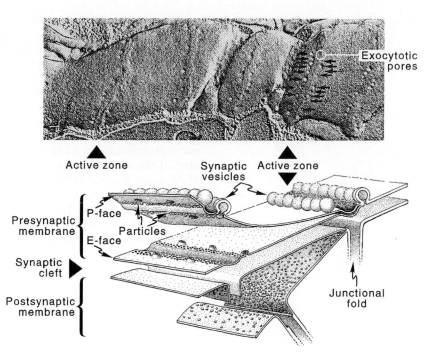

**FIG. 9.** Synaptic vesicle exocytosis captured by rapid-freezing technique. (Top) P-face of freeze-fractured replica of a stimulated and rapidly frozen nerve terminal showing images of synaptic vesicles caught in the process of exocytosis at the active zones. The junction was activated in presence of 1 m$M$ 4-aminopyridine to enhance the level of transmitter release per single presynaptic impulse. (Bottom) Three-dimensional representation of pre- and postsynaptic membranes in relation to the freeze-fracture image above. The freeze-fracture splits membrane bilayer into protoplasmic (P-face) and extracellular (E-face) leaflets of the bilayer. The protruding particles seen in P-face and the corresponding pits in the E-face are thought to be the presynaptic Ca channels. The openings in the P-face and their craterlike continuations in the E-face are the exocytotic pores formed when synaptic vesicles fuse with the plasmalemma along the parallel bars of the acitve zone. [Adapted from Heuser *et al.*, 1979 (top) and Kuffler, and Nicholls, 1977 (bottom). Micrograph produced by Dr. John Heuser, Washington University, St. Louis. Reproduced from *Journal of Cell Biology*, 1979, vol. 81, pp. 275–280, by copyright permission of Rockefeller University Press.]

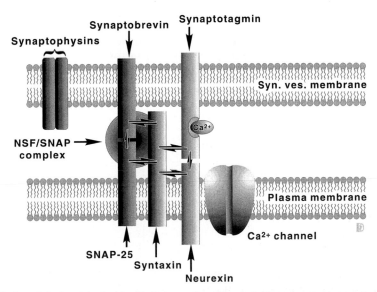

**FIG. 10.** A simplified model of vesicle docking/fusion complex. The model incorporates interactions among the proteins of synaptic vesicles (synaptobrevins and synaptotagmins), plasma membrane proteins (syntaxins, SNAP-25, neurexin), and the elements of constitutive fusion machinery (NSF, SNAP). The complex is thought to form in proximity to the plasmalemmal Ca channels, with synaptotagmin serving as the key Ca sensor. The vesicle membrane proteins synaptophysins and plasma membrane proteins physophylins (not shown) may participate in the formation of vesicle fusion pore.

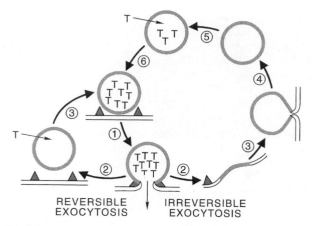

**FIG. 11.** Synaptic vesicle cycling in nerve endings. Following vesicle docking and fusion (step 1), synaptic vesicles can recycle through a short pathway (steps 2 and 3 on the left) or via the more complicated sequence of steps 2–6 on the right. The short pathway involves transient fusion between vesicle and plasma membrane, followed by fission and recharging of the emptied vesicle with new transmitter (T). In the long pathway, vesicle membrane collapses into the plasmalemma to be eventually retrieved and reformed within the terminal, and reloaded with new transmitter.

being already positioned at the active zone, release the newly formed transmitter in preference to the reserve pool. This could explain the well-documented phenomenon of the preferential release of newly synthesized transmitter. A more generally accepted idea is that following the fusion and exocytosis, vesicle membrane collapses into the presynaptic plasmalemma and moves laterally away from the active zone to be then retrieved by endocytosis and reformed via a series of transformations into functional vesicles (Heuser and Reese, 1973) (Fig. 11). It is possible that both of these processes occur, but the extent to which one predominates over the other may be a function of the rate of presynaptic stimulation and possibly other factors. The evidence for local recycling of synaptic vesicles in the nerve terminals has been provided by the observation that vesicles become labeled with high molecular weight markers such as horseradish peroxidase (HRP) or dextrans when nerve terminals are stimulated in the presence of these markers in the extracellular medium (Ceccarelli *et al.*, 1973). More recent evidence for vesicle recycling has been provided by tracking the movement of fluorescently labeled synaptic vesicles during stimulation of motor nerve terminals at the frog NMJ (Betz and Bewick, 1992).

The vesicles undergoing local recycling within the nerve terminals are progressively degraded and replaced by vesicles that are formed *de novo* in the cell body and transported to the nerve terminals by the fast axoplasmic transport. The half-life of vesicles has been estimated at 7–14 days. Evidently, synaptic vesicles can undergo numerous cycles of transmitter release–reloading before being replaced with new ones.

## V. Postsynaptic Processes

The nature of postsynaptic responses is determined by the type of receptor-gated ionic conductances that are activated in the postsynaptic membrane. At excitatory synapses, the transmitters activate receptor-gated channels that conduct cations, principally $Na^+$ and $K^+$, but exclude anions. The net current through the synaptic channels is inward and carried by $Na^+$ ions, causing a depolarization of the postsynaptic membrane, or EPSP. The EPSPs generated at the skeletal NMJ are called **end-plate potentials (EPPs)** and those at other peripheral synapses, for example, at nerve-smooth muscle junctions, are frequently referred to as **excitatory junctional potentials (EJPs).**

At inhibitory synapses, transmitters activate receptor-gated channels that conduct $Cl^-$ ions. Influx of chloride ions through the synaptic channels tends to increase the negativity of the cell interior and hyperpolarize the postsynaptic membrane. The hyperpolarizing postsynaptic potentials are called inhibitory postsynaptic potentials because they tend to move the membrane potential away from the threshold.

### A. The End-Plate Potential at the Skeletal Neuromuscular Junction

The EPP at the NMJ is generated when ACh released from the presynaptic motor nerve terminal activates the postjunctional nAChR. When recorded focally at the end plate, EPP is seen to rise rapidly, reaching its peak within 1–2 msec, and then declines exponentially with time. The amplitude and time course of EPPs decline with distance, as the recording electrode is moved away from the end plate (Fig. 12). Thus, EPP is a local, electrotonic potential whose sole function is to depolarize the end-plate region to the critical membrane potential (threshold) at which the propagating muscle AP is initiated. In normal, healthy NMJ, the EPP is always of sufficient amplitude to trigger the muscle AP, and the NMJ is said to have a high safety factor.

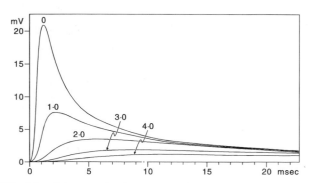

**FIG. 12.** Electrotonic character of EPPs. Recordings were obtained at the frog NMJ at the indicated distances (in millimeters) from the end plate (0 distance). Note that as the electrode is moved away from the end plate, the amplitudes of EPPs decrease and their rise time and decay are slowed. (Adapted from Fatt and Katz, 1952.)

### 1. The Molecular Structure and Cation-Gating Function of the Nicotinic Acetycholine Receptor

The nAChR is a transmembrane cation channel-forming complex. The distribution of the receptor at the end plate has been analyzed by quantitative electron microscopic autoradiography of $\alpha$-[$^{125}$I]bungarotoxin ($\alpha$-BGT) binding, a polypeptide component of Formosan krait venom, which binds with high affinity and specificity to the receptor. The receptors are concentrated at the juxtaneuronal crests of the postjunctional folds. The density of $\alpha$-BGT sites in the sternomastoid muscle in mice has been estimated at $30,000 \pm 8000 \ \mu m^{-2}$ (Fertuck and Salpeter, 1974). Because each receptor contains two toxin binding sites, there are approximately 15,000 receptors per square micrometer of the subsynaptic membrane. The density of the receptors decreases steeply with distance away from the end plate, explaining why the extrajunctional membrane is chemically inexcitable (i.e., cannot be activated by ACh).

The molecular structure of nAChR (Fig. 13) has been determined most thoroughly for the nAChR from the electric organ of the ray *Torpedo,* a tissue that is particularly abundant in the receptors. The receptor consists of four homologous polypeptide chains ($\alpha$, $\beta$, $\gamma$, and $\delta$) assembled into a pentameric complex ($\alpha_2\beta\gamma\delta$) that forms a transmembrane funnel-like pore. The channel is approximately 6.5 $\times$ 6.5 Å at its narrowest point and contains three negatively charged rings formed by glutamic and aspartic amino acid residues. The negatively charged rings serve as selectivity filters that exclude anions but allow passage of Na$^+$ and K$^+$ and other cations less than 6.5 Å in diameter. The ACh recognition site is on the $\alpha$-subunit. The quaternary structure of the nicotinic receptor in mammalian adult muscle is similar except for a replacement of the $\delta$-subunit by a homologous $\varepsilon$-subunit.

The nAChR exists in either closed (nonconducting) or open (conducting) states. The binding of two ACh mole-

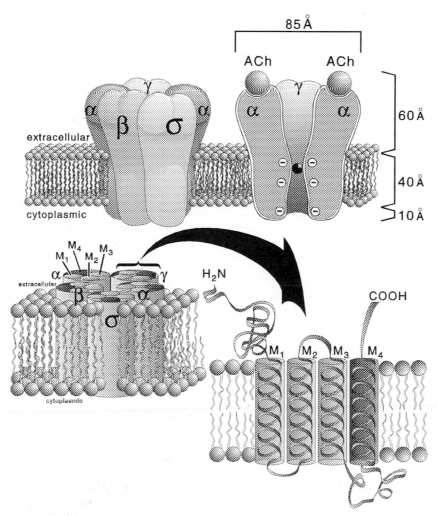

**FIG. 13.** A model of nAChR. The receptor is a transmembrane channel-forming complex composed of five subunits. Each subunit contains four membrane-spanning helical domains denoted M1, M2, M3, and M4. The M2 domains are thought to line the channel pore. ACh recognition site is on the $\alpha$-subunit and there are two $\alpha$-subunits per receptor complex. The gate, which screens out the negative charges but allows passage of cations such as Na$^+$ and K$^+$, lies within the pore. (Based on Numa, 1989; Gadzi *et al.,* 1991; modified from Kandel *et al.,* 1991.)

cules to the receptor induces a conformational change to the open state:

$$2ACh + R \leftrightarrow ACh_2R_{close} \leftrightarrow ACh_2R_{open} \quad (2)$$

Patch-clamp recordings of single-channel currents indicate that the mean channel open time is about 1 msec and the single-channel conductance is about 25 pS, allowing a transfer of approximately $1.5 \times 10^4$ Na$^+$ and K$^+$ ions/msec channel open time and producing a depolarization of the postsynaptic membrane by about 0.3 $\mu$V from its resting potential.

### 2. Synaptic Current and Equilibrium Potential

The current ($i_s$) flowing through a single ACh-activated channel is determined by the channel conductance ($\gamma_s$) and the electrochemical driving force ($V_m - E_s$) acting on the ions moving through the channel:

$$i_s = \gamma_s(V_m - E_s) \quad (3)$$

where $V_m$ and $E_s$ are membrane potential and the synaptic equilibrium potential, respectively.

In resting synapse, most of the ACh-gated channels are closed and the conductance of the postsynaptic membrane to ions is very low. When ACh is released by presynaptic impulse, it binds to its receptors in the postsynaptic membrane and opens the associated channels for a short, 1–2 msec duration. The resultant total synaptic (end-plate) current, $I_s$, is a sum of currents through all opened channels:

$$I_s = n\gamma_s(V_m - E_s) = g_s(V_m - E_s) \quad (4)$$

where $n$ is the number of channels that are activated and $g_s$ is the total synaptic (end-plate) conductance ($n\gamma_s$).

Because the channels permit the passage of both Na$^+$ and K$^+$ ions with about equal ease, the synaptic current is the sum of Na$^+$ and K$^+$ currents:

$$I_s = I_{Na} + I_K = g_{Na}(V_m - E_{Na}) + g_K(V_m - E_K) \quad (5)$$

where $g_{Na}$ and $g_K$ are the Na$^+$ and K$^+$ ion conductances, and $E_{Na}$ and $E_K$ are the Na$^+$ and K$^+$ equilibrium potentials, respectively. The direction and magnitude of ion flux is determined by the electrochemical driving forces ($V_m - E_i$). Because $g_{Na}(V_m - E_{Na}) > g_K(V_m - E_K)$ (i.e., $I_{Na} > I_K$), the net synaptic current is always inward and carried by Na$^+$ ions, resulting in membrane depolarization. As the membrane potential ($V_m$) becomes depolarized, the term ($V_m - E_{Na}$) decreases and ($V_m - E_K$) increases until equilibrium is reached, where the Na$^+$ inward current is exactly equal to the outward K current,

$$I_{Na} = -I_K \quad (6)$$

or

$$g_{Na}(V_m - E_{Na}) = -g_K(V_m - E_K) \quad (7)$$

The membrane potential $V_m$ at which this occurs is the end-plate equilibrium potential ($E_{EPP}$). By solving Eq. (7) for $V_m$, we see that $E_{EPP}$ is a weighted average of sodium and potassium equilibrium potentials:

$$E_{EPP} = [g_{Na}/(g_{Na} + g_K)]E_{Na} + [g_K/(g_K + g_{Na})]E_K \quad (8)$$

where $E_{EPP}$ is the limiting potential to which the postsynaptic membrane can be depolarized during the action of transmitter. Any further depolarization beyond this point would result in $I_K > I_{Na}$ and reversal of net end-plate current from inward to outward direction. Therefore, the $E_{EPP}$ is also called a reversal potential ($E_r$). Since $g_{Na} \sim g_K$, $E_{Na} \sim$ +60 mV, and $E_K \sim -90$ mV, the $E_{EPP}$ at vertebrate NMJ can be calculated at about $-15$ mV.

### 3. Relationship between the Synaptic Current and Depolarization of the Postsynaptic Membrane

The relationship between the synaptic current and depolarization of the adjacent muscle membrane can be analyzed in terms of equivalent electrical circuit consisting of parallel synaptic and nonsynaptic branches (Fig. 14). The synaptic branch represents the synaptic receptor-gated channels ($g_s$) in series with the synaptic battery $E_{EPP}$. The nonsynaptic branch consists of membrane capacitance $C_m$ and leakage channels ($g_m$) in series with the battery of the resting membrane potential, $E_m$. During activation of the end plate by ACh, the end-plate current ($I_s$) flows inward through the synaptic branch and outward through the parallel capacitive and resistive elements of the nonsynaptic branch as $I_c$ and $I_m$, respectively:

$$I_s = -(I_c + I_m) \quad (9)$$

At the onset of synaptic action most of the synaptic current flows through the capacitive branch because the outward driving force ($V_m - E_m$) on current flow through the nonsynaptic channels ($g_m$) is small. The deposition of positive charges on the interior of the lipid bilayer and removal of an equal number of positive charges from the outside results in the discharge of membrane capacitance and depolarization. Once the membrane capacitance is discharged to its final value, all the synaptic current exits through the leakage channels ($g_m$). Thus at the peak of synaptic activation, $I_c = 0$ and

$$I_s = -I_m \quad (10)$$

where

$$I_m = g_m(V_m - E_m) \quad (11)$$

Substituting Eqs. 4 and 11 into Eq. 10 and solving for $V_m$, one obtains the expression for the membrane potential at the peak of EPP:

$$V_m = [g_s/(g_s + g_m)]E_{EPP} + [g_m/(g_s + g_m)]E_m \quad (12)$$

The difference between $V_m$ and $E_m$ ($V_m - E_m$) is the amplitude of the EPP ($\Delta V_{EPP}$).

Equation 12 shows that the value of membrane potential $V_m$ at the peak of synaptic activation is a weighted average of $E_{EPP}$ and $E_m$, where the weighting factors are the relative magnitudes of synaptic ($g_s$) and nonsynaptic ($g_m$) conductances. During peak activation of synaptic ACh channels, $g_s \gg g_m$ and the membrane potential will tend toward the $E_{EPP}$; that is, to about $-15$ mV. However, the amplitude of the EPP will be "loaded down" by th e$g_m$ of the nonsynaptic membrane so that the actual $V_m$ will be somewhat less than the theoretical value of $E_{EPP}$. For example, at

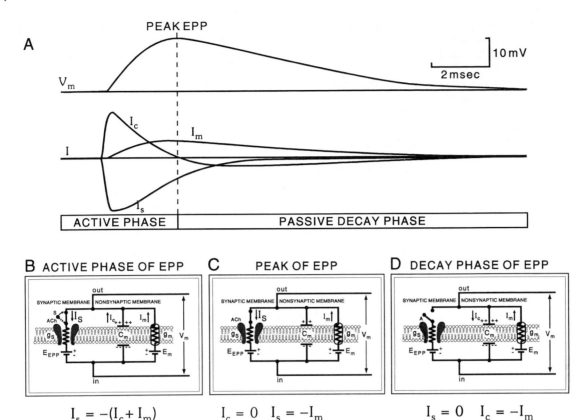

**FIG. 14.** Electrical equivalent circuit description of EPP. (A) Postsynaptic potential (EPP) and the underlying membrane currents: synaptic current $I_s$, capacitive current $I_c$, and nonsynaptic membrane current $I_m$. (B and C) Equivalent circuit representations of the three phases of end plate. (B) During the active phase the synaptic switch s is closed (synaptic channels are open) and $I_s$ flows in through the synaptic branch ($g_s$) and out as $I_c$ and $I_m$ through the nonsynaptic branch. (C) At the peak of EPP, capacitance $C_m$ has been discharged to its final (steady-state) value, $I_c = 0$ and $I_s = -I_m$. (D) During the decay phase, the membrane repolarizes back to resting potential. Synaptic channel has reclosed (switch open) and $I_s = 0$. The current through the nonsynaptic membrane consists of outward $I_m$ and inward $I_c$, which recharges the membrane capacitance and repolarizes the membrane back to the resting potential. (Modified from Suszkiw and Sperelakis, 1993; based on Kandel *et al.*, 1991.)

the vertebrate NMJ the conductance of resting muscle membrane ($g_m$) is about $5 \times 10^{-6}$ S. During activation by ACh the conductance of the end plate ($g_s$) increases to $5 \times 10^{-5}$ S. Assuming that $E_m = -90$ mV and $E_{EPP} = -15$ mV, one obtains $V_m = -23.8$ mV, which is about 7 mV more negative than the theoretical value. Nevertheless, the depolarization to $-23$ mV far exceeds that necessary threshold for muscle AP (about $-60$ to $-50$ mV); that is, transmission at healthy NMJ is very secure so that each presynaptic potential normally will always elicit a muscle AP.

### 4. Decay of End-Plate Potential and Termination of Synaptic Action of Acetylcholine

The synaptic current rises rapidly during the dynamic phase of EPP and then decays exponentially as the channels spontaneously reclose. The time constant of this process is about 1 msec, that is, the same as the average channel open time determined from statistical analysis of single-channel kinetics. The decay of EPP reflects an exponential repolarization of the membrane, that is, recharging the

membrane capacitance back to its resting membrane potential. The rate of EPP decay is determined by the resistance capacitance (RC) time constant ($\tau_m$) of the muscle membrane.

During approximately the same time, the concentration of ACh in the synaptic cleft diminishes rapidly due to AChE-catalyzed hydrolysis of ACh to inactive acetate and choline. Rapid lowering of ACh concentration in the cleft diminishes the probability of repetitive activation of receptors and effectively terminates its synaptic action. The importance of the enzymatic degradation of ACh for normal transmission is illustrated by the fact that inhibition of AChE and consequent buildup of ACh in the synaptic cleft results in increased and prolonged EPP, leading initially to multiple discharges of muscle AP and eventual block of synaptic transmission.

### 5. Genesis of Muscle Action Potential

Unlike the propagated AP, the EPP is a locally generated electrotonic potential that passively decays with time and distance according to time and length constants of the muscle membrane. Because of the cable properties of the muscle fiber with a length constant of about 1 mm (the distance at which the EPP will decay to 37% of its peak value at the end plate) the EPP essentially disappears at a distance >2 mm from the end plate (see Fig. 12). However, the local depolarization of the electrically excitable membrane in the immediate vicinity of the end plate is more than sufficient to reach the threshold potential (i.e., about −60 mV) so that a muscle potential is triggered. Because the NMJ is usually located somewhere near the middle of a muscle fiber, the APs triggered by EPPs are propagated toward each end of the fiber. This propagating wave of depolarization elicits muscle twitch (depolarization–contraction coupling).

### B. Excitatory and Inhibitory Potentials in Central Synapses

The NMJ is an example of a one-to-one excitatory synapse designed to ensure that every presynaptic AP will give rise to muscle AP and muscle contraction. This is accomplished by release of several hundreds of transmitter (ACh) quanta sufficient to generate suprathreshold depolarization of the postsynaptic muscle membrane. In contrast, activation of a single excitatory synapse in brain produces a relatively small change in the membrane potential of the postsynaptic cell so that several excitatory synaptic inputs must be active in order to bring the neuron to its threshold for firing an AP. Furthermore, the effectiveness of the excitatory synaptic inputs can be diminished or nullified by activation of inhibitory synaptic inputs. Thus, whereas the sole function of the skeletal NMJ is a secure transmission of neural command signals to the muscle for the purpose of force generation, the function of central synapses is the integration of many inputs converging onto the postsynaptic cell.

### 1. Excitatory Postsynaptic Potentials

The ionic mechanism of EPSPs at neuro-neuronal synapses is analogous to that described for EPP at the neuromuscular junction. EPSPs are generated when a transmitter released from a presynaptic terminal activates specific postsynaptic receptors to increase transiently membrane permeability to $Na^+$ and $K^+$ ions, giving rise to net inward synaptic current ($I_s$), which is carried by $Na^+$ ions and depolarizing (Fig. 15). The synaptic equilibrium or reversal potential is given by the expression already derived for the endplate, that is,

$$E_{EPSP} = E_r = (g_{Na}/g_K + g_{Na})E_{Na} + (g_K/g_K + g_{Na})E_K \tag{13}$$

The amplitude of postsynaptic potential at the peak synaptic activation can be readily obtained recognizing that the current flowing through the nonsynaptic membrane can be written (Ohm's law) as:

$$I_m = \Delta V_m/R_m \tag{14}$$

where $\Delta V_m$ is the change in the voltage of nonsynaptic membrane and $R_m$ is the cell's input resistance. Since at

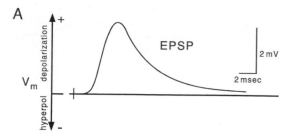

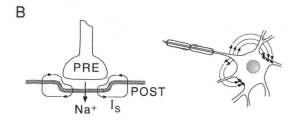

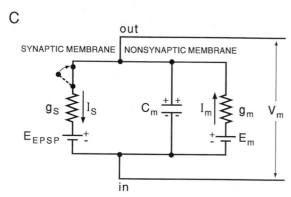

**FIG. 15.** Excitatory postsynaptic potentials. (A) EPSP is a depolarizing potential, few millivolts in amplitude and few milliseconds in duration. (B) The current that gives rise to EPSP is inward through the synaptic channels and outward through the nonsynaptic membrane. (C) Electrical equivalent circuit description of EPSP at the peak synaptic activation. Synaptic current ($I_s$) flows inward through the synaptic channels ($g_s$) and exits through the nonsynaptic leakage channels ($g_m$) as $I_M$. $E_{EPSP}$ is the synaptic battery (the equilibrium potential of EPSP) and $E_M$ is the synaptic battery of the resting membrane. Compare this electrical circuit description of EPSP with the more detailed description of EPP in Fig. 14. (Adapted from Eccles, 1994.)

peak synaptic activation $I_s = -I_m$ (eq. 10), one can further write:

$$I_{EPSP} = \Delta V_m/R_m \tag{15}$$

so that

$$\Delta V_m = \Delta V_{EPSP} = I_{EPSP} \cdot R_m \tag{16}$$

where $R_m$ is the cell's input resistance and $\Delta V_m$ is the change in the voltage across the nonsynaptic membrane, that is, $V_m - E_m$.

Expression 16 shows that the size of EPSP is determined by the magnitude of synaptic current and the cell's input resistance. The higher the $R_m$ the greater the depolarizing effectiveness of the $I_{EPSP}$. Since $I_{EPSP} = g_s(V_m - E_s)$ [Eq.

(4)], the magnitude of $I_{EPSP}$ itself is determined by the number of synaptic channels opened, that is, the synaptic conductance $g_s$ and on the membrane potential $V_m$ through its effect on the driving force ($V_m - E_s$). The EPSPs generated at any single synaptic input are small, usually in the range of only 0.5–2 mV. This is because only few transmitter quanta are released per presynaptic AP at central synapses. Consequently, several EPSPs must summate in order to depolarize the membrane to the threshold potential for discharging the propagated AP.

## 2. Inhibitory Postsynaptic Potentials

Inhibitory postsynaptic potentials are generated when a transmitter activates the receptor-gated Cl⁻ channels or, less frequently, through second messenger-mediated activation of K⁺ channels (indirectly gated transmission). The direction of current through the synaptic channels at inhibitory synapses is outward (Fig. 16) and opposite to that at the excitatory synapses. The expression for the synaptic current at inhibitory synapses ($I_{IPSP}$) is analogous to that for synaptic currents at excitatory synapses:

$$I_{IPSP} = g_s(V_m - E_{IPSP}) \qquad (17)$$

where the term $g_s$ refers to Cl⁻ or K⁺ conductance, and $E_{IPSP}$, the inhibitory synaptic equilibrium potential, is equivalent to the Cl⁻ or K⁺ equilibrium potentials, respectively. Thus, at synapses where transmitter activates receptor-gated Cl⁻ channels, the synaptic current is associated with influx of Cl⁻ ions, which will tend to move membrane potential toward the chloride equilibrium potential ($E_{Cl} = -80$ mV). At synapses where transmitter activates permeability to K⁺ ions, the synaptic current will be associated with efflux of K⁺ ions through the synaptic channels and membrane hyperpolarization toward the value of $E_K$.

The inhibitory effect of IPSPs on the cell electrical excitability (e.g., spike generation), however, is not only due to membrane hyperpolarization but also to the decrease in the cell's overall input resistance $R_m$. This can be seen from consideration of Eq. 16, which shows that the effectiveness of $I_{EPSP}$ in depolarizing the membrane is a function of membrane input resistance. The depolarizing effectiveness of $I_{EPSP}$ is reduced as activation of inhibitory synapses and opening of the synaptic Cl⁻ channels reduces the cell's $R_m$, and provides low resistance conductance pathways for current flow out of the cell, thus in effect attenuating the depolarizing effectiveness of $I_{EPSP}$.

## 3. Integration of Synaptic Activity and Genesis of Action Potential

Neurons integrate postsynaptic potentials by adding the EPSPs and subtracting the IPSPs from the membrane potential at any instant of time. It is intuitively evident that when several excitatory synapses are activated at nearly the same time, the net inward synaptic current will be the sum of inward currents produced at individual synapses, and the net EPSP (compound EPSP) will be a sum of the corresponding EPSPs elicited at several synapses. Furthermore, it is evident that when excitatory and inhibitory

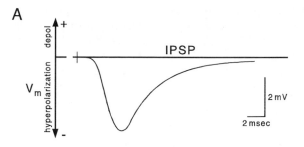

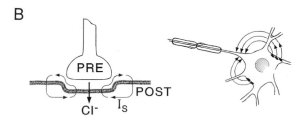

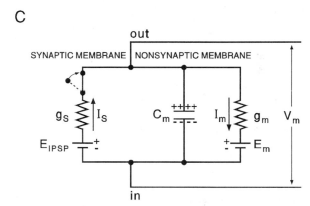

**FIG. 16.** Inhibitory postsynaptic potential. (A) IPSP is a hyperpolarizing potential. (B) IPSP is associated with an outward current through the synaptic membrane and inward through the nonsynaptic membrane. (Note that the direction of bioelectric current flow is, by convention, the movement of positive charge. Influx of Cl⁻, that is, entry of negative charge into the cell is equivalent to movement of positive charge out of the cell). (C) Electrical equivalent circuit of an inhibitory synapse describing current flow during peak IPSP. Note again that synaptic current $I_s$ is outward through synaptic membrane and returns as $I_m$ through the nonsynaptic membrane. (Based on Eccles, 1994.)

synapses are activated at about the same time, then the size of the inward current produced at excitatory synapses will be reduced proportionally to the size of the outward current at the inhibitory synapse, resulting in smaller net EPSP.

When two or more topographically separated synapses are activated nearly simultaneously so that their PSPs can add together, the process is called *spatial summation* (Fig. 17A). The effectiveness of spatial summation depends on the neuronal membrane space constant, that is, the distance at which electrotonic potentials decay to $1/e$ or 37% of their amplitude at the point of origin. Clearly, synaptic inputs separated by a distance smaller than the space constant can summate much more effectively than those that

A

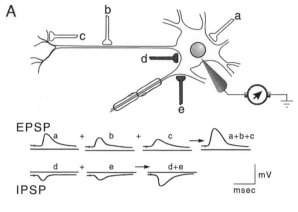

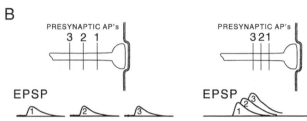

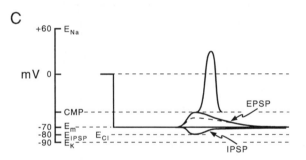

B

C

**FIG. 17.** Integration of synaptic potentials. (A) Spatial and (B) temporal summation. Postsynaptic potentials are recorded with an intracellular electrode from a neuron receiving excitatory synaptic inputs a, b, and c, and inhibitory synapses d and e. A. Activation of each synapse individually gives rise to small EPSPs or IPSPs. Activation of a, b, and c synapses nearly simultaneously gives rise to an EPSP that is the sum of EPSPs generated at each, spatially separate E synapse. Similarly, near-simultaneous activation of I synapses yields a summated IPSP. (B) When the interval between presynaptic APs exceeds the membrane time constant, each EPSP decays before the next one is evoked (left); when presynaptic potentials fire at a frequency that results in overlap between the succeeding EPSPs (right) the EPSPs add to give a summated response that is larger than any individual EPSP. (C) Integration of excitatory (EPSPs) and inhibitory (IPSPs) potentials. The diagram illustrates the relationship between resting membrane potential ($E_m$), synaptic equilibrium potentials ($E_{EPSP}$ and $E_{IPSP}$), and the equilibrium potentials for $Na^+$, $K^+$, and $Cl^-$. CMP is critical membrane potential, that is, the threshold potential at which AP is fired. Activation of E synapses will trigger an AP when the (spatially and temporally) summated EPSP reaches the CMP. Activation of I synapses generates IPSPs, which subtract from the EPSP so that the depolarizing potential (dashed waveform) is below threshold and cannot elicit an AP. (Modified from Suszkiw and Behbehani, 1993.)

are separated by a distance exceeding the space constant. In other words, the larger the membrane space constant, the greater the cell's capability to summate the PSPs generated at various locations on the neuron.

When action potentials in a single presynaptic neuron fire in rapid succession, such that the interval between the presynaptic APs is less than the duration of the postsynaptic potential (e.g., 2–10 msec), the process is referred to as *temporal summation* (Fig. 17B). The effectiveness of temporal summation depends on the membrane time constant. The membrane time constant determines the time required for an electrotonic potential to decay to $1/e$ or 37% of its peak value. The longer the time constant, the longer the duration of the PSP and thus greater the opportunity for summation of successive PSPs to occur.

A propagated AP will be triggered only if the net current is of sufficient magnitude to depolarize the neuronal membrane to the threshold (Fig. 17C). The AP is triggered usually at the axon hillock, which is the region of the neuron with the lowest threshold. Because summation of synaptic currents at the axon hillock is the principal determinant of whether or not an AP will be fired, the initial segment is referred to as the *integrative zone* of the neuron.

### 4. Termination of Transmitter Action

With the exception of the transmitter acetylcholine, which is degraded extracellularly by AChE, the synaptic actions of most other transmitters are terminated by reuptake into the presynaptic terminals and probably also the surrounding glial cells. The neurotransmitter uptake is mediated by specific sodium-dependent, high-affinity transporters (Amara and Kuhar, 1993). Following reuptake, the transmitters may be repackaged into synaptic vesicles or catabolized by intracellular degradative enzymes.

## VI. Slow Synaptic Transmission Mediated by G-Protein-Coupled Receptors

In contrast to fast synaptic potentials mediated by directly transmitter-gated channels, slow synaptic responses are mediated by a distinct family of proteins that transduce the transmitter binding into cellular responses through activation of GTP-binding regulatory proteins or G proteins. Binding of a transmitter to a specific receptor transforms the associated G protein into an active form, which modulates activity of ionic channels at some distance from the receptor either through direct interaction with the channel and/or through second messenger systems. The slow synaptic transmission may involve either an increase or decrease in postsynaptic membrane conductance due to a channel opening or channel closure, respectively. For example, activation of muscarinic receptors in heart muscle causes a G-protein-mediated opening of a certain class of voltage-gated $K^+$ channels and relatively long-lasting (seconds) membrane hyperpolarization. Similar, G-protein-modulated potassium channels K(G) are present in central neurons where they mediate slow IPSPs. In contrast, activation of a muscarinic receptor in sympathetic ganglionic neurons and certain neurons in the CNS is associated with a second messenger mediated closure of certain K$^+$ channels and membrane depolarization. The responses to transmitter activation of G-protein-coupled receptors need not be limited to modulation of ionic channels; they also involve modifications of other regulatory proteins resulting in long-term modifications of the postsynaptic cell's physiology. For a more detailed discussion of G proteins and second messengers in slow synaptic actions of neurotransmitters, the reader is referred to Hille (1992).

## VII.  Cotransmission

In addition to classical neurotransmitter, many neurons have been shown to contain and release physiologically active small peptide hormones (Table 2) that may act as cotransmitters or modulators of synaptic transmission. Synthesis, packaging, and release of conventional transmitters and neuroactive peptides are regulated differently. Unlike the classical transmitters, which are locally synthesized and packaged in synaptic vesicles in nerve terminals, peptides are synthesized and packaged into secretory granules in the cell body whence they are then transported via axoplasmic transport to the nerve terminals. The peptides are generally associated with large dense-core vesicles, which tend to undergo exocytosis outside the active zone structures, and generally require higher frequencies of stimulation than those that release conventional transmitter. In contrast to the fast synaptic actions of conventional neurotransmitter, the postsynaptic actions of peptides tend to be slower in onset and longer in duration, thus providing a background modulation for the fast synaptic transmission. Neuroactive peptides may also alter efficacy of fast synaptic transmission by modulating transmitter release via presynaptic actions.

An interesting example of cotransmission, not involving peptides, is provided by the corelease of ATP and noradrenaline from sympathethic nerve endings at the vascular smooth muscle junctions, where ATP generates fast junctional excitatory potentials and the action of noradrenaline is to generate the slow, junctional potentials. In this instance, both small molecular weight messengers are costored in and released from the same secretory vesicles and both substances serve as *bona fide* transmitters.

## VIII.  Modulation of Synaptic Transmission

The efficacy of signal transmission at the chemical synapses can be modulated by extrinsic and intrinsic factors, including the pattern of ongoing activity as well as history of previous activity. The efficacy of transmission may be altered by mechanisms that affect the dynamics of presynaptic transmitter release and/or modify the postsynaptic receptor-mediated events. These modifications may be short lasting or may persist for some time. Thus, synaptic modulation provides for fine tuning of ongoing synaptic activity as well for longer lasting changes that are likely to play an important role in learning processes.

### A.  Activity-Dependent Modulation of Synaptic Transmission

#### 1. Depression

Depression or fatigue of synaptic transmission refers to a progressive reduction in the amplitudes of postsynaptic potentials in the course of prolonged, relatively high-frequency activation of presynaptic neuron, reflecting progressive depletion of releasable transmitter store (Fig. 18). Recovery from fatigue may take from minutes to hours.

#### 2. Facilitation

Facilitation is a frequency-dependent, progressive increase in the amplitude of postsynaptic potentials evoked by closely spaced presynaptic action potentials (Fig. 18). The increase in the amplitudes of succeeding PSPs reflects a progressively larger amount of transmitter released per each nerve impulse, in the course of stimulation. The mechanism is thought to involve buildup of ionized Ca$^{2+}$ within the terminals. That is, when presynaptic neuron is stimulated at a certain frequency, the diffusion and clearance of Ca$^{2+}$ from the release sites begins to lag, and the residual Ca$^{2+}$ adds to the Ca transient evoked by the next arriving impulse. Buildup of Ca$^{2+}$ increases the probability ($P$) of transmitter quanta being released. Facilitation is a relatively short-lived process, lasting for only a few seconds.

#### 3. Post-Tetanic Potentiation

When the nerve is activated with relatively prolonged and/or high-frequency (tetanic) bursts of impulses, the quantity of transmitter is increased on subsequent stimulation, even after a relatively long intervening rest period (Fig. 18). This phenomenon is known as **post-tetanic potentiation (PTP).** In contrast to facilitation, PTP may last for minutes and sometimes hours, suggesting a long-term modification of presynaptic function secondary to an increase in cytosolic Ca$^{2+}$ levels.

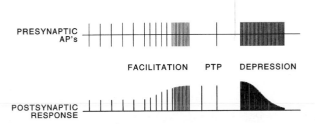

**FIG. 18.** Diagrammatic illustration of the relationship between the frequency of presynaptic APs (top) and postsynaptic responses (bottom). A presynaptic neuron firing APs at low frequency evokes PSPs that are relatively constant in amplitude. An increase in presynaptic AP frequency results in progressive increase in the amplitudes of PSP due to progressively larger amounts of transmitter released (facilitation). After a train of high-frequency (tetanic) activation, the response remains enhanced for some time and this is referred to as post-tetanic potentiation (PTP). Prolonged, high-frequency stimulation may, however, result in progressive reduction of PSPs amplitudes due to progressive depletion of transmitter available for release (depression).

### 4. Long-Term Potentiation

Long-term potentiation (LTP) refers to a long-lasting increase in EPSP following tetanic stimulation in the presynaptic neurons. It is distinguished from PTP in that the latter is a strictly presynaptic phenomenon, whereas induction and expression of LTP involves both postsynaptic and presynaptic elements. LTP was first described (Bliss and Lømo, 1973) and has been analyzed most extensively in the hippocampus. In the CA1 region, the LTP involves a special glutamate receptor subtype, the NMDA receptor. The NMDA receptor-gated channel is normally blocked by $Mg^{2+}$, but can be activated by glutamate when the postsynaptic neuron is sufficiently depolarized so that the $Mg^{2+}$ blockade of the channel is relieved. The induction of LTP appears to be associated with influx of $Ca^{2+}$ and activation of Ca-dependent protein kinases CaM kinase II and protein kinase C, and possibly other protein kinases in the postsynaptic dendritic spine, which somehow leads to increased efficacy of synaptic transmission that may last for days to weeks. It is thought that while induction of LTP involves the postsynaptic events, the maintenance of LTP may be associated with long-term increase in the probability of transmitter release, that is, a presynaptic event.

### B. Presynaptic Receptors and Transmitter Release

Receptors for neurotransmitters are not confined to the postsynaptic sites, but are also found on presynaptic nerve terminals. The modulation of transmitter release by presynaptic receptors that respond to transmitter release by another neuron is referred to as **heterosynaptic modulation.** Heterosynaptic modulation may involve either inhibition or facilitation of transmitter release. In addition, certain presynaptic "autoreceptors" recognize the cell's own neurotransmitter. In this case, the neuron's transmitter may modulate its own release by interacting with these receptors. This is called **automodulation.** For example, many cholinergic neuron terminals possess muscarinic autoreceptors, and ACh released from these terminals acts on the autoreceptors to inhibit its own release. Although, the physiological role of presynaptic receptors has been a subject of debate, it is evident that they provide a potential homeostatic mechanism that could function to fine-tune the level of transmitter release.

### 1. Presynaptic Inhibition

Presynaptic inhibition is exerted at axo-axonic synapses between a terminal of an inhibitory neuron and a terminal of an excitatory neuron (Fig. 19). Activation of this kind of axo-axonic synapse brings about reduction in the release of excitatory transmitter and, consequently, a smaller EPSP in the postsynaptic cell. Presynaptic inhibition differs from postsynaptic inhibition in that it diminishes the effectiveness of EPSP generated at the synapse without directly hyperpolarizing the postsynaptic neuron, that is, it does not produce IPSPs.

Presynaptic inhibition tends to be found at sensory inflow points where it may act to reduce or even "switch off"

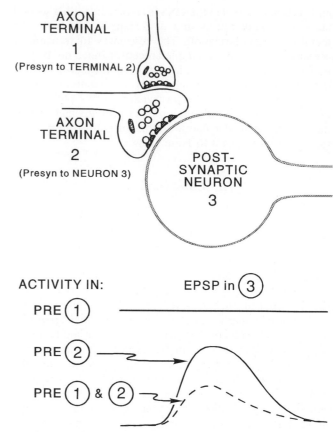

**FIG. 19.** Presynaptic inhibition. An axo-axonic synapse that mediates presynaptic inhibition is illustrated. Nerve terminal 1 is presynaptic to nerve terminal 2, which forms excitatory connection onto the postsynaptic neuron 3. Activation of 1 alone has no direct effect on neuron 3. Activation of 2 alone produces normal EPSP in the neuron 3. However, if activation of 2 is preceded by activation of 1, the EPSP is reduced due to inhibition of transmitter release from 2 consequent to activation of the inhibitory input 1 onto the excitatory neuron terminal 2.

selected inputs without affecting the responsiveness of the postsynaptic neuron to other synaptic inputs. In the spinal cord, presynaptic inhibition has been shown at axo-axonic synapses formed between spinal inhibitory interneurons and the excitatory endings of primary afferent fibers from the muscle stretch receptors, which provide powerful excitatory input to the alpha motoneurons. Activation of the interneurons mediating presynaptic inhibition in the mammalian spinal cord causes a depolarization of the terminals of the excitatory primary afferent fibers (primary afferent depolarization, or PAD). Depolarization of presynaptic terminals reduce the quantity of transmitter released, probably because the amplitude of the presynaptic spike in a partially depolarized terminal is smaller, fewer $Ca^{2+}$ channels are activated, and less $Ca^{2+}$ enters the terminal. Lasting depolarization of the terminal arborizations may also block AP invasion into the terminals. PAD appears to be mediated by GABA, as it is blocked by GABA antagonist, bicuculline. The presynaptic inhibition is of relatively long duration and may last up to several hundred milliseconds.

## C. Receptor Desensitization

Following a prolonged application of agonist, many receptors tend to exhibit decreased responsiveness to the drug. This "desensitization" is thought to reflect an agonist-induced conformational state in which receptor binds the agonist with high affinity but is unable to open the channel. Whether or not desensitization plays a role in normal physiological processes is not certain; however, prolonged exposure to drugs that act as receptor agonists can lead to desensitization. The block of neuromuscular transmission by receptor agonists such as nicotine or following poisoning with anticholinesterases, which due to inhibition of the enzyme causes persistent elevation of ACh in the synaptic cleft, may in part be related to receptor desensitization.

## D. Denervation Supersensitivity

Many peripheral effector organs develop increased sensitivity to the transmitter after being denervated, that is, chronically deprived of the motor nerves that normally innervate the tissue. Denervation supersensitivity is associated with synthesis and spread of receptors to extrasynaptic regions. The mechanism by which denervation triggers increase in the synthesis of new receptors and their spread to extrasynaptic regions is not well understood. It is thought that distribution and turnover of receptor in the muscle membrane are regulated in part by the trophic factors provided by the neuron and/or activity. Indeed, direct stimulation of the denervated muscle may slow down and sometimes even reverse the effects of denervation.

## IX. Effects of Drugs and Toxins

Chemical synaptic transmission is vulnerable to a number of drugs and toxins that may interfere with either the presynaptic or postsynaptic process. The toxicity of various agents may reflect their relative accessibility. For example, certain toxins may exert powerful effects at the peripheral synapses but due to the existence of the blood–brain barrier may be unable to penetrate and exert their action on central synapses. A brief survey of presynaptic and postsynaptic actions of selected agents is provided next. For detailed information on specific neurotransmitter systems, the reader is referred to standard textbooks on pharmacology.

## A. Presynaptic Actions

Presynaptic actions of drugs may involve interference with transmitter biosynthesis, uptake and storage in synaptic vesicles, or the transmitter release process.

### 1. Inhibitors of Transmitter Synthesis

Inhibitors of precursor uptake or the key biosynthethic enzymes will poison synaptic transmission by reducing the availability of transmitter. For example, the drug hemicholinium-3 is a potent inhibitor of the high affinity Ch transporter at the cholinergic nerve terminals. Block of choline transport will result in depressed formation of ACh. During repetitive activation, ACh stores become depleted and synaptic fatigue ensues.

### 2. Inhibitors of Transmitter Packaging

Reserpine is probably the best known example of a drug that interferes with storage of catecholamines in secretory vesicles. In reserpinized animals, sympathethic nerves are depleted of noradrenaline and adrenergic transmission is impaired. Uptake of ACh into synaptic vesicles is blocked by the drug Vesamicol ($d$,l-2-(4-phenylpiperidino)cyclohexanol). In the presence of the drug, synaptic vesicles cannot be refilled with ACh and the releasable store of vesicle-bound transmitter becomes rapidly depleted in the course of stimulation.

### 3. Interference with Depolarization-Secretion Coupling Process

Divalent metal ions, particularly heavy metals such as $Pb^{2+}$, $Cd^{2+}$, and $Hg^{2+}$, are potent inhibitors of voltage-gated Ca channels. By reducing influx of $Ca^{2+}$ these cations depress or abolish the AP-evoked transmitter release. Interestingly, several of the heavy metal ions may enter the terminals and substitute for $Ca^{2+}$ in activating transmitter exocytosis. In this instance, heavy metal cations can be said to exert dual action on transmitter release, that is, suppression of the evoked release and augmentation of spontaneous release.

### 4. Interference with Exocytosis and Vesicle Cycling

Clostridial botulinus and tetanus toxins are potent inhibitors of transmitter exocytosis. The toxins exert their action intraterminally following receptor-mediated internalization and proteolysis of synaptobrevin, syntaxin, or SNAP-25. Botulinus neurotoxin is selective for cholinergic synapses and is one of the most potent neuroparalytic agents known. In severe poisoning respiratory failure and death may result. Tetanus toxin is taken up by spinal motor nerve terminals and transported retrogradely to the spinal cord where it blocks the release of glycine at inhibitory synapses. Spread of the toxin throughout the spinal cord and brain can lead to generalized convulsions and death.

The $\alpha$-latrotoxin from black widow spider venom causes massive transmitter exocytosis and depletion of synaptic vesicles from presynaptic nerve terminals. The toxin appears to exert its effect by binding to a presynaptic plasma membrane protein neurexin.

## B. Postsynaptic Actions

Various agents may interfere with the postsynaptic processes by interacting with the transmitter binding to the receptor or with the receptor channel.

### 1. Receptor Agonists

Agents that mimic the actions of natural transmitter substance are called **receptor agonists.** Nicotine is a well-known agonist of the ACh receptor at NMJ. Indeed, designation of this receptor as nicotinic originally derives from its responsiveness to nicotine. Nicotine binds to the receptor in activates it as ACh does. However, in contrast to ACh, whose action at the receptor is terminated by AChE, nicotine is not degraded by this enzyme. Consequently, nicotine causes persistent receptor activation, which ultimately results in "depolarizing" block of neuromuscular transmission, probably due to inactivation of the muscle spiking mechanism, receptor desensitization, or both.

### 2. Receptor Antagonists

Compounds that compete with the transmitter for the receptor binding sites and prevent receptor activation are called **receptor antagonists.** An example is curare, which is a classic antagonist of the nicotinic ACh receptors at the skeletal NMJ. By preventing ACh binding to the receptor, curare causes a progressive decrease in the amplitude and shortening of EPPs. In severe curare poisoning, the EPPs are depressed to below the threshold potential and transmission is blocked. Selective antagonists exist for most transmitter receptors and their subtypes. At central synapses, the antagonists of GABA receptors, for example, bicuculline, are well-known convulsants.

### 3. Receptor Modulators

Several clinically important drugs act through modulation of the GABA receptor function. The benzodiazepines and barbiturates interact with distinct sites on the receptor to augment the GABA-activated $Cl^-$ currents. Benzodiazepines such as diazepam (Valium) or chlordiazepoxide (Librium) are used as antianxiety drugs and muscle relaxant, whereas barbiturates (e.g., phenobarbital) have anaesthetics and hypnotic effects.

### 4. Receptor Channel Blockers

A number of compounds may interfere with receptor-gated permeability changes by interfering with the channel itself. Histrionicotoxin is an example of an agent that interferes with the cationic pore of the nAChR at the NMJ. Picrotoxin, which blocks the GABA receptor-activated $Cl^-$ channel, and strychnine, which blocks the glycine receptor-activated $Cl^-$ channels, are potent blockers of respective inhibitory synapses and known convulsants.

### C. Interference with Transmitter Inactivation

As has been discussed before, efficient inactivation of synaptic transmitters is required for normal repetitive synaptic activation. Thus, any agent that interferes with transmitter inactivation will tend to prolong its synaptic action and disrupt transmission. A well-known example is provided by inhibition of acetylcholinesterase at the NMJ. When AChE is inhibited, ACh persists in the synaptic cleft for longer time and its concentration builds up. This leads to multiple receptor activation, resulting in increased amplitude and prolongation of EPPs. This in turn may initially

result in repetitive spiking and muscle spasm. Eventually, persistent receptor activation and local depolarization of the end-plate region results in block of transmission and flaccid paralysis. The mechanism probably involves receptor desensitization and/or inactivation of the spiking mechanism, as discussed earlier in the case of nicotine. The most potent anticholinesterases are organophosphorus compounds, for example, diisopropylphosphofluoridate, which phosphorylate the active serine in the esteratic site of the enzyme. Since dephosphorylation is an extremely slow process, the enzyme is inactivated essentially irreversibly.

A number of clinically important, psychotropic drugs may exert their action by augmenting synaptic activity consequent to inhibition of neurotransmitter reuptake. The antidepressant drugs fluoxetine and imipramine block the serotonin transporter and the tricyclic antidepressant desipramine inhibits the norepinephrine reuptake. The transporters for dopamine, norepinepherine, and serotonin are all targets for cocaine and may be the primary site of its action.

## X. Summary

The transmission of synaptic signals is mediated by chemical neurotransmitter substances. Neurotransmitters are synthesized in presynaptic terminals and stored in synaptic vesicles. Transmitter release is evoked by presynaptic APs, which activate influx of $Ca^{2+}$ into terminals and trigger a $Ca^{2+}$-dependent exocytosis of transmitter from synaptic vesicles into the synaptic cleft. Once released, neurotransmitters activate specific receptor-gated channels in the postsynaptic cell and elicit a transient change in the membrane permeability to cations or anions. Fast synaptic transmission is mediated by directly transmitter-gated channels (ionophoric receptors). Slow synaptic transmission is mediated by G-protein-coupled receptors. Excitatory postsynaptic potentials are associated with the transmitter-induced increase in $Na^+$ and $K^+$ conductance of the synaptic membrane, resulting in net entry of positive charge carried by $Na^+$ and membrane depolarization. Inhibitory postsynaptic potentials are associated with transmitter-activated influx of $Cl^-$ or efflux of $K^+$ results in membrane hyperpolarization. The EPSPs at the skeletal neuromuscular junction are called end-plate potentials. EPPs are always large enough to depolarize the muscle membrane to threshold and trigger muscle AP. The EPSPs generated at any single neuro-neuronal synapse are usually too small to depolarize the postsynaptic neuron to threshold. Synaptic signals converging onto a neuron are normally integrated through summation of EPSPs and IPSPs, and AP is triggered only when the resultant membrane potential reaches or exceeds the threshold for AP.

Chemical synaptic transmission is subject to modulation by intrinsic and extrinsic factors, including frequency and pattern of AP firing, which can either facilitate or depress the transmission across any given synapse. A number of neuroactive peptides are costored and coreleased with the classic, low molecular weight transmitters and modulate their synaptic signals. Many drugs, toxins, or diseases, may impact on the efficacy of synaptic transmission by interfering either with the process of transmitter release, transmitter–receptor interactions, or transmitter inactivation.

## Bibliography

Akert, K., Pfenniger, K., Sandri, C., and Moor, H. (1972). Freeze etching and cytochemistry of vesicles and membrane complexes in synapses of the central nervous system. *In* "Structure and Function of Synapses" (G. D. Pappas and D. P. Purpura, Eds.), Chap. 3, pp. 67–86. Raven Press, New York.

Almers, W. (1990). Exocytosis. *Annu. Rev. Physiol.* **52**, 607–624.

Amara, S. G., and Kuhar, M. J. (1993). Neurotransmitter transporters: Recent progress. *Annu. Rev. Neurosci.* **16**, 73–93.

Bennett, M. K., and Scheller, R. H. (1993). The molecular machinery for secretion is conserved from yeast to neurons. *Proc. Natl. Acad. Sci. USA* **90**, 2559–2563.

Bennett, M. V., Barrio, L. C., Bargiello, T. A., Spray, D. C., Hertzberg, E., and Saez, J. C. (1991). Gap junctions: New tools, new answers, new questions. *Neuron* **6**, 305–320.

Betz, W. J., and Bewick, G. S. (1992). Optical analysis of synaptic vesicle recycling at the frog neuromuscular junction. *Science* **255**, 200–203.

Bliss, T. V. P., and Lømo, T. (1973). Long-lasting potentiation of synaptic transmission in the dentate area of the anaesthetized rabbit following stimulation of the perforant path. *J. Physiol.* **232**, 331–356.

Boyd, I. A., and Martin, A. R. (1956). The end-plate potential in mammalian muscle. *J. Physiol.* **132**, 30–38.

Ceccarelli, B., Hurlbut, W. P., and Mauro, A. (1973). Turnover of transmitter and synaptic vesicles at the frog neuromuscular junction. *J. Cell Biol.* **54**, 30–38.

Couteaux, R., and Pecot-Dechavassine, M. (1970). Vesicules synaptiques et poches au niveau des "zones actives" de la jonction neuromusculaire. *C.R. Acad. Sci. Hebd. Acad. Sci. J.* **271**, 2346–2349.

delCastillo, J., and Katz, B. (1954). Quantal components of the end-plate potential. *J. Physiol.* **124**, 560–573.

Dreyer, F., Peper, K., Sandri, C., and Moor, H. (1973). Ultrastructure of the "active zone" in the frog neuromuscular junction. *Brain Res.* **62**, 373–380.

Eccles, J. C. (1964). "The Physiology of Synapses." Academic Press, New York.

Fatt, P., and Katz, B. (1952). Spontaneous subthreshold activity at motor nerve endings. *J. Physiol.* **117**, 109–128.

Fertuck, H. C., and Salpeter, M. M. (1974). Localization of acetylcholine receptor by $^{125}$I-$\alpha$-bungarotoxin binding at mouse motor endplates. *Proc. Natl. Acad. Sci. USA* **71**, 1376–1378.

Gadzi, J. C., Revah, F., Bessis, A., and Changeux, J.-P. (1991). Functional architecture of the nicotinic acetylcholine receptor: From electric organ to brain. *Annu. Rev. Pharmacol.* **31**, 37–72.

Gilman, A. G., Goodman, L. S., and Gilman, A. (Eds.). (1980). "Goodman and Gilman's The Pharmacological Basis of Therapeutics," 6th ed. Macmillan, New York.

Hall, Z. W., and Sanes, J. R. (1993). Synaptic structure and development: The neuromuscular junction. *Neuron* **72**, Suppl., 99–121.

Heuser, J. E., and Reese, T. S. (1973). Evidence for recycling of synaptic vesicle membrane during transmitter release at the frog neuromuscular junction. *J. Cell Biol.* **57**, 315–344.

Heuser, J. E., Reese, T. S., Dennis, M. J., Jan, Y., Jan, L., and Evans, L. (1979). Synaptic vesicle exocytosis captured by quick freezing and correlated with quantal transmitter release. *J. Cell Biol.* **81**, 275–300.

Hille, B. (1992). Modulation, slow synaptic action, and second messengers. *In* "Ionic Channels of Excitable Membranes" (B. Hille, Ed.), pp. 177–201. Sinauer Associates, Sunderland, MA.

Jessell, T. M., and Kandel, E. R. (1993). Synaptic transmission: Bidirectional and self-modifiable form of cell-cell communication. *Neuron* **72**, Suppl., 1–30.

Jessell, T. M., Kandel, E. R., Lewin, B., and Reid, L. (Eds.). (1993). Signaling at the synapse. Review supplement to *Cell*, **72**. *Neuron* **10**, 1–30.

Kandel, E. R., Schwartz, J. H., and Jessel, T. M. (Eds.). (1991). "Principles of Neural Science," 3rd ed. Elsevier Science, New York, and Appleton & Lange, Norwalk, CT.

Katz, B. (1966). "Nerve, Muscle, and Synapse." McGraw-Hill, New York.

Katz, B., and Miledi, R. (1967). The timing of calcium action during neuromuscular transmission. *J. Physiol.* **189**, 535–544.

Kuffler, S. W., and Nicholls, J. G. (1977). "From Neuron to Brain." Sinaur Associates, Sunderland, MA.

Llinas, R. R. (1977). Calcium and transmitter release in squid synapse. *In* "Approaches to the Cell Biology of Neurons" (W. M. Cowan and J. A. Ferrendelli, Eds.), Society for Neuroscience Symposia, Vol. II, pp. 139–169. Society for Neuroscience, Bethesda, MD.

Llinas, R. R., Gruner, J. A., Sugimori, M., McGuinness, T. L., and Greengard, P. (1991). Regulation by synapsin I and $Ca^{2+}$-calmodulin-dependent protein kinase II of transmitter release in squid giant synapse. *J. Physiol.* **436**, 257–282.

Martin, R. A. (1977). Junctional transmission II. Presynaptic mechanisms. *In* "Handbook of Physiology. The Nervous System," Vol. 1, Part 1, Chap 10. American Physiological Society.

Maycox, P. R., Hell, J. W., and Jahn, R. (1990). Amino acid neurotransmission: Spotlight on synaptic vesicles. *Trends Neurosci.* **13**, 83–87.

Miledi, R. (1973). Transmitter release induced by injection of calcium ions into nerve terminals. *Proc. R. Soc. London* **183**, 421–425.

Niemann, H., Blasi, J., and Jahn, R. (1994). Clostridial neurotoxins: New tools for dissecting exocytosis. *Trends Cell Biol.* **4**, 179–185.

Numa, S. (1989). A molecular view of neurotransmitter receptors and ionic channels. *Harvey Lect.* **83**, 121–165.

Pappas, G. D., and Purpura, D. P. (Eds.). (1972). "Structure and Function of Synapses." Raven Press, New York.

Söllner, T., Bennett, M., Whiteheart, S., Scheller, R., and Rothman, J. (1993). A protein assembly–disassembly pathway *in vitro* that may correspond to sequential steps of synaptic vesicle docking, activation and fusion. *Cell* **75**, 409–418.

Südhof, T. C. (1995). The synaptic vesicle cycle: A cascade of protein-protein interactions. *Nature* **375**, 645–653.

Südhof, T. C., and Jahn, R. (1991). Proteins of synaptic vesicles involved in exocytosis and membrane recycling. *Neuron* **6**, 665–677.

Suszkiw, J. B., and Behbehani, M. (1993). Central synapses. *In* "Physiology" (N. Sperelakis and R. O. Banks, Eds.), Chap. 9, pp. 139–148. Little, Brown, Boston.

Suszkiw, J. B., and Sperelakis, N. (1993). Neuromuscular junction of skeletal muscle. *In* "Physiology" (N. Sperelakis and R. O. Banks, Eds.), Chap. 8, pp. 123–137. Little, Brown, Boston.

Whittaker, V. P. (1987). Cholinergic synaptic vesicles from the electromotor nerve terminals of Torpedo. Composition and life cycle. *Ann. NY Acad. Sci.* **493**, 77–91.

Winkler, H., Fischer-Colbrie, R., Obendorf, D., and Schwarzenbrunner, U. (1988). Adrenergic and cholinergic vesicles: Are there common antigens and common properties *In* "Cellular and Molecular Basis of Synaptic Transmission" (H. Zimmermann, Ed.), NATO ASI Series, Vol. H21, pp. 305–314. Springer-Verlag, New York.

*Nicole Gallo-Payet and Marcel Daniel Payet*

# 41

## Excitation–Secretion Coupling

## I. Introduction

The secretory process for peptide and amine molecules begins with the synthesis, modification, and sorting of the molecules to be secreted. Synthesis occurs in the rough endoplasmic reticulum and sorting in the Golgi complex. The secretory molecules are packaged in secretory granules or vesicles, which are then transported to the cell periphery before they are released in the extracellular space by fusion with the plasma membrane. This complex process is named **exocytosis** or **reverse pinocytosis** (see Fig. 6 later in this chapter). Secretion of peptide or neurotransmitters by neurons and other secretory cells generally involves one of two processes: a constitutive or regulated process. Constitutive secretion is unregulated and closely follows the rate of synthesis of the secretory products. This form of secretion occurs in many cell types, including lymphocytes, hepatocytes, and pancreatic $\beta$ cells. In regulated secretion, fusion of the secretory granules with the plasma membrane is triggered by a specific signal such as an increase in cytosolic calcium concentration ($[Ca^{2+}]_i$). In steroid-secreting cells (e.g., adrenal cortex, ovary, testis), the process of synthesis begins with cholesterol stored in lipid droplets followed by subsequent steps occurring in mitochondria and smooth endoplasmic reticulum. Secretory vesicles are not present and secretion and/or release of steroids is tightly coupled to their synthesis. It is generally assumed that steroids are free to diffuse throughout the aqueous cytoplasm and lipid phase of the plasma membrane.

The processes of exocytosis and secretion represent a fascinating interplay between cellular components and secretory vesicles. Granule storage not only allows secretory tissues to store large amounts of secretory products in a relatively small volume, but also protects this material from intracellular degradation while providing a very efficient means for transporting and releasing fixed quantities of secretory material. Our knowledge of the various membrane components of the secretory vesicles has grown rapidly in recent years, including a better knowledge of the interactions between secretory vesicles and calmodulin, GTP-binding proteins, $Ca^{2+}$, and cytoskeletal components. Exocytosis is an all-or-none phenomenon in which $Ca^{2+}$ plays a pivotal role. $Ca^{2+}$ is required for second messenger activity, for the control of cytoskeletal dynamics, and probably for the vesicle–plasma membrane fusion process. In neurons and certain endocrine cells, the electrical activity of the cell, that is, the action potential (AP), leads to the opening of voltage-dependent calcium channels with a subsequent increase in cytosolic $Ca^{2+}$. Neurotransmitter release at synaptic and neuromuscular junctions (NMJs), peptide hormone secretion from the pituitary gland, and catecholamine release by the adrenal medulla all belong to this class. In nonexcitable cells, the triggering $Ca^{2+}$ signal is provided by release of $Ca^{2+}$ from intracellular stores after appropriate stimulation. Depletion in $Ca^{2+}$ from these intracellular pools activates an influx of $Ca^{2+}$, which is responsible for the sustained increase in $[Ca^{2+}]_i$ observed in many cell types. Moreover, this $Ca^{2+}$ influx provides $Ca^{2+}$ ions for the replenishment of internal stores.

## II. Cellular Components Involved in Excitation–Secretion Coupling

### A. Receptors, Second Messengers, and Cytoskeleton

Although the process of synthesis differs, stimulation of secretion by peptide hormones, neurotransmitters, and steroids involves the same cascade of molecular events. After specific binding to their receptors, the stimuli activate second messenger production, several cascades of phosphorylation/dephosphorylation of intracellular proteins, cytoskeleton reorganization, synthesis of new products, and release of secretory products. Cytoskeleton and $Ca^{2+}$ ion are probably the most important players involved in this excitation–secretion coupling.

The cytoskeleton forms a cytoplasmic network of fibrillar structures providing an overall framework for the cell. In molecular terms, the cytoskeleton is composed of three types of fibers—microfilaments, microtubules, and intermediate filaments—and their associated proteins (for review, see Carraway and Carothers Carraway, 1989). The cytoskeletal elements are described in Chapter 5. Therefore the brief descriptions given here on microfilaments and microtubules are aimed at understanding the mechanisms involved in excitation–secretion coupling.

Microfilaments are composed of G-actin monomers, which are assembled in linear actin polymers (F-actin) of approximately 7 nm in diameter. Polymeric actin filaments consist of two staggered, parallel rows of monomers noncovalently bound and twisted into a helix (Fig. 1A). In the living cell, actin filaments are often linked into a three-

dimensional network (Fig. 1B). They can cross-link myosin to form stress fibers (Fig. 1C) or form a very dense network at the cell periphery, the cell cortex (Figs. 1D and E). These diverse polymeric forms of actin are in dynamic equilibrium with each other and with monomeric G-actin. Status of actin in a cell and organization of the microfilaments depend on a large and diverse group of actin-binding proteins, including profilin (which binds G-actin), gelsolin and scinderin (which cap and sever F-actin), fimbrin, $\alpha$-actinin, villin, and fodrin (which cross-link and bundle microfilaments).

Microtubules are 25-nm tubular filaments composed of $\alpha$- and $\beta$-tubulin heterodimers. Polymerization, stabiliza-

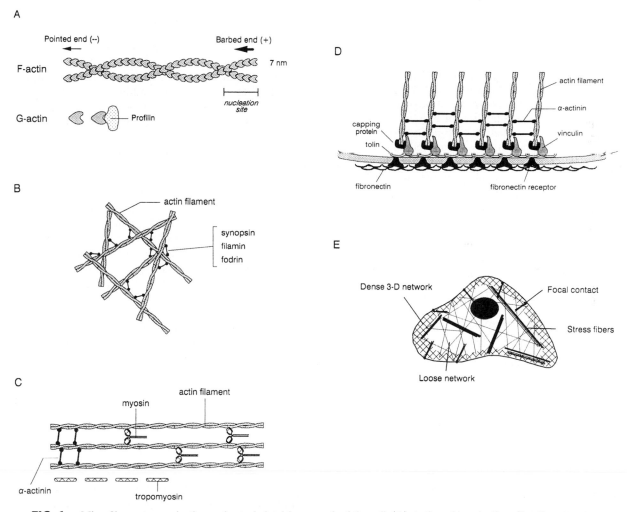

**FIG. 1.** Microfilament organization and cytoskeletal framework of the cell. (A) Actin polymerization. G-actin monomers are incorporated preferentially at the barbed end of actin filaments. Polymeric actin filaments consist of two staggered parallel rows of filaments twisted into a helix. When unpolymerized, G-actin monomers are linked to profilin, forming a profilactin complex. Microfilaments are often linked into a stiff three-dimensional network by cross-linking proteins (B) the most important of which is filamin, a long flexible molecule; synapsin and fodrin may also be found. Temporary contractile bundles of parallel actin filaments (cross-linked by $\alpha$-actinin) and myosin can be found, forming stress fibers in the cell (C). (D) A possible arrangement of a bundle of actin filaments and transmembrane linker glycoproteins at a special attachment site of the plasma membrane called *focal contact*. At this level, an especially dense network of actin filaments and associated proteins is found just beneath the plasma membrane. This network constitutes the cell cortex, which excludes large particles and organelles (E). Cytoskeleton elements are under tension and could exhibit, under appropriate stimulation, dynamic changes responsible for secretion, division, and motility.

tion, and modulation of microtubule function depend on several microtubule-associated proteins (MAPs), which adorn the tubulin-containing core of the tubules and can be purified along with microtubules.

## B. Interaction of Cytoskeletal Structures with Transmembrane Signaling

### 1. Interactions with Receptors, Transducers, and Effectors

Association of cell surface molecules with the cytoskeleton is widely believed to be one of the early consequences of cellular activation for many systems. Several studies conducted both in peptide and steroid-producing cells have shown that microfilament disruption (with cytochalasin B) or microtubule disruption (with colchicine or vinblastine) decreased or blocked secretion and second messenger production. In many examples, external signals have immediate and profound effects on the organization and activity of the microfilaments. Biochemical analysis of the G/F-actin ratio of cell extracts has shown that ligand–receptor interaction induces a rapid polymerization of actin, demonstrated by a rise in the proportion of F-actin over G-actin and by an increased interaction of F-actin with the membrane (Fig. 2A vs B and C). In several models, cell activation is associated with a rapid and transient increase in F-actin content. Moreover, a close association of actin filaments with receptors, G-proteins, $Ca^{2+}$ channels, and phospholipase C was shown. These associations may be ensured by the plecktrin homology domain (PH domain) found in all of these studies (Inglese *et al.*, 1995). Recent studies have shown that $G_{q/11}$ protein localization overlaps F-actin distribution (Figs. 2C and D). Cell activation is accompanied by a rapid translocation of $G_{q/11}$ and F-actin from the cytosol to the membrane. Recent data support the observation that association of $G_{q/11}$ with microfilaments is essential in promoting phospholipase C activation (Ibarrondo *et al.*, 1995). A close functional association between G-proteins and microtubules has also been extensively described during the past five years (Podova *et al.*, 1994). Such observations indicate that the cytoskeleton operates as a matrix improving the efficiency of the signal transduction cascade. Actin-binding proteins are in large part responsible for these cytoskeleton–membrane receptor interactions.

### 2. Interaction of Actin-Binding Proteins-Phosphoinositides

*a. Profilin.* *In vitro,* the actin-binding protein, profilin, binds to G-actin and inhibits actin nucleation and monomer addition at the pointed end of the filament (Fig. 1A). Profilactin (profilin-actin) complexes interact specifically and with high affinity to membrane phosphatidylinositol-bisphosphate ($PtdInsP_2$) and to a lesser extent with the phosphatidylinositol-phosphate (PtdInsP). In resting cells, 95% of the $PtdInsP_2$ may be complexed with profilin. The $PtdInsP_2$–profilactin interaction induces dissociation of the profilactin complexes and promotes actin polymerization

(Fig. 3A). This is because $PtdInsP_2$ can compete with actin for its binding site on profilin. $PtdInsP_2$ molecules bind to profilin with a stoichiometry of about 8 : 1. This 8 : 1 association of $PtdInsP_2$ : profilin complexes protects $PtdInsP_2$ from the cleaving action of phospholipase C. Following receptor activation, phospholipase C activity increases to a level where profilin protection can be overcome, resulting in some $PtdInsP_2$ hydrolysis. Hydrolysis of one or two of the eight $PtdInsP_2$ molecules bound to each profilin leads to a rapid decrease in $PtdInsP_2$ affinity and release of the remaining profilin-bound $PtdInsP_2$ molecules (Fig. 3B). Thus, receptor-stimulated phospholipase C activation may lead to a progressive increase in substrate availability and rate of $PtdInsP_2$ hydrolysis (for review, see Forscher, 1989).

*b. Gelsolin: A $Ca^{2+}$-Dependent F-Actin Severing Molecule.* Gelsolin is another actin-binding protein, which is $Ca^{2+}$- and $PtdInsP_2$-regulated. This protein regulates the length of actin filaments within the cells by several different mechanisms depending on cell status. Gelsolin contains two spatially separate G-actin-binding sites: a $Ca^{2+}$-sensitive site in the C-terminal domain and a PtdIns-sensitive site closer to the N terminus. In resting cells ($[Ca^{2+}]_i \approx 100$ n*M*), the bulk of gelsolin is cytoplasmic and a small fraction is membrane bound. Gelsolin has little affinity for actin under these conditions and is in an actin-free state (Fig. 3A). When the cell is activated, there is a transient rise in $[Ca^{2+}]_i$ due to $InsP_3$ action or $Ca^{2+}$ influx, which then activates gelsolin, causing a 200-fold increase in its affinity for actin. This results in rapid filament side-binding, followed by severing and capping of any free barbed filament ends, inducing a dramatic disruption of existing actin network structure in the vicinity of $Ca^{2+}$ elevation (Fig. 3B).

In the absence of PtdIns turnover, much of the $PtdInsP_2$ is likely to be tightly associated with profilin and thus unavailable to gelsolin. If gelsolin is activated under these conditions (for example, by an increase in $[Ca^{2+}]_i$, independent of the PtdIns turnover), severing of actin networks is observed, but without subsequent actin reassembly. This leads to two possible modes of gelsolin activation: (1) Calcium influx produced by activation of PtdInsP-independent pathways (i.e., voltage- or agonist-gated $Ca^{2+}$ channels) leads to actin severing and capping only, or (2) in contrast, activation of these same $Ca^{2+}$ channels concomitant with PtdInsP turnover results in severing and capping followed by actin polymerization, that is, actin remodeling.

*c. MARCKS, A Protein Kinase C Substrate That Interacts with Cytoskeleton.* MARCKS (myristoylated, alanine-rich C kinase substrate) is a specific protein kinase C (PKC) substrate that is targeted to the membrane by its amino-terminal binding domain. In resting cells, MARCKS associates with the cytoplasmic face of the membrane. In its nonphosphorylated form, MARCKS cross-links actin, favoring a rigid actin meshwork at the membrane level. Activated PKC phosphorylates MARCKS, which is released from the membrane, but still associates with actin filaments, but can no longer cross-link actin. The actin linked to MARCKS is spatially separated from membrane and more

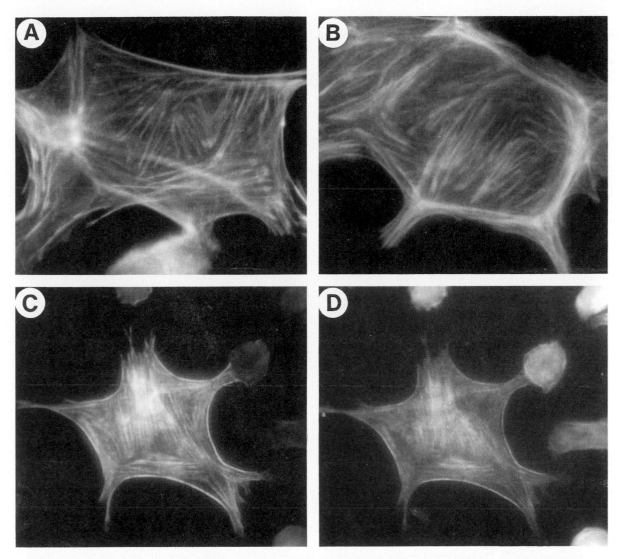

**FIG. 2.** Effect of ACTH and angiotensin II on immunofluorescence labeling of actin and Gq protein in rat glomerulosa cells. Rat glomerulosa cells were cultured for 3 days on plastic coverslips and then incubated for 1 min in HBS medium in the (A) absence or (B) presence of 100 n$M$ ACTH or (C and D) Ang II. After formaldehyde fixation and permeabilization with 0.1% Triton X-100, cells were double stained with anti-actin antibody and rhodamine (A–C) and anti-$\alpha_q$ antibody and FITC (D). All panels are shown at the same magnification of 3120×.

plastic. An increase in intracellular Ca$^{2+}$ promotes binding of calmodulin to MARCKS, inhibiting its actin cross-linking activity, which results in a less rigid actin meshwork. Thus, PKC induces a local destabilization of the actin skeleton through the phosphorylation of MARCKS (Aderem, 1992). MARCKS is phosphorylated when synaptosomes are depolarized, suggesting a role in secretion (Fig. 4).

*d. FAK and rho, Two Actin-Binding Proteins That Link F-Actin to the Membrane.* Several newly identified proteins, such as the focal adhesion molecule (p125$^{FAK}$), paxillin, or the small GTP-binding protein, rho, are also closely implicated in F-actin polymerization after hormonal stimulation. Tyrosine phosphorylation of p125$^{FAK}$ and paxillin and their association with cytoskeleton and $\beta\gamma$-subunits of G-pro-

teins have been recently identified as an early event in the action of several growth factors and G-protein-couped receptors (such as angiotensin II and vasopressin) (Inglese *et al.,* 1995). However, to date, their role has been ascribed to regulating cell adhesion, motility, or proliferation rather than secretion.

*e. Microtubule-Associated Proteins.* Recent studies have shown that some MAPs could be implicated in the secretory process (Bennett Jefferson and Schulman, 1991). MAPs are substrates for several protein kinases (PKs), including a Ca$^{2+}$/calmodulin-dependent kinase, a cAMP-dependent kinase, tyrosine kinases, and PKC. Covalent modification of these MAPs in T-cell lymphocytes may lead to an alteration in the integrity of cytoplasmic signal(s)

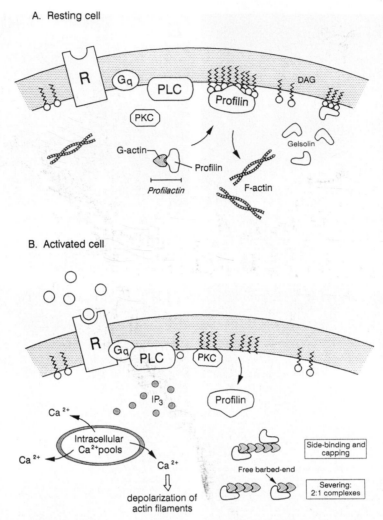

**FIG. 3.** Interaction of profilin/PtdInsP$_2$. (A) In resting cells, profilin interacts with high affinity with PtdInsP$_2$ and with low affinity to G-actin monomers. Profilin interacts with about eight molecules of PtdInsP$_2$ and protects them from hydrolysis by phospholipase C. Dissociation of profilactin complexes induces F-actin polymerization. (B) In activated cells, ligand binding to its receptor activates phospholipase C, which hydrolyzes PtdInsP$_2$ and leads to PtdInsP$_3$ and diacylglycerol (DAG). The rise in [Ca$^{2+}$]$_i$ (due to the PtdInsP$_3$ binding to intracellular pools of Ca$^{2+}$) induces F-actin depolymerization. (Adapted from Forscher, 1989.) (C) In resting cells, gelsolin is mostly cytosolic and has low affinity for actin. (D) In activated cells, gelsolin binds, severs, and caps actin filaments into 2:1 complexes to the barbed ends. [Adapted with permission from Forscher, P. (1989). *Trends Neurosci.* **12**, 468–474.]

from the membrane to the nucleus. Moreover, in the pituitary cell line GH3, MAP-1 and MAP-2 are responsible for binding secretory granules to microtubules. Substances that increase levels of cAMP in GH3 cells, such as vasoactive intestinal peptide (VIP), forskolin (which acts directly on adenylyl cyclase) or cholera toxin (which acts directly on the α-subunit of G$_s$), are known to stimulate prolactin secretion. On the other hand, increasing Ca$^{2+}$ levels in cells with K$^+$ depolarization, Ca$^{2+}$ ionophore, or the thyroid releasing hormone (TRH) also stimulates prolactin and growth hormone (GH) secretion. Both cAMP (via protein kinase A) and Ca$^{2+}$ (via Ca$^{2+}$-calmodulin PK) lead to phosphorylation of MAP-2. However, the exact link between phosphorylation of MAP-2 and secretion in

GH3 cells is not clearly understood. In endocrine pancreatic β cells, secretory granules are bound to microtubules via MAPs and in retinal neurons, MAP-2 is colocalized with serotonin-immunoreactive and tyrosine hydroxylase-immunoreactive cells. These results strengthen the probable role of MAPs in secretory processes.

### C. Actin and Actin-Binding Proteins Associated with Secretory Granules

Chromaffin cells of the adrenal medulla synthesize, store, and secrete catecholamines. These cells contain numerous electron-dense secretory granules, which discharge their contents into the extracellular space by exocytosis. The

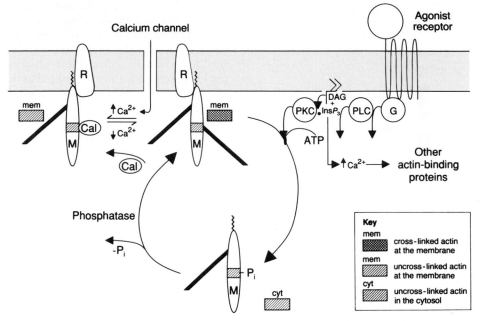

**FIG. 4.** Model indicating a possible mechanism by which MARCKS may regulate actin–membrane interaction. Agonist receptor activates PKC through a cascade involving G-proteins (G) and phospholipase C (PLC). PKC phosphorylates MARCKS (M), which is released from its membrane receptor (R). An increase in cellular $Ca^{2+}$ results in the binding of calmodulin (Cal) to MARCKS. [Reproduced with permission from Aderem, A. (1992). *Trends Biochem. Sci.* **17**, 438–443.]

subplasmalemmal area is characterized by the presence of a highly organized cytoskeletal network. F-actin seems to be exclusively localized in this area, and together with specific actin-binding proteins, forms a dense viscoelastic gel. Fodrin, vinculin, and caldesmon, three actin-binding proteins, as well as gelsolin and scinderin, two actin-severing proteins, are found in this plasmalemmal region. Because fodrin, caldesmon, and $\alpha$-actinin-binding sites exist on secretory granule membranes, actin filaments can also link to secretory granules (Figs. 1D and E).

Caldesmon is a calmodulin-dependent actin-binding protein that, at low $Ca^{2+}$ concentrations (100 nM), binds and cross-links actin filaments. The binding of caldesmon to actin filaments is inhibited in the presence of micromolar concentrations of $Ca^{2+}$. Under these conditions, caldesmon interacts reversibly with chromaffin vesicle membranes. The flip-flop regulation of caldesmon may be important for secretory vesicle function during the changes in intracellular $Ca^{2+}$ levels observed on stimulation.

Synapsin I is associated with synaptic vesicles. It is a phosphoprotein and a substrate for protein kinase A and calmodulin-dependent protein kinase II. Synapsin I also binds to spectrin and actin microfilaments and may serve as an anchor between synaptic vesicles and the cytoskeleton. The affinity of synapsin I for synaptic vesicles is decreased by phosphorylation, and neurotransmitter release is accompanied by a reversible phosphorylation of synapsin I. Therefore, synapsin I phosphorylation results in the release of synaptic vesicles from their anchorage sites on the cytoskeleton, thus allowing the vesicles to move to the active synaptic zones. Chromaffin granules can be en-

trapped in this subplasmalemmal lattice, and thus the cytoskeleton acts as a barrier preventing exocytosis.

### D. Docking and Fusion Proteins

The nature and the role of the docking complex is slowly being elucidated. Several proteins from the secretory vesicle membrane, the plasma membrane, and the bulk cytosol interact to form the core complex with docking, priming, and fusioning properties. The membrane of the synaptic vesicles contains several proteins that mediate vesicle membrane traffic such as docking, fusion, and budding. It is beyond the scope of this chapter to draw a complete list of these proteins (see Südhof, 1995). Briefly, the following play an important role in such complexes: vesicular membrane proteins, synaptobrevins (or vesicle-associated membrane proteins, VAMPS), synaptophysin, Rab 3; the plasma membrane proteins syntaxin, synaptosome-associated protein (SNAP-25; $M_r$ of 25 kDa); and the cytoplasmic proteins NSF (*N*-ethylmalimide-sensitive fusion protein), the soluble NSF attachment proteins (SNAPs) (Fig. 5). The proposed sequence for docking and priming is described hereafter (Südhof, 1995). Before and/or during docking, syntaxin is bound to Munc 18 and synaptophysin to synaptobrevin. These two complexes, syntaxin/Munc 18 and synaptophysin/synaptobrevin, must dissociate in order for the core complex to be formed. Formation of this complex is considered to be the first step of vesicle priming. Syntaxin and SNAP-25 (t-SNARE) can then bind tightly to form a high-affinity site for synaptobrevin (v-SNARE) located on the vesicle membrane; the core complex has a stochiometry

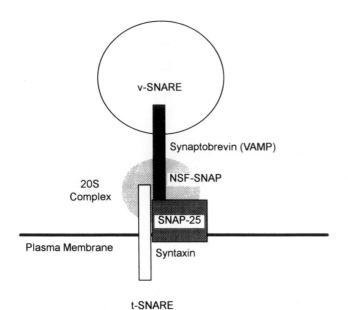

**FIG. 5.** SNARE, the receptor involved in docking and fusion. The 20S particle, which forms the core complex, contains several interacting proteins. v-SNARE, related to synaptobrevin (VAMP) and located on the vesicle, binds to t-SNARE, related to syntaxin and SNAP-25 and located on the plasma membrane, to form the SNARE or SNAP receptor. SNAPs and NSF can then bind to the receptor to complete the core complex. Numerous SNARE-related proteins, each specific for a single kind of vesicle or target membrane ensure vesicle-to-target specificity.

of $1:1:1$. The trimeric core complex serves as a receptor (SNARE) for the soluble SNAPs proteins (not related to SNAP-25). NSF will only interact with SNAPs ($\alpha$, $\beta$) bounded on the trimeric complex. The NSF-SNAP receptor forms a multisubunit particle that sediments at 20S; it may form the core of a generalized apparatus catalyzing bilayer fusion (Söllner *et al.,* 1993). NSF is a trimeric protein that cross-links multiple core complexes into a network. The core complex is then disrupted by enzymatic activity of NSF under ATP hydrolysis. Botulinum A and tetanus toxins are toxin proteases able to digest synaptobrevin, SNAP-25, and syntaxin. When entering in the nerve terminal, they irreversibly inhibit exocytosis. However, the number of docked granules is not decreased by the toxins, indicating that the primary function of the core complex is fusion and not docking. Hydrolysis of ATP by NSF is followed by ATP-independent steps (see later) sensitive to temperature, $H^+$ and $Ca^{2+}$. The last step is $Ca^{2+}$ sensitive and likely involves a $Ca^{2+}$ sensor at the site of exocytosis.

Synaptotagmins (Syt) are membrane glycoproteins found in brain secretory vesicles of which eight forms have been cloned. One of these, synaptotagmin I (Syt I), plays a pivotal role in the $Ca^{2+}$-triggered neurotransmitter release as a $Ca^{2+}$ sensor. The functional implication of multiple synaptotagmins is unknown. Syt I binds $Ca^{2+}$ cooperatively and undergoes a $Ca^{2+}$-dependent conformational change; the coefficient of cooperativity (4) is similar to that observed for $Ca^{2+}$-triggered release. Syt I also binds phospholipids as a function of $Ca^{2+}$ with high affinity (half

maximal binding, $5–6\ \mu M$). Syntaxin, a plasma membrane protein involved in the formation of the core complex interacts with Syt I for intracellular $Ca^{2+}$ ranges similar to those observed for $Ca^{2+}$-dependent release (half maximal binding, $200\ \mu M$). All of these observations suggest that Syt I interactions with phospholipids, syntaxin, and possibly other unknown proteins will advance the secretory granules from a hemifused state to a fused state on increases in intracellular calcium.

In addition to phospholipids and syntaxin, Syt I binds to neuroxins, a family of neuronal cell surface proteins and to AP-2, a protein complex involved in synaptic vesicle endocytosis. In addition to a $Ca^{2+}$ sensitivity similar to $Ca^{2+}$-triggered release, the role of Syt as a $Ca^{2+}$ sensor has been illustrated in several ways. Knockout mice for Syt I impairs $Ca^{2+}$-triggered transmitter release but release can still be obtained by $Ca^{2+}$-independent agents such as hypertonic sucrose or the excitatory neurotoxin $\alpha$-latrotoxin (the receptor for $\alpha$-latrotoxin belongs to the neuroxin family). Synaptotagmin *Drosophila* mutants show a severe but incomplete block of neurotransmission with an altered $Ca^{2+}$ dependence in some mutants. Injection of synaptotagmin peptide in squid nerve terminal inhibits release and vesicles accumulate possibly by competing for a common effector. Synaptotagmins are also able to interact with non-neuronal syntaxin, indicating that they can play a role in a variety of cell types from endocrine and immune systems.

Synaptophysin (p38) is a major integral membrane protein of small presynaptic vesicles ($M_r$ of 38 kDa). Synaptophysin is found in vesicle membranes of neurons and some neuroendocrine cells, and represents a major cytosolic site of $Ca^{2+}$ binding. The presence of phosphorylation sites for tyrosine kinase could indicate that phosphorylation modulates protein activity. The primary structure was deduced from the cDNA sequence, leading to the proposition that synaptophysin spans the membrane four times with its N- and C-terminals located in the cytoplasmic side. This transmembrane topology resembles the gap junction monomer. Furthermore, synaptophysin forms hexameric complexes in the vesicle membrane with properties similar to those of an ionic channel. Reconstituted into planar lipid bilayers, the synaptophysin channel has a conductance of 150 pS. The open probability of the channel is voltage dependent, increasing at positive voltages but no activity is observed at negative voltages (Thomas *et al.,* 1988). An antibody against synaptophysin impairs the release of catecholamine in chromaffin cells.

The annexin family includes several $Ca^{2+}$- and phospholipid-binding proteins with conserved structure. Two of these are thought to be involved in the mechanism of exocytosis. Synexin (annexin VII) is a calcium-binding protein (47 kDa) with four transmembrane domains. This protein demonstrates a voltage-dependent $Ca^{2+}$ channel activity. Synexin is able to induce the aggregation of chromaffin vesicles in the presence of $Ca^{2+}$. A model for the synexin-driven $Ca^{2+}$-dependent membrane fusion has been proposed in which synexin monomers polymerize as the $Ca^{2+}$ concentration of $Ca^{2+}$ increases. The polymerized synexin forms a hydrophobic bridge between the two membranes. Annexin II (calpactin) is located on the cytoplasmic side of the plasma membrane in chromaffin cells. Annexin II can be found in monomeric (36 kDa) and heterotetrameric ($p36_2$ $p10_2$) forms. Sites of phosphorylation for PKC, cAMP-, and calmodulin-dependent protein kinases have been localized within the N-terminal domain; phosphorylation of these sites could inactivate the protein. A channel-like activity has not been reported. A possible role for annexin II could be to link the granule with the plasma membrane (docking?) and/or to induce fusion (Burgoyne, 1991).

## III. Cellular and Molecular Events in Chromaffin, Mast Cells, and Neuronal Synaptic Vesicles

### A. Dynamic Changes in the Cytoskeletal Networks Required for Exocytosis

Secretion is a process that requires (1) the movement of secretory vesicles toward the plasma membrane, (2) the fusion of vesicles with the plasma membrane, and (3) subsequent release of secretory contents in the cell exterior. The process of secretion is mediated by contractile elements associated with the secretory vesicles or present elsewhere in the cells. Because F-actin is preferentially localized in the cortical surface of the chromaffin cell, F-actin may act as a barrier to the secretory granules, impeding their contact with the plasma membrane (Trifaró et al., 1992). Stimulation of the cell produces disassembly of actin networks and removal of the barrier (Fig. 6) Studies using fluorescent rhodamine-labeled phalloidin (a drug that stabilizes actin filaments *in vitro* and stops actin disassembly on stimulation) and actin antibodies have shown that, in resting cells, a filamentous actin network is visualized as a strong cortical fluorescent ring (Burgoyne, 1991). Cholinergic receptor stimulation produces a fragmentation of the fluorescent ring, leaving cell cortical areas devoid of fluorescence. These changes are accompanied by a decrease in F-actin associated with a concomitant increase in G-actin.

The actin-binding proteins present in the cell cortex also undergo changes on stimulation. In a resting cell, fodrin is localized in the cell cortex. On stimulation, it rearranges into patches beneath the plasma membrane. This redistribution could be related to the clearing of exocytotic sites at the plasma membrane. Scinderin is a cytosolic protein that shortens actin filament length when $Ca^{2+}$ is present in the medium (Fig. 6B). Stimulation induces both redistribution of scinderin from cytosol to cell cortex and F-actin disassembly, which precedes exocytosis. Thus, stimulation-induced redistribution of scinderin and F-actin disassembly produce subplasmalemmal areas of decreased cytoplasmic viscosity and high secretory vesicle mobility. All these processes require the presence of $Ca^{2+}$ in the extracellular medium. Therefore, only secretagogues that induce $Ca^{2+}$ entry are able to produce these effects.

F-actin network disassembly has also been observed in mast cells on stimulation and in synaptosomes on depolarization. The existence of actin-binding proteins that regulate the dynamics of actin networks strongly suggests a role for these proteins in the disassembly of actin filaments triggered by cell stimulation. Actin disassembly is closely correlated with exocytosis. Direct evidence for an actin barrier has come from the use of drugs that affect actin assembly and disassembly. Cytochalasin B and DNAse I prevent actin assembly and drive the system toward net disassembly and increased secretion in permeabilized chromaffin cells. Results from the use of toxins also support the concept of a cortical actin barrier. Botulinum C2 toxin, which ADP-ribosylates actin and inhibits actin polymerization, enhances secretion in PC12 cells. In contrast, tetanus and botulinum A toxins, which block actin disassembly, inhibit exocytosis on cholinergic stimulation.

In isolated chromaffin cells, stimulation with nicotinic agonists can result in secretion of about 30% of the total catecholamine. Electron microscopic observations show that a small number of granules lay within the exclusion zone of the cell cortex. This demonstrates the importance of changes in cortical actin to allow movement of the bulk of the granules involved in a full secretory response. Nevertheless, low levels of exocytosis, due to these granules in the cortical exclusion zone, could occur without generalized changes in cortical actin. Thus, in a physiological situation, where relatively few exocytotic events occur per stimulus, changes in cortical actin may not be necessary for the initial wave of exocytosis, but are required for the movement, into the cortical exclusion zone, of granules ready for the next stimulus. A similar picture has emerged from studies on the nerve terminal cytoskeletal phosphoprotein, synapsin I, which is believed to cross-link synaptic vesicles and release them following depolarization and phosphorylation of the synapsin I.

### B. Physical Events Associated with the Fusion of Vesicles to Plasma Membrane

In regulated exocytosis, fusion of the secretory granules with the plasma membrane is triggered by an appropriate signal. The fusion process, following the triggering signal, can be very fast such as in mammalian nerve terminals, with delays of less than 0.2 msec between the action potential and exocytosis. In some cells, however, delays of 0.2 sec (chromaffin cells) to 50 sec (mast cells) are observed. This delay is thought to be caused by the time required for production of second messengers possibly involved in exocytosis and removal of the cytoskeletal barrier, which immobilizes the vesicles. However, the physical interactions between the granule membrane and the plasma membrane remain similar whether the exocytotic delay is fast or slow. Accordingly, the following fusion events are described along general lines based on a sequence proposed by Almers (1990).

#### 1. Capacitance Jump

Each time a secretory granule fuses with the plasma membrane, the total capacitance of the cell increases by a value proportional to the surface of the new membrane added to the existing cell membrane. Assuming that biolog-

A

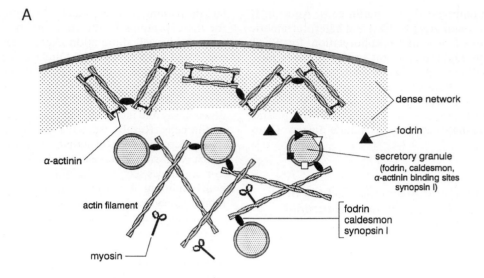

B

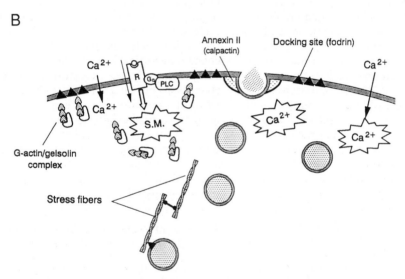

**FIG. 6.**  Dynamic changes in cytoskeleton during exocytosis. (A) In resting cells ($[Ca^{2+}]_i \approx 100$ n$M$, microfilaments form a network that is cross-linked and stabilized by fodrin and by $\alpha$-actinin. Calmodulin, scinderin, and myosin are also present in unactivated forms. (B) Activation ($[Ca^{2+}]_i \approx 1$ m$M$) induces (1) dissociation of actin from fodrin, (2) patching of fodrin along the plane of the plasma membrane (docking site), and (3) activation of scinderin and gelsolin with subsequent capping and severing of the actin microfilaments. These events result in a decrease in viscosity, which favors movement of granules toward the plasma membrane-releasing sites. Actin–myosin interactions could facilitate granule displacement in cytosol.

ical membranes have a constant specific capacitance of about 1 $\mu$F/cm$^2$ allows a simple calculation of the granule size. On fusion of a single vesicle, the capacitance value increases abruptly to a new stable value as the membrane of the vesicle and the plasma membrane become continuous and the vesicle lumen opens into the extracellular space. Figure 7 illustrates the equivalent circuitry of a resting cell (Fig. 7A) and that of a cell undergoing exocytosis (Fig. 7B). During degranulation, several granules fuse with the plasma membrane, which produces a typical staircase recording (Fig. 7C). Degranulation in three different cell types is presented in Fig. 8: human neutrophil (Fig. 8A), guinea-pig eosinophil (Fig. 8B), and horse eosinophil (Fig. 8C). Note that the amplitude of the individual step capaci-

tance is lower in human neutrophil than in horse eosinophil, reflecting the different sizes of the vesicles. A capacitance step amplitude histogram is built by measuring the step height of a large number of individual events. An example of this is illustrated in Fig. 8D for neutrophils. The capacitance step amplitudes range from 1 to 6 fF (1 to 6 $10^{-15}$ F) with a greater number of events having an amplitude of 2 fF. Assuming a spherical shape, the diameter of the granule can be calculated from the step change in capacitance, $\delta C_M$, in fF by the relation

$$\delta C_M = \pi d^2 \qquad (1)$$

The frequency distribution of the sizes of the granules obtained from the capacitance step amplitude histogram

A        B        C

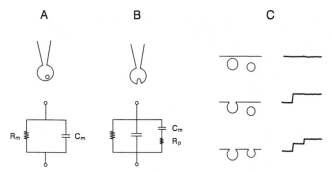

**FIG. 7.** Capacitance measurement and equivalent electrical circuit. (A) The cell is patch-clamped in the whole-cell configuration. The cell is at rest and an unfused granule is shown near the plasma membrane. The equivalent circuit is represented by membrane resistance $R_M$ in parallel with membrane capacitance $C_M$. The series resistance due mainly to the micropipette has been omitted for clarity. (B) Once the granule has fused with the plasma membrane, its capacitance $C_v$, proportional to the surface of membrane added, is added to the total capacitance of the cell. $R_p$ is the fusion pore resistance, which rapidly decreases as the pore dilates. (C) The fusion of a granule with the plasma membrane increases the capacitance by step. Granules of identical sizes induce the same increases in capacitance.

is represented in Fig. 8E for guinea-pig eosinophils. The distribution is fitted (smooth line) by the sum of two Gaussian curves with means of 520 and 590 nm (Lindau and Gomperts, 1991). This size distribution fits in well with the morphometric data obtained by direct microscopic observation of the secretory vesicles. Capacitance step measurements have been achieved in a variety of cell types including adrenal chromaffin cells, mast cells, pancreatic acinar cells, neutrophils, eosinophils, basophils, pituitary lactotrophs, and nerve terminals derived from the posterior pituitary. Capacitance step values generally range between

1 and 30 fF, corresponding to granule diameters of 0.2–1 $\mu$m.

Giant vesicles with diameters ranging between 1 and 5 $\mu$m are found in mast cells from a strain of genetically defective beige mice (strain C57BL/6J-bg$^j$/bg$^j$). In these mice, mast cells and other granulocytes are unable to limit the size of their secretory vesicles; mast cells contain 10 to 40 giant vesicles, which can easily be observed under photonic microscopy. These cells thus provide an ideal material for exocytosis studies and, for this reason, have been extensively used. The capacitance method offers the possibility to study line degranulation; time analysis of the secretory process reveals that the granules fuse sequentially, one by one, with the plasma membrane. However, in mast cells, capacitance step analysis demonstrates the presence of step values greater than 60 fF, which could not be produced by the fusion of a single vesicle. A detailed analysis of the capacitance step histogram reveals a multimodal distribution of granules size, which indicates that the larger granules could be formed by the fusion of two to five single granules with each other.

### 2. Capacitance Flickering

The pattern of the staircase increase in capacitance during degranulation demonstrates that each step builds on the previous one, indicating that the fusion event is irreversible. However, closer observation of capacitance jumps in mast cells reveals the existence of "on" and "off" steps. The "on" step is produced by the opening of a small-diameter pore, called a **fusion pore,** which adds the surface of the vesicle to that of the cell. Once opened, the fusion pore can close quickly and reopen (Fig. 9). Rapid oscillation between open and closed states gives the appearance of a flickering of the capacitance (Almers, 1990). Size distribution of capacitance steps during the flickering period shows that large steps are absent and that all steps remain in the range of values expected for single vesicle fusion events. The size of the fusion pore is proportional to its conductance. An unexpected result is that the reclosing of the pore occurs, not only in small-diameter pores (low conductance), but also in larger sized pores having a conductance of several nano-Siemens. Capacitance flickering is rather frequent in mast cells, but very few have been observed in eosinophils. This raises the question of the undetectable presence of flickering during the fusion of all vesicles before

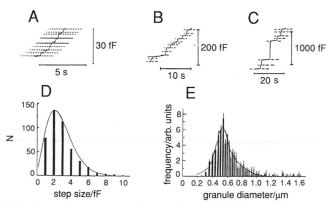

**FIG. 8.** Analysis of step capacitance. Capacitance recordings obtained from different cells: (A) human neutrophils, (B) guinea-pig eosinophils, and (C) horse eosinophils. Note the staircase appearance of the recordings, which reflects the sequential fusion of individual granules and the size of the step capacitances, which are proportional to the size of the granules. (D) Frequency distribution of step sizes for human neutrophils. (E) Distribution of granule size for guinea-pig eosinophils derived from capacitance measurement. (Reproduced from Lindau and Gomperts, 1991.)

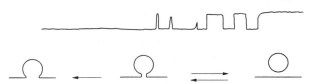

**FIG. 9.** Flickering of the membrane capacitance. Fusion of the granule is reversible and can oscillate between an unfused state and a fused state. Each time the granule fuses with the plasma membrane, the capacitance of the cell increases by step. If the fusion pore closes, the capacitance recovers its initial value; incomplete recovery indicates very brief closures exceeding the speed of the recording system. Eventually, a stable fused state can be reached.

the irreversible fused state is attained. Indeed, the initial phase of fusion is a very fast process that cannot be faithfully recorded by the speed of the recording techniques used. The earliest steps as well as short-lived flickering are certainly missed.

### 3. Fusion Pore

Freeze-fracture images of exocytosis in mast cells and neutrophils reveal the presence of narrow pores formed between the granules and the plasma membrane. The size of the pore increases as fusion progresses, allowing the release of vesicle contents. The presence of pores joining two secretory vesicles is also observed (Lindau and Gomperts, 1991). Electrically, the first opening of the fusion pore generates a brief transient current from which several parameters can be deduced.

*a. Size of Fusion Pore.* The conductance of the fusion pore can be calculated from the current transient produced by movement of the charges between two differently charged membranes, that is, the plasma membrane and vesicle membrane. The initial value of the current $I_0$ is related to the initial pore conductance $g_0$ by the relationship

$$g_0 = I_0/(E_c - E_v) \qquad (2)$$

where $E_c$ is the potential across the plasma membrane and $E_v$ the potential across the vesicle membrane.

The initial conductance of the fusion pore in mast cells was found to be 230 pS. The corresponding size diameter was calculated assuming a resistivity of 100 $\Omega$m and a pore length of 15 nm. The abrupt increase in conductance corresponds to the all-or-none opening of a pore having an inner diameter of less than 2 nm. For comparison purposes, gap junction channels have conductances that vary from 80 to 240 pS and a diameter of approximately 2 nm. It can thus be proposed that the fusion pore is a large protein spanning across two membrane thicknesses, and having a structure and function resembling that of a channel. Data obtained from electron microscopic observations reveal that the smallest pores that can be observed have diameters between 30 and 50 nm, values far higher than the values reported from conductance measurements. A rapid dilatation of the pore soon after its formation might explain this discrepancy. Once the pore has been formed, the conductance increases abruptly to a value near 250 pS. Thereafter, the pore begins to dilate, possibly by infiltration of lipid molecules between the subunits of the protein structure, followed by an increase in conductance. The pore conductance increases from 500 to 3000 pS in about 25 msec; a plateau is reached after 150 msec, corresponding to a pore diameter of more than 16 nm.

*b. Does the Fusion Pore Leak?* Once established, the diameter of the fusion pore is similar to that of the gap junction, that is, between 1.5 and 2 nm. Because gap junctions allow the intercellular passage of molecules weighing up to 1900 Da, the question can be asked as to whether granule contents can leak through the fusion pore. In guinea-pig eosinophils, the irreversible fusion of the granule with the plasma membrane is occasionally preceded by a long-lived fusion pore having a conductance of 70–250 pS. When granules were loaded with the fluorescent dye quinacrine, no release of the dye in the extracellular medium could be observed during the life of fusion pore. Only after the fusion pore had completely dilated, and the vesicle reached its irreversible state of fusion, was the dye released. These results indicate that the fusion pore is too narrow for the release of granule contents.

However, computations based on the diffusion of a small molecule such as histamine predict that a granule with a diameter of 0.8 mm should release its contents rapidly through the opening of the fusion pore. Recently, experimental proof was provided by Neher and collaborators using bovine chromaffin cells (Chow *et al.,* 1992). Secretion was measured by voltametry, while cells were studied by voltage-clamp techniques. The cells were stimulated by depolarizing the membrane from a holding potential of $-60$ mV to a step potential of $+10$ mV for 25 msec to activate the $Ca^{2+}$ channels. The amperometric signals, which represent the detection of the released catecholamine molecules by the carbon electrode (potential of 800 mV), were transient with a fast or slow rising phase and variable amplitudes. A histogram of integrals of current transient amplitude obtained on several cells showed that the mean charge transfer had a value of 0.76 pC, which is equivalent to the release of $2.36 \times 10^6$ molecules of catecholamine. Sometimes, larger events were detected presumably corresponding to multigranular exocytosis. One interesting and surprising feature was that most of the fast-rising events were preceded by a small "foot" or "pedestal," as illustrated in Fig. 10A. The mean duration

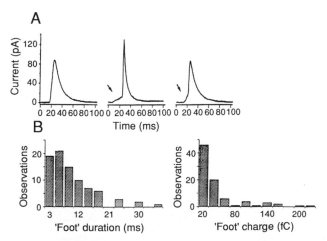

**FIG. 10.** Amperometric current recorded in chromaffin cells. (A) The amperometric signal recorded with the voltametry method shows a fast rising phase followed by a slower decrease. Occasionally, the rising phase is preceded by a pedestal or foot. The foot signal is thought to be due to the leak of catecholamine by the fusion pore before its complete dilatation. (B) Histogram of the foot signal duration (left panel) and histogram of the charge of the foot signal (right panel), with a mean of 34 fC corresponding to $1.05 \times 10^5$ molecules. [Adapted with permission from Chow *et al.* (1992). *Nature,* **356,** 60–63. Copyright 1992 Macmillan Magazines Limited.]

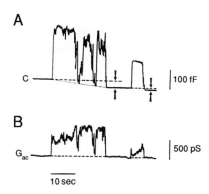
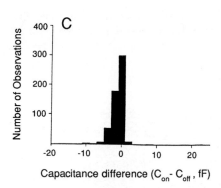

**FIG. 11.** Decrease of total membrane capacitance after fusion. (A) Capacitance and (B) conductance measurements during transient fusion of a giant secretory granule from beige mouse mast cells. Capacitance and conductance increase on fusion. The fusion pore closes twice with a decrease in conductance and capacitance. Note that the conductance returns to its initial level but that the capacitance of the cell is lower than its initial value. This indicates a reduction of the surface of the plasma membrane due to a leak of lipid molecules toward the granule membrane. (C) Histogram showing the size distribution of the capacitance difference measured after transient fusion in mast cells. (Reproduced from Monck *et al.*, 1990.)

of the foot was 8.26 msec and the mean charge was 34 fC, equivalent to $1.05 \times 10^5$ molecules (Fig. 10B). The foot was interpreted as reflecting a slow leakage of catecholamine molecules through the fusion pore formed during the early step of the fusion process. A second important result of this study was the discovery and quantification of a long latency period between the end of the stimulus and catecholamine release. The majority of the secretory events occurred 5–100 msec after the end of the electrical stimulus, which is rather long when compared to nerve endings where the delay is about 1 msec. A complex cascade of intracellular events triggered by the increase of cytosolic $Ca^{2+}$ concentration could be responsible for this long latency.

### 4. Membrane Tension as a Driving Force for Fusion

As previously described, flickering is characterized by the opening and closing of the fusion pore. In some cases, it has been shown that after a period of flickering, the capacitance of the plasma membrane declines to a value lower than its initial value (Monck *et al.,* 1990). Figure 11A shows that this decrease in capacitance is paralleled by an increase in the conductance of the fusion pore (Fig. 11B), thus establishing a relationship between the dilatation of the pore and decreased plasma membrane capacitance. Once the flickering stops, the conductance recovers its initial value but the whole-cell capacitance remains lower. These results are interpreted as reflecting a decrease of the plasma membrane surface due to a net transfer of material to the granule membrane. The difference between the "on" and "off" steps (found in one-half of the transient fusions with values between −2 to −4 fF) is proportional to the duration of contact between the plasma membrane and the secretory vesicle (Fig. 11C). The rate of cell surface area reduction is 0.16 $\mu m^2 sec^{-1}$. The transfer of membrane is facilitated by the fact that the membrane of the secretory vesicle is under tension. Upon fusion, a movement of phospholipid molecules occurs from the plasma membrane to

the granule membrane. A possible mechanism for generating tension in the granule membrane is osmotic swelling. However, fusion can proceed in isotonic, hypotonic, or hypertonic solutions with no change in the kinetics of capacitance increase.

### 5. Fusion Steps

Several lines of evidence favor the hypothesis that a pore-forming protein could be involved in the fusion process. The abrupt opening of the fusion pore with an initial conductance of 250 pS, similar to the conductance of the gap junction channel, and the occurrence of rapid flickering are the strongest arguments. A similar mechanism was analyzed during the fusion of fibroblasts expressing influenza virus hemagglutinin (a membrane-spanning fusion protein) and red blood cells. As described previously for exocytosis, the initial event is the opening of an aqueous pore having an initial conductance of 600 pS, eventually followed by a period of flickering activity (Spruce *et al.,* 1989). Prior to their fusion, secretory granules are docked to the plasma membrane. In synaptic nerve endings, docking is localized to a restricted region and, on stimulation, yields localized secretions. Docked granules have been observed in a variety of systems including chromaffin cells. In these latter cells, localized secretion was also reported depending on the applied stimulus.

Figure 12 illustrates a mechanism of fusion in five steps (a–e) proposed by Almers (1990). In this model, the docking structure and the pore-forming protein are considered to be the same entity, perhaps located in the vesicle membrane. During the docking phase, the protein inserts itself into the plasma membrane. Activation of the protein by a cytosolic messenger induces a conformational change to the open state (conductance 250 pS), with an eventual flickering before the dilatation phase by diffusion of lipid molecules between the pore subunits (conductance increases slowly by several nano-Siemens) (Fig. 12, steps f–h). Finally, the fusion becomes irreversible.

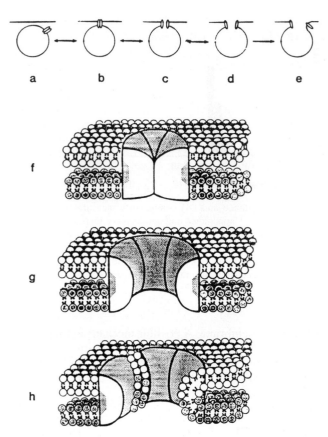

**FIG. 12.** Model for exocytosis showing an hypothetical formation of the fusion pore. Steps a–e are the five steps describing the docking of the granule and the formation of the fusion pore. Steps f–h illustrate the position of the fusion pore complex in the two lipid bilayers of the plasma membrane and the vesicle membrane. Infiltration of lipid molecules between the subunits of the fusion pore is responsible for dilatation of the pore. Stippled surfaces of the pore protein are hydrophilic; clear surfaces are hydrophobic. (Reproduced, with permission, from the *Annual Review of Physiology,* Volume 52, © 1990 by Annual Reviews Inc.)

## C. Control of Exocytosis

Exocytosis occurs in a variety of electrically excitable and nonexcitable systems in response to receptor activation. The regulatory pathways that couple stimulation and secretion vary widely among cell types. In the following section, analysis of the factors controlling exocytosis focuses mainly on two well-studied systems: the chromaffin cell from the adrenal medulla (which belongs to the class of excitable cells) and the mast cell from the immune system (a nonexcitable cell).

### 1. Effectors of Exocytosis

*a. Calcium Signaling and Sources of Calcium.* In many cells, $Ca^{2+}$ is the key signal for triggering exocytosis. In chromaffin cells, the resting $Ca^{2+}$ concentration has a value ranging between 50 and 100 n$M$. Chromaffin cells can be stimulated by two different classes of acetylcholine receptors. The nicotinic receptor has a channel-like structure consisting

of five transmembrane subunits. The binding of two acetylcholine molecules opens the channel, which leads to a large net influx of $Na^+$ ions. This influx causes a membrane depolarization, which can activate voltage-dependent $Ca^{2+}$ channels. The nicotinic receptor channel is also permeable to $K^+$ ions and, to a lesser extent, $Ca^{2+}$ ions. The muscarinic receptor belongs to the family of the seven-span transmembrane domain receptor proteins, which utilizes the G-protein cascade pathway as its signal-transducing mechanism. Stimulation of chromaffin cells with the cholinergic agonist nicotine induces a rapid rise in $[Ca^{2+}]_i$ (measured with $Ca^{2+}$ sensitive fluorescence dye) up to 1 $\mu M$; this $[Ca^{2+}]_i$ increase is followed by exocytosis. Muscarinic stimulation of chromaffin cells also increases $[Ca^{2+}]_i$ but does not induce secretion.

In a $Ca^{2+}$-free external medium, secretion is abolished regardless of the stimulus. This observation reinforces the fact that $Ca^{2+}$ influx from the external medium is crucial for secretion in chromaffin cells. Several voltage-dependent $Ca^{2+}$ channels have been described in chromaffin cells, namely, (1) the L-type dihydropyridine-sensitive channels, (2) the N-type $\varpi$-conotoxin-sensitive dihydropyridine-insensitive $Ca^{2+}$ channels, and (3) the dihydropyridine-sensitive facilitation $Ca^{2+}$ channels. Electrical depolarization, $K^+$ depolarization, and nicotinic stimulation open the $Ca^{2+}$ channels, allowing an immediate influx of $Ca^{2+}$ ions from the external medium. The video-imaging technique allows the recording of $[Ca^{2+}]_i$ with good spatial definition. As shown in Fig. 13 (right panel), the increase in $[Ca^{2+}]_i$ is restricted to the immediate vicinity of the plasma membrane after activation of $Ca^{2+}$ channels. The need for a high $Ca^{2+}$ concentration for exocytosis could imply that secretory granules are stored near the $Ca^{2+}$ channels. Recent experimental evidence confirms that when $Ca^{2+}$ channels are activated, the concentration of $Ca^{2+}$ at the secretory sites exceeds 10 $\mu M$ (Augustine and Neher, 1992).

In chromaffin cells, internal $Ca^{2+}$ release is provided by two different pools: the $InsP_3$-sensitive pool and the $Ca^{2+}$-induced $Ca^{2+}$ release pool. The activation of the PtdIns-specific phospholipase C by the $G_q$-protein-coupled receptor or by cytosolic $Ca^{2+}$ elevation induces the hydrolysis of the $PtdInsP_2$, thus generating two messengers: $InsP_3$ and DAG. The binding of $InsP_3$ to specific sites on the endoplasmic reticulum induces $Ca^{2+}$ release. $InsP_3$ receptors were also recently found in secretory granules of endocrine and neuroendocrine cells. Their activation by $InsP_3$ induces the release of $Ca^{2+}$ ions from the granules and provides a localized increase of $[Ca^{2+}]_i$. $Ca^{2+}$ ion alone or caffeine can trigger the release of $Ca^{2+}$ from the second pool. A different type of receptor, the muscarinic receptor, induces the release of $Ca^{2+}$ from the $InsP_3$-sensitive store in chromaffin cells without $Ca^{2+}$ channel activation. The observed $Ca^{2+}$ increase is low and sometimes localized to one pole of the cell (Fig. 13, left panel). Despite this $[Ca^{2+}]_i$ increase, there is little or no stimulation of secretion at the site where $[Ca^{2+}]_i$ has increased. A more uniform $[Ca^{2+}]_i$ increase can be generated by release from the $Ca^{2+}$-induced $Ca^{2+}$ release pool by caffeine or from the $InsP_3$ pool by introduction of GTP-$\gamma$S into the cell. However, exocytosis still remains unstimulated. These results conclusively dem-

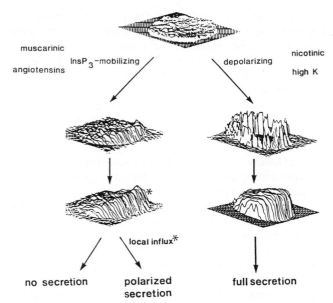

muscarinic
angiotensins

InsP$_3$-mobilizing

depolarizing

nicotinic
high K

local influx*

no secretion

polarized
secretion

full secretion

**FIG. 13.** Ca$^{2+}$ signal in a chromaffin cell in response to various agonists. The distribution of [Ca$^{2+}$]$_i$ is obtained by video imaging. This figure illustrates the fact that concentration and distribution of [Ca$^{2+}$]$_i$ are important for full secretion. The right portion of the figure shows that a depolarization of the plasma membrane, induced by high [K$^+$]$_o$ or nicotinic agonist, opens voltage-activated Ca$^{2+}$ channels with an immediate and high [Ca$^{2+}$]$_i$ increase at the periphery of the cell. Afterward, [Ca$^{2+}$]$_i$ increases in the entire cell due to release of Ca$^{2+}$ from internal stores; a full secretion is obtained. The left portion of the figure shows that release of Ca$^{2+}$ from the IP$_3$-sensitive pool is not sufficient to induce secretion although some localized secretion can be obtained in the region of the cell where [Ca$^{2+}$]$_i$ increase was the highest. (Reproduced with permission from Burgoyne, 1991.)

onstrate that, in chromaffin cells, only a considerable rise in Ca$^{2+}$ concentration in the proximity of the plasma membrane is able to trigger exocytosis. This highly localized [Ca$^{2+}$]$_i$ increase can only be achieved via the Ca$^{2+}$ channels, and not by a small increase due to release from internal stores. On the other hand, histamine is a potent stimulator of chromaffin cells that does not activate the opening of the classical voltage-dependent Ca$^{2+}$ channels. Nevertheless, the [Ca$^{2+}$]$_i$ increase is localized near the plasma membrane. Apparent contradictory results were recently published by Augustine and Neher (1992). In bovine chromaffin cells, bradykinin (which releases Ca$^{2+}$ from the InsP$_3$-sensitive pool), ionomycin (which releases Ca$^{2+}$ from internal stores), as well as dialysis of the cell with a low Ca$^{2+}$ medium (0.34 $\mu M$) are able to trigger exocytosis. Furthermore, the relationship between [Ca$^{2+}$]$_i$ and the rate of secretion is insensitive to the means by which the rise in [Ca$^{2+}$]$_i$ was brought about, that is, by a dialysis, internal release, or Ca$^{2+}$ influx through channels.

Ca$^{2+}$ alone is not able to trigger histamine secretion in mast cells. There is a requirement for a guanine nucleotide together with Ca$^{2+}$ for exocytosis. Mast cells are stimulated by antigenic binding on cell surface IgE receptors or by compound 48/80. A sudden rise in Ca$^{2+}$ concentration (released from the InsP$_3$-sensitive pool) is observed, reaching

a level of up to several micromolars. The signal is transient and [Ca$^{2+}$]$_i$ declines within several seconds to its original baseline value. Degranulation begins soon after this transient response. Two important features should be noted concerning [Ca$^{2+}$]$_i$ and secretion in mast cells: (1) The [Ca$^{2+}$]$_i$ transient is not dependent on the presence of Ca$^{2+}$ in the external medium and (2) simultaneous recording of capacitance and membrane conductance reveals that conductance is constant throughout the duration of exocytosis. These results indicate that Ca$^{2+}$ entry through Ca$^{2+}$ channels is not a required signal for exocytosis, but rather its release from internal stores. Moreover, voltage-dependent Ca$^{2+}$ channels have never been found in mast cells. Recently, a Ca$^{2+}$ current, called $I_{CRAC}$ (calcium-release activated calcium), was described in these cells (Hoth and Penner, 1992); $I_{CRAC}$ is activated by the depletion of intracellular Ca$^{2+}$ pools. The role of $I_{CRAC}$ in nonexcitable cells could be to maintain a high [Ca$^{2+}$]$_i$ and to replenish empty Ca$^{2+}$ stores after stimulation. A possible direct role in exocytosis has not yet been shown.

One conclusion is that Ca$^{2+}$ release from internal stores is sufficient for exocytosis in mast cells. However, intracellular application of InsP$_3$ induces a transient [Ca$^{2+}$]$_i$ increase but does not trigger secretion, emphasizing that the signal is more complex. Application of the nonhydrolyzable GTP analogue GTP-$\gamma$S (100 $\mu M$) inside a mast cell (via a patch pipette) induces a transient [Ca$^{2+}$]$_i$ increase, after a delay of 10–20 sec, followed by an increase in capacitance. GTP-$\gamma$S activates G-proteins, including the G$_q$ protein coupled to PLC, with a resulting production of InsP$_3$ and DAG. Ca$^{2+}$ is released from the InsP$_3$-sensitive pool by the same mechanism as with the direct application of InsP$_3$ described earlier. However, in this case, exocytosis is stimulated, leading to the conclusion that GTP-$\gamma$S had activated a G-protein that plays a crucial role in exocytosis. A possible role for G$_q$ was ruled out, however, by inhibition of PLC with neomycin. A new G protein, called G$_E$, is now thought to be involved in exocytosis (Lindau and Gomperts, 1991).

*b. Guanine Nucleotides.* As seen earlier, Ca$^{2+}$ and a guanine nucleotide are necessary and sufficient effectors to ensure secretion in mast cells. The stable analogue GTP-$\gamma$S is the commonly used guanine nucleotide, although any ligand that binds to G protein is able to stimulate secretion: GTP, ITP, XTP, GTP-$\gamma$S, GppNHp, GppCH$_2$p, and AlF$_4^-$. In chromaffin cells, two types of observations were recorded: (1) GTP-$\gamma$S increases the Ca$^{2+}$-sensitivity of secretion in permeabilized cells and (2) GTP analogues stimulate secretion in a Ca$^{2+}$-independent manner (Morgan and Burgoyne, 1990). This secretion is blocked by GDP-$\beta$S suggesting that a G$_E$-like protein could be involved. GTP analogues do not enhance the secretion induced by high Ca$^{2+}$ concentration (10 $\mu M$), which indicates that the two stimuli (GTP and Ca$^{2+}$) act on the same exocytotic pathway.

Evidence for the involvement of a GTP-binding protein in secretion has been obtained from a variety of cells: neutrophils; platelets; parathyroid; pituitary lactotroph, gonadotroph and melanotroph; insulin secreting cells; and pancreatic acinar cells. Various effects have also been observed in these cells: GTP or GTP analogues behave as effectors able to trigger a $Ca^{2+}$-independent secretion, or as modulators that increase $Ca^{2+}$ sensitivity, or more surprisingly as inhibitors that block the exocytotic pathway at a late stage. The exact nature of the G proteins is not yet known, although some analogy can be made with GTP proteins involved in vesicular traffic (Rothman and Orci, 1992).

The small Ras-like GTPases are involved in the formation, transport, and fusion of vesicles. In yeast, the small Ras-related GTP-binding protein SEC4 is required for targeting and/or fusion of vesicles to the plasma membrane. Rab 3A from the rab-family proteins has been located in the membrane of neurosecretory vesicles. More than 30 Rabs have been identified. They have a regulatory role in secretion. Under its GTP-bounded form, Rab 3A inhibits the secretion possibly by stabilizing a fusion-incompatible conformation. Under an appropriate signal, GTP is hydrolyzed to GDP with the help of GAP (a GTPase-activating protein); this exchange of GTP for GDP releases the inhibition. The fusion can then proceed to subsequent steps. Two accessory proteins are involved in the cycle: rabphilin-3A, which binds on the GTP-Rab 3A form. Rabphilin is phosphorylated by many kinases and contains binding sites for $Ca^{2+}$ and phospholipids. GDI (a GDP-dissociation inhibitor) binds to Rab 3A under its GDP form and removes it from the membrane. After dissociation of the GDI–GDP-Rab 3A complex, GTP-Rab 3A can bind again to another vesicle. Thus rabphilin-3A and GDI control the Rab 3A cycle but do not directly participate in the fusion (Südhof, 1995).

## 2. Modulators of Exocytosis

*a.* ATP.   Permeabilized cells rapidly lose their secretory response to agonists unless ATP is present in the medium. However, the sequence of exocytosis can be separated in an early phase that requires MgATP to proceed and a late phase that is MgATP independent. ATP could act at various levels of the secretory response, acting as a substrate for protein phosphorylation or modulating various kinases involved in secretion. The last ATP-requiring steps in exocytosis have been identified; they involve ATPase, NSF, and the formation of $PtdInsP_2$. NSF forms a large (20S) complex with the attachment proteins (SNAPs) and SNAP receptors (SNARE), which has been proposed to be the fusion particle (see Fig. 5). The hydrolysis of ATP by NSF is thought to produce energy for the fusion (priming step). The formation of $PtdInsP_2$ by phosphatidylinositol-4-phosphatase-5-kinase and a phosphatidylinositol transfer protein requires the presence of ATP. $PtdInsP_2$ is thought to be involved in the interaction between cytoskeleton and secretory granules (docking step). Docking and priming of secretory granules are thus ATP dependent. During vesicle transport through the Golgi, ATP hydrolysis by NSF occurs after docking, but in granule secretion no such evidence exists.

*b. Cyclic Nucleotides.*   In chromaffin cells, cAMP concentration increases after cholinergic stimulation, leading to the hypothesis that cAMP could have a role in exocytosis. In pancreatic $\beta$ cells, the cAMP-dependent protein kinase A enhances secretion. However, contradictory results have been reported on the effects of high and low cAMP concentrations on secretion modulation, that is, inhibition and potentiation, respectively. Nicotinic-induced secretion in chromaffin cells is inhibited by high cGMP concentration, whereas low concentrations potentiate the secretory response.

*c. Calmodulin.*   A calmodulin-binding protein of 65 kDa, called 65-CMBP or p65, has been found in several secretory vesicles such as synaptic, neurohypophyseal, chromaffin, platelets, and pancreatic islet granules. This protein also binds to phospholipids with high affinity. A possible role of calmodulin in exocytosis was investigated by blocking its action. The calmodulin antagonist calmidazolium inhibits secretion from intact chromaffin cells. Antibodies raised against calmodulin inhibit the $Ca^{2+}$-dependent binding of calmodulin to vesicle membrane by acting on the docking and/or fusion steps of exocytosis. However, some reports indicate that the less specific calmodulin inhibitor trifluoperazine (TFP) has no effect on permeabilized cells. At present, it appears that calmodulin is not essential for $Ca^{2+}$-dependent exocytosis, although interaction between calmodulin and cytoskeletal proteins should nevertheless be considered.

*d. Protein Kinase C.*   A role for PKC could be inferred from the fact that any antagonist that stimulates secretion also induces activation of PKC. Indeed PLC activation (by a G protein or by $[Ca^{2+}]_i$ elevation) generates two messengers: $InsP_3$ and the PKC activator, DAG. PKC has a modulating role in $Ca^{2+}$ affinity, by decreasing $Ca^{2+}$ requirements for exocytosis. Inhibition of PKC by staurosporine or through down-regulation partially inhibits secretion. However, in gonadotrophs, PKC-stimulated LH exocytosis is independent of $Ca^{2+}$. Obviously, the role of PKC in exocytosis is not yet completely understood. It is possible that PKC is not essential for $Ca^{2+}$-dependent exocytosis (modulator), but that a second $Ca^{2+}$-independent pathway could coexist in which PKC acts as an effector of secretion. Indeed, it has been recently shown in chromaffin cells that PKC acts at a late state in exocytosis before the final $Ca^{2+}$-sensitive step. The role of PKC would be to increase the size of the readily releasable pool (RRP) of secretory granules by speeding their maturation after they dock with the plasma membrane. Moreover, direct stimulation of PKC by the phorbol ester PMA (phorbol 1,2 meristate 1,3-acetate) leads to a disruption of the actin network near the plasma membrane, which increases the number of docked granules. Several proteins involved in the regulation of the cytoskeleton are substrates for PKC: annexin I, annexin II, and MARCKS (see Fig. 4).

*e. Phospholipase $A_2$ and Arachidonic Acid.*   A possible fusogen role of arachidonic acid was inferred from the fact that *in vitro* granules, aggregated by synexin and $Ca^{2+}$, fuse if arachidonic acid is added to the medium. As previously mentioned for PKC activation, arachidonic acid is produced each time secretion is stimulated. Inhibition of $PLA_2$ and arachidonic metabolism inhibits secretion of chromaffin cells due to a blockage of $Ca^{2+}$ entry.

### 3. Secretory Granule Pools

In bovine chromaffin cells, increase in $[Ca^{2+}]_i$ triggers secretion at various rates: ultrafast secretion (time constant, <0.5 sec), fast secretion (time constant, 3 sec), and slow secretion (time constant, 10–30 sec). Nevertheless, there is a weak correlation between the rate of secretion and $[Ca^{2+}]_i$ levels; the three types of responses are mainly observed for 10–50 $\mu M$, above 80 $\mu M$, and around 170 $\mu M$ $[Ca^{2+}]_i$, respectively. The time constant of the increase in membrane capacitance can be considered as a measure of the hormone released from a particular store. This indicates that the secretory granules are in various states of releasability. The ultrafast response comes from vesicles, docked to the plasma membrane and immediately available for release by an increase in $[Ca^{2+}]_i$; they belong to the immediately releasable pool (IRP). A second pool of granules is located near the membrane in a nearly releasable pool (NRP) and can be released within seconds of a rise in $[Ca^{2+}]_i$; the fast response originates from this pool. Finally, the slow response originates from vesicles from a depot store in the bulk cytoplasm (Neher and Zucker, 1993). Movement of granules from the NRP to the IRP is $Ca^{2+}$ dependent and could involve calpactin in the docking process. However, not all the docked granules are available for the last steps, that is, opening of the granule induced by $Ca^{2+}$. The fastest response, the exocytotic burst, mobilizes only one-tenth of the docked granules, which indicates that after docking, granules undergo a maturation process in the IRP. Figure 14 describes a possible sequence for exocytosis in endocrine cell. PKC and ATP hydrolysis are involved in the docking and priming steps. Thereafter, three steps have been proposed based on their sensitivity to temperature, blockage by acidification, and $Ca^{2+}$ ion (three or four ions) triggering; PKC is thought to play a role in this sequence. The pivotal role of ATP in secretion has been outlined, however, as a large pool of granules can be released in the absence of ATP by an increase in $[Ca^{2+}]_i$. This indicates first that ATP hydrolysis by NSF is an early step in the priming process and second that if the energy of hydrolysis powers fusion, it remains stored in the core complex until $[Ca^{2+}]_i$ increases (Parsons *et al.*, 1995; Gillis *et al.*, 1996).

## IV. Hormone Release in Endocrine Cells

### A. Polypeptide and Thyroid Hormones

#### 1. Secretion of Insulin

Insulin secretion by the pancreatic $\beta$ cells provides an excellent example of a cellular activity that requires direction. Insulin is packaged in secretory vesicles, which have to migrate to the plasma membrane and fuse with it to release the entrapped insulin. Both microscopic and biochemical studies have shown that secretory granules are linked to microtubules, which direct attached vesicles to the cell surface. However, a cortical band of fine microfilaments is consistently observed in $\beta$ cells. Alteration of this cell web by cytochalasin B is associated with an enhancement of glucose-induced secretion of insulin by isolated islets. This microfilamentous web plays an important role in the exocytosis of insulin secretory granules by controlling access to the cell membrane via a mechanism probably similar to that previously described for chromaffin cells. $Ca^{2+}$ appears to initiate the cascade of events by which microtubules facilitate the displacement of granules toward the cell membrane. Glucose metabolism increases intracellular concentration of ATP, which closes the ATP-sensitive $K^+$ channels, consequently inducing cell depolarization and $Ca^{2+}$ influx, while cAMP modifies the intracellular distribution of $Ca^{2+}$ by increasing the cytosolic pool at the expense of $Ca^{2+}$ bound to intracellular organelles. Protein kinase C also appears to be involved in the secretion of insulin.

#### 2. Secretion of Pituitary Hormones

Anterior pituitary cells, in their diversity and heterogeneity, provide a rich source of models for secretory function

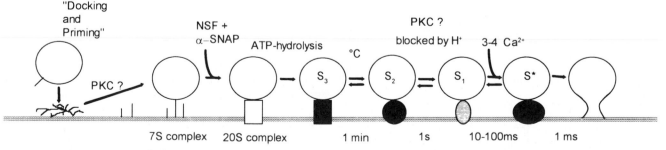

**FIG. 14.** Docking and fusion of secretory granules in endocrine cells. Docking of vesicles involves the proteins synaptobrevin and syntaxin/SNAP-25. In the first step of docking, PKC could have a role in the disruption of the actin barrier. ATP hydrolyzed by NSF is thought to produce energy for fusion. Several steps have been identified after the ATP-dependent step based on their temperature, $H^+$, and $Ca^{2+}$ dependence. PKC could be involved before the final $Ca^{2+}$-dependent step. (Adapted from Parson *et al.*, 1995; Gillis *et al.*, 1996, both copyright of Cell Press.)

(for review, see Mason *et al.,* 1988). Secretion of the pituitary hormone is controlled by both specific hypothalamic-releasing peptides and neurotransmitters. Upon binding to their receptors, these agonists activate both $Ca^{2+}$ influx by different types of $Ca^{2+}$ channels (voltage-activated channels, ligand-activated channels, second messenger–activated channels) and second messenger production, thus resulting in hormone secretion. As for insulin secretion, cytoskeletal structures are tightly associated with hormone release. Purified secretory granule membranes cosediment with microtubules, MAPs being involved in this association. This suggests that microtubules facilitate the movement of secretory granules from the Golgi apparatus to the plasma membrane, by providing tracks along which the granules can move. The granule membrane can then dissociate from the microtubules and fuse with the cell membrane, followed by exocytosis and release of the hormone into the circulation. Moreover, actin and microtubules are cross-linked by MAPs, forming three-dimensional networks. This cross-linking activity can be inhibited if MAPs are heavily phosphorylated. These observations suggest that MAPs might play an important role in the binding of secretory granules to tubulin and actin. Binding of actin to secretory granules suggests a role for actin in the final steps of exocytosis, as described previously.

### 3. Secretion of Thyroid Hormones

The first evidence for a role of the cytoskeleton in endocrine activity stems from studies of the thyroid gland. Before reaching the circulation, thyroid hormones stored as thyroglobulin must be released from their peptide form. Secretion of T3 and T4 is organized in a strictly vectorial manner as a two-way flow of biosynthetic intermediates: (1) from the periphery of the cell to the follicular lumen for synthesis and storage in the colloid, and (2) transport from follicular lumen to the periphery for secretion. Microfilaments and microtubules are both important in these processes. The first step leading to the secretion of the hormones is the movement of colloid from the interior of the follicle to the cytoplasm. Many studies have shown that colchicine and vinblastine inhibit this movement. Addition of TSH (which stimulates cAMP production and $Ca^{2+}$ influx) stimulates release of the hormones, which can be inhibited by addition of colchicine. In response to TSH, the follicular lumen membrane develops protrusions that engulf stored colloid. The protrusions or villus processes do not develop in the presence of colchicine. It is tempting to propose that microtubules are responsible for the construction of villus processes, and that the contractile properties of microfilaments facilitate the movement of colloid into the cell.

The colloid droplets (which contain thyroglobulin) enter the thyroid cell by micro- or macropinocytosis, resulting in the formation of small vesicles. These small vesicles fuse with lysosomes to form phagolysosomes, in which thyroglobulin proteolysis occurs. This results in the release and secretion of thyroid hormones into the blood by a mechanism still unknown. Iodotyrosines, (which are rapidly deiodinated) and the free iodide (which is recycled by

the gland) are also released at the same time. These vesicles are different from colloid vesicles or apical vesicles of exocytosis, which originate from pseudopod formations. Because TSH increases the amount of actin associated with lysosomes, this suggests a role for actin in intracellular movement and perhaps attachment to the membranes. Moreover, thyroid cells contain myosin, implying an involvement in the intracellular transport of structures associated with microfilaments.

### B. Stimulation of Steroid Synthesis and Secretion

Steroid hormones are synthesized from cholesterol contained in lipid droplets. The process of steroidogenesis begins in the mitochondria, where cholesterol is converted to pregnenolone from which other steroids are synthesized. Stimuli induce a rapid increase in the production and secretion of steroids, without cytoplasmic storage.

It has been shown that microfilaments are strictly required for both spontaneous and stimuli-induced corticosteroid secretions, including those utilizing the cAMP-dependent pathways (ACTH, serotonin, vasoactive intestinal peptide) and those utilizing the phosphoinositide pathway (angiotensin II, acetylcholine). They are involved in a common and probably late step of steroidogenesis (translocation of the 11-deoxycorticosterone from endoplasmic reticulum to mitochondria).

In contrast, microtubules seem involved in an early step in the mechanism of hormone action, probably at the level of their interaction with the $\alpha$-subunit of the G protein. Microtubules and microfilaments are closely associated with the plasma membrane. That colchicine and vinblastine stimulate basal steroidogenesis may be explained by the fact that most of the tubulin is linked to cholesterol, in lipid droplets. Thus, by acting on tubulin, antimicrotubular drugs could release cholesterol, causing an increase in basal steroid secretion.

In adrenocortical cells, findings suggest that intermediate filaments could facilitate or increase the transport of cholesterol to mitochondria in response to ACTH and cAMP. Figure 15 summarizes the role of the three types of cytoskeletal fibers in the steroidogenic effects of the main stimuli of corticosterone and aldosterone secretions (Feuilloley and Vaudry, 1996).

## V. Summary

The first step in the secretory process for peptide hormones and neurotransmitters involves synthesis, modification, and sorting of the molecules to be secreted. These secretory molecules are packaged in secretory granules or vesicles, which are then transported to the cell periphery where they are released in the extracellular space by fusion with the plasma membrane. This complex process is named *exocytosis.* Exocytosis is an all-or-none phenomenon in which $Ca^{2+}$ plays a pivotal role.

Interaction of cytoskeletal structures with transmembrane signaling is an important feature of excitation–secretion coupling. Actin-binding proteins are in large part

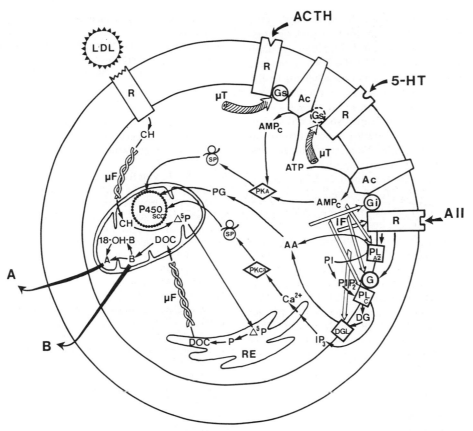

**FIG. 15.** Cytoskeleton implication in adrenal corticosteroidogenesis. Microtubules ($\mu$T) are implicated in the action of ACTH and 5-HT, probably at a level of $G_s$ protein. Microfilaments ($\mu$F) control the transfer of steroid precursors (cholesterol from lipid droplets or membrane low density lipoproteins to mitochondria, and from mitochondria to endoplasmic reticulum). Intermediate filaments (IF) appear implicated in the action of angiotensin II (AII), between hormone–receptor coupling and second messenger production. A, aldosterone; AA, arachidonic acid; Ac, adenylyl cyclase; cAMP, adenosine monophosphate; ATP, adenosine triphosphate; B, corticosterone; DAG, diacylglycerol; DOC, 11-deoxycorticosterone; 18-OH-B, 18-hydroxycorticosterone; P, progesterone; P450scc, P450scc cytochrome; $\Delta^5$P, pregnenolone; PG, prostaglandins; PI, PtdIns; PIP$_2$, PtdInsP$_2$; PKA, protein kinase A; PKCa, protein kinase Ca$^{2+}$-dependent; PLA$_2$, phospholipase A$_2$; PLC, phospholipase C; PS, protein synthesis. (Adapted from Feuilloley *et al.*, 1989.)

responsible for cytoskeleton–receptor interactions. In resting cells, more than 95% of the PtdInsP$_2$ may be complexed with profilin. This interaction promotes actin polymerization. Following receptor activation, phospholipase C activity increases to a level where profilin protection can be overcome, resulting in PtdInsP$_2$ hydrolysis. Gelsolin is another actin-binding protein, which is Ca$^{2+}$ and PtdInsP$_2$ regulated. In resting cells, the bulk of gelsolin is cytosolic. When the cell is activated, there is a transient rise in [Ca$^{2+}$]$_i$ due to InsP$_3$ action or Ca$^{2+}$ influx, which activates gelsolin, causing a 200-fold increase in its affinity for actin. This results in rapid filament side binding, followed by severing and capping, inducing a dramatic disruption of existing actin network structure in the vicinity of Ca$^{2+}$ elevation. Microtubule-associated proteins are also implicated in the transduction of secretory process. MAPs are substrates for several protein kinases, including a Ca$^{2+}$/calmodulin-dependent kinase, a cAMP dependent kinase, tyrosine kinases, and PKC.

Dynamic changes in the cytoskeletal network are required for exocytosis. The subplasmalemmal area of secretory cells is characterized by the presence of a highly organized cytoskeletal network. F-actin seems to be exclusively localized in this area and, together with specific actin-binding proteins, forms a dense viscoelastic gel. Fodrin, vinculin, and caldesmon, three actin-binding proteins, as well as gelsolin and scinderin, two actin-severing proteins, are found in this plasmalemmal region. Because fodrin, caldesmon, and $\alpha$-actinin-binding sites exist on secretory granule membranes, actin filaments can also link to secretory granules. Caldesmon and synapsin I proteins are also associated with synaptic vesicles. Synapsin I binds to spectrin and actin microfilaments, and may serve as an anchor between synaptic vesicles and the cytoskeleton. Synapsin I phosphorylation results in the release of synaptic vesicles from their anchoring site on the cytoskeleton, allowing the vesicles to move to the active synaptic zones. Chromaffin granules can also be entrapped in this subplasmalemmal lattice.

Secretion is a process that requires (1) the movement of secretory vesicles toward the plasma membrane, (2) the fusion of vesicles with the plasma membrane, and (3) subsequent release of secretory contents into the cell exterior. The process of secretion is mediated by contractile elements either associated with secretory vesicles or present elsewhere in the cells. Because filamentous actin (F-actin) is preferentially localized in the cortical surface of the chromaffin cell, F-actin may act as a barrier to the secretory granules, impeding their contact with the plasma membrane. Upon stimulation, fodrin rearranges into patches beneath the plasma membrane. Such a redistribution could be related to the clearing of exocytotic sites at the level of the plasma membrane. Scinderin is a cytosolic protein that shortens actin filament length when $Ca^{2+}$ is present in the medium. Stimulation induces both redistribution of scinderin from the cytosol to the cell cortex as well as F-actin disassembly, which precedes exocytosis. Docking of granules on the plasma membrane is an important step in exocytosis. Several proteins are involved in this process: synaptobrevin, syntaxin, and SNAP-25, which form the SNAP receptor (SNARE) (see Fig. 5). Priming of the granules is also a pivotal step in exocytosis. NSF, a part of the 20S core complex, primes the granule through ATP hydrolysis. Once primed, an increase in $[Ca^{2+}]$ triggers the release. All these processes require the presence of $Ca^{2+}$ in the extracellular medium. Therefore, only secretagogues that induce $Ca^{2+}$ entry are able to produce these effects.

In regulated exocytosis, fusion of secretory granules with the plasma membrane is triggered by an appropriate signal. Every time a secretory granule fuses with the plasma membrane, the total capacitance of the cell increases by a value proportional to the surface of the new membrane added to the existing cell membrane. Capacitance step values generally range between 1 and 30 fF, corresponding to granule diameters of 0.2–1 $\mu$m. Freeze-fracture images of exocytosis reveal the presence of narrow pores formed between the granules and the plasma membrane called *fusion pores*. The size of the pore increases as fusion progresses, allowing for the release of vesicle contents.

Control of exocytosis occurs in a variety of electrically excitable and nonexcitable systems in response to receptor activation. The regulatory pathways that couple stimulation and secretion vary widely among cell types. In many cells, calcium is the key signal for triggering exocytosis. $Ca^{2+}$ influx from the external medium is crucial for secretion but internal $Ca^{2+}$ release from internal stores also plays a pivotal role, depending on cell type. In some cases, $Ca^{2+}$ alone is not able to trigger secretion, but the presence of a guanine nucleotide together with $Ca^{2+}$ is necessary and sufficient for exocytosis.

## Acknowledgments

The authors thank their colleagues, Drs. Gilles Dupuis, Eric Rousseau, and Diego Bellabarda, and their students, Liette Laflamme and Alain Giguère, for fruitful discussions and critical comments on the manuscript.

## Bibliography

Aderem, A. (1992). Signal transduction and the actin cytoskeleton: The roles of MARCKS and profilin. *Trends Biochem. Sci.* **17**, 438–443.

Almers, W. (1990). Exocytosis. *Annu. Rev. Physiol.* **52**, 607–624.

Augustine, G. J., and Neher, E. (1992). Calcium requirements for secretion in bovine chromaffin cells. *J. Physiol.* **450**, 247–271.

Bennett Jefferson, A., and Schulman, H. (1991). Phosphorylation of microtubule-associated protein-2 in GH3 cells. *J. Biol. Chem.* **266**, 346–354.

Burgoyne, R. D. (1991). Control of exocytosis in adrenal chromaffin cells. *Biochim. Biophys. Acta* **1071**, 174–202.

Carraway, K. L., and Carothers Carraway, C. A. (1989). Membrane-cytoskeleton interactions in animal cells. *Biochim. Biophys. Acta* **988**, 147–171.

Chow, R. H., Von Rüden, L., and Neher, E. (1992). Delay in vesicle fusion revealed by electrochemical monitoring of single secretory events in adrenal chromaffin cells. *Nature* **356**, 60–63.

Feuilloley, M., and Vaudry, H. (1996). Role of the cytoskeleton in adrenocortical cells. *Endocr. Rev.* **17**, 269–288.

Feuilloley, M., Netchtaïlo, P., Delarue C., Leboulanger, F., Pelletier, G., and Vaudry, H. (1989). Quels sont les rôles du cytosquelette dans les processus de sécrétion des hormones stéroïdes? *Medicine/Sciences* **5**, 674–677.

Forscher, P. (1989). Calcium and polyphosphoinositides control of cytoskeletal dynamics. *Trends Neurosci.* **12**, 468–474.

Gillis, K. D., Mössner, R., and Neher, E. (1996). Protein kinase C enhances exocytosis from chromaffin cells by increasing the size of the readily releasable pool of secretory granules. *Neuron* **16**, 1209–1220.

Hoth, M., and Penner, R. (1992). Depletion of intracellular calcium stores activates a calcium current in mast cells, *Nature* **355**, 353–356.

Ibarrondo, J., Joubert, D., Dufour, M.-N., Cohen-Solal, A., Homburger, V., Jard, S., and Guillon, G. (1995). Close association of the α-subunits of $G_q$ and $G_{11}$ G proteins with actin filaments in $WRK_1$ cells: Relation to G protein-mediated phospholipase C activation. *Proc. Natl. Acad. Sci. USA* **92**, 8413–8417.

Inglese, J., Koch, W., Touhara, K., and Lefkowitz, R. (1995). Gβγ interactions with PH domains and ras-MAPK signaling pathways. *Trends Biochem. Sci.* **20**, 151–156.

Lindau, M., and Gomperts, B. D. (1991). Techniques and concepts in exocytosis. Focus on mast cells. *Biochim. Biophys. Acta* **1071**, 429–471.

Mason, W. T., Rawlings, S. R., Cobbett, P., Sikdar, S. K., Zobec, R., Akerman, S. N., Benham, C. D., Berridge, M. J., Cheek, T., and Moreton, R. B. (1988). Control of secretion in anterior pituitary cells—Linking ion channels, messengers and exocytosis. *J. Exp. Biol.* **139**, 287–316.

Monck, J. R., Alvarez de Toledo, G., and Fernandez, J. M. (1990). Tension in secretory granule membranes causes extensive membrane transfer through the exocytotic fusion pore. *Proc. Natl. Acad. Sci. USA* **87**, 7804–7808.

Morgan, D., and Burgoyne, R. D. (1990). Stimulation of $Ca^{2+}$-independent catecholamine secretion from digitonin-permeabilized bovine adrenal chromaffin cells by guanine nucleotide analogues, $AT_2$. *Biochem. J.* **269**, 521–526.

Neher, E., and Zucker, R. S. (1993). Multiple calcium-dependent processes related to secretion in bovine chromaffin cells. *Neuron* **10**, 21–30.

Parsons, T. D., Coorssen, J. R., Horstmann, H., and Almers, W. (1995). Docked granules, the exocytic burst, and the need for ATP hydrolysis in endocrine cells. *Neuron* **15**, 1085–1096.

Podova, J. S., Johnson, G. L., and Rasenick, M. M. (1994). Chimeric Gαs/Gαi2 proteins define domains on Gαs that interact with tubulin for β-adrenergic activation of adenylyl cyclase. *J. Biol. Chem.* **269**, 21748–21754.

Rothman, J. E., and Orci, L. (1992). Molecular dissection of the secretory pathway. *Nature* **355**, 409–415.

Söllner, T., Whiteheart, S. W., Brunner, M., Erdjument-Bromage, H.,

Geromanos, S., Tempst, P., and Rothman, J. E. (1993). SNAP receptors implicated in vesicle targeting and fusion. *Nature* **362,** 318–323.

Spruce, A. E., Iwata, A., White, J. M., and Almers, W. (1989). Patch-clamp studies of single cell-fusion events mediated by a viral fusion protein. *Nature* **342,** 555–558.

Südhof, T. C. (1995). The synaptic vesicle cycle: A cascade of protein–protein interactions. *Nature* **375,** 645–653.

Thomas, L., Hartung, K., Langosch, D., Rehm, H., Bamberg, E., Franke, W. W., and Betz, H. (1988). Identification of synaptophysin as a hexameric channel protein of the synaptic vesicle membrane. *Science* **242,** 1050–1053.

Trifaró, J.-M., Vitale, M. L., and Rodríguez Del Castillo, A. (1992). Cytoskeleton and molecular mechanisms in neurotransmitter release by neurosecretory cells. *Eur. J. Pharmacol. Mol. Pharmacol. Sect.* **225,** 83–104.

*Stanley Misler, David M. Pressel, and David W. Barnett*

# 42

# Stimulus Transduction in Metabolic Sensor Cells

## I. Introduction

The term **sensory receptor** is usually applied to a limited group of cells, located near the surface of the organism, that transduces stimuli from the external environment (e.g., photons of light, air- or waterborne molecules, pressure disturbances) into the release of a chemical transmitter, which ultimately signals neurons that project into the central nervous system (CNS). However, to maintain a "steady state," organisms must also detect changes in their internal environment. To do this, specialized **internal receptor** cells must sense changes (1) in the levels of circulating metabolic fuels (e.g., nutrient metabolites and $O_2$), (2) in the levels of metabolic wastes (e.g., $CO_2$), (3) in the ionic composition of the extracellular lymphlike fluid that bathes the cells, and (4) in the internal or lumenal tension in hollow tubular organs (e.g., blood vessels and gut).

Currently, among these internal receptor cells, those sensitive to metabolic changes are better understood than those responding to changes in ionic composition or lumenal tension. In this chapter, we examine, in some detail, sensory transduction by three types of such **metabolic sensor cells:**

1. *Pancreatic islet $\beta$ cells.* These cells sense changes in the levels of several circulating metabolites, especially glucose, and respond with secretion of the hormone insulin, which, in turn, promotes the uptake of glucose and its storage as glycogen in muscle and liver and maintains fat stores in adipocytes.
2. *Chemoreceptor cells of carotid body.* These cells sense changes in the $O_2$ and $CO_2$ content of blood, and synapse onto sensory nerve fibers, which project into the CNS and help shape respiratory drive.
3. *Vascular smooth muscle cells.* These cells sense local changes in $O_2$ tension, and respond by regulating their tone, thereby locally controlling blood flow.

These metabolic sensor cells share a common property; each makes use of $K^+$ channels specially attuned to metabolic changes to transduce stimuli. These cells may also use the ability of metabolic intermediates to modulate the activity of more conventional channels to help shape their response. A familiarity with different patch-clamp recording configurations is assumed (see Chapter 25).

## II. Stimulus–Secretion Coupling in Insulin-Secreting $\beta$ Cells

The study of how a rise in plasma glucose leads to rapid secretion of insulin by $\beta$ cells of the pancreatic islets of Langerhans has provided a fertile ground for the interaction of a wide range of disciplines in cell physiology. Since the mid-1970s, it has been widely appreciated that the release of insulin, which is stored in secretory granules, is dependent on extracellular $Ca^{2+}$ concentration $[Ca^{2+}]_o$ (the **calcium hypothesis**), requires oxidative metabolism of glucose or other metabolite fuels (the **fuel hypothesis**), and follows the onset of complex electrical activity in the $\beta$ cells (the **depolarization–secretion coupling hypothesis**). Aspects of all three hypotheses are illustrated in Fig. 1. The fuel hypothesis was originally based on three seminal findings: (1) insulin release is often proportional to glucose metabolism (uptake, phosphorylation, and ultimately the production of labeled $CO_2$ and $H_2O$ from radiolabeled glucose), (2) among other sugars only those that are metabolized by $\beta$ cells stimulate secretion, and (3) inhibitors of early steps of glycolysis and mitochondrial electron transport inhibit glucose-induced secretion (see Fig. 1A). The calcium hypothesis and the depolarization–secretion coupling hypothesis were initially prompted by findings that concentrations of extracellular $K^+$ expected to depolarize $\beta$ cells were effective in substituting for glucose in evoking insulin release provided adequate $[Ca^{2+}]_o$ was present. Shortly thereafter it was found that $\beta$ cells maintain resting potentials of about $-60$ mV at fasting glucose concentrations. Within several minutes of addition to the bathing medium of concentrations of fuel metabolites, which trigger insulin secretion, these cells depolarize and often develop bursts of action potential (AP) activity (see Fig. 1B).

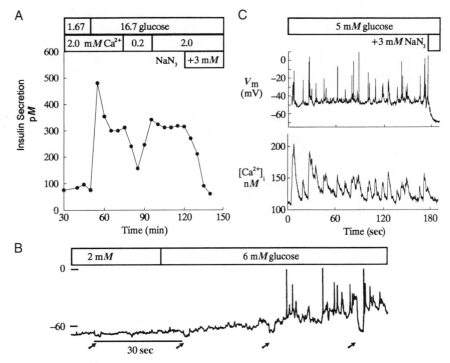

**FIG. 1.** Basic evidence of $Ca^{2+}$ and metabolic dependence of stimulus–secretion coupling in inslin-secreting $\beta$ cells of human pancreatic islets of Langerhans. (A) Insulin secretion measured from isolated intact islets by radioimmunoassay. Note the rapid onset of insulin secretion on increasing glucose from 1.67 to 16.7 m$M$, the interruption of the subsequent plateau secretion by reduction in $[Ca^{2+}]_o$ from 2.0 to 0.2 m$M$, and the termination of secretion by the addition of sodium azide (NaN$_3$), an inhibitor of electron transport by mitochondrial cytochromes. Leucine, glyceraldehyde, and acetoacetate produce similar patterns of $Ca^{2+}$-dependent metabolically dependent insulin secretion. (B) Glucose-induced electrical activity in human $\beta$ cells. $\beta$ cells progressively depolarize, undergo a reduction in resting membrane conductance ($G_m$), and generate action potentials with increasing frequency as the ambient glucose concentration is increased from 2 to 6 m$M$. The negative-going membrane potential excursions, marked with arrows, denote voltage responses to small pulses of current injection ($-5$ pA). From Ohm's law (i.e., $\Delta V = \Delta I/G_m$), an increasing voltage excursion in response to constant current pulse indicates a decrease in resting membrane conductance. (C) Simultaneous recording of electrical activity and cytosolic $Ca^{2+}$ transients from single rat $\beta$ cells. Note in 5 m$M$ glucose, the cell fires individual or short trains of action potentials. Even single action potentials are sufficient to induce a $Ca^{2+}$ transient. Addition of NaN$_3$ abruptly hyperpolarizes the cell to $-70$ mV and blocks the appearance of $Ca^{2+}$ transients. (Unpublished experiments by D. Barnett and S. Misler.)

Electrical activity is inhibited by removal of extracellular $Ca^{2+}$ or application of $Ca^{2+}$ channel blockers.

By the mid-1980s, through the analysis of membrane currents by patch-clamping and estimation of cytosolic $Ca^{2+}$ by microspectrofluorimetry of cells containing a $Ca^{2+}$-sensitive dye, it was possible to define more closely some of the ion channels underlying $\beta$-cell excitability, and to demonstrate directly a concomitant rise in cytosolic $Ca^{2+}$ with electrical activity (see Fig. 1C). Most recently, it has been possible to link more directly $Ca^{2+}$ entry and the rise in cytosolic $Ca^{2+}$ to secretion by a number of approaches. In cells that have been electropermeabilized (i.e., whose membranes have been perforated by intense electric shock), release of insulin is graded with the bath concentration of $Ca^{2+}$ (which equilibrates with the cytoplasm). In patch-clamped $\beta$ cells, depolarization-induced granule exocytosis (or fusion of the granule membrane with the plasma membrane) was measured electrically as an increase in membrane capacitance, or electrochemically as release of packets of native insulin or the false transmitter serotonin

(the latter loaded into insulin granules during bath incubation). In either case the measure of exocytosis is graded with depolarization-induced $Ca^{2+}$ entry.

On the basis of these experiments, a consensus hypothesis has emerged during the past decade for glucose-induced insulin secretion from $\beta$ cells (Fig. 2). Uptake of glucose (predominantly via a Glut-2-type transporter) provides glucose entry into glycolysis (via glucokinase, a low-affinity hexokinase) and thereafter results in mitochondrial generation of adenosine triphosphate (ATP) (step 1). This leads to the closure of ATP-sensitive K$^+$ channels here abbreviated as **K$^+$(ATP) channels** (step 2). K$^+$(ATP) channel closure results in the depolarization of the $\beta$-cell membrane (step 3). When the membrane potential reaches the threshold for activating voltage-dependent $Ca^{2+}$ and Na$^+$ channels, electrical activity and voltage-dependent $Ca^{2+}$ entry begin (step 4). The resultant rise in cytosolic $Ca^{2+}$ (step 5) triggers $Ca^{2+}$-dependent exocytotic fusion of insulin granules with the plasma membrane (step 6). Steps 4, 5, and 6 are reminiscent of the process of depolarization–secretion

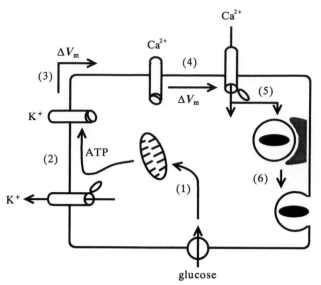

**FIG. 2.**  Schematic depiction of steps in stimulus–secretion coupling in β cells. See text for details.

coupling well studied at nerve terminals, but subtle difference are discussed later.

## A.  Role of ATP$_i$-Sensitive K$^+$ Channels in Coupling Metabolism to β-Cell Depolarization

How do fuel metabolites produce β-cell depolarization? Early patch-clamp experiments, in which β cells were recorded from in the cell-attached patch mode, revealed the presence of a K$^+$ channel, which is probably a critical link (Fig. 3). This K$^+$ channel is inwardly rectifying (i.e., passes inflow of potassium better than outflow) and shows little voltage dependence. It is open at the low ambient glucose concentrations encountered during fasting (2–3 m$M$), but rapidly closes at plasma glucose concentrations encountered after a meal (5–7 m$M$). Channel closure is usually followed closely by the onset of electrical activity. In the presence of the higher glucose concentration, the channel can be reopened by the addition to the bath of an inhibitor of glucose transport or phosphorylation, inhibitors of early steps in glycolysis, or mitochondrial electron transport (see Fig. 3A). The time course and relative change in channel activity seen in a cell-attached patch during a variety of metabolic maneuvers closely parallels the changes in whole-cell resting $G_m$. This strongly suggests that the activity of this channel tends to move $V_m$ to a level near the potassium equilibrium potential, $E_K$. With so many of these channels open at rest, their closure *en masse* is needed to make the K$^+$ conductance of the cell comparable to that of other conductance pathways operative at rest (including nonselective cation channels and anion conductance channels, with estimated equilibrium potentials of 0 or ca. $-30$ mV, respectively). This permits $V_m$ to depolarize to the threshold for activating cell electrical activity (ca. $-45$ mV).

Recalling that the metabolically regulated K$^+$ channels, recorded in the cell-attached patch configuration, are not

directly exposed to bath glucose, it is safe to assume that the channel is gated by a metabolic intermediate(s). *A priori*, cell metabolism might affect a host of metabolic intermediates, ranging from high-energy phosphate compounds (e.g., ATP and ADP) to redox equivalents to cytosolic H$^+$ or Ca$^{2+}$. Early experiments showed that the metabolite-regulated K$^+$ channel seen in the cell-attached patch was avidly gated by ATP at its inner surface after excision of the membrane patch in an inside-out excised patch configuration (Fig. 3B), and hence the name K$^+$(ATP) channel is used. Free concentrations of ATP, or its nonhydrolyzable analogs (e.g., AMP-PNP), as low as 10 $\mu M$ reduce channel activity by half. This suggests that ATP closes this channel by directly "gating" it rather than by phosphorylation. In parallel with this, in a whole-cell patch experiment that allows ATP to dialyze spontaneously out of the cell (due to the absence of ATP in the pipette), resting $G_m$ increases and the cell hyperpolarizes. Interestingly, ATP inhibition of channel activity in the excised (inside-out) patch can be partially reversed by addition of ADP to the bath. Hence, changes in the relative concentrations of ATP and ADP, which usually change in a reciprocal fashion as metabolism is altered, would appear to be an important factor in channel gating. To be sure, the activity of the K$^+$(ATP) channel is also altered by other potential intermediates, such as the NADH/NAD ratio, cytosolic H$^+$, and Ca$^{2+}$; however, the concentration changes in these other mediators needed to affect channel activity are far larger than those seen during physiological cell stimulation.

How is the metabolite-regulated K$^+$(ATP) channel gated *in situ*? Two critical questions merit consideration. First, how is the low $K_d$ value (micromolar) for ATP-induced channel closure reconciled with the millimolar concentrations of ATP in the cytosol? Several factors may mitigate this discrepancy. (1) Recent evidence suggests that ADP may exert complex effects on channel activity. ADP competition for an ATP-binding site, combined with allosteric action of ADP at an independent binding site, might adjust the half maximal effective concentration ($K_d$) of ATP, measured in the presence of ADP, closer to the millimolar range. (2) Steep cytoplasmic gradients of ATP and ADP, due to consumption of ATP and concomitant ADP production by neighboring plasma membrane ATPases, might reduce the ATP available to the channel. (There may be compartmentation of the ATP.) (3) Given the huge preponderance of ATP-sensitive K$^+$ conductance over all other contributors to resting $G_m$, near millimolar concentrations of submembrane ATP may be needed to close down a sufficient fraction of the K$^+$ channels so that the threshold voltage for electrical excitation is approached, and thereafter small changes in $G_K$ further affect $V_m$. Second, what is the source of the ATP used? The weight of evidence supports a mitochondrial source; mitochondrial inhibition totally prevents glucose-induced channel closure, whereas alternative metabolites such as ketone bodies as well as amino acids and their deamination products (e.g., leucine and ketoisovalerate), which can directly enter mitochondria, stimulate channel closure even in the face of inhibition of glycolysis. Curiously, however, glyceraldehyde, an intermediate early in glycolysis, stimulates insulin

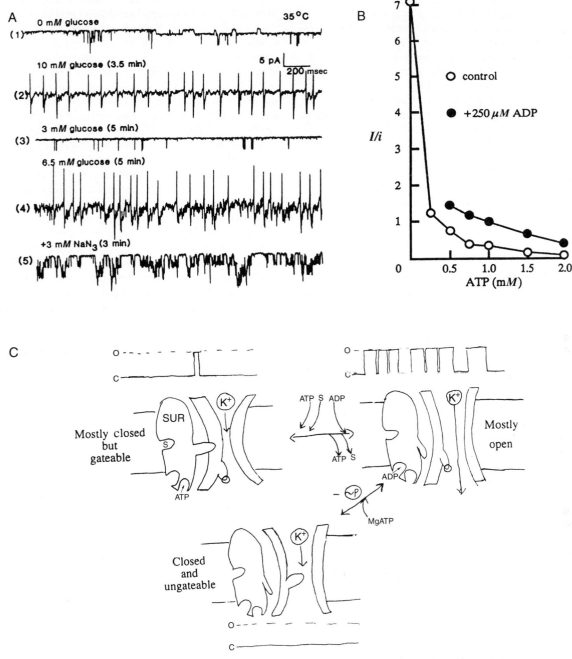

**FIG. 3.** Identifying features of metabolically regulated, ATP$_i$-inhibited K$^+$ channels [K$^+$(ATP)] in pancreatic islet $\beta$ cells. (A) Response of K$^+$(ATP) channels to metabolic changes recorded from the cell-attached patch. With the cell bathed in an extracellular solution containing 0 mM glucose, K$^+$(ATP) channel activity is evident along with the intermittent activity of a longer lived nonselective cation channel (see trace 1). On raising glucose to 10 or even 6.5 mM, K$^+$(ATP) channel activity nearly ceases and the cells fire APs with increasing frequency. However, on addition of NaN$_3$ to 6.5 mM glucose, APs disappear and K$^+$(ATP) channels open very robustly. (Reproduced from *Diabetes,* Vol. 38, p. 423, Fig. 1a, 1989.) (B) ATP/ADP sensitivity of K$^+$(ATP) channels in an inside-out excised patch; nucleotides applied to the cytoplasmic surface. Note the steep inhibition of channel activity by ATP and partial disinhibition of ATP action by ADP. (C) Model for gating of K$^+$(ATP) channel by ATP, ADP, and sulfonylureas (S). All three substances appear to bind to the SUR subunit of channel. Binding to ATP to a nucleotide-binding facet (NBF) or S to its receptor changes the confirmation of the Kir 6.2 subunit from a mostly open state to a closed but gateable state. In contrast, binding of ADP to a second NBF, particularly at low ATP concentration and in the absence of S, helps maintain the channel in a mostly open state. Dephosphorylation of a critical site, here taken as a gate within the pore, brings the channel into a closed and ungateable state from which it can only be rescued by rephosphorylation in the presence of magnesium ATP. Recent evidence suggests that the stoichometry of the channel is 4SUR:4Kir 6.2 subunits.

secretion, whereas pyruvate, which couples glycolysis and the Krebs cycle, does not stimulate secretion. Recent evidence suggests that glyceraldehyde actively suppresses $K^+(ATP)$ channel activity, whereas neither pyruvate nor inhibition of mitochondrial pyruvate transport affects $K^+(ATP)$ channel activity. More detailed experiments along these lines suggest that NADH, produced early in the glycolysis process, might be a critical intermediate in both the glycerol phosphate/dihydroxyacetone phosphate as well as the oxaloacete/malate shuttle pathways across the mitochondria. Both shuttles feed in to critical electron transport sites.

The metabolite-regulated $K^+(ATP)$ is a key locus of action by hypoglycemia (glucose-lowering) and hyperglycemic (glucose-raising) drugs and hormones that modify glucose responsiveness and insulin release by $\beta$ cells (see Fig. 3C). Perhaps most interesting among these agents are the hypoglycemic **sulfonylureas** [e.g., **tolbutamide** and **glyburide** (glibenclamide)]. (Clinically, these drugs are useful in the treatment of non-insulin-dependent diabetes mellitus, in which $\beta$ cells produce significant quantities of insulin, but appear to "tune out" the glucose stimulus, perhaps due to defective glucose transporters.) At concentrations equal to free plasma levels that enhance glucose-induced electrical activity and insulin secretion in intact animals, the sulfonylureas close $K^+(ATP)$ channels in cell-attached or excised patches alike, while leaving voltage- and calcium-gated K channels unaffected. These experiments suggest that the $K^+(ATP)$ channel or another molecule in close proximity serves as a receptor for the sulfonylureas and have led to the cloning of the $K^+(ATP)$ channel. Experiments with cloned $K^+(ATP)$ have shown that to reconstitute channel activity in naive cells is is necessary to transfect such a cell with two subunits of the native channel. The first is a subunit resembling a standard inward-rectifier-type subunit (Kir 6.2) consisting of two membrane-spanning domains separated by an H loop dipping into the bilayer and endowing the channel with its typical $K^+$ conduction properties. The second is a complex high molecular weight sulfonylurea receptor (SUR) subunit, which has two distinct nucleotide-binding facettes (NBFs) and ATPase activity and resembles an ABC-type transporter (this endows the channel with its features of gating by nucleotides and sulfonylureas). Mutations in SUR lacking one of the NBFs have been identified in persistent hyperinsulinemic, hypoglycemia of infancy, a disease where $K^+(ATP)$ channel activity is not detected and $\beta$ cells are persistently depolarized and secreting. Coexpression of the mutant SUR and Kir 6.2 fails to produce detectable $K^+(ATP)$ channels. This type of analysis may help to identify the mode of action on the channel of structurally related vasodilator drugs, such as diazoxide, that inhibit glucose-induced insulin secretion largely by opening these channels and hyperpolarizing the $\beta$ cell.

Is this finely tuned coupling of cell metabolism to $K^+(ATP)$ channel activity used as a sensory mechanism by the two other major glucose receptor cells of the body, namely, the **satiety center** of the hypothalamus and the peripheral sweet **taste receptors**? Recently, $K^+(ATP)$ channels have been sought in glucose receptive (GR) neurons

of the ventromedial nucleus whose activity is associated with satiety and suppression of food intake. In these neurons, reduction of bath glucose or addition of the glycolytic inhibitor mannoheptulose results in hyperpolarization and cessation of electrical activity accompanied by an increase in resting $G_m$; oppositely, tolbutamide produces excitation. In cell-attached patch recordings, cessation of electrical activity is associated with increased activity of a $K^+$-selective channel that is voltage insensitive. After patch excision, this channel is inhibited by ATP at its cytoplasmic surface. This channel is distinct from the $K^+(ATP)$ channel in $\beta$ cells; under identical excised patch recording conditions, the $K_i$ constant for ATP inhibition is 2.3 m$M$ (rather than 20 $\mu M$) and the single-channel conductance is 135 pS (rather than 65 pS). In contrast, in peripheral taste chemoreceptors, glucose reception may be working through G-proteins and a second messenger cascade. In these cells application of sucrose or saccharin reduces the amplitude of a voltage-dependent outward $K^+$ current. This effect can be mimicked by application of membrane-permeant analogs of cAMP. In excised patches, similar $K^+$ channels are closed by exposure of the patch to ATP and cAMP-dependent protein kinase, but not ATP alone, suggesting that this channel is not "gated" by ATP binding, but rather requires phosphorylation for closure. [Here we should remark that $K^+(ATP)$ channel activity in $\beta$ cells also benefits from phosphorylation. In the presence of Mg ATP, but not free ATP or MgAMP-PNP, channel activity fails to show its usual rundown of activity in the inside-out excised patch.]

## B. Depolarization–Secretion Coupling in $\beta$ Cells

The vast majority of evidence suggests that the link between depolarization and secretion is provided by $Ca^{2+}$ entry largely (if not exclusively) through voltage-dependent $Ca^{2+}$ channels activated during glucose-induced electrical activity. To date, a wide variety of patterns of glucose-induced electrical activity has been recorded in $\beta$ cells of various species; these range from periodic bursts of tetrodotoxin (TTX)-insensitive APs, riding on a plateau depolarization, to individual large-amplitude TTX-sensitive APs. In all cases examined, simultaneous recordings of cytosolic $Ca^{2+}$ and cell $V_m$ show that $Ca^{2+}$ entry occurs in a pulsatile fashion in phase with AP activity or continuously during a sustained plateau (see Fig. 1C). It is still an open question whether this $Ca^{2+}$ influx triggers regenerative or secondary $Ca^{2+}$ release from intracellular organelles and whether this is critical for secretion.

Figure 4 presents features of the "star" players involved in electrical activity of the $\beta$ cell. The $Ca^{2+}$ channel most often encountered is a high-voltage-activated (HVA) (or L-type) one. It is slowly inactivating, moderately sensitive to dihydropyridines (DHPs), and is enhanced by replacement of bath $Ca^{2+}$ with $Ba^{2+}$. Otherwise, it is similar to the $Ca^{2+}$ channel underlying the plateau of the cardiac AP. In cells that show clustered bursts of spikes, slow inactivation of $I_{Ca}$ makes it possible for this current to sustain plateau depolarizations on which bursts of APs may ride. When present in sufficient density, standard-

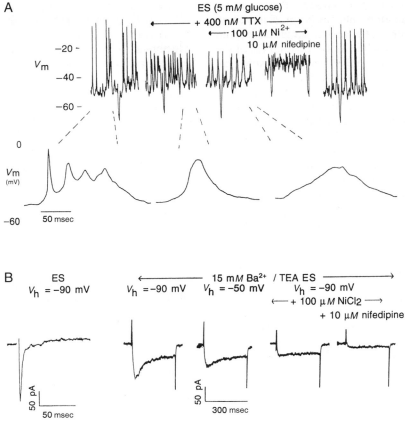

**FIG. 4.** Identification of voltage-dependent currents underlying electrical activity of a single human $\beta$ cell. (A) Perforated patch, whole-cell current clamp recording (pipette filled with high $K^+$ intracellular-like solution) from a cell bathed in extracellular-like solution containing 5 m$M$ glucose. Under control conditions, short bursts of moderate amplitude APs are seen. Addition of 1 m$M$ TTX reduces the amplitude of each AP and broadens its width. Subsequent additions of 100 $\mu M$ NiCl$_2$ and 10 $\mu M$ nifedipine reduce the frequency of initiation of the coarser appearing APs and, ultimately, produce plateau depolarizations. Expanded traces, taken from a similar experiment, demonstrate the dramatic (nearly 50-fold) reduction of maximum upstroke velocity of APs following addition of TTX and nifedipine to the bath. Changes in AP morphology produced by these agents are reversible with their removal. (B) Voltage-clamp recording from same cell. In the standard ES, depolarization from a holding potential ($V_h$) = $-90$ mV to a test potential ($V_c$) = $-20$ mV produces a rapidly activating large inward current followed by a delayed outward current. Replacement of the standard ES with a modified one containing 15 m$M$ BaCl$_2$ and 120 m$M$ tetraethylammonium chloride (TEACl) produced a smaller, more slowly inactivating inward current presumed to be carried by Ba$^{2+}$. The transient component of the latter current is reduced in amplitude by decreasing the $V_h$ from $-90$ to $-50$ mV and is blocked by 100 m$M$ NiCl$_2$. The sustained component remaining after addition of Ni$^{2+}$ is largely blocked by nifedipine. These results suggest that a Na$^+$ current and two distinct Ca$^{2+}$ currents (one low voltage activated and the other high voltage activated) contribute to the total inward current. (Reproduced from *Pflugers Archiv., Eur. J. Physiol.*, 1995, Vol. 431, Figs. 1B and C.)

voltage-activated Na$^+$ currents and rapidly inactivating, low-voltage-activated (or T-type) Ca$^{2+}$ currents "kick in" at $V_m$ values negative to threshold for HVA Ca$^{2+}$ currents. Na$^+$-dependent APs are characteristic of canine islet cells, whereas Na$^+$ currents and both types of Ca$^{2+}$ currents contribute to the initiation of electrical activity in human $\beta$ cells. Repolarization of individual APs is largely brought about by delayed rectifier K$^+$ channels, abbreviated K$^+$(DR). Despite many ideas and much effort in modeling, the mechanism(s) causing the termination of a burst of APs remains elusive. Possibilities range from the activation of miniconductance (SK) or even maxiconductance (BK) calcium-activated K$^+$ channels to slow inactivation of an

HVA Ca$^{2+}$ current coupled with a rise in K$^+$(DR), to a cyclical rise in K$^+$(ATP) channel activity.

Although $\beta$ cells appear to contain variants of most of the granule and plasma membrane proteins thought to contribute to the exocytotic machinery (SNAPs, SNAREs, synaptotagmins), to date little is known about the detailed molecular interactions mediating Ca$^{2+}$-dependent fusion of insulin granules with the plasma membrane. However, during the past decade two real-time single-cell assays of exocytosis have provided a glimpse at several general key features of the Ca dependence: timing and dynamics of the release process. One assay, **membrane capacitance tracking,** is designed to assess rapid changes in membrane surface area using patch clamp techniques. This assay assumes that the plasma membrane can be modeled as an equivalent circuit consisting of a lumped conductance and a capacitance element in parallel. Since the total $C_m$ of the plasma membrane of an isolated $\beta$ cell (roughly a spherical lipid bilayer) is directly proportional to the surface area of the plasma membrane, rapid fusion of a granule with the plasma membrane should predictably add to the total $C_m$ in stepwise fashion and this $\Delta C_m$ should be directly calculable from the surface area of the granule. In practice, small changes in $C_m$ of a voltage-clamped cell are often estimated from a complex membrane response. A sinusoidal voltage stimulus is imposed on the membrane and the resulting current is then passed through a lock-in amplifier (LIA) that can decompose this current into two components, one in phase and the other orthogonal to (90° out of phase with) the voltage sinusoid. If the electrical circuit of the cell membrane only consisted of $R_m$ and $C_m$ in parallel, $C_m$ would be directly proportional to the amplitude of the orthogonal current sinusoid. (Recall from the discussion of voltage clamping in an earlier chapter that when a membrane is charged to a new voltage the current $i_c$, flowing to charge the capacitor is $i_c = C_m(dV_m/dt)$. If $V_m = V_a \sin \omega t$, $I_c$ will be $V_a \cos \omega t$, or 90° out of phase). However, because the complete electrical circuit of the cell includes the access resistance to the cytoplasm, this is not precisely so. To compensate for this, one must either set the LIA to a special phase angle where changes in the resistance parameters do not affect the amplitude of the orthogonal current sinusoid or use circuit analysis to model the three circuit parameters. In some situations a stimulus consisting of the sum of two or more sine waves may be useful.

In the second assay, known as **amperometry,** the transient release of the secretory product is assayed as it appears near the cell surface and is electrochemically oxidized or reduced at the tip of a small-diameter carbon fiber electrode positioned at the cell's surface. The oxidation–reduction of a packet of chemical produces a spike of current. Knowing the number of oxidations–reductions per molecule and the Faraday constant, one can estimate the number of molecules whose oxidation–reduction contributes to the current spike. This estimate of granule content often compares quite favorably with estimates from other biochemical measurements made on intact granules. Last, by ramping the voltage applied to the electrode and examining the rapid current response, it is possible to obtain an electrochemical signature (or *voltammagram*) of the molecule released.

As shown by the idealized data in Fig. 5, much of the verification of these two approaches resulted from their early and often mutual application to the study of exocytosis from mast cells of the beige mouse. These cells contain giant (1–2 $\mu M$ in diameter) secretory granules loaded with readily oxidizable serotonin (5-HT). In these cells dialysis of the cell with a pipette solution containing GTP-$\gamma$S triggers slow but inexorable fusion of all of the secretory granules. Steplike jumps in capacitance, consistent with the fusion of one secretory granule, are associated, after a delay of tens of milliseconds, with amperome-

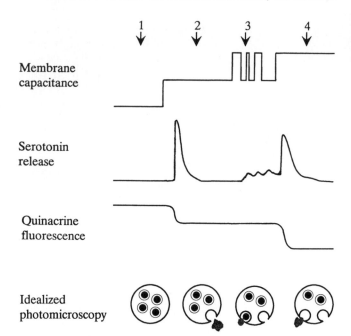

**FIG. 5.** Schematic presentation of the best available evidence validating membrane capacitance tracking and amperometry as single-cell assays or secretory granule exocytosis. In this experiment, whole-cell recording is done on the mast cell of a beige mouse, which contains fewer than twenty 1 to 2-$\mu$m diameter dense core granules. Introduction into the cytoplasm of GTP-$\gamma$S in the presence of submicromolar concentrations of Ca$^{2+}$ results in a stepwise increase in membrane capacitance. The sustained steps are accompanied by spike-like release of serotonin (5-HT) and are followed by visible extrusion of the contents of the giant secretory granule, monitored optically as the appearance of fuzzy granule contents or spectrofluorometrically as the loss from the cell of quinacrine, a dye that loads into acidic vesicles. Transient (or "flickering") capacitance increases and are accompanied by "flickery" pulses of 5-HT release; these produce no visible extrusion of granule contents. These flickers may increase in intensity until a stable fusion event is seen, and most likely they represent the formation, followed by widening or closure, of the so-called "fusion pore" connecting the vesicle interior and the extracellular solution. Fortuitously, the secretory granule membrane is densely packed with H$^+$ATPases, which causes the interior of the granule to be hyperpolarized with respect to the cytoplasm. Hence, fusion of the granule with the plasma membrane produces a measurable small burst of current, signifying the equalization of membrane potential (V$_m$) across the fusing granule with that of the plasma membrane. From the total charge crossing the membrane (Q) and the final capacitance increase ($\Delta C_m$), $\Delta V$ can be calculated as $Q/\Delta C_m$. Using the time course and magnitudes of $i$, $\Delta V$, and Ohm's law, it is possible to calculate the time course of the development of conductance across the fusion pore. One possibility is that the fusion pore begins as a protein scaffold and rapid infiltration of the scaffold by fluid lipids causes expansion.

tric spikes of 5-HT release as well microscopically visible expulsion of granule contents. These steady-state measurements of very large signals provide confidence for undertaking analysis of smaller signals evoked by transient stimuli.

In $\beta$ cells, experiments in which patch-clamp recordings (sometimes augmented by cytosolic Ca measurements) are combined with either capacitance tracking or amperometry have revealed some key features of depolarization–secretion coupling (Fig. 6). First, as expected, exocytosis is strongly dependent on $Ca^{2+}$ entry occurring during the voltage-clamp pulse. With long-duration, widely separated depolarizations, both $\Delta C_m$'s measured within 500 msec after the depolarization and the sum of amperometric events measured during the depolarization show very similar dependence on depolarization as does the $Ca^{2+}$ current. There is often a second-power dependence of $\Delta C_m$ on the $Ca^{2+}$ influx estimated by the integral of the $Ca^{2+}$ current. Second, short barrages of amperometric spikes can be seen during, and distinct $C_m$ increases can be seen after, a short train of APs is generated in response to a secretogogue, such as tolbutamide. A combination of patch-clamp recording in the cell-attached mode and amperometry demonstrates that $K^+(ATP)$ channels have closed, and electrical activity is in progress before secretion commences. Third, like insulin secretion measured by radioimmunoassay of islet perfusate, the depolarization-induced, $Ca^{2+}$-dependent $\Delta C_m$ and amperometric spikes show a temperature threshold (e.g., being absent at temperatures <25°C)

and are enhanced by a variety of maneuvers that raise the cAMP level. Fourth, as expected from experiments in permeabilized cells, exocytosis is highly sensitive to small changes in cytosolic $Ca^{2+}$. When cytosolic $Ca^{2+}$ is globally increased to only 1–2 $\mu M$ by cell dialysis, sizable exocytosis commences; and when raised in the range of ~10 $\mu M$ by photolysis of a caged Ca compound, the rates of exocytosis recorded are within several-fold of those seen after cell depolarization. Fifth, the exocytosis assayed by these single-cell approaches appears to be chiefly from the large dense core granules seen in the cells. The smallest sustained steps in capacitance that can be reliably measured in response to depolarization or cell dialysis are 1.0–1.5 fF; this roughly corresponds to that expected from the rapid fusion of a 200-nm-diameter granule with plasma membrane followed by a slow (seconds long) endocytotic process. In fact, it is this slow endocytosis that makes measurement of depolarization-induced exocytosis even possible, since capacitance cannot be measured accurately and rapidly during the actual depolarization. The charge transfer associated with the largest amperometric events is consistent with the estimated insulin or serotonin contents of the

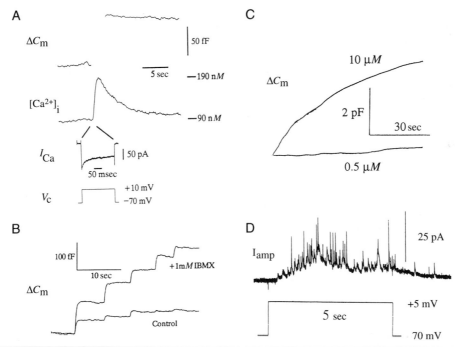

**FIG. 6.** Using capacitance tracking and amperometry to study $Ca^{2+}$-dependent depolarization–secretion coupling in $\beta$ cells. (A) A cell was preloaded with the membrane permeant $Ca^{2+}$-sensitive dye fura-2. AM is recorded from the perforated-patch variant of whole-cell recording. Note that a depolarizing pulse from −70 to +10 mV produces a peak $Ca^{2+}$ current ($I_{Ca}$) of −50 pA, and is sufficient to raise rapidly the estimated cytosolic $Ca^{2+}$ concentration and to trigger a 50-fF increase in $C_m$, corresponding to the exocytosis of 30–50 granules. The estimated cytosolic $Ca^{2+}$ increase is an average over the entire cell. The effective $Ca^{2+}$ concentrations near the membrane, which trigger exocytosis, may well be in the micromolar range, especially if one considers that $[Ca^{2+}]_i$ must be raised to >1 $\mu M$ by dialysis to provoke a much slower rate of continuous secretion (see part C). (B) Effect of phosphodiesterase inhibitor (IBMX) on secretion evoked by repetitive depolarization. Note that after treatment with IBMX, not only is initial release enhanced but release is sustained through multiple subsequent depolarizations. (C) Effect of cytosolic $Ca^{2+}$ on exocytosis. Whole-cell recordings were made using intracellular solutions containing free $Ca^{2+}$ of 0.5 $\mu M$ vs 10 $\mu M$. (D) Time course of amperometric events during continuous depolarization to a clamping potential where $Ca^{2+}$ entry is maximal. Note that the first release events are seen >200 msec after start of voltage-clamp pulse, while release events continue, albeit at a low frequency, for >2 sec after depolarization is completed. (Parts A and C are from unpublished data by D. Barnett and S. Misler. Part D is from unpublished data by Z. Zhou and S. Misler.)

granules. (These measurements are subject to the vagaries of variable 5-HT loading and subsequent leakage as well as the rundown in sensitivity of the specially coated carbon fiber electrodes used to detect insulin.)

A closer study of some of these responses reveals some other interesting aspects. *First, especially in human and canine β-cells, the kinetics of depolarization-induced exocytosis is quite complex and has interesting implications for the interplay of $Ca^{2+}$ channels and release sites.* While depolarizations of 50- to 200-msec duration provoke subsequent measurable release, brief depolarizations of 10–25 msec in duration generally do not unless they are grouped into a high-frequency train. Amperometric assay of quantal release during a sustained depolarization or a train of short depolarizations demonstrates that release begins well into the depolarization and can continue up to several seconds after its cessation. These "slow to start and stop" release kinetics are quite different from those at fast transmitting nerve terminals where the vast majority of release occurs within 1–2 msec of $Ca^{2+}$ entry (see Fig. 6d). One possibility, supported by the rare ultrastructural observations of "docked" granules, is that in β cells secretory granules are not as tightly colocalized with $Ca^{2+}$ channels as vesicles are with $Ca^{2+}$ channels at the "active zone" of secretion at nerve terminals. Hence, in β cells $Ca^{2+}$ would need to diffuse greater distances, over tens of milliseconds, and encounter more interposed buffer molecules, prior to interacting with a granule fusion site. This being so, the effective concentration of $Ca^{2+}$ needed at the release site would probably be much lower in β cells (probably $<5\ \mu M$) than in nerve terminals (probably $>50\ \mu M$). This more diffuse but higher $Ca^{2+}$-affinity mode of exocytosis of dense core granules has been studied and modeled in detail in chromaffin cells of the adrenal medulla (see Klingauf and Neher, 1997). *Second, release is highly history dependent, yet can be modulated by second messengers, offering interesting implications for the availability of readily releasable granules.* In many cells repetitive depolarizations imposed at intervals of several seconds initial release and subsequent rapid depression of release despite little or no change in depolarization-evoked $Ca^{2+}$ entry or the resultant rise in cytosolic $Ca^{2+}$. However, addition of an agent that enhances cytosolic cAMP not only enhances initial exocytosis but sustains exocytosis through repeated depolarizations without necessarily enhancing $Ca^{2+}$ entry. In concert with this is evidence that the initial rate as well as the total extent of $C_m$ increase, in response to diffusion of a set concentration of $Ca^{2+}$ from the patch pipette, is at least two- to threefold greater in the presence of $100\ \mu M$ cAMP than in its absence. This suggests that there is a finite but easily depletable release-ready granule pool whose total size can be modulated by agents known to enhance secretion.

*This basic scheme of $Ca^{2+}$ entry and diffusion-dependent exocytosis regulated by the size of a release-ready pool* provides a focal point for examining more complex physiological aspects of stimulus-secretion coupling. (1) Is glucose-induced electrical activity, often characterized by short (several second) bursts of intense electrical activity separated by longer quiescent periods, *tuned* to maximize secretion in the face of the delay release and need for pool

refilling? A burst of electrical activity might be an efficient way to ensure sufficient entry of $Ca^{2+}$ that slowly diffuses to distant release-activating sites. Pauses in $Ca^{2+}$ entry might prevent overload during periods when the release-ready pool needs to be replenished. (2) Does glucose regulate secretion beyond triggering electrical activity? In addition to triggering volleys of electrical activity, under some circumstances glucose appears to enhance the activity of $Ca^{2+}$ channels as well as stimulate the production of at least two second messengers. Arachidonic acid, generated via a $Ca^{2+}$-independent phospholipase $A_2$ upon stimulation by rising ATP levels, has possible $Ca^{2+}$ channel-enhancing as well as fusigenic properties. Cyclic ADP-ribose, generated from NAD, may enhance release of $Ca^{2+}$ from intracellular stores and "prime" secretion by providing a higher background level of $Ca^{2+}$ on which pulsatile $Ca^{2+}$ entry can build. (3) Postprandial insulin secretion appears to be "jump started" by vagally released acetylcholine (ACh) interacting with muscarinic receptors as well as insulin-stimulating peptides ("incretins") such as glucagon-related peptide-1 (GLP-1) released by endocrine cells of the gut. Does the muscarinic ACh receptor trigger G-protein-related release of $Ca^{2+}$ from cytosolic stores? Does GLP-1 largely operate by increasing cytosolic cAMP and the readily releasable pool? (4) Do other hormones or transmitters work to increase effective pool size by affecting monomeric G-proteins (GTPases such as Rab-3) that contribute to the $Ca^{2+}$-independent route for exocytosis, as well as enhance the $Ca^{2+}$-dependent route.

## III. Carotid Chemoreceptor Cells

Neural control of respiratory drive is critical in maintaining physiologic levels of plasma $O_2$, $CO_2$, and pH. In the rhythmic neuromuscular process of breathing, trains of APs, generated by pacemaker cells in the CNS, trigger motor neurons and activate periodic contractions of the inspiratory muscles (diaphragm and intercostal muscles). This action expands the chest, thereby sucking air (high $O_2$, low $CO_2$ content) into the lungs. Gas exchange consists of net $CO_2$ diffusion from capillary to adjacent lung air spaces (alveoli) and net $O_2$ diffusion in the opposite direction (from alveoli to the capillary). Relaxation of inspiratory muscles, sometimes combined with the contraction of expiratory muscles, expels the air (high $CO_2$, low $O_2$) from the lungs. A major stimulus for altering the pattern of respiratory drive is a drop in the partial pressure of $O_2$ dissolved in plasma (i.e., plasma $pO_2$); a secondary stimulus is a rise in plasma $pCO_2$ or a fall in pH. The carotid body is the organ that senses changes in plasma $pO_2$, $pCO_2$, and $pH$, and mediates changes in CNS respiratory drive. Located at the bifurcation of the carotid artery, a branch of the aorta, the carotid body consists of a central core of **chemoreceptor** (or **glomus**) **cells,** of neural crest origin. The glomus cells synapse on dendritic endings of sensory nerve fibers that comprise the carotid sinus nerve traveling into the CNS. This glomus cell core is surrounded by more superficial glial or sustentacular cells and a dense network of highly porous capillaries. The glomus cell is the transduction site of changes in plasma $pO_2$, $pCO_2$, and pH into neural activity in the afferent carotid sinus nerve. Decreases in $pO_2$, as well increases in $pCO_2$ or decreases in pH, lead to an increase in release of dopamine and probably as yet unidentified transmitters from glomus cells, as well as an increased AP frequency in the dopamine-sensitive carotid sinus nerve. This contributes to increased respiratory drive.

## A. Role of O₂-Sensitive K⁺ Channels in Transduction of Hypoxia

Two divergent views have emerged as to the mechanism for glomus cell chemotransduction of hypoxia (Fig. 7A). The first theory, let's call it **depolarization–secretion coupling,** maintains that hypoxia induces cell depolarization, resulting in opening of voltage-gated $Ca^{2+}$ channels, the firing of APs, $Ca^{2+}$ influx, and synchronized exocytotic release of dopamine. This might give rise to sufficiently large excitatory postsynaptic potentials in the dendritic region of a single sinus nerve cell to trigger propagating APs. An alternative view, let's call it **hypoxia-induced asynchronous release,** holds that hypoxia induces quantal release of transmitter in a $Ca^{2+}$-dependent manner, which is independent of voltage-gated Ca entry, by discharging $Ca^{2+}$ from intracellular stores. This increase in asynchronous

quantal release of transmitter, generating a rise in the frequency of miniature excitatory postsynaptic potentials, might enhance ongoing electrical activity in the postsynaptic carotid sinus nerve. These contrasting viewpoints have arisen from data generated with different preparations from different species.

Several lines of evidence support the classic scheme for depolarization–secretion coupling in glomus cells. First, glomi are electrically excitable sensory cells and contain HVA $Ca^{2+}$ currents, delayed rectifier $K^+$ currents, and in some cells voltage-dependent $Na^+$ currents. Second, in some isolated cell preparations glomus cells respond to hypoxia by depolarizing and generating APs. (However, in other preparations, glomus cells fire APs only in response to current injection or release of the cell from sustained hyperpolarization. These cells show no change in passive electrical activity in response to hypoxia.) Third, in glomus cells that fire in response to hypoxia, depolarization results in increases both in cytosolic $Ca^{2+}$, measured with $Ca^{2+}$-sensitive intracellular dyes, and in amperometrically measured dopamine release. Depolarization–secretion coupling requires extracellular $Ca^{2+}$ and is reduced by blockers of the HVA-type $Ca^{2+}$ channels.

A very exciting development in chemoreceptor physiology consistent with this depolarization–secretion coupling scheme is the finding that, in whole-cell recordings, low-

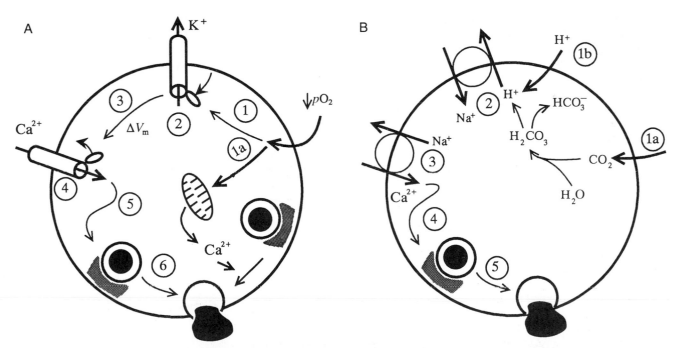

**FIG. 7.** Stimulus-secretion coupling in carotid chemoreceptor cells. (A) O₂-induced transmitter release. Key steps in classical scheme of depolarization–secretion coupling are outlined. $\downarrow pO_2$ results in decreased O₂ binding to the intracellular receptor coupling to a specific voltage-dependent $K^+$ channel (1). This results in the closure of this channel (2), cell depolarization (3), increased probability of the opening of HVA $Ca^{2+}$ channels (4), increased calcium entry (5), and the enhanced rate of fusion of dopamine-containing vesicles with the plasma membrane (6). Alternatively (see path beginning with 1a), hypoxia-induced rundown of the mitochondrial proton gradient might result in reduced ATP generation and slow release of $Ca^{2+}$ from mitochondria or other $Ca^{2+}$ stores, resulting in a slow rise in spontaneous quantal release. (B) CO₂- and H⁺-induced transmitter release. $\uparrow pCO_2$ (1) or H⁺ entry (1b) results in $\downarrow pH_i$ and stimulation of the $Na_o^+/H_i^+$ exchanger (2). The resultant influx in $Na^+$ stimulates the $Ca_o^{2+}/Na_i^+$ exchanger (3). This, in turn, increases phasic $Ca^{2+}$ influx (4) and $Ca^{2+}$-dependent exocytosis (5).

ering ambient $pO_2$ selectively and reversibly reduces the outward K current flowing through $K^+(DR)$ channels of the glomus cells. Reducing $pO_2$ from 160 to 90 mmHg reduces the peak $K^+(DR)$ current by ~30%. This effect is not dependent on the concentrations of ATP (0–3 m$M$) or $Ca^{2+}$ (<1 n$M$–0.5 $\mu M$) in the pipette. In outside-out patches of membrane, reduced $pO_2$ reversibly decreases the probability of opening ($P_o$) of a 20-pS $K^+(DR)$-type channel (Fig. 8). Early on, the relevance of this $O_2$-sensitive $K^+$ channel to hypoxia chemotransduction was questioned; the original published study suggested that *in vitro* the channel was most responsive to changes in $O_2$ over the range of $pO_2$ values between 110 and 150 mmHg while the carotid sinus nerve actually fires optimally at $pO_2$'s of less than 70 mmHg. However, the $O_2$ sensitivity of the channel can be shifted into a more physiological range by the addition of a membrane-permeant analog of cAMP, and many isolated cells in culture lose their usual ability to elevate cAMP in response to messengers. More recent data have shown modulation of the activity of these $K^+(DR)$-type channels over partial pressures of $O_2$ ranging from 20 to 150 mmHg. In whole-cell current-clamp recordings, the effect of a reduction in $pO_2$ is (1) an increase in the rate of cell depolarization on release from maintained hyperpolarization, (2) an increase in the frequency of spike activity in the resultant short train impulses, and (3) an increase in AP overshoot. Hence, in glomus cells with some intrinsic spontaneous electrical pacemaker activity, $O_2$-dependent changes in $K^+$ current could alter the frequency of APs and hence the frequency of $Ca^{2+}$ transients, thereby increasing [Ca]$_i$-dependent transmitter release onto the afferent nerve. In glomus cells with little automaticity that maintain

resting potentials of between −50 and −40 mV, closure of $K^+(DR)$-type channels could still cause steady-state depolarization "generator potential" and trigger electrical activity *de novo*. More extensive perforated-patch recordings are needed to determine the range of electrical activity patterns exhibited by these cells.

These data suggest that $O_2$ maintains the activity of a delayed rectifier-type $K^+$ channel through a novel gating mechanism. Several possibilities for the molecular mechanism of $O_2$ transduction and its relationship to channel gating have been suggested but there is no definitive evidence for any one. The first is that a heme protein, analogous to a subunit of the $O_2$-carrying protein hemoglobin, is attached to the channel, or, resembling the SUR of the $K^+(ATP)$ channels, is a functional subunit of the channel. In this way a change in configuration of the protein, on losing $O_2$, would alter channel gating. The second is the presence of an oxidase that, on reduction of cytosolic $pO_2$, produces less hydrogen peroxide ($H_2O_2$), consequently altering the concentration of redox intermediate and thereby channel conformation.

An alternative view of hypoxia-induced chemotransduction in glomus cells is that the rise in intracellular $Ca^{2+}$ necessary for dopamine secretion is due to $Ca^{2+}$ release from intracellular stores and that voltage-activated currents play only a secondary role, perhaps allowing replenishment of intracellular $Ca^{2+}$ stores. This hypothesis has arisen from data generated from both isolated glomus cells and *in situ* carotid cell bodies. Mitochondrial poisons, such as cyanide, produce a condition of *histotoxic hypoxia,* which mimics true hypoxia in stimulating carotid body nerve activity and respiratory drive. These poisons produce in-

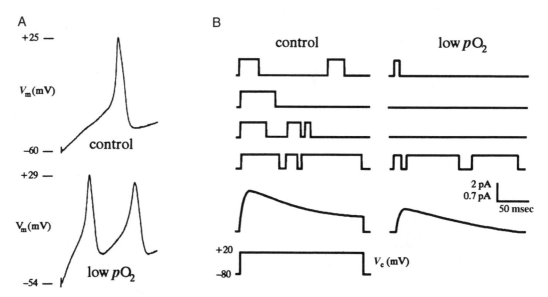

**FIG. 8.**  Origin of low $pO_2$-induced enhancement of electrical activity in carotid glomus chemoreceptors. (A) Reduction in ambient $pO_2$ results in increased frequency of spontaneous AP activity, partly due to reduced rate of repolarization of the AP. Hence, any background depolarizing current might be more effective in raising the membrane potential toward threshold for firing a spike. (Reproduced from *The Journal of General Physiology,* 1989, Vol. 93, p. 979, Fig. 14. By permission of the Rockefeller University Press.) (B) Reduction in ambient $pO_2$ reduces the open probability $P_o$ of $K^+(DR)$ channels, on depolarization from −80 to +20 mV. Lowest traces in each column represent the average of many individual channel current traces. (Idealized traces adapted from Ganfornina and Lopez Barneo, 1991.)

creases in intracellular $Ca^{2+}$, which is unaffected by pharmacologic maneuvers to abolish or enhance the AP, suggesting that increases in $Ca^{2+}$ arise from intracellular stores. Hypoxia and mitochondrial poisons produce a rise of cell $NAD(P)H$. Additionally, studies from *in situ* carotid bodies have shown that hypoxia-induced carotid sinus nerve APs are not blocked by drugs that block outward $K^+$ currents and that whole-cell glomus membrane resistance is not changed in response to hypoxia. A proposed mechanism for hypoxia inducing intracellular $Ca^{2+}$ release is that low $O_2$ decreases mitochondrial efficiency, perhaps slowing electron transfer in the respiratory chain, and leading to a decreased mitochondrial proton electrochemical gradient. A consequence of this could be the release of $Ca^{2+}$ from mitochondria or decreased ATP production, resulting in slow $Ca^{2+}$ release from other intracellular stores (e.g., the ER). For this mitochondrial hypothesis to work, it might be necessary for $O_2$-trapping properties of mitochondrial cytochrome oxidase and mitochondrial $Ca^{2+}$ storage properties in glomus cells to differ significantly from that of other tissues.

### B. Transduction of Increased $pCO_2$ and Decreased Plasma pH

Glomus cells rapidly equilibrate $pH_i$ and $pH_o$. Hence, it is probable that with acidic stimuli the cell is actually sensing a fall in $pH_i$. But how does an increase in $[H^+]_i$ result in $Ca^{2+}_o$-dependent transmitter release, especially since $H^+_i$-induced release is insensitive to blockers of HVA $Ca^{2+}$ channels and HVA $Ca^{2+}$ channels are often inhibited by reduction in $pH_i$? The proposed link between increased $[H^+]_i$ and $Ca^{2+}$ entry in promoting dopamine release is that increased $[H^+]_i$ activates a $Na^+_o/H^+_i$ exchanger, which, in turn, elevates $Na^+$ and recruits a $Na^+/Ca^{2+}$ exchanger. The activity of the latter exchanger is often augmented by cell depolarization; this may, in fact, occur because a drop in $pH_i$ reduces $K^+(DR)$-type $K^+$ current much as a drop in $pO_2$ does. An increase in background "spontaneous" quantal release of dopamine, which is not phase linked to AP activity, could increase steady-state depolarization of carotid nerve fibers and, perhaps, augment ongoing impulse activity (Fig. 7B). In considering this scheme, it should be cautioned that thus far, good evidence only exists for activation of a $Na^+/H^+$ exchanger.

To be sure, respiratory drive varies significantly according to age and physiological status. Human fetuses do not exhibit regular respiratory movements *in utero*, while newborn infants often display periodic breathing different from breathing patterns seen in older children and adults. Disease states produce abnormal breathing patterns such as Cheyne–Stokes breathing, the diamond-shaped changes in respiratory excursions followed by a period of non-breathing (apnea), seen in stroke, congestive heart failure, and chronic hypoxia. In infants, prolonged apnea despite persistent hypoxia or hypercarbia can culminate in sudden infant death syndrome (SIDS). Recent data suggest that carotid body glomus cells from animals raised under hypoxic conditions show blunted electrophysiological responses to hypoxia compared to cells from healthy animals.

Disorders of breathing such as apnea and SIDS are seen in higher frequency in infants at increased risk for chronic hypoxia due to conditions such as prematurity or second-hand cigarette smoke exposure. Studies on the mechanism of changes in carotid body chemoreception in normal development and in various disease states should be a fruitful area for further inquiry.

## IV. Vascular Smooth Muscle Cells

### A. Hypoxic Vasodilation of Coronary and Mesenteric Vessels

Smooth muscle cells of resistance arterioles are very sophisticated metabolic sensors. In fact, they encompass a complete vasodilator reflex system in a single cell. In response to a drop in ambient $pO_2$, these cells reduce their tension generation. This results in vessel relaxation and the local redistribution of $O_2$ supply to $O_2$-consuming tissue. A key to the explanation of this phenomenon is the increase in $K^+$ conductance and membrane hyperpolarization, which precedes the fall in tension. However, the joint inhibition of electrical and mechanical activity during metabolic blockage appears to be a fairly general property of vertebrate muscle including cardiac and skeletal muscle. These are worth exploring prior to returning to the case of smooth muscle.

In cardiac myocytes, depression of cellular energy levels, achieved by inhibition of energy production (e.g., substrate or $O_2$ deprivation) or by overstimulation of energy use (e.g., repetitive stimulation), all result in slow but progressive reduction in the duration of the $Ca^{2+}_o$-dependent plateau phase of the AP. This progressively reduces plasma membrane $Ca^{2+}$ entry, which is needed as a trigger for regenerative release of $Ca^{2+}$ from the sarcoplasmic reticulum. The net result is a reduction in twitch tension. All of this occurs prior to any detectable change in resting $V_m$ or reduction in $Ca^{2+}$ currents measured by voltage clamping. The critical clue to the mechanism underlying metabolic block was the finding of increased background $K^+$ conductance; this increase was blocked by intracellular injection of ATP. The search for the single-channel correlate of the $K^+$ conductance resulted in the first identification of a voltage-independent, $ATP_i$-inhibitable $K^+$ channel in the excised patch. Recalling that the plateau phase of the AP represents a time of delicate balance between outwardly directed $K^+$ current, through the slowly activating $K^+(DR)$ channel (this tends to hyperpolarize the cell), and inwardly directed $Ca^{2+}$ current, through the slowly inactivating L-type $Ca^{2+}$ (this tends to depolarize the cell), increases in background $G^+_K$ will tip the balance toward earlier repolarization. However, other mechanisms might contribute to ischemia-induced contractile failure. These include (1) altered $[Ca^{2+}]_i$ "homeostasis" as well as changes in the binding affinity of the contractile apparatus for $Ca^{2+}$ induced by changes in $pH_i$ and $ATP_i$ and (2) activation of other $K^+$ channels, such as muscarinic, G-protein-gated $K^+$ channels by metabolitic intermediates generated during hypoxia.

In skeletal muscle, failure of neuromuscular transmission and the AP, as well as inhibition of contraction, are seen with metabolic blockade or intense exercise. Here, the AP and twitch are both quite brief and not reliant on $Ca^{2+}$ entry. However, under some conditions, large increases in $K^+(ATP)$ channel activity occur. The increased background $K^+$ conductance overwhelms the effect of the generalized increase in carbon conductance evoked by acetylcholine release from the motor neuron. Hence, transmitter release produces little or no depolarization. Failure of impulse initiation and its conduction into the T-tubule blocks depolarization-activated release of $Ca^{2+}$ from the sarcoplasmic reticulum.

$K^+(ATP)$ channels in cardiac and skeletal muscle have single-channel conductance, kinetics, and ATP/ADP gating similar to those found in the $\beta$ cell. However, in contrast to those in $\beta$ cells, muscle $K^+(ATP)$ channels are not open in the absence of hypoxia or metabolic poisoning. Cardiac $K^+(ATP)$ channels respond more dramatically to changes in glycolytic activity than to changes in oxidative phosphorylation; addition of glucose in the presence of a mitochondrial inhibitor reduces channel activity. Glycolytic enzymes may be bound to the plasma membrane, thereby preferentially routing ATP to cardiac $K^+(ATP)$ channels.

Contractile failure, prior to metabolic exhaustion, appears to be adaptive for skeletal muscle. A drop in tension development by some motor units results, via spinal reflex, in recruitment of other less active motor units. The adaptive advantage is not that clear for the ventricle, which is an electrical syncytium and must contract as a unit. In the ventricle, shortening of the widely propagated AP may locally reduce the refractory period and promote reentry or rebound excitation. In addition, the period of reoxygenation may enhance arrhythymogenicity by promoting large transient inward currents ($I_{ti}$) which result in spontaneous depolarizations. Hence, it would be better for the ventricle to anticipate rather than react to metabolic deprivation. Such "anticipation" could occur through local vasodilation of coronary arterioles. Here's how.

Returning to vascular smooth muscle, it is well appreciated that resistance vessels in the coronary and mesenteric (gut) circulation display resting tone. Their myocytes maintain a resting potential of −40 to −50 mV and have an abundance of $K^+(ATP)$ and L-type (HVA) $Ca^{2+}$ channels. In fact, in smooth muscle of mesenteric artery, the $K_d$ value (micromolars) for ATP-induced closure of $K^+(ATP)$ channels is roughly double that in cardiac myocytes, in principle making $K^+(ATP)$ channels of these myocytes more sensitive to hypoxia than the $K^+(ATP)$ channels in the heart. Under these circumstances, small changes in resting potential would be expected to affect resting tone. A small depolarization of 5–10 mV caused by closure of $K^+(ATP)$ channels should result in increasing opening of HVA $Ca^{2+}$ channels and subsequent vasoconstriction (Fig. 9A), whereas a small hyperpolarization of 5–10 mV affected by addition of a $K^+(ATP)$ channel opener or a DHP $Ca^{2+}$ channel antagonist should reduce resting tone and result in vasodilation. These predictions have been borne out experimentally. A modest drop in microenvironment $pO_2$ is sufficient to reduce arteriolar oxidative metab-

olism and cytosolic ATP, and produce vasodilation attributable to the opening of smooth muscle cell $K^+(ATP)$ channels, while leaving cardiac excitation–contraction coupling unaffected. [In bulk cardiac muscle *in situ,* this preferential opening of smooth muscle $K^+(ATP)$ channels might be further augmented by the release of adenosine by active cardiac tissue; extracellular adenosine has been shown to activate $K^+(ATP)$ channels via a G-protein-dependent mechanism.] Hence, preferential hypoxia-induced arteriolar dilation, with its attendant increases in local $O_2$ delivery, may spare cardiac myocytes the risk of hypoxia-induced alterations in excitation–contraction coupling. This scheme might work for the regulation of local blood supply to the brain and gut as well as to the heart.

## B. Hypoxic Vasoconstriction of Pulmonary Vessels

In contrast to coronary, mesenteric, and cerebral vessels, pulmonary artery and its smooth muscle constrict in response to a drop in $pO_2$. This hypoxic pulmonary vasoconstriction (HPV) constitutes an adaptive response in the lung bed because it ensures that areas of the lung that are poorly oxygenated will receive less blood flow; the extra blood flow is "redirected" toward better oxygenated areas to optimize gas exchange. HPV is critical in fetal pulmonary development when the maturing airspaces are filled with secreted fluid rather than inspired air. Under these conditions HPV maintains the relatively high pulmonary vascular resistance that shunts venous return around the low flow pulmonary bed, through the ductus arteriosus of the cardiac septum, and into the left heart. With inflation of the newborn's lungs to air containing substantially higher $pO_2$, $O_2$-induced pulmonary vasodilation occurs, thereby promoting blood flow through the pulmonary circulation. In the adult, HPV is useful in maintaining moment-to-moment matching of local ventilation to perfusion; this reduces the risk of hypoxia that can occur when a portion of the lung is poorly inflated. However, when the areas of local poor ventilation are widespread or with chronic hypoxia (such as at high altitudes), chronic HPV is accompanied by smooth muscle proliferation (i.e., "work hypertrophy of muscle"). The end result is the development of increased resistance and pressure in the total pulmonary vascular bed. This *pulmonary artery hypertension* imposes an increased *afterload* for the right ventricle and a stimulus for its hypertrophy.

What cellular mechanisms support HPV? Pulmonary artery myocytes maintain a low resting tension, a $V_m$ of ca. −40 mV and a resting cytosolic $Ca^{2+}$ of <100 n$M$. They respond to progressive hypoxia (e.g., a slow fall in $pO_2$ from 150 to 15 mm Hg) with a 15-mV depolarization, which is not affected by changes in $[Ca^{2+}]_o$, followed by $Ca^{2+}_o$-dependent increases in both cytosolic $Ca^{2+}$ and tension. Given that the $V_m$ in these cells show a Nernstian relationship to $[K^+]_o$, these results suggest that membrane depolarization is due to a large decrease in $Ca^{2+}_o$-independent resting membrane conductance (e.g., $G_K^+$), while depolarization ultimately results in a small increase in $G_{Ca}^{2+}$, which is sufficient to provide $Ca^{2+}$ entry to trigger contraction.

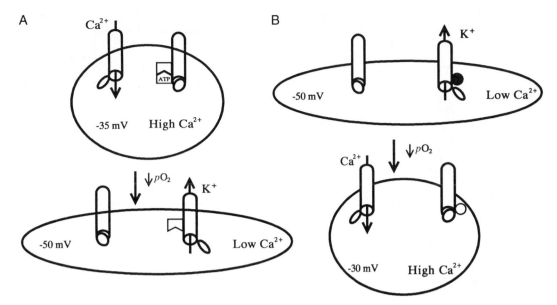

**FIG. 9.** Hypoxia and stimulus-contraction coupling in vascular myocytes. (A) Model of hypoxia-induced vasodilation in cardiac, cerebral, and mesenteric vessels. In normoxic conditions, $K^+$(ATP) channels are largely closed, voltage-dependent $Ca^{2+}$ channels are open, $V_m = -35$ mV, and cytosolic $Ca^{2+}$ is low. With a decrease in $pO_2$, cytosolic ATP levels drop, resulting in the opening of $K^+$(ATP) channels, repolarization to $-50$ mV, and closure of $Ca^{2+}$ channels. This results in a drop in cytosolic $Ca^{2+}$ and myocyte relaxation. (B) Model of hypoxia-induced vasoconstriction in pulmonary arteries. In normoxic conditions, $V_m = -50$ mV, $O_2$-sensitive $K^+$ channels are largely open, and voltage-dependent $Ca^{2+}$ channels are largely closed. Hence, cytosolic $Ca^{2+}$ levels are low and there is little resting tension. With a decrease in $pO_2$, $O_2$-sensitive $K^+$ channels close, resulting in depolarization, opening of $Ca^{2+}$ channels, $Ca^{2+}$ entry, and $Ca^{2+}$-induced contraction.

Tension is maintained as long as $Ca^{2+}$ entry through voltage-gated $Ca^{2+}$ channels exceeds $Ca^{2+}$ efflux via the $Na^+$/$Ca^{2+}$ exchanger. Whole-cell voltage-clamp experiments have provided good evidence that a voltage-activated, delayed rectifier-type $K^+$ current open at $V_m$ values positive to $-50$ mV and, hence at rest, is significantly inhibited by hypoxia. Depolarization of a few microvolts would be sufficient to open HVA $Ca^{2+}$ channels and result in vasoconstriction (see Fig. 9B). Hence, as in the carotid chemoreceptor, hypoxia-induced reduction in the activity of $K^+$(DR)-type channels in a cell with a very low background $G_m$, appears to be responsible for depolarization and stimulus-response coupling.

An important paradox that remains to be explained is how pulmonary artery myocytes manage to "hide" $K^+$(ATP) channels, known to be present in the sarcolemma, during hypoxia. In these cells, $K^+$(ATP) channel openers abort high $[K^+]_o$-induced tension increase, while sulfonylureas enhance tension generation, as they do in resistance vessels, yet hypoxia does not open these channels.

## V. Metabolic Coupling of Transport in Epithelial Cells

Evolution tends to be a conservative process. Channels used to couple stimuli to secretion and contraction in sensory and sensor-effector cells might be used in other more "vegetative" cell processes. As a fine example, it is salutary to examine an example of the use of $K^+$(ATP) channels in transcellular transport in epithelial cells.

The functional unit of the kidney, the nephron, excretes nitrogenous waste, water, and salts by a complex sequence of events. Initial urine is formed by bulk filtration of plasma, under hydrostatic pressure, through the glomerulus, a porous cuplike outgrowth of the renal tubule that encompasses a porous parallel capillary bed. This "filtrate" is turned into final urine by (1) bulk reabsorption (i.e., transtubular transport and corrective flow into capillaries) of solute and water at the proximal segment of the nephron, followed by (2) selective reabsorption of salt and water coupled with selective secretion of $K^+$ and $H^+$ at more distal nephron segments. Transtubular reabsorption of ions ($Na^+$, $Cl^-$, $HCO_3^-$, and $PO_4^{3-}$) and organic solutes (glucose, amino acids) are largely driven by the $Na^+$/$K^+$-ATPase (the $Na^+$ pump) situated at the basolateral (or peritubular) surface of the cell (i.e., facing the reabsorptive capillaries). The $Na^+$ pump regulates the electrochemical gradient for $Na^+$, which, in turn, allows passive $Na^+$ entry into the tubule cell via carrier-type mechanisms ($Na^+$/amino acid or glucose symporters or a $Na^+$/$H^+$ antiporter) or via $Na^+$ channels situated in the apical- or lumenal-facing membrane. Accompanying anions move via transcellular or paracellular routes. To maintain $K_i^+$ in the face of continuous $K^+$ entry via the pump, the cell must maintain continuous $K^+$ exit. $K^+$ channels in the basolateral membrane, and often in apical membranes, accomplish this exit. Patch-clamp

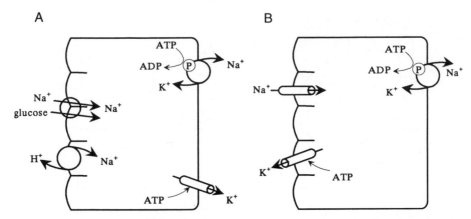

**FIG. 10.**  K$^+$(ATP) channels in renal tubular epithelial cells. (A) Role of basolateral K$^+$(ATP) channels in permitting K$^+$ exit to maintain cytosolic [K$^+$] during mass transcellular Na$^+$ transport in proximal tubule cells. (B) Role of K$^+$(ATP) channels in the exchange of Na$^+$ for K$^+$ at the apical membrane of the cortical collecting duct.

data have provided evidence for channel gating and maintenance phosphorylation by ATP of both basolateral K$^+$ channels of the proximal tubule and the apical K$^+$ channel of the distally located cortical collecting tubule. This K$^+$(ATP) channel, which differs from its counterpart in the $\beta$ cell in its conductance properties and ATP dependence, nevertheless, is inhibited by sulfonylureas. The presence of these channels in the nephron has interesting physiological implications. In the proximal tubule depicted in Fig. 10A, addition to the lumen of glucose (or alanine) stimulates apical Na$^+$ transport, basolateral Na$^+$/K$^+$ATPase activity, and ATP consumption. The resultant ATP consumption lowers cytosolic ATP levels, increases sulfonylurea-blockable basolateral conductance and single K$^+$(ATP) channel activity. This demonstrates that the activity of a metabolically regulated channel can be closely coupled to that of a Na$^+$/K$^+$ pump. In the cortical collecting duct depicted in Fig. 10B, the strategic placement in the apical membrane of Na$^+$ and K$^+$ channels, both of which show hormonal regulation, provides the opportunity for K$^+$ excretion to be graded with Na$^+$ delivery. For example, "revving up" transcellular Na$^+$ transport by hormonal (aldosterone) stimulation of the basolateral Na$^+$/K$^+$ pump reduces [Na$^+$]$_i$ and increases apical Na$^+$ conductance. Increased $g_{Na^+}$ brings apical $V_m$ toward $E_{Na^+}$ and away from $E_K$. If this is coupled with a drop in ATP production to stimulate K$^+$(ATP) channel activity, enhancement of passive exit of K$^+$ into the lumen is ensured.

## VI. Summary

This chapter focused on several types of metabolic sensor cells, namely, pancreatic islet $\beta$ cells, carotid chemoreceptor cells, and vascular smooth muscle cells, all of which transduce changes in plasma levels of metabolite fuels, $O_2$, or $CO_2$, into specific cell responses (secretion or contraction). These cells are "special" secretory and contractile cells because their stimulus-response coupling begins with metabolic stimuli changing K$^+$ channel activity. This, in

turn, results in changes in $V_m$ that trigger Ca$^{2+}$ entry and ultimately cell response.

Several K$^+$ channels involved in metabolic sensing have been identified. An ATP-sensitive K$^+$ channel, which closes with cell oxidation of nutrient metabolites, can regulate membrane potential and Ca$^{2+}$ entry. In pancreatic $\beta$ cells this channel controls stimulus-induced insulin secretion, whereas in the smooth muscle of resistance vessels it regulates contraction. An $O_2$-sensitive K$^+$ channel, which closes on cell exposure to decreased $O_2$, in part regulates transmitter release from carotid chemoreceptors and contraction of smooth muscle in pulmonary artery. The molecular biology of these channels and the complexities of their interactions with molecular species that gate are beginning to be unraveled.

These metabolic sensor cells and their metabolically regulated channels are likely to be just the tip of the iceberg as far as metabolic modulation of the passive ionic permeability of cells. Already, ATP-sensitive K$^+$ channels have been found in a variety of epithelial cells lining much of the nephron. The channels contribute to the ion recycling needed for transcellular solute transport. Other channels, such as Ca$^{2+}$ and K$^+$ channels, may be metabolically modulated by the rate of their phosphorylation, governed partly by available ATP or by novel chemical intermediates generated during transitions in cell metabolism.

## Acknowledgments

We thank David Bryant for preparing early drafts of Figs. 2, 5, and 7 through 10.

## References

Almers, W. (1990). Exocytosis. *Annu. Rev. Physiol.* **52,** 607–624.
Ashcroft, F. M., and Rorsman, P. (1989). Electrophysiology of the pancreatic $\beta$-cell. *Prog. Biophys. Mol. Biol.* **54,** 87–143.
Barnett, D. W., and Misler, S. (1995). Coupling of exocytosis to depolarization in rat pancreatic islet $\beta$-cells: effects of Ca$^{2+}$, Sr$^{2+}$ and

Ba$^{2+}$-containing extracellular solutions. *Pfluegers Arch., Eur. J. Physiol.* **430,** 593–595.

Ganfornina, M. D., and Lopez-Barneo, J. (1991). Single K$^+$ channels in membrane patches of arterial chemoreceptor cells are modulated by O$_2$ tension. *Proc. Natl. Acad. Sci. USA* **88,** 2927–2930.

Gonzalez, C., Almaraz, L., Obeso, A., and Rigual, R. (1992). Oxygen and acid chemoreception in the carotid body chemoreceptors. *TINS* **15,** 146–157.

Klingauf, J., and Neher, E. (1997). Modeling buffered Ca$^{2+}$ diffusion near the membrane: implications for secretion in neuroendocrine calls. *Biophysical J.* **72,** 674–690.

Misler, S., Barnett, D. W., Pressel, D. M., and Gillis, K. D. (1992). Electrophysiology of stimulus-secretion coupling in human $\beta$-cells. *Diabetes* **42,** 1220–1227.

Newgard, C. B., and McGarry, J. D. (1995). Metabolic coupling factors in pancreatic $\beta$-cell signal transduction. *Annu. Rev. Biochem.* **64,** 689–719.

Nichols, C. G., and Lederer, W. J. (1991). Adenosine triphosphate-sensitive potassium channels in the cardiovascular system. *Am. J. Physiol.* **261,** H1675–H1686.

Seino, S., Inagaki, N., Namba, N., and Gonoi, T. (1996). Molecular biology of the $\beta$-cell ATP-sensitive K$^+$ channel. *Diabetes Rev.* **4,** 177–190.

Tsuchiya, K., Wang, W., Giebisch, G., and Welling, P. A. (1992). ATP is a coupling modulator of parallel Na/K$^+$ATPase and K$^+$ channel activity in the renal proximal tubule. *Proc. Natl. Acad. Sci. USA* **89,** 6418–6422.

Weir, E. K., and Archer, S. L. (1995). The mechanism of acute hypoxic pulmonary vasoconstriction: The tale of two channels. *FASEB J.* **9,** 183–189.

Wollheim, C. B., Lang, J., and Regazzi, R. (1996). The exocytotic process of insulin secretion and its regulation by Ca$^{2+}$ and G-proteins. *Diabetes Rev.* **4,** 276–297.

Zhou, Z., and Misler, S. (1995). Amperometric detection of quantal secretion from patch-clamped rat pancreatic $\beta$-cells. *J. Biol. Chem.* **271,** 270–277.

Catherine E. Morris

# 43

# Mechanosensitive Ion Channels in Eukaryotic Cells

## I. Introduction

The idea of mechanically gated channels has a long history and there is abundant evidence that mechanosensory specialist cells use mechanosensitive (MS) ion channels for mechanotransduction (French, 1992; Hudspeth, 1989). Although several mechanosensory specialists have been intensively studied by electrophysiologists and molecular biologists (e.g., hair cells, crayfish stretch receptors, and, for molecular work, *Caenorhabditis elegans* mechanosensory neurons), nature has uncooperatively placed their mechanotransducing channels in structures that defy single-channel recording. Moreover, it is increasingly evident from cell biological, biophysical, and molecular work on hair cells and on a nematode model that specialized mechanotransduction involves a supramolecular assembly wherein a MS channel is connected fairly rigidly to intracellular and/ or extracellular filamentous proteins (Hamill and McBride, 1996b). It is not likely to become possible to preserve the necessary cytoarchitecture while making single-channel recordings from these mechanotransducers.

It is somewhat ironic, therefore, that for more than a decade, single-channel recordings from nonspecialized cells—bacteria, fungi, plant, and animal cells—have revealed a plethora of MS ion channels. The irony is compounded because for all of these cases except one, compelling evidence that the MS channels act as physiological mechanosensors is lacking. It is easy to point to physiological tasks requiring mechanosensors in bacteria, fungi, plant, and animal cells, but it has been frustratingly difficult to prove that these tasks are mediated by MS channels. The one satisfying exception (MS channels in hypothalamic neurons that regulate blood osmolarity) is described later in the Physiology section.

How far one should extrapolate from single MS channel data to putative physiological function is a matter of controversy (Gustin *et al.*, 1991; Morris and Horn, 1991); various workers offer substantially different perspectives, as is evi-

dent in recent reviews (Martinac, 1993; Morris, 1992; Hamill and McBride, 1995; Sackin, 1995; Sachs and Morris, 1997). An exhaustive account of MS channel pharmacology is also available (Hamill and McBride, 1996a).

## II. Three Mechanosensitive Channel Breakthroughs

Since the previous edition of this book was published, MS channel breakthroughs have occurred on three fronts: electrophysiologic, genetic, and molecular. Although this chapter focuses on eukaryotic MS channels that can be studied by single-channel recording (Fig. 1), we allude briefly to the genetic and molecular work. Substantial progress on hair cell mechanotransduction is described in Chapter 44 (see also Corey and Garcia-Anoveros, 1996).

Breakthroughs in the molecular biology of MS channels centered on two preparations, *Escherichia coli* and *C. elegans*. For the nematode, *C. elegans*, a large body of genetics by Chalfie and colleagues laid the groundwork for the subsequent molecular biology; the genetic approach continues to be crucial (reviewed by Bargmann, 1994; Tavernarakis and Driscoll, 1997). From each preparation, a MS channel has been cloned, albeit with important caveats—for *E. coli*, we do not know if the channel is mechanosensitive *in vivo*. For the *C. elegans* case, the missing piece is just the opposite—the cloning strategy was based on *in vivo* mechanoreception but functional expression of the putative channel in a heterologous system remains to be accomplished. Nevertheless, both are major findings, made all the more intriguing because they generate radically different pictures of channel mechanosensitivity.

The *E. coli* MS channel is the simplest of channels; a single small protein (~15 kDa) oligomerizes, probably as a hexamer, to form a high-conductance ion pore (1–3 nS) (Sukaharev *et al.*, 1997). Though this channel, MscL, can exhibit its mechanosensitive behavior in an artificial lipid

A

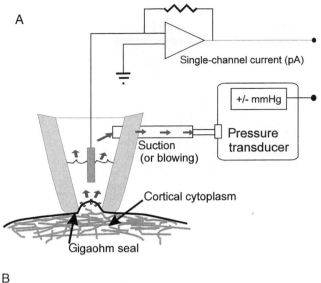

B

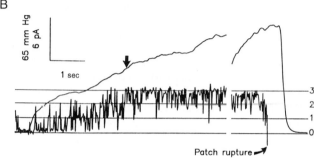

**FIG. 1.** Recording MS channels using pipette pressure to create membrane tension. (A) The most commonly used single-channel recording configuration for MS channels. The fire-polished tip is ~1 μm in diameter. A glass-to-membrane gigaohm seal forms several micrometers from the tip; underlying cytoplasm sustains some damage, both as a result of seal formation and from subsequent stimulation by suction or blowing. The patch can be excised to form an inside-out patch without abolishing mechanosensitivity of MS channels. (B) A recording of SA channel events from an *Aplysia* neuron. Suction is increased progressively, activating a maximum of three channels, whose open probability saturates near −100 mm Hg, as indicated by the arrow on the pressure trace (see Vandorpe *et al.*, 1994).

ating a detailed physical model (Fig. 2). The model features a pore, a heteromer of degenerin-like proteins, each with two hydrophobic membrane-spanning domains, as well as a chain of residues that may contribute to an aqueous cation-selective pore. A filamentous cytoskeletal element (stiffened microtubules) plus a linker molecule (a stomatin-like globular protein) and collagen in the extracellular matrix are all necessary for functional MS machinery in the sensory neurons. Given the elaborateness of the machinery, it is perhaps not surprising that heterologous expression of the degenerins—the putative channel proteins—has not yielded functional expression of a conductance. It may be inherently impossible to gate the channel mechanically minus its ancillary bells and whistles. There was, therefore, real excitement when it was realized that nematode MS channel degenerins are related by homology to mammalian epithelial Na channels. The evidence for mechanosensitivity in the latter channels (Awayda *et al.*, 1995; Achard *et al.*, 1996) is, however, still too rudimentary to be fully convincing. Nevertheless, a compelling picture is emerging for the degenerins as *C. elegans* MS channels: They appear to be part of a supramolecular entity broadly comparable to what is envisaged for the mechanotransducer channel in vertebrate hair cells. In hair cells there is cell biological and biophysical evidence that a cation channel (identity unknown), an extracellular tip link (protein unknown but visible by electron microscopy and perturbable by low calcium solutions), and actomyosin-linked elements contribute to the MS channel machinery.

The electrophysiological milestone? The excitement here is that finally there is a channel whose MS activity can be followed from the single-channel level all the way to the level of cellular and, in fact, whole organism physiology. The channel in question exhibits stretch inactivation (Figs. 3 and 4). Stretch-inactivated (SI) channels have aroused interest before, first when they were shown to coexist in growth cone membrane with stretch-activated (SA) channels (Morris and Sigurdson, 1989) and second when SI channel behavior was linked to the pathophysiology of dystrophic muscle (Franco and Lansman, 1990). But SI channel involvement in pathophysiology has never been confirmed and the molluscan SI channels proved not to act as a physiological mechanotransducer (Morris and Horn, 1991). Bourque and colleagues (reviewed in Bourque and Oliet, 1996), however, found that a SI channel mediates the osmotransducer current of hypothalamic osmosensory neurons (OSNs). Here, the picture holds together well, but it is tricky in its details. All-important is the concurrence between the single channel and macroscopic input/output relations (the whole-cell osmosensory data). Because this is somewhat convoluted, it is not given here, but is discussed later in the Physiology section. Unfortunately, this important prototype has yet to shed light on why there is not a more general concordance between single-channel and macroscopic findings. There are no notable differences in membrane density, in dynamic range, or in apparent mechanosensitivity between the osmotransduction SI channel and the many other MS channels that have, to date, *failed* to yield the expected macroscopic "mechanocurrents."

bilayer, it generally activates only at near-lytic tensions. No physiological role for the channel is known (an *E. coli* strain with the *MscL* gene knocked out had no "osmotic phenotype"). Moreover, the protein has no known eukaryotic homologs. This channel is, nevertheless, biophysically fascinating and will allow for the testing of many hypotheses about the basis for mechanosensitivity: For example, is oligomerization *per se* involved in activation of the current or in its subsequent slow adaptation? What is the role of interactions between the channel protein and bilayer lipid?

In the case of the putative *C. elegans* mechanosensory channel, genetic analysis of mutant individuals is pivotal. Although electrophysiology has not been possible, analysis of stereotyped touch sensitivity behavior (yes, whole animal behavior!) and of leaky channel-induced cell death in mutant worms has proven astonishingly effective in gener-

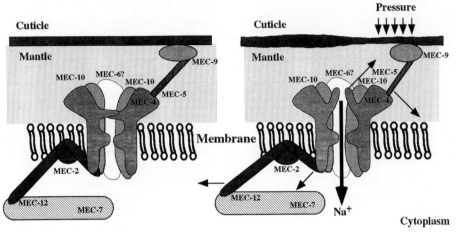

**FIG. 2.** Model for a touch transducing complex in *C. elegans* mechanosensory neurons. "MEC number" labels refer to a variety of proteins which, when mutated, can yield mechanosensory malfunctions. The degenerin MECs form the channel itself. In the absence of mechanical stimulation (left) the channel is closed. Application of a mechanical force to the body of the animal (right) stresses the assembly of interacting molecules in a way that favors the open state of the channel. (Modified from Tavernarakis and Driscoll, 1997).

## III. Stimulation of Mechanosensitive Channel Activity

MS channel currents can be elicited in patch-clamp studies of most types of cells, including a multitude of animal cells, plus wall-free cellular preparations from bacteria, fungi, and plants (Morris, 1990; Martinac, 1993). Bacterial MS channels are unique in that they retain full mechanosensitivity in artificial bilayer preparations and in liposomes (see Martinac, 1993; Hase *et al.*, 1995); for eukaryotes,

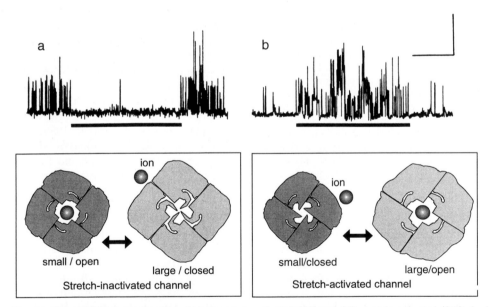

**FIG. 3.** Two versions of MS channel gating. (A) Stretch-inactivated (SI) and (B) stretch-activated (SA) single-channel events recorded from molluscan neurons. Bars: application of −40 mm Hg suction. Scale: 1 pA, 2 sec (SI) or 4 pA, 2 sec (SA). (From Morris and Sigurdson, 1989.) Below each trace is a cartoon of a SI or SA channel seen face-on, suggesting how in-plane tension (conveyed from the plasma membrane itself, or from membrane skeleton or both) could act allosterically in MS gating. At a given tension, a channel has a fixed probability of being open or closed; two-headed arrows represent stochastic equilibrium. At higher in-plane tensions, both the SI and SA channels would have higher probabilities of being in the large-diameter configuration. Whether a channel is SI or SA depends on details of a protein's "internal levers and pulleys"; the diameter of the conducting pore (the "active site" of this allosteric model) can either increase (SA) or decrease (SI) when the channel (or some tension-responsive region of the channel) flips into the larger configuration. As drawn here, tension stresses the channel's perimeter (the "other site" of this allosteric gating model), altering the transition probabilities. Tension does not literally "pull" the channel open. An example: consider two tensions at which a SA channel's open probability is, say, 0.01 and 0.4. At the higher tension the open probability increases 40× yet the channel is still 2.5× more stable *closed* than open! As discussed in the text, if the open and closed states of the channels had different inherent elasticities, this would add another level of complexity to the gating mechanism.

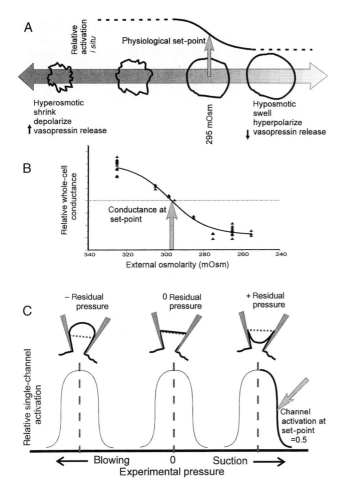

**FIG. 4.** The MS channels in osmosensory neurons; interpreting the activation curves. (A) Top, the sigmoidal activation curve for a SI channel, where hyperosmotic conditions (low tension) are shown to the left and hyposmotic conditions (high tension) to the right. The sigmoid has a negative slope, corresponding to data from osmosensory neurons. (B) Whole-cell osmosensitive conductance as a function of bath osmolarity. Data from Bourque, with permission; a Boltzmann has been fit to the data. (See Oliet and Bourque, 1996, for references.) (C) Single-channel "bell-shaped" activation curves. Illustrations show a pipette with a patch at two different experimental pressures. The dashed line indicates the membrane position for minimal membrane tension and the solid line indicates its position at zero experimental pressure. When the two do not coincide (i.e., when residual pressure in the pipette provides an offset), the bell curve is displaced. The bolded half of the right-hand bell curve would correspond to the sigmoids in parts A and B.

convincing MS channel currents have not been reported for membrane skeleton-free membrane.

To obtain recordings of MS channels in animal cells, a patch of membrane several square micrometers in area is sealed at the tip of a recording pipette and pressure (usually suction) is applied to distend the membrane (see Fig. 1). On cells that lack extracellular matrix, the membrane–pipette seal may occur with no treatment of the cell surface, but most cells require enzymic digestion to remove material (e.g., basement membrane, cell wall) that prevents direct access to the membrane. Sachs and colleagues have provided detailed visual and biophysical information (e.g., capacitances, elasticities) about the state of patch-clamped

membrane and its underlying cytoplasmic cortex (see Sokabe *et al.*, 1991). There is no tidy mechanical stimulus for patches. Although we like to imagine that membrane tension is all that changes when suction is applied, the plasma membrane is not a simple elastic solid, so the responses of membranes patches only roughly approximate those of a Hookean (elastic solid) thin-walled shell. In addition to tension changes, patches under suction pressure exhibit bilayer flow (i.e., liquid-like behavior), alterations of the structural relations between cortical cytoskeleton and the overlying membrane (i.e., viscoelastic behavior), microvesiculation, and sundry other nonideal (i.e., non-Hookean) behaviors.

The mechanism by which suction-induced membrane tension alters the open probability of MS channels is not known. It is likely that different explanations will be needed for channels in different cell types. In bacteria, experiments with amphipaths that partition asymmetrically into the bilayer's leaflets suggest that differential stress in the two leaflets of the bilayer may activate these MS channels (see Martinac, 1993). In eukaryotic cell membranes, simultaneous visual and capacitance measurements led Sokabe *et al.* (1991) to conclude that suction induces the bilayer to flow past a more unyielding protein network-like skeleton that acts as the tension-bearing (elastic) component; the channels are anchored by (and hence responsive to) unknown membrane skeletal element(s). Consistent with this view, Sheetz and colleagues (see Dai and Sheetz, 1995) measured membrane tether forces; they estimate that even the largest tensions in swelling molluscan neurons are considerably smaller than the tensions that gate most MS channels, which generally activate at near-lytic tensions (i.e., 1–10 mN/m) (Dai *et al.,* 1996). Table 1 lists mechanical quantities and attempts to put them in a helpful context for understanding the forces acting on MS channels.

## IV. Diverseness of Mechanosensitive Channels

The designation of an ion channel as mechanosensitive is empirical—it signifies only that the channel's open probability responds to membrane deformation. In most animal cells, once a gigaohm seal is obtained during cell-attached patch-clamp procedure, a suction of −10 to −100 mm Hg applied through the recording electrode will generally suffice to half maximally activate any SA MS channels present (see Table 1). Often it is difficult to saturate the effect because MS gating is elicited only as membrane tension approaches the lytic tension for the patch (e.g., Bedard and Morris, 1992; Vandorpe *et al.*, 1994). The density of MS channels is normally on the order of one channel per square micrometer of membrane. High densities such as those that can occur for ligand- and voltage-gated channels (say, $>100/\mu m^2$) have never been reported for MS channels. A cell may have only one type of MS channel [for example, the SA cation channels in *Xenopus* oocytes (Hamill and McBride, 1992)] or they may have multiple types, as indicated in Table 2. Neither excision of a patch nor removal of all cytoplasmic-side calcium abolishes the mechanosensitivity of MS channels (e.g., Vandorpe *et al.*, 1994).

**TABLE 1** Mechanical Quantities for MS Channels

**In the context of MS channels:**

1 Newton is an enormous force
  [$1N \sim = 100$ g (... ~1 apple in earth's gravity)]
1 Newton per meter is a large tension that would rupture a
  membrane
  [$1$ N/m $= 10^3$ dyne/cm (... ~1 apple dangling from a
  meter-wide banner)]
1 Newton per square meter is a very small experimental pressure
  [$1$ N/m$^2$ = 1 Pa (... ~1 apple per coffee table)]

**Forces** (see Sheetz and Dai, 1996; Sachs and Morris, 1997):

3 pN ... force generated by myosin molecule
7 pN ... force to pull membrane tether from a neuron
10–20 pN ... calculated force for activation of a "typical" MS
  channel
20 pN ... actin–gelsolin bond
50 pN ... force to pull erythrocyte membrane tethers
100 pN ... actin–actin bond
30,000 pN ... carbon–carbon bond

**Tensions**[a]

0.12 mN/m = 0.12 dynes/cm ... resting tension of plant
  protoplast membrane (see Kell and Glaser, 1993)
0.4 and 0.12 mN/m ... membrane tensions in normal and
  swollen molluscan neurons (Dai *et al.*, 1996)
1 mN/m = 1 dyne/cm = 1 pN/nm ... MS channel activation
  [yeast (Gustin, 1992); muscle (Sokabe *et al.*, 1991)]
4 mN/m ... lytic tension for plant protoplast (see Kell and
  Glaser, 1993)
2–10 mN/m ... lytic tension for lipid bilayers and cell
  membranes (see Sheetz and Dai, 1996)

**Pressures**[b]

1 kPa = 7.5 mm Hg
100 mm Hg = 13.3 kPa = 133 mbar
1 mm Hg = 1.36 cm H$_2$O
760 mm Hg = 1 atm

---

[a] The relevant membrane tension for MS channels is *in-plane tension*, not interfacial surface tension, though the units are the same. Another important quantity with the units of N/m is the spring constant (elasticity or stiffness) of Hook's law. Tension can also have the dimensions energy/volume as in $k_BT/nm^3$. Sometimes, *forces* are referred to as "tensions" (the magnitude of a force exerted, say, via a string) but "tension" that counteracts a force is a force.

[b] ±1–200 mm Hg: applied via pipettes to gate MS channels. Ideally, tension goes as Laplace's law (pressure = 1/2 × tension × radius) so the more curved a membrane, the lower the tension.

---

There are nonselective, cation-selective (i.e., mono- and divalent), and K$^+$- and Cl$^-$- or anion-selective MS channels (see Table 2). Most MS channels are SA and a few are SI channels. There are also, however, accounts of channels whose MS activity depends on the direction of membrane curvature (Bowman *et al.*, 1992; Marchenko and Sage, 1996). From their diversity, it is evident that MS channels belong to many different channel families. Many are activated by a variety of stimuli other than mechanical stress (Table 3), though some are known only from their MS gating (e.g., the SA cation channel in *Xenopus* ooctyes). The champion so far, with four types of activation, is the smooth muscle K channel reported by Kirber *et al.* (1992), although this channel may really be a ligand-gated channel in disguise. Fatty acids, liberated from the membrane by stretcth, appear to be the mediator of stretch activation

for this channel (Ordway *et al.*, 1995). This possibility needs to be considered more widely.

In addition to the *E. coli* MS channel, two other channels that exhibit weak MS gating at the single-channel level have been cloned. Even a cursory glance at this collection reveals that the requirements for MS gating must be rather catholic. The cloned channels include MscL, the nonselective channel of nano-Seimen conductance, from *E. coli* (Sukharev *et al.*, 1997); the *N*-methyl-D-aspartate (NMDA) channel, a glutamate-activated cation channel (Paoletti and Ascher, 1994); smooth muscle Ca-activated K channels (Dopico *et al.*, 1994); and a G-protein-regulated K-permeant inward rectifier, GIRK (Pleusamran and Kim, 1995). Of these, only MscL was cloned because of its mechanosensitivity. McsL forms homomultimers, probably **homohexamers** (Sukharev *et al.*, 1997), and bears no sequence resemblance to either the **heteropentameric** NMDA channel or the **heterotetrameric** GIRK.

MS channels are defined by their responses to membrane deformation but not all are equally responsive. Dynamic ranges of several orders of magnitude are not uncommon for SA channels in, say, molluscan neurons, chick muscle, and *Xenopus* oocytes (Sigurdson and Morris, 1989; Guharay and Sachs, 1984; Hamill and McBride, 1992); this compares to dynamic ranges of voltage- and ligand-gated channels. For some MS channels, however, mechanical stimulation only slightly changes open probability. For example, the NMDA-type glutamate channel increases its open probability ca. threefold in response to pipette suction (Paoletti and Ascher, 1994), and the bovine epithelial Na channel ENaC only ca. twofold (Awayda *et al.*, 1995). Eventually it would be desirable to reserve the term *mechanosensitive* for channels that recognize mechanical defor-

---

**TABLE 2** Cells with Multiple Types of MS Channels

**Amphibian kidney cells:** All demonstrable K channels in the basolateral membrane of amphibian proximal tubule cells are mechanosensitive (Cemerikic and Sackin, 1993).

**Molluscan heart** (*Lymnaea stagnalis*): SA K (see Morris, 1990) and (zero extracellular Ca only) SA Na channels (Gardiner and Brezden, 1990).

**Molluscan neurons:** (1) *Lymaea stagnalis*: SA K and SI K channels; the latter are less common and have a smaller conductance (Morris and Sigurdson, 1989). (2) *Cepaea nemoralis*: SA K channels and (excised patches only) SA Cl channels (Bedard *et al.*, 1992).

**Amphibian smooth muscle:** SA cation and SA K channels (see Kirber *et al.*, 1992).

**Chick heart cells:** Five distinct SA channels (three K selective, two cation selective) (Ruknudin *et al.*, 1993).

**Mammalian outer hair cells:** The lateral walls of guinea pig outer hair cells have a SA cation channel and one that appears K selective (Ding *et al.*, 1992).

**Mammalian osteoblast-like cells:** SA K channel, 60-pS SA cation channel, 20-pS SA channel (Davidson *et al.*, 1990).

**Crustacean stretch receptor neurons:** SA cation channels, rectifying SA cation channels (Erxleben, 1989).

***E. coli*:** Several molecularly distinct nonselective nano-Seimen conductance channels (see Sukharev *et al.*, 1997).

**TABLE 3** MS Channels with Gatings Controlled by Other Factors

**Avian SA cation channel:** Also activates with *depolarization* (see Sachs, 1992).

**Amphibian smooth muscle SA K channel:** A $Ca^{2+}$-*activated* K channel, also activated by *fatty acids* and by membrane *depolarization* (Kirber *et al.*, 1992).

**Amphibian smooth muscle SA cation channel:** The SA cation channels (or a subset of the population) is activated by *hyperpolarization* (Hisada *et al.*, 1991).

**Crayfish rectifying SA-cation channel:** Activated by *hyperpolarization* (Erxleben 1989).

**E. coli MS channel:** Also *depolarization* (see Martinac, 1993).

**Mammalian atrial SA K channels:** (1) An *arachidonic acid* and *depolarization-activated* SA K channel (Kim, 1992); (2) a distinct *ATP-sensitive* (*inhibited* by) SA K channel (Van Wagoner, 1993; but see Kim, 1992).

**Liver SA cation channel:** Identical to an *ATP-activated* (*via purinergic receptors*) channel of these cells (Bear and Li, 1991).

**Osteosarcoma SA-cation channel:** *Parathyroid hormone* acts via cytoplasmic messengers to increase the open probability and the single-channel conductance (Duncan *et al.*, 1992).

**Aplysia S(erotonin) channel:** Activated by stretch and (via *FMRFamide* receptor pathway) by *arachidonic acid metabolites* (see Vandorpe *et al.*, 1994).

**Mammalian neuron NMDA channel:** Once activated by its agonist (*NMDA* or glutamate), activity is potentiated by mechanical stimuli (Paoletti and Ascher, 1994).

**Cardiac GIRK[a]:** When *liganded by GTP*, the activity of this G-protein-regulated inward rectifier is enhanced by membrane stretch (Pleusamran and Kim, 1995).

[a] These channels are only known to be mechanosensitive when they are ligand-bound.

mation as a physiological signal. Then, whether the dynamic range was two to three orders of magnitude, the channel would qualify as MS only if mechanosensitivity mattered in the life of the cell.

Various MS channels can be activated via the ligands of second messenger systems as well as by membrane stretch. They include such the K-selective *Aplysia* S channel (Vandorpe and Morris, 1992) and cation channels in osteoblasts (Duncan *et al.*, 1992), hepatocytes (Bear and Li, 1991), and kidney cells (Verrey *et al.*, 1995). Physiologically, the *Aplysia* S channel (Vandorpe *et al.*, 1994) is activated by arachidonic acid metabolites and inhibited by an A kinase, ligands controlled via the neurotransmitters FMRF amide and serotonin, respectively. This voltage-independent K channel's physiological job is to modulate electrical excitability according to neurotransmitter signaling. Under patch-clamp conditions (cell attached or excised), however, the S channel can be activated by stretch alone (see, e.g., Fig. 1B); in fact, the maximal effect of stretch on a given patch considerably exceeds activation via second messenger paths (Vandorpe *et al.*, 1994). MS K channels akin to the *Aplysia* S channel are ubiquitous in neurons of *Aplysia* and other molluscs (Bedard and Morris, 1992; Morris, 1992) but only in the identified *Aplysia* neurons are the signaling

neurotransmitters known. *In situ* the channels do not readily activate with mechanical stimuli (Morris and Horn, 1991; Wan *et al.*, 1995), yet "SA K channel" is currently the best designation we have for these "S-like" channels. Many other channels may have been designated as "MS channels" simply because the primary physiological stimulus, perhaps a second messenger ligand, has not been identified. The *Xenopus* oocyte MS channel (Steffensen *et al.*, 1991) may be a case in point.

For all MS channels, tension must be conveyed to the channel via the surrounding lipids, via membrane skeleton, via extracellular matrix elements, or via some combination. Although these structures are universal, suction does not affect gating in all channels. The classical example is the nicotinic acetylcholine channel of chick skeletal muscle. This is a cation channel with very similar permeation properties to a cation-selective SA channel in the same membrane (and a heteropentameric structure not unlike the NMDA channel), but the acetylcholine channel's kinetics are not affected by stretch (Guharay and Sachs, 1984).

Mechanotransduction at membranes need not be mediated by MS ion channels. Recently, it was shown that expression of a heterologous metabotropic adenosine triphosphate (ATP) receptor confers mechanosensitivity on *Xenopus* oocytes. The explanation is that oocyte deformation releases ATP, which then acts on the receptor (Nakamura and Strittmatter, 1996). Here, the mechanical event that needs explanation is the stretch-induced release of ATP.

## V. Mechanosensitive Channels: Does Stretch Stimulate or Disrupt?

Because MS channels are almost ubiquitous, the question of whether their mechanosensitivity is meaningful or merely epiphenomenal is not trivial. If, *in situ*, many MS channels are not used for cellular mechanotransduction, why is MS gating in membrane patches so common? Perhaps for many channels anchored to the membrane skeleton, spurious (essentially pathological) MS gating can occur when the integrity of the cortical cytoskeleton in the channel's immediate environs is inadequate (Table 4). The assumption here is that an intact cortex absorbs mechanical loads that would otherwise be felt by the membrane skeleton and by attached channels. MS channels "designed" for use as physiological mechanotransducers would, by contrast, require an environs specifically modified to enhance mechanical loading of the channel via membrane skeleton linkages.

Patch recording is inevitably accompanied by mechanical damage to the patched membrane region. Small and Morris (1994) showed that in molluscan neurons, starting from a near-intact patch, successive rounds of mechanical stimulation augment the responsiveness of MS channels to the mechanical stimuli. *Stimulation* means "aspiration of the membrane and the underlying cytoskeleton" and it would be specious to argue that this is entirely benign. In *Xenopus* oocytes, too, MS channel behavior is damage dependent: Steady-state activation of the oocyte MS channels is enhanced whereas rapid adaptation is destroyed by patch trauma (i.e., by successive stimuli). The ability of patch-

**TABLE 4** Trauma, Pathology, Nociception, and MS Channel Activity

**Dystrophic muscle:** Lacks membrane skeleton dystrophin so spectrin may experience abnormal stress, which may be transferred to channels linked to spectrin. Dystrophic cells are fragile and leaky to calcium. MS cation channels in dystrophic muscle exhibit abnormal MS gating (Franco and Lansman, 1990). There may be a causal relationship.

**Brain ischemia:** In the presence of agonist, NMDA-gated channels are mechanosensitive. In swelling neurons, NMDA channels generate whole-cell currents, which grow larger as swelling progresses (Paoletti and Ascher, 1994). In stroke, this could exacerbate glutamate excitotoxicity.

**Nociception and mechanical trauma:** Neuronal SA channels are easier to activate following trauma (Small and Morris, 1994). Osteoclast MS channel mechanosensitivity is enhanced in chronically and intermittently strained cells (Duncan and Hruska, 1994). Sustained rather than transient responses of MS cation channels predominate after membrane trauma (Hamill and McBride, 1992), a phenomenon that could potentially be exploited for nociceptive signaling.

**Ectopic foci in demyelinated neurons:** These foci are hypermechanosensitive (see references in Waxman *et al.*, 1994). Abnormalities in membrane and cortex may increase the mechanosensitivity of normally refractory channels.

---

induced damage to facilitate sustained MS channel responses may explain much of the widespread discordance between single-channel and whole-cell recordings (Morris and Horn, 1991).

What then, is the "meaning" of channel mechanosensitivity? There is no unique answer, but an encapsulation follows: *MS channels are capable, for reasons not yet understood, of responding to mechanical stimuli by gating between open and closed states, but the channel's cellular environment can either act as a restraint, absorbing mechanical energy so the channel can attend to other stimuli, or it can enhance mechanosensitivity, yielding a mechanotransducer. Mechanosensitivity may vary with physiological and pathological conditions that perturb the cytoskeleton and membrane skeleton (e.g., cell swelling, cell division, cell growth, ischemia, mechanical trauma, nociceptive stimulation).*

Whether or not it is used by the cell, mechanosensitivity is an inherent trait in diverse integral membrane proteins. Should we be marvelling that mechanosensitivity evolved independently as a "selected for" feature in each protein? Or should we surmise that a tendency for mechanical perturbations to trigger gating transitions is a difficult-to-suppress trait for membrane-spanning enzymes? Both postulates are valid and deeper understanding will require molecular studies of mechanosensitivity in eukaryotic channels that exhibit strong mechanosensitivity (say, a dynamic range of >100-fold) and can be manipulated by molecular biology then reexamined at the single-channel level.

### VI. Linkages to the Membrane Skeleton

When MS channels were first described (Guharay and Sachs, 1984) it was clearly demonstrated that neither tu-

bulin nor *F*-actin was required for the MS gating. It was postulated that MS channels are linked into the membrane skeleton network and that when the in-parallel actin cytoskeleton is depolymerized, stress is readily transferred to the channel-linked component. What that parallel component may be has not been resolved, although it is not dystrophin since MS channels are evident in dystrophic muscle cells (Franco and Lansman, 1990). Spectrin is a candidate that has not been ruled out.

For eukaryotic systems, the bilayer may seldom experience significant stress except near lysis. Supporting this notion is the fact that the elastic constant of patches (Sokabe *et al.*, 1991) is much less than for lipid membranes. The bilayer is not, therefore, the patch's load-bearing element. That leaves the cortical cytoskeleton and the membrane skeleton. Because the elastic constant of patches is insensitive to the actin disrupter cytochalasin (Sokabe *et al.*, 1991) that leaves, by further elimination, a cytochalasin-insensitive membrane skeletal network as the only candidate. In an intact cell, one would expect cortical actin to contribute to the elasticity of the membrane–cortex region. The fact that, in patches, it does not contribute suggests that the act of patch formation disrupts actin structure. This is consistent with the finding of Small and Morris (1994) who showed that cytochalasin, like mechanical damage, augments rather than diminishes the responsiveness of SA channels to membrane stretch. Other membrane skeletal filaments, perhaps spectrin, may constitute the load-bearing elastic elements that transfer mechanical energy to the channels.

A variety of well-characterized membrane channels (and other transporters) are tethered to the membrane skeleton either to β-spectrin via ankyrin (Lambert and Bennett, 1993) or directly to α-spectrin (Rotin *et al.*, 1994). Additionally, some voltage- and ligand-gated channels are anchored to *postsynaptic density*-family proteins (Kim *et al.*, 1995). Although we surmise that eukaryotic MS channels receive mechanical energy via the membrane skeleton, there are no grounds for turning the argument around and invoking channel mechanosensitivity as a reason for linkages. Transporters may anchor to the membrane skeleton in order stay fixed at some cellular locale. In other cases, as in erythrocytes, the transporters may serve as anchor points for a membrane skeleton whose major job is to strengthen the membrane. It is probably counterproductive for many anchored membrane enzymes to allow their activity to be influenced by mechanical stress. Thus, in asking how and if specialized channel/membrane skeletal linkages might be used by MS channels to convey force to the channel, we should also remember to ask the reverse question: What features can make such linkages stress-proof?

### VII. Delay and Adaptation: Mechanically Fragile Aspects of Mechanosensitive Channel Behavior

The mechanical properties of patches are labile and history dependent (Hamill and McBride, 1992; Small and Morris, 1994). In fibroblasts stretch causes F-actin to depolymerize within 10 sec, though when the cells are left to rest briefly F-actin reorganizes, reappearing at greater than control levels, stiffening the cortex as it does so (see Glogauer *et al.*, 1997). Patch formation and stretch stimulation of cells may have qualitatively similar influences on the state of actin. Thixotropic properties of actin [in which mechanical stimulation of actin gels promotes the sol state of actin (see Heidemann and Buxbaum 1994)] may affect MS channel mechanosensitivity.

In molluscan neurons and in *Xenopus* oocytes (Hamill and McBride, 1992; Small and Morris, 1994), the behavior of SA channels following a step of applied suction has been studied in patches formed with a minimum of mechanical disruption—"gentle" patches. The responses of MS channels in the gentle patches differ dramatically from those in standard patches. The designations *gentle* and *standard* patch are operational; gentle patches inevitably becoming progressively more standard with successive mechanical stimuli.

In gentle oocyte patches, SA channels exhibit rapid adaptation; adaptation is abolished in standard patches. In gentle snail neuron patches, SA channels show a prolonged delay before activation; delay is abolished in standard patches. Changes in patch size and changes in cortical cytoskeleton integrity are probably not mutally exclusive, so the interpretation of input/output relations for MS channels can be fraught with difficulty. However, the mechanical fragility of the two dynamic phenomena—adaptation and delay—cannot be explained by increases in patch size, which may accompany a patch's progressive transformation from "gentle" to "standard."

As indicated, molluscan neuron SA channel activation in response to a first hit (a first large step of suction applied to a gentle patch) proceeds only after a delay, as if the patch were viscoelastic. The delay is substantial—>2 sec at −130 mm Hg—and changes produced by treatments are easily detected. Treatments that dramatically decrease delay include repetition of mechanical stimulation, use of larger mechanical stimuli, pretreatment with cytochalasin (a drug that promotes actin depolymerization), use of recently isolated (i.e., recently disrupted) as opposed to well-established cultured neurons, hyposmotic swelling of neurons, exposure to the sulfhydryl reagent *N*-ethylmaleimide, and elevation of intracellular Ca (Morris, 1996; Small and Morris, 1994). In addition to shortening delay, these treatments increase the activation elicited for the same stimulus. In summary, the response time is long and the extent of activation is small when the cortical cytoskeleton is intact, but not when it is compromised. Are the channels mechanosensitive at all when the cortex is perfectly intact? Probably not: for smaller first hit stimuli (say, −20 to −50 mm Hg), delays increase to tens of seconds, even though in standard patches, channels activate immediately with −20 mm Hg (Small and Morris, 1994). Delayed activation of SA channels (~30 sec) has also been reported by Kim *et al.* (1995). Although the volume-activated Cl channel in lymphocytes has not been confirmed as a SA channel, it is noteworthy that comparable delays occur during swelling-induced activation of this channel (Lewis *et al.*, 1993).

Mechanical inflation of neurons under whole-cell clamp does not activate MS current except just before the cell rapidly expands and ruptures; a catastrophic disruption of the cortex could explain both the MS currents and the ensuing membrane rupture (Wan *et al.*, 1995) (Fig. 5).

The fragility of MS channel dynamic responses can cause consternation and confusion but it also advances the possibility that substantial modification of MS responses may be possible through hormones that modify the cytoskeleton and by cell motility-related and volume-regulation-related changes in the cytoskeleton (Schweibert *et al.*, 1994). For example, vasopressin causes both actin depolymerization and SA channel activation in kidney cells (Verrey *et al.*, 1995). Another example is during the cell cycle (Bregestovski *et al.*, 1992; see discussion in Section IX 3) in which there are major cytoskeletal rearrangements that may partly account for the observations that SA channel sensitivity varies during the cycle.

Since the formation of focal adhesions is accompanied by major cytoskeletal rearrangements, it seems likely that MS channel should show altered mechanosensitivty in connection with changed states of cell adhesivity. Hints that this might be so are beginning to accumulate. Macrophages have a MS K channel whose activity, monitored by single-channel recording at the "upper" surface of the cell, increases when the "lower" surface of the cell becomes adherent to a substrate (Martin *et al.*, 1995). MS channel currents in osteoblasts subjected to chronic intermittent strain are more readily activated than those in quiescent cells (Duncan and Hruska, 1994).

Thus, it is possible that the mechanical fragility of delay is telling us about variable *in situ* states of MS channels. But what about adaptation? Here there is as yet, little insight. In phasic mechanosensory neurons, the purpose of rapid adaptation is to provide a mechanical high-pass filter, but what it might mean for amphibian oocytes, tunicate eggs, and yeast MS channels is a mystery. By contrast, the other dynamic property, delay, could in principle act like a low-pass filter; it could be useful for, say, muscle or bone cell MS channels, enabling them to ignore momentary stresses, yet to inform the cell of sustained mechanical inputs and hence modify some aspect of its metabolism (volume regulation, $Ca^{2+}$ loading, protein synthesis).

## VIII. Physiology of Mechanosensitive Channels

What establishes that a putative mechanotransducer channel performs a MS physiological (or developmental) task? Suggestions about functions of the channels are legion, but experimental evidence conclusively linking MS channel activity to physiological processes is rare. A dearth of appropriate pharmacological tools continues to hinder progress as does the difficulty of mechanically stimulating membrane in a calibrated and reproducible manner. Finding MS channels in patches does not mean that MS currents can be readily recorded from the channels in the intact cell.

### A. Criteria for Establishing That a Mechanosensitive Channel Performs a Mechanical Cellular Task

The criteria that need to be met are comparable to those that have been met by ligand-gated and voltage-gated channels with established physiological roles. The following criteria summarize a list given by Morris (1992):

1. Establish the MS channel's biophysical characteristics by recording at the single-channel level.

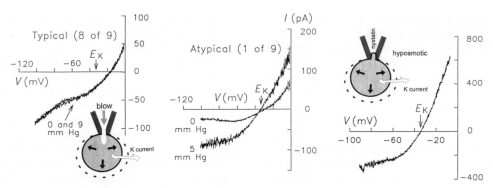

**FIG. 5.** Inflation of molluscan neurons under whole-cell clamp rarely elicits MS K currents in spite of the abundant MS channels. Macroscopic MS currents are observed only at pressures marginally less than those that rupture the plasma membrane; the left and middle graphs are ramp-clamp currents for the stated pressures (applied via the recording pipette as shown). As expected from Laplace's law, large cells can withstand only small pressures. In the same neurons, K currents are elicited by osmotic swelling (right; a difference current obtained by subtracting ramp currents before swelling from those during swelling). Are these "volume-activated" currents the direct result of swelling (mechanical) stress? One cannot say, since swelling causes an increase in membrane capacitance (possibly recruiting new channels) and it alters cytoplasmic chemistry (possibly stimulating channel activity via ligand changes) as well as increasing tension. (Modified from Wan *et al.*, 1995).

2. Establish a way of altering channel function (e.g., channel blocker or a mutant channel).
3. Demonstrate that some physiological or developmental aspect of the cell is MS.
4. Demonstrate that the MS cellular function is specifically impaired by the channel blocker (or by the mutation).
5. Make macroscopic recordings under conditions of minimal damage to the membrane–cortex region, and demonstrate MS currents that correspond to the single-channel recordings according to (a) selectivity; (b) pharmacology; (c) noise characteristics; (d) saturability at large mechanical stimuli; (e) saturating macroscopic current, which has the magnitude expected given the single-channel membrane-density; (f) expected MS current density variations—spatial and/or temporal—as predicted from any spatial or temporal variations noted by single-channel recording; (g) degree of mechanosensitivity; and (h) sensitivity to other known physiological influences (e.g., hormones, intracellular ions).

## B. A SI Cation Channel: The Basis for Hypothalamic Osmotransduction

For one MS channel, many of the criteria just listed have been met. A subpopulation of neurons in the hypothalamus detects and responds to blood osmolarity. These magnocellular OSNs possess MS cation channels that, it has been convincingly shown, underlie the osmotransduction currents that hyperpolarize OSNs during hyposmotic stimuli and depolarize them during hyperosmotic stimuli (see Figs. 4A and B). This in turn leads to respective decreases and increases in release of the osmoregulatory hormone, vasopressin, from the OSNs. Under cell-attached recording conditions with physiological solutions, the OSN channels have a conductance in the 30-pS range and reverse near $-40$ mV, sufficiently depolarized with respect to resting potential for active channels to be excitatory. The idea that MS channels

give rise to the osmotransduction currents is supported by the demonstration that gadolinium ions, a blocker of many cation channels including MS cation channels, block both the macroscopic osmosensor currents and the single-channel MS currents.

Activation curves for the OSN single-channel currents in cell-attached patches are bell shaped (centered at zero mechanical stimulus) rather than the sigmoid curve characteristic of voltage-gated channels and of other MS channels (Fig. 4C). This is not inconsequential; well-established physical explanations exist for sigmoid activation curves ("Boltzmanns"), but there is no precedent for "Gaussians" when it comes to activation curves. Does the bell shape reflect an inherent peculiarity of the channels? It appears not to. Rather, the unusual input/output relation is thought to stem from a systematic error in assessing the true pipette pressure. Although seeming to be data obtained at applied pressures symmetric about zero (see Fig. 4C, middle curve), the activation curves were obtained entirely with suction (negative pressures) (see Fig. 4C, right); the published bell curve peaks at zero simply because an offset was applied in the graph. Macroscopic currents from the same cells show a monotonic dependence on hypotonicity that, as shown here, fit well to a sigmoid (see Fig. 4B). If both single-channel and intact cell currents reflect activity of the same channels, what explains the dramatic qualitative difference in input/output relations? To answer, we need to visualize how applied tensions in patch recordings would correlate with osmotic stimuli for whole-cell recording.

When the whole-cell data (see Bourque and Oliet, 1997) are plotted with the *x* axis going from low to high osmotically induced membrane stretch (as in Fig. 4B), they fit a sigmoid Boltzmann relationship (akin to Hodgkin-Huxley h-∞) as expected for the intrinsic tension-activity relations of SI channels (Morris and Sigurdson, 1989; Lecar and Morris, 1993). The expected sigmoid can be found in the single-channel bell curve (see Fig. 4C, right) if the following two assumptions hold: (1) The MS channel is indeed a SI channel and (2) MS channel activity is recorded with net positive pressure in the pipette, even though suction is applied.

These assumptions are not merely post hoc rationalizations—they have a precedent. For snail neuron SI channels (Morris and Sigurdson, 1989), sigmoids are more common but bell-shaped activation curves too can be obtained if the appropriate pressure experimental pressure range is used. The bell corresponds to the expected h-∞-type sigmoid plus its mirror image reflected on the *zero-tension* point of the pressure axis. Because the *x* axis is the experimentally applied pressure (see Fig. 4C), not membrane tension, the true zero-tension point is not known. The bell-shape "illusion" arises because the membrane patch is sealed firmly in the electrode rim, and the patch is concave, flat, or convex depending on the effective pipette pressure. Tension sufficient to gate the channels can be achieved in both the concavely and convexly stretched patch. If residual pressure persists in the pipette after gigaseal formation, the bell-shaped curve will be centered not at the origin but elsewhere along the axis of experimentally applied pressure, according to the sign and magnitude of the residual pressure (see Fig. 4C).

To summarize, the bell shape probably need not reflect an intrinsic peculiarity of the OSN MS channel if it arises from the stimulus mechanics of patches. A bell shape is simply the SI channel congener to the U shape obtained for SA channels when the full stimulus range (pressures positive and negative to zero) is used. Presumably all of the single-channel OSN channel recordings were obtained from a starting point of positive residual (not zero) pressure in the pipette. Assuming that the channel responsible for the osmosensory currents is an SI channel, then the microscopic and macroscopic input/output relations are not at odds.

This story has just begun to unfold. What makes the OSN MS cation channels sufficiently sensitive *in situ* that they can gate in response to small osmotic perturbations (plus or minus about 20 mOsm, as illustrated in Fig. 4B)? Are the OSNs good stress detectors because, like strain gauges, they have exceptionally stiff detector components? Are there specialized membrane skeletal or cytoskeletal arrangements? Is the large size of OSNs germane to their osmosensory role, as might be expected for a cell that detects osmotic pressure as membrane tension according to the dictates of Laplace's law (see Table 1, bottom)? Do the OSNs have an unusually high water permeability? A major challenge will be to clone the MS channel and study it at the molecular level.

## C. Other Roles Proposed for Mechanosensitive Channels

In this section, I quote from a *potpourri* of papers in which mammalian or avian MS channels were tentatively linked to a wide range of cellular phenomena. The microscopic-to-macroscopic correspondences have yet to be established except for the case described earlier. There is a pressing need to establish whether or not various MS channels perform *in situ* the varied mechanical tasks suggested for them on the basis of their single-channel behavior.

Bear (1990): "It is proposed that stretch-activated channels in liver cells permit the transient influx of $Ca^{2+}$, which in turn acts to trigger changes in ion conductance or cytoskeletal components involved in **cell volume regulation** (see Bear and Li, 1991).

Suleymanian *et al.* (1995): "Stretch-activated channel blockers **modulate cell volume** in cardiac ventricular myocytes . . . support[ing] the idea that [the channels] are involved in cardiac cell volume regulation."

Puro (1991): "Stretch-activated channels may help mediate a **compensatory response of glia to swelling** . . . [because] activation of calcium-permeable stretch-sensitive channels is associated with an increase in the activity of calcium-activated potassium channels. Activation of potassium channels to produce and efflux of potassium with a subsequent loss of . . . cell water could . . . decrease glial cell volume."

Hansen *et al.* (1991): "Our results indirectly implicate stretch-activated channels in the genesis of stretch-induced [cardiac] arrhythmias and provide preliminary evidence for a potential new mode of **antiarryhthmic drug action**—blockade of stretch-activated channels."

Kent *et al.* (1989): "Sodium entry [an initial cellular response requisite to the growth-inducing activity of many substances] through stretch-activated ion channels is stimulated by deformation of the sarcolemma. . . . Streptomycin, a cationic blocker of the mechanotransducer ion channels . . . inhibited, in a dose-dependent manner, **[load-dependent] protein synthesis** otherwise observed in contracting [myocardial] muscles developing tension."

Ding *et al.* (1992): "SA channels . . . could affect the **motile response [of outer hair cells**, via] membrane potential or by allowing the entry of free $Ca^{2+}$ which could lead to [cellular length changes via] actin and myosin. SA channels could also . . . [regulate the cell's osmotic pressure] thereby influencing its **electro-osmotic response**."

Iwasa *et al.* (1991): In auditory outer hair cells "stretch-activated channels may play an important role in producing a mechanical feedback, an indispensible element in **cochlear tuning**."

Coleman and Parkington (1992): "Propagation of electrical and mechanical **activity in uterine smooth muscle**: a functional role for stretch-sensitive channels."

Davidson *et al.* (1990): "We propose that one or more of these [three] mechanosensitive ion channels [in osteoblast-like cells] is involved in the **response of bone to mechanical loading**."

Sigurdson *et al.* (1992): "Gently prodding [tissue-cultured chick heart] cells with a pipette produced a $Ca^{2+}$ influx that often led to **waves of calcium-induced calcium release spreading from the site of stimulation**. . . . The mechanical sensitivity probably arose from stretch-activated ion channels."

## IX. Other Recent Explorations of Mechanosensitive Channels

A few widely ranging examples provide some flavor of potentially exciting findings. In each case, the question testifies to the need for more information about the biological impact of the readily observed mechanosensitivity.

1. *Fungal germling of bean rust—topographical sensing?* Rust is a parasite whose germling needs to find the stomatal openings of host-plant leaves, a feat it achieves by sensing minute topographical features. Kung and colleagues (Zhou *et al.*, 1991) showed that germ tube protoplasts of the rust have SA channels that are evident in both macroscopic and single-channel recordings. Saturating MS currents were readily obtained in both types of recording. Moreover, gadolinium blocks both single-channel and whole-cell currents in protoplasts and inhibits germ tube growth and differentiation *in vivo*. The rust channels are nonselective and have a very high conductance ($>0.5$ nS), so that channel openings could profoundly alter the ionic composition of the minute volume of cytoplasm in the growing tips. Presumably fine control could be achieved *in vivo* by having exceedingly brief openings.

2. *Hyphal tips in fungi—MS channels and development?* On the basis of numbers of SA channels per patch, Levina *et al.* (1994) report variations in the density of the channels toward hyphal tips in fungi. Based on the tip-high staining pattern of filamentous actin, and on data with cytochalasin, the authors suggest that actin causes the channels to cluster, either by linking channels to the cytoskeleton or by increasing transport of channels to the tip. Alternately, regional variations in membrane mechanics may explain the data; channels may be present at uniform density but they may be less mechanosensitive in "old" membrane far from the tip and easy to stimulate in "new" membrane near the tip. This would echo the case of neuronal cytochalasin-sensitive SA channels, which become increasingly difficult to activate as the cell culture matures (Small and Morris, 1994).

3. *Fish embryo—a role in the cell cycle?* SA K channels in loach embryos (2–256 cells) show dramatic changes in activity during the cell cleavage cycle, perhaps explaining the membrane voltage oscillations. At the beginning of interphase, channels tend to be inactive and less responsive to stretch, whereas at prometaphase they are highly active. Evidence is presented showing that the activity of these channels is regulated through cAMP-dependent phosphorylation (Medina and Bregestovski, 1991).

4. *Xenopus oocytes MS channels—why the rapid adaptation?* On "gentle" patches, a pressure step elicits a transient increase in MS channel activity—the response shows "adaptation," comparable perhaps to the phasic responses of some mechansensory cells. Is adaptation a type of slow inactivation (comparable to inactivation in voltage-gated channels) or is it associated with a change in mechanosensitivity? Double-step protocols indicate that adapted channels will reactivate with stronger stimulation, suggesting that "inactivation" of the MS gates has not occurred. By elimination, the transient change seems to be associated with gating sensitivity. As described earlier, adaptation is a fragile phenomenon, critically dependent on the mechanical history of the patch. A surprising aspect of adaptation is that it is not seen at depolarizing potentials. This is also true of adaptation in yeast MS channels (see Martinac, 1993; Gustin, 1992). From the point of view of potential physiological significance, these are interesting, but puzzling findings because there would seem to be a Catch-22 operating. Adaptation occurs only at hyperpolarized potentials. *In vivo*, therefore activation of the MS cation channels would depolarize the membrane, thereby precluding adaptation. Hyperpolarizing influences working in parallel with the MS channel to override their depolarizing effect (e.g., $Ca^{2+}$-activated K channels) might, however, "undo the Catch-22."

5. *Epithelial SA Cl channels—a role in volume regulation?* In shark rectal gland cells and in renal cells, Cl channel activity depends of the state of actin (Schwiebert *et al.*, 1994; Mills *et al.*, 1994). In renal glands, filamentous actin is associated with inactive channels and depolymerized actin with active channels. Membrane stretch also stimulates channel activity. It is suggested that during swelling-induced actin depolymerization, cell swelling could be limited through activation of MS chloride channels.

6. *NMDA (glutamate) channels in mouse neurons—pathology? physiology?* Membrane tension potentiates the activity of ligand-bound NMDA channels in a variety of recording configurations (Paoletti and Ascher, 1994). By contrast, activated kainate-type glutamate channels in the same patches are not mechanosensitive. The potentiation by tension (at best, threefold) is not impressive compared to many MS channels and stretch alone will not activate the channels. However, successive stimuli do yield progressively larger effects, suggesting that NMDA channel mechanosensitivity is increased by disrupting the cytoskeleton. NMDA responses to stretch in the whole-cell mode tend to support this view, since osmosensitivity only becomes evident after prolonged dialysis. The experiments were performed with internal pH and pCa buffers, no ATP or GTP, and with $F^-$ (a blocker of most phosphatases); the simplest explanation of "run-up" of mechanosensitivity is that there is a progressive improvement in the transfer of force to the NMDA channels. If it is "real," mechanosensitivity in NMDA-activated channels could have far-ranging implications for neuronal plasticity. If it reflects pathology, it is a phenomenon worth understanding in the context of the glutamate excitotoxicity and neuronal swelling associated with stroke.

## X. Two-State Barrier Models for MS Gating

Formal models for mechanosensitive gating provide a framework for relating experimentally applied force to kinetic data. If a channel has two states and is in some sense a spring, then the mechanical features that should matter are the spring's position and its elasticity. There are various candidates for MS channel springs. Is the spring the channel itself, or is some part of it elastic? Is the spring an extracellular or intracellular tether (tip links in hair cells, strands of membrane skeleton or extracellular matrix) mechanically in series with the channel? Is it the collection of noncovalent bonds between lipid and protein at the channel/bilayer interface? Is it some combination? Whatever the identity of the spring, a two-state model assumes that force acting through the spring biases a channel's open probability; the channel would have a free-energy difference of several kT between open and closed states and the force would help surmount this barrier. When the larger state has a pore in the conducting configuration, then the channel is a SA channel, whereas when the larger state has a nonconducting pore, the channel is a SI channel.

A simple and satisfying two-state model is that developed by Hudspeth and colleagues (see Corey and Howard, 1994) for hair cell channels. It posits, on excellent empirical grounds, that extracellular tip links act like an elastic spring in series with a stochastic gate. An external force changes the spring's tension and thereby the mean position of the spring. Two extremes of position are helpful to visualize—a position in which the spring is fully relaxed and with its attached gate occluding the channel (this is the smaller state) and another position in which the spring is under maximal tension with the gate away from the now-open pore (this is the larger, "stretch-activated" state). If external force alters the mean position of the gate and thereby the channel's open probability, then, reciprocally, thermally induced opening and closing of the channel should randomly stress and relax the gating spring. Measurements of channel noise (which indicates the rates of gating transitions) and hair cell compliance (assumed to be an indicator of gate position) in stimulated hair cells support this prediction. Note that in this model, the spring's elasticity is critical to its action, but it is regarded as a fixed quantity.

The possibility that a MS channel's two major states have **different** elasticities has also been formalized. In this case (Sachs and Lecar, 1992; Lecar and Morris, 1993) the MS gate is depicted as a parabolic harmonic oscillator. At a given applied force (membrane tension) the free energies of the open and closed states depend on both mean gate position and the state-specific elasticities of the two states. Using reasonable quantities for channel transition rates and positional changes, Corey and Howard (1994) estimate, however, that differences in the second factor—elasticity—would be too small to have appreciable impact. Intriguingly, however, they point out that were the closed state significantly "softer" than the open state, the prediction is that a channel's open probability open will go through a maximum as force increases, rather than generating a sigmoid activation curve. It may be worth bearing this latter prediction in mind as further studies of SI channel mechanisms are undertaken. As discussed earlier, Oliet and Bourque (1993) had to assume, in explaining their data, that their experimental conditions generated a persistent tension offset in OSN membrane. Should it be found that there is no such offset and that the bell-shaped activation curves represent the true input/output relations of these MS channels, then the possibility of state-specific elasticities could be explored through detailed single-channel kinetics.

Given the extreme range of channels that exhibit mechanosensitivity and the likelihood that varied mechanisms are responsible, it would be pleasing but perhaps surprising if the formalism that applies for the hair cell captured the major gating characteristics of all MS channels. Some MS channels may undergo mechanosensitive open-closed transitions along a reaction coordinate whose units are not well described as a linear position (Lecar and Morris, 1993).

## XI. Summary

MS channels are ubiquitous and they are abundant. They have few unifying features other than their propensity to change their open probability when suction is applied under patch-clamp conditions. The term *MS channel* carries an implicit assumption that some susceptible gating structure is put under tension by experimental application of differential pressure across the membrane, and that the resulting tensile strain biases the channel's open probability. The identity of the MS structure(s) is a mystery for MS channels in most eukaryotic cells.

What information there is on possible mechanisms of ion channel mechanosensitivity points at diversity rather than at universality. At one extreme—the specialized receptors—cytoarchitecture is central to function. Specialized animal mechanoreceptors (e.g., hair cells in vertebrates, mechanosensory neurons in the nematode *C. elegans*) require that mechanotransducer channels be precisely positioned with respect to special extracellular and intracellular structural proteins and aligned properly with respect to structures in the rest of the cell. At the other extreme we have MS channels from bacteria—they can be reconstituted into artificial bilayers and retain their mechanosensitivity. Ought one to use the term *MS channel* to include the whole spectrum? Provided that one's minimal mechanistic model includes the notion of the channel's gating structure under tensile strain, the term should be acceptable. This includes most, but not all scenarios; an imaginary MS channel that would *not* be a MS channel would be one whose open probability changed because a mechanical deformation of the membrane induced a critical lipid-phase transition or altered the fatty acid constituents of the bilayer, thereby altering channel gating. Later on, we may wish to reserve the term for physiological mechanotransducer channels.

Ingenious ideas about how MS channels might contribute to cellular control processes are not in short supply, and work is under way on many fronts to test these ideas. Although MS channels have been on the scene for more than a decade, they have not attained a status coequal with voltage-gated and ligand-gated channels. But now that there is a prototype case (the OSN that is MS channels) wherein membrane stretch acts on a MS channel to mediate a physiologically relevant voltage change, and now that a putative MS channel is known in molecular terms from *C. elegans*, this state of affairs has begun to change.

## Bibliography

Achard, J. M., Bubien, J. K., Benos, D. J., and Warnock, D. G. (1996). Stretch modulates amiloride sensitivity and cation selectivity of sodium channels in human B lymphocytes. *Am. J. Physiol.* **39**, C224–C234.

Awayda, M. S., Ismailov, I. I., Berdiev, B. K., and Benos, D. J. (1995). A cloned renal epithelial Na⁺ channel protein displays stretch activation in planar lipid bilayers. *Am. J. Physiol.* **268**, C1450–C1459.

Bargmann, C. I. (1994). Molecular mechanisms of mechanosensation? *Cell* **78**, 729–731.

Bear, C. E. (1990). A nonselective cation channel in rat liver cells is activated by membrane stretch. *Am. J. Physiol.* **258,** C421–C428.

Bear, C. E., and Li, C. (1991). Calcium-permeable channels in rat hepatoma cells are activated by extracellular nucleotides. *Am. J. Physiol.* **261,** C1018–C1024.

Bedard, E., and Morris, C. E. (1992). Channels activated by stretch in neurons of a helix snail. *Can. J. Physiol.* **70,** 207–213.

Bourque, C. W., and Oliet, S. H. (1997). Osmoreceptors in the central nervous system. *Annu. Rev. Physiol.* **59,** 601–619.

Bowman, C. L., Ding, J. P., Sachs, F., and Sokabe, M. (1992). Mechanotransducing ion channels in astrocytes. *Brain Res.* **584,** 272–286.

Bregestovski, P., Medina, I., and Goyda, E. (1992). Regulation of potassium conductance in the cellular membrane at early embryogenesis. *J. Physiol.* [*Paris*] **86,** 109–115.

Cemerikic, D., and Sackin, H. (1993). Substrate activation of mechanosensitive, whole-cell currents in renal proximal tubule. *Am. J. Physiol.* **264,** F697–F714.

Coleman, H. A., and Parkington, H. C. (1992). Propagation of electrical and mechanical activity in uterine smooth muscle: a functional role for stretch-sensitive channels. *Jap. J. Pharm.* **58,** Suppl 2, 369P.

Corey, D. P., and Garcia-Anoveros, J. (1996). Mechanosensation and the DEG/ENac ion channels. *Science* **273,** 323–324.

Corey, D. P., and Howard, J. (1994). Models for ion channel gating with compliant states. *Biophys. J.* **66,** 1254–1257.

Dai, J., and Sheetz, M. P. (1995). Mechanical properties of neuronal growth cone membranes studied by tether formation with laser optical tweezers. *Biophys. J.* **68,** 988–996.

Dai, J., Sheetz, M. P., Herring, T., and Morris, C. E. (1996). Membrane tension in swelling and shrinking molluscan neurons. *Mol. Biol. Cell* **7,** 449a.

Davidson, R. M., Tatakis, D. W., and Auerbach, A. L. (1990). Multiple forms of mechanosensitive ion channels in osteoblast-like cells. *Pfleugers Archiv.* **416,** 646–651.

Ding, J. P., Salvi, R. J., and Sachs, F. (1992). Stretch-activated ion channels in guinea pig outer hair cells. *Hearing Res.* **56,** 19–28.

Dopico, A. M., Kirber, M. T., Singer, J. J., and Walsh, Jr., J. V. (1994). Membrane stretch directly activates large conductance $Ca^{2+}$-activated $K^+$ channels in mesenteric artery smooth muscle cells. *Am. J. Hyperten.* **7,** 82–89.

Duncan, R. L., and Hruska, K. A. (1994). Chronic, intermittent loading alters mechanosensitive channel characteristics in osteoblast-like cells. *Am. J. Physiol.* **267,** F909–F16.

Duncan, R. L., Hruska, K. A., and Misler, S. (1992). Parathyroid hormone activation of stretch-activated cation channels in osteosarcoma cells (UMR-106.1). *FEBS Lett.* **307,** 219–223.

Erxleben, C. (1989). Stretch-activated current through single ion channels in the abdominal stretch receptor organ of the crayfish. *J. Gen. Physiol.* **94,** 1071–1083.

Franco, A., Jr., and Lansman, J. B. (1990). Calcium entry through stretch-inactivated ion channels in *mdx* myotubes. *Nature* **344,** 670–673.

French, A. S. (1992). Mechanotransduction. *Annu. Rev. Physiol.* **54,** 135–152.

Gardiner, D. R., and Brezden, B. L. (1990). Ion channels in *Lymnaea stagnalis* heart ventricle cells. *Comp. Biochem. Physiol.* **96A,** 79–85.

Glogauer M., Arora, P., Yao, G., Sokholov, I., Ferrier, J., and McCulloch, C. A. G. (1997). Calcium ions and tyrosine phosphorylation interact coordinately with actin to regulate cytoprotective responses to stretching. *J. Cell Sci.* **110,** 11–21.

Guharay, F., and Sachs, F. (1984). Stretch-activated single ion channel currents in tissue-cultured embryonic chick skeletal muscle. *J. Physiol.* **352,** 685–701.

Gustin, M. C. (1992). Mechanosensitive ion channels in yeast. Mechanisms of activation and adaptation. *Adv. Compar. Environ. Physiol.* **10,** 19–38.

Gustin, M. C., Sachs, F., Sigurdson, W., Ruknudin, A., Bowman, C., Morris, C. E., and Horn, R. (1991). Single channel mechanosensitive currents. Technical comments. *Science* **253,** 800–853.

Hase, C. C., Le Dain, A. C., and Martinac, B. (1995). Purification and functional reconstitution of the recombinant large mechanosensitive ion channel (MscL) of *Escherichia coli.* *J. Biol. Chem.* **270,** 18329–18334.

Hamill, O. P., and McBride, D. W., Jr. (1992) Rapid adaptation of single mechanosensitive channels in *Xenopus* oocytes. *Proc. Natl. Acad. Sci. USA* **89,** 7462–7466.

Hamill, O. P., and McBride, D. W., Jr. (1995). Mechanoreceptive membrane channels. *Am. Sci.* **83,** 30–37.

Hamill, O. P., and McBride, D. W., Jr. (1996a). The pharmacology of mechanogated membrane ion channels. *Pharmacol. Rev.* **48,** 231–252.

Hamill, O. P., and McBride, D. W., Jr. (1996b). A supramolecular complex underlying touch sensitivity. *Trends Neurosci.* **19,** 258–261.

Hansen, D. E., Borganelli, M., Stacy, G. P., Jr., and Taylor, L. K. (1991). Dose-dependent inhibition of stretch-induced arrhythmias by gadolinium in isolated canine ventricles. Evidence for a unique mode of antiarrhythmic action. *Circ. Res.* **69,** 820–831.

Heidemann, S. R., and Buxbaum, R. E. (1994). Mechanical tension as a regulator of axonal development. *Neurotoxicol.* **15,** 65–108.

Hisada, T., Ordway, R. W., Kirber, M. T., Singer, J. J., and Walsh, J. V., Jr. (1991). Hyperpolarization-activated cationic channels in smooth muscle cells are stretch-sensitive. *Pflugers Archiv.* **417,** 493–499.

Hong, K., and Driscoll, M. (1994). A transmembrane domain of the putative channel subunit MEC-4 influences mechanotransduction and neurodegeneration in *C. elegans* [see comments]. *Nature* **367,** 470–473.

Huang, M., Gu, G., Ferguson, E. L., and Chalfie, M. (1995). A stomatin-like protein necessary for mechanosensation in *C. elegans.* *Nature* **378,** 292–295.

Hudspeth, A. J. (1989). How the ear's works work. *Nature* **341,** 397–404.

Iwasa, K. H., Li, M., Jia, M., and Kachar, B. (1991). Stretch sensitivity of the lateral wall of the auditory outer hair cell from the guinea pig. *Neurosci. Lett.* **133,** 171–174.

Kell, A., and Glaser, R. W. (1993). On the mechanical and dynamic properties of plan cell membranes: Their role in growth, direct gene transfer and protoplast fusion. *J. Theor. Biol.* **160,** 41–62.

Kent, R. L., Hoober, J. K., and Cooper, G. (1989). Load responsiveness of protein synthesis in adult mammalian myocardium: role of cardiac deformation linked to sodium influx. *Circ. Res.* **64,** 74–85.

Kim, D. (1992). A mechanosensitive $K^+$ channel in heart cells. Activation by arachidonic acid. *J. Gen. Physiol.* **100,** 1021–1040.

Kim, E., Niethammer, M., Rothschild, A., Jan, Y. N., and Sheng, M. (1995). Clustering of *shaker*-type $K^+$ channels by interaction with a family of membrane-associated guanylate kinases. *Nature* **378,** 85–88.

Kirber, M. T., Ordway, R. W., Clapp, L. H., Walsh, J. V., Jr., and Singer, J. J. (1992). Both membrane stretch and fatty acids directly activate large conductance $Ca^{2+}$-activated $K^+$ channels in vascular smooth muscle cells. *FEBS Lett.* **297,** 24–28.

Lambert, S., and Bennett, V. (1993). From anemia to cerebellar dysfunction. A review of the ankyrin gene family. *Eur. J. Biochem.* **211,** 1–6.

Lecar, H., and Morris, C. E. (1993). Biophysics of mechanotransduction. *In* "Mechanoreception by the Vascular Wall" (G. Rubanyi, Ed.). Futura Pub. Co., Mount Kisco, NY.

Levina, N. N., Lew, R. R., and Heath, I. B. (1994). Cytoskeletal regulation of ion channel distribution in the tip-growing organism. *Saprolegnia ferax. J. Cell Sci.* **107,** 127–134.

Lewis, R. A., Ross, P. E., and Cahalan, M. D. (1993). Chloride channels activated by osmotic stress in T lymphocytes. *J. Gen. Physiol.* **101,** 801–826.

Liu, J., Schrank, B., and Waterston, R. H. (1996). Interaction between a putative mechanosensory membrane channel and a collagen. *Science* **273**, 361–364.

Marchenko, S. M., and Sage, S. O. (1996). Mechanosensitive ion channels from endothelium of excised rat aorta. *Biophys. J.* **70**, A365.

Martin, D. K., Bootcov, M. R., Campbell, T. J., French, P. W., and Breit, S. N. (1995). Human macrophages contain a stretch-sensitive potassium channel that is activated by adherence and cytokines. *J. Memb. Biol.* **147**, 305–315.

Martinac, B. (1993). Mechanosensitive ion channels: Biophysics and physiology. *In* "Thermodynamics of Membrane Receptors and Channels" (M. B. Jackson, Ed.), pp. 327–352. CRC Press, Boca Raton, FL.

McBride, D. W., Jr., and Hamill, O. P. (1992). Pressure clamp: A method for rapid step perturbation of mechanosensitive channels. *Pflugers Archiv.* **421**, 606–612.

Medina, I. R., and Bregestovski, P. D. (1991). Sensitivity of stretch-activated K$^+$ channels changes during cell-cleavage cycle and may be regulated by cAMP-dependent protein kinase. *Proc. R. Soc. London B* **245**, 159–164.

Mills, J. W., Schwiebert, E. M., and Stanton, B. A. (1994). Evidence for the role of actin filaments in regulating cell swelling. *J. Exp. Zool.* **268**, 111–120.

Morris, C. E. (1990). Mechanosensitive ion channels. *J. Membr. Biol.* **113**, 93–107.

Morris, C. E. (1992). Are stretch-sensitive channels in molluscan cells and elsewhere physiological mechanotransducers? *Experientia* **48**, 852–858.

Morris, C. E., and Horn, R. (1991). Failure to elicit neuronal macroscopic mechanosensitive currents anticipated by single channel studies. *Science* **251**, 1246–1249.

Morris, C. E., and Sigurdson, W. J. (1989). Stretch-inactivated ion channels coexist with stretch-activated ion channels. *Science* **243**, 807–809.

Nakamura, F., and Strittmatter, S. M. (1996) P2Y1 purinergic receptors in sensory neurons: Contribution to touch-induced impulse generation. *Proc. Natl. Acad. Sci. USA* **93**, 10465–10470.

Oliet, S. H. R., and Bourque, C. W. (1993). Mechanosensitive channels transduce osmosensitivity in supraoptic neurons. *Nature* **364**, 341–343.

Oliet, S. H. R., and Bourque, C. W. (1996). Gadolinium uncouples mechanical detection and osmoreceptor potential in supraoptic neurons. *Neuron* **16**, 175–181.

Ordway, R. W., Petrou, S., Kirber, M. T., Walsh, J. V., Jr., and Singer, J. J. (1995). Stretch activation of a toad smooth muscle K$^+$ channel may be mediated by fatty acids. *J. Physiol.* **484**, 331–337.

Paoletti, P., and Ascher, P. (1994). Mechanosensitivity of NMDA receptors in cultured mouse central neurons. *Neuron* **13**, 645–655.

Pender, N., and McCulloch, C. A. (1991). Quantitation of actin polymerization in two human fibroblast sub-types responding to mechanical stretching. *J. Cell Sci.* **100**, 187–193.

Pleusamran, A., and Kim, D. (1995). Membrane stretch augments the cardiac muscarinic K$^+$ channel activity. *J. Membr. Biol.* **148**, 287–297.

Puro, D. G. (1991). Stretch-activated channels in human retinal Muller cells. *Glia* **4**, 456–460.

Rotin, D., Bar-Sagi, D., O'Brodovich, H., Merilainen, J., Lehto, V. P., Canessa, C. M., Rossier, B. C. and Downey, G. P. (1994). An SH3 binding region in the epithelial Na$^+$ channel ($\alpha$rENaC) mediates its localization at the apical membrane. *EMBO J.* **13**, 4440–4450.

Ruknudin, A., Sachs, F., and Bustamante, J. O. (1993). Stretch-activated ion channels in tissue-cultured chick heart. *Am. J. Physiol.* **264**, H960–H972.

Sachs, F. (1992). Stretch-sensitive ion channels: An update. *In* "Sensory Transduction" (D. Corey, Ed.), pp. 242–260. Rockefeller University Press, New York.

Sachs, F., and Lecar, H. (1991). Stochastic models for mechanical transduction. *Biophys. J.* **59**, 1143–1145.

Sachs, F., and Morris, C. E. (1997). Mechanosensitive ion channels in non-specialized cells. *Rev. Physiol. Biochem. Pharmacol.* in press.

Sackin, H. (1995). Mechanosensitive channels. *Annu. Rev. Physiol.* **57**, 333–353.

Schweibert, E. M., Mills, J. W., and Stanton, B. A. (1994). Actin-based cytoskeleton regulates a chloride channel and cell volume in a renal cortical collecting duct cell line. *J. Biol. Chem.* **269**, 7081–7089.

Sheetz, M. P., and Dai, J. (1996). Modulation of membrane dynamics and cell motility by membrane tension. *Trends Cell Biol.* **6**, 85–89.

Sigurdson, W. J., Ruknudin, A., and Sachs, F. (1992). Calcium imaging of mechanically induced fluxes in tissue-cultured chick heart: role of stretch-activated ion channels. *Am. J. Physiol.* **262**, H1110–H1115.

Small, D. L., and Morris, C. E. (1994). Delayed activation of single mechanosensitive channels in *Lymnaea* neurons. *Am. J. Physiol.* **267**, C598–C606.

Sokabe, M., Sachs, F., and Jing, Z. (1991). Quantitative video microscopy of patch clamped membranes stress, strain, capacitance, and stretch channel activation. *Biophys. J.* **59**, 722–728.

Steffensen, I., Bates, W. R., and Morris, C. E. (1991). Embryogenesis in the presence of blockers of mechanosensitive ion channels. *Dev. Growth Diff.* **5**, 437–442.

Sukharev, S. I., Blount, P., Martinac, B., and Kung, C. (1997). Mechanosensitive channels of *Escherichia coli*: the MscL gene, protein, and activities. *Annu. Rev. Physiol.* **59**, 633–657.

Suleymanian, M. A., Clemo, H. F., Cohen, N. M., and Baumgarten, C. M. (1995). Stretch-activated channel blockers modulate cell volume in cardiac ventricular myocytes. *J. Mol. Cell. Cardiol.* **27**, 721–728.

Tavernarakis, N., and Driscoll, M. (1997). Molecular modeling of mechanotransduction in the nematode *Caenorhabditis elegans*. *Ann. Rev. Physiol.* **59**, 659–689.

Vandorpe, D. H., and Morris, C. E. (1992). Stretch activation of the *Aplysia* S channel. *J. Membr. Biol.* **127**, 205–214.

Vandorpe, D. H., Small, D. L., Dabrowski, A. R., and Morris, C. E. (1994). FMRFamide and membrane stretch as activators of the *Aplysia* S-channel. *Biophys. J.* **66**, 46–58.

Van Wagoner, D. R. (1993). Mechanosensitive gating of atrial ATP-sensitive potassium channels. *Circ. Res.* **72**, 973–983.

Verrey, F., Groscurth, P., and Bolliger, U. (1995). Cytoskeletal disruption in A6 kidney cells: Impact on endo/exocytosis and NaCl transport regulation by antidiuretic hormone. *J. Membr. Biol.* **145**, 193–204.

Wan, X., Harris, J. A., and Morris, C. E. (1995). Responses of neurons to extreme osmo-mechanical stress. *J. Memb. Biol.* **145**, 21–31.

Waxman, S. G., Kocsis, J. D., Black, J. A. (1994). Pathophysiology of demyelinated axons. *In* "The Axon: Structure, Function & Pathophysiology" (S. G. Waxman, J. D. Kocsis, P. K. Stys, Eds.). Oxford University Press, New York.

Zhou, X. L., Stumpf, M. A., Hoch, H. C., and Kung, C. (1991). A mechanosensitive channel in whole cells and in membrane patches of the fungus *Uromyces*. *Science* **253**, 1415–1417.

Michael M. Behbehani

# 44

# Sensory Receptors and Transduction

## I. Introduction

Before sensory transduction occurs, the external energy associated with that sensation must be changed to a series of action potentials (APs). The sensory receptors are the first component of this translation process. They transduce information in the external world to APs. In general, there are three types of sensory receptors. Activation of one type leads to conscious perception; these are called **extroceptors.** Some examples of extroceptors are photoreceptors that lead to the sensation of vision and olfactory receptors that lead to the sensation of smell. **Proprioceptors** that encode muscle length and tension are the second type of sensory receptors. The third type of receptors includes the **interoceptors.** Activation of this class of receptors does not lead to conscious perception. Examples of this type of receptors are the sensory receptors that sense oxygen tension and those that sense blood pressure. Although there are significant differences among these types, all receptors change external energies into changes in the membrane potential that, if sufficiently large, can produce APs. Membrane potential changes that are produced by action of external energy on sensory receptors occur by opening or closing of ionic gates. In some receptors, such as the pacinian corpuscle, which encodes the sense of vibration, mechanical pressure produces opening of $Na^+$ and $K^+$ channels. In other receptors, such as vertebrate photoreceptors, absorption of light closes $Na^+$ channels. In general, transduction can occur through activation of second messenger system (such as $Ca^{2+}$, cAMP, or cGMP) or by direct coupling of the external energy to particular types of ionic channels. For example, transduction in the photoreceptors and olfactory receptors involves cGMP. In the stretch receptors, on the other hand, stretching of the cell membrane opens Na and K channels without involvement of second messenger systems. Table 1 gives a summary of the properties of several sensory receptors.

## II. Common Operations of the Sensory Transduction Process

Transduction of sensory signals involves at least four operations (Shepard, 1991): detection, amplification, discrimination, and adaptation. The first operation is detection. Sensory receptors operate at the low limit of their respective stimulus energy. For example, under a highly dark-adapted state, absorption of a single photon can activate rod photoreceptors and lead to light perception. Similarly, displacement of hair cells in the organ of Corti by a few nanometers can produce a significant change in the membrane potential and produce APs in the auditory nerve. The stimulus for which a sensory receptor is most sensitive is called the **adequate stimulus.** Although sensory receptors are highly sensitive to their adequate stimuli, the latency of their response is highly variable. In some mechanoreceptors, such as hair cells in the cochlea, the latency can be as short as 10 $\mu$sec. On the other hand, the latency of the response of olfactory neurons to odorants can be several hundred milliseconds. In general, the sensory systems that encode sensations that have high-frequency components have shorter response latency.

The second operation that is performed by sensory receptors is amplification of the signal. Several processes that lead to signal amplification have been identified. By far the most widely used mechanisms involve activation of a second messenger system in which production of a single molecule of an enzyme can cause alteration of several hundred molecules of a substrate. For example, in the photoreceptors, absorption of a few photons can produce a few molecules of phosphodiesterase, which, in turn, hydrolyze several hundred molecules of cGMP. In this fashion, the absorption effect is amplified. Other amplification mechanisms include nonlinear voltage-gated channels and positive feedback.

The third operation is discrimination among several stimuli. In some receptors the quality of discrimination

682

**TABLE 1** Properties of Five Different Types of Sensory Receptors

| Receptor type | Adequate stimulus | Transduction mechanism | Types of channels | Channel characteristics |
|---|---|---|---|---|
| Photoreceptors | Photons | Reduction in cGMP | cGMP activated | Nonselective cation, 2-pS conductance |
| Olfactory | Odorant | Increase in cGMP | cGMP activated | Nonselective cation, 2-pS conductance |
| Taste | Salt, bitter sour, sweet | Increase in cGMP or cAMP | cGMP or cAMP or tastant specific | Nonselective cation, 2–80-pS conductance |
| Touch, vibration, and osmolarity | Mechanical pressure | Stretching of the cell membrane | Stretch activated | Nonselective cation, 2–40-pS conductance |
| Nociceptors | Noxious pressure, temperature $>50°C$ or $<4°C$ | Bradykinin, prostaglandins, histamines, K, Cl, and other unknown peptides | Unknown | Unknown |
| Thermal (heat and cold) | Temperature | Unknown | Unknown | Unknown |

depends on properties of the receptor molecules. For example, in photoreceptors, color discrimination is due to the presence of three different chromophores in the cone systems. In other receptors, the physical location of the receptor determines its discrimination ability. For example, the hair cells located at the tip of the basilar membrane of the organ of Corti in the inner ear encode high-frequency auditory signals because of their physical location.

Adaptation is the fourth operation that all sensory receptors perform. Adaptation is defined as the reduction in the response of a sensory receptor to repeated or prolonged application of a stimulus. On the basis of their adapting properties, the receptors can be classified as slowly adapting or rapidly adapting. The rapidly adapting receptors encode sensations that change rapidly. For example, the sense of vibration is encoded by the rapidly adapting pacinian corpuscles. In contrast, the slowly adapting receptors encode sensations that must be transmitted continuously for proper function of a system. For example, voluntary movements require the knowledge of the position of the joints at all times. For this reason, the joint receptors adapt very slowly. The cellular mechanisms for adaptation differ for different types of receptors. For example, in the slowly adapting stretch receptors of the lobster, adaptation occurs by slow inactivation of $Na^+$ channels followed by a change in the activity of the electrogenic Na/K pump. On the other hand, in the rapidly adapting receptor of the lobster, adaptation is produced by inactivation of Na channels alone (Gestrelius and Grampp, 1983). In other receptors the changes in level of activity of second messenger systems lead to adaptation; for example, photoreceptors and olfactory receptors use these mechanisms for their adaptation.

## III. Physiology of Mechanoreceptors

Mechanoreceptors are cells that respond to stretching of their cell membrane, thereby transforming mechanical energy into nerve impulses. Some important mechanoreceptors are (1) joint receptors, which include (a) transient receptors that encode movement of the joint regardless of the direction of the movement, (b) velocity detectors that encode changes of joint angle, and (c) position detectors that signal the position of the joint in the absence of movement; (2) free nerve endings, such as those in the nasopharyngeal epithelium and in the epithelial lining of the airways from trachea to the respiratory bronchioles and those of the larynx that are involved in the cough reflex; (3) hair cells of the organ of Corti that encode auditory signals, and hair cells of the semicircular canal that encode the static position of the body and acceleration in three-dimensional space; (4) stretch receptors that encode the length and tension of the muscle; and (5) pacinian corpuscles that encode rapidly changing mechanical pressure and vibration.

All mechanoreceptors contain mechanically activated channels. The properties of these channels were first described by Guharay and Sachs in 1984, and several excellent review articles dealing with the physiology and kinetics of these channels have been written (French, 1992; Morris, 1990; Sachs, 1988). Two types of mechanosensitive channels have been observed. In one type, the stretch-activated (SA) channel, stretching the membrane increases the probability of opening. In the second type, the stretch-inactivated (SI) channel, the open probability decreases with stretch. Mechanoreceptors in the stretch receptors have a range of sensitivity between $-30$ and $-40$ mm Hg (Fig. 1). The conductance of a mechanically activated channel is highly species dependent. In yeasts, the conductance can be as high as 36 pS, and in some bacteria the conductance can be as low as 1 nS (Szabo et al., 1990; Zoratti et al., 1990). In general, mechanical stimulation causes an increase in the probability of channel opening without increasing the conductance of individual channels (Erxleben, 1989; Medina and Bregestovski, 1988; Sackin, 1989, Gustin et al., 1988; Sigurdson et al., 1987; Yang and Sachs, 1989).

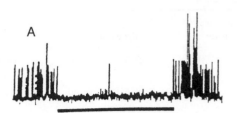 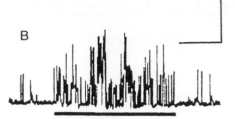

**FIG. 1.**   Time course and magnitude of currents recorded from (A) stretch-inactivated and (B) stretch-activated channels from snail neurons. Channels in these receptors are $K^+$ selective and produce outward $K^+$ current. The scale in part A is 1 pA, 2 sec and in part B, 4 pA, 2 sec. (Adapted from Morris, 1990.)

## A.  Crustacean Stretch Receptor

An excellent model of a mechanoreceptor is the crustacean stretch receptor. This receptor is relatively large, and because of its morphology, has been studied in detail. It is found in the tail muscle of crustaceans and consists of specialized muscle cells that are innervated by the dendrites of a sensory neuron. There are two types of muscle bundles: slow and fast, which differ considerably in their anatomy and in their response to stretch. The sensory neurons supplying these bundles are large and can be impaled with two microelectrodes (Brown *et al.*, 1978).

When the slow bundle is stretched moderately, the sensory neuron depolarizes and fires APs. At the beginning of the response, the firing frequency is high. In time, the firing frequency decreases to a steady-state level that is maintained as long as the muscle is stretched. In contrast to the slow bundle, the fast bundle responds to stretch by several APs and then stops firing even though the muscle is still stretched.

If the impulse production is prevented by adding a small quantity of an agent that blocks voltage-gated $Na^+$ channels, such as tetrodotoxin (TTX), to the bathing media, the effect of stretch on the membrane potential can be investigated. In this case the stretch of the fast or the slow bundle causes a depolarization of the sensory neuron. This depolarization is the generator potential.

### 1.  Ionic Basis for the Generator Potential

By inserting two microelectrodes in the sensory neuron, one to pass current and the other to record the generator potential, the reversal potential of the stretch receptor can be determined. This potential is near 0 mV, which implies that during the stretch the membrane becomes permeable to Na ions. This can be verified by replacing the Na ions in the bathing medium by choline. Since Na channels are much less permeable to choline than to Na, there will be no movement of positive charges into the cell and the generator potential almost disappears. Recent studies of stretch receptor neurons of the crayfish have shown the presence of two types of channels, both of which are permeable to mono- and divalent cations. These channels have nearly identical single-channel conductance but differ in their voltage range of activation. The current–voltage relationships for both channels are linear with slope conduc-

tance for Na = $50 \pm 7.4$ pS; for K = $71 \pm 11$ pS; and for Ca = $23 \pm 2.8$ pS. One type of channel is voltage independent, whereas the second type shows strong inward rectification (Erxleben, 1989).

The slow muscle bundle fires as long as the stimulus is applied. However, the fast bundle fires a short burst of APs only when the stimulus is applied and stops firing even though the stimulus application persists. The slow bundle is a slowly adapting cell, whereas the fast bundle is a rapidly adapting cell. In this circumstance the generator potential is the same for both bundles. Therefore, the adaptive properties of the fast bundle are attributed to that part of the membrane that produces an AP. This component of adaptation is called *accommodation*. Another factor in adaptation is the mechanical properties of the receptors. The amplitude of the generator potential is larger soon after the stimulus is applied than later during stimulation. This is attributed to mechanical factors such as slippage of the muscle fiber, which leads to unloading of the stretch receptor.

Measurements of membrane currents of the rapidly adapting stretch receptor neuron of the lobster have shown that the rapid adaptation of the receptors is due to slow Na current inactivation, which causes a regenerative process of accommodation. The increased firing frequency at the onset of the stimulus is due to inactivation of K current that occurs faster than the slow Na current inactivation at a comparable degree of membrane polarization (Edman *et al.*, 1987).

## B.  Pacinian Corpuscle

Another example of a mechanoreceptor is the pacinian corpuscle. This receptor is an ellipsoidal body made of several concentric laminae. The myelinated nerve fiber enters at one end and the final node of Ranvier occurs within the corpuscle.

In an intact corpuscle, application of pressure leads to a depolarizing generator potential that dies out rapidly; a second generator potential is recorded when the pressure is removed. Therefore, this receptor is an example of a rapidly adapting receptor. The amplitude of the generator potential increases as the magnitude of pressure increases. If the laminae of the pacinian corpuscle are removed and a pulse of pressure is applied to the receptor, a generator potential is recorded as long as the pressure is applied. This generator potential has no refractory phase and has spatial and temporal summation properties; that is, the generator responses set up by two weak stimuli delivered sequentially at one spot or two different spots on the nerve terminal add to produce a generator potential that is larger than the response of each stimulus alone.

In intact corpuscle, the laminated structure forms a mechanical filter and is responsible for the adaptive property of this receptor. However, if one records the AP in a corpuscle in which the laminae have been removed, the application of pressure produces a generator potential that lasts as long as the stimulus. However, still only one or two APs are produced at the termination of the pressure (Loewenstein, 1971). This indicates that the axon has accommodative properties.

By using two microelectrodes, one to pass current and the other to record the receptor potential, the reversal potential of the generator potential can be determined. This value is near zero, which indicates that the pressure causes an increase of the permeability of the membrane to $Na^+$. In a $Na^+$-free situation, the generator potential is abolished.

## C. Osmoreceptors

In mammals, the volume and salt concentration of the extracellular fluid (ECF) can change drastically due to hydration, dehydration, or ingestion of salt. Because cells are exposed to ECF, changes in osmolarity can have a significant effect on their function. For example, a decrease in the effective extracellular osmolarity can lead to swelling of cells and in extreme cases can cause the rupture of their membrane. For this reason, the osmolarity of the ECF is well regulated. The first step in this regulatory process is transduction of osmolarity by the osmoreceptors. The osmoreceptors are present both in the periphery and in the central nervous system. The peripheral receptors are located in the hepatic portal vein and in the region containing the mesenteric and portal vasculature that drain the upper part of the small intestine. In addition, the splanchnic mesentery and liver also contain osmoreceptors. The activities of these peripheral osmoreceptors are transmitted to the nucleus of the solitary tract and the ventrolateral medulla. The central osmoreceptors are located in the paraventricular nucleus, the median preoptic nucleus, the antroventral of the third ventricle, the subfornical organ (SFO) and organnum vasculosum lamina terminalis (OVLT) of the hypothalamus. In addition to these regions of the hypothalamus, other areas including zona inserta and parabrachial nucleus are also involved in osmoregulation.

Recently the effect of osmolarity on the membrane potential, membrane resistance, and firing pattern of neurons of OVLT and SFO has been studied. Using whole-cell voltage-clamp recording, Bourque and colleagues (1994) have shown that increasing osmolarity of the bathing solution bathing SFO neurons causes the appearance of an inward current and an increase in membrane conductance. The current/voltage relationship recorded from these cells indicated a reversal potential of −45 mV, which indicates

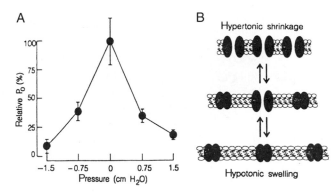

**FIG. 2.** The relationship between the opening probability of cation osmolarity-sensitive channels and pressure. (A) The opening probability–pressure curve is bell shaped, indicating that the open probability decreases as the pressure deviates from a neutral osmotic pressure. (B) A model that explains the bipolar response of osmolarity-sensitive receptors. According to this model, at zero pressure, a certain number of channels are open. Under hypotonic condition, the cell swells. This causes closing of pressure-sensitive cation channels, which leads to hyperpolarization of the cell. In contrast, under hypertonic conditions, the membrane shrinks and more cation channels are opened and the cell depolarizes. [From Bourque, C. W., Oliet, S. H. R., and Richard, D. (1994). *Frontiers Neuroendocrinol.* **15**, 231–274.]

that an increase in osmolarity causes nonselective cation channels to open.

Recordings from magnocellular neurosecretory cells in the hypothalamus also indicate that a change in osmolarity can cause opening of nonselective cationic channels. In this preparation, hypotonic stimuli have been found to reduce membrane conductance. These cells have osmolarity-sensitive channels with an average conductance of 2 pS. A decrease in the osmolarity of the bathing solution causes a decrease in the frequency of opening of these channels. In contrast, increase in the osmolarity of the bathing solution causes a significant increase in the firing frequency of these gates. That is, the gates open at a higher rate. The osmolarity-sensitive channels in these cells are highly sensitive, and can detect osmolarity changes as small as 1%.

The osmoreceptor channels in the magnocellular cells can also be activated by application of small amounts of pressure to the inside of the recording pipette, indicating that these channels are a type of mechanosensitive channel (for a complete discussion of these channels, see Chapter 43). Further characterization of these gates using positive and negative pressure indicated that the open probability has a set point for pressure such that a change in pressure (increase or decrease) causes a decrease in the open probability. An operational model of osmoreceptors is shown in Fig. 2.

## D. Hair Cell

Another mechanoreceptor we consider is the hair cell. Hair cells encode minute changes that occur when sound vibration causes a movement of the basilar membrane in the organ of Corti as well as movement of fluids in the

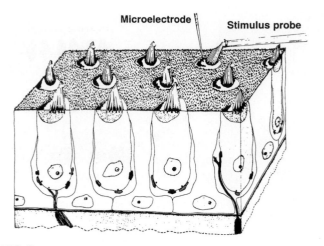

**FIG. 3.** The structure of hair cells in bullfrog. Hair cells are embedded in an epithelial sheet and their apices are covered by microvilli. Each cell is separated by connective tissue. The bending of the microvilli opens mechanically activated channels that lead to alteration of $K^+$ and $Ca^{2+}$ conductances. (From Huspeth, 1989; *Nature*, copyright 1989 MacMillan Magazines Limited.)

semicircular canal. The apical surface of every hair cell is immersed in an unusually high K solution. The junction around the hair cell separates this high K concentration from a solution that has a low K concentration. Each hair cell contains a hair bundle that projects from the apical surface of the hair cell (Fig. 3). The hair bundle is a cluster of 20–300 cylindrical processes consisting of an actin cy-

toskeleton ensheathed by plasma membrane. Within this structure, adjacent actin filaments are cross-linked by fimbrin. Because of cross-linkage, the stereocilium is much more rigid than bundles of actin, which can move against each other (Pickeles *et al.*, 1984).

In an unstimulated hair cell, 15% of the transducing ionic channels are in the open state. Flow of inward current through these channels causes the membrane potential of the cell to be near −60 mV. A force that moves the bundle toward its tall edge opens additional channels and allows influx of cations into the cell, thereby causing depolarization. A stimulus that moves the hair bundle toward the short edge of the bundle closes the channels that are normally open at the resting state and hyperpolarizes the cell. The displacement required to change the state of mechanosensitive channels corresponds to displacement of approximately 120 nm at its tip.

In the bullfrog sacculus, a filamentous link extends between the tips of neighboring cilia. A model proposed by Huspeth (1989) has suggested that these filaments act as gating springs that are connected to the gate of the transduction channel. According to this model (Fig. 4), when the hair cell is displaced, the gating spring is stretched and causes the opening of the transduction gates. There is evidence that a portion of the work that displaces a bundle goes into the gating spring. These gating springs also contribute to the stiffness that opposes displacement of the hair cells. The evidence for existence of the gating spring has come from the time course of channel gating. The response latency of hair cells is about a few microseconds, which rules out the involvement of a second messenger

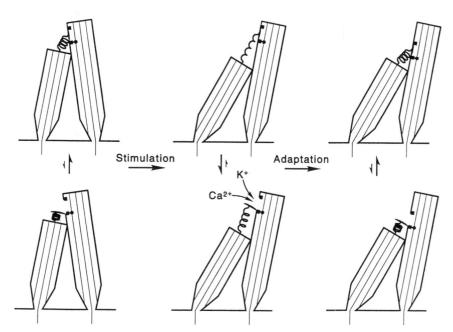

**FIG. 4.** A model proposed by Huspeth that accounts for transduction and adaptation of bullfrog hair cell. At rest (left) there is rapid transition between the open and closed states of the mechanically activated channels that are assumed to be at the apex of the receptor. In this state, the channels are mostly in the closed configuration. When the receptor is stimulated (center), the mechanical force stretches the gating spring and opens the channel that would allow entry of cations ($K^+$ and $Ca^+$). In this model the adaptation (right) occurs by altering the position of the gating spring.

system. In addition, larger step stimuli elicit electrical responses that are larger in magnitude and have a faster rise time. This behavior is consistent with a mechanical model in which the mechanical energy of the stimulus is stored in a spring attached to the channel's gate. In this fashion, the opening and closing rates of the channels will be determined by the probability that the energy content of the spring exceeds the transition-state energy for channel opening or closing.

The gating spring also can explain the adaptive properties of the hair cells. With continuous force application, the hair bundle becomes less stiff. This can occur when the tension in the gating springs is reduced. Huspeth has proposed that prolonged stimulation repositions the anchoring position of the gating spring. When a stimulus step increases the tension in the spring, the transduction channels open and allow the influx of cations including $Ca^{2+}$, which accumulates in the stereocilium and interacts with the motor protein, decreasing its upward force. The gating spring then shortens by pulling down the motor molecule. When the spring reaches its resting tension, the channels close and $Ca^{2+}$ influx reduces to its resting level. This restores the balance between the upward force produced by the protein and the downward force tension in the spring.

## IV. Summary

The sensory receptors transform external energies into changes in the membrane potential. All sensory receptors have the following mechanisms in common: detection, amplification, discrimination, and adaptation. The mechanisms that lead to changes in the membrane potential of the sensory receptors can involve either direct coupling of the stimulus and ionic channels or activation of second messenger systems. The majority of receptors use the second messenger system as a means of both detection and amplification of the signal. The second messengers that are known to be involved in transduction and amplification are cAMP and cGMP; in addition, calcium is involved as a second or third messenger in sensory receptors. All sensory receptors produce generator potential and show summation.

All mechanoreceptors have a type of channel that is mechanically activated. The conductance of this channel can be as small as 1 nS or as large as 40 pS. These channels do not require a second messenger system. The ionic channels that are operated by these receptors are TTX insensitive. The adaptation in the mechanoreceptor can occur through specialized structures, such as laminae, that cover pacinian corpuscles or through specialized ionic channels.

## Bibliography

Bourque, C. W., Oliet, S. H. R., and Richard, D. (1994). Osmoreceptors, osmoreception, and osmoregulation. *Front. Neuroendocrin.* **15**, 231–274.

Brown, H. M., Ottoson, D., and Rydqvist, B. (1978). Crayfish stretch receptor: An investigation with voltage-clamp and ion-sensitive electrodes. *J. Physiol.* **284**, 155–179.

Edman, A., Gestrelius, S., and Grampp, W. (1987). Analysis of gated membrane currents and mechanisms of firing control in the rapidly adapting lobster stretch receptor neuron. *J. Physiol.* **384**, 649–669.

Erxleben, C. (1989). Stretch-activated current through single ion channels in the abdominal stretch receptor organ of the crayfish. *J. Gen. Physiol.* **94**, 1071–1083.

French, A. S. (1992). Mechanotransduction. *Annu. Rev. Physiol.* **54**, 135–152.

Gestrelius, S., and Grampp, W. (1983). Impulse firing in the slowly adapting stretch receptor neurone of lobster and its numerical simulation. *Acta Physiol. Scand.* **118**, 253–261.

Guharay, F., and Sachs, F. (1984). Stretch-activated single ion channel currents in tissue cultured embryonic chick skeletal muscle. *J. Physiol.* **352**, 685–701.

Gustin, M. C., Zhou, X.-L., Martinac, B., and Kung, C. (1988). A mechanosensitive ion channel in the yeast plasma membrane. *Science* **242**, 762–765.

Huspeth, A. J. (1989). How the ear's works work. *Nature* **341**, 397–404.

Lowenstein, W. R. (1971). Mechano-electric transduction in the pacinian corpuscle. Initiation of sensory impulses in mechanoreceptors. In "Principles of Receptor Physiology" (W. R. Lowenstein, Ed.), pp. 269–290. Springer-Verlag, New York.

Medina, I. R., and Bregestovski, P. D. (1988). Stretch-activated ion channels modulate the resting membrane potential during early embryogenesis. *Proc. R. Soc. London Ser. B* **235**, 95–102.

Morris, C. E. (1990). Mechanosensitive ion channels. *J. Membr. Biol.* **113**, 93–107.

Pickeles, J. O., Comis, S. D., and Osborne, M. P. (1984). Cross-links between stereocillia in the guinea pig organ of corti and their possible relation to sensory transduction. *Hearing Res.* **75**, 103–112.

Sachs, F. (1988). Mechanical transduction in biological systems. *CRC Crit. Rev. Biomed. Eng.* **16**, 141–169.

Sackin, H. (1989). A stretch-activated $K^+$ channel sensitive to cell volume. *Proc. Natl. Acad. Sci. USA* **86**, 1731–1735.

Shepherd, G. M. (1991). Sensory transduction: Entering the mainstream of membrane signaling. *Cell* **67**, 845–851.

Sigurdson, W. J., Morris, C. E., Brezden, B. L., and Gardner, D. R. (1987). Stretch activation of a $K^+$ channel in molluscan heart cells. *J. Exp. Biol.* **127**, 191–209.

Szabo, I., Petronilli, V., Guerra, L., and Zoratti, M. (1990). Cooperative mechanosensitive ion channels in *Escherichia coli. Biochem. Biophys. Res. Commun.* **171**, 280–286.

Yang, X. C., and Sachs, F. (1989). Block of stretch-activated ion channels in *Xenopus* oocytes by gadolinium and calcium ions. *Science* **243**, 1068–1071.

Zoratti, M., Petronilli, V., and Szabo, I. (1990). Stretch-activated composite ion channels in *Bacillus subtilis. Biochem. Biophys. Res. Commun.* **30**, 443–450.

*Daniel C. Marcus*

# 45

# Acoustic Transduction

## I. Introduction

The detection of sound by mammals depends on a series of biological systems beginning with the collection of sound pressure waves by the external ear, followed by the mechanical transmission through the middle ear ossicles to the **cochlea** of the inner ear. The sound pressure waves in the cochlea induce motion of the basilar membrane to which the sensory organ, the **organ of Corti,** is attached. Motion of the organ of Corti with respect to another structure (tectorial membrane) causes movements of the sensory cilia (hairs) on the **hair cells** and modulation of the flow of current through these hair cells, leading to modulation of the rate of firing of the afferent auditory nerve fibers, which synapse to the base of the hair cells. The nerve fibers carry the auditory information to the brain where the signal undergoes central processing, leading to the perception of sound. Peripheral auditory processing is modulated by efferent signals originating from the brain. This chapter focuses on the cellular aspects of acoustic transduction in the auditory periphery. Much of what we know about the function of the mammalian cochlea is derived from experiments performed on preparations from the vestibular labyrinth of mammals, birds, and amphibians and from the cochlea of birds.

## II. Mammalian Inner Ear Structure

The transduction apparatus is part of an **epithelium** forming the **cochlear duct** and separating two distinct cochlear fluids, **endolymph** and **perilymph.** The composition and importance of these fluids is related later. A diagram of a cross-section of the mammalian cochlear duct is shown in Fig. 1. The organ of Corti is comprised of the sensory **inner and outer hair cells** (Fig. 11, shown later), which are surrounded by Deiters' cells, pillar cells, and Hensen's cells. A mechanically stiff apical surface of the organ of Corti, the **reticular lamina,** is formed by cuticular plates just under the apical membrane of these cells. The organ of Corti sits on the **basilar membrane,** a fibrous sheet that transmits the acoustic stimulus.

In the medial direction from the organ of Corti, the duct is comprised of inner sulcus cells and interdental cells of the spiral limbus. The **tectorial membrane** is a gelatinous, acellular structure in the cochlear lumen apparently secreted by the interdental cells. Lateral from the organ of Corti, the cochlear duct consists of outer sulcus cells, spiral prominence cells, and **marginal cells** of the **stria vascularis.** Reissner's membrane forms the remaining wall of the triangular-shaped cochlear duct and is comprised of a thin, avascular sheet of epithelial cells.

The apical and basolateral membranes of all of these cells are separated by tight junction complexes near the endolymphatic surface, which serve to join each cell to its neighbor and to complete the barrier between endolymph and the fluid bathing the basolateral membranes. The basolateral fluid is perilymph for all cell types except strial marginal cells, as described later. The cochlear duct is closed at the apex of the cochlea and is joined at the base of the cochlea via a constriction in the epithelial lumen (ductus reunions) to the vestibular system. Cellular physiologists have focused most of their attention on the strial marginal cells, which provide the energy source for the transduction process, and on the sensory hair cells themselves.

## III. Cell Physiology of Endolymph Homeostasis

### A. Composition

Even in the absence of an acoustic stimulus, the cochlea is highly active, maintaining the electrolyte composition of endolymph (Table 1) and a standing current analogous to the "dark current" of photoreceptors (Chapter 47) (Wangemann and Schacht, 1996). The transduction current from endolymph through the hair cells is carried by $K^+$ and so depends on the high concentration of that ion in endolymph. Gross changes in endolymph composition ei-

**688**

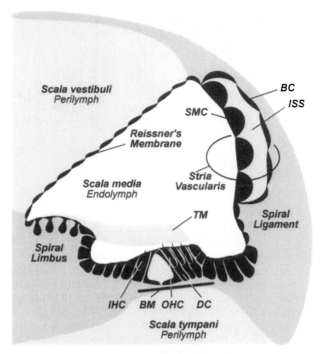

**FIG. 1.** Diagram of the epithelial cell boundary of the cochlear duct. The triangular cross-section is about 0.3–0.5 mm on each side. The lateral wall is the site of the K+-secretory stria vascularis, a tissue consisting of two barrier layers of strial marginal cells (SMCs) and basal cells (BCs), which enclose the intrastrial space (ISS). The organ of Corti sits on the fibrous basilar membrane (BM), forms most of the wall separating scala media and scala tympani, and consists of the inner hair cells (IHCs), outer hair cells (OHCs), and several supporting cell types including Deiters' cells (DCs). The sensory hairs at the apical membranes of the OHC project into the gelatinous tectorial membrane (TM), whereas for the most part those of the IHC do not. The spiral ligament is the connective tissue between the stria vascularis and the bone (not shown). (Adapted with permission from Wangemann, 1997; Kalium-Ionensekretion und Entstehung des endokochlearen Potentials in der Stria vascularis. *HNO* **45**, 205–209.)

ther by a tear in Reissner's membrane or by genetic interference in K+ secretion lead to degeneration of the hair cells with subsequent degeneration of the synapsing afferent auditory nerve (Vetter *et al.,* 1996). The dimensions of the gelatinous tectorial membrane are sensitive to the levels of K+, Na+, H+, and Ca$^{2+}$ as well as on unknown chemical factors (Shah *et al.,* 1995). Swelling has been associated with substitution of Na+ for K+ and with decreases of Ca$^{2+}$. Movements of water into and out of the "membrane" are believed to be related to the level of screening of fixed charges.

### B. Stria Vascularis

The stria vascularis has two primary known functions: secretion of K+ and generation of the lumen-positive **endocochlear potential** (EP; ca. +80 mV). The high endolymphatic K+ concentration provides the carrier of the transduction current. The driving force for that current consists

**TABLE 1** Approximate Ion Composition and Electrical Potential of Cochlear Endolymph and Perilymph[a]

| Ion | Endolymph | Perilymph |
| --- | --- | --- |
| Potassium (m$M$) | 157 | 5 |
| Sodium (m$M$) | 1 | 145 |
| Calcium (m$M$) | 0.02 | 1 |
| Chloride (m$M$) | 132 | 125 |
| Bicarbonate (m$M$) | 31 | 25 |
| pH | 7.4 | 7.3 |
| Potential (mV) | +80 | 0 |

[a] From Wangemann and Schacht, 1996.

almost exclusively of the voltage across the transduction channels in the stereocilia (Section IV) and that voltage is the sum of the intracellular potential of the hair cells and the EP. The large EP therefore heightens the sensitivity of the cochlear transduction process compared to vestibular organs, which do not have this high transepithelial electrical polarization. The cellular transport model by which the stria vascularis secretes potassium and generates the EP is shown in Fig. 2 and described next. Note that all epithelial cells that produce a vectorial transport of substances do

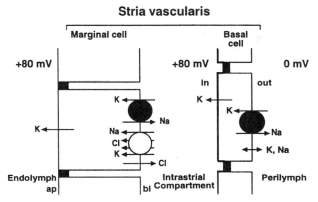

**Stria vascularis**

**FIG. 2.** Model of ion transport by the stria vascularis. The marginal cell epithelium secretes K+ into endolymph, and the basal cell layer produces the voltage (endocochlear potential) between endolymph and perilymph. The ion transport processes ascribed to the marginal cells have a strong experimental basis. The transport processes proposed in the model of the basal cells have not yet been experimentally demonstrated and may be distributed among basal cells, intermediate cells of the stria vascularis, and fibrocytes of the spiral ligament, which are all connected by gap junctions. Arrows, ion channels; open circle, ion carrier; filled circles, primary-active ion "pump." (Adapted from *Hear Res.* **90**, Wangemann, P., Comparison of ion transport mechanisms between vestibular dark cells and strial marginal cells, pp. 149–157, Copyright 1995 with kind permission of Elsevier Science–NL, Sara Burgerhartstraat 25, 1055 KV Amsterdam, The Netherlands.)

so by virtue of different membrane properties of their apical and basolateral membranes (Chapter 21).

## 1. Division of Function between Marginal and Basal Cells

The stria vascularis consists of two barriers formed by the **marginal cells** and the **basal cells.** Each barrier consists of a continuous sheet of cells joined by tight junction complexes (Fig. 2). Between these barriers is the intrastrial space with the capillary bed for which the tissue is named and a discontinuous layer of intermediate cells. The basal cells are joined via **gap junctions** (Chapter 30) to the intermediate cells and to fibrocytes in the adjacent connective tissue, suggesting a level of cooperation among these three cell types (Kikuchi *et al.,* 1995). Unlike most sheets of epithelial cells, strial marginal cells are not coupled to each other nor to other cells by gap junctions (Kikuchi *et al.,* 1995; Sunose *et al.,* 1997). The physiological significance of this functional independence of marginal cells is not known.

There is strong evidence that there is a division of function between strial marginal cells and basal cells. The basal cell barrier produces and supports the endocochlear potential and the strial marginal cells secrete K[+]. In spite of the separation of cellular function, the two processes are closely tied together through the composition of the intrastrial space. For example, inhibition of K[+] secretion by the marginal cells leads to a decline of the endocochlear potential, presumably due to a rise in intrastrial K[+] concentration and the consequent depolarization of the intrastrial membrane of the basal cells. Several lines of indirect evidence, including histochemical, biochemical, electrophysiological, and flux studies, have strongly suggested that the stria vascularis is responsible for secretion of K[+] into the cochlear lumen and for production of the EP (reviewed in Wangemann and Schacht, 1996). Recent evidence of a more direct nature is presented here.

## 2. K[+] Secretion by Strial Marginal Cell Epithelium

Active K[+] secretion in the cochlear duct has been demonstrated by flux measurements of radiolabeled K[+] introduced in either the perilymphatic or vascular space and its appearance in endolymph (Konishi *et al.,* 1978; Sterkers *et al.,* 1982). These fluxes were inhibited by transport blockers and anoxia. The stria was assumed to be the site of secretion since the flux was nearly the same when the radiotracer was added to either scala vestibuli or scala tympani; possible contributions by the spiral limbus were disregarded.

More recently, the stria was isolated from the cochlear duct, a K[+]-selective self-referencing probe (see Appendix at the end of this chapter) was placed near the tissue *in vitro* and a K[+] gradient was found directed away from the luminal surface (Wangemann *et al.,* 1995a) (Fig. 3). The marginal cell layer of the stria vascularis produces this K[+] flux using the constellation of transport processes shown in Fig. 2. The cell model is identical to that for the vestibular dark cells, the vestibular homolog of strial marginal cells (Wangemann, 1995). K[+] is taken up across the **basolateral**

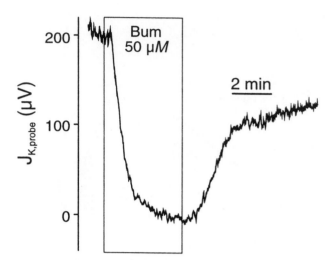

**FIG. 3.**  K[+]-secretory flux from stria vascularis measured with the K[+]-selective self-referencing probe. The ordinate reflects the difference in microvolts measured at the two positions of the probe. Flux was decreased by basolateral perfusion of the inhibitor of Na-K-Cl cotransport, bumetanide (Bum). (Adapted from *Hear Res.* **84,** Wangemann, P., Liu, J., and Marcus, D. C., Ion transport mechanisms responsible for K[+] secretion and the transepithelial voltage across marginal cells of stria vascularis *in vitro,* pp. 19–29, Copyright 1995 with kind permission of Elsevier Science–NL, Sara Burgerhartstraat 25, 1055 KV Amsterdam, The Netherlands.)

**membrane** from the **intrastrial space** fluid by two transporters: the Na,K-ATPase and the Na-K-Cl cotransporter. The first is a primary-active process, which uses the energy from cytosolic adenosine triphosphate (ATP) and the second is secondary-active and uses the large Na[+] concentration gradient between extracellular and intracellular compartments created by the Na,K-ATPase. Na[+] and Cl[–] taken up by the cotransporter are removed from the cytosol by the Na,K-ATPase and basolateral Cl[–] channels, respectively (Wangemann and Schacht, 1996).

K[+] secretion occurs passively via channels in the apical membrane (Fig. 4). These channels are of an unusual type known as **IsK** (or **min K) channels.** The essential contribution of the IsK protein to hearing and balance is most dramatically illustrated by the development of a mouse in which the *isk* gene was knocked out (Vetter *et al.,* 1996). In these mice the entire endolymphatic space collapsed (Fig. 5) after the point in embryonic development at which endolymph secretion normally commences, resulting in degeneration of the hair cells and the afferent nerves. The mice showed behavior that was characteristic of a lack of hearing and balance. Measurements of short circuit current (I$_{sc}$) under constitutive and stimulated conditions showed a lack of K[+] secretion. The marginal cells remained present and were in other respects normal. The absence of this protein clearly has its most profound effect on the function of the auditory and vestibular periphery.

The IsK channels are characterized by slow activation (over seconds) in response to depolarization of the cell membrane and have a single-channel conductance of about 14 pS under *in vivo*-like conditions (Marcus and Shen,

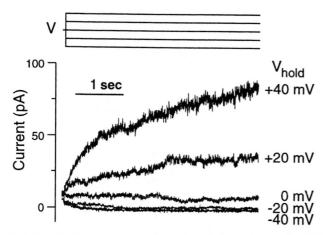

**FIG. 4.** Patch-clamp currents from the apical membrane of a strial marginal cell, on-cell configuration. Sustained depolarizations activate the current slowly over several seconds, characteristic of IsK channels. (Reproduced with permission from Shen *et al.*, 1996.)

1994; Shen *et al.*, 1996). These channels are believed to carry the transepithelial $K^+$ flux since maneuvers that alter transepithelial $K^+$ secretion, such as basolateral bumetanide or elevated bath $K^+$ concentration, have the same effects on IsK currents measured from the apical membrane of strial marginal cells and vestibular dark cells in cell-attached patches (Marcus and Shen, 1994; Wangemann, 1995; Shen *et al.*, 1997). The functional IsK channel is thought to be composed of two IsK protein subunits of only 130 amino acids with a single transmembrane domain and of another subunit, KvLQT1, which forms the $K^+$-selective channel (Takumi *et al.*, 1988; Wang and Goldstein, 1995; Barhanin *et al.*, 1996; Sanguinetti *et al.*, 1996). The IsK protein itself apparently modulates the activity of the associated channel (Vetter *et al.*, 1996; Tzounopoulos *et al.*, 1995; Ben-Efraim *et al.*, 1996). Although the IsK chan-

nel has often been studied as exogenously expressed in *Xenopus* oocytes, heart myocytes are the only cells other than strial marginal cells and vestibular dark cells where IsK channels are currently thought to be both endogenously expressed and constitutively active.

### 3. Production of Endocochlear Potential by Strial Basal Cells

An electric current directed from the stria vascularis into the cochlear duct was found based on a voltage gradient observed when an electrode was tracked within the cochlear duct *in situ* (Zidanic and Brownell, 1990). More recently, the stria was isolated from the cochlear duct and a **voltage-sensitive vibrating probe** (see Appendix at the end of this chapter) was placed near the tissue *in vitro* (Wangemann *et al.*, 1995a). An electric field gradient was found directed away from the luminal surface and the associated current was strongly reduced by bumetanide, a substance known to decrease the endocochlear potential *in vivo*. Evidence for the role of the basal cell layer in generation of the endocochlear potential was obtained from experiments in which the stria vascularis was mounted in a micro **Ussing chamber** to measure the voltage generated by the isolated tissue *in vitro* (Wangemann *et al.*, 1995a). It was found that the voltage was greater when the basal cell layer was intact compared to when the barrier properties had been mechanically compromised. In addition, under conditions when only the marginal cell layer contributed to the measurements, the voltage was nearly zero if the apical perfusion was high in $K^+$ concentration (artificial endolymph). Under symmetrical conditions, both sides of the epithelium are perfused with the same mammalian physiological saline (high $Na^+$, low $K^+$ concentrations), which ensures that the transepithelial current is due completely to active cellular transport processes and not to passive chemical driving forces across the paracellular pathway (Koefoed-Johnsen and Ussing, 1958).

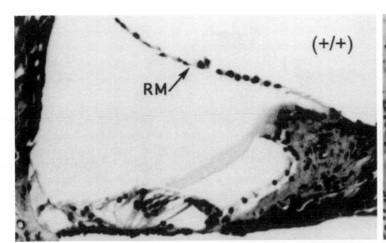
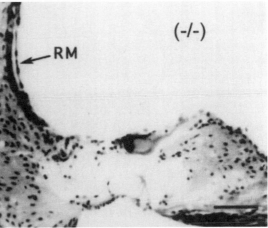

**FIG. 5.** Cross-section of the cochlear duct from mice with (+/+) and without (−/−) the *isk* gene. Absence of the isk gene led to total collapse of the duct such that Reissner's membrane (RM) was in close apposition to the stria vascularis and tectorial membrane. (Reproduced with permission from Vetter *et al.*, 1996. Copyright 1996 by Cell Press.)

The proposition that the basal cell layer (perhaps in association with the intermediate cells and fibrocytes) produces the endocochlear potential is partly based on double-barreled microelectrode measurements of electrical potential and K⁺ concentration while traversing the stria *in situ* (Salt *et al.,* 1987). In the spiral ligament, the potential was taken as zero and the K⁺ concentration was a few millimolar (Fig. 6). As the electrode was advanced, a region was found where the K⁺ concentration was also low but the potential had risen to about +80 mV. This was interpreted as being located in the **intrastrial space.** Further penetration led to a jump of the K⁺ concentration to that commonly found in endolymph with no appreciate change in the potential. The voltage gradient was therefore largest across the basal cell layer and the K⁺ gradient largest across the marginal cell layer.

The **cell model** proposed for the basal cells is shown in Fig. 2. The basolateral membrane would support little potential difference due to a large nonselective cation conductance and similar cation concentrations in the basolateral fluid (perilymph permeating the spiral ligament) and in the cytosol. The resting membrane potential of the basal cells would therefore be clamped to near zero when measured with respect to perilymph. If the membrane facing the intrastrial space has a large K⁺-selective conductance, that membrane would support a large potential difference (intrastrial space positive with respect to basal cell cytosol) due to a large K⁺ gradient between intra- and extracellular fluids. It is this K⁺ conductance of the intrastrial membrane that generates the EP. There is little biophysical evidence yet to support experimentally the basal cell model although K⁺ channels of the maxi-K⁺ type have been observed on isolated basal cells (Takeuchi and Irimajiri, 1996). Some components of the model may be located in the basal cells themselves, whereas others may be associated with the

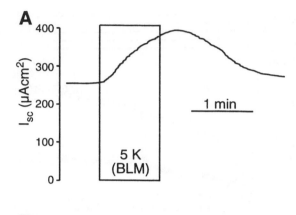

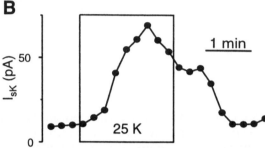

**FIG. 7.**    Dependence of (A) short circuit current ($I_{sc}$) and (B) apical IsK channel current on K⁺ concentration in vestibular dark cells, the homologs to strial marginal cells. (Adapted from Wangemann, P., Shen, Z., and Liu, J, *Hear. Res.* **100,** K⁺ induced stimulation of K⁺ secretion involves activation of the IsK channel in vestibular dark cells, pp. 201–210. Copyright 1996 with kind permission of Elsevier Science–NL, Sara Burgerhartstraat 25, 1055 KV Amsterdam, The Netherlands.)

intermediate cells and/or fibrocytes (Wangemann and Schacht, 1996).

### 4. Regulation of Ion Transport in Strial Marginal Cells and Vestibular Dark Cells

Every cell type encounters perturbations and signals by systemic variables such as osmotic strength and extracellular pH ($pH_o$), by hormones and/or neurotransmitters, and by cytosolic variables such as intracellular pH ($pH_i$). K⁺ secretion in the inner ear is altered in response to many such stimuli.

The increase in flow of K⁺ through hair cells in response to mechanical stimulation (sound in the cochlea, acceleration in the vestibular labyrinth) is known to increase the concentration of K⁺ in the perilymph surrounding the basolateral membranes of the hair cells and the nerve endings. In strial marginal cells and vestibular dark cells, the rate of the basolateral uptake of K⁺ and of its secretion through the apical IsK channel has been found to be exquisitely sensitive to the concentration of perilymphatic K⁺ (Fig. 7). This K⁺ sensitivity is believed to play a significant role in regulating the recirculation of K⁺ back into endolymph (Wangemann *et al.,* 1996b). It has been proposed that diffu-

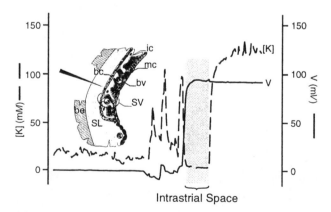

**FIG. 6.**    Recording of profiles for voltage and K⁺ concentration during penetration of the stria vascularis. A region was found where the voltage with respect to perilymph was highly positive but the K⁺ concentration was low; this region was interpreted to be the intrastrial space. [Adapted with permission from figures in Salt *et al.,* 1997, and (inset) Dallos, P. (1996). Overview: Cochlear neurobiology. *In* "The Cochlea" (P. Dallos, A. N. Popper and R. R. Fay, Eds.), pp. 1–43. Springer-Verlag, New York.]

sion of K+ from the base of the hair cells to the K+ secretory cells is facilitated by a system of supporting cells and fibrocytes connected by gap junctions (Kikuchi *et al.*, 1995).

Both **extracellular** and **intracellular acidification** transiently elevate $I_{sc}$ in vestibular dark cells with a given decrease in $pH_i$ causing a sevenfold greater increase in $I_{sc}$ than the same decrease in $pH_o$ (Fig. 8) (Wangemann *et al.*, 1995b). Fluctuations normally occur in $pH_o$ as a result of systemic acidosis and alkalosis, whereas $pH_i$ changes in response to varying metabolic activity of the cell. These challenges to the cellular homeostasis of $pH_i$ are controlled by transporters such as the Na+/H+ exchanger and H+/ monocarboxylate transporter in vestibular dark cells and strial marginal cells (Wangemann *et al.*, 1996a; Shimozono *et al.*, 1996). In addition to controlling their own cytosolic pH, cells bordering the cochlear duct regulate the pH of endolymph (Wangemann and Schacht, 1996).

Most cells respond to a hypoosmotic challenge by first swelling passively as water rushes in and then **regulating** their **cell volume** over several seconds or minutes by releasing KCl (**regulatory volume decrease; RVD**) by one or more of several pathways that are normally quiescent but activated by osmotic challenge (Chapter 19). In contrast, vestibular dark cells undergo RVD via K+ and Cl– selective pathways, which include the constitutively active IsK channel (Wangemann *et al.*, 1995c; Shiga and Wangemann, 1995; Wangemann and Shiga, 1994). Surprisingly, a hypotonic challenge also results in sustained stimulation of K+ secretion (Fig. 9).

There is no innervation of the stria vascularis, but these cells contain receptors coupled to ion transport for several local or systemic **hormones,** including catecholamines, ATP, and adrenocorticosteroid hormones (Table 2). Vasopressin (antidiuretic hormone) affects the EP and longitudinal K+ concentration gradient (Julien *et al.*, 1994; Mori

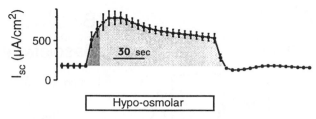

**FIG. 9.** Effect of a hypo-osmotic challenge from 300 to 150 mOsm on $I_{sc}$ in vestibular dark cells, the homologs to strial marginal cells. The dark shading indicates the increased amount of K+ secretion that is sufficient to account for cell volume regulation by release of osmolyte (K+). The subsequent sustained elevation of $I_{sc}$ (light shading) indicates that a hypotonic challenge causes an upregulation of the steady-state rate of K+ secretion. (Adapted with permission from Wangemann *et al.*, 1995c)

*et al.*, 1989), but the cells with vasopressin receptors have not yet been identified.

Perilymphatic perfusion of epinephrine causes no appreciable effect on the cochlear microphonic, a stimulus-dependent electrical response dependent on the EP (Klinke and Evans, 1977). However, basolateral perfusion *in vitro* of the β-adrenergic agonist isoproterenol caused maximal increases in $I_{sc}$ of 40–75% in strial marginal cells and vestibular dark cells (Wangemann and Liu, 1996). **β-Adrenergic receptors** are commonly coupled via G-proteins to adenylate cyclase (Chapter 8). Indeed, an increase of cytosolic cAMP in strial marginal cells and vestibular dark cells by direct stimulation of adenylate cyclase, by perfusion of a membrane-permeable **cAMP** analogue, or by inhibition of phosphodiesterases that catalyze the breakdown of cAMP all lead to an increase of $I_{sc}$ (Sunose *et al.*, 1997; Wangemann and Liu, 1996).

In isolated strial marginal cell epithelium, both apical and basolateral perfusion of micromolar **ATP** significantly alter $I_{sc}$ (Fig. 10) (Liu *et al.*, 1995). Apical ATP and analogues monotonically down-regulate $I_{sc}$, whereas basolateral ATP and analogues transiently increase $I_{sc}$ followed by down-regulation. The sequences of agonist effects are consistent with the presence of purinergic receptors of the

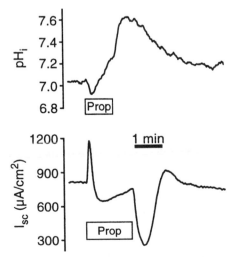

**FIG. 8.** Effect of propionate (Prop) on intracellular pH ($pH_i$) and short circuit current ($I_{sc}$) in vestibular dark cells, the homologs to strial marginal cells. Propionate caused a biphasic change in $pH_i$ that was mirrored in the $I_{sc}$. (Adapted with permission from Wangemann *et al.*, 1995b.)

**TABLE 2** Hormones and Signal Pathways Regulating IsK Current

| Hormone/signal | Stimulate | Inhibit |
|---|---|---|
| β-Adrenergic agonist basolateral | X | |
| Purinergic agonist | | |
| Apical | | X |
| Basolateral | Transient | X |
| cAMP | X | |
| Phospholipase C | | X |
| Protein kinase C | | X |
| Cytosolic Ca²⁺ | X | |
| Cytosolic pH | Transient | X |
| Cell swelling | X | |

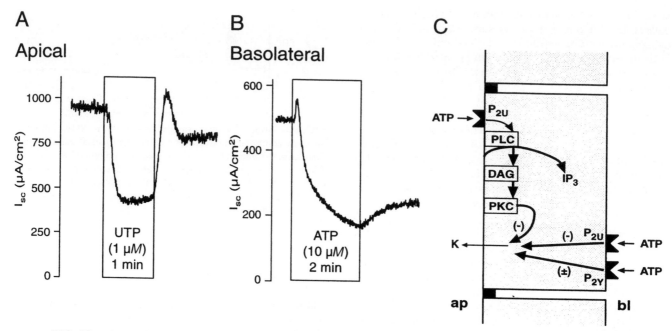

**FIG. 10.**  Regulation of K⁺ secretion by extracellular ATP in strial marginal cells. (A) Representative monophasic response of short circuit current ($I_{sc}$) to apical perfusion of UTP, agonist of the $P_{2Y2}$ ($P_{2U}$) receptor. (B) Representative biphasic response of $I_{sc}$ to basolateral perfusion of ATP, agonist for $P_2$ receptors. (C) Diagram of signal pathway between activation of apical $P_{2Y2}$ receptor and down-regulation of IsK channel and of $P_2$ receptor subtypes in basolateral membrane with effects of each on K⁺ secretion. (A and B, Adapted with permission from Liu *et al.*, 1995.)

$P_{2Y2}$ ($P_{2U}$) subtype on the apical membrane and coexisting populations of both $P_{2Y1}$ and $P_{2Y2}$ subtypes on the basolateral membrane. The transient increase in $I_{sc}$ from the basolateral side is due to activation of the $P_{2Y1}$ receptor and the subsequent decline due to both the $P_{2Y1}$ and $P_{2Y2}$ receptors. Both $P_{2Y}$ receptor subtypes are known in other cells to be coupled to G-proteins that stimulate phospholipase C (PLC). PLC catalyzes the breakdown of membrane phospholipid to produce both inositol trisphosphate (IP₃) and diacylglycerol (DAG) (Chapter 8). Although perilymphatic perfusion of high concentrations of ATP *in vivo* has little effect on the EP (Kujawa *et al.*, 1994), the basal cell layer can be expected to offer a significant barrier to diffusion from perilymph to the basolateral membrane of strial marginal cells.

It is most common that the IP₃ branch of the PLC pathway regulates ion channel activity through the cytosolic level of free $Ca^{2+}$. However, it was shown that the apical $P_{2Y2}$ receptor down-regulates K⁺ secretion primarily via the DAG–protein kinase C (PKC) branch (Marcus *et al.*, 1997) even though inositol phosphate production is increased in the cochlear lateral wall in response to $P_{2Y}$ agonists (Ogawa and Schacht, 1995). In fact, elevation of cytosolic $Ca^{2+}$ increased the IsK channel current while activation of PKC decreased the current.

Binding sites for both **glucocorticosteroid** (Ten Cate *et al.*, 1993) and **mineralocorticosteroid** (Yao and Rarey, 1996) have been demonstrated in the stria vascularis. Both corticoids control the activity of Na,K-ATPase in the stria vascularis (adrenalectomy reduced activity 60% and sys-

temic administration of either the glucocorticosteroid dexamethasone or the mineralocorticosteroid aldosterone restored activity) (Curtis *et al.*, 1993), and the increase in Na,K-ATPase with aldosterone was not dependent on major changes in blood plasma cation concentration (Ten Cate *et al.*, 1994). However, in spite of the strong dependence of the EP on strial Na,K-ATPase, a reduction of adrenocorticosteroids by adrenalectomy did not significantly reduce the EP in the presence or absence of strong acoustic stimulation (Ma *et al.*, 1995).

## IV.  Cell Physiology of Accoustic Transduction

The mammalian cochlea has an exquisite sensitivity, being able to detect sound pressure fluctuations of less than one millionth atmospheric pressure. At the threshold of hearing, the organ of Corti vibrates less than 1 nanometer. This sensitivity includes a gain of 100–1000 due to active mechanical amplification by cells of the organ of Corti, most likely the outer hair cells. In spite of the delicacy of this process, the dynamic range for the amplitude detected by the cochlea is about one million to one due to compression of the response (Patuzzi, 1996). The cellular processes responsible for transduction by the hair cells (Figs. 11 and 12) are described in this section.

### A.  Apical Membrane Channels

The mechanosensitive organelle of the sensory cells is the hair bundle, which consists of 30–300 **stereocilia** ar-

A

B

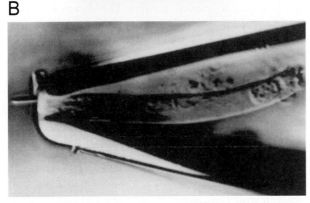

**FIG. 11.** Photomicrographs of hair cells: (A) inner hair cell and (B) outer hair cell. (Reproduced with permission from (A) Kros and Crawford, 1990, (B) Evans and Dallos, 1993. Copyright 1993 National Academy of Sciences, U.S.A.)

ranged in rows of increasing height. Motion of the hair bundle modulates the fractional open time (open probability $P_o$) of the **mechanosensitive transduction channels** near the tips of the stereocilia. Modulation of $P_o$ leads to corresponding fluctuations of the current through the apical membrane and to changes in the membrane potential of the hair cell. These changes in the membrane voltage are referred to as the **receptor potential,** the primary event to modulate synaptic transmission.

Bundle movement is believed to be the result of direct interaction of the OHC stereocilia with the tectorial membrane and of the IHC stereocilia with flowing endolymph pumped back and forth in the channel under the tectorial membrane. The cytoskeleton of the stereocilia is formed by a rigid matrix of actin filaments cross-linked by fimbrin. Deflection of the hair bundle results in a rotation of the individual cilia about a pivotal region at their base, which causes a shearing motion between the ciliary tips. Shear resulting from movement of the hair bundle in the direction of the tallest stereocilia increases $P_o$ of the transduction channels from the resting activity of about 5–15%, whereas movement of the hair bundle in the opposite direction decreases $P_o$ (Hudspeth and Gillespie, 1994). Motion in the stimulatory direction is caused by the rarefaction phase of the external acoustic pressure wave. The asymmetric position on the receptor potential–stimulus transfer function of the stereocilia under unstimulated conditions (zero displacement in Fig. 13) leads to larger changes in the receptor potential with a given stimulatory displacement

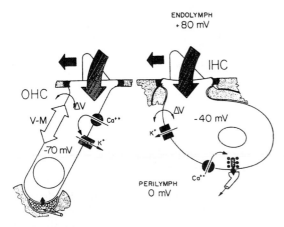

**FIG. 12.** Cellular processes of acoustic transducton in inner (IHC) and outer (OHC) hair cells. Shown are the influx of K$^+$ through the transduction channels in the stereocilia in response to acoustic stimulation (filled horizontal arrows), voltage-gated Ca$^{2+}$ channels in the basolateral membranes that initiate synaptic transmission in IHC and activate K$^+$ channels in OHC, and the voltage-driven motor elements (V-M) in the OHC. [Reproduced with permission from Dallos, P. (1992). The active cochlea. *J. Neurosci.* **12,** 4575–4585.]

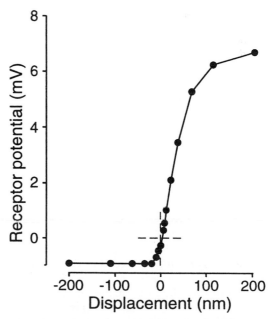

**FIG. 13.** Response function of the receptor potential of OHC to mechanical displacement of the stereocilia. The asymmetrical position of the quiescent point is indicated by dashed cross-hairs. [Adapted with permission from Russell, I. J., Koessl, M., and Richardson, G. P. (1992). Nonlinear mechanical responses of mouse cochlear hair bundles. *Proc. R. Soc. London Ser. B* **250,** 217–227.]

than for the same size inhibitory displacement. This asymmetry is a primary cause of distortion, which generates many of the nonlinear properties of the auditory system, such as distortion products and otoacoustic emissions (see later section on Reverse Transduction).

The shear between stereocilia is posited in the **gating-spring model** of hair cell function to transmit a mechanical force through an elastic element (the gating spring) to the mechanosensitive gate of each **transduction channel.** Increased tension on the spring increases $P_o$ and decreased tension reduces $P_o$. The most widely accepted hypothesis is that the fine extracellular links (called **tip links**) that run between the tips of shorter stereocilia and the sides of taller ones constitute the elastic element of the model. This hypothesis is supported by the shared sensitivity to $Ca^{2+}$ of the tip link structure, the transduction process, and the mechanical manifestation of the gating springs (Jaramillo, 1995).

The question of the **location of the transduction channels** has been approached by several means. The simplest and most widely accepted configuration is the presence of one channel at the insertional plaque of the tip link in the taller stereocilia of each pair. The entry of $Ca^{2+}$ into the cytosol of the stereocilia via transduction channels was visualized with fluorescent dyes and found to be slightly below the stereocilia tip where the tallest of the two stereocilia bridged by a tip link has its insertion point. However, $Ca^{2+}$ also enters many of the shortest stereocilia as well, suggesting that transduction channels may be located at both ends of the tip link (Denk *et al.,* 1995).

A different view is provided by evidence suggesting that the transduction channels may reside at yet another location. An antibody that recognizes an epitope on several amiloride-sensitive transport proteins binds to stereocilia not at the insertional plaque but at the point of contact between the shorter and taller stereocilia, a place below the insertion point of the tip link on the taller stereocilia (Hackney and Furness, 1995). The transduction channel is known to be inhibited by amiloride and is therefore likely the target of the antibody.

Electrophysiologic and micromechanical data from a variety of hair cells, both *in vivo* and *in vitro,* are best in accord with a model of the transduction channel in which there are transitions among one open state (O) and two closed states ($C_1$) and $C_2$) (Markin and Hudspeth, 1995):

$$C_1 \leftrightarrow C_2 \leftrightarrow O$$

and in which $Ca^{2+}$ is required at an intracellular site to promote the transition from $C_1$ to $C_2$ (Hackney and Furness, 1995). It has been proposed that there may also be a second open state (reviewed in Kros, 1996).

Tip links are thought to be maintained under tension (see next section on Adaptation), which is either increased or decreased by motion toward or away from the stimulatory direction. The time constant for response of the receptor current to a step stimulus has been found in saccular hair cells of the bullfrog to be in the range of 100–500 $\mu$s at 4°C and to become faster at higher temperatures (Corey and Hudspeth, 1983). This time course is substantially faster than can be accounted for by typical enzymatic or second messenger pathways and points to a direct mechanical coupling between the stimulus and the gate of the transduction channel.

The **conductance of the transduction channel** has been estimated from patch-clamp records to be on the order of 300 pS under *in vivo* conditions (Kros, 1996). The channel is permeable to cations including monovalent, divalent, and small organic cations such as tetraethylammonium while selecting against anions (Corey and Hudspeth, 1979). Because potassium and calcium in endolymph are above electrochemical equilibrium with respect to the hair cell cytosol, it is primarily these two cations that flow through the transduction channel. Potassium carries the bulk of the current due to its high concentration, whereas entry of $Ca^{2+}$ through the transduction channel is thought to control adaptation (see following section). The channel is known to be blocked by aminoglycoside antibiotics (multivalent cations) and amiloride (a blocker of several $Na^+$ transport processes).

### B.  Adaptation

When a sustained displacement is applied to hair cells, the transduction current is first stimulated but then relaxes in the continued presence of the stimulus with a characteristic time course, a process referred to as **adaptation** (Jaramillo, 1995; Hudspeth and Gillespie, 1994). The mechanism underlying this phenomenon is thought to be an active process in which myosin motors in the insertional plaque maintain tension within the tip link through a balance of climbing and slipping (Fig. 14). The tip link slackens when there is a deflection in the negative direction but tension is soon reestablished as the motor climbs along actin filaments within the taller stereocilium of each pair. Deflection in the positive direction increases tension in the tip links, opening the mechanosensitive transduction channel. $Ca^{2+}$ coming in through the channel is thought to bind to **calmodulin,** a myosin regulatory protein found to be concentrated in the tips of stereocilia, which in turn promotes the dissociation of myosin from the actin filaments, allowing the plaque with the motors to slip.

The **physiologic significance** of adaptation is clear for vestibular hair cells where prolonged static displacements naturally occur during which sensitivity to transient stimuli must be maintained. However, the function of adaptation in the mammalian cochlea is less obvious (Kros, 1996). First, no sustained displacements due to acoustic stimuli are expected and, second, the concentration of $Ca^{2+}$ in cochlear endolymph is about one-tenth that of vestibular endolymph, greatly reducing the expected influx through the transduction channels during stimulation. Third, the adaptation process has been estimated to have an upper frequency limit of about 160 Hz, limiting its potential usefulness as a response to acoustic stimuli. Physiological significance in the cochlea may lie more with homeostasis of the system in the face of small readjustments of the position and/or size of cells during normal changes in systemic osmolarity and local metabolic processes or during development.

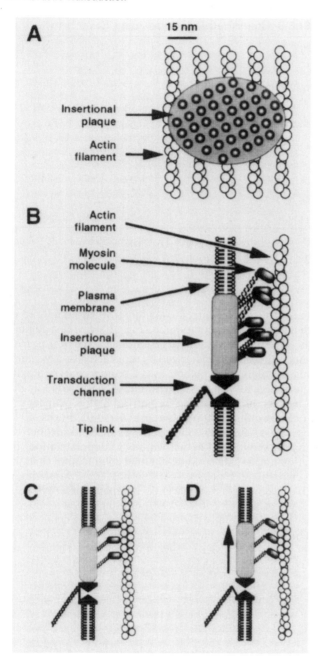

**FIG. 14.** Proposed mechanism of adaptation: (A) frontal view of insertional plaque on actin filaments; (B) side view of insertional plaque with associated mechanosensitive transduction channel in cell membrane; and (C and D) power stroke of actin–myosin complex. (Reproduced with permission from Hudspeth and Gillespie, 1994. Copyright 1994 by Cell Press.)

## C. Basolateral Membrane Channels

Although there are differences among hair cell types, they all contain a major basolateral **K⁺ conductance,** comprised of voltage-dependent and Ca²⁺-activated K⁺ channels, and **voltage-gated Ca²⁺ channels** (Kros, 1996; Fuchs, 1992). The K⁺ channels serve to shape the receptor poten-

tial initiated by the acoustically gated transduction channels and the Ca²⁺ channels to initiate transmitter release at the presynaptic sites, respectively. The presence of voltage-gated Ca²⁺ channels alone would be expected to lead to regenerative depolarization during acoustic stimulation, however, the basolateral voltage-gated K⁺ channels provide an active repolarizing influence.

The Ca²⁺ channels in hair cells appear to be of the L type, which are characterized by activation at depolarized membrane potentials, little or no inactivation, and block by dihydropyridines such as nifedipine. These channels differ slightly from the L-type current of neurons by a more negative voltage activation range and more rapid kinetics (Figs. 15A and B) (Fuchs, 1992). In **inner hair cells,** the Ca²⁺ currents are activated above −60 mV and peak near −20 to 0 mV (see Fig. 15A), a range that correlates well with that of receptor potentials found in recordings made *in vivo* and with predictions from *in vitro* experiments (Kros, 1996). These Ca²⁺ channels are therefore well poised to participate in the graded release of neurotransmitter at the basolateral synapses to type I afferent fibers in the auditory nerve. The kinetics of channel activation have been estimated to be sufficiently fast so as not to limit accurate following of nervous discharge rates to the acoustic stimulus (Kros, 1996). By contrast, the Ca²⁺ currents in **outer hair cells** are activated only above −30 mV (see Fig. 15B), a range that is far removed from the resting membrane potential near −80 mV found *in vivo* and never reached by receptor potentials, which apparently saturate at 15 mV (Kros, 1996). (The curve would be expected to shift slightly to the left at physiological perilymphatic Ca²⁺ levels, but the channel would still not likely show significant activation at voltages observed in OHC.) There is therefore no evidence at this time for a function of these voltage-gated Ca²⁺ channels in OHCs; there are also no measurements of neural activity in the type II afferent fibers in the auditory nerve that synapse exclusively to the OHCs.

The K⁺ channels in hair cells (Table 3) vary considerably among type and species. The cochlear hair cells of many nonmammalian vertebrates and the vestibular hair cells of mammals have Ca²⁺-activated K⁺ channels that act in synergy with the voltage-gated Ca²⁺ channels to **electrically resonate** at the characteristic frequency of the cell. This forms part of the basis of frequency discrimination in these species. Electrical tuning is not, however, significantly present in mammalian cochlear hair cells. Sharp tuning of mammalian IHCs *in vivo* is thought to be the result of active mechanical feedback from other cochlear elements such as OHCs (see later section on Reverse Transduction).

In inner hair cells, the K⁺ currents are activated by depolarization in the range of −60 to −20 mV, a range that correlates well with that of the resting and receptor potentials found in recordings made *in vivo* and with that of the voltage-gated Ca²⁺ channels described earlier. The resting potential of IHCs *in vivo* is about −40 mV, which reflects a compromise of the highly negative EMF from the basolateral K⁺ channels by the depolarized EMF of the nonselective cation conductance in the apical membrane (Dallos, 1986).

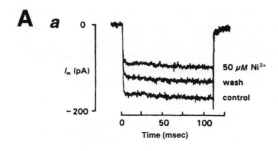

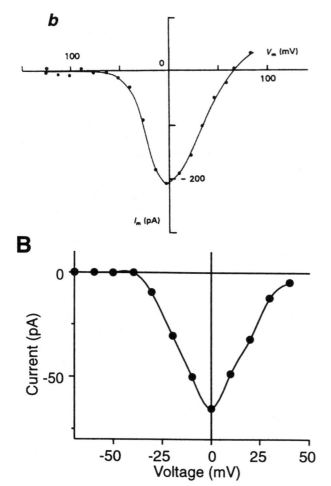

**FIG. 15.** Basolateral calcium currents (whole-cell recordings): (A) inner hair cells, representative traces (*a*) and current-voltage relationship (*b*) and (B) outer hair cells. [Reprinted from (A) Fuchs, P. A. *Prog. Neurobiol.* **39,** Ionic currents in cochlear hair cells, pp. 493–505. Copyright 1992 with kind permission from Elsevier Science Ltd, The Boulevard, Langford Lane, Kidlington OX51GB, UK. and (B) from Nakagawa, T., Kakehata, S., Akaike, N., Komune, S., Takasaka, T., and Uemura, T., *Neurosci. Lett.* **125,** Calcium channel in isolated outer hair cells of guinea pig cochlea, pp. 81–84. Copyright 1991 with kind permission of Elsevier Science–NL, Sara Burgerhartstraat 25, 1055 KV Amsterdam, The Netherlands.]

**TABLE 3**   Inner Hair Cell Potassium Channels

| Property | Fast K channel | Slow K channel |
|---|---|---|
| Activation time constant (msec) | 0.2–0.4 | 2–10 |
| Activation by membrane potential | Depol. | Depol. |
| Inactivation | No | No |
| Single-channel conductance | >200 pS | Smaller |
| Block by TEA | Yes | No |
| Block by CTX | Yes | Not tested |
| Block by 4-AP | No | Yes |
| Apamin | Not tested | No |

TEA, tetraethylammonium; CTX, charybdotoxin; 4-AP, 4-amino pyridine.

The total cell $K^+$ current was found to be composed of both a fast and slow component (Fig. 16). The fast component is carried by channels resembling *maxi K* (or *BK*) channels that are characterized by rapid activation by membrane depolarization, block by TEA and charybdotoxin, insensitivity to 4-amino pyridine (4-AP), and single-channel conductance with symmetrical high $K^+$ concentration of more than 200 pS (Gitter *et al.,* 1992; Fuchs, 1992; Kros, 1996). The slow component is a delayed rectifier $K^+$ channel exhibiting slow activation by depolarization, no inactivation, a smaller single-channel conductance than the fast component, block by 4-AP but not TEA. Surprisingly, both components are apparently independent of cytosolic $Ca^{2+}$ (Kros and Crawford, 1990) although they are highly sensitive to $Ca^{2+}$ in other types of hair cell (Fuchs, 1992).

The profile of $K^+$ currents in outer hair cells is different from that of IHCs (Kros, 1996; Fuchs, 1992). The primary $K^+$ conductance is due to $I_{K,n}$ channels, which are deacti-

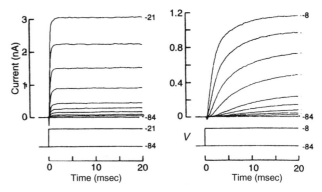

**FIG. 16.** Basolateral potassium currents (whole-cell recordings) from inner hair cells. Left panel: fast component; right panel: slow component. [Reproduced with permission from Kros, C. J., and Crawford, A. C. (1989). Components of the membrane current in guinea-pig inner hair cells. *In* "Cochlear Mechanisms. Structure, Function and Models" (J. P. Wilson and D. T. Kemp, Eds.), pp. 189–195. Plenum Press, New York.]

vated below −90 mV and fully activated at −50 mV, have a single-channel conductance of about 45 pS, are activated by $Ca^{2+}$ ($P_o$ = 0.5 at 0.2 $\mu M$), and are blocked by $Ba^{2+}$ (Mammano et al., 1995). There is also a $K^+$ current that activates at potentials more depolarized than −35 mV and with increasing cytosolic $Ca^{2+}$ activity ($P_o$ = 0.5 at 0.2 $\mu M$), has a single-channel conductance of >200 pS, is noninactivating, is inhibited by TEA, and makes only a very small contribution to the total $K^+$ conductance of the basolateral membrane of OHCs (Gitter et al., 1992; Housley and Ashmore, 1992). This channel has been found to also activate with membrane stretch and thus may play a role in mechanical feedback (see "Reverse transduction" below) (Iwasa et al., 1991). Both channels have been found to display these properties in OHCs in situ as well as in separated cells (Mammano et al., 1995).

In addition to these two $K^+$ currents, the basolateral membrane contains a nonselective cation conductance that is permeable to $Ca^{2+}$ as well as $Na^+$ and $K^+$ (Housley and Ashmore, 1992). Its function in vivo is not known but may be associated with a purinergic receptor (see later discussion). This conductance is clearly not dominant in vivo because the resting membrane potential [−70 to −80 mV (Dallos, 1986)] is not far from the Nernst potential for $K^+$.

### D. Synaptic Release of Vesicles

Depolarization by acoustic stimulation leads to $Ca^{2+}$-dependent release of neurotransmitter at the 30–40 synapses with the afferent fibers at the base of each IHC (Sewell, 1996). This process can be monitored in vitro as changes in **membrane capacitance** (Lindau and Neher, 1988). Opening of $Ca^{2+}$ channels during depolarization raises the local free $Ca^{2+}$ concentration to tens or hundreds of micromolar (Roberts, 1994), which is thought to trigger synaptic exocytosis (Chapter 40). Depolarization led to increases in membrane capacitance (up to 5% of the initial value) that continued for up to 2 sec. It was estimated that this was sufficient to exhaust more than five times the number of vesicles initially in close apposition to the plasma membrane at active zones, suggesting that hair cells are capable of rapidly replenishing vesicles at release sites (Parsons et al., 1994). Capacitance returned during repolarization toward its prestimulus level with a time constant of about 14 sec, but only in perforated-patch recordings (Chapter 25), suggesting that membrane retrieval depends on a diffusible intracellular factor.

### E. Reverse Transduction

Discrimination of frequency information in acoustic signals is performed broadly at an initial level by mechanical tuning of the basilar membrane as a function of position along the cochlea. In lower vertebrates, frequency discrimination is accomplished by tuning the electrical properties of the hair cells. The exquisite sensitivity and fine frequency discrimination of mammalian hearing depends on amplification of the incoming sound within the organ of Corti. It is widely held that the OHCs sense displacements caused by sound and feedback forces, which enhance the basilar membrane motion by reducing the inherent damping of the cochlear partition. In support of this hypothesis, OHCs show membrane potential-induced length changes at acoustic rates. This process has been termed **reverse transduction** by the **cochlear amplifier**. This property of the organ of Corti leads to an epiphenomenon called **otoacoustic emission** in which mechanical fluctuations are generated by the inner ear, resulting in sounds emanating from the ear (Brownell, 1990).

**Outer hair cells shorten and lengthen** by up to 5% when depolarized and hyperpolarized, respectively. Longitudinal motions of the OHCs are assumed to be transmitted in vivo to the whole organ of Corti by the rigid reticular lamina at the top and by the Deiters' cell bodies at the base, mechanically coupled to the basilar membrane. Shortening in vivo would cause an upward pull on the basilar membrane and/or a downward pull on the reticular lamina. The latency is less than 0.1 msec and can be driven to frequencies in excess of 22 kHz (measurement of upper frequency limited by instrumentation) (Dallos and Evans, 1995). This response is driven by membrane potential rather than ionic currents and is independent of ATP, ruling out commonly observed cytoskeletal motors (Alberts et al., 1994). It is widely believed that the motor molecules are voltage-sensitive proteins packed into the lateral membrane of the OHCs; activation of the motor proteins is thought to be associated with movements of gating charges (Fig. 17). This hypothesis is largely based on the observations that there is a congruence of the distributions of (1) densely packed particles, (2) membrane gating currents, (3) "nonlinear membrane capacitance," and (4) a force-generating membrane only on the lateral membrane between the level of the nucleus to near the cuticular plate. The presence of both gating current (Fig. 17) and nonlinear capacitance suggests that the motors act by conformational changes of the protein rather than by electro-

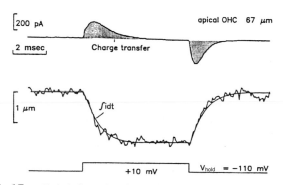

**FIG. 17.** Correlation of gating charge movements (upper panel) with displacement of stereocilia (lower panel). Integral of the gating charge has been scaled and is shown plotted over the displacement for comparison of the kinetics. (Reproduced from Ashmore, J. F. Neurosci. Res. Suppl. **12,** Forward and reverse transduction in the mammalian cochlea, S39–S50. Copyright 1990 with kind permission of Elsevier Science–NL, Sara Burgerhartstraat 25, 1055 KV Amsterdam, The Netherlands.)

static repulsion between separate elements within the plane of the lipid bilayer (Holley, 1996). Forces estimated from distortions of the organ of Corti *in situ* due to changes of membrane potential can account for the enhanced responses to sound required by the cochlear amplifier (Mammano *et al.*, 1995).

It was initially assumed that this mechanism would function by motile responses to the OHC receptor potential. It was recently shown from measurements of the amplitude and phase relationships of the receptor potential to stereociliary displacement that those data were consistent with *attenuation* rather than amplification of the stimulus at high frequencies (Preyer *et al.*, 1996). An alternate proposed source of voltage changes that might drive the lateral motors *in vivo* is extracellular potential gradients across the OHC (Dallos and Evans, 1995).

**Deiters' cells** were previously thought to offer only a passive support to the OHCs and a mechanical connection to the acoustic stimulus at the basilar membrane. Recent evidence, however, suggests that these cells have a **$Ca^{2+}$-dependent motile response.** Increases in cytoplasmic $Ca^{2+}$ caused an extension of the head of the phalangeal process away from the body by 0.5 to 1 $\mu$m within a few hundred milliseconds and the stiffness of the phalangeal process to increase by 28 to 51% (Dulon, 1995).

Sinusoidal fluctuations in membrane potential have been shown in lower vertebrate hair cells to cause **mechanical oscillations of the hair bundle,** but the operational frequency range is only from 20 to 320 Hz, possibly limited by the time constant of the basolateral membrane (Hudspeth and Gillespie, 1994). Myosin in the stereocilia is estimated to be sufficiently plentiful and powerful to account for the amplification produced by the organ of Corti. However, if the stereociliary motion were modulated by $Ca^{2+}$ rather than membrane potential, activation of stereociliar myosin by $Ca^{2+}$ would not be restricted by the membrane's time constant. It remains to be determined whether these motors can operate at the highest frequencies detected by the mammalian cochlea.

### F. Receptors

A number of receptors for neurotransmitters and neuromodulators have been identified on cochlear hair cells. Some are known to mediate synaptic transmission while the physiological significance of others has not yet been identified.

### 1. Outer Hair Cells

Receptors for several neurotransmitters and neuromodulators (Table 4) have been identified on outer hair cells by histochemical and electrophysiologic observations (Sewell, 1996). The efferent synapses on the outer hair cells release **acetylcholine** (ACh) and *γ-aminobutyric acid* (GABA) as neurotransmitters from two populations of fibers and may corelease ATP with ACh (Guinan, 1996; Sewell, 1996). There is evidence for the presence of both ionotropic and metabotropic receptors for ATP and ACh and for the ionotropic GABA$_A$ receptor (Housley *et al.*, 1995; Sewell, 1996). **Ionotropic receptors** are coupled directly to ion channels, whereas **metabotropic receptors** act via G-protein pathways that ultimately regulate ion channel activity and/or other cellular processes. It is not yet clear whether the responses to ACh occur via a single unusual receptor isoform containing $\alpha_9$-subunits or whether there are multiple isoforms of ACh receptors present in OHCs.

Activation of ACh receptors causes membrane hyperpolarization apparently through an ion channel permeable to cations including $Ca^{2+}$. The entry of $Ca^{2+}$ activates $Ca^{2+}$-dependent $K^+$ channels, leading to hyperpolarization. Ionotropic ATP receptors depolarize the hair cell by opening associated nonselective cation channels permeable to $Ca^{2+}$ as well as monovalent cations. Activation of the ionotropic GABA$_A$ receptor causes membrane hyperpolarization via a $Cl^-$-selective channel (see Chapter 39) (Plinkert *et al.*, 1993). The channels associated with the ionotropic ATP and ACh receptors admit $Ca^{2+}$ into the cell, leading to a relatively slow elongation (a process referred to as **slow motility**). This effect of $Ca^{2+}$ is mediated via calmodulin (Schacht *et al.*, 1995) and may involve phosphorylation of effector proteins by myosin light chain kinase and/or calmodulin-dependent protein kinase II (Puschner and Schacht, 1997). Activation of the GABA$_A$ receptor leads to a slow elongation of OHCs (Plinkert *et al.*, 1993) but the pathway leading to the motile response is not yet known.

In addition to their effects on slow motility, both ACh and GABA have been found to control the gain and magnitude of fast motility in OHCs (Sziklai *et al.*, 1996). ACh evokes an increase in magnitude and gain of electromotility, which is sensitive to the muscarinic blocker atropine. GABA causes the opposite effects and is sensitive to the GABA$_A$ receptor antagonist bicuculline methiodide (Sziklai *et al.*, 1996). The effects of ACh on electromotility

**TABLE 4** Effects of Outer Hair Cell Receptors on Fast and Slow Motility, Membrane Potential, and Calcium Concentration

|             | Fast | Slow | $V_c$  | [Ca]$_i$ |
|-------------|------|------|--------|----------|
| ACh         | x    | x    | Hyper  | x        |
| ATP         |      | x    | Depol. | x        |
| GABA        |      |      | Hyper  | No       |
| Substance P |      |      | Hyper  | No       |
| CGRP        |      |      |        | No, but potentiates ACh response |

were found to be likely mediated via phosphorylation by PKC of undetermined proteins (King *et al.,* 1996).

Receptors have also been found on OHCs for the peptides substance P and calcitonin gene-related peptide (CGRP). Substance P hyperpolarizes these cells by downregulating the nonselective cation conductance in the lateral wall via a pertusis toxin-insensitive G-protein (Kakehata *et al.,* 1993). CGRP had no effect on the resting cytoplasmic $Ca^{2+}$ concentration but potentiated by about threefold the increase in $Ca^{2+}$ caused by ACh stimulation in chick hair cells (Shigemoto and Ohmori, 1990).

## 2. Inner Hair Cells

Inner hair cells have been found to have both metabotropic and ionotropic receptors for extracellular ATP (Sugasawa *et al.,* 1996). At submicromolar ATP concentrations, the metabotropic receptors raise intracellular $Ca^{2+}$ concentration, which hyperpolarizes IHCs via $Ca^{2+}$-sensitive $K^+$ channels. The ionotropic receptors are nonselective cation channels activated at higher ATP concentrations and which mainly have a depolarizing effect on the IHCs. Although the source of agonist for these receptors has not yet been identified, $Ca^{2+}$-dependent release of ATP from the organ of Corti has been observed (Wangemann, 1996).

## G. Echolocation

Bats can capture flying prey by echolocation (biosonar) even among vegetation and other bats. Sounds are vocalized by the bat at frequencies near 60 kHz (the upper range of human hearing is about 20 kHz) and typically consist of a constant-frequency tone pulse for 10–100 msec followed by a shorter period of dropping frequency. The pulse is reflected from surfaces in front of the bat and is returned and heard before the end of the emitted pulse. Several characteristics of the returned sound pulse are interpreted by the bat as an acoustic image. A Doppler shift of the frequency of the returned sound from that of the emitted sound carries velocity information (the rate of closure on a target). For example, a Doppler shift from 61 to 62 kHz corresponds to 6.3 miles/hour. The delay of the echo gives the distance to the target such that a 1-msec echo delay corresponds to a target distance of 17 cm. Temporal delay acuity has been reported to be as fine as 10 nsec (Simmons *et al.,* 1990), a surprisingly short time span for recognition by a biological system. It has been suggested, however, that bats may not respond directly to processing of the time delay but rather to spectral processing, which would require only a resolution of about a few kilohertz (reviewed in Neuweiler and Schmidt, 1993, but see Simmons, 1993; Simmons *et al.,* 1996).

The fine tuning of hearing used in echolocation is mostly achieved by mechanical specializations of the cochlea, which spread out the physical mapping of the biosonar frequencies over a full half turn of the cochlea. There is also an amplifying reverberation of the acoustic wave traveling on the basilar membrane, which is due either to reflections at a discontinuity of the basilar membrane thickness or to different radial oscillation modes of the basilar membrane (Russell and Koessl, 1995). As in other mammalian cochleae, the tuning (and the bat's special reverberation) is thought to depend on active processes in the OHCs. Tuning of auditory nerve fibers, and therefore likely of basilar membrane motion, to acoustic stimuli is amazingly high in the bat with Q10 values (center frequency/bandwidth 10 dB from the tip) as high as 610 in comparison to typical values of 1 to 10 for nonecholocating animals (Russell and Koessl, 1995). Processing by central neural pathways of the detected echos has been reviewed by Suga (1989), Suga *et al.,* (1995), Simmons (1989), and Simmons *et al.,* (1996). An interesting specialization of the bats' prey has been found in the dogbane tiger moth, which apparently emits clicks when a bat approaches in order to interfere with the sonar echos, thereby "jamming" the signal and averting becoming a meal (Fullard *et al.,* 1994)!

## Summary

The cellular bases of hearing in the auditory periphery were described with emphasis on the sensory hair cells and on cells of the stria vascularis. The latter is a complex structure responsible both for the generation of a lumen-positive electrical potential and for the secretion of potassium to an unusually high level in the cochlear lumen. Transduction of sound waves into neural impulses by the hair cells relies on modulation of an electric current carried by potassium flowing through the hair cells. The ion channels, carriers, and pumps in the marginal cell membranes, which account for constitutive potassium secretion, were described in this chapter. A wide variety of extracellular and cytosolic signaling pathways that regulate the rate of potassium secretion were also described. The specializations of hair cell physiology were discussed, including the transduction channel in the sensory cilia and the highly unusual lateral membrane motors of OHCs. The description of the physiological mechanisms employed by inner ear epithelial cells was related in this chapter to the function of the organ as a whole.

## Bibliography

Alberts, B., Bray, D., Lewis, J., Raff, M., Roberts, K., and Watson, J. D. (1994). "Molecular Biology of the Cell," 3rd ed. Garland Publishing, New York.

Barhanin, J., Lesage, F., Guillemare, E., Fink, M., Lazdunski, M., and Romey, G. (1996). $K_v$LQT1 and IsK (minK) proteins associate to form the $I_{Ks}$ cardiac potassium current. *Nature* **384,** 78–80.

Ben-Efraim, I., Shai, Y., and Attali, B. (1996). Cytoplasmic and extracellular IsK peptides activate endogenous $K^+$ and $Cl^-$ channels in *Xenopus* oocytes—Evidence for regulatory function. *J. Biol. Chem.* **271,** 8768–8771.

Brownell, W. E. (1990). Outer hair cell electromotility and otoacoustic emissions. *Ear Hearing* **11,** 82–92.

Corey, D. P., and Hudspeth, A. J. (1979). Ionic basis of the receptor potential in a vertebrate hair cell. *Nature* **281,** 675–677.

Corey, D. P., and Hudspeth, A. J. (1983). Kinetics of the receptor current in bullfrog saccular hair cells. *J. Neurosci.* **3,** 962–976.

Curtis, L. M., Ten Cate, W. J., and Rarey, K. E. (1993). Dynamics of Na,K-ATPase sites in lateral cochlear wall tissues of the rat. *Eur. Arch. Otorhinolaryngol.* **250**, 265–270.

Dallos, P. (1986). Neurobiology of cochlear inner and outer hair cells: Intracellular recordings. *Near. Res.* **22**, 185–198.

Dallos, P., and Evans, B. N. (1995). High-frequency motility of outer hair cells and the cochlear amplifier. *Science* **267**, 2006–2009.

Denk, W., Holt, J. R., Shepherd, G. M., and Corey, D. P. (1995). Calcium imaging of single stereocilia in hair cells: Localization of transduction channels at both ends of tip links. *Neuron* **15**, 1311–1321.

Dulon, D. (1995). Ca$^{2+}$ signaling in Deiters' cells of the guinea-pig cochlea: Active process in supporting cells? *In* "Active Hearing" (A. Flock, D. Ottoson, and M. Ulfendahl, Eds.), pp. 195–208. Elsevier Science, New York.

Evans, B. N., and Dallos, P. (1993). Stereocilia displacement induced somatic motility of cochlear outer hair cells. *Proc. Natl. Acad. Sci. USA* **90**, 8347–8351.

Fuchs, P. A. (1992). Ionic currents in cochlear hair cells. *Prog. Neurobiol.* **39**, 493–505.

Fullard, J. H., Simmons, J. A., and Saillant, P. A. (1994). Jamming bat echolocation: The dogbane tiger moth *Cycnia tenera* times its clicks to the terminal attack calls of the big brown bat *Eptesicus fuscus. J. Exp. Biol.* **194**, 285–298.

Gitter, A. H., Froemter, E., and Zenner, H. P. (1992). C-type potassium channels in the lateral cell membrane of guinea-pig outer hair cells. *Hear. Res.* **60**, 13–19.

Guinan, J. J., Jr. (1996). Physiology of olivocochlear efferents. *In* "The Cochlea" (P. Dallos, A. N. Popper, and R. R. Fay, Eds.), pp. 435–502. Springer-Verlag, New York.

Hackney, C. M., and Furness, D. N. (1995). Mechanotransduction in vertebrate hair cells: Structure and function of the stereociliary bundle. *Am. J. Physiol.* **268**, C1–13.

Holley, M. C. (1996). Outer hair cell motility. *In* "The Cochlea" (P. Dallos, A. N. Popper, and R. R. Fay, Eds.), pp. 386–434. Springer-Verlag, New York.

Housley, G. D., and Ashmore, J. F. (1992). Ionic currents of outer hair cells isolated from the guinea-pig cochlea. *J. Physiol. (London)* **448**, 73–98.

Housley, G. D., Connor, B. J., and Raybould, N. P. (1995). Purinergic modulation of outer hair cell electromotility. *In* "Active Hearing" (A. Flock, D. Ottoson, and M. Ulfendahl, Eds.), pp. 221–238. Elsevier Science, New York.

Hudspeth, A. J., and Gillespie, P. G. (1994). Pulling springs to tune transduction: Adaptation by hair cells. *Neuron* **12**, 1–9.

Iwasa, K. H., Li, M., Jia, M., and Kachar, B. (1991). Stretch sensitivity of the lateral wall of the auditory outer hair cell from the guinea pig. *Neurosci. Lett.* **133**, 171–174.

Jaramillo, F. (1995). Signal transduction in hair cells and its regulation by calcium. *Neuron* **15**, 1227–1230.

Julien, N., Loiseau, A., Sterkers, O., Amiel, C., and Ferrary, E. (1994). Antidiuretic hormone restores the endolymphatic longitudinal K$^+$ gradient in the Brattleboro rat cochlea. *Pflugers Archiv.* **426**, 446–452.

Kakehata, S., Akaike, N., and Takasaka, T. (1993). Substance P decreases the non-selective cation channel conductance in dissociated outer hair cells of guinea pig cochlea. *Ann. NY Acad. Sci.* **707**, 476–479.

Kikuchi, T., Kimura, R. S., Paul, D. L., and Adams, J. C. (1995). Gap junctions in the rat cochlea: Immunohistochemical and ultrastructural analysis. *Anat. Embryol. (Berlin)* **191**, 101–118.

King, B. F., Wang, S. Y., and Burnstock, G. (1996). P$_2$ purinoceptor-activated inward currents in follicular oocytes of *Xenopus laevis. J. Physiol. (London)* **494**, 17–28.

Klinke, R., and Evans, E. F. (1977). Evidence that catecholamines are not the afferent transmitter in the cochlea. *Exp. Brain Res.* **28**, 315–324.

Koefoed-Johnsen, V., and Ussing, H. H. (1958). The nature of the frog skin potential. *Acta Physiol. Scand.* **42**, 298–308.

Konishi, T., Hamrick, P. E., and Walsh, P. J. (1978). Ion transport in guinea pig cochlea. I. Potassium and sodium transport. *Acta Otolaryngol. (Stockholm)* **86**, 22–34.

Kros, C. J. (1996). Physiology of mammalian cochlear hair cells. *In* "The Cochlea" (P. Dallos, A. N. Popper, and R. R. Fay, Eds.), pp. 318–385. Springer-Verlag, New York.

Kros, C. J., and Crawford, A. C. (1990). Potassium currents in inner hair cells isolated from the guinea-pig cochlea. *J. Physiol. (London)* **421**, 263–291.

Kujawa, S. G., Erostegui, C., Fallon, M., Crist, J., and Bobbin, R. P. (1994). Effects of adenosine 5′-triphosphate and related agonists on cochlear function. *Hear. Res.* **76**, 87–100.

Lindau, M. and Neher, E. (1988). Patch-clamp techniques for time-resolved capacitance measurements in single cells. *Pflugers Archiv.* **411**, 137–146.

Liu, J., Kozakura, K., and Marcus, D. C. (1995). Evidence for purinergic receptors in vestibular dark cell and strial marginal cell epithelia of the gerbil. *Auditory Neurosci.* **1**, 331–340.

Ma, Y. L., Gerhardt, K. J., Curtis, L. M., Rybak, L. P., Whitworth, C., and Rarey, K. E. (1995). Combined effects of adrenalectomy and noise exposure on compound action potentials, endocochlear potentials and endolymphatic potassium concentrations. *Hear. Res.* **91**, 79–86.

Mammano, F., Kros, C. J., and Ashmore, J. F. (1995). Patch clamped responses from outer hair cells in the intact adult organ of Corti. *Pflugers Archiv.* **430**, 745–750.

Marcus, D. C., and Shen, Z. (1994). Slowly activating, voltage-dependent K$^+$ conductance is apical pathway for K$^+$ secretion in vestibular dark cells. *Am. J. Physiol.* **267**, C857–C864.

Marcus, D. C., Sunose, H., Liu, J., Shen, Z., and Scofield, M. A. (1997). P$_{2U}$ purinergic receptor inhibits apical IsK/KvLQT1 channel via protein kinase C in vestibular dark cells. *Am. J. Physiol.* in press.

Markin, V. S., and Hudspeth, A. J. (1995). Gating-spring models of mechanoelectrical transduction by hair cells of the internal ear. *Annu. Rev. Biophys. Biomol. Struct.* **24**, 59–83.

Mori, N., Shugyo, A., and Asai, H. (1989). The effect of arginine-vasopressin and its analogues upon the endocochlear potential in the guinea pig. *Acta Otolaryngol. (Stockholm)* **107**, 80–84.

Neuweiler, G., and Schmidt, S. (1993). Audition in echolocating bats. *Curr. Opin. Neurobiol.* **3**, 563–569.

Ogawa, K., and Schacht, J. (1995). P2y purinergic receptors coupled to phosphoinositide hydrolysis in tissues of the cochlear lateral wall. *Neuroreport* **6**, 1538–1540.

Parsons, T. D., Lenzi, D., Almers, W., and Roberts, W. M. (1994). Calcium-triggered exocytosis and endocytosis in an isolated presynaptic cell: Capacitance measurements in saccular hair cells. *Neuron* **13**, 875–883.

Patuzzi, R. (1996). Cochlear micromechanics and macromechanics. *In* "The Cochlea" (P. Dallos, A. N. Popper, and R. R. Fay, Eds.), pp. 186–257. Springer-Verlag, New York.

Plinkert, P. K., Gitter, A. H., Mohler, H., and Zenner, H. P. (1993). Structure, pharmacology and function of GABA-A receptors in cochlear outer hair cells. *Eur. Arch. Otorhinolaryngol.* **250**, 351–357.

Preyer, S., Renz, S., Hemmert, W., Zenner, H. P., and Gummer, A. W. (1996). Receptor potential of outer hair cells isolated from base to apex of the adult guinea-pig cochlea: Implications for cochlear tuning mechanisms. *Auditory Neurosci.* **2**, 145–157.

Puschner, B., and Schacht, J. (1997). Slow motility in outer hair cells is regulated by calmodulin-dependent protein kinases. *Assoc. Res. Otolaryngol.* **20**, 74.

Roberts, W. M. (1994). Localization of calcium signals by a mobile calcium buffer in frog saccular hair cells. *J. Neurosci.* **14**, 3246–3262.

Russell, I. J., and Koessl, M. (1995). Measurements of the basilar membrane resonance in the cochlea of the mustached bat. *In* "Active Hearing" (A. Flock, D. Ottoson, and M. Ulfendahl, Eds.), pp. 295–306. Elsevier Science, New York.

Salt, A. N., Melichar, I., and Thalmann, R. (1987). Mechanisms of endocochlear potential generation by stria vascularis. *Laryngoscope* **97**, 984–991.

Sanguinetti, M. C., Curran, M. E., Zou, A., Shen, J., Spector, P. S., Atkinson, D. L., and Keating, M. T. (1996). Coassembly of K$_v$LQT1 and minK (IsK) proteins to form cardiac $I_{Ks}$ potassium channel. *Nature* **384**, 80–83.

Schacht, J., Fessenden, J. D., and Zajic, G. (1995). Slow motility of outer hair cells. *In* "Active Hearing" (A. Flock, D. Ottoson, and M. Ulfendahl, Eds.), pp. 209–220. Elsevier Science, New York.

Sewell, W. F. (1996). Neurotransmitters and synaptic transmission. *In* "The Cochlea" (P. Dallos, A. N. Popper, and R. R. Fay, Eds.), pp. 503–533. Springer-Verlag, New York.

Shah, D. M., Freeman, D. M., and Weiss, T. F. (1995). The osmotic response of the isolated, unfixed mouse tectorial membrane to isosmotic solutions: Effect of Na$^+$, K$^+$, and Ca$^{2+}$ concentration. *Hear. Res.* **87**, 187–207.

Shen, Z., Marcus, D. C., Sunose, H., Chiba, T., and Wangemann, P. (1997). I$_{sK}$ channel in strial marginal cells: Voltage-dependence, ion-selectivity, inhibition by 293B and sensitivity to clofilium. *Auditory Neurosci.* **3**, 215–230.

Shiga, N., and Wangemann, P. (1995). Ion selectivity of volume regulatory mechanisms present during a hypoosmotic challenge in vestibular dark cells. *Biochim. Biophys. Acta* **1240**, 48–54.

Shigemoto, T., and Ohmori, H. (1990). Muscarinic agonists and ATP increase the intracellular Ca$^{2+}$ concentration in chick cochlear hair cells. *J. Physiol. (London)* **420**, 127–148.

Shimozono, M., Liu, J., and Wangemann, P. (1996). Vestibular dark cells contain a H$^+$/monocarboxylate$^-$ cotransporter in their basolateral membrane. *J. Gen. Physiol.* **108**, 18a.

Simmons, J. A. (1989). A view of the world through the bat's ear: The formation of acoustic images in echolocation. *Cognition* **33**, 155–199.

Simmons, J. A. (1993). Evidence for perception of fine echo delay and phase by the FM bat, *Eptesicus fuscus. J. Comp. Physiol. A* **172**, 533–547.

Simmons, J. A., Dear, S. P., Ferragamo, M. J., Haresign, T., and Fritz, J. (1996). Representation of perceptual dimensions of insect prey during terminal pursuit by echolocating bats. *Biol. Bull.* **191**, 109–121.

Simmons, J. A., Ferragamo, M., Moss, C. F., Stevenson, S. B., and Altes, R. A. (1990). Discrimination of jittered sonar echoes by the echolocating bat, *Eptesicus fuscus:* The shape of target images in echolocation. *J. Comp. Physiol. A* **167**, 589–616.

Sterkers, O., Saumon, G., Tran Ba Huy, P., and Amiel, C. (1982). K, Cl, and H$_2$O entry in endolymph, perilymph, and cerebrospinal fluid of the rat. *Am. J. Physiol.* **243**, F173–F180.

Suga, N. (1989). Principles of auditory information-processing derived from neuroethology. *J. Exp. Biol.* **146**, 277–286.

Suga, N., Butman, J. A., Teng, H., Yan, J., and Olsen, J. F. (1995). Neural processing of target-distance information in the mustached bat. *In* "Active Hearing" (A. Flock, D. Ottoson, and M. Ulfendahl, Eds.), pp. 13–30. Elsevier Science, New York.

Sugasawa, M., Erostegui, C., Blanchet, C., and Dulon, D. (1996). ATP activates non-selective cation channels and calcium release in inner hair cells of the guinea-pig cochlea. *J. Physiol. (London)* **491**, 707–718.

Sunose, H., Liu, J., Shen, Z., and Marcus, D. C. (1997). cAMP increases apical I$_{sK}$ channel current and K$^+$ secretion in vestibular dark cells. *J. Membr. Biol.* **156**, 25–35.

Sziklai, I., He, D. Z. Z., and Dallos, P. (1996). Effect of acetylcholine and GABA on the transfer function of electromotility in isolated outer hair cells. *Hear. Res.* **95**, 87–99.

Takeuchi, S., and Irimajiri, A. (1996). Maxi-K$^+$ channel in plasma membrane of basal cells dissociated from the stria vascularis of gerbils. *Hear. Res.* **95**, 18–25.

Takumi, T., Ohkubo, H., and Nakanishi, S. (1988). Cloning of a membrane protein that induces a slow voltage-gated potassium current. *Science* **242**, 1042–1045.

Ten Cate, W. J., Curtis, L. M., and Rarey, K. E. (1994). Effects of low-sodium, high-potassium dietary intake on cochlear lateral wall Na$^+$,K$^+$-ATPase. *Eur. Arch. Otorhinolaryngol.* **251**, 6–11.

Ten Cate, W. J., Curtis, L. M., Small, G. M., and Rarey, K. E. (1993). Localization of glucocorticoid receptors and glucocorticoid receptor mRNAs in the rat cochlea. *Laryngoscope* **103**, 865–871.

Tzounopoulos, T., Maylie, J., and Adelman, J. P. (1995). Induction of endogenous channels by high levels of heterologous membrane proteins in *Xenopus* oocytes. *Biophys. J.* **69**, 904–908.

Vetter, D. E., Mann, J. R., Wangemann, P., Liu, J., McLaughlin, K. J., Lesage, F., Marcus, D. C., Lazdunski, M., Heinemann, S. F., and Barhanin, J. (1996). Inner ear defects induced by null mutation of the *isk* gene. *Neuron* **17**, 1251–1264.

Wang, K. W., and Goldstein, S. A. (1995). Subunit composition of minK potassium channels. *Neuron* **14**, 1303–1309.

Wangemann, P. (1995). Comparison of ion transport mechanisms between vestibular dark cells and strial marginal cells. *Hear. Res.* **90**, 149–157.

Wangemann, P. (1996). Ca$^{2+}$-dependent release of ATP from the organ of Corti measured with a luciferin-luciferase bioluminescence assay. *Auditory Neurosci.* **2**, 187–192.

Wangemann, P. (1997). Kalium-Ionensekretion und Entstehung des endokochlearen Potenials in der Stria vascularis. *HNO* **45**, 205–209.

Wangemann, P., and Liu, J. (1996). Beta-adrenergic receptors but not vasopressin-receptors stimulate the equivalent short circuit current in K$^+$ secreting inner ear epithelial cells. *J. Gen. Physiol.* **108**, 31a.

Wangemann, P. and Schacht, J. (1996). Homeostatic mechanisms in the cochlea. *In* "The Cochlea" (P. Dallos, A. N. Popper, and R. R. Fay, Eds.), pp. 130–185. Springer-Verlag, New York.

Wangemann, P. and Shiga, N. (1994). Cell volume control in vestibular dark cells during and after a hyposmotic challenge. *Am. J. Physiol.* **266**, C1046–C1060.

Wangemann, P., Liu, J., and Marcus, D. C. (1995a). Ion transport mechanisms responsible for K$^+$ secretion and the transepithelial voltage across marginal cells of stria vascularis *in vitro. Hear. Res.* **84**, 19–29.

Wangemann, P., Liu, J., and Shiga, N. (1995b). The pH-sensitivity of transepithelial K$^+$ transport in vestibular dark cells. *J. Membr. Biol.* **147**, 255–262.

Wangemann, P., Liu, J., Shen, Z., Shipley, A., and Marcus, D. C. (1995c). Hypo-osmotic challenge stimulates transepithelial K$^+$ secretion and activates apical I$_{sK}$ channel in vestibular dark cells. *J. Membr. Biol.* **147**, 263–273.

Wangemann, P., Liu, J. Z., and Shiga, N. (1996a). Vestibular dark cells contain the Na$^+$/H$^+$ exchanger NHE-1 in the basolateral membrane. *Hear. Res.* **94**, 94–106.

Wangemann, P., Shen, Z., and Liu, J. (1996b). K$^+$-induced stimulation of K$^+$ secretion involves activation of the I$_{sK}$ channel in vestibular dark cells. *Hear. Res.* **100**, 201–210.

Yao, X. F., and Rarey, K. E. (1996). Localization of the mineralocorticoid receptor in rat cochlear tissue. *Acta Otolaryngol. (Stockholm)* **116**, 493–496.

Zidanic, M., and Brownell, W. E. (1990). Fine structure of the intracochlear potential field. I. The silent current. *Biophys. J.* **57**, 1253–1268.

# Appendix: Self-Referencing Electrodes for the Measurement of Extracellular Potential and Chemical Gradients

Many physiological processes produce extracellular gradients of electric fields, ions, nutrients, and respiratory gases. Such processes include constitutive ion transport and aerobic metabolism, as well as mechanisms involved in development and neural activity. In many cases these processes can be characterized through measurements of the gradients under control and experimental conditions. In many bulk tissues, ion and oxygen fluxes can be measured with radiotracers and/or static electrodes, and ion movements in single cells can often be monitored with fluorescence microscopy and patch-clamp electrodes. There are, however, a wide variety of situations to which these techniques cannot be applied and for which the self-referencing electrode is a powerful addition to the physiologist's arsenal of techniques. Self-referencing electrodes (Smith, 1995; Smith *et al.,* 1994) are capable of making noninvasive measurements from preparations of otherwise intractable geometries (irregular shapes and/or extremely small size) and are therefore ideal for measuring fluxes from inner ear epithelia that occur in patches with dimensions on the order of 0.1 mm. Unlike most other electrophysiological approaches, including patch-clamp recordings, ion fluxes can be observed with self-referencing ion-selective probes from electroneutral as well as electrogenic transport processes.

The purpose of self-referencing the probes is to average out noise and reduce the impact of the drift normally associated with microsensing electrodes by creating a periodic signal from a source that is intrinsically static or varying only slowly and irregularly. The first such probe measurements were made of current density by detection of the associated voltage drop as the current flowed through a resistive medium, the physiological saline in which the preparation was bathed (Jaffe and Nuccitelli, 1974). High-sensitivity voltage gradient measurements are made by vi-

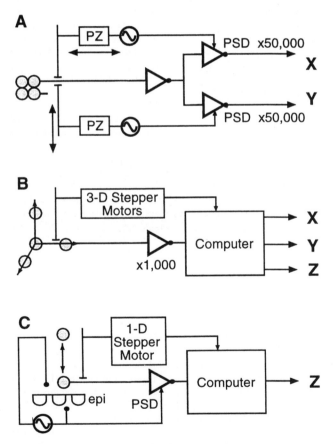

**FIG. A1.** Self-referencing probes. (A) Vibrating current probe driven in two orthogonal directions by piezoelectric (PZ) elements; signal is measured with a phase-sensitive detector (PSD). (B) Ion-selective noninvasive probe driven by stepper motors. (C) System for mapping transepithelial conductance with a combination of stepper motors and PSDs.

brating a metal electrode at frequencies between 100 and 1000 Hz, with signal analysis by a phase-sensitive detector (lock-in amplifier) for each plane of vibration (Fig. A1A) (Scheffey, 1989; Jaffe and Nuccitelli, 1974). In contrast to static electrodes, which intrinsically have a noise level and stability on the order of 1 mV, the vibrating current probe can measure signals in the *nanovolt* range. The electrode is typically made from a platinum wire that has been tapered to a point and insulated to within a few micrometers of the tip. The tip is electroplated to form a "fuzzy" surface of platinum black, which greatly increases the capacitance by expanding the area of electrical contact with the bathing solution. This large contact area gives good capacitive coupling (low impedance) between the bath and electrode at the frequencies used. The electrode can be vibrated in two orthogonal directions at different frequencies over an excursion of about 20–40 $\mu$m (Fig. A1A). The fine spatial resolution of the probe tip (down to approximately 4–8 $\mu$m with fine-tipped electrodes) can be exploited to map the spatial distribution of current-producing processes on large and robust single cells or tissues; two-dimensional vectors representing the current flows can be constructed.

A second method (Fig. A1B) was developed for the measurement of chemical gradients using more slowly responding sensors such as ion-selective microelectrodes (Kühtreiber and Jaffe, 1990). An electrode is stepped between two or more positions, residing at each for about 1 sec per position. Each position is between ~2–50 $\mu$m apart depending on the experiment. Motion of the electrode between the positions disturbs the gradient being measured, so a waiting period is introduced before the signal is collected at each position after a movement, followed by a period over which the signal is collected and subsequently

averaged. The averaged signal from each position is subtracted from that at the "origin." Either one-, two-, or three-dimensional paths can be defined with a trade-off between spatial and temporal resolution (Shi and Borgens, 1995). A three-dimensional path is shown in the figure. Noise reduction by this technique permits measurements of voltage gradients in the *microvolt* range.

A third method (Fig. A1C) combines elements of the first two and is applicable to quantifying variations in transmural epithelial conductance by observing the modulation by the tissue of an imposed sinusoidal electric field (Koeckerling *et al.,* 1993). The sensor is stepped between two positions at a slow rate as in the second method. The noise reduction of the signal at each position, however, is performed by oscillating the imposed electric field at a fixed frequency and passing the signal from the sensor through a phase-sensitive detector (lock-in amplifier) referenced to that frequency. This technique is reported to give a resolution of better than 0.5 $\mu$V.

These techniques have been applied to a wide variety of biologic systems (Nuccitelli, 1990). Early measurements focused on mapping electric currents and ion gradients around eggs and in the vicinity of plant rootlets and pollen tubes (Anderson *et al.,* 1994; Pierson *et al.,* 1994; Kochian *et al.,* 1992). They have also been applied to the measurements of currents and ion fluxes across epithelia either with heterogeneous cell populations (Breton *et al.,* 1996; Durham *et al.,* 1989) or small homogeneous domains, such as the epithelia regulating the composition of endolymph in the inner ear (Chapter 45 and Fig. A2). The probes have been found useful in studies of cell development and wound healing (Shi and Borgens, 1995; Hotary and Robinson, 1992; Borgens *et al.,* 1977) and in localizing currents in

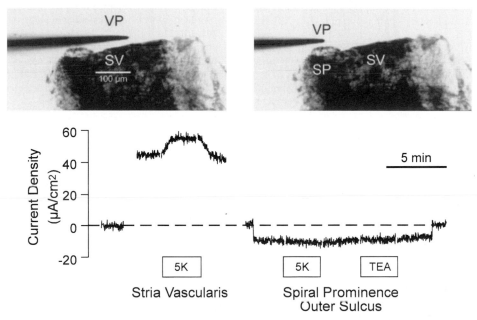

**FIG. A2.** Current density near stria vascularis (SV) and spiral prominence (SP). Upper panels: Vibrating probe (VP) near SV (*left*) and SP (*right*). Lower panel: Recording of current density showing outward (positive) current from SV and its sensitivity to change of bath $K^+$ concentration from 3.6 to 5m$M$ (*left*) and the inward (negative) current from SP and its insensitivity to elevated $K^+$ concentration and to the maxi-K channel blocker tetraethylammonium (TEA).

neurons (Duthie *et al.*, 1994). The probes can also be used in combination with other physiologic measurement techniques such as transepithelial voltage clamping, single-cell patch-clamp recording, and intracellular fluorescence microscopy (Foskett and Machen, 1985; Scheffey *et al.*, 1991; P. J. S. Smith, unpublished observations). With the recent addition of polarographic detection systems to the repertoire of self-referencing probes, measurements of minute oxygen gradients near respiring cells and tissues can now be recorded.

Resources are available for investigators interested in exploring the feasibility of applying these techniques to their own preparation. Self-referencing probe technology was originally developed at the National Vibrating Probe Facility (Dr. Lionel Jaffe, director) at the Marine Biological Laboratory (MBL), Woods Hole, Massachusetts. The successor to this group, the BioCurrents Research Center (Peter J. S. Smith, director), is a facility of the National Center for Research Resources (National Institutes of Health) and can host visiting scientists by application to the director (see the MBL web page at http://www.mbl.edu). Another laboratory providing access to probe technology is the University of Massachusetts Vibrating Probe Facility, directed by Dr. Joseph G. Kunkel. An internet web site for this facility is at http://www.bio.umass.edu/biology/kunkel/.

## Bibliography

Anderson, M., Bowdan, E., and Kunkel, J. G. (1994). Comparison of defolliculated oocytes and intact follicles of the cockroach using the vibrating probe to record steady currents. *Dev. Biol.* **162,** 111–122.

Borgens, R. B., Vanable, J. W., Jr., and Jaffe, L. F. (1977). Bioelectricity and regeneration: II. Large currents leave the stumps of regenerating newt limbs. *Proc. Natl. Acad. Sci. USA* **74,** 4528–4532.

Breton, S., Smith, P. J., Lui, B., and Brown, D. (1996). Acidification of the male reproductive tract by a proton pumping (H$^+$)-ATPase. *Nat. Med.* **2,** 470–472.

Durham, J. H., Shipely, A., and Scheffey, C. (1989). Vibrating probe localization of acidification current to minority cells of the turtle bladder. Ann. NY Acad. Sci. **574,** 486–488.

Duthie, G. G., Shipley, A., and Smith, P. J. (1994). Use of a vibrating electrode to measure changes in calcium fluxes across the cell membranes of oxidatively challenged Aplysia nerve cells. *Free Radic. Res.* **20,** 307–313.

Foskett, J. K., and Machen, T. E. (1985). Vibrating probe analysis of teleost opercular epithelium: Correlation between active transport and leak pathways of individual chloride cells. *J. Membrane Biol.* **85,** 25–35.

Hotary, K. B., and Robinson, K. R. (1992). Evidence of a role for endogenous electrical fields in chick embryo development. *Development* **114,** 985–996.

Jaffe, L. F., and Nuccitelli, R. (1974). An ultrasensitive vibrating probe for measuring steady extracellular currents. *J. Cell Biol.* **63,** 614–628.

Kochian, L. V., Shaff, J. E., Kühtreiber, W. M., Jaffe, L. F., and Lucas, W. J. (1992). Use of an extracellular, ion-selective, vibrating microelectrode system for the quantification of K$^+$, H$^+$, and Ca$^{2+}$ fluxes in maize roots and maize suspension cells. *Planta* **188,** 601–610.

Koeckerling, A., Sorgenfrei, D., and Fromm, M. (1993). Electrogenic Na$^+$ absorption of rat distal colon is confined to surface epithelium: A voltage-scanning study. *Am. J. Physiol.* **264,** C1285–C1293.

Kühtreiber, W. M., and Jaffe, L. F. (1990). Detection of extracellular calcium gradients with a calcium-specific vibrating electrode. *J. Cell Biol.* **110,** 1565–1573.

Nuccitelli, R. (1990). Vibrating probe technique for studies of ion transport. *In* "Noninvasive Techniques in Cell Biology" (J. K. Foskett and S. Grinstein, Eds.), pp. 273–310. Wiley-Liss, New York.

Pierson, E. S., Miller, D. D., Callaham, D. A., Shipley, A. M., Rivers, B. A., Cresti, M., and Hepler, P. K. (1994). Pollen tube growth is coupled to the extracellular calcium ion flux and the intracellular calcium gradient: Effect of BAPTA-type buffers and hypertonic media. *Plant Cell* **6,** 1815–1828.

Scheffey, C. (1989). Vibrating probe method for measuring local transepithelial current. *Ann. NY Acad. Sci.* **574,** 485.

Scheffey, C., Shipley, A. M., and Durham, J. H. (1991). Localization and regulation of acid–base secretory currents from individual epithelial cells. *Am. J. Physiol.* **261,** F963–F974.

Shi, R., and Borgens, R. B. (1995). Three-dimensional gradients of voltage during development of the nervous system as invisible coordinates for the establishment of embryonic pattern. *Dev. Dyn.* **202,** 101–114.

Smith, P. J. S. (1995). Non-invasive ion probes—Tools for measuring transmembrane ion flux. *Nature* **378,** 645–646.

Smith, P. J. S., Sanger, R. H., and Jaffe, L. F. (1994). The vibrating Ca$^{2+}$ electrode: A new technique for detecting plasma membrane regions of Ca$^{2+}$ influx and efflux. *Methods Cell Biol.* **40,** 115–134.

Anita L. Zimmerman

# 46

## Cyclic Nucleotide-Gated Ion Channels

### I. Introduction

Cyclic nucleotides have long been known as intracellular second messengers that regulate cell function by controlling the activity of protein kinases, which in turn control many other cellular proteins (reviewed in Greengard, 1978). However, in 1985, Fesenko and his colleagues made a startling discovery that changed our view of the physiological role of cyclic nucleotides. These investigators found that the ion channel mediating the electrical response to light in retinal rod cells was directly opened by the binding of guanosine $3',5'$-cyclic monophosphate (cGMP); no phosphorylation reaction was required. It now appears that the rod channel is a member of a special class of ion channels—the cyclic nucleotide-gated (CNG) channels.

Why would Nature directly gate ion channels with cyclic nucleotides? When a cyclic nucleotide regulates a kinase, it is also, in effect, regulating all the proteins controlled by that kinase and by substrates of the kinase. Regulating ion channels is similar in that, like kinases, ion channels have diverse physiological effects. For example, the opening of nonselective cation channels (such as those opened by cyclic nucleotides) depolarizes the cell membrane and also allows the entry of $Ca^{2+}$, another important second messenger. Membrane depolarization opens $Ca^{2+}$ channels, further increasing the entry of $Ca^{2+}$. Many cell functions are controlled by membrane potential and/or intracellular $Ca^{2+}$, including nerve impulses, muscle contraction, and the secretion of neurotransmitters and hormones. Finally, depolarization and $Ca^{2+}$ open potassium channels, which repolarize the membrane and thereby contribute to the termination of the cellular response. Thus, there are numerous possibilities for control of cell function by cyclic nucleotide-gated ion channels. Furthermore, ion channel gating and permeation are much faster than phosphorylation reactions. Thus, changes in cyclic nucleotide levels could have fast effects mediated by ion channels, followed by slower, longer lasting effects mediated by protein kinases.

### II. Physiological Roles and Locations

Since their discovery in retinal rods, CNG channels have been identified in many other types of cells (reviewed in Zimmerman, 1995; Zagotta and Siegelbaum, 1996). In particular, CNG channels have been implicated generally in sensory transduction, as they also have been found in retinal cones, olfactory cells, invertebrate photoreceptors, cochlear hair cells, and pineal gland cells. Furthermore, CNG channels may mediate the glutamate response of retinal bipolar cells. Since CNG channels have been purified, cloned, and expressed (reviewed in Kaupp, 1991), it is now possible to screen other types of tissue for CNG channels. For example, mRNA probes against the rod CNG channel have revealed its expression in cells of the heart, testes, and kidney. Furthermore, the pacemaker channel ($I_h$) in the heart sinoatrial node, and some other voltage-gated channels contain regulatory cyclic nucleotide binding sites. The CNG channels have been studied most thoroughly in rods and cones and in olfactory cells. Therefore, CNG channels from these cells are discussed in most detail here.

In rods and cones, CNG channels are key players in visual transduction (reviewed in Yau and Baylor, 1989; Lamb and Pugh, 1990; McNaughton, 1990; Detwiler and Gray-Keller, 1992; Lagnado and Baylor, 1992; Baylor, 1996). It is these channels that conduct the so-called "dark current" and whose closure generates the hyperpolarizing response to light, which decreases the secretion of glutamate onto bipolar cells at the rod–bipolar synapse. The physiological second messenger in the photoreceptors is cGMP, which is at relatively high cytosolic concentration in the dark and decreases in the light after hydrolysis by a phosphodiesterase (PDE). A similar system exists in cone visual transduction. Details of the enzyme cascade controlling the level of cGMP are given in the next section.

Rods and cones are particularly well suited to patch-clamp studies of CNG channels, since the plasma membranes of their light-sensitive "outer segments" contain essentially no other type of ion channel (although Na/K/

Ca exchange carriers are present). Furthermore, the rod outer segment plasma membrane has an extremely high density of CNG channels—hundreds per square micrometer—allowing nanoamperes of current to be recorded from a single excised patch. Such large currents are useful in studying channel block (e.g., by divalent cations). In contrast, cone outer segments have relatively low channel densities, allowing one to study single-channel kinetics in patches containing only one channel. However, rods and cones interestingly have about the same number of total CNG channels because of the much larger plasma membrane area in cone outer segments (a consequence of the characteristic infolding of this membrane). CNG channels have also been found at low density in rod inner segments, but their functional properties appear altered there.

Olfactory receptor cells use CNG channels in sensing odorants (reviewed in Shepherd, 1992; Firestein, 1991). In this system, however, there are apparently numerous receptor types; adenosine 3',5'-cyclic monophosphate (cAMP) is the physiological second messenger; and the stimulus triggers cAMP production by adenylate cyclase, rather than its degradation by a PDE. Thus, in response to an odorant, the CNG channels open, and the olfactory receptor cell depolarizes, increasing the probability of generation of an action potential. Despite these differences, there is considerable similarity between the visual and olfactory transduction systems and between their respective CNG channels.

Like rods and cones, the olfactory cell has its CNG channels concentrated in a specialized region: the olfactory cilia and ciliary knob. Although the channels have been studied in excised patches from olfactory cilia (Nakamura and Gold, 1987), such experiments are extremely difficult because of the small diameter of a cilium. Luckily, the knob is larger, and some CNG channels are also located (at lower density) in the membrane of the soma. Furthermore, whole-cell patch-clamp methods have yielded considerable information on the olfactory CNG channels.

## III. Control by Cyclic Nucleotide Enzyme Cascades

Like cyclic nucleotide-regulated protein kinases, CNG channels are sensors of the local concentration of cyclic nucleotides. Stimulus-induced changes in cyclic nucleotides are mediated by GTP-binding proteins (G-proteins). The stimulus-activated receptor interacts with a G-protein, causing it to release GDP and bind GTP and to dissociate into two components: an $\alpha$ subunit and a $\beta\gamma$ subunit complex. The $\alpha$ subunit of the G-protein, now bound with GTP, stimulates either adenylate cyclase (in olfactory receptors) or a cGMP-specific phosphodiesterase (in photoreceptors). For photoreceptors, the stimulus that activates the receptor is a photon, whereas for olfactory cells, the stimulus is a particular odorant molecule that acts as a receptor ligand. This type of enzyme cascade is diagrammed in Fig. 1 for a rod photoreceptor and in Fig. 2 for an olfactory cell. The cascade in cones is similar to that in rods, except that all the membrane-associated players

are located on the plasma membrane, since cones lack internal disks (see Yau and Baylor, 1989).

Cyclic nucleotide enzyme cascades are not fixed in their behavior. Instead, they are regulated by feedback systems, some of which involve CNG channels. For example, in rods, the $Ca^{2+}$ that enters through CNG channels has been found to modulate the cGMP cascade. The sites and mechanisms of action of $Ca^{2+}$ are not clear, and this topic is currently under intense investigation and debate. There is evidence that $Ca^{2+}$ (in association with $Ca^{2+}$ binding proteins) inhibits guanylate cyclase and inhibits the shutoff of rhodopsin (see Lagnado and Baylor, 1992; Yarfitz and Hurley, 1994). In olfactory receptors, $Ca^{2+}$ appears to link an inositol triphosphate transduction cascade with the cAMP system by its effects on adenylate cyclase via calmodulin (Anholt and Rivers, 1990) and protein kinase C (Frings, 1993).

In addition to such feedback regulatory systems, there are the standard shutoff mechanisms employed in cyclic nucleotide cascades (reviewed in Lagnado and Baylor, 1992; Yarfitz and Hurley, 1994). These include phosphorylation of the receptor (e.g., the phosphorylation of rhodopsin and its binding to arrestin), GTPase activity of the G-protein (converting it back to the GDP-bound inactive form), cessation of the stimulus, and competing hydrolysis or synthesis of the cyclic nucleotide. There are also hints that the ability of the channels to respond to the cyclic nucleotide may be modulated (see following).

## IV. Functional Properties

### A. Channel Gating

CNG channels are very sensitive detectors of the local concentration of cyclic nucleotides, and they appear designed to work in the physiological concentration range of their respective agonists. Dose–response curves (e.g., Fig. 3) for activation of rod channels by cGMP give half-saturating concentrations ($K_{1/2}$ values) ranging from about 5 to 60 $\mu M$, which is well within the expected physiological concentration range. The rather wide range of values of $K_{1/2}$ may reflect functional modulation of the channels by other factors (see below).

The form of the dose–response curve is well described by the Hill equation,

$$r/r_{max} = [cGMP]^n/(K_{1/2}^n + [cGMP]^n)$$

where $r$ is the response to cGMP (e.g., the cGMP-activated component of the membrane current measured in a patch-clamp experiment), $r_{max}$ is the maximum response (obtained with a saturating concentration of cGMP to open all channels in the patch), and $n$ is the Hill coefficient. Reported Hill coefficients have ranged between about 1.5 and 4, suggesting that several molecules of cGMP must bind to each channel to open it. Because the dose–response curve for channel activation is so steep, small changes in the concentration of cAMP or cGMP produce very large changes in channel open probability. As discussed later, each cGMP molecule may bind to one of four channel

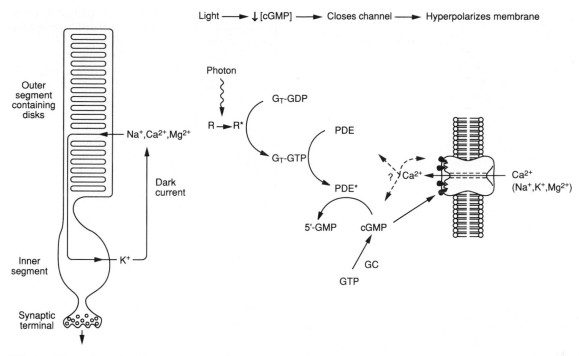

**FIG. 1.** The cyclic nucleotide cascade controlling CNG channels in rods. R, R*, rhodopsin in its inactive and active forms, respectively. $G_T$, G-protein ("Transducin"), bound to either GDP or GTP. PDE, PDE*, phosphodiesterase in its inactive and active forms, respectively. GC, guanylate cyclase. Calcium ions entering through the CNG channels are thought to modulate the function of several players in the cascade, including the channels themselves. A similar cascade operates in cones.

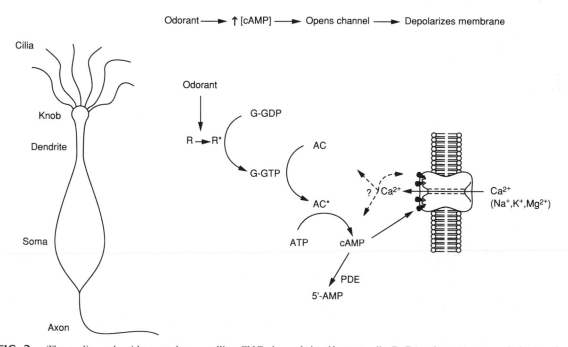

**FIG. 2.** The cyclic nucleotide cascade controlling CNG channels in olfactory cells. R, R*, odorant receptor in its inactive and active (odorant-bound) forms, respectively. G, G-protein, bound to either GDP or GTP. AC, AC*, adenylate cyclase in its inactive and active forms, respectively. PDE, phosphodiesterase. Here, as in photoreceptors, entering $Ca^{2+}$ appears to modulate the cascade, including the CNG channels.

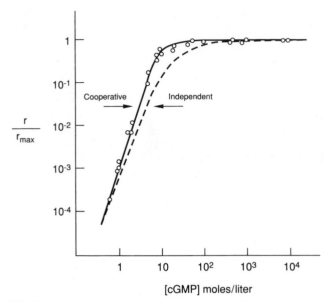

**FIG. 3.** Dose–response curve for activation of rod CNG channels by cGMP. The response, $r$, to cGMP is the cGMP-activated current obtained from excised, inside-out patches from salamander rod outer segments. The response is normalized to the value ($r_{max}$) obtained with all channels open at saturating [cGMP]. The smooth curve (Hill relation) is based on the assumption that the channels open when at least three cGMP molecules bind very cooperatively. The dashed curve assumes three independent cGMP-binding sites of equal affinity. Divalent cation concentrations were submicromolar, since these ions tend to reduce the single-channel conductance (see text). (Adapted with permission from *Nature*, Zimmerman and Baylor, 1986. Copyright 1986 Macmillan Magazines Limited.)

subunits. The good fit of the Hill equation to the data suggests that the binding of cGMP to the channel is very cooperative. That is, the data cannot be described simply by independent binding of cGMP molecules to sites of equal affinity (dashed curve, Fig. 3: $r/r_{max} = \{[\text{cGMP}]/(K_{1/2} + [\text{cGMP}])\}^n$). Instead, the binding of one cGMP apparently enhances the binding or action of others.

Activation of olfactory channels is similar to that of rod and cone channels except for relative cycle nucleotide sensitivities. Rod and cone channels are much less sensitive to cAMP than to cGMP, with a $K_{1/2}$ for activation by cAMP of about 1.5 m$M$. Native olfactory CNG channels are more sensitive to both cyclic nucleotides than are photoreceptor channels, but they are only two to five times more sensitive to cGMP than to cAMP, with most $K_{1/2}$ values for activation by cGMP in the range of 1 to 5 $\mu M$ (and a few as high as 20 $\mu M$—see Nakamura and Gold, 1987).

Although some strongly voltage-dependent channels are regulated by cyclic nucleotides, channels that are primarily activated by cyclic nucleotides are only weakly voltage dependent, with no voltage-dependent inactivation, (reviewed in Zagotta and Siegelbaum, 1996). In current/voltage ($I/V$) relations from excised patches, the voltage dependence of channel gating is most obvious at low cyclic nucleotide concentrations, where it introduces rectification (e.g., see Fig. 6D later). High concentrations of cyclic nucle-

otides overcome the voltage dependence of gating, driving the channels (by mass action) toward high open probabilities at all voltages and linearizing the $I/V$ curves. Note, however, that much of the CNG channel rectification seen in intact cells probably results from voltage-dependent channel block by $Ca^{2+}$ and $Mg^{2+}$, as discussed later. Finally, the end result of both forms of rectification is that current is relatively independent of voltage in the physiological voltage range. Thus, CNG channels are able to transduce faithfully in the face of changes in membrane potential that originate at either the transducing region or elsewhere in the cell, where there are many kinds of voltage-dependent ion channels.

Unlike the acetylcholine receptor channel, CNG channels do not appear to desensitize on continued exposure to their agonists, although some evidence to the contrary has appeared (reviewed in Yau and Baylor, 1989). Studies on the rod channel using rapid jumps in cGMP concentration or in voltage suggest that cyclic nucleotide binding and channel opening are very rapid and that under physiological conditions activation is limited by the time required for cGMP to diffuse to the channel.

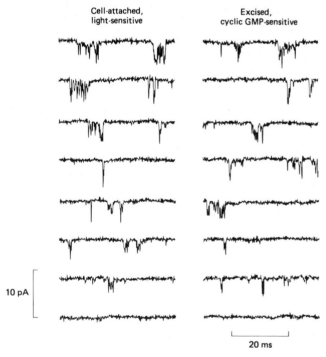

**FIG. 4.** Single-channel recordings of CNG channels in native rod outer segment membranes demonstrate such rapid gating kinetics that many open–shut transitions are poorly resolved. Channel openings give downward deflections; holding potential −148 mV. Left: Cell attached patch from toad rod outer segment; the bottom trace was obtained in a saturating light, and all others in darkness. To prevent channel block, the pipette was filled with a solution lacking $Ca^{2+}$ and $Mg^{2+}$. Right: The same patch after excision, with the intracellular surface bathed in either no cGMP (bottom trace) or 10 $\mu M$ cGMP (all other traces); both pipette and bathing solutions lacked $Ca^{2+}$ and $Mg^{2+}$. (Reproduced with permission from Matthews and Watanabe, 1987.)

Experiments to "trap" divalent transition metal cations in the rod channel have revealed evidence that the channel gate resides between an ion binding site and the extracellular mouth of the pore (Karpen et al., 1993). In these excised patch experiments, the kinetics of channel block by internal $Ni^{2+}$ were found to be independent of the channel open probability, but the kinetics of block accelerated with increasing open probability when $Ni^{2+}$ was applied to the extracellular side of the patch. These results suggest that $Ni^{2+}$ could hop in and out of the channel from the intracellular surface whether the channel gate was closed or open, but that it could not do so from the extracellular surface until the gate was open. The structural correlate of this gate remains to be determined.

Single-channel studies demonstrate that CNG channels have particularly fast open–shut transitions in the native membrane. This flickery behavior is striking in the cell-attached and excised-patch recordings obtained from toad rods by Matthews and Watanabe (1987) (Fig. 4). However, when purified and reconstituted, or cloned and expressed, the channels were initially found to have much slower gating kinetics (Fig. 5; Kaupp et al., 1989) that are more typical of other ion channels. A similar difference was found when cGMP-gated channels in the rod outer segment were compared with those that occur at low density in the inner segment (Torre et al., 1992), a region of the cell with different internal structures and functions and whose plasma membrane has a different lipid and protein composition. Although $Ca^{2+}$ and $Mg^{2+}$ produce flicker block of these channels, the flickery gating behavior persists even in the absence of these ions. Furthermore, the transitions are too fast to reflect the binding and unbinding of cyclic nucleotides, and they also apparently do not reflect block by protons. Evidence from Chen et al. (1993) suggests that the flickery gating pattern of the native rod CNG channel results partly from the presence of a channel subunit that was missing in the original reconstitution and expression studies, which identified only one kind of subunit (see further, molecular structure). These fast gating kinetics, along with the appearance of different single-channel con-

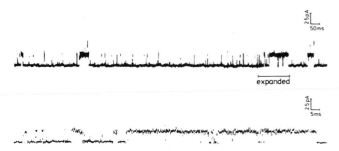

**FIG. 5.** Single-channel recordings obtained from the homo-oligomeric form of cloned bovine rod CNG channels after expression in a *Xenopus* oocyte. Gating kinetics are much slower than those demonstrated by the native channel (Fig. 4). This inside-out, excised patch was exposed to 5 $\mu M$ cGMP in a bathing solution low in divalent cations; the pipette was filled with the same solution without cGMP. (Reproduced with permission from *Nature*, Kaupp et al., 1989. Copyright 1989 Macmillan Magazines Limited.)

ductance levels, have hindered extensive kinetic analysis like that feasible for other types of channels (e.g., Magleby and Weiss, 1990).

## B. Permeation, Selectivity, and Block

Most CNG channels described so far are nonselective cation channels with no significant anion permeability. (Note, however, that some $K^+$-selective channels are regulated by direct binding of cyclic nucleotides; reviewed in Zimmerman, 1995, and Zagotta and Siegelbaum, 1996). Thus, reversal potentials for the CNG channels studied under normal ionic conditions are about +5 to +20 mV. The monovalent alkali cation permeability sequence for the rod channel has been reported to be $Li^+ \geq Na^+ \geq K^+ > Rb^+ > Cs^+$ (reviewed in Yau and Baylor, 1989). This sequence suggests that the pore contains a region of high negative charge density, that is, a "high field strength site" (Menini, 1990; Furman and Tanaka, 1990). $Ca^{2+}$ and $Mg^{2+}$ ions appear to be more permeant than the monovalent alkali cations (e.g., $P_{Ca}/P_{Na} = 12.5$), as they carry a larger share of the dark current than otherwise expected (Nakatani and Yau, 1988). Studies with organic cations suggest that the selectivity filter region of the rod channel is equivalent to a rectangle about 3.8 by 5 Å in size (Picco and Menini, 1993).

Cone and olfactory CNG channels also poorly discriminate among monovalent alkali cations, although their exact permeability sequences and ratios are not identical to those of the rod CNG channel. Strict comparisons are difficult because there is some variability in reported permeability ratios for each cell type. Relative permeabilities obtained may depend on whether the channels are studied in the intact cell, in excised patches, or in reconstitution or expression systems. The permeabilities may also depend on intracellular factors controlling functional modulation of the channels (see following). In intact rods, the relative currents carried by monovalent and divalent cations have been found to depend on the concentration of cGMP (Cervetto et al., 1988). Thus, the functional properties of CNG channels may change with the number of ligand molecules bound. If so, this may reflect a very unusual feature of these channels.

CNG channels generally behave as if they were single-file pores with one dominant ion-binding site that has a higher affinity for divalent cations than for monovalent cations (Frings et al., 1992; Picones and Korenbrot, 1992; Zimmerman and Baylor, 1992). However, some behavior indicates the presence of more than one ion-binding site, and this is confirmed by structure–function work (reviewed in Zimmerman, 1995, and Zugotta and Siegelbaum, 1996). Under physiological conditions, the channel is occupied by $Ca^{2+}$ or $Mg^{2+}$ most of the time, and these ions prevent the passage of $Na^+$ and $K^+$, which are able to pass through the pore much more rapidly. As a result of the very slow transport rate of the divalent cations, the mean single-channel conductance is extremely low—only about 0.1 pS in rods studied under physiological conditions. The channels can be blocked from either side, but under physiological conditions, they are mostly blocked by extracellular

$Ca^{2+}$ and $Mg^{2+}$. To resolve single-channel currents, one must reduce the concentration of divalent cations to the micromolar range. In the absence of divalent cations, the single-channel conductance is as high as tens of pico-Siemens. The extremely low single-channel conductance in physiological solutions gives an excellent signal-to-noise ratio for photon detection by rods, since the random openings and closings of individual channels produce only very tiny fluctuations in the dark current. The absorption of a single photon elicits the closure of hundreds of channels, giving a smooth, stereotypical waveform. If all those channels had single-channel conductances in the range of tens of pico-Siemens, the rod cell would have to contend with the consequences of a large influx of $Na^+$ and $Ca^{2+}$. Cones and olfactory cells would have a similar problem, since their transducing regions also contain many CNG channels.

In the absence of divalent cations, $I/V$ relations for CNG channels (with saturating cyclic nucleotide concentrations) are linear or nearly so (Figs. 6A and D, upper curve). Divalent cations introduce extreme nonlinearity in the $I/V$ relations (Figs. 6B and C). When the rod channel is studied in the presence of physiological concentrations of divalent cations, its $I/V$ relation is very outwardly rectified. Although some of this rectification is a consequence of the weak voltage dependence of channel gating described earlier; much of its results from channel block by $Ca^{2+}$ and $Mg^{2+}$. Thus, at negative membrane potentials in the physiological range (about $-40$ to $-80$ mV), external $Ca^{2+}$ and $Mg^{2+}$ are drawn into the channel, reducing $Na^+$ entry and giving an approximately flat $I/V$ relation over a wide range of voltage. This very low, voltage-independent conductance in the physiological voltage range allows light-induced outer segment voltage changes to travel relatively unattenuated to the inner segment to regulate synaptic transmission and also prevents the outer segment photon-sensing mechanism from fluctuating with voltage.

$I/V$ relations of the olfactory channel, with and without divalent cations, are essentially indistinguishable from those of the rod channel, but surprisingly, the $I/V$ relation for the cone channel is rather different. Whereas cone $I/V$ curves from excised patches are linear in the absence of

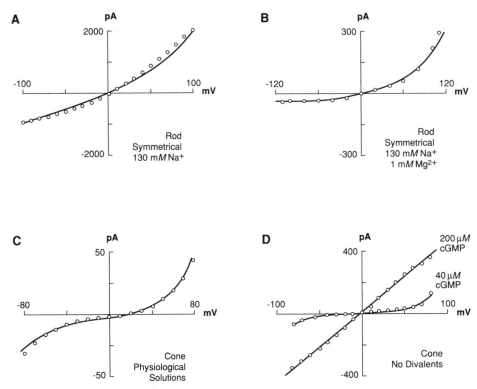

**FIG. 6.** Current–voltage relations for photoreceptor cyclic GMP-activated currents from multichannel, excised patches. (A) The rod CNG channel relation is almost linear with saturating [cGMP] and no divalent cations. Olfactory CNG channels demonstrate an essentially identical relation (not shown). (B) The addition of millimolar $Mg^{2+}$ to both sides of the membrane gives strong outward rectification. Similar outward rectification has been reported for olfactory CNG channels. (C) Cone CNG channels show both outward and inward rectification in the presence of physiological (millimolar) levels of $Mg^{2+}$ and $Ca^{2+}$. (D) The $I/V$ relation for cone CNG channels is approximately linear with saturating [cGMP] and no divalent cations (top curve). Decreasing [cGMP] produces a nonlinear $I/V$ relation (bottom curve), because gating is slightly voltage dependent. [Redrawn with permission from Zimmerman and Baylor, 1992 (A) and (B); from *Nature,* Haynes and Yau, 1985. Copyright 1985 Macmillan Magazines Limited. (C); and Picones and Korenbrot, 1992. (D) Copyright 1992 from The Rockefeller University Press.)

divalent cations, they rectify in both positive and negative directions when divalents are present (Fig. 6C). Although the functional significance of this difference between cone CNG channels and those in rod cells and olfactory receptors is not clear, structurally it may reflect a different location of the dominant ion-binding site within the cone channel. Rod channel data have been fit with a model that assumes the ion binding site is 33% of the electrical distance from the inside surface (Zimmerman and Baylor, 1992), whereas cone channel data have been fit assuming that the site is approximately in the middle of the membrane electric field (Picones and Korenbrot, 1992). At subsaturating concentrations of cyclic nucleotides (which are closer to the physiological concentrations), the cone channel $I/V$ relation is much more flat in the physiological range of membrane potential (Fig. 6D, lower curve).

In whole-cell recordings of light-regulated current measured with physiological solutions, the cone $I/V$ curve bends down at less negative potentials than does that of the rod (Fig. 7), consistent with the different voltage dependence measured in excised patches. Thus, the cone $I/V$ begins to bend down at voltages more negative than about $-30$ mV, whereas the rod $I/V$ curve does not clearly bend until about $-100$ mV. In both cells, the increasing inward current at large negative potentials probably reflects unblock of the channels; that is, as a cell is hyperpolarized from $+50$ mV toward $-100$ mV, external divalent cations are first drawn into the pore, lowering conductance, and then pushed all the way through to the inside of the cell.

Single-channel recordings of the rod channel in the absence of divalent cations have revealed at least two conductance states: one of about 25 pS and the other with a conductance about one-third as large. However, it has been suggested that the channel has at least one more, and perhaps many more, conductance levels. Because of the extremely rapid gating kinetics of this channel, numerous open–closed transitions are no doubt unresolved in the single-channel records. Thus, it is difficult to determine the exact number of distinct conductance states. It is also not clear whether these states are characterized by truly different ion transport rates or merely by different (incompletely resolved) open times, giving the appearance of different conductance levels. In the absence of divalent cations, the olfactory and cone channels demonstrate major single-channel conductances around 45 to 50 pS, also with apparent subconductance states.

A few pharmacological agents have been tested on CNG channels. Since these channels have a strong affinity for $Ca^{2+}$, various calcium channel blockers (including verapamil, dihydropyridines, and diltiazem) have been tested (reviewed for the rod channel in Yau and Baylor, 1989). Most block only weakly or not at all, but *l-cis*-diltiazem and an amiloride analogue, 3′,4′-dichlorobenzamil, have been found to block effectively from the cytoplasmic surface of the membrane, with $K_i$ values in the micromolar range. Although block by *l-cis*-diltiazem has been studied in some detail, it remains unclear whether this drug acts as a pore-occluding blocker or a modifier of channel gating (Haynes, 1992). Frings *et al.*, (1992) have found

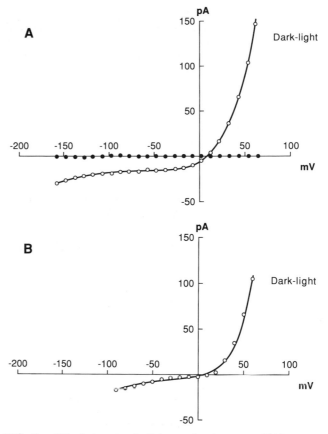

**FIG. 7.** $I/V$ relations for the light-regulated current of (A) a transducing rod and (B) cone. Rectification results from block by $Ca^{2+}$ and $Mg^{2+}$ and from a slight voltage dependence of gating. It is not clear why the rod $I/V$ relation is flat over a larger negative voltage range than is the cone relation. In each case, an outer segment was drawn into a glass pipette (suction electrode) to measure the current in dark and light, and the membrane potential was controlled by a patch-clamp electrode sealed onto the inner segment with the patch ruptured. Open symbols are the difference currents. Filled symbols in part A represent outer segment currents in a saturating light that shut down all the CNG channels; the small magnitude of these currents suggests that few if any, other types of channels exist in the outer segment. [Redrawn with permission from (A) Zimmerman and Baylor, 1992 and (B) Reproduced from *The Journal of General Physiology*, 1993, vol. 101, pp. 875–904, by copyright permission of The Rockefeller University Press.]

olfactory channels to be blocked by amiloride, D600, and diltiazem.

## V. Molecular Structure

A subunit of the bovine rod CNG channel was cloned by Kaupp *et al.* (1989), who proposed that the channel consists of multiple identical subunits, each with a cGMP-binding site located in the C-terminal region. Structurally similar olfactory CNG channels have also been cloned (reviewed in Kaupp, 1991). Studies of the rod channel highlight some of the many remaining mysteries regarding CNG channel structure. First, recent work (discussed later) suggests that all subunits are not equivalent after all. Second, although the total number of subunits per channel appears to be four (Liu *et al.*, 1996), the number of each type of subunit per channel is not known. Third, when the original channel subunit ($\alpha$) is biochemically purified or labeled *in situ* in the rod plasma membrane, it has a molecular weight of 63,000 compared with 78,000 for the cloned subunit (expressed in either COS-1 cells or *Xenopus* oocytes). The difference appears to be a segment near the N-terminus of the peptide, which may be removed by post-translational cleavage before insertion into the rod membrane (see dashed-line region in Fig. 8).

Although CNG channels are ligand-gated channels, their molecular structures more closely resemble those of voltage-gated channels. This similarity is brought out dramatically in recent work in which a deletion of two amino acids in a voltage-gated $K^+$ channel converted it into a nonselective cation channel that was blocked by divalent cations and that demonstrated the flickery gating kinetics typical of CNG channels (Heginbotham *et al.*, 1992). Citing sequence similarities, Jan and Jan (1992) have pointed out that CNG channels probably belong to a superfamily of channels including voltage-gated $Na^+$, $Ca^{2+}$, and $K^+$ channels, as well as $Ca^{2+}$-activated $K^+$ channels.

The members of this CNG/voltage-gated channel superfamily are characterized by one or more repeats of a structure containing six potential membrane-spanning segments, located between the hydrophilic N- and C-terminal regions that project into the cytosol. Also characteristic of these channels is an S4 segment that contains many basic residues and is thought to be a voltage-sensing region. Although CNG channels have S4 segments, they have only very weak voltage sensitivity. It has been proposed that glutamate residues in the vicinity of S4 contribute negative charges that may neutralize the effects of the positively

charged arginine and lysine residues of S4 that are thought to confer voltage sensitivity to the channel (Wohlfart *et al.*, 1992). The channels in this family also contain a putative pore region ("P-region") in which a crucial glutamate interacts with permeating ions (reviewed in Zimmerman, 1995, and Zagotta and Siegelbaum, 1996). Although there are subtle differences even among the various types of CNG channels, the similarities are striking and suggest a common ancestry. Figure 8 shows a current model for the CNG channel subunit that was first discovered (Bönigk *et al.*, 1993). Gordon and Zagotta (1995) have shown that the N-terminus of one subunit interacts with the C-terminus of the adjacent one in an assembled channel.

Although Kaupp *et al.* (1989) obtained functional expression of the rod CNG channel in *Xenopus* oocytes, certain properties of the expressed channel differed from those of the native channel. For example, the expressed channel was not blocked by *l-cis*-diltiazem, and it did not demonstrate the flickery single-channel gating kinetics typical of the native channel. Chen and colleagues (1993) and Körschen *et al.* (1995) have found a "missing subunit" that, when coexpressed with the CNG channel subunit, appears to restore these lost functions. Expression of the missing subunit alone, however, does not appear to give CNG channel function. Furthermore, neither the homo-oligomeric nor the hetero-oligomeric cloned channel has a cGMP affinity as high as that of the native channel. It is possible that this difference between native and cloned channels relates to modulation of the channels by intracellular factors (see following), rather than to intrinsic channel structure. How many of each subunit are needed for proper channel function and whether additional types of subunits (or modifications of subunits) are required to give full functional reconstitution remain to be determined.

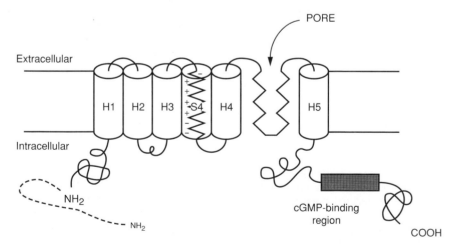

**FIG. 8.** Model for the organization of the first identified CNG channel subunit in the outer segment plasma membrane. Expression of multiple copies of this subunit produces a functional CNG channel, with one cGMP-binding site on each copy. At least one other subunit type may be required for complete function (see text). H1 through H5 and S4 are putative membrane-spanning regions. N- and C-termini are cytoplasmic. The first 92 amino acids (dashed line) on the N-terminus appear to be cleaved posttranslationally. The S4 segment is thought to be the voltage-sensing region, and the putative pore region lies between H4 and H5. Newer terminology refers to the transmembrane segments as S1 through S6. (Adapted with permission from Bönigk *et al.*, 1993. Copyright 1993 by Cell Press.)

## VI. Functional Modulation

New information suggests that CNG channels may be modulated in ways that are only beginning to be elucidated. As mentioned previously, there are hints that the ionic selectivity of the rod CNG channel may vary with the amount of cyclic nucleotide bound. There is also now evidence that the cGMP sensitivity of the rod channel may be tuned up or down by phosphorylation and reduced by diacylglycerol (reviewed in Zimmerman, 1995), as well as by calmodulin in the presence of $Ca^{2+}$ (Hsu and Molday, 1993; Gordon *et al.*, 1995). Whether other channel properties are also altered by these modulators has not been determined. Calcium seems particularly useful in modulatory mechanisms involving feedback control since its concentration in the cell is determined partly by its entry through the CNG channels. Functional modulation of the channel may explain the large variability in reported cGMP affinity and cooperativity and may play a role in some aspects of visual transduction.

The gating of olfactory CNG channels has been found to be modulated by intracellular $Ca^{2+}$ (Zufall *et al.*, 1991; Kramer and Siegelbaum, 1992). Figure 9, from Zufall *et al.* (1991), shows single-channel recordings obtained from an excised patch of olfactory dendritic membrane. When the concentration of $Ca^{2+}$ bathing the intracellular surface of the patch was increased from 0.1 to 3 $\mu M$, there was a striking increase in channel closed time, giving a decrease in open probability. Since the single-channel amplitude and mean channel open time were not affected, the mechanism of $Ca^{2+}$ action most likely involved allosteric effects on channel gating (perhaps by stabilizing one or more closed states), rather than channel block. Further evidence for channel inhibition by $Ca^{2+}$ was presented by Kramer and Siegelbaum (1992), who found that $Ca^{2+}$ shifts the cAMP dose–response curve to the right. The channel also appears to be modulated by $Ca^{2+}$-calmodulin (reviewed in Zimmerman, 1995). The functional significance of this modulation remains to be determined.

## VII. Summary

Cyclic nucleotide-gated channels are nonselective cation channels that are sensitively and directly activated by cGMP and/or cAMP. Like other cyclic nucleotide-regulated proteins, CNG channels are powerful modifiers of cell function. They are found in photoreceptors and olfactory cells, where their role in sensory transduction is well established. However, they are also beginning to be discovered in other types of cells, where their function is not always clear.

The enzyme cascades controlling cyclic nucleotide levels, and therefore CNG channel opening, have been thoroughly studied in vertebrate photoreceptors and olfactory cells. In a rod or cone, photon absorption by a pigment molecule leads to activation of a G-protein, which in turn stimulates a phosphodiesterase to hydrolyze cGMP. The resultant closure of CNG channels hyperpolarizes the cell and thereby reduces synaptic transmitter release onto the next

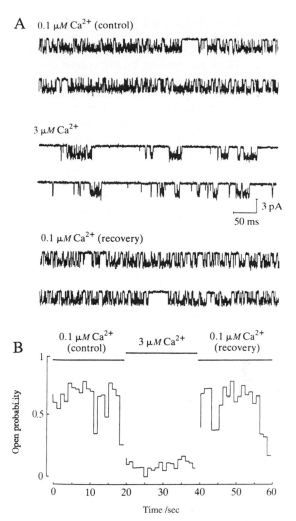

**FIG. 9.** Modulation of an olfactory CNG channel by $Ca^{2+}$. Raising the $Ca^{2+}$ concentration from 0.1 to 3 $\mu M$ at the intracellular surface of an excised patch produced a reversible increase in channel closed time, resulting in a decrease in channel open probability. The cAMP concentration was 100 $\mu M$, and the holding potential was $-60$ mV. Channel opening gives a downward deflection in current. (Reproduced with permission from Zufall *et al.*, 1991.)

neuron in the visual circuit. In an olfactory cell, an odorant receptor with bound odorant activates a G-protein, which stimulates adenylate cyclase to synthesize cAMP from ATP. The resultant opening of CNG channels depolarizes the cell, thereby increasing its action potential firing rate.

Although it appears that the primary steps in these two cascades have been determined, many questions remain, particularly regarding feedback control mechanisms in sensory adaptation. In fact, it has become clear that many of the molecular players controlling the cascades have not yet been identified, and there is recent evidence for modulation of the CNG channels themselves. Thus, CNG channel function may be modulated by phosphorylation, by calmodulin, and/or by $Ca^{2+}$, and perhaps by other factors as well. The molecular mechanisms of modulation and how many aspects of channel function are modulated remain to be determined.

CNG channels have striking gating and permeation properties. They are opened by the cooperative binding of several molecules of cGMP or cAMP, and appear to be operating in the physiological range (micromolar) of relevant cyclic nucleotide concentrations. Because the dose–response curve for channel activation is so steep, small increases in [cAMP] or [cGMP] dramatically increase CNG channel open probability. Calcium and magnesium ions permeate CNG channels very slowly, so that they block sodium and potassium transport under physiological conditions, drastically reducing the mean single-channel conductance and increasing the signal-to-noise ratio in sensory transduction.

Amino acid sequences of cloned CNG channels suggest that they belong to the same channel superfamily as the voltage-gated $Na^+$, $K^+$ and $Ca^{2+}$ channels. CNG channels were initially thought to be homo-oligomers, but it now appears that they have at least two distinct types of subunits. It is not known how many of each type of subunit make up a native channel. Our knowledge of the correlation of CNG channel structure and function is still in its infancy.

## Bibliography

Anholt, R. R. H., and Rivers, A. M. (1990). Olfactory transduction: Cross-talk between second-messenger systems. *Biochemistry* **29**, 4049–4054.

Baylor, D. A. (1996). How photons start vision. *Proc. Natl. Acad. Sci. USA* **93**, 560–565.

Bönigk, W., Altenhofen, W., Müller, F., Dose, A., Illing, M., Molday, R. S., and Kaupp, U. B. (1993). Rod and cone photoreceptor cells express distinct genes for cyclic GMP-gated channels. *Neuron.* **10**, 865–877.

Cervetto, L., Menini, A., Rispoli, G., and Torre, V. (1988). The modulation of the ionic selectivity of the light-sensitive current in isolated rods of the tiger salamander. *J. Physiol.* **406**, 181–198.

Chen, T. Y., Peng, Y-W., Dhallan, R. S., Ahamde, B., Reed, R. R., and Yau, K-W. (1993). A new subunit of the cyclic nucleotide-gated cation channel in retinal rods. *Nature* **362**, 764–767.

Detwiler, P. B., and Gray-Keller, M. P. (1992). Some unresolved issues in the physiology and biochemistry of phototransduction. *Curr. Opin. Neurobiol.* **2**, 433–438.

Fesenko, E. E., Kolesnikov, S. S., and Lyubarsky, A. L. (1985). Induction by cyclic GMP of cationic conductance in plasma membrane of retinal rod outer segment. *Nature* **313**, 310–313.

Firestein, S. (1991). A noseful of odor receptors. *Trends Neurosci.* **14**, 270–272.

Frings, S. (1993). Protein kinase C sensitizes olfactory adenylate cyclase. *J. Gen. Physiol.* **101**, 183–205.

Frings, S., Lynch, J. W., and Lindemann, B. (1992). Properties of cyclic nucleotide-gated channels mediating olfactory transduction. *J. Gen. Physiol.* **100**, 45–67.

Furman, R. E., and Tanaka, J. C. (1990). Monovalent selectivity of the cycle guanosine monophosphate-activated ion channel. *J. Gen. Physiol.* **96**, 57–82.

Gordon, S. E., and Zagotta, W. N. (1995). Subunit interactions in coordination of $Ni^{2+}$ in cyclic nucleotide-gated channels. *Proc. Natl. Acad. Sci. USA* **92**, 10222–10226.

Gordon, S. E., Downing-Park, J., and Zimmerman, A. L. (1995). Modulation of the cGMP-gated ion channel in frog rods by calmodulin and an endogenous inhibitory factor. *J. Physiol.* **486**, 533–546.

Greengard, P. (1978). Phosphorylated proteins as physiological effectors. *Science* **199**, 146–152.

Haynes, L., and Yau, K.-W. (1985). Cyclic GMP-sensitive conductance in outer segment membrane of catfish cones. *Nature* **317**, 61–64.

Haynes, L. W. (1992). Block of the cyclic GMP-gated channel of vertebrate rod and cone photoreceptors by *l-cis*-diltiazem. *J. Gen. Physiol.* **100**, 783–801.

Heginbotham, L., Abramson, T., and MacKinnon, R. (1992). A functional connection between the pores of distantly related ion channels as revealed by mutant $K^+$ channels. *Science* **258**, 1152–1155.

Hsu, Y.-T., and Molday, R. S. (1993). Modulation of the cGMP-gated channel of rod photoreceptor cells by calmodulin. *Nature* **361**, 76–79.

Jan, L. Y., and Jan, Y. N. (1992). Tracing the roots of ion channels. *Cell* **69**, 715–718.

Karpen, J. W., Brown, R. L., Stryer, L., and Baylor, D. A. (1993). Interactions between divalent cations and the gating machinery of cyclic GMP-activated channels in salamander retinal rods. *J. Gen. Physiol.* **101**, 1–25.

Kaupp, U. B. (1991). The cyclic nucleotide-gated channels of vertebrate photoreceptors and olfactory epithelium. *Trends Neurosci.* **14**, 150–157.

Kaupp, U. B., Niidome, T., Tanabe, T., Terada, S., Bönigk, W., Stühmer, W., Cook, N. J., Kangawa, K., Matsuo, H., Hirose, T., Miyata, T., and Numa, S. (1989). Primary structure and functional expression from complementary DNA of the rod photoreceptor cyclic GMP-gated channel. *Nature* **342**, 762–766.

Körschen, H. G., Illing, M., Seifert, R., Sesti, F., Williams, A., Gotzes, S., Colville, C., Müller, F., Dosé, A., Godde, M., Molday, L., Kaupp, U. B., and Molday, R. S. (1995). A 240 kDa protein represents the complete $\beta$ subunit of the cyclic nucleotide-gated channel from rod photoreceptor. *Neuron* **15**, 627–636.

Kramer, R. H., and Siegelbaum, S. A. (1992). Intracellular $Ca^{2+}$ regulates the sensitivity of cyclic nucleotide-gated channels in olfactory receptor neurons. *Neuron* **9**, 897–906.

Lagnado, L., and Baylor, D. (1992). Signal flow in visual transduction. *Neuron* **8**, 995–1002.

Lamb, T. D., and Pugh, E. N., Jr. (1990). Physiology of transduction and adaptation in rod and cone photoreceptors. *Neurosciences* **2**, 3–13.

Liu, D. T., Tibbs, G. R., and Siegelbaum, S. A. (1996). Subunit stoichiometry of cyclic nucleotide-gated channels and effects of subunit order on channel function. *Neuron* **16**, 983–990.

Magleby, K. L., and Weiss, D. S. (1990). Identifying kinetic gating mechanisms for ion channels by using two-dimensional distributions of simulated dwell times. *Proc. R. Soc. London Ser. B* **241**, 220–228.

Matthews, G., and Watanabe, S.-I. (1987). Properties of ion channels closed by light and opened by guanosine $3',5'$-cyclic monophosphate in toad retinal rods. *J. Physiol.* **389**, 691–715.

McNaughton, P. A. (1990). Light response of vertebrate photoreceptors. *Physiol. Rev.* **70**, 847–883.

Menini, A. (1990). Currents carried by monovalent cations through cyclic GMP-activated channels in excised patches from salamander rods. *J. Physiol.* **424**, 167–185.

Miller, J. L., and Korenbrot, J. I. (1993). In retinal cones, membrane depolarization in darkness activates the cGMP-dependent conductance: A model of Ca homeostasis and the regulation of guanylate cyclase. *J. Gen. Physiol.* **101**, 875–904.

Nakamura, T., and Gold, (1987). A cyclic nucleotide-gated conductance in olfactory receptor cilia. *Nature* **325,** 442–444.

Nakatani, K., and Yau, K.-W. (1988). Calcium and magnesium fluxes across the plasma membrane of the toad rod outer segment. *J. Physiol.* **395,** 695–729.

Picco, C., and Menini, A. (1993). The permeability of the cGMP-activated channel to organic cations in retinal rods of the tiger salamander. *J. Physiol.* **460,** 741–758.

Picones, A., and Korenbrot, J. I. (1992). Permeation and interaction of monovalent cations with the cGMP-gated channel of cone photoreceptors. *J. Gen. Physiol.* **100,** 647–673.

Shepherd, G. M. (1992). Toward a consensus working model for olfactory transduction. *In* "Sensory Transduction" (D. P. Corey and S. D. Roper, Eds.), pp. 20–37. Rockefeller Univ. Press, New York.

Torre, V., Straforini, M., Sesti, F., and Lamb, T. D. (1992). Different channel-gating properties of two classes of cyclic GMP-activated channel in vertebrate photoreceptors. *Proc. R. Soc. London Ser. B.* **250,** 209–215.

Wohlfart, P., Haase, W., Molday, R. S., and Cook, N. J. (1992). Antibodies against synthetic peptides used to determine the topology and site of glycosylation of the cGMP-gated channel from bovine rod photoreceptors. *J. Biol. Chem.* **267,** 644–648.

Yarfitz, S., and Hurley, J. B. (1994). Transduction mechanisms of vertebrate and invertebrate photoreceptors. *J. Biol. Chem.* **269,** 14329–14332.

Yau, K.-W., and Baylor, D. A. (1989). Cyclic GMP-activated conductance of retinal photoreceptor cells. *Annu. Rev. Neurosci.* **12,** 289–327.

Zagotta, W. N., and Siegelbaum, S. A. (1996). Structure and function of cyclic nucleotide-gated channels. *Annu. Rev. Neurosci.* **19,** 235–263.

Zimmerman, A. L., and Baylor, D. A. (1986). Cyclic GMP-sensitive conductance of retinal rods consists of aqueous pores. *Nature* **321,** 70–72.

Zimmerman, A. L., and Baylor, D. A. (1992). Cation interactions within the cyclic GMP-activated channel of retinal rods from the tiger salamander. *J. Physiol.* **449,** 759–783.

Zufall, F., Shepherd, G. M., and Firestein, S. (1991). Inhibition of the olfactory cyclic nucleotide gated ion channel by intracellular calcium. *Proc. R. Soc. London Ser. B.* **246,** 225–230.

Anita L. Zimmerman

# 47

# Visual Transduction

## I. Introduction

The miracle of vision begins when our photoreceptors absorb light that is reflected by our surroundings. However, the photoreceptor's duty does not end with photon capture. In addition, the photoreceptor must convert the energy of the absorbed photon into an electrochemical signal to be relayed to the visual cortex of the brain. This process of visual transduction involves a G-protein-mediated second messenger system which ultimately controls membrane potential and neurotransmitter release.

This chapter gives an overview of vertebrate visual transduction. Further information can be obtained from many excellent reviews on the subject (e.g., Baylor, 1996; Yarfitz and Hurley, 1994; Yau and Baylor, 1989; Lamb and Pugh, 1990; McNaughton, 1990; Detwiler and Gray-Keller, 1992; Lagnado and Baylor, 1992). Details of the cyclic nucleotide-gated ion channels that mediate the light response are covered in Chapter 46. Invertebrate visual transduction is not discussed here, but thorough descriptions are available elsewhere (e.g., Bacigalupo *et al.,* 1990; Yarfitz and Hurley, 1994; Zuker, 1996). One interesting difference between vertebrate and invertebrate photoreceptors is that the former hyperpolarize in response to light, whereas the latter depolarize.

## II. Photoreceptor Cells

The two kinds of vertebrate photoreceptors—rods and cones—are specialized for use under different conditions. The rods are employed for vision in dim light, whereas the cones are used in moderate to bright light. Thus, while a rod can reliably detect a single photon in a darkened room, its sensing mechanism shuts down completely when normal room lights are switched on. Cones, on the other hand, are not sensitive enough to detect single photons, but give us color vision and high spatial and temporal resolution when light levels are sufficiently high.

Rods and cones also occupy different parts of the retina. Cones are concentrated in the center of the retina, the fovea, which receives light from the center of our visual field. The rods, however, are almost entirely excluded from the fovea and instead dominate the peripheral retina. However, a few cones are scattered throughout the peripheral retina, and a few rods exist at the edge of the fovea. Thus, in daylight, our peripheral vision has much lower spatial resolution than our central vision because of the sparse distribution of peripheral cones. At night, our ability to locate a dim star is enhanced by looking "out of the corners of our eyes," using the abundant rods in the peripheral retina.

As shown in Fig. 1, the vertebrate photoreceptor can be divided into three regions: the outer segment, the inner segment, and the synaptic terminal. The outer segment is the region of the cell dedicated to photon capture and visual transduction. The inner segment, located closer to the front of the eye, contains the nucleus, mitochondria, and other general cell machinery. The synaptic terminal contains vesicles of neurotransmitter (thought to be glutamate) for release onto the second-order retinal neurons (the bipolar cells).

Rods and cones have very similar anatomy, except for one major feature: Rods have intracellular disks, and cones have, instead, infoldings of the plasma membrane called *sacs* (Fig. 1). The reason for this difference is not known. The membranes of the disks and sacs contain photopigment molecules and various other proteins used in visual transduction. The disks or sacs are packed very tightly together: usually about 2000 in a typical outer segment whose length is about 60 $\mu$m. This parallel array of membranes serves to align the photopigment molecules in the correct orientation for optimal absorption of light, which normally travels the length of the cell from the synaptic terminal toward the tip of the outer segment. In rods, the parallel arrangement of the disks may be maintained by cytoskeletal filaments that have been found to connect the disks to each other and to the plasma membrane (Roof and Heuser, 1982).

In both rods and cones, the outer and inner segments are connected by a narrow region called the *cilium.* In some way that is not yet understood, the cilium segregates

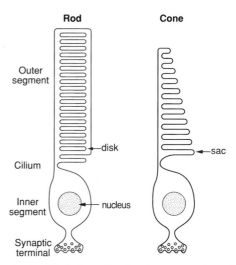

**FIG. 1.** A simplified view of rod and cone structure. The outer segment is specialized for visual transduction. Cone photopigment molecules are located on infoldings of the plasma membrane, called *sacs,* rather than in the membranes of intracellular disks as in rods.

the inner and outer segment membrane proteins. For example, whereas the inner segment plasma membrane contains many types of ion channels (Bader *et al.,* 1982; Barnes and Hille, 1989), the outer segment plasma membrane contains essentially a pure population of light-regulated cGMP-gated channels (Baylor and Nunn, 1986). It is in the ciliary region that new disks or sacs are inserted during the outer segment regeneration process that occurs daily (Bok, 1985). Old disks or sacs are removed from the tip of the outer segment by phagocytosis by the pigment epithelial cells. These very opaque cells exist at the back of the retina, with fingerlike cellular processes that hug the photoreceptor outer segments and also help to reduce light scatter. Because the cilium is such a narrow structure, outer and inner segments often break apart (and their membranes reseal) during cell isolation from the retina. Thus, many studies of visual transduction are conducted on isolated outer segments, which are somewhat simpler preparations than whole cells.

## III. Physiology of Visual Transduction

The physiological trademark of the vertebrate photoreceptor is its light-regulated "dark current" (Fig. 2; Hagins *et al.,* 1970). This current circulates between the outer and inner segments in the dark and is reduced upon light absorption. The dark current is carried into the outer segment mainly by $Na^+$ ions flowing through cGMP-gated nonselective cation channels (see Chapter 46) and out of the inner segment by $K^+$ ions flowing through voltage-gated $K^+$ channels. Since the cell has a relatively high $Na^+$ permeability in the dark, its resting potential is somewhat more positive than that for most cells (about $-40$ mV, instead of $-70$ mV). This resting depolarization tends to open voltage-gated $Ca^{2+}$ channels near the synaptic terminal, allowing

$Ca^{2+}$ entry and vesicular release of neurotransmitter in the dark. When a photon is absorbed by a rhodopsin molecule in a disk membrane, a series of reactions occurs that lead to a reduction in the guanosine $3',5'$-cyclic monophosphate (cGMP) concentration, with a resulting closure of the cGMP-gated channels in the outer segment plasma membrane (see later). The resulting reduction in the dark current (i.e., reduction in $Na^+$ permeability) causes a membrane hyperpolarization, which in turn causes closure of the voltage-gated $K^+$ channels and $Ca^{2+}$ channels and finally a decrease in neurotransmitter release.

The dark current and light response can be measured in a variety of ways, including a voltage-clamp method using intracellular microelectrodes, the whole-cell patch-clamp method, and the suction electrode method. The relatively noninvasive suction electrode method (Fig. 3) works particularly well with vertebrate photoreceptors, which are usually quite small and fragile. This method is much gentler than the other techniques, yet is able to measure the dark current without interference from inner segment currents.

Families of light responses recorded with a suction electrode from a monkey rod and cone are shown in Fig. 4. Light response amplitudes increase in a graded manner with light intensity until they reach a limiting value obtained with a "saturating light" that shuts off all the inward dark current (i.e., that closes all the cGMP-gated channels in the outer segment plasma membrane). Supersaturating light intensities cannot give larger responses, but instead give longer lasting ones. Cones are less sensitive than rods, requiring 50 to 100 times as many photons to shut down half the dark current. However, the cone responses are several times faster than those of rods, as evidenced by the shorter time to peak response. The undershoots seen in the cone responses (but not in normal rod responses) are not yet fully understood, but appear to make the cone especially sensitive to changes in illumination (Schnapf *et al.,* 1990).

Although single photon responses are too small to be measured in cones, they are about 1.0 pA in rods, and therefore have been studied in some detail there. Single photon responses are very reliable: They occur at least 80% of the time after absorption of a photon by a rhodopsin molecule, and they almost never occur in the absence of photon absorption (Baylor, 1987). A single absorbed photon gives a highly amplified response, shutting down 3–5% of the total dark current by closing a few hundred cGMP-gated channels in a narrow band (one to a few micrometers in width) of outer segment plasma membrane near the site of absorption. Like the dim flash responses in Fig. 4 (bottom traces of each set), the single photon response has a stereotypical waveform, with a slow, S-shaped rise and a very slow decay. These complex, slow kinetics reflect the complex enzyme cascade underlying the response (see later).

When membrane voltage is recorded instead of outer segment current, the waveform resembles an inverted version of the current for dim lights, but not for light levels near or beyond saturation. When many photons are absorbed, the voltage recordings (Fig. 5) show a characteristic "nose" and "plateau." These features result from a shaping

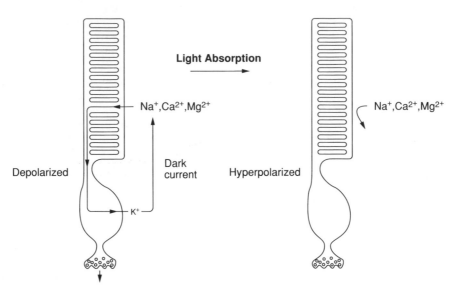

Light ⟶ ↓[cGMP] ⟶ Channels close ⟶ Hyperpolarization ⟶ ↓Transmitter release

**FIG. 2.** The absorption of light shuts down the "dark current" that circulates between the outer and inner segments of the photoreceptor. This reduction in current hyperpolarizes the cell, reducing the release of neurotransmitter.

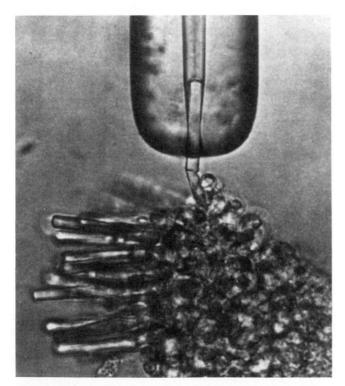

**FIG. 3.** The use of a suction electrode to record the dark current flowing into the outer segment of a rod attached to a piece of toad retina. The horizontal line just below the suction electrode is the junction between the outer and inner segments. The electrode is filled with a physiological salt solution and connected to a sensitive amplifier. (Reproduced with permission from Baylor *et al.,* 1979.)

of the response by ion channels in the inner segment. Several types of channels contribute to the waveform, but the dominant channel giving rise to this more complicated shape appears to be an inner segment, nonselective cation channel that is opened by hyperpolarization. Thus, the closure of outer segment cGMP-gated channels gives the initial hyperpolarization (beginning of the nose). This hyperpolarization opens the inner segment nonselective cation channels, which in turn depolarize the membrane, giving the transition from nose to plateau. Eventually, these channels close because of the depolarization, while at the same time the outer segment channels reopen during recovery from the light response, and the membrane potential returns to its initial dark value.

Aside from the ion channels described earlier, two other types of ion transport protein are of obvious importance to the vertebrate photoreceptor. First, $Na^+/K^+$ pumps in the inner segment are responsible for expelling the $Na^+$ that enters through the cGMP-gated channels in the outer segment. Second, $Na^+ : Ca^{2+}$, $K^+$ exchange carriers in the outer segment expel the $Ca^{2+}$ that enters through the cGMP-gated channels. The $Na^+ : Ca^{2+}$, $K^+$ carriers appear to be particularly important in light adaptation, as discussed in Section IV,B.

The ability of the visual system to detect light depends partly on its degree of adaptation to background illumination. Although much of light and dark adaptation involves pupillary responses and processes occurring in the brain and nonphotoreceptor layers of the retina, some aspects of adaptation clearly occur in the rods and cones themselves. In continuous light, there is a partial recovery of the dark current after its initial suppression when the light is switched on. This sag in the photoresponse is accompa-

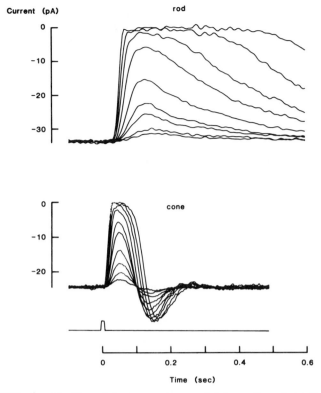

**FIG. 4.** Families of photocurrents recorded by suction electrode from a rod and a cone of the monkey, *Macaca fascicularis*. Brief light flashes were given at time zero (rectangular pulse below the current recordings). Upward deflections indicate reductions in the inward dark current. From the bottom to the top of each family, flash strengths were increased by factors of 2. Expected numbers of photoisomerizations ranged from 2.9 to 860 for the rod, and from 190 to 36,000 for the cone. Saturation of the responses (top traces in each family) occurred when all dark current was shut off. (Reproduced with permission from Baylor, 1987.)

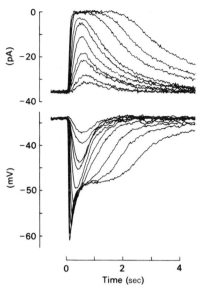

**FIG. 5.** Comparison of light-evoked changes in outer segment current with those in membrane potential for a salamander rod. The currents were measured by a suction electrode, while the membrane potential was monitored by an intracellular microelectrode. The current and voltage responses to dim flashes have similar form, but a prominent "nose" and "plateau" are seen in the voltage responses to bright flashes. 500-nm, 11-msec flashes were given at $t = 0$, and photon densities increased by factors of about 2 between 1.5 and 430 photons $\mu m^{-2}$. (Reproduced with permission from Baylor and Nunn, 1986.)

nied by a decrease in light sensitivity and an acceleration of response kinetics. Thus, the response to a light flash during continuous background light is smaller and briefer than that to an equally bright flash occurring in the dark. To make the amplitude of the flash response equal to that obtained in the dark, one must increase the flash intensity (Fig. 6). This light adaptation allows the photoreceptor to respond over a much larger range of light intensities than it otherwise could. In effect, the photoreceptor uses a nonlinear gain adjustment to partially overcome the response saturation that occurs as a result of having a finite number of cGMP-gated channels to close in the light.

Photoreceptor dark adaptation is a rather complicated set of processes. After exposure to a light bright enough to bleach all the photopigment, a photoreceptor ultimately cannot return to its dark state until nonexcited photopigment molecules have been regenerated (including reinsertion of the chromophore, see following). However, when the photoreceptor has been exposed to only moderately bright light, which leaves some photopigment unbleached, it is able to respond to light again after a briefer period of adaptation. This period is characterized by a lingering

suppression of the dark current for many minutes after the light is switched off, a decrease in light sensitivity similar to that found in the presence of background light, and an increase in outer segment current noise (Fig. 7). These features limit visual detection and may derive at least in

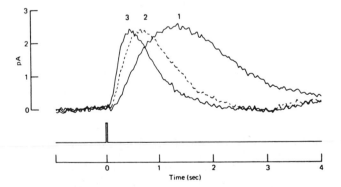

**FIG. 6.** Decreased sensitivity and accelerated flash response kinetics characteristic of adaptation to background light. These outer segment currents were obtained by suction electrode recording from a toad rod that was given a test light flash in darkness (curve 1), with a dim background light (curve 2), and with a brighter background light (curve 3). In the presence of background light, the test flash intensity had to be increased to obtain responses of approximately the same amplitude. For example, the flash intensity used to obtain curve 2 was approximately five times the intensity used to obtain curve 1. (Reproduced with permission from Baylor et al., 1979.)

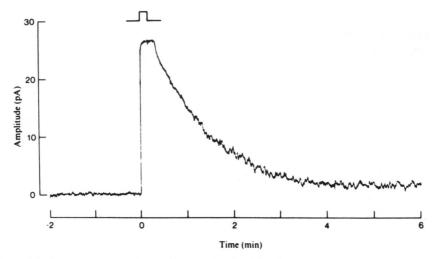

**FIG. 7.** Prolonged dark current suppression and increased noise following a strong bleach. The rectangular pulse above the response indicates when the dark-adapted rod was given a bright flash of light that bleached about 0.7% of its rhodopsin ($7.2 \times 10^5$ photons $\mu m^{-2}$, calculated to isomerize $1.8 \times 10^7$ rhodopsins). Closer examination of the lingering noise revealed a strong resemblance to that produced by the absorption of single photons. (Reprinted with permission from *Nature,* Lamb, 1980. Copyright 1980 Macmillan Magazines Limited.)

part from reversibility in rhodopsin shutoff reactions (Lamb, 1987).

## IV. Molecular Mechanisms

### A. Photopigment Activation and Shutoff

Visual pigments are integral membrane proteins called *opsins* that contain a ubiquitous light-absorbing chromophore, retinal (Fig. 8; for reviews of visual pigments, and rhodopsin in particular, see, for example, Wald, 1968; Birge, 1990; Hargrave and McDowell, 1992; Nathans,

1992). Retinal is derived from vitamin A and comes in two forms in vertebrates: retinal$_1$ and dehydroretinal, or retinal$_2$ (Dowling, 1987). Human photoreceptors use retinal$_1$. Interactions with different opsins give the retinals different spectral tuning characteristics. Thus, each type of photoreceptor has its own type of opsin, which results in a characteristic photopigment absorption spectrum, and therefore a characteristic color sensitivity of the cell. Rods contain rhodopsin (Rh) and are therefore most sensitive to blue-green light (peak wavelength around 490 nm), whereas primate cones are of three types, with characteristic peak absorbances in the short-wavelength (peak about 430 nm), middle-wavelength (peak about 530 nm), and long-wave-

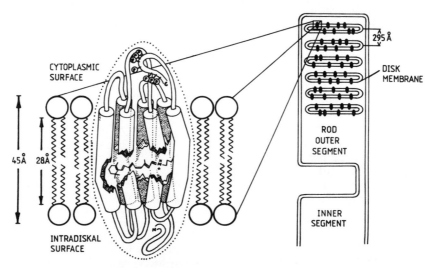

**FIG. 8.** A model for the structure of rhodopsin in the disk membrane of a rod. The light-absorbing chromophore, retinal, sits in a binding pocket near the middle of the integral membrane protein, opsin. (Reproduced with permission from Dratz and Hargrave, 1983.)

length (peak about 560 nm) regions of the visible spectrum (Fig. 9). These three types of cone are sometimes referred to as blue-, green-, and red-sensitive cones, respectively, although 560 nm actually corresponds to yellow, rather than red, light. The characteristic absorption spectra determine only the probability that a photon will be absorbed by the cell's photopigment. The photoreceptor responds in exactly the same way to any absorbed photon, independent of its wavelength.

Absorption of a photon causes isomerization of retinal from the 11-*cis* to the all-*trans* form (Fig. 10), destabilizing its position within the protein. Eventually all-*trans*-retinal hops out of its binding pocket within opsin, but a chain of conformational intermediates is generated before this dissociation occurs. For the best-studied pigment, rhodopsin, these intermediates are (in the order that follows photon absorption) photorhodopsin, bathorhodopsin, lumirhodopsin, metarhodopsin I, metarhodopsin II, and metarhodopsin III (reviewed in Hargrave and McDowell, 1992). Similar transitions are thought to occur in other visual pigments. The metarhodopsin II (meta II) conformational state activates the G-protein, transducin, as discussed later.

For the photoreceptor to reset to its dark condition after a light is switched off, the molecular players in the light response also must be switched off. Most of what we know about photopigment shutoff comes from work on rhodopsin (depicted in part of Fig. 11 and reviewed in Hargrave and McDowell, 1992; Hofmann *et al.*, 1992; Yarfitz and Hurley, 1994). The shutoff of rhodopsin's ability to activate transducin begins even before all-*trans*-retinal has left the

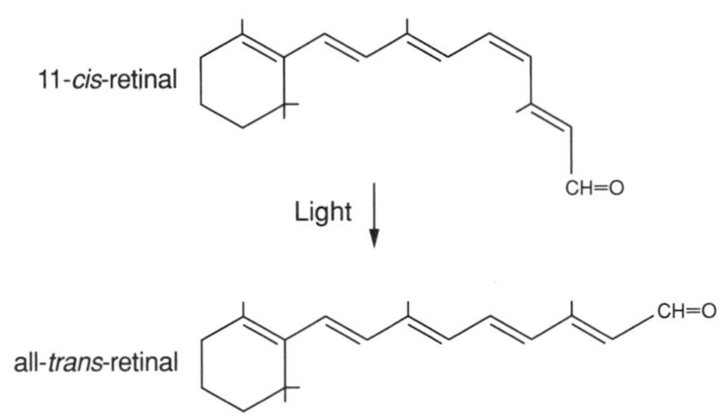

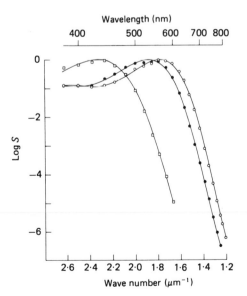

**FIG. 10.** The light-induced isomerization of retinal₁ from the 11-*cis* to the all-*trans* form. This is the first step in vertebrate visual transduction, and the only one that is a direct action of light.

protein. Rhodopsin kinase phosphorylates photoexcited rhodopsin at multiple serine and threonine residues near the C-terminal. A 48-kDa protein called *arrestin* then binds phosphorylated rhodopsin, blocking its interaction with transducin. There are recent hints for still other players in the shutoff process, but all the details have not been resolved. Once all-*trans*-retinal has separated from opsin, the photopigment cannot be reset to its photon-receptive dark state until 11-*cis*-retinal has been reinserted, arrestin has been released, and a phosphatase (serine/threonine phosphatase type 2A) has removed the phosphates from the C-terminal residues. This process takes several minutes. The new 11-*cis*-retinal that is inserted into opsin is made in the pigment epithelial cells and transported into the photoreceptors.

## B. Cyclic Nucleotide Cascade

A cyclic nucleotide enzyme cascade is used to translate the message of photon absorption into the language of the brain—electrical and chemical signals. Many publications written in the 1970s and early 1980s describe the "$Ca^{2+}$ hypothesis" (Hagins, 1972) in which $Ca^{2+}$ is proposed to be the second messenger mediating the response to light. There is now overwhelming evidence that the key second messenger is cGMP, rather than $Ca^{2+}$, and that $Ca^{2+}$ is involved in more subtle aspects of visual transduction (see later). Most of our information on the molecular mechanism of visual transduction derives from studies on rods, but the process appears to be similar in cones. A large body of biochemical and physiological evidence suggests that absorption of a single photon by rhodopsin initiates the following chain of events (diagrammed in Fig. 11):

1. Photoexcited rhodopsin (Rh*) catalyzes the replacement of GDP by GTP on transducin (T). Both Rh* and the peripheral membrane protein, T, are able to diffuse laterally in the disk membrane, where they find each other and interact. Before Rh* is shut off, it activates about 500 Ts, representing the first amplification step in the transduction process.

2. The α subunit of T, now with GTP bound, dissociates from the rest of T (the β–γ subunit complex) and binds to and removes the γ inhibitory subunit of another peripheral

**FIG. 9.** Cone spectral sensitivities, as measured by the suction electrode method. These spectra were obtained from blue , green , and red-sensitive monkey cones (squares, filled circles, and open circles, respectively). The spectral sensitivity, *S*, was derived from the cone's response to flashes of different color, and it reflects the probability of absorption of a photon of that color (or wavelength). (Reproduced with permission from Baylor, 1987.)

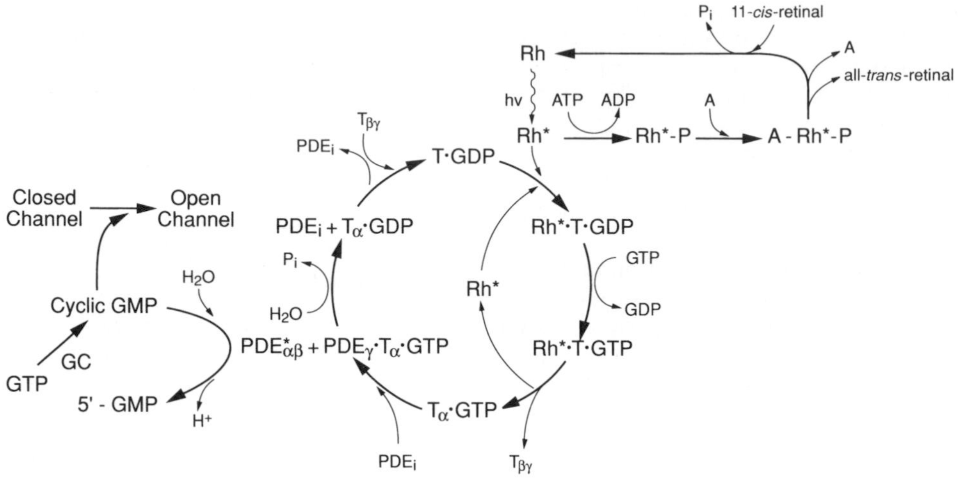

**FIG. 11.** The enzyme cascade controlling visual transduction in rods. Abbreviations: A, arrestin; channel, cGMP-gated outer-segment channel; GC, guanylate cyclase; PDE, cGMP-specific phosphodiesterase (*, active form): Rh, rhodopsin (*, photoexcited, active form); T, the G-protein, transducin. There is increasing evidence that in spite of its complexity, this diagram is quite incomplete. (Modified with permission from Stryer, 1991, and drawing on information from Hargrave and McDowell, 1992, and Hofmann *et al.*, 1992.)

disk membrane protein, cGMP-specific phosphodiesterase (PDE). Each $T\alpha$ can activate only one PDE since the catalytic portion of PDE, $PDE\alpha\beta$, is only free to work when $PDE\gamma$ is bound to $T\alpha$ (and therefore unable to reassociate with $PDE\alpha\beta$).

3. Once disinhibited, one PDE molecule ($PDE\alpha\beta$) hydrolyzes about a million cGMP molecules, leading to the closure of hundreds of cGMP-gated ion channels and thereby halting entry of over a million $Na^+$ ions during the time course of the single photon response.

The recovery of the dark current after photon absorption is accomplished by many processes, including these:

1. Shutoff of rhodopsin, as discussed earlier.
2. Inherent GTPase activity of $T\alpha$, converting $T\alpha$ back to the GDP-bound, inactive form that reassociates with $T\beta\gamma$.
3. Reassociation of $PDE\gamma$ with $PDE\alpha\beta$, returning PDE to its more inhibited, dark state.
4. Synthesis of new cGMP from GTP by guanylate cyclase (GC).
5. Reopening of cGMP-gated ion channels, as more cGMP becomes available.

Although these processes are the most established ones, there is recent evidence that other processes and factors may be involved in regulating the cGMP cascade, mostly through feedback mechanisms. Some of the factors that may play a role are regulatory cGMP-binding sites on PDE; calcium-binding proteins that may affect guanylate cyclase, PDE, rhodopsin, transducin, and/or the cGMP-gated channel; members of the inositol phosphate system whose function has not yet been established; and various kinases and phosphatases whose functions are also not yet clear. The field of regulation of the photoreceptor cGMP cascade is currently very unsettled and very intriguing. Some interesting findings on this subject are reviewed in Lagnado and

Baylor (1992), Detwiler and Gray-Keller (1992), and Yarfitz and Hurley (1994).

One aspect of visual transduction that has been proposed to be controlled by feedback regulation is light adaptation. The following argument is suggested: Coupled with calcium-binding proteins, $Ca^{2+}$ has been found to inhibit guanylate cyclase (GC) and to stimulate PDE (via inhibition of Rh phosphorylation). In the dark, when the cGMP concentration is relatively high, $Ca^{2+}$ enters the cell through the cGMP-gated channels, raising intracellular $[Ca^{2+}]$ ($[Ca]_i$). When the channels close following photon absorption, $[Ca]_i$ decreases because its influx through the channels is decreased while its extrusion by $Na^+:Ca^{2+}$, $K^+$ exchangers continues. The decreased $[Ca]_i$ would be expected to stimulate guanylate cyclase and to inhibit PDE, leading to a subsequent partial recovery of the cGMP level and a reopening of some of the channels.

While this scheme sounds very logical and appealing, some results suggest that it is very oversimplified. Detwiler and Gray-Keller (1992) have found that high $[Ca]_i$ speeds the recovery from a light response, as if facilitating the return to dark levels of [cGMP] and the consequent reopening of cGMP-gated channels. This behavior is inconsistent with the argument just presented, which predicts that high $[Ca]_i$ would *decrease* [cGMP] by stimulating PDE and inhibiting guanylate cyclase. These and other recent findings reveal that many pieces of the transduction puzzle are still missing (also see Chapter 46).

## V. Summary

The process of vertebrate vision begins with photon capture and visual transduction in the rods and cones. The rods are outstanding single photon detectors that are used in dim light, whereas the cones provide excellent visual acuity and color vision in bright light. There are slight differences in the structures of rods and cones, but these photoreceptors have the same basic design for optimal photon capture and visual transduction.

The first step in the response to light is the absorption of a photon by a visual pigment molecule. The visual pigment (or "photopigment") consists of a particular protein (an opsin) containing a ubiquitous chromophore, retinal. The differences in color sensitivity of the rods and cones result from spectral tuning of retinal by their different opsins. The photopigment absorption spectra determine the probability that a photon of a particular wavelength will be absorbed.

Visual transduction consists of a kind of signal conversion: The message of photon absorption is converted into a decrease in membrane conductance, which results in a hyperpolarization that decreases the release of neurotransmitter onto other retinal neurons. The light-induced decrease in membrane conductance is accomplished by the closure of nonselective cation channels in a specialized region of the photoreceptor called the outer segment. These channels conduct the so-called "dark current" that flows into the outer segment in the dark. The channels are kept open in the dark by cGMP, and they close in the light when the intracellular cGMP concentration falls.

The light-induced fall in [cGMP] occurs when a photoexcited visual pigment molecule activates a G-protein, which in turn activates a phosphodiesterase to hydrolyze cGMP. There are elaborate mechanisms for returning the molecular machinery of the photoreceptor to its dark state and for light and dark adaptation. Although many of the fundamental molecular steps in visual transduction have been determined, numerous mysteries still remain.

## Bibliography

Bacigalupo, J., Johnson, E., Robinson, P., and Lisman, J. E. (1990). Second messengers in invertebrate phototransduction. *In* "Transduction in Biological Systems" (C. Hidalgo, J. Bacigalupo, E. Jaimovich, and J. Vergara, Eds.), pp. 27–45. Plenum, New York.

Bader, C. R., Bertrand, D., and Schwartz, E. A. (1982). Voltage-activated and calcium-activated currents studied in solitary rod inner segments from the salamander retina. *J. Physiol.* **331**, 253–284.

Barnes, S., and Hille, B. (1989). Ionic channels of the inner segment of tiger salamander cone photoreceptors. *J. Gen. Physiol.* **94**, 718–743.

Baylor, D. A. (1987). Photoreceptor signals and vision. *Invest. Ophthalmol. Visual Sci.* **28**, 34–49.

Baylor, D. A. (1996). How photons start vision. *Proc. Natl. Acad. Sci. USA* **93**, 560–565.

Baylor, D. A., and Nunn, B. J. (1986). Electrical properties of the light-sensitive conductance of rods of the salamander *Ambystoma tigrinum. J. Physiol.* **371**, 115–145.

Baylor, D. A., Lamb, T. D., and Yau, K.-W. (1979). The membrane current of single rod outer segments. *J. Physiol.* **288**, 589–611.

Baylor, D. A., Nunn, B. J., and Schnapf, J. L. (1987). Spectral sensitivity of cones of the monkey *Macaca fascicularis. J. Physiol.* **390**, 145–160.

Birge, R. R. (1990). Nature of the primary photochemical events in rhodopsin and bacteriorhodopsin. *Biochim. Biophys. Acta* **1016**, 293–327.

Bok, D. (1985). Retinal photoreceptor-pigment epithelium interactions. *Invest. Ophthalmol. Visual Sci.* **26**, 1659–1694.

Detwiler, P. B., and Gray-Keller, M. P. (1992). Some unresolved issues in the physiology and biochemistry of phototransduction. *Curr. Opin. Neurobiol.* **2**, 433–438.

Dowling, J. E. (1987). "The Retina: An Approachable Part of the Brain." pp. 195–197. The Belknap Press of Harvard Univ. Press, Cambridge, MA.

Dratz, E. A., and Hargrave, P. A. (1983). The structure of rhodopsin and the rod outer segment disk membrane. *Trends Biochem. Sci.* **8**, 128–131.

Hagins, W. A. (1972). The visual process: Excitatory mechanisms in the primary receptor cells. *Annu. Rev. Biophys. Bioeng.* **1**, 131–158.

Hagins, W. A., Penn, R. D., and Yoshikami, S. (1970). Dark current and photocurrent in retinal rods. *Biophys. J.* **10**, 380–412.

Hargrave, P. A., and McDowell, J. H. (1992). Rhodopsin and phototransduction. *Int. Rev. Cytol.* **137B**, 49–97.

Hofmann, K. P., Pulvermuller, A., Buczylko, J., Van Hooser, P., and Palczewski, K. (1992). The role of arrestin and retinoids in the regeneration pathway of rhodopsin. *J. Biol. Chem.* **267**, 15701–15706.

Lagnado, L., and Baylor, D. A. (1992). Signal flow in visual transduction. *Neuron* **8**, 995–1002.

Lamb, T. D. (1980). Spontaneous quantal events induced in toad rods by pigment bleaching. *Nature* **287**, 349–351.

Lamb, T. D. (1987). Sources of noise in photoreceptor transduction. *J. Opt. Soc. Am. (A)* **4**, 2295–2300.

Lamb, T. D., and Pugh, E. N., Jr. (1990). Physiology of transduction and adaptation in rod and cone photoreceptors. *Neurosciences* **2**, 3–13.

McNaughton, P. A. (1990). Light response of vertebrate photoreceptors. *Physiol. Rev.* **70**, 847–883.

Nathans, J. (1992). Rhodopsin: Structure, function, and genetics. *Biochemistry* **31**, 4923–4931.

Roof, D. J., and Heuser, J. E. (1982). Surfaces of rod photoreceptor disk membranes: Integral membrane components. *J. Cell Biol.* **95**, 487–500.

Schnapf, J. L., Nunn, B. J., Meister, M., and Baylor, D. A. (1990). Visual transduction in cones of the monkey *Macaca fascicularis. J. Physiol. (London)* **427**, 681–713.

Stryer, L. (1991). Visual excitation and recovery. *J. Biol. Chem.* **266**, 10711–10714.

Wald, G. (1968). The molecular basis of visual excitation. *Nature* **219**, 800–807.

Yau, K.-W., and Baylor, D. A. (1989). Cyclic GMP-activated conductance of retinal photoreceptor cells, *Annu. Rev. Neurosci.* **12**, 289–327.

Yarfitz, S., and Hurley, J. B. (1994). Transduction mechanisms of vertebrate and invertebrate photoreceptors. *J. Biol. Chem.* **269**, 14329–14332.

Zuker, C. S. (1996). The biology of vision in *Drosophila. Proc. Natl. Acad. Sci. USA* **93**, 571–576.

# 48

## Olfactory/Taste Receptor Transduction

## I. Introduction

Many sensory receptor cells are modified epithelial cells and share properties common to epithelial tissues. These properties include the existence of two different membrane regions—**apical membrane** and **basolateral membrane.** These regions are separated by intercellular junctional complexes that also serve to bind adjacent cells. The two different regions of epithelial cells and sensory receptor cells are exposed to vastly different milieu. This is especially the case in *gustatory* and *olfactory* receptor cells. Their apical membrane tips are exposed to an external environment that can vary profoundly in chemical composition. In contrast, their basolateral membranes are bathed in a protected constant medium.

As a first generalization about olfactory and gustatory chemoreceptors, it is important to note that the apical membrane is the site where **chemosensory transduction** occurs (although there may be exceptions in taste receptor cells; see later discussion). The intercellular junctional complex restricts or confines most chemical stimulation to the specialized apical membrane. This membrane surface is exposed to an ionic medium that is quite different from that of the tissue spaces surrounding the basolateral membrane. Because chemosensory transduction is often associated with a **receptor** (or **generator) current** (that is, a translocation of ions) across the stimulated membrane, the unique ionic environment at the apical region often imposes important considerations for transduction. For example, olfactory receptor neurons utilize a **Ca-dependent $Cl^-$ conductance** that exploits the low external $Cl^-$ concentration in the mucus surrounding the cilia; this $Cl^-$ conductance generates an **inward (depolarizing) receptor current** (i.e., $Cl^-$ efflux) that supplements cation influx through odorant-activated channels. Receptor currents generated in the apical chemosensitive tips of olfactory and gustatory receptor cells flow out of the cell across the basolateral membrane.

In the case of olfactory receptors, signal transmission to the brain is a one-step process. Olfactory receptor cells communicate directly to the central nervous system (CNS) via their axons (Fig. 1). In the case of gustatory receptor cells, signal transmission involves two or more steps. Taste cells communicate via **chemical synapses** and **electrical synapses** with other taste cells, and with **sensory afferent fibers** that transmit signals to the CNS. This suggests that some degree of information processing via **lateral synaptic interactions** (i.e., cross-talk) might occur in taste buds.

## II. Taste Receptor Cells

### A. General Comments

Receptor cells for gustation are situated in tiny clusters, the taste buds, throughout the oral cavity. There are approximately 5000 taste buds in humans. Each taste bud consists of 50–100 cells. Taste buds are embedded in the lingual epithelium on papillae (**fungiform, foliate,** and **circumvallate**). Taste buds are also localized on the soft palate, uvula, epiglottis, pharynx, larynx, and esophagus.

Most of the cells in a taste bud are narrow elongate cells, extending nearly the full thickness of the lingual epithelium. The basal processes reach to the basement membrane and the apical processes stretch up to a tiny cavity in the lingual surface, the **taste pore.** This cellular morphology sets taste buds apart from the surrounding tissue and creates tiny islands of simple columnar epithelium within the larger sheet of stratified squamous lingual epithelium. Such an arrangement positions the taste cells in the electric field generated by **transepithelial ion transport** currents. This may have implications for taste transduction mechanisms (discussed later).

The apical tips of the taste cells have numerous **microvilli** that extend into the taste pore and are the primary sites of chemosensory transduction (Fig. 1). Taste cells vary in their morphological appearance and they have been categorized into groups such as **dark cells** and **light cells,** or **types I, II,** and **III** cells, based on their cytological and

**726**

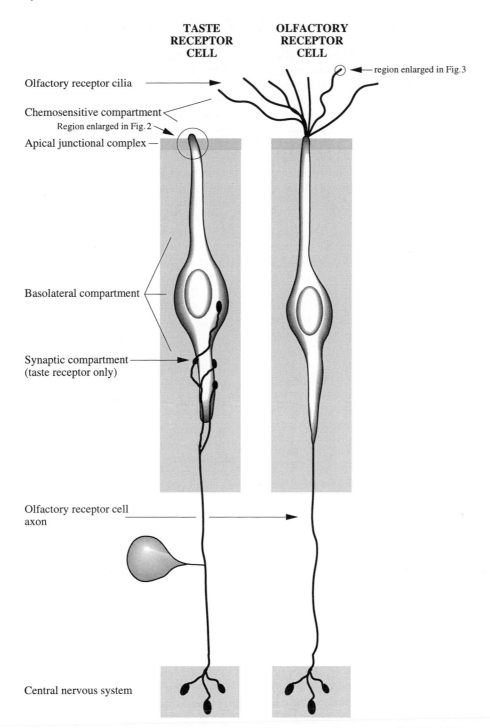

**TASTE RECEPTOR CELL**

**OLFACTORY RECEPTOR CELL**

region enlarged in Fig.3

Olfactory receptor cilia

Chemosensitive compartment

Region enlarged in Fig.2

Apical junctional complex

Basolateral compartment

Synaptic compartment (taste receptor only)

Olfactory receptor cell axon

Central nervous system

**FIG. 1.** Schematic drawing comparing taste receptor cells (left) and olfactory receptor cells (right). Three regions, or compartments, are illustrated. The apical, or chemosensitive compartment is the portion of the receptor cell extending above the apical junctional complex. The initial events in chemosensory transduction occur here. The basolateral compartment is the portion of the receptor cell below the apical junctional complex. Taste receptor cells have an additional, synaptic, compartment in the taste bud where sensory information is transmitted to other cells.

ultrastructural features. There is no consensus yet on which of these cell types are the actual chemoreceptive elements and which, if any, might be sustentacular or mucus-secreting cells. Some ultrastructural studies have reported that all taste cells possess synaptic connections and thus are believed to be chemoreceptor cells. Other studies report that only one of the cell types (type III) possesses synaptic connections and is the actual receptor cell. Recent evidence suggests that some of the cells are **neuromodulatory cells:** They release **serotonin** to modify the chemosensitivity of the taste bud (Ewald and Roper, 1994; Delay *et al.,* 1997).

Taste buds also contain a population of small ovoid **stem cells** near the base of the gustatory organ. These cells, like the adjacent nontaste epithelium, are a **renewing population.** Taste bud cells have an estimated life span of approximately 10 days. As stem cells divide and their daughter cells differentiate into mature chemoreceptive cells, they extend a long narrow process up to the taste pore. One interpretation of the morphological classes in the taste bud is that the different cell types represent different stages of maturation of taste receptor cells. Ultrastructural and physiological studies tend to support this interpretation, but the data are not yet conclusive.

A mature taste receptor cell can be thought of as having three functional regions or compartments: (1) an **apical chemosensitive membrane** where receptor currents are generated, (2) a large **basolateral membrane** across which receptor currents spread, and (3) **synaptic regions** on the basolateral membrane that transmit information to sensory axons and to adjacent cells within the taste bud. Chemosensory transduction, the focus of this chapter, is primarily concerned with the first of these compartments, the apical chemosensory membrane. However, it is important to consider the influences of the other two regions because they contribute to the gustatory signals. For example, the basolateral membrane is the site where serotonin, an important neuromodulator in taste buds, (1) acts to modify the passive (electrotonic) spread of receptor currents throughout the taste cell and (2) modulates $Ca^{2+}$ currents that are important in synaptic transmitter release.

Taste receptor cells are **excitable** and possess a wide variety of **voltage-dependent ionic conductances.** These include voltage-dependent tetrodotoxin (TTX)-sensitive $Na^+$ channels; delayed-rectifier $K^+$ channels; type A $K^+$ channels; inward-rectifier $K^+$ channels; $Ca^{2+}$-activated $K^+$ channels, $Ca^{2+}$-activated $Cl^-$ channels; and voltage-dependent $Ca^{2+}$ channels (Lindemann, 1996). Certain of these ion channels are intimately involved in the taste transduction process, as described next.

## B. Nature of Taste Stimuli

Taste stimuli consist of water-soluble chemicals usually at high concentrations (millimolar to molar). Exceptions are certain bitter-tasting substances (such as quinine, caffeine, and strychnine) and artificial sweeteners that elicit responses in the micromolar to low millimolar range. It is clear that taste cells have low stimulus sensitivity relative to the exquisite responsiveness of photoreceptors (i.e., single photons) or olfactory and pheromone chemoreceptors (i.e., picomolar stimulus concentrations, possibly single molecules). One function of taste organs is to regulate food intake. Therefore, taste receptor cells may be optimized to respond to relatively high, physiologically relevant concentrations of nutrients. The exception for bitter-tasting chemicals may be related to the fact that many deleterious substances are toxic at low concentrations. Generally, these substances elicit bitter taste, an aversive stimulus. Thus, it is understandable that bitter taste transduction is optimized

to sense low concentrations of potentially harmful chemicals.

Taste stimuli are dissolved in saliva and are delivered to the first compartment, the chemosensitive microvillar surface of gustatory receptor cells in the taste pore. Most hydrophilic stimuli are prevented from reaching the basolateral regions of the taste bud because of the **tight junctions** that seal off the apical chemosensitive membrane in the taste pore. These intercellular junctions, however, have been shown to permit the passage of certain ions, such as $Na^+$ and $Cl^-$, which can therefore penetrate into the interstitial spaces within the taste bud. It is possible that chemotransduction mechanisms for certain taste stimuli, especially $Na^+$ and $Cl^-$, exist on the basolateral membranes of taste receptor cells as well as on their exposed apical tips. Basolateral transduction mechanisms may explain **intravenous taste,** where certain chemicals that are dissolved in the blood elicit gustatory sensations. The bitter taste of some drugs injected intravenously may be an example of this phenomenon.

## C. Initiation of Taste Receptor Potentials

Unlike signal transduction in most other sensory organs, there is no single general transduction scheme that explains how taste bud cells respond to the various taste stimuli. The necessity of taste organs to sense a wide range of chemical substances, from simple ions to complex molecules, has led to the evolution of **multiple taste transduction mechanisms.** Presumably, different types of taste receptor cells possess different transduction mechanisms and respond to different taste stimuli. Recent findings from $Ca^{2+}$ imaging studies and patch-clamp recordings support this view. That is, an individual taste receptor cell responds to a limited repetoire of chemical substances. Although each taste cell itself may thus be "tuned" to a subset of chemical stimui, a taste bud contains a population of several different types of taste receptor cells. Thus, one taste bud containing up to 100 taste cells can respond to a diverse array of chemical stimuli.

Taste transduction can be divided into three basic mechanisms:

1. *Passive ion permeation through ion channels.* Some taste stimuli enter taste cells through ion channels in the apical membrane, thereby generating transmembrane receptor currents and **receptor potentials.** This is the case when the taste stimulus is itself an ion (for example, $Na^+$ or $K^+$ salts, see later discussion).
2. *Block of ion channels.* As explained later, certain ions (e.g., $Ca^{2+}$, $H^+$) *block* the passive permeation of $K^+$. This also generates a receptor potential because leak conductances for other ions such as $Na^+$, which tend to depolarize the cell, then exert a greater influence on the resting potential.
3. *Receptor-ligand binding.* Interactions occur between chemical stimuli (ligands) and membrane-bound receptors (ligand-receptor binding) that ultimately lead to receptor currents or the release of intracellular $Ca^{2+}$.

Lastly, lipophilic stimuli such as caffeine may diffuse into the cell and activate taste cells independent of these three basic mechanisms, but little is presently known about such possibilities. Taste transduction mechanisms involving ion permeation and receptor-ligand interactions are described next.

## 1. Ion Permeation and Current Flow through Apical Ion Channels

*a. Na Channels.* Sodium salts ("salty taste") are transduced in a subset of taste cells by the direct permeation of $Na^+$ through passive Na-selective channels (as opposed to TTX-sensitive, voltage-dependent Na channels). In some (but not all) taste cells, these ion channels resemble renal **amiloride-sensitive $Na^+$ channels.** Indeed, mRNA isolated from lingual epithelium shows the presence of $\alpha$, $\beta$, and $\gamma$ subunits of the amiloride-blockable Na channels found in renal and other tissues. *In situ* hybridization shows that mRNA for all three subunits is present in taste and nontaste lingual epithelium alike. Consistent with these findings, immunohistochemistry reveals the amiloride-blockable $Na^+$ channel protein in taste cells. Immunostaining in surrounding nontaste epithelium is less intense.

The apical amiloride-sensitive $Na^+$ channels are not voltage-dependent and are open at rest to allow the influx of $Na^+$, driven by this cation's electrochemical gradient. These $Na^+$ channels have small unitary conductance (1–2 pS) and are blocked by low concentrations of amiloride ($K_m = 0.2$–$1 \mu M$). The **influx of $Na^+$** at the specialized apical chemosensitive compartment of the taste cell generates a **depolarizing receptor current** in those cells that possess this channel. This current exits across the entire basolateral cell membrane, thus forming a return current loop by way of an extracellular pathway that includes the intercellular tight junctions near the taste pore. Consequently, any factors that affect the basolateral membrane conductance or the extracellular resistance will also affect the magnitude of the receptor currents.

The concentration of $Na^+$ in human saliva varies from about 3 to 63 m$M$, depending on the rate of salivary secretion. This results in an equilibrium potential for $Na^+$ ($E_{Na}$) of about $-30$ to $+50$ mV (assuming $[Na]_i$ is 10 m$M$) across the apical membrane. The resting potential is estimated to be $-60$ to $-80$ mV. Thus, there should be a steady-state influx of $Na^+$ at rest in those taste receptor cells that possess apical $Na^+$ channels. During salt (that is, NaCl) stimulation, the electrochemical gradient driving $Na^+$ influx is increased. This generates an **inward receptor current** that depolarizes the cell. The density of amiloride-sensitive $Na^+$ channels, which is under hormonal control, determines the magnitude of the influx. Specifically, arginine-vasopressin (antidiuretic hormone, ADH) and angiotensin increase the number of amiloride-blockable sodium channels, albeit via different mechanisms. Sodium conductance in taste cells may be tonically depressed by salivary $Na^+$ ("self-inhibition") but this inhibition is overridden during chemostimulation with Na salts. Accumulation of $Na^+$ within the taste cell is prevented by the action of the Na/K-ATPase distributed on the basolateral membrane compartment. Na/K-ATPase pumps excess $Na^+$ out of the cell into the interstitial spaces surrounding the taste cells. The depolarizing current produced by the $Na^+$ influx may be offset in part by the electrogenic Na/K pump (which is hyperpolarizing) and partly by the compensatory efflux of $K^+$ through basolateral $K^+$ channels (outward, hyperpolarizing $K^+$ current).

Although one might anticipate that amiloride-blockable $Na^+$ channels would be localized to the apical chemosensitive tips of the taste receptors cells in fact, immunostaining for these channels reveals their presence on apical *and* basolateral membranes. The function of the amiloride-blockable $Na^+$ channels on the basolateral membrane is not yet understood. Taste cells should be depolarized by the influx of $Na^+$ from interstitial fluid through basolateral amiloride-sensitive $Na^+$ channels.

To complicate the issue further, the taste of $Na^+$ may also be transduced via mechanisms other than by the amiloride-blockable $Na^+$ channels. For example, there also are $Na^+$ channels in taste cells that are insensitive to amiloride. These channels may consist of a related, but different, subunit composition than amiloride-blockable channels. Indeed, amiloride blocks $Na^+$ taste in some species, but not in others. In humans, the ability of amiloride to block "salty" taste *per se* is disputed, and the possible presence of amiloride-blockable $Na^+$ channels on human taste cells remains to be investigated. Also, the effects of amiloride on taste cells vary depending on the location of the taste bud: Vallate taste cells are unaffected by amiloride, whereas fungiform taste cells are exquisitely sensitive. Yet, $Na^+$ stimulates vallate and fungiform taste buds alike. The consensus to date, however, is that isoforms of amiloride-sensitive $Na^+$ channels play a dominant role in $Na^+$ taste.

In addition to their role in salt taste, the amiloride-blockable $Na^+$ channels participate in other taste qualities, particularly "sour" (discussed later). Amiloride reduces acid ("sour") taste in behavioral studies on animals and in psychophysical studies on humans.

$Na^+$ salts have an important secondary effect on taste buds apart from permeating through $Na^+$-selective ion channels in the apical tips of taste cells. The presence of $Na^+$ in saliva results in a steady transepithelial transport of $Na^+$ across the entire lingual epithelium, from mucosal (apical, saliva) to serosal (basolateral, interstitial fluid) spaces. This $Na^+$ transport generates a standing voltage across the lingual epithelium, estimated to be about 10 mV or higher (serosal side positive). Consequently, transepithelial transport imposes an electric field across taste buds that, as described earlier, represent tiny islands of simple columnar cells embedded in a stratified squamous epithelium. During NaCl stimulation, transepithelial Na transport is increased and the transepithelial voltage gradient increases, thereby increasing the electric field across the taste buds. The transepithelial voltage drives a return current through the paracellular and transcellular pathways back to the mucosal surface (Fig. 2). The net effect of the transcellular current is to hyperpolarize the basolateral membrane of taste cells, where synaptic sites are located, and reduce the potential difference across their apical membranes, where chemosensory transduction takes place.

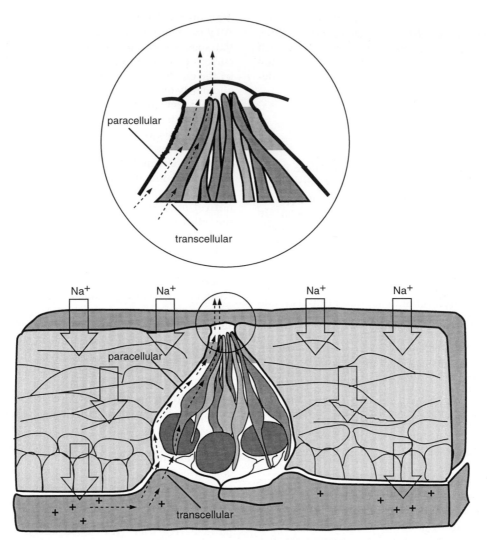

**FIG. 2.** Na$^+$ transport across the lingual epithelium can drive current loops through taste buds. Na$^+$ influx through passive apical channels is actively pumped out of the basolateral membranes of epithelial cells by electrogenic Na/K-ATPase, generating a directional flux of Na$^+$ (open arrows) and transepithelial currents (dashed lines). Current loops pass around taste cells (paracellular pathway) and through taste cells (transcellular pathway). Current through taste cells (i.e., transcellular pathway) hyperpolarizes the basolateral membrane where synapses are located, and depolarizes the apical membrane where transduction occurs. The relative contribution of current through these two pathways is determined in large part by the resistance of the junctional complex in the region of the taste pore (gray stippling in inset, top) and the resistance of the taste cell membranes. Certain taste stimuli, such as organic anions, alter the resistance of the junctional complex, and thereby change the balance of current through taste buds, possibly altering synaptic transmission and/or apical transduction.

These secondary effects oppose the direct excitation of taste receptor cells by Na$^+$ salts. Furthermore, when the electrical resistance of the paracellular current pathways is increased, for example, by the presence of organic anions (e.g., stimulation with Na$^+$ acetate), proportionately more of the transepithelial transport return current is shunted through the taste cells (transcellular pathway, Fig. 2). This even further hyperpolarizes the basolateral membrane and decreases the apical membrane potential. Thus, current loops generated by widespread transepithelial Na transport in the tongue may modulate taste cell activity, independent of any direct actions of Na$^+$ on taste receptor cells *per se*.

This modulation may have implications for salty taste and for taste mixtures when combinations of NaCl and other taste stimuli are present, as normally occurs during food intake.

*b. K Channels.* During chemostimulation with K$^+$ salts ("bitter-salty taste"), K$^+$ passively permeates through K$^+$ channels located in the apical membrane. The passive influx of K$^+$ through **apical K$^+$ channels** generates a depolarizing current. Apical K$^+$ channels are prominent in taste cells from some species; it is not known whether taste cells from mammalian taste buds also possess apical K$^+$

channels. K$^+$ may also diffuse into taste cells through K$^+$ channels located in the basolateral membrane.

**Inwardly rectifying K$^+$ channels** that open when the cell is hyperpolarized below $E_K$ are also present in taste cells, and together with the delayed rectifier and other K channels, may serve to stabilize the resting potential near $E_K$. Taste cells have a high input resistance (20–50 G$\Omega$) near their resting potential. Thus, the **receptor cell at rest is in a state of maximal sensitivity.** The inward rectifier will buffer a taste cell against influences that might tend to hyperpolarize the membrane and depress excitability. Such influences could include electrogenic Na$^+$ pumps in taste cells or transcellular currents generated by transepithelial transport (discussed earlier).

Acids ("sour taste") are transduced in a subset of taste cells by protons blocking apical K$^+$ channels. Protons also permeate the apical membrane and generate a depolarizing receptor current. Interesting, H$^+$ is believed to permeate into taste cells via amiloride-sensitive Na$^+$ channels (Fig. 3). In mixtures of Na$^+$ salts and acids, protons bind to the ion selectivity site in the channel more strongly than do Na$^+$ ions; H$^+$ permeability is less than Na$^+$ permeability. Consequently, protons interfere with the influx of Na$^+$. (This interference between Na$^+$ and H$^+$ permeation may help explain the culinary wisdom that acidifying food, for example, by adding vinegar or lemon juice, reduces its salty taste.) The fact that H$^+$ and Na$^+$ ions utilize a common ion channel pathway raises the question of how the taste of acids is distinguished from a salty (NaCl) taste. The answer may be that protons affect taste cells in additional ways independent of the amiloride-sensitive Na$^+$ channels.

For example, as already mentioned, H$^+$ also blocks K$^+$ channels in certain taste cells and this depolarizes the cells. Protons may have additional less well-characterized actions on the apical chemosensory membrane that contribute to sour taste. It remains to be determined whether different subsets of taste cells are activated by acids versus by Na$^+$ salts.

Other salts, such as those of the divalent cations, Ca$^{2+}$ and Ba$^{2+}$, are also transduced by reducing ion permeation through the apical membrane. These salts elicit a bitter taste. Ca$^{2+}$ and Ba$^{2+}$ block apical K$^+$ channels. This results in a depolarizing receptor current; that is, the resting outward repolarizing (leak) K$^+$ current that maintains the resting potential is blocked (Fig. 4).

Apparently, transduction mechanisms for salty, sour, and certain bitter substances overlap and affect many of the same ion channels. If all taste cells were to possess identical ion channels, it would be a mystery as to how taste perceptions of these substances can be differentiated. One likely answer is that different taste receptor cells may express different combinations of ion channels. Thus, different subsets of receptor cells are activated during stimulation with salts, acids, and bitter stimuli.

### 2. Membrane-Bound Receptor Events

*a. Receptor-Ligand Binding.* Certain taste stimuli interact with specific **membrane-bound receptors** rather than penetrate the plasma membrane through ion channels. Sweet tasting substances and amino acids are examples of such stimuli. For humans, the taste of amino acids varies. Some are sweet (glycine and proline), bitter (phenylalanine), or

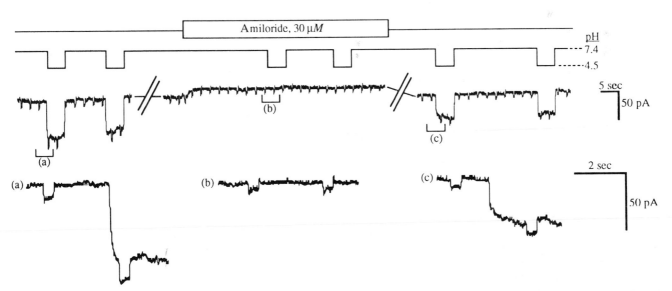

**FIG. 3.** Acid taste stimuli are transduced, in part, by proton permeation of amiloride-sensitive Na$^+$ channels. This figure shows a patch-clamp record from a taste cell in an isolated fungiform taste bud from a hamster. The cell was held at −80 mV throughout the record. Acid stimulation (2.5 mM citric acid, pH 4.5) elicits an inward current (a) due to influx of H$^+$. This response is blocked by perfusing the bath with amiloride (30 μM) (b). Washing out amiloride from the bath restores the acid response (c). The top two traces indicate the periods when amiloride and citric acid were applied. Bottom traces are expanded records to show the increases in membrane conductance during citric acid stimulation more clearly. Membrane conductance was monitored by applying a train of small, constant, hyperpolarizing pulses throughout the record. (From Gilbertson *et al.*, 1993. Copyright Cell Press.)

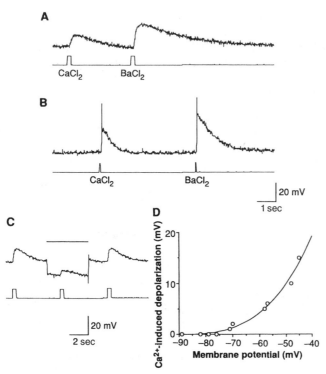

**FIG. 4.** Divalent cations, which elicit bitter taste in humans, produce depolarizing receptor potentials by blocking $K^+$ conductance. The figure shows intracellular recordings from three different taste cells in *Necturus* (A–C). Applying $Ca^{2+}$ or $Ba^{2+}$ (40–60 m$M$) focally to the taste pore produced responses, which, if large enough, elicited impulses (B). The lower traces in (A), (B), and (C) show the pulses of air pressure applied to the barrel of the stimulus micropipette containing $CaCl_2$ and $BaCl_2$. (C) Voltage dependence of the amplitude of the receptor potential induced by $CaCl_2$. Responses to $CaCl_2$ at resting potential ($-45$ mV), at a membrane potential of $-58$ mV established by passing DC current through the intracellular microelectrode (indicated by the horizontal bar), and again at resting potential. (D) Amplitude of $CaCl_2$ responses as a function of membrane potential (same cell as Part C). (Reprinted with permission from Bigiani and Roper, 1991. Copyright 1991 American Association for the Advancement of Science.)

sour (tryptophan), and others are difficult to categorize (for example, glutamate, as in monosodium glutamate). The unusual taste properties of glutamate have led many researchers to believe that this compound evokes its own unique taste, termed **umami,** that is fundamentally different from sweet, sour, salty, and bitter. Analyses based on recordings from taste receptor cells, nerve recordings, and activity of CNS neurons generally support the concept of a separate taste mechanism and taste quality for glutamate.

The most straightforward receptor mechanism is direct ligand gating of ion channels **(ionotropic channels),** similar to the action of acetylcholine at the neuromuscular junction. Arginine and proline are believed to act this way and gate nonselective cation channels, as has been shown in catfish taste cells.

More complex receptor mechanisms involve **G-protein-coupled receptors.** Taste receptor binding triggers intracellular second messenger cascades. For example, the taste of sucrose is believed to be generated in this manner. Inves-

tigators postulate the sucrose receptor is coupled to a $G_s$ protein that stimulates adenylyl cyclase (AC) and increases $[cAMP]_i$. The most plausible interpretation of studies to date is that the increase in intracellular cAMP stimulates a protein kinase that phosphorylates and blocks basolateral $K^+$ channels, thereby indirectly depolarizing the taste cell. Experimental evidence for this pathway has been gathered from electrophysiological, biochemical and $Ca^{2+}$ imaging studies on taste cells from several species. Thus, stimulating taste tissue with sucrose enhances AC activity and increases cyclic adenosine monophosphate (cAMP). Injecting cAMP intracellularly into a subset of taste cells produces a depolarization, accompanied by a decreased membrane conductance. This depolarization depends on adenosine triphosphate (ATP) and is blocked by protein kinase A (PKA) delivered through the patch pipette. In isolated patches derived from the basolateral membrane of taste cells, PKA was shown to block certain K channels. Collectively, these findings indicate a **primary role for AC and cAMP in sucrose taste transduction.**

A collection of compounds other than sucrose also elicits a preferred taste which in human studies is perceived as "sweet." Not all of these compounds are sugars, and include such diverse compounds as lead salts, certain amino acids, proteins, and artificial sweeteners (e.g., saccharin). Cross-adaptation studies and cellular analyses of taste cell responses reveal that there are multiple mechanisms, likely reflecting multiple receptors, for "sweet" taste. For example, saccharin has been shown to increase intracellular inositol trisphosphate ($IP_3$) in rat taste cells and thereby release intracellular $Ca^{2+}$. In the same taste cells, sucrose elicits $Ca^{2+}$ influx from the extracellular medium, consistent with activation of voltage-gated $Ca^{2+}$ channels. That is, there is a dual pathway in some taste cells for sucrose and saccharin, involving cAMP and $IP_3$ second messengers, respectively.

Based on the involvement of cAMP and $IP_3$ mechanisms, it is reasonable to infer that sucrose and saccharin act via heptahelical G-protein-coupled receptors.[1] To date, no receptor has been cloned or identified for the taste of sugars or nonsugar sweeteners.

Other taste stimuli, for example, certain bitter compounds, are also thought to activate membrane receptors that are coupled to G-proteins. For instance, bitter compounds such as denatonium, strychnine, and caffeine transiently generate $IP_3$ in taste cells and stimulate $Ca^{2+}$ release from internal stores, presumably via G-protein-coupled receptors other than those believed to be involved in sweet taste. The findings are consistent with the existence of bitter receptor(s) that activate phospholipase C, digesting membrane phosphatidylinositol-4-5-bisphosphate to diacylglycerol (DAG) and releasing $IP_3$ into the cytosol. An additional receptor mechanism for bitter taste has been proposed that involves a reduction of intracellular cAMP, possibly via a transducin-coupled receptor mechanism similar to that in photoreception (Kolesnikov and Margolskee, 1995; Ruiz-Avila *et al.*, 1995).

---

[1] However, there is also evidence that saccharin and other taste stimuli may penetrate the taste cell and directly activate G-proteins, bypassing any specific receptors.

Although specific membrane-bound receptors for sweet, bitter, and other tastes have not yet been identified, there is evidence that a class III metabotropic glutamate receptor transduces the taste of monosodium glutamate (*umami*). Class III metabotropic glutamate receptors are selectively stimulated by the glutamate agonist, L-2-amino-4-phosphonobutanoic acid (L-AP4). A class III metabotropic glutamate receptor, mGluR4, has recently been localized selectively to taste buds in lingual epithelium (Chaudhari *et al.*, 1996). Previously, metabotropic glutamate receptors have only been identified at synapses in the CNS and in the retina. In taste cells, this receptor appears to have been put to another use, namely, to sense free glutamate as a component of food. Activation of class III metabotropic glutamate receptors in neurons is negatively coupled to cAMP. Thus, one might predict that in taste cells, *umami* stimulation elicits a decline in intracellular cAMP. **Ionotropic (NMDA-like) glutamate receptors** may also contribute to *umami* taste. Ionotropic glutamate receptors have been identified in lingual tissue. However, unlike mGluR4, ionotropic glutamate receptors are not localized to taste buds; they are expressed in lingual epithelium bearing taste buds and epithelium free of taste buds alike. Physiological studies have revealed taste cell responses consistent with activating metabotropic and ionotropic glutamate receptors.

*b. G-Proteins.* As inferred from the preceding section, heptahelical receptors in taste are presumed to be coupled to G-proteins. A G-protein that is unique for taste, **gustducin,** has been identified in taste buds (McLaughlin *et al.*, 1992) and may be a candidate for the G-protein that couples to bitter and sweet receptors. Studies are under way to determine its specific functional role in taste transduction; gene-knockout mice lacking gustducin show deficits in bitter and sweet taste. This implies that there may be shared or interacting elements in the second messenger pathways for bitter and sweet taste stimuli. Additionally, a G-protein found in photoreceptors, **transducin,** has also been shown to be expressed in taste cells (McLaughlin *et al.*, 1993). This suggests there may be molecular mechanisms in common between photoreception and chemoreception. A likely scenario is that there are G-protein-coupled receptors in taste cells that are coupled to **phosphodiesterase** (PDE), leading to the catabolism of cAMP (cf. earlier discussion of *umami* taste). Which taste stimuli activate this pathway and how this might result in taste cell excitation remain unanswered.

Transduction mechanisms for gustatory receptor cells are summarized in Fig. 5.

### D. Termination of Taste Signals

Termination of chemostimulation in taste cells is probably due to two factors: **diffusion** of the stimulatory substance away from the chemoreceptive surfaces and receptor cell **adaptation.** Very little is known about these mechanisms. However, it is possible that a **Ca-dependent Cl⁻ conductance** that exists in taste bud cells contributes to sensory adaptation in taste (Taylor and Roper, 1994). This Cl⁻ conductance would repolarize the receptor cell even in the presence of a maintained chemical stimulus.

That is, excitatory receptor currents result in an increase in $[Ca]_i$ either by way of $Ca^{2+}$ permeation during stimulation, by influx through voltage-dependent $Ca^{2+}$ channels secondarily activated by depolarizing receptor and action potentials, or by release from intracellular $Ca^{2+}$ stores. Increased $[Ca]_i$ activates the Ca-dependent Cl⁻ conductance and thereby triggers an influx of Cl⁻ from the interstitial fluid, that is, an outward, repolarizing current. This would serve to reduce the effectiveness of a maintained excitatory stimulus.

## III. Olfactory Receptor Cells

### A. General Comments

Receptor cells for **odorants** are distributed in specialized patches of sensory epithelium embedded within the nasal respiratory epithelium. These patches comprise about 2 cm² of surface in each nostril in the human, or approximately 5% of the total nasal epithelium.

**Olfactory receptor cells** are specialized neurons. Olfactory receptor neurons extend the full thickness of the epithelium. Receptor neurons are embedded among elongate supporting cells. Olfactory receptor neurons possess apical membrane specializations for receiving chemical signals. However, unlike the microvillar surface of gustatory receptor cells, the apical tips of olfactory receptor neurons consist of a tiny knob from which 6 to 12 long cila extend (see Figure 1). The opposite, basal, pole of an olfactory receptor neuron transforms into an axon. Consequently, olfactory receptor neurons bypass any synaptic intervention or modulation in the periphery and transmit action potentials directly to the brain, and specifically to the **olfactory bulbs.** Parenthetically, this raises the interesting point that certain agents, such as drugs or toxins, can be taken up by olfactory receptor neurons and transported directly into the brain along the axons of the olfactory neurons. That is, the olfactory epithelium represents a "window" into the brain that might be exploited for pharmaceutical therapies or may be at the root of certain CNS disorders that are linked to environmental toxins.

A population of **stem cells** resides at the base of the olfactory epithelium, similar to the stratum germinativum of a stratified epithelium. Stem cells are capable of dividing and differentiating into olfactory receptor neurons. The life span of olfactory receptor neurons was originally thought to be approximately 30 days (Graziadei and Monti Graziadei, 1978), but more recent findings indicate that the mature receptor neurons may live for many months (Mackay-Sim and Kittel, 1991a,b). Mitotic activity of stem cells provides a reserve population of immature olfactory receptor neurons that can rapidly differentiate and replace mature ones if necessary, but that otherwise die.

Closely related, but separate receptor neurons exist for **pheromones.** Pheromones are airborne chemical stimuli that are key in **endocrine responses** and in shaping social and sexual interactions (e.g., territoriality, mating) among members of a species, probably including humans. Pheromone receptor neurons lie in a separate chemosensory

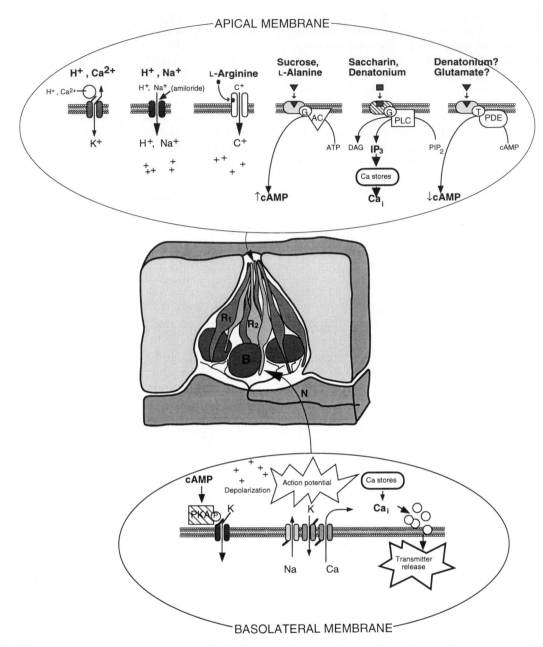

**FIG. 5.** Summary of chemosensory transduction in taste. Events at the apical membrane of taste receptor cells, top, include (shown here from left to right) $H^+$ (sour) and divalent cations (bitter) blocking $K^+$ channels; $H^+$ (sour) and $Na^+$ (salty) passing through apical, sodium-selective channels (that, in some species, are blocked by amiloride); L-arginine binding to a ligand-gated, nonselective cation channel; sucrose (sweet) and L-alanine activating G-protein-coupled receptors and stimulating AC; saccharin (sweet) and denatonium (bitter) activating G-protein-coupled receptors linked to phospholipase C, resulting in an increase in $IP_3$; lastly, denatonium and glutamate (*umami*) activating receptors that are believed to be negatively coupled to cAMP, shown here as being mediated via the G-protein transducin and PDE. Different cells are believed to possess different transduction mechanisms. Events illustrated on the basolateral membranes of taste receptor cells, bottom, include depolarization of the membrane consequent to an influx of cations at the apical membrane; reduction of $K^+$ conductance due to cAMP buildup, which also depolarizes the membrane; generation of an action potential when depolarizations reach threshold; influx of $Ca^{2+}$ ions through voltage-gated $Ca^{2+}$ channels or intracellular release of $Ca^{2+}$ from $IP_3$-gated stores; release of neurotransmitter consequent to increased $[Ca^{2+}]_i$. Abbreviations: AC, adenylyl cyclase; B, basal taste cell; $C^+$, cations; DAG, diacyl glycerol; G, G-protein; $IP_3$, inositol 1,4,5-trisphosphate; N, nerve bundle; PDE; phosphodiesterase; $PIP_2$, inositol 4,5-bisphosphate; PLC, phospholipase C; R, taste receptor cell; T, transducin.

organ, the **vomeronasal organ** (VNO). The VNO is a shallow pit of specialized epithelial cells located in the nasal cavity, anterior to the olfactory epithelium. Receptor neurons in the VNO, as in the olfactory epithelium, possess axons. However, unlike most olfactory receptor neurons, sensory neurons of the VNO possess **apical microvilli,** not cilia. The initial events of transduction are believed to occur on the microvilli. VNO sensory neurons send their axons to the **accessory olfactory bulb,** a separate portion of the olfactory bulb that is reserved for processing pheromone stimuli.

Sensory mechanisms for transducing pheromones and odorants are believed to be quite similar, both involving increases in intracellular cAMP and the opening of **cyclic nucleotide-gated (CNG) ion channels,** as discussed later. The principal differences for the purpose of this chapter are (1) olfactory receptor neurons are ciliated; VNO sensory neurons possess microvilli; (2) olfactory receptor neurons send their axons to the main olfactory bulb; VNO neurons project to a subdivision of the olfactory bulb, the accessory olfactory bulb; and (3) odorant receptors and pheromone receptors represent distinctly different classes of G-protein-coupled receptors. This chapter focuses mainly on olfactory transduction, which has been studied much more extensively to date.

The olfactory receptor neuron, like the gustatory receptor cell, can be divided into three functional compartments: (1) the chemosensitive membrane (cilia) where receptor currents are generated, (2) a conducting membrane that conveys receptor currents to the basolateral regions of the cell, and (3) an axon that propagates action potentials to the brain. The remainder of this chapter focuses on the first compartment, where odorants directly interact with the membrane.

## B. Nature of Olfactory Stimuli

Olfactory receptor neurons are generally much more sensitive to chemical stimulation than are taste receptor cells. Odorants in the range of nanomolar to micromolar concentrations can be detected. Odorants for land-dwelling animals are volatile compounds, typically diluted in large volumes of air. Odorants are only poorly soluble in aqueous solutions such as those that bathe the chemosensitive olfactory epithelium (i.e., mucous layer). The exquisite sensitivity of olfactory receptor neurons might be considered as an optimization of the sensory neuron to low concentrations of odorants.

The low concentration of odorants imposes certain constraints on the possible receptor mechanisms for olfaction. Namely, amplifying intracellular second messenger cascades are likely to be involved in transduction, as is shown to be the case later. Additionally, all three compartments of olfactory receptor neurons possess a high membrane resistance. The input resistance of olfactory receptor neurons is substantial (up to 30 GΩ). Consequently, tiny currents—some investigators believe even those generated by the opening of a single ion channel—can produce significant voltage changes sufficient to raise the membrane potential to impulse theshold. These factors contribute to

increase the sensitivity of olfactory receptor neurons to the low concentration of odorants.

Volatile odoriferous compounds partition into the mucous layer covering the olfactory epithelial surface. A specialized carrier or transport protein has been identified in the mucous layer, **odorant-binding protein** (OBP). It has been hypothesized that OBP facilitates the partitioning of odorants into the mucus. According to this hypothesis, odorants bound to OBP are carried and presented to the chemoreceptive surface on the cilia that protrude up into the mucous layer. OBP may also help rid the olfactory epithelium of odoriferous molecules (see discussion later).

## C. Initiation of Olfactory Receptor Potentials

As far as is known, olfactory transduction involves odorant binding to specific receptors localized in the ciliary membrane.

### 1. Membrane-Bound Receptors

The transduction pathways that have been characterized for olfaction to date involve G-protein-coupled receptors that are integral proteins in the ciliary membrane. When occupied by odorant, the receptor activates a G-protein. Unlike taste stimuli, olfactory stimuli do not permeate the plasma membrane through ion channels and carry charge (current). Volatile compounds, being lipophilic, probably do partition directly into the membrane, but this is not believed to be a principal mechanism in olfactory transduction.

Putative receptors for odorants and pheromones have been cloned and sequenced (Buck and Axel, 1991; Dulac and Axel, 1995). Odorant and pheromone receptors are molecules with seven transmembrane-spanning regions and members of the large G-protein-coupled receptor superfamily. There is little sequence homology between odorant and pheromone receptors; they represent different gene families. It has been estimated that there may be as many as 1000 different odorant receptor genes and 100 pheromone receptor genes. The preponderance of data suggest that any given olfactory or VNO sensory neuron expresses a single odorant or pheromone receptor, respectively. Researchers have been able to express odorant receptors in *Xenopus* oocytes and in an insect cell line (baculovirus/Sf9 system) and have shown the receptors to be activated by certain odorants. However, other laboratories attempting these experiments have not been successful. Intensive efforts, using different strategies, are presently under way to express members of the odorant receptors. It is fair to say that currently there is no consensus on the identification of a specific odorant for any of the putative odorant receptors.

### 2. G-Proteins in Olfactory Receptor Neurons

Following receptor binding, second messenger pathways are activated. Two separate G-protein pathways for olfaction exist. For one, a G-protein that is enriched in olfactory epithelium, $G_{olf}$ has been cloned and sequenced. $G_{olf}$ is believed to be critically involved in olfactory transduction. It is a member of the $G_s$ family and activates type III AC expressed in olfactory epithelium. It has long been known that exposing isolated olfactory cilia to low concentrations (in the micromolar range) of certain fruity, floral, or herbaceoous odors results in an increase in intracellular cAMP. These observations were key in establishing AC in the intracellular pathway for signal transduction in olfaction, especially for what are considered pleasant odors.

### 3. Cyclic Adenosine Monophosphate and Cyclic Nucleotide-Gated Channels in Olfactory Receptor Neurons

Cyclic AMP binds to and rapidly opens cation-selective ($Na^+$, $K^+$, $Ca^{2+}$) channels (CNG channels) from the inside of the ciliary membrane. This was demonstrated by applying cAMP and cGMP to isolated patches of membrane pulled from olfactory cilia and observing cyclic nucleotide-gated, cation-selective conductance increases. Curiously, cAMP and cGMP both activate CNG channels in olfactory cilia, although cAMP is the principal ligand for odorant activation. The cAMP-gated channels have little or no voltage dependence and have a small conductance (femtosiemens). Three or more molecules of cAMP are required to open the CNG channel. The consequence of cAMP-gated channel activity is a **depolarizing receptor current.** CNG channels from olfactory receptor neurons have been cloned and sequenced and are closely homologous with CNG channels found in photoreceptors. The principal differences between CNG channels from photoreceptors and olfactory receptor neurons is that the CNG channels in olfactory neurons are sensitive to cAMP and cGMP alike; in photoreceptors, cAMP is a much less effective ligand. It is an enigma that olfactory CNG channels are comparably sensitive to cAMP and cGMP because odorant stimulation is believed to elevate only cAMP (discussed later). A possible role for an indirect elevation of cGMP in **olfactory adaptation** may resolve this puzzle. Details of CNG channels are discussed in Chapter 46.

During odorant stimulation, $Ca^{2+}$ and $Na^+$ enter the olfactory cilium through CNG channels. This **cationic current** depolarizes the sensory cell, raising the membrane potential to threshold and initiating impulses that are conducted by the sensory neuron's axon to the olfactory bulb. The influx of $Ca^{2+}$ has several important consequences in addition to supplying inward (cationic) current. Specifically, $Ca^{2+}$ influx activates a **$Ca^{2+}$-dependent $Cl^-$ conductance** in the olfactory cilia. Because the olfactory cilia are suspended in a mucus that is low in $Cl^-$, activating the anion conductance in these cells causes an **efflux of $Cl^-$,** that is, a **depolarizing current** that complements the influx of cations. The combined action of the cAMP-gated and $Ca^{2+}$-dependent $Cl^-$ channels may be to ensure a depolarizing receptor current

even in the face of fluctuating extracellular (mucosal) cation concentrations.[2]

Equally important, the increase in intracellular $Ca^{2+}$ exerts negative feedback on the CNG channel itself. Additionally, $Ca^{2+}$ has positive and negative regulatory effects on a number of the enzymes involved in the second messenger pathways in olfactory signal transduction, particularly regarding signal adaptation. Some of these are discussed next.

### 4. Inositol Trisphosphate in Olfactory Receptor Neurons

There is evidence, albeit controversial at present, that another second messenger cascade may be activated by a different set of odorants than those that are transduced by activation of AC. The main impetus to search for alternative signal transduction pathways stems from observations that some odorants stimulate the rapid generation of $IP_3$ in olfactory receptor cilia. G-proteins other than those of the $G_s$ category, described earlier, activate the membrane-bound enzyme, **phospholipase C** (PLC). PLC hydrolyzes a phospholipid, phosphatidyl inositol 4,5-bisphosphate, in the plasma membrane of olfactory receptor neurons. The resultant byproducts, $IP_3$ and **diacyglycerol** (DAG), are powerful bioactive compounds, as discussed previously for taste transduction. However, in olfactory receptor neurons, $IP_3$ acts directly on the intracellular face of ion channels in the plasma membrane and opens $IP_3$-gated ion channels, similar to the action of cAMP on CNG channels. $IP_3$-gated, nonselective cation channels have been demonstrated in olfactory cilia immunocytochemically and in certain species, especially invertebrates, electrophysiologically. In other tissues such as taste buds (discussed earlier), $IP_3$ receptors are found on intracellular $Ca^{2+}$ storage compartments (e.g., endoplasmic reticulum) and control the release of $Ca^{2+}$ into the cytosol. No such storage sites have been demonstrated in olfactory cilia. The actions of $IP_3$ in olfactory transduction, if it is involved, are believed to be mediated by $IP_3$-gated channels in the plasma membrane.

Both mechanisms—AC → cAMP and PLC → $IP_3$—may coexist in a single receptor cell and may be activated by different odorants. This has been clearly shown to occur in *invertebrate* olfactory receptors where the two pathways mediate opposing effects—excitation ($IP_3$-gated cation channels) versus inhibition (cAMP-gated $K^+$ channels) (Ache, 1994). However, if this is also the case in mammals, it might require the expression of two or more different odorant receptors in the same cell. This would contradict the prevailing belief.

Lastly, although certain findings in vertebrate species support the involvement of $IP_3$ in olfactory signal transduc-

---

[2] As stated earlier, activating the $Ca^{2+}$-dependent anion conductance in taste receptor cells results in the opposite effect, namely an *influx* of $Cl^-$ (i.e., a repolarizing, outward current). This is because the basolateral membrane of taste cells—where the anion conductance channels are situated—is exposed to interstitial fluid. Consequently, the equilibrium potential for $Cl^-$ is believed to be near the resting potential in taste cells and opening $Cl^-$ channels will allow $Cl^-$ to enter the taste cell.

tion, the data are far from compelling. In fact, recent evidence from gene-knockout mice lacking olfactory CNG channels indicates that odorant responses to a wide selection of odors—even odors that are known to elevate intracellular $IP_3$ and complex odor mixtures—are depressed (Brunet *et al.,* 1996). The ability to inhibit odor responses by only eliminating CNG channels in olfactory cells seriously challenges the hypothesis that $IP_3$ is directly involved in mammalian olfactory transduction, at least in the simple scheme presented here.

The transduction mechanisms for olfactory receptor neurons are summarized in Fig. 6.

The transduction pathway in VNO sensory receptors is currently under investigation. Findings to date indicate a generally similar scheme involving cAMP as described previously for olfaction. Key differences include the following: type II AC is found in VNO sensory neurons, in contrast with type III AC found in olfactory receptor neurons; $G_{\alpha o}$ and $G_{\alpha i2}$ are highly expressed in VNO sensory neurons but not $G_{olf}$; only one type of olfactory CNG channel subunit (oCNC2) is expressed in VNO cells, but not another channel subunit (oCNC1) that characterizes olfactory receptor neurons (Berghard and Buck, 1996).

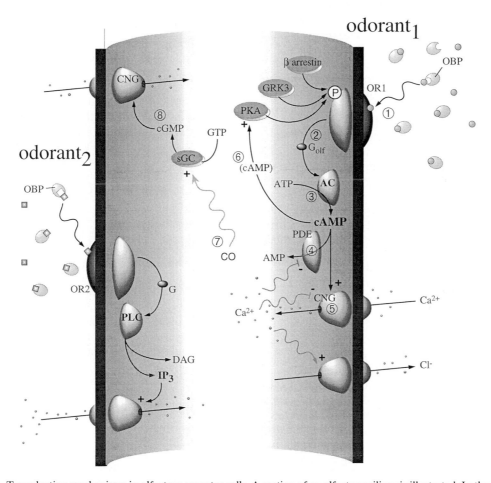

**FIG. 6.** Transduction mechanisms in olfactory receptor cells A section of an olfactory cilium is illustrated. In the mucous layer, odorants may be bound to carrier molecules, odorant binding protein (OBP). Some odorants (odorant₁) activate receptors (①) that stimulate a G-protein (②) that activates AC, converting ATP to cAMP (③). cAMP activates CNG channels, allowing an influx of $Ca^{2+}$ (also $Na^+$). cAMP is hydrolyzed by PDE (④). $Ca^{2+}$ influx opens Ca-dependent $Cl^-$ channels, allowing an efflux of $Cl^-$ (i.e., a depolarizing current) (⑤). cAMP also stimulates protein kinases that phosphorylate and desensitize the odorant receptor (⑥), thereby quenching the response. Intracellular $Ca^{2+}$ decreases the long-term responsiveness of the olfactory receptor neuron, leading to adaptation. $Ca^{2+}$ accomplishes this by exerting negative feedback on the CNG channel, positive feedback on PDE, and concentration-dependent effects on AC. These actions are shown by gray wavy lines. Odorant activation also generates carbon monoxide (⑦), which stimulates soluble guanylyl cyclase (sGC) to produce cGMP (⑧). cGMP exerts a tonic, subthreshold excitatory influence on CNG channels, allowing a small but steady influx of $Ca^{2+}$ into the olfactory receptor neuron. Other odorants (odorant₂) may activate receptors that are coupled via a G-protein to PLC. When these receptors are activated, the intracellular second messenger, $IP_3$, is formed. $IP_3$ opens channels in the olfactory receptor neuron membrane. The existence and significance of odorant-activated $IP_3$ pathways in olfactory transduction in invertebrate species is currently under debate.

## D. Termination of Olfactory Signals

Olfactory receptor excitation is terminated by a number of mechanisms. An obvious one is the unbinding and disappearance of the odorant from the chemoreceptive surface of the cilia, perhaps aided by OBP (see Fig. 6). The removal of odorants is complicated by the lipophilic nature of most odoriferous compounds; they will tend to partition into the plasma membrane and thus may linger in the vicinity of the receptors. Although the details are only now unfolding, one proposed mechanism for ridding the chemosensitive surfaces of odorants is the **enzymatic modification** of odorants by broad-spectrum biotransformation enzymes found in adjacent supporting cells. These enzymes are similar to the **detoxification enzymes** found in the liver. Olfactory epithelial supporting cells adjacent to receptor neurons contain high concentrations of the detoxifying enzymes glutathione transferase, cytochrome P-450, and UDP gluconosyl transferase (Lazard *et al.*, 1991).

Besides disappearance of the odorant, **receptor desensitization** and olfactory receptor neuron **adaptation** can also terminate signaling. Adaptation mechanisms are believed to be activated by the increase in intracellular $Ca^{2+}$ secondary to $Ca^{2+}$ influx during odorant stimulation. $Ca^{2+}$ inhibits CNG channels in olfactory cilia. Thus, an increase in $Ca^{2+}$ in the ciliary cytosol will depress odorant responses. $Ca^{2+}$ also affects the second messenger enzyme cascade. At high concentrations (millimolar range), $Ca^{2+}$ inhibits AC, thereby reducing the continued generation of cAMP during maintained stimulation. However, at nanomolar to micromolar concentrations, $Ca^{2+}$ has the oppostive effect— namely, it stimulates AC. Consequently, the net effect of $Ca^{2+}$ influx on AC activity during odor stimulation is concentration dependent. Lastly, $Ca^{2+}$ influx stimulates PDE, thereby accelerating the degradation of cAMP and reducing the odorant response.

An intriguing role for the novel gaseous neurotransmitter, **carbon monoxide** (CO) in $Ca^{2+}$-mediated adaptation has recently been proposed (Leinders-Zufall *et al.*, 1996). CO, like the related neurotransmitter nitric oxide, NO, is a highly diffusible substance and is generated during olfactory transduction. It activates **soluble guanylyl cyclase** (sGC) and leads to the production of cGMP in the stimulated olfactory sensory cell as well as in surrounding cells. At first glance, this would seem contrary to adaptation and would represent positive feedback during odorant stimulation because cGMP opens CNG channels. However, activation of CNG channels via CO → sGC → cGMP produces only a low-level, subthreshold inward current. This maintains a persistant leakage of $Ca^{2+}$ into the sensory neurons, chronically reducing CNG channel activity. The rapid transient influx of $Ca^{2+}$ through CNG channels (the immediate effect of odorant stimulation), combined with the sustained influx via cGMP activation of these channels, overall leads to a long-lasting (minutes) reduction of the sensitivity of olfactory receptor neurons after the initial excitation.[3]

---

[3] Until recently, it was believed that NO played this role in olfactory adaptation. However, evolving data utilizing blockers of CO synthesis instead now implicate CO, not NO, as the diffusible intra- and intercellular controller of CNG channel activity in olfactory receptor neurons.

As a final note, $Ca^{2+}$ influx will also activate $Ca^{2+}$-dependent ion channels in the olfactory receptor neuron, independent of its negative feedback on CNG channels. An increase in cytosolic $Ca^{2+}$ will open Ca-dependent $K^+$ channels that conduct outward repolarizing currents. Counterbalancing this, $Ca^{2+}$ activation of $Cl^-$ conductances in the olfactory cilia produces inward depolarizing currents, as described previously.

Thus, the actions of $Ca^{2+}$ are complex. The net effect of $Ca^{2+}$ entry during olfactory stimulation appears to be to reduce the responsiveness of olfactory receptor neurons perhaps as much as 20-fold, moving the stimulus–response relationship to the right.

Direct **receptor desensitization** following odorant stimulation also plays a key role in terminating odorant responses. The increase of cAMP produced by odor stimulation, along with having the short-term action of directly opening CNG cation channels, secondarily activates PKA. PKA and a specialized G-protein-coupled receptor kinase, GRK3,[4] phosphorylate odorant-bound receptors (Fig. 7). These protein kinases, together with regulatory proteins such as arrestin, render the odorant receptor inactive and quench the subsequent transduction cascade. Receptors are resensitized by the action of phosphatases. Thus, a cycle of phosphorylation/dephosphorylation controls the active state of odorant receptors and, ultimately, the responsiveness of olfactory receptor neurons.

## IV. Summary

Taste and olfaction share certain common features. For example, transduction mechanisms in both these chemical senses involve G-protein-coupled receptors and increases in intracellular cyclic nucleotides. This is the case for certain sweet taste stimuli and many odoriferous stimuli. $IP_3$ mechanisms may also play a role. In taste, $IP_3$ releases intracellular $Ca^{2+}$. In olfaction, $IP_3$ is believed to gate specific channels.

Transduction channels and integral membrane receptors appear to be concentrated on the exposed apical tips of both types of chemosensory receptor cells, although the basolateral membranes of taste cells may be additional sites for chemosensory transduction. In some regards, sensory transduction in taste and olfaction is similar to that in other modalities, notably photoreception. Photoreceptors, taste receptor cells, and olfactory receptor neurons possess ion channels that are modulated by cyclic nucleotides. Further, transducin, the G-protein in photoreceptors, is also found in taste cells.

However, there are key differences between olfaction and taste. Olfactory receptor neurons are optimized for low concentrations of chemical stimuli, taste receptor cells for higher concentrations. Furthermore, whereas an increase in cAMP opens cation channels in olfactory receptor neurons, cAMP closes $K^+$ channels in taste receptor cells (by activating PKA, leading to channel phosphorylation).

---

[4] GRK3 was formerly named $\beta$-adrenergic receptor kinase 2, or $\beta$ARK2.

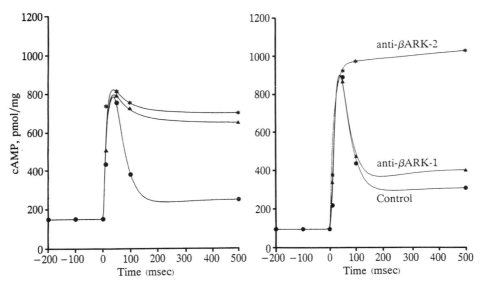

**FIG. 7.** Desensitization of odorant receptors during prolonged stimulation. Olfactory second messenger signaling is quenched by phosphorylation cascades. Left, inhibiting protein kinase activity with Walsh inhibitor or heparin markedly prolongs increases in cAMP induced by odorant stimulation in cilia preparations from the rat. ●, odorant alone (1 $\mu M$ citralva); ▲, odorant stimulation of cilia pretreated with Walsh inhibitor (3.8 $\mu M$); *, odorant stimulation of cilia pretreated with heparin (1 $\mu M$). Right, pretreating the cilia preparation with antibodies to GRK3 (previously named $\beta$ARK-2) specifically prevents desensitization of the odorant receptor and prolongs its activation (measured here as an increase in cAMP). ●, odorant alone (1 $\mu M$ citralva); ▲, odorant stimulation of cilia pretreated with anti-$\beta$ARK-1 antibodies (diluted 1 : 5000); *, odorant stimulation of cilia pretreated with anti-$\beta$ARK-2 antibodies (diluted 1 : 5000). (Modified from Schleicher *et al.,* 1993. Copyright 1993 National Academy of Sciences, U.S.A.)

An additional difference is that taste cells possess ligand-gated (ionotropic) ion channels and ion channels that are either permeated or blocked by certain chemical stimuli (especially salts). These mechanisms do not participate in olfactory transduction.

Current topics of intensive research in the peripheral transduction mechanisms for taste and olfaction include extending our knowledge of the G-protein-coupled effector cascades in the receptor cells; discovering the molecular structure of olfactory and taste receptor proteins; understanding how taste and olfactory receptor neurons distinguish among the multitude of different chemical stimuli; learning how the output from the peripheral sensory organs is encoded; and, for taste cells, determining the synaptic mechanisms that transmit signals between cells within the taste bud and on to sensory afferent fibers that innervate the taste bud.

## Bibliography

Ache, B. (1994). Towards a common strategy for transducing olfactory information. *In* "Seminars in Cell Biology" (H. Breer, Ed.), pp. 55–63, Academic Press, San Diego.

Berghard, A., and Buck, L. B. (1996). Sensory transduction in vomeronasal neurons: Evidence for Gao, Gai2 and adenylyl cyclase II as major components of a pheromone signaling cascade. *J. Neurosci.* **16,** 909–918.

Bigiani, A., and Roper, S. D. (1991). Mediation of responses to calcium in taste cells by modulation of a potassium conductance. *Science* **252,** 126–128.

Breer, H. (1994). Biology of the olfactory system. *In* "Seminars in Cell Biology" (H. Breer, Ed.). Academic Press, San Diego, pp. 1–2.

Breer, H., and Boekhoff, T. (1992). Second messenger signalling in olfaction. *Curr. Biol.* **2,** 439–443.

Breer, H., Raming, K., and Kireger, J. (1994). Signal recognition and transduction in olfactory neurons. *Biochim. Biophys. Acta* **1224,** 277–287.

Brunet, L. J., Gold, G. H., and Ngai, J. (1996). General anosmia caused by a targeted disruption of the mouse olfactory cyclic nucleotide-gated cation channel. *Neuron* **17,** 681–693.

Buck, L., and Axel, R. (1991). A novel multigene family may encode odorant receptors: A molecular basis for odor recognition. *Cell* **66,** 175–187.

Corey, D. P., and Roper, S. D. (1992). "Sensory Transduction." Rockefeller University Press, New York.

Chaudhari, N., Yang, H., Lamp, C., Delay, E., Cartford, C., Than, T., and Roper, S. (1996). The taste of monosodium glutamate: Membrane receptors in taste buds. *J. Neurosci.* **16,** 3817–3826.

Dahmen, N., and Margolis, F. L. (1992). Expression of olfactory receptors in *Xenopus* oocytes. *J. Neurochem.* **58,** 1176.

Delay, R. J., Kinnamon, S. C., and Roper, S. D. (1997). Serotonin modulates voltage-dependent calcium current in *Necturus* taste cells. *J. Neurophysiol.* in press.

DeSimone, J. A. (1991). Transduction in taste receptors. *Nutrition* **7,** 146–147.

Dhallan, R. S., Yau, K.-W. Schrader, K. A., and Reed, R. R. (1990). Primary structural and functional expression of a cyclic nucleotide-activated channel from olfactory neurons. *Nature* **347,** 184–187.

Dionne, V., and Dubin, A. E. (1994). Transduction diversity in olfaction. *J. Exper. Biol.* **194,** 1–21.

Dulac, C., and Axel, R. (1995). A novel family of genes encoding putative pheromone receptors in mammals. *Cell* **83,** 195–206.

Ewald, D. E., and Roper, S. (1994). Bidirectional synaptic transmission in *Necturus* taste buds. *J. Neurosci.* **14,** 3791–3801.

Farbman, A. (1990). Olfactory neurogenesis: Genetic or environmental controls? *Trends Neurosci.* **13,** 362–365.

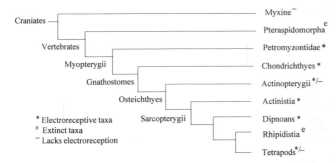

**FIG. 1.** Cladogram illustrating the distribution of electroreception among craniates. Electroreception in tetrapods is found only in nonanuran amphibians and montreme mammals.

The tuberous organs detect changes in the intensity and temporal pattern of the electric field produced across the body by the electric organ discharge (EOD), whether by the presence of items in the field of objects with different conductances than the water, or by the addition of the electric organ discharges of another individual. Such forms of electrodetection responding to alterations of the standing EOD are referred as electroreception in the **active mode** (Kalmijn, 1988). In this chapter we consider the morphology and physiology of both ampullary and tuberous electroreceptors. Space limitations preclude a discussion of the central processing of electrosensory information, which has become one of the most engrossing stories in modern neuroethology. The reader is referred to several excellent recent reviews on this topic (Bullock and Heiligenberg, 1986; Zakon, 1988; Heiligenberg, 1991).

The ability to detect geomagnetic fields directly, or *magnetoreception,* has been described in a number of different taxa (for review, see Tenforde, 1989). These studies have been primarily behavioral; no studies exist concerning the physiology of magnetoreceptors or magnetoreception, and even the behavioral studies have been difficult to replicate or confirm (Moore, 1988). The direct identification of a magnetoreceptor organ in these taxa has itself been problematic; such organs are believed to be associated with localized deposits of magnetite ($Fe_3O_4$). Localized magnetite domains have been described in bees (Gould *et al.,* 1978), salmon (Kirschvink *et al.,* 1985), tuna (Walker *et al.,* 1984), turtles (Perry *et al.,* 1981), pigeons (Walcott *et al.,* 1979, Presti and Pettigrew, 1980), dolphins (Zoeger *et al.,* 1981), and humans (Baker *et al.,* 1983), but the organs, transduction mechanisms, and innervation have yet to be positively identified. A recent description of a putative magnetoreceptor in honeybees (Hsu and Li, 1994) has been disputed (Nesson, 1995; Nichol and Locke, 1995; Kirschvink and Walker, 1995). The exception is the description of magnetotactic behavior in magnetite-containing bacteria (Blakemore, 1975; Kalmijn and Blakemore, 1978). However, magnetoreception is believed to be possible in some electroreceptive taxa via the detection of electric fields induced by an animal's movement through a geomagnetic field (see discussion later).

## II. Ampullary Electroreceptors

The wide phylogenetic distribution of ampullary electroreceptors among extant vertebrate taxa suggests that this class of electroreceptor has served important biological functions for hundreds of millions of years and has subsequently "re-evolved" several times (Fig. 1; also see Fig. 8 later). With the exception of weakly electric fishes, which also possess tuberous electroreceptors, most species with ampullary electroreceptors lack electric organs. Thus, behaviorally relevant electric field stimuli for most species are thought to originate primarily from extrinsic sources. Ampullary electroreceptors are known to be important for the detection of prey (Kalmijn, 1971; Peters and Meek, 1973; Tricas, 1982), mates (Tricas *et al.,* 1995), and orientation to local inanimate electric fields (Kalmijn, 1982; Pals *et al.,* 1982). In addition, the ampullary electroreceptor system is theoretically capable of mediating navigation by detecting electric fields induced by movement of the animal through the earth's magnetic field (Kalmijn, 1974, 1982; Paulin, 1995), which would represent a form of electroreception in the **active mode,** as discussed earlier (Kalmijn, 1988).

### A. Morphology

The ampullary electroreceptor organ in lampreys is known as an **end bud** (Fig. 2) and differs considerably in morphology from the ampullary electroreceptors found in cartilaginous and teleost fishes (Ronan and Bodznick, 1986). Each end bud consists of numerous support cells and 3 to 25 sensory cells in the epidermis that are in direct contact with the surrounding water. Individual receptor cells have numerous small microvilli on the apical surface but lack a kinocilium. Small groups or lines of end buds are distributed over the head and body surface with multiple buds being innervated by a single sensory lateral line nerve fiber (Ronan, 1986; Bodznick and Preston, 1983). Little is known about transduction in end buds, but the similarity of their responses to primitive gnathostome electroreceptors supports a homologous transduction mechanism (see discussion later). It is not known whether end buds represent the primitive electroreceptor state or whether they are a derived condition unique to the lampreys, which first appeared in their current form in the Carboniferous era considerably after the appearance of jawed fishes (Moy-Thomas and Miles, 1971).

The anatomy and physiology of primitive ampullary electroreceptor organs have been studied most extensively in the chondrichthyan fishes. Chondrichthyan fishes include the elasmobranchs (rays, skates, and sharks) and the rat fishes (Fields *et al.,* 1993), all of which possess ampullary electroreceptor organs of a similar morphology. In elasmobranch fishes the electroreceptive unit is a prominent and highly specialized structure known as an **ampulla of Lorenzini** (Fig. 3). The ampulla proper in the marine skate is composed of multiple **alveolar sacs,** which share a common lumen (Waltman, 1966). The apex of each ampulla chamber is connected by a highly insulated **marginal zone** to a single subdermal **canal,** which is approximately 1 mm in

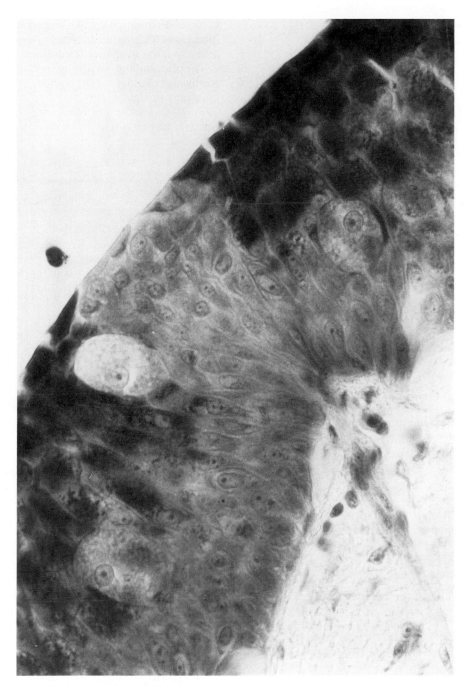

**FIG. 2.** Light photomicrograph of the electrosensory end bud of the silver lamprey. Scale bar-25 $\mu$m. (From Ronan, 1986. Copyright © 1986 John Wiley & Sons. Reprinted by permission of John Wiley & Sons, Inc.)

diameter and terminates as a small epidermal pore. The canal wall is 1–2 $\mu$m thick and composed of two layers of flattened epithelial cells, which are separated by a basement membrane to which the lumenal layer is also united by tight junctions. Both the canal lumen and the ampullary chambers are filled with a $K^+$-rich, mucopolysaccharide, jelly-like matrix that is secreted by the superficial layer and has an extremely low resistivity (Murray and Potts, 1961; Waltman, 1966). These remarkable morphological and cytological features make each canal a low-resistance core conductor that connects the internal ampullary lumen

with the surface pore. The sensory epithelium within the alveolus is composed of two cell types which form a monolayer that is approximately 15 $\mu$m thick (Fig. 4). The vast majority of the alveolar surface is formed by accessory cells that are highly resistive to transmembrane currents and are bound together by tight junctions that prevent ionic leakage across the lumenal and basal surfaces of the epithelium. Interspersed among the accessory cells are flask-shaped receptor cells (thought to be modified hair cells), which possess a single kinocilium on the apical surface and lack microvilli. This physical arrangement results

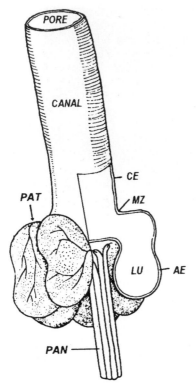

copious distribution of the ampullary pores over the cephalic surface provide an extensive array of receptors with a high degree of spatial resolution.

The morphology of the ampullary electroreceptors in freshwater elasmobranchs is thought to reflect sensory adaptations to their highly resistive environment (Kalmijn, 1974, 1982; Raschi and Mackanos, 1989). The freshwater rays, *Pomatotrygon* and *Dasyatis garouaensis*, have a hy-

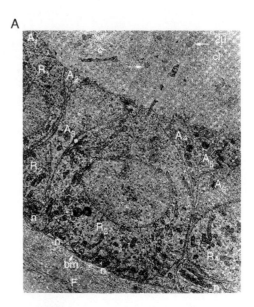

**FIG. 3.** Ampulla of Lorenzini from the marine skate, *Raja.* The ampulla proper consists of multiple alveoli formed by the alveolar epithelium (AE). A high-resistance marginal zone (MZ) connects the sensory walls of the ampulla to the high resistance canal epithelium (CE), which projects to the surface of the skin and terminates as a small pore confluent with the surrounding water. The ampulla lumen (LU) and canal are filled with a highly conductive jelly, which makes the lumen isopotential with voltages present at the pore. Myelinated primary afferent neurons (PANs) innervate the base of the ampullae and their unmyelinated primary afferent terminals (PATs) receive chemical excitation from the basal region of the sensory cells in the epithelial layer. (Modified from Waltmann, 1966.)

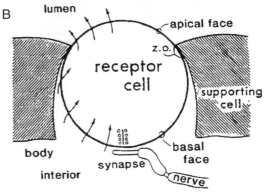

in only a small fraction of the receptor cell surface being exposed to the ampullary chamber.

The basal membrane surface of the receptor cell forms a ridge seated in a postsynaptic invagination that is separated by a distance of 100–200 Å (Waltmann, 1966). A **synaptic ribbon** about 250 Å wide and 2 $\mu$m in length is located within the presynaptic ridge. A single layer of synaptic vesicles covers the ribbon, and exocytotic release of chemical neurotransmitter contained within these vesicles depolarizes the postsynaptic membrane of the innervating fibers of the anterior lateral line nerves. Unlike the hair cell receptors of the mechanosensory lateral line and octaval systems, all ampullary electroreceptors, both primitive and derived, lack efferent innervation.

Chondrichthyan fishes typically possess hundreds of ampullae that are associated in specific **regional clusters** that are closely bound by a dense matrix of connective tissue (Fig. 5A). From these clusters the subdermal canals radiate omnidirectionally and terminate in surface pores on the head, and on the enlarged pectoral disk of rays and skates. The multiple orientations of the receptor canals, and the

**FIG. 4.** Receptor cell of the skate ampulla of Lorenzini. (A) Photomicrograph of flask-shaped receptor cells (R) and adjacent accessory cells (A) are united by tight junctions to form the alveolar epithelium. A single kinocilium projects from each receptor cell into the lumen and, together with a small portion of the apical surface, is exposed to electric stimuli. Primary afferent neurons (n) innervate the basal portion of the receptors. The basement membrane (bm) and fibroblast cells (f) lie beneath the sensory epithelium. (From Waltmann, 1966.) (B) Diagrammatic representation of the receptor cell illustrating current flow during excitation. In the case of the elasmobranch, a cathodal (negative) stimulus relative to the basal region of the receptor excites the apical surface of the cell, which causes an increase in outward current flow (arrows). This in turn causes release of chemical transmitter at the cell synapse and excitation of the primary afferent. Anodal potentials in the lumen will decrease outward current flow at the apical surface and subsequently decrease the rate of transmitter release. (Reproduced from *The Journal of General Physiology*, 1972, vol. 60, pp. 534–557, by copyright permission of The Rockefeller University Press.)

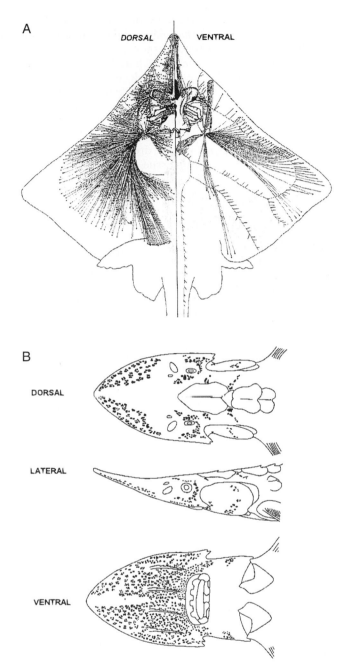

**FIG. 5.** Distribution of ampullary receptors on nonteleost fishes. (A) In the skate, *Raja laevis,* ampullary canals originate at specific regional clusters and radiate in multiple directions. These canals function to measure voltage gradients that pass across the body of the animal. (Adapted from Raschi, 1986). (B) In the sturgeon, *Scaphirhynchus,* ampullary canals are extremely short and are not arranged in clusters. Instead they measure transcutaneous potential relative to the internal potential of the animal. (From Northcutt, 1986. Copyright 1986 John Wiley & Sons. Reprinted by permission of John Wiley & Sons.)

pertrophied, thick epidermis that functions to increase transcutaneous electrical resistance. The ampullary electroreceptors are greatly reduced in size and referred to as **miniampullae** or **microampullae,** which are distributed individually across the skin rather than in clusters, and have

very short canals (about 0.3–2.1 mm long) that traverse the integument.

The anatomy and organization of ampullary electroreceptor organs in other nonteleost fishes and amphibians are generally similar to those of elasmobranch fishes, with which they are believed to be homologous. Ampullary electroreceptors in chondrostean (sturgeon and paddlefishes), cladistian (bichirs), and dipnoan fishes (lungfishes) share in most respects a similar morphology among alveoli, canals, and ampullary pores. However, the ampullary organs in these fishes are most commonly arranged as single units (as opposed to large clusters) and are located immediately beneath the skin with very short (generally <0.25 mm) and small diameter (generally <0.14 mm) canals (Fig. 5B). The abundance and distribution of receptors to the head region with the exception of the lungfishes, in which there are single ampullary electroreceptors on the head and small groups consisting of 3 to 5 ampullae scattered widely over the body (Pfeiffer, 1968; Northcutt, 1983). Ampullary electroreceptors of the head are innervated by a ramus of the anterior lateral line cranial nerve, while those on the body are innervated by a recurrent branch of the anterior lateral line nerve complex. In the bichir, *Polypterus* (Cladistia), there are about 1000 ampullae on the head region (Northcutt, 1986). Among the chondrostean fishes, electroreceptor organs in sturgeon, *Scaphirhynchus,* are arranged in about 1300 clusters of about 20 ampullae each (Northcutt, 1986), whereas in the related paddlefish, *Polyodon,* there are 50,000 to 75,000 ampullae on the elongate rostral "paddle," which are also arranged in small clusters (Nachtrieb, 1910; Jørgensen *et al.,* 1972). In the marine coelacanth, *Latimeria,* the "rostral organ" located between the eye and olfactory organ represents a complex of three principal canals that end centrally in small sensory crypts (Millot and Anthony, 1956), and is thought to be a homologous structure to the elasmobranch ampullae of Lorenzini (Bemis and Heatherington, 1982).

There is significant variability also in ampullary receptor cell morphology, particularly at the level of the apical membrane. Ampullary receptor cells in bichirs possess both a kinocilium and microvilli (Jørgensen, 1982), whereas those of chondrostean sturgeons (Teeter *et al.,* 1980) and paddlefish (Jørgensen *et al.,* 1972; Jørgensen, 1980) possess only a kinocilium as in the elasmobranchs. The receptor cells in lungfishes lack a kinocilium, but possess microvilli as in the jawless lampreys (Jørgensen, 1984). The receptor cells of the urodele amphibians (salamanders) are highly variable in morphology (Istenic and Bulog, 1984), whereas the tropical subterranean gymnophion have only microvilli (Fritsch and Münz, 1986). The primitive condition for electroreceptors is generally thought to be one possessing both kinocilium and microvilli (like other hair cells), but the reason for the loss of either kinocilium or microvilli in the various taxa and possible physiological ramifications are not known. All primitive electroreceptor cells also possess synaptic ribbons at the basal cell region, although some variation in synaptic morphology has been observed.

The fine structure of ampullary electroreceptors in teleost fishes closely resembles that of the freshwater elasmobranchs (Szabo, 1974) and other nonteleost taxa. However, these receptors are not homologous to the primitive receptors but represent a case of parallel homoplasy, presumably the result of developmental and functional constraints necessary for the detection of extrinsic electric fields and their derivation from the hair cell receptors of the lateral line. The organs are located at the level of the basement membrane of the epidermis with a very short canal (usually about 200 $\mu$m) connecting to a pore upon the skin surface. The cells of the inner walls of the ampulla and canal consist of three to five layers of flattened epithelial cells connected by tight junctions preventing current leakage across the canal wall, and the canal is filled with a conductive jelly that provides a low-resistance pathway through the lumen (Pfeiffer, 1968). The parallelism of ampullary electroreceptors among both primitive and derived groups is further demonstrated in the few existing species of electroreceptive marine teleosts. In the marine catfish, *Plotosus,* the ampullary canals have elongated, forming long subdermal tubules terminating centrally in alveolar clusters strikingly similar to the ampullae of Lorenzini in marine elasmobranchs (Obara, 1976). These teleosts can detect electric field stimuli at 80 $\mu$V/cm (Kalmijn, 1988), which is much more sensitive than the ampullary system of freshwater teleosts and approaches that of the elasmobranchs at 5 $\mu$V/cm (Kalmijn, 1982).

The ampullary receptor cells of teleosts are located in the base of the alveolus and are connected to the supporting cells via tight junctions, with only a small portion of their apical face exposed to the lumen. Teleost electroreceptor cells generally possess only microvilli; kinocilia are known only in *Xenomystus* (Notopteridae). The synaptic structure of receptors in more recently derived fishes is similar to those of the more primitive species, in which synaptic ribbons and presynaptic membrane evaginations are surrounded by a prominent postsynaptic "cup" (Szabo, 1974). Unlike primitive ampullary electroreceptor cells, ampullary receptors in teleosts may be innervated by either anterior or posterior lateral line nerves depending upon location on the body surface. Like most other nonteleost fishes the ampullae are distributed widely over the head, but differ in that they are usually distributed across the trunk in distinct patterns that are species specific.

## B. Physiology

In marine elasmobranchs such as the skate, *Raja,* the resistance of the skin is only moderately higher than that of the body tissues (Kalmijn, 1974). Imposition of a uniform extrinsic field results in a voltage gradient across the skin and internal tissues of the body. In contrast, the ampullary canals act as **core conductors** via the highly conductive environment of the conductive jelly that is bounded by the high-resistance canal walls that lead into the ampulla. The high resistance of the alveolar surface to current leakage results in the lumen being effectively isopotential with the voltage at the ampullary pore on the skin. The electroreceptor cell potential varies as a function of the voltage drop across the apical membrane surface and the basal surface, which is at a different potential due to field gradient in the fish's body. In effect, the receptor measures the

voltage drop of the field gradient along the length of the canal, and thus in a uniform field the longer the canal the greater the response of the receptor.

In freshwater elasmobranchs such as *Potamotrygon* the resistance of the skin is relatively high compared to marine species, and the resistance of the internal tissues is relatively low, presumably as a result of osmoregulatory constraints. In these fishes the internal environment is essentially at a common reference potential. Individual ampullae detect the transepidermal voltage drop between an applied external field and the internal tissue reference. Hence most freshwater elasmobranchs, as well as most other ampullary-bearing taxa, have short ampullary canals that cross only the epidermis. The majority of these taxa spend some or all of their lives in a freshwater environment.

The large size of the ampullary organ makes it possible to remove complete sensory units from the fish and conduct physiological recordings *in vitro*. Unfortunately, technical difficulties have made it impossible to obtain detailed intracellular records from single ampullary electroreceptor cells. The membrane biophysics of ampullary receptor excitation is best described for the skate, *Raja*, in which the receptor field activity was recorded as a voltage potential near the sensory epithelium within the ampulla (Obara and Bennett, 1972; Bennett and Clusin, 1978). In unstimulated electroreceptors there exists a depolarizing **bias current** caused by a standing $Ca^{2+}$ conductance across the apical cell membrane. This resting depolarization results in a steady synaptic release of neurotransmitter at the basal cell surface and depolarization of the adjacent postsynaptic afferent nerve fiber. As a result ampullary electrosensory primary afferent neurons are typically characterized by a regular discharge pattern. Modulation of the bias current by imposition of an extrinsic electric field results in changes in afferent fiber activity (Fig. 6). Presentation of an anodal stimulus hyperpolarizes the apical membrane and decreases the $Ca^{2+}$-generated bias current, inhibiting neurotransmitter release. Presentation of a relatively strong cathodal stimulus results in an increase in the $Ca^{2+}$ conductance and a depolarization of the apical membrane. This depolarization results in the activation of a large inward $Ca^{2+}$ conductance in the apical face, followed by activation of an inward $Ca^{2+}$ conductance in the basal face, which increases the release of neurotransmitter and excites the primary afferent neuron. A subsequent outward $K^+$ current in the basal face results in repolarization of the receptor (Bennett and Clusin, 1978). Thus, changes in the polarity and intensity of the electric potential at the skin pore (and apical surface of the receptor cell) over time will modulate the resting discharge pattern of the primary afferent neuron.

Cathodal excitation and anodal inhibition of the homologous ampullary receptors has also been described in nonteleost fishes and amphibians, presumably through similar mechanisms. However, the ampullary electroreceptors in teleosts exhibit a reverse response pattern in that they are excited by anodal and inhibited by cathodal stimuli. The major difference between primitive and derived electroreceptors is that the apical surface of the teleost receptor is not excitable but has a very low electrical resistance, possibly due to a greatly increased surface area provided by the

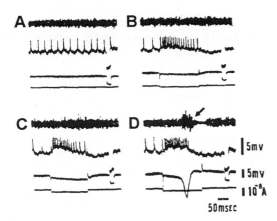

**FIG. 6.** The response of ampullary receptor currents and primary afferent neurons to cathodal stimuli. The upper traces in each panel show compound action potentials recorded from the mandibular nerve. Second traces show discharges of a single primary afferent. Third traces show the potential recorded in the lumen of the ampulla and represent the microphonic record sum for all electroreceptor cells in the ampulla. Bottom traces show the excitatory current pulse injected into the canal. (A) Spontaneous discharge activity of the primary afferent without an applied extrinsic stimulus shows a regular discharge pattern. (B) Application of a weak negative current pulse evokes an excitation of the primary afferent discharge that slowly adapts. Note that the ampullary potential follows the stimulus current, the single unit activity ceases for a brief time following end of the stimulus and that little change is evident in the compound action potentials recorded from the whole mandibular nerve. (C) Application of a slightly stronger stimulus evokes a similar response from the single unit and a weak response from the whole nerve. (D) A similar stimulus can also produce a depolarizing *receptor spike*, which excites other ampullary receptors as seen in the whole nerve recording, but has little additional excitatory effect on the single unit. Note that all primary afferent discharges appear to cease for a brief period during repolarization of the cell membrane. (Reproduced from *The Journal of General Physiology*, 1972, vol. 60, pp. 534–557, by copyright permission of The Rockefeller University Press.)

numerous microvilli. Anodal stimulation therefore results in a direct depolarization of the basal membrane, activating an inward $Ca^{2+}$ conductance across the basal membrane and resulting in an increased release of neurotransmitter. Conversely, a negative potential applied to the apical surface of the receptor will directly hyperpolarize the receptor potential at the basal membrane and decrease the rate of transmitter release (Bennett, 1971a; Bennett and Obara, 1986).

The high sensitivity of electrosensory primary afferent neurons was first established at a voltage gradient of about 1 $\mu$V/cm (Murray, 1962) and has recently been extended to near 20 nV/cm by Tricas and New (1997). The neural response to a prolonged constant current field is sustained for a duration of a few seconds before it begins to adapt back to the resting discharge rate. Prolonged, constant stimulation results in a return to resting levels and accommodation of the receptor, resulting in no change in the overall sensitivity of the receptor (Bodznick *et al.*, 1993). Work on a variety of species with both primitive and derived ampullary electrosenses shows a maximum response to sinusoidal electric fields at frequencies of 1–10 Hz (Fig. 7) (Andrianov *et al.*, 1984; New, 1990; Montgomery,

1984; Peters and Evers, 1985; Tricas *et al.*, 1995). Sensitivities of primary afferent fibers innervating ampullary electroreceptors to a sinusoidal uniform field are 0.9 spikes/sec per $\mu$V/cm for the little skate, *Raja erinacea* (Montgomery and Bodznick, 1993), 4 spikes/sec per $\mu$V/sec for the thornback guitarfish, *Platyrhinoidis triserata* (Montgomery, 1984), and 24 spikes/sec per $\mu$V/cm average for the round stingray, *Urolophus halleri* (Tricas and New, 1996). The ampullary electroreceptors of other freshwater taxa are less sensitive to field stimuli with neural thresholds in the range of 10–100 $\mu$V (Teeter *et al.*, 1980; Teeter and Bennett, 1981).

Recordings from the lateral line nerve in the behaving dogfish (Dijkgraaf and Kalmijn, 1966) and single units from unparalyzed *in vivo* preps (Akoev *et al.*, 1967a,b) show that the regular discharge of primary afferent neurons is modulated in rhythmic bursts that are in phase with the ventilatory movements of the fish. This reafferent neuromodulation is explained by the standing (DC) bioelectric field that arises from the differential distribution of ionic charges in the animal (reviewed by Kalmijn, 1974, 1988), which in the skate is a result of both diffusion potentials and osmoregulatory ion pumping at the gills (Bodznick *et al.*, 1992). The modulation of this standing field occurs as the animal opens and closes the mouth, gills, or spiracles during the ventilatory cycle, which changes the resistance pathway between the animal's internal tissues and surrounding seawater (Bodznick *et al.*, 1992). The resultant transcutaneous potential is the source of electrosensory

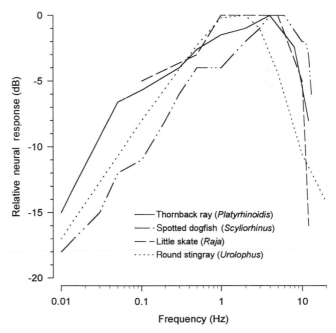

**FIG. 7.** Frequency response of ampullary electrosensory primary afferent neurons in four species of elasmobranch fishes. As a group the most sensitive frequencies range from 1 to 10 Hz, but individually some species may show distinct sensitivity peaks within this range. (Data modified for *Platyrhinoidis* from Montgomery, 1984; for *Scyliorhinus* from Peters and Evers, 1985; for *Raja* from New, 1990; and for *Urolophus* from Tricas *et al.*, 1995.)

self-stimulation or *ventilatory reafference* (Montgomery, 1984) by which a change in the internal potential of the animal (and basal regions of the ampullary receptor cells) proportionately modulates the regular discharge of all primary afferent neurons. Thus electrosensory receptors and primary afferents exhibit common mode noise, which has important implications for central processing of electrosensory information (New and Bodznick, 1990; Bodznick and Montgomery, 1992).

## III. Tuberous Electroreceptors

As is the case for derived ampullary electroreceptors, **tuberous electroreceptors** appear in separate and distantly related families of teleost fishes, indicating that they have evolved independently in each lineage (Bullock *et al.,* 1983; see also Zakon, 1986a, 1988). Among the Osteoglossomorpha, members of two families of African fishes, the Mormyridae and Gymnarchidae (superfamily Notopteroidea) (Nelson, 1994), possess both ampullary and tuberous electroreceptors; a sister group, the Xenomystinae, possesses only ampullary electroreceptors (Fig. 8) (Braford, 1986). In the superorder Ostariophysi, the siluriforms (catfishes) and South American gymnotiforms are sister groups. Catfish possess only a low-frequency ampullary system, but gymnotiforms possess both ampullary and tuberous systems. The distribution of electric senses in bony fishes supports the hypothesis that the high-frequency tuberous system is derived from the low-frequency sensitive ampullary system. Tuberous electroreceptors are found only in animals also possessing ampullary organs and weak electric organs, which have coevolved with the tuberous organs. Both taxa possessing tuberous organs have sister groups that lack tuberous organs but possess ampullary organs, which are themselves presumably derived from the neuromasts of the mechanosensory lateral line. Despite the considerable phylogenetic distance separating the two lineages, both groups show remarkable similarities in morphology and physiological response properties of the

various types of tuberous receptors, presumably the result of fundamental constraints associated with the analysis of high-frequency electric organ discharges (Zakon, 1986a, 1988).

Tuberous electroreceptor organs occur only in fishes possessing **electric organs.** These specialized organs produce weak (millivolts to a few volts), high-frequency (100–1000 Hz range) discharges, to which the receptors are tuned. The electric organ–tuberous electroreceptor system therefore represents a distinct sensory channel, isolated from the low-frequency electrical signals produced by inanimate and biological sources in aquatic environments. These electric organs are usually derived from specialized muscle (see Bennett, 1971b, and Bass, 1986, for reviews of electric organs), however in the gymnotid, *Apteronotus leptorhynchus,* a larval myogenic electric organ is replaced by a neurogenic organ during development (Kirschbaum, 1983). The temporal pattern and frequency spectrum of the EOD is typically species specific, although individual and sexually dimorphic variations exist within a given species. The EODs of all weakly electric fish species can be broadly divided into one of two categories; wave- or pulse-type EODs (Fig. 9). Wave-type EODs are quasi-sinusoidal in nature, with little if any resting interval between successive discharges. Pulse-type species produce discharges in which the interval between pulses is much longer than the duration of the pulse itself. Both wave- and pulse-type species are found in each of the lineages possessing tuberous organs, and have evolved independently within each taxon.

The EOD serves both an **electrolocation** and a **communicative** function in the animal's behavioral repertoire. The EOD generates an electric field, to which the tuberous electroreceptors are sensitive, in the water surrounding the animal. If an object of greater or lesser conductivity than the water is placed within the field, and spatial configuration of the field is altered and the resulting change in potential at the skin will be detected by the receptors (Fig. 10). Tuberous receptors allow animals to detect objects in their immediate vicinity and are especially useful in situations where other sensory systems, such as vision, are of limited usefulness in the turbid waters in which these animals generally live. In addition, the EOD produced by one animal may be detected by conspecifics; such electrocommunication plays important roles in the lives of weakly electric fishes (for reviews, see Bullock and Heiligenberg, 1986; Heiligenberg, 1977, 1991).

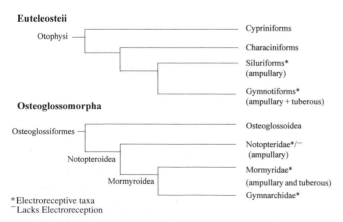

**Euteleosteii**

Otophysi —
- Cypriniforms
- Characiniforms
- Siluriforms* (ampullary)
- Gymnotiforms* (ampullary + tuberous)

**Osteoglossomorpha**

Osteoglossiformes —
- Osteoglossoidea

Notopteroidea —
- Notopteridae*/⁻ (ampullary)

Mormyroidea —
- Mormyridae* (ampullary and tuberous)
- Gymnarchidae*

*Electroreceptive taxa
⁻Lacks Electroreception

**FIG. 8.** Cladogram illustrating the distribution of electroreception in the two distantly related taxa of teleost fishes possessing ampullary electroreceptors or both ampullary and tuberous electroreceptors.

## IV. Gymnotid Tuberous Electroreceptors

### A. Anatomy

Gymnotid tuberous receptors exhibit a fairly uniform morphology despite the existence of several different functional subtypes of receptor. The organ consists of a roughly spherical chamber located in the epidermis, which communicates with the external environment via a short canal ending in an epidermal pore (Fig. 11A) (Szabo, 1965, 1974). The walls of the canal and capsule are composed of numer-

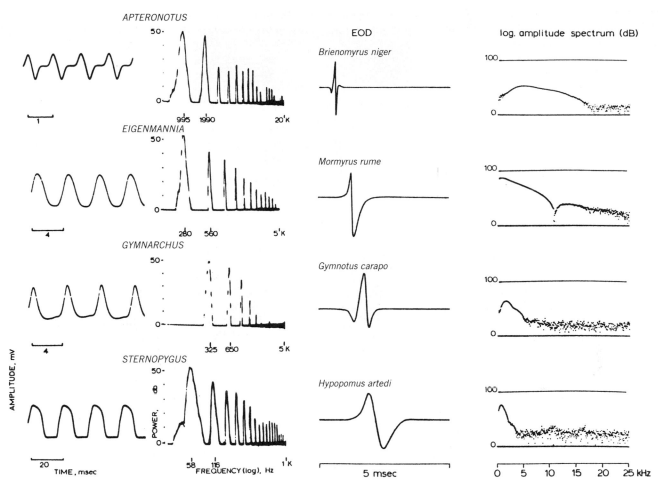

**FIG. 9.** Electric organ discharges and corresponding power spectra produced by several wave- and pulse-type weakly electric teleost fishes. EODs were recorded via a positive electrode placed in front of the head and a negative electrode behind the tail. (Reprinted from Heiligenberg, 1977, with permission.)

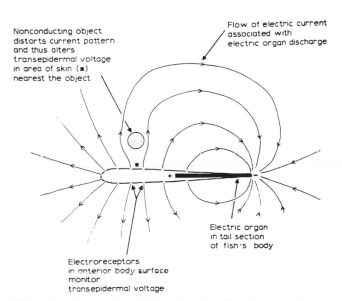

**FIG. 10.** The principle of electrolocation. The black bar depicted within the caudal portion of the fish's body indicates the position of the electric organ. (Reprinted from Heiligenberg, 1977, with permission.)

ous (up to 50) layers of extremely flattened epithelial cells (Szamier and Wachtel, 1970). The many layers of cell membranes in series result in a decreased capacitance of the canal walls and less shunting of high frequencies across the walls of the canal and capsule than in low-frequency sensitive ampullary organs (Bennett, 1971a). Typically, the canal of gymnotid electroreceptors is filled with a plug of loosely packed epithelial cells. Directly beneath the plug and extending across the lumen of the capsule is a layer of covering cells. The cells are joined to each other and to the walls of the lumen via tight junctions (Wachtel and Szamier, 1966; Szamier and Wachtel, 1970; Szabo, 1974; Zakon, 1984a). The layer of covering cells maintains a constant ionic environment within the lumen of the capsule and also provides capacitive coupling of the receptor cells to the external environment.

The receptor cells themselves are located in the base of the capsule. Unlike ampullary receptors they are attached to the base of the chamber only at the basal portion of the receptor cell's membrane, leaving most of the cell's membrane exposed to the lumen. The basal portion of the cell membrane is electrically isolated from the lumen via tight junctions with the supporting cells in the lumen

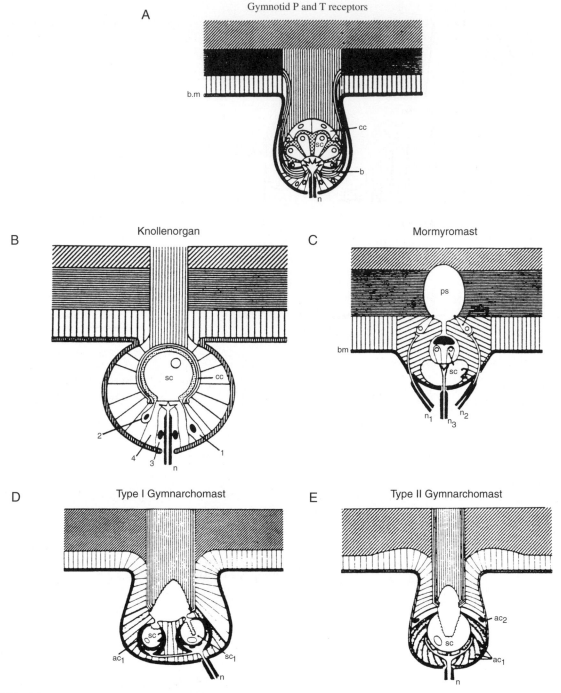

**FIG. 11.** Tuberous electroreceptor organs of (A) gymnotid, (B,C) mormyrid, and (D,E) gymnarchid weakly electric fishes. Abbreviations: $ac_1$, $ac_2$, accessory cells; b, capsule wall; bm, basement membrane; cc, covering cells; $n_{\#}$, afferent nerve fibers; ps, perisensory space; $sc_{\#}$, sensory cells. Numbers indicate different cell types within a given organ. (Reprinted from Szabo, 1974, with permission.)

(Wachtel and Szamier, 1966; Szamier and Wachtel, 1970). Generally there are 20 to 30 receptor cells per organ, however, some species may possess as many as 100 cells in certain specialized receptors (Szabo, 1965; Echague and Trujillo-Conez, 1981; Zakon, 1986b, 1987). The receptor cells are typically 20–30 $\mu$m long, with numerous apical microvilli, and large numbers of mitochondria evident in the apical region of the cell (Wachtel and Szamier, 1966; Szabo, 1974). All of the receptor cells within a tuberous organ are innervated by a single afferent nerve fiber and a given fiber may innervate either a single organ or several organs together. In the latter case the organs form a distinct *rosette*, which is the result of division of a single organ with growth. Physiological experiments support this finding: The receptive field of an afferent fiber is centered on an individual tuberous organ (Wachtel and Szamier, 1966; Bennett, 1967; Szabo, 1974; Baker, 1980; Bastian and Heiligenberg, 1980; Zakon, 1984a, 1987). Variations in the morphology of the afferent fibers and the terminal boutons have served as a basis for anatomical classification and correspond to functional differences in the organ's physiological responses. Most tuberous organs in pulse-type fish possess an afferent fiber that loses its myelin sheathing on entering the capsule and that innervates the individual receptor cells via numerous narrow branches (B units, type II of Szabo) (Wachtel and Szamier, 1966; Szabo, 1974; Zakon, 1984a). Other tuberous organs are innervated by markedly thicker fibers, which remain myelinated until in proximity to the receptor cells, where the fiber swells to form a large post-synaptic knob (M units, type of Szabo) (Szabo, 1970, 1974; Szamier and Wachtel, 1970; Srivastava, 1973; Echague and Trujillo-Conez, 1981; Zakon, 1987). The synapses of tuberous ampullary organs in gymnotids are similar in appearance to those of ampullary receptors, with a presynaptic evagination of the receptor cell surrounded by the postsynaptic membrane of the afferent fiber. As in ampullary receptors, a synaptic ribbon is also evident at the presynaptic site in the receptor cell (Echague and Trujillo-Conez, 1981).

## B. Physiology

Intracellular recordings from individual tuberous electroreceptors have yet to be successfully conducted, and a considerable amount remains to be learned about the physiology of transduction in these receptors. They appear to function, at least in principle, in a manner similar to that of teleost ampullary receptors, by measuring the voltage drop across the receptor cell's basal membrane surface between the internal potential of the receptor cell and the internal "reference" potential of the animal (Bennett, 1971a). The layer of covering cells that isolates the internal capsule of gymnotiform tuberous receptors from the external environment provides capacitance coupling between the two environments and acts as a high-pass filter (Fig. 12) (Bennett, 1971a). The numerous microvilli on the apical membrane surface greatly increase the surface area of the cell, which increases the input capacitance of the cell and decreases the input resistance. Bennett has suggested that the apical membrane face is electrically inexcitable and that the high blocking capacitance and low resistance of the apical face result in a membrane time constant permitting passage of frequencies in the range of the EOD. Bennett (1971a) has further suggested that for maximum sensitivity the resistance across the excitable basal membrane face should be higher than that across the apical membrane so that the larger voltage drop is across the former.

Tuberous receptor organs generally exhibit sharper tuning characteristics than are suggested simply by the high-pass filter characteristics suggested by its morphology

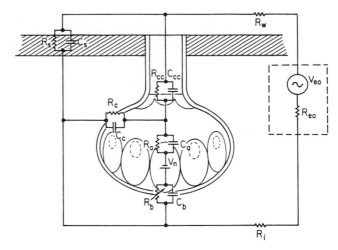

**FIG. 12.** Electrical equivalent circuits of tuberous electrical organs. Abbreviations: $C_a$, capacitance of receptor cell apical membrane; $C_b$, capacitance of receptor cell basal membrane; $C_c$, capacitance of canal and capsule wall; $C_{cc}$, capacitance of covering cells; $C_s$, capacitance of skin; $R_a$, resistance of receptor cell apical membrane; $R_b$, resistance of receptor cell basal membrane; $R_c$, resistance of canal and capsule wall; $R_{cc}$, resistance of covering cells; $R_{eo}$, internal resistance of the electric organ; $R_i$, internal resistance of fish; $R_s$, resistance of skin; $R_w$, resistance of the water; $V_{eo}$, internal voltage of the electric organ; $V_n$, resting potential of receptor cells. Resistances and capacitances of supporting cells are not indicated and are believed to be passive. (Modified from Bennett, 1967, and Zakon, 1988.)

alone. Active mechanisms on the basal membrane of the receptor cells have been suggested as the basis for the frequency response characteristics (Scheich *et al.,* 1973; Bastian, 1976; Scheich, 1977; Viancour, 1979a,b). Tuberous receptors produce oscillatory potentials on stimulation: The frequency of this "ringing" is the same as the best frequency of the cell and appears to be based on inward $Ca^{2+}$ and outward $K^+$ currents (Zipser and Bennett, 1973; Zakon, 1984b). The tuning properties of the cell could thus be a function of the kinetics of these two currents at different stimulus frequencies.

Although the general processes of transduction in tuberous organs may be said to be similar, different populations of gymnotid tuberous organs display varying responses to elements of identical stimuli. Although a number of complex and conflicting classification schemes have been suggested to identify functional types of gymnotid tuberous electrosensory organs, we follow the classification originally proposed by Hopkins (1983) and adopted by Zakon (1986a) of two general types of tuberous organs based on their responses to EOD-like stimulation. The first group may be said to be **rapid-timing units,** which show a tightly phase-locked response to presentation of the EOD signal. The response of the receptor is remarkably precise: Single spikes are phase locked to each cycle of the EOD, exhibiting very little timing "jitter" ($<100$ $\mu$sec). Rapid-timing units do not adapt and are typically sharply tuned. The second class of receptors is **amplitude-modulated units,** which demonstrate poor phase-locking of the response to the EOD but which demonstrate either an increase in the probability of a spike being produced or an increase in the number of spikes produced with increases in amplitude of the EOD stimulus. These units adapt readily and are generally more broadly tuned to lower frequencies than are rapid timing units. Thus the first category of cells encodes information on the timing of when an EOD pulse occurs, whereas the second provides information on changes in amplitude of the EOD over time. Rapid timing and amplitude-modulated units are observed in species producing both wave- and pulse-type EODs. In wave-type species such as *Eigennmania, Sternopygus,* and *Apteronotus,* rapid-timing units have been termed **T units** (timing or phase-coding) and amplitude-modulated receptors are called **P units** (probability coding) (Fig. 13) (Scheich *et al.,* 1973; Feng and Bullock, 1977; Scheich, 1977; Bastian and Heiligenberg, 1980). The principal physiological difference between these receptor types appears to be the threshold of excitation of the receptors. Both receptor subtypes demonstrate similar widths of their dynamic ranges (approximately 20 dB) and demonstrate poor phase-locking responses at stimulus intensities that are just suprathreshold, as well as demonstrating a 1:1 following of the stimulus at intensities that are of saturating intensity. However, the stimulation threshold and saturation intensities of T units are generally 15–20 dB lower than that of P units; hence at normal EOD intensities T units are saturated and following at 1:1, whereas P units are responding in their dynamic range, generating greater or lesser spike frequencies as a function of EOD amplitude (Scheich *et al.,* 1973; Feng and Bullock, 1977). Similar units are observed in species producing pulse-type EODs such as *Hypopomus* and *Gymnotus.* **Pulse marker units** (M units) are similar to T units

in their responses, producing a strongly phase-locked response to the EOD, whereas **burst duration receptors** (B units) respond to increasing EOD amplitude with monotonically increasing bursts of spikes (Hagiwara *et al.,* 1962, 1965; Hagiwara and Morita, 1963; Suga, 1967; Bastian, 1976; Hopkins and Heiligenberg, 1978; Watson and Bastian, 1979; Baker, 1980). Both types of receptors demonstrate a directional sensitivity to current flow, with optimal responses to stimuli at the optimal azimuth for transepidermal current flow (Yager and Hopkins, 1993). This directional sensitivity appears to result from the frequency response characteristics of the receptors: The directional sensitivity can be modified by altering the stimulus frequency (McKibben *et al.,* 1993). A correlation between receptor morphology and response has been suggested in both wave- and pulse-producing species. Rapid-timing units (T, M units) are those receptors possessing thick, myelinated afferent fibers that end in large synaptic sites at the receptors. Amplitude-modulated organs (P, B units) are those with long, thin unmyelinated afferents innervating the receptor cells (Szamier and Wachtel, 1970; Szabo, 1974).

A singular property of many gymnotid tuberous electroreceptors is their relatively narrow tuning characteristics. In general, the best frequency (BF) of an electrosensory animal's electroreceptor is closely correlated with the peak power of the EOD frequency spectrum (Fig. 14) (Scheich *et al.,* 1973; Bullock 1975; Bastian, 1976, 1977; Hopkins, 1976; Zakon and Meyer, 1983). Additionally, a number of species show polymorphic EOD frequency spectra. In many species there exists a distinct sexual dimorphism of the EOD, and in wave-producing species individuals appear to have unique preferred frequencies of the EOD within the frequency range of their particular species and gender (Hopkins, 1972, 1974; Bass and Hopkins, 1983; Hagedorn and Heiligenberg, 1985). The narrow tuning of these receptor's to the fish's EOD likely provides the animal with a degree of communicative and reproductive isolation from the discharges produced by different species. Furthermore, the frequency of the EOD often changes during the fish's lifetime: In *Sternopygus* the frequency of the EOD shifts downward in maturing males and increases in females. Treatment with exogenous sex steroid hormones can alter both the frequency of the EOD and the tuning of the receptors: Androgens decrease the frequency of the EOD and estrogens increase it (Meyer and Zakon, 1982; Meyer, 1983; Bass and Hopkins, 1984). Following administration of exogenous hormones the frequencies of the EOD and receptor tuning shift at about the same rate, but with the receptor tuning lagging by a period of approximately one week (Meyer and Zakon, 1982; Meyet, 1983). Administration of steroids to animals in which the EOD has been silenced can still result in retuning of the receptors, indicating that the receptors are not merely "following" a hormonally induced shift in EOD frequency. In *Sternopygus,* tuberous receptors developing in regenerating skin segments following removal of the epidermis initially demonstrate broad tuning properties, but rapidly become tuned to the BF of the fish, even if the EOD has been silenced (Zakon, 1986b).

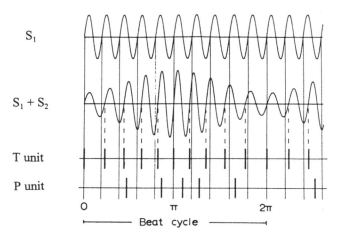

**FIG. 13.**   Schematic representation of the responses of gymnotid rapid-timing units (T units) and amplitude-modulated units (P units) to modulation of the EOD. The EOD of the fish ($S_1$) is modulated by the addition of a conspecific's EOD of a slightly different frequency ($S_1 + S_2$), creating a beat envelope. T units are tightly phase-locked to the positive zero-crossing of the modulated EOD waveform. P units are not phase-locked but exhibit an increased probability of firing as the amplitude of the EOD beat envelope increases. (Reprinted from Heiligenberg, 1986, with permission.)

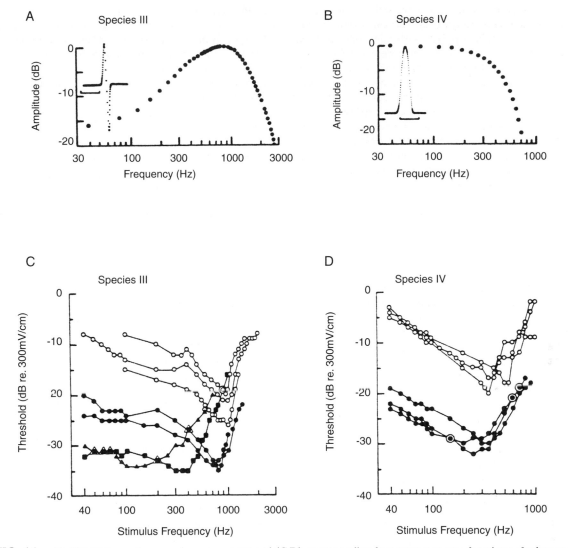

**FIG. 14.** (A,B) EOD waveforms and power spectra and (C,D) corresponding frequency response functions of tuberous electroreceptors from two species of the gymnotiform *Hypopomus*. Tuning curves with open symbols are from pulse markers (M units); those with closed symbols are from burst duration coders (B units). (From Bastian, 1977.)

## V. Momyroidea

### A. Anatomy

Despite the fact that the tuberous electroreceptors in the Mormyroidea (Mormyridae and Gymnarchidae) are not homologous to those of the gymnotids, the receptors show striking similarities in both morphology and physiological responses, presenting a striking case of parallel homoplasy. It is likely that some of the similarities are the result of receptors in both groups being derived from hair cell octavolateralis receptors (and probably from the nonhomologous ampullary receptors in both lineages) but also likely represents the physical, physiological, and developmental constraints incumbent on the evolution of a high-frequency electrosensory system for communicative and localization purposes.

Unlike gymnotids, mormyrids possess two morphologically distinct types of tuberous electroreceptor organs: Knollenorgans (or K units) and mormyromasts (Fig. 11B,C). The **Knollenorgans** are similar in morphology to gymnotid tuberous receptors, with a number of receptor cells located in a roughly spherical chamber, the walls of which consist of flattened epithelial cells (Szabo, 1965; Harder, 1968). The capsule is connected via a canal and epidermal pore to the external environment and, as in gymnotids, the canal is loosely plugged with epithelial cells. Generally, Knollenorgans possess a small number of receptor cells (1 to 10), although larger numbers are reported in some species. Each of the large (40–50 $\mu$m in diameter), spherical receptor cells themselves are individually encapsulated by a layer of covering cells within the larger capsule of the organ. The receptor cells are attached to the walls of the capsule only at their basal membrane surface, leaving most of the receptor membrane exposed to the lumen within the covering cell capsule. The apical surface of the cell membrane is densely supplied with microvilli, and numerous mitochondria are apparent just beneath the apical membrane. Unlike most teleost electroreceptors the membrane of Knollenorgans is believed to be electrically excitable: Stimulation of the afferent nerve fiber results in a spike that can be recorded from the organ. Although it has been suggested that an electrotonic synapse exists between Knollenorgan and afferent fiber (Bennett, 1971a; Steinbach and Bennett, 1971), only chemical synapses have been observed (Derbin and Szabo, 1968; Schessel et al., 1982). Although a synaptic ribbon is present, it does not extend into an evagination of the presynaptic membrane as is the case in gymnotids.

**Mormyromast** electroreceptors (D units) are more complex organs, with two distinct populations of electroreceptor cells each innervated by different afferent nerve fibers (Szabo and Wersäll, 1970). Mormyromast receptors consist of two chambers; a superficial chamber that resembles an ampullary receptor and a deep chamber that is similar in morphology to a Knollenorgan. The deep chamber (or sensory chamber) possesses three to five small (2–3 $\mu$m) receptor cells (B cells). The receptor cells are covered with microvilli and, as in other tuberous organs, are attached to the capsule only at the basal membrane surface. A narrow canal connects the deep chamber to the superficial chamber, which is filled with a jelly-like matrix. The five to seven receptor cells of the superficial chamber (A cells) are embedded within the walls of the receptors with only the apical membrane exposed to the chamber lumen in a manner similar to ampullary electroreceptors. These cells lack microvilli, but possess a curious small dense conical structure on the apical membrane. Each of the receptors in the upper chamber is innervated by one of two to four nerve fibers, each of which innervates a series of adjacent cells.

Tuberous electroreceptor organs in gymnarchids are quite different from those of their mormyrid sister group. **Gymnarchomast** receptors are divided into two morphologically distinct types (Figs. 11D and E). In both types of organs the cell bodies of the receptors are embedded in the capsule walls with the apical face exposed to the jelly-like matrix of the lumen. **Gymnarchomast type I** organs possess two distinct receptor cell morphologies, usually with one or more pairs of each type per organ. One type of receptor cell possesses a deep groove-like invagination of the apical membrane surface, the other possesses only a shallow indentation. Both types of receptor cells have apical microvilli, and all of the receptors within an organ are innervated by a single afferent nerve fiber. **Gymnarchomast type II** organs possess a single type of large receptor cell, with a characteristic deep invagination of the apical membrane and long apical microvilli. All of the sensory cells (7 to 12) in a type II receptor are innervated by a single afferent fiber (Szabo, 1965, 1974). Ultrastructural studies have demonstrated relatively high levels of $Ca^{2+}$ in the cytoplasm of the receptor cells (Djebar et al., 1995) presumably as a result of the inward depolarizing current produced at the membrane basal face.

## B. Physiology

As in gymnotids, the tuberous electroreceptors of mormyrids can be differentiated into rapid-timing units and amplitude-coding units. The broadly tuned Knollenorgans respond to presentation of EOD-like stimuli by generating a single, highly phase-locked spike per pulse (Bennett, 1967; Szabo, 1974; Hopkins and Bass, 1981). The temporal pattern of the EOD pulse is critical to the receptor response, only stimuli consisting of positive–negative transitions with rise–fall profiles similar to the animal's own EOD will reliably elicit a spike: Phase-shifted stimuli with power spectral peaks similar to the EOD are also poor at eliciting receptor responses (Hopkins and Bass, 1981). The Knollenorgans play a functional role in the communicative behavior of mormyrid fishes by detecting EODs produced by other conspecific individuals since responses of Knollenorgans to the animal's own EOD are suppressed in the central nervous system. Hopkins and Bass (1981) suggested that different classes of Knollenorgan (KI and KII) may respond to different portions of the EOD cycle; the interval between the firing of the two receptor types will thus be specific for a given EOD waveform forming a basis for conspecific recognition at the level of the receptor.

Mormyromast receptors respond to the amplitude of the EOD, producing afferent fiber spikes as a function of EOD intensity. Unlike the B units of gymnotiforms, mormyromasts typically generate fewer spikes (1 to 5) per stimulus, and both response latency and interspike intervals shift with increasing stimulus amplitude (Szabo and Hagiwara, 1967). The threshold of mormyromast receptors is relatively high, indicating that these receptors are stimulated primarily by the animal's own EOD, and very little by the attenuated EODs of conspecifies at a distance. Alteration of the electric field produced by the EOD via presentation of items of varying conductances near the fish's skin produces systematic shifts in spike latency and number in afferents innervating mormyromasts. Afferent nerve fibers innervating either of the two populations of receptor cells in the mormyromast organ project to different regions of the first-order medullary electrosensory nucleus (Bell et al., 1989) and exhibit differing responses to presentation of EOD stimuli. Fibers innervating A cells exhibit higher thresholds and smaller numbers of maximum spikes per burst per stimulus, whereas those innervating B cells have lower thresholds and higher numbers of maximum spikes per burst (Bell, 1990a,b). The dynamic range of the mormyromast response intensity function is relatively limited (6 dB) suggesting that these cells are very sensitive to slight modulations of EOD amplitude within that range (Szabo and Hagiwara, 1967). The properties of the mormyromast response to EOD stimulation suggest that EOD amplitude is coded by a **latency code,** in which the intensity of the EOD is encoded by the latency of the first spike of the response, rather than a **burst duration code,** as in the B units of gymnotids, in which the amplitude is encoded by the number of spikes generated in the burst (Szabo and Hagiwara, 1967; Bell, 1990b). Mormyromasts also exhibit extreme sensitivity to EOD phase modulation (von der Emde and Bleckmann, 1992). The different behavioral functions of Knollenorgans and mormyromasts (namely, communication vs electrolocation) in the ethogram of mormyrids is maintained throughout the central nervous system and information from the two receptor types is segregated into two parallel pathways through the neuraxis.

The gymnarchomast receptors show striking similarities to those of gymnotids producing wave-type EODs, exhibiting either rapid-timing (S units) or amplitude-coding (O units) responses to presentation of EOD-like stimuli. S units have a low threshold and narrow dynamic range. At "normal" EOD intensities the saturated S units show little temporal variation in their response to changes in stimulus intensity. O units have higher thresholds and dynamic ranges, changes in EOD amplitudes are encoded as rapidly adapting bursts of afferent fiber spikes (Bullock *et al.*, 1975).

# Bibliography

Akoev, G. N., Ilyinsky, O. B., and Zadan, P. M. (1976a). Physiological properties of electroreceptors of marine skates. *Comp. Biochem. Physiol.* **53A**, 201–209.

Akoev, G. N., Ilyinsky, O. B., and Zadan, P. M. (1976b). Responses of electroreceptors (ampullae of Lorenzini) of skates to electric and magnetic fields. *J. Comp. Physiol.* **106A**, 127–136.

Andres, K. J., and von During, M. (1988). The platypus bill. A structural and functional model of a pattern-like arrangement of cutaneous receptor receptors. *In* "Sensory Receptor Mechanisms" (A. Iggo, Ed.), pp. 81–89. World Scientific Publishing Company, Singapore.

Andrianov, G. N., Broun, G. R., and Ilyinsky, O. B. (1984). Frequency characteristic of skate electroreceptive central neurons responding to electrical and magnetic stimulation. *Neurophysiol.* **16**, 365–376.

Baker, C. (1980). Jamming avoidance behavior in gymnotoid electric fish with pulse-type discharges: Sensory encoding for a temporal pattern discrimination. *J. Comp. Physiol.* **136**, 165–181.

Baker, R. R., Mather, J. G., and Kennaugh, J. H. (1983). Magnetic bones in human sinuses. *Nature* **301**, 78–80.

Bass, A. H. (1986). Electric organs revisited: Evolution of a vertebrate communication and orientation organ. *In* "Electroreception" (T. H. Bullock and W. Heiligenberg, Eds.), pp. 13–70. John Wiley and Sons, New York.

Bass, A. H., and Hopkins, C. D. (1983). Hormonal control of sexual differentiation: Changes in electric organ discharge waveform. *Science* **220**, 971–974.

Bass, A. H., and Hopkins, C. D. (1984). Shifts of frequency tuning in electroreceptors in androgen treated mormyrid fish. *J. Comp. Physiol.* **155**, 713–724.

Bastian, J. (1976). Frequency response characteristics of electroreceptors in weakly electric fish (Gymnotoidei) with a pulse discharge. *J. Comp. Physiol.* **112**, 165–180.

Bastian, J. (1977). Variations in the frequency response of electroreceptors dependent on receptor location in weakly electric fish (Gymnotoidei) with a pulse discharge. *J. Comp. Physiol.* **121**, 53–64.

Bastian, J. and Heiligenberg, W. (1980). Neural correlates of the jamming avoidance response in *Eigenmania*. *J. Comp. Physiol.* **136**, 135–152.

Bell, C. C. (1990a). Mormyromast receptor ogans and their afferent fibers in mormyrid fish. II. Intra-axonal recordings show initial stages of central processing. *J. Neurophys.* **43**, 303–318.

Bell, C. C. (1990b). Mormyromast receptor ogans and their afferent fibers in mormyrid III. Physiological differences between two morphological types of fibers. *J. Neurophys.* **43**, 319–332.

Bell, C. C., Zakon, H., and T. E. Finger (1989). Mormyromast receptor ogans and their afferent fibers in mormyrid fish. I. Morphology. *J. Comp. Neurol.* **286**, 391–407.

Bemis, W. E., and Hetherington, T. E. (1982). The rostral organ of *Latimeria chalumnae*: Morphological evidence of an electroreceptive function. *Copeia* **1982**, 467–471.

Bennett, M. V. L. (1967). Mechanisms of electroreception. *In* "Lateral Line Detectors" (P. Cahn Ed.), pp. 313–393. Indiana University Press, Bloomington, IN.

Bennett, M. V. L. (1971a). Electroreception. *In* "Fish Physiology" (W. S. Hoar and D. S. Randall, Eds.), Vol. 5, pp. 493–574. Academic Press, New York.

Bennett, M. V. L. (1971b). Electric organs. *In* "Fish Physiology" (W. S. Hoar and D. S. Randall, Eds.), Vol. 5, pp. 347–491. Academic Press, New York.

Bennett, M. V. L., and Clusin, W. T. (1978). Physiology of the ampulla of Lorenzini, the electroreceptor of elasmobranchs. *In* "Sensory Biology of Sharks, Skates, and Rays" (E. S. Hodgson and R. F. Mathewson, Eds.), pp. 483–505. Office of Naval Research, Arlington, VA.

Bennett, M. V. L., and Obara, S. (1986). Ionic mechanisms and pharmacology of electroreceptors. *In* "Electroreception" (T. H. Bullock and W. Heiligenberg, Eds.), pp. 157–181. John Wiley and Sons, New York.

Blakemore, R. (1975). Magnetotactic bacteria. *Science* **190**, 377–379.

Bodznick, D. (1989). Comparisons between electrosensory and mechanosensory lateral line systems. *In* "The Mechanosensory Lateral Line: Neurobiology and Evolution" (S. Coombs, P. Görner and H. Münz, Eds.), pp. 655–680. Springer-Verlag, New York.

Bodznick, D., and Montgomery, J. C. (1992). Reafference in the elasmobranch electrosensory system: Medullary neuron receptive fields support a common-mode rejection mechanism. *J. Exp. Biol.* **171**, 127–137.

Bodznick, D., and Preston, D. G. (1983). Physiological characterization of electroreceptors in the lampreys *Ichthyomyzon unicuspis* and *Petromyzon marinus*. *J. Comp. Physiol.* **152**, 209–217.

Bodznick, D., Montgomery, J. C., and Bradley, D. J. (1992). Suppression of common-mode signals within the electrosensory system of the little skate, *Raja erinacea*. *J. Exp. Biol.* **171**, 107–125.

Bodznick, D., Hjelmstad, G., and Bennett, M. V. L. (1993). Accommodation to maintained stimuli in the ampullae of Lorenzini: How an electroreceptive fish achieves sensitivity in a noisy world. *Japan J. Physiol.* **43**, Suppl. 1, S231–S237.

Braford, M. R. (1986). African knifefishes: The Xenomystinae. *In* "Electroreception" (T. H. Bullock and W. Heiligenberg, Eds.), pp. 453–464. John Wiley and Sons, New York.

Bullock, T. H., Behrend, K. and Heiligenberg, W. (1975). Comparison of the jamming avoidance response in gymnotoid and gymnarchid electric fish: A case of convergent evolution of behavior and its sensory basis. *J. Comp. Physiol.* **103**, 97–121.

Bullock, T. H., Bodznick, D. A., and Northcutt, R. G. (1983). The phylogenetic distribution of electroreception: Evidence for convergent evolution of a primitive vertebrate sense modality. *Brain Res. Rev.* **6**, 25–46.

Bullock, T. H., and Heiligenberg, W. (1986). "Electroreception." John Wiley and Sons, New York.

Derbin, C., and Szabo, T. (1968). Ultrastructure of an electroreceptor (Knollenorgan) in the mormyrid fish, *Gnathonemus petersii*. *Int. J. Ultrastruct. Res.* **22**, 469–484.

Dijkgraaf, S., and Kalmijn, A. J. (1966). Versuche zur biologishen Bedeutung der Lorenzinischen Ampullen an Haifischen. *Z. Vgl. Physiol.* **47**, 438–456.

Djebar, B., Bensouilah, M., and Denizot, J.-P. (1995). Ultrastructural distribution of calcium in cutaneous electroreceptor organs of teleost fish. *Biotech. Histochem.* **70,** 81–89.

Echague, A., and Trujillo-Conoz, O. (1981). Innervation patterns in the tuberous organs of *Gymnotus carapo*. *In* "Advances in Physiological Sciences," Vol 31: "Sensory Physiology of Aquatic Lower Vertebrates" (T. Szabo and G. Czeh, Eds.), pp. 29–40. Akademiai Kiado, Budapest.

Feng, A. S., and Bullock, T. H. (1977). Neuronal mechanisms for object discrimination in the weakly electri fish, *Eigennmania viriscens*. *J. Exp. Biol.* **66,** 141–158.

Fields, R. D., Bullock, T. H., and Lange, G. D. (1993). Ampullary sense organs, peripheral, central and behavioral electroreception in chimeras (*Hydrolagus*, Holocephali, Chondrichthyes). *Brain Behav. Evol.* **41,** 269–289.

Fritzsch, B., and Münz, H. (1986). Electroreception in amphibians. *In* "Electroreception" (T. H. Bullock and W. Heiligenberg, Eds.), pp. 483–496. John Wiley and Sons, New York.

Gould, J. L., Kirschvink, J. L., and Deffeyes, K. F. (1978). Bees have magnetic resonance. *Science* **201,** 1026–1028.

Gregory, J. E., Iggo, A., McIntyre, A. K., and Proske, U. (1989). Responses of electroreceptors in the snout of the echidna. *J. Physiol. (Lond.)* **414,** 521–538.

Hagedorn, M., and Heiligenberg, W. (1985). Court and spark: Electric signals in the courtship and mating of gymnotoid fish. *Anim. Behav.* **33,** 254–265.

Hagiwara, S., Kusano, K., and Negishi, K. (1962). Physiological properties of electroreceptors of some gymnotids. *J. Neurophys.* **25,** 430–449.

Hagiwara, S., and Morita, H. (1963). Coding mechanisms of electroreceptor fibers in some electric fish. *J. Neurophys.* **26,** 551–567.

Hagiwara, S., Szabo, T., and Enger, P. S. (1965). Electroreceptor mechanisms in a high-frequency weakly electric fish, *Sternarchus albifrons*. *J. Neurophys.* **28,** 784–799.

Harder, W. (1968). Die Beziehungen zwischen Elektrorezeptoren, elektrischem Organ, Seitenlinienorganen und Nervensystem bei den Mormyridae (Teleostei, Pisces). *Z. Vgl. Physiol.* **59,** 272–318.

Heiligenberg, W. (1977). "Principles of Electrolocation and Jamming Avoidance Response in Electric Fish." Springer, New York.

Heiligenberg, W. (1986). Jamming avoidance responses. *In* "Electroreception" (T. H. Bullock and W. Heiligenberg, Eds.), pp. 613–649. John Wiley and Sons, New York.

Heiligenberg, W. F. (1991). Neural nets in electric fish. MIT Press, Cambridge.

Hopkins, C. D. (1972). Sex differences in signalling in an electric fish. *Science* **176,** 1035–1037.

Hopkins, C. D. (1974). Electric communication: Functions in the social behavior of *Eigennmania viriscens*. *Behaviour* **50,** 270–305.

Hopkins, C. D. (1976). Stimulus filtering and electroreception: Tuberous electroreceptors in three species of gymnotoid fish. *J. Comp. Physiol.* **103,** 171–207.

Hopkins, C. D. (1983). Functions and mechanisms in electroreception. *In* "Fish Neurobiology," Vol. 1: "Brain Stem and Sense Organs" (R. G. Northcutt and R. E. Davis, Eds.), pp. 215–259. University of Michigan Press, Ann Arbor, MI.

Hopkins, C. D., and Bass, A. H. (1981). Temporal coding of species recognition signals in an electric fish. *Science* **212,** 85–87.

Hopkins, C. D., and Heiligenberg, W. F. (1978). Evolutionary designs for electric signals and electroreceptors in gymnotoid fishes of Surinam. *Behav. Ecol. Sociobiol.* **3,** 113–134.

Hsu, C.-Y., and Li, C.-W. (1994). Magnetoreception in honeybees. *Science* **265,** 95–97.

Istenic, L., and Bulog, B. (1984). Some evidence for the ampullary organs in the European cave salamander *Proteus anguineus* (Urodela Amphibia). *Cell Tissue Res.* **235,** 393–402.

Jørgensen, J. M. (1980). The morphology of the Lorenzinian ampullae of the sturgeon *Acipenser ruthenus* (Pisces: Chondrostei). *Acta Zool.* **61,** 87–92.

Jørgensen, J. M. (1982). Fine structure of the ampullary organs of the bichir *Polypterus senegalus* Cuvier, 1829 (Pisces: Brachiopterygii) with some notes on the phylogenetic development of electroreceptors. *Acta Zool.* **63,** 211–217.

Jørgensen, J. M. (1984). On the morphology of the electroreceptors of the Australian and one African lungfish species. *Vidensk. Meddr. Dansk Naturh. Foren.* **145,** 77–85.

Jørgensen, J. M. (1989). Evolution of octavolateralis sensory cells. *In* "The Mechanosensory Lateral Line: Neurobiology and Evolution" (S. Coombs, P. Görner, and H. Münz, Eds.), pp. 115–146. Springer-Verlag, New York.

Jørgensen, J. M., Flock, Å., and Wersäll, J. (1972). The Lorenzinian ampullae of *Polyodon spathula*. *Z. Zellforsch.* **130,** 362–377.

Kalmijn, A. J. (1971). The electric sense of sharks and rays. *J. Exp. Biol.* **55,** 371–383.

Kalmijn, A. J. (1974). The detection of electric fields from inanimate and animate sources other than electric organs. *In* "Handbook of Sensory Physiology" (A. Fessard, Ed.), Vol. III/3, pp. 147–200. Springer-Verlag, New York.

Kalmijn, A. J. (1982). Electric and magnetic field detection in elasmobranch fishes. *Science* **218,** 915–918.

Kalmijn, A. J. (1988). Detection of weak electric fields. *In* "Sensory Biology of Aquatic Animals" (J. Atema, R. R. Fay, A. N. Popper, and W. N. Tavogla, Eds.), pp. 151–186. Springer-Verlag, New York.

Kalmijn, A. J., and Blakemore, R. (1978). The magnetic behavior of mud bacteria. *In* "Animal Migration, Navigation and Homing" (K. Schmidt-Koenig and W. T. Keeton, Eds.), pp. 347–353. Springer Verlag, New York.

Kirschbaum, F. (1983). Myogenic electric organ precedes the neurogenic organ in apteronotid fish. *Naturwissenschaften* **70,** 205.

Kirschvink, J. L., and Walker, M. M. (1995). Honeybees and magnetoreception. *Science* **269,** 1889.

Kirschvink, J. L., Walker, M. M., Chang, S. B., Dizon, A. E., and Peterson, K. A. (1985). Chains of single-domain magnetite particles in chinook salmon. *J. Comp. Physiol. A* **157,** 375–381.

Lissman, H. W. (1958). On the function and evolution of electricx organs in fish. *J. Exp. Biol.* **35,** 156–191.

Lissman, H. W., and Machin, K. E. (1958). The mechanisms of object location in *Gymnarchus niloticus* and similar fish. *J. Exp. Biol.* **35,** 451–486.

Manger, P. R., and Hughes, R. L. (1992). Ultrastructure and distribution of epidermal sensory receptors in the beach of the echidna, *Tachyglossus aculeatus*. *Brain Behav. Evol.* **40,** 287–296.

McKibben, J. R., Hopkins, C. D., and Yager, D. D. (1993). Directional sensitivity of tuberous electroreceptors: Polarity preferences and frequency tuning. *J. Comp. Physiol.* **173,** 415–424.

Meyer, J. H. (1983). Steroid influences upon the discharge frequency of a weakly electric fish. *J. Comp. Physiol.* **153,** 29–38.

Meyer, J. H., and Zakon, H. H. (1982). Androgens alter the tuning of electroreceptors. *Science* **217,** 635–637.

Millot, J., and Anthony, J. (1956). L'organe rostral de *Latimeria* (Crossoptérygien Coelacanthidè). *Ann. Sci. Nat. Zool., 11ᵉ Série,* **18,** 381–387.

Montgomery, J. C. (1984). Frequency response characteristics of primary and secondary neurons in the electrosensory system of the thornback ray. *Comp. Biochem. Physiol.* **79A,** 189–195.

Montgomery, J. C., and Bodznick, D. (1993). Hindbrain circuitry mediating common-mode suppression of ventilatory reafference in the electrosensory system of the little skate, *Raja erinacea*. *J. Exp. Biol.* **183,** 203–315.

Moore, B. R. (1988). Magnetic fields and orientation in homing pigeons: Experiments of the late W. T. Keeton. *Proc. Natl. Acad. Sci. USA* **85,** 4907–4909.

Moy-Thomas, J. A., and Miles, R. S. (1971). "Paleozoic Fishes." W. B. Saunders, Philadelphia.

Murray, R. W. (1962). The response of the ampullae of Lorenzini of elasmobranches to electrical stimulation. *J. Exp. Biol.* **39**, 119–128.

Murray, R. W., and Potts, T. W. (1961). The composition of the endolynph and other fluids of elasmobranches. *Comp. Biochem. Physiol.* **2**, 65–75.

Nachtrieb, H. F. (1910). The primitive pores of *Polyodon spathula* (Walbaum). *J. Exp. Zool.* **9**, 455–468.

Nelson, J. S. (1994). "Fishes of the World," 3rd ed. John Wiley and Sons, New York.

Nesson, M. H. (1995). Honeybees and magnetoreception. *Science* **269**, 1889–1890.

New, J. G. (1990). Medullary electrosensory processing in the little skate. I. Response characteristics of neurons in the dorsal octavolateralis nucleus. *J. Comp. Physiol.* **167**, 285–294.

New, J. G., and D. Bodznick (1990). Medullary electrosensory processing in the little skate. II. Suppression of electrosensory reafference via a common-mode rejection mechanism. *J. Comp. Physiol.* **167**, 295–307.

Nichol, H., and Locke, M. (1995). Honeybees and magnetoreception. *Science* **269**, 1888–1889.

Northcutt, R. G. (1983). The primary lateral line afferents in lepidosirenid lungfishes. *Soc. Neurosci. Abstr.* **9**, 1167.

Northcutt, R. G. (1986). Electroreception in nonteleost bony fishes. *In* "Electroreception" (T. H. Bullock and W. Heiligenberg, Eds.), pp. 257–285. John Wiley and Sons, New York.

Northcutt, R. G., Brändle, K., and Fritzsch, B. (1995). Electroreceptors and mechanosensory lateral line organs arise from single placodes in axolotls. *Dev. Biol.* **168**, 358–373.

Obara, S. (1976). Mechanisms of electroreception in ampullae of Lorenzini of the marine catfish Plotosus. *In* "Electrobiology of Nerve, Synapse and Muscle" (J. P. Reuben, D. P. Purpura, M. V. L. Bennett, and E. R. Kandel, Eds.), pp. 128–147. Raven Press, New York.

Obara, S., and Bennett, M. V. L. (1972). Mode of operation of ampullae of Lorenzini of the skate, *Raja. J. Gen. Physiol.* **60**, 534–557.

Pals, N., Valentijn, P., and Verwey, D. (1982). Orientation reactions of the dogfish, *Scyliorhinus canicula,* to local electric fields. *Neth. J. Zool.* **32**, 495–512.

Paulin, M. G. (1995). Electroreception and the compass sense of sharks. *J. Theor. Biol.* **174**, 325–339.

Perry, A., Bauer, G. B., and Dizon, A. E. (1981). Magnetite in the green turtle. *Trans. Am. Geophys. Union* **62**, 850.

Peters, R. C., and Bretschneider, F. (1972). Electric phenomena in the habitat of the catfish, *Ictalurus nebulosus* LeS. *J. Comp. Physiol.* **81**, 345–362.

Peters, R. C., and Evers, H. P. (1985). Frequency selectivity in the ampullary system on an elasmobranch fish (*Scyliorhinus canicula*). *J. Exp. Biol.* **118**, 99–109.

Peters, R. C., and Meek, F. (1973). Catfish and electric fields. *Experientia (Basel)* **29**, 299–300.

Pfeiffer, W. (1968). Die Fahrenholzschen Organe der Dipnoi und Brachiopterygii. *Z. Zellforsch.* **90**, 127–147.

Preston, D., and Pettigrew, J. D. (1980). Ferromagnetic coupling to muscle receptors as a basis for geomagnetic field sensitivity in animals. *Nature* **285**, 99–101.

Raschi, W. (1986). A morphological analysis of the ampullae of Lorenzini in selected skates (Pisces-rajoidei). *J. Morphol.* **189**, 225–247.

Raschi, W., and Mackanos, L. A. (1989). The structure of the ampullae of Lorenzini in *Dasyatis garouaensis* and its implications on the evolution of freshwater electroreceptive systems. *J. Exp. Zool.* **2**, 101–111.

Ronan, M. C. (1986). Electroreception in cyclostomes. *In* "Electroreception" (T. H. Bullock and W. Heiligenberg, Eds.), pp. 209–224. John Wiley and Sons, New York.

Ronan, M. C., and Bodznick, D. (1986). End buds: Non-ampullary electroreceptors in adult lampreys. *J. Comp. Physiol.* **158**, 9–16.

Scheich, H. (1977). Neural basis of communication in the high frequency electric fish, *Eigennmania viriscens* (jamming avoidance response). III. Central integration in the sensory pathway and control of the pacemaker. *J. Comp. Physiol.* **113**, 229–255.

Scheich, H., Bullock, T. H., and Hamstra, R. H. (1973). Coding properties of two classes of afferent fiber: High frequency electroreceptors in the electric fish. *Eigennmania. J. Neurophys.* **36**, 39–60.

Scheich, H., Langner, G., Tidemann, C., Coles, R. B., and Guppy, A. (1986). Electroreception and electrolocation in platypus. *Nature* **319**, 401–402.

Schessel, D. A., Ginzburg, R. D., and Highstein, S. M. (1982). Ultrastructural and physiological evidence for chemical synaptic transmission between the vestibular type I hair cell and its primary nerve chalice. *Anat. Rec.* **202**, 168.

Srivastava, C. B. L. (1973). Peripheral nerve ending types in tuberous electroreceptors of a high frequency gymnotid, *Sternarchus albifrons. J. Neurocytol.* **2**, 77–83.

Steinbach, A. B., and Bennett, M. V. L. (1971). Effect of divalent ions and drugs on synaptic transmission in phasic electroreceptors in a mormyrid fish. *J. Gen. Physiol.* **58**, 580–598.

Suga, N. (1967). Coding in tuberous and ampullary organs of a gymnotid electric fish. *J. Comp. Neurol.* **131**, 437–453.

Sugawara, Y. (1989). Two $Ca^{++}$ current components of the receptor current in the electroreceptors of the marine Catfish *Plotosus. J. Gen. Physiol.* **93**, 365–380.

Sugawara, Y., and Obara, S. (1989). Receptor $Ca^{++}$ current and $Ca^{++}$-gated $K^+$ current in tonic electroreceptors of the marine catfish *Plotosus. J. Gen. Physiol.* **93**, 343–364.

Szabo, T. (1965). Sense organs of the lateral line system in some electric fish of the Gymnotidae, Mormyridae and Gymnarchidae. *J. Morphol.* **117**, 229–250.

Szabo, T. (1970). Morphologische and funktionelle Aspekte bei Elektrorezeptoren. *Verh. Dtsch. Zool. Ges.* **64**, 141–148.

Szabo, T. (1974). Anatomy of the specialized lateral line organs of electroreception. *In* "Handbook of Sensory Physiology" (A. Fessard, Ed.), Vol. III/3, pp. 13–58. Springer-Verlag, New York.

Szabo, T., and Hagiwara, S. (1967). A latency-change mechanism involved in sensory coding of electric fish (mormyrids). *Physiol. Behav.* **2**, 331–335.

Szabo, T., and Wersäll, J. (1970). Ultrastructure of an electroreceptor (mormyromast) in a mormyrid fish. *Gnathonemus petersii. J. Ultrastruct. Res.* **30**, 473–490.

Szamier, R. B., and Wachtel, A. W. (1970). Special cutaneous receptor organs of fish. VI. Ampullary and tuberous organs of Hypopomus. *J. Ultrastruct. Res.* **30**, 450–471.

Teeter, J. G., and Bennett, M. V. L. (1981). Synaptic transmission in the ampullary electroreceptor of the transparent catfish, *Kryptopterus. J. Comp. Physiol.* **142**, 371–377.

Teeter, J. G., Szamier, R. B., and Bennett, M. V. L. (1980). Ampullary electroreceptors in the sturgeon *Scaphirhynchus platorynchus* (Rafinesque). *J. Comp. Physiol.* **138**, 213–223.

Tenforde, T. S. (1989). Electroreception and magnetoreception in simple and complex organisms. *Biomagnetics* **10**, 215–221.

Tricas, T. C. (1982). Bioelectric-mediated predation by swell sharks, *Cephaloscyllium ventriosum. Copeia* **1982**, 948–952.

Tricas, T. C., and New, J. G. (1997). Sensitivity and response dynamics of elasmobranch electrosensory primary afferent neurons to near threshold fields. *J. Comp. Physiol.* in press.

Tricas, T. C., Michael, S. W., and Sisneros, J. A. (1995). Electrosensory optimization to conspecific phasic signals for mating. *Neurosci. Lett.* **202**, 129–132.

Viancour, T. (1979a). Electroreceptors of a weakly electric fish. I. Characterization of tuberous receptor tuning. *J. Comp. Physiol.* **133,** 317–325.

Viancour, T. (1979b). Electroreceptors of a weakly electric fish. II. Individually tuned receptor oscillations. *J. Comp. Physiol.* **133,** 327–338.

Vischer, H. A. (1995). Electroreceptor development in the weakly electric fish *Eigennmania:* A histological and ultrastructural study. *J. Comp. Neurol.* **360,** 81–100.

von der Emde, G., and Bleckmann, H. (1992). Extreme phase sensitivity of afferents which innervate mormyromast electroreceptors. *Naturwissenschaften* **79,** 131–133.

Wachtel, A. W., and Szamier, R. B. (1966). Special cutaneous receptor organs of fish: The tuberous organs of *Eigennmania. J. Morphol.* **119,** 51–80.

Walcott, C., Gould, J. L., and Kirschvink, J. L. (1979). Pigeons have magnets. *Science* **201,** 1027–1029.

Walker, M. M., Kirschvink, J. L., Chang, S. B. R., and Dizon, A. E. (1984). A candidate magnetic sense organ in the yellowfin tuna, *Thunnus albacares. Science* **224,** 751–753.

Waltmann, B. (1966). Electrical properties and fine structure of the ampullary canals of Lorenzini. *Acta Physiol. Scand.* **66,** Suppl. 264, 1–60.

Watson, D., and Bastian, J. (1979). Frequency response characteristics of electroreceptors in the weakly electric fish, *Gymnotus carapo. J. Comp. Physiol.* **134,** 191–202.

Wu, C. H. (1984). Electric fish and the discovery of animal electricity. *Am. Sci.* **72,** 598–607.

Yager, D. D., and Hopkins, C. D. (1993). Directional characteristics of tuberous electroreceptors in the weakly electric fish, Hypopomus (Gymnotiformes). *J. Comp. Physiol.* **173,** 401–414.

Zakon, H. H. (1984a). Postembryonic changes in the peripheral electrosensory system of a weakly electric fish: Addition of receptor organs with age. *J. Comp. Neurol.* **228,** 557–570.

Zakon, H. H. (1984b). The ionic basis of the oscillatory receptor potential of tuberous electroreceptors in Sternopygus. *Soc. Neurosci. Abstr.* **10,** 193.

Zakon, H. H. (1986a). The electroreceptive periphery. *In* "Electroreception" (T. H. Bullock and W. Heiligenberg, Eds.), pp. 103–156. John Wiley and Sons, New York.

Zakon, H. H. (1986b). The emergence of tuning in newly-generated tuberous electroreceptors. *J. Neurosci.* **6,** 3297–3308.

Zakon, H. H. (1987). Variation in the mode of receptor cell addition in the electrosensory system of gymnotiform fishes. *J. Comp. Neurol.* **262,** 195–214.

Zakon, H. H. (1988). The electroreceptors: Diversity in structure and function. *In* "Sensory Biology of Aquatic Animals" (J. Atema, R. R. Fay, A. N. Popper and W. N. Tavolga, Eds.), pp. 813–850. Springer, New York.

Zakon, H. H., and Meyer, J. H. (1983). Plasticity of electroreceptor tuning in the weakly electric fish *Sternopygus dariensis. J. Comp. Physiol.* **153,** 477–487.

Zipser, B., and Bennett, M. V. L. (1973). Tetrodotoxin resistant electrically excitable responses of receptor cells. *Brain Res.* **62,** 253–259.

Zoeger, J., Dunn, J. R., and Fuller, M. (1981). Magnetic material in the head of the common Pacific dolphin. *Science* **213,** 892–894.

# SECTION
# VI

# Muscle and Other Contractile Systems

Nicholas Sperelakis

# 50

# Skeletal Muscle Action Potentials

## I. Introduction

Action potential (AP) generation and excitability in neurons were covered in an earlier chapter. Most of the general electrophysiological principles discussed there also apply to skeletal muscle fibers and so are only briefly reviewed and summarized in this chapter. In most respects, the electrogenesis of the APs in nerve axons and skeletal muscle fibers is quite similar, with the exceptions of the type of afterpotentials, saltatory propagation in myelinated axons, and high $Cl^-$ conductance in skeletal muscle fibers. Both are long cables and have very brief and fast-rising APs whose inward current is carried by $Na^+$ ion through fast $Na^+$ channels. Skeletal muscle fibers, however, have the added complexity of an extensive internal transverse (T) tubular system formed by a periodic invagination of the surface cell membrane (Fig. 1), forming an orderly three-dimensional array of tubules that propagate excitation from the cell surface into the deep interior of the fiber for purposes of excitation–contraction coupling.

The skeletal AP is considerably different from that of cardiac muscle cells, which has a very long duration with a pronounced plateau and a substantially lower rate of rise and propagation velocity. The myocardial cells are short, and there is a slight delay in propagation at each cell-to-cell junction. Like skeletal muscle muscle fibers, myocardial cells have a fast $I_{Na}$ responsible for the rapid upstroke of the AP, but the delayed rectifier $K^+$ conductance is turned on slowly and there is a substantial inward $I_{Ca}$ during the entire plateau.

The APs of smooth muscles are markedly different from those of skeletal muscle fibers, in that the resting (takeoff) potential is lower (more depolarized), the rate of rise of the APs is much slower, the AP overshoot is much less, and the AP duration (APD) is considerably longer. Propagation velocity is much slower, and the smooth muscle cells are short and small in diameter. The inward current for the APs in smooth muscle cells is primarily a slow $Ca^{2+}$ current carried through L-type

$Ca^{2+}$ channels, but some cells do possess some functioning fast $Na^+$ channels.

## II. General Overview of Electrogenesis of the Action Potential

The ion distributions and ion pumps and exchangers found in skeletal muscle fibers are similar to those of other types of cells, as described in the chapter on the resting potential (Fig. 1). The APs in vertebrate skeletal muscle twitch fibers consist of *spike* followed by *depolarizing* (*"negative"*) *afterpotential* (Figs. 2 and 3). A large fast inward $Na^+$ current, passing through fast $Na^+$ channels, is responsible for electrogenesis of the spike, which rises rapidly (400–700 V/sec). Subsequently, a small slow inward $Ca^{2+}$ current, passing through kinetically slow channels, may be involved in excitation–contraction coupling. The skeletal muscle cell membrane has at least two types of voltage-dependent $K^+$ channels (Fig. 4). One type allows $K^+$ ions to pass more readily inward than outward, the so-called *inward-going rectifier*. This channel is responsible for *anomalous rectification* (i.e., decrease in $g_K$ with depolarization). There is a quick decrease in $K^+$ conductance on depolarization and increase in $K^+$ conductance with repolarization. The second type of $K^+$ channel is similar to the usual $K^+$ channel found in squid giant axon membrane, the so-called *delayed rectifier*. Its conductance turns on more slowly than $g_{Na}$ on depolarization. This channel allows $K^+$ to pass readily outward down the electrochemical gradient for $K^+$. The activation of this channel produces the large increase in total $g_K$ that helps to terminate the AP (see Fig. 3).

The AP amplitude is about 120 mV, from a resting potential of $-80$ mV to a peak overshoot potential of about $+40$ mV (see Figs. 2 and 3). The duration of the AP (at 50% repolarization, or $APD_{50}$) ranges between 3 and 6 msec, depending on the species and temperature. The *threshold potential* ($V_{th}$) for triggering of the fast $Na^+$ channels is about $-55$ mV; thus, a *critical depolarization* of about

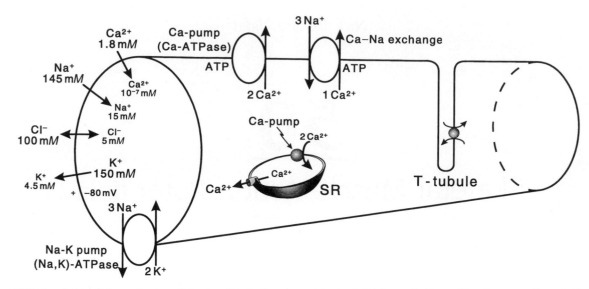

**FIG. 1.** Intracellular and extracellular ion distributions in vertebrate skeletal muscle fibers. Also shown are the polarity and magnitude of the resting potential. Arrows indicate direction of the net electrochemical gradient. The $Na^+$–$K^+$ pump and $Ca^{2+}$–$Na^+$ exchange carrier are located in the cell surface membrane. A Ca-ATPase and $Ca^{2+}$ pump, similar to that in the sarcoplasmic reticulum (SR), is located in the cell membrane.

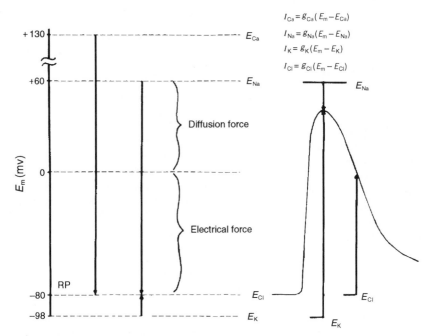

**FIG. 2.** Representation of the electrochemical driving forces for $Na^+$, $Ca^{2+}$, $K^+$, and $Cl^-$ at rest (left diagram) and during the AP in a skeletal muscle fiber (right diagram). Equilibrium potentials for each ion (e.g., $E_{Na}$) are positioned vertically according to their magnitude and sign; they were calculated from the Nernst equation for a given set of extracellular and intracellular ion concentrations. Measured resting potential is assumed to be −80 mV. Electrochemical driving force for an ion is the difference between its equilibrium potential ($E_i$) and the membrane potential ($E_m$), that is, ($E_m − E_i$). Thus, at rest, the driving force for $Na^+$ is the difference between $E_{Na}$ and the resting $E_m$; if $E_{Na}$ is +60 mV and resting $E_m$ is −80 mV, the driving force is 140 mV. The driving force is then the algebraic sum of the diffusion force and the electrical force, and is represented by the length of the arrows in the diagram. Driving force for $Ca^{2+}$ (about 210 mV) is even greater than that for $Na^+$, whereas that for $K^+$ is much less (about 18 mV). Direction of the arrows indicates the direction of the net electrochemical driving force, namely, the direction for $K^+$ is outward, whereas that for $Na^+$ and $Ca^{2+}$ is inward. If $Cl^-$ is passively distributed, then for a cell sitting a long time at rest, $E_{Cl} = E_m$ and there is no net driving force. The driving forces change during the AP, as depicted. The equations for the different ionic currents are given in the upper right-hand portion of the figure. (Adapted from Sperelakis, 1979.)

**A**

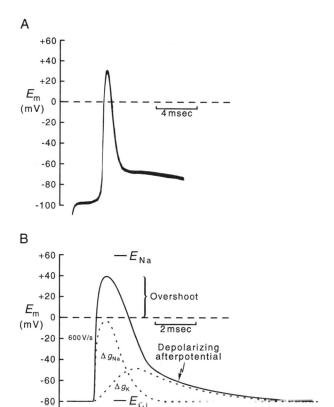

**B**

**FIG. 3.** (A) Action potential (AP) recorded with an intracellular microelectrode in a skeletal muscle fiber of frog semitendinosus muscle bathed in normal frog Ringer solution. Note the prominent depolarizing afterpotential. Shock artifact is at left of spike. (Modified from Sperelakis *et al.*, 1973.) (B) Diagrammatic representation of the relative conductance changes for Na$^+$ and K$^+$ during an AP. The rising phase of the AP is caused by an increase in $g_{Na}$, which brings the membrane potential ($E_m$) toward the Na$^+$ equilibrium potential ($E_{Na}$). The falling phase in the AP is due to the rise in $g_K$ and to the decrease in $g_{Na}$ and to an outward Cl$^-$ current. The depolarizing afterpotential is explained by the fact that the delayed rectifier K$^+$ channel is less selective for K$^+$ (30:1 over Na$^+$) than is the resting channel (100:1).

25 mV is required to reach $V_{th}$. The turn-on of the fast $g_{Na}$ (fast $I_{Na}$) is very rapid (within 0.2 msec), and $E_m$ is brought rapidly toward $E_{Na}$ (Figs. 2 and 3). There is an explosive (positive exponential initially) increase in $g_{Na}$, caused by a positive-feedback relationship between $g_{Na}$ and $E_m$.

From the current versus voltage curves, the maximum inward fast Na$^+$ current occurs at an $E_m$ of about $-20$ mV. The current decreases at more depolarized $E_m$ levels because of the diminution in electrochemical driving force as the membrane is further depolarized, even though the conductance remains high. At the reversal potential ($E_{rev}$) for the current, the current goes to zero; $I_{Na}$ then reverses direction with greater depolarization.

As $E_m$ depolarizes, it crosses $V_{th}$ for slow Ca$^{2+}$ channels located in the transverse tubules, which is about $-35$ mV. Turn-on of the Ca$^{2+}$ conductance ($g_{Ca}$) and $I_{Ca}$ is slower. The peak $I_{Ca}$ is considerably smaller than the peak $I_{Na}$.

## III. Ion Channel Activation and Inactivation

As discussed in the chapter on nerve excitability, the fast Na$^+$ channels (and the slow Ca$^{2+}$ channels) have a double gating mechanism; one gate is the *inactivation gate* (I-gate), and the second gate is the *activation gate* (A-gate). For a channel to be conducting, both the A-gate and the I-gate must be open; if either is closed, the channel is nonconducting. The A-gate is closed at the resting $E_m$ and opens rapidly on depolarization, whereas the I-gate is open at the resting $E_m$ and closes slowly on depolarization. In the Hodgkin–Huxley (1952) analysis, the opening of the A-gate requires simultaneous occupation of three negatively charged sites by three positively charged $m^+$ particles. Therefore,

$$g_{Na} = \bar{g}_{Na} m^3 h \qquad (1)$$

where $m$ is the activation variable, $h$ is the inactivation variable, and $\bar{g}_{Na}$ is the maximum conductance. A small gating current ($I_g$) has been measured that corresponds to the movement of the charged $m$ particles (or rotation of an equivalent dipole). The outward $I_g$ leads into the inward $I_{Na}$.

The fast $I_{Na}$ lasts only for 1–2 msec because of the spontaneous *inactivation* of the fast Na$^+$ channels, that is, they inactivate quickly, even if the membrane were to remain depolarized. In contrast, the slow Ca$^{2+}$ channels inactivate more slowly. Inactivation is produced in the ion channels by the voltage-dependent closing of the inactivation gate (I-gate).

The voltage dependency of inactivation is given by the $h_\infty$ versus $E_m$ curve. The Na$^+$ conductance ($g_{Na}$) at any time is equal to the maximal value ($\bar{g}_{Na}$) times $m^3 h$. Therefore, when $h = 0$, $g_{Na} = 0$, and when $h = 1.0$, $g_{Na} = \bar{g}_{Na}$ (if $m = 1.0$). At the normal resting potential, $h_\infty$ is nearly 1.0 and diminishes with depolarization, becoming nearly zero at about $-30$ mV. The maximal rate of rise of the AP (max $dV/dt$) is directly proportional to the net inward current or $I_{Na}$, which is directly proportional to $g_{Na}$ and can be expressed as

$$\max dV/dt \propto \frac{I_{Na}}{C_m} = \frac{\bar{g}_{Na} m^3 h (E_m - E_{Na})}{C_m} \qquad (2)$$

Therefore, the decrease in $h_\infty$ is the cause of decrease in max $dV/dt$. At about $-30$ mV, $h_\infty \to 0$, and there is nearly complete inactivation of the fast Na$^+$ channels. Thus, depolarization by any means (e.g., elevated [K]$_0$ or applied current pulses) decreases max $dV/dt$, and excitability disappears at about $-50$ mV.

The *slow Ca$^{2+}$ channels* behave much the same way as the fast Na$^+$ channels with respect to inactivation, with one main difference being the voltage range over which the slow channels inactivate: $-45$ to $-10$ mV for the slow channels, compared with $-80$ to $-30$ mV for the fast Na$^+$ channels. Another major difference is that the slow Ca$^{2+}$ conductance inactivates much more slowly than the fast Na$^+$ conductance; that is, they have a long inactivation time constant ($\tau_{inact}$). (The $h$ variable for the slow channel is sometimes referred to as the $f$ variable, and the $m$ variable as the $d$ variable.) The recovery process for the slow

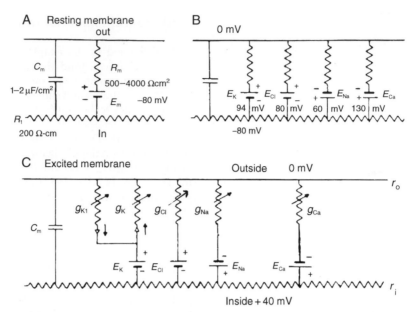

**FIG. 4.** Electrical equivalent circuits for a skeletal muscle fiber cell membrane (A and B) at rest and (C) during excitation. (A) Membrane as a parallel resistance–capacitance circuit, the membrane resistance ($R_m$) being in parallel with the membrane capacitance ($C_m$). Resting potential ($E_m$) is represented by an 80-mV battery in series with the membrane resistance, the negative pole facing inward. (B) Membrane resistance is divided into four component parts, one for each of the four major ions of importance: $K^+$, $Cl^-$, $Na^+$, and $Ca^{2+}$. Resistances for these ions ($R_K$, $R_{Cl}$, $R_{Na}$, and $R_{Ca}$) are parallel to one another and represent totally separate and independent pathways for permeation of each ion through the resting membrane. These ion resistances are depicted as their reciprocals, namely, ion conductances ($g_K$, $g_{Cl}$, $g_{Na}$, and $g_{Ca}$). Equilibrium potential for each ion (e.g., $E_K$), determined solely by the ion distribution in the steady state and calculated from the Nernst equation, is shown in series with the conductance path for that ion. Resting potential of $-80$ mV is determined by the equilibrium potentials and by the relative conductances. (C) Equivalent circuit is further expanded to illustrate that, for the voltage-dependent conductances, there are at least two separate $K^+$-conductance pathways (labeled here $g_{K1}$ and $g_K$). In series with the $K^+$ conductances are rectifiers pointing in the direction of least resistance to current flow. There is one $Na^+$ conductance pathway, the kinetically fast $Na^+$ conductance ($g_{Na}$). In addition, there is a kinetically slow pathway that allows $Ca^{2+}$ to pass through. Arrows drawn through the resistors indicate that the conductances are variable, depending on membrane potential and time. (Adapted from Sperelakis, 1979.)

$Ca^{2+}$ channels is slow compared with that for fast $Na^+$ channels.

The *$K^+$ channel* (delayed rectifier) may have only an activation gate, because it does not inactivate quickly. In the Hodgkin–Huxley analysis of squid giant axon, the A-gate opens when four positively charged $n^+$ particles simultaneously occupy four favorable positions (negatively charged sites). If $n$ is the probability that one site is occupied, then $n^4$ is the probability that all four sites are occupied. Therefore,

$$g_K = \bar{g}_K n^4 \tag{3}$$

The fourth power to which $n$ must be raised causes a delay (sigmoidal foot) in turn-on of the $K^+$ conductance.

## IV. Mechanisms of Repolarization

The skeletal AP is terminated partly by the turn-on of the $K^+$ conductance ($g_K$) (see Fig. 3). The increase in $g_K$ acts to bring $E_m$ toward $E_K$ (about $-98$ mV), since the membrane potential at any time is determined primarily by the ratio of $g_{Na}/g_K$. This type of $g_K$ channel is activated by

depolarization and turned off by repolarization. Therefore, this $g_K$ channel is *self-limiting*, in that it turns itself off as the membrane is repolarized by its action.

In addition to the $g_K$ turn-on, turnoff of $g_{Na}$ occurs (see Fig. 3) (contributing to repolarization) for two reasons: (1) spontaneous inactivation of fast $Na^+$ channels that had been activated, that is, closing of their I-gate (inactivation $\tau$ of 1–3 msec); and (2) reversible shifting of activated channels directly back to the resting state (*deactivation*), because of the rapid repolarization occurring due to the $g_K$ increase (Fig. 5). Theoretically, it would be possible to have an AP that would repolarize (but more slowly) even if there were no $g_K$ mechanism, because the $g_{Na}$ channels would spontaneously inactivate, and so the $g_{Na}/g_K$ ratio and $E_m$ would be slowly restored to their original resting values.

In skeletal muscle, there is an important third factor involved in repolarization of the AP: the *$Cl^-$ current* (see Fig. 2). The $Cl^-$ permeability ($P_{Cl}$) and conductance ($g_{Cl}$) are very high in skeletal muscle. In fact, $P_{Cl}$ of the surface membrane is much higher than $P_K$, the $P_{Cl}/P_K$ ratio being about 3–7. As discussed in the chapter on resting potential, the $Cl^-$ ion is passively distributed, or nearly so, and thus cannot determine the resting potential under steady-state

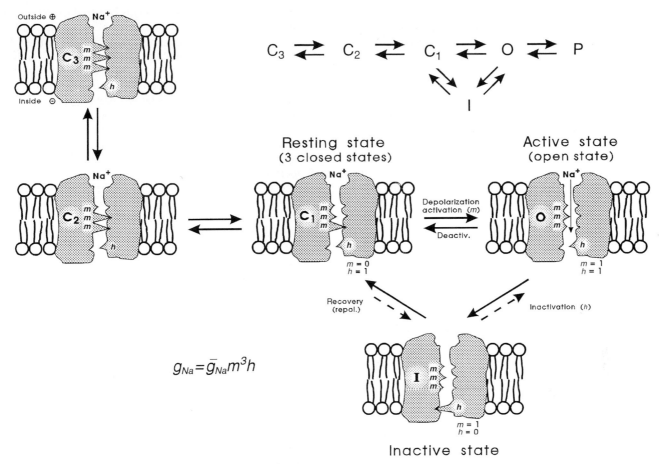

$$g_{Na} = \bar{g}_{Na} m^3 h$$

**FIG. 5.** Illustration of the hypothetical states of the fast Na⁺ channel. The three states patterned after the Hodgkin–Huxley view were modified to reflect the fact that there is evidence for three closed states. As depicted, in the most closed state ($C_3$), all three $m$ gates (or particles) are in the closed configuration. In the mid-closed state ($C_2$), two $m$ gates are closed and one is open. In the least closed state ($C_1$), one gate is closed and two are open. In the resting state, the activation gate (A) is closed and the inactivation gate (I) is open: $m = 0$, $h = 1$. Depolarization to the threshold activates the channel to the active state, the A-gate opening rapidly and the I-gate still being open: $m = 1$, $h = 1$. The activated channel spontaneously inactivates to the inactive state due to closure of the I-gate: $m = 1$, $h = 0$. The recovery process on repolarization returns the channel from the inactive state back to the resting state, thus making the channel again available for reactivation. Na⁺ ion is depicted as being bound to the outer mouth of the channel and poised for entry down its electrochemical gradient when both gates are in the open configuration. The reaction between the resting state and the active state is readily reversible, and there is some reversibility of the other reactions. The fast Na⁺ channel is blocked by tetrotodoxin (TTX) binding to the outer mouth and plugging it.

conditions. However, net Cl⁻ movements inward (hyperpolarizing) or outward (depolarizing) do affect $E_m$ transiently until reequilibration occurs and there is no further net movement. Although there is no net electrochemical driving force for Cl⁻ current ($I_{Cl}$) at the resting potential, since $E_m = E_{Cl}$, during the AP depolarization, there is a larger and larger driving force for outward $I_{Cl}$ (i.e., Cl⁻ influx), since $I_{Cl} = g_{Cl}(E_m - E_{Cl})$. In other words, the large electric field that was keeping Cl⁻ out (i.e., $[Cl]_i \ll [Cl]_o$) diminishes during the AP, and so Cl⁻ ion enters the fiber. This Cl⁻ entry is hyperpolarizing, and so tends to repolarize the membrane more quickly than would otherwise occur. That is, repolarization of the AP is sharpened by the Cl⁻ mechanism. There may be

voltage-dependent $g_{Cl}$ channels, as well as a high $g_{Cl}$ in the membrane at rest.

To further illustrate some of the preceding points on the role of Cl⁻, when skeletal muscle fibers are placed into Cl⁻-free Ringer solution (e.g., methanesulfonate substitution), depolarization and spontaneous APs and twitches occur for a few minutes until most or all of the $[Cl]_i$ is washed out. After equilibration, the resting $E_m$ returns to the original value (ca. $-90$ mV for frog skeletal muscle and $-80$ mV for mammalian, clearly indicating that Cl⁻ does not determine the resting potential and that net Cl⁻ efflux produces depolarization. Readdition of Cl⁻ to the bath produces a rapid large hyperpolarization, for example to $-120$ mV, due to net Cl⁻ influx; the $E_m$ then slowly returns to the original value ($-90$ mV) as $Cl^-$ reequilibrates, that is, redistributes itself passively. These same effects would occur in cardiac muscle, smooth muscle, and nerve, but to a lesser extent, because in these tissues $P_{Cl}$ is much lower (e.g., $P_{Cl}/P_K$ ratio is only about 0.5 in vascular smooth muscle).

The importance of the $Cl^-$ current in repolarization in skeletal muscle fibers is illustrated by one type of *myotonia* in which an abnormally low $P_{Cl}$ causes repetitive APs to occur. Because $g_{Cl}$ is abnormally low, the $Cl^-$ influx during AP repolarization is much less than normal, and so the repolarization process is slowed. This increases the duration of the AP and leads to repetitive discharge of APs in the muscle fibers. That is, the muscle fibers lose their tight control by the motor neurons, and so contraction becomes partly involuntary. For example, persons with myotonia find it difficult to release a handshake or to remove their hand from a drinking glass. There are several causes of myotonia, including genetic abnormalities in ion channels as well as drug-induced conditions. Any agent that greatly lowers $P_{Cl}$ or $g_{Cl}$ will have the same effect. It has been shown that simply decreasing $g_{Cl}$ causes repetitive firing in equivalent circuit models of skeletal muscle fibers. In addition, $K^+$ ions tend to accumulate in the lumen of the T-tubules under normal conditions (see section following on late depolarizing afterpotentials), and this accumulation is exaggerated with the prolonged APs and so tends to depolarize partially the fibers and increase their excitability. Some forms of myotonia are produced by *abnormal fast $Na^+$ channels*; namely, a small fraction of these channels do not inactivate as quickly as usual (i.e., their I-gates do not close normally), and so causes a prolonged small depolarization after the AP and repetitive discharge.

## V. Voltage-Dependent $Cl^-$ Channels

As discussed previously, the $Cl^-$ conductance ($g_{Cl}$) of skeletal muscle fibers is very high normally, and it is important for producing a sharp repolarization of the AP. As stated, one type of myotonia is caused by an abnormally low $Cl^-$ conductance, and therefore low $Cl^-$ current ($I_{Cl}$), which leads to repetitive firing of APs. Also as mentioned previously, the $P_{Cl}/P_K$ ratio is about 3–7 in skeletal muscle, thus indicating that the dominant permeability or conductance ($g_{Cl}/g_K$ ratio) is $Cl^-$.

This high $g_{Cl}$ is due to a large number of voltage-dependent gated $Cl^-$ channels, which are outwardly rectifying. These $Cl^-$ channels are located primarily on the surface sarcolemma in frog skeletal muscle, there being only a few in the T-tubules; whereas in rat or mouse, they are primarily in the T-tubules, there being very few on the surface membrane. Denervation of mammalian fibers causes $g_{Cl}$ to decrease almost to zero.

In frog, there are several subtypes of $Cl^-$ channels that have single-channel conductances ranging between 40 and 70 pS, and each channel may exhibit several subconductance states. Those that do often have a main gate that opens or closes the entire channel. In fetal mammalian fibers, $Cl^-$ channels with conductances of about 40, 60, and

300 pS have been observed. Myoballs cultured from muscle biopsies of patients having one form of myotonia had a reduced (ca. 50%) single-channel conductance for the $Cl^-$ channel, which would contribute to the myotonia (Fahlke *et al.*, 1993). In primary cultures of rat skeletal muscle, the fast $Cl^-$ channel showed a behavior consistent with six closed states and two open states (Weiss and Magleby, 1992). The $Cl^-$ channel in myoblasts and myotubes of the L6 cell line derived from rat skeletal muscle had a high conductance of about 330 pS (Hurnak and Zachar, 1992). Voltage-gated $Cl^-$ channels have also been found in the SR membrane of skeletal muscle.

Some $Cl^-$ channels described for other tissues include (1) $Ca^{2+}$-dependent $Cl^-$ channels, (2) stretch-activated $Cl^-$ channels, and (3) cyclic AMP-stimulated $Cl^-$ channels. The receptor-operated $Cl^-$ channels apparently have a G-protein (e.g., $G_s$ or $G_i$) as intermediate for coupling.

The voltage-dependent $Cl^-$ channels can be blocked relatively selectively by several methods, including acidosis and use of compounds such as the stilbene derivatives (DIDS and SITS) and 9-anthracene carboxylic (9-AC) acid. The $Cl^-$ channels in frog skeletal muscle are relatively insensitive to 9-AC acid, whereas those in adult mammalian muscle are highly sensitive. The anion selectivity sequence for some voltage-dependent $Cl^-$ channels is $I^- > Br^- > Cl^- > F^-$.

## VI. Slow Delayed Rectifier $K^+$ Current

Two types of $K^+$ delayed rectifier currents occur in skeletal muscle. A slow $I_K$ was first described by Adrian and co-workers (1970a,b) in voltage-clamped frog sartorius fibers; the slow component of outward $I_K$ reached a maximum in about 3 sec (at $-30$ mV) and declined with a time constant of about 0.5 sec (at $-100$ mV). In voltage-clamped frog toe muscle, Lynch (1978) observed that, with depolarizing clamps more positive than the threshold of $-55$ mV for activation of the outward $K^+$ current (about 5 mS/cm$^2$), most fibers had both fast and slow components of the outward $I_K$. The voltage dependences of both $K^+$ currents were shifted equally in the depolarizing direction by elevated $[Ca]_o$ or $[H]_o$, presumably due to altering the net negative outer surface charge of the membrane, thereby "hyperpolarizing" (see chapter on resting potential). Acidosis also increased the rate of turn-on of the slow delayed rectifier ($pK_a = 5.8$). The fast delayed current was relatively selectively blocked by TEA or by a sulfhydryl reagent, whereas the slow delayed current was selectively depressed by a histidine reagent. It was estimated that about 25% of the delayed rectifier channels are in the T-tubular membrane. The functional significance of the slow outward $I_K$ is unknown, although it may be partly responsible for the delayed depolarizing afterpotential (see following).

## VII. Electrogenesis of Depolarizing Afterpotentials

As mentioned previously in this chapter, the AP spike in skeletal muscle fibers is followed by a prominent *depolar-*

*izing afterpotential* (also called a "negative" afterpotential based on the old terminology from external recording) (see Fig. 3). In addition to this early depolarizing afterpotential (i.e., emerging from the spike), there is a late depolarizing afterpotential that follows a tetanic train of spikes (e.g., 10 spikes). The electrogeneses of the early and *late afterpotentials* are different. The early afterpotential is due to a conductance change, whereas the late afterpotential may be due primarily to $K^+$ *accumulation* in the T-tubules.

The *early depolarizing afterpotential* of frog skeletal fibers is about 25 mV in amplitude immediately after the spike component, and gradually decays to the resting potential in 10–20 msec. It was shown by Adrian *et al.* (1970a) that this decay results from the fact that the delayed rectifier $K^+$ channel that opens during depolarization to terminate the spike is less selective for $K^+$ (ca. 30:1, $K^+$:$Na^+$) than is the $K^+$ channel in the resting membrane (ca. 100:1). Therefore, the constant-field equation (see chapter on resting potential) predicts that the membrane should be partly depolarized when the membrane potential is dominated by this $K^+$ conductance that is turned on during the AP. Thus, the early depolarizing afterpotential is apparently due to the persistence and slow decay of this less-selective $K^+$ conductance.

The *late depolarizing afterpotential* of frog skeletal fibers may result from accumulation of $K^+$ ions in the T-tubules (Adrian and Freygang, 1962). During the AP depolarization and turn-on of $g_K$ (delayed rectifier), there is a large driving force for $K^+$ efflux from the myoplasm coupled with a large $K^+$ conductance, resulting in a large outward $K^+$ current [$I_K = g_K (E_m - E_K)$] across all surfaces of the fiber, namely, across the surface sarcolemma and T-tubule walls. The $K^+$ efflux at the fiber surface membrane can rapidly diffuse away and mix with the relatively large interstitial fluid (ISF) volume, whereas the $K^+$ efflux into the T-tubules (TT) is trapped in this restricted diffusion space. The resulting high $[K]_{TT}$ decreases $E_K$ across the T-tubule membrane, and thereby depolarizes this membrane. Because of cable properties, part of this depolarization is transmitted to the surface sarcolemma, and is recorded by an intracellular microelectrode. The $K^+$ accumulation in the T-tubules can only be dissipated relatively slowly by diffusion out of the mouth of the T-tubules and by active pumping back into the myoplasm across the T-tubule wall. Thus, the decay of the late afterpotential will be a function of these two processes.

The amplitude and duration of the late depolarizing afterpotential are a function of the number of spikes in the train and their frequency. That is, the greater the spike activity, the greater the amplitude and duration of the late afterpotential. If the train consists of 20 spikes at a frequency of 50/sec, a typical value for the amplitude of the late afterpotential in frog fibers is about 20 mV. When the diameter of the T-tubules is increased by placing the fibers in hypertonic solutions,[1] the amplitude of the late afterpotential decreases as expected because of the greater

dilution of the $K^+$ ions accumulating in the T-tubule lumen. When the T-tubular system is disrupted and disconnected from the surface membrane by the *glycerol osmotic shock* method,[2] the late afterpotentials disappear (whereas the early afterpotentials persist).

An alternative explanation for the late depolarizing afterpotential is that it may be due to the slow delayed rectifier $g_K$ change described previously (Adrian *et al.*, 1970b). The equilibrium potential for the slow $I_K$ is $-83$ mV, and the sign (direction) of the late afterpotential reverses when the fiber is depolarized below $-80$ mV. Hence, the late afterpotential may arise from the slow relaxation of a component of the $K^+$ conductance increase, which is less selective for $K^+$ than the $K^+$ channels open in resting membrane. That is, in this view, the electrogenesis of the late afterpotential would be similar to that for the early afterpotential, but a different $K^+$ channel is involved.

All depolarizing afterpotentials, regardless of whether early or late, have physiological importance because they alter excitability and the propagation velocity of the fiber. A depolarizing afterpotential should enhance excitability (lower threshold) to a subsequent AP. This is because the *critical depolarization* required to reach the *threshold potential* would be decreased. A large late depolarizing afterpotential, such as that due to $K^+$ *accumulation* in the T-tubules, can, under certain pathological conditions, trigger repetitive APs. The effect of depolarizing afterpotentials on velocity of propagation in skeletal muscle fibers involves two opposing factors: (1) the decrease in critical depolarization required and (2) the decrease in maximal rate of rise of the AP (max $dV/dt$), which is a function of the takeoff potential ($h_\infty$ vs $E_m$ curve). Therefore, what factor dominates will depend on the degree of depolarization and the shape of the $h_\infty$ curve. When frog skeletal fibers are depolarized slightly by elevating $[K]_0$, only a decrease in propagation velocity is observed (Sperelakis *et al.*, 1970).

## VIII. $Ca^{2+}$-Dependent Slow Action Potentials

*Slow APs* are recorded under conditions in which the fast $Na^+$ current is blocked by $Na^+$-deficient solution, TTX, or voltage inactivation of the fast $Na^+$ channels in high $[K]_0$. Under these conditions, the only carrier of inward current available to produce an AP is $Ca^{2+}$ ion. Spontaneously occurring slow APs were first observed in frog sartorius fibers equilibrated in $Cl^-$-free solution containing TTX (Sperelakis *et al.*, 1967). Upon addition of $Ba^{2+}$ ion (e.g., 0.5 m$M$), which is a potent blocker of $K^+$ channels and $P_K$, the fibers partially depolarize, and spontaneously discharge slowly rising (e.g., 1–10 V/sec) overshooting APs of long duration (e.g., several seconds), having a prominent plateau component (resembling a cardiac AP in shape). $Ba^{2+}$ depolarizes rapidly in $Cl^-$-free solution, because the volt-

---

[1] In hypertonic solutions, skeletal muscle fibers shrink (fiber diameter decreases) like an imperfect osmometer (with an osmotically inactive volume of about 32%), but the T-tubules swell.

[2] To produce glycerol osmotic shock, about 300 mOsm glycerol is added to Ringer solution. The glycerol rapidly permeates into the fiber interior (so the fiber shrinks only transiently) and equilibrates. But when the glycerol is washed out, there is a great hypotonic shock that disrupts the T-tubules.

age-clamping effect of the Cl⁻ distribution ($E_{Cl}$), due to the large $P_{Cl}$, is circumvented. In addition, since Cl⁻-free solution raises the resistance of the cell membrane about sevenfold, an intracellular microelectrode can better detect the potential changes occurring across the T-tubule wall because of less short-circuiting by the surface membrane.

By using two intracellular microelectrodes, one for applying intracellular current to stimulate the frog skeletal muscle fiber and the other for recording voltage a short distance away in the same fiber ($[K]_o$ of 25 mM to depolarize the fibers to about −45 mV and thereby voltage-inactivate the fast Na⁺ channels, and $[Na]_o$ reduced to zero so that there could be no inward fast Na⁺ current), application of small hyperpolarizing current pulses during the slow AP indicates that membrane resistance increases progressively during the plateau component, leading to an abrupt repolarization terminating the AP (Kerr and Sperelakis, 1982). The AP responses fatigued with repetitive stimulation. The rate of rise, overshoot, and duration of the slow APs are a function of $[Ca]_o$ (Beaty and Stefani, 1976; Vogel et al., 1978; Kerr and Sperelakis, 1982). For example, the AP duration at 50% amplitude ($APD_{50}$) was about 8 sec when $[Ca]_o$ was 6 mM. The amplitude of the slow AP plotted against log $[Ca]_o$ gave a straight line with a slope of 28 mV/decade, which is close to the theoretical 29 mV/decade (at 21°C) from the Nernst relationship for a situation in which only Ca²⁺ ion carried the inward current. The slow APs were depressed and blocked by the Ca-antagonistic and slow-channel-blocking drugs, verapamil and bepridil, with an $ED_{50}$ of about $5 \times 10^{-8}$ M. In voltage-clamp studies on frog muscle, elevation of $[Ca]_o$ increased $I_{Ca}$, and $I_{Ca}$ was depressed by the slow Ca²⁺ channel blockers D-600, nifedipine, and Ni²⁺ (Stanfield, 1977; Sanchez and Stefani, 1978; Almers et al., 1981). Detubulation by the glycerol osmotic shock method abolishes $I_{Ca}$ (Nicola-Siri et al., 1980; Potreau and Raymond, 1980).

It was shown that the slow AP arises from the T-tubular system of the skeletal muscle fiber (Vogel et al., 1978; Kerr and Sperelakis, 1982), based on their disappearance when the T-tubules were disrupted and disconnected from the surface membrane by the glycerol osmotic shock method (Eisenberg and Gage, 1969). The normal fast APs are not affected by the glycerol treatment. These results indicate that the slow Ca²⁺ channels giving rise to the slow APs are located primarily in the tubular system.

A substantial contraction of between 20 and 50% of the normal twitch tension accompanies the slow APs (Vogel et al., 1978), suggesting that the voltage-dependent Ca²⁺ channels in the tubular system may play a role during excitation–contraction (E–C) coupling. In contrast, Gonzalez-Serratos et al. (1982) concluded that the slow inward Ca²⁺ current plays no role in EC coupling, based on the finding that diltiazem did not depress the twitch or tetanic contractions. However, in dysgenic mice in which contraction of skeletal muscles is weak, the Ca²⁺ channels in the T-tubules are few or absent.

Slow APs were also recorded from mouse skeletal muscle fibers equilibrated in a solution that was Cl⁻ free, low Na⁺ (10 mM), and high K⁺ (20 mM) (Kerr and Sperelakis, 1982). As with frog muscle, the slow APs were abolished

after detubulation. The rate of rise, amplitude, and duration of the slow AP increased as a function of $[Ca]_o$, the maximal rate of rise (max $dV/dt$) being about 0.5 V/sec (in 8 mM $[Ca]_o$). The slow APs also were blocked by verapamil, bepridil, Mn²⁺ and La³⁺.

The various conformational states that the Ca²⁺ slow channels undergo during excitation are depicted in Fig. 6. These states are similar to those of the fast Na⁺ channels (see Fig. 5), except there are only two closed states.

In summary, there are voltage-dependent slow Ca²⁺ channels located primarily in the T-tubules of amphibian and mammalian skeletal muscle fibers that have properties similar to those of the slow Ca²⁺ channels found in heart muscle and smooth muscle. Thus, the slow Ca²⁺ channels may play an important role in EC coupling. This Ca²⁺ influx could trigger the release of more Ca²⁺ from the nearby TC–SR via the Ca²⁺-trigger Ca²⁺-release mechanism (Fabiato, 1982). Although contractions can still be evoked when the surface membrane is depolarized to $E_{Ca}$ to prevent Ca entry (Miledi et al., 1977), this does not exclude a possible role for the Ca²⁺ entry across the T-tubule wall because it is the $\Delta E_m$ across the T-tubule that would affect Ca²⁺ influx at this location. Because the time course of the slow AP is much longer than that of a twitch contraction, it was suggested that the inward Ca²⁺ current may play a role in K⁺ contracture, in tetanic contraction, or in long-term regulation of contraction, perhaps by increasing the Ca²⁺ concentration in the SR, and thereby increasing the amount of internal Ca²⁺ available for release on subsequent activation (Nicola-Siri et al., 1980). $[Ca]_{SR}$ does increase following tetanic stimulation (Gonzalez-Serratos et al., 1982).

## IX. Developmental Changes in Membrane Properties

The cell membranes of most excitable cells apparently pass through similar stages of differentiation during development. For example, young (2- to 3-day-old) embryonic chick heart (tubular) has few or no functional fast Na⁺ channels, but has a high density of slow (Na⁺ and Ca²⁺) channels and fires slowly rising TTX-insensitive APs. Fast Na⁺ channels then appear and progressively increase in number, reaching the maximal (adult) level at late embryonic development (e.g., day 20). The $P_{Na}/P_K$ ratio is high in young hearts due to a low $P_K$ and accounts for the low resting potential and automaticity in nearly all the cells.

Skeletal muscle fibers and neurons also undergo developmental changes in membrane electrical properties (e.g., Spector and Prives, 1977; Spitzer, 1979) (see Chapter 34). In general, fast Na⁺ channels are absent in the young less-differentiated neuronal cells, but they do possess excitability because of a large number of slow channels. The AP is TTX insensitive, slowly rising, and of long duration, resembling a slow AP in cardiac muscle. Later during development, fast Na⁺ channels make their first appearance, and both the fast Na⁺ channels and the slow channels coexist. At that period, TTX does not abolish the APs, but reduces max $dV/dt$ (i.e., slow APs remain). At a later stage,

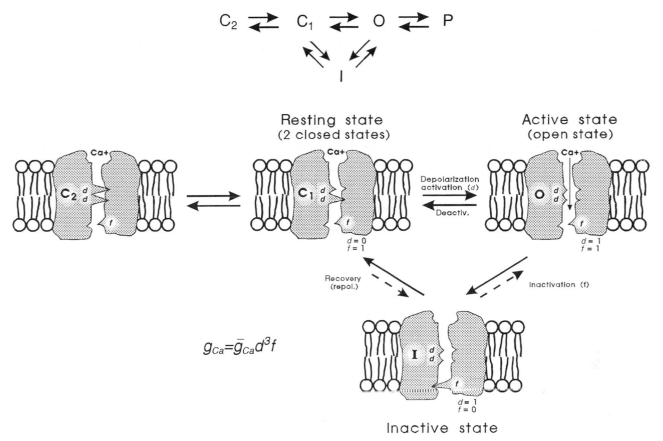

**FIG. 6.** Illustration of the four hypothetical states of the slow $Ca^{2+}$ channel. There is evidence for two closed states. As depicted, in the most closed state ($C_2$), both $d$ gates (or particles) are in the closed configuration. In the least closed state ($C_1$), one gate is closed and one is open. In the resting closed states ($C_2$, $C_1$), the activation gate (A) is closed and the inactivation gate (I) is open: $d = 0$, $f = 1$. Depolarization to the threshold activates the channel to the active state, the A-gate opening rapidly and the I-gate still being open: $d = 1$, $f = 1$. The activated channel spontaneously inactivates to the inactive state due to closure of the I-gate: $d = 1$, $f = 0$. The recovery process on repolarization returns the channel from the inactive state back to the resting state, thus making the channel again available for reactivation. $Ca^{2+}$ ion is depicted as being bound to the outer mouth of the channel and poised for entry down its electrochemical gradient when both gates are in the open configuration. The reaction between the resting state and the active state is readily reversible, and there is some reversibility in the other reactions. The slow channels behave similarly to the fast channels, except that their gates appear to move more slowly on a population basis; that is, the slow channels activate, inactivate, and recover more slowly. (Although the gates of any individual slow channel may move quickly, the stochastic behavior of the population of channels is such that their summed conductance changes slowly.) The slow channel gates operate over a different voltage range than the fast channels (i.e., less negative, more depolarized). TTX does not block the slow channels, but drugs such as nifedipine do block by binding to the channel.

the slow channels in the sarcolemma are lost (or greatly reduced in number), and the fast $Na^+$ channels progressively increase in density. The APs become fast rising and of short duration and are completely abolished by TTX. As discussed previously, some functional slow $Ca^{2+}$ channels remain in the T-tubular system.

## X. Electrogenic $Na^+$–$K^+$ Pump Stimulation

The Na,K-ATPase pump is *electrogenic* in skeletal muscle fibers (both mammalian and amphibian). The pump is electrogenic, producing a net outward current, because 3 $Na^+$ ions are pumped out to every 2 $K^+$ ions pumped in.

The *electrogenic pump potential* contribution to the resting potential (see chapter on resting potential) is about 12–16 mV in rat skeletal muscle fibers (Sellin and Sperelakis, 1978). The *net pump current* can be stimulated by increasing the number of pump sites per unit area of cell membrane or by increasing the turnover rate of each pump site. β-Adrenergic agonists (e.g., isoproterenol) rapidly (within 5 min) hyperpolarize (e.g., 7–9 mV) skeletal muscle fibers, and insulin more slowly (e.g., peak reached by 10 min) hyperpolarizes to a smaller degree (e.g., 5–7 mV) (Iannaccone *et al.*, 1989). Since cAMP also hyperpolarizes, the action of β-agonists is believed to be mediated by elevation of cAMP and phosphorylation of the Na–K pump (or an associated regulatory protein) by protein kinase A (PKA).

The action of insulin is thought to be mediated by the incorporation of spare membrane from an internal pool, which contains Na–K pumps, into the cell membrane.

The pump current ($I_p$) can be directly measured in single fibers (cultured skeletal myotubes rounded by use of colchicine, a microtubule disrupter) by doing whole-cell voltage clamp under conditions in which all the ionic conductances are blocked. When this is done, the pump current can be measured at different voltages and normalized for unit membrane capacitance and membrane area. Values of about 1 pA/pF or 1.0 $\mu$A/cm$^2$ were obtained, with a reversal or zero current potential of about $-140$ mV (Li and Sperelakis, 1994).

When [K]$_o$ is lowered below the normal physiological level, for example, from 4.5 m$M$ to about 0.1 m$M$, a large depolarization occurs in mammalian skeletal muscle fibers (Fig. 7). This depolarization is caused, in part, by inhibition of the Na–K pump current. The $K_m$ value for [K]$_o$ for the Na,K-ATPase is about 2 m$M$, and the relationship between Na, K-ATPase activity and [K]$_o$ is very steep. Therefore, inhibition of the Na–K pump occurs. The contribution of the electrogenic pump to the RP is relatively large in mammalian skeletal muscle fibers.

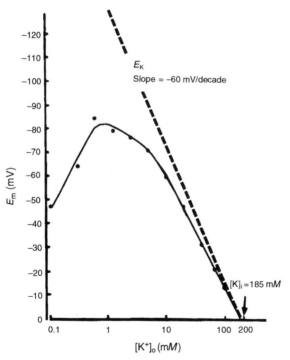

**FIG. 7.** The mean resting membrane potential ($E_m$) of normal mouse skeletal muscle plotted as a function of the extracellular K$^+$ concentration ([K]$_o$) on a logarithmic scale. The straight line drawn through the data points for 20 m$M$ [K]$_o$ and above has a slope of 50 mV/decade. Extrapolation of this line to zero potential gives the intracellular K$^+$ concentration ([K]$_i$) of 185 m$M$. The dashed line gives the calculated $E_K$ values (slope of 61 mV/decade). Note the "fold-over" of the $E_m$ curve at [K]$_o$ levels below 1 m$M$, presumably due to inhibition of the electrogenic pump potential ($V_p$) and to a decrease in $P_K$ and $g_K$ at low [K]$_o$ levels. (Reproduction from Sellin and Sperelakis, 1978.)

## XI. Slow Fibers

One type of skeletal muscle fiber, known as *slow fibers*, subserves tonic functions including posture. Slow fibers should not be confused with "slow twitch fibers." The true slow fibers do not fire APs, whereas all types of twitch fibers do. The slow fibers are usually smaller in diameter than twitch fibers, and they exhibit a less distinct myofibrillar arrangement. Slow fibers have been found in a number of vertebrate muscles, for example, in the frog rectus abdominus muscle, frog ileofibularis muscle, and mammalian extraocular muscles. It is probable that careful searching will reveal some slow fibers in other mammalian muscles.

The slow fibers have *multiple innervation* by a series of motor endplates (spaced about 1 mm apart) from a single motoneuron. As with twitch fibers, acetylcholine (ACh) is the synaptic transmitter. The force of contraction of the slow fibers is controlled by graded *end-plate potentials* (EPPs). That is, an increase of frequency of impulses in the motoneuron produces a larger EPP (by temporal summation) and this, in turn, produces a greater contraction in the vicinity of the end-plate. Since the end-plates are spaced closely together—at a distance of about one length constant—the entire fiber becomes nearly uniformly depolarized, even though there are no propagated APs. Therefore, the entire length of the slow fiber contracts almost uniformly.

The slow fibers do possess *T-tubules* and triadic junctions with the terminal cisternae of the SR (TC-SR). Therefore, the T-tubules may act as *passive conduits* in the slow fibers to bring the depolarization (produced in the surface membrane by the EPP) deep into the fiber interior. Thus, depolarization of the T-tubule occurs by their passive cable properties. This depolarization, in turn, could bring about the influx of Ca$^{2+}$ by activation of voltage-dependent slow Ca$^{2+}$ channels located in the T-tubules.

APs normally cannot be induced to occur in vertebrate slow fibers under a variety of experimental conditions. However, denervation of frog slow fibers does allow an AP-generating mechanism to appear (Miledi *et al.*, 1971). APs can be induced in slow fibers of invertebrates [e.g., crustacean skeletal muscles and horseshoe crab (*Limulus*) heart] (Fatt and Ginsborg, 1958). In the neurogenic *Limulus* heart, which normally is activated by summing excitatory postsynaptic potentials, propagating (ca. 5 cm/sec) and overshooting spontaneous APs can be rapidly induced by Ba$^{2+}$ (0.1–10 m$M$) (Rulon *et al.*, 1971). These slowly rising (ca. 1.0 V/sec) APs are resistant to TTX, and these voltage-dependent slow channels can pass Ba$^{2+}$, Sr$^{2+}$, and Ca$^{2+}$.

## XII. Conduction of the Action Potential

When the EPP, generated at the neuromuscular junction, reaches threshold for eliciting an AP in the skeletal muscle fiber, an AP is propagated down the muscle fiber in both directions from the end-plate. (In some muscle fibers, there is a second end-plate innervated by a motoneuron exiting the spinal cord at another level.) The AP is overshooting (to about +40 mV) and propagates at a constant velocity of about 5 m/sec over the surface sarcolemma. As discussed in the chapter on propagation in axons, propagation occurs by means of the *local-circuit currents* that accompany the propagating impulse. The *radial (transmembrane) currents* are triphasic: (1) The first phase is outward across the membrane and is of moderate intensity; (2) the second phase is inward and is most intensive (at the current "sink" region); and (3) the third phase is outward and is the least intensive. The first outward phase corresponds to the passive exponential foot of the AP due to the *passive cable spread* of voltage and current. The second inward phase corresponds to the large increase in $g_{Na}$ and fast inward $Na^+$ current during the later portion of the rising phase and peak of the AP. The third outward phase corresponds to the small increase in $g_K$ and net outward $K^+$ current and the less-steep repolarizing phase of the AP spike.

The *longitudinal (axial) currents*, both inside and outside the fiber, are biphasic: (1) The first phase is most intensive and is in the direction of propagation of the impulse (forward); and (2) the second phase is less intensive and is in the backward direction. The external longitudinal currents, of course, must equal the internal axial currents, but they are in the opposite directions. The internal axial currents are confined to the myoplasm, whereas the external longitudinal currents can use the entire ISF space or so-called *volume conductor* (since "current takes the path of the least resistance"). It is this latter fact that allows the *electromyogram* (EMG) to be recorded from the skin overlying an activated skeletal muscle. The amplitude of the EMG potentials becomes larger when more fibers within the muscle are activated (fiber summation), because of summation of the IR voltage drops produced by each fiber activated simultaneously. The frequency of the EMG potentials reflects the frequency and asynchrony of activation of the muscle.

The skeletal muscle fibers, composed of *myoblast cells* fused end to end and multinucleated, behave as semi-infinite cables. That is, an AP can propagate from one end of the fiber to the other, uniformly and unimpeded. The space or *length constant* ($\lambda$) of the fiber cable is about 1.5 mm for frog sartorius fibers (Sperelakis *et al.*, 1967) and about 0.76 mm for the rat EDL muscle (Sellin and Sperelakis, 1978). The length constant is the distance over which the potential impressed at one region would decay to $1/e$ ($1/2.717 = 0.37$) or 37% of the initial value. That is, in a passive cable, voltage decays exponentially with a certain length constant as given by

$$V_x = V_o \, e^{-x/\lambda} \tag{4}$$

where $V_x$ is the voltage at the distance $x$, and $V_o$ is the voltage at the origin ($x = 0$); $\lambda$ is given by

$$\lambda = \sqrt{\frac{r_m}{r_i + r_o}} \tag{5}$$

$$cm = \sqrt{\frac{\Omega \cdot cm}{\dfrac{\Omega}{cm} + \dfrac{\Omega}{cm}}} = \sqrt{cm^2}$$

Assuming that $r_o$ (the outside longitudinal resistance) is negligibly small compared to $r_i$ (this would be true for a superficial fiber in a bundle immersed in a large bath),

$$\lambda = \sqrt{\frac{r_m}{r_i}} = \sqrt{\frac{R_m}{R_i} \frac{a}{2}}$$

$$cm = \sqrt{\frac{\Omega \cdot cm}{\Omega/cm}} = \sqrt{\frac{\Omega \cdot cm^2}{\Omega \cdot cm} \frac{cm}{1}} = \sqrt{cm^2} \tag{6}$$

where $r_m$ ($\Omega \cdot cm$) and $r_i$ ($\Omega/cm$) are the membrane resistance and the internal longitudinal resistance normalized for unit length of fiber; $R_m$ ($\Omega \cdot cm^2$) is the membrane resistance normalized for both fiber radius and length; $R_i$ ($\Omega \cdot cm$) is the resistivity of the myoplasm (normalized for length and cross-sectional area); and $a$ (cm) is the fiber radius. $R_m$ is often loosely called membrane resistivity or specific resistance, but this is not accurate because for true membrane resistivity ($\rho_m$) there must be correction for membrane thickness $\delta$:

$$\rho_m = \frac{R_m}{\delta}$$

$$\Omega \cdot cm = \frac{\Omega \cdot cm^2}{cm} \tag{7}$$

For a derivation of these equations, and for the interconversions between $r_i$ and $R_i$, and $r_m$ and $R_m$, the reader is referred to the Appendix and this book.

The factors that determine active *velocity of propagation* ($\theta_a$) include the intensity of the local-circuit current, thresh-

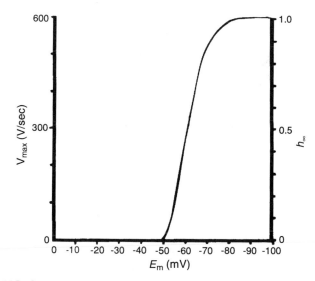

**FIG. 8.** Graphic representation of the maximal rate of rise of the action potential (max $dV/dt$) as a function of resting $E_m$ or takeoff potential. Max $dV/dt$ is a measure of the inward current intensity (membrane capacitance being constant), which is dependent on the number of channels available for activation; $h$ is the inactivation factor of Hodgkin Huxley as $g_{Na} - \bar{g}_{Na}m^3h$, where $g_{Na}$ is the $Na^+$ conductance, $\bar{g}_{Na}$ is the maximal conductance, and $m$ and $h$ are variables; $h_\infty$ represents $h$ at $t = $ infinity or steady state (practically, after 20 msec). The fast $Na^+$ channels begin to inactivate at about $-75$ mV, and nearly complete inactivation occurs at about $-30$ mV ($h_\infty$ low). Therefore, max $dV/dt$ decreases because $h_\infty$ decreases.

old potential, and the passive cable properties, $\lambda$ and $\tau_m$. As discussed previously, the greater the rate of rise of the AP, the greater the intensity of the local-circuit current, hence the greater the $\theta_a$. In addition to its dependence on the density of the fast $Na^+$ channels (determinant of the maximum $Na^+$ conductance, $g_{Na}$), the kinetic properties of the channel gating, and the threshold potential ($V_{th}$), max $dV/dt$ is determined also by the resting (takeoff) potential (related to the $h_\infty$ vs $E_m$ curve), as discussed previously (Fig. 8). In addition, because cooling decreases max $dV/dt$ ($Q_{10} \simeq 3$), $\theta_a$ is slowed accordingly. $R_m$ also is increased by cooling, the $Q_{10}$ for $R_K$ in frog sartorius fibers being about 2.8 (ion diffusion in free solution has a $Q_{10}$ of about 1.2) (Sperelakis, 1969).

In a passive cable, such as a skeletal muscle fiber, the *passive propagation velocity* ($\theta_p$) is directly proportional to the length constant[3] and inversely proportional to the time constant:

$$\theta_p \propto \frac{\lambda}{\tau_m}$$

$$\frac{cm}{sec} = \frac{cm}{sec} \tag{8}$$

$$\theta_p \propto \frac{\sqrt{\dfrac{R_m a}{R_i 2}}}{R_m C_m}$$

$$\frac{cm}{sec} = \frac{\sqrt{\dfrac{\Omega \cdot cm^2}{\Omega \cdot cm}} \cdot \dfrac{cm}{}}{\Omega \cdot cm^2 \cdot \dfrac{F}{cm^2}} = \frac{\sqrt{cm^2}}{sec} \tag{8a}$$

$$\theta_p \propto \frac{\sqrt{a}}{\sqrt{R_m R_i 2} \, C_m} \tag{8b}$$

$$\frac{cm}{sec} = \frac{\sqrt{cm}}{\sqrt{\Omega \cdot cm^2} \sqrt{\Omega \cdot cm} \, \dfrac{F}{cm^2}} = \frac{\sqrt{cm}}{cm \sqrt{cm} \, \dfrac{sec}{cm^2}} = \frac{cm}{sec}$$

Therefore, propagation velocity is directly proportional to the square root of the fiber radius ($a$), and inversely proportional to membrane capacitance ($C_m$) and to the square root of $R_i$ and the square root of $R_m$. For example, propagation velocity is greater in large-diameter muscle fibers.

The relationship between propagation velocity and membrane current density ($I_m$) is given by

$$I_m = \frac{a}{2} \frac{1}{R_i} \frac{1}{\theta^2} \frac{d^2V}{dt^2}$$

$$\frac{amp}{cm^2} = (cm)\left(\frac{1}{\Omega \cdot cm}\right)\left(\frac{1}{cm^2/sec^2}\right)\left(\frac{V}{sec^2}\right) = \frac{amp}{cm^2} \tag{9}$$

where $d^2V/dt^2$ is the second time derivative of the AP. As indicated, membrane current is proportional of $d^2V/dt^2$,

---

[3] The length constant for sinusoidally varying applied currents ($\lambda_{AC}$) is shorter than $\lambda_{DC}$, depending on the AC frequency.

whereas the longitudinal current ($I_1$) or the capacitive current ($I_C$) is proportional to $dV/dt$:

$$I_c = C_m \frac{dV}{dt} \tag{10}$$

## XIII. Excitation Delivery to Fiber Interior

### A. Conduction into the T-Tubular System

The experiments of Huxley and Taylor (1958) were the first to provide evidence that there was some structure, located at the level of the Z-lines in frog skeletal muscle fibers, which is involved in EC coupling. These investigators applied current pulses at different points along the length of the sarcomeres in isolated fibers and found that when the microelectrode tip was opposite the Z-line, graded contractions of the two half-sarcomeres occurred. The greater the current, the greater the inward spread of the contraction. In addition, they discovered that there were *sensitive spots* located around the perimeter of the fiber at the Z-line level; that is, the membrane was not uniformly sensitive. At about the same time, it was discovered by electron microscopists that transverse (T) tubules were located at the level of the Z-lines in amphibian skeletal muscle (and at the level of the A-I junctions of the sarcomere in mammalian skeletal muscle). Thus, the T-tubules probably represent the morphological conduit for the findings of Huxley and Taylor (1958). The morphological arrangement of the sarcotubular system of skeletal muscle fibers is illustrated in Fig. 9.

*Diffusion* of some substance from the surface membrane into the skeletal muscle fiber interior is much too slow to account for the relatively short latent period of about 1–3 msec between the beginning of the AP and the beginning of contraction. That is, the diameter of the fibers (mean value of about 70 $\mu$m in frog sartorius fibers) is much too large for a diffusion mechanism from the fiber surface to be responsible. *Diffusion time* (for 95% equilibration) increases by the square of the distance, and would require about 2.5 sec for a small molecule freely diffusing across a cell radius of 50 $\mu$m; estimates for $Ca^{2+}$ diffusion time are considerably longer than this (Podolsky and Costantin, 1964). Therefore, the T-tubular system serves to bring excitation deep into the fiber interior rapidly and thereby to reduce the required diffusion distance to an average value of about 0.7 $\mu$m (Sperelakis and Rubio, 1971). Disruption of the T-tubules (with their disconnection from the surface membrane by the glycerol osmotic-shock method) uncouples contraction from excitation (Eisenberg and Gage, 1969), thus further underscoring the essential role of the T-tubules in EC coupling.

Estimates of the *length constant* of the T-tubules ($\lambda_{TT}$), assuming the T-tubule membrane has about the same resistivity ($R_m$) as the surface membrane ($\lambda = \sqrt{R_m/R_i} \sqrt{a/2}$), give values of about 50 $\mu$m. Because the resistivity of the T-tubule membrane of frog muscle is probably higher than that of the surface sarcolemma because of a lower $g_{Cl}$ (Hodgkin and Horowicz, 1959; Adrian and Freygang, 1962; Sperelakis and Schneider, 1968), this would give a longer $\lambda$ value. Therefore, it is possible for the T-tubules to serve as *passive conduits* to bring the depolarization from the surface membrane (during its AP) into the fiber interior.

However, there is evidence that the T-tubules actually fire APs, that is, they *actively propagate impulses* inward, and so bring large depolarization deep into the fiber interior. The evidence for this includes the observation of a threshold for sudden initiation of localized contraction by Costantin and colleagues (1967, 1973). The T-tubule AP is sensitive to TTX and is $Na^+$ dependent, and therefore is apparently similar in nature to the surface membrane AP. By use of high-speed cinemicrography to measure sequential activation of the myofibrils in a radial direction, Gonzalez-Serratos (1971) estimated the propagation velocity of the T-tubule AP to be about 10 cm/sec. Although this velocity is about 50 times slower than propagation down the fiber longitudinally (about 5 m/sec), it is sufficient to account for the short latent period before contraction begins.

It has been proposed that the *fatigue* observed during relatively brief tetanic contractions (i.e., a progressive fall-off in the force of contraction) is due to $Na^+$ depletion in the T-tubule network, particularly in the deeper parts far from the orifice at the fiber surface (Bezanilla *et al.*, 1972). The fatigue occurs more rapidly in fibers preequilibrated in low $[Na]_0$ (e.g., 60 m$M$). It is thought that the $Na^+$ influx (the inward fast $Na^+$ current) with each AP in the T-tubule produces a progressive decline in $[Na]_{TT}$, which slows propagation velocity down the T-tubules and eventually leads to loss of excitability when $[Na]_{TT}$ drops below some critical level (e.g., 30 m$M$). $Na^+$ depletion should occur more rapidly deep in the T-tubule network of the fiber because there would be less diffusion of $Na^+$ in from the mouth of the T-tubule to replenish the $Na^+$ loss. Active Na–K pumping in the T-tubules may not occur fast enough to keep up with the $Na^+$ loss into the fiber myoplasm.

There are also voltage-dependent slow $Ca^{2+}$ channels in the T-tubule membrane, and slow APs that arise from the T-tubule can be recorded under appropriate conditions (Sperelakis *et al.*, 1967; Vogel and Sperelakis, 1978). The evidence for the existence of this type of channel and some of its properties was discussed in Section VIII. The $Ca^{2+}$ influx into the myoplasm through these $Ca^{2+}$ channels could play a role in EC coupling.

## B. Evidence for T-Tubule Communication with the Sarcoplasmic Reticulum across the Triadic Junction under Some Conditions

$Ca^{2+}$ for contraction in skeletal muscle is primarily released from the TC-SR (Winegrad, 1968), and there is an *internal cycling* of $Ca^{2+}$ ion. Changes in $[Ca]_0$ of the bathing solution take a relatively long time (e.g., 30–60 min) before exerting a large effect on the force of contraction in skeletal muscle. In contrast, in cardiac muscle, the effect of lowered $[Ca]_0$ is obvious within a few seconds, indicating that the primary determinant of the force of contraction is the $Ca^{2+}$ influx across the sarcolemma through the slow $Ca^{2+}$ channels. Therefore, in skeletal muscle, excitation propagates

actively down the T-tubules and $Ca^{2+}$ is released from the TC-SR, but it is not known how the signal is transferred from the T-tubule to the TC-SR across the triadic junction.

Electron-opaque *tracer molecules*, like horseradish peroxidase (HRP) (ca. 60 Å in diameter), enter into the T-tubules and from there can enter into some of the TC-SR of frog skeletal muscle (Rubio and Sperelakis, 1972; Kulczycky and Mainwood, 1972) (Figs. 10A and B). Exposure of the fibers to hypertonic solutions facilitates the entry of HRP into the TC-SR, so that nearly 100% of the TC-SR becomes filled (Figs. 10C and D). Thus, there may be a functional connection between the SR and the extracellular space (Sperelakis *et al.*, 1973). If so, there may be lumen-to-lumen continuity between the T-tubules and TC-SR during excitation, allowing the AP in the T-tubules to invade directly into the TC-SR to depolarize and bring about the release of $Ca^{2+}$. The depolarization of the TC-SR could activate voltage-dependent slow $Ca^{2+}$ channels, allowing $Ca^{2+}$ influx into the myoplasm down an electrochemical gradient.

If the longitudinal SR (L-SR) were electrically isolated from the TC-SR by a substantial resistance (e.g., zippering between the two SR compartments described later), this would account for the *fiber capacitance* measured being relatively low (Mathias *et al.*, 1980). The effect of this would be to remove the very large membrane surface area of the L-SR and hence greatly reduce the capacitance that would be measured.

It has been suggested that the SR is depolarized during the release of $Ca^{2+}$ in EC coupling. For example, optical signals (e.g., birefringence and fluorescence changes) can be recorded from the SR membranes during contraction (e.g., Baylor and Oetliker, 1975; Bezanilla and Horowicz, 1975). In addition, Natori (1965) demonstrated that propagation of contraction (1–3 cm/sec) triggered by electrical stimulation can occur in muscle fiber regions that had been *denuded (skinned)* of its sarcolemma, the propagation of excitation presumably occurring by means of the SR membranes.

Investigators (Sperelakis *et al.*, 1973) have demonstrated that *EC uncoupling* could be produced by exposing frog skeletal muscle fibers to $Mn^{2+}$ (1 m$M$) or $La^{3+}$ (1 m$M$) while in hypertonic solution (to facilitate entry of the blockers into the TC-SR). After the fibers were returned to normal Ringer solution, normal fast APs could be elicited, but there were no contractions accompanying them; that is, a "permanent" EC uncoupling was produced. These results were interpreted as suggesting that $Mn^{2+}$ and $La^{3+}$ entered into the lumen of the TC-SR and blocked the slow $Ca^{2+}$ channels. A similar exposure of frog sartorius fibers to $Mn^{2+}$, $La^{3+}$, or to $Ca^{2+}$-free solution blocked the caffeine-induced contracture as well (Rubio and Sperelakis, 1972). Thus, from these physiological and ultrastructural studies, it was suggested that the lumen of the SR is continuous with that of the T-tubule under conditions of the hypertonicity, and that substances can enter into the TC-SR to exert an effect on $Ca^{2+}$ release into the myoplasm.

*Compartmental analysis* of skeletal muscle has also suggested that the SR is open to the ISF. [In contrast, in

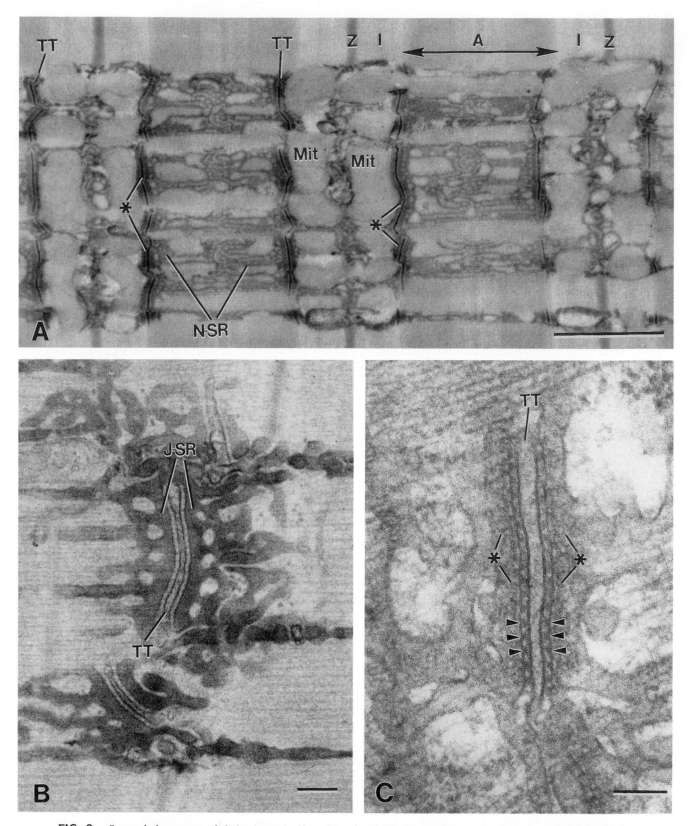

**FIG. 9.** Sarcotubular system of skeletal muscle fibers from (A,C) tibialis anterior muscle of mouse and (B) iliotibialis msucle of lizard. (A) Longitudinal section showing the sarcomere structure of several myofibrils: A-band, I-band, Z-line. The network sarcoplasmic reticulum (N-SR), also known as the longitudinal SR, appears as a torn sleeve surrounding the surface of each myofibril. The N-SR is continuous with the junctional SR (J-SR) that abuts close to the transverse tubules (TT). The TT membranes are invaginations of the cell surface membrane at the level of the A-I junctions in mammalians

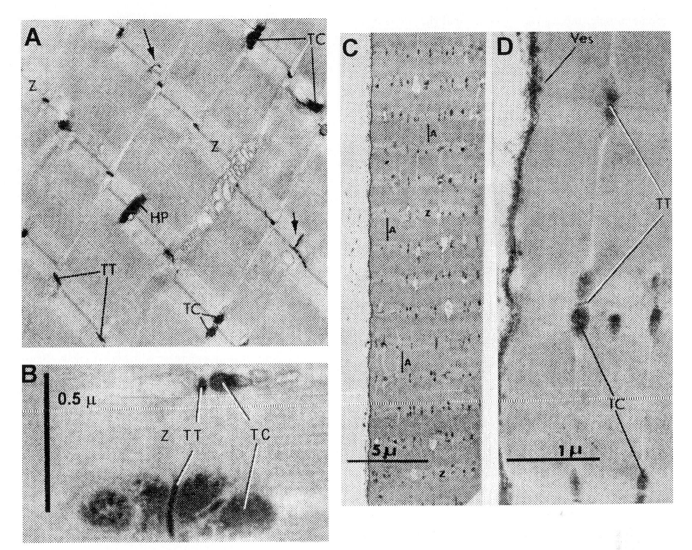

**FIG. 10.** Evidence that large molecules of horseradish peroxidase (HP) can enter into the terminal cisternae (TC) of the SR via the transverse tubules (TT) of frog sartorius fibers. Electron micrographs of longitudinal sections. (A,B) Fiber was exposed to HP under isosmotic conditions. (A) Section through several myofibrils showing presence of HP activity (as a dense electron-opaque material) in the TT and in some of the TC at triadic junctions. In amphibian muscle, the TT occur at the level of the Z-lines (Z) of the sarcomeres. Arrows point to two branches of the TT running longitudinally. (B) Higher magnification of two triads, one with both cisternae filled with HP and the other with only one cisterna filled. (C,D) Fiber was exposed to HP under hypertonic condition (3× isotonic, using NaCl), showing that almost all cisternae were filled with peroxidase. (C) Section at low magnification. (D) Portion of same section as in Part C shown at higher magnification. The surface vesicles (Ves) also became filled with HP. [Modified from Figs. 2 and 4 of Rubio, R., and Sperelakis, N. (1972). *Z. Zellforsch.* **124,** 57–72. Dr. H. Yokoshiki kindly helped in the preparation of this figure.]

---

(or at the level of the Z-line in lizards and amphibians). The J-SR and TT form the complex coupling known as a triad (*). The N-SR is continuous across the I-band, but this is obscured in this section by the presence of paired mitochondria (Mit) over the I-bands. The TT and SR are both selectively filled with osmium tetroxide precipitate, causing their profiles to be more electron opaque than the other structures. Scale bar at lower right represents 1.0 μm. (B) Higher magnification of a triad to show more detail. As shown, the triad consists of a single T-tubule sandwiched between two cisternae of the J-SR. Scale bar = 0.1 μm. (C) High magnification of a triadic junction to illustrate the array of regularly spaced junctional processes or SR foot processes (several indicated by arrowheads) that project between the TT membrane and the J-SR membrane. There are dense granules within the lumen of the J-SR cisternae (*). Scale bar = 0.1 μm. (Electron micrographs provided courtesy of Dr. Mike Forbes, University of Virginia.)

cardiac muscle, there is no evidence that the SR is open to the ISF (Rubio and Sperelakis, 1971).] For example, Conway (1957), Harris (1963), and Keynes and Steinhardt (1968) concluded that $Na^+$ in frog skeletal muscle fibers is distributed in two separate compartments. Harris (1963) suggested that the $Na^+$, $K^+$, and $Cl^-$ concentrations in one compartment (presumably the SR) were about equal to those of the ISF, and Rogus and Zierler (1973) concluded that the $Na^+$ concentration in the SR of rat skeletal muscle approximates that of the ISF. The volume of the SR compartment was 14.3% of fiber volume, and in hypertonic solution, the SR volume increased and the washout of the SR compartment was faster. Tasker *et al.* (1959) also had reported a large sucrose space of 26.5% for frog sartorius fibers.

Other researchers (Birks and Davey, 1969) have demonstrated that the volume changes of the SR of skeletal muscle in hypertonic (sucrose) and hypotonic solutions were always opposite of those occurring within the myoplasmic compartment. They concluded that sucrose must enter into the SR, pulling in water osmotically from the myoplasm, to produce the marked swelling of the SR that occurred in hypertonic solutions. Vinogradova (1968) concluded from the distribution of nonpenetrating sugars in frog sartorius muscle that the SR compartment is continuous with the ISF; the *inulin space* was 19.0% and increased in hypertonic solution and decreased in hypotonic solution and in glycerol-treated fibers (for disruption of the T-tubules).

The total [³H] *sucrose space* of frog sartorius muscles was found to be 18.0% in isotonic solution and 22.6% in twofold hypertonic solution (Sperelakis *et al.*, 1978). The relative SR volume (including the small T-tubule volume) was 12.4 and 17.0% of fiber volume, respectively. This value for SR volume of frog skeletal muscle is close to that measured by ultrastructural techniques (Peachey, 1965; Mobley and Eisenberg, 1975). Evidence that the TC-SR and L-SR may not be freely connected to one another under resting conditions comes from the observations that (1) the L-SR did not fill with HRP, whereas the TC-SR did (Rubio and Sperelakis, 1972) and (2) ther is a *"zippering"* of the membranes connecting these two components of the SR in mouse and frog skeletal muscle (Howell, 1974; Wallace and Sommer, 1975; Forbes and Sperelakis, 1979).

In $^{45}Ca$ washout experiments on frog muscles, Kirby *et al.* (1975) found three compartments, similar to the three *sucrose compartments* described before, except the half-times were about two- to threefold shorter. They suggested that the first compartment was the ISF space, the second was the T-tubule plus the TC-SR, and the third was the L-SR. Bianchi and Bolton (1974) also found a transient increase in $^{45}Ca$ efflux and a marked loss of muscle $Ca^{2+}$ from frog sartorius muscles exposed to hypertonic solutions (twice isotonicity), and suggested that hypertonicity produces transient communication between the TC-SR and the T-tubules, thus allowing their $Ca^{2+}$ to be lost to the ISF. In addition, it has been reported in a human muscle disease, polymyositis, that the T-tubules are spatially con-

tinuous with the SR, as visualized with lanthanum tracer, and that enzymes leak from the TC-SR into the T-tubules and ISF (Chou *et al.*, 1980).

Frog skeletal muscle fibers have an *osmotically inactive volume* of about 32% when placed into Ringer solution made hypertonic with sucrose or other nonpenetrating solutes; that is, fiber diameter does not shrink to the theoretical value expected if it were a perfect *osmometer* (Sperelakis and Schneider, 1968; Sperelakis *et al.*, 1970). For example, in twofold hypertonic solution, there should be a decrease in fiber volume to one-half and fiber radius to 0.707 ($1/\sqrt{2}$) of the original value. The observed change is to only 0.81 of the original diameter. Because the SR volume increases in hypertonic solution (Huxley *et al.*, 1963; Sperelakis and Schneider, 1968; Birks and Davey, 1969), it is likely that the osmotic inactive volume is due to the SR. The swollen SR would prevent the fiber volume from decreasing to one-half in twofold hypertonic solution, even if the volume of the myoplasm proper were to decrease to one-half.

In cardiac muscle, an osmotically inactive volume is not present (Sperelakis and Rubio, 1971), electron-opaque tracers do not enter the SR (Sperelakis *et al.*, 1974), and the SR volume does not increase with hypertonicity (Sperelakis and Rubio, 1971).

This brief section was included to let the student know that there are data that do not fit with currently accepted hypotheses.

## XIV. Summary

The resting potential of skeletal muscle twitch fibers is about $-80$ mV (mammalian) or $-90$ mV (amphibian), and the AP overshoots to about $+40$ mV. The maximal rate of rise of the AP is very fast, being about 500–700 V/sec, and is due to a large inward fast $Na^+$ current ($I_{Na}$), which brings $E_m$ up close to $E_{Na}$. The $APD_{50}$ is brief, being about 1–3 msec, and the falling phase of the AP is produced by several repolarizing factors: (1) $Na^+$ channel inactivation, (2) increase in the delayed rectifier $K^+$ conductance, (3) $I_{Na}$ deactivation due to some repolarization, and (4) $Cl^-$ influx (outward $I_{Cl}$).

The skeletal muscle fibers are long cables, formed by fusion of myoblast cells and therefore produce a multinucleated long myotube or fiber, and have a fast propagation velocity of about 5 m/sec. Each skeletal muscle twitch fiber is normally closely controlled by the motor innervation, there being one or two motor end-plates (neuromuscular junctions) located near the midregion of each fiber. Excitation spreads in both directions from the neuromuscular junction.

The twitch fibers undergo developmental changes similar to those in cardiac muscle and neurons. In early development, there are few or no fast $Na^+$ channels, and the AP upstroke is slow and produced by an inward current through slow $Ca^{2+}$ channels. The AP duration is also long because the delayed rectifier $K^+$ conductance is not fully developed. During subsequent development, fast $Na^+$

channels are gained, the resting potential is greater (more negative), and the AP shortens to a brief spike.

The skeletal muscle AP spike is immediately followed by a large and prominent early depolarizing afterpotential, that slowly decays over 10–20 msec. This early afterpotential is caused by the persistence and slow decay of the delayed rectifier $K^+$ conductance (that was turned on by the $Na^+$ influx-caused depolarization), which has a $Na^+ : K^+$ selectivity or $P_{Na}/P_K$ ratio higher (e.g., 1/30) than that of the resting membrane (e.g., 1/100). Since this delayed rectifier $K^+$ conductance dominates, $E_m$ is held for a time more depolarized than the normal resting potential.

After (and during) a tetanic burst (train) of AP spikes, a large prominent late depolarizing afterpotential is produced. This late afterpotential is caused by a cumulative $K^+$ accumulation in the T-tubules that acts to depolarize them due to the decrease in $E_K$, and thereby depolarize the surface sarcolemma passively. In addition, a slow component of the delayed rectifier $K^+$ conductance may persist during the train and slowly decay after the train, and as discussed previously, this $K^+$ conductance is less selective for $K^+$ than is the resting $K^+$ conductance.

The AP invades the T-tubules and propagates inward at a slow velocity and thereby serves to bring excitation deep into the fiber interior. The depolarization of the T-tubules activates slow (L-type) $Ca^{2+}$ channels that are present in them, and this serves as a critical step in EC coupling in skeletal muscle fibers. In some circumstances, the lumen of the TC of the SR may communicate directly with the lumen of the T-tubule, perhaps allowing excitation to invade directly into the TC-SR, from which $Ca^{2+}$ is released into the myoplasm to bring about contraction. The chapter on EC coupling provides a detailed discussion of a mechanism that involves the $Ca^{2+}$ channels acting as voltage sensors that somehow are coupled to and open the $Ca^{2+}$ release channels in the TC-SR (surface facing the T-tubule).

Some skeletal muscles also contain a fraction of fibers that are nontwitch slow muscle fibers, which normally do not fire APs. They are multiply innervated by the motor neuron, with numerous motor end-plates spaced about 1 mm apart along the entire length of the fiber. Graded contraction of each fiber is produced by varying the frequency of axon APs that increase the amplitude of the EPPs by temporal summation. The membrane potential change produced by the summed EPPs is carried passively into the T-tubules to bring about contraction.

## Acknowledgment

Thanks to Glenn Doerman for his assistance in creating figures for Dr. Sperelakis for this and other chapters throughout the book.

## References

Adrian, R. H., and Freygang, W. H. (1962). The potassium and chloride conductance of frog muscle membrane. *J. Physiol.* (*London*) **163**, 61–103.

Adrian, R. H., Chandler, W. K., and Hodgkin, A. L. (1970a). Voltage clamp experiments in striated muscle fibers. *J. Physiol.* (*London*) **208**, 607–644.

Adrian, R. H., Chandler, W. K., and Hodgkin, A. L. (1970b). Slow changes in potassium permeability in skeletal muscle. *J. Physiol.* (*London*) **208**, 645–668.

Almers, W., Fink, R., and Palade, P. T. (1981). Calcium depletion in frog muscle tubules: The decline of calcium current under maintained depolarization. *J. Physiol.* (*London*) **312**, 177–207.

Baylor, S. M., and Oetliker, H. (1975). Birefringence experiments on isolated skeletal muscle fibres suggest a possible signal from the sarcoplasmic reticulum. *Nature* **253**, 97–101.

Beaty, G. N., and Stefani, I. (1976). Calcium dependent electrical activity in twitch muscle fibers of the frog. *Proc. R. Soc. London* (*Biol.*) **194**, 141–150.

Bezanilla, F., and Horowicz, P. (1975). Fluorescence intensity changes associated with contractile activation in frog muscle stained with Nile Blue A. *J. Physiol.* (*London*) **246**, 709–735.

Bezanilla, F., Caputo, C., Gonzalez-Serratos, H., and Venosa, R. A. (1972). Sodium dependence of the inward spread of activation in isolated twitch muscle fibres of the frog. *J. Physiol.* (*London*) **223**, 507–523.

Bianchi, C. P., and Bolton, T. C. (1974). Effect of hypertonic solutions and glycerol treatment on calcium and magnesium movements of frog skeletal muscle. *J. Phamacol. Exp. Ther.* **188**, 536–552.

Birks, R. I., and Davey, D. F. (1969). Osmotic responses demonstrating the extracellular character of sarcoplasmic reticulum. *J. Physiol.* (*London*) **21**, 171–188.

Chou, S. M., Nonaka, I., and Voice, G. F. (1980). Anastomoses of transverse tubules with terminal cisternae in polymyositis. *Arch. Neurol.* **37**, 257–266.

Conway, E. J. (1957). Nature and significance of concentration relations of potassium and sodium ions in skeletal muscle. *Physiol. Rev.* **37**, 84–132.

Costantin, L. L., and Podolsky, R. J. (1967). Depolarization of the internal membrane system in the activation of frog skeletal muscle. *J. Gen. Physiol.* **50**, 1101–1124.

Costantin, L. L., and Taylor, S. R. (1973). Graded activation in frog muscle fibers. *J. Gen. Physiol.* **61**, 424–443.

Eisenberg, R. S., and Gage, P. W. (1969). Ionic conductances of the surface and transverse tubular membranes of frog sartorius fibers. *J. Gen. Physiol.* **53**, 279–297.

Fabiato, A. (1982). Mechanism of calcium-induced release of calcium from the sarcoplasmic reticulum of skinned cardiac cells studied with potential-sensitive dyes. *In* "The Mechanism of Gated Calcium Transport Across Biological Membranes" (S. T. Ohnishi and M. Endo, Eds.), pp. 237–255. Academic Press, New York.

Fahlke, C., Zachar, E., and Rudel, R. (1993). Chloride channels with reduced single-channel conductance in recessive myotonia congenita. *Neuron* **10**, 225–232.

Fatt, P., and Ginsborg, B. L. (1958). The ionic requirements for the production of action potentials in crustacean muscle fibers. *J. Physiol.* (*London*) **142**, 516–543.

Forbes, M. S., and Sperelakis, N. (1979). Ruthenium red staining of skeletal and cardiac muscles. *Z. Zellforsch. Cell Tissue Res.* **200**, 367–382.

Gonzalez-Serratos, H. (1971). Inward spread of activation in vertebrate muscle fibers. *J. Physiol.* (*London*) **212**, 777–799.

Gonzalez-Serratos, H., Valle-Aguilera, R., Lathrop, D. A., and del Carmen Garcia, M. (1982). Slow inward calcium currents have no obvious role in muscle excitation-contraction coupling. *Nature* **298**, 292–294.

Harris, E. J. (1963). Distribution and movement of muscle chloride. *J. Physiol.* (*London*) **166**, 87–109.

Hodgkin, A. L., and Horowicz, P. (1959). The influence of potassium and chloride ions on the membrane potential of single muscle fibers. *J. Physiol. (London)* **148**, 127–160.

Hodgkin, A. L., and Huxley, A. F. (1952). Currents carried by sodium and potassium ions through the membrane of the giant axon of Loligo, *J. Physiol. (London)* **116**, 449–472.

Howell, J. N. (1974). Intracellular binding of ruthenium red in frog skeletal muscle. *J. Cell. Biol.* **62**, 242–247.

Hurnak, O., and Zachar, J. (1992). Maxi chloride channels in L6 myoblasts. *Gen. Physiol. Biophys.* **11**, 389–400.

Huxley, A. F., and Taylor, R. E. (1958). Local activation of striated muscle fibres. *J. Physiol. (London)* **144**, 426–441.

Huxley, H. E., Page, S., and Wilkie, D. R. (1963). Appendix. An electron microscopic study of muscle in hypertonic solutions. (M. Dydynsk and D. R. Wilkie, Eds.). *J. Physiol. (London)* **169**, 312–329.

Iannaccone, S. T., Li, K.-X., Sperelakis, N., and Lathrop, D. A. (1989). Insulin-induced hyperpolarization in mammalian skeletal muscle. *Am. J. Physiol.* **256**, C368–C374.

Kerr, L. M., and Sperelakis, N. (1982). Effects of the calcium antagonists verapamil and bepridil (CERM-1978) on $Ca^{2+}$-dependent slow action potentials in frog skeletal muscle, *J. Pharmacol. Exp. Ther.* **222**, 80–86.

Keynes, R. D., and Steinhardt, R. A. (1968). The components of the sodium efflux in frog muscle. *J. Physiol. (London)* **198**, 581–599.

Khan, A. R. (1981). Influence of ethanol and acetaldehyde on electromechanical coupling of skeletal muscle fibres. *Acta Physiol. Scand.* **111**, 425–430.

Kirby, A. C., Lindley, B. D., and Picken, J. R. (1975). Calcium content and exchange in frog skeletal muscle. *J. Physiol. (London)* **253**, 37–52.

Kulczycky, S., and Mainwood, G. W. (1972). Evidence for a functional connection between the sarcoplasmic reticulum and the extracellular space in frog sartorius muscle. *Can. J. Physiol. Pharmacol.* **50**, 87–98.

Li, K.-X., and Sperelakis, N. (1994). Electrogenic Na–K pump current in rat skeletal myoballs. *J. Cell. Physiol.* **159**, 181–186.

Lynch, C., III (1978). Kinetic and biochemical separation of potassium currents in frog striated muscle. Ph.D. thesis. University of Rochester, New York.

Mathias, R. T., Levis, R. A., and Eisenberg, R. S. (1980). Electrical models of excitation-contraction coupling and charge movement in skeletal muscle. *J. Gen. Physiol.* **76**, 1–31.

Miledi, R., Stefani, E., and Steinbach, A. B. (1971). Induction of the action potential mechanism in slow muscle fibres of the frog. *J. Physiol. (London)* **217**, 737–754.

Miledi, R., Parker, R. I., and Schalow, G. (1977). Measurement of calcium transients in frog muscle by the use of arseno III. *Proc. R. Soc. London (Biol.)* **198**, 201–210.

Mobley, B. A., and Eisenberg, B. R. (1975). Sizes of components in frog skeletal muscle measured by methods of stereology. *J. Gen. Physiol.* **66**, 31–45.

Natori, R. (1965). Propagated contractions in isolated sarcolemma-free bundle of myofibrils. *Jikeidai Med. J.* **12**, 214–221.

Nicola-Siri, L., Sanchez, J. A., and Stefani, E. (1980). Effect of glycerol treatment on calcium current of frog skeletal muscle. *J. Physiol. (London)* **305**, 87–96.

Peachey, L. D. (1965). The sarcoplasmic reticulum and transverse tubules of the frog's sartorius. *J. Cell. Biol.* **25**, 209–231.

Podolsky, R. J., and Costantin, L. L. (1964). Regulation by calcium of the contraction and relaxation of muscle fibers. *Fed. Proc.* **23**, 933–939.

Potreau, D., and Raymond, G. (1980). Calcium-dependent electrical activity and contraction of voltage-clamped frog single muscle fibers. *J. Physiol. (London)* **307**, 9–22.

Rogus, E., and Zierler, K. L. (1973). Sodium and water contents of sarcoplasm and sarcoplasmic reticulum in rat skeletal muscle: Effects of anisotonic media, ouabain, and external sodium. *J. Physiol. (London)* **233**, 227–270.

Rubio, R., and Sperelakis, N. (1971). Entrance of colloidal $ThO_2$ tracer into the T-tubules and longitudinal tubules of the guinea pig heart. *Z. Zellforsch.* **116**, 20–36.

Rubio, R., and Sperelakis, N. (1972). Penetration of horseradish peroxidase the terminal cisternae of frog skeletal muscle fibers and blockade of caffeine contracture by $Ca^{++}$ depletion. *Z. Zellforsch.* **124**, 57–71.

Rulon, R., Hermsmeyer, K., and Sperelakis, N. (1971). Regenerative action potentials induced in the neurogenic heart of Limulus polyphemus. *Comp. Biochem. Physiol.* **39A**, 333–335.

Sanchez, J. A., and Stefani, E. (1978). Inward calcium current in twitch muscle fibers of the frog. *J. Physiol. (London)* **283**, 197–209.

Sellin, L. C., and Sperelakis, N. (1978). Decreased potassium permeability in dystrophic mouse skeletal muscle. *Exp. Neurol.* **62**, 609–617.

Spector, I., and Prives, J. M. (1977). Development of electrophysiological and biochemical membrane properties during differentiation of embryonic skeletal muscle in culture. *Proc. Natl. Acad. Sci. USA* **74**, 5166–5170.

Sperelakis, N. (1969). Changes in conductance of frog sartorius fibers produced by $CO_2$, $ReO_4$, and temperature. *Am. J. Physiol.* **217**, 1069–1075.

Sperelakis, N. (1979). Origin of the cardiac resting potential. *In* "Handbook of Physiology, the Cardiovascular System, Vol. 1: The Heart" (R. Berne and N. Sperelakis, Eds.), Chap. 6, pp. 187–267. American Physiological Society, Bethesda, MD.

Sperelakis, N., and Rubio, R. (1971). Ultrastructural changes produced by hypertonicity in cat cardiac muscle. *J. Mol. Cell. Cardiol.* **3**, 139–156.

Sperelakis, N., and Schneider, M. F. (1968). Membrane ion conductances of frog sartorius fibers as a function of tonicity. *Am. J. Physiol.* **215**, 723–729.

Sperelakis, N., Schneider, M. F., and Harris, E. J. (1967). Decreased $K^+$ conductance produced by $Ba^{++}$ in frog sartorius fibers. *J. Gen. Physiol.* **50**, 1565–1583.

Sperelakis, N., Mayer, G., and Macdonald, R. (1970). Velocity of propagation in vertebrate cardiac muscles as functions of tonicity and $[K^+]_0$. *Am. J. Physiol.* **219**, 952–963.

Sperelakis, N., Valle, R., Orozco, C., Martinez-Palomo, A., and Rubio, R. (1973). Electromechanical uncoupling of frog skeletal muscle by possible change in sarcoplasmic reticular content. *Am. J. Physiol.* **225**, 793–800.

Sperelakis, N., Forbes, M. S., and Rubio, R. (1974). The tubular systems of myocardial cells: Ultrastructure and possible function. *In* "Recent Advances in Studies on Cardiac Structure and Metabolism" (N. S. Dhalla and G. Rona, Eds.), Myocardial Biol., Vol. 4, pp. 163–194. University Park Press, Baltimore.

Sperelakis, N., Shigenobu, K., and Rubio, R. (1978). $^3$H-Sucrose compartments in frog skeletal muscle relative to sarcoplasmic reticulum. *Am. J. Physiol.* **234**: C181–C190.

Spitzer, N. C. (1979). Ion channels in development. *Annu. Rev. Neurosci.* **2**, 363–397.

Stanfield, P. R. (1977). A calcium dependent inward current in frog skeletal muscle fibers. *Pflugers Arch.* **368**, 267–270.

Stephenson, E. W. (1981). Activation of fast skeletal muscle: Contributions of studies on skinned fibers. *Am. J. Physiol.* **240**, C1–19.

Tasker, P., Simon, S. E., Johnstons, B. M., Shankly, K. H., and Shaw, F. H. (1959). The dimensions of the extracellular space in sartorius muscle. *J. Gen. Physiol.* **43**, 39–53.

Vinogradova, N. A. (1968). Distribution of nonpenetrating sugars in the frog's sartorius muscle under hypo- and hypertonic conditions. *Tsitologiya* **10**, 831–838.

Vogel, S., and Sperelakis, N. (1978). Valinomycin blockade of myocardial slow channels is reversed by high glucose. *Am. J. Physiol.* **235,** H46–H51.

Vogel, S., Harder, D., and Sperelakis, N. (1978). Ca$^{++}$ dependent electrical and mechanical activities in skeletal muscle. *Fed. Proc.* **37,** 517.

Wallace, N., and Sommer, J. R. (1975). Fusion of sarcoplasmic reticulum with ruthenium red. Proc. Electron Microsc. Soc. A., 33rd, Las Vegas, pp. 500–501.

Weiss, D. S., and Magleby, K. L. (1992). Voltage-dependent gating mechanism for single fast chloride channels from rat skeletal muscle. *J. Physiol. (London)* **453,** 279–306.

Winegrad, S. (1968). Intracellular calcium movements of frog skeletal muscle during recovery from tetanus. *J. Gen. Physiol.* **51,** 65–83.

*Gordon M. Wahler*

# 51

# Cardiac Action Potentials

## I. Introduction

Ultimately, the function of the heart is to pump blood to the body. The electrophysiological behavior of heart cells, individually and in concert, subserves this function. Heart cells are similar to other cell types in that their internal ionic composition is quite different from the extracellular ionic environment. For example, measurement of the intracellular versus extracellular ionic composition shows that the intracellular ionic composition is low in sodium ions ($Na^+$) and high in potassium ions ($K^+$), while the reverse is true of the extracellular ionic composition. These concentration differences, together with the selective permeability characteristics of the cell membrane, generate a potential difference of between 80 and 90 mV across the cell membrane of the resting cardiac cell, with the inside being negative with respect to the outside. Additionally, heart cells are excitable cells. That is, they are capable of generating all-or-none electrical responses known as **action potentials** (APs). The cardiac AP is caused by the complex interaction of a number of different ionic currents. This chapter reviews the characteristics of the cardiac AP, with emphasis on the major currents responsible for the various components or phases of the APs, as well as the regional differences in cardiac APs.

The electrical activity of cardiac cells has been studied for several decades by impaling the cells with high-resistance microelectrodes. The more recent developments of methods to isolate viable single adult cardiac cells (e.g., Powell *et al.*, 1980), together with the development of the patch-clamp technique (Hamill *et al.*, 1981) for recording single-channel (microscopic) currents and whole-cell (macroscopic) currents from single cells, have led to an explosion of information on the currents responsible for generation of the cardiac AP. Although most of the information on these currents has been obtained from cardiac cells isolated from experimental animals, the small number of studies on human cardiac cells (e.g., Coraboeuf and Nargeot, 1993;

Ravens *et al.*, 1996) suggests that the currents in the human heart are similar.

## II. Resting Potential

In cardiac cells at rest, the membrane is quite permeable to $K^+$ and relatively impermeable to $Na^+$ and other ions. Thus, $K^+$ flows out of the cell down the concentration gradient, resulting in rapid buildup of a negative potential inside the cell. As the electric potential buildup increases in magnitude, it becomes sufficient to counterbalance the chemical driving force generated by the concentration gradient. At this potential, called the **equilibrium potential,** the net ion flux is zero. The Nernst equation describes the relationship between the intracellular and extracellular concentrations of a single ion and its equilibrium potential. The equilibrium potential for $K^+$ ($E_K$) is calculated by the Nernst equation:

$$E_K = \frac{-RT}{zF} \ln \frac{[K^+]_i}{[K^+]_o}$$

where $T$ is the absolute temperature (in Kelvins), $z$ is the valence (or charge) of the ion (for $K^+$ it is $+1$), $R$ is the universal gas constant, F is the Faraday constant, $[K^+]_i$ is the $K^+$ ion concentration inside the cell, $[K^+]_o$ is the $K^+$ ion concentration outside the cell, and ln is the natural logarithm. For $K^+$, the intracellular and extracellular concentrations are such that $E_K$ is approximately $-90$ mV at 37°C.

Because the ventricular cell at rest is very permeable to $K^+$, and not very permeable to other ions, the cell resting potential should be close to the calculated $E_K$. Indeed, this is the case. For example, in Fig. 1, the resting potential is approximately $-87$ mV. The resting potential does not quite reach $E_K$ because there is a small, but finite, permeability to $Na^+$ ions and, in addition, there are other electrogenic ion transport systems (primarily

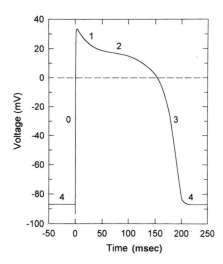

**FIG. 1.** Ventricular action potential. This is a diagrammatic representation of a typical adult mammalian ventricular AP. The five phases of the ventricular AP are labeled 0–4.

the Na,K-ATPase system) that contribute slightly to the resting potential.

## III. Currents During Phases of the Action Potential

### A. Overview

Cardiac cells have certain properties in common with other excitable cells, such as nerve and skeletal muscle cells. However, the behavior of cardiac cells also differs from the behavior of nerve and skeletal muscle cells in some important respects. The nerve and skeletal muscle APs are relatively brief, consisting primarily of a rapid depolarization phase (when the inside of the cell becomes more positive), followed immediately by a rapid repolarization phase (when the inside of the cell returns to a more negative potential). The depolarization phase of nerve and skeletal muscle cells is caused by the rapid influx of positive sodium ions into the cell, down the electrochemical gradient for $Na^+$, which makes the cell interior more positive. The subsequent repolarization is caused by the efflux of positive potassium ions from the cell, down the electrochemical gradient for $K^+$, which returns the cell interior to its original more negative resting potential. The entire process of the nerve or skeletal muscle APs is largely complete within a few milliseconds. In cardiac cells, the APs are more complex and generally much longer in duration. Thus, for example, a typical cardiac ventricular AP may be at least 100–200 msec in duration (Fig. 1). Additionally, unlike for nerve and skeletal muscle cells, APs from different regions of the heart vary substantially in shape.

The configuration of the cardiac AP can be divided into several phases. The following description describes the five phases of the AP found in ventricular cells (phases 0 through 4), and indicates the current or currents primarily responsible for each phase. The phases and primary currents in the ventricle are summarized in Table 1. Phase 0 is the upstroke of the AP; phase 1 is the early repolarization phase; phase 2 is the plateau phase; phase 3 is the primary repolarization phase; and phase 4 is the resting potential phase of the ventricular AP. Additional information about regional differences in the AP phases and currents will be also presented for sinoatrial nodal cells, atrial cells, atrioventricular nodal cells, and Purkinje fibers in Section V.

### B. Phase 0: Sodium Current

Phase 0 is the rapid upstroke of the AP. In ventricular cells, as well as in cells of the atria and His–Purkinje network, this upstroke is dependent on a fast $Na^+$ current quite similar to current responsible for the upstroke of the nerve or skeletal muscle APs. This current is designated as fast because it exhibits very rapid activation and inactivation kinetics, in contrast to other currents. Figure 2 illustrates the kinetics and voltage dependence of this current. Following a large depolarization, $I_{Na}$ reaches a peak in less than 1 msec. Following this peak activation of $I_{Na}$, the amplitude of this current spontaneously decreases. This decay of $I_{Na}$ is due to closure (inactivation) of the fast $Na^+$ channels; thus, $I_{Na}$ is nearly zero after only a few milliseconds. The fast activation and large magnitude of $I_{Na}$ has caused difficulties in accurately recording this current in voltage-clamped cardiac cells. Thus, $I_{Na}$ can generally only be recorded in adult cardiac cells at reduced $[Na^+]_0$ levels and/or at low temperatures. Alternatively, $I_{Na}$ can be studied in very small cells, such as the embryonic chick ventricular cell (e.g., see Fig. 2), because $I_{Na}$ (and other currents) are at least an order of magnitude smaller than in adult cells.

For the rapid upstroke of phase 0 to occur, the cell needs to be depolarized to the voltage necessary to open the fast $Na^+$ channels (approximately $-70$ mV). When the $Na^+$ channels begin to open, $Na^+$ flows down the electrochemical gradient into the cell. This causes the cell to depolarize further (i.e., the inside becomes more positive), which opens additional $Na^+$ channels. Therefore, once sufficient $Na^+$ channels open, the process becomes self-perpetuating, resulting in rapid depolarization (i.e., the membrane potential rapidly moves toward the equilibrium potential for $Na^+$). The voltage at which a sufficient number of $Na^+$

**TABLE 1** Phases of the Ventricular Action Potential and Primary Currents Responsible[a]

| Phase | Current |
| --- | --- |
| Phase 0 | $I_{Na}$ |
| Phase 1 | $I_{to}$ |
| Phase 2 | $I_{Ca}$  $I_K$, $I_{to}$ |
| Phase 3 | $I_K$, $I_{K1}$ |
| Phase 4 | $I_{K1}$ |

[a] $I_{Na}$ and $I_{Ca}$ are inward currents; the potassium currents ($I_{to}$, $I_K$, $I_{K1}$) are outward currents.

A

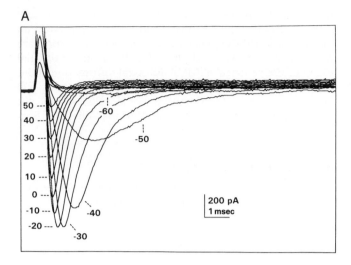

B

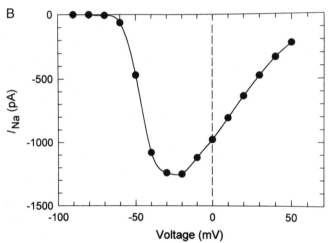

**FIG. 2.** Sodium current ($I_{Na}$) recorded from a small embryonic chick ventricular cell. In these cells $I_{Na}$ is much smaller than in adult cells, due to the very small size of these cells. Additionally, the small spherical shape of these cells and lack of T-tubules also contribute to better voltage control (A) The original $I_{Na}$ currents obtained on stepwise depolarization from −100 mV to the indicated voltages. The current peaks in less than 1 msec at very depolarized potentials, and then rapidly inactivates. (B) The current–voltage curve of the peak current at each voltage. Note that the threshold for $I_{Na}$ is approximately −70 mV and the current peaks at approximately −25 mV.

channels open to initiate the AP is the threshold for firing of the AP. As the cell begins to depolarize further beyond the threshold, $I_{Na}$ increases as more $Na^+$ channels are activated. Eventually, with even greater depolarization, $I_{Na}$ begins to decline as the membrane potential approaches $E_{Na}$. Thus, the peak $I_{Na}$ occurs between −30 to −20 mV. The membrane potential never reaches $E_{Na}$ for several reasons: (1) As the membrane potential gets closer to $E_{Na}$, the driving force for $Na^+$ influx is diminished. (2) The $Na^+$ channels close shortly after opening (beginning after about 1 msec); thus, some $Na^+$ channels are already closing during the latter part of the upstroke. (3) Repolarizing currents are beginning to activate during the latter portion of the

upstroke. Thus, the maximum positive membrane potential is approximately +35 mV (see Fig. 1). Nevertheless, the upstroke causes a substantial voltage change (110–120 mV) within 1–2 msec.

## C. Phase 1: Transient Outward Current

Phase 1 of the cardiac AP is the transient and relatively small repolarization phase that immediately follows the upstroke of the AP. The size of phase 1 repolarization varies between species and also between different regions of the heart within a given species. Thus, APs recorded from the outer (epicardial) layer of ventricular cells display a more prominent phase 1, whereas APs recorded from the inner (endocardial) layer of ventricular cells display a small phase 1 repolarization (Liu *et al.*, 1993). Phase 1 is also very large in Purkinje fibers and in atrial cells, but is largely absent in nodal cells.

Identification of the specific current or currents responsible for phase 1 repolarization has been controversial. Early experiments in which $Cl^-$ ions were replaced by impermeant anions resulted in depression of phase 1. These experiments suggested that the current responsible for phase 1 repolarization was a $Cl^-$ current. This conclusion was based on the assumption that the only effect of $Cl^-$ replacement was a reduction in the $Cl^-$ influx. It was subsequently demonstrated that $Cl^-$ replacement has a number of other effects, and it is these confounding effects which are largely responsible for the depression of phase 1, rather than a reduction in $Cl^-$ influx. It is now clear that the primary ion responsible for most of the phase 1 repolarization is $K^+$, rather than $Cl^-$ (Kenyon and Gibbons, 1979), although a small $Cl^-$ component appears to contribute to phase 1 repolarization (see later discussion).

Phase 1 repolarization is largely due to a transient outward current ($I_{to}$), which turns on rapidly with depolarization (i.e., beginning during the final portion of the AP upstroke) and is only active at very depolarized potentials. The threshold for activation is approximately −30 mV (Fig. 3). Thus, $I_{to}$ has a characteristic transient shape—the rapid activation of this current is followed by inactivation during the AP plateau. Because of its voltage dependence and time course, $I_{to}$ significantly overlaps (and opposes) the inward $Ca^{2+}$ current (which is the primary depolarizing current during the plateau phase; see later discussion).

This current ($I_{to}$) is actually composed of at least two separate currents ($I_{to1}$ and $I_{to2}$), which are carried through two physically distinct channels (Tseng and Hoffman, 1989). One of the currents ($I_{to1}$) is a $K^+$ current that is independent of the internal $Ca^{2+}$ concentration ($[Ca^{2+}]_i$) and is sensitive to the $K^+$ channel blocker 4-aminopyridine (4-AP). This component of $I_{to}$ is very similar to the $I_A$ current recorded in nerve fibers. The second component of $I_{to}$ ($I_{to2}$) is $Ca^{2+}$ dependent and less sensitive to 4-AP, but is more sensitive to another $K^+$ channel blocker, tetraethylammonium ion (TEA$^+$). There is some evidence, however, that $I_{to2}$ may be, at least in part, a $Ca^{2+}$-activated $Cl^-$ channel (Harvey, 1996). Thus, some disagreement remains about the exact nature of $I_{to2}$.

A

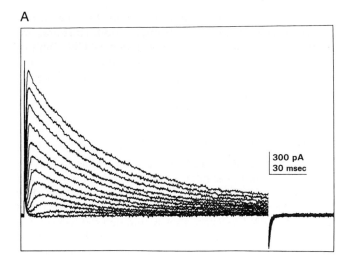

```
300 pA
30 msec
```

B

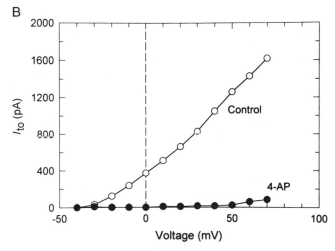

**FIG. 3.** Transient outward current ($I_{to}$) recorded from a 21-day rat ventricular cell. (A) The original currents obtained on stepwise depolarization from $-80$ mV. $I_{to}$ activates rapidly and then inactivates. Currents were recorded in the presence of tetrodotoxin to eliminate $I_{Na}$ and cadmium to eliminate overlapping $I_{Ca}$. (B) The current–voltage curve shows increasing activation of this current at voltages above approximately $-30$ mV (open circles). This current is the $Ca^{2+}$-independent, 4-aminopyridine-sensitive component of $I_{to}$ (i.e., $I_{to1}$) and is therefore readily blocked by 4-aminopyridine (4-AP, closed circles).

Under physiological conditions, the first $I_{to}$ component, $I_{to1}$, is by far the larger of the two components. Thus, irrespective of the exact nature of $I_{to2}$, efflux of $K^+$ (through $I_{to1}$ channels) appears responsible for the vast majority of phase 1 repolarization. The second component ($I_{to2}$) may become more important when intracellular levels of $Ca^{2+}$ become too high. Thus, under $Ca^{2+}$-overload conditions, $I_{to2}$ would be activated and shorten the AP duration, thereby indirectly abbreviating the duration of $Ca^{2+}$ current, resulting in a reduced $Ca^{2+}$ influx. Thus, activation of $I_{to2}$ by intracellular $Ca^{2+}$ likely acts as a negative feedback mechanism to reduce calcium overload.

### D. Phase 2: Calcium Current

Phase 2 is commonly called the plateau. It follows the early repolarization phase (phase 1), and is a period of time in which the membrane potential remains relatively constant; that is, it does not rapidly repolarize. The presence of this prominent plateau phase is responsible for the long AP duration in cardiac cells, which is the major difference between cardiac cells and nerve or skeletal muscle fibers. The plateau is caused by an approximate balance of positive inward current (which would tend to cause depolarization) and positive outward currents (which would tend to cause repolarization). The primary inward current is a $Ca^{2+}$ current. The primary outward $K^+$ current during the plateau phase of ventricular cells (particularly in latter part of the plateau), is the classic slowly activating $K^+$ current known as the delayed rectifier, which is described in greater detail under phase 3 repolarization (Section III,E). Additionally, $I_{to}$ contributes to the early plateau phase in those cells that have a substantial $I_{to}$. In addition to the $Ca^{2+}$ and $K^+$ currents, there is a small contribution of the $Na^+$ current to the plateau. Thus, although $I_{Na}$ largely inactivates within a few milliseconds after depolarization, a very small fraction of the $I_{Na}$ inactivates slowly, causing a late sodium current known as the *sodium window current*, which is sufficient to affect the plateau of the AP and, hence, AP duration (Attwell *et al.*, 1979).

The inward $Ca^{2+}$ current ($I_{Ca}$) exhibits activation and inactivation much like $I_{Na}$, but on a slower timescale (Fig. 4). This second inward current is carried by $Ca^{2+}$ and peaks within a few milliseconds, but requires a few hundred milliseconds to completely inactivate. $I_{Ca}$ is at least one order of magnitude smaller than $I_{Na}$ in a given cell. The threshold for activation of $I_{Ca}$ is approximately $-40$ mV, the current is maximal near 0 mV, and the $E_{Ca}$ is around $+100$ mV. Thus, $I_{Ca}$ is active over a more positive (depolarized) potential range than $I_{Na}$. This classical $Ca^{2+}$ current is now commonly referred to as the **L-type calcium current, $I_{Ca(L)}$**, to distinguish it from the novel transient **T-type $Ca^{2+}$ current** described in Section VI,B,2.

The L-type $Ca^{2+}$ current is central to many aspects of cardiac function. For example, it is the primary link in excitation–contraction coupling in the heart. Thus, the influx of $Ca^{2+}$ ions during the plateau of the AP is what links the electrical events of the AP to the mechanical events, namely, contraction. $Ca^{2+}$ influx via $I_{Ca(L)}$ stimulates $Ca^{2+}$ release from internal stores (which is the major source of contractile $Ca^{2+}$ in most adult mammalian hearts), replenishes the internal $Ca^{2+}$ stores available for subsequent release (for more detail, see the chapter on excitation–contraction coupling), and, to some extent, directly activates the contractile proteins. $I_{Ca(L)}$ is also very important in automaticity and conduction, due to the $Ca^{2+}$-dependent nature of the nodal APs (see Sections V,B and V,D).

$I_{Ca(L)}$ is a major regulatory site in control of cardiac electrical activity and contraction by neurotransmitters, hormones, intracellular ions, etc. Perhaps the most important regulator of $I_{Ca(L)}$ in the heart is the autonomic nervous system. Thus, release of norepinephrine from car-

A

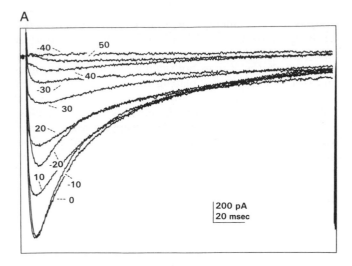

B

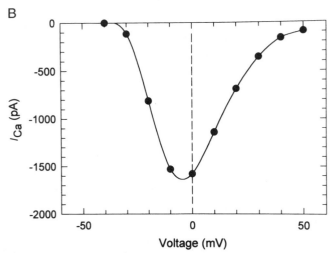

**FIG. 4.** L-type calcium current, $I_{Ca(L)}$, recorded from an adult guinea pig ventricular cell. (A) The original currents obtained on stepwise depolarization from −80 mV to the indicated voltages. $I_{Na}$ was inactivated by briefly pulsing to −40 mV prior to applying the indicated test pulse. The $Ca^{2+}$ currents show the same general shape as the $Na^+$ currents (see Fig. 2); that is, they activate on depolarizing steps and inactivate during the voltage pulse. However, both activation and inactivation are much slower than for $I_{Na}$ (note difference in timescale). (B) The current–voltage curve shows a similar shape as the $I_{Na}$ current–voltage relationship; however, both the threshold voltage and peak voltage are shifted approximately 30 mV in the depolarizing direction. $I_{Ca(L)}$ is roughly the same size as $I_{Na}$ shown in Fig. 2 only because this cell is so much larger than the embryonic chick cell used in Fig. 2. $I_{Na}$ in this guinea pig cell would be approximately 20 nA or more, and could not be voltage-clamped under these conditions.

diac sympathetic nerves or release of epinephrine from the adrenal gland stimulates the β-adrenergic receptors of the cardiac cell. A cascade of events ensues, which involves production of cyclic adenosine 3′,5′-monophosphate (cAMP) and stimulation of the cAMP-dependent protein kinase. This protein kinase directly phosphorylates the $Ca^{2+}$ channel, thereby enhancing the channel activity. The net result of this cascade is that $I_{Ca(L)}$, and thereby force

of contraction, is stimulated. For a detailed review of this process, see the chapter on phosphorylation. The parasympathetic neurotransmitter, acetylcholine (ACh), inhibits cAMP formation. Additionally, ACh stimulates cyclic guanosine 3′,5′-monophosphate (cGMP) production. Cyclic GMP is a second messenger similar to cAMP, but which reduces $I_{Ca(L)}$ (see Wahler and Dollinger, 1995).

### E. Phase 3: Delayed Rectifier Current

Phase 3 is the late or final repolarization phase following the AP plateau. It is similar to the repolarization observed in nerve and skeletal muscle cells. It is primarily caused by the imbalance of the currents that were relatively balanced during phase 2. $I_{Ca(L)}$ decreases with time (due to inactivation) and the delayed rectifier current ($I_K$) increases (due to slow activation). This eventually leads to the outward current $I_K$ overwhelming the inward current $I_{Ca(L)}$. $I_K$ is the primary repolarizing current in most ventricular preparations. The delayed rectifier activates slowly, compared to most other currents, and does not inactivate with time. Thus, $I_K$ increases gradually during sustained depolarization at voltages around the plateau level. This current is similar to the delayed rectifier in nerve cells, although slower. $I_K$ can be clearly distinguished from $I_{to}$ by slow activation, lack of inactivation, and different pharmacology. For technical reasons, $I_K$ is often greatly underestimated in patch-clamp recordings.

There are several components of $I_K$; thus, many investigators divide $I_K$ into a very slow component ($I_{Ks}$) and a more rapid component ($I_{Kr}$) (e.g., see Lindblad et al., 1996). The sensitivity of the two $I_K$ components to pharmacological agents differs, suggesting that they are carried through distinct channels. In addition, some ventricular preparations exhibit a $Ca^{2+}$-dependent $I_K$ (Tohse, 1990). This component may help to shorten the AP duration similar to the effect of the $Ca^{2+}$-dependent $I_{to}$ described earlier. However, the relative magnitude and importance of this component is still unclear.

The β-adrenergic–cAMP cascade can stimulate $I_K$, similar to its ability to stimulate $I_{Ca(L)}$ (e.g., Walsh and Kass, 1988), which would tend to shorten the AP. Thus, β-adrenergic stimulation tends to lengthen the AP duration by enhancing $I_{Ca}$ and at the same time tends to shorten AP duration by enhancing $I_K$. The overall effect of β-adrenergic stimulation on AP duration, thus, is determined by the relative contribution of the changes in these two currents (and perhaps also the contribution of $I_{Cl}$; see discussion later).

As the membrane potential continues to repolarize and approaches the resting membrane potential, largely due to the effects of $I_K$, the inwardly rectifying $K^+$ current (which is the primary determinant of the resting potential; see Section III,F), also begins to contribute to repolarization.

### F. Phase 4: Inward Rectifier Current

In ventricular cells, and most other cardiac cells, phase 4 is the resting potential. The resting potential is defined as the stable, negative potential that occurs between APs in nonspontaneous cells. It is very near $E_K$ due to a relatively high permeability of the resting ventricular cell membrane to $K^+$ ions, and a very low permeability to other ions. Thus, the resting potential of nonautomatic cardiac cells is similar to the resting potential in skeletal muscle cells. The resting potential is determined largely by a $K^+$ current known as the **inward** (or **anomalous**) **rectifier**, $I_{K1}$.

The most notable characteristic of $I_{K1}$ is that it displays inward rectification. Thus, it passes current more readily in the inward direction than in the outward direction. This characteristic is evident in Fig. 5. Of course, the outward $K^+$ current is the only one occurring physiologically, since the membrane potential does not normally hyperpolarize beyond $E_K$. In most preparations, $I_{K1}$ also demonstrates a negative slope region in the outward direction. That is, as the cell is depolarized from near its resting potential to progressively more depolarized potentials, the outward current first increases and then decreases with further depolarization. Thus, the contribution of $I_{K1}$ to repolarization tends to be less at potentials near the plateau, and greater as the membrane potential approaches the resting potential, just as $I_K$ is declining.

## IV. Additional Currents Contributing to the Action Potential

### A. Pump Current

The Na,K-ATPase pump is the primary transport system that maintains the ionic imbalance between the cell exterior and interior. That is, each pump cycle extrudes three $Na^+$ ions out of the cell and transports two $K^+$ ions into the cell, thus building up $[K]_i$ and reducing $[Na]_i$. Because of the exchange of three positive for two positive ions, each pump cycle generates a net loss of one positive charge, which generates a pump current ($I_{pump}$). At the resting potential, $I_{pump}$ is an outward current that may hyperpolarize the membrane potential slightly. In addition, $I_{pump}$ can cause considerable shortening of the action potential when $[Na]_i$ increases pathologically (Gadsby, 1984). However, the primary role of the Na,K-ATPase is not to create a small pump current, but rather to set up and maintain the ionic gradients that generate the currents responsible for the action potential.

### B. Na/Ca Exchange Current

A Na/Ca exchanger exists in the cardiac sarcolemma. Working in the "normal mode" the Na/Ca exchanger exchanges intracellular $Ca^{2+}$ for extracellular $Na^+$; thus, the exchanger is an important mechanism whereby $Ca^{2+}$ is removed from the cytoplasm. The exchanger may also work in the "reverse" mode, exchanging intracellular $Na^+$ for extracellular $Ca^{2+}$. Under these conditions, the exchanger can contribute $Ca^{2+}$ influx for excitation–contraction (EC) coupling. The exchanger transports three $Na^+$ for each $Ca^{2+}$ under most conditions, leading to the net movement of one positive charge. Thus, the exchanger generates a

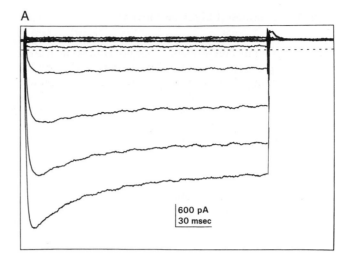

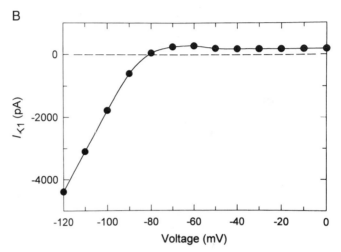

**FIG. 5.** Inwardly rectifying $K^+$ current ($I_{K1}$) recorded from an adult guinea pig ventricular cell. (A) The original barium-subtracted currents obtained on stepwise hyperpolarization and depolarization from a holding potential of $-40$ mV. The current displays the typical inward rectification. That is, the current is smaller in the outward (physiological) direction than in the inward direction. (B) The current–voltage curve for $I_{K1}$. The current also displays a negative slope region between approximately $-60$ and $-30$ mV. Thus, larger depolarizations actually result in a decrease in outward $K^+$ flux during this voltage range, such that the current at $-30$ mV is less than half as large as at $-60$ mV. The net result is that during repolarization (phase 3) the $K^+$ current increases as the membrane potential approaches the resting potential.

current that can also contribute to the action potential. The equilibrium potential for Na/Ca exchange current is generally slightly negative to 0 mV; therefore, near the resting potential the Na/Ca exchanger works in the normal mode and generates an inward current. During the initial portion of the plateau, the Na/Ca exchanger transiently works in the reverse mode and briefly generates an outward current prior to returning to the normal mode. Thus, the Na/Ca exchange current may contribute to the shape of the AP (for review, see Janvier and Boyett, 1996).

## C. cAMP-Stimulated Chloride Current

Under basal conditions, the Cl$^-$ current ($I_{Cl}$) in the heart is relatively small, and probably does not contribute a great deal to the configuration of the AP. However, when cAMP levels are stimulated (as with sympathetic nerve stimulation), a significant time-independent Cl$^-$ current develops. Activation of this current by $\beta$-adrenergic stimulation can cause a small depolarization of the resting potential and significant shortening of the action potential (Harvey *et al.*, 1990).

## D. ATP-Sensitive Potassium Current

When the oxygen supply declines, as occurs during ischemia, AP duration shortens. The shortening of the AP duration accelerates inactivation of $I_{Ca}$, thereby reducing contractility. The reduced contractility greatly decreases the energy demands of the cell, thereby sparing adenosine triphosphate (ATP). This mechanism contributes to the survival of the myocardial cell during temporary ischemia. However, in addition to this beneficial effect, regional shortening of the AP can also lead to arrhythmias.

The shortening of the AP during ischemia is caused, in large part, by activation of a novel outward K$^+$ current. The decreased oxygen supply reduces ATP levels in the cell and $I_{K1}$ is inhibited. At the same time, another K$^+$ current, $I_{K(ATP)}$ is activated (Noma, 1983). This ATP-sensitive K$^+$ current is inhibited by physiological levels of intracellular ATP (Fig. 6) and, thus, appears to contribute little to the AP configuration under conditions of adequate oxygenation. However, during inadequate oxygenation, the AP is shortened as the larger $I_{K(ATP)}$ replaces $I_{K1}$. The physical characteristics of the $I_{K(ATP)}$ channel (ATP sensitivity, degree of rectification) are altered in some pathophysiological states, such as hypertrophy (Cameron *et al.*,

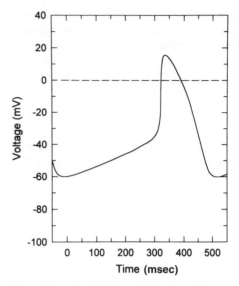

**FIG. 7.** Sinoatrial node AP. This is a diagrammatic representation of a typical SA node AP in the mammalian heart. Note the pacemaker potential and the less negative membrane potential, and the slower upstroke compared to the fast AP of ventricular cell (see Fig. 1).

1988) or diabetic cardiomyopathy (Smith and Wahler, 1996). This may be an important factor in the abnormal responses to ischemia in these conditions.

## V. Regional Differences in Action Potentials

### A. Overview

The normal pathway for electrical activation of the heart is the following: sinoatrial (SA) node, atria, atrioventricular (AV) node, bundle of His, Purkinje fibers, ventricles. The APs differ from region to region, reflecting the different roles played by the different cell types. The following description characterizes the APs potentials in each region, and indicates how each differs from the ventricular AP.

### B. Sinoatrial Node

The SA node contains specialized cells that generate APs which are quite different from the ventricular APs described earlier (Fig. 7). Unlike ventricular APs, these cells do not have a true resting potential, that is, the membrane potential between APs is not stable, but rather exhibits a slow spontaneous depolarization known as **phase 4 depolarization** or the **pacemaker potential.** Because there is no resting potential in these cells, the most negative potential the cell reaches between APs is called the **maximum diastolic potential.** This potential is less negative than the resting potential of ventricular cells, due to a lower K$^+$ permeability (caused by a lack of $I_{K1}$ in these cells). The maximum diastolic potential ranges from approximately $-55$ mV for true primary pacemaker cells to approximately $-70$ mV for transitional cells on the border between the SA node and atria. The pacemaker potential takes the nodal cell from the maximum diastolic potential to the

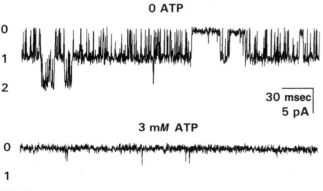

**0 ATP**

**3 m*M* ATP**

**FIG. 6.** ATP-sensitive K$^+$ current, $I_{K(ATP)}$, recorded from an inside-out patch from an adult rat ventricular cell. Single-channel currents shown were recorded in the absence of ATP (upper traces) and in the presence of physiological levels of ATP (3 m*M*, lower traces). Channel openings are downward. In the absence of ATP, channel activity is high. Thus, up to two channels are open simultaneously (number of open channel levels are indicated at the side). In the presence of 3 m*M* ATP, the channels are rarely open. In the presence of ATP, the openings are extremely brief, such that they do not appear to reach the normal open level (1).

threshold for generating an AP in these cells (approximately $-40$ mV). Thus, these cells are spontaneously active, and the slope of the phase 4 depolarization is an important determinant of the rate of AP generation and, thereby, heart rate (see Section VI,C). Once the phase 4 depolarization brings the cell to the threshold for AP firing, an AP occurs, as in ventricular cells. However, in nodal cells, the upstroke (phase 0) is quite different from the upstroke in ventricular (and nerve or skeletal muscle) cells: It is a much slower upstroke and is $Ca^{2+}$ dependent rather than $Na^+$ dependent. Thus, the upstroke in nodal cells is generated by the inward $Ca^{2+}$ current, $I_{Ca(L)}$. Because the speed of the upstroke largely determines the speed of conduction, this slow phase 0 is very important in cardiac function, since it results in a slow conduction in nodal cells. There is generally no phase 1 and a brief plateau (phase 2) in nodal cells. Phase 3 repolarization returns the cell to the maximum diastolic depolarization. Because nodal cells only repolarize to the maximum diastolic potential of approximately $-55$ to $-70$ mV, the lack of $I_{K1}$ in these cells does not significantly slow repolarization, because $I_{K1}$ contributes to repolarization primarily at potentials nearer to $E_K$.

### C. Atria

Atrial cell APs are similar in many aspects to the ventricular APs described earlier. Thus, the resting potential (phase 4) is approximately $-85$ mV, and there is a fast upstroke (phase 0) generated by $I_{Na}$. The most distinguishing feature of the atrial AP is that it has a more triangular appearance than the ventricular AP. This more triangular appearance seems to be due to a prominent phase 1 in atrial cells. Thus, in atrial cells phases 1, 2, and 3 tend to run together resulting in a triangular shape, with a distinct plateau not always apparent. This is likely due to a large $I_{to}$ in atrial cells.

### D. Atrioventricular Node

The cells of the AV node generate APs that are quite similar to the APs of the SA node. Thus, these cells fire $Ca^{2+}$-dependent APs, and also display spontaneous phase 4 depolarization (i.e., automaticity). However, the rate of the phase 4 depolarization in AV nodal cells is much slower than the rate of phase 4 depolarization in SA nodal cells. Thus, the SA node cells fire APs before the AV node cells fire, which is why the SA node cells are the normal pacemaker cells of the heart.

### E. Purkinje Fibers

Purkinje APs are in most respects similar to the ventricular APs. Thus, these cells have a negative resting potential between APs (phase 4) and a very rapid upstroke (phase 0) generated by $I_{Na}$. Purkinje fibers do differ from ventricular cells in that they have a more prominent phase 1 repolarization and a longer plateau (phase 2). The plateau is followed by a phase 3 repolarization that is virtually identical to phase 3 in ventricular cells. Cells in the bundle of His

appear to have APs similar to those in the Purkinje fibers; however, in general, bundle of His cells have not been studied in detail.

Additionally, Purkinje cells may exhibit automaticity, especially when the extracellular $K^+$ concentration is low. Thus, under some conditions, they exhibit phase 4 depolarization. Purkinje cells have all the currents found in ventricular cells, described earlier ($I_{Na}$, $I_{to}$, $I_{Ca}$, $I_K$, $I_{K1}$). Purkinje cells also have some additional currents, which are absent or very small in ventricular cells, that are related to the latent pacemaker function of these cells. These additional currents are described in Section VI,B.

## VI. Automaticity

### A. Overview

Automaticity refers to the ability of some cardiac cells to depolarize and fire repetitive APs spontaneously. Thus, as noted earlier, automatic cells (e.g., in the SA node) do not have a stable resting potential between APs, but rather have a maximum diastolic potential followed by a spontaneous phase 4 depolarization (pacemaker potential). The spontaneous depolarization to threshold generates the AP in that cell type. Following repolarization of the AP, the spontaneous phase 4 depolarization occurs again. The slope of the pacemaker potential (i.e., rate of spontaneous phase 4 depolarization) largely determines the rate of AP firing. Because the rate of firing of APs is normally fastest in the primary pacemaker cells of the SA node, this region acts as the normal pacemaker of the heart. Once this region fires an AP, the wave of depolarization is propagated to other regions of the heart, ultimately leading to contraction of the heart. Automatic cells in other regions of the heart (e.g., AV node) normally do not have the opportunity to fire spontaneously before the wave of depolarization arising from the SA node drives them to threshold.

### B. Mechanisms of Automaticity

Automaticity is a property of several cell types in the heart under physiological conditions. Under pathophysiological conditions, even normally nonspontaneous cells (e.g., ventricular cells) may exhibit automaticity. Physiologically, the most important pacemaker potential is that of the normal pacemaker cells in the SA node. However, the very small size of the SA node cells makes all the currents difficult to measure accurately; additionally, the very high input resistance (due to lack of $I_{K1}$ in the SA node) means that extremely small currents may be significant contributors to the pacemaker potential in the SA node cells. These technical difficulties have limited our understanding of the pacemaker process in SA node cells. Therefore, there is considerable controversy regarding the precise mechanism of automaticity in SA node cells (for reviews, see Baumgarten and Fozzard, 1992; Campbell et al., 1992). Because of these limitations, much of what we know about automaticity is experimentally derived from other automatic cell types (e.g., Purkinje cells) and then extrapolated to SA node cells using various mathematical models. There are limitations to this approach, since there are considerable differences between automatic cell types. Thus, for instance, the pacemaker potential range is much more depolarized in SA node cells compared to Purkinje cells, which is bound to influence the relative contribution of various currents.

## 1. Automaticity in Purkinje Fibers

Under physiological conditions, Purkinje fibers can have an extremely slow phase 4 pacemaker potential. In contrast to SA node cells, the mechanisms for automaticity in Purkinje cells are fairly well understood. As noted earlier, the Purkinje cells have all the currents that ventricular cells have, plus some additional currents not found to any significant degree in adult ventricular cells. The large $I_{K1}$ current in Purkinje cells tends to clamp the membrane potential near $E_K$ and, therefore, a large depolarizing current is needed to overcome this clamping effect. The primary current responsible for the pacemaker potential in Purkinje has some unusual properties, earning it the designation **funny current** ($I_f$).

$I_f$ is a slowly activating inward depolarizing current activated by hyperpolarization that is present in automatic cells. $I_f$ is largely absent in adult ventricular cells (or it may be present but nonfunctional in ventricular cells due to voltage inactivation; see Yu *et al.*, 1993). It is a nonselective cation current, that is, it is carried by a mixture of both $Na^+$ and $K^+$ ions. In Purkinje cells, $I_f$ is responsible for most of the depolarizing current that generates the pacemaker potential.

In addition to $I_f$, there may be a small contribution of a novel $Ca^{2+}$ current to the latter stages of the pacemaker potential in Purkinje cells (see description later of automaticity in nodal cells), as well as a small contribution of $I_{Na}$ to the final portion of the pacemaker potential in these cells. Once the Purkinje cell is depolarized to the threshold for $I_{Na}$ (approximately $-70$ mV), $Na^+$ channels will open. The inward flux of $Na^+$ may contribute to the final phase of the pacemaker potential, as the membrane potential approaches the threshold for AP generation. Thus, the Purkinje fiber will fire an AP that is generated by $I_{Na}$, as described previously for ventricular cells. The remainder of the AP is generated by essentially the same mechanisms as previously described for ventricular cells.

## 2. Automaticity in Nodal Cells

Several factors contribute to the pacemaker potential in nodal cells. Because of the very low density of $I_{K1}$ channels in nodal cells, the resting $K^+$ permeability is lower in nodal cells than in ventricular cells. The large resting $K^+$ permeability in ventricular cells generated by $I_{K1}$ tends to keep the interior of the cells negative, opposing depolarization of the cell toward threshold by "clamping" the membrane potential near $E_K$. A much smaller current is sufficient to depolarize the nodal cells, due to this much lower resting $K^+$ permeability. Thus, currents that may be too small to measure accurately using present electrophysiological techniques (small background currents or currents produced by various electrogenic transport mechanisms could produce sufficient current to affect the pacemaker potential. Because of this limitation, the analysis of the relative contribution of various currents to the pacemaker potential in nodal cells is much less clear than for Purkinje cells. Therefore, investigators often use mathematical models of nodal electrophysiology that are based largely on the characteristics of various currents measured in other automatic cell types (e.g., Purkinje fibers, sinus-venosus of amphibian hearts) in which the measurements are more readily made.

The major depolarizing current during the pacemaker potential of Purkinje cells, $I_f$, is also present in nodal cells and probably contributes to the pacemaker potential in the SA node. However, $I_f$ is an unusual depolarizing current in that it is activated by hyperpolarization. Thus, it may not be quantitatively quite as important to pacemaker activity in nodal cells as in Purkinje cells. In addition to $I_f$, the interaction between a depolarizing background current ($I_b$) and the decay of the major repolarizing current, the delayed rectifier, $I_K$ may also provide a depolarizing current capable of contributing to the pacemaker potential in nodal cells. Because the relative contribution of these currents to the pacemaker potential in nodal cells cannot be accurately determined experimentally, there has been considerable controversy over which depolarizing current ($I_f$ or $I_b$) plays the greater role in determining the slope of the pacemaker potential in these cells.

The small inward background current ($I_b$), present in nodal cells, is a cation current carried primarily by $Na^+$ ions (Hagiwara *et al.*, 1992). $I_b$ contributes a constant small depolarizing current. Due to the small size of this current, little is known about its magnitude and characteristics in mammalian SA node cells; however, indirect evidence suggests that it may be a very important component in determining automaticity (see Campbell *et al.*, 1992; Dokos *et al.*, 1996). The role of a constant $I_b$ in generating a variable pacemaker potential likely stems from the interaction of $I_b$ with a variable $I_K$. $I_K$ is the primary current responsible for repolarization in nodal cells, as in other cardiac cells. It displays no inactivation during a prolonged depolarizing pulse, but displays a slow decay on repolarization toward $E_K$. The time course of the $I_K$ decay is very slow at membrane potentials in the voltage range of the pacemaker potential in nodal cells. The depolarizing action of $I_b$ is opposed by $I_K$. Thus, the depolarization due to a constant background current $I_b$ increases progressively with time due to a gradual reduction of opposing repolarizing current, $I_K$. This is thought to be a significant factor in development of the pacemaker potential in nodal cells. Other small currents (e.g., $I_{pump}$) may also modulate the pacemaker potential in nodal cells.

The upstroke (phase 0) of nodal cells is generated by a $Ca^{2+}$ current rather than a $Na^+$ current. This $Ca^{2+}$ current is identical to the classic long-lasting L-type $Ca^{2+}$ current, $I_{Ca(L)}$, that is largely responsible for the plateau in both nodal and working cardiac cells. In addition to $I_{Ca(L)}$, nodal cells (and other pacemaker cells) have a second type of $Ca^{2+}$ current that is activated at more negative potentials and has a much more rapid inactivation (Hagiwara *et al.*, 1988). This second, rapid type of $Ca^{2+}$ current has been named the **transient,** or **T-type,** $Ca^{2+}$ current, $I_{Ca(T)}$, in contrast to the classical, slowly inactivating long-lasting, or L-type, $Ca^{2+}$ current, $I_{Ca(L)}$. This T-type $Ca^{2+}$ current is small to nonexistent in adult ventricular cells. It appears that both T- and L-type $Ca^{2+}$ currents may contribute to the latter part of the pacemaker potential in nodal cells (Doerr *et al.*, 1989). Because the T-type channels are active at more negative potentials than the L-type channels, the presence of T-channels effectively lowers the threshold for $I_{Ca}$. That is, opening of T-type channels can cause $Ca^{2+}$ influx, which then depolarizes the cells further toward the threshold for the L-type channels.

## C. Modulation of Automaticity

$I_f$ and $I_{Ca(L)}$ are both enhanced by the sympathetic neurotransmitter, norepinephrine (NE), and inhibited by the parasympathetic neurotransmitter, acetylcholine (ACh). Thus, NE increases the slope of the phase 4 depolarization, the threshold is reached sooner, and heart rate increases. In contrast, ACh decreases the slope of the phase 4 depolarization, the threshold is reached more slowly, and heart rate decreases. In addition to the effect on $I_f$ and $I_{Ca(L)}$, ACh also activates another specific $K^+$ current, $I_{K(ACh)}$, which hyperpolarizes the cell, that is, drives the maximum diastolic potential further from threshold. Thus, when $I_{K(ACh)}$ is activated, it takes a longer time to reach threshold and, also, the rate of phase 4 depolarization (slope) is decreased; thus, the heart rate is decreased. The pacemaker cells of the SA node are richly innervated by sympathetic and parasympathetic nerves. These actions of NE and ACh on $I_f$, $I_{Ca(L)}$, and $I_{K(ACh)}$ in SA node cells are the basis of the stimulating and inhibiting effect on heart rate of sympathetic or parasympathetic nerve stimulation.

## VII. Summary

APs in the heart are generated by the complex time-dependent interaction of several currents carried primarily by $Na^+$, $K^+$, and $Ca^{2+}$ ions. Some cardiac cells display automaticity, due to the presence of a cyclical spontaneous depolarization called the pacemaker potential. The cells of the SA node normally exhibit the fastest spontaneous depolarization of automatic cells in the heart and are, therefore, the normal pacemaker cells.

There are two primary types of cardiac cells. One cell type is found in the working cells of the atria, ventricles, and the specialized conduction cells of the His–Purkinje network. These cells have a high resting $K^+$ permeability between APs. The APs in these cells are generated by a fast $Na^+$ current. The prolonged duration of the AP in ventricular cells and Purkinje fibers is caused by an approximate balance of an inward $Ca^{2+}$ current and outward $K^+$ currents, which results in a prominent plateau phase. Atrial cells have a less prominent plateau.

The second type of cardiac cell is found in the SA and AV nodes. These cells have a slow $K^+$ permeability be-

tween APs and are automatic (i.e., spontaneously active). The upstroke of these APs is generated by a $Ca^{2+}$ current, which is smaller and slower than the fast $Na^+$ current. Because of this, and the relatively low density of $Ca^{2+}$ channels, conduction in nodal cells is much slower than conduction in regions having $Na^+$-dependent APs. Automaticity in nodal cells is caused by the interaction of several currents, including a hyperpolarization-activated cation current (the "funny current"), a small background $Na^+$ current, the delayed rectifier $K^+$ current, and $Ca^{2+}$ currents.

The sympathetic and parasympathetic nerves play important (and opposite) roles in regulating the force of myocardial contraction and heart rate. Thus, the sympathetic nerves stimulate the $Ca^{2+}$ current, which enhances the force of contraction of the heart. The parasympathetic nerves inhibit the $Ca^{2+}$ current, which decreases the force of contraction of the heart. The sympathetic nerves enhance the rate of spontaneous phase 4 depolarization in the SA node, resulting in an increased heart rate. The parasympathetic nerves decrease the rate of phase 4 depolarization, and also hyperpolarize SA node cells, resulting in a decreased heart rate. Thus, actions of neurotransmitters on cardiac ionic currents are a major site for modulating cardiac function.

## Bibliography

Attwell, D., Cohen, I., Eisner, D., Ohba, M., and Ojeda, C. (1979). The steady-state TTX-sensitive ("window") sodium current in cardiac Purkinje fibers. *Pflugers Archiv.* **379**, 137–142.

Baumgarten, C. M., and Fozzard, H. A. (1992). Cardiac resting and pacemaker potentials. *In* "The Heart and Cardiovascular System: Scientific Foundations" (H. A. Fozzard, H. Haber, R. B. Jennings, A. M. Katz, and H. E. Morgan, Eds.), Vol. 1, pp. 963–1001. Raven Press, New York.

Bouron, A., Potreau, D., and Raymond, G. (1991). Possible involvement of a chloride conductance in the transient outward current of whole-cell voltage-clamped ferret ventricular myocytes. *Pflugers Archiv.* **419**, 534–536.

Cameron, J. S., Kimura, S., Jackons-Burns, D. A., Smith, D. B., and Bassett, A. L. (1988). ATP-sensitive $K^+$ channels are altered in hypertrophied ventricular myocytes. *Am. J. Physiol.* **255**, H1254–H1258.

Campbell, D. L., Rasmussen, R. L., and Strauss, H. C. (1992). Ionic current mechanisms generating vertebrate primary cardiac pacemaker activity at the single cell level: An integrative view. *Annu. Rev. Physiol.* **54**, 279–302.

Coraboeuf, E., and Nargeot, J. (1993). Electrophysiology of human cardiac cells. *Cardiovasc. Res.* **27**, 1713–1725.

DiFrancesco, D. (1990). Current $I_f$ and neuronal modulation of heart rate. *In* "Cardiac Electrophysiology: From Cell to Bedside" (D. Zipes and J. Jalife, Eds.), pp. 28–35. W. B. Saunders, Philadelphia.

Doerr, T., Denger, R., and Trautwein, W. (1989). Calcium currents in single SA nodal cells of the rabbit heart studied with the action potential clamp. *Pflugers Archiv.* **413**, 599–603.

Dokos, S., Celler, B., and Lovell, N. (1996). Ion currents underlying sinoatrial node pacemaker activity: A new single cell mathematical model. *J. Theor. Biol.* **181**, 245–272.

Gadsby, D. C. (1984). The $Na^+/K^+$ pump of cardiac cells. *Annu. Rev. Biophys. Bioeng.* **13**, 373–378.

Hagiwara, N., Irisawa, H., and Kameyama, M. (1988). Contribution of two types of calcium currents to the pacemaker potential of rabbit sino-atrial node cells. *J. Physiol.* **395**, 233–254.

Hagiwara, N., Irisawa, H., Kasanuki, H., and Hosoda, S. (1992). Background current in sino-atrial node cells of the rabbit heart. *J. Physiol.* **448,** 53–72.

Hamill, O. P., Marty, A., Neher, A., Sakmann, B., and Sigworth, F. J. (1981). Improved patch-clamp techniques for high-resolution current recording from cells and cell-free membrane patches. *Pflugers Archiv.* **391,** 85–100.

Harvey, R. D. (1996). Cardiac chloride currents. *NIPS* **11,** 175–181.

Harvey, R. D., Clark, C. D., and Hume, J. R. (1990). Chloride current in mammalian cardiac myocytes. Novel mechanism for autonomic regulation of action potential duration and resting membrane potential. *J. Gen. Physiol.* **95,** 1077–1102.

Janvier, N. C., and Boyett, M. R. (1996). The role of Na–Ca exchange current in the cardiac action potential. *Cardiovasc. Res.* **32,** 69–84.

Kenyon, J. L., and Gibbons, W. R. (1979). Influence of chloride, potassium, and tetraethylammonium on the early outward current of sheep cardiac Purkinje fibers. *J. Gen. Physiol.* **73,** 117–138.

Lindblad, D. S., Murphey, C. R., Clark, J. W., and Giles, W. R. (1996). A model of the action potential and underlying membrane currents in a rabbit atrial cell. *Am. J. Physiol.* **271,** H1666–H1696.

Liu, D. W., Gintant, G. A., and Antzelevitch, C. (1993). Ionic bases for electrophysiological distinctions among epicardial, midmyocardial, and endocardial myocytes from the free wall of the canine left ventricle. *Circ. Res.* **72,** 671–687.

Noma, A. (1983). ATP-regulated K$^+$ channels in cardiac muscle. *Nature* **305,** 147–148.

Powell, T., Terrar, D. A., and Twist, V. W. (1980). Electrical properties of individual cells isolated from adult rat ventricular myocardium. *J. Physiol.* **302,** 131–153.

Rasmussen, R. L., Clark, J. W., Giles, W. R., Shibata, E. F., and Campbell, D. L. (1990). A mathematical model of a bullfrog cardiac pacemaker cell. *Am. J. Physiol.* **259,** H352–H369.

Ravens, U., Wettwer, E., Ohler, A., Amos, G. J., and Mewes, T. (1996). Electrophysiology of ion channels of the heart. *Fundam. Clin. Pharmacol.* **10,** 321–328.

Smith, J. M., and Wahler, G. M. (1996). ATP-sensitive potassium channels are altered in ventricular myocytes from diabetic rats. *Mol. Cell. Biochem.* **158,** 43–51.

Tohse, N. (1990). Calcium-sensitive delayed rectifier potassium current in guinea pig ventricular cells. *Am. J. Physiol.* **258,** H1200–H1207.

Tseng, G. N., and Hoffman, B. F. (1989). Two components of transient outward current in canine ventricular myocytes. *Circ. Res.* **64,** 633–647.

Wahler, G. M., and Dollinger, S. J. (1995). Nitric oxide donor SIN-1 inhibits mammalian cardiac calcium current through cGMP-dependent protein kinase. *Am. J. Physiol.* **268,** C45–C54.

Walsh, K. B., and Kass, R. S. (1988). Regulation of a heart potassium channel by protein kinase A and C. *Science* **242,** 67–69.

Yu, H., Chang, F., and Cohen, I. S. (1993). Pacemaker current exists in ventricular myocytes. *Circ. Res.* **72,** 232–236.

Zygmunt, A. C., and Gibbons, W. R. (1991). Calcium-activated chloride current in rabbit ventricular myocytes. *Circ. Res.* **68,** 424–437.

R. Kent Hermsmeyer

# 52

# Smooth Muscle Action Potentials

## I. Introduction

**Action potentials** (APs) were first defined in nerve fibers as propagating electrical impulses with uniform conduction velocity and amplitude. APs in smooth muscle tissues are different from those in most long, thin cells such as nerve or skeletal muscle. Perhaps it is more appropriate to refer to these transient potentials in smooth muscles as **spikes** of electricity, as a generic, more descriptive term. APs are much more uniform in amplitude (and overshoot), whereas smooth muscle AP spikes can be variable in amplitude, shape, pattern, and conduction. In skeletal muscle fibers, which are very long with an almost uniform geometry throughout most of their length, it is accurate to portray the **propagation** (conduction) of electrical impulses, or APs, as having constant velocity and uniform amplitude (Keynes, 1983). The feature of APs in nerve and skeletal muscle fibers that differentiates them from those in smooth muscle cells is the constancy of amplitude and conduction velocity.

An important difference between smooth muscle and other muscle types is the very small size of the cells and the dependence on **multicellular interaction.** Smooth muscle cells consist of an assembly of short (e.g., 100–300 $\mu$m in length) and very narrow (e.g., 3–10 $\mu$m in diameter) cells with differing biophysical properties (Table 1). Figure 1 shows an example of a living single vascular muscle cell with Nomarski optical contrast enhancement to illustrate that this is among the smallest cells found in mammals. These cells interact by multiple coordination mechanisms to produce relatively long or tonic contractions associated with the function of the blood vessels, hollow organs, and viscera in which they appear. At least some of the preparations may function through propagation of spikes by electrical coupling, while others may use chemical or mechanical mechanisms. Thus, the best generalization is that, through a variety of electrical and nonelectrical coupling mechanisms, smooth muscle cells produce **coordinated contractions or relaxations** to carry out the functions of hollow organs, such as blood vessels, urinary bladder, intestine, or uterus.

Given the function of blood vessels and internal organs for which smooth muscle cells are the contractile cells, a different classification of muscle types is appropriate. The principal classification system used for smooth muscles in the last 50 years has been the **unitary** and **multiunit** categories of Bozler (1938, 1948). This classification, referred to in many textbooks, is based on whether the sphere (or cylinder) that contains a wall of smooth muscle acts as a single unit, or whether the smooth muscle cells are separately controlled through innervation, thus forming multiple functional units in the organ. With the recently available information on physiology and pharmacology of smooth muscle cells, it is now possible to classify an array of different smooth muscle types.

Let us consider first the common features found in all smooth muscle cells. By definition, these cells lack **striations** (aligned dark and light bands at the light microscope level), and this is the principal unifying principle. Although the cells are generally small, and relatively slow in their contraction/relaxation cycle, exceptions can be found, such as the large and quick muscle cell of the ctenophore, *Beroë* (Hernandez-Nicaise *et al.*, 1980), and fast contractions and relaxations of single vascular muscle cells in mammals (Hermsmeyer, 1979). While most attempts to find other generalizable features of smooth muscle have failed, a second common feature can be recognized. Smooth muscle cells have multiple receptors, activation mechanisms, and inhibitors. They can be stimulated or inhibited by transmitters released from nerve endings, by hormones circulating or locally released, by neighboring cells or even by the muscle cells themselves, and by drugs. The **multiplicity of receptors** on a given smooth muscle cell is noteworthy because the resulting array of possibilities for integration of separate excitatory and inhibitory inputs offers diversity in end-organ responses. The additional complexity added by multi-input modulation of the activity of individual cells provides a diversity that virtually defies monolithic classification. Thus, the second generalization about smooth mus-

**TABLE 1**  Electrical Properties of Some Smooth Muscles ($\Theta = 0–15$ cm/sec[a])

| Muscle | Resting $E_m$ (mV) | Spike amplitude (mV) | $dV/dt$ max (V/sec) | Cell length ($\mu$m) | $S$ ($\mu$m) | $C_m$ $\mu$F |
|---|---|---|---|---|---|---|
| Vascular muscle | −55 | 40 | 5 | 100 | 150 | 1.9 |
| Taenia coli (GP) | −50 | 55 | 7 | 150 | 1500 | 1.7 |
| Intestinal (longitudinal) | −60 | 65 | 5 | 150 | 2000 | 1.7 |
| Uterine (pregnant) | −70 | 75 | 5 | 200 | 1500 | 1.5 |
| Vas deferens | −60 | 65 | 5 | 150 | 2100 | 1.5 |
| Beroë | −60 | 70 | 25 | 6000 | 1700 | 1.7 |

[a] Data are from Prosser *et al.* (1960), Tomita (1970), and Hernandez-Nicaise *et al.* (1980).

cle cells, after their lack of striations, is the importance of integration of multiple inputs and consequently diverse mechanisms for modulation (Table 2). Because each cell can have an array of **receptors** (high-affinity drug-binding sites) and modulations, and there are many permutations and combinations, attempts to categorize smooth muscle cells require a multicriterion system principally useful to specialists. For practical purposes of considering one property, the concept from relational databases of sorting by one or a few criteria will be most productive. Differences that will predictably appear in other parameters must be overlooked for purposes of such pragmatic classifications.

A major concept of this chapter is that the electrophysiology of smooth muscle cells is highly variable. Within an organ, and at times within even a small part of a single organ, there will be a variety of cells contracting, relaxing,

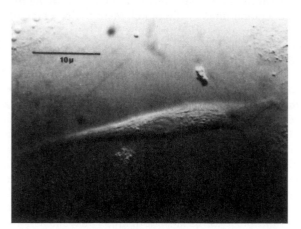

**FIG. 1.**  A single smooth muscle cell has a ribbon shape when it is isolated from other cells and attaches to a glass coverslip (part of a microscope chamber for single-cell experiments). These cells are typically 5–10 $\mu$m in diameter at their widest part and approximately 1–200 $\mu$m in length. As such, they have one of the smallest diameters of cells that exist in mammals, and have one of the highest ratios of surface area to cell volume. This very thin dimension means that the membrane is exceedingly important for smooth muscle function as it also is known to be for many nerve cells. This example is from the azygos vein (a large collecting vein near the vena cava) of a rat, and it maintains its contractile characteristic for days or weeks under primary cell culture conditions. This photomicrograph is a differential interference contrast (Nomarski) image at 1200× magnification. (Reproduced with permission from Hermsmeyer, 1979.)

and releasing signals to carry out the functions, including reduction or enlargement of the dimensions of that organ. Within a single blood vessel or visceral organ, **spontaneously active** cells exist that show pacemaker properties capable of generating spontaneous electrical spikes, which will be conducted for distances that range from a few cells to several thousand cells. In contrast, other cells will show no spontaneous electrical activity, and may even remain **quiescent** during attempts to electrically stimulate these cells, and yet can contribute to overall contractions of the organ through multiplicity of stimulatory inputs. Evidence from cellular studies is showing that chemical modulation of smooth muscle function dominates transitions from contraction to relaxation to rest, and that multiple endogenous mediators contribute.

Also within a given organ, there appear to be mechanisms for sustaining contractions for long periods of time with relatively low energy cost, a concept referred to as **smooth muscle tone.** The electrophysiological specialization for such a long-term maintenance of a **state of activation** (reduced diameter) correlates closely in some cells

**TABLE 2**  Receptor Types on Smooth Muscles[a]

| Type | Transmitters | Receptors |
|---|---|---|
| Adrenergic | Norepi, epi | $\alpha$, $\beta$ |
| Cholinergic | Acetylcholine | ACh |
| Serotonergic | Serotonin (5-HT) | 5-HT |
| Nucleotide | ATP, adenosine | ATP |
| Angiotensin | Angio | AT |
| Vasopressin | Vasopressin (ADH) | VP |
| Endothelin | Endothelin | ET |
| Prostanoid | Prostaglandins E, F, H, I | PG |
| Leukotrienes | Leukotriene series | L |
| Insulin | Insulin | I |
| Interleukin | IL series | IL |
| Histamine | Histamine | H |
| Thromboxane | Thromboxane | Tx-A$_2$ |

[a] This is a partial list of receptors that can be found on a single smooth muscle cell.

with membrane potential, and in others most strongly with chemical mediators. Electrophysiological correlations with the rapid phasic contractions of vascular smooth muscle cells have demonstrated tight temporal correlation, which implies cause and effect. These phasic contractions have characteristics nearly as quick as those of striated muscles. Because cells with fast contractions are easier to identify, they are more often studied, and have been dominant in the smooth muscle literature. Certainly, it is very important to our understanding of the function of a smooth muscle organ to characterize both (1) the electrophysiology and (2) the chemical modulation. The activation state will be highly variable because the state of the smooth muscle is regulated by innervation and by substances released from other cells, including the endothelium and the smooth muscle cells themselves.

Smooth muscle cells are not easily categorized by innervation, which is a useful criterion in skeletal muscle. Although the muscle cells are coordinated by motor nerves of the **autonomic** (involuntary) nervous system, the distribution is more typically diffuse, with release of **neurotransmitters** such as norepinephrine, acetylcholine, serotonin, angiotensin, adenosine triphosphate (ATP), substance P, calcitonin gene-related peptide (CGRP), adenosine, and still other important regulatory substances, often in a mixture and with diffusion over hundreds or even thousands of cells. The function of the nerves in smooth muscles is more often to modulate than to initiate contractions. The muscle cells are slower to contract, but maintain long-duration contractions, the amplitude of which is continuously variable, as illustrated by the depolarization of an arterial muscle by norepinephrine (Fig. 2). Smooth muscle

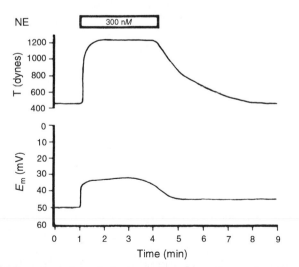

**FIG. 2.** Graded depolarization and corresponding tension allow a steady state to be maintained during 300 nM stimulation by norepinephrine in rat caudal artery. These drawings are traced from original records and illustrate the parallelism between tension and membrane potential ($E_m$) that is typical of the arteries that have been studied. Both $E_m$ and tension levels can remain relatively constant for many minutes during stimulation by vasoconstrictors. This graded depolarization and tension probably explain a major part of how tone is developed in blood vessels. (Reproduced with permission from Hermsmeyer, 1982.)

cells, many of which have intrinsic activity and are controlled through response to multiple stimuli that are only partially understood, can be modulated by stretch, changes in oxygen tension, pH osmolality, locally sensed chemical substances, ionic imbalances, and substances that are found in abundance in that organ. Nerve fibers are generally confined to the border between the outer layers in many smooth muscles, and this is particularly true in the case of blood vessels. Even in heavily innervated arteries, there is a well-defined border between the outer adventitia and the media (muscle) layer, which is the point beyond which nerve fibers do not extend. The comparatively diffuse innervation of a small artery is shown in Fig. 3.

These neurotransmitters and some vasoactive hormones released by the cells themselves **(autocrine),** or by cells nearby **(paracrine),** affect the electrical properties of vascular muscle cells. Changes in the level of **resting membrane potential** are apparently the predominant influence on contraction in most smooth muscle cells, as exemplified well by blood vessels. **Depolarization** (change in membrane potential toward 0) causes increasing states of contraction because the membrane excitation process that is characterized by an influx of depolarizing ions makes the inside of the cell less negative (its normal resting state), and results in $Ca^{2+}$ ion increases, both from influx and from release that produces contraction. On the other hand, **hyperpolarization** (change to more negative potential inside the cell) results in a decrease of intracellular free $Ca^{2+}$ and relaxation. Regulation of contraction and of function in many kinds of cells by voltage-sensitive ion channels explains why changes in membrane potential and therefore ion conductances (e.g., $Ca^{2+}$ conductance, the electrical manifestation of $Ca^{2+}$ channels) are of great importance. The action of norepinephrine on vascular muscle of the rabbit ear artery is an example showing both hyperpolarization (at 10 nM) and depolarization (at 100 nM to 10 μM) and close correlation with contraction (Fig. 4), although the same preparation had been errantly reported to contract without depolarization (Droogmans et al., 1977).

However, it is important to recognize that there are multiple mechanisms for activation of smooth muscle cells via a rich array of membrane processes. Membrane depolarization provides a most effective signal mechanism that allows for electrical **integration of activity** throughout the cell and, therefore, efficient function. Many smooth muscle cells, even those that do not show spontaneous electrical spikes, show depolarization, which appears to depend primarily on $Ca^{2+}$ channels acting to cause intracellular $Ca^{2+}$ release. However, this very revelation of $Ca^{2+}$ release as the result of $Ca^{2+}$ entry suggests an important additional mechanism that is prominent in smooth muscle cells.

An additional trigger mechanism is the direct release of intracellular $Ca^{2+}$, which can occur without membrane depolarization. The pathways identified that might allow such a direct signaling of intracellular $Ca^{2+}$ release, without opening of surface membrane $Ca^{2+}$ channels, are membrane enzymes that begin a cascade which leads to release of $Ca^{2+}$ from sarcoplasmic reticulum and other $Ca^{2+}$ storage sites. The example chosen shows that norepinephrine can also act directly on $Ca^{2+}$ release, in this case revealing a

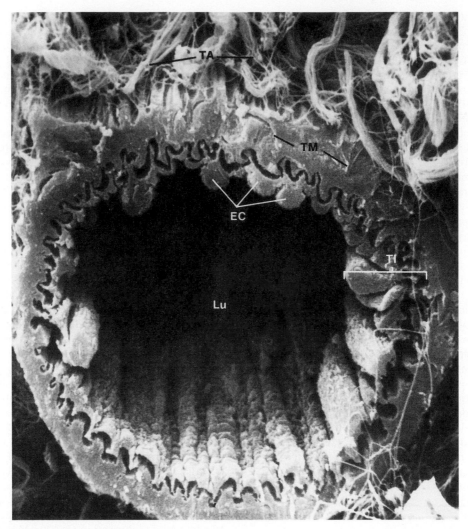

**FIG. 3.**   The scanning electron micrograph of a small mesenteric artery shows that the outer layer (tunica adventitia, TA) is composed of connective tissue fibers that hide nerve fibers, which are known to be in the adventitia as revealed by staining characteristics, and end at the outer edge of the vascular smooth muscle layer (the tunica media, TM). Endothelial cells (EC) and elastic fibers make up the inner layer (tunica intima, TI). Lu is the lumen of this artery, imaged here at 3000× magnification. (Reproduced with permission from Kessel and Kardon, 1979.)

potential independent excitation mechanism. In most cases, norepinephrine causes depolarization with entry of $Ca^{2+}$ and $Na^+$, which are the result of the permeability increase caused by norepinephrine. However, in addition, there is an enzyme, **phospholipase C,** in the plasma membrane that appears to be directly linked to norepinephrine. Activation of phospholipase C causes the conversion of the principal substituent of the membrane, phosphatidylinositol into metabolites, *inositol 1,4,5-trisphosphate* ($IP_3$), which causes localized (but highly effective) release of intracellular $Ca^{2+}$ (Berrige and Irvine, 1984). Diacylglycerol (DAG) is also formed in this phospholipase-C-induced metabolism of phosphatidyl inositol, and DAG causes the activation of **protein kinase C** (PKC). This PKC signal is a profound inducer of contraction through multiple mechanisms. Activation of PKC causes intracellular $Ca^{2+}$ release, phosphorylation of the contractile proteins and thus increased contraction for a given level of $Ca^{2+}$, and markedly

prolonged contraction. A diagram of membrane modulatory mechanisms for vascular muscle is shown in Fig. 5.

This combination of activation responses triggered by PKC results in strong, long-lasting contractions that correlate poorly, or at times not at all, with the level of depolarization, especially at later times during the contraction. $IP_3$ and PKC are examples of mechanisms that can be characterized as **pharmacomechanical coupling,** that is, changes in contraction caused by substances or drugs directly and without requisite participation of membrane potential (Somlyo and Somlyo, 1990). Another variation of the pharmacomechanical coupling hypothesis suggests that norepinephrine activates ion channels that allow $Ca^{2+}$ influx (independent of the depolarization). One postulate is that there are channels that allow only $Ca^{2+}$ to pass through which are opened by norepinephrine (Bolton, 1979). Although there have not been demonstrations of such norepinephrine-operated $Ca^{2+}$ channels (*receptor-*

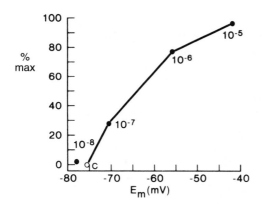

**FIG. 4.** Tension as a function of membrane potential for rabbit ear artery shows the remarkable correspondence between $E_m$ and tension. Depolarization was caused by exposure to increasing concentrations of norepinephrine. The correlation of voltage and tension is very strong over the 35 mV range of $E_m$ between $-75$ and $-40$ mV in this artery. Other arteries show a different range of membrane potentials where this correlation holds, but the principle is the general rule. (Reproduced with permission from Trappani *et al.*, 1981.)

*operated channels*), there are examples of **ligand-gated channels** in smooth muscles as well as in neurons and striated muscles. The notable examples of these are the **ATP-gated channels** of vascular muscle, the classical case of the nicotinic receptor at the neuromuscular junction in striated muscle, and the $\gamma$-aminobutyric acid (GABA)-gated channels found in neurons (Fig. 6). All of these ligand-gated channels are nonselective, opening to $K^+$ and $Cl^-$ as well as $Ca^{2+}$. These are properly considered to be ligand-gated channels (none of which is selective for $Ca^{2+}$). More details about ligand-gated ion channels can be found in Chapter 39. In any event, it is important to understand that both **electromechanical** and pharmacomechanical coupling are important in regulating the degree of contraction in smooth muscle cells.

To summarize, norepinephrine produces contraction via multiple mechanisms. One is the voltage-dependent $Ca^{2+}$ channel, which is activated by depolarization. The second is pharmacomechanical coupling, which takes the form of intracellular $Ca^{2+}$ release by multiple membrane signaling mechanisms. The pharmacomechanical coupling category includes both $Ca^{2+}$-dependent and $Ca^{2+}$-independent components, because there is $Ca^{2+}$ release and modulation via phosphorylation that may or may not involve ion channels. Thus it is possible to block norepinephrine-induced contractions by decreasing the activation of PKC (Hermsmeyer and Miyagawa, 1996). To understand how antagonists of smooth muscle contraction operate, it is important to keep both of these mechanisms in mind and to focus on signaling intracellular $Ca^{2+}$ entry and intracellular $Ca^{2+}$ release.

## II. Ionic Currents

The ion that carries the inward (depolarizing) current that causes the electrical spike found in smooth muscles

is usually calcium ($Ca^{2+}$), although there are also examples of fast sodium ($Na^+$) current contributions to the rising phase of the AP. In comparison with $Na^+$ channels, $Ca^{2+}$ channels have a slower rate of **activation** (onset of current) and a lower current density (important for conduction) by at least an order of magnitude (Fig. 7). Even more importantly, the **inactivation** (turn-off of current) of $Ca^{2+}$ channels occurs over time courses that are thousands of times as long as those found with $Na^+$ channels. An example from an arterial muscle cell is shown in Fig. 8. Thus, although activation amplitude is smaller in smooth muscles, the $Ca^{2+}$ signal accumulated over seconds or longer can be larger. Relatively slower inactivation of the $Ca^{2+}$ channels is important in this accumulation process, and can be varied to allow $Ca^{2+}$ (and contractile force) to reach higher levels, and be sustained hundreds of times longer than $Na^+$ channels. This prolonged process of inactivation allows for a diversity of function, and for custom-tailoring of the excitation event by hormones and regulatory substances, acting through variables of time and space. Additional degrees of freedom are added by other ion channels that affect membrane potential, as can be modeled by an electrical equivalent circuit for a vascular muscle cell membrane (Fig. 9).

Smooth muscle spikes show a high degree of diversity, which translates into varied excitation. This diversity is a contrast to the rapid (but inflexible) signals that result from fast $Na^+$ channels in nerve and skeletal muscle. A useful way to think of the difference might be that stimulation of skeletal muscle for activity requires the fastest (largest) electrical wiring system to allow for quickest response time, for example, major groups of skeletal muscle for whole-animal movement. In contrast, the variability in the electrical signal in smooth muscles based on $Ca^{2+}$ channels provides a range of gradation of activity for internal organs that is like a more intricate network of finer wiring and multiple inputs to a terminal. With summation of positive and negative signals and wiring diameters, there is gradation rather than speed and constancy. The electrical spikes of smooth muscle are an ideal solution to the question of electrical signaling in organs with varied purposes, which must be able to switch from one to the other as a matter of the daily routine, for example, intestinal motility.

Spikes in smooth muscles are based on $Ca^{2+}$ channels that are in some ways similar to $Ca^{2+}$ channels in other cells, and yet are dissimilar in important characteristics such as drug sensitivity. Smooth muscle cells are preferentially affected by several classes of **$Ca^{2+}$ antagonists** that are increasingly in common medical use as safe, highly effective smooth muscle relaxants (also see Chapter 37 for more details). The high sensitivity of smooth muscles to $Ca^{2+}$ antagonists is useful in a number of conditions, for example, hypertension, coronary artery disease, vasospasm, asthma, and urinary bladder dysfunction. Two types of $Ca^{2+}$ channels are recognized in smooth muscles, only one of which is sensitive to $Ca^{2+}$ antagonists. The T-type and L-type designations have been developed to recognize the voltage, kinetic, permeability, and drug sensitivity differences (Sturek and Hermsmeyer, 1986). The class of $Ca^{2+}$ channels called the T-type (for transient type) are relatively

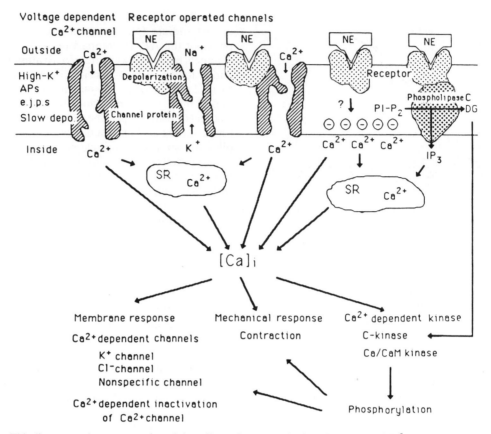

**FIG. 5.** This diagrammatic representation of the cell membrane emphasizes the sources of $Ca^{2+}$ for contraction of vascular muscle. Several major sources of $Ca^{2+}$ are shown, including (1) voltage-dependent $Ca^{2+}$ channels; (2) a ligand-gated (receptor-operated) channel that is nonspecific, having approximately equal permeabilities for $Na^+$, $K^+$, and $Ca^{2+}$; (3) a $Ca^{2+}$ channel responding directly to norepinephrine (presented here for logical completeness); (4) $Ca^{2+}$ release from the surface membrane; and (5) $Ca^{2+}$ release from intracellular sources, presumably almost entirely from the sarcoplasmic reticulum with an important signal messenger, $IP_3$, as a mediator. The key enzyme leading to production of $IP_3$ is phospholipase C (PLC), the activity of which is regulated by a GTP-binding protein. Another important intracellular messenger is PKC, which is activated by diacylglycerol (DG), also a product of the PLC pathway. When norepinephrine excites a vascular muscle cell, there is stimulation of the $\alpha$-adrenergic receptor leading to depolarization and thus the activation of the signal process, resulting in $Ca^{2+}$ release and contraction. (Reproduced with permission from Sperelakis and Ohya, 1989.)

insensitive to $Ca^{2+}$ antagonists. It is the L-type (for longer duration, or sustained) that are highly sensitive to $Ca^{2+}$ antagonists. Separation of T- and L-type currents is shown in Fig. 10. Sources for the $Ca^{2+}$ that triggers contraction consist of $Ca^{2+}$ flux through $Ca^{2+}$ channels in the **sarcolemma** (the cell surface membrane) and $Ca^{2+}$ release from intracellular stores. Intracellular $Ca^{2+}$ release is primarily from membranous structures thought to function analogously to the **sarcoplasmic reticulum** (SR), which is the prominent $Ca^{2+}$ release organelle in skeletal muscle. One portion of the SR that regulates intracellular $Ca^{2+}$ by release in response to membrane signals is located immediately (within 0.1 $\mu$m) inside the surface membrane, and functions as the superficial buffer barrier (Daniel *et al.*, 1995). Subsarcolemmal SR is important in $Ca^{2+}$ homeostasis in vascular muscle cells because $Ca^{2+}$ entering through sarcolemmal $Ca^{2+}$ channel must pass through this region of SR, giving rise to the buffer barrier concept. Intracellular stores are modulated by the $Ca^{2+}$ content of the superficial buffer barrier (Chen and van Breemen, 1993). Intracellular

$Ca^{2+}$ levels rise from a resting level, currently believed to be from 100 to 300 $\mu M$, to (locally) greater than 100 times the resting level when $Ca^{2+}$ is released from the larger central component of SR. Vasoactive substances bring about contraction by increases in $[Ca]_i$ through a combination of sarcolemmal $Ca^{2+}$ channel and SR $Ca^{2+}$ release mechanisms.

Once membrane excitation has begun, a complex set of membrane-related processes that amplify the signal commences. $Ca^{2+}$ entry as influx or release from intracellular stores (or both) triggers further release of intracellular $Ca^{2+}$ through production of inositol trisphosphates from phosphatidylinositol during the activation of PLC in the cell membrane. This process is a biochemical signal cascade, capable of causing activation of multiple intracellular processes through amplification by intracellular messengers. The key event appears to be $Ca^{2+}$ entry through ion channels, with perhaps even a single $Ca^{2+}$ channel opening (Fig. 11) beginning the messenger cascade leading to contraction. As noted earlier, some vasoactive agents are be-

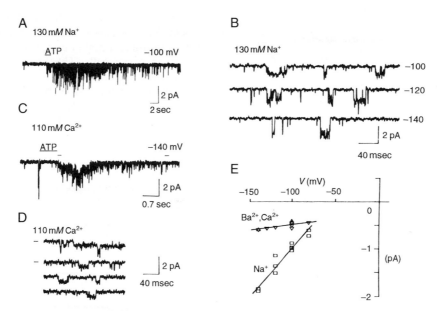

**FIG. 6.** Individual currents activated by ATP in outside-out membrane patches torn from a cell are shown here. This type of patch-clamp recording can isolate a single ion channel, to allow individual events of this single molecule in a cell membrane to be measured. (A) Currents carried by Na⁺ and (B) The single-channel events are spread out for individual inspection. Potentials at the right of each trace indicate the holding potentials and the period of application in part A is shown by the horizontal line under ATP. (C,D) The same kind of membrane patch is shown, but with $Ca^{2+}$ as the current carrier. From these records, the amplitude of current for a given potential can be determined and plotted (E). From such plots, the conductance (current/voltage) can be determined as 25 pS for the divalent ion current and 200 pS for the Na⁺ currents. (Reproduced with permission from *Nature*, **328**, 275–278. Benham and Tsien, 1987. Copyright 1987 Macmillan Magazines Limited.)

lieved to produce contraction through a direct action on the intracellular signaling processes, whether or not membrane potential is allowed to change. In cases where membrane potential is held perfectly constant, for example, during a voltage-clamp experiment, such changes in contraction caused by drugs can be unequivocally identified as pharmacomechanical coupling. Such direct activation of contractile proteins occurs in some preparations when stimulation produces contraction without changes in membrane potential. Interference with this process can be at several points, but $Ca^{2+}$ entry and release are particularly notable. $Ca^{2+}$ antagonists interfere with the $Ca^{2+}$ signal process and are most effective smooth muscle relaxants (for more details, see Chapter 37).

## III. Spontaneous Activity

In cells that are known for their spontaneous spiking activity, there is a high incidence of T-type $Ca^{2+}$ channels. Although spontaneously active smooth muscle cells have L-type channels that generate electrical spikes which are inhibited by $Ca^{2+}$ antagonists at low concentration, the hallmarks of pacemaker smooth muscle cells are prominent T-type channels that seem to be necessary for robust pacemaker function. Examples of cells showing prominent $Ca^{2+}$ currents include (1) the spontaneously active azygos vein of newborn animals; (2) the hepatic portal vein (in all animals), which is the most widely studied of the spontaneously active vascular muscle cells; and (3) the pacemaker

cells of the progesterone-dominated uterus. All of these smooth muscle cells are a bit less sensitive to $Ca^{2+}$ antagonists than quiescent counterparts, but are capable of indefatigable pacemaker activity.

It is difficult to inhibit the T-type $Ca^{2+}$ channels in smooth muscles. The only reported examples are certain inorganic ions, for example, cadmium, manganese, cobalt, or lanthanum. In each of these cases, the inhibiting cation appears to block the $Ca^{2+}$ channel, or an important regulatory site necessary for operation of the $Ca^{2+}$ channel. Thus, pacemaker activity can be effectively suppressed by these inorganic cations. However, the fact that inhibition can occur allows experimental manipulation and suggests that there may be a possibility for development of drugs acting specifically on T-type $Ca^{2+}$ channels that would effectively inhibit spontaneous activity in smooth muscles. This decrease in pacemaker function might be used to control unwanted motility, for example, in premature labor of the uterus.

Other mechanisms for smooth muscle spikes include Na⁺ channels, which can be found in certain smooth muscle cells, especially those that show prominent pacemaker activity. As stated previously, the cells of the azygos vein of newborns, the hepatic portal vein, and the progesterone-dominated uterus provide good examples of spontancity and thus correspondingly exemplify Na⁺ channel function in smooth muscles. Block of the Na⁺ channels is possible with local anesthetics (like procaine) or by the specific Na⁺ channel toxins, tetrodotoxin (TTX) and saxitoxin (STX). Further examples can be found in the chapter on ion chan-

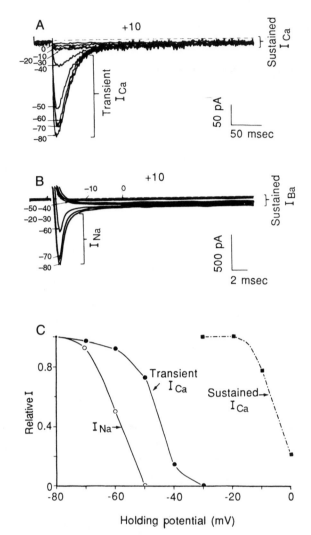

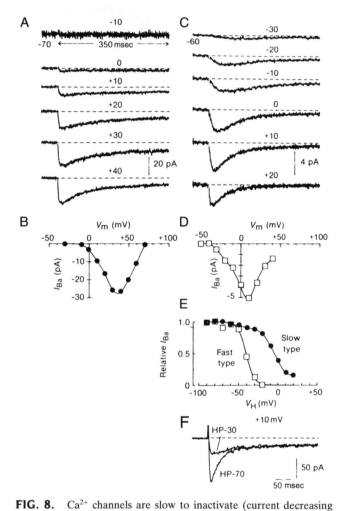

nels as targets for toxins (Chapter 36). Inhibition of smooth muscle $Na^+$ channels usually (except in the pregnant uterus) requires a higher concentration than would be effective on nerve cells. The function of $Na^+$ channels in smooth muscles that have a $Ca^{2+}$ channel pacemaker which functions independently of $Na^+$ is enigmatic. One useful way to think of a function for $Na^+$ channels in smooth muscle is the role of an adjunct to the pacemaker mecha-

**FIG. 7.**   Comparison of (A) $Ca^{2+}$ and (B) $Na^+$ currents in a single vascular muscle cell show the tremendous differences in amplitude and duration for these clearly separable ion channels. $Ca^{2+}$ currents in part A show both transient and sustained (long-lasting or L) components, as also shown (C) in the plot of amplitude as a function of holding potential. $Ca^{2+}$ currents can be activated over a relatively wide range of potentials, including values between normal resting potential, all the way up to positive potentials. In contrast, $Na^+$ channels, which may play a relatively minor or even vestigal role in these cells, are nearly completely inactivated at a normal resting membrane potential of $-50$ mV. However, it is easy to construct a mechanism for automaticity (spontaneous pacemaker-type potentials) in vascular muscle if even a small $Na^+$ current exists near resting membrane potential. Furthermore, hyperpolarization by drugs used as vasodilators often involves hyperpolarization that would allow availability of the $Na^+$ channels, and the possibility for synchronized contraction. These families of currents shown in parts A and B and the inactivation analysis shown in part C suggest how multiple excitatory mechanisms in a single vascular muscle cell can cause a variety of activation responses, and that both $Ca^{2+}$ and $Na^+$ may be important for blood vessel regulation. [Reprinted with permission from Sturek, M., and Hermsmeyer, K. (1986). Calcium and sodium channels in spontaneously contracting vascular muscle cells. *Science* **233**, 475–478. Copyright 1986 American Association for the Advancement of Science.]

**FIG. 8.**   $Ca^{2+}$ channels are slow to inactivate (current decreasing back up to near the zero line) and thus smooth muscle $Ca^{2+}$ currents have been characterized as relatively long-lasting (L-type) and can also be called sustained or slow. The important aspect of these prolonged $Ca^{2+}$ currents (A,B) of this record from mesenteric artery is that continued $Ca^{2+}$ influx occurs and, therefore, there is maintained contraction. The fast inactivating transient $Ca^{2+}$ current (T-type), which is also known as the fast-type current, inactivates within 100 msec (C,D) and is more like $Na^+$ currents that are thought to be important for pacemaker or trigger events. (E) Shows the greater range of potentials, over which the slower inactivating L-type are active. The cells that show both the T- and the L-type currents appear to be a relatively small fraction, but these cells are capable of giving both the transient and the sustained components of $Ca^{2+}$ current (F) at holding potentials of $-30$ and $-70$ mV. All of these records show inward $Ca^{2+}$ channel currents carried by $Ba^{2+}$ to illustrate how $Ca^{2+}$ entry can occur in vascular muscle cells. (Reproduced with permission from Bean *et al.*, 1986, copyright 1986, American Heart Association.)

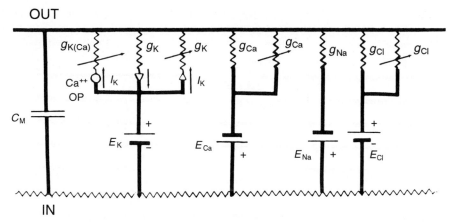

**FIG. 9.**   This is an electrical equivalent circuit for the cell membrane of a vascular muscle cell. Conductance channels are shown as having both resting and excited states by the arrows through the resistances (to indicate variable resistance over voltage and time). The conductances operate in parallel and are driven by batteries that result from the electrochemical potentials (Nernst potentials) that are generated by the ion distributions across the cell membrane. The cell membrane capacitance ($C_m$) is also shown, to complete the electrical analogy. This complex set of conductances and potential is continuously changing under modulation by voltage, time, and a number of signal substances released from nerves, nearby cells, and even the cells themselves. This complex situation allows for a wide variety of responsiveness in smooth muscle cells. (Reproduced with permission from Sperelakis and Ohya, 1989.)

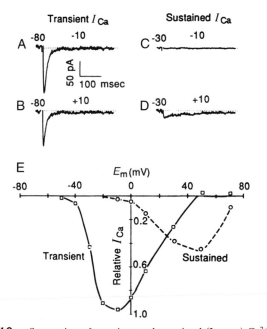

**FIG. 10.**   Separation of transient and sustained (L-type) $Ca^{2+}$ currents are shown for rat vascular muscle cell. Currents recorded from holding potentials of −80 mV (A,B) or from −30 mV (C,D) are plotted as peak currents in (E). The more positive range of potentials over which the L-type currents are active is evident in part E. This activation analysis shows that voltage can be used to separate the T and L-type $Ca^{2+}$ currents and emphasizes the individuality of these ion channels. [Reprinted with permission from Sturek, M., and Hermsmeyer, K. (1986). Calcium and sodium channels in spontaneously contracting vascular muscle cells. *Science* **233**, 475–478. Copyright 1986 American Association for the Advancement of Science.]

nism. It appears likely that in those cells having pacemaker activity, there are multiple pacemaker mechanisms. If the $Ca^{2+}$ pacemaker should fail to initiate spontaneous spikes, the $Na^+$ pacemaker is present and available for initiating spikes. Thus, it would provide a safety function for maintenance of pacemaker function.

## IV. K⁺ Channel Functions

The primary determinant of resting potential in smooth muscle, as in other excitable cells, is $K^+$ conductance. Actually, $K^+$ conductance is now known to consist of several types of $K^+$ channels, each of which contributes various aspects of modulation of excitability. These multiple types of $K^+$ channels also contribute to the variable responses to transmitters and drugs. The $K^+$ conductance has been found to be most important for producing electronegativity of smooth muscle as with most other kinds of cells. In fact, $Cl^-$ is the only other ion that has both the substantial concentrations and conductance to cause the electronegativity observed as resting membrane potential. The final component important for membrane potential in smooth muscle cells is an electrogenic $Na^+,K^+$-ATPase pump. Studies of a wide variety of smooth muscle cells have indicated that $K^+$ conductance and the electrogenic sodium pump are the primary determinants of membrane potential under a variety of resting and stimulated conditions. An example from arterial muscle is shown in Fig. 12.

It is the ratio of permeabilities of $Na^+$ to $K^+$ given to the fundamentally important formulation Goldman–Hodgkin–Katz that dominates membrane potential, namely:

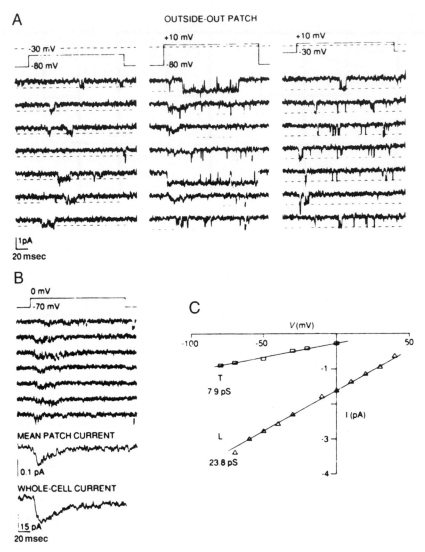

**FIG. 11.**  Single-channel currents from isolated patches of vascular muscle cell membrane show the single channel analog of whole-cell currents. (A,B) Individual T- and L-type channel openings can be resolved. When their amplitudes are plotted (C), conductances of 8 and 24 pS can be found for T- and L-type $Ca^{2+}$ channels, respectively, under these conditions of 110 m$M$ $Ba^{2+}$ as the charge carrier. Smaller conductances would be expected in physiological concentrations of extracellular $Ca^{2+}$. (Reproduced with permission from Benham *et al.*, 1987, *Circ. Res.* Copyright 1987 American Heart Association.)

$$E_{m} = \frac{RT}{F} \ln \frac{P_{Na}[Na]_{o} + P_{K}[K]_{o} + P_{Cl}[Cl]_{i}}{P_{Na}[Na]_{i} + P_{K}[K]_{i} + P_{Cl}[Cl]_{o}}$$

The value of $[K]_{i}$ can be estimated from a complete plot of $E_{m}$ versus log $[K]_{o}$ and extrapolating to the zero potential, which gives the point at which $[K]_{o} = [K]_{i}$. This electrophysiological method gives values that cluster very closely around 160 m$M$ in blood vessels and other smooth muscle types. The selectivity ratio for the $P_{Na}/P_{K}$ ratio is approximately 0.17 for a physiological $[K]_{o}$ of 4–5 m$M$. This ratio of $P_{Na}/P_{K}$ varies with the $[K]_{o}$ because the $K^{+}$ conductance is sensitive to extracellular $K^{+}$ concentration and is changing as extracellular $K^{+}$ is changing. Thus, the plot of $E_{m}$ vs $[K]_{o}$ is not a straight line, but bends substan-

tially at $[K]_{o}$ below approximately 10 m$M$, and to a lesser extent again at about 60 m$M$ (Fig. 13).

$K^{+}$ channels provide an important way in which spikes in smooth muscle can be regulated and contribute to variability in the shape and amplitude of spikes. Many drugs also affect $K^{+}$ channels in smooth muscle cells, and can thus modify membrane potential during spikes. The most important aspect of drug action on $K^{+}$ channels is through changes in resting membrane potential. In smooth muscles, as in many other cells, the probability for opening along with the equilibrium potential for $K^{+}$ ($E_{K}$, calculated from the Nernst equation) determine the resting membrane potential. Through the inactivation properties of ion channels, the resting potential provides a highly important influence

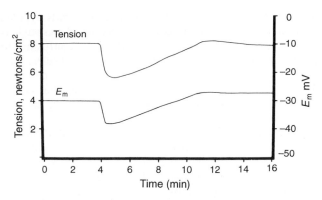

**FIG. 12.** Simultaneously recorded tension and $E_m$ were traced from oscilloscope records to show the correspondence between relaxation and underlying hyperpolarization that occur when vascular muscle cells of the caudal artery of a rat were returned to $K^+$ solution after 15 min in 0 $K^+$ solution. There is a transient hyperpolarization–relaxation due to activation of the electrogenic sodium pump, which normally contributes to membrane potential of the caudal artery. This example of hyperpolarization–relaxation, even under stimulation by a midrange dose of norepinephrine, shows the powerful modulation of contraction by membrane potential. [Reproduced with permission from Hermsmeyer, K., *et al.* (1981). Membrane-potential dependent tension in vascular muscle. *In* "Vasodilation" (P. M. Vanhoutte and I. Leusen, eds.), pp. 273–284. New York, Raven Press.]

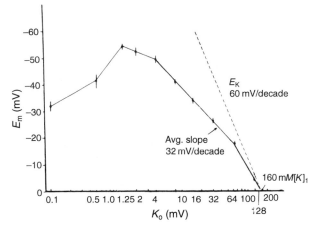

**FIG. 13.** Membrane potential as a function of extracellular $K^+$ reveals the dependence of membrane potential on $K^+$, which becomes very strong at higher $K^+$ concentrations. The theoretical relationship that would result if the membrane were selective only to $K^+$ is given by the dashed line. The potential reaches zero at the $K^+$ concentration where extracellular $K^+$ equals intracellular $K^+$, which is approximately 160 m*M*. At $K^+$ concentrations lower than this symmetrical value for the intracellular concentration, potential is more and more negative, reaching a peak of −55 mV at $K^+$ concentrations lower than physiological. From the peak negativity of approximately −55 mV, potential becomes less negative at very low $K^+$ concentrations, which is predicted by the Goldman–Hodgkin–Katz constant field equation and decreasing $K^+$ permeability at low $K^+$ concentrations. At the normal physiological concentration of 4.7 m*M* $K^+$, resting membrane potential for the vascular muscle cell of the guinea pig superior mesenteric artery is −50 mV. (Reproduced with permission from Harder and Sperelakis, 1979.)

over the availability of $Ca^{2+}$ channels and $Na^+$ channels and, thus, the capability to produce an electrical spike.

Therefore, the action of the $K^+$ channels is twofold. The first is to set the membrane potential, and thus influence the amplitude of inward currents for $Ca^{2+}$ and for $Na^+$ that will produce the electrical spike. The second action is to modify the form (shape and duration) of the electrical spike, because the $K^+$ channels allow a flow of electric current that repolarizes the cells. This current, called an **outward current,** is present even at rest, and contributes to the pacemaker mechanism and spike repolarization. If the $K^+$ channels are blocked by such drugs as tetraethylammonium (TEA) or by the cation, barium, there is (1) depolarization of the cells (usually producing contraction); (2) slowed repolarization of the spikes, which often leads to enhanced electrical activity, especially in spontaneously active organs; and (3) usually a deterioration of function as $Ca^{2+}$ overloads the cell if $K^+$ channel block continues. $K^+$ channels thus importantly influence the spikes in smooth muscle cells. When there is a large background (or resting) $K^+$ conductance, block by $K^+$ channel blockers can induce excitability (Fig. 14).

Pacemaker activity depends on $K^+$ channels in a balance of ion currents. The constant traffic of inward currents against outward currents determines whether a cell will depolarize threshold to trigger an electrical spike. A model of pacemaker activity based on $Ca^{2+}$ and $K^+$ channels is shown in Fig. 15. In pacemaker cells, the spontaneous depolarization caused by an influx of $Ca^{2+}$ and/or $Na^+$ is balanced by an outflow of $K^+$ current that tends to balance membrane potential. If there is a slight increase in outward $K^+$ channel current, hyperpolarization and inhibition of spontaneous activity occur. If $K^+$ conductance slightly decreases, the cell depolarizes sufficiently for T channels to inactivate (at about −30 mV) and suppression of pacemaker activity occurs. It is only under conditions where

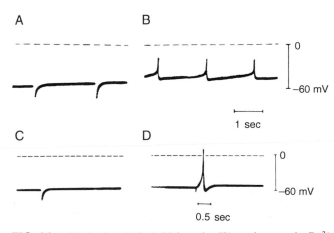

**FIG. 14.** Block of a relatively high resting $K^+$ conductance by $Ba^{2+}$ or TEA induces excitability in the form of electrical spikes. Electrical stimulation in the control condition (A) can be converted to electrical spikes by 0.5 m*M* $Ba^{2+}$ (B). In another guinea-pig superior mesenteric artery, the inexcitability (C) is converted by 5 m*M* TEA to an overshooting electrical spike (D). (Reproduced with permission from Harder and Sperelakis, 1979.)

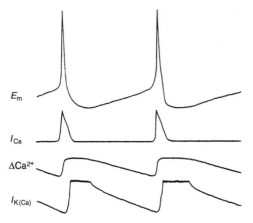

**FIG. 15.** Pacemaker potentials are modeled by this diagram of membrane potential ($E_m$), $Ca^{2+}$ current ($I_{Ca}$), intracellular $Ca^{2+}$ concentration ($\Delta Ca^{2+}$), and $Ca^{2+}$-dependent $K^+$ current ($I_{K(Ca)}$). In this hypothetical vascular muscle cell, pacemaker potentials are evident as the slow upward drift of membrane potential toward zero to an inflection point, at which the slow upward movement suddenly bends to a nearly vertical rise that is the electrical spike. The spike is brief because of inactivation of $Ca^{2+}$ channels that is evident in the return to baseline of $Ca^{2+}$ current. The increase in $Ca^{2+}$ shown by $\Delta Ca^{2+}$ results in an increase in current through the $Ca^{2+}$-sensitive $K^+$ channels shown as $I_{K(Ca)}$, which accelerates the repolarizing phase and allows the downsweep to the most negative $E_m$ at the beginning of the pacemaker depolarization. The important feature is that $K^+$ current is unstable, decreasing as the intracellular $Ca^{2+}$ is pumped into $Ca^{2+}$ stores, or out through the cell membrane, which allows the depolarization and results in opening of $Ca^{2+}$ channels that cause the spike. This cooperation of an inward current and a decaying outward current is the type of pacemaker encountered in cardiac and spontaneously active nerve cells. This pacemaker potential in smooth muscle cells allows for spontaneous activity as in other excitable cells. [From Hermsmeyer and Akbarali (1989). Reproduced with permission of S. Karger AG, Basel.]

$K^+$ conductance is in an optimum range when there is a balance of currents that tireless, repetitive cycles of spontaneous activity can occur. There must be a decaying outward $K^+$ current in the face of a persistent inward current to bring together the continuously changing balance that is the basis for rhythmic spontaneous activity seen in smooth muscles. Superimposed on this is a strongly **electrogenic $Na^+/K^+$ pump,** or the $Na^+$ pump for short, that hyperpolarizes the cell in response to increasing intracellular $Na^+$, which this transport mechanism vigorously regulates (Hermsmeyer, 1982). The degree of rhythmicity is influenced because membrane potential is changed and consequently the pacemaker mechanism. With this multiplicity of regulators, spontaneous activity is somewhat irregular in smooth muscle cells, distinguishing it from the more clocklike timing mechanism found in the pacemaker in heart.

## V. Heterogeneous Electrical Properties of Smooth Muscle Cells

The electrical properties of smooth muscles, including resting potentials and electrical spikes, have many funda-

mental similarities to other types of excitable cells. Among smooth muscles, there are common characteristics such as $Ca^{2+}$, $Na^+$, and $K^+$ channels, but it is confusing to generalize about them. In fact, smooth muscles, like striated muscles and perhaps even more like neurons, have a diverse array of electrical properties that correspond with functional differences in various organs, and even parts of an organ.

For example, the resting membrane potential of smooth muscles varies over a range from about $-70$ to $-40$ mV. This variation reflects real differences in the cells being studied. In fact, within a smooth muscle organ, there is a diversity of resting membrane potentials that distributes even within 1 mm, suggesting different populations of cells within an organ. The range of resting potentials in arteries has been reported by many investigators to span typically more than 10 mV (for the 95% confidence intervals). Studies of visceral smooth muscle show a similar range, possibly contributing to the composite characteristics of that organ. The differences are not simply due to quality of impalement, as was first supposed, for several reasons. When only the best quality recordings are considered, based on criteria such as impalement stability, input resistance of a cell being measured, freedom from tip potential or changes in tip potential, and constancy of all parameters before versus after the impalement, there is the same variability. The same $-10$-mV span for 95% confidence intervals appears. Furthermore, studies using electron probe microanalysis show a distribution of intracellular $K^+$ and $Cl^-$ that predicts the same range of resting potentials. Thus, it is best to think of an organ as composed of smooth muscle cells consisting of composite subpopulations that have a range of membrane properties that, when taken together, constitute the characteristics of that organ.

Smooth muscle cells exhibit an array of rhythms and variability in the timing of those rhythms. Pacemakers have timing mechanisms on the order of minutes or seconds, and a basic organ rhythm can be identified for various kinds of smooth muscles. Pacemaker types and modulation allow for diversity in rhythms, which emanate from electrical spikes triggered by continuously changing membrane potential. The spike fired is a membrane electrical signal conducted within a fraction of 1 sec, synchronizing contraction, and recognized by spontaneous activity. Smooth muscle pacemakers, though relatively less studied than those in heart, provide vital functions for blood vessels and visceral organs and are important for understanding regulatory mechanisms and drug actions.

One particular danger in studies of single cells is that of extrapolating observations found on subcultured cells back to the intact organs. This is, in fact, a dangerous step because cells can lose membrane (and other) characteristics with time in cell culture. It is characteristic of smooth muscle cells that when they are "passaged," there is loss of expression of membrane characteristics. It is much easier to buy subcultured cells that have uniform characteristics than it is to use primary cultures that are as complex as the tissue from which they are isolated and must be prepared each week. However, it is primary cells that have been in culture for a matter of days, and for which conditions have been optimized to keep membrane, and especially contraction–relaxation capabilities, that are most like the tissue from which they were derived. In particular, it is extremely important to evaluate data coming from cultured smooth muscle cells based on the care that has been taken to maintain the properties (usually ion channels or other membrane phenomena) that are being studied. When the primary cell culture and contractile stipulations are included, cultured smooth muscle cells are particularly useful, and many of the conflicting and contrary data are eliminated. However, it is common to find reports on subcultured cells, and thus the reader must exercise judgment about extrapolations of data. It is particularly important to justify any freshly isolated or cultured cell data.

## VI. Modulation

Electrical spikes can be modulated by transmitters, as shown by the spiking behavior induced in the carotid artery by norepinephrine, or in the progesterone-dominated uterus by oxytocin. In either case, the occurrence of electrical spikes is dependent on the right hormonal milieu and may involve several synchronous factors. The kind of electrical spike that occurs is determined not only by norepinephrine in the case of the carotid artery or oxytocin in the case of the uterine muscle, but by the other factors that are also present. For example, uterine muscle is very sensitive to cycling levels of steroid hormones (estrogen and progesterone). Electrical spikes vary sufficiently that the state of hormonal regulation of uterine muscle can be identified by the shape of electrical spikes. Greatly prolonged electrical spikes have been demonstrated in blood vessels and uterine muscle under the influence of experimental substances as well as those that can be found in the blood and extracellular fluid. For example, in response to barium, decreases in $K^+$ conductance allow prolonged spikes and the resulting tonic-type contractions. $K^+$ channels that are inhibited by intracellular ATP can contribute to long-duration enhanced electrical activity. When $K^+$ channels are blocked, a state of hyperexcitability often occurs that leads to more frequent spikes, and can cause very prolonged contractions.

These examples show how alteration of the electrical activity can be sufficient to explain varying function in smooth muscle cells. The electrical spike is directly transduced into $Ca^{2+}$ release, which triggers contraction. Spikes or changes in membrane potential that cause $Ca^{2+}$ release are major determinants of smooth muscle shortening. This has been demonstrated (Fig. 16) by recording the electrical spike and associated contraction from a single vascular muscle cell (Hermsmeyer, 1979). Our recent appreciation that both T- and L-types of $Ca^{2+}$ channels are fundamentally involved in such examples as stroke-prone genetically hypertensive rats (Self *et al.*, 1994) increases the importance of understanding these key $Ca^{2+}$ signals that are the basis for smooth muscle APs. It is likely that as the electrical signals activating smooth muscles are better understood, the variations in function, both in normal and in abnormal situations, can be better explained.

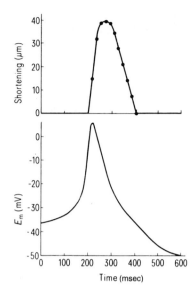

**FIG. 16.** Correspondence of contraction (upper) and membrane potential ($E_m$) (lower) show the direct correlation of membrane depolarization and contraction in an isolated single vascular muscle cell. Electrical recording was done via an intracellular micropipette electrode filled with KCl and with a 0.2-$\mu$m tip diameter. Shortening was determined by high-speed filming of the contracting cell through the microscope as a simultaneous recording. The high shortening velocity shown here occurs in specialized vascular muscle cells and is particularly useful for showing the close correlation of tension to membrane potential. (Reproduced with permission from Hermsmeyer, 1979.)

## VII. Summary

Smooth muscle cells are small and arranged in groups to form the contractile wall of hollow organs and blood vessels. In most cases, the muscle cell specialization is for slow, maintained contraction (tone) rather than speed. The AP phenomenon here is to activate by a multiplicity of stimulatory and inhibitory inputs to a diverse array of receptors on a single cell. Coordination through chemicals released by nerve fibers, secretory cells, or other smooth muscle cells is important. An array of ion conductances with slower inactivation properties allows for long-duration contractions (tone) and continuously graded regulation. Superimposed electrogenic ion transport allows for additional modulatory inputs. With the multiplicity of receptors and response mechanisms found in the deceptively small smooth muscle cell, the complexity of signaling mechanisms possible with the simple smooth muscle spike should be regarded as a compact masterpiece of multiple responsiveness.

## Bibliography

Benham, C. D., and Tsien, R. W. (1987). A novel receptor-operated $Ca^{2+}$-permeable channel activated by ATP in smooth muscle. *Nature* **328**, 275–278.

Benham, C. D., Hess, P., and Tsien, R. W. (1987). Two types of calcium channels in single smooth muscle cells from rabbit ear artery studied with whole-cell and single-channel recordings. *Circ. Res.* **61**, 1-10-1-16.

Berridge, M. J., and Irvine, R. F. (1984). Inositol trisphosphate, a novel second messenger in cellular signal transduction. *Nature* **312**, 315–321.

Bolton, T. B. (1979). Mechanisms of action of transmitters and other substances on smooth muscle. *Physiol. Rev.* **59**, 606–718.

Bozler, E. (1938). Electric stimulation and conduction of excitation in smooth muscle. *Am. J. Physiol.* **122**, 614–623.

Bozler, E. (1948). Conduction, automaticity, and tonus of visceral muscles. *Experientia* **4**, 213–218.

Chen, Q., and van Breemen, C. (1993). The superficial buffer barrier in venous smooth muscle: Sarcoplasmic reticulum refilling and unloading. *Br. J. Pharmacol.* **109**, 335–343.

Daniel, E. E., Van Breemen, C., Schilling, W. P., and Kwan, C. Y. (1995). Regulation of vascular tone: Cross-talk between sarcoplasmic reticulum and plasmalemma. *Can. J. Physiol. Pharmacol.* **73**, 551–557.

Droogmans, G., Raeymaekers, L., and Casteels, R. (1977). Electro- and pharmaco-mechanical coupling in the smooth muscle cells of the rabbit ear artery. *J. Gen. Physiol.* **70**, 129–148.

Harder, D. R., and Sperelakis, N. (1979). Action potentials induced in guinea pig arterial smooth muscle by tetraethylammonium. *Am. J. Physiol.* **237**, C75–C80.

Hermsmeyer, K. (1979). High shortening velocity of isolated single arterial muscle cells. *Experientia* **35**, 1599–1602.

Hermsmeyer, K., Trapani, A., Abel, P. W. (1981). Membrane-potential dependent tension in vascular muscle. *In* "Vasodilation" (P. M. Vanhoutte and I. Leusen, eds.), pp. 273–284. New York, Raven Press.

Hermsmeyer, K. (1982). Electrogenic ion pumps and other determinants of membrane potential in vascular muscle. *Physiologist* **25**(6), 454–465.

Hermsmeyer, K., and Akbarali, H. (1989). Cellular pacemaker mechanism in vascular muscle. *Prog. Appl. Microcirc.* **15**, 32–40.

Hermsmeyer, K., and Miyagawa, K. (1996). Protein kinase C mechanism enhances vascular muscle relaxation by the $Ca^{2+}$ antagonist, Ro 40-5967. *J. Vasc. Res.* **33**, 71–77.

Hernandez-Nicaise, M.-L., Mackie, G. O., and Meech, R. W. (1980). Giant smooth muscle cells of *Beroë*. Ultrastructure, innervation, and electrical properties. *J. Gen. Physiol.* **75**, 79–105.

Holman, M. E., and Neild, T. O. (1979). Membrane properties. *Br. Med. Bull.* **35**(3), 235–241.

Kessel, R. G., and Kardon, R. H. (1979). "Tissues and Organs: A text-atlas of scanning electron microscopy," p. 41. San Francisco, W. H. Freeman and Company.

Keynes, R. D. (1983). The Croonian Lecture. Voltage-gated ion channels in the nerve membrane. *Proc. Royal Soc. Lond. B.* **220**, 1–30.

Prosser, C. L., Burnstock, G., and Kahn, J. (1960). Conduction in smooth muscle: Comparative structural properties. *Am. J. Physiol.* **199**, 545–552.

Self, D. A., Bian, K., Mishra, S. K., and Hermsmeyer, K. (1994). Stroke-prone SHR vascular muscle $Ca^{2+}$ current amplitudes closely correlate with lethal increases in blood pressure. *J. Vasc. Res.* **31**, 359–366.

Somlyo, A. P., and Somlyo, A. V. (1990). Flash photolysis studies of excitation–contraction coupling, regulation, and contraction in smooth muscle. *Annu. Rev. Physiol.* **52**, 857–874.

Sperelakis, N., and Ohya, Y. (1989). Electrophysiology of vascular smooth muscle. *In* "Physiology and Pathophysiology of the Heart, 2nd Edition," pp 773–811. Kluwer Academic Publishers.

Sturek, M., and Hermsmeyer, K. (1986). Calcium and sodium channels in spontaneously contracting vascular muscle cells. *Science* **233**, 475–478.

Tomita, T. (1970). Electrical properties of mammalian smooth muscle. In "Smooth Muscle" (E. Bülbrin, A. F. Brading, A. W. Jones, and T. Tomita, Eds.). pp. 197–243. Williams & Wilkins. Baltimore.

Trapani, A. J., Matsuki, N., Abel, P. W., and Hermsmeyer, K. (1981). Norepinephrine produces tension through electromechanical coupling in rabbit ear artery. *Eur. J. Pharmacol.* **72**, 87–91.

*Judith A. Heiny*

# 53

## Excitation–Contraction Coupling in Skeletal Muscle

## I. Introduction

Skeletal muscle is activated to contract by a sequence of fast, electrically driven events that are collectively termed **excitation–contraction coupling** (EC coupling). EC coupling is a highly organized process of signal transduction that utilizes specialized membranes, membrane junctions, and ion channels on both the exterior and interior of the cell. This chapter describes the cellular and molecular processes that mediate electrical excitation of the outer membranes and transduce it into a signal for intracellular $Ca^{2+}$ release, focusing on vertebrate fast-twitch skeletal muscle which is the best characterized. Chapter 54 discusses molecular mechanisms for control of $Ca^{2+}$ release and reuptake by the sarcoplasmic reticulum. Chapter 55 discusses mechanisms of force generation by the contractile proteins, actin and myosin.

Skeletal muscle is the fastest contracting of the three major muscle types and is related to cardiac and smooth muscle types evolutionarily and embryonically. The proteins that mediate excitation–contraction in cardiac, smooth, and skeletal muscle share a high degree of homology. Nonetheless, each muscle type has evolved highly differentiated mechanisms of using these proteins to serve its specialized functions. Whereas cardiac and smooth muscle mediate primarily enteric contractile processes that are under autonomic and humoral control, skeletal muscle primarily mediates rapid, willed bodily movements that are under control of the central nervous system (CNS). Consequently, the defining features of skeletal muscle activation are speed and voluntary control, as needed for fine control of body movement.

## II. Overview of Excitation–Contraction Coupling

The sequence of EC coupling in a vertebrate fast-twitch skeletal muscle fiber is shown schematically in Fig. 1. Acti-vation begins when an action potential from a motor neuron arrives at the **neuromuscular junction** (NMJ), resulting in the release of the neurotransmitter **acetylcholine** (ACh) into the synaptic clefts (see Fig. 1 of Chapter 40). Binding of ACh to ACh receptors on the adjacent end plate (see Figs. 4 and 5 of Chapter 39) locally depolarizes the postsynaptic membrane to threshold for exciting an action potential on the muscle **sarcolemma.** The action potential rapidly propagates the depolarization to the entire sarcolemma by a mechanism similar to action potential generation in the nervous system (Chapter 19). The action potential also propagates into the fiber interior via a specialized system of tubular membranes **(tranverse-tubules** or **T-tubules),** which are continuous with and invaginate tranversely from the sarcolemma at periodic intervals. The T-tubules provide the conduit for the action potential to reach the fiber interior, and also bring the outer membranes into proximity with the internal **sarcoplasmic reticulum** (SR) at specialized junctions called triads which mediate rapid communication between the extracellular sarcolemma/T-tubules and the intracellular SR. At the triad junctions, **voltage-dependent $Ca^{2+}$ channels** in the T-tubular membrane [also called **voltage sensors** and **dihydropyridine receptors** (DHPRs) because of their sensitivity to the dihydropyridine class of $Ca^{2+}$ channel blocking drugs] detect the depolarization and transduce it into a signal for opening **$Ca^{2+}$ release channels** [also called the **ryanodine receptors** (RyRs) because they bind this plant alkalyoid with high affinity] on the closely opposed SR membrane. The molecular mechanism of interaction between these two distinct types of $Ca^{2+}$ channel is not completely known but it is likely that some kind of allosteric interaction is involved. It is hypothesized that membrane depolarization drives a conformational change on charged, intramembrane domains of the T-tubule $Ca^{2+}$ channel, which, in turn, drive a conformational change leading to opening of the SR $Ca^{2+}$ release channel. Whether the T-tubule $Ca^{2+}$ channel interacts with the SR $Ca^{2+}$ release channel directly or via accessory proteins as part of a

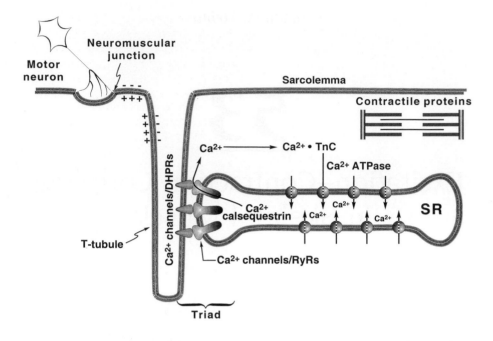

**FIG. 1.** Sequence of muscle activation and relaxation.

- ◆ **Resting [Ca]$_i$ ~0.1 μ$M$**
- ◆ **Neuromuscular transmission**
- ◆ **Action potential propagation along sarcolemma and into T-tubules**
- ◆ **Signal transduction from voltage-dependent Ca$^{2+}$ channels to Ca$^{2+}$ release channels at triad junctions**
- ◆ **Ca$^{2+}$ release from SR**
- ◆ **Cytosolic [Ca]$_i$ reaches 1–10 μ$M$**
- ◆ **Diffusion and binding of Ca$^{2+}$ ions to TnC**
- ◆ **Removal of troponin inhibition of the contractile proteins**
- ◆ **Contractile proteins shorten to generate force**
- ◆ **[Ca]$_i$ returns to resting levels via:**

   **Reuptake of Ca$^{2+}$ into SR by Ca$^{2+}$-ATPase**
   **Inactivation of Ca$^{2+}$ release channels**
   **Binding of Ca$^{2+}$ to calsequestrin**

larger macromolecular complex remains to be established. The SR, related evolutionarily to the endoplasmic reticulum, is a highly specialized membrane compartment for controlling cytosolic Ca$^{2+}$. In a resting muscle fiber, the SR actively maintains cytosolic [Ca$^{2+}$] at submicromolar levels, typically 0.1 μ$M$, by active transport mediated by Ca$^{2+}$-ATPase molecules present at high density along its length. On receiving a stimulus from the T-tubule, it rapidly releases Ca$^{2+}$ into the cytosol at rates approaching 100 μ$M$/msec to raise cytosolic [Ca$^{2+}$] to micromolar levels, typically 1–10 μ$M$. Ca$^{2+}$ efflux occurs through RyRs localized on the junctional regions of SR **(terminal cisternae)** nearest the T-tubules. Released Ca$^{2+}$ ions diffuse and bind to **troponin-C** (TnC), the regulatory subunit of troponin, thereby

removing troponin's inhibitory effect on the **contractile proteins,** actin and myosin, which then shorten to generate force (see Chapter 55). Force output is proportional to cytosolic Ca$^{2+}$ concentration (Fig. 2). EC coupling is terminated when cytosolic Ca$^{2+}$ returns to resting levels. High density **Ca$^{2+}$-ATPase** pumps on the SR membrane rapidly pump Ca$^{2+}$ back into the SR where it is largely bound to the Ca$^{2+}$ binding protein, **calsequestrin.** Additional mechanisms such as a Ca$^{2+}$-dependent inactivation of the Ca$^{2+}$ release channel/RyR and active Ca$^{2+}$ extrusion across the sarcolemma participate to turn off Ca$^{2+}$ release.

Thus, cytosolic [Ca$^{2+}$] is the central link between membrane excitation and activation of the contractile proteins (Fig. 3). Contraction is inhibited at low resting Ca$^{2+}$ levels

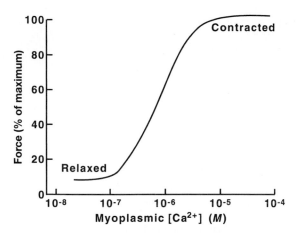

**FIG. 2.** Relationship between force and myoplasmic Ca²⁺ concentration.

and proceeds when $Ca^{2+}$ is transiently elevated in response to membrane excitation.

## III. Speed of Skeletal Muscle Activation

Activation of skeletal muscle is rapid despite the fact that skeletal muscle is one of the largest of mammalian cells. Skeletal muscle fibers are multinucleated cells that form during embryogenesis by fusion of precursor myoblasts. They are cylindrically shaped with diameters ranging from 50 to 150 $\mu M$ and lengths from millimeters to centimeters. To produce effective mechanical force, all parts of a muscle fiber must develop tension simultaneously. It was recognized early that a trigger process other than one involving a diffusion-limited second messenger must be involved (Hill, 1948). For example, a small ion entering through the outer membrane of a 100-$\mu$m muscle fiber would require 100–200 msec to diffuse and bind to putative activating sites distributed throughout the cross-

section. By comparison, a fast-twitch muscle fiber is activated in 20–30 msec. Synchronous activation of skeletal muscle is achieved via an electrical signal that propagates rapidly along the sarcolemma and into the cell interior along the specialized conducting membranes of the T-tubules. The final trigger $Ca^{2+}$ diffuses at most a few microns from its release site on the SR to the activator sites on TnC. Figure 4A shows the typical delay between membrane excitation and contraction for a frog fast-twitch muscle fiber at 18°C. Tension begins developing within 3–5 msec of the action potential and rises to a peak in 20–30 msec. Table 1 summarizes the speed of the key events following neuromuscular transmission that contribute to this delay in a frog muscle fiber at 18°C. For example, propagation of the action potential laterally along the sarcolemma from the middle of the fiber to both tendons introduces a delay of 5 msec. Signal transduction into the fiber interior, including all events from sarcolemma depolarization to activation of the innermost contractile proteins, takes 4–5 msec. Each of these events is faster in mammalian muscle at 37°C. Figure 4B shows the time

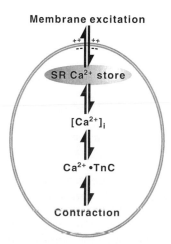

**FIG. 3.** A rise in cytosolic [Ca²⁺] links membrane excitation to activation of contractile proteins.

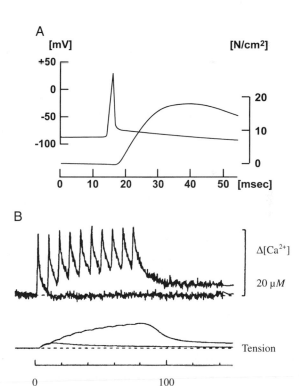

**FIG. 4.** Speed of EC coupling in skeletal muscle. (A) Speed of EC coupling in a fast-twitch frog skeletal muscle fiber at 18°C, showing the action potential (top trace) and tension (lower trace). (Modified from Hodgkin and Horowicz, 1957.) (B) Speed of EC coupling in a mouse skeletal muscle fiber at 37°C. Top traces show the myoplasmic Ca²⁺ change measured with a Ca²⁺ indicator dye. The lower trace shows tension. The fiber was stimulated at time zero with a single action potential, and with repetitive action potentials at high frequency, to simulate a muscle tetanus. (From Hollingworth *et al.*, 1996. Reproduced from *The Journal of General Physiology* by copyright permission of The Rockefeller University Press.)

**TABLE 1**   Duration of Key Steps in the Activation of a Fast Twitch
Skeletal Muscle[a]

| E–C coupling steps | Duration (msec) |
| --- | --- |
| Action potential propagation along sarcolemma | 5–10 |
| Action potential propagation to center of fiber along T-tubules | ~0.7 |
| Signal transduction at triad junction, from T-tubule depolarization to activation of RyR on SR | ~0.5 |
| Peak rate of $Ca^{2+}$ release to peak $Ca^{2+}$ binding to TnC (start of tension) | 2–3 |
| Peak myoplasmic $Ca^{2+}$ change to peak tension | 15–25 |

[a] Durations were calculated for hypothetical frog fiber of 50 $\mu$m in diameter and 5 cm in length, having a central end plate. Literature values for conduction velocity and duration of intermediate steps (Gonzales-Serratos, 1971; Vergara and Delay, 1986; Pape *et al.*, 1993) were adjusted to 18°C.

course of the intracellular $Ca^{2+}$ change and tension in a fast-twitch mammalian muscle at 37°C, stimulated with a single action potential or at high frequency to simulate a tetanus.

## IV. Membrane Architecture of Excitation–Contraction Coupling

Synchronous activation depends on effective and rapid communication between the specialized extracellular membranes of the sarcolemma/T-tubules and the intracellular SR. To achieve this, the membranes of skeletal muscle are highly organized to bring into physical proximity the proteins involved in each of the key events: membrane excitation, $Ca^{2+}$ release, and force generation (reviewed in Franzini-Armstrong and Jorgensen, 1994). Figure 5 illustrates the three-dimensional architecture of the sarcolemma, T-tubules, triad junctions, and SR membranes of a fast-twitch skeletal muscle fiber and their physical relationship to the contractile proteins. The sarcolemma and T-tubule membranes are organized to conduct the action potential rapidly to all parts of the fiber. The T-tubules comprise the majority of the sarcolemma membrane, representing 50–80% of the total sarcolemma area (Peachey, 1965). They invaginate from the sarcolemma in a transverse plane approximately twice every 1–2 $\mu M$ (at two planes per sarcomere). Within this plane they branch extensively to form a network that covers the entire cross-sectional area of the muscle cell, as shown in Fig. 6. Up to 80% of the tubular membrane is associated with the SR at triad junctions (Peachey, 1965; Dulhunty, 1984). Given this membrane architecture, no part of the sarcolemma is more than a few tenths of a micron, and thus no more than a few milliseconds conduction time, from the $Ca^+$ release channels of the SR.

The triad junctions are optimized for rapid signal transduction from the T-tubule to SR. Figure 7A shows the morphology of the triad and Fig. 7B shows the molecular organization of the key proteins of the triad junction. The T-tubule and SR membranes flatten at the junctions and face each other across a narrow gap of about 10 nm. The junctional surface of the terminal cisternae of the SR contains two rows of proteins corresponding to the $Ca^{2+}$ release channels/RyRs, which are aligned in two rows in a skewed pattern with a center-to-center spacing of about 30 nm. The RyRs are also termed **foot proteins** because of their unique appearance as dense, bridging structures in transverse electron micrographs of the triad. Each $Ca^{2+}$ release channel is composed of four identical subunits. Each subunit has a membrane-spanning domain and a large cytosolic domain that extends across the junctional gap and comes to within at most 1 nm of the T-tubule membrane. The junctional T-tubule membranes also contain a high density of proteins that are aligned with a regular periodicity in parallel rows of four-particle arrays, termed **junctional tetrads.** The tetrads have been identified as groups of four voltage sensors/DHPRs (Block *et al.*, 1988) and are arranged in a zigzag spacing that corresponds exactly to the position of every other RyR on the adjacent terminal cisternae (indicated by dotted lines in Fig. 7B).

Figure 7B was taken from a fish skeletal muscle fiber, in which tetrads line up opposite every other RyR in a 1:2 ratio. A similar arrangement of tetrads is suggested in triads and peripheral couplings of other muscles from the frog, rat, chicken, and human (Franzini-Armstrong and Jorgensen, 1994; Lamb, 1992). However, in some skeletal muscles of the frog and rabbit, the ratio of DHPRs to RyR may be more variable and can range from 0.6 to as high as 7 (Anderson *et al.*, 1994). This finding raises the question of whether tetrads of DHPRs line up opposite every other RyR in a fixed ratio in every skeletal muscle, as expected for an allosteric coupling mechanism.

## V. Molecular Mechanisms of Signal Transduction at the Triad Junctions

The essential events in signal transduction at the triad junctions are detection of the depolarization by voltage-sensing molecules in the T-tubule, communication to $Ca^{2+}$ release channels/RyRs on the SR, and $Ca^{2+}$ efflux from

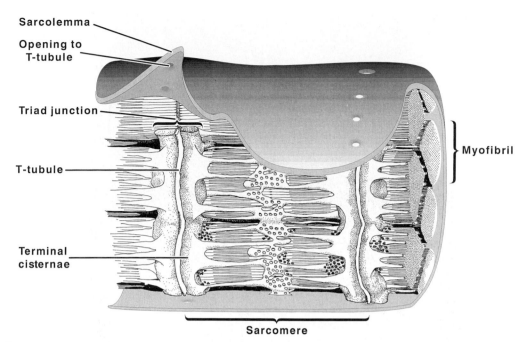

**FIG. 5.** Three-dimensional reconstruction of a longitudinal section of a skeletal muscle fiber. The sarcolemma surrounds bundles of contractile proteins called *myofibrils*. T-tubules invaginate transversely from the sarcolemma at periodic intervals and form junctions with the SR along their entire length. At the triad junctions, the longitudinally oriented SR membranes widen into sacs called *terminal cisternae*, which closely oppose the T-tubules and contain the $Ca^{2+}$ release channels/RyRs. The terminal cisternae also closely overlay the activating $Ca^{2+}$ sites on TnC. The functional unit of force generation is the sarcomere, consisting of overlapping actin and myosin filaments anchored at each end. The longitudinally oriented tubular regions of SR contain the $Ca^{2+}$-ATPase, which takes up released $Ca^{2+}$ all along the sarcomere. (Adapted from *The Journal of Cell Biology,* 1965, vol. 25, pp. 209–232 by copyright permission of The Rockefeller University Press.)

the SR. As noted, this is accomplished by the specialized interaction between two distinct $Ca^{2+}$ ion channels, the DHPR and the RyR.

The DHPR/T-tubule $Ca^{2+}$ channel is a voltage-gated ion channel belonging to the L-type superfamily of $Ca^{2+}$ channels. It is characterized by high-voltage activation, brief openings, and low single-channel ion flux. The RyR/SR $Ca^{2+}$ channel belongs to the superfamily of ligand-gated ion channels, structurally related to the $IP_3$ receptor of the ER. It is identified by its activation by $Ca^{2+}$ and adenine nucleotides, and high single-channel conductance. Thus, at the triad junction a unique intracellular signal transduction occurs whereby a voltage-dependent $Ca^{2+}$ channel on an extracellular membrane controls the opening of a separate, normally ligand-gated $Ca^{2+}$ channel on an intracellular membrane.

### A. Voltage Sensing

It was recognized early that EC coupling in fast skeletal muscle of higher vertebrates is directly controlled by membrane potential, without a requirement for extracellular $Ca^{2+}$. This contrasts with EC coupling in skeletal muscle of some lower invertebrates and cardiac muscle in which $Ca^{2+}$ influx across the sarcolemma functions as a second messenger and is absolutely required for contraction. In a classic experiment, Armstrong *et al.* (1972) demonstrated that a frog skeletal muscle fiber continues to twitch when stimulated with action potentials in the absence of extracellular $Ca^{2+}$. Similarly, voltage-clamp experiments demonstrated that a skeletal muscle fiber can develop tension provided only that the membrane is depolarized to a minimum potential and duration, termed the **mechanical**

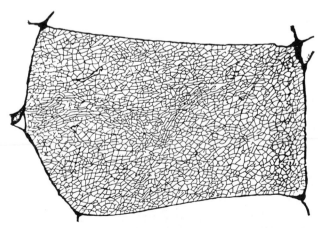

**FIG. 6.** Reconstruction of the T-tubule network over the whole cross section of a frog twitch muscle fiber. The reconstruction was made by tracing electron micrographs of transverse serial slices of muscle perfused with the electron-dense and membrane-impermeant molecule, peroxidase, which diffused into the T-tubules. (Reproduced with permission from Peachey and Eisenberg, 1978.)

A

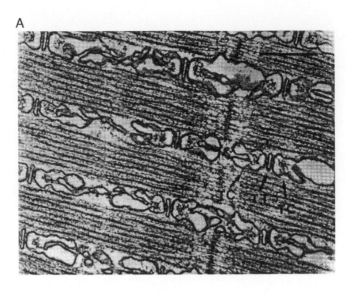

B

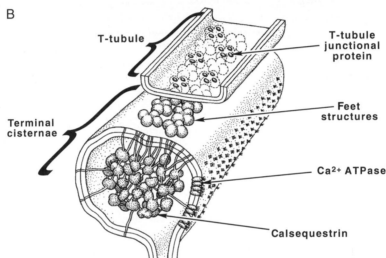

**FIG. 7.**   (A) Electron micrograph of a longitudinal section from a fish swim bladder muscle triad junction. (From Franzini-Armstrong and Peachey, 1981.) (B) Three-dimensional reconstruction of a half-triad showing the organization of the junctional membranes and proteins involved in signal transduction. T-tubule and SR membranes face each other across a gap of about 10 nm. The junctional surface of the terminal cisternae of the SR bears two rows of proteins corresponding to the $Ca^{2+}$ release channels/RyR, which project into the gap and come to within 1 nm of the T-tubule membrane. The junctional T-tubule membrane contains parallel rows of four-particle arrays, termed junctional tetrads, presumed to be DHPRs. In fast-twitch fish muscle, tetrads are arranged in a zigzag spacing that corresponds exactly to the position of every other RyR on the adjacent SR (indicated by dotted lines). Nonjunctional and longitudinal regions of the SR are continuous with the terminal cisternae and contain densely packed $Ca^{2+}$-ATPase molecules. The lumen of the SR contains calsequestrin, a high-capacity $Ca^{2+}$-binding protein. (Reproduced from *The Journal of Cell Biology,* 1988, vol. 107, p. 2587 by copyright permission of The Rockefeller University Press.)

**threshold.** That is, when the $Na^+$, $K^+$, and $Cl^-$ ion channels that mediate the action potential are blocked, all that is required to activate contraction is a suprathreshold change in membrane potential. The relationship between tension and membrane potential is shown in Fig. 8.

Subsequently, Schneider and Chandler (1973) discovered the presence in skeletal muscle membranes of mobile charges whose movement could be detected as a voltage-dependent capacity current, termed **charge movement.** Based on the similarity of the kinetics and voltage dependence of charge movement to mechanical activation (Fig. 9), they proposed that it reflected the movement of charged intramembrane domains of a molecule in the T-tubules that functioned as the essential voltage sensor for EC coupling. They further suggested that the voltage sensor might interact directly with a hypothetical $Ca^{2+}$ conducting protein on the SR to cause its opening. This initial mechanical hypothesis of EC coupling, formulated before any proteins of the triad junction had been identified, has formed the framework for subsequent investigations

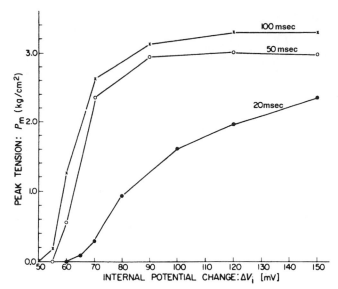

**FIG. 8.** Relationship between peak tension and membrane potential, in the absence of an action potential. The fibers were depolarized by holding the membrane potential constant for 20, 50, or 100 msec with a voltage clamp. Tension was measured simultaneously with a transducer attached to one tendon. (From Caputo *et al.*, 1984. Reproduced from *The Journal of General Physiology* by copyright permission of The Rockefeller University Press.)

into the nature of the molecules and molecular interactions involved.

Recently, several lines of evidence led to the identification of the voltage sensor as the DHPR/T-tubule $Ca^{2+}$ channel that was shown to be localized in skeletal muscle T-tubules at an extremely high density (reviewed in Melzer *et al.*, 1995; Rios and Pizarro, 1991; Beam *et al.*, 1989; Vergara and Asotra, 1987; Schneider, 1994; Lamb, 1992). A definitive identification came following the cloning of

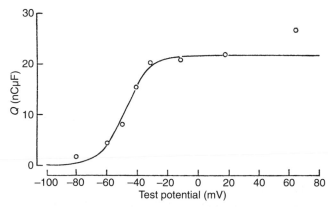

**FIG. 9.** Relationship between charge movement (nC/$\mu$F) and membrane potential (mV). Charge moved at each potential was obtained by integrating the voltage-dependent capacity current, normalized to the fiber linear capacitance near the resting potential. Symbols represent the mean values of charge moved at the "on" and "off" of the pulse. The continuous curve is a fit of the data to a two-state Boltzmann function, with these parameters: $V_{mid} = -47.7$ mV, $k = 8$ mV, and $Q_{max} = 21.5$ nC/$\mu$F. (Reproduced with permission from Chandler *et al.*, 1976.)

the skeletal muscle $Ca^{2+}$ channel (Tanabe *et al.*, 1987) and its use in experiments on dysgenic mice (reviewed in Adams and Beam, 1990). Muscular dysgenesis is a recessive lethal mutation that results in reduction of DHPR protein, lack of slow $Ca^{2+}$ currents and charge movement, and failure of EC coupling. The defect arises from a point mutation in the gene for the DHPR $\alpha_1$-subunit. In an elegant series of experiments, Tanabe *et al.* (1988) demonstrated rescue of $Ca^{2+}$ currents and EC coupling in fibers transfected with the cDNA for the $\alpha_1$-subunit of the DHPR.

Figure 10 (top) shows the inferred 2-D structure of the skeletal $Ca^{2+}$ channel (see also Chapter 26). The S4 transmembrane helix of each homologous repeat contains a large number of positively charged residues that are presumed to form the voltage sensor and generate intramembrane charge movement. The H5-H6 domains fold into the membrane to form the pore for $Ca^{2+}$ current flow. Activation of the skeletal $Ca^{2+}$ channel is slow compared with the closely related cardiac isoform. The S4 charge movements may be followed by slower, less steeply voltage-dependent conformational changes leading to opening. It is thought that the early rapid movement of S4 domains, and not the $Ca^{2+}$ current that follows after a delay of several hundred milliseconds, has been exploited evolutionarily by skeletal muscle to control RyR activation. Because of this slow kinetics of opening, $Ca^{2+}$ influx through the channel pore is not significant during a normal action potential of a few milliseconds duration. However, it may build up to measurable levels during a tetanus, when a skeletal muscle is activated repetitively at rates of 50–100 Hz. In this case $Ca^{2+}$ entry may add to and perhaps modulate intracellular $Ca^{2+}$ release (Oz and Frank, 1991).

## B. Mechanism of Interaction between the DHPR and RyR

Detection of the depolarization by the DHPR/T-tubule $Ca^{2+}$ channel is communicated by the RyR/$Ca^{2+}$ release channel on the SR, resulting in its opening. Figure 11 shows a three-dimensional reconstruction of the RyR in closed and open conformations. The molecular mechanism by which the RyR moves from a closed to the open configuration under control of the T-tubule $Ca^{2+}$ channel is the subject of intense research (reviewed in Rios *et al.*, 1991a; Rios and Pizarro, 1991; Melzer *et al.*, 1995; Ebashi, 1991; Schneider, 1994). Two basic mechanisms have been considered—mechanical and chemical. The mechanical hypothesis of EC coupling proposes that movement of charged voltage-sensing domains of the DHPR allosterically alters an activation domain of the RyR, either directly or via an intermediate protein. In this model, the triad junction exists to bring together the key proteins in a macromolecular complex, analogous to the proteins involved in electron transport on mitochondria membranes. In contrast, the chemical hypothesis proposes that movement of voltage-sensing domains of the DHPR causes the release of a chemical transmitter, perhaps $Ca^{2+}$ or $IP_3$, from the T-tubule junctional membrane (reviewed in Vergara and Asotra, 1987). In this model, the triad junction functions in a manner analogous to a chemical synapse by minimizing

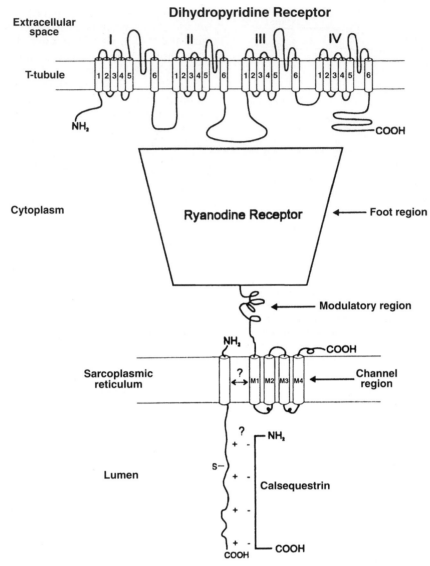

**FIG. 10.**   Key proteins of the triad junction, the DHPR (top) and RyR (bottom), showing primary sequence and presumed secondary membrane topology. For simplicity, only the $\alpha_1$-subunit of the DHPR and only one subunit of the ryanodine receptor homotetramer are shown. DHPRs are composed of one each of four subunits ($\alpha_1,\alpha_2$-$\delta,\beta,\gamma$). The DHPR $\alpha_1$-subunit contains the key functional domains including the voltage sensor, pore, and DHPR binding site. It consists of four homologous internal domains (I–IV) composed of six transmembrane $\alpha$-helices (S1–S6, drawn schematically as cylinders). Alternative models of the RyR suggest as many as 12 transmembrane domains. A question mark indicates the speculative nature of the interaction between the RyR and calsequestrin. (Adapted with permission from McPherson and Campbell, 1993.)

the diffusion distance between the two membranes. Currently, the mechanical hypothesis is favored by a large body of evidence from molecular, biochemical, and functional studies.

### C. Information from Molecular Studies

The RyR has a putative intramembrane $Ca^{2+}$ channel domain as well as a huge cytoplasmic domain that makes up most or all of the foot structure that spans the triadic gap. Thus, opening and closing of the transmembrane pore of the RyR could be remotely controlled by a physically separate domain that makes contact with the DHPR. For example, the DHPR in its resting state could exert an inhibitory influence on the RyR/$Ca^{2+}$ release channel, which could be removed on depolarization, or the RyR could be activated by conformational movements of the DHPR during depolarization. Inhibition of the RyR by the DHPR in the resting state and its removal by depolarization is attractive because of the inherent speed of this type of mechanism and because positively charged transmembrane helices in the S4 domain of the DHPR would move outward in the T-tubule membrane on depolarization, away from the RyR.

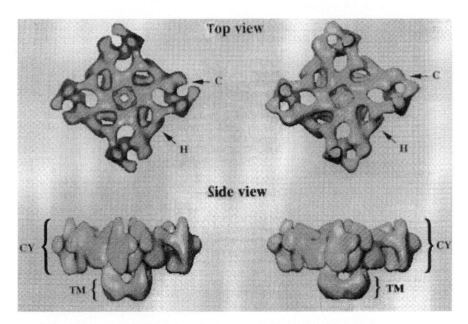

**FIG. 11.** The open and closed states of the skeletal muscle RyR/$Ca^{2+}$-release channel are shown in two different views: top, and side. The putative transmembrane (TM) and cytoplasmic (CY) parts of the protein resemble the stem and cap of a mushroom-shaped protein. The cytoplasmic side of the tetramer contains clamp-shaped (C) and handle (H) domains, which alternately touch and overlap in the open and closed conformations. The cytoplasmic sides of the tetramer are the regions of spatial superposition with the T-tubule tetrads/DHPRs. The transmembrane region of the channel appears open toward the SR, whereas in the closed state a central opening is not seen in this region. (From Orlova *et al.*, 1996. Reproduced from *The Journal of General Physiology* by copyright permission of The Rockefeller University Press.)

Results from studies with dysgenic mice have identified specific domains of the DHPR that are important for controlling activation of the RyR. The $\alpha_1$-subunits of the L-type $Ca^{2+}$ channel/DHPR of skeletal and cardiac muscle share a high degree of homology and both play a key role in triggering $Ca^{2+}$ release. However, as noted, a key functional difference is that $Ca^{2+}$ entry through the channel is required to elicit $Ca^{2+}$ release in cardiac but not in skeletal muscle. Beam and collaborators (Tanabe *et al.*, 1990) exploited this difference to identify molecular regions of the $\alpha_1$-subunit that are critical for skeletal-type EC coupling. When the skeletal muscle $\alpha_1$-subunit was expressed in dysgenic myotubes, a $Ca^{2+}$ current-independent $Ca^{2+}$ release could be elicited in response to electrical depolarization, whereas $Ca^{2+}$ entry was required to elicit $Ca^{2+}$ release when the cardiac $\alpha_1$-subunit was expressed. By constructing and expressing chimeric DHPRs composed of various cardiac and skeletal domains, Tanabe *et al.* (1990) demonstrated that the cytoplasmic loop between domains II and III is a critical determinant of skeletal-type EC coupling. When a chimeric DHPR having the cardiac protein backbone but the skeletal II–III loop was expressed (Fig. 12), contraction could be elicited by electrical stimulation even when the $Ca^{2+}$ current was blocked. This indicates that the II–III loop is involved in signal transmission from the DHPR to the RyR of the SR. Interestingly, expression of both the wild-type skeletal and cardiac DHPRs as well as the chimeric channels all led to the restoration of charge movement. This indicates that the properties of the voltage-sensing step in EC coupling do not depend on the nature

of the transmission mechanism, and likely involve separate molecular domains.

### D. Evidence from Biochemical Studies

Biochemical evidence also supports the existence of protein–protein interactions between the DHP and ryanodine receptor and suggests that the critical II–III loop of the DHPR may mediate a specific interaction with the RyR. Marty *et al.* (1994) demonstrated that the DHPR $\alpha_1$-subunit and RyR can form a complex *in vitro*. Meissner and collaborators (Lu *et al.*, 1994) demonstrated that purified cytoplasmic II–III loop peptides of the skeletal or cardiac $\alpha_1$-subunit can activate the skeletal muscle RyR/$Ca^{2+}$ release

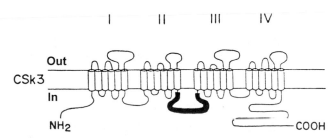

**FIG. 12.** Cardiac–skeletal chimeric DHPR composed of a cardiac backbone and a skeletal muscle II–III loop. (Reprinted with permission from *Nature*, Tanabe *et al.*, 1990, vol. 346, pp. 567–569. Copyright 1990 Macmillan Magazines Limited.)

channel by increasing open channel probability and the affinity of ryanodine binding. The peptides did not activate the cardiac muscle $Ca^{2+}$ release channel. This demonstrates that the II–III loop of cardiac and skeletal DHP $\alpha_1$-subunit interacts functionally with the skeletal $Ca^{2+}$ release channel and provides convincing evidence for a direct allosteric interaction between these proteins. The fact that the II–III peptides activated the skeletal but not the cardiac RyR also suggests that, in addition to the DHPR, the identity of the RyR may determine the type of EC coupling. In a subsequent study, Lu *et al.* (1995) showed that a cAMP-dependent phosphorylation of the DHPR $\alpha_1$-subunit may play a role in the functional interaction between the DHPR and RyR in skeletal muscle. The II–III peptide could activate the skeletal RyR only when it was dephosphorylated at a specific Ser residue.

### E. Evidence from Functional Studies

Functional studies have demonstrated that manipulations that alter T-tubule $Ca^{2+}$ channel function influence SR $Ca^{2+}$ release, and conversely that manipulations that affect $Ca^{2+}$ release alter the function of the DHPR. Such bidirectional cross-talk between the DHPR and RyR, involving both feed-forward and feedback interactions, is expected if there is a physical association between them. Experimentally, conformational transitions of the DHPR are detected as charge movement and SR $Ca^{2+}$ release is detected by monitoring myoplasmic $Ca^{2+}$ transients using optical techniques and $Ca^{2+}$-indicator dyes introduced into the myoplasm.

Rios and collaborators (Rios *et al.*, 1993; Gonzalez and Rios, 1993; Ma *et al.*, 1993) investigated interactions be-

tween the voltage sensor and the $Ca^{2+}$ release process by studying the effects of perchlorate ($ClO_4^-$) on charge movement and $Ca^{2+}$ release. This chaotropic anion is a potent and specific potentiator of EC coupling in skeletal muscle (reviewed in Rios and Pizarro, 1991; Melzer *et al.*, 1995). It shifts the voltage dependence of charge movement by 15 to 30 mV to more negative voltages, slows the kinetics of both "on" and "off" charge movements, and reduces the minimum threshold charge required to activate $Ca^{2+}$ release. Rios *et al.* (1993) demonstrated that several of the actions of $ClO_4^-$ on charge movement and $Ca^{2+}$ release could be explained, at least qualitatively, by a simple model in which the two proteins are allosterically coupled, as shown in Fig. 13. In their model, the RyR/$Ca^{2+}$ release channel is assumed to be a homotetramer that can exist in two conformations—open and closed. The four voltage-sensing subunits of the DHPR act as heterotropic ligands to modulate the kinetics and the equilibrium of this transition. Each shift of a voltage sensor subunit to the active state increases the probability that the release channel will open. In this model, $ClO_4^-$ acts primarily on the RyR/$Ca^{2+}$ release channel to increase its probability of opening and secondarily on the DHPR via allosteric feedback. As part of this feedback to the DHPR, one kinetic component of charge movement, termed $Q_\gamma$ charge, could be simulated. This latter finding complemented a related hypothesis from this group that released $Ca^{2+}$ ions in the triadic gap modify the voltage-sensing process in a positive-feedback manner to generate additional charge movement. Thus, both allosteric interactions between the RyR and the DHPR, and $Ca^{2+}$ ions released into the triadic gap may contribute to the generation of $Q_\gamma$ charge movements.

Functional studies from other laboratories (Pape *et al.*, 1995, 1996; Jong *et al.*, 1995, 1996) have also provided

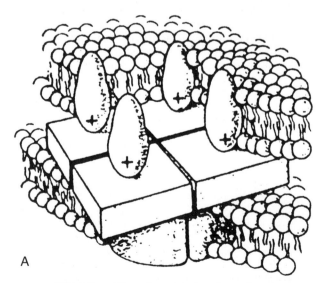

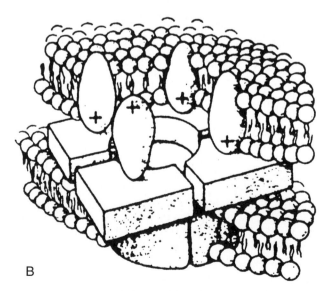

**FIG. 13.** Allosteric model of E–C coupling. Four independent voltage sensor molecules/DHPRs in the T-tubule membrane are aligned with and contact the RyR/$Ca^{2+}$ release channel, depicted as a homotetramer. Two states of the release channel (closed and open) are pictured, as well as two of five possible dispositions of the voltage sensors. (A) All S4 domains in the inactive conformation. (B) One S4 domain in the activating conformation and three S4 domains inactive. (From Rios *et al.*, 1993. Reproduced from *The Journal of General Physiology* by copyright permission of The Rockefeller University Press.)

strong evidence that $Ca^{2+}$ release can influence the DHPR and vice versa. However, the nature of the interaction may be different from the allosteric model described earlier. Chandler and collaborators (Pape *et al.*, 1995; Jong *et al.*, 1996) showed that the steep voltage dependence of $Ca^{2+}$ release is an intrinsic function of the voltage sensor in the membranes of the T-tubule system. In other words, the steep voltage dependence of $Ca^{2+}$ release is conferred directly by the voltage sensor, without additional positive feedback from $Ca^{2+}$ release. They further demonstrated (Jong *et al.*, 1995; Pape *et al.*, 1996) that SR $Ca^{2+}$ content or release or some related event directly alters the kinetics of charge movement, but does not generate a separate species of charge. Both of these findings suggest a close association between the DHPR and the SR $Ca^{2+}$ release channel.

Although the $Ca^{2+}$ current is not required for contraction in skeletal muscle, evidence that $Ca^{2+}$ release and/or the RyR can modify the slow $Ca^{2+}$ current also supports the idea of a close interaction between these proteins. Intracellular BAPTA, a rapid $Ca^{2+}$ buffer that is capable of buffering released $Ca^{2+}$ ions in the triadic space near the release sites (Jong *et al.*, 1996), eliminated the slow $Ca^{2+}$ current (Feldmeyer *et al.*, 1993). Experiments with dyspedic skeletal muscle, a recombinant model that lacks the RyR, suggest that there is reciprocal signaling between the skeletal DHPR and the skeletal RyR such that $Ca^{2+}$ release activity of RyR is controlled by the DHPR and the $Ca^{2+}$ channel activity of the DHPR is controlled by RyR. Dyspedic muscles lack EC coupling and have a 30-fold reduction in $Ca^{2+}$ current density, without change in the density of DHPRs or charge movement (Nakai *et al.*, 1996).

## VI. Summary

Excitation–contraction coupling in skeletal muscle is a fast signal transduction process by which an action potential on the sarcolemmal/T-tubule membranes is transduced into a biochemical trigger for the contractile proteins to shorten and generate force. The immediate intracellular trigger is $Ca^{2+}$ released from intracellular stores. Contraction is inhibited when cytosolic $Ca^{2+}$ is at resting, submicromolar levels, and proceeds when $Ca^{2+}$ rises to micromolar levels.

The key transduction steps than link the depolarization to $Ca^{2+}$ release from the SR occur at interior triad junctions, which bring together the sarcolemmal and SR membranes. At the triads, a sarcolemmal $Ca^{2+}$ channel senses the depolarization during the action potential and communicates it to the Ca release channels/RyRs on the SR, resulting in the release of stored $Ca^{2+}$. The molecular mechanisms of interaction between these two distinct $Ca^{2+}$ channels is not completely known, but likely involves some kind of allosteric process.

## Bibliography

Adams, B. A., and Beam, K. G. (1990). Muscular dysgenesis in mice: A model system for studying excitation–contraction coupling. [Review]. *FASEB J.* **4**, 2809–2816.

Anderson, K., Cohn, A. H., and Meissner, G. (1994). High-affinity [3H]PN200-110 and [3H]ryanodine binding to rabbit and frog skeletal muscle. *Am. J. Physiol.* **266**, C462–C4626.

Armstrong, C. M., Bezanilla, F. M., and Horowicz, P. (1972). Twitches in the presence of ethylene glycol bis(-aminoethyl ether)-*N,N'*-tetracetic acid. *Biochim. Biophys. Acta* **267**, 605–608.

Beam, K. G., Tanabe, T., and Numa, S. (1989). Structure, function, and regulation of the skeletal muscle dihydropyridine receptor. [Review]. *Ann. NY Acad. Sci.* **560**, 127–137.

Block, B. A., Imagawa, T., Campbell, K. P., and Franzini-Armstrong, C. (1988). Structural evidence for direct interaction between the molecular components of the transverse tubule/sarcoplasmic reticulum junction in skeletal muscle. *J. Cell Biol.* **107**, 2587–2600.

Caputo, C., Bezanilla, F., and Horowicz, P. (1984). Depolarization-contraction coupling in short frog muscle fibers. A voltage clamp study. *J. Gen. Physiol.* **84**, 133–154.

Chandler, W. K., Rakowski, R. F., and Schneider, M. F. (1976). A non-linear voltage-dependent charge movement in frog skeletal muscle. *J. Physiol.* **254**, 243–283.

Dulhunty, A. F. (1984). Heterogeneity of T-tubule geometry in vertebrate skeletal muscle fibres. *J. Muscle Res. Cell Motil.* **5**, 333–347.

Ebashi, S. (1991). Excitation–contraction coupling and the mechanism of muscle contraction. [Review]. *Ann. Rev. Physiol.* **53**, 1–16.

Feldmeyer, D., Melzer, W., Pohl, B., and Zollner, P. (1993). A possible role of sarcoplasmic $Ca^{2+}$ release in modulating the slow $Ca^{2+}$ current of skeletal muscle. *Pflugers Arch.* **425**, 54–61.

Franzini-Armstrong, C., and Jorgensen, A. O. (1994). Structure and development of E–C coupling units in skeletal muscle. [Review]. *Ann. Rev. Physiol.* **56**, 509–534.

Franzini-Armstrong, C., and Peachey, L. D. (1981). Striated muscle—contractile and control mechanisms. [Review]. *J. Cell Biol.* **91**, 166s–186s.

Gonzalez, A., and Rios, E. (1993). Perchlorate enhances transmission in skeletal muscle excitation–contraction coupling. *J. Gen. Physiol.* **102**, 373–421.

Gonzalez-Serratos, H. (1971). Inward spread of activation in vertebrate muscle fibres. *J. Physiol.* **212**, 777–799.

Hill, A. V. (1948). On the time required for diffusion and its relation to processes in muscle. *Proc. R. Soc. London Ser. B* **135**, 446–453.

Hodgkin, A. L., and Horowicz, P. (1957). The differential action of hypertonic solutions on the twitch and action potential of a muscle fibre. *J. Physiol.* **136**, 17–18.

Hollingworth, S., Zhao, M., and Baylor, S. M. (1996). The amplitude and time course of the myoplasmic free [$Ca^{2+}$] transient in fast-twitch fibers of mouse muscle. *J. Gen. Physiol.* **108**, 455–469.

Jong, D., Pape, P. C., and Chandler, W. K. (1995). Effect of sarcoplasmic reticulum calcium depletion on intramembranous charge movement in frog cut muscle fibers. *J. Gen. Physiol.* **106**, 659–704.

Jong, D., Pape, P. C., Giebel, J., and Chandler, W. K. (1996). Sarcoplasmic reticulum calcium release in frog cut muscle fibers in the presence of a large concentration of EGTA. *In* "Organellar Ion Channels and Transporters," pp. 256–268. The Rockefeller University Press, New York.

Lamb, G. D. (1992). DHP receptors and excitation–contraction coupling. [Review]. *J. Muscle Res. Cell Motil.* **13**, 394–405.

Lu, X., Xu, L., and Meissner, G. (1994). Activation of the skeletal muscle calcium release channel by a cytoplasmic loop of the dihydropyridine receptor. *J. Biol. Chem.* **269**, 6511–6516.

Lu, X., Xu, L., and Meissner, G. (1995). Phosphorylation of dihydropyridine receptor II–III loop peptide regulates skeletal muscle calcium release channel function. Evidence for an essential role of the beta-OH group of Ser687. *J. Biol. Chem.* **270**, 18459–18464.

Ma, J., Anderson, K., Shirokov, R., Levis, R., Gonzalez, A., Karhanek, M., Hosey, M. M., Meissner, G., and Rios, E. (1993). Effects of perchlorate on the molecules of excitation–contraction coupling of skeletal and cardiac muscle. *J. Gen. Physiol.* **102**, 423–448.

Marty, A., Robert, M., Villaz, M., De Jongh, K., Lai, Y., Catterall, W. A., and Ronjat, M. (1994). Biochemical evidence for a complex involving dihydropyridine receptor and ryanodine receptor in triad junctions of skeletal muscle. *Proc. Natl. Acad. Sci. USA* **91,** 2270–2274.

McPherson, P. S., and Campbell, K. P. (1993). The ryanodine receptor/Ca$^{2+}$ release channel. *J. Biol. Chem.* **268,** 13765–13768.

Melzer, W., Herrmann-Frank, A., and Luttgau, H. C. (1995). The role of Ca$^{2+}$ ions in excitation–contraction coupling of skeletal muscle fibres. [Review]. *Biochim. Biophys. Acta* **1241,** 59–116.

Nakai, L. J., Dirkson, R. T., Nguyen, H. T., Pessah, I. N., Beam, K. G., and Allen, P. D. (1996). Enhanced dihydropyridine receptor channel activity in the presence of ryanodine receptor. *Nature* **380,** 72–75.

Orlova, E. V., Serysheva, I. I., Van Heel, M., Hamilton, S. L., and Chiu, W. (1996). Two structural configurations of the skeletal muscle calcium release channel. *J. Gen. Physiol.* **106,** 659–704.

Oz, M., and Frank, G. B. (1991). Decrease in the size of tetanic responses produced by nitrendipine or by extracellular calcium ion removal without blocking twitches or action potentials in skeletal muscle. *J. Pharmacol. Exper. Therapeut.* **257,** 575–581.

Pape, P. C., Jong, D. S., Chandler, W. K., and Baylor, S. M. (1993). Effect of fura-2 on action potential-stimulated calcium release in cut twitch fibers from frog muscle. *J. Gen. Physiol.* **102,** 295–332.

Pape, P. C., Jong, D. S., and Chandler, W. K. (1995). Calcium release and its voltage dependence in frog cut muscle fibers equilibrated with 20 m*M* EGTA. With an Appendix: Measurement of the rapidly available buffering power of myoplasm. *J. Gen. Physiol.* **106,** 259–336.

Pape, P. C., Jong, D., and Chandler, W. K. (1996). A slow component of intramembranous charge movement during sarcoplasmic reticulum calcium release in frog cut muscle fibers. *J. Gen. Physiol.* **107,** 79–101.

Paul, R. J., Ferguson, D. G., and Heiny, J. A. (1996). Muscle physiology: Molecular mechanisms. In "Essentials of Physiology" (N. Sperelakis and R. O. Banks, Eds.), pp. 203–216. Little, Brown and Co., Boston.

Peachey, L. D. (1965). The sarcoplasmic reticulum and transverse tubules of the frog's sartorius. *J. Cell Biol.* **25,** 209–232.

Peachey, L. D., and Eisenberg, B. R. (1978). Helicoids in the T system and striations of frog skeletal muscle fibers seen by high voltage electron microscopy. *Biophys. J.* **22,** 145–154.

Rios, E., Ma, J. J., and Gonzalez, A. (1991a). The mechanical hypothesis of excitation–contraction (EC) coupling in skeletal muscle. [Review]. *J. Muscle Res. Cell Motil.* **12,** 127–135.

Rios, E., Shirokov, R., Levis, R., Gonzalez, A., Stavrovsky, J., Ma, J., Mundina-Weilenmann, C., and Hosey, M. M. (1991b). Differential effects of perchlorate on skeletal muscle EC coupling, cardiac gating currents and gating of DHP receptors in bilayers. *Biophys. J.* **59,** 201a.

Rios, E., Karhanek, M., Ma, J., and Gonzalez, A. (1993). An allosteric model of the molecular interactions of excitation–contraction coupling in skeletal muscle. *J. Gen. Physiol.* **102,** 449–481.

Rios, E., and Pizarro, G. (1991). Voltage sensor of excitation–contraction coupling in skeletal muscle. [Review]. *Physiol. Rev.* **71,** 849–908.

Schneider, M. F. (1994). Control of calcium release in functioning skeletal muscle fibers. [Review]. *Annu. Rev. Physiol.* **56,** 463–484.

Schneider, M. F., and Chandler, W. K. (1973). Voltage-dependent charge movement in skeletal muscle: A possible step in excitation–contraction coupling. *Nature* **242,** 244–246.

Tanabe, T., Takeshima, H., Mikami, A., Flockerzi, V., Takahaski, H., Kangawa, K., Kojima, M., Matsuo, H., Hirose, T., and Numa, S. (1987). Primary structure of the receptor for calcium channel blockers from skeletal muscle. *Nature* **328,** 313–318.

Tanabe, T., Beam, K. G., Powell, J., and Numa, S. (1988). Restoration of excitation–contraction coupling and slow calcium current in dysgenic muscle by dihydropyridine receptor complementary DNA. *Nature* **336,** 134–139.

Tanabe, T., Beam, K. G., Adams, B., Niidome, T., and Numa, S. (1990). Regions of the skeletal muscle dihydropyridine receptor critical for excitation–contraction coupling. *Nature* **346,** 567–569.

Vergara, J., and Delay, M. (1986). A transmission delay and the effect of temperature at the triadic junction of skeletal muscle. *Proc. R. Soc. London Ser. B* **229,** 97–110.

Vergara, J., and Asotra, K. (1987). The chemical transmission mechanism of excitation–contraction coupling in skeletal muscle. [Review]. *NIPS* **2,** 182–186.

Gerhard Meissner

# 54

# Ca²⁺ Release from Sarcoplasmic Reticulum in Muscle

## I. Introduction

In excitable cells, the release of $Ca^{2+}$ ions from intracellular membrane compartments can be triggered by an excitatory electrical signal, or it can occur via a chain of voltage-independent steps that involve agonist-induced formation of inositol 1,4,5-trisphosphate ($IP_3$) and subsequent activation of an intracellular membrane receptor/$Ca^{2+}$ channel complex, the **$IP_3$ receptor** (Berridge, 1993). The voltage-dependent mechanism, commonly referred to as **excitation–contraction** (EC) **coupling,** has been most thoroughly studied in vertebrate striated muscle.

Figure 1 illustrates the major components involved in EC coupling: a *transverse (T)-tubule* membrane system of invaginations through which muscle contraction is triggered, and an intracellular $Ca^{2+}$ storing and $Ca^{2+}$ releasing membrane system, the **sarcoplasmic reticulum** (SR), which contains an ATP-dependent $Ca^{2+}$ pump and a $Ca^{2+}$ release channel. The release channels span the narrow gap where the SR (the $Ca^{2+}$ store) and the T-tubule (the conduit of the action potential) are within ∼15 nm of each other (for reviews, see Fleischer and Inui, 1989; Franzini-Armstrong and Jorgensen, 1994). The SR $Ca^{2+}$ release channels are also known as **feet** or **junctional processes,** and as **ryanodine receptors** (RyRs) because they have the ability to bind the plant alkaloid ryanodine with high affinity and specificity. They are often viewed, at least in skeletal muscle, as directly linked to four particles located in the T-tubule membrane and presumed to represent another $Ca^{2+}$ channel (L-type) also known as the **dihydropyridine receptor** (DHPR). The relative number of RyRs and DHPRs in striated muscle varies greatly, ranging from about 10 RyRs per DHPR in cardiac muscle to a 1:1 stoichiometry in fast-twitch mammalian skeletal muscle. Accordingly, not all RyRs may be linked or closely apposed to DHPRs in striated muscle.

In addition to the $Ca^{2+}$ release channel, the SR of skeletal and cardiac muscle contains monovalent ion-selective channels: (1) a $K^+$ channel, (2) a $Cl^-$ channel, and (3) a $H^+$ ($OH^-$) permeable pathway. Movement of ions through the monovalent cation- and anion-selective channels has been proposed to compensate charge movements across the SR membrane, and thereby to support rapid $Ca^{2+}$ release and reuptake (see review by Meissner, 1983).

## II. Mechanisms of Excitation–Contraction Coupling

Different EC coupling mechanisms apparently exist in **skeletal** and **cardiac muscles** (see reviews by Fabiato, 1983; Rios and Pizarro, 1991). A distinguishing feature is that EC coupling in cardiac muscle is dependent on extracellular $Ca^{2+}$, whereas skeletal EC coupling is not (in the short run). In cardiac muscle, DHP-sensitive (L-type) $Ca^{2+}$ channels located in the surface membrane and T-tubule have been shown to mediate the influx of $Ca^{2+}$ during an action potential by functioning as voltage-dependent $Ca^{2+}$ channels (Fig. 2). The resulting rise in intracellular $Ca^{2+}$ concentration is believed to trigger the massive release of $Ca^{2+}$ by opening SR $Ca^{2+}$ release channels. This process is known as **calcium-induced $Ca^{2+}$ release** (CICR). Intracellular $Ca^{2+}$ transients have been suggested to arise as the sum of localized $Ca^{2+}$ release events called **$Ca^{2+}$ sparks.** The opening of a single T-tubule $Ca^{2+}$ channel is apparently sufficient to evoke a $Ca^{2+}$ spark by activating a functional $Ca^{2+}$ release unit, which may consist of one or more $Ca^{2+}$ release channels (Santana *et al.,* 1996).

It has been suggested that $Ca^{2+}$ also plays a major role in regulating a ryanodine-sensitive SR $Ca^{2+}$ release channel in crustacean skeletal muscle. **Mammalian smooth muscle** and **neurons** also contain $Ca^{2+}$-activated $Ca^{2+}$ release channels (Meissner, 1994).

In vertebrate skeletal muscle, a different mechanism of EC coupling appears to be present. A unique mechanism, referred to as the **mechanical coupling mechanism,** has

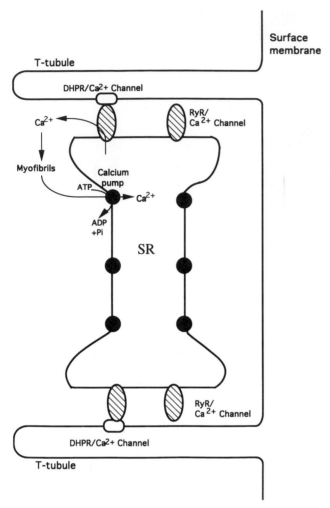

**FIG. 1.** Schematic representation of a segment of a mammalian skeletal muscle cell illustrating the major components involved in E–C coupling.

tissues, including muscles, express three RyR/Ca$^{2+}$ release channel isoforms, which are encoded by three different genes, *ryr1, ryr2,* and *ryr3.* The three RyR isoforms are also known as the skeletal muscle (RyR1), cardiac muscle (RyR2), and brain (RyR3) RyR because they were first identified and isolated from skeletal muscle, cardiac muscle, and brain, respectively (see reviews by Coronado *et al.,* 1994; Meissner, 1994; Ogawa, 1994). All three share a high-affinity binding site for [$^3$H]ryanodine and a high-conductance pathway for Ca$^{2+}$ and monovalent cations, but display isoform- and species-dependent differences in their *in vitro* regulation by Ca$^{2+}$ and other effector molecules.

## III.  Isolation of Membrane Fractions Enriched in Ca$^{2+}$ Release Channels and Ryanodine Binding Activity

The molecular properties of the RyR/Ca$^{2+}$ release channels have been most extensively studied with rabbit skeletal muscle as the source for the RyRs. Fragmentation of the SR during homogenization and subsequent fractionation by differential and density-gradient centrifugation yields a "heavy" SR vesicle fraction that is enriched in Ca$^{2+}$ release channels and [$^3$H]ryanodine binding activity and corresponds to the junctional region of the SR **(junctional SR)** (see Fig. 1). Another important advance allowing study of the coupling between T-tubule depolarization and SR Ca$^{2+}$ release in skeletal muscle (Ikemoto *et al.,* 1985) has been the isolation of membrane fractions composed of a T-tubule segment sandwiched between two junctional SR vesicles (Fleischer and Inui, 1989). These junctional complexes are known as **triads.**

Microsomal membrane fractions enriched in ryanodine-sensitive Ca$^{2+}$ release channels have been also isolated from other excitable tissues, including cardiac muscle, smooth muscle, and brain. In these cases, however, the membrane fractions are typically of a lower purity than those from skeletal muscle.

## IV.  Isolation and Structure of RyR/Ca$^{2+}$ Release Channel

The isolation and structural determination of the SR Ca$^{2+}$ release channel has been greatly facilitated by the identification of ryanodine as a channel-specific ligand. Ryanodine is a neutral plant alkaloid that is obtained from the stems of the South American shrub, Ryania speciosa, and is composed of two major compounds: ryanodine and 9,21-didehydroryanodine (Fig. 3). Ryanodine is a highly toxic compound. Its pharmacological effects have been most clearly shown in muscle, where, depending on muscle type and activity, it can cause either contracture or a decline in contractile force (Jenden and Fairhurst, 1969). Ca$^{2+}$ flux studies with isolated SR vesicles and single-channel recordings have shown that ryanodine specifically affects the SR Ca$^{2+}$ release channel. Ryanodine activates the chan-

been formulated and suggests that the SR Ca$^{2+}$ release channel is opened via a direct physical interaction with a voltage-sensing molecule in the T-tubule (see Fig. 2). Biophysical and pharmacological evidence, as well as molecular expression studies, has suggested that the T-tubule DHPR is the **voltage sensor** for EC coupling in vertebrate skeletal muscle (Rios and Pizarro, 1991). A more recent view holds that vertebrate skeletal muscle EC coupling is regulated in addition to DHPR-dependent mechanisms, by Ca$^{2+}$-dependent mechanisms (see review by Schneider, 1994). According to this view, Ca$^{2+}$ ions are initially released by SR Ca$^{2+}$ release channels that are directly linked to DHPRs. The released Ca$^{2+}$ then open Ca$^{2+}$ release channels that are not linked to DHPR (see Fig. 1). Other proteins may have an important role in the coupling of the DHPR and RyR (Meissner, 1994).

We discuss later studies carried out to determine the structural and functional features of the SR RyR/Ca$^{2+}$ release channel. These studies have shown that mammalian

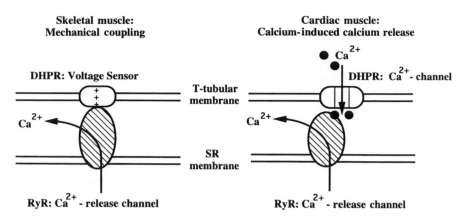

FIG. 2. Models of muscle E–C coupling. Depicted are the mechanical coupling model for vertebrate skeletal muscle, and the calcium-induced calcium release (ClCR) model for cardiac muscle.

nel at low (nanomolar) concentrations, but inhibits the channel at high (micromolar) concentrations (see Figs. 10 and 11 in a later section). Because the drug binds with high specificity and dissociates slowly from the high-affinity site of the receptor (either membrane bound or detergent solubilized), [³H]ryanodine has been found to be an ideal probe in the isolation of the RyR from a variety of tissues and species.

The RyR was first isolated from striated muscle because its role in regulating free cytoplasmic Ca²⁺ levels was originally recognized in muscle, and relatively large amounts of membranes enriched in [³H]ryanodine binding activity can be obtained from skeletal muscle and cardiac muscle. In most studies, the membrane-bound Ca²⁺ release channels were solubilized in the presence of [³H]ryanodine, using the zwitterionic detergent CHAPS and high ionic strength (1.0 $M$ NaCl); they were then purified by sequential column chromatography (Inui *et al.*, 1987). Purification of a functional receptor can also be obtained, in essentially

one step, by immunoaffinity chromatography (Smith *et al.*, 1988) or density-gradient centrifugation through a linear sucrose gradient (Lai *et al.*, 1988). Figure 4A shows the [³H]ryanodine pattern on the sucrose gradients and sedimentation profile of the proteins associated with junctional SR vesicles isolated from rabbit skeletal muscle. A single peak of bound radioactivity, comigrating with a small protein peak possessing an apparent sedimentation coefficient of 30S, is observed in the lower half of the gradients. Binding to the small protein peak is specific since no radioactivity is present in the lower half of the gradients when the membranes are incubated with an excess of cold ryanodine. A sedimentation coefficient of 30S suggests that the RyR is a very large macromolecular protein complex with a molecular weight ($M_r$) in excess of 1,000,000.

The sucrose gradient centrifugation procedure is relatively simple and straightforward. The procedure results in efficient separation of the large 30S RyR complex from the other solubilized smaller SR proteins because of its faster sedimentation rate. This method has been used to isolate a functional 30S RyR from several species and tissues including skeletal muscle, cardiac muscle, smooth muscle, and brain (Meissner, 1994).

Sodium dodecyl sulfate–polyacrylamide gel electrophoresis (SDS–PAGE) has shown that the 30S protein complexes of mammalian skeletal and cardiac muscles are composed of a high molecular weight RyR polypeptide and isoform-specific low molecular weight immunophilins (FK506 binding protein), which migrate with an apparent $M_r$ >340,000 (see Fig. 4B) and $M_r$ ~12,000 (Timerman *et al.*, 1996; not visible on the gels in Fig. 4B), respectively. Cloning and sequencing of the complementary DNA of the mammalian skeletal and cardiac muscle RyR isoforms have revealed an open reading frame of about 15 kb and encoding RyR polypeptides of $M_r$ ~560,000 (Meissner, 1994; Ogawa, 1994). Purified RyR preparations from mammalian brain, crustacean skeletal muscle, and the nematode *Caenorhabditis elegans* also displayed a single protein band of high molecular weight on SDS gels. In contrast, the presence of two immunologically distinct high molecular weight RyR protein bands (corresponding to the mamma-

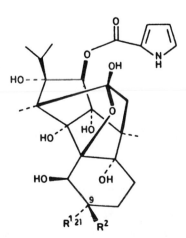

FIG. 3. Structure of ryanodine (R¹ = CH₃) and 9,21-didehydroryanodine (R¹ and R² = CH₂).

lian RyR1 and RyR3) has been described for the main skeletal muscles of chicken, frog, and fish (see Section V). The two RyR isoforms are present as discrete homo-oligomers in amphibian and avian skeletal muscles. They were shown to differ in their $Ca^{2+}$ sensitivity and to have immunological properties and an electrophoretic mobility, respectively, characteristic of the mammalian skeletal and cardiac receptor isoforms. The appearance of the RyR1 isoform alone in some very fast-contracting muscle of fish suggests that this isoform is selectively expressed when rapid contraction is required in nonmammalian vertebrate muscles (O'Brien *et al.*, 1993).

Electron microscopy has revealed that the purified mammalian RyRs have a morphology nearly identical to that of the protein bridges (junctional feet) that span the T-tubule/SR junctional gap in vertebrate skeletal muscles (Franzini-Armstrong and Jorgensen, 1994). The skeletal muscle RyR is composed of four polypeptides of $M_r$ ~560,000 as evidenced by (1) the four-leaf clover-like (quatrefoil) appearance of negatively stained samples (Fig. 5), (2) a high apparent sedimentation coefficient of 30S (see Fig. 4), and (3) cross-linking studies (Lai *et al.*, 1989). Three-dimensional reconstruction of images from electron microscopy indicates that the skeletal muscle RyR consists of a large, loosely packed cytosolic "foot" region with overall dimensions of 29 × 29 × 12 nm and a smaller

"transmembrane" region that extends ~7 nm toward the SR lumen and likely contains the $Ca^{2+}$ channel pore (Radermacher *et al.*, 1994; Serysheva *et al.*, 1995).

## V. Molecular Cloning and Expression of RyR

Current evidence indicates that mammalian tissues express three types of RyR that are encoded by three different genes. The primary structure of the skeletal muscle (RyR1), cardiac muscle (RyR2), and brain (RyR3) ryanodine receptors was determined by cDNA cloning and sequencing. The three RyR proteins are composed of about 5000 amino acid residues, and the predicted amino acid sequence between the three types of RyR is 65–70% identical (Ogawa, 1994). cDNAs encoding the amphibian homologs of mammalian RyR1 and RyR3 (Oyamada *et al.*, 1994) and avian homolog of mammalian RyR3 (Ottini *et al.*, 1996) have been isolated and sequenced. Isolation and sequencing of a gene for a RyR in the fruit fly (*Drosophila melanogaster*) (Takeshima *et al.*, 1994) revealed that there is 45–47% identity between the amino acid sequences of *Drosophila* RyR and the three mammalian RyRs.

Northern blot analysis of mRNA from a variety of mammalian tissues has indicated that the full-length skeletal isoform is abundant in fast-twitch and slow-twitch skeletal

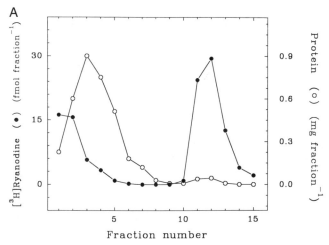

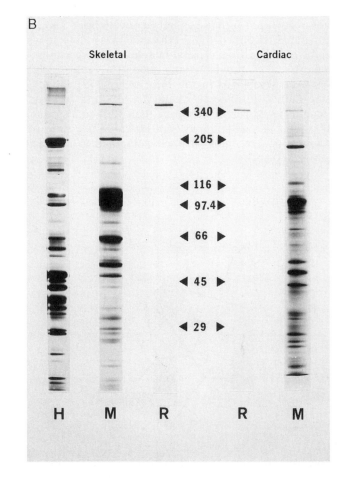

**FIG. 4.** Sedimentation profile and sodium dodecyl sulfate (SDS) gel electrophoresis of rabbit skeletal muscle and canine cardiac RyRs. (A) Heavy skeletal muscle SR vesicles were solubilized in CHAPS, centrifuged through a linear sucrose gradient, and fractionated. Fractions were analyzed for protein and ³H radioactivity. Unbound [³H]ryanodine and the majority of the solubilized proteins sedimented near the top of the gradient (fractions 1–6), whereas the [³H]ryanodine-labeled RyR comigrated with a small protein peak to the bottom of the gradient (fractions 11–13) [Modified with permission from Lai *et al.*, *Nature* **331,** 315–319 (1988). Copyright 1988 Macmillan Magazines Limited.] (B) Silver stained SDS–polyacrylamide gel of whole rabbit skeletal muscle homogenate (H), and rabbit skeletal and canine cardiac muscle heavy SR membranes (M) and purified RyRs (R). Sizes of molecular weight standards are shown (×10⁻³) [Reprinted with permission from Meissner *et al.*, *Mol. Cell. Biochem.* **82,** 59–65 (1988).]

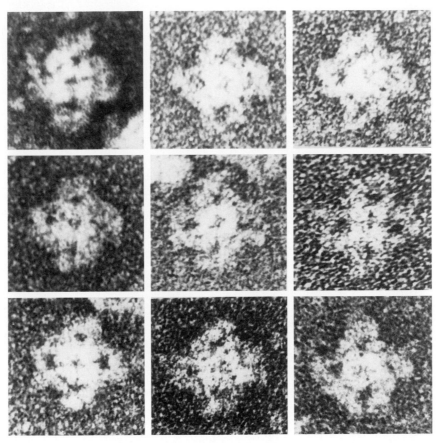

**FIG. 5.** Negative-stain electron micrograph of the purified rabbit skeletal muscle RyR. Shown is a selected panel of particles displaying the characteristic four-leaf clover (quatrefoil) structure of the 30S RyR complex. Dimensions of the quatrefoils are 34 nm from the tip of one leaf to the tip of the opposite one, with each leaf 14 nm wide. The central electron-dense region has a diameter of 14 nm with the central "hole" of a diameter of 1–2 nm. [Reprinted with permission from Lai *et al.*, *Nature* **331**, 315–319 (1988). Copyright 1988 Macmillan Magazines Limited.]

muscles, and present at low levels in the brain as a full-length and abbreviated species of ~2400 nucleotides (Takeshima *et al.*, 1993; Furuichi *et al.*, 1994; Ledbetter *et al.*, 1994; Giannini *et al.*, 1995). The cardiac isoform is expressed, in addition to the heart, at low levels in the brain. The brain RyR mRNA is expressed, in addition to the brain, in tissues containing smooth muscle such as aorta and uterus, and as a minor component in skeletal muscle and cardiac muscle (Furuichi *et al.*, 1994; Ledbetter *et al.*, 1994; Giannini *et al.*, 1995). Expression studies with the mammalian cDNAs in cultured cells and *Xenopus* oocytes have suggested that the $M_r$ 560,000 polypeptides are sufficient to form Ca²⁺ channels sensitive to ryanodine, caffeine, and Ca²⁺.

Two RyR gene-knockout mice were constructed to address the question of a functional requirement of the predominant (RyR1) and minor (RyR3) RyR isoforms for skeletal muscle EC coupling. Mutant mice lacking RyR1 died perinatally and skeletal muscle fibers failed to show a contractile response to electrical stimulation under physiological conditions. In contrast, mutant mice lacking RyR3 showed normal growth, and EC coupling appeared to be

normal. These results indicate that RyR1 but not RyR3 is essential for skeletal muscle EC coupling (Takeshima *et al.*, 1996).

Analysis of the hydropathy of the predicted amino acid sequence has provided some valuable clues regarding the disposition of the RyR in the SR membrane. Hydropathy plots of the amino acid sequences of the three mammalian RyRs have suggested that the $M_r$ ~560,000 polypeptides have two major structural regions: (1) a carboxy-terminal pore region that exhibits a high extent of similarity in amino acid sequence and traverses the membrane at least four times (hence 16 or more transmembrane segments per tetrameric RyR), and (2) a large more variable extramembrane region, which is thought to correspond to the cytoplasmic foot structure. Primary sequence predictions also suggest the presence of several phosphorylation sites and sites for binding of cytoplasmic Ca²⁺, nucleotides, and calmodulin (Meissner, 1994).

One site for phosphorylation by a calmodulin-dependent protein kinase has been identified (Ser²⁸⁴³ and Ser²⁸⁰⁹ in the skeletal muscle and cardiac muscle RyR, respectively). Phosphorylation of the site in the cardiac RyR activated

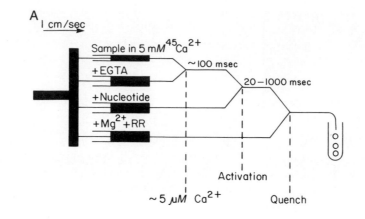

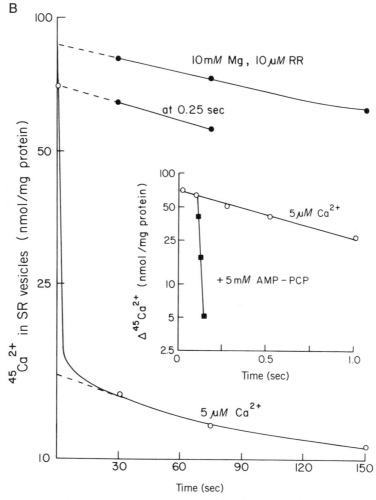

**FIG. 6.** Measurement of rapid $^{45}Ca^{2+}$ efflux rates. (A) Diagrammatic representation of rapid $^{45}Ca^{2+}$ efflux quench experiments. A major part of the rapid mixing apparatus is a ram that pushes simultaneously four syringes filled with SR vesicles and three mixing solutions. In the example depicted, vesicles are passively loaded with 5 m$M$ $^{45}Ca^{2+}$ by prolonged incubation (2 hr at room temperature) before being placed in the top syringe. In the first mixing chamber, the extravesicular $[Ca^{2+}]$ is reduced to a more physiological concentration of 5 $\mu M$ or less by dilution into a solution containing a $Ca^{2+}$ chelating agent (EGTA). A nucleotide such as ATP or AMP-PCP can then be added in a second mixing chamber to induce very rapid $^{45}Ca^{2+}$ release. In the third mixing chamber, $^{45}Ca^{2+}$ release is inhibited by the addition of $Mg^{2+}$ and ruthenium red (RR), which in combination, strongly inhibit the skeletal muscle RyR. After the final mixing step, vesicles are collected, placed on a filter under vacuum, and washed. The radioactivity remaining with the vesicles is determined by liquid scintillation counting. (B) Time course of $^{45}Ca^{2+}$ efflux from heavy rabbit skeletal muscle SR vesicles that were passively loaded with 5 m$M$ $^{45}Ca^{2+}$ by incubation at 23°C for 2 hr and then diluted into iso-osmolal, unlabeled release media. Rapid $^{45}Ca^{2+}$ efflux was inhibited by the addition of a quench solution (10 m$M$ $Mg^{2+}$ and 10 $\mu M$ RR). $^{45}Ca^{2+}$ remaining with the vesicles was determined by placing them on a filter and washing with the quench solution. Amounts of $^{45}Ca^{2+}$ initially trapped by all

the $Ca^{2+}$ release channel by apparently reversing the inhibition of the channel that is observed in the presence of **calmodulin** (Witcher *et al.*, 1991). Some progress has been also made in localizing the regulatory sites for $Ca^{2+}$, ATP, and calmodulin within the large $Ca^{2+}$ release channel complex using molecular biological and immunological approaches. Calmodulin-binding studies with synthesized RyR peptides have identified three calmodulin-binding sites of the three mammalian RyRs (Guerrini *et al.*, 1995). A point mutation (Arg615 → Cys) in the $Ca^{2+}$ release channel of porcine skeletal muscle has been shown to alter the channel's sensitivity to $Ca^{2+}$ and to be responsible for the hypersensitivity to caffeine and halothane in the rare muscle disorder known as malignant hyperthermia (Otsu *et al.*, 1994). Studies with site-directed antibodies (against 4478–4512) have provided evidence for a second $Ca^{2+}$-sensitive region in the skeletal muscle RyR (Chen *et al.*, 1993).

## VI. RyR Is a High-Conductance Ligand-Gated Channel

Function of the $Ca^{2+}$ release channel has been studied *in vitro* by three complementary techniques: (1) measurement of macroscopic $Ca^{2+}$ fluxes from passively or actively loaded SR vesicles, (2) recording from single $Ca^{2+}$ release channels incorporated into planar lipid bilayers, and (3) measurement of [³H]ryanodine binding.

### A. SR Vesicle–Ca²⁺ Efflux Measurements

In skeletal and cardiac muscles, the SR releases its $Ca^{2+}$ stores in milliseconds in response to an action potential. To measure similarly rapid $Ca^{2+}$ fluxes *in vitro*, rapid mixing and filtration devices must be employed. The released $Ca^{2+}$ can be measured using $Ca^{2+}$ indicator dyes such as fura-2 or a radioisotope of $Ca^{2+}$ ($^{45}Ca^{2+}$).

Figure 6 shows a typical $^{45}Ca^{2+}$ efflux experiment. A rapid mixing device (Fig. 6A) is used to measure $^{45}Ca^{2+}$ release by a passively loaded, heavy SR vesicle fraction derived from junctional SR. $^{45}Ca^{2+}$ efflux is slow when the vesicles are directly diluted into a medium containing two inhibitors ($Mg^{2+}$ and ruthenium red) of the $Ca^{2+}$ release channel (Fig. 6B). This allows determination of the amounts of $^{45}Ca^{2+}$ trapped by all vesicles at zero time. In release media containing micromolar concentrations of free $Ca^{2+}$ (5 $\mu M$ in Fig. 6B), the SR $Ca^{2+}$ release channel is partially activated, resulting in the release of a majority of the trapped $^{45}Ca^{2+}$ in just a few seconds. Some radioactivity remains with the vesicles for longer times (>30 sec) because not all of them contain the $Ca^{2+}$ release channel. In the figure inset, $^{45}Ca^{2+}$ release is stopped by the addition of

$Mg^{2+}$ and ruthenium red at varying time intervals (ranging from 25 to 1000 msec). In the presence of 5 $\mu M$ free $Ca^{2+}$, the vesicles release half their $^{45}Ca^{2+}$ stores within 0.7 sec, corresponding to a first-order rate constant of about 1.0 $sec^{-1}$. Figure 6B further shows that AMP-PCP [a nonhydrolyzable adenosine triphosphate (ATP) analog] increases the initial release rate about 50-fold, resulting in nearly complete release in 50 msec.

Table 1 compares the $Ca^{2+}$ release properties of skeletal muscle and cardiac muscle SR vesicles that were diluted into media containing different concentrations of $Ca^{2+}$, $Mg^{2+}$, adenine nucleotide, and calmodulin. In both membrane fractions, $^{45}Ca^{2+}$ release is activated by millimolar ATP and micromolar $Ca^{2+}$. In the presence of micromolar $Ca^{2+}$ and millimolar AMP-PCP, skeletal and cardiac vesicles release half their $^{45}Ca^{2+}$ stores in about 10 msec. Millimolar $Ca^{2+}$ and $Mg^{2+}$ and micromolar calmodulin have an inhibitory effect.

Major differences in the *in vitro* regulation of the skeletal muscle and cardiac muscle $Ca^{2+}$ release channels have been observed. At $10^{-5}$ $M$ extravesicular (cytoplasmic) $Ca^{2+}$, the half time of $^{45}Ca^{2+}$ release is about 20 msec for cardiac vesicles, as compared with 600 msec for skeletal vesicles (Table 1). In contrast, adenine nucleotides are more effective in stimulating $^{45}Ca^{2+}$ release from skeletal than from cardiac vesicles. In addition, $^{45}Ca^{2+}$ efflux from cardiac vesicles is only partially inhibited by 1 m$M$ $Ca^{2+}$ or 1 m$M$ $Mg^{2+}$, whereas the $Ca^{2+}$-activated release channel of skeletal muscle is nearly fully inhibited.

### B. Single-Channel Planar Lipid Bilayer Measurements

The RyR/$Ca^{2+}$ release channel is a cation-selective channel that displays an unusually large ion conductance for

**TABLE I**  Comparison of $^{45}Ca^{2+}$ Efflux from SR Vesicles of Rabbit Skeletal Muscle and Canine Cardiac Muscle

| Free $Ca^{2+}$ ($M$) | Additions to efflux medium ($M$) | Rabbit skeletal | Canine cardiac |
|---|---|---|---|
| | | $^{45}Ca^{2+}$ efflux ($t_{1/2}$, sec) | |
| <$10^{-8}$ | — | 8 | 25 |
| | $5 \times 10^{-3}$ ATP | 0.06 | 12 |
| ~$10^{-5}$ | — | 0.6 | 0.02 |
| | $5 \times 10^{-3}$ AMP-PCP | 0.01 | 0.01 |
| | $2 \times 10^{-6}$ Calmodulin | 1.2 | 0.1 |
| | $10^{-3}$ $Mg^{2+}$ | 15 | 0.25 |
| $10^{-3}$ | — | 10 | 0.1 |

*Note:* SR vesicles were passively loaded with 1 m$M$ $^{45}Ca^{2+}$ and diluted into efflux media containing the indicated concentrations of free $Ca^{2+}$ and other additions; $t_{1/2}$ indicates the time required for the $Ca^{2+}$-permeable vesicles to release half of their $^{45}Ca^{2+}$ content.

vesicles (87 nmol/mg protein) and not readily released by subpopulation of vesicles lacking the $Ca^{2+}$ release channel (15 nmol/mg protein) were estimated by extrapolating back to the time of vesicle dilution (broken lines). In the inset, the time course of $^{45}Ca^{2+}$ efflux from the vesicle population containing the $Ca^{2+}$ release channel was obtained by subtracting the amount not readily released (15 nmol/mg protein) [Reprinted with permission from Meissner, *Methods Enzymol.* **32**, 417–437 (1988).]

monovalent cations (~750 pS with 250 m$M$ K$^+$ as the current carrier) and divalent cations (~150 pS with 50 m$M$ Ca$^{2+}$). The existence of a large Ca$^{2+}$ conductance was originally demonstrated in single-channel recordings with native skeletal and cardiac muscle SR vesicles fused with planar lipid bilayers, and subsequently in studies with purified RyRs incorporated into lipid bilayers (Meissner, 1994). Since the Ca$^{2+}$ release channel conducts monovalent cations such as Na$^+$, K$^+$, or Cs$^+$ more efficiently than Ca$^{2+}$, many of the reported studies have been performed using monovalent cations rather than Ca$^{2+}$ as the conducting ion. As illustrated in Fig. 7, the 30S RyR complex, purified in the absence of [$^3$H]ryanodine, can be recorded in planar lipid bilayers in the presence of a symmetric 250 m$M$ KCl medium. The CHAPS-solubilized, purified RyR can be directly incorporated into a lipid bilayer; however, in our hands more reproducible results are obtained when the CHAPS-solubilized, purified RyR is first reconstituted into lipid bilayer vesicles by removal of the detergent by dialysis. The lipid vesicles are then fused with a planar lipid bilayer (Tripathy *et al.*, 1995). In most experiments, the channel complex incorporates in the bilayer with the cytoplasmic (foot) region of the channel facing the *cis* side of the bilayer (i.e., the side to which the complex is added). Regulation of the channel can therefore be conveniently studied by varying the Ca$^{2+}$, Mg$^{2+}$, or ATP concentration in the *cis* chamber of the bilayer apparatus. If desired, a Ca$^{2+}$ conducting channel can be obtained by perfusing the *trans* chamber with a CaCl$_2$ buffer. Cl$^-$ can be used in these studies because the channel does not conduct anions.

In Fig. 8, the purified skeletal muscle Ca$^{2+}$ release channel (Fig. 8A) and cardiac muscle Ca$^{2+}$ release channel (Fig. 8B) were incorporated into planar lipid bilayers in a symmetrical KCl medium. The single-channel traces show that the activity of both channels is dependent on the Ca$^{2+}$

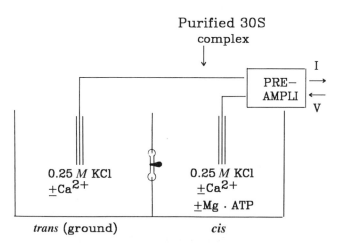

**FIG. 7.** Diagrammatic representation of planar lipid bilayer apparatus used for recording purified 30S Ca$^{2+}$ release channel complexes. The *trans* and *cis* sides of the planar lipid bilayer are equivalent to the lumenal and cytoplasmic sides of the SR membrane, respectively.

concentration at the *cis* (cytoplasmic) side of the bilayer. The activity of the skeletal muscle channel is increased by increasing *cis* Ca$^{2+}$ from a submicromolar concentration to 100 $\mu M$. Further increase in *cis* Ca$^{2+}$ to 600 $\mu M$ inhibits channel opening. The cardiac channel shows a similar Ca$^{2+}$ activation behavior. However, as seen in vesicle–Ca$^{2+}$ efflux studies (Table 1), higher (millimolar) *cis* Ca$^{2+}$ concentrations are required to inhibit the cardiac Ca$^{2+}$ release channel. Single channel recordings such as those shown in Fig. 8, as well as vesicle–ion flux and [$^3$H]ryanodine binding measurements, have suggested that the mammalian skeletal and cardiac Ca$^{2+}$ release channels possess cytosolic high-affinity (activating) and low-affinity (inhibitory) Ca$^{2+}$-binding sites.

Calcium ions may regulate the Ca$^{2+}$ release channel in additional ways. Recent studies have indicated that lumenal Ca$^{2+}$ flowing through the skeletal muscle Ca$^{2+}$ release channel has access to the cytosolic Ca$^{2+}$ activation and Ca$^{2+}$ inactivation sites (Tripathy and Meissner, 1996). Other Ca$^{2+}$-dependent mechanisms that may activate and reduce SR Ca$^{2+}$ release are calmodulin activation and inhibition of Ca$^{2+}$ release channel activity (Tripathy *et al.*, 1995), SR Ca$^{2+}$ depletion (Baylor and Hollingworth, 1988), and Ca$^{2+}$ release channel adaptation (Gyorke and Fill, 1993).

## C. Dihydropyridine Receptor–Ryanodine Receptor Interactions in Skeletal Muscle

Different coupling mechanisms exist in cardiac and skeletal muscle. As discussed earlier, a unique property of the mammalian skeletal muscle RyR is its direct linkage to the skeletal muscle DHPR/Ca$^{2+}$ channel isoform, which acts in skeletal muscle EC coupling as a voltage-sensing molecule rather than Ca$^{2+}$ channel (see Fig. 2). Compelling evidence for the different function of the DHPR in cardiac and skeletal muscle EC coupling, that is, as a Ca$^{2+}$ channel in cardiac muscle and as a voltage-sensing molecule in skeletal muscle, has been obtained through the elegant work of Tanabe and Beam. These investigators used an animal model lacking the skeletal muscle DHPR $\alpha_1$-subunit to show that one of the putative cytoplasmic loop regions (II–III region) of the $\alpha_1$-subunit plays a major role in determining the type of EC coupling that exists in muscle (Tanabe *et al.*, 1990). Evidence for a direct functional interaction between the DHPR and RyR was obtained by Lu *et al.* (1994) who showed that a peptide expressed in *Escherichia coli* and derived from the II–III loop region of the skeletal muscle DHPR $\alpha_1$-subunit activated the purified skeletal muscle Ca$^{2+}$ release channel in planar lipid bilayers (Fig. 9). In the upper trace of Fig. 9, a single skeletal muscle channel was recorded in the presence of a suboptimally activating Ca$^{2+}$ concentration in the *cis* (cytoplasmic) chamber. The middle trace shows that the addition of 48 n$M$ peptide to the *cis* chamber resulted in an increase in Ca$^{2+}$ release channel activity. Removal of the peptide by perfusion (lower trace) decreased single-channel activity to the control level showing that the activation of the channel by the peptide was reversible. Additional studies showed that a peptide derived from the putative II–III loop

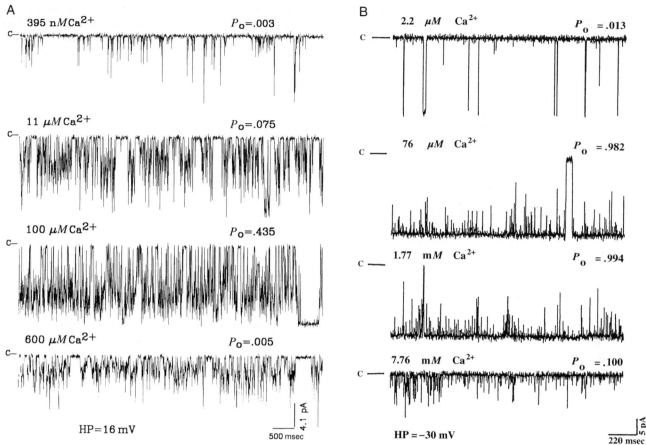

**FIG. 8.**   Single-channel recordings of skeletal muscle and cardiac muscle Ca²⁺ release channels. Shown are the effects of *cis* (cytoplasmic) free [Ca²⁺] on single-channel activity of (A) purified rabbit skeletal muscle and (B) canine cardiac muscle 30S channel complexes incorporated into a planar phospholipid bilayer in symmetric 250 mM KCl buffer. Unitary conductances are about 750 pS (with μM Ca²⁺ in the *cis* chamber and with 250 mM K⁺ as the conducting ion). Bars on the left represent the closed (c) channel, and $P_o$ the open probability of the channel.

region of the cardiac DHPR $\alpha_1$-subunit was as effective as the skeletal peptide in activating the skeletal muscle Ca²⁺ release channel. Both the skeletal muscle and cardiac muscle peptide bound to the skeletal muscle RyR but neither peptide was able to bind or activate the cardiac Ca²⁺ release channel (Lu *et al.*, 1995). These results imply that, in addition to the DHPR, the RyR has an important role in determining the EC coupling mode in muscle. The observation that phosphorylation of a serine residue abolished activation of the skeletal muscle RyR by the II–III skeletal muscle peptide suggests that phosphorylation may regulate skeletal muscle EC coupling (Lu *et al.*, 1995).

## D. Effects of Ryanodine and Caffeine

Among the large number of drugs that have been identified to affect SR Ca²⁺ release (Coronado *et al.*, 1994), ryanodine and caffeine are two of the best known because of their extensive use in the assessment of SR function in controlling cytoplasmic Ca²⁺ concentrations.

Measurements of ⁴⁵Ca²⁺ efflux have shown that in SR vesicles ryanodine has a dual effect in that nanomolar

concentrations lock the channel into an open configuration, whereas at concentrations greater than 10 μM, ryanodine completely closes the channel (Fig. 10). Single-channel recordings have allowed researchers to study directly the functional consequences of the interaction of ryanodine with its receptor. In turn, in the planar lipid bilayer measurements, the channel's highly characteristic modification by ryanodine provides a reliable means of distinguishing between the Ca²⁺ release channel and other types of ion channel currents. Figure 11A shows that several minutes after the addition of micromolar concentrations of ryanodine, the channel enters into a subconductance state with a channel open probability close to unity. In this study, a relatively high ryanodine concentration of 30 μM was used to reduce the time required to observe the otherwise very slow interaction of ryanodine with the channel. Upon the addition of millimolar ryanodine *cis*, the channel's subconductance state generally disappears and the channel enters into a fully closed state. A characteristic property of the ryanodine-modified channel states is their insensitivity to regulation by Ca²⁺, Mg²⁺, and ATP, all of which greatly affect the gating behavior of the unmodified channel.

**Control**

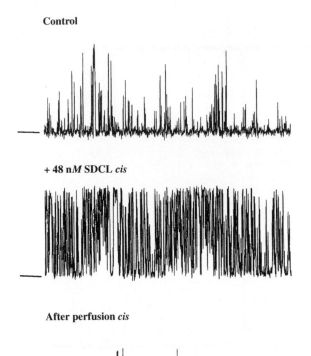

**+ 48 n*M* SDCL *cis***

**After perfusion *cis***

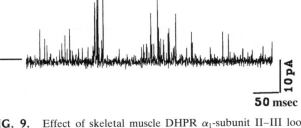

10 pA

**50 msec**

**FIG. 9.** Effect of skeletal muscle DHPR $\alpha_1$-subunit II–III loop peptide on the single-channel activity of purified skeletal muscle $Ca^{2+}$ release channel. Proteoliposomes containing the purified 30S channel complex were fused with a planar lipid bilayer (see Fig. 7). Single-channel currents, shown as upward deflections, were recorded in symmetric 0.25 *M* KCl media with 6 and 50 $\mu M$ free $Ca^{2+}$ in the *cis* and *trans* chambers, respectively. Top trace: Control, $P_o = 0.04$; middle trace: 3 min after the addition of 48 n*M* peptide to the *cis* chamber, $P_o = 0.71$; bottom trace, after perfusion of *cis* chamber with 0.25 *M* KCl buffer containing 6 $\mu M$ free $Ca^{2+}$, $P_o = 0.003$. Holding potential = 30 mV. [Reprinted with permission from Lu *et al.*, *J. Biol. Chem.* **269**, 6511–6516 (1994).]

[$^3$H]ryanodine binding studies have provided independent evidence for a complex interaction of the drug with the $Ca^{2+}$ release channel (Fig. 12) (Meissner, 1994). The existence of multiple interacting high- and low-affinity sites and the dependence of [$^3$H]ryanodine binding on $Ca^{2+}$, $Mg^{2+}$, ATP, and ionic strength have been described. As a general rule, conditions that open the channel, such as the presence of millimolar $Ca^{2+}$, millimolar ATP, or high ionic strength, have been found to favor the interaction of ryanodine with the channel by increasing the rate of association and affinity of [$^3$H]ryanodine binding.

In contrast to ryanodine, caffeine activates the SR $Ca^{2+}$ release channel by increasing channel open probability without significantly affecting single-channel conductance

**TABLE 2** Properties of Mammalian Skeletal Muscle (RyR1) and Cardiac Muscle (RyR2) $Ca^{2+}$ Release Channel Complexes

| Property | Skeletal muscle | Cardiac muscle |
|---|---|---|
| Apparent sedimentation coefficient | 30S | 30S |
| Morphology | Quatrefoil | Quatrefoil |
| High-affinity ryanodine binding ($K_D$) | n*M* | n*M* |
| Single-channel conductance | | |
| In 0.25 *M* KCl | 770 pS | 745 pS |
| In 0.05 *M* $Ca^{2+}$ (with 0.25 *M* KCl *cis*) | 145 pS | 145 pS |
| Regulation | | |
| Activation by $Ca^{2+}$ | $\mu M$ | $\mu M$ |
| Activation by m*M* ATP and m*M* caffeine | Yes | Yes |
| Inhibition by m*M* $Ca^{2+}$ or $Mg^{2+}$ | Yes | Yes |
| Modification by ryanodine | | |
| To subconducting state | Yes | Yes |
| To closed state | Yes | Not determined |

(see Fig. 11B). Another important difference is that caffeine activates the $Ca^{2+}$ release channel without loss of sensitivity to regulation by $Mg^{2+}$ and ATP. Several dimethylxanthines (1,3-(theophylline), 3,7-(theobromine), 1,7-) have been found to be more potent than caffeine (1,3,7-trimethylxanthine) in triggering the release of $Ca^{2+}$ from skeletal muscle SR membranes (Rousseau *et al.*, 1988).

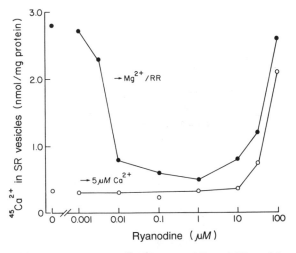

**FIG. 10.** Dependence of $^{45}Ca^{2+}$ permeability of SR vesicles on ryanodine concentration. Heavy rabbit skeletal muscle SR vesicles were incubated for 45 min at 37°C with 0.1 m*M* $^{45}Ca^{2+}$ and the indicated concentrations of ryanodine. Amounts of $^{45}Ca^{2+}$ retained by the vesicles in $Ca^{2+}$ release channel inhibiting [10 m*M* $Mg^{2+}$, 10 $\mu M$ ruthenium red (RR, ●)] and activating (5 $\mu M$ free $Ca^{2+}$, ○) media are shown. [Reprinted with permission from Meissner, *J. Biol. Chem.* **261**, 6300–6306 (1986).]

**A**

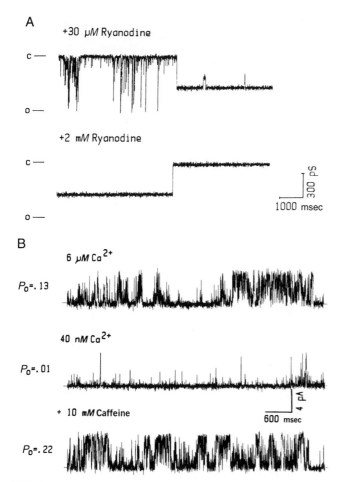

**B**

ryanodine binding site, activation by $Ca^{2+}$, and a high conductance mono- and divalent cation pathway, which can be modified by ryanodine in an essentially identical manner (see Table 2). On the other hand, the two RyR isoforms exhibit differences in their sensitivity to activation by $Ca^{2+}$ and ATP, and inhibition by $Mg^{2+}$ (see Table 1 and Fig. 8).

The functional properties of the third mammalian RyR isoform (RyR3) are less well known at present. The RyR3 isoform is expressed in various tissues including brain, cardiac muscle and skeletal muscle, but the low concentrations of RyR3 have hindered its study. A limited number of single-channel measurements has suggested that the mammalian vascular RyR exhibits a $Ca^{2+}$-gated channel activity that could be modified by caffeine (Herrmann-Frank *et al.*, 1991). [³H]Ryanodine binding studies have shown that the mammalian RyR3 in brain shares several functional characteristics with the amphibian homolog. These include a requirement of $Ca^{2+}$ for [³H]ryanodine binding and an enhanced $Ca^{2+}$ sensitivity in the presence of caffeine and 1 $M$ salt (Murayama and Ogawa, 1996).

The isolation of RyRs from nonmammalian sources including amphibian, avian, and crustacean skeletal muscles and *C. elegans* has been reported. In *Drosophila,* a gene for a RyR has been sequenced and shown to display ~50% sequence identity with the mammalian RyRs. Study of these RyRs is of interest because distinctly different patterns of regulation by $Ca^{2+}$ and other effector molecules have been observed (Meissner, 1994; Ogawa, 1994).

The physiological role of several other endogenous effector molecules and proteins implicated in the regulation of the RyR/$Ca^{2+}$ release channel needs to be better defined. These include cyclic ADP ribose (a novel $Ca^{2+}$ mobilizing agent in sea urchin eggs), lipid metabolites such as acylcarnitines (Dumonteil *et al.*, 1994), polyamines such as spermine (Uehara *et al.*, 1996), and proteins such as triadin (a

**FIG. 11.** Effects of ryanodine and caffeine on single skeletal muscle RyRs. 30S purified channel complexes were reconstituted in their soluble form into planar lipid bilayers (see Fig. 7). (A) Upper trace shows appearance of subconducting channel state with open probability ($P_o$) of ~1, following several minutes after the addition of 30 $\mu M$ *cis* ryanodine. An additional, infrequent substate is also observed. Lower trace illustrates the sudden transition from the subconductance state to a fully closed state in 1 min after addition of 2 m$M$ *cis* (cytoplasmic) ryanodine. [Reprinted with permission from Lai *et al., J. Biol. Chem.* **264**, 16776–16785 (1989).] (B) Single-channel activity was with 6 $\mu M$ free *cis* $Ca^{2+}$ (upper trace), after *cis* free $Ca^{2+}$ was decreased to 40 n$M$ by the addition of a $Ca^{2+}$ buffer (EGTA) (middle trace), and after the addition of 10 m$M$ caffeine to the 40 n$M$ free *cis* $Ca^{2+}$ medium (bottom trace). In contrast to ryanodine, caffeine activates the channel without changing its conductance. Bars on left represent the closed (c) channel. [Reprinted with permission from Rousseau *et al., Arch. Biochem. Biophys.* **267**, 75–86 (1988).]

## VII. Summary

The identification of [³H]ryanodine as a specific probe has enabled the isolation and subsequent cloning of intracellular 30S RyR complexes composed of four ~560-kDa polypeptides. The mammalian RyRs are encoded by three different genes and show a tissue-specific expression. The skeletal muscle (RyR1) and cardiac muscle (RyR2) isoforms share several properties including an apparent sedimentation coefficient of 30S, the presence of a high-affinity

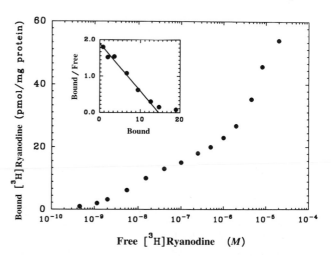

**FIG. 12.** High- and low-affinity [³H]ryanodine binding to heavy rabbit skeletal muscle SR membranes. Bound [³H]ryanodine was determined as the difference between total [³H]ryanodine and unbound [³H]ryanodine before and after removal of SR membranes by centrifugation, respectively. Scatchard analysis (inset) reveals a curvilinear slope indicating the present of both high- ($K_D = 7$ n$M$) and low-affinity sites. [Adapted from Lai *et al., J. Biol. Chem.* **264**, 16776–16785 (1989).]

$M_r$ 90,000 junctional SR protein (Fan *et al.*, 1995; Guo and Campbell, 1995) and FK506 binding protein ($M_r$ 12,000) (Timerman *et al.*, 1996).

## Acknowledgments

I would like to thank Xiangyang Lu for providing Figs. 1 and 2 and Le Xu for Fig. 8. Support from U.S. Public Health Service grants AR18687 and HL27430 is gratefully acknowledged.

## Bibliography

Baylor, S. M., and Hollingworth S. (1988). Fura-2 calcium transients in frog skeletal muscle fibers. *J. Physiol.* **403**, 151–192.

Berridge, M. J. (1993). Inositol tris-phosphate and calcium signalling. *Nature* **361**, 315–325.

Chen, S. R. W., Zhang, L., and MacLennan, D. H. (1993). Antibodies as probes for $Ca^{2+}$ activation sites in the $Ca^{2+}$ release channel (ryanodine receptor) of rabbit skeletal muscle sarcoplasmic reticulum. *J. Biol. Chem.* **268**, 13414–13421.

Coronado, R., Morrissette, J., Sukhareva, M., and Vaughan, D. M. (1994). Structure and function of ryanodine receptors. *Am. J. Physiol.* **266**, C1485–C1504.

Dumonteil, E., Barre H., and Meissner, G. (1994). Effects of palmitoyl carnitine and related metabolites on the avian $Ca^{2+}$-ATPase and $Ca^{2+}$ release channel. *J. Physiol.* **479**, 29–39.

Fabiato, A. (1983). Calcium-induced release of calcium from the cardiac sarcoplasmic reticulum. *Am. J. Physiol.* **245**, C1–C14.

Fan, H., Brandt, N. R., Peng, M., Schwartz, A., and Caswell, A. H. (1995). Binding sites of monoclonal antibodies and dihydropyridine receptor alpha 1 subunit II–III loop on skeletal muscle triadin fusion proteins. *Biochemistry* **34**, 14893–14901.

Fleischer, S., and Inui, M. (1989). Biochemistry and biophysics of excitation–contraction coupling. *Annu. Rev. Biophys. Biophys. Chem.* **18**, 333–364.

Franzini-Armstrong, C., and Jorgensen, A. O. (1994). Structure and development of E-C coupling units in skeletal muscle *Annu. Rev. Physiol.* **56**, 509–534.

Furuichi, T., Furutama, D., Hakamata, Y., Nakai, J., Takeshima, H., and Mikoshiba, K. (1994). Multiple types of ryanodine receptor/ $Ca^{2+}$ release channels are differentially expressed in rabbit brain. *J. Neurosci.* **14**, 4794–4805.

Giannini, G., Conti, A., Mammarella, S., Scrobogna, M., and Sorrentino, V. (1995). The ryanodine receptor/ $Ca^{2+}$ release channel genes are widely and differentially expressed in murine brain and peripheral tissues. *J. Cell. Biol.* **128**, 893–904.

Guerrini, R., Menegazzi, P., Anacardio, R., Marastoni, M., Tomatis, R., Zorzato, F., and Treves, S. (1995). Calmodulin binding sites of the skeletal, cardiac and brain ryanodine receptor $Ca^{2+}$ channels: Modulation by the catalytic subunit of cAMP-dependent protein kinase? *Biochemistry* **34**, 5120–5129.

Guo, W., and Campbell, K. P. (1995). Association of triadin with the ryanodine receptor and calsequestrin in the lumen of the sarcoplasmic reticulum. *J. Biol. Chem.* **270**, 9027–9230.

Gyorke, S., and Fill, M. (1993). Ryanodine receptor adaptation: Control mechanism of $Ca^{2+}$-induced $Ca^{2+}$ release in heart. *Science* **260**, 807–809.

Herrmann-Frank, A., Darling, E., and Meissner, G. (1991). Functional characterization of the $Ca^{2+}$-gated $Ca^{2+}$ release channel of vascular smooth muscle sarcoplasmic reticulum. *Pflugers Arch.* **18**, 353–359.

Ikemoto, N., Antonium, B., and Meszaros, L. G. (1985). Rapid flow chemical quench studies of calcium release from isolated sarcoplasmic reticulum. *J. Biol. Chem.* **260**, 14096–14100.

Inui, M., Saito, A., and Fleischer, S. (1987). Isolation of the ryanodine receptor from cardiac sarcoplasmic reticulum and identity with the feet structures. *J. Biol. Chem.* **262**, 15637–15642.

Jenden, D. J., and Fairhurst, A. S. (1969). The pharmacology of ryanodine. *Pharmacol. Rev.* **21**, 1–25.

Lai, F. A., Erickson, H. P., Rousseau, E., Liu, Q. Y., and Meissner, G. (1988). Purification and reconstitution of the calcium release channel from skeletal muscle. *Nature* **331**, 315–319.

Lai, F. A., Misra, M., Xu, L., Smith, H. A., and Meissner, G. (1989). The ryanodine receptor–$Ca^{2+}$ release channel complex of skeletal muscle sarcoplasmic reticulum. Evidence for a cooperatively coupled, negatively charged homotetramer. *J. Biol. Chem.* **264**, 16776–16785.

Ledbetter, M. W., Preiner, J. K., Louis, C. F., and Mickelson, J. R. (1994). Tissue distribution of ryanodine receptor isoforms and alleles determined by reverse transcription polymerase chain reaction. *J. Biol. Chem.* **269**, 31544–31551.

Lu, X., Xu, L., and Meissner, G. (1994). Activation of the skeletal muscle calcium release channel by a cytoplasmic loop of the dihydropyridine receptor. *J. Biol. Chem.* **269**, 6511–6516.

Lu, X., Xu, L., and Meissner, G. (1995). Phosphorylation of dihydropyridine receptor II–III loop peptide regulates skeletal muscle calcium release channel function. *J. Biol. Chem.* **270**, 18459–18464.

Meissner, G. (1983). Monovalent ion and calcium ion fluxes in sarcoplasmic reticulum. *Mol. Cell Biochem.* **55**, 65–82.

Meissner, G. (1994). Ryanodine receptor/$Ca^{2+}$ release channels and their regulation by endogenous effectors. *Annu. Rev. Physiol.* **56**, 485–508.

Murayama, T., and Ogawa, Y. (1996). Properties of Ryr3 ryanodine receptor isoform in mammalian brain. *J. Biol. Chem.* **271**, 5079–5084.

O'Brien, J., Meissner, G., and Block, B. A. (1993). The fastest contracting muscles of nonmammalian vertebrates express only one isoform of the ryanodine receptor. *Biophys. J.* **65**, 2417–2418.

Ogawa, Y. (1994). Role of ryanodine receptors. *Crit. Rev. Biochem. Mol. Biol.* **29**, 229–274.

Otsu, K., Nishida, K., Kimura, Y., Kuzuya, T., Hori, M., Kamada, T., and Tada, M. (1994). The point mutation $Arg^{615} \rightarrow Cys$ in the $Ca^{2+}$ release channel of skeletal muscle sarcoplasmic reticulum is responsible for hypersensitivity to caffeine and halothane in malignant hyperthermia. *J. Biol. Chem.* **269**, 9413–9415.

Ottini, L., Marziali, G., Conti, A., Charlesworth, A., and Sorrentino, V. (1996). $\alpha$ and $\beta$ isoforms of ryanodine receptor from chicken skeletal muscle are the homologues of mammalian RyR1 and RyR3. *Biochem. J.* **315**, 207–216.

Oyamada, H., Murayama, T., Takai, T., Iino, M., Iwabe, N., Miyata, T., Ogawa, Y., and Endo, M. (1994). Primary structure and distribution of ryanodine-binding protein isoforms of the bullfrog skeletal muscle. *J. Biol. Chem.* **269**, 17206–17214.

Radermacher, M., Rao, V., Grassucci, R., Frank, J., Timerman, A. P., Fleischer, S., and Wagenknecht, T. (1994). Cryo-electron microscopy and three-dimensional reconstruction of the calcium release channel/ryanodine receptor from skeletal muscle. *J. Cell Biol.* **127**, 411–423.

Rios, E., and Pizarro, G. (1991). Voltage sensor of excitation–contraction coupling in skeletal muscle. *Physiol. Rev.* **71**, 849–908.

Rousseau, E., Ladine, J., Liu, Q. Y., and Meissner, G. (1988). Activation of the $Ca^{2+}$ release channel of skeletal muscle sarcoplasmic reticulum by caffeine and related compounds. *Arch. Biochem. Biophys.* **267**, 75–86.

Santana, L. F., Cheng, H., Gomez, A. M., Cannell, M. B., and Lederer, W. J. (1996). Relation between the sarcolemmal $Ca^{2+}$ current and $Ca^{2+}$ sparks and local control theories for cardiac excitation–contraction coupling. *Circ. Res.* **78**, 166–171.

Schneider, M. F. (1994). Control of calcium release in functioning skeletal muscle fibers. *Annu. Rev. Physiol.* **56**, 463–484.

Serysheva, I. I., Orlova, E. V., Chui, W., Sherman, M. B., Hamilton, S. L., and van Heel, M. (1995). Electron cryomicroscopy and angular reconstitution used to visualize the skeletal muscle calcium release channel. *Nature Struct. Biol.* **2,** 18–24.

Smith, J. S., Imagawa, T., Ma, J., Fill, M., Campbell, K. P., and Coronado, R. (1988). Purified ryanodine receptor from rabbit skeletal muscle is the calcium-release channel of sarcoplasmic reticulum. *J. Gen. Physiol.* **92,** 1–26.

Takeshima, H., Nishimura, S., Nishi, M., Ikeda, M., and Sugimoto, T. (1993). A brain-specific transcript from the 3'-terminal region of the skeletal muscle ryanodine receptor gene. *FEBS Lett.* **322,** 105–110.

Takeshima, H., Nishi, M., Iwabe, N., Miyata, T., Hosoya, T., Masai, I., and Hotta, Y. (1994). Isolation and characterization of a gene for a ryanodine receptor/calcium channel in *Drosophila melanogaster. FEBS Lett.* **337,** 81–87.

Takeshima, H., Ikemoto, T., Nishi, M., Nishiyama, N., Shimuta, M., Sugitani, Y., Kuno, J., Saito, I., Saito, H., Endo, M., Iino, M., and Noda, T. (1996). Generation and characterization of mutant mice lacking ryanodine receptor type-3. *J. Biol. Chem.* **271,** 19649–19652.

Tanabe, T., Beam, K. G., Adams, B. A., Niidome T., and Numa, S. (1990). Regions of the skeletal muscle dihydropyridine receptor critical for excitation–contraction coupling. *Nature* **346,** 567–569.

Timerman, A. P., Onoue H., Xin, H. B., Barg, S., Copello, J., Wiederrecht, G., and Fleischer, S. (1996). Selective binding of FKBP12.6 by the cardiac ryanodine receptor. *J. Biol. Chem.* **271,** 20385–20391.

Tripathy, A., and Meissner, G. (1996). Sarcoplasmic reticulum lumenal Ca²⁺ has access to cytosolic activation and inactivation sites of skeletal muscle Ca²⁺ release channel. *Biophys. J.* **70,** 2600–2615.

Tripathy, A., Xu, L., Mann, G., and Meissner, G. (1995). Calmodulin activation and inhibition of skeletal muscle Ca²⁺ release channel (ryanodine receptor). *Biophys. J.* **69,** 106–119.

Uehara, A., Fill, M., Velez, P., Yasukochi, M., and Imanaga, I. (1996). Rectification of rabbit cardiac ryanodine receptor current by endogenous polyamines. *Biophys. J.* **71,** 769–777.

Witcher, D. R., Kovacs, R. J., Schulman, H., Cefali, D. C., and Jones, L. R. (1991). Unique phosphorylation site on the cardiac ryanodine receptor regulates calcium channel activity. *J. Biol. Chem.* **266,** 11144–11152.

Richard J. Paul

# 55

# Contractility of Muscles

## I. Introduction

The generation of force and movement by muscle is an area of physiology and biophysics that has fascinated scientist and layman alike since the dawn of scientific inquiry. The history of the study of muscle is elegantly chronicled by Dorothy Needham in *Machina Carnis* (1971). Because of its highly organized and repeating structure, skeletal muscle has proven more amenable to biophysical analysis (such as X-ray diffraction) than most biological tissues. Thus it has served as a paradigm for unraveling relationships between function and structure. This has become particularly exciting in conjunction with the techniques of molecular biology, which offer the potential for altering particular amino acids and directly testing the links between structure at nanometer resolution and function. The goal of this chapter is to discuss the nature of the mechanochemical energy conversion. Muscle is one of the most efficient energy converters known, and studies in this area couple classical enzyme kinetics and muscle mechanics.

Although we will focus on the molecular and cellular level, one should be aware of the functions and consequences of muscle activity at the organ and whole animal levels. Many important effects at the whole organism level are related simply to muscle mass. Skeletal muscle constitutes approximately 40% of human body mass. If we include cardiac and smooth muscle, the total muscle mass reaches the 50% level. One consequence of this is that approximately 30% of basal metabolism is related to muscle and as much as 90% of a person's total metabolism during strenuous exercise can be related to meeting the energy requirements of muscle. This chemical activity can in turn produce a significant heat load for the organism. Other functions that are associated with the large muscle mass are the storage and mobilization of metabolites (primarily glucose and amino acids). Also, there are significant consequences to alterations in muscle electrolyte metabolism, since it is a major storage site for ions such as $H^+$, $K^+$, and $Mg^{2+}$. Some of these ramifications, as related to molecular

processes in muscle, are presented. However, the major emphasis of this chapter is on the relationships between muscle structure and function at the cellular and subcellular levels. (Other nonmuscle motile systems relating to cilia and flagella are covered in the following two chapters.)

The study of the relationships between structure and function can be divided into two primary areas. The first area involves investigation of mechanisms underlying the generation of macroscopic force and shortening. Muscle is primarily a device for the conversion of chemical energy to mechanical energy. We first consider how force is developed and how the mechanical behavior of muscle is quantitated, a field known collectively as *muscle mechanics*. This is then integrated with the current picture of muscle structure as a first step in constructing theories of muscle function. Next, we consider the thermodynamic rules governing energy conversion and the constraints they place on models proposed for muscle contraction. We relate energetics and mechanics to the kinetics of the myosin ATPase as a basis for crossbridge cycling mechanisms. Muscle metabolism and the matching of adenosine triphosphate (ATP) demand with ATP synthesis will complete the picture of mechanochemical energy conversion.

The second major area involves study of the mechanisms underlying the regulation of muscle contraction. The control of intracellular $Ca^{2+}$ concentration, a key intracellular messenger, forms an area of study known as *excitation–contraction coupling* and is the focus of Chapters 53 and 54. The intracellular receptors transducing the $Ca^{2+}$ signal reflect the great diversity of types of muscle that have evolved in response to a wide variety of functional needs. These muscles also have much in common. All muscle contains the proteins actin and myosin, which are the locus of the mechanochemical energy conversion. Chemical energy in the form of ATP hydrolysis is the immediate driving reaction for all muscle energy transduction. Another feature common to all muscle types is that calcium ions, at micromolar concentrations, are the primary second messenger in the regulatory mechanisms. We first focus on these common aspects, using a generalized striated muscle as the model. Then, with an understanding of these common mechanisms, we consider the different muscle types.

## II. The Mechanisms of Force Production and Shortening: Muscle Mechanics

Studies of the mechanical behavior of muscle have played a central role in our understanding of muscle and also have formed an integral part of the language of muscle physiology (Hill, 1965). These studies before the late 1950s were primarily phenomenological, though they were also important in characterization of muscle performance (Jewel and Wilkie, 1958). Such studies remain important in characterization of muscle myopathies and in current mechanistic studies (for example, in describing the functional consequences of changing muscle protein isoforms in transgenic animals).

There are two arbitrary but natural divisions in studies of mechanics. The first division involves *steady-state* relationships. This information was crucial to the development of the sliding-filament theory and was extensively investigated in the 1960s (Gordon *et al.,* 1966). Studies of *transients* form a second division. They assumed a more central importance during the early 1970s (Huxley and Simmons, 1971) with the growing realization that information at the level of individual crossbridges could be gained from such mechanical studies on single fibers. These studies of muscle responses to rapid changes in mechanical constraints are even more valuable to unraveling crossbridge behavior when coupled with the recently improved temporal resolution of X-ray diffraction of muscle (Huxley, 1996).

### A. Steady-State Relations between Force and Length and the Sliding Filament Theory

A first step in muscle mechanics involves a description of muscle behavior in terms of relationships between the mechanical variables of force and length. Apparatus for transduction and recording of these variables have evolved considerably over this century but the historical apparatus are still responsible for much of the language of muscle physiology. Since it was easier to control force than length, by hanging a fixed load on the muscle, terms such as *preload* and *afterload* entered this vocabulary (see later section on force–velocity relationships). However, in view of what is now known about structure, it is conceptually easier to use length as the independent variable.

Figure 1 shows a schematic of an experimental apparatus for measurement of muscle force–length relationships. In this setup, length is controlled, and the steady-state force at various lengths is measured. In developing these relationships, we consider the performance of an isolated muscle (or single muscle cell) known as a *muscle fiber*. After mounting in the apparatus muscle length is varied, then held isometric at a specific length (constant/total length) while the steady-state force is measured. The relationship between the unstimulated force (often designated as passive force) and muscle length is designated the passive *length–tension relationship*. This relationship can be char-

acterized as an exponential spring, $F = A_1 + A_2 \exp(A_3 X)$, whose behavior is similar to that of a rubber band. This relationship can show a dependence on the direction of the imposed length changes, known as *hysteresis,* but deviations are small in a true steady state. Some form of passive force is common to all muscles but an exact anatomical assignment of the structures underlying passive force is dependent on both the type of muscle and the type of study.

The next page of this analysis involves a similar protocol but includes stimulation of the muscle to identify the parameters associated with activated muscle. With active muscle, the language that evolved reflected the state of understanding and the experimental apparatus. The response to a single electrical shock is known as a *twitch contraction.* At one time, this was believed to be some form of elemental or quantal behavior of muscle, hence its historical importance. Increasing the frequency of stimulation leads to a summation in time of the individual twitch responses, known as *temporal summation.* Beyond a certain frequency (depending on muscle type and temperature) the force response becomes a smooth fused curve, called an *isometric tetanus.* These responses to stimulation are ultimately related to the $Ca^{2+}$ handling underlying activation of the contractile proteins. For our present purposes, we consider only isometric tetani, so that the mechanical behavior of the fully activated contractile apparatus can be considered, without complications arising from behavior attributable to non-steady-state $Ca^{2+}$ signaling.

Tetanic stimulation adds an additional increment of force to the passive force present at a given length. The passive force plus active (stimulated) force, measured as a function of muscle length, is shown in Fig. 1. This relationship is known as the *total force–length curve.* The total force–length relationship varies considerably from muscle to muscle, though the component passive and active force–length relationships are qualitatively similar. The differences are largely ascribable to the relative amount of passive force developed at the length at which active force is optimal.

For the understanding of mechanism, the relationship between the additional active force generated when a muscle is stimulated and muscle length is paramount. This relationship, the *active force–length curve,* is unusual in that it decreases to zero at both long and short muscle lengths. For most materials, including polymers like rubber, force increases as length is increased. This typical behavior is also seen for the passive force as shown in Fig. 1. The observation that active force decreased at long muscle lengths was critical to eliminating theories that involved folding of continuous muscle filaments as the basis for the generation of force upon activation. To understand the relationship between active force and muscle length, it is necessary to consider muscle structure.

### B. The Structure of Muscle: Interdigitating Filament Systems

Muscle cells are composed of a filament system underlying their mechanical properties and an internal membrane system related to their control functions. The filament

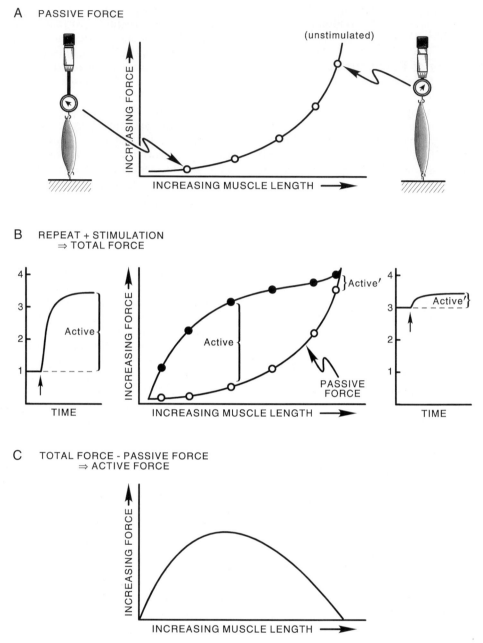

**FIG. 1.** Measurement and operational definitions of muscle force–length relationships. (A) Relation between isometric force under unstimulated conditions, passive force, and muscle length. (B) Relation between isometric force and length under stimulated conditions, total force. (C) Active force operationally defined as the difference between total force and passive force is shown as a function of muscle length.

structure repeats in both the transverse and longitudinal directions as shown in Fig. 2. A muscle fiber is composed of myofibrils whose fundamental longitudinal repeating unit is the sarcomere. The sarcomere consists of two inter-digitating filament systems—thick (14-nm) myosin-con-taining filaments and thin (7-nm) actin-containing fila-ments—which underlie the banding seen under optical microscopy, characteristic of striated or striped muscle. The optical properties of the sarcomere due to the overlapping filaments gave rise to the nomenclature for the banding

regions. The I-band contains only thin filaments and is optically *isotropic,* whereas the A-band contains both fila-ment types and is *anisotropic.* The constancy of A-band dimensions, independent of total muscle length (Huxley and Niedergerke, 1954), was a key experimental finding, leading to the concept that filament length was constant. Constant filament lengths and interdigitating filaments are best observed at the electron microscope level (Fig. 3), where interpretation of the changing banding pattern with muscle length was first elucidated.

## C. Sliding Filament Theory

We can now understand better the tenets of the sliding-filament theory (the most generally accepted theory of muscle contraction). This theory postulates that muscle force is generated by the interaction between thick and thin filaments of constant length, which are free to interdigitate and slide past one another. In this theory, the ability to generate isometric force is proportional to the extent of overlap between the filaments. This mechanical correlate of the proposed sliding-filament structure was tested in the classic work of Gordon *et al.* (1966) and is summarized in Fig. 4. The clearest correlation between the level of active isometric force and extent of filament overlap is in the region of decreasing force between sarcomere lengths of 2.2 and 3.6 $\mu$m, that is, the descending limb of the active force–length curve. The interpretation of the ascending limb of the force–length curve (e.g., 1.2 to 2.2 $\mu$m) is not as straightforward. It is complicated by possible changes in activation at short lengths. The loss of force at these lengths can also be attributed to double filament overlap and compression of the contractile elements. While inter-

digitating, sliding filaments are widely accepted, the basis of the force generation remains an active area of research (see later section on crossbridge theory).

## D. Relationships between Force, Velocity, Work, and Energy Utilization

Together with the relationships between force and length, the relationship between force and shortening velocity is a fundamental characteristic of muscle performance. Studies of work production under various conditions (to define muscle performance) have an extensive history. To eliminate considerations due to transient activation, only contractions under maximum tetanic stimulation are presented here.

With current technology, one would use an ergometer to impose a constant velocity or force, and measure the resulting force or velocity as a function of time. However, it is instructive to consider the apparatus used historically because some of the nomenclature prevalent in muscle physiology can be attributed to these protocols. Such apparatus is shown in Fig. 5. A muscle is attached to a lever system so that addition of a weight or *preload* can freely stretch the muscle. The importance of the preload, in this context, is that is sets the initial length of the muscle and, consequently, the active force, as governed by force–length relationship. A mechanical stop placed on the lever system allows additional weights to be added without changing length. The total load now on the muscle is called the *afterload,* as the muscle can support this load only after stimulation, when its force development exceeds that of the afterload. Afterloaded isotonic contractions are depicted in Fig. 5, which shows the effects of increasing afterload on muscle length as a function of time after stimulation. Although the change in length is curvilinear with time, reflecting changes in force due to altered filament overlap, the initial change in length is constant with time and thus has a steady velocity. To isolate the relationship between force and velocity from changes in force due to different degrees of thick and thin filament overlap, experiments are designed to exploit the plateau region of force–length relationship in which the crossbridge number is constant. Furthermore, muscles in which the passive force was negligible in this region (e.g., the frog semitendinosis) are commonly used to avoid complications in interpretation of force–velocity data due to passive force. The relationship between afterload and velocity under these conditions is hyperbolic (Fig. 5). Several equations have been proposed to fit this relationship between velocity ($V$) and afterload ($F$). However, the most widely used is an equation attributed to A. V. Hill and is expressed as

$$(F + a)(V + b) = a(V_{max} + b) = b(F_o + a) \qquad (1)$$

where $a$ and $b$ are constants and $a/F_o = b/V_{max}$ are dimensionless constants that determine the curvilinearity of the Hill equation and are approximately 0.25 for skeletal muscle. The mechanistic origin of this parameter is controversial. Although low values are often associated with efficient muscle performance, a generally accepted theoretical basis remains to be found. The parameter $a$ was initially thought

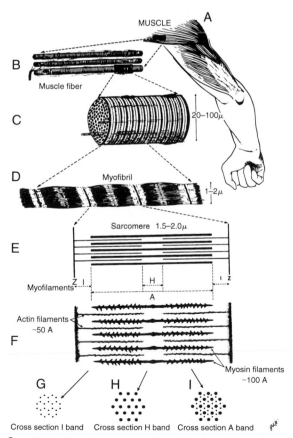

**FIG. 2.** Skeletal muscle structure. The whole muscle is constructed of fundamental repeating units. A muscle cell or fiber is composed of myofibrils, and its fundamental longitudinal repeating unit is the sarcomere. All views are longitudinal except the cross-sections of the sarcomere shown in the bottom left drawings. (From Bloom and Fawcett, 1994, *Textbook of Histology,* 11th ed. New York: Chapman and Hall.)

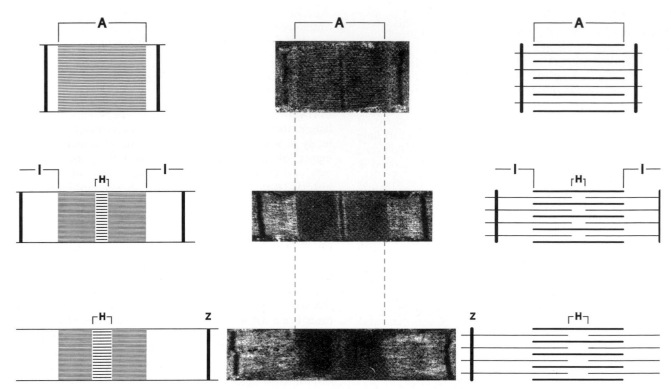

**FIG. 3.** Electron micrographs and diagrams of sarcomeres at various muscle lengths. The sliding-filament theory is based on the constancy of the thick and thin filaments. (A) A-bands; (I) I-bands; (Z) Z-bands; (H) H-zone.

to be related to a constant derived from heat measurements, called the *shortening heat.* However, subsequent experimentations proved that the shortening heat was not a constant but a function of load as well. Although other equations can fit the relationship between force and velocity, the scientific stature of Hill and the apparent link between heat and mechanical measurements led to the dominance of the Hill equation in the literature.

The Hill equation empirically provides a reasonable guide to muscle performance. A consequence of the hyperbolic nature is that the power output of muscle shows a relatively broad range around a maximum. And advantage of the Hill formulation over others is that it is symmetrical. This can be seen best in its normalized form,

$$(\xi + \theta)(\nu + \theta) = \theta(\theta + 1) \tag{2}$$

where $\xi$ is the normalized force ($F/F_o$), $\nu$ is normalized velocity ($V/V_{max}$), and $\theta = a/F_o = b/V_{max}$. A power is given by the product of force times velocity, Eq. 1 or 2 can be readily used to derive a relationship between power and load or velocity. The optimal load or velocity for maximum power output can also be derived from these equations and is equal to $\sqrt{\theta^2 + \theta} - \theta$. For skeletal muscle this equals about 0.3 times the maximum load or velocity. These empirical characteristics are useful in the design of ergonometric devices to optimize performance.

The Hill equation is used most commonly as a characteristic for testing models of muscle mechanical behavior. The most widely known of these models is the kinetic scheme

of A. F. Huxley (1957). In this model, rate constants for crossbridge formation/breakage are functions of position. This model (Fig. 6) can explain the relationship between force and velocity, but it is not adequate for transitions between states, such as those seen (see Fig. 10) when rapid step changes are imposed (see later section on tension transients). It is interesting that a major constraint, within which the parameters of Huxley's kinetic scheme were developed, was based on data of heat production as a function of shortening. With hindsight, we now know that these data probably do not simply represent ATP breakdown at the crossbridge level as Huxley assumed. Direct chemical measurement of ATP breakdown has improved substantially with development of rapid-freezing apparatus and application of NMR analysis to intact muscle. However, the dependence on mechanical conditions of the ATP hydrolysis rate attributable to crossbridges is still not known with certainty (see Section III, Muscle Energetics).

### E. Crossbridge Theory

The most prevalent theory for force generation involves the concept of cyclic interactions between the heads of myosin molecules (crossbridges) projecting from the thick filaments and the actin molecules comprising the thin filaments. Evidence from electron micrographs indicates that these crossbridges, in various conformations, bridge the gap between thick and thin filaments. Based on both micrographs and X-ray diffraction data, these projections occur at intervals of 14.3 nm. The best evidence (from mass comparisons) is that 3 myosin molecules are located at each site, with an identical repeat at about 43 nm. The structure of the thick filament is shown in Fig. 7.

The evidence for cyclic interaction is partly based on structural considerations. Striated muscle can generate force and shorten over a range of about 50–150% of its rest length. Since a muscle is composed of identical sarcomeres, each sarcomere could also operate over the same range. For a sarcomere of 2.2 $\mu$m, this operating range would be about 1.1–3.3 $\mu$m. Since the thin filaments of each half of the sarcomere move toward the center, a sarcomere shortening from 3.3 to 1.1 $\mu$m ($\Delta$ of 2.2 $\mu$m) would require a movement of thin filaments on each side of the sarcomere of one half that distance, or 1.1 $\mu$m relative to the thick filament. This distance is considerably longer than the crossbridge spacing (0.014 $\mu$m) and is, in fact, longer than a single myosin molecule (0.150 $\mu$m). It is thus difficult to envision a myosin crossbridge remaining attached to actin in a structure with interdigitating filaments of constant length and relative sliding of more than 1 $\mu$m. Hence some form of cyclic interactions between the myosin crossbridge and actin are envisioned to permit the observed degree of

shortening. For example, if the working range of a crossbridge is 10 nm, then 100 repeated cycles of attachment and detachment would be required for a relative filament movement of 1 $\mu$m.

A second basis for the crossbridge theory arises from biochemical studies on isolated muscle proteins. The myosin molecule consists of a long rodlike region important in assembly into thick filaments and a globular head region, which contains the ATPase activity. The globular head, or S1 region (Fig. 7), has dimensions consistent with the projections identified with the crossbridge in electron micrographs. Studies of the kinetics of the ATP hydrolysis by myosin, involving stopped-flow apparatus and other techniques for the measurement of rapid time courses, have provided a framework for cyclic interaction of the S1 myosin head with actin. The kinetic scheme is shown in Fig. 8. The physiological ATPase activity (known as $Mg^{2+}$-ATPase activity) of purified myosin is relatively low. But importantly, it is activated by a factor of 200-fold by actin, the principal protein of the thin filament. In the absence of ATP, purified actin and myosin bind strongly. In intact muscle, this is the cause for the stiffness of muscle in the absence of nucleotide associated with "rigor." Addition of ATP to a "rigor complex" of actin and myosin dissociates the complex by producing a weak binding state, because the binding of ATP to myosin has a higher affinity (Fig. 8, steps 3 to 4). ATP hydrolysis by myosin alters the binding characteristics and crossbridge orientation, leading to a strong binding state (Fig. 8, step 2). The biochemical cycle is completed as the products of ATP hydrolysis, first inorganic phosphate ($P_i$) and then ADP, dissociate from the myosin head. One ATP molecule is hydrolyzed per crossbridge cycle. The structural details of this cycle have recently been significantly upgraded due to the high resolution X-ray diffraction data of the myosin head by Rayment (1996) and colleagues (Rayment *et al.*, 1996). As shown in the insets in Fig. 8, these data suggest that binding of ATP to myosin alters the conformation of the myosin head and consequently its binding to actin. Correlating this crossbridge cycle with mechanical steps in force generation is a major focus of muscle physiology. To understand how the mechanical behavior of an intact macroscopic muscle can be interpreted in terms of individual crossbridge behavior, we must consider the studies of mechanical transients.

### F. Transient Mechanical Behavior and the Crossbridge Cycle

Analysis of the mechanical properties of any material generally involves the imposition of known perturbations (for example, step or sinusoidal changes in length or force) while recording the response of the system. When modeling the response, combination of two types of components (springlike and viscous elements) can account for the time course of most materials. The behavior of springlike material is characterized by an instantaneous relationship between force and length, whereas for a viscous element, a resistive force is proportional to the rate of imposed length change. Combination of these elements can approximate

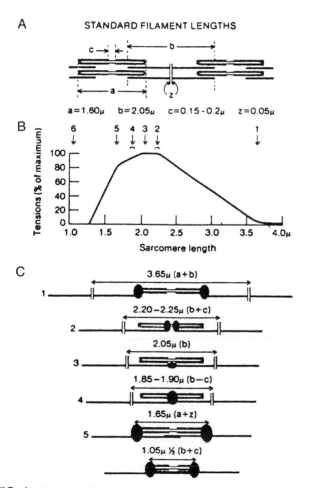

**FIG. 4.** Structural basis for the active isometric force–length relationship. (From Gordon *et al.*, 1966.)

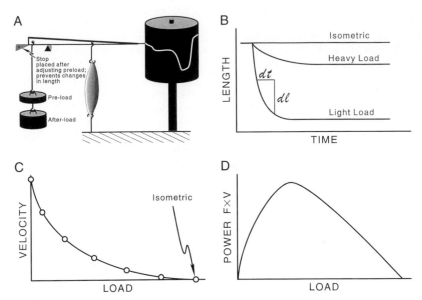

**FIG. 5.** Relationships among velocity of shortening, power output, and load. (A) The experimental apparatus used to generate afterloaded, isotonic conditions. (B) The experimental results yielded by various loads. (C and D) The derived relationships between velocity and load and power and load, respectively. $dl$ = change in length; $dt$ = change in time; $F \times V$ = force × velocity = power.

the behavior of many materials, including muscle to a degree (see Fig. 9).

The steady-state behavior of muscle can be adequately described by two parallel elements: a springlike element, representing the passive force–length characteristics; and a contractile apparatus, representing the active force characteristics. We consider the case of a skeletal muscle at

$L_o$, a length of optimal filament overlap at which little passive force exists. Thus the behavior we are describing is that of the contractile element alone. Imposition of a rapid step shortening leads to a rapid drop in active force, followed by a redevelopment of tension. This suggests that muscle behavior can be modeled by two components: a series elastic spring, which instantaneously responded to

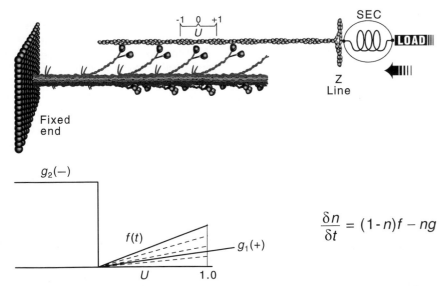

**FIG. 6.** Diagram representing a two-element model of muscle and Huxley's (1957) mathematical model. Top: Half-sarcomere containing a myosin thick filament and actin-containing thin filament, which constitutes a contractile component, coupled in series with a series elastic component (SEC). Bottom left: Functional form of rate constants for crossbridge attachment ($f$) and detachment ($g_1$ and $g_2$) as a function of $U$; the crossbridge position coordinate shown at top. Bottom right: Differential equation describing the change in number of crossbridges. Force generated in this model is equal to the number of attached crossbridges, $n$, times the force of an individual crossbridge.

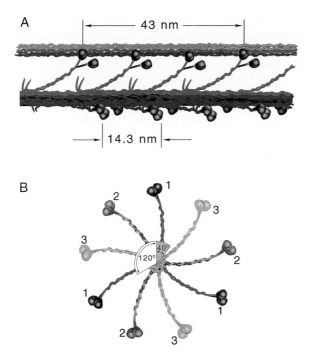

**FIG. 7.** Thick filament. (A) A rendition of the thick-filament structure based on current evidence. (B) The filament viewed on end.

shortening phase, also consistent with this two-component model.

Although this modeling can explain mechanical behavior, its value lies in its potential for associating anatomical elements with model elements. A localization for the series elastic component (SEC) has evolved and has considerably altered our view of muscle mechanics. The SEC can be characterized in terms of the extent of shortening required to discharge the maximal isometric force ($F_o$). Studies on intact, whole muscle suggested that a change in the length of muscle of 3% $L_o$ was sufficient to transiently reduce active force to zero. This elasticity is ascribed to connective elements such as tendons and similar passive structures. A 3% change in $L_o$ can be translated to a relative filament motion of 33 nm. This is significantly larger than the crossbridge spacing (14.3 nm), reinforcing the concept that such behavior is external to the contractile component and in series with crossbridges. The two-component model is useful phenomenologically and can predict behavior of many whole tissues, particularly those with large intrinsic SEC, such as some smooth muscle. However, as improvements in the temporal resolution of mechanical measurements advanced, this model was found to be inadequate for the behavior of single skeletal muscle fibers.

In the 1970s techniques for imposition of step changes greatly improved as both transducers and recording apparatus achieved millisecond resolution. At the same time, techniques for working with single muscle fibers and devices for control of sarcomere length, rather than overall muscle length, were developed. These new measurements indicated that the true SEC extent was much smaller than

the length change; and a contractile element, whose behavior could be described by a viscouslike relationship between force and velocity (see later section). Step changes in force yield a rapid, in-phase letter change followed by a slower

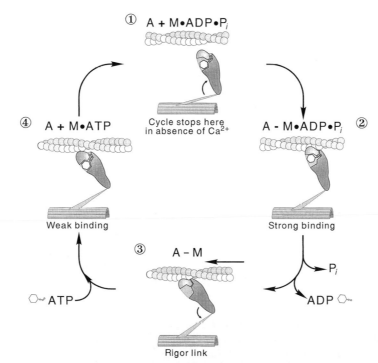

**FIG. 8.** The crossbridge cycle. This depiction represents a crossbridge cycle based on currently accepted information from muscle structure, mechanics, and biochemical kinetics of the actin-activated myosin ATPase.

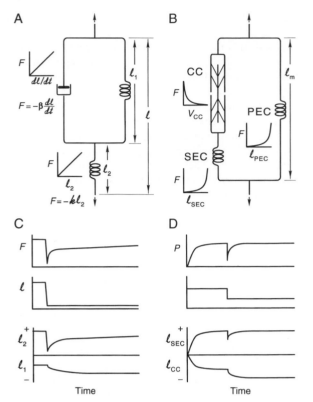

**FIG. 9.** Models forming the basis of classic analysis of muscle mechanics. (A) Three-element Voight model containing two linear spring elements ($l_1$ and $l_2$) combined with a "dashpot," a linear viscous element. (B) Three-element model of muscle containing a contractile component (CC), a parallel elastic component (PEC), and a series elastic component (SEC). Graphs indicate the behavior of each element. (C) Time course of the response of the Voight model and its individual components to imposition of a rapid step change in length. (D) Similar responses of the muscle model to a rapid step change (note that the initial muscle length here is chosen such that the PEC does not play a role in this response).

crossbridges and, consequently, independent of force. Alternatively, if the instantaneous elasticity was external to the crossbridges, it would be governed by its own relationship between force and extension. Reduction of force (and number of crossbridges) by changing the initial muscle length in this case would reduce the step shortening required. Huxley and Simmons (1971) provided evidence that showed that altering active force by changing the initial muscle length did not alter the size of the step shortening required to discharge the maximum isometric force at any initial length. These results supported the concept that the instantaneous springlike behavior is an intrinsic crossbridge property. Moreover, it also provides evidence that the crossbridges act as independent force generators. An important corollary of these studies is that the instantaneous stiffness can be used as an index of the number of crossbridges operating at any moment. This corollary has been used in numerous studies to assess the effects of various interventions on crossbridge number.

Recent mechanical (Higuchi *et al.,* 1995) and X-ray diffraction studies (Huxley, 1996) suggest that a significant fraction of the instantaneous stiffness is attributable to thin filaments. Thus one should exercise some caution in the interpretation of stiffness measurements.

To summarize, force generation is due to cyclic interaction of myosin heads projecting from the thick filament with actin of the thin filament. These crossbridges act as independent sites for force generation. Total active force is proportional to the number of activated crossbridges whose number is geometrically constrained by the extent of filament overlap. The mechanism of crossbridge force generation is still the subject of intense investigation. A current model based on biochemical kinetics and crossbridge mechanic studies is shown in Fig. 8. A shift in orientation of the crossbridge in the strong binding states (steps 3–4) is associated with the force-generating step in this model.

that previously measured in whole muscle. (Previous measurements apparently included an artifact of the slow response time of the recording system.) Current estimates of the SEC for striated muscle are less than 0.5% $L_o$, approximately 6 nm per half-sarcomere, clearly in a range to be potentially associated with crossbridges themselves. Moreover, the time courses of responses (Fig. 10) were not consistent with an instantaneous spring connected in series with a contractile element, which also was characterized by an instantaneous force–velocity relationship. Huxley and Simmons (1971) provided the first evidence that an instantaneous springlike behavior could be attributed to the crossbridges. They exploited the fact that the number of crossbridges could be varied by taking advantage of the force–length relationship. As shown in Fig. 10, if the SEC is intrinsic to the crossbridge, then the shortening step required to discharge the instantaneous elasticity is also intrinsic to the crossbridge. Thus the change in length to discharge force would be independent of the number of

## III. Muscle Energetics

Studies of muscle energetics parallel those of mechanics and biochemical kinetics and have the ultimate goal of understanding the mechanism of mechanochemical energy transduction at the crossbridge level. Historically, muscle energetics has been associated with the measurements of muscle heat production. Heat production is the first indication of the chemical reactions in muscle, and the temporal resolution of measurements of heat far exceeds that of direct chemical determinations. Moreover, heat measurements were made long before the chemical reactions were known. In fact, heat measurements and thermodynamic analysis were essential to ultimately identifying the chemical reactions underlying muscle activity. Thus it is not surprising that a tremendous literature, complete with its own terminology, has arisen. The essential principle comes from an application of the first law of thermodynamics. The change in enthalpy, $\Delta H$, is equal to the sum of heat ($Q$) plus work ($W$) of the muscle. The change in enthalpy, in turn, can be related to the sum of the changes in number

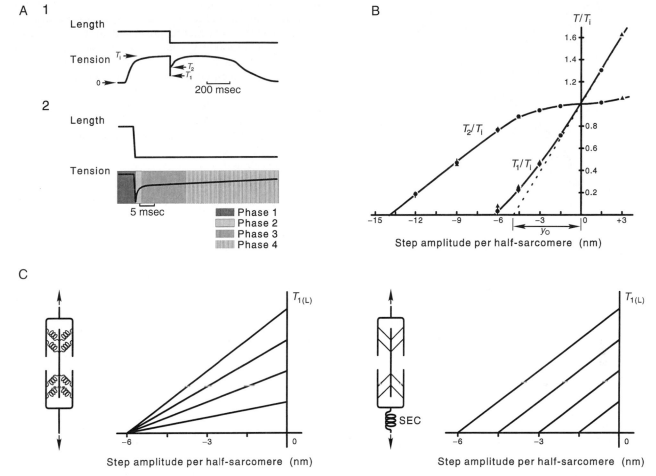

**FIG. 10.** Huxley–Simmons (1971) experiments. (A) Transient force responses to very rapid (<1 msec) step changes in length, which led to revision of classical mechanics. (B) Dependence of phase 1 and phase 2 force amplitudes on muscle length. (C) Dependence of $T_1$ if the series elasticity resided within the crossbridge itself (left), and if the series elasticity was located in a classic SEC (right) Huxley and Simmons data support left panel of part C.

of moles of each species ($\Delta n_i$) multiplied by its specific enthalpy ($h_i$):

$$\Delta H = \sum_i \Delta n_i \times h_i = Q + W \tag{3}$$

Thus, by measuring muscle heat and work some indication of the underlying chemical reactions can be obtained.

One of the first applications of this type of study was carried out by Wallace Fenn (1924) in the laboratory of A. V. Hill. A theory, known as the *viscoelastic model,* proposed that stimulation elicited a fixed or quantal extent of some unknown reaction(s), whose energy was transduced to a fixed amount of heat plus work. A muscle shortening against a load would produce work and according to this theory, less heat than a muscle under isometric conditions not producing work. Fenn's studies showed that the $Q + W$ in a work-producing contraction exceeded the heat produced in an isometric contraction (in which work is minimal). This "Fenn effect" was important in that theories of muscle energy conversion had to be configured in terms of chemical reactions that were closely coupled to mechanical events.

The studies of the nature of the chemical reactions coupled to muscle activity, both those immediately coupled to mechanical events and those involved in the subsequent recovery by intermediary metabolism, have been closely associated with muscle energetics. The history of the search for the identity of the reactions immediately coupled to muscle mechanical output parallels the history of biochemistry itself. Because of increased understanding of the biochemistry of fermentation, lactate was first believed to be this driving reaction. However, Lundsgaard (1938) showed that muscle treated with iodoacetate (which blocks glycolysis and thus lactate production) could still contract. Fiske and SubbaRow (1927) showed that inorganic phosphate ($P_i$) is liberated during contraction, arising not from a glycolytic intermediate, but from a new "phosphagen" phosphocreatine (PCr). Myosin was known to be an ATPase, but there was little change in ATP concomitant with contractile activity. The connection was made by Lohmann (1934), who discovered an enzyme (creatine kinase) that catalyzes the reversible transphosphorylation:

$$PCr + ADP \leftrightarrow ATP + Cr \tag{4}$$

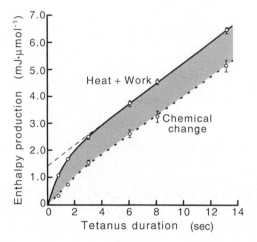

**FIG. 11.** Skeletal muscle "energy balance." Enthalpy (heat + work) production as a function of tetanus duration is given as the solid line. The expected enthalpy (chemical change times the enthalpy of PCr breakdown, 34 kJ/mole) is plotted as the dotted line. Shaded area represents the unexplained enthalpy. [Adapted from Kushmerick, M. J. 1983. Energetics of muscle contraction. *In* "Handbook of Physiology (Section 10)—Skeletal Muscle" (Peachy, Lee and Adrian, Richard H., Eds.), pp. 189–236. Oxford University, New York.]

From these observations grew the energetic schema that ATP was the immediate energy source coupled to the myosin ATPase. In intact muscle, PCr rapidly rephosphorylated ADP via this Lohmann reaction for creatine kinase reaction), such that breakdown of PCr to Cr and $P_i$ was the only net reaction occurring, and this accounted for muscle heat production during contraction. Quantitative tests of this scheme awaited the development of rapid-freezing technology and microchemical analysis of muscle extracts, which occurred during the late 1960s. These "energy balance" studies (Woledge *et al.,* 1985) tested the validity of Eq. (3). Under conditions in which metabolic resynthesis was minimized, the measured breakdown of PCr multiplied by the partial molar enthalpy (34 kJ/mole) would have been equal to $Q + W$, if this had been the only chemical reaction occurring. As shown in Fig. 11, this equality was not valid as $Q + W$ was significantly greater than that accounted for by PCr breakdown. The implication was that there was a "missing" reaction associated with an "unexplained" enthalpy production occurring during contraction. The implications of these observations were critical to the validity of the energetic dogma. Moreover, heat records to this day are still interpreted largely in terms of ATP breakdown (coupled to PCr breakdown) in most current kinetic models, starting with the classic Huxley (1957) formulation. Thus energetics studies were focused on identification of the nature of the unexplained enthalpy and a potential missing reaction associated with contraction.

In a parallel fashion, "biochemical" balance studies (Kushmerick, 1983) were also aimed at approaching the question of a "missing" reaction but with chemical rather than physical techniques. The rationale was that if PCr had been the only net reaction occurring during contraction, and if resynthesis (which is governed by theoretical biochemical stoichiometry) had occurred during recovery, then the amount of oxygen consumed during recovery would have been equal to 1/6.5 times the PCr broken down during contraction. Again as shown in Fig. 12, an imbalance was observed, with the extent of recovery metabolism associated with a resynthesis of ATP that exceeded that accounted for by PCr breakdown during contraction. Paul

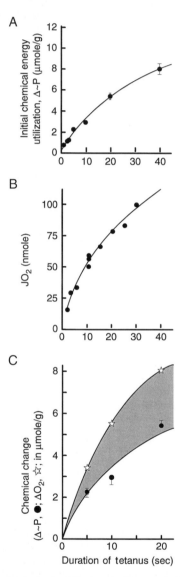

**FIG. 12.** Skeletal muscle "biochemical energy balance." (A) The initial high-energy phosphagen (PCr + ATP) breakdown during contraction as a function of tetanus duration. (B) Total recovery oxygen consumption elicited by the contraction durations plotted on the ordinate. (C) The total phosphagen resynthesized based on theoretical stoichiometry and the measured $O_2$ consumption (open stars) and the measured phosphagen breakdown during contraction (filled circles). The shaded area represents the excess of ATP synthesis over the phosphagen breakdown measured during contraction.

(1983a), using both "energy balance" techniques, showed that oxidation of substrate was the only net reaction occurring. The combinations of techniques yield data that indicated that the chemical imbalance is attributable to additional breakdown of ATP after contraction.

The resolution of the "missing" reaction question apparently resides in the discovery of the protein parvalbumin, and its role as a $Ca^{2+}$-binding site (Gillis et al., 1982). Parvalbumin binds a significant amount of the $Ca^{2+}$ released by the sarcoplasmic reticulum (SR) during contraction. This binding is associated with a substantial evolution of heat. There is also evidence that $Ca^{2+}$ is returned to the SR after contraction activity ceases, which would require activity of the $Ca^{2+}$ pump and concomitant hydrolysis of ATP. This explanation is perhaps the most widely held, but is not universal. For example, the biochemical balance studies would suggest that the postcontractile ATP breakdown is proportional to the duration of stimulus, whereas the enthalpy balances apparently indicate that the unexplained enthalpy attains a plateau value. Moreover, some unexplained enthalpy is associated with mammalian striated muscle, which does not contain parvalbumin. Nevertheless, the schema involving ATP as the immediate driving reaction, with PCr as the source of chemical energy for rapid resynthesis in intact muscle, appears valid. This is further supported by (1) thermodynamic data indicating that the free energy provided by ATP during contraction is sufficient to account for the work produced (Woledge et al., 1985) and (2) many studies using permeabilized muscle indicating that ATP is the only chemical energy source required.

## IV. Muscle Metabolism

Mechanochemical energy transformation and the breakdown of high-energy phosphagens under various conditions form one branch of study known as muscle energetics. Energy metabolism and intermediary metabolism are two terms often used to describe the other major branch of energetics, that of the study of the metabolic provision of phosphagen in support of muscle activity and the mechanisms of coordination of metabolism with contractility. Muscle mechanical activity is supported by ATP provided by both oxidative phosphorylation and glycolysis. Historically, the biochemistry of these pathways often mirrors our understanding of muscle, as this tissue was often the major model for biochemical studies. Although general metabolism and energy production were covered in Chapter 6, the historical aspects (Needham, 1971) and biochemical details are beyond the scope of this text; the treatise of Needham is highly recommended to this end.

Perhaps of most recent interest in this field is the reexamination of the theories whereby contractile ATP use is coordinated with metabolic resynthesis of ATP. The most prevalent theory is that of acceptor- or ADP-limited respiration. In the case of isolated mitochondria in the presence of substrate, their rate of oxidative phosphorylation (state III) is limited by the availability of ADP as the substrate

for rephosphorylation (Kushmerick, 1983). Because the hydrolysis of ATP concomitant with contractile activity is increased, one might anticipate that an increase in ADP could naturally couple the increased usage with synthesis. The creatine kinase reaction coupled with the amount of PCr in striated muscle buffers the ATP concentration. Under the assumption that this reaction is in equilibrium, the level of free (i.e., not bound) ADP can be calculated to be in the region of 10 $\mu M$. This is also in the range for the $K_m$ for ADP for mitochondrial oxidative phosphorylation. This is apparently the major mechanism for skeletal muscle (Kushmerick, 1983), for which significant changes in PCr and thus ADP occur during contractile activity and correlate well with the increased oxidative metabolism. On the other hand, significant increases in cardiac (Heineman and Balaban, 1990) and smooth muscle oxidative metabolism occur in the absence of changes in PCr or ATP, indicating that increases in energy metabolism can match the contractile ATP utilization in the absence of changes in ADP. There are a number of hypotheses in this area. The most prominent ones propose regulation of oxidative metabolism by $[Ca^{2+}]_i$ and mechanisms involving a substrate level "push," for example, by mobilization of substrate, leading to increased NADH levels.

A second area of growing interest on the metabolism side of muscle energetics is the effect of microcompartmentation of metabolism and its subserved function. Instead of being distributed in a random fashion in solution in the cytosol, many enzymes important in energy metabolism are apparently localized within the cell, often in close apposition to energy-dependent processes. Perhaps the most widely studied system is that of creatine kinase (Ishida et al., 1994), which is localized at the m-line of striated muscle as well as in membrane structures and the mitochondria. In permeabilized muscle, endogenous phosphocreatine alone can support contractile activity, suggesting a close relationship between the myosin ATPase and creatine kinase. Similarly, phosphocreatine has been shown to support $Ca^{2+}$ uptake by sarcoplasmic reticulum isolated from skeletal muscle. The enzymes involved with glycolysis are also localized, primarily on the thin filaments and associated with the plasmalemma and sarcoplasmic reticulum. There are increasing numbers of studies (Ishida et al., 1994) for both cardiac and smooth muscle indicating that ATP provided by glycolysis supports functions different from those dependent on oxidative metabolism. This is particularly striking in smooth muscle, in which the oxidative and glycolytic components of metabolism can be of similar magnitude (Paul, 1983b). ATP-dependent processes associated with the plasma membrane, such as the Na–K pump and ATP-dependent potassium channels, are apparently particularly dependent on ATP generated by aerobic glycolysis, whereas the ATP requirements for the force generating actin–myosin interaction are apparently more strongly correlated with oxidative metabolism. The bases and consequences of this cellular organization are not fully understood, but have been suggested to be related to either greater energy transduction efficiency or the need for independent regulation of different cellular functions.

## V. Comparative Muscle Physiology

### A. Striated Muscle

The general features just described are common to muscle function; however, there are a great variety of muscle types adapted to specialized conditions. Among skeletal muscle, perhaps the most familiar adaptation is the difference between red and white muscle. The nomenclature has a long history and many different schemes exist. Histologically and functionally three major classifications predominate, termed fast glycolytic (FG), slow oxidative (SO), and fast, oxidative–glycolytic (FOG), reflecting their gross properties. Most whole muscle contains various mixtures of these three basic muscle fibers and appears pale compared to red muscle, which is composed primarily of SO fibers. The red color is due to the presence of myoglobin, a protein that facilitates the diffusion of oxygen.

These fiber types are adapted for different power requirements. The FG fibers provide large forces that can be rapidly activated and relaxed, but are susceptible to fatigue. Their metabolism depends heavily on glycolytic production of ATP, which can respond rapidly to large changes in energy requirements associated with muscle contraction, but has relatively limited capacity. For low-level but continuous muscular activity, such as that required by postural muscles, the SO fibers are adapted with a high oxidative metabolic capacity. A lower actin-activated myosin ATPase of the SO fibers relative to FG fibers facilitates a more economical maintenance of force: however, this is accompanied by a slower velocity of shortening.

In primates, FOG fibers are relatively rare and highly specialized muscles (some ocular muscles appear to be composed primarily of this fiber type). Their twitch duration and shortening speed are intermediate between those of SO and FG fibers and are characterized by high levels of both glycolytic and oxidative metabolism. These differences in adaptation also extend to speed of activation, which is approximately three times faster in FG than SO fibers. Cardiac muscle has its own special adaptations and in many ways shares similarities with SO skeletal muscle fibers. (These are considered elsewhere in this text). A third muscle type, smooth muscle, is significantly different from striated muscle and is discussed in detail in the following section.

These gross differences in muscle fiber type have long been studied. Applying advanced molecular biology techniques, we now know that most muscle proteins exist as isoforms. Thus recent interest has focused on the isoform–function relationships as well as the regulation of the expression of these various isoforms. What is of particular interest is that the number of expressed isoforms, and indeed the theoretical possible variants, far exceed the three gross skeletal fiber types. Detailed information on isoforms is beyond the scope of this text and is available in specialized treatises (Mahdavi *et al.*, 1986; Pette and Staron, 1990).

### B. Smooth Muscle

Historically, smooth muscle has been of interest because of its specialized function. It lines the hollow organs, such as blood vessels and the gastrointestinal (GI) tract. Its structure is less organized than that of striated muscle and thus somewhat less amenable to biophysical experimentation. For many years, its properties have been simply extrapolated from striated muscle. However, it is now clear that while many similarities exist, there are also significant differences. Studies of smooth muscle have significantly increased over the past decade. This can be attributed largely to its clinical relevance in terms of the diseases of industrialized society, such as hypertension, asthma, and GI motility disorders. However, much of the current interest has arisen from the discovery that the regulatory mechanisms at the contractile filament level are radically different from those of striated muscle.

Smooth muscle can develop isometric forces per cross-sectional area that are equal to or greater than those generated by striated muscle. This is all the more surprising in that the myosin content of smooth muscle is considerably less (about one-fifth) and that of striated muscle. These large forces can be maintained with an ATP use that is 100- to 500-fold lower than the corresponding rate in skeletal muscle. The trade-off for high forces and economy of force maintenance is apparently shortening velocity, which is much slower (up to 1000-fold) than that in striated muscle. The efficiency, in terms of work per ATP, is also somewhat lower (approximately one-fifth that of skeletal muscle); however, the data here are less extensive. The key question then is how does smooth muscle accomplish these specialized functions given that its contractile apparatus uses basically similar actin and myosin components?

#### 1. Smooth Muscle Structure and Its Relationship to Function

The structure of smooth muscle is shown in Fig. 13. Smooth muscle cells are smaller than skeletal muscle, with maximum diameters of the spindle-shaped cells in the range 10–20 $\mu$m. Filament structure is less organized, with no distinct sarcomeric or banding structure. There are no Z-bands but actin filaments are organized through attachment to specialized cytoskeletal regions called dense bodies or patches. These areas contain $\alpha$-actinin, a Z-band protein, which further strengthens this analogy. It is also likely that these structures mediate transmission of force between cells. The ratio of thin to thick filaments is much higher in smooth muscle ($\sim$15:1) than in striated muscle (2:1).

This latter observation raises the question of whether structural factors could account for these functional differences. The answer clearly depends on the mechanism of contraction in smooth muscle. It is assumed, by analogy to striated muscle, that a sliding-filament mechanism is involved in smooth muscle contraction. There is very limited evidence to date but the available data are consistent with this theory. Primarily, the evidence in intact tissue is limited to a dependence of active force of length, which is qualitatively similar to that of striated muscle. The loose organization of myofilaments is not readily amenable to direct structural data regarding the extent of thin–thick filament overlap for comparison with force–length behavior. Other supportive evidence can be gained from the

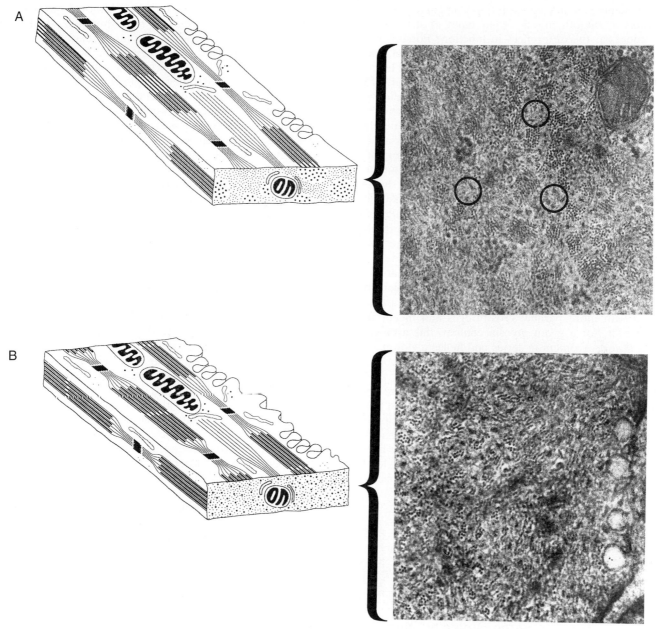

**FIG. 13.** Morphology of relaxed and contracted smooth muscle. (A) This diagram represents a portion of a relaxed smooth muscle cell, highlighting the arrangement of the actin-containing thin filaments and the myosin-containing thick filaments. This rendering is greatly simplified from the actual arrangement. From the longitudinal orientation, thin filaments arise at dense bodies and project to interact with thick filaments, forming the contractile apparatus. In the cross section, the filaments are distributed in a nonuniform manner, with thick filaments forming clusters surrounded by groups of thin filaments. This arrangement is shown in the electron micrograph (circled), which is a cross-section of a relaxed visceral smooth muscle cell. (B) A contracted smooth muscle cell. The notable differences from the relaxed cell shown in part A are that, in the longitudinal views, the thin filaments overlap considerably more of the thick filaments, drawing the dense bodies closer together and shortening the cell. In cross section, the thin and thick filaments are randomly distributed to form a uniform pattern. The electron micrograph is a cross section of a contracted visceral smooth muscle cell, which shows more uniform distribution of the thin and thick filaments. (Diagrams are modified from Heumann, 1973.)

more recent "motility assays." In these experiments, actin filaments are observed to move on a myosin-coated surface, or myosin-coated beads move on actin filament networks (Harris and Warshaw, 1993b). One can infer that folding of filaments is not essential to motion, consistent with a sliding-filament model. Moreover, the only difference between striated and smooth muscle myosin in these experiments is a slower velocity observed with smooth muscle myosin. Interestingly, the type of actin apparently has little effect (Harris and Warshaw, 1993a).

Assuming an analogous sliding-filament mechanism operates in smooth muscle, what types of structure could account for the differences in function? Force is proportional to the number of crossbridges in parallel, whereas shortening velocity and the maximal shortened length are related to the number of fundamental units in series. Models of arrangement of cells or sarcomeres yielding an increase in force and holding economy while reducing velocity are shown in Fig. 14. While increasing the number of cells in parallel (Fig. 14A) works in the appropriate direction (large force, low tension cost), it falls short of true smooth muscle behavior in that the amount of total tissue shortening is very limited. Smooth muscle can shorten to relatively short lengths; some 20–30% of initial tissue length is not uncommon. In the model of Fig. 14A, the absolute length change would be that attributable to shortening of only one cell and would not account for the shortening of any macro size tissue.

In Fig. 14B, different mechanical properties are associated with the assembly of the sarcomere. Longer myosin filaments and consequently longer sarcomeres are associated with more parallel crossbridges and higher forces. However, for a given myosin content, this arrangement has fewer sarcomers in series and hence a slower overall velocity. Moreover, if the individual myosin crossbridge ATPase is not altered, one would have a similar ATP use for both models, and thus the longer sarcomere version would have a greater *economy* force maintained per rate of ATP hydrolysis) or a lower *tension cost* (reciprocal of economy). These changes are in the direction that distinguishes smooth muscle from skeletal muscle. The question then is how much of the difference between skeletal and smooth muscle can be related to simply the "mechanical

advantage" of longer sarcomeres? Although no obvious sarcomeric structure exists for smooth muscle, the length of the myosin filament of smooth muscle relative to striated answers this question. There is evidence that some smooth muscles, particularly in invertebrates, such as the scallop or mussel, whose sarcomere lengths may be up to 10-fold greater than mammalian skeletal muscle, make use of this mechanical advantage. However, for mammalian tissues, myosin filament length has been estimated at approximately 2.2 $\mu$m (Somlyo, 1980), a figure not significantly different from that of striated myosin (1.6 $\mu$m), and differences here are not primarily ascribable to structural features. The differences in contractile properties between smooth and skeletal muscle would thus appear to reside largely in the nature of the smooth muscle myosin molecule and its intrinsically lower ATPase.

## 2. Regulation of Smooth Muscle Contractility

Regulation of contractility can be divided into (1) mechanisms for control of $[Ca^{2+}]_i$ and (2) mechanisms for transduction of the $Ca^{2+}$ signal to activation of the contractile apparatus. Regulation of $[Ca^{2+}]_i$ is considered elsewhere and the focus here is on transduction mechanisms.

Up to the mid-1970s, mammalian smooth muscle, based largely on analogy to striated muscle, was considered to be a "thin-filament" regulated system, that is, one in which the troponin is the $Ca^{2+}$ receptor and tropomyosin is a transducing component of the regulatory system (see Chapter 53, on regulation of striated muscle). Several lines of evidence led to the current, most widely accepted view that phosphorylation/dephosphorylation of the 20-kDa light chain of myosin (MLC-$P_i$) is the primary tranduction site of the $Ca^{2+}$ signal (see Hartshorne, 1987; de Lanerolle and Paul, 1991). Although tropomyosin is known to be a major smooth muscle protein, troponin is not present in smooth muscle. However, proving a protein absent is difficult and that proof came slowly. Smooth muscle actomyosin also proved considerably different from that of striated muscle in that as the purity increased, the ATPase activity of smooth muscle actomyosin decreased, the opposite of that found for striated actomyosin. This suggested that smooth muscle myosin required activation, whereas striated muscle myosin required de-inhibition. This was tested more rigorously by "competition" experiments. In these experiments, the $Ca^{2+}$ sensitivity of the actin-activated Mg-ATPase of myosin was measured using "unregulated" thin filaments (i.e., purified actin without troponin and tropomyosin). Striated muscle myosin showed little $Ca^{2+}$ sensitivity, whereas vertebrate smooth muscle myosin retained $Ca^{2+}$ sensitivity, indicating that the site of its regulation was on the myosin itself. The regulatory site was identified at the isolated protein level when it was discovered that the actin activation of the Mg-ATPase of smooth muscle myosin was dependent on phosphorylation of the 20-kDa light chain.

Several lines of evidence supported this thick filament regulatory mechanism at the cellular level. One strategy was to alter MCL–$P_i$ independent of $[Ca^{2+}]_i$ through manipulation of the kinase or phosphatase. Inhibition of

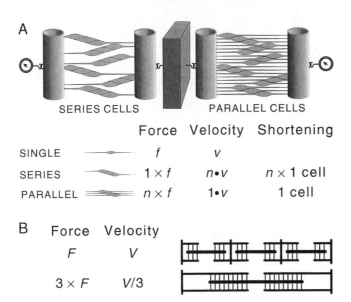

**FIG. 14.** (A) Parallel and series models of force transmission in smooth muscle-containing tissue. Parallel arrangement increases total force, but with a decrease in velocity and total shortening compared with cells coupled in series. (B) Parallel and series model at the level of the sarcomere. Both sliding filament models have the same myosin content, but it is arranged in short (top) and long sarcomeres (bottom).

MLCK by various inhibitors or use of calmodulin-binding peptides leads to relaxation and dephosphorylation in the presence of $[Ca^{2+}]_i$. Alternatively, proteolysis of MLCK yields a fragment that is constitutively active, independent of $Ca^{2+}$. This enzyme added to permeabilized smooth muscle generates a contraction, mechanically indistinguishable from that induced by $Ca^{2+}$. Pressure injection of this constitutively active MLCK into intact smooth muscle cells also leads to a contraction. This is important, for one could argue that making tissue permeable could lead to loss of regulatory elements. Taken in total these results suggest that myosin light chain phosphorylation is the major regulatory site and is necessary for activation.

Whether myosin light chain phosphorylation is sufficient as a regulatory mechanism or indeed has additional roles is open to question. During the past several years a number of experimental observations suggest that smooth muscle regulation may be more complex than initially envisioned. Figure 15 summarizes the time courses of several parameters after stimulation. Isometric force increases monotonically and then maintains a plateau value, whereas MLC-$P_i$ rapidly increases to a maximum achieved early in the contraction and then decreases. The extent of the decline in the MLC-$P_i$ is controversial, and this is likely attributable to different smooth muscle types, dependence on the nature of the stimulus, and technique for measurement of MLC-$P_i$. The decline in MLC-$P_i$ with maintained or increasing force suggests that other regulatory factors may play a role. Moreover, near maximal forces are attained at MLC-$P_i$ levels considerably lower than 100%, which further questions its role as a simple switch. The decline in MLC-$P_i$ is similar to that of $[Ca^{2+}]_i$ and, importantly, to the decline in shortening velocity. In view of these observations, Murphy and colleagues (Hai and Murphy, 1988, 1989) suggested that MLC-$P_i$ might be a regulatory of contractile velocity. They coined the expression *latch* for the state of maintained force with reduced MLC-$P_i$ and slower contraction speeds. *Latch bridges* or crossbridges in this state were initially postulated to be non- or slowly cycling, and thus they reduced the overall tissue velocity

by acting as a type of internal load on the more rapidly cycling bridges. As force retained its dependence on $Ca^{2+}$ in the latch state, it was postulated that latchbridges may be regulated by a system different from MLC-$P_i$. This is a rapidly changing and intensely investigated area.

Currently there are two major classes of theories for latch behavior, which are not necessarily mutually exclusive. Thin-filament regulatory mechanisms based on several actin-binding proteins, namely, leiotonin, caldesmon, and calponin, have been suggested. A common feature is that they are all proposed to be inhibitory, and for the latter two, the interaction with actin is sensitive to $Ca^{2+}$-calmodulin. Although all are promising in terms of the ability to inhibit myosin ATPase activity in the test tube, the relevance to intact smooth muscle has yet to be unequivocally demonstrated. As smooth muscle myosin requires activation, these systems could only be ancillary. Based on the ability of MLC-$P_i$ to activate smooth muscle in the absence of $Ca^{2+}$, it is difficult to postulate that these systems can be inhibitory in the sense that they regulate activation on an "on–off" basis. However, these systems could regulate crossbridge cycling and thus velocity, but more evidence of this is needed.

An alternative hypothesis suggested by Hai and Murphy (1989) does not require regulatory components beyond myosin light chain phosphorylation/dephosphorylation. A version of their model is shown in Fig. 16. They postulate that if a phosphorylated crossbridge were dephosphorylated while attached, its detachment rate would be significantly slower than if it were to remain phosphorylated. The kinetic scheme of Fig. 16 is sufficient to fit the available data on the time courses of force and MLC-$P_i$. It can also predict the time course of ATP utilization, which also decreases similarity to velocity, with stimulation time. However, this scheme predicts that a strongly curvilinear relationship exists between the rate of ATP utilization and force and that ATP utilization due to the "futile" cycle of myosin light chain phosphorylation/dephosphorylation is the dominant (~85%) source of ATP utilization. This view has been challenged by Paul (1990) based on the available smooth muscle energetics data. Moreover, he shows that simply a high ratio of attachment to detachment rate constants, as proposed several decades earlier by Bozler (1977), can fit the available data without invoking a special "latch" detachment cycle. Recently, Hai and Murphy (1992), using the model of Fig. 16 coupled with the kinetic scheme of Huxley (1957) for crossbridge cycling, were able to circumvent some of the objections to this model. At present, experiments specifically designed to test the energetic consequences are needed to help distinguish between these contrasting theories of smooth muscle regulation.

## VI. Summary

This chapter focuses largely on muscle structure in relation to contractility. Mechanisms of regulation of intracellular $Ca^{2+}$ and excitation–contraction coupling are treated in more detail elsewhere (Chapters 53 and 54). Muscle structure is first developed in conjunction with mechanics,

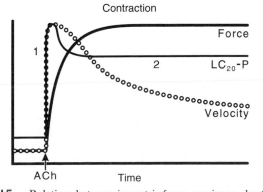

**FIG. 15.** Relations between isometric force, maximum shortening velocity, and myosin regulatory light chain phosphorylation (LC$_{20}$-P) and the duration of stimulation. Data are for tracheal smooth muscle (de Lanerolle and Paul, 1991). The decrease in velocity at times when force is maintained is the basis for the *latch bridge* theories.

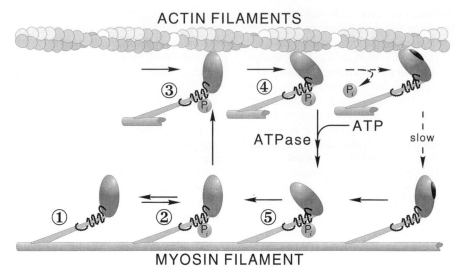

**FIG. 16.** A schematic model of the smooth muscle myosin crossbridge interaction with actin filaments is depicted. The transition from state 1 to state 2 represents the $Ca^{2+}$-calmodulin-dependent phosphorylation/dephosphorylation activation mechanism proposed for smooth muscle regulation. The cycle of interaction (states 2–5) hydrolyzed 1 ATP molecule and generates 1 quantum of tension with each pass. Only the angulated attached myosin crossbridge (state 4), however, generates isometric force. The inherent speed of the cycle is much slower in smooth muscle than in striated, with the ATP-dependent dissociation step (state 4–5) apparently being slower. The dashed arrows indicate formation of a proposed dephosphorylated actomyosin crossbridge (Hai and Murphy, 1989), which then dissociates only very slowly and may be responsible for the high holding economy of vascular smooth muscle.

leading to the formulation of the sliding-filament theory. Muscle contractility continues with further characterization of the relationships among force, velocity, work, and energy utilization. The mechanism at the molecular level, that is, the crossbridge cycle, is similarly developed, matching structural and biochemical knowledge with transient mechanical studies. The energy requirements of the crossbridge cycle lead to presentation of energy use in intact muscle, which is considered in detail in muscle energetics. Energy use then is followed by a discussion of muscle metabolism and the route for synthesis of ATP necessary for contractile activity. This chapter then shifts from considerations of muscle in general to comparative muscle physiology and behavior of different fiber types. In particular, smooth muscle is treated in detail, and its regulation, mechanical properties, and energetics are considered and contrasted to those of striated muscle.

## Bibliography

Bloom, W., and Fawcett, D. W. (1968). "A Textbook of Histology." W. B. Saunders, Philadelphia.

Bozler, E. (1977). Introduction: Thermodynamics of smooth muscle contraction. *In* "The Biochemistry of Smooth Muscle" (N. L. Stevens, Ed.), pp. 3–14. University Park Press.

Dillon, P. F., Aksoy, M. O., Driska, S. P., and Murphy, R. A. (1981). Myosin phosphorylation and the cross-bridge cycle in arterial smooth muscle. *Science* **211**, 495–497.

Fenn, W. O. (1924). The relation between the work performed and the energy liberated in muscular contraction. *J. Physiol.* **58**, 373–395.

Fiske, C. H., and SubbaRow, Y. (1927). The nature of inorganic phosphate in voluntary muscle. *Science* **65**, 401–403.

Gillis, J. M., Thomason, D., Lefèvre, J., and Kretsinger, R. H. (1982). Parvalbumins and muscle relaxation: A computer simulation study. *J. Muscle Res. Cell Mot.* **3**, 377–398.

Gordon, A. M., Huxley, A. F., and Julian, F. J. (1966). The variation in isometric tension with sarcomere length in vertebrate muscle fibres. *J. Physiol.* **184**, 170–192.

Hai, C.-M., and Murphy, R. A. (1988). Regulation of shortening velocity by cross-bridge phosphorylation in smooth muscle. *Am. J. Physiol.* **255**, C86–C94.

Hai, C.-M., and Murphy, R. A. (1989). $Ca^{2+}$, crossbridge phosphorylation, and contraction. *Annu. Rev. Physiol.* **51**, 285–298.

Hai, C.-M., and Murphy, R. A. (1992). Adenosine 5'-triphosphate consumption by smooth muscle as predicted by the coupled fourstate crossbridge model. *Biophys. J.* **61**, 530–541.

Harris, D. E., and Warshaw, D. M. (1993a). Smooth and skeletal muscle actin are mechanically indistinguishable in the in vitro motility assay. *Circ. Res.* **72**, 219–224.

Harris, D. E., and Warshaw, D. M. (1993b). Smooth and skeletal muscle myosin both exhibit low duty cycles at zero load in vitro. *J. Biol. Chem.* **268**, 14765–14768.

Hartshorne, D. J. (1987). Biochemistry of the contractile process in smooth muscle. *In* "Physiology of the Gastrointestinal Tract" (L. R. Johnson, Ed.), Vol. 1, pp. 423–482. Raven Press, New York.

Heineman, E. W., and Balaban, R. S. (1990). Control of mitochondrial respiration in the heart *in vivo*. *Annu. Rev. Physiol.* **52**, 523–542.

Heumann, H. G. (1973). Smooth muscle: Contraction hypothesis based on the arrangements of actin and myosin filaments in different states of contraction. *Philos. Trans. R. Soc. London (Biol.)* **B265**, 213.

Higuchi, H., Yanagida, T., and Goldman, Y. E. (1995). Compliance of thin filaments in skinned fibers of rabbit skeletal muscle. *Biophys. J.* **69**, 1000–1010.

Hill, A. V. (1965). "Trails and Trials in Physiology." Arnold, London.

Huxley, A. F. (1957). Muscle structure and theories of contraction. *Prog. Biophys. Chem.* **7**, 255–318.

Huxley, A. F., and Niedergerke, R. (1954). Interference microscopy of living muscle fibres. *Nature* **173**, 971–973.

Huxley, A. F., and Simmons, R. M. (1971). Proposed mechanism of force generation in striated muscle. *Nature* **233**, 533–538.

Huxley, H. E. (1966). A personal view of muscle and motility mechanisms. *Annu. Rev. Physiol.* **58**, 1–19.

Ishida, Y., Riesinger, I., Wallimann, T., and Paul, R. J. (1994). Compartmentation of ATP synthesis and utilization in smooth muscle: Roles of aerobic glycolysis and creatine kinase. *Mol. Cell. Biochem.* **133/134**, 39–50.

Jewell, B. R., and Wilkie, D. R. (1958). An analysis of the mechanical components in frog's striated muscle. *J. Physiol.* **143**, 515–540.

Krisanda, J. M., and Paul, R. J. (1984). Energetics of isometric contraction in porcine carotid artery. *Am. J. Physiol.* **246**, C510–C519.

Kushmerick, M. J. (1983). Energetics of muscle contraction. *In* "Handbook of Physiology (Section 10)—Skeletal Muscle" (Peachy, Lee and Adrian, Richard H., Eds.), pp. 189–236. Oxford University, New York.

de Lanerolle, P., and Paul, R. J. (1991). Myosin phosphorylation/dephorphorylation and the regulation of airway smooth muscle contractility. *Am. J. Physiol.* **261**, L1–L14.

Lohmann, K. (1934). Über die enzymatische Aufspaltur der Kreatinephosphorsäure; zugleich ein Beitrag zum Chemismus der Muskelkontraktion. *Biochem. Z.* **271**, 264–277.

Lundsgaard, E. (1938). The biochemistry of muscle. *Annu. Rev. Biochem.* **7**, 377–398.

Mahdavi, V., Streber, E. E., Periasamy, M., Wieczorek, D. F., Izumo, S., and Nadal-Ginard, B. (1986). Sarcomeric myocin heavy chain gene family: Organization and pattern of expression. *Med. Sci. Sports Ex.* **18**, 299–308.

Murray, J. M., and Weber, A. (1974). The cooperative action of muscle proteins. *Sci. Am.* **230**, 58.

Needham, D. M. (1971). "Machina carnis. The Biochemistry of Muscular Contraction in Its Historical Development," pp. 1–782. University Press, Cambridge.

Paul, R. J. (1983a). Physical and biochemical energy balance during an isometric tetanus and steady state recovery in frog sartorius at 0°C. *J. Gen. Physiol.* **81**, 337–354.

Paul, R. J. (1983b). Aerobic glycolysis and ion transport in vascular smooth muscle. *Am. J. Physiol.* **244**, C399–C409.

Paul, R. J. (1990). Smooth muscle energetics and theories of cross-bridge regulation. *Am. J. Physiol.* **258**, C369–C375.

Pette, D., and Staron, R. S. (1990). Cellular and molecular diversities of mammalian skeletal muscle fibers. *Rev. Physiol. Biochem. Pharmacol.* **116**, 2–47.

Rayment, I. (1996). The structural basis of the myosin ATPase activity. *J. Biol. Chem.* **271**, 15850–15853.

Rayment, I., Smith, C., and Yount, R. G. (1996). The active site of myosin. *Annu. Rev. Physiol.* **58**, 671–702.

Somlyo, A. V. (1980). Ultrastructure of vascular smooth muscle. *In* "Handbook of Physiology (Section 2)—The Cardiovascular System, (Vol. II) Vascular Smooth Muscle," pp. 33–67. American Physiological Society, Bethesda, MD.

Warshaw, D. M., Desrosiers, J. M., Work, S. S., and Trybus, K. M. (1990). Smooth muscle myosin cross-bridge interactions modulate actin filament sliding velocity in vitro. *J. Cell Biol.* **111**, 453–463.

Woledge, R. C., Curtin, N. A., and Homsher, E. (1985). "Energetic Aspects of Muscle Contraction," pp. 1–357. Academic Press, London/New York.

Edna S. Kaneshiro

# 56

# Amoeboid Movement, Cilia, and Flagella

## I. Introduction

Contractility and motility are fundamental properties of all living cells. Movement is characteristic of life, whether it be of whole organisms, individual cells, or cytoplasmic components. For a system to cause movement, a force must be exerted on a mass over some distance; thus movement requires expenditure of energy in one form or another. It appears that only four basic mechanisms (Table 1) for the generation of force have evolved, and these can be represented by the following biological systems: (1) muscle: actin–myosin (ATPase) interaction + energy from adenosine triphosphate (ATP); (2) eukaryote cilia and flagella: microtubule–dynein (ATPase) interaction + energy from ATP; (3) spasmonemes: spasmin–$Ca^{2+}$ interaction + energy from the $Ca^{2+}$ chemical potential; (4) bacterial flagella: basal body–cell membrane interaction (rotary motor) + energy from transmembrane ion gradient. Muscle contraction was discussed in the preceding chapter, so the present chapter covers nonmuscle cells. Spasmonemes and bacterial flagella are discussed in the next chapter.

## II. Amoeboid Movement and Actin-Based Systems

The discussion on actin-based systems will be restricted to cell motility involving molecules similar, or related to, actin and myosin in nonmuscle systems such as platelets. Several cellular functions are attributed to actin-based systems that are involved in pseudopod formation and in exocytosis and endocytosis in cells such as macrophages, neutrophils, and retinal pigmented epithelia (Table 2). Temperature-independent precipitation or patching of antibodies bound to cell surfaces is followed by actin-based aggregation of the patches called **capping**, which requires metabolic energy and is temperature dependent. The contractile ring in the cortex of cells undergoing cytokinesis contains actin, myosin, and other actin-binding proteins.

Thus, actin-based motility also controls the basic cellular events involved in cell proliferation.

Actin filaments may have functions other than that of cell motility. For example, stereocilia of vertebrate mechanoreceptors are filled with actin filaments. Microvilli in the vertebrate cochlea that respond to sound vibrations and the vestibular apparatus that respond to position and movement are actin-filament-based sensory receptors.

### A. Actin Polymerization

One of the most dramatic examples of rapid actin polymerization was described by Tilney in the sperm of the sea cucumber, *Thyone*. The acrosome vacuole of unstimulated sperm is filled with actin monomers and profilin. The term **profilactin** (profilamentous actin) was originally coined to describe the unpolymerized state of precursors below the acrosome vacuole; thus this term should not be confused with profilin. Upon activation, the acrosome process forms as the result of polymerization of actin monomers into filaments. Filaments about 60–70 $\mu$m long are assembled within 10 sec.

Actin monomers bind ATP, which is subsequently hydrolyzed during polymerization. Polymerization can occur by the addition of actin monomers at the plus end (fast growing) as well as at the minus end (slow growing) of actin filaments. The plus and minus ends correspond to the barbed and pointed ends, respectively, of heavy meromyosin (HMM)-decorated filaments. When associated with membranes, it is the plus end that is usually found adjacent to the membrane.

Inhibitors of actin polymerization include the cytochalasins, compounds that bind to the ends of actin filaments, resulting in the inhibition of monomer addition. At high concentrations, cytochalasins can even sever filaments. Cellular functions, such as cell locomotion, cytokinesis, and cell movements during morphogenesis, are inhibited by these alkaloid drugs. In contrast, the highly toxic compound phalloidin stabilizes actin filaments and inhibits actin depolymerization.

**TABLE I** Comparison of Actin, Tubulin, Spasmin–Centrin, and Flagellin

| Property | Actin | Tubulin | Spasmin–centrin | Flagellin |
|---|---|---|---|---|
| Molecular mass | 42 kDa | 50 kDa[b] | | 51 kDa |
| Isoforms[a] | 4 muscle ($\alpha$): cardiac, skeletal, vascular, and enteric; 2 nonmuscle: ($\beta$ and $\gamma$): cytoplasmic | | Spasmin A-18 kDa Spasmin B-20 kDa | |
| Isoelectric point[a] | | | Spasmin A-47 | |
| | $\alpha$-5.41 | $\alpha$-4.75 | Spasmin B-4.8 | |
| | $\beta$-5.44 | $\beta$-4.69 | $\alpha$ Centrin-4.9 | |
| | $\gamma$-5.47 | $\gamma$-5.55 | $\beta$ Centrin-4.8 | |
| Unpolymerized form | Monomer | Dimer | ? | Monomer |
| Nucleotide required for polymerization | ATP (1/monomer) | GTP (2/dimer) | ? | None |
| Polymer structure | Double helical filament | Hollow tube of 13 protofilaments (14 nm lumen) | Fibrous filament | Hollow cylinder (5 nm lumen) |
| Polymer diameter | 8 nm | 25 nm | 4 nm | 14–20 nm |
| Subunits/$\mu$m polymer | 370 monomers | 1600 dimers | ? | 2000 monomers |
| Subunits/turn | 15 | 13 | ? | 5.5 |

[a] Differences occur among various species.
[b] Isoforms separate mainly by differential detergent binding.

In recent years, dozens of proteins associated with actin-based contractile systems have been described in many different cell types. These actin-binding proteins sever, stiffen, nucleate, bundle, cap, cause sliding between, cross-link, polymerize, and depolymerize actin filaments, and others interact with actin monomers. Some actin-binding proteins can function in more than one capacity; for example, they can cap and sever filaments (Table 3). These molecules have at least one actin-binding site. Many, such as those involved in gel–sol transformations, also have binding sites for $Ca^{2+}$ or the signal-transducing lipid, phosphatidylinositol 4,5-bisphosphate ($PIP_2$).

Profilins are proteins that can regulate actin polymerization. In the soil amoeba *Acanthamoeba* and in yeast, profilin has two actin-binding sites. It caps the free ends of actin polymers by forming dimers with the actin monomers, thus stopping filament growth. In vertebrates, profilin has only one actin-binding site, and hence cannot cap actin filaments. Profilin and actin-depolymerizing factor inhibit actin polymerization. After an external stimulus, actin must first dissociate from such molecules before cell surface membrane extensions can proceed. The dissociation may be controlled by inositol lipids in the membrane.

**TABLE 2** Functions of Actin-Based Motility Systems

| Function | Examples |
|---|---|
| Locomotion | Vertebrate smooth, cardiac and skeletal muscles, amoeboid cells, neutrophils |
| Cytoplasmic streaming (cyclosis) | *Nitella, Physarum* |
| Feeding and endocytosis | *Amoeba*, macrophages, heliozoans, forams |
| Exocytosis and secretion | Neurotransmitter release |
| Cell proliferation | Contractile ring formed during cytokinesis |
| Reproduction | Invertebrate sperm acrosome process |
| Transport and absorption | Intestinal and kidney brush border microvilli |
| Sensory reception and transduction | Vertebrate stereocilia, squid photoreceptor |
| Tissue morphogenesis | Neurulation |

### B. Amoeboid Locomotion

#### 1. Crawling of Amoeboid Cells

The mechanism of amoeboid movement was among the earliest nonmuscle actin–myosin systems that received the attention of Mast and other early cell biologists. Studies were first done mainly on large free-living amoebae such as *Amoeba proteus*. The amoeba cell body is differentiated into three regions: the endoplasm (sol) at the core of the cell occupied by organelles; a cortical ectoplasm (gel); and a shear zone separating the ectoplasm from the endoplasm (Fig. 1). Organelles are excluded from a thin clear hyaline layer adjacent to the cell surface membrane. This layer is thicker at the pseudopodial tip, where it is called the **hyaline**

**TABLE 3**   Some Actin-Binding Proteins

| Protein | Subunit molecular mass (kDa) | Activity |
| --- | --- | --- |
| Actin-binding protein (ABP) | 250 | Gelating |
| Actin depolymerizing factor (ADF) | 19 | Monomer binding, severing |
| $\alpha$-Actinin | 90–100 | Bundling, gelating |
| $\beta$-Actinin | 34–37 | Capping |
| Actophorin–depactin | 17 | Monomer binding |
| Band 4.9 | 48–52 | Bundling |
| Caldesmon | 87 | Bundling |
| Calpactin | 36 | Bundling, membrane binding |
| Capping protein | 29–31 | Capping |
| Cofilin | 21 | Polymer binding |
| DNAse-1 | 35 | Monomer binding |
| Fascin | 58 | Bundling |
| Filamin | 250 | Gelating |
| Fimbrin | 68 | Bundling |
| Fragmin, severin | 42 | Capping, severing, nucleating |
| Gelactins I–IV | 23–28 | Gelating |
| Gelsoln | 90 | Capping, severing, nucleating |
| Myosin I | 110–130 | Sliding, membrane binding |
| Myosin II | 175–200 | Sliding |
| Ponticulin | 17 | Membrane binding |
| Profilin | 12–15 | Monomer binding, capping |
| Radixin | 82 | Capping |
| Spectrin | 220–260 | Gelating |
| Synapsin I | 76–78 | Bundling |
| Tropomyosin | 35 | Polymer binding |
| Troponin I | 23 | Polymer binding |
| Villin | 95 | Capping, bundling, severing, nucleating |
| 120-kDa protein | 120 | Gelating |

**cap.** *Amoebae* move by extension of the pseudopodium, attachment of the pseudopodial membrane to the substratum, and detachment of the membrane to the substratum, and detachment of the membrane from the substratum at the posterior uroid, thus allowing the cell to translocate itself to an anteriorad location. Concomitantly, there is a movement of solated cytoplasm in the endoplasm anteriorly into the newly forming pseudopod and movement of cytoplasm in the gel of the cortex posteriorly. Similarly, rapid cytoplasmic streaming (cyclosis) commonly occurs and is easily observed in large plant cells (such as *Nitella*) and in fungi (such as *Dictyostelium* and *Physarum*).

During locomotion, cell surface membrane is added at the forming pseudopodium and internalized at the posterior uroid of the cell. Large free-living amoebae have numerous large Golgi bodies (dictyosomes), correlating with the demand for active membrane turnover and recycling. The addition of new membrane material to the cell surface during locomotion places a high demand on the cell. Hence, when an amoeba is actively crawling along a substratum, it stops feeding and drinking (phagocytosis, pinocytosis), which are endocytotic processes known to also require actin, myosin, and ATP. Thus, amoebae apparently are unable to walk and eat at the same time.

Microinjection of the $Ca^{2+}$-sensitive bioluminescent protein aqueorin shows that $Ca^{2+}$ is elevated at the tip of extending pseudopodia and at the posterior uroid region. The pseudopod tip membrane is apparently more electrically active than the rest of the cell. Ionic currents, probably carried by $Ca^{2+}$, have been recorded by the sensitive vibrating microprobe technique (which measures extracellular currents concentrated near "hot spots").

A

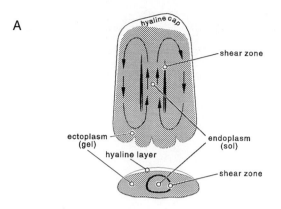

B

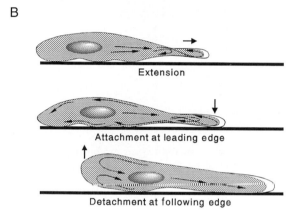

Extension

Attachment at leading edge

Detachment at following edge

C

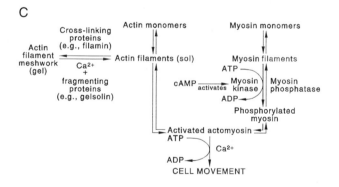

**FIG. 1.** Amoeboid locomotion and cytoplasmic streaming. (A) The cytoplasm of a large free-living amoeba such as *Amoeba proteus* is differentiated into the solated endoplasm, the gelated ectoplasm at the cortex, and the clear region next to the cell surface membrane called the *hyaline layer,* which is enlarged at the tip of the pseudopod, where it is called the hyaline cap. Organelles move in the endoplasm toward the extending pseudopod. In the central endoplasm, actin filaments are presumably short and not highly cross-linked, whereas in the ectoplasm, where organelles move posteriad, they are long and in a gelatin state. (B) Locomotion of the cell involves three events: (1) The extension of the pseudopod. This extension involves the assembly of actin filaments by addition of monomers to the plus ends next to the cell membrane. Movement of other molecules to the advancing pseudopod may be achieved by actin–myosin I interaction. (2) Attachment of the extended pseudopod membrane to the substratum is correlated with the aggregation of actin in this region. (3) The membrane at the posterior becomes detached from the substratum, thus allowing the cell to locomote. The cortical network of actin in the posterior region probably contracts by myosin II action, generating the force for the flow of cytoplasmic organelles in the solated endoplasm. (C) Regulation of actin and myosin interactions for amoeboid locomotion is schematically diagrammed. (From "Animal Physiology: Mechanisms and Adaptions" by Eckertt, Randall, and Augustine. Copyright © 1988 W. H. Freeman and Company. Reprinted with permission.)

## 2. Contractile Proteins

Ultrastructural studies of extracted amoebae indicate that the cortical region is filled with actin filaments (Fig. 2). These actin filaments can be decorated with heavy meromyosin (HMM) from vertebrate muscle. Actin associated with the cell membrane was thus recognized as a mechanism for explaining surface extensions and other cell shape changes that occur in amoeboid cells during crawling, feeding, and dividing. The first convincing biochemical indication that amoeboid movement was based on molecules and mechanisms similar to those of muscle was obtained by Korn and Pollard on the small soil amoeba, *Acanthamoeba,* which can be grown to high densities in culture. Prior to these studies, two models for amoeboid movement were hotly debated: (1) the rear contraction model (posterior

cortex contracts, pushing the cytoplasm forward) and (2) the fountain head model (molecules at the anterior fold resulting in transformation of sol to gel and pulling of endoplasm forward). With recent biochemical, molecular, and immunochemical data, these models are being integrated into a more comprehensive view of this type of cellular movement.

Contractile proteins purified from *Acanthamoeba* indicate that actin has been highly conserved in different organisms. Amoeba myosin molecules, however, were found to be much smaller and less complex than vertebrate muscle myosin, suggesting that myosin may have become more elaborate during evolution. Two forms of myosin have now been characterized in *Acanthamoeba:* (1) myosin II, which forms bipolar filaments like myosin in striated muscle; and (2) myosin I, which has a single head and a short tail and does not self-assemble into bipolar filaments as does

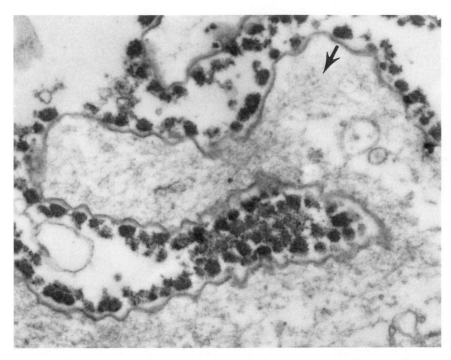

**FIG. 2.**   Electron micrograph of actin filament networks. The free-living amoeba *Chaos carolinensis* exhibits typical amoeboid locomotion. An extensive network of actin filaments is detected by electron microscopy of the gelated ectoplasm cortex under the cell surface membrane (arrow). The cell surface membrane is clearly identified by its association with the stain Alcian blue, which binds surface glycocalyx mucopolysaccharides. Magnification, ×49,000 (Courtesy of Vivian Nachmias.)

myosin II. Myosin I, which can bind to actin and to cell membranes, causes cell shape changes in many nonmuscle cells by forces generated at the cell surface (Fig. 3).

It was demonstrated that even after the cytoplasm is removed from the amoeba (by suction or by allowing the cytoplasm to move into a glass capillary tube), cytoplasmic streaming continued; thus, the machinery and energy for this movement reside within the cytoplasm. The mechanism controlling the direction and rate of cytoplasmic streaming involves the cell surface. It is now generally accepted that extensions and retractions of cell surface in various types of pseudopods of protists (as in microspikes, lamellopodia and filopodia of platelets, macrophages, and fibroblasts) is regulated by transmembrane signaling events such as those involving inositol lipids and G proteins. Stimulation of the cell surface results in the polymerization of actin by addition of monomers at the plus end attached to the cell surface membrane. Hormones and growth factors stimulate cell surface ruffling activity in tissue culture systems. Thus, it is likely that transmembrane signals activate second messengers, cyclic nucleotides, and protein phosphorylations by protein kinases.

### 3. Sol–Gel Transformation

A current model for explaining the reversible transformation of sol to gel states in free-living amoeba cytoplasm proposes that the viscous gel contains cross-linked actin filaments. Cytoplasm isolated from cells that exhibit active streaming contains high concentrations of actin-binding proteins (such as myosin) and cross-linking and severing proteins. Proteins such as α-actinin and filamin can link actin at both ends, and thereby cross-link the filaments. Cross-linking produces a gel state of the actin network in the cytoplasm (endoplasm). If highly cross-linked, the gel is more rigid and less able to contract.

Proteins such as gelsolin and severin tightly bind actin, resulting in the displacement of adjacent monomers in the filament. The displacement causes the fragmentation or disintegration of the filament, resulting in shorter actin filaments. In response to increased intracellular $Ca^{2+}$ and/ or $H^+$, the cross-links between actin filaments are broken, and the actin filaments are now present as short fragments and transform the cytoplasm from the gel to the sol state.

### 4. Movement of Cytoplasm and Organelles

Amoeboid locomotion is accompanied by cytoplasmic streaming involving reversible sol to gel transformations of actin filament networks. However, myosin causes cytoplasm to move by providing the motor for the sliding of actin filaments (see Fig. 1). In the gel state, cytoplasmic actin filaments exist in a network or mesh with phosphorylated myosin. In the sol state, the short actin filaments interact with the small myosin molecules and contraction occurs using the energy derived from ATP hydrolysis. The forces generated by the contraction in these regions are then transferred to surrounding gel regions. Myosin I provides this motor function, and myosin II may function in the generalized contraction of the cell cortex. The distribution of myosins has been described in *Acanthamoeba* by antibody binding studies. Myosin II is concentrated in the posterior uroid, whereas myosin I is concentrated in the anterior pseudopod where the cell membrane forms local ruffling and microspikes. *Acanthamoeba* myosin I can interact directly with membrane lipids, especially acidic lipids such as phosphatidylserine and inositol phospholipids, which may enable linking of the regulatory mechanism at the membrane with movement in the cytoplasm. Myosin bound to specific lipids in the membrane may be the motor for movement of the cell surface and organelles along actin filaments (see Fig. 3). The actin network of many different cell types contains actin-binding proteins, such as cross-linking filamin and the $Ca^{2+}$-sensitive severing gelsolin and fragmin. Some of these proteins bind to integral membrane proteins, some link actin with membrane-associated proteins (e.g., erythrocyte spectrin), and others (e.g., calpactin

and lipocortin) bind to membrane phospholipids. Because $PIP_2$ binds to myosin I, $\alpha$-actinin, profilin, and gelsolin, this lipid may be directly involved in transduction of signals across the membrane to the cytoplasmic actin network.

Movement of organelles involves different mechanisms: however, most are either microtubule or actin based. Organelles are moved along microtubules or actin filaments by motor molecules.

### 5. Chemotaxis

The amoeboid stage of the slime mold *Dictyostelium* exhibits chemotactic behavior. It moves up a cyclic adenosine monophosphate (cAMP) concentration gradient (Fig. 4). Upon receptor binding by the chemoattractant at the cell surface membrane, increases occur in cortical actin polymerization, cell surface ruffling, and microspike formation. When a pseudopod extends, this establishes a polarity to the cell and the direction of cytoplasmic flow. Since the attachment of the pseudopodial membrane to the substratum is required for actual crawling movement, these foci of membrane attachments may be sites of actin polymerization. It is expected that the part of the cell closest to the higher concentration of attractant molecules would have higher receptor occupancy and local transmembrane signals, ion currents, second messengers, and protein phosphorylations. Since the plus ends of actin filaments are adjacent to the cell membrane, actin polymerization may enhance myosin-activated movement for extension anteriorly as well as increase movement up a chemoattractant gradient.

### C. Contractile Ring

Cytokinesis is accomplished by the pinching of the cell into daughter cells by forces in the cell cortex. The cleavage furrow of dividing cells contains actin filaments recruited from other parts of the cell and arranged parallel to the membrane in a belt called the **contractile ring.** The location of the contractile ring is controlled by and develops in the plane bisecting the mitotic spindle. The contractile ring also contains myosin, $\alpha$-actinin, and filamin (Fig. 5). In smooth muscles and in various types of nonmuscle cells, myosin light chains are regulated by phosphorylation catalyzed by $Ca^{2+}$-calmodulin-associated myosin light chain kinases. Phosphorylation allows the myosin to interact with actin, resulting in contraction; dephosphorylation causes dissociation of actin and myosin. Similarly, the division of a cell by the contractile ring within the cell cortex involves a cascade of protein phosphorylations. After cell division, the components of the contractile ring disperse.

### D. Embryogenesis

Movement associated with whole tissues has long been recognized during a number of developmental events, such as neurulation (Fig. 6). Neurulation involves epithelial cells with apical actin filament bundles arranged parallel to the cell surface. Cells in the epithelial sheet are cemented together at the apical pole by belt desmosomes. Thus, con-

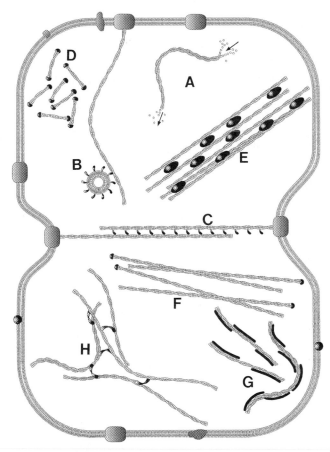

**FIG. 3.** Mechanisms of actin functions and the role of actin-binding proteins. A number of actin-binding proteins serve to change the state of actin filaments and their aggregation. (A) Polymerization of monomers into filaments and depolymerization. Depolymerization can cause movement of organelles such as chromosomes. Some actin-binding proteins serve as motor molecules in the movement of organelles. (B) A motor molecule attached to a vesicle membrane can cause movement of the organelle along actin filaments. (C) Sliding of actin filaments past each other in the contractile ring is aided by the action of myosin motor molecules during cytokinesis. Cell shape changes can also occur by motor molecules attached to the cell surface membrane interacting with membrane-attached actin filaments. (D–H) Various actin binding proteins cause fragmenting (D), bundling (E), capping (F), stiffening (G), and gelating (H) of actin filaments.

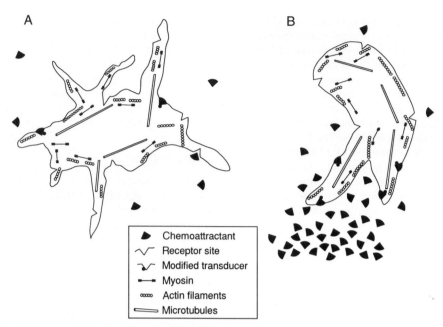

**FIG. 4.** Chemotactic responses in amoeboid cells. (A) In the absence of an attractant or repellent concentration gradient, the chemotactic cell exhibits normal random pseudopod extensions and locomotion. (B) In the presence of a gradient of chemoattractant, more receptor sites on the cell membrane facing the gradient become occupied, thus changing the conformation of the receptor or modifying the transducer. Tranducer-induced transmembrane signals result in cascade reactions involving intracellular second messengers and an increase in cyclic nucleotides, which in turn activate myosin and allow it to bind and interact with actin filaments. Cell shape changes involving microtubules may also participate. The events localized at the site of pseudopod extension cause the cell to turn and move up the attractant gradient.

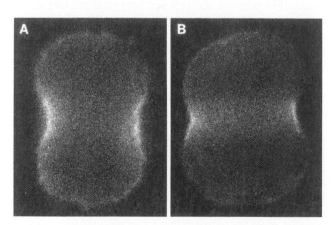

**FIG. 5.** Contractile ring in the cortex of a dividing human cell. (A) Antibodies directed against purified actin were tagged with a fluorescent label and reacted with a HeLa cell at early telophase of mitosis. The antibody bound the actin filaments concentrated in the cleavage furrow. (B) Similar to (A) but using anti-myosin II antibodies demonstrating the concentration of myosin in the contractile ring. α-Actinin and radixin are also concentrated in the cleavage furrow. (Courtesy of Pamela Maupin and Thomas Pollard.)

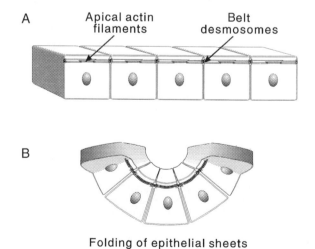

Folding of epithelial sheets

**FIG. 6.** Mechanism of epithelial folding during embryogenesis. (A) Actin filament networks are formed at the apical pole of epithelial cells in a sheet. They are aligned parallel to the apical cell membrane. Belt desmosomes serve as welding structures between the cells. (B) When the apical actin network contracts, presumably involving myosin interaction with the filaments, the entire epithelial sheet undergoes synchronous folding.

traction at the apexes of the cells and the folding of the epithelial sheet are highly coordinated. Contraction presumably results from sliding of actin filaments caused by interaction with myosin.

## E. Microvilli

Brush borders in intestinal and kidney epithelia consist of extensive 1- to 2-$\mu$m-long outfoldings of the cell surface membrane called **microvilli.** Microvilli contain bundles of actin filaments cross-linked by the actin-associated proteins fimbrin and villin (Fig. 7). Fimbrin aligns actin filaments in a parallel paracrystalline array. Villin has three actin-binding sites and, like fimbrin, also bundles actin; but at high $Ca^{2+}$ concentrations, villin serves the filament and caps the plus end at the break. The actin filaments are linked laterally to the cell membrane by myosin I associated with the $Ca^{2+}$-binding protein calmodulin. Movement of the cell membrane relative to the actin bundle may be related to the high turnover and sloughing of intestinal brush border membranes.

At the tip of microvilli, the filaments end in an electron-dense amorphous structure. The tip of the microvillus corresponds to the actin filament plus end. At the base of microvilli, the filaments are anchored in a terminal web composed of the cytoskeletal elements myosin II, tropomyosin, spectrin, and caldesmon (another actin-binding

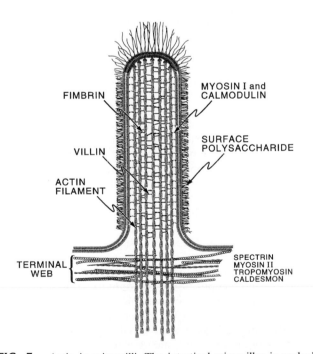

**FIG. 7.** Actin in microvilli. The intestinal microvillus is packed with parallel-bundled actin filaments. The filaments are held together by actin-binding bundling molecules such as villin and fimbrin. The filaments are also linked to the membrane through actin-binding myosin I that is associated with calmodulin. In the terminal web other actin-binding proteins, caldesmon and spectrin, tropomyosin, and myosin II anchor the actin bundle into the cytoplasm, attach the terminal web to the cell surface membrane, and generate tension to maintain the stiffness of the microvillus structure.

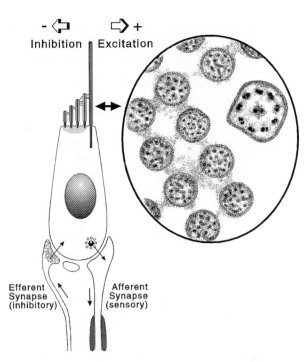

**FIG. 8.** Stereocilia and a kinocilium of a hair cell. The stereocilia are actually actin-filled microvilli. Linkers attach the tips of adjacent stereocilia. The single kinocilium is a typical cilium with a 9 + 2 axoneme pattern (inset). Mechanostimulation that bends stereocilia in different directions causes increased or decreased rates of firing of the afferent second order neuron. (Adapted from "Animal Physiology: Mechanisms and Adaptions" by Eckert, Randall and Augustine. Copyright © 1988 W. H. Freeman and Company. Reprinted with permission.)

protein). Villin is absent from the terminal web. Myosin II and tropomyosin may interact with actin in the terminal web by mechanisms similar to that in muscle. The terminal web complex may be responsible for maintaining the stiff cell surface microvillus extensions, thereby ensuring the maximum absorptive area of the epithelial cell.

## F. Sensory Reception

Among vertebrates, hair cells are found in the organs for hearing (cochlea), balance and equilibrium (semicircular canals), and the lateral line system of fish and amphibians. When present, hair cells contain one true cilium, the kinocilium. Because adult mammals lack a kinocilium in the cochlea, the kinocilium may not be critical for hair cell function. The actual mechanoreceptors of the hair cells are the other dozen or so extensions of the cell called stereocilia, which are actin filament-filled microvilli. During development, the filaments grow distally (plus end) to form the microvilli and then firmly anchor into the cytoplasm by monomer addition; this probably occurs at the minus end to form an extension of the actin bundle deep into the cytoplasm. Stereocilium length increases from one end of the cell to the other. The stereocilia are arranged in a teepee-like fashion with their tips touching the tips of neighboring stereocilia (Fig. 8). Thin bridging material links the tip of one stereocilium to what may be ion channels in the membrane at the tip of adjacent stereocilia. Displacement of the tuft in the direction of the longest stereocilia results in depolarization of the hair cell, whereas displacement toward the shortest stereocilia results in a hyperpolarizing receptor potential. Depolarizing stimuli produce greater potential changes than equivalent hyperpolarizing stimuli.

Microvilli in the squid photoreceptor contain rhodopsin (visual pigment) and a single actin filament. Because light disrupts the filament, it may function with the visual pigment in transducing light energy.

## III. Eukaryote Cilia and Flagella and Other Microtubular-Based Systems

Eukaryote cilia and flagella should not be equated to prokaryote (bacterial) flagella (described in the next chapter) nor to the stereocilia (an unfortunate misnomer). As described previously stereocilia are microvillar structures. Prokaryote flagella are extracellular organelles with their distinct substructures, biochemical composition, and mechanism of force generation. Eukaryote cilia and flagella are intracellular organelles that are surrounded by the cell surface membrane. Force generation uses the energy released from ATP hydrolysis and involves interaction between microtubules and dynein ATPases.

Cytoplasmic singlet microtubules can interlink and bind to other cytoskeletal elements. They also function like railroad tracks, guiding vesicles and other organelles to specific sites in the cell. Microtubule- and actin-based contractions (see previous discussion) are probably the most important structures governing changes in cell shape and organelle transport. However, actin and the microtubule protein tubulin differ in many respects (see Table 1).

### A. Cilia and Flagella

A swimming *Paramecium* cell, propelled by ciliary activity, would stop almost instantaneously when ciliary activity stops. This can be shown by evaluating the Reynolds number (Re), which reflects the ratio of inertial forces to viscous forces (density × length × velocity/viscosity). If Re is low ($<10^{-2}$), movement is governed by viscous forces; if it is high ($>10^2$), resistance to motion or flow of material over cilia is primarily the result of inertial forces. The Re of a *Paramecium* cell is about $10^{-1}$; hence, compared with viscous forces of the medium, inertial forces of the moving cell is small. The Re for a single cilium is about $10^{-4}$; thus inertial effects are negligible in the movement of an individual cilium. Transport of mucus by metachronal waves of ciliated epithelia (see later) has different viscomechanical features. Appropriate adaptations in ciliary length, beat patterns, and metachrony enable effective mucous movement by these tissues.

The locomotory organelles of many unicellular eukaryotic organisms, sperm cells, invertebrate metazoan larvae, and ciliated epithelia, as well as nonmotile sensory receptors, are based on common microtubular structures (Fig. 9). Cilia were first observed around 1676 by von Leeuwenhoak, who described them as "little legs" of *Paramecium*. Eukaryote cilia and flagella have similar substructures and their functions are based on similar mechanisms: thus these terms will be used interchangeably in this section. However, although both cilia and flagella are quite similar in diameter (ca. 0.2 $\mu$m), cilia are usually shorter (5–15 $\mu$m) and often occur in fairly large numbers and in proximity to each other. Cilia exhibit asymmetrical beat cycles (effective power stroke and recovery stroke), causing fluid to move parallel to the cell surface in the direction of the effective power stroke. The movement of cilia propels small organisms through the medium, or they move the medium or mucus along tissue surfaces. During the effective stroke, the cilium is relatively straight with a bend near the base. During the recovery stroke, the bend propagates from the base to the tip, thus offering minimal resistance to the medium.

In contrast to cilia, flagella are longer (up to 100–200 $\mu$m or more), often appear singly or in low numbers, and exhibit relatively symmetrical waves that propagate along their lengths (either base-to-tip or tip-to-base), causing fluid to move parallel to the flagellar long axis. Flagella occur in all the major eukaryotic phylogenetic groups, with the possible exception of some lower fungi. Nematodes, crustaceans, and adult insects lack motile cilia, but they do have sensory cilia.

The many functions attributed to these organelles (Table 4) reflect their importance in the normal physiology of cells and organisms. Because so many fundamental biological processes involve microtubule-based motility, this implies that this mechanism evolved relatively early. It has been suggested that eukaryote cilia first evolved by symbiotic relationships of prokaryotic ancestral forms, and that these organelles may have evolved from several independent events. These ideas parallel those concerning the origins of mitochondria and chloroplasts. Currently, there seems to be a consensus that mitochondria and chloroplasts evolved from prokaryotic symbiotes that became dependent on the host cell (and vice versa), concomitant with the loss or transfer of some genetic information between the organelles and the nucleus. It is more feasible to test this hypothesis for mitochondria and chloroplasts by modern molecular genetics techniques, because these organelles contain substantial amounts of DNA. However, this hypothesis is more difficult to evaluate in the case of cilia, since DNA (of the structure and composition commonly found in nuclei) has not been detected in these organelles. For decades, sporadic reports appeared in the literature presenting evidence for DNA in the basal bodies of cilia only to be followed by other reports refuting such conclusions.

### B. Structure

Familiarity with the ultrastructure of the cilium is essential for understanding how the organelle functions. The major regions of the cilium, the shaft, transition zone, and the basal body (kinetosome), are diagrammed as observed by longitudinal sections and cross-sections (see Fig. 9).

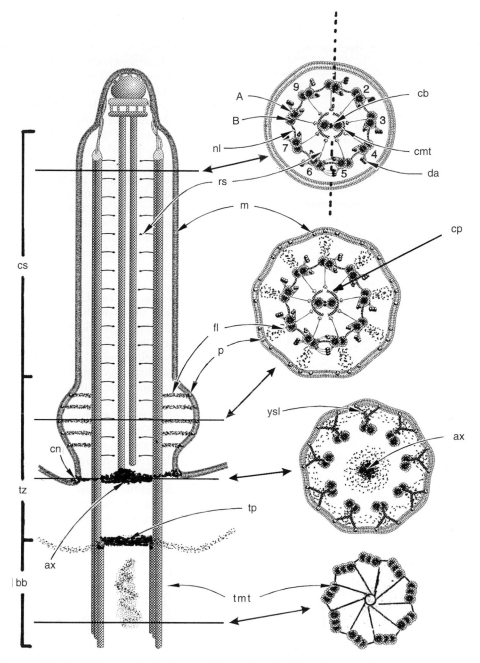

**FIG. 9.** Longitudinal and cross-sectional views at different levels of the cilium. A longitudinal section illustrates the different regions of the cilium: the basal body (bb) or kinetosome, the transition zone (tz), and the ciliary shaft (cs). The type of modification of the tip depends on species. The ciliary membrane (m) is contiguous with the cell surface membrane. The outer doublet microtubules in the ciliary shaft (shown in cross-section as viewed from base to tip) are each composed of an A (complete) microtubule and B (incomplete) microtubule. The outer dynein arms (with three large globular subunits) and the inner dynein arms (with two or three large globular subunits) are attached to the A microtubule. At the 5–6 bridge, the dynein arms are permanently linked to the adjacent doublet. Outer doublets are joined by nexin linkers (nl). Radial spokes (rs) extend centrally toward the pair of central singlet microtubules (cmt); the enlarged spoke heads are in close proximity to the pair of central microtubule projections (cp). The plane (dotted line) that bisects the central microtubule bridge (cb) and passes through the 5–6 bridge and an outer doublet (doublet No. 1) is the plane of the effective power stroke of ciliary action. In the transition zone, fibrous linkers (fl) span the outer doublets and the ciliary membrane at a level where the membrane bulges out and assumes a scalloped outline in cross-section. Within the membrane, intramembranous particles that form the ciliary membrane plaques (p) are in register with these linkers (also see Fig. 13). Proximally, at about the level of the axosome (ax) and the transverse plate (basal plate, axosome plate), another intramembranous particle array, the ciliary necklace (cn) is in register with linking material that appears Y-shaped (ysl) in cross-sections. One of the central microtubules is embedded in the axosome, whereas the other appears at a distance distal to the axosome. The transition zone ends proximally by the terminal plate of the basal body (tp). In the basal body (centriole), triplet microtubules form a characteristic pattern as viewed in cross-section.

**TABLE 4** Comparison of Axonemal Mechanisms of Movement[a]

| Sliding system[b] | Control system[c] |
|---|---|
| Trypsin-resistant | Trypsin-sensitive |
| Dynein arm–microtubule doublet interactions | Spoke-central microtubule projection or other interactions |
| Unconstrained, isotopic sliding | Constrained, anisotopic sliding |
| Single polarity (active) | Bidirectional (active vs passive) |
| Sliding uncoupled to bending | Sliding strictly coupled to bending |
| No systematic bend formation or propagation | Systematic bend formation and propagation |
| Unaffected by high $Ca^{2+}$ | Affected by high $Ca^{2+}$ |

[a] After Satir (1984).
[b] Observed under conditions in which sliding is uncoupled from the control system.
[c] Observed under conditions in which sliding is coupled to the control system.

Freeze-fracture and other ultrastructural techniques such as deep etching have helped elucidate substructures, such as those of the ciliary membrane and dynein arms.

## 1. Shaft and Axonemal Nomenclature

The common ultrastructural pattern of the axoneme (microtubules and associated cytoplasmic structures) is described as 9 + 2. That is, the axoneme is composed of 9 outer doublet microtubules and 2 central single microtubules. Other patterns exist; these are described as 9 + 0, 9 + 1, 3 + 0, 6 + 0, etc. A comparative study of sperm ultrastructure indicates that there is a correlation between species with external fertilization and the typical 9 + 2 pattern. These include sperm of coelenterates, echinoderms, and protists. On the other hand, sperm of nematodes, insects, crustaceans, molluscs, and mammals with internal fertilization have large variations in structures, including a variety of accessory structures. These less common patterns occur naturally and are stable within a given cell type in an organism; some of these patterns result from experimentally induced mutations. Afzelius has described several axonemal mutations in humans. Patients with immotile cilia syndrome may have recurrent respiratory problems, abnormal ciliary or sperm motility, infertility, and abnormal axonemes. Abnormal axonemes are also correlated with *situs inversus,* a condition characterized by displacement of organs (such as the heart) to the opposite side of the body. The abnormal body symmetry occurs by defective movements of tissues during embryonic development.

By convention, micrographs of cross-sections are published as viewed from the base looking toward the tip of the cilium (see Fig. 9). Thus, cross-sections of the shaft (with the typical 9 + 2 axonemal pattern) show inner and outer dynein arms extending clockwise from the A tubule (subfiber) toward the B tubule of the adjacent outer doublet. Convention adopted in the numbering of the outer doublets comes from their relative positions with respect to the central singlet microtubules. Furthermore, the effective stroke of the ciliary beat (1) occurs in a plane (shown in Fig. 9) perpendicular to the bridge linking the central microtubules, (2) passes through the permanent bridge between outer doublets number 5 and number 6 (5–6 bridge), and (3) passes through the outer doublet opposite the 5–6 bridge. That doublet has been designated outer doublet number 1 and numbering proceeds clockwise. Individual doublets are also designated in a more general way as doublet $N$ with its dynein arms pointing at the adjacent doublet, $N + 1$.

*a. Microtubules and Tubulins.* The central singlets and A tubules of the outer doublets are complete microtubules composed of 13 protofilaments that extend the length of the microtubule. The B tubule is incomplete and may consist of 10 or 11 protofilaments (Fig. 10). The protofilaments are observed as fibrils that splay out at the tips of damaged microtubules, indicating strong longitudinal and weak lateral bonds between subunits. Tubulin exists as a 110-kDa dimer of p$I$ 5.4 and consists of 25% $\alpha$-helical structure. The $\alpha$ and $\beta$ monomers are resolved by denaturing SDS–PAGE with migrations corresponding to proteins of about 55 kDa. Microtubule protofilaments commonly consist of subunits of $\alpha$- and $\beta$-tubulin heterodimers. The protofilament diameter is 4 nm and the tubulin dimer repeat occurs longitudinally every 8 nm. The tubulin dimer dimension determines the periodicities of axonemal structures and the mechanism

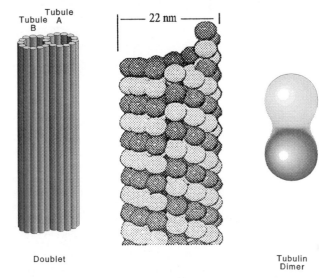

**FIG. 10.** Subunit structure of microtubules. A complete microtubule is composed of 13 protofilaments (longitudinal arrangement of dimers). In a doublet, tubule A is complete and shares three protofilaments with tubule B, which is incomplete and composed of 10–11 protofilaments. In many microtubules, the tubulin dimer occurs as a heterodimer of $\alpha$- and $\beta$-tubulin.

of ciliary movement (note below that nanometer distances between repeats of structures along the axoneme can be divided by 8). Another isoform, $\gamma$-tubulin, is found at sites called microtubule organizing centers (MTOC) and may be involved in the initiation of microtubule polymerization and determination of microtubule subunit polarity.

Growth or regeneration of axonemes occurs at the tip of cilia as demonstrated by Rosenbaum and Child. When cells of the unicellular biflagellate alga *Chlamydomonas* are deflagellated, synthesis of flagellar proteins is stimulated. Autoradiograms of deflagellated cells exposed to radiolabeled amino acids show radioactivity at the flagellar tips, indicating that newly synthesized material was added to the axoneme tips.

The distal tip of the cilium can have elaborate modifications in different cell types. In general, the central and A tubules extend further than do the B tubules (see Fig. 9). The central microtubules have caps and the A tubules contain plugs that prevent the addition of tubulin after reaching the appropriate length. The cap and plugs have linking structures that anchor the microtubules to the ciliary membrane. The cap material shares cross-reactive epitopes with kinetochores, structures associated with centromeres of chromosomes (see following), indicating a common mechanism for microtubule length changes in these two structures.

The central microtubules twist around each other in longitudinal double helices, as documented by high-voltage electron microscopy, and are joined by the central pair bridge. The central tubule projections (formerly referred to as the central sheath) are complex structures that extend from each central microtubule, some of which have a periodicity of 16 nm along the axoneme. Since the projections on the two microtubules can differ in size, this helps distinguish the one central microtubule from the other. The central microtubules may serve as a distributor that activates outer doublet sliding. It has also been suggested that the central microtubules are structures by which tubulin subunits are transported to the tip of growing flagella. There is evidence suggesting that growth of the central microtubules themselves occurs by addition at the proximal transition zone of the cilium.

*b. Dynein Arms and ATPases.* Axonemal ATPase activity was demonstrated in elegant experiments conducted by Gibbons (1965), who combined biochemical methods with ultrastructural analyses of *Tetrahymena* cilia. He extracted cilia with a low-ionic-strength buffer containing EDTA and noted that the arms on outer microtubule doublets were removed. Along with the loss of doublet arms was the loss of ATPase activity of the cilia. These armless cilia were then incubated with fractions of cilia extracts with ATPase activity, an experimental manipulation that restored the arms to the outer microtubule doublets. Based on these observations, he coined the term *dynein* (force protein) for material in the arms, and postulated that the enzyme was essential for releasing energy from ATP to generate force for ciliary motility.

Most cilia observed thus far contain two categories of arms, although those with only outer or inner arms have

been reported. The outer dynein arms consist of several polypeptides and have a longitudinal periodicity of 24 nm along the A tubule of outer doublets. Three large polypeptides (<400,000 MW) with ATPase activity have been detected and may correspond to the three heads seen in its substructure. Beat frequency of *Chlamydomonas* mutant flagella lacking outer arms is one-third that exhibited by wild-type cells. The outer arms thus provide much of the power for flagellar movement. In some cells, only inner arms with two heads have been observed. In other cells, two types of inner arms have been described: dyads and triads with two and three globular heads, respectively. These arms are arranged longitudinally in the cilia in 96-nm repeats along the A tubules, with each repeat unit composed of one triad and two dyads. The inner arms contain six large polypeptide ATPases in addition to several intermediate and low molecular weight chains. Beating of *Chlamydomonas* inner arm mutants indicates that inner arms may regulate waveform symmetry. Actin and centrin (discussed later) have been identified at the base of the inner arms. These proteins may regulate the activity of these structures.

*c. Linkers and Accessory Structures.* Nexin (interdoublet links) connect the A tubules to the adjacent outer doublet B tubules. These structures have a longitudinal periodicity of 96 nm.

Radial spokes extend from A tubules toward the center, ending in an enlargement called *spoke heads,* which may have ATPase activity. Spoke heads occur along the length of microtubules in groups of either two or three that have a longitudinal periodicity of 96 nm.

Membrane-to-microtubule linkers occur along the axoneme shaft, especially at the tips. These linkers may occur in a helical array along the length of the cilium. Links joining the outer dynein arm and the membrane have been described in some cilia.

Within the flagellar shaft, accessory structures are commonly found between the axoneme and membrane. For example, in many sperm cells, prominent dense fibers are associated with each outer doublet and a fibrous sheath is found external to the dense fibers (Fig. 11). In the midpiece, located between the principal piece (shaft) and the cell body of the sperm cell, extensive mitochondrial profiles are visible surrounding the axoneme and dense fiber structures.

In the phytoflagellate *Euglena* and in the trypanosomes, the paraxial rod is an accessory structure that runs along the length of the flagellum (Fig. 12). In *Euglena*, this structure and the paraxial ribbon are sites of mastigoneme (see following) attachment. The substructure of the paraxial rod consists of fibrous helices with a pitch of 45° and a longitudinal periodicity of 54 nm. Intermediate filament-type tektin filaments have been detected running the length of ciliary microtubules in some cilia.

*d. Ciliary Membrane.* The ciliary membrane is closely apposed to the axoneme along the length of the shaft. Unlike most of the axoneme, its growth is by insertion at the base. At the tip, modifications of the membrane can occur, such as the transient bulbous expansion that occurs when flagel-

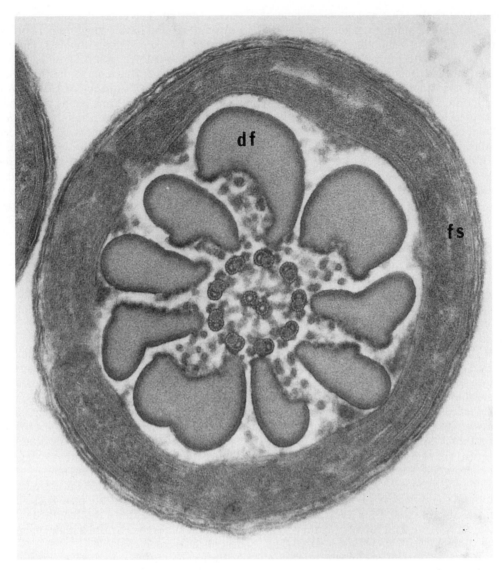

**FIG. 11.** Accessory structures typical of sperm flagella. This electron micrograph of the flagellum of a Chinese hamster spermatozoa shows dense fibers (df) and the fibrous sheath (fs), found in many sperm cells. (Courtesy of David Phillips.)

lar tips adhere (tipping) during mating of *Chlamydomonas* (Figs. 13 and 14). The outer surfaces of some cilia are decorated at their tips with fibrillar material called *crowns,* which can be elaborate in some cell types. Hemidesmosomes are forms at points along the trypanosome flagellar membrane (Fig. 15) where the parasite firmly attaches to host tissue and desmosomes link the flageller membrane with its cell surface membrane. In many flagellated protists, the site of flagellar emergence from the flagellar pocket (reservoir) is characterized by desmosomes between the flagellar membrane.

Scales and mastigonemes (flagellar hairs) decorate the flagellar exterior in some unicellular organisms (see Fig. 12). These external structures have species-specific architecture useful for taxonomic assignments. The arrangement of mastigonemes can be classified as acroneme (single hair at the tip), pantoneme (many hairs surrounding the flagel-

lum shaft), or stichoneme (hairs only on one side of the flagellum). Scales and mastigonemes, which are composed primarily of glycoproteins, are assembled in the Golgi complex (maturing from proximal, or *cis* to distal, or *trans* regions) and are seen within vesicles that eventually undergo exocytosis. Scales are easily removed from the membrane, indicating electrostatic bonding, whereas mastigonemes are firmly attached by linking material (or accessory structures) to the axoneme through the membrane.

### 2. Transition Zone

*a. Axoneme.* The proximal region of the cilium, usually at the level of the cell surface membrane, contains unique structures not present in the shaft (see Fig. 9). The outer doublet microtubules continue in this zone but here they lack dynein arms, nexin links, and radial spokes. Cilia of

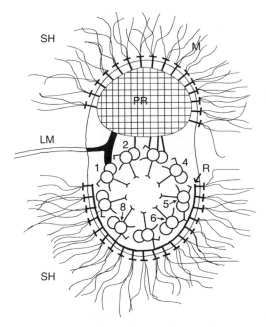

**FIG. 12.** The *Euglena* flagellum has many accessory structures. The flagellum contains a paraxial rod (PR). Two half-sheaths (SH) partially enclose the flagellar membrane; one associated with the paraxial rod, and the other attached to a paraxial ribbon (R). The half-sheaths are sites of mastigoneme (M) attachment and of anchoring to the membrane and underlying cytoplasmic structures. A unilateral row of long mastigonemes (LM) also exhibits anchoring through the membrane. (Adapted from Bouck *et al.,* 1990.)

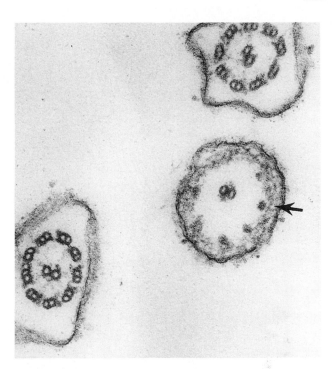

**FIG. 14.** Ultrastructure of the *Chlamydomonas* gametocyte flagellum. Electron micrographs of cross-sections of *Chlamydomonas* gametocyte flagella demonstrate that the flagellar tips are modified during activation. Unactivated flagellar tips have the A microtubule in proximity to the membrane, whereas activated flagella have fibrous material (arrow) that accumulates between the membrane and the single A microtubule. The distal region of the tip also elongates, thereby increasing it by 30%. (Courtesy of Ursula Goodenough.)

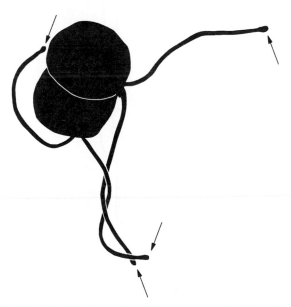

**FIG. 13.** Modifications of the *Chlamydomonas* flagellum. During mating, flagella of *Chlamydomonas* gametocytes develop bulbous tips and mating pairs adhere by their flagellar tips (Adapted from *The Journal of Cell Biology,* 1980, vol. 84, pp. 599–617, by copyright permission of The Rockefeller University Press.)

different species vary with respect to some structures: however, the proximal end of this region is normally clearly delineated by the terminal plate of the basal body (kinetosome, centriole). The number of plates across the axoneme in this zone can vary and there seems to be little consistency in the names used to refer to each one. The most distal transverse plate has been called the **basal plate** (indicating the base of the ciliary shaft) or the **axosome plate** (describing the presence of the embedded axosome in this structure).

The axosome is an amorphous structure from which one of the two central microtubules arises. The other central microtubule begins at a distance above the basal plate. A thin intermediate plate is present in some cilia. The most proximal transverse plate is the terminal plate of the basal body, which is always present regardless of whether the basal body is associated with a cilium. This plate has also been called the basal plate by some authors.

The transition zone of some flagella contains two central cylinders (proximal and distal) that have an H shape when viewed in longitudinal sections. Cross-sections of this region demonstrate fibers connecting the cylinders to the A tubules, forming a nine-pointed stellate pattern. Flagellar scission has been shown to occur along the plane of the organelle containing the nine-pointed stellate pattern (see centrin following).

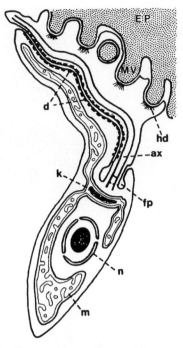

**FIG. 15.** Flagellar membrane of a parasitic protozoan. *Trypanosoma brucei* attaches to the microvillous (MV) surface of the tsetse fly salivary gland epithelium (EP). The parasite anchors by forming hemidesmosomes (hd). Desmosomes (d) are commonly observed between the flagellar membrane and flagellar pocket (fp) at the site of flagellar emergence from the cell body. Desmosomes occur along the entire length of the flagellum, linking it with the cell body. The kinetoplast (k), axoneme (ax), nucleus (n), and mitochondrion (m) are indicated. (From Vickerman and Tetley, 1990.)

*b. Ciliary Membrane.*   Unlike the shaft, the ciliary membrane bulges out in the transition zone. Intramembranous particles (IMP) are organized in two distinct arrays, the ciliary necklace and the ciliary plaques (patches) (Fig. 16). Other IMP arrays are found in various species. A single plaque usually consists of three longitudinal rows containing three to six 10-nm IMP. Nine plaques encircle the cilium in register with the nine peripheral doublets in the axoneme and are attached to the doublets by linking structures. The rectangular plaques are located distal to the axosome plate (basal plate, transverse plate). Proximal to the plaques, the ciliary necklace, consisting of one or more rows of IMP, surrounds the cilium in a scalloped pattern. The scalloped pattern corresponds to the rims of champagne-glass-shaped linkers (Y-shaped in thin sections) that attach to the microtubule doublets. The plaque and/or necklace particles may be sites of Ca²⁺ entry into the cilium or may have Ca²⁺-ATPase activity responsible for active pumping of this divalent cation out from the cilium against a concentration gradient. Studies on deciliation in the ciliate *Tetrahymena* demonstrate that the plaques are removed with the excised cilia and the necklace remains with the cell body. Thus, scission occurs where there are no microtubule–membrane links,

dyneins, nexin links, IMP, central microtubules, or radial spokes.

### 3. Basal Body

The basal body (kinetosome, centriole) is characterized by triplet microtubules connected by linkers between the A and C tubules (see Fig. 9). The A and B tubules are continuous with those of the transition zone and are joined with the C tubule in the basal body. In the proximal regions of many basal bodies, triplets, intertriplet and radial linkers form a distinctive cartwheel pattern in cross-sections of the structure.

The basal body is essential for ciliary and centriolar growth and replication, probably serving as a template. Cilia appear and grow only at sites first occupied by basal bodies. How this occurs is not known, but these structures characteristically develop at right angles to the existing basal body during replication (see preceding for discussion on DNA). Basal bodies serve to nucleate axonemes during reconstitution *in vitro;* their exclusion results in growth of only singlet microtubules. However, centrioles can appear *de novo* during spermiogenesis in several plant species and the amoeboflagellate *Naegleria.*

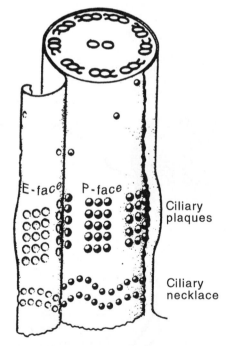

**FIG. 16.** Intramembranous particles within ciliary membranes. The membrane is fractured between the lipid bilayer revealing particles on the protoplasmic face (P-face) with imprints on the external face (E-face). Characteristic arrays called the *ciliary plaques* occur as nine groups in register with the nine outer doublets of the axoneme. Each plaque consists of three longitudinal rows of three to six particles. The ciliary necklace is proximal to the ciliary plaques and consists of two or more zigzag rows of particles, depending on species. (Adapted from Sleigh, 1989.)

Several types of accessory structures can be found associated with basal bodies: cross-striated rootlets, basal feet, microtubular ribbons and bands, kinetodesmal fibers, and microfibrillar bundles. These appear to anchor the cilium to the cell body and help maintain proper orientation or polarity of cortical structures as well as the rest of the cell. Fibrous linkers connect the cell surface membrane to basal bodies of mechanoreceptor cilia of nudibranch statocysts. This relationship suggests a mechanism of signal transduction from cilium to membrane.

## C. Mechanism of Ciliary Motility

### 1. Sliding of Outer Doublet Microtubules

Early experiments by Hoffman–Berling showed that glycerinated striated muscle (models) that had lost small, soluble molecules after damage to the cell surface membrane contracted in the presence of ATP and $Mg^{2+}$. Sperm models prepared in a similar manner began swimming upon subsequent exposure to reactivation solutions containing ATP and $Mg^{2+}$. This indicated that flagellar motility, like muscle, was also based on a system using ATP to generate force. Gibbons confirmed this in *Tetrahymena* extracted models and, as described earlier, identified ATPase activity in cilia extracts and correlated it to the presence of dynein arms on the outer doublets.

The sliding microtubule hypothesis was proposed by Satir, who noted that at the tips of straight cilia all doublets ended at the same level, but in bent cilia the outer doublets along the convex outer side did not extend as far as those on the concave inner side. This suggested that the microtubules themselves did not shorten during force generation and established the concept that the mechanism for cilia movement involved the sliding of outer microtubules past each other, analogous to sliding of thick filaments past thin filaments in striated muscles. However, unlike striated muscle, except for the 5–6 bridge, electron micrographs of cilia did not show crossbridges attached between outer doublets and dynein arms. The relatively large 10-nm gap seen between doublets posed a problem for models involving microtubule–dynein interaction. Purified dynein preparations were shown to decorate both A and B tubules of extracted axonemes in reconstitution experiments, indicating that both A and B tubules had binding sites for dynein. By then, it was known that maximal crossbridging in striated muscle occurs upon ATP depletion during rigor. Gibbons created a rigor state in flagella by first preparing models, reactivating them with ATP, and then rapidly decreasing the ATP levels by diluting the medium. Electron micrographs of these flagella in rigor showed a reduced axonemal diameter with all dynein arms pointed toward the flagellar base and attached to neighboring doublets (Fig. 17). This experiment provided strong evidence that the microtubule–dynein arm sliding hypothesis was feasible. These studies also indicated that the dynein binding site on the A tubule is ATP insensitive, whereas the dynein binding site on the B tubule is ATP sensitive.

Definitive direct evidence that microtubules can slide past each other was demonstrated by Summers and Gib-

**FIG. 17.** Flagellar axoneme in a rigor state. Dynein arms of sperm flagella bind to adjacent doublets when the preparation is depleted of ATP. (Adapted from Gibbons, 1975. Adapted with permission of the Society of General Physiologists. Copyright © 1975.)

bons (1971) who treated flagella by limited tryptic digestion that destroyed the membrane and various linkers. Upon the addition of ATP and $Mg^{2+}$, the axonemes elongated by a telescope action of sliding microtubules. The maximum length of telescoping axonemes sometimes reached nine times the original length before disintegration occurred. Trypsin-treated axonemes of *Chlamydomonas* paralyzed mutants that lack central microtubules and radial spokes also undergo sliding disintegration in reactivation solutions. Thus, sliding involves doublet–doublet interactions and does not require normal central microtubules or radial spokes.

If sliding occurs during ciliary beating and dynein arms are equivalent to crossbridges of muscle, the arms must have hinge action that generates force and the arms must undergo cyclic attachment and detachment to neighboring doublets. Longitudinal sections show dynein arms at 90° as well as at smaller angles in relationship to the microtubule, evidence that arms undergo hinge activity. The position at an angle less than 90° is described as the extended conformation of the arm. The direction in which force is applied by the arm of the A tubule upon the adjacent doublet was analyzed by Sale and Satir (1977) in trypsin-treated axonemes undergoing ATP-induced sliding. The extended conformation of dynein arms of all doublets was found pointed toward the base; thus they concluded that dynein arms of doublet $N$ attach to the ATP-sensitive binding site on the B tubules of adjacent doublets. Then, using energy released by ATP hydrolysis, the attached dynein arm swings and pushes doublet $N + 1$ in the base-to-tip direction.

### 2. Conversion of Sliding to Bend Formation

If all microtubule doublets slide past each other with each doublet $N$ pushing $N + 1$ tipward, ciliary beat or flagellar waves cannot be formed. Since every dynein arm apparently works in the same manner and all doublets slide in the same way, if all operated at the same time, the resultant forces would cancel each other's effects, resulting in no ciliary movement. Direct proof of sliding was demonstrated only when the membrane and linkers were digested. In an intact cilium, the membrane would restrict the amount of sliding, but the presence of the surrounding membrane alone does not explain the action of normal ciliary beat. Therefore, there must be regulation of local, transient resistance to, and activation of, inter-doublet sliding for coordinated activity. Separate sliding and control systems must operate in the cilium (see Table 4).

Warner and Satir reasoned that the 5–6 bridge permanently restricts sliding between these doublets and hence the effective stroke is always perpendicular to and toward it. They proposed that bend is produced in cilia by the result of local and temporary restriction to sliding (Fig. 18). They suggested that spoke head–central sheath (central microtubule projections) binding causes resistance to sliding. Along the length of the flagellum, radial spokes are perpendicular to microtubules at straight regions, but are tilted in bent regions. They suggested that at the leading edge of a propagating bend, attachment causes resistance to doublet sliding, and at the trailing edge, spoke heads

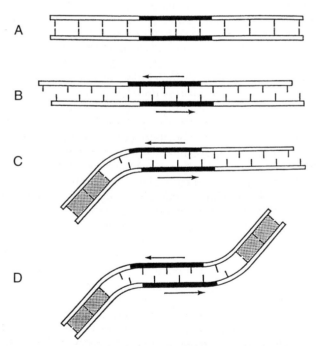

**FIG. 18.** Conversion of sliding into bending. (A) Outer doublet microtubules with associated linking structures. (B) During sliding of doublet microtubules caused by forces generated by dynein arm hinge action (darkened region), linking structures also slide freely. (C) Local and transient resistance to sliding caused by activation of linking structures in one region of the cilia (stippled area) causes bend formation during sliding in an adjacent region (darkened region). (D) Resistance to sliding (stippled region) on both sides of the active force generation site (darkened region) causes a bend in the opposite direction, resulting in the formation of a wave. (Reprinted by permission from *Nature*, vol. 265, pp. 269–270. Copyright © 1977 Macmillan Magazines Limited.)

become detached, releasing the resistance to sliding and allowing that region to become straightened. However, the presence of motile flagella of mutants without normal central microtubule or radial spoke structures implies that other structures cause resistance to sliding and ciliary bend formation. Thus, restriction of sliding probably involves nexin links or the binding of dynein arms to neighboring microtubule doublets, analogous to the permanent 5–6 bridge. The radial spoke–central microtubule interaction may control wave symmetry.

### 3. Coordination of Sliding during the Ciliary Beat Cycle

Unlike the one-dimensional sliding of thick and thin filaments of muscle, ciliary motility is achieved by movement in two or three dimensions. Although internal, local restriction of interdoublet sliding can result in bend formation, propagation of bends and the different movements in the effective and recovery strokes in the ciliary beat cycle require coordination of events in different parts of the cilium. The switch-point hypothesis was proposed by Satir (1984) based on observations made on doublet sliding displacements seen at ciliary tips during the ciliary cycle. Naitoh and Sugino (1984) verified this hypothesis by refined computer simulations of the coordinated movements of *Paramecium* cilia. According to this model, dynein arms of only doublet numbers 1, 2, 3, and 4 on one side of the cilium are active during the effective stroke (Fig. 19). During the recovery stoke, activity switches to doublet numbers 6, 7, 8, and 9 on the other side. All doublets in each half do not act synchronously, but are staggered from doublet $N$ to doublet $N + 1$. Two pauses occur in the activation of outer doublets corresponding to the sensitive switch points separating the effective and recovery strokes. Activity along the length of the cilium is also not synchronous, but is staggered and begins at the base with activity spreading tipward.

The pair of central microtubules is seen arranged in a helical fashion, and there is evidence that these continuously rotate in one direction during ciliary activity. These observations make the central microtubules attractive candidates for controlling the coordination of doublet activation during the beat cycle by functioning as a distributor. However, if indeed this is true, the obvious transducing structures in the axonemes are the radial spokes (but recall the existence of mutants with motile flagella that lack detectable radial spokes).

### 4. Modulations and Control of Ciliary Activity

Changes in beat frequency, arrest of movement, and reversal of the direction of the effective stroke (ciliary reversal) are modulations superimposed on the dynein–ATP–$Mg^{2+}$–microtubule system. These control mechanisms, which impart meaningful behavior such as taxis, mating, and the avoidance reaction, primarily involve regulation by the intraciliary level of $Ca^{2+}$, which in turn is controlled by the ciliary membrane. Neuronal, hormonal, and purely physical interactions of adjacent cilia within a group are other modulation factors. Two basic responses,

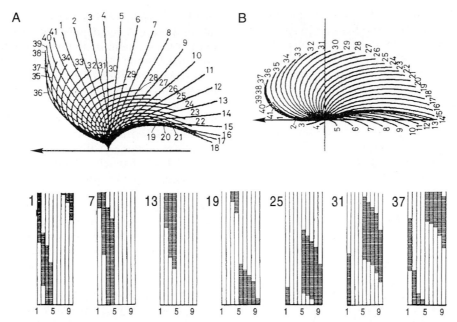

**FIG. 19.** Computer simulation of ciliary activity and the switch point hypothesis. (A) Side view of the activity of a *Paramecium* somatic cilium during one cycle of effective and recovery strokes. The position of the cilium at different times during the cycle is numbered. The arrow indicates the direction of forward swimming (toward the cell's anterior); the effective stroke is directed posteriorly. (B) Similar to (A) but showing the top view of the effective and recovery stroke of a *Paramecium* somatic cilium. The effective stroke begins at the position indicated No. 1; the cilium rapidly and relatively stiffly bends posteriorly offering high resistance to the medium. During the recovery stroke the cilium exhibits a greater bend, moves relatively slowly, and is directed more laterally, thus offering less resistance to the surrounding medium. (C) To obtain the movements of the ciliary cycle, activation of the individual outer doublets (hatched regions) changes with different positions of the cilium. During the effective stroke (e.g., positions 1, 7, 13), activation of doublets 1–4 dominates, whereas during the recovery stroke (e.g., positions 19, 25, 31, 37), activation of doublets 5–9 dominates.

beat orientation (changes in the direction of the effective stroke) and beat frequency, are modified by stimuli.

*a. Ciliary Reversal and the Avoidance Reaction.* A *Paramecium* cell usually swims about 1–2 mm/sec in a forward left spiral pattern with the effective stroke of cilia surrounding its cell body directed toward its posterior (Fig. 20). When the ciliate encounters an appropriate negative stimulus (mechanical, electrical, or chemical stimulation at the anterior end), it reverses the direction of the ciliary effective stroke. Thus the cell rapidly swims backward, away from and avoiding the stimulus source (ciliary reversal). Then, during a renormalization period (partial ciliary reversal), some cilia revert back to normal beat while others are still reversed, or individual cilia exhibit circling movements, all of which cause the cell to gyrate or spin while remaining in the same spot. After all effective strokes of cilia are directed posteriorly, the ciliate returns to forward swimming in a new direction and completes the avoidance reaction.

The resting membrane potential of *Paramecium* is about −40 mV (inside negative). Early in the 1930s, Kamada's laboratory observed that the membrane of *Paramecium* exhibited depolarizations that correlated with reversals of the effective strokes of somatic cilia, indicating electrophysiologic events at the membrane were involved in the modulation of ciliary activity. The definitive evidence that

Ca²⁺ regulated this activity was provided by Naitoh and Kaneko (1973), who found that detergent-extracted models of *Paramecium* reactivated in solutions containing ATP and $Mg^{2+}$ swam in a forward left spiral pattern, but that those reactivated in ATP, $Mg^{2+}$ plus $Ca^{2+}$ swam backward. After a series of experiments on *Paramecium*, Eckert and Naitoh formulated the $Ca^{2+}$ hypothesis for ciliary reversal (Fig. 21).

The intracellular $Ca^{2+}$ concentration is normally $<10^{-6}$ *M*, whereas extracellular concentrations are usually in the millimolar range. Upon stimulation to the anterior end of *Paramecium*, receptor potentials that activate voltage-sensitive $Ca^{2+}$ channels located in the ciliary membrane are generated. The open $Ca^{2+}$ channels allow extracellular $Ca^{2+}$ to enter the cilia by diffusion down its concentration gradient, causing a regenerative $Ca^{2+}$ current that lasts 10–20 msec. This depolarization is graded with stimulus intensity; the spike upstroke is about 10 times slower than those of neurons. The inward $Ca^{2+}$ current at depolarizing voltages has been measured in voltage-clamped cells. These voltage-sensitive $Ca^{2+}$ channels also allow $Sr^{2+}$ or $Ba^{2+}$ into the cilium, which is accompanied by all-or-none membrane action potentials and backward jerks by the cell called the **barium dance** (Fig. 22).

The location in the cilia of $Ca^{2+}$ channels was shown by experiments in which cells were deciliated, resulting in the loss of the $Ca^{2+}$ response in stimulated cells. During

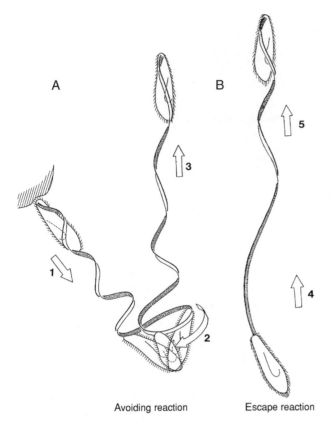

A                    B

Avoiding reaction          Escape reaction

**FIG. 20.** Locomotory behavior in *Paramecium*. (A) The avoiding reaction occurs when the cell is stimulated (e.g., bumps against a solid object with its anterior end). The cell reacts by swimming backward (1) by reversing the effective stroke of the ciliary beat, in which the effective stroke of the somatic cilia is directed toward the cell's anterior. The cell then pivots about its posterior pole during a renormalization period (2), then resumes normal front left spiral forward swimming, in which the effective power stroke of somatic cilia is redirected toward the cell's posterior (3). (B) The escape reaction is initiated by stimulation of the posterior pole of the organism. The cell responds by increasing the velocity of swimming in a front left spiral forward direction (4), then resumes its normal rate of forward swimming (5) (Adapted from Naitoh and Sugino, 1984, with permission.)

regeneration, the $Ca^{2+}$ response returned and the amplitude increased as cilia grew; thus it is probable that $Ca^{2+}$ channels are distributed along the length of the ciliary membrane.

The $Ca^{2+}$ influx through channels results in increased intraciliary $Ca^{2+}$ concentrations, which activates a switch mechanism that causes the reversal of the effective stroke. Thus, the ciliary effective stroke is directed anteriorly, and the ciliate swims backward. The latent period between stimulus and ciliary reversal lasts several milliseconds. In some ciliated epithelia (e.g., in the bivalve mussel gill), increased intracellular $Ca^{2+}$ causes beat arrest instead of ciliary reversal. Although the structural correlate of the switch mechanism has not been identified, most evidence suggests that it is located at the base of the cilium and that calmodulin or centrin may serve as the receptor.

The depolarization-activated $Ca^{2+}$ channels in the ciliary membrane are inactivated by the $Ca^{2+}$ that enters the or-

ganelle. The inactivation of the $Ca^{2+}$ channels and an outflux of $K^+$ via delayed depolarization-activated $K^+$ channels in the somatic membrane restore the membrane's resting potential. During the renormalization period of the avoidance reaction, $Ca^{2+}$-ATPases known to be present in ciliary membranes are thought to pump $Ca^{2+}$ out, thus lowering its intraciliary concentration. Several distinct $Ca^{2+}$-ATPase activities specifically found in the ciliary membrane have been characterized; these are distinct from those of the somatic membrane. Although $Ba^{2+}$ and $Sr^{2+}$ can enter cells through $Ca^{2+}$ channels, they apparently are not effectively pumped out by $Ca^{2+}$-ATPase pumps and are also not as effective as $Ca^{2+}$ in inactivating the activated $Ca^{2+}$ channels. Following the decrease of intracellular $Ca^{2+}$ to a threshold value, a switch in the control mechanism restores the direction of the effective stroke toward the posterior and the ciliate resumes forward swimming.

Hundreds of locomotory behavioral mutants of *Paramecium* that are useful for dissecting events at the membrane and at the axoneme have been isolated. Kung first described pawn mutants that fail to exhibit the avoidance reaction (pawn chess pieces are not allowed to be moved backward). In a series of studies with Eckert and Naitoh, Kung demonstrated that extracted models of the mutant

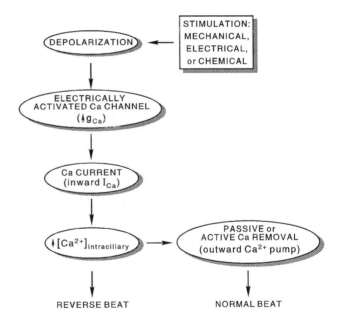

**FIG. 21.** The calcium hypothesis of reversal of the effective stroke of ciliary beat. In the avoidance reaction of a ciliated protozoan, a stimulus causes depolarization of the cell surface membrane, including the ciliary membrane. The depolarization of the membrane activates voltage-sensitive $Ca^{2+}$ channels in the ciliary membrane, resulting in the membrane increasing its $Ca^{2+}$ conductance. The inward $Ca^{2+}$ current reflects the movement of the divalent cation down its concentration gradient into the cell (cilia), resulting in increased $Ca^{2+}$ concentration at the switch mechanism in the cilium. The calcium-sensitive switch causes a modification of the ciliary beat by changing the direction of the effective stroke. During the renormalization period, $Ca^{2+}$ is removed from the switch mechanism, probably aided by outward-directed $Ca^{2+}$ pumps (ATPase) in the ciliary membrane, leading to the resumption of the normal direction of the ciliary effective stroke.

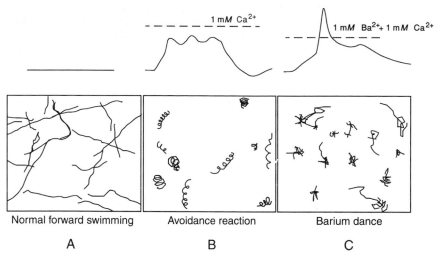

FIG. 22. Electrical properties of the cell surface membrane and swimming behavior. (A) Top: In the absence of a stimulus, the membrane exhibits a stable resting potential of about −40 mV. Bottom: Swim tracks of the organism reflect normal front left spiral forward locomotion. (B) Top: Stimulation is followed by the $Ca^{2+}$ response, a spike graded with respect to stimulus strength. Bottom: The behavioral correlate is illustrated by the swim tracks reflecting backward swimming in a tight spiral pattern. (C) Top: The ciliary membrane voltage-sensitive $Ca^{2+}$ channel also accommodates $Ba^{2+}$ and $Sr^{2+}$. In the presence of $Ba^{2+}$, all-or-none action potentials occur. Bottom: The cell exhibits a characteristic "barium dance" of rapid backward jerks.

swam backward when $Ca^{2+}$ was present in the reactivation solution, indicating that the defect was not in the axoneme. Electrophysiologic data showed no $Ca^{2+}$ spike in response to depolarizing current stimuli and no inward $Ca^{2+}$ current under voltage-clamp conditions, thus identifying the voltage-sensitive $Ca^{2+}$ channels as the lesion site in this mutant (Fig. 23). Although the channel (or gate) molecules have not yet been biochemically isolated and identified, they have been reconstituted into artificial lipid membranes by fusion with ciliary membrane vesicles, and single-channel conductances have been examined in patch-clamp experiments. Several types of channels, one of which behaves like the voltage-dependent $Ca^{2+}$ channels of intact cells, have been thus detected. Identification of the $Ca^{2+}$ channel

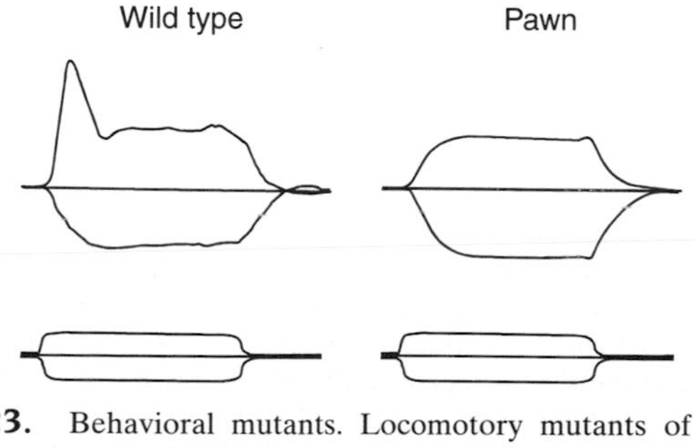

Wild type          Pawn

FIG. 23. Behavioral mutants. Locomotory mutants of various types have been correlated with ciliary and flagellar abnormalities. In *Paramecium*, a membrane-defective pawn mutant was isolated by its inability to swim backward. Electrophysiological analysis demonstrated that when stimulated, the mutant membrane did not exhibit the $Ca^{2+}$ spike. Top traces are the responses (mV) to depolarizing and hyperpolarizing stimuli; the bottom traces are the corresponding current stimuli ($I_s$). (Adapted from Kung and Eckert, 1972, with permission of the author.)

has also been approached by successful, but temporary, rescuing or curing of pawns by microinjection of wild-type cytoplasm and purified RNA.

*b. Beat Frequency and the Escape Reaction.* Beat frequency of cilia in nonstimulated *Paramecium* is about 15 Hz, which can change to 50 Hz in response to a stimulus. In the avoidance reaction, the beat frequency during ciliary reversal increases to a maximum of 40–50 Hz and then decreases with cilia beat reorientation.

*Paramecium* responds to mechanical stimulation at its posterior by increasing cilia beat frequency (ciliary augmentation), and swimming forward faster, described as the escape reaction (see Fig. 20). Beat frequency is also controlled by intraciliary $Ca^{2+}$ levels (Fig. 24). Hyperpolarization of the membrane follows posterior stimulation and probably involves $Ca^{2+}$ influx and voltage- and $Ca^{2+}$-dependent $K^+$ fluxes, which are not understood as well as those associated with the avoidance reaction. There is evidence indicating that hyperpolarization activates adenylate cyclase, leading to increased intraciliary cAMP, which in turn stimulates cAMP-dependent protein kinase(s) that phosphorylates several ciliary proteins (Fig. 25). Protein kinases may phosphorylate dyneins or other proteins associated with the arms, including a low molecular weight light chain.

Calmodulin, adenylate and guanylate cyclases, protein kinases, cAMP and cyclic GMP (cGMP), cAMP- and cGMP-dependent protein kinases, and phosphoprotein phosphatase (calcineurin) have been detected in *Paramecium* cilia. In sperm, it has been demonstrated that cAMP is required to activate flagellar motility of fully assembled but nonmotile sperm upon release from the male reproductive tract. Hence $Ca^{2+}$ regulation of cilia activity probably

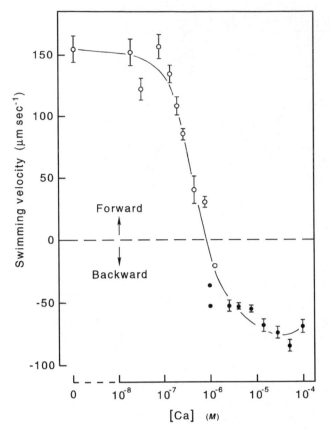

**FIG. 24.** Intraciliary calcium concentration regulates the velocity of backward and forward swimming. Correlation of swimming velocity was determined by analyzing ATP-Mg$^{2+}$-reactivated Triton-extracted models of the organism. (From Naitoh and Sugino, 1984, with permission.)

crosurgical cuts of fibrillar structures in the cortex of ciliates did not alter metachonry; waves continued over small gaps made in the cell cortex. Also, detergent-extracted models were shown to exhibit metachrony. Furthermore, it was discovered that certain protozoa harbor extracellular spirochete bacteria attached to their surfaces. As the spirochetes become closely packed, metachronal waves are seen in their movements.

### 5. Sensory Receptors

The association of cilia with sensory reception is common among animals. Usually these cilia lack dynein arms and have a 9 + 0 axoneme pattern. The following examples of

involves cascade systems involving second messengers and cyclic nucleotides; the substrates that become phosphorylated are now being examined. It has been suggested that guanylate cyclase activity regulates the avoidance reaction, but the rate of increase in cGMP may not be sufficiently rapid to account for this behavioral response.

In ciliated epithelial systems, acetylcholine and serotonin cause an increase in beat frequency, whereas epinephrine causes a decrease. Increased intraciliary Ca$^{2+}$ (experimentally induced by Ca$^{2+}$ ionophores) causes the arrest of cilia beat in mussel gills. This response also occurs *in situ* when the nerve is stimulated. The involvement of transmitter substances in the control of ciliary activity further implicates second messenger and cyclic nucleotide cascade reactions in ciliary reversal, ciliary augmentation, or both.

*c. Metachrony.* The coordinated beating of a group of cilia forming large waves over the tops of cilia on the cell surface is called a **metachronal wave** (Fig. 26). The waves result from purely physical viscous coupling between neighboring cilia transmitted through the surrounding fluid. The coupled-oscillator hypothesis developed as the result of experimental evidence refuting the idea that protozoan cells have intracellular neural systems and that ciliary coordination was controlled by such neuromotor networks. Mi-

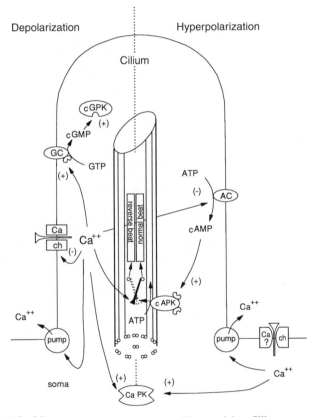

**FIG. 25.** Control of *Paramecium* ciliary activity. Ciliary reversal (avoidance reaction, depolarization) and augmentation (escape reaction, hyperpolarization) involves Ca$^{2+}$ and chemical reactions involving cascade reactions and cyclic nucleotides. Upon depolarization, ciliary voltage-dependent Ca$^{2+}$ channels open, increasing the intraciliary Ca$^{2+}$ concentration that activates the switch controlling reversal of the effective stroke and backward swimming. The Ca$^{2+}$ channels are inactivated by the increase in intraciliary Ca$^{2+}$, and the concentration is restored by sequestration and/or extrusion by pumps, resulting in the resumption of forward swimming. The Ca$^{2+}$ response activates a protein kinase (CaPK) and guanylate cyclase (GC) that causes an increase in cGMP, which in turn activates a cGMP-dependent protein kinase (cGPK). Membrane hyperpolarization is accompanied by the activation of adenylate cyclase (AC) and increased cAMP. The increase in intraciliary cAMP activates a cAMP-dependent protein kinase (cAPK) that leads to increased ciliary beat frequency. (From Preston and Saimi, 1990.)

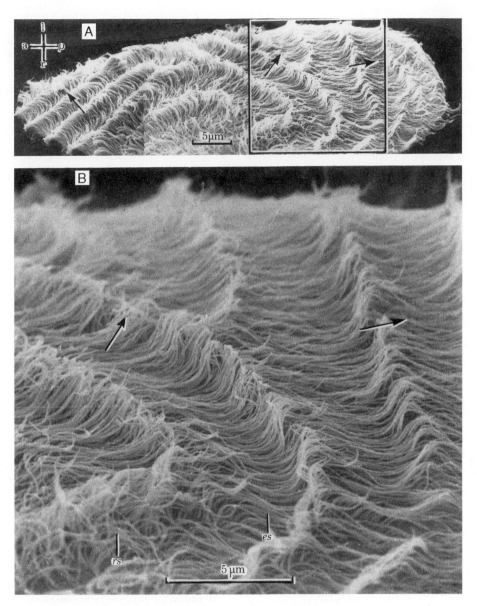

**FIG. 26.** Metachronal waves. Coordinated activity of many closely packed cilia and flagella form large metachronal waves over the surface of the cell. (A) Scanning electron micrograph of *Opalina* that was rapidly fixed to preserve the metachronal waves. (B) Enlargement in which individual flagella are resolved. The coordination of flagellar beat resulting in metachrony is attributed to physical viscous coupling interactions between neighboring flagella. (Reprinted with permission from Tamm and Horridge, 1970.)

sensory cilia demonstrate that they are commonly involved in sensory reception and transduction and may also serve additional functions in these cells.

The major diversion during metazoan evolution apparently has separated photoreceptors into two groups. The photoreceptors of most protostomes (annelids, arthropods, and molluscs), such as the cells of squid retina, have actin-based microvilli substructures, whereas most deuterostomes (echinoderms and chordates) have microtubule-based axonemes.

The vertebrate photoreceptors have a specialized region between the inner and the outer segments that contains a single short connecting cilium that is analogous in structure to the transition zone. The outer segment is a modified cilium. During development of the retina, just before the outer segment forms, the cilium grows from a basal body–centriole complex. The membrane of the distal region of the cilium expands, forming the photosensitive disk membranes. The immature cilium is differentiated into proximal and distal regions. In cross-sections the proximal region contains Y-shaped membrane–microtubule doublet linkers, and an IMP array, comparable to the ciliary necklace, is present within the membrane. This configuration is also true of the mature connecting cilium. Involvement of the distal region in the transport of opsin into the developing and mature outer segment has been indicated by immunocytochemical evidence. These regions bind anti-opsin antibodies, as does the mature connecting cilium. Thus, in addition to separating the photosensory outer segment domain of the cell from the release of synaptic transmitter by the inner segment, the connecting cilium may function in the delivery and turnover of the disk membrane.

In olfactory receptors, evidence for odorant receptors within the ciliary membrane has been provided by the loss of electrophysiologic responses after experimental deciliation of the neurons. Receptor occupancy probably causes a voltage drop across the membrane, which then spreads to the cell body. As in motile cilia, cyclic nucleotides are involved in olfactory sensory transduction.

Some cilia-based mechanoreceptors may involve the deformation of the ciliary membrane, causing the opening of ion channels such as the $Na^+$ channels in nudibranch statocysts. The change in ion conductance then elicits generator potentials in the sensory cell.

### D. Cytoplasmic Singlet Microtubules

#### 1. Occurrence of Nonaxonemal Microtubules

Nonaxonemal microtubules are widespread and occur as singlets in the cytoplasm, as they do in the mitotic apparatus of dividing cells (Table 5). They also occur as singlets linked by bridging material forming elaborate organized arrays, such as the heliozoan axopodium (Fig. 27).

#### 2. Microtubule Polymerization and Depolymerization

In some *Chlamydomonas* cells, only one flagellum becomes detached during the deflagellation treatment. Rosenbaum noted that the intact flagellum decreased its length until it was the same length as the one undergoing growth, after which the two flagella grew at the same rate, reaching their original lengths. These and other studies demonstrate that microtubule polymerization and depolymerization, as well as tubulin pools and posttranslational modifications, are highly regulated in a cell.

A dynamic and dramatic example of rapid microtubule polymerization and depolymerization can be seen in heliozoans (Fig. 28). The axopodia of these protozoan are stiff pseudopodia radiating from the cell in a beautiful starburst pattern. These structures collapse when a prey organism is taken into the cell body. This disintegration and reextension can happen so rapidly that special fixation procedures are necessary for ultrastructural visualization of the structure. The axopod contains a stiff bundle of microtubules

**TABLE 5**   Functions of Microtubular Systems

| Function | Examples |
| --- | --- |
| Cilia and flagella | |
| Locomotion | Ciliates; flagellates; flatworms; sperm cells; aquatic metzoan larvae |
| Feeding (generation of water currents that move suspended or mucus-trapped food toward the mouth and/or toward phagocytic cells) | Ciliates; flagellates; sponge choanocytes; sea anemone tentacles and gastrovascular cavities; mucociliary feeding in gastropods, bivalves, crinoids, and annelids |
| Respiration (move water over respiratory surface) | Aquatic mollusk ctenidia; sea star epidermis and tube feet |
| Excretion and osmoregulation | Protonephridial flame cells of flatworms and rotifers; ciliated funnels of annelid nephridia |
| Surface cleaning | Ciliated epithelia of vertebrate tracheal and bronchial airways; ependymal surfaces of cerebrospinal cavity; eustachian tube and middle ear |
| Mating (cell–cell recognition, gamete agglutination) | Mating-reactive flagellates and ciliates; invertebrate and vertebrate gametes |
| Transport | Passage of egg along mammalian oviduct; movement of ingested material along alimentary tracts of annelids, mollusks, echinoderms, and tunicates |
| Adhesion and anchoring | Parasitic trypanosomes and leishmanias to host tissues; cilia of mammalian respiratory tract to bacterial pathogens |
| Circulation | Coelomic fluid of invertebrates; cerebrospinal fluid of vertebrates |
| Sensory reception and transduction | Connecting cilia in photoreceptor cells; kinocilia of fish and amphibian lateral line systems; nudibranch, jellyfish and crayfish statocysts |
| Morphogenesis and development | Rotation of archenteron (*situs inversus* correlated with immotile cilia), mirror-image and other global cortical abnormalities in ciliates |
| Cytoplasmic singlet microtubules | |
| Cell division | Chromosomal and polar fibers of mitotic appaaratus of eukaryotic cells |
| Transport and intracellular trafficking–guidance | Movement of organelles toward and away from neuron soma; transport of lens precursors during development of squid eye; movement of discoid vesicles to ciliate oral region |
| Structural integrity and architecture | Cytoskeletal network; axopods of heliozoan protozoa |
| Cell shape changes | Amoeboid cells |

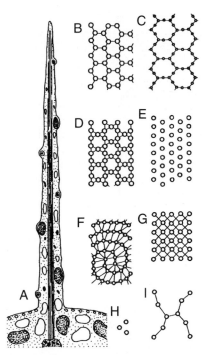

**FIG. 27.** Microtubules are important cytoskeletal elements. (A)–(I) Illustration of the diversity of microtubular arrays found in various protozoa. (From Sleigh, 1989.)

extending to the nucleus. When viewed in cross-sections, an elaborate architectural pattern is revealed that looks like two microtubule sheets rolled up together and maintained by extensive linking structures. The rapid disintegration and reextension of axopodia are controlled by temperature (cold favors depolymerization of the microtubules) and $Ca^{2+}$ levels. The processes of polymerization and depolymerization can occur in seconds, as rapidly as actin polymerization in the acrosome reaction of sea cucumber sperm described previously.

Compounds such as colchicine bind tubulin and inhibit polymerization. Colchicine is used to purify tubulin. Purified tubulin was reconstituted *in vitro* to form microtubules by Weisenberg, who found that $Ca^{2+}$ was inhibitory and that unlike actin, which requires ATP for polymerization, GTP was the nucleotide that allowed tubulin to polymerize into microtubules. In a $Ca^{2+}$-free solution, one dimer of tubulin binds two GTP molecules, one of which is hydrolyzed during polymerization. During polymerization, the other GTP serves as a cap at the plus end (fast growing), which enhances the rate of subunit addition.

### 3. Mitotic Spindle

As in cilia and flagella, mitotic spindles form by growth of microtubules nucleated at centrosomes and by addition of tubulin distally. At the end of prophase, the nuclear envelope in most cells disintegrates and three types of

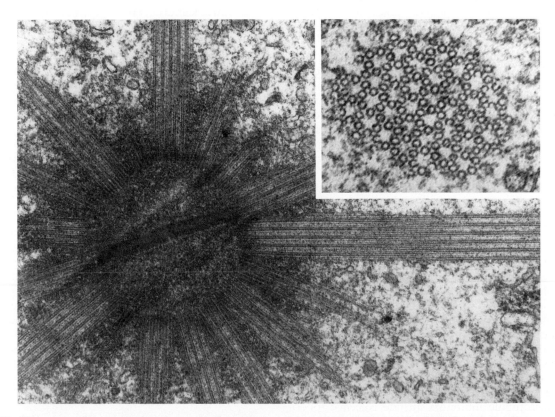

**FIG. 28.** Axopodia are microtubular organelles. Electron micrograph of a heliozoan showing microtubules that support the organism's axopod. The microtubular bundles radiate from the centroplast, which contains a centriolar structure. The inset shows a cross-sectional view. (Reproduced from Bardele, 1977, Courtesy of The Company of Biologists Ltd., Cambridge, UK.)

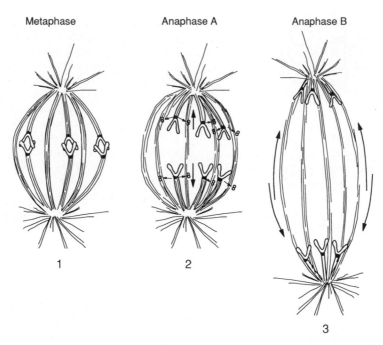

Metaphase    Anaphase A    Anaphase B

1           2           3

**FIG. 29.** Mitotic spindle activity. Microtubular bundles occur in the mitotic spindle as astral rays (1), chromosomal fibers (2), and polar fibers (3). During anaphase A, the chromosomal microtubules depolymerize at the end attached to the kinetochore, resulting in shortening of the microtubules and movement of chromosomes toward the poles. During anaphase B, the distance between the two poles increases and the entire spindle apparatus becomes elongated, presumably by sliding of polar microtubules past each other by motor molecule action.

microtubules grow from the spindle poles with plus ends distal to the poles: (1) the polar microtubules that extend from the pole in the direction of the opposite pole, (2) astral microtubules that end freely in the cytoplasm in directions away from the pole, and (3) kinetochore microtubules that attach to chromosomes (Fig. 29). The kinetochore is a trilaminar disk, conspicuous at metaphase, when chromatid pairs are seen with a constriction called the centromere. In higher eukaryotic cells, the chromosomal centromeres are associated with kinetochores. The microtubules growing from the pole bind to the kinetochore in prometaphase.

At anaphase, two movements are associated with the mitotic spindle: (1) chloral hydrate-insensitive chromosome movement toward the poles (anaphase A), and (2) chloral hydrate-sensitive separation of the poles (anaphase B) by apparent lengthening of the polar fibers. Polar movement of chromosomes in anaphase A results from depolymerization at the kinetochore. Depolymerization at the spindle pole may also occur. Yeast chromosomal fibers have only one microtubule at the kinetochore, which demonstrates that depolymerization of a single microtubule can move chromatids to the poles. The pushing apart of mitotic poles during anaphase B does not result from tubulin polymerization. Sliding of overlapping polar microtubules driven by kinesin or dynein in the central zone of the spindle may be responsible for this movement.

Other microtubule inhibitors include colcemid, podophyllotoxin, and nocodazole. Vinblastine and vincrystine inhibit microtubules by precipitation of tubulin and the formation of crystalline aggregates. Taxol, which is used as an anticancer drug, favors polymerization and stabilizes microtubules, leading to cellular accumulation of microtubules.

Not only are there isoforms of tubulin, but microtubules from various sources also differ in sensitivity to cold, cholchine, solubility at different ionic strengths, pH, and detergents. Recognition of these differences led to the identification of posttranslational modifications of tubulin. Amino acid sequences of tubulin from organisms representing broad phylogenetic backgrounds indicate that the gene for tubulin is highly conserved. Microtubule maturation is often accompanied by posttranslational modifications of tubulin, as well as the association with specific microtubule-associated proteins (MAPs). The most widespread modifications involve tubulin acetyltransferase (and deacetylase) and tubulin detyrosinase. Enzymatic acetylation occurs at the axonemal tip as microtubule subunits are added during growth of cilia. Tubulin synthesis, which is controlled by negative feedback mechanisms, is finely regulated in the cell. An increase in cellular free tubulin decreases the half-life of tubulin mRNA. The enzyme for deacetylation of tubulin is present in the cytoplasm; hence, tubulin subunits may be exchangeable between axonemal and cytoplasmic pools regulated by complex interactions of feedback loops.

## 4. Movement of Particles and Organelles along Microtubules

Placing *in vitro* reassembled microtubules on glass slides coated with purified axoneme outer arm dynein results in gliding of the microtubules along the surface. Analogous movements occur *in situ* in the pigmented granules that move along microtubules in fish chromatophores (Fig. 30). In fish chromatophore cells, the granules aggregate to the center of the cell in the presence of ATP and $10^{-7}$ $M$ $Ca^{2+}$ and disperse when $Ca^{2+}$ is decreased to $10^{-8}M$. The reticulopodial network (rete) of protozoans called **forams** have long challenged the imagination of cell biologists. Movement of particles is bidirectional on a single microtubule in the reticulopodia.

As the result of purification of microtubules by assembly and disassembly, several microtubule-binding MAPs have been detected. The high molecular weight proteins ranging in size from 200 to 400 kDa are called MAPs, whereas the smaller MAP (about 50 kDa) are designated tau proteins. In *Drosophila* eggs, at least 50 proteins bind to tubulin affinity columns; thus we can expect to learn much more about cytoplasmic movement mechanisms when the properties and functions of these proteins are elucidated. Some MAPs, such as MP2 and tau, are known to cross-link microtubules, and others are known to provide the force or motor for movement of organelles within the cell.

In some cytoplasmic systems, dynein (or other motor) activity has been visualized by the movement of polystyrene beads along microtubules isolated from the cells. In neurons, the motor for moving organelles along microtubules toward the nerve terminus (toward the plus end) has been called **kinesin,** distinguishing it from dynein, responsible for axonal transport toward the soma (toward the microtubule minus end). The discovery of these motor molecules explains the observation that extruded axoplasm of squid giant axons continues cytoplasmic streaming. Kinesin and dynein are now believed to be responsible for fast anterograde and retrograde axonal transport, respectively. A number of force-generating molecules are now being isolated from several different cell types. In many cells, two distinct motor molecules are responsible for movements along microtubules in opposite direction; however, a single motor molecule from the amoeba *Reticulomyxa*, a close relative of the forams, supports bidirectional movement of particles. Motor proteins all have two domains, a globular head and a tail. The ATPase activation of hinge action follows the binding of the motor molecule to the ATP-sensitive site of microtubules, analogous to the inter-action of motors with actin filaments. In some cells, organelle movement by actin-based mechanisms may operate in one direction, and either microtubule polymerization or sliding of microtubules acts in the opposite direction. The tail of motor molecules varies in binding specificity, depending on the cytoplasmic organelle, food particle, or other material that is moved along the microtubule or actin filament.

## VI. Summary

Two mechanisms for cell motility, actin–myosin and tubulin–dynein, use the energy from ATP hydrolysis and often complement each other during cell shape changes and organelle transport. Studies of striated muscle structure and function identified the nature of relevant molecules and the concept of sliding as a mechanism by which contraction can be achieved. Hence, amoeboid locomotion of nonmuscle cells, cytoplasmic streaming, and cell shape changes can now be explained by similar actin-based systems in which smaller motor molecules provide the generation of force. It is still not clear how actin functions in processes such as sensory reception and transduction.

Although ciliary and flagellar movement is based on a different microtubule–dynein system, it shares features with the actomyosin system; that is, motility is based on a sliding mechanism. Furthermore, motors that interact with microtubules are now known to cause such cell motility as organelle transport, similar to the actin filament system. In some cases actin filament- and microtubule-based organelle movement cannot be explained by sliding, for example, shortening of microtubules during movement of chromosomes to spindle poles. Polymerization and depolymerization of actin filaments or microtubules apparently constitute the underlying mechanism.

Three-dimensional ciliary and flagellar movements involve several levels of regulation. The generation of force is achieved by hinge action of dynein ATPases using the energy released by $Mg^{2+}$-ATP hydrolysis. Conversion of sliding to bend formation results from local, transient resistance to sliding of doublet microtubules. Reversal of the ciliary effective stroke and cessation of beat of some epithelial cila are controlled by $Ca^{2+}$ levels at the switch mechanism; the structural correlate of the switch is unknown. The intraciliary $Ca^{2+}$ level also controls beat frequency in both forward and backward swimming of ciliated protozoa. Intraciliary $Ca^{2+}$ concentration is regulated by voltage-sensitive $Ca^{2+}$ channels in the ciliary membrane that allow external $Ca^{2+}$ to enter the cilia. In *Paramecium*, stimulation of the cell's anterior results in depolarization of the cell membrane, increased intraciliary $Ca^{2+}$, and the avoidance reaction. Restoration of normal $Ca^{2+}$ levels in the cilia probably occurs by the action of outward-directed ATPase pumps. Stimulation of the cell's posterior results in hyperpolarization and the escape reaction. Second messengers and cyclic nucleotides participate in the transduction of signals during changes in the rate of ciliary beat and the direction of the effective stroke.

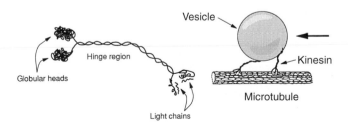

**FIG. 30.** Kinesin, a motor molecule. Kinesin is a relatively small motor molecule that interacts with microtubules. Interaction of kinesin with microtubules and membrane-bound organelles is responsible for movement of organelles within the cytoplasm. Arrow indicates the direction in which the vesicle moves relative to a microtubule. (From Bray, 1992.)

# Bibliography

Adamek, G. D., Gestland, R. C., Mair, R. G., and Oakley, B. (1984). Transduction physiology of olfactory receptor cilia. *Brain Res.* **310,** 87–97.

Bardele, C. F. (1977). Comparative study of axopodial microtubule patterns and possible mechanisms of pattern control in the centrohelidian heliozoa *Acantocystis. Raphidiophrys* and *Heterophrys. J. Cell Sci.* **25,** 205–232.

Bloodgood, R. A. (ed.) (1990). "Ciliary and Flagellar Membranes." Plenum, New York.

Bouck, B., Rosiere, T. K., and Levasseur, P. J. (1990). *Euglena gracilis:* A model for flagellar surface assembly, with reference to other cells that bear flagellar mastigonemes and scales. *In* "Ciliary and Flagellar Membranes" (R. A. Bloodgood, Ed.), pp. 65–90. Plenum, New York.

Bray, D. (1992). "Cell Movements." Garland Publ. Inc., New York.

Brokaw, C. J. (1985). Cyclic AMP-dependent activation of sea urchin and tunicate sperm motility. *Ann. NY Acad. Sci.* **438,** 132–141.

Conrad, G. W., and Schroeder, T. E. (eds.) (1990). "Mechanisms of Furrow Formation During Cell Division," Ann. NY Acad. Sci. Vol. 582. New York Acad. Sci., New York, 325 pages.

Doughty, M. J., and Kaneshiro, E. S. (1983). Divalent cation-dependent ATPase activities associated with cilia and other subcellular fractions of *Paramecium:* an electrophoretic characterization on Triton polyacrylamide gels. *J. Protozool.* **30,** 565–573.

Eckert, R., and Brehm, P. (1979). Ionic mechanisms of excitation in *Paramecium. Annu. Rev. Biophys. Bioenerg.* **8,** 353–383.

Eckert, R., Randall, D., and Augustine, G. (1988). "Animal Physiology. Mechanisms and Adaptation." Freeman, New York.

Gibbons, I. R. (1965). Chemical dissection of cilia. *Arch. Biol.* **75,** 317–352.

Gibbons, I. R. (1975). The molecular basis of flagellar motility in sea urchin spermatozoa. *In* "Molecules and Cell Movement" (S. Inoué and R. E. Stephens, Eds.), pp. 207–232. Raven Press, New York.

Inoué, S., and Stephens, R. E. (eds.) (1975). "Molecules and Cell Movement." Raven Press, New York.

Kaneshiro, E. S. (1984). Symposium—The structure and function of cilia and flagella. *J. Protozool.* **31,** 7–40.

Klumpp, S., Steiner, A. L., and Schultz, J. E. (1983). Immunocytochemical localization of cyclic GMP, cGMP-dependent protein kinase, calmodulin and calcineurin in *Paramecium tetraurelia. Eur. J. Cell Biol.* **32,** 164–170.

Kung, C., and Eckert, R. (1972). Genetic modification of electric properties in an excitable membrane. *Proc. Natl. Acad. Sci. USA* **69,** 93–97.

Lefebvre, P. A., and Rosenbaum, J. L. (1986). Regulation of the synthesis and assembly of ciliary and flagellar proteins during regeneration. *Annu. Rev. Cell Biol.* **2,** 517–546.

Mesland, D. A. M., Hoffman, J. L., Caligor, E., and Goodenough,

U. W. (1980). Flagellar tip activation stimulated by membrane adhesions in *Chlamydomonas* gametes. *J. Cell Biol.* **84,** 599–617.

Naitoh, Y., and Eckert, R. (1969). Ionic mechanisms controlling behavioral responses of *Paramecium* to mechanical stimulation. *Science* **164,** 963–965.

Naitoh, Y., and Kaneko, H. (1973). Control of ciliary activities by adenosine-triphosphate and divalent cations in Triton-extracted models of *Paramecium caudatum. J. Exp. Biol.* **58,** 657–676.

Naitoh, Y., and Sugino, K. (1984). Ciliary movement and its control in *Paramecium. J. Protozool.* **31,** 31–40.

Omoto, C. K., and Witman, G. B. (1981). Functionally significant central-pair rotation in a primitive eukaryotic flagellum. *Nature* **290,** 708–710.

Preston, R. R., and Saimi, Y. (1990). Calcium ions and the regulation of motility in *Paramecium. In* "Ciliary and Flagellar Membranes" (R. Bloodgood, Ed.), pp. 173–200. Plenum, New York.

Sale, W. S., and Satir, P. (1977). Direction of active sliding of microtubules in *Tetrahymena* cilia. *Proc. Natl. Acad. Sci. USA* **74,** 2045–2049.

Satir, P. (1984). The generation of ciliary motion. *J. Protozool.* **31,** 8–12.

Satir, P. (1968). Studies on cilia. III. Further studies on the cilium tip and a sliding filament model of ciliary motility. *J. Cell Biol.* **39,** 77–94.

Shinghoji, C., Murakami, A., and Takahashi, K. (1977). Local reactivation of Triton-extracted flagella by iontophoretic application of ATP. *Nature (London)* **265,** 269–270.

Sleigh, M. A. (Ed.) (1974). "Cilia and Flagella." Academic Press, New York.

Sleigh, M. (1989). "Protozoa and Other Protists." Arnold, London.

Summers, K. E., and Gibbons, I. R. (1971). Adenosine triphosphate-induced sliding of tubules in trypsin-treated flagella of sea urchin sperm. *Proc. Natl. Acad. Sci. USA* **68,** 3092–3096.

Tamm, S. L., and Horridge, G. A. (1970). The relationship between the orientation of the central fibrils and the direction of beat in cilia of *Opalina. Proc. R. Soc. London Sec. B* **175,** 219–233.

Vickerman, K., and Tetley, L. (1990). Flagellar surfaces of parasitic protozoa and their role in attachment. *In* "Structure and Function of Ciliary and Flagellar Surfaces" (R. A. Bloodgood, Ed.), pp. 267–304. Plenum, New York.

Warner, F. D., and Satir, P. (1974). The structural basis of ciliary bend formation. Radial spoke positional changes accompanying microtubular sliding. *J. Cell Biol.* **63,** 35–63.

Warner, F. D., Satir, P., and Gibbons, I. R. (Eds.) (1989). "Cell Movement," Vol. 1. Liss, New York.

Warrick, H. M., and Spudich, J. A. (1987). Myosin structure and function in cell motility. *Annu. Rev. Cell Biol.* **3,** 379–421.

Yanagimachi, R. (1988). Mammalian fertilization. *In* "The Physiology of Reproduction" (E. Knobil, J. Neill, L. L. Ewing, G. S. Greenwald, C. L. Markert, and D. W. Pfaff, Eds.), pp. 135–185. Raven Press, New York.

Edna S. Kaneshiro

# 57

# Centrin-Based Contraction and Bacterial Flagella

## I. Spasmonemes and Centrin-Containing Structures

The peritrich ciliates include solitary (e.g., *Vorticella*) and colonial (e.g., *Zoothamnium*) forms that have stalks, some of which contract rapidly when the organism is mechanically, chemically, or electrically stimulated (Fig. 1). Contractile peritrich stalks can contract to 30% of their original lengths in an all-or-none response. In some colonial forms, the stalks of the individual organisms contract independently, whereas in others, stimulation results in a synchronous contraction of all stalks, branches, and the trunk. The tubular stalk is formed by extracellular secretions of the cell when the nonstalked, free-swimming telotroch stage settles to the substratum. Secretion of the stalk transforms the telotroch to the sessile form.

### A. Correlation of Contractility with Spasmonemes and Myonemes

Contractility of peritrich stalks is absolutely correlated with the presence of spasmonemes; stalks of peritrichs without spasmonemes are not contractile. The spasmoneme, readily visible by light microscopy, is present in the cytoplasmic extension of the cell body within the stalk. Spasmonemes are commonly about 10 $\mu$m in diameter; however, the giant spasmoneme found in the colonial peritrich *Zoothamnium* is 30 $\mu$m in diameter and 1 mm long. The stalk also contains stiffening fibers that serve to reextend the contracted stalk. When contracted spasmonemes within stalks break, the stalk sheath rapidly recoils and the stalk straightens and extends.

Recognizing that spasmonemes differ from the actinomyosin-based contractile systems (Table 1 in Chapter 56), early workers preferred the term **spasmoneme**. The term myoneme implied too close a relationship with muscle; however, the structures in cell bodies that behave like spasmonemes are still called myonemes. Myonemes are joined to spasmonemes in peritrichs. Similar myonemes are present in stalkless ciliates such as the large ciliate *Spirostomum* (Fig. 1), which can contract to 50% of its original length, and the trumpet-shaped *Stentor*, which can contract its cell body down to 13% of its original length.

### 1. Ultrastructure

*a. Spasmonemes.* Spasmonemes have a trabecular texture containing 4-nm microfibrils arranged roughly along the long axis of the spasmoneme. By negative staining, extracted spasmoneme microfibrils show a 3.5-nm longitudinal periodicity. Numerous mitochondria and a membranous system (endoplasmic reticulum, ER) of saccules and tubules are associated with the microfibrils (Fig. 2). The ER vesicles or cisternae surround and penetrate microfibrillar bundles. This enclosed membrane system is believed to be analogous to the sarcoplasmic reticulum of muscle fibers and is thought to sequester and release $Ca^{2+}$ during spasmoneme contraction and extension. This notion is supported by electron microscopy of oxalate-treated myonemes that show deposits of calcium precipitates within the vesicles.

*b. Myonemes.* The myonemes of the stalkless, giant ciliates *Stentor* and *Spirostomum* are similar in ultrastructure to those in the cell bodies of stalked peritrich ciliates. Huang and Pitelka studied *Stentor* that had been pretreated with EGTA. The myonemes in these treated cells were relaxed and electron microscopy showed a meshwork of the 4-nm microfibrils. In cells stimulated to contract during fixation, the microfibrils coiled, forming short 10- to 12-nm-diameter tubes with a 4- to 5-nm wall consisting of 4 to 6 globular subunits (Fig. 3). However, these coiled tubular conformations are not seen in all contracted myonemes of all species, nor are they seen in spasmonemes of stalked peritrichs; therefore, this configuration may represent a supercoiled state of the microfibrils.

Myonemes in *Stentor* are associated with cross-linked microtubular bundles (kinetodesmal fibers), which was suggested by Huang and Pitelka to be the $Mg^{2+}$-ATP-requiring mechanism for reextension of the cell body. The microtubular bundles extend from ciliary basal bodies. The

**875**

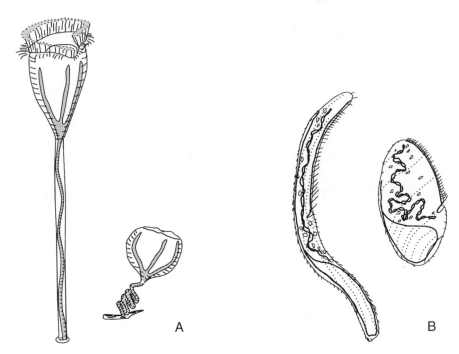

**FIG. 1.**   Contraction of *Vorticella* and *Spirostomum*. (A) Both the stalk and the cell body of *Vorticella* contract. The stalk coils during contraction by spasmoneme activity. Similar structures, called myonemes, in the cell body are responsible for the shortening of the cell body. (B) Myonemes in the cortex of *Spirostomum* cause contraction of the cell along its longitudinal axis. (Modified from *The Journal of General Physiology*, 1970, vol. 56, pp. 168–179, by copyright permission of The Rockefeller University Press.)

number of overlapping microtubules at any given anterior-to-posterior level of the cell is greater in contracted cells than in extended cells. During contraction, cross-links between microtubules detach, suggesting that this allows passive sliding of microtubules when myonemes contract. Sliding of microtubules in the opposite direction may accompany reextension and therefore may serve as the antagonist of myoneme contraction, but this idea has not been tested.

*c. The Linkage Complex.*   Allen described a structure associated with the microfibrils and called it the **linkage complex.** The linkage complex is oriented in various directions with respect to the spasmoneme or myoneme long axis. It consists of a midpiece, thin structures that cross perpendicular to that midpiece, rails that run parallel to the midpiece, and long, thin, striated microfilaments that connect the midpiece with structures at the outer cortex of the cell or stalk (Fig. 4). At the cortex, the striated filaments make close associations with basal bodies or modified basal bodies and can even be seen passing through the lumen of modified basal bodies.

## 2. Properties of Spasmonemes and Myonemes

Many of the early significant advances in the understanding of spasmoneme structure and function came from a group that included Weiss–Fogh and Amos (see, for example, Routledge *et al.*, 1976). Difficulty in obtaining large numbers of purified spasmonemes for biochemical analyses

and the lack of a suitable method for growing mass cultures of peritrichs have slowed progress in understanding the mechanism by which these organelles contract. Nonetheless, considerable information has been assimilated by some imaginative approaches used by these and other investigators.

*a. Speed of Contraction.*   Spasmoneme and myoneme contractions differ from those in the actomyosin–ATP systems in many respects. Myonemes do not bind HMM. Following a latent period of 1–4 msec, spasmonemes contract at a speed of 11–21 cm/sec, or about 200 lengths/sec, whereas the fastest known striated muscle (in mouse fingers) contracts at only 22 lengths/sec. In peritrichs, the contraction begins at the cell body–stalk junction, which has also been shown to have a lower stimulus threshold compared with distal regions. Contraction propagates at a maximum velocity of 21–60 cm/sec, in long stalks and 1.4–5.6 cm/sec in small stalks. In fatigued stalks, conduction velocity decreases to 3 cm/sec, with fatigue appearing first at the distal end. The duration of contraction is 2–20 msec. In contrast, reextension of contracted spasmonemes takes several seconds. The $Q_{10}$ of extension is 2.5, suggesting that enzymatic processes are involved during this slower reextension process.

*b. Role of $Ca^{2+}$.*   Injection of $Ca^{2+}$ into cells with myonemes results in contraction, and stimulation of cells preloaded with the $Ca^{2+}$-sensitive bioluminescent protein aequorin results in the emission of light before contraction

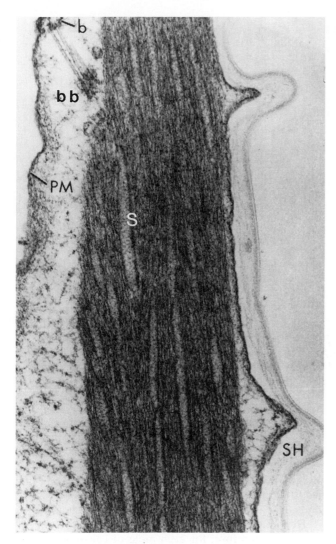

**FIG. 2.** Electron microscopy of the *Vorticella* spasmoneme. The spasmoneme in a peritrich stalk consists of 4-nm fibrils arranged parallel to the long axis of the organelle and the stalk. Within the organelle, membrane-bound saccules (S), sites of calcium sequestration, are seen interspersed between the fibrils. The stalk sheath (SH), the cell surface membrane (PM), and a modified basal body (bb) associated with short bridges (b) are indicated. (Reproduced with permission from Allen, 1973.)

(Fig. 5). Glycerinated or detergent-extracted models of peritrich stalks contract and extend repeatedly when sequentially exposed to a $Ca^{2+}$ solution and then to an EGTA solution; ATP or $Mg^{2+}$ is not required. In extracted models, $Ca^{2+}$ can be substituted by $Sr^{2+}$ and less so by $Ba^{2+}$, but $Mg^{2+}$ does not initiate contraction. The threshold for $Ca^{2+}$ is estimated at about $10^{-7}$g ions/liter. Contraction and extension of spasmoneme models can be repeated for at least 2 days at room temperature. This activity is not affected by the mercury-containing inhibitor salyrgan (mersalic acid) nor by parachloromercurobenzoate, cyanide, dinitrophenol, or fluorodinitrobenzene. It is also not affected by the detergents digitonin, saponin, and Tween 80.

*c. Optical Properties.* An important difference between spasmonemes and striated muscles lies in their optical properties. Like the A-band of muscle, spasmonemes exhibit positive birefringence (A-band, $2.3 \times 10^{-3}$; spasmonemes, $4 \times 10^{-3}$), which indicates that substructures are ordered with their long axis in line with the long axis of the fibers; therefore, spasmonemes and muscle A-bands are comparable. The positive birefringence of spasmonemes is largely due to form (substructures), rather than to intrinsic (molecular conformations) birefringence. During contraction, the birefringence of the striated muscle A-band remains the same, indicating no dramatic molecular conformational changes. However, the positive birefringence of spasmonemes decreases to almost zero during contraction (Fig. 6). This indicates that during spasmoneme contraction there is a decrease in order or folding of polymeric molecules. Since the birefringence appears to be due mainly to form birefringence, the optical properties of spasmonemes suggest a mechanism of microfibril coiling, in agreement with the observation that microfibrils in *Stentor* myonemes coil or supercoil during contraction. Unstrained *Zoothamnium* giant spasmonemes are apparently isotropic but they increase their birefringence when stretched.

*d. Energetics.* With respect to energetics, the tension developed by spasmonemes is similar to that of striated muscles operating at the optimum tension for performance of work. During isometric tension, the work done by striated muscle in a single twitch is equivalent to $10^5$–$10^6$ N m$^{-2}$ (N = $10^5$ dynes). Spasmonemes develop 4–8 $\times 10^4$ N m$^{-2}$ tension per unit cross-sectional area, and the work done in a single twitch has been calculated at 11 J/kg wet wt. Therefore, spasmonemes are equivalent to, or develop somewhat lower tension than, striated muscle. But, during contraction with different loads, striated muscles exhibit high shortening velocity with low load, which is associated with very little force and thin–thick filament cross-bridging. On the other hand, in spasmonemes (as measured by changing the viscosity of the medium), force does not vary with speed. This is similar to the energetics of a rubber band or metal spring. The calculated instantaneous power is 2.7 kW/kg wet wt, compared with that of the most energetic known muscle (insect flight muscle), which is only 0.05–0.2 kW/kg wet wt.

Spasmonemes have a high degree of passive extensibility and can be stretched to four times their resting lengths and then return to their normal lengths. At low $Ca^{2+}$ concentrations, a measurable pushing force and increased birefringence develop. If the stalk is prevented from elongating, the spasmoneme is seen elongating and bending within the stalk. This pushing force observed during spasmoneme extension is not thoroughly understood, but it may involve ATP hydrolysis, which is consistent with the presence of abundant mitochondria closely associated with spasmonemes.

Elemental analysis (calcium $K_\alpha$ peak) by electron microscopic microprobe of glycerinated *Zoothamnium* giant spasmonemes indicates that the organelle has 1.5 and 0.36 g bound calcium/kg dry wt during contraction and extension, respectively (Fig. 7). The change in chemical potential of

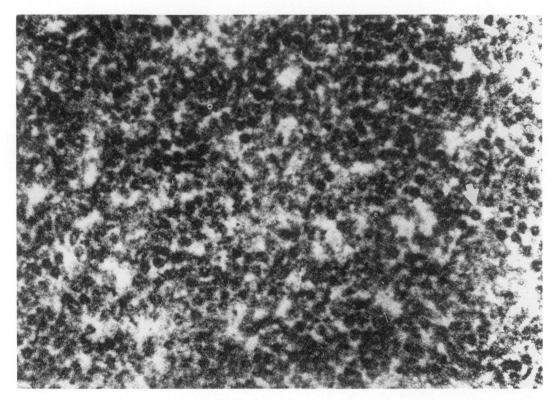

**FIG. 3.**    Contracted myoneme of *Stentor*. The 4-nm fibrils in myonemes of a contracted *Stentor* cell form supercoils. They appear doughnut-shaped when viewed by electron microscopy in cross sections of the myoneme. [Reprinted with permission from Huang, B., and Mazia, D. (1975). Microtubules and filaments in ciliate contractility. *In* "Molecules and Cell Movement" (S. Inoué and R. E. Stephens, Eds.), pp. 389–409. Raven Press, New York.]

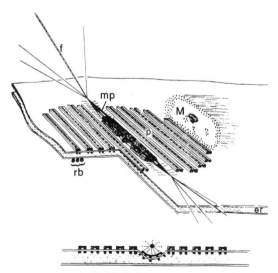

**FIG. 4.**    Linkage complex of spasmonemes and myonemes of *Vorticella*. The linkage complex is associated with the myoneme and the rough ER (er); the ribosomes of the ER are shown. The complex consists of rails (R), the midpiece (mp), and several filaments (f) that fan out from the tips of the midpiece; (rb), ribosomes. A cross-sectional view of the linkage complex is shown at the bottom. (Reproduced with permission from Allen, 1973.)

calcium ($\Delta\mu_{Ca} = RT \log_e [Ca^{2+}]_{high}/[Ca^{2+}]_{low}$) was examined to determine whether $Ca^{2+}$ alone can account for work done by spasmonemes during contraction. The value of $10^4$ J/mol calcium was calculated for a change from $10^{-8}$ to $10^{-6}$ *M*; therefore, for each mole of calcium bound to the spasmoneme, $10^4$ J of energy would be available for work. As mentioned previously, the work done in a single twitch of the spasmoneme is 11 J/kg wet wt. Therefore, to produce the estimated energy output, the spasmoneme must bind at least $11/10^4$ moles of $Ca^{2+}$, or 44 mg calcium/kg wet wt. The microprobe elemental analyses indicated values of 1.5 (contracted spasmoneme) and 0.36 (extended spasmoneme) g calcium bound/kg wet wt. Therefore, the chemical potential of calcium can account for the work done by spasmoneme contraction.

## 3. Spasmins and the Mechanism of Contraction

The major proteins of spasmonemes, spasmins, are acidic proteins with p*I* values of 4.7 and 4.8. They constitute 50–60% of the total stainable proteins of the isolated organelles separated by SDS–PAGE. Their migrations, which are not affected by $\beta$-mercaptoethanol (no —S—S— bonds), indicate that their masses are about 18 kDa (spasmin A) and 20 kDa (spasmin B). In EGTA–$Ca^{2+}$-buffered gels, their mobilities (anodally) decrease in the presence of high $Ca^{2+}$ ($10^{-6}$ *M*), suggesting a decrease in net negative charge, an increase in the Stokes radius of the molecules, or both. However, in alkaline urea gels with

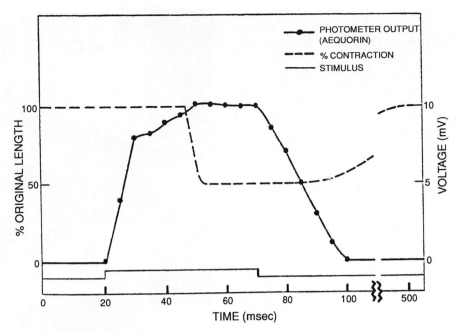

**FIG. 5.** Excitation–contraction coupling in *Spirostomum*. Temporal relationship between excitation, increase in free calcium, and the contraction of a *Spirostomum* cell preinjected with aequorin, a $Ca^{2+}$-dependent luminescent protein. Contraction begins when aequorin-induced photon emission is maximal. Cessation of electrical stimulation is followed by a decrease in photon emission and the cell relaxes. (Reproduced from *The Journal of Physiology*, 1970, vol. 56, pp. 168–179, by copyright permission of The Rockefeller University Press.)

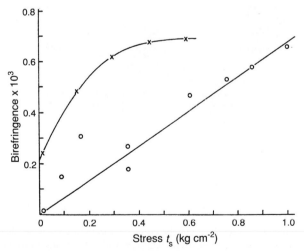

**FIG. 6.** Birefringence changes of spasmonemes. In the presence of high calcium (o) the spasmoneme birefringence is directly proportional to stress ($t_s$ = tensile stress). Thus, in the high calcium state, the spasmoneme behaves optically and mechanically as rubber. However, in the low calcium state at $10^{-8}$ $M$ concentration (X), the relationship between birefringence and stress is not the same. In the low calcium medium, the spasmoneme actively elongates and exhibits a pushing force, and the unstressed organelle becomes birefringent. [Reprinted with permission from Amos, W. B. (1975). Contraction and calcium binding in the vorticellid ciliates. *In* "Molecules and Cell Movement" (S. Inoué and R. E. Stephens, Eds.), pp. 411–436. Raven Press, New York.]

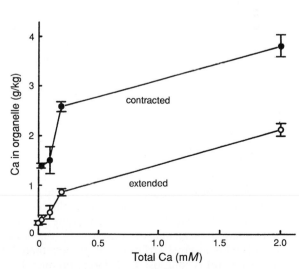

**FIG. 7.** Electron microprobe analysis of the calcium content of isolated glycerinated *Zoothamnium* giant spasmonemes. Contracted spasmonemes were prepared by treatment with $10^{-6}$ $M$ $Ca^{2+}$. Extended spasmonemes were treated with $10^{-8}$ $M$ $Ca^{2+}$, except at the ordinate, at which a 20 m$M$ EDTA solution was used. [Reprinted with permission from Amos, W. B. (1975). Contraction and calcium binding in the vorticellid ciliates. *In* "Molecules and Cell Movement" (S. Inoué and R. E. Stephens, Eds.), pp. 411–436. Raven Press, New York.]

$Ca^{2+}$, spasmin B exhibits increased mobility, which is not consistent with a decrease in net negative charge due to the binding of calcium. This observation provided strong evidence against an earlier proposed hypothesis, the electrostatic model of spasmoneme contraction. This hypothesis was based on the idea that $Ca^{2+}$ binding neutralized repulsive negative charges, resulting in the folding of polymeric molecules. Using the microprobe data for bound calcium in contracted and extended spasmonemes, it was calculated that $Ca^{2+}$ binds to spasmin with a stoichiometry of one or two $Ca^{2+}$ per molecule. Spasmins are high in serine, aspartate + asparagine, and glutamate + glutamine. Unlike troponin C and parvalbumin, spasmins lack cysteine.

Coiling of microfibrils is a plausible hypothesis for myoneme contraction. An alternative hypothesis is the entropic rubber model, which generally equates spasmonemes with material such as elastin or rubber that possesses oriented isoprene units. In this model, contraction results by the release of stored elastic energy.

### 4. Excitation–Contraction Coupling

Less examined and understood is the nature of the coupling of excitation with contraction (EC coupling). Since the activation time is on the order of 2–3 msec in *Zoothamnium,* and its giant spasmoneme is 30 $\mu$m in diameter, the signal is transduced 10 times faster than can be explained by simple diffusion of $Ca^{2+}$ through the surface. Currently, the linkage complex associated with spasmonemes appears the most likely candidate for EC coupling (see Fig. 4). Allen has suggested that the striated filament that links the midpiece of the linkage complex with cortical structures near the surface membrane (e.g., basal bodies) may transduce the electrical signal to the midpiece. The midpiece is physically coupled to the ER and hence could trigger $Ca^{2+}$ release from the cisternae. Furthermore, the rails may be involved in the movement of $Ca^{2+}$ in and out of the ER and may be the sites of ATPase activity for concentrating $Ca^{2+}$ into the ER.

### B. Striated Rootlets

Striated rootlets (rhizoplasts), well developed in unicellular organism such as flagellated algal cells (e.g., *Tetraselmis*), are organelles that are associated with the flagellar basal bodies (Fig. 8). They extend, branch, and link to adjacent basal bodies. In some cases they extend and pass along the nuclear envelope and anchor near the cell membrane opposite the flagella. The anchor site has been termed laminated oval, half-desmosome, or rhizankyra. The rootlet exhibits cross-striations of greater than 160-nm periodicity and is composed of 3- to 8-nm microfibrils aligned parallel to the long axis of the organelle. That these were contractile organelles was clearly demonstrated by fixation of *Tetraselmis* cells in 0.1–5 m*M* $Ca^{2+}$. Salisbury (1989) found that compared with organelles in cells fixed without $Ca^{2+}$, the organelle was 65% shorter, 30% wider, and had a reduced cross-striation periodicity (see Fig. 8). Other studies of striated rootlets demonstrate that regions proximal to the flagellar basal bodies have smaller periodic-

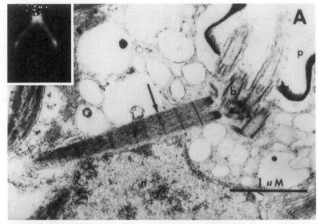

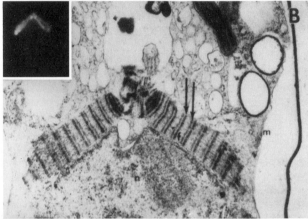

**FIG. 8.** Striated rootlet of the marine flagellate, *Tetraselmis.* Cells were fixed for electron microscopy under various conditions. The insets show the binding patterns of fluorescence-labeled anti-centrin antibodies. (A) Fixation with low $Ca^{2+}$ illustrates the normal width of the organelle and the periodicity of its cross-striations. (B) Fixation with high $Ca^{2+}$ plus ATP may be equivalent to the reextension phase after contraction. (C) Fixation with high $Ca^{2+}$ (0.1–5 m*M*) caused a dramatic shortening of the striated rootlets. The organelle in this state is characterized by increased thickness and reduced periodicity of the cross-striations. (Courtesy of Jeffrey Salisbury.)

ities than distal regions, suggesting a polarity in the organelle. However, this observation may reflect the direction in which contraction propagates. In living cells, it was observed that the rootlets undergo cyclic contractions and extensions with millimolar concentrations of $Ca^{2+}$ and ATP.

Much like spasmonemes, neither muscle HMM nor myosin S-1 fragments decorate extracted rootlets. Also, striated rootlets and other related organelles contract rapidly within less than 20 msec; reextension is slow (a few seconds to about 1 hr). Contraction is initiated by elevated intracellular $Ca^{2+}$ and is independent of ATP hydrolysis. When contracted, the microfibrils in striated rootlets assume a supercoiled configuration. Although the direct requirement for ATP in the reextension of spasmonemes is not clearly established, there is good evidence that ATP is required by striated rootlets during their reextension process.

The major proteins of striated rootlets, centrins (called caltractin in *Chlamydomonas*), make up greater than 60% of the total proteins of the isolated organelle. Centrins are low-molecular-mass (20-kDa) acidic proteins ($\alpha$ centrin, p*I* 4.9; and $\beta$ centrin, p*I* 4.8).

Metabolic radiolabeling with $^{32}PO_4$ demonstrates that the $\beta$ isoform is phosphorylated whereas the $\alpha$ isoform is not. The role of phosphorylation of these contractile proteins is currently unclear. However, it has been suggested that the phosphorylated $\beta$ and dephosphorylated $\alpha$ isoforms are correlated with the extended and contracted states, respectively. Both isoforms of centrin exhibit $Ca^{2+}$-sensitive mobilities in electrophoresis gels. However, in alkaline urea gels, the mobilities decrease in the presence of $Ca^{2+}$, similarly to the mobility of spasmins in SDS–PAGE under denaturing conditions. These observations on centrin may be related to those made with troponin C, which binds $Ca^{2+}$ in urea gels, causing an increase in mobility. This has been interpreted as a change to a more compact molecular conformation, analogous to the interaction of troponin I and troponin C when the two proteins interact and form a slower migrating complex.

## C. Occurrence and Functions of Centrin and Related Molecules

Monospecific antibodies to flagellar rootlet centrin proteins have been reported to bind to centrioles and centrosomes of mitotic spindle poles, as well as to the mitotic spindle matrix (Fig. 9). These observations, made on a wide range of cell types, revealed the ubiquitous occurrence of centrin-like molecules and inspired Salisbury (1989) to coin the term centrin for the protein. These findings indicate that centrin may have a universal function in cells, since it is found in cell division structures.

The anti-centrin antibodies also bind to a nine-pointed stellate structure in the transition zone between the flagellar shaft and the basal body of some flagella. The antibody binding occurs at the plane of scission during the deflagellation process. The stellate structure was shown to contract during $Ca^{2+}$-induced deflagellation, implicating that structure in the generation of force required for autotomy (Fig. 10). However, autotomy occurs in many cilia and flagella that normally do not have the nine-pointed stellate structure, and recently Jarvik has described autotomy in a *Chlamydomonas* mutant lacking this structure.

The centrin-related proteins are related to the E–F hand superfamily of calcium-binding proteins. Huang isolated and sequenced *Chlamydomonas* caltractin (centrin) cDNA and demonstrated that the protein contains four E–F hands. Sequence identity with calmodulin is 45–48%, and sequence identity with the yeast spindle pole *CDS31* gene is 50%. Anti-caltractin antibodies, however, do not cross-react with calmodulin.

An alternative hypothesis for the function of centrin has been proposed for the biflagellated alga *Spermatozopsis*, whose rhizoplast (rootlet) does not exhibit cross-striations. A connecting fiber joins the two basal bodies. During forward swimming, the two flagella of the cell exhibit asymmetrical breast stroke movements. The organism displays a photophobic response to light (avoidance reaction), resulting in backward swimming. During this response, the flagella undulate and the basal bodies reorient by contraction of the connecting fiber. This switch for basal body reorientation is independent of the motor that drives flagellar beating.

Anti-centrin antibodies also bind to a structure called the **paraxial rod,** which is found within the transverse flagellum of some dinoflagellates. The paraxial rod may cause a rapid and strong contraction of the dinoflagellate transverse flagellum that causes a stop reaction of the swimming cell (equivalent to an avoidance reaction). Thus, centrin may function as a modulator of motile structures.

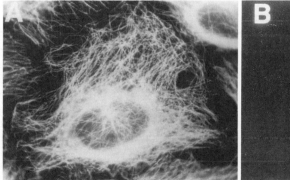
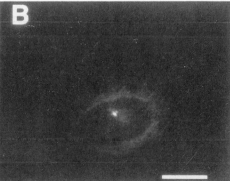

**FIG. 9.** Anti-centrin antibody binding to cells. The wide occurrence of centrin-like molecules is demonstrated by a HeLa cell double-stained with fluorescence-labeled anti-tubulin and anti-centrin antibodies. (A) Anti-tubulin binding pattern. (B) Anti-centrin binds to the centrosomes of cells from broad phylogenetic backgrounds. (Courtesy of Jeffrey Salisbury.)

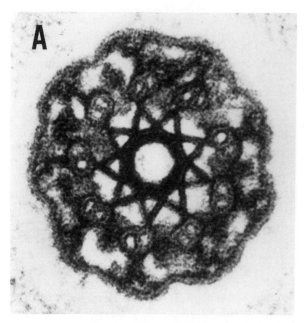
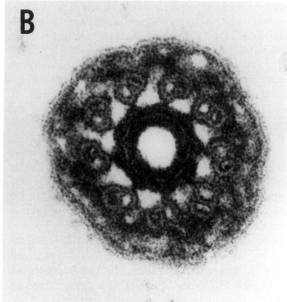

**FIG. 10.**  Transition zone of *Chlamydomonas* flagella at the nine-pointed stellate structure. (A) The configuration in a normal flagellated cell. (B) The nine-pointed stellate structures appears contracted in a cell treated with high concentrations of $Ca^{2+}$, a treatment that induces autotomy. Constriction of the stellate structure may be the direct cause of weakening at this plane leading to severing of the flagella. (Reproduced from *The Journal of Cell Biology*, 1989, vol. 108, pp. 1751–1760, by copyright permission of The Rockefeller University Press.)

## II.  Prokaryote Locomotion

Locomotion, characteristic of some bacteria, serves as a useful taxonomic criterion. Motility assays are standardly employed in identifying bacterial isolates. Two types of locomotion are recognized in prokaryotic cells: (1) flagella-driven swimming and (2) gliding. Gliding of prokaryotes and eukaryotes is discussed later in Section III.

### A.  Bacterial Flagella

Flagellated bacteria swim by the rotation of rigid helical flagella driven by motors at their bases. Bacterial flagella are thin (ca. 20 nm in diameter) extracellular organelles that extend about 15–20 $\mu$m from the surface of cells (Fig. 11). They can occur as a single flagellum (monopolar) or a bundle of several flagella (lophotrichous) at one end of the cell. When bundles occur at both poles, the pattern is called **amphitrichous.** Cells with randomly distributed flagella are **peritrichous.** In spirochetes, flagella do not protrude from the cell surface. Instead, an axial filament consisting of two sets of fibrils that extend from pole to pole, enclosed within the outer layer of the cell surface, is responsible for the flexing movements exhibited by these organisms.

### B.  Structure

When stained, bacterial flagella are visible by light microscopy, but details of their structure can be elucidated only by special electron microscopic and spectroscopic techniques. The bacterial flagellum is composed of three parts: the basal body, hook, and filament (Fig. 12).

### 1.  Basal Body

The basal body is relatively small, but is a complex structure made up of a rod and rings. The rod is analogous to a transmission shaft. Basal body rings differ in gram-negative and gram-positive bacteria, and these substructures correlate with the differences in cell wall structures. In gram-negative species such as *Escherichia coli* and *Salmonella typhimureum,* there are usually four rings: (1) the M ring resides within the cell membrane; (2) the S ring is located within the periplasmic space; (3) the P ring is in the peptidoglycan layer; and (4) the L ring is within the lipopolysaccharide-containing outer layer. The outer two rings probably serve as the motor's bushing and allow the rod to pass through the outer layers of the cell envelope. The inner two are involved in rotation of the motor. How rotation is generated is not yet clear. It has been suggested that the M ring may serve as the stator and the P ring the rotor. Alternatively, some have argued that a stator needs a large mass offering sufficient stability into which it can anchor, since the filament needs to overcome high viscous drag. The most likely structure for greater mass is the peptidoglycan layer. In this case, the rotation would then be generated with the cell membrane and the M ring directly involved in the generation of rotation. In gram-positive species that lack the outer layer, only two rings are found in the basal body: One is the equivalent of the M ring within the cell membrane, and the other, the outer ring, is associated with the teichoic acid component of the cell wall.

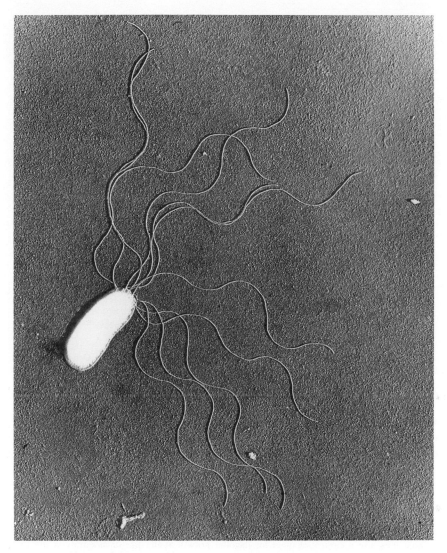

**FIG. 11.** Electron micrograph of *Pseudomonas marginalis.* Flagella emerge from one pole of the bacterial cell. (Courtesy of Arthur Kelman.)

### 2. Hook

The hook functions like an elastic universal joint between the basal body and filament. Its 90-nm length is genetically predetermined and involves hook-associated proteins (HAP) at the hook–filament junction. Mutants lacking HAP continuously secrete unassembled filament protein into the medium.

### 3. Filament

The filament, which acts like a propeller, is a rigid, brittle helical polymer of a single 51-kDa protein, flagellin. Some bacterial flagella (such as in *Caulobacter*) are composed of two types of flagellin molecules. The flagellin subunits determine the final shape of the filament structure by nonequivalent interactions between the subunits. Left-handed helices are more common than right-handed helices. Flagellin self-assembles *in vitro,* forming a cylindrical wall of about 11 subunits per two turns. A mixture of two different flagellins also self-assembles *in vitro.* Unlike that of actin and tubulin, polymerization of flagellin does not require energy from nucleotide triphosphates. Flagellins are among the most antigenic proteins known, which is an adaptation of the vertebrate immune system for effective elimination of bacterial infections.

### C. Assembly of Flagellar Substructures

Much like that observed in growing eukaryote flagellar axonemes, the assembly of the growing flagellum occurs by the addition of individual protein subunits from the proximal to distal end; the tip is the last to be added. This base-to-tip mechanism probably evolved in bacteria, since assembly occurs primarily outside the cell membrane and the export and addition of simple subunits to a single structure are easier than the export and addition of larger, more complex modular structures. Also, since most of the torsional load is at the proximal end, and because the motor

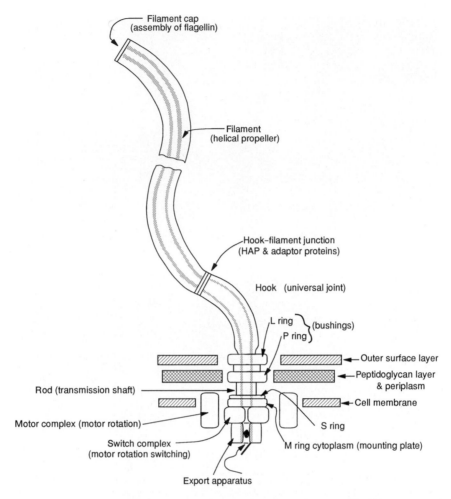

**FIG. 12.** Structure of the flagellum of a gram-negative bacterium. The filament is flagellin polymer that is rigid and helical. The hook, L ring, P ring, and rod lie distal to the cell surface membrane. The S ring and M ring are embedded in the membrane bilayer and are associated with the complexes that serve as the motor and the motor switch. The export apparatus allows molecules into a central channel for flagellar growth at the site of assembly at the tip. (Modified from MacNab, 1990, with permission of Cambridge University Press.)

rotates while the organelle is being elongated, any proximal separation in the nascent structure for insertion of new material would be mechanically difficult. The precursors, therefore, must be exported, single-file, through the growing structure; they do not reach the tip on the external surface. The information for the entire flagellum structure is probably inherent in the structure of each subunit, since self-assembly *in vitro* reconstitutes the different components very close in size and sequence to the intact organelle. Moreover, all rod, hook, HAP, and filament proteins do not have cleaved signal peptides, strongly suggesting that the subunits include topological targeting information. Assembly, therefore, is closely regulated by the sequence of synthesis and the export of the subunits. A central channel of 5-nm diameter has been detected in bacterial flagella by X-ray fiber diffraction techniques. This channel is sufficiently large to serve as a conduit for single subunits to get to the growing tip.

## D. Energy Source

The energy for rotation of the motor does not come from the hydrolysis of high-energy phosphate bonds of ATP. It is derived from the dissipation of an ionic transmembrane electrochemical gradient (ion-motive force). In most neutrophilic species, such as *E. coli*, the ionic species that has been actively pumped out of the cell by the electron transport chain in the bacterial membrane during respiration is $H^+$ (proton-motive force). The coupling of energy released in the respiratory chain requires solute and ion transport, not ATP, as an intermediate; therefore, ATP synthesis and flagellar rotation are alternative links with the respiratory chain. When $H^+$ is the ion actively pumped out, forming a transmembrane proton chemical and electrical gradient, the proton-motive force becomes available to drive flagellar rotation. Under anerobic conditions, the translocation of protons is achieved by proton ATPase activities. In marine and alkalophilic bacteria, the ion is $Na^+$, which has been pumped out of the cell by a $Na^+–H^+$ antiporter or by the electron transport chain. Since the energy can come from the electrochemical gradients of either $H^+$ or $Na^+$, the rotation of the motor must involve electrostatic interactions and events involving association and dissociation of ions.

Although no morphological structures have been identified, genetic and biochemical evidence indicates that there must be structures involved in energy transduction for converting ion-motive force to motor rotation. Since it is difficult to control specific membrane potentials and pH gradients experimentally in gram-negative bacteria, energy coupling has been best examined in gram-positive bacteria such as *Streptococcus* and *Bacillus subtilis*. It has been clearly established that rotation is coupled to the inward current of protons and that the speed of rotation is proportional to the proton-motive force. A flux of 1200 protons per revolution, independent of load or velocity, has been calculated. Membrane potentials or pH gradients of opposite polarity also drive the rotation of the motor. Gliding motility in some species of bacteria has also been shown to be powered by proton-motive force.

### E. Swimming Behavior

In general, peritrichous bacteria swim slower and rotate in a relatively straight line; those with flagella restricted to poles swim in many directions and spin rapidly. During swimming, the flagella of peritrichous bacteria rotate as a coordinated bundle (Fig. 13). In amphitrichous cells, the bundle at one end rotates clockwise (CW) and the one at the other end rotates counterclockwise (CCW) to propel the cell in one direction. The rotary nature of the flagellar motor can be visualized in tethered cells with their flagella adhered onto the substratum. Tethered cells rotate; the shorter the cell, the less the hydrodynamic drag, and thus the faster the cell rotates.

The force created by the dissipation of the transmembrane potential propels the cell at a speed of 20–80 $\mu$m/sec or $\geq$10 lengths/sec. In cells like *E. coli*, swimming is a three-dimensional random walk with gentle curves called runs. The flagellar bundle revolves CCW at a rate of 12,000 rpm. The current consumption of each flagellum has been calculated at $10^{-5}$ A. Considering that bacteria commonly live in dilute aqueous solutions, with Reynolds numbers of $10^4$–$10^5$, inertia is insignificant. That is, if the cell stopped swimming, it would come to a halt estimated at $<10^{-5}$ of its cell length. Swimming patterns are well characterized and they result from the control of flagellar activity. Runs are interrupted by brief breaks called *tumbles* or *twiddles* during which individual flagella in bundles splay out causing a pause. The flagella in monopolar cells or flagellar bundles in peritrichous cells reverse to a CW rotation and the cell swims backward.

### F. Chemotaxis

### 1. Responses to Attractants and Repellents

In the presence of attractants such as amino acids and sugars or repellents such as phenol, organic acids, and

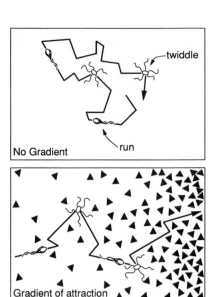

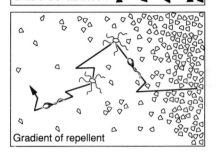

**FIG. 13.** Swim tracks illustrating chemotaxis in bacteria. Top: The cell exhibits spontaneous runs and tumbles (twiddles) with random orientations when attractant or repellent molecules are uniformly distributed in the medium. During runs, flagella form a bundle and rotate together. Tumbling results from the flagella splaying out of the bundle. Middle: In the presence of a gradient of an attractant, a random walk changes to a biased walk characterized by longer runs toward the region of higher attractant concentration [bottom: away from higher repellent concentration].

heavy metals, random walk becomes biased (see Fig. 13). This is unlike chemotaxis of *Dictyostelium* amoebae, which move directly toward or away from attractants and repellents, respectively. Taxis is possible because the flagellar motor has an all-or-none switch. The morphological identity of the motor switch has yet to be determined, but there is good genetic and physiological evidence for its existence. When *E. coli* is placed in a spatial chemical gradient of an attractant, suppression of tumbling leads to net progress toward the attractant; repellents increase tumbling. Actually, bacteria do not sense a spatial gradient; that is, they cannot detect differences in concentrations between their anterior and posterior parts. Berg found that chemotactic behavior is the result of a response to a temporal gradient that develops as the cell moves through the medium. Thus, bacteria must be able to measure the attractant concentration, store that information, and compare it with a value measured at a later time. To do this, they must have memory. The time between stimulus and response (response latency) is on the order of 200 msec. In signal transduction, the cell can integrate information from one type of receptor or it can integrate information from different receptors; the effect of an attractant (CCW flagellar rotation) is canceled by the effect of a repellent (CW flagellar rotation). Chemotactic responses are transient; thus, the cell exhibits adaptation.

### 2. Receptor Binding and Transduction of Signal to Switch

An important breakthrough in understanding bacterial chemotaxis was the discovery in Koshland's and Adler's laboratories that methylation of proteins was involved in the behavioral response. Since then, use of the powerful tools of molecular genetics and biochemistry has advanced this field at a remarkable rate. In *E. coli* and *S. typhimurium,* the components of most flagellar structures are biochemically and genetically identified; almost all flagellar genes have now been cloned and sequenced.

The chemotactic system of *E. coli* involves four receptor–transducer proteins, six cytoplasmic proteins, and three components of the flagellar switch. Different attractants bind to different receptors called Tsr, Tar, Trg, and Tap. These receptor–transducer proteins are involved in measuring past and current concentrations of the ligand and in comparing the two values. The ligand-binding site of the transducer is in the periplasmic domain. In the cytoplasmic domain the transducer molecules contain the methyl-accepting sites. Ligand binding at the periplasmic domain is fast and leads to covalent modifications of the cytoplasmic domain; the reactions involved in modification of the cytoplasmic methyl-accepting sites are slower.

In three dimensions (Fig. 14), a receptor–transducer protein is thought to contain four helical bundles in the periplasmic space and two helices that contain the ligand-binding sites. Two transmembrane helices connect the periplasmic and cytoplasmic domains. In the cytoplasm, there are four helical bundles with four to five methyl-accepting sites. Occupancy of ligand-binding sites represents the measurement of the current ligand concentration, and modification by methylation of the cytoplasmic domain is involved in storage of information about prior concentrations (Fig. 15). In the transduction cascade, the initial source of methyl groups is *S*-adenosylmethionine (SAM), and the initial source of phosphate groups is adenosine triphosphate (ATP).

In the unstimulated state the ligand-binding site of the receptor is unoccupied. In this state, the methyl-accepting sites on the cytoplasmic domain of the transducer are continuously methylated and demethylated at a steady state of about one methyl group per protein.

In the stimulated state the ligand-binding site is occupied, which leads to a bias of the switch to CCW rotation. This is translated to suppression of tumbling (longer runs) as the cell moves up the attractant gradient. The information at the binding site in the periplasmic domain is transduced through the cell membrane domain to the cytoplasmic domain, resulting in the activation of methyl-accepting sites. In the cytoplasm, phosphotransfer reactions involving the proteins CheA, CheW, CheY, CheZ,

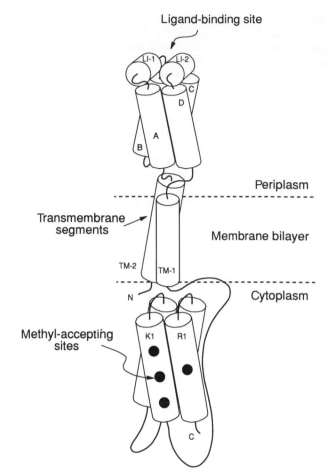

**FIG. 14.** The transmembrane receptor–transducer molecule. The receptor sites (LI-1 and LI-2) that bind ligands lie in the periplasmic space of gram-negative bacteria cell surfaces. There are segments (TM-1 and TM-2) that span the bilayer membrane and methyl-accepting sites (solid dots) are on α-helical cylindrical structures (K1 and R1) within the cytoplasm. The C- and N-termini of the polypeptide are indicated. (From Hazelbauer *et al.,* 1990; reprinted with permission of the Cambridge University Press.)

CheR, and CheB are activated. CheA is a protein kinase that is phosphorylated by ATP and then transfers phosphate groups to CheY and CheB. Phosphorylated CheY and CheB are unstable compounds. Phospho-CheY may interact directly with the flagellar motor to induce CCW rotation. Experiments using cell envelopes with depleted cytoplasmic contents demonstrate that flagellar motors are locked in the excited CCW rotation state.

Methylation of the cytoplasmic domain of the transducer offsets the effects of ligand binding, leading to adaptation; thus, the cytoplasmic components of the chemotactic system receive a null signal. The cytoplasmic protein CheR is an enzyme that catalyzes the methylation of the transducer.

Loss of the ligand on the periplasmic domain represents a negative stimulus. The effects of methylation of the cytoplasmic domain are unbalanced and cause a bias of the flagellar switch to CW rotation. The cytoplasmic domain of the transducer becomes activated for demethylation.

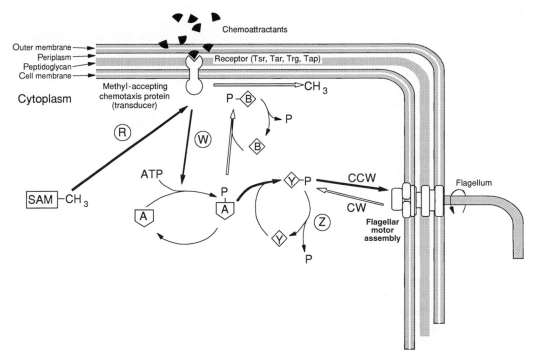

**FIG. 15.** Bacterial chemotactic behavior. The attractant molecules bind to receptors located in the periplasm of gram-negative bacteria. This induces a signal that is transduced through the region spanning the membrane bilayer to the methyl-accepting portion in the cytoplasm. Activation of the cytoplasmic portion of the receptor–transducer methyl-accepting protein results in methylation involving *S*-adenosylmethionine (SAM), ATP, and the enzyme CheR. A protein kinase (CheA) is phosphorylated, which in turn phosphorylates CheY (which interacts with the flagellar motor switch to induce CCW rotation) and CheB (an enzyme that demethylates the transducer). The phosphatase CheZ dephosphorylates CheY; dephosphorylated CheY interacts with the flagellar switch to induce CW rotation.

The cytoplasmic protein CheB is an enzyme that catalyzes the demethylation of the transducer. CheZ is a phosphatase that acts on phospho-CheY, increasing its hydrolysis rate.

The result of demethylation of the cytoplasmic domain balances the effect of ligand loss at the periplasmic domain as the transducer assumes the adapted state. Again, the cytoplasmic components of the chemotactic system receive a null signal. The system then returns to the unstimulated state in the absence of a chemotactic gradient.

## III. Gliding and Other Movements

There are several cell motility systems that may not belong to one of the four systems described in this chapter. Examples of unique movements that have been reported are described here.

### A. Gliding

Gliding of an organism is characterized by active movement in contact with a solid substratum, with no obvious locomotory organelle responsible for this movement nor any distinct change in shape of the organism.

#### 1. Bacteria

Extensive investigations for over a century have yet to bring consensus on the mechanism of bacterial gliding. It

is only agreed that gliding requires cell contact with the substratum and that extracellular slime, mucilage, or mucoid material is probably involved.

Particles such as polystyrene latex beads placed on the surface of *Cytophaga* adhere and move longitudinally in both directions, as well as clockwise and counterclockwise. Hence the entire surface of the cell is capable of gliding motility, and the force for this movement is generated in all directions lateral to the cell surface. The cell can turn at a rate of 100 rpm and move as fast as 10 $\mu$m/sec.

Gliding mutants were grouped by Pate (1988) into two types: (1) mutants with defective machinery or motors that are paralyzed or nonmotile and (2) mutants with defective systems that interact with the motor to translocate the cell over the surface. Mutants that can glide and move beads over their surfaces, but whose colonies fail to spread, may be defective in accessory systems.

All truly nonmotile mutants are resistant to phages that infect wild-type cells and have several membrane properties that differ from those of wild-type membranes. One of several major cell envelope proteins can be missing from different nonmotile strains; therefore, these may be components of the surface machinery or motor. Leadbetter has identified an interesting biochemical difference between wild-type and nonmotile mutants that do not exhibit colony spreading, gliding, or movement of latex beads over their surfaces. The mutants are deficient in a specific group

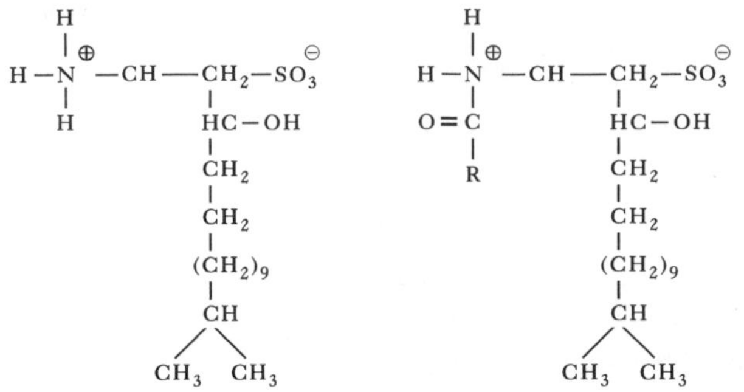

$$
\begin{array}{cc}
\underset{\text{Capnine}}{
\begin{array}{l}
\overset{\displaystyle H}{\underset{\displaystyle H}{H-\overset{\oplus}{N}}}-CH-CH_2-SO_3^{\ominus} \\
\qquad\quad | \\
\qquad\quad HC-OH \\
\qquad\quad | \\
\qquad\quad CH_2 \\
\qquad\quad | \\
\qquad\quad CH_2 \\
\qquad\quad | \\
\qquad\quad (CH_2)_9 \\
\qquad\quad | \\
\qquad\quad CH \\
\qquad\quad \diagup\;\diagdown \\
\qquad CH_3\;\;CH_3
\end{array}}
&
\underset{\textit{N}\text{-acylcapnine}}{
\begin{array}{l}
\overset{\displaystyle H}{\underset{\displaystyle O=C}{H-\overset{\oplus}{N}}}-CH-CH_2-SO_3^{\ominus} \\
\qquad\quad | \\
\qquad\quad R\qquad HC-OH \\
\qquad\qquad\quad | \\
\qquad\qquad\quad CH_2 \\
\qquad\qquad\quad | \\
\qquad\qquad\quad CH_2 \\
\qquad\qquad\quad | \\
\qquad\qquad\quad (CH_2)_9 \\
\qquad\qquad\quad | \\
\qquad\qquad\quad CH \\
\qquad\qquad \diagup\;\diagdown \\
\qquad\quad CH_3\;\;CH_3
\end{array}}
\end{array}
$$

**FIG. 16.** Gliding motility in some bacteria is correlated with the presence of sulphonolipids. Sulphonolipids are lipids not commonly found in organisms. They contain a fatty acid (R—CO—) moiety.

of unusual lipids called **sulfonolipids** (Fig. 16). He further demonstrated that restoration of normal sulfonolipids in the cell by supplementation of mutant cultures with a sulfonolipid precursor also restored all types of movement present in wild-type cells. Several other changes in surface polypeptides and carbohydrates have been observed in both classes of mutants, but it is still unclear whether these biochemical lesions are primary or secondary to the machinery at the cell surface driving gliding motility.

The fluidity of the membrane bilayer can modulate gliding, as shown by a rapid temperature downshift that quickly stops bacteria from moving. If bacteria are kept at the lower temperature, gliding resumes after an adaptation period during which the fatty acid composition of the bacteria lipids has changed with an increase in the ratio of unsaturated to saturated fatty acids, as well as an increase in the ratio of branched-chain to straight-chain fatty acids. The cell surface membrane lipids probably undergo thermal phase transition from liquid crystalline to crystalline gel, and metabolic compensations resulting in more fatty acids that increase bilayer fluidity allow the transition point to be lowered.

There is evidence that the membrane potential or proton-motive force, not ATP, is probably the source of energy for bacterial gliding. The presence of many rotary motors distributed over the entire cell surface may explain this form of locomotion. These motors are thought to be similar to the motors at the proximal end of bacterial flagella (see preceding) but are obscure because they do not have obvious external structures. Extracts of cells examined by electron microscopy reveal ring structures that resemble flagellar motors of bacteria.

## 2. Unicellular Eukaryotes

Gregarines are large protozoans that live as parasites in invertebrates. They exhibit smooth gliding locomotion but have no cilia or other external organelles that can account for this movement. The cell surface has an unusual ultrastructure that may be related to its motility, or it may simply

represent an efficient absorptive surface for nutrients. The cell surface is highly folded with parallel longitudinally arranged ridges (Fig. 17). The organism secretes copious amounts of mucus to which small particles adhere and are moved posteriad. The ridges undulate, which may cause mucus transport and/or gliding. Although myosin has been identified in the organism, it is still not clear if it functions in gliding and/or other movements such as dilating, contracting, and flexing of the cell body.

Gliding is exhibited by unicellular algae such as diatoms. It is thought that extracellular mucoid material at the suture between the two parts of the cells is responsible for movement of the cell.

Desmids, another group of unicellular algae, are composed of adjoining semicells joined by an isthmus. They exhibit end-over-end flipping, which probably involves extracellular mucoid secretions released through pores at each end of the cell. The structure of the mucilage is described as prismatic or fibrillar and hence may contain contractile proteins. Like the gregarines, the desmids' surfaces are also complex. Microfibrillar networks are present in the inner, primary layer of the cell wall, and microfibril bundles present in the external, secondary layer form flat, longitudinal ridges or ribbons.

## 3. Movement along Surfaces of Eukaryote Flagella

*Euglena, Chlamydomonas,* and other protists exhibit cellular movements that are not related to other movements involving known motility mechanisms. When *Chlamydomonas* flagella make contact with a solid substratum, it can

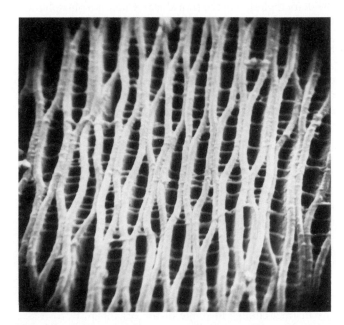

**FIG. 17.** Gliding motility in gregarine protozoa is not understood. Scanning electron microscopy of the surface of a gregarine cell suggests that deep cortical folds undulate. The undulations may move secreted mucus posteriad, thus causing the gliding locomotion of the organism. (Courtesy of Eugene Small.)

glide at a velocity of 2 $\mu$m/sec. Bloodgood demonstrated that beads move up and down the flagellar surface in a saltatory fashion. Nongliding mutants, and cells treated with local anesthetics (lidocaine), calmodulin inhibitors (trifluoperazine), and low extracellular $Ca^{2+}$ ($<10^{-6}$ $M$), fail to exhibit latex bead movement. Thus $Ca^{2+}$ is apparently required for flagellar surface movements. The motor for this movement has not been identified or located.

## B. The Rotary Axostyle

The guts of termites and wood-eating roaches are filled with many different protozoan species. One group, the devescovinid flagellates, has complex morphologies that serve as excellent landmarks for the investigator (Fig. 18). Internal landmarks include the nucleus and the Golgi apparatus, which wraps around the posterior part of the nucleus and part of the axostyle. These organelles form a complex coupled during axostyle rotation. External landmarks include an asymmetrical papilla at the tip and several different species of exosymbiotic bacteria with distinct distributions over the surface of individual devescovinid flagellate cells. Some of the devescovinid flagellates rotate by mechanisms that probably involve the flagella or other external motor mechanism. At least one species has a rotary axostyle.

The axostyle extends from the anterior to the posterior poles of the cell. Axostyles are composed of numerous microtubules arranged similar to the axostyles in heliozoan axopodia. Here the microtubules are arranged by crosslinks in a spiral pattern when viewed in cross-section (Fig. 19). Surrounding the axostyle is a sheath containing microfilaments aligned perpendicular to the axostyle axis. Outside this sheath is a region called the **girdle,** consisting of

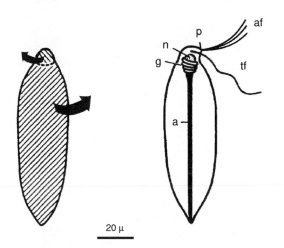

20 $\mu$

**FIG. 18.** Rotary motor of some devescovinid flagellates. The anterior pole ("head") of the cell spins in one direction relative to the cell body. The axostyle (a) in the cytoplasm of the cell also rotates in one direction with the "head." This organelle is associated with the nucleus (n), which is encircled by the Golgi complex (g) at the anterior of the cell. Three anterior (af) and one trailing (tf) flagella arise from the papilla (p). The cytoplasmic organelles and flagella serve to visualize the rotation of the anterior pole. (Adapted from Tamm, 1976, with permission.)

a finely granular inner layer and an outer layer filled with vesicles and inclusions. The outer layer of the girdle remains stationary during axostyle rotation.

Flagella that emerge from the anterior papilla exhibit normal flagellar waves. Independent of those flagellar movements, the cells exhibit another type of movement that involves the rotation of the anterior pole (head) relative to the rest of the cell (body). The parts of the cell that rotate include the cell surface membrane at the anterior pole, the flagella, the nucleus–Golgi complex, the axostyle, the sheath, and the inner layer of the girdle. Rotation rate can be as high as 1.5 sec per turn. The drive shaft is the rigid axostyle, which does not contract or undulate as do axostyles in some other flagellates. This devescovinid axostyle rotates in only a clockwise direction, looking at the cell from the anterior. If the head is tethered and prevented from moving, the body rotates. The cell surface membrane between the head and body is called the **shear zone.** Here membrane domains continuously move past each other and must therefore be highly fluid and free of ectosymbiotes. Paralysis of flagellar activity by pH shock does not stop the rotational movements, thus, the force for rotation does not come from flagellar action.

Tamm graphically demonstrated that the axostyle generates a torque along its length. Following destruction of segments along the axostyle by a laser microbeam, the posterior half continues rotating and does so at higher speeds. The head rotates slower or stops. Thus, the piece of axostyle attached to the head, which represents a load, provides less force for rotation than the intact axostyle. The axostyle is analogous to the armature of an electric motor and the surrounding cytoplasm represents the field coils. Torque is probably developed between the sheath and the inner layer of the girdle since the sheath rotates with the axostyle. The absence of a sheath and girdle is correlated with the lack of rotary motors in devescovinids.

Experiments on glycerinated cell models demonstrate that reactivation of the rotary motor requires ATP and $Mg^{2+}$; $Ca^{2+}$ apparently does not affect motor activity. It has been suggested, but not proved, that force generation may be actin-based, arranged in one direction similar to cytoplasmic streaming in *Nitella*.

## IV. Summary

The $Ca^{2+}$–spasmin/centrin system does not use ATP for contraction, but it may be responsible for reextension. The mechanism for spasmoneme and centrin-based contraction cannot be explained by sliding; coiling of 4-nm filaments or more basic molecular conformational changes may be involved in this type of contraction. Spasmins and centrins are polymers that appear to require only the binding of $Ca^{2+}$ to change their lengths. The role of phosphorylation of centrin function is currently unknown. Much less is understood about this motility system compared with the actin–myosin and tubulin–dynein systems, but the demonstration of its distribution in diverse organisms makes it highly probable that increased attention will be given to this motility system in the future.

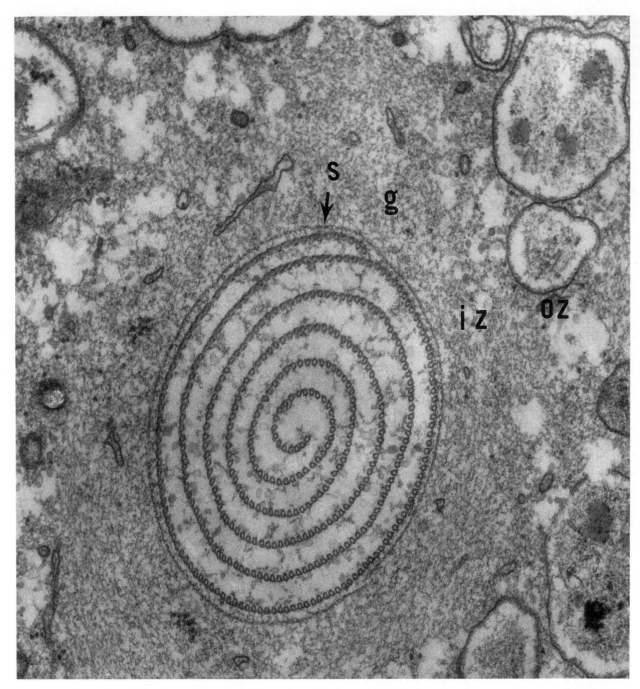

**FIG. 19.** The axopod of the devescovinid flagellate serves as a shaft for the rotary motor. Electron micrographs of a cross-section of the axopod illustrates its microtubular structure. The microtubules are organized within a sheath (s), which is surrounded by a region called the girdle (g). The granular inner zone (iz) of the girdle is distinct from the outer zone (oz), where vesicles and other inclusions are present. It is not known how the girdle–sheath interaction gives rise to rotation of the axostyle. (Courtesy of Sidney Tamm.)

The energy for the rotation of bacterial flagella is derived from a transmembrane ion gradient (e.g., proton-motive force) initially established by input of energy (e.g., pumps) to set up the gradient. Although energy from ATP hydrolysis may be involved in setting up an electrochemical gradient across the cell membrane, ATP does not directly act on the rotation of bacterial flagella. Cell motility using the energy of ion gradients has yet to be demonstrated in eukaryotic cells. The molecular genetics of bacterial flagella and proteins involved in chemotaxis are well elucidated, but only in a few species of bacteria. Unlike eukaryotic cilia and flagella, bacterial flagella are rigid helical structures that are rotated in one direction by the rotary motor associated with the cell membrane. Details on the mechanism by which the motor is driven by the ion-motive force are still unclear.

There are cell movements (e.g., gliding) that are not explained by the four recognized motility systems discussed in this and the previous chapter. It may be that many of these will be shown to be governed by one of the known mechanisms, but there remains the possibility for the discovery of novel ones.

## Bibliography

Abbanat, D. R., Leadbetter, E. R., Godchaux III, W., and Escher, A. (1986). Sulfonolipids are molecular determinants of gliding motility. *Nature* **324,** 367–369.

Allen, R. D. (1973). Contractility and its control in peritrich ciliates. *J. Protozool.* **20,** 25–36.

Amos, W. B. (1975). Contraction and calcium binding in the vorticellid ciliates. *In* "Molecules and Cell Movement" (S. Inoué and R. E. Stephens, Eds.), pp. 411–436. Raven Press, New York.

Cachon, J., and Cachon, M. (1985). Non-actin filaments and cell contraction in *Kofoidinium* and other dinoflagellates. *Cell Motil.* **5,** 1–15.

Ettiene, E. M. (1970). Control of contractility in *Spirostomum* by dissociated calcium ions. *J. Gen. Physiol.* **56,** 168–179.

Hazelbauer, G. L., Yaghmai, R., Burrows, G. G., Baumgartner, J. W., Dulton, D. P., and Morgan, D. G. (1990). Transducers: Transmembrane receptor proteins involved in bacterial chemotaxis. *In* "Biology of the Chemotactic Response" (J. P. Armitage and M. M. Lackie, Eds.), pp. 107–134. Cambridge University Press, London/New York.

Huang, B., and Mazia, D. (1975). Microtubules and filaments in ciliate contractility. *In* "Molecules and Cell Movement" (S. Inoué and R. E. Stephens, Eds.), pp. 389–409. Raven Press, New York.

Huang, B., Mengersen, A., and Lee, V. D. (1988). Molecular cloning of cDNA for caltractin, a basal body-associated Ca$^{2+}$-binding protein: Homology in its protein sequence with calmodulin and the yeast CDC31 gene product. *J. Cell Biol.* **107,** 133–140.

MacNab, R. M. (1990). Genetics, structure and assembly of the bacterial flagellum. *In* "Biology of the Chemotactic Response" (J. P. Armitage and J. M. Lackie, Eds.), pp. 77–106. Cambridge University Press, London/New York.

Manson, M. D. (1992). Bacterial motility and chemotaxis. *In* "Advances in Microbial Physiology" (A. H. Rose, Ed.), pp. 277–346. Academic Press, New York.

Melkonian, M., Beech, P. L., Katsaros, C., and Schulze, D. (1992). Centrin-mediated cell motility in algae. *In* "Algal Cell Motility" (M. Melkonian, Ed.), pp. 179–221. Chapman & Hall, New York.

Pate, J. L. (1988). Gliding motility in procaryotic cells. *Can. J. Microbiol.* **34,** 459–465.

Roberts, T. M. (1987). Fine (2–5 nm) filaments: New types of cytoskeletal structures. *Cell Motil. Cytoskel.* **8,** 130–142.

Routledge, L. M., Amos, W. B., Yew, F. F., and Weis-Fogh, T. (1976). New calcium binding contractile proteins. *In* "Cell Motility" (R. Goldman, T. Pollard, and J. Rosenbaum, Eds.), pp. 93–114. Cold Spring Harbor Laboratory Press, Cold Spring Harbor, New York.

Salisbury, J. L. (1989). Algal centrin: Calcium-sensitive contractile organelles. *In* "Algae as Experimental Systems" (A. W. Coleman, L. J. Goff, and J. R. Stein-Taylor, Eds.), pp. 19–37. Liss, New York.

Salisbury, J. L., and Floyd, G. L. (1978). Calcium induced contraction of the rhizoplast of a quadriflagellate green alga. *Science* **202,** 975–977.

Sanders, M. A., and Salisbury, J. L. (1989). Centrin-mediated microtubule severing during flagellar excision in *Chlamydomonas rheinhardii.* *J. Cell Biol.* **108,** 1751–1760.

Tamm, S. L. (1976). Properties of a rotary motor in eukaryotic cells. *In* "Cell Motility," Book C (R. Goldman, T. Pollard, and J. Rosenbaum, Eds.), pp. 949–967. Cold Spring Harbor Laboratory Press, Cold Spring Harbor, NY.

Tamm, S. L. (1978). Laser microbeam study of a rotary motor in termite flagellates. Evidence that the axostyle complex generates torque. *J. Cell Biol.* **78,** 76–92.

Thomas K. Akers

# 58

# Physiological Effects of Pressure on Cell Function

## I. Introduction

Two kinds of pressure can affect cell function. The first is a pure hydrostatic pressure and the second is the pressure of dissolved gases. Air-breathing organisms under compressed air have altered effects of their physiological function depending on the composition of the gaseous environment. The earliest attempts to understand the effects of pressure on living organisms did not distinguish between hydrostatic pressure and pressure of gases. Studies of specimens from extreme ocean depths usually involved decompression, as these were brought to the surface for examination.

There were a number of expeditions starting in the late nineteenth century, which began the study of the ocean, leading to oceanography as a new branch of science. The *Challenger* was one of the earliest British ships to be developed as a laboratory ship. The researchers crossed the Antarctic Circle, bringing up specimens from 26,850 feet and describing at least 4000 new species of animals. In the sixteenth century, Magellan made the first sounding of the ocean at 1200 feet, but did not do any biological dredging. The *Challenger* discoveries began a period of intense investigation of the ocean depths. A thorough review of studies of this kind was done by W. O. Fenn (1969).

## II. Pressure

### A. Gas Laws

To understand the role of dissolved gases in the effects of pressure on living cells, one must understand the physical laws that describe gas behavior. Boyle's law states that the pressure $(P)$ and volume $(v)$ of gas are inversely related for any given number of moles of a gas, with the product of the pressure and volume constant at any given temperature:

$$Pv = k \tag{1}$$

This law is paramount for understanding pressure changes in the volumes and pressures of body cavities.

Charles's law states that the volume of gas is directly proportional to the absolute temperature $(T)$, assuming that the pressure remains constant. This may be expressed as

$$\frac{V_1}{V_2} = \frac{T_1}{T_2} \tag{2}$$

where subscript 1 indicates initial conditions and subscript 2 the final conditions. If the gas volume remains constant, then the pressure is directly proportional to the absolute temperature. Since both Charles's law and Boyle's law describe interrelationships between temperature, pressure, and volume, these laws yield to give a general gas law, which is

$$Pv = nRT \tag{3}$$

where $n$ is the number of gas molecules, and $R$ is the gas constant (whose numerical value depends on the units used for the variables).

### B. Pressure Units

Until recently, there had been no standard way to represent pressure. Various authors have used feet of seawater (fsw), meters of seawater (msw), atmosphere absolute (ATA), bars, millimeters of mercury (mmHg), and pounds per square inch gauge (psig). Several years ago the Underseas Medical Society adopted the pressure numenclature defined by the International System of Units (SI). Therefore in papers published since 1990, pressure is expressed as a pascal (Pa = newton/meter squared), kilopascal (kPa), or megapascal (MPa). Table 1 summarizes the interconversions between these various units of pressure.

**TABLE I**  Pressure Conversion

| Atmosphere absolute (ATM ABS) (ATA) | Pounds per square inch gauge (PSIG) | Bars | Kilopascals (kPa) | Millimeters of mercury (mm Hg) | Feet of seawater (FSW) | Meters of seawater (MSW) |
|---|---|---|---|---|---|---|
| 1 | 0 | 1.0132 | 101.3247 | 760 | 0 | 0 |
| 2 | 14.6959 | 2.0264 | 202.6494 | 1520 | 33.08 | 10.13 |
| 0.98697 | 14.5037 | 1 | 100 | 750.064 | 32.646 | 10 |
| 1.068 | 1 | 0.06842 | 6.842 | 51.71 | 2.245 | 0.6842 |
| 1.0302 | 0.4823 | 0.03048 | 3.048 | 23.03 | 1 | 0.3048 |

*Note.* International System of Units (SI): pascal (Pa) = newton × meter$^2$; kilopascal (kPa) or megapascal (MPa) is used at present.

## III. Problems

Problems with dissolved gases in tissues became evident in caisson work and with dry hyperbaric facilities. When ventilation engineers were trying to provide comfortable conditions using standard principles of heat transfer and air conditioning, it became apparent that there were various problems associated with spending time in high-pressure gaseous environments. The alterations due to gaseous environments can be subdivided into several groups as follows: oxygen toxicity, inert gas narcosis, and high-pressure nervous syndrome.

### A. Oxygen and Toxicity

The first report that breathing pure oxygen caused pathological effects came from Priestley's studies. In 1775, Priestley discovered oxygen and subsequently noted that candles burned brighter and faster in oxygen than in air. He concluded that animals breathing pure oxygen would "live out too fast and the animal powers be too soon exhausted in this pure kind of air." Lavoisier, in 1783, noted that breathing pure oxygen caused pneumonia in experimental animals, and he proposed that oxygen caused pulmonary congestion.

Before 1900, Bert and Smith observed that experimental animals in hyperbaric oxygen exhibited chronic convulsions and death. Autopsies revealed blood-congested lungs. These toxic effects of hyperbaric oxygen in the presence or absence of inert gases (e.g., nitrogen) were attributed to cellular attacks at neuronal and pulmonic sites, respectively. For the past 70 years, the quest to understand oxygen toxicity has centered on Bert and Smith's initial observations that oxygen tensions above those of air cause fatal pulmonary lesions. Excellent reviews describing many hyperbaric oxygen experiments and data can be found in the literature (Bean, 1945). The need for safe space cabins and undersea environments has promoted an increase in research on oxygen effects on biological tissues in animals chronically exposed to oxygen in the range of 160–760 mm Hg. As a result, much information is available about the effect of hyperoxic exposures, with and without inert gases, on cell cultures and on animals at low altitude and at sea level pressures. Research on the clinical application of hyperoxic treatment has also expanded to include hyperoxic management of microbial infections, tumor treatment in conjunction with X-ray therapy, tissue preservation, and surgery.

Until recently, it was not known whether animals could survive exposure to an environment lacking in diluent gas. The answer came from MacHattie and Rahn (1960), who reported that mice survived for 51 days in pure oxygen at 197 mm Hg without difficulty. Several animals died within 48 hr of exposure with pulmonary atelectasis, but significant numbers were conceived, born, and raised in the absence of inert gas.

Long-term oxygen exposure at 160–450 mm Hg produces no toxic effects on growing mice, rats, and adult men. Deleterious oxygen effects have been produced by exposure to oxygen above this range. Pulmonary edema and congestion occur in animals kept at 258 mm Hg oxygen for 92 days.

Some species, such as cold-blooded animals and birds, appear to be resistant to hyperoxia. It is thought that low body temperature offers some protection to poikilotherms, whereas the complex lung structure offers protection for the birds.

Deleterious oxygen effects have been reported in lysosomal enzymes (Menzel *et al.*, 1967), lungs, and metabolic pathways (Larsen, 1968). Ross and Akers (1976) demonstrated that exposure to oxygen at 600 mm Hg for 24 days produced gross structural changes in the lung. Akers and Crittenden (1988) further demonstrated that alveoli in guinea pigs subjected to partial pressure of oxygen of 500 mm Hg for 6 days showed major changes in basic structure, including alteration in the endothelial cells and epithelial cells. There was some degree of protection when the animals were pretreated and maintained on a reserpine regimen during the pressure exposure, thus implicating the sympathetic nervous system in the pulmonary oxygen toxic effects.

Fenn (1967) suggested that most animals normally live in a critical oxygen range, so that both increases and decreases in oxygen are harmful. Attempts were made to add small quantities of nitrogen to the oxygen environment or interrupt the oxygen exposure with brief exposures of air.

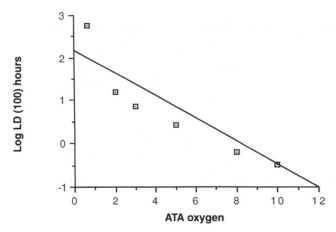

**FIG. 1.** Oxygen toxicity. Time to lethality is represented as the log of the LD 100 in hours ($r = 0.838$).

Both methods are thought to alleviate the onset of oxygen poisoning. Brauer *et al.* (1966) tried high-altitude acclimatization of rodents before exposing them to 825 mm Hg of oxygen and found that the onset of pulmonary damage was delayed and the threshold of central nervous system sensitivity to oxygen was lowered.

The inverse relationship between survival times and oxygen pressures has been validated many times. Thompson *et al.* (1970) showed that in the presence of hyperbaric inert gas, rat survival time is considerably shortened. There is a shift in the rat hyperbaric nitrogen/oxygen survival curves to the left of pure oxygen curves, which agrees with Fenn's reports that $N_2$-$O_2$ or Ar-$O_2$ shortens survival time of fruit flies (Fig. 1).

During the past 20 years, research has done much to heighten our awareness of the role of oxygen radicals in various pathophysiological processes and human diseases. This is true for hyperoxic pressure effects reviewed in the preceding paragraphs. Oxygen not only induces trauma, ischemia, or tissue injury at hyperbaric conditions, but also has been linked to many chronic diseases at much lower levels. The critical molecular components of the free radical biology and interrelations of these components that are tied to tissue injury have been studied extensively. An excellent review of this area can be found in the proceedings of an Upjohn Symposium of Oxygen Radicals in Tissue Injury (Halliwell, 1987).

## IV. Inert Gas Narcosis

Additional observations of animals and humans have indicated that pressure may enhance the narcotic effect of certain weak anesthetic agents. Indeed, if the partial pressure of nitrogen becomes high enough, it may itself produce a narcosis (Bennett, 1975).

From observations of the effects of elevated ambient pressures on animals and man, several principles have been established. It is known that oxygen is necessary in the pressurized gas, yet hyperbaric oxygen itself is toxic (Clark and Lambertsen, 1971). Thus, a diluent gas is necessary.

This diluent gas is nitrogen in the normal atmosphere. However, nitrogen at hyperbaric pressure produces a narcosis, often termed "rapture of the deep." Other inert gases have therefore been substituted for nitrogen, but many of these also produce a narcosis. The problem of nitrogen narcosis may be partially circumvented by the use of helium as the diluent gas, since it is not narcotic at normal physiological pressure. However, helium also produces problems related to pressure, termed the high-pressure neurological syndrome (Miller *et al.*, 1967).

Inert gas narcosis is an intoxication with signs and symptoms similar to those of alcoholic inebriation or anesthesia. The symptoms of this type of narcosis include euphoria, retardation of higher mental functioning, slowing of responses to sensory stimuli, amnesia, fixation of ideas, and a possible decrement in coordination and motor functioning. Junod in 1835, Green in 1861, and Bert in 1878 were the first to report the occurrence of narcosis associated with breathing of compressed air. With the use of compressed air by divers and caisson workers, it was noted that decompression back to the surface pressure resulted in a rapid cessation of the narcotic symptoms. Behnke *et al.* (1935) suggested that nitrogen ($N_2$) was producing the narcosis.

The cause of inert gas narcosis is complex. Inert gas narcosis apparently involves both the nature of the gas breathed and its pressure. Oxygen has been ruled out as the gas responsible for narcosis.

One theory of inert gas narcosis holds that carbon dioxide ($CO_2$) retention is responsible (Bean, 1950). An increased arterial pH during compression was noted and attributed to carbon dioxide. Measurement of alveolar carbon dioxide also was elevated, suggesting that ($CO_2$) might be the cause of compressed air intoxication; however, subsequent experiments failed to support the ($CO_2$) theory.

### A. Meyer–Overton Hypothesis

Since the suggestion of Behnke that $N_2$ was responsible for inert gas narcosis, the inert gas theory has been the one most widely supported. The theory takes its basis from the Meyer–Overton hypothesis relating the narcotic potency of a gas to its lipid solubility. This theory has been refined to state that narcosis may be produced by any gaseous or volatile substance that may penetrate into the cell lipids to a precise molar concentration that is specific for each gas. Based on this hypothesis, it is possible to relate the narcotic potency of a gas to its lipid solubility.

Since the Meyer–Overton hypothesis, many other theories have attempted to explain the mechanisms whereby inert gases and volatile anesthetics may induce a state of narcosis or anesthesia. These theories basically fall into two main categories: those proposing the lipid phase as the site, and those proposing the aqueous phase.

### B. Aqueous Phase Hypothesis

Theories expounding the aqueous phase were first suggested by Pauling (1961) and Miller (1961). Pauling suggested that clathrates, hydrated crystalline structures, were induced by the inert or anesthetic gas atoms. Miller, on the other hand, suggested that inert gases or anesthetics may increase the area of highly structured water surrounding the dissolved gas molecules. Both of these theories imply that the increased structuring of the aqueous phase surrounding membranes may trap ions associated with impulse conduction, thereby lowering conductance and altering membrane function. Research has provided little support for the aqueous phase as the critical site for anesthetic action.

### C. Critical Volume Hypothesis

Evidence continues to accumulate indicating that the lipid phase of membranes is the site of the narcotic action of inert gases and general anesthetics. Based on this accumulating evidence, a new hypothesis has been proposed as an extension of the Meyer–Overton hypothesis, known as the **critical volume hypothesis.** It has been calculated that an anesthetic concentration in the lipids, estimated at 0.05 $M$, will produce a thickening in the membrane of about 0.4%. It is this thickening that may disrupt membrane processes and thus produce the narcosis.

An antagonism between pressure and anesthetics has been known for some time. Narcosis in luminous bacteria, tadpoles, newts, and mice has been reversed by application of helium pressure. It is believed that this pressure reversal of anesthesia is a consequence of the pressure compressing the membranes back to the original volume, thus reducing thickness.

This theory has been extended by Stern and Frisch (1973) to suggest that it is the "free volume" in the lipids of the cell membrane that may induce the narcotic potency of inert gases. According to this theory, not only lipid solubility but also the environmental temperature, thermal expansivity, compressibility of the liquid phase, and the hydrostatic pressure are involved in the narcosis. Inert gas narcosis will occur when the concentration of the dissolved gas increases the lipid phase free volume greater than some critical value volume.

The lipid solubilities of various inert gases differ and depend on the hydrostatic pressure gradient. There is excellent agreement between the lipid solubility of the inert gases and their narcotic potencies. Based on the relative narcotic potencies, nitrogen should produce a narcosis at one-fourth the partial pressure that is needed for hydrogen. Furthermore, helium should not produce a narcosis at any partial pressure, and it apparently does not.

### V. Helium Effects

Because of the low lipid solubility of helium, it is not considered narcotic. On the contrary, at increased pressure, helium produces compression of the lipid phase instead of an expansion as seen with nitrogen. Although helium might have a small narcotic potency at certain hyperbaric pressures, the narcosis is masked by a more dominating compression effect.

Because of low tissue solubility, as well as the rapid diffusion of helium across membranes, helium has been used as a diluent gas as a replacement for nitrogen since the early 1900s. The rapid diffusion of helium should make it a better gas for rapid decompression.

Two further characteristics of helium are its low density and its high thermal conductivity. At increased pressures, the number of gas molecules occupying the same space increases, and thus the density of the inspired gas increases. This results in an increased respiratory effort required under hyperbaric conditions. The low density of helium will partially counteract this effect and relieve the excessive respiratory effort. High thermal conductivity of helium increases the convective heat transfer away from the body.

### VI. Hydrostatic Pressure

Hydrostatic pressure may be produced by increased pressure of a liquid such as water or by hyperbaric helium, which has been shown to produce compression without significant expansion of any cellular phase. High hydrostatic pressure improves the mechanical performance of striated muscle by inhibiting the catalytic activity, substrate binding, and mechanical work of three mechanisms. These include the SR calcium cycle, the SL calcium cycle, and the cross-bridge cycle. Each of these mechanisms involves an ATPase. The myosin ATPase is a cytosolic enzyme not associated with a lipid membrane. Therefore, the inhibition of ATP hydrolysis may involve more than alterations in the critical volume of membranes. Pressure inhibition of ATPase activity (particularly Na–K pump) leads to an accumulation of intracellular calcium, resulting in a pressure-induced inotropic effect (Hogan and Besch, 1993).

It was believed that helium could produce a narcosis much like that produced by nitrogen when a sufficient pressure was reached (approximately 31–43 ATA). However, Brauer *et al.* (1966) observed convulsions before any narcosis when helium was used. Likewise, Miller studied the reversal of anesthesia and observed convulsions produced by helium at high pressure. The set of symptoms produced by increased pressure is collectively termed the **high-pressure neurological syndrome** (HPNS).

### VII. High-Pressure Neurological Syndrome

Anesthetic gases antagonize the neurological effects of high-pressure nervous syndrome, just as high pressure antagonizes the effects of anesthetics. This suggests that both pressure and anesthetic agents exert their opposing effects at the same hydrophobic site.

#### A. Site and Mechanism of HPNS

The symptoms of HPNS begin with motor disturbances such as tremors. This early stage may be followed by isolated myoclonic jerks and generalized clonic or tonic convulsions. The tremor and convulsion thresholds show considerable individual variation.

The search for the causes of HPNS has led to the finding of many diverse factors. Among these are hypoxia, hypercapnia, hypothermia, and gas-induced osmosis as well as the direct effects of hydrostatic pressure. The effects of hydrostatic pressure on physical, chemical, and biological systems, however, seem to best explain the production of HPNS. Gases under high pressure may form solid hydrates, thus producing increased ordering of water molecules (Featherstone *et al.,* 1971). Pressure tends to favor ionization and alter pH; ionization leads to a decrease in volume (Johnson and Eyring, 1970). Likewise, molecular volumes may be altered by pressure, thus causing structure changes in large molecules such as enzymes. In any enzyme–substrate reaction involving an increase in volume, hydrostatic pressure would tend to be inhibitory. If such reaction resulted in a volume decrease, the result would be stimulation. These pressure-dependent changes in physical and chemical reactions may be involved in the mechanism of HPNS. Although hydrostatic pressure produces physical and chemical changes, the mechanisms by which these changes alter physiological function to produce a set of symptoms such as HPNS is still not understood.

The question of what functions within the nervous system are altered to produce the symptoms characteristic of HPNS still remains. The two mechanisms within the nervous system responsible for information transfer, namely, axonal conduction and synaptic transmission, are the most likely sites of action of pressure-induced effects. Nerve conduction has been implicated in these pressure-related changes in function. Grundfest and Cattell (1935) and Spyropoulos (1957) have demonstrated increased action potential amplitude, increased conduction velocity, and increased excitability in isolated frog nerves and giant squid axon under high pressure. However, such nerve conduction changes occur only at pressure ranges considerably above the pressure at which HPNS first occurs (Fig. 2).

## B. Membrane Effects

Based on the decreased safety factor for transfer of information across the synapse, this site could possibly be important in the etiology of HPNS. Yet, only in recent years have there been any reports concerning the effects of pressure on synaptic receptors. Akers and Carlson (1975) reported that the appearance of actions of acetylcholine and norepinephrine on rabbit duodenum was delayed by application of 30 ATA helium. They proposed that a change in the receptor molecule may be responsible. Athey and Akers (1978) demonstrated that the acetylcholine receptor complex was altered under pressure, and they proposed a model suggesting conformational changes in the receptor–ligand complex.

It has been shown that high (up to 200 ATA) hydrostatic pressure depresses (1) amplitude of excitatory junctional potentials, (2) frequency facilitation, and (3) post-tetanic potentiation. These results suggest that reduced transmitter release occurs.

The amplitude of postsynaptic potentials at a known synapse in the aplasia central nervous system can be modified by hydrostatic pressure. Postsynaptically controlled functions, including response to acetylcholine in the course of time decay of the synaptic response, remain unaffected

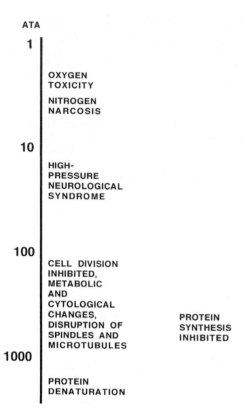

**FIG. 2.** Pressure spectrum for various cellular effects. Pressure in the Philippine Trench is approximately 1000 ATA.

by hydrostatic pressure. But, frequency facilitation post-tetanic potentiation that results from presynaptic processes is altered by pressure in a manner similar to that of agents that block transmitter release (Parmentier *et al.,* 1981).

## C. Cell Division

One effect of pressure is inhibition of cell division. The rate of cell division of *Tetrahymena* is reduced at pressures greater than 100 ATA, but gases inhibited division to a lesser degree than pure hydrostatic pressure. Hydrogen produced the same effect as hydrostatic pressure at 130 ATA or more, and the effect of helium appeared at 175 ATA. There was an inhibition of cell division in *Tetrahymena* at 60 ATA nitrogen, which was reversed by addition of 40 ATA helium, so that cell division was nearly normal at 100 ATA total pressure. The combination of 30 ATA nitrogen and 70 ATA helium, however, caused severe inhibition. Cell division of *Escherichia coli* is also inhibited by hydrostatic pressure.

## D. Subcellular Effects

Hildebrand and Pollard (1972) demonstrated the effect of high hydrostatic pressure on protein synthesis. They used polyuridylic acid to direct the synthesis of polyphenylalanine and showed that there was an apparent increase in synthesis at 100 ATA. There was a sharp reduction in synthetic rate with increased pressures up to 640 ATA (95% inhibition).

When pressure is increased in steps to 3000 psig at temperatures below 22°C, predictable reductions in spindle equilibrium occurred, demonstrated by a change in birefringence and retraction of the metaphase arrested spindles. Hydrostatic pressure of 68 ATA prevents reversal ("avoiding") motility in paramecium, an effect on the membrane processes (Otter and Salmon, 1979). Sebert *et al.* (1987) studied the effects of hydrostatic pressure on kinetic processes in fish brought up from deep depths and showed that 101 ATA applied for 3 hr caused an increase in adenosine monophosphate (AMP) and a decrease in adenosine triphosphate (ATP) in muscle. Cossins and MacDonald (1989) studied the adaptation of biological membranes to temperature and pressure using fish from the deep and demonstrated that there are adaptations that take place in the nerve cell membranes. They believed that these fish display a membrane adaptation that largely overcomes the ordering effects of pressure, and thus prevents the high-pressure nervous syndrome as seen with helium and hydrostatic pressure in other fish. Similarly, Bourns *et al.* (1988) studied the microtubule and actin organization in cultured epithelial cells exposed to pressures from 1 to 610 ATA and found that exposure to pressures of 290 ATA or greater caused cell rounding and retraction from the substrate. This response became more pronounced with increased pressure, microtubules vanished from the cells at 320 ATA, and most actin stress fibers disappeared by 290 ATA.

The red blood cell has proven to be a good experimental model for pressure studies. Paciorec (1982) showed that there was a stress on human erythrocytes after they had been exposed in the human during a dive at pressures up to 60 bar even for some time later. About 20% of the total red cells formed either echinocytes or smooth cup-shape stomatocytes. This did not occur to any of the red cells at pressures below 40 bar and did not apparently increase much in distortion from 70–350 bar. Scanning electron micrographs of erythrocytes in pulmonary capillaries revealed no effect of hyperoxia on cell shape, but copper deficiency caused erythrocyte shape changes that can be described as stomatocytic, echinocytic, or knizocytic (Akers and Saari, 1993).

## E. Effects on Seaweed

Shameel (1976) showed that there were changes in the cellular morphology of *Bryopsis plumosa* under hydrostatic pressure and temperature. Using 200–800 atm and 5–15°C applied for 5 hr on the assimilating filaments of the *B. plumosa*, the author found that cell morphology was increasingly affected by rising pressure. Aftereffects lasted up to 3 weeks. These seaweeds had physiologic limits much lower than those of other seaweeds and they seemed to be extreme baraphobes. Masuda and Albright (1978) showed that hydrostatic pressures >400 atm caused the release of cellular components of *Vibriomarinus* without fracture of the cell membrane. The leakage was believed to be due

to increased affinity of the transport system for the substrate at increased pressures.

## VIII. Anesthetics

### A. Molecular Mechanisms

A classical view of the molecular mechanism of gaseous anesthetics basically states that they operate by a dissolution of membrane lipids in the cells. This view points to a relationship between the potency of anesthetics and their lipid solubility coefficient. As indicated earlier, fluidization of the membrane is involved, with increased movement of lipid molecules that disrupt the normal lipid protein interactions, which in turn reduce the function of nerve and cell membranes. Hunt (1985) has developed a model of the mechanism of anesthetic action that proposes that there is an optimal phase relationship of membrane lipids so an ideal phase of solid gel and fluid liquid crystals exists in the region of membrane lipids. The anesthetics are known to lower the phase transition temperature of phospholipids, that is, causing melting so they disrupt the optimal phase relationship in the lipid and indirectly disrupt the function of the membranes. All the other theories ignore the phospholipid bilayer of the plasma membrane and propose that anesthetics disrupt membrane function by a direct interaction with the proteins of the membrane. In comparison to anesthetic agents, membrane proteins are very large and contain many hydrophobic regions.

### B. Similarity of Alcohol and Gas Anesthetics

Gas anesthetics and alcohols, such as ethanol, seem to work by binding to nonspecific and nonstereospecific receptors (Seeman, 1974). Gas anesthetics and alcohols, which produce an effect in the millimolar range or below, are considered to be less potent than other anesthetic drugs such as barbiturates and benzodiazepines, which bind a specific receptor, have considerable stereospecificity, and need only micromolar amounts within the membrane to achieve anesthesia (Miller *et al.,* 1972). Nitrous oxide, argon, and nitrogen gases also have similar characteristics in that they are colorless, odorless, and relatively nontoxic with brief exposure. These gaseous anesthetics are not metabolized because they enter and exit through the lungs and there are no confounding effects on different rates of metabolism. Gaseous anesthetics such as nitrogen, nitrous oxide, and argon are low-lipid-soluble gases, and for that reason, high pressure is needed to obtain sufficient concentration in the membrane for anesthesia (Fig. 3).

### C. Helium Antagonism

Helium has an antagonizing effect on a membrane opposing the effects of ethanol and anesthetics. It causes the membrane to become more rigid or orderly. The higher the pressure, the more effective the reversal of anesthesia. The reversal of anesthetics is apparently due to high pressure of the helium and is not due to pharmacological prop-

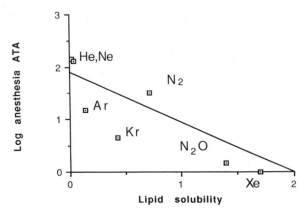

**FIG. 3.** Effect of lipid solubility of various gases on anesthetic potency. The pressure required to produce anesthesia is given in log ATA. Lipid solubility is the olive oil/gas partition coefficient at 37°C ($r = 0.763$).

erties of the anesthetic drug. High pressure is also known to act as an antagonist to ethanol withdrawal syndrome.

Brauer and Way (1970), using high pressure, were able to obtain an anesthetic concentration within a membrane of nitrous oxide, nitrogen, and argon. These gases have different effects because they have different solubilities. Argon needs 16 atm of pressure to produce loss of righting reflex (LORR) in mice, whereas 36 atm is required for nitrogen and 2 atm for nitrous oxide. High pressure acts as a vehicle for the anesthetics to enter into or dissolve in the membrane. However, there are two processes: (1) An increase in pressure allows more gas to dissolve into the membrane by binding to a protein, or membrane expansion occurs, or a combination of these two in the millimolar concentration range until eventually anesthesia occurs; and (2) as pressure increases so do the antagonistic effects, constriction and rigidness of the membrane, thus opposing and preventing anesthesia. With an extremely lipid-insoluble gas such as helium, under hyperbaric conditions, the high-pressure antagonizing effects override the anesthetic effects as a result of obtaining a very low membrane concentration. Therefore, the depressant effects of dissolved helium in the membrane at high pressure are offset by the hyperexcitability resulting from that pressure. The more lipid-soluble gases have anesthetic domination above certain pressures so that anesthesia occurs despite the antagonistic effect of the high pressure. This means that the anesthesia caused by 30 atm of nitrogen in mice is reversed by the use of helium above 100 atm. Alternatively, HPNS can be antagonized or reversed by an administration of nitrogen or nitrous oxide.

LORR due to ethanol was reversed with 6–12 atm of heliox. High pressure has been shown to reverse membrane expansion and fluidity caused by ethanol and other anesthetic agents. The higher the pressure, the more effective the reversal of anesthesia. The reversal anesthetics appears to be due to high pressure of the helium and not due to pharmacological properties of the anesthetic drug. High pressure has been demonstrated to act as an antagonist in its effect on ethanol withdrawal syndrome. When helium,

12 atm absolute, was applied during withdrawal of ethanol-dependent mice, the high pressure increased the intensity of the withdrawal (Alkana *et al.*, 1985). A whole series of genetic mouse models has been developed to study alcohol effects, such as loss of righting reflex and long sleep effects. The mechanism of LORR between ethanol and the gaseous anesthetics, nitrous oxide, argon, and nitrogen, are similar. Long sleep bred mice were significantly more hypothermic than were any of the short sleep bred mice to all three gases (Faber *et al.*, 1991).

## VIII. Summary

Hydrostatic pressure apparently affects the cell primarily by its action on protein structures within the cell. There is evidence that protein structures become deformed by the pressure, thus inhibiting their functions. Inhibition of the SR calcium cycle, the SL calcium cycle, and the cross-bridge cycle in striated muscle is involved in inhibition of various ATPases. Therefore, enzymatic function of the cell is reduced by extremely high pressure, as is transport through membranes. There is evidence that receptor proteins are altered and are delayed in undergoing conformational changes when transmitters are attached.

Cell division is inhibited in both plant and animal species. The inhibition can occur in two ways: (1) Inert narcotic gases (40 ATA) block division, but can be reversed by 60 ATA of helium pressure or hydrostatic pressure; and (2) division is blocked by helium and/or hydrostatic pressure above 175 ATA.

Protein synthesis is blocked at 640 ATA and above. Spindle microtubules become depolymerized at 200 ATA.

Membrane alterations occur at 680 ATA in paramecia, reducing their ability to avoid noxious stimuli.

Inert gases produce a general hyposensitivity of cell membranes by producing an expansion of the cell membrane, thus producing a narcosis similar to alcohol narcosis. The rate and severity of this narcosis are dependent to some extent on the lipid solubility of the gases.

Helium, on the other hand, produces a compression of cell membranes, thus producing a hypersensitivity of the cell. This becomes critical in multicellular animals when the hypersensitivity–hyperexcitability of cells produces an increased rate of conduction across synapses, thus disorganizing critically timed nervous loops. This results in the high-pressure neurological syndrome.

Oxygen in excess of 0.5 ATA produces toxic responses, most likely because of increased free radicals in the oxygen environment attacking protein ends.

## Bibliography

Akers, T. K., and Carlson, L. C. (1975). The changes in smooth muscle receptor coupling of acetylcholine and norepinephrine at high pressure. *In* "5th Symposium on Underwater Physiology" (C. J. Lambertsen, Ed.). FASEB, Bethesda, MD.

Akers, T. K., and Crittenden, D. J. (1988). Into deepest lung with drug and camera. *Neurosci. Biobehav. Rev.* **12,** 315–332.

Akers, T. K., and Saari, J. T. (1993). Hyperbaric hyperoxia exaggerates respiratory membrane defects in the copper-deficient rat lung. *Biol. Trace Element Res.* **38**, 149–163.

Alkana, R. L., Finn, D. A., Galleisky, G. G., Syapin, P. J., and Malcolm, R. D. (1985). Ethanol withdrawal in mice precipitated and exacerbated by hyperbaric exposure. *Science* **229**, 772–774.

Athey, G. R., and Akers, T. K. (1978). An analysis of frog neuromuscular function at hyperbaric pressures. *Undersea Biomed. Res.* **5**, 127–131.

Bean, J. W. (1945). Effects of oxygen at increased pressure. *Physiol. Rev.* **25**, 1–72.

Bean, J. W. (1950). Tensional changes of alveolar gas in reaction to rapid compression and decompression, and the question of nitrogen narcosis. *Am. J. Physiol.* **161**, 417–425.

Behnke, A. R., Thomson, R. M., and Motley, E. P. (1935). The psychological effects of breathing air at 4 atmospheres pressure. *Am. J. Physiol.* **112**, 554–558.

Bennett, P. B. (1975). Inert gas narcosis. *In* "The Physiology and Medicine of Diving" (P. B. Bennett and D. H. Elliott, Eds.), pp. 207–230. Williams & Wilkins, Baltimore.

Bert, P. (1878). "La Pression Barometrique." Masson, Paris.

Bourns, B., Franklin, S., Cassimeris, L., and Salmon, E. D. (1988). High hydrostatic pressure effects in vivo: Changes in cell morphology, microtubule assembly, and actin organization. *Cell Motil. Cytoskel.* **10(3)**, 380–390.

Brauer, R. W., and Way, R. O. (1970). Relative narcotic potencies of hydrogen, helium, and nitrogen and their mixtures. *J. Appl. Physiol.* **29**, 23 31.

Brauer, R. W., Johnsen, D. O., Pessotti, R. L., and Redding, R. W. (1966). Effects of hydrogen and helium at pressures to 67 atmosphere on mice and monkeys. *Fed. Proc.* **25**, 202.

Clark, J. M., and Lambertsen, C. J. (1971). Pulmonary oxygen toxicity: A review. *Pharmacol. Rev.* **23**, 37–133.

Cossins, A. R., and MacDonald, A. G. (1989). The adaptation of biological membranes to temerature and pressure: Fish from the deep and cold. *J. Bioenerg. Biomembr.* **21**, 115–135.

Faber, J. M., Akers, T. K., Belknap, J. K., and Aasen, G. H. (1991). Commonalities between gas anesthetics (nitrous oxide, nitrogen, and/or argon) and ethanol intoxication in HOT and COLD selection line mice. *Biomed. Sci. Instrum.* **27**, 127–130.

Featherstone, R. M., Hegeman, S., and Settle, W. (1971). Effects of inert gas pressures on protein structure and function. *In* "Proc. 4th Symp. on Underwater Physiol." (C. J. Lambertsen, Ed.), pp. 95–100. Academic Press, New York.

Fenn, W. O. (1967). Interactions of oxygen and inert gases in drosophila. *Resp. Physiol.* **3**, 117.

Fenn, W. O. (1969). The physiological effects of hydrostatic pressures. *In* "The Physiology and Medicine of Diving and Compressed Air Work" (P. B. Bennett and P. H. Elliott, Eds.). Ballière, Tindall & Cassell, London.

Grundfest, H., and Cattell, M. (1935). Some effects of hydrostatic pressure on nerve action potentials. *Am. J. Physiol.* **113**, 56–57.

Halliwell, B. (Ed.) (1987). Oxygen radicals and tissue injury. *Proc. of Upjohn Symposium.* FASEB, Bethesda, MD.

Hildebrand, C. E., and Pollard, E. C. (1972). Hydrostatic pressure effects on protein synthesis. *Biophys. J.* **12**, 1235–1250.

Hogan, P. M., and Besch, S. R. (1993). Vertebrate skeletal and cardiac muscle *In* "Advances in Comparative and Environmental Physiol-

ogy, Vol. 17: Effects of High Pressure on Biological Systems" (Alister Macdonald, Ed.). Springer-Verlag. New York/Berlin.

Hunt, W. A. (1985). "Alcohol and Biological Membranes." Guilford Press, New York.

Johnson, F. H., and Eyring, H. (1970). The kinetic basis of pressure effects in biology and chemistry. *In* "High Pressure Effects on Cellular Processes" (M. Zimmerman, Ed.). Academic Press, New York.

Larsen, J. A. (1968). The effect of oxygen breathing at atmospheric pressure on the metabolism of glycerol and ethanol in cats. *Acta Physiol. Scand.* **73**, 186.

MacDonald, A. G. (1984). The effects of pressure on the molecular structure and physiological functions of cell membranes. *Philos. Trans. R. Soc. London B* **304**, 47–68.

MacHattie L., and Rahn, H. (1960). Survival of mice in absence of inert gas. *Proc. Soc. Exp. Biol. Med.* **104**, 772–776.

Masuda, K. V., and Albright, L. J. (1978). Hydrostatic pressure effects upon cellular leakage and active transport by *Vibrio marinus*. *Z. Allg. Mikrobiol.* **18**, 731–740.

Menzel, D. B., Lee, S. A., Shaw, A. M., Miquel, J., and Brooskby, J. A. (1967). Lysosomal enzymes in rats exposed to 100 percent oxygen. *Aerosp. Med.* **38**, 722–725.

Miller, K. W., Paton, W. D. M., Street, W. B., and Smith, E. B. (1967). Animals at very high pressures of helium and neon. *Science* **157**, 97–98.

Miller, K. W., Paton, W. D. M., Smith, E. B., and Smith, R. A. (1972). Physicochemical approaches to the mode of action of general anesthetics. *Anesthesiology* **36**, 339–351.

Miller, S. L. (1961). A theory of gaseous anesthetics. *Proc. Natl. Acad. Sci. USA* **47**, 1515–1524.

Otter, T., and Salmon, E. D. (1979). Hydrostatic pressure reversibly blocks membrane control of ciliary motility in Paramecium. *Science* **206**, 358–361.

Paciorec, J. A., (1982). *In vitro* pressure stressed human erythrocytes. *Undersea Biomed. Res.* **9**, Suppl., A13.

Parmentier, J. L., Shrivastav, B. B., and Bennett, P. B. (1981). Hydrostatic pressure reduces synaptic efficiency by inhibiting transmitter release. *Undersea Biomed. Res.* **8**, 175–183.

Pauling, L. (1961). A molecular theory of general anesthesia. *Science* **134**, 15–21.

Ross, B. K., and Akers, T. K. (1976). Scanning electron microscopy of normoxic and hyperoxic hyperbaric exposed lungs. *Undersea Biomed. Res.* **3**, 283–299.

Sebert, P., Barthelemy, L., Caroff, J., and Hourmant, A. (1987). Effects of hydrostatic pressure per se (101 ATA) on energetic processes in fish. *Comp. Biochem. Physiol. A* **86**, 491–495.

Seeman, P. (1974). The membrane expansion theory of anesthesia: Direct evidence using ethanol and a high-precision density meter. *J. Cell Biol.* **70**, 247–251.

Shameel, M. (1976). Changes in the cellular morphology of *Bryopsis plumosa* (Bryopsidophyceae) under hydrostatic pressure and temperature. *Pak. J. Bot.* **8**, 103–110.

Spyropoulos, C. S. (1957). Response of single nerve fibers at different hydrostatic pressures. *Am. J. Physiol.* **189**, 214–218.

Stern, S. A., and Frisch, H. L. (1973). Dependence of inert gas narcosis on lipid "free volume." *J. Appl. Physiol.* **34**, 366–373.

Thompson, R. E., Nielsen, T. W., and Akers, T. K. (1970). Synergistic oxygen-inert gas interactions in laboratory rats in a hyperbaric environment. *Aerospace Med.* **41**, 1388–1392.

Anthony L. Gotter, Marcia A. Kaetzel, and John R. Dedman

# 59
## Electrocytes of Electric Fish

## I. Introduction

**Electric fish** such as the marine **electric ray** (genus *Torpedo*) and the freshwater **electric eel** (*Electrophorus electricus*) are capable of generating powerful electrical discharges that can be measured in the water surrounding these animals. These fish use the production of bioelectricity as an effective mechanism to stun prey and ward off predators. Electrical discharges are generated by electric cells, called **electroplax** or **electrocytes,** that produce end-plate potentials and action potentials that are remarkably similar to the membrane potentials of neurons and myocytes. In fact, the membrane receptors, ion channels, and ATPases responsible for electric tissue electrophysiology are biochemically and functionally identical to those of mammalian muscle and nerve. For this reason, electrocytes have been used extensively as a specialized and appropriate model system for the study of excitable cell membrane electrophysiology and biochemistry. Due to the specialized nature of electric tissue, it has also been used as an enriched source of membrane proteins for biochemical studies. Previous chapters have described in detail the generation of acetylcholine (ACh)-mediated muscle end-plate potentials and the propagation of action potentials of nerve and muscle (see Chapters 23, 24, 27, 40, and 50). This chapter examines the anatomy and cellular morphology that electric fish have evolved in order to produce powerful electrical discharges. An electrophysiological and biochemical comparison is made between the electrocytes of the freshwater electric eel and the marine electric ray. The major contributions that electric tissue has made to the understanding of the electrophysiology and biochemistry of excitable membranes is also reviewed.

The shocking sensations produced by electric fish were undoubtably experienced by mankind long before the recording of scientific phenomena. Some of the first recorded reports of unusual effects produced by electrical discharges of electric fish were of the Nile river catfish, *Malapterus electricus.* Nile river fishermen reported unpleasant sensations when handling live *Malapterus,* or even the water-soaked nets containing the fish. Godigno, a seventeenth century Jesuit father, noted that dead fish could be induced to move when a live *Malapterus* was thrown among them (Grundfest, 1957). At the time when Ben Franklin and other investigators were experimenting with static electricity of the Leyden jar, the electric eel provided insight into the basic conductive properties of electricity. In 1775, John Walsh conducted numerous experiments, one of which involved 10 people holding hands in a circle where the first and last "subjects" touched the opposite ends of a moderately sized eel. All 10 people received a severe shock. The relative conductivities of various materials including glass, wood, silk, brass chains, and iron rods were then determined by holding these materials between two of the investigators and noting the severity of the electrical discharge. Although these experiments were likely to be very convincing to Walsh and his assistants, others doubted the electrical nature of the discharge from *Electrophorus* and *Torpedo.* The bioelectric nature of the discharge had not gained widespread acceptance until Du Bois-Raymond demonstrated that nerve and muscle were electrogenic (Grundfest, 1957). Since that time, the usefulness of *Electrophorus, Torpedo,* and other electric fish as models for excitable membranes has been realized.

## II. Anatomy of *Electrophorus* and Mechanism of the Electrical Discharge

Powerful electric fish possess a specialized anatomy and cellular morphology devoted to the production of electrical discharges. The electric eel is an excellent example of this specialization. It has been well characterized on the cellular and biochemical level and is used here to describe the production of bioelectricity. Figure 1A depicts the location of the electric organs within *Electrophorus*. The viscera are crowded into the rostral 20% of the animal; the remaining 80% is comprised predominantly of electric tissue and swimming muscles. The electric organs are confined to the ventral portion of this caudal region, whereas most of the

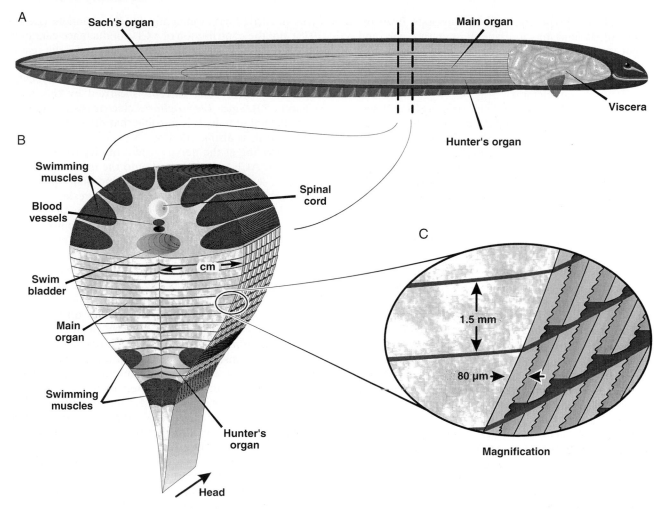

**FIG. 1.** Anatomy of the electric eel. (A) Diagram illustrating the anatomical orientation of electric organs. (B) A section through the middle portion of the eel, drawn such that the anterior surface is nearest the reader. (C) Columns of electrocytes extend the length of the electric organ. In this panel, the flatter, caudal surface of each electrocyte would be innervated by numerous electromotor neurons (not shown).

swimming muscles, major blood vessels, and spinal cord are situated in the dorsal one-third (Fig. 1B). The eel also possesses a tubular swim bladder that extends the length of the fish, and is positioned dorsal to the main electric organ and ventral to the spinal cord. The central nervous system consists of a small brain typical of teleost fish and spinal cord that extends down the length of the animal. The electrical discharge is coordinated in the central control nucleus in the medulla. Axons from these neurons of the brain project caudally and synapse on neurons of the electromotor nucleus of the spinal cord. **Electromotor neurons** radiate into the electric organ, innervating individual electrocytes (Bennett and Sandri, 1989). To generate the whole-animal electrical discharge, each electrocyte of the entire electric organ must be stimulated simultaneously. In other words, action potentials reaching proximal electrocytes of the electric organ must be delayed to varying degrees relative to more distal regions. Neurons innervating proximal electrocytes are smaller in diameter and conduct action potentials more slowly. Some of these neurons

wind their way to these rostral electrocytes, thereby slowing stimulation of this part of the electric organ, aiding in synchronous activation (Bennett, 1971). The delay may also occur in the electromotor nucleus of the spinal cord where central neurons synapse on electromotor neurons. Presumably, synaptic transmission is slower to electromotor neurons innervating proximal portions of the electric organ, whereas faster signaling occurs between neurons that innervate electrocytes near the tail (Szabo, 1961).

As depicted in Fig. 1, *Electrophorus* has three well-defined electric organs. The main organ is the largest and is responsible for voluntarily generating powerful high-voltage discharges. The main organ extends from behind the peritoneal cavity of the viscera, down the tail of the eel where it eventually gives rise to Sach's organ. This organ, along with Hunter's organ, generates repetitive low-voltage discharges and is thought to be involved in electrolocation of objects in the eel's environment. In the cross-sectional view of Fig. 1B, Hunter's organ is seen to be partially delineated from the main organ by two columns of skeletal muscles. Electric tissue develops from these columns of skeletal muscle tissue in immature eels, and is thought to arise from embryonic myocyte precursor cells (Keynes, 1961). As shown later, the membranes of electrocytes are biochemically and functionally very similar to skeletal muscle sarcolemma.

**Electrocytes** of both the main electric organ and Sach's organ are large ribbon-shaped cells. Each electrocyte extends laterally from the midline of the electric organ to the skin, a distance of up to 4 cm. They have a width of up to 1.5 mm and thickness of 80 $\mu$m. As seen in Fig. 1C, electrocytes are positioned one after another along their flat axis to give rise to long rectangular columns of cells running in the longitudinal axis of the eel. These columns are delineated and electrically insulated from one another by **connective tissue septa,** which help to maintain the physical structure of the electric organ. This stacked arrangement is a common feature of electric organs in electric fish, and enables greater voltages to be produced (see below). When viewed under light microscopy, electrocytes are seen as multinucleated syncytiums, similar to the skeletal muscle myocytes from which they are derived. Electrocytes in cross-section are seen to have one relatively flat posterior membrane relative to the other more undulated anterior membrane. The flat caudal membrane is innervated by ACh-releasing electromotor neurons that form synapses that are morphologically similar to motor end plates of skeletal muscle cells (Chapter 40).

Immunofluorescent localization microscopy and electrophysiological experiments have led to the understanding of how the electrical discharge is generated at the cellular level. Electrocytes use membrane receptors and ion channels polarized to either the innervated or noninnervated membrane in order to produce **transcellular potentials** that give rise to the discharge of the electric organ (Fig. 2A). The caudal innervated membrane contains acetylcholine receptors (AChRs), inward and outward rectifying $K^+$ channels, a high density of voltage-gated $Na^+$ channels, and only trace amounts of $Na^+,K^+$-ATPase. This membrane is both chemically and electrically excitable. That is, this surface of the electrocyte produces **action potentials** (APs) in response to artificial stimulation with AChR agonists or direct electrical stimulation. The noninnervated membrane, on the other hand, does not respond to these manipulations, since it has no AChRs or voltage-gated $Na^+$ channels. Instead, this membrane contains a high concentration of $Na^+,K^+$-ATPase and ion channels responsible for maintaining the **resting potential** of $-70$ to $-85$ mV. Evidence suggests that the resting current of *Electrophorus* electrocytes is carried predominantly by $K^+$ channels (Lester, 1978), but the contribution of a $Cl^-$ conductance cannot be excluded (Nakamura *et al.,* 1965). In *Torpedo*, the resting current has been shown to be carried at least partially by $Cl^-$ (Miller and White, 1980). In skeletal muscle myocytes, 30–70% of the resting current is carried by this anion (Chapter 50). Since *Electrophorus* electric tissue is derived from skeletal muscle, it is possible that the resting current of these cells is also carried by $Cl^-$.

APs arriving at the nerve termini of electromotor neurons cause ACh to be released onto the innervated membrane of electrocytes (Fig. 2A). **End-plate potentials** (EPPs) produced by AChRs surpass the threshold for $Na^+$ channel activation, and trigger the production of APs that propagate very short distances between electromotor junctions and have overshoots of $+40$ to $+65$ mV. Meanwhile, the potential of the noninnervated membrane remains at the resting value of up to $-85$ mV, due to the abundance of resting current channels and the absence of voltage-gated $Na^+$ channels. When the AP peaks on the innervated membrane, a net **transcellular potential** of up to 150 mV results across the electrocyte ($+65$ mV of the innervated membrane minus $-85$ mV of the noninnervated membrane). This transcellular potential difference is accompanied by a net flow of positive current moving in the innervated membrane to noninnervated membrane direction (left to right as shown in Fig. 2). Insulating septa that form electrocyte columns effectively insulate the extracellular regions on either side of the electrocyte. These connective tissue structures prevent current from flowing around the outside of the electrocyte, which would short circuit the transcellular potential difference (Fig. 2B). This arrangement allows each electrocyte to act as a simple battery having an electrical potential of up to 150 mV. Since each electrocyte of a column is stimulated simultaneously, the potentials of each electrocyte battery within a column summate to generate a large voltage, as predicted by Ohm's law. In Fig. 2B, the potentials of three electrocytes summate to give a potential of 450 mV. In large eels where the potentials of many thousands electrocytes are summated, the net electric discharge can reach 700 V.[1] Connective tissue septa channel the current down the longitudinal axis of the electric organ toward the head of the eel. The current leaves the eel through low-resistance regions of the skin,

---

[1] A discharge of this magnitude requires that at least 4700 electrocytes be stimulated simultaneously. That is, 4700 electrocytes × 0.15 V per electrocyte = 705 V. This situation is analogous to a flashlight where more batteries aligned in series produce a brighter light source.

---

**FIG. 2.** Diagrammatic representation of electrocytes. The left surface of each cell represents the posterior innervated membrane. (A) At rest, both the innervated and noninnervated membrane exhibit a potential of $-85$ mV. When stimulated, activated AChRs generate EPPs, triggering $Na^+$ channel-mediated APs peaking at $+55$ mV on the innervated membrane. The noninnervated membrane contains no voltage-gated $Na^+$ channels and maintains the $-85$-mV resting potential. The result is a transcellular potential difference of approximately 150 mV. (B) Because each cell is stimulated simultaneously, electrocyte transcellular potentials summate. The potentials of three electrocytes culminate to produce 450 mV. Currents generated by stimulated electrocytes flow down electrocyte columns in the posterior to anterior direction. The circuit is closed by current flowing out the head of the eel, through the water, and back into the tail region.

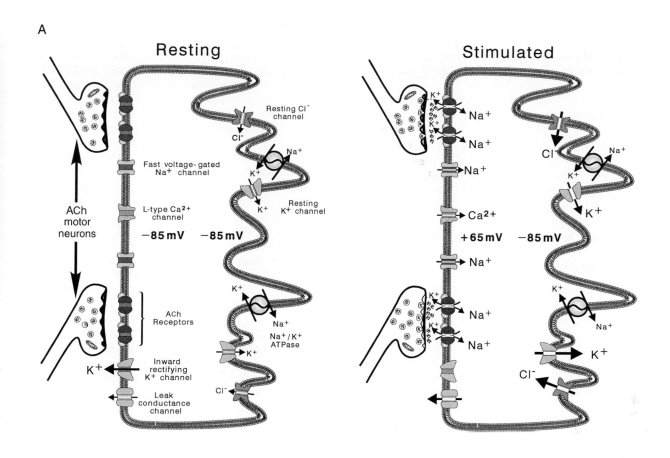

A

**Resting**

Resting Cl⁻ channel

Fast voltage-gated Na⁺ channel

L-type Ca²⁺ channel

ACh motor neurons

−85 mV    −85 mV

Resting K⁺ channel

ACh Receptors

Na⁺/K⁺ ATPase

K⁺

Inward rectifying K⁺ channel

Leak conductance channel

**Stimulated**

+65 mV    −85 mV

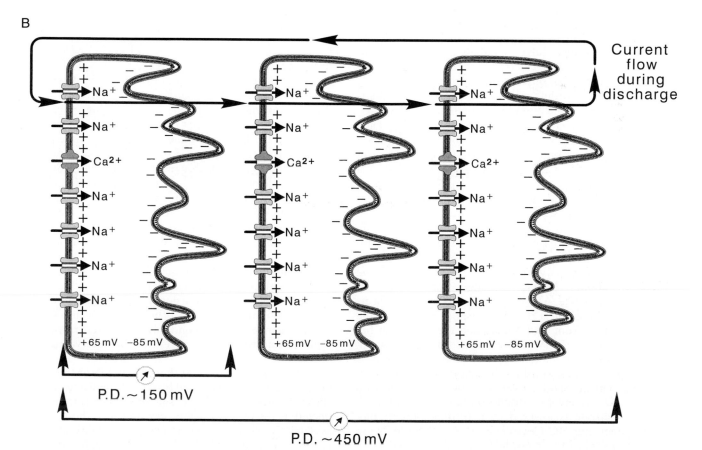

B

Current flow during discharge

+65 mV   −85 mV      +65 mV   −85 mV      +65 mV   −85 mV

P.D. ~ 150 mV

P.D. ~ 450 mV

and is conducted through the water producing an electric field, the magnitude of which diminishes with the square of the distance from the animal. Objects, like other fish or the human hand, experience a potential difference in this electric field, and currents sufficient to excite muscles, nerves, and sensory endings flow through them producing a stunning sensation. The circuit of the discharge is closed by current flowing through the skin of the tail region of the eel back into electrocyte columns from which the electromotive force (EMF) originated (Bennett, 1971).

## III. Electrocyte Membrane Electrophysiology

### A. Membrane and Extracellular Potentials

Electrocyte membranes contain many of the same protein elements found in myocytes and neurons. In fact, the individual APs of the innervated membrane of *Electrophorus* electrocytes are quite similar to those of other excitable cells. However, electric cells are different, in that potential changes are polarized to a particular membrane resulting in the generation of transcellular potentials and an asymmetric flow of current. Also, to produce whole-animal electrical discharges having the maximum possible voltage or current output, electrocytes have evolved to express exaggerated amounts of key excitable membrane proteins.

In electric tissue preparations where the flat innervated electrically excitable surface of electrocytes is exposed, APs can be triggered with extracellular stimulating electrodes situated close to the membrane surface. Membrane potentials and **transcellular potentials** can then be measured through recording electrodes lowered across the innervated membrane or through the entire cell, respectively (Fig. 3). As the recording electrode approaches the innervated membrane and a stimulus is applied, a negative deflection is recorded (Fig. 3A). Because the recording electrode measures the potential difference between the region just outside the membrane compared to the reference electrode placed in the bath, the AP here is recorded as a negative deflection. After the electrode is advanced through the innervated membrane, an AP that has propagated from the site of stimulation to the recording electrode is detected. The characteristics of electrocyte membrane potentials vary considerably from cell to cell, but are typically similar to those measured on the myocyte sarcolemma or neuronal axolemma. As seen in Fig. 3B, a typical resting potential is about −75 mV, and ranges from −65 to −85 mV. The electrocyte AP seen in Fig. 3B is typical in its 3.5-msec duration and +50-mV overshoot. Generally, the duration ranges from 2 to 4 msec and the overshoot between +35 and +65 mV. These values for the overshoot are considerably larger than that of other excitable cells and are due to an extraordinarily high density of voltage-gated $Na^+$ current, and a relatively low level of outward rectifying $K^+$ current (see discussion later). When the recording electrode is lowered even further until it completely penetrates the electrocyte, transcellular APs are recorded. Because the recording electrode is once again in the extracellular space, the resting potential here is measured as 0 mV. However, the interstitium where the recording electrode is positioned is electrically insulated by connective tissue septa from the reference electrode located in the bath solution. When the electrocyte is stimulated, an AP is recorded that is identical to the intracellular AP, except that it initiates at 0 mV. It has a peak equal to the total amplitude of the intracellular AP, in this case, about 125 mV. This transcellular AP arises because the noninnervated inexcitable membrane does not fire an AP that would cancel out the spike of the innervated membrane. Insulating connective tissue septa prevent the potential difference from being short-circuited around the outside of the electrocyte. If the stimulus is increased such that the electrocyte beneath the recording electrode is also stimulated, a negative deflection in the extracellular potential is recorded just behind the 125-mV transcellular AP (Fig. 3C, right). This negative deflection corresponds to the resulting AP of the lower electrocyte, which is stimulated later than the electrocyte on the surface. In the intact electric organ, these electrocytes would be stimulated simultaneously by the eel's nervous system such that these APs would be occurring at the same time. In this way, the transcellular potentials summate to yield a powerful electrical discharge.

Ionic currents responsible for *Electrophorus* electrocyte resting potentials, as well as EPPs and APs are similar to other excitable cells, and the reader is referred to previous chapters where their mechanisms have been described in detail. However, some differences between electrocyte membranes and those of neurons and myocytes are worthy of mention. The noninnervated membrane of electrocytes has a very low resistance of about 0.1 $\Omega/cm^2$, which is one to two orders of magnitude less than that typically found for nerve or muscle (Nakamura *et al.,* 1965). Physiologically, the eel needs this high $K^+$ and $Cl^-$ current to clamp the noninnervated membrane at the resting potential in order to set up transcellular potentials like those observed in Fig. 3C. On the other hand, the innervated membrane at rest has a resistance of 3–6 $\Omega/cm^2$ (Nakamura *et al.,* 1965). On stimulation, this resistance decreases due to a large voltage-activated $Na^+$ current. In fact, Shenkel and Sigworth (1991) were able to measure macroscopic $Na^+$ currents in excised patches of the innervated membrane corresponding to a density of as much as 1300 channels/ $\mu m^2$. This high density of $Na^+$ channels is accompanied by relatively few outwardly rectifying $K^+$ channels. Because of this distribution of channels in the innervated membrane, the electrocyte AP has a large overshoot that nearly reaches the $Na^+$ equilibrium potential, $E_{Na}$. The repolarization phase is therefore due primarily to the inactivation of $Na^+$ channels, and secondarily to delayed rectifier $K^+$ channels and the resting current of the innervated membrane.

Skeletal muscle-like EPPs that trigger APs on the surface of electrocytes are generated by AChR-mediated currents. Basically, macroscopic and single-channel currents conducted by the eel AChR are very similar to those seen on skeletal muscle sarcolemma. The permeability of the eel receptor to both $Na^+$ and $K^+$ is nearly equal since the reversal potential is the midpoint between $E_{Na}$ and $E_K$

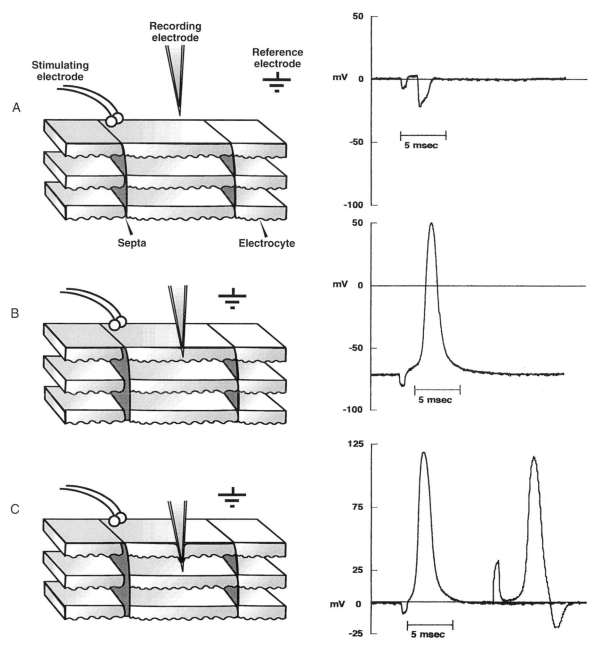

**FIG. 3.** Extracellular potentials, APs, and transcellular potentials of *Electrophorus* electrocytes in an electric tissue slice preparation. Diagrams to the left depict a slice of electric tissue in cross-section where electrocytes are oriented such that the innervated membrane is uppermost. Columns of electrocytes run in the vertical direction and are delineated by insulating connective tissue septa. Potentials recorded by the recording electrode in the indicated positions are shown on the right. (A) The recording electrode near the innervated membrane records a negative deflection in the extracellular potential. (B) The recording electrode penetrating the innervated membrane records an intracellular AP. (C) Transcellular potentials measured after the recording electrode has penetrated the entire electrocyte. The second potential recording on the right shows the negative deflection of the electrocyte beneath the recording electrode when a higher intensity stimulus is applied.

(Sheridan and Lester, 1977; Pasquale *et al.*, 1986). In other words, the peak of the EPP moves toward a value of approximately $-10$ mV in order to activate $Na^+$ channels for an AP. Eel AChRs elicit single-channel opening events that are similar to the receptor from mammalian sources in that their mean open time is dependent on membrane potential, temperature, and the AChR agonist used. Single-channel conductance's through individual receptors do not depend on the ligand used. However, these preparations of the eel AChR are different from other excitable cells in that single-channel open times can be fitted to a single exponential compared to the more complex distributions found for other sources of the receptor. This indicates that the eel expresses only one isoform of each of the receptor subunits yielding a receptor with a single unique conductance.

### B. Equivalent Circuits

From what is known of electrocyte electrophysiology, equivalent circuits can be derived that explain the production of transcellular potentials that arise during the AP. Because electrocytes are large flat cells comprised of essentially two parallel membranes having a uniform potential across their entire surfaces, whole-cell potentials can be described with two equivalent circuits pertaining to the innervated and noninnervated membranes connected by a resistor representing the resistance of the cytoplasm ($R_{cyt}$). At rest, the permeability of both membranes to $K^+$ and $Cl^-$ is high so that their equilibrium potentials are expressed more than that of $Na^+$, resulting in an $E_m$ of about $-85$ mV (for a detailed description, see Chapter 13). The equivalent circuit for the electrocyte at rest can then be reduced to that seen in Fig. 4A (right), where both membranes have composite $E_m$ values that drive an outward flow of positive current. Notice that the equivalent circuits for both membranes are mirror images of one another, both having a symmetrical outward movement of current that cancels to give a transcellular potential of 0 mV. At the peak of the AP, however, the permeability of the innervated membrane to $Na^+$ increases dramatically, such that the membrane potential is influenced primarily by $E_{Na}$. The composite $E_m$ of the innervated membrane has now reversed its polarity, so that current across this membrane moves inward (Fig. 4B, right). Because the noninnervated membrane has no $Na^+$ conductance, the polarity of the potential here is the same as at rest, and a net outward current continues to flow. Now, the total driving force for both membranes is in the same direction, so that current flows from the innervated membrane to and through the noninnervated membrane. If one considers that the $R_{cyt}$ and $R_{m(non)}$ are negligible compared to $R_{m(inn)}$, the equivalent circuit for the electrocyte can be reduced to a resistor and battery in series, where the resistance is equal to $R_{m(inn)}$ and the battery represents the composite potentials of $E_{m(inn)} + E_{m(non)}$. During stimulation, the potential across this unit, then, equals the transcellular potential having a magnitude equal to that of the amplitude of the innervated membrane's AP, or approximately 150 mV.

In the electric organ of *Electrophorus*, electrocytes are stacked one after another in very long columns. Electrically, this arrangement is represented by many electrocyte resistor-battery units connected in series as shown in Fig. 5A. Each unit contributes an additional 150 mV to the overall electrical discharge. However, Kirchhoff's first law dictates that the current measured at every point along an unbranching leg of a circuit, such as a series of batteries, is constant.[2] In other words, the value of the overall current

[2] Specifically, Kirchoff's first law states that the current entering a point along a circuit is equal to the sum of the currents of all the branches leaving that point. Therefore, if the circuit is unbranching, then the current at every point is constant.

output of a column of electrocytes does not depend on the number of cells in series, but on the electrocyte that has the greatest resistance to the flow of current. (For a comprehensive description of Ohm's law and Kirchoff's laws applied to biological equivalent circuits, see Sperelakis, 1979.) In order for an electric organ to increase the current of an electrical discharge, it must have additional electrocytes arranged in parallel. This amplification is accomplished in electric fish by having numerous electrocyte columns situated alongside one another. The electric eel, being a long slender animal, has fewer electrocyte columns arranged in parallel relative to some other electric fish. Because of this anatomical arrangement, the eel produces discharges of very high voltage with less current compared to other electric fish, such as *Torpedo,* the electric ray.

The electric ray is a marine elasmobranch having two large electric organs positioned laterally on either side of its flattened head (Fig. 6). Hexagonal-shaped columns of electrocytes run in the vertical direction, and conduct current up away from the ocean floor, around the edge of the lateral fin and back into the bottom of the electric organ. *Torpedo* has a short, yet wide electric organ accommodating large numbers of electrocyte columns situated next to one another. As shown in Fig. 5B, Kirchoff's law predicts that electrocyte resistor-battery units in parallel produce electrical discharges of low voltage and high current, proportional to the total number of electrocyte units in parallel. This is indeed the case with the electric ray where discharges of up to 16 A and 60 V, totaling up to 1 kW, have been measured (Grundfest, 1960).

### IV. Comparative Physiology of *Electrophorus* and *Torpedo*—Models for Mammalian Excitable Cells

#### A. *Electrophorus*

##### 1. Na⁺ Channel

Because electric tissue is specialized for membrane excitability and carries out its function with membrane proteins common to mammalian tissues, electrocytes of electric fish provide excellent models for mammalian excitable cell membranes. Compared to electrocytes of *Torpedo* and other electric fish, *Electrophorus* electrocytes express more of the membrane proteins common to mammalian excitable tissues and therefore provide a more general model for excitable membranes. For example, both voltage-gated $Na^+$ and AChR-mediated currents have been measured from the innervated membranes of these cells. Early preparations took advantage of the large size of these cells by sealing single electrocytes over windows in Lucite chambers (Schoffeniels, 1961). Because the resistance of the noninnervated membrane is negligible relative to the innervated membrane, electrocytes in this configuration are treated as a single membrane without having to thread a space clamping electrode down the middle of the cell. With this method, Nakamura *et al.* (1965) were able to measure $Na^+$ currents responsible for the rising phase of eel electrocyte APs, and to construct a current–voltage ($I$–$V$) relation

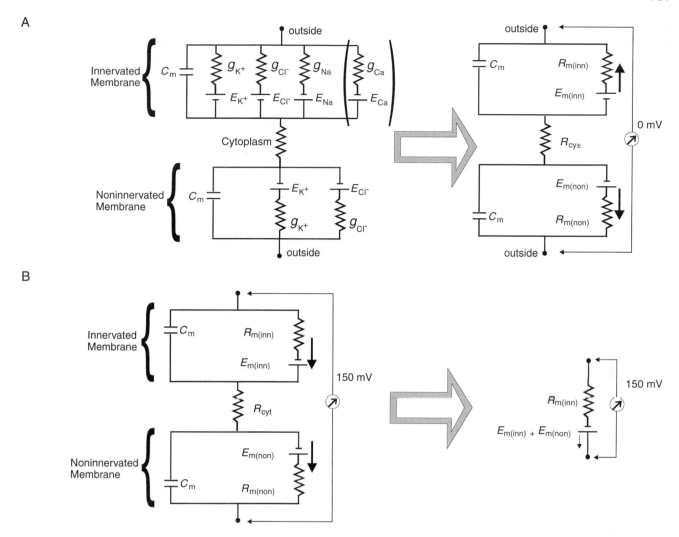

**FIG. 4.** Electrical equivalent circuit diagrams for both the innervated and noninnervated membranes of an electrocyte. The innervated membrane is uppermost in both parts A and B. Both membranes are represented by parallel resistance–capacitance circuits, and are connected by the cytoplasmic resistance, $R_{cyt}$. $C_m$ is the membrane's capacitance. The conductance's for $K^+$, $Cl^-$, $Na^+$, and $Ca^{2+}$ across the membranes are represented by $g_K$, $g_{Cl}$, $g_{Na}$, and $g_{Ca}$, respectively, and are inversely proportional to the resistance of the membrane for these ions. Nernst potentials for each of these ions across the membranes are $E_K$, $E_{Cl}$, $E_{Na}$, and $E_{Ca}$. The $Ca^{2+}$ leg of the circuit is in brackets, since the existence of a selective $Ca^{2+}$ conductance in the electrocyte has not yet been demonstrated. (A) At rest, the innervated membrane reduces to a circuit where $R_{m(inn)}$ represents the total membrane resistance of this face of the cell, and is a composite of the resistance's of the membrane to each of the ions listed to the left. The $E_{m(inn)}$ and $E_{m(non)}$ are the resting membrane potentials for the innervated membrane and noninnervated membrane, respectively. At rest, $E_K$ and $E_{Cl}$ are expressed the most, since the conductance of the membrane to both these ions is greatest. In this state, the circuit diagrams are mirror images of one another, no net current flows across the cell and the transcellular potential is 0 mV. (B) At the peak of the AP, the conductance of the membrane to $Na^+$ (and possibly $Ca^{2+}$) increases dramatically, and $E_{Na}$ (and $E_{Ca}$) is expressed more than $E_K$ and $E_{Cl}$. The polarity of the $E_{m(inn)}$ battery is now reversed, such that there is a net flow of positive current in the direction of the noninnervated membrane, and a transcellular 150-mV potential results. Given that $R_{cyt}$ and $R_{m(non)}$ are negligible compared to $R_{m(inn)}$, the equivalent circuit can be reduced to the "electrocyte unit" on the right.

for the *Electrophorus* Na$^+$ channel. These researchers also described an abundant inward rectifying K$^+$ current that has an *I–V* relation similar to those seen in myocytes. However, since delayed K$^+$ currents are not seen in these preparations, the repolarization of the eel electrocyte AP is apparently due to Na$^+$ channel inactivation. Much of what is known about the electrogenesis of the AP of elec-

trocytes and other excitable cells has come from studies such as this. The innervated membrane of *Electrophorus* electrocytes was also **patch-clamped** to examine macroscopic Na$^+$ currents in order to determine minute charge movements associated with channel opening, as well as to determine the relative permeability of the channel to K$^+$ relative to Na$^+$. Upon membrane depolarization, the Na$^+$

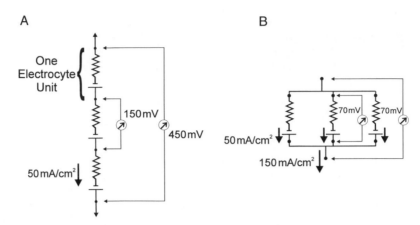

**FIG. 5.** Electrocyte units in series and in parallel. (A) When connected in series, the potentials of each 150-mV electrocyte unit summate, while the current remains constant. In this case, three electrocyte units are shown to summate their potentials to yield a total of 450 mV. This tends to be the case in the electric organ of *Electrophorus* where transcellular potentials of 150 mV and currents of 50 mA/cm² are measured. (B) The currents of electrocytes in parallel summate while the potential remains constant. Three electrocyte units each having a current of 50 mA/cm² are shown to produce a 150 mA/cm² current when arranged in parallel. This is the case in *Torpedo,* where many electrocytes are situated next to one another in parallel. The transcellular potential produced by *Torpedo* electrocytes is 70 mV (see text).

channel undergoes a conformational change that is associated with the movement of positively charged amino acids within the protein. This movement of positive charge is thought to be a consequence of the opening of the channel gate, and can be measured in a population of Na⁺ channels as a small outward current. Opening of the eel Na⁺ channel is associated with the movement of approximately 1.5 charges upon channel opening, a value similar to nerve and muscle preparations (Shenkel and Bezanilla, 1991; for a comprehensive description of gating mechanisms, see Hille, 1992). On examining the selective permeability of the eel Na⁺ channel for this ion relative to K⁺, Shenkel and Sigworth (1991) found $P_{Na}/P_K$ ratios of 8 to 43. This range represents a substantial variation that was even seen in membrane patches taken from the same cell. Given that *Electrophorus* electrocytes are known to express just one isoform of the channel protein, these results suggested that this variability might arise from posttranslational modifications such as glycosylation or even **phosphorylation.** This possibility was substantiated by experiments that showed the eel Na⁺ channel to be modulated by exogenously applied protein kinase A (Emerick *et al.,* 1993; also see Chapter 32 of this book). These studies contributed to our basic understanding of the electrophysiological function of the Na⁺ channel protein.

The first Na⁺ channel to have ever been purified and later sequenced, came from electric tissue of *Electrophorus.* The purified 260-kDa protein is heavily glycosylated and consists of a single functional α-subunit (Agnew *et al.,* 1978; Miller *et al.,* 1983). Na⁺ channels from mammalian brain and muscle express additional β-subunits. Since it has been suggested that these auxiliary subunits may regulate channel gating, preparations of the eel Na⁺ channel are advantageous in that they eliminate the possibly complicating influence of β-subunits. An *Electrophorus* electric tissue cDNA library was used to clone and sequence the

channel for the first time (Noda *et al.,* 1984). Its structure includes four homologous repeats that each contain six transmembrane-spanning helices. The fourth transmembrane segment contains a cluster of positively charged amino acids thought to be involved in gating of the channel. Movement of these positive amino acids during opening of the channel are thought to be responsible for the minute currents measured in patch-clamp experiments such as those discussed earlier. Compared to sequences of Na⁺ channels of mammalian muscle and brain, the eel Na⁺ channel shows the greatest homology with the muscle protein, as expected since electrocytes develop ontogenetically from myocytes. Both of these channels, however, lack a 202 amino acid segment located between the first and second homologous repeat domain of the brain Na⁺ channel (for complete discussions of Na⁺ channel purification, structure, and diversity, see Chapter 26 in this book and Hille, 1992).

### 2. Acetylcholine Receptor

As mentioned in a previous section, the large size of eel electrocytes has facilitated their dissection in order to measure single-channel AChR conductances. Other excitable cells express numerous isoforms of the subunits that make up the receptor, and produce single-channel recordings that show complex mean open-time distributions and variable conductance values. The eel AChR yields more homogeneous values owing to its simplified subunit composition (Pasquale *et al.,* 1986). The simple mean open-time distributions for the eel AChR were found for both the main electric organ as well as Sach's organ, suggesting that AChR diversity evolved in order to meet varying physiological needs of myocytes and neurons. The eel receptor also desensitizes to a lesser extent in the presence of sustained concentrations of agonists (Pallotta

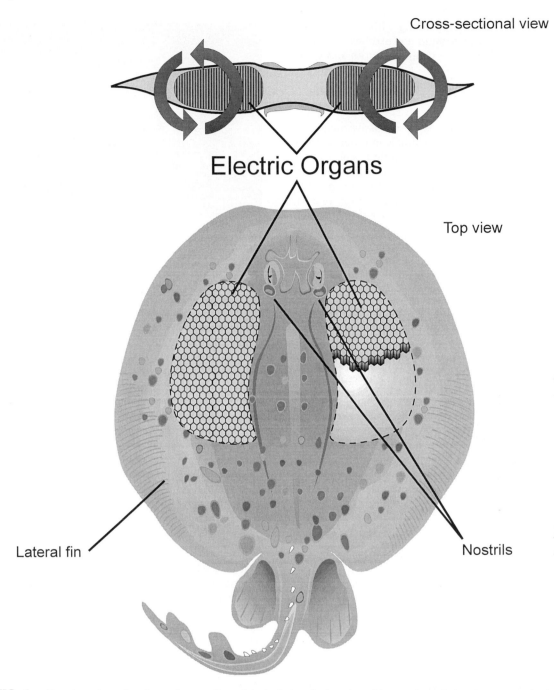

**FIG. 6.** Drawing of an electric ray (genus, *Torpedo*), depicting the location of its two lateral electric organs. The inset shows the direction of the flow of current around the fish while producing an electrical discharge.

and Webb, 1980). Simple subunit composition and lack of agonist-induced desensitization make the eel AChR advantageous for electrophysiological and biochemical studies.

Recognizing the cholinergic nature and the specialization of *Electrophorus* electrocytes, biochemists utilized electric tissue as a source for some of the first purifications of AChRs. Using various separation techniques, including differential centrifugation to separate membrane fractions

and affinity chromatography to selectively purify the AChR from other membrane proteins, a 260-kDa macromolecule was isolated (Olsen *et al.,* 1972; Biesecker, 1973). Not knowing *a priori* that the receptor is a pentameric protein made up of $\alpha$-, $\beta$-, $\gamma$-, and $\delta$-subunits (in a respective ratio of $2:1:1:1$) the initial isolation and identification of the peptides that make up the whole receptor were arduous. After some debate, the 44-kDa $\alpha$-subunit was established as the ligand-binding portion of the protein. The 50-

to 65-kDa $\beta$-, $\gamma$-, and $\delta$-subunits along with the $\alpha$-subunit are arranged symmetrically around a central axis to make up the ion channel pore of the protein (Karlin and Cowburn, 1973; Chang, 1974).

### 3. Na⁺,K⁺-ATPase

As we have seen, electrocytes express massive quantities of membrane receptors and ion channels in order to carry out their specialized function. To maintain resting potentials in the face of currents associated with EPPs and APs that dissipate Na⁺ and K⁺ gradients, electrocytes need to express large amounts of **Na⁺,K⁺-ATPase.** For this reason, *Electrophorus* electric tissue has been used as a source for the purification of this enzyme for structure–function studies. The eel protein, like that from other tissues consists of a 94-kDa $\alpha$-subunit and a glycosylated 47-kDa $\beta$-subunit. During purification, the ATPase is solubilized from electrocyte membranes with various detergents. To analyze its functional characteristics, the protein was reconstituted into liposomes where its ATP-driven translocation of radiolabeled Na⁺ and K⁺ has been found to be similar to preparations of native electrocyte membranes containing the ATPase (Yoda *et al.,* 1984). These preparations of the eel protein have been invaluable in determining the reaction mechanisms of the ATPase involved in its function. Drugs that target and inhibit different partial reactions of the protein's translocation mechanism have also been investigated and used to examine the pump's function in cell physiology. Drugs, such as the cardiac glycoside, digoxin, have also been used clinically specifically to inhibit Na⁺,K⁺-ATPase function in the heart. This treatment dissipates the membrane Na⁺ gradient, which indirectly augments intracellular $Ca^{2+}$ within cardiac myocytes. This results in an increase in the heart's force of contraction, which can alleviate some forms of heart disease. (See Chapter 16; for literature review, see Lingrel and Kuntzweiler, 1994).

### 4. Calmodulin

*Electrophorus* electric tissue also expresses large quantities of the calcium-binding protein **calmodulin.** (For a detailed discussion of $Ca^{2+}$-binding protein function, see Chapter 9.) In fact, calmodulin makes up roughly 2%, by weight, of electrocyte protein (Munjaal *et al.,* 1986). Once again, electric tissue was used as a source for the purification of this 17-kDa soluble protein (Childers and Siegel, 1975). Unlike the membrane proteins discussed earlier, the function of calmodulin within electrocytes remains elusive, even though some of the functions of this $Ca^{2+}$-mediator protein in intracellular signaling mechanisms is well documented in other electrically excitable cells. Figure 7 shows the intracellular location of calmodulin within electrocytes. Calmodulin is present throughout the cytoplasm of electrocytes, but is particularly concentrated near both the innervated and noninnervated membranes. In light of this membrane localization along with the fact that calmodulin is so abundant in this tissue specialized for membrane excitability, a role for this protein in membrane function is

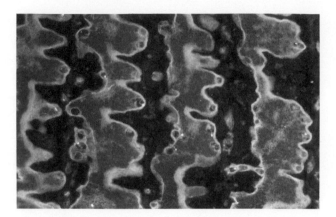

**FIG. 7.** Immunofluorescent localization of calmodulin within main organ electrocytes. Paraffin-embedded 4-$\mu$m sections were probed with anti-calmodulin sheep antibodies. The location of primary antibodies was visualized with fluorescein-labeled rabbit anti-sheep secondary antibodies, and photographed using epifluorescence microscopy.

likely. Determining the role of calmodulin in electrocyte function will undoubtedly lend insight into the role of this protein in membrane function of other excitable cells.

### B. *Torpedo*

#### 1. Comparative Electrophysiology

The marine electric ray (genus, *Torpedo* and various species: *marmorata, californica, nobilianae, occidentalis*) can produce high-amperage electrical discharges by virtue of numerous electrocyte columns arranged in parallel as discussed briefly earlier. In this way it differs from *Electrophorus,* which produces high-voltage discharges owing to numerous electrocytes arranged in series. The basic arrangement of *Torpedo* electrocytes within electric organ columns is remarkably similar to that of *Electrophorus,* considering that these two fish belong to different orders and the existence of electric tissue in both orders of fish represents convergent evolution. Although *Torpedo* electrocytes are smaller and pancake shaped (10–30 $\mu$m $\times$ 5 mm in diameter) relative to wafer-shaped *Electrophorus* electrocytes, they are still stacked one after another in columns delineated by electrically insulating connective tissue septa. *Torpedo* electrocytes also display membrane polarity similar to that depicted in the diagrams of Fig. 2.

Some basic differences in the membrane biochemistry and electrophysiology exist between electrocytes of these two fishes, however. Like the eel, *Torpedo* electrocytes can be stimulated to produce EPPs in response to nervous stimulation. That is, these electrocytes are chemically excitable. *Torpedo* electrocytes, however, do not fire APs in response to EPPs or artificially applied electrical stimuli and are therefore are electrically inexcitable. This is due to a lack of voltage-dependent Na⁺ channels on the innervated membrane that would generate and propagate APs.

Instead, these cells are richly innervated by ACh-releasing electromotor neurons, and have an abundance of postsynaptic AChRs. These ligand-gated channels conduct EPPs of 5-msec duration that peak just below 0 mV, halfway between $E_{Na}$ and $E_K$. Like *Electrophorus* electrocytes, the noninnervated membrane of these cells has a large resting current, partially carried by $Cl^-$. Transcellular potentials measured across *Torpedo* electrocytes then, have an amplitude equal to that of the EPP, or 70 to 85 mV. The equivalent circuits diagrammed in Figs. 4A and B also apply to these electrocytes, except that the potential produced by each electrocyte unit equals 70 to 85 mV, instead of 150 mV for *Electrophorus* electrocytes.

## 2. Acetylcholine Receptor

Without voltage-gated $Na^+$ channels to propagate an AP, EPPs decay exponentially with the distance traveled from the **electromotor end-plate.** However, very little EPP decay is actually measured on the innervated membrane of *Torpedo* electrocytes, because the density of end-plates is so great. In fact, one might describe the innervated membrane of these cells as one large electromotor end plate. These cells, therefore, provide a very specialized model for the motor end-plate. Like *Electrophorus* electric tissue, *Torpedo* electric tissue has been used for the purification of the AChR, which has supplied a wealth of knowledge about the biochemical properties of the receptor, as described in the previous section. The first sequences ever to be determined for each of the subunits of the receptor were obtained by screening *Torpedo* electric tissue cDNA libraries (Noda *et al.,* 1982, 1983a,b; Claudio *et al.,* 1983). The mRNAs encoding each of the receptor subunits were together injected into *Xenopus* oocytes in order for the functional protein to be expressed on the membrane of these cells. Interestingly, an increase in the ACh-induced conductance could be measured after microinjection (Mishina *et al.,* 1984). By knowing the sequence of the receptor's subunits, a three-dimensional model of the receptor has been constructed and continually amended in light of ongoing biochemical research. (A model of the AChR appears in Chapter 40, Synaptic Transmission). These studies established the basic protein structure of the nicotinic AChR, the findings of which have only been slightly modified to apply to the receptor of mammalian muscle and nerve.

## 3. Acetylcholinesterase

AChE is an important enzyme found in the postsynaptic membrane of cholinergic synapses of neurons, motor end-plates of myocytes, and the electromotor end-plates of electrocytes. It catalyzes the hydrolysis of ACh to choline and acetate, thereby terminating ligand-gated activation of the AChR. Some pesticides and chemical warfare agents contain **anti-AChE agents** that cause acute and chronic alterations in central nervous system and neuromuscular function. Anti-AChE drugs have also been developed to alleviate symptoms of glaucoma, Alzheimer's dementia, and myesthenia gravis—diseases marked by attenuated postsynaptic AChR density or compromised ACh release (for review, see Millard and Broomfield, 1995). Obviously, great care is needed when administering these drugs because overmedication can cause side effects similar to exposure to harmful anti-AChE agents.

Since the *Torpedo* electrocyte represents an exaggerated cholinergic system, it was used as a source for the purification and subsequent structural analysis of AChE. It is an 80-kDa protein that self-associates into tetramers, octamers, and dodecamers, and is anchored in the postsynaptic membrane through a phospholipid linkage (Parker *et al.,* 1978; Ratman *et al.,* 1986). *Torpedo* electric tissue provided massive enough quantities of AChE for the protein to be crystallized for subsequent X-ray diffraction studies. The resulting diffraction pattern obtained from X-rays shot through these crystals was analyzed to construct a three-dimensional structure of the protein, localizing atoms within the enzyme to within 2.8 Å (Sussman *et al.,* 1991). These experiments with the *Torpedo* enzyme will continue to be invaluable to the development of new drugs aimed at the treatment of cholinergic diseases and for therapies for individuals exposed to toxic anti-AChE agents.

## 4. $Cl^-$ Channel

Perhaps the most dramatic contribution that electric tissue has made to recent membrane biochemistry and physiology has been toward elucidating the structure and function of $Cl^-$ channels. Recall that in order for electrocytes to produce large transcellular potentials, the noninnervated membrane must have a tremendous resting current. This current clamps the noninnervated membrane potential at highly negative resting potentials even while the innervated membrane depolarizes dramatically. In *Torpedo* electrocytes, the resting current is carried at least partially by $Cl^-$. Electric tissue of the electric ray, therefore, has been used to isolate the channel protein for physiological studies, as well as a source for mRNA used in cloning and sequencing of the channel.

In the course of developing the planar lipid bilayer method for measuring ion channel conductance's, Miller and White (1980) found that vesicles derived from the noninnervated membrane of *Torpedo* electrocytes contained $Cl^-$ channels having novel **"double-barreled" gating kinetics.** With depolarization, individual $Cl^-$ channel complexes acted as two channels with two separate, but equal conductances. When one channel of the complex was open, the other was more likely to be subsequently activated as well. Single-channel recordings showed periods of inactivity until one channel of the complex was opened, after which a second equal conductance would superimpose on the first. Purification of the *Torpedo* $Cl^-$ channel confirmed that the protein was a homodimer consisting of two 90-kDa polypeptides. When the purified protein was incorporated into planar lipid bilayers for single-channel recording, the same double-barreled gating kinetics were observed (Middleton *et al.,* 1994).

The first $Cl^-$ channel ever to be sequenced came from *Torpedo* electric tissue and has greatly expanded the field of $Cl^-$ channel molecular biology. Using the expression cloning technique, the mRNA responsible for the *Torpedo* Cl⁻ conductance was identified, and its corresponding cDNA sequenced. The encoded protein was predicted to consist of 805 amino acids, and to have a molecular weight similar to that of the purified protein (Jentsch *et al.,* 1990). When mRNA for this channel, termed ClC-0, was injected into *Xenopus* oocytes, Cl⁻ conductances having double-barreled gating kinetics were expressed (Bauer *et al.,* 1991). Recognizing that electric tissue is a model for skeletal muscle membranes, Steinmeyer *et al.,* (1991b) screened a rat muscle cDNA library with oligonucleotide sequences derived from the *Torpedo* Cl⁻ channel. In this way, the sequence for the major Cl⁻ channel of mammalian skeletal muscle (called ClC-1) was obtained. It was later found that genetic aberrations in the mammalian ClC-1 gene result in symptoms of skeletal muscle myotonia (Steinmeyer *et al.,* 1991a). These findings confirmed the results of Bryant and Morales-Aguilera (1971) that showed this disease to be associated with compromised Cl⁻ conductance (see Chapter 38). Once the sequences were known for both ClC-0 and ClC-1, investigators began screening libraries derived from virtually every mammalian tissue. Numerous Cl⁻ channel sequences have now been determined, and have been implicated in various physiological functions from neuronal membrane excitability to epithelial solute transport (for a review, see Fong and Jentsch, 1995). More Cl⁻ channels having even greater diversity and function will undoubtedly be uncovered in the future.

## V. Summary

Both the freshwater electric eel and the saltwater electric ray produce extraordinarily powerful electrical discharges with membrane ion channels, receptors, and pumps common to other excitable cells. These fish have separately evolved a specialized anatomy and cellular morphology designed for this function. Because of their specialized membrane asymmetry, APs and EPPs generated on the innervated membrane are not reproduced on the noninnervated membrane, thereby setting up an asymmetrical flow of current across the cell. The arrangement allows transcellular potentials to be generated, which is essentially the basis for the generation of bioelectricity within the electric organs of these fish. Connective tissue septa that delineate columns of electrocytes prevent transcellular potentials from being short circuited around the outside of individual electrocytes, and also channel the resulting current along the electric organ.

The membrane potentials used by electrocytes to produce transcellular potentials are remarkably similar to those of other excitable cells, such as myocytes and neurons. The electrophysiology, therefore, can be explained by currents conducted through ligand-gated receptors and channels having known characteristics. Currents that give rise to electrocyte membrane potentials can even be represented by equivalent circuits similar to those of other excitable cells. However, two major differences exist between electrocytes and other excitable cells: (1) Electrocytes express exaggerated amounts of key excitable membrane pro-

teins, such as the Na⁺ channel of *Electrophorus* and the AChR of *Torpedo*. These proteins that exist in high density tend to produce greater currents and peak potentials than what is customarily seen on other excitable cells. (2) Membrane proteins are polarized to particular sides of the cell to facilitate the production of transcellular potentials. In the past, however, researchers have taken advantage of these differences to utilize these fish as useful model systems.

*Electrophorus* electrocytes provide a general model system for excitable cells such as neurons and myocytes, since they contain common membrane receptors, channels, and ATPases. They are also large and easy to dissect in order to perform potential recording, voltage-clamp analysis, and patch-clamp measurements. Since they express large quantities of proteins such as the Na⁺ channel, the Na⁺,K⁺-ATPase, AChR, and calmodulin, eel electric tissue has been used as a source for the purification of these proteins for molecular and functional analysis.

*Torpedo* electrocytes, on the other hand, are richly innervated with ACh-releasing electromotor neurons and are electrically inexcitable. Therefore, they provide a very specialized model for the motor end plate. Because of the exaggerated cholinergic nature of *Torpedo* electric tissue, it has been used as a rich protein and mRNA source for the AChR and AChE. The expression of Cl⁻ channels on the noninnervated membrane of these cells has led researchers to use this tissue as a source for ClC-0 protein and mRNA as well.

Investigations with electric tissue of both the electric eel and the electric ray have opened wide avenues of study in electrophysiology, protein biochemistry, and clinical research. Electrophysiological techniques have been used, refined, and in some cases developed while using electrocytes as model systems. Like the squid giant axon, these cells have been instrumental in defining and confirming the ionic currents responsible for excitable cell membrane potential changes. Biochemically, *Electrophorus* and *Torpedo* electric tissue has supplied abundant quantities of key excitable membrane proteins that exist in only trace amounts in mammalian tissues. Since electric tissue develops from skeletal muscle, the biochemical properties and three-dimensional structures of these proteins are similar, if not identical, to those of mammalian skeletal muscle and other excitable cells. These discoveries will continue to further our understanding of the mechanisms by which membrane potentials of excitable cells are generated and regulated, as well as for the understanding and the treatment of disease.

## Bibliography

Agnew, W. S., Levinson, S. R., Brabson, J. S., and Raftery, M. A. (1978). Purification of the tetrodotoxin-binding component associated with the voltage-sensitive sodium channel from *Electrophorus electricus* electroplax membranes. *Proc. Natl. Acad. Sci. USA* **75,** 2606–2610.

Bauer, C. K., Steinmeyer, K., Schwartz, J. R., and Jentsch, T. J. (1991). Completely functional double-barreled chloride channel expressed from a single *Torpedo* cDNA. *Proc. Natl. Acad. Sci. USA* **88,** 11052–11056.

Bennett, M. V. L. (1971). Electric organs. *In* "Fish Physiology" (W. S. Hoar and D. J. Randall, Eds.), pp. 347–491. Academic Press, New York.

Bennett, M. V. L., and Sandri, C. (1989). The electromotor system of the electric eel investigated with horseradish peroxidase as a retrograde tracer. *Brain Res.* **488,** 22–30.

Biesecker, G. (1973). Molecular properties of the cholinergic receptor purified from *Electrophorus electricus*. *Biochemistry* **12,** 4403–4409.

Bryant, S. H., and Morales-Aguilera, A. (1971). Chloride conductance of normal and myotonic goat fibres and the action of monocarboxylic aromatic acids. *J. Physiol. (London)* **219,** 367–382.

Chang, H. W. (1974). Purification and characterization of acetylcholine receptor-I from *Electrophorus electricus*. *Proc. Nat. Acad. Sci. USA* **71,** 2113–2117.

Childers, S. R., and Siegel, F. L. (1975). Isolation and purification of a calcium-binding protein from electroplax of *Electrophorus electricus*. *Biochim. Biophys. Acta* **455,** 99–108.

Claudio, T., Ballivet, M., Patrick, J., and Heinemann, S. (1983). Nucleotide and deduced amino acid sequences of *Torpedo californica* acetylcholine receptor γ subunit. *Proc. Natl. Acad. Sci. USA* **80,** 1111–1115.

Emerick, M. C., Shenkel, S., and Agnew, W. S. (1993). Regulation of the eel electroplax Na channel and phosphorylation of residues on amino- and carboxyl-terminal domains by cAMP-dependent protein kinase. *Biochemistry* **32,** 9435–9444.

Fong, P., and Jentsch, T. J. (1995). Molecular basis of epithelial Cl channels. *J. Membrane Biol.* **144,** 189–197.

Grundfest, H. (1957). The mechanisms of discharge of the electric organs in relation to general and comparative electrophysiology. *Prog. Biophys.* **7,** 3–74.

Grundfest, H. (1960). Electric organ. *McGraw-Hill Encycl. Sci. Technol.* **8,** 427–433.

Hille, B. (1992). In "Ionic Channels of Excitable Membranes" (B. Hille, Ed.). Sinauer Associates, Sunderland, MA.

Jentsch, T. J., Steinmeyer, K., and Schwarz, G. (1990). Primary structure of *Torpedo marmorata* chloride channel isolated by expression cloning in *Xenopus* oocytes. *Nature* **348,** 510–514.

Karlin, A., and Cowburn, D. (1973). The affinity-labeling of partially purified acetylcholine receptor from electric tissue of *Electrophorus*. *Proc. Natl. Acad. Sci. USA* **70,** 3636–3640.

Keynes, R. D. (1961). The development of the electric organ in *Electrophorus electricus*. *In* "Bioelectrogenesis" (C. Chagas and A. Paes De Carvalho, Eds.), pp. 14–19. Elsevier, New York.

Lester, H. (1978). Analysis of sodium and potassium redistribution during sustained permeability increases at the innervated face of *Electrophorus* electroplaques. *J. Gen. Physiol.* **72,** 847–862.

Lingrel, J. B., and Kuntzweiler, T. (1994). Na$^+$,K$^+$-ATPase. *J. Biol. Chem.* **269,** 19659–19662.

Middleton, R. E., Pheasant, D. J., and Miller, C. (1994). Purification, reconstitution, and subunit composition of a voltage-gated chloride channel from *Torpedo* electroplax. *Biochemistry* **33,** 13189–13198.

Millard, C. B., and Broomfield, C. A. (1995). Anticholinesterases: Medical applications of neurochemical principles. *J. Neurochem.* **64,** 1909–1918.

Miller, C., and White, M. M. (1980). A voltage-dependent chloride conductance channel from *Torpedo* electroplax membrane. *Ann. N.Y. Acad. Sci.* **80,** 534–551.

Miller, J. A., Agnew, W. S., and Levinson, S. R. (1983). Principal glycopeptide of the tetrodotoxin/saxitoxin binding protein from *Electrophorus electricus:* Isolation and partial chemical and physical characterization. *Biochemistry* **22,** 462–470.

Mishina, M., Kurosaki, T., Tobimatsu, T., Morimoto, Y., Noda, M., Yamamoto, T., Terao, M., Lindstrom, J., Takahashi, T., Kuno, M., and Numa, S. (1984). Expression of functional acetylcholine receptor from cloned cDNAs. *Nature* **307,** 604–608.

Munjaal, R. P., Conner, C. G., Turner, R., and Dedman, J. R. (1986). Eel electric organ: Hyperexpressing calmodulin system. *Molec. Cell. Biol.* **6,** 950–954.

Nakamura, Y., Nakajima, S., and Grundfest, H. (1965). Analysis of spike electrogenesis and depolarizing K inactivation in electroplaques of *Electrophorus electricus*, L. *J. Gen. Physiol.* **49,** 321–349.

Noda, M., Takahashi, H., Tanabe, T., Toyosato, M., Furutani, Y., Hirose, T., Asai, M., Inayama, S., Miyata, T., and Numa, S. (1982). Primary structure of α-subunit precursor of *Torpedo californica* acetylcholine receptor deduced from cDNA sequence. *Nature* **299,** 793–797.

Noda, M., Takahashi, H., Tanabe, T., Toyosato, M., Kikyotani, S., Furutani, Y., Hirose, T., Takashima, H., Inayama, S., Miyata, T., and Numa, S. (1983a). Structural homology of *Torpedo californica* acetylcholine receptor subunits. *Nature* **302,** 528–532.

Noda, M., Takahashi, H., Tanabe, T., Toyosato, M., Kikyotani, S., Hirose, T., Asai, M., Takashima, H., Inayama, S., Miyata, T., and Numa, S. (1983b). Primary structures of β- and δ-subunit precursors of *Torpedo californica* acetylcholine receptor deduced from cDNA sequences. *Nature* **301,** 251–255.

Noda, M., Shimizu, S., Tanabe, T., Takai, T., Kayano, T., Ikeda, T., Takahashi, H., Nakayama, H., Kanaoka, Y., Minamino, N., Kangawa, K., Matsuo, H., Raftery, M. A., Hirose, T., Inayama, S., Hayashida, H., Miyata, T., and Numa, S. (1984). Primary structure of *Electrophorus electricus* sodium channel deduced from cDNA sequence. *Nature* **312,** 121–127.

Olsen, R. W., Meunier, J.-C., and Changeux, J.-P. (1972). Progress in the purification of the cholinergic receptor protein from *Electrophorus electricus* by affinity chromatography. *FEBS Lett.* **28,** 96–100.

Pallotta, B. S., and Webb, G. D. (1980). The effects of external Ca$^{++}$ and Mg$^{++}$ on the voltage sensitivity of desensitization in *Electrophorus* electroplaques. *J. Gen. Physiol.* **75,** 693–708.

Parker, K. K., Chan, S. L., and Trevor, A. J. (1978). Purification of native forms of eel acetylcholinesterase: Active site determination. *Arch. Biochem. Biophys.* **187,** 322–327.

Pasquale, E. B., Udgaonkar, J. B., and Hess, G. P. (1986). Single-channel current recording of acetylcholine receptors in electroplax isolated from the *Electrophorus electricus* main and Sachs' electric organs. *J. Membrane Biol.* **93,** 195–204.

Ratman, M., Sargent, P. B., Sarin, V., Fox, J. L., Nguyen, D. L., Rivier, J., Criado, M., and Lindstrom, J. (1986). Location of antigenic determinants on primary sequences of subunits of nicotinic acetylcholine receptor by peptide mapping. *Biochemistry* **25,** 2621–2632.

Schoffeniels, E. (1961). The flux of cations in the single isolated electroplax of *Electrophorus electricus* (L.). *In* "Bioelectrogenesis" (C. Chagas and A. Paes De Carvalho, Eds.), pp. 147–165. Elsevier, New York.

Shenkel, S., and Bezanilla, F. (1991). Patch recordings from the electrocytes of *Electrophorus*. Na channel gating currents. *J. Gen. Physiol.* **98,** 465–478.

Shenkel, S., and Sigworth, F. J. (1991). Patch recordings from the electrocytes of *Electrophorus electricus*. Na currents and $P_{Na}/P_K$ variability. *J. Gen. Physiol.* **97,** 1013–1041.

Sheridan, R. E., and Lester H. A. (1977). Rates and Equilibria at the acetylcholine receptor of *Electrophorus* electroplaques. A study of neurally evoked postsynaptic currents and of voltage-jump relaxations. *J. Gen. Physiol.* **70,** 187–219.

Sperelakis, N. (1979). Origin of the cardiac resting potential. *In* "Handbook of Physiology." Vol. 1 "The Cardiovascular System" (R. M. Berne and N. Sperelakis, Eds.), pp. 187–267. American Physiological Society, Bethesda, MD.

Steinmeyer, K., Klocke, R., Ortland, C., Gronemeier, M., Jockusch,

H., Grunder, S., and Jentsch, T. J. (1991a). Inactivation of muscle chloride channel by transposon insertion in myotonic mice. *Nature* **454,** 304–308.

Steinmeyer, K., Ortland, C., and Jentsch, T. J. (1991b). Primary structure and functional expression of a developmentally regulated skeletal muscle chloride channel. *Nature* **354,** 301–304.

Sussman, J. L., Harel, M., Frolow, F., Oefner, C., Goldman, A., Toker, L., and Silman, I. (1991). Atomic structure of acetylcholinesterase from *Torpedo californica:* A prototypic acetylcholine-binding protein. *Science* **253,** 872–879.

Szabo, T. H. (1961). Anatomo-physiologie des centres nerveux specifiques de quelques organes electriques. *In* "Bioelectrogenesis" (C. Chagas and A. Paes De Carvalho, Eds.), pp. 185–201. Elsevier, New York.

Walsh, J. (1775). Experiments and observations on the *Gymnotus electricus,* or electric eel. *Philos. Trans.* **65,** 94–101.

Yoda, A., Clark, A. W., and Yoda, S. (1984). Reconstitution of $(Na^+ + K^+)$-ATPase proteoliposomes having the same turnover rate as the membranous enzyme. *Biochim. Biophys. Acta* **778,** 332–340.

Michael Levandowsky and Thomas E. Gorrell

# 60

## Physiological Adaptations of Protists

## I. Introduction

This chapter introduces the reader to the great diversity to be found in the physiology of the **single-celled eukaryotes, or protists.** These include a variety of groups, some **autotrophic** or plant-like, some **phagotrophic** or **osmotrophic** and thus animal-like, and many with a combination of these traits.

First, a word about terminology and classification. Traditionally, these organisms comprised the **algae** and the **protozoa.** The early classifications of protozoa divided them into three groups based on their locomotion (**ciliates, flagellates, amoebae**) and a parasitic group (**sporozoa**). The algae were classified largely on the basis of pigments (**red algae, brown algae, golden-brown algae, green algae**) and obvious structural differences (**cryptophytes, dinoflagellates**). Many of the flagellated groups appeared in both classifications. These characteristics are clearly important, and the early classifications and terminology tend to persist in informal usage. However subsequent work with the electron microscope, biochemical advances, and molecular approaches have changed many of our views on phylogenetic relationships, and revised classification schemes have been proposed (Margulis, 1974; Cavalier-Smith, 1993).

In this chapter, the term **Protista,** or simply **protists,** is used to describe all of these groups. Figure 1 gives an informal indication of current thinking about relationships among protistan and other eukaryotic groups, as indicated by molecular evidence, particularly rRNA homology (Sogin, 1989, 1991).

The molecular evidence suggests that many protist lines probably originated at least one billion years ago. Thus the divergence between different protistan groups is probably at least comparable to that between, for example, the animal, plant, and fungal kingdoms.

Although they are certainly phylogenetically diverse, the protists do have a certain physiological unity, based on common problems confronted by unicellular organisms. While, in general, they are typical eukaryotic cells and follow the principles described elsewhere in this book, a number of evolutionary solutions have appeared here that are not present in the cells of multicellular organisms. An example is the **light antenna** ("eyespot") present in many photoautotrophic flagellates. This structure, which appears to have evolved independently in several protist groups, is not found in the multicellular animals and plants.

A feature of protists, just starting to be appreciated, is the apparent ease with which many cells "capture" and use physiological units of other cells. A prime example is the sequestration and use of **prey organelles** such as **chloroplasts** by phagotrophic protists. Indeed, the incorporation of symbionts is a pervasive feature of protistan physiology, occurring in most if not all groups. There is a range of endosymbiotic phenomena, ranging from the temporary uptake, endocytosis, and use of foreign cells or organelles, to established symbioses in which the endosymbiont has become a required part of the host cell. A dramatic example of this was the appearance of an intracellular bacterial infection in a laboratory culture of *Amoeba*. The infecting bacterium, initially deleterious to the host cell, was transformed from an **endoparasite** to an obligate **endosymbiont** over the course of a few years' culture in the laboratory, and the host and its endosymbiont are now mutually dependent (Jeon, 1987, 1995). This nearly universal tendency of protists to form temporary or permanent endosymbiotic complexes is a feature that the physiologist comes to expect, and a major theme of this chapter.

The relationships shown in Fig. 1 come from molecular and other evidence relating to the basic genome in these groups. However, if one considers plastids and other organelles, it is clear that many of the groups shown as separate are also linked by past endosymbiotic events. Thus, some (but not all) of the dinoflagellates are photosynthetic, and it is clear that their chloroplasts were obtained endosymbiotically from various chromophyte groups. Euglenids, on the other hand, though distantly related to the dinoflagellates, apparently got their chloroplasts endosymbiotically from a green algal source. These and other complexities of protistan evolutionary relationships do not appear in simple branching "family tree" diagrams such as Fig. 1.

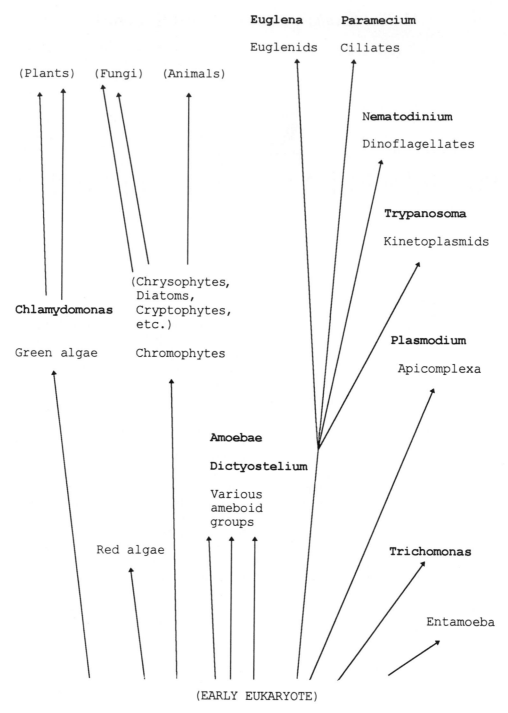

**FIG. 1.**   An informal diagram showing probable phylogenetic relations of major groups of protists. Organisms discussed in this chapter are in boldface. For a more rigorous presentation, visit http://phylogeny.arizona.edu/tree/phylogeny.html.

Also included in this chapter are descriptions of some striking protistan structures whose functions are not clear. Thus, this chapter catalogs the many types of **extrusomes,** microprojectiles that are found in many protistan groups, but whose function is in most cases still a matter of speculation.

A comprehensive review of protistan physiology is not given, because that would duplicate much of the treatment of basic cell physiology in other chapters. Rather, the focus is on unique or unusual features that have evolved in this group, and that are not found in the cells of multicellular organisms. These features are adaptations to the various free-living and parasitic niches filled by this very diverse group of organisms. In particular, emphasis is given here to understudied phenomena.

## II. Biophysical Constraints of Scale: The Example of Filter-Feeding

The dimensions of most protists and their appendages, the values of relative velocity in water, and the viscosity of water all combine to yield **Reynolds numbers**[1] much smaller than one. Thus, inertial forces are generally completely excluded as a factor in protistan biophysics. As an illustration of the practical implications of this, **filter-feeding** in flagellates and ciliates is examined. This has been worked out theoretically and experimentally by Fenchel (1986a,b) for a number of species, and his discussion is followed here.

Filter-feeding is widespread among protists that consume bacteria, other protists, or other particles. Suspended food particles may be "captured" passively by diffusion, where the particle reaches the protist by chance, through Brownian motion or swimming, and sticks to an adhesive surface. This happens with many protistan cells with ameboid tentacles (e.g., heliozoa). Particles may be caught raptorially by direct interception and ingestion during active swimming (many dinoflagellates, euglenids, ciliates). Finally they may be obtained by filtering a current of water, which is often produced by the protist. This latter method is examined here.

Following Fenchel, let $U(x)$ be an **uptake function,** where $x$ is the particle concentration. **Clearance,** $F$, is defined as the volume of water cleared per unit time. The relation between these two quantities is given by $F = U(x)/x$, where $U$ is assumed to be nearly proportional to $x$ at low particle concentrations, but becomes saturated at high concentrations. It is assumed that the rate of retention of particles by the filter is proportional to concentration, and that ingestion of retained particles takes a finite time, $t$, during which no other particles are ingested. These assumptions lead to a hyperbolic equation:

$$U(x) = F_m x(1 - Ut) \tag{1}$$

Substituting variables and rearranging yields the equation:

$$U(x) = U_m x/(U_m/F_m + x) \tag{2}$$

where $U_m = 1/t$ (the maximum rate of ingestion as $x \to \infty$), $F_m$ is the maximum clearance realized as $x \to 0$, and $U_m/F_m$ is a constant (dimension $L^{-3}$). From Eq. 2, we can see that the kinetics of filter-feeding is formally identical to the familiar Michaelis–Menten kinetics of enzymology,

---

[1] Reynolds number is defined as $R = dv/n$, where $d$ is a spatial dimension, $v$ is velocity, and $n$ is the kinematic viscosity. Essentially, it can be thought of as the ratio of inertial to viscous forces acting on a body moving through a fluid. Thus, for a large, fast-moving organism in water, such as a fish, the ratio is much greater than one and inertial forces dominate. For microbes, on the other hand, the ratio is usually much less than one, and they live in a world dominated by viscous forces.

with the constant $U_m/F_m$ corresponding to the half-saturation constant.

Equation 2 has been verified experimentally for protistan uptake of latex particles and of bacterial food, and $F_m$ varies greatly with the particle size (Fig. 2). When the size for which $F_m$ is maximal is determined experimentally, that size is assumed to be retained with an efficiency approaching 100%; thus, that value of $F_m$ can be taken as a direct measure of the rate of flow of water through the filter. This can then be compared for different flow fields of water and for different species' filter designs.

To understand the biological significance of clearance, it is useful to express it as volume-specific clearance, i.e.,

$$F_m/(\text{cell volume})$$

As an example of the application of this theory, consider a situation in which the volume fraction of bacteria in a seawater sample is $10^{-6}$ (e.g., $2 \times 10^6$ cells/mL $\times 0.5 \ \mu m^3$ cell volume). A typical heterotrophic flagellate with a specific clearance of $10^5$/hr (it filters $10^5$ times its volume of water per hour) will need 10 hr to ingest its own volume of bacteria. With a 50% growth efficiency, it would thus be able to divide every 20 hr. Growth efficiency, or **yield,** of protists is nearly invariant over a large range of growth rates (Fenchel, 1986a,b). For balanced (i.e., log phase) growth,

$$u(x) = U(x)Y \tag{3}$$

where $u(x)$ is the instantaneous growth rate at a food concentration $x$, and $Y$ is a yield constant. Thus, and as verified experimentally, $u(x)$ and $U(x)$ have similar functional forms, so that growth data can be used to estimate clearance.

Fenchel (1986b) also investigated the flow field generated by various filter-feeding flagellates and ciliates both theoretically and empirically. It is here that the assumption

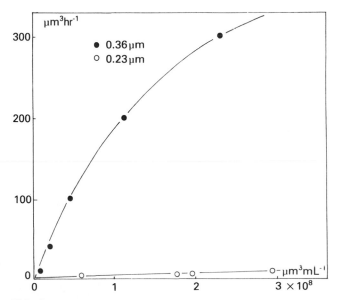

**FIG. 2.** Volume uptake of two sizes of latex beads as functions of environmental concentration by *Cyclidium* and with the data fitted to Eq. 2. (Adapted with permission from Fenchel, 1986a,b.)

of low Reynolds number is used, in calculating the theoretical flow field produced in the medium relative to the cell by ciliary motion. An example is shown in Fig. 3, where the predicted theoretical flow field is compared to the observed paths of suspended particles through the filter apparatus of a ciliate. From these flow lines one can see that a larger volume is sampled than might be expected simply from the area of the filter itself.

## III. Nutrition and Excretion

### A. Endocytosis, Digestion, and Defecation

As with other cells, protists take in nutrients and other materials by a variety of passive and active mechanisms. Of particular interest here are specialized processes of phagocytosis and digestion not found elsewhere. Many ciliates and some flagellates have complex feeding structures. The best known case is *Paramecium,* the subject of many studies by Allen and Fok (Allen, 1984; Fok and Allen, 1990). This cell has a fixed **cytopharynx** (or **buccal cavity**), a tube-shaped cavity where food particles are taken in. At the end of this is the **cytostome,** where the **food vacuoles** (FV) (also **phagosomes,** or **digestive vacuoles**) are formed. The cytostomal membrane, which pinches off to form the

FV, is distinct from the cell membrane antigenically and in structure. During formation of the FV it grows by fusion with disk-shaped vesicles, which are transported to it along microtubular ribbons. The movement is thought to be due to cytoplasmic dynein in a microtubule-based motor. The vesicles are essentially recycled membrane from earlier FVs, returning to the cytostome by a defined path.

After forming, the FV follows a defined path through the cytoplasm. As it progresses, vesicles called **acidosomes** also move along the cytopharyngeal microtubular ribbons and fuse with the vacuole just after it pinches off, lowering the pH of its interior. Whether the acidosomes are filled with acid or simply deliver proton pumps is not resolved yet. The low pH kills the prey and favors the activity of digestive enzymes. The release of the FV from the cytostome and its subsequent movement involve actin. As the acidosomes fuse with it, the original membrane of the FV pinches off to form small vesicles that recycle back to the cytostome, so that the FV is totally reconstituted. Next, lysosomes migrate from the golgi apparatus and fuse synchronously with the FV, bringing acid hydrolases, including acid phosphatase. The membrane may be completely replaced again at this stage. Digestion takes about 20 min. At the end of this phase, vesicles form and are pinched off from the FV, probably to be used again in lysosomes.

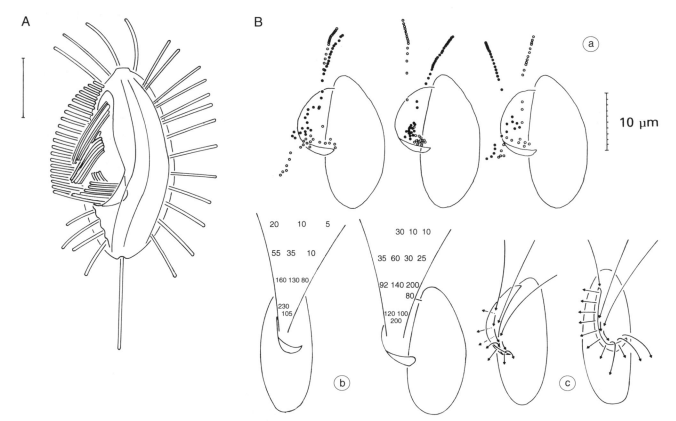

**FIG. 3.** (A) Diagrammatic rendition of *Cyclidium,* showing filtration apparatus. Bar is 5 μm. (B) *a.* Position of 1.1-μm latex particles at 0.02-sec intervals along six flow lines during filtration by *Cyclidium.* *b.* Schematic presentation of the critical flow lines and approximate velocities (μm/sec). *c.* Schematic presentation of the flow lines. (Reprinted with permission from Fenchel, 1986a.)

The remaining FV membrane binds to microtubules that guide it to the **cytoproct,** or **cytopyge,** where the FV membrane fuses with the cell membrane to form a pore. Its contents (undigested residue), are then egested (**defecation**) and the FV collapses. The cytoproct (cytopyge) is a specific location where the **plasmalemma,** a layer containing infraciliature and other structures, has a gap allowing the food vacuole to reach the cell membrane and fuse with it. After defecation the membrane is recycled in discoidal vesicles to the cytostome, completing the cycle. This last step appears to involve an actin-based system that may be calcium regulated.

## B. The Contractile Vacuole

Protists without cell walls that live in hypotonic media (freshwater species) have **contractile vacuoles** (CV), which periodically excrete fluid. In the best studied case, the ciliate *Paramecium,* this consists of a central vacuole, a surrounding complex of **ampullae** and a network, or **spongiome,** of tubules (Fig. 4). Certain tubules in this complex are decorated with peg-like elements that are vacuolar-type proton pumps. These are found in both the cellular slime mold *Dictyostelium* and the ciliate *Paramecium* (Heuser *et al.,* 1993; Allen, 1997).

Excretion occurs through a cycle. During **diastole,** the vacuole forms and grows by the fusion of smooth-membrane vesicles. During this period fluid travels from the spongiome to the CV. In **systole,** the vacuole membrane fuses with the cell membrane at one site to form a pore and the fluid is excreted. As this happens the vacuole contracts as its membrane fragments and forms vesicles again. Actin has not been detected near the CV, and contraction is thought to be due to cellular pressure after the appearance of the pore opening to the exterior. At one time it was thought that the connections of the CV to the spongiome were interrupted during systole, preventing backflow.

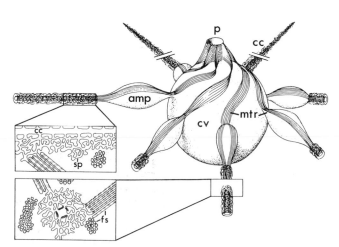

**FIG. 4.** Contractile vacuole complex in *Paramecium,* showing ampullae (amp), collecting canal (cc), contractile vacuole (cv), pore (pv), spongiomal tubules (sp), fluid segregation organelles (fs), and microtubular ribbons (mtr). (Adapted with permission from Hausmann and Hülsmann, 1996.)

However recent work indicates that the connections persist throughout the cycle. The narrowness of the tubular connexions present great resistance, so that considerable pressure would be required for a rapid backflow during systole.

A major function of the CV is clearly osmoregulation. The cycle ceases in cells placed in a hypertonic medium, resuming after a time when the cell adapts and increases internal tonicity. Fluid in the CV is high in K, Na relative to the cytoplasm, suggesting also a role in maintenance of ionic balance.

## IV. Energetic Adaptations: Fermentative Microbodies

Although aerobic protists have mitochondria similar to those in other cells, the earliest protists are thought to have lacked mitochondria. Many modern species, both parasitic and free-living, are either **microaerophiles** or **anaerobes** (facultative or obligate), and other organelles have evolved in response to energy requirements in these groups.

Studies of parasitic protists revealed a pair of organelles: **glycosomes** and **hydrogenosomes.** Glycosomes, named for the presence of several glycolytic enzymes in a microbody-like organelle, have been found in the **kinetoplastida,** a group that includes many important parasites (*Trypanosoma, Leishmania*) as well as free-living organisms. Hydrogenosomes are hydrogen-producing organelles found in the parasitic trichomonad flagellates, as well as certain ciliates and fungi.

Both hydrogenosomes and glycosomes belong to the category of organelles called **microbodies.** Most microbodies have a single surrounding membrane, and function quite differently from mitochondria and chloroplasts (the "energy organelles"). **Peroxysomes** and **glyoxysomes** are classic examples of microbodies in higher plants and animals, as well as in many protists. Glycosomes and hydrogenosomes, however, perform functions that resemble more closely the energy organelles. Each facilitates formation of adenosine triphosphate (ATP) by a fermentative process. Glycosomes contain the first seven enzymes of glycolysis (Opperdoes, 1987), while hydrogenosomes metabolize pyruvate to $H_2$, $CO_2$ and acetate (Müller, 1993). Glycosomes and hydrogenosomes function in separate groups of organisms for energy production and use organic molecules as terminal electron acceptors. ATP formation occurs in these microbodies by a **substrate-level phosphorylation** mechanism, in which a phosphorylated intermediate of glycolysis serves to phosphorylate adenosine diphosphate (ADP). Both types of organelle show unusual aerobic and anaerobic modes of metabolism. Moreover, the glycosomes are found in parasites that have a complex life cycle, including separate stages adapted to life in the bloodstream of vertebrates and in insect tissues. Both glycosomes and hydrogenosomes are of considerable interest as potential targets for development of new drugs for serious parasitic infections—e.g., African sleeping sickness, Chagas' disease, leishmaniasis, trichomoniasis. As yet, no DNA has been detected in these microbodies.

## A. Glycosomes

Glycosomes were initially discovered in 1977 in the bloodstream form of the cattle parasite, *Trypanosoma brucei,* and later in other members of the kinetoplastida. The latter are a group of mostly parasitic flagellates that have a single mitochondrion with a characteristic DNA-enriched structure, the **kinetoplast.** Besides the bloodstream form, trypanosomes can have several other morphological forms, including the **procyclic** form found in the invertebrate host. Procyclics are often used in comparative studies because they are easy to culture. Bloodstream forms (also known as slender forms) can be obtained by harvesting from infected vertebrate hosts, or in recent times by **axenic** culture (pure culture, without other cell types) (Hirumi and Hirumi, 1989). Information about glycosomal function is derived mainly from studies of overall reactions and the subcellular location of enzymes. Much remains to be learned about specific transport mechanisms.

The bloodstream form of the parasite shows high rates of glucose catabolism and enrichment of glycolytic enzymes in the glycosome (Opperdoes, 1987). Under aerobic conditions, pyruvate is the sole product of the cells, but oxygen is consumed by the **promitochondria,** a process quite distinct from that found in normal mitochondria (see Chapter 6). Without oxygen, equal amounts of pyruvate and glycerol are formed. The glycosome represents about 5% of the cell protein, and 90% of this protein consists of enzymes of the first seven steps of glycolysis (i.e., up to the formation of 3-phosphoglycerate) (Fig. 5). This phosphoglycerate then leaves the glycosome, forming a separate cytoplasmic pool for the subsequent production of pyruvate via phosphoenolpyruvate and formation of adenosine triphosphate

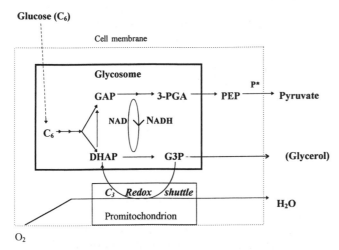

**FIG. 5.** Summary of glycosomal metabolism in bloodstream forms of *Trypanosoma brucei.* Redox shuttle: glycosomally produced glycerol-3-phosphate (G3P) is oxidized in the promitochondrion to dihydroxyacetone phosphate (DHAP) and this oxidized three-carbon unit returns to the glycosomes. The P* denotes where net formation of ATP occurs under aerobic conditions. Under anaerobic conditions glycerol is produced in equal amounts to pyruvate. Additional abbreviations: glyceraldehyde-3-phosphate, GAP; 3-phosphoglycerate, 3-PGA; phosphoenolpyruvate, PEP. (Adapted with permission from Fairlamb and Opperdoes, 1986.)

(ATP). Most interestingly, under aerobic conditions the amount of ATP consumed in activating the six-carbon sugars is balanced by formation of 2 mol of 3-phosphoglycerate. Thus, the glycosome under these conditions does not provide net formation of ATP for the cell, despite the considerable turnover of ATP.

The excretion of pyruvate by the cell leaves unanswered the question of how the glycosome oxidizes reduced nicotinamide adenine dinucleotide (NADH), a rather typical problem for a fermentative pathway. Studies of enzyme activities and their subcellular distribution lead to the conclusion that an unusual $C_3$ redox shuttle connects the glycosome with the mitochondrion of the cell (see Fig. 5). In the shuttle, DHAP reduction to glycerol-3-phosphate results in the regeneration of NAD from NADH for the pathway leading to pyruvate production. The glycerol-3-phosphate moves to the mitochondrion, where an alternative oxidase regenerates the dihydroxyacetone phosphate to resupply the glycosome. The transport systems involved have not been characterized and the oxidase has not been purified. The mitochondrion in bloodstream forms is regarded as a promitochondrion since it lacks several components of the mitochondrion. Bloodstream forms do not obtain ATP by the alternative oxidase.

For bloodstream forms under aerobic conditions with an inhibitor of the oxidase [e.g., salicyl hydroxamic acid (SHAM), a metal chelator], or under anaerobic conditions, glucose consumption occurs at a rate comparable to that of aerobic cells. Cells now produce glycerol and pyruvate at equal rates. Glycerol is a direct product of glycosomal metabolism via a glycerol kinase. It has been suggested that this formation of glycerol leads to ATP formation, despite the unfavorable thermodynamic potential. ATP formed by isolated organelles, however, has not yet been identified.

While glycosomes do function in bloodstream forms to degrade glucose, several other functions also occur in the organelle, depending on the species and the stage of the life cycle. Procyclic forms rely on respiration by a complete mitochondrion. Glycosomes function in $CO_2$ fixation. In addition, they show key enzyme activities for both fatty acid oxidation and ether lipid peroxidation. These enzyme activities, along with the presence of catalase (in a few strains) and protein sequences of glycosomal enzymes, suggest a close relationship between glycosomes and peroxysomes (Opperdoes, 1987; Sommer and Wang, 1994).

The glycosome is unique for the presence of the glycolytic enzymes. How this compartmentation benefits the cell is not clear, however. Several studies have sought to define a novel regulatory mechanism. One regulated step would be the final stage in pyruvate production: Fructose 2,6-bisphosphate is produced by the cells and stimulates pyruvate kinase (which is used in pyruvate formation). In most other eukaryotic cells fructose 2,6-bisphosphate typically stimulates phosphofructokinase. More detailed studies of transport processes are needed to help in understanding the role of compartmentation of glycolysis in cells.

## B. Hydrogenosomes

**Hydrogenosomes** have been found in several unrelated groups, including parasitic flagellates and free-living protists. Besides these protozoa, at least one rumen fungus has hydrogenosomes (Müller, 1993; Yarlett *et al.*, 1986). They were first discovered in 1973 in the cattle parasite *Tritrichomonas foetus* and in the human parasite *Trichomonas vaginalis*. Hydrogenosomes represent about 5% of the cell protein in the trichomonads. No mitochondria are present in these cells, and they lack the ATPase characteristic of oxidative phosphorylation (Müller, 1993).

The organelle was named for its ability to produce hydrogen gas while oxidizing pyruvate, produced by cytosolic glycolysis. A significant amount of carbon flows through the organelle. Net ATP formation occurs by a substrate level phosphorylation step (Fig. 6). The specific pathway of ATP formation varies with the organism. In *Trichomonas*, acetate production occurs while the cell transfers coenzyme A to succinate for subsequent ATP formation by a succinate thiokinase. This pathway has been demonstrated with hydrogenosome-enriched fractions that were incubated with pyruvate, succinate, ADP, and phosphorus. Succinate accumulates in cultures of *T. foetus* but not in *T. vaginalis*. In rumen ciliates pyruvate oxidation produces ATP via acetylphosphate, but fatty acids such as butyrate are also produced.

The production of hydrogen from pyruvate represents a highly atypical feature for a eukaryotic cell. This property is more characteristic of anaerobic bacteria, such as clostridia. Clostridia produce hydrogen by a pathway in which proteins with iron-sulfur clusters in the active site catalyze the oxidation of pyruvate, the production of hydrogen, and the transfer of electrons between two enzymes. Trichomonads were found to have a pathway similar to clostridia except the ferredoxin had a single [2Fe-2S] cluster rather than the two [4Fe-4S] clusters present in clostridia. Interestingly, two other well-studied parasitic protists, *Entamoeba histolytica* and *Giardia lamblia*, lack both hydrogenosomes and mitochondria but have the pyruvate enzyme and a clostridial-type ferredoxin. These are sometimes considered to be relicts of an early, bacteria-like eukaryote stage, the **Archezoa** (Cavalier-Smith, 1993).

Removal of pyruvate by the hydrogenosome eliminates its use as an electron acceptor for reoxidizing glycolytically reduced NAD. It was discovered recently that *T. vaginalis*

oxidizes the NADH under anaerobic conditions by the production of glycerol. In contrast to trypanosomes, trichomonads lose energy by producing glycerol. Trichomonads have a glycerol-3-phosphatase rather than a glycerol kinase. Consequently, when these cells produce glycerol they lose one pyruvate. The amount of glycerol produced varies between species, with *T. vaginalis* producing considerably greater quantities than *T. foetus*, due to the latter's production of succinate.

Trichomonads prefer growth under anaerobic or micro-aerophilic conditions, but do have the ability to consume oxygen (Müller, 1993). Likewise, the isolated hydrogenosomes consume oxygen, but in contrast to whole cells this activity is rapidly lost. The cytosol of trichomonads contains a soluble NADH oxidase that directly reduces oxygen to water; this would protect the oxygen-labile components of the hydrogenosome, such as the two iron-sulfur enzymes. Oxygen does not change the rate of glucose consumption, in contrast to aerobic eukaryotic cells, and cells gain no ATP by oxygen reduction. (Oxygen stops hydrogen production by intact cells, but acetate and $CO_2$ production continue.) Contrary to expectations oxygen does not inhibit production of glycerol by *T. vaginalis*.

The presence of the hydrogenosomal pathway is not essential for growth of trichomonads (Kulda *et al.*, 1993), since several strains of *T. foetus* and *T. vaginalis* have been obtained by long-term culture in the presence of metronidazole. This drug is reductively activated to a cytotoxic form by the oxidation of pyruvate. The cells grow at a decreased rate and show increased rate of glucose consumption, however.

Free-living ciliates with hydrogenosomes show an unusual symbiotic relationship by having established methanogenic bacteria as endosymbionts (see Fig. 12, later, Section V,B). The hydrogenosomes in these ciliates have an unusual membrane since it forms invaginations and thus the organelle structurally resembles a mitochondrion. The methane production pulls reoxidation of NADH against a standard midpoint potential difference of $-100$ mV by (presumably) removing $H_2$ produced via NADH.

The presence of hydrogenosomes in such diverse anaerobic organisms as flagellates, free-living and symbiotic ciliates, and rumen fungi provides for many variations on the hydrogenosome. Trichomonad hydrogenosomes have now been shown to have a double membrane. Additional functions for hydrogenosomes such as accumulation of Ca and formation of a transmembrane potential have also been reported for some species (Müller, 1993).

The hydrogenosome can be regarded as a fermentative analog of a mitochondrion, since both organelles degrade pyruvate and form ATP as a major function. It is believed to have originated in an endosymbiotic event.

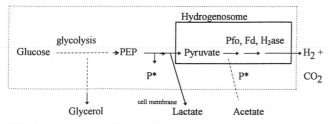

**FIG. 6.** Summary of *T. vaginalis* hydrogenosomal metabolism. Glycolysis occurs by the Embden–Meyerhoff pathway in the cytoplasm. Oxidation of pyruvate occurs in the hydrogenosome by the sequential action of pyruvate : ferredoxin oxidoreductase (Pfo), ferredoxin (Fd), and hydrogenase ($H_2$ase). Net formation of ATP is designated by P* and phosphoenolpyruvate by PEP. The dashed line indicates that more than one step occurs in the formation of acetate and of ATP. (Adapted with permission from Müller, 1993.)

## V. Sensory Adaptations, Membrane Potentials, and Ion Channels

Like other cells, protists exhibit sensitivity and respond in various ways to environmental stimuli. The basic physiological and biochemical mechanisms, as they are known, are similar to those found in metazoan systems. **Paramecium,** for example, has been referred to informally as a "swimming neuron." In this section several examples of photoreceptors and gravireceptors are examined, and then recent work on the underlying transduction mechanisms is discussed.

## A. Photoreceptors

Ambient light is a source of energy to autotrophic protists. In addition, for motile species, it can serve as a directional signal. Many flagellates, and probably all the motile photosynthetic species, exhibit a positive phototaxis. For this purpose a directional receptor is needed. Directional receptors, or "eyespots," fall into two main categories: (1) receptors with opaque screens, which detect direction by the screen's shadow, and (2) receptors with **antennae** based on interference and diffraction to detect direction, essentially the same optical principle as that seen in a reflection hologram. In addition to these directional detectors, one group of predatory dinoflagellates has developed a system of **ocelloids** with intracellular lenses. In this section we discuss the antennae and the ocelloids, two unique but understudied systems.

### 1. Receptors with Light Antennae

The interferometric **light antenna** is found in a phylogenetically diverse array of flagellates, including many genera of green algae, dinoflagellates, and cryptophytes (Foster and Smyth, 1980; Melkonian and Robenek, 1984; Smyth *et al.,* 1988). In the cases where experiments have been done, these antennae appear to be selective with regard to wavelength and light direction. The presence of essentially similar organelles in a wide array of groups considered on many other grounds, including molecular homology, to be phylogenetically diverse raises a question: Are these all cases of convergent evolution? If not, have these organelles spread through endosymbiotic events?

The antennae operate through an interference mechanism similar to that which produces structural color in iridescent objects. The essential structure consists of a number of parallel reflective layers that act as partial mirrors. These are spaced evenly, separated by a distance of a quarter of the wavelength to be selected. Thus, incident light passes through a series of partially reflecting pigment layers spaced at quarter-wavelength intervals. Normally incident light of the appropriate wavelength is reflected back at each partial mirror. These reflected light rays are in phase with each other and with the incoming radiation, leading to positive reinforcement of this light signal in front of the eyespot, where the receptor pigment is located. The basic principle is illustrated in Fig. 7. Light from other angles, and light of other wavelengths also passes through this filter, but the reflections are out of phase with each other, and the resultant interference minimizes their contribution.

The most intensive studies of light antennae were done with the green flagellate *Chlamydomonas* (Fig. 8) (Foster and Smyth, 1980), but the antennae are also found in dino-

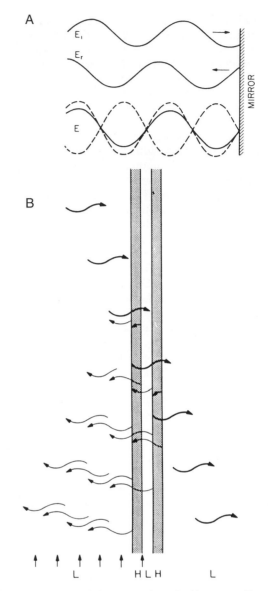

**FIG. 7.** Reflection. (A) Perfect mirror. Incident wave $E_i$, reflected wave $E_r$, and the sum $E = E_i + E_r$, a standing wave; the dashed lines show the extrema. (B) Interference reflector. A segment of a wave one wavelength long at successive instants, showing reflections from partial mirrors spaced quarter-wavelengths apart. The reflected waves are in phase. (From Foster and Smythe, 1980.)

flagellates, chrysophytes, and many other light-responsive flagellates.

### 2. Intracellular Lenses in Dinoflagellates

Perhaps the most complex and surprising photoreceptors are the ocelloids (ocelli), eyespots found in a family of nonphotosynthetic dinoflagellates, the Warnowiaceae. A typical ocelloid consists of a lens, a retinoid, and an opaque pigment cup (Fig. 9). The pigment cup, or **melanosome,** consists of a wall of pigment granules, and surrounds the paracrystalline **retinoid.** In front of this is the lens, or **hyalosome,** consisting of a peripheral corneal zone and a central crystalloid body. The entire organelle is approximately 24 $\mu$m long and 15 $\mu$m wide (Greuet, 1978, 1987).

Francis (1967) measured the refractive index of the lens in *Nematodinium* by three methods, and found it to be approximately 1.52. He determined the focal plane by tracing rays parallel to the optic axis and found that light was focused in the retinoid layer. The field of view in these species was determined to be about 30 deg, but may be wider in other species. From these observations we conclude that the ocelloids could probably act as directional light receptors.

Given the presence of **nematocysts** (see Section VI,C) in these same dinoflagellate species, and the fact that they are predators rather than autotrophs, it has been suggested that the ocelloids might act as "range finders," leading to discharge of nematocysts when the contrast of the focused image on the retinoid is maximal. While certainly plausible, such behavior has not been reported yet. Unfortunately, these species have not been cultured, and there appear to be no studies of ocelloid function, so that all we have at this time are speculations based on the structure.

### B. Gravity Receptors in Ciliates

Many single-celled organisms orient with respect to the earth's gravitational field while swimming. In some, such as the well-studied ciliate *Paramecium,* this is thought to be a purely hydrodynamic effect due to the shape and the distribution of mass in the cell: As it swims, the center of mass pulls the rear of the cell downward, orienting it so that it swims upward (Roberts, 1981). In this case, then, there seems to be no need for a gravity receptor. However, many protists can migrate either up or down, switching between a positive and a negative geotaxis in response to external signals or to a circadian or a tidal rhythm, suggesting a more complex type of response.

In particular, a group of ciliates containing the genera *Loxodes* and *Remanella* has a gravity response that de-

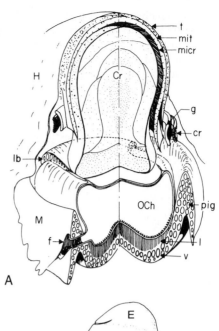

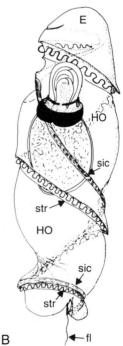

**FIG. 9.** (A) Diagram of structure of *Nematodinium* ocellus, showing interior structure: crystalline body (Cr), ocelloid channel (f), constricting ring (cr), periocelloid gallery (g), hyalosome (H), lamellae of retinal body (l), microcrystalline layer (micr), mitochondrion (mit), ocelloid chamber (OCh), pigmentary ring (pig), basal plate (lb), microtubular layer (t), vesicular layer of the retinal body (v). Magnification: 3900×. (B) Position of ocellus in the *Nematodinium* cell, showing parts of the cell: episome (E), hyposome (HO), intercingular sulcus (sic), cingular sulcus with transverse flagellum (str), posterior flagellum (fl). (From Greuet, 1978.)

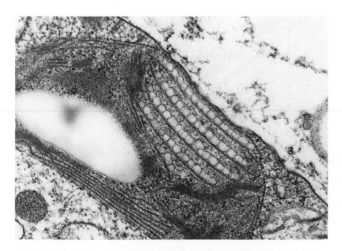

**FIG. 8.** Sectional view of the eyespot of *Chlamydomonas,* showing reflecting pigment layers forming a quarter-wavelength stack. (From Foster and Smyth, 1980.)

pends on the dissolved oxygen level. In nature, these ciliates collect at the interface between anaerobic and aerobic zones in the water column. In oxygen-containing water they swim downward, while in an anoxic environment they swim upward. They all contain a characteristic organelle, the Müller body, which acts as a gravity receptor. This is a fluid-filled vesicle in which a membrane-covered mineral body is suspended (Fig. 10). The mineral body, consisting of a strontium or barium salt, acts as a statolith, and its position on the interior surface of the vesicle appears to inform the cell of its orientation in the gravitational field (Fenchel and Finlay, 1986). The vesicle is anchored to the system of connected cilia of the cell, and it is suggested, by analogy to other mechanoreceptors, that oriented mechanical stress on the cell membrane at this point of connection may lead to a depolarization, which in turn would affect ciliary beating and determine the orientation of the

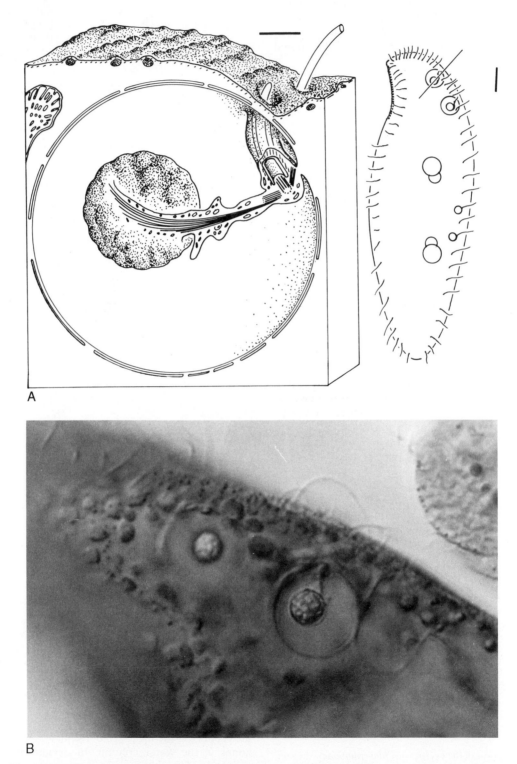

**FIG. 10.** (A) Müller body, gravity receptor. Drawing showing position in *L. striatus* cell (scale bar: 10 $\mu$m), and a three-dimensional reconstruction from serial sections (scale bar: 1 $\mu$m). (B) Muller body in living cell of *L. striatus*. (From Fenchel and Finlay, 1986.)

swimming motion (Fenchel and Findlay, 1986) (see also Section VIII).

### C. Sensory Transduction: Membrane Potentials, Ion Channels, and Intracellular Components

#### 1. Membrane Potentials, Calcium, and Behavior

As with many aspects of protistan physiology, membrane potentials are best known in the large ciliate *Paramecium.* A classic feature of ciliate behavior is the **avoidance reaction,** in which the cell stops, swims backward for a short time, then swims forward again, usually in a somewhat different direction. Early workers observed prolonged backward swimming in cells subjected to various stresses: high temperatures, sudden changes in pH or osmotic pressure, exposure to solvents and other deleterious chemicals. The common denominator in all these proved to be that the cell membrane became "leaky" or permeable to $Ca^{2+}$ and other ions. Using the methods developed by Szent–Gyorgy and others for muscle tissues, *Paramecium* cells were exposed to detergents such as Triton X-100 to render the cell membrane permeable. Such model-extracted cell preparations would swim forward if provided with ATP and $Mg^{2+}$ in a medium containing $[Ca^{2+}] < 10^{-7}$ *M.* At higher $Ca^{2+}$ concentrations the cells swam backward (Naitoh and Kaneko, 1973).

Electrophysiological studies with intracellular electrodes in normal cells have given the following general picture: During forward swimming, a $Ca^{2+}$-dependent membrane potential is maintained. Appropriate mechanical, chemical, or electrical stimuli cause an action potential in which calcium channels open in the membrane, causing depolarization and backward swimming. The membrane potential is restored by a calcium pump after a short time and forward swimming resumes. Cells are depolarized by repellents (stimuli that cause backward swimming) and hyperpolarized by attractants (stimuli that inhibit backward swimming).

#### 2. Behavioral Mutants

Kung and his colleagues obtained a number of behavioral mutants in which swimming behavior could be correlated with abnormal electrophysiological properties of the *Paramecium* membrane. Examples include the following:

1. *Pawn* (several types), in which voltage-gated $Ca^{2+}$ channels are affected, depolarization does not occur, and the cell cannot swim backward
2. *Pantophobiac,* where $Ca^{2+}$-dependent $K^+$ currents are affected, and prolonged responses occur to all stimuli
3. *Paranoiac* (several types), affecting $Ca^{2+}$-dependent $Na^+$ channels. Prolonged responses occur in $Na^+$ solutions

Many other mutants are known (Saimi and Kung, 1987).

#### 3. Ion Channel Types and Membrane Excitation

Electrophysiological and behavioral genetic studies have revealed eight distinct types of ion channels in *Paramecium,* and at least three types of membrane excitation that govern behavioral responses, as follows (Hinrichsen and Schultz, 1988):

1. Two mechanically induced currents: a depolarizing, $Ca^{2+}$-based current in the cell's anterior and a hyperpolarizing, $K^+$-based current in the posterior
2. A rectifying $K^+$ current that can cause regenerating hyperpolarization during the action potential
3. A second $Ca^{2+}$-dependent $K^+$ current, which may play a role in maintaining the resting potential

#### 4. Second Messengers and Transduction Pathways

Second messengers and internal biochemical events following mechanical or chemical stimuli have been the subject of study in various protists. In the case of the *Pantophobiac* mutants of *Paramecium,* for example, it was found that normal behavior could be restored by microinjection of wild-type calmodulin (CaM). This led to the discovery that these mutants were specific point mutations leading to amino acid substitutions at specific CaM sites. Further studies showed that the $Ca^{2+}$–CaM complex regulates calcium-dependent $Na^+$ channels by direct interaction and is also required for the functioning of $K^+$ channels.

Progress at this level has been greatest with the cellular slime mold *Dictyostelium discoideum.* In this organism there appear to be several chemosensory transduction pathways, similar in general to that found in animal cells such as leukocytes but differing in some aspects. Figure 11 shows a model summarizing current information and theory for this species. Greater diversity is expected to be found as more protistan groups are examined (Van Houten, 1992; Van Haastert, 1991; Newell *et al.,* 1990).

## VI. Incorporation of Physiological Units from Other Cells

Symbiotic relationships, including cellular endosymbiosis, are widespread in biology, including in multicellular organisms (e.g., the algal symbionts of corals and certain flatworms). In the protists, this tendency appears to be accentuated, and many new physiological opportunities have been produced by endosymbiotic combinations. Thus, many protists have prokaryotic or eukaryotic endosymbionts. In some cases these associations appear to be more or less permanent, as in the green ciliate *Paramecium bursaria,* which always has an algal *Chlorella* endosymbiont. Sometimes the association is more intimate, and parts of another cell have been incorporated into the host cell permanently. There are also more transient associations, in which the host "cultures" and uses all or part of an ingested cell in its cytoplasm for a period of time. Examples of all of these possibilities are presented in this section.

### A. Intracellular Capture and Culture of Foreign Organelles

Early reports of photosynthesis by certain natural populations of chlorophyll-containing ciliates suggested that

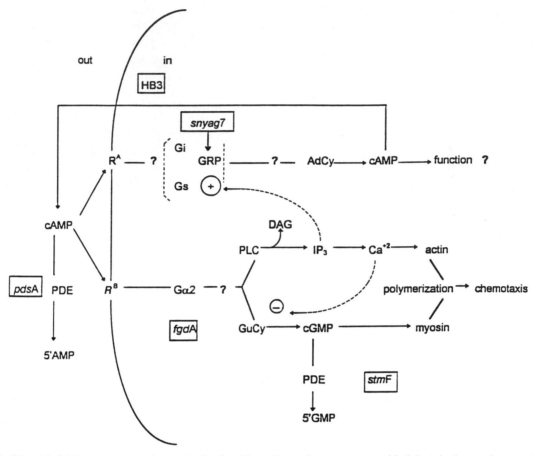

**FIG. 11.** Model for sensory transduction in *D. discoideum*. Boxes denote mutants with defects in the nearby step of the pathway. Abbreviations: cAMP, cyclic adenosine monophosphate; cGMP, cyclic guanosine monophosphate; PDE, phosphodiesterase; $R_A$, $R_B$, receptors; PLC, phospholipase C; IP$_3$, inositol trisphosphate; G, G-proteins; GRP, GTP-reconstituting protein; GuCy, guanylyl cyclase; AdCy, adenylyl cyclase; DAG, diacylglycerol. (Adapted with permission from Van Haastert, 1991.)

these putatively heterotrophic organisms might harbor algal symbionts. Electron microscopy, however, revealed that some ciliates contain isolated fragments of algal cells, which appear to be functional. The photosynthetic marine ciliate *Mesodinium rubrum,* which sometimes forms "red tides" (massive blooms that color the sea), contains endosymbiotic algal organelles in the form of chloroplast–mitochondrial complexes (Taylor *et al.,* 1971). Other species of planktonic ciliates have been found to contain intact algal chloroplasts, pyrenoids, and even eyespots. Indeed, a significant proportion of the planktonic ciliates in various marine habitats has been found to contain "captured" chloroplasts, which remain functional within the ciliate for extended periods, eventually being consumed by the host (Stoecker *et al.,* 1987).

## B. Xenosomes: Bacterial Endosymbionts

The term **xenosome** was used by Soldo (1987) to describe certain bodies in marine ciliates. These were later identified as endosymbiotic bacteria, and Corliss (1985) suggested that the term be defined to include all DNA-containing, membrane-bound intracellular bodies or organisms. By this definition, mitochondria and chloroplasts would be included. In practice, however, the term has been used largely for bacterial endosymbionts of marine ciliates.

### 1. Bacterial Endosymbionts in Ciliates

Some endosymbiotic bacteria, such as the *Kappa* and *Alpha* particles in *Paramecium* and the *Omicron* particles in *Euplotes,* are classical objects of study (Görtz 1996). More recently there have been a number of studies by Soldo and colleagues of the xenosomes from marine ciliates (Soldo, 1987; Soldo *et al.,* 1992). Originally found in the species *Parauronema acutum,* they are infective to 12 strains of this species, and also to the phylogenetically distant species *Uronema marinum,* but not to strains from five other marine ciliate genera, or the freshwater genera *Paramecium* and *Tetrahymena.* They are toxic to some marine ciliates, such as *Uronema nigricans,* and the toxic effect is abolished by proteolytic enzymes.

These xenosomes contain DNA, RNA, proteins, and lipids in amounts typical of small bacteria, and are selectively destroyed by a number of antibiotics, including penicillin, ampicillin, tetracycline and chloramphenical, but not neomycin or cycloheximide. They are small gram-negative rods, present in numbers ranging from 100 to 200 per host cell, and divide in synchrony with the host. When released by gentle mechanical rupture of the host they swim with darting motions or spin like propellers. They average two flagella per cell, inserted at random sites in the cell wall. Within the host cytoplasm they have a fairly typical gram-negative double outer membrane and single inner membrane, with a layer of peptidoglycan between these. They are not enclosed in vacuoles, but contact the host cytoplasm directly.

Xenosome chromosomal DNA is multicopy, consisting of 9–14 circularly permuted duplex molecules of about 515 kb. In addition there are several plasmids. Analysis of restriction sites revealed that all the adenines of GATC sequences in the plasmids are methylated, whereas those in chromosomal sequences are not, suggesting that there are two replicons, one controlling chromosomal and the other plasmid replication.

Surprisingly, 30% of protein in xenosomes is from the host cell. This is, however, consistent with the relatively small genome. Assuming 1.3 kb of DNA is needed to encode a protein of average $M_r$ 66,000, then the xenosome can encode fewer than 400 proteins, less than one-tenth the number for a typical free-living species such as *Escherichia coli.*

Xenosomes *in vitro* consume oxygen at a very low rate, comparable to that observed in Rickettsia and *Kappa* symbionts, and preferentially use succinate as an energy source. Symbiont-containing ciliates, on the other hand, consume oxygen at a rate 20–30% higher than symbiont-free ciliates, and host glycogen is consumed at a significantly higher rate in the latter than in the former.

The affinities and origins of these organisms present an intriguing problem. Ribosomal RNA analysis of marine ciliate xenosomes indicates some homology to the mu and pi xenosomes of *P. aurelia,* but little or no homology to known free-living species, suggesting that these endosymbionts may have a long history of association with ciliates (Soldo *et al.,* 1992).

## 2. Methanogenic Endosymbionts in Anaerobic Ciliates

Most of the free-living anaerobic ciliates examined have endosymbiotic and/or ectosymbiotic bacteria (Fenchel *et al.,* 1977), and the role of some of the former have been studied in some detail. In particular, a number of species harbor methanogenic bacteria. In the cases where these have been investigated, the bacteria are intimately associated with a series of hydrogenosomes. The arrangement of bacteria and organelles has been compared to a stack of coins, with bacteria and hydrogenosomes alternating, and the latter on the ends (Fig. 12). This complex is highly organized in some species, and possibly a permanent feature of the host cell. In the ciliate *Plagiopyla frontata* the bacteria divide synchronously with the host. The number of methanogens per host cell (about 3000) remains constant until a late stage in the host cell cycle, when it doubles. Thus, the ciliate apparently controls reproduction of the symbiont to maintain a stable population density.

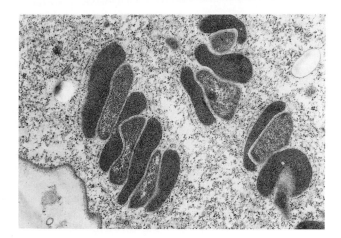

**FIG. 12.** Hydrogenosomes and methanogen symbionts in the ciliate *P. frontata.* The darker bodies are the hydrogenosomes and the lighter bodies are the symbionts. Magnification: 20,000×. (From Fenchel and Finlay, 1991.)

When the methanogen inhibitor 2-bromoethanesulfonic acid was used to inactivate the methanogenic symbionts of three ciliate species, *P. frontata, Metopus contortus,* and *M. palaeoformus,* growth rate and yield were reduced in the first two species, but not in the last. It is suggested that the energetic advantage conferred by the symbiont in the first two species may be due to the secretion of organic material by the bacteria (Fenchel and Finlay, 1991). The advantage to the methanogens of consumption of $H_2$ and acetate from the adjacent hydrogenosomes seems clear. This could be a significant advantage, especially in marine or other sulfate-rich environments, where free-living methanogenic bacteria would compete with the more efficient sulfate-reducing bacteria for $H_2$.

## 3. Bacterial Endosymbionts in *Amoeba proteus*

Many endosymbionts have been reported in large free-living amebas, such as *Amoeba proteus* (Jeon, 1995). Particularly dramatic is the case of a bacterial endosymbiont that has been studied from its initial appearance as a contaminant in laboratory culture, through the coevolution of host and endosymbiont, to a mutually dependent symbiotic relationship (Jeon, 1987). The bacterium, which first appeared in the culture in 1966, is termed the **X-bacterium.** It infected a culture of the D strain of *A. proteus,* which already contained other symbiont-like particles of unknown origin. The X-bacterium is described as a gram-negative rod, with an ultrastructure similar to *E. coli,* but is not otherwise identified taxonomically as yet. Initially most of the infected amoebae died, but a few survived, and the X-bacteria gradually lost virulence. The number of bacteria per ameboid cell, originally greater than 100,000, stabilized at about 42,000. Within a few years, the host cell, now called the xD strain, became dependent on the presence of the endosymbiont. X-bacteria can be transferred into other D strain cells by microinjection or by induced phagocytosis. The bacteria are enclosed in host-generated vesicles, or

**symbiosomes,** and when observed in freeze-fracture preparations are found to be embedded in a matrix of fibrous material. The symbiosomes do not fuse with lysosomes, and during infection the X-bacteria seem to be somewhat resistant to lysozymes, since about 10% of them avoid digestion in the phagolysosomes. Two kinds of plasmids were found in X-bacteria, and isolated X-bacteria treated with ethidium bromide or acridine orange for 3 hr failed to infect amoebae.

Several molecules produced by the symbiont have been studied. One, the Xd29 protein, appears to be a peripheral membrane protein that is constantly shed into the host cytoplasm, passing readily through the symbiosome membrane. Symbiont-produced lipopolysaccharides (LPS) have also been identified and were shown by immunostaining to be present on the cytoplasmic side of the symbiosomes. Injected antibodies to the LPS abolished the fusion-avoiding properties of symbiosomes, causing them to fuse with lysosomes. A 96-kDa protein from the symbiont is also present on the symbiosome membrane, and is suspected of playing a role in preventing lysosomal fusion. X-bacteria contain a large amount of 67-kDa heat-shock protein (HSP, GroEx), but since there are no free-living X-bacteria cultures for comparison, it is not known whether this is an indication of stress. In other intracellular infective bacteria (e.g., *Legionella*) the GroEx protein in the intracellular bacteria is more than seven times greater than in free-living cells. The complete nucleotide sequence of the *GroEx* operon of the X-bacteria was determined and it has a high degree of homology with those of other endoparasitic or symbiotic bacteria, such as *Legionella* and *Coxiella.*

Some polypeptide bands detected by gel electrophoresis of the amoebae cytosol are no longer present after prolonged endosymbiosis. One protein that disappears after symbiosis is a SAM (S-adenosyl-L-methionine) synthetase-like protein. A genomic library of *A. proteus* has been prepared in an effort to detect genetic changes as a result of long-term symbiosis with X bacteria.

The following changes are reported in experimentally infected cells:

1. Accelerated cell growth during the initial phase of experimental infection for up to 12 months
2. Increased sensitivity to starvation
3. Newly acquired temperature sensitivity above 26°C, one degree higher than their optimum growth temperature
4. Increased sensitivity to overfeeding
5. Increased sensitivity to crowding in culture vessels
6. A symbiont-synthesized 29-kDa protein in the host cytoplasm
7. Development of dependence on the symbiotic X-bacteria

Clearly, the analysis of this evolving system is only beginning. With regard to future research directions, Jeon (1995) lists the following unanswered questions:

1. Why are X-bacteria dependent on amoebae?
2. Why are host amoebae dependent on their symbionts for survival?

3. How do symbiosomes avoid fusing with amoebic lysosomes?
4. Are any chromosomal and/or plasmid genes of X-bacteria involved in rendering the bacteria resistant to digestion by lysosomal enzymes?
5. What are the precursor bacteria from which the X-bacteria arose? What are the basic differences between them and the X-bacteria?
6. Is there a permanent genomic change of amoebae caused by symbiosis?

## VII. Structures with Unknown Functions

Next, emphasizing the opportunities for future research, a number of more or less prominent structures are examined that are not understood at the basic level of function (though some have been the subject of interesting speculations).

### A. Rhoptries

A large group of medically important parasitic protists, the **Apicomplexa** (formerly the **Sporozoa**), is defined by a structure called the **apical complex.** This group includes such important parasites as *Plasmodium,* which causes malaria. All the apicomplexa are obligate intracellular parasites at some stage in their life cycle, and the apical complex is thought to be an instrument of invasion. A prominent part of this complex are the **rhoptries,** secretory organelles containing lipids and proteins (Fig. 13) that originate in

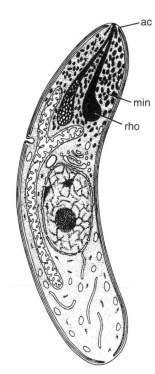

**FIG. 13.** Diagram of apicomplexid cell, showing rhoptries (rho), micronemes (min), and apical complex (ac). (Adapted from Chobotar and Scholtyseck, 1982.)

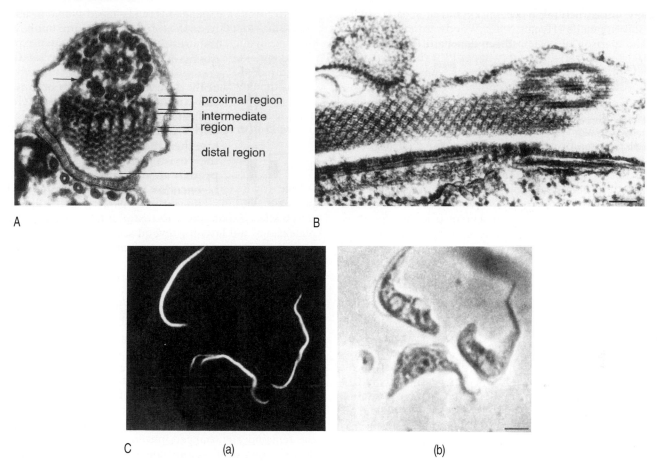

proximal region

intermediate region

distal region

A

B

C          (a)

(b)

**FIG. 14.** (A) Cross-section of the flagellum and paraflagellar rod of *Trypanosoma brucei* (scale bar: 74 nm). (B) Longitudinal section (scale bar: 256 nm). (C) Immunofluorescent staging of the PFR with monoclonal antibody ROD-1. Fluorescence picture (a) and phase image (b) (scale bar: 5.5 um). (From Sherwin and Gull, 1996.)

the **Golgi system,** filled with enzymes. For many years the general assumption has been that rhoptry and microneme secretions must play a role in the invasion of the host cell, but it has proven difficult to identify actual function (Sam-Yellowe, 1996). In some apicomplexid species the parasite is contained in a **parasitophorous vacuole** after invasion, but in others the vacuole disappears and the parasite is in direct contact with the host cytoplasm. Rhoptries contain dense protein granules, and epitopes corresponding to these have been identified in the host cell membrane, its cytoskeleton, and also in the parasitophorous vacuole membrane, where present. Their role, however, is not clear.

### B. Apicoplastids

*Plasmodium* and other parasitic apicocomplexids have a structure that appears to be homologous to the chloroplast of green algae (Köhler *et al.,* 1997). An organelle surrounded by four membranes contains a 35-kb circular fragment of DNA, which is shown by cluster analysis to be most closely related to DNA in green algal chloroplastids. Localization of the fragment was done by an *in situ* hybridization technique. The organelle divides by binary fission

and is introduced into daughter cells early in replication. The genome is transcribed, and transcription products have been identified. The presence of four membranes enclosing this organelle suggests it originated endosymbiotically, following ingestion of an algal protist that contained a plastid.

The function of this organelle, which has been named the **apicoplastid,** is not known, but it clearly must have a function. It has been found in all apicomplexids examined so far. Photosynthesis is the usual function of chloroplasts, but this is improbable in this group of obligate parasites. Other functions performed by chloroplasts include biosynthesis of amino acids and fatty acids, assimilation of nitrate, and starch storage. Plastids are found in a number of non-photosynthetic algal species.

### C. The Paraflagellar Rod

The **paraflagellar rod** (PFR) is found in parasitic kinetoplastids and also in the free-living euglenids and dinoflagellates. These three groups of flagellates are considered to be distantly related (see Fig. 1). The PFR is a complex, lattice-like structure running parallel to the flagellum (Fig. 14). In both euglenids and kinetoplastids there have been many ultrastructural, biochemical, and molecular studies of this organelle (Hyams, 1982; Woodward *et al.*, 1994).

Comparison of proteins from flagella of *Euglena* (with PFR) and *Chlamydomonas* (without PFR) indicated a pair of major proteins as PFR components. Subsequently, similar protein pairs were found in various kinetoplastid species. Cross-reactivity of kinetoplastid and euglenid PFR was established, using antibodies to PFR of the kinetoplastid *Crithidia*. In addition to the two major proteins, a number of minor protein components have been found.

Regarding function, in the euglenids the PFR is located adjacent to the eyespot, and is therefore thought to be involved in photoreception in some way. In the parasitic kinetoplastids, current opinion favors a role in attachment to host cells during infection. Direct evidence for these rather different roles is lacking however[2] (Bastin *et al.*, 1996).

### D. Extrusomes

A feature of many protists is the ability to extrude pre-formed structures. **Extrusomes** are vesicles that contain some organized substance or apparatus, which usually changes its form when released to the exterior in exocytosis. In some cases these organelles are clearly related to the feeding activities of the cell, while in others they may be a defensive adaptation, but in many cases their function is as yet unknown. They are extruded in response to chemical, mechanical, or electrical stimuli. While some metazoa eject structures, such as the **cnidocysts** (also called **nematocysts**) of **Hydra** and other coelenterates, such functions are usually performed by differentiated cells. In the unicellular protists specialized organelles have evolved instead.

The terminology and organization of this section follows that of Hausmann and Hülsmann (1996). The classification is based largely on morphology, and it is not clear whether, for example, *trichocysts* in ciliates and flagellates are homologous or represent parallel evolutionary developments. In the case of the ejectosomes, a truly remarkable question of homology arises: The Kappa particles in the ciliate *Paramecium* (bacterial endosymbionts, see earlier discussion) contain **R-bodies,** which are morphologically indistinguishable from the **ejectosomes** of certain flagellates (see later section), and these have even been observed to unroll and form tubes, as ejectosomes do. Did the ejectosome originate as part of a bacterial endosymbiont?

### 1. Spindle Trichocysts

These are probably the best studied of the extrusomes. Spindle trichocysts of *Paramecium,* a favorite demonstration of introductory biology courses, are found in the cortex of the cell, just under the plasma membrane. In the resting state, the trichocyst is a spindle-shaped or rhomboid paracrystalline protein body. In response to various chemical and/or mechanical stimuli, this unfolds in a few millisec-

onds to form an expanded thread-shaped filament, about eight times longer than the former resting form but having the same diameter. The driving force for this sudden expansion is not known. It is independent of ATP, but $Ca^{2+}$ is required.

It is thought that the function of trichocysts is to repel predators, though the evidence for this is not very clear. Trichocysts are found in ciliates, and in dinoflagellates and other flagellate groups (Hausmann and Hülsmann, 1996).

### 2. Mucocysts

Like trichocysts, **mucocysts** are found just under the plasma membrane, and consist of paracrystalline filamentous bodies. Expansion to the exterior takes place in three dimensions and lasts for several seconds. They are found in various ciliates, flagellates, and the ameboid actinopods, where they may be responsible for the sticky surface used in capturing food organisms. Otherwise, they are thought to have a protective function. They resemble the cortical granules of sea urchin eggs, which are involved in the formation of the fertilization membrane.

### 3. Discobolocysts

**Discobolocysts** are found in certain flagellates (**chrysophytes** and others). In the intracellular resting state they are almost spherical, with a disk in the part next to the plasma membrane. On ejection, the disk is unaltered, but the remainder is changed into long filamentous material. The function of these organelles is not known.

### 4. Toxicysts

These are found in ciliates and some phagotrophic flagellates. They function in the capture of prey, and perhaps in defense against predators. The toxicyst capsule contains a long tube, which during extrusion is either telescoped or everted. This enters the prey and is used to inject a toxin, which kills or paralyzes the prey. In the suctorian ciliates, they discharge on contact, and serve also to hold the prey until it can be taken in and consumed. The nature of the toxin(s) does not appear to have been investigated.

### 5. Rhabdocysts

These are rod shaped and occur in one group of ciliates (the **karyorelictids**). As with toxicysts they discharge telescopically—an event that has been compared to the discharge of an arrow from a blowpipe. Their function is unknown.

### 6. Ejectosomes

These occur in certain flagellates (**cryptophytes** and **prasinophytes**). In the intracellular resting state they are like tightly coiled ribbons. Extruded, they unroll and form very long tubes. This is said to be an escape reaction. The remarkable similarity of this organelle with the R-bodies in the *Kappa* particle endosymbionts of *Paramecium* was

---

[2] Interestingly, the only group of kinetoplastids that lacks a PFR is a group of monogenetic parasites that have bacterial endosymbionts of the genus *Bordetella.*

noted earlier, and raises the possibility of an endosymbiotic origin for this organelle.

### 7. Epixenosomes

Epixenosomes occur tightly bound to the outer surface of the plasma membrane in certain ciliates. They, like the ejectosomes, contain a tightly coiled band that unrolls and forms a tube. They contain DNA, however, and thus are epibionts, not organelles, and on the basis of several structural features it has been suggested that they may represent a primitive type of organism somewhere between the prokaryotes and the eukaryotes. In any case they are found in all specimens of different species of the genus *Euplotidinium,* from different geographic regions. Thus, this appears to be a tight, ancient symbiotic relation.

### 8. Nematocysts

These are found in certain dinoflagellates, and are thought to function in predation, though this has not been observed. As with toxicysts, they strongly resemble the **cnidocysts** (also known as nematocysts) of multicellular coelenterates such as Hydra. They are capsules containing a coiled tube, which evaginates on extrusion.

### E. Crystalline Bodies

Many protists contain crystals, usually of calcium and phosphorus salts, and with small amounts of magnesium, chloride, or organic material in some cases. Classically, these were considered simply waste products, since they seemed to be somewhat dependent on diet. More recently it has been suggested they may be reservoirs of ions needed in metabolism. They are often formed when the cells are grown in optimal conditions, and when removed experimentally tend to be replaced quickly.

## VIII. Protistan Responses to Gravity and to Gradients of Oxygen and Light: An Example from Physiological Ecology

In this section an example is presented of the integration of protists in their environment, based on some of the physiological capacities discussed in earlier sections. The example is taken from a study of a nutrient-rich pond (Berninger *et al.,* 1986). Figure 15 shows the distribution of two kinds of ciliates in relation to several critical physicochemical factors. The situation depicted is typical of summer, when temperatures are high and there is not much wind. In these circumstances, bacterial metabolism depletes oxygen in the lower part of the water column, and in the absence of mixing by wind, the pond becomes stratified. The depth profiles of two ciliate populations are shown. The first consists of three species of microaerophilic zoochlorellae-bearing ciliate. These remain in a low-oxygen zone, where major predators have difficulty following them. Their lower limit is set by the light requirement of

their symbionts, which produce the oxygen that enables them to survive in this zone. The second group consists of a species of the ciliate *Loxodes,* mentioned earlier (Section V,B) in connection with the gravity receptor, or Müller body.

*Loxodes* responds to three interacting factors: oxygen, blue light, and the force of gravity. As noted earlier, in the dark, cells accumulate in regions of low oxygen tension (ca. 5% of saturation). If then exposed to light, they swim into the dark or into anaerobic water. When exposed to high $O_2$ and light simultaneously, after initial episodes of backward swimming (the **avoiding reaction**), they swim downward (a **positive geotaxis**). If placed in anaerobic water, especially in the dark, they respond by swimming upward (**negative geotaxis**). The end result of all this is that, in the pond, they accumulate by geotaxis in a zone in the **oxycline** (oxygen gradient) that is optimal. Typically this is in the dark, just below the ciliates with zoochlorellae symbionts. The cells are sensitive to very low light levels, so that levels as low as 10 W m$^{-2}$ will cause them to swim down into anaerobic water. *Loxodes* can use nitrate as a terminal electron acceptor, which is unusual in ciliates. Meanwhile, by swimming down into the anaerobic zone, the cells avoid most of their potential predators—zooplankters, planktonic larvae, juvenile fish—which are restricted to the aerobic zone.

Figure 16 summarizes what is known or suspected regarding the physiological basis for this adaptive behavior. Obviously, much remains to be done before the physiological basis for some of the arrows in the figure is understood. Photoreception remains to be worked out in detail, but it is clear that a blue light receptor is involved, possibly a flavin, and one product of its excitation in the presence of oxygen is superoxide radical (Finlay and Fenchel, 1986). However, *Loxodes* has only low levels of superoxide dismutase and catalase. Thus, superoxide, or a product of its dismutation, for example, hydrogen peroxide, might be the internal signal for oxygen perception, binding to cytochrome oxidase or perhaps reducing another component of the electron transport system (ETS), such as cytochrome *c*. A drawback of this speculation is that the ETS is in the mitochondrion, while the pigment granules are in the cell membrane. Thus, a change in the ETS might influence the membrane potential of the mitochondrion, but it is not clear how this signal might be transmitted to the cell membrane, to influence ciliary beating and change the swimming behavior (Finlay *et al.,* 1986).

In any case the linked responses to light, oxygen tension, and gravity seem very understandable given the ecology of this organism. What is attractive in this system is the possibility of linking the organism's physiology and its ecology.

## IX. Summary: Protistan Diversity

The theme of this chapter has been the physiological diversity that has evolved among the many protistan groups, as seen in some of the better studied examples. However, even these examples are as yet only poorly un-

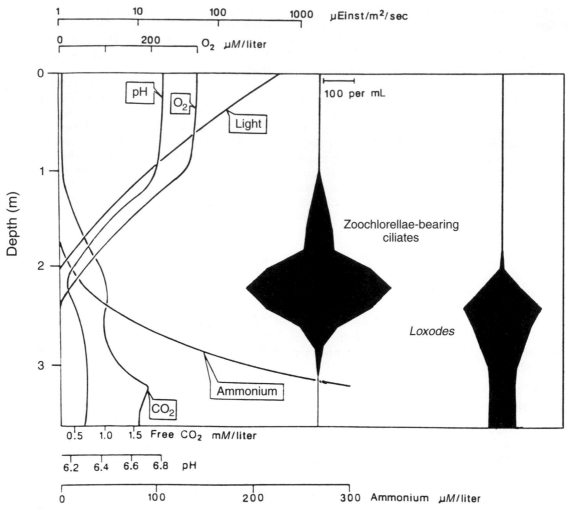

**FIG. 15.** Vertical distribution of microaerophilic zoochlorellae (algal symbiont)-bearing ciliates, and the ciliate *Loxodes* in a small productive pond, with profiles of some relevant physical and chemical factors. (Adapted from Berninger *et al.,* 1986, and Fenchel and Finlay, 1986.)

derstood, and many groups remain unstudied or little studied. The diversity of known physiological systems and mechanisms is likely to increase enormously as these groups are explored. Let us consider this at several different levels of organization.

## A. Molecular Diversity

There is great diversity at the level of molecular genetics, not treated in this chapter. Ciliates, for example, practice a slight variation on the otherwise universal genetic code: The universal stop codons, TAA and TGA, instead code for glutamine in some (but not all) ciliates. (It has been suggested this may serve as a barrier to viruses, which have not been detected in ciliates.) The presence of two nuclei in ciliates is also different: The **micronucleus** serves as the genetic archive, for storage and recombination of the genetic pattern, while the **macronucleus** is the regulator of cellular function. In trypanosomes and other kinetoplastids, the **kinetoplast** is a large network of DNA associated with a single giant mitochondrion. This **kinetoplast DNA** (kDNA) consists of a network of about 50 **maxicircles,** which carry the mitochondrial genes, and about 10,000 **minicircles.** It was noticed that many genes in the maxicircles appeared nonfunctional, since they lacked conventional punctuation and contained frameshifts. However, mRNA with functional coding was produced. It turned out that the minicircle DNA was used to produce "guide RNA," which edited the pre-mRNA from the maxicircles. Since this discovery, similar RNA editing has been found in other organisms, including a mammalian nucleus.

Indeed, protists have been a fruitful source of discovery in fundamental biology. Early in the century it was noted that some ciliates could divide approximately 50 times asexually, and then had to undergo sexual genetic recombination, anticipating Heyflick's epochal discovery of a similar mortality in mammalian cell lines by several decades. The mechanism involved here, change in length of the telomeres with each division, was first discovered in a ciliate, *Tetrahymena* (Blackburn, 1992; Blackburn and Greider,

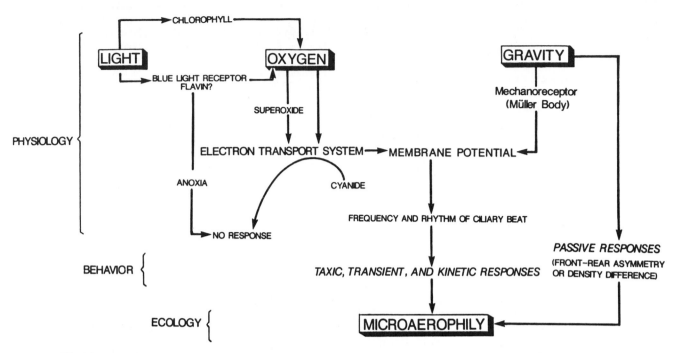

**FIG. 16.**   Model of the probable physiological and behavioral responses to the three cardinal factors (light, oxygen, and gravity) controlling microaerophily in ciliated protozoa. (Adapted with permission from Finlay, 1990.)

1995). Studies with this ciliate also revealed for the first time that RNA could be an enzyme (Cech, 1987).

### B. Organellar Diversity

Figure 1, based largely on nucleic acid homologies, gives a view of the evolutionary sequence in the development of protists, based to a large degree on molecular evidence, especially rRNA homology. However, the ever-branching treelike structure in Fig. 1 may be misleading. For example, the earliest eukaryotes, thought to be represented by certain relict contemporary anaerobic organisms such as the flagellate *Trichomonas,* lacked both mitochondria and chloroplasts, while another early group, the red algae (Rhodophyta), has these organelles but lacks the axoneme and thus the eukaryotic flagellum. If one assumes, as experts on these groups do, that the absence of these organelles is a primitive feature (i.e., not the result of secondary evolutionary losses), then logic indicates that either (1) these organelles evolved more than once or (2) an endosymbiotic event occurred, in which a cell containing mitochondria and chloroplasts combined with one having the eukaryotic flagellum with an axonemal structure. (Note that many biologists think that mitochondria and chloroplasts, and perhaps also flagella, originated from symbiotic prokaryotes.)

As seen in this chapter, though, endosymbiotic events do occur often among protists. Protists are not necessarily prisoners of their phylogenetic past. If a useful invention appears in one evolutionary line, it can cross to another by a symbiotic transfer. Many dinoflagellates are photosynthetic, but ultrastructural evidence indicates that their chloroplasts originated in other evolutionary lines (the crypto-

phytes and the chrysophytes). The ciliates, distant relatives of the dinoflagellates, also lack chloroplasts, but, as mentioned in this chapter, planktonic species often "capture" and use chloroplasts from their algal prey, and in some cases these symbioses appear to have become more or less permanent.

In summary, Fig. 1 reflects the phylogeny of sRNA and other nucleic acids in the cell, but, in organellar organization, one evolutionary line can borrow or take from another. At this level, a flowchart is more appropriate than a tree.

### C. Cellular Diversity

Finally, protistan cells can vary in space and time. Some have complex life cycles, or can form multicellular structures in which there is specialization and differentiation. Perhaps most well known here are the cellular slime molds. At one stage these function as individual small amoebae. Under certain conditions, however, the individual cells send chemical signals to each other and come together to form multicellular "slugs." These move about for a time and then metamorphose into a more or less complex multicellular reproductive structure. Curiously, in one group of ciliates essentially the same thing occurs—presumably a parallel evolution.

This chapter started with the description of filter-feeding by protists—an important ecological process that is illustrative of the problems faced by cells that are also free-living organisms. This was followed by a more detailed discussion of feeding: phagotrophy and digestion in the well-studied case of the ciliate *Paramecium.* Since mitochondria in the protists, though varying somewhat in structure, are essentially similar to those in other cells, their metabolism was not examined here. Instead the unique glycosomes and hydrogenosomes of anaerobic and microaerophilic protists were discussed.

Returning to the view of protists as organisms, several sensory organelles unique to these cells were examined, followed by a consideration of their sensory transduction. Here there are some exciting similarities and differences with vertebrate and invertebrate neurons and sensory cells. In particular, a calcium potential dominates in those cells that have been studied, and calmodulin-based pathways are prominent aspects of transduction. This is a very new area.

The widespread occurrence of endosymbiosis in protists was illustrated with well-documented examples. Next, to emphasize the opportunities for breaking new ground with protistan physiology, a number of striking organelles of unknown function were described. Finally, returning to the original emphasis on the cells as organisms, an example from the physiological ecology of a freshwater ciliate in a high-nutrient freshwater pond was described.

The record suggests that, by studying the diversity of protistan adaptations, cell physiologists can find new answers to old questions, and also find new questions.

## Acknowledgments

We thank Nigel Yarlett, Tom Fenchel, F. J. R. Taylor, and Kenneth Foster for helpful discussions.

## Bibliography

Allen, R. D. (1984). *Paramecium* phagosome membrane: From oral region to cytoproct and back again. *J. Protozool.* **31,** 1–6.

Allen, R. D. (1997). Membrane tubulation and proton pumps. New ideas in cell biology. *Protoplasma* **189,** 1–8.

Bastin, P., Matthews, K. R., and Gull, K. (1996). The paraflagellar rod of Kinetoplastida: Solved and unsolved questions. *Parasit. Today* **12,** 302–307.

Berninger, U.-G., Finlay, B. J., and Canter, H. M. (1986). The spatial distribution and ecology of zoochlorellae-bearing ciliates in a productive pond. *J. Protozool.* **33,** 557–563.

Blackburn, E. H. (1992). Telomerases. *Annu. Rev. Biochem.* **61,** 113–129.

Blackburn, E. H., and Greider, C. W. (Eds.) (1995). "Telomeres." Cold Spring Harbor Laboratory Press, Cold Spring Harbor, New York.

Cavalier-Smith, T. (1993). Kingdom Protozoa and its 18 phyla. *Microbiol. Rev.* **57,** 953–994.

Cavalier-Smith, T., and Lee, J. J. (1985). Protozoa as hosts for endosymbioses and the conversion of symbionts into organelles. *J. Protozool.* **32,** 376–379.

Cech, T. (1987). The chemistry of self-splicing RNA and RNA enzymes. *Science,* **236,** 1532–1539.

Chobotar, W., and Scholtyseck, E. (1982). Ultrastructure. *In* "The Biology of the Coccidia." (D. M. Hammond and P. L. Long, Eds.), pp. 10–37. University Park Press, Baltimore.

Corliss, J. O. (1985). Concept, definition, prevalence and host interactions of xenosomes (cytoplasmic and nuclear endosymbionts). *J. Protozool.* **32,** 373–376.

Fairlamb, A. H., and Opperdoes, F. R. (1986). Carbohydrate metabolism in African trypanosomes, with special reference to the glyco-

some. *In* "Carbohydrate Metabolism in Cultured Cells" (M. J. Morgan, Ed.), pp. 183–224. Plenum, New York.

Fenchel, T. (1986a). "Ecology of Protozoa." Springer-Verlag, New York.

Fenchel, T. (1986b). Protozoan filter feeding. *Progr. Protistol.* **1,** 65–113.

Fenchel, T., and Finlay, B. J. (1986). The structure and function of Müller vesicles in loxodid ciliates. *J. Protozool.* **33,** 69–76.

Fenchel, T., and Finlay, B. J. (1991). Endosymbiotic methanogenic bacteria in anaerobic ciliates: Significance for the growth efficiency of the host. *J. Protozool.* **38,** 18–22.

Fenchel, T., and Finlay, B. J. (1995). "Ecology and Evolution in Anoxic Worlds." Oxford University Press, Oxford, UK.

Fenchel, T., Perry, T., and Thane, A. (1977). Anaerobiosis and symbiosis with bacteria in free-living ciliates. *J. Protozool.* **24,** 154–163.

Finlay, B. J. (1990). Ecology of free-living protozoa. *Adv. Microb. Ecol.* **11,** 1–36.

Finlay, B. J., and Fenchel, T. (1986). Physiological ecology of the ciliated protozoan *Loxodes. Rep. Freshwater Biol. Ass.* **54,** 73–96.

Finlay, B. J., Fenchel, T., and Gardner, S. (1986). Oxygen perception and $O_2$ toxicity in the freshwater ciliated protoozoon *Loxodes J. Protozool.* **33,** 157–165.

Fok, A. K., and Allen, R. D. (1990). The phagosome–lysosome membrane system and its regulation in *Paramecium. Int. Rev. Cytol.* **123,** 61–94.

Foster, K. W., and Smyth, R. D. (1980). Light antennas in phototactic algae. *Microbiol. Rev.* **44,** 572–630.

Francis, D. (1967). On the eyespot of the dinoflagellate, *Nematodinium. J. Exp. Biol.* **47,** 495–501.

Gallo, J. M., and Schrevel, J. (1985). Homologies between paraflagellar rod proteins from trypanosomes and euglenoids revealed by a monoclonal antibody. *Eur. J. Cell Biol.* **36,** 163–168.

Görtz, H. D. (1996). Symbiosis in ciliates. *In* "Ciliates" (K. Hausmann and P. Bradbury, Eds.), pp. 441–462. Gustav Fischer, Stuttgart.

Greuet, C. (1978). Organization ultrastructurale de l'ocelloide de *Nematodinium.* Aspect phylogenetique du photorecepteur de Peridiniens Warnowiidae Lindemann. *Cytobiologie* **17,** 114–136.

Greuet, C. (1987). Complex organelles. *In* "The Biology of Dinoflagellates." (F. J. R. Taylor, Ed.), pp. 119–142. Blackwell, Oxford.

Hausmann, K. and Hülsmann, N. (1996). "Protozoology." Georg Thieme Verlag, Stuttgart, Germany.

Heuser, J., Zhu, Q., and Clarke, M. (1993). Proton pumps populate the contractile vacuoles of *Dictyostelium* amoebae. *J. Cell Biol.* **121,** 1311–1327.

Hinrichsen, R. D., and Schultz, J. (1988). *Paramecium:* A model system for the study of excitable cells. *Trends Neur. Sci.* **11,** 27–32.

Hirumi, H., and Hirumi, K. (1989). Continuous cultivation of *Trypanosoma brucei* bloodstream forms in a medium containing a low concentration of serum protein without feeder cell layers. *J. Parasitol.* **75,** 985–989.

Hyams, J. (1982). The *Euglena* paraflagellar rod: Structure, relationship to other flagellar components and preliminary biochemical characterization. *J. Cell Sci.* **55,** 199–210.

Jeon, K. W. (1987). Change of cellular pathogens into required cell components. *Ann. NY Acad. Sci.* **503,** 359–371.

Jeon, K. W. (1995). The large, free-living amoebae: Wonderful cells for biological studies. *J. Euk. Microbiol.* **42,** 1–7.

Joiner, K. A. (1991). Rhoptry lipids and parasitophorous vacuole formation: a slippery issue. *Parasitol. Today,* **7,** 226–227.

Köhler, S., Delwiche, C. F., Denny, P., Tilney, L. G., Webster, P., Wilson, R. J. M., Palmer, J. D., and Roos, D. S. (1997). A plastid of probable green algal origin in Apicomplexan parasites. *Science* **275,** 1485–1489.

Kulda, J., Tachezy, J., and Čerkasovova, A. (1993). *In vitro* induced anaerobic resistance to metronidazole in *Trichomonas vaginalis. J. Euk. Microbiol.* **40,** 262–269.

Margulis, L. (1974). The five kingdom classification and the origin and evolution of cells. *Evol. Biol.* **7**, 45–78.

Melkonian, M., and Robenek, H. (1984). The eyespot apparatus of flagellated green algae: a critical review. *Progr. Phycol. Res.* **3**, 193–268.

Michaels, P. A. M., and Opperdoes, F. R. (1991). The evolutionary origin of glycosomes. *Parasitol. Today* **6**, 105–109.

Müller, M. (1993). The hydrogenosome. *J. Gen. Microbiol.* **139**, 2879–2889.

Naitoh, Y., and Kaneko, H. (1973). Control of ciliary activities by adenosine triphosphate and divalent cations in Triton-extracted models of *Paramecium caudatum. J. Exp. Biol.* **58**, 657–676.

Newell, P., Europe-Finner, G., Liu, B., Gammon, G., and Wood, C. (1990). Chemotaxis of *Dictyostelium. Soc. Gen. Microbiol. Symp.* **46**, 273–297.

Opperdoes, F. R. (1987). Compartmentation of carbohydrate metabolism in trypanosomes. *Ann. Rev. Microbiol.* **41**, 127–151.

Roberts, A. M. (1981). Hydrodynamics of protozoan swimming. *In* "Biochemistry and Physiology of Protozoa" 2nd ed., Vol. 4, pp. 5–66. Academic Press, New York. (M. Levandowsky and S. H. Hutner, Eds).

Saimi, Y., and Kung, C. (1987). Behavioral genetics of *Paramecium. Ann. Rev. Genet.* **21**, 47–65.

Sam-Yellowe, T. Y. (1996). Rhoptry organelles of the apicomplexa: Their role in host cell invasion and intracellular survival. *Parasitol. Today* **12**, 308–316.

Sherwin, T., and Gull, K. (1989). The cell cycle of *Trypanosoma brucei brucei:* timing of event markers and cytoskeleton modifications. *Phil. Trans. Roy. Soc. London Ser B* **323**, 575–588.

Smyth, R. D., Saranak, J., and Foster, K. W. (1988). Algal visual systems and their photoreceptor pigments. *Progr. Phycol. Res.* **6**, 254–286.

Sogin, M. L. (1989). Evolution of eukaryotic microorganisms and their small subunit ribosomal RNAs. *Am. Zool.* **29**, 487–497.

Sogin, M. L. (1991). Early evolution and the origin of eukaryotes. *Curr. Opin. Gen. Devel.* **1**, 457–463.

Soldo, A. T. (1987). *Parauronema* and its xenosomes: A model system. *J. Protozool.* **34**, 447–451.

Soldo, A. T., Brickson, S. A., and Vazquez, D. (1992). The molecular biology of a bacterial endosymbiont. *J. Protozool.* **39**, 196–198.

Sommer, J. M. and Wang, C. C. (1994). Targeting proteins to the glycosomes of African trypanosomes. *Ann. Rev. Microbiol.* **48**, 105–138.

Stoecker, D. K., Michaels, A. E., and Davis, L. H. (1987). Large proportion of marine planktonic ciliates found to contain functional chloroplasts. *Nature* **326**, 790–792.

Taylor, F. J. R., Blackbourne, D. J., and Blackbourne, J. (1971). The redwater ciliate *Mesodinium rubrum* and its "incomplete symbionts": A review including new ultrastructural observations. *J. Fish. Res. Bd. Can.* **28**, 391–407.

Van Haastert, P. (1991). Sensory transduction in eukaryote cells. *Eur. J. Biochem.* **195**, 289–303.

Van Houten, J. (1992). Chemosensory transduction in eukaryotic microorganisms. *Ann. Rev. Physiol.* **54**, 639–663.

Woodward, R., Carden, M. J., and Gull, K. (1994). Molecular characterisation of a novel, repetitive protein of the paraflagellar rod in *Trypanosoma brucei. Mol. Biochem. Parasitol.* **67**, 31–39.

Yarlett, N., Orpin, C. G., Munn, E. A., Yarlett, N. C., and Greenwood, C. A. (1986). Hydrogenosomes in the rumen fungus *Neocallimastix patricarium. Biochem. J.* **236**, 729–739.

Dennis W. Grogan

# 61

# Physiology of Prokaryotic Cells

## I. Introduction

When the electron microscope first revealed structural detail within cells, a clear-cut distinction between two very different classes of cells emerged. Plant, animal, fungal, and protozoan cells were always found to contain a variety of internal membranous structures called **organelles,** the most prominent of which was the nucleus. In contrast, bacterial cells were never found to contain a nucleus. Plants, animals, fungi, and protozoa were thus collectively termed **eukaryotes,** that is, having a true nucleus, whereas bacteria were termed **prokaryotes,** reflecting a presumably primitive form of life predating eukaryotic cells.

A summary of prokaryotic physiology must emphasize two seemingly dissonant themes. On the one hand, the most thoroughly understood living organisms are prokaryotes. The cellular structures and functions of *Escherichia coli* have been defined in greater molecular detail than those of any eukaryote. On the other hand, this detailed knowledge does not predictably translate into an understanding of other prokaryotes. Prokaryotes do not constitute a naturally coherent group of organisms. They are ancient life-forms, and have diverged to cover a tremendously broad spectrum of metabolic and molecular alternatives. Putting prokaryotic physiology in perspective thus requires an appreciation of the distant relationships among these apparently simple organisms.

When compared at the level of ribosomal RNA sequence, all cells belong to one of three discrete natural groupings. Two of these groupings, **bacteria** and **archaea,** consist of prokaryotes that have diverged from each other about as much as either has diverged from eukaryotes. The bacteria consist of several major sublineages (Table 1), but for many purposes they are simply divided into two groups based on their response to the **Gram stain.** Bacterial species that stain in this procedure are termed **gram-positive** and have a different cell-surface architecture from bacteria that fail to stain, that is, from **gram-negative** bacteria (Section II,E). The Gram reaction is thus a useful way to distinguish

two fundamentally different cell types among the bacteria. The archaea are typically divided into three groups according to metabolic properties: the extreme thermophiles, the methanogens, and the extreme halophiles. Table 1 illustrates how various groups of bacteria, archaea, and eukaryotes compare to each other by the criterion of ribosomal RNA sequence analysis.

The relationships summarized in Table 1 provide the necessary perspective for discussing physiological features of prokaryotic cells. They imply, for example, that bacterial cells differ greatly from eukaryotic cells at the molecular level, which has been borne out by decades of research. However, at least two eukaryotic organelles, the **mitochondrion** and the **chloroplast,** seem to represent bacteria that entered into symbiotic relationships with a progenitor of modern eukaryotic cells (Schwartz and Dayhoff, 1978). Thus, the composite nature of the eukaryotic cell results in some very close analogies between eukaryotic organelles and bacterial cells. Molecular taxonomy also implies that archaea differ greatly from bacteria, despite the fact that both groups are prokaryotic. This is consistent with the available data, but the full extent of these differences remains to be discovered.

This chapter provides an overview of some physiological aspects of prokaryotic cells. It presents the prokaryotic world in a concentric fashion. *Escherichia coli,* which belongs to the "purple" group (named after its pigmented, photosynthetic members) of gram-negative bacteria, is described in greatest detail. Gram-negative bacteria that differ from *E. coli* in significant respects are discussed where appropriate. In turn, gram-positive bacteria and other structurally distinct bacteria are contrasted with the gram-negative bacteria. Finally, archaea, insofar as their physiological features have been established, are discussed in comparison to all bacteria.

## II. Prokaryotic Cytology

The prokaryotic cell is the smallest and simplest form of life capable of an independent existence. From the eukaryotic perspective, it is remarkable for what it lacks. A typical prokaryotic cell measures less than 1 $\mu$m in diameter (Fig. 1). It has no cytoskeleton, no identifiable mitotic apparatus, and no internal membranous organelles, yet it carries out all the basic functions of life, including nutrient uptake, catabolism, energy conversion, biosynthesis, genetic regulation, genome replication, cell division, and adaptive responses to environmental stimuli. The spatial and functional organization required for these processes occurs at the level of molecular assemblies, and the larger of these assemblies is visible in the electron microscope.

All prokaryotes have a **nucleoid,** a **cytoplasm,** and a **cytoplasmic membrane.** In addition to these three essential structures, most bacteria and archaea have a **cell wall** of some type. Gram-negative bacteria have an additional membrane, outside the cell wall, called the **outer membrane** (Fig. 2). The following discussion of prokaryotic cytology centers on gram-negative cells, which include

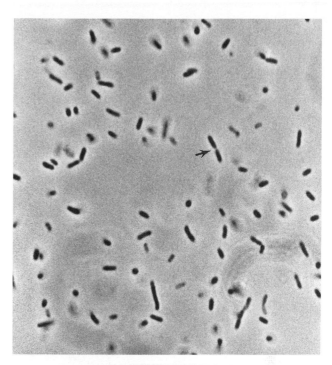

**FIG. 1.** *Escherichia coli* cells suspended in growth medium. The average width of the cells is about 0.5 $\mu$m. The arrowhead shows site of constriction (septation) in a cell undergoing division. This light micrograph (phase contrast) illustrates the limited structural information obtainable by optical methods due to the small size of prokaryotic cells.

**TABLE 1** Major Groups of Organisms According to Molecular Taxonomy

    I. Bacteria
        A. Deinococci
            1. *Thermus*
            2. *Deinococcus*
        B. Green, nonsulfur bacteria
        C. Cyanobacteria
        D. "Purple" bacteria
            1. *Escherichia*
            2. *Azotobacter*
            3. *Thiobacillus*
            4. *Desulfovibrio*
        E. Gram-positive bacteria
            1. *Clostridium*
            2. *Lactobacillus*
            3. *Streptococcus*

    II. Archaea
        A. Thermophiles
            1. *Sulfolobus*
            2. *Pyrodictium*
            3. *Thermoproteus*
            4. *Archaeoglobus*
        B. Methanogens
        C. Extreme halophiles
            1. *Haloferax*
            2. *Halobacterium*

    III. Eukaryotes
        A. Microsporidia
        B. Flagellates
        C. Slime molds
        D. Ciliates
        E. Green plants
        F. Fungi
        G. Animals

This scheme summarizes the relatedness of major groups of organisms according to ribosomal RNA comparisons (Woese *et al.,* 1990). All groups at the same level in the outline have comparable evolutionary status; several major groups have been omitted for the sake of clarity.

all of the basic cellular structures normally found in bacteria.

## A. Nucleoid

The prokaryotic chromosome consists of a single circular DNA of several million base pairs. This DNA is neither contained within a nucleus nor evenly dispersed throughout the cytoplasm. It occurs instead as a defined region, called the **nuclear region** or **nucleoid,** which occupies about 10% of the cell's internal volume. In electron microscopy of thin sections, the nucleoid appears as a convoluted region devoid of ribosomes near the center of the cell (Fig. 3). The nucleoid replicates and partitions into daughter cells during growth and cell division, and bears a functional analogy to the eukaryotic nucleus.

Although the *E. coli* cell is a short rod about 1 $\mu$m long, its DNA is about 1 mm, or $10^3$ $\mu$m in length. This illustrates the extent to which the bacterial DNA must be condensed and folded to form the nucleoid. Cations, including $Mg^{2+}$, spermidine and other polyamines, and small "histone-like" proteins help compact the DNA *in vivo.* Longer range organization occurs in the form of 50–100 topologically constrained domains or "loops." Lysis of cells under appropriate conditions (nonionic detergent and high salt concentration) releases nucleoids in apparently intact form, and allows the loops to be visualized by electron microscopy (Drlica, 1987).

## B. Cytoplasm

All other components of the cell interior are collectively termed the **cytoplasm.** Often described as a fluid (hence the equivalent term cytosol), the cytoplasm is highly concentrated and probably quite viscous. It consists primarily of metabolic enzymes and ribosomes, reflecting the central importance of rapid growth for prokaryotic organisms. An *E. coli* cell adapted to growth in simple glucose medium at 37°C contains about 19,000 ribosomes. The ribosomes, transfer RNA, and various translation factors together account for nearly half of the total cell mass; this high ribosomal content enables the *E. coli* cell to reproduce itself every 40 min under these conditions (Neidhardt *et al.,* 1990).

Its lack of internal compartmentalization (see Fig. 3) contributes to a prokaryote's ability to grow and regulate its gene expression quickly. Prokaryotic cells have no diffusional barriers segregating the sources of energy, raw mate-

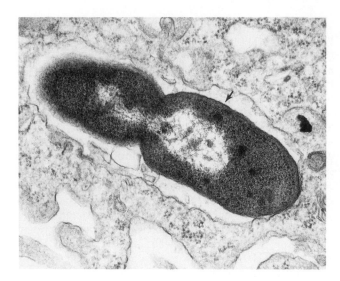

**FIG. 3.**    Electron micrograph (ultrathin section) of a bacterial cell. An *E. coli* cell is shown that has been engulfed by a macrophage and is in the late stages of cell division. The nucleoid is visible as a clear zone in each of the nascent daughter cells. The arrowhead shows a region in which the three-layer structure of the gram-negative cell envelope can be seen. (Photograph courtesy of A. J. Mukkada.)

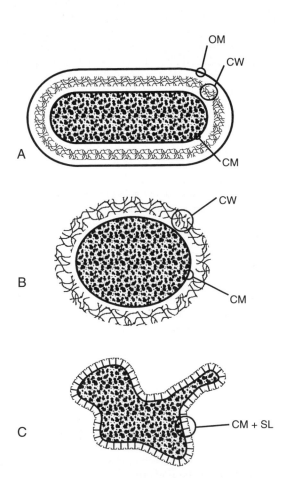

**FIG. 2.**    Basic cell structures of prokaryotes. The diagrams summarize the three most common types of prokaryotic cell structure. (A) A gram-negative bacterium has a thin cell wall sandwiched between the cytoplasmic and outer membranes (OM). (B) A gram-positive bacterium has a thick cell wall (CW) immediately outside the cytoplasmic membrane (CM). (C) Many archaea have only an S-layer (SL) composed of glycoprotein subunits as a structural support for the cytoplasmic membrane, although others (not depicted) have cell walls (Section II,H,5).

rial, and sequence information needed for DNA, RNA, and protein synthesis, and the small dimensions of the cell ensure rapid availability of precursors through simple diffusion. Furthermore, transcription and translation are temporally and spatially coupled in bacteria; a series of ribosomes begins "reading" the mRNA before the mRNA itself has been completed and released from the transcription complex. This coupling has functional significance beyond simply allowing rapid growth, as demonstrated by the phenomenon called **attenuation.** This form of genetic regulation uses disruption of the normally tight coupling between ribosome and RNA polymerase to abort the transcription of certain amino acid biosynthetic genes when the supply of those amino acids is adequate (Yanofsky and Crawford, 1987).

## C. Cytoplasmic Membrane

The prokaryotic cell interior is contained by a unit membrane called the **cell (cytoplasmic) membrane** (CM; see Fig. 2). This is the only lamellar structure common to all prokaryotic cells; it essentially defines the chemical boundary between the cell and its environment (see Fig. 1). The molecular composition of the bacterial CM generally conforms to the fluid mosaic model of other biological membranes (see Chapter 5). The phospholipid bilayer of the CM is intrinsically permeable to gases and water but impermeable to ionic or large polar molecules. This chemical insulation is necessary; without it the cell could not maintain an energized state and retain metabolites, for example (see discussion later). However, the cell also requires provision for chemical exchange with its surroundings, including nutrient uptake and environmental sensing. A large number of different integral and peripheral proteins provide the necessary communication across the phospholipid bilayer; bacterial cytoplasmic membranes are about 70% protein by weight. The phospholipid bilayer with its embedded proteins represents a nonspecific diffusional barrier breached by solute-specific pumps and gates. With this combination, the prokaryotic cell can maintain transmembrane gradients and control transmembrane fluxes of critical solutes.

## D. Cell Wall

With few exceptions (Section II,D,4), bacteria have a **cell wall** immediately outside the cytoplasmic membrane (see Fig. 2). The cell wall is composed of a polymer, called **peptidoglycan** or **murein,** that is unique to bacteria. Peptidoglycan has a dimeric amino-sugar repeating unit and $\beta$-1,4 glycosidic bonds. The resulting glycan chains are cross-linked at frequent intervals by peptide bridges (Fig. 4). This cross-linked network results in one huge, baglike macromolecule of high tensile strength that completely envelopes the mechanically fragile cell membrane (see Fig. 2).

The function of the cell wall derives from the fact that nearly all prokaryotic cells live in environments whose water activity greatly exceeds that of the cytoplasm. The

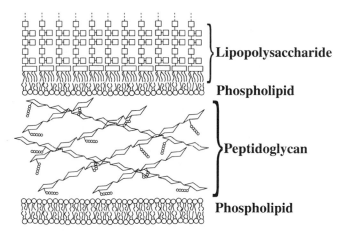

External environment

Bacterial cytoplasm

**FIG. 4.** Molecular organization of the cell envelope of gram-negative bacteria. The outer membrane (OM) consists of one leaflet of lipopolysaccharide (LPS) and one of phospholipid. Each LPS molecule has a large polysaccharide chain (linked squares) whose precise structure depends on the bacterial species. The cell wall is composed of peptidoglycan, whose repeating unit consists of a dimer of two modified sugars: *N*-acetyl muramic acid and *N*-acetyl glucosamine. Each repeating unit has a short peptide (circles) attached to the *N*-acetyl muramic acid, the peptides of adjacent glycan chains become joined by enzymes in the periplasm to form the periodic cross-links found in the mature peptidoglycan polymer. The cytoplasmic membrane (CM) is composed of phospholipid. In addition to lipid, both the OM and CM contain protein species, which are not depicted in the figure.

cell wall provides the mechanical resistance necessary to counteract the resulting osmotic (turgor) pressure (see Chapter 19). This can be as high as 20 atm in some bacteria. As a result, the bacterial cell wall preserves the integrity of the CM, determines cell shape, and plays a major role in cell division. Direct evidence of these functions in the bacterial cell can be seen in the effects of antibiotics that specifically disrupt the normal synthesis of peptidoglycan (required for the cell wall structure). Rod-shaped bacteria growing in media containing low concentrations of these antibiotics elongate but fail to divide. At moderate antibiotic concentrations, bulges (rather than constrictions) form at the normal site of bacterial cell division. At high antibiotic concentrations, the bacteria lose their rodlike shape, swell (into spheres), and lyse (Schwarz *et al.*, 1969).

## E. Outer Membrane

Gram-negative bacteria have a second lipid bilayer immediately outside the cell wall, called the **outer membrane** (OM). Thus, in contrast to the wall of a gram-positive cell, which is exposed to the external environment (see Fig. 2), the wall in a gram-negative cell is sandwiched between two membranes (see Figs. 2 and 4). The protected compartment between the cytoplasmic and outer membranes is called the **periplasm.**

The biochemical composition of the bacterial OM differs from that of the CM with respect to both lipid and protein constituents. With respect to lipid, the OM is a hybrid membrane. Its inner leaflet, which faces the periplasm, incorporates the same phospholipids as does the CM. The outer leaflet, however, consists of much higher molecular weight class of lipid known as **lipopolysaccharide** (LPS; see Fig. 4). With respect to protein, the OM differs from the CM in having a relatively simple protein composition dominated by only a few protein species (Nikaido and Vaara, 1985).

These two compositional features underly the primary physiological function of the OM, which is to shield the gram-negative cell from toxic hydrophobic compounds. An intact LPS leaflet greatly impedes diffusion of soluble hydrophobic (i.e., lipophilic) molecules across the OM. This presumably reflects the permeability properties of the dense array of cooperatively interacting charged and polar sugar residues at the membrane surface, through which nonpolar solutes must pass (see Fig. 4). Experimental evidence of the importance of this screening function can be seen in the greater sensitivities of gram-positive bacteria, as well as of LPS-depleted gram-negative cells, to toxic dyes, detergents, and hydrophobic antibiotics (Nikaido and Vaara, 1985).

As in the case of the CM, the chemical insulation from the environment provided by the OM cannot be absolute. The necessary exchange of nutrients is mediated by the few, abundant protein species of the OM. These proteins, termed **porins,** trimerize to form relatively nonspecific solute channels, or **pores,** through the OM. The pores are narrow; they exclude solutes with molecular weights of about 700 Da and above, and tend to further discriminate against anionic solutes (Nikaido and Vaara, 1985).

## F. Appendages

A bacterium may have either of two basic types of external filaments anchored (attached) to the CM. **Flagella** are helical protein filaments that extend several cell lengths (2–20 $\mu$m). Depending on the bacterial species, they may be single, numerous, clustered, dispersed over the cell, or absent altogether. Each flagellum is rotated by a protein complex at its base, and acts like a ship's propeller, pushing the cell through its liquid medium (see Chapter 57). Thus, bacterial flagella have a cellular function (i.e., motility) analogous to that of eukaryotic cilia and flagella, but no structural, molecular, or mechanistic homology to them.

**Pili** and **fimbriae** are structurally similar to each other and distinct from flagella. Both pili and fimbriae consist of straight protein fibers that protrude less than about 1 $\mu$m from the cell, giving it a bristled appearance in the electron microscope. Both types of appendages appear to mediate examples of cell-specific attachment of a bacterium to another cell. Pili attach a "donor" bacterium to a recipient cell for the subsequent transfer of DNA (conjugation), whereas fimbriae may attach a pathogenic bacterium to the cells lining the urinary tract, for example.

Flagella, pili, and fimbriae form via spontaneous self-assembly of small protein subunits. Bacterial flagella grow by adding protein subunits to the distal tip. This requires the subunits (monomers) to travel from the cytoplasm to the growing tip of the flagellum, several cell lengths away. This is accomplished by using the hollow core of the flagellum itself as a conduit. Monomers synthesized in the bacterial cytoplasm are secreted into the core of the flagellum; they diffuse to the distal tip and add to the regular helical array that makes up the flagellar filament. In contrast, pili and fimbriae appear to polymerize at the base of the fiber, where it attaches to the CM (Neidhardt *et al.*, 1990).

## G. Capsule

Many bacteria secrete hydrophilic polymers (usually polysaccharides) that form a gelatinous matrix around the cell. If this matrix remains attached to the cell and forms a defined zone, it is called a **capsule.** Capsules inhibit the ingestion of bacteria by phagocytes (in the human body) or by protozoa (in natural environments), and may help other bacteria survive temporary desiccation or starvation. Secreted polysaccharide also aids the nonspecific adherence of bacteria to wetted solid surfaces. This can lead to the formation of **biofilms** hundreds of micrometers thick on such surfaces. Bacteria in biofilms tend to tolerate much higher concentrations of antibiotics than populations of suspended cells; since these films can form on surfaces of heart valves or implanted devices, they pose serious health risks in certain situations (Costerton *et al.*, 1987).

## H. Alternative Cell-Surface Architectures

### 1. Gram-Positive Bacteria

The gram-positive bacteria lack an outer membrane and have thicker cell walls than do gram-negative bacteria. These cell walls are chemically more complex; in addition

**TABLE 2** Properties of Gram-Positive and Gram-Negative Bacteria

| Property | Gram-positive bacteria | Gram-negative bacteria |
|---|---|---|
| Cell wall: | | |
|   Chemical constituents | Peptidoglycan, teichoic, lipoteichoic, and teichuronic acids | Peptidoglycan |
|   Typical thickness | 20–80 nm | 2–3 nm |
| Outer membrane | Absent | Composed of LPS and phospholipid |
| Intrinsic resistance to detergents, dyes, and certain antibodies | Low | High |
| Dessication resistance | High | Low |
| Examples | *Clostridium* *Lactobacillus* | *Escherichia* *Thiobacillus* |

to peptidoglycan, they contain various anionic polymers (teichoic, lipoteichoic, and teichuronic acids). The resulting cell wall does not readily decolorize in the Gram stain procedure. Table 2 summarizes the major physiological differences between gram-positive and gram-negative bacterial cells.

### 2. "Acid-Fast" Bacteria

Some environmentally and clinically important bacteria (*Mycobacterium* and *Nocardia* spp.) have a unique type of cell envelope that includes very high molecular weight lipids called **mycolic acids.** These waxy lipids are complexed to an additional polysaccharide layer surrounding the cell wall of an otherwise gram-positive cell. X-ray diffraction data suggest that the mycolic acids form a very thick lipid membrane analogous to the OM of gram-negative cells (Nikaido *et al.*, 1993). The resulting cell envelope makes the cells very difficult to stain and destain.[1] The extraordinary inertness and impenetrability of these cell envelopes to solutes probably explains the other notable properties of *Mycobacterium* and *Nocardia* spp., including slow growth rates and a general resistance to antibiotics and to the immune system (Nikaido *et al.*, 1993).

### 3. Surface Layers

Both gram-positive and gram-negative bacteria may bear regular "paracrystalline" arrays of protein subunits on their surfaces. These **surface (S-) layers** give whole cells or cell envelope preparations a cobblestone pattern when metal-shadowed for electron microscopy. A number of functions have been proposed for these protein layers,

---

[1] Cells of these bacteria stained by heating in carbol-fuschin dye fail to decolorize even in acidic phenol, yielding the empirical designation "acid fast."

including sieving effects, mediation of certain cell–cell interactions, and modification of the cell's net surface charge. Bacterial S-layers are not essential for bacteria growing under laboratory conditions, however. As a result, bacteria tend to lose the ability to produce S-layers during repetitive cultivation in the laboratory (Messner and Sleytr, 1992).

### 4. Wall-Less Bacteria

A few bacteria, most notably *Mycoplasma* spp., have no peptidoglycan layer (cell wall). These cells have relatively strict requirements for the composition and osmolarity of their growth medium. As expected from the absence of a bacterial cell wall (see previous discussion), the cells have no defined shape under normal conditions, and when transferred into hypotonic medium, the cells swell and lyse.

### 5. Archaea

All archaeal cell membranes differ from bacterial cell membranes with respect to their lipids. Archaea contain only *sn*-(2,3 di-O-alkyl)-glycerol membrane lipids, which occur in no other known organism. The hydrocarbon chains of these lipids are C20 or C40 saturated isoprenoids attached to the glycerol backbone via ether linkages. Archaea do not have a second lipid membrane analogous to the OM. Some archaea (a few species of methanogens and extreme halophiles) have a cell wall, but the constituent polymer is not peptidoglycan.

Most archaea have some type of S-layer attached to the CM; its location is analogous to a polymer cell wall, which suggests an analogous function (see Fig. 1). Archaeal S-layers differ from bacterial S-layers (Section II,H,3) in several respects. Unlike bacterial S-layers, (1) their subunits are usually glycosylated, (2) they seem to be essential and are not lost during laboratory cultivation, (3) they are usually anchored to the cytoplasmic membrane (Fig. 5), and (4) they can define cell shape (Baumeister *et al.,* 1989). Note also that some archaea, such as *Thermoplasma* spp., lack both an S-layer and a cell wall; these cells are osmotically sensitive.

## III. Metabolic Strategies

The prokaryotic world is metabolically diverse. Many prokaryotes depend on metabolic processes unknown in eukaryotes.

### A. Autotrophy

Many prokaryotes can derive all of their carbonaceous cell material from $CO_2$. This property, termed **autotrophy,** requires a source of energy. Several different families of bacteria (collectively called **photoautotrophs**) use light as the energy source, but others, the **chemoautotrophs,** derive energy by mediating the oxidation and reduction of inorganic compounds in their environments.

The photoautotrophic bacteria exhibit great diversity with regard to the biochemistry of photosynthesis. Among

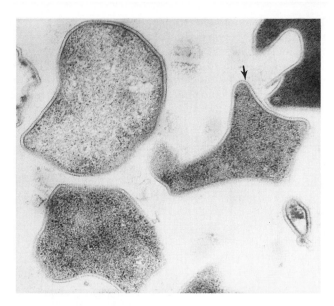

**FIG. 5.** Cells of *Sulfolobus acidocaldarius,* an archaeon from geothermal environments (Section VII). The ultrathin sections show the irregular cell shape and regular repeating structure of the glycoprotein S-layer. The S-layer is held at a fixed distance from the cytoplasmic membrane by some type of spacer (Baumeister *et al.,* 1989); this results in a "picket fence" boundary around cells or cell ghosts (Grogan, 1996a) in thin sections.

them, only the **cyanobacteria** carry out **oxygenic photosynthesis.** Cyanobacteria contain **chlorophyll b** and have two connected **photosystems,** one of which (PS2) oxidizes water to supply electrons to the other (PS1; see Chapter 63). This is the scheme also used by plant chloroplasts, to which cyanobacteria are closely related (Schwartz and Dayhoff, 1978).

The other photoautotrophic bacteria conduct **anaerobic (anoxygenic) photosynthesis.** They have only one photosystem, which can operate in a cyclic manner to produce a proton potential (Section IV,B). In these cases, the electron equivalents for the reduction of $CO_2$ come from $H_2S$, $H_2$, or dissolved organic compounds, depending on the type of bacterium and the resources available to it. In these latter organisms, photoautotrophic growth represents only one among several metabolic options, and production of photosynthetic pigments occurs only in the absence of oxygen. The photosynthetic pigments of the anoxygenic photoautotrophs, called **bacteriochlorophylls,** differ from cyanobacterial and plant chlorophylls in having intense absorbance maxima in the infrared region of the spectrum (White, 1995).

Chemoautotrophs include aerobic organisms that oxidize $H_2$, $CO$, $NH_4^+$, $NO_2^-$, elemental S, $H_2S$, or $Fe^{++}$, using $O_2$. Anaerobic chemoautotrophs derive energy from the reduction of $CO_2$ to $CH_4$ and of $SO_4^=$ to $H_2S$.

Among autotrophic prokaryotes, three metabolic pathways have been identified by which $CO_2$ is "fixed" (i.e., reduced and incorporated into some common intermediary metabolite): (1) the **Calvin cycle,** (2) the **acetyl-CoA pathway,** and (3) the **reductive tricarboxylic acid cycle** (White,

1995). Most photoautotrophic bacteria (and all green plants) use the Calvin cycle: Ribulose bis-phosphate is carboxylated to yield two molecules of 3-phosphoglycerate, which are rearranged in a complex series of reactions to regenerate ribulose bis-phosphate. The acetyl-CoA pathway, used by methanogenic archaea and sulfate-reducing bacteria, involves the differential reduction of two $CO_2$ molecules; one is reduced to a methyl group and another is reduced to form a carbonyl unit, which is condensed to the methyl group. The resulting acetyl unit is transferred to coenzyme A (CoA) for assimilation into cell material. The reductive tricarboxylic acid (TCA) pathway occurs among the "green" photosynthetic bacteria, for example. It begins with carboxylation of phospho-*enol*-pyruvate to yield oxaloacetate. The transformations of the TCA cycle then proceed in reverse, culminating in the cleavage of citrate by an adenosine triphosphate (ATP)-dependent citrate lyase to yield acetyl CoA and oxaloacetate.

### B. Nitrogen Fixation

Only prokaryotic cells can convert $N_2$ into ammonia. This capability is reasonably widespread among prokaryotic groups; it occurs in various aerobic, anaerobic, facultatively anaerobic bacteria (i.e., those able to grow either aerobically or anaerobically), heterotrophic and autotrophic bacteria, and in methanogenic archaea. A complex enzyme, **nitrogenase,** carries out the $N_2$ reduction. It is composed of an iron-containing component and a molybdenum- and iron-containing component, and utilizes ATP plus a reduced low-potential ferredoxin or flavoprotein as the source of electrons. Nitrogenases are intrinsically oxygen sensitive, and $N_2$-fixing prokaryotes use a variety of strategies to protect these enzymes. Facultative anaerobes such as *Klebsiella* spp. produce nitrogenase only under anaerobic conditions. Filamentous cyanobacteria, which generate $O_2$ during photosynthesis, sequester nitrogenase in specialized cells called **heterocysts,** in which the oxygenic photosystem II does not operate. *Azotobacter* spp. are obligate aerobes that can fix nitrogen; they appear to scavenge intracellular $O_2$ by maintaining very high respiration rates.

## IV. Energetics of Bacterial Cells

With a few notable exceptions, the endergonic cellular processes of bacteria are driven by being enzymatically coupled to the hydrolysis of ATP. Life thus requires a steady resupply of this energy currency. An *E. coli* cell growing aerobically using glucose regenerates ATP in two ways: **substrate-level phosphorylation** and **chemiosmotic coupling.**

### A. Substrate-Level Phosphorylation

As in the eukaryotic cytosol, soluble enzymes in the *E. coli* cytoplasm convert 1 mol of glucose to 2 mol of pyruvic acid via the Embden–Meyerhof–Parnas pathway ("glycolysis"), which involves phosphorylation of 2 mol of adenosine diphosphate (ADP) to form ATP and the reduction of 2 mol $NAD^+$ to NADH. If the cell has no exogenous electron acceptors available (Section IV,B), the 2 mol ATP represents the cell's sole energy harvest, whereas the reduced cofactor (NADH) represents unusable electrons. If the electrons are not transferred to some other molecule, the oxidized cofactor ($NAD^+$) will not be regenerated and a second mole of glucose cannot be metabolized. Bacteria (and eukaryotic cells) solve the latter problem by transferring the electrons to pyruvate or its metabolites, thereby forming various end products that the cell excretes; examples include lactic, acetic, propionic, or butyric acids, and ethanol, butanol, or acetone. The overall process of substrate-level phosphorylation coupled to reduction of the resulting metabolites is called **fermentation.**

### B. Chemiosmotic Coupling

The general features of this mechanism resemble those of the mitochondrion, which are described elsewhere in this volume (see Chapters 5 and 6). The process has two distinct stages, each of which involves vectoral enzymatic processes at the cytoplasmic membrane. The first stage utilizes the electrons taken from carbon compounds during glycolysis and the TCA cycle by transferring them to some other electron acceptor (such as $O_2$) via a series of enzymatically catalyzed oxidation–reduction reactions. This series of strongly exergonic reactions, known as respiratory electron transport, is coupled at certain points to the extrusion of protons from the cytoplasm, creating a **proton potential** or **proton-motive force** (PMF) across the CM. The second stage of chemiosmotic coupling converts the PMF into ATP; it uses an $F_1F_0$ **proton-translocating ATPase** to couple the entry of three protons from outside the cell to phosphorylation of an ADP molecule (Maloney, 1987). The yield of ATP per glucose molecule is many times higher for a bacterium that is respiring than for one that is fermenting.

Respiration can take many forms in prokaryotic cells; in other words, oxygen represents only one of several terminal electron acceptors that prokaryotes can use to form a PMF. Table 3 lists the major types of **anaerobic respiration** carried out by prokaryotes. For example, *E. coli* can use four electron acceptors other than oxygen (Table 3), but it produces the necessary oxidoreductases only under certain conditions, according to a complex regulatory hierarchy. This hierarchy ensures that only the energetically most favorable electron acceptor is used should more than one be available (Gunsalus, 1992).

As just outlined, the PMF of a respiring bacterium represents an energetic intermediate in the formation of ATP by the oxidation of carbon compounds. It should be emphasized, however, that the PMF appears to be a basic energetic feature of all bacterial cells whether or not they are respiring. The **lactic acid bacteria,** for example, have no electron-transport chain and cannot respire; they generate ATP only by substrate-level phosphorylation. They have an $F_1F_0$ ATPase, however, and use it in the reverse sense of respiring bacteria, that is, to maintain a PMF at the expense of ATP hydrolysis. This may reflect the fact that

**TABLE 3**   Respiratory Strategies of Prokaryotes

| Electron acceptor | Reduced product | Organism |
| --- | --- | --- |
| Oxygen | Water | *E. coli,* other aerobes |
| Nitrate | Nitrite | *E. coli,* other bacteria |
| Fumarate | Succinate | *E. coli,* other bacteria |
| Dimethyl sulfoxide | Dimethyl sulfide | *E. coli,* other bacteria |
| Trimethylamine oxide | Trimethylamine | *E. coli,* other bacteria |
| Sulfate | Hydrogen sulfide | *Desulfovibrio, Archaeoglobus*[a] |
| Sulfur | Hydrogen sulfide | *Thermoproteus*[a] |
| Carbon dioxide | Methane | *Methanobacterium*[a] |

[a] Archaea; all others are bacteria.

certain basic prokaryotic processes use the PMF directly as their energy source. For example, *E. coli* transports several nutrients (see Section V,C) and rotates its flagella by direct coupling to $H^+$ flux into the cytoplasm through the appropriate protein complexes located in the CM.

As discussed in Chapter 6, Energy Production and Metabolism, the PMF consists of two components: the electrical potential ($\Delta\Psi$) and the chemical potential ($\Delta pH$). In those bacteria which live at pH values near 7, both components contribute significantly to the PMF. Although the small size of prokaryotic cells precludes direct electrical measurement of these potentials, they can be estimated by chemical probes. The $\Delta\Psi$ can be measured by the fluorescence yield of triphenylmethyl- or tetraphenylphosphonium ions, or by the equilibrium distribution of radioactive $K^+$ across the CM in the presence of valinomycin. The $\Delta pH$ can be estimated by the distribution of radioactive weak acids such as acetic or benzoic acids across the CM, and the overall PMF can be independently estimated by the maximal accumulation of lactose by cells able to transport, but not metabolize, this sugar (Maloney, 1987).

According to these methods, which generally agree, a typical *E. coli* cell in medium at pH 6.5 has a $\Delta\Psi$ of about 100 mV (inside negative) and a $\Delta pH$ that corresponds to an additional 100 mV (Maloney, 1987). This relationship does not extend to prokaryotes which live at extremes of pH, however. Extreme acidophiles, for example, maintain an electrical potential of opposite polarity as the normal $\Delta\Psi$, that is, inside positive. This helps counteract the very large $\Delta pH$ (which can be more than 4 pH units) across the cytoplasmic membrane. In contrast, extreme alkaliphiles, which grow at external pH values of 10–12, have a negative $\Delta pH$ across the CM and, accordingly, a very large $\Delta\Psi$ (White, 1995).

## V. Solute Transport

Prokaryotes have efficient transport systems for a wide range of nutrients and inorganic ions. The reason for these systems is straightforward: Prokaryotes must scavenge nu-

trients from their environments in order to survive and grow. Table 4 lists the basic transport mechanisms used by bacterial cells.

### A. Facilitated Diffusion

This mechanism is rare among bacterial transport systems. Although mechanistically simple and energetically cheap, it is also relatively ineffective. For example, *E. coli* is known to take up glycerol by facilitated diffusion. It has a membrane channel specific for glycerol and can grow rapidly on high concentrations of glycerol as the sole carbon source. At low external concentrations, however, the rate of transport becomes growth limiting, because the difference in free glycerol concentration across the CM limits the rate of transport that can be achieved (see Chapter 12).

### B. Group Translocation

This represents an elaboration of facilitated diffusion in which the solute becomes chemically modified on entering the cytoplasm. The chemical modification, or **group translocation** (typically phosphorylation or phosphoribosylation of a neutral sugar or nucleobase) prevents exit of the modified solute through the same carrier. The neutral species actually transported by the channel remains at a lower steady-state concentration inside the cell than outside, but the modified solute can reach high concentrations, due to the investment of chemical energy in the modification process. In most cases, the chemical modification represents a necessary step in metabolizing the solute. Thus, the "trapping reaction" doubles as a metabolic step.

### C. Active Transport

Active transport in bacteria is defined by the following features (Neidhardt *et al.,* 1990): (1) Energy is expended, (2) the unmodified solute is transported, (3) accumulation occurs against a solute concentration gradient, and (4) specific steric recognition occurs between the solute and a membrane-bound transport protein. Two types of active transport are distinguished from each other by the immediate source of energy used (see Table 4).

The first type requires the hydrolysis of ATP or its equivalent. In these systems, the membrane-bound carriers typically consist of several subunits, one of which contains an ATP-binding site (Furlong, 1987). Perhaps the most notable feature of these transporters in gram-negative bacteria is the participation of a nonmembrane (i.e., soluble) protein, specific for the transport system and normally located in the periplasm. Transport requires these solute-binding proteins. This can be demonstrated by a cold hypotonic shock treatment of gram-negative bacterial cells; the small periplasmic proteins are released from the cells with concomitant loss of transport capability. Studies of bacterial mutants and of cytoplasmic membrane vesicles confirm that the membrane-bound carrier interacts specifically, not with external solute in its free form, but with the solute complexed to its cognate binding protein (Furlong, 1987).

**TABLE 4**   Solute Transport Systems in Bacteria

| Basic type | Mechanistic features | Examples in *E. coli* |
| --- | --- | --- |
| I. Facilitated diffusion | Selective channel | OM pores |
| | No net accumulation of transported species | Glycerol uptake |
| II. Group transfer | Chemical modification of transported species | Phosphotransferase system for sugars |
| III. Active transport | Solute interacts specifically with a cognate, membrane-bound carrier | |
| | Transported species accumulates against a concentration gradient | |
| A. ATP-driven | Requires ATP hydrolysis | Histidine uptake |
| B. PMF-driven | Coupled to ion flux | Lactose uptake |

The second type of active transport utilizes the PMF directly. The solute enters via symport with a proton, or in some cases, a Na$^+$ ion. Conversely, certain ions, such as Ca$^{2+}$, are pumped out of the cells by proton-coupled antiport systems (Maloney, 1987).

## VI. Signal Transduction

Prokaryotic cells must routinely adapt to changing environmental conditions in order to optimize their growth or even to survive. Prokaryotic cells generally respond to environmental change by changing their overall cellular composition (Neidhardt *et al.*, 1990). This, in turn, requires changes in gene expression, the rapidity of which is made possible by the relatively large biosynthetic capacity of the growing prokaryotic cell, its streamlined organization, and a very short half-life of messenger RNA, which in *E. coli* (under standard conditions) averages 1.3 min.

Studies elucidating such regulatory responses in bacteria have revealed an impressive array of molecular mechanisms for controlling gene expression. These regulatory mechanisms are elegant and diverse, and their discussion lies outside the scope of this chapter. However, one example warrants mention because its mechanistic aspects appear to be shared among a variety of environmental responses in bacteria, and because it involves transmembrane signaling. Two other notable regulatory responses relate to general cellular stress in bacteria.

### A. "Two-Component" Regulatory Systems

When gram-negative bacteria exhaust their normal supply of inorganic phosphate, they activate the transcription of several families of genes, whose products help the cell scavenge phosphate (Wanner, 1993). The process involves the specific interaction and phosphorylation of two proteins, named PhoR and PhoB after their respective gene designations in *E. coli*.

The phosphorylation state of the PhoB protein determines expression of the various genes of the phosphate-starvation response. The phosphorylated form of PhoB (here designated "*PhoB") activates transcription of these genes, whereas the unphosphorylated form does not. The relative abundance of *PhoB is in turn determined by the

conformational state of a CM protein, PhoR (Fig. 6). In its "activated" conformation ("*PhoR"), this protein acts as a PhoB kinase; in its "inactive" conformation, PhoR acts as a *PhoB phosphatase. Ultimately, the conformational state of PhoR is determined by the availability of

PHOSPHATE STARVATION

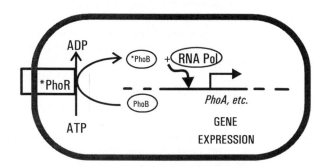

PHOSPHATE ABUNDANCE

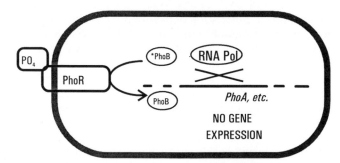

**FIG. 6.**   Basic features of a two-component regulatory system. A simplified version of phosphate-regulated gene expression is shown. The PhoR protein senses the extracellular availability of phosphate, either directly through an externally exposed domain, or indirectly via interaction with the high-affinity phosphate transporter (Wanner, 1993; Neidhardt *et al.*, 1990). When the external phosphate concentration is low, PhoR becomes "activated"; it autophosphorylates at the expense of intracellular ATP and then phosphorylates the PhoB protein. When the external phosphate concentration is high, PhoR becomes "deactivated"; in this form it removes phosphate groups from phosphorylated PhoB protein. PhoB is a soluble transcription factor for a specific set of genes, but stimulates transcription only when phosphorylated.

**TABLE 5** Examples of Two-Component Regulatory Systems

| Mnemonic | Environmental stimulus | Cellular response | Component I | Component II |
|---|---|---|---|---|
| *Pho*sphate | Low external [$PO_4$] | Induction of various $PO_4$-metabolizing enzymes | PhoR | PhoB |
| *N*itrogen *r*egulation | Low external [$NH_4^+$] | Induction of glutamine synthetase | $NR_I$ | $NR_{II}$ |
| *O*uter *m*embrane *p*rotein | External osmolarity | Regulation of alternate porin synthesis | EnvZ | OmpR |
| *A*erobic *r*espiratory *c*ontrol | Oxygen concentration | Regulation of central metabolic enzymes | ArcB | ArcA |
| *N*it*r*ate *r*eductase | External $NO_3^-$ present, oxygen absent | Induction of nitrate reductase | NarX | NarL |
| *Che*motaxis | Temporal gradients of attractants and repellants | Change in swimming interval | CheA | CheY |

inorganic phosphate in the external medium. PhoR can sense the absence of phosphate–periplasmic binding protein complexes in the periplasm and assumes its active conformation. The resulting *PhoR phosphorylates PhoB, forming *PhoB, which in turn stimulates the transcription of *phoA* and other phosphate-scavenging genes by the bacterial RNA polymerase (see Fig. 6). The result is a rapid increase in certain enzymatic activities as a means of adapting to a specific change in the environment.

Protein pairs resembling PhoR/PhoB mediate a wide variety of regulatory responses of bacteria to various types of environmental change (Tables 5 and 6). The PhoR protein, a transmembrane signal transducer, typifies the **component I** of these systems, which is often termed the **sensor-kinase.** With a few exceptions, this is a cytoplasmic membrane protein that senses the status of some environmental parameter and accordingly adopts either of two conformations. In its "active" conformation, component I has two characteristic phosphotransferase activities: (1) It phosphorylates one of its own histidine residues, using ATP, and (2) it transfers the phosphate to an aspartyl residue on its cognate component II. These two activities result in the net transfer of a phosphate group from ATP to the cognate **component II.** This second component, also called the **response regulator,** acts in most cases as an activator of transcription for a certain set of genes, but only in its phosphorylated form.

The similarities in interactions and biochemical properties of various components I and II (see Table 6) are underscored by regions of sequence homology. Although they respond to diverse signals, all components I have two highly conserved sequence motifs at their carboxy termini. Similarly, all components II have sequence homology near their amino termini, and various subfamilies may share regions of additional homology elsewhere in the amino acid sequence. The conserved regions of components I presumably recognize and interact with the conserved regions of components II as an essential part of the signal transmission process. The unconserved regions among the components I and II presumably define their specificity for the external stimulus and for the cellular response, respectively.

## B. The Stringent Response

Bacteria have additional regulatory mechanisms which help them survive life-threatening stress. One of these, the stringent response, mediates basic metabolic changes needed to cope with sudden starvation. When cells growing in complete medium are suddenly deprived of exogenous amino acids, they quickly and specifically halt the transcription of ribosomal- and transfer-RNA genes. The benefit of this response lies in the need of a starving cell to halt the production of new ribosomes at a time when the cell has simultaneously too few biosynthetic resources to meet

**TABLE 6** General Features of Two-Component Regulators

| Feature | Component I | Component II |
|---|---|---|
| Cellular location | Membrane | Cytoplasm |
| Cellular function | Sense and signal an external condition | Control gene expression |
| Biochemical activities | Autophosphorylation at histidine residue<br>Phosphorylation of cognate component II<br>May dephosphorylate cognate component II | Binding to regulatory regions of genes<br>Stimulating transcription of specific genes<br>May inhibit transcription of other specific genes |
| Sequence homology | Two highly conserved regions near carboxyl end | One highly conserved region near amino end (subfamilies have additional regions of homology near carboxyl end) |

its future growth demands and too many ribosomes (Cashel and Rudd, 1987).

The mechanism by which transcription responds to general starvation is uniquely bacterial and involves the ribosome. When unchanged tRNAs bind to the ribosomes (a sign of severe amino acid depletion), a ribosome-associated protein is stimulated to form an unusual nucleotide, **ppGpp,** from ATP and guanosine diphosphate (GDP). This **guanosine tetraphosphate** appears to interact with the bacterial RNA polymerase and specifically inhibit its transcription of rRNA and tRNA genes (Cashel and Rudd, 1987).

## C. The SOS Response

DNA-damaging agents such as UV light or mitomycin C trigger the transcription of various genes in *E. coli* and related bacteria, leading to a series of cellular changes collectively called the **SOS response**[2] (Walker, 1987). The cellular functions of the genes differ; some mediate DNA repair, some allow error-prone DNA synthesis past a lesion, and some trigger induction of latent bacterial viruses (prophages), which can actually kill the cell. Most SOS genes have one feature in common: Their transcription is normally repressed by the protein product of the *lexA* gene.

Induction of the SOS response following DNA damage has a unique molecular mechanism. The LexA protein, which is responsible for silencing the various SOS genes under normal conditions, proteolytically cleaves itself following DNA damage. This cleavage only occurs with the assistance of a specially activated form of a protein not normally involved in proteolysis: the major DNA recombinase, encoded by the *recA* gene. The effect of LexA proteolysis is the relief of repression of SOS genes (Walker, 1987). The cause of conversion of RecA protein into a form that facilitates this unusual cleavage remains unclear, but it may be triggered by single-stranded DNA or nucleotides derived from damaged DNA.

# VII. Prokaryotes Living in Extreme Environments

The physiological diversity of prokaryotes is dramatically illustrated by those organisms that not only tolerate chemical and physical extremes, but actually require them for normal cellular function. This heterogenous collection of prokaryotes includes many new species of archaea. These have been only discovered in recent years, in part, because their natural habitats are so unusual.

The **extreme halophiles** are archaea that normally grow in concentrated brines. Well-studied species include the rod-shaped *Halobacterium salinarum* (= *halobium*) and the irregularly shaped *Haloferax volcanii*. *Halobacterium salinarum* grows best in media containing 3–4 *M* NaCl and will not grow at salt concentrations of less than 1.5 *M*. The cell maintains an internal K$^+$ concentration of up to 5 *M* and contains Cl$^-$ as the major counterion. Accordingly, *H. salinarum* enzymes function best in extremely high salt

concentrations, and most denature at salt concentrations below 2 *M* (Kushner, 1985). The native structure of these halophilic enzymes tends to incorporate an unusually high number of acidic amino acid residues on the protein surface. Other notable features of the *H. salinarum* cell include small, gas-filled protein vesicles, called **gas vacuoles,** in the cytoplasm, and patches of a special retinal protein, **bacteriorhodopsin,** in the cytoplasmic membrane. The gas vacuoles make the cells buoyant, keeping them near the surface of brine ponds, and thus in contact with two sources of energy: O$_2$ and light. Oxidation of organic compounds in the environment normally supports growth, but when oxygen becomes limiting, extra energy can be supplied by bacteriorhodopsin. This protein functions as a light-driven ion pump. When the cell is illuminated, it extrudes protons from the cytoplasm; this contributes to the cell's PMF and thus, to the production of ATP, when O$_2$ is in short supply (Kushner, 1985).

**Acidophiles** require low pH for cell growth; they have been identified among the bacteria and archaea. Gram-negative bacteria of the genus *Thiobacillus* derive energy from the oxidation of reduced sulfur compounds. The oxidized end product is often H$_2$SO$_4$, so that these bacteria acidify their environment in the normal course of their metabolism. At least one species, *Tb. ferrooxidans,* can also oxidize Fe$^{2+}$, whose auto-oxidation proceeds slowly at low pH values. Some *Thiobacillus* cells grow optimally at about pH 2 but maintain a cytoplasmic pH near 6.5. This very large $\Delta$pH is mitigated by a $\Delta\Psi$ of opposite polarity (i.e., inside positive) (White, 1995).

A moderately thermophilic archaeon, *Picrophilus oshimae,* represents the most extreme acidophile known to date. Recently isolated from the soil of a Japanese geothermal field, this organism grows optimally in dilute sulfuric acid at about pH 0.9. *Picrophilus oshimae* cells can grow at pH 0 but not at pH values above 3.5. At pH values greater than 5, the cells lyse (Schleper *et al.,* 1995).

Bacteria that require low temperatures are termed **psychrophiles,** and typically occur in marine, arctic, and antarctic habitats. Enzymes of these organisms may denature, and the cells may die, at room temperature. In several cases, the minimum temperature for growth has not been determined, but lies below about $-10°C$. The membrane phospholipids of these bacteria contain unusually high proportions of unsaturated and short-chain fatty acids (Morita, 1975).

Some bacteria exhibit extreme resistance to ionizing radiation. *Deinococcus radiodurans* was first discovered in a heavily gamma-irradiated can of food that subsequently spoiled. The radiation doses necessary to kill 63% of a population of *D. radiodurans* are about 1800 krad of gamma radiation and about 900 J/m$^2$ of 254-nm UV. These doses can be compared to those required to kill 63% of a population of *E. coli,* that is, about 5 krad of X radiation and 30 J/m$^2$ 254-nm UV (Gutman *et al.,* 1993; also see Chapter 68, Effects of Ionizing Radiation). The extremely high radiation resistance of *D. radiodurans* has been attributed to DNA repair enzymes, multiple copies per cell of the circular chromosome, and a particularly active DNA recombinase. The recombinase allows a *D. radiodurans*

---

[2] The term *SOS response* is based on analogy to a distress signal.

cell to reconstruct an intact chromosome from at least 100 individual fragments formed by double-stranded breaks (Gutman *et al.*, 1993).

Prokaryotes that grow optimally at 80°C or higher temperatures are termed **extreme thermophiles** or **hyperthermophiles** (Stetter *et al.*, 1990). This group consists almost entirely of archaea isolated from geothermal habitats such as hot springs, fumaroles, and submarine thermal vents. Aerobic hyperthermophiles include *Sulfolobus* spp., which require acidic conditions as well as high temperatures. The combination of high temperature and low pH required for optimal growth (about 80°C and pH 3) is extremely effective at denaturing proteins. The *Sulfolobus* cell helps protect its cytoplasmic proteins, which appear to be generally heat stable but not acid stable (Grogan, 1996b) by maintaining its cytoplasmic pH at about 6 (Schäfer *et al.*, 1990). Physical properties of the unique lipids of *Sulfolobus* spp. probably help maintain the resulting $\Delta$pH. The ether-linked phytanyl chains of these lipids span the entire membrane, forming a stable monolayer. Such membranes exhibit very low rates of proton leakage at high temperature, compared to ester-linked phospholipid bilayers (Van De Vossenberg *et al.*, 1995).

The most thermophilic organisms known are anaerobic archaea isolated from submarine vents, where hydrostatic pressure permits liquid water to be superheated. *Pyrodictium occultum*, for example, grows optimally at about 105°C (Stetter *et al.*, 1990). Obviously, the very existence of cells with such growth requirements raises fundamental questions regarding their need to stabilize nucleic acids, proteins, and other molecules necessary for all life. The enzymes of most hyperthermophiles appear to be intrinsically thermostable. The available sequence comparisons of corresponding enzymes from mesophiles and extreme thermophiles suggest that just a few interactions of the amino acid residues, determined in turn by the primary structure of the protein itself (see Chapter 2), can stabilize an enzyme tremendously. In contrast to the intrinsic stabilization of proteins, the DNA of these cells appears to be stabilized against thermal denaturation by extrinsic factors, that is, other molecular species. Examples include polyamines and small, highly basic, DNA-binding proteins. When purified and added to purified DNA, these cellular components dramatically increase the melting (i.e., denaturation) temperature of DNA *in vitro* (Reddy and Suryanarayana, 1989).

## VIII. Summary

Prokaryotes represent the smallest and simplest living cells. They inhabit environments ranging from the dry valleys of Antarctica to hydrothermal vents; they mediate the global cycling of elements, cause disease, and provide some of the best defined and most manipulatable systems available for elucidating cellular structure and function in molecular terms. Molecular sequence comparisons and biochemical properties show that prokaryotes have diverged at very basic levels. They represent two distinctly different groups, the bacteria and the archaea.

The "minimal cell," that is, that basic set of structures common to all prokaryotic cells, consists of a nucleoid and cytoplasm enclosed in a cytoplasmic membrane. However, most free-living prokaryotes also have some type of cell wall, and many bacteria, operationally identified by their staining properties (gram-negative) have a second lipid membrane outside the cell wall. The apparently simple substructures of prokaryotic cells are functionally complex. They mediate all of the cell's diverse functions, including nutrient uptake, catabolism, energy conversion and conservation, biosynthesis, growth, secretion, genome replication, cell division, and regulation of gene expression in response to environmental change.

Prokaryotes encompass a wide range of metabolic capabilities. Several strategies of $CO_2$ fixation and photosynthesis are represented among bacteria and archaea, of which only one, that of the cyanobacteria, has been adopted by green plants. Some prokaryotes can fix $N_2$, and some can "respire" anaerobically, using electron acceptors other than $O_2$ in the environment.

Bacteria depend on solute-specific transport proteins located in the CM to take in nutrients and to move inorganic ions into and out of the cell. Those proteins involved in active transport display a high affinity for the external solute and expend energy to accumulate it to high intracellular levels. A common type of active transport in gram-negative bacteria utilizes special proteins that bind the solute in the periplasm and deliver it to the membrane-bound transporter. CM protein complexes also appear to mediate several bacterial responses to environmental change. Functional and structural homologies define a family of "two-component regulators" of bacterial gene expression. Typically, the first component resides in the CM; it senses environmental change and ultimately modifies the second component in response. The second component is a cytoplasmic transcription factor, active in only one of two phosphorylation states. By this and other streamlined mechanisms, bacteria change their macromolecular compositions in response to external conditions. This form of adaptation can be accomplished rather quickly in growing prokaryotic cells, which have a high biosynthetic capacity relative to cell mass.

The metabolic and functional diversity of prokaryotic cells becomes apparent in extreme environments. Certain bacteria require low external pH or low temperature to survive and grow, and others can cope with very high levels of DNA damage. Archaea dominate environments of extremely high salt concentration or extremely high temperature. Because these archaea differ radically from *E. coli* and other well-studied bacteria, pinpointing the relevant physiological and biochemical properties of their cells remains a challenging area of research.

## Bibliography

Baumeister, W., Wildhaber, I., and Phipps, B. (1989). Principles of organization in eubacterial and archaebacterial surface proteins. *Can. J. Microbiol.* **35**, 215–227.

Caldwell, D. R. (1995). "Microbial Physiology and Metabolism." William C. Brown Publishers, Dubuque, IA.

Cashel, M., and Rudd, K. E. (1987). The stringent response. *In* "Escherichia coli and *Salmonella typhimurium:* Cellular and Molecular Biology" (F. C. Neidhardt, Ed.), pp. 1410–1438. American Society for Microbiology, Washington, DC.

Costerton, J. W., Cheng, K.-J., Geesey, G. G., Ladd, T. I. Nickel, J. C., Dasgupta, M., and Marrie, T. J. (1987). Bacterial biofilms in nature and disease. *Annu. Rev. Microbiol.* **41,** 435–464.

Drlica, K. (1987). The nucleoid. *In* "Escherichia coli and *Salmonella typhimurium:* Cellular and Molecular Biology" (F. C. Neidhardt, Ed.), pp. 91–103. American Society for Microbiology, Washington, DC.

Dutton, P. L. (1986). Energy transduction in anoxygenic photosynthesis. *In* "Photosynthesis III: Photosynthetic Membranes and Light-Harvesting Systems" (L. A. Staehelin and C. J. Arntzen, Eds.), pp. 197–237. Springer-Verlag, Berlin.

Furlong, C. (1987). Osmotic-shock-sensitive transport systems. *In* "Escherichia coli and *Salmonella typhimurium:* Cellular and Molecular Biology" (F. C. Neidhardt, Ed.), pp. 768–796. American Society for Microbiology, Washington, DC.

Grogan, D. W. (1996a). Isolation and fractionation of cell envelope from the extreme thermoacidophile *Sulfolobus acidocaldarius. J. Microbiol. Meth.* **26,** 35–43.

Grogan, D. W. (1996b). Organization and interactions of cell envelope proteins of the extreme thermoacidophile *Sulfolobus acidocaldarius. Can. J. Microbiol.* **42,** 1163–1171.

Gunsalus, R. P. (1992). Control of electron flow in *Escherichia coli:* Co-ordinated transcription of respiratory pathway genes. *J. Bacteriol.* **174,** 7069–7074.

Gutman, P. D., Fuchs, P., Ouyang, L., and Minton, K. W. (1993). Identification, sequencing, and targeted mutagenesis of a DNA polymerase gene required for the extreme radioresistance of *Deinococcus radiodurans. J. Bacteriol.* **175,** 3581–3590.

Kushner, D. J. (1985). The Halobactericeae. *In* "The Bacteria," Vol. VIII, "Archaebacteria" (J. R. Socatch and L. N. Ornston, Eds.), pp. 171–214. Academic Press, Orlando.

Maloney, P. (1987). Coupling to an energized membrane: Role of ion-motive gradients in the transduction of metabolic energy. *In* "Escherichia coli and *Salmonella typhimurium:* Cellular and Molecular Biology" (F. C. Neidhardt, Ed.), pp. 222–243. American Society for Microbiology, Washington, DC.

Messner, P., and Sleytr, U. (1992). Crystalline bacterial cell-surface layers. *Adv. Microb. Physiol.* **33,** 212–275.

Morita, R. Y. (1975). Psychrophilic bacteria. *Bacteriol. Rev.* **89,** 144–167.

Neidhardt, F. C., Ingraham, J. L., and Schaechter, M. (1990). "Physiology of the Bacterial Cell." Sinauer Associates, Sunderland, MA.

Nikaido, H., and Vaara, M. (1985). Molecular basis of bacterial outer membrane permeability. *Microbiol. Rev.* **49,** 1–32.

Nikaido, H., Kim, S-H., and Rosenberg, E. Y. (1993). Physical organization of lipids in the cell wall of *Mycobacterium chelonae. Mol. Microbiol.* **8,** 1025–1030.

Raetz, C. (1987). Biosynthesis of lipid A in *Escherichia coli. In* "Escherichia coli and *Salmonella typhimurium:* Cellular and Molecular Biology" (F. C. Neidhardt, Ed.), pp. 498–503. American Society for Microbiology, Washington, DC.

Reddy, T. R., and Suryanarayana, T. (1989). Archaebacterial histone-like proteins. *J. Biol. Chem.* **264,** 17298–17308.

Schäfer, G., Anemüller, S., Moll, R., Meyer, W., and Lübben, M. (1990). Electron transport and energy conservation in the archaebacterium *Sulfolobus acidocaldarius. FEMS Microbiol. Rev.* **75,** 335–348.

Schleper, C., Pühler, G., Holz, I., Gambacorta, A., Janekovic, D., Santarius, U., Klenk, H-P., and Zillig, W. (1995). *Picrophilus* gen. nov., fam. nov: A novel heterotrophic, thermoacidophilic genus and family comprising archaea capable of growth around pH 0. *J. Bacteriol.* **177,** 7050–7059.

Schwartz, R. M., and Dayhoff, M. O. (1978). Origin of prokaryotes, eukaryotes, mitochondria, and chloroplasts. *Science* **199,** 395–403.

Schwarz, U., Asmus, A., and Frank, H. (1969). Autolytic enzymes and cell division of *Escherichia coli. J. Mol. Biol.* **41,** 419–429.

Stetter, K. O., Fiala, G., Huber, G., Huber, R., and Segerer, A. (1990). Hyperthermophilic microorganisms. *FEMS Microbiol. Rev.* **75,** 117–124.

Van De Vossenberg, J., Ubbink-Kok, T., Elferink, M., Driessen, A., and Konings, W. N. (1995). Ion permeability of the cytoplasmic membrane limits the maximum growth temperature of bacteria and archaea. *Mol. Microbiol.* **18,** 925–932.

Walker, G. C. (1987). The SOS response of *Escherichia coli. In* "Escherichia coli and *Salmonella typhimurium:* Cellular and Molecular Biology" (F. C. Neidhardt, Ed.), pp. 1346–1357. American Society for Microbiology, Washington, DC.

Wanner, B. L. (1993). Gene regulation by phosphate in enteric bacteria. *J. Cell. Biochem.* **51,** 47–54.

Westerhoff, J. V., and Welch, G. R. (1992). Enzyme organization and the direction of metabolic flow: Physicochemical considerations. *In* "Current Topics in Cellular Regulation" (E. R. Stadtman and P. B. Chock, Eds.), Vol. 33, Academic Press, New York.

White, D. (1995). "The Physiology and Biochemistry of Prokaryotes." Oxford University Press, New York.

Woese, C. R., Kandler, O., and Wheelis, M. (1990). Towards a natural system of organisms: Proposal for the domains Archaea, Bacteria, and Eucarya. *Proc. Natl. Acad. Sci. USA* **87,** 4576–4579.

Yanofsky, C., and Crawford, I. P. (1987). The tryptophan operon. *In* "Escherichia coli and *Salmonella typhimurium:* Cellular and Molecular Biology" (F. C. Neidhardt, Ed.), pp. 1453–1472. American Society for Microbiology, Washington, DC.

# SECTION

# VIII

# Plant Cells, Photosynthesis, and Bioluminescence

*Richard G. Stout*

# 62

# Plant Cell Physiology

## I. Introduction

Superficially, plants and animals seem to be completely different forms of life. Sedentary flowering plants may seem alien to motile warm-blooded mammals like ourselves. However, if we look at the cellular level, we see that animals and plants are not so different and unrelated. Plant cells and animal cells are connected evolutionarily, sharing a common ancestor, but separated by over half a billion years of evolution.

Despite this length of time, plant and animal cells are remarkably similar with regard to fundamental cellular processes such as chemiosmotic adenosine triphosphate (ATP) synthesis, DNA replication, and protein synthesis. Membranes and membrane-bound organelles, such as mitochondria, nuclei, Golgi, and endoplasmic reticulum (ER), appear similar in both plant and animal cells, although differences can be detected at the biochemical and molecular levels. The differences between plant and animal cells become more evident at the higher order cellular functions, such as cell replication, transport, defense, and intercellular communication.

## II. Plant Cell Ultrastructure

The major structural features of plant cells are illustrated in Fig. 1. Throughout the plant's life, new cells are typically produced in localized regions of relatively high mitotic activity called **meristems.** Meristematic regions are usually found at the tips of shoots and roots and at the base of branch shoots. Compared with meristematic cells, mature plant cells are distinguished by a large central vacuole. Along with the vacuole, the plastids (including chloroplasts) and the cell wall are the main structural features of plant cells that distinguish them from animal cells. The cell wall affects virtually every aspect of plant cell physiology, including development, transport, defense, and osmoregulation (Burgess, 1985; Tolbert, 1980). Morphological and physiological specialization of plant cells occurs primarily through modifications of the cell wall and plastids.

## A. Cell Wall Structure and Biosynthesis

The walls of higher plant cells, like most eukaryotic walled organisms, consist of a skeleton of interwoven microfibrils embedded in an amorphous matrix (Bacic *et al.,* 1988). Cellulose ($\beta$1,4-linked poly-D-glucose) is the microfibrillar component of plant cell walls (Fig. 2A). Cellulose molecules are linear ribbonlike polymers of 500 to 2000 glucose residues. Cellulose microfibrils generally consist of 40 to 70 cellulose molecules, running in parallel, overlapping and oriented with the same polarity. The linear nature of the cellulose molecule is due to the $\beta$1–4 linkage between the glucose units. Extensive hydrogen bonding occurs among the linear cellulose chains, promoting the formation of a highly ordered crystalline aggregate.

The cellulose microfibrils are cross-linked in the plant cell wall by the matrix polysaccharides, the hemicelluloses and the pectins. The hemicelluloses (Fig. 2B) are a mixture of branched heterogeneous polymers of glucose, xylose, galactose, arabinose, and other sugars. Structurally, the hemicelluloses are typified by a backbone of $\beta$1–4-linked glucose or xylose, from which relatively short side chains extend. The long backbone of the hemicellulose molecules may hydrogen bond to the surface of a cellulose microfibril, with the side chains protruding perpendicular to the surface of the microfibril.

A heterogeneous complex family of branched polymers, the pectins (Fig. 2C), is enriched in galacturonic acid residues. Their major function is probably to cross-link cellulose microfibrils by attaching to the microfibril-bound hemicelluloses (Fig. 3).

In addition to hemicelluloses and pectins, the cell wall matrix contains both structural glycoproteins and enzymes. The structural glycoproteins are analogous to the animal structural protein collagen. Like collagen, they are fibrous proteins containing the rarely found amino acid hydroxyproline, along with relatively high levels of serine, alanine, and threonine. One hydroxyproline-rich glycoprotein in cell walls is extensin. Woven into and covalently bound to the polysaccharide components of the plant cell wall, extensin serves to reinforce structurally the wall (Varner

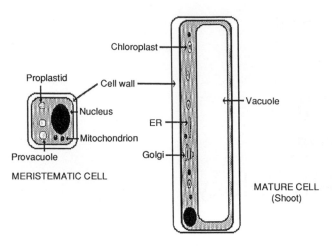

**FIG. 1.** Plant cells.

and Lin, 1989). The rodlike extensin molecules may also be cross-linked to one another (via tyrosine residues), catalyzed by peroxidases present in the cell wall.

The relative proportion of these cell wall components depends on the type of cell wall. Dividing and growing plant cells contain primary cell walls, which are thinner and less rigid than secondary cell walls, typically synthesized by older maturing plant cells after growth is finished. Such cells may construct a thick and rigid secondary wall to provide structural support, such as in stems. Secondary cell walls usually contain a higher proportion of cellulose and less pectin than primary walls. Lignins (polymers of aromatic alcohols) are common matrix components of secondary cell walls. Lignins contribute to cell wall rigidity and greatly increase resistance to cell wall degradation. Cork cell walls, for example, contain very high levels of lignins. High levels of pectins are found in the middle lamella, the region between the walls of adjacent plant cells, and act to glue the cells together.

Cellulose microfibrils, synthesized by enzymes in the plasma membrane, appear in the electron microscope to be pressed against the outer surface of the plasma membrane. A microfibril can sometimes be traced along the membrane surface until it terminates (originates?) at a structure called a **terminal complex** or **rosette.** Many plant cell biologists think these are cellulose synthases (Delmer and Amor, 1995). Multiple cellulose chains may be synthesized by a single "rosette" and are deposited into the cell wall space (Fig. 4). These terminal complexes presumably are laterally mobile in the fluid plasma membrane, and as cellulose microfibrils are made, the complexes are pushed around the surface of the plasma membrane, as shown in Fig. 5 (Giddings and Staehelin, 1991). The cellulose synthases likely use cytoplasmic UDP-glucose as the building blocks for cellulose.

The deposition of cellulose microfibrils at the cell surface is not random. The morphology of a plant cell is determined by the orientation of these cellulose microfibrils in the cell wall. Cellulose synthase complexes may somehow be guided as they move laterally through the lipid bilayer

matrix of the plasma membrane by one or two microtubule "tracks" (see Fig. 5).

As shown in Fig. 4, the matrix polysaccharides and structural proteins are synthesized within the Golgi membrane system, packaged into vesicles by the Golgi apparatus, and deposited into the cell wall space via vesicular fusion to the plasma membrane (Hayashi, 1989; Carpita *et al.,* 1996).

## B. Cell Wall Functions

Cell walls play roles in addition to serving as scaffolding for the society of plant cells that compose a living plant. For example, cell walls provide a physical barrier to insects and microbial pathogens.

In addition to acting as a suit of armor, the cell wall allows plants to take up water. Plant cells take up water passively, that is, by osmosis. Thus, to remain hydrated, the cell's water potential must be lower (more negative) than its surroundings. Plant cells must be hypertonic to their immediate environment. In this condition, the cells would eventually swell and burst without rigid cell walls. Since the cell wall prevents swelling, a relatively small amount of water uptake leads to a large increase in hydrostatic pressure (turgor pressure) within the cell. As turgor pressure increases, the cell approaches osmotic equilibrium with its surroundings (Taiz and Zeiger, 1991).

Cell turgor pressure is critical to plants. For example, it is responsible for the distention of leaves. Moreover, reversible changes in cell turgor pressure allow the regulation of gas exchange between the leaf and the environment through pores on the leaf epidermis called **stomata.** Stomata consist of two kidney-shaped cells, called guard cells, that change shape with changes in their turgor pressure. This reversible change in shape leads to the opening and closing of the pores. The most important role of plant cell turgor pressure is, however, to drive cell enlargement. Without plant cell turgor pressure, plants would not be able to grow.

## C. Developmental Aspects of the Cell Wall

Plant cell division cannot occur without the formation of a new cell wall. In plants, the final phase of cell division (cytokinesis) is characterized by the formation of a cell plate between daughter cells. Components of the cytoskeleton, primarily microtubules, may serve to direct the orientation of cell plate formation along the plane of cell division.

Soon after a plant cell divides, it begins to enlarge. Most of a plant's size comes from the growth of its cells. In general, plant cells enlarge much more than animal cells. Plant cells may grow in volume by 10- to 50-fold. Most of this increase in volume is due to water uptake into the vacuole (see Fig. 1).

For uptake to occur, either the osmotic solute concentration in the cell must increase or the cell's turgor pressure must decrease. In growing plant cells, the turgor pressure decreases because of a special process called **stress relaxation.** Because water is not compressible, only a slight

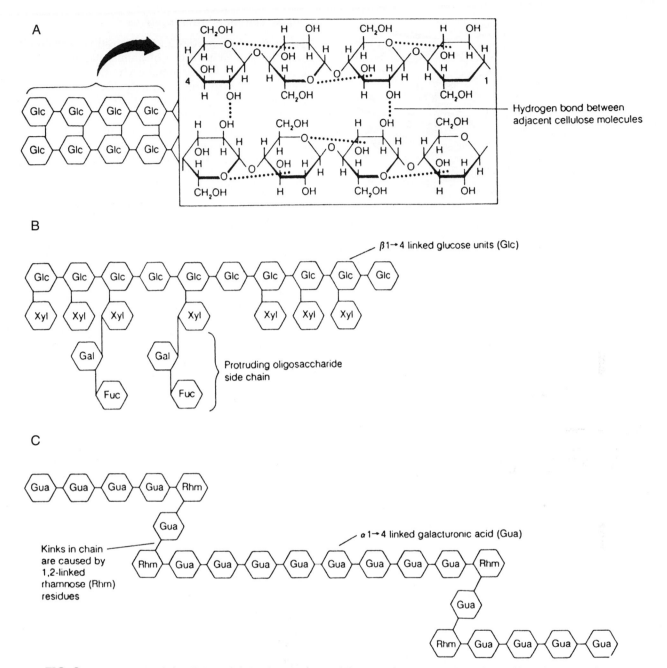

**FIG. 2.** Structures of (A) cellulose, (B) hemicellulose, and (C) pectin. (From Taiz and Zeiger, 1991; *Plant Physiology,* Copyright © 1991 Benjamin/Cummings Publishing Company and Dr. Eldon Newcomb, Department of Botany, University of Wisconsin, Madison, WI.)

"give" in the cell wall is sufficient to reduce the cell's turgor pressure and, simultaneously, the stress on the wall. With this drop in turgor pressure, the cell's water potential is lowered (more negative) relative to its surroundings, and water uptake occurs. For plant cell growth to be sustained, the cell strikes a balance between water uptake (tends to increase turgor pressure) and cell wall extension (tends to decrease turgor pressure). During this growth phase, the plant cell accumulates enough solutes to maintain the osmotic concentration as the cell volume increases.

In most nongrowing plant cells, the cell wall is rigid, except for a slight elasticity. For the cells to enlarge, the cell wall must irreversibly extend in response to the cell's turgor pressure. It must become less rigid throughout, that is, "loosened." This increased wall extensibility depends on the rate of complex biochemical processes in the cell wall. Several mechanisms have been proposed for this. One is called the acid growth theory (Rayle and Cleland, 1992). According to this hypothesis, the growing plant cell actively extrudes protons into the cell wall space (probably via

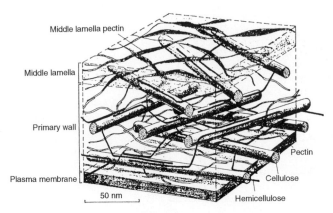

**FIG. 3.** A simplified illustration of how the three main components of the primary cell wall of higher plants may be spatially arranged in three dimensions. (From McCann and Roberts, 1991, by permission.)

## D. Plant Cell Membrane Systems

The physical properties and composition of plant membranes are similar to those of animal membranes. Plant and animal cells also resemble each other when it comes to fundamental membrane-related processes. For example, the plasma membrane of plant cells is constructed, replenished, and recycled as it is in animal cells. Golgi-derived vesicles fuse with the plasma membrane, adding lipids and proteins to it in addition to depositing their contents in the cell wall region. These vesicles contain structural glycoproteins and a wide range of cell wall polysaccharides. These macromolecules are synthesized by ER and Golgi enzymes. (In contrast, the Golgi apparatus in animal cells synthesizes, packages, and secretes mostly glycoproteins.)

The plant cell plasma membrane is recycled in a manner reminiscent of animal cells. Coated pits have been observed in plant cells, and evidence suggests that macromolecular components of the plasma membrane are recycled via endocytosis, eventually being deposited in the central vacuole (Robinson and Depta, 1988).

ATPase proton pumps in the plasma membrane). The resulting lowered pH may activate enzymes called expansins (Taiz, 1994; Carpita *et al.*, 1996) present in the cell wall that cleave load-bearing bonds. "Acid growth" is augmented by the active deposition of new cell wall material via exocytosis.

The cell wall also determines the morphology of the cell during growth and differentiation. This is due to the orientation of the cellulose microfibrils around the cell (see Fig. 5; Section II,A). For example, a horizontal barrel-hoop arrangement of the microfibrils will result in vertically elongated cells typical of shoots and roots.

## E. Vacuoles

In some mature plant cells, the central vacuole may constitute more than 90% of the cell volume. Meristematic and other immature cells may have many small vacuoles called provacuoles (see Fig. 1), which are probably derived from the Golgi apparatus. They coalesce as the cell grows to maturity and eventually give rise to the large central vacuole.

The vacuole is bounded by a lipid bilayer membrane called the **tonoplast,** and it contains inorganic ions, metabo-

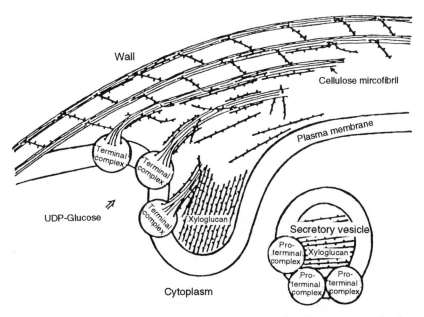

**FIG. 4.** A model of hemicellulose (xyloglucan) exocytosis in the plant cell wall, accompanied by deposition of terminal complexes in the plasma membrane. (Reproduced, with permission, from the *Annual Review of Plant Physiology and Plant Molecular Biology,* vol. 40. © 1989 by Annual Reviews, Inc.)

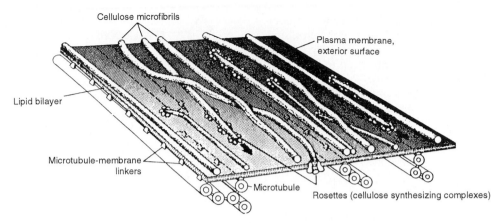

Cellulose microfibrils

Plasma membrane, exterior surface

Lipid bilayer

Microtubule-membrane linkers

Microtubule

Rosettes (cellulose synthesizing complexes)

**FIG. 5.** A model depicting microtubule-directed cellulose microfibril deposition during primary cell wall formation. (From Staehelin and Giddings, 1982. Copyright © 1982 John Wiley and Sons, Inc. Reprinted by permission of John Wiley & Sons, Inc.)

lites (i.e., organic acids, sugars, amino acids), phenolic and alkaloid compounds, and hydrolytic enzymes. The acidic vacuolar pH (4 to 6) is often several pH units below the cytoplasmic pH (ca. pH 7.0).

The vacuole is the cell's principal storage site. Vacuoles also may (1) help to defend the cell, (2) play a role in the sexual reproduction of plants, (3) function as a cytoplasmic homeostat, (4) recycle macromolecular components of the cell, and (5) most important, enable cells to grow.

As a large compartment separated from the cytoplasm, the vacuole serves as both a short- and long-term storage site for metabolic compounds or inorganic ions such as $Na^+$ that may be potentially harmful to cytoplasmic functions. Organic acids such as malate may be temporarily stored in the vacuole, depending on the rate of photosynthesis. Toxic compounds produced by cells are permanently stored in the vacuole. These noxious compounds may discourage insects and other herbivores from eating such plant tissues. Vacuoles may also help to attract beneficial insects. Colorful pigments such as anthrocyanins may be stored in the vacuole of flower petal cells, attracting insect pollinators.

The metabolic and signal transduction processes in the cell cytoplasm are very sensitive to changes in the $H^+$ and $Ca^{2+}$ concentration. The vacuole serves as both source and sink for these two ions. Transport systems located in the tonoplast (discussed later in Section IV) actively mediate this dynamic ebb and flow of $H^+$ and $Ca^{2+}$.

Sometimes referred to as the cell's garbage dump, the vacuole is better described as the cell's recycling center, behaving like a giant lysosome. The hydrolytic enzymes in the vacuole digest proteins, RNA, lipids, etc., that have outlived their usefulness. The resulting breakdown products may be reused by the cell or, in the case of dying cells, may be transported to other parts of the plant.

As previously mentioned, plant cells can grow much larger than animal cells because of the vacuole. The vacuole is much less complex and protein-rich than the cytoplasm, which costs more to support in terms of energy and mineral input (particularly nitrogen). For most plants, nitrogen is in short supply in the soil. Thus, by increasing their volume through water uptake primarily into the vacuole, plant cells can enlarge without having to support a larger, nitrogen-rich cytoplasm.

### F. Plastids

Chloroplasts are only one example of a group of plant cell organelles called plastids (Burgess, 1985). Other plastids include proplastids, amyloplasts ("starch-containing"), and multicolored chromoplasts (found in fruits and flowers). The membrane system within the plastids, as well as their contents, is what defines the different types of plastids. Immature undifferentiated plastids, the proplastids (four to five times smaller than chloroplasts) are most commonly found in meristematic cells. These proplastids give rise to the other plastids.

The nature of the plastids found in a cell is determined by the cell type and the developmental stage of the tissue. Most plastid proteins (ca. 90%) are encoded in the cell nucleus. These genes are translated by cytoplasmic ribosomes, and the resulting polypeptides are targeted to the plastids and actively imported by them. Differential expression of these nuclear genes determines the kind of plastids present in a given cell (Mullet, 1988).

### G. The Cytoskeleton

The plant cell cytoskeleton appears to be a dynamic structure composed of microtubules, actin filaments, and intermediate filaments (Lloyd, 1991). As previously discussed, the orientation of cortical microtubules may direct the deposition of cellulose microfibrils in the cell wall (see Section II,A). The orientation of the filaments that constitute the cytoskeleton may be affected by light, gravity, and plant hormones.

In large cells, cortical actin filaments may help stir the cell's cytoplasm. This phenomenon, called **cytoplasmic streaming,** is commonly observed as a unidirectional streaming of cytoplasm in large vacuolated cells. In such large cells, the rate of diffusion may be too slow to accommodate normal metabolism. Cytoplasmic organelles, such as mitochondria and chloroplasts, have been observed traveling along bundles of actin filaments at up to 75 $\mu$m/sec in giant algal cells (Williamson, 1986).

The cytoskeleton also directs intracellular vesicular traffic. As previously mentioned, microtubules direct Golgi-derived vesicles to migrate to the plane of cell division and deposit their contents there during cell plate formation.

### H. Plasmodesmata

Most plasmodesmata are established during the final stage of cell division. Strands of tubular ER become entrapped in the developing cell plate, possibly leading to the formation of hundreds of tiny conduits in the wall separating the two new daughter cells. Under special circumstances, secondary plasmodesmata (having multiple branches united via a central cavity) may develop in nondividing cells.

As illustrated in Fig. 6, plasmodesmata are structurally complex (Robards and Lucas, 1990). They are lined with plasma membrane, thus linking the plasma membranes of adjacent cells. A single plasmodesma contains an axial component called the **desmotubule**, which passes through the plasmodesma and often connects the ER of adjacent cells. Indeed, the desmotubule may be an extension of the ER. In between the desmotubule and the plasma membrane is a sleeve of cytoplasm. The two neck regions of a plasmodesma may be ringed by nine or ten "5-nm subunits," which may govern aperture size (analogous to a sphincter).

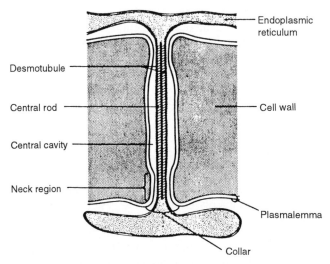

**FIG. 6.** Diagram of the components of a simple plasmodesma as seen in longitudinal section. [Reproduced from Robards, A. W. (1976). Plasmodesmata in higher plants. *In* "Intercellular Communication in Plants: Studies on Plasmodesmata" (B. E. S. Gunning and A. W. Robards, Eds.), p. 31. Springer-Verlag, New York/Berlin, with permission.]

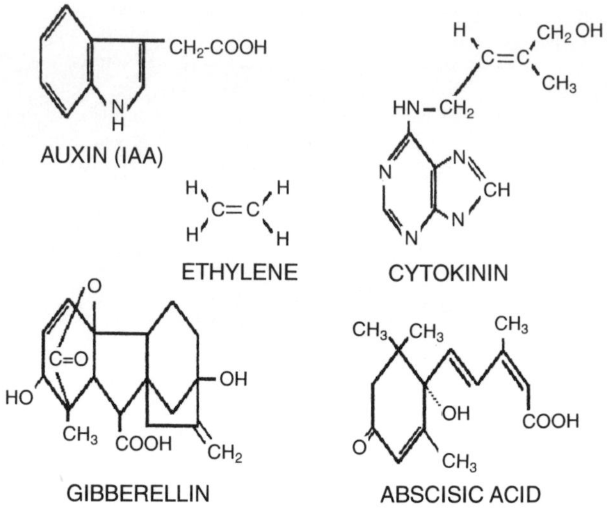

**FIG. 7.** Structures of plant hormones.

## III. Cell-to-Cell Communication

### A. Intercellular Transport via Plasmodesmata

Plasmodesmata probably mediate transport between adjacent plant cells, as gap junctions facilitate transport between animal cells (Robards and Lucas, 1990). Most plasmodesmata exclude molecules with a molecular weight of 700–900 or higher, comparable to molecular size exclusion limits of gap junctions. Gap junctions and plasmodesmata, however, possess no structural similarities.

### B. Plant Hormones

Like animals, higher plants depend on chemical signals for intercellular communication. Five plant hormones, auxin, cytokinin, gibberellin, abscisic acid, and ethylene, play many roles in plant growth and development (Davies, 1988; Taiz and Zeiger, 1991). They are all relatively small molecules (Fig. 7), which allows them easy passage through cell walls and plasmodesmata, and they are effective at low concentrations. Plant hormones often act at points remote from their site of synthesis. Recent evidence suggests that plants may also have steroid hormones (Hooley, 1996).

Each of these hormones may affect several different physiological processes during the course of a plant's life (see Table 1). A hormone may elicit a particular biological response depending on several factors. These include its point of action within the plant, the stage of development of the tissue, the concentration of the hormone, and the presence or absence of one or more of the other hormones.

### C. Defensive Signals

Plant cells have the ability to defend themselves against attack by herbivores or pathogens. As noted previously, vacuoles may contain metabolites, such as alkaloids, as constitutive chemical deterrents to attack. Cells also have inducible defensive responses, including (1) the biosynthe-

**TABLE I** Primary Physiological Roles of Plant Hormones

| Hormone | Main functions |
| --- | --- |
| Auxin | Promotes stem elongation, bud dormancy, and fruit development; stimulates formation of vascular tissue; promotes cell division |
| Ethylene | Promotes fruit ripening and lateral growth in stems |
| Cytokinin | Promotes cell division; delays sensescence |
| Gibberellin | Promotes shoot elongation, flowering, and seed germination |
| Abscisic acid | Inhibits growth; promotes stomatal closure; promotes desiccation tolerance |

One of these bacteria, *Rhizobium,* forms a symbiotic relationship with leguminous plants such as soybean, pea, and alfalfa (Long, 1989). This symbiosis provides the plant with a valuable mineral nutrient, namely, nitrogen. Plant cells cannot directly use nitrogen from the atmosphere. Plants must rely on mineral nitrogen (primarily $NO_3^-$) from the soil, and nitrate availability often limits plant growth and productivity. Some plants can obtain nitrogen derived from atmospheric $N_2$ by forming symbiotic relationships with prokaryotic organisms that "fix" nitrogen. This ability to incorporate or "fix" atmospheric $N_2$ into organic form is limited to only certain prokaryotes, including some cyanobacteria and bacteria in the genus *Rhizobium.* The plant

sis of antibiotic compounds, collectively known as phytoalexins, toxic to some microorganisms; (2) the activation of genes coding for protease inhibitor proteins that inhibit the digestive serine proteases of microbial pathogens or insect herbivores; and (3) the increased deposition of extensin and ligins into the cell wall region to fortify the wall.

These inducible defensive responses can be elicited locally and systematically by signal molecules produced at the site of pathogen or insect attack. Such wounding results in the rapid accumulation of phytoalexins and protease inhibitors, not only in the wounded tissues but also in neighboring and distal unwounded parts of the plant, thereby indicating that a signal, or signals, is released from wounded cells to travel throughout the plant.

Specific cell wall fragments (oligosaccharides) of fungi and higher plants (Fig. 8) can induce defensive responses in cells (Ryan and Farmer, 1991). Because oligogalacturonides have been shown to move only a relatively short distance from a wound site, cell wall fragments may function chiefly as localized intercellular wound signals. Other, more mobile signals have also been reported.

An 18-amino-acid polypeptide, systemin (perhaps the first polypeptide hormone found in plants) is able to induce the synthesis of wound-inducible protease inhibitor proteins in several plant species (McGurl *et al.,* 1992). Systemin is relatively mobile within the plants tested to date.

Abscisic acid (ABA), ethylene, salicylate (structurally related to aspirin), and methyl jasmonate are other possible defensive signal molecules in plants. Both salicylate (Raskin, 1992) and methyl jasmonate, a lipid-derived molecule found in many higher plants, trigger the production of protease inhibitors (Farmer and Ryan, 1990). Because of their volatile nature, methyl jasmonate and ethylene may function not only in intercellular communication but also in plant–plant communication (Staswick, 1992).

### D. Interactions between Plants and Other Organisms

Complex interactions exist between plants and other living organisms. A glimpse of this complexity is provided by the current understanding of how plant cells interact with two soil-borne bacteria.

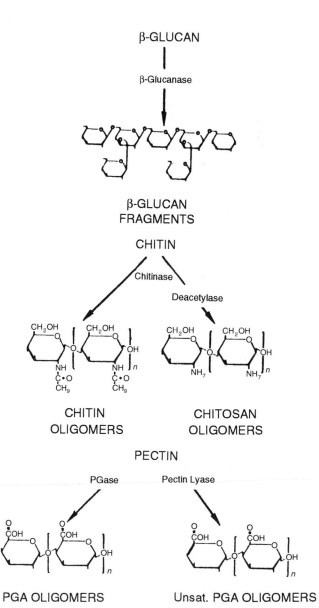

**FIG. 8.** Enzymatic degradation of fungal and plant cell walls produces oligosaccharides that may function as localized defensive signals in plants. (Reprinted with permission from Ryan, 1988. Copyright 1988 American Chemical Society.)

receives a usable form of nitrogen from the *Rhizobium,* and the bacteria derive energy in the form of carbohydrates from the plant.

Legumes that thrive in nitrogen-poor soil likely have small, pea-size nodules that are packed with symbiotic nitrogen-fixing *Rhizobium* on their roots. The steps involved in establishing such a relationship are numerous and complex. The coordinated expression of both bacterial and plant genes is critical for the successful formation and maintenance of this symbiosis. Recent advances in the molecular genetics of *Rhizobium* have revealed that most symbiotic genes are located on a large plasmid called the Sym plasmid. Sym plasmid genes involved in host-specific infection and nodule formation are referred to as *nod* genes. Some of these bacterial *nod* genes may be required for the production of a signal that leads to the expression in the host plant of a family of proteins, called nodulins, involved in various stages of nodule formation.

The release into the soil of chemical signals produced by the plants may be the first step toward symbiosis. For example, alfalfa root cells release flavones and other related compounds that preferentially attract the species *R. meliloti.* These compounds may not only mediate this chemotactic response, but also regulate bacterial genes (Geurts and Franssen, 1996). One of these flavones, luteolin, induces the expression of *nod* genes in *R. meliloti.* It apparently does so by activating a 33-kDa protein, constitutively expressed in these bacteria and coded for by the *nodD* gene (Fisher and Long, 1992).

Though little is known about the biochemical functions of the *nod* genes, they are suspected of being responsible for the production of signal molecules critical for the establishment of legume/*Rhizobium* symbiosis. Some of these Nod factors may mediate plant–microbe recognition, and others may cause the early events of nodule formation.

The other soil-borne bacterium, *Agrobacterium tumifaciens,* causes crown gall disease in plants. It transfers genes to infected plant cells, transforming them into tumor cells. The transformed cells grow and divide abnormally to form large galls on the plant.

*Agrobacterium* infects the plant only at wound sites. Ironically, wounded plant cells release a chemical that signals *Agrobacterium* to activate the infection process. This signaling molecule specifically activates *Agrobacterium* genes that mediate the infection process. These *vir* (virulence) genes, as well as the DNA transferred to the cell (T-DNA), are part of a large plasmid of *Agrobacterium,* the Ti (tumor-inducing) plasmid. The T-DNA portion of the Ti plasmid is copied within *Agrobacterium,* transferred to the cell, and integrated into the cell nuclear genome. The expression of the T-DNA in the cells results in their transformation into tumor cells (Beijersgergen *et al.,* 1992).

The ability of *Agrobacterium* to transfer and stably integrate T-DNA into cells is the basis for most plant genetic engineering. Recombinant DNA technology can be used to replace tumor-causing genes from within the T-DNA with "desirable" genes. Then the T-DNA can be used as a vector for the genetic transformation of higher plants.

## IV. Membrane Transport

### A. Electrogenic Proton Pumps

Unlike animal cells, the cells of higher plants, fungi, and bacteria can live in an environment of little nutritional value. The ability to accumulate nutrients such as mineral ions, sugars, and amino acids, which may be present extracellularly at only minute levels, may be explained in two ways. First, these organisms produce a substantial resting membrane potential, and second, they rely on the ubiquitous proton, rather than $Na^+$, to generate this potential.

The primary active transport mechanisms in cell membranes generate electrochemical $H^+$ gradients to drive secondary active transport. The average plasma membrane potential in cells ranges from $-120$ to $-180$ mV (cytoplasmic side negative), with the electrogenic pump component being $-60$ to $-120$ mV. Electrogenic $H^+$ transport occurs not only in plants but also in bacteria and fungi. Some animal cells, such as those of the gastric epithelium, also possess proton pumps. In plant cells, this electrogenic component is produced by proton-translocating ATPases ($H^+$-ATPases), which pump protons from the cytoplasm, across the plasma membrane, into the cell wall space (Michelet and Boutry, 1995).

Recent efforts to clone and sequence plant plasma membrane $H^+$-ATPase genes have revealed differences from and similarities to animal P-type ATPases. The enzyme has eight membrane-spanning regions, with both the C- and the N-terminal regions present on the cytoplasmic side of the plasma membrane. The sequences of several plant P-type $H^+$-ATPases, except for certain hydrophilic regions, do not show great overall homologies with analogous enzymes in animal cells. One conserved region (containing aspartate) may be associated with the formation of the phosphorylated intermediate; another may be the binding site for ATP (Sussman and Harper, 1989).

There are multiple forms of the P-type $H^+$-ATPase genes in plants, and their expression may be cell- or tissue-specific. There is evidence that the plasma membrane $H^+$-ATPase exists as multigene families in oat, tobacco, and tomato. In *Arabidopsis thaliana,* characterized by a relatively small genome, at least 10 putative isoforms of plant P-type $H^+$-ATPase have been identified. At least one of these isoforms is expressed principally in roots.

How plasma membrane $H^+$-ATPases are regulated is not known. Regulation of the $H^+$-ATPase may involve a putative autoinhibitor domain located within the C-terminal region. There are several reports that $Ca^{2+}$-activated protein kinases may be involved. The nature of the phospholipids that surround the $H^+$-ATPases in the plasma membrane may also modulate proton-pumping activity.

The plasma membrane may contain components of a redox chain involved in $H^+/e^-$ transport across the membrane (Crane and Barr, 1989). Although it is not clear whether this system is electrogenic, it likely increases $H^+$ extrusion through exporting negative charge, thus stimulating the $H^+$-ATPases by reducing the "backpressure" of the charge gradient on the proton pumps. This redox system may be modulated by light.

A proton-motive force also drives transport at the vacuolar membrane (tonoplast), though this membrane potential (−90 mV, cytoplasmic side negative) is typically less than that of the plasma membrane. The primary active transport system in the tonoplast is a V-type H⁺-ATPase that pumps protons into the vacuole (Sze *et al.,* 1992). These V-type H⁺-ATPases are distinguished in several ways from the P-type ATPases in the plasma membrane. Structurally more complex, the vacuolar H⁺-ATPases are large (450- to 650-kDa) oligomeric proteins consisting of 8 to 13 subunits. A comparison of vacuolar H⁺-ATPases from several plant species indicates considerable variation in subunit composition. As with the plasma membrane H⁺-ATPases, the cell genome may contain multiple genes coding for the individual subunits of the vacuolar H⁺-ATPases. Regulation of the vacuolar V-type H⁺-ATPase is poorly understood. Factors that affect the H⁺-ATPases include $Ca^{2+}$ concentration and cytoplasmic pH.

These factors also affect another type of electrogenic proton pump in the tonoplast. A proton-translocating pyrophosphatase (PPase) may significantly contribute to the electrogenic component of the vacuolar membrane potential. Pyrophosphate (PPi) may be generated by a variety of metabolic reactions in cells.

## B. Ion Channels

The application of the patch-clamp technique has demonstrated the presence of channels conducting K⁺, Cl⁻, $Ca^{2+}$, and organic ions in the plasma membrane and vacuolar membrane of plant cells (Hedrich and Schroeder, 1989; Tester, 1990). In addition, recent advances in the cloning, sequencing, and expression of putative K⁺ channel gene have revealed some similarities to K⁺ channels found in animal nerve cells.

Present at relatively high concentrations in cells (typically 75–100 mM), K⁺ activates cytoplasmic enzymes and serves as a major osmotic solute. In most plant cells, K⁺ passively enters the cytoplasm down the relatively large electrical gradient across the plasma membrane (inside negative). Major pathways for this K⁺ uptake by cells are probably K⁺-influx channels. These are activated by membrane potentials more than −100 mV, generated by the action of the plasma membrane H⁺-ATPases.

Some cells, particularly Characean algae such as *Nitella* and *Chara,* display action potentials (APs), that is, a transient depolarization of the membrane potential (Davies, 1987). The AP may occur in response to mechanical stimulation. In *Chara,* the rising phase of the AP is caused by efflux of Cl⁻.

K⁺ efflux is mediated by channels that are also regulated by the membrane potential. In the case of K⁺-efflux channels, however, K⁺ efflux is activated by the depolarization of the membrane potential to values more positive than −40 mV. The physiological roles for the K⁺ efflux channels include rapid turgor/volume regulation in stomatal guard cells (see Section II,B) and motor tissue cells in plants as well as repolarization of APs in both higher plants and algae.

Plant cell K⁺ channels differ both physiologically and biophysically from animal cell K⁺ channels. In plant cells, the two types of K⁺ channels mediate the long term K⁺ transport into and out of the cells. As in animal cells, K⁺ channels help to reset the membrane potential close to the equilibrium potential for K⁺ after the induction of short-term potential changes by other channels. Biophysically, plant voltage-dependent K⁺ channels activate 10 to 100 times more slowly than those of animal cells. Despite these functional differences, plant and animal K⁺ channels may have some structural similarities.

Several other plasma membrane ion channels have been reported in higher plants (Fig. 9). These include anion channels (particularly Cl⁻) and nonselective stretch-activated channels (Tyerman, 1992). The latter may function as turgor sensors and mechanosensors (Cosgrove and Hedrich, 1991). Signal-regulated $Ca^{2+}$-permeable ion channels may also be present in both the plasma membrane and the tonoplast (see separate discussion of $Ca^{2+}$-transport later).

Patch-clamp studies of the vacuolar membrane indicate the presence of voltage-dependent ion channels that are modulated by cytosolic-free $Ca^{2+}$. One is the SV-type channel [the kinetics of activation of these currents are slow, hence the term "slow-vacuolar" (SV-type) currents] and the other is the FV-type channel, characterized by relatively "fast" kinetics. The SV-type channels are permeable to both cations and anions and are activated by an increase in cytoplasmic $Ca^{2+}$ (>0.3 $\mu M$). The gating of the $Ca^{2+}$-dependent channels depends on both $Ca^{2+}$ concentration and voltage across the membrane. The FV-type channels may be chiefly responsible for the physiological passage of anions into the vacuole. Another ion channel in the vacuolar membrane, distinct from the SV and FV channels, may transport malate. An abundant anion in vacuoles, malate serves as the major anionic solute involved in charge balance with inorganic cations such as K⁺ and $Ca^{2+}$. The tonoplast may also contain water channels (Chrispeels and Maurel, 1994).

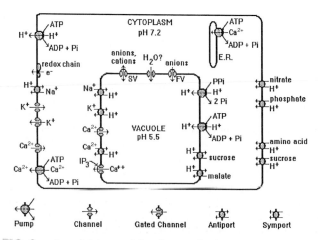

**FIG. 9.** A model summarizing the membrane transport systems in plant cells.

## C. Carriers

In addition to ion-selective (or nonselective) channels that create pores in the lipid bilayer, plant cell membranes have carriers or porters. Carrier-mediated transport of sugars, amino acids, and ions across the plasma membrane and tonoplast is coupled to proton transport down the $H^+$ gradients across these membranes. These proton cotransporters may be either symporters ($H^+$ and solute travel in same direction) or antiporters ($H^+$ and solute exchange). In the plasma membrane, $H^+$-coupled symporters may transport sucrose, amino acids, and anions such as nitrate or phosphate. $H^+$-coupled antiporters may include $Na^+$ carriers in the plasma membrane and $Na^+$, $Ca^{2+}$, $Cd^{2+}$, and sucrose carriers in the vacuolar membrane.

Sugar/proton cotransport carriers may exist in both the plasma membrane and the tonoplast. The "superfamily" of sugar transporter genes reported for bacteria and mammalian cells supports the idea that cell sugar transporters may be part of this family. Nitrogen-containing substances, such as nitrate and amino acids, are probably taken up by cells by proton-coupled symporters. Electrical measurements and genetic studies support the idea of carrier-mediated $NO_3^-$ uptake at the plasma membrane. An $H^+/NO_3^-$ antiporter may also exist in the vacuolar membrane. The proton-coupled uptake of inorganic phosphate at the plasma membrane has also been proposed on the basis of electrical measurements.

## D. $Ca^{2+}$ Transport Mechanisms

In plant cells, as in animal cells, $Ca^{2+}$ is an important second messenger in intracellular signal transduction (Leonard and Hepler, 1990). For this and other reasons, the regulation of $Ca^{2+}$ levels in the cytoplasm is critical for cell physiology. Although extracellular and intravacuolar $Ca^{2+}$ concentrations may be in the millimolar range, the cytoplasmic concentration of free $Ca^{2+}$ is about 10,000 times lower (100 n$M$ in resting conditions). Obviously, transport mechanisms must exist to maintain such steep $Ca^{2+}$ concentration gradients.

$Ca^{2+}$-ATPases may be associated with both the plasma membrane and the ER. The plasma membrane $Ca^{2+}$-ATPases act as primary $Ca^{2+}$-efflux pumps, moving $Ca^{2+}$ into the cell wall space against a large electrochemical gradient. Primary $Ca^{2+}$-ATPases pump $Ca^{2+}$ into the lumen of the ER, which may be a major site of intracellular storage. These $Ca^{2+}$-ATPases are P-type ATPases and are stimulated by calmodulin (similar to the $Ca^{2+}$-ATPase of red blood cells). The $Ca^{2+}$-ATPase is a dimer, consisting of two identical polypeptides of 130–140 kDa each.

A secondary active transport mechanism for $Ca^{2+}$ may exist on the vacuolar membrane. In mature higher plant cells, the most important intracellular pool of $Ca^{2+}$ is the vacuole. Calcium ion transport into the vacuole occurs against both a chemical and an electrical potential difference. The $H^+$ gradient across the tonoplast may drive $Ca^{2+}$ uptake into the vacuole via $H^+/Ca^{2+}$ antiporters.

$Ca^{2+}$ channels also mediate transmembrane $Ca^{2+}$ movement. $Ca^{2+}$ actively pumped out of the cell can passively reenter the cell through $Ca^{2+}$ channels in the plasma membrane. $Ca^{2+}$ influx channels in the plasma membrane have been implicated in the establishment (and maintenance?) of cell polarity. For example, in the zygote of the brown alga *Fucus*, asymmetry in the environment (e.g., low light, gravity, sperm entry, or low pH) leads to localization (or local activation) of $Ca^{2+}$-influx channels and $Ca^{2+}$-efflux pumps in the cell membrane. This leads to a $Ca^{2+}$ flux out at one side of the cell and in at the opposite side. This intracellular $Ca^{2+}$ gradient may promote structural asymmetries, such as cytoskeletal orientation and cell wall deposition, leading to cell polarity.

There are also at least two $Ca^{2+}$ channels in the vacuolar membrane. One is activated (gated) by inositol 1,4,5-trisphosphate ($IP_3$) and may increase cytoplasmic free Ca through releasing $Ca^{2+}$ from the vacuolar store. A voltage-dependent $Ca^{2+}$ channel has also been recently reported in vacuoles. This channel may be activated after the release of $Ca^{2+}$ into the cytoplasm by the $IP_3$-stimulated $Ca^{2+}$-release channels. $IP_3$-stimulated $Ca^{2+}$ release from the vacuole may polarize the vacuolar membrane potential which, in turn, would activate the inward-rectifying $Ca^{2+}$ channels. These channels may assist the $Ca^{2+}$-ATPases and the $H^+/Ca^{2+}$ antiporters in reestablishing the low cytoplasmic resting levels of calcium. As discussed in the following section, changes in the cytoplasmic-free $Ca^{2+}$ concentration may play a critical role in stimulus–response coupling.

# V. Signal Perception and Response

## A. Photoperception

Light has profound effects on many aspects of plant physiology, especially development. Sunlight, absorbed primarily by the green pigment chlorophyll, is the energy source for the life-giving process of photosynthesis. By activating specific nuclear genes coding for proteins involved in chlorophyll biosynthesis and chloroplast development, sunlight also acts as the environmental cue to developing cells to begin photosynthesis. Light does not directly alter gene expression, of course, but acts through a mediator, a photoreceptor known as phytochrome (von Arnim and Deng, 1996). In addition to mediating the light-triggered greening of plant tissue, phytochrome is involved in many other light-sensitive processes, including seed germination, shoot elongation, circadian rhythms, leaf development, and flower induction.

Phytochrome has been well characterized. The native phytochrome molecule is a dimer of approximately 240 kDa. Each monomer (ca. 120 kDa) contains a single covalently bound chromophore, an open-chain tetrapyrrole. A water-soluble protein, phytochrome can exist in either of two conformations. The conversion from one form to the other is triggered by specific wavelengths of light and is fully reversible. The Pr form of phytochrome absorbs red light (650–680 nm) and is converted to the Pfr form, which absorbs light of longer wavelengths, that is, far-red light (710–740 nm). In most cases, Pfr (produced by red light) is the active form of phytochrome. The Pfr form initiates a transduction process that leads to altered expression of specific genes. The resulting differential gene expression culminates in modified cell growth or physiology appropriate to the prevailing light environment.

The intracellular signal transduction mechanism probably involves $Ca^{2+}$. Since there is some evidence that the Pfr form associates with membranes (and the Pr form is soluble), Pfr may act by altering the cytoplasmic $Ca^{2+}$ concentration through interaction with one or more of the membrane-bound $Ca^{2+}$-transport mechanisms discussed earlier.

Phytochrome is involved in a wide range of light-sensitive processes. From recent cloning and sequencing studies of phytochrome, the smallest known plant genome (*Arabidopsis*) was found to contain as many as five different phytochrome genes. This supports earlier evidence for the existence of more than one type of phytochrome in higher plants. Different phytochrome isoforms may mediate different red light-sensitive responses.

Another class of photoreceptors in plant cells is activated by blue light. Blue light is involved in the control of a variety of physiological and developmental processes. The most well studied of these include phototropism (e.g., the growing of a shoot toward light), stomatal opening, and inhibition of shoot elongation. The blue-light photoreceptors that mediate these responses are likely plasma membrane-bound flavoproteins.

Blue light apparently directly activates a kinase moiety on a plasma membrane-bound flavoprotein to cause rapid phosphorylation of a class of plasma membrane proteins (Kaufman, 1993). This may represent an early step in the signal transduction chain for phototropism and for the blue-light inhibition of shoot elongation. Thus, at least one blue-light photoreceptor in cells may be a photoactivated protein kinase and may initiate a phosphorylation cascade culminating in the inhibition of elongation.

Evidence also indicates that a protein phosphorylated by a blue-light-mediated photoreaction may be associated with cytoskeletal elements. This suggests that the cell cytoskeleton may play a functional role in the inhibition of cell growth by blue light.

## B. Gravity Perception

Both shoots and roots correct their direction of growth if they are displaced from the vertical. The growing regions of a plant, particularly the zones of cell elongation in primary roots and shoots, are particularly responsive to gravity, though roots and shoots respond in opposite ways. When a young shoot is displaced from the vertical, it will curve upward after several hours. Conversely, a root in the same situation will curve downward. Both responses are a result of unequal rates of cell elongation on the upper side of the organ compared with the lower side. As the curving root or shoot regains a vertical orientation, the elongation rates of the cells become equal.

Gravity perception by higher plants probably involves the movement of free-falling bodies within specialized cells (Volkmann and Sievers, 1979). These gravity-sensitive

bodies are called **statoliths.** In gravity-sensing cells (statocytes), the statoliths are probably starch grains located within amyloplasts, which are specialized plastids. The amyloplasts sediment through the cytosol and come to rest on the lowermost part of the cell (Fig. 10). This result somehow redirects the transport of the plant growth substance auxin, leading to unequal elongation growth across the root or shoot (McClure and Guilfoyle, 1989; Hasenstein and Evans, 1988).

### C. Mechanosensory Mechanisms

Most plants display altered development as a result of mechanical stimulation. For example, plants growing in windy landscapes tend to be shorter, with thicker stem diameters, than the same species sheltered from the wind. To be shorter and sturdier in windy environments is obviously an adaptive advantage. This change in plant growth and development in response to mechanical stimulation is called **thigmomorphogenesis.** In the laboratory, thigmomorphogenesis can be simulated by touch. These touch responses are different from touch-induced rapid movements in plants (e.g., the Venus flytrap or the "touch-sensitive" *Mimosa*) in that the plant movements occur through rapid and reversible turgor pressure changes in motor cells. Thigmomorphogenesis involves irreversible alterations in cell development. Recent reports of mechanosensitive plasma membrane ion channels (Cosgrove and Hedrich, 1991), known to exist in animal cells, may offer a clue to how mechanical stimulation is perceived by plant cells. Patch-clamp analyses of stomatal guard cells (see Section II,B) have revealed at least three types of stretch-activated (SA) ion channels (anion efflux, $K^+$ efflux, and $Ca^{2+}$ influx) in the plasma membrane. These SA channels are gated, presumably by strain on the membrane, and may act primarily in the regulation of guard cell volume

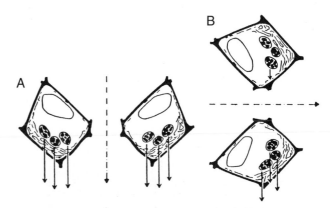

**FIG. 10.** A diagram illustrating the reorientation of amyloplasts in statocytes in response to the direction of gravity (denoted by the solid arrows). The dashed arrows point to the root tip. (A) Vertical orientation of the root. (B) Horizontal orientation of the root. [Reproduced from Volkmann, D., and Sievers, A. (1979). Graviperception in multicellular organs. *In* "Encyclopedia of Plant Physiology, New Series" (W. Haupt and M. E. Feinleib, Eds.), Vol. 7, pp. 573–600. Springer-Verlag, Berlin/New York, with permission.]

and turgor (see Section II,B). SA channels in actively grow-ing cells may also serve as growth-rate sensors so that cell wall expansion can be coordinated with the plasma membrane material via Golgi-derived vesicular fusion. Stretch-activated $Ca^{2+}$-influx channels in the plasma membrane may modulate the process of cell enlargement through changes in cytoplasmic $Ca^{2+}$ concentration. This $Ca^{2+}$ signal would be amplified by a variety of $Ca^{2+}$-modulated proteins, including calmodulin (CaM).

As shown in Fig. 11, activated CaM may amplify the signal in several ways: first, through CaM-modulated protein kinases that subsequently phosphorylate other en-zymes; second, through alterations in the microtubules and microfilaments as well as through increased production of the plant hormone ethylene, increasing the lateral growth of cells (e.g., leading to shorter, thicker stems). Finally, $Ca^{2+}$-activated CaM also may stimulate the expression of calmodulin genes, serving as a positive feedback mecha-nism. Changes in cytosolic $Ca^{2+}$ may also activate anion efflux channels in the plasma membrane, which may ac-count for action potentials (see Section IV,B) that have been observed in algae and higher plants following me-chanical stimulation (Braam and Davis, 1990a).

### D. Receptor Proteins

As previously discussed, chemical signals, such as plant hormones and cell wall fragments, affect cell development and physiology. Most hormone receptor proteins in animal cells are located on the cell surface, that is, in the plasma membrane, the exception being steroid hormone receptors, which are located intracellularly. Similarly, most of the putative receptor proteins identified in plant cells are asso-ciated with the plasma membrane.

Although putative receptors exist for most of the plant hormones (see Fig. 7), the two most well characterized are for auxin and ethylene. A probable auxin receptor has been located in elongating cells. This 22-kDa protein is apparently bound to the plasma membrane, with the auxin-binding site on the extra-protoplasmic surface. Physiologi-cal evidence suggests that the auxin receptor may somehow modulate ion transport activity at the plasma membrane

(Venis and Napier, 1995). Plants sense ethylene by a pro-tein kinase cascade (Bleeker and Schaller, 1996).

Specific receptors for oligosaccharide signal molecules, such as beta-glucans (see Section III,C), may also reside in the plasma membrane (Cheong and Hahn, 1991). The receptors for these and other extracellular effector mole-cules that elicit defensive responses in cells (e.g., systemin, methyl jasmonate, and salicylic acid) presumably trigger signal transduction pathways, culminating in the transcrip-tional activation of defense-related genes.

### E. Signal Transduction Mechanisms

Many of the signal perception mechanisms discussed previously invoke the participation of $Ca^{2+}$, inositol tris-phosphate ($IP_3$), and protein kinases in intracellular signal-ing (Palme, 1992). Two of the most important intracellular second messengers in animal cells are cyclic AMP (cAMP) and $Ca^{2+}$. Although there is little reliable evidence that cAMP functions as such in plants, there is ample experi-mental evidence that $Ca^{2+}$ does. Changes in cytosolic $Ca^{2+}$ occur during stimulus–response coupling in higher plants. Calcium-modulated proteins, including calmodulin and calcium-dependent protein kinases, are present in plant cells. Several CaM-regulated proteins have been identified, notably, $Ca^{2+}$-ATPase, NAD kinase, and protein kinase. The recent identification of annexins in plants, which are involved in the regulation of exocytosis in animal cells, provides evidence that $Ca^{2+}$-binding proteins may serve to link increases in cytosolic $Ca^{2+}$ with the stimulation of exocytosis in plant cells (Xu et al., 1992).

At least two breakdown products of phosphatidylinositol 4,5-bisphosphate ($PIP_2$), $IP_3$, and diacylglycerol (DAG), serve as intracellular signaling molecules in animal cells. These products of $PIP_2$ hydrolysis are known to exist in plants (Trewavas and Gilroy, 1991), but unequivocally demonstrating their participation in signal transduction has been difficult. The enzyme that hydrolyzes $PIP_2$, namely, phospholipase C, has been localized to the cytoplasmic surface of plasma membranes. Thus, the enzymatic basis for signal transduction via $PIP_2$ hydrolysis is apparently present in cells (Roberts and Harmon, 1992). A conceptual-ization of events in cells involving stimulus-evoked turn-over of phosphatidylinositol in the plasma membrane and the resulting $Ca^{2+}$ release is summarized in Fig. 12.

In animal cells, GTP-binding regulatory proteins (G pro-teins) serve as intermediates between plasma membrane receptors and the enzymes or ions channels they affect. G-protein-linked receptors typically activate a chain of events that alter the intracellular concentration of second messengers such as cAMP, $Ca^{2+}$, and $IP_3$. They also trans-duce extracellular signals into membrane-mediated events, including regulation of ion channels. GTP-binding proteins have been reported to exist in plants. Antibodies to animal G proteins have been shown to cross-react with plasma membrane proteins from a variety of plants, and a G-protein gene has been isolated from Arabidopsis. Al-though there is evidence that a G protein may regulate inward $K^+$-channel current in stomatal guard cells (Fairly-Grenot and Assman, 1991), G proteins have yet to be

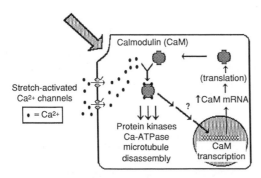

**FIG. 11.** A hypothetical model of mechanosensory perception and response at the cellular level in plants. (Modified from Braam and Davis, 1990b, by permission.)

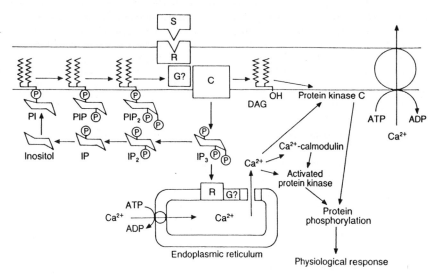

**FIG. 12.** A summary of events in plant cells involving the stimulus-induced turnover of plasma membrane-bound phosphatidylinositol, leading to an increase in cytoplasmic $Ca^{2+}$ concentration. S, stimulus; R, receptor; G, G-protein; C, phospholipase C; DAG, diacylglycerol; $IP_3$, inositol 1,4,5-trisphosphate; $IP_2$, inositol 1,4-bisphosphate; IP, inositol 1-phosphate; PI, phosphatidylinositol; PIP, phosphatidylinositol 4-phosphate; $PIP_2$, phosphatidylinositol 4,5-bisphosphate. The G-protein may stimulate the gating of ion channels. (From Trewavas and Gilroy, 1991, by permission.)

shown to mediate second messenger production in plant cells.

## VI. Summary

The cell wall, plastids, and the vacuole are the three main structural features that distinguish plant cells from animal cells. Consisting of cellulose microfibrils embedded in a polysaccharide and protein matrix, the cell wall of higher plants both supports and protects the protoplast. Chloroplasts serve as the photosynthetic organelles in green plant tissues, while other types of plastids function in carbohydrate storage (amyloplasts) and in reproduction (chromoplasts). Most mature plant cells are characterized by a large central vacuole, which permits plant cells to enlarge to a greater degree than animal cells and serves as an intracellular storage area for metabolites and for hydrolytic enzymes involved in the breakdown and recycling of cellular components. In contrast to these three structural features, plant and animal cell membrane and cytoskeletal systems display fundamental similarities.

Plant cells use chemical signals to communicate with each other as well as with other organisms. Such molecules may travel from cell to cell via plasmodesmata, small conduits through the cell walls of adjacent cells, which normally allow the passage of relatively small molecules (<800 kDa). The coordination and regulation of plant development and physiology are primarily accomplished by only five different plant hormones. These, as well as other chemical signals such as specific oligosaccharides and oligopeptides, can trigger systematic defensive responses in plants. Complex interactions, such as differential gene expression,

between microbial symbionts or pathogens and plant cells are mediated by chemical signals produced by both the plant and the microorganisms.

Plant cells maintain a substantial resting potential ($-120$ to $-180$ mV) across their plasma membrane primarily through electrogenic, proton-translocating ATPases. This primary active transport mechanism drives most of the other secondary transport systems in plasma membrane. An electrochemical proton gradient, maintained by proton-transport ATPases and pyrophosphatases, also drives transport at the vacuolar membrane. Gated channels for potassium and calcium ions, for example, are present in both the plasma and vacuolar membranes. Many of the transport proteins in plant cells structurally resemble analogous proteins in animal cells.

Plants can perceive subtle changes in light quality and quantity through at least two different kinds of photoreceptors, phytochrome (red/far-red light) and a blue-light photoreceptor. Gravity perception is probably achieved through falling bodies in the cytoplasm called statoliths. Mechanical stimulation may be mediated by plasma membrane bound, stretch-activated ion channels. Both chemical and environmental signal responses in plant cells may be mediated by second-messenger systems similar to those found in animal cells, that is, calcium ions, phosphoinositides, and G-proteins.

### Bibliography

Bacic, A., Harris, P. J., and Stone, B. A. (1988). Structure and function of plant cell walls. *In* "Biochemistry of Plants: A Comprehensive Treatise" (J. Preiss, Ed.), Vol. 14, pp. 298–371. Academic Press, San Diego.

Beijersgergen, A., Dulk-Ras, A. D., Schilperoort, R. A., and Hooy-kaas, P. J. J. (1992). Conjugative transfer by the virulence system of *Agrobacterium tumefaciens*. *Science* **256,** 1324–1327.

Bleeker, A. B., and Schaller, G. E. (1996). The mechanism of ethylene perception. *Plant Physiol.* **111,** 653–660.

Braam, J., and Davis, R. W. (1990a). Rain-, wind-, and touch-induced expression of calmodulin and calmodulin-related genes in *Arabidopsis*. *Cell* **60,** 357–364.

Braam, J. and Davis, R. W. (1990b). The mechanosensory pathway in *Arabidopsis*: touch-induced regulation of expression of calmodulin and calmodulin-related genes and alterations of development. *Curr. Top. Plant Biochem. Physiol.* **9,** 85–100.

Burgess, J. (1985). "An Introduction to Plant Cell Development." Cambridge University Press, London/New York.

Carpita, N., McCann, M., and Griffing, L. R. (1996). The plant extracellular matrix: News from the cell's frontier. *Plant Cell* **8,** 1451–1463.

Cheong, J.-J., and Hahn, M. G. (1991). A specific, high-affinity binding site for the hepta-beta-glucoside elicitor exists in soybean membranes. *The Plant Cell* **3,** 137–147.

Chrispeels, M. J., and Maurel, C. (1994). Aquaporins: The molecular basis of facilitated water movement through living plant cells. *Plant Physiol.* **105,** 9–13.

Cosgrove, D. J., and Hedrich, R. (1991). Stretch-activated chloride, potassium, and calcium channels coexisting in plasma membranes of guard cells in *Vicia faba* L. *Planta* **186,** 143–153.

Crane, F. L., and Barr, R. (1989). Plasma membrane oxidoreductases. *Crit. Rev. Plant Sci.* **8,** 273–307.

Davies, A. (1987). Action potentials and multifunctional signals in plants: A unifying hypothesis to explain disparate wound responses. *Plant Cell Environ.* **10,** 623–631.

Davies, P. J. (Ed.). (1988). "Plant Hormones and Their Role in Plant Growth and Development." Kluwer, Dordrecht, The Netherlands.

Delmer, D. P., and Amor, Y. (1995). Cellulose biosynthesis. *Plant Cell* **7,** 987–1000.

Fairly-Grenot, K., and Assman, S. M. (1991). Evidence for G-protein regulation of inward K⁺-channel current in guard cells of Fava Bean. *The Plant Cell* **3,** 1037–1047.

Farmer, E. E., and Ryan, C. A. (1990). Interplant communication: Airborne methyl jasmonate induces the synthesis of proteinase inhibitors in plant leaves. *Proc. Natl. Acad. Sci. USA* **87,** 7713–7716.

Fisher, R. F., and Long, S. L. (1992). *Rhizobium*-plant signal exchange. *Nature* **357,** 655–660.

Geurts, R., and Franssen, H. (1996). Signal transduction in *Rhizobium*-induced nodule formation. *Plant Physiol.* **112,** 447–453.

Giddings, T. H., Jr., and Staehelin, L. A. (1991). Microtubule-mediated control of microfibril deposition: A re-examination of the hypothesis. *In* "The Cytoskeletal Basis of Plant Growth and Form" (C. W. Lloyd, Ed.), pp. 85–100. Academic Press, London.

Hasenstein, K. H., and Evans, M. L. (1988). Effects of cations on hormone transport in primary roots of *Zea mays*. *Plant Physiol.* **86,** 890–894.

Hayashi, T. (1989). Xyloglucans in the primary cell wall. *Annu. Rev. Plant Physiol. Plant Mol. Biol.* **40,** 139–168.

Hedrich, R., and Schroeder, J. I. (1989). The physiology of ion channels and electrogenic pumps in higher plants. *Annu. Rev. Plant Physiol. Plant Mol. Biol.* **40,** 539–569.

Hooley, R. (1996). Plant steroid hormones emerge from the dark. *Trends Genet.* **12,** 281–283.

Kaufman, L. S. (1993). Transduction of blue-light signals. *Plant Physiol.* **102,** 333–337.

Leonard, R. T., and Hepler, P. K. (Eds.). (1990). "Calcium in Plant Growth and Development." American Society of Plant Physiologists, Rockville, MD.

Lloyd, C. W., Ed. (1991). "The Cytoskeletal Basis of Plant Growth and Form." Academic Press, London.

Long, S. R. (1989). Rhizobium-legume nodulation: Life together in the underground. *Cell* **56,** 203–214.

McCann, M. C., and Roberts, K. (1991). Architecture of the primary cell wall. *In* "The Cytoskeletal Basis of Plant Growth and Form" (C. W. Lloyd, Ed.), pp. 109–130. Academic Press, London.

McClure, B. A., and Guilfoyle, T. (1989). Rapid redistribution of auxin-regulated RNAs during gravitropism. *Science* **243,** 91–93.

McGurl, B., Pearce, G., Orozoco-Cardenas, M., and Ryan, C. A. (1992). Structure, expression, and antisense inhibition of the systemin precursor gene. *Science* **255,** 1570–1573.

Michelet, B., and Boutry, M. (1995). The plasma membrane H⁺-ATPases. *Plant Physiol.* **108,** 1–6.

Mullet, J. E. (1988). Chloroplast development and gene expression. *Annu. Rev. Plant Physiol. Plant Mol. Biol.* **39,** 475–502.

Palme, K. (1992). Molecular analysis of plant signalling elements: Relevance of eukaryotic signal transduction models. *Int. Rev. Cytol.* **132,** 223–283.

Raskin, I. (1992). Salicylate, a new plant hormone. *Plant Physiol.* **99,** 799–803.

Rayle, D. L., and Cleland, R. E. (1992). The acid growth theory of auxin-induced cell elongation is alive and well. *Plant Physiol.* **99,** 1271–1274.

Robards, A. W. (1976). Plasmodesmata in higher plants. *In* "Intercellular Communication in Plants: Studies on Plasmodesmata" (B. E. S. Gunning and A. W. Robards, Eds.), p. 31. Springer-Verlag, New York/Berlin.

Robards, A. W., and Lucas, W. J. (1990). Plasmodesmata. *Annu. Rev. Plant Physiol. Plant Mol. Biol.* **41,** 369–419.

Roberts, D. M., and Harmon, A. C. (1992). Calcium-modulated proteins: targets of intracellular calcium signals in higher plants. *Annu. Rev. Plant Physiol. Plant Mol. Biol.* **43,** 375–414.

Robinson, D. G., and Depta, H. (1988). Coated vesicles. *Annu. Rev. Plant Physiol. Plant Mol. Biol.* **39,** 53–99.

Ryan, C. A. (1988). Oligosaccharides as recognition signals for the expression of defensive genes in plants. *Biochemistry* **27,** 8879–8883.

Ryan, C. A., and Farmer, E. E. (1991). Oligosaccharide signals in plants: A current assessment. *Annu. Rev. Plant Physiol. Plant Mol. Biol.* **42,** 651–674.

Staehelin, L. A., and Giddings, T. H. (1982). Membrane control of microfibrillar order. *In* "Developmental Order: Its Origin and Regulation" (S. Subtelney and P. B. Green, Eds.), p. 144. Alan R. Liss, New York.

Staswick, P. E. (1992). Jasmonate, genes, and fragrant signals. *Plant Physiol.* **99,** 804–807.

Staves, M. P., Wayne, R., and Leopold, A. C. (1992). Hydrostatic pressure mimics gravitational pressure in characean cells. *Protoplasma* **168,** 141–152.

Sussman, M. R., and Harper, J. F. (1989). Molecular biology of the plasma membrane of higher plants. *The Plant Cell* **1,** 953–960.

Sze, H., Ward, J. M., and Lai, S. (1992). Vacuolar H⁺-translocating ATPases from plants: Structure, function, and isoforms. *J. Bioenerg. Biomembr.* **24,** 371–381.

Taiz, L. (1994). Expansins: Proteins that promote cell wall loosening in plants. *Proc. Natl. Acad. Sci. USA* **91,** 7387–7389.

Taiz, L., and Zeiger, E. (1991). "Plant Physiology." Benjamin/Cummings, Redwood City, CA.

Tester, M. (1990). Plant ion channels: Whole-cell and single-channel studies. *New Phytol.* **114,** 305–340.

Tolbert, N. E. (Ed.) (1980). "The Plant Cell." Academic Press, New York.

Trewavas, A., and Gilroy, S. (1991). Signal transduction in plant cells. *Trends Genet.* **7,** 356–361.

Tyerman, S. D. (1992). Anion channels in plants. *Annu. Rev. Plant Physiol. Plant Mol. Biol.* **43,** 351–373.

Varner, J. E., and Lin, L.-S. (1989). Plant cell wall architecture. *Cell* **56,** 231–239.

Venis, M. A., and Napier, R. M. (1995). Auxin receptors and auxin binding proteins. *Crit. Rev. Plant Sci.* **14,** 27–47.

Volkmann, D., and Sievers, A. (1979). Graviperception in multicellular organs. *In* "Encyclopedia of Plant Physiology, New Series" (W. Haupt and M. E. Feinleib, Eds.), Vol. 7, pp. 573–600. Springer-Verlag, Berlin/New York.

von Arnim, A., and Deng, X.-W. (1996). Light control of seedling development. *Annu. Rev. Plant Physiol. Plant Mol. Biol.* **47,** 215–243.

Williamson, R. E. (1986). Organelle movement along actin filaments and microtubules. *Plant Physiol.* **82,** 631–634.

Xu, P., Lloyd, C. W., Staiger, C. J., and Drobak, B. K. (1992). Association of phosphatidylinositol-4-kinase with the plant cytoskeleton. *The Plant Cell* **4,** 941–951.

Darrell Fleischman

# 63

# Photosynthesis

## I. Introduction

### A. Historical Aspects

It must have been apparent from the earliest times that animals can survive only by eating plants or other animals, whereas plants seem to subsist on nothing but sunshine and a little water and earth. Julius Mayer, who introduced the concept of conservation of energy, suggested in 1845 that green plants can capture the energy of light and store it in a chemical form. His insight was correct. Almost all of the free energy used by living organisms is derived from energy of sunlight, which has been stored in photosynthetically produced molecules.

Clues about how plants might store free energy were obtained early (Loomis, 1960). The English philosopher Joseph Priestly began a series of experiments in 1771 that demonstrated that plants produce a gaseous substance, oxygen, which animals require for respiration. By the end of the nineteenth century it was recognized that oxygen is produced only by the green parts of the plant and that its production requires light, carbon dioxide, and water. The first visible products of photosynthesis were grains of starch, a polymer of glucose.

We now know that photons are captured by **chlorophyll,** the green pigment found in leaves (Fig. 1). Plants use the energy of the photons to extract electrons from water (leaving oxygen as a by-product), and use them to convert inorganic molecules such as carbon dioxide to organic molecules such as glucose:

$$6CO_2 + 6H_2O \xrightarrow{\text{Light}} C_6H_{12}O_6 + 6O_2$$

The products of this process contain 686 kcal more free energy per mole of glucose formed than do the reactants, that is,

$$\Delta G^{\circ\prime} = 686 \text{ kcal/mol}$$

Nitrogen and sulfur also can be fixed, that is, converted from inorganic to organic form, by the photosynthetic system. Nitrogen and sulfur contributed by inorganic starting materials [including dinitrogen ($N_2$), nitrate, and sulfate] are incorporated into amino acids and other organic molecules. In this chapter we discuss in detail only carbon fixation.

### B. Photosynthetic Microorganisms

Higher plants are not the only organisms that perform **photosynthesis.** Algae do so as well, as do many bacteria, including the **cyanobacteria** (blue-green algae) and the **purple** and **green photosynthetic bacteria,** heliobacteria and *Chloroflexus* (see Chapter 61). As much as a third of the earth's photosynthesis is performed by such microorganisms in the oceans. Algae and cyanobacteria use water as their source of electrons, and evolve oxygen, as higher plants do. Most other photosynthetic bacteria cannot derive electrons in this manner, and instead extract electrons from donors such as elemental sulfur, sulfide, thiosulfate, organic acids, alcohols, and hydrogen. The bacterium *Halobacterium halobium* performs photosynthesis in a quite different way. Rather than chlorophyll, it uses **bacteriorhodopsin,** a carotenoid-containing protein resembling the visual pigment rhodopsin, to capture the energy of light in a process that does not involve electron transfer (Stoeckenius, 1978).

### C. Evolution of Photosynthesis

It is believed that the earliest photosynthetic bacteria appeared more than 3.5 billion years ago, at a time when the earth's atmosphere contained almost no oxygen. Like present-day purple and green photosynthetic bacteria, these primordial bacteria could not use water as an electron donor. Then, perhaps as early as 3.6 billion years ago, there appeared a bacterium that, in the light, could generate an oxidant strong enough to oxidize water (Blankenship, 1992). The new bacterium had access to a seemingly limitless source of electrons—water. The bacterium formed oxygen as the by-product of water oxidation. At first the oxygen may have been consumed in chemical reactions with reductants such as $Fe^{2+}$, but by two billion years ago

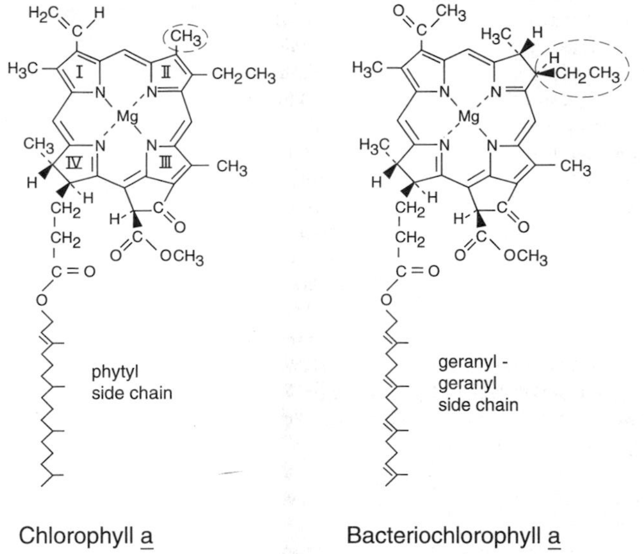

Chlorophyll <u>a</u>                    Bacteriochlorophyll <u>a</u>

**FIG. 1.** The structures of chlorophyll *a* (Chl *a*) and bacteriochlorophyll *a* (BChl *a*). Differences between the structures of the various forms of chlorophyll contribute to differences between the absorption spectra of chlorophyll proteins, allowing them to capture light in different regions of the spectrum. The conjugated system in BChl has one fewer double bond (in ring II) than that of Chl. Single resonance forms of the conjugated systems of double bonds are shown. In Chl *b*, —HC=O replaces the circled methyl substituent on ring II of Chl *a*. In BChl *b*, =CH—CH$_3$ replaces the circled substituents on ring II of BChl *a*. In some bacteria, phytyl rather than geranylgeranyl serves as the side chain of bacteriochlorophyll *a* or bacteriochlorophyll *b*. Other forms of chlorophyll are found in some organisms. Pheophytins resemble the respective chlorophylls, but the Mg$^{2+}$ is replaced by two protons.

it was accumulating in the atmosphere. The oxygen gave rise to the ozone layer, and so permitted the development of higher organisms by shielding them from ultraviolet radiation.

Once oxygen was present, respiration and oxidative phosphorylation became possible. Using oxygen as an electron acceptor, bacteria could reoxidize photosynthetically synthesized molecules and regain some of the energy stored in them:

$$C_6H_{12}O_6 + 6O_2 \rightarrow O_2 + 6H_2O$$

$$\Delta G^{\circ\prime} = -686 \text{ kcal/mol}$$

Many of the proteins photosynthetic bacteria used for photosynthesis could also be used for oxidative phosphorylation. Several lines of evidence suggest that purple photosynthetic bacteria were the ancestors of nonphotosynthetic bacteria that perform oxidative phosphorylation, and that the mitochondria of eukaryotic cells evolved from such bacteria, which had formed symbiotic associations with primitive eukaryotic cells (Woese, 1987). Thus, it is no coincidence that the mechanisms of photosynthesis and respiration are similar in many respects. Much of our understanding of oxidative phosphorylation has come from studies of photosynthesis, which have exploited

the experimental advantages of a system that can be driven by light.

In algae and higher plants, photosynthesis takes place in organelles known as **chloroplasts.** There is evidence that chloroplasts evolved from oxygen-evolving photosynthetic bacteria, such as cyanobacteria, which had formed symbiotic associations with eukaryotes. This process appears to have occurred several times.

## II. Chloroplasts

In higher plants, most chloroplasts are found in the **mesophyll cells** of leaves (Fig. 2). The chloroplast consists of a collection of flattened membranous vesicles called **thylakoids,** which are enclosed in an **envelope** formed by a double membrane (Fig. 3). The outer membrane contains channels formed by the protein **porin** and is freely permeable to substances whose molecular mass is below about 10 kDa. The inner membrane is selectively permeable and contains a number of metabolite transporters, which mediate the interaction between metabolism in the plant cytosol and within the chloroplast (Flügge and Heldt, 1991). The volume surrounding the thylakoids is known as the **stroma.** The thylakoid membranes contain the pigments that capture light and, like the cristae of mitochondria, they contain an electron transport system coupled to an adenosine triphosphate (ATP) synthase. The soluble enzymes responsible for carbon dioxide assimilation are found in the stroma. In many chloroplasts of land plants and green algae, the thylakoid membranes are stacked tightly together in some regions to form structures called **grana,** which look like green grains under the light microscope. The grana are connected by nonstacked thylakoid extensions, the **stroma lamellae.**

Chloroplasts still retain genes, which are located on a chromosome found in the stroma, for many of the proteins involved in photosynthesis and for transfer and ribosomal RNA. Proteins coded by chloroplast genes are synthesized on ribosomes located in the stroma. Genes for other chloroplast proteins are now located on nuclear chromosomes of the plant cell. Remarkably, many multisubunit proteins of the chloroplast thylakoid, including the ATP synthase, contain both subunits coded on nuclear genes and synthesized on cytosolic ribosomes and subunits coded on chloroplast genes and synthesized on stromal ribosomes. How the plant coordinates the synthesis of these subunits and the assembly of the proteins presents an intriguing research problem. Chloroplast proteins that are synthesized in the plant cytosol are initially formed with amino terminal extensions, termed **transit peptides,** which allow them to bind to receptors on the outer envelope membrane and then enter the chloroplast. The process requires guanosine triphosphate (GTP), ATP, and molecular chaperones (Schnell, 1995). After the proteins have entered the chloroplast stroma, the transit peptides are removed by a highly specific protease. Proteins destined for the membrane or lumen of the thylakoid are formed with an additional presequence that is proteolytically cleaved after it has allowed the protein to enter the thylakoid membrane or lumen.

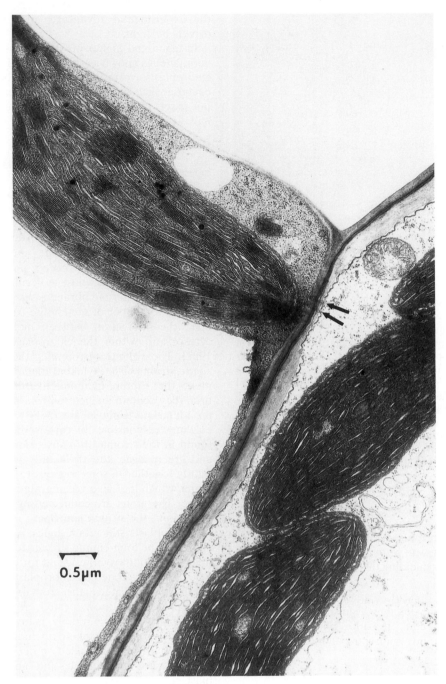

**FIG. 2.**  Transmission electron micrograph of corn (*Zea mays*) chloroplasts. Upper left: chloroplast within mesophyll cell. Lower right: chloroplast within bundle sheath cell. Arrows indicate plasmadesmata through which metabolites are exchanged between the cells. The mesophyll cells lie near the leaf surface. The bundle sheath cells lie near the leaf vascular bundles. Note the grana in the mesophyll chloroplast. (Micrograph courtesy of Iain M. Miller.)

## III.  Biochemistry of Carbon Assimilation

### A.  The Reductive Pentose Phosphate Cycle

To determine the sequence of reactions involved in the conversion of carbon dioxide to sugars, in the early 1950s

Melvin Calvin and his associates exposed algae to $^{14}CO_2$ and light for different periods. After treatment with boiling alcohol, the carbon compounds that had been formed were separated by paper chromatography. The manner in which the distribution of $^{14}C$ among the carbon atoms of the various carbon compounds changed with time revealed the

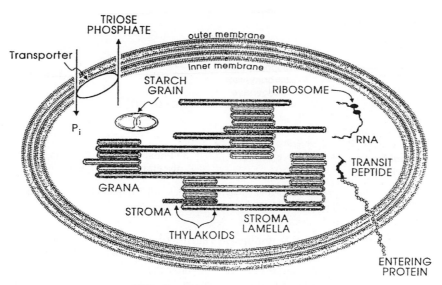

**FIG. 3.** Diagram of a chloroplast.

operation of a cyclic process, which has become known as the **reductive pentose phosphate cycle,** or **Calvin cycle** (Fig. 4).

In the first step, $CO_2$ combines with ribulose 1,5-bisphosphate to form two molecules of 3-phosphoglycerate. The reaction is catalyzed by ribulose 1,5-bisphosphate carboxylase/oxygenase, often referred to as "rubisco." This enzyme has a rather low affinity for $CO_2$. To compen-

sate, it is present in extremely high concentrations in the chloroplast, and is probably the most abundant protein on earth. Next, the 3-phosphoglycerate is phosphorylated to form 1,3-bisphoglycerate, which in turn is reduced to glyceraldehyde 3 phosphate by NADPH. In the remaining steps of the cycle, ribulose 1,5-bisphosphate is regenerated by a pathway that includes a complicated series of rearrangements catalyzed by transketolases and aldolases.

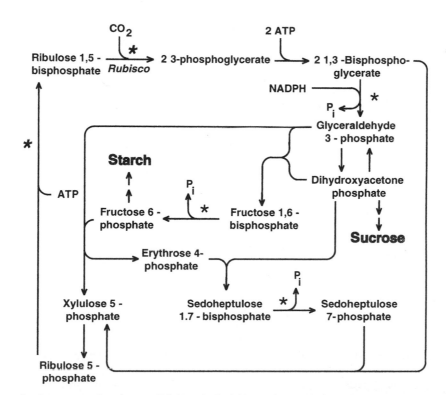

**FIG. 4.** The reductive pentose phosphate or Calvin cycle. Stoichiometries are shown only for the first two steps. Erythrose, xylulose, and sedoheptulose are sugars that contain four, five, and seven carbons, respectively. Stars indicate steps catalyzed by light-regulated enzymes.

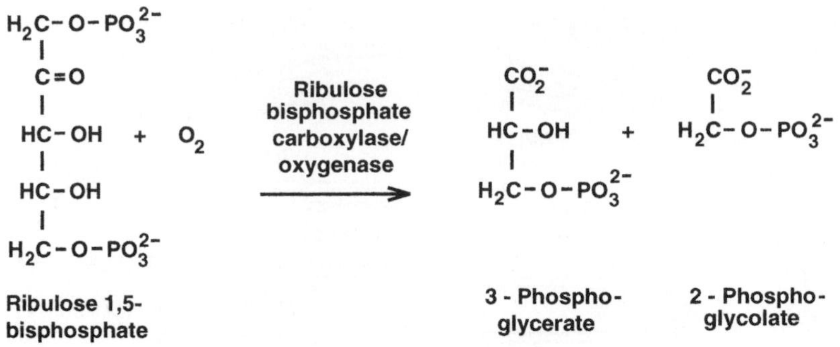

$$
\begin{array}{c}
H_2C-O-PO_3^{2-} \\
| \\
C=O \\
| \\
HC-OH \quad + \quad O_2 \\
| \\
HC-OH \\
| \\
H_2C-O-PO_3^{2-}
\end{array}
\quad
\xrightarrow[\text{oxygenase}]{\substack{\text{Ribulose} \\ \text{bisphosphate} \\ \text{carboxylase/}}}
\quad
\begin{array}{c}
CO_2^- \\
| \\
HC-OH \quad + \quad \begin{array}{c} CO_2^- \\ | \\ H_2C-O-PO_3^{2-} \end{array} \\
| \\
H_2C-O-PO_3^{2-}
\end{array}
$$

**Ribulose 1,5-bisphosphate**                                   **3 - Phospho-glycerate**        **2 - Phospho-glycolate**

**FIG. 5.**   The oxygenation reaction catalyzed by ribulose 1,5-bisphosphate carboxylase.

Intermediates in the cycle, including fructose 6-phosphate, glyceraldehyde 3-phosphate, and dihydroxyacetone phosphate, serve as precursors for the starch, sucrose, and amino acids that are the final products of photosynthesis.

ATP is required for the conversion of 3-phosphoglycerate to 1,3-bisphosphoglycerate and for the conversion of ribulose 5-phosphate to ribulose 1,5-bisphosphate. NADPH is required for the reduction of 1,3-bisphosphoglycerate to glyceraldehyde 3-phosphate. For each molecule of $CO_2$ fixed, three molecules of ATP and two molecules of NADPH are needed. It is the function of the light-driven reactions in the thylakoid membranes to furnish this ATP and NADPH.

## B. Photorespiration and C₄ Plants

Not only does ribulose 1,5-bisphosphate carboxylase/oxygenase have a low affinity for $CO_2$, it also catalyzes a competing reaction, in which $O_2$ rather than $CO_2$ is added to ribulose 1,5-bisphosphate (Fig. 5). The products are 3-phosphoglycerate and phosphoglycolate. This reaction seems to serve no useful purpose. Many plants grow much faster in a $CO_2$-enriched atmosphere in which the carboxylation reaction can compete more effectively with the oxygenation reaction. It appears that in the two billion years since plants began to fill the atmosphere with oxygen while removing $CO_2$ from it, they have been unable to modify the enzyme so that its affinity for $CO_2$ is increased or its affinity for $O_2$ is decreased significantly. Molecular biologists are now trying to accomplish that task.

Part of the carbon appearing in phosphoglycolate is rescued. In a series of reactions occurring in **peroxisomes** and in mitochondria, two molecules of phosphoglycolate are converted to one molecule of glycerate, which is returned to the chloroplast and rephosphorylated. But one carbon atom is lost as $CO_2$, and ATP and $O_2$ are consumed. This process is known as **photorespiration.** It is not coupled to ATP formation, and the net result is waste of ATP and fixed carbon.

A number of plants, including corn, sugarcane, and crabgrass, partially avoid photorespiration by concentrating $CO_2$ in the cells that contain the Calvin cycle enzymes, so that carboxylation can compete more effectively with oxygenation. In these plants, the Calvin cycle enzymes are located in the chloroplasts of the **bundle sheath cells,** which

surround the vascular bundles deep within the leaves (Fig. 6). $CO_2$ is first captured in the mesophyll cells, which lie near the leaf surface, by carboxylation of phosphoenolpyruvate (PEP):

Phosphoenolpyruvate + $CO_2$

$$\xrightarrow{\text{PEP carboxylase}} \text{oxaloacetate} + P_i$$

The carboxylation is catalyzed by PEP carboxylase, an enzyme that has a high affinity for $CO_2$ and does not catalyze an oxygenation reaction. Oxaloacetate is next reduced to malate (or transaminated to form aspartate in some

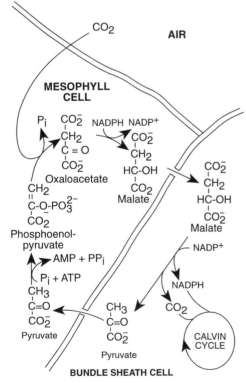

**FIG. 6.**   The mechanism used by C₄ plants to concentrate $CO_2$ and NADPH in bundle sheath cells. $CO_2$ is captured in the mesophyll cell by carboxylation of phosphoenolpyruvate. Malate is transferred to the bundle sheath cells, where $CO_2$ is released and NADPH is formed to enter the Calvin cycle. Pyruvate returns to the mesophyll cells.

plants). The malate or aspartate is transferred to the bundle sheath cells through fibers known as **plasmadesmata.** Some of these can be seen at the arrows in Fig. 2. Malate is oxidatively decarboxylated to pyruvate in the bundle sheath cells, in a reaction that also generates NADPH. The $CO_2$ that is released enters the Calvin cycle in the bundle sheath cells. Pyruvate returns to the mesophyll cells, where it is again transformed to PEP. The net result is the transfer of $CO_2$ and NADPH to the bundle sheath cells.

The $CO_2$-concentrating mechanism consumes energy, since ATP is converted to adenosine monophosphate (AMP) and phosphate. Nevertheless, plants that use it are often referred to as efficient plants because avoidance of photorespiration more than compensates for this expense. Such plants are usually known as $C_4$ plants because of the involvement of 4-carbon acids. Many $C_4$ plants are native to tropical areas, where the $C_4$ pathway is especially advantageous since high temperatures and bright sunlight encourage photorespiration.

## IV. Formation of Adenosine Triphosphate

### A. Light-Driven Electron Transport and Adenosine Triphosphate Synthesis

Carbon dioxide fixation requires ATP and NADPH. It seemed reasonable to suspect that the role of light is to provide the energy necessary for their formation. Photosynthetic membranes contain electron transport chains much like those of mitochondria, and light can drive elec-

tron transport along the chains (Figs. 7 and 8). As in mitochondria, ATP formation is coupled to the electron transport.

How can electron flow through the electron carriers shown in Fig. 7 cause ATP formation? Mitchell (1961) suggested that electron transport and ATP formation are both coupled to the movement of protons across membranes. His **chemiosmotic hypothesis** helps us understand the reason for the arrangement of the electron carriers within photosynthetic membranes (Mitchell, 1979).

### B. The Chemiosmotic Hypothesis

Mitchell's proposal was beautifully simple. He pointed out that ATP formation from ADP and $P_i$ is a dehydration reaction:

$$ADP + P_i \leftrightarrows ATP + H_2O$$

Any process that could remove water from a compartment containing ADP and $P_i$ would shift the equilibrium to the right and cause ATP formation. For example, suppose such a compartment were separated from an aqueous phase having a high pH by a membrane permeable to protons but not to hydroxide ions, and from an aqueous phase having a low pH by a membrane permeable to hydroxide ions but not to protons (Fig. 9). As water dissociated into its component $H^+$ and $OH^-$ ions, they would diffuse across the respective membranes, depleting the water in the compartment and pulling the reaction toward ATP synthesis. ATP formation would be driven by the difference in pH

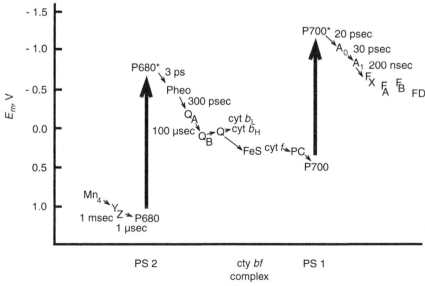

**FIG. 7.** The electron transport system of higher plant chloroplasts. Midpoint redox potentials ($E_m$) of the components are indicated. P680 and P700 are chlorophyll $a$ dimers that, when excited by light (large arrows), transfer an electron to the acceptors pheophytin $a$ (Pheo) and chlorophyll $a$ ($A_0$), respectively. Typical electron transfer times are shown for the electron transfer steps, which are indicated by arrows. $Mn_4$, a complex containing four manganese nuclei that accepts electrons from water; $Y_Z$, tyrosine; $Q_A$, $Q_B$, Q, plastoquinone; cyt $b_H$ and cyt $b_L$, the high and low potential hemes, respectively, of cytochrome $b_6$; FeS, the 2 Fe-2S center of the Rieske protein; cyt $f$, cytochrome $f$; PC, plastocyanin, a water-soluble copper-containing protein; $A_1$, phylloquinone; $F_X$, $F_A$, $F_B$, 4 Fe-4 S clusters; FD, ferredoxin, a water-soluble protein containing a 4 Fe-4 S cluster.

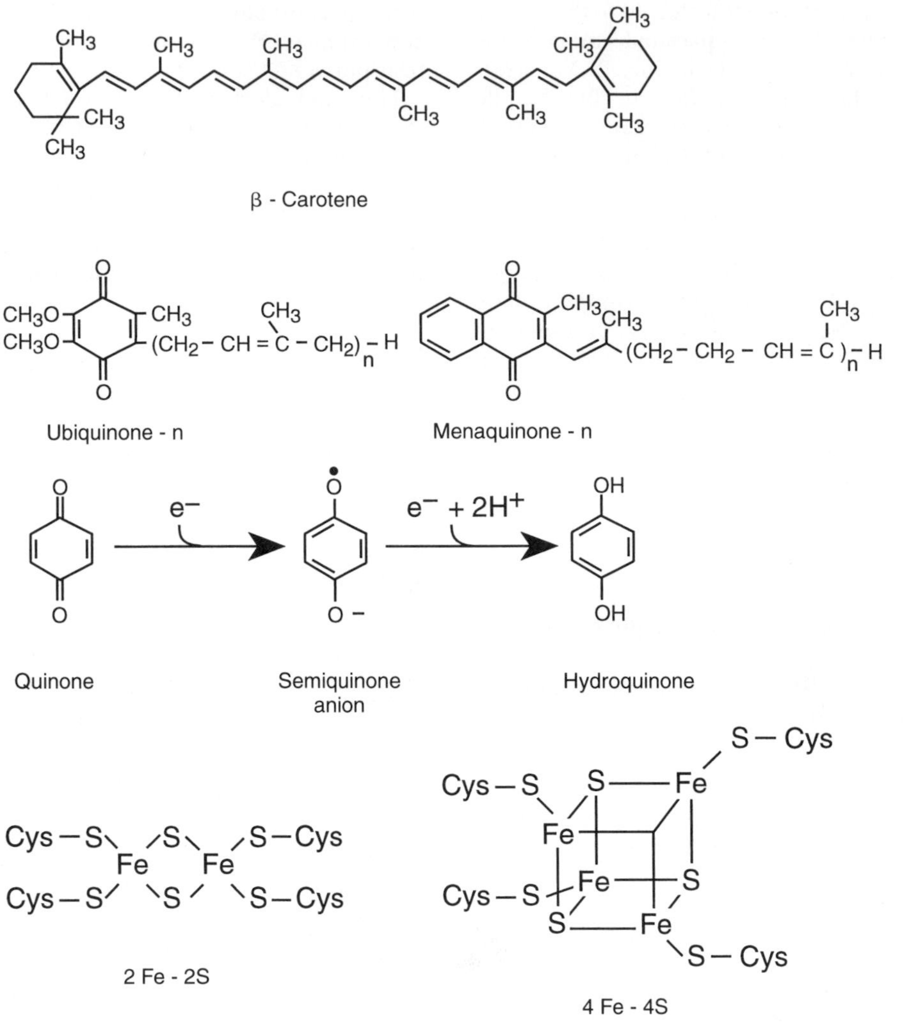

β - Carotene

Ubiquinone - n                          Menaquinone - n

Quinone                    Semiquinone                  Hydroquinone
                              anion

2 Fe - 2S

4 Fe - 4S

**FIG. 8.** Structures of some of the molecules and cofactors involved in photosynthesis. Other carotenoids and quinones are found in many photosynthetic organisms. The quinone, semiquinone anion, and hydroquinone forms of the quinone six-membered ring are shown.

across the membrane. How might the pH difference (often called a pH gradient) be established?

Some electron carriers, such as cytochromes and iron-sulfur proteins, cannot cross membranes. Others, such as quinones, are soluble in membranes and, as illustrated in Fig. 8, may bind protons after they have accepted electrons.

Mitchell proposed that the electron carriers are arranged asymmetrically in the membrane, as in the lower part of Fig. 9. Carriers that bind protons would alternate in the electron transport chain with those that do not. Electron transfers that result in $H^+$ binding (A → B) would occur on one side of the membrane (the stromal side in Fig. 9), while electron transfers that result in $H^+$ release (B → C) would occur on the opposite side (the luminal side). In one direction (from the stromal side to the luminal side), electrons and protons would always move together across the membrane. In Fig. 9, BH serves as the carrier of an electron and a proton. Electrons would always move alone across the membrane in the opposite direction (C → D).

In this way electron transport would be accompanied by the net transfer of protons across the membrane. Such electron transport is sometimes described as **vectorial,** since it has a defined direction within the membrane, unlike electron transfer that occurs in solution.

### C. Storage of Free Energy in an Electrochemical Gradient

When an uncharged molecule moves from a region where its concentration is $C_1$ to a region where its concentration is $C_2$ there is a free energy change

$$\Delta G = RT \ln C_2/C_1$$

per mole transferred.

$$\Delta G = \Delta H - T \Delta S$$

The free energy change is due to the change in entropy, $\Delta S$, that accompanies the change in distribution of the

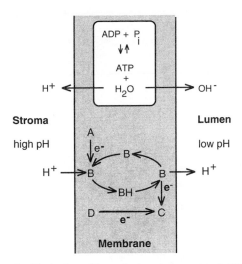

**FIG. 9.** An illustration of how electron transport and ATP formation might be coupled in chloroplasts, based on the ideas presented by Mitchell in 1961. The compartment containing ADP, $P_i$, and ATP is separated from the stroma by a membrane permeable only to $H^+$ ions and from the thylakoid lumen by a membrane permeable only to $OH^-$. A, B, C, and D are electron carriers that cannot move through the membrane. B is a molecule, such as a quinone, which becomes protonated when it accepts an electron, diffuses across the membrane, and releases the proton when it is reoxidized.

molecules. When a charged particle such as a proton moves between regions whose electrical potentials differ, there is an additional free energy change,

$$\Delta G = zF \, \Delta\psi$$

per mole transferred, where $z$ is the number of charges on the molecule, $F$ is Faraday's constant (23 kcal/V mol), and $\Delta\psi$ is the electrical potential difference in volts. Such a difference in both concentration and electrical potential is often called an **electrochemical gradient.** There may be an electrochemical proton gradient between the chloroplast stroma and the thylakoid lumen. This gradient is created by the transfer of protons and electrons across the membrane during electron transport. For the transfer of protons ($z = +1$) from the stroma to the lumen

$$\Delta G = RT \ln([H^+]_{lumen}/[H^+]_{stroma}) + F(\psi_{lumen} - \psi_{stroma})$$

$$\Delta G = -RT \, \Delta pH + F \, \Delta\psi$$

This free energy difference is sometimes known as **proton-motive force** and abbreviated $\Delta p$, $\Delta\mu_{H+}$ or PMF. Protons can be transported across the thylakoid membrane against their electrochemical gradient because the transfer is coupled to exergonic electron transfer. The proton translocation shown in Fig. 9 can occur if the sum of $\Delta G$ for proton translocation and $\Delta G$ of electron transfer from A to C is less than zero.

### D. Phosphorylation after an Artificially Imposed pH Gradient

Two experiments involving photosynthetic material played a critical role in convincing scientists that the chemi-

osmotic hypothesis must be taken seriously. Chloroplasts will synthesize ATP even if they are illuminated in the absence of ADP and $P_i$ and then left in the dark for several minutes before these substrates are added. A relatively stable "high-energy intermediate" has been formed in the light and can later drive ATP formation in the dark. Andre Jagendorf and Ernest Uribe learned that the illumination step was most efficient when the pH of the medium was about 4, while the ATP synthesis step was most efficient at pH 8. Therefore they routinely illuminated the chloroplasts in medium buffered at pH 4 with succinate, then increased the pH to 8 before adding ADP and $P_i$. But ATP was formed by this procedure even if the chloroplasts were not illuminated! The quick increase in pH had created a proton gradient across the thylakoid membranes; the lumen pH had remained near 4, while the external pH had increased to 8. The simplest explanation was that the proton gradient had driven the ATP formation.

### E. Bacteriorhodopsin

*Halobacterium halobium* is a red bacterium found in lakes that have a high salt content. It can form ATP in the light in a quite unusual way. The membrane of the bacterium contains purple domains in which molecules of the transmembrane protein bacteriorhodopsin are packed together in a hexagonal lattice. Each molecule of bacteriorhodopsin contains a molecule of all-*trans* retinal attached as a Schiff base to a lysine residue. Upon absorbing light, the retinal is isomerized to the 13-*cis*-isomer, initiating a cycle of events resulting in the translocation of a proton from the inside to the outside of the cell and regeneration of all-*trans* retinal (Khorana, 1988). The electrochemical proton gradient that is established is used to drive ATP formation.

ATP synthesis in mitochondria is catalyzed by a membrane-associated ATP synthase. Photosynthetic ATP formation is catalyzed by a similar protein. Ephraim Racker and Walter Stoeckenius incorporated purified bacteriorhodopsin into artificial phospholipid vesicles. When the vesicles were illuminated, protons were pumped into the vesicles. Protons were pumped inward rather than outward because the bacteriorhodopsin incorporated into the membranes in an orientation that was opposite to that in the bacterial membrane. When purified mitochondrial ATP synthase was also incorporated into the vesicles, light caused ATP to be formed from ADP and $P_i$. This experiment showed that the mitochondrial ATP synthase can catalyze ATP formation in the absence of any other mitochondrial protein, including those of the electron transport chain. Only an electrochemical proton gradient is required.

### F. ATP Synthase

The chloroplast ATP synthase has a structure similar to that of the mitochondrial ATP synthase. It consists of a transmembrane domain ($CF_0$), composed of four subunits and containing a proton channel, and a catalytic domain ($CF_1$), composed of five subunits and located on the stromal surface of the thylakoid membrane (Nalin and Nelson,

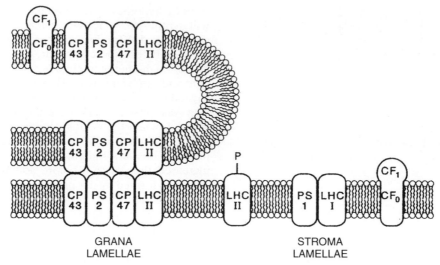

GRANA
LAMELLAE

STROMA
LAMELLAE

**FIG. 10.**  Distribution of photosynthetic complexes between the grana and stroma lamellae. Photosystem 2 (PS 2) is found primarily in the grana lamellae, whereas photosystem 1 (PS 1) and ATP synthase are primarily in stroma lamellae and unappressed granal membranes. When PS 2 activity exceeds PS 1 activity, LHC II complexes become phosphorylated and move from the grana lamellae to the stroma lamellae.

1987). The ATP synthases are found in the stroma lamellae and in the unappressed membranes of the grana, but rarely in appressed granal membranes (Fig. 10). ATP synthesis is thought to be coupled to the transfer of three protons through the ATP synthase from the lumen to the stroma; thus ATP formation is thermodynamically possible when $\mu_{H+}$ is more than three times $\Delta G$ of ATP formation.

The ATP synthase is intricately regulated, presumably to prevent the reversal of ATP synthesis in the dark. It is activated by the presence of a sufficiently large $\Delta\mu_{H+}$. Until it is activated it will neither synthesize nor hydrolyze ATP. Reduction of a $CF_1$ disulfide bridge by reduced thioredoxin, which is formed when chloroplasts are illuminated, enhances the efficiency of the activated ATP synthase when $\Delta\mu_{H+}$ is small. The mechanism by which ATP synthesis is coupled to proton efflux is still incompletely understood.

## V.  Photosynthetic Electron Transport

### A.  The Interaction of Light with Molecules

A light wave consists of an electric field and a magnetic field whose directions are perpendicular to each other. The wave moves through space at a velocity

$$v = nc$$

where $n$ is the refractive index of the medium through which the light is moving and $c$ is the speed of light in a vacuum, $3.0 \times 10^8$ m/sec (Fig. 11). The wavelength of the light, $\lambda$, is the distance between the crests of the waves. Visible light has a wavelength between 400 and 700 nm. The frequency of the light, $\nu$, is the frequency with which crests of the waves pass a given point in space, and

$$\nu = c/\lambda$$

According to the principles of quantum theory, light can be absorbed or emitted only in discrete units called **quanta.** The energy, $E$, of such a quantum depends on the frequency of the light according to the equation

$$E = h\nu$$

where $h$ is Planck's constant, $2.86 \times 10^{-37}$ kcal sec.

Just as the electrons of atoms occupy atomic orbitals having discrete energies, the electrons in molecules occupy molecular orbitals having discrete energies. Molecular orbitals may extend over several nuclei. Each molecular orbital can contain two electrons of opposite spin. Absorption of light can promote (excite) an electron to an orbital of higher energy (Fig. 12) if the energy difference between the orbitals is equal to the energy of a quantum of the light. Several competing processes may occur after light is absorbed. The excited electron may drop back to its ground molecular orbital. As it does so, the excitation energy may

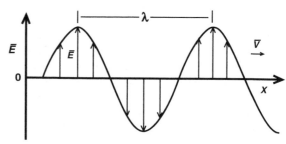

**FIG. 11.**  The electric field of a light wave. The magnetic field (not shown) is perpendicular to the electric field and to the direction of motion of the wave.

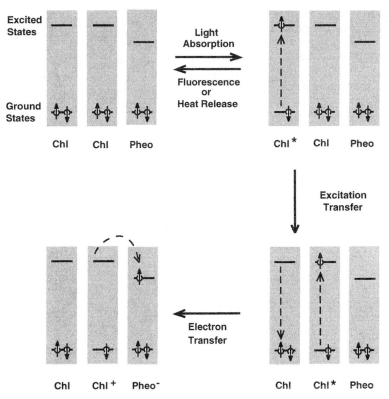

**FIG. 12.** Light absorption and emission, excitation transfer, and photochemical electron transfer. Chlorophyll (Chl) and pheophytin (Pheo) molecules are used as examples. Circles with upward and downward arrows represent electrons having opposite spins. Molecules have many ground and excited orbitals. Only one of each is shown. Chl*, electronically excited Chl; Chl⁺, oxidized Chl; Pheo⁻, reduced Pheo.

be reemitted as light (**fluorescence**), it may be lost as heat, or it may be transferred to a neighboring molecule, causing excitation of one of the neighbor's electrons to a higher orbital. Finally, the excited electron may be transferred to an empty orbital of a neighboring molecule. When this occurs, part of the energy of the absorbed light is conserved as chemical free energy of the electron donor–acceptor pair.

The tendency of a biological molecule to donate electrons is usually expressed as its midpoint redox potential at pH 7, $E_{m7}$. The **midpoint potential,** $E_m$, is the electrical potential difference that would exist between the standard hydrogen half cell and a solution in which the oxidized and reduced forms of the molecule are present in equal concentrations. In the standard hydrogen half cell, the pH is 0 and the partial pressure of $H_2$ is 1 atm. The $E_{m7}$ values for members of the chloroplast electron transport system are indicated in Fig. 7. A molecule in which an electron has been excited to a higher orbital has a more negative $E_m$ than does the molecule in its ground state, by an amount about equal numerically to the energy difference between the ground and excited orbitals (in electron volts). Thus, it is a better electron donor. Note the difference between the $E_m$'s of P680 and P680* (and between P700 and P700*) in Fig. 7.

The standard free energy change of an electron transfer is related to the difference between the $E_m$ values of the electron donor and acceptor,

$$\Delta G^{\circ\prime} = -nF\,\Delta E_m$$

where $n$ is the number of electron equivalents transferred per mole. Thus, excitation of a molecule lowers the free energy change accompanying the transfer of an electron to another molecule. This equation can be used to calculate the free energy changes accompanying the light-induced charge separations and the subsequent electron transfers shown in Fig. 7.

Especially lucid explanations of the photochemical processes involved in photosynthesis have been given by Clayton (1970, 1980).

## B. The Presence of Two Photochemical Reactions in Chloroplasts

The most careful measurements of the **quantum yield** of oxygen evolution indicate that at least eight light quanta must be absorbed for each molecule of oxygen that is evolved. Because the oxidation of two water molecules to form one $O_2$ molecule requires the removal of only four electrons,

$$2H_2O \rightarrow O_2 + 4H^+ + 4e^-$$

it appeared that two quanta are necessary for the extraction of each electron from water. Robert Emerson illuminated the green alga *Chlorella* with light of different wavelengths and observed the rates of $O_2$ evolution. He found that when light of wavelength less than 680 nm and light of wavelength greater than 680 nm were given together, the rate of $O_2$ evolution was greater than the sum of the rates observed when the algae were illuminated with light of each wavelength separately; that is, the two wavelengths acted synergistically. This phenomenon, which has come to be known as **Emerson enhancement,** suggested that efficient $O_2$ evolution requires the cooperation of a system that absorbs long-wavelength light and a system that absorbs short-wavelength light. L. N. M. Duysens found that 680-nm light caused the oxidation of cytochrome $f$, whereas 562-nm light caused its re-reduction. In 1960 R. Hill and F. Bendall suggested the **Z scheme,** in which a photosystem driven by short-wavelength light (**photosystem 2, PS 2**) and a photosystem driven by long-wavelength light (**photosystem 1, PS 1**) are connected in series. Light absorbed by PS 2 would cause the transfer of an electron from $H_2O$ to a chain of electron carriers between the photosystems (cytochrome $f$ is one of these). Light absorbed by PS 1 would cause the transfer of the electron from this chain to $NADP^+$ to form NADPH:

$$H_2O \rightarrow PS\ 2 \rightarrow cytochrome\ f\ (etc.) \rightarrow PS\ 1 \rightarrow NADP^+$$

Absorption of two quanta, one by each photosystem, would be required for the transfer of an electron all the way from $H_2O$ to $NADP^+$. Cytochrome $f$ would be reduced by PS 2 and oxidized by PS 1. The complete Z scheme as it is now understood is presented in Fig. 7.

The light-absorbing pigments associated with PS 1 and PS 2 have somewhat different absorption spectra. The PS 2 pigments have very little absorption beyond about 680 nm, so light in this wavelength region drives PS 1 almost exclusively.

### C.  Primary Electron Donors and Light-Harvesting Pigments

The absorption spectra of a typical plant cell and of Chl $a$ and Chl $b$ are shown in Fig. 13. Bessel Kok noticed that when chloroplasts are illuminated, there is a small decrease in their optical absorbance ("bleaching") at 700 nm. He suggested that the change was due to the light-induced oxidation of a chlorophyll molecule, resulting in the loss of its optical absorbance at that wavelength. It was suggested that the primary light reaction is the transfer of an electron from a special chlorophyll molecule, named **P700,** to an electron acceptor (see Fig. 12). In bacteria, illumination caused cytochrome oxidation (detected as a change in optical absorbance at wavelengths where cytochrome absorbs light) as well as bacteriochlorophyll oxidation. To determine whether bacteriochlorophyll or cytochrome oxidation occurs first, William Parson illuminated bacteria with very short flashes from a Q-switched laser. Immediately after the flash, he observed oxidation of bacteriochlorophyll. The bacteriochlorophyll became reduced again in a few milliseconds and, as it did, the cytochrome became oxidized. The chlorophyll had been oxidized first; the cytochrome had then donated an electron to the oxidized chlorophyll.

The laser flash had been so brief that it ended before the electron lost by the bacteriochlorophyll had been re-

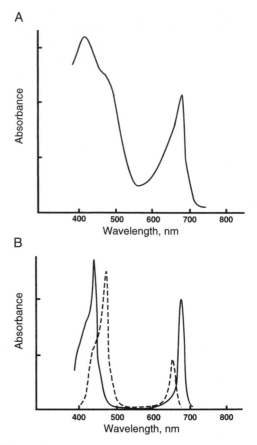

**FIG. 13.**  (A) Absorption spectrum of a typical plant chloroplast. (B) Absorption spectra of Chl $a$ (solid line) and Chl $b$ (dashed line). Each absorbance band corresponds to excitation of chlorophyll from the ground state to a different excited state. In the chloroplast both Chl $a$ and Chl $b$ absorb light between 400 and 500 nm and between 600 and 700 nm. In addition, $\beta$-carotene absorbs light between 400 and 500 nm.

placed by an electron from the cytochrome, so the flash had caused the transfer of only a single electron from the bacteriochlorophyll. Such **single turnover flashes** have been a powerful tool for the study of photosynthetic electron transport. Lasers capable of producing flashes femtoseconds ($10^{-15}$ sec) in duration have made it possible to measure the rates of early electron transfer steps, some of which occur in picoseconds ($10^{-12}$ sec).

P700 is the primary electron donor in PS 1. The primary donor of PS 2 absorbs at 680 nm, and is known as **P680.** Oxidized chlorophyll contains an unpaired electron (see Fig. 12). Such molecules can be studied by the technique of **electron paramagnetic resonance,** in which the absorption of microwaves by the unpaired electrons is measured. Study of P680 and P700 by this technique, as well as other experiments, suggested that they are *dimers* of chlorophyll. After they have lost an electron, the remaining unpaired electron is shared between the two chlorophyll molecules that comprise the dimers.

The details of many biophysical techniques, including optical spectroscopy and magnetic resonance, that are employed in photosynthesis research are described in Amesz and Hoff (1996).

Most of the chlorophyll in cells serves only to capture light and does not participate in photochemistry. Each photochemically active chlorophyll dimer (P680, P700) is associated with an aggregate of several hundred light-harvesting chlorophyll molecules, known as a **photosynthetic unit.** After any chlorophyll molecule in the unit absorbs a quantum of light, the excitation energy migrates among the chlorophyll molecules in the aggregate (Figs. 12 and 14). When the excitation reaches the photochemically active dimer, electron transfer occurs. A typical photosynthetic unit contains about 300 chlorophyll molecules. **Beta-carotene** and other carotenoids also serve as light-harvesting pigments, as do tetrapyrrole pigments known as **phycobilins** in cyanobacteria and some algae. The pigments are attached to protein molecules. Such accessory pigments allow the capture of light in regions of the spectrum where chlorophyll does not absorb light strongly.

## D. The Structure of Photosynthetic Reaction Centers

It has been possible to isolate many of the oligomeric protein complexes involved in photosynthesis by dissolving the membranes in detergent and purifying the complexes by chromatography and other techniques. The complexes that contain the photochemically active chlorophyll dimers are referred to as **reaction centers.** It has become increas-

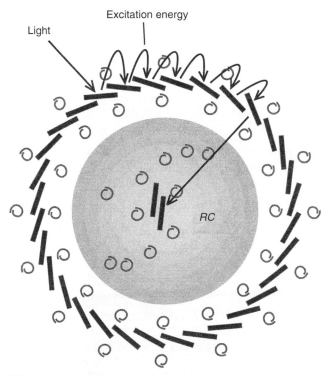

**FIG. 14.** Light absorption and the transfer and capture of excitation energy. Light is absorbed by any chlorophyll molecule (dark rectangles) in a large array. Excitation energy is transferred among the chlorophyll molecules and finally captured by a special chlorophyll pair in the reaction center (RC). The figure is based on the structure of the bacterial light harvesting complex. Protein dimers to which BChl is bound form a ring around the reaction center. (Higher plant light-harvesting pigments have a quite different structure.)

ingly apparent that PS 1, PS 2, and the reaction centers of photosynthetic bacteria resemble each other much more closely than had been suspected (Nitschke and Rutherford, 1991; Büttner *et al.*, 1992). They may have evolved from a common ancestor (Blankenship, 1992). PS 1 bears an especially strong resemblance to the reaction centers of the green bacteria and heliobacteria (Olson, 1996), while PS 2 resembles those of the purple bacteria and *Chloroflexus.* Roderick Clayton and Dan Reed, and George Feher succeeded in isolating reaction centers from purple bacteria. Hartmut Michel, Johann Deisenhofer, and Robert Huber devised a way to crystallize reaction centers from the purple bacterium *Rhodopseudomonas viridis* and determined their structure by x-ray diffraction (Deisenhofer and Michel, 1989; Deisenhofer and Norris, 1993). Many of the current ideas about the structure and function of PS 1 and PS 2 are based on the belief that their structures probably resemble that of the *Rp. viridis* reaction center. In addition, the reaction center structure provides important insights into the general properties to be expected of integral membrane proteins, which have been extremely difficult to crystallize. The *Rp. viridis* reaction center consists of four subunits (Fig. 15). Subunits L and M each contain five $\alpha$-helices that cross the membrane. The cofactors that participate in electron transport are bound to the L and M subunits. Their arrangement is sketched in Fig. 15 and can be seen in three dimensions in Fig. 16. The primary electron donor, P985, is indeed a dimer, in this case of bacteriochlorophyll *b*. The reaction center also includes two additional molecules of bacteriochlorophyll *b*; two of bacteriopheophytin *b*; one molecule each of menaquinone, ubiquinone, and carotenoid; and a nonheme ferrous iron ion. Surprisingly, the reaction center has approximate twofold symmetry. It has been shown that electrons move only down the L side of the reaction center after light has been absorbed. The electrons move from P985 to the pheophytin associated with subunit L, to $Q_A$ (menaquinone), and finally to $Q_B$ (ubiquinone). Why they move only along this pathway, what function the other cofactors may have, and whether the bacteriochlorophyll *b* monomer associated with the L subunit participates in electron transfer are currently being investigated. The reaction center also includes a cytochrome subunit containing four hemes, which is located on the periplasmic surface (facing the outside of the cell) of the membrane. The hemes donate electrons to oxidized P985. There is also an H subunit, which forms a sort of cap on the cytoplasmic side of the L and M subunits and is anchored to the membrane by a single-membrane-spanning helix. Its function is not known. All aspects of bacterial photosynthesis are discussed in Blankenship *et al.*, 1995.

## E. The Structure and Function of the Chloroplast Electron Transport System

We can now summarize the way the chloroplast electron transport system is believed to operate (Fig. 17; Barber, 1992; He and Malkin, 1996; Nugent, 1996; Ort *et al.*, 1996). The components have redox potentials that allow electrons to move forward along the chain; they are arranged in the membrane and oriented in such a way that forward electron

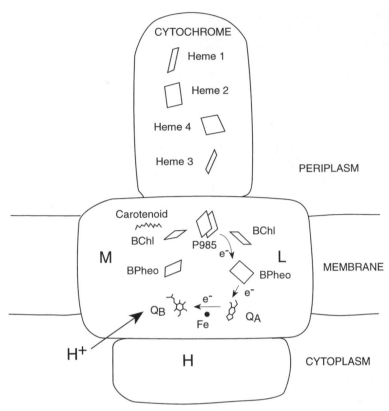

**FIG. 15.**    The arrangement of the cofactors in the *Rp. viridis* reaction center. The reaction center has approximate twofold (C$_2$) symmetry, but electrons move only along the path indicated. After Q$_B$ accepts two electrons, two protons from the cytoplasmic side of the membrane are bound. P985, the photochemically active bacteriochlorophyll *b* dimer; BChl, bacteriochlorophyll *b*; BPheo, bacteriopheophytin *b*; Q$_A$, menaquinone; Q$_B$, ubiquinone; Fe, ferrous iron.

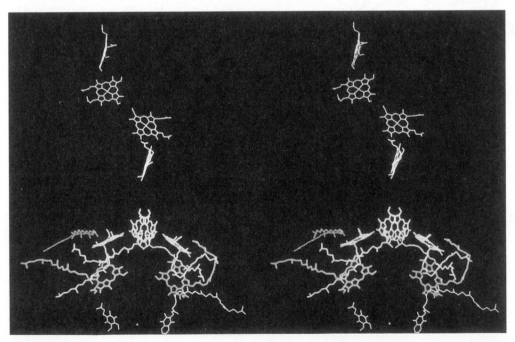

**FIG. 16.**    Stereo drawing of the *Rp. viridis* reaction center cofactors. To view the drawing in three dimensions, focus on a spot midway between the figures and allow their images to become superimposed. [From *Embo J.* **8,** 2149–2170 (1989). © 1989 The Nobel Foundation.]

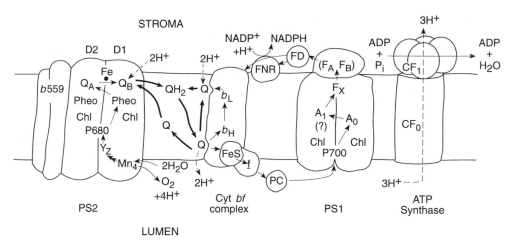

**FIG. 17.** Electron and proton transport and ATP formation in chloroplasts. Symbols are defined in the legend of Fig. 7.

transfer is fast and electron flow results in translocation of protons across the membrane.

PS 2 contains two subunits, D1 and D2, whose amino acid sequences partially resemble those of purple bacterial reaction center subunits L and M. Like the L and M subunits, D1 and D2 bind the electron transport cofactors. PS 2 contains several additional proteins that are not discussed here.

Formation of one $O_2$ molecule requires the removal of four electrons from two molecules of $H_2O$. Kok suggested that each photochemical act removes one electron from an oxygen-evolving complex (OEC). When four electrons have been removed, oxygen molecules are released (Fig. 18). The OEC contains four manganese nuclei. Determining its structure is a major objective of current research. The following sequence of events is now believed to occur (see Fig. 8). Upon excitation, P680 transfers an electron to a special pheophytin $a$ molecule (pheophytin $a$ is similar to Chl $a$, but the $Mg^{2+}$ is replaced by two protons). An electron is then transferred to $P680^+$ from $Y_Z$, a tyrosine residue, which is part of D1. Next, $Y_Z^+$ receives an electron from the OEC, leaving a positive charge stored in the OEC. This sequence of events is repeated until four positive

charges have been stored in the OEC and $O_2$ is released. During the process, four protons are released into the chloroplast lumen.

The electron that had been transferred to pheophytin $a$ moves to $Q_A$ (plastoquinone) and then to $Q_B$ (also plastoquinone) to form a plastosemiquinone anion. The rates of these electron transfers are shown in Fig. 8. After a second quantum is absorbed, a second electron moves down the chain to $Q_B$, reducing it to $Q_B^{2-}$. $Q_B^{2-}$ then binds two protons from the stroma to form plastohydroquinone. The structures of quinone, semiquinone anion, and hydroquinone are shown in Fig. 8; the mechanism that may be involved in proton binding is discussed by Shinkarev and Wraight (1993) and Okamura and Feher (1995). At this point two protons have been taken up from the stroma and four protons have been released into the lumen. The herbicides atrazine and diuron act by competing with plastoquinone for binding to the $Q_B$ site on D1.

The plastohydroquinone is released from its binding site on PS 2 and diffuses within the membrane (as $QH_2$ in Fig. 17) until it reaches the cytochrome $bf$ complex. Here it binds to a site near the luminal surface of the membrane.

At this point a series of reactions is initiated that will result in the transfer of additional protons across the membrane. A possible mechanism of the proton translocation, which was suggested by Mitchell, is known as a **Q cycle** (Hope, 1993). The bound $QH_2$ donates an electron to a 2 Fe-2 S center known as the **Rieske Fe-S center.** The electron moves from there to cytochrome $f$. The cytochrome $bf$ complex also contains two cytochrome $b$-type hemes, $b_H$ and $b_L$. $QH_2$ as a reductant is not strong enough to donate an electron to these hemes, but the plastosemiquinone that was formed when $QH_2$ transferred the electron to the Fe-S center is, and does so. In effect, the transfer of the first electron from $QH_2$ to the Fe-S center has driven the transfer of the second electron to $b_L$. $QH_2$ has been oxidized to Q, releasing two protons into the lumen. A second $QH_2$ molecule is oxidized by a similar mechanism, and a second electron is donated to the hemes. The two heme electrons are donated to a plastoquinone

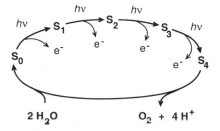

**FIG. 18.** Storage of positive charges in the OEC during a series of flashes. $S_0, S_1, S_2, S_3,$ and $S_4$ represent states of the OEC containing 0, 1, 2, 3, and 4 stored positive charges, respectively. $O_2$ is released after $S_4$ is formed.

(Q) molecule, which is bound near the stromal surface. It then binds two protons from the stroma. In the net process occurring at the cytochrome $bf$ complex, one $QH_2$ has been oxidized, two protons have been removed from the stroma, and four protons have been released into the lumen.

The electrons that were transferred to cytochrome $f$ move to **plastocyanin,** a soluble, copper-containing protein. The plastocyanin diffuses within the lumen until it reaches the PS 1 complex, where it transfers the electron to $P700^+$. P700 is a chlorophyll $a$ dimer.

Along with several other proteins, PS 1 contains a protein dimer, the PsaA and PsaB proteins, to which electron transfer cofactors are bound. The crystal structure of PS 1 reveals that the PsaA-PsaB dimer has approximate twofold rotational symmetry, as does the purple bacterial reaction center (Krauss $et\ al.$, 1993). Excited P700 donates an electron to $A_0$, a chlorophyll $a$ monomer (remember that the acceptor in PS 2 is pheophytin $a$) (Chitnis, 1996). From $A_0$ the electron moves to $A_1$, which is phylloquinone, then through the [4 Fe-4 S] cluster $F_X$ to the [4 Fe-4 S] clusters $F_A$ and $F_B$, and finally to **ferredoxin,** a soluble [4 Fe-4 S] protein. From ferredoxin the electron is transferred to $NADP^+$ by way of the flavoprotein **ferredoxin-NADP oxidoreductase** (FNR). On reduction the NADP binds a proton in the stroma. Protons return from the lumen to the stroma through the ATP synthase and ATP is formed.

When the demand for ATP is greater than the demand for NADPH, some of the ferredoxin molecules transfer their electrons back to oxidized quinone in the membrane, by a mechanism that is not well understood, so that they can be used to generate more ATP. This process is known as **cyclic photophosphorylation** (Bendall and Manasse, 1995).

## VI. Regulation of Photosynthesis

Virtually every step in the photosynthetic process is tightly regulated. The precision of this regulation is illustrated in Fig. 19. During illumination of spinach leaves with light whose intensity was varied sinusoidally, the net rate of $CO_2$ fixation perfectly paralleled the light intensity except between hours 4 and 10 when it was saturated. As the light intensity changed, the rates of the reactions leading to $CO_2$ uptake changed in such a way that efficiency of the use of the light remained constant. The final output of photosynthesis, export of fixed carbon from the leaves, remained almost perfectly constant throughout the cycle.

If photosynthesis is to work efficiently, PS 1 and PS 2 must transfer electrons at the same rate. Chloroplasts regulate the rates of the photosystems by regulating the transfer of excitation energy to P680 and P700. Both PS 1 and PS 2 contain chlorophyll molecules bound to proteins that are a permanent part of the PS 1 and PS 2 complexes. CP 43, CP 47, and LHC I shown in Fig. 10 are among these. Chlorophyll is also bound to the PsaA and PsaB proteins of PS 1. Such chlorophyll is known as **antenna chlorophyll.** Chloroplast thylakoids also contain a protein known as

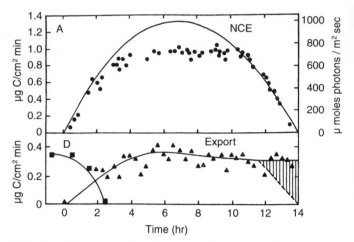

**FIG. 19.**   Illustration of how rigorously photosynthesis is regulated. Spinach leaves were illuminated with sinusoidally varying light to simulate the variation of light intensity during the day. Top: Net $CO_2$ exchange. Solid line, light intensity; closed circles, net uptake of $CO_2$. Bottom: Export of fixed carbon from the leaf. Triangles, newly fixed carbon; squares, carbon originating from stored starch. (From Servaites $et\ al.$, 1989, by permission of the American Society of Plant Physiologists.)

**LHC II** (for light-harvesting complex II), to which are typically attached eight molecules of chlorophyll $a$, seven of chlorophyll $b$, and one or two of xanthophyll per peptide (Allen, 1992). Under some conditions LHC II is tightly attached to PS 2, to which it transfers excitation energy. If PS 2 is working faster than PS 1, the electron carriers between PS 2 and PS 1 become overreduced. When this happens, a kinase is activated. The kinase catalyzes the phosphorylation of some of the LHC II. The phosphorylated molecules dissociate from PS 2 and diffuse into the stroma lamellae (Fig. 10). As a result, they no longer transfer excitation energy to PS 2, and so its rate slows. It is suspected that some of the phosphorylated LHC II may associate with PS 1, which is found predominantly in the stroma lamellae, and transfer excitation energy to it instead. The kinase activity is believed to be regulated by the oxidation state of the cytochrome $bf$ complex and perhaps also by the oxidation state of plastoquinone. Once the electron carriers between PS 2 and PS 1 are no longer overreduced, the kinase is inactivated and a phosphatase removes the phosphate from the LHC II. It returns to the grana lamellae and reassociates with PS 2.

A regulatory mechanism that deserves special mention involves the protein **thioredoxin,** which serves a signal that it is daytime. Thioredoxin contains a disulfide bridge that can be reduced by ferredoxin in a reaction catalyzed by ferredoxin-thioredoxin reductase. In the light, ferredoxin is reduced by PS 1 and in turn reduces thioredoxin. Thioredoxin then activates a number of Calvin cycle enzymes, and increases the sensitivity of the ATP synthase to $\Delta\mu_{H^+}$, by disulfide exchange. As a result the Calvin cycle and the ATP synthase are active when there is enough light to drive electron transport.

## VII. Summary

Plants, algae, and photosynthetic bacteria capture the energy of sunlight and store it in molecules such as starch and sucrose. Such photosynthetically synthesized molecules are the source of almost all of the free energy used by living organisms.

Light is first absorbed by antenna and light-harvesting pigments such as chlorophyll and $\beta$-carotene, raising them to electronically excited states. The excitation energy is then transferred to a special pair of chlorophyll molecules in the photosynthetic reaction center.

Excitation energy is converted to chemical energy in the reaction center as the excited special chlorophyll pair transfers an electron to an acceptor, which is either a pheophytin molecule or a chlorophyll molecule. The electrons proceed along an electron transport chain. The photosynthetic reaction centers of some purple photosynthetic bacteria have been isolated and crystallized and their structures determined by X-ray diffraction.

In higher plants, photosynthesis takes place in organelles called chloroplasts. Chloroplasts contain two photosystems, each having a reaction center. The photosystems are connected in series, forming an electron transport system known as the Z scheme. As electrons move along this electron transport chain from water to $NADP^+$, protons are bound on the stromal side of the chloroplast thylakoid membrane and released on the luminal side, forming a proton gradient. As the protons move back across the membrane through the ATP synthase, ATP is formed.

The ATP and NADPH are used for the synthesis of carbohydrates in a cyclic series of reactions known as the reductive pentose phosphate cycle.

## Bibliography

Allen, J. F. (1992). Protein phosphorylation in regulation of photosynthesis. *Biochim. Biophys. Acta* **1098**, 275–335.

Amesz, J., and Hoff, A. J. (Eds.) (1996). "Biophysical Techniques in Photosynthesis." Kluwer Academic Publishers, Dordrecht.

Barber, J. (Ed.) (1992). "The Photosystems: Structure, Function and Molecular Biology." Elsevier, Amsterdam.

Bendall, D. S., and Manasse, R. S. (1995). Cyclic photophosphorylation and electron transport. *Biochim. Biophys. Acta* **1229**, 23–38.

Blankenship, R. E. (1992). Origin and early evolution of photosynthesis. *Photosynthesis Research* **33**, 91–111.

Blankenship, R. E., Madigan, M. T., and Bauer, C. E. (Eds.) (1995). "Anoxygenic Photosynthetic Bacteria." Kluwer Academic Publishers, Dordrecht.

Büttner, M., Xie, D., Nelson, H., Pinther, W., Hauska, G., and Nelson, N. (1992). Photosynthetic reaction center genes in green sulfur bacteria and in photosystem 1 are related. *Proc. Natl. Acad. Sci. USA* **89**, 8135–8139.

Chitnis, P. R. (1996). Photosystem I. *Plant Physiol.* **111**, 661–669.

Clayton, R. K. (1970). "Light and Living Matter, Volume 1: The Physical Part." McGraw-Hill, New York.

Clayton, R. K. (1980). "Photosynthesis: Physical Mechanisms and Chemical Patterns." Cambridge University Press, Cambridge.

Deisenhofer, J., and Michel, H. (1989). The photosynthetic reaction centre from the purple bacterium *Rhodopseudomonas viridis*. *EMBO J.* **8**, 2149–2169.

Deisenhofer, J., and Norris, J. R. (Eds.) (1993). "The Photosynthetic Reaction Center," Vols. I and II. Academic Press, San Diego.

Flügge, U., and Heldt, H. W. (1991). Metabolite translocators of the chloroplast envelope. *Annu. Rev. Plant Physiol. Plant Mol. Biol.* **2**, 129–144.

Geiger, D. R., Servaites, J. C., and Shieh, W.-J. (1992). Balance among parts of the source-sink system: A factor in crop productivity. *In* "Crop Photosynthesis: Spatial and Temporal Determinants" (N. Baker and H. Thomas, Eds.), pp. 155–192. Elsevier, Amsterdam.

He, W.-Z., and Malkin, R. Photosystem II and Photosystem I. *In* "Photosynthesis: A Comprehensive Treatise" (A. S. Raghavendra, Ed.), Cambridge University Press, Cambridge (in press).

Hope, A. B. (1993). The chloroplast *bf* complex: A critical focus on function. *Biochim. Biophys. Acta* **1143**, 1–22.

Khorana, H. G. (1988). Bacteriorhodopsin, a membrane protein that uses light to translocate protons. *J. Biol. Chem.* **263**, 7439–7442.

Krauss, N., Hinrichs, W., Witt, I., Fromme, P., Pritzkow, W., Dauter, Z., Betzel, C., Wilson, K. S., Witt, H. T., and Saenger, W. (1993). Three-dimensional structure of system I of photosynthesis at 6 A resolution. *Nature* **361**, 326–337.

Loomis, W. E. (1960). Historical introduction. *In* "Encyclopedia of Plant Physiology" (W. Ruhland, Ed.), Vol. V, pp. 85–114. Springer-Verlag, Berlin.

McDermott, G., Prince, S. M., Freer, A. A., Hawthornewaite-Lawless, A. M., Papiz, M. Z., Cogdell, R. J., and Isaacs, N. W. (1995). Crystal structure of an integral membrane light-harvesting complex from photosynthetic bacteria. *Nature* **374**, 517–521.

Mitchell, P. (1961). Coupling of phosphorylation to electron and hydrogen transfer by a chemi-osmotic mechanism. *Nature* **191**, 144–148.

Mitchell, P. (1979). Keilin's respiratory chain concept and its chemiosmotic consequences. *Science* **206**, 1148–1159.

Nalin, C. M., and Nelson, N. (1987). Structure and biogenesis of chloroplast coupling factor $CF_0CF_1$-ATPase. *Curr. Topics Bioenerg.* **15**, 273–294.

Nitschke, W., and Rutherford, A. W. (1991). Are all the different types of photosynthetic reaction centers variations of a common structural theme? *Trends Biochem. Sci.* **16**, 241–245.

Nugent, J. H. A. (1996). Oxygenic photosynthesis. Electron transfer in photosystem I and Photosystem II. *Eur. J. Biochem.* **237**, 519–531.

Okamura, M. Y., and Feher, G. (1995). Proton-coupled electron transfer reactions of $Q_B$ in reaction centers from photosynthetic bacteria. *In* "Anoxygenic Photosynthetic Bacteria" (R. E. Blankenship, M. T. Madigan, and C. E. Bauer, Eds.), pp. 577–594. Kluwer Academic Publishers, Dordrecht.

Olson, J. M. (1996). Iron-sulfur-type reaction centers. Introduction. *Photochem. Photobiol.* **64**, 1–4.

Ort, D. R., Yocum, C. F., and Heichel, I. F. (Eds.) (1996). "Photosynthesis: The Light Reactions." Kluwer Academic Publishers, Dordrecht.

Schnell, D. J. (1995). Shedding light on the chloroplast protein import machinery. *Cell* **83**, 521–524.

Servaites, J. C., Fondy, B. R., Li, B., and Geiger, D. R. (1989). Source of carbon for export from spinach leaves throughout the day. *Plant Physiol.* **90**, 1168–1174.

Shinkarev, V., and Wraight, C. A. (1993). Electron and proton transport in the acceptor quinone complex of reaction centers of phototropic bacteria. *In* "The Photosynthetic Reaction Center" (J. Deisenhofer and J. R. Norris, Eds.), Vol. I, pp. 194–255. Academic Press, San Diego.

Stoeckenius, W. (1978). Bacteriorhodopsin. *In* "The Photosynthetic Bacteria" (R. K. Clayton and W. R. Sistrom, Eds.). Plenum Press, New York.

Woese, C. R. (1987). Bacterial evolution. *Microbiol. Rev.* **51**, 221–271.

J. *Woodland Hastings*

# 64

# Bioluminescence

## I. Introduction

Although the emission by living organisms of light that is visible to other organisms is a rather rare occurrence, and in that sense a curiosity of nature, it has several different and fascinating functions. Bioluminescence is also a unique tool for investigating and understanding numerous different basic physiological processes, both cellular and organismic. It allows one to explore in one system a gamut of questions that confront biologists, ranging from gene expression and its regulation to enzymology, bioenergetics, physiology, function, ecology, and evolution (Hastings, 1976, 1983). Lastly, luciferases and associated proteins have recently been developed for use as reporters of gene expression. Such proteins may be visualized noninvasively from the same cell *in vivo,* and over an extended time course, for example, during development (Hastings *et al.,* 1997; Chalfie and Kain, 1997).

The phenomenon is not only rare; in the different groups that do emit light, the biochemical and physiological mechanisms responsible for it are very different, as are their specific functional roles. Indeed, bioluminescence is *not* an evolutionarily conserved function; in the different groups of organisms the genes and proteins involved are mostly unrelated and evidently originated and evolved independently. How many times this may have occurred is difficult to say, but it has been estimated that present-day luminous organisms come from as many as 30 different evolutionarily distinct origins (Harvey, 1952; Herring, 1978; Hastings, 1983; Hastings and Morin, 1991).

## II. Physical and Chemical Mechanisms

Bioluminescence does not come from or depend on light absorbed by the organism. It derives from an enzymatically catalyzed chemiluminescence, a highly exergonic reaction in which chemical energy is transformed into light energy (McCapra, *in* Herring, 1978; Campbell, 1988; Wilson, 1985, 1995). Thus in the reaction of sub-stance A with substance B, one of the reaction products is formed in an electronically excited state (D*), which then emits a photon ($h\nu$).

$$A + B \rightarrow C + D*$$
$$D* \rightarrow D + h\nu$$

Chemiluminescence is a special case of the more general phenomenon of luminescence, in which energy is specifically channeled to a molecule; excited state production is not dependent on the temperature of the molecule. Other kinds of luminescence include fluorescence and phosphorescence, in which the excited state is created by the prior absorption of light, or triboluminescence and piezoluminescence, involving crystal fracture and electric discharge, respectively. The color is a characteristic of the excited molecule, independent of how it was excited.

Luminescence is contrasted with incandescence, in which excited states are produced by virtue of the thermal energy. An example is the light bulb, in which a filament is heated, and the color of the light depends on the temperature ("red hot" reflecting a lower temperature than "white hot"). The energy ($E$) of the photon is related to the color or frequency of the light, and is given by the equation $E = h\nu$, where $h$ is Planck's constant and $\nu$ the frequency. In the visible light range, $E$ is very large in relation to most biochemical reactions. Thus, the energy released by a mole of photons ($6.02 \times 10^{23}$) in visible wavelengths is about 50 kcal, which is much more than the energy from the hydrolysis of a mole of ATP—about 7 kcal. A visible photon is thus able to do a lot of work (e.g., photosynthesis) or a lot of damage (mutation; photodynamic action, which can kill). Conversely, it takes a highly exergonic reaction to create a photon.

A question of fundamental importance, then, is what kind of chemical process possesses enough energy, and evidently in a single step (an important point), to "populate" an excited state. A clue is the fact that chemiluminescences in solution generally require oxygen, which in its reaction with a substrate, forms an organic peroxide. The energy from the breakdown of such peroxides—which can generate up to 100 kcal per mole—is ample to account for a product in an electronically excited state.

A model of the reaction mechanism in such chemiluminescent reactions is referred to as chemically initiated electron exchange luminescence (CIEEL) (Schuster, 1979; Catalani and Wilson, 1989). In this mechanism, peroxide breakdown involves electron transfer with chemiexcitation. It is initiated by an electron transfer from a donor species (D) to an acceptor (A), which is the peroxide bond in this case. After electron transfer the weak O—O bond cleaves to form products B and $C^-$. The latter is a stronger reductant than $A^-$, so the electron is transferred back to D with the concomitant formation of a singlet excited state $D^*$ and emission. Thus,

$$A + D \rightarrow A^- + D^+ \qquad D^* \rightarrow h\upsilon + D$$
$$\downarrow$$
$$B + C^- \qquad C$$

The mechanisms of different bioluminescence reactions can possibly be accommodated within this general scheme.

A useful way to think of chemiluminescence is to regard it as the reverse of a photochemical reaction, in which the excited state created by the absorption of a photon gives rise to chemical species capable of further reaction. But such species can also react reversibly to repopulate the excited state. Photosynthesis is a good example: the primary chemical states formed in photosynthesis are comparable to $C^-$ and $D^+$, which may be the penultimate states in bioluminescence.

In the primary step of photosynthesis, the energy of the excited state of chlorophyll (Chl*) gives rise to an electron transfer, with the consequent formation of a primary oxidant and a primary reductant (Govindjee et al., 1986). With A as the electron acceptor, these steps can be represented as

$$\text{Chl} + h\upsilon \rightarrow \text{Chl}^*$$
$$\text{Chl}^* + A \rightleftarrows \text{Chl}^+\cdot + A^-\cdot.$$
$$\underbrace{\qquad}_{\text{stable oxidant}} \underbrace{\qquad}_{\text{stable reductant}}$$

Most of these redox species give rise to stable products and ultimately to $CO_2$ fixation. However, some species recombine and reemit a "delayed light." This is essentially the reverse of the reaction, with the formation of the singlet excited state of chlorophyll and its subsequent emission of a photon.

In bioluminescence, the substrates and enzymes, though chemically different in different organisms, are all referred to as luciferin and luciferase (DeLuca, 1978). These are thus generic terms, and to be correct and specific, each should be identified with the organism (Tables 1 and 2), thus firefly luciferin or bacterial luciferase. The luciferases, as well as the structures and reaction intermediates of the different known luciferins, will be discussed in connection with the individual groups of organisms.

## III. Luminous Organisms: Abundance, Diversity, and Distribution

Bioluminescence is indeed rare as measured by the total number of luminous species, but it is phylogenetically diverse, being found in more than 13 phyla (Herring, in Herring, 1978; Herring, 1987); only the major ones are considered here (Table 1). These include bacteria, unicellular algae, and fungi, as well as animals ranging from jellyfish, annelids, and mollusks to shrimp, fireflies, echinoderms, and fish. Luminescence is unknown in higher plants and in vertebrates above the fish (Cormier, in Herring, 1978). It is also absent in several invertebrate phyla. In some phyla or taxa, a substantial population of the genera are luminous (e.g., ctenophores, ~50%; cephalopods, >50%; echinoderms and annelids, ~4%). Commonly, all members of a luminous genus emit light, but in some cases there are both luminous and nonluminous species.

The fact that the enzymes and substrates, as well as the physiological and functional aspects of bioluminescence, differ in the several major taxa is indicative of their independent evolutionary origins. In fact, there are chemically different systems found in different taxa within some phyla, so the total number of evolutionarily independent groups may be 30 or more (Hastings, 1983). Fewer than half of these have been studied in detail, and knowledge of the luciferins and luciferases is available for only about 10.

Although luminescence is prevalent in the deep sea (Herring, 1985a), it is not associated especially with organisms that live in total darkness. There are no known luminous species either in deep freshwater bodies, such as Lake Baikal, Russia, or in the darkness of terrestrial caves. There are luminous dipteran larvae (Arachnocampa) that live near the mouths of caves in New Zealand and Australia (Fig. 1), but they also occur in culverts and the undercut banks of streams, where there is considerable daytime illumination. Although insect displays of bioluminescence are among the most spectacular, bioluminescence is relatively rare in the terrestrial environment (<0.2% of all genera). Some other terrestrial luminous forms are millipedes, centipedes, earthworms, and snails, but in none of these is the display very bright.

For reasons that are still obscure, bioluminescence is most prevalent in the marine environment (Kelly and Tett, in Herring, 1978); it is greatest at midocean depths (200–1200 m), where daytime illumination fluxes range between ~$10^{-1}$ and $10^{-12}$ $\mu$W cm$^{-2}$. In these locations bioluminescence in fish may occur in over 95% of the individuals and 75% of the species; similar percentages are tabulated for shrimp and squid. The midwater luminous fish Cyclothone is considered to be the most abundant vertebrate on the planet. Where high densities of luminous organisms occur, their emissions can exert a significant influence on the communities and may represent an important component in the ecology, behavior, and physiology of these organisms. Above and below midocean depths, luminescence decreases to <10% of all individuals and species. It may be somewhat higher (~20%) at abyssal depths, whereas among coastal species, less than 2% are bioluminescent.

**TABLE I**    Luminous Organisms: Biochemical Mechanisms and Biological Functions

| Type of organism | Representative genera | Luciferins and other factors (Emission max, nm) | Displays and functions |
|---|---|---|---|
| Bacteria | Photobacterium Vibrio Xenorhabdus | Reduced flavin and long-chain aldehyde (475–535) Some with accessory emitters | Steady bright glow Autoinduction of luciferase Function as symbionts |
| Mushrooms | Panus, Armillaria Pleurotus | Unknown (535) | Steady dim glow; function unknown |
| Dinoflagellates | Gonyaulax Pyrocystis Noctiluca | Linear tetrapyrrole pH change (470) | Short (0.1 sec) bright flashes function to frighten or deter |
| Cnidaria Jellyfish Hydroid Sea pansy | Aequorea Obelia Renilla | $Ca^{2+}$, coelenterazine Imidazo pyrazine nucleus (460–510), some accessory emitters | Bright flash or train of flashes; function to frighten or deter |
| Ctenophores | Mnemiopsis Beroe | $Ca^{2+}$, coelenterazine (460) | Bright flashes; function to frighten or deter |
| Annelids Earthworms Marine polychaetes Syllid fireworm | Diplocardia Chaetopterus Odontosyllis | *N*-Isovalyeryl-3 amino propanal Unknown Unknown (480) | Cellular exudates, sometimes very bright; function to divert and to deter; others unknown |
| Molluscs Limpet Clam Squid | Latia Pholas Heteroteuthis | Aldehyde Clam luciferin, but structure is unknown, $Cu^{2+}$ | Exuded luminescence in all three; photophores and symbiotic bacteria in some squid; diversion, decoy |
| Crustacea Ostracod Shrimp (euphausids) Copepods | Vargula Meganyctiphanes | Imidazolopyrazine nucleus (465) Linear tetrapyrrole (470) | Squirts enzyme and substrates; diversion, decoy Photophores; camouflage |
| Insects Coleopterids (beetles) Firefly Click beetles Railroad worm | Photinus, Photuris Pyrophorus Phengodes, Phrixothrix | Benzothiazole, ATP, $Mg^{2+}$ Similar chemistry in all coleoptera | Flashes, specific kinetic patterns Communication: courtship, mating |
| Diptera (flies) | Arachnocampa | Biochemistry unknown | Lure to attract prey |
| Echinoderms Brittle stars Sea cucumbers | Ophiopsila Laetmogone | Biochemistry unknown Biochemistry unknown | Trains of rapid flashes; frighten, divert predators Unknown |
| Chordates Tunicates | Pyrosoma | Organelles evolved from bacteria (480–500) | Brilliant trains of flashes stimulated by light and other factors |
| Fish Cartilaginous Bony Ponyfish Flashlight fish Angler fish Midshipman | Squalus Leiognathus Photoblepharon Cryptopsaras Porichthys | Biochemistry unknown Symbiotic luminous bacteria (~490) Symbiotic luminous bacteria (~490) Symbiotic luminous bacteria (~490) Self luminous, Vargula type luciferin, nutritionally obtained | Unknown Camouflage, ventral luminescence To attract and capture prey Camouflage? courtship display? |
| Midwater fish | Cyclothone Neoscopelus Tarletonbeania | Self-luminous, biochemistry unknown Self-luminous, biochemistry unknown Self-luminous, biochemistry unknown | Many photophores, ventral and lateral Photophores: lateral, on tongue Sexual dimorphism; males have dorsal (police car) photophores |

**TABLE 2**  Luciferases

|  | kDa | E.C. No. |
|---|---|---|
| Bacterial | ~80 | 1.14.14.3 |
|  | ($\alpha$, 41; $\beta$, 39) |  |
| Dinoflagellate | ~135 |  |
| Coelenterate | ~35 | 1.13.12.5 |
| Mollusk |  | 1.14.99.21 |
| Firefly | ~60 | 1.13.12.7 |
| Crustacean | ~68 | 1.13.12.6 |

## IV. Functions of Bioluminescence

The functional importance of bioluminescence and its selection in evolution are believed to be based largely on its being detected by another organism; the response of that organism then favors in some way the luminous individual

**FIG. 1.**  Luminous dipteran larvae (*Arachnocampa*) on the ceiling of a cave in New Zealand.

(Buck, in Herring, 1978; Hastings, 1983; Buck, 1938; Herring, 1990). In higher animals luminescence is generally controlled neurally (Anctil, 1987).

Bioluminescence is interesting biologically because it is a clear and well-documented example of a function that, while not metabolically essential, confers an advantage on the individual. Although bioluminescence has evidently arisen independently many times, it may also have been lost many times in different evolutionary lines, particularly where it was not a truly important function.

Bioluminescence may be thought of as a bag of tricks: The light can be used in different ways and for different functions. Most of the perceived functions of bioluminescence may be classified under three main rubrics: defense, offense, and communication (Table 3).

Important defensive strategies associated with bioluminescence are to frighten, to serve as a decoy, to provide camouflage, and to aid in vision. Organisms may be frightened or diverted by flashes, which are typically bright and brief (0.1 sec); light is emitted in this way by many organisms, and experimental studies confirm that flashes can indeed frighten (Morin, 1983).

On the other hand, a glowing object in the ocean often appears to attract feeders or predators. Although a luminous organism would evidently be at risk by virtue of this attraction, the fact can be used defensively if an organism creates a decoy light to attract the predator, and then slips off under the cover of darkness. This is exactly what quite a number of organisms do. A luminous squid in darkness squirts luminescence instead of ink; ink would be useless in such a case. Some organisms sacrifice more than light; in scaleworms and brittle stars, a part of the body may be automized (broken off) and left behind as a luminescent decoy to attract the predator. In these cases the body part flashes while still attached but glows after detachment, exemplifying that a flash deters whereas a glow attracts.

A unique method for evading predation from below is to camouflage the silhouette by emitting light that matches the color and intensity of the downwelling background light. By analogy with countershading in reflected light, this has been called counterillumination. Imagine a plane in the sky during the day. If it could emit light from its bottom surface matching the sky behind, it would be invisible from below. Actually, it is not necessary for the entire surface to emit light; emission by only a part would mean that the object would no longer look like a plane. This can be called **disruptive illumination,** and many luminous marine organisms, including fish, apparently use this to aid in escaping detection (McFall-Ngai and Morin, 1991). Another novel defensive strategy has been dubbed the burglar alarm: Dinoflagellates flash when grazed upon, which may enhance predation on the grazers, and thus reduce grazing on dinoflagellates (Abrahams and Townsend, 1993).

There are also several ways in which luminescence can aid in predation. Several of these, such as helping in vision, may be of value for both offense and defense. For example, flashes, which are more typically used defensively, can be used offensively to temporarily startle or blind prey. A

**TABLE 3**   Functions of Bioluminescence

| Function | Strategy | Method |
|---|---|---|
| Defense | Frighten, startle | Bright, brief flashes |
| | Decoy, diversion | Glow, luminous cloud, sacrificial lure |
| | Camouflage | Ventral luminescence during the day, disrupting or concealing the silhouette seen from below |
| Offense | Frighten, startle | Bright flash may temporarily immobilize prey |
| | Lure | Glow to attract, then capture prey |
| | Vision | To see and capture prey |
| Communication | Courtship, mating | Specific flashing signals; patterns of light emission recognized by opposite sex |
| Dispersal, propagation | Glow to attract feeders | Bacteria ingested by feeder pass through gut tract alive and are therefore dispersed |

glow can also be used offensively; it can serve as a lure. The organism is attracted to the light but is then captured by the organism that produced the light. Camouflage may also be used offensively, allowing the luminous predator to approach its prey undetected. Vision is certainly useful offensively; prey may be seen and captured under conditions that are otherwise dark, as practiced by the flashlight fish *Photoblepharon* (Fig. 2).

Communication involves information exchange between individual members of a species, and luminescence is used for this in several organisms, including annelids, crustaceans, insects, squid, and fishes. The most common such use of light is for courtship and mating, as in fireflies (Buck and Buck, 1976; Lloyd, 1977, 1980). But there are numerous examples in the ocean (Herring, 1990). In the syllid fireworm *Odontosyllis* a truly extraordinary display occurs as the animals engage in mating, which occurs just post-twi-

light a few days after the full moon. Readily observed in many parts of the world (e.g., Bermuda), the females come to the surface and swim in a tight circle. A male streaks from below and joins the female, with eggs and sperm shed in the ocean in a luminous circle. Another example occurs over shallow reefs in the Caribbean: male ostracod crustaceans produce complex species-specific trains of secreted luminous material—ladders of light—which attract females (Morin and Cohen, 1992).

There remain some luminous organisms for which it is difficult to determine what the function of light emission may be. In the luminous fungi (Wassink, in Herring, 1978) (Fig. 3), both the mycelium and mushrooms emit a continu-

**FIG. 2.**   The flashlight fish (*Photoblepharon*), showing the light organ (harboring luminous bacteria) just below the eye.

**FIG. 3.**   Luminous mushrooms.

ous light, day and night, but it is typically very dim. Moreover, the mycelium itself is almost never exposed: it is underground or inside a decaying tree. The role of luminescence in this and other such cases remains to be understood.

## V. Bacterial Luminescence

### A. Occurrence and Functions

Luminous bacteria (Fig. 4) occur ubiquitously in the oceans and can be isolated from most seawater samples from the surface to depths of 1000 m or more. A primary habitat where most species abound is in association with another (higher) organism, dead or alive, where growth and propagation occur. Planktonic forms are readily isolated, but they do not grow in seawater, as it is a poor growth medium, so they may be viewed as having overflowed from primary habitats (Nealson and Hastings, 1991).

The most exotic specific associations involve specialized light organs (e.g., in fish and squid) in which a pure culture

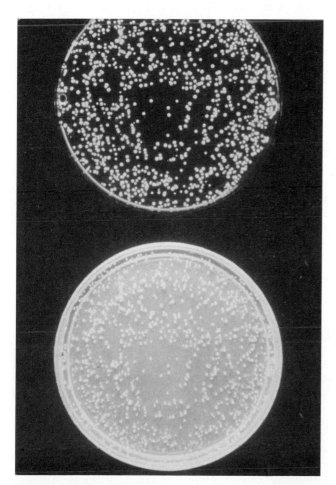

**FIG. 4.** Luminous bacteria, photographed by their own light (top) and by room light (bottom). Note the dark mutant near the center.

of luminous bacteria is maintained at a high density and at high light intensity. In teleost fishes, some 11 different groups carrying such bacteria are known (Figs. 2 and 5). Exactly how symbioses are achieved—the initial infection, exclusion of contaminants, nutrient supply, restriction of growth but bright light emission—is not understood (Hastings *et al.*, 1987). In such associations, the host receives the benefit of the light and may use it for one or more specific purposes; the bacteria in return receive a niche and nutrients.

Direct (nonspecific) associations include parasitizations and commensals. Intestinal bacteria in marine animals, notably fish, are often luminous, and heavy pigmentation of the gut tract is often present, presumably to prevent the light from betraying the location of the fish to predators. Indeed, the light emission in these cases is thought to benefit the bacteria more directly, for example, by attracting predators and thereby promoting bacterial dispersion and propagation. Luminous bacteria growing on a substrate, be it a parasitized crustacean, the surface of a dead fish, or a fecal pellet, can produce a light bright enough to attract other organisms, presumably to feed on the material. Attraction to promote dispersal and thus propagation of bacteria may thus be viewed as yet another function of bioluminescence although, because the luminous organism must survive ingestion, it is probably limited to bacteria.

Terrestrial luminous bacteria are rare. The best known are those harbored by nematodes that are parasitic on insects such as caterpillars. The nematode carries the bacteria as symbionts and injects them into the host along with its own fertilized eggs. The bacteria grow and the developing nematode larvae feed on them. The dead but now luminous caterpillar (Fig. 6) attracts predators, which serves to disperse the nematode offspring, along with the bacteria.

Luminous bacteria are also examples of organisms that can exploit—or at least survive in—different habitats, for example, in a light organ or in the planktonic environment. This versatility also includes the capacity to turn the luminescent system on and off, at both physiological and genetic levels (Hastings, 1987). Where advantageous, it is expressed at high levels; where not, the genes are repressed and energy is conserved.

### B. Biochemistry

Luminous bacteria typically emit a continuous light, usually blue-green. When strongly expressed, a single bacterium may emit $10^4$ or $10^5$ photons sec$^{-1}$. The system is biochemically unique and is diagnostic for a bacterial involvement in the luminescence of a higher organism, as endosymbionts, for example. The pathway itself (Fig. 7) constitutes a shunt of cellular electron transport at the level of flavin, and reduced flavin mononucleotide is the substrate (luciferin) that reacts with oxygen in the presence of bacterial luciferase to produce an intermediate peroxy flavin (Meighen and Dunlap, 1993; Baldwin and Ziegler, 1992; Hastings *et al.*, 1985). This intermediate then reacts with a long-chain aldehyde (tetradecanal) to form the acid and the luciferase-bound hydroxy flavin in its excited state.

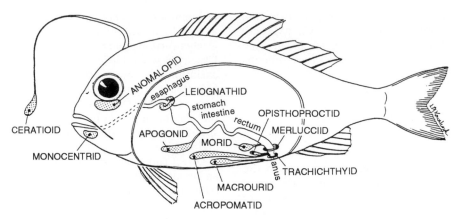

**FIG. 5.** The "ichthylicht." A diagrammatic fish is used to indicate the approximate locations, sizes, and configurations of the light organs in the several groups of luminous fishes that culture luminous bacteria as a source of light for the organ.

Although there are two substrates in this case, the flavin can claim the name luciferin on etymological grounds, since it forms (bears) the emitter. The bioluminescence quantum yield has been estimated to be about 30%.

The enzyme is an external flavin monoxygenase (EC

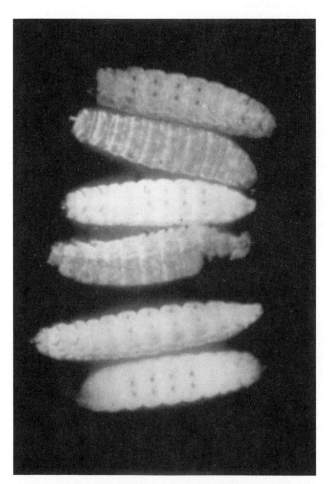

**FIG. 6.** Luminous caterpillars; caused by parasitic luminous bacteria.

1.14.14.3). Curiously, no other enzymes of this general type have been found to emit light, even at very low quantum yields. The light-emitting steps have been modeled in terms of an electron exchange mechanism (see Section II), and the experimental evidence is consistent with this.

There are enzyme systems that serve to maintain the supply of myristic aldehyde, and genes coding for these enzymes are part of the *lux* operon (Meighen, 1991; Fig. 8). The luciferases themselves are homologous heterodimeric ($\alpha$-$\beta$) proteins (~80 kDa) in all species. They possess a single active center per dimer, mostly associated with the $\alpha$-subunit. Structurally, they appear to be relatively simple; that is, no metals, disulfide bonds, prosthetic groups, or nonamino acid residues are involved.

The luciferase and the mechanism of the bacterial reaction have been studied in great detail. An interesting feature of this luciferase reaction is its inherent slowness: At 20°C the time required for a single catalytic cycle is about 20 sec. The luciferase peroxy flavin itself has a long lifetime; at low temperatures (0 to −20°C) it has been isolated, purified, and characterized. It can be further stabilized by aldehyde analogs such as long-chain alcohols and amines, which bind at the aldehyde site.

### C. Regulation of Bacterial Luminescence

Typically, bacteria are unable to regulate emission on a fast time scale (msec, sec), as in organisms that emit flashes. However, bacteria are generally able to control the development and expression of luminescence at both physiological and genetic levels. The most unique of these mechanisms is "autoinduction," in which the transcription of the luciferase and aldehyde synthesis genes of the *lux* operon is regulated by a gene product of the operon itself. A substance produced by the cells called autoinducer (Fig. 9) is a product of the *lux* I gene (Nealson and Hastings, 1991). The ecological implications are evident: in planktonic bacteria, a habitat where luminescence has no value, autoinducer cannot accumulate, and no luciferase synthesis occurs. However, in the confines of a light organ, high autoinducer levels are reached and the luciferase genes

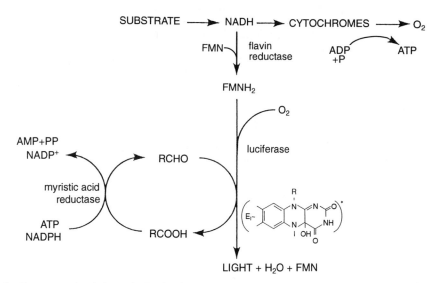

**FIG. 7.** The luciferase reaction in bacteria. In the electron transport pathway, adenosine triphosphate (ATP) is generated; luciferase shunts electrons at the level of reduced flavin ($FMNH_2$) directly to molecular oxygen. In the mixed function oxidation with long-chain aldehyde, hydroxy-FMN is produced in its excited state (*) along with long-chain acid. The FMN product is reduced again and recycles and the aldehyde is regenerated from the acid.

are transcribed. Interestingly, it has recently been discovered that autoinduction-type mechanisms similarly control the expression of other specific genes in bacteria (Fuqua *et al.,* 1994).

There are a number of other control mechanisms that serve to regulate the transcription of the *lux* operon, including glucose (catabolic repression), nutrient levels, iron, and oxygen. Each of these factors represents a different mechanism for the physiological control of gene expression, and each has implications concerning the ecology of luminous bacteria and the function of their luminescence. But there is also control at the genetic level: in some species of bacteria there are "dark" (very, very dim) mutants that may arise spontaneously (see Fig. 4). In these the synthesis of the luminescent system fails altogether to occur, irrespective of conditions, and this is an inheritable property. However,

the *lux* genes are not lost and revertants occur. Thus, the organism can compete under conditions where luminescence is not advantageous, yet be able to produce luminous forms and populate the appropriate habitat when and where it is found. Indeed, the bacterial *lux* genes may occur in many bacterial strains but not be expressed. This means that there may be many more potentially luminous bacteria than would be deduced from the luminescence of colonies on plates.

## VI. Dinoflagellate Luminescence

### A. Occurrence and Function

Dinoflagellates occur ubiquitously in the oceans as planktonic forms, and contribute substantially to the so-

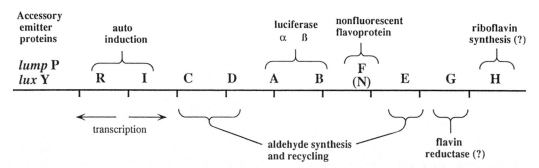

**FIG. 8.** Organization of the *lux* genes in *Vibrio fischeri*. The operon on the right, transcribed from the 5′ to the 3′ end, carries genes for synthesis of autoinducer (*lux* I), for luciferase α and β peptides (*lux* A and B), and for aldehyde production (*lux* C, D, and E). The operon on the left encodes for a gene (*lux* R), which encodes for a receptor molecule that binds autoinducer; the complex controls the transcription of the right operon. Other genes, *lux* F (N), G and H (right), are associated with the operon but with still uncertain functions; genes for accessory emitter proteins also occur (left).

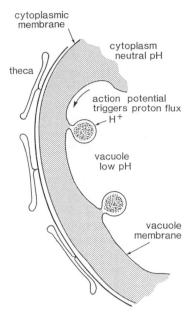

**FIG. 9.**   Structure of autoinducer from *Vibrio fischeri*.

called "phosphorescence" commonly seen at night (especially in summer) when the water is disturbed (Hastings, in Govindjee *et al.,* 1986). They occur primarily in surface waters and many species are photosynthetic. In the phosphorescent bays (e.g., in Puerto Rico and Jamaica), high densities of a single species (*Pyrodinium bahamense*) usually occur. The so-called "red tides" are blooms of dinoflagellates.

About 6% of all dinoflagellate genera contain luminous species, but since there are no luminous dinoflagellates among the freshwater species, the proportion of luminous forms in the ocean is higher. As a group, dinoflagellates are important as symbionts, notably for contributing photosynthesis and carbon fixation in animals, but unlike bacteria, no luminous dinoflagellates are known from symbiotic niches.

Since dinoflagellates are stimulated to emit light when predators (e.g., crustacea) are active, predators on the crustacea might be alerted, resulting in a reduced predation on dinoflagellates (Abrahams and Townsend, 1993). Predation on dinoflagellates may also be impeded more directly, since the flash could startle or divert the predator. The response time to stimulation (msec) is certainly fast enough to have this effect.

## B.  Biochemistry and Cell Biology

Luminescence in dinoflagellates is emitted from many small ($\sim$0.5 $\mu$m) cortical structures, identified as a new type of organelle, termed the **scintillon** (flashing unit) (Hastings and Dunlap, in DeLuca and McElroy, 1986). They occur as outpocketings of the cytoplasm into the cell vacuole, like a balloon, with the neck remaining connected (Fig. 10). Scintillons contain only dinoflagellate luciferase and luciferin (with its binding protein), other cytoplasmic components being somehow excluded. Ultrastructurally, they can be identified by immunolabeling with antibodies raised against the luminescence proteins (Nicolas *et al.,* 1987), and visualized by their bioluminescent flashing after stimulation, as well as by the fluorescence of luciferin (Fig. 11). Dinoflagellate luciferin is a novel tetrapyrrole related to chlorophyll (Fig. 12).

Activity can be obtained in extracts made at pH 8 simply by shifting the pH from 8 to 6; it occurs in both soluble and particulate (scintillon) fractions, suggesting that during extraction some scintillons are lysed, whereas others seal off at the neck and form closed vesicles. With the scintillon fraction, the *in vitro* activity occurs as a flash ($\sim$100 msec), very close to that of the living cell, and the kinetics are independent of the dilution of the suspension. For the soluble fraction, the kinetics are dependent on dilution, as

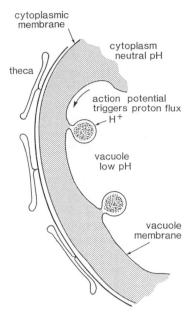

**FIG. 10.**   Scintillons of dinoflagellates represented as organelles formed as cytoplasmic outpocketings hanging in the acidic vacuole.

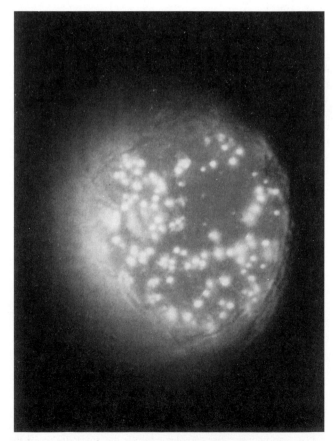

**FIG. 11.**   A *Gonyaulax* cell visualized by fluorescence microscopy, showing scintillons by the fluorescence of dinoflagellate luciferin.

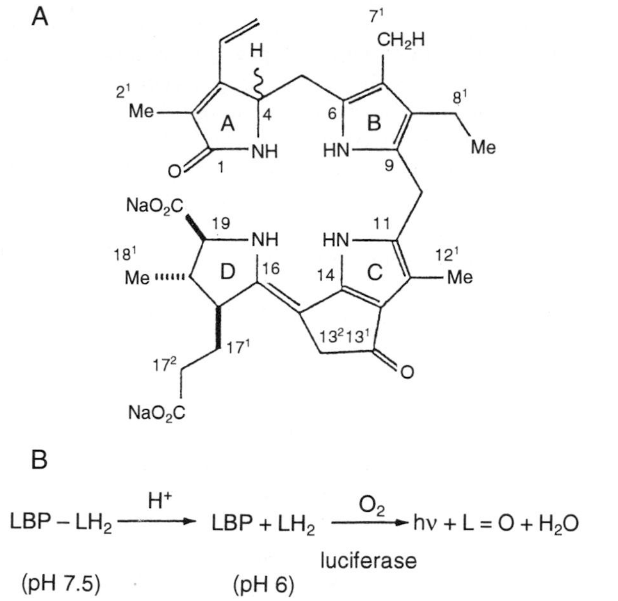

**A**

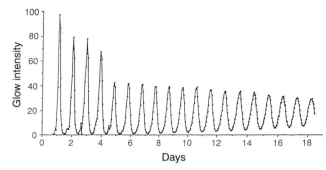

**B**

$$LBP - LH_2 \xrightarrow[\text{(pH 7.5)}]{H^+} LBP + LH_2 \xrightarrow[\text{luciferase}]{O_2} h\nu + L = O + H_2O$$

(pH 6)

**FIG. 12.** (A) Structure of dinoflagellate luciferin, a tetrapyrrole (fluorescence maximum, ~475 nm). (B) Reaction steps for *Gonyaulax* bioluminescence. Luciferin ($LH_2$) is attached to luciferin binding protein (LBP) at pH 8, but is free at pH 6 and thus able to be oxidized by luciferase; the product has a carbonyl at position $13^2$.

**FIG. 13.** The circadian rhythm of the steady glow of bioluminescence in *Gonyaulax*. A culture grown in a 24-hr light–dark cycle was transferred at zero time to constant conditions (19°C; dim white light); measurements were made about once every hour for 18 days. The average period length was 22.8 hr.

in an enzyme reaction. A distinctive feature is that at pH 8 the luciferin is bound to a protein (luciferin binding protein), which thereby prevents it from reaction with luciferase, but it is free at pH 6 (Fig. 12B).

The flashing of dinoflagellates *in vivo* is postulated to result from a transient pH change in the scintillons, triggered by an action potential in the vacuolar membrane which, while sweeping over the scintillon, opens ion channels that allow protons from the acidic vacuole to enter (see Fig. 10).

### C. Control of Dinoflagellate Luminescence: The Circadian Clock

The composition of the medium and nutrient conditions apparently have little effect on the development and expression of bioluminescence in dinoflagellates. However, in *Gonyaulax polyedra* and some other dinoflagellates, luminescence is regulated by day–night light–dark cycles and an internal circadian biological clock mechanism (Morse *et al.,* 1990). The flashing response to mechanical stimulation is far greater during the night than during the day, and a steady low-level emission (glow) exhibits a peak toward the end of the night phase. The regulation is attributed to an endogenous mechanism; cultures maintained under constant conditions (light, temperature) continue to exhibit rhythmicity for weeks (Fig. 13), but with a period that is not exactly 24 hr; it is only about (circa) one day (diem), thus the origin of the term.

The nature of this circadian clock remains one of the real enigmas in physiology. In humans and other higher animals, where it regulates the sleep–wake cycle and many other physiological processes, the mechanism involves the nervous system (Hastings *et al.,* 1991). But it also occurs in plants and unicellular organisms, including *Euglena, Chlamydomonas,* and *Paramecium.* In the case of *G. polyedra* it is known that daily changes occur in the cellular concentrations of luciferase, luciferin, and its binding protein; the proteins are synthesized and destroyed each day. Hence, the biological clock exerts control at a very basic level, by controlling gene expression.

## VII. Coelenterates and Ctenophores

### A. Occurrence and Function

Luminescence is common and widely distributed in these groups (Cormier, in Herring, 1978; Herring, in Herring, 1978). In the ctenophores (comb jellies), luminescent organisms constitute over half of all genera, whereas in the coelenterates (cnidaria) the figure is about 6%. Luminous hydroids, siphonophores, sea pens, and jellyfish, among others, are well known. The organisms are mostly sessile or sedentary, and upon stimulation emit light as flashes (Fig. 14A). Bioluminescence is absent in sea anenomes and corals.

Hydroids occur as plantlike growths, typically adhering to rocks below low tide level in the ocean. When they are touched, a sparkling emission is conducted along the colony; repetitive waves from the origin may occur. Luminous jellyfish (such as *Pelagia noctiluca*) are well known; the bright flashing comes from photocytes along the edge of the umbrella at the base of the tentacles. *Aequorea,* a hydromedusan that is very abundant during the summer in the ocean off the northwest United States (San Juan Islands region), has been the material used for much of the research on the biochemistry of the system (Shimomura, 1985). The sea pansy, *Renilla,* which occurs near shore on sandy bottoms, has also figured importantly in the elucidation of the biochemistry of coelenterate luminescence (Cormier, 1981).

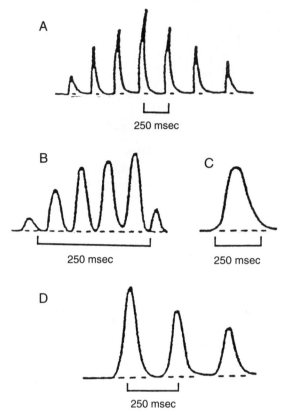

250 msec

250 msec          250 msec

250 msec

**FIG. 14.** Bioluminescent flashes. (A) A train of flashes in the hydrozoan *Obelia geniculata* following a single electrical stimulus. Spontaneous flashes from three species of fireflies: (B) *Photinus eva-nescens* (C) *P. marginellis*, and (D) *P. versicolor*.

## B. Biochemistry, Cell Biology, and the Control of Flashing

Photocytes occur as specialized cells located singly or in clusters in the endoderm. They are commonly controlled by epithelial conduction in hydropolyps and siphonophores and by a colonial nerve net in anthozoans. The light may be emitted as one to many flashes per stimulus. The putative neurotransmitter involved in luminescence control in *Renilla* is adrenaline or a related catecholamine.

The luciferin from coelenterates, coelenterazine, possesses an imidazolopyrazine skeleton (Fig. 15A). It is notable for its widespread phylogenetic distribution, but whether the reason is nutritional or genetic (hence, possible evolutionary relatedness) has not yet been elucidated. In some cases (e.g., *Renilla*), the sulfated form of luciferin may occur as a precursor or storage form and is convertible to active luciferin by sulfate removal with the cofactor 3′,5′-diphosphadenosine. The active form may also be sequestered by a $Ca^{2+}$-sensitive binding protein, analogous to the dinoflagellate binding protein. In this case $Ca^{2+}$ triggers the release of luciferin and flashing.

Another, more novel type of control of the reaction occurs in other cnidaria (e.g., *Aequorea*); this involves a luciferase–peroxyluciferin intermediate poised for the completion of the reaction. The photoprotein aequorin (Shimomura, 1985; Charbonneau *et al.,* 1985), isolated from

the jellyfish *Aequorea* (in the presence of EDTA to chelate calcium), emits light simply upon the addition of $Ca^{2+}$, which is presumably the trigger *in vivo* (Blinks *et al.,* 1982; Cormier *et al.,* 1989). This luciferin and the luciferase (EC 1.13.12.5) react with oxygen to form the peroxide in a calcium-free compartment (the photocyte), where it is stored. An action potential allows $Ca^{2+}$ to enter and bind to the protein, changing its conformational state and allowing the reaction to continue, but without the need for free oxygen at this stage. An enzyme-bound cyclic peroxide, a dioxetanone, is a postulated intermediate; it breaks down with the formation of an excited emitter, along with a molecule of $CO_2$. Coelenterate luciferase possesses homology with calmodulin (Lorenz *et al.,* 1991).

In several coelenterates the light emission occurs at a longer wavelength *in vivo* than *in vitro*. This is attributed to energy transfer from the excited luciferase-bound emitter to the fluorophore of the accessory green fluorescent protein (GFP), now widely used as a reporter of gene expression (Chalfie and Kain, 1997; Hastings *et al.,* 1997).

It had been reported in the early literature that coelenterates could emit bioluminescence without oxygen. The explanation is now evident: the animal contains the luciferase-bound peroxyluciferin (analogous to the bacterial flavin peroxide) in a stored and stable state, and only calcium is needed for the light-emitting step.

## VIII. Fireflies

### A. Occurrence and Function

Only about 100 genera of insects are classed as luminous out of a total of approximately 70,000 insect genera (Lloyd, in Herring, 1978). But when it occurs, the luminescence is impressive, most notably in the many species of beetles, fireflies, and their relatives. Fireflies themselves possess ventral light organs on posterior segments (Fig. 16). The South American railroad worm, *Phrixothrix*, has paired green lights on the abdominal segments and red head lights, while the click and fire beetles, *Pyrophorini*, have both running lights (dorsal) and landing lights (ventral). The dipteran cave glow worm (see Fig. 1) exudes beaded strings of slime from its ceiling perch, serving to entrap minute flying prey, which are attracted by the light emitted by the animal.

The variety of different fireflies, with their different habitats and behaviors, is impressive. The major function of light emission in fireflies is for communication during courtship, in which one sex emits a flash as a query, to which the other responds, usually in a species-specific pattern (Case, 1984; Lloyd, 1977, 1980). Some flashing patterns are shown in Figs. 14B–14D. Two signal system types have been distinguished. In the first, one sex (usually female) is stationary and emits light or a flashing signal and the other sex is attracted to it. In the second, one sex (usually the flying male), emits a species-specific flash, while the other emits a species-specific flash response. The time delay between the two may be a signaling feature; for example, it is 2 sec in some North American species. But the flashing pattern (e.g., trains distinctive in duration and/or intensity)

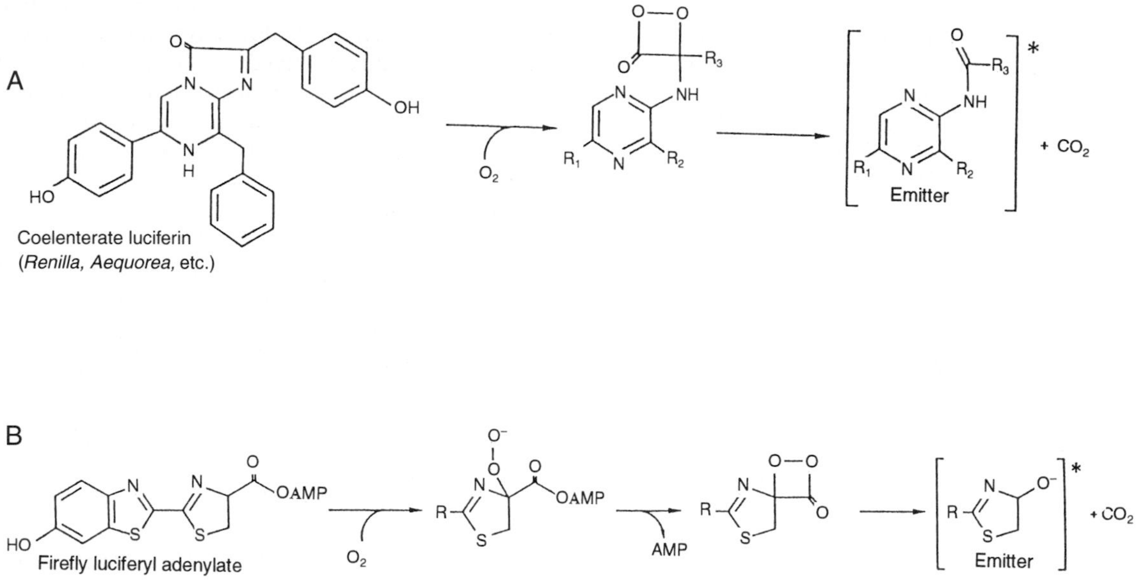

**A**

Coelenterate luciferin
(*Renilla, Aequorea*, etc.)

Emitter

$+ CO_2$

**B**

Firefly luciferyl adenylate

$O_2$

AMP

Emitter

$+ CO_2$

**FIG. 15.** (A) The structure of coelenterate luciferin (coelenterazine) and the light-emitting reaction, showing the postulated cyclic peroxide intermediate and excited state in the light emitting reaction. (B) The structures of firefly luciferin, showing the postulated cyclic peroxide intermediate and excited state.

**FIG. 16.** Ventral light organs of fireflies.

is also important in some cases, as is the kinetic character of the individual flash (duration; onset and decay kinetics). In some species, flickering occurs in the flashes, sometimes at frequencies higher than those detectable by the human eye ($\sim$40 Hz).

In other cases, the communication patterns and their interpretations are not as clear or as readily classified. Fireflies in Southeast Asia are particularly noteworthy in this respect, especially the synchronous flashing of *Pteroptyx* spp. These fireflies form congregations of many thousands in single trees, where the males produce an all-night-long display, with flashes every 1–4 sec, dependent on species (Buck and Buck, 1976). This appears to serve to attract females to the tree.

### B. Biochemistry

The firefly system was the first in which the biochemistry was well characterized. It had been known since before 1900 that cell-free extracts could continue to emit light for several minutes or hours, and that after the complete decay of the light, emission could be restored by adding a second extract, prepared using boiling water to extract the cells (then cooled). The enzyme luciferase was assumed to be in the first (cold water) extract (with all the luciferin substrate being used up during the emission), whereas since the enzyme was denatured by the hot water extraction, some substrate was left intact. This was referred to as the **luciferin–luciferase reaction,** and it was already known in the first part of this century that luciferins and luciferases from the different major groups would not cross-react, indicative of their independent evolutionary origins (Harvey, 1952).

In 1947 it was discovered that the addition of adenosine triphosphate (ATP), a "high-energy" intermediate, to an "exhausted" cold water extract resulted in an enormous bioluminescence response. This response showed that luciferin had not actually been consumed in the cold water extract; ATP could not be the emitter, since it did not have the appropriate fluorescence. For sometime, ATP was believed to be providing the energy for light emission. But, as noted already, the energy available from ATP hydrolysis is only about 7 kcal per mole, whereas the energy of a visible photon is 50 kcal per mole or more. It was soon discovered that firefly luciferin, later shown to be a unique benzothiazole (Fig. 15B), was still present in large amounts in the exhausted cold water extract and that it was ATP that was consumed, but available in the hot water extract. ATP was shown to be required to form the luciferyl adenylate intermediate, which then reacted with oxygen to form a cyclic luciferyl peroxy species, which broke down to yield $CO_2$ and an excited state of the carbonyl product (McElroy and DeLuca, in Herring, 1978). Luciferase catalyzes both the luciferin activation and the subsequent steps leading to the excited product.

In reactions in which luminescence has decreased to a low level (this may continue for days), it was long ago found that emission is greatly increased by coenzyme A, but the reason for this was obscure. The discovery that long-chain acyl-CoA synthetase (EC 6.2.1.3) has homologies with firefly luciferase (EC 1.13.12.7) both explains this observation and indicates the evolutionary origin of the gene.

Firefly luciferase has been cloned and expressed in other organisms, including *Escherichia coli* and tobacco. In both cases, luciferin must be added exogenously; tobacco plants "light up" when the roots are immersed in luciferin (Ow *et al.,* 1986). There are some beetles in which the light from different organs is a different color. The same ATP-dependent luciferase reaction with the same luciferin occurs in the different organs, but the luciferases are slightly different, coded by different (but homologous) genes (Wood *et al.,* 1989). They are presumed to differ with regard to the site that binds the luciferin, which could thereby alter the emission wavelength.

### C. Cell Biology and Regulation of Flashing

The firefly light organ comprises a series of photocytes arranged in a rosette, positioned radially around a central tracheole, which supplies oxygen to the organ. The organ itself comprises a series of such rosettes, stacked side by side in many dorso ventral columns. Photocyte granules or organelles containing luciferase have been identified with peroxisomes on the basis of immunochemical labeling.

It is still not known how flashing is controlled in fireflies (Case and Strause, in Herring, 1978). Although flashing is initiated by a nerve impulse that travels to the light organ, most of the nerve terminals in the light organ are not on photocytes but on special tracheolar cells, which control the supply of oxygen. This accounts for the considerable time delay between the arrival of the nerve impulse at the organ and the onset of the flash. The possibility that the flash is somehow directly triggered by an action potential thus seems unlikely. Also, none of the ions typically gated by membrane potential changes ($Na^+$, $K^+$, and $Ca^{2+}$) ap-

pear to be likely candidates for controlling luminescence chemistry. An alternate theory is that the flash is controlled by the availability of oxygen, which is required in the luminescence reaction. A comparison of different species shows a strong positive relationship between the extent of the tracheal supply system in the adults and the flashing ability. On the other hand, the mechanism must account for the rapid kinetics, complex waveforms, multiple flashes, and high-frequency flickering, all of which seem unlikely to be regulated by a gas in solution. However, although oxygen might diffuse slowly, it reacts very rapidly chemically in this system. The half rise time of luminescence with the anaerobic enzyme system (luciferase–luciferyl adenylate) is about 10 msec. The mechanism is still being actively investigated.

## IX. Other Organisms: Other Chemistries

The four systems already described are known best, but several others have been studied in some detail. Several of these are briefly described below.

### A. Mollusks

Snails (gastropods), clams (bivalves), and cephalopods (squid) have bioluminescent members (Young and Bennett, 1988). The squid luminous systems are by far the most numerous and diverse, in both form and function, rivaling the fishes in these respects. As is also true for fishes, some squid use symbiotic luminous bacteria, but most are self-luminous, indicating that bioluminescence had more than one evolutionary origin within the class.

Many squid possess photophores, which may be used in spawning and other interspecific displays (communication). Photophores are compound structures with associated optical elements such as pigment screens, chromatophores, reflectors, lenses, and light guides. They may emit different colors of light and are variously located near the eyeball, on tentacles, on the body integument, or associated with the ink sac or other viscera. In some species, luminescence intensity has shown to be regulated in response to changes in ambient light, indicative of a camouflage function.

Along the coasts of Europe a clam, *Pholas dactylus,* inhabits compartments that it makes by boring holes into the soft rock. When irritated, these animals produce a bright cellular luminous secretion, squirted out through the siphon as a blue cloud. This animal and its luminescence have been known since Roman times, and the system was used by DuBois in his discovery and description of the luciferin–luciferase reaction in the 1880s (see Section VIII). Well ensconced in its rocky enclosure, the animal presumably uses the luminescence to somehow deter or thwart would-be predators.

The *Pholas* reaction has been studied extensively, but the structure of the luciferin, which involves a protein-bound chromophore, remains unknown. The luciferase is a copper-containing large (>300-kDa) glycoprotein (Henry *et al.,* 1975). It can serve as a peroxidase with several alternative substrates, indicating the involvement of a peroxide in the light-emitting pathway; the superoxide ion is apparently involved in the reaction.

There are luminous species in several families of gastropods; a New Zealand pulmonate limpet, *Latia neritoides,* is notable as the only known luminous eukaryote that can be classed as a truly freshwater species. It also secretes a bright luminous slime (green emission, $\lambda_{max} = 535$ nm), whose function may be similar to that of the *Pholas* emission. Its luciferin is an enol formate of an aldehyde, but the emitter and products in the reaction are unknown; in addition to its luciferase ($M_r$ ~170 kDa; EC 1.14.99.21), a "purple protein" ($M_r$ ~40 kDa) is required, but only in catalytic quantities, suggesting that it may be somehow involved as a recycling emitter.

## B. Annelids

The annelids also include many luminous species, both marine and terrestrial (Herring, in Herring 1978). *Chaetopterus* are marine polychaetes that construct and live in U-shaped tubes in sandy bottoms; they also exude luminescence upon stimulation, but the chemistry of the reaction has eluded researchers. Other marine polychaetes include the Syllidae, such as the Bermuda fireworm mentioned earlier, and the polynoid worms, which shed their luminous scales as decoys. Extracts of the latter have been shown to emit light upon the addition of superoxide ion.

More but still limited knowledge is available concerning the biochemistry of the reaction in terrestrial earthworms, some of which are quite large, over 60 cm in length (Wampler, in DeLuca and McElroy, 1981). Upon stimulation they exude coelomic fluid from the mouth, anus, and body pores. This exudate contains cells that lyse to produce a luminous mucus, emitting in the blue-green region. However, the exudate from animals deprived of oxygen does not emit, but will do so after the admission of molecular oxygen to the free exudate. In *Diplocardia longa* the cells responsible for emission have been isolated; luminescence in extracts involves a copper containing luciferase ($M_r$ ~300 kDa), and the luciferin (*N*-isovaleryl-3-amino-1 propanal). The *in vitro* reaction requires $H_2O_2$, not free $O_2$.

## C. Crustaceans

Many crustaceans are luminescent (Herring, 1985b). The cypridinid ostracods such as *Vargula* (formerly *Cypridina*) *hilgendorfii* are small organisms that possess two glands with nozzles from which the luciferin and luciferase (EC 1.13.12.6) are squirted into the seawater, where they react and produce a spot of light, useful either as a decoy or for communication.

Cypridinid luciferin and its reaction have differences and similarities to those of the coelenterazine system (Cormier, in Herring, 1978). The luciferin in both is a substituted imidazopyrazine nucleus that reacts with oxygen to form an intermediate cyclic peroxide, which then breaks down to yield $CO_2$ and an excited carbonyl. However, the cypridinid luciferase gene has been cloned and appears to have no homologies with the gene for the corresponding coelenter-

ate proteins, and calcium is not involved in the cypridinid reaction. The two different luciferases reacting with similar luciferins have apparently had independent evolutionary origins, indicative of convergent evolution at the molecular level.

Euphausiid shrimp possess compound photophores with accessory optical structures and emit a blue ventrally directed luminescence. The system is unusual because both luciferase and luciferin cross-react with the dinoflagellate system. This cross taxon similarity indicates another possible exception to the rule that luminescence in distantly related groups had independent evolutionary origins. The shrimp might obtain luciferin nutritionally, but the explanation for the occurrence of functionally similar proteins is not evident. One possibility is lateral gene transfer; convergent evolution is another. Analyses of gene structures for homologies should provide insight into this question.

## D. Fish

Bioluminescence in fish is highly diverse and occurs in both teleost (bony) and elasmobranch (cartilaginous) fish. Partly because animals have not been so readily available, relatively little is known about their physiology and biochemistry, but many have been described (Herring, 1982).

As noted in the section on bacteria, many fish obtain their light-emitting ability by culturing luminous bacteria in special organs (see Figs. 2 and 5), but most are self-luminous. Self-luminous species include *Porichthys,* the midshipman fish, so-called because of the array of its photophores distributed linearly along the four pairs of lateral lines, much as are the buttons on a military uniform. Because it occurs close to shore, it has been the object of considerable study, and more is known about the physiological control of luminescence in *Porichthys* than in any other fish. Biochemically it is less well characterized, but its luciferin and luciferase cross-react with the cypridinid ostracod crustacean system already described. This was an enigma until it was discovered that Puget Sound fish have photophores but are nonluminous; however, they can emit if injected with cypridinid luciferin or fed the animals, showing that luciferin may be obtained nutritionally. Did the luciferase in this fish originate independently to make use of the available substrate, or was the ability to synthesize luciferin lost secondarily from a complete system? If the latter, this would be analogous to the loss of the ability to synthesize vitamins in mammals.

Open sea and midwater species include the sharks (elasmobranchs), some of which may have several thousand photophores. The teleosts include the gonostomatids such as *Cyclothone,* with simple photophores, and the hatchet fish, having compound photophores with elaborate optical accessories; emission is directed exclusively downward, indicative of a camouflage function of the light.

Fish often possess different kinds of photophores located on different parts of the body, particularly ventrally and around the eyes, evidently with different functions. One interesting arrangement, known in both midwater squid and myctophids, makes use of a special photophore positioned to shine on the eye or a special photoreceptor.

Its intensity parallels that of the other photophores, so it provides information to the animal concerning its own brightness, thus allowing it to match the intensity of its own counterillumination to that of the downwelling ambient light. Another clear case of functional use is in *Neoscopelus;* in addition to the many photophores on the skin, they also occur on the tongue, allowing it to attract prey to just the right location.

Sexual dimorphism is also frequent in fish luminescence. Males and females of the myctophid *Tarletonbeania* were originally thought to be different species. Only one (now known to be the male) has caudal luminous organs, and the occurrence of these fish was known only from stomach contents of predator fish; only the female could be captured in nets, but never together with a male. The apparent explanation is that when a predator attacks, the males dart off in all directions with their dorsal lights flashing, like a police car, leading the predators on a wild chase (and sometimes getting caught), but leaving the females, who remain in place, safe from the predator in the cover of darkness, yet easy to catch in a net.

A number of self-luminous fish eject luminous material; in the searsid fishes this is cellular in nature, but it is not bacterial, and its biochemical nature is not known. Such animals may also possess photophores.

## X. Applications of Bioluminescence

Instrumentation for measuring light emission is very sensitive and free of background and interference characteristic of many other analytical techniques (Wampler, in Herring, 1978; Kricka and Whitehead, 1984). A typical photon counting instrument can readily detect an emission of about $10^4$ photons sec$^{-1}$, which corresponds to the transformation of $10^5$ molecules sec$^{-1}$ (or $6 \times 10^6$ min$^{-1}$) if the quantum yield is 10%. (Those bioluminescent reactions that have been measured have quantum yields of this or higher; the firefly yield is about 90%.) A substance at a concentration of $10^{-9}$ $M$ could readily be detected in a single cell 10 $\mu$m in diameter by this technique.

For these and other reasons, luminescence has come into widespread use (Campbell *et al.,* 1994; Hastings *et al.,* 1997). Luminescent tags have been developed that replace radioactivity with as high a sensitivity. Since the different luciferase systems have different specific chemistries, quantitative determinations of many different specific substances can be accomplished. One of the first and still widely used assays uses firefly luciferase to detect ATP. Many different enzymes use or produce ATP, so their activities may be followed using this assay. With bacterial luciferase, any reaction that produces or utilizes NAD(H), NADP(H), or long-chain aldehyde, either directly or indirectly, can be coupled to this light-emitting reaction.

A photoprotein from the scaleworm has been used to detect superoxide ion; the purified photoprotein aequorin is widely used to detect intracellular $Ca^{2+}$ and its changes under various experimental conditions (e.g., during muscle contraction). The protein is relatively small, nontoxic, and may be readily injected into cells, reporting $Ca^{2+}$ over the range of $3 \times 10^{-7}$ to $10^{-4}$ $M$. The luminescence of bacteria has also been used as a very sensitive test for oxygen, making use of the fact that the $K_m$ for $O_2$ in that reaction is extremely low. An oxygen electrode incorporating luminous bacteria has been developed recently.

Luciferase genes have also been exploited as reporters for many different specific genes. Such systems are noninvasive and nondestructive, and the relevant activity can be measured in the living cell and in the same cell over the course of time. Recent studies of circadian rhythms have made use of the expression of the bacterial *lux* gene for the study of rhythms in cyanobacteria (Kondo *et al.,* 1995), the firefly *luc* gene for rhythms in higher plants (Millar *et al.,* 1995), and the *aequorin* gene for tracking intracellular calcium rhythms in *Arabidopsis* (Johnson *et al.,* 1995). As indicated earlier, the *gfp* gene is in wide use in many different types of applications (Chalfie and Kain, 1997).

These many diverse applications illustrate the fact that studies of bioluminescence, sometimes thought to be of little importance, have contributed to knowledge in ways not foreseen at the time the investigations were being made. Basic knowledge is a powerful tool in more than one respect.

## XI. Summary

The emission of visible light by living organisms is an unusual phenomenon, both in terms of its relative rarity and with respect to the biochemical and regulatory mechanisms involved. But where it does occur, bioluminescence is sometimes spectacular and can usually be inferred to have functional importance—a consequence of the fact that another organism detects and responds to the light.

The uses of the light may be classified under three headings: defense, offense, and communication. Light may be used defensively to startle or frighten (flashes), to divert predators, as a decoy, or to provide camouflage. Offensively, light may be used as a lure, to attract and convert would-be predators into prey. Communication occurs in courtship and mating displays.

Biologically, the single most striking aspect of bioluminescence is its wide and diverse phylogenetic distribution and the independent evolutionary origin of different systems. The ability to emit light occurs in some 13 different phyla, and apparently originated many times, perhaps 30 or more. This is reflected not only in the gene and protein structures, but also in its biological, biochemical, and functional diversity, as well as its sporadic phylogenetic distribution.

Another unusual and unexplained fact is that bioluminescence is primarily a marine phenomenon. Although there are terrestrial forms, it is virtually absent in fresh water; only one such species is known. It is also not confined to or especially prevalent in animals that live in complete darkness (caves, the deep ocean).

Bioluminescence is an enzymatically catalyzed chemiluminescence, a chemical reaction that emits light. The enzymes involved are all referred to generically as luciferases, somewhat unfortunately, because they are not conserved evolutionarily, and are thus structurally different in different groups of organisms. The genes coding for several of the luciferases have been cloned and sequenced, confirming that they possess no homologous regions in common. The substrates, generally called luciferins, react with molecular oxygen to form intermediate luciferase-bound peroxides, which break down to give a product in an excited state, which subsequently emits light.

In the marine environment, luminous bacteria are ubiquitous as planktonic forms in seawater, and they are also responsible for the light emission of many species of higher organisms, usually as symbionts. Terrestrial forms are not common, but do occur as symbionts in nematodes, as an agent in the nematode's parasitization of insects. All species use the same biochemical system, a shunt of the electron transport pathway, in which reduced flavin and aldehyde are oxidized by molecular oxygen to give a luciferase-bound flavin intermediate in an excited state. Luminous species are versatile with respect to the alternate uses of light emission in different situations, the exploitation of alternate habitats, and the capacity to turn the synthesis of the system on and off at both the physiological and genetic levels. The genes involved may be widely distributed and mobile.

Dinoflagellates are unicellular algae; in the ocean these organisms are largely responsible for the sometimes brilliant sparkling "phosphorescence" seen at night when the water is disturbed, and also for "red tides." Their luminescent flashing originates from novel cellular organelles called scintillons, formed as spherical outpocketings of the cytoplasm into the vacuole. They contain dinoflagellate luciferase and luciferin (a novel reduced linear tetrapyrrole), the latter bound to a second protein. Flashing is triggered by an action potential in the vacuolar membrane that causes a transient pH change in the scintillon, releasing the luciferin from its binding protein. Luminescence in some dinoflagellates is controlled by a cellular circadian biological clock, which causes the synthesis and destruction of the components to occur on a daily cycle.

There are many luminous comb jellies, hydroids, siphonophores, sea pens, and jellyfish. Upon stimulation, light is generally emitted as brief bright flashes or trains of flashes. The luminescence originates in specialized cells called photocytes, triggered by a conducted epithelial or nerve action potential. The luciferin is a substituted imidazole called coelenterazine, which, the name notwithstanding, occurs in several other phyla; its oxidation results in light emission. The reaction, catalyzed by coelenterate luciferase, is regulated by $Ca^{2+}$; this is a case where the evolutionary origin of the luciferase is known: it bears homology with calmodulin.

Fireflies typically emit light as flashes, which are used as species-specific signals for communication in courtship. The light organ is a complex structure with photocytes arranged in a rosette pattern, invested with tracheoles to transport the required molecular oxygen directly to the cells. Indeed, it may be that oxygen ultimately regulates flashing, though a nerve impulse initiates the process. The firefly reaction is unique in having a requirement for ATP, which serves to "activate" the luciferin (a benzothiazole).

The luciferyl adenylate is thus the "true" substrate that reacts with oxygen, forming an intermediate cyclic peroxide whose breakdown results in light emission. Firefly luciferase also shows homologies with another protein, namely, long-chain acyl-CoA synthetase.

Other major luminous groups include the mollusks (snails, clams, squid), annelid worms (both marine and terrestrial), crustacea (shrimp and ostracods), echinoderms (brittle stars, starfish, sea cucumbers), and fish, both cartilaginous (sharks) and teleost (bony fishes). Of all the groups, fish and squids have the greatest variety of luminous systems; some make use of symbiotic luminous bacteria as a source of light, whereas others are self-luminous. Luminous organisms are most abundant at midwater depths (500–1000 m) in the open ocean.

## Bibliography

Abrahams, L. V., and Townsend, L. D. (1993). Bioluminescence dinoflagellates: A test of the burglar alarm hypothesis. *Ecology* **74**, 258–260.

Anctil, M. (1987). Neural control of luminescence. *In* "Nervous Systems in Invertebrates" (M. A. Ali, Ed.), pp. 573–602. Plenum, New York.

Baldwin, T., and Ziegler, M. (1992). The biochemistry and molecular biology of bacterial bioluminescence. *In* "Chemistry and Biochemistry of Flavoenzymes" (F. Müller Ed.), Vol. 3, pp. 467–530. CRC Press, Boca Raton, FL.

Blinks, J. R., Wier, W. G., Hess, P., and Prendergast, F. G. (1982). Measurement of $Ca^{2+}$ concentrations in living cells. *Prog. Biophys. Mol. Biol.* **40**, 1–114.

Buck, J. B. (1938). Synchronous rhythmic flashing of fireflies. II. *Q. Rev. Biol.* **63**, 265–289.

Buck, J. B., and Buck, E. (1976). Synchronous fireflies. *Sci. Am.* **234**, 74–85.

Campbell, A. K. (1988). "Chemiluminescence: Principles and Applications in Biology and Medicine." Verlag Chemie, Weinheim.

Campbell, A. K., Kricka, L. J., and Stanley, P. E. (1994). "Bioluminescence and Chemiluminescence: Fundamentals and Applied Aspects." John Wiley & Sons, Chichester.

Case, J. (1984). Firefly behavior and vision. *In* "Insect Communication" (T. Lewis, Ed.), pp. 195–222. Royal Entomological Society of London, published for them by Harcourt Brace Jovanovich, New York.

Catalani, L. H., and Wilson, T. (1989). Electron transfer and chemiluminescence. Two inefficient systems: 1,4-Dimethoxy-9,10 diphenylanthracene peroxide and diphenoylperoxide. *J. Am Chem. Soc.* **111**, 2633–2639.

Chalfie, M., and Kain, S. (Eds.) (1997). "GFP: Green Fluorescent Protein Strategies and Applications." John Wiley & Sons, New York.

Charbonneau, H., Walsh, I., McCann, R., Prendergast, F., Cormier, M., and Vanaman, T. (1985). Amino acid sequence of the calcium-dependent photoprotein aequorin. *Biochemistry* **24**, 6762–6771.

Cody, C. W., Prasher, D. C., Westler, W. M., Prendergast, F. G., and Ward, W. W. (1993). Chemical structure of the hexapeptide chromophore of the *Aequorea* green-fluorescent protein. *Biochemistry* **32**, 1212–1218.

Cormier, M. J. (1981). *Renilla* and *Aequorea* bioluminescence. *In* "Bioluminescence and Chemiluminescence" (M. DeLuca and W. D. McElroy, Eds.), pp. 225–233. Academic Press, New York.

Cormier, M. J., Prasher, D. C., Longiaru, M., and McCann, R. O. (1989). The enzymology and molecular biology of the $Ca^{2+}$-activated photoprotein, aequorin. *Photochem. Photobiol.* **49**, 509–512.

DeLuca, M. (1978) (Ed.). *Methods Enzymol.* **57.**

DeLuca, M., and McElroy, W. D. Eds., (1981). "Bioluminescence and Chemiluminescence." Academic Press, New York.

DeLuca, M., and McElroy, W. D. (1986). *Methods Enzymol.* **133.**

Fuqua, W. C., Winans, S. C., and Greenberg, E. P. (1994). Quorum sensing in bacteria: The LuxR-LuxI family of cell density-responsive transcriptional regulators. *J. Bact.* **176,** 269–275.

Govindjee, Amesz, J., and Fork, D. C. (1986). "Light Emission by Plants and Bacteria." Academic Press, New York.

Grober, M. S. (1988). Brittle star bioluminescence functions as an aposematic signal to deter crustacean predators. *Anim. Behav.* **36,** 493–501.

Harvey, E. N. (1952). "Bioluminescence." Academic Press, New York.

Hastings, J. W. (1976). Bioluminescence. *Oceanus* **19,** 17–27.

Hastings, J. W. (1983). Biological diversity, chemical mechanisms and evolutionary origins of bioluminescent systems. *J. Mol. Evol.* **19,** 309–321.

Hastings, J. W. (1987). Why luminous bacteria emit bright light—and sometimes do not. *In* "Perspectives in Microbial Ecology," (F. Megusar and M. Ganter, Eds.). pp. 249–252. Ljubljana, Yugoslavia.

Hastings, J. W., and Morin, J. G. (1991). Bioluminescence. *In* "Neural and Integrative Animal Physiology" (C. L. Prosser, Ed.), pp. 131–170. Wiley-Interscience, New York.

Hastings, J. W., Potrikus, C. J., Gupta, S., Kurfürst, M., and Makemson, J. C. (1985). Biochemistry and physiology of bioluminescent bacteria. *Adv. Microb. Physiol.* **26,** 235–291.

Hastings, J. W., Makemson, J., and Dunlap, P. V. (1987). How are growth and luminescence regulated independently in exosymbionts? *Symbiosis* **4,** 3–24.

Hastings, J. W., Boulos, Z., and Rusak, B. (1991). Circadian rhythms. *In* "Neural and Integrative Animal Physiology" (C. L. Prosser, Ed.), pp. 435–546. Wiley-Interscience, New York.

Hastings, J. W., Kricka, L., and Stanley, P. (Eds.) (1997). "Bioluminescence and Chemiluminescence: Molecular Probes." John Wiley & Sons, Chichester.

Henry, J. P., Monny, C., and Michelson, A. M. (1975). Characterization and properties of *Pholas* luciferase as metalloglycoprotein. *Biochemistry* **14,** 3458–3466.

Herring, P. J. (Ed.) (1978). "Bioluminescence in Action." Academic Press, New York.

Herring, P. J. (1982). Aspects of bioluminescence in fishes. *Oceanogr. Mar. Biol. Annu. Rev.* **20,** 415–470.

Herring, P. J. (1985a). How to survive in the dark: Bioluminescence in the deep sea. *In* "Physiological Adaptations of Marine Animals" (M. S. Laverack, Ed.), Soc. Exp. Biol. Symp. No. 39, pp. 323–350. The Company of Biologists Ltd., Cambridge, UK.

Herring, P. J. (1985b). Bioluminescence in the crustacea. *J. Crustacean Biol.* **5**(4), 557–573.

Herring, P. J. (1987). Systematic distribution of bioluminescence in living organisms. *J. Biolumin. Chemilumin.* **1,** 147–163.

Herring, P. J. (1990). Bioluminescent communication in the sea. *In* "Light and Life in the Sea" (P. J. Herring, A. K. Campbell, M. Whitfield and L. Maddock, Eds.), pp. 245–264. Cambridge University Press, London/New York.

Johnson, C. H., Knight, M., Kondo, T., Masson, P., Sedbrook, J. Haley, A., and Trewavas, A. (1995). Circadian oscillations of cytosolic and chloroplastic free calcium in plants. *Science* **269,** 1863–1865.

Kondo, T., Strayer, C., Kulkarni, R., Taylor, W., Ishiura, M., Golden, S., and Johnson, C. (1993). Circadian rhythms in prokaryotes: Lucif-erase as a reporter of circadian gene expression in cyanobacteria. *Proc. Natl. Acad. Sci. USA* **90,** 5672–5676.

Kricka, L., and Whitehead, T. P. (Eds.) (1984). "Analytical Applications of Bioluminescence and Chemiluminescence." Academic Press, New York.

Lloyd, J. E. (1977). Bioluminescence and communication. *In* "How Animals Communicate" (T. A. Sebeok, Ed.), pp. 164–183. Indiana University Press, Bloomington.

Lloyd, J. E. (1980). Firefly signal mimicry. *Science* **210,** 669–671.

Lorenz, W. W., McCann, R. O., Longiaru, M., and Cormier, M. J. (1991). Isolation and expression of a cDNA encoding *Renilla reniformis* luciferase. *Proc. Natl. Acad. Sci. USA* **88,** 4438–4442.

McFall-Ngai, M., and Morin, J. W. (1991). Camouflage by disruptive illumination in leiognathids, a family of shallow-water, bioluminescent fishes. *J. Exp. Biol.* **156,** 119–137.

Meighen, E. A. (1991). Molecular biology bacterial bioluminescence. *Microbiol Rev.* **55,** 123–142.

Meighen, E. A., and Dunlap, P. V. (1993). Physiological, biochemical and genetic control of bacterial bioluminescence. *Adv. Microb. Physiol.* **34,** 1–67.

Millar, A. J., Carré, I. A., Strayer, C. A., Chua, N.-H., and Kay, S. A. (1995). Circadian clock mutants in Arabidopsis identified by luciferase imaging. *Science* **267,** 1161–1163.

Morin, J. (1983). Coastal Bioluminescence: Patterns and functions. *Bull. Mar. Sci.* **33,** 787–817.

Morin, J. G., and Cohen, A. C. (1991). Bioluminescencent displays, courtship and reproduction in ostracodes. In "Crustacean Sexual Biology" (R. Bauer and J. Martin, Eds.), pp. 1–16. Columbia University Press, New York.

Morse, D., Fritz, L., and Hastings, J. W. (1990). What is the clock? Translational regulation of circadian bioluminescence. *Trends Biochem. Sci.* **15,** 262–265.

Nealson, K., and Hastings, J. W. (1991). The luminous bacteria. *In* "The Prokaryotes" 2nd ed., Volume I, Part 2, Chap. 25 (A. Balows, H. G. Trüper, M. Dworkin, W. Harder and K. H. Schleifer, Eds.), pp. 625–639. Springer-Verlag, New York.

Nicolas, M.-T., Nicolas, G., Johnson, C. H., Bassot, J.-M., and Hastings, J. W. (1987). Characterization of the bioluminescent organelles in *Gonyaulax polyedra* (dinoflagellates) after fast-freeze freeze fixation and antiluciferase immunogold staining. *J. Cell Biol.* **105,** 723–735.

Ow, D. W., Wood, K. V., DeLuca, M., de Wet, J. R., Helinski, D. R., and Howell, S. H. (1986). Transient and stable expression of the firefly luciferase gene in plant cells and transgenic plants. *Science* **234,** 856–859.

Schuster, G. B. (1979). Chemiluminescence of organic peroxides. *Acc. Chem. Res.* **12,** 366–373.

Shimomura, O. (1985). Bioluminescence in the sea: Photoprotein systems. *Soc. Exp. Biol. Symp.* **39,** 351–372.

Wilson, T. (1985). Mechanism of chemiluminescence. *In* "Singlet Oxygen" (A. Frimer, Ed.), Vol. 2, pp. 37–57. CRC Press, Boca Raton, FL.

Wilson, T. (1995). Comments on the mechanisms of chemi- and bioluminescence. *Photochem. Photobiol.* **62,** 601–606.

Wood, K. V., Lam, Y. A., Seliger, H. H., and McElroy, W. D. (1989). Different cDNAs elicit bioluminescence of different colors. *Science* **244,** 700–702.

Young, R. E., and Bennett, T. M. (1988). Cephalopod luminescence. *In* "The Mollusca" (M. R. Clarke and E. R. Trueman, Eds.), Vol. 12, pp. 241–251. Academic Press, New York.

# SECTION

# IX

## Cell Division and Programmed Cell Death

Andrés A. Gutiérrez

# 65

# Regulation of Cell Division in Higher Eukaryotes

## I. Introduction

The cells of a living organism must choose either one of three pathways: live and reproduce, live without reproducing, or die. Immortality could only happen in the case that ever-dividing cells (e.g., cancer cells) could count on an endless nutrient supply; but since this is an unprobable situation, they eventually die too. Interestingly, cell division has been found to be tightly controlled in association with other fundamental events in the cell's life: **differentiation, senescence,** and **cell death** (e.g., apoptosis). Deregulation of these processes has been involved in the origin of pathological conditions, such as malignancy and autoimmune disorders. How and by which pathways the cell cycle is controlled have been central questions since the **cell theory** was proposed more than a century ago.

Research has been mainly oriented toward the phase of the cycle in which the cell divides into two replicas of itself, that is, **mitosis.** In contrast, the other part of the cell cycle, known as **interphase,** remained largely unexplored. Late in the 1960s, the biochemical description of the **maturation promoting factor** (MPF) in amphibian eggs became the first clue as to what controls the cell cycle passage from interphase to mitosis (reviewed by Norbury and Nurse, 1992). It is now known that this factor is a dimer composed of a **protein kinase unit, Cdc2,** and its **regulatory subunit** known as **cyclin.** Further studies revealed the existence of an homologous protein kinase, Cdc28, in budding yeast. However, in this particular case, the kinase was responsible for regulating all the phases of the cell cycle by binding to different cyclins at different periods of time.

From these initial observations, a large number of proteins have recently been cloned and characterized for each one of the phases. These proteins include various **cyclins, Cyclin-Dependent Kinases (CDKs), protein kinase inhibitors** and **transcription factors** such as the products of **tumor suppressor genes,** p53 and pRB. All of these molecules have been shown to interact in critical ratios to determine the passage from one phase to another in the cell cycle. We could conceive this process as smooth oscillations of proteins that must reach threshold concentrations for performance. The final goal is that these signals occur in a specific spatiotemporal sequence to ensure a high-fidelity transmission of genetic information.

Unfortunately, as the list of participant molecules keeps enlarging, so does the complexity of understanding how the cell cycle is controlled. Nevertheless, the cell cycle machinery of the eukaryotic cells studied to date, from yeasts to humans, has been well conserved throughout evolution. This similarity has greatly facilitated the characterization of, for example, presumed human cyclin genes in reconstitution assays by its ability to rescue mutant strains of defective yeasts. Other powerful biochemical and genetic approaches have also contributed to fully characterize these genes and their products: (1) from transcriptional control to posttranslational modifications, (2) from biochemical assays to crystallographic analysis of the interactions between cyclins and CDKs, or (3) from physiological responses in cultured cells to those occurring in transgenic mice (Table 1). Moreover, by using standard chemical kinetic theory, a mathematical model of the M phase in cell-free extracts from *Xenopus* eggs has already been developed (Tyson *et al.,* 1996).

Considering the limitation of space and the overwhelming amount of information available, this chapter deals exclusively with novel concepts on the regulation of the mitotic cell cycle of higher eukaryotic cells. To complement this information, recent reviews on each subject have been mentioned throughout the chapter. Finally, neither the meiotic cell cycle nor studies in other eukaryotic cells (e.g., yeasts) are analyzed in detail. A comprehensive review on these topics can be found elsewhere (Murray and Hunt, 1993; Kleckner, 1996).

## II. General Overview

The **mitotic cell cycle** of higher eukaryotic cells consists of four phases (Fig. 1A): (1) **gap phase 1,** or **G₁ phase,** when the cell grows and prepares to synthesize DNA;

**1003**

**TABLE 1** Common Strategies to Identify and Characterize Key Molecules of the Cell Cycle

**Gene cloning strategies**
    Yeast two hybrid screens
    Positional cloning
    Subtractive hybridization

***In Vitro* assays**
    Kinase (phosphorylating) activity
        Substrates: Histone H1, pRB,[a] p107
    Immunoprecipitations of complexes (e.g., cyclin–CDK–CDK inhibitors)
    Ubiquitination assays
    Fluorescence *in situ* hybridization (FISH)
    Crystallographic analysis

***In Vivo* assays**
    Reconstitution assays in defective mutant yeasts and insects
    Cells in culture
        DNA synthesis–inhibitory activity (e.g., thymidine incorporation)
        Cell cycle analysis (e.g., FACS[b])
        Genetic overexpression and down-regulation (e.g., Ab[c] and antisense)
    Transgenic mice
        Gain of function or knock-out

---

[a] pRB, retinoblastoma protein.
[b] FACS, fluorescence-activated cell sorter.
[c] Ab, antibody.

(2) **S phase,** when DNA synthesis takes place and replicates the whole set of chromosomes; (3) a second gap period, **G$_2$ phase,** in which the cell prepares for division; and (4) the mitotic, or **M phase,** in which the actual division of the original cell into two daughter cells takes place. Mitosis also consists of various phases: (1) **prophase,** when chromosome condensation takes place; (2) **prometaphase–metaphase,** in which two members of each pair of sister chromatids attach to microtubules; (3) **anaphase,** when sister chromatids are separated and move along the microtubules; and (4) **telophase,** with the reformation

of nuclei and the decondensation of the chromosomes (Fig. 1B).

At the exit from mitosis, and depending on the circumstances, the new daughter cells (1) reenter the G$_1$ phase and keep cycling, (2) remain temporarily quiescent in a stage named the **G$_0$ phase,** or (3) are terminally differentiated and unable to reenter the cell cycle.

To ensure the high-fidelity transmission of genetic information, eukaryotes have developed a complicated network of molecules that act in a finely coordinated manner during the cell cycle. Central participants of the cell cycle machinery are **CDK complexes.** The latter contain a catalytic subunit, the CDK, and a regulatory subunit, a cyclin. Once the complexes have been activated, they are able to phosphorylate diverse substrates and, directly or indirectly, control gene expression and the progression of the cell cycle. As a feedback mechanism, cyclin–CDK complexes are negatively regulated by two different families of **CDK inhibitors:** (1) CDK-inhibitor proteins, **Cip/Kip,** and (b) Inhibitors of kinases, **Ink** (see Section III,B).

Moreover, the cell counts on at least three "checkpoints" to safeguard the normal progression of the cell cycle: at G$_1$/S, at G$_2$/M, and at the exit of mitosis (Pines, 1994). The term "checkpoint" was originally coined by Hartwell and Weinert (1989) in relation to the gene products that negatively regulate the cell cycle in response to DNA damage in yeasts. In their experiments in yeasts, cell cycle arrest occurred in response to chromosome damage by X-rays, and this arrest was mediated by the expression of the *RAD 9* gene. Thus, the term was used in the sense of a "border" that could not be crossed in the presence of DNA damage. In this case, checkpoints could be considered monitors of the physical integrity of chromosomes that coordinate cell cycle transitions. However, other authors have also used the term "checkpoint" in relation to biochemical pathways that ensure that the initiation of particular cell cycle events is dependent on the successful completion of others.

To help the reader through this general overview, Fig. 2 shows the most important mechanisms that control the mitotic cell cycle in higher eukaryotic cells. Tables 2 and 3 list the number of amino acid residues, the NCBI accession

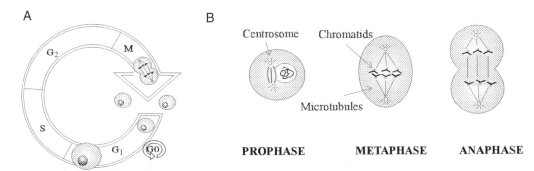

**FIG. 1.** Phases of cell cycle: (A) Interphase consists of three phases known as G$_1$, S, and G$_2$, in which the cell prepares for cell division. The latter takes place during mitosis. (B) Phases of mitosis. See text for details.

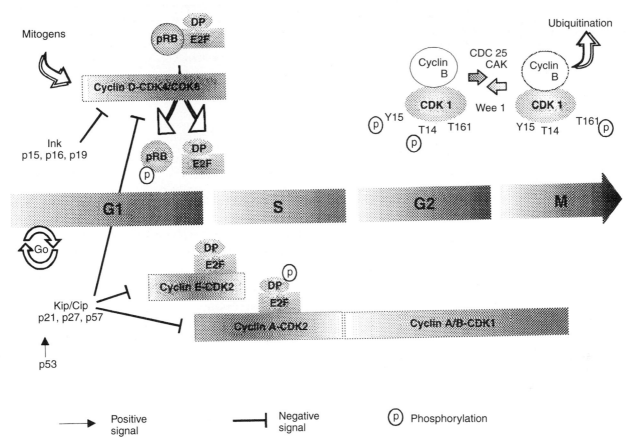

**FIG. 2.** Regulation of the mitotic cell cycle in higher eukaryotic cells. The sequential formation, activation, and inactivation of various cyclin–CDK complexes govern the progression of the cell cycle. The timing of the events herein depicted and the patterns of interaction among the different members of this machinery are representative and may differ for some models. Negative regulation of the cell cycle occurs at checkpoints $G_1/S$, $G_2/M$, and exit of mitosis. CDK, cyclin-dependent kinase; E2F-DP, transcription factor; pRB, retinoblastoma protein; Ink, Inhibitors of kinase; Kip/Cip, CDK-kinase inhibitor protein; CAK, CDK-activating kinase; Cdc25, dual specificity phosphatase; Wee1 phosphorylase; T14, Thr 14; Y 15, Tyr 15; T161, Thr 161. Of note, pRB has more than a dozen phosphorylation sites on either serine or threonine residues. For further details, see Section II.

numbers,[1] and the main activities of the participants in charge of regulating this process.

## A. Regulation of $G_1$/S Phase Transition

The driving force that makes cells cycle comes from an intricate network of negative and positive signals (Müller *et al.*, 1993). When the ratio of growth stimulatory signals to growth inhibitory signals increases, the cell cycle is "switched on." For simplicity, these signals have been referred to as **mitogens** in Fig. 2.

Mitogens induce the synthesis of **D-type cyclins** (D1, D2, D3), which are short-lived proteins ($t_{1/2} < 25$ min). The highest cyclin D levels are reached in the late $G_1$ phase, and they rapidly decay after the cell has moved into

the S phase. Thereafter, D cyclins are no longer needed for the completion of the cell cycle (Sherr, 1995). The importance of these cyclins is that they become transducers of signals from the environment by acting as regulator subunits for cyclin-dependent kinases CDK2, CDK4, and CDK6.

As previously mentioned, the activation of CDKs regulates progression through critical transitions in the cell cycle (Morgan, 1995; Nigg, 1995). Two distinct types of CDKs control entry into S phase: CDK4/CDK6 and CDK2. In addition to cyclin binding, CDK4 and CDK2 require phosphorylation by a **CDK-activating-kinase** (CAK) to become fully active. In contrast, inactivation of either kinase activity leads to cessation of proliferation and withdrawal from the mitotic cycle.

As D cyclins accumulate, they associate with CDK4 and CDK6, and form active holoenzyme complexes, that is, cyclin D–CDK4 and cyclin D–CDK6. These complexes are critical because they facilitate the exit from $G_1$ by phosphorylating a master regulatory protein of the cell cycle at the $G_1$/S transition: the **retinoblastoma protein**

---

[1] NCBI accession numbers refer to the loci of the protein sequences at the National Center for Biotechnology Information, NIH, USA. These sequences are compiled from different databases (e.g., GenBank, EMBL, PIR), and the complete list can be obtained from the NCBI-NIH server: http://www.ncbi.nlm.nih.gov.

**TABLE 2** Human Cyclins and Cyclin-Dependent Kinases

| | aa[a] | Accession[b] | Binds | Synonyms/functions |
|---|---|---|---|---|
| **Cyclin** | | | | |
| A | 432 | 116169 | CDK1, CDK2<br>p107 | $G_2/M$ specific cyclin<br>$G_2/M$ cyclins are abruptly destroyed at mitosis |
| B | 433 | 116176 | CDK1 | $G_2/M$ specific cyclin<br>MPF = cyclin B–CDK 1 complex |
| C | 303 | 1117984 | CDK2, CDK1 | CCNC gene. $G_1/S$ specific cyclin |
| D1 | 295 | 116152 | CDK4<br>CDK6 | Cyclin PRAD1 encoded by the bcl-1 linked oncogene<br>CCND gene [11q13].[c] Control the $G_1/S$ transition |
| D2 | 289 | 231741 | CDK2, CDK4/6 | Control the $G_1/S$ transition |
| D3 | 292 | 231743 | CDK2, CDK4/6 | Control the $G_1/S$ transition |
| E | 395 | 116154 | CDK2<br>CDK3 | CCNE gene<br>Nuclear protein essential for the $G_1/S$ transition |
| F | 786 | 1082313 | ? | CCNF gene [16p13.3]. Nuclear protein<br>Activity during interphase with peak levels in $G_2$ |
| G1 | 295 | 1236913 | ? | Damage inducible genes (p53 dependent) |
| G2 | 344 | 1236915 | ? | Damage inducible genes (p53 dependent) |
| H | 323 | 1090760 | CDK7 | CAK (MO15)-regulatory subunit along with MAT1 |
| I | 377 | 1183162 | ? | High-constant levels in postmitotic tissues |
| **Other CDK regulators** | | | | |
| p35 | 307 | 1090763 | CDK5 | Neuronal specific–noncyclin regulator of CDK5 |
| PCNA | 261 | 129694 | CDK2, CDK4<br>CDK6 | Proliferating cell nuclear antigen<br>Auxiliary protein of DNA pol $\delta$<br>Controls eukaryotic DNA replication |
| p50 | 378 | 1421821 | CDK4 | p50/Cdc 37 protein kinase homolog<br>Targets hsp90 and both stabilize CDK4 |
| CKS1 | 79 | 461746 | CDK2 | p9. CDK regulatory subunit 1 |
| CKS2 | 79 | 461747 | CDK2 | p9. CDK regulatory subunit 2<br>CKS1 and 2 bind to the catalytic subunit of CDKs<br>Involved in CDK oligomerization ($\cong$6 kinase subunits) |
| MAT1 | 309 | 1089848 | CDK7<br>H | *Menage-à-trois 1:* CDK7-cyclin H assembly factor<br>p36–RING finger protein |
| Wee1 | 646 | 1085293 | CDK1 | Tyrosine protein kinase. Phosphorylates Cdc 2<br>Negative regulator of entry into mitosis at $G_2/M$ |
| **CDKs**[d] | | | | |
| CDK1 | 297 | 115922 | A, B<br>p9 | p34-Cdc 2. T loop contains Thr 161<br>Phosphorylates the C-terminus of RNApol II<br>Regulates the formation of the mitotic spindle<br>Essential for entry into S phase and mitosis |
| CDK2 | 298 | 116051 | A, E, D2, D3<br>E1A<br>Skp1/Skp2 | p33. T loop contains Thr 160<br>Adeno. E1A associated kinase<br>Maximum activity during S and $G_2$ phases |
| CDK3 | 305 | 231726 | E | Putative role in $G_1$ exit by activating E2F 1, 2, and 3 |
| CDK4 | 303 | 1168867 | D1–D3<br>p50/Hsp90 | PSJK3. Phosphorylates pRB<br>Essential for progression at $G_1/S$ transition<br>Constant levels throughout the cell cycle |
| CDK5 | 291 | 543973 | p35<br>D ? | PSSALRE. Homology to the CDK1/Cdc 28 family<br>Role of CDK 5/p35 in neuronal differentiation |
| CDK6 | 326 | 266423 | D1–D3 | PLSTIRE. Phosphorylates pRB<br>Essential for progression at $G_1/S$ transition |
| CDK7 | 346 | 468789 | MAT1<br>H | CDK-activating kinase, CAK (MO15) = CDK7-cyclin H-MAT-1. Phosphorylates: <u>CDK 2</u>-(cyclin E), <u>CDK 4</u>-(cyclin D) and Thr 161 in <u>CDK 1</u> kinase |
| CDK8 | 464 | 1345718 | D-type and C | Interacts with D type and C cyclins |

[a] aa, Number of amino acid residues.

[b] Accession, NCBI locus.[1]

[c] Brackets indicate chromosomal location (when available).

[d] Serine/threonine kinases that belong to the Cdc 2/Cdc 28 family.

**TABLE 3**  Human CDK Inhibitors and Pocket Proteins[a]

| Protein | aa[b] | Accession[c] | Synonyms/functions[d] |
|---------|------|-------------|----------------------|
| **Ink4 family** | | | |
| p16 | 148 | 1168868 | CDK4-inhibitor A (CDK4I) (p16-Ink4A) [9p21] <br> Multiple tumor suppressor 1 (MTS1) <br> Inhibits CDK4/6 interaction with cyclins D and their function <br> Negative regulator of the cell cycle at $G_1$ |
| p15 | 138 | 639716 | CDK4-inhibitor B (p14-Ink 4B) (p15-Ink 4B) (MTS2) [9p21] <br> Effector of TGF$\beta$-induced $G_1$ arrest <br> Inhibits CDK4/6 interaction with cyclins D and their function |
| p18 | 168 | 1168870 | CDK6 inhibitor (p18-Ink 6; Ink 4C). Weak interaction with CDK4 <br> Arrests cell cycle with a correlated dependence on pRB <br> Contain ANK repeat [1p32] |
| p19 | 166 | 1418221 | CDK inhibitor specific to CDK4 and CDK6 (p19-Ink4D) <br> Inhibitor of cyclin D-dependent kinases [19p13] |
| **Cip/Kip family** | | | |
| p21 | 164 | 729143 | CDK inhibitor 1 (CDKN1A, Cip1, Sdi1, Waf1, Cap20) <br> Potent "universal" inhibitor of G1 CDKs <br> Induction by DNA damage (p53 dependent), IFN, hypoxia, terminal differentiation, and senescence |
| p27 | 198 | 745448 | CDK inhibitor p27Kip1 [12p–12p13.1] <br> Potent inhibitor of G1 CDKs <br> N'-terminal (CDK–binding/inhibitory domain) similar to p21 <br> Inactivates cyclin E/A-CDK2 and cyclin D/CDK4 complexes <br> Mediates extracellular antimitogenic signals |
| p57 | 315 | 790248 | p57Kip2. Inhibitor of G1 CDKs [11p15.5] <br> Not a significant suppressor in adult cells <br> N'-terminal (inhibitory domain) similar to p21 |
| **Pocket proteins** | | | |
| p105 | 928 | 292421 | pRB, Retinoblastoma tumor suppressor gene [13q14] <br> Blocks cells in G1 phase. Phosphorylated by cyclin D-CDK4/6 <br> Inhibits E2F-DP activity and RNA pol I–III transcription <br> Binds various cellular transcription factors and viral proteins |
| p107 | 1068 | 449005 | RB-related p107 protein <br> Phosphorylated by cyclins A/E-CDK2 complexes and D kinases <br> Similar but nonredundant pRB actions |
| p130 | 1139 | 627364 | RB related p130 protein [16q12.2–13] <br> Same as in p107 |

[a] Human CDK inhibitors are classified in one of two families: Ink4 or Cip/Kip. Pocket proteins include the retinoblastoma suppressor gene and its related proteins, p107 and p130.

[b] aa refers to the number of amino acid residues of each protein.

[c] Accession refers to the NCBI locus.[1]

[d] Functions are also given the names by which the same protein might be known and the chromosome mapping in brackets.

(p105Rb). Indeed, cyclin D–kinase activity is the most important mechanism by which pRB is phosphorylated in mid to late $G_1$ phase. pRB is the product of a tumor suppressor gene and a member of a family referred to as **pocket proteins,** which includes p107 and p130. Like pRB, p107 and p130 are phosphorylated by various CDKs (Weinberg, 1995). When pRB proteins are phosphorylated by CDKs, they become "inactive" and release normally bound transcription factors known as **E2F-DP heterodimers.** The latter, in turn, mediate the progression toward S phase. Because pRB phosphorylation occurs 1–2 hr before the start of the S phase, a vast array of S-phase genes

is induced by the E2F-DP heterodimers during this period (La Thangue, 1994; Weinberg, 1996).

But the activation of CDKs (and their effects) does not depend only on cyclin abundance and on the phosphorylation state of their kinase subunits. As mentioned earlier, their activity is also controlled by the presence of negative regulators that belong to either one of two families: Inks or Cip/Kips (Pines, 1994; Sherr and Roberts, 1995). These two families of kinase inhibitors differ in structure and specificity. Three members of the **Cip/Kip family** have now been described, and they are known as **p21, p27,** and **p57,** according to their protein molecular weight. These molecules are considered "universal" inhibitors, since they can bind to and inhibit the kinase activity of various CDK2, CDK4, and CDK6 complexes. In this respect, the Cip/Kip family can be considered more promiscuous than the Ink family, since the latter only binds to cyclin D–CDK4 and cyclin D–CDK6 complexes. Members of the **Ink family** include proteins **p15, p16, p18,** and **p19.**

In addition to these events, the $G_1/S$ transition requires the presence of **cyclin E–CDK2** and **cyclin A–CDK2 complexes.** Just like cyclin D–kinase activity (cyclin D–CDK4/6), cyclin E–CDK2 complexes also participate in pRB phosphorylation. The levels of cyclin E–CDK2 complexes correlate with their kinase activity, that is, they are first observed in late $G_1$, peak at the $G_1/S$ transition, and disappear in early S. Thus, their activity precedes the S-phase role of cyclin A.

**Cyclin A** appears around the $G_1/S$ transition, and their levels gradually increase and peak at one of the phases of mitosis known as prophase. Then, cyclin A is abruptly degraded by ubiquitin-mediated proteolysis at the metaphase–anaphase transition of mitosis (Section III,D). This cyclin is found in the nucleus, and its expression occurs earlier than that of cyclin B. In contrast to the latter, cyclin A binds both CDK2 and CDK1 kinases. Thus, cyclin A–CDK2 complexes are formed as the synthesis of cyclin A progresses, reaching maximum concentration during S phase. In fact, these are not binary but quaternary complexes, because p107 proteins associate with cyclin A, whereas E2F-DP heterodimers associate with the CDK2 unit of these complexes. The important concept is that cyclin A–CDK2 holoenzymes are important in S-phase progression.

Novel families of cyclins have been recently described (i.e., F, G, and I), but their functions have not been fully characterized (see Section III,A,1).

Once cells enter into the S phase, they become fully committed to dividing, and the cell cycle continues without the presence of further external signals. The process of DNA synthesis, chromatin assembly, chromosome replication and telomerase activity is beyond the scope of this chapter and further information can be found elsewhere (Murray and Hunt, 1993; Felsenfeld, 1996; Wuarin and Nurse, 1996; Roth and Allis, 1996; Shay, 1996).

## B. Regulation of G₂/M Phase Transition

The $G_2/M$ transition is one of the best characterized phases of the cell cycle. In this phase, **cyclin A/B–CDK1** (i.e., cyclin A/B–Cdc2) complexes become the dominant players. The role of cyclin A has already been described, so only that of cyclin B is referred to in this section. The expression of this cyclin is transcriptionally regulated throughout the cell cycle. As shown in Fig. 2, it appears in S phase (mainly in the cytoplasm), and it is destroyed along with cyclin A at metaphase. Interestingly, cyclin B is imported into the nucleus just before nuclear envelope breakdown.

As cyclin B accumulates, cyclin B–CDK1 complexes are formed. Originally described as **maturating promoter factors,** these complexes require further posttranslational modifications to become fully active (Solomon, 1994; Tyson *et al.,* 1996). In higher eukaryotes, this activation requires phosphorylation of CDK1 at Thr 161, and dephosphorylation at Tyr 15–Thr 14 (see Fig. 2 and Table 2). The inhibitory phosphorylation at Thr 14 and at Tyr 15 is carried out by a **kinase, Wee1,** while a dual specificity **phosphatase, Cdc25 homolog,** carries out their dephosphorylation. Interestingly, both enzymes also require up-regulatory phosphorylation by a protein kinase that is active at M phase, probably CDK1. As the cell enters into mitosis, Wee1 decreases its kinase activity, while Cdc25 increases its phosphatase activity. At this time, the **CDK-activating kinase (CAK)** exerts its up-regulatory activity by phosphorylating the Thr 161 residue in the T-loop of CDK1.

Once the cyclin A/B–CDK1 complexes are active, they promote M-phase progression. It is during this phase that the actual division of the cell into two replicas of itself takes place. Evidence from years of descriptive and experimental work has established the existence of different phases in mitosis: prophase, prometaphase–metaphase, and transition to anaphase. Thus, there is an overwhelming number of studies on diverse events that occur during each one of these phases, including membrane partitioning (Warren, 1993), mitotic chromosome condensation (Hirano, 1995), and mitotic spindle formation and chromosome movement (McIntosh, 1991; Miyazaki and Orr-Weaver, 1994). One of the areas of major interest at present is the control and function of the mitotic spindle. This spindle consists of a dynamic array of microtubules and associated proteins, and it is responsible for segregation of chromosomes during mitosis. Various **microtubule motor proteins** (e.g., **kinesins**) have been found within the spindle, so they are generally considered as controllers of the segregation process. However, the importance of kinesins and related proteins in this process remains undetermined (reviewed in Walczak and Mitchison, 1996). Interestingly, the product of the **tumor suppressor gene,** p53, has also been implicated in a **spindle checkpoint** that controls the ploidy of the cells. In fact, cells lacking p53 display high rates of aneuploidy and gene amplification (Cross *et al.,* 1995).

## C. Exit from Mitosis

At the end of the cell cycle, during metaphase, cyclin A and B levels decrease abruptly through **ubiquitin-mediated proteolysis.** This mitotic cyclin proteolysis is essential for the exit from mitosis. As described later, cyclin A and B contain a motif within the N'-termini, the "destruction box," which confers regulation by ubiquitination pathways. Once the cyclin–CDK1 complexes are disrupted by destruction of their cyclin subunits, CDK1 monomers are rapidly and completely dephosphorylated (Murray, 1995).

Although the disruption of cyclin A/B–CDK1 complexes has been largely recognized as a major event for the exit from mitosis, recent studies have shown that anaphase entry is controlled by yet another mechanism in which cyclin destruction is not required, that is, **sister chromatid separation** at **kinetochores** (Murray, 1995). This process seems to be orchestrated by an **E3 holoenzyme** composed of two subunits, CDC16 and CDC27, which conducts the ubiquitination of proteins whose degradation is needed for sister chromatid separation (Tugendreich *et al.*, 1995). Therefore, ubiquitination and the subsequent proteolysis of these proteins by **proteasomes** seem to be important events for progression from metaphase to anaphase. Interestingly, this E3 complex also mediates cyclin destruction during $G_1$ phase (see Section III,D).

## III. Participants of the Cell Cycle

### A. Cyclins and Cyclin-Dependent Kinases

As mentioned previously, CDKs are recipients of **growth signals** that regulate passage through sequential cell cycle transitions. In its monomeric form, CDKs are inactive. Thus, CDKs are regulated by association with *positive* regulatory subunits (cyclins), with *negative* regulators known as CDK inhibitors, and by phosphorylation. The balance of these factors control CDK activity, and the levels of all of these proteins are tightly controlled both transcriptionally and posttranslationally. The latter is mainly achieved by ubiquitin-mediated proteolysis.

The complexity of the cell cycle machinery in higher eukaryotes is unparalleled in other organisms. For example, while the cell cycle events are triggered by a single protein kinase subunit, Cdc28, in budding yeast, mammalian cells require at least seven different CDKs to achieve the same task. Furthermore, Cdc28 associates with cyclins Cln and Clb, while mammalian CDKs bind to and form complexes with various types of cyclins (i.e., A, B, C, D, F, G, H, and I). In comparing the activity of these cyclins, cyclin A is homologous to Clb5/6 and cyclin B to Clb1/2; cyclins D and E are homologous to Cln3 and Cln1/2, respectively.

So, to have a quick reference of most molecules involved in this process, Table 2 shows the protein sequences, NCBI accession numbers, and functions of most cyclins and CDKs known to date. The phylogenetic trees of these and other cell cycle regulators are also shown in Fig. 3.

#### 1. Cyclins

As mentioned before, the cyclins are a group of well-conserved proteins through evolution and among different species (see Fig. 3). Cyclins A to I are structurally related to each other by 11.5–60.6% similarity. This is mainly apparent at the so-called **"cyclin box,"** a 100–150 amino acid region that is the putative site for interaction with CDKs. Cyclins also contain a **"destruction box"** (RxxLxxxxN) in the N′-terminal region, or a consensus sequence **"PEST"** at the C′-terminus (single-letter amino acid code; Table 4). The former is important for timing-triggered degradation, for example, in cyclins A, B, F, and I. The latter is rich in proline, glutamic acid, aspartic acid, serine, and threonine residues, which are necessary for the rapid turnover of these proteins, for example, in cyclins C, D, E, F, G2, and I.

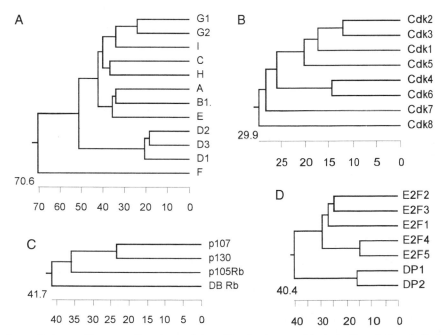

**FIG. 3.** Phylogenetic trees of human cell cycle regulators. The dendograms illustrate the relative similarity between the members of four different groups of human cell cycle regulators. The scale beneath the tree measures the distance between sequences and the units indicate the number of substitutions events, that is, horizontal branch lengths are inversely proportional to the degree of similarity between the sequences. The multiple sequence alignment of these proteins was created by the Megalign program from the DNASTAR package, using the Clustal method with PAM250 residue weight table (licensed to GGA, INNSZ, Mexico). Except for E2F-DP proteins, the sequence accession numbers used for this analysis are shown in Tables 2 and 3. (A) Cyclins, (B) CDKs, (C) pocket proteins, and (D) E2F and DP transcription factors: E2F1, 181918; E2F2, 1082847; E2F3, 738758; E2F4, 1095443; E2F5, 1095444; DP1, 424317; DP2, 604479; DP2, 604479.

**TABLE 4** Amino Acid Code

| | |
|---|---|
| A | alanine |
| B | aspartate or asparigine |
| C | cystine |
| D | aspartate |
| E | glutamate |
| F | phenylalanine |
| G | glycine |
| H | histidine |
| I | isoleucine |
| K | lysine |
| L | leucine |
| M | methionine |
| N | asparagine |
| Q | glutamate |
| R | arginine |
| S | serine |
| T | threonine |
| U | selenocysteine |
| V | valine |
| W | tryptophan |
| Y | tyrosine |
| Z | glutamate or glutamine |
| X | any |

Because cyclin oscillations are critical throughout the cycle, they must be finely regulated by various mechanisms, including (1) the transcriptional control of mRNA production, (2) by specific proteolysis mediated by special ubiquitin ligases that turn on and off at particular times in the cell cycle (see discussion later), and (3) by external events such as a diminution of nutrients (in yeasts), or growth factors (in mammalian cells).

The functions of these molecules have already been described in different phases of the cell cycle, and they are summarized in Table 2. Some others are described in relation to CDKs and CDK inhibitors. However, the recent description of three new types of cyclins (i.e., F, G, and I) deserves a brief comment. **Cyclin F** is a nuclear protein ubiquitously expressed in human tissues. Its expression dramatically oscillates during the cell cycle, peaking in $G_2$ phase, and decreasing prior to cyclin B destruction in mitosis. It is presumed that cyclin F controls cell cycle transitions through pathways that are different from those reported for other cyclins, but this remains to be established (see Table 2).

Another group of human cyclins is composed by **cyclins G1 and G2,** whose expression differs in various tissues. Cyclin G1 levels can be found constitutively expressed either throughout the cell cycle or exclusively elevated during $G_1$ phase, depending on the type of cell under study. In contrast, cyclin G2 levels oscillate during the cell cycle, with peak expression in late S phase (Horne *et al.*, 1996).

The function of G cyclins is unknown, but they seem to be transcriptionally activated by the **tumor suppressor protein p53** in the case of DNA damage. In contrast, it is known that down-regulation of G1 cyclins (with antisense molecules) inhibits the growth and survival of osteosarcoma cells.

Finally, **cyclin I** mRNA is primarily expressed in human postmitotic tissues. Because cyclin I is constantly expressed during the cell cycle, it is possible that their functions are independent of the cell cycle control (Table 2).

### 2. Cyclin-Dependent Kinases

Proteins with putative CDK activity have been identified and isolated through their ability to complement mutant strains of yeast or insects defective for a particular CDK. These reconstitution assays test the capacity of a particular gene to replace the activity of the defective mutant. For example, CDK1, CDK2, and CDK3 genes have been found to complement and rescue Cdc28 mutants of *Saccharomyces cerevisiae*.

All members of this group belong to a family of serine/threonine protein kinases that can phosphorylate histone H1 or other substrates *in vitro* (Nigg, 1995). CDK1 to CDK8 are related to one another by 20.3–75.5% similarity, and the phylogenetic tree of these proteins is shown in Fig. 3. Interestingly, CDK1, CDK2, and CDK3 share a **PSTAIRE motif** at their N′-terminal region, while the rest of the CDKs differ in this motif, that is, CDK4, PISTVRE; CDK5, PSSALRE; CDK6, PLSTIRE; CDK7, NRTALRE; and CDK8, SMSACRE (single-letter amino acid code, Table 4). In CDK2, for example, the PSTAIRE motif has been found in a helix known to be important for recognition of cyclin A. In the inactive state, the helix is displaced in CDK2 so the residues for binding ATP are wrongly disposed. Once cyclin A binds to the catalytic cleft of CDK2, it induces conformational changes in its activation segment and PSTAIRE helix. These changes, in turn, activate the kinase function of CDK2 (Johnson *et al.*, 1996). The significance of these motifs in other CDKs is under study.

Just as in the case of cyclins, the activity of CDKs is also tightly controlled by various mechanisms: (1) cyclin abundance, that is, synthesis rate; (2) levels of cyclin–CDK complexes formed, that is, activation; (3) presence of CDK inhibitors; and (4) by the state of CDK subunit phosphorylation state. The latter might be, for example, increased (activatory) or decreased (inhibitory) depending on the balance of CAK and Wee1 activities, respectively (see Fig. 1). The intricate network has been recently reviewed by Nigg (1995) and Morgan (1995), and the following paragraphs describe some novel findings on this subject.

*a. CDK1 (Cdc2).* Although the mechanisms of activation of Cdc2 are now well understood and its *in vitro* substrates recognized (e.g., histone H1), the identification of its *in vivo* substrates has remained elusive. Certainly, CDK1 plays an important role by triggering, directly or indirectly, various processes at the onset of mitosis (Morgan, 1995). For example, it is known that CDK1 can activate/phosphorylate the C′-terminus of RNA pol II. Moreover, recent studies in HeLa cells have shown that cyclin B–CDK1 complexes are also able to activate/phosphorylate a protein, Eg5, required for the efficient separation of centrosomes and the assembly of the bipolar mitotic spindles. Of relevance, this process occurs in a cell-cycle-dependent manner, with CDK1 phosphorylating a Ser residue during S phase and another critical one, Thr 297, during M phase (Blangy *et al.*, 1995).

*b.* CDK2. As mentioned in Section II,A, the activity of cyclin A–CDK2 complexes peaks at late $G_1$ and S phases in both normal and malignant cells. These complexes promote the passage through the S phase. For example, in normal fibroblasts, the binary complexes are really quaternary ones, composed by cyclin A–CDK2 bound to **proliferating cell nuclear antigen** (PCNA) and the CDK inhibitor, p21 (Section III,B,2,a). In malignant cells, as the tumor suppressor gene p53 is lost, so is the level of p21 protein found in these complexes. This could explain why, in this type of cells, a substantial fraction of cyclin A–CDK2 complexes associate to other proteins of 9, 19, and 45 kDa (i.e., $p9^{Cks1/Cks2}$, $p19^{Skp1}$, $p45^{Skp2}$). Binding of p19 to cyclin A–CDK2 complexes requires p45, and these proteins are essential for kinase activity (Zhang *et al.*, 1995). Moreover, the p45 protein bound to cyclin A–CDK2 complexes is required for execution of DNA replication in both normal and malignant cells. Because p45 levels have been found greatly increased in various tumoral cells, this could be another important player in the tumorigenic process.

*c.* CDK3. CDK3 has been proposed as a regulator of $G_1$ exit. Little is known about the pathway(s) by which this CDK might mediate this effect. One hypothesis is that CDK3, through its binding to the DP subunit of the transcription factors E2F-1 and E2F-2, might be licensing events necessary for S-phase entry (Hofmann and Livingston, 1996). However, the mechanism by which the activation of E2F factors takes place and its exact timing, whether this occurs prior to or after the release of E2F from the pRb, remain to be elucidated.

*d.* CDK4 *and* CDK6. CDK4 and CDK6 both bind selectively to D-type cyclins during $G_1$ phase (Table 2). As mentioned in Section II,A, the cyclin D–CDK4/CDK6 complexes are the main regulators that phosphorylate pRB at the $G_1$/S phase. Both CDKs are specifically downregulated by Ink proteins and, nonselectively, by Cip/Kip inhibitors. In addition, CDK4 phosphorylation by CAK upregulates its functions. CAK is composed of three subunits: CDK7, MAT1, and cyclin H, and it is reviewed in Section III,A,2,f.

*e.* CDK5. CDK5 stands apart from the rest of the known CDKs. Apparently, CDK5 does not play an important role in the control of the cell cycle in terminally differentiated cells. Its levels are low in most adult tissues, except in the brain where it is associated to a **noncyclin activator,** p35. Intriguingly, the CDK5/p35 kinase has been found to be involved in neuronal differentiation during embryogenesis and in the control of cytoskeletal functions in post mitotic neurons (Nikolic *et al.*, 1996). The way by which p35 activates CDK5 is unknown. However, recent studies in embryo brain extracts and Mv1Lu cells have demonstrated that p35 can inhibit the recognition of CDK5

by the p27 inhibitor (Lee *et al.*, 1996). This "protective" function could partially explain the positive regulatory activity that p35 exerts on CDK5.

*f.* CDK7. Formerly known as MO15, CDK7 is the kinase subunit of the CAK complex (Morgan, 1995). The other subunits of this complex are cyclin H and MAT1 (*menage-à-trois*). The name *MAT1* was given to this protein because it actually acts as the assembly factor between CDK7 and cyclin H. However, the activation of the CAK complex is not complete without the phosphorylation of CDK7 in the T-loop (Thr 170). Once that CAK becomes active, it phosphorylates CDK1, CDK2, and CDK4 at their Thr residues in the T-loop (i.e., Thr 160/161), up-regulating the functions of these molecules. Thus, a single CAK activity can control major cyclin kinase activities throughout the cell cycle (Morgan, 1995).

### 3. Crystallographic Studies

Standing as landmarks in our understanding of cyclin–CDK interactions are the recent determinations of the crystal structures of cyclin A, CDK2, and that of the partially activated human cyclin A–CDK2 complex (for a review, see Johnson *et al.*, 1996).[2] These studies have revealed how cyclin A-binding induces the conformational changes in CDK2 that allow the formation of a catalytic center. In brief, cyclin A binding to CDK2 produces structural arrangements at the activation segment and at the PSTAIRE helix of the kinase subunit. At the same time, these changes expose and allow the phosphorylation of a Thr 160 residue, which is the homologous T-loop-phosphorytable Thr 161 in CDK1. As a result, the phosphorylated cyclin A–CDK2 complex has a 17-fold increase in activity in comparison to the nonphosphorylated form. Surprisingly, as the complexes become active, they also become susceptible to inhibition by Cip/Kip inhibitors. This mechanism, therefore, could act as the negative feedback of cyclin A–CDK2 activation. It is impossible to predict if all CDKs are activated as CDK2 does, but this could indeed be the case.

### B. Cyclin–CDK Inhibitors

Recent work has identified two families of proteins that act as negative regulators of CDKs: **inhibitors of CDK 4/6** (Ink4) and **CDK-inhibitor proteins** (Cip/Kip). The former specifically target the CDK4 and CDK6 kinases and bind to these proteins preventing their interaction with cyclins D and/or inhibiting the kinase activity of formed complexes. In contrast Kip binds preferentially to cyclin–CDK complexes and either prevents their activation by CDK-activating kinase (CAK) or inhibits their kinase activity. In addition, and unlike Inks, Kips are promiscuous and interact with most $G_1$–CDKs (see Fig. 2). For example, they can effectively target CDK2 in complexes with cyclins A and E, as well as CDK4 and CDK6 in complexes with cyclins D1, D2, and D3. Therefore, Cip/Kip family mem-

---

[2] The crystal structure of PCNA, complexed with a 22-residue peptide derived from the C-terminus of the protein p21, has also been described by Gulbis *et al.*, 1996.

bers have been referred to as "universal" CDK inhibitors (for a review, see Sherr and Roberts, 1995). A summary of the names, NCBI accession numbers, and functions of these inhibitors is given in Table 3.

## 1. Ink4 Family

**p16**[Ink4] (p16) was the first member of the Ink family to be described. This inhibitor was initially observed as a CDK4-associated protein in human cells, and subsequently found to be a specific cyclin D–CDK4/6 inhibitor. Since the original description of p16 in 1993, three other members of this family, **p15**[Ink4b] (p15), **p18**[Ink4c] (p18), and **p19**[Ink4d] (p19), have been isolated (Sherr, 1995). They all share structural and biological properties, although each one has a different transcriptional control. For example, by using two different promoters, the same locus Ink4a gives rise to two different transcripts: p16[Ink4a] and p19[ARF] (Quelle *et al.*, 1995). Both transcripts have common exons (E2 and E3) but differ at the 5′ exon: E1$\alpha$ for p16[Ink4a] and E1$\beta$ for p19[ARF] (note that the latter is not the same as p19[Ink4d]). Interestingly, the locus that encodes p15, Ink4b, is in proximity to the Ink4a locus of chromsome 9p21.

A characteristic of this family is that all of its members specifically bind to CDK4 and CDK6. The only exception is p18, which binds weakly to CDK4. It is also known that p15, p16, and p18 have ankyrin repeats (Ank repeats), a protein interaction motif, that could mediate the action of these inhibitors. It is considered that Ink proteins arrest the cell cycle through these interactions.

Another characteristic of Ink proteins is that they are inducible. For example, p15 is induced by TGF$\beta$, while E2F proteins and terminal differentiation induce p16. However, the ectopic overexpression of either one of these proteins arrests the cell cycle at G$_1$. Intriguingly, p19[ARF] overexpression arrests the cycle in both the G$_1$ and G$_2$ phases, but apparently through mechanisms not involving direct inhibition of known cyclin–CDK complexes.

Another important aspect of this family is that p16 and p18 rely on the presence of functional pRB in order to arrest the cycle (this fact is unknown for p19). Since the ability of Ink proteins to arrest G$_1$ progression is restricted to pRB positive cells, this implies that loss of these proteins, like loss of pRB, could predispose toward tumor development.

In fact, this hypothesis has proved to be right in the case of p16, which is also known as **Multiple Tumor Suppressor gene 1** (MTS1). First, many human cancers have been shown to harbor deletions and mutations at the Ink4a locus (Table 5). These abnormalities encode for functional aberrant p16 proteins that apparently allow unrestricted cell proliferation. On the other hand, Ink4a $-/-$ (null) mice have been found to develop tumors at an early age (Table 6), and this fact clearly corroborates the role of p16 as a tumor suppressor gene (Serrano *et al.*, 1996). Since the p15 locus is found in physical proximity to the p16 locus, it has also been considered as a tumor suppressor gene and it was named **multiple tumor suppressor gene 2** (MTS2). However, its role in tumorigenesis has not been established

**TABLE 5**  Cell Cycle Players in Tumorigenesis

| Gene | Genetic changes | Human tumors |
|------|-----------------|--------------|
| p16[a] | Homozygous deletions | NSC lung cancer,[b] ALL,[c] renal, bladder, prostate, head, and neck cancers |
|  | Point mutation + small deletions | Pancreas, esophagus, biliary tract, f. melanoma[d] |
|  | Methylation | Breast, colon |
| D1 | Rearrangement, amplification/over-expression | Parathyroid adenoma; B-cell lymphoma; breast, colorectal, squamous cell cancers |
| pRB[a] | Germline mutations | Hereditary retinoblastoma |
|  | Mutation/deletion of both alleles | Sporadic retinoblastoma, osteosarcoma, SC lung and breast cancers |
| p53[a] | Mutation/deletion | Wide range of tumors |

[a] Tumor suppressor genes.
[b] NSC, non-small cell.
[c] ALL, acute lymphocytic leukemia.
[d] f. melanoma, familial melanoma.

as in the case of p16 and p19[ARF] mutations, which coexist in up to 50% of human tumors.

## 2. Cip/Kip Family

The three members of the Cip/Kip family are **p21, p27,** and **p57** (see Table 3). As a general characteristic, these proteins can inhibit all the cyclin–CDK complexes essential for G$_1$ progression and S-phase entry. Indeed, when either one of these inhibitors is ectopically overexpressed in cells *in vitro*, these arrest in G$_1$.

It has been recently found that the sequence WNFDFXXXXPLEGXXXWXXV is a common amino acid motif at the N′-terminal region of p21 and p27. This sequence seems to be the active CDK-binding/inhibitory domain of both proteins (Nakanishi *et al.*, 1995). A more detailed analysis of this region in p21 has revealed that (1) the amino acids 42 to 71 are important for growth inhibition, (2) the minimum amino acid sequence required for DNA synthesis–inhibition spans amino acids 22–71, and (3) amino acids 49 to 71 are involved in actual binding/inhibition of CDK2 activity. Of note, this sequence similarity in members of the Cip/Kip family is not present in p16 or in p15, so there must be at least two distinct mechanisms for CDK inhibition.

In contrast, Cip/Kip inhibitors are more divergent at their carboxy-terminal region. Nonetheless, both p21 and p27 share a PCNA-binding domain by which they could inhibit its replicative activity. The function of the carboxy-terminal region of p57 remains to be elucidated.

*a. p21 and the p53 Tumor Suppressor Gene.* The p21 gene was first cloned as an inhibitor of DNA synthesis that was overexpressed in terminally nondividing senescent human fibroblasts (SDI1), and later as a p53 transactivated gene (WAF1) and a CDK-interacting protein (CIP1,p21) that inhibited cyclin-dependent kinase activity. Now it is known that p21 is the only member of this family induced by p53. Other inducers of this inhibitor include various growth factors (e.g., TGF $\beta$) and transcription factors (e.g., Myo D and C/EBP $\alpha$).

p21 is considered a universal inhibitor, and cells might use it in various ways: (1) in proliferating cells in $G_1$ phase, it may contribute to cell cycle transitions timing or $G_1$ arrest; (2) in S phase, it may block proliferating cell nuclear antigen, PCNA, and slow down DNA synthesis to facilitate repair processes (Gulbis et al., 1996; Umar et al., 1996); and (3) in development, it may contribute to cell cycle arrest in terminally differentiating cells. Finally, the ectopic expression of p21 in human diploid fibroblasts and mouse NIH3T3 has been shown to inhibit DNA synthesis and, in tumor cells, growth inhibition.

It has been shown that p21 contains specific cyclin-binding motifs that are important for its activity (Chen, J., et al., 1996). Immunoprecipitations with antibodies against different cyclins and CDKs have demonstrated that p21 preferentially associates with cyclin–CDK complexes in vivo. In normal fibroblasts, for example, it can be found associated with CDK2 bound to cyclins A and E, and with CDK4 or CDK6 bound to D-type cyclins. Furthermore, the majority of cyclin–CDK molecules in proliferating, nontransformed cells are assembled in heterodimeric complexes containing stoichiometric amounts of both PCNA and p21. In contrast, p21 usually disappears from these complexes in transformed cells apparently due to the lack of p53 antioncogene function (see Section III,A,2,b). This is because p21 is a p53 inducible inhibitor (its promoter has two consensus sites for p53). Thus, if p53 function is lost in tumor cells, p21 expression is also stopped. On the contrary, low DNA precursor pools, radiation, hypoxia, and other DNA damaging agents induce p53-mediated $G_1$ arrest through the transcriptional activation of p21 (Linke et al., 1996). According to the current cell cycle model, DNA damage in p53 +/+ cells induces p21 activation and inhibits cyclin–kinase activity. These changes, in turn, block pRB phosphorylation and the expression of S-phase genes by keeping E2F bound to pRB. Thereafter, the cells follow one of two pathways: (1) repair DNA damage and continue in the cell cycle or (2) die. This mechanism, then, is an essential regulator that coordinates cell cycle progression with transcriptional activity, DNA repair mechanisms, and cell death (e.g., apoptosis).

But p21 can also be induced by p53-independent pathways. For example, p21 is elevated in the postmitotic state in terminally differentiated cells, during embryogenesis, and in senescent cells. In these cases, p21 elevation is not necessarily p53 dependent. For example, p21 is elevated in developing mouse embryo muscle by the transcriptional activator of the myogenic Myo D, and during the growth and differentiation of mouse preadipocytes by c/EBP$\alpha$ (Timchenko et al., 1996).

While p21 levels increase reversibly in quiescent cells, they irreversibly increase in senescent cells. The exact mechanism by which p21 exerts its inhibitory effects in senescence has not yet been established. However, one putative target of p21 could be inhibition of the SAPK group of MAP kinases. Since SAPK phosphorylation of transcription factors is important in stress-activated signaling cascades, p21 could be the converging point for the regulation of cellular stress, cell cycle, tumor suppression, and senescence. It has been suggested that this inhibition may reside in the N'-terminal domain of p21 too (Shim et al., 1996).

*b. p27 Inhibitor.* This protein was first identified as a CDK inhibitor present in contact-inhibited cells and in mink lung epithelial cells made quiescent by the antimitogenic cytokine transforming growth factor beta (TGF$\beta$). In the latter case, the cells contained cyclin E–CDK2 complexes associated with a protein of M 27 K. Further studies demonstrated that p27 cooperates with p15 to cause the $G_1$ arrest induced by TGF$\beta$. In other models, p27 also mediates $G_1$ arrest induced by serum ablation and cAMP. In contrast, it can be down-regulated by growth-stimulatory signals, such as interleukin-2 (IL-2) and adenovirus E1A proteins.

It is now known that p27 predominantly inhibits cyclin D–CDK4 kinase activity and also inactivates cyclin E/A–CDK2 complexes. Through the former action, p27 inhibits pRB phosphorylation/inactivation, the master control at the $G_1$/S transition. Thus, if p27 expression is downregulated with antisense molecules, the fraction of cells in S phase increases and that in $G_1$ decreases.

Recent evidence from three independent groups has confirmed the importance of p27 in cell growth control (Fero et al., 1996, and reviewed in Raff, 1996). In these studies, both copies of the p27 gene were inactivated in mice (p27 nullizygous mice, −/−) by targeted gene disruption, and in all cases the animals grew more rapidly and bigger than the controls. Since these abnormalities could not be attributed to other factors (e.g., endocrinopathy), p27 inactivation was implicated as the main responsable of these results. Thus, p27 probably limits cell proliferation in various tissues by inhibiting $G_1$ CDK activities (Table 6).

*c. p57 Inhibitor.* As previously mentioned, the p57 inhibitory domain is homologous to the one in the N-terminal region of p21. This inhibitor, as well as the other members of this family, induces $G_1$ arrest. Nonetheless, its contribution for the maintenance of the nonproliferative state in adult tissues is marginal. However, due to the identification of imprinting of the p57 gene on chromosome 11p15, it has been considered a putative tumor suppressor gene in embryonal carcinomas and in a cancer prone syndrome known as Beckwith–Weidmans (BWS). The importance of this finding resides in being the first cell cycle regulatory protein controlled by imprinting, but its tumor suppressor activity remains to be proven (Matsuoka et al., 1996).

### C. Pocket Proteins and E2F-Transcription Factors

#### 1. Pocket Proteins

This family includes three members: the **retinoblastoma tumor suppressor protein** (p105RB), and the related proteins **p107** and **p130**. A fourth member has already been

**TABLE 6**　Transgenic Mice

| Gene | Phenotype |
| --- | --- |
| p53 −/− | Viable and normal development<br>Females with exencephaly (some)<br>T-cell lymphomas and sarcomas<br>Mammary tumors (Donehower *et al.*, 1995) |
| E2F1 −/− | Viable and fertile<br>Testicular atrophy<br>Exocrine gland dysplasia<br>Thymic and lymph node enlargement<br>Reproductive tract sarcomas, lung adenocarcinomas<br>　and lymphomas (Field *et al.*, 1996; Yamasaki *et al.*,<br>　1996) |
| pRB −/− | Lethal phenotype (<16th embryonic day)<br>Neuronal cell death–abnormal erythropoiesis (Lee *et al.*, 1992) |
| +/− | Viable and fertile<br>Pituitary and thyroid tumors<br>Bronchial epithelium hyperplasia (Jacks *et al.*, 1992) |
| D1 −/− | Viable and fertile/dwarfism<br>Severe retinopathy<br>Hypoplasia of mammary glands (Fantl *et al.*, 1995) |
| ++/++ | Selective overexpression in stratified epithelia<br>Normal differentiation and function<br>Epidermal hyperproliferative response<br>Severe thymic aplasia (Robles *et al.*, 1996) |
| ++/++ | Selective overexpression in mammary gland<br>Abnormal proliferation/adenocarcinoma (Wang *et al.*, 1994) |
| p21 −/− | Viable and normal development<br>Nontumorigenic (Deng *et al.*, 1995) |
| ++/++ | Selective overexpression in hepatocytes<br>Halts postnatal liver development and regeneration<br>　(Wu *et al.*, 1996) |
| p27 −/− | Normal prenatal development and viable<br>Enlarged body size/female sterility<br>Hyperplasia of thymus, adrenals, and gonads<br>Pituitary tumors and disorganization of retina (Fero<br>　*et al.*, 1996) |
| p16 −/− | Viable<br>Extramedullary hematopoiesis<br>Soft tissue sarcomas and lymphomas (Serrano *et al.*, 1996) |

described in *Drosophila,* RBF, but its human homolog has not been identified yet (Du *et al.*, 1996).

Pocket proteins share many structural and functional features. They display high sequence homology in two discontinuous regions (CR1 and CR2) that make up the "pocket domain" (i.e., E1A/T binding domain). This region is particularly important in complex formation and modulation of the activity of several cellular and viral proteins. For example, all three proteins bind E2F-DP heterodimers at the pocket region and promote transcriptional repression. As shown in Fig. 4, the pRB pocket also interacts with (1) other transcription factors such as MyoD,

BRG1, UBF; (2) nuclear RNA-polymerases pol I, pol II, and pol III; (3) viral proteins such as HPV-16, simian virus large T antigen and adenovirus E1A; and (4) chaperone molecules, for example, hsp 75 (Chen, C.-F., *et al.*, 1996). Furthermore, pRB protein contains a HPV E7 binding domain at the carboxy-terminal that is able to inhibit bindxing of E2F to pRB with the sole union of E7. Thus, viral proteins, whether bound to the pocket or to the E7 domain, can displace and modify the function of cellular proteins normally bound to the pocket. The consequence of these interactions can be, for example, that free E2F-DP heterodimers become transcriptionally active and promote cell cycle progression.

The mechanism by which pocket proteins regulate the cell cycle progression and arrest the cells in $G_1$ phase is not fully understood. First, depending on the cell under study, these proteins exhibit different growth–suppressive properties and exert control at various phases of the cell cycle. Thus, despite acting in a similar way, they are not functionally redundant. As described later, pocket proteins also have preferential binding to certain transcription factors, for example, pRB binds to E2F1, E2F2 and E2F3, while p107 and p130 bind to E2F4 and E2F5.

The activation/inactivation of the RB family members is also puzzling. Both pocket protein–phosphorylation and pocket protein–E2F complex formation are cell cycle dependent. As mentioned earlier, pRB is phosphorylated by cyclin D–CDK4/6 and cyclin E–CDK2 complexes in mid to late $G_1$ phase. Once that pRB is phosphorylated/inactivated, it dissociates from E2F-DP heterodimers and allows cell cycle progression. Similarly, p107 and p130 stably interact *in vivo* with cyclin E/A–CDK2 complexes that appear in late $G_1$ phase. These complexes seem to be responsible for the existence of various phosphorylated forms of p107 and p130 throughout the different phases of the cell cycle. However, there is evidence that their phosphorylation by cyclin E/A–CDK2 complexes does not necessarily lead to dissociation of pocket proteins from E2F-DP heterodimers (i.e., transcriptional activation). For example, cyclin E/A–CDK2 complexes phosphorylate p107–E2F4 complexes but this phosphorylation does not dissociate p107 from E2F4. Instead, only cyclin D–kinase activity seems to be able to dissociate them, resulting in the relief of the p107 $G_1$ exit block (Xiao *et al.*, 1996). Thus, the timing of activation of the various complexes formed by pocket proteins and the activities of these complexes are numerous and intricate. This information is beyond the scope of this chapter and it has recently been updated by Weinberg (1995).

However, it is important to highlight the tumor suppressor activity of pRB. Mutations of pRB (e.g., deletions, duplications, point mutations) are frequently found in human tumors, and the inactivation of both copies of the gene appears to be the rate-limiting event in the genesis of retinoblastoma and osteosarcoma. Moreover, pRB nullizygous transgenic mice (pRB −/−) are not viable, whereas adult mice heterozygous (RB +/−) are highly predisposed to pituitary and thyroid carcinomas (Table 6). These findings have reinforced the essential role of pRB in growth control and development and, despite the fact that p107

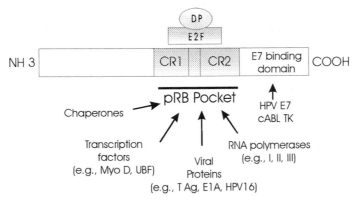

**FIG. 4.** RB protein domains and regulators. The pRB protein is the master regulatory protein at the G₁/S transition. Its main function is to inhibit the activity of the E2F family of transcription factors. However, other important participants of the cell cycle, such as polymerases, transcription factors, and chaperones, also interact with it. In contrast, mutations and binding of viral proteins can inactivate the pRB gene and its protein, respectively.

and p130 exert related functions, mutations of these proteins have never been described in human tumors. For this reason, it is believed that p107 and p130 only serve in a limited fashion in the inhibition of tumor formation. An emerging concept is that the tumor suppressor functions of pRB not only correlate with the suppression of E2F-DP heterodimers, but also with the inhibition of RNA pol I, pol II, and pol III transcriptional activity (White *et al.*, 1996). Since the transcription by RNA pol III is higher in the S and G₂ phases, its activity must certainly be important in protein synthesis and cell growth. Therefore, if pol activities are also repressed by pRB, this could explain its critical role in arresting unrestrained proliferation and tumorigenesis.

## 2. E2F and DP (DRTF1) Proteins

**E2F-DP heterodimers** are a group of cell-cycle-regulated transcription factors that coordinate the expression of a variety of viral and cellular genes (La Thangue, 1994). While E2F-DP activity has been traditionally considered an essential step for the G₁/S transition, studies have also suggested a role for these heterodimers in abnormal growth proliferation and in the control of programmed cell death (Lam and La Thangue, 1994). This is because E2F-DP binding sites are present in the promoters of a wide number of genes that encode proteins necessary for these processes.

Structural analyses have shown that these heterodimers consist of two distinct proteins: E2F and DP (i.e., E2F-DP). The **E2F subunit** is encoded by a family of genes consisting of at least five members (i.e., E2F1–E2F5). These proteins are related to one another by 20.9–61.4% similarity (see Fig. 3). Despite being highly conserved and structurally related proteins, E2F1, E2F2, and E2F3 prefer liaisons with pRB, while E2F4 and E2F5 associate with other pocket proteins, p107 and p130 (Weinberg, 1995, 1996). Another important difference is that E2F1, E2F2, and E2F3 present a cyclin A binding domain, which is nonexistent in other E2F proteins (see discussion later). Intriguingly, the formation of diverse E2F-DP–pocket protein complexes occurs

at different times during the cell cycle (Lam and La Thangue, 1994).

However, two human **DP proteins** have been identified as heterodimeric partners of E2F: DP1 and DP2. DP1 protein (410 aa) results from the ubiquitous expression of a single message. Instead, the DP2 gene encodes at least five DP2 RNAs from which three DP2-related proteins are expressed *in vivo* (55, 48, and 43 kDa) (Rogers *et al.*, 1996). DP proteins are conserved proteins in their DNA-binding domains, which, in the carboxy-termini, contain the **DEF box.** This domain is essential for the dimerization of DP with E2F proteins. It is worth mentioning that there is already a DP3 gene cloned form mouse, but its human homolog has not yet been described.

Despite both type of phosphoproteins, DP1 and DP2, being unable to bind pRB directly (they do not have an E2F-like pRB-binding domain), either one can first form transcriptionally active complexes with E2F1, E2F2, and E2F3 through the DEF box and then associate to pRB.

Late in G₁ phase, cyclin D–CDK4/6 complexes phosphorylate retinoblastoma proteins, and these, in turn, release E2F-DP heterodimers from their "pockets." Thus, E2F-DP becomes transcriptionally active and binds to promoters in diverse type of genes whose products control the cell cycle progression. The latter include (1) transcription genes: dihydrofolate reductase (DHFR), polymerase alpha, thymidine kinase (TK); (2) CDK activity: cyclins D, E, A, and CDK1; (3) nuclear oncogenes: c-Myc, N-myc, B-myb; (4) pocket proteins; and (5) the E2F1 gene itself. Of note, it has also been found in *Drosophila melanogaster* embryos that cyclin E acts as an upstream activator of E2F, and that E2F-directed transcription cannot proceed in cyclin E mutated strains (Duronio *et al.*, 1996).

Later in the cycle, during S phase, the function of DP1 is partially controlled by phosphorylation. Thus, it has been observed that cyclin A–CDK2 complexes initially bind to a cyclin A-binding domain in the amino-terminal region of E2F (see Fig. 2). Subsequently, the kinase activity of this complex phosphorylates DP1 and inhibits the E2F-DP activity. Interestingly, when these DP1 heterodimers are phosphorylated *in vitro* by cyclin A–CDK2 complexes (but not cyclin E–CDK2), its E2F1–DNA binding and transactivation activities are suppressed. This could explain why some promoters that are transcriptionally activated by E2F during $G_1$ are found inactive during S phase.

However, when E2F1 mutants—defective in cyclin A–CDK2 binding—are expressed in fibroblasts, these cells undergo S phase delay/arrest followed by regrowth or apoptosis depending on the transactivation activity of DNA bound-E2F (Krek *et al.*, 1995). The connection between a continuous E2F activity and apoptosis has also been shown in *Drosophila* imaginal disks where the overexpression of E2F activates programmed cell death (Asano *et al.*, 1996). Since the current model considers E2F as an up-regulator of cell proliferation, these results were completely unexpected.

The findings in **E2F1 null mice** (E2F1 −/−) have been puzzling too. These animals not only were viable but, contrary to what was expected, they developed tumors. For this reason, the E2F gene has been proposed as the first of a class that could exert oncogene and antioncogene functions (Weinberg, 1996). Many theories have been developed in trying to understand those confusing results. Nevertheless, it will certainly take some years to properly understand what exactly the E2F-DP heterodimers do in the cell cycle.

From these facts, it is obvious that normal cell cycle progression requires a fine balance between cyclin–CDK activity, transcriptional control, and DNA replication.

## D. Protein Ubiquitination and Destruction

As already stated, cyclin abundance is often rate limiting for CDK activity, and by controlling rates of cyclin synthesis and degradation the cell is able to regulate cell cycle transitions. However, cyclins are not the only type of cell cycle regulators whose activity is controlled by proteolysis. In any case, this process is generally carried out by **ubiquitin-mediated proteolysis** (Fig. 5).

In this proteolytic system, ubiquitin is conjugated with target proteins through the sequential activation of three enzymes: (1) An **ubiquitin-activating enzyme,** E1, uses ATP to form a thioester bond between itself and ubiquitin; (2) E1 transfers ubiquitin to an **ubiquitin-conjugating enzyme,** E2; and finally (3) ubiquitin is transferred to the substrate directly by an E2 enzyme, or by E2 acting together with an **ubiquitin-protein ligase,** E3. Once the protein has been ubiquitinated, it is prone to ATP-dependent proteolysis by the **proteasome.** The latter is a large protease (2000 kDa) also known as the 26S proteasome complex. This complex results from the interaction of a 20S proteasome with the proteasome activator 700 (i.e., PA700). It is through this association that the complex obtains ubiquitin dependency, although many other substrates of the ubiquitin system have been described.

Similarly, E2 is an enzyme composed by at least two subunits, Ubc4 and Ubc9, while E3 is composed of three subunits, CDC16, CDC23, and CDC27, and is regulated by phosphorylation (Murray, 1995). E3 is the same holoenzyme previously referred to as *anaphase-promoting complex* (APC) or 22S cyclosome.

Currently, three major cell cycle transitions are known to require the degradation of specific proteins by the ubiquitin-proteasome pathway: (1) entry into S phase, (2) separation of sister chromatids, and (3) exit from mitosis.

### 1. Proteolysis and Entry into S Phase

As already mentioned, G1 cyclin activity is rate limiting for cell division in mammals and requires tight posttranslational control (i.e., proteolysis). In budding yeast, a molecule known as Cdc53 has been recently shown to ubiquitinate G1 cyclins *in vivo* and it also binds to the E2-ubiquitin-conjugating enzyme. Thus, Cdc53 could be regulating G1 cyclin activity by promoting proteolysis. Because Cdc53 homologs have also been identified in higher eukaryotic cells, it has been suggested that Cdc53 might play a role in $G_1$ phase similar to that of APC in M phase, that is, the former destroying G1 cyclins while APC would be in charge of destroying mitotic cyclins. This is yet to be demonstrated.

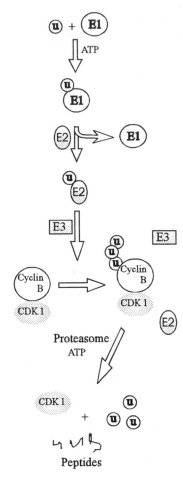

**FIG. 5.** Ubiquitin-mediated proteolysis. The ubiquitination and subsequent proteolysis of mitotic cyclin B is shown. However, the same pathway is used for the degradation of other type of proteins during the cell cycle. See Section III,D for further details.

On the other hand, there is a **licensing model** in which an essential replication factor, a **licensing factor,** is destroyed within the nucleus during each S phase and can only access chromatin—and be replenished from the cytoplasmic stores—when the nuclear membrane is disrupted during mitosis. A licensing factor in yeasts could, indeed, be a complex of proteins that includes MCM, Orc, and Cdc6/Cdc18. The phosphorylation/activation of this replicative complex occurs while the $G_1$ chromatin is in a permissive state for replication. Following this activation, a specific proteolytic machinery (including CDC16 and CDC27) degrades Cdc6/Cdc18 and inactivates the complex during S phase (Wuarin and Nurse, 1996).

### 2. Proteolysis and Separation of Sister Chromatids

This process has already been mentioned in Section II,C (Tugendreich *et al.,* 1995).

### 3. Proteolysis and Exit from Mitosis

As previously mentioned, cyclin A/B–CDK1 complexes are important during mitosis, and the cyclins in these complexes undergo ubiquitin-mediated proteolysis at the metaphase–anaphase transition. The ubiquitination of these proteins is carried out by the APC complex (CDC16 and CDC27) and the proteolysis is carried out by the proteasome (see Fig. 5). Once they are degraded, the CDK1 and the ubiquitin molecules are released intact (King *et al.,* 1995).

## IV. Transgenic Mice

Once a putative molecule has been cloned and shown to be of relevance in the cell cycle in *in vitro* assays, the final goal is to test its potential regulatory activity in *in vivo* conditions. This has been achieved by the creation of transgenic mice for each one of the molecules of interest. For this purpose, pluripotential embryonic stem (ES) cells in culture are genetically modified prior to its return to mouse, where they will contribute to all tissues of a mouse including the germ line. ES cells are modified to either (1) carry and overexpress foreign DNA for **gain of function** or (2) gene target, to create loss of function mutations in the mouse, that is, **knockout mice.**

A partial list of the transgenic mice specifically created for evaluating the activity of cell cycle regulator proteins is shown in Table 6.[3] Interestingly, although the phenotype of some of these chimeric animals has fully corroborated the putative activity of certain regulators in *in vitro* experiments (e.g., pRB, p53), in other cases the phenotype has significantly differed from the expected outcomes (e.g., EF2-1, cyclin D1, p21). Some of these findings have already been discussed in relation to specific proteins in Section III, and pertinent notes are added in this one.

---

[3] Further information regarding available transgenic mice can be accessed at TBASE, http://www.gdb.org/Dan/tbase/tbase.html.

Except for the **pRB nullizygous mice** (pRB $-/-$), none of the rest of the knockout mice present a lethal phenotype. Indeed, pRB is not necessary for cell division or for differentiation up to day 13 or 14 of gestation. However, this lethal phenotype can be partially potentiated in pRB $-/-$, p107 $-/-$ mice. In contrast, **pRB heterozygous** animals (pRB $+/-$) are viable and develop tumors of the thyroid and pituitary glands after 8 months of age. This finding is in agreement with its role as a tumor suppressor gene, but these heterozygous animals do not develop retinoblastomas or osteosarcomas as do humans.

Interestingly, pRB heterozygous mice (pRB $+/-$) and **p27 nullizygous** mice (p27 $-/-$) share the feature of generating pituitary tumors. This has given further support to the concept of an integrated pRB–cyclin D–CDK4–p27 pathway. Furthermore, p27 $-/-$ mice grow faster and present with generalized organomegaly in comparison to their controls. Since these effects could not be attributed to another factor (e.g., endocrinopathy), p27 has been implicated as an essential regulator of growth in multiple cell types (Raff, 1996).

As expected, **p53 nullizygous** (p53 $-/-$) and **p16 nullizygous** (p16 $-/-$) mice present normal development and tumor formation early in life. These findings are in agreement with their tumor suppressor activities *in vitro,* and with their known genetic abnormalities in human cancers. However, it was also thought that **p21 nullizygous** mice (p21 $-/-$) were going to develop tumors, but this was not the case. Nonetheless, both p53 $-/-$ thymocytes and p21 $-/-$ embryonic fibroblasts are (1) significantly deficient in their ability to arrest in $G_1$ in response to DNA damage, and (2) both type of cells achieve a high saturation density due to a significant alteration in growth control *in vitro*. In contrast, p21 overexpression in liver halts cell cycle progression, and postnatal liver development and regeneration, suggesting a critical role of p21 in normal development.

Another unexpected finding was the one in **mice lacking E2F1** (i.e., E2F1 $-/-$). These animals were viable, developed normally, and presented a broad spectrum of tumors. Thus, although overexpression of E2F1 in tissue culture cells promotes growth proliferation and oncogenesis, the lack of this gene in E2F1 $-/-$ mice can also induce oncogenesis *in vivo*. For this reason, the E2F1 gene has been considered the first of a class acting both as an oncogene and as a tumor suppressor gene (Weinberg, 1996).

Puzzling too was the case of cyclin D1 nullizygous mice. It is known that the sole overexpression of cyclin D1 deregulates cell proliferation and induces tumorigenesis in the tissues where it is selectively expressed. Thus, cyclin D1 has been considered an essential component in promoting cell proliferation and **cyclin D1 $-/-$ mice** were expected to present a lethal phenotype. However, these animals developed to term and showed a reduced body size. They also lacked normal development in retina and in mammary gland. It was concluded, then, that (1) D1 kinase activity is important for $G_1$ progression, but that it is dispensable for the proliferation of the majority of tissues in mammals; and (2) that other D-type cyclins could partially replace cyclin D1 activity in these animals.

## V. Summary

The mitotic cell cycle of higher eukaryotes consists of four phases: $G_1$, S, $G_2$, and mitosis. In the first three phases, the cell prepares for cell division, which actually takes place during mitosis. To guarantee the high-fidelity transmission of genetic information, eukaryotic cells have developed a complicated network of molecules that act in a finely coordinated manner. Central participants in this process include cyclins, CDKs, CDK inhibitors, and transcription factors, such as the products of tumor suppressor genes (e.g., pRB, p53). Cyclins bind to CDKs and act as regulatory subunits of these catalytic kinases. The activation of these cyclin–CDK complexes, and their effects, depend on various factors: (1) cyclin abundance, (2) phosphorylation state of their kinase subunits, and (3) the presence of negative regulators known as CDK inhibitors.

Once CDKs become active, they phosphorylate diverse substrates that, directly or indirectly, control gene expression and the progression of the cell cycle. Nine groups of cyclins (i.e., A type to I type) and eight types of CDKs (i.e., 1, 2, etc.) are known to date, some of which have various isoforms. Established interactions include (1) cyclin D–CDK4 and cyclin D–CDK6 during $G_1$ phase; (2) cyclin E–CDK2 and cyclin A–CDK2 at the $G_1$/S transition and into S phase; and (3) cyclin A–CDK1 and cyclin B–CDK1 from $G_2$ until mitosis.

Cyclin D–CDK4/CDK6 complexes phosphorylate pRB protein, the product of the tumor suppressor gene pRB and the master regulatory protein of cell cycle at the $G_1$/S transition. Once phosphorylated, pRB protein liberates E2F-DP heterodimers, which are active transcription factors that promote the passage from $G_1$ to S phase and allow the progression of the cell cycle. For this transition to occur, the activities of cyclin E/A–CDK2 complexes are also required. Once in $G_2$ phase, cyclin A/B–CDK1 complexes appear and promote M-phase progression. Interestingly, these complexes dissociate at metaphase (i.e., one of the phases of mitosis) due to cyclin degradation via ubiquitin-mediated proteolysis. As previously discussed, this process is essential for the exit from mitosis, but proteolytic mechanism also controls other phases of the cell cycle.

As a negative feedback, CDK inhibitors down-regulate the function of CDK complexes. These inhibitors belong to either one of two families, Ink4 or Kip/Cip. Members of the former family include proteins p15, p16, p18, and p19 and they exclusively inhibit cyclin–CDK4/CDK6 complexes. In contrast, the Kip/Cip family proteins, p21, p27, and p57, are considered universal inhibitors since they can down-regulate the activities of cyclin–CDK4/6 and cyclin–CDK2 complexes. All of these inhibitors are very important in the control of the cell cycle during various physiological (e.g., embryogenesis, terminal differentiation, senescence) and pathological conditions (e.g., p53-mediated $G_1$ arrest through the transcriptional activation of p21 in the presence of DNA damage). Although this model is supported by evidence from years of descriptive and experimental work, it is still a simplified scenario that must not be taken as a paradigm. Novel proteins and interactions

are constantly found and unexpected results, such as those in transgenic mice, have made clear that what it is found in *in vitro* conditions is not necessarily what occurs *in vivo*.

## Bibliography

Asano, M., Nevins, J. R., and Wharton, R. P. (1996). Ectopic E2F expression induces S phase and apoptosis in *Drosophila* imaginal discs. *Genes Dev.* **10**, 1422–1432.

Blangy, A., Lane, H. A., d'Herin, P., Harper, M., Kress, M., and Nigg, E. A. (1995). Phosphorylation by p34 cdc2 regulates spindle association of human Eg5, a kinesin-related motor essential for bipolar spindle formation in vivo. *Cell* **83**, 1159–1169.

Brugarolas, J., Chandrasekaran, C., Gordon, J. I., Beach, D., Jacks, T., and Hannon, G. J. (1995). Radiation-induced cell cycle arrest compromised by p21 deficiency. *Nature* **377**, 552–557.

Chen, C.-F., Chen, Y., Dai, K., Chen, P.-L., Riley, D. J., and Lee, W.-H. (1996). A new member of the hsp90 family of molecular chaperones interacts with the retinoblastoma protein during mitosis and after heat shock. *Mol. Cell. Biol.* **16**, 4691–4699.

Chen, J., Saha, P., Kornbluth, S., Dynlacht, B. D., and Dutta, A. (1996). Cyclin-binding motifs are essential for the function of p21$^{CIP1}$. *Mol. Cell. Biol.* **16**, 4673–4682.

Cross, S. M., Sanchez, C. A., Morgan, C. A., Schimke, M. K., Ramel, S., Idzerda, R. L., Raskind, W. H., and Reid, B. J. (1995). A p53-dependent mouse spindle checkpoint. *Science* **267**, 1353–1356.

Deng, C., Zhang, P., Harper, J. W., Elledge, S. J., and Leder, P. (1995). Mice lacking p21$^{CIP1/WAF1}$ undergo normal development, but are defective in G1 checkpoint control. *Cell* **82**, 675–684.

Donehower, L. A., Godley, L. A., Marcelo Aldaz, C., Pyle, R., Shi, Y. P., Pinkel, D., Gray, J., Bradly, A., Medina, D., and Varmus, H. E. (1995). Deficiency of p53 accelerates mammary tumorigenesis in Wnt-1 transgenic mice and promotes chromosomal instability. *Genes Dev.* **9**, 882–895.

Du, W., Vidal, M., Xie, J.-E., and Dyson, N. (1996). RBF, a novel RB-related gene that regulates E2F activity and interacts with cyclin E in Drosophila. *Genes Dev.* **10**, 1206–1218.

Duronio, R. J., Brook, A., Dyson, N., and O'Farrel, P. H. (1996). E2F-induced S phase requires cyclin E. *Genes Dev.* **10**, 2505–2513.

Fantl, V., Stamp, G., Andrews, A., Rosewell, I., and Dickson, C. (1995). Mice lacking cyclin D1 are small and show defects in eye and mammary gland development. *Genes Dev.* **9**, 2364–2372.

Felsenfeld, G. (1996). Chromatin unfolds. *Cell* **86**, 13–19.

Fero, M. L., Rivkin, M., Tasch, M., Porter, P., Carow, C. E., Firpo, E., Polyak, K., Tsai, L.-H., Broudy, V., Perlmutter, R. M., Kaushansky, K., and Roberts, J. M. (1996). A syndrome of multiorgan hyperplasia with features of gigantism, tumorigenesis and female sterility in p27$^{Kip1}$-deficient mice. *Cell* **85**, 733–744.

Field, S. J., Tsai, F.-Y., Kuo, F., Zubiaga, A. M., Kaelin, W. G., Livingston, D. M., Orkin, S. H., and Greenberg, M. E. (1996). E2F-1 functions in mice to promote apoptosis and suppress proliferation. *Cell* **85**, 549–561.

Gulbis, J. M., Kelman, Z., Hurwitz, J., O'Donnell, M., and Kuriyan, J. (1996). Structure of the C-terminal region of p21$^{WAF1/CIP1}$ complexed with human PCNA. *Cell* **87**, 297–306.

Hartwell, L. H., and Weinert, T. A. (1989). Checkpoints: Controls that ensure the order of cell cycle events. *Science* **246**, 629–634.

Hirano, T. (1995). Biochemical and genetic dissection of mitotic chromosome condensation. *Trends. Biochem. Sci.* **20**, 357–361.

Hofmann, F., and Livingston, D. (1996). Differential effects of cdk2 and cdk3 on the control of pRB and E2F function during G1 exit. *Genes Dev.* **10**, 851–861.

Horne, M. C., Goolsby, G. L., Donaldson, K. L., Tran, D., Neubauer, M., and Wahl, A. F. (1996). Cyclin G1 and cyclin G2 comprise a new family of cyclins with contrasting issue-specific and cell-cycle regulated expressions. *J. Biol. Chem.* **271**, 6050–6051.

Jacks, T., Fazeli, A., Schmitt, E. M., Bronson, R. T., Goodell, M. A., and Weinberg, R. A. (1992). Effects of and *Rb* mutation in the mouse. *Nature* **359**, 295–300.

Johnson, L. N., Noble, M. E. M., and Owen, D. J. (1996). Active and inactive protein kinases: Structural basis for regulation. *Cell* **85**, 149–158.

King, R. W., Peters, J. M., Tugendreich, S., Rolfe, M., Hieter, P., and Kirschner, M. W. (1995). A 20S complex containing CDC27 and CDC16 catalyzes the mitosis-specific conjugation of ubiquitin to cyclin B. *Cell* **81**, 279–288.

Kleckner, N. (1996). Meiosis; how could it work? *Proc. Natl. Acad. Sci. USA* **93**, 8167–8174.

Krek, W., Xu, G., and Livingston, D. M. (1995). Cyclin A-kinase regulation of E2F-1 DNA binding function underlies suppression of an S phase checkpoint. *Cell* **83**, 1149–1158.

La Thangue, N. B. (1994). DRTF1/E2F: An expanding family of heterodimeric transcription factors implicated in cell-cycle control. *Trends Biochem. Sci.* **19**, 108–114.

Lam, E. W.-F., and La Thangue, N. B. (1994). DP and E2F proteins: Coordinating transcription with cell cycle progression. *Curr. Opin. Cell. Biol.* **6**, 859–866.

Lee, E. Y.-H. P., Chang, C. Y., Hu, N., Wang, Y.-C. J., Lai, C.-C., Herrup, K., Lee, W.-H., and Bradley, A. (1992). Mice deficient for Rb are nonviable and show defects in neurogenesis and hematopoiesis. *Nature* **359**, 288–294.

Lee, M.-H., Nikolic, M., Baptista, C., Lai, E., Tsai, L.-H., and Massague, J, (1996). The brain specific activator allows Cdk 5 to escape inhibition by p27Kip1 in neurons. *Proc. Natl. Acad. Sci. USA* **93**, 3259–3263.

Linke, S. P., Clarkin, K. C., DiLeonardo, A., Tsou, A., and Wahl, G. M. (1996). A reversible, p53-dependent $G_0/G_1$ cell cycle arrest induced by ribonucleotide depletion in the absence of detectable DNA damage. *Genes Dev.* **10**, 934–947.

Matsuoka, S., Thompson, J. S., Edwards, M. C., Barletta, J. M., Grundy, P., Kalikin, L. M., Harper, J. W., Elledge, S. J., and Feinberg, A. P. (1996). Imprinting of the gene encoding a human cyclin-dependent kinase inhibitor, p57KIP2, on chromosome 11p15. *Proc. Natl. Acad. Sci. USA* **93**, 3026–3030.

McIntosh, J. R. (1991). Spindle fiber action and chromosome movement. *Annu. Rev. Cell. Biol.* **7**, 403–426.

Miyazaki, W. Y., and Orr-Weaver, T. L. (1994). Sister-chromatid cohesion in mitosis and meiosis. *Annu. Rev. Gen.* **28**, 167–187.

Morgan, D. O. (1995). Principles of CDK regulation. *Nature* **374**, 131–134.

Müller, R., Mumberg, D., and Lucibello, F. C. (1993). Signals and genes in the control of cell-cycle progression. *Biochem. Biophys. Acta* **1155**, 151–179.

Murray, A. (1995). Cyclin ubiquitination: The destructive end of mitosis. *Cell* **81**, 149–152.

Murray, A., and Hunt, T. (1993). "The Cell Cycle." W. H. Freeman, New York.

Nakanishi, M., Robetorye, R. S., Adami, G. R., Pereira-Smith, O. M., and Smith, J. R. (1995). Identification of the active region of the DNA synthesis inhibitory gene p21 Sdi1/Cip1/WAF1. *EMBO J.* **14**, 555–563.

Nigg, E. A. (1995). Cyclin-dependent protein kinases: Key regulators of the eukaryotic cell cycle. *Bioassays* **17**, 471–480.

Nikolic, M., Dudek, H., Kwon, Y. T., Ramos, Y. F. M., and Tsai, L. H. (1996) The cdk5/p35 kinase is essential for neurite outgrowth during neuronal differentiation. *Genes Dev.* **10**, 816–825.

Norbury, C., and Nurse, P. (1992). Animal cell cycles and their control. *Annu. Rev. Biochem.* **61**, 441–470.

Quelle, D. E., Zindy, F., Ashmun, R. A., and Sherr, C. J. (1995). Alternative reading frames of the INK4a tumor suppressor gene

encode two unrelated proteins capable of inducing cell cycle arrest. *Cell* **83**, 993–1000.

Raff, M. C. (1996). Size control: The regulation of cell numbers in animal development. *Cell* **86**, 172–175.

Reed, S. I. (1992). The role of p34 kinases in the G1 to S-phase transition. *Annu. Rev. Cell Biol.* **8**, 529–561.

Robles, A. I., Larcher, F., Whalin, R. B., Murillas, R., Richie, E., Gimenez-Conti, I. B., Jorcano, J. L., and Conti, C. J. (1996). Expression of cyclin D1 in epithelial tissues of transgenic mice results in epidermal hyperproliferation and severe thymic hyperplasia. *Proc. Natl. Acad. Sci. USA* **93**, 7634–7638.

Rogers, K. T., Higgins, P. D. R., Milla, M. R., Phillips, R. S., and Horowitz, J. M. (1996). DP-2, a heterodimeric partner of E2F: Identification and characterization of DP-2 proteins expressed *in vivo. Proc. Natl. Acad Sci. USA* **93**, 7594–7599.

Roth, S. Y., and Allis, C. D. (1996). Histone acetylation and chromatin assembly: A single escort, multiple dances? *Cell* **87**, 5–8.

Sabbatini, P., Lin, J., Levine, A. J., and White, E. (1995). Essential role for p53-mediated transcription in E1A-induced apoptosis. *Genes Dev.* **9**, 2184–2192.

Serrano, M., Lee, H.-W., Chin, L., Cordon-Cardo, C., Beach, D., and DePinho, A. (1996). Role of the *INK4a* locus in tumor suppression and cell mortality. *Cell* **85**, 27–37.

Shay, J. W. (1996). Telomerase activity in human cancer. *Curr. Opin. Oncol.* **8**, 66–71.

Sherr, C. J. (1995). D-type cyclins. *Trends Biochem. Sci.* **20**, 187–190.

Sherr, C. J., and Roberts, J. M. (1995). Inhibitors of mammalian G1 cyclin-dependent kinases. *Genes Dev.* **9**, 1149–1163.

Shim, J., Lee, H., Park, J., Kim, H., and Choi, E. J. (1996). A non-enzymatic p21 protein inhibitor of stress-activated protein kinases. *Nature* **381**, 804–807.

Solomon, M. J. (1994). The function(s) of CAK, the $p34^{cdc2}$-activating kinase. *Trends Biochem. Sci.* **19**, 496–500.

Timchenko, N. A., Wilde, M., Nakanishi, M., Smith, J. R., and Darlington, G. J. (1996). CCAAT/enhancer-binding protein α (C/EBP α) inhibits cell proliferation through the p21 (WAF-1/CIP-1/SDI-1) protein. *Genes Dev.* **10**, 804–815.

Tugendreich, S., Tomkiel, J., Earnshaw, W., and Hieter, P. (1995). CDC27Hs colocalizes with CDC16Hs to the centrosome and mitotic spindle and is essential for the metaphase to anaphase transition. *Cell* **81**, 261–268.

Tyson, J., Novak, B., Odell, G. M., Chen, K., and Thron, C. D. (1996). Chemical kinetic theory: Understanding cell-cycle regulation. *Trends Biochem. Sci.* **21**, 89–96.

Umar, A., Buermeyer, A. B., Simon, J. A., Thomas, D. C., Clark, A. B., Liskay, R. M., and Kunkel, T. A. (1996). Requirement for PCNA in DNA mismatch repair at a step preceding DNA resynthesis. *Cell* **87**, 65–73.

Walczak, C. E., and Mitchison, T. J. (1996). Kinesin-related proteins at mitotic spindle poles: Function and regulation. *Cell* **85**, 943–946.

Wang, T. C., Cardiff, R. D., Zukerberg, L., Lees, E., Arnold, A., and Schmidt, E. V. (1994). Mammary hyperplasia and carcinoma in MMTV-cyclin D1 transgenic mice. *Nature* **369**, 669–671.

Warren, G. (1993). Membrane partitioning during cell division. *Annu. Rev. Biochem.* **62**, 323–348.

Weinberg, R. A. (1995). The retinoblastoma protein and cell cycle control. *Cell* **81**, 323–330.

Weinberg, R. A. (1996). E2F and cell proliferation: A world turned upside down. *Cell* **85**, 457–459.

White, R. J., Trouche, D., Martin, K., Jackson, S. P., and Kouzarides, T. (1996). Repression of RNA polymerase III transcription by the retinoblastoma protein. *Nature* **382**, 88–90.

Wu, H., Wade, M., Krall, L., Grisham, J., Xiong, Y., and VanDyke, T. (1996). Targeted *in vivo* expression of the cyclin dependent kinase inhibitor p21 halts hepatocyte cell-cycle progression, postnatal liver development and regeneration. *Gene Dev.* **10**, 245–260.

Wuarin, J., and Nurse, P. (1996). Regulating S phase: CDKs, licensing and proteolysis. *Cell* **85,** 785–787.

Xiao, Z.-X., Ginsberg, D., Ewen, M., and Livingston, D. M. (1996). Regulation of the retinoblastoma protein-related protein p107 by G1 cyclin associated kinases. *Proc. Natl. Acad. Sci. USA* **93,** 4633–4637.

Yamasaki, L., Jack, T., Bronson, R., Goillot, E., Harlow, E., and Dyson, N. J. (1996). Tumor induction and tissue atrophy in mice lacking E2F-1. *Cell* **85,** 537–548.

Zhang, H., Kobayashi, R., Galaktionov, K., and Beach, D. (1995). p19^Skp1 and p45^Skp2 are essential elements of the cyclin A-CDK2 S phase kinase. *Cell* **82,** 915–925.

*Christopher D. Heinen, Kira Steigerwald, Donna McLaren, and Joanna Groden*

# 66

## The Cancer Cell

## I. Introduction

Understanding the process of neoplastic transformation implies an understanding of how the physiology of the transformed cell has altered in comparison to its normal counterpart. Although sometimes obscure in their beginnings, the characteristics of transformed cells eventually announce themselves by a generalized loss of responsiveness to the signals that normally promote growth arrest, differentiation, or sometimes cell death. The study of how these normal processes are altered during neoplastic transformation will enable the development of newer and better therapies to treat a variety of tumors.

One approach to defining the physiological changes that occur in the transformed cell is through the use of genetics. Such techniques have led to the identification of genes that are altered during tumorigenesis and **disease genes**—genes whose aberrant alleles are responsible for defined clinical syndromes that predispose to cancer (Fig. 1). These genetic clues have introduced us to the pathways and processes that are normally required for growth control and have allowed the definition of physiological systems within the cell.

It is widely accepted that as a cell becomes neoplastic, it acquires mutations in a variety of genes and that these mutations lead to both losses and gains of function. These genes encode the regulatory components of systems normally required to maintain the cell in equilibrium with its environment. Through mutation of one or more of these genes, the transformed cell acquires the ability to evade normal inhibitory signals from the environment and to become dysplastic or ultimately invasive and/or metastatic. Such physiological systems that are targets of mutation include those that regulate the cell cycle and the transcription of genes by signal transduction pathways, that maintain genomic integrity, that control the ability of cells to interact with their neighbors, or that induce programmed cell death or apoptosis under specific circumstances.

## II. The Cell Cycle

The life of a growing cell is cyclical in nature. It progresses through four major phases during which specific cellular functions are performed. In S phase, the cell replicates its genome producing two sets of chromosomes. The cell ultimately divides into two daughter cells during M phase, with each daughter inheriting one set of chromosomes. Following M, and prior to S, is a growth phase, $G_1$, in which the cell grows and prepares for S phase. Similarly, following S and prior to M is a second growth phase, $G_2$, in which the cell grows further and prepares for M. This cycle is tightly controlled, ensuring that one S phase is completed prior to every M phase.

The cell commits to one complete cycle at a specific point during the $G_1$ phase. Beyond this point, referred to as START in yeast or the restriction point in mammalian cells, the cell will no longer be influenced by external signals until it has completed the full cycle. Prior to START, the cell receives information from its external environment in the form of positive regulators such as growth factors, or negative regulators such as contact inhibition or other signals that induce differentiation and exit from the cell cycle.

These positive and negative signals are communicated to the cell cycle machinery of the nucleus through a cascade of proteins that begins at the cell surface. Cell-surface receptors receive external signals and set off a wave of protein–protein interactions in the cytoplasm that eventually conduct signals to the nucleus. The result is transcription of specific genes involved in cell cycle regulation. A transformed cell may often produce a positive growth signal independent of external regulators. This self-stimulation occurs via three different mechanisms: (1) autocrine production of growth factors, (2) production of altered growth factor receptors that function as activated receptors even in the absence of ligand, and (3) constitutive activation of components in growth factor signal transduction cascades (reviewed in Grunicke, 1990). The malignant cell will continue to grow and divide, ignoring external cues to stop cycling and begin differentiation.

An example of the third mechanism of self-stimulated growth involves mutations of the *ras* family of genes. The *ras* proto-oncogenes (*H-ras, K-ras,* and *N-ras*) belong to

*1021*

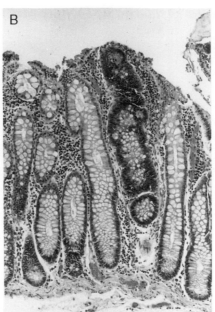
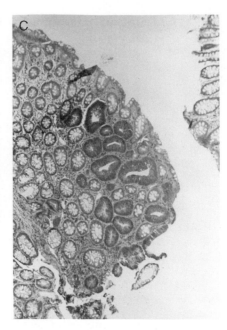

**FIG. 1.** Disruptions of the tumor suppressor gene *APC* are associated with abnormal proliferation of the colonic epithelium. With the identification of genes involved in hereditary cancers, researchers have begun the study of different physiological processes in normal cells, such as those that control cell growth. When the *APC* gene is mutated in the germline, numerous adenomas form in the colon and rectum of the affected individual. (A) A colon from a patient with adenomatous polyposis coli (APC) that is carpeted with numerous adenomas. (B) A hematoxylin-and-eosin (H&E)-stained section of the colon from a different APC patient, illustrating the change in the epithelium of a single-crypt adenoma as compared to the normal crypts surrounding it. A second *APC* mutation has most likely occurred in these cells that have become neoplastic. (C) An H&E-stained section from a larger adenoma, showing the increase in epithelial dysplasia. (Courtesy of Cecelia Fenoglio-Preiser, M.D., University of Cincinnati.)

a superfamily of low molecular weight guanosine triphosphate (GTP)-binding proteins implicated in signal transduction pathways for cellular growth and differentiation (reviewed in Bokoch and Der, 1993). The active, GTP-bound form of Ras contains a weak GTPase activity, which converts GTP to guanosine diphosphate (GDP) and consequentially deactivates Ras. This GTP hydrolysis activity is enhanced $10^5$-fold by interactions with GTPase-activating proteins or GAPs. The Ras–GAP interaction can be considered an off-switch in most cells, shutting down the signal transducted via Ras. Ras–GAP interactions can be disrupted by specific point mutations of the *ras* gene that occur in human tumor cells. These mutations exclusively affect codons 12, 13, and 61 of the Ras proto-oncogenes. Disruption of these amino acids leads to a constitutively active form of the Ras protein, which ultimately affects signal transduction. In addition, alterations in the Ras signaling pathway can occur by the alteration of interacting proteins needed for the down-regulation of the Ras growth-promoting signal.

Studies of families with the autosomal dominant disorder neurofibromatosis type I lead to the mapping and positional cloning of this disease gene, known as *Nf1*. Individuals who inherit one mutated *Nf1* gene develop numerous neurofibromas and are afflicted with other abnormalities of growth and development. Analysis of the *Nf1* gene product identified a region with significant homology to the GAP proteins; this region also has been shown to catalyze the Ras–GTP to Ras–GDP reaction (Xu *et al.*, 1990). It is now believed that the *Nf1* gene product is necessary for the down-regulation of the Ras–GTP growth-promoting signal in some specific cell types and that its loss leads to a generalized up-regulation of a positive growth signal. It is unknown how the different GAPs interact to mediate Ras signaling, although numerous researchers have investigated the ability of the Ras pathway to up- and down-regulate growth.

In classic experiments, transfection of rodent fibroblasts with an oncogenic form of *ras* resulted in cells with a transformed phenotype that grew independently of serum growth factors, lost contact inhibition, and caused tumors when injected into athymic mice (Goldfarb *et al.*, 1982; Land *et al.*, 1983; Parada *et al.*, 1982). Gene transfer of *Nf1* into melanoma cells has been shown to down-regulate the Ras growth signal (Johnson *et al.*, 1994). In a different cell type, however, the rat pheochromocytoma cell line PC12, adding a constitutively activated Ras, leads to differentiation of these cells, not transformation (Bar-Sagi and Feramisco, 1985). Thus, components of one signal transduction pathway may contribute to different pathways, resulting in opposite outcomes.

Cells induced to differentiate exit the cell cycle and enter a quiescent state referred to as $G_0$. The decision to enter $G_0$ or to proceed through the cell cycle is determined by a series of cell cycle control proteins. The major proteins of the cell cycle machinery are a family of kinases called the cyclin-dependent kinases (CDKs). These kinases are activated by pairing with partner cyclin proteins, which then are able to phosphorylate downstream targets and propel the cell through the cell cycle. An additional class of cell cycle proteins, the cyclin–CDK inhibitors (CDIs) inhibit progression through the cell cycle by interfering with the cyclin activation of a CDK.

In addition to regulating the transition at START, different members of these protein families control transitions prior to S and M phases. These transition points are referred to as **cell cycle checkpoints.** Disruptions of these checkpoints were first identified in specific tumor types and in cells infected with tumor viruses. In addition, tumor suppressor proteins were shown to regulate these checkpoints, of which the retinoblastoma gene product or RB is the best example (Fig. 2).

The identification of the RB tumor suppressor stemmed from studies of a pediatric tumor of the eye, retinoblastoma, known to occur in either a sporadic or familial form. The work of Knudson (1971) first suggested that the familial form was due to the inheritance of one mutant allele at a locus and that the differences in age of onset and tumor number in sporadic versus familial cases was due to the frequency with which the second allele acquired an inactivating mutation in familial cases. Sporadic cases were due to the acquisition of two mutations at the same locus. This paradigm has been applied to understanding the function of most tumor suppressors associated with inherited predisposition to cancer, where loss of function at both alleles of a locus is associated with loss of growth control in a cell and where the inheritance of one mutant allele accelerates the incidence of tumor formation. Subsequent cloning of the gene and the analysis of its gene product showed that RB was a nuclear protein that was phosphorylated in a cell cycle-specific manner.

It is now known that the RB tumor suppressor plays a key role at the START checkpoint. During $G_1$, the RB protein is hypophosphorylated and arrests cells in $G_1$ by binding members of the E2F family of transcription factors, which are essential in the progression of cells into S phase. Following phosphorylation of RB, the E2F proteins are released and cells begin to cycle (reviewed in Hinds and Weinberg, 1994). This phosphorylation is mediated by the D-type cyclins in association with their partner CDKs (CDK4 and CDK6) (reviewed in Hunter and Pines, 1994). Inhibiting this reaction is the CDI, p16[Ink4], which binds to CDK4 or CDK6 and prevents the binding of cyclin D. The CDIs represent an additional level at which progression of the cell cycle can be regulated (reviewed in Hunter and Pines, 1994).

Another well-characterized cell cycle checkpoint occurs at the $G_1$/S transition, where cells will arrest in response to DNA damage. This delay is mediated by the p53 tumor suppressor, germline mutations of which lead to an increase in childhood sarcomas and/or the familial cancer disorder known as Li–Fraumeni syndrome. Experiments by Kuerbitz *et al.* (1992) reveal a cell cycle block at $G_1$ in cells with normal p53 in response to radiation-induced DNA damage. The $G_1$ block is not present, however, when the experiment is repeated with cells that lack functional p53. When DNA damage is detected, the p53 protein is upregulated at the $G_1$/S boundary (Kastan *et al.,* 1991), which leads to an upregulation of the CDI p21 (reviewed in Sherr, 1994). The p21 protein can interfere with a number of $G_1$ cyclin–CDK complexes as well as the proliferating cell nuclear antigen (PCNA) that functions in DNA replication and repair. This pauses the cell cycle progression, presumably to correct the damaged DNA prior to replication. Failure to arrest the damaged cells would result in the propagation of any DNA mutations during S phase (reviewed in Sherr, 1994).

In addition to the block at $G_1$/S, an arrest in response to DNA damage occurs at the $G_2$/M transition, although this checkpoint remains functional in the absence of normal p53. The proteins involved in the $G_2$/M checkpoint are less well understood. It is known that the complex of cyclin B and p34[CDC2] is necessary for entry into M phase; however, the mechanism by which the cell would be arrested at this checkpoint remains unclear. Interestingly, the APC tumor suppressor protein, mutated in one type of familial predisposition to colon cancer, is phosphorylated by p34[CDC2] at the $G_2$/M boundary, suggesting that it may function in some type of regulatory checkpoint in late $G_2$ or M phase of the cell cycle (Trzepacz *et al.,* 1997).

### III. Genome Stability

The likelihood of developing a tumor might seem rare when considering the requirement for multiple mutations in tumor development and the slow rate of mutation in human cells, estimated at $1.4 \times 10^{-10}$ mutations/nucleotide/cell generation. Therefore, researchers have hypothesized that the mutation rate of tumor cells must exceed that of normal cells (Loeb *et al.,* 1974). Therefore, tumors might require early mutational events that not only confer a growth advantage to the cell, but also lead to an increase in genomic instability.

Studies of *p53* and *Ha-ras* have revealed increased levels of genomic instability when these genes are mutant. Using

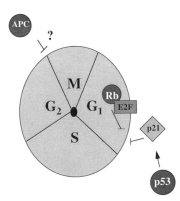

**FIG. 2.** Tumor suppressors control cell cycle checkpoints. The four major phases of the cell cycle are shown in this figure. Tumor suppressors such as RB and p53 regulate a cell cycle checkpoint during $G_1$. Changes in the phosphorylation status of APC in M phase may point to a role for APC as a checkpoint regulator in $G_2$. Mutations of these tumor suppressor genes result in the loss of checkpoint regulators and, thus, uncontrolled growth.

cells from Li–Fraumeni patients, which carry one mutation of *p53,* it was observed that gene amplification occurred at a much higher frequency in cells that lose the remaining *p53* allele than those with at least one functional *p53* gene (Livingstone *et al.,* 1992; Yin *et al.,* 1992). In addition, NIH-3T3 cells that have been stably transfected with activated *Ha-ras* demonstrate an increase in the frequency of chromosome breaks and rearrangements (Denko *et al.,* 1994).

Additional candidates for genes involved in maintaining genome stability are those involved in DNA replication or repair. The strongest evidence for this has been the discovery of a link between hereditary nonpolyposis colon cancer (HNPCC) and genes involved in the mismatch repair system (reviewed in Kinzler and Vogelstein, 1996). Molecular analyses have identified four genes responsible for HNPCC: *hMLH2, hMLH1, PMS1,* and *PMS2.* The products of these four genes are involved in the human mismatch repair system, which identifies and corrects DNA replication errors. A feature of HNPCC tumors is the frequent presence of instability at small repeat sequences called **microsatellites.** This microsatellite instability results from slippages of the DNA polymerase when it replicates these tracts of DNA that lead to deletion or insertion of additional repeat units (reviewed in Kinzler and Vogelstein, 1996).

Microsatellite instability has been detected in a variety of tumor types and also in the earliest detectable stages of colon tumor development, indicating that mutations of the DNA mismatch repair system are early events in tumorigenesis (Shibata *et al.,* 1994; Augenlicht *et al.,* 1996; Heinen *et al.,* 1996). Microsatellite instability has been identified in the regenerating colonic mucosa of ulcerative colitis patients (Brentnall *et al.,* 1996; Heinen *et al.,* 1997). In addition, microsatellite instability has been reported by Brentnall *et al.* in normal pancreatic tissue from patients with chronic pancreatitis. Interestingly, an increased risk of cancer also exists for patients with these chronic inflammatory diseases, suggesting again that genomic instability can be a very early event in the tumor predisposition associated with these somatic disorders.

Finally, the chromosome breakage syndromes such as ataxia telangiectasia, Bloom's syndrome, Fanconi anemia, Werner's syndrome, and xeroderma pigmentosum are autosomal recessive diseases that are characterized by increases in both chromosome instability and cancer predisposition. The discovery of the disease genes for these disorders has led to the identification of numerous proteins necessary for cells to replicate or repair their DNA and include helicases, kinases that respond to DNA damage, and proteins necessary for nucleotide excision repair. Mutation of these genes and the genes of the DNA mismatch repair system act in a recessive manner in cells (similar to that of tumor suppressors), although their absence does not directly affect growth regulation. Mutation of these genes results in a mutator phenotype within the cell, which leads to an increased frequency of further mutation, the targets of which may then be those genes directly involved in growth control (Fig. 3).

Therefore, inherited predisposition to human cancer can

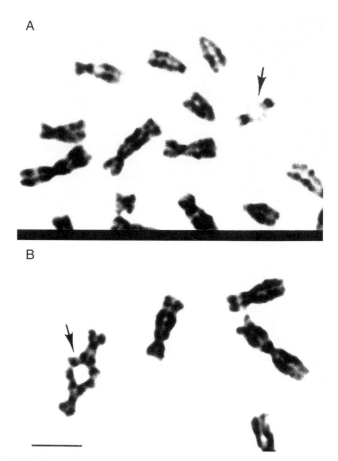

**FIG. 3.** Symmetrical quadriradial configurations (Qrs) from a Bloom's syndrome lymphocyte are examples of genomic instability. Arrows point to Qrs in G-banded chromosome preparations. (A) A Qr that is due to an exchange between the long arms of chromosome 19. (B) A Qr that is due to an exchange between the long arms of chromosome 9 (scale bar: 5 μm). The increased genomic instability in Bloom's syndrome cells may explain the increased predisposition to cancer in these patients. (Courtesy of Steven Schonberg, Ph.D., and James German, M.D., The New York Blood Center.)

occur by one of two genetic mechanisms. The first is by the inheritance or acquisition of a germline mutation in one of the genes involved in a growth control pathway for a particular cell type. The second is by the inheritance or acquisition of a germline mutation in a gene that controls the ability of a cell to replicate or repair its DNA and whose disruption consequently is associated with an increased mutation frequency throughout the genome. The disruption of such genes can be recognized clinically by an increased incidence of a particular tumor type in affected individuals or by an increased incidence of many tumor types in affected individuals.

## IV. Cell Adhesion and Motility

A cell that has lost growth control and progressed to a neoplasm eventually becomes restricted from further growth due to the size of the cell mass and the lack of nutrients in the immediate area. There are three possible outcomes: The neoplasm remains a benign tumor and growth ceases; the neoplasm induces angiogenesis such that vascularization allows the neoplasm to grow into a larger tumor; or cells of the neoplasm undergo further mutations that allow them to invade surrounding normal tissue and metastasize to secondary sites. Such cells acquire alterations in adhesive properties and motile properties, and will often inappropriately express proteases.

Cellular adhesion plays a fundamental role during development and is required for maintenance of tissue integrity in the adult. Almost all cells in the human body are in contact with surrounding cells, either of the same tissue type or a different type, and/or surrounding extracellular matrix (ECM). Cellular adhesion can also be considered in two subsets: cell–cell adhesions and cell–matrix adhesions.

The specificity of cellular adhesion is at the molecular level. Cell adhesion molecules (CAMs) are molecules that facilitate adhesion between a cell and another substrate (typically a cell surface or matrix). Adhesion between identical molecules is termed **homophilic adhesion** whereas adhesion between distinct molecules is termed **heterophilic adhesion.** As discussed later, CAMs can perform more than just an adhesive role (reviewed in Kirkpatrick and Peifer, 1995). The CAMs are categorized according to their binding specificities and are arranged into families that share homology and functional domains. The four major families are the immunoglobulin superfamily, the cadherins, the integrins, and the selectins (Fig. 4). These molecular families are increasing in size, although they are becoming less distinct because functional domains are often shared between more than one family of molecules. For example, the protein tyrosine phosphatases have been shown to function as adhesion receptors and as signaling molecules (Brady-Kalnay and Tonks, 1995). CAMs contain an extracellular domain, a transmembrane domain, and a cytoplasmic domain. Regulation of CAM function can occur at the level of protein expression, conformational alteration of the protein, or biochemical modification of the CAM. This regulation is crucial in determining when a cell is adhesive.

During development, many embryonic cells down-regulate certain adhesion molecules prior to their migration through the embryonic tissue. Once the cells arrive at their destination, those same molecules are up-regulated again, in order for cells to coalesce and form mature tissue. For example, it has been well demonstrated that neural crest cells, a transient population of migratory embryonic cells, decrease expression of neuronal CAM (N-CAM) just prior to their exit from the neural tube (Akitaya and Bronner-Fraser, 1992). As cell–cell adhesion is reduced, the neural crest cells can separate from one another and, following specific guidance cues, migrate through the developing embryo. Once these cells reach their destination, N-CAM is upregulated again and the cells readhere in order to develop adult tissue, such as the dorsal root ganglia of the peripheral nervous system. Function-blocking antibodies against N-CAM can inhibit neural crest cell migration,

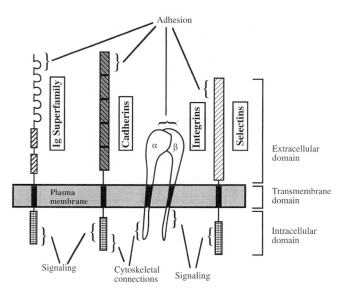

**FIG. 4.** Structure and function of cell adhesion molecules. This figure diagrams the four major families of CAMs: the immunoglobulin (Ig) superfamily, the cadherins, the integrins, and the selectins. Note that each CAM molecule contains an extracellular domain, a transmembrane domain, and a cytoplasmic domain. While there are common functional domains found *within* each family, note the common functions found *between* families. CAMs serve as adhesion molecules by way of extracellular domains. The CAMs also transduce signals from the outside of the cell to the inside of the cell via activation of regions in the cytoplasmic domain. These regions then can bind cytoskeletal elements thereby forming a line of communication between the outside of the cell and the motile apparatus.

demonstrating that it not only is involved in cell–cell adhesion, but also in the cell–matrix adhesion required for migration (Bronner-Fraser *et al.,* 1992). Many of the early studies on neural crest cell migration can be considered a template for current studies on metastasis. The basic questions are these: What controls the detachment of cells from an originating population? What mechanisms are involved in their invasion of surrounding tissue and/or migration? What determines the final resting place where a neural crest cell will differentiate or a tumor cell will proliferate and form a new tumor mass?

A classic example of a CAM whose function is modified in many tumors is E-cadherin. E-cadherin is decreased in malignant carcinomas, conferring invasive potential to those cells. Alternatively, molecules that regulate cadherin function, for example, the catenins, can be absent due to mutational inactivation or down-regulation, giving rise to the same outcome (Morton *et al.,* 1993; Sommers *et al.,* 1994; Breen *et al.,* 1995; Vermeulen *et al.,* 1995).

In order for a cell to migrate, however, it also must secrete proteases to digest the local ECM. Many types of brain tumors with an invasive nature express increased levels of membrane-type matrix metalloproteinases (MT-MMP), suggesting an important role for these enzymes in invasion. *In vivo* studies of malignant astrocytomas demonstrated elevated levels of MT-MMP mRNA and protein in tumor cells compared to normal brain tissue. Low-grade gliomas that do not invade surrounding normal brain tissue do not express increased levels of MT-MMPs (Yamamoto *et al.,* 1996). Conversely, tissue inhibitors of metalloproteinases (TIMPs) are negative regulators of MMPs. Their down-regulation correlates with increased invasiveness of glioblastomas and astrocytomas (Mohanam *et al.,* 1995). Other proteases, such as cathepsin B, produced by some gliomas, also have been shown to contribute to a more invasive phenotype (Sivaparvathi *et al.,* 1995).

In addition to activating proteases, a cell needs to engage its motile apparatus. CD44 is a transmembrane glycoprotein that binds the ECM component collagen extracellularly and cytoskeletal components intracellularly. The cytoplasmic domain of CD44 interacts with the cytoskeletal protein ankryin. This binding can be modified biochemically by phosphorylation, acylation, and binding of GTP. Intracellular regulation of CD44 controls the adhesive properties of the extracellular domain of CD44, forming a line of communication between cell adhesion and the cell motility apparatus. Some metastatic carcinomas express isoforms of CD44 that do not interact with ankryin in the correct manner, allowing the cells to lose their spatial restriction (Bourrguignon *et al.,* 1995).

Additionally, the presence of antiadhesive molecules such as proteoglycans, mucins, laminins, and thrombospon-dins has been well documented in the developing nervous system (Chiquet-Ehrismann, 1995), where they prohibit the migration of certain cell types. These molecules may play a role in restricting invading tumor cells to a more limited terrain. On the other hand, if such antiadhesive molecules are inappropriately down-regulated, invasion might be more severe.

The combined effects of cellular adhesion, protease secretion, and motility must be considered when studying tumor progression to metastasis. The events involved in the progression to the metastatic state can be considered a recapitulation of normal developmental events; the difference being that the changes a developing cell undergoes are preprogrammed whereas the changes a cancer cell undergoes are the result of random mutation (Fig. 5). In teleological terms, the tumor cells are undergoing similar changes to the normal cells, given that they have equivalent objectives: to migrate to a distant site, settle there, begin to divide, and populate the new area.

## V. Apoptosis

Apoptosis is a programmed cell death necessary for proper development and cell turnover (Kerr *et al.,* 1972). It is functionally and morphologically distinct from necrosis, or cell death due to tissue injury. During apoptosis

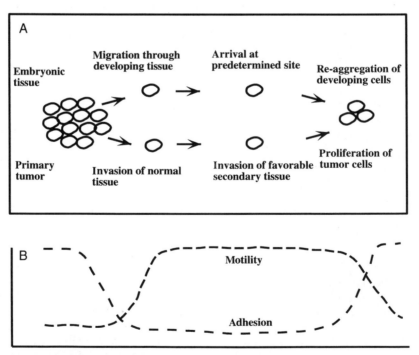

**FIG. 5.**   Tumor metastasis is a recapitulation of development. This figure illustrates the parallels between cellular events that take place during development and during metastasis. (A) A population of cells that represents embryonic cells in the upper portion of the figure or primary tumor cells in the lower portion. In both scenarios, the migrating cell will undergo similar changes in the relative levels of adhesion and motility (depicted in B). In order for the cell to detach from its original mass, it must down-regulate adhesion molecules. In order for the cell to migrate, it must up-regulate molecules involved in motility. Once the embryonic cell reaches its destination, or the tumor cell finds itself in a suitable environment, the opposite occurs: Adhesion is up-regulated and motility is down-regulated.

chromatin compacts into sharply delineated masses associated with the nuclear envelope; nuclear and cellular outlines become convoluted (Fig. 6). The cell and the nucleus then convolute extensively, forming discrete, membrane-enclosed fragments containing intact organelles. These cell fragments are phagocytosed by surrounding cells and degraded inside lysosomal vacuoles. No tissue inflammation is present and apoptosis can occur on a single-cell basis without affecting the status of neighboring cells. Necrosis is direct cell lysis triggered by cell membrane damage or high levels of toxicity and usually involves many cells in an inflamed area. Chromosomes condense, however, the condensed chromatin forms irregular clumps inside the nucleus. All organelles appear swollen, as the membranes of the cell and organelles begin to break down. The resulting cellular debris is removed by mononuclear phagocytes that have migrated from small blood vessels and capillaries during the inflammatory response.

During development of multicellular organisms, subsets of cells undergo apoptosis as an obligate part of the developmental process (Sanders and Wride, 1995). During normal vertebrate development, apoptosis has been observed as early as the blastocyst stage, in both the uterine epithelium and the blastocyst during the implantation process. Later in development, apoptosis is essential for normal nervous system development, loss of the tail bud in the human embryo, and proper development of vertebrate limbs, which begin as paddle-shaped outgrowths and assume their final shape by a balance of both cell growth and cell death (reviewed in Hincliffe, 1981).

In the adult, apoptosis is important in the immune system for elimination of autoimmune T cells in the thymus. In addition, apoptosis maintains a constant cell number in continuously renewing cell populations such as haemato-poietic and colonic epithelial cells. Thus, the size of these cell populations is determined not only by the rate of cell proliferation, but also by the rate of cell death. When the apoptotic mechanism is disrupted, the ability to remove unwanted cells is lost. The down-regulation of apoptosis is associated with a number of different tumor types, such as chronic myeloid leukemia (CML), many human lymphomas (Vaux *et al.*, 1988), and in the transformation of normal epithelium to carcinoma in the colon (Bedi *et al.*, 1995) and prostate (Denmeade *et al.*, 1996). Interestingly, the level of apoptosis in some tumors of the breast is increased (Wu, 1996), although the mechanism of this type of deregulation is not understood.

A better understanding of the apoptotic process has been gained by studying the genes affected in some of these tumor types. For example, a cytogenetic hallmark of CML is a balanced translocation, t(9;22), known as the Philadelphia chromosome, which results in a chimeric *bcr-abl* gene. This chimeric gene encodes a fusion protein specific to CML cells. Normally, the *bcr* gene is ubiquitously expressed and encodes a 160-kDa protein of unknown function (Hariharan and Adams, 1987; Stam *et al.*, 1987), while the *abl* gene encodes a protein with homology to the tyrosine kinase family of proteins (Konopka and Witte, 1985). Studies of the fusion protein BCR-ABL reveal that its expression suppresses apoptosis *in vitro;* inhibition of BCR-ABL expression with antisense nucleotides reverses this effect. Since most studies have found the relative rates of cell proliferation not increased in CML (Strife and Clarkson, 1988), the primary mechanism by which BCR-ABL increases cell number in CML appears to be through the suppression of apoptosis.

In a second example, non-Hodgkin's B-cell lymphoma is associated with a t(14;18) translocation in the majority of cases. These translocations also disrupt normal apoptotic function while joining the *bcl-2* gene with the immunoglobulin heavy-chain locus. The result is transcriptional deregulation of the *bcl-2* gene. Normally, the role of Bcl-2 in the cell is to suppress apoptosis (Vaux *et al.*, 1988). Increased levels of Bcl-2 in a clonal population of cells such as a lymphoma leads to an increase in cell number caused by a decrease in cell death. Additionally, increased levels of Bcl-2 *in vitro* confer resistance to killing by some glucocorticoids (Miyashita and Reed, 1992), indicating that the apoptotic status of a tumor can dictate which chemotherapeutic regimes are more effective.

Apoptosis also plays a role in the maintenance of consistent epithelial cell populations. Therefore, loss of apoptotic function can be involved in the formation of carcinomas. The antihormone-induced mammary tumor cell death may be apoptotic in nature (Wu, 1996) as is the loss of androgen-dependent prostatic cancer cells following surgical androgen ablation. Also, a loss of apoptotic activity is associated with the transformation of normal colorectal epithelium to carcinoma (Bedi *et al.*, 1995). The progression of breast cancer may be associated with high levels of Bcl-2 expression (Wu, 1996).

Gross rearrangements of *bcl-2*, like the lymphoma-related *bcl-2* translocations, have not been observed in

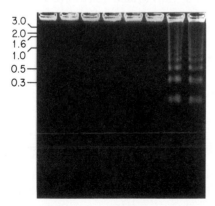

**FIG. 6.** DNA laddering is a characteristic of a cell population undergoing apoptosis. This figure shows an ethidium bromide-stained agarose gel containing DNAs from a transformed cell line that is dependent on interleukin-3 (IL-3) for growth. When IL-3 is removed from the culture medium, the cell line undergoes programmed cell death or apoptosis. Evidence for this process is seen in lanes 7 and 8, with degradation of DNA in a laddering pattern. Size markers are represented on the left in kilobases. (Courtesy of David Askew, Ph.D., University of Cincinnati.)

carcinomas, suggesting that alternative mechanisms for dysregulation of apoptosis exist. A candidate for such dysregulation is the alteration of p53 function that is observed in a large number of carcinomas. The importance of p53 in apoptosis was discovered by experiments using thymocytes from *p53* knockout mice. Comparisons of these thymocytes with control thymocytes, when treated with ionizing radiation, revealed that the *p53* −/− cells were extremely resistant to the effects of radiation. The control cells, meanwhile, underwent massive apoptosis at doses as low as 1 Gy (Lowe *et al.*, 1993; Clark *et al.*, 1993). p53 is now believed to function in response to DNA damage, either by arresting the cell cycle at the $G_1/S$ boundary to facilitate repair or by its ability to initiate apoptosis and eliminate cells that have acquired DNA damage. The loss of p53 function in many cancers is consistent with the idea that a tumor must acquire mutations during tumorigenesis that inhibit the apoptotic pathway.

In addition, normal p53 is a transcription factor that has been shown to down-regulate the expression of *bcl-2 in vitro* via a 195-bp segment in the 5′ untranslated region of *bcl-2*. p53 also up-regulates expression of *bax*, which encodes a Bcl-2-related protein that is a dominant inhibitor of Bcl-2 (Miyashita and Reed, 1995). The *bax* gene is presumably regulated at the level of expression via p53 binding at the *bax* promoter region, which contains four p53-binding motifs (Miyashita and Reed, 1995). Additionally, radiation-induced DNA damage causes increased *bax* expression by p53 (Miyashita *et al.*, 1994).

Colorectal carcinomas are usually characterized by mutations in both copies of the *p53* gene. In addition, the earliest mutation in the formation of most of these carcinomas is mutation of the *APC* gene. *APC* mutations lead to an increase in the number of epithelial cells from an intestinal crypt, either from a decrease in apoptosis or by an increase in cell growth. Gene transfer studies have been designed to address this question and have shown that the introduction of *APC* to tumorigenic cells in culture can lead to a decrease in cell number by an increase in apoptosis (Morin *et al.*, 1996). Such experiments suggest that the down-regulation of multiple apoptotic pathways can occur in some tumor types and that the APC tumor suppressor may be involved in a different apoptotic pathway than p53. APC may indeed be part of a normal developmental process in the colonic crypts, where apoptosis of the lumenal epithelial cells is constantly occurring.

## VI. Summary

Proteins that regulate the cell cycle have been identified through the study of inherited diseases that predispose to cancer and by genetic analyses of different tumor types. Many of the rate-limiting steps of such growth control pathways have been revealed by finding the genes mutated in tumors and in subsequent study of their gene products. Consequently, researchers have begun to dissect the numerous pathways of protein interactions that control the ability of a cell to replicate its genome and to divide in attempts to understand the normal mechanics of the cell cycle, and how these mechanisms have broken down in tumors.

We also have seen that the processes of regulated DNA replication and cell division are intimately involved with the maintenance of the genome. Gene mutations have been identified that increase genomic instability and allow cells to accumulate further mutations at a rapid rate. Such a mutator phenotype may be an essential characteristic of many tumors, facilitating the number of mutations that a cell must acquire before becoming tumorigenic.

Malignant tumors are characterized by the ability to invade surrounding normal tissues. Many of these cells then may metastasize to new sites through subsequent changes in the expression of genes that regulate cell adhesion and motility. These events seem to echo normal processes of embryonic development where cells are programmed to migrate to distant sites. Studies of the genes involved in development, therefore, have led to new information about the metastatic potential of tumor cells.

Finally, the unregulated growth of a tumor may result from the inability of the tumor cells to induce appropriate cell death. Such apoptotic pathways frequently are disrupted by mutations of specific genes in tumors. Mechanisms for deleting unwanted or aberrant cells from these tumors are lost, thus adding to the tumor mass.

As cells replicate their DNA and divide, they must coexist and cooperate with neighboring cells to compose different tissues. Therefore, the growth and life span of each cell is tightly regulated, with a precise balance between cell division and its inhibition. When the regulatory systems that maintain this balance are disrupted, the scales tilt in favor of growth with a subsequent increase in cell number. Such altered cells ultimately produce clonal populations of cells, or what we see clinically as tumors. An understanding of the systems that regulate appropriate cell growth is developing from the identification of genes and their gene products that are altered in tumors. Comparative studies of normal and transformed cells have revealed many of the proteins and processes involved in maintaining the normal physiology of the cell. Ultimately, this information will lead to new therapies that can modulate the systems that are disrupted in tumors.

## Bibliography

Akitaya, T., and Bronner-Fraser, M. (1992). Expression of cell adhesion molecules during initiation and cessation of neural crest cell migration. *Dev. Dyn.* **194**, 12–20.

Augenlicht, L. H., Richards, C., Corner, G., and Pretlow, T. P. (1996). Evidence for genomic instability in human colonic aberrant crypt foci. *Oncogene* **12**, 1767–1772.

Bar-Sagi, D., and Feramisco, J. R. (1985). Micro-injection of the *ras* oncogene protein into PC12 cells induces morphological differentiation. *Cell* **42**, 841–848.

Bedi, A., Pasricha, P., Akhtar, A., Barber, J., Bedi, G., Giardiello, F., Zehnbauer, B., Hamilton, S., and Jones, R. (1995). Inhibition of apoptosis during development of colorectal cancer. *Cancer Res.* **55**, 1811–1816.

Bokoch, G. M., and Der, C. J. (1993). Emerging concepts in the *ras* superfamily of GTP-binding proteins. *FASEB* **7**, 750–759.

Bourrguignon, L. Y., Iida, N., Welsh, C. F., Zhu, D., Krongrad, A., and Pasquale, D. (1995). Involvement of CD44 and its variant isoforms in membrane-cytoskeleton interaction, cell adhesion and tumor metastasis. *J. Neurooncol.* **26**, 201–208.

Brady-Kalnay, S. M., and Tonks, N. K. (1995). Protein tyrosine phosphatases as adhesion receptors. *Curr. Opin. Cell. Biol.* **7**, 650–657.

Breen, E., Steele, G., Jr., and Mercurio, A. M. (1995). Role of the E-cadherin/adlpha-catenin complex in modulating cell–cell and cell–matrix adhesive properties of invasive colon carcinoma cells. *Ann. Surg. Oncol.* **2**, 378–385.

Brentnall, T. A., Chen, R., Lee, J. G., Kimmey, M. B., Bronner, M. P., Haggit, R. C., Kowdley, K. V., Hecker, L. M., and Byrd, D. R. (1995). Microsatellite instability and K-ras mutations associated with pancreatic adenocarcinoma and pancreatitis. *Cnacer Res.* **55**, 4264–4267.

Brentnall, T. A., Crispin, D. A., Bronner, M. P., Cherian, S. P., Hueffed, M., Rabinovitch, P. S., Rubin, C. E., Haggitt, R. C., and Boland, C. R. (1996). Microsatellite instability in nonneoplastic mucosa from patients with chronic ulcerative colitis. *Cancer Res.* **56**, 1237–1240.

Bronner-Fraser, M., Wolf, J. J., and Murray, B. A. (1992). Effects of antibodies against N-Cadherin and N-Cam on the cranial neural crest and neural tube. *Dev. Biol.* **153**, 291–301.

Chiquet-Ehrismann, R. (1995). Inhibition of cell adhesion by anti-adhesive molecules. *Curr. Opin. Cell. Biol.* **7**, 715–719.

Clarke, A. R., Purdie, C. A., Harrison, D. J., Morris, R. G., Bird, C. C., Hooper, M. L., and Wyllie, A. H. (1993). Thymocyte apoptosis induced by p53-dependent and independent pathways. *Nature* **362**, 849–852.

Denko, N. C., Giaccia, A. J., Stringer, J. R., and Stambrook, P. J. (1994). The human *Ha-ras* oncogene induces genomic instability in murine fibroblasts within one cell cycle. *Proc. Natl. Acad. Sci. USA* **91**, 5124–5128.

Denmeade, S. R., Lin, X. S., and Isaacs, J. T. (1996). Role of programmed (apoptotic) cell death during the progression and therapy for prostate cancer. *Prostate* **28**, 251–265.

Goldfarb, M., Shimizu, K., Perucho, M., and Wigler, M. (1982). Isolation and preliminary characterization of a human transforming gene from T24 bladder carcinoma cells. *Nature* **296**, 404–409.

Grunicke, H. (1990). Signal transduction mechanisms in cancer. *Biochem. Soc. Trans.* **18**, 67–69.

Hariharan, I. K., and Adams, J. M. (1987). cDNA sequencc for human *bcr*, the gene that translocates to the *abl* oncogene in chronic myeloid leukaemia. *EMBO J.* **6**, 115–119.

Heinen, C. D., Shivapurkar, N., Tang, Z., Groden, J., and Alabaster, O. (1996). Microsatellite instability in aberrant crypt foci from human colons. *Cancer Res.* **56**, 5339–5341.

Heinen, C. D., Noffsinger, A. E., Belli, Straughen, J., Fischer, J., Groden, J., and Fenoglio-Preiser, C. M. (1997). Regenerative lesions in ulcerative colitis are characterized by microsatellite mutation. *Genes, Chromosomes Cancer* **19**, 170–175.

Hincliffe, J. R. (1981). Cell death in embryogenesis. *In* "Cell Death in Biology and Pathology" (I. D. Bewn and R. A. Lockshin, Eds.), pp. 35–78. Chapman & Hall, London.

Hinds, P. W., and Weinberg, R. A. (1994). Tumor suppressor genes. *Curr. Opin. Genet. Dev.* **4**, 135–141.

Hunter, T., and Pines, J. (1994). Cyclins and cancer II: Cyclin D and CDK inhibitors come of age. *Cell* **79**, 573–582.

Johnson, M. R., DeClue, J. E., Felzmann, S., Vass, W. C., Xu, G., White, R., and Lowy, D. R. (1994). Ncurofibromin can inhibit Ras-dependent growth by a mechanism independent of its GTPase-accelerating function. *Mol. Cell Biol.* **14**, 641–645.

Kastan, M. B., Onyekwere, O., Sidransky, D., Vogelstein, B., and Craig, R. W. (1991). Participation of p53 protein in the cellular response to DNA damage. *Cancer Res.* **51**, 6304–6311.

Kerr, J. F. R., Wyllie, A. H., and Currie, A. R. (1972). Apoptosis: Basic biological phenomenon with wide-ranging implications in tissue kinetics. *Br. J. Cancer* **26**, 239–257.

Kinzler, K. W., and Vogelstein, B. (1996). Lessons from hereditary colorectal cancer. *Cell* **87**, 159–170.

Kirkpatrick, C., and Peifer, M. (1995). Not just glue: Cell–cell junctions as cellular signaling centers. *Curr. Opin. Genet. Dev.* **5**, 56–65.

Knudson, A. G. (1971). Mutation and cancer: Statistical study of retinoblastoma. *Proc. Natl. Acad. Sci. USA* **68**, 820–823.

Konopka, J. B., and Witte, O. N. (1985). Detection of c-abl tyrosine kinase activity *in vitro* permits direct comparison of normal and altered gene products. *Mol. Cell. Biol.* **5**, 3116–3123.

Kuerbitz, S. J., Plunkett, B. S., Walsh, W. V., and Kastan, M. B. (1992). Wild-type p53 is a cell cycle checkpoint determinant following irradiation. *Proc. Natl. Acad. Sci. USA* **89**, 7491–7495.

Land, H., Parada, L. F., and Weinberg, R. A. (1983). Tumorigenic conversion of primary embryo fibroblasts requires at least two cooperating oncogenes. *Nature* **304**, 596–602.

Livingstone, L. R., White, A., Sprouse, J., Livanos, E., Jacks, T., and Tlsty, T. D. (1992). Altered cell cycle arrest and gene amplification potential accompany loss of wild-type p53. *Cell* **70**, 923–935.

Loeb, L. A., Springgate, C. F., and Battula, N. (1974). Errors in DNA replication as a basis of malignant change. *Cancer Res.* **34**, 2311–2321.

Lowe, S. W., Schmitt, E. M., Smith, S. W., Osborne, B. A., and Jacks, T. (1993). p53 is required for radiation-induced apoptosis in mouse thymocytes. *Nature* **363**, 847–849.

Miyashita, T., and Reed, J. C. (1992). *Bal-2* gene transfer increases relative resistance of S49.1 and WEHI7.2 lymphoid cells to cell death and DNA fragmentation induced by glucocorticoids and multiple chemotherapeutic drugs. *Cancer Res.* **52**, 5407–5411.

Miyashita, T., and Reed, J. C. (1995). Tumor suppressor p53 is a direct transcriptional activator of the human *bax* gene. *Cell* **80**, 293–299.

Miyashita, T., Harigai, M., Hanada, M., and Reed, J. C. (1994). Identification of a p53-dependent negative response element in the bcl-2 gene. *Cancer Res.* **54**, 3131–3135.

Mohanam, S., Wang, S. W., Rayford, A., Yamamoto, M., Sawaya, R., Nakajima, M., Liotta, L. A., Nicolson, G. L., Stetler-Stevenson, W. G., and Rao, J. S. (1995). Expression of tissue inhibitors of metalloproteinases: Negative regulators of human glioblastoma invasion *in vivo*. *Clin. Exp. Metastasis* **13**, 57–62.

Morin, P. J., Vogelstein, B., and Kinzler, K. W. (1996). Apoptosis and APC in colorectal tumorigenesis. *Proc. Natl. Acad. Sci. USA* **93**, 7950–7954.

Morton, R. A., Ewing, C. M., Nagafuchi, A., Tsukita, S., and Isaacs, W. B. (1993). Reduction of E-cadherin levels and deletion of the alpha-catenin gene in human prostate cancer cells. *Cancer Res.* **53**, 3585–3590.

Parada, L. F., Tabin, C. J., Shih, C., and Weinberg, R. A. (1982). Human EJ bladder carcinoma oncogene is homologue of Harvey sarcoma virus ras gene. *Nature* **297**, 474–478.

Sanders, E. J., and Wride, M. A. (1995). Programmed cell death in development. *Int. Rev. Cytol.* **163**, 105–173.

Sherr, C. J. (1994). G1 phase progression: Cycling on cue. *Cell* **79**, 551–555.

Shibata, D., Peinado, M. A., Ionov, Y., Malkhosyan, S., and Perucho, M. (1994). Genomic instability in repeated sequences is an early somatic event in colorectal tumorigenesis that persists after transformation. *Nature Genet.* **6**, 273–281.

Sivaparvathi, M., Sawaya, R., Wang, S. W., Rayford, A., Yamamoto, M., Liotta, L. A., Nicolson, G. L., and Rao, J. S. (1995). Overexpression and localization of cathepsin B during the progression of human gliomas. *Clin. Exp. Metastasis* **13**, 49–56.

Sommers, C. L., Gelmann, E. P., Kermler, R., Cowin, P., and Byers, S. W. (1994). Alterations in beta-catenin phosphorylation and plakoglobin expression in human breast cancer cells. *Cancer Res.* **54,** 3544–3552.

Stam, K., Heisterkamp, N., Reynolds, F. H., and Groffen, J. (1987). Evidence that the Ph gene encodes a 160,000-dalton phosphoprotein with associated kinase activity. *Mol. Cell. Biol.* **7,** 1955–1960.

Strife, A., and Clarkson, B. (1988). Biology of chronic myelogenous leukemia: Is discordant maturation the primary defect? *Semin. Hematol.* **25,** 1–19.

Suzuki, H., Harpaz, N., Tarmin, L., Yin, J., Jiang, H-Y., Bell, J. D., Hontanosas, M., Groisman, G. M., Abraham, J. M., and Meltzer, S. J. (1994). Microsatellite instability in ulcerative colitis-associated colorectal dysplasias and cancers. *Cancer Res.* **54,** 4841–4844.

Trzepacz, C., Lowy, A. M., Kordich, J. J., and Grodin, J. (1997). Phosphorylation of the tumor suppressor APC by the cyclin-dependent kinase p34$^{cdc2}$. *J. Biol. Chem.,* in press.

Vaux, D. L., Cory, S., and Adams, J. M. (1988). *Bcl-2* gene promotes haemopoietic cell survival and cooperates with *c-myc* to immortalize pre-B cells. *Nature* **335,** 440–442.

Vermeulen, S. J., Bruyneel, E. A., Bracke, M. E., De Bruyne, G. K., Vennekens, K. M., Vleminckx, K. L., Berx, G. J., van Roy, F. M., and Mareel, M. M. (1995). Transition from the noninvasive to the invasive phenotype and loss of alpha-catenin in human colon cancer cells. *Cancer Res.* **55,** 4722–4728.

Wu, J. (1996). Apoptosis and angiogenesis: Two promising tumor markers in breast cancer. *Anticancer Res.* **16,** 2233–2240.

Xu, G. F., Lin, B., Tanaka, K., Dunn, D., Wood, D., Gesteland, R., White, R., Weiss, R., and Tamanoi, F. (1990). The catalytic domain of the neurofibromatosis type 1 gene product stimulates *ras* GTPase and complements *ira* mutants of *S. cerevisiae. Cell* **63,** 835–841.

Yamamoto, M., Mohanam, S., Sawaya, R., Fuller, G. N., Seiki, M., Sato, H., Gokaslan, Z. L., Liotta, L. A., Nicolson, G. L., and Rao, J. S. (1996). Differential expression of membrane-type matrix metalloproteinase and its correlation with gelatinase A activation in human malignant brain tumors *in vivo* and *in vitro. Cancer Res.* **56,** 384–392.

Yin, Y., Tainsky, M. A., Bischoff, F. Z., Strong, L. C., and Wahl, G. M. (1992). Wild-type p53 restores cell cycle control and inhibits gene amplification in cells with mutant p53 alleles. *Cell* **70,** 937–948.

*Agustin Guerrero and Juan Manuel Arias*

# 67

# Apoptosis

## I. Introduction

**Apoptosis** is a program for cell deletion triggered by either physiological or abnormal signals. Apoptosis or **programmed cell death** has been the subject of intense work in recent years for many reasons. First, it has been recognized that the amount of cells in a given tissue depends on the balance between cell death and cell division. Second, important biochemical and molecular advances have been developed in the elucidation of the pathways regulating and triggering cell death. Third, it has been found that the underlying cause of different diseases are alterations in the cell death machinery; one example is cancerous growth that could be originated by both an increased rate of cell division or reduced cell death. Alterations in this balance are also present in degenerative disorders as well, illness that could arise from a reduced production of cells or an enhanced rate of cell death. Furthermore, physiological cell death is an essential process in the ontogeny of multicellular organisms. In this respect it has been proposed that normal cell death is involved in different embryonic processes such as the development of the central nervous system, the lumina of tubular structures, or the shaping of limbs. It has been suggested that physiological cell death is programmed particularly during development, implying that some internal molecular clock is involved in triggering cell death regardless of the surrounding signals. However, there are also an abundance of cases where physiological cell death could be triggered by either external signals or the absence of trophic factors. It is quite possible that both internal and external signals are involved in regulating the fate of a given cell. Interestingly, a picture is emerging where signals for cell growth and cell death are so closely intertwined that they probably represent different manifestations of a common theme. Examples of that are the **nerve growth factor (NGF)** or the **T-cell receptor (TCR),** both of which could either induce growth and differentiation or kill a cell. The mechanism for that, and more importantly how the same signal or receptor could produce such a divergent result, is just one of many areas of study in cell death.

## II. Morphological Characterization of Cell Death

Physiological cell death was described initially based on morphological grounds only, and was given the name **apoptosis** (Kerr *et al.,* 1972). This was defined as a physiological cell death occurring sporadically in tissues, that did not generate an inflammatory reaction, and could be either programmed or induced, but it was morphologically different from necrosis. It was also proposed that apoptosis was an active process, because messenger RNA (mRNA) and protein synthesis might be required for cell death to occur. At that time, it was appreciated that apoptosis could be a key element in tissue homeostasis by balancing cell division with cell death. It was also suggested that apoptosis could be important in ontogeny and in some pathological processes (Kerr *et al.,* 1972).

Strictly speaking then, apoptosis depicts only the morphological changes of dying cells in a tissue. Therefore, to name a given cell death event apoptosis, it has to fulfill the morphological characterization described later. This together with the absence of *in vitro* models (that is, before the appearance of molecular and biochemical markers of apoptosis) retarded elucidation of the mechanisms involved in physiological cell death. Such models for studying apoptosis *in vitro* have emerged by combining morphological, biochemical, and molecular approaches to the point that apoptosis can be studied in cell-free extracts now. Despite this progress, there is still an incipient classification of the different types of cell deletion. Necrosis and apoptosis are two different expressions of cell death that are clearly recognized, but there may be more forms of cell deletion.

### A. Necrosis

**Necrosis** is considered to be a nonphysiological cell death. This occurs apparently as a consequence of an extremely toxic stimulus or a massive cell injury. Cell death by necrosis is characterized by cellular swelling, most importantly of the mitochondria, and the absence of morphological alterations in the nucleus. The integrity of the plasma membrane is lost at the end, producing leakage of the internal contents into the extracellular milieu and vice versa. The disappearance of the plasma membrane as a selective barrier seems to be a differentiating characteristic in necrosis. Presumably, this is why necrosis produces a very strong inflammatory reaction, which is not the case for apoptosis (Wyllie *et al.,* 1980). The fact that the same noxious stimulus could produce either necrosis or apoptosis depending on the concentration and the duration of exposure has made difficult differentiation of apoptosis from necrosis (Slater *et al.,* 1995; Trump and Berezesky, 1995). This is further complicated because *in vitro* models cannot be used to show one of the key features of apoptosis, that is, being imperceptible by the organism.

## B. Apoptosis

Apoptosis was described as a particular set of transformations at the microscopic level associated with cell death (Kerr *et al.,* 1972). Apparently, one of the earliest steps a cell takes when it is committed to die in a tissue is to cease communicating with its neighbors. This is evident as the dying cell detaches from the adjacent ones and rounds up (Wyllie *et al.,* 1980; Ruoslahti and Reed, 1994). Condensation of the chromatin at the nuclear membrane and fragmentation of the nucleus are manifested. The cell shrinks due to cytoplasmic condensation, probably in response to cross-linking of proteins and loss of water. Organelles preserve their normal ultrastructure except for the endoplasmic reticulum, which becomes slightly dilated. An increase in the activity of the cell surface is another event characteristic of apoptosis, where the plasma membrane becomes ruffled and blebbed. Eventually, this activity separates the cell in a number of membrane-bound fragments of different size, termed **apoptotic bodies** (Kerr *et al.,* 1972). They could contain cytoplasm only or fragments of nucleus and other organelles depending on the size of such bodies. Membrane integrity of apoptotic bodies is not lost before being readily engulfed by either surrounding macrophages or, most frequently, the healthy neighboring cells. An intact plasma membrane that impedes the contact of the cytoplasm with the immune system seems to be the reason for the absence of inflammatory reaction in apoptosis, making this process practically imperceptible to the organism.

The morphological hallmark of apoptosis is the condensation of the nuclear chromatin in either crescents around the periphery of the nucleus or a group of condensed spherical fragments. These changes can be easily seen with permeable fluorescent DNA-intercalating dyes, like Hoeschst 33342, by fluorescence microscopy. It has been suggested that these nuclear changes are due to the activation of a $Ca^{2+}$- and $Mg^{2+}$-dependent endonuclease. This enzyme produces the characteristic ladder pattern of DNA fragmentation (Wyllie *et al.,* 1984), which is now considered one of the biochemical hallmarks of apoptosis (Fig. 1).

Apoptosis seems to be an active process, implying that the cell participates in its own demise ("suicide program").

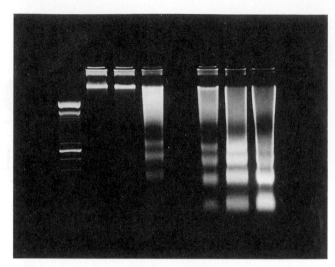

**FIG. 1.** Typical ladder pattern of DNA degradation in apoptosis. Cells were exposed to ionomycin (a $Ca^{2+}$ ionophore) for 14 hr in the presence (lanes 2–4) or the absence (lanes 5–7) of serum. DNA was extracted and electrophoresed in a 2% agarose gel and stained with ethidium bromide. Lanes 2 and 5 are in the absence of ionomycin and show the effect of serum removal. Lanes 3 and 6 are DNA from cells incubated with 1 $\mu M$ ionomycin and lanes 4 and 7 with 10 $\mu M$ ionomycin. Lane 1 is the molecular weight markers. Notice how DNA is fragmented in multiples of approximately 200 bp generating the typical ladder pattern associated with apoptosis.

This is, in principle, an important difference with necrosis where the cell is just a victim of a strong harmful condition. Initially, it was suggested there was a strict requirement for RNA and protein synthesis (Kerr *et al.,* 1972; Wyllie *et al.,* 1984). However, recent studies aimed at reevaluating the role of protein synthesis in thymocyte apoptosis found their inhibition does not always block DNA fragmentation (Chow *et al.,* 1995). Furthermore, the cytoplasmic changes typical of apoptosis can also be seen in enucleated cells, suggesting that physiological cell death is regulated by a cytoplasmic factor. This also argues that the characteristic nuclear changes in apoptosis and the nuclear activity do not determine the fate of the cell (Jacobson *et al.,* 1994). The apoptosis machinery appears to be already present in the cytoplasm of healthy cells but it is not functional. Energy is required in apoptosis, most probably to generate those signals that will start the death machinery but not to keep it working (Jacobson *et al.,* 1993, 1994).

Physiological cell death could be considered a programmed event mainly during ontogeny, meaning that some kind of internal molecular clock determines the time when a given cell should die. This is ultimately believed to be associated with specific genes being either turned on or deactivated to ignite the cell death machinery. This has originated a strong effort in characterizing the battery of genes involved in this process, as is reviewed later. However, it seems that apoptosis is not always programmed and important signal transduction events are triggered by external signals. Programmed cell death can also be considered in more general terms to mean that there is a "program" or a sequence of steps associated with cell death, known as the **execution phase.**

## III. Regulation of Programmed Cell Death

Signal transduction in apoptosis is one of the rapidly evolving areas in the study of physiological cell death. It is evident now that there is a cell suicide mechanism or program that involves receptors, regulators, and effectors. Some of the key questions in apoptosis include these: Is there a single checkpoint where all different inducers of cell death converge? What are the nature and mechanisms of physiological regulators of cell death? How does a cell know when to die? Unraveling the mechanisms involved in controlling cell death will be an important step for coping with the pathological consequences of alterations in the apoptosis program.

### A. Inducers of Apoptosis

Apoptosis can be induced by a large variety of conditions that *a priori* do not seem to have anything in common (Fig. 2). Interestingly, for some inducers, the same signal could induce either cell death or proliferation on the same type of cell. It is not clear how this is achieved, but it has been suggested that the difference could be either due to the nature of the signal transduction pathway involved or the intensity of the signal. Cell death occurs in response to specific messages; however, cell proliferation has been shown to be affected by some of the same signals that induce cell death. In other cases a signal could produce

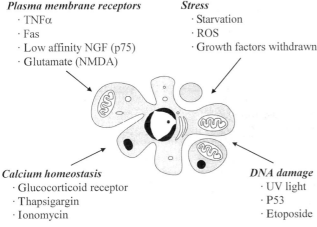

*Plasma membrane receptors*
· TNFα
· Fas
· Low affinity NGF (p75)
· Glutamate (NMDA)

*Stress*
· Starvation
· ROS
· Growth factors withdrawn

*Calcium homeostasis*
· Glucocorticoid receptor
· Thapsigargin
· Ionomycin

*DNA damage*
· UV light
· P53
· Etoposide

**FIG. 2.** Inducers of apoptosis. A schematic separation of inducers of apoptosis based on similar mechanisms for inducing cell death. However, there is an important overlap among the different groups. For instance, reactive oxygen species (ROS) seem to have an effect by increasing $Ca^{2+}$ influx, and the opposite (i.e., induction of ROS by increased $[Ca^{2+}]_i$) has also been reported.

either apoptosis or necrosis; this is apparently related to the level of stress imposed by the noxious signal. Lower levels will induce apoptosis where the cell is still able to activate its own demise, but for higher insults, necrosis will be the result of an uncontrollable degradation due to a drastic depletion of energy (Samali *et al.,* 1996). Examples of this are oxidative stress or excitotoxicity (see discussion later). *In vitro*, apoptosis is always followed by secondary necrosis, that is, plasma membrane disintegration.

We have subdivided the different inducers of apoptosis into four different groups (see Fig. 2), mainly to facilitate discussion of the different transducing mechanisms involved; however, keep in mind that many of them overlap quite importantly.

### 1. Plasma Membrane Receptors

Plasma membrane limits a cell and it is endowed with a multitude of receptors that recognize external signals and transduce their messages by fostering the generation of second messengers. These molecules in turn trigger a new cell behavior by activating different effector cascades inside the cell. Apoptosis is not the exception to that picture of signal transduction. Different plasma membrane receptors are involved in triggering cell death. Three members of the family of **tumor necrosis factor receptors** (TNFRs), that is, Fas, TNF-R1, and p75 (a low-affinity version of the NGF receptor) are well identified as plasma membrane receptors coupled to apoptosis. However, more integrants of this kind of cell death receptor are being continuously discovered, suggesting this family is bigger than previously thought (Chinnaiyan *et al.,* 1996).

Receptors of TNFR family are characterized by having extracellular cysteine-rich regions and a single membrane-spanning domain, their activation requires dimerization or a higher degree of association, and their transduction mechanism is based on protein–protein interaction. In a broad sense, TNFR family is more related to the receptors for trophic factors than those coupled to G-proteins, highlighting the similarity between cell death and cell growth. Nevertheless, the transduction mechanism by the TNFR family is different to receptors for growth factors. Fas is activated by Fas ligand (FasL), a protein expressed in the plasma membrane of different activated T cells (Nagata and Golstein, 1995). Once the receptor is activated a chain of interacting proteins is generated, which is mediated by coupler proteins (FADD or MORT1) with the final activation of specific proteases (FLICE or MACH1) associated with cell death (Muzio *et al.,* 1996; Boldin *et al.,* 1996). The TNFR family receptors that were thought to be specifically involved in cell death, under some circumstances, can also increase cell proliferation. It has been proposed that this occurs because the activation of the transcription regulator nuclear factor κB (NF-κB) could be up-regulating genes encoding apoptosis-inhibitory proteins (Wang *et al.,* 1996a). Actually, the capability to kill a cell by TNF-R1 is better expressed by combining TNF with inhibitors of transcription or blockers of protein synthesis, in agreement with the capacity of TNF-R1 to activate NF-κB. Interestingly, besides TNF-R1, ionizing radiation and cancer chemotherapeutic compounds also activate NF-$\kappa$B. This might explain their failure in eliminating cancer cells (Wang *et al.*, 1996a). NGF is a well-known inducer of neuronal survival during development and it does that by activating a specific tyrosine kinase receptor (*trk*A). Surprisingly, NGF can also produce cell death, in this case a different type of receptor is involved known as p75. This is a low-affinity receptor for NGF without an intrinsic tyrosine kinase activity that belongs to the TNFR family of cell death receptors (Frade *et al.*, 1996).

**Integrins** form another family of plasma membrane proteins that have being associated with apoptosis of normal epithelial or endothelial cells. When these cells are detached from the extracellular matrix or forced to grow in suspension, they undergo apoptosis. This effect can be specifically prevented by attaching these cells with anti-integrin antibodies, suggesting that integrins produce apoptotic signals when they are not interacting with the extracellular matrix (Ruoslahti and Reed, 1994).

Interestingly, all these studies have documented the existence of plasma membrane receptors linked to apoptosis, but also that the same receptors can alter cell proliferation. This shows that cell death and cell division are closely intertwined, suggesting the presence of common points in these two supposedly divergent events.

## 2. DNA Damage

Apoptosis can also be seen as a safeguard mechanism because it is involved in deleting damaged cells. *p53* is a tumor suppressor gene normally activated by DNA damage and is the most frequently mutated gene in human tumors (Yonish-Rouach, 1996). Activation of *p53* blocks cell cycle in $G_1$ phase and, it is presumed, this gives time to the DNA repairing machinery to restore DNA before replication. On activation of *p53* the cell could undergo either growth arrest or apoptosis. Growth arrest induced by *p53* depends on its role as transcriptional activator, which is located in the N-terminus of the protein. On the other hand, apoptosis induced by DNA damage, which has been shown to depend strictly on *p53* (Lowe *et al.*, 1993), does not seem to require gene expression (Caelles *et al.*, 1994; Haupt *et al.*, 1995). It seems the ability of *p53* to induce cell death or growth arrest resides in different parts of the peptide sequence. Similarly to the TNFR family, the induction of cell death by *p53* does not require protein synthesis. This is in agreement with the cell death machine being present in the cytoplasm of healthy cells but under an active repression.

## 3. Stress

Withdrawal of growth factors or serum deprivation has been shown to induce cell death by apoptosis. However, this effect depends on the stage of the cell cycle, because there are points where cells can be arrested and survive (Farinelli and Greene, 1996). This seems to be linked to the presence of cyclin-dependent kinase (CDK) inhibitors, like p21 or p16 (Wang and Walsh, 1996). The former produces growth arrest at the $G_1$ phase of the cell cycle, allowing the cell to survive. This could be explained as if the presence of these inhibitors impedes the interaction of conflicting signals (proliferation in harsh conditions) that otherwise would eventually trigger apoptosis.

Studying the mechanism of action of Bcl-2 (an important blocker of apoptosis, discussed later), it was found that this protein protects cells against oxidative insults. This raised the possibility that reactive oxygen species (ROS) could be important mediators of cell death. However, it was shown later that apoptosis can occur in the absence of ROS, implying that they are not universal signals for apoptosis. Nevertheless, the effect of different inducers of apoptosis can be blocked by antioxidants and ROS scavengers (Slater *et al.*, 1995). Further, Bcl-2 protects cells from the lethal effects of $H_2O_2$ or *t*-butyl hydroperoxide, two well-known inducers of oxidative stress (Korsmeyer *et al.*, 1995). Interestingly, mutations in a Cu/Zn superoxide dismutase, an enzyme that catalyzes the dismutation of the toxic superoxide anion ($O_2^-$) to $O_2$ and $H_2O_2$, have been associated with loss of motor neurons in some cases of familial amyotrophic lateral sclerosis (Rosen *et al.*, 1993). All of these studies point to the importance of ROS as inducers of cell death and argues for the relevance of having mechanisms of protection against oxidative insults.

## 4. $Ca^{2+}$ Homeostasis

Calcium ions have been involved practically since the inception of apoptosis. In the beginning, researchers thought that the activation of a $Ca^{2+}$- and $Mg^{2+}$-dependent endonuclease committed the cell to apoptosis (Wyllie, 1980; Wyllie *et al.*, 1984). This enzyme is the best candidate to be producing DNA fragmentation in the characteristic ladder pattern that it is widely associated with apoptosis (see Fig. 1). It is now recognized that a DNA ladder is not an unequivocal indicator of apoptosis because other forms of DNA degradation have been described for this type of cell deletion (Samali *et al.*, 1996). Patterns of DNA degradation described in apoptosis are single-strand nicks, generation of large DNA fragments of 50 to 300 kbp and the typical 180–200 bp internucleosomal fragmentation that makes the characteristic ladder pattern when electrophoresed on an agarose gel (see Fig. 1). DNA degradation is blocked when the $Ca^{2+}$ buffer capacity of the cell is increased either by overexpressing $Ca^{2+}$-binding proteins like Calbindin $D_{28k}$ or by introducing to the cell $Ca^{2+}$ chelators like Bapta (Dowd, 1995). The requirement for $Ca^{2+}$ of DNA degradation can be due to one or more of the different $Ca^{2+}$-dependent processes described in apoptosis. As indicated previously, there is evidence for the activation of a $Ca^{2+}$- and $Mg^{2+}$-dependent endonuclease (Montague and Cidlowski, 1996). Other candidates could be $Ca^{2+}$-regulated proteases associated with the nuclear scaffold; these enzymes will degrade nuclear lamins, proteins involved in the nuclear architecture and also in anchoring DNA with a 20–50 kbp periodicity (McConkey, 1996).

$Ca^{2+}$-dependent apoptosis is the consequence of a sustained elevation in $[Ca^{2+}]_i$, which is due to a continued $Ca^{2+}$ influx through the plasma membrane (Dowd, 1995).

However, the nature of the $Ca^{2+}$ channels involved and their regulation are not known. Inducers of apoptosis like ionomycin and thapsigargin are well known for their ability to deplete internal $Ca^{2+}$ stores; other inducers such as dexamethasone and $H_2O_2$ have been shown to lower $Ca^{2+}$ content in internal stores. A reduction in the amount of $Ca^{2+}$ stored activates what are called store-operated $Ca^{2+}$ (SOC) channels, also known as depletion-activated channels (Berridge, 1995). Accordingly, it has been proposed that inducers of apoptosis increase $[Ca^{2+}]_i$ via SOC channels by depleting internal $Ca^{2+}$ stores. The poor pharmacological and molecular characterization of SOC channels has precluded the corroboration of this hypothesis.

Although SOC channels might play a critical role in apoptosis, there is also evidence for the involvement of other types of $Ca^{2+}$ channels in this process. It has recently been shown an increment in the expression of a $Ca^{2+}$-permeable channel (the type 3 of inositol 1,4,5-trisphosphate receptor) in the plasma membrane of lymphocytes undergoing apoptosis (Khan *et al.*, 1996). Furthermore, when the expression of this channel is inhibited, so is apoptosis for this type of cell (Khan *et al.*, 1996). This clearly demonstrates that the expression of $Ca^{2+}$-permeable channels is a requisite for lymphocytes to undergo apoptosis. In neurons, cell death has been associated with another kind of $Ca^{2+}$-permeable channel, the NMDA type of glutamate receptors. The increased activity of this channel produces a $Ca^{2+}$ overload, which has been associated with cell death by excitotoxicity. This condition can result in either necrosis or apoptosis depending on the metabolic state of the cell (Ankarcrona *et al.*, 1995).

As discussed earlier, oxidative stress can induce apoptosis, and the same type of insult has been shown to activate directly $Ca^{2+}$-permeable, nonselective cation channels in endothelial cells (Koliwad *et al.*, 1996), linking $Ca^{2+}$ influx and apoptosis induced by reactive oxygen species. Apoptosis in prostate cells is also dependent on $Ca^{2+}$ influx (Furuya *et al.*, 1994) where a new type of $Ca^{2+}$-permeable channel, not operated by internal $Ca^{2+}$ stores, has recently been identified and associated with apoptosis (Gutierrez *et al.*, 1997). Interestingly, cationic channels not permeable to $Ca^{2+}$, encoded by *mdeg* genes, produce cell death by swelling, characteristic of necrosis (Ellis *et al.*, 1991; Waldmann *et al.*, 1996). This suggests that cationic channels do not produce apoptosis unless they are permeable to $Ca^{2+}$. From these limited studies on the role of $Ca^{2+}$ channels in apoptosis, it appears that the nature of the $Ca^{2+}$ channel activated is not critical, but its capacity to increase $[Ca^{2+}]_i$ is. These channels must be kept open for a long period of time to produce a sustained increase in $[Ca^{2+}]_i$ (Dowd, 1995).

A rise in $[Ca^{2+}]_i$ is a widely used second messenger for different physiological events (Fig. 3). It is up to the different $Ca^{2+}$-binding proteins to transduce correctly $Ca^{2+}$-rising external messages in specific cell behaviors. In this regard, different groups have seen an increment in calmodulin expression induced by cell death agonists, despite a general reduction in protein synthesis (Furuya *et al.*, 1994; Dowd, 1995). Calmidazolium, a potent inhibitor of calmodulin, blocks apoptosis in different systems (Dowd, 1995) indicat-

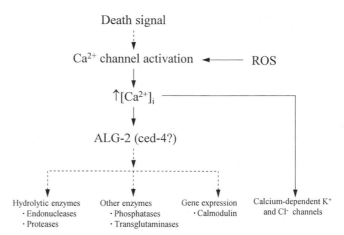

**FIG. 3.** Working model of the role of $Ca^{2+}$ as second messenger in apoptosis. Death signals, like reactive oxygen species (ROS), could directly activate $Ca^{2+}$-permeable channels or indirectly do so by depleting internal $Ca^{2+}$ stores. An increase in $[Ca^{2+}]_i$ is recognized by apoptosis-specific $Ca^{2+}$-binding proteins that activate different kinds of cell death effectors. Solid lines indicate direct activation that has been documented. Dashed lines represent an indirect pathway that is still undefined.

ing that this $Ca^{2+}$-binding protein could be transducing the elevated $[Ca^{2+}]_i$ in an active cell death program. Ced-4 is a protein essential for the expression of apoptosis in the worm *C. elegans*, its mechanism of action is not known yet; however, it has been proposed ced-4 could be a $Ca^{2+}$-binding protein (Yuan and Horvitz, 1992). Recently, ALG-2 has been identified, a new kind of $Ca^{2+}$-binding protein in thymocytes (Vito *et al.*, 1996). This protein has two canonical $Ca^{2+}$-binding domains, and apoptosis triggered by different, unrelated inducers is blocked when the expression of ALG-2 is specifically abrogated by antisense oligonucleotides. The expression of this protein is not limited to thymocytes since it is also present in hepatic and kidney cells, suggesting that it could be involved in mediating cell death in those cells as well.

Such elevated $[Ca^{2+}]_i$ could be activating different $Ca^{2+}$-dependent enzymes like endonucleases, proteases, and tissue transglutaminase (Fig. 3), which are considered general effectors in cell death (see following discussion). In this regard, activation of **calcineurin,** a $Ca^{2+}$-dependent phosphatase, readily induces cell death in the presence of low levels of serum. Furthermore, the overexpression of a constitutively active form of calcineurin does induce apoptosis without a $Ca^{2+}$ increase, implying that the activation of this phosphatase is enough to trigger programmed cell death (Shibasaki and McKeon, 1995).

Bcl-2 is a membrane protein that blocks apoptosis, as discussed later. This protein modifies regulation of $[Ca^{2+}]_i$ in different types of cells. The main effect of the overexpression of Bcl-2 is the preservation of the endoplasmic reticulum $Ca^{2+}$ pool (Baffy *et al.*, 1993; Distelhorst *et al.*,

1996; Reynolds and Eastman, 1996). This protein also reduces $Ca^{2+}$ influx through the plasma membrane (Lam *et al.*, 1994). Bcl-2 seems to block preferentially the apoptosis-associated rise in intranuclear $[Ca^{2+}]$ suggesting this is how Bcl-2 inhibits DNA degradation (Marin *et al.*, 1996). Whether this effect is directly due to Bcl-2 or is just a consequence of blocking cell damage is not clear yet.

$Ca^{2+}$ is a very important second messenger in a variety of physiological processes, the evidence reviewed here suggests that it is also involved in inducing cell death. Whether this occurs because $Ca^{2+}$ is directly activating $Ca^{2+}$-dependent enzymes involved in the execution path of apoptosis (see later), or indirectly because alterations in $Ca^{2+}$ regulation block cell growth, which in turn triggers cell death, is not clear yet. Understanding how Bcl-2 affects regulation of $[Ca^{2+}]_i$ and how proteins like ALG-2 transduce cell death will undoubtedly shed light on the role of $Ca^{2+}$ in apoptosis.

Calcium and apoptosis could be indirectly related by the functional state of the endoplasmic reticulum. Depletion of internal $Ca^{2+}$ pools by irreversible blocking endoplasmic reticulum $Ca^{2+}$ pumps with thapsigargin, for instance, will impair the role of luminal $Ca^{2+}$-sensitive chaperonin proteins (Gill *et al.*, 1996). Depletion of nuclear $Ca^{2+}$ stores stops traffic between the nucleus and the cytoplasm, blocking protein synthesis (Perez-Terzic *et al.*, 1996). Therefore, it is plausible that depletion of internal $Ca^{2+}$ stores associated with the endoplasmic reticulum or the nuclear membrane could somehow trigger a warning signal activating the apoptosis program, making the increase in $[Ca^{2+}]_i$ a consequence of store depletion and not necessarily a signal for apoptosis.

## B. Regulators of Apoptosis

Apoptosis is finely regulated and the discovery of genes involved in this task has made the study of them one of the quickly developing areas in program cell death. The importance of these genes, for instance the *bcl-2* family, has been widely demonstrated. Nevertheless, there are other proteins unrelated to the *bcl-2* family that seem to modulate apoptosis as well. The role of $Ca^{2+}$ as second messenger in cell death is still under intense research (see Section III,A,4). Concomitantly with this work, there are reports of other second messengers, like ceramide, in programmed cell death.

### 1. Bcl-2 Family of Proteins

Bcl-2 is a membrane protein considered a **proto-oncogene.** It was found to be expressed in up to 85% of human follicular B-cell lymphomas. This was the consequence of a t(14;18) chromosomal translocation that put *bcl-2* (18q21) under the transcriptional control of the Ig H-chain locus (14q32). This explains its abundance in lymphoma cells (Korsmeyer *et al.*, 1995).

It was later discovered that Bcl-2 produces lymphomas without increasing the rate of cell proliferation, as was expected for an oncogene, but by slowing down cell death (Korsmeyer *et al.*, 1995). Bcl-2 became then the foundation

of a new family of regulators of cell death. The importance of this family was more evident when realized that *ced-9* (an inhibitory gene of cell death in the worm *C. elegans*) is a functional and structural homolog of *bcl-2*. This suggests that programmed cell death has been highly conserved in evolution from invertebrates to humans (Häcker and Vaux, 1995).

Bcl-2 is a 25-kDa membrane protein localized to the membranes of mitochondria, nuclear envelope, and endoplasmic reticulum. Bcl-2 insertion in intracellular membranes requires the hydrophobic carboxy-terminus of the protein. This is called Bcl-2α, which is the most abundant isoform. There is a splice variant (Bcl-2β) that lacks the anchoring region but that binds to Bcl-2α isoform. Deletion of the hydrophobic region of Bcl-2α reduces but does not abolish its function. This suggests that the active site is located in its cytoplasmic part and the hydrophobic C-terminus targets the protein to the different organelles. It also implies that the location of Bcl-2 plays a role in protecting cells from apoptosis.

Most of the organs in Bcl-2 deficient mice (bcl-2 $-/-$) complete development quite normally, indicating that other proteins must be involved in regulating cell death (Korsmeyer *et al.*, 1995). It is now evident that Bcl-2 is only one member of a still growing family of regulators of apoptosis (Table 1). Remarkably, this family comprises proteins that are present in some viruses as well. The Bcl-2 family of proteins is divided in two main groups, agonists and antagonists, based on their ability to favor or block apoptosis, respectively (Table 1).

Bcl-2 is the archetypal antagonist. *In vitro,* this protein is able to block or retard apoptosis triggered by a wide and diverse variety of inducers (Table 2), although *in vivo* its function seems to be more limited as indicated earlier. This is not the case for another antagonist, Bcl-$x_L$. This protein is highly expressed in central nervous system, kidney, and marrow. In this case *bcl-x* knockout embryos were not viable. Massive apoptosis was evident in the immature nervous system and in hematopoietic cells. This is in agreement with the abundance of Bcl-$x_L$ in those same cells, implying a very important role for the *bcl-x* gene in development (Motoyama *et al.*, 1995). Interestingly, long (Bcl-$x_L$) and short (Bcl-$x_S$) forms of Bcl-x have been identified. The former is an antagonist but the latter counteracted the antagonistic action of Bcl-2, suggesting that this form produces cell death.

Bax was the first protein identified to be an agonist of apoptosis; this protein was isolated in association with Bcl-2. Similarly, Bax is an integral membrane protein of 21 kDa that shares 21% identity and 43% similarity with Bcl-2. Bcl-2 homology is clustered in three regions termed BH domains. They are found in the following order, BH3, BH1, and BH2, from the amino to the carboxy end of the protein (see Table 1). Bax-deficient mice show alterations due to uncontrolled accumulations of only lymphocytes and germ cells (Knudson *et al.*, 1995). Two conclusions arise from this study. First, Bax is effectively an agonist of cell death. Second, there must be more agonists of cell death, since only few cells were affected by the absence of Bax (see Table 1).

**TABLE 1** Presence of Bcl-2 Sequence Homologies (BH) and Transmembrane Domain for Agonist and Antagonist of Apoptosis

| Protein | BH1 | BH2 | BH3 | TD | Cell death | Ref. |
|---------|-----|-----|-----|-----|-----------|------|
| Bcl-2 | + | + | + | ± | − | 1, 12 |
| Bcl-X$_L$ | + | + | + | + | − | 6, 12 |
| Ced-9 | + | + | −$^a$ | + | − | 6, 8, 12 |
| BHRF1 | + | + | − | + | − | 4, 6, 11 |
| Bad | + | + | − | − | + | 5, 6, 12 |
| Bak | + | + | + | + | +>− | 3, 7, 8 |
| Bax | + | + | + | ± | + | 5, 9, 11 |
| Bcl-X$_S$ | − | − | + | + | + | 6, 12 |
| Bik | − | − | + | + | + | 2 |
| Bid | − | − | + | − | + | 10 |

*Note:* The presence (+) or absence (−) of Bcl-2 sequence homology regions (BH1 to BH3) is shown based on the references listed in the rightmost column. The transmembrane domain (TD) is either present (+) or not (−); (±) represents splice variants lacking the TD. Cell death column indicates whether the protein is an agonist (+) or an antagonist (−) of apoptosis.

$^a$ The presence of this domain is not clear (see Refs. 8 and 12).

*References:* (1) Boise *et al.*, 1995; (2) Boyd *et al.*, 1995; (3) Chittenden *et al.*, 1995a; (4) Chittenden *et al.*, 1995b; (5) Davis, 1995; (6) Häcker and Vaux 1995; (7) Kiefer *et al.*, 1995; (8) Muchmore *et al.*, 1996; (9) Oltvai *et al.*, 1993; (10) Wang *et al.*, 1996b; (11) Yin *et al.*, 1994; (12) Zha *et al.*, 1996. Complete references are found at the end of the chapter.

**TABLE 2**

| Apoptosis inducers | Reference |
|--------------------|-----------|
| Physical | |
| Gamma and UV radiation | 1 |
| Heat shock | 6 |
| Chemical | |
| Chemotherapeutic drugs | 2 |
| Reactive oxygen species | 6 |
| Azide | 6 |
| Cell death receptors | |
| TNF | 6 |
| Fas | 1 |
| CD3/TCR | 7 |
| Dexamethasone | 2 |
| Growth factor withdrawal | |
| Glucose | 6 |
| IL-3 | 4 |
| Androgens | 5 |
| Neurotrophic factors | 1 |
| Others | |
| Calcium | 3 |
| p53 | 6 |
| c-Myc | 1 |
| Viruses | 6 |

This table is not intended to be an exhaustive list of apoptosis inducers, but to highlight their diversity in nature. (1) Boise *et al.*, 1995; (2) Cidlowski *et al.*, 1996; (3) Dowd, 1995; (4) Hockenbery *et al.*, 1990; (5) Raffo *et al.*, 1995; (6) Reed, 1994; (7) Sentman *et al.*, 1991. Complete references are found at the end of the chapter.

BH1 domain is crucial in the antagonistic action of the Bcl-2 family of proteins and BH3 has been associated with their agonistic effect (see Table 1). Point mutations in a highly conserved glycine in BH1 completely abrogate the antagonistic action of Bcl-2 and Bcl-x$_L$. The hypothesis that BH3 is involved in triggering cell death stems from the existence of pure agonists of apoptosis that have only BH3, like Bcl-x$_S$, Bik, and Bid (see Table 1). However, BH3 is also present in antagonists such as Bcl-2 and Bcl-x$_L$, indicating that there must be other factors besides these domains to explain their mechanism of action.

The three-dimensional structure of a monomer of Bcl-x$_L$ has been resolved (Muchmore *et al.*, 1996). This study showed that BH1, BH2, and BH3 form a hydrophobic groove inside of which the highly conserved glycine of BH1 is located. Apparently, this hydrophobic cleft allows the interaction with Bax, which has been suggested to be necessary for the antagonistic actions of Bcl-2 and Bcl-x$_L$. Accordingly, increasing the side chain of the highly conserved glycine in BH1 would generate a steric hindrance that would diminish the strength of the interaction with Bax. Although homodimerization and heterodimerization among the different proteins of the Bcl-2 family is well documented, the importance of this interaction on their physiological roles is not clear-cut (McDonnell *et al.*, 1996). Thus, Muchmore *et al.*, (1996) have highlighted the structural similarity between Bcl-x$_L$ and the membrane insertion domain of diphtheria toxin, insinuating that proteins of Bcl-2 family could be making pores as well

Understanding how the Bcl-2 family of proteins works will become a cornerstone in the study of apoptosis. Although Bcl-2 is a very powerful blocker of oxidative stress and many inducers of apoptosis produce an oxidative insult (Korsmeyer *et al.*, 1995), this protein can also block cell death in the absence of ROS. This implies that Bcl-2 function is wider than just a ROS scavenger. Whatever the Bcl-2 mechanism of action is, it must be downstream and very close to a common effector in the execution phase of programmed cell death. This is because Bcl-2 is able to abrogate or retard the effect of widely diverse inducers of apoptosis or even necrosis that *a priori* do not have anything else in common (see Table 2).

### 2. Other Regulators of Apoptosis

The Bcl-2 family of proteins is the most studied one in terms of regulation of apoptosis, yet it appears it is not the only family of proteins involved in regulating programmed cell death. For instance, the gene encoding the **neuronal apoptosis inhibitory protein** (NAIP) does not show any resemblance to *bcl-2,* yet it still effectively protects neurons from apoptosis. Deletions of NAIP gene have been correlated with some cases of spinal muscular atrophy, a fatal autosomal recessive disorder characterized by an important neuronal cell death (Chinnaiyan and Dixit, 1996).

Studies on the mechanism of action of **tumor necrosis factor (TNF)** found that a sphingomyelinase is activated during TNF-induced apoptosis. This enzyme is a sphingomyelin-specific type C phospholipase that generates phosphocholine and **ceramide** by hydrolysis of sphingomyelin. The use of cell-permeable ceramides suggested that natural ceramide could be involved in triggering cell death, among other cellular responses, probably by modulating the activity of different transcription factors (Hannun, 1996). Subsequently, it has been demonstrated that activation by TNF-R1 of plasma membrane neutral sphingomyelinase is not linked to the dead domain of this receptor. However, the dead domain of TNF-R1 is required for the indirect activation of an acidic sphingomyelinase. In conclusion, it appears as TNF-R1 activates both neutral and acidic sphingomyelinases; ceramide produced by the former is involved in cell survival, but the one produced by the latter is linked to apoptosis (Testi, 1996). Ceramides are insoluble, so they are restricted to those membranes where they are generated. That, combined with segregation of the effector system sensitive to ceramide, could explain the paradoxical roles of ceramide in cell physiology.

**Phosphorylation/dephosphorylation** events are central in controlling cell function. It is conceivable that these events are also critical in apoptosis; however, the evidence has been contradictory so far. For example, activation of protein kinase C (PKC) blocks apoptosis in some cases and the inhibition of this kinase induces cell death. This is in agreement with the very well-documented role of PKC as a key regulator of cell proliferation. Nonetheless, there are also reports that activation of PKC with phorbol esters induce apoptosis (Lavin *et al.,* 1996). It is possible that variability in the rate of down-regulation of PKC, due to its sustained activation with phorbol esters, could be the reason for cell death also being associated with the apparent activation of PKC. Alternatively, some members of the PKC family could have an antagonistic effect on cell death, while others promote apoptosis. In any event, the role of kinases and phosphatases in apoptosis is still far from being understood and more information is needed to generate a coherent scheme for these enzymes in apoptosis.

### C. Effectors in Apoptosis

Apoptosis is characterized by specific morphological changes in the nucleus, plasma membrane, and endoplasmic reticulum, among others. These changes are very similar in different types of cells, implying that a conserved program exists that produces those modifications when activated during the execution phase of cell death. The apoptosis program involves a series of enzymes that are responsible for such morphological changes; once they are activated, the fate of the cell is deadly irreversible. Many different enzymes and proteins are in this group, albeit endonucleases and proteases have attracted much of the attention, probably because their effects are also irreversible.

### 1. Endonucleases

Alterations in the nuclear morphology are one of the most important changes associated with apoptosis (see Section II,B) and one of the goals has been to understand the biochemical events underlying such changes. It was suggested that the activation of a $Ca^{2+}$- and $Mg^{2+}$-dependent endonuclease produces the nuclear modifications already described (Wyllie, 1980; Wyllie *et al.,* 1984). The DNA degradation in apoptosis seems to result from changes in the DNA accessibility and in the specific activation of endonucleases, most likely by proteolytic processing. The initial stage apparently involves the action of an endonuclease that produces single-strand nicks in regions where large chromosome domains are attached to the nuclear matrix. The second phase is the generation of DNA fragments varying from 300 to 50 kb which are evident only by using pulsed-field gel electrophoresis. These patterns of fragmentation seem to be tightly associated with apoptosis. NUC18, a $Ca^{2+}$- and $Mg^{2+}$-dependent endonuclease has been proposed to be the active enzyme in this stage (Montague and Cidlowski, 1996). The third phase is the characteristic DNA ladder (see Fig. 1) that is generated by cleavage at the linker regions connecting the nucleosomes within a DNA loop. The third stage even when is a good diagnostic element of apoptosis, its absence, does not negate apoptotic cell death since stage one or two could be the only ones present. This means that different enzymes are activated in each one of the different stages of DNA degradation. Interestingly, internucleosomal DNA fragmentation seems to be occurring only in repeat elements of structural DNA (Walker and Sikorska, 1994), implying that DNA degradation would not necessarily affect transcription immediately.

Calcium chelation, zinc, and aurintricarboxilic acid block DNA fragmentation as well as apoptosis, clearly indicating the importance of DNA degradation as a biochemical marker of apoptosis. On the other hand, the observation that bacterial micrococcal nucleases are capable of producing DNA ladder fragmentation and the characteristic nuclear morphology associated with apoptosis in isolated thymic nuclei has made difficult the identification of specific endonucleases activated in apoptosis by a reconstitution approach. This observation also stresses the importance of using more than a single test to define apoptosis.

## 2. Proteases

The discovery that *ced-3* (an essential gene for cell death in the worm *C. elegans*) was similar to an interleukin-1β-converting enzyme (ICE) highlighted the importance of **proteases** in the execution phase of apoptosis. Those proteases shown to be involved in apoptosis are substrate-specific, nonlysosomal enzymes that have either cysteine or serine in their active site.

The role and regulation of cysteine proteases in apoptosis is currently under intense research. Conversely, we know much less for serine-dependent proteases in cell death. Granzyme B is a serine protease specific for aspartic acid P1 residue, which seems to be required for the cytotoxic T-cell induced apoptosis (Chinnaiyan and Dixit, 1996). It has been shown in vitro that granzyme B can activate another protease, CPP32, which is an important effector in apoptosis (discussed later). Presumably the introduction by perforin of granzyme B into targeted cells will trigger apoptosis by activation of CPP32. Another enzyme activated in apoptosis is a $Ca^{2+}$-dependent serine protease involved in the degradation of the nuclear scaffold (McConkey, 1996), and it is one of the candidates producing the characteristic apoptotic nuclear alterations.

To date, all cysteine proteases described to be involved in apoptosis belong to the ICE/Ced-3 family characterized by cleaving at aspartic residues in P1 (Chinnaiyan and Dixit, 1996). Again, these enzymes show substrate specificity based on determined peptide sequences: ICE recognizes tyr-val-ala-asp so it generates interleukin-1β from its precursor by cleaving after aspartate. Another enzyme, CPP32/Yama/apopain, recognizes asp-glu-val-asp (Nagata, 1996), cleaving and inactivating poly(ADP-ribose) polymerase, one of the enzymes involved in DNA repair. Similarly to endonucleases, the reconstitution approach cannot be used to test for the role of a given protease in apoptosis. This arises from the finding that ectopic expression of general degradative proteases, such as trypsin or chymotrypsin, also produces apoptosis although there is no evidence these enzymes are involved in programmed cell death. However, the use of permeable and specific inhibitory tetrapeptides together with some viral proteins (specific inhibitors of the ICE/Ced-3 family of proteases) substantiates the role of the ICE-like family of proteases in the execution phase of apoptosis. In general, these enzymes are present as inactive proenzymes that activate on proteolysis, in some cases by autocatalysis. Eventually, cascades of proteases will be unraveled that will probably be both specific for a cell type and also for the inducer of apoptosis.

Changes in the shape and volume of the cell and nucleus are very prominent characteristics of apoptosis. These represent important modifications in the cytoskeleton (Zhivotovsky *et al.*, 1996). Indeed, proteolytic modification of

different cytoskeletal proteins have been documented: α-fodrin, actin, and vimentin, among others. Proteins with roles in nuclear function and shape are also substrates for apoptosis-activated proteases. Enzymes involved in DNA repair like poly(ADP-ribose) polymerase and the DNA-dependent protein kinase are also cleaved by proteases. Proteins associated with the matrix attachment region of the nucleus, like lamins and the U1 small ribonucleoprotein, and many others like histone H1 and topoisomerase I and II are other examples of proteins cleaved during apoptosis. The specific cleavage of these proteins conceivably disrupts the normal nuclear integrity and function preparing the cell for a complete shutdown.

It is clear that the regulation of proteases in apoptosis is a key event that will show significant progress in the near future and that should become an important tool for manipulating the outcome in apoptosis.

## 3. Transglutaminase

Limited cytoplasmic leakage is another characteristic of apoptosis, which seems to be due to a shell formed around the cell. It has been proposed that tissue transglutaminase (tTG) could be involved in generating such a stabilizing structure. This enzyme is a cytoplasmic, $Ca^{2+}$-dependent, protein-glutamine γ-glutamyltransferase that cross-links proteins leading to their polymerization (this probably confers cyto-architectural stability). A substantial increase in tTG mRNA occurs on induction of apoptosis in different types of cells (Fesus *et al.*, 1996; Piacentini, 1995). Overexpression in some cell lines of tTG cDNA increased the susceptibility to apoptosis and, conversely, transfection with antisense tTG cDNA increased resistance to apoptotic stimuli by diminishing expression of tTG. Although these experiments suggest that tTG can induce apoptosis, further evidence suggests that the role of tTG in apoptosis, albeit not universal, is that of an effector and not a regulator of cell death. Interestingly, overexpression of tTG cDNA can transform necrotic cell death in typical apoptosis, stressing the importance of this enzyme in the expression of programmed cell death (Fesus *et al.*, 1996).

## 4. Other Effectors

Significant reductions in cell volume are associated with apoptosis. This could be carried out either by budding of the endoplasmic reticulum or alterations in the cell volume regulatory mechanisms. The former involves the release in the extracellular space of endoplasmic reticulum-derived vesicles that were not allowed to condensate. The latter may be due to an increased efflux of KCl in response to an elevated activity of $Ca^{2+}$-dependent $K^+$ and $Cl^-$ channels, among other ion transporters.

Programmed cell death is characterized by "silent" removal of apoptotic cells from the organism. Alterations in the plasma membrane signal neighboring cells and surrounding phagocytes for engulfing and deletion of the dying cell. There are different mechanisms for the recognition of target cells by phagocytes (Hart *et al.*, 1996). One of these mechanisms is the **externalization of phosphatidylserine.**

Phosphatidylserine (PS) is a negatively charged phospholipid that is asymmetrically present in the internal leaflet of the plasma membrane in normal conditions. This asymmetry is rapidly lost by externalization of PS on induction of apoptosis. Inhibition of ICE-like proteases abolishes the reduction in PS asymmetry, placing the activation of proteases upstream to the externalization of PS. This also means that proteins are responsible for keeping lipid asymmetry in normal cells. In this regard, cytoskeletal proteins such as $\alpha$-fodrin or enzymes, like translocases, have been implicated in regulating this asymmetry and they could be the targets of apoptosis proteases.

A plasma membrane receptor present in phagocytes is involved in PS recognition; the best candidate is CD68 or macrosialin. However, it seems that this protein is only part of a more complex structure involved in the recognition of apoptotic and senescent cells.

## IV. Roles of Physiological Cell Death

It is now recognized that not all physiological cell death occurs by apoptosis. Different forms of cell death could arise because there is one single program (highly conserved from worms to mammals, i.e., ICE/Ced-3 family of proteases) with different downstream effectors. The other possibility is the existence of more than one program. The latter seems to be supported by the observation that ciliary ganglion neurons can die by either apoptosis or not, depending on the stimulus (Schwartz *et al.,* 1993). The different functions of physiological cell death are considered here, regardless of whether it is manifested as apoptosis or not.

Physiological cell death is presumably present in multicellular organisms only, with prominent roles in development and homeostasis (see Fig. 4). Accordingly, it has been proposed that cell death can be divided into five different categories depending on the roles of the dying cell in the organism (Ellis *et al.,* 1991). First, we consider **futile cells,** those that have no apparent function. In general, they are removed in very early stages of development, and it has been proposed they represent evolutionary vestiges. Second, we have **surplus** or **redundant cells.** One case is the nervous system, where an excess of cells is produced and this is suggested to help innervation of target sites, those neurons that failed making contact are removed later. Third, **mistaken** or **damaged cells** occur when cells develop improperly or sustain genetic damage, respectively. Fourth, **obsolete cells** are those that have completed their function, either in development or homeostasis of a given organ. Fifth, **harmful cells** include, for example, the negative selection of autoreactive T lymphocytes in the thymus. These are cells of the immune system that, unless deleted, represent a risk of autoimmune diseases.

Alterations of the cell death program will inevitably generate important disorders, due to the relevance of cell deletion in development and homeostasis in multicellular organisms (Thompson, 1995). Diseases such as cancer, autoimmune disorders, and viral infections are some examples where diminished cell death has been implicated. However, neurodegenerative disorders, such as Alzheimer's

*Cell division*

**Neoplasia**   **Homeostasis**   **Degeneration**

*Cell death*

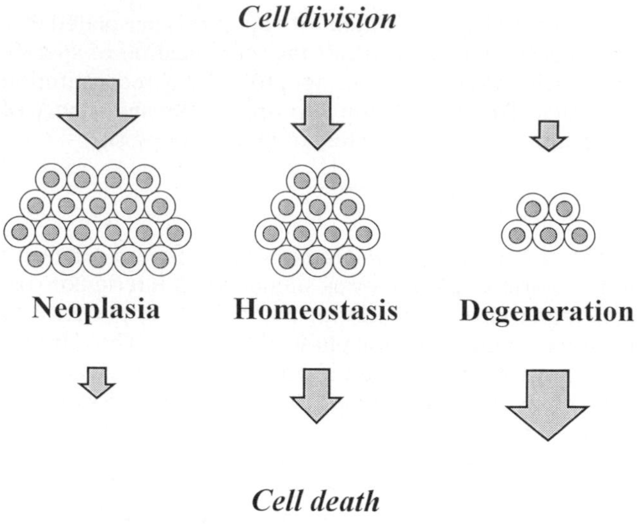

**FIG. 4.** Homeostasis depends on the balanced rates of cell division and cell death. The combination of an increased cell division with a reduced cell death in the generation of "proliferative" disorders is depicted. The opposite is present in degenerative diseases. The scheme is trying to stress the idea that both processes work in concert to produce dramatic changes in the homeostasis of a tissue. However, it is possible to have alterations in cell death independently of cell division.

and Parkinson's diseases, as well as osteoporosis and AIDS, are cases where an accelerated rate of cell deletion has been implicated in their pathophysiology.

## V. Frontiers in the Study of Apoptosis

Given the importance of cell death in both normal and altered functions, its control will have important repercussions in medicine. Currently, apoptosis has become synonymous with regulation of proteolytic enzymes. Identifying those mechanisms involved in controlling proteases in apoptosis and defining the diversity and isoforms of the execution phase will permit the design of strategies for therapeutically regulating cell death.

Bcl-2 is a pluripotent antagonist of apoptosis *in vitro.* However, we do not know how this is achieved. The fact that diverse and unrelated inducers of apoptosis are blocked by Bcl-2 suggests that there is a single, intracellular homeostasis sensor that activates the execution path on alterations in cell physiology. This is probably the target being repressed by the different antagonists of cell death. Identification of this homeostasis sensor will be an important step in understanding signal transduction in programmed cell death.

## VI. Summary

Physiological cell death, apoptosis, and programmed cell death may not be synonyms strictly speaking. However, the wealth of information on cell deletion emphasizes the importance of a cellular machinery, which, for the most part, is constitutively present in the cytoplasm and readily executes a cell death program. This program seems to be the same regardless of the nature of the inducer triggering apoptosis; however, it is also apparent that there is redundancy in this machinery, that is, different proteases cleaving the same substrate. This redundancy could open the possibility of exploiting specificity for the cell involved as well as the inducer of apoptosis. Therefore, understanding the regulation of proteolytic enzymes in apoptosis is a necessary step in order to design therapeutical approaches to eliminate cancer cells specifically or stop cell death in degenerative disorders.

RNA transcription and protein synthesis do not appear to be a key characteristic for the action of the execution machinery in apoptosis, as opposed to previous thinking. This indicates that cell death effectors are constitutively present but repressed. Nevertheless, other stages of cell death, like signaling or determination, could be under transcriptional or translational control. Thus, the role of protein synthesis might be more evident in physiological conditions as opposed to *in vitro* studies.

Although we do not understand how Bcl-2 works yet, this protein has revolutionized our understanding of cell death mechanisms already. First, we know that the so-called "proliferative" disorders could also originate by a reduction in cell death besides an increment in cell proliferation (see Fig. 4). Second, the finding that Bcl-2 can partially protect against necrotic insults in addition to apoptosis implies that both processes could have more in common than previously thought and the difference may be due to the intensity of the cell death signal.

Interestingly, the two ends, cell death and proliferation, seem to come in close contact. This is evident when we consider the relationship between apoptosis and cell cycle and also the paradoxical effect of some inducers of cell death in cell proliferation. This has made difficult the understanding and identification of apoptosis-specific regulatory mechanisms. Nevertheless, important advances have been carried out in the molecular identification of the cell death machinery, for regulators and effectors as well. The similarity of this program from worms to mammals has allowed the combination of forces for a faster understanding of cell deletion and it also highlights the importance of apoptosis in cell function.

## Acknowledgments

We thank Jorge Cerbón and Marco A. Meraz for critical reading of the manuscript.

## Bibliography

Ankarcrona, M., Dypbukt, J. M., Bonfoco, E., Zhivotovsky, B., Orrenius, S., Lipton, S. A., and Nicotera, P. (1995). Glutamate-induced neuronal death: A succession of necrosis or apoptosis depending upon mitochondrial function. *Neuron* **15**, 961–973.

Baffy, G., Miyashita, T., Williamson, J. R., and Reed, J. C. (1993). Apoptosis induced by withdrawal of interleukin-3 (IL-3) from a IL-3-dependent hematopoietic cell line is associated with repartitioning of intracellular calcium and is blocked by enforced Bcl-2 oncoprotein production. *J. Biol. Chem.* **268**, 6511–6519.

Berridge, M. J. (1995). Capacitative calcium entry. *Biochem. J.* **312**, 1–11.

Boise, L. H., Gottschalk, A. R., Quintáns, J., and Thompson, C. B. (1995). Bcl-2 and Bcl-2 related proteins in apoptosis regulation. *Curr. Top. Microbiol. Immunol.* **200**, 107–121.

Boldin, M. P., Goncharov, T. M., Goltsev, Y. V., and Wallach, D. (1996). Involvement of MACH, a novel MORT1/FADD-interacting protease, in Fas/APO-1 and TNF receptor-induced cell death. *Cell* **85**, 803–815.

Boyd, J. M., Gallo, G. J., Elangovan, B., Houghton, A. B., Malstrom, S., Avery, B. J., Ebb, R. G., Subramanian, T., Chittenden, T., Lutz, R. J., and Chinnadurai, G. (1995). Bik, a novel death-inducing protein shares a distinct sequence motif with Bcl-2 family of proteins and interacts with viral and cellular survival-promoting proteins. *Oncogene* **11**, 1921–1928.

Caelles, C., Helmberg, A., and Karin, M. (1994). p53-dependent apoptosis in the absence of transcriptional activation of p53-target genes. *Nature* **370**, 220–223.

Chinnaiyan, A. M., and Dixit, V. M. (1996). The cell-death machine. *Curr. Biol.* **6**, 555–562.

Chinnaiyan, A. M., O'Rurke, K., Yu, G. L., Lyons, R. H., Garg, M., Duan, D. R., Xing, L., Gentz, R., Ni, J., and Dixit, V. M. (1996). Signal transduction by DR3, a death domain-containing receptor related to TNFR-1 and CD95. *Science* **274**, 990–992.

Chittenden, T., Flemington, C., Houghton, A. B., Ebb, R. G., Gallo, G. J., Elangovan, B., Chinnadurai, G., and Lutz, R. J. (1995a). A conserved domain in Bak, distinct from BH1 and BH2, mediates cell death and protein binding functions. *EMBO J.* **14**, 5589–5596.

Chittenden, T., Harrington, E. A., O'Connor, R., Flemington, C., Lutz, R. J., Evan, G. I., and Guild, B. (1995b). Induction of apoptosis by the Bcl-2 homologue Bak. *Nature* **374**, 733–736.

Chow, S. C., Peters, I., and Orrenius, S. (1995). Reevaluation of the role of *de novo* protein synthesis in rat thymocytes apoptosis. *Exp. Cell Res.* **216**, 149–159.

Cidlowski, J. A., King, K. L., Evans-Storm, R. B., Montague, J. W., Bortner, C. D., and Hughes, F. M. (1996). The biochemistry and molecular biology of glucocorticoid-induced apoptosis in the immune system. *Recent Prog. Horm. Res.* **51**, 457–491.

Davis, A. M. (1995). The Bcl-2 family of proteins and the regulation of neuronal survival. *Trends Neurosci.* **18**, 355–358.

Distelhorst, C. W., Lam, M., and McCormick, T. S. (1996). Bcl-2 inhibits hydrogen peroxide-induced ER $Ca^{2+}$ pool depletion. *Oncogene* **12**, 2051–2055.

Dowd, D. R. (1995). Calcium regulation of apoptosis. *In* "Advances in Second Messenger and Phosphoprotein Research" (A. R. Means, Ed.), Vol. 30, pp. 255–281. Raven Press, New York.

Ellis, R. E., Yan, J., and Horvitz, H. R. (1991). Mechanisms and functions of cell death. *Annu. Rev. Cell Biol.* **7**, 663–698.

Farinelli, S. E., and Greene, L. A. (1996). Cell cycle blockers mimosine, ciclopirox, and deferoxamine prevent the death of PC12 cells and postmitotic sympathetic neurons after removal of trophic support. *J. Neurosci.* **16**, 1150–1162.

Fesus, L., Madi, A., Balajthy, Z., Nemes, Z., and Szondy, Z. (1996). Transglutaminase induction by various cell death and apoptosis pathways. *Experientia* **52**, 942–949.

Frade, J. M., Rodriguez-Tebar, A., and Barde, Y. A. (1996). Induction of cell death by endogenous nerve growth factor through its p75 receptor. *Nature* **383**, 166–168.

Furuya, Y., Lundmo, P., Short, A. D., Gill, D. L., and Issacs, J. T. (1994). The role of calcium, pH, and cell proliferation in the programmed (apoptotic) death of androgen-independent prostatic cancer cells induced by thapsigargin. *Cancer Res.* **54**, 6167–6175.

Gill, D. L., Waldron, R. T., Rys-Sikora, K. E., Ufret-Vincenty, C. A., Graber, M. N., Favre, C. J., and Alfonso, A. (1996). Calcium pools, calcium entry, and cell growth. *Biosci. Rep.* **16**, 139–157.

Gutierrez, A., Garcia, L., Mas-Oliva, J., and Guerrero, A. (1997). $Ca^{2+}$ permeable channel activated during apoptosis of a prostatic cancer cell line. *Biophys. J.* **72**, A271.

Häcker, G., and Vaux, D. L. (1995). A sticky business. *Curr. Biol.* **5**, 622–624.

Hannun, Y. A. (1996). Functions of ceramide in coordinating cellular responses to stress. *Science* **274**, 1855–1859.

Hart, S. P., Haslett, C., and Dransfield, I. (1996). Recognition of apoptotic cells by macrophages. *Experientia* **52**, 950–956.

Haupt, Y., Rowan, S., Shaulian, E., Vousden, K. H., and Oren, M. (1995). Induction of apoptosis in HeLa cell by trans-activation deficient p53. *Genes Dev.* **9**, 2170–2183.

Hockenbery, D., Nuñez, G., Milliman, C., Scheiber, R. D., and Korsmeyer, S. J. (1990). Bcl-2 is an inner mitochondrial membrane protein that blocks programmed cell death. *Nature* **348**, 334–336.

Jacobson, M. D., Burne, J. F., King, M. P., Miyashita, T., Reed, J. C., and Raff, M. C. (1993). Bcl-2 blocks apoptosis in cells lacking mitochodrial DNA. *Nature* **361**, 365–369.

Jacobson, M. D., Burne, J. F., and Raff, M. C. (1994). Programmed cell death and Bcl-2 protection in the absence of nucleus. *EMBO J.* **13**, 1899–1910.

Kerr, J. F. R., Wyllie, A. H., and Currie, A. R. (1972). Apoptosis: A basic biological phenomenon with wide ranging implications in tissue kinetics. *Br. J. Cancer* **26**, 1790–1794.

Khan, A. A., Soloski, M. J., Sharp, A. H., Schilling, G., Sabatini, D. M., Li, S. H., Ross, C. A., and Snyder, S. H. (1996). Lymphocyte apoptosis: Mediation by increased type 3 inositol 1, 4, 5-trisphosphate receptor. *Science* **263**, 503–507.

Kiefer, M. C., Brauer, M. J., Powers, B. C., Wu, J. J., Umansky, S. R., Tomei, L. D., and Barr, P. J. (1995). Modulation of apoptosis by the widely distributed Bcl-2 homologue Bak. *Nature* **374**, 736–739.

Knudson, C. M., Tung, K. S., Tourtellote, W. G., Brown, G. A., and Korsmeyer, S. J. (1995). Bax-deficient mice with lymphoid hyperplasia and male germ cell death. *Science* **270**, 96–99.

Koliwad, S. K., Kunze, D. L., and Elliot, S. J. (1996). Oxidant stress activates a non-selective cation channel responsible for membrane depolarization in calf vascular endothelial cells. *J. Physiol.* **491**, 1–12.

Korsmeyer, S. J., Yin, X. M., Oltvai, Z. N., Veis-Novack, D. J., and Linette, G. P. (1995). Reactive oxygen species and the regulation of cell death by Bcl-2 gene family. *Biochim. Biophys. Acta* **1271**, 63–66.

Lam, M., Dubyak, G., Chen, L., Nuñez, G., Miesfeld, R. L., and Distelhorst, C. W. (1994). Evidence that Bcl-2 represses apoptosis by regulating endoplasmic reticulum-associated $Ca^{2+}$ fluxes. *Proc. Natl. Acad. Sci. USA* **91**, 6569–6573.

Lavin, M. F., Watters, D., and Song, Q. (1996). Role of protein kinase activity in apoptosis. *Experientia* **52**, 979–994.

Lowe, S. W., Schmitt, E. M., Smith, S. W., Osborne, B. A., and Jacks, T. (1993). p53 is required for radiation-induced apoptosis in mouse thymocytes. *Nature* **362**, 847–849.

Marin, M. C., Fernandez, A., Bick, R. J., Brisbay, S., Buja, L. M., Snuggs, M., McConkey, D. J., Eschenbach, A. C., Keating, M. J., and McDonnel, T. J. (1996). Apoptosis suppression by *bcl-2* is correlated with the regulation of nuclear and cytosolic $Ca^{2+}$. *Oncogene* **12**, 2259–2266.

McConkey, D. J. (1996). Calcium-dependent, interleukin 1β-converting enzyme inhibitor-insensitive degradation of lamin $B_1$ and DNA fragmentation in isolated thymocyte nuclei. *J. Biol. Chem.* **271**, 22398–22406.

McDonnell, T. J., Beham, A., Sarkiss, M., Andersen, M., and Lo, P. (1996). Importance of Bcl-2 family in cell death regulation. *Experientia* **52**, 1008–1017.

Montague, J. W., and Cidlowski, J. A. (1996). Cellular catabolism in apoptosis: DNA degradation and endonuclease activation. *Experientia* **52**, 957–962.

Motoyama, N., Wang, F., Roth, K. A., Sawa, H., Nakayama, K., Negishi, I., Senju, S., Zhang, Q., Fuji, S., and Loh, D. Y. (1995). Massive cell death of immature hematopoietic cells and neurons in Bcl-x-deficient mice. *Science* **267**, 1506–1510.

Muchmore, S. W., Sattler, M., Liang, H., Meadows, R. P., Harlan, B. S., Yoon, H. S., Nettesheim, D., Chang, B. S., Thompson, C. B., Wong, S. L., Ng, S. C., and Fesik, S. W. (1996). X-ray and NMR structure of human Bcl-$x_L$, an inhibitor of programmed cell death. *Nature* **381**, 335–341.

Muzio, M., Chinnaiyan, A. M., Kischkel, F. C., O'Rourke, K., Shevchenko, A., Ni, J., Scaffidi, C., Bretz, J. D., Zhang, M., Gentz, R., Mann, M., Krammer, P. H., Peter, M. E., and Dixit, V. M. (1996). FLICE, a novel FADD-homologous ICE/CED-3-like protease, is recruited to the CD95 (Fas/APO-1) death-induced signaling complex. *Cell* **85**, 817–827.

Nagata, S. (1996). Apoptosis: Telling cells their time is up. *Curr. Biol.* **6**, 1241–1243.

Nagata, S., and Golstein, P. (1995). The Fas death factor. *Science* **267**, 1449–1455.

Oltvai, Z. N., Milliman, C. L., and Korsmeyer, S. J. (1993). Bcl-2 heterodimerizes *in vivo* with a conserved homolog, Bax, that accelerates programmed cell death. *Cell* **74**, 609–619.

Perez-Terzic, C., Pyle, J., Jaconi, M., Stehno-Bittel, L., and Clapham, D. E. (1996). Conformational states of the nuclear pore complex induced by depletion of nuclear $Ca^{2+}$ stores. *Science* **273**, 1875–1877.

Piacentini, M. (1995). Tissue transglutaminase: A candidate effector element of physiological cell death. *Curr. Top. Microbiol. Immunol.* **200**, 163–175.

Raffo, A. J., Perlman, H., Chen, M. W., Day, M. L., Streitman, J. S., and Buttyan, R. (1995). Overexpression of Bcl-2 protects prostate cancer cells from apoptosis *in vitro* and confers resistance to androgen depletion *in vivo*. *Cancer Res.* **55**, 4438–4445.

Reed, J. C. (1994). Bcl-2 and the regulation of programmed cell death. *J. Cell Biol.* **124**, 1–6.

Reynolds, J. E., and Eastman, A. (1996). Intracellular calcium stores are not required for Bcl-2-mediated protection from apoptosis. *J. Biol. Chem.* **271**, 27739–27743.

Rosen, D. R., Siddique, T., Patterson, D., Figlewicz, D. A., Sapp, P., Hentati, A., Donaldson, D., Goto, J., O'Regan, J. P., Deng, H. X. et al. (1993). Mutations in Cu/Zn superoxide dismutase gene are associated with familiar amyotrophic lateral sclerosis. *Nature* **362**, 59–62.

Ruoslahti, E., and Reed, J. C. (1994). Anchorage dependence, integrins, and apoptosis. *Cell* **77**, 477–478.

Samali, A., Gorman, A. M., and Cotter, T. G. (1996). Apoptosis—the story so far. *Experientia* **52**, 933–941.

Schwartz, L. M., Smith, S. W., Jones, M. E. E., and Osborne, B. A. (1993). Do all programmed cell deaths occur via apoptosis? *Proc. Natl. Acad. Sci. USA* **90**, 980–984.

Sentman, C. L., Shutter, J. R., Hockenber, D., Kanagawa, O., and Korsmeyer, S. J. (1991). Bcl-2 inhibits multiple forms of apoptosis but not negative selection in thymocytes. *Cell* **67**, 879–888.

Shibasaki, F., and McKeon, F. (1995). Calcineurin functions in $Ca^{2+}$-activated cell death in mammalian cells. *J. Cell Biol.* **131**, 735–743.

Slater, A. F. G., Nobel, C. S. I., and Orrenius, S. (1995). The role of intracellular oxidants in apoptosis. *Biochim. Biophys. Acta* **1271**, 59–62.

Testi, R. (1996). Sphingomyelin breakdown and cell fate. *Trends Biochem. Sci.* **21,** 468–471.

Thompson, C. B. (1995). Apopotosis in the pathogenesis and treatment of disease. *Science* **267,** 1456–1462.

Trump, B. J., and Berzesky, I. K. (1995). Calcium-mediated cell injury and cell death. *FASEB J.* **9,** 219–228.

Vito, P., Lacaná, E., and D'Adamio, L. (1996). Interfering with apoptosis: $Ca^{2+}$-binding protein ALG-2 and Alzheimer's disease gene ALG-3. *Science* **271,** 521–525.

Waldmann, R., Champigny, G., Voilley, N., Lauritzen, I., and Lazdunski, M. (1996). The mammalian degenerin MDEG, an amiloride-sensitize cation channel activated by mutations causing degeneration in *Caenorhabditis elegans. J. Biol. Chem.* **271,** 10433–10436.

Walker, P. R., and Sikorska, M. (1994). Endonuclease activities, chromatin structure, and DNA degradation in apoptosis. *Biochem. Cell Biol.* **72,** 615–623.

Wang, J., and Walsh, K. (1996). Resistance to apoptosis conferred by Cdk inhibitors during myocyte differentiation. *Science* **273,** 359–361.

Wang, C. Y., Mayo, M. W., and Baldwin, A. S., Jr. (1996a). TNF- and cancer therapy-induced apoptosis: Potentiation by inhibition of NF-κB. *Science* **274,** 784–787.

Wang, K., Yin, X. M., Chao, D. T., Milliman, C. L., and Korsmeyer, S. J. (1996b). BID: A novel BH3 domain-only death agonist. *Genes Dev.* **10,** 2859–2869.

Wyllie, A. H. (1980). Glucocorticoid-induced thymocyte apoptosis is associated with endogenous endonuclease activation. *Nature* **284,** 555–556.

Wyllie, A. H., Kerr, J. F. R., and Currie, A. R. (1980). Cell death: The significance of apoptosis. *Int. Rev. Cytol.* **68,** 251–306.

Wyllie, A. H., Morris, R. G., Smith, A. L., and Dunlop, D. (1984). Chromatin cleavage in apoptosis: Association with condensed chromatin morphology and dependence on macromolecular synthesis. *J. Pathol.* **142,** 67–77.

Yin, X. M., Oltavi, Z. N., and Korsmeyer, S. J. (1994). BH1 and BH2 domains of Bcl-2 are required for inhibition of apoptosis and heterodimerization with Bax. *Nature* **369,** 321–323.

Yonish-Rouach, E. (1996). The p53 tumour suppressor gene: A mediator of a G1 growth arrest and of apoptosis. *Experientia* **52,** 1001–1007.

Yuan, J., and Horvitz, H. R. (1992). The *Caenorhabditis elegans* cell death gene *ced-4* encodes a novel protein and is expressed during the period of extensive programmed cell death. *Development* **116,** 309–320.

Zha, H., Aimé-Sempé, C., Sato, T., and Reed, J. C. (1996). Proapoptotic protein Bax heterodimerizes with Bcl-2 and homodimerizes with Bax via a novel domain (BH3) distinct from BH1 and BH2. *J. Biol. Chem.* **271,** 7440–7444.

Zhivotovsky, B., Burgess, D. H., and Orrenius, S. (1996). Proteases in apoptosis. *Experientia* **52,** 968–978.

Eric J. Hall and James G. Kereiakes

# 68

# Effects of Ionizing Radiations on Cells

## I. Introduction

Life on earth has developed with an ever-present background of radiation. It is not something new, invented by the wit of man; radiation from natural sources has always been here in the form of, for example, radioactivity in the ground and in food, as well as cosmic rays from outer space. An issue often debated is whether life has evolved in spite of the potential deleterious effects of radiation—the winner in a constant battle—or whether the ability of radiation to cause mutations has been a vital factor in the continued upward evolution of biological species. No one is sure at present which is the case, and it is probable that the answer will never be known with any certainty.

What is new, what is man-made, is the extra radiation to which we are subjected from medical X-rays in the hospital or dentist's office, from journeys in high-flying aircraft, from the fallout of nuclear weapons testing, from nuclear reactors built to generate electrical power, and from low-level radioactive waste. There can be no denying that a man-made component of radiation is being continuously added to the natural background level. This is a cause of great concern to the public, and must ultimately implicate the whole of society because of the critical choices and issues involved. However, the use of radiation has become an integral part of modern life. From the X-ray picture of a broken limb to the treatment of cancer, the medical applications of radiation are now accepted as commonplace, and X-ray facilities are available in every community hospital.

## II. Types of Radiation

Radiation is ionizing if it is able to disrupt the chemical bonds of molecules of which living things are made and thus cause biologically important changes. Visible light, radiowaves, and the radiant heat from the sun are also forms of radiation, but are not able to produce damage in this way by ionization, though, of course, they too can cause biological effects if larger amounts are involved.

Ionizing radiations come in several varieties. First, there are rays, such as **X-rays** or **gamma rays.** These represent energy transmitted in a wave without the movement of any material, just as heat and light from the sun cross the vast emptiness of space to reach the earth. X- and gamma rays do not differ from one another in nature or in properties; the only difference between the two is where they originate. X-rays in general, are made in an electrical device, such as may be seen in any dentist's office, whereas gamma rays are emitted by unstable or radioactive isotopes. The electromagnetic spectrum is illustrated in Fig. 1.

Other types of ionizing radiations are fast-moving particles of matter. Some carry a charge of electricity, some do not. The penetrability of different types of radiation is illustrated in Fig. 2.

**Neutrons** are the only uncharged particles of any consequence, and they are an important form of ionizing radiation because they are generally associated with atomic bombs and nuclear reactors. Neutrons are particles with a mass similar to that of the proton, but they carry no electrical charge. Because they are electrically neutral, they are very penetrating in material of all sorts, including living tissues. Neutrons constitute one of the fundamental particles of which the nucleus of most atoms is built. Neutrons are emitted as a by-product when heavy radioactive atoms such as uranium undergo fission, that is, split up to form two smaller atoms. They can also be produced artificially by large accelerators in physics research laboratories.

**Electrons** are small negatively charged particles, which are found in all normal atoms. Electrons are often given off when radioactive materials break down and decay, in which case they are called **beta rays.** They can also be produced artificially in the laboratory and accelerated in electrical devices.

**Protons** are positively charged particles found in the nucleus of every atom. They have a mass approximately equal to that of the neutron and are almost 2000 times heavier than an electron. Protons are not usually given off

**1044**

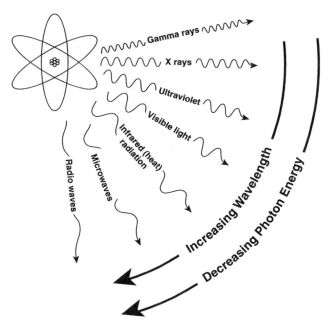

**FIG. 1.** Illustrating the electromagnetic spectrum. X-rays and gamma rays have the same nature as visible light, radiant heat, and radiowaves; however, they have a shorter wavelength and consequently a larger photon energy—more energy per "packet."

by radioactive isotopes on earth, but they are found in great abundance in outer space and may constitute a hazard to astronauts.

**Alpha particles** are nuclei of helium atoms; that is, helium atoms with the planetary electrons stripped off. An alpha particle consists of two protons and two neutrons stuck together. They have a net positive charge and are relatively massive. These particles are commonly emitted when heavy radioactive isotopes, such as uranium or radium, decay and break down.

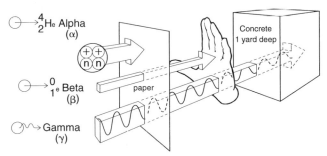

**FIG. 2.** Illustration of some of the different types of ionizing radiations that were labeled $\alpha$, $\beta$, and $\gamma$ before their true nature was known or understood. Gamma ($\gamma$) rays form part of the electromagnetic spectrum and are of the same nature as heat or light. They can be very penetrating and pass through thick barriers. Beta ($\beta$) rays are comprised of a stream of electrons, tiny negatively charged particles. Beta rays can pass through a hand, but unless they are of very high energy, they can be stopped by a modest barrier. Alpha ($\alpha$) rays are relatively massive positively charged particles: they are in fact helium nuclei and each is made up of two neutrons and two protons in close association. In general, alpha rays can be stopped by a thin barrier—even a sheet of cardboard.

**Heavy ions** are the nuclei of any atoms that are stripped of their planetary electrons and are moving at high speed. Ions of almost all of the known elements are present in space and constitute one of the problems of space flight. These particles move with great speed and have enormous energy, and it is virtually impossible to design spacecraft to fully protect their occupants from all of the heavy ions.

## III. Interactions of Radiation with Matter

When such radiations pass through living things, they give up energy to the tissues and cells of which all biological material is made. The energy is not spread out evenly, but is deposited or dumped very unevenly in discrete "packets." A lot of energy is given to some parts of some cells, and little, if any, to others.

This uneven pattern of energy deposition accounts for the special consequences of ionizing radiation. The total amount of energy involved may be small; however, some cells of the living material may be adversely affected because it is deposited so unevenly. The smallness of the amount of energy involved may be illustrated in a number of ways. For example, the dose of X-rays that would undoubtedly kill a person if given to the whole body may be compared with heat energy; it would be less than that absorbed by drinking a cup of warm coffee, or sunbathing for a few minutes on a hot day. Energy in the form of heat is absorbed uniformly and evenly, and much greater quantities of energy in these forms are required to produce damage in living things.

The situation may be illustrated by an example that, although trivial, contains the essence of the difference between ionizing and nonionizing radiations. Imagine throwing a kilogram of material at a running rabbit. One may choose to throw sand, in which case millions of particles would make up one kilogram, or one may choose to throw a single rock weighing one kilogram. The same total amount of energy would be involved in throwing either projectile a given distance. In the case of the sand, the energy would be divided into such small individual packets that no damage would result to the rabbit from the impact of any of them. If a single rock were chosen instead, the chance of scoring a hit at all would be greatly reduced, but biological damage would undoubtedly follow in the event that a hit occurred. Ionizing radiations constitute large discrete "packets" of energy. To absorb a dose of ionizing radiation is to be hit by a rock, not by many particles of sand!

Radiations may be directly or indirectly ionizing. All of the charged particles previously discussed including $\alpha$ particles and heavy ions are **directly ionizing**; that is, provided the individual particles have sufficient kinetic energy, they can directly disrupt the atomic structure of the absorber through which they pass and produce chemical and biological changes. Electromagnetic radiations (X- and gamma rays) are **indirectly ionizing.** They do not produce chemical and biological damage themselves, but when they are absorbed in the material through which they pass they give up their energy to produce fast-moving electrons.

The process by which X-ray photons are absorbed depends on the energy of the photons concerned and the chemical composition of the absorbing material. At high energies, the Compton process predominates, while at lower energies the photoelectric process is more important; but in either case the end result is the same: The photon energy is converted into kinetic energy of a fast-moving secondary electron. In practice, when an X-ray beam is absorbed by tissue, a vast number of photons interact with a vast number of atoms to produce a large number of fast electrons, many of which can ionize other atoms of the absorber, break vital chemical bonds, and initiate the chain of events that ultimately is expressed as biological damage.

Neutrons are uncharged particles. For this reason they are highly penetrating compared with charged particles of the same mass and energy. They are indirectly ionizing and are absorbed by elastic or inelastic scattering. Fast neutrons differ basically from X-rays in the mode of their interaction with tissue. **X-ray photons** interact with the **orbital electrons** of atoms of the absorbing material and set in motion **fast electrons. Neutrons,** however, interact with the **nuclei** of atoms of the absorbing material and set in motion **fast recoil protons, $\alpha$ particles,** and **heavier nuclear fragments.**

In the case of intermediate fast neutrons, **elastic scattering** is the dominant process. The incident neutron collides with the nucleus of an atom of the absorber (which in the case of tissue is predominantly hydrogen). Part of its kinetic energy is transferred to the nucleus and part is retained by the deflected neutron, which may go on to make further collisions. The recoil protons that are set in motion lose energy by excitation and ionization as they pass through the biological material.

At energies above about 5 MeV, **inelastic scattering** begins to take place and assumes increasing importance as the neutron energy rises. The neutron may interact with a carbon nucleus to produce three $\alpha$ particles, or with an oxygen nucleus to produce four $\alpha$ particles. These are the so-called "spallation" products, which become very important at higher energies. The $\alpha$ particles produced in this way represent a relatively modest proportion of the total absorbed dose, but they are densely ionizing and have an important effect on the biological characteristics of the radiation.

## IV. Measuring Radiation

The amount or quantity of radiation, or as it is usually called, the **dose,** is measured in terms of the energy absorbed in the tissues. The unit of absorbed dose is the Gray (Gy) defined to be an energy absorption of 1 joule per kilogram.

In some instances, the dose is very much less than one Gray and then the units milligray (mGy) or microgray ($\mu$Gy) are used. There are 1000 mGy or one million $\mu$Gy in 1 Gy. To give an idea of the way in which these units are used, two examples are quoted from later sections. A total body dose of 5 Gy to a human being would most probably be fatal; the average natural background radiation to which we are exposed is about 3 mGy per year.

Equal doses of different types of radiation do not necessarily produce equal biological effects. For example, 0.5 Gy of neutrons is more effective than 0.5 Gy of X-rays. In general, X-rays, gamma rays, and electrons are least effective for a given dose, while heavy ions are the most damaging. Neutrons fall somewhere between.

For general discussions of radiation effects, a different quantity is used, namely, **equivalent dose,** which is the absorbed dose (in Gray) multiplied by a factor known as the **radiation weighting factor** ($W_r$), which allows for the relative effectiveness of the particular type of radiation involved. If dose is measured in Gray, the equivalent dose will be in Sievert (Sv). X-rays and gamma rays are regarded as the standard, and for these types of radiation Gray and Sievert are interchangeable. A dose of 1 Gy of X-rays is, by definition, 1 Sv. Alpha particles, however, are roughly 20 times as effective as X-rays for the same absorbed dose and are therefore assigned a radiation weighting factor ($W_r$) of 20. Consequently, a dose of 1 Gy of $\alpha$ particles represents an effective dose of 20 Sv. In some instances, the equivalent dose may be much less than 1 Sv or $\mu$Sv, in which case smaller subunits are used. There are 1000 millisieverts (mSv) and 1 million microsieverts ($\mu$Sv) in 1 Sievert.

Next, there is the **collective equivalent dose.** This is obtained by multiplying the equivalent dose by the number of individuals exposed. Thus, for example, in the "incident" at Three Mile Island, an estimate of the collective equivalent dose received by the 2 million people who lived within 50 miles turned out to be 32 person-sieverts. Some individuals received up to 1 mSv, others less than 0.01 mSv.

## V. DNA Damage and Chromosome Breaks

Many lines of evidence support the view that DNA is the principal target for the biological effects of ionizing radiations. For example, using a polonium-tipped microneedle that emits short-range $\alpha$ particles, it has been shown that irradiating the cytoplasm of the cell to huge doses in excess of 250 Gy does not affect proliferation, whereas small doses to the nucleus prove to be lethal. Further, the incorporation of the short-range beta emitter tritiated thymidine into DNA is lethal, and incorporation of halogenated pyrimidines into DNA results in increased radiosensitivity, which is directly proportional to the amount incorporated.

The interaction of ionizing radiation with DNA has classically been described as resulting from a *direct* or an *indirect* interaction. In the direct interaction, a photon is absorbed by the medium, resulting in a secondary fast-moving electron that directly hits and breaks the DNA strand. In the indirect interaction the fast electron hits a water molecule (tissue is mostly water) producing a hydroxyl radical (OH·). A free radical is an atom or molecule with an unpaired electron in the outer orbit, a state that is associated with a high degree of chemical reactivity. This free radical then diffuses to the DNA causing a strand

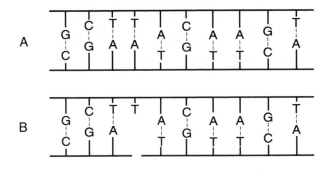

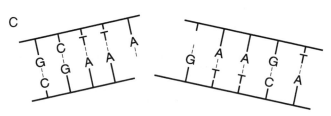

**FIG. 3.** Diagrams of single- and double-strand DNA breaks caused by radiation. (A) Two-dimensional representation of the normal DNA double helix. The base pairs carrying the genetic code are complementary (i.e., adenine pairs with thymine, guanine pairs with cytosine). (B) A break in one strand is of little significance because it is readily repaired, using the opposite strand as a template. (C) If breaks occur in both strands and are directly opposite or separated by only a few base pairs, this may lead to a double-strand break where the chromatin snaps into two pieces. (Courtesy of John Ward, 1993.)

break. Experiments using free radical scavengers have led to the estimation that two-thirds of DNA damage from photon irradiation results from the indirect action. For $\alpha$ particles and neutrons the balance shifts in favor of the direct effect, which becomes dominant.

DNA consists of two strands that form a double helix, with each strand composed of deoxynucleotides, the sequence of which must be complementary, A pairing with T, and C with G. A break in a single strand, caused by either the direct or indirect effects discussed earlier, has little biological consequence since the break is rapidly repaired using the opposite strand as a template. However, if breaks occur in both strands that are directly opposite, or separated by only a few base pairs, then a double-strand break (DSB) occurs and the piece of chromatin snaps in two. This is illustrated in Fig. 3.

There is good evidence to believe that most of the important biological effects of ionizing radiations are a direct consequence of the rejoining of two DSBs. Figure 4 illustrates two ways in which a DSB in each of two chromosomes may rejoin. In the lower half of the figure, the two broken chromosomes rejoin in such a way that a dicentric and an acentric fragment are formed. This exchange type aberration is lethal to the cell; the fragment with no centromere will be lost during cell division, whereas the aberrant chromosome with two centromeres will make a normal mitosis impossible. Aberrations of this type represent the principal mechanism whereby radiation kills cells. Note that breaks in *two* chromosomes are necessary for an exchange type aberration. The two breaks can be produced by a single electron track (see Fig. 5) in which case the probability of an interaction will be proportional to dose—that is, doubling the dose doubles the yield of aberrations. However, the two chromosome breaks may result from two separate electrons, in which case the probability of an interaction increases with the (dose)$^2$—that is, doubling the dose makes an aberration *four* times as likely.

In practice, both possibilities occur, so that the yield of chromosomal aberrations is a linear-quadratic function of dose, that is,

$$\text{Aberration yield} = \alpha D + \beta D^2 \qquad (1)$$

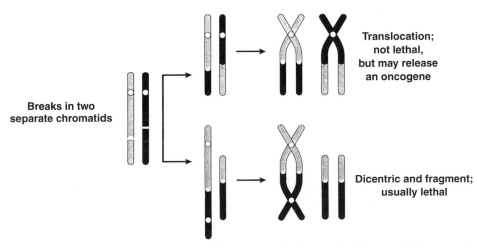

**FIG. 4.** Most biological effects of radiation are due to the incorrect joining of breaks in two chromosomes. For example, the two broken chromosomes may recombine to form a dicentric (a chromosome with two centromeres) and an acentric fragment (a fragment with no centromere). This is a lethal lesion resulting in cell death. Alternatively, the two broken chromosomes may exchange broken ends. This is called a symmetrical translocation. It does not lead to the death of the cell, but in a few special cases activates an oncogene by moving it from a quiescent to an active site.

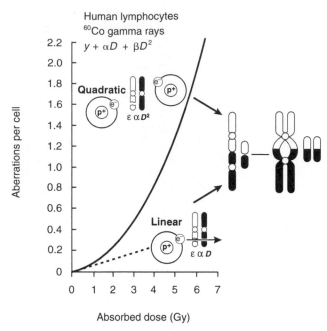

**FIG. 5.** The frequency of chromosomal aberrations (dicentrics and rings) is a linear-quadratic function of dose because the aberrations are the consequence of the interaction of two separate breaks. At low doses, both breaks may be caused by the same electron; the probability of an exchange aberration is proportional to the square of the dose ($D^2$). (Redrawn from Hall, 1994.)

This linear-quadratic relationship is ubiquitous in radiation biology for all biological effects of gamma and X-rays. Scoring dicentric aberrations can be used as a biological dosimeter. Suppose someone is suspected of being accidentally exposed to a large dose of radiation. A sample of blood can be taken, the peripheral lymphocytes stimulated to divide, and the incidence of dicentric aberrations scored. The lowest dose that can be assessed conveniently is about 25 cGy.

Referring back to Fig. 4 it is possible for the breaks in the two chromosomes to rejoin as shown in the upper half of the figure. This leads to a symmetrical translocation that is quite compatible with cell viability. However, it represents a rearrangement of genetic material, and in a few instances leads to a malignant change. For example, Burkitt's lymphoma and some types of leukemia result from a translocation that moves an oncogene from a quiescent to an active chromosomal site. This is one of the most likely mechanisms for radiation-induced carcinogenesis.

## VI. Cell Survival Curves

### A. Clonogenic

A **cell survival curve** describes the relationship between the radiation dose and proportion of cells that survive. What is meant by "survival"? Survival is the opposite of death! For differentiated cells that do not proliferate, such as nerve, muscle, or secretory cells, death can be defined

as the loss of a specific function. For proliferating cells, such as hematopoietic stem cells or cells growing in culture, loss of the capacity for sustained proliferation—that is, loss of **reproductive integrity**—is an appropriate definition. This is sometimes called **reproductive death** and a cell that survives by this definition is said to be **clonogenic** because it can form a clone or colony.

This definition is generally relevant to the radiobiology of whole animals and plants and their tissues. It has particular relevance to the radiotherapy of tumors. For a tumor to be eradicated, it is only necessary that cells be "killed" in the sense that they are rendered unable to divide and cause further growth and spread of the malignancy.

### B. Why Cells Die

The classic mode of cell death following exposure to radiation is "mitotic death." Cells die while attempting to divide because of chromosomal damage, such as the formation of a ring or a dicentric that causes loss of genetic material and prevents the clean segregation of DNA into the two daughter cells. Death does not necessarily occur at the first mitosis following irradiation; cells can often manage to complete several divisions, but death is inevitable if chromosomal damage is severe.

The other form of cell death is programmed cell death or **apoptosis.** This is an important form of cell death during the development of the embryo, and is implicated, for example, in the regression of the tadpole tail during metamorphosis. Is it also important in many facets of biology including cell renewal systems and hormone-related atrophy. Apoptosis is characterized by a sequence of morphological events; cells condense and the DNA breaks up into pieces before the cells are phagocytosed and removed. In radiation biology, apoptosis is a dominant mode of radiation-induced cell death in cells of lymphoid origin, but of variable importance in most other cell types.

### C. The Shape of a Cell Survival Curve

Most cell survival curves have been obtained by growing cells *in vitro* in petri dishes. Many cell lines have been established from malignant tumors and from normal tissues taken from humans or laboratory animals.

If cells are seeded as single cells, allowed to attach to the surface of a petri dish, and provided with culture medium, and appropriate conditions, each cell will grow into a macroscopic colony that is visible by eye in a period of a few weeks. If, however, the same number of cells is placed into a parallel dish and exposed to a dose of radiation just after they have attached to the surface of the dish, some cells will grow into colonies indistinguishable from those in the unirradiated dish, but others will form only tiny abortive colonies because the cells die after a few divisions. This is illustrated in Fig. 6. The **surviving fraction** is the number of macroscopic colonies counted in the irradiated dish divided by the number on the unirradiated dish. This process is repeated so that estimates of survival are obtained for a range of doses; surviving fraction is plotted on a logarithmic scale against dose on a linear scale in Fig. 7.

**Control** **4 Gy**

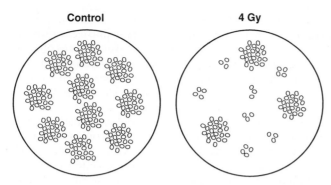

**FIG. 6.** In the left-hand dish, 10 cells were seeded and grew up into 10 macroscopic colonies, each containing hundreds or thousands of cells. The same number of cells were seeded in the right-hand dish but it was exposed to a dose of 4 Gy of X-rays as soon as the cells had attached. In this case, 3 cells grew into colonies indistinguishable from those on the unirradiated dish, while three formed abortive colonies. The surviving fraction is 3/10, or 0.3. In practice, of course, much larger numbers of cells are used to achieve better precision.

Qualitatively, the shape of the survival curve can be described in relatively simple terms. At "low doses" for sparsely ionizing radiations, such as X-rays, the survival curve starts out straight on the log-linear plot with a finite initial slope; that is, the surviving fraction is an exponential function of dose. At higher doses, the curve bends. This bending or curving region extends over a dose range of a few Gray. At very high doses the survival curve often tends to straighten again; in general, this does not occur until doses are in excess of 10 Gy. By contrast, for densely ionizing (high linear energy transfer) radiations, such as $\alpha$ particles or low-energy neutrons, the cell survival curve is a straight line from the origin; that is, survival approximates to an exponential function of dose.

The linear quadratic model is currently the model of choice for cell survival curves and is fitted to the data in Fig. 7. This model assumes that there are two components to cell killing by radiation, one that is proportional to dose and one that is proportional to the square of the dose. The notion of a component of cell inactivation that varies with the square of the dose goes back to the early work with chromosomes in which many chromosome aberrations are clearly the result of two separate breaks, as previously described.

By this model the expression for the cell survival curve is where $S$ is the fraction of

$$S = e^{-\alpha D - \beta D^2} \qquad (2)$$

where $S$ is the fraction of cells surviving a dose $D$, and $\alpha$ and $\beta$ are constants. The components of cell killing that are proportional to dose and to the square of the dose are equal when:

$$\alpha D = \beta D^2 \qquad (3)$$

or

$$D = \alpha/\beta \qquad (4)$$

This is an important point that bears repeating: The linear and quadratic contributions to cell killing are equal at a dose that is equal to the ratio of $\alpha/\beta$. This is illustrated in Fig. 7.

### D. Survival Curves for Normal Tissues *in Vivo*

A great deal of ingenuity has been shown in devising systems to determine clonogenic survival curves for the cells of normal tissues *in vivo*. A compilation of many of these is shown in Fig. 8. There is a range of radiosensitivity with bone marrow stem cells being the most radiosensitive; these cells are characterized by a survival curve that has little shoulder, probably because these cells are prone to die an apoptotic rather than a mitotic death.

### E. Radiosensitivity of Various Organisms

Figure 9 is a compilation of survival data to compare the radiosensitivity of various organisms. Its purpose is to illustrate that mammalian cells are exquisitely radiosensitive to radiation compared with microorganisms. Bacteria

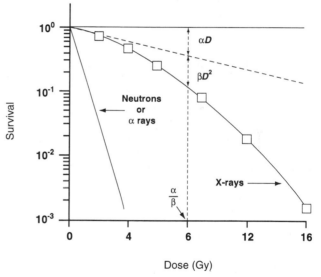

**FIG. 7.** Shape of survival curve for mammalian cells exposed to radiation. The fraction of cells surviving is plotted on a logarithmic scale against dose on a linear scale. For $\alpha$ particles or low-energy neutrons (said to be densely ionizing) the dose–response curve is a straight line from the origin (i.e., survival is an exponential function of dose). The survival curve can be described by just one parameter, the slope. For X- or gamma rays (said to be sparsely ionizing), the dose–response curve has an initial linear slope, followed by a shoulder; at higher doses the curve tends to become straight again. The experimental data for X-rays are fitted to a linear-quadratic function. There are two components of cell killing: One is proportional to dose ($\alpha D$); the other is proportional to the square of the dose ($\beta D^2$). The dose at which the linear and quadratic components are equal is the ratio $\alpha/\beta$. The linear-quadratic curve bends continuously but is a good fit to experimental data for the first few decades of survival. (Redrawn from Hall, 1994.)

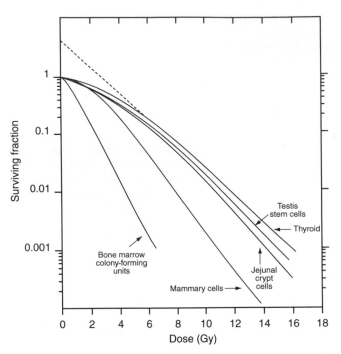

**FIG. 8.** Summary of survival curves for clonogenic assays of cells from normal tissues. The bone marrow colony-forming units, together with the mammary and thyroid cells, represent systems in which cells are irradiated and assayed by transplantation into a different tissue in recipient animals. The jejunal crypt and testis stem cells are examples of systems in which cells are assayed for regrowth *in situ* after irradiation.

and yeast are much more resistant to radiation. *Escherichia coli* B/r is a resistance mutant of *E. coli* that has particularly efficient repair mechanisms. Top of the league is *Micrococcus radiodurance;* that is no measurable killing even after a dose of 100 Gy! This dramatic variation of radiosensitivity is largely a function of DNA content—a large DNA content leads to radiosensitivity. This figure also illustrates why a dose of tens of thousands of Gray is required when radiation is used to sterilize surgical devices; common bacteria are *very* resistant.

## VII. Sensitivity and Phase of the Cell Cycle

### A. The Cell Cycle

Mammalian cells propagate by mitosis. When a cell divides, two daughter cells are produced, each of which carries a chromosome complement identical to that of the mother cell. After an interval of time has elapsed, each of the daughter cells may undergo a further division. The time between successive divisions is known as the **mitotic cycle time** or, as it is commonly called, the **cell cycle time.**

When a population of dividing cells is observed with a conventional light microscope, the only event in the entire cell cycle that can be distinguished is mitosis itself. Just before the cell divides to form two daughter cells, the

chromosomes (which are diffuse and scattered in the cell in the period between mitoses) condense into clearly distinguishable forms. In addition, in monolayer cultures of cells, just before mitosis, the cells round up and become loosely attached to the surface of the culture vessel. This whole process of mitosis—in preparation for which the cell rounds up, the chromosome material condenses, the cell divides into two, and then stretches out again and attaches to the surface of the culture vessel—lasts only about 1 hr. The remainder of the cell cycle, interphase, occupies all of the intermitotic period. No events of interest can be identified with a conventional microscope during this time.

Because cell division is a cyclic phenomenon, repeated in each generation of the cells, it is usual to represent it as a circle, as shown in Fig. 10. The circumference of the circle represents the full mitotic cycle time for the cells (T); the period of mitosis is represented by M. The remainder of the cell cycle can be further subdivided by the use of **autoradiography.** This technique was first introduced by Howard and Pelc (1953) and has revolutionized the study of cell biology.

The basis of the technique is to feed the cells thymidine, a basic building block for making a new set of chromosomes, which has been labeled with radioactive tritium. Cells that are actively synthesizing new DNA as part of the process of replicating their chromosome complement will incorporate the radioactive thymidine. The surplus radioactive thymidine is then flushed from the system, and the preparation of cells is coated with a very thin layer of nuclear (photographic) emulsion. Beta particles from cells that have incorporated radioactive thymidine pass through the nuclear emulsion and produce a latent image. When the emulsion is subsequently developed and fixed, the area through which a $\beta$ particle has passed appears as a black spot. It is then a comparatively simple matter to view the preparation of cells and to observe that some of the cells have black spots or "grains" over them, which indicates that they were actively synthesizing DNA at the time radioactive thymidine was made available. Other cells do not have any grains over their nuclei; this is interpreted to mean that the cells were not actively making DNA when the radioactive label was made available to them. When this is done, it becomes obvious that cells incorporate thymidine, that is, make DNA, only during a discreet part of the cycle, labeled S in Fig. 10. The interval between mitosis and S was called by Howard and Pelc "The first gap in activity," or $G_1$; the interval between S and M represents the second gap or $G_2$. Because of the problems related to disposal of radioactive waste, tritiated thymide has been largely replaced by 5-bromodeoxyuridine, which is not radioactive and is identified by fluorescence.

All proliferating mammalian cells, whether in culture or growing normally in a tissue, have a mitotic cycle consisting of mitosis (M), followed by $G_1$, a period of DNA synthesis (S), and $G_2$, after which mitosis occurs again. The overall length of the cell cycle may vary from about 10 hr for a hamster cell grown in culture to hundreds of hours for stem cells in some self-renewal tissues. This is due almost entirely to a dramatic variation in the length of $G_1$. The remaining components of the cell cycle M, S, and $G_2$, vary

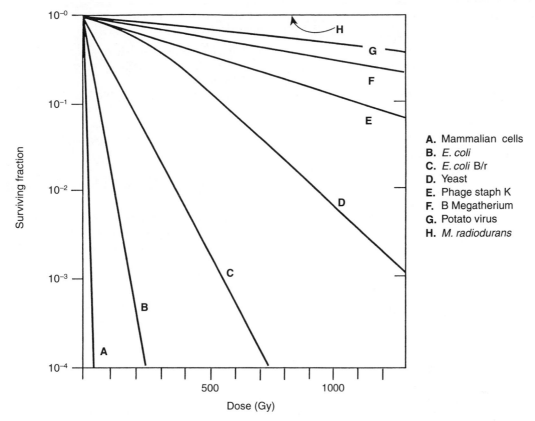

**FIG. 9.** A comparison of the radiosensitivy of various organisms. Mammalian cells are exquisitely sensitive compared with bacteria and yeast, principally because of their larger DNA content.

comparatively little between different cells in different circumstances.

## B. Synchronously Dividing Cells

A study of the variation of the radiosensitivity with the position or age of the cell in the cell cycle was only made possible by the development of techniques to produce synchronously dividing cell cultures—populations of cells in which all of the cells occupy the same phase of the cell cycle at a given time. The most satisfactory way to produce a synchronously dividing cell population is to use the **mitotic harvest** technique, first described by Terasima and Tolmach (1963). This technique can only be used for cultures that grow in monolayers attached to the surface of the growth vessel. It exploits the fact that when such cells are close to mitosis, they round up and become loosely attached to the surface. If at this stage the growth medium over the cells is subjected to gentle motion (by shaking), the mitotic cells become detached from the surface and float in the medium. If this medium is then removed from the culture vessel and plated out into new petri dishes, the population consists almost entirely of mitotic cells. Incubation of these cell cultures at 37°C then causes the cells to move together synchronously in step through their mitotic cycle. By delivering a dose of radiation at various

times after the initial harvesting of mitotic cells, one can irradiate cells at various phases of the cell cycle.

## C. Radiosensitivity and the Cell Cycle

Using the mitotic harvest technique, complete survival curves at a number of discrete points during the cell cycle were measured by Sinclair. The results are shown in Fig. 11. Survival curves are shown for mitotic cells (M), for cells in $G_1$, and $G_2$, and for cells in early and late S. It is at once evident that the most sensitive cells are those in M and $G_2$, which are characterized by a survival curve that is steep and has no shoulder. At the other extreme, cells in the latter part of S phase exhibit a survival curve that is less steep, but the essential difference is that the survival curve has a very broad shoulder. The other phases of the cycle, such as $G_1$ and early S, are intermediate in sensitivity between the two extremes.

The experiments of Sinclair and Morton (1966), shown in Fig. 11, were performed with a rapidly dividing cell having a very short $G_1$. In other cell lines with a long $G_1$, which may be closer to the situation for most proliferating stem cells *in vivo*, there is a second relatively radioresistant period in early $G_1$.

The following is a summary of the main characteristics

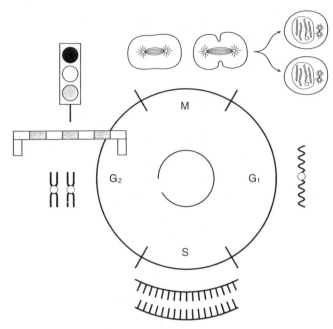

**FIG. 10.** The stages of the mitotic cycle for actively growing mammalian cells (M, mitosis; S, DNA synthetic phase; $G_1$ and $G_2$, "gaps" or periods of apparent inactivity between the major discernible events in the cycle). Also shown are the site of action and function of the molecular checkpoint gene. Cells exposed to any DNA damaging agent, including ionizing radiation, are arrested in $G_2$ phase. The function of the pause in cell cycle progression is to allow for a check of chromosome integrity before the complex task of mitosis is attempted. Cells in which the checkpoint gene is inactivated are more sensitive to killing by gamma rays or ultraviolet light.

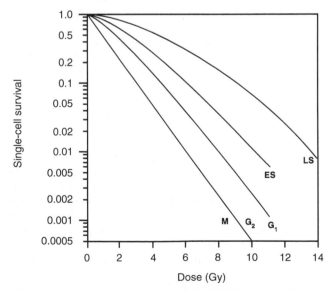

**FIG. 11.** Cell survival curves for Chinese hamster cells at various stages of the cell cycle. The survival curve for cells in mitosis is steep and has no shoulder. The curve for cells in late S phase is shallower and has a large initial shoulder. $G_1$ and S phases are intermediate in sensitivity. (Redrawn from the data of Sinclair and Morton, 1966.)

of the variation of radiosensitivity with cell age in the mitotic cycle:

1. Cells are most sensitive at or close to mitosis.
2. Resistance is usually greatest in the latter part of the S phase.
3. If $G_1$ has an appreciable length, a resistant period is evident early in $G_1$, followed by a sensitive period toward the end of $G_1$ phase.
4. $G_2$ is usually sensitive, perhaps as sensitive as M.

## VIII. Molecular Checkpoint Genes

Cell cycle progression is controlled by a family of genes known as **molecular checkpoint genes.** In radiation biology the most important appears to be the block in the $G_2$ phase of the cycle where cells are temporarily halted after a dose of radiation (see Fig. 10). The importance of this function is to halt cells and allow chromosomal damage to be repaired before the complex task of mitosis is attempted. The checkpoint gene has been cloned and sequenced in some strains of yeast. Mutant cell lines in which the checkpoint gene is defective are very sensitive to radiation or, in fact, to any DNA damaging agent.

## IX. Repair of Radiation Damage

Radiation damage to mammalian cells may be (1) **lethal damage,** which is irreversible, is irreparable, and by definition, leads irrevocably to cell death; or (2) **sublethal damage,** which under normal circumstances can be repaired unless additional sublethal damage is added (e.g., from a second dose of radiation) with which it can interact to form lethal damage. **Sublethal damage repair** is the operational term for the increase in cell survival that is observed when a given radiation dose is split into two fractions separated by a time interval.

Figure 12 shows data obtained in a split-dose experiment with cultured Chinese hamster cells. Figure 12A refers to cells that are not moving through the cycle because they were maintained at 24°C, a nonphysiological temperature, between doses.

A single dose of 15.58 Gy leads to a surviving fraction of 0.005. When the dose is divided into two equal fractions, separated by 30 min, the surviving fraction is already appreciably higher than for a single dose. As the time interval is extended, the surviving fraction increases until a plateau is reached at about 2 hr, corresponding to a surviving fraction of 0.02. This represents about four times as many surviving cells as for the dose given in a single exposure. A further increase in the time interval between the dose fractions is not accompanied by any additional increment in survival. The increase in survival in a split-dose experiment is due to the repair of sublethal radiation damage. This illustrates the phenomenon of repair without the complication of movement of cell through the cell cycle.

Figure 12B shows the results of the parallel experiment in which cells were exposed to split doses while being

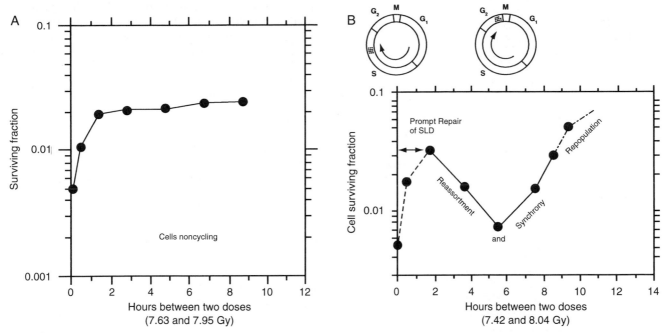

**FIG. 12.** (A) Survival of Chinese hamster cells exposed to two fractions of X-rays and incubated at room temperature for various time intervals between the two exposures. (B) Survival of Chinese hamster cells exposed to two fractions of X-rays and incubated at 37°C for various time intervals between the two doses. The survivors of the first dose are predominantly in a resistant phase of the cell cycle (late S). When the interval between doses is about 6 hr, these resistant cells will have moved to the $G_2$/M phase, which is sensitive. (Redrawn from Elkind *et al.,* 1965.)

maintained at their normal growing temperature of 37°C. The pattern of repair seen in this case differs from that observed for cells kept at room temperature. In the first few hours prompt repair of sublethal damage is again evident, but at longer intervals between the two split doses the surviving fraction of cells decreases again. An understanding of this phenomenon is based on the age–response function described earlier. When an asynchronous population of cells is exposed to a large dose of radiation, more cells are killed in the sensitive than in the resistant phases of the cell cycle. The surviving population of cells, therefore, tends to be partly synchronized. In Chinese hamster cells most of the survivors from a first dose are located in the S phase of the cell cycle. If about 6 hr are allowed to elapse before a second dose of radiation is given, this cohort of cells will progress around the cell cycle and will be in $G_2$ or M, a sensitive period of the cell cycle, at the time of the second dose. If the increase in radiosensitivity in moving from late S to the $G_2$/M period exceeds the effect of repair of sublethal damage, the surviving fraction will fall.

The pattern of repair shown in Fig. 12B is, therefore, a combination of three processes occurring simultaneously. First, there is the prompt repair of sublethal radiation damage. Second, there is progression of cells through the cell cycle during the interval between the split doses, which has been termed **reassortment.** Third, there is an increase in surviving fraction due to cell division, or repopulation, when the interval between the split doses is 10–12 hr be-

cause this exceeds the length of the cell cycle of these rapidly growing cells.

A simple experiment, performed *in vitro,* illustrates the three "R"s of radiobiology: **repair, reassortment,** and **repopulation.** It should be emphasized that the dramatic dip in the split-dose curve at 6 hr, caused by reassortment, and the increase in survival by 12 hr, because of repopulation, are seen only for rapidly growing cells. Hamster cells in culture have a cycle time of only 9 or 10 hr. The time sequence of these events would be longer in more slowly proliferating normal tissues *in vivo.* Repair of sublethal radiation damage has been demonstrated in just about every biological test system for which a quantitative endpoint is available.

Figure 13 shows that when a dose of radiation is split into several fractions, each separated by a time interval sufficiently long for sublethal damage to be repaired, more cells survive than for the same total dose given in a single fraction, because the shoulder of the curve must be repeated with each fraction. In general, there is a good correlation between the extent of repair of sublethal damage and the size of the shoulder of the survival curve. This is not surprising, since both are manifestations of the same basic phenomenon: the accumulation and repair of sublethal damage. Some mammalian cells are characterized by a survival curve with a broad shoulder, and split-dose experiments then indicate a substantial amount of sublethal damage repair. Other types of cells show survival curves with a minimal shoulder, and this is reflected in more lim-

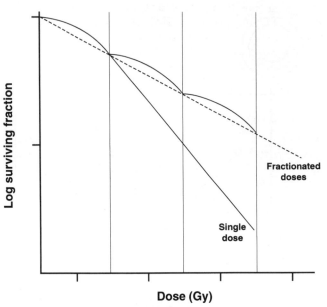

**FIG. 13.** The effect of fractionation when radiation is delivered in a series of fractions. With a time interval between fractions sufficiently long for sublethal damage to be repaired, fewer cells are killed than for the same dose delivered in a single exposure.

ited repair of sublethal damage. In the terminology of the linear-quadratic ($\alpha/\beta$) description of the survival curve, it is the quadratic component ($\beta$) that causes the curve to bend and results in the sparing effect of a split dose.

## X. The Mechanism of Sublethal Damage Repair

As previously mentioned, there is a good correlation between cell killing and the production of asymmetrical chromosomal aberrations, such as dicentrics. This in turn is a consequence of an interaction between two (or more) double-strand breaks in the DNA. On this interpretation the repair of sublethal damage is simply the repair of DSBs. When a dose is split into two parts separated by a time interval, some of the DSBs produced by the first dose are rejoined and repaired before the second dose. The breaks in two chromosomes that must interact to form a dicentric may be formed by (1) a single track breaking both chromosomes or (2) separate tracks breaking the two chromosomes. The component of cell killing that results from single-track damage will be the same whether the dose is given in a single exposure or fractionated. The same is not true of multiple-track damage. If the dose is given in a single exposure (i.e., two fractions with $t = 0$ between them), all breaks produced by separate electrons can interact to form dicentrics. On the other hand, if the two dose fractions, $D/2$, are separated by (for example) 3 hr, breaks produced by the first dose may be repaired before the second dose is given. Consequently, there will be fewer interactions between broken chromosomes to form dicentrics and more cells will survive. On this simple interpreta-

tion, the repair of sublethal damage reflects the repair and rejoining of DSBs before they can interact to form lethal lesions. This interpretation readily accounts for the repair of radiation damage in cells where mitotic death dominates. In cells that die an apoptotic death, survival tends to be an exponential function of dose, that is, the survival curve is straight with no shoulder on the usual semilog plot. In this case there is no sparing from a split-dose exposure.

### A. Repair and Radiation Quality

For a given biological test system the size of the initial shoulder on the acute survival curve and, therefore, the amount of sublethal damage repair indicated by a split-dose experiment varies with the type of radiation used. The shoulder is largest for X-rays, smaller for neutrons, and nonexistent for densely ionizing $\alpha$ particles, where survival is an exponential function of dose. The amount of sublethal damage repair follows the same pattern; largest for X-rays, smaller for neutrons, and nonexistent for $\alpha$ particles.

### B. The Dose-Rate Effect

For x- or gamma ray doses, rate is one of the principal factors that determines the biological consequences of a given absorbed dose. As the dose rate is lowered and the exposure time extended, the biological effect of a given dose is generally reduced. Continuous low dose-rate irradiation may be considered to be an infinite number of infinitely small fractions; consequently the survival curve under these conditions would also be expected to have no shoulder and to be shallower than for single acute exposures.

The magnitude of the dose-rate effect from the repair of sublethal damage varies enormously between different types of cells. Cells characterized by a survival curve for acute exposures that has a small initial shoulder usually exhibit a modest dose-rate effect. This is to be expected, since both are expressions of the cell's capacity to accumulate and repair sublethal radiation damage. This is generally true of cells for which apoptosis is an important mechanism of cell death. Cell lines characterized by a survival curve for acute exposures, which has a broad initial shoulder, exhibit a dramatic dose-rate effect. An example is shown in Fig. 14 for Chinese hamster cells; in this cell line, mitotic death dominates and apoptosis is unimportant.

## XI. The Oxygen Effect

### A. Nature of Oxygen Effect

Many chemical and pharmacologic agents that modify the biological effect of ionizing radiations have been discovered. None is simpler than oxygen, and none produces such a dramatic effect. The oxygen effect was observed by Schwarz (1910), who noted that the skin reaction produced on his forearm by a radium applicator was reduced if the applicator was pressed hard onto the skin. In England,

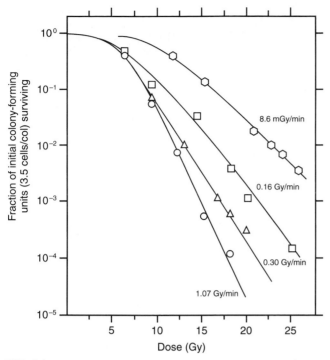

**FIG. 14.** Dose–response curves for Chinese hamster cells (CHL-F line) grown *in vitro* and exposed to cobalt-60 gamma rays at various dose rates. At high doses a substantial dose rate effect is evident even between 1.07, 0.3, and 0.16 Gy/min. The decrease in cell killing becomes even more dramatic as the dose rate is further reduced. (Redrawn from Bedford *et al.,* 1973.)

Mottram, in the 1930s, explored the question of oxygen in detail and was the first to discuss the possible importance of the oxygen effect in the radiotherapy of human cancer.

Survival curves for mammalian cells exposed to X-rays in the presence and absence of oxygen are illustrated in Fig. 15. The ratio of hypoxic to aerated doses needed to achieve the same biological effect is called the **oxygen enhancement ratio** (OER). For sparsely ionizing radiations, such as X- and gamma rays, the OER at high doses has a value of between 2.5 and 3. The OER has been determined for a wide variety of chemical and biological systems with different endpoints, and its value for X-rays always tends to fall in this range.

### B. Mechanism of Oxygen Effect

In practical terms, oxygen must be present during irradiation for its sensitizing effect to be observed. In fact, sophisticated experiments have shown that oxygen still sensitizes if it is added a few *microseconds* after a brief pulse of radiation. This is the clue to the mechanism of the oxygen effect since microseconds corresponds to the lifetime of the free radicals formed when X-rays interact with water.

Oxygen is essential to "fix" the damage to DNA resulting from radiation-produced free radicals. In the absence of oxygen, radical induced damage is readily repaired, but if oxygen is present the damage is made permanent, that is, "fixed." The word "fix" here is used in the European sense

of making something permanent, as in fixing a film, not in the American sense of repairing, as in fixing a flat tire.) Consequently, oxygen modifies only that component of radiation damage mediated by free radicals—the indirect damage. Oxygen has no effect on the direct damage caused by ionization of DNA.

### C. Oxygen Effect for Different Types of Radiation

The oxygen enhancement ratio for X- or gamma rays is approximately 3. It is this large because two-thirds of the damage produced by X-rays is mediated by free radicals. Densely ionizing $\alpha$ particles kill cells primarily by direct action, that is, by ionization of the DNA, a process not dependent on oxygen. Consequently, the OER for low energy (2-MeV) $\alpha$ particles is 1.0. The OER for neutrons has an intermediate value of about 1.6, because while the direct action is dominant there is still a contribution from free radical damage.

In summary, the oxygen effect is large and important in the case of sparsely ionizing radiations, such as X-rays; is absent for densely ionizing radiations, such as $\alpha$ particles; and has an intermediate value for fast neutrons.

### D. Concentration of Oxygen Required

Only a small amount of oxygen is required to produce the dramatic and important oxygen effect characteristic of X-rays. An oxygen concentration of 3 mm Hg (about 1/2%)

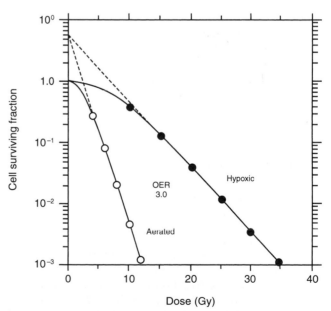

**FIG. 15.** Survival curves for mammalian cells exposed to X-rays under aerated and hypoxic conditions produced by passing a stream of pure nitrogen over the cells. These data are typical of many in the literature. Oxygen is a dose-modifying factor; that is, at all levels of cell survival, the dose required under hypoxic conditions is three times greater than that required under aerated conditions to produce the same biological effect. This ratio of doses is known as the oxygen enhancement ratio (OER).

results in a radiosensitivity halfway between complete hypoxia and 100% oxygen. By a concentration of 30 mm Hg, characteristic of most normal tissues in the body, the radiosensitivity of cells is indistinguishable from that in air or even in 100% oxygen.

### E. Oxygen Effect in Radiotherapy of Human Cancer

Malignant tumors tend to outgrow their blood supply so that areas of necrosis (dead cells) are a common feature of most cancers. Oxygen diffusing from blood vessels has a limited range because it is used up by rapidly growing and respiring cancer cells. Consequently there are pockets of cells distant from capillaries that are hypoxic—still with sufficient oxygen to be viable, but at a low enough oxygen concentration to be resistant to killing by X-rays. This diffusion-limited hypoxia is known as "chronic" hypoxia. Regions of "acute" hypoxia develop in tumors as a result of the temporary closing of a particular blood vessel. If this blockage were permanent, the cells downstream would, of course, eventually die and would be of no further consequence. There is, however, good evidence that tumor blood vessels open and close in a random fashion so that different regions of the tumor become hypoxia intermittently. At the moment when a dose of radiation is delivered, a proportion of the tumor cells may be hypoxic, and consequently resistant to killing by X-rays. The problem of hypoxia, both chronic and acute, is largely overcome in practice by delivering radiotherapy in a large number of small fractions, typically 30 to 40, over a period of 6–8 weeks. Other attempts to eliminate the problem of hypoxia cells has been to use neutrons rather than X-rays (because of their smaller OER) and to develop chemicals that specifically sensitize or kill hypoxic cells.

## XII. Radiation Quality and Biological Effects

### A. Deposition of Radiant Energy

When radiation is absorbed in biological material, ionizations and excitations occur that are not distributed at random but tend to be localized along the tracks of individual charged particles in a pattern that depends on the type of radiation involved. For example, photons of X-rays give rise to fast electrons, particles carrying unit electric charge and having very small mass; neutrons, on the other hand, give rise to recoil protons, particles again carrying unit electric charge but having a mass nearly 2000 times greater than that of the electron. Alpha particles carry two electric charges on a particle four times as heavy as a proton. The charge-to-mass ratio of $\alpha$ particles therefore differs from that for electrons by a factor of about 8000. As a result, the spatial distribution of the ionizing events produced by different particles vary enormously. For X-rays, the primary ionizing events are well separated in space and for this reason X-rays are said to be "sparsely ionizing." The ionizing events produced along the track of a low-energy $\alpha$ particle form a dense column, and for this reason $\alpha$ particles are referred to as "densely ionizing."

### B. Linear Energy Transfer

**Linear energy transfer (LET),** a term introduced by Zirkle (1940), is the energy transferred per unit length of the track. The special unit usually used for this quantity is kiloelectron volt per micron (keV/$\mu$m) of unit density material. The International Commission of Radiological Units (1962a, 1962b) defined this quantity as follows: The linear energy transfer ($L$) of charged particles in medium is the quotient of $dE/dl$, where $dE$ is the average energy locally imparted to the medium by a charged particle of specified energy in traversing a distance of $dl$. That is,

$$L = dE/dl \tag{5}$$

Since most radiations in practice consist of a wide spectrum of energies, LET can only be an average quantity. Typical LET values for commonly used radiations are listed in Table 1. Note that for a given type of charged particle, the higher the energy, the *lower* the LET and therefore the lower its biological effectiveness. For example, gamma rays and X-rays both give rise to fast secondary electrons; therefore, 1.1-MV cobalt-60 gamma rays have a lower LET than 250-kV X-rays and are less effective biologically by about 10%. By the same token, 150-MeV protons have a lower LET than 10-MeV protons and are, therefore, slightly less effective biologically.

### C. Relative Biological Effectiveness

Figure 16 illustrates the survival curves obtained for three different types of radiation, namely, X-rays, 15-MeV neutrons, and $\alpha$ particles. As the LET increases from about 2 keV/$\mu$m for X-rays up to 150 keV/$\mu$m for $\alpha$ particles, the survival curve changes in two important respects. First, the survival curve becomes steeper. Second, the extrapolation number tends toward unity; that is, the shoulder of the curve becomes progressively smaller as the LET increases.

It is evident from Fig. 16 that *equal* doses of different types of radiation do not result in *equal* biological effects. One Gray of $\alpha$ particles produces more cell killing than 1 Gy of neutrons, which in turn produces more killing than 1 Gy of X-rays. The key to the difference lies in the pattern of energy deposition at the microscopic level. In comparing different radiations it is customary to use X-rays as the standard. The formal definition of **relative biological effec-**

**TABLE 1** Typical LET Values

| Radiation | Track avg. | LET (keV/$\mu$m) | Energy avg. |
|---|---|---|---|
| Colbalt-60 X-rays | | 0.2 | |
| 250-kV X-rays | | 2.0 | |
| 10-MeV protons | | 4.7 | |
| 150-MeV protons | | 0.5 | |
| 14-MeV neutrons | 12 | | 100 |
| 2.5-MeV $\alpha$ particles | | 166 | |
| 2-GeV Fe ions | | 1000 | |

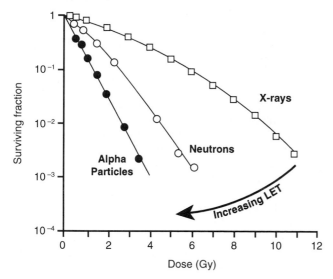

**FIG. 16.** Survival curves for cells of human origin exposed to 250-kVp X-rays, 15-MeV neutrons, and 4-MeV $\alpha$ particles. As the LET of the radiation increases, the slope of the survival curves gets steeper and the size of the initial shoulder gets smaller. (Redrawn from Broerse *et al.*, 1968.)

tiveness (RBE) is as follows: The RBE of some test radiation ($r$) compared with X-rays is defined by the ratio $D_x/D_r$, where $D_x$ and $D_r$ are, respectively, the doses of X-rays and the test radiation required for equal biological effect.

To determine the RBE of some test radiation, say, $\alpha$ particles, one chooses a biological system in which the effect of radiation may be scored quantitatively, and one compares $\alpha$ particles with the standard radiation, that is, X-rays. The data in Fig. 16 are an example. To achieve a surviving fraction of $10^{-2}$ requires a dose of about 9.5 Gy of X-rays or 2.5 Gy of $\alpha$ particles. The RBE of $\alpha$ particles relative to X-rays is the ratio of doses required to produce the same biological effect, that is, 9.5/2.5 = 4.6. Neutrons have a somewhat lower RBE; the dose to result in a surviving fraction of $10^{-2}$ in 4.3 Gy, so that the RBE is about 9.5/4.3 or 2.2.

It is important to note that, because the survival curves for X-rays (low LET) and neutrons or $\alpha$ particles (high LET) have a different shape, with the survival curve for X-rays having an appreciable shoulder, RBE does not have a unique value. At lower doses, corresponding to higher levels of survival, the RBE will assume a larger value; the maximum RBE will correspond to the ratio of the initial slopes for X-rays and for the test radiation.

### D. Relative Biological Effectiveness for Different Cells and Tissues

Even for a given total dose, the RBE varies significantly according to the tissue or endpoint used to measure it. Survival curves for mammalian cells exposed to x-rays have a large but variable shoulder region, while for high LET radiations the shoulder region is reduced or eliminated. As a consequence, the RBE will be different for each cell

line. In general, cells characterized by an x-ray survival curve with a large shoulder, indicating that they can accumulate and repair a large amount of sublethal radiation damage, will show a large RBE for neutrons or $\alpha$-particles. Conversely, cells for which the x-ray survival curve has little if any shoulder will exhibit small RBE values for $\alpha$-particles or neutrons.

### E. Relative Biological Effectiveness as a Function of Linear Energy Transfer

Figure 16 shows data for just three types of radiation to illustrate the way in which the shape of the survival curve changes as the density of ionization (LET) increases. Studies have been performed with many types of radiation covering a wide range of LET values. Figure 17 is a plot of the RBE as a function of LET. As the LET increases, the RBE increases slowly at first, and then more rapidly as the LET increases beyond 10 keV/$\mu$m. Between 10 and 100 keV/$\mu$m, the RBE increases rapidly with increasing LET and in fact reaches a maximum of about 100 keV/$\mu$m. Beyond this value for the LET, the RBE falls to lower values.

The LET at which the RBE reaches a peak is much the same (about 100 keV/$\mu$m) for a wide range of mammalian cells, from mouse to human, and is the same for all biological endpoints. It reflects the "target" size and is related to the DNA content, which is similar for all mammalian cells. The reason why radiation with an LET of 100 keV/$\mu$m is optimal in terms of producing a biological effect can be understood from the upper panel of Fig. 17. At this density

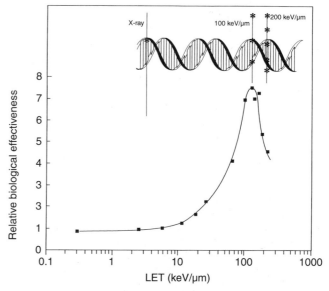

**FIG. 17.** Variation of RBE with LET. RBE rises to a maximum at an LET of about 100 keV/$\mu$m and subsequently falls for higher LET values. The upper panel illustrates that radiation with an LET of 100 keV/$\mu$m is most biologically effective because the average separation between ionizing events coincides with the diameter of the DNA double helix (2 nm); consequently a single-track of this quality can produce double-strand breaks with greatest efficiency. (Lower panel redrawn from Barendsen, 1968.)

of ionization, the *average* separation between ionizing events just about coincides with the diameter of the DNA double helix (20 Å or 2 nm). Radiation with this density of ionization is most likely to cause a double-strand break by the passage of a single charged particle, and DSBs are the basis of most biological effects, as discussed earlier. This is illustrated in the top panel of Fig. 17. In the case of X-rays, which are more sparsely ionizing, the probability of a single track causing a double-strand break is low, and in general more than one track will be required to produce a double-strand break. As a consequence, X-rays have a low biological effectiveness. At the other extreme, much more densely ionizing radiations (with a LET of 200 keV/$\mu$m, for example) will readily produce DSBs but energy will be "wasted" because the ionizing events are too close together. Since RBE is the ratio of *doses* to produce equal biological effect, this more densely ionizing radiation will have a lower RBE than the optimal LET radiation. The more densely ionizing radiation will be just as effective *per track,* but less effective per unit dose. It is possible, therefore, to understand why RBE reaches a maximum value in terms of the production of DSBs, since the interaction of two DSBs to form an exchange type aberration is the basis of most biological effects. Radiation having this optimal LET includes neutrons of a few hundred kiloelectron volts, as well as low-energy protons and $\alpha$ particles.

## F.  Radiation Weighting Factors

It is evident from the preceding discussion that different types of radiation differ in their biological effectiveness per unit of absorbed dose. The complexities of RBE are too difficult to apply in specifying dose limits in radiation protection; it is necessary to have a simpler way to consider differences in biological effectiveness of different radiations. One cGy of neutrons, for example, is more hazardous than 1 cGy of X-rays. The term **radiation weighting factor** ($W_r$) has been introduced for this purpose. The quantity produced by multiplying the absorbed dose by the weighting factor is called the **equivalent dose.** When dose is expressed in Gray (Gy), the equivalent dose is in Sievert (Sv). Radiation weighting factors are chosen by the International Commission on Radiological Protection (1991), based on a consideration of experimental RBE values, biased for biological endpoints relevant to radiation protection at low dose and low dose rate. There is a considerable element of judgment involved. The $W_r$ is set at unity for all low LET radiations (X-rays, gamma rays, and electrons), with a value of 20 for maximally effective neutrons and $\alpha$ particles. Thus, using this system, an absorbed dose of 0.1 Gy of radiation with a radiation weighting factor of 20 would result in an equivalent dose of 2 Sv.

## XIII.  Radio Protectors

In the late 1940s it was discovered that either cysteine or cysteamine could protect mice from the lethal effects of total body X-irradiation if administered in large amounts prior to exposure. The structures of these compounds are:

$$SH-CH_2-CH \Big\langle {}^{NH_2}_{COOH} \qquad \text{cysteine} \qquad (6)$$

$$SH-CH_2-CH_2-NH_2 \qquad \text{cysteamine} \qquad (7)$$

Animals receiving these radioprotectors could tolerate almost double the radiation dose. The ratio of the radiation dose to produce the same biological effect in the presence and absence of the drug is known as the **dose reduction factor** (DRF). Many similar compounds have been tested and found to be effective as radioprotectors. The most efficient tend to have certain structural features in common: a free SH group (or potential SH group) at one end of the molecule and a strong basic function such as amine or guanidine at the other end, separated by a straight chain of two or three carbon atoms. Sulfhydryl compounds are efficient radioprotectors against sparsely ionizing radiations, such as X- or gamma rays. The mechanism of action involves the scavenging of free radicals.

X-ray photons give up their energy to produce fast electrons as described earlier. These electrons may hit the DNA and cause a break directly (direct action) or interact with a water molecule to form a hydroxyl radical (OH·), which diffuses to the DNA and causes a break (indirect action). The protective effect of sulfhydryl compounds stems from their ability to "scavenge" free radicals. This process is illustrated in Fig. 18.

The protective effect of sulfhydryl compounds tends to parallel the oxygen effect, being maximal for sparsely ionizing radiations (e.g., X- or gamma rays) and minimal for densely ionizing radiations (e.g., low-energy $\alpha$ particles). It might be predicted that with effective scavenging of all free radicals the largest possible value of DRF would equal the oxygen enhancement ratio, with a value of 2.5 to 3.0.

It is not surprising that the discovery in 1948 of a compound that offered protection against radiation excited the interest of the U.S. Army, since the memory of Nagasaki and Hiroshima was vivid in the years immediately after World War II. Although cysteine is a radioprotector, it is also toxic and induces nausea and vomiting at the dose levels required for radioprotection. Consequently, the Walter Reed Army Hospital in Washington, D.C., synthesized more than 3000 compounds in an attempt to find the perfect radioprotector, one that would protect against radiation without debilitating side effects. At an early stage the important discovery was made that the toxicity of the compound could be greatly reduced if the sulfhydryl group was covered by a phosphate group. This is illustrated in Table 2. The $LD_{50}$ of the compound in animals can be doubled and the protective effect in terms of the DRF greatly enhanced if the SH group in cysteamine is covered by a phosphate. This tends to reduce systemic toxicity. Once in the cell, the phosphate group is stripped, and the SH group begins scavenging free radicals. Many compounds have been synthesized, but only two have been put to practical use. The structure and effectiveness of these is summarized in Table 3. The first compound, WR-638, called Cystaphos, was said to be carried routinely in the field pack of Russian infantry in Europe during the era of the Cold War for use

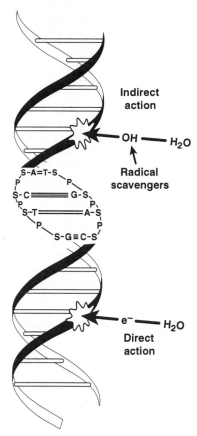

**FIG. 18.** Photons of X-rays either interact directly with DNA to produce a strand break (direct action) or indirectly via the production of a free radical (indirect action). Radioprotectors that are radical scavengers "intercept" the indirect action. (Redrawn from Hall, 1994.)

in the event of a nuclear conflict. Its usefulness would have been largely psychological, since the compound was carried as a tablet to be administered orally, when in fact these sulfhydryl compounds break down in stomach acid and are only effective when administered intravenously or intraperitoneally. A further factor, of course, is that such compounds will protect only from sparsely ionizing radiation; consequently, they would offer little protection against the prompt release of neutrons produced by the detonation of a nuclear device. They would be effective only against the gamma rays from the resulting fallout.

**TABLE 2** Effect of Adding a Phosphate-Covering Function on the Free Sulfhydryl of β-Mercaptoethylamine

| Drug | Formula | $LD_{50}$ in mice | DRF |
|------|---------|------------------|-----|
| MEA | $NH_2-CH-CH_2-SH$ | 343 (323–364) | 1.6 at 200 mg/kg |
| MEA-PO₃ | $NH_2-CH_2-CH-SH_2PO_3$ | 777 (700–864) | 2.1 at 500 mg/kg |

**TABLE 3** Two Protectors in Practical Use

| Compound | Structure | Drug dose (mg/kg) | DRF 7 days | DRF 30 days |
|----------|-----------|-------------------|-----------|-------------|
| WR-638 | $NH_2CH_2CH_2SPO_3HNa$ | 500 | 1.6 | 2.1 |
| WR-2721 | $NH_2(CH_2)_3NHCH_2CH_2SPO_3H_2$ | 900 | 1.8 | 2.7 |

The second compound, WR-2721, now known as amifostine, is perhaps the most effective of those synthesized in the Walter Reed series (Kligerman *et al.,* 1992). It gives good protection to the blood-forming organs, as can be seen by the DRF for 30-day death in mice, which approaches the theoretical maximum value of 3. It was probably the compound carried by the U.S. astronauts on their trips to the moon to be used if a solar event occurred. On these missions, when the space vehicle left earth's orbit and began coasting toward the moon, the astronauts were committed to a 14-day mission, since they did not have sufficient fuel to turn around without first orbiting the moon and using its gravitational field. If there had been a major solar event in that period, the astronauts would have been exposed to a shower of high-energy protons, resulting in an estimated dose of several hundred Gray. The availability of a radioprotector with a DRF of between 2 and 3 would have been very important in such a circumstance. As it turned out, no major solar event ever occurred during any manned lunar mission.

Amifostine also has a potential in radiotherapy. One application would be for local topical use, to reduce for example the mucosal reaction that occurs during radiotherapy of head and neck cancers. Another possibility is to protect all normal tissues by delivering the drug systemically; there is some evidence that the drug gets into tumors more slowly than into normal tissues, so that if the radiation is delivered within minutes of the drug, a differential effect is obtained with more protection of the normal tissues than of the tumor.

## XIV. Summary

Life on earth has always been exposed to radiation, but there is now an additional exposure from medical X-rays, nuclear power, and journeys in high-flying aircraft. Of concern are ionizing radiations, so called because they have a sufficient photon energy to knock electrons out of the atoms through which they pass, break chemical bonds, and cause a variety of biological changes.

Ionizing radiations come in a variety of types. X-rays are usually generated in an electrical device. Gamma rays are emitted by a radioactive material during decay. X- or gamma rays are electromagnetic waves, similar to radiant heat and visible light, but having a shorter wavelength. Neutrons are uncharged particles, often associated with nuclear fission. High-energy high-z charged particles are a hazard to astronauts in long-duration space flights.

The principal target for all biological effects of ionizing radiations is the DNA. The basic lesion is a double-strand break, caused when a charged particle breaks both strands of the DNA double helix, causing the chromatin strand to snap. Chromosomal aberrations represent one of the earliest detectable biological changes; if DSBs occur in two separate chromosomes, the broken ends may rejoin in incorrect and bizarre ways. These rearrangements can often be seen at the first metaphase after irradiation.

Following irradiation, cells may die from one of two mechanisms:

1. Mitotic cell death, due to the failure of cells to cope with severely damaged chromosomes, or
2. apoptosis, or programmed cell death, similar to the way that cells die and are removed during embryogenesis

The quantity of radiation, or dose, is measured in terms of the energy absorbed per unit mass. The current mass is the Gray (Gy) defined to be an energy absorption of one joule per kilogram.

A number of factors influence the biological effects of a given absorbed dose of radiation. First, the response of cells to radiation depends critically on their position in the division cycle at the time that they are exposed. In general, cells are most radiosensitive when they are close to mitosis. Second, cells vary greatly in their ability to repair radiation damage. In addition, a given dose of radiation is much less effective when spread out over a long period of time, because much of the radiation damage can be repaired during a prolonged exposure. Third, in the case of X- and gamma rays (but less so for neutrons and $\alpha$ particles), the biological consequences of a given dose can be modified by chemical means. For example, cells are more radiosensitive if molecular oxygen is present than if it is absent. On the other hand, compounds that scavenge free radicals are often found to be radioprotectors, and such compounds were used by the astronauts during lunar missions.

Fourth, the biological effect of a given absorbed dose depends on the "quality" of the radiation. X- and gamma rays are said to be "sparsely ionizing" because the ionizing events are well separated along the tracks of the electrons set in motion. Consequently, they are less effective than neutrons and $\alpha$ particles that are said to be "densely ioniz-

ing" because the ionizing events are clustered together and more likely to produce a DSB in a chromosome.

## Bibliography

Barendsen, G. W. (1968). Responses of cultured cells, tumors and normal tissues to radiations of different linear energy transfer. *Curr. Top. Radiat. Res.* **4,** 293.

Bedford, J. S., and Mitchell, J. B. (1973). Dose-rate effects in synchronous mammalian cells in culture. *Radiat. Res.* **54,** 316–327.

Broerse, J. J., Barendsen, G. W., and Van Kersen, G. R. (1968). Survival of cultured human cells after irradiation with fast neutrons of different energies in hypoxic and oxygenated conditions. *Int. J. Radiat. Biol.* **13,** 559–572.

Elkind, M. M., Sutton-Gilbert, H., Moses, W. B., Alescio, T., and Swain, R. B. (1965). Radiation response of mammalian cells in culture: V. Temperature dependence of the repair of x-ray damage in surviving cells (aerobic and hypoxic). *Radiat. Res.* **25,** 359–376.

Hall, E. J. (1994). "Radiobiology for the radiologist." (4th ed.) J. B. Lippincott Company, Philadelphia.

Howard, A., and Pelc, S. R. (1953). Synthesis of deoxyribonucleic acid in normal and irradiated cells and its relation to chromosome breakage. *Heredity* **6,** Suppl., 261.

International Commission on Radiological Protection. (1991). "Recommendations," Report No. 60. Pergamon Press, New York.

International Commission on Radiological Units and Measurements. (1962a). "Radiation Quantities and Units," Report 10a, Handbook '84. National Bureau of Standards, Washington, DC.

International Commission on Radiological Units and Measurements. (1962b). "Physical Aspects of Irradiation," Report 10b, Handbook 85. National Bureau of Standards, Washington, DC.

Kligerman, M. M., Liu, T., Liu, Y., Scheffler, B., He, S., and Zhang, S. (1992). Interim analysis of a randomized trial of radiation therapy of rectal cancer with/without WR-2721. *Int. J. Radiat. Oncol. Biol. Phys.* **22,** 799–802.

Mottram, J. C. (1936). Factor of importance in radiosensitivity of tumors. *Br. J. Radiol.* **9,** 606–614.

Schwarz, W. (1910). *Wein. Klin. Wochenschr. Nr.* **11S,** 397.

Sinclair, W. K., and Morton, R. A. (1966). X-ray sensitivity during the cell generation cycle of cultured Chinese hamster cells. *Radiat. Res.* **29,** 450–474.

Terasima, R., and Tolmach, L. J. (1963). X-ray sensitivity and DNA synthesis in synchronous populations of HeLa cells. *Science* **140,** 490–492.

Zirkle, R. E. (1940). The radiobiological importance of the energy distribution along ionization tracks. *J. Cell. Comp. Physiol.* **16,** 221.

Nicholas Sperelakis

# Appendix
# Review of Electricity and Cable Properties

## I. Introduction

To understand many aspects of cell physiology and bio-physics and the electrophysiology of excitable cells, an appreciation of some elementary principles of electricity must be obtained. The purpose of this section is to give a brief review of the most relevant aspects of electricity and electric circuits that pertain to the cell membrane.

## II. Definition of Circuit Elements and Ohm's Law

The simple electrical circuit schematized in Fig. A-1 consists of a battery, resistor, ammeter, switch, and copper wire. Closing the switch allows electrical current to flow from the positive side of the battery, through the resistor and ammeter, and back to the negative side of the battery. Electrical current is considered, by convention, as flowing from positive to negative through the external circuit, even though the electrons—the negatively charged particles—actually flow in the opposite direction, that is, from negative to positive. One can consider that holes, that is, the absence of an electron in a space that previously contained one, flow in the opposite direction of the electrons, that is, from positive to negative around the external circuit. Within the battery, the flows are just the opposite; namely, electrons flow from positive to negative, and current flows from negative to positive.

In the circuit of Fig. A-1, the relationship between the applied voltage, the total resistance, and the current that flows is given by **Ohm's law.** This relationship, first formulated by George S. Ohm, states that the current ($I$) that flows is equal to the applied voltage ($V$) divided by the resistance ($R$):

$$I = \frac{V}{R} \tag{1}$$

$$\text{amperes} = \frac{\text{volts}}{\text{ohms}}$$

An ampere (A) of current is the movement past a point of one coulomb (coul) of electrical charge per second (1 A = 1 coul/sec). The unit of resistance is the ohm ($\Omega$), which may be operationally defined as that value of resistance through which there is a current of 1 A when the applied potential is 1 V ($R = V/I$). Resistance is a measure of the opposition to current flow; namely, for a given applied voltage, the greater the resistance the less the flow of current (inverse proportionality). The volt can be operationally defined from Ohm's law as that potential difference (PD) necessary to produce 1 A of current flow through a resistance of 1 $\Omega$:

$$V = IR \tag{1a}$$

Two points are at a PD of 1 V when it takes 1 joule (J) of energy to transport 1 coul of charge from one point to the other, given as

$$1 \text{ V} = 1 \text{ J/coul} \tag{2}$$

Note that energy equals voltage times the charge:

$$\text{joules} = \text{volts} \times \text{coulombs} \tag{2a}$$

(Thus the electron-volt unit of atomic physics is a unit of energy, the charge on an electron being $1.60 \times 10^{-19}$ coul.)

A coulomb of charge is equal to that represented by $6.24 \times 10^{18}$ electrons or univalent ions (reciprocal of $1.60 \times 10^{-19}$ coul/electron).

$$1 \text{ coul} = 6.24 \times 10^{18} \text{ univalent ions} \tag{3}$$

Multiplication of the charge on an electron ($q_e$) by Avogadro's number ($N_A$) (the number of molecules per mole) gives the number of coulombs in an equivalent (eq) of charge or mole of univalent ions

$$F = q_e \times N_A \tag{4}$$

$$\frac{\text{coul}}{\text{eq}} = \left(1.60 \times 10^{-19} \frac{\text{coul}}{\text{univalent ion}}\right)\left(6.02 \times 10^{23} \frac{\text{ions}}{\text{eq}}\right)$$

$$\cong 96,500 \frac{\text{coul}}{\text{eq}}$$

This derived value is known as the **Faraday constant** ($F$),

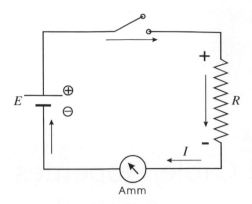

**FIG. A-1.** A simple electrical circuit containing a battery ($E$) and resistance ($R$). When the switch is closed, electrical current ($I$) flows in the direction indicated (arrows), from the positive pole of the battery to the negative pole in the external circuit, and can be measured by the ammeter (Amm) placed in series with the resistor. Current through the battery flows from the negative pole to the positive pole. The relationship between, $E$, $R$, and $I$ is given by Ohm's law: $I = E/R$.

used in electrochemistry for equations such as the Nernst equation.

Multiplying the Faraday constant by the valence of the ion (eq/mol) then gives the number of coulombs per mole:

$$zF = \frac{eq}{mol} \times \frac{coul}{eq} = \frac{coul}{mol} \qquad (5)$$

Conductance ($G$) is the reciprocal of the resistance:

$$G = \frac{1}{R} \qquad (6)$$

The greater the resistance, the lower the conductance, and vice versa. Therefore, Ohm's law can also be written as

$$I = \frac{V}{R} = \frac{1}{R}V = GV \qquad (7)$$

This states that the current is a product of conductance and voltage. This form is useful for electrophysiology, since conductances in parallel simply add. The unit of conductance is the mho (ohm spelled backwards) or $\Omega^{-1}$ or siemens (S).

## III. Resistors and Conductances in Series and in Parallel

### A. Resistors in Series

For resistors in series (Fig. A-2A), the total resistance ($R_T$) is a simple sum of the values of the resistances in series:

$$R_T = R_1 + R_2 + R_3 + \ldots \qquad (8)$$

If all the resistors are of the same value ($R_1 = R_2 = R_3$), then the total resistance is

$$R_T = R_1 N \qquad (9)$$

where $N$ is the number of resistors in series. Because myelin consists of many wrappings of membrane around the axon membrane (membranes in series), it should be obvious why myelination effectively raises membrane resistance.

### B. Resistors in Parallel

For resistors in parallel (Fig. A-2B), the reciprocal of the total resistance ($1/R_T$) is equal to the sum of the reciprocals of the resistances:

$$\frac{1}{R_T} = \frac{1}{R_1} + \frac{1}{R_2} + \frac{1}{R_3} + \cdots \qquad (10)$$

If all the resistors are the same value ($R_1 = R_2 = R_3$), then the total resistance is equal to any one resistance divided by the total number of resistors in parallel:

$$R_T = \frac{R_1}{N} \qquad (11)$$

If there are only two resistances in parallel, then Eq. 10 can be reduced algebraically to

$$\frac{1}{R_T} = \frac{1}{R_1} + \frac{1}{R_2} = \frac{R_1 + R_2}{R_1 R_2} \qquad (11a)$$

A

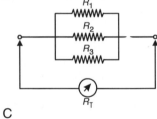

B

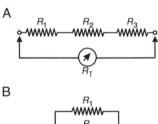

C

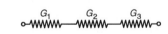

D

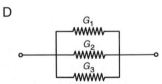

**FIG. A-2.** (A) Resistances placed in series—the total resistance ($R_T$) is the sum of the resistances: $R_T = R_1 + R_2 + R_3$. (B) Resistances placed in parallel—the reciprocal of the total resistance is equal to the sum of the reciprocals of the individual resistances: $1/R_T = 1/R_1 + 1/R_2 + 1/R_3$. (C) Reciprocal of the total conductance ($1/G_T$) of conductances in series is equal to the sum of the reciprocals of the individual conductances: $1/G_T = 1/G_1 + 1/G_2 + 1/G_3$. (D) Total conductance of conductances in parallel is equal to the sum of the conductances: $G_T = G_1 + G_2 + G_3$.

Taking the reciprocal gives

$$R_T = \frac{(R_1 \times R_2)}{(R_1 + R_2)} \qquad (11b)$$

Equation 11b states that the total resistance of two resistors in parallel is equal to the product of the resistances divided by the sum (often called the rule of product over the sum). Note that $R_T$ is always less than the lowest $R$.

## C. Conductances in Series

Conductances in series (Fig. A-2C) add like resistances in parallel; namely, the reciprocal of the total conductance $(1/G)$ is equal to the sum of the reciprocals of each conductance $(g)$:

$$\frac{1}{G} = \frac{1}{g_1} + \frac{1}{g_2} + \frac{1}{g_3} \qquad (12)$$

## D. Conductances in Parallel

Conductances in parallel (Fig. A-2D) add like resistances in series; namely, the total conductance $(G)$ is simply equal to the sum of each conductance:

$$G = g_1 + g_2 + g_3 + \dots \qquad (13)$$

That this is true is demonstrated by substituting $G = 1/R$ into this equation and obtaining

$$\frac{1}{R_T} = \frac{1}{R_1} + \frac{1}{R_2} + \frac{1}{R_3} + \dots \qquad (13a)$$

which is identical to the equation for resistances in parallel. It is convenient to use conductances in the study of the electrical properties of membranes, since the conductance pathways for the various species of ions are in parallel to each other, and therefore can simply be added to give the total conductance.

## IV. Kirchhoff's Laws

Kirchhoff's two laws may be stated in the following manner: (1) The current going toward a point is equal to that leaving it. For example, in a simple circuit composed of three resistors in series and a battery in series (Fig. A-3A), the current is constant throughout the circuit; that is, an ammeter placed anywhere in the circuit gives the same reading. In a simple, parallel circuit composed of a network of three resistors in parallel connected to a series battery (Fig. A-3B), the current is not constant throughout. The sum of the current in each branch is equal to that in the main branch; in other words, the sum of the currents converging toward a branch point is equal to that leaving it:

$$I_T = i_1 + i_2 + i_3 \qquad (14)$$

(2) The algebraic sum of the voltage drops $(IR)$ across each of the resistors in any closed network is equal to the source electromotive force (emf), e.g., battery. For the series circuit shown in Fig. A-3C,

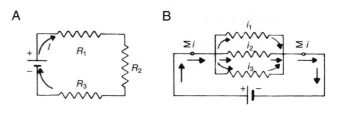

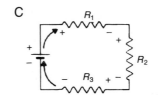

**FIG. A-3.** Diagrammatic representation of Kirchhoff's two laws: (1) Current approaching a point is equal to that leaving the point. Thus in the simple series circuit (A) the current is constant throughout, and in the simple parallel circuit (B), the sum of the currents in each leg is equal to the total current $(I_T)$: $I_T = i_1 + i_2 + i_3$. (2) The sum of the voltage drops $(IR)$ in the external circuit must equal the source emf. Thus in the simple series circuit (C), the sum of the $IR$ drops is equal to the source emf (battery voltage): source emf = $IR_1$ + $IR_2$ + $IR_3$.

$$\text{source emf} = IR_1 + IR_2 + IR_3 \qquad (15)$$

The $IR$ or voltage drop across each resistor is always in the polarity depicted; namely, the proximal side with respect to the direction of the current flow is always positive, and the distal side is negative. Current is always considered to flow from positive to negative in the external circuit. By convention, a positive potential is higher (or greater) than a negative one; hence, the potential is progressively dropped or lowered around the external circuit.

## V. Nature of Capacitors

A capacitor is a device that can store electric charge or electricity. It consists of two parallel plates (conductors) separated by a dielectric. All effective dielectrics have a high electrical resistance, but the reverse is not necessarily true; namely, all high resistors are not necessarily effective dielectrics (e.g., air). The dielectric constant $(\varepsilon)$ of a material is a relative index of its ability to function as an efficient dielectric compared to a vacuum, which has an $\varepsilon$ value of 1.000. The dielectric constants of some selected material are given in Table A-1. The greater the dipolar nature of the molecules, the greater the dielectric constant. For example, water is an efficient dielectric because of its highly dipolar nature. The dipoles must be free to flip-flop and orient with changes in the electric field, much like the water dipole orients around small charged ions (e.g., $Na^+$, $K^+$) to form the hydration layers. Polar phospholipids in the cell membrane should have a higher dielectric constant than less polar oils.

It is seen from Coulomb's law, for the force (F, in dynes) of interaction between point charges, that a large dielectric constant reduces the force of interaction between electrical

**TABLE A-1**   Dielectric Constants of Some Materials

| Material | Dielectric constant ($\varepsilon$) |
| --- | --- |
| Vacuum | 1.00000 |
| Air | 1.0006 |
| Oils | 3–5 |
| Water (distilled) | 80 |
| Cell membrane | 3–6 |
| Solid insulators | 2–10 |
| Paraffin wax | 2–3 |
| Quartz | 4–5 |
| Mica | 6–7 |
| Glass | 5–7 |
| Thin layers of oxidized metal | High |

charges, that is, the dielectric shields or buffers (see Section IX):

$$F = \frac{q_1 q_2}{\varepsilon d^2} \tag{16}$$

where $q_1$ and $q_2$ are the magnitudes of the charges (in statcoulomb), $d$ is the distance between the charges, and $\varepsilon$ is the dielectric constant. The distance squared is in Coulomb's law because of the geometrical factor: the surface area of a sphere equals $4\pi a^2$ or $\pi d^2$, in which $a$ is the radius and $d$ is the diameter. Therefore, the surface area over which the total force is distributed increases with the square of the radius or diameter of the sphere. From Eq. 16, a statcoulomb may be defined as that charge which, when placed 1 cm from an equal and like charge in a vacuum, repels the charge with a force of 1 dyne. A dyne is the force necessary to give a mass of 1 g an acceleration of 1 cm/sec/sec. Because the gravitational constant is 980 cm/sec$^2$, 1 dyne is equal to approximately 1 mg of force (1 g wt/980).

The magnitude of a capacitance ($C$) is equal to

$$C = \frac{\varepsilon A}{d} \frac{1}{4\pi k} \tag{17}$$

in which $A$ is the area of each plate (in cm$^2$), $d$ is the distance (in cm) between the plates (i.e., thickness of the dielectric), and $k$ is a constant ($9.0 \times 10^{11}$ cm/F). The unit of capacitance is the farad (F), named after Michael Faraday. Thus, the greater the dielectric constant, the greater the capacitance (Eq. 17).

The greater the area, the greater the capacitance (Eq. 17). Therefore, the total capacitance of several capacitors in parallel should be the simple sum (see Section VI,A) because the area of the plates is effectively increased. For example, the capacitance of the tuning capacitor (dial) of a radio is varied by altering the degree of overlap between two sets of interdigitating plates: the more the overlap, the greater the capacitance, because the area ($A$) is greater; air is the dielectric in this case. (Commercial capacitors

may use substances such as air, oil, paper, and mica for the dielectric; the area can be made very large by many wrappings of double sheets of aluminum foil separated by paper or mica.) The area term is often incorporated with the capacitance to give the specific capacitance ($C/A$), that is, the capacitance normalized to unit area (e.g., 1 cm$^2$):

$$\frac{C}{A} = \frac{\varepsilon}{d} \frac{1}{4\pi k} \tag{17a}$$

The shorter the distance between the plates, the greater the capacitance (Eq. 17). Therefore, the cell membrane has a very high capacitance, because the distance between the plates is only about 3.5–7.0 nm or 35–70 Å (the nonpolar phospholipid tail length of the membrane bilayer). For example, if the measured specific capacitance of the cell membrane ($C_m$) is 1.0 $\mu$F/cm$^2$, and assuming a dielectric constant of 5, then from Eq. A-17a

$$
\begin{aligned}
d &= \frac{\varepsilon}{(C/A)} \frac{1}{4\pi k} \\
&= \frac{5}{1 \times 10^{-6} \text{ F/cm}^2} \frac{1}{(12.56)(9.0 \times 10^{11} \text{ cm/F})} \\
&= 44.2 \times 10^{-8} \text{ cm} \\
&= 44 \text{ Å} \\
&= 4.4 \text{ nm}
\end{aligned}
\tag{17b}
$$

All capacitors have a breakdown voltage; that is, at more than a critical voltage level, the dielectric material cannot withstand the PD impressed across it. Therefore the capacitor breaks down, and electric current can now flow through the dielectric material from one plate to the other. The capacitor can no longer function as a capacitance, because the two plates are almost shorted to one another. A typical breakdown voltage of a commercial radio capacitor may be 200 V. In contrast, the capacitance of the nerve or muscle membrane has a breakdown voltage of only about 200 mV. Because the two plates of the membrane capacitor are much closer (about 50 Å or 0.5 × 10$^{-6}$ cm), however, the voltage gradient necessary for breakdown is 200 mV/0.5 × 10$^{-6}$ cm or 400,000 V/cm. At a resting potential of −80 mV, the voltage gradient is about 160,000 V/cm. Thus, the cell membrane is indeed a very effective capacitor and dielectric.

## VI. Capacitors in Parallel and Series

### A. Capacitors in Parallel

For capacitances in parallel (Fig. A-4A), the total capacitance ($C_T$) is a simple sum of the capacitances

$$C_T = C_1 + C_2 + C_3 + \ldots \tag{18}$$

The reason for this should be clear from Eq. 17, which demonstrates that capacitance is directly proportional to the area of the plates; hence placing capacitors in parallel effectively increases the area of the plates, as can be deduced from Fig. A-4A. If the number ($N$) of capacitors in parallel all have the same value ($C_1 = C_2 = C_3$), then the total capacitance is

$$C_T = C_1 N \tag{19}$$

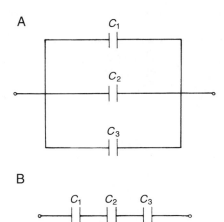

**FIG. A-4.** (A) Capacitors placed in parallel—the total capacitance is the sum of the individual capacitances: $C_T = C_1 + C_2 + C_3$. (B) Capacitors placed in series—the reciprocal of the total capacitance is equal to the sum of the reciprocals of the individual capacitances: $1/C_T = 1/C_1 + 1/C_2 + 1/C_3$.

Thus, capacitors in parallel add like resistors in series. Invaginations of the skeletal muscle fiber or ventricular myocardial cell surface membranes to form transverse (T) tubules increase the total measured fiber capacitance, because the increased membrane area effectively adds capacitors in parallel.

### B. Capacitors in Series

For capacitances in series (Fig. A-4B), the reciprocal of the total capacitance ($1/C_T$) equals the sum of the reciprocals of each capacitance:

$$\frac{1}{C_T} = \frac{1}{C_1} + \frac{1}{C_2} + \frac{1}{C_3} + \cdots \tag{20}$$

If there are several capacitors in series that are all of equal value ($C_1 = C_2 = C_3 = \ldots$), then the total capacitance is equal to any capacitance divided by $N$:

$$C_T = \frac{C_1}{N} \tag{21}$$

If there are only two capacitors in series, regardless of values, the total capacitance equals the product over the sum:

$$C_T = \frac{C_1 C_2}{C_1 + C_2} \tag{20a}$$

Note that $C_T$ is always less than the lowest $C$. Because myelin consists of many (20–200) wrappings of membrane around the axon membrane, it should be obvious why myelination reduces the total capacitance of a nerve fiber. Capacitors in series effectively increase the distance between the plates ($d$), and thereby lower the total capacitance (see Eq. 17).

### VII. Capacitive Reactance

Capacitors have almost infinite resistance to direct current (DC). Their impedance ($Z$, in $\Omega$) or so-called "AC

resistance" to sinusoidal alternating current (AC) is less, however, and is a function of the frequency ($f$) of the AC. The higher the frequency, the lower the capacitive reactance ($X_C$, in $\Omega$). The relationship is given by

$$X_C = \frac{1}{2\pi f C} \tag{22}$$
$$= \frac{1}{\omega C}$$

in which $C$ is capacitance and the angular velocity ($\omega$) in radians/sec is equal to $2\pi\,f$. For DC, $f = 0$ and $X_C = \infty$. At $f = \infty$, $X_C = 0$. For a given $f$, the larger $C$ is, the lower $X_C$ is.

For a nerve or muscle membrane, if $C$ is $1.0\ \mu\text{F/cm}^2$ and $f$ is 1000 cycles/sec, then

$$X_C = \frac{1}{(6.28)(1{,}000/\text{sec})(1 \times 10^{-6}\ \text{F/cm}^2)} \tag{22a}$$
$$= 0.159 \times 10^3 \frac{\text{sec} \cdot \text{cm}^2}{\text{F}}$$
$$= 159\ \Omega \cdot \text{cm}^2$$

### VIII. Membrane Impedance

#### A. Parallel RC Circuit

Because the cell membrane has a capacitive component (because of the continuous phospholipid bilayer matrix) in parallel with the resistive component (because of protein ion channels embedded in the lipid bilayer matrix), membrane impedance ($Z$) should diminish as the AC frequency is increased (Fig. A-5A). The relationship between frequency and relative impedance for a typical cell membrane is given in Fig. A-5B. When $f$ is extremely low, $X_C$ is extremely high, and $Z$ approaches $R$. As $f$ is increased, $X_C$ diminishes, and more and more of the current passes through $C$. Current does not actually pass through the dielectric of $C$, but it appears so because of the phenomenon of induced charge that occurs (see Section IX).

When $f$ is extremely high, $X_C$ approaches 0, and therefore $Z$ approaches 0. This is analogous to two resistors in parallel, one of whose impedance decreases as a function of the applied frequency. In diathermy, a very high frequency AC is used to heat damaged tissues without producing excitation of the nerve and muscle membranes; this is possible because of the capacitive component of the membrane, which causes $Z_m$ to be extremely low and which does not allow a significant membrane potential change to be produced.

At the frequency at which $X_C = R$, however, $Z$ is not equal to $0.5 \times R$, but is actually $0.707 \times R$, as determined by the following equation for a resistor and capacitor in parallel (note its similarity to that for two resistors in parallel, except for the squared function):

$$\left(\frac{1}{Z}\right)^2 = \left(\frac{1}{X_C}\right)^2 + \left(\frac{1}{R}\right)^2 \tag{23}$$

$$Z = \sqrt{\frac{X_C^2 R^2}{X_C^2 + R^2}} \tag{23a}$$

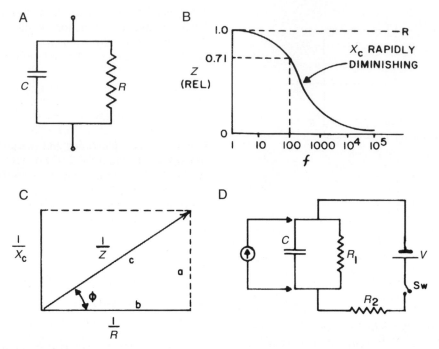

**FIG. A-5.** (A) Parallel resistance–capacitance ($RC$) network model of the cell membrane. (B) Impedance ($Z$) of such a parallel resistance–capacitance network decreases as the sinusoidal frequency ($f$) is increased: a plot of $Z$ vs log frequency is an inverse sigmoidal shape. Impedance decreases because the capacitive reactance ($X_C$) decreases according to $X_c = 1/2\pi fC$. (C) Parallelogram method of obtaining the impedance of a parallel resistance–capacitance circuit. Parallelogram constructed of $1/X_C$ against $1/R$ gives the vector that is equal to $1/Z$; the phase angle ($\phi$) between current and voltage is also obtained. ($\phi$ approaches 90° as resistance ($R$) approaches infinity; i.e., the circuit becomes purely capacitive.) Also, as can be reasoned, the larger the capacitance (for a given resistance), the lower the $X_C$ for a given $f$ and hence the larger the $\phi$. Thus the circuit behaves as the most purely capacitive ($\phi \to 90°$), the larger the capacitance and the larger the resistance. (D) Circuit diagram illustrates why current leads the voltage (by a certain phase angle) in a resistance–capacitance circuit. The instant the switch (Sw) is closed the frequency appears to be infinite, and so $X_C$ is zero; therefore no voltage appears across the capacitance, although the (induced) current through the capacitance is maximum; that is, current leads the measured voltage.

The impedance ($Z$) and the phase angle ($\phi$) between current and voltage can also be measured graphically by the parallelogram method, as illustrated in Fig. A-5C. In this method, the reciprocal of the capacitive reactance is plotted on the ordinate against the reciprocal of the resistance on the abscissa, and the parallelogram is completed by drawing lines parallel to the two axes. The diagonal line is equal to the reciprocal of the impedance, and the angle between the diagonal and the abscissa gives the phase angle ($\phi$) between voltage and current. In a purely resistive circuit, the current and voltage are exactly in phase ($\phi = 0°$). In a purely inductive circuit, current lags the voltage by 90°, whereas in a purely capacitive circuit, current leads the voltage by 90°. In a resistance–capacitance circuit, parallel or series, $\phi$ is between 0° and 90°, depending on the relative values of $R$ and $X_C$.

The reason that current leads the voltage in a resistance–capacitance circuit may be discerned from Fig. A-5D. At the instant the switch is closed, the frequency appears to be infinite (step change), and so $X_C = 0$. Therefore, there can be no potential difference across the capacitor, and the entire voltage is dropped across $R_2$. Thus, the current through the capacitor is maximum, while the voltage across

the capacitor is minimum, that is, the current leads the voltage by 90°.

The identity of the parallelogram method with Eq. 23 is given by the Pythagorean theorem of trigonometry,

$$c^2 = a^2 + b^2 \tag{24}$$

where $c$ is the length of the hypotenuse of a right triangle, and $a$ and $b$ are the lengths of the other two sides. By substituting for $a$, $b$, and $c$ from Fig. A-5C, the following expression, which is identical to Eq. 23 is obtained:

$$\left(\frac{1}{Z}\right)^2 = \left(\frac{1}{X_C}\right)^2 + \left(\frac{1}{R}\right)^2 \tag{23}$$

### B. Series RC Circuit

For a series resistance–capacitance network (Fig. A-6A), the corresponding analysis is

$$Z^2 = X_C^2 + R^2 \tag{25}$$

or

$$Z = \sqrt{X_C^2 + R^2} \tag{25a}$$

A

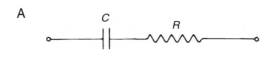

B

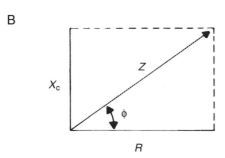

**FIG. A-6.** (A) Series resistance–capacitance (*RC*) network. (B) Parallelogram method of obtaining the impedance (*Z*) of a series resistance–capacitance circuit. Parallelogram constructed with capacitative reactance (*X*$_C$) on the ordinate and resistance (*R*) on the abscissa gives the vector which is equal to *Z*. The phase angle (*φ*) between current and voltage approaches 90° as *R* approaches zero or *X*$_C$ approaches infinity (*f* → 0, or the capacitance has a lower and lower value).

which is similar to having two resistors in series, except for the squared function. When $X_C = R$, then $Z = 1.414R$ (not $2 \times R$). At $f = \infty$, $X_C = 0$, and $Z = R$. At $f = 0$ (DC), $X_C = \infty$, and $Z = \infty$. The parallelogram analysis is shown in Fig. A-6B.

## IX. Capacitive Charge and Capacitive Current

The total charge (*Q*, in coulombs) stored on the plates of a capacitor is equal to the product of the capacitance and applied voltage:

$$Q = CV \qquad (26)$$
$$\text{coul} = (\text{F})(\text{V})$$

Thus, if a certain charge is placed on a capacitor (air dielectric) and then the distance between the plates is decreased so that capacitance increases, it follows that the PD across the capacitor must decrease in proportion. It also follows that because an increase in dielectric constant increases capacitance, an increase in dielectric constant must also decrease the PD for a given charge separation; that is, it reduces the force acting between the separated charges.

The electric charge stored in a capacitor can be used to do work. Kinetic energy is needed to charge a capacitor, the charged capacitor representing potential energy (like a battery), and discharge of the capacitor converts the potential energy back into kinetic energy. From Eq. 26 and because electric current flow (*I*) is a charge flowing past a point per unit of time ($I = Q/t$), the capacitive current flow ($I_C$, in amps) during charge or discharge of a capacitor is equal to the capacitance (in Farads) times the rate of change of the PD:

$$I_C = C\frac{dV}{dt} \qquad (27)$$
$$\frac{\text{coul}}{\text{sec}} = \text{F}\frac{\text{volts}}{\text{sec}}$$

Thus for a given capacitance, the greater the rate of change of voltage, the greater the capacitive current is. In a myocardial cell membrane, for example, the maximum rate of change in PD across the membrane during the rising phase of the AP is relatively large (e.g., 100 mV/0.5 msec or 200 V/sec); therefore capacitive current is also relatively large. There is little or no change in the value of the membrane capacitance during an AP.

As stated in Section VIII, current does not actually flow through the capacitor, but only appears to do so because of the phenomenon of induced charge. When electrons flow into one plate of the capacitor, the negative charges on this plate repel the electrons from the other adjacent plate, since like charges repel one another. Thus as diagrammed in Fig. A-7, a flow of electrons into the lower plate of the capacitor leads to a flow of electrons out of the upper plate. Hence it appears as though the flow of electrons was directly through the dielectric. The deficiency of electrons in the upper plate makes the upper plate become positive, whereas the lower plate is negatively charged.

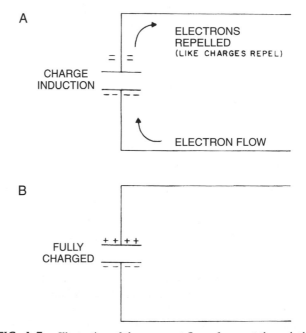

**FIG. A-7.** Illustration of the apparent flow of current through the dielectric of a capacitor. Current does not actually flow through the capacitor but appears to because of the phenomenon of charge induction. (A) When one plate of the capacitor is made negative, this plate repels the electrons from the other plate nearby, because like charges repel. (B) Thus the second plate has an electron deficiency that gives the plate a positive charge.

## X. Membrane Time Constant

When current is through a simple resistance or series of resistors, the PD across the network is developed instantaneously. This is not true for a biological membrane or for the parallel resistance–capacitance network shown in Fig. A-8A. Instead, the PD builds in a nearly exponential (negative) manner until the final maximal value is attained, as illustrated in Fig. A–8B. The slow charging occurs because of the presence of the capacitance. The time it takes for

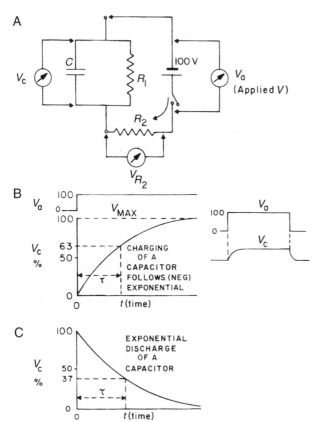

63% of the final PD to be reached is, by definition, the time constant ($\tau$). The equation describing this PD buildup is

$$V_t = V_{max}(1 - e^{-t/\tau}) \tag{28}$$

where $V_t$ is the voltage across the network at any time ($t$) and $V_{max}$ is the maximum or final PD when the steady state is reached. When $t = \tau$, then $e^{-t/\tau}$ becomes

$$e^{-1} = \frac{1}{e} = \frac{1}{2.717} = 0.37 \tag{28a}$$

Therefore,

$$V_t = V_{max}(1 - 0.37) \tag{29}$$
$$= 0.63 V_{max}$$

For all other values of $t$, Eq. 28 is solved by taking the logarithms and obtaining

$$V_t = V_{max} - \text{antiln}\left[\ln V_{max} - \frac{t}{\tau}\right] \tag{30}$$

$$V_t = V_{max} - \text{antilog}\frac{2.303 \log V_{max} - \dfrac{t}{\tau}}{2.303} \tag{31}$$

On discharge of the resistance–capacitance network, the voltage across the capacitor ($V_C$) decreases exponentially, as indicated in Fig. A-8C. Again, the time it takes for 63% of the final PD to be reached (which is to the level of 37% of the initial value) is the time constant ($\tau$). The equation describing the decay in PD as a function of time (analogous to that for the decay in voltage as a function of distance along a cable) is

$$V_t = V_{max}\,e^{-t/\tau} \tag{32}$$

where $V_{max}$ is the maximum or initial PD at $t = 0$. When $t = \tau$,

$$V_t = 0.37 V_{max} \tag{32a}$$

For all other values of $t$, Eq. 32 is solved by taking the logarithms

$$\ln V_t = \ln V_{max} - \frac{t}{\tau} \tag{33}$$

$$V_t = \text{antiln}\left[\ln V_{max} - \frac{t}{\tau}\right] \tag{34}$$

$$V_t = \text{antilog}\frac{2.303 \log V_{max} - \dfrac{t}{\tau}}{2.303} \tag{34a}$$

**FIG. A-8.** (A) Exponential (negative) charge and discharge of a capacitor for the parallel resistance–capacitance ($R_1 C$) circuit in series with a second resistance ($R_2$). Applied voltage is from a 100-V battery upon closing of the switch and is measured by voltmeter $V_a$. Voltmeter $V_C$ measures the voltage across the capacitor, and voltmeter $V_{R2}$ measures the voltage across $R_2$. (B) Upon closing the switch, $V_a$ suddenly jumps from 0 to 100 V. Potential recorded by $V_C$, however, slowly increases exponentially; the time constant ($\tau$) is the time it takes for $V_C$ to build to 63% $(1 - e^{-1})$ of its final (or maximal) value. The instant that the switch is closed, the frequency effectively is infinite, and so capacitive reactance ($X_C$) is 0; therefore no voltage appears across the parallel resistance–capacitance network, although the induced current through the capacitor is maximum. The entire voltage drop of the battery thus appears across $R_2$. After a long time, that is, when the circuit reaches steady state, $f = 0$ and $X_C$ is infinite; therefore the voltage across the parallel resistance–capacitance network is equal to $(R_1/R_1 + R_2) \times 100$ V. This charging of a capacitor can be viewed as a compartment filling with electrons and obeying first-order kinetics. (C) Upon opening of the switch, the capacitor discharges exponentially through $R_1$. Time constant is the time it takes for discharge to 37% $(e^{-1})$ of the initial (maximal) value.

In the circuit illustrated in Fig. A-8A, upon closing the switch, the capacitor ($C$) charges exponentially with a time constant about equal to $R_1 C$ (if $R_1$ is much greater than $R_2$). At the first instant after the switch is closed, there is a step change in the applied potential that appears as an infinite frequency. Therefore, because $X_C = 1/2\pi fC$, $X_C \cong 0$. According to Kirchhoff's laws pertaining to the relative voltage drops across series impedances, the voltage drop is in proportion to the relative impedance (i.e., the higher impedance has the greater voltage drop across it), and the sum of all voltage drops must equal the source emf. Thus, all the emf of the battery should be dropped across $R_2$ ($V_{R_2} = 100$ V) and none across $R_1$ ($V_c = 0$ V).

At the final steady-state value (plateau of applied voltage pulse), the applied voltage pulse is steady, and therefore the frequency is zero; hence $X_C \cong \infty$, and from analysis of the impedance of a parallel resistance–capacitance network, $Z = R_1$. At this time, nearly all of the battery emf should be dropped across $C$ (or $R_1$) because $R_1 \gg R_2$. In this condition, the current continues to flow through $R_1$ and $R_2$.

At all intermediate instants, $V_C > 0$ V but $<100$ V. The capacitor charges exponentially. One way to view why a capacitor charges and discharges exponentially is to assume that the negative plate of a capacitor is a compartment that can fill up with electrons. If so, the rate of fill-up (charging) at any moment should be a function of the actual degree of fill-up at that moment, according to the laws of first-order kinetics. As the concentration builds, the rate of further buildup becomes slower and slower (a negative exponential). The process reaches equilibrium when the maximum number of electrons (maximum density or concentration of electrons) that can be stored on the plate for a given impressed voltage is reached.

Similarly, the rate of emptying of the compartment at any instant depends only on the number of electrons available at that instant. As concentration decays, the rate of further decay becomes slower and slower. The time to build up or to decay to one-half (50%) of the final value is called the **half-time,** $t_{1/2}$. At one half-time, the rate should be one-half what it was initially, because only half of the original number of electrons are now available. At the second half-time, the rate should be one-half what it was at the first half-time or one-fourth of what it was initially, and so on. About seven half-times are necessary to reach more than 99% of the final steady-state value.

The **time constant** is the time required to build to 63% of the final steady-state value or to decay to 37% (100% − 63%) of the initial value. About five time constants are necessary to reach more than 99% of the final value. The time constant and half-time values for a particular exponential process can be interconverted by the relationship

$$\tau = \frac{t_{1/2}}{\ln 2} = \frac{t_{1/2}}{0.693} = 1.44 t_{1/2} \tag{35}$$

Thus, if the membrane time constant were 1.0 msec, the half-time is 0.69 msec. The time constant (in seconds) and first-order **rate constant** ($k$, in $\text{sec}^{-1}$) are reciprocally related as

$$k = \frac{1}{\tau} \tag{36}$$

In a nerve or skeletal muscle fiber cable, the change in membrane potential during the application of electrotonic constant-current pulses is actually an **error function,** because the longitudinal resistance of the intracellular fluid is not negligible. An error function curve builds more rapidly than does an exponential curve. The equivalent time constant for such an error function curve corresponds to a buildup to 84% of steady-state value, rather than to 63%.

## XI. Specific Resistance and Specific Capacitance

### A. Specific Resistance

The resistance ($R$) actually measured for any material depends on the nature of the material (its specific resistance or **resistivity,** $\rho$) and on the amount and shape of the material used. If discussion here is confined to uniform elongated cylinders or cubes of material (Fig. A-9A), then

$$R = \rho \frac{L}{A_X} \tag{37}$$

$$\Omega = (\Omega \cdot \text{cm}) \left( \frac{\text{cm}}{\text{cm}^2} \right)$$

In Eq. 37, $A_X$ is the cross-sectional area ($\pi a^2$) and $L$ is the length. As seen from this equation, the greater the resistivity of the material, the greater the resistance measured for any given length or cross-sectional area of material. The resistivity, by definition, is equal to the resistance of a material when the material is of unit cross-sectional area (1 cm$^2$) and of unit length (1 cm), hence of unit volume (1 cm$^3$). The units of resistivity are in $\Omega \cdot \text{cm}$. Rearranging Eq. 37 gives

$$\rho = R \frac{A_X}{L} \tag{37a}$$

$$\Omega \cdot \text{cm} = \Omega \frac{\text{cm}^2}{\text{cm}}$$

Mammalian Ringer solution has a resistivity of about 50 $\Omega \cdot \text{cm}$. The resistivity of such electrolyte solutions varies with the concentration of each of the salts present, and

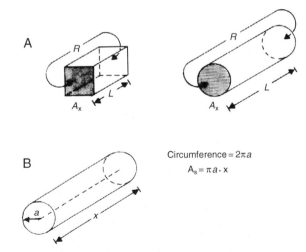

**FIG. A-9.** Resistivity measurements on a material. (A) For material such as myoplasm or axoplasm, the longitudinal resistance ($R$) depends on the resistivity of the material ($R_i$), the cross-sectional area ($A_x$ or $\pi a^2$, where $a$ is the radius), and length ($L$): $R = R_i L / \pi a^2$. Box shape and cylindrical shape are illustrated. (B) For skin material such as cell membrane, the transverse resistance ($R$) depends on the so-called "resistivity" of the membrane ($R_m$) and on the surface area ($A_s$ or $2\pi a \cdot x$): $R = R_m / 2\pi a x$.

with the mobilities of the various species of ions (e.g., $K^+$, $Na^+$, and $Ca^{2+}$). The resistivity (reciprocal of conductivity or specific conductance) of solutions decreases with temperature elevation (whereas that of metallic conductors increases).

## B. Resistivity of Axoplasm of Myoplasm

The notation is confusing because physicists use $\rho$ to denote resistivity, whereas physiologists use $R$. Thus $R_i$ is the resistivity of the internal cytoplasm (usual values of 100–300 $\Omega \cdot$ cm for axoplasm and myoplasm), and $R_o$ is the resistivity of the outside or external solution, which is normally the interstitial fluid (value of about 50 $\Omega \cdot$ cm, comparable to that of Ringer solution). Hence, we are forced to use $r$ (instead of $R$) to represent the absolute resistance (in $\Omega$), and Eq. 37a becomes

$$R_i = r\frac{A_X}{L} \tag{38}$$
$$\Omega \cdot \text{cm} = \Omega \, \frac{\text{cm}^2}{\text{cm}}$$

It is convenient to substitute $r_i$ for $r/L$,

$$R_i = r_i A_X$$
$$\Omega \cdot \text{cm} = (\Omega/\text{cm})(\text{cm}^2) \tag{39}$$

where $r_i$ is the internal resistance normalized for unit length, but not for unit cross-sectional area. That is, $r_i$ is numerically equal to the internal resistance of a 1-cm length of a given cable, such as a skeletal muscle fiber or nerve axon. Because $A_X = \pi a^2$, Eq. 39 can be modified to

$$R_i = r_i \pi a^2 \tag{39a}$$

which is the final useful form of the equation when making biological measurements (see Table A-2).

## C. Specific Resistance and Resistivity of the Cell Membrane

In dealing with the resistance of the cell surface membrane, we may proceed starting from the basic Eq. 37a and rearranging it to give

$$\rho L = RA_X \tag{37b}$$

Because the length being considered is the thickness of the membrane ($\delta$) and because the area being considered is the surface area ($A_S$) of the cylinder (Fig. A-9B), we must make the appropriate substitutions into Eq. 37b. Letting $\delta = L$,

$$\rho\delta = RA_X \tag{40}$$

Letting $r = R$ and $A_S = A_X$

$$\rho\delta = rA_S \tag{41}$$

Because $A_S$ = circumference ($2\pi a$) times length ($x$) = $2\pi ax$,

$$\rho\delta = r2\pi ax \tag{41a}$$

Letting $R_m = \rho\delta$,

$$R_m = r2\pi ax \tag{42}$$

Rearranging gives

$$R_m = (rx)(2\pi a) \tag{42a}$$

Letting $r_m = rx$,

$$R_m = r_m(2\pi a) \tag{43}$$
$$\Omega \cdot \text{cm}^2 = (\Omega \cdot \text{cm})(\text{cm})$$

where $r_m$ is the membrane resistance normalized for unit length, but not for unit surface area. That is, $r_m$ is numerically equal to the transverse membrane resistance of a 1-cm length of a given cable (e.g., frog sartorius fiber). The $r_m$ is converted to $R_m$ by multiplying by the circumference. These interconversions are summarized in Table A-2.

Although $R_m$ is often called the membrane resistivity or specific resistance, the term is not accurate, because as just seen, the units for $R_m$ are in $\Omega \cdot \text{cm}^2$, not $\Omega \cdot$ cm. That is, the true resistivity has not been normalized for the unit length (thickness of 1 cm). Thus, if $R_m$ is 1000 $\Omega \cdot \text{cm}^2$ and $\delta$ is 100 Å ($10^{-6}$ cm), then the true membrane resistivity ($\rho_m$) would be

$$\rho_m = \frac{R_m}{\delta} \tag{44}$$
$$= \frac{1000 \, \Omega \cdot \text{cm}^2}{1 \times 10^{-6} \, \text{cm}}$$
$$= 1 \times 10^9 \, \Omega \cdot \text{cm}$$

This resistivity value makes the cell membrane an effective insulator indeed, comparable to the best commercial insu-

**TABLE A-2**  Summary of the Interconversions Used in the Measurements of Properties of Biological Cables

| Orientation, with respect to fiber axis | Actual value measured | Operation | | Normalized for unit length of fiber | | Operation | | Normalized for radius of fiber ($A_x$ or circumference) | |
|---|---|---|---|---|---|---|---|---|---|
| Longitudinal | $r(\Omega)$ | $\div \times$ | $=$ | $r_i$ | $(\Omega/\text{cm})$ | $\times$ | $\pi a^2$ | $= R_i$ | $(\Omega \cdot \text{cm})$ |
| Transverse | $r(\Omega)$ | $\times \times$ | $=$ | $r_m$ | $(\Omega \cdot \text{cm})$ | $\times$ | $2\pi a$ | $= R_m$ | $(\Omega \cdot \text{cm}^2)$ |
| Transverse | $c(\text{F})$ | $\div \times$ | $=$ | $c_m$ | $(\text{F}/\text{cm})$ | $\div$ | $2\pi a$ | $= C_m$ | $(\text{F}/\text{cm}^2)$ |

*Note: $A_x$, cross-sectional area; $r$, resistance; $x$, fiber length; $r_i$, internal resistance per unit length of fiber; $a$, fiber radius; $R_i$, internal resistivity; $r_m$, membrane resistance for a unit length of fiber; $R_m$, membrane resistivity; $c$, capacitance; $c_m$, membrane capacitance per unit length of fiber; $C_m$, membrane specific capacitance.*

lators. The cell membrane, however, is extremely thin (the animal cannot afford the space to have a thick insulation (e.g., 1 cm thick) for each nerve and muscle fiber); therefore, the actual nerve cable has a great deal of energy loss (cable decrement). Note that $r_m$ is not the true resistivity even though it has the proper units.

## D. Specific Capacitance

In dealing with the capacitance of the cell surface membrane, we may proceed in a similar manner using an analysis analogous to that for $R_m$ given earlier, and taking into account the fact that the total resistance of resistors added in parallel becomes smaller, whereas the total capacitance of capacitors added in parallel becomes larger. Therefore, the measured capacitance must be divided by the surface area—rather than multiplied, as for membrane resistivity—to obtain the specific capacitance. Thus, the units for membrane specific capacitance ($C_m$) are $F/cm^2$, compared to $\Omega \cdot cm^2$ for membrane specific resistance ($R_m$). In addition, as with $R_m$, $C_m$ is not prorated for 1-cm thickness, but is given for the actual thickness of the membrane; hence, the units are $F/cm^2$ (and not $F/cm$). Thus, the equation analogous to that for membrane resistivity is

$$C_m = \frac{c}{A_S} = \frac{c}{2\pi a x} \tag{45}$$

where $c$ is the absolute capacitance in Farads.

Letting $c_m = c/x$,

$$C_m = \frac{c_m}{2\pi a} \tag{46}$$

$$\frac{F}{cm^2} = \frac{F/cm}{cm}$$

where $c_m$ (measured in $F/cm$) is the membrane capacitance normalized for unit length but not for unit surface area. The $c_m$ is converted to $C_m$ ($F/cm^2$) by dividing by the circumference (Table A-2). Thus, $C_m$ is normalized both for unit length of the fiber and for unit circumference (radius).

## XII. Biological Cable Decrement

Nerve and muscle fibers make relatively poor cables, in comparison with commercial cables (such as a transoceanic submarine cable). There is a great deal of energy loss in the transmission of information along these biological cables. If the cable remained passive (no active impulse), the signal would become greatly attenuated and distorted after traveling only a relatively short distance, e.g., 1 mm. Hence biological signal transmission in long fibers is an active energy-consuming process, the energy input being distributed along the entire length of the fiber. The biological cable is poor for the following reasons: (1) $R_i$ is approximately $10^7$ times that for copper wire; (2) $R_m$ is approximately $10^{-6}$ times that for an effective insulator such as rubber. (But, as stated in the preceding section, the true resistivity of the cell membrane is comparable to that of

an efficient commercial insulator; that is, the cell membrane would be as efficient as a commercial insulator if both were equally thick); and (3) the $C_m$ is high.

The biological cable, however, is the best possible, given the biological circumstances and requirements. In addition, the nerve cable has been vastly improved by an evolutionary development in the vertebrates (namely, **myelination**); myelination increases $R_m$ and decreases $C_m$ by more than 100-fold. There are two consequences that result from improvement of the nerve cable by myelination: (1) The velocity of propagation is greatly increased; and (2) the energy cost of signaling is greatly decreased.

A cable consists of two conductors arranged in parallel (usually concentrically) separated by an insulating material, for example, a copper wire in the center core surrounded by insulation and a sleeve of copper, as illustrated in Fig. A-10A. The circuit diagram for an ideal cable is given in Fig. A-10B. In a submarine cable, the ocean can serve as the outer conductor. In the biological cable, the tissue fluids serve as the ocean, providing a fairly effective outside conductor. The inside conductor is not as effective, however, as discussed previously (because of high $R_i$ and small cross-sectional area); hence the internal resistance is not negligible. Therefore the circuit diagram shown in Fig. A-10C more closely approximates the biological cable.

Furthermore, efforts are made to minimize any stray (unwanted) capacitance in the submarine cable so that AC signals are not attenuated and distorted. As stated in the previous paragraph, the biological cable has a relatively large capacitance (but that capacitance is greatly reduced in myelinated fibers). Hence the biological cable is more accurately represented by the circuit diagram (Fig. A-10D), which includes the large $c_m$, the large $r_i$, and the relatively low $r_m$. These resistance and capacitance parameters are distributed along the length of the cable.

A potential applied at one end of a cable diminishes exponentially with distance ($x$), with a certain **length** (or **space) constant** ($\lambda$) as given by the equation

$$V_x = V_o e^{-x/\lambda} \tag{47}$$

where $\lambda$, by definition, is the distance to which the potential has fallen to $1/e$ or 37% of the initial value. This exponential decay can be determined experimentally using a cable network or can be calculated using Ohm's and Kirchhoff's laws as the principal analytic aids.

Equation 47 is for an infinitely long fiber cable [i.e., one whose length is at least 10 times the length (space) constant $\lambda$]. In such cases, there is an exponential fall of potential in both directions from the site of current injection ($x = 0$). Cable properties are discussed in more detail in Chapter 23. However, if the fiber is short (i.e., a **truncated cable,** shorter than the hypothetical $\lambda$), then Eq. 47 does not apply. Instead the following equation applies:

$$V_x = V_0 \frac{\cosh\left(\dfrac{L-x}{\lambda_h}\right)}{\cosh\left(\dfrac{L}{\lambda_h}\right)} \tag{48}$$

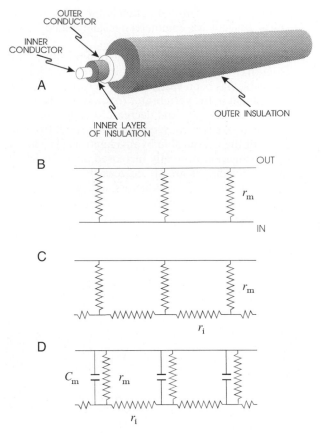

**FIG. A-10.** Diagrams illustrating cables. (A) Cable composed of copper wire coated with insulation and surrounded concentrically by a copper sleeve, with insulation over the latter. (B) Simple circuit diagram for an idealized cable, which consists of two good conductors separated by a material of high resistance ($R$). (C) For a biological cable, such as nerve fiber or skeletal muscle fiber, the inner conductor is not zero or low resistance (as would nearly be true for a copper wire), and therefore a resistance must be drawn. Even in the biological cable, however, the transverse membrane resistance ($r_m$) is much higher than the longitudinal internal resistance ($r_i$). In the biological cable, the outside resistance ($r_o$) can be drawn as a straight line, since $r_o$ is relatively low in resistance compared with $r_i$, because the external current makes use of the entire interstitial fluid space (i.e., many resistors in parallel). (D) Biological cable with the large membrane capacitance ($C_m$) included.

where cos*h* represents the hyperbolic cosine, $L$ is the length of the cell, $x$ is the distance from the site where current is injected (e.g., at one end of the fiber) and $V_x$ is measured, $\lambda_h$ is the hypothetical length constant, and $V_0$ is the voltage change at the site of current injection ($x = 0$). This equation indicates that when $x = 0$, the cos*h* term becomes 1.0, and so $V_x = V_0$. When $x$ is not zero, but is at the middle or far end of the fiber, then $V_x$ is less than $V_0$, in accordance with Eq. 48. The actual values calculated from this equation are given in Table A-3.

Cardiac muscle cells and smooth cells are short, so they behave as truncated cables. If there are no gap junction channels (connexons) between contiguous cells lying end to end, then Eq. 48 applies. The cable is considered to be "open" at both ends, that is, terminated in nearly infinite

resistance (like an open electrical wall switch), because of the high-resistance cell membrane at the ends. Applied current is considered to reflect from the high-resistance end, thus causing the voltage falloff over the length of the cell to be very small as indicated in Table A-3. The reader is referred to Appendix 1 of Chapter 23 for evidence and references showing that there is actually very little voltage decay within a single myocardial cell. If there were many connexon channels providing low-resistance pathways between the cells, then Eq. 48 would not apply.

The values given in Table A-3 are for a cylindrical myocardial cell having a length ($L$) of 150 $\mu$m, and $V_0$ is assumed to be 100 mV. The $V_x$ values were calculated at two points in the cell: at midpoint ($x = 75 \mu$m) and at the far end ($x = 150 \mu$m). The calculations were done for a wide range of assumed $\lambda_h$ values. As can be seen, the larger the $\lambda_h$ value, the less voltage decay along the length of the cell.

To measure $\lambda_h$ requires two microelectrodes to be placed within one cell at a known distance apart, and the voltage decay measured between the two electrodes. The first electrode is used to apply current ($I_0$) and to measure simultaneously the voltage change ($V_1$ or $V_0$) by use of a balanced Wheatstone bridge circuit. From these data, $\lambda_h$ can be calculated. However, the procedure is difficult and inaccurate because of bridge imbalance that often occurs due to the microelectrode resistance changing with applied current, pressure, and intracellular ionic milieu; in addition, the exact interelectrode distance of the microelectrode tips is hard to determine, and the placing of two microelectrodes into one short cell often damages the cell.

Because of these difficulties, it is simpler and more accurate to measure the input resistance of an isolated single cell, and from this, calculate the resistance of the cell membrane (the resistance of the myoplasm is negligibly low). Values for $R_{in}$ of 10–80 M$\Omega$ have been reported (Sperelakis *et al.,* 1960; Tarr and Sperelakis, 1964; Sperelakis and Lehmkuhl, 1966; Josephson *et al.,* 1984). Then the membrane specific resistance $R_m$ can be calculated using the measured surface area of the cell. $R_m$ is the product of $R_{in}$ and the surface area ($A_s$) of the cell ($A_s = \pi 2aL$):

**TABLE A-3** Calculations of Voltage Falloff at Two Distances within a Myocardial Cell Behaving as a Truncated Cable and for Which Eq. 48 Applies

| $\lambda_h$ ($\mu$m) | $V_x$ (mV) | |
|---|---|---|
| | $x = 75 \mu$m | $x = 150 \mu$m |
| 150 | 73.1 | 64.8 |
| 200 | 82.7 | 77.2 |
| 300 | 91.4 | 88.7 |
| 500 | 96.8 | 95.7 |
| 1000 | 99.2 | 98.9 |
| 3000 | 99.9 | 99.9 |

*Note:* Calculations for cell length ($L$) of 150 $\mu$m, $V_0$ of 100 mV, and assumed hypothetical $\lambda_h$ values ranging between 150 and 3000 $\mu$m.

$$R_m = R_{in} (\pi 2 a L) \tag{49}$$

For an $R_{in}$ of 30 M$\Omega$:

$$R_m = (30 \times 10^6 \, \Omega)(7.54 \times 10^{-5} \, cm^2)$$
$$= 2260 \, \Omega \cdot cm^2$$

For a higher or lower $R_{in}$ value, $R_m$ becomes proportionally higher or lower.

The resistivity of the myoplasm $R_i$ is a relatively constant value, and so can be assumed to be 200 $\Omega \cdot cm$. Therefore, from the measured radius ($a$) of the cell, the $\lambda_h$ value can be calculated using the following relationship:

$$\lambda_h = \sqrt{\frac{R_m}{R_i} \frac{a}{2}} \tag{50}$$

This is the same as Eq. 3 of Chapter 23. Table A-4 gives the calculated $\lambda_h$ values for a myocardial cell with a radius of 8 $\mu$m for various assumed $R_m$ values ranging between 100 $\Omega \cdot cm^2$ and 30,000 $\Omega \cdot cm^2$. The calculations are given for two different assumed $R_i$ values. As expected, $\lambda_h$ is a direct function of the square root of $R_m$. Values of $\lambda_h$ of 320–375 $\mu$m have been reported (Josephson *et al.,* 1984). If so, then $R_m$ must be about 500 $\Omega \cdot cm^2$ (see Table A-4), and there is only about a 10% decay of potential over the length of the cell (see Table A-3).

## XIII. Inductance, Inductive Reactance, and Oscillations

### A. Nature of Inductance

An inductor consists of a coil of wire with multiple turns. When current in any conductor is changed in magnitude, there is an opposing emf induced in any neighboring conductor, for example, in the adjacent turns of the coil. The greater the rate of change of the current, the larger the opposing emf induced. The opposing emf is in the opposite polarity of the applied voltage, and therefore acts to oppose the change in current. This is why an inductor is also known as a choke, because the inductor reacts against AC, causing

**TABLE A-4** Calculations of the Hypothetical Length Constant $\lambda_h$ from Eq. 50 for a Myocardial Cell Having Radius of 8 $\mu$m and for Various Assumed $R_m$ Values

| $R_m$ ($\Omega cm^2$) | $\lambda_h$ (mm) | |
| | $R_i = 200 \, \Omega cm$ | $R_i = 100 \, \Omega cm$ |
| --- | --- | --- |
| 100 | 0.141 | 0.200 |
| 300 | 0.245 | 0.346 |
| 500 | 0.316 | 0.447 |
| 1,000 | 0.447 | 0.632 |
| 3,000 | 0.775 | 1.095 |
| 10,000 | 1.414 | 2.000 |
| 30,000 | 2.449 | 3.464 |

*Note:* Calculations made for two different assumed $R_i$ values.

a high impedance (or high so-called "AC resistance") to rapidly varying currents. When the current through the coil is increasing, the **induced emf** opposes the current flow, whereas when the current is decreasing, the induced emf facilitates and prolongs the current. Thus, inductance ($L$) tends to keep current constant when the voltage is varied.

The induced emf is due to the magnetic field that surrounds a wire that is carrying current. When current increases, the magnetic field expands, and when current decreases, the magnetic field shrinks. Thus, if the conductor is in the form of a coil, the expanding and shrinking magnetic field cuts across the neighboring turns of the wire. Whenever a moving magnetic field cuts across a conductor, an electromotive force (emf) is generated in the conductor. [This is the principle of the electric generator, in which the magnetic field is usually fixed (permanent magnet) and the conductor (rotor armature) is moved.] An expanding field induces the emf in one polarity (opposing), and the shrinking field induces the emf in the opposite polarity (facilitating). For example, a large voltage and current (spark) is produced when a circuit containing an inductance is opened (broken).

The phenomenon of induction of an opposing emf and a facilitating emf in a simple coil is known as **self-inductance**. The self-inductance of any given coil is a constant value, depending on the number of turns in the coil. A coil has a self-inductance of 1 H when 1 V of opposing emf is caused by a $dI/dt$ of 1 A/sec:

$$L = \frac{V}{dI/dt} \tag{51}$$

Inductance is analogous to mechanical inertia or a mass that resists an increase or decrease in velocity. The energy (in joules) stored in the magnetic field of an inductance is equal to $0.5LI^2$ (analogous to 0.5 mass $\times$ velocity$^2$). **Mutual inductance** occurs between two closely spaced parallel coils, for example, in a transformer.

### B. Inductive Reactance and Impedance

In contrast to capacitive reactance, the **inductive reactance** ($X_L$, in ohms) varies directly with frequency and magnitude of the inductance

$$X_L = 2\pi f L$$
$$= \omega L \tag{52}$$

Hence, for a given inductance, the higher the frequency, the greater the reactance. The inductive reactance is zero for DC ($f = 0$). The unit of inductance is the Henry (H), named for the research by Joseph Henry published in 1831. In a purely inductive circuit, the current lags the voltage by a phase angle of 90°. The current through the inductance ($I_L$) is equal to the applied voltage ($V$) divided by $X_L$ (according to Ohm's law).

The impedance for series and parallel resistance–inductance circuits is similar to those for the corresponding resistance–capacitance circuits. For a series resistance–inductance circuit

$$Z^2 = X_L^2 + R^2 \tag{53}$$

For a parallel resistance–inductance circuit

$$\left(\frac{1}{Z}\right)^2 = \left(\frac{1}{X_L}\right)^2 + \left(\frac{1}{R}\right)^2 \tag{54}$$

The same parallelogram analysis also applies, except by convention $X_L$ is drawn downward (instead of upward, as in the case of $X_C$) to represent the fact that the current lags the voltage.

## C. Resonance and Oscillations

If an inductance is placed in parallel with a capacitance (parallel inductance–capacitance or so-called **"tank circuit"**), then the impedance follows a peculiar pattern as the AC frequency is increased. As $f$ is increased, $X_L$ increases, whereas $X_C$ decreases; hence $Z$ passes through a rather sharp maximum (approaches infinite), and the current is minimum (approaches zero), at some intermediate frequency called the **resonant frequency** ($f_0$). In a series inductance–capacitance circuit, at the resonant frequency, $Z$ is minimum and the current is maximum. These facts are used in many electronic tuning circuits and filter circuits.

The resonant frequency for either the parallel or the series inductance–capacitance circuit may be calculated from

$$f_0 = \frac{1}{2\pi}\sqrt{\frac{1}{LC}} \tag{55}$$

At low frequencies, $X_C > X_L$ and current leads voltage; at high frequencies, $X_L > X_C$ and current lags voltage. At $f_0$, the current is in phase with the applied voltage and $X_C = X_L$. Thus,

$$\frac{1}{\omega C} = \omega L \tag{56}$$

or

$$\omega^2 = \frac{1}{LC} \tag{56a}$$

and

$$\omega = \sqrt{\frac{1}{LC}} \tag{56b}$$

which is the same as Eq. 55, since $\omega = 2\pi f$. If a significant resistance (in series with the inductive branch, as for the excitable membrane) is present in the parallel $(L + R)C$ circuit, then a damping factor ($R^2/L^2$) must be introduced, and $f_0$ is now given by

$$f_0 = \frac{1}{2\pi}\sqrt{\frac{1}{LC} - \frac{R^2}{L^2}} \tag{57}$$

One characteristic of the parallel inductance–capacitance tank circuit is that when suitable energy is supplied to such a circuit, oscillations in potential tend to occur at the resonant (or natural) frequency. In effect, at one instant, all the energy is stored in the magnetic field around the inductor; this field collapses when the current through the inductor goes to zero. This effect induces an

emf that drives current in the opposite direction and charges the capacitor. When fully charged, all the energy is stored as electrical charge on the plates of the capacitor. When the capacitor is fully charged and the current in the circuit goes to zero, the inductance now appears as a zero resistance ($f = 0$) across the capacitor, causing the capacitor to discharge into the inductor. This changing current induces a magnetic field in the inductor, and the cycle is repeated over and over. The oscillations produced are sinusoidal in nature. A small amount of energy needs to be put into the system; otherwise the oscillations dampen out because of the small resistance of the components and the resultant power ($I^2R$) loss as heat.

Oscillations of membrane potential, namely, the pacemaker potentials, are known to occur in many nerve and muscle cells under natural conditions and can be made to occur in all nerve and muscle membranes under experimental conditions. The nerve membrane (e.g., squid giant axon) has been demonstrated by Cole, Mullins, and others to also contain an inductive component (see Cole, 1968). Estimates of the **apparent inductance** for biological membranes ($L_m$) range from 0.5 to 10 H · cm². This inductance does not have a simple physical molecular basis, as do the resistance and capacitance components; hence it is described as an apparent inductance only. The $L_m$ seems to be intimately associated with the $K^+$ resistive component of the membrane ($R_K$), as depicted in the circuit diagram of Fig. A-11. As indicated by the arrow through the $K^+$ resistance, $R_K$ varies as a function of the membrane potential ($E_m$);

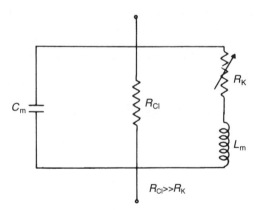

**FIG. A-11.** Circuit diagram (K. C. Cole circuit) depicting the apparent inductance of biological membranes ($L_m$). $L_m$ is in parallel to the membrane capacitance ($C_m$) and is due to the peculiar behavior of the $K^+$ resistance ($R_K$); that is, there is no physical basis for the inductance as there is for the capacitance (namely, the lipid bilayer matrix). When $Cl^-$ resistance ($R_{Cl}$) is high compared to $R_K$, as is true for most myocardial cells and neurons, the parallel inductance–capacitance circuit tends to oscillate at a resonant frequency ($f_0$): $f_0 = 1/2\pi \sqrt{1/LC}$. Arrow through $R_K$ indicates that it is a variable resistance, depending on membrane potential. When the membrane is depolarized, $R_K$ increases (because of the $K^+$ channels responsible for so-called "anomalous rectification"). Because $I_K = (E_m - E_K)/R_K$ the increase in driving force ($E_m - E_K$) for outward $K^+$ current ($I_K$) is counterbalanced by the increase in $R_K$, the latter acting to hold the $K^+$ current constant. This is behavior typical of an inductance and gives rise to the apparent membrane inductance.

**TABLE A-5**  Range of the Electromagnetic Spectrum

| | Frequency | (Hz) | Wavelength (m) |
|---|---|---|---|
| Longwave | $1 \times 10^4$ | (10 kHz) | 30,000 (30 km) |
| Radio broadcast | | | |
| AM | $1 \times 10^6$ | (1,000 kHz) | 300 (0.3 km) |
| Shortwave | $1 \times 10^7$ | (10 MHz) | 30 |
| FM | $3 \times 10^8$ | (300 MHz) | 1.0 |
| TV broadcast | $3 \times 10^8$ | (300 MHz) | 1.0 |
| Microwave | $3 \times 10^{10}$ | (30 GHz) | 0.01 (1.0 cm) |
| Light | | | |
| Infrared | $3 \times 10^{12}$ | | $1 \times 10^{-4}$ (0.1 mm) |
| Visible | $3 \times 10^{14}$ | | $1 \times 10^{-6}$ (1.0 $\mu$m) |
| Ultraviolet | $3 \times 10^{16}$ | | $1 \times 10^{-8}$ (0.01 $\mu$m; 10 nm) |
| X-rays | $3 \times 10^{18}$ | | $1 \times 10^{-10}$ (1.0 Å; 0.1 nm) |
| Gamma rays | $3 \times 10^{20}$ | | $1 \times 10^{-12}$ (0.01 Å) |

*Source: Now You're Talking,* edited by L. D. Wolfgang, J. Kearman, and J. P. Kleinman. Published by the American Radio Relay League, Newington, CT, 1996. Frequencies listed are for about the midrange of each spectrum. The corresponding wavelengths were calculated from the following equation, where $c$ is the velocity of light, $f$ is the frequency, and $L$ is the wavelength. The sample calculation is for an AM radio wave of 1000 kHz.

$$L = \frac{c}{f} = \frac{\text{m/sec}}{\text{cycles/sec}} = \frac{\text{m}}{\text{cycle}}$$

$$= \frac{300,000,000 \text{ m/sec}}{1000 \text{ kHz}} = \frac{3 \times 10^8 \text{ m/sec}}{1 \times 10^6 \text{ cycles/sec}} = 3 \times 10^{-2} \text{ m}$$

that is, there is $K^+$ rectification. In $K^+$ anomalous rectification, when the membrane is depolarized, thereby increasing the driving force $(E_m - E_K)$ for outward $K^+$ current, $g_K$ decreases; this decrease in $g_K$ tends to keep the outward $K^+$ current constant, since $I_K = g_K (E_m - E_K)$, thereby mimicking the behavior of an inductance.

## XIV. Electromagnetic Spectrum

The electromagnetic spectrum from radiowaves and light waves through gamma rays is summarized in Table A-5. As can be seen, the higher the frequency of the electromagnetic wave, the shorter its wavelength. The equation relating wavelength ($L$) to frequency ($f$) and to relativistic velocity ($c$) is given in the footnote to Table A-5. Note the similarity of this relationship to that for calculating the wavelength of a nerve impulse discussed in Chapter 23. The information given in this table has relevance to Fig. 1 of Chapter 68.

## Bibliography

Cole, K. S. (1968). "Membranes, Ions, and Impulses." University of California Press, Berkeley.

Josephson, I. R., Sanchez-Chapula, J., and Brown, A. M. (1984). A comparison of calcium currents in rat and guinea pig single ventricular cells. *Circ. Res.* **54,** 144–156.

Sperelakis, N. (1979). Origin of the cardiac resting potential. *In* "Handbook of Physiology, The Cardiovascular System, Vol. 1: The Heart (R. M. Berne and N. Sperelakis, Eds.), Chap. 6, pp. 187–267, American Physiological Society, Bethesda, MD.

Sperelakis, N. (1995). "Electrogenesis of Biopotentials." Kluwer Academic Publishers, New York.

Sperelakis, N., and D. Lehmkuhl. (1966). Ionic interconversion of pacemaker and nonpacemaker cultured chick heart cells. *J. Gen. Physiol.* **49,** 867–895.

Sperelakis, N., Hoshiko, T., and Berne, R. M. (1960). Non-syncytial nature of cardiac muscle: Membrane resistance of single cells. *Am. J. Physiol.* **198,** 531–536.

Tarr, M., and Sperelakis, N. (1964). Weak electrotonic interaction between contiguous cardiac cells. *Am. J. Physiol.* **207,** 691–700.

# Index